FISHES OF ARKANSAS

FISHES OF ARKANSAS

2ND EDITION

Henry W. Robison and Thomas M. Buchanan

THE UNIVERSITY OF ARKANSAS PRESS | FAYETTEVILLE | 2020

Copyright © 2020 by The University of Arkansas Press

ISBN: 978-1-68226-103-3
eISBN: 978-1-61075-673-0

23 22 21 20 19   5 4 3 2 1

Designer: April Leidig

∞ The paper used in this publication meets the minimum
requirements of the American National Standard for
Permanence of Paper for Printed Library Materials
Z39.48-1984.

Montage on half title courtesy of Garold Sneegas

Library of Congress Cataloging-in-Publication Data
Names: Robison, Henry W., author. | Buchanan, Thomas M.,
    1943– author.
Title: Fishes of Arkansas / Henry W. Robison and Thomas M.
    Buchanan.
Description: 2nd edition. | Fayetteville : The University of
    Arkansas Press, 2019. | Includes bibliographical references
    and index.
Identifiers: LCCN 2019000731| ISBN 9781682261033 (cloth :
    alk. paper) | ISBN 9781610756730 (paperback : alk. paper)
Subjects: LCSH: Freshwater fishes—Arkansas—Identification. |
    Freshwater fishes—Arkansas. | Fishes—Arkansas—
    Identification. | Fishes—Arkansas.
Classification: LCC QL628.A8 R63 2019 | DDC 597.17609767—
    dc23
LC record available at https://lccn.loc.gov/2019000731

DEDICATION

We dedicate this book to those whose love, nurturing, guidance, and sacrifice shaped our early lives, our professional lives, and our married lives.

HWR — To my family, especially my mother, Margie Frances Shiver, who lovingly supported me in all my pursuits, my sweet grandmother, Leila Marie ("Doll") Robison, my grandfather, Henry Avon Robison, my dearest aunt, Montene Griffin, who steadfastly encouraged me throughout her life, and my stepfather, Col. Morris E. Shiver, who jointly made me the person I am today by nurturing and supporting my love of the natural world, my joy of learning, and my dedication to purpose.

To my major professors, George L. Harp, who first taught me scientific rigor and the fascinating complexity of the aquatic world, and Rudolph J. Miller who gently taught me to appreciate fish behavior and natural history, instructed me in the science and the art of taxonomy, and transformed my initial love of fishes and biology into my dream of becoming an ichthyologist.

To my dear wife, Catherine, my son Patrick, daughter Lindsay, and my grandchildren, Ryan, Katy, Callie, Reid, Hayes, Chloe, and Edie, for their combined love, continuing patience, unceasing support, and occasional understanding of "Papa's" obsession with the fishes of Arkansas. All these wonderful individuals have so richly blessed my life and provided me countless hours of happiness and joy!

TMB — To all my family, but especially to my mother, Velmagene Thomas Gray, my grandmother, Lilian Goodale Thomas, my grandfather, A.F. Thomas, Sr., my uncle, A. F. Thomas, Jr., my aunts, Vivian Thomas Robertson, Venici Thomas Mann, and Mary Ellen Thomas Halbert, and my stepfather, Leo Gray who collectively molded my early years, gave me an appreciation for nature, and always encouraged me in my endeavors.

To my major professors, Kirk Strawn, who first encouraged me to study fishes, and Clark Hubbs, who shaped my development as an ichthyologist.

To my wife of forty-eight years, Lauren Marie Buchanan, my daughter, Jennifer, and my grandchildren, Sheridyn, Madelyn, and Ryan, for their love, encouragement, support, and patience during my work on this project.

We both make a special dedication to our fathers who gave their lives for this country during World War II: Lt. Henry W. Robison, fighter pilot, who died in a training mission over the Gulf of Mexico in 1944, and Leo B. Buchanan who died in combat in Normandy, France in 1944.

CONTENTS

List of Figures xiii

List of Tables xxi

Foreword xxiii

Preface to the Second Edition xxv

Acknowledgments xxvii

List of Abbreviations xxxiii

1. Arkansas Fishes: Diversity, Derivation, and Zoogeography

Worldwide Fish Diversity 3

North American Fish Diversity 4

Arkansas Fish Diversity 5

Derivation of the North American Fish Fauna 7

    North American Fishes with a Marine Origin 8

    North American Fishes with a North American Origin 8

    North American Fishes with a Eurasian Origin 9

    North American Fishes with a Central American Origin 9

    North American Fishes with a South American Origin 10

Ages of North American Fish Families 10

Derivation of the Arkansas Fish Fauna 10

    Northern Component 10

    Great Plains Component 11

    Eastern North American Component 11

    Central and South American Component 11

    Archaic, Relict Component 11

    Endemic Component 11

Zoogeographic Concepts and North American Fish Zoogeography 11

    Center of Origin/Dispersal 12

    Vicariance 12

Arkansas Fish Zoogeography 13

    Introduction 13

    Distribution of Arkansas Fishes by River System 14

        White River System 15

        Arkansas River System 16

        Ouachita River System 17

        Red River System 17

        St. Francis River System 18

        Mississippi River System 18

    Distribution of Arkansas Fishes by Ecoregion 25

    Arkansas Fish Faunal Regions 26

2. History of Ichthyology in Arkansas

Age of Discovery: 10,000 BC–AD 1850 37

The Early Years: 1850–1900 38

Classic Period: 1900–1960 41

Resurgence: 1960–1987 43

Post First Edition Modern Period: 1988–2019 45

3. The Environmental Setting

Terrestrial Ecoregions 49

Aquatic Ecosystems 56

Environmental Alterations 61

    General Overview 61

    Agricultural Activities 62

    Stream Channelization 63

    Construction of Dams and Navigation Systems 66

    Gravel Mining 71

    Accidental Chemical Spills 71

    Forestry Activities 72

    Industrial Pollution 73

    Municipal Pollution 74

    Groundwater Contamination and Withdrawal 75

    Roadbuilding Activities 77

4. Classification and Systematics

What Is a Fish? 79

Classification Basics 79

Higher Categories of Fish Classification 79

Linnaean System of Classification and Nomenclature 80

Common Names 81

Modern Evolution-Based Systems of Classification 82

    Evolutionary Taxonomy 82

    Cladistics—Phylogenetic Systematics 82

Characters 83

What Is a Species? 84

    Biological Species Concept (BSC) 84

    Phylogenetic Species Concept (PSC) 84

    Evolutionary Species Concept (ESC) 85

The Subspecies Concept 85

Molecular Techniques 86

    Early Applications 86

    Phylogeography 86

    Recent Applications 86

    DNA Barcoding 87

    Characters: Molecular Markers 88

    Phylogenetic Methods and Markers 89

5. Conservation of Arkansas Fishes

Fishes in Trouble 91

Rarest Fishes 93

Introduced Species 93

    Categories of Introduced Fishes 93

Commercial Fishery 96

6. Organization of the Chapters Containing Species Accounts

Order, Family, and Genus Accounts 99

Species Accounts 99

    Scientific Name and Common Name 99

    Photographs 100

    Distribution Map 100

    Characters 100

    Life Colors 101

    Similar Species 101

    Variation and Taxonomy 101

Distribution 101

Habitat and Biology 101

Conservation Status 101

7. How to Identify a Fish

Keying Out Arkansas Fishes 103

Dichotomous Keys 103

Counts and Measurements 104

Key to Arkansas Fish Families 106

8. Order Petromyzontiformes—Lampreys

Family Petromyzontidae *Lampreys* 114

    Genus *Ichthyomyzon* Girard, 1858—
River Lampreys 115

    Genus *Lampetra* Bonnaterre, 1788—
Brook Lampreys 122

    Genus *Lethenteron* Creaser and Hubbs, 1922—
Northern Brook Lampreys 124

9. Order Acipenseriformes—
Sturgeons and Paddlefishes

Family Acipenseridae *Sturgeons* 128

    Genus *Acipenser* Linnaeus, 1758—
Holarctic Sturgeons 129

    Genus *Scaphirhynchus* Heckel, 1836—
Shovelnose Sturgeons 133

Family Polyodontidae *Paddlefishes* 140

    Genus *Polyodon* Lacépède, 1797—
North American Paddlefishes 140

10. Order Lepisosteiformes—Gars

Family Lepisosteidae *Gars* 147

    Genus *Atractosteus* Rafinesque, 1820a—
Alligator Gars 149

    Genus *Lepisosteus* Lacépède, 1803—
Longnose Gars 154

11. Order Amiiformes—Bowfins

Family Amiidae *Bowfins* 161

    Genus *Amia* Linnaeus, 1766—Bowfins 162

12. Order Hiodontiformes—Mooneyes

Family Hiodontidae *Mooneyes* 167

    Genus *Hiodon* Lesueur, 1818a—Mooneyes 168

13. Order Anguilliformes—Eels

Family Anguillidae *Freshwater Eels* 173

    Genus *Anguilla* Schrank, 1798—Freshwater Eels 174

14. Order Clupeiformes—Herrings

Family Clupeidae *Herrings* 179

    Genus *Alosa* Linck, 1790—River Herrings 180

    Genus *Dorosoma* Rafinesque, 1820a—
Gizzard Shads 185

15. Order Cypriniformes—
Minnows, Carps, Loaches, and Suckers

Family Cyprinidae *Carps and Minnows* 192

    Genus *Campostoma* Agassiz, 1855—Stonerollers 203

    Genus *Carassius* Jarocki, 1822—Goldfishes 221

Genus *Chrosomus* Rafinesque, 1820a—
Redbelly Daces 222

Genus *Ctenopharyngodon* Steindachner, 1866—
Grass Carps 225

Genus *Cyprinella* Girard, 1856—Satinfin Shiners 228

Genus *Cyprinus* Linnaeus, 1758—True Carps 243

Genus *Erimystax* Jordan, 1882—Slender Chubs 246

Genus *Hybognathus* Agassiz, 1855—
Silvery Minnows 249

Genus *Hybopsis* Agassiz, 1854—Bigeye Chubs 257

Genus *Hypophthalmichthys* Bleeker, 1860—
Bigheaded Carps 261

Genus *Luxilus* Rafinesque, 1820a—
Highscale Shiners 268

Genus *Lythrurus* Jordan, 1876a—
Finescale Shiners 276

Genus *Macrhybopsis* Cockerell and Allison, 1909—
Blacktail Chubs 284

Genus *Mylopharyngodon* Peters, 1881—
Black Carps 294

Genus *Nocomis* Girard, 1856—Chubs 296

Genus *Notemigonus* Rafinesque, 1819—
Golden Shiners 301

Genus *Notropis* Rafinesque, 1818a—True Shiners 303

Genus *Opsopoeodus* Hay, 1881—
Pugnose Minnows 353

Genus *Phenacobius* Cope, 1867b—
Suckermouth Minnows 356

Genus *Pimephales* Rafinesque, 1820a—
Bluntnose Minnows 358

Genus *Platygobio* Gill, 1863—Flathead Chubs 368

Genus *Pteronotropis* Fowler, 1935—Flagfin Shiners 370

Genus *Scardinius* Bonaparte, 1837—Rudds 374

Genus *Semotilus* Rafinesque, 1820a—
Creek Chubs 375

Family Catostomidae *Suckers* 377

Genus *Carpiodes* Rafinesque, 1820a—
Carpsuckers 382

Genus *Catostomus* Lesueur, 1817b—
Finescale Suckers 388

Genus *Cycleptus* Rafinesque, 1819—Blue Suckers 390

Genus *Erimyzon* Jordan, 1876b—Chubsuckers 394

Genus *Hypentelium* Rafinesque, 1818c—
Hog Suckers 398

Genus *Ictiobus* Rafinesque, 1820a—Buffalofishes 400

Genus *Minytrema* Jordan, 1878b—
Spotted Suckers 407

Genus *Moxostoma* Rafinesque, 1820a—
Redhorse Suckers 409

16. Order Siluriformes—Catfishes

Family Ictaluridae *North American Catfishes* 425

Genus *Ameiurus* Rafinesque, 1820a—
Bullhead Catfishes 430

Genus *Ictalurus* Rafinesque, 1820a—
Forktail Catfishes 436

Genus *Noturus* Rafinesque, 1818d—Madtoms 440

Genus *Pylodictis* Rafinesque, 1819—
Flathead Catfishes 465

17. Order Osmeriformes—Smelts

Family Osmeridae *Smelts* 469

Genus *Osmerus* Linnaeus, 1758—Rainbow Smelts 469

18. Order Salmoniformes—Trout, Salmon, and Whitefish

Family Salmonidae *Trouts and Salmons* 473

Genus *Oncorhynchus* Suckley, 1861—
Pacific Trouts and Salmons 475

Genus *Salmo* Linnaeus, 1758—
Atlantic Trouts and Salmons 479

Genus *Salvelinus* Richardson, 1836—Chars 482

19. Order Esociformes— Pikes and Mudminnows

Family Esocidae *Pikes and Mudminnows* 487

    Genus *Esox* Linnaeus, 1758—Pikes 488

    Genus *Umbra* Kramer, 1777—Mudminnows 495

20. Order Percopsiformes—Trout-Perches

Family Aphredoderidae *Pirate Perches* 499

    Genus *Aphredoderus* Lesueur, 1833— Pirate Perches 499

Family Amblyopsidae *Cavefishes* 502

    Genus *Troglichthys* Eigenmann, 1899a— Ozark Cavefishes 503

    Genus *Typhlichthys* Girard, 1859— Southern Cavefishes 506

21. Order Mugiliformes—Mullets

Family Mugilidae *Mullets* 511

    Genus *Mugil* Linnaeus, 1758—Mullets 511

22. Order Atheriniformes—Silversides

Family Atherinopsidae *New World Silversides* 515

    Genus *Labidesthes* Cope, 1870—Brook Silversides 516

    Genus *Menidia* Bonaparte, 1836— Atlantic Silversides 522

23. Order Cyprinodontiformes— Killifishes, Topminnows, Livebearers, and Toothcarps

Family Fundulidae *Topminnows* 527

    Genus *Fundulus* Lacépède, 1803—Topminnows 528

Family Poeciliidae *Livebearers* 541

    Genus *Gambusia* Poey, 1854—Gambusias 542

24. Order Anabantiformes— Labyrinth Fishes

Family Channidae *Snakeheads* 547

    Genus *Channa* Scopoli, 1877—Snakeheads 548

25. Order Scorpaeniformes— Mail-Cheeked Fishes

Family Cottidae *Sculpins* 553

    Genus *Uranidea* DeKay, 1842— Uranidea Sculpins *554*

26. Order Perciformes— Perches and Relatives

Family Moronidae *Temperate Basses* 561

    Genus *Morone* Mitchill, 1814—Striped Basses 562

Family Centrarchidae *Sunfishes* 573

    Genus *Ambloplites* Rafinesque, 1820a— Rock Basses 575

    Genus *Centrarchus* Cuvier, 1829—Fliers 582

    Genus *Lepomis* Rafinesque, 1819—Panfishes 584

    Genus *Micropterus* Lacépède, 1802—Black Basses 609

    Genus *Pomoxis* Rafinesque, 1818c—Crappies 619

Family Elassomatidae *Pygmy Sunfishes* 624

    Genus *Elassoma* Jordan, 1877b—Pygmy Sunfishes 625

Family Percidae *Perches and Darters* 627

    Genus *Ammocrypta* Jordan, 1877b—Sand Darters 634

    Genus *Crystallaria* Jordan and Gilbert, 1885— Crystal Darters 638

Genus *Etheostoma* Rafinesque, 1819—
Etheostoma Darters 640

Genus *Nothonotus* Putnam, 1863—Lined Darters 712

Genus *Perca* Linnaeus, 1758—Perches 717

Genus *Percina* Haldeman, 1842—
River Darters and Logperches 720

Genus *Stizostedion* Rafinesque, 1820a—
Pikeperches 752

Family Sciaenidae *Drums and Croakers* 757

Genus *Aplodinotus* Rafinesque, 1819—
Freshwater Drums 758

Family Cichlidae *Cichlids and Tilapias* 760

Genus *Oreochromis* Günther, 1889—Tilapias 762

Appendix 1. Arkansas Fish Collections 769

Appendix 2. Preservation of Specimens 771

Appendix 3. Aids to Identification of Fishes 773

Appendix 4. Pre-1988 Distribution Maps for Arkansas
Fishes 779

Appendix 5. Checklist of Species and List of Higher
Taxonomic Categories of Arkansas Fishes 821

Glossary 827

Literature Cited 841

Index to Common and Scientific Names 951

FIGURES

1.1. General origins of the major families of North American freshwater fishes. 9

1.2. Directional zoogeographic affinities of 178 native freshwater fishes in Arkansas. 14

2.1. Charles Frederic Girard (1822–1895). 39

2.2. David Starr Jordan (1851–1931). 40

2.3. Charles Henry Gilbert (1859–1928). 41

2.4. Seth Eugene Meek (1859–1914). 41

2.5. Carl L. Hubbs (1894–1979). 42

2.6. John D. Black (1908–1996). 43

2.7. Neil H. Douglas (1932–). 44

3.1. Arkansas Counties. 50

3.2. Physiographic regions and six ecoregions (Natural Divisions) of Arkansas. 50

3.3. Environmental Protection Agency Level IV Ecoregions of Arkansas. 51

3.4. Shaded relief map showing six major ecoregions of Arkansas. 52

3.5a. Boston Mountains, Newton County, Arkansas (Ozark Mountains Ecoregion). 52

3.5b. Ozark Mountains near Sam's Throne, Arkansas (Ozark Mountains Ecoregion). 52

3.6. Buffalo River, Newton County, Arkansas. Ozark Mountains Ecoregion. 52

3.7a. Large spring in the Springfield Plateau. 52

3.7b. White River in Springfield Plateau, Arkansas. 53

3.8. Calcareous bluffs in the Salem Plateau. 53

3.9. Arkansas Valley Ecoregion. 53

3.10. Magazine Mountain, Arkansas Valley Ecoregion. 53

3.11a. Ouachita Mountains Ecoregion. 54

3.11b. Ouachita Mountains, Perry County, Ouachita Mountains Ecoregion. 54

3.12a. Athens Piedmont Plateau, Ouachita Mountains Ecoregion. 54

3.12b. Athens Piedmont Plateau, Cossatot River, Ouachita Mountains Ecoregion. 54

3.13a. Mississippi Alluvial Plain Ecoregion. 55

3.13b. Lake in the Mississippi Alluvial Plain Ecoregion. 55

3.14a. West Gulf Coastal Plain Ecoregion. 55

3.14b. Mature Beech-Holly forest in West Gulf Coastal Plain Ecoregion. 55

3.15a. Stream on Crowley's Ridge, Crowley's Ridge Ecoregion. 56

3.15b. Loess soils of Crowley's Ridge Ecoregion. 56

3.16. Natural lakes greater than 500 surface acres. 57

3.17. Major rivers and streams of Arkansas. 58

3.18. River Systems of Arkansas. 58

3.19. Red River near Texarkana, Arkansas. 59

3.20a. Upper Ouachita River. 59

3.20b. Lower Ouachita River. 59

3.21a. Arkansas River, Conway County, Arkansas. 59

3.21b. Dam on the Arkansas River. 59

3.22a. White River at Calico Rock, Arkansas. 60

3.22b. White River near Oil Trough, Arkansas. 60

3.23. St. Francis River near Forrest City, Arkansas. 60

3.24. Mississippi River. 60

3.25. Lake Chicot, largest oxbow lake in Arkansas in Chicot County. 60

3.26. Ditch in Cross County, Arkansas, Mississippi Alluvial Plain. 64

3.27. Channelized stream (Dark Corner, Arkansas) in Mississippi Alluvial Plain. 64

3.28. Channelized stream (same stream as Fig. 3.27) in Mississippi Alluvial Plain. 64

3.29. Channelization of Cache River, Mississippi Alluvial Plain. 64

3.30. Reservoirs greater than 500 surface acres. 68

3.31. Tailwaters below Norfork Dam. 69

3.32. Mississippi River with barge traffic. 70

5.1. Leopard Darter, *Percina pantherina*. 92

5.2. Yellowcheek Darter, *Nothonotus moorei*. 92

6.1. Arkansas localities sampled for fishes between 1988 and 2019. 100

7.1. Petromyzontidae. 106

7.2. Anguillidae. 106

7.3. Amblyopsidae. 106

7.4. Polyodontidae. 106

7.5. Acipenseridae. 106

7.6. Lepisosteidae. 106

7.7. Ictaluridae. 107

7.8. Osmeridae. 107

7.9. Salmonidae. 107

7.10. Aphredoderidae. 107

7.11. Top: Position of anus in front of pelvic fins in Aphredoderidae (Pirate Perch). Bottom: Typical position of anus behind pelvic fins as in Centrarchidae (Bluegill). 107

7.12. Amiidae. 107

7.13. Channidae. 108

7.14. Poeciliidae (male and female). 108

7.15. Fundulidae. 108

7.16. Esocidae. 108

7.17. Clupeidae. 108

7.18. Hiodontidae. 109

7.19. Cyprinidae. 109

7.20. Catostomidae. 109

7.21. Atherinopsidae. 109

7.22. Cottidae. 109

7.23. Mugilidae. 109

7.24. Moronidae. 110

7.25. Elassomatidae. 110

7.26. Cichlidae. 110

7.27. Centrarchidae. 110

7.28. Sciaenidae. 111

7.29. Percidae. 111

8.1. Ammocoete larval form. 114

8.2. Oral discs of lampreys. 115

8.3. *Ichthyomyzon castaneus*, Chestnut Lamprey. 115

8.4. *Ichthyomyzon castaneus* mouth. 116

8.5. *Ichthyomyzon castaneus* mouth and *I. gagei* mouth with teeth. 116

8.6. *Ichthyomyzon castaneus* head fastened on a rock. 117

8.7. *Ichthyomyzon gagei*, Southern Brook Lamprey. 118

8.8. Oral disc of *Ichthyomyzon gagei*, Southern Brook Lamprey. 119

8.9. *Ichthyomyzon unicuspis*, Silver Lamprey. 120

8.10. Oral disc of *Ichthyomyzon unicuspis* with teeth. 120

8.11. *Ichthyomyzon unicuspis* feeding on an adult Paddlefish. 121

8.12. *Lampetra aepyptera*, Least Brook Lamprey. 122

8.13. Oral disc of *Lampetra aepyptera*, Least Brook Lamprey. 122

8.14. *Lethenteron appendix*, American Brook Lamprey. 124

8.15. *Lethenteron appendix* mouth. 125

9.1. Undersides of heads of sturgeons. 129

9.2. *Acipenser fulvescens*, Lake Sturgeon. 129

9.3. Head of *Acipenser fulvescens*, Lake Sturgeon. 130

9.4. *Acipenser oxyrinchus*, Atlantic Sturgeon. 132

9.5. *Scaphirhynchus albus*, Pallid Sturgeon. 133

9.6. Coloration of three live Arkansas sturgeons compared side-by-side. 134

9.7. *Scaphirhynchus platorynchus*, Shovelnose Sturgeon. 137

9.8. *Scaphirhynchus platorynchus* barbels on underside of head. 138

9.9. *Scaphirhynchus platorynchus* showing dorsal, lateral, and ventral scutes. 138

9.10. *Polyodon spathula*, Paddlefish. 141

9.11. Paddlefish with mouth open to show large size and gill arches. 141

9.12. Eye and spiracle of Paddlefish, *Polyodon spathula*. 142

9.13. Adult Paddlefish. 143

9.14. Juvenile Paddlefish. 143

9.15. Paddle of Paddlefish supported by dorsal and ventral sheets of numerous and densely packed interdigitating stellate bones. 144

9.16. Paddlefish taken from the Mississippi River. 146

10.1. Alligator Gar being hunted by bow and arrow in the 1950s on the White River in Arkansas. 149

10.2. Tops of heads of gars. 150

10.3a, b. *Atractosteus spatula*, Alligator Gar. 150

10.4. *Atractosteus spatula* tooth rows. 151

10.5. *Lepisosteus oculatus*, Spotted Gar. 154

10.6. *Lepisosteus oculatus* and reflection. 155

10.7. *Lepisosteus osseus*, Longnose Gar. 156

10.8. *Lepisosteus platostomus*, Shortnose Gar. 159

11.1. Gular plate of *Amia calva*. 162

11.2. *Amia calva*, Bowfin. 162

11.3. Juvenile (70 mm TL) Bowfin showing orange and black colors, Altamaha River system, Georgia. 163

11.4. Male Bowfin in breeding coloration. 165

12.1. *Hiodon alosoides*, Goldeye. 168

12.2. Anal fin shapes in *Hiodon*. 170

12.3. *Hiodon tergisus*, Mooneye. 170

13.1. *Anguilla rostrata*, American Eel. 174

13.2. Leptocephalus larva. 176

14.1. Lateral view of *Dorosoma* heads showing shape and relative lengths of upper and lower jaws. 180

14.2. Lateral view of *Alosa* heads showing relationship of upper and lower jaws. 180

14.3. *Alosa alabamae*, Alabama Shad. 181

14.4. *Alosa chrysochloris*, Skipjack Herring. 183

14.5. *Dorosoma cepedianum*, Gizzard Shad. 185

14.6. *Dorosoma petenense*, Threadfin Shad. 188

15.1. Breeding pair of *Pteronotropis hubbsi*, Bluehead Shiner, from southern Arkansas. 194

15.2. Key characters of some minnow genera. 195

15.3. Lateral view of head of *Cyprinus carpio* showing two barbels. 195

15.4. Lateral view of head of *Nocomis biguttatus* showing barbel. 196

15.5. Lateral view of head of *Semotilus atromaculatus* showing minute, concealed barbel. 196

15.6. Characteristics of some minnow mouths (lateral views). 196

15.7. Cyprinid gut (intestinal) coiling and looping patterns and peritoneal melanism. 197

15.8. Heads of *Campostoma* breeding males showing tubercle patterns. 197

15.9. Ventral views (diagrammatic) of body cavities in minnows, as they would appear if lower body-wall were cut away. 199

15.10. Basioccipital bones in four species of *Hybognathus*. 199

15.11. Wedge-shaped caudal spot in *Notropis greenei*. 199

15.12. Upper lip shapes in *Notropis potteri* and *N. blennius*. 200

15.13. Head and anterior lateral line scales of *Notropis volucellus* and *Notropis girardi*. 201

15.14. Cephalic lateral line system of *Notropis volucellus* and *N. buchanani*. 201

15.15. Orientation of dorsal fin relative to origin of pelvic fins in minnows. 202

15.16. Ventral view of head showing chin pigmentation in *Notropis percobromus* and *Notropis atherinoides*. 202

15.17. Extension of the maxillary bone behind the angle of the mouth in *Hybopsis amnis*. 203

15.18a. *Campostoma oligolepis*, Largescale Stoneroller breeding male. 206

15.18b. *Campostoma oligolepis*, Largescale Stoneroller female. 206

15.19. Dorsal and ventral views of nuptial male stoneroller heads. 209

15.20. *Campostoma plumbeum*, Plains Stoneroller. 211

15.21. *Campostoma plumbeum* grazing scars on rock. 213

15.22a. *Campostoma pullum*, Finescale Stoneroller male. 214

15.22b. *Campostoma pullum*, Finescale Stoneroller female. 214

15.23. Frontal views of head of *Campostoma pullum* male from Village Creek, Cross County, Arkansas, showing tubercle pattern. 215

15.24. *Campostoma spadiceum*, Highland Stoneroller. 217

15.25. *Campostoma spadiceum*. A—Nuptial male from Glover River, McCurtain County, Oklahoma, collected 20 March 2007; B—Close-up of caudal peduncle of same male to show detailed coloration. 218

15.26. Coloration of various size *Campostoma spadiceum* from Glover River, McCurtain County, Oklahoma, 27 July 2004. 219

15.27. *Campostoma spadiceum* nuptial male with tubercles. Mill Creek, Clark County, Arkansas. 220

15.28. Tubercle patterns of nuptial males of *Campostoma plumbeum* and *Campostoma spadiceum*. 220

15.29. *Carassius auratus*, Goldfish. 221

15.30a. *Chrosomus erythrogaster*, Southern Redbelly Dace male. 223

15.30b. *Chrosomus erythrogaster*, Southern Redbelly Dace female. 223

15.31. *Ctenopharyngodon idella*, Grass Carp. 224

15.32. Pharyngeal teeth of Grass Carp. 226

15.33. *Cyprinella camura*, Bluntface Shiner. 229

15.34. *Cyprinella galactura*, Whitetail Shiner. 231

15.35a. *Cyprinella lutrensis*, Red Shiner male. 233

15.35b. *Cyprinella lutrensis*, Red Shiner female. 233

15.36. *Cyprinella spiloptera*, Spotfin Shiner. 236

15.37. *Cyprinella venusta*, Blacktail Shiner. 239

15.38. *Cyprinella whipplei*, Steelcolor Shiner. 241

15.39. *Cyprinus carpio*, Common Carp. 243

15.40. *Erimystax harryi*, Ozark Chub. 246

15.41. *Erimystax x-punctatus*, Gravel Chub. 247

15.42. *Hybognathus hayi*, Cypress Minnow. 250

15.43. Ventral view of head showing relative snout length in *Hybognathus hayi* and *Hybognathus nuchalis*. 251

15.44. *Hybognathus nuchalis*, Mississippi Silvery Minnow. 252

15.45. *Hybognathus placitus*, Plains Minnow. 254

15.46. *Hybopsis amblops*, Bigeye Chub. 257

15.47. *Hybopsis* tubercles. 258

15.48. *Hybopsis amnis*, Pallid Shiner. 259

15.49. *Hypophthalmichthys molitrix*, Silver Carp. 262

15.50. Ventral keels of bigheaded carps. 263

15.51. *Hypophthalmichthys nobilis*, Bighead Carp. 265

15.52. *Luxilus cardinalis*, Cardinal Shiner. 268

15.53. *Luxilus chrysocephalus*, Striped Shiner. 270

15.54. *Luxilus chrysocephalus* male in breeding colors. 270

15.55. Dorsal views of *Luxilus chrysocephalus* subspecies. 271

15.56. *Luxilus pilsbryi,* Duskystripe Shiner. 272

15.57. Duskystripe Shiners in spawning coloration (29 April 2015), Yocum Creek, Carroll County, Arkansas. 273

15.58. *Luxilus zonatus,* Bleeding Shiner. 274

15.59. *Lythrurus fumeus,* Ribbon Shiner. 277

15.60. *Lythrurus snelsoni,* Ouachita Mountain Shiner. 279

15.61a. *Lythrurus umbratilis,* Redfin Shiner male. 281

15.61b. *Lythrurus umbratilis,* Redfin Shiner female. 281

15.62. Coloration variation in *Lythrurus umbratilis* populations from the Ouachita River drainage. 282

15.63. *Macrhybopsis gelida,* Sturgeon Chub. 284

15.64. *Macrhybopsis hyostoma,* Shoal Chub. 286

15.65. *Macrhybopsis meeki,* Sicklefin Chub. 289

15.66. *Macrhybopsis storeriana,* Silver Chub. 292

15.67. *Mylopharyngodon piceus,* Black Carp. 294

15.68. *Nocomis asper,* Redspot Chub. 297

15.69. Dorsal and lateral views of head of nuptial male *Nocomis asper.* 297

15.70. *Nocomis biguttatus,* Hornyhead Chub. 299

15.71. Dorsal and lateral views of the head of a nuptial male *Nocomis biguttatus.* 300

15.72a. *Notemigonus crysoleucas,* Golden Shiner (golden form). 302

15.72b. *Notemigonus crysoleucas,* Golden Shiner (silver form). 302

15.73. *Notropis atherinoides,* Emerald Shiner. 304

15.74. *Notropis atrocaudalis,* Blackspot Shiner. 306

15.75. *Notropis bairdi,* Red River Shiner. 308

15.76. *Notropis blennius,* River Shiner. 310

15.77. *Notropis boops,* Bigeye Shiner. 312

15.78. *Notropis buchanani,* Ghost Shiner. 314

15.79. *Notropis chalybaeus,* Ironcolor Shiner. 316

15.80. Heads and mouths of *Notropis chalybaeus* and *Opsopoeodus emiliae.* 317

15.81. *Notropis girardi,* Arkansas River Shiner. 318

15.82. *Notropis greenei,* Wedgespot Shiner. 321

15.83. *Notropis maculatus,* Taillight Shiner. 322

15.84. *Notropis nubilus,* Ozark Minnow. 324

15.85. *Notropis ortenburgeri,* Kiamichi Shiner. 326

15.86. *Notropis ortenburgeri* with inset showing chin pigmentation. 327

15.87a. *Notropis ozarcanus,* Ozark Shiner. 328

15.87b. *Notropis ozarcanus* breeding male, Buffalo River, Searcy County, Arkansas. 328

15.88. *Notropis percobromus,* Carmine Shiner. 329

15.89. *Notropis percobromus* in breeding coloration in northeastern Oklahoma. 330

15.90. *Notropis perpallidus,* Peppered Shiner. 332

15.91. *Notropis potteri,* Chub Shiner. 334

15.92. *Notropis sabinae,* Sabine Shiner. 336

15.93. *Notropis shumardi,* Silverband Shiner. 338

15.94. *Notropis stramineus,* Sand Shiner. 340

15.95. *Notropis suttkusi,* Rocky Shiner. 342

15.96. *Notropis telescopus,* Telescope Shiner. 344

15.97. *Notropis texanus,* Weed Shiner. 346

15.98. *Notropis texanus* anal fin. 347

15.99. *Notropis volucellus,* Mimic Shiner. 348

15.100. Postdorsal pigmentation in *Notropis volucellus* and *N. wickliffi.* 350

15.101. *Notropis wickliffi,* Channel Shiner. 351

15.102. *Opsopoeodus emiliae,* Pugnose Minnow. 354

15.103. *Phenacobius mirabilis,* Suckermouth Minnow. 356

15.104. *Pimephales notatus,* Bluntnose Minnow. 359

15.105. *Pimephales* heads. 360

15.106. *Pimephales notatus* male tending eggs attached to the underside of a large rock in a Flint Hills stream, Kansas. 361

15.107. *Pimephales notatus* male rubbing eggs attached to the underside of a large rock in a Flint Hills stream, Kansas. 361

15.108. *Pimephales promelas,* Fathead Minnow. 361

15.109a. Rosy Reds (*Pimephales promelas*) baitfish school. 363

15.109b. Rosy Red bait minnow (*Pimephales promelas*). 363

15.110. *Pimephales tenellus,* Slim Minnow. 364

15.111. *Pimephales vigilax,* Bullhead Minnow. 366

15.112. *Platygobio gracilis,* Flathead Chub. 368

15.113. *Pteronotropis hubbsi,* Bluehead Shiner. 371

15.114. *Pteronotropis hubbsi* "flag male" in spawning coloration. 372

15.115. *Scardinius erythrophthalmus,* Rudd. 374

15.116. *Semotilus atromaculatus,* Creek Chub. 375

15.117. *Moxostoma carinatum,* River Redhorse, in White River near Cotter, Arkansas. 377

15.118. Diagrammatic views of the heads of *Carpiodes* and *Ictiobus.* 379

15.119. Lateral views of the heads of the *Ictiobus* species. 380

15.120. Carpsucker lips. 380

15.121. Characteristics of the lips of some suckers. 380

15.122. Frontal views of heads of two suckers. 381

15.123. Molarlike pharyngeal teeth of *Moxostoma carinatum.* 381

15.124. Ventral views of lips of *Moxostoma* species. 381

15.125. Intestinal coiling patterns of ictiobine suckers. 382

15.126. *Carpiodes carpio*, River Carpsucker. 382

15.127. *Carpiodes cyprinus*, Quillback. 384

15.128. *Carpiodes velifer*, Highfin Carpsucker. 386

15.129. *Catostomus commersonii*, White Sucker. 388

15.130. *Cycleptus elongatus*, Blue Sucker. 391

15.131. Head of nuptial male Blue Sucker. 391

15.132. *Cycleptus elongatus* small juvenile, 32 mm SL, Mississippi River, Tennessee. 392

15.133. *Erimyzon claviformis*, Western Creek Chubsucker. 394

15.134. Head of *Erimyzon claviformis* male with 3 breeding tubercles on each side. 395

15.135. *Erimyzon sucetta*, Lake Chubsucker. 396

15.136. *Hypentelium nigricans*, Northern Hog Sucker. 398

15.137. *Ictiobus bubalus*, Smallmouth Buffalo. 401

15.138. *Ictiobus cyprinellus*, Bigmouth Buffalo. 403

15.139. *Ictiobus niger*, Black Buffalo. 405

15.140. *Minytrema melanops*, Spotted Sucker. 407

15.141. *Moxostoma anisurum*, Silver Redhorse. 409

15.142. *Moxostoma carinatum*, River Redhorse. 411

15.143. *Moxostoma duquesnei*, Black Redhorse. 414

15.144. *Moxostoma erythrurum*, Golden Redhorse. 416

15.145. Nuptial male *Moxostoma erythrurum* displaying over a spawning site in the Verdigris River, Chase County, Kansas. 417

15.146. *Moxostoma lacerum*, Harelip Sucker. 418

15.147. *Moxostoma pisolabrum*, Pealip Redhorse. 420

15.148. *Moxostoma poecilurum*, Blacktail Redhorse. 423

15.149. *Moxostoma poecilurum* juvenile. 423

16.1. Characteristic caudal and adipose fins in catfishes. 427

16.2. Pectoral fin spines of four *Ameiurus* species showing degree of serration. 428

16.3. Lateral views of heads of catfishes. 428

16.4. Pectoral fin spines showing degree of posterior serration in 13 *Noturus* species. 428

16.5. Dorsal view of one side of head showing internasal pores of *Noturus exilis* and *N. nocturnus*. 429

16.6. Premaxillary tooth patch in upper jaw of catfishes. 429

16.7. *Ameiurus melas*, Black Bullhead. 430

16.8. *Ameiurus natalis*, Yellow Bullhead. 432

16.9. *Ameiurus nebulosus*, Brown Bullhead. 434

16.10. *Ictalurus furcatus*, Blue Catfish. 436

16.11. *Ictalurus punctatus*, Channel Catfish. 438

16.12. *Noturus albater*, Ozark Madtom. 441

16.13. *Noturus eleutherus*, Mountain Madtom. 443

16.14. *Noturus eleutherus* in gravel. 444

16.15a. *Noturus exilis*, Slender Madtom. 445

16.15b. White River form of *Noturus exilis* with blackened margins of medial fins. 445

16.16. *Noturus flavater*, Checkered Madtom. 447

16.17. *Noturus flavus*, Stonecat. 449

16.18. *Noturus gladiator*, Piebald Madtom. 451

16.19a. *Noturus gyrinus*, Tadpole Madtom, slender-bodied form. 453

16.19b. *Noturus gyrinus*, Tadpole Madtom, humpbacked form. 453

16.20. *Noturus lachneri*, Ouachita Madtom. 455

16.21. *Noturus maydeni*, Black River Madtom. 457

16.22. *Noturus miurus*, Brindled Madtom. 458

16.23. *Noturus nocturnus*, Freckled Madtom. 460

16.24. *Noturus phaeus*, Brown Madtom. 462

16.25. *Noturus taylori*, Caddo Madtom. 464

16.26. *Pylodictis olivaris*, Flathead Catfish. 465

16.27. Juvenile Flathead Catfish. 466

17.1. *Osmerus mordax*, Rainbow Smelt. 470

18.1. *Oncorhynchus clarkii*, Cutthroat Trout. 475

18.2. *Oncorhynchus mykiss*, Rainbow Trout. 477

18.3. *Salmo trutta*, Brown Trout. 480

18.4. *Salvelinus fontinalis*, Brook Trout. 482

18.5. *Salvelinus namaycush*, Lake Trout. 484

19.1. Heads of pikes. 488

19.2. *Esox americanus*, Redfin Pickerel. 489

19.3. *Esox lucius*, Northern Pike. 491

19.4. *Esox masquinongy*, Muskellunge. 492

19.5. *Esox niger*, Chain Pickerel. 494

19.6. *Umbra limi*, Central Mudminnow. 496

20.1. *Aphredoderus sayanus*, Pirate Perch. 500

20.2. Jugular position of anus and urogenital opening in *Aphredoderus sayanus*. 500

20.3. Sensory papillae on cavefish caudal fins. 503

20.4. *Troglichthys rosae*, Ozark Cavefish. 503

20.5. *Typhlichthys eigenmanni*, Salem Plateau Cavefish. 507

20.6. *Typhlichthys* sp. *cf. eigenmanni*, Ghost Cavefish from Alexander Cave, Stone County. 508

21.1. *Mugil cephalus*, Striped Mullet. 512

22.1. Dorsal view of silverside heads. 516

22.2. Midlateral stripe of *Labidesthes* species. 516

22.3. Morphological differences in posttemporal process of *Labidesthes* species. 516

22.4. *Labidesthes sicculus*, Brook Silverside. 517

22.5. *Labidesthes vanhyningi*, Hardy Silverside. 520

22.6. *Menidia audens*, Mississippi Silverside. 522

23.1. *Fundulus blairae*, Western Starhead Topminnow. 529

23.2a. *Fundulus catenatus*, Northern Studfish male. 530

23.2b. *Fundulus catenatus*, Northern Studfish female. 530

23.3a. *Fundulus chrysotus*, Golden Topminnow male. 533

23.3b. *Fundulus chrysotus*, Golden Topminnow female. 533

23.4. *Fundulus dispar*, Starhead Topminnow. 535

23.5. *Fundulus notatus*, Blackstripe Topminnow. 537

23.6. *Fundulus olivaceus*, Blackspotted Topminnow. 539

23.7a. *Gambusia affinis*, Western Mosquitofish male. 542

23.7b. *Gambusia affinis*, Western Mosquitofish female. 542

23.8. Gonopodium of male *Gambusia affinis*. 543

23.9. Pregnant female *Gambusia affinis*. 543

23.10. Gonopodia of *Gambusia affinis* and *G. holbrooki*. 544

24.1. *Channa argus*, Northern Snakehead. 548

24.2. Gular region of Northern Snakehead. 549

24.3. Northern Snakehead mouth. 549

25.1. Knobfin Sculpin on stream bottom in White River. 554

25.2. *Uranidea carolinae*, Banded Sculpin. 555

25.3. *Uranidea immaculata*, Knobfin Sculpin. 557

25.4. Knobfin Sculpin in Cotter Spring, White River at Cotter, Arkansas. 558

26.1. Anal fin spine lengths and size of second anal spine in *Morone* species. 562

26.2. Tooth patches on tongue of *Morone* species. 563

26.3. *Morone americana*, White Perch. 563

26.4. *Morone chrysops*, White Bass. 565

26.5. *Morone mississippiensis*, Yellow Bass. 568

26.6. *Morone saxatilis*, Striped Bass. 570

26.7. *Morone chrysops* × *M. saxatilis* hybrid. 572

26.8. Gill rakers in *Lepomis* species. 575

26.9. Differences in flexibility of opercles in sunfishes. 575

26.10. Roof of mouth showing palatine teeth of *Lepomis auritus*. 575

26.11a. *Ambloplites ariommus*, Shadow Bass. 576

26.11b. *Ambloplites ariommus*, Shadow Bass juvenile. 576

26.12. *Ambloplites constellatus*, Ozark Bass. 578

26.13. *Ambloplites rupestris*, Rock Bass. 580

26.14. *Centrarchus macropterus*, Flier. 582

26.15. Juvenile *Centrarchus macropterus*. 583

26.16. Small juveniles of *Lepomis* species. 584

26.17. *Lepomis auritus*, Redbreast Sunfish. 585

26.18. *Lepomis cyanellus*, Green Sunfish. 587

26.19. *Lepomis gulosus*, Warmouth. 589

26.20. *Lepomis humilis*, Orangespotted Sunfish. 591

26.21a. *Lepomis macrochirus*, Bluegill male. 593

26.21b. *Lepomis macrochirus*, Bluegill female. 593

26.22. Coppernose Bluegill, Fanning Springs, Suwannee River drainage, Florida. 595

26.23. *Lepomis marginatus*, Dollar Sunfish. 597

26.24. *Lepomis megalotis*, Longear Sunfish. 599

26.25. Male *Lepomis megalotis* from the White River in breeding coloration. 601

26.26. *Lepomis microlophus*, Redear Sunfish. 603

26.27a. *Lepomis miniatus*, Redspotted Sunfish male. 605

26.27b. *Lepomis miniatus*, Redspotted Sunfish female. 605

26.28. *Lepomis symmetricus*, Bantam Sunfish. 607

26.29. *Micropterus dolomieu*, Smallmouth Bass. 609

26.30. *Micropterus dolomieu* juvenile with tricolored caudal fin. 610

26.31. *Micropterus punctulatus*, Spotted Bass. 613

26.32. *Micropterus punctulatus* juvenile with tricolored caudal fin. 614

26.33. *Micropterus salmoides*, Largemouth Bass. 616

26.34. *Micropterus salmoides* juvenile. 616

26.35. *Pomoxis annularis*, White Crappie. 619

26.36. *Pomoxis nigromaculatus*, Black Crappie. 622

26.37. Blacknose Crappie, *Pomoxis nigromaculatus*. 623

26.38. *Elassoma zonatum*, Banded Pygmy Sunfish. 625

26.39. Differences in scales on the belly and width between bases of the pelvic fins in darters. 629

26.40. Snout differences between adult *Percina macrolepida* and *P. caprodes/P. fulvitaenia*. 630

26.41. Gill membrane connections of darters. 630

26.42. Differences in snout pigmentation between *Etheostoma chlorosoma* and *E. nigrum*. 631

26.43. Darter frena. 632

26.44. Palatine teeth in roof of mouth of *Etheostoma stigmaeum*. 633

26.45. *Ammocrypta clara*, Western Sand Darter. 634

26.46. *Ammocrypta vivax*, Scaly Sand Darter. 636

26.47. *Crystallaria asprella*, Crystal Darter. 639

26.48a. *Etheostoma artesiae*, Redspot Darter male. 642

26.48b. *Etheostoma artesiae*, Redspot Darter female. 642

26.49a. *Etheostoma asprigene*, Mud Darter male. 644

26.49b. *Etheostoma asprigene*, Mud Darter female. 644

26.50a. *Etheostoma autumnale*, Autumn Darter male. 646

26.50b. *Etheostoma autumnale*, Autumn Darter female. 646

26.51a. *Etheostoma blennioides*, Greenside Darter male in breeding coloration. 648

26.51b. *Etheostoma blennioides*, Greenside Darter, nonbreeding pigmentation. 648

26.52a. *Etheostoma caeruleum*, Rainbow Darter male in breeding coloration. 651

26.52b. *Etheostoma caeruleum*, Rainbow Darter female. 651

26.53. *Etheostoma chlorosoma*, Bluntnose Darter. 654

26.54a. *Etheostoma clinton*, Beaded Darter male in breeding coloration. 655

26.54b. *Etheostoma clinton*, Beaded Darter female. 655

26.55a. *Etheostoma collettei*, Creole Darter male. 657

26.55b. *Etheostoma collettei*, Creole Darter male in breeding coloration. 657

26.55c. *Etheostoma collettei*, Creole Darter female. 657

26.56a. *Etheostoma cragini*, Arkansas Darter male. 660

26.56b. *Etheostoma cragini*, Arkansas Darter female. 660

26.57a. *Etheostoma euzonum*, Arkansas Saddled Darter male. 662

26.57b. *Etheostoma euzonum*, Arkansas Saddled Darter female. 662

26.58. *Etheostoma flabellare*, Fantail Darter. 664

26.59. *Etheostoma flabellare* eggs on underside of rock. 666

26.60. *Etheostoma fragi*, Strawberry Darter. 667

26.61. *Etheostoma fusiforme*, Swamp Darter. 669

26.62. *Etheostoma gracile*, Slough Darter. 670

26.63. *Etheostoma histrio*, Harlequin Darter. 672

26.64. *Etheostoma microperca*, Least Darter. 674

26.65. *Etheostoma mihileze*, Sunburst Darter. 676

26.66. *Etheostoma nigrum*, Johnny Darter. 678

26.67. *Etheostoma pallididorsum*, Paleback Darter. 681

26.68. *Etheostoma parvipinne*, Goldstripe Darter. 683

26.69. *Etheostoma proeliare*, Cypress Darter. 685

26.70a. *Etheostoma pulchellum*, Plains Darter male. 687

26.70b. *Etheostoma pulchellum*, Plains Darter female. 687

26.71. *Etheostoma* sp. *cf. pulchellum* 1 (Red Belly Form). 689

26.72. *Etheostoma* sp. *cf. pulchellum* 2 (Blue Belly Form). 692

26.73. *Etheostoma radiosum*, Orangebelly Darter. 694

26.74a. *Etheostoma* sp. *cf. spectabile*, Ozark Darter. 696

26.74b. Ozark Darters in Cotter Spring, Arkansas. 696

26.75a,b. *Etheostoma squamosum*, Plateau Darter. 698

26.76. *Etheostoma stigmaeum*, Speckled Darter. 701

26.77. *Etheostoma teddyroosevelt*, Highland Darter. 703

26.78. *Etheostoma uniporum*, Current Darter. 705

26.79a. *Etheostoma whipplei*, Redfin Darter male. 706

26.79b. *Etheostoma whipplei*, Redfin Darter female. 706

26.80a. *Etheostoma zonale*, Banded Darter breeding male. 709

26.80b. *Etheostoma zonale*, Banded Darter female. 709

26.81. *Nothonotus juliae*, Yoke Darter. 712

26.82a. *Nothonotus moorei*, Yellowcheek Darter male. 715

26.82b. *Nothonotus moorei*, Yellowcheek Darter female. 715

26.83. *Perca flavescens*, Yellow Perch. 718

26.84. *Percina brucethompsoni*, Ouachita Darter. 721

26.85. *Percina caprodes*, Logperch. 723

26.86. *Percina copelandi*, Channel Darter. 725

26.87. *Percina evides*, Gilt Darter. 728

26.88. *Percina fulvitaenia*, Ozark Logperch. 730

26.89. *Percina macrolepida*, Bigscale Logperch. 732

26.90. *Percina maculata*, Blackside Darter. 735

26.91. *Percina nasuta*, Longnose Darter. 737

26.92. *Percina pantherina*, Leopard Darter. 739

26.93. *Percina phoxocephala*, Slenderhead Darter. 741

26.94. *Percina sciera*, Dusky Darter. 743

26.95. *Percina shumardi*, River Darter. 746

26.96. *Percina uranidea*, Stargazing Darter. 748

26.97. *Percina vigil*, Saddleback Darter. 750

26.98. *Stizostedion canadense*, Sauger. 753

26.99. *Stizostedion vitreum*, Walleye. 755

26.100. *Aplodinotus grunniens*, Freshwater Drum. 758

26.101. *Oreochromis aureus*, Blue Tilapia. 763

26.102. *Oreochromis mossambicus*, Mozambique Tilapia. 764

26.103. *Oreochromis niloticus*, Nile Tilapia. 766

A3.1. Structural features of fishes often used for identification. 773

A3.2. The major head bones of a generalized bony fish. 773

A3.3. Kinds of caudal fins in fishes. 774

A3.4. Hypothetical anal fins showing how fin rays are counted. 774

A3.5. First gill arches showing differences in number and shape of gill rakers. 774

A3.6. Gill membranes of fishes in relation to ventral body wall. 774

A3.7. Pharyngeal arches of two suckers and a minnow on a horizontal surface with teeth projecting upward. 775

A3.8. Areas and appendages of fishes. 775

A3.9. Some caudal fin shapes in fishes. 775

A3.10. Areas and parts of the head, including the canals of the cephalic lateralis system. 776

A3.11. Types of scales. 776

A3.12. Counting lateral line scales and scale rows above and below the lateral line. 776

A3.13. Predorsal and anterodorsal scales. 776

A3.14. Counts of median fin rays. 776

A3.15. *Chrosomus* anal fin. 777

A3.16. *Noturus* anal fin. 777

A3.17. Anterior views of pharyngeal arches of cyprinids showing teeth configurations. 777

A3.18. *Chrosomus erythrogaster* pharyngeal teeth. 777

A3.19. Dorsal view of pharyngeal teeth of minnow with arches rotated outward so teeth are visible. 777

A3.20. Dorsal view of pharyngeal teeth of Common Carp, *Cyprinus carpio*, with arches rotated outward so teeth are visible. 777

A3.21. Dorsal view of pharyngeal teeth of Grass Carp, *Ctenopharyngodon idella*, with arches rotated outward so teeth are visible. 778

A3.22. Stoneroller (*Campostoma*) with operculum removed to show gill arches, gill rakers, and gill filaments. 778

A3.23. Stoneroller (*Campostoma*) pharyngeal arches showing relative position of pharyngeal teeth and basioccipital pad. 778

TABLES

Table 1. Top States Ranked by Native Freshwater Fish Species Richness. 5

Table 2. Arkansas Fish Families with Number of Genera, Native Species, and Introduced Species. 6

Table 3. Valid Fish Species Described from Arkansas. 7

Table 4. Species Richness of River Drainages of Arkansas by Total Number of Species and by Number of Species per Unit Area (Number per 1,000 Square Miles). 15

Table 5. Distribution of Native and Nonindigenous Fishes of Arkansas by River System. 19

Table 6. Distribution of Native and Nonindigenous Fishes of Arkansas by Ecoregion. 27

Table 7. Natural Lakes of Arkansas. 57

Table 8. River Systems of Arkansas. 58

Table 9. Reservoirs of Arkansas. 67

Table 10. Arkansas Fishes in Trouble. 94

Table 11. Nonnative Fish Species Introduced into Arkansas. 95

Table 12. Native Fish Species Intentionally Stocked in Arkansas. 97

Table 13. Comparison of Meristic and Other Morphological Characters of the Four Species of *Campostoma* in Arkansas. Data are from Arkansas Populations Only. 207

Table 14. Comparison of Some Characters of Five Forms of *Lepomis megalotis* in Arkansas. Data are Mainly from Arkansas Populations, but Ouachita River Information Includes 16 Specimens from Louisiana. 600

FOREWORD

When I began sorting fish from my first collections on Piney Creek in north Arkansas, there was far less help for a novice than is available to students today. In August and again in December 1972, I had sampled fish communities at 15 sites on Piney Creek (Izard County) for my MS thesis at Arkansas State University. But I was sorting the collections in the Warrenton, Missouri, high school classroom where I taught, completely isolated from help by my major professor, George Harp. Keep in mind that in those days there was no internet, no email, no Facebook, no computer connections of any kind, and that the only way to reach George was by expensive long-distance phone calls, or by US mail. So, with only a limited introduction to "fishes" from one previous class in natural history of the vertebrates, I was slogging through trying to identify about 6,000 minnows, darters, sunfish, suckers, and other taxa that I had collected.

To aid with identifications I had at my disposal the key to "Fishes," by George Moore, in *Vertebrates of the United States* by W. F. Blair et al., published in 1968, including Moore's identification scheme for every species of fish known in the United States at the time. I also relied heavily on the old paperback *Checklist of the Fishes of Missouri* that Bill Pflieger published in 1968, and without which I would not have survived. Near the end of my sorting efforts I also discovered Frank Cross's 1967 *Handbook of Fishes of Kansas,* and I got a dim photocopy of the typewritten dissertation "The Distribution of the Fishes of Arkansas," by John D. Black, which he had loaned me personally. to make matters more complicated, Dr. Black's dissertation included many "manuscript names,"apparently the influence of his mentor, Carl Hubbs, and those names, which were never incorporated into the formal literature, were sometimes more confusing than helpful to a novice. And Dr. Black's dissertation contained no keys for identification. Then I somehow obtained a copy of the brand new *Key to the Fishes of Arkansas,* published in 1973 by Tom Buchanan (which I deeply wished he had published a year earlier!). But as of the early 1970s there was no single reference to turn to for information on identification,

distribution, and basic biology of fishes in Arkansas. The stage was set for such a badly needed volume!

So, that was pretty much that for aids to identification of fishes or information on their biology in the Ozark streams of north Arkansas in 1972 or 1973. Pflieger's comprehensive *The Fishes of Missouri* did not appear until 1975, and the first edition of *Fishes of Arkansas* by Henry W. Robison and Thomas M. Buchanan was still years away, to be published in 1988. When it did appear, the first edition of *Fishes of Arkansas* was a very welcome and masterful account of the biology, natural history, and of course identification and distribution, for all the fishes in Arkansas, as their taxonomy was known at the time. The first edition represented a massive amount of work by Tom and by Henry, with the former concentrating on studies in western Arkansas, and the latter crossing the entire state many times, usually at his own expense and with volunteer helpers, to make more than 2,000 collections of Arkansas fishes. Imagine for a moment the magnitude of work that went into that first edition, with both authors teaching heavy loads at their respective institutions, yet finding time to take students, friends, colleagues, and helpers of all sorts into the field under all manner of conditions to document in great detail the distribution of fishes throughout the state. That book was a tribute to their tenacity, and a welcome aid to all students, professionals, amateur naturalists, and fisherpersons in the state of Arkansas and elsewhere. It set the tone for Arkansas ichthyology for a generation, and I have used it extensively in my classes and in the museum at the University of Oklahoma, where I was curator of fishes in the years since the first edition appeared.

Now we have in hand a very welcome and much-needed second edition that builds upon and greatly enhances the first edition. The first edition was masterful at its publication date, but much has changed in classification, taxonomy, and nomenclature of fishes since 1988. Much new has appeared in the peer-reviewed literature that greatly informs knowledge about Arkansas fishes since publication of the first edition. The first edition, for example, retained the genus *Notropis* for all shiner minnows in Arkansas, in accord with the accepted taxonomy at the time. It was not until 1991 that the Fifth Edition American Fisheries Society *Checklist* formally recognized all the genera in which various "*Notropis*" were placed, including *Cyprinella*, *Lythrurus*, and *Luxilus*. Likewise, "*Hybopsis*" in 1988 included most of the "chubs" in Arkansas, some of which are now separated into other genera like *Macrhybopsis* and *Erimystax*. Such changes in taxonomy, all now incorporated into the second edition, are neither trivial nor the work of misanthropic academics bent on making "knowing fish" more challenging. Instead, such changes recognize and properly reflect the most recent phylogenetic hypotheses available to express correctly the complex evolutionary relationships among the fishes of Arkansas, and their positions in the larger worldview of evolution of fishes, writ large.

A second major advance in the second edition of *Fishes of Arkansas* is inclusion of the most recently recognized species of fishes within the state, based on discovery of entirely new kinds of fish, separation of former species into two or more species as knowledge about their relationships improved, or as species from outside have been introduced into or invaded the state of Arkansas. They have also appropriately deleted from consideration a number of species that briefly appeared in Arkansas but have not been found in the state for decades. And they report appropriately on the new arrivals, or the expansion, of numerous exotic species, the impact of which may not be known for years to come—but this helps to alert us all to their presence. All these many important advances in knowledge are incorporated into

the second edition. The second edition also goes far beyond the first in providing the reader with a comprehensive view of fishes of the world, and the ways fishes in Arkansas fit into the "big picture," as well as a detailed review of the zoogeographic relationships of Arkansas fishes to regions to the east, west, north, and south, with thoughtful discussion of how these findings relate to some earlier concepts. Finally, the introductory chapter to the new edition provides, for those in need of rapid summaries, massive tables documenting the distribution of Arkansas fishes among all major river drainages, and, separately, among recognized ecoregions. These tables, clearly the product of an incredible effort by the authors to track down all the primary literature, are alone worth the price of the book for professionals in need of rapid information, and for students seeking a baseline for ecological projects.

In summary, Henry Robison and Tom Buchanan have produced a masterful, scholarly, and much-needed second edition to their *Fishes of Arkansas* that will be a tremendous addition to the bookshelf of any professional, fisheries student, or amateur naturalist interested in Arkansas fishes. This book is the product of many decades of dedicated work by the authors, from extensive "dirty boots" work in the field to extensive research of the literature. This excellent volume will be *the* authoritative tome on the fishes of the State of Arkansas for many decades.

William J. Matthews
Professor Emeritus, Department of Biology,
University of Oklahoma

PREFACE TO THE SECOND EDITION

In the more than thirty-two years since the first edition of *Fishes of Arkansas* was published in 1988, much additional knowledge about the fishes of our state has accumulated. In addition, many technological advancements have revolutionized our daily lives and greatly improved our ability to revise this book, not the least of which is the widespread use of personal computers, cell phones, and digital photography. The reader of this second edition will find that we made extensive revisions to the first edition. We updated and greatly increased the amount of information on identification, systematics, distribution, life history, conservation status, and other aspects of the biology of Arkansas fishes, and we significantly altered the structure of the book. Topics are now organized into chapters with figures numbered consecutively by chapter, and the chapters with species accounts are now organized phylogenetically by order. All current species distribution maps and tables are now numbered consecutively across chapters throughout the book. In addition to expanded information in the species accounts, including much new information specific to Arkansas populations, new chapters and sections on biodiversity and zoogeography, classification and systematics, and how to identify a fish are provided. The glossary, list of abbreviations, and literature cited sections have been greatly expanded. A new distribution map with county lines, stream drainages, and distribution points in contrasting colors showing fish collection localities from 1988 through 2019 is given in each species account. The species distribution maps from the first edition are now in Appendix 4 in the second edition, allowing a comparison of recent with historical distribution for each species. In Appendix 4, we also provide new maps showing the pre-1988 distributions (when known) for species not described or known from Arkansas when the first edition was published. The 2019 checklist of Arkansas species along with the higher taxonomic categories pertaining to Arkansas fishes are presented in Appendix 5. New to each species account is our opinion of the state conservation status based on current information about the species in

Arkansas. We also added molecular information regarding systematic relationships to most species accounts. Our manuscript for the second edition has undergone far more extensive peer review than the manuscript for the first edition. Each chapter received input from one or more of our colleagues (see acknowledgments).

The amount of taxonomic change since 1988 has been astonishing. Since the first edition, 32 species have been added to the Arkansas checklist, while 16 species were removed because of not having been collected in Arkansas since 1988 (2 species) or technically removed because of taxonomic changes to their scientific names (14 species). Nineteen of the added species are the result of new species descriptions (7 species), elevation of formerly recognized subspecies to full species status (8 species), or undescribed species we consider valid species (4 species); 14 species (including introduced and native forms) were added by discovery in the state's waters, and we added a species account for the extinct *Moxostoma lacerum* to emphasize its historical occurrence in the Arkansas fish fauna. In addition, approximately 26% of the species included in the first edition have undergone a change in scientific and/or common name or family placement since 1988. In most taxonomic matters, we accepted the most recent American Fisheries Society designations of common and scientific names of fishes (Page et al. 2013), but 22 species (see checklist and individual species accounts) in our second edition differ in scientific and/or common name from species listed in the 2013 AFS list. We recognize 14 species as having a different scientific or common name from the AFS list, 4 undescribed species not on the AFS list for which we provide species accounts, and 4 species (*Etheostoma clinton*, *Etheostoma teddyroosevelt*, *Percina brucethompsoni*, and *Labidesthes vanhyningi*) described after the most recent AFS list was published. Two introduced species, *Ameiurus catus* (Map A4.110) and *Micropterus coosae* (Map A4.175), included in the first edition have been removed from the second edition because they have not been collected in Arkansas since at least 1988. The 14 species no longer listed among the Arkansas fish fauna because of taxonomic changes are *Campostoma anomalum, Hybopsis aestivalis, Hybopsis dissimilis, Notropis rubellus, Erimyzon oblongus, Moxostoma macrolepidotum, Noturus stigmosus, Typhlichthys subterraneus, Menidia beryllina, Lepomis punctatus, Etheostoma punctulatum, Etheostoma spectabile, Percina ouachitae,* and *Cottus hypselurus.* Changes in the abundance and distribution have also occurred for some state fish species—for example, Asian carps and their dispersal—and we know much more about our big-river fishes, a group poorly known in 1988.

At least 33 native Arkansas species currently have an unsettled and/or controversial taxonomic status. We evaluated the existing morphological and genetic evidence, often consulting with colleagues, and made our best judgement on how to assign a scientific name to each. Sometimes we took a conservative approach (16 species) and used the recognized AFS scientific name (Page et al. 2013), but in other cases we decided on a bolder approach (17 species) and used a genus name or specific epithet different from that of Page et al. (2013). The 16 species for which we used the conservative name are *Nocomis biguttatus* (instead of *N.* sp. *cf. biguttatus* for White River populations), *Notropis percobromus* (instead of *N.* sp. *cf. percobromus* for upper Ouachita River populations), *Notropis volucellus* (instead of *N.* sp. *cf. volucellus*), *Notropis wickliffi* (instead of *N.* sp. *cf. wickliffi*), *Fundulus catenatus* (instead of *F.* sp. *cf. catenatus*), *Ambloplites ariommus* (instead of *A. rupestris* for upland populations), *Lepomis megalotis* (instead of *L.* sp. *cf. megalotis* for Little River populations), *Micropterus salmoides* (used herein to include *M. s. floridanus*), *Etheostoma blennioides* (instead of *E.* sp. *cf. newmanii*), *Etheostoma caeruleum* (instead of *E.* sp. *cf. caeruleum*), *Etheostoma euzonum* (instead of *E. erizonum* for Current River populations), *Etheostoma fusiforme* (instead of *E. barratti*), *Etheostoma microperca* (instead of *E.* sp. *cf. microperca* for Ozark populations), *Percina copelandi* (instead of *P.* sp. *cf. copelandi* for Ouachita River drainage populations), *Percina evides* (instead of *P.* sp. *cf. evides*), and *Percina phoxocephala* (instead of *P.* sp. *cf. phoxocephala* for Red River drainage populations). There are 17 species for which we felt evidence was sufficient to use a scientific name different from the name designated by the AFS Committee. They are *Campostoma plumbeum* and *Campostoma pullum* (instead of *Campostoma anomalum*), *Troglichthys rosae* (instead of *Amblyopsis rosae*), *Typhlichthys eigenmanni* (instead of *Typhlichthys subterraneus*), *Typhlichthys* sp. *cf. eigenmanni*

(instead of *Typhlichthys subterraneus*), *Uranidea carolinae* (instead of *Cottus carolinae*), *Uranidea immaculata* (instead of *Cottus immaculatus*), five species of the Orangethroat Darter complex, *Etheostoma pulchellum, E.* sp. *cf. pulchellum* 1, *E.* sp. *cf. pulchellum* 2, *E.* sp. *cf. spectabile* in the White River, and *E. squamosum* (instead of *Etheostoma spectabile*), *Nothonotus juliae* (instead of *Etheostoma juliae*), *Nothonotus moorei* (instead of *Etheostoma moorei*), *Percina fulvitaenia* (instead of *Percina caprodes fulvitaenia*), *Stizostedion canadense* (instead of *Sander canadensis*), and *Stizostedion vitreum* (instead of *Sander vitreus*). Additional information about our taxonomic decisions can be found in the species accounts.

As with the first edition, the main purposes of this new edition are to serve as a reference for identifying the native and introduced fish species known to occur in Arkansas and to summarize the known biological information for those species. To acquaint the reader with the fishes found in Arkansas, we provide tested taxonomic keys, new color photographs (sometimes two or more fish photographs in a single species account), improved distribution maps, and current pertinent information on the biology of each species. While the book will be most useful to professional ichthyologists, fisheries biologists, and state and federal government managers of aquatic resources, we have attempted to appeal to a wider audience, including all biologists, undergraduate and graduate students, amateur naturalists, environmental consultants, anglers, and anyone else who might be interested in the state's fishes.

Even though we have filled many gaps in our knowledge since 1988, there is still much to be learned about Arkansas's fishes. There are undoubtedly new species to be discovered, new distributional records to be found, and new interpretations of the positions of various taxa in the systematic hierarchy, and much information is still needed about the natural history of our state's nongame species (Matthews 2015). Populations of rare and endangered species must be frequently monitored, and there will always be a need to monitor fish populations throughout the state to detect changes in fish community structure and changes in status of species with populations currently considered stable. We hope readers enjoy learning about the fishes of the "Natural State" as much as we enjoyed the fieldwork, laboratory work, and writing of this book. Our fishes are among the most important aquatic resources in Arkansas, and constant vigilance will be required to protect and maintain fish biodiversity.

ACKNOWLEDGMENTS

Many persons can be thanked for their generosity and unfailing assistance in compiling this second edition of *Fishes of Arkansas*. The contributions of these friends and colleagues inspired and stimulated us to do our best in providing this new edition. Assembling the amount of detailed information necessary for this book was almost overwhelming, and we are certain that our friends, colleagues, acquaintances, and others made the difference in achieving our goals. Without their input, this new edition would not have been possible. Below, we acknowledge persons, agencies, and institutions contributing to the production of this edition. We hope we have not inadvertently overlooked anyone who made a contribution. If we have, we extend our utmost apologies to anyone forgotten.

For partial funding of the second edition of *Fishes of Arkansas*, we are especially grateful to the Arkansas Game and Fish Commission and the United States Fish and Wildlife Service Wildlife Conservation and Restoration Programs. A grant from those agencies enabled the authors to travel to museums to examine previously collected Arkansas material, gather additional distributional data on the fishes of Arkansas, update the computer database of Arkansas fishes, and finish the manuscript for the second edition. We thank Southern Arkansas University and the University of Arkansas–Fort Smith for use of facilities and for excellent administrative cooperation with our research efforts during the past forty-five years or so. Other agencies that also provided invaluable information and assistance were the USDA Forest Service, Ouachita National Forest, Ozark National Forest, Arkansas Department of Environmental Quality, the Arkansas Soil and Water Conservation Commission, and the U.S. Army Corps of Engineers Waterways Experiment Station at Vicksburg, Mississippi.

Special thanks are owed to Dr. David A. Neely, Tennessee Aquarium, who graciously reviewed and edited the entire manuscript. Dave is a consummate, skilled, and superb ichthyologist with enormous field experience in the fishes of multiple continents, and he brought fresh and interesting ideas to the manuscript as well as a devotion to the task. The manuscript also benefited from Dave's wide knowledge of fishes and his unique background as an artist as well as ichthyologist. In addition, Dave kept the authors apprised of several major papers published after the review process had started and several current developments in ichthyology occurring during the review process, to make sure the manuscript was as up-to-date as possible upon publication. Dave also contributed all the original line drawings of fishes and fish structures. His artistic talent is showcased in more than 54 illustrations throughout this book.

A tremendous source of encouragement and information on ichthyology, taxonomy in general, taxonomic problems of fishes, and fish biology was provided by our dear friend and ichthyologist extraordinaire, Wayne Starnes, North Carolina State Museum. Wayne was a constant resource, confidant, and friend during the many years of the writing of this second edition. His wise counsel and many suggestions greatly benefitted the second edition. In addition, we thank our generous and erudite friend Robert E. Jenkins for his encyclopedic knowledge of the *Moxostoma* suckers and their behavior and ecology. Numerous emails and long phone calls clarified many issues related to sucker taxonomy and reproductive biology. Bob freely gave of his time and wide experience with redhorses and other fishes when asked. In addition, Bob graciously allowed us to use several of his unpublished ideas and pertinent observations made during his decades of studies on redhorse spawning behavior and ecology. Specifically, HWR is eternally indebted to REJ for his long years of mentoring, encouragement, and direction in the study of the Arkansas ichthyofauna in general. For his unwavering and enthusiastic support, we are extremely grateful to Jan J. Hoover, U.S. Army Corps of Engineers Waterways Experiment Station in Vicksburg, Mississippi, who time and again furnished valuable literature citations and copies of literature from his personal files, loaned important references books to HWR, and supplied numerous important photographs of various fish species, habitats, and people during production of this edition. Jan was also available for a myriad of questions involving big-river fishes, collecting methodology, and general information about fishes, including several historical references unknown to the authors.

We continue to owe a special debt of gratitude to Neil H. Douglas, professor emeritus of the University of Louisiana at Monroe, our most delightful and encouraging friend, mentor, and collector extraordinaire. Neil and his large cadre of former graduate students systematically documented many of the stream-fish faunas of Arkansas for the first time, making our job easier, and allowed us to use their thesis information and NLU museum collection data liberally. Neil's museum was a gold mine of ichthyological knowledge concerning Arkansas and the southeastern region of the United States, but the NLU fish collection has now been dispersed to other institutions (Appendix 1). We are very deeply indebted to Michael J. Blum and Fernando Alda, Tulane University, for discussions and clarification of the *Campostoma* taxonomic problems in Arkansas. Mike and Fernando were gracious in sharing unpublished data about the various forms of *Campostoma* in Arkansas, including genetic data and distributional information. Mike's advice and counsel are greatly appreciated, especially during the latter days of compiling this manuscript. A special acknowledgment is extended to Jack Killgore of the U.S. Army Corps of Engineers Waterways Experiment Station in Vicksburg, Mississippi, for making its extensive 25-year Arkansas fish collection database available for our use. More than 400 fish collections were made in Arkansas by the Fish Ecology Team, Environmental Laboratory, Engineer Research and Development Center, USACE, and persons primarily responsible for those collections, in addition to Jack Killgore, were Jan Hoover, Todd Slack, Steven G. George, Catherine Murphy, and Neil H. Douglas. Special thanks are extended to Thomas Turner (University of New Mexico), Steven Norris (California State University Channel Islands), Benjamin Keck (University of Tennessee), and Richard Broughton (University of Oklahoma) for supplying information on modern systematic methods used in ichthyology, including current molecular techniques. Discussion with these individuals concerning modern technology used in current ichthyology studies is greatly appreciated.

Reviewers are very special people who willingly volunteer or are cajoled into reviewing manuscripts. They take these duties on with all their other various job-related responsibilities and are still able to provide such a wonderful service to those of us trying to write books and scientific papers, spreading knowledge about a particular field or research question. We owe the following individuals (listed alphabetically) our deepest gratitude for reviewing the various family chapters or parts of a chapter (in parentheses): Ginny L. Adams (Cottidae, Mugilidae, Channidae), Reid Adams (Lepisosteidae), Henry L. Bart, Jr. (*Ictiobus* and *Carpiodes*), Bruce H. Bauer (*Lepomis marginatus* and *L. megalotis*), Michael J. Blum (*Campostoma*), John Bruner (genus *Stizostedion*), Robert C. Cashner (Fundulidae), Patrick A. Ceas (*Etheostoma spectabile* complex in Arkansas), Dave Coughlan (Hiodontidae, Salmonidae, Osmeridae, Mugilidae, and Sciaenidae), David Eisenhour (*Macrhybopsis hyostoma*), David A. Etnier (Cyprinidae), Mike Freeze (Asian carps), Charles J. Gagen (Atherinopsidae, Poeciliidae, Sciaenidae), G. O. Graening (*Typhlichthys* sp. *cf. eigenmanni*), John L. Harris (*Erimystax* subgenus), Sam D. Henry (Cichlidae), Jan Hoover (Acipenseridae, Polyodontidae, Lepisosteidae, and Amiidae), Robert Hrabik (*Notropis volucellus* and *Notropis wickliffi* accounts), Robert E. Jenkins (*Moxostoma* accounts), Rebecca Blanton Johansen (*Noturus exilis, Etheostoma flabellare*), Benjamin P. Keck (Classification and Systematics, Percidae), Andrew Kinziger (Cottidae), Bernard Kuhajda (*Scaphirhynchus*), Steve E. Lochmann (Moronidae, Centrarchidae), F. Douglas Martin (Fundulidae, Moronidae, Centrarchidae, Elassomatidae), Samuel Martin (*Chrosomus*), Chris T. McAllister (History of Arkansas Ichthyology), Michael J. Miller (Anguillidae), Catherine Murphy (*Scaphirhynchus*), David A. Neely (entire manuscript), Matthew L. Niemiller (Amblyopsidae), Steven M. Norris (Classification and Systematics), Bill Posey (Commercial Fishery), Jeffrey W. Quinn (Acipenseridae, Lepisosteidae, Amiidae, Anguillidae, Polyodontidae), Fred C. Rohde (Clupeidae, Esocidae, Aphredoderidae, Atherinopsidae), Harold L. Schramm, Jr. (Environmental Alterations), Todd Slack (*Percina copelandi*), Joe Stoeckel (Ictaluridae), Renn Tumlison (History of Arkansas Ichthyology), Thomas Turner (Classification and Systematics), and Jeffrey Williams (Salmonidae).

We wish to especially thank our colleagues who provided valuable locality data, information about the systematics of a particular group, electronic copies of their publications or theses (or publications of others in their possession), literature sources, line drawings, or their own collection records; identified museum specimens; loaned or donated specimens; or provided other materials. In alphabetical order, they are Ginny Adams, Reid Adams, Susan Adams, Jonathan W. Armbruster, Allison Asher, Henry L. Bart, Jr., Bruce Bauer, Andy C. Bently, Michael Blum, the late Herbert Boschung, Richard E. Broughton, John Bruner, Richard T. Bryant, Robert C. Cashner, Mollie Cashner, Don Cloutman, Matthew B. Connior, Walt Courtenay, Casey A. Cox, Rob Criswell, Betty Crump, Michael H.

Doosey, Neil H. Douglas, Anthony Echelle, David R. Edds, David J. Eisenhour, Carole Engle, William N. Eschmeyer, David A. Etnier, Brook L. Fluker, Joe Mike Fowler, Byron Freeman, Mary C. Freeman, Mike Freeze, Charles J. Gagen, Johnnie L. Gentry, Anna George, Steven G. George, Carter R. Gilbert, George L. Harp, John L. Harris, Marty Harvill, Eric Hilton, Christopher W. Hoagstrom, Jan Hoover, Robert Hrabik, Clark Hubbs, Robert E. Jenkins, Rebecca Blanton Johansen, Carol Johnston, Ray Katula, Jack Killgore, Andrew Kinziger, Bernard Kuhajda, Nicholas Lang, William G. Layher, William LeGrande, Steve Lochmann, John Lundberg, Samuel Martin, Edie Marsh-Matthews, William J. Matthews, Richard L. Mayden, Chris T. McAllister, Maurice (Scott) Mettee, Michael J. Miller, Rudolph J. Miller, Catherine Murphy, Thomas Near, David A. Neely, Joseph Nelson, Matthew L. Niemiller, Steven Norris, Jennifer Ogle, Patrick O'Neil, David Ostendorf, Lawrence M. Page, Tim Patton, the late Jimmie Pigg, Kyle R. Pillar, William L. Pflieger, Jeff Quinn, Morgan Raley, Fred Rohde, Nelson Rios, Stephen T. Ross, William Roston, Mark Sabaj, Jeremiah M. Salinger, Christopher Scharpf, J. R. Shute, the late Thomas Simon, Todd Slack, Wayne C. Starnes, Nathan Stone, Chad Thomas, Matt Thomas, Bruce A. Thompson, Gary Tucker, Renn Tumlison, Thomas Turner, William Voiers, Brian Wagner, Melvin Warren, David Werneke, James D. Williams, and Jim Wise.

The following persons provided museum records, loaned specimens, or identified specimens at our request: Henry L. Bart, Jr. and Justin Mann (Tulane University Museum of Natural History), Matt E. Roberts and Matthew D. Wagner (Mississippi Museum of Natural Science), Edie Marsh-Matthews (University of Oklahoma Sam Noble Museum of Natural History), William J. Matthews (University of Oklahoma), Adam E. Cohen and F. Douglas Martin (Texas Natural History Collections), Kevin W. Conway (Texas A&M University Biodiversity Research and Teaching Collections), Nancy McCartney (University of Arkansas Collections Facility), Lawrence M. Page and Robert H. Robins (Florida Museum of Natural History), Scott Payton and Andrew Simons (University of Minnesota Bell Museum of Natural History), Wayne C. Starnes (North Carolina Museum of Natural Sciences), Anthony A. Echelle (Oklahoma State University Collection of Vertebrates), Andrew Bentley (University of Kansas Biodiversity Institute & Museum of Natural History), and Chris Taylor (Illinois Natural History Survey).

Arkansas State University faculty and graduate students have continued to elucidate the ichthyofauna of the state.

For access to the information in their care and for their continual encouragement, we gratefully acknowledge two ASU former faculty, George L. Harp and the late John K. Beadles. Dr. Harp provided the entire ASU Museum fish catalog for our use in the Arkansas Fishes Database which forms the basis of the distributional maps used herein. Specific information from previous ASU theses was also continually provided by Dr. Harp to HWR on multiple occasions as was information concerning individual fish collections and specimens in the ASU Fish Collection. The gracious input and important assistance of Dr. Harp is deeply appreciated. We are also grateful to Brook L. Fluker, curator of fishes of the Arkansas State University Museum, for continuing the long-standing ASU tradition of providing valuable information to us from the ASU Fish Collection.

The contributions of the Arkansas Game and Fish Commission to our study of Arkansas's fishes cannot be overemphasized. Over the past thirty years since the publication of the first edition (1988), numerous employees of the AGFC have freely given their assistance in collecting fish all over the state. Many of them gave up weekends, holidays, and vacation time to increase our knowledge of Arkansas's fishes. Many of them have also gone out of their way to preserve unusual or interesting specimens or to otherwise provide valuable information.

Steve Filipek and Brian K. Wagner played an invaluable role in moving the second edition of this book forward and were instrumental in getting the grants needed to complete the initial work. In addition, Brian helped develop the Arkansas Fishes Database originally begun by HWR, assisted in the development of the computerized state maps so helpful to this project, and made valuable research contributions to our knowledge of threatened fishes in northwestern Arkansas. Ken Shirley, now retired from the Mountain Home District, was especially helpful in securing certain fish specimens for photography and study and in making us aware of important records of specimens collected by him during the past ten years. Ken personally made sure that HWR had fresh sucker specimens from Crooked Creek and the White River for much-needed photographs for the second edition. Ken's assistance and cooperation were invaluable and will always be remembered, especially his years of fieldwork with TMB. Drew Wilson and Les Claybrook deserve special recognition for their years of tireless fieldwork at our request, especially the ten years devoted to sampling the Red River and its associated floodplain habitats. For more than forty years, Drew worked with TMB (more than with any other biologist) to sample fishes in eastern and southwestern Arkansas.

Bob Limbird is legendary for his persistence and determination in the field, and his years of collecting fishes with TMB produced some of our more important data. Bob could always be counted on to preserve samples of fishes at our request when we could not be present to do so. Special acknowledgment is also due Sam Barkley and Sam Henry, without whose help we would not have been able to sample the Mississippi River and other areas of northeastern Arkansas. Jeff Quinn deserves special recognition for many contributions to our fieldwork. Jeff has collected fishes with us throughout the state (often accommodating our requests for help on short notice) and has played an important role in gaining access to desired collecting sites, in utilizing a variety of collecting techniques, and in preserving specimens for our inspection when we could not be present.

Others who have worked extensively in the field with TMB during the past thirty years were Jim Ahlert, Diana Andrews, Eric Brinkman, Tom Bly, Colton Dennis, Jeff Farwick, Ralph Fourt, Brett Hobbs, Ron Moore, Mark Oliver, Phil Penny, Carl Perrin, Jerry Smith, Brett Timmons, Don Turman, Brian Wagner, and Stuart Wooldridge. Without exception, over the years all fisheries biologists throughout the state have altered their schedules to help us with fieldwork whenever we requested it. In addition to AGFC biologists listed in the first edition of this book and above, other former and current AGFC biologists who have participated in fieldwork with us are Mike Armstrong, Mike Bivin, Darrell Bowman, Randall Bullington, Tim Burnley, Casey A. Cox, Shawn Hodges, Steve Filipek, April Layher, Frank Leone, Myron Means, Noah Moses, Paul Port, Jeremy Risley, Fred Shells, and Stan Todd. The following state fish hatchery managers also provided valuable information or access to specimens: Don Brader (Andrew H. Hulsey SFH), Tommy Laird (C. B. Craig SFH), and Jason Miller (Joe Hogan SFH). Jeff Quinn not only reviewed several chapters and helped with fieldwork, but also supplied us with numerous important publications on sturgeons, Paddlefish, and Freshwater Eels. Bill Posey furnished data on sturgeon and Paddlefish roe harvest. Other former or present Arkansas Game and Fish Commission personnel who provided information, were cooperative in granting collecting permits, or otherwise assisted in the publication of this book were Diana Andrews, Eric Brinkman, Justin Homan, Andrea Russenberger, John Stein, Andrew Yung, William E. Keith, former Chief of Fisheries; and Scott Henderson, former Chief of Fisheries. We also wish to thank Mike Freeze, Keo Fish Farms of Lonoke, Arkansas, for generously providing a series of Black Carp and Grass Carp for our photographic use. In addition, we greatly appreciate Stan Speight, Margaret Elli, Stewart Carleton, and Brad Holiman, Arkansas State Parks, for their help in finding locations for collection of fish specimens needed for study.

We also thank our many colleagues in the USDA Forest Service who ably helped collect specimens, provided data, and assisted in many ways to speed the production of this second edition. A most special thanks to Betty Crump, an outstanding USDA Forest Service aquatic biologist, who assisted HWR on too many fish collecting trips in the Ouachita National Forest to count and was an unfaltering colleague when we needed her help with a number of fish problems or specimens to photograph or otherwise study. Betty's perpetual positive outlook, contagious enthusiasm for the project, professionalism, skill in electroshocking, and helpfulness at every turn proved key to getting this second manuscript finished. Thanks also go to Forest Service biologists and personnel: Richard Standage, Rhea Whalen, Gene Leeds, Louie Leeds, Keith Whalen, Jobi Brown, Terry McKay, Duane Rambo, Robert Bastarache, Jessica Wakefield, and Brian Pounds. These folks generously assisted the collections of fishes on U. S. Forest Service lands and endeavored to make sure we were able to collect target species for color photographs for this edition. We also thank William G. Layher and April Layher for securing fish specimens for photographs. Their time and assistance will never be forgotten.

A number of people inspired HWR with their vast knowledge of Arkansas natural history and their willingness to share their experiences with various fishes, streams, rivers, mountainous areas of the Natural State, or simply accompanied HWR on a multitude of fieldtrips across the state in search of fishes include the late Don Crank, the late Dr. Neil Compton, Bruce and Lana Ewing (Mena), Dr. George L. Harp (Jonesboro), John Harris (Scott), Brad Holleman (Needmore), Joe Kremers (Clarksville), Bill and April Layher (Pine Bluff), Gene Leeds (Lamar), Louie Leeds (Clarksville), Bill Matthews and Edie Marsh-Matthews (OU–Norman, OK), the late Dan Marsh, Chris T. McAllister (Broken Bow, OK), Rudy and Helen Miller (OSU–Stillwater, OK), the late Barney Sellers, Dr. John Simpson (Hot Springs), and Dr. Renn Tumlison (Henderson State University). Chris McAllister introduced HWR to the excitement of fish parasitology as we journeyed across the entire state during the past fifteen years in search of new species and vital information regarding the fish parasites of state fishes, especially the nongame species so poorly known with regard to their parasites. All the above individuals are greatly appreciated for their time, talents, information, and the joy with which they shared their respective expertise with HWR and TMB. Our appreciation is also extended to Charles and Lynne King of Pocahontas for allowing HWR to collect fish on their property. We also thank commercial fishermen Jim Cunningham of Fulton, Arkansas and Dwight Ferguson of Black Rock, Arkansas for obtaining specimens of Blue Suckers and other species for us.

We are particularly indebted to David A. Neely and Uland Thomas for generously providing their digital images of fishes for use in the second edition. From 2010 to 2017 this photographic team of Dave and Uland traveled thousands of miles with HWR all over Arkansas in search of fresh specimens, especially breeding males in full nuptial coloration for color photographs. In addition, photographic trips into Oklahoma, New Mexico, Tennessee, and Texas were taken to assure fresh specimens were collected of fishes currently rare or no longer occurring in Arkansas. We faced drought, high water, tornados, flooded big rivers, muddy back roads, and occasional venomous snakes, but we persevered! These wonderfully talented gentlemen (Dave and Uland) graciously allowed use of their impressive fish images to enhance this second edition. They spent untold hours setting up and taking down their photo tanks in the field beside rivers and streams, often under difficult circumstances, and in inhospitable weather conditions so we would have photographs of fresh and colorful fish specimens. After hundreds of hours in the field collecting and photographing fish specimens, they then spent long difficult hours editing thousands of digital images. These indefatigable collectors labored long and hard in every aquatic habitat in the state that supported fishes in an effort to secure the best possible Arkansas fish specimens for photographs. All their collecting effort was accomplished long after the original moderate grant monies were exhausted. Their total dedication to this project, extreme professionalism, tenacity, and collecting zeal were a source of constant inspiration to HWR in making sure the best photographs were available for publication in this second edition. In addition, our dear friend and colleague Renn Tumlison kindly loaned more than 135 images of Arkansas fishes for use in this second edition that he has made over the course of his fieldwork in Arkansas during the past twenty years. Renn specifically took photos requested by HWR of various fish anatomy, fish structures, and color variations of state fishes. He made available his collection of fishes at Henderson State University to HWR for study whenever the need arose as well as participated in numerous discussions with HWR involving behavioral and ecological questions relating to Arkansas fishes. A number of other photographers kindly agreed to allow use of their fish photographs in this edition or discussed their methods of photographing fish, chief among them Garold W. Sneegas, Chad Thomas, Richard T. Bryant, Wayne C. Starnes, Jan Hoover, Dante Fenolio, Isaac Szabo, Rob Criswell, and Konrad Schmidt. In addition, William J. Matthews, Edie Marsh-Matthews, Patrick O'Neil, Scott Mettee, Jack Killgore, Steven G. George, Robert Hrabik, Matt Thomas, David Ostendorf, David Eisenhour, Fritz Rohde, John Harris, Thomas L. Taylor, Paul Schumann, William L. Pflieger, Michael J. Miller, Zachary S. Randall, Robert H. Robins, Brook L. Fluker, and Mark Harness graciously loaned digital images of fish for use in this project. Also extremely important were the Arkansas landscape digital images of streams, rivers, lakes, swamps, mountains, delta lands, and other ecological areas of the state kindly provided by William Rainey, Keith Sutton, Brent Baker, Ken Buoc, Paul Otto, George Simpkins, Jack Killgore, Jan Hoover, Dero Sanford, Craig M. Fraiser, and Dustin Lynch. The high level of skill of all these photographers is duly noted and greatly appreciated. Special thanks are extended to Joe Tomelleri for allowing us to publish for the first time his original colorful drawing of the extinct Harelip Sucker, *Moxostoma lacerum*. Last, but not least, we gratefully acknowledge the historical photographs provided by Kraig Adler, Brooks M. Burr, digitized and archived by the Florida Museum of Natural History, Edward Pister, Caleb D. McMahan, The Field Museum of Natural History, Chicago, the late Laura Hubbs, Truman State University, Kirksville, Missouri, and Jan Hoover.

Special thanks and appreciation are extended to several people at my (HWR) former institution, Southern Arkansas University, who played a key role in the eventual publication of *Fishes of Arkansas*, notably, Donna McCloy, SAU interlibrary loan specialist librarian, who steadfastly obtained my (HWR) every interlibrary loan request during the past eight years with prompt delivery of research papers needed, even after I retired from SAU. Ms. McCloy even photocopied papers in the SAU library for me when conditions dictated immediate action regarding the particular article in question. My deepest and most sincere appreciation is extended to her for her constant professionalism, dedication, tenacity, hard work in securing obscure titles, interest in my work, and promptness with which she delivered all my requested papers. Even the most trivial of requests in oftentimes obscure journals were graciously granted. She never failed in securing important titles and I deeply appreciate her and her total professionalism. I (HWR) also had the complete and unfailing support of my former chair, James Rasmussen, SAU chairperson of biological sciences, during the numerous years of arduous fieldwork necessary for data gathering. I sincerely thank Dr. Rasmussen for his encouragement, confidence, trust, and total support during the long years of preparation of this second edition. Finally, I thank Dr. David Rankin, past president of SAU, who for more than forty years has remained a dear and loyal friend, staunch supporter, and trusted colleague. His full and steadfast support of my research throughout my career at SAU is gratefully acknowledged and genuinely appreciated. In addition, I would be remiss without thanking Mrs. Christa Marsh, SAU, who aided the early production of this second edition. Christa ably assisted me in many facets of data collection, specimen processing, fieldwork, typing, interlibrary loan requests, and a multitude of additional tasks of book production in its earliest iteration. Thanks also to my SAU colleague James T. (Tim) Daniels, who always supported my scientific efforts and constantly loaned me various biology and natural history books from his personal library as well as encouraged my continual pursuit of the Arkansas ichthyofauna. I (HWR) also thank Bob Pradaxay and all the good folks at FedEx in North Little Rock for printing thousands of pages of material for me.

We also thank the many students and colleagues at our two respective institutions without whose help in the field and enthusiasm this second edition would not have been possible. I (HWR) wish to personally recognize and thank these former Southern Arkansas University students for assistance on the second edition: Patrick H. Robison, Lindsay M. (Robison) Fowler, Jan Rader, Christa Marsh, Lindsey Fowler, Josh Talley, John Harris, Ken Ball, Darryl Koym, and Nick Covington. Persons (mostly students) who aided TMB in the collection of fishes since the first edition (in addition to previously mentioned AGFC personnel) are Frederick Breuer, Mary Burgess, Tonya Chronister, Gaylene Douglas, Tom Duncan, Richard Evans, Bill Freeman, Derrick Gant, Jason Gilkey, John Gold, Derek Gordon, Sabrina Hardcastle, Chad Hargrave, Patrick Harrison, Kyler B. Hecke, Laura Hough, Laura Hudson, Joe Hyland, Cheryl Jackson, Kim Lankford, Jane Lowry, David Mayo, Cindy Moore, Bill Newman, Lana Newman, Josh Nichols, Tim Patton, Elizabeth Phillips, Brandy Ree, Jade Ryles, Jeff Shaver, A. J. Spires, Josh D. Steckwell, Dustin R. Thomas, Kirby Williams, and Emily Wilson. I also want to thank my aunt and uncle, Mary Ellen and Jim Halbert of Heber Springs, and my aunt Vivian Robertson of Hot Springs,

who on numerous occasions provided a place for me to stay while collecting fish in their areas of the state. I am especially indebted to the former chancellor of the University of Arkansas–Fort Smith, Joel R. Stubblefield, who understood the value of scientific inquiry, encouraged and supported all my endeavors, and established a novel program (Scholar-Preceptor Program) to involve undergraduate students in my work with fishes. Special gratitude is extended to our former Biological Sciences Laboratory Coordinator at the University of Arkansas–Fort Smith, Alyce J. Spires, who for thirty years made sure that I had all equipment and supplies necessary to carry out my field and laboratory activities, and who often made useful suggestions. I also thank my good friend and colleague for more than twenty-five years at UAFS, Ragupathy Kannan, who patiently listened to my ideas, complaints, and observations, and gave suggestions, support, and constructive comments. I also thank Myron Rigsby and Daiho Uhm of the UAFS Department of Mathematics, who guided me through various probability calculations pertaining to Arkansas fish characters. I (TMB) am also grateful to Don Clover, Daxton S. Dupire, Eric A. White, Tim Smith, Katie R. Puckett, and Elizabeth Rowland, Aquatic Biologists with the City of Fort Smith Wastewater Management Team, for allowing me to participate in electrofishing samples in the Lee Creek Reservoir watershed, for demonstrating fish collection techniques to my ichthyology and ecology classes, for saving fish specimens for me to examine, and for always accommodating my requests for help in sampling fishes.

We wish to thank our former major professors, George L. Harp and Rudolph J. Miller (HWR), and Kirk Strawn and Clark Hubbs (TMB) for igniting and encouraging our interest in studying fishes and for guiding our earliest ichthyological pursuits. Their influence, dedication to task, and sterling examples continue to be strongly reflected even in this second edition.

Finally but most importantly, we thank our families for putting up with all the stolen hours away from them while collecting, writing, photographing, and doing the multitude of jobs involved in the production of this second edition. To the HWR family—Catherine, Patrick, Lindsay, Ann, Joe, and the grandkids, Ryan, Katie, Callie, Reid, Hayes, Chloe, and Edie; to the TMB family—Marie, Jennifer, Sheridyn, Madelyn, and Ryan, our undying love and eternal gratitude.

ABBREVIATIONS

ADEQ	Arkansas Department of Environmental Quality (formerly ADPCE)
ADPCE	Arkansas Department of Pollution Control and Ecology
AFD	Arkansas Fishes Database
AFLP	Amplified Fragment Length Polymorphism
AFS	American Fisheries Society
AGFC	Arkansas Game and Fish Commission
ANSP	Academy of Natural Sciences at Philadelphia
ARV	Arkansas River Valley
ASU	Arkansas State University
AUM	Auburn University Museum of Natural History
BSC	Biological Species Concept
cf.	Latin *conferro*: compare
CFS	cubic feet per second
COI	cytochrome *c* oxidase subunit I
CU	Cornell University
CUMV	Cornell University Museum of Vertebrates
DDT	dichloro-diphenyl-trichloroethane
DNA	deoxyribonucleic acid
DO	dissolved oxygen
EDC	endocrine disrupting chemical
e.g.	Latin *exempli gratia*: for example
EPA	Environmental Protection Agency
ESC	evolutionary species concept
FL	fork length
GPS	Global Positioning System

GSI	Gonadosomatic Index (or Gonosomatic Index)
ha	hectare
HSU	Henderson State University
HUC	hydrologic unit code
HWR	Henry W. Robison
IBI	Index of Biotic Integrity
ICZN	International Commission on Zoological Nomenclature
INHS	Illinois Natural History Survey
IRI	Index of Relative Importance
IUCN	International Union for Conservation of Nature
JTU	Jackson turbidity unit
LMBV	Largemouth Bass Virus
KU	University of Kansas
KUI	University of Kansas Biodiversity Institute
LMR	Lower Mississippi River
MMNS	Mississippi Museum of Natural Science
MMR	Middle Mississippi River
mtDNA	mitochondrial DNA
mya	million years ago
NCSM	North Carolina State Museum
nDNA	nuclear DNA
NLU	Northeast Louisiana University (now University of Louisiana Monroe)
NOR	nucleolus organizer region
NPDES	National Pollutant Discharge Elimination System
NTU	nephelometric turbidity unit

OKMNH	Sam Noble Oklahoma Museum of Natural History		TL	total length
PCBs	polychlorinated biphenyls		TMB	Thomas M. Buchanan
PCR	polymerase chain reactions		TNHC	Texas Natural History Collection
pers. comm.	personal communication		TTW	Tennessee-Tombigbee Waterway
pers. obs.	personal observation		TU	Tulane University
ppb	parts per billion		UAFC	University of Arkansas Fish Collection
ppm	parts per million		UAFS	University of Arkansas–Fort Smith
PSC	phylogenetic species concept		UF	Florida Museum of Natural History
rkm	river kilometer		ULM	University of Louisiana Monroe
RFLP	restriction fragment length polymorphism		UMMZ	University of Michigan Museum of Zoology
SAU	Southern Arkansas University		USACE	United States Army Corps of Engineers
SD	standard deviation		USDAFS	United States Department of Agriculture Forest Service
SL	standard length		USFDA	United States Food and Drug Administration
SNP	single nucleotide polymorphisms			
sp.	species (singular)		USFWS	United States Fish and Wildlife Service
spp.	species (plural)		USNM	United States National Museum
sp. *cf.*	species *conferro*: compare		VHS	viral hemorrhagic septicemia
TCDD	2, 3, 7, 8-tetrachlorodibenzo-p-dioxin		WRNWR	White River National Wildlife Refuge
TCWC	Texas A&M University Biodiversity Research and Teaching Collection		ybp	years before present
			YOY	young-of-the-year

FISHES OF
ARKANSAS

CHAPTER 1

Arkansas Fishes

Diversity, Derivation, and Zoogeography

Arkansas is an environmentally diverse landscape with complex terrestrial and aquatic ecosystems, including forests, prairies, mountains, lowlands, rivers, lakes, springs, swamps, and subterranean environments. The "Natural State" is home to the largest or second-largest (depending on the number of undescribed species counted) native freshwater fish fauna of states west of the Mississippi River. To more fully understand the fishes of Arkansas, we first summarize some aspects of worldwide fish diversity, then the diversity, derivation, and zoogeography of the North American fish fauna, followed in each case by the diversity, derivation, and zoogeography of the Arkansas fish fauna.

Worldwide Fish Diversity

Fishes are the largest group of craniates on Earth and were the first animals with a backbone to appear in the fossil record. Living fishes comprise approximately 85 orders, 536 families, 4,655 genera, and about 32,000 species (Nelson et al. 2016). Fishes constitute slightly more than one-half the total number of approximately 60,000 recognized living vertebrate species (with the tetrapods accounting for 28,000 species) (Nelson et al. 2016). The actual number of extant fish species has been conservatively projected at close to 32,500 (Eschmeyer and Fong 2017), and it may be much greater than that (based on the rate at which new species are currently being described). Of the 536 fish families

with living species recognized by Nelson et al. (2016), the nine largest (most species rich) families, each with more than 400 species, contain approximately 33% of all species (10,560). Those nine families, in descending order of number of species are Cyprinidae, Gobiidae, Cichlidae, Characidae, Loricariidae, Baltoridae, Serranidae, Labridae, and Scorpaenidae. Interestingly, about 66.9% of the species in the largest families are freshwater fishes, whereas only about 43% of all fish species occur exclusively or almost exclusively in freshwater (Nelson 2006). Sixty-four families are monotypic, containing only one species.

The oldest fossils identifiable as "fishes" date to the Early Cambrian (about 540–530 mya). Those Early Cambrian jawless vertebrates (agnathans) lack bone but have a well-preserved soft anatomy and were first found in Yunnan, China (Janvier 1999; Shu et al. 1999, 2003). Although sharks have a fossil record from the Devonian (Ginter 2004), the ancestry of the Chondrichthyes is not supported by fossils. The oldest chondrichthyian fossil remains may be scales or dermal denticles of Late Ordovician age (about 455 mya); fossils of shark teeth go back to the Early Devonian (about 418 mya), and the oldest intact shark fossil is about 409 million years old (Early Devonian) (R. F. Miller et al. 2003; Nelson 2006). Although bony fishes are known from the Devonian, modern bony fishes (Division Teleostei) appeared in the Lower Mesozoic (Middle or Late Triassic, 245–202 mya), and representative forms of most major groups (i.e., orders or divisions) of modern fishes were present by at least the Middle Mesozoic (Jurassic), 202–145 mya (Nelson 2006; Helfman et al. 2009; Ross 2013). Two of the five extant classes of fishes are represented in Arkansas, the jawless lampreys (Petromyzontida) with five species, and the ray-finned fishes (Actinopterygii) accounting for more than 95% of the species in Arkansas. However, it is possible for a third class (Chondrichthyes) to occur in Arkansas. Bull Sharks (*Carcharhinus leucas*) are known to ascend rivers around the world for considerable distances, and a record of a Bull Shark from the Mississippi River of Illinois in 1937 (Thomerson et al. 1977) indicates that it must have traveled through that river in Arkansas to get to Illinois. There are also records of Bull Sharks from the Atchafalaya and Red rivers in Louisiana, more than 160 and 190 miles inland, respectively (Gunter 1938; Etnier and Starnes 1993).

Fishes live in almost every conceivable type of aquatic habitat on Earth (Nelson 2006). Carl L. Hubbs is noted for frequently reminding students, "Where there is water, there are fish." Fishes are found at elevations up to 5,200 meters in Tibet, and in South America's Lake Titicaca, the world's highest large lake (3,812 m in elevation). Fishes inhabit the world's deepest lake, Lake Baikal (1,620 m) (Kozhova and Izmest'eva 1998) and extend down to 7,000 m below the ocean's surface. Some freshwater fishes live in total darkness in subterranean (hypogean) environments confined to caves in North America, Tibet, China, and India, and many marine fishes live below the euphotic zone of the ocean in darkness. Some species live their entire lives in one small spring (e.g., Devil's Hole Pupfish in Nevada), but the American Eel exhibits a round-trip migration of thousands of miles to spawn in the Sargasso Sea near Bermuda in the Atlantic Ocean. Fishes are divided into freshwater forms and marine forms, with approximately 225 species being diadromous, that is, capable of living in and moving between both freshwater and marine environments. Freshwater fishes make up about 43% of all fish species, but liquid freshwater accounts for only 0.01% of all water on earth. Therefore, nearly half of all fish species live in less than 1% of the world's water supply (Helfman et al. 2009), an indication of the greater diversity of habitats found in freshwater. A typical freshwater fish has only 15 km^3 available to it. The greatest diversity of freshwater fishes occurs in the tropics, but the southeastern United States has also been an evolutionary center for some groups of fishes. Approximately 56% of fish species live in saltwater and 0.7–1% move between freshwater and the sea during their life cycles (Cohen 1970; Nelson et al. 2016).

Fishes exhibit great diversity in behavior, life history characteristics, and size. A variety of species can produce electricity, sound, light, and even venom. While most fishes are ectotherms, some sharks and scombrids (e.g., tunas and mackerels) exhibit endothermy for at least part of the body. Most fishes have external fertilization, but many species use internal fertilization, and females of some species provide nutrients to the developing embryos (there are varying degrees of viviparity). While most fishes are gonochoristic (having fixed sexual patterns), hermaphroditism is known from at least 14 families of teleosts (Moyle and Cech 2004). A few species are parthenogenetic. Fishes range in size from the smallest fish, a 7.9 mm (0.35 inch) mature cyprinid (*Paedocypris progenetica*) from southeastern Asia (Kottelat et al. 2006; Mayden and Chen 2010) to the gigantic 12 m Whale Shark (*Rhincodon typus*). Some fishes are brilliantly colored (e.g., many darters [Percidae] and many coral reef fishes), while other fishes are drab or silvery in appearance the entire year (e.g., the Gizzard Shad). Some fishes barely live 10 weeks (some African killifishes and Great Barrier Reef pygmy gobies) while others, such as some sturgeons, live nearly 150 years (Helfman et al. 2009). A few species, such as lampreys, eels, and the Brook Silverside, have only a single spawning event during the life cycle (semelparity), but most fishes normally reproduce for two or more seasons (iteroparity).

North American Fish Diversity

North America has the world's most diverse temperate freshwater fish fauna. Burr and Mayden (1992) defined North America as the entire North American continent, with the southern boundary located at 18° north latitude on the Atlantic Coast (Rio Papaloapan drainage) and 16° north latitude on the Pacific Coast (Rio Verde/Atoyac drainage) of Mexico. With this definition, the total number of native freshwater fish species in North America is 1,061 (G. R. Smith et al. 2010; Ross 2013; Ross and Matthews 2014). The North American fish fauna is primarily a fauna of flowing water, with relatively few contemporary species unique to lakes (Ross 2013). Unfortunately, approximately 40% of fishes in North America are listed as endangered, threatened, or imperiled, with an estimated 61 species considered extinct or extirpated from their natural habitats (Jelks et al. 2008).

The 1,061 North American species are contained in 201 genera, 50 families, and 24 orders (Burr and Mayden 1992; G. R. Smith et al. 2010). Families with the greatest number of species include minnows (Cyprinidae, 297 species), perches (Percidae, 186 species), suckers (Catostomidae, 71 species),

Table 1. Top States in the U.S. Ranked by Native Freshwater Fish Species Richness.

EAST OF THE MISSISSIPPI RIVER	
Tennessee	301 species (Etnier and Starnes 1993; Page and Burr 2011; Layman and Mayden 2012)
Alabama	295 species (Boschung and Mayden 2004)
Georgia	265 species (Brett Albanese and Chris Skelton, pers. comm.)
Kentucky	242 species (D. J. Eisenhour and B. M. Burr, pers. comm. 2017)
Mississippi	220 species (including freshwater, freshwater entering estuaries, and diadromous species, Ross 2001; Matthew Wagner, pers. comm. 2018)
Florida	121 freshwater species plus 53 marine species that commonly enter freshwaters (Robins et al. 2018)
WEST OF THE MISSISSIPPI RIVER	
Missouri	219 species (Robert Hrabik, pers. comm., 2019)
Arkansas	219 species
Texas	191 primarily freshwater species (F. D. Martin, pers. comm.)
Oklahoma	177 species (primarily Miller and Robison 2004)
Louisiana	166 primarily freshwater species, 48 marine species that ascend freshwater rivers, and 36 diadromous species (H. L. Bart, Jr., M. H. Doosey, and K. R. Piller, pers. comm. 2018)

livebearers (Poeciliidae, 69 species), and North American catfishes (Ictaluridae, 46 species). Those five families comprise 62% of the North American fishes, and when 10 additional families are included, 15 families comprise 90% of the North American freshwater fish fauna (Ross 2013). Fish diversity is greatest in eastern North America, especially in the southeastern United States, where more than half (662 native species) of the fish species live (Warren et al. 2000; G. R. Smith et al. 2010). The southeastern United States is so rich in species that it has been referred to as a "piscine rainforest" (Warren and Burr 1994).

If we confine the analysis of the North American fish fauna to just the United States and Canada, the number of native freshwater fish species falls to 831 (Page and Burr 2011). Another 58 species of fishes from other parts of the world have become established in North America, and 20 marine species are encountered frequently enough in freshwater to bring the total to 909 species currently known from the freshwaters of the United States and Canada (Page and Burr 2011). Of the 536 families of fishes recognized worldwide, 34 families (6%) are represented by one or more species native to freshwater lakes and streams of the United States and Canada, and another 11 families have marine species that occasionally enter our rivers (Page

and Burr 2011). Eight additional families are represented by introduced exotic fishes. Table 1 lists states with the greatest native fish diversity by number of freshwater species.

Arkansas Fish Diversity

Arkansas is inhabited by a rich and diverse fish fauna of 243 species distributed in 81 genera (68 native and 13 introduced), 28 families, and 19 orders that occupy various aquatic habitats (Table 2). Twenty-four of those species are nonnative introductions in Arkansas (Table 11, Chapter 3). Of the 24 introduced fish species, 22 species (plus 6 hybrid combinations) were intentionally introduced into Arkansas. Two species, the Rainbow Smelt (*Osmerus mordax*) and White Perch (*Morone americana*), made their way into Arkansas after introduction into states to the north. The native fish fauna of Arkansas comprises 219 fish species classified in 68 genera contained in 24 families. This figure also includes one native species, the Harelip Sucker (*Moxostoma lacerum*), that became extinct in the 1890s throughout its range. The native ichthyofauna of Arkansas represents about 20% of the freshwater fishes occurring in North America. The richness and diversity are the combined result of several factors, including a sufficiently long geological history

Table 2. Arkansas Fish Families with Number of Genera, Native Species, and Introduced Species.

Family	Genera	Native Species	Introduced Species	Total Species
Petromyzontidae	3	5	0	5
Acipenseridae	2	4	0	4
Polyodontidae	1	1	0	1
Lepisosteidae	2	4	0	4
Amiidae	1	1	0	1
Hiodontidae	1	2	0	2
Anguillidae	1	1	0	1
Clupeidae	2	4	0	4
Cyprinidae	24	65	7	72
Catostomidae	8	19	0	19 (1 Extinct)
Ictaluridae	4	19	0	19
Osmeridae	1	0	1	1
Salmonidae	3	0	5	5
Esocidae	2	3	2	5
Aphredoderidae	1	1	0	1
Amblyopsidae	2	3	0	3
Mugilidae	1	1	0	1
Atherinopsidae	2	3	0	3
Fundulidae	1	6	0	6
Poeciliidae	1	1	0	1
Channidae	1	0	1	1
Cottidae	1	2	0	2
Moronidae	1	2	2	4
Centrarchidae	5	17	2	19
Elassomatidae	1	1	0	1
Percidae	7	53	1	54
Sciaenidae	1	1	0	1
Cichlidae	1	0	3	3
Total	**81**	**219**	**24**	**243**

of favorable climates and habitats, a lack of glacial ice during Pleistocene glacial advances, a fortuitous centrally located geographic position, and a wide variety of aquatic habitats due to the varied physiography of the state.

The greatest diversity of Arkansas fishes is concentrated in five families: Cyprinidae (72), Percidae (54), Centrarchidae (19), Ictaluridae (19), and Catostomidae (19). Together these five families comprise 183 species or 76% of the state's fish fauna. In contrast, eight families (Polyodontidae, Amiidae, Anguillidae, Aphredoderidae, Mugilidae, Poeciliidae, Sciaenidae, and Elassomatidae) are represented in the Arkansas native fish fauna by only one species each.

Fish are very important economically to the people of Arkansas. According to the Arkansas Game and Fish Commission, resident and nonresident total fishing expenditures equaled $427 million in 2011. Trout fishing accounted for $129 million or 26% of the total expenditures. Of the visitor sample of 12,000 survey respondents in 2011, 7% reported participating in fishing while in Arkansas. Joanne Hinson, Director of Tourism Research and Information, Arkansas Department of Parks and Tourism, informed us that total travel and tourism expenditures by 23.2 million visitors to Arkansas in 2013 totaled $5.9 billion. Of the 9,000 respondents to the 2013 survey, 10% included fishing as an activity during their trips to Arkansas. In addition, Arkansas is the number one baitfish producing state in the nation, with annual sales of more than $20 million. Catfish farming in the United States has grown rapidly since it began in the 1960s. In 1963 Arkansas was the first state to produce farm-raised catfish on a commercial basis. Today, catfish farming is the largest aquaculture industry in the United States, producing 600 million pounds of catfish from 165,000 pond water acres in 2005. Importantly, the farm-raised catfish industry has the highest economic value of any aquaculture industry in the US, with $450 million in annual production. Commercially, Arkansas ranks third after Mississippi and Alabama in the production of Channel Catfish, with annual sales of nearly $13 million. The next highest valued aquaculture industry in the country is trout culture, with $74 million in annual production. See Chapter 5 for additional information on the commercial fishery of Arkansas. While game, commercial, and baitfishes are of great economic importance to Arkansas, most of the fishes of Arkansas are not directly used for food or recreation and have no measurable economic value. However, the ecological value of these native species is becoming increasingly apparent as we learn more about the structure and functioning of aquatic ecosystems. All native species have specific ecological roles and contribute to the diversity and stability of aquatic environments. Many native species are highly sensitive to environmental disturbances and are important indicators of environmental quality.

Arkansas fish species vary tremendously in size, appearance, reproductive behavior, feeding habits, and habitat.

Diversity is especially apparent in size. The smallest Arkansas fish is the Least Darter (*Etheostoma microperca*), which attains a maximum total length of about 1.8 inches (46 mm). The Alligator Gar is easily the largest state species with an 8-foot, 3-inch (2,515 mm), 350-pound (159 kg) individual taken in the 1930s from Arkansas. Twenty-eight of the 219 native fish species in Arkansas were originally described from specimens collected from the state's waters between 1856 (by Charles Girard) and 2014 (by Robison et al. 2014) (Table 3). Seven of those species are endemic to Arkansas: Ouachita Madtom (*Noturus lachneri*), Caddo Madtom (*N. taylori*), Beaded Darter (*Etheostoma clinton*), Strawberry Darter (*E. fragi*), Paleback Darter (*E. pallididorsum*), Yellowcheek Darter (*Nothonotus moorei*), and Ouachita Darter (*Percina brucethompsoni*).

Derivation of the North American Fish Fauna

North American freshwater fish families have a variety of origins, including archaic groups that originated in Pangaea (Fig. 1.1) (Ross and Matthews 2014).

Five areas have been proposed as the origins of the North American freshwater fish fauna. In deceasing order of importance, they are (1) the marine environment, (2) North America itself which includes Laurasian-Pangaean lineages (relict fauna), (3) Central America, (4) South America, and (5) Eurasia (Burr and Mayden 1992).

Table 3. Valid Fish Species Described from Arkansas.

Originally described as	Current name	Type locality	Author(s) and date
CYPRINIDAE			
Alburnops blennius	*Notropis blennius*	Arkansas River at Fort Smith	Girard 1856
Alburnops shumardi	*Notropis shumardi*	Arkansas River at Fort Smith	Girard 1856
Alburnus umbratilis	*Lythrurus umbratilis*	Sugar Loaf Creek*, 20–22 miles S of Fort Smith, Sebastian County	Girard 1856
Cyprinella whipplii	*Cyprinella whipplei*	Sugar Loaf Creek*, 20–22 miles S of Fort Smith, Sebastian County	Girard 1856
Dionda spadicea	*Campostoma spadiceum*	Fort Smith, Arkansas	Girard 1856
Exoglossum mirabile	*Phenacobius mirabilis*	Arkansas River at Fort Smith	Girard 1856
Notropis hubbsi	*Pteronotropis hubbsi*	Locust Bayou, Calhoun County	Bailey and Robison 1978
Notropis perpallidus	*Notropis perpallidus*	Saline River 5 miles N of Warren, Bradley County	Hubbs and Black 1940
Notropis pilsbryi	*Luxilus pilsbryi*	White River system, Rogers, Benton County	Fowler 1904
Notropis snelsoni	*Lythrurus snelsoni*	Mt. Fork River, W of Hatfield, Polk County	Robison 1985
ICTALURIDAE			
Noturus lachneri	*Noturus lachneri*	Middle Fork of Saline River, 11.2 miles N of Mountain Valley, Garland County	Taylor 1969
Noturus maydeni	*Noturus maydeni*	Strawberry River, 4 miles N of Evening Shade, Sharp County	Egge in Egge and Simons 2006
Noturus nocturnus	*Noturus nocturnus*	Saline River, Benton, Saline County	Jordan and Gilbert 1886
Noturus taylori	*Noturus taylori*	South Fork of Caddo River, 1.6 km SE of Hopper, Montgomery County	Douglas 1972
CENTRARCHIDAE			
Ambloplites constellatus	*Ambloplites constellatus*	Buffalo River, near Rush, Marion County	Cashner and Suttkus 1977
Bryttus humilis	*Lepomis humilis*	Sugar Loaf Creek*, Sebastian County	Girard 1858

Table 3. (*continued*)

Originally described as	Current name	Type locality	Author(s) and date
ELASSOMATIDAE			
Elassoma zonata	*Elassoma zonatum*	Little Red River, Judsonia, White County	Jordan 1877
PERCIDAE			
Etheostoma clinton	*Etheostoma clinton*	Caddo River at St. Hwy. 182, 3.2 km N of Amity, Clark County	Mayden and Layman in Layman and Mayden 2012
Etheostoma (Ulocentra) histrio	*Etheostoma histrio*	Poteau River near Hackett, Saline River at Benton, Ouachita R. at Arkadelphia	Jordan and Gilbert in Gilbert 1887
Etheostoma moorei	*Nothonotus moorei*	Devils Fork, Little Red River, near Drasco, Cleburne County	Raney and Suttkus 1964
Etheostoma pallididorsum	*Etheostoma pallididorsum*	Caddo River, near Black Springs, Montgomery County	Distler and Metcalf 1962
Etheostoma spectabile fragi	*Etheostoma fragi*	Spring Creek (Strawberry River dr.), Sharp County	Distler 1968
Poecilichthys euzonus	*Etheostoma euzonum*	Buffalo River, St. Joe, Searcy County	Hubbs and Black 1940
Poecilichthys whipplii radiosus	*Etheostoma radiosum*	Sugar Loaf Creek, trib. to Caddo River, Hot Spring County	Hubbs and Black 1941
Hadropterus shumardi	*Percina shumardi*	Arkansas River near Fort Smith	Girard 1859
Etheostoma (Cottogaster) uranidea	*Percina uranidea*	Ouachita River, Arkadelphia, Clark County	Jordan and Gilbert in Gilbert 1887
Hadropterus nasutus	*Percina nasuta*	Middle Fork of Little Red River, near Leslie, Searcy County	Bailey 1941
Percina brucethompsoni	*Percina brucethompsoni*	Ouachita River at St. Hwy. 88 at Pine Ridge, Montgomery County	Robison, Cashner, and Near 2014

* The type locality for some species described by Girard was "Sugar Loaf Creek, Arkansas." There has been considerable confusion about the location of Sugar Loaf Creek. Black (1940) could not find this creek on any map and after interviewing local residents tentatively concluded that it was a tributary of the James Fork of the Poteau River which rises on the north side of Sugar Loaf Mountain near Hartford in Sebastian County, Arkansas. Black believed it was possible to secure the variety of fishes reported by Girard at a locality on this creek four miles northeast of Hartford. However, recent maps show Sugar Loaf Creek as a direct tributary of the Poteau River arising to the northwest of Hartford and flowing into Oklahoma to join the Poteau River.

North American Fishes with a Marine Origin

Burr and Mayden (1992) suggested that half of the 50 North American fish families had a marine origin. Families of marine derivation that have a long independent freshwater existence include Cottidae (sculpins), Sciaenidae (drums and croakers), Gadidae (codfish), and Atherinopsidae (silversides). Cyprinodontidae (pupfishes), Fundulidae (topminnows), and Petromyzontidae (lampreys) are largely freshwater in distribution and possibly origin, but also have marine or brackish water species. In addition, Salmonidae (trouts) and Moronidae (temperate basses) have several species that are anadromous, and Anguillidae (freshwater eels) is a family with catadromous species.

North American Fishes with a North American Origin

The archaic or relict bony fishes of North America make up one of the most interesting and characteristic elements of the North American freshwater fish fauna (Gilbert 1976). Four families of archaic fishes inhabit North America, including Amiidae (bowfins), Lepisosteidae (gars), Polyodontidae (paddlefishes), and Acipenseridae (sturgeons). All have fossil histories dating back 75 million years to the Cretaceous or earlier.

Seven fish families are endemic to North America: Ictaluridae (North American freshwater catfishes, with about 40 species); Centrarchidae (sunfishes and black

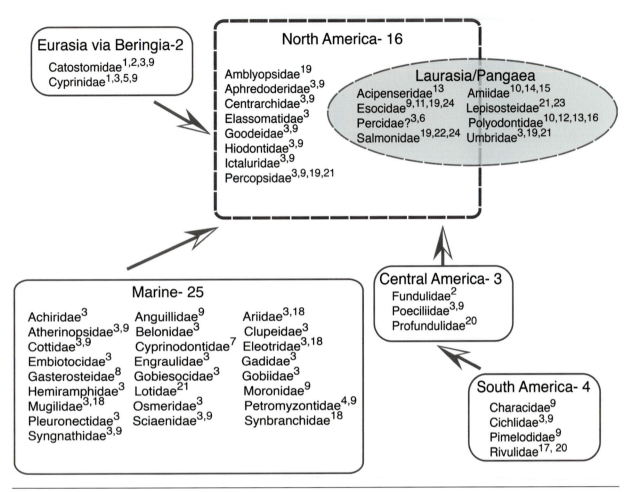

Figure 1.1. General origins of the major families of North American freshwater fishes. Families listed as North American include those of Laurasian-Pangaean origin because of the general uncertainty in determining exact locations. References: (1) Berra (2001), (2) Briggs (1986), (3) Burr and Mayden (1992), (4) Cavender (1986), (5) Cavender (1991), (6) Collette and Bănărescu (1977), (7) Echelle and Echcelle (1992), (8) Foster et al. (2003), (9) Gilbert (1976), (10) Grande (1984), (11) Grande (1999), (12) Grande and Bemis (1991), (13) Grande and Bemis (1996), (14) Grande and Bemis (1998), (15) Grande and Bemis (1999), (16) Grande et al. (2002), (17) Hrbek and Larson (1999), (18) Miller and Smith (1986), (19) Moyle and Cech (2004), (20) Parenti (1981), (21) Patterson (1981), (22) Smith and Stearly (1989), (23) Wiley (1976), (24) Wilson and Williams (1992). Photo by Stephen T. Ross from *Freshwater Fishes of North America, Volume 1.*

basses, 33 species), Goodeidae (goodeids, 35–40 species), Amblyopsidae (cavefishes, 9 species), Hiodontidae (mooneyes, 2 species), Percopsidae (trout-perches, 2 species), and Aphredoderidae (pirate perches, 1 species). Five of the seven families (all except Goodeidae and Percopsidae) are represented today in Arkansas.

North American Fishes with a Eurasian Origin

Families originating in Europe or Asia include Cyprinidae (minnows and carps, with about 297 North American species), Percidae (true perches, about 186 species), Catostomidae (suckers, about 71 species), and Esocidae

(pikes and mudminnows, 8 North American species). Each of these families is believed to have migrated to North America across the Bering Land Bridge, which has united the Asian and North American continents at various times in the past (Gilbert 1976).

North American Fishes with a Central American Origin

Poeciliidae (livebearers) clearly have a Central American origin (Gilbert 1976). Derived from the Cyprinodontidae (pupfishes), poeciliids are salt-tolerant and can inhabit salt or brackish water as well as freshwater. North American members of the family Fundulidae (topminnows) apparently are derived from Central American ancestors (Briggs 1986). The Cyprinodontidae also originated in Central America.

North American Fishes with a South American Origin

Fish families of South American origin inhabiting North America include Cichlidae (cichlids), Characidae (tetras), and Pimelodidae (Neotropical catfishes); however, none has penetrated far into North America (Gilbert 1976). There are no species native to Arkansas in those families. Numerous cichlids have been introduced into areas of the United States, including Arkansas (see Cichlidae family and species accounts).

Ages of North American Fish Families

The diverse fish fauna of North America was derived over a span of hundreds of millions of years, from as early as the Late Paleozoic through the Pleistocene, and continuing into the present as populations respond to changing environmental conditions (Ross 2013). The ages of the individual North American freshwater families vary. According to Ross and Matthews (2014), data for estimating faunal ages are available for 27 families and subfamilies (from phylogenies calibrated by geological events, fossils, or molecular data). The oldest known extant family is Petromyzontidae (lampreys) which dates from the Paleozoic. Ancient groups like the Acipenseridae (sturgeons), Polyodontidae (paddlefishes), Amiidae (bowfins), Lepisosteidae (gars), and Esocidae (pikes and mudminnows) date back to the Cretaceous of the Mesozoic. All other North American fish families date from the Cenozoic. The Hiodontidae, Clupeidae, Ictaluridae, Catostomidae, Cyprinidae, Percopsidae, Salmonidae, Moronidae, Centrarchidae, and Aphredoderidae all have occupied North America since the Paleogene of the Early Tertiary. In the Neogene of the Late Tertiary, the Goodeidae, Poeciliidae, Percidae, Cichlidae, Fundulidae, Cyprinodontidae, Atherinopsidae, Cottidae, Gasterosteidae, and Sciaenidae appeared. The oldest fossils of Percidae in North America date from the Pleistocene, but Carlson et al. (2009), using a fossil-calibrated molecular phylogeny of darters, provided evidence that the separation of darters from other percids occurred much earlier (19.8 mya in the Paleogene). Near and Keck (2005) estimated that within the darter genus *Nothonotus*, the most recent common ancestor dates to 18.5 mya. Ross and Matthews (2014) speculated that percids likely occurred in North

America at least by the Early Miocene (about 23 mya). By the Early Miocene (23–16 mya), 78% of the 27 major families were present in North America.

Derivation of the Arkansas Fish Fauna

Our knowledge of the developmental history of the Arkansas fish fauna is based on the present-day distribution patterns of the state fishes and what is known of environmental conditions and drainage relationships in eastern North America in the geologic past. Unfortunately, we have little in the way of fish fossil evidence to aid our understanding of past events in Arkansas. The derivation of Arkansas fish families was included in the previous section on origin of North American families; therefore, this section emphasizes the derivation of lower taxa, primarily species. As Pflieger (1971) noted, a basic tenet of this approach is the assumption that the habitat preferences and environmental tolerances of present-day fish species do not differ significantly from those of their immediate ancestral stocks. In support of this assumption, R. R. Miller (1965) observed that most if not all fishes known from Pleistocene deposits are morphologically indistinguishable from existing species. If we assume that their ecological requirements have likewise undergone little change during the same period, then we can have greater confidence in inferences concerning the fish faunas of past geologic ages based on what is known about environmental conditions that existed at that time (Pflieger 1971). In addition, we must assume that the theory of geographic (allopatric) speciation (Mayr 1942) is a valid mechanism for forming new species.

The present-day fish fauna of Arkansas is a mixture of species derived from several geographic regions. Making up the Arkansas fish fauna is (1) a northern component, (2) a western, Great Plains component, (3) an eastern North American component, (4) a small component from Central/South America, (5) an archaic relict indigenous component, and (6) an endemic Arkansas component of fishes.

Northern Component

Pflieger (1971) investigated the zoogeography of Missouri fishes and listed species with affinities to certain areas. Since many Missouri species also occur in Arkansas, we can list the affinities of many Arkansas fish species based on Pflieger's work. Arkansas fishes with northern affinities include Silver Lamprey (*Ichthyomyzon unicuspis*), Least Brook Lamprey (*Lampetra aepyptera*), American Brook Lamprey (*Lethenteron appendix*), Lake Sturgeon (*Acipenser fulvescens*), Goldeye (*Hiodon alosoides*), Southern

Redbelly Dace (*Chrosomus erythrogaster*), Spotfin Shiner (*Cyprinella spiloptera*), Hornyhead Chub (*Nocomis biguttatus*), Largescale Stoneroller (*Campostoma oligolepis*), Creek Chub (*Semotilus atromaculatus*), Bigeye Chub (*Hybopsis amblops*), Striped Shiner (*Luxilus chrysocephalus*), Wedgespot Shiner (*Notropis greenei*), Ozark Minnow (*N. nubilus*), Carmine Shiner (*N. percobromus*), Mimic Shiner (*N. volucellus*), Channel Shiner (*N. wickliffi*), White Sucker (*Catostomus commersonii*), Northern Hog Sucker (*Hypentelium nigricans*), Silver Redhorse (*Moxostoma anisurum*), Black Redhorse (*M. duquesnei*), Harelip Sucker (*M. lacerum*), Pealip Redhorse (*M. pisolabrum*), Slender Madtom (*Noturus exilis*), Stonecat (*N. flavus*), Piebald Madtom (*N. gladiator*), Brindled Madtom (*N. miurus*), Redfin Pickerel (*Esox americanus*), Smallmouth Bass (*Micropterus dolomieu*), Rainbow Darter (*Etheostoma caeruleum*), Fantail Darter (*E. flabellare*), Least Darter (*E. microperca*), Johnny Darter (*E. nigrum*), Logperch (*Percina caprodes*), Ozark Logperch (*P. fulvitaenia*), Channel Darter (*P. copelandi*), Gilt Darter (*P. evides*), Slenderhead Darter (*P. phoxocephala*), Walleye (*Stizostedion vitreum*), Sauger (*S. canadense*), Banded Sculpin (*Uranidea carolinae*), and Knobfin Sculpin (*U. immaculata*).

Great Plains Component

Arkansas fishes having affinities with western areas or the Great Plains include the Pallid Sturgeon (*Scaphirhynchus albus*), Plains Stoneroller (*Campostoma plumbeum*), Red Shiner (*Cyprinella lutrensis*), western subspecies of Blacktail Shiner (*C. venusta venusta*), Plains Minnow (*Hybognathus placitus*), Sturgeon Chub (*Macrhybopsis gelida*), Shoal Chub (*M. hyostoma*), Sicklefin Chub (*M. meeki*), Ribbon Shiner (*Lythrurus fumeus*), Red River Shiner (*Notropis bairdi*), Arkansas River Shiner (*N. girardi*), Silverband Shiner (*N. shumardi*), Sand Shiner (*N. stramineus*), Suckermouth Minnow (*Phenacobius mirabilis*), Flathead Chub (*Platygobio gracilis*), Quillback (*Carpiodes cyprinus*), Orangespotted Sunfish (*Lepomis humilis*), Mississippi Silverside (*Menidia audens*), Arkansas Darter (*Etheostoma cragini*), Orangbelly Darter (*E. radiosum*), Plains Darter (*E. pulchellum*), and Bigscale Logperch (*Percina macrolepida*).

Eastern North American Component

A third group of state fishes has affinities with the Eastern Appalachian Mountains and includes the Whitetail Shiner (*Cyprinella galactura*), Ozark Shiner (*Notropis ozarcanus*), Telescope Shiner (*N. telescopus*), Northern

Studfish (*Fundulus catenatus*), Ozark cavefish (*Troglichthys rosae*), Salem Plateau Cavefish (*Typhlichthys eigenmanni*), Ghost Cavefish (*T.* sp. *cf. eigenmanni*), Greenside Darter (*Etheostoma blennioides*), Arkansas Saddled Darter (*E. euzonum*), Banded Darter (*E. zonale*), Yoke Darter (*Nothonotus juliae*), Yellowcheek Darter (*N. moorei*), and Longnose Darter (*Percina nasuta*).

Central and South American Component

This small group of fishes with connections to Central and South America include the Western Mosquitofish (*Gambusia affinis*) and the topminnows (6 species of *Fundulus*). North American fundulids apparently are derived from Central American ancestors (Briggs 1986).

Archaic, Relict Component

An archaic, relict fish faunal component derived from ancestors existing since the Cretaceous includes Paddlefish (*Polyodon spathula*), Bowfin (*Amia calva*), and Alligator Gar (*Atractosteus spatula*).

Endemic Component

An Ozark Mountains endemic group of fishes includes the Cardinal Shiner (*Luxilus cardinalis*), Duskystripe Shiner (*L. pilsbryi*), Bleeding Shiner (*L. zonatus*), Ozark Madtom (*Noturus albater*), Checkered Madtom (*N. flavater*), Black River Madtom (*N. maydeni*), Ozark Bass (*Ambloplites constellatus*), Plateau Darter (*Etheostoma squamosum*), and Ozark Darter (*Etheostoma* sp. *cf. spectabile*). The Ouachita Mountain endemic fishes are the Ouachita Mountain Shiner (*Lythrurus snelsoni*), Kiamichi Shiner (*Notropis ortenburgeri*), Peppered Shiner (*N. perpallidus*), Rocky Shiner (*N. suttkusi*), Ouachita Madtom (*Noturus lachneri*), Caddo Madtom (*N. taylori*), Beaded Darter (*Etheostoma clinton*), Paleback Darter (*E. pallididorsum*), Ouachita Darter (*Percina brucethompsoni*), and Leopard Darter (*P. pantherina*).

Zoogeographic Concepts and North American Fish Zoogeography

Moyle and Cech (2004) commented that the study of fish zoogeography is both fascinating and frustrating. It is frustrating because the explanation of fish distribution requires putting together knowledge from many areas of ichthyology (i.e., ecology, physiology, systematics, and paleontology), and because so much of our knowledge in those areas is fragmentary or incomplete. Therefore, any attempt to explain fish distribution patterns is bound to contain gaps that must be bridged with educated guesses until better information becomes available (Moyle and Cech 2004). The use of new information and tools has revolutionized the study of fish zoogeography, in part because ichthyologists have been among the leaders in developing these new approaches (Mooi and Gill 2002). There are two major paradigms of modern zoogeography to explain distributional patterns: (1) centers of origin/dispersal explanations, and (2) vicariance explanations.

Center of Origin/Dispersal

Organisms have a natural tendency to disperse within and between areas of suitable habitat. We see numerous examples of this tendency in fishes as well as in other vertebrates and invertebrates. In the dispersal model, organisms are assumed to have migrated across preexisting barriers and one taxon (or more) founded a new population to achieve a disjunct distribution (Jenkins and Burkhead 1994). This model was supported by early zoogeographers such as Alfred Russel Wallace (1876) and Phillip Darlington (1957). Among ichthyologists, John Briggs (1974, 1995) was a major proponent of the importance of dispersal as a primary mechanism of zoogeography (Ross 2013).

Vicariance

Vicariance is an alternative model that explains the existence of closely related taxa or biota in different geographical regions as the result of separation by the formation of natural barriers to dispersal. The vicariance model basically holds that organisms have been passively transported by movement of tectonic plates or have been separated by other geological processes (Ross 2013). Ross (2013) points out that if this has occurred, then several or more taxa should share common distribution patterns, where the distribution of each taxon may be referred to as a "track." The investigator then searches for common patterns of distribution or generalized tracks among various taxa. The vicariance model emphasizes patterns generated by many (but not necessarily closely related) taxa. Croizat (1958; Croizat et al. 1974) was an early vicariance biogeographer. Modern ichthyologists who have strongly supported vicariance biogeography include Gareth Nelson, Norman Platnick, Edward Wiley, and Donn Rosen.

In the dispersal model, the barrier is older than at least one of the isolated populations, and the age of the barrier

is older than the disjunction in range (because the organism crossed the barrier to achieve its current disjunct distribution). In the vicariance model, the populations predate the age of the barrier, and ages of the barrier and the disjunction in the range are the same (Ross 2013). While the debate continues, Ross (2013) believes that a synthesis of views is required to understand the distributional patterns of fishes.

Biogeographic regions are geographical areas that correspond to boundaries of resident biotic components (Matamoros et al. 2016). For the study of worldwide patterns of fish zoogeography, it is informative to divide the world into the six zoogeographic regions of Darlington (1957): (1) the African Region, which consists of the African continent south of the Sahara Desert; (2) the Neotropical Region, consisting of South and Central America; (3) the Oriental Region, which consists of the Indian subcontinent, Southeast Asia, most of Indonesia, and the Philippines; (4) the Palearctic Region, which includes Europe and Asia south to the Himalayas; (5) the Nearctic Region, which is North America south to central Mexico; and (6) the Australian Region, comprising Australia, New Zealand, New Guinea, and the smaller islands of the same region. The fish fauna of each region has distinctive elements that reflect its isolation from (as well as its connections to) other regions (Moyle and Cech 2004). Matamoros et al. (2016) noted that the classifications of biogeographic regions are hierarchical and include the categories of realm, region, dominion, province, and district. They further noted that the level of province is the focus of many biogeographical studies because it represents a geographical region with a relatively homogenous faunal composition and is believed to reveal means of faunal assembly and evolution.

Our focus in this section is the Nearctic or North American region, which consists of North America south to the southern edge of the Mexican Plateau. The Nearctic can be further divided into three broad ichthyological subregions: (1) the Arctic-Atlantic subregion, consisting of all drainages into the Arctic and Atlantic oceans as well as the Gulf of Mexico south to the Rio Panuco; (2) the Pacific subregion comprising all drainages along the Pacific Coast south to the Mexican border, and all interior drainages west of the Rocky Mountains; and (3) the Mexican subregion consisting of most of Mexico (except the drainages flowing into the upper Gulf of Mexico and related interior drainages) south to the northern boundary of Central America. Abell et al. (2000) called the three subdivisions of the Nearctic region ecoregions and broke them down further into 76 subregions.

The zoogeography of North American fishes is the

best documented of any continent because of two important works (Lee et al. 1980; Hocutt and Wiley 1986) and the fascination of North American ichthyologists with the subject (Moyle and Cech 2004). Approximately 1,061 fish species belonging to more than 200 genera and 56 families are native to the three Nearctic subregions (Burr and Mayden 1992; Lundberg et al. 2000). If the Mexican subregion is excluded because of its transitional nature, then approximately 916 fish species in 30 families are in the two other subregions combined (Moyle and Cech 2004). The families Cyprinidae (34%), Percidae (18%), Catostomidae (8%), Poeciliidae (8%), Goodeidae (5%), Fundulidae (4%), Ictaluridae (5%), and Centrarchidae (3%) contain most of the species. Diadromous families such as the Salmonidae, Osmeridae, and Petromyzontidae make up another 6% of the fishes, while the remainder of the fish fauna comprises freshwater representatives of marine families like Cottidae (3%) and euryhaline marine species from another 15 families.

The Arctic-Atlantic ichthyological subregion (of the Nearctic region) is subdivided into 17 ichthyological provinces (Moyle and Cech 2004). The Mississippi province (which includes Arkansas) consists of all the United States and Canada drained by the Mississippi-Missouri river systems as well as many western Gulf Coast drainages. The Mississippi province is by far the richest area of North America, with 375 species (40% endemic) in 31 families. Freshwater dispersants make up most of the species, predominately Cyprinidae (32%), Percidae (30%), Centrarchidae (5%), Castostomidae (6%), and Ictaluridae (6%). Diadromous fishes and freshwater representatives of marine families are poorly represented in this province. With the high percentage (40%) of endemic freshwater dispersants, it is apparent that the Mississippi-Missouri drainage system is an ancient one (Moyle and Cech 2004).

Within the Mississippi province, the Mississippi River was a center of fish evolution (Robison 1986), and the lower Mississippi River was an important refuge during Pleistocene glacial times. During times of glacial retreat, species reoccupied areas once covered by continental glaciers. Its role in the development of the North American fish fauna is indicated by the families Centrarchidae, Percidae (Etheostomatinae), Ictaluridae, Cyprinidae, and Esocidae, which are most abundant and diverse in the mainstem Mississippi River and its tributaries (Moyle and Cech 2004). Its role as a refuge has not been confined just to the Pleistocene, because it remains a refuge today for many relict Chondrostei and Neopterygii, including teleosts. The chondrosteans are the sturgeons and paddlefish, and ancient neopterygians of the Mississippi River drainage include six species of gars (Lepisosteidae) and Amiidae (Bowfin). Including one gar species in Central America, gars and Bowfin are the only extant representatives of the fishes that were abundant in freshwater during much of the Mesozoic. Relict teleosts inhabiting the Mississippi include the Mooneye and Goldeye (Hiodontidae), *Cycleptus elongatus* (Blue Sucker), the Elassomatidae (pygmy sunfishes), Percopsidae (trout-perches), Amblyopsidae (cavefishes), and Aphredoderidae (Pirate Perch) which are confined to North America.

Arkansas Fish Zoogeography

Introduction

When we think of the distribution of an organism, we usually think of where the organism occurs geographically. However, an organism has three types of distribution: (1) a geographical distribution, (2) an ecological distribution, and (3) a temporal distribution (evolutionary history). What is known about the temporal distribution, or distribution through geological time of Arkansas fishes based on the fossil record is here generally provided in the family accounts, but unfortunately, adequate information about most fish species through geological time is lacking. Therefore, this section focuses mainly on the geographic and, to a lesser extent, on ecological distributional patterns of Arkansas fishes.

Early distributional studies of the fishes of Arkansas were conducted in the late 1800s and early 1900s, but John D. Black was the first ichthyologist to study the distribution of Arkansas fishes in detail. In his dissertation, "Distribution of the Fishes of Arkansas," Black (1940) used 339 collections from 268 locations in Arkansas made during 1938–1939 and all earlier material known to him to initially describe fish faunal regions of Arkansas. Black emphasized differences among major river systems but provided no quantitative data (Matthews and Robison 1988). Later, Mayden (1985a) reviewed the distribution and possible origins of Ouachita Mountains upland fishes in Arkansas and Oklahoma and compared Ozark and Ouachita distributions (presence-absence) of some species. Cross et al. (1986) and Robison (1986) provided general overviews comparing fishes by river system in Arkansas, but did so in a very broad geographical context. Schwemm (2013) used mtDNA and microsatellite variation to assess phylogeographic history of small-bodied fishes of the Ouachita Highlands. Matthews and Marsh-Matthews (2015) compared fish distributional patterns in Oklahoma and western Arkansas before and after the period of dam building (see Chapter 3 for conclusions).

Black (1940) concluded that the Arkansas fish fauna has its greatest affinities with fish faunas of the Great Plains and eastern North America, especially the Ohio Valley. Later, in a series of papers, Mayden (1985a, 1987a; Mayden et al. 1992b) compared the distributional relationships of the upland fishes of Arkansas to the Appalachian Highlands to the east, defining the Central Highlands biogeographic track, linking Ozark, Ouachita, and the Appalachian Highlands as an evolutionary unit. A comprehensive treatment of zoogeography of Arkansas fishes was published by Matthews (1998) in *Patterns in Freshwater Fish Ecology.* Matthews (1998) originally examined 168 native Arkansas species (the actual number was 178; William Matthews, pers. comm.) but excluded species for which he could not determine affinity (i.e., species for which Arkansas was almost in the center of their range). Matthews also excluded Arkansas endemics from his analysis. For the fish fauna of Arkansas, Matthews (1998) used eight 45° compass or clock directions (indicating direction of species affinity) to plot native fish species distributional affinities (Fig. 1.2).

We reexamined Matthews's data using 178 native fish species. In our analysis, 46 species (25.8%) had distributional affinity to the north, including the upper Mississippi–Missouri River basin, 55 species (30.9%) were allied with the northeast (eastern Mississippi River–Ohio River basin tributaries), 22 species (12.4%) had range or distributional affinities to the southeast, including the rich Gulf Coast fauna (Swift et al. 1986), and 19 (10.7%) had affinities to the east. Only 12 species (6.7%) had affinities to the northwest (i.e., Kansas, central Great Plains), and relatively few had affinities to the west (9 species), southwest (6 species), or south (9 species) (Oklahoma, Texas, and Louisiana drainages, respectively). The fish faunas of the latter four regions are less species rich (Conner and Suttkus 1986; Cross et al. 1986; Robison 1986) than those of the upper Mississippi and Ohio river basins (Robison 1986; Burr and Page 1986). The most surprising aspect of our analysis and that of Matthews (1998) was the small number of fish species with affinity to the east (19 species—10.7%) given Black's (1940) assessment of a strong relationship of the upland Ozark fish fauna (north Arkansas) to the upland fish fauna of the Appalachian Mountains east of the Mississippi River. It is apparent that some upland-restricted fish species such as *Cyprinella galactura, Notropis telescopus,* and *Fundulus catenatus* connect the Ozark Uplands and the Appalachians, but Matthews (1998) pointed out that these fishes do not make up a preponderance of the fishes of Arkansas. Matthews (1998) also stated that if the goal is to understand the overall fish fauna of Arkansas, it may be more practical, for ecological purposes, to consider the directional

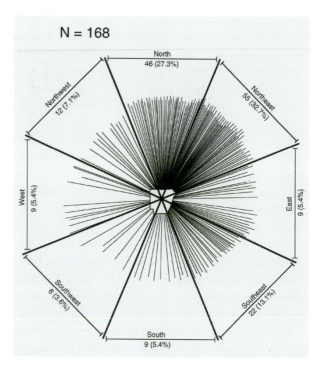

Figure 1.2. Directional zoogeographic affinities of 178 native freshwater fishes in Arkansas. *Courtesy of W. J. Matthews.*

affinities of all species, rather than to focus on a few species (or sister-species pairs [Mayden 1985a, 1987a; Mayden et al. 1992b]) that connect two regions with zoogeographic tracks of evolutionary affinity. The best approach may be to take both approaches, look for evolutionary tracks *sensu* Mayden (1987a, b; Mayden et al. 1992b), and determine the preponderance of distributional affinity for all the species.

The following analyses of Arkansas fish distribution by river system and ecoregion are based on more than 5,000 collections of fishes by the authors and colleagues who have worked with the ichthyofauna of Arkansas for many years and have made their collection data available to us.

Distribution of Arkansas Fishes by River System

Although all river drainages in Arkansas are part of the large Mississippi River basin, the state can be subdivided into six main river systems for a better explanation and understanding of fish distributional patterns: White River, Arkansas River, Ouachita River, Red River, St. Francis River, and Mississippi River.

Table 4 shows the historical species richness of fishes in the six river drainages in Arkansas by total number of species and by the number of species per unit area (number/1,000 sq. miles). For the most part, the number of

Table 4. Historical Species Richness of River Drainages of Arkansas by Total Number of Native Species and by Number of Native Species Per Unit Area (number per 1,000 square miles).

River Drainage	Area (mi²)	Total Native Species	No. of Native Species per 1,000 mi²
Arkansas	11,100	152	13.7
Mississippi	4,755	108	22.7
Ouachita	11,548	136	11.8
Red	4,433	131	29.6
St. Francis	4,200	115	27.4
White	17,143	176	9.7

native species in a river is directly related to the size of the area drained; however, the number of species per unit area drained does not follow a similar pattern. The White River has the highest number of native fish species documented within Arkansas at 176 species, but it has the lowest number of species per unit area drained (10.3). In contrast, the Mississippi River drainage (as defined herein) has the lowest species richness (108) in Arkansas but 22.7 native fish species per 1,000 sq. miles.

The following analysis of the historical distribution of Arkansas fishes by river system (Table 5) does not address relative abundance within each river system.

WHITE RIVER SYSTEM

The White River drains 27,765 square miles of northern Arkansas and southern Missouri. It is the largest river drainage in Arkansas, draining 17,143 sq. mi. within the state (Table 4). The large Black River drainage empties into the White River system near Newport, Arkansas, and forms the northeastern part of the White River system. The richest fish fauna of the state occurs in the White River system, with 176 native fish species (Table 5). Another 20 nonnative fish species have been introduced into the White River partly because of its cold water from spring-fed tributaries and cold-water release below the impoundments on the mainstem White River. Introduced species such as the Rainbow Trout (*Oncorhynchus mykiss*) and Brown Trout (*Salmo trutta*) do quite well in the system, and for several years a Brown Trout caught in the Little Red River held the world record for that species. The combination of native and introduced species brings to 195 the total number of extant fish species in the White River system. The large total is due in part to the system's considerable length, relationships with a more northern fauna, and extreme diversification of habitats (Black 1940). It is the largest stream drainage in which the main stem has its headwaters, transition zone,

lowland portion, and mouth within Arkansas (38% of the White River drainage is in Missouri).

The Harelip Sucker, *Moxostoma lacerum*, the only Arkansas fish species known to have become extinct, formerly occurred in the White River system. Most interesting among the 195 fish species are the 15 Arkansas species endemic to the White River system, including the Ozark Chub (*Erimystax harryi*), Ozark Shiner (*Notropis ozarcanus*), Duskystripe Shiner (*Luxilus pilsbryi*), Bleeding Shiner (*L. zonatus*), Ozark Madtom (*Noturus albater*), Checkered Madtom (*N. flavater*), Black River Madtom (*N. maydeni*), Ozark Bass (*Ambloplites constellatus*), Knobfin Sculpin (*Uranidea immaculata*), Autumn Darter (*Etheostoma autumnale*), Arkansas Saddled Darter (*E. euzonum*), Strawberry Darter (*E. fragi*), Current Darter (*E. uniporum*), Yoke Darter (*Nothonotus juliae*), and Yellowcheek Darter (*N. moorei*). In addition, other currently known undescribed fish endemics of the White River system include the Ozark Darter (*Etheostoma* sp. cf. *spectabile* (Ceas and Burr 2002; Pat Ceas, pers. comm.), a divergent clade of the Slender Madtom (*Noturus exilis*) (Blanton et al. 2013), and a newly discovered but undescribed form of *Nocomis* sp. cf. *biguttatus* (Echelle et al. 2014).

The White River is quite intriguing since it appears to be a very old river (i.e., has persisted in its present form for a long span of geological time) and consists of three very different ecological regions. Black (1940) first commented that the White River was rather abruptly divided into three regions: lower, middle, and upper. The large lower part of the White River below Batesville and Black Rock is mostly a lowland river with a slightly silty, large river appearance. The floodplain meanders, and oxbow lakes that characterize a large river are absent upstream of Batesville on the White River proper and slightly above Black Rock, Arkansas, on the Black River. There is a short ecotone (transitional zone) in the main White River between

Batesville and Newport, but from Newport downstream, the fish fauna is decidedly lowland with fish species such as *Cyprinella venusta, Hybognathus hayi, Notropis buchanani, Aphredoderus sayanus, Lepomis macrochirus,* and *Centrarchus macropterus* abundant in the river. Also present are the larger river fishes such as *Polyodon spathula, Amia calva, Atractosteus spatula, Ictiobus* spp., and *Carpiodes* spp. The White River above Batesville is a clear, swift stream of medium size at about Cotter, Arkansas, but most of the mainstem flow today at this locality is regulated by tailwater releases from large dams. The Buffalo River is the major tributary entering the middle section of the White River. This section also makes up an important part of the Arkansas trout fishery. Native fishes commonly found in the middle section include Plains Stoneroller (*Campostoma plumbeum*), Largescale Stoneroller (*C. oligolepis*), Whitetail Shiner (*Cyprinella galactura*), Bigeye Chub (*Hybopsis amblops*), Ozark Chub (*Erimystax harryi*), Striped Shiner (*Luxilus chrysocephalus*), Duskystripe Shiner (*L. pilsbryi*), Hornyhead Chub (*Nocomis biguttatus*), Bigeye Shiner (*Notropis boops*), Ozark Minnow (*N. nubilus*), Ozark Shiner (*N. ozarcanus*), Carmine Shiner (*N. percobromus*), Telescope Shiner (*N. telescopus*), Bluntnose Minnow (*Pimephales notatus*), Northern Hog Sucker (*Hypentelium nigricans*), Northern Studfish (*Fundulus catenatus*), Knobfin Sculpin (*Uranidea immaculata*), Ozark Bass (*Ambloplites constellatus*), Longear Sunfish (*Lepomis megalotis*), Smallmouth Bass (*Micropterus dolomieu*), Greenside Darter (*Etheostoma blennioides*), Rainbow Darter (*E. caeruleum*), Banded Darter (*E. zonale*), and Logperch (*Percina caprodes*). Finally, an upper part of the White River system occurs above the reservoirs as the diverse tributaries reach their headwater origins in the Ozark Mountains, where spring outflows from the karst mountainous regions produce clear, cold, waters. The fish fauna indicates these physicochemical changes as fishes like *Chrosomus erythrogaster, Luxilus pilsbryi, Semotilus atromaculatus, Uranidea immaculata* and *U. carolinae,* and numerous darter species (*Etheostoma* and *Percina*) can be found.

White River fishes with northern affinities include *Catostomus commersonii, Chrosomus erythrogaster, Nocomis biguttatus, Campostoma oligolepis, Etheostoma caeruleum, Percina evides, Uranidea carolinae,* and *U. immaculata.* Interestingly, none of those species occur in the mountainous areas of the Ouachita River or Red River systems farther south in Arkansas. Other forms with northern affinities, but not so sharply limited to cold waters are also found in the Arkansas, Ouachita, and Red river systems, including *Hypentelium nigricans, Semotilus atromaculatus, Notropis volucellus, Noturus miurus, Micropterus dolomieu, Etheostoma microperca, E. nigrum,* and *Stizostedion vitreum.*

ARKANSAS RIVER SYSTEM

The Arkansas River bisects Arkansas as it flows from a general northwest to southeast direction for 310 miles through the center of the state, draining 11,100 sq. mi. (Table 4) before it empties into the Mississippi River near Arkansas Post. Tributaries from both the southern Ozark (Mulberry River, Illinois Bayou, Big Piney Creek) and northern Ouachita Mountains (Poteau River, Petit Jean River, Fourche LaFave River, Maumelle River) dilute the turbidity and salinity carried by the Arkansas River as it enters Arkansas from Oklahoma. Farther downstream a few lowland tributaries such as Bayou Meto add to the flow. One hundred fifty-two native species of fishes have been documented from the Arkansas River system as it flows through Arkansas (Table 5). Fifteen species are nonnative introductions. This total of 167 fishes makes the Arkansas River the second-richest river system in the state. Several species occur only in the Arkansas River system, including *Notropis girardi, Pimephales tenellus tenellus, Etheostoma mihileze, E. squamosum,* and the Red Belly form of the *E. spectabile* complex (*E.* sp. *cf. pulchellum* 1).

Black (1940) recognized the Arkansas River as a great dispersal channel for fishes, and it has been an important corridor for both upstream and downstream fish movements. Black (1940) viewed the Arkansas River fish fauna as transitional between the faunas of the uplands and lowlands, but also as containing representatives of the Great Plains fauna. He noted the pattern of many lowland species extending up the Arkansas River Valley into Oklahoma. Representatives from the Ozark Mountains and the Ouachita Mountains penetrate the Arkansas River Valley from the north and south respectively, while Mississippi Alluvial Plain forms (*Hybognathus nuchalis, Fundulus chrysotus, Morone mississippiensis*) push upstream from the southeast and some Great Plains fishes (*Hybognathus placitus, Notropis girardi*) extend downstream from the west. Its intermediate character in terms of geographic position and fish fauna, makes it difficult to classify. Fenneman (1938) considered it physiographically a separate portion of the Ozark Upland physiographic province, but eastern, western, and southern boundaries of the Arkansas River Valley are not pronounced (Black 1940). The highest mountain in Arkansas (Mount Magazine, 2,853 feet) rises directly from the valley floor, and to the south, the valley grades into a rugged and sharply dissected region. The Poteau River is a major tributary of the Arkansas River and penetrates well into the Ouachita Mountains almost to the town of Eagleton, Arkansas, at the foot of Rich Mountain. The lower Poteau River possesses many of the characters of a lowland stream where it meets the Arkansas River at Fort Smith, and most of its tributaries are relatively warm (Black 1940).

While it has been a great dispersal channel for some fishes, the Arkansas River is also a barrier to movement for other fishes. It has been an especially effective barrier to movement of Ozarkian fishes into the Ouachita Mountains. No sculpins (*Uranidea*) or Southern Redbelly Dace (*Chrosomus erythrogaster*) are found in the Ouachita Mountains, although some of the rivers and streams are cold enough to support them and do not show great differences physicochemically from the Ozark streams and rivers. *Chrosomus erythrogaster* was reported in the Illinois Bayou (Robison et al. 2013), which enters the Arkansas River from the north. Interestingly, this dace does not occur south of the Arkansas River in any northward flowing Ouachita Mountain tributaries to the Arkansas River.

OUACHITA RIVER SYSTEM

The Ouachita River originates at the bases of Rich and Black Fork mountains in the Ouachita Mountains of Arkansas and flows 362 miles southeast from the mountains onto the Coastal Plain as it winds its way south toward its eventual junction with the Red River at Alexandria, Louisiana. Within Arkansas, the Ouachita River drains 11,548 square miles (Table 4) and contains 136 native species of fishes (Table 5). Ten fish species represent nonnative introductions. This total of 146 fishes makes the Ouachita River the third-most speciose river system in the state. Part of the reason for the high species richness is that the Ouachita River is the second-largest drainage in the state that has most of its headwaters, the entire main stem transitional zone between upland and lowland stream, and most of the lowland portion of the main stem within Arkansas. Six endemic fish species are known from the Ouachita River, including the Peppered Shiner (*Notropis perpallidus*), Ouachita Madtom (*Noturus lachneri*), Caddo Madtom (*N. taylori*), Paleback Darter (*Etheostoma pallididorsum*), Beaded Darter (*E. clinton*), and the Ouachita Darter (*Percina brucethompsoni*) (Robison and Allen 1995; Robison et al. 2014). In addition, an undescribed form of *Lythrurus* sp. cf. *umbratilis* inhabits the upper Ouachita River.

RED RIVER SYSTEM

One hundred thirty-one native species of fishes are known from the Red River system in Arkansas (Table 5), with 80 native species historically known from the Red River main stem (Buchanan et al. 2003). The Red River flows for approximately 135 miles (217 km) through the extreme southwestern portion of Arkansas draining 4,433 sq. miles of Arkansas (Table 4) before it continues south into Louisiana to join the Atchafalaya River. The Red River formerly flowed directly into the Mississippi River in Louisiana, but the flood of 1927 and the subsequent construction of levees diverted its flow southward into the Atchafalaya River (Douglas 1974). The Arkansas portion of the Red River has not been impounded, but the lower portion of the Red River downstream from US Route 71 has been modified by man-made levees, revetted banks, and wing dikes. Upstream from U.S. 71 to the Oklahoma state line, there are few channel modification structures (Buchanan et al. 2003). The most important tributaries of the Red River system in Arkansas are the Little River, McKinney Bayou, and Sulphur River. Its largest Arkansas tributary, the Little River drainage, flows south from the southern Ouachita Mountains and comprises the Mountain Fork, Rolling Fork, Cossatot, and Saline rivers which harbor many of the fishes known from the Red River system in Arkansas. In addition to 131 native species, there are 7 nonnative introductions (Table 4), making a total of 138 fishes and ranking the Red River fourth in species richness in the state. Six fish species endemic to the Red River drainage are known from the Red River system in Arkansas, including *Notropis bairdi*, *N. suttkusi*, *Lythrurus snelsoni*, *Percina pantherina*, and two undescribed darters (not included in Table 5), *Percina* sp. cf. *phoxocephala* and *Percina* sp. cf. *copelandi* (Robison et al. 2014; Todd Slack, pers. comm.). In Arkansas, the distribution of the Plains Darter, *Etheostoma pulchellum* (as currently diagnosed) is confined to the Red River drainage (although it is more widely distributed in other drainages in other states). The native distribution of Bigscale Logperch in Arkansas is also presumably confined to the Red River although it occurs throughout the Arkansas River main stem, probably as the result of introduction (Buchanan and Stevenson 2003). Other Arkansas species found only in the Red River drainage (in Arkansas) are Blackspot Shiner (*Notropis atrocaudalis*), Chub Shiner (*N. potteri*), and Western Starhead Topminnow (*Fundulus blairae*). Black (1940) concluded that "the Red River is the least suited to minnows of all the rivers in Arkansas." Our data on cyprinid species richness support that statement to some extent. There are 14 native minnow species historically known from the Red River main stem in Arkansas, compared with 21 from the Arkansas River and 23 from the Mississippi River main stems. However, the Red River is only about one-third as long in Arkansas as the other two large rivers.

ST. FRANCIS RIVER SYSTEM

The St. Francis River system drains the smallest region of the state, only 4,200 sq. mi. of northeast Arkansas (Table 4). Meek (1896) provided an excellent early account of the St. Francis River system in Arkansas. Today it is primarily a turbid, lowland stream as a result of extensive channelization in the 1950s, and it bears little physical resemblance to the stream that Meek described in 1896. There is little or no vegetation, and substrates consist largely of mud, sand, and, very infrequently, small gravel areas near sandbars. Black (1940) attributed the relative lack of species inhabiting the St. Francis River to the uniformity of habitats in the Arkansas portion of the system and commented on the intimate relationship of the fish fauna of the St. Francis River with the fishes of the White River. This river system has only 115 documented native fishes from the Arkansas portion (Table 5), partly because of its short length within the state and its basic uniform lowland character. Seven nonnative species have been introduced into the system, making a total of 122 fish species reported for the St. Francis River system in Arkansas. Many of its tributaries drain heavily agricultural areas of northeast Arkansas that have been affected by pollution and other human modifications such as channelization. No fishes living in the Arkansas portion of the St. Francis River system are endemic to the system, but the Arkansas distribution of the Finescale Stoneroller, *Campostoma pullum*, is confined almost entirely to that drainage. Meek's (1896) capture of *Umbra limi* in the St. Francis River system was the only record for this species in Arkansas for more than 100 years. It was not found again in Arkansas until the late 1990s, when a specimen was taken from a stream immediately below a bait minnow farm leaving it suspect as a recent bait release. Typically, the St. Francis River is dominated by lowland species such as *Lepisosteus platostomus*, *Amia calva*, *Cyprinella venusta*, *Hybognathus nuchalis*, *Macrhybopsis hyostoma*, *Notropis atherinoides*, *Pimephales vigilax*, *Ictalurus punctatus*, *Gambusia affinis*, and *Percina shumardi*.

Black (1940) noted the presence of the Redfin Shiner subspecies *Lythrurus umbratilis cyanocephalus* and hypothesized that it was evidence that the St. Francis fish fauna had an origin distinct from that of the fauna of the waters to the west. This subspecies occurs in the St. Francis River and the Cache River (White River drainage), but is replaced by *Lythrurus umbratilis umbratilis* in other parts of the White River system to the west.

MISSISSIPPI RIVER SYSTEM

All waters of Arkansas in recent geologic time flow into the Mississippi River; however, for our analysis, we delineate and restrict the Mississippi River drainage in Arkansas to the mainstem Mississippi River and its adjacent, direct lowland tributaries (excluding the large rivers) and oxbows, which drain 4,755 sq. miles (Table 4). Lake Chicot, the largest oxbow lake in Arkansas, is included within this system. The Mississippi River system, which forms the eastern border of Arkansas, harbors only 108 native species of fishes (Table 5), making it the least speciose river system in the state. In addition, 10 nonnative species have been introduced, making a total of 118 species known from this system in Arkansas. No endemic fish species are known from the Arkansas portion of the Mississippi River system. Several large river species inhabit the mainstem Mississippi River that do not occur anywhere else in Arkansas, including *Macrhybopsis gelida*, *M. meeki*, *Notropis stramineus*, *Platygobio gracilis*, *Noturus gladiator*, and a small-eyed form of *Noturus flavus*.

The topographic feature of eastern Arkansas known as Crowley's Ridge played an important role in fish distribution in the past. The Mississippi River once flowed in the channel now occupied by the Cache River on the western side of Crowley's Ridge, which runs north and south. The old Ohio River followed a course on the eastern side of Crowley's Ridge in what is currently the bed of the St. Francis River, joining the Mississippi River below Helena, Arkansas (Call 1891; Branner 1891; Robison 1986). Erosion by the Mississippi and Ohio rivers cut away the Tertiary gravel deposited over this entire region by outwash from Pleistocene glaciers to the north and left Crowley's Ridge as a river divide. Later, the rivers shifted their courses eastward, with the Ohio River shifting eastward to the present position of the Mississippi River channel, and the Mississippi River breaking through Crowley's Ridge near the present-day Arkansas-Missouri state line to flow into the abandoned channel of the Ohio River. The Mississippi River later shifted eastward to intercept the Ohio River farther upstream near the present junction of those rivers, with the St. Francis River then occupying its current position, the abandoned Mississippi River channel east of Crowley's Ridge (Black 1940). Black speculated that the Ohio Valley fauna was introduced into Arkansas by these river changes, allowing the Tennessee Upland fauna, now

Table 5. Historical Distribution of Native and Nonindigenous Fishes of Arkansas by River System.

Species	Arkansas	White	St. Francis	Mississippi	Ouachita	Red
PETROMYZONTIDAE						
Ichthyomyzon castaneus	X	X	X	X	X	X
Ichthyomyzon gagei	X	X		X	X	X
Ichthyomyzon unicuspis		X				
Lampetra aepyptera		X				
Lethenteron appendix		X			X	
ACIPENSERIDAE						
Acipenser fulvescens	X	X	X	X		
Acipenser oxyrinchus					X	
Scaphirhynchus albus			X	X		
Scaphirhynchus platorynchus	X	X	X	X		X
POLYODONTIDAE						
Polyodon spathula	X	X	X	X	X	X
LEPISOSTEIDAE						
Atractosteus spatula	X	X	X	X	X	X
Lepisosteus oculatus	X	X	X	X	X	X
Lepisosteus osseus	X	X	X	X	X	X
Lepisosteus platostomus	X	X	X	X	X	X
AMIIDAE						
Amia calva	X	X	X	X	X	X
HIODONTIDAE						
Hiodon alosoides	X	X	X	X	X	X
Hiodon tergisus	X	X		X	X	
ANGUILLIDAE						
Anguilla rostrata	X	X	X	X	X	X
CLUPEIDAE						
Alosa alabamae	X	X			X	
Alosa chrysochloris	X	X	X	X	X	X
Dorosoma cepedianum	X	X	X	X	X	X
Dorosoma petenense	X	X	X	X	X	X
CYPRINIDAE						
Campostoma oligolepis	X	X	X			
Campostoma plumbeum	X	X				
Campostoma pullum		X	X	X		
Campostoma spadiceum	X				X	X
Carassius auratus	I	I	I	I	I	
Chrosomus erythrogaster	X	X	X			
Ctenopharyngodon idella	I	I	I	I	I	I
Cyprinella camura	X					
Cyprinella galactura		X				
Cyprinella lutrensis	X	X	X	X		X

Table 5. (*continued*)

Species	Arkansas	White	St. Francis	Mississippi	Ouachita	Red
Cyprinella spiloptera	X	X				
Cyprinella venusta	X	X	X	X	X	X
Cyprinella whipplei	X	X			X	X
Cyprinus carpio	I	I	I	I	I	I
Erimystax harryi		X				
Erimystax x-punctatus	X	X			X	
Hybognathus hayi	X	X	X	X	X	X
Hybognathus nuchalis	X	X	X	X	X	X
Hybognathus placitus	X			X		X
Hybopsis amblops	X	X				
Hybopsis amnis	X	X	X	X	X	X
Hypophthalmichthys molitrix	I	I	I	I	I	
Hypophthalmichthys nobilis	I	I	I	I	I	I
Luxilus cardinalis	X					
Luxilus chrysocephalus	X	X	X		X	X
Luxilus pilsbryi		X				
Luxilus zonatus		X				
Lythrurus fumeus	X	X	X	X	X	X
Lythrurus snelsoni						X
Lythrurus umbratilis	X	X	X	X	X	X
Macrhybopsis gelida				X		
Macrhybopsis hyostoma	X	X	X	X	X	X
Macrhybopsis meeki				X		
Macrhybopsis storeriana	X	X	X	X	X	X
Mylopharyngodon piceus	I			I		
Nocomis asper	X				X	
Nocomis biguttatus		X				
Notemigonus crysoleucas	X	X	X	X	X	X
Notropis atherinoides	X	X	X	X	X	X
Notropis atrocaudalis						X
Notropis bairdi						X
Notropis blennius	X	X	X	X		
Notropis boops	X	X			X	X
Notropis buchanani	X	X	X	X	X	X
Notropis chalybaeus	X	X	X		X	X
Notropis girardi	X					
Notropis greenei	X	X				
Notropis maculatus	X	X	X	X	X	X
Notropis nubilus	X	X				
Notropis ortenburgeri	X				X	X
Notropis ozarcanus	X	X				

Table 5. (*continued*)

Species	Arkansas	White	St. Francis	Mississippi	Ouachita	Red
Notropis percobromus		X			X	
Notropis perpallidus					X	
Notropis potteri						X
Notropis sabinae		X	X			
Notropis shumardi	X	X	X	X		X
Notropis stramineus				X		
Notropis suttkusi						X
Notropis telescopus		X				
Notropis texanus	X	X	X	X	X	X
Notropis volucellus	X	X	X	X	X	X
Notropis wickliffi		X		X		
Opsopoeodus emiliae	X	X	X	X	X	X
Phenacobius mirabilis	X		X			X
Pimephales notatus	X	X	X	X	X	X
Pimephales promelas	X	X	X	X	X	X
Pimephales tenellus	X	X	X		X	X
Pimephales vigilax	X	X	X	X	X	X
Platygobio gracilis				X		
Pteronotropis hubbsi					X	X
Scardinius erythrophthalmus				I		
Semotilus atromaculatus	X	X	X	X	X	X
CATOSTOMIDAE						
Carpiodes carpio	X	X	X	X	X	X
Carpiodes cyprinus	X	X	X	X	X	
Carpiodes velifer	X	X		X	X	X
Catostomus commersonii	X	X				
Cycleptus elongatus	X	X	X	X		X
Erimyzon claviformis	X	X	X		X	X
Erimyzon sucetta	X	X	X	X	X	X
Hypentelium nigricans	X	X			X	
Ictiobus bubalus	X	X	X	X	X	X
Ictiobus cyprinellus	X	X	X	X	X	X
Ictiobus niger	X	X	X	X	X	X
Minytrema melanops	X	X	X	X	X	X
Moxostoma anisurum		X				
Moxostoma carinatum	X	X			X	X
Moxostoma duquesnei	X	X			X	X
Moxostoma erythrurum	X	X	X	X	X	X
Moxostoma lacerum		E				
Moxostoma pisolabrum	X	X	X	X		
Moxostoma poecilurum					X	

Table 5. (*continued*)

Species	Arkansas	White	St. Francis	Mississippi	Ouachita	Red
ICTALURIDAE						
Ameiurus melas	X	X	X	X	X	X
Ameiurus natalis	X	X	X	X	X	X
Ameiurus nebulosus	X	X	X	X	X	X
Ictalurus furcatus	X	X	X	X	X	X
Ictalurus punctatus	X	X	X	X	X	X
Noturus albater		X				
Noturus eleutherus		X		X	X	X
Noturus exilis	X	X				I
Noturus flavater		X				
Noturus flavus				X		
Noturus gladiator				X		
Noturus gyrinus	X	X	X	X	X	X
Noturus lachneri					X	
Noturus maydeni		X				
Noturus miurus	X	X	X		X	X
Noturus nocturnus	X	X	X	X	X	X
Noturus phaeus						X
Noturus taylori					X	
Pylodictis olivaris	X	X	X	X	X	X
OSMERIDAE						
Osmerus mordax				I		
SALMONIDAE						
Oncorhynchus clarkii		I				
Oncorhynchus mykiss	I	I			I	
Salmo trutta	I	I			I	
Salvelinus fontinalis		I				
Salvelinus namaycush		I				
ESOCIDAE						
Esox americanus	X	X	X	X	X	X
Esox lucius	I	I			I	I
Esox masquinongy		I			I	
Esox niger	X	X	X		X	X
Umbra limi				X		
APHREDODERIDAE						
Aphredoderus sayanus	X	X	X	X	X	X
AMBLYOPSIDAE						
Troglichthys rosae	X	X				
Typhlichthys eigenmanni		X				
Typhlichthys sp. *cf. eigenmanni*		X				

Table 5. (*continued*)

Species	Arkansas	White	St. Francis	Mississippi	Ouachita	Red
MUGILIDAE						
Mugil cephalus	X	X		X	X	X
ATHERINOPSIDAE						
Labidesthes sicculus	X	X			X	X
Labidesthes vanhyningi	X	X	X	X	X	X
Menidia audens	X	X	X	X	X	X
FUNDULIDAE						
Fundulus blairae						X
Fundulus catenatus	X	X			X	X
Fundulus chrysotus	X	X	X	X	X	X
Fundulus dispar	X	X	X	X	X	
Fundulus notatus	X	X	X	X	X	X
Fundulus olivaceus	X	X	X	X	X	X
POECILIIDAE						
Gambusia affinis	X	X	X	X	X	X
CHANNIDAE						
Channa argus		I	I	I		
COTTIDAE						
Uranidea carolinae	X	X				
Uranidea immaculata		X				
MORONIDAE						
Morone americana	I					
Morone chrysops	X	X	X	X	X	X
Morone mississippiensis	X	X	X	X	X	X
Morone saxatilis	I	I	I	I	I	I
CENTRARCHIDAE						
Ambloplites ariommus	X	X	X		X	X
Ambloplites constellatus		X				
Ambloplites rupestris	I					
Centrarchus macropterus	X	X	X	X	X	X
Lepomis auritus	I	I				
Lepomis cyanellus	X	X	X	X	X	X
Lepomis gulosus	X	X	X	X	X	X
Lepomis humilis	X	X	X	X	X	X
Lepomis macrochirus	X	X	X	X	X	X
Lepomis marginatus	X	X	X	X	X	X
Lepomis megalotis	X	X	X	X	X	X
Lepomis microlophus	X	X	X	X	X	X
Lepomis miniatus	X	X	X	X	X	X
Lepomis symmetricus	X	X	X	X	X	X

Table 5. (*continued*)

Species	Arkansas	White	St. Francis	Mississippi	Ouachita	Red
Micropterus dolomieu	X	X			X	X
Micropterus punctulatus	X	X	X	X	X	X
Micropterus salmoides	X	X	X	X	X	X
Pomoxis annularis	X	X	X	X	X	X
Pomoxis nigromaculatus	X	X	X	X	X	X
ELASSOMATIDAE						
Elassoma zonatum	X	X	X	X	X	X
PERCIDAE						
Ammocrypta clara		X	X		X	X
Ammocrypta vivax	X	X	X		X	X
Crystallaria asprella		X	X		X	X
Etheostoma artesiae					X	X
Etheostoma asprigene	X	X	X	X	X	X
Etheostoma autumnale		X				
Etheostoma blennioides	X	X			X	
Etheostoma caeruleum		X	X			
Etheostoma chlorosoma	X	X	X	X	X	X
Etheostoma clinton					X	
Etheostoma collettei					X	X
Etheostoma cragini	X					
Etheostoma euzonum		X				
Etheostoma flabellare	X	X				
Etheostoma fragi		X				
Etheostoma fusiforme	X	X	X	X	X	X
Etheostoma gracile	X	X	X	X	X	X
Etheostoma histrio	X	X	X	X	X	X
Etheostoma microperca	X					
Etheostoma mihileze	X					
Etheostoma nigrum	X	X			X	X
Etheostoma pallididorsum					X	
Etheostoma parvipinne		X	X		X	X
Etheostoma proeliare	X	X	X	X	X	X
Etheostoma pulchellum						X
Etheostoma sp. cf. pulchellum 1	X					
Etheostoma sp. cf. pulchellum 2	X	X				
Etheostoma radiosum					X	X
Etheostoma sp. cf. spectabile		X				
Etheostoma squamosum	X					
Etheostoma stigmaeum	X	X	X		X	X
Etheostoma teddyroosevelt	X	X				

Table 5. (*continued*)

Species	Arkansas	White	St. Francis	Mississippi	Ouachita	Red
Etheostoma uniporum		X				
Etheostoma whipplei	X	X				
Etheostoma zonale	X	X			X	
Nothonotus juliae		X				
Nothonotus moorei		X				
Perca flavescens		I		X		
Percina brucethompsoni					X	
Percina caprodes	X	X	X	X	X	X
Percina copelandi	X				X	X
Percina evides		X				
Percina fulvitaenia	X	X				
Percina macrolepida	I	I				X
Percina maculata	X	X	X		X	X
Percina nasuta	X	X	X			
Percina pantherina						X
Percina phoxocephala	X	X				X
Percina sciera	X	X	X	X	X	X
Percina shumardi	X	X	X	X	X	X
Percina uranidea		X			X	
Percina vigil		X	X		X	
Stizostedion canadense	X	X	X	X	X	
Stizostedion vitreum	X	X			X	X
SCIAENIDAE						
Aplodinotus grunniens	X	X	X	X	X	X
CICHLIDAE						
Oreochromis aureus	I	I				
Oreochromis mossambicus		I				
Oreochromis niloticus		I		I		I
Total Native Species	152	176	115	108	136	131
Total Introduced Species	15	20	7	10	10	7
Total Species	167	195 extant, 1 extinct	122	118	146	138

remotely separated from the Ozarks, to come into contact with the fauna of the Ozark Uplands.

Distribution of Arkansas Fishes by Ecoregion

Ecoregions are similar to physiographic regions or provinces in name and distribution, but Arkansas ecoregions (as used herein), in addition to having a distinctive combination of geography and topography, also exhibit distinctive fish assemblages. Topographically, Arkansas is readily divisible into the Interior Highlands (uplands) and the Gulf Coastal Plain (lowlands). Within these two geographic divisions, six ecoregions are generally recognized: Ozark Mountains, Arkansas Valley, Ouachita Mountains, West Gulf Coastal Plain, Mississippi Alluvial Plain, and Crowley's Ridge (Fig. 3.2). These ecoregions have been characterized in Chapter 3. Table 6 lists the fishes historically known from each ecoregion and provides a quick comparison of species among ecoregions.

The ecoregions have existed long enough to play an important role in the evolution of the fishes of Arkansas. The richest ecoregions based on number of fish species are the Ozark Mountains ecoregion, with 162 extant native species, and the Ouachita Mountains ecoregion, with 138 native species (Table 6). The Mississippi Alluvial Plain (136), West Gulf Coastal Plain (138), and Arkansas River Valley (130) ecoregions are all similar in native fish species richness (Table 6). The least species-rich ecoregion is Crowley's Ridge, which contains only 57 native fish species; however, that is to be expected given its small area compared with the other ecoregions of Arkansas. Non-native introductions are highest in the Ozark Mountains ecoregion (15) and Mississippi Alluvial Plain (15). The Ouachita Mountains ecoregion has 11 nonnative fish species, the Arkansas River Valley has 9, the West Gulf Coastal Plain has 7, and only 2 nonnative fish species have been introduced into Crowley's Ridge.

Arkansas Fish Faunal Regions

On a small scale (e.g., within a state), a fish faunal region is identified by its distinctive fish assemblage resulting from a combination of physiographic, ecological, and evolutionary influences. Keith (1987) and Bennett et al. (1987) attempted to identify distinctive fish assemblages by ecoregion by determining the characteristic species composition of selected least disturbed reference streams in six Arkansas ecoregions. Using the five largest Arkansas fish families (Cyprinidae, Catostomidae, Centrarchidae, Ictaluridae, and Percidae), Keith (1987) found that the Delta and Gulf Coastal Plain ecoregions were distinctively dominated by the Centrarchidae. The Arkansas River Valley ecoregion was also dominated by Centrarchidae, with Cyprinidae only slightly less dominant. In the Boston Mountain ecoregion, Percidae dominated, followed closely by Cyprinidae and Centrarchidae. The Ozark Highland ecoregion was strongly dominated by Cyprinidae followed by Centrarchidae and Percidae. Similarly, the Ouachita Mountains ecoregion fish communities were dominated by Cyprinidae although not as distinctively as in the Ozark Highlands. The average number of species collected per sample site was very similar among all ecoregions; however, the total number of species collected per ecoregion was as follows: Arkansas River Valley 75, Gulf Coastal Plain 66, Ouachita Mountains 61, Boston Mountains 60, Ozark Highland 60, and Delta 51.

The first comprehensive quantitative analysis of the distribution of the fishes of Arkansas to identify distinctive faunal groupings was by Matthews and Robison (1988), who used a multivariate analysis of the distribution of native stream fishes based on 2,323 collections between 1968 and 1988 by the authors and fish collections of other ichthyologists. Matthews and Robison defined 101 drainage units covering the entire state and permitting a more detailed assessment of extant fish distribution patterns. Principal component analysis identified four different (and mutually exclusive) faunal groupings of Arkansas fishes: PC I included a group of fishes characteristic of lowlands but extending well into some upland streams, PC II contained taxa restricted to or most typical of the uplands in north or southwest Arkansas, PC III was composed of true lowland forms characteristic of the Gulf Coastal Plain, and PC IV included species distinctive of low-gradient big rivers and/or with western affinities.

A complementary detrended correspondence analysis by Matthews and Robison (1988) permitted a different definition of fish faunal patterns within Arkansas, clarifying patterns of fish distribution corresponding to natural northwest-to-southeast gradients in environmental conditions. Five fish faunal regions emerged from that analysis. Region 1 is an Ozark upland faunal region largely within the White River drainage, including most of the Black River system. This faunal region also includes the Neosho-Illinois system in which faunal similarities are strong between the two drainages. Fishes such as the Cardinal Shiner (*Luxilus cardinalis*), which is closely related to the Duskystripe Shiner (*Luxilus pilsbryi*), the Ozark Minnow (*Notropis nubilus*), and the Largescale Stoneroller (*Campostoma oligolepis*) cross over into the Neosho-Illinois drainage, supporting the concept of a prehistoric connection between the White and the Neosho-Illinois drainages first proposed by Black in 1940. Region 2 includes the southern Ozark Mountains (i.e., streams tributary to the Arkansas River) and most of the Ouachita Mountains. Specifically, within this region are some of the easternmost streams draining the Ozark Uplift, the middle White River below Batesville, the Little Red River, most of the tributaries of the Arkansas River draining the extreme southern flanks of the Ozark Uplift, and almost all streams of the Ouachita Uplift in western Arkansas south of the Arkansas River. Region 3 is in part an ecotone between the uplands and lowlands and thus includes many of the southward-draining streams of the Western Gulf Coastal Plain below the Fall Line, including the lower Ouachita River, the lower Little Missouri River, and the middle and lower Saline River. Pflieger (1971) also identified an

Table 6. Historical Distribution of Native and Nonindigenous Fishes of Arkansas by Ecoregion.

Species	Ozark Mountains	Arkansas Valley	Ouachita Mountains	Mississippi Alluvial Plain	West Gulf Coastal Plain	Crowley's Ridge
PETROMYZONTIDAE						
Ichthyomyzon castaneus	X	X	X	X	X	
Ichthyomyzon gagei	X		X	X	X	
Ichthyomyzon unicuspis	X			X		
Lampetra aepyptera	X					
Lethenteron appendix	X		X	X		
ACIPENSERIDAE						
Acipenser fulvescens				X		
Acipenser oxyrinchus					X	
Scaphirhynchus albus				X		
Scaphirhynchus platorynchus	X	X	X	X	X	
POLYODONTIDAE						
Polyodon spathula	X	X	X	X	X	
LEPISOSTEIDAE						
Atractosteus spatula		X	X	X	X	
Lepisosteus oculatus	X	X	X	X	X	X
Lepisosteus osseus	X	X	X	X	X	X
Lepisosteus platostomus	X	X		X	X	X
AMIIDAE						
Amia calva	X	X	X	X	X	X
HIODONTIDAE						
Hiodon alosoides	X	X	X	X	X	
Hiodon tergisus	X	X	X	X	X	
ANGUILLIDAE						
Anguilla rostrata	X	X	X	X	X	
CLUPEIDAE						
Alosa alabamae	X	X	X			
Alosa chrysochloris	X	X	X	X	X	
Dorosoma cepedianum	X	X	X	X	X	X
Dorosoma petenense	X	X	X	X	X	X
CYPRINIDAE						
Campostoma oligolepis	X					X
Campostoma plumbeum	X	X				
Campostoma pullum				X		X
Campostoma spadiceum	X	X	X		X	
Carassius auratus	I	I	I	I	I	
Chrosomus erythrogaster	X					X
Ctenopharyngodon idella	I	I	I	I	I	I

Table 6. (*continued*)

Species	Ozark Mountains	Arkansas Valley	Ouachita Mountains	Mississippi Alluvial Plain	West Gulf Coastal Plain	Crowley's Ridge
Cyprinella camura	X	X				
Cyprinella galactura	X					
Cyprinella lutrensis		X	X	X	X	X
Cyprinella spiloptera	X					
Cyprinella venusta	X	X	X	X	X	X
Cyprinella whipplei	X	X	X	X	X	
Cyprinus carpio	I	I	I	I	I	I
Erimystax harryi	X					
Erimystax x-punctatus	X		X			
Hybognathus hayi			X	X	X	X
Hybognathus nuchalis	X	X	X	X	X	X
Hybognathus placitus		X		X		
Hybopsis amblops	X	X		X		
Hybopsis amnis		X	X	X	X	
Hypophthalmichthys molitrix		I	I	I	I	
Hypophthalmichthys nobilis		I		I	I	
Luxilus cardinalis	X	X				
Luxilus chrysocephalus	X	X	X	X	X	X
Luxilus pilsbryi	X					
Luxilus zonatus	X					
Lythrurus fumeus		X	X	X	X	X
Lythrurus snelsoni			X			
Lythrurus umbratilis	X	X	X	X	X	X
Macrhybopsis gelida				X		
Macrhybopsis hyostoma	X	X	X	X	X	
Macrhybopsis meeki				X		
Macrhybopsis storeriana		X	X	X	X	
Mylopharyngodon piceus				I		
Nocomis asper	X		X			
Nocomis biguttatus	X					
Notemigonus crysoleucas	X	X	X	X	X	X
Notropis atherinoides	X	X	X	X	X	X
Notropis atrocaudalis					X	
Notropis bairdi					X	
Notropis blennius		X		X		
Notropis boops	X	X	X		X	
Notropis buchanani		X	X	X	X	
Notropis chalybaeus				X	X	
Notropis girardi		X				

Table 6. (*continued*)

Species	Ozark Mountains	Arkansas Valley	Ouachita Mountains	Mississippi Alluvial Plain	West Gulf Coastal Plain	Crowley's Ridge
Notropis greenei	X	X	X			
Notropis maculatus	X	X		X	X	
Notropis nubilus	X	X				
Notropis ortenburgeri			X			
Notropis ozarcanus	X					
Notropis percobromus	X		X		X	
Notropis perpallidus			X		X	
Notropis potteri					X	
Notropis sabinae	X			X		
Notropis shumardi		X		X	X	
Notropis stramineus				X		
Notropis suttkusi			X			
Notropis telescopus	X					
Notropis texanus	X	X		X	X	
Notropis volucellus	X	X	X	X	X	
Notropis wickliffi	X			X		
Opsopoeodus emiliae	X	X	X	X	X	X
Phenacobius mirabilis		X	X	X	X	
Pimephales notatus	X	X	X	X	X	X
Pimephales promelas	X	X	X	X	X	X
Pimephales tenellus	X	X	X	X	X	
Pimephales vigilax	X	X	X	X	X	X
Platygobio gracilis				X		
Pteronotropis hubbsi				X	X	
Scardinius erythrophthalmus	I			I		
Semotilus atromaculatus	X	X	X	X	X	X
CATOSTOMIDAE						
Carpiodes carpio	X	X	X	X	X	
Carpiodes cyprinus	X	X	X	X	X	
Carpiodes velifer	X	X		X	X	
Catostomus commersonii	X					
Cycleptus elongatus	X	X		X	X	
Erimyzon claviformis	X	X	X	X	X	X
Erimyzon sucetta		X		X	X	X
Hypentelium nigricans	X	X	X		X	
Ictiobus bubalus	X	X	X	X	X	
Ictiobus cyprinellus	X	X	X	X	X	X
Ictiobus niger	X	X	X	X	X	
Minytrema melanops	X	X	X	X	X	X

Table 6. (*continued*)

Species	Ozark Mountains	Arkansas Valley	Ouachita Mountains	Mississippi Alluvial Plain	West Gulf Coastal Plain	Crowley's Ridge
Moxostoma anisurum	X					
Moxostoma carinatum	X	X	X		X	
Moxostoma duquesnei	X	X	X		X	
Moxostoma erythrurum	X	X	X	X	X	
Moxostoma lacerum	E					
Moxostoma pisolabrum	X	X		X		
Moxostoma poecilurum			X	X	X	
ICTALURIDAE						
Ameiurus melas	X	X	X	X	X	X
Ameiurus natalis	X	X	X	X	X	X
Ameiurus nebulosus	X	X	X	X	X	
Ictalurus furcatus	X	X	X	X	X	X
Ictalurus punctatus	X	X	X	X	X	X
Noturus albater	X					
Noturus eleutherus	X		X	X	X	
Noturus exilis	X	X	I			
Noturus flavater	X					
Noturus flavus				X		
Noturus gladiator				X		
Noturus gyrinus	X	X	X	X	X	X
Noturus lachneri			X			
Noturus maydeni	X					
Noturus miurus	X	X	X	X	X	
Noturus nocturnus	X	X	X	X	X	
Noturus phaeus					X	
Noturus taylori			X			
Pylodictis olivaris	X	X	X	X	X	X
OSMERIDAE						
Osmerus mordax				I		
SALMONIDAE						
Oncorhynchus clarkii	I					
Oncorhynchus mykiss	I	I	I	I		
Salmo trutta	I		I			
Salvelinus fontinalis	I					
Salvelinus namaycush	I					
ESOCIDAE						
Esox americanus	X	X	X	X	X	X
Esox lucius	I		I			
Esox masquinongy	I		I			
Esox niger	X	X	X	X	X	

Table 6. (*continued*)

Species	Ozark Mountains	Arkansas Valley	Ouachita Mountains	Mississippi Alluvial Plain	West Gulf Coastal Plain	Crowley's Ridge
Umbra limi				X		
APHREDODERIDAE						
Aphredoderus sayanus	X	X	X	X	X	X
AMBLYOPSIDAE						
Troglichthys rosae	X					
Typhlichthys eigenmanni	X					
Typhlichthys sp. *cf. eigenmanni*	X					
MUGILIDAE						
Mugil cephalus				X	X	
ATHERINOPSIDAE						
Labidesthes sicculus	X	X	X	X		
Labidesthes vanhyningi		X		X	X	X
Menidia audens	X	X	X	X	X	
FUNDULIDAE						
Fundulus blairae			X		X	
Fundulus catenatus	X		X		X	
Fundulus chrysotus		X	X	X	X	
Fundulus dispar	X	X	X	X	X	X
Fundulus notatus	X	X	X	X	X	
Fundulus olivaceus	X	X	X	X	X	X
POECILIIDAE						
Gambusia affinis	X	X	X	X	X	X
CHANNIDAE						
Channa argus				I		
COTTIDAE						
Uranidea carolinae	X					
Uranidea immaculata	X					
MORONIDAE						
Morone americana		I				
Morone chrysops	X	X	X	X	X	
Morone mississippiensis	X	X	X	X	X	
Morone saxatilis	I	I	I	I	I	
CENTRARCHIDAE						
Ambloplites ariommus	X	X	X	X	X	
Ambloplites constellatus	X					
Ambloplites rupestris	I					
Centrarchus macropterus	X	X	X	X	X	
Lepomis auritus	I		I			
Lepomis cyanellus	X	X	X	X	X	X
Lepomis gulosus	X	X	X	X	X	X

Table 6. (*continued*)

Species	Ozark Mountains	Arkansas Valley	Ouachita Mountains	Mississippi Alluvial Plain	West Gulf Coastal Plain	Crowley's Ridge
Lepomis humilis	X	X	X	X	X	X
Lepomis macrochirus	X	X	X	X	X	X
Lepomis marginatus		X		X	X	
Lepomis megalotis	X	X	X	X	X	X
Lepomis microlophus	X	X	X	X	X	X
Lepomis miniatus	X	X	X	X	X	
Lepomis symmetricus		X	X	X	X	
Micropterus dolomieu	X	X	X	X	X	
Micropterus punctulatus	X	X	X	X	X	X
Micropterus salmoides	X	X	X	X	X	X
Pomoxis annularis	X	X	X	X	X	X
Pomoxis nigromaculatus	X	X	X	X	X	X
ELASSOMATIDAE						
Elassoma zonatum	X	X	X	X	X	
PERCIDAE						
Ammocrypta clara	X			X	X	
Ammocrypta vivax	X	X	X	X	X	
Crystallaria asprella	X		X	X	X	
Etheostoma artesiae			X	X	X	
Etheostoma asprigene		X		X	X	X
Etheostoma autumnale	X					
Etheostoma blennioides	X	X	X		X	
Etheostoma caeruleum	X					X
Etheostoma chlorosoma	X	X	X	X	X	X
Etheostoma clinton			X			
Etheostoma collettei			X		X	
Etheostoma cragini	X					
Etheostoma euzonum	X					
Etheostoma flabellare	X	X	X			
Etheostoma fragi	X					
Etheostoma fusiforme		X		X	X	
Etheostoma gracile		X	X	X	X	X
Etheostoma histrio	X	X	X	X	X	
Etheostoma microperca	X					
Etheostoma mihileze	X	X				
Etheostoma nigrum	X	X	X		X	
Etheostoma pallididorsum			X			
Etheostoma parvipinne			X	X	X	X
Etheostoma proeliare	X	X	X	X	X	X
Etheostoma pulchellum			X		X	

Table 6. (*continued*)

Species	Ozark Mountains	Arkansas Valley	Ouachita Mountains	Mississippi Alluvial Plain	West Gulf Coastal Plain	Crowley's Ridge
Etheostoma sp. *cf. pulchellum 1*	X	X	X			
Etheostoma sp. *cf. pulchellum 2*	X	X	X			
Etheostoma radiosum			X		X	
Etheostoma sp. *cf. spectabile*	X					
Etheostoma squamosum	X					
Etheostoma stigmaeum	X		X	X	X	
Etheostoma teddyroosevelt	X	X	X			
Etheostoma uniporum	X					
Etheostoma whipplei	X	X	X	X		
Etheostoma zonale	X	X	X		X	
Nothonotus juliae	X					
Nothonotus moorei	X					
Perca flavescens	I			X		
Percina brucethompsoni			X			
Percina caprodes	X		X	X	X	X
Percina copelandi	X	X	X		X	
Percina evides	X					
Percina fulvitaenia	X	X	X	X		
Percina macrolepida		I		I	X	
Percina maculata	X	X	X	X	X	X
Percina nasuta	X	X	X			
Percina pantherina			X			
Percina phoxocephala	X	X			X	
Percina sciera	X	X	X	X	X	
Percina shumardi	X	X	X	X	X	
Percina uranidea	X		X	X	X	
Percina vigil	X		X	X	X	
Stizostedion canadense	X	X	X	X	X	
Stizostedion vitreum	X	X	X		X	
SCIAENIDAE						
Aplodinotus grunniens	X	X	X	X	X	
CICHLIDAE						
Oreochromis aureus				I		
Oreochromis mossambicus				I		
Oreochromis niloticus				I	I	
Total Native Species	163	130	138	136	138	57
Total Introduced Species	15	9	11	15	7	2
Total Species	177 extant, 1 extinct	139	149	151	145	59

ecotonal Ozark border fish faunal group in Missouri. Region 3 also includes parts of the lower Black and Current rivers and tributaries draining eastward from Crowley's Ridge in northeast Arkansas. Additionally, several tributaries of the Arkansas River and the westward-draining Poteau River system are in fish faunal Region 3. Region 4 is a true lowland faunal region and is more centered in streams of northeast Arkansas, within the Mississippi Alluvial Plain (Delta), but this faunal region also includes some major south-draining streams of the western Gulf Coastal Plain in southern Arkansas, interspersed between streams that are included in Region 3. Finally, Region 5 is a Big River region. Fish species characteristic of the five faunal regions were specified by Matthews and Robison (1998) and are discussed in more detail below. Correlation and multiple regression analyses showed that fish distributions were significantly related to environmental variables; however, comparison of upland fish faunas among major river drainages within the state showed that a strong component of historical zoogeography also influenced patterns in fish distribution (Matthews and Robison 1988).

Faunal Region I of Matthews and Robison (1988) is an Ozark upland fish faunal region located largely within the White River drainage but also includes the Black River system. Fishes typical of the White River drainage of Arkansas are American Brook Lamprey (*Lethenteron appendix*), Largescale Stoneroller (*Campostoma oligolepis*), Southern Redbelly Dace (*Chrosomus erythrogaster*), Whitetail Shiner (*Cyprinella galactura*), Ozark Chub (*Erimystax harryi*), Hornyhead Chub (*Nocomis biguttatus*), Ozark Minnow (*Notropis nubilus*), Ozark Shiner (*N. ozarcanus*), Telescope Shiner (*N. telescopus*), Duskystripe Shiner (*Luxilus pilsbryi*), Bleeding Shiner (*L. zonatus*), Black Redhorse (*Moxostoma duquesnei*), Slender Madtom (*Noturus exilis*), Checkered Madtom (*N. flavater*), Ozark Madtom (*N. albater*), Black River Madtom (*N. maydeni*), Northern Studfish (*Fundulus catenatus*), Ozark Bass (*Ambloplites constellatus*), Smallmouth Bass (*Micropterus dolomieu*), Autumn Darter (*Etheostoma autumnale*), Rainbow Darter (*E. caeruleum*), Arkansas Saddled Darter (*E. euzonum*), Fantail Darter (*E. flabellare*), Strawberry Darter (*E. fragi*), Current Darter (*E. uniporum*), Ozark Darter (*E. sp. cf. spectabile*), Yoke Darter (*Nothonotus juliae*), Yellowcheek Darter (*N. moorei*), Gilt Darter (*Percina evides*), Longnose Darter (*P. nasuta*), Knobfin Sculpin (*Uranidea immaculata*), and Banded Sculpin (*U. carolinae*).

The second faunal region of Matthews and Robison (1988) includes the southern Ozark Mountains (i.e., streams tributary to the Arkansas River and parts of the White River drainage) and most of the Ouachita

Mountains. Fishes restricted to or most typical of Faunal Region 2 include the Southern Brook Lamprey (*Ichthyomyzon gagei*), Plains Stoneroller (*Campostoma plumbeum*), Highland Stoneroller (*C. spadiceum*), Bigeye Chub (*Hybopsis amblops*), Gravel Chub (*E. x-punctatus*), Striped Shiner (*Luxilus chrysocephalus*), Wedgespot Shiner (*Notropis greenei*), Peppered Shiner (*N. perpallidus*), Ouachita Mountain Shiner (*Lythrurus snelsoni*), Duskystripe Shiner (*Luxilus pilsbryi*), Carmine Shiner (*Notropis percobromus*), Bluntnose Minnow (*Pimephales notatus*), Slim Minnow (*P. tenellus*), Golden Redhorse (*Moxostoma erythrurum*), Mountain Madtom (*Noturus eleutherus*), Ouachita Madtom (*N. lachneri*), Caddo Madtom (*N. taylori*), Northern Hog Sucker (*Hypentelium nigricans*), Golden Redhorse (*Moxostoma erythrurum*), Northern Studfish (*Fundulus catenatus*), Smallmouth Bass (*Micropterus dolomieu*), Scaly Sand Darter (*Ammocrypta vivax*), Beaded Darter (*Etheostoma clinton*), Arkansas Darter (*E. cragini*), Least Darter (*E. microperca*), Sunburst Darter (*E. mihileze*), Johnny Darter (*E. nigrum*), Paleback Darter (*E. pallididorsum*), Plains Darter (*E. pulchellum*), Redbelly Form (*E. sp. cf. pulchellum* 1), Blue Belly Form (*E. sp. cf. pulchellum* 2), Orangebelly Darter (*E. radiosum*), Plateau Darter (*E. squamosum*), Highland Darter (*E. teddyroosevelt*), Banded Darter (*E. zonale*), Logperch (*Percina caprodes*), Ouachita Darter (*P. brucethompsoni*), Channel Darter (*P. copelandi*), and Leopard Darter (*P. pantherina*).

Faunal Region 3 of Matthews and Robison (1988) includes the ecotone between uplands and lowlands. Typical fishes that may be found in this faunal region include Chestnut Lamprey (*Ichthyomyzon castaneus*), Longnose Gar (*Lepisosteus osseus*), Spotted Gar (*L. oculatus*), Blacktail Shiner (*Cyprinella venusta*), Steelcolor Shiner (*C. whipplei*), Redfin Shiner (*Lythrurus umbratilis*), Bigeye Shiner (*Notropis boops*), Kiamichi Shiner (*N. ortenburgeri*), Sabine Shiner (*N. sabinae*), Black Redhorse (*Moxostoma duquesnei*), River Redhorse (*M. carinatum*), Western Creek Chubsucker (*Erimyzon claviformis*), Mountain Madtom (*Noturus eleutherus*), Brindled Madtom (*N. miurus*), Freckled Madtom (*N. nocturnus*), Brown Madtom (*N. phaeus*), Flathead Catfish (*Pylodictis olivaris*), Spotted Sucker (*Minytrema melanops*), Longear Sunfish (*Lepomis megalotis*), Redspotted Sunfish (*L. miniatus*), Spotted Bass (*Micropterus punctulatus*), Crystal Darter (*Crystallaria asprella*), Scaly Sand Darter (*Ammocrypta vivax*), Mud Darter (*Etheostoma asprigene*), Redspot Darter (*E. artesiae*), Bluntnose Darter (*E. chlorosoma*), Creole Darter (*E. collettei*), Harlequin Darter (*E. histrio*), Cypress Darter (*E. proeliare*), Plains Darter (*E. pulchellum*), Speckled Darter (*E. stigmaeum*), Logperch (*Percina caprodes*), Channel Darter (*P. copelandi*), Blackside Darter (*P. maculata*), Slenderhead Darter (*P. phoxocephala*), Dusky Darter (*P. sciera*), Stargazing Darter (*P. uranidea*), Saddleback Darter (*P. vigil*) and Walleye (*Stizostedion vitreum*).

Faunal Region 4 of Matthews and Robison (1988) is the lowland fish faunal region (excluding the large rivers). Fishes typical of the true lowland region of Arkansas are the Shortnose Gar (*Lepisosteus platostomus*), Goldeye (*Hiodon alosoides*), Cypress Minnow (*Hybognathus hayi*), Pallid Shiner (*Hybopsis amnis*), Blacktail Shiner (*Cyprinella venusta*), Blackspot Shiner (*Notropis atrocaudalis*), Ironcolor Shiner (*N. chalybaeus*), Taillight Shiner (*N. maculatus*), Weed Shiner (*N. texanus*), Mimic Shiner (*N. volucellus*), Pugnose Minnow (*Opsopoeodus emiliae*), Bluehead Shiner (*Pteronotropis hubbsi*), Bullhead Minnow (*Pimephales vigilax*), Black Buffalo (*Ictiobus niger*), Blacktail Redhorse (*Moxostoma poecilurum*), Lake Chubsucker (*Erimyzon sucetta*), Highfin Carpsucker (*Carpiodes velifer*), Tadpole Madtom (*Noturus gyrinus*), Flathead Catfish (*Pylodictis olivaris*), Grass Pickerel (*Esox americanus vermiculatus*), Chain Pickerel (*E. niger*), Pirate Perch (*Aphredoderus sayanus*), Golden Topminnow (*Fundulus chrysotus*), Starhead Topminnow (*F. dispar*), Western Starhead Topminnow (*F. blairae*), Western Mosquitofish (*Gambusia affinis*), Banded Pygmy Sunfish (*Elassoma zonatum*), Flier (*Centrarchus macropterus*), Warmouth (*Lepomis gulosus*), Orangespotted Sunfish (*L. humilis*), Dollar Sunfish (*L. marginatus*), Redear Sunfish (*L. microlophus*), Redspotted Sunfish (*L. miniatus*), Bantam Sunfish (*L. symmetricus*), Black Crappie (*Pomoxis nigromaculatus*), Mud Darter (*Etheostoma asprigene*), Swamp Darter (*E. fusiforme*), Slough Darter (*E. gracile*), Goldstripe Darter (*E. parvipinne*), and Speckled Darter (*E. stigmaeum*).

Faunal Region 5 of Matthews and Robison (1988) consists of the large rivers including all segments of the Mississippi and Red rivers within Arkansas, the Arkansas River, and the lower White River. It is obvious that the large rivers of Arkansas have a fish fauna largely distinct from that of other streams, including many of the medium-sized rivers (Matthews and Robison 1988). Pflieger (1971) similarly identified a big-river faunal group in Missouri as did Burr and Warren (1986b) in Kentucky and Miller and Robison (2004) in Oklahoma. Large river assemblages have a high percentage of large fish species, and many species are characterized by streamlined, more terete bodies with sharper entering wedges, long and falcate fins, a flat ventral surface, an arched dorsal contour, slender caudal peduncles, and smaller scales (Hubbs 1941). Among the species restricted to or most typical of big rivers are the Shovelnose Sturgeon (*Scaphirhynchus platorynchus*), Pallid Sturgeon (*S. albus*), Paddlefish (*Polyodon spathula*), Alligator Gar (*Atractosteus spatula*), Goldeye (*Hiodon alosoides*), Mooneye (*H. tergisus*), Skipjack Herring (*Alosa chrysochloris*), Shoal Chub (*Macrhybopsis hyostoma*), Sicklefin Chub (*M. meeki*), Sturgeon Chub (*M. gelida*), Silver Chub (*M. storeriana*), Flathead Chub (*Platygobio gracilis*), Red Shiner (*Cyprinella lutrensis*), River Shiner (*Notropis blennius*), Chub Shiner (*N. potteri*), Silverband Shiner (*N. shumardi*), Blue Sucker (*Cycleptus elongatus*), Smallmouth Buffalo (*Ictiobus bubalus*), River Carpsucker (*Carpiodes carpio*), Quillback (*C. cyprinus*), Blue Catfish (*Ictalurus furcatus*), Mississippi Silverside (*Menidia audens*), White Bass (*Morone chrysops*), White Crappie (*Pomoxis annularis*), River Darter (*Percina shumardi*), Sauger (*Stizostedion canadense*), and Freshwater Drum (*Aplodinotus grunniens*).

Overall, it is apparent that the distribution of native stream fishes in Arkansas is a product of zoogeographic chance (i.e., river system) as well as numerous interconnected environmental factors that correspond to the northwest-to-southeast gradient from uplands to lowlands within the state (Matthews and Robison 1988).

CHAPTER 2

History of Ichthyology in Arkansas

M oyle and Cech (2004) provided an overview of the history of ichthyology in western civilization, tracing its beginnings from the work of Aristotle and focusing on the major contributors from the 1500s to modern times. Ichthyology as a science began with Francis Willoughby (1635–1672) in England and Peter Artedi (1705–1735) in Sweden (Boschung and Mayden 2004). Peter Artedi is considered by many "the father of ichthyology." George S. Myers (1964, 1979) provides a brief history of ichthyology for North America to the year 1850, and Carl L. Hubbs (1964, 1979) provides an informative overview of the history of ichthyology in the United States from 1850 to 1964. An excellent historical synopsis of European and North American ichthyology is given by Pietsch and Grobecker (1987), and a compilation of the contributions of women to ichthyology appears in Balon et al. (1994). Pietsch and Anderson (1997) provide a rich account of collection building in ichthyology and herpetology during the past 300 years. Some recent important discoveries are reviewed in Lundberg et al. (2000), and Boschung and Mayden (2004) present an informative history of the science of ichthyology. Our brief treatment of the history of Arkansas ichthyology presented herein focuses mainly on the people who made major contributions to the study of Arkansas fishes and on major trends in the study of this state's fishes. We do not attempt to list all contributors to this state's ichthyological knowledge.

There are rich archaeological and historical records dealing with fishes in Arkansas. The archaeological record of the area that was to become Arkansas begins soon after the arrival of the earliest migratory Paleo-Indians, runs through Hernando de Soto's early invasion (AD 1541–1542) of the Native Americans' land, and continues to the beginning of the period of permanent European settlement (about AD 1820). The historical record begins with de Soto's journal, spans the period of the early government railroad surveys, the classic stream surveys and initial documentation of our state ichthyofauna, and covers modern fishery management techniques as well as recent investigations. As might be expected, our knowledge of the fishes of Arkansas has not accumulated at a uniform rate but has come from a classic period of early activity, followed by a slowdown in ichthyological pursuits, and finally a resurgence of workers and subsequent publications dealing with the fishes of Arkansas that continues to this day.

Age of Discovery: 10,000 BC–AD 1850

Indigenous peoples have lived in Arkansas for at least 12,000 years. The first human inhabitants of Arkansas, known as Paleo-Indians, are presumed to have been hunters, fishers, and gatherers of wild foods (Morse and Morse 1983). B. D. Smith (1978) and Limp and Reidhead (1979) emphasized the importance of fishes as an aboriginal protein source. Archaeological evidence in the form of fish remains and fishing tackle from several sites in Arkansas confirms the extensive use of fishes in the diet of the early Arkansas Native Americans by 5000 BC. Several fish species turn up consistently in middens, including Bowfin (*Amia calva*), gars (*Lepisosteus*), buffalofishes (*Ictiobus*), catfishes (*Ictalurus*), sunfishes (*Lepomis*), basses (*Micropterus*), and Freshwater Drum (*Aplodinotus grunniens*). Smith (1986) stated that by the time of the Mississippian Native Americans (AD 1000–1541), backwater fish species and migratory waterfowl of oxbow lakes consistently contributed at least half the animal protein consumed by tribes living in Arkansas. There is even some evidence that early Native Americans practiced a type of primitive fish farming (Frank Schambach, pers. comm.). Some archaeologists think that the thousands of large "borrow pits" resulting from the building of mounds at ceremonial centers may have been deliberately stocked with fishes once they had filled with water.

It has long been known that fishes were used by Native Americans as effigies in their pottery making. At a site known as Cedar Grove in Lafayette County, the remains of a late Caddo farmstead (1650–1750) have been excavated, revealing the use of whole fish as offerings. In one grave, a whole gar was found lying across the forehead of the corpse (Trubowitz 1984). In two other graves, whole fish offerings (gar and Gizzard Shad) were found across the faces. These were high-status graves (Frank Schambach, pers. comm.), and the offering was not meant to serve as a warning or to be taken in a demeaning manner. The offerings suggest that certain fish species had totemic significance in addition to their importance as food.

In June 1541, the first Europeans set foot in Arkansas under the leadership of Hernando de Soto, the famous Spanish conquistador. In their quest for gold, these adventurers crossed the Mississippi River from the east into southern Phillips County below the present city of Helena in eastern Arkansas, where they encountered an Indian village between the St. Francis and Mississippi rivers. The first written records about Arkansas fishes came from de Soto's journalist, who wrote,

> There was a fish called the bagres; the third part of it was head, and it had on both sides the gills and along the sides great spikes like very many sharp awls. Those that were in the lakes were as big as pikes; there were some of an hundred and of an hundred and fifty pounds weight, and many of them were taken with the hook. . . . There was another fish called a peel-fish; it had a snout of a cubit [18 inches] long, and at the end of the upper lip it was made like a peel . . . and all of them

[all the species of fishes described] had scales, except the bagres and the peel-fish (Hempstead 1890: 35–36).

Renn Tumlison (pers. comm.) concluded that since both species were scaleless, and *bagres* in Spanish means catfish, the fish referred to as bagres were probably Flathead Catfish, and the peel-fish was the Paddlefish, which attains a weight of up to 180 pounds and is almost scaleless.

One hundred thirty-two years passed before Europeans again visited Arkansas. In 1673, the French explorers Jacques Marquette and Louis Joliet arrived at the mouth of the Arkansas River while exploring the Mississippi River. Sutton (1986) quotes Marquette:

> "We met from time to time monstrous fish, which struck so violently against our canoes that we took them to be large trees!"

Soon after La Salle claimed all the Louisiana Territory for France in 1632, his lieutenant Henri de Tonti, "the father of Arkansas," established Arkansas Post. Arkansas Post, formed in 1686, is Arkansas's oldest permanent settlement and the first of its kind in the lower Mississippi Valley. It served as a trading center and opened the Mississippi River area to commercial hunting and trapping. In 1721, the French explorer Jean-Baptiste Bénard de la Harpe wrote, "The [Arkansas] river continues to widen . . . it is abounding with fish . . . ," thus providing us with an early first-hand account of the bountifulness of this large river. President Thomas Jefferson's acquisition of the Louisiana Purchase in 1803 prompted a flood of settlers into what is now Arkansas. Later in 1819 Arkansas became a territory separate from Missouri and statehood followed in 1836.

Although the famous Swedish naturalist and "father of taxonomy" Carolus Linnaeus (1707–1778) never collected fishes in Arkansas or described them from specimens that came from Arkansas, he described several fish species that occur in Arkansas, including the Longnose Gar (*Lepisosteus osseus*), Bowfin (*Amia calva*), and Redbreast Sunfish (*Lepomis auritus*). Another species described by Linnaeus is the Striped Mullet (*Mugil cephalus*), a marine species that frequently enters Arkansas waters.

The Early Years: 1850–1900

With the interest of the country increasingly turning to the west, the government decided to survey this vast, uncharted expanse for a feasible route to the Pacific Ocean for the relatively new transportation miracle, the railroad. The government surveys played an important role in early Arkansas ichthyology. Systematic investigation of the fish fauna of Arkansas began in 1852 when army personnel

Figure 2.1. Charles Frederic Girard (1822–1895), first ichthyologist to undertake the scientific study of Arkansas fishes. *Photograph courtesy of Kraig Adler, Cornell University.*

under the leadership of Captain Randolph Barnes Marcy surveyed the Louisiana Territory. Fishes collected during the expedition, which began at Fort Smith (Moore 1973), were sent to the Smithsonian Institution, where they were studied by Charles Frederic Girard (1822–1895) (Fig. 2.1) and Spencer Fullerton Baird (1823–1887).

Girard, a French pupil of the famous Louis Agassiz in Switzerland and later his colleague in America, was the most prolific North American ichthyologist in the 1850s (Boschung and Mayden 2004). Girard alone, or with S. F. Baird, described about 700 species in 278 genera, mostly from the Mexican Boundary Pacific and Railroad surveys. In his first paper on Arkansas fishes, Girard (1856) gave no collection dates for specimens sent to him and merely stated, "The fishes were collected at different times and periods by several naturalists and surgeons attached to the various surveys undertaken within the five years past." Moore (1973) identified the personnel credited with collecting specimens as Dr. George G. Shumard, surgeon; H. B. Mollhausen, artist; and Lieutenant E. G. Beckwith, all under the command of Lieutenant Whipple. Charles F. Girard's initial report (1856) on the cyprinid and catostomid

fishes collected during these surveys marks the beginning of the scientific study of Arkansas fishes. Girard reported 21 species of fishes from the Arkansas River near Fort Smith, 20 of which were described as new to science, although few of Girard's species remain valid today. Later, in a detailed review of Oklahoma ichthyology, Moore (1973) concluded that most of Girard's species were collected in what is now Oklahoma and not within Arkansas's present political boundaries. After Girard's initial 1856 report, a short paper by Girard (1858b) described *Calliurus formosus* (= *Lepomis cyanellus*) and *Bryttus humilis* (= *Lepomis humilis*) from "the fresh waters of Arkansas." *Calliurus formosus* was later shown to be a synonym of *Lepomis cyanellus*, invalidating the Arkansas description.

Two years after Girard's initial effort, his major work on Arkansas fishes (Girard 1858a) was published in which he reported on collections made by the Pacific Railroad Survey in Arkansas, Indian Territory, Texas, and Kansas, describing and redescribing previously reported forms. Unfortunately, there are numerous inaccuracies in this first major treatment of the Arkansas fish fauna, which ascribed many records to Arkansas that have since been determined to have come from other states (Moore 1973). Typical is the case of the original description of the River Darter, *Hadropterus shumardi* (= *Percina shumardi*) from the Arkansas River "near Fort Smith" by Girard (1859) but actually taken from Oklahoma (Moore 1973).

Almost 20 years elapsed before further works on Arkansas fishes were published. David Starr Jordan (1851–1931) (Fig. 2.2), "the father of American ichthyology" and the preeminent ichthyologist of North America during the late 19th and early 20th centuries, entered the Arkansas picture with four publications in 1877 (1877a, b, c, d).

Jordan described the Banded Pygmy Sunfish, *Elassoma zonatum,* from the Little Red River at Judsonia, White County, from two specimens sent to him by Professor H. S. Reynolds, and the Pirate Perch, *Asternotremia mesotrema* (= *Aphredoderus sayanus*), from the same location, in addition to listing two other species (Jordan 1877b). No detailed chronicle of Jordan's collections in Arkansas at Judsonia and later at Mammoth Spring ever appeared, although the data were used to portray distribution patterns of several species in Jordan's later general works, such as Jordan and Evermann (1896).

Oliver Perry Hay (1846–1930), the first curator of ichthyology at The Field Museum of Natural History in Chicago, Illinois, contributed in a small way to our knowledge of Arkansas fishes. Although known primarily for his work along the Gulf Coast, Hay collected 22 fish species in shallow pools on the Arkansas side of the Mississippi

Figure 2.2. David Starr Jordan (1851–1931). Photo from the collection of Brooks M. Burr, digitized and archived by the Florida Museum of Natural History.

River across from Memphis, Tennessee, in June 1881 (Hay 1882). That he also later collected elsewhere in Arkansas is supported by six specimens of the now federally threatened Leopard Darter, *Percina pantherina* (USNM 101183) in the U.S. National Museum Collection taken by him from Arkansas in 1884. Unfortunately, no record of his other collections in Arkansas ever appeared. Hay would go on to gain fame as a herpetologist and vertebrate paleontologist.

In September of 1884, David Starr Jordan and his students, Charles Henry Gilbert, Joseph Swain, and Seth Eugene Meek, working under the auspices of the U.S. National Museum and the U.S. Fish Commission, collected in Arkansas, the Indian Territory, and Texas. The group collected near Eureka Springs (White and Kings rivers), Fort Smith (Arkansas River, Poteau River, James Fork, Lee Creek), Benton (Saline River), Arkadelphia (Ouachita River), and Fulton (Red River) (Jordan and Gilbert 1886). At the James Fork of the Poteau and the Poteau River at Slate Ford they collected with "a fine-meshed seine of large

size" (Cross and Moore 1952). It is solely from Jordan and Gilbert (1886) that we know of the presence of the now extinct Harelip Sucker, *Moxostoma lacerum* (formerly *Lagochila lacera*) in Arkansas. They listed the sucker as "not rare" in a collection from the White River near Eureka Springs in the summer of 1884. They also described the Freckled Madtom, *Noturus nocturnus*, from Arkansas but did not designate an official type locality. Instead, they stated "The best specimens obtained (USNM 36461) from the Saline River at Benton." Jordan and Gilbert (Gilbert 1887) described the Harlequin Darter, *Etheostoma histrio*, the Stargazing Darter, *Etheostoma uranidea* (= *Percina uranidea*), and the Saddleback Darter, *Etheostoma ouachitae* (= *Percina vigil*) from Arkansas. Also listed was *Etheostoma zonale*.

En route to Texas in July 1888, one of the most important descriptive ichthyologists in the United States, Charles Henry Gilbert (1859–1928) (Fig. 2.3), stopped briefly to collect from a small tributary of the Poteau River, about seven miles west of Waldron, Arkansas. He reported 10 species, adding three forms to the state total (Gilbert 1889); however, he reported *Notropis heterodon*, a misidentification of *Notropis boops*. Gilbert's *Etheostoma caeruleum lepidum* was probably a misidentification of the undescribed *E.* sp. *cf. pulchellum* 1, and *Etheostoma microperca* was probably *E. proeliare* (Moore 1973). Gilbert is best known as a pioneer ichthyologist who made significant contributions to our knowledge of the natural history of the western United States. He was an original member of the "Indiana School" of natural history in the 1850s and was closely associated with D. S. Jordan, first as a student and later as an assistant, colleague, and friend (Dunn 1997).

The single greatest contributor to knowledge of Arkansas fishes before 1900 was Seth Eugene Meek (1859–1914), who had accompanied Jordan to Arkansas earlier in 1884. Meek (Fig. 2.4) could rightfully be called "the father of Arkansas ichthyology."

As professor of biology and geology at the Arkansas Industrial University (University of Arkansas), Meek explored the fishes of the state from 1889 to 1893 under the auspices of the U.S. Fish Commission and the University of Arkansas. Meek's investigations were the most significant of the early works and remain the foundation from which all subsequent workers on Arkansas fishes have launched their studies. Meek would later gain additional fame as the first compiler of a book on Mexican freshwater fishes, and he coauthored the first book on the freshwater fishes of Panama.

During July and August 1889, Meek spent six weeks investigating the clear streams of the Ozark region of

Figure 2.3. Charles Henry Gilbert (1859–1928), student of David Starr Jordan and pioneer ichthyologist who made significant contributions to the study of the natural history of the western United States and coauthor with Jordan of important papers on Arkansas fishes. *Photo from the collection of Brooks M. Burr, digitized and archived by the Florida Museum of Natural History.*

Figure 2.4. Seth Eugene Meek (1859–1914), author of the first attempt to compile a list of the fishes of Arkansas, published in 1894. *Photograph courtesy of The Field Museum, Z92927.*

Missouri and Arkansas. In his report on that work, Meek (1891) described two new subspecies, neither of which is currently recognized. Two reports by Meek appeared in 1894 containing much of the same information. They were primarily summaries of all known records of Arkansas fishes, although he had made some collections in 1891 and 1892. His major work, "A Catalog of the Fishes of Arkansas," (1894a) was the first attempt to compile a list of the fishes known to occur in Arkansas. HWR first stumbled upon this classic treatise as an undergraduate at Arkansas State University in the basement of the ASU library when looking through the Arkansas Geological Survey Reports. In this catalog, Meek discusses 137 forms, although several are incorrect, others are unsubstantiated, and many have since been synonymized as a result of recent nomenclatorial changes. Appearing later, Meek's 1894b paper omitted

unverified forms and listed 132 forms; however, misidentifications and misunderstandings of synonymy continued to plague the accuracy of his works.

Meek's 1896 publication was based on collections from the Arkansas and St. Francis rivers, adding five new forms to the state faunal list. The earliest and only Arkansas record for more than 100 years for the Central Mudminnow, *Umbra limi*, was included. This report marks the last of Meek's works in Arkansas and brought to a close the second period in the study of ichthyology in the state.

Classic Period: 1900–1960

During the third period, 1900–1960, mostly sporadic collections were made from various regions of Arkansas. Dr. Henry W. Pilsbry, a leading conchologist, traveled through Arkansas in April 1903 collecting fishes in the White River system at Rogers, Benton County, and in the Arkansas River system near Hartford, Sebastian County. Later Pilsbry sent Henry Weed Fowler (1878–1965) at the Philadelphia Academy of Sciences 10 fish species from Arkansas, including one undescribed form. Fowler had earlier studied under David Starr Jordan and rose to become the curator of fishes at the Academy of Natural Sciences in Philadelphia from 1940 to 1965. Fowler (1904) honored Dr. Pilsbry by naming the new Arkansas fish species *Notropis pilsbryi* (= *Luxilus pilsbryi*) from the types collected at Rogers, Arkansas. Later, Fowler (1906) presented an annotated list of the "more generalized percoid fishes" housed in the collections of the Academy of Natural Sciences of Philadelphia, reviewing several centrarchids and percids, some previously collected from Arkansas by Meek, Gilbert, and Jordan. Although widely traveled, Fowler never collected in Arkansas. He is probably best remembered as the last ichthyologist who attempted to single-handedly revise the taxonomy of the fishes of the world (Smith-Vaniz and Peck 1997).

In a cursory report, Gudger (1925) reviewed all known Arkansas fishes, although he reported many records that were actually from Oklahoma. During the summer of 1927, an expedition led by A. I. Ortenburger from the University of Oklahoma Museum of Zoology and under the auspices of the Oklahoma Biological Survey (OBS) collected from localities in eastern Oklahoma and western Arkansas from the Red, Ouachita, and Arkansas river drainages. Fishes from the OBS survey were sent for identification to the renowned ichthyologist Carl Leavitt Hubbs (Fig. 2.5) at the University of Michigan.

The results of those collections were published by Hubbs and Ortenburger (1929b), marking the entrance of Professor Hubbs, one of Charles H. Gilbert's last students, onto the Arkansas scene. This 1929 report corrected and clarified many errors in nomenclature and added significantly to the knowledge of the taxonomy of Arkansas fishes. Jordan et al. (1930) published the landmark checklist of the fishes and fishlike vertebrates of North and Middle America. Back in Arkansas, Gudger (1932) documented the Shovelnose Sturgeon, *Scaphirhynchus platorynchus,* from the Arkansas River, in addition to listing four other ganoid species.

Under the direction of Carl Hubbs, John D. Black (1908–1996) (Fig. 2.6), a doctoral student at the University of Michigan, became the next important contributor to the knowledge of Arkansas fishes. Black was raised in the Ozark Mountains at Winslow, Arkansas, and between 1922 and 1935 prior to earning his bachelor's degree at

Figure 2.5. Carl L. Hubbs (1894–1979), prolific ichthyologist and author of many papers on Arkansas fishes. *Photograph courtesy of Laura Hubbs.*

the University of Kansas, published 17 papers on birds in Arkansas (James and Neal 1986). In the summers of 1938 and 1939, Black made 315 fish collections from 163 different localities in every major drainage system in the state. This was the most extensive collecting effort in Arkansas up to that time. When Black began his Arkansas fieldwork in 1938, the number of fish species known from Arkansas was 133. When he finished his collecting in 1939, the number of Arkansas species had grown to 165. Black (1940) also correctly predicted several additional species that would eventually be found in Arkansas. It is difficult to imagine how Black accomplished this major effort given the condition of the highways in Arkansas at that time. This fieldwork culminated in Black's (1940) unpublished doctoral dissertation, "The Distribution of the Fishes of Arkansas." Unfortunately, Black's work also introduced much confusion into the study of Arkansas fishes because of his frequent use of manuscript names, many of which were never published by Black or Hubbs. Despite its drawbacks, no serious student of Arkansas fishes should neglect this detailed work that includes important historical distribution maps of state fishes.

Following Black's (1940) treatise on Arkansas fishes, information on state fishes appeared in a variety of publications dealing notably with cyprinid and etheostomatine species (Hubbs and Moore 1940; Hubbs and Black 1940a, c, 1941; Bailey 1941; Hubbs and Black 1947; Moore and

Figure 2.6. John D. Black (1908–1996), whose studies of Arkansas fishes in 1938 and 1939 culminated in an unpublished doctoral dissertation on the fishes of Arkansas in 1940. *Photograph courtesy of Truman State University, Kirksville, Missouri.*

Rigney 1952; Moore and Reeves 1955; Hubbs and Crowe 1956). Two important stream surveys by Moore and Paden (1950) and Cross and Moore (1952) on the Illinois and Poteau rivers, respectively, added greatly to our knowledge of the fishes of western Arkansas.

Resurgence: 1960–1987

The year 1960 marks the beginning of a period of resurgence in ichthyological investigations in Arkansas, stimulated by the arrival of Dr. Kirk Strawn at the University of Arkansas, where Seth Meek had labored years earlier. Dr. Strawn and his students were especially active in fish surveys of the White River system (Keith 1964; Cashner 1967; Brown 1967; Cashner and Brown 1977). These works form the foundation of documented fish distribution in northwestern Arkansas and provide a comparative platform from which to view the many human perturbations dealt to the White River ecosystem.

Important papers appeared in the 1960s describing new fishes from Arkansas. Distler and Metcalf (1962) described the Paleback Darter, *Etheostoma pallididorsum* from the upper Caddo River in Montgomery County, Arkansas, and eminent Cornell University ichthyologist Edward C. Raney and his equally prominent student Royal D. Suttkus described the Yellowcheek Darter, *Etheostoma moorei* (= *Nothonotus moorei*), from the Little Red River system. The Yellowcheek Darter was named in honor of Dr. George A. Moore, the dean of Oklahoma ichthyologists and former doctoral student of Carl L. Hubbs (Raney and Suttkus 1964). Distler's (1968) classic monograph on the *Etheostoma spectabile* complex describes the Current River Orangethroat Darter, *Etheostoma spectabile uniporum*, and the Strawberry River Orangethroat Darter, *Etheostoma spectabile fragi*, as new subspecies from Arkansas. A year later, Birdsong and Knapp (1969) described the Creole Darter, *Etheostoma collettei*, as a new species from southern Arkansas. The Smithsonian Institution was again active in Arkansas ichthyology as William Ralph Taylor's (1969) landmark monograph on the madtom catfish genus *Noturus* appeared. This long-awaited treatment of 23 *Noturus* species included 10 new species. Nine species treated in Taylor's monograph occur in Arkansas, and three new Arkansas madtom species were described: the Ouachita Madtom, *Noturus lachneri,* Checkered Madtom, *Noturus flavater,* and Ozark Madtom, *Noturus albater.*

The 1970s was a period of intense ichthyological activity in Arkansas that produced detailed studies of Arkansas fishes. First, another well-known Smithsonian Institution ichthyologist, Ernest Lachner, together with Cornell graduate student Robert E. Jenkins described the Redspot Chub, *Nocomis asper* from the Arkansas River drainage (Lachner and Jenkins 1971). Neil H. Douglas (1972) described the Arkansas endemic Caddo Madtom, *Noturus taylori,* from the upper Caddo River, naming this catfish after the Smithsonian madtom catfish authority, William Ralph Taylor. Several years later, Bailey and Robison (1978) described the unique cyprinid *Notropis* (= *Pteronotropis*) *hubbsi,* the Bluehead Shiner, from the Coastal Plain of southern Arkansas. Robison and Harris (1978) reported the habitat and zoogeography of the state endemic Caddo Madtom, *Noturus taylori.*

Arkansas ichthyology is especially indebted to Dr. Neil H. Douglas (Fig. 2.7) and his graduate students from the University of Louisiana at Monroe (formerly Northeast Louisiana University) who provided the impetus for a revival of the investigation of the fishes of southern Arkansas, an area almost wholly neglected by systematic sampling in earlier years by visiting ichthyologists.

Figure 2.7. Neil H. Douglas (1932–), University of Louisiana at Monroe, systematically collected Arkansas fishes and guided multiple graduate theses on the fishes of various river systems of Arkansas. *Photograph courtesy of Jan Jeffrey Hoover, USACE.*

Under his tutelage, many important theses were completed which helped build the foundation of our knowledge of Arkansas fishes. Douglas's students produced surveys of the Caddo River (Fruge 1971), the Saline River (Reynolds 1971; Stackhouse 1982), the Buffalo River (Guidroz 1975), the Cossatot River (Etheridge 1974), the Ouachita River (Raymond 1975; Harris 1977), Bayou Bartholomew (Thomas 1976), the Little Missouri River (Meyers 1977; Loe 1983), the Saline River of southwestern Arkansas (Johnson 1978), the Rolling Fork River, (Corkern 1979), and the Spring River (Winters 1985). The fish collection amassed by Dr. Douglas and his students was the largest repository of Arkansas fishes both by species and by numbers of individuals, but the Arkansas specimens in that collection were relocated to ASU in 2017. The second-largest collection of Arkansas fishes is housed at Tulane University, largely through the work and diligence of Dr. Royal D. Suttkus. Suttkus and his students made many collections in Arkansas, and when the University of Arkansas was considering disposing of its fish collection in the 1960s, much of which included uncataloged material stored in quonset huts on the Fayetteville campus, Dr. Suttkus obtained most of this material for the Tulane collection.

During the period of resurgence (1960–1988), Arkansas State University took an active role in surveying the fishes of streams in the northern portion of the state under the direction of Dr. George L. Harp and Dr. John K. Beadles. Both supervised several master's theses on fishes from a variety of streams in northern and northeastern Arkansas. In addition, a variety of published papers have

resulted from these graduate theses and investigations by ASU personnel, including Robison and Harp (1971), Strawberry River; Jenkins and Harp (1971), Big Creek; Jackson and Harp (1973), Big Creek; Fowler and Harp (1974), Jane's Creek; Matthews and Harp (1974), Piney Creek (Izard County); Green and Beadles (1974), Current River; Bounds and Beadles (1976), Fourche Creek; Yeager and Beadles (1976) and Bounds (1977), Cane Creek; Johnson and Beadles (1977), Eleven Point River; Fulmer and Harp (1977), Crowley's Ridge; Frazier and Beadles (1977), Sylamore Creek; Bounds et al. (1977), Randolph County; Mauney and Harp (1979), Cache River and Bayou DeView; and Carter (1984), Mississippi River; Hilburn (1987), Strawberry River; Harvill (1989), St. Francis River; and Holt and Harp (1993), Village Creek. Other ASU thesis projects included systematic work on the stoneroller genus *Campostoma* by Sewell (1979a).

Also important to Arkansas ichthyology in the 1970s were stream surveys by Cloutman and Olmsted (1976), Washington County fishes; Olmsted et al. (1972), Mulberry River; Buchanan (1976), Arkansas River Navigation System; Cloutman and Olmsted (1974), Cossatot River; Robison and Beadles (1974) and Robison (1979), Strawberry River; Geihsler et al. (1975), Illinois River; Robison (1975b), lower Ouachita River system; Robison (1976), Petit Jean River; Dewey and Moen (1978), Caddo River; Harris and Douglas (1978), upper Ouachita River; Robison and Winters (1978), Moro Creek; Jeffers and Bacon (1979), Ten Mile Creek; Buchanan et al. (1978), Gulpha Creek; and upper Saline River (Polk and Howard counties) by Sewell (1981). New distributional records for state fishes were published by Robison (1974c, 1975b). The first keys specifically for Arkansas fishes were compiled by Buchanan (1973b) and checklists of fishes were presented by Buchanan (1973a) and Robison (1975a).

Two of the most significant contributions to Arkansas ichthyology during the 1980s were the description of the Ouachita Mountain Shiner, *Notropis snelsoni* (= *Lythrurus snelsoni*), from the Little River system of Arkansas and Oklahoma (Robison 1985), and the description of the Ozark Sculpin, *Cottus hypselurus,* formerly known as *Cottus bairdi,* from the White River system of Arkansas and Missouri (Robins and Robison 1985). Arkansas populations of *Cottus hypselurus* are currently known as *Uranidea immaculata*. Stream surveys published in the 1980s include Carter and Beadles (1980) on Rock Creek; Sewell (1981) on the western Saline River; Baldwin (1983) on Little Red River; Ponder (1983) on Terre Noir Creek; and Robison et al. (1983) on Antoine River. Importantly, Paige et al. (1981) reported the second record of *Typhlichthys subterraneus* from Arkansas. Lindsey et al. (1983) updated a previous survey of the Poteau River in Oklahoma and Arkansas by Cross and Moore (1952) by re-collecting in Arkansas in upstream areas of the Poteau watershed. Carter and Beadles (1983a) reported the first specimens of *Macrhybopsis meeki* in Arkansas from the Mississippi River 15 miles south of the Missouri-Arkansas state line. Baker and Armstrong (1987) inventoried the Spring River and added several species to the fishes known from that river system. Keith et al. (1987) rediscovered the Suckermouth Minnow (*Phenacobius mirabilis*) in Arkansas in Little Bay Ditch (St. Francis River drainage) in Craighead County. Stream ecology papers dealing with Arkansas fishes are generally lacking during the 1980s; however, papers by Dewey (1981), Frietsche (1982), Matthews (1982), and Matthews (1986a, b) on upland streams provided a new impetus for ecological studies among state workers.

Several papers dealt with the zoogeography of Arkansas. Mayden (1985a) reviewed distributions and possible origins of Ouachita Upland fishes in Arkansas and Oklahoma. Robison and Harp (1985) reviewed the distribution and status of the Ouachita Madtom, *Noturus lachneri*. Cross et al. (1986) and Robison (1986) provided general overviews comparing fishes by river systems in Arkansas in a broad geographical context. Keith (1987) investigated Arkansas fish distribution in selected least disturbed reference streams within six Arkansas ecoregions, while Rohm et al. (1987) evaluated the aquatic ecoregion classification of streams in Arkansas.

Post First Edition Modern Period: 1988–2019

In 1988, a comprehensive compilation of information on Arkansas fishes was published, titled *Fishes of Arkansas* (Robison and Buchanan 1988). Publication coincided with the beginning of a new era of ichthyological study in this state, spurred by several developments. One of the most important developments has been the emergence of the World Wide Web and the use of personal computers in general. The main contributions of these new technologies are obvious, including greatly improved access to information and easier, faster communication with colleagues. Other new or improved technologies were developed, including new methods in molecular biology that permit an unprecedented ability to study mitochondrial and nuclear genes (see Chapter 4); the advent of GPS instruments for rapidly and precisely pinpointing localities by latitude and longitude; and the development of the Missouri trawl (Herzog

and Hrabik 2013; Herzog et al. 2005, 2009) for sampling small fishes from some of the most inaccessible habitats of large rivers. Another trend was the dramatic increase in the number of people in Arkansas (as well as outside the state) carrying out investigations of Arkansas fishes, including more ichthyologists not only at state and federal agencies, but also at universities across the state. In 1988, only two universities in Arkansas (University of Arkansas in Fayetteville and Arkansas State University in Jonesboro) offered postgraduate degrees based on ichthyological projects. By 2015, two additional universities (Arkansas Tech University and University of Arkansas at Pine Bluff) offered masters and doctoral degrees in this area, and three others (Henderson State University, University of Central Arkansas, and University of Arkansas at Little Rock) offered masters degrees. The post-1988 upsurge in ichthyological investigation has produced a large volume of publications in scientific journals, many unpublished master's theses, and some doctoral dissertations. There have been more studies on life history aspects, more studies to determine the status of rare and endangered fish species, and more papers on new distributional information. Another important trend during this period has been greater emphasis on collecting fishes from the historically least-sampled environment in Arkansas: the large rivers. Buchanan et al. (2003) surveyed the fishes of the Red River, and extensive sampling was carried out in the Mississippi, Red, and White rivers by the U.S. Army Corps of Engineers Waterways Experiment Station in Vicksburg, Mississippi (data provided by K. J. Killgore, USACE). The Arkansas River Navigation System was sampled by the Arkansas Game and Fish Commission, TMB (2005, and unpublished data reported in our species accounts), and by graduate students from Arkansas Tech University.

The year 1988 was significant in the study of Arkansas fishes, not only because of the publication of the first edition of *Fishes of Arkansas*, but also because a paper by Matthews and Robison (1988) provided a detailed multivariate analysis of the distribution of Arkansas fishes based on thousands of personal collections of Arkansas fishes by HWR from 101 drainage units. Such an analysis of a large dataset on Arkansas fishes had never been attempted. More than 3,000 collections of fishes from Arkansas were used by Matthews and Robison (1988), Matthews et al. (1992), and Matthews and Robison (1998) to analyze the distributions of Arkansas fishes and the geological, climatological, and water quality correlates they described across the state. Maps taken from Robison and Buchanan (1988), in part, formed the basis for an in-depth assessment of fish distribution, diversity, and conservation status for hydrologic units in the Ozark-Ouachita Highlands Assessment led by USDA Forest Service (Standage 1999; Warren and Hlass 1999; Warren and Tinkle 1999).

The publication of Mayden's (1989) monograph on the cyprinid genus *Cyprinella* had a major impact on the taxonomy of Arkansas cyprinids. Mayden presented hypotheses about relationships based primarily on cladistic analysis of osteology, tuberculation, and body and fin pigmentation and in the process recommended elevating several subgenera of the cyprinid genus *Notropis* to genus level. Not since the proposals of the eminent American ichthyologist Reeve Bailey (1951) had such an extensive overhaul of North American cyprinid nomenclature been proposed in a single work (Snelson 1991). Robison and Buchanan (1993) reviewed nomenclature changes and composition of the Arkansas fish fauna since the appearance of Robison and Buchanan (1988).

William J. Matthews' groundbreaking 1998 *Patterns in Freshwater Fish Ecology* used numerous examples from the Arkansas fish fauna which Matthews had studied for more than 40 years. This work focused on (1) the structure of local fish assemblages, that is, those species found at a single location in freshwater lakes or streams, and on (2) the effects of those fishes in their communities. Particularly informative was Matthews's presentation of an analysis of the zoogeographic affinities of the native fishes of Arkansas and specifically of the fishes of Piney Creek in Izard County. Matthews also used data from the multi-year collections of fishes from the Strawberry River made by HWR to test the "continuum" versus "zonation" hypotheses for stream fishes. In addition, Hoover and Killgore (1998) studied fish communities in forested wetlands of the southern United States, including the Cache River in Arkansas. Matthews and Marsh-Matthews (2016, 2017) provided fresh analyses of the short- and long-term dynamics of stream fish communities based on four decades of data from streams in Arkansas, Oklahoma, and Virginia.

During the 1988–2019 period, 11 species (not including newly described or undescribed species) were reported from Arkansas for the first time: *Ichthyomyzon unicuspis* (Robison et al. 2011b), *Mylopharyngodon piceus* (Nico et al. 2005), *Notropis wickliffi* (Robison and Buchanan 1994), *Scardinius erythrophthalmus* (Robison and Buchanan 1993), *Morone americana* (Buchanan et al. 2007), *Percina macrolepida* (Buchanan et al. 1996), *Perca flavescens* (Buchanan et al. 2000), *Oreochromis aureus* (Buchanan et al. 2000), *Oreochromis mossambicus* and *Oreochromis niloticus* (reported herein), and *Channa argus*.

The decade following the turn of the 21st century was one of renewed vigor in the study of Arkansas fishes. Fishes of the Ouachita Mountains became a prime subject of ecological research by U.S. Forest Service personnel as well as other workers. Taylor (2000) studied 74 upland localities in the Little River system of southeastern Oklahoma and southwestern Arkansas to compare riffle and pool communities. Taylor and Warren (2001) investigated how local abundance of fishes was associated with immigration and extinction rates in the Alum Fork of the Saline River system and Little Glazypeau Creek of the Ouachita River system, and Williams et al. (2003) studied the structure of regional fish faunas in these same Ouachita Mountain stream systems. Williams et al. (2002) evaluated large-scale effects of timber harvesting on stream systems in the Ouachita Mountains of Arkansas. Killgore and Hoover (2001) evaluated the fish species composition relative to dissolved oxygen concentrations in Mercer Bayou, a vegetated impoundment of the Sulphur River (Red River drainage) in southwestern Arkansas. Petersen (2004) and Petersen and Justus (2005) surveyed the fishes of the Buffalo River. Robison (2005b) surveyed the fishes of the Pine Bluff Arsenal, and Robison (2006b) compiled a list of fishes in the White River drainage.

One of the most encouraging aspects in the continuing study of fishes in Arkansas was the landmark formation of the Arkansas Fishes Database (AFD), a computerization of more than 3,500 collections of fishes in Arkansas begun by H. W. Robison from U. S. Forest Service streams and later expanded to include the entire state (Robison et al. 2004). The AFD originally grew out of discussions between HWR and Melvin Warren of the U.S Forest Service on the need for synthesis of state fish distribution data that was available in many venues but not available in a single computerized system. The Forest Service was interested in establishing a computerized database of fishes of streams in Forest Service lands, and HWR was initially funded to computerize his thousands of fish records from those lands. This database was subsequently expanded to include fish collection data from other ichthyologists and collections housed at the University of Louisiana Monroe, Tulane University, and Arkansas State University. Following initial construction of the AFD, support from the Arkansas Game and Fish Commission was forthcoming, and HWR continued to expand and update the database. This new database is a Geographical Information System (GIS) compatible fish research database for the entire state of Arkansas to document historical and present fish distributions (Robison et al. 2004). In addition, this digital database allows identification of unique ecological or taxonomic fish community assemblages and centers of fish diversity within and across drainages of Arkansas. Historical changes in stability and persistence of community assemblage patterns and historical trends in species distributions can be easily associated with land use. Finally, the database provides for geospatial tools to assess conservation status of individual Arkansas fish species. The Arkansas Fishes Database is currently maintained by the Arkansas Game and Fish Commission.

Important nomenclatural changes occurred in 2010. First, Cashner et al. (2010) redescribed the Ouachita Mountain and Arkansas River Valley populations of stonerollers as *Campostoma spadiceum*, resurrecting an older name from the synonymy of *Campostoma anomalum*. Shortly thereafter, Mayden (2010) analyzed the *Etheostoma punctulatum* populations of the Ozarks and recognized two new species in Arkansas, *Etheostoma autumnale* (Autumn Darter) and *E. mihileze* (Sunburst Darter), concluding that the true *E. punctulatum* did not occur in Arkansas. Next, Kinziger and Wood (2010) analyzed *Cottus hypselurus* populations in northern Arkansas and southern Missouri and concluded that *C. hypselurus* was made up of two distinct species. They described the Arkansas (and some Missouri) populations as a new species, the Knobfin

Sculpin, *Cottus immaculatus*, with the true *Cottus hypselurus* (now *Uranidea hypselura*) limited to Missouri.

An analysis of the systematics of the *Etheostoma stigmaeum* complex by Layman and Mayden (2012) resulted in the description of five new species, including two new species in Arkansas. The first, the Beaded Darter, *Etheostoma clinton*, is a state endemic occurring only in the upper Caddo River and the upper Ouachita River, and the second new species, *Etheostoma teddyroosevelt*, inhabits the White and Arkansas river systems in Arkansas. William J. Matthews et al. (2014) published the latest in a 40+ year ecological study (begun in 1972) of the fishes of Piney Creek, a White River tributary, in the Ozark Mountains of northern Arkansas. The most recent additions to the state ichthyofauna were made by Robison et al. (2014), who published the long-anticipated description of the Ouachita Darter, *Percina brucethompsoni*, a percid species related to *Percina nasuta* that had been known for a few decades but never formally described, and *Labidesthes vanhyningi* (Werneke and Armbruster 2015).

The Environmental Setting

Terrestrial Ecoregions

Arkansas is situated geographically between the parallels of 33° and 36°30' north latitude and 89°39' and 94°37' west longitude, placing it in the midsection of the southcentral United States. The state is bounded on the north by Missouri; on the east by the Mississippi River, which separates it from Mississippi and Tennessee; on the south by Louisiana; and on the west by Oklahoma and Texas. Approximately 250 miles (402 km) from north to south and 235 miles (378 km) from east to west with a total area of 53,255 square miles, Arkansas ranks 26th in size among the states. There are 75 counties in Arkansas (Fig. 3.1).

Arkansas is readily divisible into two broad geographic regions based on topography, geological substrate, and dominant vegetation: (1) the Interior Highlands (uplands), and (2) the Gulf Coastal Plain (lowlands) (Fig. 3.2).

The Interior Highlands, reaching elevations of more than 2,800 feet (853.4 m), were originally formed by uplifting, folding, and faulting processes in the Pennsylvanian (300 mya), whereas the Gulf Coastal Plain was the result of inundation of this area by the Gulf of Mexico during the Cretaceous, with resulting deposition of unconsolidated sediments. Within these two basic geographic divisions, most workers generally recognize six physiographic regions or physiographic provinces based on geography and topography, but we use the term ecoregion to denote

that these regions are also characterized by distinctive fish assemblages. The six ecoregions in Arkansas (Fig. 3.3) are Ozark Mountains, Arkansas Valley, Ouachita Mountains, West Gulf Coastal Plain, Mississippi Alluvial Plain, and Crowley's Ridge. To most ecologists, the term ecoregion is synonymous with the terms natural region and natural division, and the Arkansas Natural Heritage Commission (ANHC) and the Environmental Protection Agency (EPA) consider these terms the same (Woods et al. 2004).

The following abbreviated descriptions and characterizations of the ecoregions of Arkansas (Fig. 3.4) are based on information in Croneis (1930), Fenneman (1938), Thornbury (1965), Foti (1974, 1976), Pell (1983), Robison (1986), Howard (1989), Woods et al. (2004), and Gentry et al. (2013).

The major northern division of the Interior Highlands is the Ozark Mountains ecoregion, which occupies most of northwestern and northcentral Arkansas and is characterized by rugged, flat-topped mountains, long, deep valleys, steep cliffs and ledges, and clear, spring-fed streams (Fig. 3.5).

This ecoregion comprises three plateaus of different ages, originally uplifted as a single unit with few folds and faults and severely dissected by many swift rivers, enhancing the ruggedness of the region. Local relief of more than 750 feet (229 m) is not uncommon. Most of this area has been above

Figure 3.1. Arkansas Counties.

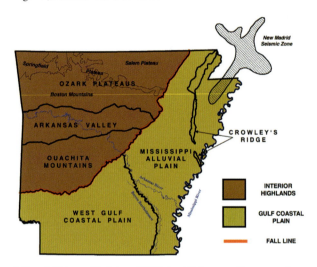

Figure 3.2. Physiographic regions and six ecoregions (natural divisions) of Arkansas. *Courtesy of Gentry et al. (2013).*

sea level continuously for the nearly 300 million years since the Pennsylvanian. Elevations range from 250 to 2,450 feet (76–747 m) above sea level. Principal rock types are Ordovician limestone and dolomite, Pennsylvanian sandstone, and Pennsylvanian and Ordovician shales. Soils are primarily residual, and vegetation is mostly upland hardwood forests of white oak, red oak, and hickory. The three distinct topographic subdivisions of the Ozark Mountains are as follows: the Salem Plateau, the Springfield Plateau, and the Boston Mountains. The Salem Plateau (the oldest) lies primarily in the northeastern portion of the ecoregion, the Boston Mountains (the youngest) occupy the southern portion, and the Springfield Plateau extends

in a narrow band between them. The Boston Mountains have the highest elevation, are the youngest geologically (Pennsylvanian), and are the most extensively eroded (dissected), and therefore include the most rugged terrain of the Ozark Mountains in Arkansas (Fig. 3.6).

Gorges and ravines 500 to 1,200 feet (152–366 m) deep are common. Elevations range from 1,900 to 2,578 feet (579–786 m) above sea level at the headwaters of the Buffalo River in southwestern Newton County. Surface rock is sandstone and shale of the Atoka Formation. Soils are sandy and clay loams, generally acid, medium textured, and moderately permeable. The elevations of the Springfield Plateau are intermediate between those of the other two subdivisions, creating gently rolling hills ranging from 1,000 to 1,500 feet (305–457 m) above sea level in the northwestern corner of the state (Fig. 3.7). Solution valleys and caves are common.

This subdivision is highly dissected on its northern border, the Eureka Springs Escarpment, and near its southern terminus, whose erosional remnants stand 1,000 to 1,500 feet (305–457 m) above the plateau surface. Surface rocks are generally limestone and insoluble chert of the Boone Formation (Mississippian) with some infrequent sandstone and shale. Soils are quite thin and strewn with rocky rubble. The Salem Plateau is the largest as well as the lowest of the Ozark Plateaus, ranging from 250 to 1,250 feet (76–381 m) in elevation throughout its rolling landscape. It is situated mainly east and north of the Springfield Plateau. Steep slopes are common only near the White River (Fig. 3.8).

Surface rocks of the Salem Plateau (the oldest in the Arkansas portion of the Ozarks) are largely limestone and Cotter dolomite (Ordovician) with occasional sandstone outcrops. Soils are rocky, dry, and thin. Because this plateau is relatively level or gently rolling, many areas have been converted to pastures and residential building lots.

The Arkansas Valley ecoregion (also called the Arkoma Basin) is a deep synclinal trough that lies between the Ozark Mountains and the Ouachita Mountains and acts in many ways as a transitional zone between the two (Fig. 3.9).

The northern boundary of this ecoregion is the Boston Mountains, and the Fourche Mountains represent the southern limit. Geologically, the valley contains the youngest Paleozoic rocks in Arkansas. It is generally characterized by rolling plains 500–600 feet (152–183 m) above sea level and 25 to 30 miles (40–48 km) wide, which extend from Searcy in White County west to Fort Smith in Sebastian County. Within this rolling lowland are isolated, flat-topped synclinal mountains (Mount Nebo, Petit

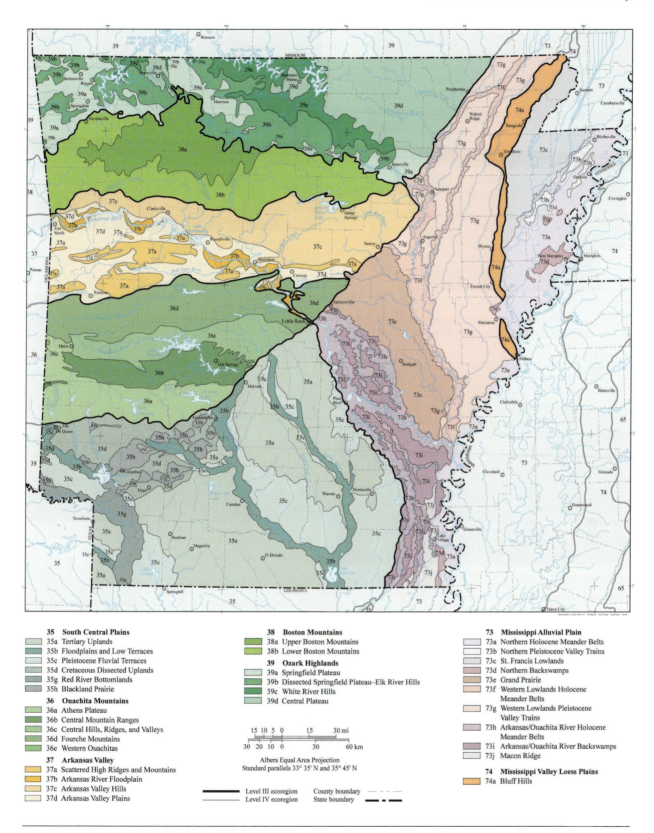

35 South Central Plains
35a Tertiary Uplands
35b Floodplains and Low Terraces
35c Pleistocene Fluvial Terraces
35d Cretaceous Dissected Uplands
35g Red River Bottomlands
35h Blackland Prairie

36 Ouachita Mountains
36a Athens Plateau
36b Central Mountain Ranges
36c Central Hills, Ridges, and Valleys
36d Fourche Mountains
36e Western Ouachitas

37 Arkansas Valley
37a Scattered High Ridges and Mountains
37b Arkansas River Floodplain
37c Arkansas Valley Hills
37d Arkansas Valley Plains

38 Boston Mountains
38a Upper Boston Mountains
38b Lower Boston Mountains

39 Ozark Highlands
39a Springfield Plateau
39b Dissected Springfield Plateau–Elk River Hills
39c White River Hills
39d Central Plateau

15 10 5 0   15   30 mi
30 20 10 0   30   60 km

Albers Equal Area Projection
Standard parallels 33° 35' N and 35° 45' N

Level III ecoregion   County boundary
Level IV ecoregion   State boundary

73 Mississippi Alluvial Plain
73a Northern Holocene Meander Belts
73b Northern Pleistocene Valley Trains
73c St. Francis Lowlands
73d Northern Backswamps
73e Grand Prairie
73f Western Lowlands Holocene Meander Belts
73g Western Lowlands Pleistocene Valley Trains
73h Arkansas/Ouachita River Holocene Meander Belts
73i Arkansas/Ouachita River Backswamps
73j Macon Ridge

74 Mississippi Valley Loess Plains
74a Bluff Hills

Figure 3.3. Environmental Protection Agency Level IV Ecoregions of Arkansas. *From Woods et al. (2004).*

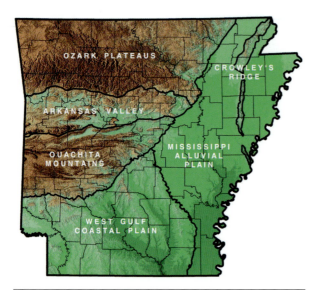

Figure 3.4. Shaded relief map showing six major ecoregions of Arkansas. *Courtesy of Gentry et al. (2013).*

Figure 3.5a. Boston Mountains, Newton County, Arkansas (Ozark Mountains Ecoregion). *William Rainey.*

Figure 3.5b. Ozark Mountains near Sam's Throne, Arkansas (Ozark Mountains Ecoregion). *Keith Sutton.*

Figure 3.6. Buffalo River, Newton County, Arkansas, Ozark Mountains Ecoregion. *Dero Sanford.*

Figure 3.7a. Large spring in the Springfield Plateau, Arkansas. *Chris T. McAllister.*

Jean Mountain) rising as high as 1,500 feet (457 m). The Arkansas Valley ecoregion is the site of the highest elevation and greatest vertical relief in the state of Arkansas, Magazine Mountain (2,800 feet [853 m]) (Fig. 3.10).

Although the Arkansas River flows eastward through this ecoregion before turning southward between Conway and Little Rock, its bottomlands should not be confused with the Arkansas Valley ecoregion, as the river has apparently always occupied only a small part of the Valley.

The Ouachita Mountains ecoregion is the southern division of the Interior Highlands, comprising a belt about 60 miles (965 km) wide and 120 miles (193 km) long (Fig. 3.11).

Unlike their northern counterpart, the Ozark Mountains, the Ouachita Mountains form a series of long ridges running east and west. Relatively wide valleys, each drained by a river or stream, separate the long, narrow ridges. Only the southernmost portion of the Ouachitas can be

Figure 3.7b. White River in the Springfield Plateau, Arkansas. *Brent Baker, Arkansas Natural Heritage Commission.*

Figure 3.8. Calcareous bluffs in the Salem Plateau. *Craig M. Fraiser.*

Figure 3.9. Arkansas Valley Ecoregion. *Keith Sutton.*

Figure 3.10. Magazine Mountain, Arkansas Valley Ecoregion. *William Rainey.*

considered an uplifted plateau. The rocks of the Ouachita ecoregion are mainly Paleozoic sedimentary sandstone and shale ranging in age from Cambrian or Ordovician through Pennsylvanian. Those strata were warped, twisted, and folded under tremendous pressure millions of years ago by tectonic events associated with continental collisions. Some of the most interesting rocks in the state occur here, including Arkansas Stone, Ouachita Stone (novaculite), and quartz crystal. Elevations range from 250 feet (76 m) along the Arkansas River to more than 2,600 feet (793 m) on Rich Mountain (2,681 feet [818 m]). Shortleaf pine, upland hardwood, and bottomland hardwood forests predominate in the Ouachita Mountains. Soils are derived from shale and sandstone with recent alluvium in the bottomlands of the main rivers.

The Ouachita Mountains ecoregion can be divided into three subregions: the Fourche Mountains to the north,

Figure 3.11a. Ouachita Mountains Ecoregion. *William Rainey.*

Figure 3.11b. Ouachita Mountains, Perry County, Ouachita Mountains Ecoregion. *Brent Baker, Arkansas Natural Heritage Commission.*

Figure 3.12a. Athens Piedmont Plateau, Ouachita Mountains Ecoregion. *Brent Baker, Arkansas Natural Heritage Commission.*

Figure 3.12b. Athens Piedmont Plateau, Cossatot River, Ouachita Mountains Ecoregion. *Keith Sutton.*

the Novaculite Uplift or Central Ouachita Mountains in the center, and the Athens Piedmont Plateau on the south (Fig. 3.12).

The Fourche Mountains occupy a belt approximately 25 miles (40 km) wide extending from Little Rock west to the Oklahoma state line near Mena. These mountains have rugged, narrow ridges, rising more than 2,600 feet (793 m) above sea level (Blue Mountain 2,623 feet [800 m]), decreasing in height to 250 feet (76 m) eastward near Little Rock. Atoka and Jackfork sandstone, along with Johnsville and Stanley shales, are the principal bedrock types. The Central Ouachitas begin near Little Rock and extend westward to southern Polk County as a spindle-shaped area. The novaculite that occurs here is a dense, siliceous rock found only in Arkansas, Oklahoma, and two outcrops in central and western Texas. The oldest surface rocks in Arkansas (Cambrian shales) are found in this subdivision. Some of the most rugged topography in Arkansas

Figure 3.13a. Mississippi Alluvial Plain Ecoregion. *Photograph by Paul Otto.*

Figure 3.14a. West Gulf Coastal Plain Ecoregion. *Brent Baker, Arkansas Natural Heritage Commission.*

Figure 3.13b. Lake in the Mississippi Alluvial Plain Ecoregion. *Keith Sutton.*

Figure 3.14b. Mature Beech-Holly forest in West Gulf Coastal Plain Ecoregion. *Photograph courtesy of the Arkansas Natural Heritage Commission.*

occurs in this natural division, especially in the western regions in the Cossatot Mountain group where the ridges are closely crowded and elevations of 1,500 to 2,000 feet (457–610 m) are common. Soils are shallow and generally stony. The Athens Piedmont Plateau, a belt 8 to 18 miles (13–29 km) wide, extends from near the Ouachita River in Hot Spring County west to Oklahoma. Near its northern boundary, there are elevations of more than 1,000 feet (305 m). Novaculite ridges in northern Howard and Pike counties slope to 400 feet (122 m) on the southern border. Cretaceous and Early Tertiary deposits separate the Coastal Plain from the Paleozoic formations of the Interior Highlands.

The Coastal Plain is a relatively flat region lying south and east of the Interior Highlands. It is separated into three ecoregions: the Mississippi Alluvial Plain (Delta); a southwestern portion, the West Gulf Coastal Plain; and Crowley's Ridge, which bisects the Delta. The Mississippi Alluvial Plain ecoregion, occupying much of the eastern

third of Arkansas (about 8 million acres), is a level plain that slopes toward the south (Fig. 3.13).

It is covered with recent alluvium and terrace-deposits at elevations from 100 to 300 feet (30–91 m) above sea level. Soils are the result of the Ohio, Arkansas, and Mississippi rivers carrying material, including glacial deposits, from the north and west. Recently transported alluvium covers most of the area. Extensive wetlands once characterized this area, but by 1970 only 20% of the original forested area remained (Holder 1970). Wetland draining and intensive agricultural development of this rich area continue today. The Mississippi Alluvial Plain is subdivided into the plains (predominantly ancient river terraces or soils derived from loess) and the bottomlands (predominantly recent floodplains).

In southwestern Arkansas, the West Gulf Coastal Plain ecoregion is a large rolling area inundated by the Gulf of Mexico 50 million years ago, but now eroded by southward and southeastward flowing streams (Fig. 3.14).

Figure 3.15a. Stream on Crowley's Ridge, Crowley's Ridge Ecoregion. *Keith Sutton.*

Figure 3.15b. Loess soils of Crowley's Ridge Ecoregion. *Keith Sutton.*

(windblown soils) and rises 100 to 250 feet (30–76 m) above the broad alluvial plain surrounding it. Because of the nature of the soils, the Ridge is subject to extensive erosion resulting in ravines and gullies (Fig. 3.15). Its vegetation is more closely related to the tulip tree-oak forest of Tennessee than to the oak-hickory forest of the Ozarks.

Aquatic Ecosystems

Arkansas has a wealth of aquatic resources, including 600,000 acres of lakes and more than 20,000 miles of streams (Robison and Smith 1984). Within this realm a myriad of natural standing and running water environments exist, including cave streams, springs, upland streams, upland rivers, lowland streams, lowland rivers, big rivers, meander scar (oxbow) lakes, alluvial swamps, and sinkhole ponds. The largest natural lakes (500 surface acres or more) in Arkansas are listed in Table 7 and Figure 3.16.

Arkansas is drained entirely by the Mississippi River; however, the overall state drainage may be conveniently subdivided into five smaller drainage basins, the Red, Ouachita, Arkansas, White, and St. Francis rivers. The Mississippi River, with adjacent tributary streams and oxbows, can be included as a sixth basin within the state (Table 8; Figs. 3.17, 3.18).

The Red River is a turbid plains stream farther west in Oklahoma and drains only 4,433 square miles of extreme southwestern Arkansas. Because of the fine silt that enters the stream from the western portions of Oklahoma and Texas as well as Arkansas, the river within Arkansas is almost continuously turbid (although construction of Lake Texoma in the 1940s reduced turbidity downstream of Denison Dam to some extent) (Fig. 3.19).

It has been divided into two subdivisions, Southwestern Arkansas and Southcentral Arkansas. Elevations range from 200 to 700 feet (61–213 m) above sea level. The distinctive Southwestern Arkansas subdivision consists primarily of Cretaceous sediments, including sand, clay, gravel, and marl with some sandstone, chalk, and limestone. Terrain is level to hilly with elevations reaching 800 feet (244 m) along the northern limits. Most of the West Gulf Coastal Plain ecoregion is in the Southcentral Arkansas subdivision, an area of poorly defined soils of noncalcareous sand, silt, and clay.

Crowley's Ridge ecoregion is a long, narrow, low ridge ranging from 1 to 10 miles (1.6–16 km) wide, originating in southern Missouri and extending southward 150 miles (240 km) to Helena, Arkansas. Interestingly, the Ridge ecoregion represents an ancient divide between the Mississippi and Ohio rivers when the Mississippi flowed west of the ridge and the Ohio flowed east of it (Robison 1986). Crowley's Ridge was originally deposited as loess

Table 7. Natural Lakes in Arkansas*: Approximately 500 Acres or More at Normal Levels.

Lake	Acres	Classification[†]
1. Beulah	800	Private (½ Ark., ½ Miss.)
2. Big Lake	6,500	National Wildlife Refuge Lake
3. Chicot	5,000	Private, USACE & AGFC
4. Council	1,075	Private (½ Ark., ½ Miss.)
5. Faulkner	600	Private
6. Ferguson	2,025	Private
7. Grand	1,400	Private
8. Grassy	2,000	Private
9. Horseshoe	2,500	Private
10. Island 66 (Desoto Lake)	1,100	Private (½ Ark., ½ Miss.)
11. Lee	1,800	Private (½ Ark., ½ Miss.)
12. Old River–Brooks Griffin (Mellwood)	1,000	Private (½ Ark., ½ Miss.)
13. Old River	700	Private
14. Old Town	1,200	Private
15. Paradise	900	Private (½ Ark., ½ Miss.)
16. Porter	500	Private
17. Swan	500	Private
18. Tunica (Island 58)	3,900	Private (½ Ark., ½ Miss.)
19. Wapanocca	1,800	National Wildlife Refuge
20. Whittington	4,000	Private (½ Ark., ½ Miss.)

*Almost all of Arkansas's natural lakes are meander scar lakes (oxbows) representing former river channels. Today most of these lakes are entirely cut off from their associated rivers during normal water levels, but during periods of high water most may be inundated by the river. A few of these lakes such as Island 66 (DeSoto), Mellwood, Ferguson, and Whittington are permanently connected to the Mississippi River by narrow channels, except during periods of extremely low water levels. A number of the natural lakes have artificial water level control structures such as levees, gates, and pumping plants. Wapanocca and Big Lake are the only large natural lakes that are not oxbows but were created by earthquakes.

[†]The state owns the lakebeds of all natural lakes formed by legally defined navigable streams, but private ownership of the surrounding land may exclude public access to any natural lake, including Chicot, the largest oxbow.

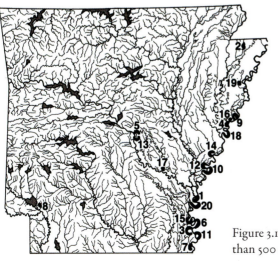

Figure 3.16. Natural lakes greater than 500 surface acres.

Table 8. River Systems of Arkansas.

System	Arkansas length (miles)	Total length (miles)	Arkansas drainage area (sq. mi.)	Total drainage area (sq. mi.)	Volume of flow	Proportion of state's total land
Red River	135	1,209	4,433	69,213	18%	11%
Ouachita River	362	542	11,548	25,999	11%	21%
Arkansas River	310	1,450	11,100	160,533	38%	25%
White River	625	720	17,143	27,765	29%	34%
St. Francis River	140	475	4,200	8,400	5%	9%
Mississippi River	321	2,400	4,755	1,245,000		

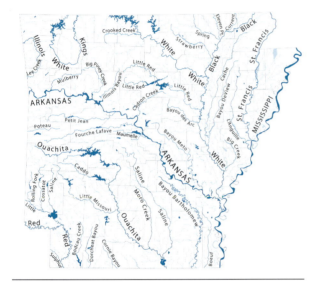

Figure 3.17. Major rivers and streams of Arkansas.

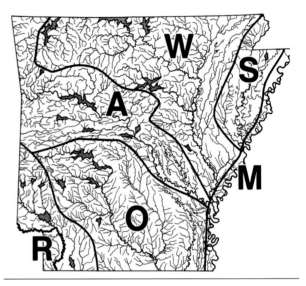

Figure 3.18. River Systems of Arkansas.

M—Mississippi   S—St. Francis
W—White   O—Ouachita
A—Arkansas   R—Red

The Arkansas portion of the Red River is unimpounded, but there are many bank and channel stabilization structures downstream of U.S. Route 71. Major tributaries in Arkansas are the Little River (with its three upland river tributaries: Rolling Fork, Cossatot, and Saline rivers), McKinney Bayou, Sulphur River, Bodcau Bayou, and Bayou Dorcheat.

The Ouachita River flows 362 miles through southern Arkansas and drains 11,548 square miles in two different ecoregions of the state: the Interior Highlands of the Ouachita Mountains and the Coastal Plain ecoregion south of Arkadelphia (Fig. 3.20).

From the upper mountainous regions, the swift-flowing, clear, spring-fed river flows eastward before turning southeasterly to flow onto the gently rolling Coastal Plain.

Major tributaries are the Caddo River, Little Missouri River, Moro Creek, Saline River, Bayou Bartholomew, Big Cornie Creek, and Bayou de L'Outre.

The Arkansas River passes through the center of the state in a general northwest to southeast direction for 310 miles and drains 11,100 square miles. Suspended silt provides a continuous turbid appearance to the river, but turbidity has been drastically reduced by the construction of the Arkansas River Navigation System (Fig. 3.21).

Waters are slightly saline as a result of the river crossing large, natural salt beds in Oklahoma and Kansas, but salinities have been steadily decreasing for the past 40 years. Mountain tributaries flowing out of the Ozark Mountains to the north and the Ouachita Mountains to the south have a diluting effect on the salinity as the river enters Arkansas.

Figure 3.19. Red River near Texarkana, Arkansas. *Keith Sutton.*

Figure 3.21a. Arkansas River, Conway County, Arkansas. *Brent Baker, Arkansas Natural Heritage Commission.*

Figure 3.20a. Upper Ouachita River in Arkansas. *Henry W. Robison.*

Figure 3.20b. Lower Ouachita River in Arkansas. *Keith Sutton.*

Figure 3.21b. Dam on the Arkansas River. *Keith Sutton.*

Figure 3.22a. White River at Calico Rock, Arkansas. *Henry W. Robison.*

Figure 3.23. St. Francis River near Forrest City, Arkansas. *Keith Sutton.*

Figure 3.22b. White River near Oil Trough, Arkansas. *Keith Sutton.*

Figure 3.24. Mississippi River. *Jan Jeffrey Hoover, USACE.*

Major tributaries are Clear Creek, Mulberry River, Illinois Bayou, Piney Creek, Cadron Creek, Bayou Meto, Petit Jean River, Fourche LaFave River, and Maumelle River.

The White River originates in northwestern Arkansas and flows northeasterly for a short distance into Missouri before turning southeasterly to reenter Arkansas and ultimately join the Arkansas River near its mouth (Fig. 3.22).

The White River has the largest drainage basin of any river system in the state, with 17,143 square miles in Arkansas. Major tributaries are the Black River system: which includes the Strawberry, Spring, Eleven Point, and Current rivers; and the Kings, James, Buffalo, North Fork, Little Red, and Cache rivers; Bayou DeView, Des Arc Bayou, Bayou La Grue, and Big Creek.

The long, narrow St. Francis River basin drains 4,200 square miles of Arkansas. About 100 miles of the St. Francis River are in Arkansas (Fig. 3.23), and another 40 miles form the Arkansas-Missouri Bootheel border.

Principal tributaries of the St. Francis River are Big Slough, Little River, Tyronza River, and L'Anguille River.

Figure 3.25. Lake Chicot, largest oxbow lake in Arkansas in Chicot County. *Keith Sutton.*

The largest river, the Mississippi, forms the eastern boundary of Arkansas, and the St. Francis, White, and Arkansas rivers are its major tributaries (Fig. 3.24). There are only few direct small stream tributaries in Arkansas. Numerous oxbow lakes, including the largest in the state, Lake Chicot in Chicot County, are a part of the Mississippi River system (Fig. 3.25).

Environmental Alterations

General Overview

Human impact on the aquatic environments of Arkansas has been substantial. Although aquatic resources abound in Arkansas, virtually all natural waters in the state have been influenced in some way by human activity. Some of the effects of man's influence have been only temporary because of the natural self-purifying and self-healing mechanisms of aquatic ecosystems; however, other effects of human activity have been long lasting and some are permanent. Lipton and Strand (1997) summarized the economic effects of pollution in fish habitats, and Ensign et al. (1997) discussed factors influencing stream fish recovery following large-scale disturbance. The National Wetlands Priority Conservation Plan identified Arkansas as one of 19 states that experienced significant decreases in wetlands (mostly in the Delta ecoregion) from 1954 to 1974 (ADEQ 2016). Today, only a fraction of the original wetland areas of Arkansas remain, making the impact of any additional loss more significant.

Alteration of aquatic environments began before settlement of the state by Europeans almost two centuries ago, and changes have accelerated as the population has grown in recent decades. Smith et al. (1984) traced the land use history of Arkansas from the first European explorer (Hernando de Soto in 1541) to the mid-1980s. Early naturalists' descriptions of Arkansas in the 1800s (Nuttall 1999) bear little resemblance to conditions today, particularly in the lowlands. Young (2011) used historical documentation along with modern measurements to explain trends in fish habitat following European settlement in the Illinois Bayou watershed of Arkansas. He concluded that reduced wetted stream width associated with incision of the stream valley since 1845 resulted in decreased opportunity to deposit sediment on the Illinois Bayou floodplain. Subsequent forest management practices (including fire suppression) resulted in tree densities almost three times those of presettlement levels, likely exacerbating stream drying during summer months resulting from increased evapotranspiration rates. Young (2011) further noted that land-use activities that reduce surface flows in summer are expected to adversely affect fish habitat in the Illinois Bayou and similar streams. Illinois Bayou is the only known stream in the United States where two species of the Orangethroat Darter complex occur syntopically.

In 1976, 33% of the streams in Arkansas did not meet the federal water quality goals for "fishable/swimmable" waters (ADPCE 1976). By 1984, Arkansas had only managed to "hold its own" in regard to those water quality standards (ADPCE 1984). Using somewhat different standards, the Environmental Protection Agency's 2008–2009 National Rivers and Streams Assessment classified 57% of the state's sampled river miles as exhibiting "poor" condition based on total nitrogen, total phosphorous, and/or salinity. In 2008, the Arkansas Department of Environmental Quality (ADEQ) classified 41% of the waters assessed in Arkansas as impaired for all parameters measured, including nutrients. The 2010, 2012, and 2014 ADEQ reports classified 38%, 37%, and 40% as impaired, respectively (at least one measured parameter was not met, ADEQ 2016). In general, the highest water quality is found in streams of the sparsely populated Boston Mountains ecoregion of the Ozark Uplands, much of which is located within the Ozark National Forest, with streams of the Ouachita Uplands running a close second. However, water quality is deteriorating in the many streams of the Springfield Plateau of the Ozark Highlands ecoregion. In that region, land use practices and a rapidly increasing human population have contributed to increased water contamination from infrastructure development as well as surface erosion from construction activities. The karst topography in this area has exacerbated these effects. This region also has some of the highest livestock production in the state. Concentrated animal wastes provide a potential source of contaminants, such as nitrates and phosphates, for surface and subsurface waters. Petersen et al. (2014) used a fish index of biotic integrity and several other fish metrics associated with feeding preference, spawning preference, and tolerance of habitat degradation to assess the effects of increasing urban and agricultural land use in the Illinois River basin in Benton and Washington counties. Those metrics indicated some disturbance of the fish communities, especially at urban sites. The highly agricultural Mississippi Delta has the poorest water quality, and parts of the West Gulf Coastal Plain and Arkansas River Valley have severe water quality problems. Poor water quality in the Delta is caused by nonpoint source agricultural runoff combined with the fact that the majority of the streams in this region are a network of extensively channelized drainage ditches (ADEQ 2016).

The effects of environmental alterations on the fishes of Arkansas are difficult to accurately assess. To our knowledge, only one native Arkansas species, the Harelip Sucker, *Moxostoma lacerum,* has become extinct. Its extinction apparently occurred by 1900, long before our awareness of extensive environmental modification. Three other native species, *Cyprinella camura, Notropis bairdi,* and *Notropis girardi,* although still present in neighboring states, have not been collected in Arkansas in more than 60 years and may be victims of habitat alteration. Many other fishes are rare or endangered in Arkansas and could be extirpated by future environmental perturbations. Considering the extent of aquatic habitat alteration in Arkansas, it is amazing that more fish species have not been eliminated from the state. It is clear that the ranges of many species have been drastically reduced and fragmented, and the fish community structures of many natural waters have been significantly altered. Large rivers in particular provide diverse habitats capable of supporting many different fish species (H. L. Schramm, Jr., pers. comm.), but there is little information on fish community changes induced by human activity in large rivers in Arkansas. Parks et al. (2014) documented historical changes in fish assemblage structure in three large rivers of Iowa, and it is possible that similar changes have occurred in the large rivers of Arkansas. In that study, there was significant change in species composition, with the most prominent declines occurring in backwater species with phytophilic spawning strategies.

Anthropogenic disturbances to aquatic ecosystems have often resulted in fish kills in all parts of the state. Martin (1970) reported that in the early 1950s fish kills covering more than 50 miles (80 km) of stream were yearly occurrences in the Ouachita River between Camden, Arkansas, and Monroe, Louisiana. The most frequent cause of fish kills was low concentration of dissolved oxygen (DO). Although natural events may result in low DO, most fish kills caused by low DO are due to human-made inflows of organic matter such as sewage, feedlot effluent, and other agricultural wastes. Agricultural chemicals and industrial wastes have also accounted for a sizeable percentage of kills in Arkansas. From 1976 through 1983, an average of more than 11 fish kills per year were reported in the state, with a high of 18 reported in 1981. For a recent four-year period (2010–2013) an average of 10 reported fish kills per year occurred in Arkansas, with 16 kills occurring in 2011 (ADEQ, pers. comm.). Undoubtedly, many kills go unreported each year. In comparison, some states have many more reported fish kills; for example, California reported 50 in 1975, Ohio reported 70 that same year, New York reported 90+ in 2009, and Maryland reported 95 in 2012.

It is beyond the scope of this book to attempt to address every known type and example of adverse human impact on Arkansas fishes, but major categories of activities and actions (some of which overlap) causing environmental alterations are (1) agricultural activities, (2) stream channelization, (3) construction of dams and navigation systems, (4) gravel mining, (5) accidental chemical spills, (6) forestry activities, (7) industrial pollution, (8) municipal pollution, (9) groundwater contamination and depletion, and (10) road-building activities. Although often less obvious than physical and chemical habitat alteration, introductions of nonnative fishes can also have severe direct and indirect effects on the native fishes. Impacts of nonnative fishes have been inadequately studied in Arkansas. Fish faunas in the United States have become more similar through time because of widespread introductions of a group of cosmopolitan species (Rahel 2000, 2002). Rahel reported that on average, pairs of states have 15.4 more species in common now than before European settlement of North America. The 89 pairs of states that formerly had no species in common now share an average of 25.2 species. Cucherousset and Olden (2011) summarized the literature reporting ecological impacts of nonnative freshwater fishes.

Agricultural Activities

Agriculture in Arkansas has had a profound influence on the state's fishes. Hernando de Soto in 1541 reported extensive farming by Native Americans in eastern and southwestern Arkansas. Ensuing settlement of the state by Europeans and slaves of African descent in the 1700s and early 1800s further expanded farming activities. This expansion continued to modern times, but after World War II the state's agriculture changed significantly from small rural farm operations to large-scale farming. Today, farms make up 45% of the total land area of Arkansas, with 9 million acres in crop production (not including commercial forestry acreage) and 3 million acres in pasture land and other agricultural uses (ADEQ 2016). Most of the cropland acreage is concentrated in the eastern third of the state, with the remainder located primarily along the Arkansas and Red river valleys.

Fish habitat in the Mississippi Delta has been severely degraded by agricultural runoff containing silt, pesticides, herbicides, excess nutrients, and organic material. Sediment is Arkansas's largest single water pollutant by volume, with an estimated annual yield of 104 million tons. About half the sediment originates on agricultural land, and most is due to sheet and rill erosion. Most native Arkansas fishes are intolerant of turbidity and silt. Silt can cause suffocation by clogging the gills, but it can also affect fish populations by smothering spawning beds, reducing photosynthesis by blocking sunlight (causing a drop in dissolved oxygen), and destroying the habitat for fish food organisms such as insect larvae (Owen 1985). In the Chattahoochee River, Georgia and Alabama, there was a significant positive relationship between agricultural land use and instream sediment, and stream depth heterogeneity decreased significantly with increased sediment (Walser and Bart 1999). In that study, mainstream reaches draining agricultural lands had significantly lower levels of fish diversity than forested reaches. Excessive turbidity and siltation due to agriculture have become increasingly severe in some Arkansas streams containing populations of rare or threatened fish species or species of very restricted range. For example, the Strawberry and Black rivers, which constitute the entire range of *Etheostoma fragi* and a large part of the range of *Etheostoma uniporum* respectively, had turbidity and phosphorous levels much above the criterion for impaired water bodies (ADEQ 2016). In the same report, East Fork Cadron Creek with an undescribed species of the Orangethroat Darter clade, also exceeded the turbidity criterion for impaired water bodies. Four years after implementing best-management practices in the upper watershed of the Strawberry River in an attempt to alleviate the impacts of cattle grazing, water quality was not improved (Brueggen-Bowman et al. 2015). Following the implementation of best-management practices, all study sites showed significant increases in at least one water quality parameter, and the concentration of *E. coli* for three of the sampling locations exceeded the maximum allowable concentration.

Pesticides and herbicides have often caused fish kills throughout the Delta region. Many agricultural chemicals are persistent in the environment and their accumulation in food chains is well documented. It was discovered in 1979 that sediments and fishes in Bayou Meto were contaminated with dioxin (2,3,7,8-tetrachlorodibenzo-*p*-dioxin [TCDD]) for more than 150 miles (250 km) downstream from Jacksonville, Arkansas. Subsequently, a total ban on commercial fishing and a limited ban on sport fishing were placed on this stream. During the mid-1980s, TCDD concentrations ranged as high as 2,500 ppt in sediments and 1,900 ppt in fishes, but by 1991, TCDD levels in sediments and fishes had declined to 276 ppt and 296 ppt, respectively, immediately below the point source (Johnson et al. 1996). In that study, concentration of TCDD in food fish fillets exceeded the USFDA allowable levels (25 ppt) for more than 12 miles (20 km) below the point source. Recent sampling in Bayou Meto found that dioxin was still present in the fish tissues, and the upper segments of Bayou Meto were still under a fish consumption advisory (ADEQ 2016). The source of the dioxin has been eliminated and the contamination is being allowed to diminish through natural attenuation.

Although there have been hundreds of reported fish kills in Arkansas during the past four decades, it is often difficult to pinpoint the direct causes of the kills resulting from agricultural activities. Furthermore, few studies have been done in this state to document the reestablishment of fish populations in the affected areas. One such study was of a fish kill in Mud Creek, Washington County, Arkansas in 1971, caused by an unknown pesticide used to rid cattle of ticks (Olmsted and Cloutman 1974). The effects of the chemical were short lived, and all the damage was done as the substance moved downstream. Twenty-nine species of fishes (100% mortality) were eliminated from the 0.9-mile (1.5 km) kill area of this small upland creek. Repopulation of Mud Creek began almost immediately after the pesticide dissipated, and after 1 year most of the species had repopulated the kill zone at population levels similar to those prior to the kill.

Fertilizers and organic wastes from farms contribute to the pollution of natural waters. Feedlots occupy more than 12,164 acres in the state and, along with poultry houses, often introduce harmful runoff to nearby streams. Oxygen-demanding organic wastes reduce the levels of dissolved oxygen, often causing fish kills. The Illinois River, which originates in northwestern Arkansas, was designated a scenic river by the state of Oklahoma. In 1987, wastewater discharge by the city of Fayetteville, Arkansas, was identified as a major source of excessive phosphorus levels in that river. Oklahoma sued Arkansas to stop this discharge, and the suit went to the U.S. Supreme Court in 1992, which ruled that the upstream state must enforce the water quality rules of the downstream state. An agreement was reached between the two states, with Arkansas agreeing to reduce phosphorus output from its wastewater treatment plants in northwestern Arkansas by 75% over the next 10 years, although it did not address poultry-farm runoff. In 2010 another lawsuit was filed by the state of Oklahoma claiming that phosphorus from poultry litter produced by six poultry companies in northwest Arkansas continued to pollute the Illinois River watershed in Arkansas and Oklahoma. The EPA began monitoring Illinois River water quality in 2010 and planned to develop a strategy to ensure that the amount of phosphorus entering the watershed does not exceed 0.037 mg/L of water.

Stream Channelization

Many of Arkansas's natural streams have been converted to ditches for flood control, to increase the amount of arable land, or, to a lesser extent, for navigation (Fig. 3.26).

The channelization of streams and the subsequent draining of their associated wetlands has had devastating effects on fish populations, particularly in the lowlands. The Lower Mississippi River has been channelized and shortened by 142 miles (229 km) but remains undammed;

Figure 3.26. Ditch in Cross County, Arkansas, Mississippi Alluvial Plain. *Keith Sutton.*

Figure 3.27. Channelized stream (Dark Corner, Arkansas) in Mississippi Alluvial Plain. *Dustin Lynch, Arkansas Natural Heritage Commission.*

Figure 3.28. Channelized stream (same stream as Fig. 3.27) in Mississippi Alluvial Plain. *Dustin Lynch, Arkansas Natural Heritage Commission.*

Figure 3.29. Channelization of Cache River, Mississippi Alluvial Plain. *Keith Sutton.*

its natural floodplain has been decreased about 90% by levee construction begun in 1927 (Fremling et al. 1989). H. L. Schramm, Jr. (pers. comm.) pointed out that the levees in the free-flowing reach (including the segment bordering Arkansas) are set back from the river, and a substantial area of active floodplain remains (unlike the levees constructed close to the Mississippi River in the impounded reach north of the mouth of the Ohio River).

Most channelized streams in eastern Arkansas are little more than straight, shallow ditches having few or no deep pools and steep banks (Fig. 3.27; Fig. 3.28; Fig. 3.29).

These streams, which have often lost connectivity to floodplains, have little instream fish habitat such as natural obstructions or other cover essential for a diverse and productive fish community (Buchanan 1975). The removal of the natural forest canopy from the floodplains of streams causes a rise in water temperatures and greater bank erosion with resulting siltation that is further intensified by increased agricultural activity, often right up to the stream banks. Channelized streams must undergo periodic maintenance (deepening) in order to be efficient drainage ditches.

Numerous studies on the effects of stream channelization on fishes have been conducted in the United States, and some examples of recent studies are those of Lau et al. (2006), Beugly and Pyron (2010), and Knight et al. (2012). There has been little study of the effects of stream channelization in Arkansas, and detailed studies of any given stream both before and after channelization have not been made. Published information on stream channelization in Arkansas can be found in the following references: Holder (1970) documented the extent and effects of the disappearing wetlands of eastern Arkansas; Alexander (1973) discussed the general environmental effects of a proposed plan by federal agencies to "develop" the basin of the upper White River; and Dale (1975) provided an environmental evaluation report on various completed channelization projects in the Village Creek Basin in Randolph, Jackson, and Lawrence counties. Mauney and Harp (1979) studied the effects of channelization on fish populations of Cache River and Bayou DeView. Johnson (2011) sampled fishes at 42 sites in the Big Piney Creek watershed in a highly agriculturalized area of eastern Arkansas in Monroe and Lee counties. The results of that study suggested that the fish fauna once present in Big Piney Creek prior to agricultural ditching can still be found in agricultural ditches, even after a complete change in land cover.

Keith (pers. comm. 1985) made a biotic and abiotic comparison of a channelized stream and an unchannelized stream in the Mississippi Delta of eastern Arkansas. Both streams were tributaries of L'Anguille River and were similar in physiographic location, size of watershed, and measured flows. Brushy Creek in Cross and Poinsett counties had been channelized with the removal of instream and riparian vegetation, and cultivation extended to the edge of the stream channel. Second Creek in St. Francis County had not been channelized, and there was very little channel alteration or removal of instream and riparian vegetation. The fish populations of the two creeks were very different, with Second Creek having many more species present, a 56% greater fish biomass, greater diversity, and species composition more typical of a relatively undisturbed deltaic watershed. The fish population of Brushy Creek was characteristic of a creek lacking suitable instream habitat and having high turbidity. Keith's study supports what other investigators have found outside Arkansas regarding the effects of stream channelization on fishes: (1) *Reduced abundance*—There are fewer numbers and less biomass of fishes in channelized streams, not only because of the destruction of fish habitat, but also because of reduction of the food supply. Often one or more trophic levels of food chains are completely eliminated. Also, the average size of fishes in channelized streams is smaller. In the Embarras River, Illinois headwaters, the meandering reach had greater morphological variability, both over time and space, than the adjacent channelized reach, and also contained more fish and larger individuals than the channelized reach, suggesting that increased geomorphological complexity resulted in increased fish abundance and total biomass (Frothingham et al. 2001). (2) *Shifts in relative abundance of species*—Channelization completely changes the community structure of a stream. In general, gamefish populations are more severely affected than rough fish or forage fish, the latter two categories becoming

relatively more abundant. Groen and Schmulbach (1978) found significantly higher harvest rates of sport fishes in unchannelized sections of the Missouri River than in the channelized sections. This reflected real differences between the fish populations of the study areas. Sauger, Channel Catfish, and White Bass were the most abundant species in the unchannelized river, while Common Carp, Channel Catfish, and Freshwater Drum were the most abundant fishes in the channelized section. The shift toward forage species appears to be greater in the smaller channelized streams that often become intermittent or dry in the summer than in larger, more permanent channelized streams. Keith also noted that primary consumers were more abundant and top carnivores less abundant in the channelized Brushy Creek in eastern Arkansas. In the upper Yazoo Basin, Mississippi, Knight et al. (2012) found that the shallow depth and lack of woody debris in channelized streams provided a selective advantage for smaller species of fish that could use available shoreline habitats as protection from the current (as opposed to larger species requiring deeper water and more cover). More than 95% of the biomass comprised species reaching an adult length smaller than 300 mm, and the lotic omnivorous fishes that dominated the biomass were species more commonly associated with smaller streams. The channel straightening led to channel incision, bank failure, and over-widening that provided habitats too shallow to support a community of fishes typical of natural northern Mississippi riverine systems (Knight et al. 2012). (3) *Reduced species richness and diversity*—Fewer species occur in channelized streams as well as fewer numbers of them. Tarplee et al. (1971), in a study of North Carolina streams, compared the difference in mean species diversity (generally considered a good indicator of biotic integrity) between natural and channelized streams and concluded that the overall quality of streams was reduced by 27.5% following channelization. (4) *Elimination or reduction in abundance of rare or threatened species*—All strictly lowland species of fish have had their numbers and distributional areas greatly reduced in Arkansas because of the widespread channelization of streams and draining of wetlands in the Coastal Plain lowlands. No species of fish, as far as is known, has yet been completely eliminated from the state by channelization activities, but if continued, these activities may pose a serious threat in the future. One sizeable "oasis" of relatively undisturbed lowland habitat left in eastern Arkansas is the 128,000-acre White River National Wildlife Refuge (WRNWR). Even this sanctuary is threatened by proposed extensive channel modification throughout the lower White River basin. Buchanan (1997) documented the remarkable species richness of Indian Bayou, a stream on the WRNWR. Encouraging developments in recent years have been the creation and expansion of the Cache River National Wildlife Refuge on the Cache River and Bayou DeView in eastern Arkansas, and the creation of new Arkansas Game and Fish Commission wildlife management areas in the Coastal Plain lowlands.

Construction of Dams and Navigation Systems

The topography of Arkansas and the abundance of runoff water have led to the construction of many reservoirs, primarily in upland areas. Arkansas ranks high among the 50 states in the number of reservoirs greater than 500 acres at average pool level with 60 such reservoirs with a total surface area of 356,269 acres (Table 9 and Fig. 3.30). Seventeen of the reservoirs were constructed by the U.S. Army Corps of Engineers and impound 316,030 acres. In addition, there are 265,473 acres of lakes on private lands, including 2,470 lakes greater than 5 surface acres (totaling 56,642 acres).

A small percentage of the lakes on private lands are natural oxbow lakes (Table 7 and Fig. 3.16).

Several benefits of damming a free-flowing stream are highly visible and desirable: the creation of a large reservoir fishery and associated recreational opportunities, flood control, hydroelectric power, water supply, and navigation. However, the impoundment of a reservoir always has some negative impact on native fish populations. The negative effects are often not immediately apparent.

When a stream is impounded, some of the native stream fishes survive and flourish in the reservoir created. This is particularly true of Bluegill, Largemouth Bass, and Channel Catfish, but other stream species survive in small populations, and some become rare or are eliminated. Habitat homogenization through reservoir construction contributes to biotic homogenization as local riverine faunas are replaced with broadly tolerant, cosmopolitan species (Rahel 2002). Fitz (1968) compared pre- and postimpoundment fish populations in the area of the Clinch River included in Melton Hill Reservoir in Tennessee. Twelve of the 47 species found before impoundment were not found in the impounded reservoir. Timmons et al. (1977) surveyed the fishes of the Chattahoochee River before and after impoundment of West Point Reservoir in Alabama and Georgia. Fifty-three species were found in the reservoir area prior to impoundment. Immediately after impoundment, 11 of those species were not collected, 5 more species disappeared during the first year, and only 37 of those 53 species were found in the reservoir 2 years after impoundment. Similar trends have been documented

in Arkansas. A 1962–1963 preimpoundment survey of the Beaver Reservoir drainage described an assemblage of warmwater fish species, predominantly Cyprinidae, Ictaluridae, Centrarchidae, and Percidae (Keith 1964). All postimpoundment surveys of the Beaver Reservoir tailwaters found dramatically reduced species richness and diversity. Postimpoundment surveys conducted in 1965–1966 (Brown et al. 1967 and Bacon et al. 1968b) documented immediate changes in the fish assemblage, with stonerollers accounting for 45–50% and five species of darters accounting for 41–42% of the fish collected. Thirty years after impoundment, Quinn and Kwak (2003) found that the tailwater fish assemblage was composed almost entirely of coldwater species. Knobfin Sculpin, *Uranidea immaculata*, and four species of introduced trout (Salmonidae) accounted for 98% of the fish assemblage by number, and 77% of the fluvial-specialist species that were present in historical collections were not found. Quinn and Kwak (2003) further concluded that short-term monitoring following impoundment was inadequate to determine the effects of dams on lotic fish assemblages and suggested that long-term postimpoundment monitoring is necessary to determine when a fish assemblage has stabilized. In another study of downstream effects, Bonner and Wilde (2000) found that the magnitude of postimpoundment changes in the fish assemblage of the Canadian River, Texas, was related to the degree of discharge decline after impoundment. Downstream from Lake Meredith, mean annual discharge decreased by 76% more than 30 years after impoundment, and the historic mainstem fish assemblage of the Canadian River (which had been an intermittent stream before impoundment of Lake Meredith) was almost completely replaced by species that formerly were restricted to tributary streams.

Hubbs and Pigg (1976) calculated that reservoirs in Oklahoma constituted 52% of the total hazards to threatened fishes of that state. Oklahoma, like Arkansas, is a state with a predominantly riverine-adapted aquatic biota, and the establishment of lacustrine environments favors those native fishes adapted to oxbows more than those adapted for currents. A great volume of literature exists on various aspects of reservoir fisheries, and individual studies have focused on the effects of reservoirs on rare or endangered species and the overall reduction in biodiversity resulting from loss of current velocity and habitat heterogeneity characteristic of lotic environments. However, there have been few, if any, long-term studies of the small, stream-adapted species occurring in impoundments in a geographic area as large as Arkansas. Buchanan (2005) studied 66 Arkansas impoundments to determine the distribution of small,

Table 9. Reservoirs of Arkansas*: Approximately 500 Acres or More at Normal Levels.

Reservoir	Acres	Classification
1. Ashbaugh 1981	500	AGFC
2. Atkins 1955	750	AGFC
3. Balboa 1987	1,000	Private (Hot Springs Village)
4. Banfield's	640	Private duck club
5. Bear Creek 1938	625	Forest Service Lake
6. Beaver 1964	28,220	Corps of Engineers
7. Beaver Fork 1957	900	Conway City Lake
8. Blue Mountain 1947	2,910	Corps of Engineers
9. Bois D'Arc 1961	650	AGFC
10. Brewer 1983	1,100	Conway City Lake
11. Bull Shoals (Ark. portion) 1951	37,300	Corps of Engineers
12. Cane Creek 1986	1,700	AGFC
13. Catherine 1923	1,940	Corporate-owned (AP&L)
14. Calion 1956	500	AGFC managed, Union County owned
15. Camp 9	800	Private
16. Charles 1963	645	AGFC
17. Claypool	1,300	Private
18. Columbia 1986	2,950	Columbia County. Rural Development Authority
19. Conway 1949	6,700	AGFC
20. Crown 1971	640	Corporate-owned (Horseshoe Bend)
21. Dardanelle 1964	34,300	Corps of Engineers
22. DeGray 1972	13,400	Corps of Engineers
23. DeQueen 1977	1,680	Corps of Engineers
24. Dierks 1975	1,360	Corps of Engineers
25. Erling 1956	7,100	Corporate-owned (International Paper Company)
26. Felsenthal† 1985	14,500	Corps of Engineers
27. Fort Smith 2007	1,390	City of Fort Smith
28. G. S. Rodgers Jr. 1945	537	Private
29. Georgia Pacific 1963	1,700	Corporate-owned (G.P. Corp.)
30. Gillham 1976	1,370	Corps of Engineers
31. Grand Marias†	900	Corps of Engineers
32. Greer's Ferry 1963	31,500	Corps of Engineers
33. Greeson (Narrows) 1950	7,260	Corps of Engineers
34. Halowell 1957	630	AGFC
35. Hamilton 1931	7,460	Corporate-owned (AP&L)
36. Harris Brake 1955	1,300	AGFC
37. Hinkle 1971	960	AGFC
38. Huckleberry Creek 1995	509	City of Russellville
39. John Hampton	1,068	Private
40. Lee Creek 1992	634	City of Fort Smith
41. Loch Lomond 1982	499	Private (Bella Vista)

Table 9. (*continued*)

Reservoir	Acres	Classification
42. Maumelle 1958	8,900	City of Little Rock
43. Merrisach 1969	2,000	Corps of Engineers
44. Millwood 1963	30,000	Corps of Engineers
45. Monticello 1997	1,520	City of Monticello
46. Nimrod 1942	3,600	Corps of Engineers
47. Norfork (Ark. portion) 1943	20,600	Corps of Engineers
48. Ouachita 1952	40,100	Corps of Engineers
49. Overcup 1964	1,025	AGFC
50. Ozark 1969	10,600	Corps of Engineers
51. Peckerwood 1942	4,000	Private
52. Pine Bluff 1962	500	AGFC
53. Poinsett 1969	550	AGFC
54. Sequoyah 1959	500	City of Fayetteville
55. Swepco (Flint Creek) 1977	530	Corporate-owned (SWEPCO)
56. Table Rock (Ark. portion) 1957	1,000	Corps of Engineers
57. W. G. Alexander	500	Private
58. White Oak, Lower 1960	1,080	AGFC
59. White Oak, Upper 1960	630	AGFC
60. Winona 1948	1,240	City of Little Rock

*Ten navigation pools created by the construction of the Arkansas River Navigation System range in surface area from 140 acres (Pool 1) to 10,600 acres (Pool 2) in Arkansas and are somewhat reservoir-like; however, these navigation pools (sometimes referred to as flow-through reservoirs) are not true reservoirs and are therefore excluded from the list. Only Dardanelle and Ozark on the main navigation system and Merrisach off pool 2 are considered true reservoirs of this system in Arkansas.

†Originally a series of small natural lakes that are now impounded beneath one human-made lake formed by the construction of Lock and Dam Number 6 on the Ouachita River at Felsenthal.

‡A lake formed by the construction of a system of locks and dams on the Ouachita River near Jonesville, Louisiana. Felsenthal Lock and Dam actually reduced the size of Grand Marias from about 1,300 acres to 900 acres because that lock and dam was built on part of Grand Marias. Grand Marias Lake is now a part of Felsenthal Lake.

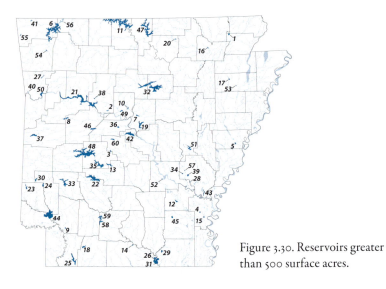

Figure 3.30. Reservoirs greater than 500 surface acres.

nongame fish species in man-made lentic environments. Eighty-five small fish species distributed among 13 families were found in the impoundments. Some small species apparently maintained breeding populations in reservoirs, others occurred in low numbers and may not have maintained breeding populations, and some species occurred sporadically, probably as stragglers from nearby tributary streams. Fifty-nine percent of the small fish species native to Arkansas were found in the 66 impoundments. Based on number of specimens collected and the number of reservoirs in which a species was found, it was estimated that less than 50% of the small native Arkansas fish species can maintain breeding populations in reservoirs.

Sometimes populations of rare or endangered species can be eliminated by impoundments. For example, the range of the endemic, federally endangered Yellowcheek Darter *(Nothonotus moorei)* was drastically reduced by the impoundment of Greers Ferry Reservoir, and the threatened Leopard Darter, *Percina pantherina,* suffered a range reduction caused by the construction of Gillham Reservoir on the Cossatot River. In addition to range reduction, reservoirs often fragment a species into isolated populations that are unable to exchange genetic material. This is particularly problematic for the approximately 50% of the small native Arkansas fish species that are unable to tolerate reservoir conditions even for the short time needed to move from one tributary stream to another (Buchanan 2005). Schwemm (2013) studied the role of large reservoir construction on genetically effective population sizes of the Leopard Darter. Results from mitochondrial and microsatellite DNA indicated that contemporary genetically effective population sizes were four to five orders of magnitude lower than historic values, and Schwemm concluded that the associated estimate of time since the decline was consistent with dam construction as the causative factor. Schwemm (2013) also emphasized the need for managers to implement a program of artificial gene flow among populations in such situations. Other uncommon Arkansas fishes that have suffered significant range reduction and fragmentation because of the construction of impoundments are the Ouachita Madtom (*Noturus lachneri),* Caddo Madtom (*Noturus taylori*), Yoke Darter (*Nothonotus juliae*), Gilt Darter (*Percina evides*), Longnose Darter (*Percina nasuta*), and Ouachita Darter (*Percina brucethompsoni*). A popular game species that had its range reduced in the state by reservoir construction is the Smallmouth Bass, *Micropterus dolomieu*. Although in many areas north of Arkansas there is a northern strain of Smallmouth Bass that thrives in both lentic and lotic environments, in Arkansas this species commonly inhabits clear, permanent streams, a

Figure 3.31. Tailwaters below Norfork Dam. *Keith Sutton.*

habitat type that diminishes with the construction of each new impoundment. It should be noted that populations of Smallmouth Bass persist in a few large Arkansas reservoirs.

Another effect of impoundments is the blocking of the migrations of certain fish species. In other areas of the United States, this has reduced the populations of economically important species, including many salmonids. However in Arkansas this blocking of migrations has not posed a significant problem for the vast majority of our native species. Two Arkansas species that may have had their migrations adversely affected by dams are the Alabama Shad, *Alosa alabamae*, and the American Eel, *Anguilla rostrata* (Hitt et al. 2012).

Not only does the construction of a reservoir have negative effects on native fishes, but the tailwaters released from the dams (Fig. 3.31) may also create problems downstream (Edwards 1978).

The tailwaters from large dams, such as those on the upper White River, are emitted from the hypolimnions of the reservoirs and are cold year-round, rarely exceeding 73–75°F (22.8–23.9°C) in the summer. Because of this, the nature of a stream is completely altered for many miles below the dam and few native warmwater fishes can survive in such an environment. Mathis (1970) concluded that part of the decline of commercial fishing on the White River during the 1950s and 1960s could be attributed to the construction of the large headwater dams. The cold tailwaters from Norfork Dam (completed in 1944) and the larger Bull Shoals Dam (completed in 1951) eliminated the warmwater fishery as far downstream as Lock and Dam No. 3 near Mountain View, and severe damage to the warmwater fish assemblage was evident as far downstream as Lock and Dam No. 1 at Batesville. This has created most of the trout fishery in the state. Mathis also noted that more subtle but lasting damage to the White River ecosystem occurred

Figure 3.32. Mississippi River with barge traffic.
K. J. Killgore, USACE.

farther downstream because of the reduced frequency and severity of flooding: (1) more land along the river was cleared and used for crops as the flood dangers subsided, and (2) the old-river oxbow lakes in the basin no longer flooded, resulting in a decline in fishing.

Arkansas's largest river, the Mississippi, has sufficient natural depth and width to be navigable by large commercial barges carrying various cargoes (Fig. 3.32).

No locks and dams have been built along the Arkansas portion of the river, but the U.S. Army Corps of Engineers has constructed levees, dikes, revetments, and other channel stabilization structures to maintain a dependable navigation channel. Although the portion of the Mississippi River along Arkansas is free-flowing, there are other rivers in Arkansas that are navigable because of the construction of lock-and-dam systems on them. The Arkansas River was made navigable to Tulsa, Oklahoma, in the early 1970s by the construction of a series of 17 locks and dams, 12 of which are in Arkansas. A small portion of the lower White River is included in this system. An additional lock and dam was constructed where the Arkansas River Navigation System connects to the Mississippi River in 2004. A 9-foot (2.7 m) deep navigation channel was completed on the Ouachita River in 1985 as far upstream as Camden, and proposals for further developing the navigation capacities of the Red and lower White rivers in Arkansas are under consideration. There are also plans to deepen the Arkansas River navigation channel to 12 feet. The construction of navigation systems comprising lock and dams is considered here because the run-of-the-river navigation impoundments have many of the same general negative effects on native fish faunas as those previously listed for reservoirs. Navigation systems also have unique environmental impacts. It is necessary to remove large quantities of sand and gravel from

the navigable rivers by dredging to maintain a sufficiently deep navigation channel. Some of the impacts of dredging were addressed by Lagasse et al. (1980) and Lagasse (1986). Sedimentation is also a significant threat to off-channel habitats in impounded river systems. In many rivers modified for navigation, including the Arkansas River, the channel is locked in its course by engineering structures (such as dikes and revetments) to control flow patterns. Because no new habitat is created postimpoundment, sedimentation of off-channel habitats in these rivers results, over time, in a loss of aquatic habitat diversity and area (Schramm et al. 2008). Increasing attention has recently been given to the effects of towboats on fishes inhabiting navigation channels. Main channel environments are used by a variety of fish species (Dettmers et al. 2001a, b), and Gutreuter et al. (2003, 2006, 2009) estimated mortality rates of adult fish from entrainment through towboat propellers and provided evidence that persistent disturbance by commercial navigation altered the relative abundance of channel-dwelling fishes in a large river. Hoover et al. (2005) and Killgore et al. (2011) also quantified the extent of fish entrainment and injury through towboat propellers on the Mississippi and Illinois rivers.

Schramm et al. (2008) assessed habitat changes in the 296-mile (477 km) portion of the Arkansas River Navigation System within Arkansas from 1973 to 1999. They found that the total aquatic area declined by 9% from 42,404 to 38,655 ha and that aquatic habitat losses were 1–17% among pools. Greatest habitat losses occurred in diked secondary channels (former secondary channels with flow reduced by rock dikes) and backwaters adjacent to the main channel. Most of the area of dike pools (aquatic habitat downstream of rock dikes), diked secondary channels and adjacent backwaters were less than 0.9 m deep. It is difficult to accurately assess specific impacts on native Arkansas fishes caused by the construction and maintenance of the navigation systems in the state because preconstruction fish assessments are lacking. Buchanan (1976) discussed some of the possible negative effects on native fishes caused by the construction of the Arkansas River Navigation System. Pennington et al. (1985) discussed the response of fishes to revetment placement in the Mississippi River.

There are, however, some changes in the fish fauna of the Arkansas River that appear related to the construction of the navigation system, based on long-term data from Arkansas Game and Fish Commission rotenone population sampling and from more than 40 years of Arkansas River sampling by TMB. Two lowland species have increased their ranges and abundance westward throughout the river in Arkansas; *Morone mississippiensis* and *Fundulus chrysotus* have become common throughout the Arkansas River, whereas those species were not commonly found upstream from Conway prior to construction of the navigation pools. Three species known from the Arkansas River near Fort Smith in the early 1970s have not been collected from the western portion of the river in Arkansas for almost three decades: *Atractosteus spatula*, *Hiodon alosoides*, and *Notropis shumardi* (although all three species become increasingly more common downstream from Conway). The Bigscale Logperch, *Percina macrolepida*, which was introduced into the Arkansas River basin in western Oklahoma in the 1950s, was present in the Arkansas River near Fort Smith and throughout the Arkansas portion of the river at least as early as 1974 (Buchanan and Stevenson 2003). It has spread throughout the Arkansas River Navigation System in the state where it occurs sympatrically with the Ozark Logperch, *Percina fulvitaenia*. It is currently the dominant logperch species throughout the mainstem Arkansas River, and Buchanan and Stevenson (2003) concluded that the construction of the navigation system increased habitats favored by *P. macrolepida* at the expense of *P. fulvitaenia*. Another nonnative species, *Morone americana*, has invaded Arkansas from Oklahoma via the Arkansas River Navigation System (Buchanan et al. 2007).

Gravel Mining

The removal of gravel from the banks and beds of streams causes direct habitat degradation and greatly accelerates siltation problems within streams. Mining-induced changes to the geomorphic structure of a stream can significantly affect fish habitat and abundance (Meador and Layher 1998), and Hayer and Irwin (2008) found that gravel mining in four Alabama streams lowered the detection probabilities for 78% of the fish species. The damage to riverine ecosystems is often extensive, sometimes affecting large areas of watersheds, and the damage often persists for decades or longer. In Arkansas, gravel is taken from stream channels in ever-increasing amounts wherever it is available in the state, despite the existence of alternatives to instream gravel mining that are far less damaging to the environment and often more cost efficient in their operation. In the late 1990s, the Illinois River had 38 gravel mining sites, and Kings River and Crooked Creek each had 11 sites. Brown et al. (1998b) studied the impacts of gravel mining on fish in three Ozark Plateau gravel-bed streams in Arkansas and found that mining significantly altered the geomorphology, fine-particle dynamics, turbidity, and biotic communities.

Stream channel form was altered by increased bankfull widths, lengthened pools, and decreased riffles in affected reaches. Total densities of all fish in pools and gamefish in pools and riffles were reduced by the gravel mines, and silt-sensitive species of fish were less numerous downstream from the mines. Brown et al. (1998b) further concluded that attempts to mitigate or restore streams impacted by gravel mining may be ineffective because the disturbance results from changes in physical structure of the streambed over distances of kilometers upstream and downstream of mining sites. There are data that suggest that grade control structures can stop upstream movement of a headcut, and some western stream studies indicate stabilizing the riparian zone will result in a stream becoming narrower and deeper; that is, regaining prealteration or predisturbance width and depth (H. L. Schramm, Jr., pers. comm.).

Accidental Chemical Spills

Accidental spills of chemicals toxic to fishes have become increasingly common in recent years in Arkansas. Sometimes these spills result from accidents at industrial sites, construction sites, or other localities, but spills are also commonly associated with transportation accidents such as train derailments and truck accidents. An astonishing variety of substances are transported within the state, including radioactive wastes, pesticides, fuel, and many other chemicals. Sometimes a chemical spill is locally contained before much damage is done, but often chemicals flow into natural waters and adversely affect fish populations. In 1980, 8.5% of the fish kills reported in the United States were caused by transportation accidents (Owen 1985). Even though natural waters have self-purifying abilities and can sometimes rebound quickly from isolated chemical spills, the fish populations often recover rather slowly. The greatest danger, of course, is that a spill might eliminate populations of rare or endangered species or otherwise adversely affect unique communities.

Chemical spills are such a common occurrence in Arkansas that space would not permit a listing of the known spills for even just the past few years. A few examples (mostly from information provided by ADEQ and AGFC) will suffice to illustrate the nature of the problem. On 9 November 1976, a truck leaking the insecticide phorate (an organophosphorus compound) stopped on the Interstate 40 bridge over McKinney Creek in Franklin County. Phorate is a Restricted Use Pesticide among the most poisonous chemicals commonly used for pest control (Walker and Keith 1992). A considerable amount of phorate leaked into the creek, causing a massive fish kill in that creek and farther downstream in Horsehead Creek, a distance of 12 miles (19.3 km) from the original spill. Lower Horsehead Creek had been one of the few reported historical collection localities in Arkansas for the Suckermouth Minnow, *Phenacobius mirabilis.* The phorate finally became diluted to nontoxic levels after entering Dardanelle Reservoir. Water samples taken two weeks after the spill still contained 25 ppb of phorate 9 miles (14.5 km) downstream from the spill site. This was five times the level toxic to fishes. The odor of phorate persisted near the spill area for well over 1 year. A permanent new stream channel was dug to divert McKinney Creek around the highly contaminated area containing the bottom sediments of phorate. No fishes repopulated the affected downstream areas for several weeks. Five months after the fish kill, four species (*Campostoma spadiceum, Notropis boops, Cyprinella whipplei,* and *Lepomis macrochirus*) inhabited the new channel, and only four species (*Notropis boops, Cyprinella whipplei, Lythrurus umbratilis,* and *Pimephales notatus*) were collected farther downstream in Horsehead Creek. In contrast, a collection of fishes from an unaffected area of the nearby Dirty Creek had 19 species of fishes. Eighteen months after the spill, the new channel contained 9 species and the affected area of Horsehead Creek had 15 species (data provided by TMB).

One of the most disastrous chemical spills in Arkansas occurred on 25 December 1978, when a tank car from a derailed freight train landed in the Saline River just south of Benton, spilling crotonaldehyde. The chemical moved downstream in a fairly compact slug killing a high percentage of all fish species. By 30 December, it had killed fishes for 86 miles (138.4 km) downstream from the spill site before being neutralized by the application of large amounts of potassium permanganate. The Arkansas Game and Fish Commission estimated 158,749 pounds (72,009 kg) of fish were killed. Counts of the various species killed and subsequent sampling by gill nets and electrofishing of the 86-mile (138.4 km) kill area revealed that a total kill of all fishes occurred for 31 miles (49.9 km) downstream from the spill site, an overall kill of 90% occurred along the next 37 miles (59.5 km), and a 70% kill occurred in the final 18 miles.

An intentional large-scale attempt to eradicate the exotic Northern Snakehead, *Channa argus,* from the Big Piney Creek watershed in Monroe and Lee counties, Arkansas, with a massive application of rotenone in 2009 provided an opportunity to study the recovery of a lowland fish community after chemical eradication. Johnson (2011) sampled fishes at 17 sites in that watershed for more than a year after the eradication and found that although fish populations appeared to be recovering at most sites, sustained changes in density by reproductive guild and trophic structure were still present in summer 2010.

Many chemical spills occur on a much smaller scale. On 25 July 2010, an employee of Rivercliff Golf Course in Bull Shoals, Arkansas, applied chlorpyrifos, a commonly used pesticide, to the entire golf course and poured 15 to 20 gallons of leftover chemical on the ground directly adjacent to a tributary creek of the White River. He told authorities that he did not think the remaining pesticide would have any effects. More than 6,000 fish and 600 crayfish were killed in the small creek. Containment efforts were made to ensure that trout populations downstream in the White River were not able to ingest any of the dead organisms. Another small-scale incident occurred on 26 August 2010, when an outside contractor hired by the University of Arkansas at Fayetteville to repair and clean the roof of a building on campus did not seal off drains from the roof. The roof drains emptied into a storm drain that emptied into Mullins Creek near the university. As a result, a chemical spill leaked into Mullins Creek and killed dozens of fish.

Some of the recent fish kills in the state were due to spillage of petroleum products. This can be a serious environmental problem because these substances contain highly toxic components and can remain in an ecosystem for a relatively long time. McKee (1956) summarized the detrimental effects of petroleum products on fishes. On 29 March 2013, an underground pipeline operated by ExxonMobil burst, pouring at least 200,000 gallons of tar sands oil across a residential landscape in the town of Mayflower, Arkansas. Despite cleanup efforts, some of the oil reached nearby Lake Conway. The effects on that aquatic ecosystem have yet to be determined.

Forestry Activities

More than half the land area (53.4%) of Arkansas is covered by forests, although virgin and old growth forests are essentially gone (Smith et al. 1984). There are approximately 18 million acres of forests in the state; however, not all this acreage is managed for timber production (ADEQ 2016). About 14% (2.5 million acres) of the forests are National Forests managed by the U.S. Forest Service for timber production and recreation. The heaviest concentration of forests is in the southcentral part of the state. Western and northern Arkansas also have extensive forested regions, and the eastern third of the state has the least amount of forest habitat. Silviculture is the dominant land use in two of the state's ecoregions: the Boston Mountains and the Ouachita Mountains (ADEQ 2016). Forested watersheds provide important benefits to fishes by protecting lands from erosion and by stabilizing stream flow and water temperature, so it is no coincidence that streams of the Boston Mountains and Ouachita Mountains have the highest water quality in the state. However, certain forestry practices are harmful to fishes. One of the main types of disturbance is construction and maintenance of the forest road network and water crossing structures (discussed in the "Roadbuilding Activities" section below). Another forestry activity is clear-cutting, the practice of harvesting all the trees in a stand of timber in one cutting. Clear-cutting is the dominant forest-harvesting method in the United States and can have devastating effects on the fishes in nearby streams if measures are not taken to reduce the impacts. Maintenance of riparian forest buffer areas is essential to fish habitat preservation. It contributes to retention of the physicochemical quality of water and provides shelter and structure to fish habitat. In clear-cut areas, water runoff and erosion are greatly accelerated, and the increased turbidity and siltation negatively affect most fish species (Boschung and O'Neil 1981). Deforestation of riparian zones in 12 stream segments of the Little Tennessee River drainage of Georgia and North Carolina led to shifts in structure of stream fish assemblages (Jones et al. 1999). Alfermann and Miranda (2013) found that change in forest representation in the land cover, both in the lakeshore and the catchment, was the most important environmental factor associated with centrarchid assemblages in 53 floodplain lakes of the Mississippi alluvial valley. In a study of the effects of clear-cutting on stream fish assemblages in the Little River watershed, Oklahoma, Rutherford et al. (1992) found that r-selected fishes (small, short-lived species) may respond quickly to clear-cutting perturbations, whereas K-selected species (large, long-lived) exhibit a delayed response and recovery. That study further suggested that the effects of clear-cutting in the Little River system are limited to temporary changes in local fish assemblage structure. Hlass et al. (1998) used a modified Index of Biotic Integrity (IBI) to characterize the fish assemblages and evaluate the biotic integrity of four forested streams in the lower Ouachita Mountains ecoregion of Arkansas. That study found that differences in IBI scores were related to differences in physical and chemical characteristics of the streams, including the varying intensities of forest management. There were significant differences among the four streams between reference and even-aged treatments and between even-aged and uneven-aged treatments. Turbidity and total suspended solids were inversely related to IBI scores.

The use of chemicals toxic to fishes is also a problem in forests. In the 1970s and 1980s, the U.S. Forest Service advocated the intensive use of herbicides and fertilizers in young forests to kill hardwood trees and encourage the faster growth of desired pine species. In 1978, more than 140,000 acres of timber in Arkansas were sprayed with 2, 4, 5-T, a herbicide containing small amounts of dioxins. Dioxins are toxic to fish, cause adverse developmental effects, and can accumulate in fish tissues, making them a hazard for human consumption (King-Heiden et al. 2012). The runoff from such chemicals into streams and lakes poses a hazard to fish populations, but no studies were done in Arkansas to document effects on fishes.

Industrial Pollution

Today, Arkansas's economy and primary land uses are based mainly on agriculture and forestry. However, the state's industrial development has accelerated in recent decades, resulting in increased pressures on aquatic environments. Native fish populations have been adversely affected by a wide variety of industrial activities. Often, the harmful effects on fishes result from discharges of toxic chemicals into natural waters. Almost all waters of Arkansas are subject to receiving a wide variety of discharges from industries that hold federal permits that allow such discharges into streams. The large rivers of the state have long been viewed as convenient, acceptable channels for disposal of many types of substances. The high concentration of reported recent fish kills in the Arkansas River Valley is, in part, a reflection of the industrial activity in that region.

Fish often become biological concentrators of harmful chemicals that travel through food chains, thereby posing hazards to man. For example, dangerous polychlorinated biphenyls (PCBs) were first detected in Lake Pine Bluff in 1981, and high levels were again found in 1987, prompting a ban on human consumption of fish from that lake. In 2016, fish consumption advisories issued throughout the state warned against human consumption of top carnivores (mostly Largemouth Bass) in certain rivers and lakes because of high concentrations of mercury in fish tissues (ADEQ 2016). The source of the mercury contamination was usually not known, and much of it may have been of natural origin.

A number of industries in Arkansas engage in activities that are potentially harmful to fishes. The mining of nonrenewable resources occurs in every ecoregion, with the fossil fuels of petroleum, natural gas, and coal accounting for almost half the value of the state's total annual mineral market. The Gulf Coastal Plain ecoregion of southern

Arkansas has historically exhibited site specific impacts due primarily to the extraction, storage, transport, and processing of petroleum products, brine, bromine, barite, gypsum, bauxite, gravel, and other natural resources (ADEQ 2016). For example, fish kills have frequently occurred in Lake June in Lafayette County after heavy rains flushed acid drainage from nearby oil fields into the lake.

In addition to fossil fuels, at least 15 nonmetallic minerals and 2 metals are mined in the state. Stone is the most commonly mined nonmetal. Gravel and sand mining are sometimes carried out in streambeds, resulting in tremendous turbidity and sedimentation downstream from the mining site. Fish populations in streams affected by stone mining often recover a short time after the mining is stopped; however, the effects of the mining of other substances are sometimes long lasting. In 1980, a barite mining and processing plant was opened near Caddo Gap in Montgomery County. During its two years of operation, large quantities of silt and sulfates entered the South Fork of the Caddo River with devastating effects on the fish. Because the company had not complied with water pollution control permits or with the reclamation plan it had filed, fish were killed throughout the South Fork of the Caddo River and in its tributary, Back Valley Creek. A population of the rare Caddo Madtom, *Noturus taylori,* was among the fishes wiped out. The mining operation was abandoned in 1982, but high levels of sulfates and total dissolved solids were found in the affected stream in 1985. By 2014, sulfate levels were within the normal range, but the South Fork of the Caddo River did not meet water quality standards for chlorides, copper, lead, or zinc (ADEQ 2016).

Pulp and paper mills in Arkansas produce large quantities of organic wastes that can cause oxygen depletion in natural waters. Even though this industry has seen great advances in pollution control in recent years, its effluents still occasionally result in fish kills. In 1982, a fish kill attributed to pulp mill effluents occurred in the Arkansas River near Pine Bluff. Recent studies in other states have focused on the ability of some chemicals found in pulp mill effluents to act as endocrine disruptors (EDCs). Evidence of masculinization, feminization, and disruption of reproduction in fish downstream from pulp and paper mills has been documented for some time (Manning 2005). Because there are so many chemicals released in the pulping of wood, it is difficult to isolate the ones causing these effects. Hewitt et al. (2002) found that reverse osmosis treatment of the effluent from a pulp mill reduced the EDC effects on the Mummichog, *Fundulus hetroclitus,* but could not identify specific chemicals that appeared to be causing the effects. Larsson and Forlin (2002) provided evidence that masculinization of fish near a pulp mill was due to some chemical in the mill effluent by showing that the effects were not apparent after a short-term shutdown of the mill.

Great amounts of heat are produced by several industries that use water for cooling. After being used, the heated water is usually returned to the nearest waterway, raising its temperature. In Arkansas, electric power plants produce most of the waste heat. This thermal pollution, although not yet a serious problem in Arkansas, has the potential of adversely affecting fishes in the future as more power plants are built. Cole (1971) discussed the complex biological effects of heated water on fishes. A temperature increase of only 3–4°C can sometimes affect fish and other aquatic life. With increased water temperature, metabolism increases and organisms require more oxygen; but there is less oxygen available at higher temperatures. As temperatures rise, nuisance plants and rough fishes may flourish while more sensitive species lose the capacity to reproduce and then die out. Nuclear power plants produce more thermal pollution (per megawatt generated) than fossil fuel plants. They also routinely discharge radioactive wastes. These low-level radioactive wastes do not exceed the limits set by the Nuclear Regulatory Commission, but little is known about the effects, if any, of these wastes on the fish fauna. Arkansas presently has one nuclear power plant with two nuclear reactors located near Russellville. In April 1985, 3,262 gallons (12,347 L) of radioactive wastes were accidentally spilled into Lake Dardanelle (Arkansas River), raising the radioactivity of this water to one-tenth the allowable level.

Radioactive wastes have entered Arkansas waters from Oklahoma via accidental spills and discharges of low-level radioactive material into the Arkansas River from a uranium processing plant at Gore, Oklahoma, approximately 40 miles (64.4 km) upstream from Fort Smith. The plant, which was shut down in 1993, annually discharged an average 2.5% of the total uranium allowed by federal permit (about 5,000 kg/year) into the river. However, in 1984, it exceeded federal permit limits on 13 occasions. The groundwater at the plant site remained contaminated with uranium and thorium in 2016.

Municipal Pollution

Nationwide, municipal wastes ranked a close second to industrial contaminants as river polluters in one federal study (McCaull and Crossland 1974). Wynes and Wissing (1981) found marked differences in the fish community of the Little Miami River, Ohio, above and below the area of influx of domestic sewage. Sharp increases in nutrient levels from sewage plant effluents resulted in downstream changes in the species composition of fish and macroinvertebrate communities. The dominant species at upstream sites were darters, and the Central Stoneroller, *Campostoma anomalum,* was the dominant species downstream. Fish distribution was affected by water quality and was positively correlated with food densities.

Wastewater from municipal sewage treatment plants is a major source of water pollution in Arkansas. Sediments, infectious organisms, detergents, human excrement, and other organic materials are all components of municipal sewage. Sewage treatment plants (which often receive domestic, industrial and/or agricultural waste) release a complex mixture of natural and synthetic chemicals into the aquatic environment, following their partial or complete biodegradation during the treatment process (Jobling and Tyler 2003). It is estimated that 60,000 human-made chemicals are in routine use worldwide, and most of these enter the aquatic environment (Shane 1994). In many towns, industrial plants also discharge their wastes into the sewage systems. To minimize the potentially harmful effects of these pollutants on aquatic ecosystems, municipal sewage is processed at sewage treatment plants before it is discharged into the state's streams. Most cities in Arkansas provide at least primary treatment for their sewage, and many also provide secondary treatment, but some raw sewage is still often discharged into streams. Combined sewers, which transport both domestic sewage and storm water, frequently cause the discharge of untreated sewage into rivers after heavy rains.

Every river in Arkansas receives sewage effluent of some type. The most important harmful effects on fish are usually caused by the oxygen-demanding organic wastes. The processes involved in the breakdown of organic wastes have been summarized by Bylinsky (1971). The quality of wastewater is often measured in terms of its biochemical oxygen demand (BOD) or the amount of dissolved oxygen needed by bacteria for decomposing the wastes. Problems occur when the demand for dissolved oxygen exceeds the available supply. Large quantities of organic pollutants such as sewage alter the balance; bacteria feeding upon the pollutants multiply and consume the oxygen, and organic debris accumulates; anaerobic areas develop, and species of fishes sensitive to oxygen deficiency can no longer survive. The chemical, physical, and biological characteristics of a stream are altered and the fish community is sometimes drastically changed.

In recent years there has been increasing concern about the effects of endocrine disrupting chemicals (EDCs) on fish populations. Endocrine disruptors are synthetic (or sometimes natural) chemicals that can affect the health of organisms by either mimicking or blocking the action of natural hormones or by interfering with the processes for making, excreting or delivering natural hormones to their site of action. A wide range of human-made substances are thought to cause endocrine disruption, including pharmaceuticals (such as synthetic hormones used in contraceptive pills and in hormone replacement therapy), dioxin and dioxin-like compounds, PCBs, DDT and other pesticides, plasticizers such as bisphenol A, alkylphenol ethoxylates (chemicals widely used as surfactants in detergents), and a number of other chemicals (Manning 2005). Although there are different ways for these chemicals to enter the environment through a variety of point and nonpoint source discharges, research from a number of countries suggests that the presence of estrogenic materials in municipal effluents is a relatively common phenomenon (Ankley and Johnson 2004; Ankley et al. 2009). The effects of EDCs on fishes range from subtle changes in the physiology and sexual behavior to permanently altered sexual differentiation and impairment of fertility (Jobling and Tyler 2003). Some of the pioneering work from the United Kingdom documented the presence of feminized male fish in rivers downstream from municipal wastewater treatment plants, and also associated this response with the occurrence of steroidal estrogens and some types of industrial chemicals (alkylphenols) (Desbrow et al. 1998; Purdom et al. 1994; Routledge et al. 1998). Since then, a wide range of endocrine-related effects have been documented in fish both in the field and in the laboratory, and further work is needed to clarify the long-term implications at population and community levels.

Municipal landfills also have the potential for harming fish populations in nearby streams. Surface water runoff may flush toxic substances into streams; however, the greatest threat from landfills is groundwater pollution, discussed below.

Groundwater Contamination and Withdrawal

A portion of the rain that falls on the land gradually seeps downward through the soil and collects in aquifers to form groundwater. In some areas, groundwater is essentially a nonrenewable resource because of its extremely slow recharge rate or because it is being withdrawn faster than recharge can replenish it. It is also a major regulator of natural stream flow in most areas of Arkansas, and its quantity and quality can have great impact on fish populations. Groundwater reserves hold 60–70 times more water than is present in lakes and streams at any one time, but heavy withdrawal of groundwater causes less water to be available to streams. Shallow freshwater aquifer systems are found throughout Arkansas, and groundwater is used for industrial, municipal, agricultural, and domestic purposes. Until fairly recently, no one would have thought that a water-rich state like Arkansas would have severe problems with excessive groundwater withdrawal. Groundwater use in Arkansas has more than doubled since 1985, resulting in water level declines in many areas of the state (ADEQ 2016). The most severely depleted aquifers in the state are the Alluvial aquifer (depleted for agricultural irrigation) in eastern Arkansas and the Sparta aquifer (depleted for municipal and industrial use) in southern and eastern Arkansas. Water levels have declined substantially in those aquifers, and large "cones of depression" have developed across broad areas (ADEQ 2016).

Intentionally or unintentionally, humans have often treated groundwater as a dump for their wastes, but it is not exposed to the same natural purification systems that recycle and cleanse surface waters. Hazardous industrial chemicals such as metal ions, phenols, tar residues, brines, and exotic organics may contaminate groundwater through accident, carelessness, or wastewater discharge (McGauhey 1971). For example, petroleum production in the Gulf Coastal Region of southern Arkansas has caused large increases in chloride and sodium concentrations in the groundwater. Leaching from solid waste landfills can pollute the groundwater with a variety of substances. Ten Arkansas sites were on the EPA's 1987 list of the nation's most hazardous waste sites, and the 2014 EPA list included 15 Arkansas sites. In comparison, the numbers of 2014 EPA sites in states surrounding Arkansas were Missouri 37, Oklahoma 15, Texas 64, Louisiana 26, Mississippi 11, and Tennessee 28.

Agricultural activities contribute pesticide residues, nutrients applied as fertilizers, and other chemicals to the groundwater. In addition, the chemical products of biodegradation of organic wastes such as those from feedlots and poultry houses generally move quite freely into the groundwater. McGauhey (1971) considered the most serious aspect of groundwater quality from a human standpoint to be the buildup of dissolved solids because of public health and water quality hazards and its diminished suitability for use in crop irrigation.

Little is known about the effects of groundwater withdrawal and contamination on fishes. In Arkansas the fishes most obviously and directly at the mercy of the groundwater quality are the three species of cavefishes that live permanently in the groundwater. These species live in underground streams in the mountainous regions of northwest and northcentral Arkansas, and the pollution of their aquatic habitat poses life-threatening danger to them. But there is also a tendency to forget about the intimate relationship between the surface and subsurface systems. They are often treated as if they were two separate, unrelated entities. Problems with the groundwater tend to be most severe in northwestern Arkansas because of the types of rock and soil near the surface (Steele 1985). In this karst area, much groundwater flow discharges as springs and seeps into nearby streams, and movement of contaminants within the karst aquifer system has a more pronounced effect on both surface and subsurface water quality because of the rapidity and higher degree of groundwater-surface interaction relative to other geologic settings in the state (ADEQ 2016). Because of the karst topography in this predominantly limestone region, especially in the Springfield Plateau, the groundwater receives little filtration (due to lack of development of a mature soil profile and the presence of sinkholes); therefore, it is highly susceptible to contamination. Sinkholes often provide a direct route for surface contaminants to rapidly enter the groundwater. Recognizing the fragile nature of these underground ecosystems, the Arkansas Department of Transportation rerouted the construction of a U.S. Route 71 bypass away from its original route, which would have traversed the recharge area of Cave Springs Cave in Benton County containing the largest known population of the threatened Ozark Cavefish. A revealing study by Graening and Brown (2000a) illustrates the susceptibility of cave ecosystems to pollution. Those authors found that the water quality of Cave Springs Cave steadily declined over two decades and was often not fit for human consumption or even for primary contact recreation. Microbes, and particularly fecal bacteria, were at unnaturally high densities in the cave stream, especially during rain storms. Concentrations of nutrients were also elevated in the cave stream water, and heavy metals were present at toxic levels. These metals were also accumulating in the cave sediments and in tissues of the cave organisms. The source of those contaminants was from the cave stream's recharge zone, a 15 square mile area that captures polluted surface water from several drainage basins and focuses the water into the cave spring by many underground channels. This recharge area has many pollution sources, especially from applications of confined animal wastes and municipal sewage sludge to pastures, and the increasing numbers and decreasing effectiveness of septic systems (Graening and Brown 2000a, b). Graening and Brown considered the cave's fauna to be stressed.

Roadbuilding Activities

Trombulak and Frissell (2000) reviewed the scientific literature on the ecological effects of roads and found support for the general conclusion that roads are associated with negative effects on biotic integrity in both terrestrial and aquatic ecosystems. The construction of many kinds of roads in Arkansas has created a massive transportation network throughout the state. Four interstate highways, 10 U.S. highways, and numerous state and county highways have been built, but thousands of small, unpaved roads have made even the most remote areas of the state accessible by vehicle. One need only look at any county highway map to become aware of the extent of this system. In the entire United States there is an average of one mile of road for every square mile of land, and interstate highways require around 24 acres of land per mile. The effects of road building on fishes in Arkansas have received little attention, but to its credit, the Arkansas State Highway and Transportation Department takes great pains to minimize environmental damage near its construction sites. There is little that can be done to change road density, but how roads are built can significantly affect the flow of streams and rivers and therefore, greatly influence fish habitat. Schaefer et al. (2003) studied the effects of natural versus man-made barriers on the federally threatened Leopard Darter. At a road crossing site, all documented movement was in a downstream direction, and at least two darters traversed culverts in the low-water bridge. Laboratory studies of movement across several types of culverts suggested that culverts significantly decrease the probability of Leopard Darter movement among habitat patches. Warren and Pardew (1998) studied road crossings as barriers to small-stream fish movement in the Ouachita Mountains, and

Ryles (2012) studied the effects of consecutive road crossings on fish movement and community structure in a Ouachita Mountain stream.

To summarize environmental alterations in Arkansas, it is important to realize that the actions/activities that alter fish habitats have social and economic value and provide necessary products. Some positive steps have been taken to minimize the damage to fish habitat, and special attention has been directed toward preserving endangered and threatened species. Federal and state agencies have intensified their efforts to monitor aquatic habitats, identify the critical fish species and habitats requiring protection, and develop action plans to improve environmental quality. At the state level, the AGFC and ADEQ have led the way in these endeavors. At the federal level, the EPA, USFWS, and USDAFS have been at the forefront. On a positive note, Matthews and Marsh-Matthews (2015) compared 381 collections made in Oklahoma and western Arkansas from 1975 to 2009 with 86 collections from the same region by A. I. Ortenburger from 1925 to 1927 and found that fish distributional patterns appear to have changed little at the regional level, based largely on analysis of the 25 species that were most commonly found in Ortenburger's collections. Eighty-one of the 95 species collected in 1925–1927 were found in 1975–2009, and species association patterns were similar. Fifteen percent of the species found in 1925–1927 were not found in 1975–2009, but with a more intensive sampling effort in the latter period, 47 species were recorded that were not found in 1925–1927. An examination of the same data across Hydrologic Units (HUCs) for the 25 most common species in Ortenburger's collections indicated that even at the smaller scale of HUCs, fishes present in collections made before the era of dam building were mostly still present in those drainages. The future conservation of Arkansas fishes will require continuous monitoring of fish assemblages throughout the state to detect changes in fish distributions and habitats.

CHAPTER 4

Classification and Systematics

What Is a Fish?

Surprisingly, this is not an easy question to answer because the term "fish" is one of convenience and does not represent a taxonomic unit or ranking. Heiser (2009) presented two reasons for the difficulty of clearly defining what seems at first glance to be a straightforward noun. First, several distinct evolutionary lineages, presumed to be derived from a common unknown ancestor (but with little fossil record), survive today, having diverged greatly through evolutionary processes. Secondly, fishes are the most diverse and species-rich living vertebrates. In the past, the word "fish" was often used by laypersons to describe everything from a crayfish, shellfish, jellyfish, or starfish to a Blue Catfish or a Largemouth Bass. Ichthyologists, however, recognize that the organisms grouped together as fishes do not form a natural group. They are paraphyletic rather than monophyletic; therefore, the term "fish" does not merit taxonomic rank. Even though lacking status as a taxon, we may still define fishes as poikilothermic, aquatic chordates with appendages (when present) developed as fins, whose chief respiratory organs are gills, and whose bodies are usually covered with scales (Berra 2001; Helfman et al. 2009). It is a convenient category for aquatic organisms as diverse as hagfishes (which are not even vertebrate animals), lampreys, sharks, rays, lungfishes, sturgeons, gars, and the more derived ray-finned fishes (Helfman et al. 2009). It should

be noted that "fish" is singular and plural when referring to a single species, whereas "fishes" is plural when referring to more than one species.

Classification Basics

Classification is the organization or grouping of things into categories. *Taxonomy* is the branch of science that has produced a formal system for naming and grouping species to communicate the diversity of life. Taxonomy is part of the broader science of *systematics* in which studies of variation among populations are used to reveal the evolutionary relationships of species or higher taxa such as families or orders (Hickman et al. 2017). The best classification is the most natural one that best represents the phylogeny (evolutionary history) of an organism and its relatives (Helfman et al. 2009). Lundberg and McDade (1990) presented a summary of fish systematics, and Betancur-R. et al. (2017) reviewed the main criteria used by ichthyologists during the past 50 years to define their classification criteria.

Higher Categories of Fish Classification

The Animal Kingdom is divided into approximately 35 phyla, with fishes comprising the major portion of phylum Chordata and subphylum Craniata. Most recent classifications have adopted the name Craniata (e.g., Page et al. 2013) to replace the older subphylum name Vertebrata because hagfishes (long included in Vertebrata) lack vertebrae. Some recent classifications divide the subphylum Craniata into two superclasses: Agnatha (jawless chordates, the hagfishes and lampreys) and Gnathostomata (jawed chordates), but we favor separation of the hagfishes and lampreys at a higher taxonomic level (Appendix 5). We prefer the classification of Nelson et al. (2016) which divides subphylum Craniata into two infraphyla: Myxinomorphi, the hagfishes, and Vertebrata, all vertebrate animals. The jawless vertebrates then become superclass Petromyzontomorphi (lampreys), and all other vertebrates, including tetrapods form superclass Gnathostomata. The two infraphyla of Craniata contain the following five classes of fishes that have uncertain evolutionary relationships with one another: Myxini (hagfishes), Petromyzontida (lampreys), Chondrichthyes (cartilaginous fishes), Sarcopterygii (lobe-finned fishes), and Actinopterygii (ray-finned fishes). We do not support the inclusion of tetrapods in class Sarcopterygii even though the two share a common ancestry. There are 78 species of hagfishes (Nelson et al. 2016) that are all marine, and Sarcopterygian fishes live in marine (two species) and freshwater environments (four species occur in Africa, and one species each in South America and Australia). Chondrichthyans (sharks) live mainly in marine waters (there are a few freshwater species) and seldom enter freshwater. Ray-finned fishes (Actinopterygii) comprise most of the bony fishes. Older classifications used the name Osteichthyes as a class containing all bony fishes, but most recent classifications do not recognize Osteichthyes as a valid class. However, Nelson et al. (2016) supported returning Sarcopterygii and Actinopterygii to class Osteichthyes as subclasses. Even those ichthyologists who don't support recognition of class Osteichthyes still use "bony fishes" as a term of convenience to refer to fishes with endochondral bone. Some recent authors divide class Actinopterygii into three subclasses, and that is the classification we use herein. The three subclasses are: Cladistia (bichirs, formerly placed in Chondrostei), Chondrostei (paddlefishes and sturgeons), and Neopterygii (gars, Bowfin, and teleosts). The major clade of neopterygians is the division Teleostei (recognized by Wiley and Johnson 2010 and Betancur-R. et al. 2017 as infraclass Teleostei, and by Nelson et al. 2016 as subdivision Teleostei in division Teleosteomorpha), the modern bony fishes with approximately 30,720 described species representing 96% of all living fishes. It should be noted that Nelson et al. (2016) used a different system of designating classes, subclasses, and several other lower taxa from the one we favor. It has been estimated that there are more than 5,000 undescribed species of fishes, and in the several years prior to 1998, approximately 200 new species of fishes were described each year, about half of which were teleosts. Between 2006 and 2016, 3,890 new species of fishes were described (Nelson et al. 2016). The higher taxonomic categories (family and above) of Arkansas fishes used herein are presented in Appendix 5.

Linnaean System of Classification and Nomenclature

In the 18th century, Carolus Linnaeus introduced our current hierarchical system of naming and classifying species. The tenth edition of Linnaeus' important book, *Systema Naturae*, published in 1758, marks the beginning of modern classification and the starting point for taxonomic nomenclature for all animals. Even though Linnaeus's system was not based on evolutionary relationships (Linnaeus predated Darwin by almost 100 years), many of its features, such as the two-word naming system (binomial nomenclature) and hierarchical arrangement of taxa, are compatible with evolutionary theory and continue to be used in systematics (Reece et al. 2015). Each species can have only one valid scientific name. The first word of the scientific name, the genus, is always capitalized. The second word is called the species epithet and is never capitalized. For example, the scientific name of the Largemouth Bass is *Micropterus salmoides*. *Micropterus* is the genus (plural: genera) name and *salmoides* is the specific epithet. If a subspecies category is recognized, a third descriptive name is used, e.g. *Micropterus salmoides floridanus*. Scientific names are commonly Latin or Latinized words, but more recently, new names have been rooted in languages as diverse as Greek, Spanish, Cherokee, and even Sanskrit. All genus and species names are italicized or in a contrasting font. In the Linnaean taxonomic hierarchy for animals, genera are grouped into families whose names always end in *-idae*. The Zoological Code of the ICZN does not prescribe uniform name endings above the family level, and Wiley and Johnson (2010) proposed the adoption of standard endings for those taxa. Families are grouped into orders, and ichthyologists have long accepted the uniform ending of *-iformes*. Orders are grouped into classes, and Wiley and Johnson (2010) proposed the adoption of *-ii* as the standard ending for a class name. Names for the larger groupings like classes, orders, and families are capitalized but not italicized. Diminutive forms of family names and order names are frequently used (e.g., centrarchid = Centrarchidae, cyprinid = Cyprinidae, cypriniform = Cypriniformes, and perciform = Perciformes). The term *taxon* (plural taxa) is used to refer to any category within the classification system hierarchy. For example, the Common Carp (*Cyprinus carpio*) and the Ozark Minnow (*Notropis nubilus*) both belong to the same taxon at the family level (Cyprinidae); however, they belong to different taxa at the genus level. The term *alpha taxonomy* is generally used to refer to the study of the basic unit of classification, the species.

Taxonomy also includes the rules of nomenclature that govern the use of taxonomic category names. Animals are assigned to taxonomic categories according to the rules of the International Code of Zoological Nomenclature established by the International Commission on Zoological Nomenclature (ICZN). The purposes of the Code are to ensure uniqueness, universality, and stability. An important part of the ICZN rules is the Rule (or Principle) of Priority that states that only scientific names published in 1758 or later are valid. Names published prior to 1758 are invalid. Following the Rule of Priority, the correct name is known as the senior synonym, the earliest species name, and all later names become junior synonyms. Some common and/or widespread fish species such as the Greenside Darter, *Etheostoma blennioides*, have long lists of junior synonyms.

A new species is officially recognized only when a name and formal description are published and made available to the scientific community (Wheeler and Pennak 2013). Species and subspecies are based on type specimens, and higher taxa are based on type taxa. Primary types include the *holotype*, the single specimen upon which the description of a new species is based. The *type locality* is the place where the holotype was collected. Secondary types include *paratypes*, which are additional specimens used in the descriptions of a new species. *Topotypes* are specimens taken from the same locality as the primary type (holotype) and therefore, are useful in understanding variation of the population that included the specimen upon which the description was based. Occasionally after the publication of the original description, one of the syntypes is selected and designated through publication to serve as "the type," called a *lectotype* (e.g., see the *Lepomis humilis* species account). Combinations of these terms can be made such as *paratopotype*, that is, other specimens from the type locality used in the original description of the species. If all primary types of a species are known to have been destroyed subsequent to the original description, a new specimen called the *neotype* is selected to serve as the type specimen. The scientific name is sometimes followed by the name of the author or authors who originally described the species and the year

that description was published. For example, the Common Carp has the scientific name *Cyprinus carpio* Linnaeus, 1758, indicating that Linnaeus was the describer (author) and 1758 was the year his description was published.

Although Linnaeus never personally collected fishes in Arkansas, or described species from Arkansas specimens, he described several species (in addition to *Cyprinus carpio*) that occur in the state, including *Lepisosteus osseus* (Longnose Gar), *Amia calva* (Bowfin), *Carassius auratus* (Goldfish), *Scardinius erythrophthalmus* (Rudd), *Salmo trutta* (Brown Trout), *Esox lucius* (Northern Pike), *Lepomis auritus* (Redbreast Sunfish), and *Oreochromis niloticus* (Nile Tilapia). Therefore, Linnaeus does share a connection with the Natural State. Another fish Linnaeus described is *Mugil cephalus* (Striped Mullet), a marine species that frequently enters state waters.

Scientific names continue to be modified as our knowledge and understanding of the biological relationships among fishes expand (C. R. Gilbert and Williams 2002). Such changes often cause much consternation and annoyance for many scientists and nonscientists. Helfman et al. (2009) provided four primary situations in which scientific names change: (1) splitting what was considered a single species into two or more species; (2) lumping two species that were considered distinct into one species; (3) changes in classification (e.g., a species is placed in a different genus); and (4) an earlier name is discovered and becomes the valid name by the Rule of Priority. A well-known example of a combination of these aspects can be seen in the change in name of the Rainbow Trout, one of the best-known fishes on Earth. For years (and in the first edition of this book) the Rainbow Trout had the scientific name of *Salmo gairdneri*. This long-held name changed in 1988 when G. R. Smith and Stearley (1989) changed the scientific name to *Oncorhynchus mykiss* because of a new generic classification as well as a lumping of species previously considered distinct.

Common Names

Common names are different in concept from scientific names. They often vary regionally and even within the same region of the country. Even within a state, people often use different common names for the same species, and, conversely, the same common name is often applied to more than one species, further complicating matters. Common names, unlike scientific names, are not covered by a set of rules set forth by the ICZN. For example, *Amia calva*, the Bowfin, a common fish of the lowland areas of

Arkansas, is called grinnel, grindle, dogfish, cypress trout, and mudfish in different parts of the country. The Bowfin may be called a cypress trout in Louisiana and southern Arkansas or a grinnel in northeastern Arkansas.

To reduce confusion and establish standardized common names for North American fishes, the American Fisheries Society (AFS) for several decades has published a list of accepted common and scientific names agreed upon by a joint committee of the AFS and the American Society of Ichthyologists and Herpetologists. The list is updated about every 10 years (see Page et al. 2013 for the most recent version). The AFS list has gained widespread acceptance, and ichthyologists and fisheries biologists usually adhere to it when dealing with common names of fishes. In instances in this book where we disagree with the committee, we explain our rationale for choosing a common or scientific name that differs from that used by the AFS committee. Taxonomic disagreements are rather common among ichthyologists and do not indicate personal animosity between colleagues. We follow Nelson et al. (2002, 2016) and Page et al. (2013) in treating common names as proper nouns and capitalizing them in this second edition of *Fishes of Arkansas*.

Modern Evolution-Based Systems of Classification

As long as there are active, creative ichthyologists, there will be major disagreements in our classification in the foreseeable future. Fish classification is in a dynamic state, and the student pursuing ichthyology will find that all groups can be reworked.—J. S. Nelson (2006)

Since Darwin, the goal of systematics has gone beyond simple organization to have classification reflect evolutionary relationships (Reece et al. 2015). The concepts and methodologies underlying how evolutionary relationships are studied, and how best to portray the resultant phylogenies, have been areas of considerable debate and advancement over the last several decades (Ross 2013). For an overview, see Mayden and Wiley (1992) and Mayden and Wood (1995). Currently, there are two major approaches to classification that have dominated the area of systematics: (1) evolutionary systematics and (2) phylogenetic systematics (or cladistics). A third approach, phenetics (or numerical taxonomy) used somewhat in the 1950s–1980s, has lost favor among most current taxonomists because it rejects phylogeny.

Evolutionary Taxonomy

Traditional evolutionary taxonomists group species into higher taxa using the joint criteria of common descent and adaptive evolution (Hickman et al. 2017). Each taxon has a single evolutionary origin and occupies a distinctive adaptive zone. The broader the adaptive zone occupied by a group of organisms, the higher the rank given to the corresponding taxon. Evolutionary taxa must have a single evolutionary origin, and must show unique adaptive features as shown on a phylogenetic tree. This philosophy recognizes monophyletic or paraphyletic groups, but recognition of paraphyletic taxa requires that our taxonomies distort patterns of common descent. Hickman et al. (2017) concluded that if we want our taxa to constitute adaptive zones, we compromise our ability to present common descent effectively.

Cladistics—Phylogenetic Systematics

Phylogenetic systematics, or cladistics, emphasizes common descent exclusively in grouping species into higher taxa. Cladistics is a system based on the presumed phylogenetic relationships and evolutionary histories of groups of organisms instead of being based solely on shared features. Cladistic theory had its origin in the 1950s, based on the work of the German entomologist Willi Hennig (1913–1976). Hennig (1966) united groups of organisms into clades based on morphology and genetic characters (traits that are heritable from parents to offspring). Each clade is a branch from a tree of life with an ancestor and all its descendants. Hennig's system assumes that a character shared by a group of organisms, but not by other groups, indicates that they have a closer recent ancestor in common and share the same evolutionary history. Although controversial when first proposed, cladistics has moved to the forefront of modern classification theory, spurred mainly by advances in technology of DNA analysis, particularly in the 1990s. Hennig's approach was further developed by Niles Eldredge and Joel Cracraft (1980), Edward O. Wiley (1981), and many others (Ross 2013). Readers are also referred to Patterson (1982b), Janvier (1984), C. L. Smith (1988), and Etnier and Starnes (1993) for more thorough discussions of cladistics.

A clade is a group of species (or other taxa) containing an ancestor and all its descendant taxa. Such an inclusive group of ancestor and descendants, whether a genus, family, or some broader taxon, is said to be monophyletic. The cladistic approach involves a search for shared derived characters called synapomorphies that diagnose monophyletic groups or clades. Clades reflect the branching pattern of evolution and can be used to construct cladograms (a diagram showing phylogenetic relationships). Character states ancestral for a taxon are called plesiomorphic, and shared ancestral characters among species, called symplesiomorphies, do not provide data useful in constructing phylogenetic classifications, because ancestral characters may be retained in a wide variety of distantly related taxa (Helfman et al. 2009). Taxa that are considered derived, as well as ancestral taxa, may possess symplesiomorphies. Sometimes it is difficult to determine whether a character is ancestral (found in the ancestor of the clade) or derived (a new character found only in the descendents of the ancestor), because convergent evolution can result in organisms possessing structurally analogous, but not evolutionarily related characters (termed homoplasies). Unique specialized characters that are present in only one clade of the group are known as autapomorphies. Autapomorphies are important in describing a particular taxon, but not useful for constructing a phylogenetic tree or revealing information about relationships.

In cladistics, the systematist seeks to resolve which two taxa of a group of three or more are each other's closest genealogical relatives (Nelson 2006). Those species or groups of species that are derived from the same most recent common ancestor are called sister species or sister groups. A cladogram or dichotomously branching diagram is constructed in which paired lineages (sister groups) are recognized based on shared derived character states (synapomorphies), with a particular derived character state being termed apomorphic. It is hypothesized that sister groups share a derived character because they descended from an immediate ancestor from which this derived feature evolved (Boschung and Mayden 2004). Plesiomorphies are ancestral states and do not indicate the existence of sister groups. The sister group possessing more synapomorphic character states relative to the other is considered the derived group, while the other is the ancestral one. Each is considered to have the same taxonomic rank.

Ichthyologists have traditionally used phylogenetic trees (based on evolutionary taxonomy) to depict hypotheses about the evolutionary history of fish species (i.e., to depict the record of diversification over time). In the cladistics system, evolutionary relationships are shown hierarchically in the branching cladograms. Each node or point of divergence in a cladogram is based on a derived character and indicates evolutionary divergence from a common ancestor. The endpoints of a tree usually represent individual species, and every node and its descendant branches and subbranches constitute a clade. Cladograms are hypotheses of evolutionary relationships, and the data can often be used to construct a cladogram in more than one way. In order to choose the best possible cladogram, taxonomists usually adopt the Principle of Parsimony; that is, they choose the cladogram involving the least number of steps or character transformations to explain the observed relationship between the groups. Support for cladistic hypotheses increases with the number of synapomorphies used in a study (studies based on few characters can be misleading), and with the number of studies, based on different characters, that come to the same or similar conclusions (Ross 2013). However, even the best cladogram (or phylogenetic tree) represents only the most likely hypothesis based on the available evidence at the time. As ichthyologists accumulate new data, hypotheses may be revised and new trees drawn.

It is important to note that both modern philosophies of classification seek to determine phylogenetic relationships (therefore, we often refer to phylogenetic systematics simply as cladistics), and both use the same sources of phylogenetic information. The differences between the two philosophies arise largely from how the data are interpreted. Both philosophies accept monophyletic groupings and reject polyphyletic groups. Evolutionary taxonomy accepts paraphyletic groups, but cladistic philosophy rejects them, seeking to identify monophyletic lineages exclusively. See Hickman et al. (2017) for more detailed information on the two philosophies. Cladistic and evolutionary taxonomy continue to coexist with the Linnaean system of classification. Both modern philosophies of systematics provide important ways of investigating the past and determining how closely related different species are, and the Linnaean classification system provides a concise way of identifying and organizing familiar groups and their names (Burnie and Wilson 2011). It should be noted that there are biologists today who advocate abandoning the traditional Linnaean categories of classification (Zachos 2011). It is also common today to see rank-free cladograms presented in the scientific literature. In phylogenetics, ranks are considered devices to express the relative position of clades in hierarchies. The only taxa directly comparable in a biological manner are sister taxa and species (Wiley and Johnson 2010).

Characters

Characters are features of organisms that vary among species. The formal taxonomy of animals that we use today was established using the principles of evolutionary taxonomy and has been revised in recent decades using the principles of cladistics. Initially, the introduction of cladistic principles replaced paraphyletic groups with monophyletic subgroups while leaving the remaining taxonomy mostly unchanged. Mayden and Wiley (1992) reviewed various types of characters used in taxonomic studies of fishes. Traditionally, characters were morphological features, but characters were eventually considered to be any feature the taxonomist used to study variation within and among species. Today, morphological characters such as meristic (countable), morphometric (measurable), and others are still important, but cytological, ecological, behavioral, and especially molecular (e.g., identifying and sequencing nuclear and mitochondrial genes) characters have become equally important. Those characters are used by both modern philosophies of systematics to reconstruct nested hierarchical relationships among taxa that reflect the branching of evolutionary lineages through time. The fossil record also provides estimates of the ages of evolutionary lineages that can be integrated into phylogenetic trees by fossil calibration analysis. Comparative studies and the fossil record jointly permit us to reconstruct a phylogenetic tree (based on evolutionary systematics) or a cladogram (based on cladistics theory) representing the evolutionary history of the animal kingdom.

What Is a Species?

No matter what the methods and data source utilized in systematics studies of fishes or other organisms, the interpretation of that data among different workers is always subject to controversy.—Etnier and Starnes (1993)

Species are the fundamental units of classification. In Darwin's time, Thomas Huxley in 1859 posed the question, "Just what is a species?" Biologists ever since have continued to debate this deceptively simple question. Numerous attempts have been made to define a species, and Mayden (1997, 1999) outlined more than 20 different species concepts that have been used in the past. While our concepts of species have become more sophisticated today, the diversity of different concepts and disagreements surrounding their use are as evident now as in Darwin's time (Hickman et al. 2017). Our goal in this section is not to resolve this long-lasting debate, but rather to define the three most widely used modern species concepts and point out some modern areas of disagreement among ichthyologists on how a species should be defined.

Biological Species Concept (BSC)

The BSC was the most widely used species concept throughout most of the 20th century. It was inspired by Darwinian evolutionary theory and formulated by Theodosius Dobzhansky and Ernst Mayr (Hickman et al. 2017). Mayr (1966) defined the BSC as a group of actually or potentially interbreeding natural populations that are reproductively isolated from other such groups. A species was an interbreeding natural population of individuals having common descent and sharing intergrading characteristics. Members of the reproductive community of a species are also expected to share common ecological properties. Reproductive isolation is the key element of the BSC. Most biologists in the mid-20th century were taught only the BSC, and HWR remembers a professor declaring that the BSC held for nine-tenths of all organisms in nature and that was good enough! Since that time, studies of variation in molecular characters have proven valuable in identifying geographical boundaries of reproductive communities. Molecular analyses have sometimes revealed the existence of cryptic or sibling species (Bickford et al. 2007) that are very different genetically but too similar in morphology to be diagnosed as separate species by morphological characters alone. An example of a species in Arkansas that does not fit the BSC definition was the description by Egge and Simons (2006) of the Black River population of the Ozark Madtom, *Noturus albater*, as a new cryptic species, *Noturus maydeni*, based entirely on genetic evidence.

Critics of the BSC contend that because the ability of species to interbreed has not always been found to reflect the genealogical descent of species, reproductive isolation has limited significance in the definition of a species (Boschung and Mayden 2004). Reproductive isolation is especially difficult to perceive in nature for freshwater fishes isolated in different drainages, and evidence of intergradation is seldom rigorously tested in practice (Piller et al. 2008).

Phylogenetic Species Concept (PSC)

The PSC was defined by the eminent ornithologist Joel Cracraft as an irreducible (basal) grouping of organisms diagnosably distinct from other such groupings and within which there is a parental pattern of ancestry and descent. The PSC emphasizes most strongly the criterion of common descent and includes asexual and sexual groups (Hickman et al. 2017); therefore, a phylogenetic species is a single population lineage with no detectable branching. If one adheres strictly to cladistic systematics, the Phylogenetic Species Concept is ideal because only this concept guarantees strictly monophyletic units at the species level.

Evolutionary Species Concept (ESC)

Probably most ichthyologists today support the ESC, which defines a species as a single lineage of ancestor-descendant populations that maintains its identity from other such lineages and that has its own evolutionary tendencies and historical fate (Wiley 1978; Wiley and Mayden 2000). Although this sounds almost identical to the definition of the Phylogenetic Species Concept, the two concepts differ mainly in that the PSC emphasizes recognizing as separate species the smallest groupings of organisms that have undergone independent evolutionary change (Hickman et al. 2017). The ESC places greater emphasis on recognizing the variability across the populations of a species. It groups into a single species geographically disjunct populations that demonstrate some phylogenetic divergence but are judged similar in their evolutionary tendencies. Future gene exchange and merging are possible.

The differences among the three modern species concepts as briefly defined above have sometimes led to disagreements in assigning fish populations to species. The BSC proponents often complain that the recognition of every demonstrably genetic variant population as a separate species does not take into account the natural genetic variability of wide-ranging species and provides little or no information on reproductive isolation. Conversely, ESC proponents argue that a zone of intergradation between two closely related species does not automatically negate their recognition as separate species. If evidence (often using a range of morphological and molecular characters) supports their specific distinctness outside the zone of intergradation, then that is usually considered sufficient for specific recognition. Ichthyologists have long known about the occurrence of hybrids between fish species in nature, but the full extent of hybridization has only recently become apparent. Recent molecular studies, especially those based on nucleotide sequences of mitochondrial DNA, have shown that hybridization and subsequent introgression between species are much more widespread and common than previously known (Ray et al. 2008; Bossu and Near 2009; Keck and Near 2009, 2010; Near et al. 2011). This is especially true for sunfishes and darters, and we have come to realize that reproductive isolation alone (as required by the BSC) is not a sufficient criterion for determining the specific distinctness of many pairs of fish species.

The BSC has been criticized for recognizing species based on more subjective decisions than either the ESC or PSC. In actual practice, all three modern species concepts involve some degree of subjectivity, and some ichthyologists today only half-jokingly refer to Regan's 1926 statement that "a species is what a competent taxonomist says it is." The continuing debate over which species concept is best or which one should be used exclusively should be seen as positive. It shows that the field of systematics has acquired unprecedented activity and importance in biology, and as Hickman et al. (2017) pointed out, "we cannot predict which concepts of species will remain useful 10 years from now." Most of the fish species recognized in this book would be valid by the principles of the BSC, but there are some instances where we have supported species recognition based largely on the ESC.

The Subspecies Concept

Geographic population variants are sometimes regarded by biologists as microgeographic races or subspecies of a species. This phenomenon is common in species that have large ranges. Mayr and Ashlock (1991) defined a subspecies as an aggregate of phenotypically similar populations of a species inhabiting a geographic subdivision of the range of that species and differing taxonomically from other populations of that species. Subspecies of a species can and do interbreed, and a zone of intergradation is considered by proponents of the Biological Species Concept as evidence that these geographic variants do not warrant recognition as full species. Even when interbreeding is evident, studies of DNA sequences often show large amounts of genetic divergence among populations (Hickman et al. 2017). Proponents of the Phylogenetic and Evolutionary Species Concepts use this as evidence for recognizing these variant populations as separate species, and many fish populations that were formerly recognized as subspecies have been elevated to full species status based on genetic differences (or a combination of morphological and genetic differences).

Recognition of taxa below the species level has been controversial, and many biologists and ichthyologists today do not recognize subspecies since this taxon is even less clearly defined than a species (Near and Koppelman 2009). Piller et al. (2008) considered the subspecies rank a category of taxonomic convenience rather than a meaningful biological entity. Those authors contended that darter (Percidae) taxonomy is undergoing a paradigm shift with the result that no new subspecies of darters have been described since 1979. Formal recognition of subspecies has lost favor with taxonomists largely because subspecies recognition is often based on minor differences in appearance that do not necessarily diagnose evolutionarily distinct units (Hickman et al. 2017). The American Fisheries Society discontinued listing subspecies in the 6th edition of *Common and Scientific Names of Fishes from the United States, Canada, and Mexico* (Nelson et al. 2004). Delisting of subspecies continued in the 7th edition (Page et al. 2013), but the authors recognized that subspecies, with their own evolutionary history in allopatry, have importance in evolutionary inquiry and may be given special protective status and recognized in studies of biodiversity. We herein list previously recognized subspecies as reported in Page and Burr (2011) when applicable to species found in Arkansas. Subspecies are discussed in the individual species accounts under the Variation and Taxonomy subsection.

Molecular Techniques

Early Applications

With the elucidation of the molecular basis of inheritance, biological macromolecules assumed an increasingly important role in studies of fish systematics. Nucleic acids (DNA and RNA), proteins, and chromosomes provide a broadly applicable set of heritable markers to examine genetic structure of populations and estimate relationships among taxa (Hillis et al. 1996). White (1973) showed that variation in chromosome structure and number provided a valuable source of genetic markers within and between species. Zuckerkandl and Pauling (1962) compared amino acid sequences of proteins to provide the first indications of a molecular clock, and that concept has been expanded in recent years and used extensively to estimate phylogeny (Goodman et al. 1987; Near et al. 2011).

Ichthyologists first began using molecular techniques to study the products of genes (proteins) before methods were developed to study DNA-level variation in the genes themselves. Compared with initial morphology-based studies on fish evolution, early genetic investigations of fishes were focused on smaller time scales of evolution, and investigations centered on genetic differentiation among populations and among closely related species based on multilocus allozyme data obtained from protein electrophoresis (Chen and Mayden 2010). Allozymes are enzymes (proteins) isolated from tissues. Studies typically used allozymes from varying tissues such as liver, eyes, and muscle. For each allozyme, the DNA locus coding for the enzyme is assumed to be the same across the group of organisms under consideration. Analysis of proteins using isozyme electrophoresis was another widely used early approach in molecular systematics (Avise 1974). Allozymes were used relatively sparingly for estimation of phylogenetic relationships compared with their use in studies of population genetic structure and gene flow. Studies of allozymes peaked in 1998 and dropped off considerably in the past decade because we now have molecular methods for studying nucleotide sequences of genes directly. An example of a study using allozymes to estimate phylogeny is found in Wood and Mayden (1997), who used this technique to estimate phylogenetic relationships among selected darter (Percidae) subgenera.

Phylogeography

In 1987, Avise et al. (1987) introduced the term "intraspecific phylogeography" for the use of molecular data to reconstruct population histories in relation to geography. The essence of their approach was a three-stage process whereby (1) molecular data are obtained from individuals sampled from geographically distinct populations; (2) those data are then used to generate a tree showing genealogical relationships among individuals; and (3) the geographic distribution of individuals is compared by superimposing the geographic location of each individual on the tree (Wiley and Hagen 1997). Avise et al. (1987) argued that patterns of concordance between genealogy and geography should reflect historical events responsible for the current distribution of an organism. The fundamental assumption is that molecular data preserve a record of genealogy that is independent of the historical pattern of dispersal or vicariance among populations (Wiley and Hagen 1997). In Arkansas, this type of analysis has been used on percid fishes by Turner et al. (1996) in a study of life history and comparative phylogeography of darters (Percidae) of the Central Highland region.

Recent Applications

Recent advances in the manipulation and analysis of nucleic acids have led to the widespread study of DNA and, to a lesser extent, RNA variation. Sequences assayed have come from the nucleus, the mitochondrion, and the chloroplast in plants. The 1990s saw an explosion of research in the field of molecular phylogenetics or molecular systematics (Brusca and Brusca 2003). This field infers phylogenies from molecular data that consist primarily of long sequences of the four nucleotides that make up the information encoded in DNA molecules. Theoretically, sequences retrieved from the genes of closely related species should differ only slightly, whereas sequences from the same areas of those genes from more distantly related species should have accumulated more differences (Brusca and Brusca 2003). Although relatively straightforward, the deep relationships in the Tree of Life have proven difficult to reconstruct with both morphological and molecular data, and anyone attempting to estimate a phylogeny from DNA sequence data should read Maddison (1997) (Ben Keck, pers. comm.). Scientists have sequenced more than 153 billion bases of DNA from thousands of species (Reece et al. 2015). This enormous database (e.g., GenBank) has fueled a boom in the study of phylogeny and clarified many evolutionary relationships of animals. Betancur-R. et al. (2013) published the first explicit phylogenetic classification of bony fishes based on a molecular phylogeny, and Betancur-R. et al. (2017) updated that version. Wiley and Johnson (2010) cautioned that molecular analysis of teleost relationships is still in its infancy in terms of gene sampling and taxon sampling. Those authors pointed out that studies with different genes often yield different results, leading to opposing conclusions.

Applications of modern molecular methods in ichthyology became widespread in the 1990s with polymerase chain reaction (PCR) amplification of targeted sequences in the genome, increasing automation, and decreasing costs of DNA sequencing, and other technologies, such as DNA microsatellite genotyping and SNPs (single nucleotide polymorphisms). In fishes, information on whole genomes and other data (e.g., anatomy, development, ontologies), particularly with existing (zebrafish, medaka) and emerging (cichlids, sticklebacks) model fish species, provide a new comparative context and data for existing fields (Mabee et al. 2007). Chen and Mayden (2010) also pointed out that the integration of these data raises the possibility of a new era of ichthyological research that includes genome-wide (or phylogenomic) studies (Volff 2005).

The original goal of the phylogenomic approach was to determine how similar gene sequences isolated from different genomes are related by descent, as well as to understand the evolution of individual gene trees (Chen and Mayden 2010). Maddison (1997) emphasized that a tree estimated from a single gene reflects the evolutionary history of that gene locus and does not necessarily reflect the evolutionary history of the species. The phylogenomic approach consists of using large-scale or multiple-loci sequence data obtained from intensive sequencing efforts or available genomic databases to reconstruct a better supported evolutionary history of organisms (Chen et al. 2004). Phylogenomics has become increasingly popular in molecular systematics for assembling the Tree of Life of extant organisms.

DNA Barcoding

One of the recent techniques for identifying animal species is DNA barcoding (Teletchea 2009, 2010). DNA barcoding offers a standardized method for identifying species using a short mitochondrial DNA sequence from the COI gene present in all animals to provide a "barcode" for comparing taxa. The mitochondrial gene encoding cytochrome c oxidase subunit 1 (COI), which contains about 650 nucleotide base pairs, is a standard barcode region for animals (Hickman et al. 2017). Although DNA sequences of COI usually vary among individuals of the same species, the variation is not extensive, so that variation within a species is much smaller than differences between species. DNA barcoding complements taxonomy, molecular phylogenetics, and population genetics (Hajibabaei et al. 2006, 2007). It has gained wide acceptance leading to public databases of DNA barcodes, such as the Barcode of Life and GenBank, where the mitochondrial COI gene sequences (barcodes) for species are stored (Hebert et al. 2003). These databases provide a central location where gene sequences for identifying species can be readily and easily accessed (Narain et al. 2013). This process is particularly useful in instances when reproductively isolated taxa are morphologically indistinguishable. Because the DNA sequences of a species are acquired over time, the DNA barcode developed for any species could be used to separate cryptic species. Such cryptic diversity occurs across the animal kingdom. Most cryptic species are, as Bickford et al. (2007) described, in "morphological stasis"—exhibiting limited or no morphological change due to selection, adaptation, and/or environmental condition. In one study, DNA barcoding found as many as nine undescribed species embedded within a single known species of skipper butterfly (Hebert et al. 2004). In an unusual example in fishes, individuals thought to belong to two different species were found to be the male and female of a single species (Byrkjedal et al. 2007). Since nearly 10% of the North American freshwater fish fauna is thought to remain undescribed (Warren et al. 2000; Butler and Mayden 2003), DNA barcoding may play a major role in the future in identifying and resolving new taxonomic problems.

There was a Fish Barcode of Life Campaign in 2009 (Ward et al. 2009), and April et al. (2011) generated a standard reference library of mtDNA sequences (DNA barcodes) derived from expert-identified museum specimens for 752 North American freshwater fishes. This established a barcode reference library for more than 80% of the named freshwater fish species in North America. Those authors used the survey of outstanding genetic diversity as an independent calibration of current taxonomic resolution within the North American fish fauna to reveal key areas of uncertainty where discrepancies between the genetic data and morphologically based taxonomy occur.

DNA barcoding has proven to be a powerful new tool in controlling traffic in endangered species or stolen specimens. The DNA barcode library provides an identification system with many applications, including the identification of fish parts or remnants, such as fish filets, sushi, smoked fish, caviar eggs, and larvae that are not recognizable using morphological observations (P. J. Smith et al. 2008; Wong and Hanner 2008; Victor et al. 2009). DNA barcoding can also facilitate tracking of exotic invasive species through water samples (Ficetola et al. 2008) and aid in food web reconstruction through gut content or fecal sample analysis (Kaartinen et al. 2010). Barcoding was used to identify larval fish species captured in plankton samples (Victor et al. 2009; Loh et al. 2014), and it is now being used by more than 50 countries to protect their biodiversity (Knowlton 2010).

Realistically, there are some disadvantages of DNA barcoding (DeWalt 2011). For example, with conserved primers there is the possibility of nuclear genome integrations into the mtDNA sequence, confounding the potential to clearly identify species (Narain et al. 2013). It should be noted that mtDNA introgression (i.e., hybridization) can strongly affect the accuracy of DNA barcodes because they are based on a single mtDNA gene that is inherited through the female parent. For example, it is possible to get an individual that is a Bluegill but has a barcode that indicates an Orangespotted Sunfish. The presence of pseudogenes in the mtDNA and its inconsistent evolutionary rate among lineages are also disadvantages in relying on COI as the sole marker for taxonomic identifications (Chu et al. 2009). DNA barcoding also does not resolve the controversies among different species concepts. However, the advantages of DNA barcoding are far greater than the disadvantages (Narain et al. 2013) for applications described above.

Characters: Molecular Markers*

RFLP, AFLP, Microsatellites (Microsats), and SNPs: These techniques and others came into favor in the early 2000s, and all are different ways of detecting mutations and recording those as discrete characters. RFLP (restriction

fragment length polymorphism) and AFLP (amplified fragment length polymorphism) are techniques that exploit variations in homologous DNA sequences. RFLP and AFLP use restriction enzymes that cut genomic DNA into smaller, various sized pieces (fragments). Both techniques focus on differences between samples of homologous DNA molecules that have different nucleotide sequences and are recognized by particular restriction enzymes. Both also use a related laboratory technique whereby differences in fragment sizes can be illustrated. Each restriction enzyme cuts genomic DNA only where it encounters a certain nucleotide sequence. Microsatellites are short tandemly repeated base sequences (usually one to six nucleotides) found in the genomes of many organisms. Although some mechanisms have been proposed, the generation of microsatellites is not fully understood. DNA microsatellites (usually tandem repeats or short repeats like CTCTCTCT) and SNPs (single nucleotide polymorphisms) are usually identified by first obtaining DNA sequence data for a locus from a representative sample of the organisms under investigation, finding an area of the sequence that is variable within that sample, and then characterizing the nucleotide sequence (by RFLPs or direct sequencing) or fragment length (in base pairs) by electrophoresis. For instance, an SNP could be one population with the base sequence ATTTC while the other population sequence has ATCTC. Specific PCR primers and restriction enzymes are developed first to replicate the sequence and then to cut the DNA at that point in the sequence for one of those, say, ATCTC. Then by electrophoretic separation, the individuals with the ATCTC will be identified by fragments that did not travel as far on the gel. SNPs are also being employed in next-generation sequencing, but the identification process is different. RFLP, AFLP, DNA microsatellites, and SNPs are relatively inexpensive tools for studying genetic variation in populations of the same species or closely related species. RFLP and AFLP studies seem to be declining, but microsatellites and SNPs are still heavily used for studies in population genetics and evolutionary ecology (genotype by environment interactions).

DNA sequences—traditional Sanger sequencing: Traditional Sanger sequencing starts with isolating DNA from a specimen and using Polymerase Chain Reaction (PCR) to amplify specific loci (using targeted primers).

---

* If the following section seems too technical for the scope of this book, our purpose was to present information about some modern methods of DNA analysis that are often unfamiliar even to many ichthyologists. We do not presume to cover all aspects of molecular methods and theory in this brief summary. It is almost impossible today to read a scientific journal article on fish systematics without encountering at least some of the terminology found below. We do not consider ourselves experts on molecular techniques and have relied heavily on the input of others who are experts in that field.

Amplicons from PCR are then run through one or more PCRs with labeled (radioactive or fluorescent) nucleic acids. This product is then run through the sequencing machine that records the sequence based on identifying the labels. The cost of Sanger sequencing has dropped dramatically over the past 15 years, and the quality and length of reads have increased. In the early 2000s, isolating the DNA from tissue, PCR, and sequencing the nearly ubiquitous (for vertebrates) mitochondrial gene *cytb* at 1140 base pairs for one individual took about 5 days and $40; today it takes at most 2 days and $20. Sequencing the entire mitochondrial genome is common today and not prohibitively expensive. In ichthyology, phylogenies from single mitochondrial DNA loci began appearing in the literature in the mid-1990s, and the number of publications and number of loci used have increased ever since. Next-generation sequencing technologies may eventually make traditional Sanger sequencing obsolete, but, for now, its low cost, ease of analysis, and utility for answering certain questions probably mean that Sanger sequencing will be used for many years to come. Examples of darter phylogenies estimated from Sanger sequenced DNA loci include Song et al. (1998), Sloss et al. (2004), Near et al. (2011), and Near and Keck (2012). Other papers on smaller taxonomic groups in darters and other North American fishes include Clements et al. (2012) and Nagle and Simons (2012).

DNA sequences—next generation sequencing methods: DNA sequencing technology is advancing so rapidly that many names are often used for essentially the same type of data production. Generally, next-generation sequencing (NGS) methods produce large amounts of data, often measured in terabytes instead of kilobytes as with Sanger sequencing data. A traditional Sanger sequenced dataset for 100 taxa and a modest number of loci may include approximately 1.5 million base pairs of sequence. For the same 100 taxa, the various NGS methods can result in anywhere from 50 million to almost 1 billion base pairs of sequence data. RADseq (or RADtag and other names) is a method in which genomic DNA is cut using restriction enzymes, much as for RFLPs, AFLPs, and SNPs, and then amplified with PCR. Then only certain sized fragments are isolated, usually 300–500 base pair fragments, and these fragments are labeled with a unique tag for that individual. RADseq is usually used when a reference genome is not available for assembly and annotation of NGS data for an organism. In such cases, a (relatively) closely related model species can be used for annotation purposes, with some caveats. SNPs are appropriate for population genetics studies, and there are now some new computer programs that can estimate phylogenies from

these data, but in studies of broad taxonomic coverage, we run into issues of low confidence in the assumed homology of the loci. Another new NGS method is targeted genome capture, in which about 500 loci that have been annotated to a reference genome are amplified for each individual, and each amplified locus is 1,000 to 3,000 base pairs depending on the sequencing platform. This type of NGS is better suited for phylogenetics because of the confidence in homology of the loci across individuals, and the reads are long enough that models of molecular evolution can be estimated. There are other methods that amplify 100 to 300 base pair fragments from throughout the genome without any reduction in the types of loci captured, and the data from these can be used to reconstruct the entire genome of an organism. These could certainly be used in phylogenetics, but the computational power required to first build the genome and then model the evolution of the genome to reconstruct the phylogeny is massive and restricts its application to studies of a few species at this time.

Phylogenetic Methods and Markers

The following books help explain phylogenetic methods and markers: Hillis et al. (1996), Page and Holmes (1998), Avise (2004), Felsenstein (2003), Knowles and Kubatko (2010), and Wiley and Lieberman (2011). In addition, discussions with Benjamin Keck (pers. comm.) and correspondence with Richard Broughton (pers. comm.) and Thomas Turner (pers. comm.) helped immensely in preparing this section. The basic premise for all phylogenetic methods is that some form of information contained in the organisms of interest can be used to infer relationships, or common ancestry. The information gathered from the organisms is varied and has evolved as technology progresses. Similarly, the methods used to infer the relationships from these different sources have changed with advances in evolutionary theory, genetic technologies, algorithms, and computational capabilities.

When the first edition of *Fishes of Arkansas* was published more than 30 years ago, ichthyologists were still using primarily morphological features, mainly meristics and morphometrics, to study fishes and to describe newly discovered species. Since then, a revolution has taken place in methodology, and a host of new techniques have been developed at the molecular level to study fishes. Another example of the relatively recent use of molecular techniques in fish systematics is the fifth edition of *Fishes of the World*, which was the first edition of that series to rely heavily on molecular information to classify the world's fishes

(Nelson et al. 2016). One of the most important points is that the rate of progress in methods and theory is increasing quickly, and any book that attempts to describe modern methodology will be out of date relatively shortly after its publication. For instance, in the early 2000s a phylogeny estimated from one mitochondrial gene locus would easily warrant publication in a respectable journal; from 2006–2010 those same journals would usually expect a study to include 5 to 20 loci; and more recently the expectation is a combination of mitochondrial and nuclear genes often involving 20 or more loci or data from a next-generation sequencing method (Ben Keck, pers. comm.).

There is still disagreement about the relative value of molecular versus morphological data, the constancy rates of molecular evolution, the neutrality and polarity of molecular variants, the type of data that should be collected, the various philosophical approaches to analyzing data, and the meaning of homology in relation to molecular characters (Hillis et al. 1996). Most recent systematic studies have integrated both morphological and molecular methods. Near et al. (2017) emphasized the importance of using a combination of morphological and molecular evidemce in systematic studies and pointed out that it is impossible to reconstruct meristic and morphometric data for individual specimens from summary tables that appear in published systematic studies. Those authors advocated depositing raw morphological data used in species descriptions in publicly available accessible databases (similar to the molecular database GenBank). Where we have made taxonomic decisions in this book that conflict with those of the AFS Committee on Common and Scientific Names, we have tried to base those decisions on a combination of morphological and molecular information.

CHAPTER 5

Conservation of Arkansas Fishes

Fishes in Trouble

Freshwater ecosystems provide vital resources for humans and are the sole habitat for an extraordinarily rich, endemic, and sensitive biota (Strayer and Dudgeon 2010). Unfortunately, the health and vitality of the planet's freshwater habitats and resources are being undermined at an unprecedented rate (Stiassny 1996; Vitousek et al. 1997). Species and whole communities are being lost in some cases even before they can be identified and long before they can be studied (Stiassny 1996). In North America, extinctions of 3 genera, 27 species, and 13 subspecies of fishes were documented between 1889 and 1989 (Miller et al. 1989). Freshwater makes up only 0.01% of the Earth's water and approximately 0.8% of Earth's surface, yet this tiny fraction of global water supports at least 100,000 species, almost 6% of all described species on Earth (Dudgeon et al. 2006). Freshwater habitats are also hotspots in the sense that they support approximately one-third of vertebrate species (Strayer and Dudgeon 2010). Walsh et al. (2011) considered loss of biodiversity one of the greatest impending environmental crises. Freshwater fisheries around the world are seriously overexploited, and large-bodied freshwater fishes are in global decline (Allan et al. 2005; Dudgeon et al. 2006). Globally, perhaps 10,000 to 20,000 freshwater species are extinct or imperiled because of human activities (Strayer 2006). Homogenization of fish assemblages is a real and serious threat to freshwater

biodiversity worldwide (Meffe and Carroll 1997; Vitousek et al. 1997; Scott and Helfman 2001). Rahel (2000) analyzed homogenization in North American fish assemblages and concluded that homogenization among state faunas is extensive and occurs primarily because of deliberate introductions of a small group of relatively cosmopolitan species that are useful to humans.

There are 700 fish taxa currently considered imperiled in North America's inland waters, representing 133 genera in 36 families (Jelks et al. 2008). The number imperiled represents a 92% increase over a nearly 20-year period dating to 1989 (Williams et al. 1989). Of the nearly 9,000 species of freshwater fishes worldwide, 1,800 (20%) were extinct or in serious decline by the 1990s (Moyle and Leidy 1992). Of the approximately 1,229 described species of freshwater fishes inhabiting North America (Page et al. 2013), 40% are imperiled or extinct (Jelks et al. 2008), and 350 (35%) need some type of protection. Fifty-seven taxa of North American fishes (39 species and 18 subspecies) have become extinct in the past century (Burkhead 2012). Factors responsible for these extinctions included habitat alteration (73% of extinctions), introduced species (68%), chemical alteration or pollution (38%), hybridization (38%), and overharvesting (15%). The modern extinction rate of freshwater fishes is nearly three orders of magnitude greater than the prehistoric background rate based on the fossil record (Burkhead 2012). All modern fish extinctions occurred since 1900, and the rate following World War II was distinctly higher, likely due to indirect effects of the postwar baby boom. Those indirect effects include demographic shifts from rural to urban areas, increased construction of large and small dams, increased alteration of natural water bodies (e.g., channelization, pollution), and other consequences of economic growth and industrial expansion (Burkhead 2012).

In Arkansas, as elsewhere, the delicate balance of nature has been upset by human perturbation of the environment, and many fish species need protection. The federal government protects organisms through the Endangered Species Act of 1973 (Public Law 93–205), which provides a program for the protection of species considered in jeopardy of extinction. The Act requires the listing of species based on specified criteria, prohibits the removal of any cited species, and encourages the preservation of habitats. Three fish species in Arkansas, the Arkansas River Shiner (*Notropis girardi*), the Ozark Cavefish (*Troglichthys rosae*), and the Leopard Darter (*Percina pantherina*) (Fig. 5.1), have been given a federal status of Threatened. Two Arkansas species, the Pallid Sturgeon (*Scaphirhynchus albus*) and the Yellowcheek Darter (*Nothonotus moorei*) (Fig. 5.2), are listed as Endangered.

At any given time, one or more Arkansas species are usually under review by the U.S. Fish and Wildlife Service. In a review of imperiled North American fishes, Jelks et al. (2008) considered 36 Arkansas taxa imperiled (23 vulnerable, 10 threatened, 3 endangered, and 1 extinct). The extinct fish, the Harelip Sucker, *Moxostoma* (= *Lagochila*) *lacerum*, was last collected from Arkansas in 1884 from the White River near Eureka Springs by David Starr Jordan. This sucker was the most widespread species to disappear from North America. It occurred in eight states and likely occurred in three others (Jenkins and Burkhead 1994; D. Neely, pers. comm.). Its demise was probably due to extensive sedimentation of streams as the result of poor farming practices (Jenkins and Burkhead 1994). The fine sediments likely caused the decline of the Harelip Sucker's prey, freshwater gastropods, which are highly sensitive to deposition of fine sediments (Johnson et al. 2013).

Earlier attempts to clarify the status of certain fishes in Arkansas were made by Buchanan (1974) and Robison (1974b). Robison and Buchanan (1988) later reviewed threatened fishes in Arkansas. Our assessment of the current conservation status of each fish species historically known to inhabit Arkansas is provided at the end of each species account. Our assessment refers to the status of a species in Arkansas, regardless of its status outside the state. Table 10 is our list of 75 extant species and 3 subspecies (approximately 35% of the extant Arkansas fish fauna)

Figure 5.1. Leopard Darter, *Percina pantherina*. *Uland Thomas.*

Figure 5.2. Yellowcheek Darter, *Nothonotus moorei*. *Uland Thomas.*

in need of protection and one extinct species in Arkansas. Our five categories and their definitions used to denote the conservation status of Arkansas fishes believed to be in trouble are as follows:

Endangered (E) A fish whose prospects for survival in Arkansas are in immediate jeopardy; in danger of extirpation and/or extinction throughout all or a significant portion of its range in Arkansas. This category also includes fishes now listed or being considered for inclusion on the *List of Endangered Fauna of the United States* as provided under the Endangered Species Act of 1973 (Public Law 93–205).

Threatened (T) A fish likely to become endangered within the foreseeable future throughout all or a significant part of its range in Arkansas. This includes forms whose numbers have decreased beyond the limits of normal population fluctuation or documented range contraction, but are not yet considered endangered. Also included are fishes listed under the provisions of Public Law 93–205.

Special Concern (SC) A fish that should be frequently monitored in Arkansas (a) because it exists in only one or few small geographic areas and/or is rare (low population density) over a relatively broad range; (b) because its existence may become endangered due to the destruction, drastic modification, or severe curtailment of the habitat; (c) because certain characteristics or requirements make it especially vulnerable to specific pressures; (d) because there is a documented historic decrease in range or demonstrable decrease in population size; or (e) for other reasons identified by experienced researchers.

Status Undetermined (U) A fish that knowledgeable ichthyologists have suggested as possibly threatened or endangered in Arkansas but data are insufficient to accurately determine its status.

Extinct or Extirpated (Ex) A fish that has become extinct throughout its range, or a fish which, though formerly recorded in Arkansas, probably no longer exists in the state, as determined by historical documents and/or specimens.

Rarest Fishes

Ten extant species documented as occurring in Arkansas are the rarest fishes known for the Natural State. There are no records since 1988 (dates of last known collection given) for the following: *Acipenser oxyrinchus* (18 July 1956), *Cyprinella camura* (four pre-1960 samples), *Cyprinella spiloptera* (23 May 1972), *Notropis bairdi* (pre-1950), *Notropis girardi* (1939), *Platygobio gracilis* (July and December 1980), and *Noturus gladiator* (1976). For *Umbra limi* and *Phenacobius mirabilis* there is one record each since 1988, and there are two records for *Noturus phaeus*. All the above species have been documented by museum records and specimens.

Introduced Species

Ten percent of the fish species known from Arkansas today are not native to the state. Humans have always transported and introduced animals and plants they found useful (Diamond 1999), and fishes are no exception. Introductions of fishes occur as a deliberate result of human endeavor or as a by-product of human activity (Helfman 2007). Several reasons are usually behind such introductions. The most common reasons for introducing species are to enhance aquaculture, improve sport or commercial fishing, augment and improve imperiled wild stocks, promote ornamental species, and practice biological control (Welcomme 1988; Lever 1996). Such introductions are often known by a multitude of names, including nonnative, nonindigenous, introduced, alien, exotic, transplanted, translocated, allochthonous, invasive, feral, and/or biological pollutant (Helfman 2007).

Categories of Introduced Fishes

Because of the confusing array of terminology applied to this subject, we herein use the categories defined as follows:

Native Species—A species that has historically occurred in Arkansas, and its occurrence in the state is not the result of an introduction.

Transplanted Species—A species native to North America but introduced into Arkansas.

Native Transplant—A species native to Arkansas that has also been intentionally stocked in this state.

Exotic Species—A species not native to North America that has been introduced into Arkansas. Such species are often referred to as alien or nonindigenous

Aquaculture, the tropical fish trade, and deliberate introductions to enhance sport or commercial fishing have all been implicated in the spread of exotic species (Barton 2007). The introduction of the Common Carp (*Cyprinus carpio*) in the early 1800s is a prime example of an exotic species having undesirable consequences. In some instances, sport fisheries have benefited by the introduction of exotic species such as the Brown Trout and Striped Bass. A chronology of introductions of exotic fishes into the United States is provided by Nico and Fuller (1999), and Vitule et al. (2009) give examples of freshwater fish introductions that have resulted in catastrophic ecological consequences. The Black Carp (*Mylopharyngodon piceus*), one of the most recent introductions in Arkansas, was originally introduced into commercial catfish ponds for snail control. It currently poses a threat to mollusk species native to the Mississippi drainage (Fuller et al. 1999; Strong and Pemberton 2000).

Buchanan et al. (2000) identified 25 species of fishes not native to Arkansas that have been accidentally or intentionally introduced into the state's waters, and today the known list of exotic and transplanted species has grown to 31 (Table 11). At least six hybrid combinations have been stocked in Arkansas, including *Esox lucius* × *E. masquinongy*—tiger muskie, *Salvelinus fontinalis* × *Salmo trutta* (Brook Trout × Brown Trout)—tiger trout, *Morone saxatilis* × *M. chrysops*—hybrid striper, *Stizostedion canadense* × *S. vitreum*—saugeye, and *Ictalurus furcatus* × *Ictalurus punctatus*—Blue Catfish × Channel Catfish hybrid. Two species, the Rainbow Smelt (*Osmerus mordax*) and White Perch (*Morone americana*) occur in Arkansas accidentally as a result of their transplantation in states north or west of Arkansas. Most introduced fishes were brought into the state to provide fishing opportunities, but five of the species (all large cyprinids), including the Grass Carp, *Ctenopharyngodon idella,* Common Carp, *Cyprinus carpio,* Silver Carp, *Hypophthalmichthys molitrix,* Bighead

Table 10. Arkansas Fishes in Trouble. The designated conservation status for a species refers only to its status in Arkansas (regardless of its status outside Arkansas) as determined by HWR and TMB.

Endangered Species in Arkansas (15 species and 1 subspecies)
Scaphirhynchus albus, Pallid Sturgeon
Alosa alabamae, Alabama Shad
Notropis atrocaudalis, Blackspot Shiner
Hybognathus placitus, Plains Minnow
Phenacobius mirabilis, Suckermouth Minnow
Pimephales tenellus tenellus, Western Slim Minnow
Platygobio gracilis, Flathead Chub
Noturus phaeus, Brown Madtom
Umbra limi, Central Mudminnow
Troglichthys rosae, Ozark Cavefish
Typhlichthys eigenmanni, Salem Plateau Cavefish
Typhlichthys sp. *cf. eigenmanni*, Ghost Cavefish
Etheostoma cragini, Arkansas Darter
Etheostoma microperca, Least Darter
Nothonotus moorei, Yellowcheek Darter
Percina pantherina, Leopard Darter
THREATENED SPECIES (14 SPECIES, 1 SUBSPECIES)
Acipenser fulvescens, Lake Sturgeon
Acipenser oxyrinchus desotoi, Gulf Sturgeon
Lythrurus snelsoni, Ouachita Mountain Shiner
Macrhybopsis meeki, Sicklefin Chub
Notropis ortenburgeri, Kiamichi Shiner
Notropis perpallidus, Peppered Shiner
Notropis suttkusi, Rocky Shiner
Noturus lachneri, Ouachita Madtom
Noturus taylori, Caddo Madtom
Moxostoma anisurum, Silver Redhorse
Etheostoma clinton, Beaded Darter
Etheostoma fragi, Strawberry Darter
Etheostoma pallididorsum, Paleback Darter
Etheostoma pulchellum, Plains Darter
Percina phoxocephala, Slenderhead Darter
SPECIAL CONCERN (39 SPECIES, 1 SUBSPECIES)
Ichthyomyzon unicuspis, Silver Lamprey
Lampetra aepyptera, Least Brook Lamprey
Lethenteron appendix, American Brook Lamprey
Scaphirhynchus platorynchus, Shovelnose Sturgeon
Atractosteus spatula, Alligator Gar
Campostoma pullum, Finescale Stoneroller
Erimystax harryi, Ozark Chub
Nocomis asper, Redspot Chub

Notropis chalybaeus, Ironcolor Shiner
Notropis ozarcanus, Ozark Shiner
Notropis potteri, Chub Shiner
Notropis sabinae, Sabine Shiner
Pteronotropis hubbsi, Bluehead Shiner
Pimephales tenellus parviceps, Eastern Slim Minnow
Catostomus commersonii, White Sucker
Cycleptus elongatus, Blue Sucker
Erimyzon sucetta, Lake Chubsucker
Moxostoma poecilurum, Blacktail Redhorse
Noturus eleutherus, Mountain Madtom
Noturus flavater, Checkered Madtom
Noturus flavus, Stonecat
Noturus gladiator, Piebald Madtom
Noturus maydeni, Black River Madtom
Mugil cephalus, Striped Mullet
Fundulus blairae, Western Starhead Topminnow
Etheostoma euzonum, Arkansas Saddled Darter
Etheostoma fusiforme, Swamp Darter
Etheostoma mihileze, Sunburst Darter
Etheostoma nigrum, Johnny Darter
Etheostoma parvipinne, Goldstripe Darter
Etheostoma sp. *cf. pulchellum* 2, Blue Belly Form
Etheostoma squamosum, Plateau Darter
Etheostoma teddyroosevelt, Highland Darter
Etheostoma uniporum, Current Darter
Nothonotus juliae, Yoke Darter
Percina brucethompsoni, Ouachita Darter
Percina evides, Gilt Darter
Percina nasuta, Longnose Darter
Percina uranidea, Stargazing Darter
Percina vigil, Saddleback Darter
EXTIRPATED FROM THE STATE (4 SPECIES)
Cyprinella camura, Bluntface Shiner
Cyprinella spiloptera, Spotfin Shiner
Notropis bairdi, Red River Shiner
Notropis girardi, Arkansas River Shiner
EXTINCT (1 SPECIES)
Moxostoma lacerum, Harelip Sucker
UNDETERMINED (3 SPECIES)
Acipenser oxyrinchus, Atlantic Sturgeon
Macrhybopsis gelida, Sturgeon Chub
Notropis stramineus, Sand Shiner

Table 11. Nonnative Fish Species Introduced into Arkansas.

EXOTIC SPECIES	
Family Cyprinidae—Carps and Minnows	*Carassius auratus*, Goldfish *Ctenopharyngodon idella*, Grass Carp *Cyprinus carpio*, Common Carp *Hypophthalmichthys molitrix*, Silver Carp *Hypophthalmichthys nobilis*, Bighead Carp *Leuciscus idus*, Ide *Mylopharyngodon piceus*, Black Carp *Scardinius erythrophthalmus*, Rudd *Tinca tinca*, Tench
Family Characidae—Characins	*Colossoma* or *Piaractus*, Unidentified Pacus
Family Salmonidae—Trouts and Salmons	*Salmo trutta*, Brown Trout
Family Channidae—Snakeheads	*Channa argus*, Northern Snakehead
Family Cichlidae—Cichlids and Tilapias	*Astronotus ocellatus*, Oscar *Oreochromis aureus*, Blue Tilapia *Oreochromis mossambicus*, Mozambique Tilapia *Oreochromis niloticus*, Nile Tilapia *Tilapia zillii*, Redbelly Tilapia
TRANSPLANTED SPECIES	
Family Clupeidae—Herrings	*Alosa sapidissima*, American Shad
Family Ictaluridae—North American Catfishes	*Ameiurus catus*, White Catfish
Family Osmeridae—Smelts	*Osmerus mordax*, Rainbow Smelt
Family Salmonidae—Trouts and Salmons	*Oncorhynchus clarkii*, Cutthroat Trout *Oncorhynchus mykiss*, Rainbow Trout *Oncorhynchus nerka*, Sockeye Salmon *Salvelinus fontinalis*, Brook Trout *Salvelinus namaycush*, Lake Trout
Family Esocidae—Pikes and Mudminnows	*Esox lucius*, Northern Pike *Esox masquinongy*, Muskellunge *Esox lucius* × *E. masquinongy*, hybrid tiger muskie
Family Moronidae—Temperate Basses	*Morone americana*, White Perch *Morone saxatilis*, Striped Bass *Morone saxatilis* × *M. chrysops*, hybrid striper
Family Centrarchidae—Sunfishes	*Ambloplites rupestris*, Rock Bass *Lepomis auritus*, Redbreast Sunfish *Lepomis macrochirus mystacalis*, Coppernose Bluegill (or Florida Bluegill) *Micropterus coosae*, Redeye Bass *Micropterus salmoides floridanus*, Florida Largemouth Bass
Family Percidae—Perches and Darters	Perca flavescens, Yellow Perch

Carp, *Hypophthalmichthys nobilis*, and Black Carp, *Mylopharyngodon piceus*, were introduced largely for vegetation control.

Some introduced fishes have established breeding populations in Arkansas. Among the exotics, the Common Carp has long been established statewide, and some breeding populations of Brown Trout also exist in Arkansas.

The Grass Carp was established in the lower Mississippi River by the late 1970s (Conner et al. 1980), and reproduction has been verified in various parts of Arkansas. The disturbing scenarios and media coverage of the explosion by populations of Silver Carp (*Hypophthalmichthys molitrix*) and Bighead Carp (*H. nobilis*) in the Mississippi River have led to fears of these cyprinids gaining access into the

Great Lakes. While no specimens of either Silver Carp or Bighead Carp have yet been collected in the Great Lakes, researchers have discovered environmental DNA of those fishes in the waters of the Great Lakes. More recently, another Asian carp, the Black Carp (*Mylopharyngodon piceus*), has been discovered in Arkansas. Although Goldfish (*Carassius auratus*) are collected occasionally in Arkansas, most are probably the result of bait releases, and no wild breeding populations are known in the state. The Rudd was introduced into the state in Horseshoe Lake in Crittenden County (Robison and Buchanan 1993), Spring River (Rasmussen 1998), and later in Lonoke County, Arkansas. Another recently established exotic is the Northern Snakehead, *Channa argus*, which occurs in the lower White and St. Francis river drainages.

An exotic species that will almost certainly eventually be documented from Arkansas is the Round Goby, *Neogobius melanostomus* (order Perciformes: family Gobiidae). The Round Goby was first reported in the United States from the Great Lakes in 1990 and quickly spread throughout all five Great Lakes within a few years (Fuller et al. 2018). This Eurasian native was introduced into the Great Lakes from the Black Sea through the release of freighter ballast. There are large populations in lakes Erie and Ontario, and it spread throughout the Illinois River of Illinois by 2010. Robert Hrabik (pers. comm.) informed us that the Round Goby was collected near the mouth of the Illinois River in Illinois by D. P. Herzog on 21 August 2018 and from the Mississippi River of Missouri by USFWS biologists on 22 August 2018. It may have already reached Arkansas waters. Distributional and biological information for *N. melanostomus* can be found in Fuller et al. (2018).

Aquarium releases have apparently been responsible for several isolated records of exotic fishes from Arkansas. There are five angling reports of pacus (*Colossoma* or *Piaractus*): Benton County 1992, Pulaski County 1992 and two records in 1995, and Carroll County in 2006. There is even an established state angling record for pacu of 7 pounds, 1 ounce (3.2 kg) from Lakewood Lake in Pulaski County on 22 July 1995. There is also an unverified angling record of one Oscar, *Astronotus ocellatus*, from a lake near Hot Springs in the early 1990s. Although no vouchered specimens are available for those apparent aquarium releases, their records are reported from Arkansas on the USGS Nonindigenous Aquatic Species website, along with a 1992 record for Redbelly Tilapia, *Tilapia zillii*. We have not included the pacus, Oscar, or Redbelly Tilapia in the keys or species accounts because of their sporadic occurrence and probable inability to become established in the state. Other introduced species not included in the keys or

accounts because there are no recent records from Arkansas are American Shad (*Alosa sapidissima*), Ide (*Leuciscus idus*), Tench (*Tinca tinca*), White Catfish (*Ameiurus catus*), and Redeye Bass (*Micropterus coosae*).

Among the transplanted fishes, the Rainbow Trout, Striped Bass, Rock Bass, Redbreast Sunfish, and Yellow Perch appear to have established self-sustaining populations in Arkansas. The White Perch may also be established. Additional information about the introduced species can be found in the individual species accounts.

A number of fishes native to Arkansas have been stocked intentionally or accidentally in various parts of the state (Table 12). Most of the native transplants are game species cultured by the Arkansas Game and Fish Commission. Some, including Channel Catfish, Bluegill, Redear Sunfish, and Largemouth Bass, have been distributed statewide. Stocking of the Smallmouth Bass has occurred on a more limited basis. Threadfin Shad, Fathead Minnow, and Mississippi Silverside have been stocked to provide forage species for desirable gamefish. The Golden Shiner and Fathead Minnow have been widely distributed in Arkansas via bait release.

Commercial Fishery

The economic value of Arkansas fishes is determined not only by the sport fishing industry, but also by commercial fish-farming operations (aquaculture) and by a wild commercial fishery. In 1979, baitfishes were produced on 51% of the farmed waters and were valued at more than $16 million (Fiegel and Freeze 1981). The principal bait species in order of importance were Golden Shiner, *Notemigonus crysoleucas*, Fathead Minnow, *Pimephales promelas*, and Goldfish, *Carassius auratus*, and they are still the predominant bait species currently raised in Arkansas. In 2007, there were 21,965 acres of ponds producing baitfish, and the total baitfish value was estimated to be $21.5 million; in 2013, there were 21,695 acres of ponds producing an estimated $19 million of baitfish (Nathan Stone, pers. comm.).

Food fishes were produced on 22.9% of the intensively farmed waters in 1979–1980 and were valued at more than $16.5 million. Several species were raised for food, but the most commonly raised fishes were Channel Catfish, *Ictalurus punctatus*, Blue Catfish, *I. furcatus*, Bigmouth Buffalo, *Ictiobus cyprinellus*, and Rainbow Trout, *Oncorhynchus mykiss*. Catfish farming has become an increasingly important part of aquaculture in Arkansas in recent decades, but it can also be a very volatile industry. Between 2007 and 2012, many catfish farmers either significantly reduced production or completely stopped raising

Table 12. Native Fish Species Intentionally Stocked in Arkansas (Native Transplants).

Family Clupeidae—Herrings	*Dorosoma petenense*, Threadfin Shad
Family Cyprinidae—Carps and Minnows	*Notemigonus crysoleucas*, Golden Shiner *Pimephales promelas*, Fathead Minnow
Family Ictaluridae—North American Catfishes	*Ameiurus nebulosus*, Brown Bullhead *Ictalurus furcatus*, Blue Catfish *Ictalurus punctatus*, Channel Catfish
Family Esocidae—Pikes and Mudminnows	*Esox niger*, Chain Pickerel
Family Atherinopsidae—New World Silversides	*Menidia audens*, Mississippi Silverside
Family Moronidae—Temperate Basses	*Morone chrysops*, White Bass
Family Centrarchidae—Sunfishes	*Ambloplites constellatus*, Ozark Bass *Lepomis cyanellus*, Green Sunfish *Lepomis macrochirus*, Bluegill *Lepomis microlophus*, Redear Sunfish *Lepomis macrochirus* × *L. cyanellus*, hybrid bream *Micropterus dolomieu*, Smallmouth Bass *Micropterus salmoides*, Largemouth Bass *Pomoxis annularis*, White Crappie *Pomoxis nigromaculatus*, Black Crappie
Family Percidae—Perches and Darters	*Stizostedion canadense*, Sauger *Stizostedion vitreum*, Walleye *Stizostedion canadense* × *S. vitreum*, saugeye

the fish, primarily because the cost of commercial food rose dramatically. Catfish acreage in the state decreased from about 40,000 acres in 2007 to approximately 10,000 acres in 2012. However, by 2012, the total annual value of the Arkansas aquaculture industry had grown to $167 million, and Arkansas ranked second among aquaculture-producing states and led the nation in baitfish production, raising more than 80% of all baitfish in the United States (Carole R. Engle, pers. comm.). The baitfish category also includes small species that are sold to aquarium markets, and increasing numbers of Arkansas fish farmers have begun to raise ornamental fish species, including koi and fancy goldfish. The most important sport fish produced for stocking recreational ponds include Largemouth Bass, Bluegill, Black Crappie, hybrid bream (most commonly a Bluegill × Green Sunfish cross), and Redear Sunfish (Southern Regional Aquaculture Center Fact Sheet FSA9055). Arkansas also surpasses all other states in the production of fingerling Largemouth Bass, hybrid Striped Bass fry, and Asian carps, and reportedly has the world's largest baitfish farm (Lonoke County), Largemouth Bass farm (Monroe County), Goldfish farm (Lonoke County), and hybrid Striped Bass hatchery (Lonoke County) (Carole R. Engle, pers. comm.). Largemouth Bass are also raised in Arkansas for sale as food. The primary markets

are ethnic markets in cities on the East and West coasts. Hybrid Striped Bass fry are exported to fish farms in Taiwan, Israel, and Europe. The Arkansas Game and Fish Commission has approved 81 species (68 native) distributed among 19 families for aquaculture use in the state.

A commercial fishery for wild fish has existed in Arkansas for almost 150 years, but documented official catch records for all commercial species are available only from the mid-1970s into the late 1980s. The comprehensive statewide monitoring program for commercial fisheries ended around 1988 and, except for a roe monitoring program (Paddlefish, sturgeon, and Bowfins), has not been resumed (Bill Posey, per. comm.). During the 1800s and early 1900s there were no records and little regulation of commercial fishing. During those times, game and nongame species alike were taken commercially by a variety of methods. Newspaper articles and magazine accounts reported a drastic decline in the fishery resources of Arkansas by the late 1800s. However, this trend was not referred to by Seth Meek or any other ichthyologists collecting in Arkansas during that time. The Arkansas General Assembly in 1873 was concerned enough to enact a law making it a misdemeanor to use poison to take fish and, in 1885, passed laws further limiting "market fishing." Eventually, commercial fishing came under the jurisdiction of the Arkansas Game and Fish Commission, and regulations were established limiting the species of fishes that could be taken commercially and specifying the type of fishing gear to be used.

In comparison to neighboring states, Arkansas had, by far, the most extensive wild commercial fishery in the 1980s. By 1985, more than 17 million pounds (7,711,200 kg) of fishes worth $7.5 million were being harvested annually (Patterson 1985). Nationwide, as late as the 1970s, only Michigan had a greater commercial catch of freshwater fishes than Arkansas. The complete yearly harvest and the status of the commercial fishery in Arkansas have been tabulated between 1975 and 1985 and summarized in the Arkansas Game and Fish Commission's annual reports (AGFC, Little Rock).

The state's wild commercial fishery is confined primarily to the larger lowland lakes and main stems of the Arkansas, White, Red, and Mississippi river systems. On rare occasions, commercial fishing is permitted in a few reservoirs. Hoop nets, gill nets, and trammel nets are the main types of fishing gear used (Crawford and Freeze 1982), but long lines (trot and/or snag lines), fiddler nets (small mesh hoop nets), and slat traps are also sometimes used.

During the years of comprehensive data tabulation, buffalofishes (*Ictiobus bubalus, I. cyprinellus,* and *I. niger*) annually comprised the greatest biomass of species taken commercially in Arkansas, followed by catfishes (primarily *Ictalurus punctatus, I. furcatus,* and *Pylodictis olivaris*), carps (*Cyprinus carpio*), gars (primarily *Lepisosteus osseus* and to a lesser extent *L. platostomus* and *Atractosteus spatula*), Freshwater Drums (*Aplodinotus grunniens*), and carpsuckers (mainly *Carpiodes carpio,* but possibly a small number of *C. cyprinus* and *C. velifer*) contributed significantly to the annual commercial catch. Grass Carp (*Ctenopharyngodon idella*), Paddlefish (*Polyodon spathula*), Bowfin (*Amia calva*), and Shovelnose Sturgeon (*Scaphirhynchus platorynchus*) made up small percentages of the annual commercial catch. The variable nature of the wild commercial fishery is related to several factors, one of the most important being the number of commercial fishermen licensed from one year to another. From the late 1970s into the early 1980s, the number of commercial fishermen in Arkansas steadily increased, as did their total catch. Annual fluctuations in total catch appeared to be closely related to changes in the numbers of fishermen.

It is not possible to detect long-term trends in Arkansas's commercial fishery from the 10 years of reliable data available; however, certain short-term trends are discernible. In the 1980s Grass Carp made up an increasingly higher percentage of the total commercial catch (although

still a very small percentage) than in the late 1970s. Also noteworthy was the development of a modest sturgeon and Paddlefish fishery in the 1980s. Because of the drastic decline of Eurasian sturgeon fisheries in recent decades and the subsequent termination of international caviar export quotas from that fishery, there has been a dramatic increase in the commercial fishery for Paddlefish, sturgeon, and Bowfin to help meet the demand for caviar. Arkansas is one of eight states to permit commercial harvest of Paddlefish. In addition, the flesh from Paddlefish and sturgeon is sold in fish markets or to processing facilities where it may be smoked and sold retail. However, Jansen (2012) found that Shovelnose Sturgeon were not abundant in the Arkansas River upstream of Pool 1 and indicated that commercial fishing had the potential to adversely affect sturgeon populations in that river. The Mississippi River is currently closed to the harvest of sturgeon for Arkansas commercial fishers (Posey 2006). Today a substantial fishery for Paddlefish exists in the Arkansas, Mississippi, and White rivers. In 2002 the Arkansas Game and Fish Commission began a study of Paddlefish populations in the upper three navigation pools of the Arkansas River in Arkansas. In 2002, 1,066 Paddlefish were captured, tagged and released in the Ozark Pool. Based on recapture records, a population estimate of the number of catchable Paddlefish in Ozark Pool was approximately 5,000 individuals in the 10,600-acre pool (Jeff Quinn, pers. comm.). Most of the following information on the Arkansas roe harvest is from Posey (2006, and pers. comm. 2014). A total of 26,176 pounds of roe were harvested in the 2005–2006 harvest season (November through May) from 9,375 fish. An additional 3,152 fish were harvested for the fresh food markets, where it was sold as "boneless catfish" or smoked Paddlefish, or sent to an international market for resale. Of this roe harvest, Paddlefish provided 25,050 pounds while sturgeon and Bowfin composed 824 and 70 pounds, respectively. Bowfins accounted for a very small percentage of the roe harvest from 2002 to 2013, ranging from 12 to 500 pounds. Between 2006 and 2013, the total Arkansas roe harvest ranged from a low of 9,266 pounds in 2009–2010 to a high of 26,249 pounds in 2006–2007. Arkansas waters in which the harvest of roe currently occurs, listed in decreasing order of roe harvested between 2002 and 2013 with percentage of the total roe harvest for those years in parentheses, are the Arkansas (31.8%), Mississippi (29.5%), White (21.9%), Ouachita (9.2%), Black (3%), St. Francis (0.5%), and Little (0.3%) rivers. Fish farms accounted for 3% of the harvest, and other waters, such as a few oxbow lakes, made up the remaining 1%. A total of 174,013.6 pounds (87.01 tons) of caviar were harvested within that time period.

CHAPTER 6

Organization of the Chapters Containing Species Accounts

Order, Family, and Genus Accounts

In this second edition, we have added brief descriptions of the 19 orders of fishes represented by Arkansas species, and the chapters containing species accounts are now organized by order. An order account is given at the beginning of each chapter, and an account for each Arkansas family in the order is given when the family is discussed. Also new to this edition are accounts for all genera. We feel these changes enable the reader to better understand the taxonomic placement of the individual species of Arkansas fishes within the framework of the overall classification of fishes.

Each order account provides basic information on distinguishing characters, worldwide distribution, fossil history, and number of families, genera, and species currently recognized within the order. Much of the information for the ordinal accounts was taken from Etnier and Starnes (1993), Boschung and Mayden (2004), Nelson (2006), and Nelson et al. (2016), and orders are presented in phylogenetic sequence according to Page et al. (2013), except for Anabantiformes (Appendix 5). Arkansas orders are arranged phylogenetically to reflect the general evolutionary history of fishes from the most ancestral fishes (Petromyzontiformes) through the most derived ray-finned fishes (Perciformes).

Family names and the sequence of families within orders follow Page et al. (2013), except for Elassomatidae

(Appendix 5). For each family, information is provided on distinguishing family characters, number of genera and species in the family, the number of species in that family occurring in Arkansas, fossil history, worldwide distribution, and selected biological information pertinent to that family. For all families, genus accounts are also provided. Genera and species within genera in a family are arranged alphabetically by scientific name, first by genus and then by species. For each family containing more than a single species, a dichotomous key is provided for identifying all Arkansas species in that family.

Species Accounts

This book includes accounts for 243 species of fishes. In each species account, the basic categories of information presented are as follows:

Scientific Name and Common Name

The scientific and common names of Arkansas fish species used in the second edition for the most part follow *Common and Scientific Names of Fishes from the United States, Canada, and Mexico*, 7th edition (Page et al. 2013). In some instances, we disagreed with Page et al. (2013) in designating genus, species, and/or common name. We also follow Nelson et al. (2002) and Page et al. (2013) in capitalizing the common names of fishes. The first item that appears in the species account is the scientific name of the species, followed by the name of the person who originally described the species, for example, *Lythrurus snelsoni* (Robison). If the author's name is in parentheses, it means the genus name of the original description of this species has been changed. In the previous example, this species was originally described as *Notropis snelsoni* Robison. Later, this shiner was placed in a different genus, *Lythrurus*. Beneath each scientific name is the accepted common name according to Page et al. (2013).

Photographs

One or more color photographs are provided for each species, showing the species in breeding (nuptial) coloration or in its normal, nonbreeding appearance. Images of male and female are provided for some species, particularly those species having strong sexual dimorphism. Nearly all photographs in this second edition are new, color, digital images of live Arkansas fish specimens, rather than preserved museum specimens. Digital photographs were taken of live fish specimens using photographic techniques developed by Mark Sabaj-Perez, David A. Neely, and Uland Thomas. Sabaj-Perez (2009) described these most interesting techniques. Every attempt was made to secure Arkansas specimens to photograph. When that was not possible, we used photographs of specimens from surrounding states. For one species, the extinct *Moxostoma lacerum*, an artist's illustration was used. Most fish photographs used were those by David A. Neely and Uland Thomas, whose beautiful images of fishes made an important contribution to this edition. When fishes were captured in the field, live specimens were photographed separately by Dave and Uland, who used slightly different methods to take their photos, including use of different backgrounds against which fishes were photographed. Dave typically used a black background, while Uland used either a blue or gray background. This resulted in two different "looks" from which to choose the photographs for this book.

Distribution Map

We provide a state geographic distribution map for each species showing the known collection localities from 1988 through 2019, plus a generalized inset map of the overall native and/or introduced range of the species in the contiguous United States. Maps from the first edition of this book (with modifications to reflect taxonomic changes since 1988) showing older distribution records prior to 1988 for all species are presented in Appendix 4, so that

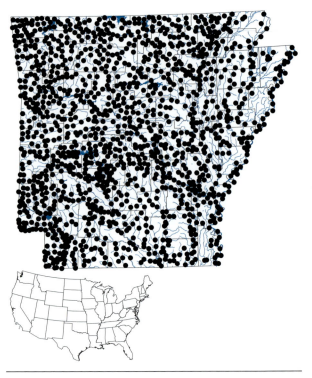

Figure 6.1. Arkansas localities sampled for fishes between 1988 and 2019.

historical changes in known distributions can be determined. The distributions of the fishes of Arkansas since 1988 are based on approximately 2,640 samples from our personal collections, those of colleagues, the referenced literature, and Arkansas specimens housed in several museums, especially Tulane University (TU), University of Louisiana at Monroe (NLU), Arkansas State University (ASUMZ), and the University of Arkansas–Fort Smith (UAFS) (see Appendix 1 for additional collections containing Arkansas specimens). Figure 6.1 shows localities sampled between 1988 and 2019.

Characters

This section describes the most useful morphometric, meristic, pigmentation, and other morphological characters for identification of the species. Such morphological characters as shape of head, mouth, and fins are provided. Morphometrics include such things as ratios of snout length, eye diameter, body depth, and body length to some other morphological feature, while meristic characters include such countable entities as number of lateral line scales, number of spines in the dorsal fin, number of pharyngeal teeth, and various other scale and fin element counts. Lengths and weights are given in English (usually inches and pounds) as well as metric units (usually millimeters and kilograms). Because of inherent variation within a species in morphometrics (due primarily to allometric growth) and meristics, ratios and counts are often given as a range from low to high. Average size in Arkansas and maximum reported size attained by the species are also given. Arkansas angling records, if any, are also listed.*

Life Colors

Color descriptions of each species are given for living specimens in nonbreeding condition. In addition, breeding coloration is described for those species exhibiting such coloration. It should be remembered that coloration often varies greatly. In addition to individual color variation, there may be sexual dichromatism between male and female and coloration differences between life stages, such as juveniles and adults. A good example of the latter is the difference in the coloration between juvenile and adult Fliers (*Centrarchus macropterus*).

Similar Species

Species most similar to the fish in question are compared, diagnosed, and differentiated to aid the reader in quickly distinguishing between a species discussed and other species likely to be confused with it, especially similar sympatric species. Even though there is some redundancy in the information in this section across species accounts, we feel that the Similar Species section is necessary to facilitate the identification process and to avoid, as much as possible, the need to frequently thumb through pages of the book in search of comparative information.

Variation and Taxonomy

The original describer(s) of a species is listed along with the type locality and scientific name used (if the original name is different from the current scientific name). Because species exhibit geographic variation, this section gives known variation of the species in Arkansas (and often elsewhere throughout the species range) and presents the current taxonomic status of state populations including sister species relationships if known. Pertinent current taxonomic literature is also provided. New to this edition is information

on genetic analyses where available. For many species, there have been recent studies of variation in mitochondrial and nuclear DNA that sometimes conflict with traditional morphological interpretations of phylogeny. Where taxonomic controversies exist, we have attempted to examine the existing evidence and give our opinion of the correct current usage. Therefore, our recognition of species (and sometimes genera) does not always agree with Page et al. (2013), particularly in the family Percidae.

Distribution

The total species range and the Arkansas distribution for each species are described. Much of the information on species range is from Page and Burr (2011).

Habitat and Biology

The topics in this section are arranged in the following order: habitat characteristics, feeding biology, and reproductive biology. Macro- and microhabitat preferences of Arkansas species are described to indicate the preferred habitat of the fish. It should be noted that fishes are sometimes collected in habitats other than their preferred one. Known aspects of the biology of our state fishes are discussed, including information on general behavior, feeding and reproductive behavior, and general comments on age at maturity, maximum size, growth, and longevity. This information was obtained from our collection experiences and observations of a combined 100+ years sampling Arkansas streams and rivers, as well as a survey of the scientific literature, including several other state fish books. While we used information about a species across its known geographic range, we attempted to present and discuss information specific to Arkansas populations in greatest detail.

Conservation Status

The conservation status in Arkansas is provided for each species. This is our opinion of its status in this state, regardless of its status outside Arkansas (see Chapter 5 for our definitions of status categories). We also note if a species is federally Threatened or Endangered.

---

* Typically, one or more new state angling records are established each year, but some angling records have persisted for decades. Readers should access the Arkansas Game and Fish Commission website to obtain the most recent list of Arkansas angling records.

CHAPTER 7

How to Identify a Fish

Keying Out Arkansas Fishes

Accurate identification of fishes is essential for those in the management of natural resources, academia (e.g., ichthyologists and ichthyology students), the angler seeking an identification of his or her catch, and for anyone interested in learning more about fishes. Although many individuals, particularly avid fishermen, can identify common Arkansas fishes such as Channel Catfish, Bluegill, crappies, or the various basses, the most reliable method of identifying a fish is to use the dichotomous keys specifically designed for Arkansas fishes. Mastery of the keys requires some study, and repetition of their use facilitates this mastery. Although identification of fishes becomes easier with practice and experience, identification skills erode over time if not used frequently.

Identifying fishes requires making counts of meristic features, measurements of body parts, and, occasionally, dissection of specimens. A few pieces of equipment are necessary: a dissecting microscope, a ruler or Vernier calipers, a scalpel, scissors, a single-edged razor blade, forceps, dissecting needles, and a pair of dividers are the most commonly used tools.

Dichotomous Keys

The keys used herein are dichotomous because they consist of paired or contrasting statements (couplets) arranged in a sequence and preceded by a number so that the user must make a choice. To use a key, start with the first couplet and decide which of the two statements in the couplet best fits the specimen you are keying out. For example, beginning with couplet 1 in the family key, both contrasting statements (1A and 1B) should be read while examining the specimen. At the end of the statement chosen, a number will direct you to the next couplet. By following the key to the next number and repeating the selection process, the specimen will be keyed out and an identification completed by a process of elimination. We provide keys to the 28 families of Arkansas fishes and keys to individual species within families. First, the reader must use the family key to establish the correct family identity of the specimen, and then proceed to species keys for that family. When keying an unknown specimen using the family keys, always begin with the first couplet. If your specimen lacks pelvic fins (1A), then go to couplet 2. If it has a sucking mouth, lacks pectoral fins, and has 7 gill openings (2A), it is a lamprey of the family Petromyzontidae, and one should proceed to the key to lampreys. If the specimen has jaws and pectoral fins (2B), proceed to couplet 3 and decide between the two alternatives (3A and 3B) and select the one that agrees best with the specimen. If the specimen has well-developed eyes and some skin pigment (3A), then it belongs to family Anguillidae; however, if the specimen lacks both eyes and pigment (3B), it belongs to the family Amblyopsidae (cavefishes) and can be identified by using the key for that family.

When one starts at the first couplet and 1B applies to your specimen, then proceed to couplet 4 and continue through the key until the correct family is located. The illustrations of representative members of each family accompanying the family key should aid in determining whether the specimen in hand generally resembles a typical member of the family. The same procedure through the consecutive couplets (always beginning with couplet 1) should be used to identify the species in each family. After assigning a name to the specimen by working through the key, compare the specimen with the photograph of that species, in addition to consulting the distinguishing characters given in the species accounts. If still in doubt as to the identity of the species, it should be taken to an ichthyologist at one of the state's universities or sent to a professional ichthyologist at a state or federal agency for identification. Ichthyologists are often asked to identify a fish from a photograph that was emailed to them. Sometimes a positive identification can be made from the photograph, but it is often not possible to be certain of the identification. It is always best to examine the actual specimen whenever possible.

Counts and Measurements

Certain counts and measurements are important in the identification of fish species. In addition to Robison and Buchanan (1988), we mainly used Etnier and Starnes (1993), Jenkins and Burkhead (1994), Ross (2001), Hubbs and Lagler (2004), and Miller and Robison (2004) to derive the following definitions of standard counts and measurements for fishes (consult the glossary for definitions of other terms):

abdominal length: In Atherinopsidae, the distance from the symphysis of the coracoids to the anterior insertion of the pelvic fin.

body depth: Greatest vertical distance in a straight line (at deepest part of body) from the midline of the dorsal surface to midline of the ventral surface. Fins and fleshy or scaly structures associated with fins are not included.

body width: Greatest distance in a straight line from one side of the body to the other.

branchiostegal ray number: Count of rays on one side, including short anteriormost ones that may be wholly or partially concealed.

breast scale rows: See scale rows on breast.

caudal peduncle depth: The least, straight-line vertical distance (dorsal to ventral) of the caudal peduncle.

caudal peduncle scale rows: Count of scale rows around the most slender portion of the caudal peduncle by zigzagging back and forth around an imagined line around this area.

caudal ray number: Count of all rays in caudal fin in catfishes, including rudimentary ones. In other fishes, only principal rays (branched and unbranched) are counted.

circumferential scales (or body circumferential scales): See **scale rows around body**.

diagonal (transverse) scale rows: Counted in a diagonal row beginning just above the first anal fin element and extending anterodorsally to the base of the dorsal fin. The count includes the lateral line scale.

dorsal and anal spine numbers: Unless otherwise specified, all spine elements are counted.

eye length (diameter): Greatest horizontal distance across eyeball between orbital rims.

fork length: The distance from the fork of the caudal fin to the anteriormost tip of the head.

gill raker number: Count of all rakers on the first gill arch on the right or left side, including the rudimentary ones. The gill arch is divided by an angle into a lower and upper limb, and usually a single count represents the total number on both limbs. Sometimes a count of the gill rakers on the lower limb of the first arch is made. A gill raker in the angle of the arch is counted as being on the lower limb of the arch.

head length (HL): Distance from tip of snout or upper lip to posteriormost tip of opercular membrane.

lateral line scale count (see also **scales in lateral row**): Count begins with the first lateral line scale behind the head that touches the pectoral girdle; the count extends backward along the lateral line, or along scale rows when pores are absent, to the base of the caudal fin (hypural plate). Scales posterior to crease formed by hypural plate are not counted.

lateral row tooth count (in lampreys): Lateral row tooth counts are made on one side of the specimen beginning with the tooth just lateral or anterolateral to the supraoral teeth; this count includes the circumoral cusp(s), each counted separately, in the row.

length of paired fins (pectoral and pelvic): Distance from middle of fin base to tip of longest ray.

myomere counts (in lampreys): This count is particularly important in the identification of lamprey species; myomeres are vertically oriented body muscles separated by myosepta (seen externally as grooves). The most anterior myomere counted is the one whose posterior myoseptum passes distinctly and entirely behind the groove surrounding the base of the fringed margin of the last gill opening, so as to leave a definite (though often narrow) band of muscle between the groove and the myoseptum.

The most posterior myomere counted is that whose lower posterior angle lies in part or wholly above the anterior end of the cloacal slit (it can be located by probing). Initial determination of the placement of the last myomere is preferred, and then count forward (Jenkins and Burkhead 1994).

number of anal rays and dorsal rays: Count of all rays when rudimentary rays grade gradually in length into developed rays, such as in the Salmonidae, Esocidae, and Ictaluridae. When rudimentary rays are few and distinctly set apart in length from the developed rays, such as in Catostomidae and Cyprinidae, only the developed (principal) rays are counted.

number of rays in paired fins (pectoral and pelvic): Count of all rays in these fins, including rudimentary ones.

pharyngeal tooth count (suckers and minnows): Count of the teeth on the fifth pharyngeal arch on each side of head. It is often necessary to examine the teeth on the pharyngeal arches of suckers and minnows. The arches can be removed from the fish by first cutting and lifting the operculum back away from the underlying gills. The pharyngeal arch lies behind the last gill-bearing arch and can be partially exposed by pushing the gill arches anteriorly. The arch is then removed by first severing the muscles holding its upper and lower ends and then gently pulling it out with forceps. Practice is necessary to properly remove the pharyngeal arches from fishes of different sizes, and each person should develop his or her own special technique to see the teeth clearly. It is often necessary to clean a pharyngeal arch after its removal. This can be accomplished by soaking the arch for a few minutes in a 1:1 solution of household laundry bleach and water. Each pharyngeal arch has from one to three rows of teeth. In species having one row of teeth on each arch, a tooth formula such as 0,4–4,0 is used to indicate that only one row of teeth is present and that the left arch has four teeth in its one row, and the right arch has four teeth in its row. For those species having two rows of teeth on each arch, a tooth formula such as 1,4–4,1 is used to indicate that the left arch has one tooth in its outer row and four teeth in its inner row, and the right arch has four teeth in its inner row and one tooth in its outer row. The pharyngeal arches of small fishes are often fragile and teeth may be missing or broken. In this situation, the tooth socket should be counted. A dissecting microscope is often a necessity for removing the arch and counting the teeth. After the teeth have been counted, the arch should be put back into the gill cavity and the operculum replaced.

pored lateral line scales (number): Count of all lateral line scales containing a pore or opening.

predorsal length: Distance of a straight line from the tip of the snout or upper lip to the dorsal fin origin.

predorsal scale count: A count of all scales that wholly or in part touch the middorsal line from the occiput to the origin of the dorsal fin.

scale rows above the lateral line: Normally, this refers to scales counted in a diagonal row on one side of the body starting with the scale nearest the base of the first dorsal fin element (but not including the scale on the dorsal midline) and extending posteroventrally to, but not including, the lateral line scale. For *Campostoma* species, this count (also referred to as body circumferential scales above the lateral line) is the number of scales on the upper part of the body from lateral line to lateral line (but not including the lateral line scales) just anterior to the dorsal fin.

scale rows below the lateral line: Counted in a diagonal row on one side of the body beginning with the scale nearest the base of the first anal fin element (but not including the scale on the ventral midline) and extending anterodorsally to, but not including, the lateral line scale.

scale rows around body (circumferential scales): The number of horizontal scale rows around the trunk at its greatest depth, usually in front of the dorsal fin. The count begins with the scale immediately in front of the dorsal fin origin, and all scale rows are counted around body until returning to the dorsal fin origin (but be careful not to recount the first scale at the dorsal origin).

scale rows on breast (such as in *Ambloplites*): Count is made by counting the scale at the lower base of the left pectoral fin insertion, counting downward and forward to the ventral midline and then upward and backward to the lower insertion of the right pectoral fin.

scale rows on cheeks: Count of all scale rows that cross an imaginary line extending from the eye to the angle of the preopercle bone.

scales in lateral row (number): Count of a series of scales along the side of the fish in the normal position of a lateral line. This count, also called lateral scale count, is used when a lateral line is incomplete or absent.

snout length: Distance from tip of snout or upper lip to front edge of orbit.

standard length (SL): Distance in a straight line from the anteriormost part of snout or upper lip to base of hypural plate.

thoracic length: The distance from the anterior insertion of the pelvic fin to the urogenital opening. Used to separate species of *Labidesthes*.

total length (TL): Greatest distance in a straight line

from the anteriormost tip of the snout or upper lip to the posteriormost tip of the caudal fin when its rays are squeezed together.

upper jaw length: Distance from anteriormost tip of jaw (premaxillary) to posteriormost point of jaw (maxilla).

ventral scute counts (in clupeids): Scutes are counted along the ventral midline from the throat to and including the scute whose point is at or slightly posterior to the anterior pelvic fin insertion (prepelvic count), and from the next posterior scute to the anus (postpelvic count).

width of gape: The greatest distance in a straight line from the posterior edge of one mandibular joint to the posterior edge of the other mandibular joint.

Key to Arkansas Fish Families

1A   Pelvic fins absent.   2
1B   Pelvic fins present.   4
2A   Pectoral fins absent; jaws absent, mouth an oval suction cup; 7 gill openings on each side of head; single median nostril.

Petromyzontidae *Lampreys* Page 114

Figure 7.1. Petromyzontidae

2B   Pectoral fins present; jaws present; 1 gill opening on each side of head; nostrils paired.   3
3A   Eyes present; body long and slender, snakelike; body heavily pigmented with brown and yellowish-brown.

Anguillidae *Freshwater Eels* Page 173

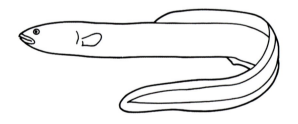

Figure 7.2. Anguillidae

3B   Eyes absent; body not snakelike; body unpigmented or with minute pigment specks.

Amblyopsidae *Cavefishes* Page 502

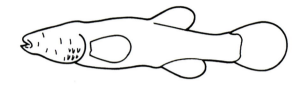

Figure 7.3. Amblyopsidae

4A   Mouth entirely behind front of eye; caudal fin deeply forked, with backbone turned upward and extending nearly to tip of upper lobe of fin.   5
4B   Mouth at least partially ahead of eye; caudal fin forked or not, but if forked, with backbone not turned upward and not extending into upper lobe of fin.   6
5A   Snout long and paddle-shaped, with 2 tiny barbels on lower surface; body without bony plates.

Polyodontidae *Paddlefishes* Page 140

Figure 7.4. Polyodontidae

5B   Snout shovel-shaped or conical, with 4 large barbels on lower surface; body with several rows of bony plates.

Acipenseridae *Sturgeons* Page 128

Figure 7.5. Acipenseridae

6A   Body with a continuous armorlike sheath of hard, platelike scales; snout a bony, strongly toothed beak.

Lepisosteidae *Gars* Page 147

Figure 7.6. Lepisosteidae

6B Body without scales or with flexible scales that overlap like shingles; snout not a bony, strongly toothed beak.
7

7A Adipose fin present. 8

7B Adipose fin absent. 10

8A Scales absent; 8 large barbels on front of head near mouth; pectoral fin with a strong, sharp spine at front.
Ictaluridae *North American Catfishes* Page 425

Figure 7.7. Ictaluridae

8B Scales present (sometimes small and easily overlooked); no barbels near mouth; pectoral fin without a spine. 9

9A Lower jaw strongly projecting beyond upper jaw; usually 70 or fewer scales in lateral line; without a thin membranous flap at pelvic fin origin; enlarged teeth on jaws and tongue. Osmeridae *Smelts* Page 469

Figure 7.8. Osmeridae

9B Jaws equal or lower jaw only slightly projecting beyond upper jaw; usually more than 70 scales in lateral line; a thin membranous flap present at pelvic fin origin; teeth not conspicuously enlarged on jaws and tongue.
Salmonidae *Trouts and Salmons* Page 473

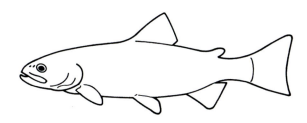

Figure 7.9. Salmonidae

10A Anus in front of pelvic fins in adults, jugular in position. Aphredoderidae *Pirate Perches* Page 499

Figure 7.10. Aphredoderidae

10B Anus behind pelvic fins, just in front of anal fin. 11

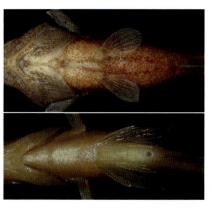

Figure 7.11. Top: Position of anus in front of pelvic fins in Aphredoderidae (Pirate Perch). Bottom: Typical position of anus behind pelvic fins as in Centrarchidae (Bluegill).

11A Dorsal fin single, spines absent or with 1 stiff saw-toothed spine. 12

11B Dorsal fin separated into 2 distinct parts, or single and with 4 or more stiff spines. 20

12A Length of dorsal fin base more than half of total fish length; dorsal fin rays 42 or more. 13

12B Length of dorsal fin base less than half of total fish length; dorsal rays 40 or fewer. 14

13A Gular plate present between lower jaws; anal fin short with fewer than 12 rays; top of head naked.
Amiidae *Bowfins* Page 161

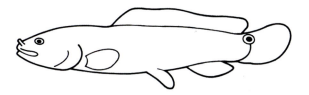

Figure 7.12. Amiidae

13B   Gular plate absent; anal fin long with 20 or more rays; top of head with enlarged scales.

Channidae *Snakeheads* Page 547

Figure 7.13. Channidae

14A   Scales present on cheek and opercle.   15
14B   Cheek and opercle naked.   17
15A   Dorsal fin base entirely or almost entirely behind anal fin base; scales in lateral series usually 30 or fewer; anal fin of male slender and rodlike.

Poeciliidae *Livebearers* Page 541

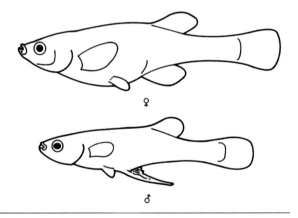

♀

♂

Figure 7.14. Poeciliidae (male and female)

15B   Dorsal fin base almost entirely over or in front of anal fin base; scales in lateral series usually more than 30; anal fin of male rounded, not slender and rodlike.   16
16A   Mouth superior; premaxillaries protractile, separated from snout by a deep transverse groove; origin of pelvic fins closer to tip of snout than to caudal fin base.

Fundulidae *Topminnows* Page 527

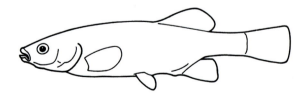

Figure 7.15. Fundulidae

16B   Mouth terminal; premaxillaries not protractile, lacking a transverse groove between snout and upper lip; origin of pelvic fins closer to caudal fin base than to tip of snout.

Esocidae *Pikes and Mudminnows* Page 487

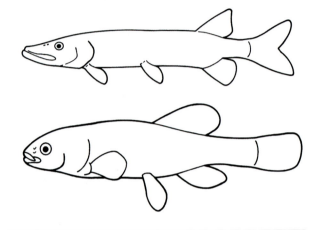

Figure 7.16. Esocidae—2 shapes

17A   Anal fin with 18 or more rays; keel present on midline of belly; small flaplike projection (axillary process) present at upper margin of pelvic fin base; gill opening extending forward on throat to beneath eye.   18
17B   Anal fin with 16 or fewer rays; usually no keel on midline of belly; axillary process absent; gill opening not extending forward on throat to beneath eye.   19
18A   Lateral line absent; keel on midline of belly with sharp, saw-toothed projections; dorsal fin far in front of anal fin.   Clupeidae *Herrings* Page 179

Figure 7.17. Clupeidae

18B Lateral line present; keel on midline of belly smooth, without sharp, saw-toothed projections; dorsal fin at least partly over anal fin.

Hiodontidae *Mooneyes* Page 167

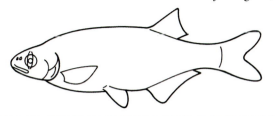

Figure 7.18. Hiodontidae

19A Dorsal fin with 8 or 9 rays, or if longer, with a stiff, saw-toothed spine at front; fewer than 9 pharyngeal teeth on each side.

Cyprinidae *Carps and Minnows* Page 192

Figure 7.19. Cyprinidae—2 shapes (carp and minnow)

19B Dorsal fin with 10 or more rays, the first dorsal ray never a stiff, saw-toothed spine, but flexible at its tip; 20 or more pharyngeal teeth on each side.

Catostomidae *Suckers* Page 377

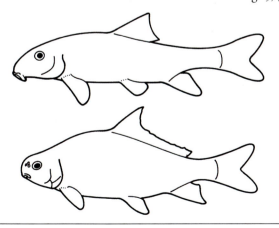

Figure 7.20. Catostomidae—2 shapes

20A Pelvic fins abdominal, their bases closer to front of anal fin than to rear margin of gill cover; spinous dorsal fin separate from soft dorsal fin and with 5 or fewer slender, flexible spines.

Atherinopsidae *New World Silversides* Page 515

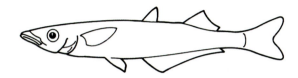

Figure 7.21. Atherinopsidae

20B Pelvic fins thoracic or subthoracic, their bases closer to the rear margin of gill cover than to front of anal fin; spinous dorsal fin separate or not from soft dorsal fin, but if separate, with stiff spines.　　　21

21A Body without scales; dorsal spines soft and flexible; anal spines absent.　　　Cottidae *Sculpins* Page 553

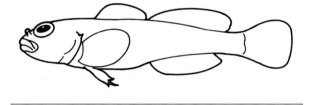

Figure 7.22. Cottidae

21B Body at least partially scaled; dorsal spines hard and stiff; 1 or more anal fin spines present.　　　22

22A Anal fin with 3 or more spines.　　　23

22B Anal fin with 1 or 2 spines (if 2, and pelvic axillary processes present, go to 23).　　　27

23A Spinous dorsal and soft dorsal fins widely separated; dorsal fin spines 4; adipose eyelids present; pelvic axillary process present.　　　Mugilidae *Mullets* Page 511

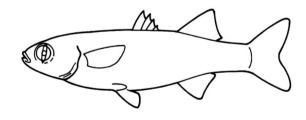

Figure 7.23. Mugilidae

23B Spinous dorsal and soft dorsal fins well connected or only slightly separated; dorsal fin spines 5 or more (if 4, then caudal fin rounded, not notched); adipose eyelids absent; pelvic axillary process absent.   24

24A Spinous dorsal and soft dorsal fins separate or only slightly connected; rear of operculum with a sharp spine; preopercle strongly saw-toothed.
Moronidae *Temperate Basses* Page 561

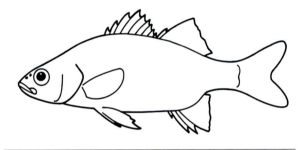

Figure 7.24. Moronidae

24B Spinous dorsal and soft dorsal fins well connected with, at most, a deep notch between them; rear margin of operculum without a sharp spine; preopercle usually smooth, sometimes weakly saw-toothed.   25

25A Spinous dorsal fin with 4 or 5 spines; lateral line absent; anal fin with 5 or 6 soft rays.
Elassomatidae *Pygmy Sunfishes* Page 624

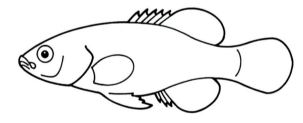

Figure 7.25. Elassomatidae

25B Spinous dorsal fin with 6–18 spines; lateral line present; anal fin with 8 or more soft rays.   26

26A A single nasal opening on each side of snout.
Cichlidae *Cichlids and Tilapias* Page 760

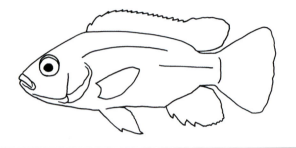

Figure 7.26. Cichlidae

26B Two nasal openings on each side of snout.
Centrarchidae *Sunfishes* Page 573

Figure 7.27. Centrarchidae (3 body shapes)

27A Lateral line extending out onto caudal fin to tips of middle rays; second anal spine stout and much longer than first; soft dorsal fin base much longer than spinous dorsal fin base and with more than 23 rays.

Sciaenidae *Drums and Croakers* Page 757

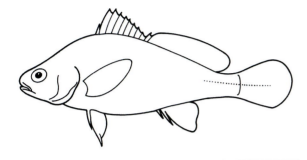

Figure 7.28. Sciaenidae

27B Lateral line present or absent, but never extending out onto the caudal fin; second anal spine, if present, slender and not much longer than the first; soft dorsal fin base not longer than spinous dorsal fin base and with fewer than 23 rays.

Percidae *Perches and Darters* Page 627

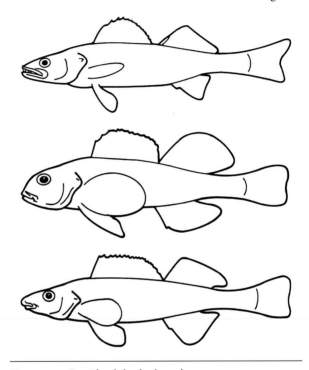

Figure 7.29. Percidae (3 body shapes)

CHAPTER 8

Order Petromyzontiformes
Lampreys

The lampreys (class Petromyzontida, order Petromyzontiformes) represent one of two surviving groups of the agnathan (jawless) craniates. The hagfishes (class Myxini, order Myxiniformes) form the second group. It is now generally accepted that the two living clades of agnathans (formerly grouped together as cyclostomes) are not sufficiently closely related to be placed in the same infraphylum and that the lampreys are sister to the gnathostomes (jawed vertebrates) (Boschung and Mayden 2004). Nelson et al. (2016) separated the two agnathan clades into different infraphyla, thereby separating the hagfishes from the lampreys (and other vertebrate animals) at a higher taxonomic level than most previous classifications. We agree with the separation into different infraphyla (see our Appendix 5).

Based on fossil discoveries in China in 1999, the earliest fossil agnathans (but not petromyzontiforms) date back approximately 550 million years to the Early Cambrian. Genetic evidence also supports an evolutionary origin of greater than 500 million ybp (Smith et al. 2013). Most agnathan groups were extinct by the close of the Devonian, almost 250 million years later (Carroll 1988). The fossil record for Petromyzontiformes is meager, with two representatives known from the Carboniferous (Bond 1996). Extant petromyzontiforms are found predominantly in the Northern Hemisphere and a few species occur in the Southern Hemisphere. Most species are freshwater forms.

The distribution of lampreys and the hypothesized ancient origin of the lineage are both consistent with vicariant events associated with the breakup of Pangaea and subsequent events in Laurasia and Gondwana (Boschung and Mayden 2004). It is generally recognized that the extant lampreys constitute three main groups, at either the subfamily (Bailey 1980) or family levels (Hubbs and Potter 1971). Nelson et al. (2016) recognized four families (one known only from fossils) and 10 genera with 46 extant species. Forty-two species occur in the Northern Hemisphere and are placed in the Petromyzontidae (Renaud 1997). The four species occurring in the Southern Hemisphere have been separated into either the monospecific Geotriidae or the monogeneric Mordaciidae (Potter 1980a). *Ichthyomyzon* and *Petromyzon* form the basal sister group in the Northern Hemisphere clade (Gill et al. 2003). Eighteen species of petromyzontiform fishes feed parasitically as adults in either marine or freshwater environments, and the other 28 species are nonparasitic and feed only during the larval stage (Potter and Gill 2003).

The order is characterized by 2 semicircular canals, 7 pairs of external lateral gill openings, eyes developed in adults, a single median nostril opening between the eyes with a pineal eye behind, body naked and eel-like, no bone, no paired fins, and tail diphycercal in adults. Teeth are present on the oral disc and tongue. Dorsal and ventral roots of spinal nerves are separated.

Family Petromyzontidae *Lampreys*

Lampreys belong to an ancient group of jawless fishes that are descendants of the first vertebrates and earliest true fishes, the ostracoderms, which date back to the Late Cambrian, approximately 500 million years ago. Family Petromyzontidae has eight genera and 42 species (Nelson et al. 2016), with more than half of lamprey species considered imperiled (Mateus et al. 2012). In North America, this family is represented by six genera and 23 species (Page et al. 2013). Three genera (*Ichthyomyzon*, *Lethenteron*, and *Lampetra*) and five species of lampreys occur in Arkansas. Petromyzontids first appeared in the Pennsylvanian (Bardack and Zangerl 1968, 1971; Janvier and Lund 1983). Some lamprey species are marine, living in the ocean for most of their adult life, but returning to spawn in freshwater streams. Other lampreys, including all five Arkansas species, are strictly freshwater. Petromyzontidae was monophyletic based on analysis of mitochondrial cytochrome *b* sequences (Lang et al. 2009).

Lampreys have fewer distinguishing morphological features for identifying species than the bony fishes. Taxonomy of lampreys has traditionally relied on myomere counts, dorsal fin shape, dentition patterns, and trophic characters, but molecular characters are now playing an important role in untangling the complicated systematic relationships. Part of the difficulty in identifying lamprey species stems from most lamprey genera containing "paired species," in which the larvae are morphologically similar or indistinguishable but, following metamorphosis, one species becomes parasitic while the other bypasses the adult feeding phase (nonparasitic) and rapidly becomes sexually mature. In Arkansas, *I. castaneus* and *I. gagei* are considered paired species.

Although occasionally mistakenly called "lamprey-eels," Arkansas lampreys are easily distinguished from true eels (and all other Arkansas fishes) by a combination of the following characters: absence of true jaws; a single median nostril anterior to the eyes; a cartilaginous skeleton; no scales; no paired fins; and seven pairs of gill pouches. Male lampreys are smaller than females and possess a long, cylindrical urogenital papilla, while females lack a papilla.

The life cycle of all lampreys includes both an extended larval stage (which is difficult to identify to species) and a relatively brief adult stage. Adults ascend flowing streams during the late winter or early spring to spawn in shallow pits excavated by spawning individuals near the upper ends of gravelly riffles. After spawning, adults do not feed and die shortly thereafter. Eggs hatch into larvae called ammocoetes. Ammocoetes resemble adults in body form but

A—Ammocoete body form.

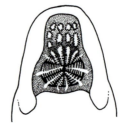

B—Ammocoete hood covering mouth.

Figure 8.1. Ammocoete larval form.

lack the typical sucking disc, rasping teeth, and eyes of the adult, and have a hood covering the mouth (Fig. 8.1).

Ammocoetes burrow into soft-bottomed, silty sand and detrital areas of streams, where they remain rather sedentary and strain organic materials and microscopic organisms from the water and soft sediments. After 1–6 years, depending on the species, the larvae undergo transformation into adults, usually during the fall. In nonparasitic species, the larvae are larger than the adults because length is lost during the transformation process. Potter (1980a) summarized the biology of ammocoetes.

Two parasitic (*Ichthyomyzon castaneus* and *I. unicuspis*) and three nonparasitic lampreys (*Ichthyomyzon gagei*, *Lethenteron appendix*, and *Lampetra aepyptera*) occur in Arkansas. Adult parasitic lampreys possess a well-developed buccal funnel (disc-like mouth) with numerous horny teeth and a rasping tongue (Fig. 8.2A). Because of these adaptations, parasitic lampreys are able to attach to their hosts and feed on blood and body fluids. Adults continue to grow as they feed and generally are longer than nonparasitic lampreys. Parasitic species have a higher fecundity (10,000–20,000+ ova) than nonparasitic lampreys (500–3,000) (Etnier and Starnes 1993). Nonparasitic lampreys possess few well-developed teeth and a less-developed oral disc (Fig. 8.2C). After transforming into adults, nonparasitic lampreys have a vestigial digestive tract and do not feed. After metamorphosis, adult lampreys spawn in the following spring. The impact of parasitic lampreys on their host species populations has been little studied, and the impact may vary by locality. Parasitic lampreys are not common enough in Arkansas to adversely affect populations of game or commercial fish species.

The identifying characters presented in the following key and species accounts pertain to adult lampreys, unless otherwise noted.

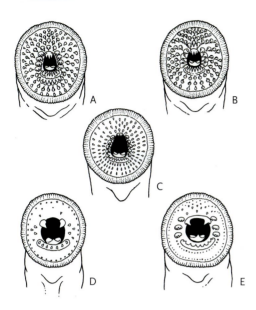

A—*Ichthyomyzon castaneus*
B—*Ichthyomyzon unicuspis*
C—*Ichthyomyzon gagei*
D—*Lampetra aepyptera*
E—*Lethenteron appendix*

Figure 8.2. Oral discs of lampreys.

Key to the Lampreys

1A Dorsal fin divided into two distinct parts.   2

1B Dorsal fin possibly slightly notched, but never divided into two parts.   3

2A Myomeres 54–60; 5 inches (127 mm) maximum length.
    Lampetra aepyptera Page 122

2B Myomeres 66–75; 6 inches (152 mm) or more maximum length.   *Lethenteron appendix* Page 124

3A Expanded oral disc narrower than head; teeth in posterior field of oral disc somewhat degenerate; no functional gut in adults; 6 inches (152 mm) maximum size.
    Ichthyomyzon gagei Page 118

3B Expanded oral disc wider than head; teeth in posterior field of oral disc not degenerate; functional gut present in adults; larger than 6 inches.   4

4A All disc teeth of innermost circle with 1 point; myomeres usually 49–52.
    Ichthyomyzon unicuspis Page 120

4B Some teeth of innermost circle with 2 points; myomeres usually 52–55.
    Ichthyomyzon castaneus Page 115

Genus *Ichthyomyzon* Girard, 1858— River Lampreys

The genus *Ichthyomyzon*, with six species, is characterized by a continuous single dorsal fin in adults and ammocoetes, unicuspid or bicuspid lateral circumoral teeth, disc teeth occur in radiating rows, and the supraoral lamina not broadly expanded. Myomeres range from 49 to 62. The genus is limited to eastern North America. Arkansas has three species of *Ichthyomyzon* lampreys: *I. castaneus*, *I. gagei*, and *I. unicuspis*. Both of our parasitic lamprey species gave rise to nonparasitic species, with *I. castaneus* giving rise to *I. gagei* and *I. unicuspis* giving rise to the more northern *I. fossor* (Hubbs and Trautman 1937). The monophyly of *Ichthyomyzon* was supported in an analysis of the cytochrome *b* gene (Lang et al. 2009). Neave et al. (2007) concluded that the lack of distinguishing charcters makes existing taxonomic keys misleading for identifying *Ichthyomyzon* lamprey larvae. Pigmentation patterns, morphometric characters, body shape, and myomere counts varied significantly among species, but were inadequate for use as diagnostic characters because of high intraspecific variation and overlapping ranges.

Ichthyomyzon castaneus Girard
Chestnut Lamprey

Figure 8.3. *Ichthyomyzon castaneus*, Chestnut Lamprey. *Uland Thomas.*

CHARACTERS   A brownish fish with no jaws, paired fins, or scales. A single median nostril is present, eyes are small, and there are 7 external gill openings (Fig. 8.3). Dorsal fin low, continuous, and slightly notched, but not separated into two distinct parts. The cup-shaped oral disc of adults of this parasitic lamprey is wider than the head when expanded and contains well-developed, horny teeth in radiating rows. Circumoral teeth 4 on either side of mouth opening, usually bicuspid; teeth elsewhere in oral disc usually have single cusps (Fig. 8.2A; Fig. 8.4). Transverse lingual lamina linear to weakly bilobed. Length of oral disc going into TL about 12.3 (10.3–16.2) times. Myomeres usually 52–56 (49–58). Mature males with urogenital papilla often extending beyond ventral margin of the body; females with prominent postanal fold and body greatly distended from behind gill region to the cloaca

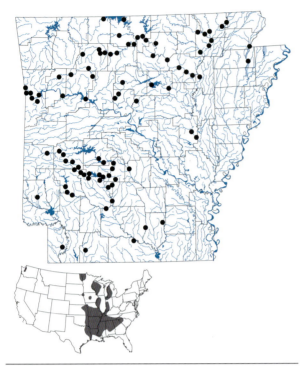

Map 1. Distribution of *Ichthyomyzon castaneus*, Chestnut Lamprey, 1988–2019.

Figure 8.4. *Ichthyomyzon castaneus* mouth. *Renn Tumlison.*

Figure 8.5. *Ichthyomyzon castaneus* mouth and *I. gagei* mouth with teeth. *Renn Tumlison.*

(A. J. Smith et al. 1968). Adults with well-developed intestine. Mature adults may reach 10–13.6 inches (254–345 mm) in length. Maximum total length is 15 inches (380 mm) (Carlander 1969). The largest known Arkansas specimen is about 12 inches (305 mm). Larvae reach a length of 6 inches (152 mm), with metamorphosis probably occurring between 100 and 130 mm (Hardisty and Potter 1971a).

LIFE COLORS   Adults brownish to olivaceous or gray dorsally, grading to lighter olive yellow on belly. Sides may be mottled or grayish. After spawning, adults become blue black. Ammocoetes generally a paler brown or gray with a cream-colored belly. Ammocoetes usually with unpigmented lateral line pores, but this character is variable (Neave et al. 2007). Lateral line pores are usually dark in large adults.

SIMILAR SPECIES   The Chestnut Lamprey is most similar to *I. unicuspis* but differs in having 52–55 myomeres (vs. 49–52 myomeres) and some disc teeth of innermost circle with two points (vs. all disc teeth with one point). Chestnut Lampreys are also similar to *I. gagei* but in adults, the oral disc in *I. castaneus* is larger, going into TL 16 or fewer times (vs. 17 or more times), and *I. castaneus* has well-developed teeth (vs. degenerate teeth in *I. gagei*) (Fig. 8.5). From *Lampetra* and *Lethenteron* species, *I. castaneus* differs in having a continuous dorsal fin (vs. dorsal fin

separated into two parts), and enlarged teeth (vs. reduced dentition).

VARIATION AND TAXONOMY   *Ichthyomyzon castaneus* was described by Girard (1858a) from Galena, Illinois, not Galena, Minnesota, as listed in some of the earlier literature. It is probably derived from *I. unicuspis* (Hubbs and Potter 1971) and is hypothesized to be the stem species to *I. gagei* (Boschung and Mayden 2004). A genetic analysis by Lang et al. (2009) also recovered *I. castaneus* as sister to *I. gagei*.

DISTRIBUTION   The main body of its range is the lower Mississippi River basin from southern Indiana across central Missouri into eastern Kansas south to Louisiana, and in Gulf Slope drainages from the Mobile Bay basin, Georgia and Alabama, west to the Sabine River, Texas. North of that range it occurs more sporadically in the St. Lawrence–Great Lakes, Hudson Bay, and upper Mississippi River basins from Manitoba, Canada, southward through western Illinois. In Arkansas, Chestnut Lampreys are found in the White, St. Francis, Arkansas, Red, and Ouachita river systems (Map 1). Occasionally this parasitic species is taken in reservoirs, and Buchanan (2005) reported it from Bull Shoals, Greers Ferry, and DeGray reservoirs, and from Pool 13 of the Arkansas River Navigation System. Although older records exist (MapA4.3), there are surprisingly no recent Arkansas records from the Mississippi River, the Illinois River drainage, or the upper White River above Beaver Reservoir. Robison et al. (2006) updated Arkansas lamprey records, McAllister et al. (2010b) reported the first record from the Strawberry River, and Tumlison and Robison (2010) collected specimens from the Saline River and Moro Creek (Ouachita River system). Connior et al. (2011) reported the southwesternmost record in Arkansas from the Sulphur River in Miller County, Connior et al. (2012) collected the first record from Bayou Dorcheat in Columbia County, and Robison et al. (2013) collected two specimens from the mainstem White River at Batesville, Independence County.

HABITAT AND BIOLOGY   Juvenile and adult Chestnut Lampreys inhabit medium to large rivers and reservoirs where generally an abundance of large fishes are present to feed upon. In Wisconsin, adults were more common in medium-sized streams, while the related *I. unicuspis* occupied larger rivers (Becker 1983), but *I. unicuspis* is not common enough in Arkansas to detect habitat segregation, if any, from *I. castaneus*. Ammocoetes, in contrast, live in smaller streams in swifter water in fine substrates, as well as in slower water with vegetation (Scott and Crossman 1973). In the Cossatot River, Howard County, ammocoetes have been collected from side backwater regions of reduced current over silt, detritus, and decaying leaves and sunken woody debris. Although widely distributed in Arkansas, populations of Chestnut Lampreys are not large. Salinger (2016) sampled 12 Arkansas streams in a search for *I. castaneus* and found only 8 individuals in 3 of those streams.

Adults are parasitic carnivores and feed on fishes by attaching the disc-like mouth to the body of a host species such as buffalo, carp, or even gamefishes such as White Bass (Fig. 8.6).

Other host species include Chain Pickerel, River Redhorse, Blacktail Redhorse, Golden Redhorse (Robison et al. 2013), Channel Catfish, Spotted Sucker, Green Sunfish, Ozark Bass, and Largemouth Bass (Hall and Moore 1954; Mayden et al. 1989; Pflieger 1997; Connior et al. 2012). Cook (1959) includes catfish, buffalo, and Paddlefish as

Figure 8.6. *Ichthyomyzon castaneus* head fastened on a rock. *Renn Tumlison.*

hosts. Salinger (2016) recorded an instance of parasitism on Spotted Bass. Cochran and Jenkins (1994) reported smaller fishes (130–150 mm, 5–6 in. TL) are also parasitized, and we have observed that adult Chestnut Lampreys maintained in aquaria readily attack small fishes. Cochran and Jenkins (1994) reasoned that such small fishes are more common upstream when lampreys metamorphose into the adult stage, and their scale and skin layers allow easier penetration by the smaller lampreys. In Wisconsin, the adults feed from April to October, with the greatest feeding activity in June and July (Becker 1983; Cochran 2014b). Some adults remain attached to hosts during winter months (Cochran et al. 2003). When attached, the horny tongue rasps a hole in the host while an anticoagulant is injected into the wound to keep the blood and body fluids flowing. Ammocoetes have a sievelike apparatus in the mouth and are filter feeders, presumably on plankton (mainly diatoms and desmids) and possibly some detritus. Feeding activities of the Chestnut Lamprey in natural streams of Arkansas apparently have little impact on host species populations. Of the 2,166 fish specimens collected from 12 Arkansas streams, only 11 individuals (0.5%) from 7 streams showed evidence of attack by *I. castaneus*, and 7 of those were the nonnative *Cyprinus carpio* (Salinger 2016). Salinger also found low incidence of parasitism by lampreys on trout in Norfork National Fish Hatchery raceways.

During the late winter and spring, adults migrate to medium-sized streams and sometimes small headwater streams to spawn. Spawning behavior was described by Case (1970). Adults create a depression in a coarse gravel stream bottom, then systematically lengthen this nest by continuing to excavate at the upper end of the nest while filling the downstream portion with stones. Males use the oral disc to attach to the head of the female while wrapping the caudal fin around the anterior part of the female's body. Males stroke the female's body, and this is followed by quivering of the pair. Case (1970) observed up to five males attached to each other in this manner. Becker (1983) reported up to 50 individuals spawning in a single nest 1 m long by 0.6 m wide by 5 cm deep. In Arkansas, spawning has been observed in late May in the Antoine and Caddo rivers over nests in coarse gravel substrate (Robison et al. 1983). Three individuals (1 male and 2 females) were observed spawning in the Antoine River in a swift riffle, 12–16 inches deep, over a gravel bottom on 19 May 1982. Water temperature was 19°C. In April 1996, 12 individuals were observed spawning in the Caddo River in an excavated gravel nest approximately 0.1 meter from shore in 45.7 cm of water (Robison et al. 2006). Water temperature was 16.7°C. Chestnut Lampreys were also observed by HWR spawning on 24 April 2003 in the Buffalo River near Hasty, Arkansas. Simon (1999b) classified lampreys as lithophils—rock and gravel spawners that do not guard their eggs, and referred to them as nonguarders and brood hiders. Adults build nests in gravel areas of streams and die shortly after spawning (Carlander 1969). Estimated fecundity ranges from 10,000 to 20,000 ova per female (Beamish and Thomas 1983); however, an 11.2-inch (284 mm) female collected in Oklahoma contained about 42,000 eggs (Hall and Moore 1954). Fecundity increases with adult size, ranging from 10,144 to 18,563 oocytes; mature ovarian oocytes average 1.34 mm (0.05 in.) in diameter (Beamish and Thomas 1983). Eggs hatch in about 2 weeks. After hatching, the larvae drift downstream, where they burrow into stream banks and stable bars having an abundance of silt, sand, and organic debris. For the next 1–5 years or longer they are filter feeders before transforming into the adult form at about 100–150 mm TL in the fall of the year (Hardisty and Potter 1971a). Ovaries begin to show development in ammocoetes larger than 88 mm (3.5 in.). The juvenile stage of this species lasts from less than a year to 1.5 years (Hall 1963). Pflieger (1997) reported that adults live about 18 months but actively feed for only about 5 months in the middle part of the adult life span. Boschung and Mayden (2004) reported a life expectancy of 5–7 years of which one-third is spent as an adult.

CONSERVATION STATUS The Chestnut Lamprey is uncommon but widely distributed in Arkansas. We do not believe it is currently in need of special protection, but because of its small population sizes, it should be monitored periodically to detect possible declines in its range.

Ichthyomyzon gagei Hubbs and Trautman
Southern Brook Lamprey

Figure 8.7. *Ichthyomyzon gagei*, Southern Brook Lamprey. *Uland Thomas.*

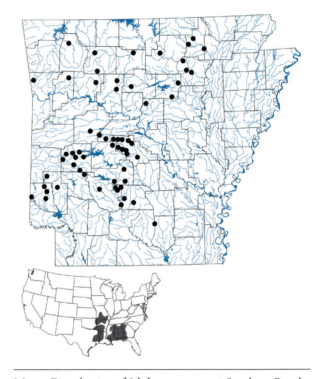

Map 2. Distribution of *Ichthyomyzon gagei*, Southern Brook Lamprey, 1988–2019.

CHARACTERS A nonparasitic lamprey without jaws, paired fins, or scales, but possessing 7 gill openings, a single median nostril, and poorly developed eyes (Fig. 8.7). Dorsal fin shallowly notched, but not separated into two distinct parts. Teeth degenerate. Transverse lingual lamina linear or weakly bilobed (Fig. 8.2C). Oral disc of the adult is small, and when expanded, the oral disc is narrower than the head; length of oral disc going 17.2–26.3 times into TL (Fig. 8.8). Intestine degenerate in adult. Myomeres 49–59, usually 52–54. Females usually larger than males. Larvae reaching a length of 7 inches (178 mm). Adults usually reach 4.7–6.5 inches (80–166 mm) TL. Maximum total adult length is 6.5 inches (166 mm) (Knapp 1951). Largest adult Arkansas specimen examined was 5.7 inches (145 mm) from Illinois Bayou.

Figure 8.8. Oral disc of *Ichthyomyzon gagei*. *Uland Thomas.*

LIFE COLORS Body color generally brownish to olivaceous dorsally, grading to white or yellow on the ventral surface and fins.

SIMILAR SPECIES *I. gagei* resembles *I. castaneus;* however, it is smaller, 5–7 inches or less TL as adults (vs. adult size usually 10–12 inches), has a degenerate intestine in the adult stage (vs. a functional intestine), has the posterior disc teeth poorly developed (vs. well developed) (Fig. 8.2), and the oral disc is smaller, its length going into TL usually 17 or more times (vs. usually 16 or fewer times). The same features distinguish *I. gagei* from *I. unicuspis.* The ammocoetes are quite difficult to separate; however, *I. gagei* tends to have less pigmentation below the eye and in front of the gill region, and usually has well-pigmented lateral line organs. Also, *I. gagei* has rapid sexual development at the end of the larval period; late-stage ammocoetes greater than 100 mm often have well-developed gonads. Ross (2001) considered the presence of developed gonads in the late ammocoete stage of *I. gagei* to be the best method of distinguishing larval forms of *I. gagei* and *I. castaneus.* *Ichthyomyzon gagei* differs from *Lampetra aepyptera* and *Lethenteron appendix* in having the dorsal fin with a shallow notch and not divided into two distinct parts (vs. dorsal fin deeply notched, with two distinct parts).

VARIATION AND TAXONOMY *Ichthyomyzon gagei* was described by Hubbs and Trautman (1937) from Dry Prong, a tributary of the Little River, Grant Parish, Louisiana. It is believed to be derived from the parasitic species *I. castaneus.* Although Robison (1974c) found little morphological variation in Arkansas specimens of *I. gagei*, Mayden et al. (1992b) suggested that *I. gagei* is probably a complex of different taxa. Genetic analysis recovered *I. gagei* as sister to *I. castaneus*, supporting previous morphological interpretations (Lang et al. 2009).

DISTRIBUTION *Ichthyomyzon gagei* occurs mainly in the lower Mississippi River basin from western Kentucky, southern Missouri, and eastern Oklahoma south through Mississippi and Louisiana, and in Gulf Slope drainages from the Florida panhandle west to Galveston Bay, Texas. Disjunct populations are found in the upper Mississippi River basin of Wisconsin and Minnesota. In Arkansas, the Southern Brook Lamprey occurs widely throughout upland streams of the White River system, upper Arkansas River tributarries such as Lee Creek, Mulberry River, and Petit Jean River, the Ouachita River, Saline River, and Little River (Map 2). Older records were collected by HWR from clear lowland streams in Columbia County (Map A4.4). We have no post-1987 records for Benton and Washington counties.

HABITAT AND BIOLOGY The adult Southern Brook Lamprey inhabits clear, permanent-flowing, small to medium-sized streams, where it is usually found in current over a rock or gravel substrate. Ammocoetes tend to be found where there is less current in decaying leaf packs and associated debris along marginal areas and pools of small streams, where they burrow into the mud and organic detritus.

Ammocoetes may spend several years as filter feeders, feeding on plankton, particularly diatoms, and organic detritus (Moshin and Gallaway 1977) in quiet areas at stream margins. In that study, feeding occurred mainly on diatoms and detritus in spring and summer, but in fall and winter, zooplankton became a significant food item. Adults do not feed following the larval interval but rely on stored energy to sustain their metabolic requirements, including gonadal maturation, through to the completion of spawning, after which they die (Moshin and Gallaway 1977). The digestive tract becomes reduced to a thin threadlike projection in the short-lived adult stage.

Adults move to clear tributary streams in the spring (late March to May), where they construct nests 6–8 inches (15–20 cm) wide and about 2 inches (5 cm) deep in sand and gravel and spawn when water temperatures reach 59–75.2°F (15–24°C). A single ripe female may release 1,000–3,200 eggs depending on her size. Females spawn with several males, sometimes in more than one nest (Boschung and

Mayden 2004). Eggs are 0.60–1.00 mm in diameter, and newly hatched larvae are 5 mm in length. Ripe females and males were collected from Ten Mile Creek (eastern Saline River system), Saline County, on 8 April 1974. From 5 to 20 adults may be active in a single nest with nest-building and spawning interspersed (Miller and Robison 2004). Adults die within 2–26 days after spawning (Dendy and Scott 1953). In Alabama, the larval stage lasts 3 years (Dendy and Scott 1953). Beamish and Thomas (1984) found that metamorphosis in Alabama begins in late August to early September and requires approximately six months to complete.

CONSERVATION STATUS The Southern Brook Lamprey is uncommon but widely distributed in Arkansas. Even though it has apparently declined in Benton and Washington counties, we do not believe it is currently in need of special protection.

Ichthyomyzon unicuspis Hubbs and Trautman
Silver Lamprey

Figure 8.9. *Ichthyomyzon unicuspis*, Silver Lamprey.
Rob Criswell.

CHARACTERS A parasitic lamprey with a shallow-notched dorsal fin, but not divided into two distinct fins; dorsal fin connected to a short, round caudal fin (Fig. 8.9). When expanded, the mouth disc is wider than the head. Disc teeth are well developed; the four teeth in the innermost circle immediately lateral to the mouth each with a single point, similar to the rest of the teeth in the disc; the dorsalmost tooth in the inner circle often has two points. (Fig. 8.2B; Fig. 8.10). Supraoral lamina with 2 (rarely 3) narrowly separated teeth. Transverse lingual lamina bilobed. Trunk myomeres usually 49–52 (46–55). Becker (1983) described sexual dimorphism in adults as follows: Male with middorsal ridge always present but not prominent; urogenital papilla occasionally extending beyond ventral margin of body. Female with prominent postanal fold, and body greatly distended from behind gill region to the cloaca. Adults reach 14–15 inches (356–381 mm) TL, and in Illinois there is a record of a 15.6-inch (395 mm) specimen (Starrett et al. 1960). Ammocoetes reach 7 inches (178 mm).

LIFE COLORS Adults are yellowish tan on the dorsum and sides, and yellow or lighter on the belly and fins. Sides sometimes mottled, occasionally slate-colored.

Map 3. Distribution of *Ichthyomyzon unicuspis*, Silver Lamprey, 1988–2019.

Figure 8.10. Oral disc of *Ichthyomyzon unicuspis* with teeth. *Konrad Schmidt.*

Spawning adults darken to blue to blue gray on the sides and almost black or bluish on the lower surface by completion of spawning. Ammocoetes are yellowish.

SIMILAR SPECIES The Silver Lamprey is most similar to *I. castaneus* but differs in usually having 49–52 myomeres (vs. 52–55 myomeres) and all lateral disc teeth of innermost circle with 1 point (vs. lateral teeth of innermost circle with 2 points). It differs from *Lampetra* and *Lethenteron* species in having the dorsal fin notched but not divided into two parts (vs. dorsal fin divided into two separate parts).

VARIATION AND TAXONOMY   *Ichthyomyzon unicuspis* was described by Hubbs and Trautman (1937) from Swan Creek, 4.8 km above its confluence with the lower Maumee River at Toledo, Lucas County, Ohio. It is the most ancestral parasitic member of its genus. Hubbs and Trautman (1937) and Hubbs and Potter (1971) considered it similar to the form that gave rise to *I. castaneus*. In a genetic analysis, Lang et al. (2009) recovered *I. unicuspis* as sister to the Northern Brook Lamprey, *I. fossor*. Based on genetic evidence, introgressive hybridization occurred between *I. unicuspis* and *I. fossor* in the Lake Huron basin (Docker et al. 2012).

DISTRIBUTION   *Ichthyomyzon unicuspis* occurs in the St. Lawrence–Great Lakes and upper Mississippi River basins of southern Canada south to Tennessee and northern Arkansas. Disjunct populations are found in the Hudson Bay drainage of Canada, the Missouri River of Nebraska and Iowa, and the lower Mississippi River of Mississippi. The Silver Lamprey is known in Arkansas from only two localities on the White River (MMNS 43754, MMNS 57161, and NLU 75186) and one locality on the Buffalo River (NLU 23654) (Robison et al. 2011b) (Map 3). The White River specimens, originally brought to the attention of HWR, were attached to Paddlefish. Although all Missouri records are reported only from the Mississippi River (Pflieger 1997), and there is one record from that river near Vicksburg, Mississippi, there are no records from the Mississippi River in Arkansas.

HABITAT AND BIOLOGY   Silver Lamprey adults inhabit large rivers and lakes throughout most of its range, but *I. unicuspis* has been found in Arkansas only in medium-sized rivers. Spawning and larval development occur in streams of medium size (Pflieger 1997). Becker (1983) noted that in Wisconsin, its distribution is complementary to that of *I. castaneus* and suggested that competition might explain this mutual avoidance of the two species. Ammocoetes are found in sand and mud areas.

During the larval period *I. unicuspis* feeds on algae, diatoms, and protozoans (Harlan and Speaker 1956). Adult Silver Lampreys are known to parasitize sturgeons, Paddlefish (Fig. 8.11), Northern Pike, Common Carp, suckers, White Bass, and catfishes (Hardisty and Potter 1971b; Becker 1983). Cochran and Lyons (2004) noted that Silver Lampreys have been most strongly associated with host species that are relatively large and have naked skin or small scales (e.g., Paddlefish, sturgeon, Lake Trout, catfishes, and esocids). Vladykov (1985) found 61 *I. unicuspis* on a single sturgeon (*Acipenser fulvescens*) in the St. Lawrence River, Canada. Some authors have questioned the ability of

Figure 8.11. *Ichthyomyzon unicuspis* feeding on an adult Paddlefish. *Konrad Schmidt.*

I. unicuspis to successfully form puncture wounds in hosts with large, heavy scales such as the invasive Common Carp. Cochran and Lyons (2004) recorded attachments to Common Carp and even Longnose Gar in the field, and the lampreys had created wounds that appeared to penetrate between adjacent scales. An attack on the host is generally made from the side, and occasionally from the rear. Most attachments by captive Silver Lampreys occurred dorsally on their hosts, and Cochran and Lyons (2004) cited studies showing that dorsal attachments are more common on benthic hosts and ventral attachments are more common on pelagic hosts. Adults are especially active at night, and 81.5% of attacks occurred at night in laboratory observations. In Wisconsin, most active parasitic feeding occurred between June and September (Becker 1983), and Cochran et al. (2003) found that some individuals remained attached to hosts in winter months. Silver Lampreys possess electroreceptors that may help in locating and attaching to the host (Bodznick and Preston 1983).

Spawning occurs in northern areas of its range in April, May, and June in clear, medium to large rivers over gravel, and occasionally over sand (Becker 1983; Cochran and Lyons 2004). Adults migrate upstream into smaller headwater streams prior to spawning when water temperature reaches 10°C (50°F) or more. Nests are constructed on sandy and gravelly riffles in medium-sized streams with moderate gradients (Trautman 1957). Spawning has been observed at depths between 23 and 79 cm, and Cochran and Lyons (2004) found that mean spawning depth was greater when another lamprey species occurred in a spawning aggregation with the Silver Lamprey. Females are generally larger than males. A sexually mature female (326 mm, 65.5 g) collected 8 April had an ovarian weight of 9.71 g and contained an estimated 27,400 eggs 0.8–1.0 mm in diameter (Becker 1983). Spawning aggregations were observed at a water temperature of 18.2°C in areas of gravel, cobble, and scattered boulders in the Oconto River, Wisconsin (Cochran and Lyons 2004). Spawning generally occurred on the upstream side of ridges or bars of gravel or cobble, just upstream of where the water surface was broken during the transition from run to riffle-like habitat. Most aggregations in the Oconto River included both Sea Lampreys and Silver Lampreys, and the number of Silver Lampreys in aggregations ranged from 1 to more than 15. Cochran and Lyons (2004) observed Silver Lampreys spawning in the same nests as Sea Lampreys. Adults die after spawning. The eggs hatch into burrowing larvae within 5 days at a developmental temperature of 18.4°C (65°F) (A. J. Smith et al. 1968). Scott and Crossman (1973) reported that the larval period lasts 4–7 years. Metamorphosis begins in August, and adults live 12 or 13 months (Becker 1983).

CONSERVATION STATUS   The Silver Lamprey is rare in Arkansas and is at least a species of special concern.

Genus *Lampetra* Bonnaterre, 1788— Brook Lampreys

The brook lampreys are characterized by a broad supraoral lamina with a cusp at each end, a feature not present in other lamprey groups (Etnier and Starnes 1993). In addition, a wide notch separates the dorsal fin into two distinct lobes. Posteriorly, the second dorsal lobe is separated from the caudal fin by a wide notch. The expanded oral disc is no wider than the head in ammocoetes and adults. Teeth are not in radiating rows and are typically fewer and less developed than in *Ichthyomyzon*. This genus contains 11 species, one of which, *L. aepyptera*, occurs in Arkansas.

Lampetra aepyptera (Abbott) Least Brook Lamprey

Figure 8.12. *Lampetra aepyptera*, Least Brook Lamprey. *Uland Thomas.*

CHARACTERS   A nonparasitic lamprey with a deeply notched dorsal fin divided into two separate parts (Fig. 8.12). Mouth disc distinctly narrower than the head when expanded in both adult and ammocoete. A pair of prominent widely separated supraoral teeth present, but in general, disc teeth are poorly developed and not in radiating rows (Fig. 8.2D; Fig. 8.13) in adults. Eyes prominent in adults. Myomeres usually 50–61 between the last gill

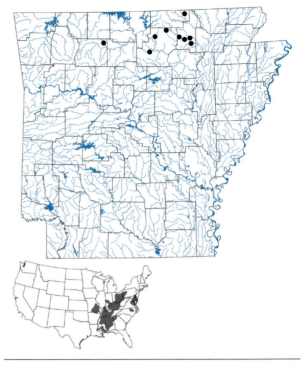

Map 4. Distribution of *Lampetra aepyptera*, Least Brook Lamprey, 1988–2019.

Figure 8.13. Oral disc of *Lampetra aepyptera*. *Uland Thomas.*

opening and anus. Smallest Arkansas lamprey with adult size about 3.5–5 inches (89–127 mm). Largest Arkansas specimen 4.9 inches (125 mm). Maximum size 7 inches (180 mm). Ammocoetes reach a length of 5.9 inches (150 mm).

LIFE COLORS   Body coloration varying from olive greenish gray to black above, lighter gray below, with a yellow or creamy venter and fins. Usually with a dark blotch anterior to gill openings (Branson 1970). Buccal cavity pigmented.

SIMILAR SPECIES   *Lampetra aepyptera* is similar to *Lethenteron appendix;* however, it has 50–61 myomeres (vs. 63–70 in *L. appendix*), poorly developed (degenerate) or lacking teeth on the posterior portion of the disc (vs. well-developed disc teeth in *L. appendix*), and a pigmented buccal cavity (vs. unpigmented in *L. appendix*). It differs from *Ichthyomyzon* species in having a dorsal fin divided into two distinct lobes (vs. dorsal fin with a shallow notch, but not divided into two separate parts).

VARIATION AND TAXONOMY   *Lampetra aepyptera* was described by Abbott (1860) as *Ammocoetes aepyptera* from the Ohio River near Portland, Meigs County, Ohio. Hubbs and Potter (1971) recognized the genus *Okkelbergia* Creaser and Hubbs for this species, but it was later returned to *Lampetra* (Vladykov and Kott 1976; Bailey 1980; and Potter 1980b). Docker et al. (1999), in a molecular phylogenetic analysis of this genus, resolved *Lampetra* as monophyletic; however, Boschung and Mayden (2004) stated that *Lampetra* was paraphyletic, with *L. aepyptera* being sister to *L. fluviatilis*, and this clade is sister to a clade containing *Lethenteron camtschaticum* (= *L. japonicum*) plus *L. appendix*. Lang et al. (2009) used genetic evidence to support returning *Lampetra aepyptera* to the monotypic genus *Okkelbergia*. The Least Brook Lamprey is probably a complex of divergent taxa (Mayden et al. 1992b). Martin and White (2008) studied genetic variation and recognized 11 clades corresponding to different drainages and/or locations within a drainage. Walsh and Burr (1981) reported limited morphological variation in a Kentucky study of *L. aepyptera*, but described a neotenic population. Kott et al. (1988) reported that the length of the urogenital papilla, expressed as a proportion of branchial length, is taxonomically useful in distinguishing among male lampreys of nonparasitic species. Cochran and Sneen (1995) investigated the effect of preservation on urogenital papilla length in *L. aepyptera*.

DISTRIBUTION   The Least Brook Lamprey occurs in Atlantic Slope drainages from Pennsylvania to North Carolina, the Mississippi River basin from Pennsylvania to southern Missouri and northern Arkansas south to southern Mississippi, and in Gulf Slope drainages from the Mobile Bay basin of Georgia to the Pearl River, Mississippi. To date, fewer than 40 individuals of *L. aepyptera* have been collected in Arkansas (Harp and Matthews 1975; Sewell et al. 1980b; McAllister et al. 1981; Robison et al. 2006, 2013). All specimens have been taken in northern Arkansas in the White and Black river systems (Buffalo River, Piney and Sylamore creeks, Strawberry River, and Spring River) (Map 4).

HABITAT AND BIOLOGY   The Least Brook Lamprey typically inhabits riffles and raceways of smaller streams than other Arkansas lampreys; therefore, headwater streams and spring runs with clean gravel riffles are primary habitats. Ammocoetes live initially in deposits of silt, mud, and fine sand, later drifting downstream to inhabit coarser substrates in creeks. The ammocoetes of *L. aepyptera* appear to be habitat specialists. D. M. Smith et al. (2011) studied selection and preference of benthic habitat by small and large ammocoetes and found that fine sand was selected with a significantly higher probability than any other substrate. Therefore, availability of fine sand habitat may limit distributions and population sizes.

Like all nonparasitic lampreys, feeding apparently occurs only in the ammocoete stage. Jenkins and Burkhead (1994) speculated that they feed on planktonic organisms and possibly particulate detritus.

Trautman (1957) suggested that spawning occurs in streams smaller than 15 feet (4.6 m) wide when water temperatures reach 50°F (10°C). The urogenital papilla in male lampreys becomes elongate as the spawning season approaches (Cochran and Sneen 1995). Spawning has been observed in southern Missouri in mid-March, and adults that had not finished spawning have been collected as late as mid-April (Pflieger 1997). Adults move into clear waters of small streams where the substrate consists primarily of clean sand and small pebbles (Boschung and Mayden 2004). Etnier and Starnes (1993) provided a description of spawning in *L. aepyptera*. They observed 10–12 lampreys in a pit constructed at the head of the spawning riffle. Nest construction appears to be an individual task in *L. aepyptera*, unlike other lampreys that engage in communal nest construction (Brigham 1973). Small, shallow nests of stones are constructed on the crest of a shallow riffle in slow-flowing streams about 15–20 cm deep by both sexes using their mouth discs and body movements to excavate. Boschung and Mayden (2004) reported that a Least Brook Lamprey anchors itself to a rock and, with rapid undulating motions of the posterior half of the body, fans the sand and pebbles from underneath, thereby eventually forming an oval depression about 3 cm deep and about 15 × 30 cm side to side, with the long axis situated in the direction of the current. F. C. Rohde (pers. comm.) observed both sexes moving stones as large as 2.5 cm in diameter and sometimes two individuals pulling simultaneously. Spawning typically involves pairs of lampreys, or one female accompanied by two males. A male attaches his oral disc to the head of a female and then curves his body around the female and opposes his cloacal papilla near her cloaca (Etnier and Starnes 1993). The male then contracts his body convulsively, as the female vibrates rapidly. This act is repeated a number of times. A single female may spawn with several males and/or spawning groups in a single nest and may deposit as few as 500 to as many as 3,800 eggs 0.63–1.36 mm in diameter (Seversmith 1953; Etnier and Starnes 1993). Eggs hatch in 4 days at a temperature of 65–72°F (18.3–22.2°C). Newly hatched larvae 0.2 inches (95 mm) TL are swept downstream and burrow into bottom substrate in low velocity areas (Hardisty and Potter 1971a; Jenkins and Burkhead 1994; Moyle and Cech 2004). By October of their first year, ammocoetes reach a length of about 1.4 inches (36 mm) (Pflieger 1997). The larval period may last 3–5.5 years (Rohde et al. 1976; Seversmith 1953; Walsh and Burr 1981). In Tennessee, the larval stage is probably 4–6 years, with metamorphosis occurring in autumn (Etnier and Starnes 1993). McAllister et al. (1981) reported size and fecundity of 13 mature *L. aepyptera* collected from Sylamore Creek, Arkansas. They reported a mean total length of 4.7 inches (119.2 mm). Ova counts ranged from 824 to 1,624 (x̄ = 1,306), and egg diameter ranged from 0.84 to 1.24 mm.

CONSERVATION STATUS   The Least Brook Lamprey is uncommon in Arkansas. It is imperiled in Mississippi (Ross 2001) and abundant in Tennessee (Etnier and Starnes 1993). Its small range in Arkansas has changed little since 1987, and we consider it a species of special concern.

Genus *Lethenteron* Creaser and Hubbs, 1922—Northern Brook Lampreys

The genus *Lethenteron* contains six or eight species, depending on authority. One species is parasitic (Renaud 2011; White 2014). As currently recognized, the genus *Lethenteron* is found in North America, Asia, and southern Europe. Controversy exists over the species composition of this genus, and Lang et al. (2009) and White (2014) provided genetic evidence that *Lethenteron* is not monophyletic. A single species, *Lethenteron appendix*, occurs in Arkansas. This nonparasitic lamprey prefers slightly larger streams than *Lampetra aepyptera*. Genus characters are presented in the species account.

Lethenteron appendix (DeKay)
American Brook Lamprey

Figure 8.14. *Lethenteron appendix*, American Brook Lamprey. *Patrick O'Neil, Geological Survey of Alabama.*

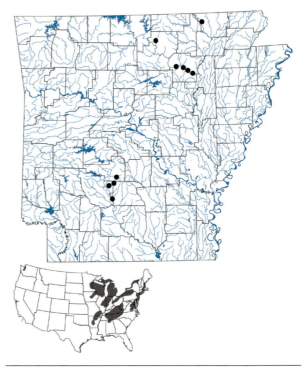

Map 5. Distribution of *Lethenteron appendix*, American Brook Lamprey, 1988–2019.

CHARACTERS   A nonparasitic lamprey with a deeply notched dorsal fin divided into two separate parts (Fig. 8.14). Oral disc, when expanded, almost as wide as the head. Oral disc length going into TL approximately 23 (16.6–37.4) times. Some teeth in the marginal fields of the oral disc are moderately well developed, usually with three pairs of bicuspids. Supraoral lamina with two widely separated teeth (Fig. 8.2E; Fig. 8.15). Teeth mostly degenerate; circumoral teeth 3 on either side of the mouth opening, all bicuspid. No lateral teeth. Myomeres 66–75. Dorsal fins of male are high and are separated by a sharp notch. A long, threadlike genital papilla distinguishes the breeding male from the breeding female, which has low dorsal fins separated by a broad notch; the female also has a prominent

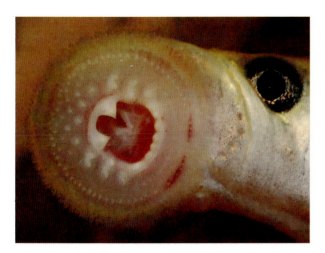

Figure 8.15. *Lethenteron appendix* mouth. *Renn Tumlison.*

anal fin fold (Becker 1983). Total lengths of up to 7.9 inches (200 mm) were reported by Pflieger (1997) and Etnier and Starnes (1993). Rohde et al. (1994) provided a range of 3.9–7.1 inches (99–180 mm) TL with a mean of 4.9 inches (125.5 mm). Maximum reported adult size is 13.75 inches (350 mm) TL (Page and Burr 2011). Larvae may reach 8.5 inches (216 mm).

LIFE COLORS   Body color varies from gray to almost black above, grading to tan or white below. Dorsal fin tinged with yellow, and caudal fin has considerable melanin proximally, lessening to gray distally. Spawning adults slate gray to grayish brown. Ammocoetes brownish on back and sides, lighter on venter. Buccal cavity unpigmented.

SIMILAR SPECIES   Adults differ from *Ichthyomyzon* species in having the dorsal fin divided into two parts (vs. dorsal fin possibly slightly notched, but never divided into two parts). It differs from *Lampetra aepyptera* in having 66 or more myomeres (vs. 60 or fewer myomeres).

VARIATION AND TAXONOMY   *Lethenteron appendix* was originally described by DeKay (1842) as *Petromyzon appendix* from Providence, Rhode Island, and the Hudson River, New York. The American Brook Lamprey has, over the past 80 years or so, been referred to as *Petromyzon wilderi, P. lamottenii, Entosphenus appendix, Lampetra appendix, L. wilderi, L. lamottenii, L. lamottei, Lenthenteron lamottenii,* and perhaps other names. In the first edition of this book, it was recognized as *Lampetra appendix.* Gill et al. (2003) found *Lampetra* (if *Lethenteron* is included) to be paraphyletic, and subsequent genetic analyses supported transferring the American Brook Lamprey from *Lampetra* to *Lethenteron* (Lang et al. 2009; White 2014). A phylogenetic analysis by Docker et al. (1999) supported the hypothesis of Hubbs and Potter (1971), Vladykov and Kott

(1979), and Potter (1980a) that *L. appendix* is derived from the parasitic *Lampetra camtschaticum* = *L. japonicum*), an anadromous species from Alaska and Canada (but see comments under *L. aepyptera*). White (2014) studied variation at two mitochondrial DNA regions among populations of the American Brook Lamprey to assess the phylogeographic history of the species. He found little variation among populations throughout its range, including 9 specimens from Arkansas (3 individuals from the White River system and 6 specimens from the Ouachita River system). Because White found little differentiation among populations, he suggested that very recent colonization and range extensions of *L. appendix* had occurred into the Great Lakes region. In addition, White (2014) rejected the hypothesis of monophyly in *L. appendix*.

In Arkansas, the American Brook Lamprey was known to occur only in the northern part of the state at a few localities in the White River system, primarily in the Ozarks, until Tumlison and Tumlison (1999) discovered this species in L'Eau Frais Creek (a tributary to the Ouachita River) in Clark and Hot Spring counties. Since the known populations were quite disjunct (about 200 km apart), it was possible that the newly discovered populations were taxonomically distinct. Tissue samples from 3 specimens collected on 20 February 1999 from the White River population (Piney Creek in Izard County) and 6 specimens collected 22 February 2007 from the L'Eau Frais Creek population in Clark County were sent to Matthew M. White to determine the degree of genetic divergence (Connior et al. 2011; Renn Tumlison, pers. comm.). Results indicated that the two disjunct populations in Arkansas are nearly genetically identical—only one substitution was found for about every 700 base pairs (M. M. White, pers. comm. to Renn Tumlison). Apparently the isolated L'Eau Frais Creek population of *L. appendix* should not be treated as a distinct taxon (Connior et al. 2011; White 2014).

DISTRIBUTION   *Lethenteron appendix* occurs in Atlantic Slope, St. Lawrence–Great Lakes, and Mississippi River basins from southern Quebec, Canada west to Minnesota and south to Virginia, northern Alabama, and southern Arkansas. In Arkansas, the American Brook Lamprey occurs in the White River system in northern Arkansas (Ward et al. 1999), primarily in the Ozarks, with one older record from the lower White River (Map A4.6). It is also found disjunctly in L'Eau Frais Creek, a tributary of the Ouachita River in Clark and Hot Spring counties (Tumlison and Tumlison 1999; Connior et al. 2012), and Jeff Quinn (pers. comm.) collected this species from the mainstem Ouachita River in Dallas County near Sparkman on 28 October 2011 (Map 5).

HABITAT AND BIOLOGY *Lethenteron appendix* is a nonparasitic brook lamprey that generally inhabits cool, clear, small to medium-sized streams, living in gravelly raceways and riffles. An exception to this is a single specimen collected from Escronges Lake, an oxbow lake of the lower White River on the White River National Wildlife Refuge on the Coastal Plain. In Wisconsin, it was most common in headwater streams over sand and small gravel, occasionally occurring over silt, clay, large gravel, and rubble (Becker 1983). It occurred at average stream widths of 5.9 (2–15) m and at depths of 0.7 (0.1–1.5) m. The larval stage, like the adult, is found in fast riffles and clean gravelly raceways of large creeks and small rivers (P. W. Smith 1979; Mundahl et al. 2006). It prefers slightly larger streams than *Lampetra aepyptera* (Etnier and Starnes 1993). This species is sensitive to various types of pollution, including turbidity (Becker 1983).

Feeding occurs only in the ammocoete stage and consists largely of small algae (principally diatoms and desmids) taken from the sediment and sometimes from the water column (Moore and Beamish 1973). Presumably, the adults do not feed. According to Boschung and Mayden (2004), size of food is an important limiting factor for a filter feeder, but there appears to be some selectivity of food items.

American Brook Lampreys spawn from March to April in Missouri (Pflieger 1997) when water temperatures reach about 65°F (18.3°C) (Becker 1983). Spawning occurs in Arkansas from at least early February through March. Gravid females of *L. appendix* were collected by Harp and Matthews (1975) on 17 February from Piney Creek (White River system) in northern Arkansas. Tumlison and Tumlison (1999) found a gravid female with mature eggs visible through the body wall in L'Eau Frais Creek in southern Arkansas on 31 January at a water temperature of 5°C. Additional gravid females were found on 24 and 25 February, the latter with 200–300 eggs visible through the abdomen. By 2 March, one female in L'Eau Frais Creek had only about half the eggs yolked, while two other females collected on that date had no yolked eggs. However, on 2 March, one male and two females were captured on a nest. Tumlison and Tumlison (1999) concluded that the reproductive season for the L'Eau Frais Creek population began in February and was completed by March. Seagle and Nagel (1982) estimated fecundity at 2,883 for gravid females; however, estimates range from 1,700 to 3,800 eggs. Eggs are 0.89–1.19 mm in diameter. Docker and Beamish (1991) reported an absolute fecundity in eight populations of this species as ranging from 503 to 5,900, while relative fecundity was 250–1,124 eggs per gram of total body weight. They noted that in populations where spawning occurs at a small

body size, females produce comparatively fewer but larger eggs. Becker (1983) described spawning of the American Brook Lamprey in Wisconsin, where males selected a site on the edge of a gravel area and constructed oval nests averaging 6–7 inches (152–178 mm) long, 4–5 inches (102–127 mm) wide, and 1–2 inches (25–51 mm) deep. Mundahl and Sagan (2005) studied the spawning ecology of *L. appendix* in Minnesota. Males arrive at the spawning site first in groups of 10–12 to begin excavation of nests. Nests were constructed in gravel and cobble substrates just upstream of riffles, spaced at an average density of 3 nests per m². The typical nest was 16 cm in diameter in water 31 cm deep, with a bottom current velocity of 14 cm/s. The nest was excavated to a depth of 4 cm below the stream bottom. Nests tended to be larger with larger spawning groups, deeper water, and slower current velocities. Pebble moving by *L. appendix* involves rapid and vigorous undulations and quivering of the body to provide a clean bed for the eggs (Boschung and Mayden 2004). Breder and Rosen (1966) reported that spawning may be a community affair for as many as 25–40 adults, and there is frequent movement among adjacent nests. During spawning, the male attaches his oral disc to the head of the female and then arches his trunk and tail to form a half-loop, bringing their cloacas together (Boschung and Mayden 2004). Alternative reproductive behaviors involving satellite males were reported by Cochran et al. (2008). In that study, 50% of matings in nests with at least three lampreys included a satellite male. Adults normally die after spawning, and Robison et al. (2013) reported finding a spent, dead male in the mainstem White River near Batesville on 12 March 2012. Newly hatched larvae are 2.6 mm TL and reach 38 mm by the end of the first year. Ammocoetes live 5–7.5 years but require at least 4.5 years to complete their development (Seagle and Nagle 1982). External signs of transformation begin in August and continue through October. Tumlison and Tumlison (1999) captured three male ammocoetes on 22 February 1997 that exhibited incomplete development of teeth and small anal fins, indicating that they were still in the process of transforming from ammocoetes to adults. The two lesser-developed specimens were larger than mature males, consistent with observations by Eddy and Underhill (1974) that ammocoetes are larger than recently metamorphosed adults. Two northern Arkansas adults measured 5.9 inches (139 mm) and 5.6 inches (143 mm) TL (Harp and Matthews 1975). The mean length of 13 L'Eau Frais Creek adults was 4.8 inches (122.2 mm) TL (7 males, $\bar{x}$ = 122 mm, and 6 females, $\bar{x}$ = 122.5 mm). Sex ratio was 1.2:1.

CONSERVATION STATUS We consider the American Brook Lamprey of special concern in Arkansas.

CHAPTER 9

Order Acipenseriformes
Sturgeons and Paddlefishes

The Acipenseriformes are an ancient lineage containing both fossil and extant taxa. This order of ancestral chondrosteans traces its lineage back about 250 million years, with an extinct suborder known from the Lower Triassic (Grande and Bemis 1996). Because they have changed little morphologically, indicating a slow evolutionary rate, they are sometimes referred to as living fossils. Extant acipenseriform diversity is represented by 27 species in two families: Acipenseridae (sturgeons) and Polyodontidae (paddlefishes) distributed in large temperate and arctic rivers of Eurasia and North America, presumably reflecting their Mesozoic distribution on the ancient northern continent of Laurasia (Bemis and Kynard 1997). The two extant families have fossil records dating back to the Mesozoic, with Polyodontidae known from the Early Cretaceous and Acipenseridae known from the Late Cretaceous (Hilton and Grande 2006). Those families diverged quite early in their evolution, and could be recognized as distinct groups by the Late Cretaceous (Doroshov 1985; Carroll 1988). The sturgeons evolved as benthic, sometimes sedentary, opercular pumping fishes, while paddlefishes became pelagic, constantly swimming ram-ventilators.

Acipenseriform species have thick skin with comparatively few and unspecialized mucous cells making them less slimy than many other fishes (Weisel 1975; 1978). Order characteristics include a largely cartilaginous skeleton, vertebrae lacking centra, the upper jaw not articulating directly with the cranium, and the notochord persisting into the adult stage. Only dermal bones of the head and pectoral girdle are ossified. The caudal fin is heterocercal, and a gular plate is absent. The intestine has a spiral valve. Grande and Bemis (1991) provided derived characters for this order. Robinson and Ferguson (2004) summarized three major aspects of the genetics of Acipenseriformes as follows: (1) Their karyotypes consist of an extremely large number of chromosomes, which for some species is as high as 500. (2) All are polyploid, and the order exhibits the highest level of polyploidy known in fishes. (3) The order is monophyletic and is the sister group of all extant Neopterygii.

Some interesting life history attributes of order Acipenseriformes include spawning possibly occurring several times within a season, females not spawning annually, and all species spawning in freshwater (Bemis and Kynard 1997). Suitable spawning sites (e.g., gravel or rocks containing many crevices) are critical for reproductive success, as are suitable water depth and moderate velocity (high flows, either natural or releases from impoundments, preclude or reduce spawning success). Spawning sites may be used year after year, suggesting homing capabilities in this group (Boschung and Mayden 2004).

Family Acipenseridae *Sturgeons*

The sturgeons represent one of the most ancestral surviving groups of bony fishes. They are a robust group of fishes whose form has withstood the test of evolutionary time and environmental pressure (Murphy et al. 2007b). Their conservative life history traits of slow maturation, slow respiration, slow metabolism, and infrequent reproduction have served them well over geologic time scales (Secor et al. 2002). The earliest members of this group probably evolved in the Lower Jurassic approximately 200 million years ago (Bemis et al. 1997). Although rare in the fossil record, the earliest undoubted fossil *Acipenser* is from the Upper Cretaceous beds of Montana about 100 million years ago (Cavender 1986; Hilton and Grande 2006). It is a Holarctic family with most species found in central and eastern Europe. The sturgeon family is comprised of four genera and 25 species occurring in the Northern Hemisphere (Page and Burr 2011; Nelson et al. 2016). Acipenseridae is subdivided into four subfamilies: Acipenserinae and Scaphirhynchinae, each with two recognized genera, plus Pseudoscaphirhynchinae and Husinae (Bemis et al. 1997; Birstein and Bemis 1997; Nelson et al. 2016). Eight species of sturgeons in two genera (*Acipenser* and *Scaphirhynchus*) are known in North America (Bemis and Kynard 1997; Page et al. 2013). Sturgeons are the largest freshwater fishes, and life spans of larger species may exceed 150 years (Etnier and Starnes 1993). The largest sturgeon species in North America is the White Sturgeon, *Acipenser transmontanus*, a Pacific Coast species that reaches 6.1 meters (Burr and Mayden 1992). Four of the eight North American sturgeons are marine species that spawn in freshwater.

Sturgeons are characterized by an elongated body that is pentagonal in cross section; an extended, hard snout; a ventral, protrusible mouth posterior to four barbels; a largely cartilaginous skeleton; a heterocercal tail; and a spiral valve in the intestine. Heavy scutes are arranged in five rows (one dorsal, two lateral, two ventrolateral) separated by thick leathery skin with denticles and scalelike plates (Weisel 1978).

Patterson (1982b) and Lauder and Liem (1983) considered the sturgeons to be monophyletic and sister to the fossil family Chondrosteidae. Grande and Bemis (1991) also considered Acipenseridae to be monophyletic, but sister to Polyodontidae. Findeis (1997) revised the classification of sturgeons using morphological characters to include the subfamily Husinae with *Huso* and subfamily Acipenserinae. The monophyly of these subfamilies has been questioned based on DNA sequence data and combined DNA and morphological data (Birstein et al.

2002). Dillman et al. (2007) recognized the two traditional subfamilies: Acipenserinae and Scaphirhychinae. Those authors supported monophyly for the genera *Scaphirhynchus* and *Pseudoscaphirhynchus*, but not for the subfamily Scaphirhychinae, nor any species within these genera.

All species of sturgeons inhabiting Arkansas are habitat specialists, mature late in life (>7 years), and do not spawn every year (1–3-year intervals). They are adapted to life in turbid, fast-flowing environments, spawn in swift mainstem, open-channel habitats, and their reproduction is cued by seasonal changes in river flow (Phelps et al. 2016). Because of these characteristics, they are sensitive to anthropogenic changes in hydrology. Sturgeon populations can be managed in their riverine environments through a combination of habitat preservation, habitat restoration, flow regulation, stocking of hatchery-reared fish, and regulation of harvest (Phelps et al. 2016).

Sturgeons, among the most important commercial fishes in the world (Becker 1983), were once of considerable economic importance in some areas of the United States because of the demand for their oil, their eggs for caviar, and their flesh for smoked steaks (P. W. Smith 1979). They have been historically overfished and are currently still taken illegally in spite of regulations in place to protect them (Auer 2004). In addition, the worldwide stocks of sturgeons are dwindling because of dams, pollution, and their slow growth rates which render them easily exploited (Becker 1983). In the free-flowing Mississippi River (i.e., below the Missouri River), sturgeons with pronounced anomalies (e.g., misshapen rostra, reduced fins, strictures and scars, missing tails) are not uncommon and are, in part, associated with trash in the river (e.g., rubber bands, O-rings) and with commercial fishing (i.e., from trot-line injuries) (Murphy et al. 2007a). The Pallid Sturgeon was listed as federally endangered in 1990. A modest sturgeon roe fishery exists today in the White River near Augusta, Arkansas for one species, *Scaphirhynchus platorynchus,* and commercial fishermen only rarely encounter them in their nets.

Key to the Sturgeons

1A Spiracles present; caudal peduncle short, not completely covered with bony plates; lower lip with 2 lobes (Fig. 9.1A); barbels not fringed; upper lobe of caudal fin without a filament (*Acipenser*).   2

1B Spiracles absent; caudal peduncle long and completely covered with bony plates; lower lip with 4 lobes (Figs. 9.1B, 9.1C); barbels fringed; upper lobe of caudal fin with a long filament (which may be abbreviated or absent in older fish) (*Scaphirhynchus*).   3

2A Postdorsal and preanal shields unpaired; plates between anus and anal fin 1 or 2; jugal bone triangular; viscera black.   *Acipenser fulvescens* Page 129

2B Postdorsal and preanal shields occurring in pairs; plates between anus and anal fin 3–9; jugal bone L-shaped; viscera white.   *Acipenser oxyrinchus* Page 132

3A Belly mostly naked; base of outer barbels posterior to base of inner barbels; dorsal fin rays usually 37 or more; anal rays usually 24 or more.
                          Scaphirhynchus albus Page 133

3B Belly mostly covered with small plates; base of outer barbels even or anterior to base of inner barbels; dorsal rays 36 or fewer; anal rays 23 or fewer.
                  Scaphirhynchus platorynchus Page 137

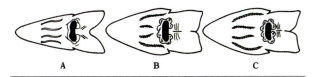

Figure 9.1. Undersides of heads of sturgeons. *Smith (1979)*.

A—*Acipenser fulvescens/oxyrinchus*
B—*Scaphirhynchus albus*
C—*S. platorynchus*

Genus *Acipenser* Linnaeus, 1758— Holarctic Sturgeons

The Holarctic genus *Acipenser*, containing 17 species, is characterized by a more or less robust snout, barbels without fringes, presence of a spiracle, and two lobes on the lower lip. The caudal peduncle is short, round in cross-section, and not completely covered by bony plates, and there is no caudal fin filament. Of the five species of *Acipenser* that inhabit North America, only the Lake Sturgeon, *Acipenser fulvescens*, has been documented in Arkansas in recent decades. An older Arkansas record exists for the Atlantic Sturgeon, *A. oxyrinchus*.

Acipenser fulvescens Rafinesque
Lake Sturgeon

Figure 9.2. *Acipenser fulvescens*, Lake Sturgeon. *David A. Neely*.

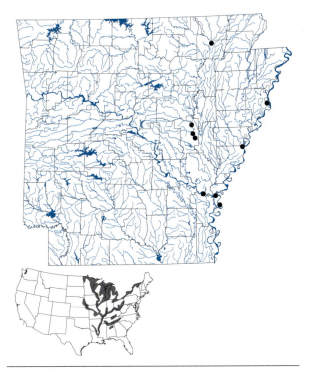

Map 6. Distribution of *Acipenser fulvescens*, Lake Sturgeon, 1988–2019.

CHARACTERS   A large sturgeon with a pointed, short, conical snout, and a robust body (Fig. 9.2). Lower lip with two lobes along posterior margin. Spiracle present at anterior end of the groove continuous with gill slit. Caudal peduncle thick, fleshy, relatively short, and only partially covered laterally by bony plates; the plates are sharp in young individuals and become very blunt or disappear altogether in large adults. Upper lobe of caudal fin without a filament, except in smallest young. Dorsal rays 35–40; anal rays 25–30. Gill rakers short and blunt, 25–40. Nostril opening closest to eye smaller than eye. Barbels on lower snout four, smooth, not fringed, and subequal (Fig. 9.1A; Fig. 9.3). Dorsal plates 9–17; lateral plates 29–42; postdorsal and preanal shields in single rows. Viscera black. Maximum size about 8 feet (2,438 mm) and over 300 pounds (136.08 kg).

LIFE COLORS   The body of this benthic fish is grayish olive to brown on back and sides, with a white to yellow-white belly. Many individuals are sharply bicolored. Fins colored the same as the area of the body they are near. Juveniles smaller than 30 cm have black blotches scattered over the head and body; blotches fade with age (Boschung and Mayden 2004).

SIMILAR SPECIES   The Lake Sturgeon superficially resembles the two species of *Scaphirhynchus* in Arkansas, *S. albus* and *S. platorynchus*. *A. fulvescens* differs from both

Figure 9.3. Head of *Acipenser fulvescens*, Lake Sturgeon. *Steven G. George, USACE.*

Scaphirhynchus species in having a spiracle (vs. spiracle absent), a lower lip with only two lobes (vs. lower lip with four lobes), smooth barbels (vs. barbels feathery with fringe projections), and a short caudal peduncle incompletely covered with bony plates (vs. a long, narrow caudal peduncle completely covered with bony plates).

VARIATION AND TAXONOMY   *Acipenser fulvescens* was originally described by Rafinesque (1817a) from Lake Erie. Boschung and Mayden (2004) reported that species relationships of *Acipenser* have been examined mainly using molecular data and the genus is not monophyletic. Those authors also considered the placement of *A. fulvescens* within the genus problematic. Birstein and DeSalle (1998) considered *A. fulvescens* the sister group to all other *Acipenser* (and *Huso*), excluding the *A. oxyrinchus–A. sturio* clade. Relationships within *Acipenser* based on morphological and karyological data have shown sister-species relationships between the White and Green sturgeons and between Shortnose and Lake sturgeons, with the Atlantic Sturgeon basal to these species (Artyukhin 1995; J. R. Brown et al. 1996; Choudhury and Dick 1998; Krieger et al. 2000). Kuhajda (2014) found that mitochondrial DNA analyses of only North American species supported these same relationships. These sister-species pairs were not recovered when other species were included in the analysis, but each species of the pair typically occurred within the same clade (Birstein and DeSalle 1998; Ludwig et al. 2001; Birstein et al. 2002).

DISTRIBUTION   *Acipenser fulvescens* occurs in the St. Lawrence–Great Lakes, Hudson Bay, and Mississippi River basins from Quebec to Alberta, Canada, south to Alabama and Louisiana. It was formerly known from the Mobile Bay drainage (Coosa River) of Alabama. Meek

(1894a) first listed the Lake Sturgeon under the name *Acipenser rubicundus* (Lesueur) as occurring in Arkansas but gave no localities or collection data. Black (1940) also provided no documentation of *A. fulvescens* from Arkansas, but predicted that it would be found in this state. Two valid records were reported by Robison and Buchanan (1988) from the Mississippi River: (1) river mile 625 near Mellwood (Phillips County), where a commercial fisherman caught a 55-inch (1,397 mm), 31-pound (14.1 kg) specimen in a hoop net in October 1981, and (2) a pre-1960 report from Mississippi County. Crump and Robison (2000) reported an overlooked newspaper account of another very large sturgeon taken from the Caddo River at Glenwood in 1945. It was assumed to be a Lake Sturgeon because of its size. The newspaper photograph did not allow positive identification, and the specimen is not available for examination. A sturgeon authority consulted by J. W. Quinn agreed that a positive identification could not be made from the photograph but was adamant that it was not a Lake Sturgeon and was possibly a hoax. Several records for Lake Sturgeon in Arkansas have been documented since the first edition of this book (Map 6). A 30-pound (13.6 kg) specimen taken on 19 May 1989 from the mouth of the White River (specimen was frozen by Ken Shirley (AGFC) but subsequently lost by a taxidermist in Fort Smith). A 47-pound (21.34 kg) specimen from the White River at DeValls Bluff caught in March 1992 was examined at a fish market by Ken Shirley (Buchanan et al. 1993). One specimen from the White River near Des Arc was photographed and released in February 2008. One specimen was caught at the confluence of the St. Francis and Mississippi rivers in a hoop net in February 2012 (photographed and released). A 4-foot (approx. 1,200 mm), 33-pound (14.97 kg) specimen was snagged by an angler in July 2014 below Dam 2 and the hydroelectric plant on the Arkansas River near Pendleton, Arkansas. A commercial fisherman reported that another specimen was caught in the St. Francis River in January 2012 about 5 miles upstream from the specimen caught in February 2012. A 7+ foot (2,134 mm), 100+ pound (45+ kg) specimen was caught by a commercial fisherman in the Mississippi River of Crittenden County on 27 September 2016 (identified from a photograph). The fisherman reported that he had captured more than 20 Lake Sturgeons in the Mississippi River in gill and hoop nets in recent years. The first record of a Lake Sturgeon from the Black River was taken by a commercial fisherman on 16 January 2019 near Powhatan in Lawrence County, Arkansas (Allison Asher, pers. comm.; identification verified by us from photographs).

HABITAT AND BIOLOGY   The Lake Sturgeon is one of the largest North American freshwater fishes. This benthic species lives on mud and gravel substrates and in the shallow shoal areas of lakes and the deepest portions of large rivers (Harkness and Dymond 1961; C. L. Smith 1985). Most Lake Sturgeons are caught at depths of 4.6–9.2 m (Scott and Crossman 1973). Prior to 1900, a substantial Lake Sturgeon commercial fishery existed in the Missouri and Mississippi rivers in Missouri (Pflieger 1997). Lake Sturgeon movements involve a spring migration to potential spawning areas, a post-spawning dispersal to feeding grounds, and a fall migration to overwintering sites (McKinley et al. 1998). Spawning migrations can be extensive (Wilson and McKinley 2004), as much as 128 km, and Auer (1996) concluded that this species is capable of traveling 1,000–1,800 km. A specimen caught in July 2014 in the Arkansas River had been tagged in Missouri by the Missouri Department of Conservation in 2008 (at 31 inches [787.4 mm] and 8 pounds [3.6 kg]) and traveled 673 miles (1,083 km) in the Mississippi River and 25 additional miles up the Arkansas River before being captured by an angler below the hydroelectric plant near Pendleton, Arkansas.

The Lake Sturgeon is anatomically adapted for feeding on bottom-living organisms as it is equipped with a ventral, protrusible, tubelike mouth (Harkness and Dymond 1961; Scott and Crossman 1973). In Ohio, it fed chiefly over a clean bottom of sand, gravel, and rocks and avoided soft, muddy bottoms (Trautman 1981). Priegel and Wirth (1971) observed feeding behavior in an aquarium. When searching for food, it swims near the bottom with the ends of its barbels dragging lightly over the substrate. When the barbels touch food, it reacts instantly by rapidly protruding the tubular mouth and sucking in substrate materials containing food. The food items are separated from the soft bottom materials and swallowed, and the debris is ejected through the gills. Young Lake Sturgeons feed on zooplankton (crustaceans) until around 178–203 mm TL, when they transition to a benthic-oriented diet (Eddy and Underhill 1974). Foods of adults include benthic organisms such as mollusks, aquatic insect nymphs and larvae (especially mayfly nymphs), worms, amphipods, crayfish, small fish, and fish eggs. They are opportunistic predators (Chiasson et al. 1997), and the diet is dependent on the availability of different prey organisms (Harkness and Dymond 1961).

Acipenser fulvescens spawns in the spring from mid-April to early June at water temperatures between 55.4°F and 64.4°F (13–18°C) at depths of 3–16.4 feet (1–5 m) in swift waters (Harkness and Dymond 1961; Gruchy and Parker

1980). Lake Sturgeon exhibit an autumn pre-spawn migration to staging areas within smaller tributaries (Bruch and Binkowski 2002). In the main channel of the river adjacent to the spawning sites, individuals begin "porpoising behavior" up to 14 days before actual spawning when water temperatures increase to 6.6–16°C (Bruch and Binkowski 2002). During this behavior, individuals quickly surface while swimming upstream, sticking only their heads out of the water but occasionally jumping entirely out of the water. Spawning occurs during both day and night (Harkness and Dymond 1961). Bruch and Binkowski (2002) provided a detailed description of spawning behavior in Wisconsin. Males move onto the spawning site 1–2 days before the arrival of the females and begin cruising. Females then move onto the spawning site and active spawning ensues. Male behaviors during and between spawning bouts include pounding the female with the tail, nosing the female, and emitting a dull, thunderous sound. Typically, one male initiates a spawning bout with an ovulating female. Seconds before the spawning bout begins, males near the female become very excited and attempt to move into position on either side of the female. Generally, two to eight males release sperm in the vicinity of a female during the somewhat violent courtship. Rapid dispersal from the spawning site occurs following the conclusion of spawning. Fecundity ranges from 49,000 to 667,000 and is among the highest of all freshwater species in North America (Harkness and Dymond 1961; Priegel and Wirth 1971; Scott and Crossman 1973). LaHaye et al. (1992) reported that spawning occurred primarily in areas with coarse gravel substrates. Eggs adhere to rocks, gravel, logs, or any other clean object they contact. Eggs hatch in 5–8 days at 16–17°C. Growth is rapid during the protracted juvenile stage, and young reach lengths of about 125 mm during their first summer and fall. Maturity in northern populations takes 18–27 years for females, with males maturing in 12–15 years at 29.5–39.4 inches (750–1,000 mm) (Scott and Crossman 1973; Bruch et al. 2001). Females spawn at intervals of 4–6 years, and males spawn annually or in alternate years. Lake Sturgeons live to be much older than other North American freshwater fishes, with maximum age estimates of 154 years (Etnier and Starnes 1993). In northern populations, females live longer (about 80 years) than males (about 55 years), but centenarians (120–150 years) have been documented.

CONSERVATION STATUS   The Lake Sturgeon reaches the southern limits of its range in Mississippi (Ross 2001) and is very rare in Arkansas. Although it has declined in numbers over most of its range and was considered a threatened species by Miller (1972), it is still common in a few Wisconsin rivers (Becker 1983). In Wisconsin, hook-and-line angling for sturgeon is still permitted during a limited open season, and spearing is also allowed. The Lake Sturgeon was listed as threatened by Williams et al. (1989) and Warren et al. (2000). It is listed as extirpated, endangered, threatened, or of special concern in 12 states (Leonard et al. 2004). Once abundant throughout its range, severe overfishing in the late 1800s and early 1900s decimated most populations (Auer 2004; Bogue 2000). Few healthy populations remain today and anthropogenic factors, including hydroelectric dams that obstruct upstream access to historic spawning grounds and degrade downstream habitats, continue to hamper most conservation and restoration efforts (Peterson et al. 2007). In Arkansas, we consider the Lake Sturgeon to be at least threatened.

Acipenser oxyrinchus Mitchill
Atlantic Sturgeon

Figure 9.4. *Acipenser oxyrinchus*, Atlantic Sturgeon. Little Missouri River specimen caught on 18 July 1956 near Chidester, Arkansas. *Photo by B. G. Crump.*

CHARACTERS   A large sturgeon with a long, V-shaped snout. Mouth ventral and small, its width less than 55% of the distance between the eyes. Lower lip with 2 lobes; barbels without accessory fringes. Spiracles present. Body plates (scutes) large and prominent. Preanal plates occurring in pairs; postdorsal plates 6–9, mostly in pairs. Usually 3–9 plates (mostly in pairs) between anus and anal fin. Dorsal plates 7–16. Lateral plates 24–35. Ventral plates 6–14. Gill rakers 17–27. Jugal bone flat with two arms forming an L-shape. A fontanelle present in skull. Dorsal rays 38–46. Anal fin origin under middle of dorsal fin base; anal rays 26–28. Viscera white and unpigmented; peritoneum white. The only verified Arkansas specimen was 6 feet 6 inches (1,943 mm) TL, and weighed about 135 pounds (61.2 kg). Maximum reported size was 14 feet

(4.27 m) TL and about 814 pounds (369 kg) (Boschung and Mayden 2004).

LIFE COLORS   The body is olive-green, bluish-gray, or brown above, grading to white on the belly. Fins gray to blue-black.

SIMILAR SPECIES   *Acipenser oxyrinchus* most closely resembles *A. fulvescens*, but differs from that species in having the postdorsal and preanal plates occurring in pairs (vs. postdorsal and preanal plates unpaired), gill rakers 17–27 (vs. 25–40), viscera white and unpigmented (vs. viscera dark), plates between anus and anal fin 3–9 (vs. 1 or 2), jugal bone flat with two arms forming an L-shape (vs. jugal bone triangular), mouth small, its width less than 55% of the distance between the eyes (vs. mouth large, its width 66–93% of distance between eyes), and a fontanelle present in skull (vs. fontanelle absent). It can be distinguished from *Scaphirhynchus* species in having spiracles (vs. spiracles absent), 2 lobes on lower lip (vs. 4 lobes), and caudal peduncle incompletely armored (vs. completely armored).

VARIATION AND TAXONOMY   *Acipenser oxyrinchus* was described by Mitchill (1815) from "New York" (no types known). Boschung and Mayden (2004) noted that some subsequent authors incorrectly emended Mitchill's original spelling of *oxyrinchus* to *oxyrhynchus*. This species has also sometimes been referred to as *Acipenser sturio*. Vladykov (1955) described Gulf of Mexico populations as a subspecies of *A. oxyrinchus*, *A. o. desotoi*, the Gulf Sturgeon. There is controversy about the taxonomic status of the Gulf Sturgeon. Some do not distinguish it as a separate species from the Atlantic Sturgeon, *A. oxyrinchus* (Page et al. 2013). However, some authors treat the previously recognized subspecies, *A. oxyrinchus desotoi* from the Gulf of Mexico as a full species, *A. desotoi* (Robins et al. 2018).

DISTRIBUTION   The Atlantic Coast subspecies (*A. o. oxyrinchus*) occurs along the Atlantic Coast from the Hamilton River in Labrador, Canada to mid-eastern Florida, with reports of its occurrence from northeastern South America, Bermuda, and even northern Europe (Baltic Sea basin) (Boschung and Mayden 2004; Page and Burr 2011). The Gulf of Mexico subspecies (*A. o. desotoi*) occurs along the Gulf Coast from Tampa Bay, Florida to Lake Pontchartrain, with an occasional record as far west as the Rio Grande, Texas. There are numerous records from freshwater rivers along the Gulf of Mexico. The only record of this species from Arkansas is from the Little Missouri River (Ouachita River drainage). The single distribution point on the pre-1988 Lake Sturgeon distribution Map A4.7 (in Appendix 4) has now been reidentified as representing *A. oxyrinchus*. Our information on the occurrence of *Acipenser oxyrinchus* in Arkansas was compiled and provided by

J. W. Quinn. Records of the anadromous Gulf Sturgeon, *Acipenser oxyrinchus desotoi*, occurring inland at latitudes approaching those of southern Arkansas are known from Alabama and Mississippi. A Lake Sturgeon record reported in Robison and Buchanan (1988) from the Little Missouri River (Ouachita River drainage) near Chidester, Arkansas was caught on 18 July 1956 and was on display in the Arkansas Game and Fish Commission Building in Little Rock (Fig. 9.4). Jeffrey W. Quinn made a detailed morphological examination of that specimen and provided conclusive evidence that its correct identity is *Acipenser oxyrinchus desotoi*, the Gulf Sturgeon. Quinn's identification was based on seven characters and was confirmed by sturgeon authorities, Rob DeVries, Eric Hilton, Martin Hochleithner, and Paul Vecsei. An attempted genetic analysis by Brian Kreiser was unsuccessful because not enough DNA could be obtained from the mounted specimen.

HABITAT AND BIOLOGY *Acipenser oxyrinchus* is known from freshwater, brackish, and marine habitats. Details of habitat preferences are poorly-known, especially for the marine environment where it occurs in shallow waters over the continental shelf. In freshwater habitats in Florida, sturgeon microhabitats in late spring and summer averaged 8.4 m deep, with an average current speed of 64 cm/s (Ross 2001). Substrates were sand or gravel. Major areas of congregation were usually downstream from a large spring.

Most feeding of adults occurs in the marine environment. This benthic feeder consumes mollusks, polychaete worms, and a variety of crustaceans (Boschung and Mayden 2004). Little or no feeding by adults occurs during migration and spawning. In fresh water, the young feed mainly on immature insects such as midge larvae and mayfly nymphs, and also on isopods and bivalve mollusks.

Acipenser oxyrinchus is anadromous, with adults ascending freshwater rivers to spawn. Atlantic Coast populations ascend rivers from February (Georgia) to June (Gulf of Maine). Gulf of Mexico adults ascend rivers in spring, from February through May, and go back out to sea in fall (Ross 2001). Temperature has the stongest influence on upstream movement, and spawning occurs in current over rubble or cobble substrates, sometimes below waterfalls (Wooley and Crateau 1985). Large females have high fecundity, sometimes producing 3.75 million eggs. Young sturgeons spend 1 to 6 years in fresh water before moving to the sea. Maximum age for the Gulf subspecies is at least 42 years (Ross 2001).

CONSERVATION STATUS The USFWS recognizes the Gulf subspecies as threatened (Robins et al. 2018). The status of *A. oxyrinchus* in Arkansas is undetermined.

Genus *Scaphirhynchus* Heckel, 1836 — Shovelnose Sturgeons

The Shovelnose Sturgeons of the genus *Scaphirhynchus* typically live in large rivers of central North America. The flattened, shovel-shaped snout, slim body with well-defined and sharply keeled scutes, and long, narrow, fully armored caudal peduncle are well suited for a fish inhabiting swift water (Phelps et al. 2016). The body shape of *Scaphirhynchus* minimizes hydrodynamic lift and its rough surfaces maximize friction with the river bottom, allowing the sturgeon to hold station in fast-flowing water and avoid being swept downstream (Webb 1994). The three species of this genus differ from those of *Acipenser* in having a broad, depressed, shovel-like snout with feathery rather than smooth rostral barbels, in having four rather than two papillose lobes on the lower lip, and in lacking a spiracle (Etnier and Starnes 1993). *Scaphirhynchus* is supported as monophyletic based on molecular (Dillman et al. 2007) and morphological (Mayden and Kuhajda 1996) data, but the relationship of *Scaphirhynchus* to other sturgeons is unclear. Two species, *S. platorynchus* and *S. albus*, occur in Arkansas. The third, *Scaprhirhynchus* species, *S. suttkusi*, is endemic to the Mobile River basin.

Scaphirhynchus albus (Forbes and Richardson)
Pallid Sturgeon

Figure 9.5. *Scaphirhynchus albus*, Pallid Sturgeon. *David A. Neely.*

CHARACTERS A pale sturgeon with a broad, flattened, shovel-shaped snout, and an elongate body tapering into a long, slender, depressed, and completely armored caudal peduncle (Fig. 9.5). Upper lobe of caudal fin often produced into a long filament. Spiracle absent. Eye small. Squamation on the ventral surface is weak and reduced (Mayden and Kuhajda 1996); belly naked or with a few small plates at all ages (although tiny denticles may be present). Usually with nine postdorsal plates. Mouth transverse. Upper lip with four lobes projecting anteriorly; lower lip with four papillose lobes projecting posteriorly (Fig. 9.1B). Oral lobes of the Pallid Sturgeon are spaced farther apart than those of the Shovelnose Sturgeon, giving the mouth of the Pallid a thin-lipped appearance. The branchiostegal connection is

Map 7. Distribution of *Scaphirhynchus albus*, Pallid Sturgeon, 1988–2019.

Figure 9.6. Coloration of three live Arkansas sturgeons compared side by side. Top to bottom: *Acipenser fulvescens*, *Scaphirhynchus albus*, and *S. platorynchus*. Steven G. George, USACE.

narrow and the posterior edge of the branchiostegal membranes is subhorizontal (Mayden and Kuhajda 1996). The bases of the outer barbels are slightly posterior to the bases of the inner barbels; length of inner barbels about half the length of the outer barbels. Inner barbel length going into head length 6.3–8 times. Dorsal fin with 35–43 rays (Mayden and Kuhajda 1996; Kuhajda et al. 2007; Murphy et al. 2007b). Pectoral fins large and rounded, with 46–56 rays. Pelvic fin rays 30–34. Anal fin rays 23–28. Maximum size about 5.5 feet (167 cm) TL and 68 pounds (31 kg) (Carlander 1969; Lee 1980a; Kallemeyn 1983). In the lower Mississippi River, individuals smaller than 3.3 feet (100 cm) TL are more typical (Killgore et al. 2007a, b).

LIFE COLORS The Pallid Sturgeon, as the name implies, is frequently lighter in pigmentation than the Shovelnose Sturgeon (Fig. 9.6), but coloration may range from white to butterscotch to dark brown. Side and ventral areas more of a grayish white. Fins similar in color to adjacent body parts. Principal rows of scutes slightly darker than surrounding body.

SIMILAR SPECIES *Scaphirhynchus albus* is similar to *S. platorynchus* but differs in having a proportionately larger head, wider mouth, thinner lips, shorter inner barbel (about half the length of outer), outer barbel base posterior to inner barbel base (vs. even with or anterior to inner

barbel base), barbels weakly fringed (vs. strongly fringed), very few or no small plates on belly (vs. belly completely or mostly covered with plates), dorsal rays usually 37 or more (vs. fewer than 37), anal rays usually 24 or more (vs. fewer than 24), gill rakers stiff and peglike (vs. gill rakers malleable and fanlike), and it achieves a larger maximum size (Bailey and Cross 1954; Kuhajda et al. 2007). The lateral and ventrolateral plates are smaller in *S. albus*, causing the space between the plate rows to be larger than in *S. platorynchus* (Kuhajda et al. 2007; Murphy et al. 2007b). Lateral-ventrolateral interspace at the tenth lateral plate was the highest-loading character along the ordination axis of greatest separation of the two species (Catherine Murphy, pers. comm.).

VARIATION AND TAXONOMY Forbes and Richardson (1905) described the Pallid Sturgeon as *Parascaphirhynchus albus* from nine specimens collected from the Mississippi River near the mouth of the Illinois River at or near Grafton, Illinois. Bailey and Cross (1954) assigned it to genus *Scaphirhynchus* and redescribed it as *S. albus* based on 17 specimens from eight localities. Biochemical systematics of *S. albus* and *S. platorynchus* were studied by Phelps and Allendorf (1983), who found no interspecific differences at 37 loci examined electrophoretically despite the several distinct morphological differences. However, genetic frequency differences between the two species based on mitochondrial and nuclear DNA were subsequently documented (Campton et al. 2000; Tranah et al. 2001; Ray et al. 2007; Eichelberger et al. 2014), despite the lack of diagnostic genetic markers. Wills et al. (2002) developed an index for distinguishing the two species based on meristics and morphometrics. Some morphological variation is present in *S. albus*, and latitudinal variation in morphology was found in Mississippi River populations (Mayden and Kuhajda 1996; Murphy et al. 2007b). Ray et al. (2007) and Schrey and Heist (2007) also found substantial genetic variation and geographic substructure in *S. albus*. Hybridization between *S. albus* and *S. platorynchus* has been reported by Carlson et al. (1985) and Schrey et al. (2007, 2011), but specimens of intermediate phenotypes have also been identified as either Pallid Sturgeons or Shovelnose Sturgeons based on genetic data (Ray et al. 2007) and morphometric data (Murphy et al. 2007b). Interestingly, *S. albus* specimens from the lower Mississippi River are morphologically more similar to *S. platorynchus* than are *S. albus* from the Upper Missouri River based on morphometric data (Murphy et al. 2007b).

DISTRIBUTION   The range of *Scaphirhynchus albus* is confined almost entirely to the main channels of the Missouri River and the middle and lower Mississippi River from Montana through Louisiana. There are also records from some of the larger tributaries. The exact historical range is not precisely known, and its distribution has now been fragmented by large dams and reservoirs. Arkansas records for Pallid Sturgeon come only from the Mississippi River, lower Arkansas River, and the lower St. Francis River in northeastern Arkansas (Map 7). The first vouchered record of *S. albus* was taken in April 1988 from the Mississippi River at river mile 665 near Helena in Phillips County (Buchanan et al. 1993). It measured 1,090 mm TL and weighed 3.75 kg. The specimen was freeze-dried and placed in the fish collection of the University of Arkansas–Fort Smith. A single specimen was taken on 8 May 1994 in the St. Francis River, five miles upstream from its confluence with the Mississippi River. It was a tagged, hatchery-reared specimen that had been released in Missouri as part of a restoration effort. Between 2001 and 2014, 81 specimens of Pallid Sturgeon were caught in the Mississippi River at 17 localities in or near Arkansas boundary waters by biologists from the U.S. Army Engineer Research and Development Center at Vicksburg, Mississippi (Killgore et al. 2007a, b, and data provided by Catherine Murphy).

HABITAT AND BIOLOGY   The Pallid Sturgeon is primarily an inhabitant of large, turbid rivers where it lives in a strong current over a firm sandy or gravelly bottom (Pflieger 1997). It is sometimes taken over a shifting sand substrate. Like other lotic, benthic fishes, juvenile *S. albus* use their pectoral fins and overall body morphology to maintain station against water velocity without swimming (Adams et al. 1999). Gravel substrates may be used more heavily in winter and spring (Koch et al. 2012), and gravel and rock substrates may also be used for feeding

and spawning (Hoover et al. 2007). It is found mainly in main channel, channel border, and secondary channel habitats with connection to the main channel (Jordan et al. 2016). Carlson et al. (1985) collected Pallid Sturgeons in sluggish areas along sandbars on the insides of bends and behind wing dams frequented by Shovelnose Sturgeons, and in faster current areas less frequented by the latter species. In the Platte River, Nebraska, it preferred water depths greater than 4 feet (1.2 m) with swift currents around 3 feet/s (0.9 m/s) at the edges of sandbars (Hrabik et al. 2015). Kallemeyn (1983) reported it inhabits areas of rapid current and prefers turbid water conditions that conceal it from prey species (Mayden and Kuhajda 1996). In the LMR, *S. albus* strongly selected island tip and natural bank habitats, and, to a lesser degree, revetted bank habitat (Herrala et al. 2014). It exhibited negative selection for the expansive main channel habitat. *Scaphirhynchus* spp. YOY were most frequently collected in the LMR in areas with a relatively fast bottom velocity (0.5–0.7 m/s) associated with channel sandbars, rootless dikes, and wing dikes along banklines, or in tributaries where bottom velocity was slower (≤0.2 m/s) (Ridenour et al. 2011). Annual population estimates in the lower Missouri River varied from 4.0 to 7.3 fish per rkm for wild and 8.4 to 18.4 fish per rkm for hatchery-reared *S. albus* (Steffensen et al. 2017). Dams and reservoirs act as barriers to sturgeon movement and have inundated and rendered unsuitable large portions of the species' former range (Jordan et al. 2016). Those authors noted that population estimates for the MMR range from 1,600 to 4,900; there are no population estimates available for the LMR (downstream from the mouth of the Ohio River).

Pallid Sturgeons are benthic or drift-feeding browsers. The larvae transition from yolk sac to exogenous food approximately 11–20 days after hatching, depending on water temperature, and observations in a hatchery environment indicate that zooplankton and small invertebrates are important first foods (Jordan et al. 2016). Young-of-the-year *Scaphirhynchus* spp. in the LMR apparently occupied sandy-bottomed habitats where they fed primarily on benthic macroinvertebrates, especially chironomids (Harrison et al. 2014). Subadult and adult foods consist of small fishes and immature aquatic insects (Cross 1967; Carlson et al. 1985), and there is a pronounced ontogenetic shift toward piscivory as body size increases (Hoover et al. 2007; Grohs et al. 2009). Hard surfaces are critical habitat for feeding in the Mississippi River. This habitat is provided by naturally occurring gravel deposits and by artificial surfaces such as stone dikes (Payne and Miller 1996). In the lower Mississippi River, prey characteristic of hard substrates (e.g., shad, Silver Chubs, Freshwater Drum, hydropsychid caddisflies) predominate over those from soft substrates (e.g., burrowing mayflies, larval chironomids), suggesting that natural gravel and possibly riprap are important feeding surfaces (Hoover et al. 2007).

The Pallid Sturgeon life history requires long, continuous reaches of free-flowing river for successful upstream spawning migrations and subsequent downstream distribution of free-embryo, larval, and juvenile life stages (Jordan et al. 2016). Like other sturgeons, it is long-lived, matures late, and exhibits intermittent spawning cycles (Steffensen et al. 2017). Those authors noted that historic migration and drift pathways have been fragmented by river modifications that altered the natural temperature, turbidity, and flow regime and negatively affected spawning areas (as well as food resources). Spawning occurs between March and July, with fish in Arkansas probably spawning early in that range. Spawning apparently occurs over coarse substrate (boulder, cobble, gravel) or bedrock in deep water with relatively fast, converging flows, and is related to environmental stimuli such as photoperiod, water temperature, and flows (Jordan et al. 2016). Details of spawning behavior are not fully known, but one gravid female and seven males were observed to form a spawning aggregation in the Yellowstone River in a current of 700–900 cm/s and a water temperature of 20°C (Fuller et al. 2007). Johnston and Phillips (2003) reported that *S. albus* produces sound during the breeding season and suggested these sounds may be used in aggregating individuals on the spawning grounds. Estimates of age at first spawn in the Mississippi River basin range from a high of 15 years (Keenlyne and Jenkins 1993) to as low as 8 (George et al. 2012). Fecundity is related to body size, with the largest upper Missouri River fish producing as many as 170,000 eggs (Keenlyne et al. 1992). In the southern part of its range, body size is smaller and fecundity ranges from 43,000 to 58,000 eggs (George et al. 2012). The acceptable temperature range for incubating Pallid Sturgeon eggs was 12–24°C, with an optimal range of temperature for embryo survival of 17–18°C (Kappenman et al. 2013). There is extreme latitudinal variation in life history characteristics of Pallid Sturgeon. In the upper Missouri River, Pallid Sturgeon are larger, slower to mature, more fecund, but longer lived than those in the Lower Mississippi River (George et al. 2012). The oldest individual caught by Killgore et al. (2007) in the LMR was 21 years, but life spans of 41 years have been reported in the upper Missouri River. Potential maximum age is probably greater than 41 years.

The high proportion of intermediate morphological characters in Mississippi River specimens of *S. albus* and *S. platorynchus* has been attributed to hybridization (Carlson et al. 1985; Keenlyne et al. 1994; Hrabik et al. 2007; Schrey and Heist 2007). Concern over misidentification of *Scaphirhynchus* specimens in the Mississippi River as hybrids is supported by Murphy et al. (2007b), who compared the identifications of 41 lower Mississippi River *S. platorynchus* and *S. albus* using principal component analysis (PCA) with a character index relying on anatomical ratios only (mCI) and a character index using both anatomical ratios and meristic characters (CI). The PCA identified two morphological intermediates, that is, putative hybrids; whereas the CI and mCI identified 37% and 73%, respectively, of the specimens as hybrids (Murphy et al. 2007b). Unfortunately, the occurrence, extent, and significance of hybridization in *Scaphirhynchus* spp. populations have yet to be fully resolved (USFWS 2007; Long and Nealis 2011).

Historical records of larval and small (smaller than approximately 300 mm FL) *Scaphirhynchus* spp. from the Mississippi River are rare, as are records of juvenile (<700 mm) *S. albus* (Hartfield et al. 2013). However, records of young-of-the-year (YOY) *Scaphirhynchus* spp. collected in the Mississippi River are increasing (Hrabik et al. 2007; Phelps et al. 2010b) in part due to the improvements of the trawls designed to catch small fishes in large rivers. It appears *S. albus is* naturally recruiting in the LMR (Herrala et al. 2014), and abundance of YOY sturgeon in that river appears to be regulated by river stage, with longer durations of high water promoting higher levels of abundance (Phelps et al. 2010a). Identification of *Scaphirhynchus* spp. <300 mm FL has proven difficult (Bailey and Cross 1954; Kuhajda et al. 2007) and is further complicated by a protracted larval development period and three larval stages. Protolarva is the early life stage from hatching to development of the first median fin rays at 21–26 mm total length (TL), mesolarva is the morphological stage from the appearance of first median fin rays through the appearance of last caudal fin rays at 50–60 mm TL in *S. platorynchus* and >81 mm TL in *S. albus*, and metalarva is the stage from the appearance of the last caudal fin rays through complete disappearance of the preanal fin fold at about 200 mm TL. (Snyder 2002).

The Pallid Sturgeon is known to migrate long distances. In 1994, a commercial fisherman from Marianna, Arkansas, caught a tagged specimen (PS4964) of the Pallid Sturgeon in the St. Francis River in eastern Arkansas that turned out to be one of the first group of Pallid Sturgeon artificially propagated and released in Missouri as part of the Missouri Department of Conservation Pallid Sturgeon Recovery Program (Kim Graham, pers. comm. to Steve Filipek). This sturgeon individual was originally released with a group of about 1,000 tagged individuals near New Madrid, Missouri, on 8 March 1994. It was caught three months later in the St. Francis River, a distance of about 216 miles (347 km) from the point of release. The sturgeon specimen was released 5 miles upstream in the St. Francis River. This represents the longest migration known to date of this species.

CONSERVATION STATUS   The Pallid Sturgeon was federally listed as an endangered species throughout its range in the United States in 1990 (Federal Register 1990). A recovery plan was approved in 1993 (Dryer and Sandvol 1993) and revised in 2014 (USFWS). The revised plan recognizes the primary threats to the Pallid Sturgeon as habitat loss and modification, altered flow regimes, entrainment, contaminants, nonnative species, low population size, recruitment failure, and hybridization with the Shovelnose Sturgeon (Jordan et al. 2016). Extensive habitat modifications within the Mississippi River have presumably interfered with the reproductive isolating mechanisms between the endangered *S. albus* and the sympatric *S. platorynchus* (Kuhajda et al. 2007). Destruction and alteration of spawning habitats have contributed greatly to the demise of the Pallid Sturgeon. In the upper Missouri River, Pallid Sturgeon have not successfully recruited in decades, and perpetuation of this species in that area is accomplished through captive spawning of wild-caught fish with subsequent release of their offspring (Saltzgiver et al. 2012). Because of the paucity of records in Arkansas, this species has long been considered rare and endangered (Robison 1974b; Buchanan 1974). Forbes and Richardson (1905) reported that only one in 500 *Scaphirhynchus* from the Mississippi River, Illinois, was *S. albus*. Carlson et al. (1985) found *S. albus* comprised 2.5% of sturgeons collected in the Mississippi River in Missouri. Etnier and Starnes (1993) reported anecdotal evidence from a commercial fisherman who estimated that about one in five *Scaphirhynchus* in his catches in the Mississippi River in Tennessee was *S. albus*. Killgore et al. (2007b) indicated that latitudinal variation in the Pallid Sturgeon to Shovelnose Sturgeon ratio varied from 1:6 to 1:77 in the Mississippi River. We continue to recognize the Pallid Sturgeon as endangered in Arkansas.

Scaphirhynchus platorynchus (Rafinesque)
Shovelnose Sturgeon

CHARACTERS   A small sturgeon with an elongate body and a broad, flattened head with four strongly fringed barbels on the underside of the snout. Bases of outer barbels are in line with or ahead of inner barbels; length of inner

Figure 9.7. *Scaphirhynchus platorynchus*, Shovelnose Sturgeon. *Uland Thomas.*

Map 8. Distribution of *Scaphirhynchus platorynchus*, Shovelnose Sturgeon, 1988–2019.

barbels much more than half the length of the outer barbels (Fig. 9.8). Snout long and spade-shaped. Spiracle absent. Upper lip with four lobes projecting anteriorly; lower lip with four papillose lobes projecting posteriorly (Fig. 9.1C). Eye small, its diameter going 16–35 (x̄ = 22) times into head length. Body with longitudinal rows of bony plates (Fig. 9.9) that converge on the completely armored, slender caudal peduncle. Dorsal plates 13–19; plates posterior to anal fin base 7–11; lateral plates anterior to dorsal fin origin 23–31. Isolated denticles 0.4–2.0 mm in greatest length occur in the skin between dorsal and lateral scutes, with fewer and smaller denticles occurring between lateral and ventral scutes. Belly between the lateral plates is covered with rhomboid plates 1.5–2.0 mm (Weisel 1978). Dorsal rays usually 31–34 (29–36); anal rays usually 19–21 (18–23). Caudal fin strongly heterocercal and produced into a long filament that is often broken off in adults. This filament is

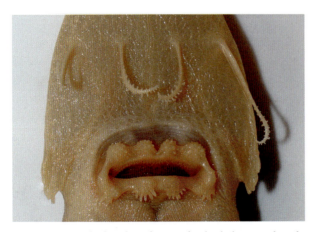

Figure 9.8. *Scaphirhynchus platorynchus* barbels on underside of head. *Renn Tumlison.*

Figure 9.9. *Scaphirhynchus platorynchus* showing dorsal, lateral, and ventral scutes. *Renn Tumlison.*

supported at its base by the notochord and contains vertebrae, nerves, and lateral line canals and pores, suggesting that it functions as a sense organ in young fish and is used in flow orientation (Weisel 1978). Maximum length is about 43 inches (1,080 mm) (Page and Burr 2011), and maximum weight about 13.7 pounds (6.2 kg). Most individuals rarely reach 30 inches (762 mm) TL. The Arkansas state angling record is 5 pounds, 0 ounces (2.3 kg) from the Spring River in 2008.

LIFE COLORS Dorsal and lateral surfaces olive or brownish, ranging to gray (Fig. 9.6). Ventral area white to cream-colored. No bold markings on body or fins. Many adults are strongly bicolored.

SIMILAR SPECIES *Scaphirhynchus platorynchus* differs from *S. albus* in having a proportionately smaller head, narrower mouth, inner barbels more than half the length of outer barbels (vs. inner barbels less than half the length of outer barbels), bases of outer barbels even with or anterior to bases of inner barbels (vs. bases of outer barbels posterior to bases of inner barbels), dorsal rays usually 30–36 (vs.

usually 37–43), anal rays usually 18–23 (vs. usually 24–29), belly mostly scaled (vs. mostly naked), papillae on leading edge of barbels are complex and branched (vs. simple and unbranched), papillae on lip lobes long, thick, and branched (vs. papillae simple and unbranched to almost absent), gill rakers fanlike and malleable (vs. gill rakers peg-like and stiff), and smaller in maximum size, not exceeding 6–7 kg (vs. adults often exceeding 10 kg). In addition, the lateral and ventrolateral plates are smaller in *S. albus*, causing the space between the plate rows to be larger than in *S. platorynchus* (Kuhajda et al. 2007; Murphy et al. 2007b). *Scaphirhynchus platorynchus* differs from *Acipenser* species in having no spiracle (vs. spiracle present), a broad, flattened, shovel-shaped snout (vs. a pointed, short, conical snout), feathery rostral barbels (vs. smooth), four papillose lobes on the lower lip (vs. two), and in having a long flattened caudal peduncle completely covered with bony plates (vs. a short, rounded peduncle only partially covered by bony plates).

VARIATION AND TAXONOMY *Scaphirhynchus platorynchus* was described by Rafinesque (1820a) as *Acipenser platorynchus*. No specific type locality was given, but Rafinesque listed "Ohio, Wabash, and Cumberland rivers, seldom reaching as high as Pittsburgh." Bailey and Cross (1954) provided taxonomic comparisons of this species and the closely related *S. albus*. Williams and Clemmer (1991), Mayden and Kuhajda (1996), and Kuhajda (2002) studied geographic variation in this species. Hybridization between *S. platorynchus* and *S. albus* may occur in the Missouri and Mississippi rivers (Carlson et al. 1985), and genetic data have both supported (Schrey and Heist 2007; Schrey et al. 2007, 2011) and failed to support hybridization (Ray et al. 2007). There is evidence that the White River population of the Shovelnose Sturgeon may be genetically and morphologically unique (Jeff Quinn, pers. comm.). Further study is needed to determine the taxonomic status of the White River form. The Shovelnose Sturgeon is apparently more closely related to *S. suttkusi* than either of those species is to *S. albus* (Mayden and Kuhajda 1996; Krieger et al. 2008).

DISTRIBUTION *Scaphirhynchus platorynchus* is known from large rivers of the Mississippi River basin from western Pennsylvania to Montana, and south to Louisiana. It also formerly occurred in the upper Rio Grande of New Mexico. In Arkansas, the Shovelnose Sturgeon generally occupies only the larger rivers of the state (Map 8). It is abundant in the Mississippi River and the lower White River, and modest populations exist in the St. Francis and Red rivers.

Arkansas River populations are sporadically distributed and not very large (Jansen 2012) and may have declined since 1988. It is conspicuously absent from the Ouachita

River in Arkansas but occurs in that river in Louisiana. There are about 1,990 records for this species from the Mississippi and lower White rivers taken between 2001 and 2014 by U.S. Army Engineer Research and Development Center biologists (data provided by K. J. Killgore and Catherine Murphy). AGFC biologists collected more than 679 Shovelnose Sturgeon from the White River for their stock assessment, with 240 individuals caught upstream of Newport, Arkansas (Jeff Quinn, pers. comm.).

HABITAT AND BIOLOGY   The Shovelnose Sturgeon occupies shallow areas and deep channels of larger rivers, inhabiting sandbars or stable substrates in areas of strong current over sand and gravel bottoms. Keenlyne (1997) and Curtis et al. (1997) found this sturgeon in main channels or in long main channel borders, in pools downstream from sandbars, and in association with wing dams. In the Mississippi River, sand was the predominant substrate when fish were located (Curtis et al. 1997). Quist et al. (1999) suggested that bottom water velocity is the most important factor in Shovelnose Sturgeon habitat selection. Hurley et al. (1987) found that *S. platorynchus* frequented areas of moderate to fast current in waters 2–7 m deep and was relatively sedentary. It moves occasionally, as much as 12 km in 1 day, primarily in May and July, with some homing behavior indicated. It is tolerant of high turbidity. In the Mississippi River, it is most common in the tailwaters below wing dams and other structures that accelerate the water flow (Becker 1983). Optimal habitats for age 0 *Scaphirhynchus* spp. in the MMR were island tips upstream of the main channel and channel border areas behind wing dikes (Sechler et al. 2012b).

This sturgeon is a benthic invertivore, raking the bottom with sensitive barbels to feed opportunistically on a variety of aquatic insects (Diptera, Trichoptera, Ephemeroptera), terrestrial insects, mollusks, crustaceans, and worms (Held 1969; Modde and Schmulbach 1977; Lee 1980b; Becker 1983; Carlson et al. 1985; Hoover et al. 2007). Trautman (1981) reported that in Ohio it fed over clean sand and gravel bottoms of chutes and bars, or wherever there was considerable current and a clean bottom; it seemed to congregate wherever there were large quantities of small clams and snails. Braaten and Fuller (2007) studied diet composition of larval and young-of-the-year fish in the upper Missouri River and reported that fish begin exogenous feeding by 16 mm (0.62 in.) TL, and individuals 16–140 mm (0.62–5.5 in.) fed exclusively on Diptera and Ephemeroptera. In the middle Mississippi River (MMR), this sturgeon was an opportunistic feeder with most of the diet consisting of immature benthic insects (Seibert et al. 2011). Dominant prey items of adults throughout all seasons were

Chironomidae, Hydropsychidae, Ephemeridae, and an exotic amphipod (Corophiidae). In the lower Mississippi River, adults fed comparably on prey from hard substrates (hydropsychid caddisworms) and soft substrates (burrowing mayflies, amphipods, larval dipterans) (Hoover et al. 2007). Sechler et al. (2012a) studied the effects of river stage height and water temperature on diet composition of age 0 *Scaphirhynchus* spp. in the MMR. River stage height varied among years and seasons, and water temperature varied among seasons but not among years. Macroinvertebrate taxa (primarily ephemeropterans, dipteran pupae, and chironomids) in the diet differed among size classes of sturgeon, but not seasons or years. Enhancement of areas with flow and substrates that facilitate the production and availability of chironomids and ephemeropterans is critical for the recruitment of age 0 *Scaphirhynchus* sturgeon in large rivers (Sechler et al. 2012b).

Shovelnose Sturgeons spawn every 2 or 3 years in tributary streams or along the borders of main river channels over hard bottoms (Keenlyne 1997). They are nonguarders, open substrate spawners, and lithopelagophils—rock and gravel spawners with pelagic embryos (Simon 1999b). Adults migrate upstream from April to early July to spawn over rocky substrates in channels of large rivers at water temperatures of 67–70°F (19.5–21.1°C). Coker (1930) reported runs of this sturgeon in the Mississippi River to be best when the river is low in spring and poor when it is high. Frequency of spawning is influenced by available food supply and the ability to store adequate fat to produce gametes (Keenlyne 1997). *Scaphirhynchus platorynchus* produces sounds associated with the breeding season (Johnston and Phillips 2003), but the exact context of the sound production is unknown. These sounds may be used in aggregating individuals on the spawning ground. Mean fecundity estimates for gravid females in Missouri, Illinois, and Indiana populations ranged from 18,000 to 30,000 or 15–22 oocytes per gram of fish with positive relationships with fork length and weight (Stahl 2008; Stahl et al. 2009). Mean oocyte diameter is 2.58 mm. In the LMR, winter endoscopy data suggest that males mature around age 7 and 563 mm FL, females at age 8 and 474–586 mm FL (Divers et al. 2009, 2013). In the middle Mississippi River, minimum age at first maturation was 8 years for males and 9 years for females (Tripp et al. 2009). Inter-gender individuals have been documented (Divers et al. 2009). This condition has been linked by other researchers to legacy pesticides and contaminants from birth control pills. Larval fish drift in the water column for up to 12 days after hatching before settling out of the water column to begin the benthic phase of life (Kynard et al. 2002; Braaten and

Fuller 2007). Allen et al. (2007) investigated the influence of substrate type, water depth, light, and relative water velocity on microhabitat selection in juvenile Shovelnose Sturgeon in an artificial stream system and reported on overall selection for sandy, deep, or heavily shaded habitats.

The construction of dams on major rivers throughout its range has contributed to the decline of this species by blocking access to traditional spawning areas and by the elimination of its required lotic environment. Sprague (1960) documented the decline in catch and condition of this sturgeon over a five-year period in the Missouri River after the impoundment of Lewis and Clark Lake. While there has been a general decline in population levels of *S. platorynchus* in Arkansas during this century, it is still fairly common in the state's big rivers. In the early 1980s, its population in the White River near Augusta was large enough to support the establishment of a modest sturgeon roe fishery. From 1980 to 1985, an average of 31,983 pounds (14,507 kg) of sturgeon per year was taken commercially in Arkansas. More recent commercial catch data are not available.

CONSERVATION STATUS Shovelnose and Pallid sturgeons are closely related and ontogenetic and latitudinal variation are substantial and can confound identifications even by experienced biologists (Murphy et al. 2007b). The USFWS was prompted to list the Shovelnose Sturgeon as threatened (to prevent accidental commercial harvest of *S. albus*) in 2010 because of its similar appearance to Pallid Sturgeon (USFWS 2010). This listing applies only to environments where the two species coexist. In Arkansas, this is mainly in the Mississippi River proper, although there are records for *S. albus* in the St. Francis River and lower White River. Because populations may have decreased in recent years, we consider the Shovelnose Sturgeon a species of special concern.

Family Polyodontidae *Paddlefishes*

The American Paddlefish, *Polyodon spathula*, of the Mississippi River system and the Chinese Paddlefish, *Psephurus gladius*, of the Yangtze River in eastern Asia are the only living species in this family and together form subfamily Polyodontinae. It is possible that the Chinese Paddlefish is extinct (Phelps et al. 2016). Adult specimens of *P. gladius* were last documented in 2002 (1 adult female) and 2003 (1 adult female), and an intensive search of the Yangtze River from 2006 to 2008 failed to find any specimens (Pough et al. 2013). *Polyodon spathula* was introduced into China in 1988 in an attempt to develop an aquaculture

industry (Hrabik et al. 2015). This primitive bony fish family dates back to the Cretaceous (75–70 mya). The Late Cretaceous fossils indicate that the family was formerly Holarctic (Berra 1981). More recent Eocene fossils (56–34 mya) are also known.

Genus *Polyodon* Lacépède, 1797— North American Paddlefishes

Lacépède established the genus *Polyodon* in 1797, thereby removing the Paddlefish from the shark genus *Squalus*. Its sharklike appearance is somewhat intimidating even though it is a harmless plankton feeder. Its closest living relative, the Chinese Paddlefish, has a swordlike snout and is piscivorous. Because of the largely naked body, many persons mistake it for a catfish, and in Arkansas it is often called a spoonbill catfish. The genus characters are those presented in the species account below.

Its abundance in Arkansas may have decreased during the 20th century, but commercial fishermen still regularly net it from large rivers. It currently constitutes an important part of the commercial catch. Pflieger (1997) reported that near the turn of the 20th century the Paddlefish ranked as one of the most important commercial fishes in the Mississippi Valley with a total harvest of nearly 2.5 million pounds (1.134 million kg) in 1899. Concerns about population sustainability and declines were documented shortly thereafter throughout its range (Stockard 1907; Wagner 1908; Alexander 1914). An excellent food fish with few bones, it is also a source of caviar, a delicacy created by preserving the eggs in special salts. Typically, the largest female Paddlefish can yield about 20 pounds of eggs or more. At a recent retail price of $35 an ounce, it is easy to see the value of this resource. *Polyodon spathula* was introduced into Europe in 1974, presumably to replace declining sturgeon populations as a source of caviar (Lenhardt et al. 2011; Hrabik et al. 2015). In the early 1980s a modest Paddlefish roe fishery existed on the lower White River of Arkansas. The roe fishery has expanded substantially in recent years in the Arkansas, Mississippi, and White rivers. Most of the U.S. commercial harvest of Paddlefish in the early 2000s was from Arkansas, Kentucky, and Tennessee (Quinn 2009). Quinn also reported that the annual Paddlefish harvest from the Arkansas River alone was 37,000 kg, and Arkansas was one of only 6 states in 2006 allowing commercial harvest. Two of the six states allowing harvest (Alabama and Mississippi) have developed more stringently regulated Paddlefish fisheries since 2006, resulting in an increase in Paddlefish harvest in Arkansas

(Jeff Quinn, pers. comm.). Controlled commercial harvest in Pool 13 and Ozark Pool of the Arkansas River in 2003, 2004, and 2006 produced exploitation estimates between 19% and 40% during the 5- to-10-day seasons (Quinn et al. 2009). There is also a sport fishery for Paddlefish in Arkansas. Although this species will not take bait, it is frequently snagged by fishermen when the rivers rise (usually in late winter through spring) below the 11 large navigation dams on the Arkansas River, below Beaver Dam on the White River, the Batesville Dam on the White River, Lake Sequoyah Dam on the White River, below Millwood Dam on the Little River, and below three dams on the Ouachita River (Hansen and Paukert 2009). Mims et al. (2009) summarized the known information on propagation and culture of Paddlefish.

Polyodon spathula (Walbaum)
Paddlefish

Figure 9.10. *Polyodon spathula*, Paddlefish. *David A. Neely.*

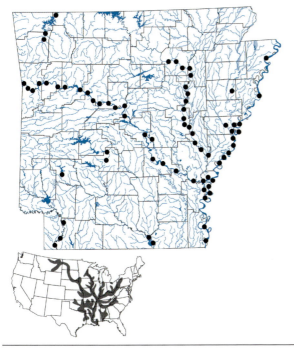

Map 9. Distribution of *Polyodon spathula*, Paddlefish, 1988–2019.

CHARACTERS   A large fish with an extremely long, paddle-shaped snout that is about one-third the body length in small individuals (Fig. 9.10). The paddle or rostrum varies in shape from elliptical to spatulate or even tapered, but it is rhomboid in cross-section with a four-sided endoskeletal mesh comprising small interconnected stellate bones (J. J. Hoover, pers. comm.). Caudal fin strongly heterocercal, especially in juveniles, the vertebral column extending far into its upper lobe. The tip of the upper caudal lobe may include a notch similar to those seen in some sharks; the epicaudal notch is clearly visible in most juveniles and is often present in adults (J. J. Hoover, pers. comm.). Mouth very large (Fig. 9.11), opening straight forward beneath the snout. Jaws apparently toothless (except in young), but teeth may be present in fish up to 20 inches, after which, they become edentulous (Adams 1942). Some adults have tiny embedded teeth that may not be easily seen but can be felt by gently rubbing a probe along the surface of the jaw; the teeth can sometimes be seen when the jaws are dried. Eyes small but well developed (Fig. 9.12).

Two minute barbels (3–4 mm) present on underside of snout in front of the mouth. Body mostly naked except for a patch of rhomboid scales along the upper lobe of the caudal fin, lunate-shaped fulcra on the caudal peduncle, denticles on pectoral girdle, and minute scales buried in the skin of the back and sides (Weisel 1975). Rear margin of

Figure 9.11. Paddlefish with mouth open to show large size and gill arches. *Jan Jeffrey Hoover, USACE.*

Figure 9.12. Eye and spiracle of *Polyodon spathula*.
Jan Jeffrey Hoover, USACE.

operculum produced into a long, pointed flap. Gill membranes broadly joined across throat. Gill rakers very long, slender, and numerous. A small spiracle is present and is usually easily seen. Intestine with a spiral valve that has 6–7 turns, with the last turn more than 2 cm from the vent. Skeleton almost entirely cartilaginous. There is evidence that the Paddlefish once reached sizes greater than 6.6 feet (2 m) TL and 154 pounds (70 kg) (Phelps et al. 2016). Most adults taken in Arkansas weigh less than 44 pounds (20 kg) and are smaller than 4 feet (121.9 cm) in length; average mass ranges from 32 to 35.1 pounds (14.5–15.9 kg) (Quinn 2009). The most recent Arkansas angling record from Beaver Lake in 2015 is 105 pounds (47.6 kg).

LIFE COLORS   Body usually dark bluish gray or slate gray except for the underside, which is white or silvery. Young-of-year fish from turbid water may be translucent or pink. Aquacultured fish raised indoors may be completely black. Operculum and rostrum with dark blue-black spots (ampullae). Head with dark star-shaped spots.

SIMILAR SPECIES   The long paddlelike snout distinguishes it from all other Arkansas fishes. It is most closely related to the members of the sturgeon family and can further be distinguished from sturgeons by the lack of bony plates on the body (vs. bony plates present). The nearly naked body further distinguishes the Paddlefish from gars which have a completely scaled body.

VARIATION AND TAXONOMY   Walbaum (1792) described the Paddlefish as a new shark species, *Squalus spathula*, due in part to its heterocercal tail and largely cartilaginous skeleton. Lacépède (1797) removed it from the shark genus and established the genus *Polyodon*. Early naturalists Rafinesque, Lesueur, and Kirtland believed there were multiple species of paddlefish, in part because of variation in rostrum size, shape, and dentition. The

prevailing modern view is that this variation is predominantly ontogenetic and that genus *Polyodon* is monospecific. Carlson et al. (1982) studied genetic variation in the Paddlefish and concluded that there were no major differences in populations. More recent data do not support this conclusion. Epifanio et al. (1996) provided evidence for genetically differentiated populations in the Mobile Basin, and possibly for the White and Arkansas river populations as well. In addition, significant differences in morphology of juvenile Paddlefish from geographically isolated basins are associated with hydrologic regimes (Hoover et al. 2009b). A study of five polymorphic DNA microsatellite loci throughout the species range by Heist and Mustapha (2008) found that nearly all pairwise estimates of genetic heterogeneity among geographic samples were significant, as was the overall test of genetic heterogeneity. Those authors also interpreted the distribution of genetic variation in Paddlefish as indicative of historically high levels of gene flow among populations that have more recently become isolated by habitat alteration (e.g., dams). Schwemm et al. (2014b) described 13 diploid microsatellite markers for the Paddlefish and found that a Neosho River, Oklahoma population had 11 loci with 2–6 alleles per locus.

All Acipenseriformes have a large number of chromosomes, and the Paddlefish has 120. A possible tetraploid origin for the species has been supported (Dingerkus and Howell 1976). Shelton and Mims (2012) found that female sex determination is heterogametic rather than the previously reported homogamety.

DISTRIBUTION   The range of *Polyodon spathula* is the Mississippi River basin from southwestern New York to Montana and south to Louisiana, and Gulf Slope drainages from Mobile Bay, Alabama to Galveston Bay, Texas. It was formerly known from the Great Lakes basin but has probably been extirpated there (Page and Burr 2011). In Arkansas, the Paddlefish is known from the Red, Ouachita, Arkansas, White, St. Francis, and Mississippi rivers and their major tributaries in the state, and it has been stocked in Beaver Lake in northwestern Arkansas (Map 9).

HABITAT AND BIOLOGY   Formerly much more widespread and common in Arkansas than it is today, the Paddlefish has declined in abundance as a result of extensive damming of streams, channelization, draining of wetlands, and other habitat alterations. Its natural habitats are the large, low-gradient rivers where adults often occupy backwater areas having rich growths of plankton. In Ozark Pool of the Arkansas River, Paddlefish avoided shallow, backwater habitats and preferred deeper, impounded creek channel habitats (Donabauer et al. 2009). It is sometimes

Figure 9.13. Adult Paddlefish. *Uland Thomas.*

Figure 9.14. Juvenile Paddlefish. *Konrad Schmidt.*

abundant in reservoirs that provide favorable feeding habitat but often little or no reproductive habitat. Many reservoir populations must be maintained by stocking. Adult Paddlefish (Fig. 9.13) prefer to feed in areas of reduced velocity.

Ideal conditions are created by constructed wing dams and naturally occurring sandbars which create eddies and scour holes with reduced current (Southall and Hubert 1984; Moen et al. 1992). In Ozark Pool of the Arkansas River Navigation System, Paddlefish selected tributary mouth habitat in all seasons, but much movement to tailwater habitat occurred in the spring (Donabauer et al. 2009). Limited movement occurred between navigation pools of the Arkansas River (Leone 2010). Population characteristics varied considerably among three adjacent navigation pools of the Arkansas River (Leone et al. 2012). Paddlefish from the most lentic and heavily fished pool, Lake Dardanelle, had the lowest catch rate, grew fastest, and had the highest mean condition factor, weight, fecundity, and mortality (67%). Paddlefish from the most lotic and least intensively fished impoundment, Pool 13 near Fort Smith, had the highest catch rate, slowest growth, and the lowest mean condition, weight, fecundity, and total annual mortality (53%) (Leone et al. 2012). Paddlefish are occasionally found in lowland oxbow lakes that are periodically flooded by adjacent large rivers. Clark-Kolaks et al. (2009) studied 11 floodplain lakes in the lower White River and concluded that Paddlefish were more likely to be found in long narrow floodplain lakes that connected to the river early in the year. Those authors also found that Paddlefish catch per unit effort increased as lake surface area, dissolved oxygen level, and variability in the start date of connection increased. Both juvenile and adult Paddlefish used White River floodplain lakes. Today most of the oxbow lakes used by Paddlefish in Arkansas are found along the Mississippi and White rivers. Many formerly flooded oxbows, such as those along the lower St. Francis River, are no longer seasonally connected to a river because of the ditching of those rivers. A Paddlefish population requires many habitats to sustain the population over time (Gerken

and Paukert 2009), and each stage of life (feeding, overwintering, spawning of adults, development of eggs and larvae, and the growth of juveniles) requires a distinct habitat. A variety of habitats are occupied within large rivers, with Paddlefish being found in the main channel, main channel borders, secondary channels, and backwater areas depending on life stage (Zigler et al. 1999, 2003). Slow-flowing water is required for Paddlefish growth and development, but spawning requires habitat with fast-flowing water and gravel substrates (Rosen et al. 1982). The distance between these suitable habitats often necessitates long-distance movements by Paddlefish (Wilson and McKinley 2004; Firehammer and Scarnecchia 2006), and Paddlefish travel extensively within river systems (Moen et al. 1992; Zigler et al. 1999; Jennings and Zigler 2000, 2009). Juvenile Paddlefish (Fig. 9.14) smaller than 155 mm eye-to-fork length were positively rheotactic and exhibited sustained swimming at water velocities up to 40 cm/s (Hoover et al. 2009a). Paddlefish are mobile throughout the year, and extensive movements are not exclusively related to spawning migrations (Zigler et al. 1999). An individual tagged in Keystone Lake, Oklahoma, in 1998 was recaptured in 2016 below Eufaula Dam, Oklahoma (Long et al. 2017). It had traveled downstream approximately 235 km, passing through three dams before moving upstream to Eufaula Dam. Stell et al. (2018) documented even longer movements. Four Paddlefish that had been tagged in Mhoon Lake, Mississippi, a remote oxbow lake, were recaptured after 8–24 months in the Missouri, Mississippi, and Ohio rivers, having traveled 1,408–2,433 km through multiple waterways.

Hintz et al. (2017) noted that in contrast to many other fishes, Paddlefish begin life as particulate feeders and later become filter feeders. The earliest stages of Paddlefish appear to actively feed on individual prey items such as zooplankton and insects until they reach about 120–250 mm TL when the gill rakers develop enough to be used as a filter (Ruelle and Hudson 1977; Rosen and Hales 1981; Michaletz et al. 1982). Young-of-the-year Paddlefish in Lewis and Clark Lake, South Dakota, fed mainly at night in open water near the surface on zooplankton and aquatic and terrestrial insects (Ruelle and Hudson 1977). In contrast, in the lotic environment of the Middle Mississippi River, age 0 Paddlefish foraged primarily on benthic macroinvertebrates, mainly trichopterans, hemipterans, and amphipods (Hintz et al. 2017). Larger juvenile Paddlefish use electroreception for locating individual zooplankton prey items (Miller 2004; Wilkens and Hofmann 2007). Juveniles may actively select larger species such as *Daphnia* when they are preying on individual zooplankton (Michaletz et al. 1982). In laboratory experiments where young Paddlefish were provided a mixture of organisms, age 0 individuals consumed primarily macroinvertebrates, while age 1 Paddlefish filtered mainly zooplankton (Hintz et al. 2017). After transition to filter feeding, Paddlefish presumably forage exclusively by filtering water through their gills (Miller 2004). Some variation in adult feeding methods and diets have been reported. Larger Paddlefish are primarily ram suspension filter feeders (Sanderson et al. 1994). The Paddlefish swims with its mouth open, straining mostly zooplankton and occasionally small insects, insect larvae, and small fish (Ruelle and Hudson 1977; Rosen and Hales 1980; George et al. 1997). In the Missouri River (Rosen and Hales 1981), in Kansas (Cross and Collins 1995), and in Louisiana (N. A. Smith et al. 2009), adult Paddlefish fed primarily on crustacean zooplankton (mainly copepods and cladocerans). Some studies indicate that they are indiscriminate filter feeders, ingesting all material greater than 0.25 mm long strained from the water column. In one study, detritus and sand made up more than 50 percent of the stomach contents by volume (Rosen and Hales 1981). In polyculture ponds, Paddlefish were preyed upon extensively by large Channel Catfish (Tidwell and Mims 1990). In a natural environment, large Blue Catfish would be more likely predators than Channel Catfish. In the White River, Paddlefish are found frequently with lamprey wounds and scars (J. J. Hoover, pers. comm.).

The long, paddlelike snout is complex in structure and lightweight (Fig. 9.15) and is not used for rooting in the bottom mud as some early naturalists and many fishermen believed.

The Paddlefish is well supplied with sense organs and nerves (Nachtrieb 1910; Norris 1923; Weisel 1975), and some have speculated that they may aid the fish in locating concentrations of plankton. It is now known that the Paddlefish snout is covered with thousands of tiny electroreceptors (visible to the naked eye as dark gray spots) (estimated by Nachtrieb in 1910 to be between 50,000 and 70,000, but reported by Phelps et al. 2016 as "more than

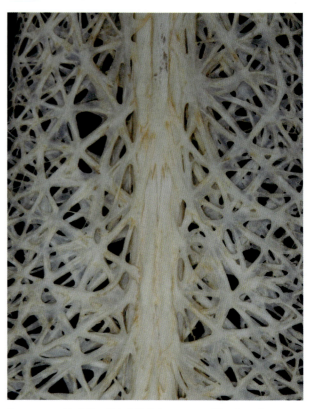

Figure 9.15. Paddle of Paddlefish supported by dorsal and ventral sheets of numerous and densely packed interdigitating stellate bones. *Jan Jeffrey Hoover, USACE.*

40,000"), similar to the ampullae of Lorenzini of sharks (Jorgensen et al. 1972; Wilkens et al. 2002). It uses the electroreceptors to detect patches of zooplankton by sensing electrical fields (Russell et al. 1999; Wilkens and Hofmann 2007). This allows feeding at night and in turbid waters. The electroreceptive rostrum also enables detection of metal objects which the Paddlefish avoids (Gurgens et al. 2000). Hoover et al. (2009b) also believed that the rostrum conferred hydrodynamic advantages of increased lift with reduced drag. Despite the documented role of the rostrum as an important sense organ, adult Paddlefish can survive with damaged, anomalous, or even missing rostra, feeding on the same foods as other Paddlefish (Gannon and Howmiller 1973). However, such individuals may be less robust than Paddlefish with intact rostra (Rosen and Hales 1980). In the lower Mississippi River, fish with fractured, incomplete, deflected, or bent rostra have been observed. Jeff Quinn (pers. comm.) also reports routinely finding many Paddlefish with missing paddles and observing many fish with damaged paddles as a result of passing through dam turbines.

While the eye is not large relative to its body size, it is well developed (Sillman and Dahlin 2004). The retina contains both rods and cones, so it can see under dim light situations and bright light conditions. The eye is devoid of a fovea centralis, the region specialized for more acute vision. There is a large and prominent pineal photoreceptor dorsally between the two eyes. This "third eye" lacks a lens but is pigmented and occupies a cavity just below the surface of translucent cartilage of the rostrum (Garman 1898).

Prior to spawning, Paddlefish (mostly but not exclusively males) develop tubercles on the dorsal surface of the rostrum, head, and body. Stockard (1907) reported that Paddlefish begin to spawn in Arkansas in mid-April. Donabauer et al. (2009) found that spawning occurred from late March to early April in Ozark Pool of the Arkansas River in the tailwaters of Trimble Lock and Dam. Gravid females migrated a median distance of 33 km in the spring of 2004 and 32 km in the spring of 2005 when water temperatures reached 10°C, and spawning occurred at water temperatures between 13°C and 19°C (Donabauer 2007). Purkett (1961) described Paddlefish reproductive biology in the Osage River, Missouri, where spawning occurred in early spring when the river was high and muddy. Spawning occurs in Alabama from March to mid-May (Boschung and Mayden 2004). The adults move upstream into swift currents over gravel bars when the water temperature reaches about 60°F (15.7°C). Spawning takes place in midstream over the gravel substrate, and the adhesive eggs stick to the first object they touch, normally stones on the stream bottom. The number of eggs/kg of body weight is a very stable metric across populations (Sharov et al. 2014). Large females may spawn more than 500,000 eggs in a season but may not spawn every year. In large reservoirs, such as Kentucky Lake, Tennessee, and Grand Lake, Oklahoma, spawning may occur annually. In the Arkansas River, Paddlefish appear to skip-spawn. Leone et al. (2012) found that both absolute and relative fecundity varied interannually in Pool 13 of the Arkansas River, and fecundities of fish of similar lengths from Ozark Pool and Pool 13 were lower than those of fish in Lake Dardanelle (probably due to food availability or reservoir productivity). Fecundity ranged between 15,000 and 16,000 eggs/kg of body weight in the Arkansas River (Leone et al. 2012). After hatching, the fry are swept downstream out of the shallows into deep pools. There is no parental care of eggs or larvae. The larval Paddlefish lacks the paddle of the adult, and the early juvenile stage has teeth in its jaws (Hogue et al. 1976). The paddle (rostrum) begins to develop rapidly about three weeks after hatching (Bemis and Grande 1992). The greatest abundance of larvae in the

Tennessee and Cumberland rivers of Tennessee occurred in April and May, respectively (Wallus 1986).

Paddlefish growth is highly variable among waters and regions (Sharov et al. 2014). In some areas of its range, Paddlefish grow about 50 inches (1,270 mm) during the first 10 years of life, after which the growth rate slows to about 2 inches (51 mm) a year. In Arkansas, they are more likely to grow to 36 inches in the first 10 years of life, and a 1,270 mm fish is more likely to be 16–20 years old. In the Arkansas River, females grew faster and attained a larger maximum size than males, and females reached maturity as early as 7 years of age (Leone et al. 2012). In Grand Lake, Oklahoma, male fish typically matured at age 6 or 7, and females matured at age 8 or 9; the five stages of the life span (immature, maturing, growth and reproduction, prime reproduction, senescence to death) are compressed into a period of 15–20 years, and the prime reproduction period occurs from ages 12 to 16 years for females (Scarnecchia et al. 2011). Despite a life span of 40–50 years reported for Yellowstone-Sakakawea Paddlefish in North Dakota and Montana (Scarnecchia et al. 2007), the life span in Arkansas appears to be similar to the life span in Oklahoma. Twenty-seven Paddlefish from lower White River oxbow lakes ranged in age from 3 to 19 years, with a mean age of 12 years (Clark-Kolaks et al. 2009). The maximum age found in three Arkansas River navigation pools was 16 years (Leone et al. 2012). In summary, Paddlefish life-span stages are more protracted and therefore more obvious in longer-lived northern stocks and more compressed and less obvious in shorter-lived southern stocks (Scarnecchia et al. 2011).

CONSERVATION STATUS   Paddlefish populations have apparently declined over the past 100 years throughout much of its range (K. Graham 1997), and it is considered extirpated in New York, Pennsylvania, Maryland, and North Carolina. The Paddlefish is currently considered vulnerable throughout its range by the American Fisheries Society and the IUCN, but the USFWS considers it secure in the United States. Declining and imperiled populations are all at the periphery of the range (Phelps et al. 2016). Bettoli et al. (2009) indicated that Paddlefish populations in most states within the Mississippi River basin were relatively stable. Currently stable populations of Paddlefish exist in the navigation pools of the Arkansas River and lower White River, and a commercial fishery exists for this species in those rivers as well as in the Mississippi River (Fig. 9.16).

A small percentage of the Paddlefish harvest also comes from the Ouachita, Little, Black, and St. Francis rivers

Figure 9.16. Paddlefish taken from the Mississippi River. *Jan Jeffrey Hoover, USACE.*

and from a few oxbow lakes. Paddlefish populations in Arkansas are vulnerable to overharvesting (Donabauer et al. 2009), and the AGFC has imposed strict regulations on this fishery. Despite multiple studies of Paddlefish biology, no estimates of sustainable and optimum exploitation rates have been developed (Sharov et al. 2014). Because of successful natural reproduction, stocking of Paddlefish has been discontinued in Arkansas in recent years (Grady and Elkington 2009). Beaver Lake was frequently stocked to reestablish fish to that area, but stocking has been discontinued even though natural reproduction has not been observed. Understanding how habitat alteration affects different life history stages is critical for the conservation and management of Paddlefish populations. Invasive species also pose a threat to Paddlefish populations. The most likely threat is from filter-feeding Bighead Carp (*Hypophthalmichthys nobilis*) and Silver Carp (*H. molitrix*) because they consume similar food resources and can also change the plankton community to one that cannot be as efficiently used by Paddlefish (Pegg et al. 2009). Arkansas currently has some of the most stable populations of this species, and we consider this species currently secure in Arkansas.

CHAPTER 10

Order Lepisosteiformes
Gars

This ancient neopterygian order was once widely distributed but today occupies a much smaller range. Fossils of this order are known in Europe from the Cretaceous to the Oligocene and in Africa and India from the Cretaceous (Wiley 1976). Cavender (1986) summarized the fossil record of lepisosteiform fishes in North America from the Cretaceous to the Recent. All extant species are placed in a single family, Lepisosteidae. Classification of the gars above the family level has been controversial for decades. Gars and fossil Semionotidae have been previously recognized in the same order, either Semionotiformes or Lepisosteiformes (Nelson 1994); however, Nelson (2006) and Nelson et al. (2016) subsequently placed them in separate orders based on the work of Grande and Bemis (1998). Wiley (1976) investigated the systematics and biogeography of fossil and recent gars and concluded that morphological data supported the monophyly of gars and their recognition as a sister group to a clade formed by Amiiformes and Teleostei. An alternate hypothesis wherein gars and amiids form a monophyletic group, sister to teleosts, was presented by Burr and Mayden (1992). Gars and Bowfins share several morphological features, such as a close correspondence between many caudal fin rays and their bony supports, and similarities in the track of ventral branches of the spinal nerve roots (features not shared with teleosts). On the other hand, the Bowfin shares some features with teleosts that are not seen in gars

(see Family Amiidae account), and we follow Nelson et al. (2016) in placing gars and bowfins in separate divisions (see our Appendix 5).

Family Lepisosteidae *Gars*

Lepisosteids have been present on the North American continent for approximately 100 million years (Wiley 1976; Wiley and Schultze 1984; Grande 2010) and represent a unique component of the extant fish fauna of North and Central America, and Cuba (Suttkus 1963; Lee et al. 1983). Gars are easily recognized as long, slender fishes with cylindrical bodies covered with rhomboid-shaped, plate-like ganoid scales. These scales articulate with one another via a peg-in-groove projection providing these heavily armored fish with sufficient flexibility for fast swimming. The snout is produced into a distinctly elongated beak with many sharp teeth. The head and snout are hard and bony. The dorsal and anal fins are set far back on the body, and the caudal fin is rounded and abbreviate-heterocercal. Internally, the intestine has a spiral valve. The vertebrae of gars are different from those of almost all other living fishes, having opisthocoelous centra (convex anteriorly and concave posteriorly) much like those of reptiles.

There are seven living species of this primitive bony fish family contained in two genera, *Lepisosteus* and *Atractosteus* (Nelson et al. 2016). Five of the species occur in the United

States, and four of those species are found in Arkansas. The phylogenetic relationships of the seven living gar species have been well resolved in the species tree inferred from mitochondrial and nuclear DNA sequences (Wright et al. 2012). The completely resolved phylogenies inferred from the molecular data strongly support previously published morphological phylogenies. The living forms are restricted to fresh and brackish waters in North and Central America. Occasionally gars are found in saltwater.

All gars are voracious predators that slowly stalk their prey or wait motionless until the prey comes close enough to seize. Scavenging behavior has been observed in some gars (Goodyear 1967). Although they feed primarily on other fishes, there are reports of predation on other vertebrate classes. Gars, using a lie-in-wait strategy, seize passing fish sideways with a sudden lunge, impaling them on their sharp teeth. The prey is then turned around, usually head-first, and swallowed whole. Gars generally spawn in mid to late spring in a similar pattern, but some interspecific and intraspecific variations in patterns have been reported. Large numbers of individuals congregate in shallow water during the breeding season, with each female accompanied by one to several males. During the spawning act, there is much thrashing in the water, and the large, adhesive eggs are scattered over the substrate and then abandoned. The newly hatched larvae have an adhesive structure on the undersurface of the end of the snout which they use to attach themselves to objects on or above the bottom (Suttkus 1963). The larvae, although capable of swimming, remain inactive until the yolk sac is absorbed, which may take up to three weeks.

Young gars possess a characteristic dorsal caudal filament at the end of the upturned vertebral column (Agassiz 1856; Wilder 1876; Carpenter 1995). Its large central core is an extension of the notochord, above which lies the dorsal nerve cord, lateral of which is a large muscle bundle, and below which are blood vessels and cartilage. The filament has prominent dorsal and ventral keels. Unlike the caudal filament of river sturgeons, which is thin, fragile, immotile, and presumably sensory (Weisel 1978), the caudal filament of gars is robust and motile and provides an important and distinctive form of locomotion for much of the early life history of the fish. This filament vibrates at a range of speeds. It is sometimes still, can be elevated or depressed, curved to one side or the other, and functions as a precise and sensitive propeller (Carpenter 1975). Movements of the filament are almost imperceptible and capable of driving the fish in almost any direction, including backwards, enabling the young gar to mimic the appearance of a slowly drifting twig. The filament appears to grow proportionately

until (some) gars are 150 mm TL or greater. It is subsequently reduced to a degenerate filament lying dorsally along the basal portion of the large caudal fin when the fish are 300 mm TL. At this time, it is only feebly and occasionally employed, and by the time fish are larger, it has disappeared (Wilder 1876).

Gars have less gill surface area than teleosts of comparable size (Landolt and Hill 1975) and, under experimental conditions, avoid low levels of dissolved oxygen (<4 mg/L) and preferentially occupy waters with moderate levels of oxygen (6–12 mg/L), especially in warm water (Hill et al. 1973). They can survive in waters that are very low in dissolved oxygen because of a large, vascularized, lunglike swim bladder connected to the esophagus. Louis Agassiz (1856) was one of the first authors to mention air gulping by gars. When oxygen levels are low, particularly in summer, they rise to the surface and gulp air to supplement gill respiration. With no access to atmospheric air, oxygen tensions greater than 3 mg/L are survivable, but oxygen levels below 2.5 mg/L are lethal (Renfro and Hill 1970). Rates of gulping air depend on physical and biological factors. Young gars (80–120 mm TL) take 20–30 breaths/hour at extreme hypoxia, 3–8 breaths/hour at normoxia, and fewer than 2 breaths/hour at high levels of dissolved oxygen (Hill et al. 1972). Young gars (86–360 mm TL) take 2–4 breaths/hour in warmer water (70 °F) and fewer than 0.5 breaths/hour in cooler water (60 °F) (Renfro and Hill 1970). Young gars are more likely to gulp air when in the presence of conspecifics (Hill 1972). Gulps by one fish (or the presence of a model) are likely to induce gulps by another gar in fewer than 2 seconds. Socially facilitated gulping by small gars may reduce predation by birds.

Historically, gars have had the reputation of being a pest to recreational anglers, a menace to fisheries, and poor table fare for consumers, so much so that efforts (some state agency-sanctioned) were made to reduce their numbers (Scarnecchia 1992). Those efforts included statutes making it illegal to return a gar alive to the water and elaborate (and expensive) methods of destruction (see Gowanloch 1933 for an example of an extensive diatribe against gars). Although seldom used as food in most parts of the United States, gars are harvested commercially in the deep South, and their flesh is firm and mild flavored and quite tasty (J. J. Hoover, pers. comm.). Potter (1923) reported that when soaked in brine overnight, baked gars are quite palatable. Many Arkansans know this, as Alligator Gar meat sold for more than $1.00 a pound in the early 1980s, when available, in eastern Arkansas fish markets. In recent years, gar meat has sold at retail for $3.50 to $6.00 per pound (Eric Brinkman, pers. comm.). In the 1970s and 1980s, gars formed an important part of the commercial fishery in Arkansas, ranking only behind the buffalofishes, catfishes, carp, and drum in pounds taken per year. From 1975 to 1985, an average of 802,130 pounds (363,846 kg) of gar per year were caught commercially in Arkansas (commercial fishing catches for gar have not been compiled in Arkansas since the late 1980s).

Gar eggs and embryos are toxic, but it is unclear to what extent this toxicity applies to humans. Several studies documented mortality of crayfish and rodents fed gar roe or injected with an extract of the roe (Netsch and Witt 1962; Fuhrman et al. 1969; Burns et al. 1981), but Ostrand et al. (1996) found that potential fish predators readily ate the embryos and grew well on an exclusive diet of gar eggs. There are anecdotal accounts of the toxicity of gar eggs to humans. One such account reports that early in his career the prominent ichthyologist George Moore ate some gar eggs to see what would happen and became ill, subsequently having his stomach pumped at a Stillwater, Oklahoma, hospital (Neil Douglas, pers. comm.). After that experience, Dr. Moore always warned his students to never eat gar eggs.

Overall, many anglers consider gars a nuisance and many kill any that are caught. In most natural waters and in large impoundments, gars play an important role in preventing overpopulation of many other species and in maintaining a proper natural balance and should be conserved rather than exterminated (Scarnecchia 1992). Alfaro et al. (2008) reviewed the biology, ecology, and physiology of different gar species as a basis for their domestication, mass production of larvae for repopulation attempts, and for the culture of commercial-size gar.

Key to the Gars*

1A Snout long and narrow, its least width going into snout length (Fig. 10.2A) more than 10 times; width of upper jaw at nostrils less than eye dameter.

Lepisosteus osseus Page 156

1B Snout short and broad, its least width going into snout length fewer than 10 times; width of upper jaw at nostrils greater than eye diameter.   2

2A Snout short and very broad, its width at nostrils 1.5 or more times eye diameter (Fig. 10.2B); distance from tip of snout to corner of mouth shorter than rest of head; snout width going 4.5 or fewer times into snout length;

Figure 10.1. Alligator Gar being hunted by bow and arrow in the 1950s on the White River in Arkansas. *Keith Sutton.*

gill rakers large, with 59 or more on first gill arch; adults with 2 rows of large teeth on each side of upper jaw (outer row on chain of interorbital bones and inner row along palatine bone).

Atractosteus spatula Page 150

2B Snout of moderate length and breadth, its width at nostrils 1.0–1.5 times eye diameter; distance from tip of snout to corner of mouth longer than rest of head; snout width going 5 or more times into snout length; gill rakers small with 35 or fewer on first gill arch; adults with single row of large teeth along infraorbital bones (outer row), and with no inner row of teeth on palatine bones.   3

3A Head and anterior part of body with large, dark spots (Fig. 10.2C); scales in lateral series 54 or 55; scale rows around body 32–38.   *Lepisosteus oculatus* Page 154

3B Head and anterior part of body usually without dark spots (Fig. 10.2D); scales in lateral series 59–64; scale rows around body 38–44.

Lepisosteus platostomus Page 159

Genus *Atractosteus* Rafinesque, 1820a— Alligator Gars

Atractosteus was described by Rafinesque in 1820 as a sub-genus of *Lepisosteus*, where it remained for most of the 20th century. It was elevated to genus status by Wiley (1976), a

---

* Because gars have an abbreviate-heterocercal caudal fin, lateral line scales are counted all the way to the base of the caudal fin rays in the lateral line scale row.

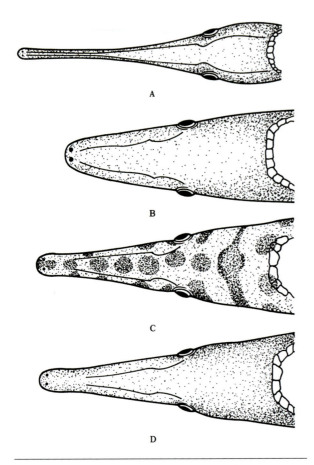

Figure 10.2. Tops of heads of gars. *Smith (1979).*

A—*Lepisosteus osseus*
B—*Atractosteus spatula*
C—*Lepisosteus oculatus*
D—*Lepisosteus platostomus*

designation that was eventually widely accepted and sup-ported by additional morphological and genetic evidence. The genus was considered monophyletic based on morpho-logical characters (Wiley 1976), and analyses of mitochon-drial and nuclear DNA also resolved *Atractosteus* as mono-phyletic (Wright et al. 2012). *Atractosteus* contains three living species distributed in North and Central America and Cuba, but the fossil record shows that this genus for-merly occurred in South America, Europe, and Africa as well (Nelson et al. 2016). The genus is characterized by 59–81 gill rakers and osteological characters (Wiley 1976). A single species is found in Arkansas.

Atractosteus spatula (Lacépède)
Alligator Gar

Figure 10.3a. *Atractosteus spatula*, Alligator Gar.
David A. Neely.

Figure 10.3b. *Atractosteus spatula*, Alligator Gar.
David A. Neely.

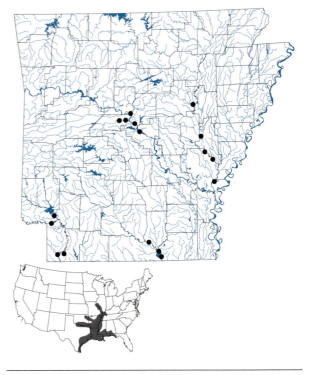

Map 10. Distribution of *Atractosteus spatula*, Alligator Gar, 1988–2019.

CHARACTERS   A large, heavy-bodied gar with a short, broad snout; snout width at nostrils going fewer than 4.6 times into snout length. Snout shorter than rest of head (Fig. 10.3); snout length going into body length 4.7–6.4 times. Large teeth in upper jaw in 2 rows on each side;

Figure 10.4. *Atractosteus spatula* tooth rows. *Renn Tumlison.*

teeth in inner row (palatine teeth) larger than teeth in outer row (Fig. 10.4). Body covered with ganoid scales. Tail abbreviate-heterocercal. Dorsal and anal fins far back on body. Dorsal fin rays 7–10; anal fin rays usually 7–10; pectoral fin rays 11–15. Predorsal scales 49–54. Scales in a diagonal row from front of anal fin to midline of back 23–32. Lateral line scales, counted to the base of the caudal fin rays in the lateral line scale row, 58–62. Gill rakers 59–66 and laterally compressed. Largest species of fish in Arkansas and one of the largest freshwater fishes in North America. The Arkansas state angling record is 215 pounds (97.5 kg) from the Arkansas River in 1964, but a specimen weighing 240 pounds (108.9 kg) was taken from the White River near St. Charles in 2004. Older unofficial records of much larger specimens exist. Burton (1970) published a photograph of an 8-foot, 3-inch (2.5 m), 350-pound (158.8 kg) Alligator Gar taken from the St. Francis River near Forrest City in the 1930s. Robins et al. (2018) reported a maximum length of 12 feet (3.7 m). The current world record is 327 pounds caught in a gill net in Mississippi in 2011.

LIFE COLORS   Back and sides olive green, grading to white or yellow on the undersides. Dorsal, caudal, and anal fins often with oval black spots; all fin rays usually dark brown. Body usually unspotted in large individuals, but specimens smaller than 30 cm TL often have dark spots on body. Moore et al. (1973) described coloration of post-yolk-sac young (24–103 mm TL) as follows:

> The dorsum was jet black with vivid white areas sharply delimited from the black. The lower sides and venter were gray. The head dorsum had a white area shaped like a spear, evidently the forerunner of the prominent, light middorsal stripe.

SIMILAR SPECIES   Many persons mistakenly refer to our other more abundant gar species as Alligator Gars. The Alligator Gar is distinguished from all other gars by its short, broad snout, the distance from the tip of snout to corner of mouth shorter than distance from corner of mouth to back of operculum (vs. distance from tip of snout to corner of mouth greater than distance from corner of mouth to back of operculum) and in having 59–66 gill rakers (vs. 35 or fewer rakers on first arch). The Alligator Gar further differs from the Longnose Gar in having 2 rows of large teeth in upper jaw (vs. 1 row of large teeth); however, this character will not allow accurate separation of the Alligator Gar from the Shortnose and Spotted gars as often reported in taxonomic keys. The young can be distinguished from the young of other gars by a light-colored middorsal stripe extending from tip of snout to dorsal fin origin (vs. dark middorsal stripe present).

VARIATION AND TAXONOMY   *Atractosteus spatula* was described by Lacépède (1803) as *Lepisosteus spatula* with no type locality given [North America]. It has been referred to by various scientific names, but a morphological analysis of fossil and Recent gars by Wiley (1976) placed the Alligator Gar (along with two other living and several fossil species) in *Atractosteus*. This placement was supported by morphological (Grande 2010) and molecular (Wright et al. 2012) analyses. The latter authors resolved *A. spatula* as sister to *A. tristoechus*. We agree with Wiley (1976) and most recent authors (Page et al. 2013) in retaining the Alligator Gar in the genus *Atractosteus*. Moyer et al. (2009a) isolated 17 polymorphic microsatellite loci from *A. spatula*. These loci possessed 2–19 alleles and observed heterozygosities of 0 to 0.974. Although hybridization in nature between *A. spatula* and *Lepisosteus* species has not been documented, Herrington et al. (2008) reported that aquarium spawning between a female *L. osseus* and a male *A. spatula* produced four hybrid offspring. Wright et al. (2012) found no genetic evidence for a history of introgressive hybridization among gar species.

DISTRIBUTION   The Alligator Gar occurs in the Mississippi River basin from southwestern Ohio and Illinois south to the Gulf of Mexico, and in Gulf Slope drainages from the Florida panhandle to Veracruz, Mexico. In Arkansas, it is found primarily in the largest rivers, the Arkansas, lower White, lower Ouachita, and Red rivers (Map 10). It is also found in the lower Little and Sulphur rivers of the Red River drainage. A population in the lower Fourche LaFave River (Arkansas River drainage) was studied by Inebnit (2009), Kluender (2011), Adams et al. (2013), and Kluender et al. (2017). The Alligator Gar was formerly known from the lower portion of the St. Francis River and the lower (eastern) Saline River, but there are no recent records from those areas. Surprisingly, there are almost no records for this species from the Mississippi River of Arkansas, and extensive sampling of that river from the mid-1990s to 2012 by the Waterways Experiment Station in Vicksburg, Mississippi, produced no new records (data supplied by K. J. Killgore).

HABITAT AND BIOLOGY   Primarily an inhabitant of the sluggish pools, backwaters of large rivers, and oxbow lakes connected to large rivers (such as Lake Chicot), the Alligator Gar sometimes enters brackish and salt waters along the Gulf Coast (Suttkus 1963). Radio-tagged fish in the Fourche LaFave drainage, Arkansas, occurred more frequently in the main channel prior to the spawning season, but proportional habitat use was highest in flooded tributaries during the spawning season (Kluender 2011; R. Adams et al. 2013; Kluender et al. 2017). Those investigators also found that when gars used main channel habitat, the microhabitats selected were along channel margins characterized by low velocity (~6.0 cm/s), shallow depth (~2.4 m), and complex instream and riparian, overhung structure. Brinkman (2008) reported that *A. spatula* was found in portions of the Red River that were relatively deep (1.8–9.1 m) compared with the average depth (typically <1.0 m) of the river. Solomon et al. (2013) found that radio-tagged juveniles stocked in a Missouri floodplain lake exhibited high site fidelity, while five fish exhibited long-distance movements. In the lower Trinity River, Texas, 83% of tagged Alligator Gar had linear home ranges smaller than 60 km during 22 months of study (Buckmeier et al. 2013). Home range size varied by season, with the smallest home ranges occurring in winter.

Until the 2000s, the biology of *A. spatula* had been little studied, but life history studies are now available for Mexico (Garcia de Leon et al. 2001), Oklahoma (Brinkman 2008), Louisiana (DiBenedetto 2009), and Arkansas (Inebnit 2009; Kluender 2011; R. Adams et al. 2013). Most of the following information on feeding and reproductive biology has been taken from those sources, particularly the studies involving Arkansas populations. It feeds mainly on fishes, but is also known to eat other small vertebrates (Raney 1942). In estuarine waters of Louisiana, it frequently eats blue crabs (Suttkus 1963). In studies in Mississippi and Louisiana, adults fed mostly on fish (about 80% by volume), but also ingested crustaceans (mainly shrimp), vegetation, bones, stones, plastic objects, hooks, and fishing line (Goodyear 1967; DiBenedetto 2009). Juveniles in Lake Texoma fed mainly on insects and *Gambusia affinis* (Echelle and Riggs 1972). The primary food of YOY *A. spatula* in Lake Sam Rayburn, Texas, was small Gizzard Shad (Toole 1971). In that same reservoir, Seidensticker (1987) found that adults were almost exclusively piscivorous, feeding mainly on Gizzard Shad, Freshwater Drum, and Channel Catfish. Other fishes

consumed were sunfish, suckers, White Bass, buffaloes, Largemouth Bass, Spotted Gar, crappies, and Common Carp. The primary food of large gars in a Mexican reservoir was Largemouth Bass (Garcia de Leon et al. 2001). The digestive tract of larval gars was completely formed 5 days after hatching, at the beginning of exogenous feeding (Mendoza et al. 2002). Artificial feeds were well accepted by the larvae and resulted in growth rates similar to gar larvae that were fed natural prey.

Suttkus (1963) reported that spawning occurs from April to June in Louisiana, and in southcentral Louisiana, spawning occurred from March through May (DiBenedetto 2009). Spawning occurred in Oklahoma in May (May and Echelle 1968), and spawning dates in Lake Texoma corresponded to rising pool elevations and water pulses of tributaries (Snow and Long 2015). Spawning usually occurs in conjunction with seasonal flooding, and preferred spawning habitat over most of its native range consists of flooded backwater areas and floodplains (Brinkman 2008; Inebnit 2009). Inebnit (2009) documented that Alligator Gars used tributary backwaters of the Fourche LaFave River, Arkansas, for spawning and nursery habitat. He found that spawning occurred from mid-May to mid-June in 2007 and 2008 during the peak of a flood event when main channel water temperatures ranged from 22°C to 28°C. More than 1,000 larvae were observed following the successful spawn of 2007. While the young of *Lepisosteus* species were found in all three macrohabitat types studied, the young of *Atractosteus* were found only in tributary backwater habitat (Inebnit 2009). The earliest date for the collection of *Atractosteus* young was 5 June 2007, and young of both gar genera were collected when floodplain water temperatures were 25–30°C. The earliest observed spawning was 23 May 2010 (R. Adams et al. 2013). Inebnit (2009) provided the following observations of a 3.5-hour spawning event at a water temperature of 29.5°C:

> Spawning bouts consisted of a group of 2 to 4 individuals thrashing, for only a few seconds, in less than 1 meter of water and in proximity with some type of vegetation (e.g., *Forestiera acuminata*, *Cephalanthus occidentalis*, *Hibiscus* sp., and/or *Carex* sp.). Furthermore, these groups were observed swimming together and making passes repeatedly in the same area. These swimming passes consisted of making a circle through deeper water and eventually coming back to the same shallow area to begin another spawning bout.

The spawned eggs were attached to recently flooded vegetation, and some were attached to floating woody debris. Similar observations were made on 3 June 2013, but spawning aggregations as large as approximately 15 individuals were observed, and eggs were found primarily in shallow water (<1.0 m) attached to flooded wetland vegetation (*Carex* sp.) (R. Adams and T. Inebnit, pers. comm.). Eggs hatched in 48–72 hours at water temperatures of 27.5–30°C, and approximately 5–7 days after the spawn date, the larvae were no longer attached to vegetation (Inebnit 2009). In August 2007, Inebnit found three size classes of juveniles present (90–250 mm TL) in the Fourche LaFave River, suggesting that multiple successful spawning events occurred that year. Fecundity in Gulf Coast populations ranged from 782 to 557,390 eggs per female ($\bar{x}$ = 4.1 eggs/g body weight) (Ferrara 2001). Simon and Wallus (1989) described early development of specimens from Lake Texoma, Oklahoma. At 15 mm TL, no yolk sac or attachment organ on the snout remained evident; at 20–32 mm, fin formation occurred, and scale development occurred by about 65 mm TL. They collected larvae in late May from a vegetated backwater slough, and juveniles (100 mm TL) were taken in late July from a shallow embayment area in association with plants. Brinkman (2008) collected multiple year classes of juvenile *A. spatula* from Lake Texoma between April and November in backwater areas and coves with woody vegetation and debris. During summer 2013, age 0 Alligator Gar in Lake Texoma grew an average of 5.49 mm and 1.15 g per day (Snow and Long 2015). Maximum reported age (from otoliths) was 50 years for females and 26 years for males (Ferrara 2001).

The Alligator Gar is now uncommon in Arkansas after a drastic decline during the last 40 years. Black (1940) reported it abundant in the lower White, Arkansas, and Mississippi rivers. At one time, there were significant sport and commercial fisheries for this gar in the state, but longtime commercial fishermen readily attest to its marked decline. In the White and Arkansas rivers (before impoundments and navigation channels were constructed), the species was avidly sought by fishermen who buried a lead weight in a ball of fluffed nylon and cast for them. When the gar snapped at this lure, its teeth became entangled in the nylon and the fight was on. Commercial fishermen report seeing Alligator Gars less frequently today, but they remark that when one is sighted, it is usually a big one. Recent confirmed records from the state are few. Whenever a specimen is taken by an angler, bow fisher, or commercial fisherman, it is usually prominently reported in the local newspaper. In addition to the recent records from the Fourche LaFave River (Inebnit 2009; R. Adams et al. 2013), some noteworthy Alligator Gar records from Arkansas

over the past four decades are: a 125-pound (56.7 kg) specimen taken from the Arkansas River near Fort Smith in 1975, an 8-foot (2,438 mm) 206-pound (93.4 kg) gar netted in the Arkansas River near Little Rock in 1977, a 122-pound (55.3 kg) gar taken in the Arkansas River near Little Rock in December 1985, and a small specimen taken by a commercial fisherman in a gill net from the Arkansas River upstream from the mouth of Palarm Creek in Faulkner County on 4 May 2012. Six Alligator Gar were collected during a 1979–1980 study of the lower Mississippi River in southern Arkansas and northern Louisiana (Beckett and Pennington 1986). Recent fish surveys of large rivers in Arkansas by Layher and Phillips (2000) and Buchanan et al. (2003) yielded 2 Alligator Gar from the Red River, 3 from the Ouachita River, and 1 from the lower White River. Extensive sampling of the Black, St. Francis, and Mississippi rivers by Layher in 2006 and 2007 produced no additional specimens of Alligator Gar, but Layher found 14 *A. spatula* in a subsequent study of the lower Ouachita River. A 50-pound specimen was caught in Cook's Lake (a lower White River oxbow) on 21 March 2014. The most recent state record of 8 feet, 2 inches and 240 pounds was taken in a hoop net by a commercial fisherman from the White River at St. Charles on 28 July 2004.

CONSERVATION STATUS Even though populations still exist in the lowland portions of large rivers, the Alligator Gar has declined in range and abundance in Arkansas. Kluender et al. (2017) emphasized the importance of connectivity between main river channel and floodplain habitats for this floodplain-obligate river species and noted that an intact, heterogeneous riparian zone creates essential microhabitat for successful spawning. Snow and Long (2015) hypothesized that the aging of reservoirs may benefit Alligator Gar because the floodplains originally lost through impoundment of the watershed may now be coming back through a prolonged process of sedimentation. The altered river-reservoir interface can then function as an alternative spawning habitat. The Alligator Gar is an Arkansas Game and Fish Commission Species of Greatest Conservation Need. Clay (2009) described aquaculture methods for propagating this species, and successful stocking programs to reestablish or supplement natural populations exist in several states (Snow et al. 2018). A project is currently underway to develop biologically feasible and economically cost effective ways for the AGFC, and other state or federal hatcheries, to produce Alligator Gar for reintroduction into suitable habitats (Steve Lochman, pers. comm.). Boschung and Mayden (2004) cautioned that a thorough geographic analysis of morphological and molecular traits is needed to identify any possible evolutionarily significant natural populations before proposed captive propagation and transplantation efforts are carried out. We consider the Alligator Gar a species of special concern in Arkansas.

Genus *Lepisosteus* Lacépède, 1803— Longnose Gars

Described in 1803, *Lepisosteus* was considered monophyletic and sister to *Atractosteus* by Wiley (1976) and Grande (2010) based on morphological evidence. A molecular analysis also resolved *Lepisosteus* as monophyletic and separate from *Atractosteus* (Wright et al. 2012). We refer to this genus as Longnose Gars because *L. osseus* is the type species of the genus and also because in all *Lepisosteus* species, the upper jaw is longer than the rest of the head. The genus is further characterized by a total gill raker count of 14–33 on the first gill arch and by osteological features (Wiley 1976). This North American genus contains four living species, three of which occur in Arkansas. Fossils of *Lepisosteus* (primarily Cretaceous and Eocene) are known from Europe, India, North America, and South America (Nelson et al. 2016).

Lepisosteus oculatus Winchell
Spotted Gar

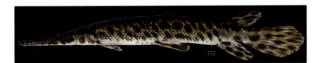

Figure 10.5. *Lepisosteus oculatus*, Spotted Gar. *David A. Neely.*

CHARACTERS   An elongate fish with a prominent, broad snout; snout width at nostrils going into snout length 5–7 times (Fig. 10.2C). Top of head, pectoral and pelvic fins, and usually the body with dark spots. Dorsal and anal fins far back on body. Body covered with ganoid scales. Tail abbreviate-heterocercal. Upper jaw slightly longer than rest of head. Large teeth in upper jaw in a single row on each side, but there is also an inner row of smaller teeth (developed to varying degrees) on each side. Dorsal fin rays 8–10, anal rays usually 8–10; pectoral rays usually 11 or 12. Lateral line scales, counted to the base of the caudal fin rays in the lateral line scale row, 53–59. Scales in a diagonal row from front of anal fin to midline of back

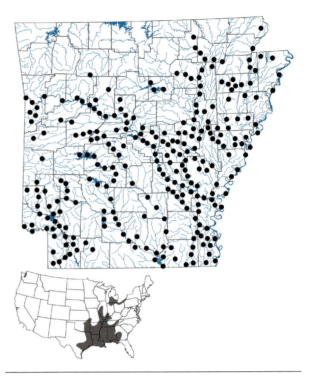

Map 11. Distribution of *Lepisosteus oculatus*, Spotted Gar, 1988–2019.

17–20. Predorsal scales 44–54. Gill rakers 15–24. One of the smallest species of gars, reaching a maximum length of around 44 inches (1.1 m) and a weight of 6–12+ pounds (2.7–5.4 kg). The Arkansas state rod and reel angling record from the Little Red River in 2018 is 7 pounds, 15 ounces (3.6 kg). The Arkansas unrestricted tackle record is 12 pounds, 5 ounces (5.6 kg) from Lake Conway in 2017.

LIFE COLORS   Back and sides olive with round, black spots on fins, head, snout, and usually on the body. Undersides white; underside of head freckled. Specimens from darkly stained waters in the lowlands have uniformly darkened bodies. Suttkus (1963) described the distinctive markings of the juveniles as follows:

> The median dorsal stripe is very broad and dark brown, often breaking up into a single row of spots posteriorly. The dark lateral band is nearly straight on its upper margin, with a narrow reddish brown stripe above it; the lateral band often breaks up into a row of spots posteriorly. The ventral surface is usually a chocolate color.

Golden (xanthic) adults have been reported from populations in Alabama and Mississippi (McIlwain and Waller 1972; Powell 1991).

SIMILAR SPECIES   The Spotted Gar is similar to the other three species of gars in Arkansas, but differs from

Figure 10.6. *Lepisosteus oculatus* and reflection.
Renn Tumlison.

them in having round, black spots on top of head and snout (vs. spots usually absent from top of head). Specimens from turbid waters may lack spots. If spots are absent, it can be distinguished from the Shortnose Gar by its 53–59 lateral scale rows (vs. usually 60–63 lateral scale rows). Specimens lacking spots on top of head can easily be distinguished from *L. osseus* by the shorter snout, its width at nostrils going into snout length 5–7 times (vs. snout width going into snout length 11 or more times in large Longnose Gar juveniles and adults). Simon and Wallus (1989) provided additional features for separating the young (50 mm or less TL) of all gar species found in Arkansas.

VARIATION AND TAXONOMY   *Lepisosteus oculatus* was described by Winchell (1864) as *Lepidosteus oculatus* from Duck Lake, Calhoun County, Michigan. Based on morphological analyses, Suttkus (1963), Wiley (1976), and Grande (2010) concluded that *L. oculatus* is the sister species of the Florida Gar, *L. platyrhincus*. This designation was supported in the molecular phylogenies of Wright et al. (2012). Moyer et al. (2009a) isolated nine polymorphic microsatellite loci from *L. oculatus*.

DISTRIBUTION   The Spotted Gar occurs in Great Lakes drainages (Lake Erie and southern Lake Michigan) in Canada, Michigan, Ohio, and Indiana; the Mississippi River basin from Illinois to the Gulf Coast; and in Gulf Slope drainages from the Apalachicola River, Florida, to the San Antonio River, Texas. In Arkansas, it is the second-most widely distributed gar species, occurring throughout most rivers of the Coastal Plain and extending northwestward through the Arkansas River Valley (Map 11). It penetrates somewhat into the streams of the Ouachita Upland but is almost absent from the Ozarks.

HABITAT AND BIOLOGY   In Arkansas, *Lepisosteus oculatus* is most abundant in quiet, clear waters that have much aquatic vegetation or standing timber. It is a common inhabitant of oxbow lakes and river backwaters and occurs in reservoirs throughout most of Arkansas. Pflieger (1997) considered it to show a greater affinity for aquatic vegetation than other gars, and that agrees with our observations of this species in Arkansas. Snedden et al. (1999) noted that Spotted Gars showed increased movement into floodplain habitats in Louisiana as water levels rose. It is a common summer inhabitant of the fresher parts of estuaries in Louisiana (Suttkus 1963), and it is apparently less tolerant of turbidity than other gars. Robertson et al. (2008) found strong habitat partitioning in the Brazos River drainage, Texas, between Spotted and Longnose gars, in which 98% of Spotted Gars were captured in oxbow habitats and 84% of Longnose Gars were captured in the river channel.

Lepisosteid jaw length differences have been linked to each species' diet, and the prey of *L. oculatus* comprises a greater variety of invertebrates and fewer fishes than the diet of *L. osseus* (Goodyear 1967; Tyler and Granger 1984). However, Robertson et al. (2008) found high diet overlap between Spotted and Longnose gars in the Brazos River drainage. Walker et al. (2013b) studied food habits of Spotted and Shortnose gars during flooding of an Arkansas River tributary and reported that the diet of Spotted Gars consisted of fish (74%), crayfish (26%), aquatic insects (11%), and terrestrial insects (9%); the Shortnose Gar consumed a higher percentage of insects and amphibians and a lower percentage of fishes. Tyler and Granger (1984) found that 82% of the Spotted Gars examined had empty stomachs, which agrees with several other studies (Scott 1938; Hunt 1953). In Oklahoma, Tyler and Granger (1984) found that fish constituted 68% and invertebrates 32% (16% crayfish and 16% insects) of food recovered. Fish ingested included Bluegills and Freshwater Drum. Large gars fed on larger food items. Echelle and Riggs (1972) found that the dominant food items in Lake Texoma, Oklahoma, were Mississippi Silversides and shad. In the Barataria Estuary, Louisiana, diets of Spotted Gars varied seasonally, with fish making up most of the diet in spring and summer, and insects predominating in the fall diet (Manley 2012). Crayfish were low in abundance in the diet during all seasons, and 45.5% of the stomachs examined were empty (Manley 2012). The fry feed on mosquito larvae and small crustaceans, but fish soon begin to appear in their diet; fish make up 90% of the diet of adults, and the remaining 10% consists of freshwater shrimp, crayfish, and insects.

Spotted Gars spawn in shallow water in the spring. Redmond (1964) observed spawning in Missouri in late April in flooded timber. Echelle and Riggs (1972) found that Spotted Gars spawned in quiet, weedy backwaters of Lake Texoma, Oklahoma, from early April through May, over dead vegetation and algal mats. Frenette and Snow (2016) emphasized the great importance of inundated vegetation or woody debris as a spawning substrate. In Lake Lawtonka, Oklahoma, spawning occurred from 22 April to 10 June, with the most intense spawning activity in mid-May (Tyler and Granger 1984). Based on histological analysis and GSI values, Spotted Gars spawned in southern Louisiana from March through May as water levels and temperatures rose (O. A. Smith 2008). *Lepisosteus oculatus* was sexually dimorphic in southern Louisiana (Love 2002). Females were significantly longer and had longer snouts than males when mass and age were considered. Love (2002) hypothesized that the greater size of females enabled them to produce larger and more numerous eggs but did not discuss the possible selective advantage of the snout length dimorphism. Typical spawning behavior was characterized by a large female swimming slowly through well-vegetated shallow areas, closely accompanied by 3–5 smaller males (Tyler and Granger 1984). During actual spawning, there was much thrashing and splashing, and eggs were apparently released when the female made quick jerking movements. The demersal, adhesive eggs are scattered over the substrate. Similar details of spawning behavior were confirmed by Frenette and Snow (2016) in Oklahoma. Spawning occurred on 3 May at a water temperature of 23°C. A single female led a group of males in a circular pattern, making several passes before moving toward inundated branches. After the female led the group into the branches, the group burst forward, vigorously thrashing and often breaking the water surface. Several thousand eggs were released during the spawning event. Males reach sexual maturity at 2 or 3 years of age, and females matured in their third or fourth year (Redmond 1964). Mean total fecundity in southern Louisiana was 6,493, but apparently not all eggs were spawned by the females (O. A. Smith 2008). Fecundity in Gulf Coast populations ranged from 284 to 5,587 eggs per female (x̄ = 0.1 eggs/g body weight) (Ferrara 2001). Eggs hatched in 6 days at 22.2°C. Gray et al. (2012) found that turbidity reduced hatching success. After hatching, vegetated areas serve as a nursery area for the young. The larvae are approximately 9.8 mm at hatching, the snout adhesive organ has largely disappeared by 17.6 mm TL, the yolk sac is absorbed by 24 mm TL, and the jaws are well differentiated and toothed by 33 mm TL (Boschung and Mayden 2004). Average length at the end of the first year was 25 cm in Tennessee (Simon and Wallus 1989). Tyler (1994) documented the first record of albinism in the genus *Lepisosteus*, a gravid 619 mm TL female

L. oculatus from Oklahoma weighing 827 g. Goodyear (1967) reported that maximum age was at least 18 years.

CONSERVATION STATUS   The Spotted Gar is a species of special concern at the periphery of its range and is considered critically imperiled in Canada (Glass et al. 2012). Loss of spawning habitat is considered a widespread threat. In Arkansas, this species is widely distributed and its populations are secure.

Lepisosteus osseus (Linnaeus)
Longnose Gar

Figure 10.7. *Lepisosteus osseus*, Longnose Gar. *Uland Thomas*.

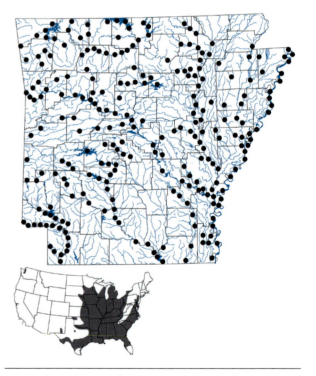

Map 12. Distribution of *Lepisosteus osseus*, Longnose Gar, 1988–2019.

CHARACTERS   An elongate fish with an extremely long and slender snout; snout width at nostrils going into snout length 11.5 or more times in large juveniles and adults (Fig. 10.2A). Snout of adult usually more than twice as long as the rest of the head. Head length going into TL 3.0–3.6 times. Body depth going into TL 12–18 times. Large teeth in upper jaw in a single row on each side. Fins spotted, body often with a few scattered spots, but top of head and snout without spots. Body covered with ganoid scales. Tail abbreviate-heterocercal. Dorsal and anal fins far back on body. Dorsal fin rays 6–9; anal rays usually 7 (8–10); pectoral rays 10–13. Lateral line scales, counted to the base of the caudal fin rays in the lateral line scale row, usually 57–63. Scales in a diagonal row from front of anal fin to middle of back 17–20. Predorsal scales 47–55. Gill rakers 14–31. A large species commonly reaching 3 feet (0.9 m) or more in length. Maximum length usually around 4.5 feet (1.37 m) and maximum weight around 55 pounds (13.9 kg), but Black (1940) reported a 6-foot, 8-inch (2 m) specimen from near Lake Village, Arkansas. The Arkansas state rod and reel angling record from Taylor Old River Lake in 2005 is 35 pounds, 12 ounces (16.2 kg). The unrestricted tackle record from the Arkansas River in 2011 is 54 pounds (24.5 kg).

LIFE COLORS   Back and sides brown or olive, grading to yellowish white on undersides. Dorsal, caudal, and anal fins with black oval spots; all fins usually tinged with orange or yellow. Body sometimes with small, indistinct spots, especially in specimens from clear water (Pflieger 1997). Young with a conspicuous black stripe along the side extending from snout through eye to base of caudal fin; the upper margin of the stripe either scalloped or fused with a narrow reddish-brown interrupted stripe or row of spots above (Suttkus 1963). A narrow, dark middorsal stripe present. Undersides of small young chocolate-colored, becoming lighter with age. Black (melanistic) adults have been documented from several states, including Florida, Virginia, and Oklahoma (Woolcott and Kirk 1976; Pigg 1998). In the Virginia population, melanistic individuals composed 11/196 (5.6%) of the Longnose Gars sampled and were equitably represented by males 560–749 mm TL and females 658–865 mm TL.

SIMILAR SPECIES   It differs from other gars in its long, slender snout; least snout width going into snout length 11.5 times (vs. snout width going into snout length fewer than 10 times). Simon and Wallus (1989) provided additional features for separating the young of all four gar species in Arkansas.

VARIATION AND TAXONOMY   *Lepisosteus osseus* was described by Linnaeus (1758) as *Esox osseus* from "Virginia." It was placed in genus *Lepisosteus* by Lacépède in 1803 and referred to as *L. gavialis*, a synonym of *L. osseus*. Various intrageneric relationships have been proposed. Based on morphology, Suttkus (1963) considered *L. osseus* sister to all currently recognized extant *Lepisosteus* species. Subsequent morphology-inferred phylogenies by Wiley (1976) and Grande (2010) concluded that *L. platostomus*

was the sister species of all extant *Lepisosteus*, with the next node resolving *L. osseus* as the sister species of a clade containing *L. oculatus* and *L. platyrhincus*. Molecular analyses by Wright et al. (2012) resolved *Lepisosteus osseus* and *L. platostomus* as sister species. Moyer et al. (2009a) isolated eight polymorphic microsatellite loci from *L. osseus*.

DISTRIBUTION   *Lepisosteus osseus* is the most widely distributed gar species, occurring in Atlantic Slope drainages from New Jersey to central Florida; St. Lawrence–Great Lakes drainages; the Hudson Bay drainage of North Dakota; the Mississippi River basin from Minnesota to Pennsylvania and south to the Gulf Coast; and in Gulf Slope drainages from central Florida to the Rio Grande drainage of Texas and Mexico. The Longnose Gar is Arkansas's most widespread and abundant gar and is found statewide in all river drainages. It is also common in reservoirs and is the only gar species that occurs throughout the Ozarks (Map 12).

HABITAT AND BIOLOGY   The Longnose Gar is a versatile species that occurs in the Coastal Plain lowlands in rivers, bayous, oxbow lakes, and swamps, and in the Ozark and Ouachita Uplands in creeks, rivers, and impoundments, typically in sluggish pools and backwaters. It is far more common in upland areas of the state than other gar species. In coastal regions of the United States it frequently enters brackish estuarine areas but is not as tolerant of saltwater as the Alligator Gar (Suttkus 1963). McGrath et al. (2013) considered it one of the few resident, euryhaline predators in the tidal rivers of Virginia. Robertson et al. (2008) found strong habitat partitioning between Longnose and Spotted gars in the Brazos River drainage, Texas, with 84% of Longnose Gars found in the main river channel and 98% of Spotted Gars captured in oxbow habitats. Young gars are most commonly found in shallow water but tend to move to deeper water as they grow. In Lake Texoma, young gars spent much time resting motionlessly close to submerged or overhanging objects near the shore during daylight hours (Echelle and Riggs 1972). At night, they were commonly found swimming actively in shallow, open water. McGrath et al. (2016) reported that while the Longnose Gar shares one trait (large eggs) with equilibrium strategists, most of its life history traits (large adult size, delayed maturation, and high fecundity) are closer to the periodic strategist side of the demographic continuum.

In Missouri, *L. osseus* fed almost entirely on fish throughout life, except briefly after hatching (Netsch 1964). In Oklahoma, Tyler et al. (1994) found a high incidence of empty stomachs (94 of 142 stomachs were empty) as did Diana (1966) and Scott (1938). However, of the gar stomachs that contained food, 98% contained fish and 2% had crayfish and some insects. Netsch and Witt (1962), Goodyear (1967), and Crumpton (1971) reported similar results. Larger gar feed on bigger prey items (Tyler et al. 1994). Tyler et al. (1994) reported 69% of the piscine prey was forage fish in Oklahoma, and Diana (1966) and Bonham (1941) also found that forage fish composed the primary diet of this species in other parts of the country. Variation in the relative importance of prey taxa in Longnose Gar diets was observed between habitats and years in the Brazos River drainage (Robertson et al. 2008). During a wet year, shad were the most important prey, followed by catfish and minnows. During a dry year, mayflies and catfish were the most important prey items while shad importance decreased. Diet overlap between Longnose and Spotted gar was high in the Brazos River drainage: 72.7% during the dry year and 90.1% during the wet year (Robertson et al. 2008). *Lepisosteus osseus* has the longest snout of all its congeners. Such elongation results in a low mechanical advantage and a high transmission of motion to the jaws (Kammerer et al. 2006), enabling individuals to open and close their jaws quickly and thereby facilitating the rapid lateral slashing capture of fast-moving prey items such as fishes. Young gar fed mostly on minnows, but older ones primarily ate Gizzard Shad. Echelle (1968) found that young Longnose Gar in the initial feeding stages (0.7–0.8 inches [17–21 mm] TL) in Lake Texoma, Oklahoma, used diverse foods, including microcrustaceans, insects, and fish. After the initial feeding stage, young Longnose Gar of all sizes fed predominantly on young-of-the-year fishes, with *Menidia audens* representing 84% of the diet. In South Carolina estuaries, there was an ontogenetic shift in prey composition from low trophic level benthic prey (fundulids) to higher trophic level pelagic prey (mainly clupeids) as Longnose Gar grew to 400–600 mm SL (Smylie et al. 2015).

Adults spawn in smaller tributaries in spring. Spawning occurs in Missouri from early May to mid-June (Pflieger 1997). In Tennessee, peak spawning occurs in May (Etnier and Starnes 1993). In Wisconsin, spawning peaked at water temperatures of 67–70°F (19.4–21.1°C) (Becker 1983). Spawning adults gather over gravelly or weedy areas. The males outnumber the females. One female usually spawns with 2–4 males at irregular intervals in a ritual of circling in shallow water with splashing and convulsive movements (Rohde et al. 2009). Matthews et al. (2012) described spawning activities along a rocky shoreline of Lake Texoma, Oklahoma. Several females about 1 m in length continually swam alongside large rocks in shallow water at the shoreline, accompanied by as many as 4

or 5 smaller males. Fertilized eggs were abundant, adhering to rocks in the shallow water. A large female can produce more than 77,000 eggs during a spawning season. Fecundity of Gulf Coast populations ranged from 8,438 to 80,137 eggs per female ($\bar{x}$ = 0.8 eggs/g body weight) (Ferrara 2001). The adhesive eggs become attached to the substrate and sometimes buried in it. Longnose Gar in Lake Texoma also spawned over algal mats and bare rock along wind-swept shorelines and rocky points (Echelle and Riggs 1972). Johnson and Noltie (1996) found that lake dwelling Longnose Gar entered tributary streams to spawn in Missouri. The spawning migration began in early April and ended in late May and was positively correlated with stream flow and water level. Goff (1984) discovered Longnose Gar eggs in nests of Smallmouth Bass in Ontario, with male bass providing brood care for the eggs and the larvae of both species. The eggs hatch in 3–9 days depending on the temperature (6 days at 68°F [20°C]). The newly hatched fry attach themselves vertically to submerged objects by an adhesive organ on the snout. After absorption of the yolk sac (about 9 days after hatching) the young no longer hang vertically but are capable of resting motionless in a horizontal position at any depth. At this time, they are able to take their first aerial breath and begin feeding on micro-crustaceans and insect larvae. Tyler et al. (1994) reported a sex ratio of 0.9:1 in 138 individuals examined. Netsch and Witt (1962) found that males in Missouri attained sexual maturity at 3 or 4 years of age and the females in about 6 years. Females also attain greater total lengths than do like-aged males (Netsch and Witt 1962; Klaassen and Morgan 1974; Johnson 1994; Johnson and Noltie 1997; McGrath and Hilton 2011) and have greater masses, pelvic girths, anal fin heights (anal fin length), and anal girths (Johnson 1994). Females also grow faster and live longer than males. In Missouri, females lived more than 30 years (Netsch and Witt 1962); in brackish and marine environments of South Carolina, maximum life span was 17 years for males and 25 years for females (Smylie et al. 2016). Sexes are dimorphic, and McGrath (2010) and McGrath and Hilton (2011) reported differences in head and anal-fin shape between males and females. Sexual dimorphism in head measurements includes males having significantly wider heads and midsnouts relative to their body sizes than females. This may help to attract females and defend preferred spawning sites. The function of the relatively longer anal fin base in male *L. osseus* remains speculative.

CONSERVATION STATUS This is a widely distributed, common species, and its populations in Arkansas are secure.

Lepisosteus platostomus Rafinesque

Shortnose Gar

Figure 10.8. *Lepisosteus platostomus*, Shortnose Gar. *Uland Thomas.*

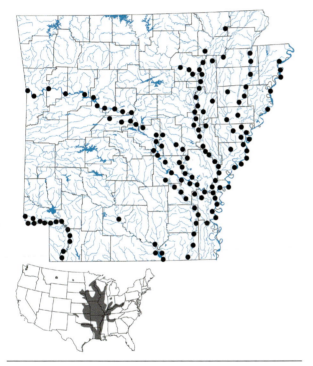

Map 13. Distribution of *Lepisosteus platostomus*, Shortnose Gar, 1988–2019.

CHARACTERS An elongate fish with a moderately short, broad snout (Fig. 10.8); snout width at nostrils going into snout length 4.6–7.1 times (Fig. 10.2D). Head length going into TL 3.8–4.4 times. Body depth going into TL 9–12 times. Large teeth in upper jaw in a single row on each side, but many specimens also have well-developed teeth in an inner row on each side. Fins with a few spots, body without spots. Tail abbreviate-heterocercal. Dorsal and anal fins far back on body. Dorsal fin rays usually 8 or 9; anal fin rays usually 8 or 9. Lateral line scales, counted to the base of the caudal fin rays in the lateral line scale row, usually 60–64. Scales in a diagonal row from front of anal fin to midline of back 20–23. Predorsal scales 50–55. Gill

rakers 27–33. Arkansas's smallest species of gar, seldom exceeding 2 feet (610 mm) in length or 5 pounds (2.3 kg) in weight. The maximum reported size is 19 pounds (8.6 kg) from Nebraska. The Arkansas unrestricted tackle record is 11 pounds, 12 ounces (5.3 kg) from Cypress Bayou in 2016.

LIFE COLORS Back and sides olive green; undersides white. Dark spots usually few and confined to unpaired fins; paired fins usually without spots. Body rarely with spots, but specimens from clear water may have body spots. Young with a fairly broad dark brown middorsal stripe and a brown lateral stripe.

SIMILAR SPECIES The Shortnose Gar differs from the Spotted Gar in lacking spots on top of head (vs. top of head with distinct dark spots). From the Alligator Gar, the Shortnose differs in distance from tip of snout to corner of mouth greater than the distance from corner of mouth to back of operculum (vs. distance from tip of snout to corner of mouth less than the distance from corner of mouth to back of operculum). It differs from the Longnose Gar in having a shorter snout, its width at nostrils going into snout length 7.1 times or less (vs. snout width going into snout length more than 10 times). Simon and Wallus (1989) provided additional features for separating the young of all four gar species in Arkansas.

VARIATION AND TAXONOMY *Lepisosteus platostomus* was described by Rafinesque (1820a) from the Ohio River (no other locality data given). Based on morphology, Suttkus (1963) considered *L. platostomus* sister to a clade containing *L. oculatus* and *L. platyrhincus*. Subsequent morphology-inferred phylogenies by Wiley (1976) and Grande (2010) resolved *L. platostomus* as the sister species of all other *Lepisosteus*, with the next node resolving *L. osseus* as the sister species of a clade containing *L. oculatus* and *L. platyrhincus*. However, molecular-inferred species trees resolved *L. osseus* and *L. platostomus* as sister species (Wright et al. 2012).

DISTRIBUTION *Lepisosteus platostomus* occurs in the Lake Michigan drainage of Wisconsin; in the Mississippi River basin from southern Ohio, Wisconsin, and Montana south to Louisiana; and in Gulf Slope drainages of Louisiana (Calcasieu and Mermentau rivers). Although found in all moderate to large rivers of Arkansas, the Shortnose Gar primarily occupies the Coastal Plain lowlands and extends up the Arkansas River Valley (Map 13). It is most abundant in the Arkansas, lower White, and Mississippi rivers and is present in the St. Francis and Ouachita river drainages, where it is common in Bayou Bartholomew. It occurs throughout the Red River of Arkansas (Buchanan et al. 2003).

HABITAT AND BIOLOGY   There is less information on the biology of *L. platostomus* than for any of the other gar species inhabiting Arkansas. More tolerant of silt and high turbidity than other gars, the Shortnose Gar is most common in the state's largest rivers, where it is often found in currents over a sandy bottom; however, it is also found in quiet backwaters and pools of the large rivers with sand and mud bottoms. Although this species reportedly prefers oxbow lakes and reservoirs in some areas (Boschung and Mayden 2004), this does not appear to be the case in Arkansas, where it is almost entirely restricted to rivers. Shortnose Gars were common throughout all Arkansas River navigation pools in annual AGFC rotenone population samples. Adult *L. platostomus* move in large schools both before and after the spawning season (Coker 1930).

Like other gars the Shortnose Gar feeds primarily on fishes, but adult gars also regularly consume crayfish and insects. Its diet in the Fourche LaFave River, Arkansas, during flooding consisted (by frequency of occurrence) of fish (79%), terrestrial insects (36%), aquatic insects (22%), and amphibians (15%) (Walker et al. 2013b). In that study, Spotted Gar diet consisted of a higher percentage of fish and lower percentages of aquatic and terrestrial insects and amphibians than the diet of Shortnose Gar. Crayfish were not an important component of the Shortnose Gar diet but composed 26% of the Spotted Gar diet. In Lake Texoma, Oklahoma, foods consisted mainly of minnows and Mississippi Silversides, with some consumption of young sunfishes and Freshwater Drum (Echelle and Riggs 1972). The first foods of Shortnose Gar are microcrustaceans and small insect larvae; fish are added to the diet when young gar reach about 1.25 inches (32 mm) (Lagler et al. 1942).

Spawning occurs from May to July in northern parts of its range at temperatures of 66–74°F (19–24°C) in shallow backwaters. In Lake Texoma, Oklahoma, spawning occurred from mid-April through May (Echelle and Riggs 1972). Spawning occurred at water depths of 1–3 feet (0.3–0.9 m) in Illinois (Carlander 1969). Spawning behavior is similar to that of other gars and involves much thrashing in shallow water by multiple individuals, with a single female often spawning simultaneously with two or more males (Potter 1926). A 9-pound (4.1 kg) female contained 36,460 eggs (Potter 1926). The greenish-yellow, adhesive eggs are scattered over vegetation or other submerged objects in shallow water. Shortnose Gar larvae were found as early as 22 April 2008 in three macrohabitat types (main channel border, adjacent floodplain, and tributary backwater) in the Fourche LaFave River drainage, Arkansas (Inebnit 2009). However, they were more commonly found in adjacent floodplain and flooded tributary backwater habitats when floodplain water temperatures were between 22°C and 31°C. Sexual maturity is reached at about 3 years.

Becker (1983) indicated that further study is warranted on the potential of this species to play an important role in maintaining a balanced fish population in some waters. Becker considered the Shortnose Gar an excellent food fish when baked or smoked.

CONSERVATION STATUS   This is the least common member of its genus in Arkansas because it is restricted mainly to large rivers and associated habitats. It may have declined some in abundance in recent decades, but its populations are currently secure.

CHAPTER 11

Order Amiiformes
Bowfins

The order Amiiformes, with only one currently recognized extant species, is an ancient group of neopterygian fishes, considered distinct since the Triassic more than 180 million years ago (Patterson 1973). There is some evidence that there may be multiple undescribed species in North America, and ongoing genetic analyses may clarify this (D. Neely, pers. comm.). Patterson and Longbottom (1989) summarized the fossil record of amiiform fishes and noted that the fossil beds of Europe, Israel, Brazil, Africa, and North America have yielded 10 genera varying in age from the Jurassic to the Eocene. The occurrence of amiiform fossils in North America is extensive from Early Cretaceous to Recent (Cavender 1986). Today, there is only a relict distribution of a formerly much more widespread order (Patterson 1982a).

Family Amiidae *Bowfins*

Although there are a number of fossil amiid species from five continents, the Bowfin is the only surviving member of this ancient bony fish family. Fossils similar to *Amia*, dating to the Mesozoic of more than 100 million years ago and persisting into the Middle Tertiary 30 mya, have been found in Europe and the United States (Grande 1980; Wilson 1982; Cavender 1986). A more recent summary of the fossil record is found in Nelson et al. (2016), including a Miocene fossil *Amia* from Japan. The oldest fossil amiids

(most of which have been interpreted to have inhabited marine environments) are of Jurassic age (Maisey 1996; Grande and Bemis 1998, 1999; Forey and Grande 1998). The recent report of an Eocene marine amiid from Senegal in West Africa contradicts existing hypotheses that marine amiids were generally absent after the Cretaceous–Paleogene boundary, having been replaced by freshwater taxa (O'Leary et al. 2012). Bowfins, which are somewhat more derived than the Acipenseridae, have cycloid scales reinforced with ganoin, an abbreviate-heterocercal tail, a long spineless dorsal fin, a spiral valve in the intestine that is reduced in comparison to other ganoids and sharks (Hopkins 1895; Hilton 1900), and a primitive skeleton that is partly bone and partly cartilage. The skull has a gular plate (Fig. 11.1) between the lower jaws, a feature not found in any of our freshwater fishes, but occurring in some marine fishes such as the Tarpon and Ladyfish (family Elopidae).

The Bowfin is similar in appearance to teleosts and possesses amphicoelous vertebrae like those of teleosts. Wiley (1976) regarded Amiidae as the sister group to all higher bony fishes (teleosts), but Nelson (1969) considered gars (a nonteleost group) the closest relatives. Etnier and Starnes (1993) pointed out that *A. calva* retains the modified heterocercal caudal fin structure found in the gars, but the scales are of a more derived type, somewhat cycloid-like, but with parallel bony ridges. Its unique position in evolution and

161

Figure 11.1. Gular plate of *Amia calva. David A. Neely.*

fish systematics makes the Bowfin useful as specimens for teaching and display (Lee 2004), and it is sold by biological supply firms for those purposes.

Because of their predatory nature, Bowfins are often caught in the lowlands of Arkansas by fishermen seeking bass or other fishes. When hooked, it is an excellent fighter. Although not generally considered a premium food fish, the Bowfin when properly cleaned, then smoked or fried, rivals traditional gamefishes (Miles 1913; Gowanloch 1933). Because of this, it is taken by commercial fishermen in Arkansas. From 1975 to 1985, an average of 74,843 pounds (33,949 kg) of Bowfins per year were taken commercially in the state. In recent years, it has made up a small part of the roe harvest for making "Cajun caviar." Koch et al. (2009) reported that unprocessed Bowfin roe sold for $55 per kg ($25 per pound) in the upper Mississippi River basin, and its value in the lower Mississippi River basin was $80 per kg ($36 per pound). Becker (1983) noted that the Bowfin is frequently part of the fish fauna in many excellent sport and panfish waters in Wisconsin and attributed the quality of such waters in part to the presence of this species. Likewise, Scarnecchia (1992) advocated management and conservation of Bowfins (along with gars) to maintain balanced fish communities and to provide diversity in fishing opportunities for anglers.

The Bowfin is known by a wide variety of common names (Lee 2004). Its appearance inspires names like dogfish, scaled ling (a ling is a type of cod), blackfish, and cabbage pike (possibly because of the green fins in breeding individuals). Its habitat is reflected in the names mudfish, beaverfish, poisson de marais (fish of the marsh), and cypress trout. Its soft flesh is indicated by the name cottonfish. The Bowfin is also known as grindle, grinnel, John A. Grindle, and lawyer (J. J. Hoover, pers. comm.).

Genus *Amia* Linnaeus, 1766—Bowfins

This monotypic genus was established by Linnaeus in 1766 from specimens collected in South Carolina. Genus characters are the same as those of the species presented below.

Amia calva Linnaeus
Bowfin

Figure 11.2. *Amia calva*, Bowfin. *David A. Neely.*

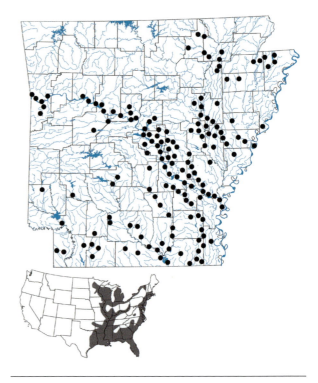

Map 14. Distribution of *Amia calva*, Bowfin, 1988–2019.

CHARACTERS   A moderate-sized, elongate, cylindrical fish with a large mouth, an abbreviate-heterocercal tail, and a long, low dorsal fin having more than 40 rays (Fig. 11.2). A large bony gular plate is present between the lower jaws. Upper jaw extending backward beyond rear margin of eye. Head length going into TL 3.8–4.9 times. Numerous sharp canine teeth present on upper and lower jaws. Body covered with cycloid scales; head naked. Nostrils prominent, each with a barbel-like flap. Swim bladder lunglike and joined to gut. Spiral valve in intestine has 4 or 4.5 turns, with the last turn about 1 cm from the vent. Lateral line complete with 65–70 scales. Fins without spines. Pectoral, pelvic, and caudal fin rounded. Dorsal fin rays 42–53; anal fin rays 9–12; pectoral rays 16–18. Pelvic fins abdominal, with 9 rays. Branchiostegal rays 10–13. The female is generally larger than the male. Size commonly 18–24 inches (457–610 mm) TL and weighing around 4 pounds (1.8 kg). Maximum length about 43 inches (1,090 mm) (Page and Burr 2011). Maximum weight about 20 pounds (9.1 kg). The Arkansas state record is 17 pounds, 5 ounces (7.9 kg) from Desha County in 1977.

LIFE COLORS   Back and upper sides a dark olive green with dark mottlings and reticulations on body and some fins. Undersides a lighter green. Upper base of caudal fin with a large black spot (ocellus), especially prominent in the young; in adult males, the black spot is bordered in orange or yellow; in adult females it is much less distinct or absent. Dorsal and caudal fins of breeding males are bright green with black bands. Other fins bright green with black margins. Young with prominent black edging to the fins, black stripes on the head, reddish fins, and no red encircling the spot at the caudal fin base (Fig. 11.3).

SIMILAR SPECIES   A distinctive fish not easily confused with any other native Arkansas species; however, the recently introduced Northern Snakehead (*Channa argus*) appears superficially similar to Bowfins (Courtenay and Williams 2004). The Bowfin can be distinguished from the Northern Snakehead by a shorter anal fin, having 9–12 rays (vs. anal fin long, usually with 31 or 32 rays) and pelvic fins that are at midbody (vs. pelvic fins closer to head). It can be further distinguished from snakeheads and all other fishes in Arkansas by its large gular plate (vs. gular plate absent).

VARIATION AND TAXONOMY   *Amia calva* was described by Linnaeus (1766) from Charleston, South Carolina. Genetic variation was studied by Bermingham and Avise (1986) and Avise et al. (1987), who separated populations in Gulf of Mexico tributaries west of the Escambia River from populations from the Escambia River eastward, based on differences in mitochondrial DNA. Although currently considered the only species in its genus, *A. calva* exhibits genetic variability in North America and new species may eventually be carved from it (D. Neely, pers. comm.).

DISTRIBUTION   The Bowfin is native to the St. Lawrence River–Great Lakes and Mississippi River basins from southern Canada west to northern Minnesota and south to the Gulf of Mexico. It also occurs in Coastal Plain drainages of the Atlantic and Gulf slopes from

Figure 11.3. Juvenile (70 mm TL) Bowfin showing orange and black colors, Altamaha River system, Georgia. *Richard T. Bryant and Wayne C. Starnes.*

Massachusetts to the Colorado River drainage, Texas. In Arkansas, it is found in all major drainages of the Coastal Plain lowlands and westward throughout the Arkansas River Valley (Map 14).

HABITAT AND BIOLOGY   *Amia calva* is primarily a lowland species and is widespread in sluggish rivers, oxbow lakes, bayous, and swamps. It prefers quiet, clear waters having aquatic vegetation, logs, or submerged debris. Pflieger (1997) noted that it avoids swift currents and excessive turbidity. It is most active at night, moving into shallower water to feed. During the day, it remains in deep water or in heavy vegetation but will occasionally seek food then. Its distinctive mode of swimming, dubbed ribbon-fin locomotion, occurs via longitudinal undulations of its long dorsal fin and is unique among North American freshwater fishes (Jagnandan and Sanford 2013). Preferred temperatures, determined experimentally, are high (>28.8–32°C) with pronounced diel variation (warmer temperatures during the day and cooler temperatures at night) (Reynolds et al. 1978). Under those conditions, there were no diurnal-nocturnal differences in activity. Midwood et al. (2018) studied seasonal movement and habitat selection in Canada. Bowfins showed high site fidelity in areas characterized by high, stable water temperatures and submerged vegetative cover. Bowfin residency increased with vegetative cover and was highest for large fish during fall and winter months. Several of the Bowfins were mobile during spring and summer months and moved distances of 5.2–12.9 km.

The swim bladder of the Bowfin functions as a lung, and under low-oxygen conditions the fish surfaces and gulps air to supplement gill respiration. It is not an obligate air breather, but when water temperatures exceed approximately 10°C, it begins to supplement its gill respiration by breathing air (Horn and Riggs 1973; Hedrick et al. 1994). In addition to its functional lung, Bowfins have specially modified gills in which lamellae of adjacent filaments are fused to one another, creating a honeycomb-like surface for gas exchange (Bevelander 1934). These sieve plates are presumed to have greater efficiency in absorbing oxygen in hypoxic water and to possibly operate in atmospheric respiration, since such gills would not collapse when the fish is out of water. Bowfins are reportedly capable of estivation under certain conditions, surviving by breathing air while lying in muddy burrows during periods of drought (Gowanloch 1933; Dence 1933; Neill 1950). Their respiratory adaptations (modified gills and lunglike swim bladder) make this tenable, but Bowfins, especially in cooler weather, would face physiological challenges from accumulations of carbon dioxide, ammonia, and associated waste products which could prove fatal (McKenzie and Randall 1990).

Microcrustaceans composed the first food of Bowfins in Alabama, and a shift in diet to small insects occurred at approximately 40 mm TL (Frazer et al. 1989). In Wisconsin, young Bowfins 45–70 mm TL fed on damselfly nymphs, chironomid larvae, and microcrustaceans (Becker 1983). Bowfins shifted to a largely piscivorous diet at about 100 mm TL, and adults fed mainly on fishes. The percentage of fish in the adult diet varies considerably in different areas. Pflieger (1997) reported that in southeastern Missouri fish represented about 65% of the diet of the adult Bowfin, with crayfish making up most of the remainder. It fed primarily on Gizzard Shad, followed in order by Golden Shiners, bullheads, and sunfishes. In two North Carolina rivers, the diet consisted mainly of crayfish and grass shrimp (Ashley and Rachels 1999), and in the Atchafalaya River drainage, Louisiana, it fed mainly on crayfish during periods of high water and on fish during periods of low water (Dugas et al. 1976). Adult Bowfins in southern Michigan fed principally on Yellow Perch and sunfishes (Lagler and Hubbs 1940), supporting the common misconception that bowfins are a threat to recreational (game) fisheries. However, field and laboratory studies indicate that Bowfins feed mainly on small individuals (Scarnecchia 1992) and on fish that are behaviorally impaired from stress, disease, parasites, or emaciation (Herting and Witt 1967). Impacts on gamefish populations, if any, are likely to be minor. Mundahl et al. (1998) reintroduced adult Bowfins (initial density approximately 32 fish/ha) into Lake Winona, Minnesota, from 1984 to 1986 to evaluate their effectiveness in controlling overabundant, stunted Bluegills, *Lepomis macrochirus*, and other centrarchids in a system with extensive macrophyte beds. Lack of natural reproduction by Bowfins, their rapid decline in numbers after stocking, and their low rate of sunfish consumption probably explain why Bowfins were ineffective in controlling the Lake Winona Bluegill population. Bowfins also preferred Fathead Minnows, *Pimephales promelas*, and crayfish, *Orconectes virilis*, over

sunfish in laboratory prey choice trials (Mundahl et al. 1998). Bowfins are often caught by anglers on natural baits and sometimes on artificial lures.

In Iowa, the Bowfin attained sexual maturity at 2 (males) or 3 years (females) of age (Koch et al. 2009). Spawning occurs in Missouri from early April into June (Pflieger 1997), in southeastern Louisiana from February to early March when water temperatures exceed 14°C (Davis 2006), and spawning has been observed in eastern Arkansas in early April. Reighard (1903) found that spawning activity occurred at water temperatures between 12°C and 26°C. Optimum temperature for nest construction and spawning is 16–19°C (61–66°F) (Scott and Crossman 1973). The male constructs a depression-like nest in shallow, weedy areas by fanning with the caudal fin and removing vegetation by mouth (Reighard 1903). Several males may make nests close together by fanning out the mud substrate. Spawning often occurs at night over the nest bottom and the adhesive eggs become attached to roots, vegetation, gravel, or other objects on the floor of the nest. A receptive female enters the nest and lies on the bottom while the male circles above her for several minutes. The male soon sinks to the bottom and lies beside the female. Both fish vibrate and eggs and milt are released over the nest bottom. Females are typically larger than males, and one male may spawn with several females. Nests often contain 2,000–5,000 eggs. Four- to 5-pound (1.8–2.3 kg) females produce 23,000–64,000 eggs, and each female may spawn in more than one nest. In the Mississippi River of Iowa, fecundity of mature females varied from 9,498 to 110,086 (Koch et al. 2009). The newly hatched black young have an adhesive organ at the tip of the snout with which they anchor themselves to objects in the nest until the yolk sac is absorbed (Reighard and Phelps 1908). The male actively guards the nest containing the developing eggs and yolk sac young against intruders (Doan 1938) and, in one case, a nest-guarding male launched itself from the water at humans on shore (Kelly 1924).

After the young develop into free-swimming fry, they remain together as a compact swarm and are herded and guarded by the male in the vicinity of the nest for a few weeks until they are about 4 inches (102 mm) long. Total parental care by the male (defense of embryos and young) may last up to two months. Cahn (1927) described the intensity of the defense of the young by the male, and noted that the male will attack anything that threatens the schooling mass of young. Because the young and adult males have a prominently developed caudal ocellus, it has been speculated that the ocellus plays a role in species recognition. It may act as a visual cue allowing the young

Figure 11.4. Male Bowfin in breeding coloration.
Patrick O'Neil, Geological Survey of Alabama.

to recognize and stay close to one another and to recognize and stay close to the male parent (J. J. Hoover, pers. comm.). HWR observed a swarm of young Bowfins estimated to contain about 1,000 individuals in a farmer's field following a flood event. Unfortunately, the entire swarm was left trapped as the high waters receded. Large swarms of several thousand young are also documented in the scientific literature (e.g., Schneberger 1937a) and in museum collections (Neil Douglas, pers. comm.). The Bowfin is the only nonteleostean bony fish in Arkansas to show such intensive parental care, and possibly only the cichlids and bullhead catfishes among Arkansas teleosts exhibit the same degree of aggressive and prolonged parental care as the Bowfin. In the upper Mississippi River of Iowa, total annual mortality of age 4 and older fish was approximately 35% (Koch et al. 2009). In Louisiana, estimated annual Bowfin mortality was 58% (Davis 2006). Maximum life span in Tennessee was around 10 years (Etnier and Starnes 1993). In the Mississippi River of Iowa, the oldest females were 13 years old, and the oldest males were 9 years old (Koch et al. 2009). Carlander (1969) reported that some individuals may live 30 years or more in captivity.

CONSERVATION STATUS This lowland species is widely distributed and common in Arkansas. The Bowfin provides a modest part of the commercial fishery in Arkansas, but in recent years, its unfertilized eggs (roe) have been harvested for use as a low-cost substitute for sturgeon and Paddlefish caviar. We consider its populations secure.

CHAPTER 12

Order Hiodontiformes
Mooneyes

The Goldeyes and Mooneyes were traditionally classified with the Osteoglossiformes (Bony Tongues), but Li and Wilson (1996) separated them into their own order, the Hiodontifomes. That separation was followed by Wiley and Johnson (2010), Page et al. (2013), Nelson et al. (2016), and Betancur-R. et al. (2017). Some morphological synapomorphies that define the Hiodontiformes are as follows: (1) the dermosphenotic bone is triradiate, (2) the opercle bone is an irregular parallelogram, (3) the dorsal arm of the posttemporal bone is more than twice as long as the ventral arm, (4) the hyomandibular bone is double-headed, (5) nasal bones are tubular and strongly curved, and (6) the pelvic fin has 7 rays. Study of the fossil hiodontid genus *Yanbiania* suggests that hiodontids separated from other osteoglossomorphs early and belong to a separate order. *Yanbiania* was originally considered an early member of the Hiodontidae based on features from the parasphenoid, the basihyal tooth plate, and the caudal skeleton (Li 1987). The order contains one family, Hiodontidae, which is indigenous to North America.

Family Hiodontidae *Mooneyes*

This primitive teleost family of freshwater fishes consists of four genera, one extant and three extinct. The extant genus, *Hiodon*, contains only two living species, the Goldeye (*H. alosoides*) and the Mooneye (*H. tergisus*), both of which are found in Arkansas. The mooneyes are North American in origin and distribution, with the earliest known fossils from the Paleocene (65–53 mya, Cavender 1986). Lauder and Liem (1983) reported fossil relatives of the North American genus *Hiodon* from China. The genus *Hiodon* is sister to the fossil genus *Eohiodon*, and several species of *Eohiodon* are known from Eocene deposits in western North America (Nelson et al. 2016). Other fossil hiodontid genera have been found in the Cretaceous of China (Li and Wilson 1996; Li et al. 1997).

Greenwood (1973) regarded hiodontids and their relatives to be closely related to clupeiform fishes, with these groups together forming the sister group to all other bony fishes above the level of the sturgeons, gars, bowfins, lungfishes, and a few other primitive groups. Goldeyes and Mooneyes superficially resemble clupeids. Both *Hiodon* species are generally silvery, slab-sided fishes with large forward-placed eyes (eye diameter greater than the length of the snout), a rounded snout, large cycloid scales on the body, adipose eyelids, pelvic axillary processes, a scaleless head, a complete but indistinct lateral line, a moderately long anal fin (anal fin base much longer than dorsal fin base), a connection between the swim bladder and the ear that consists of diverticula from the swim bladder that enter the cranium (Bond 1996), and a well-developed, forked caudal fin. They are sometimes called "toothed herrings" because of their superficial resemblance to herrings and the prominent teeth on the jaws, roof of the mouth, and the bony tongue. Mooneyes differ from true herrings by the obvious lack of the row of spiny scutes along the midline of the belly (but an untoothed, fleshy keel is present), a much more elongate body, the presence of a lateral line, and in having the dorsal fin base over the anal fin base (rather than entirely in front of the anal fin base in herrings). The soft-rayed mooneyes possess a large swim bladder and a single pyloric caecum. Eggs are ovulated directly into the body cavity and are not conveyed through oviducts (oviducts are present in most bony fishes) (Becker 1983).

Key to the Mooneyes

1A Dorsal fin rays 9 or 10; dorsal fin origin posterior to anal fin origin; iris golden; dorsal fin base approximately one-third as long as anal fin base; ventral keel long, extending from anus to bases of pectoral fins.
Hiodon alosoides Page 168

1B Dorsal fin rays 11 or 12; dorsal fin origin anterior to anal fin origin; iris silvery; dorsal fin base approximately half as long as anal fin base; ventral keel short, extending from anus to bases of pelvic fins.
Hiodon tergisus Page 170

Genus *Hiodon* Lesueur, 1818a—Mooneyes

The genus *Hiodon* contains two species and was first described by Lesueur in 1818. Genus characters are the same as those given for the family. Adaptations for feeding under low light conditions include large eyes with a well-developed guanine reflective layer (tapetum lucidum) in back of the retina (Moore and McDougal 1949). *Hiodon* species are not widely considered good food fishes. In some areas, Goldeyes are considered excellent food fish when smoked but are unpalatable when fried (Becker 1983). The Mooneye is of even less importance as a food fish.

Hiodon alosoides (Rafinesque)
Goldeye

CHARACTERS A moderately elongate, compressed, deep-bodied species having a large, terminal mouth (upper jaw extending backward beyond middle of eye) with well-developed canine teeth on the jaws and tongue, a blunt, rounded snout, and forward-placed, large eyes (Fig. 12.1). Eye diameter going 3.6–4.0 times into head length. Eyes partly covered anteriorly and posteriorly with adipose eyelids. Pelvic axillary process present. A fleshy keel extends forward from the anus nearly to the pectoral fin bases; keel

Figure 12.1. *Hiodon alosoides*, Goldeye. *Uland Thomas.*

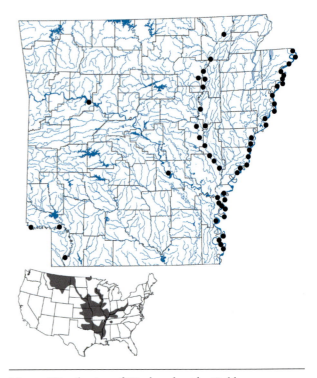

Map 15. Distribution of *Hiodon alosoides*, Goldeye, 1988–2019.

without sharp, saw-toothed projections. Dorsum nearly straight from occiput to dorsal fin origin. Dorsal fin origin behind the origin of the anal fin. Dorsal fin usually with 9 or 10 rays, rarely 11. Anal fin long with 29–34 rays; base of anal fin covered with 1–3 rows of small scales. Pectoral fin extending backward as far as pelvic fin origin; pectoral rays 11 or 12. Pelvic rays 7. Gill rakers 15–17, short and knoblike. Scales cycloid. Lateral line complete with 57–62 scales. Head scaleless. Caudal fin forked. In mature males, the anal fin has a greatly expanded, round anterior lobe; in females, the anal fin is concave. Maximum size 20 inches (51 cm) (Page and Burr 2011) and around 4 pounds (1.8 kg) (Hrabik et al. 2015). The Arkansas angling record from the Current River in 2017 is 1 pound, 10 ounces (0.74 kg).

LIFE COLORS Body with green to blue dorsum grading to silvery sides and milky white belly; entire body with a silvery sheen. Fins colorless or sometimes yellow to pink. Iris of eye yellow to gold.

SIMILAR SPECIES The Goldeye superficially resembles the shads or herrings; however, it differs in having well-developed canine teeth on the jaws and tongue (vs. canine teeth absent), a lateral line (vs. no lateral line), and a fleshy keel on the belly extending from the anus to beneath the pectoral fins (vs. a scaled keel). From the Mooneye, the Goldeye differs in usually having 9 or 10 dorsal rays (vs. 11 or 12), and in having the dorsal fin origin slightly behind the origin of the anal fin (vs. dorsal fin origin distinctly anterior to anal fin origin in the Mooneye).

VARIATION AND TAXONOMY *Hiodon alosoides* was described by Rafinesque (1819) as *Amphiodon alosoides* from the Ohio River (probably falls of the Ohio River at Louisville, Kentucky). There has been no study of phylogeographic variation, and no subspecies are recognized (Hilton et al. 2014).

DISTRIBUTION The Goldeye occurs in James Bay tributaries of Ontario and Quebec, Canada, and in the Arctic and Mississippi River basins from Northwest Territories, Canada, to western Pennsylvania and south to Louisiana. In Arkansas, it is restricted to the large rivers of the state and is most abundant in the lower White River and throughout the Mississippi River (K. J. Killgore, pers. comm.) (Map 15). The Goldeye is surprisingly absent from the Ouachita River in Arkansas. It occurs sporadically in the Red River (Buchanan et al. 2003) and in the Arkansas River, where it is recently known mainly from the lower portion in Arkansas. During the past three decades, it has declined precipitously in the upper Arkansas River in western Arkansas (compare Map A4.16 with current distribution map), and we know of only one post-1988 record from that area, a specimen taken from Dardanelle Reservoir in Yell County during a 1996 AGFC rotenone population sample. There are no records from the two uppermost navigation pools in Arkansas since 1 September 1979 (Ozark Pool) and 1 September 1983 (Pool 13). It is much more abundant in the lower Arkansas River, and in April 2012, anglers were observed cast-netting Goldeyes below Dam 2 for use as bait (Jeff Quinn, pers. comm.). Jeff Quinn (pers. comm.) and Jimmy Barnett collected the first known record (taken by electrofishing) from the Cache River in Woodruff County on 3 November 2016. Beadles (1979) reported this species from the Black River at Black Rock in Lawrence County, and there is only one post-1988 record from that drainage.

HABITAT AND BIOLOGY The Goldeye occurs in medium-sized to large rivers in Arkansas, but in northern parts of its range it is abundant in Missouri River reservoirs. An important commercial fishery for Goldeye existed in Lake Oahe, North and South Dakota, in the 1960s (Kennedy and Sprules 1967). In Mississippi, approximately 75% of the collections of Goldeyes were from rivers, as opposed to oxbow lakes and reservoirs (Ross 2001), and we have no records of its occurrence in reservoirs in Arkansas (except for spotty records from Arkansas River navigation pools). A small population persists in Lake Texoma, Oklahoma (Riggs and Bonn 1959; Lienesch et al. 2000). This schooling species prefers moderate to swift current over a firm sand substrate. In the Scioto River, Ohio, it was most abundant in the deeper pools where there was considerable current (Trautman 1981). In Nebraska, it was caught in nets at median depths of approximately 2 feet (0.91 m), with mean water column velocities of 1.8 feet/s (0.82 m/s) (Hrabik et al. 2015). It is more tolerant of turbidity than the Mooneye. The Goldeye is primarily a nocturnal species. Its eyes, having only rods and no cones, are adapted to dim light conditions and to turbid habitats (Becker 1983). It is found near the surface at night but seeks deeper water by day.

The Goldeye diet appears to be variable over its range. It is an active, fast-swimming, carnivorous fish that takes much of its food near the surface, often at night (Pflieger 1997). Age 0 Goldeyes in Canada consumed mainly calanoid copepods and cladocerans (Donald and Kooyman 1977). In the Missouri River, age 0 individuals 9.3–19.1 mm TL fed primarily on dipteran larvae and pupae, followed in decreasing order of importance by trichopteran larvae, ephemeropteran nymphs, copepods, cladocerans, coleopteran larvae, odonate nymphs, terrestrial invertebrates, hemipterans, plecopterans, and ostracods (Starks and Long 2017). The most important food items of Goldeye in North Dakota included cladocerans, chironomid larvae and pupae, and aquatic Hemiptera (Evenhuis 1970; Miller and Nelson 1974). In contrast, fish and both terrestrial and aquatic insects dominated the diet of Goldeye in Lewis and Clark Lake, South Dakota (D. H. Johnson 1963). It often congregated wherever small fishes were abundant in the Ohio River, such as in the swift waters below dams (Trautman 1981). Hoopes (1960) reported mayfly naiads (*Hexagenia*) composed 56% of stomach contents of Goldeye in the Mississippi River (Iowa) while *Potamyia flava* larvae made up about 19%, and immature Zygoptera and Odonata along with fish formed the bulk of the remaining contents. Adult terrestrial beetles were found in the guts of Goldeyes from the Mississippi River of Arkansas (Tumlison et al. 2017).

Goldeyes spawn in the early spring over shallow, gravelly shoal areas either in flowing water or shallow inshore areas of lakes at water temperatures of 50–55°F (10–12.8°C). Spawning occurs in Arkansas from late April to mid-June, based on collection of larvae from the Mississippi River (Beckett and Pennington 1986). In northern areas of its range, spawning occurs from May (just after the ice breaks up) to the first week of July (Becker 1983). Goldeyes migrate upstream in the spring prior to spawning (Eddy and Underhill 1974). Simon (1999b) considered this species a nonguarding, open substratum spawner (a lithophil—rock and gravel spawner) with pelagic, free embryos. Free embryos are pelagic by positive buoyancy or movement (Balon 1981; Simon 1999b). In Missouri, spawning aggregations move to shallow water around flooded vegetation in the Mississippi River (pers. obs., David A. Neely). Details of their spawning behavior are not known, but spawning is believed to occur in midwater at night (Wallus et al. 1990). Males are distinguished from females by the curvature of the anal fin margin, which is strongly sigmoid (convex anteriorly) in males but concave or nearly straight in females (Cross 1967) (Fig. 12.2).

An individual female (305–381 mm) may produce 5,700–25,000 semibuoyant, nonadhesive eggs (about 4 mm in diameter after fertilization) that float until hatching occurs (Battle and Sprules 1960). Eggs hatch in about 2 weeks, and the 7 mm larvae then float to the surface. Battle and Sprules (1960) and Wallus et al. (1990) provided a description of egg and larval development. Larval Goldeyes first appeared in May collections in the lower Mississippi River of Arkansas (Beckett and Pennington 1986). Recruitment of Goldeye larvae ceased during June sampling. Goldeye larvae were collected in the main channel, along revetted and natural banks, in a temporary secondary channel, and in dike fields, but were absent from backwater areas. No larval Goldeyes were ever taken past late June. This species is somewhat migratory during spawning season (Trautman 1981; Wallus et al. 1990). Males reach sexual maturity at 2–9 years while females mature in 3–10 years. The maximum life span is at least 14 years in some areas (Scott and Crossman 1973).

CONSERVATION STATUS   Impoundments on large rivers have jeopardized the Goldeye throughout much of its range (Boschung and Mayden 2004). In Arkansas the Goldeye is currently considered common in the Mississippi and lower White rivers, but it has apparently declined elsewhere in the state. Meek (1894a, b) listed the Goldeye as common in the Arkansas River at Mulberry and Fort Smith. In the 1970s and early 1980s, TMB frequently observed anglers using minnows to catch Goldeyes in the Arkansas River in July and August near Fort Smith, but Goldeyes have not been found there in at least three decades. We do not feel that it currently warrants a conservation status in Arkansas, but its populations should be monitored periodically to detect additional future declines.

Hiodon tergisus Lesueur
Mooneye

Figure 12.3. *Hiodon tergisus*, Mooneye. *Uland Thomas.*

CHARACTERS   Body elongate with dorsal fin origin slightly in front of anal fin origin; dorsal rays usually 11 or 12 (10–14). A fleshy belly keel extends forward from anus to base of pelvic fins; keel without sharp, saw-toothed projections (Fig. 12.3). Pelvic axillary process present. Snout bluntly rounded. Mouth with small, sharp teeth on upper and lower jaws, roof of mouth, and tongue. Upper jaw in adults extending backward just short of middle of eye. Eyes large, eye diameter going into head length 2.8–3.6 times. Adipose eyelids present. Tip of pectoral fin not extending posteriorly as far as pelvic fin origin. Gill rakers 15–17, short and knoblike. Pectoral rays 13–15; pelvic rays 7; anal rays 26–29. Base of anal fin usually covered by 2 or 3 rows

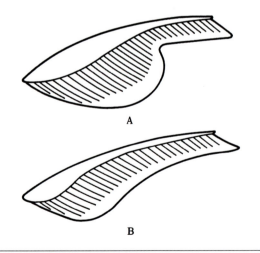

Figure 12.2. Anal fin shapes in *Hiodon. Becker (1983).*

A—Male
B—Female

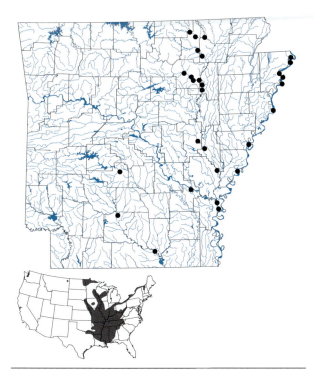

Map 16. Distribution of *Hiodon tergisus*, Mooneye, 1988–2019.

of small scales. Scales cycloid. Head scaleless. Lateral line complete, with 52–57 scales. Branchiostegals 7–9, usually 8. In mature males, the anal fin has a greatly expanded, round anterior lobe; in females, the anal fin is concave. Adults caught in Arkansas are usually 11–15 inches (279–381 mm) long and weigh 12 ounces to 2 pounds (0.3–0.9 kg). Maximum reported size 19 inches (47 cm) (Page and Burr 2011) and 2.65 pounds (1.2 kg) (Trautman 1981). There is no Arkansas state angling record.

LIFE COLORS Body greenish to silvery with faint steel-blue luster. Sides silvery grading to white below. Fins clear to dusky. Eye silvery with no yellow or golden pigment.

SIMILAR SPECIES *Hiodon tergisus* is similar to *H. alosoides* but has a deeper body, dorsal body outline gently curved (vs. straight), origin of the dorsal fin in front of the anal fin origin (vs. front of dorsal fin distinctly posterior to front of anal fin), dorsal fin with 11 or 12 rays (vs. 9 or 10), a somewhat larger eye with silvery iris (vs. a golden iris), and the fleshy ventral keel is shorter, extending forward from the vent to the pelvic fins (vs. fleshy ventral keel extending forward from the vent to the pectoral fins). The Mooneye can be distinguished from all clupeids by the presence of a lateral line (vs. lateral line absent), and by a fleshy, scaleless, ventral keel (vs. ventral keel scaled and with a saw-toothed margin).

VARIATION AND TAXONOMY *Hiodon tergisus* was described by Lesueur (1818a) from Lake Erie at Buffalo,

New York, and the Ohio River at Pittsburgh, Pennsylvania. Several synonyms exist because specimens of this species were described as a new species more than once (Smith 1979). Geographic variation of the Mooneye has not been studied, and no subspecies have been recognized (Hilton et al. 2014).

DISTRIBUTION The Mooneye occurs in the St. Lawrence–Great Lakes, Hudson Bay, and Mississippi River basins from Quebec to Alberta, Canada, south to Louisiana, and west to southeastern Oklahoma. It is also found in Gulf Slope drainages from Mobile Bay, Alabama, to Lake Pontchartrain, Louisiana. In Arkansas, it occupies large rivers, including the Arkansas, lower White, Black, Little Red, Strawberry, Spring, Current, and Ouachita rivers. It is sometimes found in the more turbid Mississippi River (Map 16). We have no records of this species from the Red River drainage of extreme southwestern Arkansas, but older records exist from the Mountain Fork and Little rivers of that drainage in Oklahoma (Miller and Robison 1973). Older records are also known from the upper Ouachita and Saline rivers (Map A4.17), and McAllister et al. (2009c) and J. W. Quinn (pers. comm. 2019) documented Mooneye specimens from the lower Ouachita River in Ouachita and Union counties. There are also older records from the upper White River in Benton and Washington counties, but there are no post-1988 records from those counties.

HABITAT AND BIOLOGY The Mooneye inhabits large, clear streams, rivers, and lakes and seems less tolerant of turbid waters than the Goldeye (Miller and Robison 2004), although it is more common in the Mississippi River of Arkansas than previously thought. In some areas outside Arkansas, it is reported to occur in impoundments (Page and Burr 2011). In Mississippi, 94% of the collections of this species were made in rivers (Ross 2001), and we have no records in Arkansas from reservoirs or oxbow lakes. It is most often taken in current over a firm bottom in deep water. It has declined in numbers in many parts of its range and is not common today in Arkansas. Habitat fragmentation, altered flow, and water-quality regimes resulting from dam construction, land-use activities, and the introduction of exotic species have probably all played a role in its decline in Arkansas.

The Mooneye feeds predominantly on aquatic insects (nymphs, larvae, and pupae), but crayfish, mollusks, and small fish are also consumed (Glenn 1975b). In Iowa, the diet also consisted mainly of immature aquatic insects, predominantly ephemeropterans, trichopterans, and simuliids (Coker 1930). Boesel (1938) found that Mooneyes up to 39 mm TL fed on zooplankton (entomostracans), and individuals 57–133 mm TL fed on insects, mostly those that had fallen into the water. Mayfly nymphs, midge and caddisfly larvae, and amphipods were also consumed. Other reported foods include immature plecopterans, dipterans, and odonates (Hoopes 1960). The structure of its eye indicates that it is probably a sight feeder, feeding most actively under low-light conditions at night or near dusk. Young Mooneyes feed more in the water column, but adults feed mainly near the surface (Glenn 1975b). Young-of-the-year smaller than 65 mm TL fed mainly on immature caddisflies, mayflies, and midges in June, and by mid-July, corixids became important in the diet (Glenn 1978). In Ohio, it fed mostly in swift waters such as those below dams (Trautman 1957).

Spawning occurs in March in the Current River in Missouri (Pflieger 1997) and in April and May in Wisconsin (Becker 1983), with adults migrating to clear tributary streams to spawn over rocks and swift gravel shoals. Gravid females were collected in Alabama between 13 March and 1 May (Katechis et al. 2007). Spawning probably occurs in Arkansas from late April to mid-June. Boschung and Mayden (2004) reported that Mooneyes were collected in the spring over habitats that appeared to be conducive to spawning such as clear, flowing water, over rocky or coarse substrates. Wallus and Buchanan (1989) suggested that Mooneyes undergo spring spawning migrations to flowing areas of the impounded Tennessee River system. Glenn and Williams (1976) reported that spawning began after 8 May and was completed by 12 June in the Assiniboine River, Canada. Males mature at 3 years of age and females at 5 years (Glenn 1975a). Wallus and Buchanan (1989) reported fecundity estimates for the Mooneye ranging from 3,037 to 7,773 eggs per female in the Tennessee and Cumberland river systems, and Katechis et al. (2007) reported fecundity of 4,956 to 8,912 ova per female in the lower Tallapoosa River, Alabama. Mooneye eggs are buoyant and nonadhesive and develop as they drift in the current (Boschung and Mayden 2004); therefore, this species probably requires moderate discharges and stable flow conditions for successful egg development. Mean diameter of eggs in the Tallapoosa River ranged from 2.16 to 2.77 mm, similar to the 2.0–2.5 mm egg diameter reported for Mooneyes in the Tennessee and Cumberland rivers (Wallus and Buchanan 1989). Mooneyes are probably complete spawners, releasing all their eggs at one time (Katechis et al. 2007). Beckett and Pennington (1986) reported Mooneye larvae first appeared in larval fish samples in the lower Mississippi River in Arkansas in May. Recruitment ceased during June. Larvae were found along revetted and natural banks in the main channels, in a temporary secondary channel, and in dike fields, but were absent from backwaters. No larvae were collected past late June. In Wisconsin, Mooneyes reached 5.5–8.4 inches (140–213 mm) during the first year of life (Becker 1983). A collection of 27 young-of-the-year Mooneyes from the Black River in Jackson County, Arkansas, on 1 August 1973 by TMB averaged 3.7 inches (96 mm) TL. Adults as small as 9.5 inches (241 mm) are sometimes mature (Coker 1930), and 6 or 7 years are required to reach a length of 13 inches (330 mm). Maximum life span is probably not more than 8 or 9 years (Katechis et al. 2007).

CONSERVATION STATUS   The Mooneye appears to be imperiled throughout much of its range today (Hilton et al. 2014). Katechis et al. (2007) suggested Mooneyes may also be negatively affected by the presence of predator species, specifically Striped Bass that have been landlocked because of dam construction (Boschung and Mayden 2004). They believed that Mooneyes may be vulnerable to Striped Bass predation, especially during the spawning season. The species may have declined somewhat in Arkansas, but stable populations remain in the lower White River and in the Mississippi River. We tentatively consider this species stable. It is probably not as common in Arkansas as the Goldeye, and its populations should be periodically monitored to detect any future declines.

CHAPTER 13

Order Anguilliformes
Eels

This large order of elopomorph fishes (subdivision Elopomorpha) contains 8 suborders with 19 families, 159 genera, and about 938 species (Nelson et al. 2016). Most eels live in shallow tropical or subtropical marine habitats, and only a few species of a few families other than Anguillidae ever occur in fresh water. Anguilliformes species are characterized by a very elongate body and no pelvic fins. The pectoral fins are also lacking in some species. The pectoral girdle, if present, is unattached to the skull, and the dorsal and anal fins are usually very long and are confluent with the caudal fin (if a caudal fin is present). The gill openings are usually narrow and placed back from the concealed edge of the operculum, but the gill region is elongate and the gills themselves are displaced posteriorly. Gill rakers and pyloric caeca are absent. Teeth are present on the maxilla bordering the mouth. The two premaxillae, the vomer (usually), and the ethmoid, are united into a single bone. There are 6–49 branchiostegal rays. A swim bladder is present, oviducts are absent, fin spines are lacking, and scales are usually absent, or if present, are cycloid and embedded. As in all members of the teleost subdivision Elopomorpha, there is a leptocephalus larval stage. The planktonic larvae are thin, transparent, and leaflike and actively swim and/or possibly drift passively with currents (M. J. Miller, pers. comm.). The leptocephalus larva has a small, round caudal fin that is continuous with the dorsal and anal fins.

Family Anguillidae *Freshwater Eels*

The family Anguillidae ranges from the Atlantic temperate waters to the Indo-Pacific regions and includes 22 species in the genus *Anguilla* and one recently described species in the genus *Neoanguilla* from Nepal (Nelson et al. 2016). Interestingly, the leptocephalus larva of anguillid eels, with its elongate, leaflike, compressed, transparent body, was originally described as a separate species because it showed such little resemblance to the adults.

The fossil record of eels extends back to the Late Cretaceous, but there are no fossil records of eels in North America (Cavender 1986). Eels are superficially similar to lampreys, but the two are not closely related. Although the term "lamprey-eel" is sometimes used by anglers, the only thing eels and lampreys have in common is a long, snakelike body. Eels differ from lampreys in having well-developed jaws, minute imbedded cycloid scales, and pectoral fins. The morphological relationships among families of eels have been difficult to determine, but genetic studies have indicated that the family Anguillidae is derived from eels that are mesopelagic and not from eels with more similar body forms like conger eels (Inoue et al. 2010). Freshwater eels are catadromous and exhibit one of nature's most incredible migrations between freshwater and saltwater environments (see the *A. rostrata* species account). However, Tsukamoto et al. (1998) discovered that some

173

adults of catadromous species from both the Pacific and Atlantic oceans apparently never enter freshwater. Eels are sometimes encountered in Arkansas by anglers, and commercial fishermen occasionally take them in nets, seines, or on trotlines. The freshwater eel population in Arkansas has been reduced by the construction of dams on virtually all the state's major rivers that impede their upstream migration.

Genus *Anguilla* Schrank, 1798— Freshwater Eels

The genus contains 22 species (Nelson et al. 2016). Characters include a very elongate body, scales absent or small and embedded, and lack of pelvic fins. All freshwater eels in North America belong to a single species, *Anguilla rostrata*, the American Eel.

Anguilla rostrata (Lesueur) American Eel

Figure 13.1. *Anguilla rostrata*, American Eel. *Uland Thomas.*

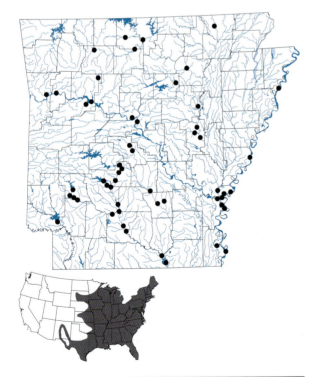

Map 17. Distribution of *Anguilla rostrata*, American Eel, 1988–2019.

CHARACTERS An elongate, slightly compressed, serpentine fish without pelvic fins, not resembling any other Arkansas fishes except lampreys, and then only superficially (Fig. 13.1). Hinged lower jaws are present with well-developed enameled teeth. There are two nostrils and only one gill slit on each side. Anterior nostrils are tubular and near the snout tip, and the posterior nostrils are round and in front of the eyes. Gill opening a short, narrow slit immediately in front of pectoral fin. Eyes well-developed. Mouth large; slightly oblique. Snout depressed and blunt. Lower jaw protruding beyond upper jaw. Dorsal, caudal, and anal fins continuous. Caudal fin bluntly rounded. Dorsal fin origin far behind pectorals; dorsal rays 240–244. Pectoral fins present; pectoral rays 14–20. Anal fin rays 200–206. Branchiostegals do not overlap. Scales cycloid, extremely small, mosaic (separated from one another or meeting at their margins), and embedded in the skin, making them difficult to see. Scales do not develop until individuals are 2–3 years old. Lateral line well developed. Adults found in freshwaters are generally 15–30 inches (381–762 mm) long and 3 pounds (1.4 kg) or less in weight. Maximum length around 60 inches (1,524 mm) (Page and Burr 2011), and maximum weight about 8 pounds (3.63 kg) (Ross 2001). Eels caught in Arkansas (293 specimens) by Cox (2014) ranged from 9.3 to 37 inches (236–940 mm) TL ($\bar{x}$ = 17.7 inches [449 mm]). The Arkansas state angling record from the Spring River in 2015 was 5 pounds, 4 ounces (2.38 kg) and 38 inches (965 mm) TL.

LIFE COLORS Coloration varies with life history stage. Body color in freshwater usually yellowish to brownish or greenish above to whitish below. Fins similar in color to adjacent body parts. Seagoing adults develop an overall silver or bronze luster and are called silver eels or bronze eels (Boschung and Mayden 2004). Eels collected from the Ouachita River in September 2017 by Jeff Quinn (pers. comm.) exhibited a range of colors including yellow, dark gray (silver), and dark brown.

SIMILAR SPECIES No other Arkansas fish species possesses the serpentine body form of the American Eel except perhaps lampreys. Eels are easily separated from all lampreys in having jaws (vs. no jaws), paired pectoral fins (vs. no pectoral fins), two nostrils (vs. one nostril), and only one gill slit on each side (vs. 7 gill slits on each side).

VARIATION AND TAXONOMY *Anguilla rostrata* was described by Lesueur (1817a) as *Muraena rostrata* from Cayuga Lake, New York. The American Eel is morphologically similar to, but genetically distinct from, the European Eel, *A. anguilla* (Lee 1980c; Inoue et al. 2010). A relatively modest effort has been devoted to the study of the population genetics of the American Eel (Cote et al. 2013). Avise et al. (1986) found that *A. rostrata* comprised a single genetically homogeneous population, based on the analysis of a single maternally inherited mitochondrial DNA locus, later confirmed by Wirth and Bernatchez (2003). Cote et al. (2013) examined 18 microsatellite gene loci of the American Eel from 32 sampling localities and concluded that all measures of differentiation were essentially zero, and there was no evidence for significant spatial or temporal genetic differentiation. Therefore, the panmixia hypothesis was supported for this species.

DISTRIBUTION   The catadromous American Eel spawns in the Atlantic Ocean and ascends streams in North America and probably the northern margin of South America. In freshwater, it occurs in Atlantic, Great Lakes, Mississippi River, and Gulf of Mexico basins from Newfoundland and Labrador, Canada, west to South Dakota and south to South America. In Arkansas, it is most common in the Ouachita River and lower White River (Map 17). Generally, it is found in the larger rivers in the state. Old AGFC rotenone records on onionskin paper show that eels were common in the Little Red River drainage prior to impoundment of Greers Ferry Lake, and a fair number of eels have been caught in Mercer Bayou in Miller County (Jeff Quinn, pers. comm.). McAllister et al. (2009c) and Tumlison and Robison (2010) presented new distribution records of eels in Clark, Hot Spring, Little River, and Ouachita counties.

HABITAT AND BIOLOGY   American Eels inhabit estuarine and freshwater habitats during their continental phase, a diversity of habitat that, coupled with their migratory plasticity, has enabled them to occupy the most extensive latitudinal range (more than 10,000 km of continental coastline approximately between the latitudes of 7° and 55° north) of any fish in the Americas (Helfman et al. 1987; Edeline 2007). *Anguilla rostrata* frequents large to moderately large streams and rivers but is occasionally found in smaller streams and reservoirs. Typically, eels prefer areas of vegetation and undercut banks over a soft-bottom substrate of mud or sand. They are also found among boulders, riprap, and root wads. In a life history study in Arkansas, 61% of the eel specimens were caught downstream of dams in the lower White and Ouachita rivers (Cox et al. 2016). In that study, the Ouachita River sites had substrata of varying size. Large woody debris was common, and water clarity was high, providing visibility to the streambed at most sites. The main channel of the lower White River study area was characterized by deep, turbid waters and high amounts of large woody debris (Cox et al. 2016). Becker (1983) reported that eels avoid cool-water habitats. Eels are common in the lower Arkansas River, where they

are sometimes unintentionally caught by anglers using live bait. Cox (2014) found that eel abundance below Norrell Dam of the Arkansas River Navigation System was linked to river flow in the lower White River. Most eels were captured during high spring flows; other captures were usually associated with spikes in river flow. Eel abundance below Norrell Dam also appeared to be related to water temperature. All but two eels were collected in water temperatures of 18–27°C. Tumlison and Robison (2010) reported the capture of 35 eels, ranging in length from 205 to 450 mm, in March after a drawdown of DeGray Lake caused reduced flow in the Caddo River. Upstream movement of eels has been linked to increased river flow, low lunar illumination, and temperature (Hammond 2003; Hildebrand 2005; Schmidt et al. 2009). It is likely that some combination of those environmental factors triggers upstream migration in freshwater riverine environments (Welsh and Liller 2013).

The carnivorous American Eel is reportedly a food generalist, acting as a bottom feeder eating a variety of aquatic insects, crustaceans, small fish, and other organisms. Young eels smaller than 16 inches (400 mm) feed primarily on aquatic insect larvae and small crustaceans, and larger eels feed mainly on fish and crayfish (Ogden 1970; Facey and LaBar 1981). Casey Cox (pers. comm.) provided the following data on food habits of eels in Arkansas. Eels from 236 to 840 mm TL fed most heavily on crayfish by percent volume and percent frequency of occurrence (approximately 60–70% of stomachs examined). Detritus was found in 18–38% (three length categories) of stomachs, and benthic macroinvertebrates other than crayfish (amphipods, isopods, ephemeropteran nymphs, megalopterans, and odonates) in 8–16%. Mollusks were eaten only by medium (303–499 mm TL) and large (505–840 mm TL) eels (1% and 8%, respectively), and fish were consumed only by large eels (3% of stomachs). Boschung and Mayden (2004) reported that eels larger than 400 mm TL also feed as scavengers on freshly dead animals, but other authors had previously concluded that eels prefer living rather than dead organisms (Scott and Crossman 1973; Becker 1983). Eels have been known to attack fish trapped in gill nets (Scott and Crossman 1973) and have been observed feeding on dead livestock that has washed into a stream (Harlan and Speaker 1956). Eels remain hidden by day and feed most actively at night. In its oceanic environment, the leptocephalus larva (Figure 13.1) has teeth and actively feeds on particulate organic matter termed "marine snow" (M. J. Miller 2009; Miller et al. 2011, 2013).

Without a doubt, the catadromous American Eel exhibits one of the most remarkable life histories of any Arkansas species. *Anguilla rostrata* mixes its entire, sexually mature

Figure 13.2. Leptocephalus larva. *Michael J. Miller.*

adult population every year by reproducing in the Atlantic Ocean south of Bermuda; therefore, the species forms a single homogeneous population (Wirth and Bernatchez 2003; Prosek 2010). Adults, after several years of living in freshwater, migrate downstream to the ocean and find their way to their ancestral spawning area in the Sargasso Sea between Bermuda and the Bahamas to spawn in the upper few hundred meters of the ocean. After spawning, all adults die. Tsukamoto et al. (2011) reported that spawning of other eel species probably occurs at less than 300 m depth in the surface layer and inferred that the American Eel also spawns at a similar depth. Schmidt (1923) was the first to provide evidence that spawning occurred in the Sargasso Sea by catching small *A. rostrata* larvae there, and this was subsequently confirmed by other studies of larval distribution (Kleckner and McCleave 1988). Breeding occurs from late winter (February) into early summer. Fecundity has been reported as high as 22 million eggs per female (Barbin and McCleave 1997; Tremblay 2009). The spawning behavior is cloaked in mystery and the actual event has not been observed.

Boschung and Mayden (2004) listed the morphological and physiological stages of the *A. rostrata* life cycle: fertilized egg > leptocephalus larva > glass eel > elver > yellow eel > silver eel > death after spawning. Only two of these stages are found in Arkansas, the yellow eel stage and individuals metamorphosing into the sexually mature silver eel stage. Casey Cox (pers. comm.) provided the following summary information on the occurrence of silver eels in Arkansas. In his study, yellow eels were the most common eel found (out of 295 eels examined). Some of the eels had GSI values greater than 2%, which was the threshold used by Tremblay (2009) to determine whether eels were at the beginning of the silver stage. GSI values in Arkansas ranged from 0.01% to 4.8% and rate of GSI increase differed significantly among river systems, suggesting different energy allocation strategies (Cox 2014). Cox could not determine whether the Arkansas eels were fully metamorphosed silver eels because of the lack of documentation of downstream migration. However, some of the larger eels with higher GSI values were caught on trotlines, indicating that they

were still actively feeding. Silver eel digestive tracts atrophy at some point during their downstream migration before they reach the ocean; therefore, it is assumed that fully mature silver eels don't feed. Cox examined guts of all eels caught, and none showed signs of digestive tract atrophy. Some of the ovaries of large eels examined by Cox contained eggs that were very well developed, and he concluded that a small number of the large, apparently mature eels were possibly silvers. If those large individuals weren't fully matured into the silver stage, they were very close. The average age of eels in Arkansas with GSI values similar to mature silver eels reported elsewhere (Tremblay 2009) was 9.4 years (Cox 2014). In 2017 and 2018, Jeff Quinn (pers. comm.) provided the following additional information on the occurrence of silver eels in Arkansas and documented downstream migration. Large eels (>680 mm TL) taken from the Ouachita River on 5 September 2017 exhibited characteristics similar to those described for silver eels elsewhere, including coloration, eye index, pectoral fin index, Fulton Condition Factor, and lateral line pore size. Nine eels from the Ouachita River at Grigsby Ford in Hot Spring County were fitted with radio transmitters on 5 September 2017 and released. On 9 November 2017, 7 of the 9 eels were still present at the Grigsby Ford site. Between 24 and 29 December 2017, four of the tagged eels passed through Felsenthal Lock and Dam in Union County near the Arkansas-Louisiana state line, indicating that downstream migration was underway. On 27 February 2018, another tagged eel passed through Felsenthal Lock and Dam. The two migration periods appeared to be related to high flow events on the Ouachita River.

After spawning in the Sargasso Sea, the eggs of *A. rostrata* hatch into planktonic leptocephalus larvae (Fig. 13.2) that are transparent and ribbonlike and drift with the Atlantic Ocean currents. Most eels are dispersed along the East Coast of North America by the Gulf Stream, and fewer eels are apparently dispersed into the Gulf of Mexico and to northern South America by other currents (Tesch 2003). It takes the larvae approximately 200 days to reach the continental shelf (Wang and Tzeng 2000). The marine leptocephalus stage metamorphoses into the glass eel phase usually prior to arriving at the continental shelf. Glass eels enter coastal rivers and streams during late winter and spring. The transparent glass eels then change into the fully pigmented elver stage when about 50–64 mm in length. The elver stage remains in the marine environment of estuaries or the backwaters of coastal rivers and streams or moves directly upstream in early spring. There is no evidence that elver stage eels ever reach as far inland as Arkansas (M. J. Miller, pers. comm.; C. A. Cox, pers.

comm.). Haro and Krueger (1991) suggested that elvers travel only limited distances in freshwater. Metamorphosis into the yellow eel stage occurs at about 100 mm TL, long before the eels arrive in Arkansas. Some of the eel population remains in the marine environment throughout life (Lamson et al. 2006). It has long been thought that male yellow eels remain near the coastal regions in brackish or freshwater, and only the females ascend the freshwater streams of the continent for hundreds of miles, where they remain for 5–20 or more years before returning to the sea. Some evidence, however, indicates that sex ratios may vary in some subpopulations of the American Eel's inland range, and males do move into inland waters in some drainages (Oliveira 1999). Sex ratios may be related to population density, with greater densities associated with a predominance of males (Krueger and Oliveira 1999). In large watersheds, eel abundance decreases with distance upstream, and only females are found far inland. Cox (2014) examined gonads of 226 Arkansas eels, and all were females. In the White River, eels appeared to have a greater minimum size upon arrival than those in the Ouachita River, allowing them to reach a GSI of 1.5 up to four years earlier in the White River (Cox et al. 2016). Those authors also reported that downstream migration of eels appeared to occur five years earlier in the White River population than in the Ouachita River population and speculated that the difference may be attributed to increased river fragmentation by dams in the Ouachita River basin. Males are generally smaller than females, and both sexes develop large eyes just before migration. In the southeastern United States, females tend to remain in inland waters for about 9 years, while males are present in freshwater for about 5 years (Rohde et al. 2009). Once maturation begins, both sexes begin the migration back to the ocean during the fall to midwinter. During this time, they become fully metamorphosed into the silver eel stage. Boschung and Mayden (2004) reported that female silver eels in the southeastern United States ready for migration to the spawning area are an average 8.6 years old, weigh 448 g, and are 584 mm TL, while male counterparts are 5.5 years old, weigh 69 g, and are 329 mm TL. In Arkansas, eels in the Ouachita River basin had a mean TL of 411.6 mm (N = 238), White River eels had a mean TL of 591.7 mm (N = 39), and Arkansas River eels had a mean TL of 683.5 mm. Mean estimated ages of the eels from the three river basins were 7.2, 5.8, and 9.4 years, respectively. Eels in Arkansas appear to mature relatively quickly compared with other inland subpopulations. Cox (2014) found that eels arrived in Arkansas at age 3 or 4 and left at ages 9 to 18. Maximum age of adult eels is more than 20 years, and silver eels in inland portions of the

St. Lawrence River were on average approximately 20 years old (Tremblay 2009).

Helfman et al. (1987) observed that American Eel lengths and ages at metamorphosis to the silver phase increased with latitude for female eels but not for male eels and hypothesized that females employ a size-maximizing strategy of migrating when an optimal size has been reached. In contrast, males employ a time-minimizing strategy and migrate when they achieve some minimum size. In both sexes, the minimum size at migration must be sufficient to enable maturation of the gonads and for reaching the spawning area (Jessop 2010). Oliveira (1999) proposed a more flexible female eel strategy that optimizes size at migration with respect to local growth conditions by balancing increased prereproductive mortality at older, larger sizes with increased fecundity at larger size. Therefore, it appears that each sex has a distinct life history strategy. Jessop (2010) examined latitudinal variability in length and age at maturity and growth rate for *A. rostrata* along the Atlantic Coast of North America. He found maturing (silver phase) female lengths and ages increased with increasing latitudes (and distance) from the Sargasso Sea spawning site, as did male ages, but not lengths. Female and male growth rates declined with increasing latitudes south of 44° north. Therefore, there is latitudinal differentiation among American Eel subpopulations. Jessop (2010) found the temperature-size rule (increase in body size at lower temperatures) evidently applies to American Eel females, but not to males. Winemiller and Rose (1992) concluded that temperate-zone anguillid eels have a periodic life history strategy. Individuals delay maturation to attain a size sufficient to enhance adult survival during periods of suboptimal environmental conditions such as, winter, variable growth conditions, and to attain high batch fecundity to take advantage of temporal and spatial patterns in conditions favorable for the survival and growth of larvae (Jessop 2010).

Worldwide, eels represent a fishery of probably hundreds of millions of dollars, but only a small percent of that occurs in the United States (Ross et al. 1984; Michael J. Miller, pers. comm.). Although not considered a food fish in Arkansas, they are delicious when grilled Japanese style in sweet soy sauce. The blood is reported to be a neurotoxin when fresh and can cause infection in humans (Herald 1961).

CONSERVATION STATUS Jessop (1997) and Haro et al. (2000) presented evidence for a decline in the populations of American Eels at regional and continental scales. Potential factors that may be influencing the decline include barriers to migration, habitat loss and alteration, hydro turbine mortality, oceanic conditions, overfishing, parasitism, and pollution. Dams on major rivers have severely limited the extent of upstream migration and greatly affected the distribution of the American Eel. However, adults are sometimes known to move overland on rainy nights to get around low falls and dams. It is estimated that only 16% of historical American Eel habitat has unobstructed connectivity to the sea (Fisheries and Oceans Canada 2014). The U.S. Fish and Wildlife Service was petitioned during 2005 and 2010 to list the species as threatened (Prosek 2010), and again more recently, but found in each review that the species is not threatened with extinction. The American Eel has declined in some Atlantic Coast drainages such as the St. Lawrence River drainage at the northern margin of the species range. Because of that decline and the uncertainty surrounding the causes of declines in anguillid eels, the American Eel, along with the European Eel and Japanese Eel, are now listed as endangered by the IUCN (Jacoby et al. 2015). Its status within the Mississippi River basin is poorly understood (Quinn et al. 2012), but it may be declining there as well (Jeff Quinn, pers. comm.). Today, eels are probably most common in Arkansas in the Ouachita River basin, followed by the lower White River (Cox et al. 2016). Cox (2014) found eels to be locally abundant in high-quality habitat in the Ouachita, Saline, and lower Caddo rivers in Arkansas. In 2012 and 2013, Cox (2014) collected 295 *A. rostrata* from those rivers (and the lower Arkansas River), indicating that Arkansas is still a valuable production area for this species. Statewide, however, eels have experienced a decline due to the construction of dams (which block or impede migration) on many of our waterways. Even though it is apparently more common in Arkansas than previously thought, the American Eel should probably be considered a species of special concern because of its overall pattern of decline.

CHAPTER 14

Order Clupeiformes

Herrings

The order Clupeiformes comprises 5 families, 92 genera, and 405 species (Nelson et al. 2016). Classification of these fishes above the order level remains controversial. Mitochondrial and nuclear DNA analyses, as well as morphological analyses often produce different groupings of clupeiforms and other orders at the level of superorder, cohort, and subdivision. About half the species in Clupeiformes are found in the Indo-West Pacific, almost one-quarter are in the Western Atlantic, and only 79 species occur primarily in freshwater (Nelson et al. 2016). Most species are adapted for living in well-lighted surface waters, where they school and feed on plankton. The Clupeidae and Engraulidae (the herrings and anchovies) are by far the largest and most individually numerous families in the order.

A specialized character that identifies Clupeiformes as a monophyletic group is the ear-swim bladder connection (Patterson and Rosen 1977; Grande 1982a, b). In clupeiform fishes, the forward extensions of the swim bladder form two large vesicles that enter the skull and terminate within ossified bullae (expansions of the prootic and pterotic bones). This type of ear-swim bladder connection is unique to the Clupeiformes (Greenwood et al. 1966; Greenwood 1973; Lauder and Liem 1983). Another diagnostic feature of living clupeiforms is a unique type of caudal skeleton, where the urostyle is made up of the centrum of the terminal vertebrae and the first uroneural; the first hypural is separated from the urostyle (Gosline 1960; Greenwood et

al. 1966; Whitehead 1985). Clupeiforms are further characterized by one or more abdominal scutes (each an unpaired element that crosses the ventral midline of the body), a compressed body with cycloid scales, no scales on the head, no spines in the fins, no adipose fin, abdominal pelvic fins, nonprotrusible jaws, and a pneumatic duct connecting the swim bladder to the gut. In addition, in Arkansas species, the acousticolateralis system is limited to the opercle and subopercle of the gill cover (i.e., there are no lateral line pores) and there are numerous, long, closely set gill rakers that filter plankton from the water.

Marine clupeiform fossils are known from the Lower Cretaceous, but the oldest freshwater clupeid, *Knightia vetusta*, dates from the Middle Paleocene in Montana. One family, Clupeidae, inhabits Arkansas.

Family Clupeidae *Herrings*

The herrings represent a large family of silvery, slab-sided, mostly schooling fishes of worldwide distribution with many marine, several freshwater, and some anadromous species. The earliest fossils of clupeids are from the Cretaceous of Europe, about 120 mya (Grande 1985). Worldwide, Clupeidae includes about 64 genera and 218 species, approximately 57 of which are freshwater species (Nelson et al. 2016). Herrings are small fishes, with most species smaller than 1 foot (300 mm) TL, but some reach 2.5 feet (760 mm). Despite their generally small size, several species are valuable food fishes. Clupeids, including the Atlantic Herring, the Pacific Sardine, and the menhaden of the Atlantic Coast and the Gulf of Mexico are of worldwide economic importance either as food fishes, or as sources of protein concentrate, fertilizer, or oil (Berra 1981). None of the four Arkansas species of clupeids is an important food fish. The rare Alabama Shad, *Alosa alabamae,* is edible, and the Skipjack Herring, *Alosa chrysochloris,* (commonly caught on hook and line) is considered a food fish in some states. The Gizzard Shad, *Dorosoma cepedianum,* and the Threadfin Shad, *D. petenense,* are not edible but are important forage species that provide food for larger predatory fishes, especially in reservoirs.

Fishes of this family are characterized by a ventral keel having sharp, saw-toothed projections. The silvery-colored herrings have no adipose fin and no lateral line. They are compressed, deep-bodied fishes having easily shed cycloid scales, a naked head, adipose eyelids (transparent eye coverings with vertical slits), a deep, small, flaplike projection (axillary process) at the upper margin of the pectoral and pelvic fin bases, and a deeply forked caudal fin. Most are schooling fishes that feed on planktonic organisms, grow rapidly, and have a short life span. The long, slender larval forms are quite different from adults in appearance.

All Arkansas clupeids are spring and early summer spawners. No nest is constructed; the eggs are deposited at random and no parental care is provided. Herrings most closely resemble the mooneyes (Hiodontidae) but can be distinguished from them by the dorsal fin, which is far in front of the anal fin base (vs. dorsal fin base over or behind front of anal fin base), and by the lack of a lateral line (vs. lateral line well developed).

Key to the Herrings

1A Dorsal fin with 14 or fewer principal rays; last ray of dorsal fin of adult greatly elongated; front of dorsal fin over or behind pelvic fin origin; mouth small, upper jaw length going into head length more than 3.2 times.   2

1B Dorsal fin with 16 or more principal rays; last ray of dorsal fin never elongated; front of dorsal fin in front of pelvic fin origin; mouth large, upper jaw length going into head length fewer than 3.2 times.   3

2A Anal rays 29–35; mouth subterminal or inferior, with lower jaw not projecting beyond tip of snout (Fig. 14.1A); upper jaw with a deep notch at its center; caudal fin not yellow in life; scales in lateral series usually 55 or more.   *Dorosoma cepedianum* Page 185

2B Anal rays 20–27; mouth terminal, with lower jaw projecting beyond tip of snout (Fig. 14.1B); upper jaw

without notch; caudal fin often yellow in life; scales in lateral series usually 50 or fewer.

Dorosoma petenense Page 188

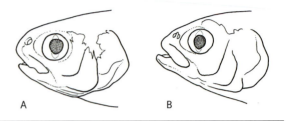

Figure 14.1. Lateral view of *Dorosoma* heads showing shape and relative lengths of upper and lower jaws.

A—*Dorosoma cepedianum*

B—*Dorosoma petenense*

3A Lower jaw projecting well beyond tip of upper jaw (Fig. 14.2A); lower jaw with dark pigment confined to anterior tip; teeth on tongue in 2–4 rows; gill rakers on lower half of first arch usually fewer than 30.

Alosa chrysochloris Page 183

3B Lower jaw equal to or projecting only slightly beyond tip of upper jaw (Fig. 14.2B); lower jaw with dark pigment along much of its length; teeth on tongue in a single median row; gill rakers on lower half of first arch usually more than 30.   *Alosa alabamae* Page 181

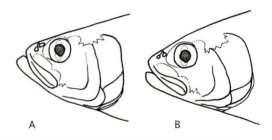

Figure 14.2. Lateral view of *Alosa* heads showing relationship of upper and lower jaws.

A—*Alosa chrysochloris*

B—*Alosa alabamae*

Genus *Alosa* Linck, 1790—River Herrings

This mostly anadromous genus, described by Linck in 1790, contains six North American species and several other species distributed along European and North African coasts and elsewhere in the Mediterranean region. The American Shad (*A. sapidissima*), native to the East Coast of North America, was successfully introduced along the West Coast. There was an unsuccessful attempt to introduce

A. sapidissima into Arkansas more than a century ago (Buchanan et al. 2000). Genus characters include an elongate, compressed body, a relatively large mouth, upper jaw usually extending backward past middle of eye, last dorsal fin ray not elongate, and 9 pelvic fin rays. There are two species of *Alosa* native to Arkansas.

Alosa alabamae Jordan and Evermann
Alabama Shad

Figure 14.3. *Alosa alabamae*, Alabama Shad. *Uland Thomas.*

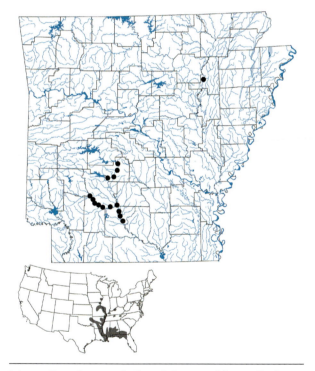

Map 18. Distribution of *Alosa alabamae*, Alabama Shad, 1988–2019.

CHARACTERS   A streamlined, slab-sided fish with a large terminal mouth, the lower jaw not projecting noticeably beyond upper jaw (Fig. 14.2B; Fig. 14.3). Adults lack teeth on the jaws; jaw teeth present only in young. Tongue with a single median row of small teeth. Black pigment distributed along most of lower jaw. Body depth going

into SL about 3.5–4.0 times. First gill arch with 30 or more gill rakers along lower half. Dorsal fin insertion is anterior to the pelvic fin base. Axillary process present at base of pectoral fin and pelvic fin. Dark humeral spot present behind operculum. Lateral line absent; scales in lateral series 48–58, usually more than 50. Usually 20–22 prepelvic scutes ($\bar{x}$ = 21.4); usually 13–16 postpelvic scutes ($\bar{x}$ = 15.2). Last ray of dorsal fin not elongated into a slender filament; principal dorsal rays 14–16 ($\bar{x}$ = 15.5). Principal anal rays 18–20 ($\bar{x}$ = 18.8). Maximum length around 20.4 inches (510 mm); maximum weight around 3 pounds (1.4 kg). The Arkansas state angling record is 2 pounds, 13 ounces from the Ouachita River in 1997.

LIFE COLORS   Back and upper side an iridescent blue green. Lower side silvery grading to silvery white on underside. Humeral spot a dark blue black. Fins clear to dusky.

SIMILAR SPECIES   *Alosa alabamae* is similar to the more common Skipjack Herring but differs from that species in having upper and lower jaws of approximately equal length (vs. lower jaw projecting well beyond upper jaw), having black pigment along most of the length of the lower jaw (vs. black pigment confined to tip of lower jaw), usually more than 50 lateral scales (vs. usually fewer than 50 lateral scales), a dusky to dark humeral spot (vs. humeral spot absent), and 30 or more gill rakers on lower limb of first gill arch (vs. 24 or fewer gill rakers on lower limb of first gill arch). The Alabama Shad differs from the Gizzard Shad and Threadfin Shad in having 14–16 principal dorsal rays (vs. usually fewer than 14 dorsal rays) and in lacking a long filamentous dorsal ray (vs. last dorsal ray greatly elongated).

VARIATION AND TAXONOMY   *Alosa alabamae* was described by Jordan and Evermann *in* Evermann (1896) from the Black Warrior River at Tuscaloosa, Alabama. This species was formerly referred to as *Alosa ohioensis*, which is considered a junior synonym of *A. alabamae*. Berry (1964) considered *A. alabamae* to form a sister pair with *A. sapidissima* of the Atlantic drainages. This grouping was generally supported by a mitochondrial DNA analysis (Bowen et al. 2008).

DISTRIBUTION   *Alosa alabamae* occurs in the Gulf of Mexico and ascends freshwater streams in Gulf and Mississippi River drainages from the Suwannee River, Florida, west to the Mississippi River, Louisiana. It ascends the Mississippi River drainage as far north as Iowa and has been reported historically in that basin from the Tennessee, Missouri, White, Arkansas, Ouachita, and Red rivers. The Alabama Shad is rare in Arkansas. Three pre-1900 records are known: the Ouachita River near Hot Springs (1 specimen), the Ouachita River at Arkadelphia (3 specimens), and the Mulberry River at Mulberry

(1 specimen). Twelve juvenile specimens collected in 1972 were reported by Stackhouse (1982) from the Saline River in Ashley County, and Loe (1983) found 3 specimens at the juncture of the Little Missouri and Ouachita rivers in 1982. Buchanan et al. (1999) collected 44 juveniles from the Little Missouri and Ouachita rivers in July 1997 and 1998 and also documented a 1.3 kg adult taken on an artificial lure on 4 April 1997 in the Ouachita River below Remmel Dam. Additional specimens were collected by T. M. Buchanan, D. Turman, C. Dennis, S. Wooldridge, and B. Hobbs in the Little Missouri and Ouachita rivers in 1999, including 15 adults below Remmel Dam (Map 18). The first records for this species from the White River were reported by Buchanan et al. (2012), and Robison et al. (2013) collected 34 juvenile *A. alabamae* from the mainstem Ouachita River at Tate's Bluff, Ouachita County on 22 October 2012. Four specimens were vouchered (AUM 59817) and photographed, and the remaining 30 specimens were released unharmed at the site. Five juveniles were collected from the same locality on the Ouachita River on 25 October 2014 (R. Hrabik, pers. comm.).

HABITAT AND BIOLOGY The Alabama Shad is Arkansas's only anadromous species, with the adults spending most of their lives in the ocean, where they feed and grow. They ascend freshwater rivers to spawn. Juveniles were found in July and August in Arkansas in moderate to swift current (0.47–1.14 m/s) in clear water at depths of 0.55–1.3 m at water temperatures of 28–32°C (Buchanan et al. 1999). The substrate at most collecting sites was gravel; a few collection sites had heavy growths of filamentous green algae, *Anacharis*, and *Potamogeton*. In the Pascagoula River, Mississippi, juvenile shad exhibited an ontogenetic habitat shift; smaller age 0 Alabama Shad favored sandbar habitat in June, but larger juveniles occupied deeper water in bank and open channel habitats (Mickle et al. 2010). Most life history information is from studies of adults and juveniles in freshwater habitats, and very little is known about its biology in marine environments of the Gulf of Mexico. Only four specimens were reported in collections from the marine environment away from estuaries prior to 2015. Mickle et al. (2015) reported the first molecularly verified marine record, an adult female with well-developed ovaries collected on 28 March 2013 from the marine waters of Mississippi. When ready to spawn, the Alabama Shad migrates inland, ascending the major Gulf Coast drainages in the spring. It formerly ascended the Mississippi River and its major tributaries including the Red, Ouachita, Arkansas, Missouri, Ohio, and Tennessee rivers for considerable distances (Burgess 1980a). In the late 1800s it was common enough during its spring spawning runs to

support a limited commercial fishery in the Mississippi River system (Mills 1972) and was reportedly a highly regarded food fish (Pflieger 1997).

Alosa alabamae drastically declined throughout its range during the 20th century, especially in recent decades. The construction of numerous dams severely limited the number of streams available for spawning migrations, and Burr and Warren (1986b) suggested that increased siltation and dredging also contributed to its decline. Declines in abundance have also been documented in coastal drainages, such as the Pearl River system of Louisiana and Mississippi (Gunning and Suttkus 1990), and in Alabama and Florida (Mettee and O'Neill 2003; Boschung and Mayden 2004). Buchanan et al. (1999) found that adult *A. alabamae* still ascended the Ouachita River system from the Mississippi River and spawned successfully in the Little Missouri and mainstem Ouachita rivers of Arkansas, despite the completion in 1985 of a series of four locks and dams on the Ouachita River (two in Louisiana and two in Arkansas). Young et al. (2012) also found that *A. alabamae* readily passed through the navigation lock of Jim Woodruff Lock and Dam on the Apalachicola River, Florida. Fish passage construction in the Apalachicola-Chattahoochee-Flint river system coincided with subsequent increases in the abundance and population estimates of adult Alabama Shad, and juveniles are now common in areas where they have historically been absent (Schaffler et al. 2015). It is apparently no longer able to ascend the Arkansas River through the system of locks and dams to spawn. Extensive sampling by several methods since the completion of the Arkansas River Navigation System in 1970 has yielded no specimens of *A. alabamae*. The recent records from the White River in Arkansas consisted of three juveniles collected on 2 August 2006 by William G. Layher (Buchanan et al. 2012).

In Arkansas, juvenile *A. alabamae* fed exclusively on benthic invertebrates in the Ouachita River (Buchanan et al. 1999). Juveniles approximately 73 mm TL fed mainly on Trichoptera, Diptera, and Plecoptera; individuals 113 mm TL fed mainly on Amphipoda, larval Diptera, and Trichoptera. No fish remains or plant materials were found in any juveniles, but juveniles in Alabama reportedly fed on young Threadfin Shad and Gizzard Shad (Boschung and Mayden 2004). Pflieger (1997) also reported that juveniles fed on small fishes as well as aquatic insects. In coastal drainages, diets of shad smaller than 50 mm SL consisted of a dark, almost black material labeled as unidentifiable organics; juveniles larger than 50 mm SL fed almost exclusively on insects (Mickle et al. 2013). That study also found that juvenile shad diets differed between drainages, with the Apalachicola River diets dominated by terrestrial insects, and the Pascagoula River diets including both terrestrial and aquatic insects. Adults feed mainly in saltwater, where fishes presumably are an important part of the diet, and most authors report that the adults do not feed during spawning migrations. Populations that move farther inland in the Mississippi River system may necessarily feed more than those that ascend the shorter Gulf drainages in Florida, and at least some feeding by adults apparently occurs in Arkansas, because it is occasionally taken by hook and line.

Much of the available life history information on *A. alabamae* is from the study of populations ascending coastal drainages from Mississippi to Florida (Laurence and Yerger 1966; Mills 1972; Ingram 2007; Mickle et al. 2010, 2013). Adults were formerly found in the Mississippi River near Keokuk, Iowa, from early May to late July (Coker 1930). We have found ripe adults in Arkansas in 1998 and 1999 in April and early May, where they congregated below Remmel Dam on the Ouachita River. In a Florida study, all spawning males were between 1 and 3 years old and females were 2–4 years old. (Ingram 2007). The estimated population size of migrating Alabama Shad in the Apalachicola River, Florida, was 25,935 in 2005, 2,767 in 2006, and 8,511 in 2007 (Ely et al. 2008). Spawning occurs in Florida at temperatures of 66.2–71.6°F (19–22°C) in moderate current over coarse sand and gravel bottoms (Laurence and Yerger 1966; Mills 1972). Eggs are fertilized as spawning pairs swim side by side. During spawning, the females scatter thousands of demersal, nonadhesive eggs over the substrate. The fertilized eggs settle into the gravel substrate, where embryonic development occurs. Females contain from 40,000 up to 250,000 eggs. It is not known whether adults that successfully reach Arkansas in spring to spawn are able to return to the ocean for future inland spawning migrations. Apparently, up to one-third of the Florida populations spawn in multiple years, but Ingram (2007) found no spawning marks on the scales of *A. alabamae* from the Apalachicola River. The juveniles remain in tributary streams until late summer or fall and then move downstream toward the ocean after reaching a length of 75–125 mm. However, Mickle et al. (2010) found no evidence that juveniles in the Pascagoula River, Mississippi, moved toward the Gulf of Mexico in their first year, and juveniles have been found in the Ouachita River in Arkansas as late as 25 October. Maximum life span is around 4 years (Laurence and Yerger 1966).

CONSERVATION STATUS This species has a federal status of special concern. The Ouachita River drainage (and possibly the White River) and a few drainages in Missouri

are the last remaining known noncoastal spawning areas for Alabama Shad, and its continued survival as part of the Arkansas fauna depends on protecting its critical spawning and nursery habitats. We consider it endangered in Arkansas.

Alosa chrysochloris (Rafinesque)
Skipjack Herring

Figure 14.4. *Alosa chrysochloris*, Skipjack Herring. *Uland Thomas.*

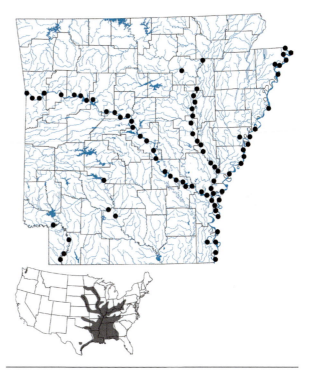

Map 19. Distribution of *Alosa chrysochloris*, Skipjack Herring, 1988–2019.

CHARACTERS A slender herring, with a laterally compressed body and a large, terminal mouth, the upper jaw reaching to below the middle of the eye. Lower jaw projecting well beyond upper jaw; black pigment on lower jaw confined to tip (Fig. 14.2A; Fig. 14.4). Caudal fin deeply forked. Body depth going 3.4–4.3 times into SL. Jaw teeth present and conspicuous at all ages. Small teeth on tongue in 2–4 rows. Axillary process present at base of pectoral fin and pelvic fin. First gill arch usually with 20–24 gill rakers along lower half. Dorsal fin insertion directly over or slightly in front of pelvic fin origin. Last ray of dorsal fin not elongated into a slender filament; dorsal rays 15–17; anal rays usually 15–18. Lateral line absent. Scales in lateral series usually 41–51 in Arkansas populations, but different counts are sometimes reported from other areas. Prepelvic ventral scutes 18–20; postpelvic scutes 15–18. Adults in Arkansas are usually smaller than 14 inches (356 mm) and 0.5 pound (0.2 kg). Maximum size about 21 inches (533 mm) and 3.75 pounds (1.7 kg). The Arkansas state angling record is 2 pounds, 10 ounces (1.2 kg) from Dardanelle Reservoir in 2004.

LIFE COLORS Back an iridescent bluish green, becoming silvery white laterally and ventrally. The blue-green coloration on the back and upper side ends abruptly, not gradually shading onto the silver side. Occasionally with a longitudinal row of 1–9 small dorsolateral dark spots extending from upper angle of gill-cleft backward along upper sides; spots near gill-cleft usually the largest. No dark humeral spot behind operculum. Fins generally clear, but large individuals may have dusky dorsal and caudal fins.

SIMILAR SPECIES Resembles the rare Alabama Shad, but differs from it in having a projecting lower jaw (vs. jaws approximately equal, lower jaw projecting only slightly beyond upper jaw), black pigment on lower jaw confined to tip (vs. black pigment distributed along lower jaw), fewer than 30 gill rakers on lower half of first arch (vs. more than 30), and scales in lateral series 41–51 (vs. 55–60 lateral scales). It differs from Gizzard Shad and Threadfin Shad in having 16 or more dorsal rays (vs. fewer than 14) and in not having the last dorsal ray produced into a long, thin filament (vs. last dorsal ray of adults greatly elongated into a filament).

VARIATION AND TAXONOMY *Alosa chrysochloris* was described by Rafinesque (1820a) as *Pomolobus chrysochloris* from the "Ohio River" (no additional details about type locality). Berry (1964) considered *A. chrysochloris* and *A. mediocris* of the Atlantic Coast to form a species pair, but Bowen et al. (2008), in a mitochondrial DNA analysis, concluded that *A. chrysochloris* was basal to all other North American *Alosa*.

DISTRIBUTION *Alosa chrysochloris* occurs in the Hudson Bay and Mississippi River basins from central Minnesota to southwestern Pennsylvania and south to the Gulf of Mexico. It is also found in Gulf Slope drainages from the Florida panhandle west to San Antonio Bay, Texas. In Arkansas, Black (1940) reported that the Skipjack

Herring occurs in "all the larger rivers in the state, and in some of moderate size as well." It is currently known from the Mississippi, lower St. Francis, lower White, Arkansas, lower Ouachita, and Red rivers (Map 19).

HABITAT AND BIOLOGY The Skipjack Herring is a freshwater species that occasionally wanders into brackish and salt waters along the coast. It is fairly common in the large rivers in Arkansas, where it inhabits the open waters, often in swift current. It is also found in the backwaters, and Beckett and Pennington (1986) and Sanders et al. (1985) reported that it was most common in backwater areas of the lower Mississippi and lower Arkansas rivers. Although it is reported to occur in lakes and reservoirs in some areas (Etnier and Starnes 1993; Ross 2001), this does not appear to be the case in Arkansas (except for the flow-through navigation pools of the Arkansas River). Several authors reported that it is rather intolerant of continuous high turbidity, but it is common in the rather turbid Mississippi River. Its numbers have reportedly declined in the upper Mississippi River of Iowa because of the construction of navigation locks and dams, but it appears to be more common today in the Arkansas River than it was before completion of the navigation system in the early 1970s because of the deepened channel and reduced turbidity. Pigg et al. (1991) also reported an increase in the number of Skipjack Herring in the Arkansas River in Oklahoma.

The adult Skipjack Herring is reported to be a schooling fish and feeds primarily on minnows and other small fishes which it actively pursues. It is known as the Skipjack because of its frequent leaps into the air to capture jumping minnows (Trautman 1957). Young smaller than 40 mm TL feed mainly on microcrustaceans, but juveniles add dipterans and other aquatic insects to the diet and quickly become piscivorous as they grow. Coker (1930) examined gut contents of 150 specimens from the Mississippi River of Iowa. Half the specimens with food in the gut had eaten fish, mainly minnows; other fish consumed included Mooneyes, Gizzard Shad, and other unidentified fishes. Fewer than half the guts with food contained insects and larvae, mainly mayflies and caddisflies. Etnier and Starnes (1993) reported that adults are often observed feeding on shad at the surface in reservoirs. Trautman (1957) described feeding activity in Ohio as follows:

The species fed in large, swiftly swimming schools which forced the huge schools of Emerald and Mimic shiners to crowd together near the water's surface. Once the minnows were closely crowded together the Skipjack dashed in among them, forcing the minnows to rise to the water's surface where they could be captured readily. The species often congregated in large numbers in the swift waters below the dams of the Ohio River where they preyed upon the immense numbers of minnows segregated there.

Little is known of its reproductive biology. It was once believed to be anadromous like the Alabama Shad, with adults present in rivers only during the summer months (P. W. Smith 1979). However, it is now known to be entirely a freshwater species, but it is somewhat migratory in the spring, traveling upstream and often assembling below dams in swift current. Coker (1930) believed that spawning occurred from early May to early July in the upper Mississippi River of Iowa. It spawns from early March to late April in Florida (Wolfe 1969) and in Alabama (Mettee et al. 1996). Spawning probably occurs in the current in the main channel over coarse sand and gravel bottoms. Wallus and Kay (1990) described egg and larval development. We found YOY Skipjack Herring in the Arkansas River near Fort Smith in early June and in the Red River near Garland in late June. Females may have as many as 300,000 eggs. The eggs hatch in about 58 hours at 17.2°C (Wallus and Kay 1990). The young grow rapidly and reach 5–6 inches (130–150 mm) TL by the end of the first growing season (Coker 1930). Some individuals live at least 4 years. It is not a good food fish but is often caught by fishermen on the Arkansas River. Becker (1983) noted that in the upper Mississippi River, the Skipjack Herring previously served as an important host for the glochidia of several species of clams.

CONSERVATION STATUS   Populations of this species are stable if not secure. It is locally common in some areas of the Arkansas and Mississippi rivers.

Genus *Dorosoma* Rafinesque, 1820a— Gizzard Shads

Described by Rafinesque in 1820, this genus is endemic to the Western Hemisphere and contains five species in North America and Central America. *Dorosoma* is unique among clupeids in having pharyngeal pockets that aid in the concentration and subsequent swallowing of its largely planktonic food. Genus characters also include an oblong body, a small mouth with upper jaw not extending backward past middle of eye, numerous pyloric caeca, a ventral keel with serrate scutes, last dorsal ray elongate (in most species), and pelvic rays 8. In addition, *Dorosoma* species have a long and complex digestive system that includes a muscular, gizzardlike stomach (Nelson and Rothman 1973) from which the well-known Arkansas forage fish,

the Gizzard Shad, gets its common name. Two species of *Dorosoma* are native to Arkansas.

Dorosoma cepedianum (Lesueur)
Gizzard Shad

Figure 14.5. *Dorosoma cepedianum*, Gizzard Shad. *Uland Thomas.*

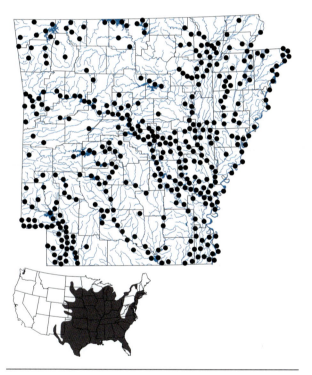

Map 20. Distribution of *Dorosoma cepedianum*, Gizzard Shad, 1988–2019.

CHARACTERS   A silvery, slab-sided fish with a thin, deeply compressed, oblong body, and a blunt snout that overhangs the small subterminal mouth that is toothless in adults (Fig. 14.5). Lower jaw short, not projecting beyond tip of snout and fitting into upper jaw (Fig. 14.1A). Upper jaw with a deep notch at its center. Body depth going about 2.3–3.1 times into SL. The stomach is very muscular and gizzardlike. Dorsal fin origin above or behind pelvic fin origin. Last ray of dorsal fin elongated into a thin filament (not developed in very small young); dorsal rays 10–13. Anal fin long, with 29–35 rays. Pectoral rays 12–17. Pelvic rays usually 8 (7–10). Axillary process present at base of pectoral fin and pelvic fin. Caudal fin deeply forked. Lateral line absent. Scales in lateral series 59–67. Ventral scutes: 17–20 prepelvic and 10–14 postpelvic. Gill rakers long and comb-like, with about 190 on first arch. Adults usually 12–16 inches (305–406 mm) long and weigh less than 1 pound (0.5 kg). Maximum size about 20.5 inches (521 mm) and 4.5 pounds (2.04 kg). Surprisingly, there is an Arkansas rod and reel angling record of 2 pounds, 14 ounces (1.3 kg) from the White River in 1992, and an unrestricted tackle record of 3 pounds, 7 ounces (1.5 kg) from DeGray Lake in 2013.

LIFE COLORS Back and upper sides bluish green, grading to silvery white laterally and ventrally. Upper sides with 6–8 horizontal dark streaks. A large dark purple-blue shoulder spot present just behind upper end of operculum in young and small adults, but becoming faint or absent in large adults. Fins dusky and without yellow colors. There are no black specks on the chin or floor of the mouth.

SIMILAR SPECIES Most like the Threadfin Shad, but differs from it in having 29–35 anal rays (vs. 20–27), 59–67 lateral scales (vs. 42–48), the upper jaw projecting well beyond the lower jaw and with a deep notch at its center (vs. lower jaw projecting slightly beyond upper jaw), chin and floor of mouth not speckled with black pigment (vs. chin and floor of mouth with black pigment), and no yellow pigment in fins in life (vs. all fins except dorsal with yellow pigment). It can be distinguished from the Alabama Shad and the Skipjack Herring by its long filamentous last dorsal ray (vs. last dorsal ray not elongated into a filament), and fewer than 14 dorsal rays (vs. 16 or more). Meristic and morphometric characteristics for distinguishing *D. cepedianum* and *D. petenense* larvae were provided by Bulak (1985) and Holland-Bartels et al. (1990).

VARIATION AND TAXONOMY *Dorosoma cepedianum* was originally described by Lesueur (1818a) as *Megalops cepediana* from Chesapeake and Delaware bays (specimens were obtained from markets in Baltimore and Philadelphia). It was formerly placed in a different family, Dorosomatidae, until Miller (1950) treated it as a subfamily of Clupeidae. Miller (1960) studied morphometric and meristic variation in *D. cepedianum* throughout its range and did not recognize subspecies; however, he placed the Gizzard Shad in the subgenus *Dorosoma*. Subspecies sometimes recognized by earlier workers are no longer considered valid (Boschung and Mayden 2004). Hatfield et al. (1982) examined genetic variation in Gizzard Shad from the Mississippi River near Eudora, Arkansas, and found that

the shad did not maintain genetically different subpopulations within the three types of river habitat that were sampled; however, significant heterogeneity was observed between the Mississippi and Ohio river populations. It is known to hybridize with *D. petenense* (Minckley and Krumholz 1960; Shelton and Grinstead 1972), and hybrids are not uncommon (Etnier and Starnes 1993). Stephens (1985) described the canals and canal branching pattern of the lateral line system of *D. cepedianum*.

DISTRIBUTION *Dorosoma cepedianum* is widely distributed in eastern North America, occurring in the St. Lawrence–Great Lakes, Mississippi River, and Atlantic and Gulf Slope drainages from Quebec, Canada, to central North Dakota and south to southern Florida and Mexico. In Arkansas, it is found statewide in all river drainages and reaches its greatest abundance and size in large rivers and impoundments (Map 20).

HABITAT AND BIOLOGY The Gizzard Shad occurs in a wide variety of habitats and is one of Arkansas's most versatile species. It is typically rare in small high-gradient creeks, but it is usually found in medium-sized creeks and rivers in the mountains as long as ample pool habitat is present. It was a commonly collected species in the upper White River prior to impoundment of Beaver Reservoir (Keith 1964). It occurs in clear as well as in turbid waters but is most abundant in waters where the fertility and productivity are high. DiCenzo et al. (1996) found that Gizzard Shad populations in eutrophic Alabama reservoirs had high densities and reduced growth rates compared with shad populations in oligo-mesotrophic reservoirs. Allen et al. (2000) found that probabilities of Gizzard Shad occurrence in Florida lakes increased with increasing chlorophyll *a* and lake surface area. It is primarily a pelagic species, usually traveling in large schools in the open waters. It prefers deep, calm water but is commonly found in strong currents as well. Along the coast it sometimes enters brackish or salt water. Although the Gizzard Shad is tolerant of turbidity, Pflieger and Grace (1987) reported that an increase in its populations was associated with decreased turbidity of the Missouri River caused by construction of upstream reservoirs. It often overpopulates reservoirs, sometimes making up more than 80% of the biomass. During flooding of the Brazos River, Texas, adult shad moved into oxbow lakes from the main channel (Zeug et al. 2009). Young shad provide excellent food for most native gamefishes, but a study of competition between larval Gizzard Shad and larval Bluegills found that zooplankton availability and Bluegill abundances were consistently low during years when Gizzard Shad dominated reservoir fish assemblages (Garvey and Stein 1998). The adults grow too large to be used by most predators. The Arkansas Game and Fish Commission employs various management techniques, such as winter drawdowns, to control their populations in certain impoundments. Another attempted shad control measure in reservoirs has been the stocking of large predatory species such as Striped Bass, Northern Pike, Muskellunge, and Walleye.

Larval Gizzard Shad feed primarily on protozoans and unicellular algae, but they soon add copepods, cladocerans, zooplankton eggs, rotifers, and small aquatic insect larvae to their diet (Armstrong et al. 1998). In a Nebraska reservoir, larval shad <30 mm TL consumed large amounts of zooplankton from different groups, and after reaching 30 mm, consumed more algae and detritus than zooplankton (Sullivan 2009). The adults feed on phytoplankton as well as zooplankton and particulate matter filtered from the water by their long gill rakers (Drenner et al. 1982), but they also graze over the bottom, ingesting detritus, sand, and bottom ooze (Baker and Schmitz 1971; Baker et al. 1971; Gido 2001). By foraging on sediments, Gizzard Shad also transport nutrients into the water column of reservoirs through excretion. This flux of nutrients can contribute substantially to the total nutrient budget of the reservoir during periods of low inflow from tributaries (Gido 2002). Sampson et al. (2009) found considerable dietary overlap between Gizzard Shad and two Asian carp species (Silver and Bighead carp) in Mississippi River backwater lakes. Cross (1967) reported that Gizzard Shad occupy progressively deeper water as they grow older and tend to increasingly consume organisms associated with the bottom sediments rather than those found near the surface. Stable isotope analysis in an Ohio lake revealed that during periods of greater Gizzard Shad biomass, adult Gizzard Shad derived most of their carbon from sediment detritus; when Gizzard Shad biomass was low, zooplankton biomass increased, and all sizes of Gizzard Shad derived most of their carbon from zooplankton (Schaus et al. 2002). In Lake Texoma, Oklahoma, stocking of Gizzard Shad in high densities in experimental enclosures significantly reduced abundance of chironomids and ostracods, but percent organic matter, algal abundance, and abundance of other macroinvertebrates in sediments did not differ significantly between areas with and without large shad (Gido 2003). Adult Gizzard Shad in Lake Texoma consumed only detritus and algae, and it is likely that disturbance of benthic sediments by shad caused the observed reduction in chironomid abundance, rather than consumption or competition for resources.

In most *D. cepedianum* populations, spawning occurs at water temperatures between 50 and 70°F (10–21°C)

(R. R. Miller 1960), but spawning may continue up to about 80°F (27°C). Spawning occurs in Arkansas from early April through May and possibly into June. Baglin and Kilambi (1968) reported that shad in Beaver Reservoir have one spawning period that extends from mid-April through May. Spawning takes place at the surface, usually in shallow backwaters or near shore; for example, along riprap in reservoirs. Most spawning occurs at night during a rise in water temperature. Females and males swimming together in a school shed eggs and sperm. The adhesive eggs sink to the bottom, becoming attached to whatever they touch. The following aspects of spawning behavior in a stream (Buncombe Creek, Oklahoma) on 27–28 March at a water temperature of 62.5°F were taken from Shelton (1972). Spawning was observed in a riffle and a pool. The spawning aggregation consisted of 40–50 adults, and most spawning occurred in the pool with the fish oriented upstream and maintaining their position in the current. A female was usually near the front of the aggregation. The female would then leave the group accompanied by 1–3 males swimming slightly below or behind her. The rate of swimming accelerated rapidly as they approached the deposition site. If deposition was in the riffle area or directly on the bottom, the female tilted 20–45° to one side. Males crowded close at this time and presumably released milt. Deposition on vegetation or other material off the bottom was accomplished by swimming more or less directly toward the substrate and abruptly turning as it was reached. There is no parental care of eggs and young. Kilambi and Baglin (1969a) found that females from Beaver and Bull Shoals reservoirs contained from 20,000 to around 170,000 mature ova depending on size and age. The oldest females were 6 years old, and average ovum diameter ranged from 0.64 to 0.67 mm. Greater fecundities (up to 543,910 oocytes for fish up to 366 mm SL) have been reported elsewhere (Bodola 1966). Eggs hatch in 95 hours at 61°F (16.1°C), or 36 hours at 80°F (26.7°C). Larval Gizzard Shad are slender and transparent and do not acquire the typical adult shape until they reach 1.25 inches (31.8 mm) TL. Larval densities in the surface waters of Beaver Reservoir reached 1,955/m³ during mid-May (Netsch et al. 1971). Gizzard Shad mature when 2–3 years old (Marcy et al. 2005). The larvae form a large part of the ichthyoplankton in Arkansas's large rivers. They were a dominant component of the drift in the upper Mississippi River from June to July (Holland and Sylvester 1983), and larval Gizzard Shad and Freshwater Drum composed 93.5% of the ichthyoplankton in the lower Mississippi River of Louisiana (Gallagher and Conner 1983). Maximum age is 10 years.

Gizzard Shad are very sensitive to rapid temperature changes and low oxygen content. It is a fragile species that dies quickly after handling and is therefore not an optimal live baitfish (although some people use it for live bait by providing a round live well with aeration). It is frequently used in Arkansas as dead bait for trotlines. It is not a good food fish and is usually only snagged accidentally by hook and line. It is a valuable forage species as long as it does not overpopulate its environment.

CONSERVATION STATUS   This is a widespread and abundant species, often overpopulating reservoir habitats in Arkansas. Its populations are secure.

Dorosoma petenense (Günther)
Threadfin Shad

Figure 14.6. *Dorosoma petenense*, Threadfin Shad. *Uland Thomas.*

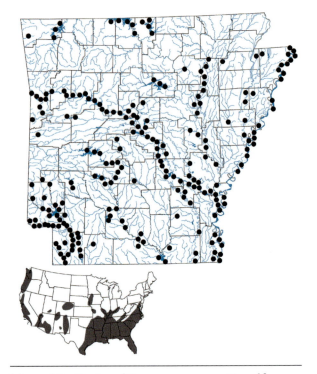

Map 21. Distribution of *Dorosoma petenense*, Threadfin Shad, 1988–2019. Inset shows native and introduced ranges in the contiguous United States.

CHARACTERS   A relatively small, silvery clupeid with a thin, deeply compressed body, a moderately pointed snout, a small terminal mouth without teeth (in adults), and a deeply forked caudal fin (Fig. 14.6). Lower jaw projecting slightly beyond upper jaw (Fig. 14.1B). Ventral edge of maxilla without a notch. Dorsal fin origin above or behind the pelvic fin origin. Last ray of dorsal fin elongated into a thin filament in adults; dorsal rays 11–14. Anal fin with 17–27 (usually 20–25) rays. Pectoral rays 12–17. Pelvic rays 7 or 8. Axillary process present at base of pectoral fin and pelvic fin. Lateral line absent. Scales in lateral series 40–48. Ventral scutes: 15–18 prepelvic and 10–12 postpelvic. Adults rarely exceed 5 inches (127 mm) in length. Maximum TL about 9 inches (230 mm).

LIFE COLORS   Back and upper sides bluish green, grading to silvery white laterally and ventrally. A round, black shoulder spot present just behind upper end of operculum; spot persisting in adults and often noticeably smaller than the shoulder spot of the Gizzard Shad. Chin and floor of mouth are sprinkled with black pigment. Caudal fin usually distinctly yellow; other fins, except dorsal, with some yellow coloring. Dorsal fin dusky. Iris of eye silvery.

SIMILAR SPECIES   It differs from the Gizzard Shad in having 20–27 anal rays (vs. 29–35), 42–48 lateral scales (vs. 59–67), snout slightly pointed, with lower jaw projecting beyond upper jaw (vs. snout rounded and fleshy, and lower jaw not projecting beyond upper jaw), chin and floor of mouth speckled with black pigment (vs. chin and floor of mouth without black pigment), caudal fin yellow in life (vs. no yellow coloration in fins). It differs from the Alabama Shad and the Skipjack Herring in its elongated filamentous last dorsal fin ray (vs. last dorsal ray not long and filamentous), and fewer than 14 dorsal fin rays (vs. 16 or more). Larval *D. petenense* are very similar to larval *D. cepedianum*, and Bulak (1985) provided meristic and other morphological features for distinguishing the two. Stephens (2016) described and compared the canals and canal branching patterns of the lateral line system on the head of *D. petenense* and *D. cepedianum*.

VARIATION AND TAXONOMY   *Dorosoma petenense* was described by Günther (1867) as *Meletta petenensis* and named after its type locality, Lake Peten, Guatemala. It is in the subgenus *Signalosa* (Nelson and Rothman 1973) and is occasionally known to hybridize with *D. cepedianum*, particularly in altered environments (Minckley and Krumholz 1960). Miller et al. (2005) noted that *D. petenense* is in need of a thorough systematic study. Ongoing systematic study (by Kyle Piller) may eventually restrict the name *D. petenense* to southern populations with a different name applied to populations in the United States (D. Neely, pers. comm.).

DISTRIBUTION   The native range of *Dorosoma petenense* is not precisely known because of widespread introductions. It is probably native to Gulf Slope and lower Mississippi River tributaries from Florida west and south to Central America (Guatemala and Belize). It has been widely introduced north of that range into the Mississippi River drainage to Ohio and central Illinois, and into Atlantic Slope drainages from Chesapeake Bay south. In western United States, it has been introduced into the Colorado River drainage of Arizona, Nevada, and California and in other Pacific drainages of California. In Arkansas, the Threadfin Shad is considered native to Coastal Plain lowland streams and throughout the Arkansas River (Map 21). It has been introduced into reservoirs in the Ozarks and the Ouachitas.

HABITAT AND BIOLOGY   The Threadfin Shad is a subtropical and southern temperate fish, preferring warm waters. Strawn (1965) reported that the ultimate lower lethal temperature of Arkansas Threadfin Shad was 41°F (5°C). Because of this, low winter temperatures below 40°F (4.5°C) often cause die-offs in northern Arkansas and occasionally farther south. Even though it is a warmwater fish, it apparently exhibits surprisingly low critical thermal maximum temperatures. Threadfin Shad maintained at 15, 20, and 25°C displayed mean critical thermal maximum temperatures of 26.5, 30.9, and 33.3°C, respectively (Monirian et al. 2010). It is primarily an inhabitant of moderate to large rivers, where it is more abundant in currents than the Gizzard Shad. It is a desirable forage fish because of its small adult size, and it is often stocked in reservoirs for that purpose (Burgess 1980b). It is a pelagic, schooling species usually found in reservoirs in the upper 5 feet (1.5 m) of water (Pflieger 1997). Along the coast, it sometimes enters brackish and salt waters. Netsch et al. (1971) studied the distribution of young Threadfin Shad in Beaver Reservoir, and Houser and Dunn (1967) found that Threadfin Shad in Bull Shoals Reservoir occupied a sharply limited layer of water between the surface and the thermocline during most of the year. Shad density decreased with increasing depth, but the horizontal distribution was remarkably uniform over the lake. The age 0 Threadfin Shad standing crop was estimated at 42 pounds/acre (49 kg/ha). In a study of 60 Florida lakes, Threadfin Shad density and biomass increased with increasing chlorophyll *a* and decreased with increasing zooplankton density, and lake surface area was the best predictor of the presence of *D. petenense* (Allen et al. 2000).

In Alabama impoundments, *D. petenense* larvae fed on the same foods as the larvae of *D. cepedianum*: zooplankton eggs, rotifers, and copepod nauplii and adults (Armstrong et al. 1998). In a study of Threadfin Shad food habits in Lake Chicot, a Mississippi River oxbow lake in southeastern Arkansas, R. V. Miller (1967) found that the bulk of its diet was made up of phytoplankton (cyanobacteria, diatoms, and green algae) and zooplankton (fish larvae, dipteran larvae, water mites, and the eggs of microcrustaceans) which it strained from the water with long gill rakers. Its food habits did not change as it increased in size. Baker et al. (1971) found that Threadfin Shad in Beaver and Bull Shoals reservoirs in northwestern Arkansas ingested large quantities of organic detritus, sand, and mud as well as phytoplankton and zooplankton. Although its food habits were very similar to those of the Gizzard Shad in these two reservoirs, consistent with mouth orientation, the Threadfin Shad was more limnetic than the Gizzard Shad when feeding and ingested less detritus than that species (Baker and Schmitz 1971). McMahon and Tash (1979) reported that the Threadfin Shad uses chemical cues in addition to vision to locate concentrations of plankton, thereby permitting feeding in turbid water and at night.

Spawning occurs in the spring when water warms to 70°F (21.3°C), and may continue at intervals for several months (Pflieger 1997). It spawned from mid-May to mid-June in Beaver Reservoir (Netsch et al. 1971). Spawning occurs in schools from dawn to shortly after sunrise in shallow water near shore (Lambou 1965). One or more females are often accompanied by several males as the group swims near the surface. The slightly adhesive eggs are scattered over plants and other submerged objects and are left unguarded. Shelton (1972) provided a detailed description of spawning behavior in Lake Texoma and noted that in contrast to *D. cepedianum*, spawning of *D. petenense* was not difficult to observe because of the extensive surface activity. The surface disturbance was quite obvious, and the sound that resulted from the splashing was audible for a considerable distance. Hatching occurs in about 3 days at 26.5 °C (Burns 1966). Larval Threadfin Shad are slender and transparent and do not acquire the typical adult shape until they reach about 1 inch (25 mm) TL. Young shad formed schools near the water surface in Beaver Reservoir, with densities of 750–1,000 individuals/m^3 (Netsch et al. 1971). In warm climates, individuals hatching early in the season may mature and spawn by the end of their first summer. Kilambi and Baglin (1969b) found that fecundity of Threadfin Shad in Beaver and Bull Shoals reservoirs ranged from around 3,000 to 25,000 eggs per female depending on length and weight. They further reported that two distinct size groups of ova were present, indicating multiple spawning during a season. Sexual maturity was reached in Beaver Reservoir at 1 year of age and in Bull Shoals at 2 years, but most spawned in their second year. Bryant and Houser (1968) found that Threadfin Shad introduced into Bull Shoals Reservoir in 1961 had slower growth and a longer life span than more southern populations. Total length averaged 2.1 inches (53 mm) after one growing season, and 4.9 inches (124 mm) (males) and 5.3 inches (135 mm) (females) after three growing seasons. The males in Bull Shoals Reservoir lived 3 years and the females lived 4 years.

CONSERVATION STATUS This species is more common and widespread in the southern half of Arkansas. Populations in northern Arkansas often experience winter die-offs but usually recover during the summer months. The Arkansas populations are secure.

CHAPTER 15

Order Cypriniformes
Minnows, Carps, Loaches, and Suckers

The order Cypriniformes is the most diverse monophyletic freshwater clade of fishes on Earth (Nelson 2006; Mayden et al. 2007, 2008, 2009). It comprises fishes that are almost completely restricted to freshwaters and estimated to number more than 3,400 species in 6 families with poorly defined subfamilies and/or tribes (Miya et al. 2008). Nelson et al. (2016) estimated the number of cypriniform species at 4,205 and recognized 13 extant families with about 489 genera, and Eschmeyer and Fong (2017) estimated more than 4,400 species. This enormously diverse order was considered monophyletic based on the kinethmoid bone, a median bone between ascending processes of the premaxillae involved in the unique protrusion mechanism of the upper jaw, and other derived skeletal synapomorphies (Fink and Fink 1981; Lauder and Liem 1983; Dimmick and Larson 1996; Wiley and Johnson 2010). Miya et al. (2008), using mitochondrial genome sequences, confirmed monophyly of the Cypriniformes and found four major clades: Cyprinidae, Catostomidae, Gyrinocheilidae, and Balitoridae + Cobitidae, with the latter two loach families reciprocally paraphyletic. Betancur-R. et al. (2017) also confirmed monophyly of Cypriniformes based on molecular evidence. Cypriniformes contains species and clades that display an amazing diversity of morphologies, natural histories, and body sizes (Winfield and Nelson 1991; Nelson 2006). The diversity, widespread distribution on four continents, and ease of care and propagation, combined with rapidly growing knowledge of phylogenetic relationships (Mayden et al. 2007, 2008; Chen and Mayden 2009) make Cypriniformes particularly appropriate as a model clade of freshwater fishes for detailed investigations of the evolutionary and coevolutionary origins of attributes, behaviors, and biogeography within a phylogenetic context (Mayden and Chen 2010). Cypriniformes is the basal sister group to other orders of series Otophysi, representing as a group the closest relative to an equally diverse group inclusive of order Characiformes plus order Siluriformes (Boschung and Mayden 2004). The characins were formerly placed in the order Cypriniformes as a suborder (Howes 1991), but there is no evidence from shared derived characters that characins and minnows are sister groups (Roberts 1973; Fink and Fink 1981).

Nelson et al. (2016) recognized 13 cypriniform families—Cyprinidae, Psilorhynchidae, Catostomidae, Botiidae, Vaillantellidae, Cobitidae, Gyrinocheilidae, Balitoridae, Gastromyzontidae, Nemacheilidae, Barbuccidae, Ellopotomatidae, and Serpenticobitidae. All families except Cyprinidae and Catostomidae are mostly small, elongate African and Eurasian fishes. Cyprinidae (carps and minnows) occur in the freshwaters of North America, Africa, and Eurasia, with the greatest diversity in southeastern Asia. Cyprinidae, as traditionally recognized, accounts for about 80% of all cypriniform species, while

Catostomidae (suckers) are Holarctic in distribution, with greatest diversity occurring in North America. Tan and Armbruster (2018) divided Cypriniformes into the suborders Gyrinocheiloidei, Catostomoidei, Cobitoidei, and Cyprinoidei.

Family Cyprinidae *Carps and Minnows*

There has been increasing support among ichthyologists in recent years for major changes in the classification of family Cyprinidae, and several papers published since 2010 have supported splitting Cyprinidae into new families. Tan and Armbruster (2018) summarized the recent taxonomic history of Cyprinidae and proposed a new classification of family-level groupings in order Cypriniformes. Based on recent molecular evidence, those authors elevated family Cyprinidae *sensu lato* to suborder Cyprinoidei and recognized 12 clades of Cyprinidae as families. According to the new classification, Arkansas species in family Cyprinidae *sensu lato* are now placed in three families: Cyprinidae *sensu stricto* (carps, barbs, and allies), Leuciscidae (minnows of Europe, Asia, and North America), and Xenocyprididae (Asian carps, culters, and allies). All native Arkansas minnows are in family Leuciscidae, but there are exotic Arkansas species in families Cyprinidae *sensu stricto* (*Carassius* and *Cyprinus*) and Xenocyprididae (*Ctenopharyngodon*, *Hypophthalmichthys*, and *Mylopharyngodon*). Although some ichthyologists believe that acceptance of the proposed classification changes is inevitable, we feel that it is premature to incorporate those changes into this second edition. Any eventual general acceptance of this revised cyprinid taxonomy lies well beyond the publication timeframe of this book, and we herein retain the most widely accepted historical family-level classification of the minnows (all in family Cyprinidae *sensu lato*). We emphasize that the proposed changes in family-level classification have no effect on our treatments in this book of lower level cypriniform taxonomy or accounts of biology, ecology, distribution, and status.

Cyprinidae, as traditionally recognized, is the largest and most widely distributed freshwater family of modern fishes in the world, with about 3,006 species in 367 genera occurring in North America (northern Canada to southern Mexico), Africa, and Eurasia (Nelson et al. 2016). With the possible exception of the largely marine family Gobiidae, Cyprinidae is the largest family of vertebrates (Nelson 2006). Of the 831 fish species native to the United States and Canada, more than 300 are cyprinid species in more than 50 genera, accounting for approximately 30%

of the North American fish fauna (Johnston 1999). A few species of Cyprinidae are used in genetic and developmental research, and many are important aquarium and food fishes. The earliest definite cyprinid fossils are of Eocene age from Asia (Nelson 2006), and fossil evidence indicates that cyprinids have inhabited North America for more than 31 million years (Cavender 1991). Cavender hypothesized that cyprinids were absent from North America in the Eocene, a time when other otophysans such as catostomids and ictalurids were present.

Success of the Cyprinidae has been attributed to a combination of factors, including a well-developed sense of hearing (due to the Weberian apparatus), the "fear scent" (schreckstoff) released by cyprinids when injured, pharyngeal teeth located on the 5th ceratobranchial arch, and high fecundity. The Weberian apparatus is a synapomorphic hearing specialization found in members of the second-largest superorder of fishes, the Ostariophysi (which includes two orders found in Arkansas, Cypriniformes and Siluriformes). The Weberian apparatus is a bony connection for transmission of vibrations from the swim bladder to the inner ear (Holt and Johnston 2011). This series of three ossicles allows for greater hearing sensitivity and a high-frequency range (Popper and Fay 2010). In addition, all cyprinids lack a well-defined stomach, and teeth are absent from the jaws and oral cavity. Pharyngeal teeth allow cyprinids to make chewing motions against a plate formed by a bony process of the skull. Pharyngeal tooth number, structure, and arrangement are often species-specific and are frequently used to identify species.

High diversity combined with widespread distribution have made cyprinids one of the most difficult groups of fishes to understand taxonomically and systematically (Mayden 1989, 1991), and resolving North American cyprinid relationships has proven especially difficult (Cuhna et al. 2002; Simons et al. 2003).

Although sister-group relationships at the levels of subfamilies and tribes within Cyprinidae received increased attention in the 1990s and revisions were proposed, studies prior to 2004 lacked adequate taxon and character sampling (Tao et al. 2010). Chen et al. (1984) divided Cyprinidae into 10 subfamilies grouped into 2 major series (Leuciscinae and Barbinae) with 6 and 4 subfamilies within each series, respectively. Cavender and Coburn (1992) hypothesized different phylogenetic positions for the rasborins and Tincinae. New morphological and molecular analyses subsequently appeared using greatly enhanced character and taxon sampling, as well as more sophisticated techniques. Saitoh et al. (2006) described six major clades of Cyprinidae and resolved two reciprocally monophyletic groups congruent with the results recognized by Cavender and Coburn (1992). Wang et al. (2012) classified Cyprinidae into two monophyletic groups identified as Leuciscinae and Cypininae and recognized eight additional groups within these clades. The larger studies of Bufalino and Mayden (2010a, b, c), and Tao et al. (2010) increased the number of different clades proposed and improved our knowledge of the relationships among constituent members, but some areas remained contentious and unresolved. He et al. (2008) studied the phylogenetic relationships of the 12 previously recognized monophyletic subfamilies within the Cyprinidae using nuclear DNA sequences. Cyprinidae were resolved as a monophyletic group. Cavender and Coburn (1992) used phylogenetic systematics of the Cyprinidae to recognize several subfamilies worldwide including the Cyprininae and Leuciscinae. Nearly all the approximately 300 North American cyprinids are members of the subfamily Leuciscinae (Mayden 1991; Nelson 1994; Simons et al. 2003) which is divided into the tribes Abramini (*Notemigonus* and many Eurasian genera) and Phoxini. Most North American Phoxini are further divided into three major clades: shiner, chub, and western clades. Yang et al. (2010) investigated the phylogenetics of the tribe Cyprinini *sensu stricto* within the subfamily Cyprininae which includes the carps (e.g., Koi) of the genus *Cyprinus* and the Crucian carps (e.g., Goldfish). Leuciscinae was diagnosed on the basis of two characters: pharyngeal arch with one or two rows of teeth and the loss of the suborbital-infraorbital connection.

Much attention has been given to cyprinid classification at and below the genus level. Mayden (1989) published a phylogenetic classification of primarily eastern North American cyprinids based on morphology, including both osteological and external characters, and resurrected nine genera. Mayden (1989) elevated several taxa from synonymy within *Notropis* (*Lythrurus, Luxilus, Cyprinella, Pteronotropis*) and *Hybopsis* (*Erimystax, Macrhybopsis, Extrarius,* and *Platygobio*). Coburn and Cavender (1992) published the first phylogenetic analysis of all North American phoxinin taxa at the genus level based on osteological and external characters and recognized three major clades: a shiner clade, sister to a chub clade, plus a western clade. Simons and Mayden (1997, 1998, 1999) examined the relationships of North American phoxinin genera using molecular characters and reported three clades of North American phoxinins: a western clade, a creek chub-plagopterin clade, and an OPM clade (open posterior myodome—a cavity in the posterior part of the cranium

that is the site of origin of the lateral rectus muscles). Subsequently, Simons et al. (2003) analyzed 83 cyprinid species in 43 genera and recovered the North American clade as monophyletic. Within the North American phoxinins, three major clades were recognized: the western clade, the creek chub-plagopterin clade, and the OPM clade (Simons and Mayden 1998). However, Simons et al. (2003) reported that a clear understanding of the relationships within and among genera of the North American phoxinins is still elusive.

Limited basic life history information exists for many cyprinid species in North America (Whittier et al. 2000). In North American rivers and lakes, minnows are an important link in the aquatic trophic web (Stewart and Watkinson 2004), yet the difficulty in identification of these small species and their relatively small economic value have made minnows an understudied group (Whittier et al. 2000). As a visit to almost any curated fish collection would confirm, no ichthyologist is immune to at least an occasional misidentification of cyprinid specimens. In Arkansas, 72 cyprinid species in 24 genera have been documented, representing about 32% of the state's fish fauna. Of those 72 species, 65 species in 18 genera are considered native to the state, while 7 cyprinid species in 6 genera have been introduced into Arkansas waters. Native genera include *Campostoma, Chrosomus, Cyprinella, Erimystax, Hybognathus, Hybopsis, Luxilus, Lythrurus, Macrhybopsis, Nocomis, Notemigonus, Notropis, Opsopoeodus, Phenacobius, Platygobio, Pimephales, Pteronotropis,* and *Semotilus.* Introduced cyprinid genera include *Carassius, Ctenopharyngodon, Cyprinus, Hypophthalmichthys, Mylopharyngodon,* and *Scardinius.* Professional biologists and students alike often have difficulty in identifying cyprinid species. It is important to remember when identifying cyprinids that a suite of characters must be used to assist in the identification of genus and species, rather than depending on a single character. It is rare that a single character will reveal the correct genus of a specimen in question, but there are exceptions where that scenario works, for example, the cartilaginous ridge in the lower lip identifies the genus *Campostoma.*

Native Arkansas cyprinids share the following diagnostic characteristics: (1) the jaws are toothless, but they have well-developed toothed pharyngeal arches; (2) all fins are without true spines; (3) there are usually 8 (9 or 10 in two species) principal rays in the dorsal fin and 16 or fewer principal rays in the anal fin; (4) there are 19 principal caudal fin rays; (5) the body is covered by cycloid scales; (6) the head is devoid of scales; (7) an adipose fin is absent; and

(8) there is no true stomach. While these characters hold true for native minnows, two of the introduced Eurasian species (Common Carp, Goldfish) have 17 or more rays in the dorsal fin and possess a stout, saw-toothed spine at the dorsal and anal fin origins.

North American phoxinins are found in a wide range of habitats and exhibit a wide range of morphologies adapting them to those habitats (Simons et al. 2003). Native cyprinids also have a variety of trophic adaptations (algivores, herbivores, planktivores, piscivores, and omnivores), often having specialized oral morphologies for obtaining food (Simons et al. 2003). Most Arkansas cyprinids feed on a variety of invertebrates and vegetation, but at least one (*Notropis potteri*) is partially a piscivore. Arkansas minnows are of great value as forage fishes and as ecological indicators of water quality. Some species such as the Golden Shiner, Goldfish, and Fathead Minnow are in great demand by sport fishermen as bait species, and some of the largest commercial minnow farms in the world are in central Arkansas near Lonoke. The introduced Common Carp (and to a lesser extent, the Grass Carp) forms a significant part of the annual commercial catch of fishes in Arkansas.

All cyprinids are egg layers, and most do not guard their eggs. Reproductive behavior is also very diverse, as cyprinids exhibit a variety of courtship displays culminating in spawning over unprepared substrate, in rock crevices, in cavities beneath rocks, or on large gravel nests. Johnston and Page (1992) divided reproductive strategies of cyprinid fishes into two main groups: (1) those that prepare the substrate for receiving fertilized eggs (a derived condition) and (2) those that make no preparation of the substrate (ancestral condition). Most cyprinid fishes are in the latter category, including species that broadcast their eggs, hide them in crevices of logs or rocks (crevice spawners), or use nests constructed by other fish species (Boschung and Mayden 2004). Those Arkansas cyprinids that prepare the substrate do so in one of four ways: (1) pit building, characteristic of *Campostoma* and *Luxilus*; (2) pit-ridge building, characteristic of *Semotilus*; (3) mound-building, characteristic of *Nocomis*; and (4) cavity nesting, characteristic of the egg-clustering species of *Opsopoeodus* and *Pimephales*. Species that spawn in the nests constructed by other minnows

and sunfishes are called nest associates (Johnston 1994a, b; Johnston and Page 1992). Cashner and Bart (2010) used genetic analyses to identify to species cyprinid eggs collected from a spawning aggregation substrate. Johnston (1994a) hypothesized that associates benefit most from parental care provided by the host and least from physical protection by the nest itself. In Arkansas, the chub genus *Nocomis* builds nests more than 1 meter in diameter and up to one-third of a meter in height (Johnson and Page 1992). In addition, some cyprinids, such as *Pteronotropis hubbsi* (Fig. 15.1), are nest associates and spawn over other cyprinid or noncyprinid fish nests such as those of centrarchids, and parental care is dutifully carried out by the nest-constructing species (Fletcher and Burr 1992).

The males of several species of Arkansas minnows become quite colorful during the breeding season, including *Luxilus pilsbryi*, *Luxilus zonatus*, *Lythrurus umbratilis*, and *Phoxinus erythrogaster*. In most species, the males develop horny protuberances known as breeding tubercles on various parts of their anatomy, particularly the snout, head, dorsum, and some fins. These tubercles have at least four functions: sex recognition, driving other males away during aggressive encounters, aiding in nest construction, and aiding in clasping the female during spawning. Tubercle patterns are useful for identifying certain species, subgenera, and genera.

Figure 15.1. Breeding pair of *Pteronotropis hubbsi*, Bluehead Shiner, from southern Arkansas. *Richard T. Bryant and Wayne C. Starnes.*

Key to the Carps and Minnows

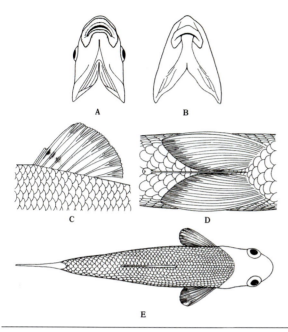

Figure 15.2. Key characteristics of some minnow genera. *Smith (1979)*.

A—Underside of head of *Campostoma*, showing cartilaginous edges of jaws.
B—Underside of head of *Phenacobius*, showing fleshy lobes on mandibles.
C—Dorsal fin of *Pimephales*, showing thickened anterior ray.
D—Ventral keel of *Notemigonus*.
E—Dorsal view of *Pimephales*, showing crowded predorsal scales.

1A Dorsal fin elongate and having more than 15 soft rays; a strong, hardened serrate ray at origin of dorsal and anal fins. 2
1B Dorsal fin with fewer than 14 soft rays; usually no hardened ray at origin of dorsal and anal fins, if hardened ray is present, its posterior margin not serrate. 3
2A Two pairs of barbels present at corners of mouth (Fig. 15.3); lateral line scales 35–38.
Cyprinus carpio Page 243
2B Barbels absent; lateral line scales 26–32.
Carassius auratus Page 221

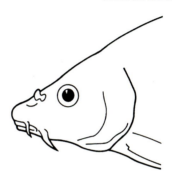

Figure 15.3. Lateral view of head of *Cyprinus carpio*, showing two barbels. *Moore (1968)*.

3A Anal fin situated far to the rear, distance from anal fin origin to caudal base going into distance between tip of snout and anal fin origin more than 2.5 times; pharyngeal teeth molariform or with prominent parallel grooves. 4
3B Anal fin situated in normal position, distance from anal fin origin to caudal base going into distance between tip of snout and anal fin origin fewer than 2.5 times; pharyngeal teeth lacking grooves. 5
4A Body coloration silvery-white or olivaceous; pharyngeal teeth with prominent parallel grooves; some fins darkly pigmented. *Ctenopharyngodon idella* Page 225
4B Body coloration blackish, blue-gray, or dark brown; pharyngeal teeth molariform without prominent parallel grooves; all fins darkly pigmented.
Mylopharyngodon piceus Page 294
5A Lateral line complete, with 85 or more scales (if 85 or more and lateral line incomplete, 2 dark lateral bands present, and ventral keel absent between anus and pelvic fins, then go to couplet 7). 6
5B Lateral line complete or incomplete, but with fewer than 85 scales in lateral series. 7

6A Ventral keel extending only from vent to pelvic fin base; gill rakers comblike; sides with numerous small, irregularly shaped dark blotches.

Hypophthalmichthys nobilis Page 265

6B Ventral keel extending from vent to isthmus; gill rakers fused into a porous plate; sides plain, without dark blotches.   *Hypophthalmichthys molitrix* Page 262

7A Barbels present (at posterior angle of the jaw or placed forward from the angle of the jaw on the maxilla) (Fig. 15.4).   8

7B Barbels absent.   18

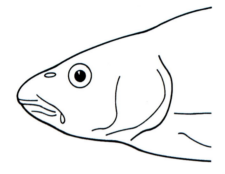

Figure 15.4. Lateral view of head of *Nocomis biguttatus*, showing barbel. *Moore (1968).*

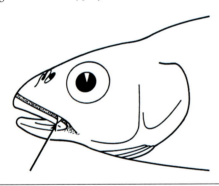

Figure 15.5. Lateral view of head of *Semotilus atromaculatus* showing minute, concealed barbel. *Miller and Robison (2004).*

8A Barbel minute, flat, concealed in groove above the maxilla, and difficult to locate (Fig. 15.5); barbel not terminal.

Semotilus atromaculatus Page 375

8B Barbel larger and terminal (at angle of jaws).   9

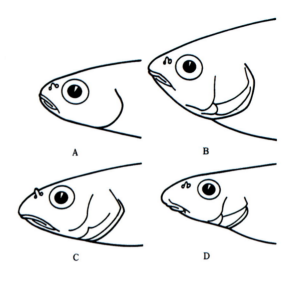

Figure 15.6. Characteristics of some minnow mouths (lateral views). *Cross (1967).*

A—Mouth terminal and oblique, upper and lower jaws equal. *Notropis boops* illustrated.
B—Mouth nearly terminal, oblique, but lower jaw included (shorter than upper jaw, closing within it). *Cyprinella camura* illustrated.
C—Mouth subterminal, scarcely oblique, lower jaw included (shorter than upper jaw and closing within it). *Notropis dorsalis* illustrated.
D—Mouth ventral and nearly horizontal. Note barbel projecting from groove at corner of mouth (barbel absent in A, B, and C). *Erimystax x-punctatus* illustrated.

9A Distance from dorsal fin origin to caudal base less than distance from dorsal fin origin to snout tip; mouth terminal (Fig. 15.6A); snout barely projecting beyond upper lip; anal fin rounded.   *Nocomis* 10

9B Distance from dorsal fin origin to caudal base equal to or greater than distance from dorsal fin origin to snout tip; mouth subterminal or inferior (Fig. 15.6D); snout projecting beyond upper lip; anal fin falcate or straight-edged, not rounded.   11

10A Breeding males with nuptial tubercles or tubercle spots restricted to the head (rarely with a few tubercles on nape), absent from the body; confined to the White River system.   *Nocomis biguttatus* Page 299

10B Breeding males with nuptial tubercles or tubercle spots present on nape and sides of body as well as on the head; found only in the Illinois River system and disjunctly in the upper Ouachita River system.

Nocomis asper Page 297

11A Body with scattered black spots, dark x-shaped markings, or a row of dark blotches.   12

11B Body without scattered black spots or dark x-shaped markings. 14

12A Body with numerous small, black scattered spots; scales in lateral line usually fewer than 40; upper jaw extends past front of eye.

Macrhybopsis hyostoma Page 286

12B Body with numerous x-shaped markings or a row of dark blotches; scales in lateral line usually more than 40; upper jaw does not extend past front of eye.

Erimystax 13

13A Lateral line scales usually 46–49; midside with a row of dark blotches; middorsal stripe composed of a series of alternating light and dark areas.

Erimystax harryi Page 246

13B Lateral line scales usually 40–43; body with numerous scattered x-shaped markings; middorsal stripe uniformly colored.

Erimystax x-punctatus Page 247

14A Lateral line scales usually more than 40; eye small, its diameter going into head length more than 4.5 times. 15

14B Lateral line scales usually fewer than 40; eye large, its diameter going into head length fewer than 4 times. 17

15A Body scales with distinct keels; snout length about equal to distance from back of eye to rear margin of gill cover; anterior rays of dorsal fin not extending beyond posterior rays when fin is flattened.

Macrhybopsis gelida Page 284

15B Body scales without keels; snout length much less than distance from back of eye to rear margin of gill cover; anterior rays of dorsal fin extending beyond posterior rays when fin is flattened. 16

16A Breast scaled; pectoral fin only slightly sickle-shaped and shorter, its tip not reaching behind pelvic fin base; head width greater than its depth; pharyngeal teeth 2,4–4,2.

Platygobio gracilis Page 368

16B Breast naked; pectoral fin prominently sickle-shaped and longer, its tip reaching behind pelvic fin base; head width less than its depth; pharyngeal teeth 1,4–4,1. *Macrhybopsis meeki* Page 289

17A Caudal fin uniformly colored; dark lateral band usually distinct; distance from dorsal fin origin to caudal base equal to or less than distance from dorsal fin origin to snout tip. *Hybopsis amblops* Page 257

17B Caudal fin with lower lobe dark with white margin; lateral band indistinct or absent; distance from dorsal fin origin to caudal base greater than distance from dorsal fin origin to snout tip.

Macrhybopsis storeriana Page 292

18A Lower jaw with hardened inner cartilaginous ridge; intestine coiled spirally around the swim bladder (Fig. 15.7A). *Campostoma* 19

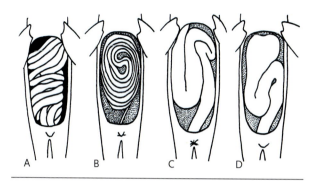

Figure 15.7. Cyprinid gut (intestinal) coiling and looping patterns and peritoneal melanism. *Jenkins and Burkhead (1994).*

A—*Campostoma*, highly coiled, looped around swim bladder.
B—*Hybognathus*, highly coiled in clock-spring pattern.
C—*Nocomis*, moderately coiled, with anterior loop.
D—Most minnows, S-shaped.

18B Lower jaw without hardened inner cartilaginous ridge; intestine not coiled spirally around the swim bladder. 22

19A Gill rakers on first gill arch usually 25 or fewer; least width of skull between eyes about equal to distance from back of eye to upper end of gill opening; breeding males usually without black pigment in anal fin and without tubercles on inner side of each nostril.

Campostoma oligolepis Page 206

19B Gill rakers on first gill arch usually more than 25; least width of skull between eyes usually less than distance from back of eye to upper end of gill opening; breeding males with black pigment in anal fin and with 1–3 tubercles on inner side of each nostril. 20

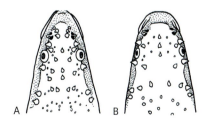

Figure 15.8. Heads of *Campostoma* breeding males showing tubercle patterns.

A—*Campostoma* species with 1–3 tubercles on inner side of each nostril.
B—*Campostoma oligolepis* without tubercles on inner side of each nostril.

20A Pharyngeal teeth in 2 rows, with 1 or 2 very thin delicate teeth in minor row on one or both sides (1,4–4,1 or 2,4–4,1); median and paired fins of adults and juveniles red or red orange throughout most of the year; nuptial males with numerous small tubercles on predorsal region usually arranged into parallel, longitudinal rows; dark pigmentation of base of caudal fin of nuptial adults vertically extended in a narrow band across 4–10 rays.

Campostoma spadiceum Page 217

20B Pharyngeal teeth in a single row, 0,4–4,0; median and paired fins of adults and juveniles without red or red-orange coloration throughout the year; nuptial males with larger tubercles, few in number, and scattered over the predorsal region, the tubercles not forming longitudinal rows on nape; dark pigmentation at base of caudal fin of nuptial adult in shape of a spot, wedge, or V. 21

21A In Arkansas, found in the White and Illinois-Neosho drainages of the Ozarks; scales smaller, sum of lateral line and circumferential scales usually 90 or more (84% of specimens; x̄ = 92.5); breeding males highly tuberculate at all sizes, often with tubercles on all fins, head, nape, and sides.

Campostoma plumbeum Page 211

21B In Arkansas, confined almost entirely to streams draining Crowley's Ridge; scales larger, sum of lateral line and circumferential scales usually less than 90 (55% of specimens; x̄ = 88.7); breeding males (except for largest specimens) usually with tubercles confined mainly to top of head and with 0–3 tubercles on anterior nape, but tubercles usually absent from fins and sides. *Campostoma pullum* Page 214

22A Belly between pelvic fins and anal fin with a sharp keel (Fig. 15.2D); anal rays 10 or more. 23

22B Belly between pelvic fins and anal fin without a sharp keel; anal rays usually fewer than 10. 24

23A Belly with a sharp, scaleless keel; dorsal rays 7–9; pharyngeal teeth 5–5; yellowish fins.

Notemigonus crysoleucas Page 302

23B Belly with a scaled keel; dorsal rays 9–11; pharyngeal teeth 3,5–5,3; blood-red fins.

Scardinius erythrophthalmus Page 374

24A Second unbranched ray of dorsal fin (Fig. 15.2C) usually thickened, short, and separated from the third ray by a membrane (first ray extremely small, usually overlooked without dissection); predorsal area broad and flattened with scales small and crowded, dorsal scales noticeably smaller than scales of upper sides (Fig. 15.2E). *Pimephales* 25

24B Second unbranched ray of dorsal fin slender and closely joined to the third ray; predorsal area not broad or flattened, with scales about same size as those on upper sides. 28

25A Body stout, greatest body depth going into standard length about 3.2–4.0 times; lateral line usually incomplete; basicaudal spot absent or indistinct; breeding males with 3 rows of tubercles on snout.

Pimephales promelas Page 361

25B Body slender, greatest body depth going into standard length about 3.9–5.1 times; lateral line complete; basicaudal spot distinct; breeding males with 1–3 rows of tubercles on snout. 26

26A Peritoneum black; mouth more ventral, upper lip overhung by a fleshy snout; intestine with several loops. *Pimephales notatus* Page 359

26B Peritoneum silvery; mouth terminal, upper lip not overhung by fleshy snout; intestine with a single S-shaped loop. 27

27A Basicaudal spot vertically elongated; well-defined lateral stripe; upper lip considerably thickened at midline; scales above lateral line usually 6; body slender, body depth going into SL 4.5–5.1 times; crosshatching distinct (pigment concentrated along scale margins); nuptial tubercles 11–13, in 3 rows; no dark crescent-shaped mark on snout between nostril and upper lip.

Pimephales tenellus Page 364

27B Basicaudal spot wedge-shaped; lateral stripe indistinct or absent; upper lip only slightly thickened at midline; scales above lateral line usually 7 or 8; body stouter, body depth going into SL 3.9–4.3 times; crosshatching indistinct (pigment dispersed on scales); nuptial tubercles 9, in 2 rows; a dark, crescent-shaped mark present on snout between nostril and upper lip.

Pimephales vigilax Page 366

28A Intestine long, with several coils that are transverse or spiraled (Fig. 15.9B); peritoneum black; pharyngeal teeth in 1 row. 29

28B Intestine short, with 1 or 2 loops (Fig. 15.9A); peritoneum various; pharyngeal teeth in 1 or 2 rows. 33

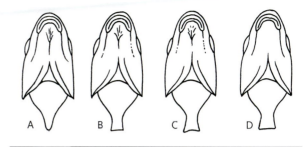

Figure 15.9. Ventral views (diagrammatic) of body cavities in minnows, as they would appear if lower body wall were cut away. *Cross (1967)*.

A—Intestine short, with a single S-shaped loop; peritoneum silvery.

B—Intestine long, looped across body cavity; peritoneum dark (usually black).

Figure 15.10. Basioccipital bones in four species of *Hybognathus*.

A—*Hybognathus placitus*   C—*H. nuchalis*
B—*H. argyritis*   D—*H. hayi*

29A Scales extremely small, more than 70 in lateral series; pharyngeal teeth 5–5.
Chrosomus erythrogaster Page 223

29B Scales large, fewer than 70 in lateral series; pharyngeal teeth 0,4–4,0. 30

30A Lateral line marked with 2 rows of dark spots; pharyngeal teeth hooked.
(Notropis, in part) Notropis nubilus Page 324

30B Lateral line not outlined by rows of dark spots; pharyngeal teeth not hooked. *Hybognathus* 31

31A Eye diameter greater than snout length; snout rounded, not easily seen from below; tip of upper lip about level with middle of eye; edges of scales on forward part of side prominently dark-edged, forming a distinct diamond-shaped pattern.
Hybognathus hayi Page 250

31B Eye diameter less than snout length; Snout more pointed, extending noticeably beyond mouth; front of upper lip below middle of eye; edges of scales on forward part of side evenly pigmented, not prominently dark-edged, not forming a distinct diamond-shaped pattern. 32

32A Eye smaller, its diameter going into head length about 5–6 times; posterior process of the basioccipital bone narrow and rodlike, not expanded (Fig. 15.10A); scale rows below lateral line 15 or more.
Hybognathus placitus Page 254

32B Eye large, its diameter going into head length about 4 times; posterior process of the basioccipital bone thin and expanded (Fig. 15.10B); scale rows below lateral line fewer than 15. *Hybognathus nuchalis* Page 252

33A Lower lip developed into a fleshy lobe on each side (Fig. 15.2B); mouth suckerlike.
Phenacobius mirabilis Page 356

33B Lower lip normally formed, without a prominent lobe on each side; mouth not suckerlike. 34

34A Dorsal fin rays 9 or more. 35

34B Dorsal fin rays 8. 36

35A Anal rays 9 or 10; body slab-sided; lateral line absent or incomplete, with only 2–9 pored scales.
Pteronotropis hubbsi Page 371

35B Anal rays 8; body terete; lateral line present and complete or incomplete, but always with more than 10 pored scales, pored scales usually extending at least onto caudal peduncle. *Opsopoeodus emiliae* Page 354

36A Caudal fin base with a distinct round or wedge-shaped spot very clearly set off from the lateral band. 37

36B Caudal fin base without distinct round or wedge-shaped spot, or, if present, spot appears as a rectangular extension of the lateral band or as an indiscrete splash. 39

37A Basicaudal spot wedge-shaped and small (Fig. 15.11); fins in breeding males not highly colored.
Notropis greenei Page 321

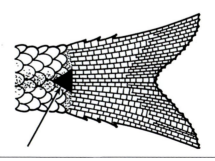

Figure 15.11. Wedge-shaped caudal spot in *Notropis greenei*. *Moore (1968)*.

37B Basicaudal spot large and round; fins in breeding males yellow or reddish. 38

38A Basicaudal spot with a small black triangle at upper and lower edge of caudal base; posterior interradial membranes of dorsal fin not blackened; pharyngeal teeth 0,4–4,0. *Notropis maculatus* Page 322

38B Basicaudal spot large, round, without small triangle at upper and lower edge of caudal base; posterior interradial membranes of dorsal fin somewhat blackened; pharyngeal teeth 1,4–4,1.
Cyprinella venusta Page 239

39A Dorsal fin with dark speckles on 1 or more membranes between fin rays, often forming a dusky blotch in posterior portion of fin, or with small spot near base of first few interradial membranes. 40

39B Dorsal fin with dark speckles absent or confined to margins of fin rays. 45

40A Body slender, its depth going into standard length more than 5 times; tip of upper lip on a level with or below middle of eye; scales in forward part of lateral line with rear margins slightly scalloped.
Notropis ozarcanus Page 328

40B Body deeper, its depth going into standard length fewer than 5 times; tip of upper lip on a level with upper part of eye; scales in forward part of lateral line with rear margins rounded. 41

41A Dorsal fin membranes uniformly pigmented, not forming a distinct posterior black spot; body deep, its depth usually going fewer than 3.5 times into standard length. *Cyprinella lutrensis* Page 233

41B Dorsal fin with dark pigment concentrated on posterior membranes forming a distinct black blotch in adults; body not markedly deep, its depth usually going into standard length more than 3.5 times. 42

42A Anal rays usually 8; dark lateral stripe on caudal peduncle narrow and prominent, centered below pores of lateral line. *Cyprinella spiloptera* Page 236

42B Anal rays usually 9; dark lateral stripe on caudal peduncle broad and faint, and centered on pores of lateral line. 43

43A Caudal fin base with 2 broad white patches; lateral line scales 39–41. *Cyprinella galactura* Page 231

43B Caudal fin base with a single narrow white patch or without white patch; lateral line scales 35–39. 44

44A Caudal fin base with a narrow milky white patch; snout blunt and head outline sharply decurved; lateral line scales 35–37. *Cyprinella camura* Page 229

44B Caudal fin base without milky white patch; snout sharply pointed and head outline not decurved; lateral line scales 37–39. *Cyprinella whipplei* Page 241

45A Upper lip distinctly swollen posteriorly; lower lip thickened medially (Fig. 15.12A); mouth large; pharyngeal teeth 2,4–4,2. *Notropis potteri* Page 334

45B Upper and lower lips thin, and little thickened; mouth size variable; pharyngeal teeth variable. 46

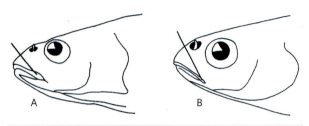

Figure 15.12. Upper lip shapes in:

A—*Notropis potteri*

B—*N. blennius*. Ross (2001)

46A Pharyngeal teeth 0,4–4,0. 47

46B Pharyngeal teeth 1,4–4,1 or 2,4–4,2. 56

47A Anal fin rays usually 7 (occasionally 8). 48

47B Anal fin rays 8–10. 51

48A Eye small, its diameter less than snout length; front of upper lip slightly below lower margin of eye; lower surface of head distinctly flattened.
Notropis sabinae Page 336

48B Eye larger, its diameter equal to or greater than snout length; front of upper lip on same level or above lower margin of eye; lower surface of head not distinctly flattened. 49

49A Lateral band conspicuous, ending in a rectangular caudal spot joined with and as wide as lateral band (in young there may be a narrow clear hiatus between the lateral band and the caudal spot).
Notropis atrocaudalis Page 306

49B Lateral band not as conspicuous and not ending in a rectangular caudal spot (if spot is present, it appears as an indiscrete splash on the basal portion of the middle caudal rays). 50

50A Nape fully scaled; mouth horizontal or nearly so; pectoral fins rounded, not falcate.
Notropis stramineus Page 340

50B Nape often naked or partly scaled; mouth more oblique, often about 45° from the horizontal; pectoral fins long, pointed, and falcate (especially in males).
Notropis bairdi Page 308

51A Anal fin rays 8. 52

51B Anal fin rays 9 or 10 (rarely 8). 55

52A Anterior lateral line scales not conspicuously elevated, their exposed surfaces about as long as high (Fig. 15.13B). *Notropis girardi* Page 318

52B Anterior lateral line scales conspicuously elevated, their exposed surfaces much higher than long (Fig. 15.13A). 53

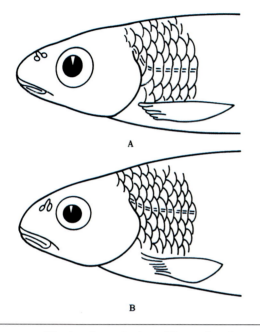

Figure 15.13. Head and anterior lateral line scales of:

A—*Notropis volucellus*
B—*Notropis girardi. Miller and Robison (2004)*

53A Infraorbital canal complete (Fig. 15.14A); peritoneum speckled with dark pigment; body coloration normal, with lateral stripe on caudal peduncle; triangular caudal spot; pigmentation of scale borders intense near middorsal line and present though diminishing downward on entire dorsum. 54

53B Infraorbital canal incomplete (Fig. 15.14B); peritoneum silvery; body coloration quite pallid; pigmentation of scale borders less intense and often absent just above lateral line. *Notropis buchanani* Page 314

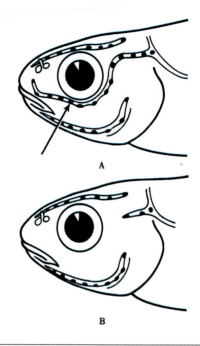

Figure 15.14. Cephalic lateral line system of:

A—*Notropis volucellus;* note the complete infraorbital canal (arrow)
B—*Notropis buchanani. Reno (1966)*

54A Dorsolateral scales with melanophores more or less concentrated along the scale margins, and the interior of the scale is unpigmented or sparsely sprinkled with melanophores; usually a noticeable predorsal dark spot anterior to the dorsal fin insertion; nuptial males have more prominent tubercles on the snout and lachrymal areas than on top of the head; widespread.

Notropis volucellus Page 348

54B Dorsolateral scales with melanophores more or less evenly distributed across the scales, especially scales anterior to the dorsal fin insertion; predorsal dark spot absent or very light; in nuptial males, snout and lachrymal tubercles are not well developed, while those on top of head (interorbital area) are more prominent; restricted to large rivers.

Notropis wickliffi Page 351

55A Lateral band well developed anteriorly and the row of scales above it largely without pigment; eye diameter greater than snout length.

Notropis ortenburgeri Page 326

55B Lateral band poorly developed anteriorly and the row of scales above it well pigmented, presenting a diamond-shaped appearance; eye diameter equal to or less than snout length.

Cyprinella lutrensis Page 233

56A Origin of dorsal fin placed well behind pelvic fin base (Fig. 15.15A). 57

56B Origin of dorsal fin not definitely behind pelvic fin base, but sometimes near the end of the pelvic fin base (Fig. 15.15B). 62

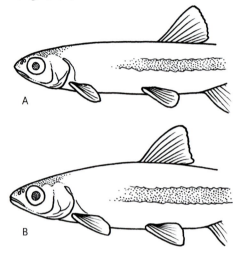

Figure 15.15. Orientation of dorsal fin relative to origin of pelvic fins in minnows.

A—Origin of dorsal fin posterior to vertical from insertion of pelvic fin.
B—Origin of dorsal fin anterior to insertion of pelvic fin.

57A Lateral line strongly decurved; lateral line scales usually more than 38; predorsal scales small and closely crowded, usually in 25 or more rows. 58

57B Lateral line gently decurved; lateral line scales 38 or fewer; predorsal scales not noticeably small and crowded, in fewer than 25 rows. 60

58A Concentration of black pigment at origin of dorsal fin; body of adult deep, its greatest depth going into standard length fewer than 4 times; dorsum usually with chevron-like markings most evident in adults; lateral band poorly developed.
Lythrurus umbratilis Page 281

58B No concentration of black pigment at origin of dorsal fin; body of adult slender, its greatest depth going 4 or more times into standard length; dorsum without chevron-like markings; lateral band more intense, extending onto snout. 59

59A Anal fin rays usually 10 (9–11); body slender, terete; breeding colors red on head; confined to Little River system. *Lythrurus snelsoni* Page 279

59B Anal fin rays usually 11 or 12 (10–13); body deep, compressed; breeding colors yellowish on fins; widespread. *Lythrurus fumeus* Page 277

60A Snout long and sharply pointed, its length greater than eye diameter; chin not sprinkled with dark pigment (Fig. 15.16A) or chin pigment sometimes present; anal rays 9 or 10; dorsal fin rounded. 61

60B Snout rather blunt, its length not greater than eye diameter; chin sprinkled with dark pigment (Fig. 15.16B); anal rays 10–13; dorsal fin pointed.
Notropis atherinoides Page 304

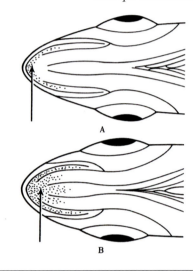

Figure 15.16. Ventral view of head showing chin pigmentation in:

A—*Notropis percobromus*
B—*Notropis atherinoides*

61A Body slender, narrower at dorsal fin and caudal peduncle; lateral line usually straight or only slightly decurved 1 or 2 scale rows; lip pigmentation light; chin region clear or with scattered melanophores; lateral stripe variable, always thicker caudally than anteriorly; lateral line well outlined by melanophores for at least half its length; Ozark and Ouachita mountains regions. *Notropis percobromus* Page 329

61B Body more robust, deeper at dorsal fin and caudal peduncle; lateral line deeply decurved 5 or 6 scale rows; lip pigmentation intense, always some pigment on chin region, typically a dense median patch present; lateral line not outlined by melanophores (or weakly outlined for a short distance) but often continuous with lateral band; lateral stripe thick, nearly equal in width from caudal base to opercular cleft; Little River drainage of Ouachita Mountain region.
Notropis suttkusi Page 342

62A Peritoneum with intense black pigment overlying the silver. 63

62B Peritoneum never with intense black pigment, but silvery and with or without scattered melanophores. 68

63A Anal fin rays 8; pharyngeal teeth 1,4–4,1 (occasionally lacking lesser row teeth on either side).
Notropis boops Page 312

63B Anal fin rays 9–11; pharyngeal teeth 2,4–4,2. 64

64A Eye very large, its diameter about equal to distance from rear of eye to rear margin of gill cover; distance from tip of snout to front of dorsal fin greater than distance from front of dorsal fin to caudal base; anal fin rays usually 10 or 11 (9–11).
Notropis telescopus Page 344

64B Eye smaller, its diameter less than distance from back of eye to rear margin of gill cover; distance from tip of snout to front of dorsal fin less than distance from front of dorsal fin to caudal base; anal fin rays usually 9.
Luxilus 65

65A Lateral band indistinct, not extending forward on head to snout tip, without a narrow stripe above and running parallel to it; body deep and compressed.
Luxilus chrysocephalus Page 270

65B Lateral band prominent, extending forward on head to snout tip with narrow stripe above and running parallel to the lateral band; body slender and not compressed. 66

66A Rear margin of gill opening bordered by dark bar; lateral band becoming abruptly narrower just behind gill opening and not touching lateral line beneath dorsal fin; area between lateral line and lateral stripe on middle part of body unpigmented.
Luxilus zonatus Page 274

66B Rear margin of gill opening not bordered by a dark bar; lateral band not becoming abruptly narrower just behind gill opening and touching lateral line beneath dorsal fin; area between lateral line and lateral stripe on middle part of body pigmented. 67

67A Lateral band extending downward only to the lateral line; 24–26 total circumferential scales; mandibular tubercles usually more than 6; gill rakers 7; confined to White River system. *Luxilus pilsbryi* Page 272

67B Lateral band quite broad, extending 1–2 scales below the lateral line; 26 or 27 total circumferential scales; mandibular tubercles usually 0–3; gill rakers 8 or 9; confined to Arkansas River system tributaries.
Luxilus cardinalis Page 268

68A Interior of mouth heavily pigmented with black.
Notropis chalybaeus Page 316

68B Interior of mouth without black pigmentation. 69

69A Anal fin rays 9–11; pigment of dorsum forming no regular pattern but with scattered, stellate melanophores. *Notropis perpallidus* Page 332

69B Anal fin rays 7–9; pigment of dorsum in a regular pattern without stellate melanophores. 70

70A Maxillary bone extending far behind the angle of the mouth (Fig. 15.17); snout very blunt and overhanging the upper lip. *Hybopsis amnis* Page 259

70B Maxillary bone not extending far behind the angle of the mouth; snout not especially blunt nor overhanging the upper lip. 71

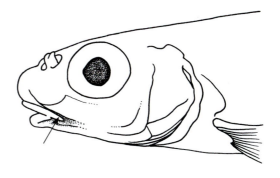

Figure 15.17. Extension of the maxillary bone behind the angle of the mouth in *Hybopsis amnis* (arrow). *Miller and Robison (2004).*

71A Anal fin rays 8 or 9; dorsal fin long and pointed, the length of longest ray contained 2 times or fewer in predorsal length. *Notropis shumardi* Page 338

71B Anal fin rays usually 7; dorsal fin short and blunt, the length of longest ray contained more than 2 times in predorsal length. 72

72A Lateral band prominent, extending forward onto head; distance from tip of snout to front of dorsal fin less than distance from front of dorsal fin to base of caudal fin. *Notropis texanus* Page 346

72B Lateral band not prominent, becoming indistinct toward head; distance from tip of snout to front of dorsal fin about equal to distance from front of dorsal fin to base of caudal fin. *Notropis blennius* Page 310

Genus *Campostoma* Agassiz, 1855— Stonerollers

The genus *Campostoma* encompasses North American herbivorous minnows generally referred to as stonerollers. Stonerollers are often the most abundant fishes in upland streams across much of central and eastern North America between southern Canada and central Mexico (Lee et al. 1980). Stonerollers share several characters considered to be adaptations to herbivory, including a keratinized ridge on the lower lip, a black peritoneum, and an elongated gut, which in most species is extensively coiled around the swim bladder (Cope 1869). Amazingly, intestine length is typically about 18 inches (460 mm) in a 5-inch (130 mm) individual (Page and Burr 2011). The genus also exhibits other shared attributes, including an elongate, thick body that is round in cross-section, no barbels, protractile premaxillae, and dorsal fin origin over or slightly behind pelvic fin origin. Breeding males usually have white lips and bright red eyes (Page and Burr 2011). There have been several attempts to resolve phylogenetic relationships of the genus *Campostoma* with other North American cyprinid genera. Cavender and Coburn (1985) hypothesized that the closest affinities were between *Campostoma* and the primarily Mexican genus *Dionda* Girard. Mayden (1989) agreed and considered *Nocomis* the sister group of those two genera. However, subsequent more inclusive molecular phylogenetic analyses recovered only *Campostoma* as sister to the chubs of genus *Nocomis* (Simons and Mayden 1999; Schönhuth and Mayden 2010).

The distinctiveness of stoneroller minnows was first recognized by Agassiz (1855), who erected the genus *Campostoma* to include *Rutilus anomalous* (described by Rafinesque from Kentucky in 1820), *Chondrostoma pullum* Agassiz (1854), and five other previously described species (Blum et al. 2008). Girard (1856) subsequently expanded the genus with several additional taxa. Blum et al. (2008) pointed out that many of these early descriptions were short, unclear, inconsistent, or not specific enough to permit accurate identification to species. Hubbs and Ortenburger (1929a) retained *C. ornatum* (primarily a Mexican species) as a valid species and lumped all other known *Campostoma* into *C. anomalum*. This recognition of only two *Campostoma* species was widely accepted for the next 40 years. Cashner et al. (2010) gave a detailed chronology of the complex nomenclatural history of stonerollers in general, and stonerollers in eastern Oklahoma and western Arkansas in particular.

In Arkansas, the taxonomic picture of *Campostoma* is complex. In his dissertation on Arkansas fishes, Black (1940) assigned all *Campostoma* in Arkansas to *C. anomalum*, recognizing the subspecies *plumbeum*, *pullum*, and *oligolepis*. Historically, all stonerollers in Arkansas were identified as the Central Stoneroller, *Campostoma anomalum*, or subspecifically as *Campostoma anomalum pullum*, until Pflieger (1971) elevated *C. anomalum oligolepis* to full specific status (*C. oligolepis*). Burr and Smith (1976) and Burr and Cashner (1983) provided additional morphological evidence that *C. oligolepis* is distinct from all other *C. anomalum* forms. Those authors noted that in the

Ozarks, *Campostoma oligolepis* and what was then called *C. a. pullum* often occur sympatrically with no evidence of hybridization. Etnier and Starnes (1993) pointed out that through most of the 20th century there had been a tendency among ichthyologists to ignore stonerollers as boringly common, ubiquitous, not very handsome, and of little taxonomic interest. That view began to change with the publications of Pflieger (1971) and Burr and Smith (1976). An electrophoretic analysis confirmed the validity of *C. anomalum* and *C. oligolepis* as distinct species (Buth and Burr 1978). Shortly thereafter, Burr and Cashner (1983) recognized *Campostoma* populations from the Apalachicola, Altamaha, and Tennessee river drainages of Tennessee, Georgia, and Alabama as a new species, *C. pauciradii*. The next new species was carved from *Campostoma anomalum* when *C. a. pullum* was elevated to full specific status by Pflieger (1997). Pflieger provided little in the way of substantiation for this elevation (no formal diagnosis was given), and it received mixed acceptance. Boschung and Mayden (2004) listed *C. pullum* as one of five known *Campostoma* species, and D. A. Etnier (pers. comm.) also indicated that *Campostoma pullum* (Agassiz) should be recognized as a valid species based in part on unpublished research on stoneroller systematics by D. A. Etnier and H. T. Boschung. Page et al. (2013) did not support recognition of *C. pullum*, mainly because a formal description was lacking.

The next new *Campostoma* species was described in 2010. As early as the 1980s, W. J. Matthews and H. W. Robison independently noticed that stonerollers in the Ouachita Mountains have reddish fins unlike their northern Arkansas counterparts. Both suggested that the Ouachita populations might be an undescribed species (Robison and Buchanan 1988). Cashner et al. (2010) subsequently described that form as a new species, resurrecting the name *Campostoma spadiceum* for those red-finned stoneroller populations in the Ouachita Mountains, the Arkansas River Valley, and the western Ozarks.

Therefore, from a single widespread, rather plain stoneroller species (*C. anomalum*) in 1970, Page et al. (2013) recognized four species: *Campostoma anomalum* (Central Stoneroller), *C. oligolepis* (Largescale Stoneroller), *C. pauciradii* (Bluefin Stoneroller), and *C. spadiceum* (Highland Stoneroller) in addition to the long-recognized *C. ornatum* (Mexican Stoneroller), for a total of five *Campostoma* species. The status and description of *C. pauciradii* and *C. spadiceum* are apparently well-accepted, whereas the current treatments of *C. ornatum*, *C. anomalum*, and *C. oligolepis* remain contentious because there may be undescribed species hiding under those names (Blum et al. 2008). The most widely distributed of the species recognized by Page et al. (2013) is *C. anomalum*, which exhibits some morphological variation but exhibits even greater genetic variation. There is also genetic evidence of cryptic evolution in the genus coupled with morphological convergence possibly due to parallel evolution of phenotypes (Blum et al. 2008). New cryptic species will likely be carved from what is now generally recognized as *C. anomalum* largely using genetic criteria. Many workers today take a conservative approach to *Campostoma* taxonomy and recognize three subspecies of *C. anomalum*: *C. a. anomalum*, *C. a. pullum*, and *C. a. michaux*. In a genus-wide study of variation across the mitochondrial cytochrome *b* gene region, Blum et al. (2008) recovered nine major clades that could be recognized as distinct taxa to provisionally resolve differences among prior systematic accounts of *Campostoma* evolutionary history. One of the nine lineages (*Campostoma spadiceum*) was subsequently described by Cashner et al. (2010), and further work on *C. ornatum* indicated that the genus likely encompasses additional evolutionary lineages (in addition to the lineages listed by Blum et al. 2008) that warrant species recognition (Schönhuth et al. 2011). Cuhna et al. (2002), Mayden et al. (2006a), and Saitoh et al. (2006) supported the cytochrome *b* gene as an excellent source of character information for phylogeny reconstruction of cypriniform species. Even though cytochrome *b* sequences represent a single gene, results of analyses of notropin phylogeny using that gene were consistent with previous studies employing morphological and/or alternative molecular characters (Mayden et al. 2006a).

Campostoma anomalum as currently recognized is clearly not monophyletic and consists of several lineages (Blum et al. 2008; Page et al. 2013). When some of those lineages are eventually elevated to full specific status, the name *C. anomalum* will be applied only to populations to the northeast of Arkansas in the Ohio River drainage (exclusive of the Tennessee and Cumberland drainages), Potomac River drainage, and parts of the Susquehanna River basin (Blum et al. 2008). Because new species will undoubtedly be described from *C. anomalum*, we believe the continued use of *C. anomalum* for any Arkansas populations of *Campostoma* should be abandoned. While the total ranges of the various *Campostoma anomalum* lineages have not yet been precisely determined, we believe that the geographic distributions of the *C. anomalum* lineages inhabiting Arkansas are rather clearly defined in this state. Blum et al. (2008) and M. Blum (pers. comm.) provided information supporting the genetic distinctness of the Arkansas lineages, and two of the lineages are also distinguishable from one another and the other *Campostoma*

species in Arkansas by a combination of morphological features. Because scientific names are available in the literature for two of the three *C. anomalum* lineages in Arkansas, we discontinue the conservative approach of using *C. anomalum* for any Arkansas populations of *Campostoma*. We believe it is appropriate to recognize two lineages as full species under the names *C. plumbeum*, Plains Stoneroller, and *C. pullum*, Finescale Stoneroller, as suggested in Blum et al. (2008). *Campostoma plumbeum* and *C. pullum* are both widely distributed outside Arkansas, and we do not herein provide a rangewide diagnosis for those species. Michael Blum and coworkers intend to publish formal species descriptions (M. Blum, pers. comm.).

Taxonomically, the nested position of *Campostoma ornatum* among the *Campostoma anomalum pullum* clades in the mtDNA genealogy provides support for elevating all three *C. a. pullum* lineages in Arkansas to specific rank (Blum et al. 2008). However, based on information provided by M. Blum (pers. comm.), we refer two of the lineages to a Great Plains clade, *Campostoma plumbeum*, because the nomenclatural priority of *C. plumbeum* has been reserved for a *Campostoma* lineage distributed across the Great Plains (Cross 1967), and because the topotypic material used in the Blum et al. (2008) genetic study (considering the description of *Dionda plumbea* by Girard in 1856 synonymous with *C. anomalum*) grouped with other individuals from Great Plains populations.

If one accepts the arguments for referring to *C. pullum* and *C. plumbeum* as full species, the question then is how to treat the two genetic lineages of *C. plumbeum* referred to by Blum et al. (2008) as *C. plumbeum* "Plains" and *C. plumbeum* "Ozarks." We are not ready to treat the genetically distinct Ozark clade of *C. plumbeum* as a full species separate from the Plains clade even though the two clades are as genetically distinct as currently recognized species of *Campostoma*. There is no available name in the literature for the Ozark clade, and we defer judgment regarding the status of Ozark populations until the ongoing studies of Michael Blum and coworkers on stonerollers are completed. We believe those studies will provide a rangewide diagnosis to support recognition of *C. plumbeum* and *C. pullum*, as well as for differentiating between the two *C. plumbeum* clades (M. Blum, pers. comm.).

We herein recognize the following four species of *Campostoma* in Arkansas: (1) *C. plumbeum*. Both lineages of *C. plumbeum* occur in Arkansas. *Campostoma plumbeum* "Plains" is found in the northwestern part of the state in Benton and Washington counties, with a total distribution that extends across the Great Plains and into the desert southwest, while *C. plumbeum* "Ozarks" is more broadly distributed in the state with a total range that spans the uplands of the Ozark Plateau, but in Arkansas is restricted to the White River drainage; (2) *C. pullum*. Within Arkansas, the true *Campostoma pullum* (as recognized by Blum et al. 2008) occurs only in northeastern Arkansas in streams draining Crowley's Ridge (St. Francis and White river drainages), with one record from the Mississippi River main stem; (3) *C. oligolepis*. This species occurs throughout the White and Illinois-Neosho drainages, and in the St. Francis River drainage; and (4) *C. spadiceum*. It is widely distributed in Arkansas, Red, and Ouachita river drainages. Table 13 compares various morphological features of the four Arkansas *Campostoma* species, based on data gathered from Arkansas populations.

Campostoma oligolepis Hubbs and Greene
Largescale Stoneroller

Figure 15.18a. *Campostoma oligolepis,* Largescale Stoneroller breeding male. *Uland Thomas.*

Figure 15.18b. *Campostoma oligolepis,* Largescale Stoneroller female. *Uland Thomas.*

CHARACTERS A terete minnow with a ventral, U-shaped horizontal mouth and the characteristic cartilaginous ridge on lower jaw of the genus *Campostoma*. Intestine extremely long and coiled around the swim bladder; peritoneum black. Barbels absent. Nape relatively flat in predorsal profile, usually without a noticeable arch. Interorbital width (least bony width of skull between eyes) approximately equal to distance from back of eye to upper end of gill opening (must be measured for accurate

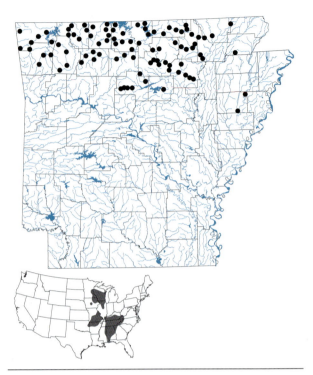

Map 22. Distribution of *Campostoma oligolepis,* Largescale Stoneroller, 1988–2019.

determination). Gill rakers on first arch 25 or fewer in 90% of Arkansas specimens (range 20–28). Pharyngeal teeth usually 0,4–4,0, but often 1,4–4,1, 1,4–4,0, or 0,4–4,1. Predorsal scales usually 20 or fewer (15–22). Lateral line complete, usually with 39–53 scales ($\bar{x}$ = 48.1). Body circumferential scales usually 31–40. Sum of lateral line and circumferential scale counts less than 88 in 87% of Arkansas specimens (73–92, $\bar{x}$ = 83.7). Body circumferential scales above lateral line 12–17. Breast and belly fully scaled. Dorsal fin origin over pelvic fin insertion; dorsal fin rays 8. Anal rays 7; pectoral rays 14–19; pelvic rays 8 or 9. Head length going into SL 3.84–4.72 times. Eye diameter going into snout length 1.05–2.85 times; eye diameter going into head length 3.24–6.58 times. Snout globose, extending noticeably beyond upper lip in ventral view. Snout length going into head length 2.14–3.08 times; snout length going into SL 8.95–13.87 times. Fins small, rounded, and clear except for a dark crescent on the dorsal fin of adults. Breeding males lack tubercles along inner margin of nostrils (Fig. 15.19A). Tubercles present on top of head, chin, branchiostegals, nape, upper sides, sometimes on lower sides (especially posteriorly), and usually on dorsal, anal, and pectoral fins. Breeding females do not develop tubercles. Maximum size in Arkansas from the Buffalo River, 7.0 inches (177 mm) TL.

LIFE COLORS Body color generally olivaceous or

Table 13. Comparison of Meristic and Other Morphological Characters of the Four Species of *Campostoma* in Arkansas. Data are from Arkansas populations only. The number of specimens examined (*N*) for each species applies to all characters except pharyngeal teeth and number of gill rakers.

Character	*C. oligolepis* N = 136	*C. plumbeum* N = 79	*C. pullum* N = 115	*C. spadiceum* N = 66
Maximum size (mm SL)	155	135	105	141
Pharyngeal teeth	Usually 0,4–4,0 but frequently 1,4–4,0; 0,4–4,1; 1,4–4,1	0,4–4,0	0,4–4,0	1,4–4,0; 1,4–4,1; or 2,4–4,2 or combinations thereof
Gill rakers on first arch	23.8 (20–28); usually 25 or fewer	33.4 (27–43); usually more than 25	32.5 (29–37); usually more than 25	30.1 (22–36); usually more than 25
Dorsal fin rays	8	8	8	8
Anal fin rays	7	7	7	7
Pectoral fin rays	16.4 (14–19)	17.4 (15–20)	17.2 (15–19)	17.2 (15–19)
Predorsal scales	19.2 (15–22)	22.8 (19–29)	20.6 (15–25)	22.2 (19–26)
Circumferential scales	35.8 (31–40)	40.4 (36–47)	38.8 (30–43)	39.7 (36–44)
Circumferential scales above lateral line	15.1 (12–17)	16.3 (13–19)	15.5 (12–17)	16.7 (15–19)
Lateral line scales	48.1 (39–53)	52.0 (47–60)	49.3 (42–57)	54.3 (49–61)
Sum of LL and circum. scales (mean, range, and distribution)	83.7 (73–90) ≤ 85 ... 66.2% ≤ 86 ... 80.9% ≤ 87 ... 86.8% ≤ 88 ... 93.4% ≤ 89 ... 97.8%	92.5 (84–105) ≥ 88 ... 98.7% ≥ 89 ... 96.2% ≥ 90 ... 84.4% ≥ 91 ... 72.2% ≥ 92 ... 27.8%	88.7 (84–96) ≥ 85 ... 94.8% ≥ 86 ... 86.1% ≥ 87 ... 78.3% ≥ 88 ... 67.0% ≥ 89 ... 96.2% ≥ 90 ... 36.5% ≥ 91 ... 20.9% ≥ 92 ... 14.8%	94.0 (85–104) ≥ 85 ... 100% ≥ 86 ... 98.5% ≥ 87 ... 98.5% ≥ 88 ... 95.5% ≥ 89 ... 90.9% ≥ 90 ... 84.8% ≥ 91 ... 83.3% ≥ 92 ... 74.2%
Crescent-shaped row of 1–3 tubercles along inner margin of nostril	Absent	Present	Present	Present
Tubercles on nape in breeding males	23.2 (5–46) Not in longitudinal rows	21.6 (1–42) Not in longitudinal rows	0.4 (0–25) Not in longitudinal rows	24.1 (0–57) Usually in longitudinal rows
Tubercles on side of body in breeding males (ant. to D insertion; post. to D insertion)	25.6 (3–48); 46.2 (13–82)	21.6 (0–32); 34.8 (0–70)	1.0 (0–28); 0 (0–60)	10.1 (0–40); 1.9 (0–9)
Large tubercles on top of head	13.5 (6–17)	11.9 (0–23)	10.7 (4–29)	10.5 (5–25)
Tubercles present on fins (% specimens) Dorsal Caudal Anal Pectoral Pelvic	73 0 18 100 0	67 67 44 78 44	7 7 0 7 0	45 18 55 91 64

Table 13. (*continued*)

Character	*C. oligolepis* N = 136	*C. plumbeum* N = 79	*C. pullum* N = 115	*C. spadiceum* N = 66
Black band in anal fin of breeding male	Absent or slight in a few individuals	Present	Present	Present
Predorsal profile	Relatively flat	Noticeably arched at nape	Usually noticeably arched at nape	Often arched at nape
Dark pigmentation at base of caudal fin of nuptial adults (if present)	In the shape of a spot, wedge, or V, often with distal extensions on the fin rays	In the shape of a spot, wedge, or V, often with distal extensions on the fin rays	In the shape of a spot, wedge, or V, often with distal extensions on the fin rays	Vertically extended in a narrow band across 4–10 rays
Eye diam. into snout length	1.76 (1.05–2.85)	1.67 (0.93–2.63)	1.37 (1.00–2.58)	1.75 (1.07–2.88)
Eye diam. into head length	4.51 (3.24–6.58)	4.35 (3.23–5.97)	3.91 (3.15–4.94)	4.40 (3.07–5.98)
Snout length into head length	2.58 (2.14–3.08)	2.65 (2.14–3.49)	2.89 (1.68–3.61)	2.55 (2.08–2.91)
Snout length into SL	11.01 (8.95–13.87)	11.15 (7.83–13.98)	12.22 (10.06–16.55)	10.75 (8.08–13.10)
Head length into SL	4.27 (3.84–4.72)	4.20 (3.58–4.85)	4.23 (3.76–4.66)	4.22 (3.78–4.09)
Snout (viewed from side or above)	Globose and extending well beyond upper lip	More pointed, extending slightly (if at all) beyond upper lip	More pointed, extending slightly (if at all) beyond upper lip	More pointed, extending slightly (if at all) beyond upper lip
Interorbital width (least bony width of skull between eyes)	Approximately equal to distance from back of eye to upper end of gill opening	Less than distance from back of eye to upper end of gill opening	Less than distance from back of eye to upper end of gill opening	Less than distance from back of eye to upper end of gill opening
Interorbital width going into distance from back of eye to upper end of gill opening	0.87–1.12 times ($\bar{x}$ = 1.00)	1.00–1.55 times ($\bar{x}$ = 1.31)	1.17–1.48 times ($\bar{x}$ = 1.31)	1.06–1.44 times ($\bar{x}$ = 1.20)

dark tan dorsally grading to white ventrally, but adults are often more sharply bicolored from snout tip to caudal base. Venter usually whitish, without pigment. Sides often with scattered dark scales above lateral line (and sometimes below). Fins clear, except for dark crescent in the dorsal fin of adults. Breeding males develop black and pink-red colors; iris of eye flushed with orange or red. Dorsal fin with a heavy black crescent. Anal fin of nuptial males without a black medial band, but a small percentage of individuals have some discernable dark pigmentation in middle of fin.

Dark pigmentation at the base of the caudal fin is in the shape of a spot or wedge, but never vertically extended into a narrow band. Pelvic fins, anal fins, and ventral area with a reddish cast. Breeding females usually develop a black band in the dorsal fin. Young lighter in coloration, often with a dark midlateral stripe.

SIMILAR SPECIES   When tuberculate males are not available, professional ichthyologists and fisheries biologists as well as students often have a great deal of difficulty distinguishing *C. oligolepis* from the other three species of

Campostoma in Arkansas. Young individuals are especially difficult to distinguish morphologically. Cataloged lots containing both *C. oligolepis* and another *Campostoma* species can usually be found in any curated fish collection. The Largescale Stoneroller is often syntopic with *C. plumbeum* and *C. pullum* but is not known to occur in Arkansas with *C. spadiceum*. Breeding males are readily distinguished from all other Arkansas *Campostoma* species in lacking tubercles along the inner margin of each nostril (vs. a row of 1–3 tubercles present along inner margin of each nostril) (Fig. 15.19A, B), and in lacking a black band in the anal fin (vs. black band present in anal fin). Nonbreeding individuals of *Campostoma oligolepis* differ from the other three Arkansas *Campostoma* species in usually having 25 or fewer gill rakers on the first gill arch (vs. usually more than 25), width of skull between eyes about equal to distance from back of eye to upper end of gill opening (vs. less than), snout more globose and extending well beyond upper lip (vs. snout more pointed and extending only slightly beyond upper lip) (Fig. 15.19C, D), and predorsal profile relatively flat, with nape little arched (vs. nape usually noticeably arched). The Largescale Stoneroller further differs from *C. plumbeum* and *C. spadiceum* (but not *C. pullum*) in sum of lateral line and body circumferential scales usually less than 90 (vs. usually greater than 90). Other more subtle distinguishing features of *C. oligolepis* described by Burr and Smith (1976) and Burr and Cashner (1983) include a more slender body shape, body coloration of adults often darker and more sharply bicolored, a wider gape, and less melanin in all fins than in *C. plumbeum*, *C. pullum*, and *C. spadiceum*.

VARIATION AND TAXONOMY *Campostoma oligolepis* was described by Hubbs and Greene (1935) as a subspecies of *Campostoma anomalum*, *C. a. oligolepis*, from the Little Rib River, a tributary of the Wisconsin River, Marathon County, Wisconsin. Black (1940) reported *C. a. pullum* and *C. a. oligolepis* in Arkansas and provided data showing the two forms were readily recognizable. Bailey (1956) noted that because *C. a. oligolepis* and *C. a. pullum* were sympatric, they should be considered separate species rather than subspecies. Pflieger (1971) studied the forms of *Campostoma* occurring in Missouri and elevated *C. anomalum oligolepis* to specific status. Most subsequent authors supported Pflieger's action, but Sewell et al. (1980a) questioned the validity of *C. oligolepis*. Burr and Smith (1976) studied variation in this species and determined that although local variation was discernible in some Ozark streams, no intermediates of *C. a. pullum* and *C. oligolepis* were found where the two species are sympatric. Fowler and Taber (1985) found microhabitat differences and food habit

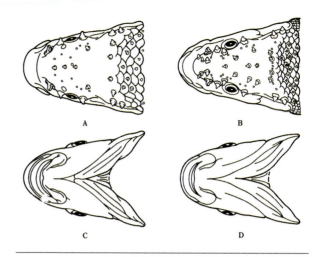

Figure 15.19. Dorsal and ventral views of nuptial male stoneroller heads. *Burr and Smith (1976).*

Dorsal:
A—*Campostoma oligolepis* (note lack of crescent-shaped row of 1–3 tubercles along inner margins of nostrils and the fewer tubercles on top of the head).
B—*Campostoma pullum* (note presence of crescent-shaped row of tubercles along inner margins of nostrils and more tubercles on top of the head).
Ventral:
C—*Campostoma oligolepis* (note larger, more globose snout, walleyed appearance, and wider gape than in *C. pullum*).
D—*Campostoma pullum.*

differences between syntopic populations of *C. a. pullum* and *C. oligolepis*. Blum et al. (2008) provided evidence that *C. oligolepis*, as currently recognized, is not monophyletic based on analysis of mtDNA. Those authors found that *C. oligolepis* from southeastern drainages are genetically distinct and concluded that only populations from the Ozarks and the upper Mississippi River basin should be recognized as *C. oligolepis*.

Arkansas populations of *C. oligolepis* have higher scale counts than those in northern populations in Illinois, Minnesota, Iowa, and Wisconsin (Burr and Smith 1976), but not higher than those east of the Mississippi River from Kentucky through Alabama (Boschung and Mayden 2004). The taxonomy of *C. oligolepis* is currently unsettled, and it is possible that populations in at least two of the disjunct areas of distribution will eventually be recognized as separate species. We found that variation among Arkansas populations occurs in some meristic and morphometric features but is especially evident in pharyngeal tooth counts. In the Illinois River drainage of Benton and Washington counties, most individuals have a tooth count of 0,4–4,0, but 35% of the specimens examined had a tooth in the minor row on one or both sides (1,4–4,0; 0,4–4,1; or 1,4–4,1). A similar tooth pattern prevailed in most White River drainage populations, with 37% of specimens having a minor row tooth on one or both sides. An exception to that pattern was found in the population in the South Fork of the Little Red River in Van Buren County, where 80% of the *C. oligolepis* specimens had a minor row tooth on one or both sides.

DISTRIBUTION *Campostoma oligolepis* occurs in three somewhat disjunct areas: (1) the upper Mississippi River and Lake Michigan drainages from Minnesota and Wisconsin south to eastern Iowa and central Illinois; (2) Ozark Upland streams of Missouri, northern Arkansas, and eastern Oklahoma; and (3) the Green, Cumberland, Tennessee, and Mobile Bay drainages from Kentucky to Virginia and North Carolina south through Alabama. In Arkansas, *C. oligolepis* is found mainly in the White River system in the Ozark Mountains, with additional records from the Illinois River in Benton and Washington counties (Burr et al. 1979) (Map 22). Pflieger (1971, 1997) reported *C. oligolepis* from the headwaters of the St. Francis River in southeastern Missouri, and Fulmer and Harp (1977) reported it from 11 sites in streams draining Crowley's Ridge in Arkansas (10 sites in the St. Francis River drainage, and 1 site in the Cache River drainage). We confirmed the presence of *C. oligolepis* in the Cache River drainage (a White River tributary) of Crowley's Ridge in Greene County, and our data indicate that it also occurs in Crowley's Ridge streams in the St. Francis River drainage as Fulmer and Harp (1977) reported. However, we have not been able to confirm its presence in the St. Francis River drainage based on tuberculate males, with one exception. One tuberculate male with *C. oligolepis* characters was collected from Village Creek in Cross County in May 2017. We examined *Campostoma* specimens from two of the localities reported for *C. oligolepis* by Fulmer and Harp (1977) and identified some individuals (based on scale counts) as *C. oligolepis*. *Campostoma oligolepis* occurs sympatrically with *C. plumbeum* throughout most of its range in Arkansas. Buchanan (2005) reported it from four Arkansas reservoirs, where more than 1,000 specimens were taken during rotenone population samples.

HABITAT AND BIOLOGY Smith (1979) reported that the habitat of *C. oligolepis* was similar to that of *C. anomalum*, except that it was less tolerant of turbidity, reduced flow, and silt. Burr and Smith (1976) noted that *C. oligolepis* prefers faster water and larger riffles and that its distribution corresponds closely with that of *Notropis nubilus*. The restricted habitat associations of *C. oligolepis* suggest that it is more specialized (Rakocinski 1980) and

better adapted to faster water (Rakocinski 1977). Becker (1983) stated that in Wisconsin the Largescale Stoneroller occurs primarily in pools associated with riffles in medium- to large-sized streams and is seldom found in small streams. In Arkansas, we have frequently collected *C. oligolepis* from large, swift riffles in large streams and rivers of the Ozarks, where it is usually more abundant than *C. plumbeum*. In the smaller streams and headwaters in the Ozarks, *C. oligolepis* becomes less common and *C. plumbeum* is the dominant stoneroller species; however, the two species are commonly found together in both large and small streams in Arkansas.

Fowler and Taber (1985) studied food habits of *C. oligolepis* and found that foods consisted primarily of nonmotile diatoms, green and bluegreen algae, and detritus. In Wisconsin, it fed in large schools near the bottom mainly on algae scraped from submerged objects (Becker 1983). That agrees with the results of gut content examination by TMB for Arkansas specimens. In Arkansas, many specimens had ingested sand, and there was little evidence that animals were either intentionally or incidentally eaten during grazing activities. Becker (1983) found small amounts of tendipedid larvae in gut contents and concluded that those insect larvae were living in the algae and were eaten incidentally. Boschung and Mayden (2004) reported that in addition to vegetation, *C. oligolepis* in Alabama also fed on rotifers, cladocerans, copepods, and midge larvae. The data suggest that *C. oligolepis* feeds in swifter riffle sections than *C. plumbeum*.

Pflieger (1997) reported that spawning has been observed in May in Missouri and that the breeding habits are like those of *C. pullum*. Boschung and Mayden (2004) noted that reproduction in Alabama occurs in the very early spring when the waters warm to about 16°C, and South and Ensign (2013) found reproductively active individuals in January in an urbanized system in Georgia, one month earlier than reproductive individuals were found in a nonurbanized system. The latter authors also discovered that reproductive maturity occurs at a larger size for females in the urbanized system. Based on data gathered by TMB, *C. oligolepis* is, by far, the earliest-breeding *Campostoma* species in Arkansas. Tubercle development in males begins in October, and ripe males and females were found in the Buffalo River on 28 January. On that date, females ranging from 115 to 124 mm TL contained an average of 1,178 ripe ova. Reproduction in Arkansas continues through May, but the ovaries of most females were spent by mid-May. Ripe females were found as late as 23 May in the White River of Washington County, but most females on that date had regressed ovaries. Burr and Smith (1976) speculated that nuptial tubercle differences between *C. oligolepis* and *C. pullum* were probably related to breeding habits and might indicate differences in spawning behavior. They further suggested that ecological isolation of the two species is maintained throughout the spawning season, allowing *C. oligolepis* to use fast, deep, rocky riffles and *C. pullum* the quieter, shallow headwater areas. Although those hypotheses have not been thoroughly tested, Boschung and Mayden (2004) provided considerable information on the reproductive biology of *C. oligolepis* in Alabama and confirmed the many similarities with *C. pullum*. Additional study is needed of the reproductive biology of the two species in sympatry. Cloutman (1976) reported parasites of White River *C. oligolepis* in northern Arkansas.

CONSERVATION STATUS The Largescale Stoneroller is widespread and abundant throughout the upland streams of northern Arkansas, and its populations are secure. Burr and Smith (1976) noted that it is less ecologically labile than *C. pullum* and more susceptible to human modification of habitat than that species. Therefore, even though it is currently abundant and in no apparent danger, biologists sampling fishes in Ozark streams should be vigilant to detect any signs of future population declines.

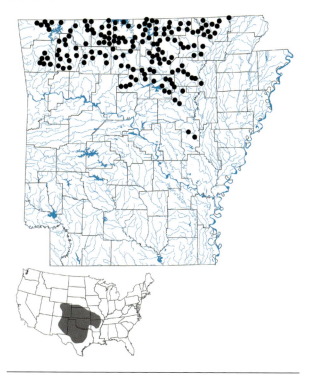

Map 23. Distribution of *Campostoma plumbeum*, Plains Stoneroller, 1988–2019.

Campostoma plumbeum (Girard)
Plains Stoneroller

Figure 15.20. *Campostoma plumbeum*, Plains Stoneroller. David A. Neely.

CHARACTERS A heavy-bodied minnow with a blunt head and rounded snout above a markedly inferior, U-shaped horizontal mouth with a cartilaginous lower lip. Snout not projecting noticeably beyond upper lip in ventral view. Intestine of this herbivore extremely long, with transverse loops, and coiled around the swim bladder. Lining of peritoneum black. All fins short and rounded. Barbels absent. Nape usually noticeably arched in predorsal profile (more strongly arched in the *C. plumbeum*-Plains form than in the *C. plumbeum*-Ozarks form). Gill rakers 27–43. Pharyngeal teeth 0,4–4,0. Dorsal fin origin approximately

over insertion of pelvic fins. Dorsal fin rays 8; anal rays 7; pectoral rays usually 16–20; pelvic rays usually 8. Lateral line complete, usually with 47–60 small scales ($\bar{x}$ = 52). Circumferential scales around body 36–47 ($\bar{x}$ = 40.4). Sum of lateral line and body circumferential scales 90 or more in 85% of Arkansas specimens (84–105, $\bar{x}$ = 92.5). Predorsal scales more than 20 in 91% of Arkansas specimens (19–29). Breast and belly fully scaled. Snout length going into head length about 2.65 times (2.14–3.49); snout length going into SL 7.83–13.98 times. Head length going into SL 3.58–4.85 times. Eye small, eye diameter going into head length 3.23–5.97 times; eye diameter going into snout length 0.93–2.63 times. Breeding males with tubercles extensively developed over top of head, sometimes on side of head, on chin and branchiostegals, upper body, lower body posteriorly, and along some rays of dorsal, caudal, and pectoral fins and frequently on anal and pelvic fins. A crescent-shaped row of 1–3 tubercles present along inner margin of each nostril of breeding male. Tubercles on nape randomly arranged. Number of large tubercles on top of head ranges from 0 to 23. Tubercles present on fins range from 44% (pelvic fins) to 78% (pectoral fins) of nuptial males. Largest Arkansas specimen examined was 6.2 inches (157 mm) TL.

LIFE COLORS Body color variable: brownish, olivaceous, or gray above grading to white ventrally; specimens at some localities are quite silvery in overall coloration. Body usually not distinctly bicolored. Sides are marked by irregular patches of black or brown spots where scales have been lost. Fins clear, except for black band in dorsal and anal fins of adults and breeding males. Breeding males are dark slate gray except ventrally; sides with a brassy sheen, belly white to yellowish. Dorsal fin orange-colored basally and with a broad black slash medially. Dorsal, anal, and pelvic fins with black pigment (R. J. Miller 1962). Caudal fin with basal suffusion of orange surrounded by diffuse dark pigment, dusky distally; dark pigmentation at the base of the caudal fin in the shape of a spot or wedge, but never vertically extended into a narrow band. Anal fin often orange basally, its membranes with black blotches medially; rays and distal edge of anal fin creamy or yellow. Pelvic fins with black slashes on membranes, yellowish distally. Pectoral fins dusky, suffused with yellow. Nuptial adults in New Mexico had orange banding on dorsal and anal fins and red orange at the insertion of the paired fins (Sublette et al. 1990). Cleithrum and opercular membrane dark (Cross 1967). Young often have a dark stripe along midside.

SIMILAR SPECIES *Campostoma plumbeum* is very similar morphologically to the other three Arkansas species of *Campostoma*. The following differentiation is based on meristic, morphometric, nuptial coloration, and tuberculation patterns of adult and nuptial male specimens collected in Arkansas and may not apply to other areas of the geographic range. *Campostoma plumbeum* differs from *C. oligolepis* in the sum of the lateral line scales and scale rows around the body just in front of the dorsal fin, usually 90 or more (vs. usually fewer than 90), gill rakers 27 or more (vs. usually 25 or fewer), the least width of the skull between the eyes is usually less than the distance from the back of the eye to the upper end of the gill opening (vs. the two measurements usually about equal), nuptial males with a crescent-shaped row of 1–3 tubercles medial to each nostril (vs. no tubercles medial to nostrils), a prominent black band in the anal fin of adult males 3 inches (75 mm) or more in SL (vs. no black band in anal fin of breeding males), predorsal profile is arched at nape in lateral view (vs. predorsal profile usually not prominently arched), and snout more pointed with little noticeable projection in ventral view (vs. snout more globose, extending well beyond upper lip). *Campostoma plumbeum* differs from *C. spadiceum* in pharyngeal teeth 0,4–4,0 (vs. pharyngeal teeth in 2 rows) (see *C. spadiceum* species account for additional differences). *Campostoma plumbeum* differs from *C. pullum* in having smaller scales, the sum of lateral line and circumferential scales usually more than 90 in 72% of specimens (vs. sum usually 90 or less in 73% of specimens), usually having

more than 21 predorsal scales (vs. usually 21 or fewer), in having more tubercles on the nape in breeding males, with an average of 21.6 (1–42) (vs. average of 0.4 [0–3]), presence of tubercles on fins in *plumbeum* (vs. usually a lack of tubercles on fins in *pullum*), and geographic location, as *C. plumbeum* inhabits the White River system and Illinois River drainage in the northcentral and northwest portion of the state, while *C. pullum* in Arkansas occurs almost exclusively in streams draining Crowley's Ridge.

VARIATION AND TAXONOMY *Campostoma plumbeum* was described by Girard in 1856 as *Dionda plumbea* from the headwaters of the Canadian River, probably Ute Creek, Quay or Harding County, New Mexico. What we herein recognize as *C. plumbeum* is commonly included in *C. anomalum pullum* by some workers (Page et al. 2013), or treated by others (Pflieger (1997; Boschung and Mayden 2004) as part of the species *C. pullum* (see genus *Campostoma* account). Blum et al. (2008) reported that mtDNA topologies did not support the recognition of subspecies of what is commonly called *C. anomalum*, but instead supported the separation of *C. anomalum* into multiple taxa as originally suggested by Etnier and Starnes (1993). Blum et al. (2008) found evidence from the mtDNA topologies that at least six distinct evolutionary lineages within *C. anomalum* warrant species rank. Based on genetic data, our morphological study of Arkansas populations, and communications with Michael Blum and Fernando Alda, we recognize two of the genetically identified lineages of Blum et al. (2008) as *C. plumbeum* (*sensu* Girard 1856), the Plains Stoneroller. The two lineages of *C. plumbeum* are *C. plumbeum* "Plains", occupying most of the species range, and *C. plumbeum* "Ozarks", restricted to the Ozark Mountains of Missouri and Arkansas. The Ozarks form is genetically distinguishable from the Plains form and may eventually be recognized as a separate species (M. Blum, pers. comm.).

DISTRIBUTION The total range of *Campostoma plumbeum* is not precisely known, but it is widely distributed throughout the southern Great Plains, mostly west of the Mississippi River from southern Nebraska and eastern Colorado south through eastern New Mexico and central Texas and eastward into southern Missouri and northern Arkansas (Map 23). In contrast, the distribution of the Plains Stoneroller within Arkansas is well known. It occurs throughout the Ozark Mountains in the Illinois, White, and Black river systems. Occasional stragglers have been reported from lowland areas of large rivers, such as the lower White River, and Buchanan (2005) found *C. plumbeum* in three Arkansas Reservoirs (Bull Shoals, Norfork, and Greers Ferry).

HABITAT AND BIOLOGY   The Plains Stoneroller inhabits small, generally clear streams with gravel, rubble, or exposed bedrock substrates. It is often the most abundant species in small, clear upland streams, where it is frequently observed in large schools in shallow water habitats. Cross (1967) found this species occupying pools throughout much of the year in Kansas. Permanent flow was not essential to its persistence. Deacon (1961) found this stoneroller to be mainly sedentary over short periods of time when stream conditions were relatively stable, and Metcalf (1959) reported rapid dispersal when flow commenced in stream channels that had previously been dry. This stoneroller is occasionally found in small numbers in upland impoundments. Cross and Collins (1995) described how the construction of impoundments, land use practices, and widespread reduction in stream flows in central and western Kansas over a 25-year period affected the distribution and abundance of this species:

> Headwater streams in the state have dried, and some river channels have acquired characteristics of their tributaries: greatly reduced but more stable flows, clearer water, and stable streambeds. Under these conditions, light penetrates to the firm substrate, allowing development of the algal films needed by Central Stonerollers. Therefore, this species has increased in channel segments of the Arkansas, Cimarron, Smoky Hill, and Solomon rivers where moderate flow persists in Kansas. In those channel segments, Central Stonerollers [= *C. plumbeum*] have replaced the Plains Minnow [*Hybognathus placitus*], an ecological analogue that predominated when flows of the larger plains rivers fluctuated widely, were usually turbid, and had riverbeds consisting mainly of loose, shifting sand.

The cartilaginous ridge of the lower jaw is used to scrape the soft diatomaceous and algal mats coating rocks in pools and slow raceways. Matthews et al. (1986) described the feeding behaviors of *C. plumbeum* and *C. oligolepis* in a tributary of Baron Fork Creek (Illinois River drainage), Oklahoma. The behaviors used to remove algal felts from hard substrates included "swiping" (the head is thrust downward and sideways, with the lower jaw scraping algae from the rock in the sideways motion), "shoveling" (pushing the lower jaw against the surface and swimming vigorously forward), and "nipping" (small, rapid bites with the body at a 45° angle to the substrate). This species can have important, conspicuous effects on the distribution of algae and invertebrates in small streams (Power and Matthews 1983; Power et al. 1985, 1988a, b; Matthews et al. 1986; Gelwick and Matthews 1992, 1997). *Campostoma*

may also have important trophic cascade consequences for standing crops or taxonomic composition of the attached algal flora (as well as fauna) by controlling pool-to-pool distribution of attached algae, or by maintaining algae at low standing crops despite high primary productivity (Power and Matthews 1983; Power et al. 1985). Pennock and Gido (2017) studied density dependence across a range of densities and resource abundance in stream mesocosms and found that mean growth of individuals was negatively associated with stocking biomass. Surprisingly, increases in fish biomass also led to increased primary productivity. Stonerollers make distinctive grazing scars when feeding upon low growth forms of attached algae on substrates in Ozark streams of Arkansas and Oklahoma (Matthews et al. 1986). These grazing scars are typically rectangular, 3–4 mm wide by 5–20 mm long, and are readily distinguishable from scars left on rocks by other common grazers in Ozark streams such as snails and herbivorous chironomids (Fig. 15.21). Other foods eaten include zooplankton, aquatic insects, and plant tissue. Plains Stonerollers consumed an estimated 27% of their body weight daily during September (Matthews 1987a). Nearly all the material consumed (mostly diatoms) is digested overnight, probably aided by bits of ingested minerals (Fowler and Taber 1985).

Spawning in Arkansas begins in late March to mid-April when water temperatures exceed 58°F (14.6°C) and continues through May. Ripe females collected on 9 May from Little Sugar Creek in Benton County by TMB averaged 102 mm TL and contained an average of 679 mature ova. Ripe eggs were approximately 1.5 mm in diameter. Males develop tubercles and other reproductive characteristics in winter (Cross 1967). Tuberculate males begin to dig nests in shallow gravel-bottomed areas when water temperatures reach 60°F (15.6°C) in Kansas (Cross 1967). R. J. Miller (1967) observed nest-building and breeding activities of this species in Spavinaw Creek in eastern Oklahoma on 13 May

Figure 15.21. *Campostoma plumbeum* grazing scars on rock. *Renn Tumlison.*

at a water temperature of 14°C and an air temperature of 16°C. *Campostoma* species are pit diggers, unlike the other mound and ridge pebble-nest building species (*Semotilus* and *Nocomis*). Males have a thick external mandibular epidermis with extensive keratin, anchored and supported by numerous strands of dermal collagen (McQuire et al. 1996). Typically, the male prepares a nest by digging into the substrate with the snout, pushing stones aside, and carrying small stones out of the spawning pit area. HWR observed a breeding male forming a spawning pit by pushing and bulldozing with the external skin surface of the mandibles rather than by lifting and transporting stones in the mouth. Recently constructed nests are often conspicuous as light-colored patches in shallow riffle areas. Ripe females then enter the nest and spawning occurs with eggs falling into the gravel interstices to develop. Stonerollers occasionally use the nests of the Hornyhead Chub (*Nocomis biguttatus*) and the Creek Chub (*Semotilus atromaculatus*) (R. J. Miller 1964). After hatching, the young school and feed along vegetated stream margins and in quiet, warm backwater stream regions (Miller and Robison 2004). Sexual maturity is reached at age 1 or 2 (Pflieger 1997). Richardson et al. (2013) documented the piscicolid fish leech *Cystobranchus klemmi* on *C. plumbeum* in Arkansas and Oklahoma.

CONSERVATION STATUS The Plains Stoneroller is widespread and abundant in northcentral and extreme northwestern Arkansas, and its populations are secure.

Campostoma pullum (Agassiz)
Finescale Stoneroller

Figure 15.22a. *Campostoma pullum*, Finescale Stoneroller male. *Uland Thomas.*

CHARACTERS A round-bodied, terete minnow with a blunt, somewhat rounded (but not globose) snout protruding slightly, if at all, above an inferior, broadly crescent-shaped mouth with a cartilaginous lower lip (Fig 15.2A). Snout does not project markedly beyond upper lip in ventral view. Intestine of this herbivore extremely long, with transverse loops, and coiled around the swim

Figure 15.22b. *Campostoma pullum*, Finescale Stoneroller female. *Uland Thomas.*

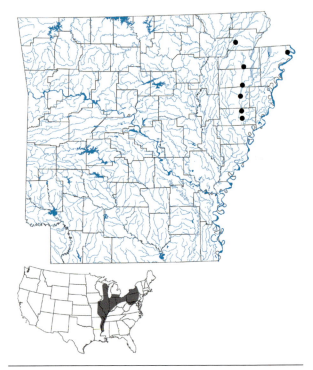

Map 24. Distribution of *Campostoma pullum*, Finescale Stoneroller, 1988–2019.

bladder. Peritoneum black. All fins short and rounded. Nape often noticeably arched. Barbels absent. Frenum absent. Pharyngeal teeth 0,4–4,0. Least bony width of skull between eyes less than distance from back of eye to upper end of gill opening. Gill rakers numerous, usually 29–37. Dorsal fin origin approximately over insertion of pelvic fins. Dorsal fin rays 8; anal rays 7; pectoral rays usually 15–19; pelvic rays usually 8. Scales rather large in Arkansas populations. Lateral line complete, usually with 45–54 scales ($\bar{x}$ = 49.3). Circumferential scales around body 30–43 ($\bar{x}$ = 38.8). Circumferential scales above the lateral line 12–17. Sum of lateral line and body circumferential scales 87 or more in 78% of Arkansas specimens (84–96, $\bar{x}$ = 88.7). Predorsal scales usually 19–21 (15–25). Breast and belly fully scaled. Head length going into

SL 3.76–4.66 times. Eyes relatively larger than in other Arkansas *Campostoma* species, eye diameter going into head length 3.15–4.94 times; eye diameter going into snout length 1.00–2.58 times. Snout relatively shorter than in other Arkansas *Campostoma* species; snout length going into head length 1.68–3.61 times; snout length going into SL 10.06–16.55 times. Breeding tubercles less extensively developed than in all other Arkansas *Campostoma* species, except in largest males (Fig. 15.23). A crescent-shaped row of 1–3 tubercles is present along the inner margin of each nostril of a breeding male. Tubercles confined mainly to top of head, especially along the occiput. Tubercles usually absent on nape, but if present, are few in number and are confined to anterior portion of nape. Tubercles generally absent from sides, but there may be a few on anterior side above the lateral line. Fins mostly without tubercles. This is Arkansas's smallest *Campostoma* species, with largest adults probably not exceeding about 4.5 inches (113 mm) TL. The name *pullum* is probably derived from the Latin *pullus* (the young of an animal) used by Agassiz (1854) in the belief that it represented the smallest species in its genus (Etnier and Starnes 1993). Maximum size may be much larger outside Arkansas, but the largest Arkansas specimen we examined (out of 160 specimens from 11 collections) was 4.4 inches (113 mm) TL, 4.1 inches (105 mm) SL, from Village Creek in Cross County (St. Francis River drainage).

LIFE COLORS   Body coloration variable: brownish, olivaceous, or gray above, grading to white ventrally; specimens occasionally quite silvery. Sides of adults usually with scattered dark flecks indicative of replacement scales (no pattern is formed). Many individuals have a dark predorsal stripe; postdorsal stripe absent or weakly developed. Scales of upper sides heavily pigmented, and lateral line scales are often the darkest lateral scales; dark pigment usually extends onto the bases of the middle three caudal rays. Scales of lower sides (below lateral line) brown and sprinkled with melanophores. Some individuals have scattered scales on upper and lower sides heavily sprinkled with melanophores causing them to stand out as dark scales. A dark crescent is present in dorsal fin and anal fin of adults. Breeding males dark slate gray with sides having a brassy sheen and whitish belly. Dorsal fin orange basally and with a broad black slash medially. Dorsal, anal, and pelvic fins with black pigment. Caudal fin with basal suffusion of orange surrounded by diffuse dark pigment, dusky distally. Anal fin often orange basally, its membranes with black blotches medially; rays and distal edge of anal fin creamy or yellow. Dark pigmentation in base of caudal fin of breeding male, if present, in the form of a spot or wedge (not vertically extended in a vertical band). Pelvic fins with black

Figure 15.23. Frontal views of head of *Campostoma pullum* male from Village Creek, Cross County, Arkansas, showing tubercle pattern. *Uland Thomas.*

slashes on membranes. Pectoral fins dusky, suffused with yellow. Some melanophores present in membranes and along rays of dorsal and caudal fins. Young and juveniles often have a dark lateral stripe and a small basicaudal spot.

SIMILAR SPECIES   The three other Arkansas species of *Campostoma* are similar to *Campostoma pullum*, but *C. pullum* is known to be syntopic only with *C. oligolepis*. Characters distinguishing *C. pullum* from *C. oligolepis* include least width of the skull between the eyes less than the distance from the back of the eye to the upper end of the gill opening (vs. skull width about equal to distance from back of eye to upper end of gill opening in *C. oligolepis*), gill rakers usually more than 25 (vs. usually 25 or fewer), nape usually noticeably arched (vs. nape relatively flat, with little arching), snout more pointed and extending only slightly if at all beyond upper lip (vs. snout globose and extending well beyond upper lip), nuptial males with a crescent-shaped row of 1–3 tubercles medial to each nostril (vs. no tubercles medial to nostrils), and a prominent black band in the anal fin of adult males 3 inches (75 mm) or more in SL (vs. no black band in anal fin of breeding males). *Campostoma pullum* differs from *C. spadiceum* in pharyngeal teeth 0,4–4,0 (vs. teeth in 2 rows), and in several other characters given in the *C. spadiceum* species account. *Campostoma pullum* in Arkansas differs from *C. plumbeum* in having lower scale counts, the sum of lateral line and circumferential scales usually 90 or less in 63% of specimens (vs. usually more than 90 in 72% of specimens), and in breeding males much less tuberculate, with most tubercles confined to the head region (vs. breeding tubercles numerous and widely distributed on head, body, and fins). Geographic location is most useful for distinguishing the two species, as they are not sympatric in Arkansas. In Arkansas, *C. pullum* occurs almost entirely in streams draining Crowley's Ridge (mainly St. Francis River system), whereas *C. plumbeum* inhabits the White River system and Illinois River drainage in the northcentral and northwestern portions of the state.

VARIATION AND TAXONOMY   *Campostoma pullum* was originally described by Agassiz (1854) as *Chondrostoma pullum* from Burlington, Des Moines County, Iowa. The *Campostoma* species that we herein recognize as *C. pullum* represents only part of the populations currently referred to as *C. anomalum pullum* by Page et al. (2013) or as *C. pullum* by Pflieger (1997) and Boschung and Mayden (2004). Blum et al. (2008) found evidence from mtDNA topologies that at least six distinct evolutionary lineages within *C. anomalum* warranted full species rank, including a form referred to as *C. pullum* that was very different in total range from the *C. pullum* referred to by Pflieger (1997), Etnier and Starnes (2001), and Boschung and Mayden (2004). We do not attempt a rangewide diagnosis for *C. pullum*, but with a formal description of the species by M. Blum and coworkers pending, we provide information about that species in Arkansas. *Campostoma pullum* is distinguishable genetically and (at least in Arkansas) by a combination of morphological features from other *Campostoma* species. As suggested to us by M. Blum (pers. comm.), we use the scientific name already available in the literature (*C. pullum*) along with the common name, Finescale Stoneroller.

It should be noted that the common name is a misnomer when applied to Arkansas populations. *Campostoma pullum* in Arkansas has comparatively large scales and is closer in most scale counts to *C. oligolepis* than to either

C. plumbeum or *C. spadiceum*. The occurrence of some specimens with low scale counts could indicate the inadvertent inclusion of some *C. oligolepis* specimens in our *C. pullum* counts, because the two species apparently occur syntopically in some Crowley's Ridge streams. However, the distribution of our counts for 13 tuberculate male *C. pullum* from Crowley's Ridge agrees closely with our counts for all other specimens identified by us as *C. pullum*. In addition, Pflieger (1971) reported sum of lateral line and circumferential scale counts as low as 84 for some St. Francis River specimens of *C. pullum* in Missouri. The considerable overlap in this sum between Arkansas specimens of *C. pullum* and *C. oligolepis* makes this character less useful for distinguishing the two species in this state. Using Pflieger's (1997) criterion of 87 or fewer lateral line plus circumferential scales for separating *C. oligolepis* from other *Campostoma* species in Missouri, we estimate that there is approximately a 76% probability (based on Bayes' theorem) that a *Campostoma* specimen from Crowley's Ridge with 87 or fewer scales is *C. oligolepis*. Tuberculate males are required for a higher percent accuracy in separating the two species. The low scale counts and other features of *C. pullum* in Arkansas (Table 13) could indicate introgression with *C. oligolepis*, a morphologically aberrant population of *C. pullum*, unintentional inclusion of some *C. oligolepis* specimens in the *C. pullum* counts, or even a taxonomically distinct form. A combined morphological and genetic study is needed to clarify the taxonomic status of Crowley's Ridge *Campostoma* populations. In northern Illinois, occasional interbreeding between *C. pullum* and *C. oligolepis* occurs, but the two species generally remain reproductively isolated there (Rakocinski 1977, 1980). Rakocinski (1984) provided evidence that postmating isolation through asymmetric hybrid inviability was important in maintaining reproductive isolation of the two species.

DISTRIBUTION   The total range of *C. pullum* is not precisely known but is primarily east of the Mississippi River. It is currently known from Great Lakes drainages of Michigan (Thornapple River of Lake Michigan drainage) and Ohio (Beaver Creek of Lake Erie drainage), and Atlantic Slope drainages of Pennsylvania (Delaware River) and New York (Susquehanna River); it is most widely distributed in the Mississippi River basin from Indiana, Illinois, eastern Wisconsin and southeastern Iowa south through eastern Missouri, southwestern Kentucky, eastern Arkansas, and western Mississippi (M. Blum and F. Alda, pers. comm.). Within Arkansas, *C. pullum* is limited mainly to streams draining Crowley's Ridge (St. Francis and Cache river drainages). There is a single record from the Mississippi River in Arkansas (Map 24).

HABITAT AND BIOLOGY *Campostoma pullum* inhabits small, generally clear streams with gravel, rubble, or exposed bedrock substrates. It is often the most abundant species in small, clear upland streams, where it is frequently observed in large schools in shallow water habitats. Burr and Smith (1976) reported that *C. pullum* is more ecologically labile than *C. oligolepis*, with which it is often sympatric in northern populations, but *C. pullum* showed a greater preference for smaller riffles and to some extent quieter water habitats than *C. oligolepis*. In Arkansas, it is found in small, short, gravel-bottom streams draining Crowley's Ridge, and it apparently attains the smallest maximum size of all *Campostoma* species in this state. We noted that *C. oligolepis* populations on Crowley's Ridge also exhibit smaller body size than elsewhere in Arkansas. In the few streams where it occurs on Crowley's Ridge, it is found only at localities having substantial gravel and cobble substrates and some current. It avoids areas of those streams having sand, silt, and clay substrates. It is absent from many Crowley's Ridge streams that apparently have suitable habitat because of an abundance of predators at those sites, primarily *Lepomis cyanellus* and *Ameiurus natalis*.

The U-shaped cartilaginous ridge of the lower jaw of *C. pullum* is used to scrape the algal and soft diatomaceous mats that coat the surfaces of the rocks in pools and slower raceways. *Campostoma anomalum* (and probably *C. pullum*) has dermal papilla with long strands of collagen extending beneath and supporting the external keratinized mandibular epidermis (McQuire et al. 1996). Collagen with keratin formed a rigid lip on the anterior margin of the mandible, making an efficient digging and scraping tool for removing algae and other organisms from the rocks. Feeding is accomplished as an individual stoneroller moves its head from side to side, using the cartilaginous edge of the lower jaw to scrape algae from the substrate, or as it pushes its lower jaw across the rocks while swimming forward (Matthews et al. 1986). There is little information on feeding biology specific for *C. pullum*, but an examination of gut contents of specimens from three Crowley's Ridge localities collected in May and June by TMB revealed a higher ingestion of immature aquatic insects than observed in the other three Arkansas *Campostoma* species.

In Arkansas, spawning of *C. pullum* usually begins in March when water temperatures exceed 58°F (14.6°C) and continues through May and sometimes into early June. Crowley's Ridge streams are small, short, and shallow, and water temperatures closely track air temperatures. Peak spawning varies with climatic conditions; that is, warm winters and early springs result in earlier spawns. In 2017 after a warm winter, spawning was largely finished in

Crowley's Ridge populations by mid-April. Ripe individuals were found by TMB and Kirk Strawn in Crow Creek in St. Francis County on 6 April 1963, and by TMB and Drew Wilson in Village Creek in Cross County on 25 March 1975 and 6 June 1985. Ripe females ranged from 55 to 62 mm TL and contained from 140 to 210 mature ova. Ripe ova were 1.2–1.5 mm in diameter. R. J. Miller's (1962) excellent description of reproductive activities of *Campostoma pullum* is abbreviated here. Tuberculate males begin to dig nests in shallow gravel-bottomed areas with slow to moderate current by pushing small stones out of the spawning pit area. A breeding male forms a spawning pit by pushing and bulldozing with the external skin surface of his mandibles rather than by lifting and transporting stones (Miller 1962; HWR pers. obs.). Video recordings of eastern *C. anomalum* digging pits in Virginia corroborate Miller's statement that inner mandibular regions rarely contact the substrate (Maurakis and Woolcott 1989). Spawning activities commence when ripe females enter the nest. Eggs are spawned into the gravel interstices, where they develop. Young individuals usually school and feed along vegetated stream margins and backwater areas of streams.

CONSERVATION STATUS Although locally common in a few streams on Crowley's Ridge, *Campostoma pullum* is not widely distributed in Arkansas. Fulmer and Harp (1977) reported *C. anomalum* [*C. pullum*] from only 2 of 40 streams sampled on Crowley's Ridge. There is also an older record of one specimen from Tuni Creek in St. Francis County collected in 1965 (University of Florida, UF64336). In 2017, we sampled 10 streams on Crowley's Ridge where stonerollers were historically known and found *C. pullum* in 3 of those streams. Because of its small range in Arkansas and susceptibility of its habitat to disturbance, we consider it a species of special concern if not threatened.

Campostoma spadiceum (Girard)
Highland Stoneroller

Figure 15.24. *Campostoma spadiceum*, Highland Stoneroller. *Uland Thomas.*

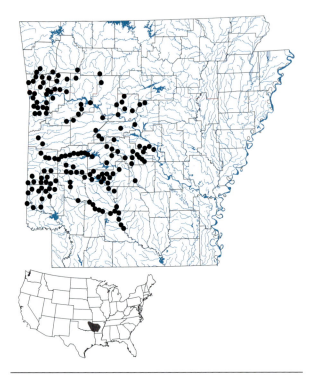

Map 25. Distribution of *Campostoma spadiceum*, Highland Stoneroller, 1988–2019.

Figure 15.25. *Campostoma spadiceum*.

A—Nuptial male from Glover River, McCurtain County, Oklahoma, collected 20 March 2007 (OKMNH 79128). B—Close-up of caudal peduncle of same male to show detailed coloration. *Photos by W. J. Matthews.*

CHARACTERS   A heavy-bodied minnow with a blunt head and rounded snout slightly overhanging an inferior, U-shaped horizontal mouth with a cartilaginous lower lip (Fig. 15.2A). Intestine extremely long, with transverse loops, and coiled around the swim bladder. Lining of peritoneum black. Fins short and rounded. Barbels absent. Nape often noticeably arched. Least bony width of skull between eyes less than distance from back of eye to upper end of gill opening. Pharyngeal teeth in 2 rows at least on one side and usually on both sides, most commonly 1,4–4,1, 1,4–4,0, or 2,4–4,1 but sometimes in other combinations thereof within a population. Dorsal fin origin approximately over insertion of pelvic fins. Dorsal fin rays 8, anal rays 7, pectoral rays 15–19, pelvic rays usually 8. Gill rakers on first arch usually more than 25 (22–36). Lateral line complete, with 49–61 small scales ($\bar{x}$ = 54.3). Body circumferential scales usually 36–44; circumferential scales above the lateral line usually 15–19. Sum of lateral line scales plus circumferential scales 90 or more in 85% of Arkansas specimens (85–104, $\bar{x}$ = 94). Predorsal scales more than 20 in 85% of Arkansas specimens (19–26). Head length 21.0–25.3% SL and going into SL 3.78–4.09 times. Snout length 0.8–1.1% SL; snout length going into SL 8.08–13.10 times; snout length going into head length 2.08–2.91 times. Caudal peduncle length 23.0–28.6% SL. Eyes small, eye diameter 3.3–5.7% SL. Eye diameter going into snout length 1.07–2.88 times; eye diameter going into head length 3.07–5.98 times. In nuptial males, there is a crescent-shaped row of 1–3 tubercles along the inner margin of each nostril, 0–57 tubercles present on nape, and 0–10 tubercles on anterolateral sides of body. Predorsal tubercles of adult males are small and numerous (100–120) and usually organized into parallel, longitudinal rows (especially in the larger tuberculate males) (Fig. 15.25A; Fig. 15.28). Tubercles may be present on all fins. Females lack tubercles. Most specimens are smaller than 5 inches (128 mm) TL. Largest Arkansas specimen examined was 6.5 inches (165 mm) TL from the Mulberry River in Franklin County.

LIFE COLORS   Body color silver gray to gray above grading to cream ventrally with caudal, dorsal, anal, and usually pectoral and pelvic fins red to reddish orange in males, females, and juveniles of all sizes (Fig. 15.26). The most intense red coloration is usually in the dorsal and caudal fins; pectoral and pelvic fins tend to have the weakest intensity of red, with the red pigment often restricted to the distal fin margin or a few of the lower rays. The Highland Stoneroller is unique among *Campostoma* species in having red or red-orange coloration in median and paired fins throughout much of the year, especially from April through November (Cashner et al. 2010). Geographic variation in fin coloration occurs, with the most intense red coloration in individuals in the Little and Ouachita river systems, while the red coloration is less intense in specimens from the Arkansas River drainage. Reddish coloration is most intense in the summer (outside the breeding season) but often persists to some degree throughout the year (Fig. 15.26). Sides may be marked by irregular patches of black or brown spots. Cashner et al. (2010) noted that nuptial coloration is completely different from the typical warm-season coloration. Body coloration of breeding males is rust to orange on the dorsolateral surface, usually with a gold wash along the ventrolateral surface. The orange pigment extends from the caudal peduncle to the edge of the operculum, and the gold wash extends from the caudal peduncle to the pectoral fins. The eye of the nuptial male is bright red, but the color is lost quickly after the breeding season and in preservation. Breeding males have a dark band of pigment medially through the dorsal fin and a black band in the basal one-half of the anal fin, well separated from the body. Breeding females are very silvery, with a tinge of orange in the paired fins. The caudal fin of breeding adults has a basal dark band usually much higher than wide, and not extending distally along medial rays (Fig. 15.25B). The dark caudal pigmentation (unique among Arkansas *Campostoma* species) is vertically extended in a narrow band across 4–10 caudal rays. Nuptial colors of adult males begin to fade at approximately the time the non-nuptial, red or red-orange coloration of all fins begins to intensify in both sexes. Small young often have a dark stripe along the midside.

SIMILAR SPECIES   The Highland Stoneroller is similar to the three other stonerollers in Arkansas. *Campostoma spadiceum* is distinguished from all other Arkansas stonerollers by the unique red or red-orange coloration in the median and paired fins of males, females, and juveniles throughout most of the year, especially from April through November (vs. red pigment absent). *C. spadiceum* further differs from contiguous populations of *C. plumbeum* and *C. pullum* in breeding males having small, numerous tubercles on the predorsal region, with tubercles from the insertion of the dorsal fin to the occiput ranging from 100 to 200 and usually organized into discrete longitudinal rows (Fig. 15.27; Fig. 15.28) (vs. tubercles larger, fewer in number, and somewhat scattered over the nape and back). Also, dark pigmentation at the base of the caudal fin (Fig. 15.25B) of nuptial adult specimens of *C. spadiceum* is vertically extended in a narrow band across 4–10 rays (vs. dark pigmentation at the base of the caudal fin in the shape of a spot, a wedge, or a V, often with a distally directed extension on the fin rays in the other Arkansas *Campostoma*

Figure 15.26. Coloration of various size *Campostoma spadiceum* collected from the Glover River, McCurtain County, Oklahoma. 27 July 2004 (OKMNH 78852). Photographed after approximately 15 minutes in 10% formalin. *Photos by W. J. Matthews.*

species), and the presence of one or occasionally two very thin delicate minor row pharyngeal teeth on one or both pharyngeal arches (vs. 0,4–4,0 in *C. plumbeum* and *C. pullum*).

Characters distinguishing *C. spadiceum* from the similar *C. oligolepis* include the sum of the scales in the lateral line plus the number of scale rows around the body just in front of the dorsal fin is usually 90 or more (vs. usually fewer than 90), and the least width of the skull between the eyes is usually less than the distance from the back of the eye to the upper end of the gill opening (vs. skull width usually approximately equal to distance from the back of the eye to the upper end of the gill opening). In addition, nuptial males of *C. spadiceum* typically possess a crescent-shaped row of 1–3 tubercles medial to each nostril and a black band is prominent in the anal fin of adult males 3 inches (75 mm) or more in SL (vs. no tubercles medial to each nostril and no black band in anal fin in *C. oligolepis*). Additionally, the predorsal profile is often arched at nape in lateral view (vs. not prominently arched in *C. oligolepis*), and the snout shape is flatter with no noticeable projection in ventral view (vs. snout globose and extending well beyond the upper lip).

Figure 15.27. *Campostoma spadiceum* nuptial male with tubercles, Mill Creek, Clark County, Arkansas. *Renn Tumlison.*

Figure 15.28. Tubercle patterns of nuptial males.

Above: *Campostoma plumbeum* (OKMNH 78848, 96 mm SL).

Below: *Campostoma spadiceum* (OKMNH 70531, 101 mm SL). *Photos by W. J. Matthews.*

VARIATION AND TAXONOMY *Campostoma spadiceum* was originally described by Girard (1856) as *Dionda spadicea* from "Fort Smith, Arkansas." It was later assigned to the genus *Campostoma* and placed into the synonymy of *C. anomalum* by Hubbs and Ortenburger (1929b). Cashner et al. (2010) redescribed the species based on morphological characters and resurrected the name *Campostoma spadiceum* from the synonymy of *C. anomalum*. They also provided the common name, Highland Stoneroller. Although the collection locality reported by Girard (1856) was not very precise, Cashner et al. (2010) noted that any one of several small creeks in the Fort Smith vicinity could have been the type locality, and Mill Creek in metropolitan Fort Smith was easily within walking distance of the fort where the collectors were encamped. Blum et al. (2008) provided molecular evidence for recognizing *C. spadiceum* as a distinct species. For further information about its taxonomic history, see genus *Campostoma* account. Schanke (2013) amd Schanke et al. (2017) found that *C. spadiceum* populations in streams in the Ouachita Mountains experienced greater reductions in gene flow along intermittent streams than along perennial streams; however, the greatest reduction in gene flow between subpopulations usually coincided with the presence of permanent instream structures.

DISTRIBUTION Originally known only from Arkansas and Oklahoma (Cashner et al. 2010), a record of *C. spadiceum* was also found in northeastern Texas (F. D. Martin, pers. comm.). In Arkansas, the Highland Stoneroller occurs in portions of the Red, Ouachita, and Arkansas river drainages (Map 25). In the Red River basin, it is found in the Little, Mountain Fork, Rolling Fork, Cossatot, and western Saline river systems. In the Ouachita River, it occurs in the Little Missouri, Caddo, Ouachita, and eastern Saline rivers, and in the Arkansas River basin it inhabits the Poteau, Mulberry, Petit Jean, and Fourche LaFave rivers, and most of the smaller tributaries north and south of the Arkansas River from Fort Smith to the Fourche Creek drainage in Little Rock in Pulaski County. It is apparently absent from the Illinois River drainage of Benton and Washington counties, where it is replaced by *C. plumbeum* and *C. oligolepis*. It is not known to be syntopic with any other *Campostoma* species in Arkansas.

HABITAT AND BIOLOGY The Highland Stoneroller inhabits primarily upland habitats of small, stony-bottomed headwater creeks to small rivers with relatively clear water and substantial base flow (Cashner et al. 2010). In the small, clear, upland streams of the Ouachita Mountains, it is often the most abundant species. *Campostoma spadiceum* typically occurs in flowing microhabitats, avoiding sluggish lowland streams and prairie streams, and does not seem to prefer backwater habitats, although it has been found below the Fall Line in Arkansas. It showed strong directional movement into pools during periods of stream drying, but survival rates were not significantly greater in pools than in riffles (Hodges and Magoulick 2011). Like the more widespread *C. plumbeum*, it seems ecologically labile and is somewhat tolerant of disturbed environments. In Fort Smith, at least four creeks that have been channelized, widened, and often with segments converted into concrete flumes, have populations of *C. spadiceum*. In those creeks, stonerollers can often be observed grazing over bedrock and concrete substrates having thick algal growths. Buchanan (2005) found the Highland Stoneroller in 17 Arkansas impoundments (reported as *C. anomalum*).

This stoneroller is algivorous and typically occurs in schools and shoals of several dozen to several hundred individuals, often with the body oriented upstream and grazing actively on attached algae on stony bottoms in modest to substantial currents (Cashner et al. 2010). Its feeding behavior is similar to the description given in detail by Matthews et al. (1986) for *C. plumbeum*. Matthews (1998) and Matthews et al. (2004) found that the feeding activities of *Campostoma anomalum* [= *C. spadiceum*] had direct or indirect effects on at least 20 important structural or functional properties of small stream ecosystems. *Campostoma spadiceum* can cause differences in algal community composition or productivity and can affect uptake and dynamics of organic matter, invertebrate community composition or life history, movement of materials in streams, carbon-nitrogen ratios, and standing crops of bacteria. This stoneroller also plays a major role in predator-driven cascades and can alter how stream ecosystems function overall.

Spawning in Arkansas begins in March and continues at least through mid-May when water temperatures exceed 58°F (14.6°C). Large nuptial males are most often found in swift water in larger riffles or chutes, whereas females may use slightly less-swift habitats (Cashner et al. 2010). Ripe females were taken by TMB and Brett Hobbs from Hot Spring Creek in Garland County on 31 March, from a swift riffle in Buffalo Creek in Scott County, Arkansas, on 10 April 2015 at a water temperature of 20°C by TMB, Jeff Quinn, and Brian Wagner, and from the Mulberry River in Franklin County by Jeff Quinn on 5 May. A ripe female 85 mm TL from Scott County contained 971 mature ova; average ovum diameter was about 1.5 mm. The largest mature female from the Mulberry River was 126 mm TL and contained 3,842 ripe ova. Juveniles generally use slower-current habitats than adults. Cashner et al. (2010) pointed out that there has been no definitive study to date of breeding behavior, food habits, and general life history of *C. spadiceum*. Richardson et al. (2013) documented the piscicolid fish leech *Cystobranchus klemmi* from *C. spadiceum* in Arkansas.

CONSERVATION STATUS   The Highland Stoneroller is widespread and abundant in mountainous regions of Arkansas and its populations appear secure.

Genus *Carassius* Jarocki, 1822—Goldfishes

Jarocki (1822) originally described *Carassius* as a subgenus of *Cyprinus*, and it was later elevated to genus status. This introduced Eurasian genus includes two species, only one of which, *C. auratus*, the Goldfish, occurs in Arkansas. These deep-bodied, coarse-scaled fishes are widely used in the aquarium trade and as baitfishes. Lonoke County, Arkansas, is one of the largest producers of Goldfish in the bait industry, with approximately 2.5 million pounds of Goldfish produced in 2014 (estimate provided by Nathan Stone). Genus characters are discussed in the species account below.

Carassius auratus (Linnaeus)
Goldfish

Figure 15.29. *Carassius auratus*, Goldfish. *Uland Thomas.*

Map 26. Distribution of *Carassius auratus,* Goldfish, 1988–2019. Inset shows introduced range (outlined) in the contiguous United States.

CHARACTERS   A stout, deep-bodied fish with a long, low dorsal fin, large scales, and a hard serrate spine (formed from fused rays) at the dorsal and anal fin origins. Mouth terminal, oblique. Barbels absent. Gill rakers on first gill arch 37–43. Pharyngeal teeth 0,4–4,0 on heavy arch. Lateral line straight and complete, with 26–32 scales. Usually 8 predorsal scales. Dorsal fin rays 15–19. Anal rays (including hard spine) 6 or 7; pectoral rays 14–17; pelvic rays 8–10. Intestine long, with several loops, about 1.5 times TL. Peritoneum dusky to black. Males with fine breeding tubercles on opercles, back, and pectoral fins. Males also have longer pectoral and pelvic fins than females. Size usually 5–12 inches (120–300 mm) TL. There are differing reports of maximum size, but it is probably around 16 inches (410 mm) TL and 4.5 pounds (2.0 kg).

LIFE COLORS Life color varies widely from dusky bronze, olive or gray green, to black or pink to gold with occasional black spots. Established populations revert to their natural wild coloration of a slaty or brownish-olive dorsum with a brassy sheen, sides lighter and often yellowish, and a yellow-white or white venter.

SIMILAR SPECIES The Goldfish differs from the Common Carp in lacking barbels (vs. barbels present), 26–32 lateral line scales (vs. 35–39), and pharyngeal teeth in one row 0,4–4,0 (vs. teeth in 3 rows 1,1,3–3,1,1). It differs from all other Arkansas cyprinids in having a stiff, serrate spine at dorsal fin origin (large Grass, Bighead, Black, and Silver carp have a stiff, nonserrate spine).

VARIATION AND TAXONOMY *Carassius auratus* was described by Linnaeus (1758) as *Cyprinus auratus* from "rivers of China and Japan." Some confusion exists regarding the taxonomy of this species. Most recent authors recognize two subspecies in its native range: *C. a. auratus* (Asia) and *C. a. gibelio* (eastern Europe). Others consider the Goldfish to be a subspecies of the Crucian Carp, *Carassius carassius* (i.e., *C. c. auratus*). Populations of Goldfish in the United States are morphologically and taxonomically diverse due to widespread and repeated introductions from many points of origin in Asia and Europe (Nico et al. 2016b). Howells (1992) reported that some workers believe that most Goldfish found in U.S. waters are actually Crucian Carp × Goldfish hybrids. Hybridization of *C. auratus* with *Cyprinus carpio* has been reported in some parts of the United States, but no hybrids are known from Arkansas.

DISTRIBUTION The Goldfish is native to Asia and adjacent Europe and was originally imported into this country primarily as an aquarium species in the late 1600s (Courtenay et al. 1984). It is now established in much of the United States into southern Canada. This exotic species has also been introduced as a baitfish in Arkansas. The state distribution is spotty, with records potentially occurring from any drainage (Map 26).

HABITAT AND BIOLOGY The Goldfish prefers shallow, warm ponds and pool areas of streams with dense submerged vegetation; however, it is tolerant of turbidity and pollution and may establish locally abundant populations in such stressed streams. No reproducing populations are known to have become established in Arkansas. Because of its extensive use as a bait species in Arkansas, the Goldfish has been found in all parts of the state; however, its sporadic occurrence in fish collections indicates that its survival in Arkansas waters after bait releases is low.

Although generally a plant and detritus feeder, it will also eat small benthic invertebrates. Jenkins and Burkhead (1994) considered it an omnivore that fed primarily on phytoplankton, macrophytes, aquatic insects, and small fishes. Boschung and Mayden (2004) also considered it an opportunistic omnivore, utilizing whatever food is available, ranging from aquatic vegetation to larval and adult insects, crustaceans, clams, snails, and worms. A palatal organ in the roof of the mouth is used for taste and touch during feeding (Lagler et al. 1977).

The Goldfish has a prolonged breeding season in the late spring and summer. Spawning continues throughout the summer in Wisconsin as long as water temperatures remain above 60°F (15.6°C) and the fish are not overcrowded (Becker 1983). Spawning begins in Mississippi in early March in shallow weedy areas where females scatter thousands of adhesive eggs over vegetation, roots, and other fixed objects (Ross 2001). A female may be accompanied by several males who fertilize the eggs as they are released. The adhesive eggs hatch in 3 days or more, depending on water temperature. Reported fecundities range from 4,000 to 400,000 eggs depending on the size of the female. Sexual maturity occurs at 1–3 years of age, and the normal life span is 6 or 7 years. The maximum observed life span is 30 years.

The Goldfish is an important part of the aquaculture industry in Arkansas. In 1982, more than 478,800 pounds (217,206 kg) of Goldfish worth $1.07 million were produced commercially in the state for bait. By 2014, Goldfish production was estimated at 2.5 million pounds valued at $7.5 million (Nathan Stone, pers. comm.). Only the production of Golden Shiners and Fathead Minnows exceeds that of Goldfish.

CONSERVATION STATUS This introduced species does not warrant a conservation status in Arkansas.

Genus *Chrosomus* Rafinesque, 1820a— Redbelly Daces

The genus *Chrosomus* has experienced several changes in taxonomic placement during the past 90 years. Jordan (1924) recognized two species in the genus *Chrosomus*, *C. erythrogaster* and *C. oreas*. Bănărescu (1964) merged the European genus *Phoxinus* with the Nearctic *Chrosomus*, and Joswiak (1980) used karyotype variation and chromosomal complements to support Bănărescu's taxonomy.

Howes (1985) provided a complete synonymy and taxonomic history in a revision of *Phoxinus* and restricted *Phoxinus* to three subgenera. The nominate subgenus (*Phoxinus*) with one species occurs in Eurasia. North American species of *Phoxinus* were classified into two subgenera: *Chrosomus* with six species and subgenus *Pfrille* with one species. Mayden et al. (2006a) resolved relationships of the genus *Phoxinus* as a non-natural grouping; North American species and subgenera (*Chrosomus* and *Pfrille*) formed a monophyletic group separate from the Eurasian subgenus *Phoxinus*. Strange and Mayden (2009) found that the genus *Phoxinus* as construed by Howes (1985) was not monophyletic. They further reported that the subgenera *Chrosomus* and *Pfrille* form a monophyletic group (as did Mayden et al. 2006a) that has affinities with the western clade of North American cyprinids (Simons and Mayden 1998; Simons et al. 2003). Strange and Mayden (2009) removed *Chrosomus* from synonymy with *Phoxinus* and assigned all seven North American species to genus *Chrosomus* (supported by Page et al. 2013). We also follow this taxonomic placement.

The genus *Chrosomus* is characterized by very small scales, giving the species a smooth metallic appearance; lateral line incomplete, usually ending at about the middle of the body; and branchiostegal membranes connected to the isthmus. Anal fin rays 8; pelvic fin rays 8. Pharyngeal teeth 0,5–5,0. Breeding individuals with a brilliant red belly. This genus contains seven species. A single species, *Chrosomus erythrogaster*, the Southern Redbelly Dace, occurs in Arkansas.

Chrosomus erythrogaster (Rafinesque)
Southern Redbelly Dace

Figure 15.30a. *Chrosomus erythrogaster*, Southern Redbelly Dace male. *Uland Thomas.*

CHARACTERS A distinctive, terete minnow having two conspicuous dusky lateral stripes: a broad one just below the midline of the side, and a narrower one situated above, with a light-colored area between the two (Fig. 15.30). Head

Figure 15.30b. *Chrosomus erythrogaster*, Southern Redbelly Dace female. *Uland Thomas.*

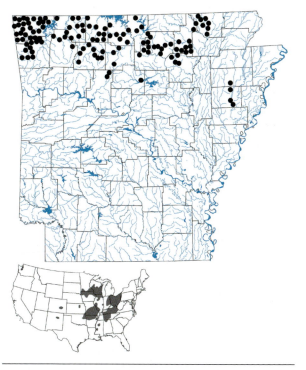

Map 27. Distribution of *Chrosomus erythrogaster*, Southern Redbelly Dace, 1988–2019.

small, its length going 3.8–4.2 times into SL. Snout length in large adults greater than diameter of eye; eye diameter going into head length 3.8–4.9 times. Mouth small, moderately oblique. Barbels absent. Pharyngeal teeth 0,5–5,0 (Fig. A3.18). Gill rakers short and conical. Scales minute. Lateral line incomplete, usually ending at about middle of body, with 70–95 scales in lateral series. Breast scaled. Dorsal, pelvic, and anal fin rays 8; pectoral fin rays 15–17. All fins small and rounded. Dorsal fin origin posterior to pelvic fin origin. Intestine long with several loops; peritoneum black. Breeding males with small tubercles scattered over the head, body, and fins. Females sometimes with weakly developed tubercles. Maximum size 3.5 inches (90 mm) TL.

LIFE COLORS Body color olivaceous above, with

scattered dark spots above the upper dark lateral stripe. Breeding males in Ozarks with bright red ventral coloration, bright yellow fins, and a pearly silver spot at the fin origins. The two dark, uninterrupted lateral stripes become black and are separated by a yellowish band. Breeding colors of Crowley's Ridge populations somewhat different and less intense than those of Ozark populations. In males, the underside of body an intense salmon pink, including lips and axils of pectoral fins; underside of caudal peduncle yellow and area between pelvic fins yellow; pectoral, pelvic, and anal fins bright yellow; some yellow in dorsal fin. Breeding females similar in overall coloration to males, but without pink coloration on venter. Breeding colors in Crowley's Ridge populations are retained for a very short time.

SIMILAR SPECIES The Southern Redbelly Dace can be distinguished from all native Arkansas cyprinids in having 70 or more lateral scales (vs. fewer than 70) and two continuous dark lateral bands (vs. one lateral band or none). Small juveniles resemble the young of several other cyprinids, particularly stonerollers and Creek Chubs. They differ from stonerollers in lacking the cartilaginous ridge of the lower jaw (vs. ridge present) and from Creek Chubs in having a smaller mouth and in lacking a barbel (vs. tiny barbel present in groove of upper lip).

VARIATION AND TAXONOMY *Chrosomus erythrogaster* was originally described by Rafinesque (1820a) as *Luxilus erythrogaster* from "Kentucky River." See genus *Chrosomus* account for information on changes in taxonomic placement during the past 90 years. Etnier and Starnes (1993) believed that *C. erythrogaster* is a complex of undescribed forms. Samuel Martin (who is currently studying the systematics and phylogeny of *Chrosomus*) supported that view and furnished the following information on *C. erythrogaster*. There are four lineages of *C. erythrogaster*, three of which reside in Arkansas: an Arkansas/Illinois drainage lineage, a White River lineage, and a Missouri River lineage. Interestingly, several of the eastern White River tributaries in Arkansas (e.g., Black River) contain populations assigned to the Missouri River lineage.

DISTRIBUTION *Chrosomus erythrogaster* occurs in the Great Lakes and Mississippi River basins from New York to Minnesota and south to northwestern Alabama, northern Arkansas, and northeastern and southern Oklahoma. Isolated populations occur in the Mississippi Embayment of Tennessee, Mississippi, and Arkansas, the Kansas River system, Kansas, and the upper Arkansas River drainage in Colorado and New Mexico. In Arkansas, it occurs across the northern part of the state in the White, Black, and Illinois river systems and disjunctly in a few streams (St. Francis River drainage) of Crowley's Ridge (Map 27). Fulmer and Harp (1977) found it in only one of 40 streams sampled on Crowley's Ridge, and Robison and Buchanan (1988) reported it from five Ridge streams. It was historically especially abundant in Village Creek in Cross County. It is still found in Village Creek and currently occurs in small numbers in three additional streams in Cross (2 streams) and Poinsett (1 stream) counties. Another isolated disjunct population was discovered in the Middle Fork of Illinois Bayou (an Arkansas River tributary) in Pope County, marking the easternmost report of *C. erythrogaster* in an Arkansas River drainage direct tributary (Robison et al. 2013). Voucher specimens from Pope County are in the Arkansas Tech University Fish Collection (Charles Gagen, pers. comm.).

HABITAT AND BIOLOGY The Southern Redbelly Dace requires clear, cool, gravel-bottom, spring-fed brooks and small headwater streams, where it is abundant in quiet pools or faster-flowing sections. It is sometimes the only fish species present in small Ozark springs. In Arkansas, it is most frequently found in waters having heavy growths of watercress, but it is not always found in or near vegetation. On Crowley's Ridge, it occurs in streams almost devoid of aquatic vegetation. In Mississippi, it occupied clear, cool (summer water temperatures of 17–24°C) streams over gravel, pebble, and sand substrates, and sites where it was most abundant usually had several long, slow-flowing pools (Ross 2001). Dace were generally absent from stream reaches with greater proportions of mud, large boulders, and silt, greater stream depth, and higher water temperatures (Slack et al. 1995). Habitat of *C. erythrogaster* in Alabama was similar to what we have observed in Arkansas. Boschung and Mayden (2004) reported that it occurs in springs or small, cool streams having abundant aquatic vegetation. In Alabama, the typical streams where it occurs are only a few meters wide, with undercut banks and long pools alternating with shallow riffles.

It is a schooling species and is often seen foraging over the bottom. Phillips (1969) and Settles and Hoyt (1976) reported that it feeds principally on algae (filamentous chlorophytes and diatoms) and organic detritus, but aquatic insects are also consumed. Fish smaller than 35 mm SL fed most heavily on diatoms, and fish greater than 35 mm SL consumed more immature aquatic insects (midges, mayflies, stoneflies, and caddisflies). Feeding occurred mostly during the day, with dace nibbling or sucking materials from the substrate (Phillips 1969; Settles and Hoyt 1976). In the Illinois River, Oklahoma, it was an omnivore, with most individuals having both aquatic invertebrates and large amounts of periphyton in their guts (Felley and Hill 1983). In a study of the effects of nutrient loading and grazing by *C. erythrogaster* in artificial outdoor stream mesocosms, Kohler et al. (2011) found that algal composition of dace diets was correlated with available algae, but there were proportionally more diatoms present in dace guts. In that study, 33% of dace guts contained invertebrate parts, but the diet was overwhelmingly dominated by diatoms and green algae.

Breeding generally occurs in Arkansas from April through June in swift, shallow riffle areas. HWR and Chris McAllister collected three females containing ripe eggs from Calico Creek at Calico Rock in Izard County on 7 July 2015, suggesting this species may prolong spawning under favorable environmental conditions (Tumlison et al. 2016). A long spawning season from March to July was reported in Kentucky (Settles and Hoyt 1978). Females spawning early in the season were in their second year. Females between 41 and 61 mm SL contained from 140 to 681 mature ova. Spawning occurred in Kansas when water temperatures were 50–60°F (10–15°C) (Cross and Collins 1995). Large spawning aggregations have been observed over shallow riffles. Breeding individuals do not construct nests or maintain territories, but they often use the nests of *Nocomis* and *Campostoma* species, and possibly other minnows. Spawning behaviors of *Chrosomus* species were compared by Hatcher et al. (2017). According to Smith (1908), the male wraps his body around the female during oviposition and holds her against the gravel with the aid of his breeding tubercles. Very often a female is flanked on either side by a male, and the trio spawns simultaneously. Although there is no noticeable fighting among the males, the largest and most active males seem to be most successful in spawning with females. The fertilized eggs sink into the gravel substrate where they develop. Young grow rapidly and are sexually mature in 1 year. Maximum life span is usually 2 years with a few individuals living into the autumn of the third year (Settles and Hoyt 1976). Walker (2011) and Walker et al. (2013a) studied movement patterns of *C. erythrogaster* in an Ozark headwater stream in Arkansas. Richardson et al. (2013) documented the piscicolid fish leech *Cystobranchus klemmi* from *C. erythrogaster* in Carroll County, Arkansas.

CONSERVATION STATUS The Southern Redbelly Dace is abundant in appropriate habitat in the mountains of northern Arkansas, and its populations there are currently secure. Populations on Crowley's Ridge are probably more vulnerable to extirpation and should be closely monitored to detect changes.

Genus *Ctenopharyngodon* Steindachner, 1866—Grass Carps

This monotypic Asian cyprinid genus contains the widely introduced Grass Carp, *C. idella*. This genus appeared during the Pliocene, 5–2 mya (Cavender 1991). *Ctenopharyngodon* is superficially similar to the Asian genus *Hypophthalmichthys*, which includes two introduced species in Arkansas. Although earlier believed to be closely related to the bigheaded carps, Howes (1981) placed the Grass Carp in a separate subfamily along with *Mylopharyngodon*. Tan and Armbruster (2018), however, placed *Ctenopharyngodon* in subfamily Xenocyprinae along with *Hypophthalmichthys* and *Mylopharyngodon*. Genus characters are included in the species account below.

Ctenopharyngodon idella (Valenciennes)
Grass Carp

Figure 15.31. *Ctenopharyngodon idella*, Grass Carp. *David A. Neely.*

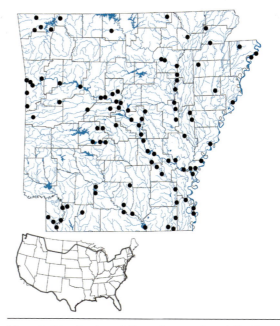

Map 28. Distribution of *Ctenopharyngodon idella*, Grass Carp, 1988–2019. Inset shows introduced range (outlined) in the contiguous United States.

CHARACTERS   A large, elongate, stout-bodied, blunt-headed, pale gray cyprinid introduced throughout the state and established in Arkansas rivers; sometimes referred to as the "white amur" (Fig. 15.31). Eyes of adults small and somewhat ventrolaterally situated; juveniles have proportionally larger eyes than adults. Fins without stiff serrated spines at origins. Dorsal fin origin over or in front of pelvic fin origin. Dorsal fin usually with 8 (rarely 9) rays. Anal fin located far posteriorly with 8–10 rays. Anal fin closer to caudal fin than in native minnows, the distance from front of anal fin base to caudal base going into the distance from anal base forward to snout tip more than 2.5 times. Pectoral rays 19–22. Pharyngeal teeth 2,4–4,2 or 2,4–5,2; teeth in the principal row with deep parallel grooves (Fig. 15.32). Gill rakers small, not comblike, usually 15 or 16. Mouth terminal, nonprotractile, with thin unspecialized lips; upper and lower jaws about equal in length. Barbels absent. Lateral line complete, with 35–47 ($\bar{x}$ = 43.0) scales. Scale rows above lateral line 6 or 7. Scales are large and dark-edged. Caudal peduncle short and thick. No abdominal keel. Intestine long with several loops, its length 1.5–3.0 times longer than the total fish length. Breeding males develop tubercles on the pectoral fins. Maximum length about 4 feet (1.25 m) (Guillory 1980; Page and Burr 2011) and 100 pounds (45.5 kg) in its native range. Specimens as large as 40.8 kg and 47.6 kg have been reported from Texas (Kolar et al. 2005) and Missouri, respectively. A 39.8 kg, 1,356 mm TL specimen was taken from Grand Lake in Oklahoma (Long and Nealis 2011). Specimens from Arkansas waters commonly weigh 15–20 pounds (6.8–9.1, kg). The Arkansas state angling record from Lake Wedington in 2004 is 80 pounds (36.3 kg) and 52 inches (1.3 m) TL.

LIFE COLORS   The dorsum is gray to green or olive-brown, becoming silvery on the sides and off-white on the belly. Scales of dorsum and sides having prominent dark edges, giving a crosshatched effect. Head often darker than rest of body.

SIMILAR SPECIES   The Grass Carp differs from all other Cyprinidae found in Arkansas (except the Black Carp) in having the anal fin situated far posteriorly, the distance from front of anal fin base to caudal base going more than 2.5 times into the distance from the anal base forward to tip of snout (vs. fewer than 2.5 times in other cyprinids). It further differs from both the introduced Bighead Carp and Silver Carp in having fewer than 50 lateral line scales (vs. more than 85), in lacking a ventral keel between vent and pelvic fin base, and in having 9 anal rays (vs. 13). It is very similar to the Black Carp in body shape and size, position and shape of fins, and position and size of eyes. It differs from Black Carp in having a more silvery

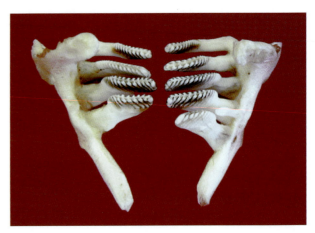

Figure 15.32. Pharyngeal teeth of Grass Carp. *Chad Thomas.*

body coloration (vs. a gray-brown [but not black] body coloration) and in having long, slender pharyngeal teeth with deep parallel grooves (vs. pharyngeal teeth molarlike). It differs from *Cyprinus carpio* in lacking barbels.

VARIATION AND TAXONOMY   *Ctenopharyngodon idella* was described by Valenciennes *in* Cuvier and Valenciennes (1844) as *Leuciscus idella* from China. Although some workers have suggested a sister group relationship between *C. Idella* and the Barbel Chub, *Squaliobarbus curriculus*, of Asia, others have supported different relationships for *Ctenopharyngodon*. He et al. (2008) found that *C. idella* was most closely related to *Mylopharyngodon piceus*. Li et al. (2009) supported a close relationship between *Ctenopharyngodon* and the bigheaded carps, *Hypophthalmichthys*.

DISTRIBUTION   The Grass Carp is native to eastern Asia from the Amur River of eastern Russia and China south to the West River of southern China. It has been reported from 45 states in the United States as a result of introduction. In Arkansas, records for this introduced species are scattered across the state from impoundments to the large rivers (Map 28).

HABITAT AND BIOLOGY   The Grass Carp is typically an inhabitant of large rivers, although it can be raised in ponds and rice fields. During most of the year, it prefers quiet, shallow waters with submerged vegetation (Bain et al. 1990). It is also found in deep holes in riverbeds during the winter (Cudmore and Mandrak 2004). Guillory and Gasaway (1978) and Guillory (1980) found that it is tolerant of a wide range of environmental conditions such as water temperatures of 0–35°C (32–95°F), salinities up to 10 ppt, and oxygen concentrations as low as .0005 ppt. Galloway and Kilambi (1984) reported a temperature preference of 95°F (35°C). Similar temperature preferences and tolerances of hybrid carp (female Grass Carp × male Bighead Carp) were reported by Kilambi and Galloway (1985), and a comparison of early developmental stages and adults of Grass Carp and the same hybrid combination was made by Kilambi and Zdinak (1981a). This native of eastern Asia was imported to the United States in 1963 by the Bureau of Sport Fisheries and Wildlife to the Fish Farming Experimental Station in Stuttgart, Arkansas (Stevenson 1965). The Grass Carp was originally introduced into the United States in an effort to control nuisance plant growth, but in some ecosystems it may negatively impact both aquatic plants and invertebrate foods that provide beneficial habitat for desired species (Stevenson 1965; Taylor et al. 1984; Chilton and Muoneke 1992; Fuller et al. 1999; Cole 2006). In the mid-1960s the Arkansas Game and Fish Commission began raising Grass Carp at the state fish hatchery in Lonoke. The first stocking of this species was in 1968 in Lake Greenlee near Brinkley. In the next several years it was introduced into lakes all over Arkansas as well as in Alabama and Oregon (Greenfield 1973). By 1978, more than 100 Arkansas lakes had been stocked with Grass Carp. By the early 1970s, Grass Carp were being caught by commercial fishermen in the Arkansas, Mississippi, and lower White rivers of the state. In 1976, 25 tons of Grass Carp were taken by commercial fishermen (less than 1% of the total state commercial catch) and by 1984, 68 tons were caught annually. There are no data for recent commercial catches.

The Grass Carp is basically a herbivore; however, it also occasionally consumes animal material such as insects and some small fish when vegetation is unavailable (Greenfield 1973; Chilton and Muoneke 1992). Larval Grass Carp first feed on zooplankton (primarily rotifers and small copepods) for 2–4 days after hatching, and then switch to larger zooplankton (primarily cladocerans) before assuming a largely herbivorous diet at around 100 mm TL (Stanley et al. 1978). Adults feed on a variety of plants (Kilambi 1980) consuming filamentous algae, aquatic vascular plants, and terrestrial plant material (Pflieger 1978). Because of its dietary habits it has only very limited potential as a gamefish. An evaluation of its catchability was made in Tennessee ponds stocked with Grass Carp (Wilson and Cottrell 1979). In more than 427 hours fishing with four types of bait, only nine *C. idella* were hooked and two landed. We have found the quality of Grass Carp flesh for eating to be highly variable. Sometimes the flavor is good, but often it is tainted with a strong algal flavor. Kilambi (1980) and Kilambi and Zdinak (1980) studied food consumption and growth in *C. idella* in Arkansas. W. R. Robison (1978) and Kilambi and Robison (1979) reported

the effects of controlled temperature and stocking density on food consumption, food conversion efficiency, and growth of Grass Carp.

In its native China, spawning occurs from April until mid-August at water temperatures between 67°F and 87°F (19.4–30.6°C). Turbulent areas at the confluence of rivers or below dams are focal points of reproduction (Stanley et al. 1978). Hargrave and Gido (2004) identified the Red River as having suitable environmental conditions for spawning of this species, and Shelton and Snow (2017) provided evidence that *C. idella* has become established in Lake Texoma, with spawning occurring in the Red and Washita rivers. The reproductive strategy of the Grass Carp is that of a nonguarder, pelagophil (Simon 1999b; Balon 1981). After a sudden rise in water level (>1.2 m within a 12-hour period usually after heavy rains) and after the appropriate water temperature is reached, spawning commences (Greenfield 1973; Chilton and Muoneke 1992). This carp is an open substratum spawner and broadcast spawner discharging ova and sperm in large numbers (Simon 1999b). During spawning, males generally outnumber females by two to one. Fish swim into the strongest currents in midstream where chasing occurs, and the male then pushes his head against the belly of the female and leans to one side. This is believed to be the time of egg and sperm release when fertilization occurs (Chilton and Muoneke 1992). Large females can produce more than 1 million semibuoyant eggs that float until hatching (Berg 1964; Stanley 1976). Hatching occurs in 26–28 hours at 75.2°F (24°C) (Stanley 1976). It was originally believed that the minimum current velocity necessary to support Grass Carp eggs in the water column until hatching was 2 feet (0.6 m) per second, but Leslie et al. (1982) demonstrated that the eggs were adequately transported in Econfina Creek, Florida, by a current of 9 inches (0.23 m) per second. Drift distance of the embryos until hatching reportedly varies between 28 and 100 km, depending on temperature and current velocity (Shelton and Snow 2017). Within 2 days after hatching, larvae move from flowing water to areas of quiet water having aquatic or submerged terrestrial vegetation (Stanley et al. 1978). Growth is rapid and the Grass Carp can reach weights of up to 5 kg in less than 2 years (C. L. Smith 1985). Females mature at 580–670 mm (22.8–26.4 in.) SL, and males mature an average of 1 year earlier at 510–600 mm (20.1–23.6 in) SL (Shireman and Smith 1983). Artificial propagation of the Grass Carp in pond culture in Arkansas was studied by Bailey and Boyd (1970). Breeding populations have been reported from a number of states, and there is ample evidence of its successful reproduction in the wild in Arkansas. We found ripe males and females in

April in the Arkansas River below Dardanelle Reservoir, and Conner et al. (1983) cited the collection of larvae as evidence of natural reproduction in the lower Mississippi River. Don Turman (pers. comm.) reported finding Grass Carp larvae throughout the lower Ouachita River, especially near Hampton, and TMB, Chad Hargrave, and Les Claybrook collected a 28 mm YOY specimen from the Sulphur River in Miller County on 12 August 1996. Shelton and Snow (2017) summarized known records of juvenile Grass Carp from Lake Texoma and the Red River, Oklahoma. Cudmore and Mandrak (2004) reported an individual in North Dakota that was about 33 years old.

CONSERVATION STATUS   As an introduced species in Arkansas, the Grass Carp has no conservation status. The effects of Grass Carp on native species are not clear, and the literature is full of contradictory studies on its interactions with other species. Shireman and Smith (1983) summarized potential negative effects of Grass Carp on natural fish communities and concluded that the impacts of its introduction are importantly related to such variables as stocking rate, macrophyte abundance, and community structure of the ecosystem.

Genus *Cyprinella* Girard, 1856— Satinfin Shiners

This genus currently contains 32 species. *Cyprinella* species are stream dwellers, typically inhabiting lotic environments, but they may also occur in lentic habitats (Mayden 1989). Adams et al. (2003) demonstrated oral grasping in *Cyprinella* as an adaptation for maintaining position in water velocities exceeding aerobic swimming abilities. *Cyprinella* was formerly treated as a subgenus of *Notropis* and contains moderate-sized to large species associated with more current than most *Notropis* species (Etnier and Starnes 1993). Mayden (1989) listed a combination of 34 derived characters by which *Cyprinella* could be distinguished from other genera of cyprinids. Many distinguishing characters are osteological (Boschung and Mayden 2004) and are not presented here. More easily distinguished genus characters include a strongly compressed body; origin of the dorsal fin slightly posterior to vertical from pelvic fin origin; free margins of lateral scales usually taller than wide, appearing elevated, especially anteriorly near lateral line; body circumferential scale rows 22–29; and circumpeduncle scales typically 14. Dorsal and pelvic rays are modally 8, and there are 7–10 anal rays. In Arkansas *Cyprinella* species, barbels are absent and pharyngeal teeth are 1,4–4,1 except in *C. lutrensis* which typically has a 0,4–4,0 tooth-count west of the Mississippi River.

Large antrorse tubercles are present on top of the head, and the tubercular pattern is discontinuous between top of head and snout, except in *C. camura* and *C. galactura*. Pigmentation in Arkansas *Cyprinella* is present on the free edges of the scales above and below the lateral line, producing a diamond-shaped pattern; dark pigment is present in the posterior interradial membranes of the dorsal fin (except in *C. lutrensis*). *Cyprinella* species are crevice spawners (Johnston and Page 1992) that spawn high in the water column and in the open. This produces intrasexual competition among males who attempt to monopolize the crevices (Heins 1990). Successful males gain access to females, and sexual selection has produced strong sexual dimorphism in *Cyprinella* species. Most *Cyprinella* are fractional spawners, with small complements of eggs intermittently spawned over a protracted period. All members of *Cyprinella* are believed to produce sounds (Delco 1960; Winn and Stout 1960; Stout 1975; Phillips and Johnston 2008a) during aggressive encounters in the breeding season.

There are several different interpretations of the relationship of *Cyprinella* to other cyprinid genera. *Cyprinella* was identified as monophyletic and sister to *Luxilus*, and this clade in turn sister to *Lythrurus* based on morphological characters (Mayden 1989). *Cyprinella* was resolved as a monophyletic group by Mayden et al. (2006a), but support for this clade was not strong. Mayden (1989) recognized two clades within *Cyprinella*: the *lutrensis* clade consisting of 10 species, and the *whipplei* clade made up of 17 species. Coburn and Cavender (1992) considered *Cyprinella* the sister group to a clade including *Pimephales* and *Opsopoeodus*; this clade in turn was sister to *Luxilus* (Boschung and Mayden 2004). Simons and Mayden (1997) reported that *Cyprinella* was more closely related to *Notropis*, but Simons and Mayden (1999) found *Cyprinella* more closely related to the western genus *Agosia* than to eastern cyprinids, and in a clade that included *Notropis* and *Hybognathus*. Broughton and Gold (2000) examined species relationships in *Cyprinella* and recognized 30 species in the genus. Dimmick (1993) found that *Cyprinella* was monophyletic based on allozyme analysis. Schönhuth and Mayden (2010) analyzed phylogenetic relationships in *Cyprinella* based on mitochondrial and nuclear gene sequences. Much remains to be done with the systematics of this genus, and there is not a consensus among ichthyologists regarding phylogenetic relationships.

The genus *Cyprinella* is represented in all major drainages across Arkansas, but there is a notable lack of recent records from the Illinois River drainage in Benton and Washington counties. Arkansas species of *Cyprinella* are

C. camura, C. galactura, C. lutrensis, C. spiloptera, C. venusta, and *C. whipplei.* The two most widely distributed species of *Cyprinella* in Arkansas are *C. venusta* and *C. whipplei. Cyprinella lutrensis* is restricted largely to the three largest rivers in the state where it is common, and *C. galactura* is widespread in the upper White River drainage. *Cyprinella camura* and *C. spiloptera* are the only *Cyprinella* species known from the Illinois River drainage (the most recent record is approximately 60 years old), and neither of these species has been found anywhere in Arkansas in more than 45 years.

Cyprinella camura (Jordan and Meek)
Bluntface Shiner

Figure 15.33. *Cyprinella camura,* Bluntface Shiner. *David A. Neely.*

Map 29. Distribution of *Cyprinella camura,* Bluntface Shiner, 1988–2019.

CHARACTERS   A deep-bodied (body depth going 3.8–5.0 times into SL), laterally compressed shiner with small eyes, a relatively blunt snout, and a slightly subterminal, oblique mouth (Fig 15.33). Eye dameter going into head length 2.4–4.5 times and into snout length 0.6–1.5 times. Smaller individuals have proportionally much larger eyes (relative to head length) than large individuals. Front of dorsal fin base slightly closer to caudal fin base than to tip of snout. Dorsal rays 8; anal fin rays usually 9 (8–10). Pharyngeal teeth 1,4–4,1. Gill rakers 8–10. Lateral line is decurved and complete, usually with 36 or 37 (35–38) scales. Predorsal scales 13–17. Breast and belly are fully scaled. Pectoral rays 14–16. Intestine short, peritoneum silvery but sprinkled with dark speckles. Breeding males develop an enlarged dorsal fin that is rounded posteriorly. Scattered tubercles on breeding males on head, snout, and between eye and snout, with no hiatus between tubercles on head and snout. Tuberculation is also extensive on the body and fins. Maximum size usually around 4.6 inches (117 mm), but Etnier and Starnes (1993) reported a maximum size of 5.9 inches (150 mm) TL.

LIFE COLORS   Dorsum olive, sides silver, middorsal stripe broad. Depigmentation (white area) at the caudal base is not as well defined as it usually is in *C. galactura,* but rather forming a light vertical bar at the base of the caudal rays and flaring dorsally and ventrally. The white caudal bar is not always evident, often causing this species to be easily confused with *C. whipplei.* Dorsal fin of large young and adults well pigmented on all membranes, with a black blotch especially prominent on posterior two or three membranes. Dusky stripe along side of caudal peduncle broad and faint, poorly defined above and centered on lateral line (most obvious in preserved specimens and usually not observed in live fish). Breeding males are bright steel blue along back and upper sides, and have salmon-colored to red dorsal and caudal fins. Pflieger (1997) reported that breeding males also have a rosy bar behind the head similar to that of the Red Shiner.

SIMILAR SPECIES   The Bluntface Shiner is most similar to *C. spiloptera, C. lutrensis, C. whipplei,* and *C. galactura.* It can be differentiated from the first two by its 9 or 10 anal rays (vs. usually 8). From *C. whipplei,* it differs in usually having a narrow, vertical white patch over the caudal base (vs. white patch absent), and fins of breeding males are salmon-pink (vs. yellow). Etnier and Starnes (1993) were unable to differentiate between *C. camura* and *C. whipplei* with certainty in Tennessee unless adults with tubercles or tubercle scars were available. Breeding males differ from males of *C. whipplei* in having no separation between the tubercles on the snout and the interorbital tubercles (vs. a distinct hiatus between the snout tubercles and interorbital tubercles in *whipplei*), and red dorsal and caudal fins (vs. a blackened dorsal fin and yellow caudal fin). Etnier and Starnes (1993) reported that juveniles are almost indistinguishable from the occasionally sympatric *C. whipplei*. *Cyprinella camura* differs from *C. galactura* in having a single narrow, vertical light patch at caudal fin base (vs. two prominent broad white patches), and in usually having 35–39 lateral line scales (vs. 39–41).

VARIATION AND TAXONOMY   *Cyprinella camura* was described by Jordan and Meek (1884) as *Cliola camura* from the Arkansas River at Fort Lyon, Colorado. Because this locality is well west of the known range of the species, Gilbert (1978) considered a more likely type locality to be the Neosho River system (Arkansas River drainage) of eastern Kansas. Gibbs (1961) found that western Arkansas River populations and eastern lower Mississippi River populations represented separate races, but recognition of subspecies was unwarranted. Gibbs assigned specimens of *C. camura* from Arkansas to the Arkansas race of this species and suggested a relationship with *C. galactura*. LeDuc (1984) studied eastern and western populations of *C. camura* using morphological and electrophoretic data, and used electrophoretic data to investigate relationships among *camura*, *galactura*, and *whipplei*. The data indicated a closer relationship between *camura* and *galactura* than between either of those species and *whipplei*. Mayden (1989) proposed a close relationship with a group consisting of *whipplei* and Atlantic Slope species, *analostomus* and *chloristia*.

DISTRIBUTION   The range of *C. camura* is divided into two distinct parts. West of the Mississippi River, it is historically known from the Arkansas River drainage in eastern Kansas, southwestern Missouri, northwestern Arkansas, and eastern Oklahoma. East of the Mississippi River, it occurs in the Mississippi Embayment from Kentucky to Louisiana. In Arkansas, the Bluntface Shiner is extremely rare, having been found only in four pre-1960 collections in the Arkansas River drainage (three records from the Illinois River drainage, Washington County, and one from Illinois Bayou, Pope County) (Map A4.30). The known Arkansas records are as follows: (1) UMMZ 123431, Washington County, Illinois River, 2 miles NE Prairie Grove, 1 July 1939, 8 specimens; (2) TU 52898, Washington County, Illinois River, 18 October 1950, 6 specimens; (3) CUMV 24362, Pope County, Illinois Bayou 4.2 miles W of Russellville, 16 August 1953, 4 specimens; and (4) CUMV 41576, Washington County, Clear Creek, 1 mile E of confluence with Illinois River at Savoy, 22 August 1959,

1 specimen. Earlier, Meek (1894a) incorrectly reported specimens from several Arkansas streams; however, these were later identified as *N. venustus* (= *C. venusta*) by C. L. Hubbs (Black 1940). The lack of recent records of *C. camura* from Arkansas (Map 29) is somewhat surprising, because there are recent records from southeastern Kansas, southwestern Missouri, and northeastern Oklahoma. Robert Hrabik (pers. comm.) informed us that it has been collected from eight localities in Jasper, Newton, and McDonald counties, Missouri, in recent years and is considered common at a few of the collection sites. There are 46 lots of *C. camura* from northeastern Oklahoma in the Sam Noble Museum of Natural History, but 45 of those lots are pre-2000 records. It was also reported as common in streams of southeastern Kansas (Cross and Collins 1995).

HABITAT AND BIOLOGY   Western populations of the Bluntface Shiner typically inhabit moderately swift, clear, upland stream sections and prefer areas of relatively high gradient and clean gravel or rubble substrates, whereas populations east of the Mississippi River inhabit sandy tributaries throughout their range (Farr 1996). It is more common in states surrounding Arkansas. In Missouri, adults are most often found near riffles in a noticeable current in moderately large rivers (Pflieger 1997; Wilkinson and Edds 2001), while the young usually inhabit nearby pools or stream margins with clear water (Cross and Cavin 1971; Fuselier and Edds 1996; Metcalf 1959). It is a common species in small and medium-sized streams in the Flint Hills of southeastern Kansas, where it prefers moderately fast, clear water and avoids streams with low gradients and mud or sand bottoms (Cross and Collins 1995). In Tennessee, it is found in the headwater portions of primarily lowland river systems, where it occurs in moderate to swift current over sand substrates (Etnier and Starnes 1993). In Mississippi, younger individuals were commonly found in slower pools or edges of riffles, often over a sand substrate (Ross 2001). The Bluntface Shiner is less tolerant of turbidity than the Red Shiner. It also has less tolerance of oxygen stress, initially gasping at the surface and then leaping from the water just before losing equilibrium (Cross and Cavin 1971).

D. A. Etnier (pers. comm.) believes that this species is ecologically very similar to *Cyprinella whipplei* and may have been gradually replaced in Arkansas by *C. whipplei*, with which it seems unable to compete, wherever the two have become syntopic. This could account for the lack of recent collections of *C. camura* in Arkansas River Valley tributaries from Fort Smith to Little Rock, where *C. whipplei* is common. Black (1940) noted that *C. camura* replaces *C. whipplei* and *C. galactura* in the Illinois-Neosho drainage, and we have no records of the latter two species from that drainage in Benton and Washington counties. If *C. camura* still occurs in Arkansas, it would most likely be found in the two northwesternmost counties.

Although foods are unknown, this shiner probably feeds on aquatic macroinvertebrates taken from the water column or occasionally from the streambed or water surface. Pflieger (1997) speculated that its life history features are probably not greatly different from those of *C. whipplei*.

In Oklahoma, *C. camura* spawns from May to August, with nuptial males captured as late as August 13 (Miller and Robison 2004). In Mississippi, spawning occurs from late March through mid-August (Farr 1996). In Kansas, most females spawned from late July through August (Cross and Cavin 1971). Spawning occurred in Kansas at water temperatures of about 80°F (26°C) or slightly cooler (Cross and Collins 1995). It is a typical crevice spawner as are the other members of the genus *Cyprinella* (D. C. Heins, pers. comm.) and deposits its eggs within a crevice or similar space which affords some protection from predation (Mayden 1989). Clean, flowing water is needed for reproductive success; silt accumulation or low dissolved oxygen can kill the eggs (Johnston 1999). Males temporarily defend spawning sites and display for females, both visually and through acoustical signals (Johnston 1999). Adults do not guard the eggs after a male abandons his spawning territory. Largest eggs in late June through July had mean diameters of 0.82–0.88 mm (Cross and Cavin 1971). Females reach sexual maturity between 36 and 40 mm SL, and the number of eggs produced increases with size of fish (Farr 1996). Counts of mature oocytes from females 36.4–57.9 mm SL ranged from 76 to 370. Mean clutch size decreased from 249 eggs in early April to 101 in mid-August. In Kansas, populations in the Cottonwood River were composed of two age groups: Age 0 fish reached 45 mm (1.8 inches) SL by the end of their first summer of growth, and age 1 fish reached 75 mm (3.0 inches) SL at the end of a second summer of growth (Cross and Cavin 1971). A few individuals might live through a second winter to age 2 (Cross and Cavin 1971).

CONSERVATION STATUS   The Bluntface Shiner is one of the rarest fishes in Arkansas. There are no verified records from the state since 1959, and Robison (1974b) and Buchanan (1974) considered it rare and endangered in Arkansas in the early 1970s. We believe it may be extirpated from the state. An intensive search for this species in the Illinois River drainage is needed to determine whether it still exists in Arkansas.

Cyprinella galactura (Cope)
Whitetail Shiner

Figure 15.34. *Cyprinella galactura*, Whitetail Shiner. *Uland Thomas.*

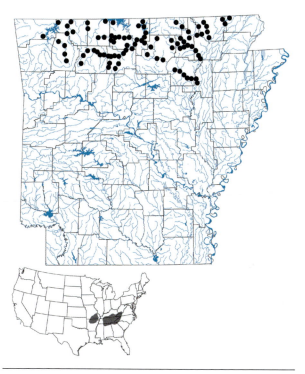

Map 30. Distribution of *Cyprinella galactura*, Whitetail Shiner, 1988–2019.

CHARACTERS   A large, slab-sided shiner with small eyes, a subterminal, slightly oblique mouth, and two broad conspicuous creamy-white patches at the base of the caudal fin (Fig 15.34). The two white patches may coalesce into one wide vertical patch. Body depth going into SL 4.6–6.9 times. Eye small in adults, its diameter going into head length 3.0–5.3 times and into snout length 0.8–1.5 times; smaller fish have proportionately much larger eyes relative to head length. Dorsal rays 8. Anal rays typically 9 (8–10). Pectoral rays 14–18. Lateral line decurved and complete, usually with 39 or 40 (38–42) scales. Predorsal scale rows 18 (17–19). Body circumferential scales usually 22–27. Breast and belly are completely scaled. Pharyngeal teeth 1,4–4,1. Gill rakers 9–11. Front of dorsal fin base slightly closer to caudal base than to tip of snout. Intestine short; peritoneum silver, but thickly sprinkled with dark speckles. Breeding males develop extensive tuberculation on the body and fins. Nuptial tubercles on dorsum of head are continuous to snout tip (no hiatus), nape tubercles are abruptly much smaller than those on head, and belly scales have small tubercles in the highest males (Etnier and Starnes 1993). Maximum size 6 inches (152 mm) TL, and 3–5-inch (76–127 mm) adults are common in Arkansas.

LIFE COLORS Body silver with two prominent white patches of pigment at caudal base. Dorsal fin usually well pigmented on all membranes, the last two or three membranes noticeably darker than the rest. Caudal fin lightly pigmented (except for depigmented basal patches), the pigment along outer edges of rays often very dark, adding to the contrast of the depigmented areas. Pectoral fin rays usually with melanophores lining the leading rays. Pelvic and anal fins clear. Anal fin base well pigmented. Breeding males develop iridescent steel-blue back and silvery blue sides. The snout is red, and they have salmon-pink pectoral, dorsal, and caudal fins. Other fins are milky white in color.

SIMILAR SPECIES *Cyprinella camura* also has a depigmented area at the caudal base similar to the white patches of *C. galactura*. See *C. camura* account for differences between the two species. Juvenile specimens of *C. galactura* may be confused with *C. whipplei* and *C. spiloptera*, but *C. galactura* has a subterminal mouth, while both *C. whipplei* and *C. spiloptera* have terminal mouths. *Cyprinella galactura* differs from all other *Cyprinella* species in Arkansas in having two white patches at caudal base.

VARIATION AND TAXONOMY *Cyprinella galactura* was described by Cope (1868a) as *Hypsilepis galacturus* from the Holston River system, Virginia. Gibbs (1961) found slight differences between Ozark populations and populations east of the Mississippi River, but did not recognize subspecies. Both Gibbs (1961) and LeDuc (1984) concluded the closest affinities of *galactura* were with *C. camura*; however, Mayden (1989) regarded *C. venusta* as its sister species. Based on analysis of the cytochrome *b* gene, Mayden et al. (2006a) recognized two clades within *C. galactura* and did not support a sister relationship with *C. camura*.

DISTRIBUTION The range of *C. galactura* is divided into two distinct parts east and west of the Mississippi Embayment. East of the Mississippi River, it occurs in the Cumberland and Tennessee river drainages from Tennessee to Virginia and south to North Carolina and northern Alabama, and in Atlantic Slope drainages (Savannah and Santee rivers) from North Carolina to Georgia. West of the Mississippi River, it is found in the St. Francis and White river drainages in Arkansas and Missouri. In Arkansas, the Whitetail Shiner is common and locally abundant in the White River system (Map 30). Although it is known from the St. Francis River drainage in Missouri, there are no known records from the St. Francis River in Arkansas. There are older records from the Little Red River above Greers Ferry Lake (Map A 4.31), but there are no post-1988 records from that drainage.

HABITAT AND BIOLOGY The Whitetail Shiner inhabits medium to large, clear, high-gradient streams with silt-free gravel, rubble, or boulder-strewn substrates. It is most abundant in swift runs or flowing pools, and it avoids small headwater streams. Buchanan (2005) found it in small numbers in two White River reservoirs (Bull Shoals and Norfork) near creek mouths. Underwater observations reveal that *C. galactura* is an active swimmer frequenting the middle to upper part of the water column, and existing in small groups that quickly move in if the substrate is stirred or food (crayfish, clams) is broken up and allowed to drift in the current.

Outten (1958) found that principal foods in North Carolina were terrestrial and aquatic insects, primarily mayflies, dipterans, and coleopterans, but larval fishes were also occasionally eaten. In that study, the diet also included oligochaetes, mites, and some plant material. It is a visual feeder that feeds on both drift and substrate items (Jenkins and Burkhead 1994). In Arkansas, the Whitetail Shiner is almost entirely insectivorous. TMB examined gut contents of 58 specimens collected from February through September from Baxter, Izard, Marion, Newton, and Sharp counties. Guts from all samples contained only insect remains. The diet in the Buffalo River in February consisted of immature insects, predominantly ephemeropteran nymphs. From May through September, *C. galactura* fed at all levels in the water column on a wide variety of terrestrial insects and immature aquatic insects. During those months, specimens from the Buffalo, Spring, Strawberry, and White rivers ingested mainly adult dipterans, coleopterans, hymenopterans, and odonates; the diet also consisted of about the same amount of immature aquatic insects, primarily ephemeropteran and plecopteran nymphs, and coleopteran and midge larvae. Some guts also contained sand grains.

Pflieger (1997) reported that this shiner spawns over an extended period from early June to mid-August in Missouri, and a similar spawning period from May to

August was reported in Virginia, North Carolina, and Tennessee (Jenkins and Burkhead 1994). A long spawning season also occurs in Arkansas, extending from at least late May into August. We have collected heavily tubercled males from mid-June to early July in Marion and Fulton counties. HWR collected two heavily tuberculate males and three gravid females from Crooked Creek, Marion County, on 23 July 2014 (Tumlison et al. 2015), and TMB found tuberculate males as late as 21 August in Baxter County. Ripe females collected by TMB in late May in Moccasin Creek (Izard County) and the Strawberry River (Sharp County) ranged from 71 to 94 mm TL and contained from 222 to 686 mature ova; ripe females from Baxter County contained approximately 313 mature ova on 3 August. Sexual maturity is reached in the second or third summer, and maximum life span is 3 or 4 years.

As a crevice spawner, its spawning habits are like those of *C. whipplei*. Male Whitetail Shiners established dominance hierarchies around crevice territories. Males performed lip locks during escalated aggression prior to hierarchy establishment, a behavior previously unreported in cyprinids (Phillips and Johnston 2008a). Females deposit eggs beneath the bark of submerged logs and in crevices of bedrock or on the sides and bottoms of rocks (Outten 1961; Pflieger 1997). *Cyprinella galactura* acoustic signals were first described by Phillips and Johnston (2008a), and Phillips and Johnston (2008b) found significant levels of divergence between agonistic and courtship signals. *Cyprinella galactura* courtship signals had higher dominant frequencies than the agonistic signals. Lower frequency signals are known to travel further in water than high frequency signals (Bradbury and Vehrencamp 2011); therefore, it was proposed that courtship calls may be used for short-range communication of males toward potentially receptive females, while agonistic calls may be intended for longer-range communication with conspecifics as well as heterospecifics. Phillips and Johnston (2009) examined signal structure and behavioral context in *C. galactura*. Males produced short knocks (autapomorphy) and chirps and rattles (synapomorphy) while they were stationary in front of their crevices, and this may serve to attract females to the area

CONSERVATION STATUS The Whitetail Shiner is a common and widely distributed upland cyprinid species in Arkansas, and we believe that its populations are currently secure.

Cyprinella lutrensis (Baird and Girard)
Red Shiner

Figure 15.35a. *Cyprinella lutrensis*, Red Shiner male. *Uland Thomas.*

Figure 15.35b. *Cyprinella lutrensis*, Red Shiner female. *David A. Neely.*

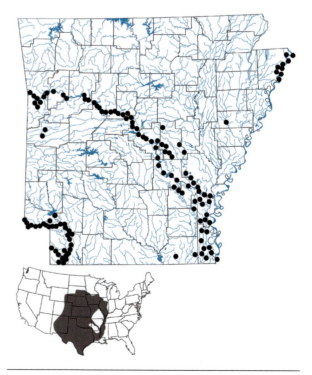

Map 31. Distribution of *Cyprinella lutrensis*, Red Shiner, 1988–2019.

CHARACTERS   A very deep-bodied, laterally compressed shiner with small eyes (eye diameter usually going 3.0–4.2 times or more into head length and 0.8–1.2 times into snout length), a terminal oblique mouth, usually 9 anal rays (occasionally 8 or 10), and diamond-shaped scales (Fig. 15.35). Snout bluntly rounded. Body depth going into SL 3.0–3.6 times. Pharyngeal teeth usually 0,4–4,0. Gill rakers 9–12. Lateral line slightly decurved and complete, with 32–37 scales. Predorsal scales 13–18. Body circumferential scales usually 24–29. Breast fully scaled; belly scalation variable, but scales may be embedded or absent anteriorly. Front of dorsal fin base about equidistant between tip of snout and base of caudal fin, and located over or slightly behind pelvic fin origin. Dorsal fin rounded, with 8 rays. Adults and large young with all membranes of dorsal fin more or less uniformly pigmented with fine dark specks, not concentrated into a dusky or black blotch on last 2 or 3 dorsal membranes. Intestine short, with a single S-shaped loop; peritoneum silvery, with many dark speckles. In breeding males, the head and lachrymal areas have prominent tubercles (described by Koehn 1965) except for a hiatus just anterior to the nares. The tubercles form 2 rows between the eyes. Sides of head without tubercles. Middorsal areas posterior to head have large tubercles, with 2 or 3 tubercles per scale, decreasing in size toward dorsal fin origin. Tubercles prominent on pectoral, pelvic, and anal fin rays. Some females may develop small tubercles. Maximum length 3.5 inches (90 mm) TL.

LIFE COLORS   Breeding males with all fins except the dorsal bright red orange, the sides a metallic blue, and a well-developed purple scapular bar above the pectoral fin behind the operculum; dorsal fin blackened. Middorsal stripe well developed. Top of head red. Large young and adults with all dorsal fin membranes more or less uniformly pigmented with fine specks, never with a posterior black blotch. Nonbreeding individuals with olive to bluish-green dorsum becoming bluish-silver laterally, and white or cream ventrally. Anal and caudal fins often with some pale red coloration.

SIMILAR SPECIES   Cyprinella lutrensis is similar to C. whipplei, but has a much deeper body with body depth going into SL 3.0–3.6 times (vs. 3,9–4,8 times), a blunter snout (vs. more pointed), and no dark blotch on the posterior rays of the dorsal fin (vs. dark blotch present at posterior base of dorsal fin). It also resembles specimens of C. venusta that lack the characteristic well-developed dark basicaudal blotch. C. lutrensis can be distinguished from those aberrant C. venusta specimens by its 9 anal rays (vs. 8) and its 0,4–4,0 pharyngeal teeth (vs. 1,4–4,1). The

0,4-4,0 tooth count distinguishes C. lutrensis from all other Cyprinella species in Arkansas.

VARIATION AND TAXONOMY   Cyprinella lutrensis was described by Baird and Girard (1853b) as Leuciscus lutrensis from Otter Creek, a tributary of the North Fork of the Red River, either Kiowa County or Tillman County, Oklahoma (not Arkansas as listed in some early reports). Matthews (1987a) studied geographic variation in Cyprinella lutrensis throughout its range and suggested that several subspecies exist. All Arkansas specimens are the nominate form Cyprinella l. lutrensis. Matthews noted that its complex taxonomic past includes at least 25 junior synonyms. Matthews (1995) studied nuptial coloration throughout its range north of Mexico and found that Red Shiner populations from the central Great Plains formed a core group based on male chromatics. More isolated populations in central and west Texas and New Mexico were more distinctive in nuptial coloration. Mayden (1989) placed this species in the C. lutrensis species group with nine other primarily western species. Mayden et al. (1992a) considered it a polytypic species that likely includes undescribed species, and Broughton and Gold (2000) provided evidence that C. lutrensis, as currently constituted, is a paraphyletic grouping.

DISTRIBUTION   Cyprinella lutrensis occurs in the Mississippi River basin from Wisconsin and eastern Indiana west to Wyoming and south to Louisiana, and in Gulf Slope drainages from west of the Mississippi River to Rio Grande drainages in Texas and northern Mexico. It has been widely introduced into other areas of the United States (1987b). In Arkansas, the Red Shiner is mainly confined to the Red, Arkansas, St. Francis, and Mississippi rivers, with a few records from the lower portions of some of their tributaries (Map 31). There are a few older records from the lower White River (Map A4.32). It was considered abundant in the Red River (Buchanan et al. 2003) and was common in the navigation pools of the Arkansas River (Buchanan 2005). In 2018, it was the most abundant minnow in collections by TMB from the Arkansas River at Fort Smith. Surprisingly, it is conspicuously absent from the Ouachita River and, in recent decades, the White River system.

HABITAT AND BIOLOGY   The Red Shiner is a habitat generalist that occupies sluggish streams and large rivers over a wide variety of bottom types (sand or mixed sand-silt-gravel) and is tolerant of high turbidities and siltation. It inhabits quiet waters as well as those with much current, but it is only rarely found in smaller streams in Arkansas. It seems to fare poorly in high-gradient, clear streams with relatively high species diversity. Cross and Collins (1995) noted that Red Shiners are most numerous in Kansas where few other kinds of fishes occur (where conditions are unfavorable for the more specialized, habitat-limited species). Those authors also reported that where other species have been depleted in Kansas by human-induced habitat changes, Red Shiners have increased in abundance. In the clear lowland tributaries of the large silty rivers in Arkansas it is replaced by *C. venusta*, but both species occur together in the main stems of the largest rivers (Mississippi, Arkansas, and Red rivers). Because the distribution of the Red Shiner was complementary to those of other *Cyprinella* species in Missouri, Pflieger (1997) suggested that competition from those species was an important factor limiting Red Shiner distribution. It is noteworthy, however, that the Red Shiner has apparently replaced some populations of other *Cyprinella* species in Illinois and Indiana (P. W. Smith 1979). In Great Plains streams, the Red Shiner is a true ecological generalist, an adaptive strategy that probably evolved during repeated exposures to desertification or extended aridity (Noel Burkhead, pers. comm.). It is adapted to a wide range of environmental conditions, including seasonal intermittent flows, degraded habitats, poor water quality, and natural physicochemical extremes. Cross and Collins (1995) considered it a remarkably hardy, adaptable fish that is ubiquitous in Kansas, where it inhabits all types of waters. Yu and Peters (2002) quantified diel and seasonal aspects of habitat use by Red Shiners in the Platte River, Nebraska. In that study, Red Shiners selected depths of less than 30 cm during the night and fine substrate during the day in summer, and avoided depths from 30 to 60 cm during the night and coarse substrates during the day in summer. Red Shiners also consistently selected slow water (<30 cm/s) and avoided fast currents (>60 cm/s) during both day and night in summer and fall. Current velocity was considered a major habitat factor affecting its distribution in the lower Platte River (Yu and Peters 2002). Matthews and Hill (1977, 1979a, b) and Matthews (1985a) documented the fairly broad ecological tolerances of this species in Oklahoma, reporting a tolerance of pH between 5 and 10, salinity of up to 10 ppt, dissolved oxygen as low as 1.5 ppm, and thermal shock of T+10 to T−21°C. Matthews (1986b) reported geographic variation in thermal tolerance.

Hale (1963) studied food habits of the Red Shiner in Lake Texoma, Oklahoma, and found that it is omnivorous, feeding mostly on aquatic and terrestrial insects including hymenopterans, trichopterans, dipterans, hemipterans, coleopterans, and ephemeropterans as well as algae and a few crustaceans. She also found similarities in the diets of *C. lutrensis* and *C. venusta*, indicating possible competition between the two cyprinids; however, *C. lutrensis* took more of its food from the surface (terrestrial insects) than *C. venusta*. Cross and Collins (1995) reported that the Red Shiner in Kansas feeds throughout the water column, where it is mainly carnivorous (primarily microcrustaceans and insects) but also consumes some algae.

Pflieger (1997) reported that spawning of *C. lutrensis* occurs repeatedly over an extended period from late May to early September in Missouri. In Oklahoma, Farringer et al. (1979) found a breeding season of April to September. In Texas and Oklahoma, spawning peaks are in May and August, or mid-June and late July in Kansas (Cavin 1962; Farringer et al. 1979). Spawning occurs in Arkansas from at least May through August, with at least one peak occurring in June. While all other members of the genus *Cyprinella* spawn only in crevices (e.g., cracks or seams in rock, or bark fissures along submerged logs), *C. lutrensis* exhibits a wider range of spawning behaviors. It has been documented as a crevice spawner, a broadcast spawner scattering its eggs over gravel, rocks, or vegetation, and a nest associate spawning in the nests of sunfishes (Minckley 1959, 1972; Vives 1993; Pflieger 1997). Cross (1967) reported "groups of more than 50 nuptial males surrounding the end of a large stump that was polished by the cleaning activity of these fish." Females produce species-specific sounds when in the presence of breeding males, and males respond positively to these "breeding calls" (Delco 1960). The ability to spawn over a variety of substrates may contribute to the ability of *C. lutrensis* to occupy a range of habitats (Mayden and Simons 2002). Spawning occurs at water temperatures between 15°C and 29°C (60–85°F) (Cross and Collins 1995). Ross (2001) reported crevice-spawning behavior as follows:

> Territorial males make slow passes along the spawning crevice, undulating their bodies and extending their red fins. Periodically males swim out toward females and attempt to lead them back to the spawning site. Females may "inspect" the spawning site hundreds of times before actually releasing eggs. When eggs are finally released, the pair makes a pass over the spawning crevice and the fertilized eggs are "sprayed" into the crevice.

Becker (1983) reported that females usually have 485–684 eggs each. Individuals may survive through two winters and breed in two successive years. Marsh-Matthews et al. (2002) showed that Red Shiners hatched in May can spawn successfully by late August, confirming age-0 individuals are capable of spawning. Females can be sexually mature at 24 mm (0.9 inches) SL, and males at 29 mm (1.1 inches) or less, but most reproduction is by age 1. Life span of this shiner is about 3 years (Larimore and Carlander 1971).

Hybridization between the Red Shiner and the Blacktail Shiner has been reported in several areas over its range, and was first noted in Arkansas in the lower Arkansas River by Buchanan (1976). Extensive introgression is apparently occurring in that river between Little Rock and Pine Bluff (John A. Baker, pers. comm.). Matthew Dugas (pers. comm.) reported that more than half the *C. lutrensis* specimens collected in the Arkansas River at Conway were *C. lutrensis* × *C. venusta* hybrids. We have also found these species hybridizing in Crow Creek (St. Francis River drainage) in St. Francis County. Johnston (1999) commented on the hybrid swarms that have been formed with congeners, especially when *C. lutrensis* contacts other species such as *C. venusta*, *C. spiloptera*, and *C. camura* (Hubbs and Strawn 1956; Page and R. L. Smith 1970; P. W. Smith 1979), most of which are correlated with habitat degradation (e.g., loss of habitat segregation, lack of visual cues). Johnston hypothesized that the sounds produced by most species of *Cyprinella* (Delco 1960; Stout 1975; Phillips and Johnston 2008a, b, 2009) during spawning act as premating isolating mechanisms. Hybrids may be formed if individuals of *C. lutrensis* are attracted to the calls of other species and spawn with congenerics. In addition, physical properties of the calls may be altered in degraded habitats if the preferred spawning habitat is unavailable (Forrest et al. 1993), and a breakdown of this premating isolating mechanism may occur (Johnston 1999). Noel Burkhead (pers. comm.) reported a hybrid swarm (*Cyprinella lutrensis* × *Cyprinella venusta*) at nearly every site where Red Shiners occurred in the upper Coosa River, Alabama.

CONSERVATION STATUS The Red Shiner is tolerant of altered or fluctuating habitats (Cross 1967) and appears to be locally abundant in large river habitats in Arkansas. Although the Red Shiner has reportedly displaced native *Cyprinella* in a few areas (Smith 1979), there is little evidence to suggest that it is a good competitor. Rather, because of its broad ecological tolerances, it seems able to take advantage of opportunities to repopulate habitats from which other species have been eliminated. It can survive and even become abundant in highly disturbed habitats. Cross and Collins (1995) discussed several adaptations of the Red Shiner that work together to account for its success in plains streams and its ability to thrive in disturbed environments where few other fishes can succeed. Matthews and Marsh-Matthews (2007) documented the loss or sharp decline of the Red Shiner in six of seven creeks

that are direct tributaries of Lake Texoma. Those authors attributed its loss to severe droughts and the inability of the Red Shiner to repopulate the creeks because of large numbers of centrarchids and other piscivores in the reservoir. In Arkansas, the Red Shiner is common and widespread in large rivers and currently in no danger.

Cyprinella spiloptera (Cope)
Spotfin Shiner

Figure 15.36. *Cyprinella spiloptera*, Spotfin Shiner. *Uland Thomas.*

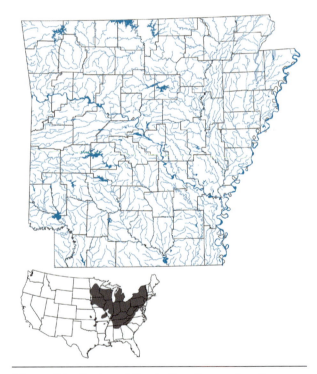

Map 32. Distribution of *Cyprinella spiloptera*, Spotfin Shiner, 1988–2019.

CHARACTERS A moderately slab-sided, small-eyed shiner (eye diameter going into head length 3.0–4.5 times and into snout length 0.9–1.3 times) possessing the characteristic diamond-shaped pigment outlines of the body scales (Fig 15.36). Snout long and pointed, its length usually greater than eye diameter. Eye size becomes proportionately smaller in larger individuals. Body depth going into SL 4.1–5.5 times. Mouth terminal or slightly subterminal and strongly oblique. Dorsal fin rays 8. Anal rays usually 8 (91% of Arkansas River specimens), rarely 9 (9% of Arkansas River drainage specimens). Pectoral rays usually 14 or 15 (13–16). Pharyngeal teeth 1,4–4,1. Gill rakers 7–10. Lateral line slightly decurved and complete, usually with 35 or 36 (34–38) scales. Predorsal scales 12–17. Body circumferential scales about 26–28. Scales around caudal peduncle modally 14. Dorsal origin slightly posterior to pelvic fin origin and closer to caudal base than to tip of snout. Dorsal fin with black blotch on last 2 or 3 membranes. Intestine short, with a single S-shaped loop; peritoneum silvery with dark speckles. The dorsal fin of breeding males is only slightly enlarged, and the distal margin is nearly straight. Breeding tubercles present on top of head, on lower jaw, and a few on scales above anal fin. A single row of tubercles on the rami of lower jaw. Breeding females usually without tubercles, but sometimes with scattered tubercles on top of head and dorsum. Maximum size 4.8 inches (122 mm).

LIFE COLORS   Body color silver with bluish cast to back and sides, caudal fin faintly pale yellowish. A narrow, prominent olivaceous lateral band on the caudal peduncle, sharply defined above and centered below the lateral line. No basicaudal spot or white patch at caudal base. A well-developed middorsal stripe extends from the back of the head to the base of the caudal rays. Scales of upper sides outlined with melanophores, creating a diamond-shaped pattern. Sides of head, lower sides, and venter are silvery white. Anterior interradial membranes of the dorsal fin lack melanophores, except in breeding males. Large young and adults with a posterior dusky or black blotch on last 2–3 membranes of dorsal fin. Breeding males steely blue with dark blue lateral band and scapular bar; snout olive yellow. Entire dorsal fin blackened; pelvic, anal, and caudal fins yellow, and all fins become milky.

SIMILAR SPECIES   Cyprinella spiloptera differs from C. whipplei, C. camura, and C. lutrensis in having 8 anal rays (vs. 9) and from C. venusta in lacking a prominent black basicaudal spot (vs. large spot usually well developed). Additionally, from C. whipplei, C. spiloptera has the melanophores in the anterior part of dorsal fin restricted to the ray margins (vs. melanophores present on all membranes of dorsal fin), a more pronounced dusky lateral band, modally 14 pectoral rays (vs. modally 15), and in distal margin of dorsal fin straight (vs. dorsal fin elongate with its distal margin rounded in breeding male). Snout of breeding male C. spiloptera is olive yellow, while the snout of the C. whipplei male is brick red. Trautman (1957) considered

the absence or scarceness of dark speckling on the membranes of the anterior half of the dorsal fin of C. spiloptera (except in breeding adults) to be the primary difference between that species and C. whipplei. See C. whipplei account for additional differences.

VARIATION AND TAXONOMY   Cyprinella spiloptera was described by Cope (1867a) as Photogenis spilopterus from the St. Joseph River, Michigan. Gibbs (1957a) recognized two subspecies, C. s. spiloptera in the eastern United States and Canada, and C. s. hypsisomata in the midwestern United States. Schaefer and Cavender (1986) reviewed rangewide variation in this species and did not recognize C. s. hypsisomata. Mayden (1989) considered spiloptera to be the sister species to the lineage containing all the eastern species (i.e., excluding the lutrensis group) of Cyprinella. Mayden et al. (2006a), based on analysis of the cytochrome b gene, found that C. spiloptera consisted of two clades that formed a clade sister to C. whipplei. Boschung and Mayden (2004) pointed out that even though the reproductive strategies of C. spiloptera and C. whipplei are very similar, they rarely hybridize in nature. Those authors speculated that perhaps behavioral, visual, or acoustic signals act as isolating mechanisms. Hybridization between C. spiloptera and C. lutrensis in Illinois was documented by Page and R. L. Smith (1970).

DISTRIBUTION   Cyprinella spiloptera occurs in Atlantic Slope drainages from the St. Lawrence River drainage of Canada south to the Potomac River drainage of Virginia; most of the Great Lakes drainages; and the Hudson Bay and Mississippi River basins from Ontario, Canada, to southeastern North Dakota and south to eastern Oklahoma and northern Alabama. This is a very rare species in Arkansas, with only three older records from widely separated localities considered valid (Map A4.33): (1) Benton County, Illinois River near the Oklahoma state line (Gilbert and Burgess 1980c), (2) Pope County, Illinois Bayou, 2 miles NW of Russellville, 4 specimens collected by A. H. Chaney and party on 4 November 1955 (UMMZ 177203), and (3) Izard County, Strawberry River near Wiseman, 5 specimens (3 specimens deposited in ASU Museum, 2 specimens in SAU Museum) collected on 23 May 1972, identified by George A. Moore and reported by Beadles (1974) (Map 32). We are aware of at least 10 other reported records for this species from Arkansas, but we have not verified any of those records. There are 20 vouchered records of C. spiloptera from eastern Oklahoma in four curated collections (KU, OKMNH, TU, and UMMZ), mostly from the 1940s and 1950s (we have not examined those specimens). The most recent Oklahoma record was 2 specimens from Cherokee County, collected by Jimmie

Pigg on 10 June 1992 (OKMNH 53008). There is one Oklahoma record from Lee Creek in Sequoyah County, 1 mile NW of Short, collected on 6 August 1971 (KUI 33455), and another Oklahoma record from the Poteau River in Le Flore County, 3 miles N of Howe, collected on 19 April 1948 (KUI 36715). It is possible that C. *spiloptera* could occur in the Arkansas portions of Lee Creek and the Poteau River, but no records currently exist from those streams. Surveys of the Poteau River by Cross and Moore (1952) and Lindsey et al. (1983) produced no records of *C. spiloptera*. It is also possible that waifs could occur in the upper Mississippi River of Arkansas because there are records from the lower Mississippi River in southeastern Missouri.

HABITAT AND BIOLOGY The Spotfin Shiner is found in moderate current over mixed gravel, sand, and silt substrates in medium to large streams and small rivers of moderate to high gradient. In Kentucky, it is usually found in or adjacent to rocky riffles in swift current or around cover in the form of logs and boulders (Burr and Warren 1986b). Hrabik et al. (2015) noted a preference for clear, cool water in Nebraska, where it has become more abundant in the Missouri River in the decades following construction of dams. It is a midwater species found in loose feeding aggregations during the day. Trautman (1981) documented habitat preferences for this species that indicated tolerance of a wide range of habitats. It can survive well in low-gradient and turbid water, is unaffected by siltation, and is typical of pools with current (Mueller and Pyron 2009). Horwitz (1982) also considered it somewhat tolerant of siltation and increased turbidity. Becker (1983) believed that it can apparently tolerate considerable siltation and domestic and industrial pollution. He concluded that it had expanded its range in Wisconsin since the 1920s because of environmental alterations caused by land use practices. It seems more tolerant of turbidity than the similar Steelcolor Shiner and is often found in larger streams than that species. Some authors reported that the Spotfin Shiner is most frequently found in pools or slow runs over a sand substrate (Gibbs 1957b; Horwitz 1982; Mayden 1989). Ross (2001) collected large numbers of Spotfin Shiners in Pickwick Reservoir (Tennessee River drainage) near the mouths of tributary streams. The Red Shiner is a competitor of the Spotfin Shiner and has displaced that species wherever the two have become syntopic in Illinois (Page and R. L. Smith 1970). Interestingly, Trautman (1981) described the Spotfin Shiner as aggressive.

Foods consist of aquatic and terrestrial insects, plant material, and small fishes, and food habits differ by region, size of fish, and season. Cross and Collins (1995) reported that the Spotfin Shiner in Kansas feeds mainly on small insects and crustaceans, usually captured as "drift" at the surface or in midwater. Starrett (1950a) found that the diet in Iowa consisted of aquatic insect nymphs and larvae (mainly ephemeropterans, trichopterans, and dipterans), but relatively high volumes of terrestrial insects were eaten during the summer months. Important food items in a Wisconsin stream were terrestrial and aquatic stages of caddisflies and mayflies, along with small crustaceans (amphipods and copepods) (Mendelson 1975). In that study, midge larvae made up most of the diet in November and December, and amphipods were important in the spring. Feeding occurred throughout midmorning and afternoon and peaked in the evening (Becker 1983). Ontogenetic changes in diet were characterized in an Indiana stream by Whitaker (1977). Individuals smaller than 20 mm SL fed primarily on midge larvae, pupae, and adults; fish between 20 and 40 mm SL added bryozoans to the diet; and fish larger than 40 mm SL consumed mainly caddisfly larvae and adults. Bottom ooze was important in the diet during fall, and carpetweed seeds contributed to the diet in winter and spring.

Hankinson (1930), Pflieger (1965), and Gale and Gale (1977) described spawning habits of the Spotfin Shiner, and most of the following reproductive information is from those sources. Spawning occurs from early June through late August and into early September. Most spawning fish are in their third summer, but some spawn in their second summer. Spawning sites are usually located near riffles and always in habitats with at least some current. Freshly submerged logs are preferred for spawning; spawning has also been reported in crevices in roots, rocks, and man-made objects. In swift water, the spawning crevices are located on the downstream side of the log, while in slow current the spawning crevices are on the side of the log facing the current. Male Spotfin Shiners defended territories that included one or more spawning crevices in the rocky substrate. Territories were about 20 inches (50 cm) or more in diameter. Agonistic behavior between rival males included grabbing the ventral fins of their opponents to drag them away from the spawning site. Attacks were frequently preceded by an erected-fin display. Courting males made display passes in which they swam slowly, sometimes undulating, along crevices where the eggs were to be deposited. If a female did not approach the crevice, the male would swim to a group of females and attempt to drive one of them to the crevice. The male circled the females as they approached the crevice. Occasionally two or more females accompanied the male on prespawning and spawning passes. Eggs were generally deposited in crevices longer than 0.8 inches (2 cm) on boulders or on the back of submerged logs. Spawning sessions lasted less than 1 hour with 156–896 eggs spawned. Spotfin Shiners are fractional spawners, spawning up to 12 times at intervals of 1–7 days (Gale and Gale 1977). Etnier and Starnes (1993) reported a single female deposited 7,474 eggs over an 8-hour period in an aquarium. Maximum egg diameter is 1.2 mm (Becker 1983). Eggs hatch in 5 days at 22°C (71°F) (Gale and Gale 1977). Males are larger than females (Pyron 1996b). Breeding adult *C. spiloptera* make sounds during courtship that probably help individuals recognize others of their own species (Winn and Stout 1960). In Wisconsin, fish grew to total lengths at ages of 0–3 of 25–30 mm (1.0–1.2 in), 45–59 mm (1.8–2.3 in), 55–67 mm (2.2–2.6 in.), and 70–86 mm (2.8–3.4 in), respectively (Becker 1983). Maximum reported life span is about 5 years (Starrett 1951).

CONSERVATION STATUS The Spotfin Shiner is exceedingly rare in Arkansas, and no recent records are known. Because it is one of the more difficult Arkansas cyprinids to identify correctly, it is easily confused with *C. whipplei*, and it may go unnoticed in collections. We consider it possibly extirpated in Arkansas.

Cyprinella venusta Girard
Blacktail Shiner

Figure 15.37. *Cyprinella venusta*, Blacktail Shiner. *Uland Thomas.*

CHARACTERS A large bluish-silver, deep-bodied, slab-sided shiner with a large distinctive squarish black blotch at the caudal base (Fig. 15.37). Body depth going into SL 3.6–5.0 times, with large fish being deeper bodied than small fish. Snout pointed. Head small, head length going into SL 3.7–4.6 times. Mouth large, slightly subterminal. Eyes small to moderate in size, eye diameter going into snout length 0.7–1.4 times and into head length 2.7–4.5 times; eye size becomes proportionately smaller in large fish. Dorsal rays 8; anal rays modally 8 (7–9). Pharyngeal

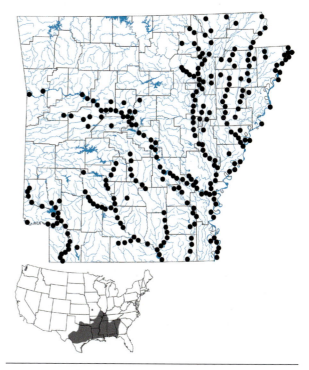

Map 33. Distribution of *Cyprinella venusta*, Blacktail Shiner, 1988–2019.

teeth 1,4–4,1 (rarely 0,4–4,0). Gill rakers 8–11. Lateral line slightly decurved and complete, usually with 36–38 (34–39) scales. Predorsal scales 15–20. Body circumferential scales 26–30. Dorsal fin origin slightly behind pelvic fin origin, closer to caudal base than to tip of snout. Intestine short; peritoneum silver with dark speckles. Breast and belly completely scaled. Breeding males with numerous tubercles on head, in front of and below the eye, along the lower jaw, on dorsal and lateral body scales, and on dorsal and anal fin rays. Tubercles often develop on some pectoral and pelvic rays. Usually 2.5–4.5 inches (63.5–114.3 mm) TL. Maximum reported size 7.5 inches (191 mm) TL (Page and Burr 2011).

LIFE COLORS Back and upper sides olive to blue green with a silvery lateral band (dark and best-developed posteriad in preserved specimens). A scapular bar is weakly developed or absent. Basicaudal blotch usually large and dark, but individuals are occasionally found in which the distinctive blotch is faint or absent. Dorsal fin well pigmented with posterior fin membranes dusky; other dorsal membranes thickly dusted with fine dark specks. As in other members of the genus *Cyprinella*, the sides have large, diamond-shaped scales outlined in black. Breeding males become bright steel blue dorsally and laterally, the caudal fin becomes dusky yellow, the pectoral fins are clear to yellow, and the pelvic and anal fins become milky yellow.

SIMILAR SPECIES   The Blacktail Shiner is similar to *Cyprinella spiloptera*, *C. whipplei*, and *C. lutrensis*. It differs from those species (and most other Arkansas cyprinids) in having the prominent black basicaudal spot (vs. spot absent). The caudal spot may be faint in specimens from turbid waters. Specimens lacking a distinct caudal spot may be distinguished from *C. lutrensis* in having 8 anal rays (vs. 9), 1,4–4,1 pharyngeal teeth (vs. 0,4–4,0), and usually fewer than 36 lateral line scales (vs. usually 36 or more). It further differs from *C. whipplei* in typically having 8 anal rays (vs. 9).

VARIATION AND TAXONOMY   *Cyprinella venusta* was described by Girard (1856) from the Rio Sabinal, at Sabinal, Uvalde County, Texas. Mayden (1989) documented 10 different names that have been used for *Cyprinella venusta*, and Boschung and Mayden (2004) reviewed its taxonomic history. Gibbs (1957c) recognized three subspecies of the Blacktail Shiner: *Cyprinella v. venusta* (Mississippi River basin and drainages to the west), *Cyprinella v. cercostigma* (Gulf drainages [except Mobile Bay] east of Mississippi River), and *C. v. stigmatura* (upper Alabama and Tombigbee river systems). Arkansas is inhabited by the nominate form, *C. v. venusta*. Kristmundsdottir and Gold (1996) and Gilbert (1998) supported the elevation of at least two clades to full specific status. Boschung and Mayden (2004) concluded that while it is possible that three or more subspecies or species may eventually warrant recognition within what is currently recognized as *C. venusta*, such changes must be supported by additional molecular and morphological evidence. Mayden (1989) considered *C. venusta* and *C. galactura* sister species, and Mayden et al. (2006a) recognized *C. venusta* as sister to a clade comprising *C. whipplei* and *C. spiloptera*.

DISTRIBUTION   The Blacktail Shiner occurs in lower Mississippi River drainages from Southern Illinois to Louisiana, and in Gulf Slope drainages from the Suwannee River of Georgia and Florida, west to the Rio Grande drainage of Texas. It is introduced into the Missouri River drainage of Missouri. In Arkansas, it is common in all large Coastal Plain rivers but extends into a few streams in the Ozarks such as the Strawberry, Black, and middle White rivers, with older records from the Little Red River (Map 33 and Map A4.34). It is found throughout the Arkansas River but becomes increasingly rare upstream from Faulkner County. It occurs throughout the Red River in Arkansas but was considered uncommon (Buchanan et al. 2003).

HABITAT AND BIOLOGY   The Blacktail Shiner is a schooling species that inhabits medium to large streams and rivers, sluggish ditches over sand bottoms, and occasionally oxbow lakes (e.g., Lake Chicot) that are sparsely vegetated. Even though it prefers some current, it is also found in some upper Arkansas River Valley impoundments. Buchanan (2005) reported it from 12 Arkansas impoundments. It occurs in moderately clear to very turbid waters. In Mississippi, it is usually found in the middle of the water column in pools and runs, usually where there is moderate to swift current over sand or gravel substrates (Ross 2001). Shallow littoral areas of sand and gravel were important nursery areas for late larval stages (Brenneman 1992). Casten (2006) studied life history features in four Alabama streams that differed in amount of environmental disturbance and concluded that *C. venusta* has the ability to alter life history parameters in harsh environments, which may contribute to its persistence in some disturbed streams. It is tolerant of relatively high water temperatures but selectively avoids areas having extreme temperatures (Ross 2001). It is not tolerant of low dissolved oxygen levels (Matthews 1987b), possibly contributing to its preference for flowing-water habitats.

The Blacktail Shiner is largely a diurnal invertivore but has sometimes been considered an omnivore. It feeds both on benthic organisms and on drift (Simon 1999a). Foods are primarily terrestrial insects and plant material (Hambrick and Hibbs 1977), with feeding most active just before sunset. Hale (1963) reported that *C. venusta* is an omnivore, with its diet consisting primarily of aquatic and terrestrial insects, and with some algae and crustaceans also ingested. She also found similarities in the diets of *C. venusta* and *C. lutrensis*, indicating competition between the two cyprinids, but noted that *C. venusta* takes more food items from the stream bottom than *C. lutrensis*. Intraspecific competition for food apparently increases with increased population density. Dekar et al. (2014) found that *C. venusta* stocked at high densities in a stream mesocosm exhibited an increase in diet breadth (compared with those stocked at low densities) that included lower-quality resources. Felley and Felley (1987) noted seasonal changes in diet in Louisiana. During the rainy months, surface prey made up 30% of the diet and benthic prey 54%, with organic detritus composing the remaining 16%. In dry months, surface prey made up only 12% of the diet, and benthic prey made up 15%, midwater prey 2%, and organic detritus 71%. Baker and Foster (1994) observed Blacktail Shiners following foraging Northern Hog Suckers to feed on food items stirred into the current by the suckers.

The Blacktail Shiner is a crevice spawner (Pflieger 1997). In Missouri, spawning occurs from June to August with eggs deposited in crevices of submerged objects (Pflieger 1997). Eggs may sometimes be deposited under small boulders or large cobble. In Mississippi, a long spawning season occurs from March through early October, with clutch sizes ranging from 139 to 459 (Heins and Dorsett 1986). Spawning over the long breeding season occurred at water temperatures between 19°C and 29°C. Sexual maturity was reached in some females as small as 32 mm SL, and all females greater than 42 mm SL were sexually mature. Spawning crevices are usually located in flowing water at current speeds of approximately 30 cm/s (Ross 2001), but reservoir populations apparently have lower preferred crevice current speeds (sometimes as low as 7 cm/s) (Baker et al. 1994). Heins (1990) observed breeding behavior in aquaria. Large males are territorial and defend selected spawning crevices from other males, with considerable fighting occurring. Heins (1990) and Rabito and Heins (1985) suggested that males may actually ejaculate sperm into the spawning crevice during their prespawning contact runs prior to the appearance of the female. Heins (1990) also believed the purpose of these males swimming alone through the crevice was to indicate the spawning area to females. He observed both large and small nonterritorial males (sneaker males) making spawning runs through the crevices when they were not defended and suggested that the nonterritorial males were also depositing sperm. Spawning females produce sounds to which males respond, permitting males to recognize females of their own species. Although able to distinguish their own sounds from those of the related Red Shiner males, hybridization sometimes occurs (see species account for *C. lutrensis*). Machado et al. (2002) studied microgeographical variation in ovum size in relation to streamflow and found a strong, positive correlation between egg size and stream discharge (variable mean annual runoff) among populations of *C. venusta*. They hypothesized that this variation may have resulted from selection for larger egg size; therefore, larger offspring size may be related to the greater environmental fluctuation or a greater number of floods in streams with higher runoff. Sexual maturity is usually attained at age 1, and maximum life span is 4.5 years (Littrell 2006). Tumlison et al. (2016) reported the parasitic copepod *Lernaea cyprinacea* from *C. venusta* from the Arkansas River in Desha County.

CONSERVATION STATUS The Blacktail Shiner is widespread in Arkansas and its populations are secure.

Cyprinella whipplei Girard
Steelcolor Shiner

Figure 15.38. *Cyprinella whipplei*, Steelcolor Shiner. *Uland Thomas.*

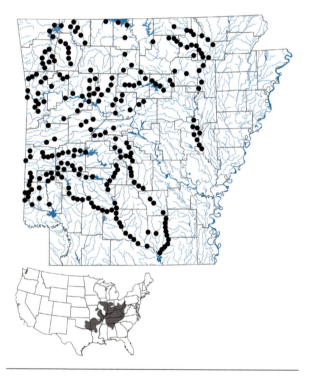

Map 34. Distribution of *Cyprinella whipplei*, Steelcolor Shiner, 1988–2019.

CHARACTERS A large, deep-bodied (body depth going 3.7–4.3 times into SL), slab-sided shiner with a terminal, oblique mouth, pointed snout, and small eyes, eye diameter going 3.2–4.4 times into head length and 0.8–1.3 times into snout length (Fig. 15.38); smaller individuals have proportionately much larger eyes relative to HL. Head length going into SL 4.0 to 4.5 times. Dorsal rays 8; anal rays modally 9 (rarely 8 or 10). Pectoral rays usually 14–16 (13–17). Pharyngeal teeth 1,4–4,1. Gill rakers 10–12. Lateral line decurved and complete, with 37–39 scales. Predorsal scales usually 16 or 17; body circumferential scales usually 24–26. Breast and belly fully scaled. Dorsal fin situated behind the pelvic fin origin, its origin closer to caudal base than to tip of snout. Sides with diamond-shaped outlines of scales. Lateral band on the caudal peduncle extending well above the midline and fading gradually on its dorsal margin and centered on lateral line. Anterior interradial membranes of the dorsal fin sprinkled with melanophores; black blotch present on last 2 or 3 membranes (blotch is faint or absent in small young). Intestine short; peritoneum silvery, usually heavily sprinkled with black pigment. Dorsal fin of nuptial males greatly enlarged and flaglike with its distal edge convex. Breeding males with fine tubercles on head and snout and extremely tiny tubercles on the nape. Tuberculation is also extensive on body and fins. Tubercles absent in females. Maximum size 6.3 inches (160 mm), but generally smaller than 4.5 inches (114 mm).

LIFE COLORS   Back and sides a metallic silver to blue gray. No pigment along anal base or midventral caudal peduncle. No distinct caudal spot. Adults with dark blotch on last 2 or 3 membranes of dorsal fin. Breeding males with steel blue on back and sides, snout pink or red, dorsal fin blackened, and other fins lemon yellow fringed with white.

SIMILAR SPECIES   The Steelcolor Shiner closely resembles *C. spiloptera* but differs in having 9 anal rays (vs. 8). Additional characters for separating these very similar species were provided by Etnier and Starnes (1993):

> In fresh specimens *spiloptera* has a faint wash of yellowish pigment on the caudal fin; this fin is clear in *whipplei*. In *spiloptera* the lateral stripe on the caudal peduncle extends little above the horizontal myoseptum, has a rather definite dorsal margin, and the oblique myosepta within the lateral stripe appear as narrow white lines; in *whipplei* the lateral stripe on the peduncle extends well above the midline and fades more gradually dorsad, and pale myosepta are not visible within the stripe. In *spiloptera* (except nuptial males) the interradial membranes of the dorsal fin lack pigment except for the spot on the posterior membranes; in *whipplei* all dorsal fin interradial membranes are liberally sprinkled with melanophores in juveniles and adults. Nuptial males of *whipplei* have greatly enlarged dorsal fins, and nape tubercles are abruptly smaller than those on the head; in *spiloptera* the dorsal fin does not enlarge, and anterior nape tubercles grade gradually in size with those of the head.

Cyprinella whipplei differs from *C. lutrensis* in having a black posterior dorsal fin blotch (vs. dorsal fin blotch absent), from *C. venusta* in lacking a discrete basicaudal spot (vs. spot present), and from *C. camura* in lacking the clear white band at the caudal base (vs. white band usually present), having 38 or 39 lateral line scales (vs. usually 36 or 37 in *C. camura*), dorsal and caudal fins not red (vs. dorsal and caudal fins with some red in *C. camura*), and in having a gap between snout and head tubercles. Gibbs (1961) concluded that *C. whipplei* and *C. camura* are not distinguishable by meristic or morphometric characters and can be reliably separated only by caudal pigmentation and tubercle pattern of breeding males.

VARIATION AND TAXONOMY   Originally described by Girard (1856) as *Cyprinella whipplii*, it was subsequently placed in genus *Notropis*, and then returned to *Cyprinella* as *C. whipplei*. The type locality was given by Girard as Sugar Loaf Creek, a tributary of the Poteau River, believed to be in Sebastian County, Arkansas. There is some uncertainty about the precise location of the type locality, and it may actually be in Oklahoma (Gilbert 1978). Gibbs (1963) studied variation but did not recognize subspecies. Both Gibbs (1963) and Mayden (1989) hypothesized the closest relatives of *C. whipplei* are *C. analostana* and *C. chloristia* of the Atlantic Slope drainages plus *C. camura*. Mayden et al. (2006a) recovered *C. whipplei* as sister to *C. spiloptera*. There is some variation in coloration of the peritoneum in Arkansas populations. In most populations, the peritoneum is heavily sprinkled with dark chromatophores, but in specimens from the Middle Fork of the Little Red River, the peritoneum is only lightly sprinkled with chromatophores. Individuals from the Little River have a completely silvery peritoneum or have only a few scattered (3–5) chromatophores.

DISTRIBUTION   The Steelcolor Shiner occurs in the Mississippi River basin from West Virginia and Ohio to Missouri and eastern Oklahoma and south to northern Louisiana and Alabama; it is largely absent from the Mississippi Embayment. It is also known from one Gulf Slope drainage, the Black Warrior system (Mobile Bay drainage) of Alabama. In Arkansas, it is confined principally to and widely distributed in the upland regions of the Ozarks and Ouachitas; however, in the Ouachita and Saline rivers, *C. whipplei* extends far onto the Gulf Coastal Plain below the Fall Line (Map 34).

HABITAT AND BIOLOGY   The Steelcolor Shiner is most common in large to medium-sized upland streams with permanent flow, clear water, and gravel bottoms. It is generally absent from the large mainstem rivers (Arkansas, Mississippi, and Red). It is frequently found in small streams (but not extreme headwaters) with permanent flow and is most commonly collected in current immediately below riffles or in the riffles, but it also occurs in pools with some current. Most streams where it is found have heavy growths of water willow along the margins. It is rather intolerant of turbidity and siltation. In the Illinois River, Oklahoma, it was most common in pools and slower water over sand and gravel substrates; average current speed in that habitat was 6–8 cm/s (0.2–0.26 feet/s) (Felley and Hill 1983). Despite its preference for current, it is easily kept in aquaria and does well under those artificial conditions if some water flow is provided. Buchanan (2005) reported *C. whipplei* from six Arkansas reservoirs.

Cyprinella whipplei is a sight feeder on material drifting with the current, but it also occasionally feeds on benthic invertebrates (Pflieger 1997). Foods include terrestrial insects found near the surface and small invertebrates found at varying depths. Substrate food items include immature insects, small crustaceans, mites, and earthworms. In the Illinois River, Oklahoma, *C. whipplei* fed mainly on terrestrial invertebrates during summer and fall, and added organic detritus to the diet in winter and spring (Felley and Hill 1983). In Indiana, foods included caddisfly larvae and adult dipterans during summer, but it switched to bottom ooze and midge larvae and pupae in fall months (Whitaker 1977). Seeds of terrestrial plants are sometimes ingested when food is scarce. This cyprinid has been caught by angling in Arkansas: two specimens were caught on an artificial rooster tail lure in Beaver Lake on 11 August 2015 by J. M. Walker (identified by TMB from photograph).

Spawning occurs from early June until mid-August, with the eggs deposited under loose bark or in crevices of submerged logs and tree roots in moderate to swift current (Pflieger 1965). Bark separating from a newly submerged tree trunk seems to be an especially favored site (Ross 2001). Spawning crevices may be located on the top or bottom surfaces of a log. The spawning site is usually in the middle to upper part of the water column. Males are highly aggressive during the breeding season. They establish poorly defined territories over the breeding site, and when a female approaches the spawning crevice, she is met by one or more males, with one male usually succeeding in driving the others away. The male is eventually able to press the female against the log where both vibrate rapidly. This spawning act is repeated two or three times farther up the log. Often three or four pairs of fish spawn over the same log simultaneously. Maturity is reached by the second summer (at 49 mm SL for males and 38 mm for females), but most spawning adults are usually in their third year. Only a few individuals survive into their fourth summer.

CONSERVATION STATUS   The Steelcolor Shiner has declined in some areas near the margins of its range. Some Oklahoma populations have shown steady declines

(Rutherford et al. 1987). Burkhead and Jenkins (1991) considered it threatened in Virginia. Ross (2001) considered it of special concern in Mississippi, as did Boschung and Mayden (2004) in Alabama. Its populations in Arkansas are widespread and stable, and it does not appear to be in any danger in this state.

Genus *Cyprinus* Linnaeus, 1758—True Carps

The Eurasian genus, *Cyprinus*, with about 24 species, is characterized by barbels present posteriorly on the upper jaw, scaled breast, lateral line scales 32–41, dorsal and anal fins preceded by a serrate spine and several anterior rudiments, dorsal fin rays 15–23, anal fin rays 4–6, pectoral fin rays 14–17, pelvic fin rays 8 or 9, and gill rakers 21–27. Pharyngeal teeth are molariform and 1,1,3–3,1,1. Coloration is usually brassy to yellowish orange, including the lower fins. Numerous varieties exist, such as genetic mutants (e.g., mirror carp, Israeli carp), and ornamental koi with orange, red, yellow, black, and white varieties. A single introduced species, *Cyprinus carpio*, inhabits Arkansas.

Cyprinus carpio Linnaeus
Common Carp

Figure 15.39. *Cyprinus carpio*, Common Carp. *Chad Thomas.*

CHARACTERS   A large, stout, laterally compressed cyprinid species with a long dorsal fin and an arched back (Fig. 15.39). The dorsal fin is highest anteriorly, but becomes shorter and more rounded posteriorly. The first ray of the dorsal and anal fins is modified into a large, hard, serrated spine. Caudal fin moderately forked. Head and eye small; eyes placed fairly high on head. Snout long. Mouth of moderate size, somewhat protractile, with 2 barbels on each side of upper jaw. Molarlike pharyngeal teeth in 3 rows 1,1,3–3,1,1 (Fig. A3.20). Gill rakers 21–27. Scales large and cycloid. Lateral line complete, with 32–41 scales. There have been genetic mutations that resulted in some forms of this species having a few large scales (e.g., mirror carp) and others

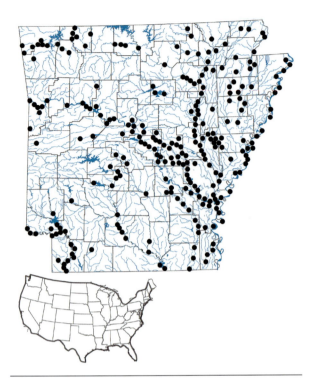

Map 35. Distribution of *Cyprinus carpio*, Common Carp, 1988–2019. Inset shows introduced range (outlined) in the contiguous United States.

completely lacking scales (leather carp). Dorsal fin rays usually 18–21 (15–23). Anal rays 4–6; pectoral rays 14–17; pelvic rays 8 or 9. Peritoneum dusky. Breeding males develop fine tubercles on head and pectoral fins. Adults usually 12–30 inches (305–762 mm) long and weigh 1–20 pounds (0.5–95 kg); maximum length 40–48 inches (1,016–1,219 mm), maximum weight 40–60 pounds (18–27 kg). The Arkansas state angling record from Lake Hamilton in 1985 is 53 pounds (24.0 kg).

LIFE COLORS   Body color is olivaceous to slate gray above, grading to greenish, bronze, or brassy gold laterally with a yellowish-white venter. Each scale is dark-edged with a black spot at the base. Fins typically are orange, golden, or light olive in color. Males becoming darker during the breeding season. There are variously colored ornamental varieties of carp known as koi which are red, orange, black, and white and can be quite expensive in the aquarium trade

SIMILAR SPECIES   The combination of 2 barbels at each end of the upper jaw, a stiff, serrated spine at the dorsal fin origin, and the dark-edged scales distinguishes the Common Carp from all other native and introduced cyprinids in Arkansas. The above combination also separates the Common Carp from all suckers.

VARIATION AND TAXONOMY   *Cyprinus carpio* was described by Linnaeus (1758) from "Europe." Several genetic strains of *C. carpio* are recognized as separate varieties. The normally scaled Common Carp with its body covered with thick cycloid scales is by far the most abundant variety found in Arkansas waters. The "leather carp" is a scaleless form and is quite rare in the state. The "mirror carp" has huge scales scattered over the body and is also rare. The offspring of the mirror carp and the leather carp are mostly fully scaled (Becker 1983). Mirror and leather carp constitute less than 2% of wild carp populations. Wohlfarth and Laharan (1963) discussed the genetic basis for these varieties.

There is some confusion regarding the status of a fourth carp variety known as the "Israeli carp," formerly stocked in Arkansas by the Game and Fish Commission to control aquatic vegetation. Most authors of state fish books have failed to mention this variety of carp, and one author (P. W. Smith 1979) reported that the "so-called Israeli carp" is morphologically indistinguishable from the ordinary carp. We disagree with that assertion and believe that the fast-growing cultured form known as the Israeli carp is quite readily distinguishable morphologically from the other three varieties of carp found in Arkansas waters. The Israeli carp most closely resembles the mirror carp in having the greatly enlarged, scattered, mirrorlike scales, but the Israeli carp has a distinctly different body shape, being much deeper bodied and thicker across the back than the wild mirror carp, which has a body conformation like the ordinary scaled variety (William E. Keith, pers. comm.). The predorsal area of the Israeli carp is strongly arched and the occipital region is noticeably concave. Additional information on Israeli carp was provided by Mayo Martin of the U.S. Fish and Wildlife Service Fish Farming Experimental Station at Stuttgart, Arkansas, and by Rom Moav, Balfour Hepher, and Shimon Tal of Israel. Israeli biologists originally crossed cultured carp varieties from Yugoslavia and the Netherlands to produce the "Israeli carp" variety. They suggested that this variety should more properly be called "improved European carp." They agree that this form is morphologically and probably physiologically distinct from any of the wild carp varieties. The Israeli carp was originally brought into the United States by fisheries biologists at Auburn University in the 1950s. The Arkansas Game and Fish Commission obtained its initial stock of Israeli carp from Auburn in the late 1950s.

Japanese breeders have selectively bred at least 19 different brightly colored varieties of carp known as koi. Koi come in a great variety of color combinations including red, white, blue, orange, yellow, and black and are usually kept as ornamental specimens in backyard ponds. In 1982 the aquaculture industry in Arkansas produced 17,000 pounds (7,711 kg) of koi valued at $85,000.

Cyprinus carpio hybridizes with the Goldfish, and the hybrids are fertile. In Ohio, 30–90% of the Common Carp–Goldfish catch were hybrids, and introgressive hybridization had occurred (Trautman 1981). The AGFC maintains state fishing records for Common Carp, the normally scaled form (state record: 36 pounds [16.3 kg] from Lake Conway in November 1985), and for Israeli carp (state record: 53 pounds [24 kg] from Lake Hamilton in March 1985).

DISTRIBUTION   Balon (1995) proposed that *C. carpio* evolved in the Caspian Sea, spread to the Black and Aral seas, then moved eastward throughout mainland Asia. It spread westward to the Danube River. This Eurasian native was first imported to North America in 1831 and liberated in the Hudson River, New York (Courtenay et al. 1984). In the late 1870s the Common Carp was quickly disseminated across the country (Jenkins and Burkhead 1994) for recreational and food purposes (Panek 1987). It was introduced into Arkansas in the 1880s, and by the 1890s its introduction was considered a serious mistake because of numerous negative effects on invaded ecosystems (Jenkins and Burkhead 1994; Weber and Brown 2009). Large commercial catches were reported in the upper Mississippi River basin by the early 1900s (Forbes and Richardson 1920; Hildebrand and Schroeder 1928). Black (1940) considered it abundant throughout the lowlands of Arkansas. Regardless of its effects on the aquatic ecosystem, it is now well established throughout Arkansas (Map 35).

HABITAT AND BIOLOGY   The invasive Common Carp inhabits a variety of aquatic habitats throughout Arkansas and is found in reservoirs, upland streams, and lowland rivers and streams. *Cyprinus carpio* becomes most abundant in soft-bottomed, weedy pools of streams, although it probably adapts to a wider variety of conditions than almost any native North American fish (Becker 1983). It is tolerant of all bottom types and is found in clear or turbid waters.

This omnivorous species eats insect larvae, crustaceans, organic sewage, plant material, and a variety of other organisms. In Iowa, aquatic insects were the most important items in the diet, but seeds and leafy plant parts were also intentionally consumed (Moen 1953). Feeding occurs most heavily at dawn and dusk, and Pflieger (1997) considered taste more important than sight in locating food. Common Carp are generally considered a nuisance or detrimental species because of their habit of rooting in the bottom during their aggressive feeding behavior, thereby causing increased turbidity. Such turbidity decreases light penetration, lowering productivity of algae and aquatic macrophytes. In addition, silt from the feeding activities may suffocate eggs of other fishes.

Many persons believe that the presence of Common Carp even in small numbers is harmful to gamefish populations. Competition does exist between young Largemouth Bass and Common Carp of all ages for the available food (Becker 1983). However, Common Carp become most abundant and a nuisance in waters where physical and/or water quality degradation of habitat has already occurred and where such degradation has substantially reduced native fishes, particularly centrarchids and other predators. The abundance of Common Carp in many Arkansas waters is probably due to changed ecological conditions such as high stream temperatures and increased siltation, which favor the Common Carp and are detrimental to gamefish populations. Therefore, Common Carp are probably a symptom and not a cause of deteriorating aquatic environments.

Spawning generally lasts from at least April through June at water temperatures between 65°F and 75°F (18.3–23.9°C). Pflieger (1997) reported that ripe males have been taken in Missouri as late as September, indicating that spawning may occur into early fall. Common Carp congregate in small groups typically consisting of one pair, or of a female and 2–4 males, or of one female with 6 or 7 males (Becker 1983). Each spawning group gathers in a shallow (12 inches [305 mm] or less), heavily vegetated area often over temporary floodplains. Females are prolific, scattering thousands of adhesive eggs over debris and vegetation on the bottom. No parental care is given. Fecundity is high but variable, depending on the size of the female. Anywhere from 50,000 to 620,000 eggs are discharged by a single female. Age and growth of Common Carp from Beaver Reservoir were studied by Kilambi and Robison (1978). Similar to native fishes, early life history stages of carp are zooplantivorous and later undergo an ontogenetic diet shift to benthic invertebrates (Britton et al. 2007; Rahman et al. 2009; Weber and Brown 2013). Howell et al. (2014) found that resource competition between age 0 carp and native fishes is most likely to occur during the early summer. Young individuals mature in 3 years. Although Common Carp are typically shunned as food fish because they are extremely bony, the meat of carp is excellent if properly prepared. Smoking and baking are primary methods of cooking them. Common Carp formed an important part of the wild commercial fishery of Arkansas in 1988, ranking behind only the buffalofishes and catfishes in number of pounds taken annually. Although current commercial fishing records are not available, it is likely still an important component of the commercial catch in Arkansas.

CONSERVATION STATUS The Common Carp is widespread and abundant in Arkansas.

Genus *Erimystax* Jordan, 1882— Slender Chubs

The North American cyprinid genus *Erimystax* is widely distributed in the Mississippi River basin and comprises five species. Those species were formerly included as a subgenus in the genus *Hybopsis*, but *Erimystax* was elevated to generic status by Mayden (1992). Mayden (1989) hypothesized that *Erimystax* is the sister group of *Phenacobius*, an assessment subsequently supported by analysis of 12S and 16S mtDNA sequences (Simons and Mayden 1999; Simons et al. 2003). Mayden (1989) also presented a phylogenetic treatment of the relationships within *Erimystax* and proposed a monophyletic group, sister to *Erimystax x-punctatus*, consisting of *E. cahni*, *E. insignis*, and *E. dissimilis* (*E. dissimilis harryi* was then recognized at the subspecific level). Simons (2004) investigated the phylogenetic and phylogeographic relationships of the members of *Erimystax* using sequences of the mitochondrial cytochrome *b* gene. Genetic distances among *Erimystax* species were high, suggesting that all speciation events occurred during the Miocene.

Characters defining this genus include a barbel at each corner of the mouth, 0,4–4,0 pharyngeal teeth, 7 anal fin rays, a hardened pad on the cheek area of breeding males, nuptial tubercles are minute or absent on top of head, scales lack basal radii, and a lack of bright breeding coloration (Hubbs and Crowe 1956; Etnier and Starnes 1993). Dimmick (1988) examined barbel structure of four species of *Erimystax* (*E. harryi* was not included) and found that the barbel ultrastructure was similar in all. All members of this genus require stream habitat with gravel or rocky substrate and little or no sedimentation (Simons 2004). Harris (1986) provided a complete synonymy and taxonomic history of *Erimystax* and described morphological variation supporting recognition of *Erimystax harryi* as distinct from *E. dissimilis*, an action supported by subsequent molecular data (Simons 2004). Two species of *Erimystax* occur in Arkansas: *E. harryi* and *E. x-punctatus*.

Erimystax harryi (Hubbs and Crowe)
Ozark Chub

Figure 15.40. *Erimystax harryi*, Ozark Chub. *Uland Thomas.*

Map 36. Distribution of *Erimystax harryi*, Ozark Chub, 1988–2019.

CHARACTERS A slender, large-eyed, small-barbeled minnow with a series of dark roundish blotches laterally, and a stripe along midline of back consisting of a series of alternating light and dark areas (Fig. 15.40). Snout blunt and rounded, projecting beyond upper lip. Mouth small, horizontal, with a small conical barbel on each side; upper jaw not extending past front of eye. Eye diameter less than snout length. Front of dorsal fin base closer to tip of snout than to base of caudal fin. Dorsal rays 8; anal rays 7. Lateral line scales usually 46–49 (43–52). Intestine long, with several loops; peritoneum black. Pharyngeal teeth 0,4–4,0. Breeding males with tubercles on head and front part of body. Maximum size about 4.7 inches long (120 mm) (Harris 1986), but TMB found a 5.1-inch (150 mm) TL individual in the Buffalo River in Marion County.

LIFE COLORS   Body color greenish yellow dorsally, with silver sides, grading to white on the belly. A series of dark oval blotches along the midsides. Middorsal stripe composed of alternating golden and dusky areas. There are no bright breeding colors.

SIMILAR SPECIES   *Erimystax x-punctatus* is similar, but *E. harryi* generally has 44–50 scales in the lateral line (vs. 38–43) and a series of dark spots along the midside, sometimes forming a dark lateral band in preserved specimens (vs. side with scattered x-shaped markings). In *E. harryi*, the middorsal stripe is composed of alternating light and dark areas (vs. a uniformly colored middorsal stripe in *E. x-punctatus*). It is distinguished from Arkansas *Hybopsis* species in having more than 42 lateral line scales (vs. fewer than 40) and 0,4–4,0 pharyngeal teeth (vs. 1,4–4,1).

VARIATION AND TAXONOMY   The Ozark Chub was originally described as a subspecies, *Hybopsis dissimilis harryi*, by Hubbs and Crowe (1956) of subgenus *Erimystax* from the White River in Barry County, Missouri. Harris (1986) reviewed the synonymy and taxonomic history of subgenus *Erimystax* and used morphological characters to support elevation of the Ozark subspecies to full specific status as *Hybopsis harryi*. Subgenus *Erimystax* was removed from genus *Hybopsis* and elevated to generic status by Mayden (1992). This designation was widely accepted and subsequently supported by genetic evidence (Simons 2004; Mayden et al. 2006a). Based on analysis of the cytochrome *b* gene, Mayden et al. (2006a) resolved *E. harryi* as sister to *E. dissimilis*.

DISTRIBUTION   The Ozark Chub occurs only in the upper White River system of Arkansas and Missouri, and in the St. Francis River in Missouri. In Arkansas, it formerly occurred throughout the upper and middle portions of the White River (including the Kings and Buffalo rivers) to its confluence with the Black River, in the Black River drainage (Spring, Eleven Point, and Current rivers), and in the Little Red River (Map A4.37). It is not known from the St. Francis River drainage in Arkansas. There are no recent records from the upper White River above Beaver Lake, or from the Little Red River (Map 36), and it may be extirpated from those areas. Strange and Burr (1997) examined mtDNA variation in the context of a Pleistocene vicariance hypothesis to explain the distribution of *E. dissimilis* and *E. harryi*. They interpreted interspecific patterns to be a result of either a pre-Pleistocene vicariant event or a Pleistocene dispersal of *E. dissimilis* from Appalachia to the Ozarks. The former hypothesis was supported by molecular DNA sequence data (Simons 2004).

HABITAT AND BIOLOGY   The Ozark Chub inhabits large, swift streams and rivers over clean gravel or rock bottoms with little or no sedimentation. It is very intolerant of turbidity and siltation and does not adapt to reservoirs. Pflieger (1997) found it most often near riffles or in pools with noticeable current, and such is the case in our experience in Arkansas. It is usually found close to the substrate in swift current, and Pflieger (1997) noted that it is often found with *Campostoma oligolepis*, *Cyprinella galactura*, and *Luxilus zonatus*.

Davis and Miller (1967) suggested that the Ozark Chub feeds primarily by sight, assisted by external taste buds. Harris (1986) stated that it is primarily a benthic herbivore/detritivore with plant material (primarily filamentous algae) and periphyton (detritus, diatomaceous algae, and bacteria) making up slightly more than 90% of the annual dietary biomass. Examination of gut contents by TMB revealed that it fed largely on algae and detritus in the Buffalo River in January and added a small amount of benthic immature insects to the diet in spring and summer months in the Spring River.

Spawning in the Buffalo River, Arkansas, occurs from early or mid-April to late May and appears to be initiated by high water levels and warming temperature. Spawning occurred in the riffles in 15.7–23.6 inches (40–60 cm) of water over clean, small to medium gravel substrate when water temperatures reached more than 60°F (>15°C) and river discharge was at or near the annual maximum. Maximum age for Ozark Chubs in the Buffalo River was 42 months.

CONSERVATION STATUS   Pflieger (1997) reported that *E. harryi* has declined drastically in the White, Black, and St. Francis rivers of Missouri following reservoir construction. It has also declined in Arkansas and may have been extirpated from the upper White River above Beaver Lake. The Ozark Chub is still widespread in its large-stream habitat in the middle White River drainage. Because of its decline and generally small population sizes, we agree with the AGFC that it is a species of special concern in Arkansas.

Erimystax x-punctatus (Hubbs and Crowe)
Gravel Chub

Figure 15.41. *Erimystax x-punctatus*, Gravel Chub. *Uland Thomas*.

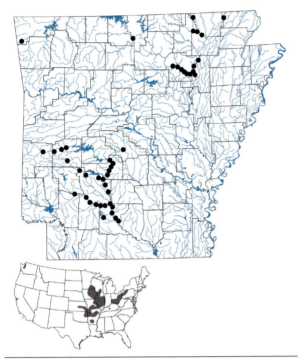

Map 37. Distribution of *Erimystax x-punctatus*, Gravel Chub, 1988–2019.

CHARACTERS   A slim, terete, barbeled, moderately large minnow with scattered x-shaped markings over the back and sides (Fig. 15.41). Head moderately large, head length going into SL nearly 4 times; preorbital length greater than postorbital head length, except in smallest young. Eye large, eye diameter going 3–3.3 times into head length, but eye diameter less than snout length. Snout blunt and rounded, projecting well beyond upper lip. Mouth small and inferior. Maxillary barbel small but conspicuous. Pharyngeal teeth 0,4–4,0, slender, and strongly hooked at tip. Lateral line complete, with 38–45 scales. Dorsal fin origin in front of pelvic fin origin. Dorsal fin falcate, with 8 rays. Anal rays 7. Pectoral fin rays 13–16. Gill rakers short and conical. Intestine long (longer than TL) with several loops. Peritoneum black. Head of breeding male covered with many minute tubercles; breast and scales on anterior half of body sometimes with tiny tubercles; larger tubercles present on pectoral fin rays 2–9. Maximum size 4 inches (102 mm).

LIFE COLORS   Color is olive above with few to many dark X- or W-shaped markings on the upper half of the body and sometimes a dusky lateral stripe. Sides silver, belly white. A small, black spot usually present at caudal base. Middorsal stripe uniformly colored, not composed of light and dark alternating areas. There are no bright breeding colors.

SIMILAR SPECIES   It differs from *Erimystax harryi* in usually having 40–43 lateral line scales (vs. 46–49), and in body with numerous scattered X-shaped markings (vs. body without X-shaped markings). The body color pattern also distinguishes *E. x-punctatus* from all other barbeled minnows in Arkansas. It further differs from *Hybopsis* species and *Macrhybopsis hyostoma* in having 40 or more lateral line scales (vs. fewer than 40), and from *M. storeriana* in having 0,4–4,0 pharyngeal teeth (vs. 1,4–4,2).

VARIATION AND TAXONOMY   *Erimystax x-punctatus* was described by Hubbs and Crowe (1956) as *Hybopsis x-punctata* from the Gasconade River, 12.9 km south of Richland, Pulaski County, Missouri. According to Harris (1986), two subspecies of *E. x-punctatus* are present in Arkansas. The eastern subspecies, *E. x-punctatus trautmani*, occurs in the White River system, and the western nominate form, *E. x. x-punctatus*, inhabits the Ouachita River system and the Illinois River system (Arkansas River drainage). Simons (2004) reviewed the phylogenetic relationships of the members of *Erimystax* and found that *E. x-punctatus* exhibited more haplotype variation than other species in the genus. He tentatively recognized four groups distributed in the following basins: Arkansas River, Ouachita River, Ozark/Upper Mississippi River, and Ohio River. Simons's study did not include the White River populations recognized by Harris (1986) as *E. x. trautmani*. Simons (2004) found that the remaining haplotypes, which represent the nominate subspecies *E. x. x-punctatus*, do not constitute a monophyletic group and suggested the apparent paraphyly of *E. x. x-punctatus* might be an artifact of sampling (he had only one specimen from the Ouachita River and one from the Arkansas River). It is apparent that additional samples from the Ouachita, Arkansas, and Missouri River tributaries are needed to resolve the status of these regional variants (Simons 2004). Analysis of the cytochrome *b* gene by Mayden et al. (2006a) resolved *E. x-punctatus* as consisting of two clades forming a clade sister to *E. cahni*.

DISTRIBUTION   The Gravel Chub occurs in the Great Lakes drainage of southern Ontario, Canada (Thames River system), in the Mississippi River basin in the upper Ohio River drainage, and from southern Wisconsin and Minnesota south to eastern Oklahoma and southern Arkansas. In Arkansas, this chub is localized in portions of the White River, Illinois River (Arkansas River drainage), and Ouachita River systems (Map 37). Simons (2004) reported that shallow divergence and wide geographic distribution in *E. x-punctatus* suggest post-Pleistocene dispersal.

HABITAT AND BIOLOGY   The Gravel Chub inhabits swift, deep riffles and raceways in medium to large-sized, clear, gravel-bottomed streams where it lives on or near the bottom. This species seems locally abundant in large downstream areas where gradient is less and the water is warmer, but it is rather intolerant of siltation. In the Illinois River drainage of northeastern Oklahoma, it most commonly occurred among or under rubble in deep riffles in fast-moving current (Becker 1983). It does not adapt to reservoirs. Moore and Paden (1950) suggested that the specific microhabitat of the gravel chub may be beneath rocks in the riffle areas, where the effects of swift water would be reduced.; however, in Arkansas it is normally taken over open gravel with few large rocks. Trautman (1957) noted that it avoided rooted aquatic vegetation, algae, and aquatic mosses.

Harris (1986) reported that the dominant foods of this chub are periphyton (detritus, diatomaceous algae, and bacteria), Trichoptera, plant material, Ephemeroptera, gastropods, and other miscellaneous aquatic insects. Our observations support those of Harris that it is an omnivore. Davis and Miller (1967) suggested that *E. x-punctatus* feeds by using its long sensitive snout to probe under rocks and in crevices. Taste buds on the barbels are unusually large (sometimes 100 microns in length) and project downward at oblique angles to give the barbels a branched appearance. Fewer than 10 taste buds occur on a barbel.

Harris (1986) indicated that spawning occurs in the Ouachita River from March to early May, with peak spawning in April. Based on examination of gonads by TMB, the spawning season in Arkansas extends a little longer than indicated by Harris. Ripe females from the Spring River in Lawrence County on 18 May ranged from 55 to 80 mm TL and contained between 125 and 342 mature ova. Ripe females were found as late as 31 May in the Caddo River (Ouachita River drainage) in Clark County, indicating that spawning may extend into June. Those females averaged 91 mm TL and contained an average of 385 ripe ova. Peak spawning occurred in April in Kansas at water temperatures around 60°F (15°C), generally at a depth of 2–3 feet (60–90 cm) (Cross and Collins 1995). Spawning occurred at the riffle heads in 15.7–23.6 inches (40–60 cm) of water over clean, small to medium gravel substrate. Harris (1986) found that most, if not all, Ouachita River Gravel Chubs begin spawning at age 2 when more than 2.0 inches (>50 mm) TL. Breeding adults are usually 2.5–3.8 inches (64–97 mm) (Trautman 1957).

CONSERVATION STATUS   The Gravel Chub is widespread and common in Arkansas. It is most abundant in the Ouachita River drainage and in the White River between Batesville and the mouth of the Black River. It was probably uncommon in the upper White River drainage even before the construction of the large reservoirs, based on lack of many older records and its absence from the unimpounded Buffalo River. Its populations are probably currently secure.

Genus *Hybognathus* Agassiz, 1855— Silvery Minnows

The genus *Hybognathus* is composed of a group of slender, silvery, herbivorous minnows with a long, coiled intestine and a black peritoneum. Species are further characterized by having no barbels and no hard, cartilaginous edge on the lower jaw. All species of *Hybognathus* have 8 rays in the dorsal, pelvic, and anal fins, 14–16 pectoral rays, 9–12 gill rakers with length of the longest raker 1.5–2.5 times its basal width, and well-scaled breasts. Lateral line scale rows 34–41. Pharyngeal teeth 0,4–4,0 without terminal hooks. The dorsal fin is usually pointed and falcate, with the tip of the anterior ray extending well past the tip of the posterior rays when the fin is depressed. Mouth structure is unusual due to the enlargement of the anterior suborbital (lachrymal) bone. Complex gut structure and black peritoneum are trophic adaptations for feeding on organic detritus that contains diatoms, filamentous algae, decaying organic matter, and small insects. Hlohowskyj et al. (1989) analyzed the pharyngeal filtering apparatus in all species. Papillae are distributed over upper and lower pharyngeal epithelia, including some gill rakers, in a pattern suggesting an accessory filtering system. The extensive modifications of the pharyngeal mucosa in *Hybognathus* seem to distinguish it from other herbivorous genera of North American cyprinids such as *Campostoma* (Hlohowskyj et al. 1989). There are no prominent breeding colors.

A close relationship was suggested between *Hybognathus* and *Campostoma* and the western genus *Dionda* in studies by Schmidt (1994), Coburn and Cavender (1992), Mayden et al. (1992a), and Cook et al. (1992). Coburn (1982b) previously suggested that *Hybognathus* was the sister group to *Pimephales*, but *Pimephales* was later hypothesized to be related to *Cyprinella* (Cavender and Coburn 1986; Coburn and Cavender 1992). Mayden (1989) placed *Hybognathus* with the chub clade. However, molecular variation strongly supported *Hybognathus* being more closely related to *Notropis*, *Agosia*, and *Cyprinella* (Simons and Mayden 1999). The taxonomic treatment of the species of *Hybognathus* has been controversial (Schmidt 1994), but the monophyly of the genus is well substantiated. Schmidt (1994) noted that monophyly of the genus was supported by at least three synapomorphies including elongated anterior processes of the urohyal, enlarged epibranchials, and organization of pharyngeal papillae. Schmidt also recognized seven species, including the three species of *Hybognathus* documented for Arkansas (*H. hayi*, *H. nuchalis*, and *H. placitus*). Moyer et al. (2009b), using a combination of mitochondrial and nuclear genes, corroborated previous studies on the monophyly of the seven species of *Hybognathus*, and monophyly of the genus was weakly supported.

In addition to the three known Arkansas *Hybognathus* species, *H. argyritis* is known from contiguous Missouri. Etnier and Robison (2004) reported an unusual 39 mm SL (UT 44.10001) specimen of *Hybognathus* collected in the lower White River at the Desha/Arkansas county line with a basioccipital process characteristic of *H. argyritis*. Close examination of this specimen revealed its eye size was larger (3.2 mm) than that of a syntopic 41 mm SL *H. nuchalis* specimen (3.0 mm). However, further comparison with a 40 mm SL *H. argyritis* specimen from Missouri River mile 16.4 made it clear (eye diameter 2.1 mm) that the Arkansas specimen was not a typical specimen of *H. argyritis*. After comparing this unusual specimen with both *H. nuchalis* and *H. argyritis*, Etnier and Robison (2004) concluded that it was either a member of a large-eyed southern race of *H. argyritis* or a member of a completely unknown species. Persons collecting fishes in the lower White River should be aware of the possible presence of this enigmatic *Hybognathus*.

Hybognathus hayi Jordan
Cypress Minnow

Figure 15.42. *Hybognathus hayi*, Cypress Minnow. *Uland Thomas.*

CHARACTERS   A silvery minnow with a rather deep, compressed body and an angular profile (Fig. 15.42). Body deepest and widest near the dorsal fin origin. Mouth terminal, large, and oblique, the front of upper lip above the middle of the eye. Snout broadly rounded, scarcely visible when viewed from below. Pharyngeal teeth 0,4–4,0, with oblique grinding surfaces, without hooks. Eye large, its diameter going 3.4–3.8 times into head length; eye diameter greater than snout length. Head short, head length

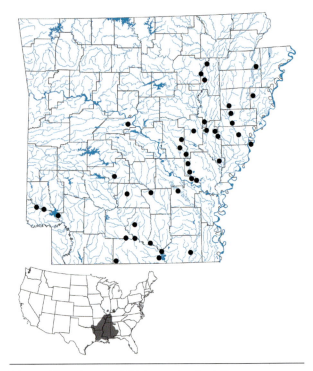

Map 38. Distribution of *Hybognathus hayi*, Cypress Minnow, 1988–2019.

going into SL 4–4.5 times. Dorsal fin origin slightly anterior to pelvic fin origin. Dorsal fin is pointed and falcate with 8 rays, the tip of the anterior ray extending past the tips of the posterior rays when the fin is depressed. Anal fin falcate with 8 rays; pectoral rays 14–16. Lateral line slightly decurved anteriorly and complete with 35–41 scales when lateral area is fully scaled; however, lateral scales are deciduous, usually resulting in an incomplete or interrupted lateral line in most preserved specimens. Predorsal scale rows 14–16. Breast completely scaled. Basioccipital bone intermediate in size and shape between those of *H. nuchalis* and *H. placitus*, its posterior margin nearly straight (Fig. 15.10D). Intestine long and coiled; peritoneum black. We have not examined tuberculate males and can find no information on tubercle patterns other than the reference by Warren and Burr (1989) to "the simple tubercle pattern of males." It is generally a smaller species than *H. nuchalis* and *H. placitus*. Maximum size 4.5 inches (114 mm).

LIFE COLORS   Body color dark olive green dorsally and silver to white on sides and belly. Scales on dorsolateral region showing distinct diamond-shaped patterns as a result of crosshatching. The first, and sometimes part of the second, scale row below the lateral line outlined with melanophores on the anterior portion of the body (most noticeable in preserved specimens). A narrow and diffuse predorsal stripe present; dorsal stripe more pronounced posterior to the dorsal fin. Melanophores on the anterior part of lateral band are small and only slightly larger than those on the upper parts of the sides and back. No nuptial coloration develops.

SIMILAR SPECIES *Hybognathus hayi* resembles some members of the genera *Notropis* and *Cyprinella* in external appearance because of a similar diamond-shaped pigmentation pattern of its scales; however, the surfaces of the pharyngeal teeth are flat and the intestine is elongate, coiled, and more than twice the length of the body in all *Hybognathus* species. *Notropis* and *Cyprinella* species have strongly hooked pharyngeal teeth, and the intestine is less than two times the length of the body (except in *Notropis nubilus*). It further differs from *Cyprinella* species in having a black peritoneum (vs, silvery peritoneum, sometimes with dark speckles), 8 anal rays (vs. 9), and pharyngeal teeth 0,4–4,0 (vs. 1,4–4,1). *Hybognathus hayi* is frequently confused with *H. nuchalis*. It differs from both *H. nuchalis* and *H. placitus* in having more conspicuous scale borders, producing a diamond pattern and crosshatched appearance; its broadly rounded snout, not easily seen from below (Fig. 15.43A); and a more oblique and nearly terminal mouth. It further differs from *H. nuchalis* in that melanophores on anterior part of the lateral band are small and only slightly larger than those on adjacent parts of the upper side and back (vs. melanophores on anterior part of lateral band large, and noticeably larger than those on upper parts of sides and back), middorsal stripe narrow and lighter in front of dorsal fin (vs. broader and darker), and in usually lacking a complete lateral line in Arkansas populations as a result of scale loss (vs. lateral line complete at least on one side of the body in 84% of *H. nuchalis* specimens).

VARIATION AND TAXONOMY *Hybognathus hayi* was described by Jordan (1885a) from the Pearl River, Hinds County, Jackson, Mississippi. Fingerman and Suttkus (1961) provided a detailed morphological description. It differs osteologically from other *Hybognathus* species by having a large opening of the myodome posteriorly, a character that has received extensive attention as it relates to more general issues of North American cyprinid relationships (Coburn 1982b; Mayden 1989; Coburn and Cavender 1992). This large opening is unique among herbivorous chubs and resembles the opening of notropin minnows (Schmidt 1994). In addition, *H. hayi* has a reduced mastigophory process of the pharyngeal pad of the basioccipital, which is the point of insertion of a ligament that extends forward to the ventral crest on the shaft of the parasphenoid. This condition resembles that found in *Campostoma*, but the process in stonerollers is much more robust. Hlohowskyj et al. (1989) found *H. hayi* to be

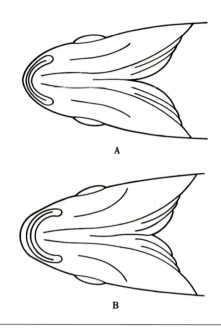

Figure 15.43. Ventral view of head showing relative snout length in:

A—*Hybognathus hayi*
B—*Hybognathus nuchalis*. Smith-Vaniz (1968)

divergent from other *Hybognathus* in the shape and number of pharyngeal papillae. Schmidt (1994) studied the phylogenetic relationships of the genus *Hybognathus* and indicated a well-corroborated clade within *Hybognathus* that included *H. hayi*, *H. nuchalis*, *H. amarus*, and *H. argyritis* which he called the *H. nuchalis* species group. Monophyly of this clade is supported by modifications of the pharyngeal process of the basioccipital. He also reported that *H. placitus* is the sister group to the *H. nuchalis* species group.

Lateral body scales of *H. hayi* in Arkansas are far more deciduous than those of *H. nuchalis*, and it is extremely rare to find a collection of preserved specimens from this state with a high percentage of individuals having a complete lateral line. In those individuals having a completely scaled lateral region, the lateral line is complete; however, most individuals in a sample usually have an incomplete or interrupted lateral line because of lost scales (as indicated by empty scale pockets, but sometimes possibly due to failure of scale development in an area). TMB examined more than 200 *H. hayi* from 6 localities in Arkansas and found that only 15.6% of the specimens possessed a complete lateral line on at least one side of the body. More than 84% of more than 300 *H. nuchalis* examined (from White, Arkansas, and Red river drainages) had a complete lateral line on at least one side.

DISTRIBUTION *Hybognathus hayi* occurs in the lower Mississippi River basin from southwestern Indiana (Ohio River system) and southern Illinois to Louisiana, and in Gulf Slope drainages from the Florida panhandle west to the Sabine River, Texas. In Arkansas, as elsewhere, it is confined to the lowland streams of the Coastal Plain (Map 38). Tumlison and Robison (2010) reported additional specimens of *H. hayi* from Union County, and Connior et al. (2012) reported the first records of this species in the Arkansas portion of the Little River system.

HABITAT AND BIOLOGY Fingerman and Suttkus (1961) reported that this minnow occupies mud-bottomed, quiet backwaters of streams and lagoons in Louisiana. In Arkansas, we have taken individuals from cypress-lined oxbows over a bottom of mud and soft detritus and from bayous and sloughs. It is often collected syntopically with the related *H. nuchalis* in this state but is usually found in quieter waters than that species. In Illinois, Cypress Minnows were found only at depths of about 0.6–1.3 m over a substrate of sand overlain with 45–60 cm of soft mud and detritus (Warren and Burr 1989). In Alabama, it abounds in oxbows and backwaters of low-gradient, medium to large rivers where there are accumulations of logs, brush, other debris, and detritus (Boschung and Mayden 2004). The Cypress Minnow apparently does not adapt well to Arkansas reservoirs. It was found in only one Arkansas impoundment (Felsenthal), but more than 6,000 specimens were taken in Felsenthal during rotenone population sampling (Buchanan 2005).

Etnier and Starnes (1993) speculated that the complex gut structure and black peritoneum were trophic adaptations for feeding on organic ooze containing diatoms, filamentous algae, decaying organic matter, and probably small insects. The pharyngeal filtering apparatus was analyzed by Hlohowskyj et al. (1989), who noted the elaborate development of papillae and taste buds along the pharynx and margins of some gill rakers. Warren and Burr (1989) believed that the food habits of *H. hayi* in Illinois were similar to those of *H. nuchalis*. Intestines of Illinois specimens contained sand, detritus, algae, and other vegetable matter. TMB examined gut contents of 61 Arkansas specimens collected in summer and fall from Arkansas, Jefferson, and Monroe counties, and found that its diet conformed to the diet predicted by Etnier and Starnes (1993). Gut contents were dominated by organic ooze, detritus, and algae. Filamentous algae and diatoms made up only a small portion of the contents. Insect remains were present in very small amounts in most specimens, indicating that insects were probably ingested incidentally.

Hybognathus hayi spawns from late December to early March in the Pearl River drainage, Mississippi (Mettee et al. 1996). Spawning occurs later in more northern areas of its range. In Illinois, females began to ripen in early March and were fully gravid in April (Warren and Burr 1989). Those authors believed that adult Cypress Minnows migrate prior to mid-April from deep portions of the Cache River, Illinois, into shallow adjacent wetlands or shoreline areas, then spawn and return to deep waters. Warren and Burr (1989) further believed that shallow coves or flooded wetlands are required for successful spawning by this species. In that Illinois study, mature females were 1 year old, ranged in SL from 57 to 64 mm, and contained between 1,500 and 2,500 eggs. Males also matured by age 1 and at about 55 mm SL; maximum life span was judged to be about 3 or 4 years. Judging from the simple tubercle pattern of males and the habitat in which ripe adults were taken, Warren and Burr (1989) speculated that spawning occurred communally, with the eggs being scattered over the substrate and with no parental care of eggs and young. Young-of-the-year Cypress Minnows were found in mid-May in Illinois, and by late August to early September, two size classes were evident: young-of-the-year, which averaged 56.5 mm SL, and individuals 1+ years of age that averaged 90.3 mm SL.

CONSERVATION STATUS The Cypress Minnow has declined drastically near the periphery of its range in Illinois, Indiana, Kentucky, Missouri, Oklahoma, and Florida (Burr and Mayden 1982a). Range fragmentation was believed to be important in its decline in northern areas of its range (Warren and Burr 1989). Black (1940) considered it common in the lowland portions of every drainage in Arkansas. It has undoubtedly declined to an undetermined extent in Arkansas, but stable populations are still widely scattered through the lowlands of the state. Some of the most stable populations are found on the White River National Wildlife Refuge. We consider its populations in Arkansas currently secure, but there is a need for future monitoring of this species.

Hybognathus nuchalis Agassiz
Mississippi Silvery Minnow

Figure 15.44. *Hybognathus nuchalis*, Mississippi Silvery Minnow. *David A. Neely.*

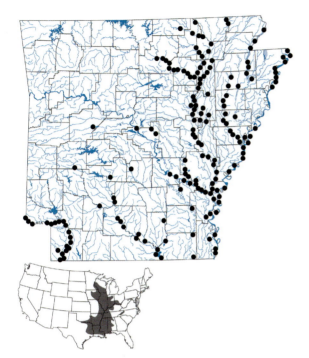

Map 39. Distribution of *Hybognathus nuchalis*, Mississippi Silvery Minnow, 1988–2019.

CHARACTERS   A large, streamlined, silvery minnow with a subterete body shape and a rounded profile (Fig. 15.44). Intestine long and coiled; peritoneum black. Becker (1983) described the intestine as "coiled like a watch spring, with elongated stomach almost the length of peritoneal cavity; ventral to the stomach are a series of alternating loops and coils, about 5.4–5.8 TL." Head small. Snout slightly rounded and projects anteriorly beyond the upper lip by about twice the thickness of the upper lip (best observed in ventral view). Mouth subterminal and nearly horizontal. Barbels absent. Eye small, its diameter less than snout length in most Arkansas populations; eye diameter going into head length about 3.5–5.0 times; diameter of eye greater than width of mouth opening. Pharyngeal teeth 0,4–4,0 with oblique grinding surfaces. Dorsal fin origin slightly anterior to pelvic fin origin and usually slightly closer to tip of snout than to caudal fin base. Dorsal fin with distinctly pointed tip; dorsal fin rays 8. Anal rays 8; pectoral rays 14–17. Dorsolateral scales appear rounded, not diamond-shaped. Predorsal scale rows 14–16. Circumferential scales 22–31. Lateral line slightly decurved anteriorly and usually complete (84% of Arkansas specimens), with 34–40 scales; 16% of specimens have an incomplete or interrupted lateral line due to scale loss. Breast completely scaled. Basioccipital process wider than long and with a distinctly concave rear margin (Fig. 15.10C). Breeding males with small tubercles on the top of the head, and smaller tubercles on the cheeks, opercles, branchiostegal areas, and fins. Females sometimes with small tubercles on the head and on body scales (Etnier and Starnes 1993). Maximum size is 7 inches (180 mm) TL (Page and Burr 2011).

LIFE COLORS   Body color tan to brown dorsally, with silvery sides grading to a whitish belly. Fins unpigmented. Middorsal dark stripe broader than dorsal fin base and especially dark anterior to dorsal fin. Breeding males with light yellow along the sides and lower fins. Melanophores on the anterior part of the lateral band large, noticeably larger than those on the upper part of the sides and back.

SIMILAR SPECIES   See *H. hayi* account for features distinguishing the three Arkansas *Hybognathus* species. The most similar sympatric species is *Notropis blennius*, but *H. nuchalis* can be distinguished from that species by its 8 anal rays (vs. 7 anal rays), 0,4–4,0 pharyngeal teeth (vs. 2,4–4,2), black peritoneum (vs. silvery peritoneum), long coiled intestine (vs. a simple S-shaped intestine), a more rounded snout (vs. a more pointed snout), an oblique groove paralleling the upper jaw (vs. lacking the additional oblique groove paralleling the upper jaw just anterior to the lachrymal bone), and with melanophores on the scale row below the lateral line (vs. lacking melanophores on the scale row below the lateral line).

VARIATION AND TAXONOMY   *Hybognathus nuchalis* was described by Agassiz (1855) from the Mississippi River at Quincy, Adams County, Illinois. The taxonomic treatment of members of *Hybognathus* has been controversial due to a lack of externally defining characters and has led to extensive synonymizing and weakly documented differentiation (Schmidt 1994). Except for *H. hankinsoni* and *H. hayi*, all the species now recognized have at one time been synonymized with *H. nuchalis* (Hubbs and Ortenburger 1929b; Hubbs and Lagler 1947; Al-Rawi and Cross 1964). A thorough rangewide systematic study of *H. nuchalis* is needed.

DISTRIBUTION   *Hybognathus nuchalis* occurs in the lowlands of the Mississippi River basin from Ohio to Minnesota and south to Louisiana, and in Gulf Slope drainages from Mobile Bay, Alabama, west to the Brazos River, Texas. In Arkansas, the Mississippi Silvery Minnow primarily occupies Coastal Plain streams but extends up the Black River system to the clear Ozark streams (Strawberry, Current, and Spring rivers) (Map 39). It does not occur in the upper White River system above Batesville. In eastern Arkansas it is common and abundant in the St. Francis River system and is one of the most abundant cyprinids in the Mississippi River. Westward this species is uncommon in the Arkansas River upstream of Little Rock (although there are records as far upstream as Pope County), and it is no longer found in that river near Fort Smith (Buchanan

1976; TMB, previously unpublished data). Southward it inhabits the Red and Ouachita river systems but avoids the upper Ouachita Mountains. It is found throughout the Red River in Arkansas but was considered uncommon in that river (Buchanan et al. 2003). It was found in six Arkansas impoundments, and a large population occurs in Lake Nimrod (Buchanan 2005).

HABITAT AND BIOLOGY The Mississippi Silvery Minnow is Arkansas's most abundant and widely distributed *Hybognathus* species and is found in a variety of lowland habitats, including large and medium-sized rivers, bayous, and oxbow lakes. It occurs in a broader range of habitats than *H. hayi* (Fingerman and Suttkus 1961) and is often abundant in both quiet water and in current over a sandy substrate or a mud and detritus bottom. It has a greater preference for current than *H. hayi*. It was formerly much more abundant in the Red River in the 1970s, where it could be seined by the hundreds. In the late 1990s and early 2000s, it was considered uncommon in the Red River (Buchanan et al. 2003). It is fairly tolerant of turbidity and is often common in channelized rivers in northeastern Arkansas such as the St. Francis River.

This schooling species is a generalized herbivore, feeding on diatoms, filamentous algae, and other plant material it ingests from soft bottoms. In Wisconsin, foods were green algae, blue-green algae, and diatoms (Becker 1983). Items found in greatest amounts were numerous diatom species, *Oscillatoria* species, and filamentous algae fragments, along with sand grains and much organic debris. Hlohowskyj et al. (1989) analyzed the pharyngeal filtering apparatus of this species in detail and concluded that the pharyngeal taste buds were arranged in a pattern that suggested a filtering apparatus for trapping diatoms and other small food items. In addition to the food items listed above, it also reportedly feeds on bottom ooze consisting of sand, silt, decaying plant materials, diatoms, various algal forms, shed exoskeletons from larval insects, and fungal material (Whitaker 1977).

Information on reproductive biology is sketchy. Spawning probably occurs in late spring and summer. Adults in spawning condition have been collected in Illinois in early June (Forbes and Richardson 1920). Spawning occurs in Wisconsin from late April to July (Becker 1983). The Mississippi Silvery Minnow is a broadcast spawner (Adams and Hankinson 1928). Its spawning behavior is probably similar to that of the eastern species *H. regius* reported by Raney (1939). Males congregate in backwater areas and two males accompany each female that enters the area to spawn. Nonadhesive eggs are scattered over the silt substrate usually over aquatic vegetation. Larger females produce 6,000–7,000 eggs per spawn. Eggs

hatch in 1 week. It is believed males mature in 2 years while some females may reach sexual maturity at the end of their first year. Males are usually darker than females. Becker (1983) reported a life span of 3 years.

CONSERVATION STATUS Etnier et al. (1979) and Etnier and Starnes (1993) attributed the decline of *Hybognathus nuchalis* in Tennessee to the construction of mainstream reservoirs. Its apparent decline in the upper Arkansas River of Arkansas may be related to the construction of the Arkansas River navigation pools, but it is still common in the navigation pools downstream from Little Rock. Larger, free-flowing rivers are important in some phase of the life history of the Mississippi Silvery Minnow. Despite apparent declines in some areas of Arkansas, we consider its populations currently secure.

Hybognathus placitus Girard
Plains Minnow

Figure 15.45. *Hybognathus placitus*, Plains Minnow. *David A. Neely.*

Map 40. Distribution of *Hybognathus placitus*, Plains Minnow, 1988–2019.

CHARACTERS   A large, terete, silvery minnow with a short head and blunt snout (Fig. 15.45). Head bluntly triangular (Sublette et al. 1990); head width considerably greater than distance from tip of snout to back of eye (Hubbs et al. 1991). Mouth thin-lipped, subterminal and nearly horizontal. Barbels absent. Eye very small, eye diameter going into head length 4.4–5.5 times; diameter of eye much less than width of mouth opening. Dorsal fin origin over pelvic fin origin. Dorsal rays 8; anal rays 8; pectoral rays 16 or 17. Lateral line scales 36–39. Predorsal scale rows usually 18–22 (17–23). Pharyngeal teeth 0,4–4,0. Intestine long and coiled, more than twice the length of the body; peritoneum black. Basioccipital process narrow and peglike with posterior end straight (Fig. 15.10A). Ostrand et al. (2001) described sexual dimorphism in this species, the most prominent differences being the longer dorsal fin rays, larger head, and caudal peduncle of the male, while the female is deeper bodied and has a relatively longer trunk from the pelvic fin origin to the anal vent. Breeding males with tubercles on head, jaws, and pectoral fins, with smaller, less numerous tubercles on other fins and on anterodorsal scales; tubercles on pectoral fin rays are curved toward the body (erect in *H. nuchalis*) (Etnier and Starnes 1993). Maximum size about 5.12 inches (130 mm) (Page and Burr 2011).

LIFE COLORS   Body color tan to brown above with silvery sides and a white belly (very similar to that of *H. nuchalis*). No lateral band. Middorsal stripe well developed. Fins colorless

SIMILAR SPECIES   This species resembles *H. nuchalis* in most features; however, it differs from *H. nuchalis* and *H. hayi* in having a peglike basioccipital process (vs. basioccipital process broad and concave or broad and straight) (Fig. 15.10). It further differs from *H. nuchalis* in having a much smaller eye in adults, eye diameter going 4.4–5.5 times into head length (vs. adult eye diameter going into head length 3.5–4.0 times); diameter of eye less than width of mouth opening (vs. eye diameter greater than width of mouth opening); predorsal scales usually 17–22 (vs. usually 14–16 predorsal scales); and with 15 or more scale rows below the lateral line, counted across the belly just anterior to the pelvic fin bases up to but not including the lateral line scale row (vs. 15 or fewer in *H. nuchalis*). *Hybognathus placitus* further differs from *H. hayi* in having snout length greater than eye diameter (vs. snout length less than eye diameter), front of upper lip below middle of eye (vs. front of upper lip about level with middle of eye), and scales on anterior side not prominently dark-edged and not forming a distinct diamond-shaped pattern (vs. scales on forward part of side prominently dark-edged and forming a distinct diamond-shaped pattern).

VARIATION AND TAXONOMY   *Hybognathus placitus* was described by Girard (1856) from the Arkansas River, near Fort Makee (= Fort Atkinson), near Cimarron, Gray County, Kansas. Specific status of *H. placitus* has been justified by the presence of a rodlike process of the basioccipital (Bailey and Allum 1962; Niazi and Moore 1962). This morphology was treated by maximum parsimony analysis as derived and autapomorphic within *Hybognathus* (Schmidt 1994). Variation was studied in *H. placitus* by Al-Rawi and Cross (1964), but no subspecies were recognized. Modifications of the parasphenoid are also considered derived for *H. placitus*. In *H. placitus*, the parasphenoid is more heavily ossified than in other members of *Hybognathus*. Schmidt (1994) studied the phylogenetic relationships of the genus *Hybognathus* and proposed that the *H. placitus* group is sister to the *H. nuchalis* species group.

DISTRIBUTION   *Hybognathus placitus* is primarily a species of the large Great Plains rivers where it is widely distributed in the Missouri, Arkansas, Red, Brazos, and Colorado river drainages from Montana and North Dakota south to New Mexico and Texas; it is also known from a short segment of the Mississippi River main stem from the mouth of the Missouri River downstream to northeastern Arkansas. In Arkansas, the following nine older records for *H. placitus* have been reported (Map A4.41): (1) Mississippi River, Mississippi County, 1 specimen taken between Barfield and Hickman by J. D. Black on 8 August 1939 (UMMZ 128575), (2) Mississippi River, Mississippi County, 1 specimen taken 2.2 miles northeast of Butler on 8 October 1968 by R. D. Suttkus (TU 54530), (3) Arkansas River, Crawford County, 1 specimen from Lee Creek, 3 miles northwest of Van Buren, collected on 5 July 1927 by the Oklahoma Museum Expedition (OKMNH 7849), (4) Arkansas River at Fort Smith, Sebastian County, 11 specimens (no date given) collected by S. E. Meek (FMNH 2209) and reported by Black (1940), (5) Arkansas River, Sebastian County, 78 specimens from backwater pool off river, 1.5 miles NE of Barling, collected by J. D. Black on 25 June 1938 (UMMZ 123308), (6) Arkansas River, near mouth of Piney Creek, Logan County (possibly Johnson County), 522 specimens collected by J. D. Black on 23 July 1939 (UMMZ 128400), (7) Arkansas River at Dardanelle, Yell County, 4 specimens collected by R. D. Suttkus on 2 May 1955 (TU 15647), (8) Arkansas River at State Hwy 1, Arkansas County, 21 specimens collected by R. D. Suttkus on 23 October 1959 (TU 22431); and (9) Red River at Spring Bank Ferry, Miller County, 3 specimens collected on 8 July 1939 by J. D. Black. Black (1940) corrected several erroneous reports of *H. placitus* from Arkansas, and we are aware of a few reports of that species from the 1950s and

1960s that we consider questionable. The Plains Minnow probably occurs today in Arkansas only as occasional stragglers. The tenth and most recent Arkansas record is a single specimen from the Arkansas River at Fort Smith (UAFS 0165) taken on 28 July 1993 during unusually high summer discharge of 60,000 cfs (Buchanan and Robison 1994) (Map 40). There are 13 cataloged lots of *H. placitus* in the Sam Noble Oklahoma Museum of Natural History from the Red River in Oklahoma downstream from Dennison Dam. Four of those lots (OMNH 48660, 49602, 53908, and 80590) were collected approximately 10 miles upstream from the Arkansas state line between 1992 and 2010. The 2010 collection contained 13 specimens; therefore, it is possible that *H. placitus* could occasionally be found in the Red River of Arkansas.

HABITAT AND BIOLOGY   The Plains Minnow is an abundant species in the Great Plains, where it inhabits the main channels of shallow, turbid, silt and sand-bottomed streams and medium to large rivers. In such streams in Kansas, it is most abundant where sediments accumulate in shallow backwaters, gentle eddies, and along the deeper edges of sand "waves" that are formed on the shifting substrate by the action of the current (Cross and Collins 1995). Pflieger (1997) reported that in Great Plains streams it often occurs in association with *Cyprinella lutrensis*, *Hybognathus argyritis*, *Macrhybopsis storeriana*, *Notropis atherinoides*, *N. stramineus*, and *Platygobio gracilis*. It is one of the most abundant species throughout much of Oklahoma, but it has not been documented in Arkansas in more than 25 years. It has high thermal, low dissolved oxygen, and high salinity tolerances (Ostrand and Wilde 2001).

In western rivers, this species occurs in very large schools. Its long and coiled intestine indicates a herbivorous diet, and Miller and Robison (2004) speculated that it feeds on benthic microflora such as algae and diatoms. Cross and Collins (1995) considered it partly herbivorous, feeding in calm, shallow backwaters along the bottom on the thin layer of diatoms and other algae, but also ingesting small animals. Based on the unique feeding apparatus found in all *Hybognathus* species, Hlohowskyj et al. (1989) suggested that foods consisted primarily of algae and organic matter obtained from the bottom ooze.

Hybognathus placitus has an extended spawning season throughout its range from April to September. Lehtinen and Layzer (1988) listed a spawning period in Oklahoma from April to July based on ovum diameters and gonadosomatic indices (age 1 and age 2 fish were reproducing). Miller and Robison (2004) reported that most spawning in Oklahoma occurred in May and June with a secondary peak in August, and spawning coincided with high or receding flows. Population age structure suggested

that multiple spawnings occur (Lehtinen and Layzer 1988; Sliger 1967). Abrupt rises (flood flow) may stimulate spawning (Lehtinen and Layzer 1988; Cross and Collins 1995). Sliger (1967) collected *H. placitus* eggs from the Cimarron River in Oklahoma only when the river was swollen by recent heavy rains. The Plains Minnow is a pelagic broadcast spawner of nonadhesive semibuoyant eggs (Platania and Altenbach 1998). Eggs are reported to gently bounce along the bottom during most of their development. Eggs of *H. placitus* have been collected from April to August in the Cimarron River in Oklahoma, with a major peak in May and June and a secondary peak in midsummer (Taylor and Miller 1990). Spawning behavior of *H. placitus* was described by Platania and Altenbach (1998). Spawning was initiated when a male aligned laterally on the right side of the female in a head-to-head orientation, with his head and abdomen slightly lower and tail slightly higher than the female. Both fish then swam rapidly and in tandem before the male curved his head under her vent and wrapped his caudal fin and peduncle over the back of the female who twisted sideways and curved her body. The male then rapidly brought his head toward his tail, appearing to squeeze the female's midsection. Simultaneously, the female released the first eggs and continued to release eggs as the male returned to a normal posture. Both fish then oriented head-to-head facing downward and momentarily drifted together. The actual duration of the spawning event (wrap-and-release) was 66.6 msec. In individuals collected from the Cimarron River, the number of mature ova (ranging in size from 1.10 to 1.40 mm) ranged from 417 to 4,134 ($\bar{x} = 817$) in fish 51–87 mm SL. While only a small number of individuals in the population may live to age 2, the much greater fecundity of those individuals compensates for their lack of numbers (Taylor and Miller 1990). Growth is rapid, with juveniles reaching a length of 28–43 mm by early September of their first year (Pflieger 1997). Taylor and Miller (1990) reported the sex ratio was 1:1 with both sexes similarly sized. Sexual maturity is reached at 45–50 mm SL.

CONSERVATION STATUS   The Plains Minnow is rare in Arkansas and probably occurs occasionally as stragglers in the Arkansas, Mississippi, and Red rivers. William J. Matthews (pers. comm. 2003) reported declines of this species in the large rivers of Oklahoma, Cross and Collins (1995) noted that it had declined precipitously over most of its range in Kansas, Hrabik et al. (2015) reported a downward trend in abundance in Nebraska, and Pflieger (1997) noted that it has undergone a drastic decline in abundance throughout its Missouri range, reducing the likelihood of stragglers from those states entering Arkansas. Winston et al. (1991) believed the construction of Altus Dam caused its extirpation from the North Fork of the Red River in

Oklahoma. Surprisingly, a nonnative population introduced into the Pecos River of New Mexico in the early 1960s flourished and largely replaced the native *H. amarus* within 10 years of its introduction (Hoagstrom et al. 2010). Older records indicate that *H. placitus* formerly occurred throughout the Arkansas River in Arkansas, but we now consider it endangered in this state.

Genus *Hybopsis* Agassiz, 1854—Bigeye Chubs

The genus *Hybopsis* has had a turbulent and often tangled systematic history, and the constitution of the genus differs depending upon authors (Mayden 1989; Coburn and Cavender 1992). Etnier and Starnes (1993) presented a thorough review of the taxonomic history of *Hybopsis*. Almost all barbeled minnows were previously placed in the genus *Hybopsis*, but barbeled minnows are currently separated into several genera. Mayden et al. (1992a) recognized 21 species in *Hybopsis*, but by 2013 only 6 species were generally recognized in that genus (Page et al. 2013). For a more comprehensive treatment of the nomenclatural history of this genus, see Boschung and Mayden (2004). Those authors listed the genus characters as follows: body forms varying from moderately robust to slender and from terete to relatively deep and compressed; head typically long; snout long and overhanging the mouth; maxillary barbels present in a single pair (in all species except *H. amnis*, which lacks barbels); eyes usually supralateral; interorbital area narrow; anal rays 7 or 8; pharyngeal teeth 1,4–4,1, those in the lesser row weak and sometimes missing; breeding colors present in some species. Skeletal characters uniting members of the genus *Hybopsis* were detailed by Mayden (1989). Two species of *Hybopsis* occur in Arkansas, *H. amblops* and *H. amnis*.

Hybopsis amblops (Rafinesque)
Bigeye Chub

Figure 15.46. *Hybopsis amblops*, Bigeye Chub.
David A. Neely.

CHARACTERS   A slender, barbeled minnow with large eyes and a well-developed dark lateral stripe (Fig. 15.46). Mouth small and inferior with a tiny barbel at the posterior tip of the jaw. Snout blunt and rounded, overhanging the mouth. Pharyngeal teeth usually 1,4–4,1. Eye large, its diameter slightly greater than snout length and going 3–3.5

Map 41. Distribution of *Hybopsis amblops*, Bigeye Chub, 1988–2019.

times into head length. Lateral line complete, with 35–38 scales; lateral line sometimes appears to be interrupted because of lost scales. Predorsal scale rows usually 13 or 14. Front of dorsal fin base about midway between tip of snout and caudal fin base. Dorsal rays 8; anal rays 8; pectoral fin rays 13 or 14; pelvic fin rays 8. Vertebrae 36–38. Dorsal and anal fins moderately to extremely falcate. Breast well scaled. Intestine short; peritoneum silvery. Nuptial males develop small tubercles that cover the top and posterior sides of the head, anterior dorsolateral scales, and the dorsal surfaces of the outer 7–10 pectoral rays (Etnier and Starnes 1993) (Fig. 15.47A). Females sometimes with tiny tubercles present on head and predorsal scales. Maximum size 3.5 inches (90 mm). Largest Arkansas specimens (90 mm) were from the White River in Marion County.

LIFE COLORS   Body color olive above, silver laterally, white below, with a prominent black lateral band extending from base of caudal fin forward onto snout. Lateral line marked by dark dots anteriorly. Faint melanophores present on edges of dorsal scales. Fins largely colorless. Dense melanophores present on middle 4 or 5 caudal rays and along first ray of pectoral fin; no melanophores on anal and pelvic fins. Bright nuptial coloration does not develop.

SIMILAR SPECIES   *Hybopsis amblops* resembles *H. amnis* and *N. volucellus*. Neither of these has barbels, as large an eye, or dorsal and anal fins as falcate as *H. amblops*. The Bigeye Chub might be confused with some of the

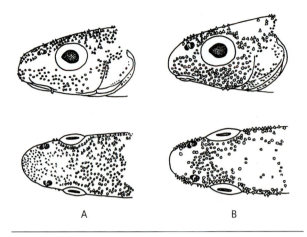

Figure 15.47. *Hybopsis* tubercles.

  A—*H. amblops*
  B—*H. amnis*

other barbeled minnows. It differs from *Macrhybopsis hyostoma*, the two Arkansas *Erimystax* species, and *Platygobio gracilis* in having pharyngeal teeth 1,4–4,1 (vs. 0,4–4,0 in *M. hyostoma* and *Erimystax* species, and 2,4–4,2 in *P. gracilis*). It can be distinguished from *Macrhybopsis gelida* and *M. meeki* by its large eye and 35–38 lateral line scales (vs. eye small and lateral line scales 39 or more). It differs from *M. storeriana* (another relatively large-eyed, barbeled minnow) in usually having a prominent dark midlateral stripe (vs. midlateral stripe faint or absent), lower lobe of caudal fin without a dark and white ventral border (vs. lower lobe of caudal fin usually dark with a white ventral border), and distance from tip of snout to front of dorsal fin equal to or greater than distance from front of dorsal fin to base of caudal fin (vs. distance from tip of snout to front of dorsal fin much less than distance from front of dorsal fin to base of caudal fin).

VARIATION AND TAXONOMY *Hybopsis amblops* was described by Rafinesque (1820a) as *Rutilus amblops* from the falls of the Ohio River at Louisville, Kentucky. Early authors referred to it by various names, including *Ceratichthys amblops*, *Semotilus amblops*, and *Erinemus hyalinus* (Smith 1979). Clemmer (1971) and Clemmer and Suttkus (1971) placed five species in the *Hybopsis amblops* species group: *H. amblops*, *H. rubrifrons*, *H. lineapunctata*, *H. winchelli*, and *H. amnis*. Later, Mayden (1989) included *Hybopsis hypsinotus* in this species group as the most basal member based on morphological characters. Shaw et al. (1995) reexamined the group using allozyme evidence and reported an undescribed species of *H. winchelli*. Grose and Wiley (2002) investigated the phylogenetic relationships among the seven members of the *H. amblops* species group

in a total evidence cladistic analysis using previously published morphological characters and partial DNA sequence data from regions of three mitochondrial genes. Their analysis supported a monophyletic *H. amblops* species group (*sensu* Mayden 1989), but relationships of *H. amblops* to other species were not clearly resolved. In an analysis of the cytochrome *b* gene, Mayden et al. (2006a) resolved *H. amblops* as sister to *H. winchelli*. We follow Jenkins and Burkhead (1994), Grose and Wiley (2002), and Page et al. (2013) in retaining the monophyletic *amblops* species group in the genus *Hybopsis*.

Burr and Page (1986) suggested that populations of *H. amblops* were originally separated in the Pleistocene into multiple refugia east and west of the Mississippi River by glacial advance and subsequently expanded into previously glaciated regions following glacial retreat. Berendzen et al. (2008a) studied the phylogeography of *Hybopsis amblops* and did not find strong support for the monophyly of the *H. amblops* species group. They found five strongly supported monophyletic groups within *H. amblops* identified as clades I–V. Clades I and II contained haplotypes from the Ozark Highlands. These haplotypes composed two monophyletic groups: clade I found in tributaries of the Arkansas River draining the western Ozarks, and clade II occurring in drainages of the northern and southern Ozarks. The division of the two clades predates the capture of the rivers of the Plains Region by the Old Ancestral Arkansas River. The modern Arkansas River and its associated lowlands presumably formed a barrier to gene flow in highland fishes (Mayden 1985a), maintaining isolation of populations of *H. amblops* in the western Ozarks from those in the northern and southern Ozarks. Clade II is Late Pleistocene in origin (Berendzen et al. 2008a).

DISTRIBUTION *Hybopsis amblops* occurs in Great Lakes drainages from New York to Michigan, and in the Mississippi River basin from New York to eastern Illinois and south to Georgia and Alabama (Tennessee River drainage) in the east and to northern Arkansas and northeastern Oklahoma in the west. In Arkansas, the Bigeye Chub is widely distributed throughout the clear streams of the middle White River system, but there is only one recent record from the upper White River drainage above Beaver Reservoir. It formerly occurred at several localities in the mountainous, southward-draining tributaries of the upper Arkansas River in western Arkansas (Map A4.42), but it has drastically declined in that drainage. The only recent records were obtained by Petersen et al. (2014) from the Baron Fork of the Illinois River in Washington County in 2012, and by TMB and Jeff Quinn from the Illinois River in Benton County on 2 August 2016. Records from the lower White River drainage of the Coastal Plain were found in 2002 in White and Woodruff counties (data from USACE provided by K. J. Killgore, with identity confirmed by Neil Douglas) (Map 41). Buchanan (2005) reported this species in Bull Shoals Reservoir.

HABITAT AND BIOLOGY   In Arkansas, the Bigeye Chub occupies clear, highly oxygenated pool areas with noticeable current and gravelly raceways of medium to large-sized streams of the Ozarks, where it occurs over silt-free gravel or rock bottoms and sometimes over shallow, moderately flowing sandy areas. This chub tends to avoid headwater streams and large rivers. It is often associated with aquatic vegetation (Clemmer 1980a). This species cannot survive where siltation is extreme. For example, in Illinois, *H. amblops* has completely disappeared where it was once widespread. It has also disappeared from many other northern parts of its range (Pflieger 1997; Trautman 1957). Davis and Miller (1967) suggested that it cannot survive in high-gradient areas that are cleanly scoured by water action. Based on the size of the pores and complete-ness of the canals of the lateralis system, Reno (1969) sug-gested a probable specialization for clear water. Boschung and Mayden (2004) considered the presence of *H. amblops* an indication of excellent water quality. Those authors also noted that it is highly susceptible to siltation and intolerant of reservoirs.

Hybopsis amblops is a visual feeder, eating small inverte-brates from clean, sandy bottoms (Davis and Miller 1967). Boschung and Mayden (2004) noted that its large eyes and extremely low numbers of cutaneous taste buds are adapta-tions for visually detecting food. Etnier and Starnes (1993) examined the gut contents of 10 specimens from Tennessee and found midge larvae, large mayfly and stonefly nymphs (*Stenonema*, *Ephemera*, and *Isoperla*), and a caddisfly (*Hydropsyche*). Jenkins and Burkhead (1994) reported that adults also fed on microcrustaceans in summer. TMB examined gut contents of 49 specimens collected between May and August from Benton, Madison, Marion, and Sharp counties and confirmed that the Bigeye Chub in Arkansas is a benthic insectivore. The diet was dominated by insect larvae, pupae, and nymphs (primarily midges and other immature dipterans, ephemeropteran nymphs, trichopterans, and odonates). More than 56% of the guts examined contained sand grains, and microcrustaceans (mainly ostracods) were also consumed in small amounts.

Pflieger (1997) collected breeding adults in Missouri in June, indicating a late spring or early summer spawn-ing season, but little is known of its reproductive biol-ogy. Examination of gonads by TMB indicated that peak spawning in Arkansas occurs in May and June,

with some reproductive activity continuing through July. Ripe females collected from the South Fork of the Spring River in Sharp County on 17 May and the White River in Madison County on 23 May ranged from 52 to 75 mm TL and contained between 126 and 392 mature ova. Ripe ova were approximately 1 mm in diameter, and most ripe females in May contained large numbers of smaller devel-oping ova. Pflieger (1997) indicated Bigeye Chubs reach about 50 mm TL after the first summer of life and are about 65–80 mm TL in the second summer. Etnier and Starnes (1993) reported sexual maturity occurs at about 55 mm TL.

CONSERVATION STATUS   Pflieger (1997) reported a drastic decline in abundance and distribution of *Hybopsis amblops* in Missouri, and we have noted some declines in Arkansas as well (compare our 1988 distribution map in Appendix 4 with our current map). The Bigeye Chub has apparently declined drastically in abundance in the Upper White River drainage above Beaver Reservoir. It is never an abundant species in collections, but it is widespread, com-monly collected, and seems to be in no imminent danger elsewhere in the White River drainage. It may be extir-pated from most Arkansas River drainages in the state, except for the Illinois River. We judge its populations to be secure only in the middle portion of the White River drainage in Arkansas.

Hybopsis amnis (Hubbs and Greene)
Pallid Shiner

Figure 15.48. *Hybopsis amnis*, Pallid Shiner. *Uland Thomas.*

CHARACTERS   A slender, slightly compressed, frag-ile shiner of moderate size with a large eye; eye diameter going 2.7–3.4 times into head length (Fig. 15.48). Upper jaw unique among native cyprinids, extending posteriorly beyond the corner of the mouth and nearly one-third hid-den by the suborbital bone when the mouth is closed (Fig. 15.17). Mouth very small, subterminal, and horizontal; rear margin of upper jaw not reaching front of eye; front (cen-ter) of upper jaw even with or lower than lower rim of orbit. Barbels absent. Snout blunt and rounded, extending well

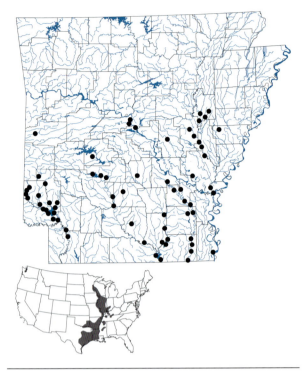

Map 42. Distribution of *Hybopsis amnis*, Pallid Shiner, 1988–2019.

beyond the upper lip. Pharyngeal teeth 1,4–4,1. Dorsal fin high and pointed with anterior rays much elongated; dorsal fin origin situated over pelvic fin origin and closer to tip of snout than to caudal fin base. Dorsal fin rays 8. Dorsal, anal, and pectoral fins with concave margins, the tip of the depressed dorsal fin reaching to or past the middle of the caudal fin base (Etnier and Starnes 1993). Anal rays 8 (7–9); pectoral rays usually 12 (13–15); pelvic rays 8. Scales highly deciduous. Lateral line complete; scales large in lateral line, but not especially elevated; lateral line scales 34–37. Predorsal scale rows 13–15. Breast scalation varies from naked anteriad to naked with 1 or 2 scales at pectoral fin base. Intestine short with a single S-shaped loop. Peritoneum silvery, sometimes with faint speckles. Nuptial tubercles of males largest on the cheeks, lachrymal area, and top of head between the orbits (Hubbs 1951). Pectoral fins with tubercles. Maximum reported size about 2.75 inches (70 mm). Largest Arkansas specimen 2.83 inches (72 mm) TL from Lake Hamilton, Garland County.

LIFE COLORS Coloration variable in Arkansas. Often, as the common name suggests, it is a very pallid, silvery fish with little or no noticeable pigmentation. Other specimens have a pale olive dorsum, silvery sides, and whitish-silver belly; fins largely unpigmented. The basal portion of the medial caudal rays and interradial membranes are heavily pigmented. Rays of remaining fins are lined with scattered melanophores, but interradial membranes are unpigmented. Lips and chin are unpigmented. Scales above dusky to dark lateral band are faintly dark-edged. The degree of lateral band development is extremely variable from entirely absent to dark and prominent. When well developed, the lateral band extends from the base of the caudal fin along midside and forward in front of eye around tip of snout. Individuals from clear waters have iridescent blue-green sides and often an orange tinge in the caudal fin in summer months.

SIMILAR SPECIES The Pallid Shiner's unique upper jaw, one-third of which extends past angle of mouth when mouth is opened, readily separates it from all other minnows. It most closely resembles *Hybopsis amblops* and several other species of barbeled minnows, but differs from them in lacking barbels (vs. barbels present). It is also similar to *Notropis volucellus* with which it is often syntopic in Arkansas and can be distinguished from that species in having pharyngeal teeth 1,4–4,1 (vs. 0,4–4,0), and in not having elevated anterior lateral line scales (vs. anterior lateral line scales elevated). The Pallid Shiner differs from *Lythrurus fumeus* in having a blunt, rounded, protruding snout and 8 anal rays (vs. 10 or 11).

VARIATION AND TAXONOMY *Hybopsis amnis* was described by Hubbs and Greene *in* C. L. Hubbs (1951) as *Notropis amnis* from the Mississippi River, 1.7 km north of Prairie du Chien, Crawford County, Wisconsin. It remained in genus *Notropis* for most of the next four decades. Clemmer (1971) and Clemmer and Suttkus (1971) placed *Hybopsis amnis* along with four other species (*H. amblops*, *H. rubrifrons*, *H. lineapunctata*, and *H. winchelli*) in the *Hybopsis amblops* species group. Mayden (1989) added *Hybopsis hypsinotus* to this species group as the most basal member based on morphological characters. After 1990, the placement of *amnis* in genus *Hybopsis* was accepted by some authors (Etnier and Starnes 1993; Page et al. 2013) and not accepted by others (Coburn and Cavender 1992; Ross 2001). Hubbs (1951) separated *H. amnis* into two subspecies: *H. a. amnis*, a northern form found in the upper Mississippi Valley, and *H. a. pinnosa* Hubbs and Bonham, a southwestern subspecies. In Arkansas, *H. a. pinnosa* is found in the Arkansas (Poteau) and Red river systems, while in northeastern Arkansas, intergrades between the two recognized subspecies occur in the St. Francis and lower White river drainages. Clemmer (1971, 1980b) disputed the validity of these subspecies because of the broad overlap in the Mississippi Valley.

DISTRIBUTION *Hybopsis amnis* occurs in the Mississippi River basin from Wisconsin and Minnesota south to Louisiana, and in Gulf Slope drainages from southwestern Louisiana west to the Guadalupe River, Texas. In Arkansas, the Pallid Shiner is found sporadically throughout the Coastal Plain province from the Black, lower White, St. Francis, lower Arkansas, Poteau, Little, Red, Saline, and Ouachita rivers (Map 42). McAllister et al. (2009a) reported specimens from 25 of 75 Arkansas counties from all major rivers of the Gulf Coastal Plain and Mississippi Alluvial Plain except the St. Francis River. Tumlison et al. (2016) reported the collection of 7 adult *H. amnis* from Big Creek (St. Francis River drainage) in Lee County, documenting its continued existence in that drainage. There are no recent records from the Black River.

We correct a recent erroneous report by McAllister et al. (2009a, 2010b) of a new record for *Hybopsis amnis* from the lower Strawberry River, Lawrence County, Arkansas. After subsequent examinations and comparisons by one of those authors (WCS), the single specimen (NCSM 47146) was reidentified as *H. amblops*, a species previously known from the Strawberry River system (Robison and Buchanan 1988).

HABITAT AND BIOLOGY   *Hybopsis amnis* occurs mainly in small to medium-sized, streams of the Coastal Plain. It is also common in some reservoirs and oxbow lakes in Arkansas. Buchanan (2005) reported it from eight Arkansas impoundments where it was often abundant (approximately 3,700 specimens taken). It appears to prefer sand and mud substrates and slow-moving waters. Becker (1983) believed that the delicate fins and skin and the moderately streamlined build are adaptations for slow currents. Although it has been found in both clear and moderately turbid waters, Pflieger (1997) reported that *H. amnis* is intolerant of excessive siltation associated with changing land use practices. This could account for its absence from recent collections in the lowlands of northeastern Arkansas. C. L. Hubbs (1951) noted that southern populations of *H. amnis* are found in smaller rivers and creeks, while northern populations typically occur in large rivers. It is rarely found in Arkansas's three largest and most turbid rivers. In the Kankakee River, Illinois, Kwak (1991) reported its preferred habitat was shallow areas with little or no current and moderately clear waters that were well oxygenated in summer. Juveniles occurred in shallower and more turbid waters than adults, and both juveniles and adults were closely associated with habitat parameters most similar to those of *Pimephales vigilax*.

Little is known about its feeding or spawning habits (Kwak 1991). Gut contents of 58 specimens collected in July and August in Garland, Perry, and Monroe counties in Arkansas were examined by TMB. The diet in Lake Hamilton (Garland County) was dominated by microcrustaceans, predominantly copepods. The remainder of the diet in that lake consisted of immature aquatic insects, primarily ephemeropterans, trichopterans, and dipteran larvae. The guts also contained much organic debris. In the Arkansas River (Perry County) and a White River oxbow lake (Monroe County), the summer diet consisted almost entirely of immature aquatic insects, with microcrustaceans, dipteran pupae, detritus, and annelids also eaten in small amounts.

Clemmer (1980b) reported that *H. amnis* breeds in late winter and early spring in the southern part of its range and that ripe adults have been taken as early as March in the Mississippi River in Arkansas. In Illinois, length-frequency distributions indicated that at least two distinct size classes existed during July and August (Kwak 1991). Kwak speculated that in most areas floodplain access may be important for successful spawning or survival of young. Kwak's data suggested that juvenile (age 0) Pallid Shiners attained a TL of up to 36 mm (1.42 in.) by August and that longer individuals collected at the same time were age 1. The largest individuals collected (53 mm and 54 mm [2.09 and 2.43 in.] TL) may represent a third age class (age 2). Becker (1983) reported lengths of 34 and 49 mm TL at ages 1 and 2 and a life span of at least 3 years in Wisconsin. Smith (1979) noted that it is one of the least-known American fishes.

CONSERVATION STATUS   Hubbs (1951) noted that the Pallid Shiner is a rare species throughout much of its range. Many northern populations have long been considered greatly reduced or extirpated (Becker 1983; Pflieger 1997; Smith 1979). It is currently more common in the southern parts of its range, but even there it is sporadically distributed and rarely abundant. Burr and Warren (1986b) reported the Pallid Shiner may be on the decline in the Midwest, and Page and Burr (2011) referred to it as generally rare and declining. McAllister et al. (2010b) provided an updated status of the Pallid Shiner in Arkansas and reported it to be stable and quite localized. We consider its populations currently secure in Arkansas.

Genus *Hypophthalmichthys* Bleeker, 1860 — Bigheaded Carps

This East Asian cyprinid genus contains three species, two of which have been introduced to central North America with potentially disastrous ecological consequences (see Bighead Carp species account). Li et al. (2009), based on analysis of complete mitochondrial genomes of Silver Carp and Bighead Carp, supported the view that the two species are congeneric. Populations of both species have exploded in the Mississippi River and other large rivers of the Mississippi River Valley. The genus *Hypophthalmichthys* is undocumented in the fossil record and like many carps and minnows, it is of comparatively recent origin (Chapman et al. 2016).

Characters include a terminal mouth, 85 or more small scales in the lateral line, a deep body, the eye in an unusual position on the anteroventral region of the head, a smooth midventral keel on abdomen, 8 dorsal fin rays, dorsal fin origin posterior to vertical from pelvic fin origin, a forked homocercal tail, and pharyngeal teeth molariform, 0,4–4,0, and opposing a grinding surface on the basioccipital area. Gill rakers are extremely close-set for filtering mid-water food particles, which are further consolidated and ground by the pharyngeal apparatus (Etnier and Starnes 1993). A specialized epibranchial (or suprabranchial) organ (Wilamovski 1972) is present on the rear palate area and presumably functions in propelling food particles into the gill strainers. The gut is long and convoluted. These fishes are pelagic planktivores and reach sizes of 27 kg (60 lbs.) or more. Both species introduced into North America are found in Arkansas.

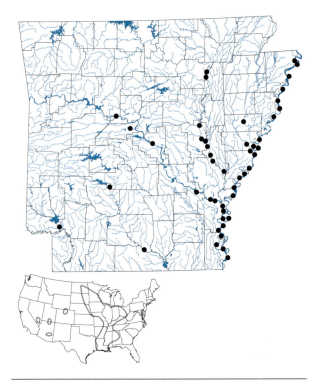

Map 43. Distribution of *Hypophthalmichthys molitrix*, Silver Carp, 1988–2019. Inset shows introduced range (outlined) in the contiguous United States.

Hypophthalmichthys molitrix (Valenciennes)
Silver Carp

Figure 15.49. *Hypophthalmichthys molitrix*, Silver Carp. *Uland Thomas.*

CHARACTERS   A deep-bodied, laterally compressed, very large, silvery cyprinid (Fig. 15.49). Scales very tiny, cycloid, resembling those of trout. Lateral line decurved and complete, usually with more than 90 (83–130) scales. Scales above lateral line 27–38, and scales below lateral line 14–19. Head and opercle scaleless. Head length going into SL more than 3.0 times. Mouth relatively large, upturned, without teeth in jaws. Eyes situated far forward along the midline of the body and projecting somewhat downward. Fins of small specimens without spines. In large specimens, front of pectoral fin with a stiff, hard spine having fine serrae on posterior margin; dorsal fin with moderately stiff, nonserrate, spinelike ray at origin; anal fin with a slightly stiffened, nonserrate spine (hardened ray) at origin. Dorsal fin origin behind pelvic fin origin; dorsal rays 8. Anal fin falcate, with 11–13 rays. Pectoral fin, when depressed, does not reach the front of the pelvic fin base; pectoral rays 15–19. Vertebrae number 36–39 (Li et al. 1990). A smooth ventral keel extends from base of anal fin to isthmus (Fig. 15.50). Gill rakers extremely numerous and thin, fused into a spongelike porous plate. Pharyngeal teeth 0,4–4,0, moderately long and bluntly rounded, slightly concave on the grinding surface (Li et al. 1990), and having striations on the grinding surface that are visible with magnification (Chapman et al. 2016). Intestine very long with many loops, its length fewer than 7 times to more than 13 times the body length, depending on season or principal foods (phytoplankton or zooplankton, Ke et al. 2008). Maximum size around 60 pounds (27.2 kg), but adults more commonly reaching 10–20 pounds (4.5–9.1 kg). The Arkansas state angling record, from the Arkansas River in 1995, is 39 pounds, 4 ounces (17.8 kg). The Arkansas unrestricted tackle record, from the Arkansas River in 2010, is 61 pounds, 0 ounces (27.7 kg).

LIFE COLORS   Back and upper sides olivaceous grading to silver below the lateral line. Belly white. Fins unpigmented. Small young are silver.

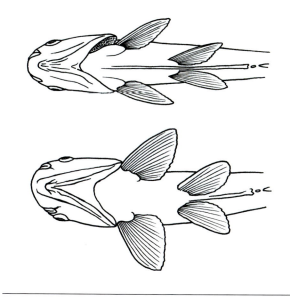

Figure 15.50. Keels of bigheaded carps.

Top—*Hypophthalmichthys molitrix*, Silver Carp
Bottom—*Hypophthalmichthys nobilis*, Bighead Carp

SIMILAR SPECIES   The Silver Carp is most similar to the Bighead Carp, but it can be distinguished from that species by the ventral keel extending from vent to isthmus (vs. keel extending from vent only to base of pelvic fins in Bighead) (Fig. 15.50), more laterally compressed body, and by its lack of small irregular dark blotches on sides (vs. body less compressed and with dark blotches on sides). It differs from all other Arkansas Cyprinidae (except *Chrosomus erythrogaster*) in having tiny scales, usually with more than 90 in lateral line. The small silvery young resemble shad but can be distinguished from all clupeids by the presence of a lateral line (vs. lateral line absent) and 15 or fewer anal rays (vs. usually more than 16).

VARIATION AND TAXONOMY   *Hypophthalmichthys molitrix* was described by Valenciennes *in* Cuvier and Valenciennes (1844) as *Leuciscus molitrix* from China. Howes (1981) studied the anatomy and phylogeny of the genus *Hypophthalmichthys* (see *H. nobilis* account).

DISTRIBUTION   (Map 43) The Silver Carp is native to Pacific drainages of eastern Asia from the Amur River of Russia south through China and possibly to northern Vietnam. It has been introduced into at least 88 countries and territories for aquaculture and for the control of nuisance plankton (Kolar et al. 2007). This invasive species is currently known from 15 states in the U.S. It was first introduced into Arkansas in 1973 by a private fish farmer (Henderson 1976) primarily for use in biological control of phytoplankton and vegetation and for aquaculture (Kolar

et al. 2007). At the time of its introduction, the Clean Water Act strongly urged the use of biological control agents to reduce the amount of pesticides and herbicides in the environment. By the mid-1970s, the Silver Carp was raised at six state, federal, and private facilities in Arkansas, and by the late 1970s it had been stocked in four municipal sewage lagoons in the state. In the 1970s and 1980s, the AGFC conducted extensive research on using Silver and Bighead carp to reduce nutrients in sewage lagoons in Arkansas. Similar research was conducted on swine waste lagoons by Homer Buck in Illinois, and additional research was conducted in other states (Mike Freeze, pers. comm.). Escape of Silver Carp from research facilities, aquaculture facilities, sewage lagoons, and reservoirs followed, and this species has now become well established in much of the Mississippi River basin (Fuller et al. 1999). Reports of Silver Carp in Arkansas's natural waters first came from the White and Arkansas rivers by 1980 (Freeze and Henderson 1982) and later from the Mississippi River (Carter and Beadles 1983b). Arkansas commercial fishermen reported catching 166 Silver Carp at seven different sites in 1980 and 1981 (Freeze and Henderson 1982). Jeff Quinn (pers. comm.) reported that the Silver Carp has now been documented from the Little River system in southwestern Arkansas (Red River drainage) and from the Ouachita River. Patton and Tackett (2012) discovered this carp in the Red River in Oklahoma, and it has also been reported as abundant in the lower St. Francis River in 2013 (Jeff Quinn, pers. comm.).

HABITAT AND BIOLOGY   The Silver Carp's history and use in Arkansas are closely intertwined with those of the Bighead Carp, and it is morphologically and biologically similar to that species. In its native range, adults are lentic, open-water fish moving about in the euphotic zones of large lowland rivers. In the Mississippi River, Silver Carp selected wing-dike areas of moderate flow (about 0.3 m/s) and elevated chlorophyll *a* (Calkins et al. 2012). In that study, no Silver Carp occurred in areas where flow was absent, and the main channel was also avoided. Silver Carp are easily startled and actively avoid nets by scattering and jumping. In the larger rivers of North America, Silver Carp are notorious for their habit of jumping out of the water individually (to a maximum height of 2.4 m) or en masse when frightened by a boat motor or other noise, sometimes actually jumping into a boat and injuring its occupants. However, this is considered a rare behavior in their native China (Chapman et al. 2016).

The Silver Carp, like the Bighead Carp, is a pelagic planktivore, straining plankton from the water. It is more highly specialized than that species for straining the smaller phytoplankton because of its unique, spongelike gill rakers that form a porous plate. It is capable of straining organisms as small as 4 microns in diameter and is apparently very efficient at ingesting Chlorophyta (green algae) and Cyanobacteria (bluegreen algae). Diet differences have been reported in different areas and at different seasons, making it difficult to predict specific effects of Silver Carp feeding activities on a given community. Rotifers formed the major part of the diet in the Mississippi River and Illinois River backwater lakes (Sampson et al. 2009). In that study, dietary overlap was greatest with Gizzard Shad followed by Bigmouth Buffalo, but there was little dietary overlap with Paddlefish. Williamson and Garvey (2005) found that phytoplankton was the most important food item in the middle Mississippi River. Minder and Pyron (2018) also found that Silver Carp fed almost exclusively on phytoplankton in the Wabash River, Indiana, with Cyanobacteria and Chlorophyta dominating the diet in spring, and Chlorophyta and diatoms dominating the summer diet. Silver Carp diet in the Wabash River overlapped with two native fishes, *Dorosoma cepedianum* and *Ictiobus cyprinellus*. In the Mississippi River, macrozooplankton and detritus were rarely found in Silver Carp guts (Calkins et al. 2012). Food items that are ingested are ground by the blunt, rounded pharyngeal teeth against a cartilaginous plate disrupting the cell walls and allowing enzymatic digestion of the algal cell contents (Xie 1999). Xie concluded that without this mechanical grinding of the food, Silver Carp are unable to effectively digest algae. The presence of phytoplankton in Silver Carp gut contents smaller than their filtering net meshes led Xie (1999) to conclude that the secretion of mucus may play an important role in collection of the smallest food particles.

In its native range, the Silver Carp requires large rivers for successful reproduction. In Asia, spawning takes place between April and July in large rivers such as the Yangtze River (Chen et al. 2007). Silver Carp reproduce in riverine areas with relatively high current velocities (>0.7 m/s), high turbidity, and water temperatures greater than 18°C (22–24°C), often moving long distances prior to spawning. Spawning habits are similar to those of the Bighead Carp and Grass Carp, and current is required to float the developing eggs, which are negatively buoyant and may float long distances downstream (Verigin et al. 1978; DeGrandchamp et al. 2007; Garvey 2007; Kolar et al. 2007). Spawning occurs in groups. During spawning, females release eggs in the water column, where they are fertilized by many males. Henderson (1979) concluded that Silver Carp could not spawn successfully in a lentic (pond) environment, but Kolar et al. (2007) demonstrated successful in vitro

embryonic development in static, sediment-laden plastic bags. A long breeding season occurred in the Wabash River, Indiana, with DNA-confirmed drifting eggs found from early May into September at water temperatures between 18.5°C and 29.7°C (Coulter et al. 2013). That Indiana study also provided evidence that Silver Carp spawn over a wider range of hydrological conditions, over a more protracted time period, and in smaller streams than reported from the large rivers in their native range. Deters et al. (2013) also confirmed spawning by Silver Carp during periods of low discharge and declining hydrograph in the Missouri River. Therefore, introduced populations exhibit plastic spawning traits that may facilitate invasion and establishment in a wider range of river conditions than previously thought (Coulter et al. 2013). Sexual maturity was reached at age 2 in the middle Mississippi River, and 5 age classes were present (Williamson and Garvey 2005). Silver Carp seem to be incremental spawners, spawning a portion of the annual production of the eggs in many separate spawning events over a longer period (Papoulias et al. 2006; Chapman et al. 2016). Fecundity estimates of Silver Carp in the Yangtze River in China ranged from an average of 396,000 eggs per individual for fish between 65 and 70 cm TL to 1.65 million eggs per individual for fish between 85 and 90 cm TL (Chapman et al. 2016). Silver Carp eggs are steel-gray or tawny with a yolk diameter between 1.3 and 1.5 mm. Most water-hardened eggs are between 4.2 and 6.2 mm (Yi et al. 2006). Average fecundity in the Mississippi River was 156,312 eggs (Williamson and Garvey 2005). Eggs hatch in 26–28 hours at a water temperature of 22.3–22.5°C (Kočovský et al. 2012). The Silver Carp is known to use floodplain habitats as juveniles (Abdusamadov 1987), which may facilitate dispersal into waters within the floodplain such as oxbow lakes (Patton and Tackett 2012). Varble et al. (2007) found that in the lower Mississippi River, nurseries may occur in the main channel and in oxbow habitats, but significant nurseries also occur on the wetland floodplain, and in some cases, are remote from the main channel spawning areas. Spawning has now been documented in the Mississippi River of Arkansas. A 27 mm TL specimen was taken from the Mississippi River in Mississippi County on 26 July 1999 by TMB and Sam Barkley. Maximum life span is around 20 years in the wild.

Like the Bighead Carp, *H. molitrix* has some potential for use in wastewater treatment plants and probably less potential than that species as a food fish. Its possible effects on natural environments and on the native fish fauna are probably similar to those documented for Bighead Carp.

CONSERVATION STATUS   This invasive species is thought to deplete plankton stocks for native larval fishes and mussels (Laird and Page 1996). It may also be a direct competitor with adults of some native species that feed on plankton such as Paddlefish and Gizzard Shad (Chen et al. 2007; Minder and Pyron 2018). This introduced species is established in Arkansas.

Hypophthalmichthys nobilis (Richardson)
Bighead Carp

Figure 15.51. *Hypophthalmichthys nobilis*, Bighead Carp. *David A. Neely.*

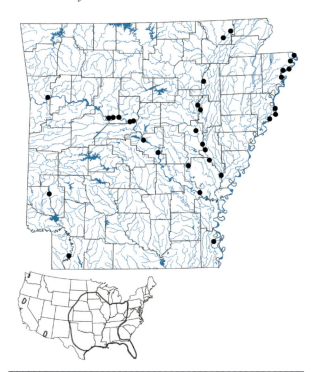

Map 44. Distribution of *Hypophthalmichthys nobilis*, Bighead Carp, 1988–2019. Inset shows introduced range (outlined) in the contiguous United States.

CHARACTERS   A deep-bodied, somewhat laterally compressed, very large cyprinid with a large head (Fig. 15.51); head length going into SL fewer than 3.0 times. Scales very tiny, cycloid, resembling those of trout. Lateral line strongly decurved anteriorly and complete, with 85–100 ($\bar{x}$ = 92.4) scales. Scale rows above lateral line 24–29 and 15–24 scale rows below lateral line. Head and opercle scaleless. Mouth large, terminal without teeth in jaws; lower jaw projecting beyond upper jaw. The premaxilla and protruding mandible form rigid bony lips. Eyes situated far forward along midline of body and projecting downward. Fins of small specimens without spines; large specimens have a heavy, stiff, nonserrate spine (hardened ray) at origin of pectoral fin, a moderately stiff nonserrate spine at dorsal fin origin, and a slightly stiffened spine at anal origin. Dorsal fin origin behind pelvic fin origin; soft dorsal rays 8 (rarely 9). Anal fin falcate, soft anal rays 13 (rarely 14); pelvic rays 7 or 8; pectoral rays 16–21. Pectoral fin, when depressed, extends posteriorly past the front of the pelvic fins. Vertebrae number 37–44. A smooth ventral keel extends from vent forward to pelvic fin base (Fig. 15.50). Gill rakers long, comblike and close-set, not fused into a porous plate. Pharyngeal teeth in a single row, 0,4–4,0, moderately long and bluntly rounded; teeth have a spoonlike shape and a shallowly concave grinding surface. Intestine long and highly convoluted, its length 3–5 times longer than the total fish length. Maximum size at least 110 pounds (50 kg), but adults more commonly reaching 10–20 pounds (4.5–9.1 kg). The Arkansas state angling record, from the Arkansas River in 2007, is 103 pounds, 8 ounces (47 kg).

LIFE COLORS   Back and upper sides dark gray grading to cream color or off-white on lower sides and belly. Many dark gray to black irregularly shaped blotches scattered over entire body. Young silvery, not developing characteristic blotches and coloration until about 8 weeks old.

SIMILAR SPECIES   Similar to the Silver Carp, but differs from that species in having a shorter ventral keel, extending from vent to pelvic fin base (vs. keel extending from vent to isthmus) (Fig. 15.50), and in having numerous scattered small, black, irregular blotches on sides of large juveniles and adults (vs. scattered black blotches absent). It differs from all other Arkansas Cyprinidae (except *Chrosomus erythrogaster*) in having very tiny scales on body, with more than 84 in lateral line. It further differs from the introduced Grass Carp in having 13 anal rays (vs. 9), a ventral keel between vent and pelvic fins, and in having the dorsal fin origin distinctly behind the pelvic fin origin. The small silvery young can be distinguished from shad and other clupeids by the presence of a lateral line (vs. lateral line absent) and 14 or fewer anal rays (vs. usually more than 16).

VARIATION AND TAXONOMY   *Hypophthalmichthys nobilis* was originally described by Richardson (1845) as *Leuciscus nobilis* from China. Oshima (1919) established the genus *Aristichthys* for *nobilis* believing that differences in gill raker form, pharyngeal dentition, and abdominal keel length between that species and *Hypophthalmichthys molitrix* were sufficient to warrant separate generic placement. Gosline (1978) believed there were clear indications of common derivation for *Hypophthalmichthys* and *Aristichthys* based on a trilobed swim bladder as one indication of this relationship. A trilobed swim bladder has such a mosaic distribution throughout the Cyprinidae that it has no value as a character indicating shared common ancestry (Howes 1981). For example, the third "lobe" in *Hypophthalmichthys* is a mere constriction or "tail" of the swim bladder and is variously developed in individual fish. However, Howes's study of the anatomy and phylogeny of *Hypophthalmichthys* supported the inclusion of Bighead Carp. Li et al. (2009) also provided genetic evidence of the congeneric status of Bighead and Silver carp.

DISTRIBUTION   (Map 44) The Bighead Carp is native to southern and central China, but introduced populations have become established in at least 74 countries and territories, mainly in Europe, Central Asia, and North America (Kolar et al. 2007). In North America, it is established in the Missouri and Mississippi river basins from Kentucky to South Dakota and south to Louisiana (Page and Burr 2011). It was first brought to Arkansas in the early 1970s by private fish farmers. By the mid-1970s, it was being raised at six state, federal, and private facilities, and by the late 1970s it had been stocked in four municipal sewage lagoons. Bighead Carp were first reported from the state's natural waters (Arkansas River near Pine Bluff) in 1986 (John Hogue, pers. comm.). Since then, the Bighead Carp has spread throughout the Mississippi River, White River, Arkansas River, and Red River drainages in Arkansas, and by 1991 it had spread to the Mississippi River of Missouri and Illinois (Tucker et al. 1996). While both the Silver and Bighead carp occur in relatively high abundance in the lower reaches of the Arkansas River in Arkansas, only the Bighead Carp has been documented as far upstream as Dardanelle reservoir (Jeff Quinn, pers. comm.), and TMB collected juveniles from the Arkansas River in Crawford County (Quinn and Limbird 2008). Unfortunately, this species has been documented to move through lock and dam systems in relatively short periods of time (Garvey 2007); therefore, its spread upstream is simply a matter of time. In 2018, a specimen was documented from the Missouri portion of Bull Shoals Reservoir of the upper White River (R. Hrabik, pers. comm.).

HABITAT AND BIOLOGY   Most information on the Bighead Carp in Arkansas and its possible role in pond fish culture is from Freeze and Henderson (1982), and Henderson (1976, 1977, 1979, 1983). In 1974 the AGFC began evaluating its biological potential. In its native range, it occupies primarily large rivers and associated floodplain backwaters and lakes. It is an open-water fish, moving about in the euphotic zones of large lowland rivers. Preferred habitat types reported from North America were confluences with tributaries, backwaters, regions below dams, side channels, and channel borders (Kipp et al. 2011). During normal spring flows in the Mississippi River, larval densities were greater in open river reaches than in impounded ones (Lohmeyer and Garvey 2009).

The problems presented by this alien species are many. Kolar et al. (2007) provided a thorough review of potential ecological impacts. There are documented ecological impacts of the Bighead Carp on the habitat and water quality of streams. These impacts include changes in nutrient contents, sediment resuspension, increased turbidity, reduced dissolved oxygen, and increased nitrogen. In addition, there are some documented impacts on trophic dynamics, including increased algal blooms (through trophic cascades associated with removing select algal species). Effects may also include changes in the benthic community, inhibition of sight-feeding predators caused by turbidity, changes in the size structure of the zooplankton community (Cooke et al. 2009), and alteration of the benthic macroinvertebrate community through high levels of excrement deposition and organic enrichment of the substrate. Additionally, there may be competitive interactions with other fishes. Patton and Tackett (2012) reported that when introduced into India, both Silver and Bighead carp dominated the commercial catch within 10 years and were negatively correlated with abundance of native planktivores. Similar instances of competitive displacement by Silver and Bighead carp may have occurred in other countries where they have been introduced, including the United States in the Mississippi River basin. Planktivorous fishes and especially planktivorous fry are the most affected. In the Mississippi River basin, there is concern that this species may reduce abundance of planktivores such as Gizzard Shad, Threadfin Shad, and Emerald Shiner, all important forage foods for other fishes and birds. Schrank et al. (2003) provided evidence that age 0 Bighead Carp have the potential to negatively affect the growth of age 0 Paddlefish. In addition, Silver and Bighead carp may affect other fish species through spatial displacement, and have been implicated in the spread of communicable fish diseases (Patton and Tackett 2012). Interestingly, even though Silver Carp are easily startled and notorious for their habit of jumping from the water when a motorboat passes, Bighead Carp are languid and rarely jump from the water (Chapman et al. 2016).

The Bighead Carp is a low trophic level filter feeder, straining plankton organisms from the water with its long, comblike gill rakers (Cremer and Smitherman 1980). In its native range, it inhabits fertile bodies of water which support large plankton populations. It often feeds at the water surface and the white mandible is often visible during feeding, assisting in identification and location of feeding fish (Kolar et al. 2007). Although it is as efficient in filter feeding as any of our native species, it is not quite as highly specialized as the Silver Carp for filtering the tiniest phytoplankton. It tends to consume more zooplankton than the Silver Carp. Although typically zooplanktivorous, it can also be very opportunistic, consuming a variety of prey items (Kolar et al. 2007). Food found in its intestine consisted mainly of large clumps of bluegreen algae, zooplankton, and aquatic insect larvae and adults. Some small Chlorophyta and diatoms have also been noted. After ingestion, the strained food organisms are passed to the pharynx, where the blunt, rounded pharyngeal teeth grind the food against a heart-shaped cartilaginous plate. Larval and juvenile Bighead Carp feed on both zooplankton and phytoplankton (Cooke et al. 2009). Dietary overlap in Mississippi River backwater habitats was greatest with Gizzard Shad and Bigmouth Buffalo (Sampson et al. 2009).

Spawning habits of Bighead Carp are similar to those of the Grass Carp and Silver Carp. A free-flowing river with sufficient current to float the semibuoyant eggs until hatching is required. Unobstructed river stretches of about 50–100 km (Kolar et al. 2007; Kipp et al. 2011) are required for successful spawning (depending on velocity and temperature). In its native range, spawning events are triggered by rising floodwaters and water temperatures, and the adults migrate upstream in response to high spring flows. Spawning occurs in China from April to August when rivers rise after heavy rains (Berry and Low 1970). The optimum spawning temperature, usually reached in April in most Arkansas waters, is 71.6–75.2°F (22–24°C), and spawning has been reported between 18°C and 30°C. Larvae were abundant in the Missouri River from May through July (Schrank et al. 2001). Multiple spawning events have been recorded throughout the summer in the United States to as late as October (Kipp et al. 2011). Bighead Carp have high fecundity. In the Yangtze River in China, fecundity of a Bighead Carp female weighing 18.5 kg was 1.1 million eggs (Chang 1966). Bighead Carp eggs are faint yellow in color with a yolk diameter of about 1.6 mm (Chapman et al. 2016). Water-hardened eggs are usually between 5.0 and 6.7 mm. Embryos develop while drifting in the current, and the eggs are only slightly heavier than water, which allows them to be kept suspended in the current by turbulence until hatching (Chapman et al. 2016). In the Mississippi River, young-of-the-year Bighead Carp are often taken in very shallow wetlands, while adult fish can be found in nearly every habitat available, using primarily low-velocity waters when not spawning (Kolar et al. 2007). A 23 mm TL juvenile was taken from the Mississippi River in Mississippi County on 26 July 1999, and specimens 21 and 39 mm TL were taken near that locality on 25 July 2000 by TMB and Sam Barkley. Bighead Carp have never been known to spawn naturally in ponds or other lentic environments in the state, despite numerous attempts to achieve natural spawning in state fish hatchery ponds. Reproduction at hatcheries is achieved through hormonal injections. Males of Bighead Carp usually mature 1 year earlier than females. In the lower Mississippi River, gravid females have been found that are less than 2 years old (J. J. Hoover, pers. comm.).

The Bighead Carp has been successfully grown in sewage oxidation lagoons and fertilized fish ponds in combination with the Silver Carp. It apparently has some potential as an agent for removing excessive nutrients and algae from wastewater. Because of its rapid growth, it could also have some potential as a food fish. Its flavor has been judged good to very good (Henderson 1976). Age and growth studies of the Bighead Carp in the Mississippi and Missouri rivers (Schrank and Guy 2002; Nuevo et al. 2004a, b) reported no specimens greater than 1 m long or older than 7 years. However, Long and Nealis (2011) reported a 1,356 mm TL, 39.8 kg Bighead Carp from Grand Lake (Arkansas River drainage) in Oklahoma estimated to be 9 years old based on estimates from three different structures (pectoral ray, branchiostegal ray, and otolith). Maximum reported age in the U.S. for all populations is typically 7 years (Tsehaye et al. 2013). Maximum age appears to be between 16 and 20 years (Kolar et al. 2005), but fish of this age have not been documented in the United States (Long and Nealis 2011). The oldest Bighead Carp reported in the United States is an 11-year-old specimen 1,316 mm TL and weighing 49.7 kg (Hoover et al. 2015). Bighead Carp in the LMR may spawn in the spring and possibly again in late summer. Fecundity estimates exceeded 1.8 million eggs. Number of eggs documented in the United States at about 0.8 million (Schrank and Guy 2002), but Hoover et al. (2015) estimated a fecundity of 1.9–2.7 million eggs (40% of which were ripe) in a 109.6-pound (49.7 kg) individual taken on 24 February 2014 from a Mississippi River oxbow lake in northwestern Mississippi. Mature eggs average 1.43 mm in diameter.

CONSERVATION STATUS This introduced Asian species merits no status.

Genus *Luxilus* Rafinesque, 1820a— Highscale Shiners

Luxilus was formerly considered a subgenus of *Notropis* (Gilbert 1964). It was elevated to genus level by Mayden (1989) and consists of nine described species (Page et al. 2013). There is no consensus on the monophyly of *Luxilus*, despite several analyses. Morphological and molecular data produced different results depending on the number of taxa sampled (Mayden 1989; Coburn and Cavender 1992; Powers and Gold 1992; Dowling and Naylor 1997; Mayden et al. 2006a). There are four species of *Luxilus* in Arkansas.

The genus *Luxilus* is diagnosed as usually having a compressed body, a large mouth, and dorsal fin directly above or slightly behind the pelvic fin origin. Height of anterior lateral line scales and adjacent scales is slightly greater than the exposed scale width in three of four Arkansas *Luxilus* species. Lateral line scales usually 37–40; predorsal scales 13–18; circumferential scales 24–30; and circumpeduncular scales usually 13–16. Anal rays 9, sometimes 10. Pectoral rays usually 15. Pharyngeal teeth 2,4–4,2. Peritoneum black or heavily speckled with black. Preorbital tubercles retrorse. Breeding males with red and pink colors.

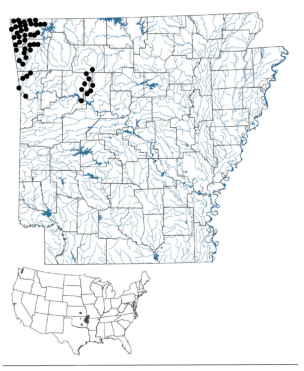

Map 45. Distribution of *Luxilus cardinalis*, Cardinal Shiner, 1988–2019.

Luxilus cardinalis (Mayden)
Cardinal Shiner

Figure 15.52. *Luxilus cardinalis*, Cardinal Shiner. *Uland Thomas.*

CHARACTERS   A large, relatively deep-bodied, slightly compressed shiner with a moderately large head, head length going into SL about 4 times. Eyes large, eye diameter going into head length 3.4–4.0 times. A prominent dark lateral band that does not become abruptly narrower just behind the gill opening; the lateral band touches the lateral line beneath the dorsal fin. A large, terminal oblique mouth (Fig. 15.52). Dorsal rays 8; anal rays 9 (rarely 8 or 10); pectoral rays 16–19. Predorsal scale rows 15–17. Pharyngeal teeth 2,4–4,2. Gill rakers usually 8 or 9. Lateral line slightly decurved anteriorly, with 39–44 ($\bar{x}$ = 41.5) scales; height of

anterior lateral line scales and adjacent scales only slightly greater than exposed scale width in adults. Total body circumferential scales usally 26 or 27. Dorsal fin origin nearly over pelvic fin origin, closer to tip of snout than to caudal fin base. Intestine short; peritoneum black. Breeding males highly tuberculate. Tuberculation is similar to that of *L. zonatus*, but usually only 0–3 mandibular tubercles. Pectoral fin rays 2–8 with a single row of large, erect tubercles forming 2 rows, with generally 1 tubercle per segment distally. Maximum size about 4.8 inches (123 mm).

LIFE COLORS   Body dark olive above with a broad, iridescent greenish-yellow middorsal stripe from occiput to caudal fin. Sides silvery with a broad, nonconstricted dusky lateral stripe extending 1–2 scales below lateral line and 2 scales above and running the length of the body; dusky lateral stripe most evident in breeding males and in preserved specimens. Basicaudal spot absent. A thin dark olive stripe above the dark lateral band separated from it by a narrow iridescent greenish-yellow stripe; belly pale and silvery. Cleithral pigmentation light or absent. Fins colorless except in breeding males. Breeding males with a rich, deep crimson red on the underside of the head and body and on all fins. Red coloration is well developed on lower sides and on base of anal fin. All fins with red pigment throughout. Lateral band becoming intensely black and wider than usual. Snout a powder blue. Some coloration develops in breeding females.

SIMILAR SPECIES   The Cardinal Shiner is most similar to *Luxilus pilsbryi* and *L. zonatus*. From *L. pilsbryi*, it differs in having more total circumferential scales, 26 or 27 (vs. 24–26); generally fewer mandibular tubercles 0–6, usually 0–3 (vs. 5–10, but usually more than 6); and gill rakers 8 or 9 (vs. 7). The Cardinal Shiner differs from *L. zonatus* in lacking dorsolateral stripes on dorsum (vs. stripes present); cleithral pigment absent (vs. present); predorsal scales 15–17 (vs. 14 or 15); crimson-red pigment strongly developed on face and lower sides of body, snout blue, and all fins with red pigment extending to base (vs. red pigment diffuse or absent on face, lower sides, and caudal peduncle, and all fins unpigmented or diffuse red, with unpigmented basal and distal bands); and lateral band broad, not constricted anteriorly, extending 1–2 scale rows below lateral line (vs. lateral band narrow, constricted anteriorly, never extending below lateral line). It can be distinguished from *L. chrysocephalus* in dark lateral stripe extending forward on head to tip of snout (vs. lateral stripe not extending to tip of snout), and scales in anterior part of lateral line with height only slightly greater than exposed width (vs. height much greater than exposed width).

VARIATION AND TAXONOMY   The Cardinal Shiner was described by Mayden (1988a) as *Notropis cardinalis* (subgenus *Luxilus*) from populations formerly identified as the Duskystripe Shiner, *Notropis pilsbryi* (now *Luxilus pilsbryi*). The type locality is Five Mile Creek south of the Kansas state line, Ottawa County, Oklahoma. Mayden (1989) elevated the subgenus *Luxilus* to generic status. Mayden (1988a) found that body size and number of mandibular tubercles varied geographically over its small range. Individuals from the upper Arkansas River and disjunct populations from the Flint Hills region in central Kansas attain a much smaller adult body size and have fewer mandibular tubercles. *Luxilus cardinalis* is generally placed in the *L. zonatus* species group, but Dowling and Naylor (1997) did not identify this group as monophyletic based on molecular evidence.

DISTRIBUTION   *Luxilus cardinalis* is found in only four states, Arkansas, Missouri, Kansas, and Oklahoma, where it occurs in Arkansas River drainages. Isolated populations in the Red River drainage in southeastern Oklahoma may be introduced (Page and Burr 2011), and there are no records known from the Red River drainage in Arkansas. In Arkansas, it is found only in the Arkansas River drainage in the northwestern part of the state (Illinois-Neosho drainages) and in clear, north bank tributaries (Big Piney Creek, Lee Creek, Clear Creek) of the Arkansas River proper downstream to Pope County (Map 45). All records from Clear Creek in Crawford County are pre-1980 (Map A4.46), and it may be extirpated from that drainage. Waifs of this turbidity-intolerant species have occasionally been found in the Arkansas River (Buchanan 1976; 2005); therefore, its absence from south bank Arkansas River tributaries is somewhat surprising.

HABITAT AND BIOLOGY   The Cardinal Shiner inhabits small, clear, gravel-bottomed upland streams or moderate to small rivers. It is a schooling species, occupying raceways and swift, deep riffles and pools with moderate current. It is usually absent from extreme headwaters and apparently requires permanent flow. In the Illinois River, Oklahoma, larger *L. cardinalis* were often found in upstream areas in flowing water over riffles, while young individuals were more prevalent in downstream slow-water habitats (Felley and Hill 1983). This shiner does not adapt well to reservoirs. Species associates often include *Campostoma plumbeum*, *Nocomis asper*, *Notropis percobromus*, and *Notropis nubilus*.

In Oklahoma, McNeely (1987) reported that Cardinal Shiners consumed equal amounts of invertebrates and plant material. Food habits were studied in a relict population in the Neosho River basin, Kansas (Alexander and Perkin 2013), and the following information on diet is from that source. The diet included aquatic invertebrates (occurring in 93.7% of individuals and constituting 29.3% of the diet), terrestrial invertebrates (76.8%, 9.5%), and algae and plant material (70.5%, 4.8%). Diet diversity increased as Cardinal Shiner size increased, so that aquatic invertebrates made up the majority of the diet among fish greater than 80 mm TL (Alexander and Perkin 2013). The dominant aquatic invertebrates eaten were dipterans, ephemeropterans, and odonates, while the dominant terrestrial invertebrates included hymenopterans and spiders.

Spawning occurs in or immediately below moderate to swift riffles. R. J. Miller (1967) observed Cardinal Shiners spawning in May over a *Semotilus atromaculatus* nest in 4.5 feet (1.4 m) of clear water in a moderately swift portion of a gravel-bottomed stream in eastern Oklahoma. Water temperature was 57.2°F (14°C). Individuals in brilliant breeding color shifted back and forth over the nest in a group of about 20–30, maintaining a position of about 1 or 2 inches (25–51 mm) above the bottom and moving to the center of the pit when possible. This species has been observed spawning in the nests of stonerollers, but occasionally males excavate their own pitlike depressions on clear gravelly riffles. Moore and Paden (1950) observed a spawning aggregate of this species in the Illinois River in May, noting that males roll from side to side, typical of the courtship behavior observed in another member of *Luxilus*, *L. cornutus* (R. J. Miller 1964); however, no clasping was observed. Spawning occurs in Arkansas from April to June and probably depends on water temperatures. Because it commonly shares spawning pits with other species of minnows, hybrids are sometimes found.

CONSERVATION STATUS Even though the Cardinal Shiner is limited to a small portion of northwestern Arkansas (Arkansas River drainage), it appears in no danger as it is common and sometimes abundant.

Luxilus chrysocephalus Rafinesque
Striped Shiner

Figure 15.53. *Luxilus chrysocephalus*, Striped Shiner. *Uland Thomas.*

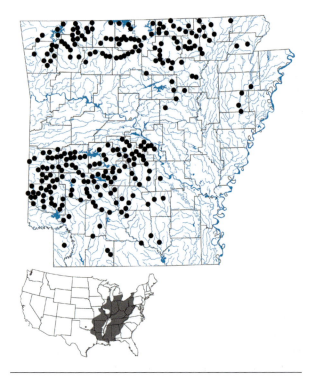

Map 46. Distribution of *Luxilus chrysocephalus*, Striped Shiner, 1988–2019.

CHARACTERS A large, silvery, slab-sided shiner with moderately large eyes (eye diameter going 3.2–4 times into head length), and a large, terminal, oblique mouth (Fig 15.53). Body depth of adult going into SL 3.5–5.0 times. Body circumferential scales at dorsal fin origin 24–29. Predorsal scales usually 13–17 (13–21). Dorsal fin origin over pelvic fin origin; dorsal rays 8. Anal fin usually with 9 (8–10) rays. Lateral line complete and decurved with 37–41 scales. Scales in and near anterior portion of lateral line deep and narrow, scale height much greater than exposed scale width, and marked by crescent-shaped vertical bars. Pharyngeal teeth 2,4–4,2. Intestine short, peritoneum black. Breeding males with tubercles on snout, head, lower jaw, and pectoral fins. Small tubercles are present on dorsum. Maximum size about 7 inches (178 mm), but generally adult SL ranges from 2.5 to 4 inches (65–100 mm) (Gilbert 1980b).

LIFE COLORS Dorsum light olive, sides silvery to green grading to silvery white on the belly. Tip of chin densely pigmented. Midline of dorsum with broad dark stripe. Lateral stripe indistinct, most obvious in preserved specimens, and not extending forward on head to tip of snout. Scales with dark crescents. Three faint to dark horizontal lines on each side of back running from behind the head and converging at the prominent middorsal stripe behind the dorsal fin to form large Vs. Predorsal scales sharply outlined with melanophores. Fins colorless except in breeding males. Body coloration of breeding males a mixture of pink, gold, and black. All fins edged with pink and often with a yellow or gold base. Dorsal fin with a wide submarginal black band. Breeding females silvery, but sometimes with faint pink tinges in fins (Fig. 15.54).

SIMILAR SPECIES The Striped Shiner is superficially similar to the Telescope Shiner, but differs from it in having a smaller eye, eye diameter less than distance from back of eye to rear margin of gill cover in adults (vs. eye larger, its diameter about equal to distance from back of eye to rear margin of gill cover), usually 9 anal rays (vs. 9–11), anterior lateral line scales with scale height much greater than exposed scale width (vs. less deepened anterior lateral line

Figure 15.54. *Luxilus chrysocephalus* male in breeding colors. *Renn Tumlison.*

scales), and lateral line not outlined by dark pigment (vs. a well-pigmented lateral line canal). *Luxilus chrysocephalus* is often sympatric with *L. cardinalis*, *L. pilsbryi*, and *L. zonatus* in Arkansas and differs from those species in the lateral stripe indistinct along midside and not extending forward on head to tip of snout (vs. lateral stripe dark and prominent along midside and extending forward on head to tip of snout), and anterior lateral line scales with scale height much greater than exposed scale width (vs. anterior lateral line scales with scale height only slightly greater than exposed scale width).

VARIATION AND TAXONOMY   *Luxilus chrysocephalus* was described by Rafinesque (1820a) from "Kentucky." Gilbert (1978) designated a neotype and restricted the type locality to a creek about 9.7 km south-southwest of Danville, Lincoln County, Kentucky. The taxonomic status of the sibling species, the Common Shiner, *Luxilus cornutus*, and the Striped Shiner, *Luxilus chrysocephalus*, has been the subject of much debate over the years. These forms seem to act as separate species in certain parts of their range (Gleason and Berra 1993) and as subspecies in other areas. In Arkansas, only the Striped Shiner form occurs. This shiner has been previously considered both a subspecies (R. J. Miller 1968; Resh et al. 1971; Miller and Robison 1973) and a full species (Gilbert 1961, 1964; Rainboth and Whitt 1974; Buth 1979; Dowling and Moore 1984, 1985a, b, 1986; Dowling et al. 1989). Arguments on both sides are persuasive. An electrophoretic analysis of 42 populations from the area of sympatry revealed that gene flow between these species (introgressive hybridization) occurs, but in 41 of the 42 populations, allele frequencies indicated that at least partial reproductive isolation was being maintained (Dowling and Moore 1984). In supporting full species recognition, they pointed out that since evolution is a continuous process, it is not surprising that we occasionally see taxa that are still involved in the process of speciation. Although reproductive isolation between *L. cornutus* and *L. chrysocephalus* is imperfect (Gilbert 1980c), most of the evidence indicates that these taxa have diverged beyond the level of subspecies and should be considered separate species (Dowling and Moore 1984). R. J. Miller (1968) earlier reached this same conclusion, that these taxa are beyond the level of subspecies; however, he argued that since there was not complete reproductive isolation, the two taxa could not be considered full species and must be relegated to the next level in the taxonomic hierarchy, the subspecies. Etnier and Starnes (1993) presented a thorough discussion of this knotty problem and supported the specific status of *L. chrysocephalus*. A molecular analysis by Dowling and Naylor (1997) indicated that *L. cornutus* was more closely

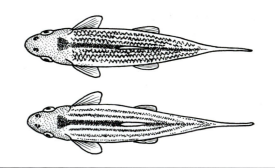

Figure 15.55. Dorsal views of *Luxilus chrysocephalus* subspecies.

Top: *L. c. chrysocephalus*
Bottom: *L. c. isolepis*

related to *L. albeolus* than to *L. chrysocephalus*. We agree with Etnier and Starnes (1993), Page and Burr (2011), and Page et al. (2013) in recognizing *Luxilus chrysocephalus* as a full species.

Two well-defined subspecies of the Striped Shiner occur in Arkansas, *L. c. chrysocephalus* Rafinesque and *L. c. isolepis* (Hubbs). These forms do not come into contact in Arkansas, as *chrysocephalus* inhabits the northern portion of the state in the Arkansas, White, and St. Francis drainages, and *isolepis* occupies the Red and Ouachita drainages in southern Arkansas. In Arkansas, the most prominent external feature distinguishing the two subspecies of *L. chrysocephalus* involves the 3 parallel, dark dorsolateral lines on either side of the prominent middorsal stripe. *Luxilus c. chrysocephalus* has rather wavy, uneven irregular lines, whereas *L. c. isolepis* has smooth, straight lines (Fig. 15.55). In addition, *L. c. chrysocephalus* usually has 14–16 predorsal scales, while *L. c. isolepis* usually has 13 or 14. Dowling et al. (1992), based on mitochondrial DNA, contended that *isolepis* may not even be a monophyletic group as currently recognized, further confusing the status of this taxon. Boschung and Mayden (2004) predicted that the two subspecies will likely be considered full species in the future.

DISTRIBUTION   *Luxilus chrysocephalus* occurs in the Great Lakes and Mississippi River basins from New York to Wisconsin and south to Alabama, Louisiana, and eastern Texas, and in Gulf Slope drainages from the Mobile Bay basin, west to the Sabine River, Louisiana. In Arkansas, it is rather widespread and locally abundant in the Ozark Uplands, Ouachita Mountains, and certain areas of the Coastal Plain in southwestern Arkansas and is common in the streams of Crowley's Ridge (Map 46). It formerly occurred sporadically in Arkansas River Valley streams (Map A4.47), but the only recent records are from

Fourche Creek in Pulaski County. There are also recent records from the Illinois River system (Arkansas River drainage) in northwestern Arkansas. It has been recorded from three Arkansas reservoirs (Buchanan 2005).

HABITAT AND BIOLOGY This large shiner seems to prefer small to moderate-sized streams with permanent flow, clear water, and rocky or gravel bottoms. It is found in some current, but avoids strong current (except when spawning). It usually avoids strong riffles and deep soft-bottomed pools, and it is often found in schools at the foot of riffles and in shallow, hard-bottomed pools with some current (Smith 1979). In some small Ouachita Mountain headwater streams, large adults may be the largest fish in a stream. It is also found in larger streams, such as the middle portion of the White River in Arkansas, and Burr and Warren (1986b) considered it an ecologically labile species that occupies a variety of riverine habitats. Shields et al. (1995) reported that *L. chrysocephalus* is intolerant of degraded conditions. Although it usually avoids calm water, it was found in Bull Shoals, Norfork, and Millwood reservoirs (Buchanan 2005).

It actively forages in the water column but also feeds on benthic material (Ross 2001). Hambrick and Hibbs (1976) found that it consumed nearly equal parts plant material and terrestrial arthropods during the day, but there was greater consumption of plant material at night. Foods consist primarily of terrestrial and aquatic insects, filamentous and unicellular algae, and aquatic invertebrates. Important invertebrate components of the diet include mayflies, termites, dragonflies, caddisflies, beetles, and adult midges (Ross 2001). During floods, large individuals forage on the inundated floodplain (Slack 1996). Boschung and Mayden (2004) reported a seasonal shift in diet from small aquatic prey such as small crayfish and mayfly nymphs during most of the year to filamentous algae and chironomid larvae in the winter. Forbes and Richardson (1920) noted that it will take a hook baited with an earthworm or grasshopper.

The Striped Shiner spawns in Mississippi from May to October, with peak larval abundance occurring from May to July (Brenneman 1992). In Alabama, spawning begins in April when water temperatures reach 16–18°C (Boschung and Mayden 2004). In the Missouri Ozarks, spawning occurs from late April to mid-June (Pflieger 1997). The spawning mode of this shiner is pit-building (Mayden and Simons 2002). Spawning occurs in shallow water over clean gravel from late April to late June in Arkansas (HWR pers. obs.; Tumlison et al. 2017) when the water temperatures reach above 60°F (15.6°C). Males are aggressive toward one another. The male excavates a craterlike depression in current in the gravel by removing stones with his mouth

and pushing aside gravel with his snout, but this species frequently spawns over the nests of the Hornyhead Chub or other nest-building species. Smith (1979) noted that the communal use of nests by more than one species results in the frequent production of hybrids between *Luxilus chrysocephalus* and species of *Notropis* and other cyprinid genera. After fertilization, the eggs hatch in 152–160 hours at temperatures of 13–15°C (Yeager 1979). Simmons and Beckman (2012) assessed the validity of age estimates for the Striped Shiner in Missouri. They found faster growth rates for males. Marshall (1939) reported maximum ages of Striped Shiners to be 4–5 years based on scales; however, Simmons (1999) found maximum age to be 6 years using otoliths.

CONSERVATION STATUS The Striped Shiner has reportedly declined in Wisconsin and Missouri. This shiner is common and widespread and in no danger in Arkansas.

Luxilus pilsbryi (Fowler)
Duskystripe Shiner

Figure 15.56. *Luxilus pilsbryi*, Duskystripe Shiner. *David A. Neely.*

CHARACTERS A large, slightly compressed shiner with a moderately large head, head length going 3.6–4.2 times into SL. Eyes moderately large, but variable in size, eye diameter usually going into head length 3–4 times in adults and often fewer than 3 times in juveniles. Eye diameter in adults usually distinctly less than distance from back of eye to rear margin of gill cover. A prominent, dark lateral band extends from tip of snout to caudal base and does not become abruptly narrower just beind gill opening; the lateral band touches the lateral line beneath the dorsal fin. Mouth large, terminal, and oblique (Fig. 15.56). Dorsal rays 8; anal rays 9. Pharyngeal teeth 2,4–4,2. Gill rakers usually 7. Lateral line slightly decurved anteriorly with 39–44 scales; height of anterior lateral line scales and adjacent scales only slightly greater than exposed scale width in adults. Scales around body at dorsal fin origin usually 24–26. Dorsal fin origin nearly over pelvic fin origin, closer to tip of snout than to caudal fin base. Intestine short;

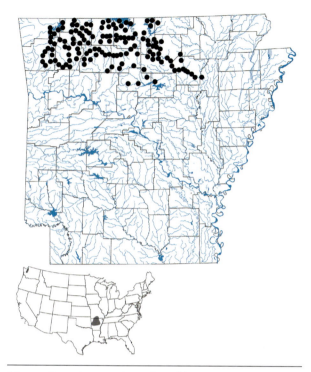

Map 47. Distribution of *Luxilus pilsbryi*, Duskystripe Shiner, 1988–2019.

Figure 15.57. Duskystripe Shiners in spawning coloration (29 April 2015), Yocum Creek, Carroll County, Arkansas. *Isaac Szabo.*

peritoneum black. Breeding males highly tuberculate, with a tuberculation pattern similar to *L. zonatus*. Maximum size 4.8 inches (122 mm).

LIFE COLORS   Body color olivaceous above with a prominent broad, middorsal iridescent or brown stripe from occiput to caudal fin. Sides silvery with a dusky lateral stripe running the length of the body; lateral stripe most evident in breeding males and in preserved specimens; lateral stripe does not become abruptly narrower behind gill opening. Basicaudal spot absent. A narrow, dark or bronze line above the lateral band is separated from the broad lateral stripe by a thin iridescent golden stripe; belly, underside of head, and lower caudal peduncle white. Breeding males (Fig. 15.57) develop some red coloration on the underside of the head and body and on all fins, but red pigment is diffuse or absent on lower sides and base of anal fin. Lateral band becoming intensely black. Tip of snout in breeding males blue.

SIMILAR SPECIES   The Duskystripe Shiner is most similar to the Cardinal Shiner but differs in having 24–26 total circumferential scales (vs. 26 or 27); mandibular tubercles 5–10, but usually more than 6 (vs. 0–6, but usually 0–3); 7 gill rakers (vs. 8 or 9), and a lateral band extending downward only to the lateral line and not as broad as in the Cardinal Shiner, in which the lateral band is wider

and extends downward 1–2 scales below the lateral line. It is similar to *L. zonatus* but lacks the dark cleithral bar along the rear margin of the gill cover (vs. dark cleithral bar present); the dorsolateral scales are better outlined by melanophores, forming zigzag lines, converging posteriorly between adjacent scale rows (vs. lacking converging zigzag lines); predorsal scales usually 14 or 15 (vs. 15–17), and the lateral band never constricted immediately behind the gill cover as in *L. zonatus*. Breeding males of *L. zonatus* are less extensively red than *L. pilsbryi* males, the red pigment being concentrated on the crest of the head, lips, margin of preopercle, axil of pectoral fin, and central part of all fin membranes. Breeding *L. pilsbryi* males have a powder blue snout with more intense red pigment on face than *L. zonatus*. *Luxilus pilsbryi* is similar to *Notropis telescopus*, with which it is often syntopic, but differs from that species in lateral line scales usually 39–44 (vs. 35–38), body circumferential scales usually 24–26 (vs. 20–22), and anal rays usually 9 (vs. usually 10 or 11, rarely 9). Difference in eye size has also been used to separate *L. pilsbryi* and *N. telescopus*, but that character is not completely reliable. *Luxilus pilsbryi* shows variation in eye size among populations and often between juveniles and adults. In large adult *L. pilsbryi*, eye diameter usually goes into head length 3 times or more (vs. fewer than 3 times in adult *N. telescopus*). However, in juvenile *L. pilsbryi*, eye diameter often goes into head length fewer than 3 times due to allometric growth.

VARIATION AND TAXONOMY   Originally described in 1904 as *Notropis pilsbryi* by H. W. Fowler from "Rogers, White River basin, Arkansas." The Duskystripe Shiner was considered a subspecies of *Notropis* (*Luxilus*) *zonatus* by Hubbs and Moore (1940) and subsequent workers until the 1960s. Gilbert (1964) reviewed the systematics of the subgenus *Luxilus* and elevated both subspecies to full specific status (*Notropis pilsbryi* and *N. zonatus*). Menzel and Cross (1977) recommended returning *N. zonatus* and *N. pilsbryi* to subspecies status based on an early biochemical analysis. However, Buth (1979) and Buth and Mayden (1981) favored continued recognition of these forms as full species, also based on biochemical properties. The subgenus *Luxilus* was elevated to genus level by Mayden (1989). We follow Buth, Pflieger (1997), Gilbert (1980d), and other recent authors (e.g., Page and Burr 2011; Page et al. 2013) in treating the two as separate species in the genus *Luxilus*. *Luxilus pilsbryi* is known to hybridize with *Chrosomus erythrogaster* (Robison and Miller 1972).

DISTRIBUTION   *Luxilus pilsbryi* is endemic to the White River system of Arkansas and Missouri, but most of its range is within Arkansas. In Arkansas, it is found in all White River tributaries and the main stem White River downstream to its confluence with the Black River, and in the upper part of the Little Red River above Greers Ferry Lake (Map 47). It is replaced upstream in the Black River drainage by *Luxilus zonatus*.

HABITAT AND BIOLOGY   The Duskystripe Shiner inhabits upland streams with swift riffles and gravel-bottomed pools with moderate current. Pflieger (1997) noted that the habitat of *L. pilsbryi* differs from that of *L. cardinalis* only in that *L. pilsbryi* occurs more frequently in headwater streams. Matthews et al. (1978) found this species most frequently in deep pools, deeper portions of the channel, and at the bases of riffles. Our collections throughout the Ozark Mountains found *L. pilsbryi* in this same habitat. It was found in small numbers in the four large White River impoundments in Arkansas (Buchanan 2005). It is a schooling species and is the most abundant minnow at many localities in upland Ozark streams (HWR, pers. obs.). Pflieger (1997) reported that the most common species associates of *L. pilsbryi* were *Notropis nubilus*, *Campostoma pullum* (= *C. plumbeum*), *Luxilus chrysocephalus*, *Notropis telescopus*, and *Nocomis biguttatus*.

The most commonly consumed foods by this sight feeder are aquatic insects. Chironomid larvae and pupae and mayfly nymphs were the principal food resources, with hemipterans and beetles occasionally taken, along with terrestrial insects, spiders, green algae, cyanobacteria, and unidentified plant material (Matthews et al. 1978). Matthews (1977) reported an unusual case of sand ingestion by this species in Piney Creek, Izard County, Arkansas. In Piney Creek, adults tend to feed below large gravel-rubble riffles approximately 70 cm deep (Matthews et al. 1978). During the day they are abundant in nearby pools (approximately 1 m deep) swimming into the current in schools apparently feeding by sight on drift items.

Matthews et al. (1978) found that large females were gravid in April in Piney Creek, Arkansas, but males had not yet attained maximum breeding colors and no spawning activities were noted. Spawning has been observed in Missouri in May and June (Pflieger 1997). Our observations indicate that spawning occurs in or immediately below moderate to swift riffles in Arkansas from late April through June. Details of spawning are similar to those described for the Cardinal Shiner (then considered to be *Notropis pilsbryi*) by R. J. Miller (1967), Moore and Paden (1950), and HWR (pers. obs.). Matthews (1977) reported that the population in Piney Creek was dominated by age 0 and age 1 individuals with numerous age 2 fish present. Scale age assessments indicated that some of the largest individuals were in their fourth year. In Piney Creek in July, young-of-year individuals ranged from 23 to 40 mm (modal size = 30 mm) and dominated the population. By December, Matthews et al. (1978) found that age 0 fish ranged from 29 to 50 mm (modal = 42 mm), and the modal size for age 1 fish increased from 60 to 65 mm. Individuals of age 0 exhibited little growth from January to April. Mean size was greatest in April. Simmons and Beckman (2012) assessed the validity of age estimates for the Duskystripe Shiner in Missouri and determined that males had faster growth rates than females. They found a maximum age of 6 years based on otoliths.

CONSERVATION STATUS   The Duskystripe Shiner is widespread and abundant throughout the White River system, and its populations are secure.

Luxilus zonatus (Agassiz)
Bleeding Shiner

Figure 15.58. *Luxilus zonatus*, Bleeding Shiner. *Uland Thomas*.

CHARACTERS   A moderately deep-bodied, slightly compressed shiner with large eyes (eye diameter going into head length about 3.5 times), large fins, and a terminal, oblique mouth (Fig. 15.58). Head moderate in size, head length going into SL about 4 times. A prominent

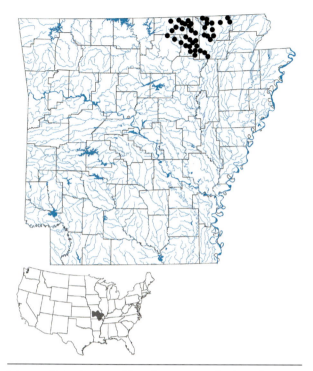

Map 48. Distribution of *Luxilus zonatus*, Bleeding Shiner, 1988–2019.

dark lateral band becomes abruptly narrower just behind gill opening and not touching lateral line beneath dorsal fin. Anal rays 9 (rarely 8 or 10). Pharyngeal teeth 2,4–4,2. Gill rakers generally 7. Lateral line scales 38–43; height of anterior lateral line scales and adjacent scales only slightly greater than exposed scale width in adults. Circumferential scales 26. Predorsal scales usually 14 or 15. Origin of dorsal fin closer to tip of snout than to caudal base. Dorsal fin origin over pelvic fin origin. Peritoneum black. Tubercles of breeding males largest on head. Mandibular tubercles 5–10 (usually more than 6). Pectoral fin rays 2–6 of breeding males with a single row of numerous fine tubercles basally and 2 rows of 1–4 (usually 3 or 4) tubercles per segment beyond branches. Maximum size 4.8 inches (122 mm).

LIFE COLORS   Body color olive above with a broad iridescent or brown middorsal stripe and silvery white below. Midside with a prominent dark black stripe that narrows anteriorly and extends forward to tip of snout. Midlateral stripe extending dorsally two scale rows above lateral line and ventrally to near, but generally not touching, lateral line below dorsal fin. Lateral stripe paralleled above by a less distinct secondary dorsolateral stripe separated by an iridescent gold stripe. Middorsal stripe prominent, intensely dark and broader than base of dorsal fin. Prominent crescent-shaped dark cleithral bar along rear margin of gill opening. Fins colorless in nonbreeding

individuals. Breeding males with intense dark markings. Head with red pigment dorsally and laterally, but weakly developed ventrally. Snout and lips of breeding adults red. Caudal fin red from base to broad, clear distal band. Dorsal, pelvic, pectoral, and anal fins with a broad medial red band and a broad clear band distally; fins clear to lightly tinted with red basally.

SIMILAR SPECIES   *Luxilus zonatus* is similar to *L. pilsbryi* and *L. cardinalis* but differs from both in having a strongly developed, crescent-shaped cleithral bar along the rear margin of the operculum (vs. cleithral bar absent), distinct dorsolateral stripes on the dorsum (vs. absent or poorly developed), generally 14 or 15 predorsal scales (vs. 15–17), lateral band constricted anteriorly and never extending below the lateral line (vs. a nonconstricted lateral band extending to or below lateral line), and a red snout in breeding adults (vs. blue snout). Sometimes confused with the syntopic *Notropis telescopus* (see *L. pilsbryi* account for differences between all Arkansas *Luxilus* and *N. telescopus*).

VARIATION AND TAXONOMY   *Luxilus zonatus* was originally described as *Alburnus zonatus* by Agassiz in 1863 (*in* Putnam 1863) from specimens collected from the Osage River, presumably in Missouri (Gilbert 1978). It was later assigned to genus *Notropis*, and Hubbs and Ortenburger (1929b) placed *N. pilsbryi* in synonymy with *N. zonatus*. Gilbert (1964) and Buth (1979) used morphological and allozyme evidence to support the specific distinctness of *N. zonatus* and *N. pilsbryi*. The subgenus *Luxilus* (containing *N. zonatus*) was elevated to generic status by Mayden (1989).

DISTRIBUTION   The Bleeding Shiner is endemic to the Ozarks of Missouri and Arkansas, where it occurs in Ozark-draining tributaries of the Missouri River and in the Meramec, Little, St. Francis, and Black rivers of the Mississippi River basin. Most of its range is in Missouri, and in Arkansas, it is known only from the Black River drainage where it is one of the dominant upland stream fishes in streams of that system (Map 48).

HABITAT AND BIOLOGY   The Bleeding Shiner inhabits permanent, small to medium-sized upland streams and small rivers where it is an abundant species frequenting deep riffles and pools with moderate current over gravel bottoms. It is intolerant of prolonged turbidity. While the adults are invariably found in current, the young are found in quieter waters. It is a schooling species often found in midwater along with several other cyprinids.

Pflieger (1997) reported that *L. zonatus* is a sight feeder that feeds on insects and other small invertebrates found floating on the surface of the water or drifting in the current. TMB examined gut contents of 60 specimens collected from May through October from 11 localities in Fulton, Randolph, and Sharp counties. It is primarily an insectivore, feeding at all levels in the water column, based on the presence of winged insects, benthic immature insects, and sand grains in the gut contents. The diet was dominated by percent frequency of occurrence by ephemeropteran and trichopteran nymphs. Terrestrial insects also contributed importantly to the diet, including dipterans, coleopterans, odonates, and ants. Other food items consumed in smaller amounts included plecopterans, fish and invertebrate embryos, filamentous algae, dipteran pupae, odonate nymphs, coleopteran larvae, and hemipterans. The Bleeding Shiner was taken from the Spring River in 2019 by a fisherman using an artificial lure (imitation trout egg) (B. Timmons, pers. comm.).

In Arkansas, spawning occurs from late April to early July. TMB found ripe females and tuberculate males with large testes in the Spring River in Sharp County on 17 May. Ripe females averaged 73 mm TL and contained an average of 336 mature ova. Females with mature eggs have been taken as late as 8 July in Sharp County (Tumlison et al. 2017). Pflieger (1997) found most reproduction was by age 2 individuals and a few individuals exceeded 3 years of age. Larger male *L. zonatus* are more colorful, acquire and hold preferred territories, and have more spawning activity than small, less colorful conspecific males (Chambers 1971). Males excavate small, pitlike depressions in clean, gravelly riffles, where females lay eggs. Spawning is often over the gravel nests of the Hornyhead Chub or stonerollers. Spawning occurs in groups of 100 or more individuals. Pflieger (1997) described the spawning behavior:

> Typically, all individuals in the spawning group are oriented with their heads upstream, and the males occupying pits are tilted downward at an angle so that the head is lower than the tail. The females remain above and behind the males until ready to spawn, at which time they dip down next to one of the dominant males. During the spawning act the male flips the female into a vertical position and briefly clasps her by throwing his body into a strong U-shaped curve. When released from the spawning embrace, the female shoots vertically towards the surface before she recovers and rejoins the spawning group.

In Missouri, Pflieger (1997) estimated age at maturity to be age 2. While males have faster growth rates than females, Chambers (1971) reported a maximum age (based on scales) of 3 years.

CONSERVATION STATUS Although its range in Arkansas is small, the Bleeding Shiner is widespread and abundant throughout the Black River system in this state. Its populations are currently secure, but future monitoring is warranted.

Genus *Lythrurus* Jordan, 1876a— Finescale Shiners

The genus *Lythrurus*, with 11 species, is characterized by very small scales on the nape (typically 21 or more predorsal scale rows, partially embedded and often nonimbricate); a slender, terete to deep and compressed body; snout acute to bluntly rounded; mouth large, terminal and oblique; dorsal fin origin well behind pelvic fin origin; breast scaled; dorsolateral area variably scaled or naked; exposed part of scales on side of body not notably taller than wide; lateral line usually complete; lateral scales rows 36–50; circumferential scales usually 29–34; circumpeduncular scales usually 14; dorsal rays 8; anal rays 10–13, rarely 9 or 14; pectoral rays usually 12–15, rarely 16; and pelvic rays 8. Pharyngeal teeth are in 2 rows, 2,4–4,2 (occasionally 1,4–4,2 or 2,4–4,1). Peritoneum silvery, spotted with melanophores. Breeding tubercles of males usually well developed on parts of head, nape, and dorsal and pectoral fins. Multiple pectoral fin tubercles per fin ray segment are moderately large, but do not form a shagreen (except in *L. fumeus* and *L. snelsoni*). Breeding colors usually red except in *L. fumeus* (yellow). Urogenital papilla of breeding females enlarged and protruding posteriorly to about anal fin origin. Most species are small to moderate size (less than 60 mm SL) and inhabit medium-sized streams.

Snelson (1972; 1973; 1980) reviewed the systematics of the members of *Lythrurus* when it was a subgenus of *Notropis*. Later revisions by Mayden (1989) and Coburn and Cavender (1992) recognized *Lythrurus* as a genus. Phylogenetic relationships of *Lythrurus* to other North American cyprinid genera remain unclear. Mayden (1989) considered *Lythrurus* to be monophyletic and sister to a clade including *Cyprinella* and *Luxilus*. Combined morphological and molecular analyses strongly supported monophyly of *Lythrurus* (Mayden et al. 2006a). Coburn and Cavender (1992) considered *Lythrurus* to be the sister group to a clade in which *Cyprinella* is the sister to a clade inclusive of *Pimephales* and *Opsopoeodus*. Species relationships within *Lythrurus* have been evaluated using morphological and molecular characters (Snelson 1972; 1973; Stein et al. 1985; Mayden 1989; Wiley and Siegel-Causey 1994; and Schmidt et al. 1998). Boschung and Mayden (2004) presented an in-depth discussion of relationships of the species of *Lythrurus*. The only study of phylogenetic relationships among all 11 species of *Lythrurus* was that of Pramuk et al. (2007) using four mitochondrial genes. Those authors found that all currently recognized species of *Lythrurus* are genetically distinct. Their results also confirmed and extended earlier studies recovering two clades within *Lythrurus*, the *L. umbratilis* clade with five species, and the *L. bellus* clade with six species (including *L. fumeus* and *L. snelsoni* in Arkansas).

Lythrurus fumeus (Evermann)
Ribbon Shiner

Figure 15.59. *Lythrurus fumeus*, Ribbon Shiner. *Uland Thomas.*

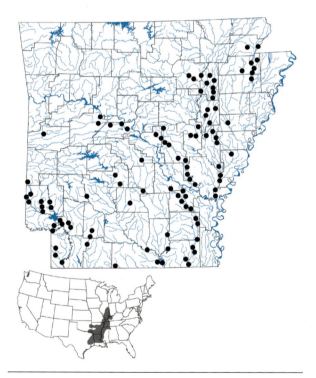

Map 49. Distribution of *Lythrurus fumeus*, Ribbon Shiner, 1988–2019.

CHARACTERS   A moderately small, slender, silver shiner with a small head, large eyes (eye diameter going into SL 3.0–3.2 times), a pointed snout, and a terminal, oblique mouth (Fig. 15.59). Eye diameter greater than snout length. Dorsal rays 8; anal rays 11–13. Pectoral fin rays 12–15.

Pelvic fin rays 8. Gill rakers 7–10. Pharyngeal teeth usually 2,4–4,2. Vertebrae 35–38. Lateral line sharply decurved with 38–42 scales (modally 39 or 40). Dorsal fin origin well behind origin of pelvic fins. Front of dorsal fin base closer to caudal base than to front of eye. Predorsal scale rows 22–26 (usually 24 or more). Body circumferential scales 28–33. Dorsal fin slightly rounded, its first principal ray shorter than second and third. Chin thickly sprinkled with dusky pigment. Intestine short, peritoneum silvery with dark speckles. Nuptial tuberculation was described by Snelson (1973). Small tubercles on dorsal surface of head most prominent from interorbital area back to occiput. Lower jaw with lateral row of tubercles that project to the side, and often with additional tubercles scattered over the ventral surface; these tubercles are smaller than largest ones on top of head. Pectoral fins with a dense shagreen of tiny tubercles on the dorsal surfaces of rays 1–8 or 9, best developed on rays 2–5, where 20 or more tubercles occur per fin ray segment proximal to major branching of rays (Etnier and Starnes 1993). Other fins lack tubercles. Nape and anterior lateral body scales with marginal row of 5–8 tubercles per scale, largest on nape. Breast and belly scales with somewhat larger and blunter tubercles, about 2 or 3 per scale. Maximum size 2.7 inches (68 mm); most adults around 2 inches (51 mm).

LIFE COLORS   Body color olive yellow with a diffuse, dusky, lateral stripe that is more intense posteriorly and more diffuse anteriorly; stripe most prominent in preserved specimens. Sides silver and belly white. Fins generally colorless, but all dorsal and caudal fin rays are bordered by pigment. No anterior basidorsal spot. Lips and chin dusky. Males develop yellow breeding colors on rays of dorsal and caudal fins and along midlateral and anterodorsal areas of body. No melanin deposited in interradial membranes of fins of nuptial males.

SIMILAR SPECIES   *Lythrurus fumeus* is similar to *Notropis percobromus*, *N. atherinoides*, *Lythrurus umbratilis*, and *L. snelsoni*. It can be distinguished from *N. percobromus* and *N. atherinoides* by the smaller scales in front of the dorsal fin with 22 or more (vs. fewer than 22), the lower and more diffuse lateral stripe, and the sharply decurved lateral line. It is often sympatric with and occasionally confused with *Notropis atherinoides* and further differs from that species in dorsal fin slightly rounded, its first principal ray shorter than the second and third (vs. dorsal fin pointed, its first principal ray longer than second and third), lateral stripe is 2–3 scale rows wide and extends anteriorly almost or fully down to the lateral line (vs. lateral stripe faint and narrow, 1 or 2 scale rows wide and more prominent on the caudal peduncle and fading abruptly anteriorly on the trunk), and body circumferential scales 28–33, modally 30–32 (vs. 22–28 circumferential scales). It is most similar to *L. umbratilis* and *L. snelsoni* but can be distinguished from the former by the absence of a prominent spot at dorsal fin origin, tuberculation of breeding males, higher anal ray count (usually 11 or 12 vs. 10), absence of dark dorsolateral chevrons in larger specimens, and yellow breeding colors (vs. red). From *L. snelsoni*, *Lythrurus fumeus* differs in having an anal ray count of 11 or 12 (vs. 10), yellow breeding color (vs. red), shagreen tuberculation, and a stouter body (vs. a more elongate and terete body in *L. snelsoni*). *Lythrurus fumeus* is syntopic with *Notropis suttkusi* in the Little River drainage and can be distinguished from that species by anal rays usually 11 or 12 (vs. usually 9 or 10), usually 22 or more predorsal scales (vs. fewer than 22), and breeding males with yellow coloration (vs. red or orange).

VARIATION AND TAXONOMY *Lythrurus fumeus* was described by Evermann (1892) as *Notropis fumeus* from Hunter Creek, San Jacinto River drainage, about 14 km west of Houston, Harris County, Texas. Snelson (1973) reviewed the taxonomic history of this species. He suggested that *L. fumeus* was somewhat intermediate between the subgenera *Lythrurus* and *Notropis*, primarily on the basis of the extremely small and dense pectoral fin tubercles, a character shared by *Notropis* but not by other *Lythrurus*. Mayden (1989) placed *Lythrurus* (including *L. fumeus*) in a lineage remote from *Notropis* (*sensu stricto*). The Ribbon Shiner is placed in *Lythrurus* based on its dorsal fin origin situated behind the pelvic fin origin, tiny, crowded predorsal scales, and enlarged urogenital papilla in breeding females. Based on morphological and genetic evidence, *L. fumeus* was placed in the *L. bellus* clade along with *L. snelsoni*, *L. alegnotus*, *L. roseipinnis*, and *L. atrapiculus* (Pramuk et al. 2007).

DISTRIBUTION *Lythrurus fumeus* occurs in the Mississippi River basin from central Illinois through Louisiana west to eastern Oklahoma and Texas, and in Gulf Slope drainages from Lake Pontchartrain, Louisiana, to the Navidad River, Texas. In Arkansas, it is primarily an inhabitant of lowland streams and rivers of the Coastal Plain but extends up the Arkansas River Valley to the Petit Jean and Poteau rivers (Map 49).

HABITAT AND BIOLOGY The Ribbon Shiner is found primarily in small to moderate-sized Coastal Plain streams characterized by low to moderate gradient and tannin-stained waters over sand, firm mud, clay, silt, or detritus substrates (Snelson 1973). Quiet pools or backwater areas are its preferred microhabitats, but it is also found in slight to moderate current. In Village Creek (Neches River drainage), Texas, most *L. fumeus* in summer samples were collected in sandbank habitats in slow to moderate currents; during fall, most *L. fumeus* were captured from backwater habitats (Moriarty and Winemiller 1997). Some investigators reported that the Ribbon Shiner is tolerant of turbidity and the associated ecological conditions of creeks and ditches flowing through agricultural areas (Smith 1979; Snelson 1980), but that does not appear to be the case in Arkansas. Moriarty and Winemiller (1997) also considered it intolerant of environmental disturbance. Although associated with Coastal Plain habitats in Arkansas, it is rarely found in large rivers.

Observations by HWR while snorkeling revealed that the Ribbon Shiner is a midwater, schooling species that often feeds from the surface. Felley and Felley (1987) reported seasonal differences in water column feeding in southern Louisiana. During the wet season, 64% of the prey volume was taken from the surface, with 34% consisting of benthic animal prey. During the dry season (when food was more limited), surface prey accounted for only 37% of the diet, with midwater and benthic prey comprising 8% and 21%, respectively. The greatest change during the dry season was increased ingestion of organic detritus, which made up 33% of the diet. TMB examined gut contents of 63 specimens collected from May through August from Bradley, Clay, Drew, Hempstead, and Yell counties in Arkansas and confirmed that *L. fumeus* is primarily an insectivore that feeds throughout the water column during spring and summer. Foods consumed in order of importance were immature aquatic insects (ephemeropteran nymphs, larval coleopterans, dipteran larvae and pupae, odonate nymphs, and hemipterans) and terrestrial insects (primarily dipterans and coleopterans). Microcrustaceans (copepods, cladocerans, and ostracods) were also frequently consumed, especially by smaller individuals. Some individuals had ingested small amounts of sand and detritus.

There is surprisingly little information on the reproductive biology of this somewhat common fish. Based on the occurrence of tuberculate males, spawning in Tennessee extends from May through early August (Etnier and Starnes 1993). In Village Creek, Texas, *L. fumeus* showed length-frequency distributions consistent with late spring-early summer (May-June) spawning (Moriarty and Winemiller 1997). Females with eggs were taken in Missouri in mid-July (Pflieger 1997). The Ribbon Shiner spawns in Arkansas from late May through July, and possibly into August. Males in yellow breeding colors have been collected by HWR as early as 24 May from Bayou Dorcheat in Columbia County. Ripe females from Clay County taken on 22 May by TMB ranged from 43 to 48 mm TL and contained between 180 and 226 nearly ripe ova. Smaller egg sizes were present, indicating multiple spawning. Ripe females collected on 12 July in Hempstead County by TMB averaged 56 mm TL and contained approximately 400 nearly ripe ova. Most individuals examined from August collections had regressed gonads, but some males had enlarged testes.

CONSERVATION STATUS   The Ribbon Shiner is a common and sometimes abundant species across the Coastal Plain of Arkansas, and its populations are secure.

Lythrurus snelsoni (Robison)
Ouachita Mountain Shiner

Figure 15.60. *Lythrurus snelsoni*, Ouachita Mountain Shiner. *David A. Neely.*

Map 50. Distribution of *Lythrurus snelsoni*, Ouachita Mountain Shiner, 1988–2019.

CHARACTERS   A small, slender, terete, silvery shiner with a bluntly rounded snout, terminal, slightly oblique mouth, and moderately large eyes (Fig. 15.60). Dorsal rays 8. Anal rays modally 10 (9–11). Pharyngeal teeth 2,4–4,2.

Front of dorsal fin base closer to caudal fin base than to tip of snout. Dorsal fin origin far behind pelvic fin origin. Lateral line complete and decurved, with 38–45 scales. Scales around caudal peduncle modally 14 (13–15); body circumferential scales usually 29–31. Intestine short; peritoneum silvery, sprinkled with dark speckles. Breeding males with small erect tubercles on head, dorsum, and pectoral rays; tubercles small and forming a shagreen. Adults seldom exceed 2 inches (51 mm SL).

LIFE COLORS   Breeding males with red coloration dorsally on the head from top of the snout to the occiput, and on chin and anterior third of gular area (Robison 1985). Fins colorless. Iris of eye washed with red. Anterior basidorsal spot absent. No chevron markings developed over myosepta of upper anterior part of body. Melanin absent in fin interradial membranes of breeding males. Body color in nonbreeding males pale green to olive. Sides silver. Scales with pigmented borders, sometimes with scattered, rather large melanophores lining the edges, rendering the outlines of each scale rather indistinct. Venter white. Lips darkly pigmented. Chin pigment dispersed over anterior portions of the dentaries and gular area. Dusky lateral stripe present. In females, all fins colorless or colored like males. Iris of eye with slight wash of red pigment.

SIMILAR SPECIES   *Lythrurus snelsoni* is closely related to *L. fumeus* and *L. umbratilis* but is distinguished from those two by the following characters: from *L. fumeus*, *L. snelsoni* differs in having generally 10 anal rays (vs. usually 11 or 12); red breeding colors (vs. yellow); and a slimmer, more elongate body (vs. deeper body). From *L. umbratilis*, *L. snelsoni* differs in lacking a predorsal spot and chevrons (vs. predorsal spot and chevrons usually present), typically 10 anal rays (vs. usually 11), in having a slim, terete body (vs. a deeper, compressed body), and small head tubercles (vs. large erect head tubercles) (See Robison 1985 for further elaboration of the differences among these three very similar species).

VARIATION AND TAXONOMY   Originally described by Robison (1985) as *Notropis snelsoni*, the type locality of *Lythrurus snelsoni* is the Mountain Fork River at State 246 bridge, 2.98 miles (4.8 km) west of Hatfield, Polk County, Arkansas. We continue to use the common name Ouachita Mountain Shiner, assigned by Robison in the original description, rather than the new name "Ouachita Shiner" used in Page et al. (2013). We feel Ouachita Mountain Shiner better describes the small, distinctive range of this shiner in the western Ouachita Mountains (especially since it is absent from the Ouachita River drainage). Mayden et al. (2006a) found *L. snelsoni* more closely related to a clade containing *L. fumeus* than to *L. umbratilis*. Based on morphological and genetic evidence, Pramuk et al. (2007) placed *L. snelsoni* in the *L. bellus* clade and also found that it is more closely related to *L. fumeus* than to *L. umbratilis*.

DISTRIBUTION *Lythrurus snelsoni* is endemic to the Little River system (Red River drainage) of Arkansas and Oklahoma. In Arkansas, *L. snelsoni* is confined to the upper portions of the Mountain Fork, Cossatot, and Rolling Fork rivers above the Fall Line (Map 50). Its absence from the adjacent Saline River (Little River system) is puzzling, but a somewhat similar pattern is exhibited by the Leopard Darter. A discriminant function analysis of habitat characteristics predicted the potential presence of *L. snelsoni* at three sites in the Rolling Fork River and one site in the Saline River, indicating that suitable habitat probably exists for this species in those rivers (Taylor 1994).

HABITAT AND BIOLOGY In contrast to most species of the genus *Lythrurus*, *L. snelsoni* frequents small to medium-sized streams from 19.7 to 65.6 feet (6–20 m) wide. It prefers pool regions of clear, high-gradient streams, congregating especially at stream margins lined with water willow and avoids current (Robison 1985). Taylor (1994) concluded that small streams with good pool development tended to have the greatest abundance of *L. snelsoni*. Wagner et al. (1987) found that *L. snelsoni* occurred over a narrower range of substrates than did *N. perpallidus*, and the two species have allopatric distributions in Arkansas. Taylor and Lienesch (1995) found this shiner to be strongly associated with high-elevation streams with boulder substrates. In that study, it was most abundant at sites with well-developed pool/riffle development providing high variability in current speed. Mean stream depth was also positively associated with abundance of *L. snelsoni*, but mean stream width was negatively associated with abundance. Taylor and Lienesch (1996) considered *L. snelsoni* and *L. umbratilis* to exhibit a parapatric distribution pattern. While we do not believe that the two species exhibit a completely parapatric pattern in Arkansas, we agree with Taylor and Lienesch that *L. snelsoni* and *L. umbratilis* are segregated by differences in preferred microhabitat. Ten habitat variables separated microhabitats occupied by the two species (Taylor and Lienesch 1996). Elevation had the strongest influence in separating microhabitats, with *L. snelsoni* occupying high-elevation streams and *L. umbratilis* occupying lower-elevation streams. Stream size was of no importance in distinguishing between the preferred microhabitats, with both species occupying a variety of stream sizes. Other noteworthy differences in microhabitats in upland streams of the Little River system included preferences by *L. snelsoni* for waters of lower conductivity, lower percent gravel substrates (reflecting its prefenence for pools), a higher per cent of small and large boulders, lower mean current velocity, and higher percent presence of algae than exhibited by *L. umbratilis* for those features. A 1993 fish survey of Little River tributaries produced specimens of *L. snelsoni* at 10 of 30 Arkansas sites sampled (Taylor 1994). In Oklahoma, according to W. L. Matthews (pers. comm.), this species has been collected in the upper part of reservoirs (e.g., Broken Bow Reservoir on the Mountain Fork River), and Buchanan (2005) found it in small numbers in Gillham Reservoir on the Cossatot River.

This shiner is a surface and water column feeder with a diet that includes chironomids, simuliids, and mayfly larvae (Miller and Robison 2004; HWR, pers. obs.). TMB examined gut contents of 66 specimens collected from the Cossatot and Mountain Fork rivers in Polk County between May and August and confirmed that it feeds mainly at or near the water surface. All guts examined contained food, with terrestrial insects dominating the diet during those months. Major food items were adult dipterans, ants, coleopterans, and other winged insects. Insects typically found in the upper part of the water column were also ingested in large amounts, including dipteran pupae and adult hemipterans. Approximately 9% of the guts contained benthic organisms, mainly ephemeropteran and plecopteran nymphs.

Reproduction occurs in Arkansas from May to late July. Tuberculate males have been taken by HWR as early as 26 May from the type locality, and gravid females have been collected at that site into mid-July. TMB found tuberculate males and ripe females in the Cossatot River as early as 5 May, and tuberculate males in the Mountain Fork River as late as 19 July. Ripe females collected in early May from the Cossatot and Mountain Fork rivers ranged from 40 to 55 mm TL and contained between 109 and 308 ripe ova. The following field notes taken by G. A. Moore and F. B. Cross on 30 May 1948, below the dam on Mountain Fork River, Oklahoma, provide some insight into spawning behavior of *L. snelsoni*:

Many males were swimming at the surface about us as we stood in waist-deep water. Occasionally a female, seemingly disinterested in the males, swam among them. Upon observing a female, a male would give pursuit and nudge her side. If another male came near, the first male would dash out to drive his rival away and then quickly return to the female. Although this behavior was repeated many times, we never observed actual spawning and assumed that we were observing a courtship pattern.

HWR observed this same courtship behavior of *L. snelsoni* on 2 June 1991 at the type locality in the Mountain Fork River, Arkansas.

CONSERVATION STATUS Because of its restricted distribution, the Ouachita Mountain Shiner is vulnerable to large-scale disturbances such as clear-cutting and reservoir construction which have occurred within its limited range. Taylor (1994) recommended a rangewide status of special concern, but we regard the Ouachita Mountain Shiner as threatened in Arkansas. Periodic monitoring of its populations should be conducted.

Lythrurus umbratilis (Girard)
Redfin Shiner

Figure 15.61a. *Lythrurus umbratilis*, Redfin Shiner male. *Uland Thomas.*

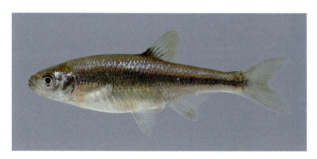

Figure 15.61b. *Lythrurus umbratilis*, Redfin Shiner female. *Uland Thomas.*

CHARACTERS A silver, deep-bodied compressed shiner with a small head (head length going into SL 4.2 times), moderately large eyes (eye diameter going into head length 3.2–3.8 times), and a terminal, oblique mouth (Fig. 15.61). Body depth going into TL 4.1–5.5 times. Dorsal fin origin slightly posterior to pelvic fin origin; dorsal rays 8. Anal fin rays usually 10 or 11 (9–12). Pectoral fin rays 13–15. Origin of pelvic fins nearer to caudal fin base than to tip of snout; pelvic fin rays 8. Anterior dorsolateral scales small and crowded, with 19–32 predorsal scales (usually more than 25). Lateral line complete and distinctly decurved

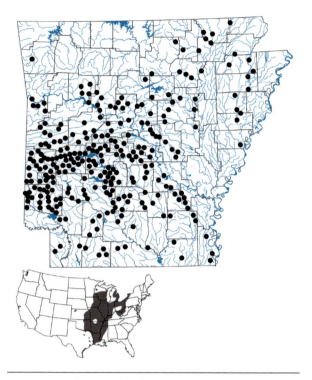

Map 51. Distribution of *Lythrurus umbratilis*, Redfin Shiner, 1988–2019.

with 37–55 scales (lateral line scale counts vary by drainage). Body circumferential scales usually 36 or 37 (32–38). Pharyngeal teeth 2,4–4,2. Intestine short with a single S-shaped loop; peritoneum silvery, sprinkled with dark, stellate speckles. Nuptial males have large tubercles below the eye. Small tubercles in males are scattered over the head, body, and fins. Maximum size 3.5 inches (88 mm) TL.

LIFE COLORS Body color pale olive yellow with fine black speckles. Sides silver with faint dusky lateral stripe posteriorly in some populations. Chin thickly sprinkled with dusky pigment. In adults, dark chevron markings develop on the anterior dorsolateral part of the body. A conspicuous small black spot present at front of dorsal fin base (base of first 2 or 3 rays). Breeding males with a powder-blue head; back and upper sides blue or black, and fins black, red, or a combination of the two. The two subspecies that occur in Arkansas differ in breeding coloration. In western Arkansas River Valley streams, *Lythrurus umbratilis umbratilis* males have black fins (except the caudal fin, which is red or reddish orange), develop a conspicuous scapular bar, have a poorly developed anterior basidorsal spot, and the top of the head is cream-white or without contrasting color. Breeding males of the eastern subspecies *L. u. cyanocephalus* have bright red fins, lack a discrete scapular bar, have a well-developed anterior basidorsal spot, and the top of the head is gray blue to bright blue. Additionally, the dark spot at the base of the first few dorsal fin rays is weakly developed in the western subspecies and well developed in the eastern subspecies.

SIMILAR SPECIES   The Redfin Shiner is most similar to *Lythrurus fumeus* and *L. snelsoni*. It differs from *L. fumeus* in having an anterior basidorsal spot (vs. no spot), chevron markings on upper anterior part of body (vs. no chevron markings), lower anal ray count (usually 10 vs. usually 11 or 12), and reddish breeding colors (vs. yellowish color). From *L. snelsoni*, it can be distinguished by the presence of the anterior basidorsal spot (vs. spot absent), chevrons (vs. no chevrons), caudal peduncle circumferential scales usually 16 (vs. usually 14), body circumferential scales usually 36 or 37 (vs. usually 29–31), and large erect head tubercles (vs. small erect tubercles). It is superficially similar to *Notropis suttkusi*, and *N. atherinoides* but differs from both by the smaller scales in front of the dorsal fin (more than 25 in *umbratilis* vs. fewer than 25 in the other two species).

VARIATION AND TAXONOMY   This species was originally described by Girard (1856) as *Alburnus umbratilis*. Type locality is Sugar Loaf Creek, a tributary of the Poteau River, Sebastian County, Arkansas. Snelson and Pflieger (1975) redescribed the Redfin Shiner and its subspecies in the central Mississippi River basin and designated a lectotype from the same type locality given by Girard (1856). Snelson and Pflieger recognized two subspecies (based primarily on tuberculation and pigmentation), *L. u. umbratilis* and *L. u. cyanocephalus*, both of which occur in Arkansas. The majority of the state is inhabited by *L. u. cyanocephalus*, except the extreme western Arkansas River system. The more western nominate form, *L. u. umbratilis*, enters Arkansas only in the Arkansas River drainage in northwestern Arkansas and near Fort Smith, with the most downstream locality being the Mulberry River near the Franklin-Crawford county line The nominate subspecies, *L. u. umbratilis*, differs from *L. u. cyanocephalus* in having many black specks behind the gill cover coalescing into a large cleithral bar in large males, and in breeding males having the cheeks and opercles covered with nuptial tubercles (vs. few black specks behind gill cover and few or no tubercles on opercle in breeding male). Populations regarded as intergrades inhabit the tributaries of the Arkansas River drainage between the towns of Delaware (eastern Logan County) and Ozark (central Franklin County). A noteworthy population of *L. u. cyanocephalus* was discovered in Cadron Creek, Faulkner County. Specimens from Cadron Creek are generally less robust, and most average measurements are lower than for

Figure 15.62. Coloration variation in *Lythrurus umbratilis* populations from the Ouachita River drainage. *Renn Tumlison*.

Upper figure—*L. u. cyanocephalus*—Arkadelphia area
Middle figure—*L. u. cyanocephalus*—Arkadelphia area
Lower figure—*L. sp. cf. umbratilis*—upper Ouachita River

other populations. The greatest differences are in fin size and body depth.

A third population of *Lythrurus umbratilis*, possibly taxonomically distinct, was discovered by Wayne C. Starnes (pers. comm.) in the upper Ouachita River in 1980. This distinctive form lacks all dark fin pigmentation (Fig. 15.62). Based on fin coloration, apparent intergrades between this novel form and *L. u. cyanocephalus* have been found in the Ouachitas along the Fall Line, with only *L. u. cyanocephalus* occurring below the Fall Line. Raley et al. (2004) examined the genetics of this distinctive form and presented cytochrome *b* data. Raley et al. (2005) hypothesized likely mitochrondrial introgression between *Lythrurus umbratilis cyanocephalus* and *Lythrurus snelsoni*. A combined morphological and genetic analysis of *L. umbratilis* populations in Arkansas is needed to clarify the systematics of this species in the state, particularly the upper Ouachita River form which is probably an undescribed species.

Mayden (1989) hypothesized that *L. ardens* is the sister species to *L. umbratilis*. An analysis of the cytochrome *b* gene by Mayden et al. (2006a) found a sister relationship between *L. umbratilis* and *L. lirus*, with both species forming a clade sister to *L. ardens*. Based on morphological and genetic evidence, Pramuk et al. (2007) recognized an *L. umbratilis* clade consisting of *L. umbratilis*, *L. lirus*, *L. fasciolaris*, *L. ardens*, and *L. matutinus*.

DISTRIBUTION   *Lythrurus umbratilis* is found in the Great Lakes and Mississippi River basins from New York and southern Ontario, Canada, west to Minnesota and south to Louisiana. It is also found in Gulf Slope drainages west of the Mississippi River in Louisiana to the Trinity and San Jacinto rivers of Texas. In Arkansas, the Redfin Shiner is almost statewide in appropriate habitats but is conspicuously absent from the White River system above Batesville (Map 51).

HABITAT AND BIOLOGY   Although occurring in a variety of habitats in Arkansas, the Redfin Shiner is generally found in small to medium-sized streams of low to moderate gradient with relatively clear warm water and, most often, in sluggish pools lined with water willow (*Justicia americana*) over gravel or sand substrates. An undescribed form currently in *L. umbratilis* is common in streams above the Fall Line in the Ouachita Mountains where it inhabits permanent pools in rocky or gravelly creeks having high gradient and low or intermittent flow. In the Coastal Plain lowlands, true *L. umbratilis* is found in sluggish creeks and ditches having vegetation and substrates of clay, mud, and detritus. It also occurs in several small, high-gradient, gravel-bottomed streams of Crowley's Ridge. The Redfin Shiner is a schooling species usually found in the mid to upper part of the water column. It appears to be intolerant of heavy siltation and high turbidity (Cross and Collins 1995), but Becker (1983) noted that in Wisconsin, it was encountered only occasionally in clear water and more frequently in turbid water. Our observations in Arkansas agree more closely with those of Cross and Collins (1995) that it avoids highly turbid environments. It also avoids the largest rivers of the state. Buchanan (2005) reported this species from five reservoirs (DeSoto, Dierks, Gillham, Millwood, and Pineda) in the southern half of the state.

The Redfin Shiner is primarily a surface feeder, consuming both aquatic and terrestrial insects and other available small animal life (Becker 1983). Pflieger (1997) also reported that it feeds largely by sight, taking most of its food from the surface. In Indiana, it fed on a variety of food items, mostly of terrestrial origin, including ants, adult midges and other dipterans, coleopterans, hemipterans, and homopterans (Whitaker 1981). Forbes and Richardson (1920) reported that zooplankton was also included in its diet. It occasionally feeds on filamentous algae and bits of higher plants (Eddy and Underhill 1974). Its diet changed seasonally in Louisiana, consisting of 57% surface prey and 40% benthic invertebrate prey during spring and early summer (wet season), and becoming more varied during the dry seasons (59% surface animal prey, 14% midwater prey, 14% benthic prey, and 13% detritus) (Felley and Felley 1987).

Spawning occurs in Kansas from May to July at water temperatures of 21°C (70°F) or higher (Cross and Collins 1995). Tuberculate males have been found in Arkansas during an extended spawning season from late April to August. We have found spawning of *L. u. umbratilis* occurring as early as 17 April in small creeks near Fort Smith. Young-of-the-year were collected as early as 10 July. Males with intense breeding colors and females with mature eggs have been collected from the Ouachita River in Garland County on 21 June, 2 July, and 21 July (Tumlison et al. 2017). M. M. Matthews and Heins (1984) reported a comparable reproductive season in Mississippi. Balon (1975) listed *L. umbratilis* as a representative of the phytolithophilous guild of nonguarding fishes. The Redfin Shiner has been classified as both a broadcast spawner and a nest associate. The formation of very dense spawning aggregations is a distinct feature of its reproductive behavior (Hunter and Hasler 1965). Hunter and Wisby (1961) noted an interesting spawning association between the Redfin Shiner and Green Sunfish, and *Lythrurus umbratilis* also deposits its eggs over the nests of other sunfish species, including Longear Sunfish and Orangespotted Sunfish. Male sunfish guarding their own nests seem to pay little attention to these intruders. The advantage of this strategy is that the sunfish nest provides a clean, silt-free substrate for development of the eggs, and the male sunfish prevents other fishes from entering the nest and eating the Redfin Shiner eggs. Hunter and Hasler (1965) and Pflieger (1997) described reproductive behavior as observed over sunfish nests. Male Redfin Shiners gathered over sunfish nests in numbers ranging from 1 to more than 100. A male defends a small territory against other males. When ready to spawn, a female enters a cluster of males and is immediately approached by a male who takes a position beside her. Spawning occurs as the two fish swim high over the nest, and the eggs drop into the sunfish nest where they complete their development. Trautman (1957) observed spawning over sandy substrates in sluggish current, even though gravel substrates with swift current were present in the same stream. Clutch sizes of females 34–52 mm SL ranged from 219 to 887 mature ova (Matthews and Heins 1984). *Lythrurus umbratilis* is relatively short-lived, maturing at about 1 year. Matthews and Heins (1984) indicated a maximum age of 2 years and considered *L. umbratilis* to be a relatively *r*-selected species based on life history traits.

CONSERVATION STATUS The population of the western subspecies in the Illinois River drainage of northwestern Arkansas may have declined because there is only one post-1988 record reported from that drainage. The Redfin Shiner is widespread and common elsewhere in Arkansas, and its populations appear to be secure.

Genus *Macrhybopsis* Cockerell and Allison, 1909—Blacktail Chubs

Species of this genus have had a tangled taxonomic history (Boschung and Mayden 2004). Bailey (1951) lumped barbeled minnows of various genera in *Hybopsis*, but Mayden (1989) found *Hybopsis* to be polyphyletic and elevated *Macrhybopsis* to generic status and included the former *Hybopsis storeriana* in *Macrhybopsis*. *Macrhybopsis storeriana* was placed in a clade with *H. meeki*, *H. gelida*, and *Platygobio gracilis*, and the genus *Extrarius* was resurrected to include members of the *aestivalis* complex (Mayden 1989). Using morphology, Coburn and Cavender (1992) placed *Extrarius* in the expanded genus *Macrhybopsis* and determined *Macrhybopsis* to be monophyletic and sister to *Erimystax*. The classification of Coburn and Cavender (1992) was accepted by Mayden et al. (1992a) and is generally followed today (Gilbert et al. 2017). Analyses of mitochondrial and nuclear genes supported monophly for the genus *Macrhybopsis* and a sister relationship to *Platygobio gracilis* (Echelle et al. 2018).

Page et al. (2013) included 8 species in *Macrhybopsis*, 4 of which (*M. gelida*, *M. hyostoma*, *M. meeki*, and *M. storeriana*) occur in Arkansas. Gilbert et al. (2017) described four new species of *Macrhybopsis* from Gulf Slope drainages east of the Mississippi River, bringing the number of species in the genus to 12, of which 9 are in the *Macrhybopsis aestivalis* complex. Echelle et al. (2018) provided genetic evidence supporting monophyly for the wide-ranging *M. aestivalis* complex. Characters of the genus include an elongated, relatively slender body, head flattened ventrally, and snout conical to blunt. There are 1 or 2 barbels at the corners of the jaws. Lateral line scale rows 34–50; predorsal scales 14–24. Anal rays 8; pectoral rays 13–18. Pharyngeal teeth 1,4–4,1 or 0,4–4,0 or a combination thereof. Nuptial tubercles present on pectoral fins. Enlarged nasal capsule; elongate branchiostegal rays. Coloration dusky or silvery. Caudal fin of most species with a darkened lower lobe and a white ventral border.

Species of this genus are mainly benthic occupants of swift, main-channel habitats of large rivers (Branson

1963; Davis and Miller 1967; Reno 1969). All species avoid upland areas. *Macrhybopsis* species are summer broadcast spawners in medium to large rivers and have semibuoyant eggs that drift downstream until hatching. After a pelagic larval stage, they become benthic insectivores (Albers and Wildhaber 2017). Taste buds and lateral line neuromasts are abundant and well developed in members of this genus, and the eyes are relatively large. Moore (1950) reported that *Macrhybopsis* species probably possess the most highly developed cutaneous sense organs of any of the North American cyprinids.

Macrhybopsis gelida (Girard)
Sturgeon Chub

Figure 15.63. *Macrhybopsis gelida*, Sturgeon Chub. *Richard T. Bryant and Wayne C. Starnes.*

Map 52. Distribution of *Macrhybopsis gelida*, Sturgeon Chub, 1988–2019.

CHARACTERS A slender, small-barbeled minnow with a small, horizontal mouth, a long, fleshy snout extending far forward of the mouth (thus the name Sturgeon Chub), a pustulose gular region, and rounded pectoral fins (Fig. 15.63). Eyes very small, eye diameter much less than snout length. Snout length about equal to distance from back of eye to rear margin of operculum. Many of the dorsal and lateral scales with distinct, longitudinal, ridgelike projections (keels), a unique character in native minnows. Head depressed, sturgeonlike, with minute sensory buds dorsally and large sensory papillae ventrally. Pharyngeal teeth 1,4–4,1. Dorsal and anal fins nearly straight-edged, not falcate. Anterior dorsal rays not extending past posterior rays when depressed. Pectoral fins not reaching pelvic fin origins. Caudal fin deeply forked. Lateral line scales 39–45. Predorsal scale rows 18–20. Dorsal rays 8; anal rays 8 (rarely 7). Pectoral fin rays 15–17 (13–17). Breast and belly scaleless. Intestine short; peritoneum silvery with scattered speckles. Although it exhibits little sexual dimorphism, males develop small tubercles behind the gills during spawning periods. Adults are usally 2–2.8 inches (50–70 mm) SL. Maximum size formerly reported as 3 inches (77 mm) TL, but Steffensen et al. (2014) reported adults exceeding 3.9 inches (100 mm) TL from the channelized Missouri River.

LIFE COLORS Body color pallid, brown dorsally with silvery sides and venter, sides without definite markings. Lower lobe of the caudal fin darker than the upper lobe except along the ventral margin, which is a milky white color. No chromatic breeding coloration in either sex.

SIMILAR SPECIES The Sturgeon Chub can be distinguished from all other chubs (and all native cyprinids) in Arkansas by the distinct keels on many of its body scales (vs. body scales without keels). It further differs from *M. meeki* in having shorter pectoral fins, the tips not reaching the pelvic fin origins (vs. pectoral fin tips reaching pelvic fin origins), and from *Platygobio gracilis* in having anterior dorsal rays not extending past posterior rays when fin is depressed (vs. extending well past), and in having 1,4–4,1 pharyngeal teeth (vs. 2,4–4,2).

VARIATION AND TAXONOMY *Macrhybopsis gelida* was originally described by Girard (1856) as *Gobio gelidus* from the Milk River, a tributary of the upper Missouri River, Montana. The Sturgeon Chub was subsequently assigned to *Ceratichthys* (*C. gelidus*) by Jordan and Gilbert (1883), and then *Hybopsis* by Jordan and Evermann (1896). It was placed in *Macrhybopsis*, then back to *Hybopsis* (Bailey 1951), and finally returned to *Macrhybopsis* by Mayden (1989) where it has remained. No subspecies are recognized. Analysis of mitochondrial and nuclear DNA produced conflicting results about the relationships of *M. gelida* to other species (Echelle et al. 2018).

DISTRIBUTION *Macrhybopsis gelida* occurs in the Missouri River basin from Montana and Wyoming to its confluence with the Mississippi River in central Missouri, and south in the Mississippi River to southern Louisiana. In Arkansas, this rare chub is known only from the mainstem Mississippi River (Map A4.53), where there are only two older records (Robison and Buchanan 1988). D. A. Etnier (pers. comm.) collected *M. gelida* from the Mississippi River in Tipton County, Tennessee, adjacent to southern Mississippi County, Arkansas. Carter (1984) subsequently reported a single specimen of *M. gelida* from Arkansas in Mississippi County. It also occurs in the Mississippi River in Missouri and Louisiana. Herzog (2004), using a Missouri trawl, collected hundreds of specimens from the Mississippi River of southeastern Missouri in 2000 and 2001. We have no post-1988 records for this species in Arkansas (Map 52).

HABITAT AND BIOLOGY The Sturgeon Chub inhabits continuously and highly turbid, warm, medium-sized to big rivers primarily in the Great Plains. It is found primarily in shallow areas (<2 m) of strong current with sand and, perhaps more frequently, a gravel bottom (Jenkins 1980a). However, bottom-trawling in deep, main channel habitats in the Missouri and Mississippi rivers also produced specimens of *M. gelida* (Rahel and Thel 2004b). This chub is highly specialized for swift, turbid water with its streamlined form, reduced eyes, and great development of taste buds over the head, body, and fins. The keeled scales may help the fish stabilize and orient itself in swift currents or serve as a means of detecting currents (Johnsgard 2005). In the Mississippi River of Missouri, Herzog (2004) found that 68% of the total catch per unit effort of Sturgeon Chub was at water depths less than 2.0 m. In that study, nearly 90% were captured in water velocities between 0.4 and 0.8 m/s. In the Missouri River, chubs <25 mm used habitats associated with L- and wing-dike structures, while chubs >25 mm were more associated with channel sandbar habitats (Ridenour et al. 2009). In North Dakota, Sturgeon Chubs were found at a wider range of depths and velocities than other chub species, with most captured at depths of 2–5 m and in current velocities from 0.5 to 1.0 m/s (Welker and Scarnecchia 2006). It is generally a rare species throughout its range, but it is most common in large turbid streams of the northern Great Plains having gravel bottoms (Cross 1967). This type of substrate is not extensive in the Mississippi River of Arkansas, but small patches of fine to medium gravel are sometimes found there. Perkin and Gido (2011) studied stream fragmentation thresholds in four

Macrhybopsis species and estimated a minimum threshold fragment of 297 river km required by the Sturgeon Chub for maintaining populations.

This chub is presumably a benthic taste feeder and is highly specialized for swift, turbid water with its streamlined form, reduced eyes, and great development of taste buds over the head, body, and fins. Its food habits are largely unknown, but Cross and Collins (1995) surmised that it eats larval insects. Etnier and Starnes (1993) reported chironomids and baetid mayflies in one specimen from the Yellowstone River. Other studies have also found benthic insects (mostly unidentifiable) in gut contents (Rahel and Thel 2004b). Juvenile diets in the Missouri River were dominated by midge pupae, and diet overlap was greatest with *M. hyostoma* (Starks et al. 2016).

Information on its reproductive biology has gradually accumulated in recent years. Spawning was confirmed to occur in the Missouri River from early May to late August (Sparks et al. 2016). Albers and Wildhaber (2017) confirmed that *M. gelida* is a pelagic, multiple batch spawner and individual females can spawn multiple times over months at water temperatures between 18.3°C and 22.7°C. Fecundity estimates range from 2,000 to 5,310 eggs per female (Albers 2014a). Hatch date histograms were congruent with the belief that *M. gelida* spawns periodically from June to September (Starks et al. 2016). Albers and Wildhaber (2017) also found that it produces nonadhesive eggs that, once water hardened, are semibuoyant and require current to remain in suspension. A drifting larval stage subsequently occurs. Young-of-year attain about 1.1–1.4 inches (28–36 mm) TL at the end of the first summer (Pflieger 1997) with sexual maturity reached at about 43 mm.

CONSERVATION STATUS *Macrhybopsis gelida* is a rare inhabitant of the mainstream Mississippi River in Arkansas, and there are no recent records from the state. We list its status in Arkansas as undetermined, but it should probably be considered an endangered species in this state.

Macrhybopsis hyostoma (Gilbert)
Shoal Chub

Figure 15.64. *Macrhybopsis hyostoma*, Shoal Chub. *Uland Thomas.*

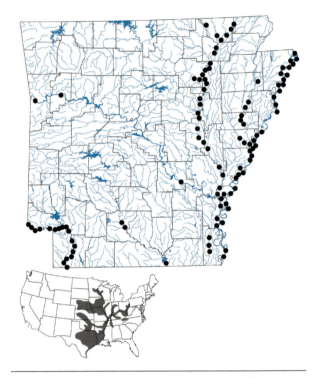

Map 53. Distribution of *Macrhybopsis hyostoma*, Shoal Chub, 1988–2019.

CHARACTERS A small, streamlined, pale, speckled minnow adapted for life on the bottom in swift waters (Fig 15.64). Head moderately rounded. Eye variable in size, generally small and suboval, but large and oval in some populations. Snout fleshy, protruding far beyond the mouth and moderately flattened ventrally. Mouth inferior, horizontal; lips moderately fleshy, not thickened posteriorly. Large taste buds expanded into papillae on gular area. Maxillary barbels 1 or 2 on each side and conspicuous; when 2 barbels are present on each side of mouth, the posterior barbel is less than the orbit length, and the anterior barbel is less than 50% of orbit length. Anterior part of body deeper and wedge-shaped; posterior part slim and tapered. Pharyngeal teeth 0,4–4,0. Gill rakers absent or rudimentary. Lateral line complete, usually with 35–38 scales. Predorsal scales usually 12–17. Caudal peduncle scale rows modally 12. Breast mostly scaleless. Nape fully scaled. Dorsal fin slightly falcate, its origin over pelvic fin origin. Dorsal rays 8. Anal fin slightly falcate; anal rays modally 8. Pectoral rays 13–16. Pelvic rays 7 or 8. Intestine short; peritoneum silvery ventrally, with some dark mottling dorsally. Pectoral fins short, their tips not reaching bases of pelvic fins in adult males. Pectoral rays 2–10, thickened in large nuptial males, and with rows of conical, antrorse uniserial tubercles. Tiny tubercles rarely present on rays of dorsal and pelvic fins in large (>50 mm SL) nuptial males in peak condition (Eisenhour 2004). Pinion and Conway (2018) reported the presence of unculiferous platelike tubercles across the surface of the head and much of the body and fins in males and females. Those authors also noted that stalk-like structures along the anterior edge of the paired and median fins are taste buds. Maximum size 3 inches (76 mm). Females attain a larger size than males; largest female 61.6 mm SL, 76 mm TL; largest male 53.6 mm SL, 66 mm TL (Eisenhour 2004).

LIFE COLORS   Color pale green or gray dorsally to silvery white below, almost translucent in life. Lateral stripe usually absent in Arkansas specimens. Lateral stripe, if present, usually best developed on caudal peduncle and more diffuse anteriorly. Random, medium-sized melanophores on the dorsolateral surface of the body produce a speckled appearance. Concentrations of small melanophores on margins or submargins of the dorsolateral scales weakly (turbid streams) to strongly (clear streams) expressed. The number of dorsal melanophores is highly variable within a population, with some individuals lacking melanophores. Sides with a silvery lateral stripe. No breeding colors.

SIMILAR SPECIES   *Macrhybopsis hyostoma* differs from all other chubs in having small black spots scattered over the back and sides (vs. random spots absent). *Erimystax harryi* has a midlateral row of dark blotches, and *E. x-punctatus* has numerous scattered x-shaped markings on body. The Shoal Chub further differs from *Erimystax* species in having fewer than 40 lateral line scales (vs. more than 40), and in the upper jaw extending past front of eye (vs. upper jaw not extending past front of eye). It is sometimes syntopic with the other three Arkansas species of *Macrhybopsis* and *Platygobio gracilis* but differs from those species in having 0,4–4,0 pharyngeal teeth (vs. 1,4–4,1 in the other *Macrhybopsis* species, and 2,4–4,2 in *P. gracilis*).

VARIATION AND TAXONOMY   *Macrhybopsis hyostoma* was described by Gilbert (1884) as *Nocomis hyostomus* from the East Fork of the White River at Bedford, Lawrence County, Indiana. Gilbert et al. (2017) reviewed the taxonomic history of the *Macrhybopsis aestivalis* complex, which includes *M. hyostoma*. The small, barbeled chubs previously known as *Hybopsis aestivalis* underwent changes in classification between 1997 and 2004 (Eisenhour 1997, 1999, 2004; Underwood et al. 2003). The western members of this species complex were formerly treated as a single, wide-ranging, morphologically plastic species, *M. aestivalis*; however, Miller and Robison (1973) had suspected for some time that this complex comprised more than one species. Subsequent phylogenetic analysis of the complex led to the breakup of the genus *Hybopsis*, with the *aestivalis*

complex initially placed in the genus *Extrarius* (Mayden 1989), and then in *Macrhybopsis* (Coburn and Cavender 1992; Dimmick 1993; Simons and Mayden 1999; Eisenhour 2004). Eisenhour (1997, 1999, 2004) provided evidence that the *Macrhybopsis aestivalis* complex west of the Mississippi River consisted of five species based on morphological variation. Only one of those species, *M. hyostoma*, occurs in Arkansas. A 4-barbeled species, *M. tetranema*, is historically known from the upper Arkansas River drainage from central Oklahoma upstream. *Macrhybopsis hyostoma* occurs syntopically with *M. tetranema* in the Arkansas River drainage in central Oklahoma. Another 4-barbeled form, *M. australis*, occurs only in the upper Red River drainage above Lake Texoma. Genetic evidence supported the specific distinctness of *M. australis* and *M. tetranema* (Echelle et al. 2018). *Macrhybopsis australis* and *M. hyostoma* are syntopic in the middle portion of the Red River in Oklahoma, and analysis of allozymic variation demonstrated differences in allele frequencies where the species occurred together (Underwood et al. 2003). The remaining two species of the *aestivalis* complex west of the Mississippi River are *M. marconis*, largely confined to streams of the Edwards Plateau in central Texas, and *M. aestivalis*, now considered endemic to the Rio San Fernando drainage and Rio Grande basin in Mexico and Texas. Gilbert et al. (2017) described four new *Macrhybopsis* species from east of the Mississippi River from populations formerly included in the *M. aestivalis* complex and suggested that further study could reveal additional nameworthy taxa.

Eisenhour (2004) redescribed *M. hyostoma* and clarified its range as occurring in streams of the West Gulf Slope and Mississippi River basin. The Shoal Chub differs from other *Macrhybopsis* species in a complex suite of characters including morphometry, modal conditions for several meristic characters, and pattern of tuberculation in breeding males (Eisenhour 1999, 2004), as well ecological differences in macrohabitat (Underwood et al. 2003). *Macrhybopsis hyostoma*, as currently recognized, exhibits considerable geographic variation in morphology, and taxonomists have frequently commented on the incredible plasticity of this species. A group of populations called the "Southern Plains" group consists of populations from the Red and Arkansas river basins that are characterized by morphological adaptations to turbid river conditions (Eisenhour 2004). Specimens are pallid overall with a poorly defined or absent lateral stripe and little or no submarginal dorsolateral scale pigmentation. They have high mean scale counts, few mean infraorbital pores, long barbels, small eyes, and often a second pair of barbels. Specimens from the lower Red and Arkansas rivers have shorter barbels, larger eyes, and darker pigmentation than specimens from the middle sections of those river basins. Another population, referred to by Eisenhour (2004) as the "Central Coastal Plain" group, consists of populations from the Sabine, Calcasieu, Ouachita, White, and St. Francis river drainages. Specimens from these populations have large eyes and dark pigmentation but tend to be intermediate between the "Ohio" group and specimens from the Southern Plains in barbel length and number, and infraorbital pore counts. This group is also characterized by relatively low lateral-line scale and vertebral counts. Based on examination of barbel number in Arkansas specimens of *M. hyostoma* by TMB, 43% of the specimens from the Red River had 2 barbels and 57% had 4 barbels. For specimens from the Arkansas River near Fort Smith, 60% had 2 barbels and 40% had 4 barbels. All specimens from the White, St. Francis, and Mississippi rivers had 2 barbels, but only a few specimens were examined from each of those drainages.

A phylogenetic analysis of selected western populations using allozyme data did not demonstrate a monophyletic *M. hyostoma* (Underwood et al. 2003). However, those authors suggested that the apparent paraphyly of *M. hyostoma* could possibly be attributed to secondary contact and introgressive hybridization of *M. hyostoma* with *M. australis* and *M. tetranema*. Eisenhour (2004) believed that given the uncertain and possibly reticulate evolutionary history among populations considered to represent *M. hyostoma*, further taxonomic subdivision of *M. hyostoma* was premature. He considered *M. hyostoma* a "catchall" taxon that grouped eastern clearwater forms with all other undescribed forms including western turbid-water forms. Despite recent studies clarifying its taxonomic status and distribution, Echelle et al. (2018) suggested that *M. hyostoma*, as currently recognized, still includes undescribed cryptic species.

DISTRIBUTION   *Macrhybopsis hyostoma* is the most wide-ranging and morphologically variable member of the *M. aestivalis* complex. As currently recognized, *M. hyostoma* is distributed from the Great Lakes drainage of Lake Michigan in Wisconsin (Wolf River); the Mississippi River basin from Ohio east to West Virginia, west to southern Minnesota, south to the Tennessee River in the east and the Red River of Louisiana and Texas in the west; and in Gulf Slope drainages from western Louisiana to the Lavaca River, Texas. In Arkansas, the Shoal Chub is confined primarily to the larger rivers of the Coastal Plain and the Arkansas River Valley (Map 53). Its range in the state includes the Mississippi, lower White, St. Francis (McAllister et al. 2010a), Arkansas, Ouachita, and Red rivers. It was found throughout the Red River in Arkansas

but was considered uncommon (Buchanan et al. 2003). It is common in the Mississippi River (TMB, unpublished data), and was the dominant cyprinid in collections in 2013 from the lower St. Francis River (Jeff Quinn, pers. comm.). It generally avoids the uplands in Arkansas.

Jordan and Gilbert (1886) reported the Speckled Chub as abundant in the Arkansas River at Fort Smith as did Meek (1896), and Black (1940) collected it from that river in Logan County. Surveys of the Arkansas River by Buchanan (1976) and by Sanders et al. (1985) failed to produce any specimens. Luttrell et al. (1999) found that this species had drastically declined in the Arkansas River basin, and it was believed to be extirpated from the Arkansas River in Arkansas. Numerous collections at Fort Smith between 1971 and 1999 (including monthly sampling below Trimble Lock and Dam from 1996 to 1999) produced no Shoal Chubs; however, monthly sampling of that site from 2000 through 2006 yielded new records for *M. hyostoma* on four occasions (TMB, unpublished data). McAllister et al. (2012b) reported the first specimens from the lower Arkansas River in Desha County, and Brown et al. (2016) discovered *M. hyostoma* in a small direct tributary of the Arkansas River just south of the lock and dam at Ozark, Franklin County, Arkansas. Thirty-five specimens were collected from this tributary, marking the first record of this species in Arkansas from a habitat other than a large river environment.

HABITAT AND BIOLOGY   The Shoal Chub inhabits swift-flowing waters in large rivers over coarse, clean sand, or gravelly raceways with strong current (Eisenhour 1997, 2004; Luttrell et al. 1999). *M. hyostoma* is highly adapted for bottom dwelling in a turbid stream by a depressed body, rather elongate snout, subterminal mouth, 2–4 well-developed barbels, reduced eyes, a well-developed olfactory area of the brain, and large pectoral fins (Moore 1950; Metcalf 1966; Davis and Miller 1967; Reno 1969). Becker (1983) also noted its preference for slightly turbid to turbid waters. This chub inhabits the Arkansas River, which had high turbidity prior to construction of the Arkansas River Navigation System. In the Mississippi River of southeastern Missouri, *M. hyostoma* was the most abundant chub species collected by trawling, with more than 800 specimens taken (Herzog 2004). Hrabik et al. (2015) noted that its populations are subject to strong year-to-year fluctuations in abundance.

The Shoal Chub feeds in turbid streams using external taste buds located on the head, body, and fins. Reno (1969) commented that it was an extremely plastic species, apparently quite labile in response to altered environmental conditions. Fishes in turbid environments rely primarily on an abundance of cutaneous taste buds for location of food, while those in clear waters probably depend more on visual sensitivity. Davis and Miller (1967) observed feeding behavior in captivity that strongly suggests this chub swims over the bottom with barbels in contact with the substrate until the taste buds on the body, fins, and barbels are stimulated by substances emanating from food. Hubbs and Ortenburger (1929a) suggested that barbel development was a recent compensatory adaptation for reduced vision in turbid habitats. Age 0 fish fed mostly on midge pupae in the Missouri River (Starks et al. 2016). Larval Shoal Chubs begin feeding 2–3 days after hatching, with most food taken from the bottom or as it falls through the water (Becker 1983). The postlarval diet consists mostly of immature aquatic insects, primarily trichopterans, hemipterans, odonates, and coleopterans. In Iowa, foods were primarily dipterans and small crustaceans, and some plant material (Starrett 1950a). In Texas, C. S. Williams (2011) reported that food items of adults in summer (by % weight) consisted of aquatic insects, including Trichoptera (24%), Chironomidae (17%), Ephemeroptera (7%), and unidentified aquatic insects (14%), sand, and detritus. During the spring and fall, aquatic insects occurring most often (% occurrence) in the diet included Chironomidae (45%), Trichoptera (38%), and Ephemeroptera (13%). Terrestrial insects were an uncommon food item. In juvenile Shoal Chubs, detritus made up the largest percent by weight of food items, followed by sand, aquatic insects, and small crustaceans (ostracods).

Spawning occurs from May through August when temperatures exceed 70°F (21.1°C) (Cross and Collins 1995). It is probably a flood-pulse spawner like *M. tetranema* (Bottrell et al. 1964). This chub is a broadcast spawner and eggs are deposited in deep water with swift current at midday or within an hour or two thereafter (Bottrell et al. 1964). Fertilized eggs develop as they drift in the current. Eggs hatch in about 25–28 hours. In the Arkansas River of Oklahoma, eggs have been found from 14 May to 28 August. TMB examined three ripe females collected by William G. Layher from the St. Francis River in Cross County, Arkansas, on 18 May. Those females were 49, 53, and 54 mm TL and contained 785, 491, and 662 mature ova, respectively; average mature ovum diameter was 0.95 mm. Williams (2011) reported that estimated mean clutch size for mature Shoal Chubs 39–70 mm TL was 198.75 and clutch size ranged from 34 to 680 oocytes. Clutch size was positively related to total length of mature females. In St. Francis River specimens, TMB found that one or two batches of smaller egg sizes were present in the ovaries, indicating the probability of multiple clutches. Williams

(2011) also found that females produced multiple cohorts of oocytes during a single reproductive season. Mean monthly GSIs for female Shoal Chubs were elevated (>3.5%) and mature ovaries were present from April through October, with some females remaining mature through November. The smallest female Shoal Chub exhibiting gonadal maturation was a 39 mm individual collected in early September. The Shoal Chub is a short-lived species, seldom surviving more than 1.5 years (Becker 1983).

CONSERVATION STATUS The Shoal Chub is uncommon in the Arkansas, Red, and Ouachita rivers, but is more abundant in the lower St. Francis, lower White, and Mississippi rivers in Arkansas. In the lower Mississippi River of Arkansas, *M. hyostoma* was collected by the thousands in the 2000s (K. J. Killgore and J. J. Hoover, pers. comm.), and collections in that river by TMB have always included many specimens. In a study by the U.S. Army Corps of Engineers, of the 56,630 fishes collected in the lower Mississippi River, Shoal Chubs made up 27% of that total (Schneider 2015). Extirpations of *M. hyostoma* in the Arkansas River basin probably were due to habitat fragmentation resulting from the construction of numerous reservoirs, channelization, and the anthropogenic disturbances to the hydrology of the streams/rivers and concomitant isolation of many populations. The species is extirpated from about 55% of its historic range in the Arkansas River basin (Luttrell et al. 1999). Its decline in the Arkansas River may also be associated with reduced turbidity caused by the construction of the Arkansas River Navigation System. Despite some apparent reductions in range, we consider its populations secure in Arkansas.

Macrhybopsis meeki (Jordan and Evermann)
Sicklefin Chub

Figure 15.65. *Macrhybopsis meeki*, Sicklefin Chub. *Uland Thomas.*

CHARACTERS A pallid, barbeled chub with a terete body, small eyes, a rounded, bulbous snout, a pustulose gular region, slender caudal peduncle, and falcate dorsal, pectoral, and anal fins (Fig. 15.65). Snout not depressed, projecting slightly beyond small, horizontal mouth. Upper

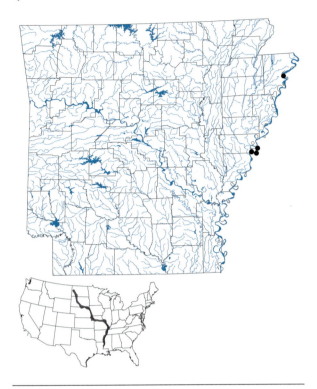

Map 54. Distribution of *Macrhybopsis meeki*, Sicklefin Chub, 1988–2019.

jaw extends posteriorly just past the anterior margin of eye. A single pair of maxillary barbels conspicuous. Eye small, its diameter going into head length 8 times or more; eyes placed high on head. Lower lobe of the caudal fin darkly pigmented, with a distinctive white ventral margin. Breast and belly scaleless. Dorsal fin origin directly over pelvic fin base. Dorsal fin distinctly falcate; dorsal rays 8. Pectoral fins very long and extremely falcate, their tips extending well behind pelvic fin origins; pectoral fin rays 16 or 17. Pelvic fins somewhat falcate, their tips reaching posteriorly to anal fin base; pelvic rays 8. Anal fin slightly falcate; anal rays 8. Body scales smooth, not keeled. Lateral line complete and straight, with 43–50 scales. Predorsal scale rows 22–24. Pharyngeal teeth 1,4–4,1. Intestine short, with a single S-shaped loop; peritoneum silvery with scattered dark speckles. Taste buds on fins are sometimes mistaken for nuptial tubercles in breeding males; however, the taste buds are restricted to the interradial fin membranes, whereas tubercles are on the fin rays (Davis and Miller 1967). Maximum size 4 inches (102 mm) TL.

LIFE COLORS   Body color pallid, yellowish green or tan dorsally, often with many dark brown and silver specks, and silvery white on the sides and belly. Lower lobe of the caudal fin darkly pigmented, with a narrow white ventral edge; other fins unpigmented. No breeding colors attained.

SIMILAR SPECIES   It differs from *M. gelida* in lacking keels on body scales, in having long pectoral fins with tips extending well behind pelvic fin origins, and in having anterior dorsal rays extending beyond posterior rays when fin is flattened. It differs from *Platygobio gracilis* in lacking scales on breast (vs. breast scales present), in pectoral fins with tips extending well behind pelvic fin origins, and in having 1,4–4,1 pharyngeal teeth (vs. 2,4–4,2). The Sicklefin Chub can be distinguished from *H. amblops* and *Macrhybopsis storeriana* by its small eye and in usually having more than 42 lateral line scales (vs. 42 or fewer lateral line scales). It differs from *M. hyostoma*, *Erimystax harryi*, and *E. x-punctatus* in having an unpigmented body lacking spots or dark markings (vs. dark spots and markings present), and in having pharyngeal teeth 1,4–4,1 (vs. 0,4–4,0).

VARIATION AND TAXONOMY   *Macrhybopsis meeki* was described by Jordan and Evermann (1896) as *Hybopsis meeki* from the Missouri River at St. Joseph, Missouri, and named in honor of Seth E. Meek, an early ichthyologist at the University of Arkansas.

DISTRIBUTION   *Macrhybopsis meeki* occurs in the Missouri River from North Dakota to its mouth, and in the Mississippi River from the mouth of the Missouri River to southern Mississippi and Louisiana. In Arkansas, it is confined to the mainstem Mississippi River (Map 54). The first Arkansas record for this species was collected and identified from the river near Blytheville, Mississippi County, on 29 December 1980 by T. M. Buchanan, F. A. Carter, and S. Henry (2 juvenile specimens—25 mm SL and 22 mm SL respectively, ASUMZ 9366) and reported by Carter and Beadles (1983a). Buchanan, Carter, and Henry also collected the first known specimens from Tennessee waters: three specimens (UAFS 0231) from the Mississippi River in Lauderdale County, Tennessee, along the east bank of the river across from Barfield, Arkansas, on 29 December 1980. In 2006 and 2008, this species was collected by trawling on three occasions from the Mississippi River near Mhoon Bend adjacent to Lee County, Arkansas, by biologists from the Vicksburg Waterways Experiment Station (data provided by K. J. Killgore). Six specimens of *M. meeki* were collected from the Mississippi River at Sans Souci Landing south of Osceola, Mississippi County on 16 October 2015 with a Missouri Trawl (Tumlison et al. 2016).

HABITAT AND BIOLOGY   In the southern portion of its range, this obligate riverine species occurs in fast water in the main channels of large, warm, turbid rivers (Mississippi River) over a bottom of firm sand or fine gravel (Davis and Miller 1967; Smith 1979). It is rarely taken in areas with soft substrates (Hrabik et al. 2015). This chub is most common over sand substrates in depths greater than

2 m (6.6 feet), with bottom velocities averaging 0.47–0.90 m/s (1.5–3.0 feet/s) (Albers 2014b). In the western portion of its range, it is most common in river segments with higher turbidity (>80 NTU), summer water temperatures of 14–26°C (57–79°F), conductivities of 402–830 uS/cm, and variable annual flows but low, stable flows in late summer (Dieterman and Galat 2004; Welker and Scarnecchia 2004, 2006). Adults prefer faster velocities with a higher percentage of gravel and lower percentage of silt relative to juveniles (Dieterman 2000). It apparently prefers deeper water than the often sympatric Sturgeon Chub, and is more commonly taken by trawling rather than by seining. Apparent minimum stream length necessary for persistence of the Sicklefin Chub is 301 river km (Dieterman and Galat 2004). In the Mississippi River of southeastern Missouri, 69% of *M. meeki* were captured at water depths greater than 4.0 m, and more than 50% were captured at water velocities between 0.4 and 0.6 m/s (Herzog 2004). Its fusiform shape, taste bud patterns, and small eyes are adapted for living in large turbid rivers with high velocity habitats (Dieterman and Galat 2005). Other adaptations for highly turbid waters include external taste buds on the dorsal and pectoral fins, abundant lateral line neuromasts, and numerous sensory papillae in the gular region and within the buccal cavity (Moore 1950; Davis and Miller 1967; Reno 1969). In the lower Missouri River, most Sicklefin Chubs occurred along shorelines associated with sandbars, and most fish were likely age 0 juveniles (19–40 mm TL) (Grady and Milligan 1998; Dieterman and Galat 2004). These findings suggest that shallow, riverine shorelines provide critical nursery habitat for Sicklefin Chubs. Shorelines with gradually sloping banks are important nursery habitats for many obligate-riverine fishes because they provide reduced current velocity, shallow water depths, greater light penetration, higher water temperatures, and increased primary production (Schiemer et al. 1995; Scheidegger and Bain 1995; Humphries and Lake 2000).

Based on observations of sensory morphology by Davis and Miller (1967), the Sicklefin Chub is presumably a benthic taste feeder. The body is covered with taste buds that aid in locating food, and there are taste buds in the mouth cavity, causing some to speculate that food items may be ingested and sorted in the mouth, with inedible foods being spit out (Pflieger 1997). Reigh and Elsen (1979) reported black fly pupae and other insects from the stomachs of *M. meeki*. Age 0 individuals fed mostly on midge pupae in the Missouri River, but *M. meeki* had a more general diet compared with *M. hyostoma* and *M. gelida* (Starks et al. 2016).

Young-of-the-year specimens collected from the lower Missouri River in July led Pflieger (1997) to suggest a spring spawning period, but subsequent investigations indicated that most spawning occurs in summer. Albers and Wildhaber (2017) reported a long spawning season extending from March through August in Nebraska and Missouri. In the Missouri and Yellowstone rivers, *M. meeki* attained sexual maturity as early as age 2 (but most matured at age 3), and gravid females containing from 7 to 1,561 mature ova were found in July and August at water temperatures ranging from 16.7°C to 25.4°C (Dieterman et al. 2006; Grisak 1996). Albers and Wildhaber (2017) found that the number of eggs released during each spawn (104–577 eggs) was lower than the previous report on the total number of oocytes in the ovaries. The following information on spawning behavior in an aquarium is from Albers and Wildhaber (2017):

> Prior to spawning, males swam quickly around the bottom of the tank and nudged one another and engaged in circular swimming with other males and occasionally with the female in addition to male-to-male posturing. Females usually swam slowly around the top sides of the tank and prior to spawning, moved to the bottom of the tank swimming among the males. Spawning occurred when a male aligned laterally on the right or left side of a female in a head-to-head orientation and nudged her to swim in a circle, with the female occupying the inner circle. . . . This circular swimming positioned the fish for the embrace; most of the time embraces occurred at the end of a circular swim. A spawning embrace involved the male wrapping his caudal fin and peduncle over the back of the female between the dorsal and caudal fins, the female arched her body laterally and, combined with the pressure from the male, both eggs and sperm were released. . . . Successful embraces during spawning could occur with different individual males or simultaneously with multiple males that could be either smaller or larger than the female.

Albers and Wildhaber (2017) were the first to confirm that *M. meeki* reproduces without any parental care, produces eggs that are nonadhesive and semibuoyant after water hardening, and has a drifting larval stage. Those workers were also the first to induce laboratory spawning of age 1+ *M. meeki* with and without hormonal injections. Temperature treatments (23°C) alone were successful in inducing spawning. Albers and Wildhaber also confirmed that Sicklefin Chubs are multiple batch spawners and individual females can spawn multiple times over months.

Larvae swam vertically in the water column after hatching, where they would swim up and then passively drift downward, rarely resting on the bottom. A change to horizontal swimming occurred before yolk-sac absorption was complete. Etnier and Starnes (1993) reported specimens collected from the Missouri River on 24 May were 43–50 mm (1.75–2 inches) TL after presumably completing nearly 1 year of growth. In a 3-year study of *M. meeki* in the Missouri River, Herman et al. (2008) found average adult length ranged from 3.6 to 10.1 centimeters (1.4–4.0 inches) with the average adult weight ranging from 0.6–6.2 grams (0.02–0.2 oz.). Mean back calculated lengths at-last-annulus for 2004 were 31 mm at age 1, 49 mm at age 2, and 65 mm at age 3. One 83 mm specimen was determined to be age 4. This chub is a relatively short-lived species with a small percentage of the population reaching age 4 (Pflieger 1997; Herman et al. 2008). In its big-river habitat, *M. meeki* serves as prey for juvenile Pallid Sturgeon (Gerrity et al. 2006). Maximum age for the Sicklefin Chub is 4 years.

CONSERVATION STATUS   The Sicklefin Chub is a rare inhabitant of Arkansas waters, found only in the mainstem Mississippi River. As a fluvial specialist (Dieterman and Galat 2004), this species is likely extirpated from highly altered portions of a river. The Sicklefin Chub is classified as globally rare and imperiled by seven of the eight states along the mainstem Missouri River (Dieterman and Galat 2004). It has suffered a 50% decline in its historical range (Stukel 2001) and is currently a candidate for listing as a threatened species by the USFWS. Dieterman and Galat (2004) found four variables significantly associated with presence/absence of *M. meeki*: (1) distance from upstream impoundment; (2) flow constancy; (3) mean turbidity; and (4) percent annual flow occurring in August. We consider the Sicklefin Chub threatened in Arkansas.

Macrhybopsis storeriana (Kirtland)
Silver Chub

Figure 15.66. *Macrhybopsis storeriana*, Silver Chub. *Uland Thomas*.

Map 55. Distribution of *Macrhybopsis storeriana*, Silver Chub, 1988–2019.

CHARACTERS   A moderately large, pallid minnow with a relatively short head (Fig. 15.66). Head length going into TL 4.8–6.1 times. Snout blunt and rounded and protruding slightly beyond mouth; mouth small, nearly horizontal and subterminal. Small conical barbel at posterior tip of mouth on each side (often inconspicuous until mouth is opened). Eye large, going into head length 3.0–4.3 times, its diameter nearly equal to snout length. Pharyngeal teeth 1,4–4,1. Gill rakers short, 4–7 total. Dorsal and anal fins weakly falcate. Pectoral fins not falcate. Dorsal fin far forward, with origin at or slightly anterior to pelvic fin origin; front of dorsal fin base much closer to tip of snout than to base of caudal fin. Dorsal fin rays 8. Tip of anterior rays of depressed dorsal fin not extending beyond tips of posterior rays in adults (may extend slightly past posterior rays in juveniles). Lateral line complete, either straight or with a broad arch and with 35–42 scales. Predorsal scale rows 14–16. Body circumferential scales 26–29. Anal fin rays 8 (7–9), pectoral rays 17 (15–18), pelvic rays 8. Breast naked or occasionally scaled posteriorly; belly fully scaled. Intestine short, with a single S-shaped loop; peritoneum silvery with scattered melanophores. Breeding males with 2–8 large, uniserial tubercles confined to pectoral fin rays 2–10 and small granular or pointed tubercles on top of the head (Etnier and Starnes 1993), but Cross (1967) considered the head structures to be sensory buds, not tubercles. Maximum size 9.1 inches (231 mm) TL.

LIFE COLORS   Color pale or olive dorsally, silver on sides and white on the belly. Dorsal scales faintly edged with small melanophores. Lower lobe of caudal fin darker than upper lobe and with a distinct white margin.

SIMILAR SPECIES   The Silver Chub is similar to *Notropis shumardi*, *N. blennius*, and *Hybognathus nuchalis*; however, its barbel, ventral mouth, large eye, falcate fins, and the darkly pigmented lower lobe of the caudal fin will help differentiate it from those species. It can be distinguished from *Platygobio gracilis* in having an unscaled breast (vs. scaled). Small specimens can be distinguished from *M. hyostoma* in lacking scattered black chromatophores on the body (vs. scattered chromatophores present), in having less conspicuous maxillary barbels, and in having 1,4–4,1 pharyngeal teeth (vs. pharyngel teeth 0,4–4,0). It differs from *M. meeki* in having 35–42 lateral line scales (vs. 43–50) and in having nonfalcate pectoral fins (vs. strongly falcate).

VARIATION AND TAXONOMY   *Macrhybopsis storeriana* was described by Kirtland (1845) as *Leuciscus storerianus* from the Ohio River, presumably from the vicinity of Cleveland, Ohio (Gilbert 1998). J. C. Williams (1963) and Trautman (1981) reported variation in this chub between Lake Erie and Ohio River populations; however, no study has examined rangewide variation. Analysis of the ND2 gene revealed two markedly divergent clades (Echelle et al. 2018), and those authors suggested that *M. storeriana* harbors undescribed cryptic species. Jenkins and Lachner (1971) hypothesized *M. storeriana* was closely related to *Platygobio*. Mayden (1989) placed *M. storeriana* in a clade with *M. gelida*, *M. meeki*, and *Platygobio gracilis*; however, Simons and Mayden (1999) considered *M. storeriana* to be more closely related to *M. aestivalis* than to *Platygobio*. Earlier, Dimmick (1993) aligned *M. storeriana* and *P. gracilis*.

DISTRIBUTION   *Macrhybopsis storeriana* is widely distributed from the Lake Erie drainage in the east; the Hudson Bay drainage (Red River drainage) of Manitoba, Canada, and Minnesota; the Mississippi River basin from Pennsylvania and West Virginia west to Minnesota and Nebraska and south to Texas and Alabama; and in Gulf Slope drainages from Mobile Bay to Lake Pontchartrain, Louisiana, with an isolated population in the Brazos River, Texas. In Arkansas, the Silver Chub occurs in the larger rivers of the state, including the Mississippi, Red, Ouachita, Arkansas, White, and St. Francis rivers (Map 55).

HABITAT AND BIOLOGY   The Silver Chub inhabits large, sandy-bottomed streams and rivers, but it is rarely abundant at most locations in Arkansas. It is common throughout the Arkansas River in the state (Buchanan 1976, 2005), where it is seldom collected during daylight hours by scining; however, trawling samples produce numerous diurnal records. It apparently inhabits deep water with moderate to swift current during the daytime, but it moves shoreward at dusk to feed. We have collected it at depths of 20–40 inches (0.5–1 m) in current in fairly large numbers at numerous Arkansas River localities at night. A dawn collection from the Red River west of Texarkana in August 1984 yielded 43 specimens, and Buchanan et al. (2003) considered it abundant in the Red River. Recent sampling of the Mississippi River in Arkansas revealed that it is common in that river as well (data from Vicksburg Waterways Experiment Station provided by K. J. Killgore). Becker (1983) reported that the Silver Chub was encountered most frequently in Wisconsin at depths of 24 inches (0.6 m) or more. McKenna and Castiglione (2014) reported it is generally found in open water over a variety of bottom types in Lake Erie but was not associated with submerged vegetation. This species seems to be quite tolerant of silty, turbid streams. According to W. J. Matthews (pers. comm.), it is also common in some reservoirs, for example, Lake Texoma (Red River), Oklahoma, and Buchanan (2005) found it in considerable numbers (about 1,400 specimens) in six Arkansas River navigation pools. In the northern part of its range it often occurs in lakes, and is the only *Macrhybopsis* species commonly found in that environment. Pflieger (1997) reported that it inhabits the quiet backwaters of large streams; however, in Arkansas the Silver Chub is invariably most abundant in moderate to swift current in the large rivers. We have collected it only in very small numbers in Arkansas River backwater habitat.

Davis and Miller (1967) observed feeding behavior in captivity that strongly suggests this chub swims over the bottom with barbels in contact with the substrate until the taste buds on the body, fins, and barbels are stimulated by substances emanating from food. Davis and Miller (1967) suggest the Silver Chub locates its food by sight as well as by taste as the relatively large eyes suggest. Simon (1999a) considered it a planktivore/invertivore. Major food items during the first year of life included microcrustaceans (copepods and cladocerans) and midge larvae and pupae (Ross 2001). Adult foods consist of caddisflies, mayflies, amphipods, fingernail clams, beetles, and dipterans. Adults in Lake Erie fed mainly on mayflies (Kinney 1954), and subsequent studies by McKenna and Castiglione (2014) and Kočovský (2019) reported that Lake Erie populations consumed mollusks (including invading Zebra Mussels) and also fed heavily on *Hexagenia* spp. mayflies. Carbon isotope analysis identified *Hexagenia* spp. as the primary source of carbon in Silver Chubs, and zebra mussels had mostly replaced native bivalves and gastropods (compared with earlier studies in Lake Erie) in the diet (Kočovský 2019).

Macrhybopsis storeriana spawns in April or May in Kansas (Cross 1967), and from April through June in Alabama (Mettee et al. 1996); however, spawning habits have not been described. Kinney (1954) observed Silver Chubs moving into shallow water in the spring, which he assumed to be a thigmotaxic response or associated with spawning activity. Large spawning aggregations were reported in the lower Pearl River, Mississippi, in February and March by R. D. Suttkus (Mettee et al. 1996). J. C. Williams (1963) reported that spawning in the Ohio River Basin occurs from mid-May to early August. Females and, presumably males, do not spawn in their first year (age 0). Mature eggs vary from 0.9 to 1.7 mm in diameter in age 1 and age 3 females, respectively. Williams (1963) found the ovaries of a 123 mm SL female from the Ohio River constituted 24.8% of her body weight. In Alabama, Boschung and Mayden (2004) reported females with nearly mature eggs were observed in mid-April at a water temperature of 15°C; however, fish did not appear ready to spawn until the water temperature reached 21°C. In Lake Erie and in Wisconsin, spawning begins in June at a water temperature of 19°C (66°F) and peaks at a water temperature of 23°C (73°F) (Kinney 1954; Becker 1983). Larvae 7.5 mm TL were collected in Lake Erie in late June and early July at depths of 18–20 m (Williams 1963). Estimates of growth are 80–130 mm TL at age 1 and 130–160 mm at age 2. Silver Chub YOY between 21 and 27 mm TL were taken from the Red River in Miller County, Arkansas, on 17 July 2000 by TMB and Drew Wilson. Sexual maturity is reached in 1 year, and the maximum life span is 3 years for males and 4 years for females (Kinney 1954). Few individuals live longer than 3 years (Gilbert 1980a).

CONSERVATION STATUS Black (1940) considered the Silver Chub "common in all suitable waters" in Arkansas. Although declining in some other areas of its distribution (McKenna and Castiglione 2014; Kočovský 2019), the Silver Chub is still a common fish of the larger rivers of Arkansas (although it is uncommon in the Ouachita River), and its populations are secure.

Genus *Mylopharyngodon* Peters, 1881— Black Carps

Mylopharyngodon is a monotypic Asian cyprinid genus containing a single species, *M. piceus*. It is introduced into Arkansas, and the characters for the genus are reported in the species account below.

Mylopharyngodon piceus (Richardson)
Black Carp

Figure 15.67. *Mylopharyngodon piceus*, Black Carp. *Uland Thomas.*

Map 56. Distribution of *Mylopharyngodon piceus*, Black Carp, 1988–2019. Inset shows introduced range (outlined) in the contiguous United States.

CHARACTERS Body stout, elongate and almost cylindrical, only slightly compressed (Fig. 15.67). Head somewhat pointed and slightly flattened. Eyes small, mouth small and terminal to subterminal. No barbels. Scales large and cycloid. Gill rakers are short and rather thick, usually with 18–21 (14–23) on first gill arch. Gill membranes are attached to the isthmus. Peritoneum dark. Intestinal length about 1–2 times TL. Dorsal and anal fins lacking strongly serrated spines, but first ray is often somewhat stiffened in adults. Dorsal fin slightly anterior to origin of the pelvic fins; dorsal fin closer to base of caudal fin than to tip of snout; dorsal rays 7 or 8. Pectoral fin rays 15–18; pelvic fin rays usually 8. Anal fin positioned more posteriorly than in other Arkansas cyprinids, except for the Grass Carp; anal fin rays usually 8 (7–9). Caudal fin forked, deeply emarginate. Lateral line complete, slightly decurved, extending along the middle of tail. Lateral line scales 39–46. Pharyngeal teeth smooth (no grooves or hooks), molarlike, and extremely strong, and in a single row (2 rows in a few specimens). Main row with 4 or 5 large molariform teeth without serrations. Diploid chromosome number is 48; triploids have 72 chromosomes. Prior to spawning, breeding males develop nuptial tubercles on the head, operculum, and pectoral fins. Females do not develop tubercles. Bardach et al. (1972: 81) stated that it is the largest of the Chinese carps. Maximum size of the Black Carp is more than 2 m (6.6 feet) TL and 70 kg (154 pounds) or more in weight (Nico and Neilson 2018). The largest known U.S. specimen was 52.2 kg (115 pounds) and 1.6 m (5.25 feet) TL, taken from the Mississippi River south of Cape Girardeau, Missouri, on 8 February 2018.

LIFE COLORS Coloration generally blackish or bluish-gray on body, and all fins are blackish-gray deepening to black toward the edges. Scales are large with dark edges, giving the fish a crosshatched appearance. Ventral surface of head and abdomen whitish.

SIMILAR SPECIES The Black Carp is superficially similar to Grass Carp in overall body shape and size, in the position and shape of the fins, in the position and size of the eyes, and in having very large scales, but it can be distinguished from Grass Carp by color of body and fins and by the pharyngeal teeth. Black Carp have molariform teeth without grooves of any kind, while Grass Carp have long and serrated or deeply grooved teeth. Grass Carp have a more cylindrical body form and differ in body coloration. Black Carp have a black, blue-gray, or dark brown color with darkly pigmented fins, while Grass Carp are olivaceous or silvery white, or olive brown above, and silvery below with most fins dark. It is further distinguished from the two Asian carp species in the genus *Hypophthalmichthys* by its large scales, with 39–46 scales in the lateral line (vs. small scales with more than 85 scales in the lateral line in Bighead and Silver carps). It is distinguished from the Common Carp by its lack of maxillary barbels.

VARIATION AND TAXONOMY The Black Carp was originally described by Richardson (1846) as *Leuciscus piceus* from Canton, China. It was later moved to the monotypic genus *Mylopharyngodon*. Bănărescu (1964) and Biro (1999) reviewed the systematics of *Mylopharyngodon*, including the controversy surrounding the placement in the subfamily Cyprininae versus Leuciscinae where recent Chinese researchers placed this species.

DISTRIBUTION The natural range of Black Carp includes most major Pacific drainages of eastern Asia from about 22°N to about 51°N latitude (Nico et al. 2005). Currently known from six states in the United States, several records of the Black Carp are known from Arkansas. On 5 April 2005, a commercial fisherman caught a specimen of the Black Carp from the White River (river mile 129), just north of De Valls Bluff (Mike Armstrong, pers. comm.). Unfortunately, the fish was sold at a fish market before state biologists could obtain the specimen and determine ploidy; however, photographs of the fish, including one showing the pharyngeal teeth, confirmed the identity (Nico et. al. 2005). There are records of Black Carp from the Mississippi River as far north as Illinois and numerous reports from Louisiana. In 2016, several specimens were found in the Mississippi River of Arkansas in Mississippi and Chicot counties (Map 56).

HABITAT AND BIOLOGY In its native range, this species can be found in rivers, streams, and lakes, but it requires large rivers for spawning. In Arkansas, it is found in the wild only in large rivers. The Black Carp has been imported on several occasions into the United States and is currently found scattered among private and state aquaculture facilities, primarily in the southeastern United States, including Arkansas (Nico et al. 2005). The first Black Carp were included with a shipment of imported Grass Carp from Asia in 1973 sent to a fish farm in Arkansas. This original stock was given to the Arkansas Game and Fish Commission for research purposes; however, these all eventually died. A second introduction of Black Carp occurred in the 1980s by fish farmers in Arkansas, Mississippi, and Missouri for the purposes of biological control of snails in farm ponds and as a food fish. Additional importations occurred in the mid-1980s and early 1990s.

Larvae and small juveniles feed almost entirely on small invertebrates such as zooplankton and aquatic insects. The diets of larger juvenile and adult Black Carp become more benthic oriented, consisting largely of snails and bivalve mollusks, although crayfish and other benthic arthropods are sometimes consumed (Nico et al. 2005). Aquaculturists view the Black Carp as a useful biological tool to control snail-borne parasites that infect farm-raised fishes. The

Black Carp has also been suggested as a possible biological control for the introduced Zebra Mussel, but there is no experimental evidence that it would be effective for that purpose. Nico et al. (2005) noted that because Black Carp do not have jaw teeth and their mouths are relatively small, it is unlikely that they are capable of breaking apart Zebra Mussel rafts. Conservationists fear the Black Carp will prey heavily on native mollusks, thereby hastening the decline of numerous native mussels and snails, 71.7% of which are already threatened or endangered or of special concern (Williams et al. 1993).

Concern was expressed about the possible impact of Black Carp on endangered mollusks in the early 1990s. In 2003, Black Carp began to be caught by commercial fishermen in Illinois, Louisiana, and eventually in the White River in Arkansas. There is some support among fisheries biologists for allowing sterile triploid Black Carp to be used for aquaculture purposes because they are incapable of reproducing. However, the effectiveness of Black Carp in significantly reducing snail populations in aquaculture ponds indicates that any Black Carp occurring in the wild (even triploid forms) may cause significant declines in certain native mollusk populations in North American streams and lakes (Nico et al. 2005). Because the life span of Black Carp is reportedly more than 15 years, sterile triploid Black Carp in the wild would be expected to persist many years and therefore have the potential to cause harm to native mollusks by way of predation (Nico et al. 2005).

Nico et al. (2005) summarized what is known about reproductive biology, and the following information comes from that source. It requires large rivers to reproduce. Reproduction takes place in late spring and summer when water temperatures and/or water levels rise. A large female Black Carp full of eggs was taken from the Mississippi River on 18 February 2018 (R. Hrabek, pers. comm.). The Black Carp is a broadcast spawner, and females are capable of releasing hundreds of thousands of eggs into flowing water, which then develop in the pelagic zone (Nico et al. 2005). After fertilization, the eggs become semibuoyant and are carried by the current until they hatch, usually in 1–2 days, depending on water temperatures. The yolk sac is absorbed in 6–8 days. They become sexually mature at 4–6 years, after which they migrate back to their spawning grounds. Life span is probably greater than 15 years. Fecundity has been reported to be 400,000 to 3 million eggs. It is not known with certainty if Black Carp are reproducing in the wild in the United States, because no larvae or small juveniles have been caught in the wild to date; however, because Black Carp have been regularly taken in the Mississipi River for 15 years, it is probably already established in Arkansas.

CONSERVATION STATUS Confirmed records of this introduced Asian carp are increasingly common from Arkansas. Nico et al. (2005) considered it on the verge of becoming established in the United States. There is probably little that can be done to eradicate it.

Genus *Nocomis* Girard, 1856—Chubs

The North American genus *Nocomis* consists of seven species of medium-sized (up to 20 cm SL) minnows long included in the genus *Hybopsis*, based primarily on the presence of a maxillary barbel. *Nocomis* was initially used as a genus name by Girard (1856), but it was subsequently recognized as a subgenus of *Hybopsis* by Jordan and Evermann (1896). Taxonomic controversy about the generic status of *Nocomis* continued until the studies of Lachner and Jenkins (1967, 1971) and Jenkins and Lachner (1971) provided conclusive evidence supporting its generic status. All *Nocomis* species occur primarily in clear, gravel-bottomed upland streams (Echelle et al. 2014), and Cross and Collins (1995) considered *Nocomis* species especially sensitive to environmental changes. Mayden (1989) hypothesized phylogenetic affinities with *Campostoma*, *Couesius*, or *Semotilus*, and a more recent analysis suggested a sister relationship with *Campostoma* (Nagle and Simons 2012). Three morphologically defined species groups are traditionally recognized (Lachner and Jenkins 1971; Jenkins and Lachner 1971; Lachner and Wiley 1971): (1) the *N. leptocephalus* group with three described species in southeastern United States, (2) the *N. micropogon* group of three species in northeastern United States, and (3) the *N. biguttatus* group of three species in drainages of the Great Lakes and the Mississippi and Ohio rivers (Echelle et al. 2014). All *Nocomis* are robust, cylindrical fishes with orange-red fins, large scales that are not crowded anterior to the dorsal fin, a complete lateral line with 37–44 scales, 7 anal rays, 8 dorsal rays, 8 pelvic fin rays, 15–19 pectoral fin rays, and a large, horizontal and slightly terminal mouth (Etnier and Starnes 1993). Eyes are generally small and situated near the dorsal border of the head. The males attain a larger size than the females. Arkansas has two allopatric species: *Nocomis asper* and *N. biguttatus*. Both species develop large breeding tubercles on the head. Small uniserial tubercles develop on dorsal surfaces of pectoral fin rays 2–7 or 8.

All *Nocomis* species are nest builders, with males constructing large nests and defending the shallow depressions in gravel areas during late spring. Males carry pebbles in their mouths to line the shallow nests, and other cyprinid species also use the nests for spawning. Peoples and Frimpong (2013) provided evidence that *Nocomis* species and their nest associates appear to receive a net benefit from the association, suggesting a mutualistic relationship. Pendleton et al. (2012) delineated groups of strong and weak nest associates of *Nocomis*. Strong nest associates showed significant geographic range overlap with *Nocomis*, and weak associates did not. Those authors also found that nest association strength was related to rarity. Strong nest associates held rare classifications based on geographic extent, habitat breadth, or local abundance. Conversely, all weak nest associates reflected common classifications. Conservation efforts for rare or imperiled nest associates should include protection of their host species.

Nocomis asper Lachner and Jenkins
Redspot Chub

Figure 15.68. *Nocomis asper*, Redspot Chub. *David A. Neely.*

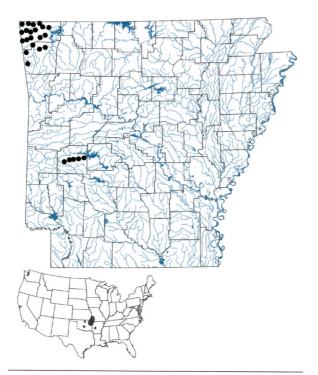

Map 57. Distribution of *Nocomis asper*, Redspot Chub, 1988–2019.

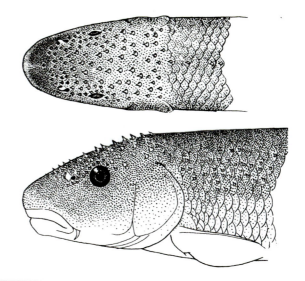

Figure 15.69. Dorsal and lateral views of the head of a nuptial male *Nocomis asper*. *Lachner and Jenkins (1971).*

CHARACTERS   A large, robust, cylindrical, barbeled chub with a large head, slightly subterminal mouth, and moderately small eyes (eye diameter much less than snout length) (Fig. 15.68). Barbel almost hidden at the corner of the mouth. Dorsal rays 8; anal rays 7. Lateral line scales 39–44. Pharyngeal teeth 1,4–4,1, rarely 1,4–4,0. Intestine short; peritoneum black. Body circumferential scales average 34 (31–37). Caudal peduncle scales about 18 (16–20). Prior to spawning, breeding males develop head tubercles distributed from about the midportion of the snout to the occiput (Lachner and Jenkins 1971) (Fig. 15.69). Nuptial tubercles are larger and mainly antrorse posteriorly on the head; smaller antrorse or erect tubercles are on the forehead and snout. Tubercles present on nape and laterally on body; body tubercles much smaller than head tubercles, and located centrally or subcentrally on exposed portions of scales. Lateral body scales, when tuberculate, often have more than 1 tubercle per scale: 75% have 1 tubercle per scale, 23% have 2 tubercles per scale, and about 2% have 3 tubercles per scale. Tubercles seldom occurring on second scale row below lateral line. Breeding females often have tiny light spots on some body scales in a position corresponding to the tubercles in males (Pflieger 1997). Maximum size 10 inches (254 mm).

LIFE COLORS   Body with olive dorsum, light olive sides, and a white belly. Pectoral and pelvic fins pinkish yellow; caudal fin pinkish olive. Adults with a conspicuous red spot behind the eye in both sexes, providing the common name Redspot Chub. Young individuals have a dark lateral band and a well-developed black caudal spot.

SIMILAR SPECIES *Nocomis asper* is morphologically very similar to *N. biguttatus*, although the two species are never sympatric. To reliably differentiate the two species, tuberculate males are needed. In *N. asper*, tubercles or tubercle spots are present laterally on the body of adults and are visible in both sexes, while in *Nocomis biguttatus*, nuptial tubercles or tubercle spots are absent on the body. Specimens of *N. asper* smaller than 4 inches (102 mm) generally have no body tubercle spots, but head tuberculation is well developed (Lachner and Jenkins 1971). It differs from *Campostoma* species in having a maxillary barbel (vs. barbels absent) and in lower jaw lacking a cartilaginous shelf (vs. shelf present). It can be distinguished from *Semotilus atromaculatus* in lacking a black anterior basidorsal spot (vs. spot present), having 1,4–4,1 pharyngeal teeth (vs. 2,5–4,2), and in having normal-sized scales in front of the dorsal fin (vs. scales small and crowded).

VARIATION AND TAXONOMY *Nocomis asper* was long included in *N. biguttatus*. It was described as a separate species by Lachner and Jenkins (1971), who designated the type locality as Big Spring Creek, 8 km south of Locust Grove, Mayes County, Oklahoma. Anthony A. Echelle (pers. comm.) found that the Ouachita River population of *Nocomis asper* in Arkansas and the Blue River population in Oklahoma are identical to those in Ozark tributaries of the Arkansas River (see Variation and Taxonomy section of *N. biguttatus* account). Echelle et al. (2014) suggested a possible sister relationship between *N. asper* and the newly discovered and undescribed cryptic White River form based on the nuclear S7 gene.

DISTRIBUTION *Nocomis asper* occurs in only four states, Arkansas, Missouri, Kansas, and Oklahoma, where it is found mainly in Arkansas River drainages in the Ozarks. There are isolated populations in the Red River drainage of Oklahoma (Blue River) and the Ouachita River drainage of Arkansas. In Arkansas, the Redspot Chub is confined mainly to upland tributaries of the Arkansas River system draining the Ozark Mountains ecoregion of Benton and Washington counties. The disjunct populations in the Ouachita River system of Arkansas include one pre-1988 record from the upper Little Missouri River above Lake Greeson (Map A4.57), and several older (Douglas and Harris 1977) as well as recent records from the South Fork of the Ouachita River (Map 57).

HABITAT AND BIOLOGY The Redspot Chub inhabits upland, clear, spring-fed streams with gravel bottoms. It requires a steady flow of water and prefers deep runs and pools. It is most common in small streams that have aquatic vegetation along their margins (Cross and Collins 1995). In the Illinois River, Oklahoma, it occurred in both fast and slow water environments (Felley and Hill 1983). Young-of-the-year were always found in vegetation or other cover, and adults were found more often in open, current-influenced habitats.

Nocomis asper feeds at the surface or in midwater (Davis and Miller 1967) on a variety of insect larvae and adults, crustaceans and other invertebrates, and occasional plant material (Miller and Robison 2004). Foods in Kansas included ephemeropterans, crayfish, snails, and other small invertebrates (Cross and Collins 1995). In the Illinois River, Oklahoma, it fed exclusively on benthic invertebrates throughout the year (Felley and Hill 1983).

The Redspot Chub is a late spring spawner in Arkansas from mid-May to mid-June, with peak activity in May. Spawning occurred in Spavinaw and Flint creeks, Oklahoma, on 7 June at a water temperatures of 21°C and 21.5°C, respectively (Lachner and Jenkins 1971). At the spawning sites, the streams were 20–40 feet (6.1–12.2 m) wide and had substrates of 70–95% small to large gravel, with the remainder of the substrate being sand and rubble. Most nests were in moderately flowing water at the heads of riffles and in pools. *Luxilus cardinalis* and *Notropis nubilus* were observed spawning over *N. asper* nests (Lachner and Jenkins 1971). Spawning behavior in *N. asper* is similar to that described for *N. biguttatus* by Maurakis et al. (1991) and Vives (1990). Sabaj (1992) divided reproductive activities of members of *Nocomis* into six sequential categories: (1) interim—male behavior between spawns; (2) approach—female behavior directed toward the interim male; (3) alignment—behavior affecting the precise orientation of the spawning pair over the substrate; (4) run—synchronized movement of the aligned pair over the substrate; (5) clasp—the momentary flexure of the male's body about an axis determined by the female's position at the end of her run; and (6) dissociation—pair separation immediately after the clasp. Maurakis and Roston (1998) provided a detailed description of spawning behavior of *N. asper* in Missouri in May in which the male begins spawning activity by first excavating a saucer-shaped concavity and then within a week builds a mound. Males construct a large mound of stones 8–12 inches (203–305 mm) high and 20–40 inches (0.5–1 m) in diameter (R. J. Miller 1964, 1967). After mound construction, the male uses his jaws to excavate a spawning pit on the upstream slope of the nest and engages in pit fanning using his anal fin and pit posturing interim behaviors like those described for *N. biguttatus* (Maurakis et al. 1991). The next step in spawning is taken by the female as she approaches the spawning pit from downstream where she has been hovering over the rear of the nest. The female then moves into the spawning pit beneath the tail of the postured male. She moves forward to his extended pelvic fins until her snout is either directly below or slightly ahead of his pectoral girdle. She aligns her body parallel to the long axis of the pit and presses the ventral surface of her body to the substrate of the pit and immediately begins her run of 2–3 cm. In response, the male tilts his body sagitally toward the female as she quivers her tail and moves slightly upstream, whereupon the male responds by accompanying her forward motion with rapid tail beats. With her ventral portion continuing to conform to the topography of the spawning pit, she then moves forward to the upstream slope of the pit, gapes, and retroflexes (i.e., the immediate pitching of her head vertically into the water column) while rolling the anterior portion of her body away from the male, thereby placing her dorsum in contact with his anterior flank. Interestingly, females occasionally perform this same behavior without a male in the pit. As the female retroflexes, the male initiates his clasp by turning his head toward the female and curving his posterior flank over her back and driving it into her side between her pectoral and pelvic fin girdles. His body contracts into a semicircle around hers. At the height of the clasp (lasting up to 1 second) the male's vent is pressed to the dorsolateral surface of the female's caudal peduncle as her vent remains in contact with the substrate on the upstream rim of the pit. Eggs are shed at this moment. Dissociation is accomplished as the male's body relaxes after contraction during the clasp. He drifts downstream and resumes interim behavior. The female simultaneously rises vertically into the water column and regains horizontal equilibrium. She then either moves downstream of the nest or drifts just downstream of the male and initiates another approach. After spawning with several females in the pit, the male covers it with pebbles from surrounding substrate, and excavates another pit where he continues spawning behavior (i.e., pit fanning and spawning). The nest attracts several other cyprinid species that also spawn over it (R. J. Miller 1964, 1967). Nest associates in Ozark populations include *Campostoma plumbeum*, *Notropis nubilus*, *Chrosomus erythrogaster*, *Luxilus cardinalis*, and *Notropis percobromus*.

CONSERVATION STATUS The Redspot Chub is uncommon in Arkansas, but its populations in northwestern Arkansas seem stable. The population previously known from the upper Little Missouri River may have been extirpated. Because most of its range in Arkansas is in Benton and Washington counties, which have experienced accelerated environmental degradation associated with increasing human population growth, we consider the Redspot Chub a species of special concern.

Nocomis biguttatus (Kirtland)
Hornyhead Chub

Figure 15.70. *Nocomis biguttatus*, Hornyhead Chub. *Uland Thomas.*

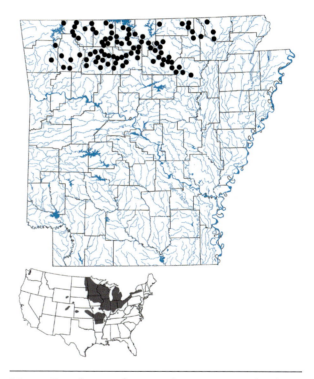

Map 58. Distribution of *Nocomis biguttatus*, Hornyhead Chub, 1988–2019.

CHARACTERS A robust, barbeled minnow with a large head, moderately small eyes (eye diameter much less than snout length), and a large, nearly terminal mouth (Fig. 15.70). Barbel small, not generally visible until mouth is opened. Pharyngeal teeth 1,4–4,1. Dorsal rays 8; anal rays 7. Circumferential scales average 35 (33–37) in Arkansas populations. Scales around caudal peduncle usually 18–20. Lateral line scales usually 40–45. Intestine short; peritoneum black. Breeding males with tubercles restricted to head (Fig. 15.71). Maximum size 10.2 inches (260 mm).

LIFE COLORS Brownish-olive dorsum and upper sides, with lower sides and belly white. Fins orange. Young

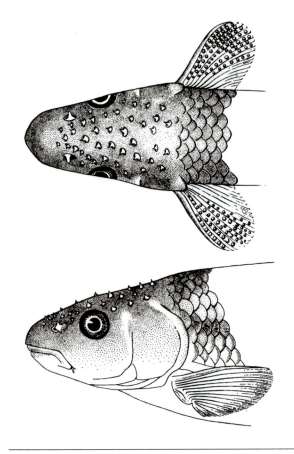

Figure 15.71. Dorsal and lateral views of the head of a nuptial male *Nocomis biguttatus. Lachner and Jenkins (1971).*

specimens with a bright orange caudal fin and a dark lateral band terminating in a discrete black caudal spot. Markings fade with age. Breeding males with a prominent red spot behind the eye, but females without the red spot (the area may be brassy-colored in breeding females).

SIMILAR SPECIES   *Nocomis biguttatus* is most similar to *Nocomis asper*, and many specimens can be separated only by locality. The Hornyhead Chub occurs only in the White River system in Arkansas, and *N. asper* is found only in the Arkansas River drainage with a disjunct population in the upper Ouachita River system. Breeding specimens are easier to separate, with *N. biguttatus* lacking tubercles on the side of the body (vs. tubercles on the body, with lateral body scales often having 2 or 3 tubercles per scale). Postorbital red spot is well developed only in adult males of *N. biguttatus* (vs. red spot conspicuously developed in adult males and females of *N. asper*). Characters distinguishing *Nocomis* species from *Campostoma* species and *Semotilus atromaculatus* are given in the *N. asper* account.

VARIATION AND TAXONOMY   *Nocomis biguttatus* was described by Kirtland (1840b) as *Semotilus biguttatus*

from Yellow Creek, a tributary of the Mahoning River, upper Ohio River basin, Ohio. It has at times been placed in the genera *Ceratichthys* and *Hybopsis* (Smith 1979). Based on tuberculation, morphology, and color of breeding males, Lachner and Jenkins (1971) stated that the *biguttatus* group is more closely related to the *micropogon* group than to *Nocomis leptocephalus*. Using reproductive behavioral characters, Maurakis et al. (1991) determined that *N. leptocephalus* and *N. biguttatus* form a monophyletic group based on two synapomorphies, spawning pit excavation and covering of eggs with pebbles after spawning. Nagle and Simons (2012) found *Nocomis effusus* to be the recently divergent (400,000 ybp) sister to *N. biguttatus*. Echelle et al. (2014) investigated the molecular systematics of the *Nocomis biguttatus* species group (*N. biguttatus*, *N. asper*, and *N. effusus*) and found that the White River form of *N. biguttatus* is apparently an undescribed cryptic species based on an analysis of the mitochondrial cyt*b* gene and two nuclear genes (S7 intron 1 and a portion of the gene for growth hormone, GH) (Echelle et al. 2014). Monophyly for independent DNA sequences (mtDNA and two nuclear genes) qualifies the White River form as a species under the phylogenetic and evolutionary species concepts (Mayden 1997). In addition, Echelle et al. (2014) found that levels of divergence for the nuclear genes are similar to or greater than those for the three traditionally recognized species of the *N. biguttatus* species group. Earlier, Lachner and Jenkins (1971) treated the White River as a separate geographic unit to allow assessment of intergradation between *N. asper* and *N. biguttatus*; however, there was no evidence of intergradation and no morphological trait was diagnostic of the White River population. Because the White River form now appears to be a separate, cryptic, undescribed species based on genetic evidence, the *Nocomis biguttatus* population in the Black River drainage is apparently the only true *N. biguttatus* in Arkansas. We refrain from recognizing the undescribed White River form as a separate species until further analysis results in a formal systematic description for it.

Nocomis specimens from the White River apparently reach larger body size (>200 mm SL) than other populations of *N. biguttatus* (<165 mm SL) and show modal shifts toward larger scale counts than all members of the *N. biguttatus* species group, except *N. effusus* (Lachner and Jenkins 1971b). In addition, head length in the White River form is shorter than in *N. asper* and other populations traditionally grouped under *N. biguttatus* (Echelle et al. 2014). Lachner and Jenkins (1971) noted that *N. asper* and the White River form represent the extremes of head length in the *N. biguttatus* species group.

DISTRIBUTION   As currently recognized, the Horny-head Chub occurs in one Atlantic Slope drainage (the Hudson River drainage), the Great Lakes drainages, the Hudson Bay basin, and in the Mississippi River basin from Ohio to North Dakota and south to northern Arkansas. Isolated western populations occur in Nebraska, Wyoming, Colorado, and Kansas. In Arkansas, it is confined to the White River system, including the Black River (Map 58). Although distributed throughout the White River system of the Ozark Mountains, the Hornyhead Chub is never as abundant as other upland minnows with which it occurs syntopically. Older records exist from the Little Red River above Greers Ferry Lake (Map A4.58), but there are no post-1988 records from that drainage.

HABITAT AND BIOLOGY   In Arkansas, the Horny-head Chub occurs mainly in medium-sized to large clear streams and rivers of the uplands having gravel substrates. Streams with permanent flow are required, and it some-times occurs in small streams having current, clear water, and substrates of gravel, rubble, and some sand. Lachner and Jenkins (1971) reported that the preferred streams gen-erally have a moderate balance of riffles and pools, but long pools of slack water often exist. It avoids turbid waters and silt-laden substrates and is often found in mixed schools with other minnows. Ecological segregation is noted in this species, as adults are typically collected near riffles although not in the swifter portions of the riffles. Young individuals seem to prefer more slack-water areas with abundant water willow, *Justicia americana*, marginally lining the region. *Nocomis* (both Arkansas species) often occur with schools of *Campostoma*, e.g., 1 or 2 *Nocomis* with 50+ stonerollers.

Nocomis biguttatus reportedly feeds at all water depths within the stream, where it feeds by sight on aquatic insects and algae and other plant material (Lachner 1950; Davis and Miller 1967). Lachner (1950) reported food habits of *N. biguttatus* and revealed that 50% by volume of the adult diet was filamentous algae and vascular plants, with the remain-der being animal material, primarily insects. He suggested that plant material was probably taken accidentally with animal material. Juveniles and adults feed mostly on a wide variety of benthic insects, with smaller amounts of crustaceans, mollusks, annelids, and fishes included in the diet (Jenkins and Lachner 1980). Age 0 fish feed mostly on aquatic vegetation, but diatoms, cladocerans, and aquatic macroinvertebrate larvae (e.g., chironomids) are also eaten (Scott and Crossman 1973). Mature fish (age 2–4) often consume insect larvae, annelids, crayfish, fish, and espe-cially snails (Scott and Crossman 1973). Angermeier (1982) found that in Jordan Creek, Illinois, Hornyhead Chubs

ingested a wide range of sizes of invertebrates and had a strong preference for aquatic prey items over terrestrial prey items. Angermeier (1982) also found that the diet varied seasonally. During March and April, the diet consisted primarily of chironomids (58%) and simuliids (15%); dur-ing May and June, the diet consisted of simuliids (57%) and helicopsychids (11%); from August through October, the diet consisted of elmids (37%) and chironomids (25%); from October through January, the diet shifted to chirono-mids (56%) and clams (15%).

In Arkansas, spawning in this nest-building minnow begins in late April and extends to late June. The male constructs and guards a dome-shaped nest consisting of a mound of pebbles by carefully selecting stones and trans-porting them in his mouth for a distance of up to 15–20 feet (4.6–6.1 m). Often, individual stones are larger than the head of the male chub. Vives (1990) reported that stones for the mound are collected primarily from the adjacent streambed away from the nest. Nests are usually conspicuous circular areas from 1 to 3 feet (0.3–0.9 m) in diameter and as much as 1 foot (0.3 m) in height. Nests taper from the upstream area to the downstream area. Spawning is accomplished when females enter the nest, which is guarded by a single male. Spawning behavior of this species is similar to that of *Nocomis asper* (see *N. asper* account for details) (Maurakis et al. 1991; Vives 1990). The female deposits several hundred eggs over the mound of stones. Several other species are attracted to the nest area for spawning purposes, including sunfishes, darters, and then minnows, that eat the eggs. Occasional interspecific encounters occur between similar-sized males of different species. Vives (1988) first documented interspecific parallel swims for *N. biguttatus* and *Luxilus cornutus*. The normal life span is 2 or 3 years.

CONSERVATION STATUS   The Hornyhead Chub is a widespread but uncommon species in Arkansas. We con-sider its populations currently secure.

Genus *Notemigonus* Rafinesque, 1819— Golden Shiners

The native North American genus *Notemigonus* has no close affinities with other North American cyprinids (Etnier and Starnes 1993) and has been considered most closely related to Eurasian minnows (Gill 1907; Cavender and Coburn 1989). *Notemigonus* has been placed in the subfamily Leuciscinae and aligned with the leuciscin lin-eage. Characters of this monotypic genus are given in the species account below.

Notemigonus crysoleucas (Mitchill)
Golden Shiner

Figure 15.72a. *Notemigonus crysoleucas*, Golden Shiner (golden form). *Uland Thomas.*

Figure 15.72b. *Notemigonus crysoleucas*, Golden Shiner (silver form). *Uland Thomas.*

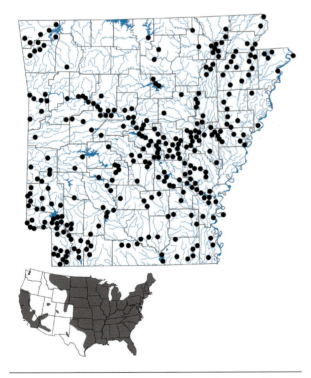

Map 59. Distribution of *Notemigonus crysoleucas*, Golden Shiner, 1988–2019. Inset shows native and introduced ranges in the contiguous United States.

CHARACTERS   A large, deep-bodied, slab-sided minnow with 0,5–5,0 pharyngeal teeth, a small head, and a pronounced scaleless midventral keel extending from the bases of the pelvic fins to the anus (Fig. 15.72). Head length going into TL 5.0–5.8 times. Snout pointed. Mouth small, strongly oblique (almost vertical). Eye going into head length 3.0–4.2 times. Lateral line complete and sharply decurved with 46–54 scales. Dorsal fin origin distinctly posterior to pelvic fin origin. Dorsal fin falcate, usually with 8 (7–9) rays. Anal fin falcate, usually with 11–14 rays. Pectoral fin rays 13 or 14; pelvic rays usually 9 (8–10). Gill rakers long and slender, with 17–19 rakers on lower half of first gill arch. Intestine short, with a single S-shaped loop; peritoneum silvery, but sprinkled with dark speckles. Breeding males develop distinct tubercles on the lower jaw, with fine tubercles covering the head, body, leading rays of the median fins, and dorsal surfaces of the pectoral fins (Burkhead and Williams 1991). Maximum size 14.5 inches (368 mm) TL from St. Johns River, Florida (McLane 1955). Maximum weight about 1.5 pounds (0.68 kg).

LIFE COLORS   Coloration variable. Body color usually a uniform silvery to brassy color. In other individuals, the dorsum is olive green with lower body portions silvery. In others, body color golden to olive; sides golden with a yellowish belly. Fins light olive or yellow. Scale bases are dark, particularly in large adults. Young with a dark lateral band. Young are sometimes difficult to identify because they closely resemble *Notropis* species.

SIMILAR SPECIES   The combination of a usually golden color, strongly oblique mouth, numerous anal rays, ventral keel, and 0,5–5,0 pharyngeal teeth separates the Golden Shiner from all other native Arkansas cyprinids. It differs from the very similar, introduced *Scardinius erythrophthalmus* in dorsal fin usually with 8 rays (vs. usually 10 or 11), lateral line scales usually 46–54 (vs. 36–45), and pharyngeal teeth 0,5–5,0 (vs. 3,5–5,3).

VARIATION AND TAXONOMY   *Notemigonus crysoleucas* was described by Mitchill (1814) as *Cyprinus crysoleucas* from New York (presumed to be in the vicinity of New York City). The Golden Shiner is unusual among cyprinids in that its anal ray counts vary greatly throughout its wide range. Anal ray counts vary from 8 to 19, but counts below 11 or above 15 are unusual (Boschung and Mayden 2004). This species has been described under several different names. In the early 1900s, two subspecies were recognized, but subspecies are not recognized today. *Notemigonus* is morphologically close to the introduced European cyprinid, Rudd (*Scardinius erythrophthalmus*), with which it is known to hybridize (Burkhead and Williams 1991).

A suggested relationship to European minnows was substantiated cladistically by Cavender and Coburn (1992) and Coburn and Cavender (1992).

DISTRIBUTION *Notemigonus crysoleucas* is considered native to Atlantic Slope drainages from Canada to Florida, Great Lakes drainages, Hudson Bay drainages, the Mississippi River basin from Canada south to the Gulf of Mexico, and Gulf Slope drainages from Florida to Texas. It has been widely introduced elsewhere in the United States. This native Arkansas species occurs statewide because of its widespread use as a bait minnow (Map 59). Not surprisingly, it was one of the most widespread and abundant small fish species found in Arkansas reservoirs (Buchanan 2005).

HABITAT AND BIOLOGY The Golden Shiner prefers well-vegetated, standing waters of ponds, reservoirs, lakes, and sloughs and quiet pools of low-gradient streams and small rivers having mud or sand bottoms. It is seldom found in riffles or in pools with current, and it is a good indicator of pollution or modification of the habitat when it outnumbers other species at a site (Smith 1979). It would not be surprising to find it in any of the state's waters. This schooling minnow inhabits a variety of areas not favored by other species because of its tolerance for turbidity and pollution. It also has one of the highest lethal temperature tolerances (near 40°C) of any North American fish (Alpaugh 1972).

It is an opportunistic sight feeder usually found from midwater to the surface, but occasionally benthic organisms are also eaten (Keast and Webb 1966). Feeding behavior was described by Ehlinger (1989), who found that it was largely a crepuscular feeder that employed two feeding modes. In one feeding mode, Golden Shiners individually selected large zooplankters by approaching them from below; the second feeding mode was employed when there were high densities of small zooplankton and involved pumping large volumes of water through the mouth and over the gill rakers to filter out the zooplankton. Foods of this species consist of zooplankton (cladocerans and copepods), plant materials, amphipods, mollusks, mites, fish eggs, and terrestrial insects (Keast and Webb 1966; Keast 1985a). There is conflicting information about its use of phytoplankton for food (Hill and Cichra 2005). Stone et al. (1997) stated that it has some capacity for masticating algae with its pharyngeal teeth but has little enzymatic ability to break down plant cell walls. Although some studies have reported phytoplankton in the diet, Stone et al. (1997) questioned its ability to effectively use algae as a major food source.

In Arkansas, spawning usually begins in mid-April, reaches a peak in May, and tapers off through June as water temperatures rise. Spawning generally occurs from early morning to noon. Stone et al. (1997) reported that natural spawning may also occur in September and October in aquaculture ponds in Arkansas. During the spawning season there are occasionally four or five distinct spawning peaks (Mansueti and Hardy 1967). The Golden Shiner does not construct a nest and provides no parental care. Spawning occurs between the temperatures of 68°F and 80°F (20–27°C) over submerged vegetation, filamentous algae, or woody debris as one or two males chase a female. Eggs are broadcast and adhere to submerged vegetation and debris. It readily uses artificial spawning mats in aquaculture ponds (Stone et al. 1997). In Minnesota, Golden Shiners were observed spawning in Largemouth Bass nests in a lake (Kramer and Smith 1960). Schools of 25–100 shiners deposited eggs in bass nests 1–2 days after the bass had spawned. The eggs were guarded by the male bass until they hatched. Golden Shiners are also nest associates of *Amia calva* (Katula and Page 1998), *Lepomis macrochirus* (DeMont 1982), *Lepomis cyanellus* (Pflieger 1997), and *Lepomis punctatus* (Carr 1946). Golden Shiners are fractional spawners with individuals reproducing several times a year (Stone et al. 1997). Fecundity varies with size of the female, ranging from around 2,000 to 200,000 eggs. The eggs hatch in 4 days at 75–80°F (23.9–26.7°C) (Becker 1983). It is a rather long-lived minnow, with a maximum life span of around 5 years in nature and 10 years in an aquarium (Becker 1983).

The Golden Shiner is the most popular and widely used bait minnow in Arkansas. It is used especially for crappie and bass fishing, and Arkansas is one of the leading states in the commercial pond culture of this species. In 1982, private fish farms in the state produced 5.8 million pounds (2.6 million kg) of Golden Shiners worth $13.8 million (Crawford 1982). By 2011, estimates of Golden Shiner production were 3.6 million pounds valued at $8.6 million (Nathan Stone, pers. comm.).

CONSERVATION STATUS This native cyprinid is widely introduced as a baitfish throughout the state and is in no danger in Arkansas.

Genus *Notropis* Rafinesque, 1818a— True Shiners

The North American genus *Notropis* is of particular interest to ichthyologists because of its large species richness. Depending on authority, the genus contains 86–90 species divided into three subgenera, *Notropis* with 20 species, *Alburnops* with 8 species, *Hydrophlox* with 5 species, and at least three species groups: *Notropis texanus* species group with 8 species, *Notropis volucellus* species group with 10 species, and *Notropis dorsalis* species group with 6 species. There is an additional group of approximately 24 species whose relationships within *Notropis* are currently unresolved (Swift 1970; Bortone 1989; Mayden 1991; Warren et al. 1994; Raley and Wood 2001; Wood et al. 2002; Cashner et al. 2011). Some molecular phylogenetic analyses of *Notropis* include those of Bielawski and Gold (2001), Raley and Wood (2001), Schönhuth and Doadrio (2003), and Berendzen et al. (2008b), with most of the earlier work using mtDNA sequences (cyt*b*) or allozymes. Cashner et al. (2011) used both mtDNA and nuclear DNA markers in a phylogenetic analysis of the subgenus *Hydrophlox*.

The genus *Notropis* as currently recognized is not monophyletic, and additional lineages within *Notropis* are likely to be recognized as separate genera when reasonable confidence concerning their monophyly develops (Mayden et al. 2006a). The most important morphological features for distinguishing the genus *Notropis* are bony elements of the skull (Mayden 1989), and Boschung and Mayden (2004) stated that otherwise, there is hardly a single character unique to the genus. *Notropis* remains a large genus, and 25 species occur in Arkansas. Arkansas *Notropis* species are characterized by a complete lateral line with 35–40 scales (except *N. maculatus*, which has an incomplete lateral line with only 8–10 pored scales), dorsal fin rays 8, anal fin rays 7–12, premaxilla protractile, four pharyngeal teeth in the greater row, mouth not greatly modified and lacking barbels, intestine short and S-shaped, with a single loop (except *N. nubilus*). Most species have a silvery or speckled peritoneum (except *N. boops*, *N. nubilus*, and *N. telescopus*) and are silvery in life with red and yellow breeding colors in many species.

Notropis atherinoides Rafinesque
Emerald Shiner

Figure 15.73. *Notropis atherinoides*, Emerald Shiner. *Uland Thomas.*

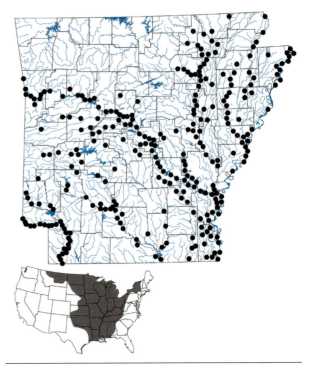

Map 60. Distribution of *Notropis atherinoides*, Emerald Shiner, 1988–2019.

CHARACTERS   An elongate, silvery shiner (Fig. 15.73) with a moderate-sized head, head length going 4 times into SL. Body slender and laterally compressed; body depth going into TL 5–7 times. Eyes moderately large, eye diameter of adults equal to or exceeding snout length and going into head length 2.5–3.3 times. Mouth terminal and slightly oblique; posterior end of upper jaw extending backward to or nearly to anterior margin of eye. Snout shape varies from slightly pointed to rounded. Pharyngeal teeth 2,4–4,2. Anal fin pointed, somewhat falcate, usually with 10 or 11 (10–13) rays. Pectoral fin rays 14–16. Dorsal fin high and pointed with 8 rays, its origin situated well behind pelvic fin origin: front of dorsal fin base closer to base of caudal fin than to tip of snout. Lateral line somewhat decurved and complete, with 35–40 scales. Scales in front of dorsal fin 18–21. Circumferential body scales 22–24, usually with 13 scales above the lateral line. Breast and belly fully scaled. Diagonal scale rows typically 19 or 20. Intestine short with a single S-shaped loop; peritoneum silvery with many dark speckles. Lips with black pigment. Chin sprinkled with dark pigment (Fig. 15.16B). Breeding males with minute tubercles on pectoral rays 2–10 and tiny tubercles on the head. Females often develop tubercles on the head and pectoral fins (Snelson 1968). Generally smaller than 3 inches (76 mm) TL, but 4-inch (102 mm) adults are occasionally collected in Arkansas. Maximum size 127 mm (5 inches) (Flittner 1964).

LIFE COLORS Body silvery with greenish-yellow dorsum; dorsal scales have pigmented margins. Sides silvery, with silver midlateral band; belly silvery white. Fin membranes unpigmented, but some fin rays are lined with melanophores. Lateral line pores not outlined or weakly outlined by black pigment. Weak lateral stripe about 1.5 scales wide is diffusely stippled anteriorly, becoming more prominently developed posteriorly on the caudal peduncle and most evident in freshly preserved specimens. Melanophores virtually absent below the lateral stripe. No brilliant spawning colors develop.

SIMILAR SPECIES It is most similar to *Notropis percobromus* but differs from that species in having large chromatophores on anterior half of chin (vs. small chromatophores on chin usually confined to outside margin of chin), head length in adult less than 25% SL (vs. greater than 25% SL), and snout short and blunt, its length less than two-thirds the distance from posterior margin of eye to posterior margin of head (vs. snout elongated and sharp, its length greater than two-thirds the distance from posterior margin of eye to posterior margin of head). See *N. percobromus* account for additional differences. The Emerald Shiner differs from *Lythrurus umbratilis* and *L. fumeus* in having fewer than 25 predorsal scales (vs. more than 25; see *L. fumeus* account for additional differences). *Lythrurus umbratilis* also has a distinct spot at the origin of the dorsal fin that is lacking in *N. atherinoides*. It is also superficially similar to *N. shumardi* but differs from that species in having a more slender body, the dorsal origin far behind pelvic fin origin, and in having 10 or 11 anal rays (vs. a deeper body, dorsal fin origin just above pelvic fin origin, and usually 8 or 9 anal rays). *Notropis atherinoides* closely resembles *N. suttkusi* with which it is syntopic in the Little River. See species account for *N. suttkusi* for differences.

VARIATION AND TAXONOMY *Notropis atherinoides* was described by Rafinesque (1818a) from Lake Erie. The Emerald Shiner is the type species of the subgenus *Notropis*. Humphries and Cashner (1994) noted a longstanding controversy concerning the taxonomic status of Great Plains versus Mississippi Valley forms of the Emerald Shiner and the frequent confusion of *N. atherinoides* with *N. percobromus* during the 20th century. Coburn 1982a studied anatomy and phylogeny. Mayden and Matson (1988) studied the relationships of *N. atherinoides* and *N. stilbius* and the *N. rubellus* species complex. Wood et al. (2002) considered *N. stilbius* most closely related to members of the *N. rubellus* species group. Mayden et al. (2006a),

in an analysis of the cytochrome *b* gene, resolved *N. atherinoides* as most closely related to *N. oxyrhynchus*, with those two species sister to a clade containing *N. amoenus* and *N. stilbius*.

DISTRIBUTION *Notropis atherinoides* occurs in Atlantic Slope drainages (St. Lawrence and Hudson rivers) of Canada and New York, the Arctic basin (Mackenzie River drainage of Canada), the Mississippi River basin from southern Canada south to the Gulf of Mexico, and Gulf Slope drainages from Mobile Bay, Alabama, to the Trinity River, Texas. In Arkansas, this common minnow occurs in appropriate habitats (all major rivers) in most parts of the state but is practically absent from the Ozark Mountains ecoregion (Map 60).

HABITAT AND BIOLOGY The Emerald Shiner is an inhabitant of medium-sized to large rivers and streams with clear to turbid waters flowing over sandy substrates. It occurs in some Arkansas impoundments and oxbow lakes becoming locally abundant, and Buchanan (2005) reported it from 12 Arkansas impoundments. It is reportedly tolerant of turbidity and low oxygen levels, but population increases in different parts of its range have been associated with decreases in turbidity. Cross and Collins (1995) believed that it was declining in abundance in Kansas due to dewatering in the Arkansas River and the effects of regulated flow in the Kansas River. It is one of the most abundant species throughout the Arkansas and Mississippi rivers in the state, and it was the most abundant species found in the Red River of Arkansas (Buchanan et al. 2003). Baker et al. (1991) reported its abundance in the lower Mississippi River. Over a 50-year period, the Emerald Shiner increased in abundance in the Missouri River from 0.1% to 28.5% of small fishes, and Pflieger (1997) attributed this increase to changes in turbidity and other factors that favored sightfeeding species. Reduced turbidity in the Arkansas River Navigation System may be partly responsible for its current numerical dominance in that system (although it is also abundant in the more turbid Red River). The Emerald Shiner had a surprisingly low critical thermal maximum temperature of 37.6°C in Oklahoma, compared with other Great Plains cyprinids (Matthews and Maness 1979). It is a schooling species usually found from midwater to the surface. Trautman (1957) noted that it avoids rooted aquatic vegetation. Although occasionally used as bait in some areas, the Emerald Shiner requires high oxygen levels, loses its scales easily, and does not survive well in bait buckets (Becker 1983; Hrabik et al. 2015).

Larval Emerald Shiners first feed on small zooplankton (rotifers, copepod nauplii, and cyclopoid copepods), and at about 12 mm TL, cladocerans are added to the diet (Siefert 1972). Algae were found in fish of all sizes, and calanoid copepods were found in fish 13 mm and larger. Fuchs (1967) reported a gradual change in young-of-the-year food habits as fish smaller than 1.6 inches (40 mm) fed primarily on algae, whereas larger fish fed chiefly on zooplankton (cladocerans and copepods) and, to a lesser extent, aquatic insects. The adult diet may vary at different times and localities, with some populations remaining largely planktivorous throughout life and other populations reportedly consuming terrestrial insects, aquatic insect larvae and pupae, or filamentous algae (Flittner 1964; Fuchs 1967; Campbell and MacCrimmon 1970; Hartman et al. 1992). Jenkins and Burkhead (1994) considered the relatively long, slender gill rakers of *N. atherinoides* an adaptation for planktivory, and Pflieger (1997) speculated that increased availability of microcrustaceans (in part as a result of plankton flushed out of upstream reservoirs) after impoundment of the Missouri River may have contributed to an increase in Emerald Shiners in that river.

Spawning in Arkansas occurs in late spring and early summer from April through June after water temperatures exceed 72°F (22.2°C). Ripe females were taken in Union County on 22 April (Tumlison et al. 2017) and from the Arkansas River near Fort Smith by TMB on 5 May and 20 May. Spawning occurs in Lake Erie offshore at night in depths of 7–20 feet (2.1–6.1 m) over a variety of substrates in spawning schools that may include millions of fish (Flittner 1964). A pelagic, broadcast spawner, females may produce 1,000 to more than 5,000 eggs each. Flittner (1964) described spawning behavior as follows:

> The shiners first appear about one to two feet below the surface milling and darting rapidly and erratically in a circular path. The smaller males appear to pursue larger females for a few seconds at a time. As these pairs swim about in a 10 to 20-foot circle, the male overtakes the female and presses closely on either the right or the left side in what appears to be an interlocking of pectoral fins. The pair gyrates briefly, and then slows down as the female arches her side upward, stops for an instant, rolling over further, and eggs are released and fertilized at the instant of rolling.

The eggs are nonadhesive and sink to the substrate. Hatching occurs in 24–32 hours. The newly hatched larvae remain on the substrate for 3 or 4 days, but after they become free swimming, they school near the surface. Populations of Emerald Shiners are characterized by high mortality (in some years) causing drastic changes in age-class structure. Maximum age is just over 4 years in

northern populations (Becker 1983). All older fish are usually females.

CONSERVATION STATUS The Emerald Shiner is a common and widespread species throughout Arkansas, and it often occurs in sufficient numbers to be an important forage species. Its populations in this state are secure.

Notropis atrocaudalis Evermann
Blackspot Shiner

Figure 15.74. *Notropis atrocaudalis*, Blackspot Shiner. *David A. Neely.*

Map 61. Distribution of *Notropis atrocaudalis*, Blackspot Shiner, 1988–2019.

CHARACTERS A robust shiner with a thick, rounded body, a blunt snout, small head (head length going about 4 times into SL), fairly large eyes (eye diameter going into head length 3.2 times), and a small, distinct black spot at the caudal base (Fig 15.74). Dark lateral stripe present. Mouth slightly subterminal, small, and somewhat oblique, slightly overhung by snout. Pharyngeal teeth variable, usually 0,4–4,0; but sometimes 1,4–4,1; or 2,4–4,2 (Hubbs et al. 1991). Intestine short, with a single S-shaped loop. Lateral line straight and complete, with 35–40 scales. Dorsal fin rays 8. Anal rays 7. Dorsal fin origin distinctly in front of pelvic fin origin. Maximum size 3.46 inches (88 mm) TL.

LIFE COLORS   Body with a conspicuous black lateral band, about as wide as the pupil of the eye, extending around the snout to the end of the caudal peduncle, where it terminates in a distinct, rectangular black spot the same depth as the lateral band. Scales above the lateral band outlined with dark pigment to form almost horizontal streaks on the dorsolateral area. Scales on the dorsum usually darker than those just above the lateral band. Lower sides and belly white. Some dark pigment present at base of anal fin extending onto fin rays.

SIMILAR SPECIES   Resembles *Hybopsis amnis* but differs from it in lacking the unique upper jaw structure and in having 7 anal rays (vs. 8). It superficially resembles *Notropis boops*, *Notropis chalybaeus*, and *Notropis texanus*, other Arkansas cyprinids that possess a dark lateral band. It differs from *N. boops* and *N. chalybaeus* in having 7 anal rays (vs. 8), and pharyngeal teeth usually 0,4–4,0 (vs. pharyngeal teeth 1,4–4,1 in *N. boops*, and 2,4–4,2 in *N. chalybaeus*). It differs from *N. texanus* in having pharyngeal teeth usually 0,4–4,0 (vs. 2,4–4,2).

VARIATION AND TAXONOMY   *Notropis atrocaudalis* was described by Evermann (1892) as *Notropis cayuga atrocaudalis* from the Neches River, about 22.5 km east of Palestine, Anderson County, Texas. It was sometimes confused with *Notropis amnis*, and it was also considered a subspecies of *Notropis heterolepis* until Hubbs (1951) properly diagnosed *N. atrocaudalis* and pointed out its specific distinctness. There has been no definitive systematic study of *Notropis atrocaudalis*, and systematic relationships are unclear.

DISTRIBUTION   The Blackspot Shiner has a small range, occurring in only four states: Arkansas, Oklahoma, Texas, and Louisiana, in the Red River drainage of the Mississippi River basin, and in Gulf Slope drainages from the Calcasieu River, Louisiana, to the Brazos River, Texas. In Arkansas, this lowland species is restricted to tributaries of the Little and Red river systems in five counties (Miller, Little River, Sevier, Howard, and Hempstead) of the southwestern part of the state (Map 61).

HABITAT AND BIOLOGY   The Blackspot Shiner is a rare species in Arkansas, restricted to clear smaller streams in the lower reaches of the Little River system and Red River tributaries. Evans and Noble (1979) reported that

this species was most abundant in headwater streams and greatly decreased in abundance downstream. Specimens reported from the Sulphur River in southwestern Miller County were collected from a cool, clear spring with slight flow over a mud and sand bottom located approximately 16.4 feet (5 m) from the main river on 22 August. Although sometimes associated with vegetation (Pigg 1977), *N. atrocaudalis* is more often found where no vegetation occurs (Bean et al. 2010).

Analysis of food habits revealed that this species is an insectivore consuming primarily aquatic insects, including immature dipterans, ephemeropterans, trichopterans, and coleopterans. The abundance of aquatic insects, lack of terrestrial insects, and high frequency of substrate material in stomachs led Bean et al. (2010) to conclude that this species is a benthic feeder (Heins and Clemmer 1975; Wilde et al. 2001). Insectivorous fish have significant top-down effects in streams by reducing densities of benthic grazing invertebrates (Hargrave 2006; Katano et al. 2006). Because *N. atrocaudalis* is often abundant where it is found in Texas, Bean et al. (2010) hypothesized that this species likely has a strong role in structuring the stream community.

The following information on reproduction is from Bean et al. (2010). Temporal patterns in ovarian development, GSI, and oocyte-diameter frequency distributions suggest that *N. atrocaudalis* spawns over a protracted period from March through June, with some individuals possessing mature ova as late as August. Three distinct oocyte size classes were present in individual mature females, strongly suggesting the production of multiple egg clutches over an extended spawning season. Four age groups (ages 0, 1, 2, and 3) were found in Texas, with an estimated maximum life span of 3 years. Age 3 fish reached a maximum total length of 88 mm, and age 0 fish were first collected in April. Growth was rapid, with some individuals reaching a total length of 56 mm in their first summer. Typically, females mature by the beginning of their second spring. Bean et al. (2010) found developing ovaries in individuals as small as 47 mm in July, indicating that some early spawned or fast-growing individuals may reach sexual maturity at age 0.

Bean et al. (2010) concluded that *N. atrocaudalis* is a species characterized by early maturation, relatively short life span, extended spawning periods, and downstream drift of eggs or larvae, traits that are common to stream fishes in variable systems (Heins and Rabito 1986; Matthews et al. 1978). Such life history traits allow rapid dispersal and recolonization after droughts and floods, which are of common occurrence in the areas inhabited by this species.

CONSERVATION STATUS   The Blackspot Shiner is a rarely encountered cyprinid in Arkansas. Because of its very small range in the state with few known localities of occurrence, we consider it endangered in Arkansas.

Notropis bairdi Hubbs and Ortenburger
Red River Shiner

Figure 15.75. *Notropis bairdi*, Red River Shiner. *David A. Neely.*

Map 62. Distribution of *Notropis bairdi*, Red River Shiner, 1988–2019. Inset shows native and introduced ranges.

CHARACTERS   A robust, slightly compressed, small to medium-sized, pale-colored shiner with a broad head, conical snout, and a large, nearly terminal, oblique mouth (upper jaw about 1.5 times as long as eye diameter and extending back to or behind front of eye) (Fig 15.75). Eyes small, round, and high on head; eye diameter going into body depth about 4 times. Back arched, body deepest under nape. Caudal peduncle deeper than in most of the other pale-colored riverine shiners. Dorsal fin origin in front of pelvic fin origin. Dorsal fin rays 8. Anal fin rays 7. Pectoral fins falcate in adult males, tip of pectoral fin often

extending backward to pelvic fin base; pectoral rays usually 15. Lateral line not decurved, either straight or with a broad arch, and often with several interruptions; lateral line scales 32–37. Pharyngeal teeth 0,4–4,0. Most specimens with a small area on the anterior nape devoid of scales, a condition unique to Arkansas species of *Notropis*; when nape scales are present, they are crowded in front of the dorsal fin. Breast partly scaled. Intestine short, with a single S-shaped loop. Peritoneum silvery with black speckles. Breeding males with small tubercles on top of head. Most individuals are smaller than 2.5 inches (64 mm) TL, but Cross and Collins (1995) reported a maximum TL of 3.5 inches (80 mm).

LIFE COLORS   Body color is generally tan to gray, grading to silver on lower half of body. Dorsolateral scales outlined by pigment. A thin predorsal stripe is present; middorsal stripe conspicuously interrupted in base of dorsal fin, producing a dark dash at base of dorsal fin (Hubbs et al. 1991). Lateral line scales not outlined by pairs of dark dashes or dots.

SIMILAR SPECIES   *Notropis bairdi* can be separated from *N. girardi*, *N. volucellus*, *N. wickliffi*, and *N. buchanani* by its 7 anal rays (vs. 8), and its 0,4–4,0 pharyngeal tooth count separates it from *N. blennius* and *N. potteri*, which have 2 rows of pharyngeal teeth. It further differs from *N. girardi* in having breast and nape partly scaled (vs. fully scaled), usually 15 pectoral rays (vs. 14), less falcate fins (vs. more falcate), and mouth nearly terminal (vs. snout slightly overhanging mouth). In Arkansas, *N. bairdi* is most likely to occur with *N. potteri* and can be further distinguished from that species by usually having a naked anterior nape (vs. anterior nape scaled), in lacking a swollen posterior portion of the upper lip (vs. posterior portion of the upper lip enlarged), and in lacking a medially thickened lower lip (vs. lower lip medially thickened). From *N. stramineus* it differs in having a larger mouth, smaller eye, deeper body, lack of pigmented dashes above and below the pores of the lateral line scales, and more crowded scales before dorsal fin.

VARIATION AND TAXONOMY   *Notropis bairdi* was described by Hubbs and Ortenburger (1929a) from the Red River, 10–14.5 km southwest of Hollis, Harmon County, Oklahoma. This shiner is usually considered most closely related to the Arkansas River Shiner (*N. girardi*) and the Smalleye Shiner (*N. buccala*). The three shiners form a close group whose precise relationships to other *Notropis* members were unclear (Gilbert 1980e). An alternative interpretation was provided by Mayden et al. (2006a), who supported placement of *N. bairdi* in genus *Alburnops*, with *bairdi* the basal sister group to the *Alburnops* clade.

Surprisingly, *N. girardi* was not recovered as a member of the *Alburnops* clade. Other Arkansas species assigned to the genus *Alburnops* by those authors were *Notropis blennius*, *N. chalybaeus*, *N. potteri*, *N. shumardi*, and *N. texanus*.

DISTRIBUTION   As the common name indicates, the Red River Shiner is endemic to the Red River drainage from the Texas panhandle and western Oklahoma downstream to southwestern Arkansas. It has been introduced into the Arkansas River system, where it has become established in three counties in southern Kansas downstream through central Oklahoma (Map 62). It was first collected from the Arkansas River drainage (Cimarron River, Oklahoma) in 1976 (Marshall 1978), and by 1979 it was established in the Cimarron River (Felley and Cothran 1981). Cross et al. (1985) believed that the introduction into the Arkansas River drainage occurred between 1964 and 1972. Luttrell et al. (1995) reported the first records of Red River Shiner from the Salt Fork of the Arkansas River, the mainstem Arkansas River, and from the South Canadian River in Oklahoma, taken by seine in 1994. Those authors also found museum records documenting its occurrence in the Arkansas River of Oklahoma in Tulsa County as early as 1982, and in the South Canadian River in 1978. This western shiner is extremely rare in Arkansas. One hundred specimens were collected in 1925 in the Red River in Oklahoma by A. I. Ortenburger, just 4 miles upstream from the Arkansas state line (OMNH 5947). Within Arkansas, it is known from only two older records from the Red River: UMMZ 128214 (2 specimens) collected by J. D. Black and R. Y. Black on 8 July 1939 from the Red River at Spring Bank Ferry at State Highway 160, Lafayette County (Black 1940), and UMMZ 170013 (4 specimens) taken on 18 August 1940 by R. M. Bailey and M. E. Davis from the Red River at U.S. Route 67 at Fulton, Miller-Hempstead County line. It has not been documented in Arkansas in approximately 80 years, despite extensive Red River sampling from 1995 to 2003 by Buchanan et al. (2003), leading to speculation that the earlier records represent waifs from more western populations. A population of *N. bairdi* persisted in the Red River of Oklahoma downstream from Lake Texoma into the mid-1990s, and specimens were collected near DeKalb, Texas, 18 km (10 miles) upstream from the Arkansas state line as recently as 1995 (OMNH 53812, data from Oklahoma Department of Environmental Quality fish samples provided by Jimmie Pigg). Additional specimens (138) were collected from the Red River near Arthur City, Texas in 1993 and 1998 (OMNH 54011, 69129). Therefore, it is possible that future records for the Red River Shiner might be obtained in Arkansas. No records are known for this species in Louisiana (Douglas

1974). There is no documentation that *N. bairdi* introduced into the Arkansas River drainage in Oklahoma has extended its range downstream in that drainage into Arkansas, but Luttrell et al. (1995) urged close monitoring of the introduced populations to detect further spread in the drainage.

HABITAT AND BIOLOGY   The Red River Shiner is primarily an inhabitant of the larger turbid rivers and streams of the Red River drainage in western Oklahoma, where it is usually associated with sand-bottomed habitats. It most commonly occurs in broad, shallow channels of the Red River main stem over bottoms of silt and shifting sand. In western Oklahoma, it is found in streams with highly fluctuating flows that often have high summer temperatures and high salinities. Echelle et al. (1972) found it at salinities ranging from 0.4 to 21.7 ppt, and it was abundant at one site with a salinity of 20 ppt. It occurred in shallow depressions in the current, along the edges of deeper pools, and in deeper areas of shallow backwaters. Introduced populations in the Cimarron River were found in Kansas in the main channel over a sand substrate, a mean depth of 7.8 inches (198 mm), and a current of 1.64 ft./s (0.5 m/s) (Cross et al. 1983). In the Cimarron River, Oklahoma, it occurred mainly in backwaters (Felley and Cothran 1981). In Oklahoma, it has rapidly displaced the related *Notropis girardi* (Luttrell et al. 1999).

Cross and Collins (1995) reported that it feeds mainly on terrestrial insects that fall into the water, and on organisms exposed by movement of the sand substrate or washed downstream. Echelle et al. (1972) examined gut contents of individuals 40–50 mm SL and found invertebrate remains and some substrate material, suggesting that some food was obtained from the bottom.

Group spawning probably occurs in summer, with the eggs drifting downstream during development. Hubbs and Ortenburger (1929a) collected hundreds of ripe or nearly ripe adults between June 16 and 26. According to Cross et al. (1983), females mature at 39 mm TL in Kansas, but Hubbs and Ortenburger (1929b) reported that females as small as 24 mm were full of ripe eggs in Oklahoma. Fecundity is apparently high, with females 42–54 mm SL containing from 1,520 to 3,314 well-developed ova; egg diameter ranged from 0.71 to 0.76 mm (Cross et al. 1983).

CONSERVATION STATUS   *Notropis bairdi* is most abundant in the upper parts of the Red River drainage of western Oklahoma downstream to the mouth of the Washita River in southeastern Oklahoma. It decreases drastically downstream from the mouth of the Washita River, but small populations apparently persist in the Red River of Oklahoma downstream of Denison Dam.

Winston et al. (1991) believed that the construction of Altus Dam caused its extirpation from the North Fork of the Red River in southwestern Oklahoma. An extremely rare species in Arkansas, the Red River Shiner is endangered and possibly extirpated in this state.

Notropis blennius (Girard)
River Shiner

Figure 15.76. *Notropis blennius*, River Shiner. *David A. Neely.*

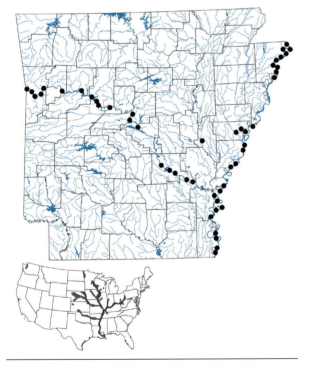

Map 63. Distribution of *Notropis blennius*, River Shiner, 1988–2019.

CHARACTERS   A robust moderately deep-bodied, pale-colored minnow (Fig. 15.76) with a large oblique, slightly subterminal mouth and 7 anal rays. Snout extending slightly past upper lip. Head large, its length going into SL 3.6–4.0 times. Tip of the upper jaw not reaching the level of the middle of the eye. Eye moderate in size, eye diameter going into head length 4.0–4.5 times, but eye diameter never as great as length of upper jaw (about three-fourths the length of the upper jaw). Pharyngeal tooth count variable, usually 2,4–4,2, but sometimes 1 or 2,4–4,1 or 2. Pectoral rays usually 15 or 16 (14–17). Pelvic rays usually 8 (7–9). Lateral line complete; lateral line scales 34–38. Predorsal scales usually fewer than 17. Diagonal scale rows usually 15 or 16 (14–18). Body circumferential scales 24–28. Intestine short; peritoneum silvery with faint speckles. Vertebrae 34–37. Breast covered to partially covered with exposed or embedded scales. Belly fully scaled. Front of dorsal fin base about equidistant between tip of snout and caudal base. Dorsal fin origin over pelvic fin origin. Dorsal rays 8. Tip of depressed dorsal fin approximately even with anal fin origin. Breeding males with tiny tubercles on snout, top of head, anterior rays (rays 2–7) of pectoral fins, and on front edges of dorsal and anal fins. Maximum size 5.2 inches (132 mm) in northern part of range (Trautman 1957), but the largest Arkansas specimens are seldom over 2.8 inches (70 mm).

LIFE COLORS   Body color pallid, varying from olive to straw dorsally with silver sides. A well-defined, broad middorsal stripe, which surrounds the dorsal fin base and continues to the caudal fin base. Milky white to silver ventrally. No basicaudal spot. Lateral band not prominent, but better developed posteriorly and in preserved specimens. The body lacks pigmentation below the lateral band. Fins colorless except for faint melanophores outlining some fin rays. No chromatic breeding coloration develops.

SIMILAR SPECIES   *Notropis blennius* can be separated from *N. girardi*, *N. volucellus*, *N. wickliffi*, and *N. buchanani* by its 7 anal rays (vs. 8), and from *N. bairdi* and *N. stramineus* by its 2 rows of pharyngeal teeth (vs. teeth in 1 row, 0,4–4,0). It is most similar to *N. potteri* but differs in lacking the swollen posterior end of the upper lip of *N. potteri* (Fig. 15.12) and in having a deeper body, body depth going into SL about 4.0 times in *N. blennius* (vs. going about 4.4 times into SL in *N. potteri*). See the species account for *N. potteri* for additional differences. It is superficially similar to *Hybognathus* species but differs in having 7 anal rays (vs. 8), a short intestine (vs. a long, coiled intestine), scaleless breast (vs. scaled), and a larger mouth.

VARIATION AND TAXONOMY   *Notropis blennius* was originally described from the Arkansas River at Fort Smith by Girard (1856) as *Alburnops blennius*. *Alburnops* was later considered a subgenus of *Notropis*, and *N. blennius* is the type species of the subgenus *Alburnops*. The River Shiner has had a confusing and tangled taxonomic history, mainly because other species (*Notropis stramineus*, *N. buchanani*, *N. volucellus*) were included under that name and because it was known by other synonyms. It was redescribed by Jordan (1878a) as *Episema jejuna* and subsequently placed in *Notropis* (*N. jejunus*) by Forbes (1884).

Forbes's *jejunus* eventually became *blennius*. Suttkus and Clemmer (1968) studied morphological variation and concluded that recognition of subspecies was unwarranted. They hypothesized its closest relatives were *Notropis edwardraneyi* of the Mobile Basin and less so *N. potteri* of the Red River system. Mayden (1989) did not resolve the relationships of these three species, but *N. potteri* was presumed closest to *N. bairdi* and *N. buccula*. Based on analysis of the cytochrome *b* gene, Mayden et al. (2006a) proposed elevation of subgenus *Alburnops* to generic status.

DISTRIBUTION   *Notropis blennius* occurs in the Hudson Bay basin of Canada to Minnesota, the Great Lakes drainage of Lake Michigan in Wisconsin, and in the Mississippi River basin from Wisconsin and Minnesota east to Pennsylvania, west to Colorado, and south to the Gulf of Mexico in Louisiana. In Arkansas, the River Shiner is limited to the larger rivers, including the Arkansas and the Mississippi rivers (Map 63). There are also a few recent records from the lower White and St. Francis rivers. Suttkus and Clemmer (1968) discussed reports of *N. blennius* from the Red River. A record of a single specimen from that river above Lake Texoma was considered a probable bait release, and the only other verified record was a single specimen from the lower Red River in Louisiana; however, four lots labeled *N. blennius* from the Red River in Choctaw and McCurtain counties, Oklahoma, collected between 1993 and 2010 are housed in the Sam Noble Oklahoma Museum of Natural History (OMNH 50513, 52774, 76000, and 80586). There are no valid records from the Red River in Arkansas, and Page and Burr (2011) showed *N. blennius* as occurring only in the lower Red River in Louisiana in close proximity to the Mississippi River. This shiner is common and locally abundant in Arkansas.

HABITAT AND BIOLOGY   This schooling midwater shiner is a big-river species, common and abundant in the main channels over sandy bottoms of the largest rivers. It is found throughout two of Arkansas's three largest rivers, the Arkansas and Mississippi rivers, and is replaced in the Red River by *Notropis potteri*. It is usually found in currents and mostly avoids quiet backwaters. We have noted, as did Trautman (1957), that the River Shiner occupies deeper water during daylight hours but moves into shallow waters at night. It seems less tolerant of turbidity than the Emerald Shiner in Illinois (Smith 1979), but it is common in the large turbid rivers of Arkansas. Pflieger (1997) reported an increase in River Shiners in the Missouri River since the 1940s, and a similar increase may have occurred in the Arkansas River after it was impounded into flow-through navigation pools (although local population reductions may have occurred).

Becker (1983) found River Shiners to be primarily insect feeders, but Hudson and Buchanan (2001) reported that *N. blennius* was omnivorous in the Arkansas River, consuming a wide variety of food items. Detritus dominated the diets of adults and juveniles in terms of volume and frequency of occurrence, and aquatic insects and algae were also important items. Occasionally, terrestrial insects were ingested when seasonally available. Small seeds were commonly found in gut contents. Animal foods ingested in decreasing order of importance included dipteran larvae and pupae, coleopterans, trichopterans, terrestrial insects, ephemeropterans, copepods, cladocerans, and oligochaetes. Protozoans, rotifers, and fish remains were present in only small amounts. Adult River Shiners fed on a wider variety of food categories than juveniles in all seasons except winter. Greatest feeding activity occurred shortly before dark.

Notropis blennius is classified as a late spawner (Starrett 1951), spawning during the summer over gravel and sandbars (Trautman 1957). Miller (1979) suggested that minnows like *Notropis blennius* which inhabit large turbid rivers of the Great Plains spawn periodically during much of the summer, often following heavy rains, a strategy that would permit early development to take place while eggs and early prolarvae are being carried along by strong currents. As the floods ebb, larvae can move into quiet havens where the water gradually clears and permits them to feed visually. Hudson and Buchanan (2001) found that *N. blennius* has a moderately long spawning season from June through August in Arkansas. Matthews and Heins (1984) concluded that protracted spawning may be adaptive to a variable environment. Water temperatures in the Arkansas River during the breeding season ranged from 26°C to 31°C, while discharge during that time varied from 142 to 1,558 m³/s. Nuptial tubercles were developed in Arkansas River males from May to October. Tubercles occurred along the rays of the pectoral fin and the pattern closely resembled that reported for Tennessee and Missouri specimens (Etnier and Starnes 1993; Pflieger 1997), but differed substantially from the pattern reported by Scott and Crossman (1973) for northern populations of this species. Arkansas females contained only 97–667 ripe eggs, but Becker (1983) reported 2,630–3,039 ripe eggs in Wisconsin River Shiners. Hatch and Elias (2002) reported clutch sizes ranging from 436 to 2,754 in the upper Mississippi River. A bimodal peak in GSI values and mean ovum diameters in June and August indicate the likelihood of multiple spawnings (Hudson and Buchanan 2001). In the upper Mississippi River, the ovarian cycling schedule also supported the hypothesis that the River Shiner is a multiple clutch spawner (Hatch and Elias 2002). Although nothing is known of the actual spawning behavior of *N. blennius*, it is thought to be a broadcast spawner (Starrett 1951).

Young-of-the-year individuals (14–20 mm) were collected during the breeding season in late June and in late July and were probably 15–30 days old. Length-frequency distributions indicated that *N. blennius* reaches a maximum age in Arkansas of about 2 years (Hudson and Buchanan 2001). Three age groups (0, 1, and 2) were present in the Arkansas River only during summer months. No discernible differences in sex ratio were found; however, females attained greater SL than males. The largest female was 69 mm, while the largest male measured 58 mm SL. Tennessee populations of *N. blennius* also reached a smaller maximum size than that reported for northern populations (Etnier and Starnes 1993). In the upper Mississippi River, mature females ranged in size from 48.3 to 87.9 mm SL (Hatch and Elias 2002). Sexual maturity is reached by males at age 1, while females mature at age 2 or older (Trautman 1957; Carlander 1969; Becker 1983). In Arkansas the smallest sexually mature males and females were 35 mm (1.4 inches) SL (Hudson and Buchanan 2001). Life span is probably 2 years.

Hudson and Buchanan (2001) concluded that *N. blennius* is a short-lived, *r*-strategist with a protracted spawning period during which a female may spawn more than once, and its generalist food habits are important life history attributes contributing to its continued success in altered environments of the Arkansas River.

CONSERVATION STATUS The River Shiner is commonly found in appropriate habitat throughout the Arkansas River despite the construction of 13 locks and dams during the 1960s to 1971. It may have experienced some local declines because it is now uncommon in Pool 13 (the pool that includes its type locality at Fort Smith). It is still common in the Mississippi River, and we consider its populations secure.

| *Notropis boops* Gilbert
| Bigeye Shiner

Figure 15.77. *Notropis boops*, Bigeye Shiner. *Uland Thomas.*

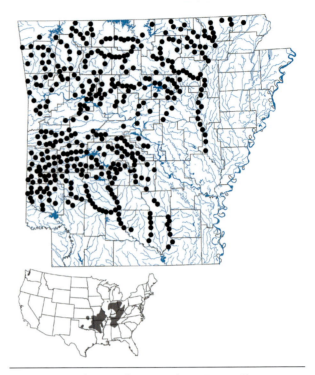

Map 64. Distribution of *Notropis boops*, Bigeye Shiner, 1988–2019.

CHARACTERS A moderately slim minnow with a black lateral band and an unusually large eye (eye diameter much longer than snout length and going into head length 2.0–2.8 times) (Fig. 15.77). Anal rays 8. Pharyngeal teeth 1,4–4,1. Lateral line complete; lateral line scales 33–39. Pectoral fin rays 13–17. Mouth terminal, oblique, and large, upper jaw extending posteriorly past front of eye. Predorsal scales usually 12–15. Diagonal scale rows usually 15 (13–16). Dorsal fin slightly falcate; dorsal rays 8. Dorsal origin directly above the pelvic fin origin and about equidistant between tip of snout and base of caudal fin. A conspicuous black lateral band with a clear band (about 1 scale row wide) directly above it. Scales above this clear band are outlined in black. Black band continuing through the eye around the snout at the top of the upper jaw. Intestine short; peritoneum black in ventral part of body cavity. Breast and nape completely scaled. Breeding males with numerous fine tubercles on the head, upper body, and anterior rays of dorsal, anal, and pectoral fins. Maximum size 3.6 inches (91 mm).

LIFE COLORS Lateral line outlined with dark pigment (especially on anterior half of body), forming a pattern of paired dots or dashes. Body color olivaceous yellow, with a silver side grading to silvery white on belly. Thin dark stripe along midline of back. A distinct black lateral stripe about 1–1.5 scale rows wide running from tip of snout and lower jaw to caudal base. Caudal spot highly variable in size, shape, and degree of development, often vague or lacking. Fin membranes clear, but rays of caudal and dorsal fins outlined by melanophores. No bright breeding colors develop.

SIMILAR SPECIES   The Bigeye Shiner is similar to a few other Arkansas shiners having a black lateral stripe. It differs from the Kiamichi Shiner in having 8 anal rays (vs. 9–11), in lacking the abundant black pigment on the chin area found in *N. ortenburgeri*, and in having a pharyngeal tooth formula of 1,4–4,1 (vs. 0,4–4,0). It is distinguished from *Notropis texanus* by its 8 anal rays (vs. 7) and a tooth formula of 1,4–4,1 (vs. 2,4–4,2). In the Ozark Upland and the Fourche LaFave River it is often sympatric with *Notropis greenei*, which it resembles in overall body coloration and eye size, but it differs from that species in having a black peritoneum (vs. peritoneum silvery but sprinkled with dark speckles), and in having 1,4–4,1 pharyngeal teeth (vs. 2,4–4,2). The caudal spot of *N. boops* is variable in size, shape, and degree of development and is not a completely reliable feature for separating it from *N. greenei*, which usually has a distinctly triangular caudal spot. *Notropis boops* differs from *N. atrocaudalis* in having 8 anal rays (vs. 7) and 1,4–4,1 pharyngeal teeth (vs. usually 0,4–4,0) and from *N. chalybaeus* in having a complete lateral line (vs. incomplete), 1,4–4,1 pharyngeal teeth (vs. 2,4–4,2), and in lacking black pigment inside the mouth (vs. inside of mouth heavily sprinkled with black pigment).

VARIATION AND TAXONOMY   *Notropis boops* was originally described by Gilbert (1884) from Salt Creek, Brown County, Indiana, and Flat Rock Creek, Rush County, Indiana. Black (1940) noted that many different names were applied to this species in early collections from Arkansas and gave a list of known erroneous reports. Burr and Dimmick (1983) redescribed the Bigeye Shiner from examination of specimens from throughout its range. No significant geographic variation was found in any of the 20 morphological characters studied; therefore, no subspecies were recognized. Swift (1970) indicated *N. xaenocephalus* and *N. scabriceps* of the *N. texanus* species group were the closest relatives. Mayden (1989) also supported placement of *N. boops* in the *texanus* species group of the subgenus *Notropis*. Both Wiley and Mayden (1985) and Mayden (1989) considered *N. xaenocephalus* the sister species of *N. boops*. Mayden et al. (2006a), based on analysis of the cytochrome *b* gene, provisionally retained *boops* in genus *Notropis* and resolved *N. boops* as sister to *N. chihuahua*.

DISTRIBUTION   *Notropis boops* is found in the Great Lakes drainage of Ohio (Lake Erie), and in the Mississippi River basin from central Ohio to eastern Kansas and south to northern Louisiana and southern Oklahoma. In Arkansas, this minnow is one of the most widespread upland species in the state (Map 64). Black (1940) referred to it as "One of the most characteristic fishes of the upland region, abounding at the ends of long holes in the gravelly mountain streams. . . . This is an exceedingly common species." It inhabits all streams of the Interior Highlands and generally avoids the more turbid streams of the Coastal Plain. The lower Saline River is an exception, as a good population extends all the way to the mouth of this river. It is also known from the Ouachita River south into Louisiana (Douglas 1974).

HABITAT AND BIOLOGY   The Bigeye Shiner prefers gravel and rock-bottomed pools lined with water willow in clear, moderate to high-gradient streams and rivers, where it is often one of the most abundant fishes. The pools usually have slow current. Pflieger (1997) noted that it usually avoids strong current and continuously cool water. In the Illinois River, Oklahoma, it was found mainly in pools during most of the year but moved into shallow, fast riffles with larger substrate particles during the spring (Felley and Hill 1983). Buchanan (2005) reported it from 12 Arkansas reservoirs where it was almost always associated with rocky substrates adjacent to water willow. Schaefer (2001) concluded that riffles may act as barriers to movement among pools in natural streams.

Trautman (1957) reported that *N. boops* feeds extensively at the surface on small insects, often jumping into the air to capture them; however, we observed this species feeding throughout the water column. Other foods include aquatic insects, organic detritus, and algae, especially in summer and fall (Felley and Hill 1983). It is often observed in schools feeding near the upper ends of pools having some current (Gorman 1988; Ross 2001). In Arkansas, the Bigeye Shiner is largely an opportunistic insectivore that ingests a wide range of prey. It apparently prefers to feed at the water surface or in the upper part of the water column but will also readily feed on benthic prey. TMB examined gut contents of 89 specimens collected from May into September from Benton, Clark, Crawford, Franklin, Garland, Polk, and Van Buren counties in Arkansas. The variety of foods eaten varied by locality, but the diet was generally dominated by terrestrial insects (primarily dipterans) and immature insects that occur most frequently in near-surface environments (dipteran larvae and pupae). Immature benthic insects were also commonly ingested at most localities and included midge larvae, ephemeropteran nymphs, plecopterans, odonates, coleopterans, trichopterans, and megalopterans. Microcrustaceans (primarily copepods) were eaten in small amounts. Filamentous algae and detritus were important food items in the Mulberry River in May and in Lee Creek in September.

Spawning occurs from late April into August in Oklahoma (Lehtinen and Echelle 1979) and from June to August in Missouri (Pflieger 1997) and Kansas (Cross and Collins 1995). Spawning in Arkansas occurs typically from May into August. Tuberculate males taken in Arkansas during those months were in their second year of growth. HWR and Chris T. McAllister collected *N. boops* females with ripe orange eggs from the Mulberry River, Franklin County, on 6 July 2015 (Tumlison et al. 2016). TMB found females with submature eggs in the Mulberry River on 5 May; ripe females were found in the Fourche LaFave River on 10 May, the Caddo River on 31 May, Middle Fork of the Little Red River on 9 June, Lake Hamilton on 11 July, and the Illinois River on 2 August. Females collected from Lee Creek on 6 September contained only atretic eggs. Females with mature eggs ranged from 63 to 85 mm TL and contained from 480 to 1,267 ripe ova. Adults move from pools into shallow, fast riffles with large substrate particles when ready to spawn; individual females likely spawn more than one clutch per season (Lehtinen and Echelle 1979). Although spawning behavior has not been documented, it probably broadcasts its eggs similar to some other members of the *N. texanus* species group. Smith (1979) indicated sexual maturity might be reached after the end of the first year's growth, and in Brier Creek, Oklahoma, it reached sexual maturity during its second summer at about 1.7 inches (43 mm) TL; few individuals survived to spawn a third summer (Lehtinen and Echelle 1979).

CONSERVATION STATUS Although currently an abundant, widespread, primarily upland species in Arkansas, populations of *N. boops* have been severely decimated in Ohio and Illinois due to excessive siltation from poor agricultural practices (Burr and Dimmick 1983; Page and Burr 2011). It is rather intolerant of siltation and continuous high turbidity (Smith 1979), and it is a species of special concern in Alabama (Boschung and Mayden 2004) and Mississippi (Ross 2001). Its populations in Arkansas are currently secure.

Notropis buchanani Meek
Ghost Shiner

CHARACTERS A small, pale, very slab-sided shiner with large eyes (eye diameter going 2.6–4.0 times into head length and 0.7–0.9 times into snout length), and a small, slightly oblique mouth (Fig. 15.78). Back arched; greatest body depth at dorsal fin origin, strongly tapering to a thin

Figure 15.78. *Notropis buchanani*, Ghost Shiner. *David A. Neely.*

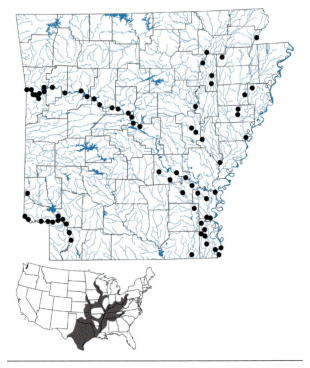

Map 65. Distribution of *Notropis buchanani*, Ghost Shiner, 1988–2019.

caudal peduncle. Body depth going into SL fewer than 3.9 times. Anal fin rays 8. Pectoral fin rays 13–15. Pharyngeal teeth 0,4–4,0. Breast naked or partly scaled. Snout short, rounded. Infraorbital canal absent or occasionally incomplete. Lateral line complete and straight; lateral line scales highly elevated, especially anteriorly, scale height often 4 times scale width; 30–35 scales in lateral line. Predorsal scales 12–15. Tips of pelvic fins reaching to or beyond front of anal fin. Dorsal fin very high and pointed. Dorsal fin rays 8. Front of dorsal fin base about equidistant between tip of snout and caudal base and slightly behind pelvic fin origin. Caudal peduncle long, its length going about 3.8–4.3 times into SL. Intestine short with a single S-shaped loop, its length about equal to SL; peritoneum silvery, without dark speckles. Breeding males with small tubercles over most of head, anterior part of body, and along dorsal surfaces of pectoral fin rays 2–7; tubercles largest on snout and top of head (Pflieger 1997). Maximum size 2.6 inches (66 mm), but most Arkansas specimens are smaller than 2 inches (51 mm) TL.

LIFE COLORS   Body color pallid and almost devoid of pigment in turbid waters, giving rise to the common name, Ghost Shiner. Some pigmentation usually occurs on the top of the head, faintly outlining the dorsal scales, and along the anal fin base. The first 10–12 anterior lateral line pores are sometimes outlined above and below by dark dots. A bright silvery lateral stripe evident. Fins colorless, but margins of fin rays sometimes have a few scattered melanophores. No breeding colors develop.

SIMILAR SPECIES   *Notropis buchanani* is similar to *N. volucellus* and *N. wickliffi* but differs from those species in having a deeper body, body depth going into SL about 3.5 times (vs. body depth going about 4–4.5 times into SL), in having the infraorbital canal absent or occasionally incomplete (vs. complete in *volucellus* and *wickliffi*), and in having very little pigment on the dorsal scales (vs. dorsal scales more prominently pigmented).

VARIATION AND TAXONOMY   *Notropis buchanani* was originally described from a small creek near Poteau, Le Flore County, Oklahoma (Arkansas River drainage), by S. E. Meek (1896), who named this species after Dr. John L. Buchanan, then president of Arkansas Industrial University (University of Arkansas). It was for many years considered a subspecies of *N. volucellus*, and Bailey (1951) elevated it to a full species. Mayden (1989) suggested *Notropis buchanani* was associated with the *N. volucellus* species group because it shares some characters with those species; however, Etnier and Starnes (1993) believed that differences in nuptial tubercle patterns suggested that these affinities should be reevaluated. Mayden et al. (2006a), based on analysis of the cytochrome *b* gene, recognized *N. buchanani* as sister to *N. volucellus*, with those species forming a clade sister to *N. ozarcanus*.

DISTRIBUTION   *Notropis buchanani* is found in Great Lakes drainages of southern Canada and Michigan (Lake Erie and Lake Huron), in the Mississippi River basin from Pennsylvania westward to Minnesota and Wisconsin, south to northern Alabama and west to Texas, and in Gulf Slope drainages from the Calcasieu River, Louisiana, to the Rio Grande of Texas and Mexico. In Arkansas, it is common in sluggish streams of the Coastal Plain and ranges northwestward through the Arkansas River Valley. It was found throughout the Red River in Arkansas but was considered uncommon there (Buchanan et al. 2003). It is found throughout the St. Francis, Black (McAllister et al. 2009d), and lower White rivers in Arkansas, and there are only a few recent records from the Mississippi River (Map

65). There are a few older records from the Ouachita River drainage, but the only recent records from that drainage are from Bayou Bartholomew.

HABITAT AND BIOLOGY   In Arkansas, the Ghost Shiner is a schooling species generally occupying medium to large, warm, sluggish streams and rivers with high turbidities. It is surprisingly uncommon in two of Arkansas's three largest rivers, the Mississippi and Red rivers. It can be common in reservoirs, and Buchanan (2005) considered it common in Arkansas River navigation pools. It is a midwater species frequenting protected backwaters and large pools away from strong currents over a bottom of silt and sand, and its numbers may have increased in the Arkansas River after the construction of the McClellan-Kerr Arkansas River Navigation System. It is sometimes found over a substrate of silt and detritus. In Kansas, it is found in gentle eddies adjacent to strong currents in the main channels of rivers, in pools where small, intermittent creeks join rivers, and alongside the lower part of gravel bars in the main stem of rivers where the direction of flow is reversed (Cross and Collins 1995). Trautman (1957) reported that in Ohio the Ghost Shiner was usually found in association with submerged aquatic vegetation.

Williams (2011) studied Ghost Shiner food habits in Texas and found that by percent weight, food items of adults consisted primarily of detritus, sand, and aquatic insects, including trichopterans (5%), dipterans (3%), hemipterans (3%), unidentified insects (14%), and <1% plecopterans, ephemeropterans, and coleopterans. His study also showed that occurrence of detritus was consistent across seasons and among adults and juveniles, occurring in 89% of adults and 85% of juveniles, and making up 60% by weight of adult and juvenile Ghost Shiner food items. Gut contents of 46 specimens, 40–55 mm TL, collected by seine from the Arkansas River near Fort Smith from March through July, were examined by TMB. By percent frequency of occurrence, the diet was dominated by small crustaceans (65%), mainly copepods, cladocerans, and amphipods; however, the diet by percent volume was dominated by immature aquatic insects, primarily ephemeropterans, dipteran larvae and pupae, and trichopterans. Feeding apparently occurred on the substrate and in the water column in the Arkansas River. Forty-four percent of the guts examined contained sand grains, and 22% contained strands of filamentous algae. There was little or no detritus in the guts.

Spawning occurs in Oklahoma from late April through early August (Miller and Robison 2004). A prolonged spawning season lasts from May through September in Texas (Williams 2011), and spawning occurs in Missouri from late April to early June over sluggish riffles composed of sand or fine gravel (Pflieger 1997). Based on examination of gonads by TMB, spawning occurs in the Arkansas River in Crawford and Yell counties from late May at least through mid-July. Arkansas River specimens had small gonads and contained no mature ova from March to early May. Tuberculate males were first found in Arkansas in early May, and ripe females were found only in June and July. Females from 45 to 50 mm TL contained from 189 to 479 mature ova. Ovaries of ripe females in June contained smaller developing egg sizes, but ovaries in mid-July did not contain smaller egg sizes. The semibuoyant eggs are broadcast in the water column to drift downstream (Williams 2011). In Texas, Williams (2011) found most spawning fishes were in their second year of a 3-year life span. Female Ghost Shiners exhibited elevated GSIs (>2.5%) and mature ovaries from May through September. The smallest female to exhibit mature ovaries was a 31 mm individual collected in August 2005. Mean clutch size was 232.6 (SD = 193.5; n = 27) and ranged from 32 to 952 oocytes (TL range: 31–49 mm). Clutch size was positively related to total length of mature females.

CONSERVATION STATUS   The Ghost Shiner is widespread in Arkansas, and was sometimes found in moderate numbers (>200 individuals in a sample) in AGFC rotenone population samples in the Arkansas River navigation pools. TMB collected 862 Ghost Shiners in a single seine sample from the Arkansas River in Crawford County on 7 April 2006. We consider its populations secure in Arkansas.

Notropis chalybaeus (Cope)
Ironcolor Shiner

Figure 15.79. *Notropis chalybaeus*, Ironcolor Shiner. *Uland Thomas.*

CHARACTERS   A small, moderately robust shiner with large eyes (eye diameter greater than snout length and going into head length about 3 times), a small head, a short, blunt snout, and a dark lateral band bordered above by a light-colored band (Fig. 15.79). Body deepest at dorsal fin

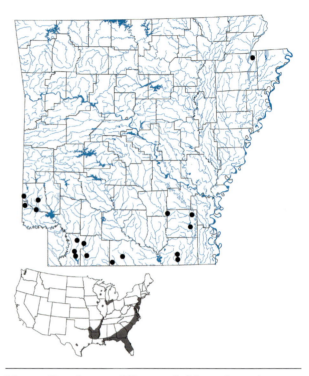

Map 66. Distribution of *Notropis chalybaeus*, Ironcolor Shiner, 1988–2019.

origin, tapering to a slender caudal peduncle. Body depth going into SL 3.7–4.5 times. Mouth terminal, moderately large and oblique. Inside of mouth heavily sprinkled with dark pigment as well as chin and lips. Barbels absent. Belly scaled; breast naked or partly scaled. Pharyngeal teeth 2,4–4,2. Lateral line incomplete, moderately decurved anteriorly, and usually with 10 or more unpored scales; lateral scale rows 31–36. Infraorbital canal interrupted. Dorsal fin rays 8. Front of dorsal fin base much closer to tip of snout than to caudal base. Dorsal fin origin over or slightly behind pelvic fin origin. Dorsal and anal fins slightly falcate. Anal fin rays 8; pectoral rays 11–13. Intestine short with a single S-shaped loop; peritoneum silvery with dark speckles. Breeding males with tubercles ringing the lower jaw and combining with a well-developed group of 8–12 tubercles at tip of symphysis. Other tubercles are found around the eyes, snout, some dorsolateral scales, and along pectoral rays 2–7. Breeding females may develop weak tubercles on the lower jaw and along pectoral rays. Maximum size 2.5 inches (64 mm).

LIFE COLORS   Body color olive yellow dorsally, white ventrally, with a prominent dark lateral band almost the diameter of the eye extending from a small, irregular caudal spot forward across the head and around the snout on both lips. Caudal spot attached to midlateral stripe. Fins fairly large and unpigmented. Dark pigment faintly outlining anal fin base and extending posteriorly along ventral edge of the caudal peduncle. Dorsolateral scales outlined with melanophores to form a diamond-shaped scale pattern. Roof and floor of mouth heavily sprinkled with dark pigment. Breeding males develop a bright orange stripe above the dark lateral band; orange spots are typically present above and below the black caudal spot with an orange cast on the caudal fin.

SIMILAR SPECIES   *Notropis chalybaeus* is similar to *Pteronotropis hubbsi*, *N. atrocaudalis*, *N. boops*, *N. texanus*, and *N. ortenburgeri*, all of which have a dark lateral band. It differs from *P. hubbsi* in having 8 anal and dorsal fin rays (vs. 9–11 anal rays and 9–11 dorsal rays). It differs from *N. atrocaudalis*, *N. texanus*, and *N. ortenburgeri*, in having a diagnostically pigmented floor and roof of mouth and 8 anal rays (vs. largely unpigmented floor and roof of mouth and anal rays 7 in *N. atrocaudalis* and *N. texanus* and 9–11 in *N. ortenburgeri*). It differs from *N. boops* in having 7 anal rays (vs. 8), a silvery peritoneum, 2,4–4,2 pharyngeal teeth, and a pigmented inside of mouth (vs. a black peritoneum, 1,4–4,1 pharyngeal teeth, and lacking pigment inside mouth). It differs from *Opsopoeodus emiliae* in several features, including mouth large and not strongly upturned (vs. mouth small and extremely oblique) (Fig. 15.80).

VARIATION AND TAXONOMY   *Notropis chalybaeus* was described by Cope (1867a) as *Hybopsis chalybaeus* from a tributary of the Schuylkill River near Conshohocken, Montgomery County, Pennsylvania. Swift (1970) and Mayden (1989) placed *Notropis chalybaeus* in the *texanus* species group. Based on analysis of the cytochrome *b* gene, Mayden et al. (2006a) proposed placing *N. chalybaeus* in the genus *Alburnops*.

DISTRIBUTION   The largest portion of the range of *Notropis chalybaeus* is along the Atlantic and Gulf slopes from the Hudson River, New York, south around Florida to the Sabine River of Texas and Louisiana, with a disjunct Gulf Slope population in the San Marcos River, Texas. It is also found in the Mississippi River basin from the northern half of Louisiana to southeastern Missouri, with disjunct populations farther north in Iowa, Illinois, Indiana, and Wisconsin, and in the Lake Michigan drainage of southern Michigan and northern Indiana. In Arkansas, the Ironcolor Shiner occurs infrequently only in the Coastal Plain below the Fall Line (Map 66). Between 1960 and 1987, it was recorded from 35 localities in Arkansas, in the White, Arkansas, Ouachita, and Red river drainages, but between 1988 and 2017, it was found at only 17 sites. It is more common in southern Arkansas than in eastern Arkansas.

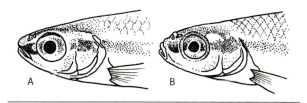

Figure 15.80. Heads and mouths of:

A—*Notropis chalybaeus*
B—*Opsopoeodus emiliae*

HABITAT AND BIOLOGY   In Arkansas, this lowland, schooling shiner inhabits middle water column depths in acid, tannin-stained, clear, sluggish Coastal Plain streams and rivers of low to moderate gradient with abundant aquatic vegetation over clean sand bottoms (Robison 1977a). Although Marshall (1947) indicated that it occurs in a wide diversity of stream habitats in Florida, that does not appear to be the case in Arkansas. Streams where *N. chalybaeus* was most common in southern Arkansas were of small to moderate size (10–30 feet [3–9.1 m] wide), approximately 3–4 feet (0.9–1.2 m) deep, with sand, mud, silt, or detritus substrates, and with submerged and emergent vegetation. It occurs in greatest numbers in streams having dense riparian vegetation of trees that form a canopy over the water. Big Creek (Red River drainage) in Columbia County where *N. chalybaeus* was fairly common was characterized physicochemically by water temperatures ranging from near 0°C in winter to 25°C in summer, dissolved oxygen values from 4.2 to 6.8 mg/L, pH of 6.2–6.7, carbon dioxide from 6.5 to 8.5 mg/L, BOD of 0.6–6.0 mg/L, chloride 11.2–26.0 mg/L, dissolved solids 93–150 mg/L, turbidity less than 25 JTU, and discharge generally less than 65 cfs (Robison 1977a). In Mississippi, it occurred in aggregations, often at the upper ends of pools, in water 60–90 cm deep, with a moderate to sluggish current, and sand, mud, silt, or detrital substrates (Ross 2001). It usually avoids still-water habitats (Swift et al. 1977), but Buchanan (2005) reported it from Lake Erling, an impoundment of Bodcau Bayou in Lafayette County.

Robison (1977a) listed the common cyprinid stream associates of *N. chalybaeus* in Arkansas as *Lythrurus fumeus*, *L. umbratilis*, and *Notemigonus crysoleucas*. In Little River County, it is also syntopic with *Notropis atrocaudalis*. The Ironcolor Shiner shows an almost completely complementary lowland distribution in Arkansas to *Notropis texanus*. Although both are lowland species, they are only rarely found in the same stream, much less at the same locality in this state. Whether this pattern is a result of competitive exclusion, the extensive habitat alteration that has occurred in the lowlands, or other factors is not known. Page and Burr (2011) reported that in other parts of its range, it is often collected with *N. texanus*.

Notropis chalybaeus is a sight feeder consuming a variety of drifting invertebrates that are swept downstream by currents or deposited from the terrestrial environment. Goldstein and Simon (1999) classified *N. chalybaeus* as an opportunistic invertivore forager. Aquatic insects are the principal item in the diet (Marshall 1947). In a study of a relict spring population in southern Texas, Perkin et al. (2012a) found prey items were dominated by aquatic insects, including Diptera (16% by weight), Ephemeroptera (13%), and Odonata (5%); terrestrial insects made up 9% and vegetation/algae constituted 3% by weight. Leckvarcik (2001) reported small crustaceans (cladocerans) constituted more than 50% of prey items in Pennsylvania. Insect prey items provide greater energy availability relative to small crustaceans for insectivorous cyprinids such as *N. chalybaeus* (Marshall 1947).

Spawning in Florida extends from early or mid-April to late September (Marshall 1947), while in the northern portions of its range (Illinois and Pennsylvania), this shiner is known to spawn for about 2–3 months (Becker 1983; Leckvarcik 2001). Spawning season increases to approximately 4 months in northern Florida and 5 months in southern Florida (Swift 1970; Boschung and Mayden 2004). In Texas, Perkin et al. (2012b) found a protracted 10-month spawning season ranging from March through December, during which multiple clutches were produced. In Arkansas, spawning occurs in the summer months (June-August) with highly tuberculate males taken in late June. No nest is constructed. Males pursue females until the female is ready to spawn. A spawning pair swims vent to vent, and upon separation, the eggs are released and fertilized (Marshall 1947). The adhesive sinking eggs stick to gravel or sand substrate or vegetative strands and particles in pool habitats (Marshall 1947; Robison 1977a; Simon 1999b, Leckvarcik 2001). The number of eggs per female in the San Marcos River was approximately 43–95 (Perkin et al. 2012b), while in a Pennsylvania population, the number per female was about 43–121 (Marshall 1947). Mean mature oocyte diameter was about 0.8 mm and the number of mature oocytes per clutch ranged from 46 to 326. Eggs hatch in just over 2 days at a temperature of about 62°F (16.7°C), and the young school near the surface soon after hatching (Marshall 1947). Leckvarcik (2001) suggested reproductive seasonality was initiated by photoperiod and terminated by declining water temperatures. Eighteen young-of-the-year, ranging from 0.6 to 0.8 inches (15.7–20 mm.) SL, were collected by HWR on 18 September from a heavily vegetated area of Bayou Dorcheat, Columbia County. The Ironcolor Shiner is a short-lived species which reaches sexual maturity at 1 year (36 mm SL) (Perkin et al. 2012b). Those authors reported four age groups (age 0–3) with a maximum life span of just over 3 years. Age 0 individuals reached a maximum length of 35 mm, age 1 reached 46 mm, age 2 reached 50 mm, and age 3 reached 52 mm. Overall ratio of male to female *N. chalybaeus* was 0.5:1 and differed from the expected 1:1 sex ratio.

CONSERVATION STATUS *Notropis chalybaeus* is a species of special concern in Texas (Hubbs et al. 1991), and Warren et al. (2000) considered its populations in the southern United States vulnerable. The Ironcolor Shiner is an uncommon inhabitant of Arkansas waters. It may have declined in this state since 1988 and should be considered a species of special concern.

Notropis girardi Hubbs and Ortenburger
Arkansas River Shiner

Figure 15.81. *Notropis girardi*, Arkansas River Shiner. *Uland Thomas.*

CHARACTERS   A small, robust shiner with a small dorsally flattened head (head length going about 4.3 times into SL), a rounded snout, and a small subterminal mouth (Fig. 15.81). Eye small, eye diameter going into head length 4.5 times or more. Dorsal rays 8; anal rays 8; pectoral rays usually 14. Pharyngeal teeth 0,4–4,0. Dorsal fin high, anterior rays extending past posterior rays when depressed. Dorsal fin origin closer to tip of snout than to caudal base. Pectoral fins long and falcate in males, rounded in females. Breast and nape fully scaled. Lateral line scales 33–37. Predorsal scales 14–17. Intestine short, forming S-shaped loop; peritoneum silver. Breeding males with 2–4 rows of tubercles on pectoral fins. Maximum size 2.65 inches (65 mm) (Wilde 2002).

LIFE COLORS   Body color pallid, tan dorsally, silver laterally, grading to white on the belly. Dorsolateral scales lightly outlined by melanophores. Anterior lateral line pores faintly outlined by dark pigment. A small black chevron usually present at the caudal base. Without a distinct middorsal stripe. Breeding males without bright colors.

Map 67. Distribution of *Notropis girardi*, Arkansas River Shiner, 1988–2019. Inset shows native and introduced ranges.

SIMILAR SPECIES   *Notropis girardi* is similar in appearance to *N. stramineus*, *N. blennius*, *N. sabinae*, and *N. bairdi*, from which it is separable in having 8 anal rays (vs. 7) and from *N. volucellus*, *N. wickliffi*, and *N. buchanani* in having the exposed portions of the lateral line scales about the same size and shape as adjacent scales (vs. anterior lateral line scales elevated, 2 to 3 times higher than long, and larger than adjacent scales on side).

VARIATION AND TAXONOMY   *Notropis girardi* was described by Hubbs and Ortenburger (1929a) from the Cimarron River, 4.9 km northwest of Kenton, Cimarron County, Oklahoma. This species has been placed in the subgenus *Notropis* (Coburn and Cavender 1992; Bielawski and Gold 2001). It is usually considered most closely related to the Red River Shiner (*N. bairdi*) and the Smalleye Shiner (*N. buccula*), but Mayden et al. (2006a) found that *N. girardi* was most closely related to *N. wickliffi* (based on analysis of the cytochrome *b* gene).

DISTRIBUTION   This Great Plains form is an Arkansas River drainage endemic historically known from western Arkansas to western Kansas, western Oklahoma, the Texas panhandle, and northeastern New Mexico (Map 67). It is now extirpated from 80% of its historical range and is currently found in only two isolated fragments of the Arkansas River basin (Wilde 2002). Within its native

range, it is apparently now almost entirely restricted to the Canadian River in New Mexico, the Texas panhandle, and Oklahoma between the Texas state line and Lake Eufaula (Pigg et al. 1999; Bonner 2000). One record was documented by Cross (1970) in the Red River drainage of Oklahoma, and a nonnative population introduced into the Pecos River, New Mexico, in 1978 has become established in that river (Hoagstrom and Brooks 2005). *Notropis girardi* was probably historically one of the rarest fishes in Arkansas. Hubbs and Ortenburger (1929b) collected 10 adults from the Arkansas River in Oklahoma, 5.5 miles southwest of Fort Smith, on 4 July 1927 and predicted that it would eventually be found in Arkansas. This species was subsequently taken only once in Arkansas over a sandbar in the Arkansas River at the mouth of Piney Creek, Logan County, on 23 July 1939 (UMMZ 128394, 7 specimens; Black 1940) (Map A4.67). Buchanan (1976) did not collect it in extensive sampling of the Arkansas River in 1974, and with the McClellan-Kerr Navigation System having altered the Arkansas River so drastically by 1971, this species probably has been extirpated from the state. It was thought to be extirpated in Kansas, but 6 specimens were found in the Arkansas River at Wichita in 1999. It occurred in only a few places in Oklahoma in the 1980s (Pigg 1991). Numerous fish collections from the Arkansas River at Fort Smith by TMB during the past 48 years have failed to produce any specimens of *N. girardi*.

HABITAT AND BIOLOGY   The Arkansas River Shiner is found in the main channels of broad, sandy-bottomed streams of the Arkansas River drainage (Miller and Robison 2004). Cross (1967) described its optimum habitat as follows:

> In those streams (where it occurs), the steady, shallow flow forms series of unstable sand-ridges, analogous to dunes in a wind-swept terrestrial environment. Arkansas River Shiners commonly lie in the lee of these transverse sand-ridges, face into the current, and feed on organisms that are exposed by movement of the sand or are washed downstream. *N. girardi* is uncommon in quiet pools or backwater, and almost never enters tributaries having deep water and bottoms of mud or stone.

Bonner (2000) also found that *N. girardi* typically occurred at shallow depths (seasonal means from 17.9 to 20.7 cm) and in slow current (30.9–40.1 cm/s). Bonner also reported that the main channel of the Canadian River was the primary habitat location for Arkansas River Shiners throughout the year. Matthews and Hill (1980) reported that *N. girardi* in the South Canadian River of central

Oklahoma showed adaptability in habitat use, changing microhabitats as environmental conditions changed. Polivka (1999) described this shiner as wide-ranging and a species that does not demonstrate a substantial or consistent level of habitat selection. In the South Canadian River, numbers were greatest in bank, island, and sand ridge habitats. Greater numbers of individuals occurred in microhabitats defined by depth down to 50 cm (19.7 in) and current speeds of 0–50 cm/s, although faster current speeds were used with high frequency. In the summer, selection of deeper water was apparent with juveniles found in shallow, slow-flowing backwater habitat types more than adults. Habitats adjacent to sand ridges were also important to juveniles and adults.

Wilde et al. (2001) studied food habits of *N. girardi* in the Canadian River of New Mexico and Texas, and the following information comes from that source. It is a generalist feeder, with terrestrial and aquatic insects making up 28% of the diet by weight. Detritus, plant materials, and silt represented 26%, 6%, and 40%, respectively. Feeding occurred over the substrate, but the presence of terrestrial coleopterans and hymenopterans indicated that it also fed in the water column on drifting invertebrates. About 76% of the shiners had food in their digestive tracts, and empty tracts were encountered more frequently in the fall and winter than in spring and summer. Aquatic insects ingested included coleopterans, megalopterans, odonates, plecopterans, trichopterans, and dipterans.

Platania and Altenbach (1998) characterized *N. girardi* as a pelagic broadcast spawner with nonadhesive, semibuoyant eggs. Pelagic-spawning cyprinids dispense gametes into pelagic zones of flowing streams. Immediately following spawning, water enters cell membranes osmotically and causes eggs to swell and become semibuoyant (slightly negatively buoyant) (Bottrell et al. 1964). The semibuoyant eggs remain suspended within the water column at current velocities above 0.01 m/s, drift for 24–48 hours before hatching, and then drift for an additional 2–3 days as developing larvae. During the drift period, individuals presumably become displaced great distances downstream (140 km) from parent localities before complete development of a swim bladder and the onset of exogenous feeding (allowing larval individuals to exit the drift) (Moore 1944; Platania and Altenbach 1998). The extent to which larval individuals continue to drift is largely unknown (Durham and Wilde 2008a). Therefore, large river fragments (100 km) are required by drifting eggs and larvae (collectively referred to as ichthyoplankton, *sensu* Dudely and Platania 2007) to allow time to develop before being deposited in impounded downstream habitats. Downstream transport

is of particular concern because high mortality rates occur among ichthyoplankton deposited within downstream reservoirs, due to suffocation within anoxic sediments or predation from lacustrine species (Platania and Altenbach 1998; Dudely and Platania 2007). Cross (1967) reported that adults of *N. girardi* in Kansas are ripe by June, but spawning does not occur until the rivers are filled by runoff from heavy rains. Moore (1944) believed that spawning occurred in July, usually following heavy rains, in the main stream channel. This species appeared to be in peak reproductive condition from May through July (Moore 1944; Polivka and Matthews 1997; Wilde 2005). This broadcast spawner has multiple asynchronous spawns in a single season (Bonner 2000; Wilde 2005; Bestgen et al. 1989). Platania and Altenbach (1998) suggested that this species spawns in the upper-water column during high flows, which would allow the eggs to remain in suspension during the 10–30-minute nonbuoyant period that occurs immediately after fertilization and before the filling of the perivitelline space with water (water hardening). Estimated total number of well-developed ova ranged from 1,012 to 3,246 in females 38–49 mm SL, with egg diameters of 0.68–0.70 mm (Cross et al. 1983). Once filled with water, the semibuoyant egg should remain suspended in the water column if current is present. If there is no flow, the eggs sink to the bottom, where they may be covered with silt and die. Durham and Wilde (2008b) found that *N. girardi* made up 41% of the larval fish drift in the Canadian River, Texas, in 2000, and 59% in 2001, with greatest drift densities occurring in June in both years. Moore (1944) also reported that hatching occurs and the larvae are capable of swimming within 3 or 4 days after spawning. When larvae are capable of horizontal movement, they move out of the main channels where food is scarce. Very small young are frequently taken from backwater pools and the quiet water at the mouths of tributaries where the deep mud affords a rich bottom fauna and where plankton is more abundant (Moore 1944). Three age groups were present in the Canadian River, Texas, with few individuals living past 2 years (Bonner 2000). In Oklahoma, Moore (1944) reported *N. girardi* lives at least 3 years. This species is best adapted for plains streams having great variability of flow, which flood unpredictably in the warm months and are almost intermittent at other times. This definitely does not describe the environment of the Arkansas River in Arkansas since the construction of the Arkansas River Navigation System. While there is still high water in the Arkansas River almost every spring, the great extremes in level seen before the construction of the navigation system do not occur today, and extremely low flows never occur (a navigation channel 9 feet (2.7 m) deep is maintained, and there are plans to increase the navigable depth to 14 feet). These altered conditions coupled with the construction of large reservoirs upstream on the Arkansas River system in Oklahoma (such as Lake Eufaula) make it very unlikely that a reproducing population of *N. girardi* could now exist in Arkansas.

CONSERVATION STATUS   The Arkansas River Shiner has been extirpated from most of its former range in Oklahoma and Kansas (Pigg 1999). Cross et al. (1983) provided evidence that reduced stream flows were largely responsible for the decline of *N. girardi* in Kansas but did not discount the possibility that displacement by the introduced *N. bairdi* may have played a secondary role in its decline. The U.S. Fish and Wildlife Service listed the Arkansas River Shiner (*Notropis girardi*) as Threatened under the Endangered Species Act in November 1998 (USFWS 1998). In Arkansas, this federally threatened shiner has disappeared from the Arkansas River and should be considered extirpated from the state.

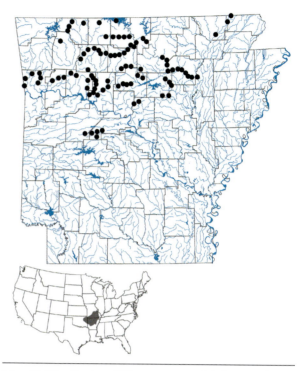

Map 68. Distribution of *Notropis greenei*, Wedgespot Shiner, 1988–2019.

Notropis greenei Hubbs and Ortenburger
Wedgespot Shiner

Figure 15.82. *Notropis greenei*, Wedgespot Shiner. *David A. Neely.*

CHARACTERS   A slender, silvery minnow with a moderately large head, head length going into SL 3.8 times. Eyes large and upwardly directed; each eye usually with a nipplelike projection extending from the front edge of the dark pupil. Eye diameter going into head length about 2.8 times. Mouth large, subterminal, and slightly oblique (Fig. 15.82). Caudal fin base with a conspicuous black wedge-shaped spot. Dorsal fin origin over pelvic fin origin; dorsal rays 8. Anal rays 8 (sometimes 9), pharyngeal teeth 2,4–4,2. Lateral line complete, with 35–38 scales; lateral line scales are large, but not elevated. Front of dorsal fin base equidistant between snout tip and caudal base. Intestine short; peritoneum silvery, but sprinkled (sometimes thickly) with dark speckles. Breeding males with small tubercles on head, front of body, and along pectoral fin rays. Maximum size usually 2.5–3.0 inches (64–76 mm). Largest Arkansas

specimen was 3.3 inches (84 mm) TL from the Mulberry River.

LIFE COLORS   Body color olive dorsally, sides silver, belly white. Scales on upper three-fourths of body outlined with pigment. Lateral band dusky, fading anteriad. Lateral line pores outlined by melanophores anteriorly. Dusky middorsal stripe is wide in front of dorsal fin and thin behind dorsal fin. Wedge-shaped spot at caudal base usually dark and prominent, but occasionally faint and poorly defined.

SIMILAR SPECIES   In Arkansas, *Notropis greenei* is often found with *N. boops*, *N. nubilus*, *Luxilus pilsbryi*, *L. cardinalis*, *N. percobromus*, and *N. telescopus* and can be distinguished from those species by its distinctly wedge-shaped spot at the base of the caudal fin. *Notropis greenei* further differs from *N. boops* and *N. nubilus* in having a silvery peritoneum and 2,4–4,2 pharyngeal teeth (vs. black peritoneum and teeth 1,4–4,1 and 0,4–4,0, respectively). It differs from *Luxilus pilsbryi*, *L. cardinalis*, and *N. telescopus* in having 8 anal rays and a silvery peritoneum (vs. 9 or more anal rays and a black peritoneum), and from *N. percobromus* in having 8 anal rays and front of dorsal fin base equidistant between snout tip and caudal fin base (vs. 9 or 10 anal rays and front of dorsal fin base closer to caudal base than to tip of snout).

VARIATION AND TAXONOMY Hubbs and Ortenburger (1929b) described *Notropis greenei* from the Elk River (tributary to Neosho River), 11.3 km north of Grove, Delaware County, Oklahoma. There has been no definitive systematic study of *Notropis greenei*, and systematic relationships are uncertain.

DISTRIBUTION The Wedgespot Shiner occurs in only three states: Arkansas, Missouri, and Oklahoma, where it is found in Ozark Upland tributaries of the Mississippi, Missouri, White, and Arkansas rivers. In Arkansas, this shiner is relatively common, though scattered, throughout the Ozark Mountains ecoregion in White River and north bank Arkansas River tributaries. It is less common in south bank Arkansas River tributaries but occurs in the upper and lower Fourche LaFave River in the Ouachita Mountains (Map 68). Although known from the Illinois River (Arkansas River drainage) in Oklahoma, there are surpisingly no records from the Illinois River in Arkansas.

HABITAT AND BIOLOGY The Wedgespot Shiner inhabits pools and runs of clear, medium-sized upland streams. It typically avoids small headwater creeks and is intolerant of reservoir conditions. This shiner occupies pools with moderate current over firm, gravel bottoms or areas near riffles, where it schools in midwater, often with other shiners. We have frequently found it in runs below riffles in Lee Creek and Clear Creek in Crawford County, where it is often found with *Campostoma spadiceum*, *Luxilus cardinalis*, and *Notropis boops*. Felley and Hill (1983) also noted that *N. greenei* was a characteristic species of the fastest currents in the Illinois River, Oklahoma.

Notropis greenei is an insectivore during spring, summer, and into early fall, feeding almost entirely on immature aquatic stages, supplemented by small numbers of terrestrial insects and sometimes other aquatic invertebrates. Gut contents of 86 specimens collected from April into September from Crawford, Franklin, Perry, Searcy, and Van Buren counties were examined by TMB. The diet was dominated by percent frequency of occurrence and percent volume in all months and at all localities by mid- and late-instar stages of ephemeropterans. Other aquatic stages of insects consumed in smaller amounts included plecopterans, coleopterans, simuliids, hemipterans, and midges. Approximately 8% of the individuals examined had ingested terrestrial insects, primarily ants and winged dipterans. Other items found in only a single specimen each included invertebrate eggs, isopods, sand, and filamentous algae.

Pflieger (1997) reported that adults in spawning condition have been collected in Missouri from late May to late August with spawning occurring over gravel riffles in swift current. Moore and Paden (1950) reported individuals in spawning condition on August 15 in Oklahoma and hypothesized that *N. greenei* has a long summer breeding period. Those observations and hypotheses do not agree with our reproductive information for Arkansas populations. Based on examination of gonads of specimens collected in multiple years from April through September by TMB, the reproductive season begins in Arkansas by late April, peaks from mid-May to mid-June, and is over by early July. Ripe females collected during those months ranged from 46 to 78 mm TL and contained between 71 and 447 maturing to ripe ova. Atretic eggs began to appear in the ovaries by late June, and ovaries and testes were atrophied by mid-July. Specimens in reproductive condition in August reported by Moore and Paden (1950) and Pflieger (1997) may have been age 0 fish that had attained sexual maturity in their first year.

CONSERVATION STATUS Although reservoir construction has reduced its range, the Wedgespot Shiner is still a common inhabitant of upland streams in Arkansas. Page and Burr (2011) reported that it is fairly common but declining in some areas. It is widely distributed in Arkansas, and even though it does not occur in large numbers at any locality, we believe its populations are secure. Future monitoring of populations in Arkansas is recommended.

Notropis maculatus (Hay)
Taillight Shiner

Figure 15.83. *Notropis maculatus*, Taillight Shiner. David A. Neely.

CHARACTERS A small, slender shiner (body depth going into SL more than 5 times) with a small head, moderately large eyes, a rounded snout that extends slightly beyond the small, slightly oblique mouth, and a prominent round, black spot at caudal fin base (Fig. 15.83). Caudal fin deeply forked. Mouth does not extend posteriorly to anterior margin of eye. Dorsal fin origin behind pelvic fin origin; front of dorsal fin base slightly closer to tip of snout than to caudal fin base. Dorsal and anal fins falcate. Dorsal rays 8; anal rays 8. Pectoral rays 13–15. Pharyngeal

Map 69. Distribution of *Notropis maculatus*, Taillight Shiner, 1988–2019.

teeth 0,4–4,0. Lateral line incomplete and short, with only 8–10 pored scales anteriorly; lateral scale rows 34–39. Predorsal scale rows 15–17. Breast and belly scaled. Intestine short, peritoneum silvery, usually with a few scattered dark speckles. Breeding males with small tubercles on head and upper surfaces of pectoral fins. Maximum size 2.8 inches (72 mm). Largest Arkansas specimen examined was 2.6 inches (65 mm) from Calion Lake.

LIFE COLORS   Important diagnostic characteristics are pigmentary, including a dark lateral stripe extending from the snout to the caudal peduncle terminating in a large, black basicaudal spot. Stripe on snout broken, tip of snout light-colored. Clear areas just above and below the caudal spot followed by small dark triangles on the margins of the caudal fin adjacent to the end of the caudal peduncle. Dorsolateral scales outlined by melanophores, creating a crosshatched pattern on back and sides. A black, vertically elongated blotch (most prominent in males) along front of dorsal fin. Breeding males with red to pink body and with red pigment on snout and iris. Median and pelvic fins red with clear tips and bases. In nonbreeding fish, red coloration is restricted to the base of the caudal fin, resembling a small taillight, and giving rise to the common name for this species.

SIMILAR SPECIES   The Taillight Shiner resembles *Notropis atrocaudalis* in having a round, black spot at the caudal fin base, but differs from that species in having 8 anal rays (vs. 7) and in having an incomplete lateral line with only 8–10 pored scales (vs. lateral line complete). It differs from *N. chalybaeus* in having a distinctly round caudal spot (vs. caudal spot irregularly shaped), in lacking dark pigment inside of mouth (vs. dark pigment present), and in having 0,4–4,0 pharyngeal teeth (vs. 2,4–4,2). It occurs syntopically with *Opsopoeodus emiliae*, but differs from that species in having a horizontal or slightly upturned mouth (vs. mouth very oblique, almost vertical), 0,4–4,0 pharyngeal teeth (vs. 0,5–5,0), and 8 anal rays (vs. 9).

VARIATION AND TAXONOMY   *Notropis maculatus* was described by Hay (1881) as *Hemitremia maculata* from the Chickasawhay River and tributaries, near Enterprise, Clarke County, Mississippi. A morphological analysis by Gilbert and Bailey (1962) suggested a close relationship between *N. maculatus* and *Opsopoeodus emiliae*. Mayden (1989) aligned *O. emiliae* and *N. maculatus* with the *N. volucellus* species group. A molecular analysis by Simons et al. (2003) recovered *Notropis maculatus* as sister to a clade containing *N. asperifrons* and *N. chrosomus*.

DISTRIBUTION   Primarily a southeastern species of the Atlantic and Gulf coastal regions and the lower Mississippi River basin, the Taillight Shiner occurs on the Coastal Plain from the Cape Fear River, North Carolina, to the Red River drainage, Texas, and northward in the Mississippi Embayment to southern Illinois. In Arkansas, it occupies the low-gradient streams below the Fall Line in the Coastal Plain where it is historically known from all rivers (Map 69).

HABITAT AND BIOLOGY   The Taillight Shiner occupies shallow, tannin-stained waters of low-gradient streams, sloughs, and lakes, including oxbows and swamps of the Coastal Plain, particularly in less disturbed portions (Robison 1978a). It is often found in the oxbow lakes associated with large rivers. Although it was not found in the mainstem Red River (Buchanan et al. 2003), it was commonly found in Red River oxbows (unpublished data, TMB). It is common in some oxbow lakes and associated vegetated borrow pits of the Mississippi River. In Louisiana, Douglas (1974) reported this species from quiet bayous, oxbows, and large lakes, but less often in rivers and streams. In Arkansas, it is most commonly found in backwater areas over substrates of decomposing vegetation, silt, and soft mud. Gravel substrates are generally avoided. It typically inhabits acidic waters with pH readings of 6.1–6.9 (Robison 1978a). Buchanan (2005) found it in eight Arkansas impoundments, sometimes in large numbers. *Notropis maculatus* is a midwater schooling species, often found in or near luxuriant aquatic vegetation

(*Myriophyllum, Nuphar, Typha*). Many of the lowland streams where *N. maculatus* occurs have a moderate to heavy canopy of bottomland hardwoods; however, it also frequents areas in the middle of oxbow lakes away from any cover where water temperatures are often higher than those in vegetated areas near the shore (Robison 1978a). Streams of the Ouachita River basin in southern Arkansas where *N. maculatus* occurred were characterized by the following physicochemical features: pH 6.1–6.9, dissolved oxygen 5.3–7.0 mg/L, total solids 105–162 mg/L, dissolved solids 97–141 mg/L, and turbidity 25–130 JTU.

Foods consist of algae, rotifers, cladocerans, ostracods, and insects (Cowell and Barnett 1974; Beach 1974). Individuals smaller than 40 mm TL feed mainly on cladocerans, while larger fish feed mainly on copepods, ostracods, chironomids, and other midge larvae. Large amounts of algae were consumed in summer and fall. TMB examined gut contents of 61 specimens from Clark, Crittenden, Hempstead, St. Francis, and Union counties in Arkansas and found that it fed mainly on midge larvae (chironomids and others) and microcrustaceans (copepods and cladocerans). Gut contents also included detritus, some filamentous algae, and other small, immature insects (coleopterans, odonates, and hemipterans).

In Arkansas, males in breeding color have been taken from April to mid-June. Females in Arkansas populations are larger than males. HWR collected three nuptial males in Calion Lake, Union County, Arkansas, on 11 June (Connior et al. 2011). Examination of gonads by TMB also supported a long breeding season in Arkansas, extending from mid-April into at least mid-July. Ripe males and females were collected as early as 14 April in Crittenden County, and as late as 23 July in Union County (but most females on 23 July contained regressed ovaries with atretic eggs). In Alabama, spawning occurs from March through June, and in Kentucky, Burr and Page (1975) indicated a March to May spawning period with young-of-the-year collected in April. Females produced up to 450 eggs (average 246 mature ova per female). Fecundity in Arkansas specimens examined by TMB was highest in April, with females containing between 156 and 550 ($\bar{x}$ = 293) ripe ova, decreased through June, and was lowest near the end of the spawning season in July (an average of 120 ripe ova per female). In Florida, the breeding season was more protracted, with ripe males and females found from March to early October at water temperatures of 23–32°C (73.4–89.6°F) (Beach 1974). In that study, the demersal, adhesive eggs hatched in 60–72 hours at 23°C (73.4°F). Chew (1974) reported this species to be a broadcast spawner and a nest

associate of other fishes such as *Micropterus salmoides*. Cowell and Barnett (1974) reported that *N. maculatus* in Florida lives only 1 year. Sexual maturity is reached in 6–9 months at about 40 mm (1.6 inches) TL. Maximum life span in Kentucky also is short, apparently less than 2 years (Burr and Page 1975).

CONSERVATION STATUS The Taillight Shiner is a widespread but uncommon fish in the Coastal Plain streams of Arkansas. Its populations are currently secure.

Notropis nubilus (Forbes)
Ozark Minnow

Figure 15.84. *Notropis nubilus*, Ozark Minnow. *Uland Thomas.*

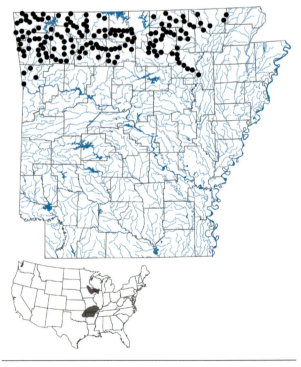

Map 70. Distribution of *Notropis nubilus*, Ozark Minnow, 1988–2019.

CHARACTERS   A slim, terete minnow with a moderately large eye (eye diameter going into head length 3.4–3.9 times and equal to or greater than snout length), a small head, and a subterminal, slightly oblique mouth (Fig. 15.84). Corner of upper lip slightly swollen. Barbels absent. Snout blunt, extending beyond mouth. Pharyngeal teeth 0,4–4,0. Most distinctive features include the long, coiled intestine (at least twice the length of the body) and black peritoneum. Lateral line complete, nearly straight, with 33–37 scales. Lateral line margined above and below with dark dashes and a dusky silvery or dark lateral stripe prominent. Dorsal fin origin over pelvic fin origin; front of dorsal fin base approximately equidistant between snout tip and caudal base. Dorsal rays 8. Anal rays 8. Pelvic rays usually 8. Breeding males tuberculate, with tubercles on head, body, and all fins except caudal. Breeding females somewhat tuberculate. Maximum length about 3.7 inches (94 mm) from Washington County, Arkansas (Becker 1983).

LIFE COLORS   Body color olive yellow, belly silvery white. Sides silvery with a prominent dark lateral band from snout to caudal peduncle; lateral band often obscuring lateral line pigmentation. Tiny wedge-shaped basicaudal spot usually evident. Upper and lower lips pigmented. Dorsum with prominent golden middorsal stripe (dusky in preserved specimens). Dorsolateral scales outlined in, and sometimes sprinkled with, dark pigment. Fins colorless. In breeding males, the lower half of the body is yellow or pinkish orange, underside of head orange, and all fins become pink to reddish orange. M. Cashner (pers. comm.) noted that the fins of *N. nubilus* remain transparent with light washes of color usually not extending beyond the basal half. Females somewhat yellow when spawning.

SIMILAR SPECIES   *Notropis nubilus* resembles *Notropis boops*, *Luxilus pilsbryi*, *L. cardinalis*, and *L. zonatus*. It schools with those species and can be distinguished from them by its long, coiled intestine and its 0,4–4,0 pharyngeal teeth (vs. short intestine and pharyngeal teeth 1,4–4,1 or 2,4–4,2). The long, coiled intestine also separates *N. nubilus* from all other species of *Notropis* in Arkansas.

VARIATION AND TAXONOMY   *Notropis nubilus* was described by Forbes *in* Jordan (1878a) as *Alburnops nubilus* from the Rock River at Oregon, Ogle County, Illinois. The Ozark Minnow was formerly placed in the genera *Hybognathus* and *Dionda*, primarily because of its elongate intestine and black peritoneum. Swift (1970) allied *Dionda nubila* with other members of *Notropis* under the subgenus *Hydrophlox* based on breeding coloration, tuberculation on the body and head, large uniserial tubercles on pectoral rays, scalloped dorsolateral scales, low circumferential

body scale count, and a sharp predorsal line; however, gut morphology, breeding biology, and distribution apparently did not support this placement (Glazier and Taber 1980; Fowler et al. 1984). Fowler et al. (1984) observed that a similarity of spawning activity and egg characters between the Ozark Minnow and *Dionda episcopa* along with similar gut morphology suggested that these species may be more closely related to each other than either is to *Notropis*. In addition, the pharyngeal teeth of the Ozark Minnow are definitely not *Notropis*-like. Schönhuth and Mayden (2010) included *N. nubilus* in their analysis of *Cyprinella*, but placement of *N. nubilus* within *Hydrophlox* was not tested. Cashner et al. (2011) tested the monophyly of eight members of the subgenus *Hydrophlox* with one mtDNA and three nuclear DNA markers. They found *Hydrophlox sensu lato* was not monophyletic, but the monophyly of a core clade composed of five species (including the type species) was well supported. Their data indicated *Notropis nubilus* was not a member of *Hydrophlox*, with *N. nubilus* being basal to all *Notropis* included in their study. Cashner et al. (2011) redefined *Hydrophlox* to include five nominal taxa: *N. rubricroceus*, *N. lutipinnis*, *N. chlorocephalus*, *N. chiliticus*, and *N. chrosomus* and concluded that morphological and behavioral variations also supported the removal of *N. nubilus* from *Hydrophlox*. We follow Page et al. (2013) in continued placement of the Ozark Minnow in *Notropis*.

DISTRIBUTION   *Notropis nubilus* has a disjunct distribution in the Mississippi River basin, occurring in the upper Red Cedar River, Wisconsin, Mississippi River tributaries from southeastern Minnesota through northeastern Iowa to northern Illinois, and in Ozark Upland drainages (Mississippi, Missouri, White, and Arkansas river drainages) of Missouri, Kansas, Oklahoma, and Arkansas. In Arkansas, the Ozark Minnow is abundant in upland streams of the White River system across north central Arkansas and in the Arkansas River system in Benton and Washington counties (Illinois-Neosho drainage). It is also currently known from two Arkansas River Valley tributaries in Crawford County (Lee Creek and Clear Creek) (Map 70). A single pre-1900 record exists from Illinois Bayou in Pope County (Black 1940) (Map A4.70). In Arkansas, as in Missouri (Pflieger 1997), its distribution parallels those of *Luxilus cardinalis*, *L. pilsbryi*, and *L. zonatus*, and it is invariably found with one of those species. Berendzen et al. (2010) used analysis of genetic variation to test hypotheses on the origin of the current disjunct distribution of *N. nubilus*. They found evidence of a northern expansion into the Paleozoic Plateau from a Southern Ozarkian refugium during the Late Pleistocene. This advance was later followed by subsequent extirpation of intervening populations caused by loss of suitable habitat, producing the current disjunct distributional pattern.

HABITAT AND BIOLOGY The Ozark Minnow is found in clear, spring-fed streams of the Ozarks over gravel and rubble substrates, where it is most abundant in pools usually having some current. It is also commonly found in backwater areas near riffles. Although it occurs in streams of various sizes, it is most abundant in moderate-sized creeks and small rivers with permanent flow. It is intolerant of siltation and turbidity, and it is also intolerant of reservoir environments.

Foods of the Ozark Minnow in Wisconsin included microscopic green algae, blue-green algae, diatoms, and other plant material (Becker 1983). Cahn (1927) reported that digestive tract contents consisted almost entirely of algae (*Spirogyra*, *Zygnema*, and *Closterium*), with occasional entomostracans and small insect larvae. Attempts to maintain Ozark Minnows in a laboratory aquarium on a diet of dry meal, *Daphnia*, and chironomid larvae were unsuccessful (Becker 1983). The fish became progressively thinner, lost their color completely, and began dying off one or two per day. In Spring Creek, Oklahoma, Ozark Minnows fed primarily on plant material and detritus, and on minor amounts of small invertebrates (McNeely 1987). *Notropis nubilus* commonly eats algae in the Ozark streams of Arkansas (Matthews et al. 1986; HWR, pers. obs.).

The onset of spawning in *N. nubilus* is closely associated with water temperature (Fowler et al. 1984). Glazier and Taber (1980) found evidence for spawning in Missouri from late April to early June; YOY were first collected in May. Spawning occurs in Arkansas from April to early July. This species often spawns over the nests of *Nocomis biguttatus*. Spawning aggregations of several hundred individuals form when water temperature reaches about 62.6°F (17°C) (Fowler et al. 1984). Spawning occurs at a water temperature of 62.6–64.4°F (17–18°C) at depths of 0.4–3.9 inches (1–10 cm) over riffles with a fine gravel bottom. Current may be slow to moderately swift. No site preparation was noted. Actual spawning commenced when 2–10 fish approached the spawning site and remained there for a few seconds until approached by 10–20 additional fish from the aggregation. All fish then began vigorous vibrating motions against the substrate and each other. Spawning acts last about 1 minute. Females greatly outnumber males. Glazier and Taber (1980) postulated a life span of less than 30 months for this species. Becker (1983) reported a 3.7-inch (93 mm) 4-year-old female Ozark Minnow from Washington County, Arkansas.

CONSERVATION STATUS The Ozark Minnow is widespread and abundant across the northern portion of the state and appears in no immediate danger.

Notropis ortenburgeri Hubbs
Kiamichi Shiner

Figure 15.85. *Notropis ortenburgeri*, Kiamichi Shiner. *David A. Neely.*

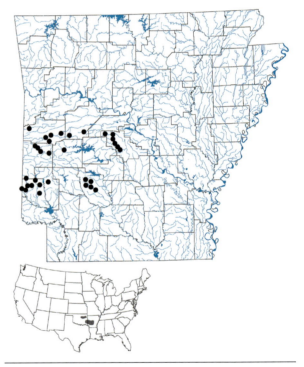

Map 71. Distribution of *Notropis ortenburgeri*, Kiamichi Shiner, 1988–2019.

CHARACTERS A small, slim, silvery shiner with a small head (head length going into SL 4.0–4.3 times), large eye (eye diameter going about 2.7 times into head length), dark lateral band, and extremely oblique mouth (Fig. 15.85). Well-developed pigment on the chin an important diagnostic feature (Fig. 15.86). Upper jaw extending backward nearly to the front of the eye; upper lip slightly above the level of the center of the pupil. Greatest body depth occurring at dorsal fin origin and strongly tapering to a narrow

Figure 15.86. *Notropis ortenburgeri* with inset showing chin pigmentation. *Garold W. Sneegas.*

caudal peduncle. Dorsal rays 8; anal rays 9–11 (usually 9 or 10). Pharyngeal teeth 0,4–4,0. Lateral line complete or nearly so, usually with 35–37 (32–37) scales. Usually 13 predorsal scales. Dorsal fin origin slightly behind pelvic fin origin, nearer to snout tip than to caudal base. Intestine short, peritoneum silvery. Gill rakers long and slender, the longest extending about to the base of the second raker below; gill rakers 12–15. Maximum size about 2.3 inches (58 mm).

LIFE COLORS   Dorsum olive, sides silvery with a relatively well-developed black lateral band extending from eye around snout to caudal peduncle, where it is perceptibly wider at its terminus. Chin darkly pigmented. No caudal spot. Scales on back dark-edged. Fins colorless.

SIMILAR SPECIES   Although superficially similar to *N. boops, N. ortenburgeri* differs in having a 0,4–4,0 tooth count (vs. 1,4–4,1 in *N. boops*), an anal ray count of 9 or 10 (vs. 8); a more oblique mouth; and a conspicuous black patch on tip of chin (absent in *N. boops*). It differs from the allopatric *Notropis texanus* in having 0,4–4,0 pharyngeal teeth (vs. 2,4–4,2) and 9 or 10 anal rays (vs. 7). It differs from *N. atrocaudalis* in having 9 or 10 anal rays (vs. 7) and in lacking a distinct basicaudal spot (vs. small, distinct spot present). It can be distinguished from *N. chalybaeus* by its 0,4–4,0 pharyngeal teeth (vs. 2,4–4,2), from *Pteronotropis hubbsi* by its 8 dorsal rays (vs. 9 or 10) and a complete lateral line (vs. incomplete, with only 2–9 pored scales), and from *Opsopoeodus emiliae* by its 8 dorsal rays (vs. 9) and 0,4–4,0 pharyngeal teeth (vs. 0,5–5,0).

VARIATION AND TAXONOMY   *Notropis ortenburgeri* was described by C. L. Hubbs (*in* Ortenburger and Hubbs 1927) from the Mountain Fork River, 16 km southeast of Broken Bow, McCurtain County, Oklahoma. There has been no definitive systematic study of *Notropis*

ortenburgeri, and systematic relationships are uncertain. Suttkus and Bailey (1990) considered *N. ortenburgeri* and *N. melanostomus* to be sister species.

DISTRIBUTION   *Notropis ortenburgeri* is found only in Arkansas and Oklahoma in the Ouachita, Red, and Arkansas river drainages. Robison (2005a) reviewed the historical distribution of the Kiamichi Shiner throughout its range. In Arkansas, the Kiamichi Shiner is confined to western Arkansas south of the Arkansas River in the Poteau, Fourche LaFave, Petit Jean, Ouachita, upper Saline, and Little river systems (Robison 1980a, 2001, 2005a) (Map 71). It is most common in the upper Ouachita River and two of its tributaries, the Saline River and Terre Noire Creek. North of the Ouachita River drainage, this species is rare in the Arkansas River drainage, and to the southwest it is also less common.

HABITAT AND BIOLOGY   The Kiamichi Shiner is rare in Arkansas and inhabits pools over gravel, rubble, or boulder-strewn substrates in small to moderate-sized clear upland streams of moderate gradient (Robison 2005a). Black (1940) considered its habitat similar to that of *Notropis greenei*, stating that it favors the ends of pools near the beginning or end of the riffles where food is no doubt easily secured. Beds of *Justicia americana* tend to predominate along the edges of pool areas and often there is a slight flow through the pools. *Notropis ortenburgeri* seems intolerant of turbidity and ecological perturbations caused by ditching in agricultural areas and modification of the watershed by clear-cutting (Robison 2005a). It is also apparently intolerant of reservoirs. McAllister et al. (2013) reported blackspot disease in the Kiamichi Shiner in Arkansas.

There is little information on the life history of the Kiamichi Shiner. TMB examined gut contents of 20 individuals collected from Perry and Scott counties in May, August, and October. Based on this small sample size, *N. ortenburgeri* is primarily an insectivore. Most feeding activity took place on or near the substrate, but some insects (including winged terrestrial forms) were ingested at or near the water surface. Foods consisted mainly of immature insects, primarily ephemeropteran nymphs and dipteran larvae and pupae. Detritus, microcrustaceans, and diatoms were ingested in small amounts.

The reproductive biology of *N. ortenburgeri* is not known. TMB found ripe females and males with enlarged testes on 10 and 14 May in Scott County.

CONSERVATION STATUS   Robison (2005a) reviewed the conservation status of the Kiamichi Shiner in Oklahoma and Arkansas and recommended a status of threatened based on the lack of recent collections. We consider it threatened in Arkansas.

Notropis ozarcanus Meek
Ozark Shiner

Figure 15.87a. *Notropis ozarcanus*, Ozark Shiner. *Uland Thomas.*

Figure 15.87b. *Notropis ozarcanus* breeding male, Buffalo River, Searcy County, Arkansas. *Henry W. Robison.*

Map 72. Distribution of *Notropis ozarcanus*, Ozark Shiner, 1988–2019.

CHARACTERS   A slender, silvery shiner with large eyes, a rounded and blunt snout projecting slightly beyond upper lip, and a small, nearly horizontal mouth (Fig. 15.87). Body depth going into SL more than 5 times. Dorsal rays 8. Anal rays 8 (rarely 7 or 9). Pharyngeal teeth 0,4–4,0. Lateral line complete, with 34–38 scales, with anterior scales much deeper than those in adjacent areas; rear margins of anterior lateral line scales slightly scalloped. Barbels absent. Intestine short; peritoneum silvery with scattered dark speckles. Dorsal fin origin slightly behind pelvic fin origin and closer to caudal base than to tip of snout. Breeding males without tubercles, but first ray of pectoral fin is thickened and roughened. Maximum size 2.8 inches (71.1 mm).

LIFE COLORS   Dorsum coloration pale yellow with sides silver and belly silvery white. Scales on dorsum and upper side prominently dark-edged. Anterior lateral line scales outlined by paired dark dots or dashes. Breeding males with black fins. Dorsal fin with a small, dusky blotch near base of the first 2 or 3 rays. All other fin membranes often dusted with fine dark specks in adults.

SIMILAR SPECIES   The Ozark Shiner superficially resembles members of the genera *Erimystax* and *Hybopsis* because of its blunt, rounded snout, but it can easily be distinguished from all chub species by its lack of maxillary barbels. It differs from *Pimephales* species in lacking a broad, flattened back in front of the dorsal fin and in lacking tiny, crowded predorsal scales (vs. back flattened and with crowded predorsal scales). *Notropis ozarcanus* is sometimes found with *Hybopsis amblops*, *Notropis sabinae*, and *N. volucellus*, all of which have rather rounded snouts. It differs from *H. amblops* in having elevated anterior lateral line scales and pharyngeal teeth 0,4–4,0 (vs. 1,4–4,1); from *N. sabinae* in having 8 anal rays (vs. 7) and a large eye, eye diameter going into head length fewer than 3.5 times (vs. eye diameter into head length more than 3.5 times); and from *N. volucellus* in having front of dorsal fin base closer to base of caudal fin than to tip of snout (vs. front of dorsal fin base equidistant between tip of snout and caudal base), and in having dorsal fin with dark pigment on one or more membranes between fin rays, often forming a small dusky blotch near base of first 2 or 3 rays (vs. dorsal fin with dark speckles absent or confined to margins of fin rays).

VARIATION AND TAXONOMY   *Notropis ozarcanus* was described by Meek (1891) from the North Fork of the White River, south of Cabool, Missouri. In an analysis of the mitochondrial cytochrome *b* gene, Mayden et al. (2006a) recovered *N. ozarcanus* as sister to a clade composed of *N. buchanani* and *N. volucellus*.

DISTRIBUTION   Endemic to the Ozark Uplands of northern Arkansas and southern Missouri, the Ozark

Shiner occurs in the White River and Black River systems in Arkansas (Map 72) and was found disjunctly in the Illinois River (Arkansas River system) by Burr et al. (1979) (Map A4.72). It is extirpated from the upper St. Francis River drainage in Missouri (Page and Burr 2011). *Notropis ozarcanus* is most abundant in the Buffalo River in Arkansas and in the Current River in Missouri (Pflieger 1997). Robison (1997) found no records of this species from the upper White River above or below Beaver Reservoir. However, sampling in White River headwater tributaries in the early 2000s by biologists from the Vicksburg Waterways Experiment Station yielded several new records, including 250 specimens from the Middle Fork of the White River in Washington County in 2004 (data provided by K. J. Killgore). Other records of Ozark Shiners obtained from that sampling included the Kings River drainage (60 specimens from two sites in 2004) and Crooked Creek (1 specimen in 2004). Robison et al. (2013) collected 2 specimens from the Spring River at Many Islands, Fulton County. Despite these recent records, fragmentation of its small range by reservoirs poses a threat to continued survival of existing populations.

HABITAT AND BIOLOGY   The Ozark Shiner prefers high-gradient stream sections below riffles in slight to moderate current in large streams and rivers. Pflieger (1997) noted that it is commonly found in association with *Luxilus chrysocephalus*, *Cyprinella galactura*, *Luxilus pilsbryi*, *Notropis percobromus*, and *N. telescopus*. It is intolerant of reservoirs, and the large White River impoundments have undoubtedly reduced its range appreciably.

There is little information on food habits. TMB examined gut contents of 26 specimens collected from the Buffalo River in February, and 8 specimens collected from the Spring River in May. The diet in February was dominated by immature benthic insects. Ephemeropteran nymphs made up the bulk of the diet, and trichopteran larvae, dipteran pupae, odonate nymphs, and sand grains were also ingested. The diet in May consisted of more immature insects from the water column but included benthic forms as well.

Little is known about the reproductive biology of this midwater schooling species. Individuals in spawning condition have been found in the Buffalo River in Arkansas from late May to August. We collected ripe females from the Spring River on 17 May, and ripe ova averaged approximately 1 mm in diameter.

CONSERVATION STATUS   Due to loss of populations, small population size (Robison 1997), and range fragmentation, the Ozark Shiner is a species of special concern in Arkansas.

Notropis percobromus (Cope)
Carmine Shiner

Figure 15.88. *Notropis percobromus*, Carmine Shiner. *Uland Thomas.*

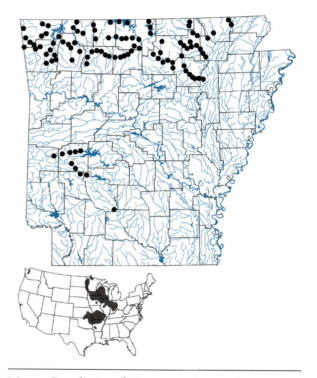

Map 73. Distribution of *Notropis percobromus*, Carmine Shiner, 1988–2019.

CHARACTERS   A very slender (body depth going into SL more than 4 times), somewhat compressed, silvery minnow with a long, sharply pointed snout (snout length longer than eye diameter), and a large terminal and slightly oblique mouth (Fig. 15.88). Barbels absent. Eye moderate in size, eye diameter going into head length 2.8–4.0 times. Anal rays typically 10 (occasionally 9); pelvic rays 8. Pharyngeal teeth, 2,4–4,2. Dorsal fin somewhat rounded with 8 rays. Dorsal fin origin posterior to pelvic fin origin and closer to the caudal base than to snout tip. Lateral line complete; lateral line scales 36–38, marked with dark dots anteriorly. Predorsal scale rows fewer than 25 (usually 17–21). Usually 25 or 26 scales around body at dorsal fin

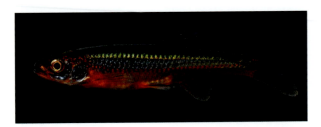

Figure 15.89. *Notropis percobromus* in breeding coloration in northeastern Oklahoma. *Fritz Rohde.*

origin. Scale rows around caudal peduncle usually 13 or 14. Intestine short, with a single S-shaped loop; peritoneum silvery but with many dark speckles. Breeding males with small tubercles present on head, body, and with a single tubercle per segment of the pectoral fin rays. Breeding females often with tubercles weakly developed. Maximum size about 3.5 inches (89 mm).

LIFE COLORS   Dorsum olive with sides silvery white with a thin iridescent green stripe above lateral line. Belly silvery white. Lateral band dusky. Anterior lateral line scales outlined by paired dark dots or dashes (most evident in preserved specimens). Chin not sprinkled with pigment, but lips are dark. Fins clear. Breeding males (Fig. 15.89) with red or orange pigment on head and body. Dark middorsal stripe narrow but well developed, especially posteriorly. Dorsolateral scales outlined by melanophores. Fins colorless. Base of dorsal fin pinkish in adults throughout the year.

SIMILAR SPECIES   *Notropis percobromus* is most similar to *Lythrurus umbratilis*, *Notropis atherinoides*, and *N. suttkusi*, and all four species are similar in having the dorsal fin origin posterior to the pelvic fin origin and closer to the caudal base than to the snout tip. The Carmine Shiner differs from *L. umbratilis* in having larger anterodorsal scales (fewer than 25) and absence of a dark spot at origin of dorsal fin (vs. 26 or more scales on midline of back in front of dorsal fin and usually with a dark spot at front base of dorsal fin in *L. umbratilis*). It can be distinguished from *N. atherinoides* by its darker lateral stripe, chin not sprinkled with dark pigment (vs. chin thickly sprinkled with dusky pigment), longer and more sharply pointed snout, dorsal fin short and somewhat rounded, its first principal ray slightly shorter than second and third (vs. dorsal fin rather high and pointed, its first principal ray longer than second and third), black edging of the lateral line pores extending past the middle of the body, and greater scattering of pigment on dorsolateral scales (vs. lateral line pores not outlined by dark pigment and pigment concentrated on scale margins in *N. atherinoides*). *Notropis percobromus* differs from *N. suttkusi* in having a larger snout and upper jaw; a more slender body; less dense pigmentation on scales, lips, and gular area; lateral line pigment outlined by melanophores rather than continuous with the lateral band; and the lateral line straight or decurved 1–2 scale rows (vs. decurved 5–6 scale rows in *suttkusi*).

VARIATION AND TAXONOMY   *Notropis percobromus* was described by Cope (1871) as *Alburnellus percobromus* from St. Joseph, Missouri. It was under the synonymy of *Notropis rubellus* for many years. Prior to 1978, it was often confused with *Notropis atherinoides*, and until 1994, *N. rubellus* was recognized as a single widespread species exhibiting geographic variation in morphological characters (Berendzen et al. 2008b). Gilbert (1978), Humphries and Cashner (1994), and Wood et al. (2002) discussed some of the troubled taxonomic history of this group. Wood et al. (2002) investigated the *N. rubellus* complex using phylogenetic analyses of 37 presumptive gene loci (allozyme analysis) in 33 populations from throughout its range. They concluded that the old polytypic *N. rubellus* should be divided into three distinct species, and they supported Humphries and Cashner's (1994) recognition of *N. suttkusi* as a fourth species of the *N. rubellus* complex. Berendzen et al. (2008b) also supported recognition of *N. suttkusi* and redefined the geographic distributions of *N. percobromus*, *N. rubellus*, and *N. micropteryx* in addition to recognizing three new clades that likely represented undescribed forms. The *Notropis rubellus* complex currently contains seven clades: *N. rubellus*, *N. percobromus*, *N. micropteryx*, *N. suttkusi*, and three undescribed forms allied with the subgenus *Notropis*. The true *Notropis rubellus* is no longer recognized in Arkansas but occurs in northeastern states from the Great Lakes and upper Ohio River drainages to Atlantic Slope drainages. Two members of the *Notropis rubellus* complex inhabit Arkansas, *N. percobromus* and *N. suttkusi* (and possibly a third undescribed form in the Ouachita River system).

In the upper Ouachita River in Arkansas, confusion reigned supreme regarding the classification of the *N. rubellus*-like form residing there. Black (1940) originally attributed the occurrence of the *rubellus*-like shiner in the Ouachita River to a bait bucket introduction. Humphries and Cashner (1994) were unable to confidently classify Ouachita River populations of the *N. rubellus* group. Some of the specimens from the Ouachita River were morphologically similar to *N. suttkusi* in some characters but more like *N. rubellus* in others. Humphries and Cashner (1994) suggested that the upper Ouachita River *Notropis percobromus* (formerly *N. rubellus*) may be of mixed ancestry involving *N. suttkusi* and *N. percobromus*, based on morphological characters and later on preliminary data from a genetic study by Wood et al. (2002). They suggested that either one or perhaps both species may have formerly existed in the Ouachita River and subsequently underwent introgressive hybridization. An alternative hypothesis was that one species may have been native but rare to the Ouachita River, and a recent bait-bucket introduction of the other species resulted in a breakdown of their genetic boundaries (Humphries and Cashner 1994). Wood et al. (2002) did not support either hypothesis of genetic mixing between species but found that *Notropis percobromus* from the upper Ouachita River was a diagnosable, divergent monophyletic clade that was consistently recognized as more closely related to *N. suttkusi* than to any populations of *N. percobromus*. Unique, derived alleles supported the independent evolution of Ouachita populations, and levels of heterozygosity did not reflect former genetic interchange between species. They considered *N. percobromus* from the Ouachita River substantially different from populations of *N. suttkusi* and other populations of *N. percobromus*. Wood et al. (2002) further suggested that the intermediate morphology referred to by Humphries and Cashner (1994) for Ouachita River *N. percobromus* was likely the result of a retained ancestral clinal variation in morphology and did not reflect recent introgression. Within the Interior Highlands clade, a sister group relationship was resolved between *N. percobromus* from the Ozarks and *N. percobromus* from the Ouachita River plus *N. suttkusi*. Based on analysis of the mitochondrial cytochrome *b* gene, Mayden et al. (2006a) found that *N. percobromus* was divided into two clades, with those clades forming a clade sister to *N. rubellus*. Berendzen et al. (2009) used traditional meristic and morphometric methods to test molecular hypotheses of cryptic diversity within the *N. rubellus* complex. They concluded that additional analyses are needed to determine whether the Ouachita River population is an undescribed species (Clade F in Berendzen et al. 2009), or is contained within the *N. suttkusi* clade. We herein recognize the upper Ouachita River populations as *N. percobromus*, pending resolution of their taxonomic status.

DISTRIBUTION As currently recognized, *Notropis percobromus* occurs in three disjunct areas: the Hudson Bay and upper Mississippi River drainages from Manitoba, Canada, south to central Illinois and central Indiana; Ozark Highland tributaries of the Mississippi, Missouri, White, and Arkansas rivers in southern Missouri, northern Arkansas, northeastern Oklahoma, and southeastern Kansas; and the Ouachita River system of southern Arkansas. In Arkansas, this shiner is widely distributed across the northern part of the state throughout the upper White River system and the Illinois River. It occurs

disjunctly in the Ouachita River system of southern Arkansas in the Ouachita River above Lake Ouachita, throughout the Caddo River, and one older record (Map A4.73) and one recent record from near the mouth of the Little Missouri River (Map 73).

HABITAT AND BIOLOGY The Carmine Shiner is an abundant, schooling inhabitant of clear streams with moderate to high gradient over gravel and rock bottoms, usually in pools with current or in areas just below riffles. In the Illinois River, Oklahoma, it was almost always found in flowing water (Felley and Hill 1983). It is not found in extreme headwaters and is usually found in water about 3 feet (0.9 m) deep. It is intolerant of siltation and turbidity. In Kansas, it prefers streambeds of limestone where the fish move in schools in the open water of pools, well above the stream bottom (Cross and Collins 1995). In Wisconsin, the Carmine Shiner occurred most frequently in moderate to swift riffles and pools in clear water at depths of 0.6–1.5 m over substrates of gravel (29% frequency), rubble (21%), sand (15%), mud (13%), silt (10%), clay (5%), boulders (4%), and bedrock (2%) (Becker 1983). Although not considered a lake-adapted species, it was found in small numbers in Bull Shoals and Norfork reservoirs (Buchanan 2005).

This schooling species feeds throughout the water column. Foods (probably located by sight) consist mainly of algae, diatoms, both aquatic and terrestrial insects, and other small invertebrates. Young Carmine Shiners feed largely on diatoms and other algae, but after the first year of life, the diet is dominated by insects (Pfeiffer 1955; Reed 1957). In Kansas, aquatic insects formed the major portion of the diet (Cross and Collins 1995), and in the Illinois River, Oklahoma, it was exclusively predaceous (Felley and Hill 1983). In one Wisconsin stream, foods consisted entirely of insects, primarily dipterans (culicids and chironomids), ephemeropterans, hemipterans, and trichopterans (Becker 1983).

In Arkansas this shiner spawns from mid-April to late June when water temperatures rise to 70°F (21.1°C) or higher (R. J. Miller 1964). *Notropis percobromus* characteristically lays its eggs over the gravel nests of other fishes, notably the Hornyhead Chub, Chestnut Lamprey, and the stoneroller minnows (*Campostoma* spp.). Spawning occurred in the Illinois River, Oklahoma, over riffles in large compact schools (Felley and Hill 1983). Cross and Collins (1995) also reported that schools spawn in riffles or eddies adjacent to gravel bars. The adhesive eggs fall to the bottom and become lodged in crevices in the gravel (Pflieger 1997). Hatching occurs in about 2.5 days. Because other cyprinids often spawn at the same time and use the same nests, hybrids are common. Life span is 3 years.

CONSERVATION STATUS The Carmine Shiner is widespread and locally abundant across northern Arkansas. It is less common in the Ouachita River drainage. We consider its populations secure.

Notropis perpallidus Hubbs and Black
Peppered Shiner

Figure 15.90. *Notropis perpallidus*, Peppered Shiner. *Uland Thomas.*

Map 74. Distribution of *Notropis perpallidus*, Peppered Shiner, 1988–2019.

CHARACTERS A small, pale, slender shiner with a small head (head length going 4.0–4.4 times into standard length), large eyes (eye diameter greater than snout length and going into head length 2.8 times), a pointed snout, and large melanophores widely scattered over dorsum and sides down to and occasionally below the lateral line (Fig. 15.90). Dorsal fin origin behind pelvic fin origin. Dorsal rays 8; anal rays 9 or 10; pectoral rays 12–14. Gill rakers moderately long, usually 8 or 9 on lower limb of first gill arch. Pharyngeal teeth usually 2,4–4,2 (65%), but count

somewhat variable, occasionally 1,4–4,1 (12%), 1,4–4,2 (15%), or 2,4–4,1 (4%). Scales large; lateral line complete, usually with 33 or 34 (32–35) scales. Predorsal scales usually 15–19. Circumferential scales usually 22–25. Scales around caudal peduncle usually 12 or 13. Infraorbital canal usually interrupted, but it is uninterrupted in about one-third of specimens. Breast and anterior portion of belly naked; body otherwise covered with thin, caducous scales. Swim bladder large and with two chambers. Intestine very short. Peritoneum silvery with a few scattered, large melanophores midventrally. Most diagnostic features are pigmentary. Breeding males tuberculate over head, dorsum, snout, chin, orbital areas, and pectoral fins; most scales on anterior third of body above lateral line margined along posterior edges with 2–4 minute tubercles. Breeding females without tubercles. One of the smallest species in its genus, maximum size about 2 inches (50 mm) TL.

LIFE COLORS Body color extremely pallid, almost translucent (milky white in preserved specimens), and no silvery guanine is visible. Scales on the dorsum and dorsolateral areas faintly outlined with tiny melanophores; aggregations of melanophores tend to form a diffuse mid-dorsal stripe. The common name is derived from very large, dark melanophores scattered on the dorsum, dorsolateral regions, and laterally, producing a "peppered" appearance. Concentrations of dark pigment are also present on top of head, snout, on both lips, chin, and along the anal fin base. Dorsal rays faintly bordered by dark pigment; anal and pelvic fins unpigmented. The distinctive caudal fin pigmentation was described by Snelson and Jenkins (1973) as follows:

> Upper and lower procurrent rays and first one or two principal rays of lower caudal lobe unpigmented. Melanin borders heaviest along outer rays of dorsal and ventral lobes. Faint pigment lines median rays of fin except for depigmented "window" over basal third of inner (lower-most) rays of upper lobe.

SIMILAR SPECIES Easily distinguished from all other Arkansas *Notropis* by its unique color pattern of huge melanophores randomly scattered over the dorsum and sides. The only other Arkansas cyprinid with large, scattered melanophores on the body is *Macrhybopsis hyostoma*. *Notropis perpallidus* can be distinguished from *M. hyostoma* by the absence of maxillary barbels (vs. barbels present), 9 or 10 anal rays (vs. 8 anal rays), and pharyngeal teeth in 2 rows (vs. teeth 0,4–4,0). Hubbs and Black (1940a) considered it superficially similar to *N. ortenburgeri*. In addition to color pattern, it differs from *N. ortenburgeri* in having pharyngeal teeth 2,4–4,2, 1,4–4,1, or combinations thereof (vs. pharyngeal teeth 0,4–4,0). It further differs from *N. texanus*, *N. boops*, and *N. chalybaeus* in having 9 or 10 anal rays (vs. 7 or 8).

VARIATION AND TAXONOMY   The type locality for *Notropis perpallidus* is the Saline River, 8.1 km north of Warren, Bradley County, Arkansas (Hubbs and Black 1940a). The original description was based on examination of only two specimens, and Snelson and Jenkins (1973) redescribed the species and established the presently used common name based on examination of 189 specimens.

DISTRIBUTION   *Notropis perpallidus* is endemic to the Ouachita and Red River drainages of southern Arkansas and southeastern Oklahoma. In Arkansas, this uncommon cyprinid is known only from the Ouachita River drainage in the Saline, Antoine, Caddo, Little Missouri, and upper Ouachita rivers (Robison 2006a) (Map 74). The remainder of its small range is in the Red River drainage of southeastern Oklahoma (Kiamichi and Little river systems, with older records from the Muddy Boggy River) (Pigg 1977; Miller and Robison 2004; Robison 2006a). It is not known from the Little River drainage of Arkansas even though there are several records from that drainage in Oklahoma. All records of this species from the Caddo and Little Missouri rivers are pre-1988 (Map A4.74), and we are not aware of any recent records from those drainages.

HABITAT AND BIOLOGY   The Peppered Shiner inhabits pool regions 2–4 feet (0.6–1.2 m) deep (or deeper) in moderate-sized, warm, clear rivers and is rarely found in small streams. Collections by Loe (1983) in Terre Noir Creek and Bell Creek (a first order stream) in Clark County demonstrated that it occasionally occurs in small streams. Typically, it occurs in the lee of islands and other obstructions away from the main current (Snelson and Jenkins 1973). It has been collected over a variety of substrates. The type specimens were taken from "a very silty, weedless backwater of the Saline River" (Hubbs and Black 1940a). Snelson and Jenkins (1973) reported that it was found in slow or quiet water 2–4 feet deep, usually downstream from riffles and shoals over a predominantly gravel-rubble substrate, with scattered boulders and bedrock outcrops. Finnel et al. (1956) found it in cut-off pools with sand-silt bottoms and in another area having a bedrock bottom and swift current. In a more detailed analysis of microhabitat utilization, Wagner et al. (1987) found that substrate type was relatively unimportant in determining the distribution of *N. perpallidus*, but that depth and current were very important. They reported that this species tends to occupy slow-moving water deeper than 20 inches (0.5 m) where current speeds are less than .01 ft./s (0.3 cm/s). It occurs both above and below the Fall Line in the Ouachita River

drainage and is frequently associated with beds of water willow which may play an important part in providing cover. It appears to be very susceptible to environmental disturbance, and Wagner et al. (1987) suggested that reservoir construction has contributed to the decline of extant populations. It is apparently intolerant of reservoirs and was not found in any Arkansas impoundments within its small range (Buchanan 2005).

Snelson and Jenkins (1973) considered the raptorial teeth, short intestine, and silvery peritoneum of *N. perpallidus* indicative of a carnivorous diet. The intestinal contents also indicated that feeding occurs on benthic, midwater, and surface organisms, including terrestrial forms. Food consisted largely of aquatic and terrestrial insects. Dipterans composed 89% of the total identified insect items and formed 47–100% of the insect items found in individual specimens. The list of food items found by Snelson and Jenkins (1973) included cladocerans, water mites, ephemeropteran nymphs, thysanopteran winged adults, trichopteran larvae, Tendipedidae larvae, and unidentified dipteran pupae and adults.

All information on reproductive biology of *N. perpallidus* is from Snelson and Jenkins (1973). A prolonged breeding season apparently occurs from late May to early August in Arkansas. Sexual maturity was reached at age 1, and all age 2 and 3 specimens of both sexes were mature. All males taken in early April lacked tubercles, but a few of the larger ones had slightly enlarged testes. The April females showed some evidence of ripening. All late May through early August males were tuberculate, and gravid females were found from May through August. The earliest date of capture of a YOY specimen (18.6 mm) was 6 August. Surprisingly for such a small cyprinid, maximum reported life span for the Peppered Shiner was around 4 years (Snelson and Jenkins 1973).

CONSERVATION STATUS   The Peppered Shiner was referred to by Jordan and Gilbert in 1886 as "A little fish very abundant in the Saline River." Almost a century later it was listed in Arkansas as rare (Robison 1974b) and threatened (Williams et al. 1989). Robison (2006a) reviewed the conservation status of this shiner throughout its range in Arkansas and Oklahoma and documented the Peppered Shiner in only two of the river systems in Arkansas from which it had been historically collected. This species is currently not abundant at any locality. The largest single series collected in Arkansas consisted of 21 specimens (CUMV 52317) from the Ouachita River at U.S. Route 270 in Montgomery County in 1967. A few larger series are known from Oklahoma (collected in the 1970s and 1980s), but there are currently no reports of large populations in that state. The low population densities and rather patchy distribution of the Peppered Shiner are reasons for concern over its future status. Robison (2006a) recommended a conservation status of threatened based on small population size and low densities, and coupled with its possible extirpation from the Caddo and Little Missouri River drainages, we agree that a threatened status is warranted.

Notropis potteri Hubbs and Bonham
Chub Shiner

Figure 15.91. *Notropis potteri*, Chub Shiner. *David A. Neely.*

Map 75. Distribution of *Notropis potteri*, Chub Shiner, 1988–2019.

CHARACTERS A moderately large shiner with a moderately compressed body (body depth going into SL 3.7–4.8 times), with greatest body depth just in front of dorsal fin. Head broad and somewhat flattened above and below. Eyes small, high on head, and somewhat directed upward; eye diameter going 4 or more times into head length. Snout wide (when viewed from above). A large, moderately oblique mouth (upper jaw length going about 2.5 times into head length) (Fig. 15.91). Posterior portion of upper lip distinctly swollen (enlarged), and the lower lip is medially thickened (medially enlarged dentary bones). Dorsal fin long and pointed, with first rays reaching beyond rest of fin when depressed. Dorsal fin origin over (but sometimes in front of) pelvic fin origin. Dorsal rays 8. Anal rays 7. Pectoral fin rays 14–17. Pharyngeal teeth 2,4–4,2. Lateral line complete at least to caudal peduncle; lateral line scales 33–37. Intestine short, with a single S-shaped loop. Predorsal scales 14–17. Body circumferential scales usually 26–29, with modally 13 scales below the lateral line. Maximum size is 4.33 inches (110 mm).

LIFE COLORS Body coloration a dusky yellowish olive or tan above and silver below. A thin dusky middorsal stripe is present. A faint lateral band underlining the silvery band. A faint, diffuse spot occasionally evident at the caudal base.

SIMILAR SPECIES The Chub Shiner is similar to *Notropis bairdi*, *Notropis blennius*, and *Notropis stramineus*. It is distinguished from *N. bairdi* and *N. stramineus* by 2 rows of teeth in the pharyngeal arch (2,4–4,2 vs. 0,4–4,0) and by the absence of the dark streak in the middle of the caudal base. It is similar to *N. blennius* in scale and fin-ray counts, but it can be distinguished from that species by its swollen posterior portion of the upper lip on each side (vs. upper lip not swollen posteriorly) and slimmer body, body depth going about 4.4 times into SL (vs. body depth going about 4 times into SL in *N. blennius*). It further differs from *N. blennius* in usually having 13 or more circumferential scales below the lateral line (vs. modally 11 circumferential scales below the lateral line) and in having a smaller eye, eye diameter going into head length 4.2 to 5.5 times (vs. eye diameter going into head length 3.0–3.4 times). Hubbs and Bonham (1951) gave the following description of pigmentation differences between *N. potteri* and *N. blennius*:

> *Notropis potteri* often has an intervening scarcely pigmented area between the lateral line pores with their associated pigmentation and the more dorsal band of large chromatophores. In *N. blennius* the pigmentation is continuous from the back, down the sides to the lateral line row of scales and often on the scale row below the lateral line.

VARIATION AND TAXONOMY *Notropis potteri* was described by Hubbs and Bonham (1951) from Waco Creek (Brazos River drainage), near Waco, McClennan County, Texas. Suttkus and Clemmer (1968) presented meristic and morphometric data for *N. potteri* and compared that species with the morphologically similar *N. blennius*. Those authors also reported remarkable ontogenetic changes in *N. potteri*. Based on analysis of the cytochrome *b* gene, Mayden et al. (2006a) placed *N. potteri* in genus *Alburnops* and considered it the sister species of *A. (Notropis) shumardi*.

DISTRIBUTION   The Chub Shiner is found in the Red River drainage from Texas through Oklahoma, Arkansas, and Louisiana; in the Mississippi River of Louisiana near the mouth of the Red River; and throughout much of the Brazos River drainage of Texas. In Arkansas, it is confined to the main channel of the Red River where it is locally abundant (Buchanan et al. 2003) (Map 75). Whether it is a native species or an introduced species in the Red River has been the subject of debate. Hubbs and Bonham (1951) considered its occurrence outside the western Gulf Slope drainages to be the result of bait bucket introductions in the Red River after the construction of Lake Texoma. Other workers dispute this claim, suggesting that the Chub Shiner is likely native to streams outside the Gulf Slope drainages (Miller 1953; Hall 1956; Suttkus and Clemmer 1968; Conner 1977). Perkin et al. (2009) reported that unpublished museum records were discovered that along with zoogeographic data now suggest that the Chub Shiner is native, rather than introduced, to the Red River drainage. We agree that *Notropis potteri* is native to the Red River drainage of Arkansas.

HABITAT AND BIOLOGY   The Chub Shiner is rather common in the large, turbid, Red River, over clean or silt-laden sand and gravel substrates along the shore. Schramm (2017) reported this shiner in lotic areas of channel borders, including secondary channels and sloughs and relatively shallow sandbar and bank habitats. Miller and Robison (2004) reported that it appeared to do well in both silt-laden and relatively clear streams and was a fairly common resident of Lake Texoma. In some parts of its range, it is found in smaller tributaries of rivers, but in Arkansas it is restricted to the main channel of the Red River. Buchanan et al. (2003) found it to be distributed throughout the Red River in Arkansas, and it was the second-most abundant species collected (exceeded in numbers only by *Notropis atherinoides*). Suttkus and Clemmer (1968) noted consistent rigid lateral projection of the pectoral fins of this species, an adaptation for living close to the bottom of the river in current. It is tolerant of high salinities, and Echelle et al. (1972) found it in Oklahoma in salinities ranging from 4.2 to 19.7 ppt.

The Chub Shiner is unique among *Notropis* species in its dietary habits (Perkin et al. 2009). In the Red River of

Oklahoma and Louisiana (and Arkansas by our observations), the Chub Shiner is primarily a benthic invertivore, but it is also considered piscivorous, which is unique for the genus *Notropis*. Felley (1984) reported that its food in Oklahoma and Louisiana consisted (percent by volume) of 63% benthic invertebrates, 13% fish, 8% open-water invertebrates (mostly cladocerans), and 16% substrate particles (probably ingested incidentally). Perkin et al. (2009) found that *N. potteri* consumed primarily fish and aquatic insects (Coleoptera, Trichoptera) in the Brazos River, Texas. In their study, the Chub Shiner showed an ontological diet shift from aquatic insects to piscine prey, which is common for piscivores (Fraser and Cerri 1982; Keast 1985a; Juanes et al. 2002). Morphological and behavioral characteristics of the Chub Shiner facilitate piscivorous feeding (Perkin et al. 2009). Gape sizes of Chub Shiners are larger than those of other sympatric cyprinids. Gape size was more similar to that of the Creek Chub, another piscivorous cyprinid that also exhibits ontological shifts in diet (Fraser and Cerri 1982). In addition, the medially enlarged dentary bones and posteriorly enlarged upper lip of the Chub Shiner (Hubbs and Bonham 1951; Douglas 1974; Robison and Buchanan 1988) are distinguishing morphological characteristics of the species (Miller and Robison 2004). Such enlarged and strong upper and lower jaws are advantageous for a firm bite when consuming large prey such as fish (Gosline 1973; Porter and Motta 2004; Hulsey and Garcia de Leon 2005). Behaviorally, piscivores pursue and ingest the whole body of their piscine prey headfirst (Simon 1999b; Juanes et al. 2002; Porter and Motta 2004). J. S. Perkin (pers. comm.) observed that Chub Shiners held in laboratory aquaria often pursue and ingest piscine prey such as Western Mosquitofish. While other *Notropis* species may occasionally and opportunistically consume fish as prey (Starrett 1950a; Felley 1984), the findings of Perkin et al. (2009) further support the observations that the Chub Shiner targets piscine prey and that fish constitute a significant portion of its diet (Goldstein and Simon 1999).

Perkin et al. (2009) reported that male Chub Shiners taken in March, May, and September in the Brazos River, Texas had elevated GSIs. A single developing female was collected in May and a sexually mature female was taken in August, indicating a long breeding season. Based on examination of gonads by TMB, *N. potteri* has a long breeding season in Arkansas, extending from late April through at least late August. Ripe females were found in the Red River of Arkansas in Lafayette County on 26 April, in Little River County on 25 July, and in Miller County on 14 and 23 August. Ripe females were 58–92 mm TL and contained between 172 and 1,962 mature ova. Females from April through August contained smaller egg sizes that appeared to be developing. Chub Shiners in the Brazos River exhibited three age groups (ages 0, 1, and 2) with a maximum life span of 2.5 years (Perkin et al. 2009). Age 0 individuals reached a maximum length of approximately 45 mm, age 1 individuals grew to approximately 70 mm, and age 2 individuals exhibited little growth before mortality in late spring (Perkin et al. 2009). There is no information on spawning behavior or other aspects of reproductive biology.

CONSERVATION STATUS Warren et al. (2000) considered this species stable across its range, but it is considered a species of special concern in Texas (Hubbs et al. 2008). There have been declines documented in the Brazos River population in Texas from which it was originally described (Perkin et al. 2009). Winston et al. (1991) found this species susceptible to habitat fragmentation and attributed its extirpation in the North Fork of the Red River, Oklahoma, to the construction of the Lake Altus dam. However, in Arkansas the Chub Shiner is locally abundant and common in the Red River along its 135-mile length through the state (Buchanan et al. 2003). Perkin et al. (2009) urged a study of the reproductive ecology of the locally abundant population of the Chub Shiner in the Red River of Arkansas. Its populations in the state are currently secure. A proposed project to extend the Red River Navigation System upstream from Shreveport, Louisiana, to Denison Dam (Oklahoma-Texas), could have a profound adverse impact on populations of *N. potteri* in the Red River. Because of its small range in Arkansas and susceptibility of its populations to sudden declines, we consider the Chub Shiner a species of special concern in this state.

Notropis sabinae Jordan and Gilbert
Sabine Shiner

Figure 15.92. *Notropis sabinae*, Sabine Shiner.
David A. Neely.

CHARACTERS A small, straw-colored shiner with a small eye (eye diameter going into head length 3.6–4 times and into snout length 0.9–1.7 times), a rather large, subterminal, almost horizontal mouth, and a slightly arched back (Fig. 15.92). Snout bluntly rounded, projecting beyond upper lip; snout length going into head length 2.4–3.9

Map 76. Distribution of *Notropis sabinae*, Sabine Shiner. 1988–2019.

times. Undersurface of head distinctly flattened. Eyes directed slightly upward. Body deepest at dorsal fin origin; body depth going into SL 3.6–4.9 times. Barbels absent. Upper jaw reaching beyond front of eye. Dorsal rays 8. Anal rays 7. Pectoral rays usually 14 or 15 (13–16). Pharyngeal teeth 0,4–4,0 (rarely 1,4–4,1). Lateral line complete, with 31–37 scales. Dorsal fin origin in front of pelvic fin origin. Front of dorsal fin base closer to tip of snout than to caudal fin base. Predorsal scale rows usually 13 or 14. Body circumferential scales 21–26, usually with 5 or 6 scale rows above and 6 rows below the lateral line. Intestine short; lining of peritoneum silvery, with a few faint speckles. Breeding males with large tubercles on snout, preorbital, and lachrymal areas, and with minute tubercles on pectoral, pelvic, dorsal, and anal fins; females may have some tubercle development on snout and side of head. Breeding females have an enlarged urogenital papilla which is absent in males. Maximum size 2.5 inches (64 mm).

LIFE COLORS Body straw-colored with a narrow middorsal line and dorsolateral scales outlined with dark pigment. Sides silvery. Ventral scales unpigmented. Fins colorless. A poorly developed lateral band present posteriorly. Anterior lateral line pores outlined with dark dashes or dots. Breeding coloration is poorly developed in both sexes, particularly females (Heins 1981). A faint lemon-yellow color may develop in the dorsal and caudal rays of the males, being strongest anterobasally; white pigmentation is variably present in these and other fins, particularly the pectoral and anal fins.

SIMILAR SPECIES The Sabine Shiner roughly resembles *N. stramineus* in coloration and general body shape, but differs from that species in having undersurface of head distinctly flattened (vs. not flattened), upper jaw extending behind front of eye (vs. upper jaw not extending past front of eye), and front of dorsal fin base much closer to tip of snout than to base of caudal fin (vs. front of dorsal fin base about equidistant between tip of snout and base of caudal fin). It differs from *Hybopsis amnis, Notropis buchanani, N. volucellus,* and *N. wickliffi* in having 7 anal rays (vs. 8).

VARIATION AND TAXONOMY *Notropis sabinae* was described by Jordan and Gilbert (1886) from the Sabine River, 8 km south of Longview, Gregg County, Texas. It is a member of the *Notropis longirostris* species group of subgenus *Alburnops* according to Swift (1970) and Heins and Baker (1992). Analysis of the cytochrome *b* gene supported retention of *N. sabinae* in a *longirostris* clade of a more restricted genus *Notropis* (Mayden et al. 2006a).

DISTRIBUTION The Sabine Shiner occurs in three disjunct areas: the White and St. Francis river drainages of the lower Mississippi River basin in Missouri and Arkansas; the Big Black and Yazoo river drainages of the lower Mississippi River basin in Mississippi; and the lower Red River drainage in Louisiana, and Gulf Slope drainages from the Calcasieu River, Louisiana, to the San Jacinto River, Texas. The distribution of *Notropis sabinae* in Arkansas is spotty, with specimens having been taken from streams in the White River system, including the Eleven Point and Spring rivers (McAllister et al. 2010b), and Piney Creek in Izard County (Matthews et al. 2014), the lower White River (an older record from the White River Refuge and a record near its confluence with the Black River), as well as the Black, Strawberry, and Current rivers (Tumlison et al. 2015) (Map 76). It was historically known from the St. Francis River drainage in Arkansas (Clay and Craighead counties, Map A4.76), but it has not been found in that drainage since 1900. There are also no recent records from the middle portion of the White River in Izard and Independence counties. Arkansas and Missouri populations are widely disjunct from remaining populations in the Coastal Plain drainages of Texas and Louisiana, and the populations reported from the Big Black and Yazoo river drainages of Mississippi (Suttkus 1991).

HABITAT AND BIOLOGY The Sabine Shiner has a spotty occurrence in streams and rivers with fine, silt-free, sand substrates, where it is found in relatively shallow runs and riffles with slight to moderate current (Williams and Bonner 2006). In Arkansas, it is most commonly found along sandbars in the Black River. It is a bottom-dwelling species that uses habitat with sandy substrate and moderate current velocities, with abundance ranging from rare to common (Gilbert 1978; Moriarty and Winemiller 1997; Ross 2001). Williams and Bonner (2006) documented spatial segregation of adult and juvenile fish, with juveniles more abundant over shallow, silt-laden bedrock and adults collected or observed over deeper, sandy areas. In Village Creek, Texas, it occurred almost exclusively in riffle and sandbar habitats (Moriarty and Winemiller 1997). Smaller individuals occurred in shallow riffles and large individuals were found most frequently in deeper water over sandbars.

Williams and Bonner (2006) examined the digestive tracts of *N. sabinae* in Texas and found aquatic insects were the dominant food organism by weight, followed by detritus, sand, silt particles, unidentified aquatic insects, terrestrial insects, crustaceans, plant material, and Collembola and Hydracarina. Dipterans were the most abundant aquatic insects consumed, followed by ephemeropterans, odonates, plecopterans, and trichopterans. Terrestrial insects included adult dipterans, coleopterans, and hemipterans. Detritus was abundant in digestive tracts from October through February, and aquatic insects were most abundant from March through September.

In Sabine River tributaries of Texas and Louisiana, *N. sabinae* has an extended spawning season from early April through late September (and possibly into October), during which multiple clutches of eggs are spawned by age 0 to age 2 females (Heins 1981). Williams and Bonner (2006) also found that individual females produced multiple cohorts of distinct oocyte sizes throughout the spawning season. Ripe females 35–48 mm SL contained from 113 to 423 mature ova; mature ovum diameter ranged from 0.63 to 0.81 mm (Heins 1981). A tuberculate male was taken by TMB from the White River in Arkansas County on 21 August at a water temperature of 73°F (22.8°C). Spawning by age 0 individuals has been suggested in *N. sabinae* (Williams and Bonner 2006). Individuals reach a maximum total length of 52 mm during their first summer. Spawning habitat and breeding behavior are unknown, but it probably spawns in midstream with the fertilized eggs drifting in the current during development. Williams and Bonner (2006) reported that downstream drift of eggs or larvae was suggested by the first appearance of age 0 fish at downstream sites and their absence at upstream sites until early fall. Heins and Baker (1989) noted that this pattern of downstream displacement is known for other stream fishes and is suggested by the data presented by Heins (1981) for *N. sabinae*. Downstream drift of eggs or larvae is a common and effective dispersal mechanism among fish species inhabiting variable stream systems (Bestgen et al. 1989; Platania and Altenbach 1998). The Sabine Shiner is a relatively short-lived species with maximum age about 2–2.5 years. Heins (1981) suggested that *N. sabinae* exhibits life history traits characteristic of an *r*-selected species.

CONSERVATION STATUS　Although Warren et al. (2000) considered populations of the Sabine Shiner to be stable, it was designated a species of conservation concern in Louisiana and Texas by the U.S. Forest Service and U.S. Fish and Wildlife Service (Williams and Bonner 2006) and a species of special concern in Mississippi (Ross 2001). In Arkansas the Sabine Shiner is rare and limited to only two disjunct areas within the state. We consider it a species of special concern.

Notropis shumardi (Girard)
Silverband Shiner

Figure 15.93. *Notropis shumardi*, Silverband Shiner. Chad Thomas.

CHARACTERS　A pale, moderately deep-bodied (body depth going into SL about 4.2–5.5 times), slab-sided fish with a deep caudal peduncle, a small head (head length going into SL 4.2 times), a large eye (eye diameter greater than snout length and going 3.2–3.4 times into head length), and a short, rounded snout (Fig. 15.93). Anal fin pointed with 8 or 9 (8–10) rays. Pharyngeal teeth 2,4–4,2. Lateral line complete, with 34–37 scales. Predorsal scale rows about 15 or 16 (14–17). Breast scaled. Mouth terminal and distinctly oblique; front (center) of upper jaw rising to level of middle of eye. Dorsal fin unusually high and pointed, with 8 rays; front rays of dorsal fin extend well past rear rays when the fin is depressed. Tip of depressed dorsal fin reaches far past the anal fin origin. Origin of dorsal fin over insertion of pelvic fin, its origin slightly closer to tip of snout than to caudal base. Intestine short, peritoneum silvery. Breeding males with small tubercles over head and pectoral rays. Maximum length about 3.4 inches (85 mm).

LIFE COLORS　Body color pale olive yellow dorsally with silvery sides, white belly, and a prominent silvery

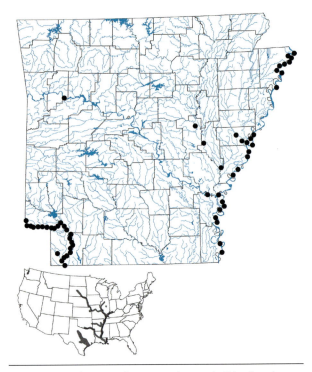

Map 77. Distribution of *Notropis shumardi*, Silverband Shiner, 1988–2019.

lateral band (dusky in preserved specimens). Fins colorless. Predorsal stripe thin. Upper and lower lips with dark pigment; chin lacking dusky pigment. Breeding males without brilliant colors.

SIMILAR SPECIES　*Notropis shumardi* is similar to *Notropis atherinoides*, *N. blennius* and *Hybognathus nuchalis*, all of which are often syntopic with it. It differs from *N. blennius* in having 8 or 9 anal rays (vs. 7), and jaw very oblique, front of upper lip rising to the level of the middle of the eye (vs. jaw more horizontal, center of the upper jaw below the middle of the eye, and usually nearer to the lower margin of the eye). Both species have pointed dorsal and anal fins, but in *N. shumardi* the tip of the depressed dorsal fin reaches far past the anal fin origin (vs. the tip of the depressed dorsal fin is about even with the anal fin origin in *blennius*). In species of *Hybognathus* there is considerable pigment below the lateral line on the anterior half of the body; both *N. shumardi* and *N. blennius* lack the pigment in this region. It can be distinguished from *N. atherinoides* in having 8 or 9 anal rays (vs. 10 or more), in having the dorsal fin origin just above the pelvic fin origin (vs. distinctly behind it), and in having a deeper body, with body depth going into SL 4.2–5.5 times in adults (vs. body depth going into SL 5.1–7.0 times).

VARIATION AND TAXONOMY　Described as *Alburnops shumardi* by Girard (1856), the type locality is the Arkansas River near Fort Smith, Arkansas. It was subsequently referred to as *Notropis illecebrosus*, and Gilbert and Bailey (1962) changed the name of this shiner to *N. shumardi* and synonymized *N. brazosensis* with the latter. In Arkansas, some variation is apparent as Arkansas River and Mississippi River populations tend to have 9 anal rays (79% and 81% of those populations, respectively), while Red River populations usually have 8 anal rays (74%). Coburn (1982b) hypothesized that *N. candidus* is the closest relative to *N. shumardi*. Based on analysis of the cytochrome *b* gene, Mayden et al. (2006a) returned *N. shumardi* to its former genus, *Alburnops*. We follow Page et al. (2013) in continuing to assign this species to genus *Notropis*.

DISTRIBUTION   *Notropis shumardi* is found in the large rivers of the Mississippi River basin (Mississippi, Missouri, Illinois, lower Ohio, and Red rivers) from South Dakota and central Illinois to the Gulf of Mexico, and in Gulf Slope drainages of Texas, from the Trinity River to the Lavaca River, with one record from the Pearl River, Louisiana-Mississippi. In Arkansas, the distribution of the Silverband Shiner is spotty (Map 77). It was previously known only from the three largest rivers of the state—the Arkansas, Mississippi, and Red rivers—but it is common today only in the latter two. It was found in small numbers in the lower White and St. Francis rivers in the early 2000s (Vicksburg Waterways Experiment Station data provided by K. J. Killgore). Buchanan et al. (2003) reported that it occurs throughout the Red River in Arkansas. It has declined drastically in the upper Arkansas River during the past three decades and has apparently disappeared from its type locality in the Arkansas River near Fort Smith, where it was still common in the 1970s. It was last collected near Fort Smith in 1983, and the last record of this species upstream from the lock and dam at Dardanelle, Arkansas (river mile 248) was in 1988. Despite intensive sampling of the upper Arkansas River, including monthly sampling at Fort Smith between 1996 and 2005, no recent records have been obtained. McAllister et al. (2012b) reported the first collection of this shiner from the lower Arkansas River in Desha County, making this only the third known record from the Arkansas River below Dardanelle Reservoir.

HABITAT AND BIOLOGY   This schooling shiner usually prefers areas of moderate to swift current in the main channels of large rivers along sand or gravel bars. It is rarely found in the tributaries of the large rivers or in oxbow lakes in Arkansas. In the Mississippi River, it is most commonly found over sand or gravel substrates in the swifter currents along the upstream sides of point bars or along the sides of towheads adjacent to the main channel (Ross 2001). It often schools with *Notropis atherinoides* and

N. blennius (Beckett and Pennington 1986), and it is usually syntopic with *N. atherinoides* and *N. potteri* in the Red River of Arkansas. Gilbert and Bailey (1962) cited its tolerance of extremely turbid conditions and high sediment loads. Hrabik et al. (2015) concluded that alterations to the Missouri River (dams, channelization, reduced turbidity and sediment loads) may have greatly reduced populations of the Silverband Shiner. Its disappearance from the upper Arkansas River in Arkansas occurred during the first two decades after the construction of the Arkansas River Navigation System. It is currently common in Arkansas only in the unimpounded Mississippi and Red rivers.

In Texas, C. S. Williams (2011) reported that the digestive tracts of adult *N. shumardi* contained primarily aquatic insects. By percent weight, the diet included Trichoptera (15%), Ephemeroptera (12%), Diptera (5%), Plecoptera (2%), and Hemiptera (<1%), detritus, algae, terrestrial insects, sand, and 1% crustaceans. During the summer, aquatic insects were the most common food items in adults, with a peak in seasonal occurrence (89%). Trichoptera (caddisflies) were the most common aquatic insects, occurring in 28% of adult *N. shumardi*. While terrestrial insects were present, they were uncommon (7%). Aquatic insects composed the greatest percent by weight of juvenile *N. shumardi* food items, followed by detritus, sand, crustaceans, and algae.

Pflieger (1997) collected females full of eggs in Missouri in mid-August, indicating a summer spawning season. We collected specimens in breeding condition on 23 April from the Red River of Arkansas. Suttkus (1980) reported that ripe specimens were collected from the Mississippi River of Louisiana from late June to early August at water temperatures ranging from 78.8°F to 84.2°F (26–29°C); breeding aggregations were observed over hard sand to fine gravel substrates in water 3–6 feet (1–2 m) deep in strong current. In Texas, Williams (2011) reported elevated monthly GSIs (>3%), and mature ovaries were present in *N. shumardi* from April through September. The smallest female to exhibit gonadal development was a mature 41 mm individual collected during August. *Notropis shumardi* females produce multiple cohorts of oocytes during a single reproductive season. Oocyte diameters consisted of three general size groups including small (<0.3 mm) and large (>0.6 mm) vitellogenic oocytes and mature oocytes (Williams 2011). Mean clutch size for mature *N. shumardi* was 387.3 and ranged from 144 to 750 oocytes in individuals 47–76 mm TL. Clutch size was positively related to total length of mature females. Williams (2011) found 3 age classes for the Silverband Shiner during each year of sampling. Age 1 was the most abundant age class until the reproductive season, when age 0 fish abundance increased. Few individuals enter their second summer, and maximum age was 2.5 years. Age 0 fish exhibited rapid growth, reaching total lengths equal to 45–65% of maximum length during the first summer and fall. Maximum life span is apparently around 3+ years.

CONSERVATION STATUS   The Silverband Shiner is a big-river inhabitant commonly encountered and widespread only in the Red and Mississippi rivers in Arkansas, where its populations are apparently secure. The Nature Conservancy ranks this shiner as S2 in the state. A proposed extension of the Red River Navigation System upstream to Denison Dam (Oklahoma-Texas) would likely adversely affect populations of this species in the Red River (similar to what occurred in the Arkansas River after the completion of the Arkansas River Navigation System). Although it does not currently merit a conservation status of concern, it is clearly susceptible to environmental alteration of its big-river habitat in Arkansas.

Notropis stramineus (Cope)
Sand Shiner

Figure 15.94. *Notropis stramineus*, Sand Shiner. *Uland Thomas.*

CHARACTERS   A pallid, robust shiner with moderately large eyes directed laterally. Eye diameter equal to or greater than snout length and going into head length 3.5 times or less. Head short, head length going 3.5–4 times into SL. Mouth slightly subterminal and oblique, upper jaw not reaching past front of eye (Fig. 15.94). Snout pointed when viewed from above. Dorsal rays 8; anal rays 7. Pharyngeal teeth 0,4–4,0. Lateral line complete and decurved, with 33–37 scales (scales not elevated); lateral line pores outlined by pairs of small dark dots or dashes. Predorsal scale rows usually 14 or 15. Dorsal fin rounded, its origin over or slightly behind pelvic fin origin. Breast naked to covered with embedded scales. Intestine short with a single S-shaped loop; peritoneum silvery with scattered dark speckles. Nuptial males with small tubercles on dorsal and lateral portions of head, most prevalent on dorsal surface posterior to nares (Etnier and Starnes 1993). Maximum size 3.2 inches (82 mm).

Map 78. Distribution of *Notropis stramineus*, Sand Shiner, 1988–2019.

LIFE COLORS   Body pallid to straw-colored; silvery sides grading to white belly. Sides with a faint lateral band appearing posteriorly on the caudal peduncle. A discrete middorsal stripe is expanded into a wedge-shaped blotch in front of the dorsal fin origin and does not encircle the dorsal fin base. Occasionally a tiny basicaudal spot is present. Dark crescent-shaped markings between nostrils. Lateral line pores outlined with paired melanophores. Fins colorless. No chromatic breeding colors.

SIMILAR SPECIES   Similar species include *Notropis volucellus*, *N. wickliffi*, *N. buchanani*, and *N. blennius*. *Notropis stramineus* differs from the first three species in having 7 anal rays (vs. 8), lack of elevated anterior lateral line scales (vs. lateral line scales elevated), a discrete middorsal dark stripe that ends in a wedge-shaped blotch in front of the dorsal fin (vs. middorsal stripe, if present, not expanded as a wedge-shaped spot), and a pointed rather than rounded snout. *N. stramineus* differs from *N. blennius* in having 0,4–4,0 teeth (vs. 2,4–4,2 teeth), and upper jaw length less than eye diameter (vs. greater than eye diameter). It differs from *N. sabinae* in having the undersurface of head not distinctly flattened (vs. distinctly flattened), upper jaw not extending past front of eye (vs. jaw extending past front of eye), and front of dorsal fin base about equidistant between tip of snout and base of caudal fin (vs.

front of dorsal fin base much closer to tip of snout than to base of caudal fin).

VARIATION AND TAXONOMY   Originally described as *Hybognathus stramineus* by Cope (1865a), the Sand Shiner was usually referred to as *Notropis blennius* (along with several other poorly known minnows) until well into the 20th century. Hubbs (1926) assigned it to *N. deliciosa*, until Suttkus (1958) found that *deliciosa* was a junior synonym of *N. texanus* and that *stramineus* was the next available oldest name. Bailey and Allum (1962) recognized two distinct subspecies, a western form, *N. s. missouriensis*, and an eastern form, *N. s. stramineus*; the latter subspecies occurs in Arkansas. Tanyolac (1973) studied variation in *N. stramineus* throughout its range and confirmed the validity of these subspecies. However, a rangewide phylogeographic analysis based on genetic variation rejected the validity of subspecies and revealed the existence of five evolutionary groups, possibly cryptic species, within the nominal *N. stramineus* (Pittman 2011). A Discriminant Function Analysis of 29 morphological features showed that morphological divergence was concordant with the reported genetic divergence (Pittman 2011). In the 1990s, this species was usually referred to as *Notropis ludibundus* after Mayden and Gilbert (1989) showed that *ludibundus* was an older available name than *stramineus*. However, a subsequent appeal to the ICZN to retain *stramineus* in the interest of stability was successful. Based on analysis of the cytochrome *b* gene, Mayden et al. (2006a) placed *N. stramineus* in the genus *Miniellus*, but we follow Page et al. (2013) in continuing to place this species in genus *Notropis*.

DISTRIBUTION   *Notropis stramineus* is a widely distributed minnow, occurring in the St. Lawrence–Great Lakes, Hudson Bay, and Mississippi River basins from southern Quebec to eastern Saskatchewan, Canada, south to Tennessee and Texas, and west to eastern Montana, Wyoming, Colorado, and New Mexico. It also occurs in Gulf Slope drainages from the Colorado River, Texas, to the Rio Grande of Texas, New Mexico, and Mexico. The Sand Shiner is extremely rare in Arkansas, having been collected only from the main channel of the Mississippi River (Map 78). David A. Etnier collected a single specimen (UT44.2276) on 30 January 1981, just upstream from the I-40 bridge in Crittenden County, Arkansas (Map A4.78). A second specimen (UT44.2269) was collected the same day from the Tennessee side of the river. Additional records were obtained from two nearby localities from the Mississippi River in Mississippi County by TMB and Sam Barkley on 26 July 2000. There are five records of *N. stramineus* from the Red River of Bryan, McCurtain,

and Choctaw counties, Oklahoma, between 1990 and 2009 (OMNH 46834, 49222, 49607, 69130, and 79747), with two of the specimens collected about 10 miles upstream from the Arkansas state line. Therefore, it is possible that this species could occasionally be found in the Red River of Arkansas even though Buchanan et al. (2003) did not find this species in a Red River survey.

HABITAT AND BIOLOGY   The Sand Shiner is a widespread species, inhabiting sand-bottomed, clear to turbid streams and rivers and is rarely found in upland areas. It is most abundant over most of its range in shallow, sandy pools of medium-sized creeks having permanent flow and moderately clear water (Pflieger 1997; Miller and Robison 2004). Mueller and Pyron (2009) reported that this species prefers sand or gravel substrate, is absent or rare in low gradient waters, is decreased in areas with erosion and siltation, and is abundant in pools with considerable current. In Arkansas, *N. stramineus* has been found only in current along sandbar habitats in the Mississippi River. Trautman (1957) reported that it was seldom found among rooted aquatic vegetation. It normally schools in midwater reaches or near the bottom (HWR, pers. obs. in Oklahoma).

Foods consist primarily of a variety of aquatic insects, but some plant material is also consumed. Pflieger (1997) noted its rather generalized food habits in Missouri, and Starrett (1950a) found it to be omnivorous in Iowa, feeding primarily on immature aquatic insects and bottom ooze, but also ingesting adult insects, small crustaceans, and plant material. The fry and adults fed heavily on microflora (bottom ooze) during summer, and the adults fed primarily on immature and adult aquatic insects during other seasons. The main insects consumed by adults were ephemeropteran nymphs and adults, trichopteran larvae, various dipterans, and corixids. In Ohio, it fed on detritus and small benthic and drifting invertebrates, including midge larvae and small mayfly nymphs (Gillen and Hart 1980). Decreased turbidity has been shown to negatively impact feeding on terrestrial and aquatic insects by Sand Shiners (Bonner and Wilde 2002).

Spawning occurs in Oklahoma from late spring throughout the summer. Sand Shiners are broadcast spawners with demersal, adhesive eggs scattered over clear gravel and sand in flowing waters (Platania and Altenbach 1998; Miller and Robison 2004; Mueller and Pyron 2009). In Kansas, *N. stramineus* spawned from May through August, with a peak in August when water levels were low and water temperatures were high (Summerfelt and Minckley 1969). Platania and Altenbach (1998) observed spawning in an aquarium whereby a male chased a female who broadcast demersal, adhesive eggs. Spawned eggs rapidly settled in the interstices of the gravel substrate of the aquarium and failed to become buoyant. Cochran (2014a) observed three spawning events of *N. stramineus* in an aquarium at 20°C. Prior to spawning, individuals actively chased each other before suddenly stopping and turning in a vertical direction as the eggs were released. Summerfelt and Minckley (1969) hypothesized that spawning in the hot, dry portion of the summer at water temperatures of 69.8–98.6°F (21–37°C), apparently enhances the survival of the larvae in rivers where spring conditions can be characterized by flooding and extreme flow fluctuations. The spawning behavior and demersal, adhesive eggs of *N. stramineus* suggest that spawning by this species is probably not correlated with flow spikes (Platania and Altenbach 1998). Spawning is also known to occur in lakes in sandy shallows and creek mouths (Hubbs and Cooper 1936). Mature eggs averaged 650–747 per female in the Rio Grande drainage (Platania and Altenbach 1998), but females in the Pomme De Terre and Loutre rivers, Missouri, contained 626–2,660 eggs (Tanyolac 1973). In Iowa, Sand Shiner populations were characterized by relatively low mean age (53% of all fish sampled were age 1, 30% age 2, 15% age 0, and 2% age 3), fast growth, high mortality (total annual mortality varied from 35 to 92.3%), and high recruitment variability (C. D. Smith et al. 2010). Maximum life span in Kansas was 3 years (Tanyolac 1973).

CONSERVATION STATUS The Sand Shiner is of undetermined status, but it is probably at least threatened in Arkansas waters.

Notropis suttkusi Humphries and Cashner
Rocky Shiner

Figure 15.95. *Notropis suttkusi*, Rocky Shiner. *David A. Neely.*

CHARACTERS A slender compressed shiner with a moderately long head (head length going about 3.7 times into SL) and a terminal, oblique mouth (Fig. 15.95). Dorsal fin rounded, not high and pointed. Dorsal rays 8, anal rays 9 or 10 (occasionally 11), and 8 pelvic rays. Lateral line complete and deeply decurved (at least 5 or 6 scale rows), usually with 34–37 (33–39) scales. In preserved specimens, a dark lateral band maintains approximately the same width from the shoulder girdle to the caudal base in adults. Lateral

Map 79. Distribution of *Notropis suttkusi*, Rocky Shiner, 1988–2019.

line pores usually not distinctly outlined by dark dashes (anterior pores sometimes weakly outlined), but dark pigment below lateral stripe usually extending well past lateral line as far ventrad as pelvic fin origin. Body circumferential scales usually 22–26. Predorsal scales usually 18–22 (16–24). Scales around caudal peduncle usually 14 or 15 (13–16). Pharyngeal teeth, 2,4–4,2. Intestine short; peritoneum silvery. Tuberculation is well developed in breeding males. The top of the head from the occiput to the posterior margin of the nares and orbital rim, upper third of opercle, cheek, lachrymal, and branchiostegals are covered evenly with relatively small, closely set, erect tubercles. Tubercles become finer between nares, forming a triangular patch in this area. Very fine, pointed tubercles occur between the larger cephalic tubercles and cover the tip of the snout, lips, and breast region (Humphries and Cashner 1994). On the dorsum, tubercles begin at the dorsal fin origin and extend anteriorly to the occiput. Three to 6 curved tubercles occur on each predorsal scale and project from the posterior margin. Tuberculate scales extend postdorsally to the caudal base, and laterally tubercles continue down 4 or 5 scale rows above the lateral line. Tubercles on pectoral fin rays limited to 1 tubercle per fin ray segment. There is some tubercle development in females on head and body scales, but not on the pectoral rays. Typically, females are larger than males and are less intensely pigmented. Maximum size about 2.5 inches (66 mm).

LIFE COLORS   Dorsum olive and sides silvery white with a thin iridescent green stripe above lateral line. Belly silvery white. Lateral band dusky. Lips darkly pigmented, especially the upper lip. Chin pigment scattered but present on one-third to all of chin region; many individuals with a dense median pigment patch on chin. Fins clear. Dark middorsal stripe narrow but well developed, especially posteriorly. Dark brown or black melanophores on dorsolateral scales evenly distributed over the scale surface and so dense (especially in males and adults) that scale margins are rarely discernible. Fins colorless. Breeding males develop a raspberry or lilac cast over much of the body. After preservation, life colors change quickly to a distinct orange or orange red. On the body, an orange wash extends slightly above the lateral stripe and over much of the upper head and gular region (Humphries and Cashner 1994). The eye is distinctly orange, and body with a vivid orange cast covering the entire venter and lateral surface below the lateral stripe from the caudal base to the cleithrum. A striking red-orange band occurs in the base of all fins and is well developed in the axial region of the pectoral fin and extends along the cleithrum. Both pectoral and caudal fin rays and membranes have a distinct orange cast and other fins with a lighter orange coloration.

SIMILAR SPECIES   Notropis suttkusi is most similar to Lythrurus umbratilis, L. fumeus, Notropis atherinoides, and Notropis percobromus, and is syntopic with the first three species. All five species have the dorsal fin origin posterior to the pelvic fin origin, and closer to the caudal base than to the snout tip. N. suttkusi can be distinguished from L. umbratilis and L. fumeus by the larger anterodorsal scales and the absence of a dark spot at the origin of the dorsal fin (vs. anterodorsal scales small in both and black spot usually present at dorsal origin in umbratilis). Body coloration of N. suttkusi generally darker overall than that of N. atherinoides (especially in males). From N. atherinoides, N. suttkusi can be distinguished by the darker lateral stripe (vs. lateral stripe diffusely stippled anteriorly, becoming more prominent on the caudal peduncle), the black pigment below the lateral stripe extending downward to or below the lateral line and ventrad past the middle of the body (vs. melanophores virtually absent below the midlateral stripe), a more rounded dorsal fin (vs. dorsal fin more pointed, with the anterior rays longer than posterior rays when fin is depressed), breeding males with red and orange coloration on head and lower body and with 1 tubercle per pectoral fin ray segment (vs. breeding males without bright colors and with 3–5 tubercles per pectoral fin ray

segment), and the greater scattering of pigment on dorsolateral scales (vs. pigment concentrated on scale margins) (pigment characters are best observed in preserved specimens). From N. percobromus, N. suttkusi differs in having a shorter snout and upper jaw (vs. a longer snout and upper jaw), a deeper, more robust body (vs. a more slender body), denser pigmentation on scales, lower lip, and gular region (vs. less pigmentation on scales, lower lip, and gular region), lateral line pores not outlined by melanophores but dark pigment below lateral band extending downward past lateral line anteriorly (vs. lateral line pores outlined by melanophores and dark pigment below lateral stripe usually extending downward no further than lateral line), lateral stripe nearly the same width from the caudal fin base to the shoulder girdle in adults (vs. lateral stripe appreciably narrower anteriorly), and lateral line decurved 5 or 6 scale rows (vs. lateral line straight or decurved only 1 or 2 scale rows in N. percobromus).

VARIATION AND TAXONOMY   Notropis suttkusi was described by Humphries and Cashner (1994). The type locality was designated as the Little River at Cow Crossing, 25 km east of Idabel, McCurtain County, Oklahoma. It was formerly included in Notropis rubellus, and prior to 1994, N. rubellus was considered a single widespread, polytypic species (Berendzen et al. 2008b). The description of N. suttkusi was based on multivariate analyses of both meristic and morphometric characters. Humphries and Cashner (1994) were unable to confidently classify upper Ouachita River specimens of the N. rubellus complex. Some specimens from the upper Ouachita River were morphologically similar to N. suttkusi for some characters but were more like rubellus for others (for additional details about this unresolved taxonomic problem, see the Notropis percobromus account). Mayden et al. (2006a) found that N. suttkusi, N. percobromus, and N. rubellus formed a well-supported clade based on analysis of the cytochrome b gene. Schwemm et al. (2014a) described 10 microsatellite markers for N. suttkusi. That genetic analysis yielded 3 to 23 alleles per locus, with mean observed and expected heterozygosities of 0.679 and 0.729, respectively.

DISTRIBUTION   Notropis suttkusi is endemic to the Red River drainage (Blue, Kiamichi, and Little rivers) of the Ouachita Mountains in southwestern Arkansas and southeastern Oklahoma, with most of its range in Oklahoma. In Arkansas, it is found in the lower Rolling Fork, Cossatot, Saline, and Little rivers (Map 79).

HABITAT AND BIOLOGY   The Rocky Shiner inhabits clearwater rivers and streams of moderate to high gradient with gravel and rubble substrates (Humphries and Cashner 1994). It is usually found in moderate current near the periphery of riffles. Pratt (2000) reported gravel substrates at all sites occupied by *N. suttkusi* in Oklahoma, but found slight variations in substrate at different sites. In the Little River, individuals were found over a substrate of fine gravel, in the Kiamichi River over a substrate of gravel and cobble, and in the Blue River over a substrate of bedrock with some gravel and cobble. It sometimes occurs in habitats with low to moderate turbidity and moderate flow. The average current in the Blue River where *N. suttkusi* was found was 0.29 m/s (0.18–0.39 m/s), and in the Kiamichi River, average current was 0.19 m/s (0.03–0.39 m/s). *Notropis suttkusi* tends to use deep pools (depth >2 m) as thermal refugia during late spring and summer months (Pratt 2000). It apparently avoids extreme headwater environments, and a 1993 survey of 30 localities in Little River tributaries in Arkansas produced no specimens (Taylor 1994). Its distribution appears to be complementary to that of *Lythrurus snelsoni*, which occurs predominantly in headwater environments.

There is very little information on the feeding biology of the Rocky Shiner. Gut contents of 10 specimens collected by TMB from the Little River in July primarily contained insect remains. It apparently feeds throughout the water column, but most food items were probably taken at or near the water surface. The diet was dominated by adult, winged dipterans, coleopterans, and odonates, but ephemeropteran nymphs and other immature insect remains were also present in some guts.

In Oklahoma, Pratt (2000) studied the reproductive season, size at first reproduction, and seasonal gonadal development in two populations, and most of the following information on reproductive biology comes from that source. Females tend to be larger than males during the breeding season. Reproductive season begins in late March and continues to early August in Oklahoma (Pratt 2000) and Arkansas (HWR, pers. obs.). This differs from the reproductive season of *Notropis percobromus*, which begins in mid-April, peaks in May and June, and ends in early July (Pflieger 1997). The presence of breeding coloration indicates that males are allocating energy toward reproduction in *N. suttkusi*. In Oklahoma, males from the Blue River began to show breeding color in late February, but Kiamichi River males did not show breeding color until March. Breeding coloration persists throughout spring and early summer in both Oklahoma and Arkansas populations (HWR, pers. obs.). By August in the Kiamichi River, males smaller than 40 mm SL did not exhibit breeding color, but males greater than 40 mm SL continued to show breeding color. Males do not show breeding color from September through December. Pratt (2000) found young-of-the-year

in the Blue River in Oklahoma in September. Females had latent eggs in January and February in Oklahoma. In March, females had eggs ranging from latent to late maturing. By April females had late maturing and mature eggs, by May/June females had early to late maturing eggs, and by August females showed evidence of reabsorption of eggs. In the Blue River, an increase in gonad mass began in March, while the increase in gonad mass began in April in the Kiamichi River. This increase in energy investment also coincided with egg stage development. Ripe females and tuberculate males were found in the Little River on 12 July by TMB. Average clutch size was 298 ova in females averaging 55 mm TL. Some females had apparently finished spawning and contained no mature ova.

CONSERVATION STATUS   Because of its limited distribution and alteration of its habitat in the Little River system, we consider the Rocky Shiner threatened (if not endangered) in Arkansas.

Notropis telescopus (Cope)
Telescope Shiner

Figure 15.96. *Notropis telescopus*, Telescope Shiner. *David A. Neely.*

CHARACTERS   A slender shiner with large eyes and a terminal, oblique mouth (Fig. 15.96). Eye diameter greater than snout length and usually going into head length 3 or fewer times. In adults, eye diameter about equal to distance from back of eye to rear margin of gill cover. Dorsal rays 8. Anal rays 9–11 (usually 10). Pectoral rays 14–16. Pharyngeal teeth 2,4–4,2. Lateral line complete; lateral line scales 35–38. Predorsal scale rows 13–16; circumferential scales usually 20–22; circumpeduncular scales modally 12. Breast scaled. Dorsal fin origin behind pelvic fin origin; front of dorsal fin base slightly closer to caudal fin base than to snout tip. Intestine short; peritoneum black. Nuptial males have small granular tubercles covering entire head, but tubercles are more dense on top of head, mandibles, lips, and gular area than on the operculum and cheeks. Scale and pectoral fin tubercles are small and numerous (Etnier and Starnes 1993). Maximum size 3.75 inches (94 mm) TL.

LIFE COLORS   Dorsum olive with a black middorsal stripe and conspicuous dark edging on scales. Middorsal

Map 80. Distribution of *Notropis telescopus*, Telescope Shiner, 1988–2019.

stripe intensely dark and broad in front of dorsal fin and becoming narrower behind dorsal fin. Two or three faint to dark wavy stripes on each side along back and upper side meet those of other side behind the dorsal fin on the caudal peduncle (similar to *Luxilus chrysocephalus*); the converging dorsolateral lines form a V- or U-shaped pattern when viewed from above. Sides silvery. Venter silvery white. Fins colorless. No basicaudal spot. Pores of lateral line prominently outlined by dark dashes or dots anteriorly. Melanophores weakly developed on chin. No chromatic breeding colors.

SIMILAR SPECIES   The Telescope Shiner is superficially similar to *Luxilus chrysocephalus*, *L. pilsbryi*, *L. cardinalis*, and *L. zonatus*, but differs from those species in usually having 10 or 11 anal rays (vs. usually 9), 35–38 lateral line scales (vs. 39–44), and in having front of dorsal fin base slightly closer to caudal fin base than to tip of snout (vs. front of dorsal fin base closer to tip of snout than to caudal base). In general, *N. telescopus* has a larger eye than *Luxilus* species, but morphometric ratios involving eye diameter are not completely reliable for distinguishing those species (see *Luxilus pilsbryi* species account for more information). *Notropis telescopus* differs from *N. percobromus* in having a black peritoneum (vs. silver), a much larger eye, eye diameter usually going into head length 3 times or less

(vs. eye diameter of *N. percobromus* usually going into head length more than 3 times), chin weakly sprinkled with dark pigment (vs. chin without dark pigment), and dorsal fin origin over or slightly behind pelvic fin origin (vs. dorsal origin distinctly behind pelvic fin origin). It differs from *N. greenei* in lacking a small wedge-shaped spot at caudal base and in usually having more than 9 anal rays (vs. usually 8). From *Notropis boops*, it differs in usually having 10 or 11 anal rays (vs. 8) and 2,4–4,2 pharyngeal teeth (vs. 1,4–4,1).

VARIATION AND TAXONOMY   *Notropis telescopus* was described by Cope (1868a) as *Photogenis telescopus* from the Holston River and its tributaries in Virginia. This shiner was long considered a subspecies of *Notropis ariommus*; however, Gilbert (1969) compared variation between Ozark populations and populations east of the Mississippi River and concluded that *N. ariommus* was a distinct species and that recognition of subspecies within *N. telescopus* was unwarranted. Coburn (1982a) included this species in the subgenus *Notropis*. Wood et al. (2002) suggested that *N. telescopus* was closely related to *N. semperasper*, *N. atherinoides*, *N. stilbius*, and the *N. rubellus* species group. Mayden et al. (2006a) questioned the placement of *telescopus* within the genus *Notropis*, but suggested no alternative placement.

DISTRIBUTION   *Notropis telescopus* has a disjunct range. East of the Mississippi River, it occurs in Atlantic Slope drainages of the upper Santee River, North Carolina and South Carolina, and in the Mississippi River basin in the Cumberland and Tennessee river drainages from Virginia west to northern Alabama and eastern Tennessee, and is introduced into the New, Yadkin, and Catawba river drainages of Virginia, North Carolina, and South Carolina. West of the Mississippi River, it occurs in the Little, St. Francis, and White river drainages of the Mississippi River basin in Missouri and Arkansas. In Arkansas, the Telescope Shiner is confined to the White and Black river systems (Map 80). It appears to have declined in the White River drainage above Beaver Lake, and only two post-1988 records are known from that area.

HABITAT AND BIOLOGY   This schooling shiner is most abundant in clear water near runs or riffles with moderate current and flowing pools over a gravel or rocky substrate in streams and small to medium rivers. It requires streams with permanent flow, such as the Buffalo River and main stem of the middle White River, and it avoids headwaters. *Notropis telescopus* is usually found in mixed schools at the bases of riffles, often near water willow, especially with *Luxilus pilsbryi*, *L. zonatus*, and *N. percobromus*. It is intolerant of reservoirs.

The large eyes of the Telescope Shiner indicate that it is a sight feeder. Based on gut contents of a few adults taken in spring and summer in Tennessee, Etnier and Starnes (1993) reported that two-thirds of the food ingested consisted of terrestrial and surface insects, and one-third immature benthic insects (trichopterans, ephemeropterans, and plecopterans). In Virginia, New River specimens fed largely on simuliids and chironomids (Jenkins and Burkhead 1994). TMB examined gut contents of 65 specimens collected from Fulton, Izard, Newton, Randolph, Sharp, and Washington counties, Arkansas, between March and November. The diet in Arkansas was similar to that reported by Etnier and Starnes (1993). It fed at all levels in the water column, but most food was obtained at or near the surface. In spring and summer, the diet was dominated by terrestrial insects, primarily dipterans and ephemeropterans. Other items important in the diet were dipteran pupae and odonate nymphs. Other benthic food items included ephemeropteran nymphs, midge larvae, trichopteran larvae, and coleopteran larvae. Less food was ingested during October and November, and benthic forms were more important in the diet at that time (although some terrestrial insects were also consumed).

N. telescopus can be characterized as having a life history strategy intermediate between opportunistic (early maturity, small clutches, low survivorship) and periodic (late maturity, large clutches, low survivorship). It is a relatively small fish with seasonal spawning, small eggs, and several spawning bouts per season (Holmes et al. 2010). The periodic strategy is typical of larger cyprinids in which females are larger than males with little or no coloration difference. Though *N. telescopus* is a small cyprinid with early maturity and small clutches, it shares two of the periodic strategy features in having little external phenotypic difference between sexes and apparently larger females (Holmes et al. 2010). In many fishes, females are of greater body size than males of comparable age, allowing females to increase their relative reproductive output because the associated increase in body volume provides more room for the development and storage of more numerous and larger eggs (Parker 1992).

Pflieger (1997) reported breeding adults were collected from the Current River in mid-April in Missouri. In Tennessee, Etnier and Starnes (1993) reported observations of the species spawning in mid-April through mid-June. HWR collected breeding specimens as early as 4 April and as late as 20 June in Arkansas in Sylamore Creek (White River drainage, Stone County), and TMB found ripe females in the Strawberry River on 18 May. In Alabama, Holmes et al. (2010) found that *N. telescopus* entered spawning season in April, peaked in June, declined in July,

and was completed by August. They found that spawning began in relatively cool water at 11°C in April, and oocyte maturation declined as water temperatures peaked in July at 24°C. Telescope Shiners are thought to be pelagic broadcast spawners (Holmes et al. 2010), and spawning seems to be correlated with elevated stream flows; that is, stream flow is highest in early spring just as *N. telescopus* enters spawning condition with lowest discharge in July and August (B. Stallsmith, pers. comm.). These same hydrologic conditions occur in Sylamore Creek, where HWR collected spawning individuals from early April to late June.

CONSERVATION STATUS The Telescope Shiner is common and widespread in the White River system of northern Arkansas, but it is usually not found in large numbers at any locality. We consider its populations secure.

Notropis texanus (Girard)
Weed Shiner

Figure 15.97. *Notropis texanus*, Weed Shiner. *Uland Thomas.*

CHARACTERS A terete, straw-colored shiner with a prominent dark lateral stripe from snout tip to caudal base (Fig. 15.97). Snout blunt. Mouth moderate in size, terminal, and slightly oblique. Barbels absent. Eye large, its diameter equal to or slightly greater than snout length; eye diameter going into head length 2.8–3.3 times. Anal rays 7. Pectoral fin rays 12–16. Pharyngeal teeth usually 2,4–4,2. Lateral line complete, with 33–38 scales. Predorsal scale rows 14–16. Circumferential scales usually 23–27. Dorsal fin rays 8. Dorsal origin over or just slightly anterior to pelvic fin origin, much closer to tip of snout than to caudal base. Breast naked; belly fully scaled. Intestine short with a single S-shaped loop; peritoneum silvery, but sprinkled with dark specks. Breeding males with tubercles on head, snout, lower jaw, and pectoral fin rays, with the largest tubercles on snout tip, lower jaw, lachrymal area, and dorsal margin of the orbits. Arkansas specimens rarely more than 2.5 inches (63.5 mm) TL. Maximum size 3.4 inches (87.5 mm) TL (Swift 1970).

LIFE COLORS Dorsum olive yellow with dark-edged scales. Venter silvery white. Conspicuous dark lateral stripe

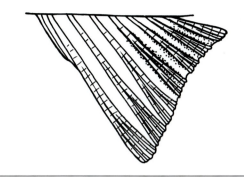

Figure 15.98. *Notropis texanus* anal fin.

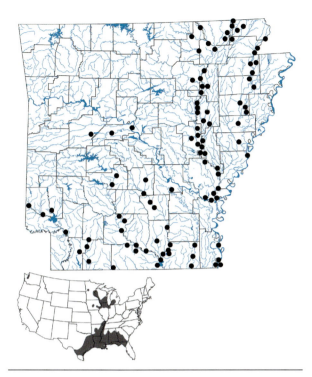

Map 81. Distribution of *Notropis texanus*, Weed Shiner, 1988–2019.

bordered above by a broad, light-colored zone (about 1.0–1.5 scales wide) in which scales are not dark-edged. Lateral black stripe extending anteriorly across operculum, through eye, and onto snout. Lips black. Small, dark, nearly round basicaudal spot present and connected to or barely separated from lateral stripe. Prominent middorsal stripe present in front of dorsal fin, but stripe becomes narrower and inconspicuous behind dorsal fin. Posterior rays of anal fin often margined with melanophores (Fig. 15.98). Breeding males with rosy pigment on the median and pectoral fins, the pale area above the lateral stripe, the top of the head, and the snout (Swift 1970; Etnier and Starnes 1993).

SIMILAR SPECIES   The Weed Shiner is very similar in appearance to *N. boops* and *N. ortenburgeri*. It is rarely sympatric in Arkansas with those two species and can be distinguished from them by its 7 anal rays (vs. 8 or more) and its pharyngeal tooth formula of 2,4–4,2 (vs. 1,4–4,1 or 0,4–4,0). It differs from the similar *N. chalybaeus* in having 7 anal rays (vs. 8), a complete lateral line (vs. incomplete), and in lacking dark pigment inside the mouth (vs. dark pigment present on floor and roof of mouth). It differs from *N. atrocaudalis* in having 2,4–4,2 pharyngeal teeth (vs. 0,4–4,0).

VARIATION AND TAXONOMY   *Notropis texanus* was described by Girard (1856) as *Cyprinella texana* from Salado Creek, near San Antonio, Bexar County, Texas. During the first half of the 20th century, this species was referred to by several names, but its taxonomic history and status were clarified by Suttkus (1958). The occurrence of 7 anal rays is one of the most consistent characteristics of Arkansas populations of the Weed Shiner. Anal ray counts from 283 Arkansas specimens revealed that 94% (266) had 7 rays, 5.6% (16) had 8 rays, and 0.4% (1) had 6 rays. Swift (1970) studied the systematics of *N. texanus* but recognized no subspecies. Boschung and Mayden (2004) noted that specimens in northern parts of its range were consistently deeper in body shape and more compressed, and mature at a smaller size than southern populations. Swift (1970), Coburn (1982a), and Mayden (1989) considered the *Notropis texanus* species group to contain that species plus *N. chalybaeus* and *N. petersoni*. Sister group relationships of *N. texanus* are not known. Mayden et al. (2006a) supported placement of *N. texanus* in genus *Alburnops*.

DISTRIBUTION   *Notropis texanus* occurs in the Lake Michigan, Hudson Bay, and Mississippi River basins from Manitoba, Canada, Wisconsin, and Michigan south to the Gulf of Mexico, and in Gulf Slope drainages from the Suwannee River of Georgia and Florida to the Nueces River, Texas, but it is most common in the southern part of its range. In Arkansas, the Coastal Plain is the area most typically occupied by the Weed Shiner, and it barely penetrates the lower reaches of typical Ozarkian streams such as the Strawberry, Spring and Black rivers (Map 81). This shiner is sporadic in its lowland occurrence and is seldom abundant. It is most common today throughout the lower White River, with a number of recent records from the St. Francis (Connior et al. 2012), Ouachita, and Saline rivers. There are also a few records from the Arkansas (McAllister et al. 2009c) and Mississippi rivers and from Red River tributaries and oxbows in extreme southwestern Arkansas. A somewhat disjunct large population occurs in Lake Nimrod and in the lower Fourche LaFave River.

HABITAT AND BIOLOGY   The Weed Shiner is most abundant in clear lowland streams of small to moderate size with sand and mud bottoms, slight current, and aquatic vegetation. It is rarely found in the large rivers of Arkansas (Arkansas, Mississippi, and Red) but occurs in small numbers in the lower White River and in a few oxbow lakes associated with large rivers. The few records from large rivers are from backwater areas with little or no current. Buchanan (2005) reported it from five Arkansas impoundments. In Village Creek, Texas, *N. texanus* was associated primarily with sandbar habitat, but as discharge increased during winter, it was found in backwaters and flooded riffles (Moriarty and Winemiller 1997). *Notropis texanus* is a microhabitat generalist (Baker and Ross 1981) but primarily occupies the lower midportion of the water column and is a schooling species (Becker 1983). Ross (2001) noted that despite the common name, it is rarely found at the microhabitat level in association with aquatic vegetation, even though on a coarser scale it may inhabit vegetated streams. Becker (1983) concluded that the Weed Shiner is extremely sensitive to environmental deterioration or changes.

Foods consist of plant debris, including filamentous algae, and animal material (Becker 1983). Ross (2001) reported that gut contents in Mississippi included aquatic insect remains, detritus, and sand grains. In Louisiana, the diet differed between wet and dry seasons, with organic debris making up 81% by volume during the dry season, and with a broader diet during the wet season that included surface prey (20%), midwater prey (5%), benthic animal prey (39%), and organic detritus (35%) (Felley and Felley 1987). Gut contents of 63 Arkansas specimens collected in March, July, August, and November from Hempstead, Monroe, Sevier, and Yell counties were examined by TMB. The diet in Arkansas was similar to that described for other states. The Weed Shiner consumed a variety of foods but was largely an insectivore, feeding at all levels in the water column. The most important foods by percent volume and frequency were immature insects, including ephemeropteran nymphs, chironomids and other midges, other dipteran larvae, pupae, and terrestrial adults, coleopterans, plecopterans, and odonates. Microcrustaceans were occasionally ingested, including cladocerans, copepods, and ostracods. Gut contents also included small amounts of rotifers, detritus, sand, and filamentous algae.

In Illinois, Smith (1979) reported collecting males in late August with dense but minute tubercles and females distended with eggs. Spawning occurs in June and July in Wisconsin (Becker 1983). Heins and Davis (1984) reported a protracted reproductive season in Mississippi extending from March through September, and Heins and Rabito (1988) found that the breeding season of Gulf Coast populations extended from March through August, with some breeding possibly occurring as early as February and as late as October. Water temperatures during the reproductive season ranged from 14°C to 25.5°C. The Weed Shiner also has a protracted breeding season in Arkansas, extending from at least late March to mid-August. Ripe females collected by TMB from Indian Bay in Monroe County on 26 March ranged from 52 to 60 mm TL and contained from 324 to 699 ripe ova ($\bar{x} = 478$). Ripe females were also found on 13 July, but ovaries of females collected on 21 August from Monroe County were regressed. Males and females reach sexual maturity at age 1, at sizes as small as 30 mm SL. Ross (2001) reported that multiple clutches of eggs are produced during the spawning season, with clutches averaging 500–700 eggs. Bresnick and Heins (1977) reported mean standard lengths of 37, 44, and 54 mm were attained at ages 1, 2, and 3, respectively. Maximum life span was 3 years or slightly longer.

CONSERVATION STATUS   The Weed Shiner is fairly widely distributed in Arkansas. It is commonly encountered in a few drainages, and we consider its populations secure.

Notropis volucellus (Cope)
Mimic Shiner

Figure 15.99. *Notropis volucellus*, Mimic Shiner.
David A. Neely.

CHARACTERS   A small, relatively chubby shiner with a fairly small head (head length going into SL 3.5–4 times) and moderately large eyes (eye diameter going into head length approximately 2.5 times in White River populations) (Fig. 15.99). Snout rounded, snout length going into head length about 3.3 times (White River). Mouth small, subterminal, and moderately oblique. Infraorbital canal complete. Fins moderately large. Dorsal fin origin over pelvic fin origin. Dorsal and anal rays 8. Pectoral fin rays 13–17. Pharyngeal teeth 0,4–4,0. Lateral line complete, nearly straight, and with 32–38 scales. Anterior scales in lateral line 2–3 times higher than wide (exposed surface) and higher than scales in adjacent rows. Breast usually

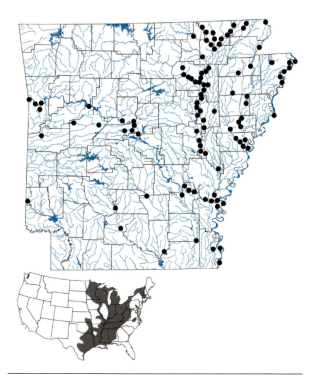

Map 82. Distribution of *Notropis volucellus*, Mimic Shiner, 1988–2019.

naked but may have some partially embedded scales. Body circumferential scales 17–22. Scale rows above lateral line 9–11; scale rows below the lateral line 3–9. Predorsal scales 13–16. Intestine short; peritoneum silvery with scattered large, dark chromatophores. Median fins falcate. Tips of pelvic fins not reaching front of anal fin base. Breeding males with minute tubercles over top of head, branchiostegals, and along rays of pectoral fin (Becker 1983). Maximum size about 3 inches (76 mm), but Arkansas specimens rarely exceed 2.3 inches (58 mm) TL. Largest Arkansas specimen examined was 2.36 inches (60 mm) TL from Vache Grasse Creek in Sebastian County.

LIFE COLORS   Body coloration basically pale, straw-colored above, sides silver, and venter silvery white. Lateral band diffuse, not extending onto head; degree of lateral band development variable in Arkansas specimens (see Variation and Taxonomy section below). Middorsal stripe absent in most specimens, but sometimes a line of melanophores (1–3 melanophores wide) present on back in front of dorsal fin. Dark blotch usually evident at front of dorsal fin base. No middorsal stripe posterior to dorsal fin. A small, triangular basicaudal spot frequently present. Scales on dorsum conspicuously outlined by melanophores. Anterior portion of lateral line often outlined by dark dots or dashes. Fins clear. No chromatic breeding colors.

SIMILAR SPECIES   *Notropis volucellus* differs from *N. buchanani* in having an uninterrupted infraorbital canal (vs. absent or interrupted in *N. buchanani*), less extremely elevated lateral line scales, tips of pelvic fins not reaching front of anal fin base (vs. tips reaching front of anal fin base), and more pigment on body. It is most similar to *N. wickliffi*, and because both species show considerable variation in pigmentation and other features, they are very difficult to distinguish from one another. It is common to find cataloged lots in museums with both species in the same jar. Furthermore, characters reported from other areas for distinguishing the two species (such as development of the pre- and postdorsal stripes, body depth, caudal peduncle depth) do not work for separating Arkansas specimens. Robert Hrabik (pers. comm.) provided the following information on diagnostic characters and suggested that the most useful (primary) characters are observable from the dorsal view. There is usually a noticeable predorsal dark spot anterior to the dorsal fin insertion in *N. volucellus*, but there is some variation in this character (vs. predorsal dark spot absent or very light in *N. wickliffi*). Dorsolateral scales of *N. volucellus* have melanophores more or less concentrated along the scale margins and the interior of the scale is unpigmented or sparsely sprinkled with melanophores (vs. *N. wickliffi* with melanophores more or less evenly distributed across the scales, especially scales anterior to the dorsal fin insertion). Overall, the melanophores are darker in *N. volucellus* and lighter in *N. wickliffi*. Degree of development of pre- and postdorsal stripes is highly variable in both species and is not reliable for distinguishing the two species. *Notropis volucellus* lacks a well-developed predorsal stripe, but *N. wickliffi* specimens also seldom have one. A postdorsal stripe is absent in *N. volucellus* and in most *N. wickliffi*, but a postdorsal stripe is present in a small percentage of *N. wickliffi* (Fig. 15.100). When viewing a specimen from the side, some crosshatching is usually observed above the part of the lateral line outlined by dark dots in *N. volucellus*, while there is usually no crosshatching evident in *N. wickliffi*. *Notropis volucellus* appears to have a smaller eye than *N. wickliffi*, but the eye diameter into head length morphometric is almost identical in the two species (approximately 2.4 times) because *N. wickliffi* has a longer head than *N. volucellus*. *Notropis wickliffi* has a longer, more pointed snout than *N. volucellus*, snout length going into head length fewer than 3 times in *N. wickliffi* and more than 3 times in *N. volucellus* (at least in syntopic White River drainage populations). In addition, nuptial males of *N. volucellus* have more prominent tubercles on the snout and lachrymal areas than those on top of the head, whereas the reverse is true in *N. wickliffi*.

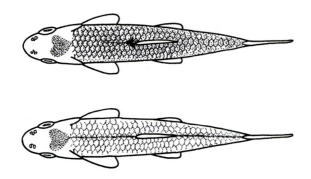

Figure 15.100. Postdorsal pigmentation in *Notropis volucellus* (top) and *N. wickliffi* (bottom).

VARIATION AND TAXONOMY   *Notropis volucellus* was described by Cope (1865a) as *Hybognathus volucellus* from the Detroit River at Grosse Isle, Wayne County, Michigan. It has at times been placed in the genera *Hybopsis* and *Alburnops*. It was also long confused with *Notropis blennius*, and Hrabik (1996) reviewed its tangled taxonomic history. Gong (1991) and Gong and Cavender (1991) studied systematics of *Notropis volucellus* and *N. wickliffi*. Hrabik (1997) studied meristic and morphometric variation in the upper Mississippi River. It exhibits substantial morphological variation across its large range from Canada to Texas and eastward, and there is variation among populations in Arkansas. *Notropis volucellus* remains a complex of several undescribed species, despite a few systematic studies. Additional rangewide morphological and genetic analyses are needed to resolve the systematics of this polytypic taxon. Mayden and Kuhajda (1989) described *N. cahabae* as a new species from the *N. volucellus* species group and elevated the formerly recognized subspecies *N. volucellus wickliffi* to full specific status. Mayden and Kuhajda (1989) also provided a population-level phylogeny of *N. volucellus* across its geographic range and noted genetic differences among drainages. They considered *N. volucellus* sister to *N. cahabae* plus *N. wickliffi*. Mayden et al. (2006a) presented a different interpretation in an analysis of the cytochrome *b* gene. Those authors recovered *N. volucellus* as sister to *N. buchanani*, with those species forming a clade sister to *N. ozarcanus*. Surprisingly, *N. volucellus* was not as closely related to *N. wickliffi* as the morphological evidence suggested.

Morphologically, at least two forms of *volucellus*-like fish inhabit Arkansas waters in addition to *Notropis wickliffi*. What is currently called *N. volucellus* in Arkansas may include more than one species, and Arkansas forms are almost certainly specifically distinct from the population of *N. volucellus* at its type locality in Michigan. Variation exists across Arkansas in *N. volucellus* in several features, such as pigmentation of the lateral line, development of a lateral band, pre- and postdorsal pigmentation, eye size, and body shape. Specimens from the Arkansas River and its tributary creeks near Fort Smith are often more strongly pigmented than those in other populations, usually having a narrow, distinct lateral band that is best developed on the caudal peduncle and somewhat more diffuse (but still obvious) anteriorly. There are scattered melanophores in a line (2 or 3 melanophores wide) on the back in front of the dorsal fin that do not form a distinct stripe. Scales on the back behind the dorsal fin have melanophores on the distal one-third of the scale, but do not form a postdorsal stripe. The anterior lateral line scales are prominently outlined by paired dots. Black (1940) referred to the form in western Arkansas (Poteau River and adjacent tributaries of the Arkansas River) as *Notropis volucellus rivalis*, the Neosho Mimic Shiner, and Hubbs and Bonham (1951) called it an "unnamed creek subspecies." Pflieger (1971, 1997) noted that this form occurs in the Neosho River drainage of southwestern Missouri and may represent an undescribed species. Additional study is needed to determine the taxonomic status of *N. volucellus* populations in western Arkansas.

DISTRIBUTION   *Notropis volucellus* is found in St. Lawrence–Great Lakes, Hudson Bay, and Mississippi River basins across Canada from Quebec to Manitoba south to the Gulf of Mexico; in a few Atlantic Slope drainages from Connecticut and Massachusetts (where it is presumably introduced) to North Carolina; and in Gulf Slope drainages from Mobile Bay, Georgia and Alabama, to the Nueces River, Texas. In Arkansas, it is widely distributed in the White, Black, St. Francis, Mississippi, Saline, Ouachita, and Arkansas rivers. There is a single record from the Red River drainage of southwestern Arkansas (Map 82).

HABITAT AND BIOLOGY   The Mimic Shiner is a schooling minnow living in midwater or at the surface. It is found in a variety of habitats but generally occurs in medium to large streams and rivers in current over gravel or hard substrates. It can also be found in lakes, primarily in the northern part of its range (Black 1945) and is known from both vegetated and unvegetated habitats. Etnier and Starnes (1993) noted that it occurs in some large reservoirs in Tennessee, but in Arkansas, the only populations reported from impoundments are in Blue Mountain Lake and in Arkansas River navigation pools (Buchanan 2005). In Missouri, it is locally abundant in clear, vegetated, and flowing Bootheel ditches (Hrabik 1996). It avoids small streams and is largely replaced in the Mississippi River of Arkansas by *N. wickliffi*, although it occurs in small numbers in that river. In the big rivers it occurs over sandy bottoms and is frequently taken near creek mouths. The Mimic Shiner is probably most abundant in Arkansas in the lower White River drainage, where it may occur syntopically with *N. wickliffi*.

The Mimic Shiner fed mainly on small crustaceans, terrestrial and aquatic insects, and some algae in Indiana and Minnesota (Black 1945; Moyle 1973). Olmstead et al. (1979) reported Mimic Shiners were active insectivores found mostly in the benthic areas of the littoral zone. Various authors have reported diel and seasonal shifts in food habits (Ross 2001). During the morning hours in lakes, it feeds heavily on midwater organisms such as cladocerans, but by midday it moves out of the central water column to feed at the surface (on emerging dipterans and terrestrial insects) or along the bottom on midge larvae and mayfly nymphs, cladocerans, algae, and detritus (Black 1945; Johnson and Dropkin 1991; Ross 2001). Terrestrial insects form an important part of the diet in spring and summer but decline in importance during fall and winter (Moyle 1973). Gut contents of 87 Arkansas Mimic Shiners collected from March through November from Craighead, Crawford, Independence, Monroe, Sebastian, and White counties were examined by TMB. That examination revealed that Arkansas populations are primarily insectivores throughout the year. It fed predominantly on immature benthic insects, but some feeding occurred throughout the water column, including the consumption of emerging adult insects and terrestrial insects near the water surface in summer months. Contrary to some reports from other areas, almost no consumption of microcrustaceans was documented. The diet in Arkansas was dominated by chironomids and other midges, other larval and pupal dipterans, ephemeropterans, plecopterans, and coleopterans. Other insects consumed in small amounts included odonates, ants, hemipterans, megalopterans, and trichopterans. Other items found in 2–17% of the guts were sand, detritus, filamentous algae, diatoms, and seeds.

Adults in spawning condition have been collected in Missouri from early June to late July (Pflieger 1997). Etnier and Starnes (1997) reported tuberculate males from late May through early October, indicating a prolonged spawning period in Tennessee. A long spawning period was also confirmed in Alabama, extending from mid-April to early August (Oliver 1986). Prolonged spawning occurs in Arkansas, with ripe females and tuberculate males known from mid-April into early August. Examination of gonads by TMB revealed that spawning occurred as early as late April in Vache Grasse Creek in Sebastian County, where ripe females between 48 and 52 mm TL contained from 246 to 336 mature ova. The presence of smaller egg sizes in the ovaries indicated multiple spawning by females. In Alabama, fecundity ranged from 74 to 386 eggs in fish 36.4–45.1 mm SL (Oliver 1986), and in northern populations, some females contained as many as 960 eggs (Becker 1983). Ripe females were found in the White River by Jeff Quinn at Batesville on 15 May, and by TMB on July 27 and 29. Ovaries of females collected in late July contained very few smaller eggs, indicating that spawning was winding down. The exact breeding sites and spawning habits of Mimic Shiners in streams are unknown, but TMB found adults in breeding condition in pools with some current in Vache Grasse Creek at a water temperature of 20°C. Black (1945) suggested Mimic Shiners are nocturnal spawners in Indiana, using deep water areas among dense weed beds. Moyle (1973) reported that *N. volucellus* was a broadcast spawner, spawning in large schools over vegetation in lakes. Middleton et al. (2013) reported Mimic Shiners exhibit indeterminate growth, and gender influences growth patterns. Age 1 females ranged from 29 to 51 mm TL and age 1 males ranged from 30 to 46 mm TL based on 884 individuals collected from Lake Erie. Age 2 females ranged from 57 to 61 mm TL, and age 2 males ranged from 54 to 56 mm TL. Life span is probably 3 years with sexual maturity reached in the first year.

CONSERVATION STATUS   The Mimic Shiner is widespread and common in Arkansas, and its populations are secure.

Notropis wickliffi Trautman
Channel Shiner

Figure 15.101. *Notropis wickliffi*, Channel Shiner. *David A. Neely.*

CHARACTERS   A small shiner with a fairly small head and moderately large eyes (eye diameter going approximately 2.4 times into head length in White River drainage populations) (Fig. 15.101). Snout rounded; longer than in *N. volucellus*; snout length going into head length approximately 2.9 times. Mouth small, subterminal, and moderately oblique. Fins moderately large. Dorsal fin origin over pelvic fin origin. Predorsal scale rows 12–15.

Map 83. Distribution of *Notropis wickliffi*, Channel Shiner, 1988–2019.

Body circumferential scales 21–24; scale rows above lateral line usually 11; scales below lateral line 9–11. Anal rays 8. Pectoral fin rays 14–16. Pharyngeal teeth 0,4–4,0. Lateral line complete, nearly straight, and with 33–36 scales. Anterior scales in lateral line elevated, higher than long (exposed surface) and higher than scales in adjacent rows. Breast naked or with a few scales. Infraorbital canal complete. Intestine short; peritoneum silvery with scattered dark speckles. Median fins falcate. Tips of pelvic fins not reaching front of anal fin base. Nuptial tubercle distribution is similar to that of *N. volucellus*, but there are differences in size of tubercles between the two species. In *N. wickliffi*, snout and lachrymal tubercles are not well developed, while those on top of the head (interorbital area) are more prominent (relative tubercle sizes are reversed in *N. volucellus*) (Etnier and Starnes 1993). Adults usually 1.5–2.8 inches (38–71 mm) TL, with maximum size around 3.1 inches (78 mm).

LIFE COLORS Body coloration basically pale, straw-colored above, sides silvery, white below with a diffuse lateral band not extending onto head. Predorsal stripe weak or absent (Fig. 15.100). Usually no middorsal stripe posterior to dorsal fin, but stripe is present in a small percentage of specimens (large individuals may have a thin postdorsal stripe). A small, triangular basicaudal spot is frequently

present. Predorsal blotch at base of dorsal fin absent or very faint. Dorsolateral scales more or less evenly pigmented with melanophores; crosshatching not evident. Fins clear. No chromatic breeding colors.

SIMILAR SPECIES *Notropis wickliffi* is most similar to *N. volucellus* and *N. buchanani* (see *N. volucellus* species account for main differences). B. R. Kuhajda (pers. comm.) provided the following additional information for distinguishing those species: *N. wickliffi* always has a naked or almost naked breast versus a fully scaled to almost naked breast in *N. volucellus*. *N. wickliffi* has a shorter and thicker body, including a shorter and thicker caudal peduncle, and also has larger eyes than *N. volucellus*. Scales on dorsum of caudal peduncle of *N. wickliffi* are uniformly pigmented to give an overall smoky-gray appearance; in *volucellus*, the center of the scale is lightly pigmented, the scale edge is heavily pigmented, and the color is more strawlike. Development of pre- and postdorsal fin stripe on back is variable and not reliable. Ross (2001) gave modal meristic features for separating the two species in Mississippi: *N. wickliffi* has modally 33 pored lateral line scales (vs. 36 in *N. volucellus*), usually 22–24 circumferential scales (vs. 19–22), and typically 15 pectoral rays (vs. 14 pectoral rays). *Notropis wickliffi* differs from *N. buchanani* in having a complete infraorbital canal (vs. absent or incomplete).

VARIATION AND TAXONOMY *Notropis wickliffi* was originally described as a subspecies of *N. volucellus*, *N. v. wickliffi*, by Trautman (1931) from the Miami River at its confluence with the Ohio River, Hamilton County, Ohio. Jenkins (1976) and Gilbert (1978) listed *N. wickliffi* as a distinct species but provided no justification for its recognition. It was elevated to species level by Mayden and Kuhajda (1989), who provided morphological and electrophoretic evidence for its elevation. Etnier and Starnes (1993) and many subsequent workers also recognized it as a distinct species, and a combined morphological and molecular analysis by Mayden et al. (2006a) further supported its specific distinctness. Despite its unsettled taxonomy, it was recognized by Page et al. (2013) as a valid species, a designation that we support. Gong (1991) and Gong and Cavender (1991) studied systematics of *Notropis volucellus* and *N. wickliffi*. Eisenhour (1996) studied the systematics and distribution of *N. wickliffi* in Illinois, and Hrabik (1997) studied meristic and morphometric variation in the upper Mississippi River. Mayden and Kuhajda (1989) considered *N. wickliffi* and *N. volucellus* to form a sister group to *N. cahabae*. Based on analysis of the cytochrome *b* gene, Mayden et al. (2006a) found that *N. girardi* was the closest relative of *N. wickliffi*.

Notropis wickliffi, as currently recognized, is a variable species. Descriptions of diagnostic features vary regionally and do not seem to refer to the same fish (Page and Burr 2011). Arkansas populations currently assigned to *N. wickliffi* may eventually be split off as a separate species.

DISTRIBUTION The range of *Notropis wickliffi* is uncertain because of confusion with *N. volucellus*, and because it is probably a complex of species. As currently recognized, it is distributed in the Ohio, Mississippi, and Tennessee rivers and their larger tributaries from Pennsylvania to Missouri and south to Alabama, Mississippi, and Louisiana. Black (1940) reported the first Arkansas record of *N. wickliffi* (then known as *Notropis volucellus wickliffi*) taken on 8 August 1939 from the Mississippi River between Barfield and Hickman in Mississippi County. No additional records were known until Robison and Buchanan (1994) reported it from the Current River. B. R. Kuhajda (pers. comm.) collected this species in the Current (Clay/ Randolph counties) and White (Independence County) rivers, and Etnier and Starnes (1993) reported *N. wickliffi* from three localities in the Mississippi River. McAllister et al. (2009d) reported seven new records for *N. wickliffi* from Desha, Mississippi, and Phillips counties of Arkansas, including one collection of 146 specimens taken with minifyke nets off sloping sandbars in the Mississippi River below Helena. It has been found in the lower Arkansas River and is now known to occur throughout the Mississippi River in Arkansas (Map 83). Because it is commonly confused with *N. volucellus*, *N. wickliffi* is likely more widely distributed in Arkansas than currently known.

HABITAT AND BIOLOGY The Channel Shiner is a schooling minnow living at the surface or in midwater regions. It is restricted mainly to large rivers in quiet water to moderate current habitats over substrates ranging from silt to gravel (Etnier and Starnes 1993), but it occasionally occurs in tributaries of large rivers. Trautman (1957) reported that it was common in large pools with no visible current, but at times it congregated in immense schools in the swift currents below dams. It is most abundant in the Mississippi River and largely replaces *N. volucellus* in that river in Arkansas. In the Mississippi River of Arkansas, it is commonly found over a sand substrate in current and can easily be seined along sandbars. Its most common species associates in the Mississippi River were *Notropis buchanani*, *N. atherinoides*, *N. shumardi*, and *N. blennius* (Hrabik 1996). In Arkansas, it is also commonly found with *Hybognathus nuchalis*.

There is little information on its biology. Gong (1991) reported that *N. wickliffi* feeds primarily on cyanobacteria,

diatoms, and immature aquatic insects. Examination of gut contents of 26 specimens collected by TMB, Sam Barkley, and Sam Henry from the Mississippi River of Arkansas in March, July, and August also supported an omnivorous diet. Gut contents were dominated by organic material, filamentous algae, and immature aquatic insect remains. Some individuals ingested sand, and there was much unidentifiable detritus.

Notropis wickliffi apparently spawns in the summer as Trautman (1931) reported spawning in Ohio from June to August, and Etnier and Starnes (1993) reported tuberculate specimens from those same months in Tennessee. During the spawning season in Ohio, it was commonly taken in large numbers over gravel bars (Trautman 1957). Ripe females were collected in late July from the Mississippi River by TMB and Sam Barkley. Eight juveniles taken at the same time averaged 26 mm TL, indicating that some spawning occurred a few weeks earlier. Ripe females ranged from 52 to 65 mm TL and contained from 200 to 511 large ova. In Ohio, YOY were 0.8–1.6 inches (20–41 mm) TL in October, and age 1 individuals were 1.2–2.5 inches (30–64 mm) TL (Trautman 1957).

CONSERVATION STATUS The Channel Shiner is more widespread than previously believed (D. A. Etnier, pers. comm.), and with reports of recent collections in Arkansas (TMB unpublished data; Vicksburg Waterways Experiment Station data provided by K. J. Killgore; McAllister et al. 2009d), we believe its populations are currently stable.

Genus *Opsopoeodus* Hay, 1881— Pugnose Minnows

Controversy over the recognition of this monotypic genus has existed for many years. Over the past 50 years, different studies have supported its placement in three genera: *Notropis* (Gilbert and Bailey 1972; Dimmick 1987; Mayden 1989), *Pimephales* (Mayden et al. 2006a), and *Opsopoeodus* (Campos and Hubbs 1973; Amemiya and Gold 1990; Cavender and Coburn 1992). Based on morphology and breeding behavior, Coburn and Cavender (1992) and Page and Johnston (1990a) indicated that *Opsopoeodus* was closely related to *Pimephales* and *Cyprinella*. We follow Blanton et al. (2011) and Page et al. (2013) in continuing to recognize the genus *Opsopoeodus*. Genus characters are given in the species account below.

Opsopoeodus emiliae Hay
Pugnose Minnow

Figure 15.102. *Opsopoeodus emiliae*, Pugnose Minnow. *Uland Thomas.*

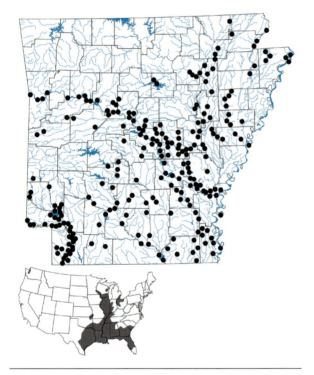

Map 84. Distribution of *Opsopoeodus emiliae*, Pugnose Minnow, 1988–2019.

CHARACTERS   A distinctive, small, moderately compressed, delicate-looking minnow with a small head, a distinctive, small, sharply upturned, vertical mouth, and a moderately large eye; eye diameter going into head length 3–3.3 times (Fig. 15.102; Fig. 15.80B). Dorsal fin with 9 rays, its origin over or slightly behind pelvic fin origin. Dorsal fin acutely pointed and high. Snout short and blunt, its length less than eye diameter. Barbels generally absent, but a barbel is occasionally present on one or both sides of the mouth. Anal rays 8. Pectoral fin rays usually 14 or 15. Scales edged with black and appear diamond shaped. Pharyngeal teeth 0,5–5,0, strongly hooked, and with oblique serrated cutting surfaces. Lateral line usually incomplete or interrupted, but about one-third of individuals in some Arkansas populations with a complete lateral line; usually 36–40 scales in lateral series. Dark lateral band extends from opercle to base of caudal fin. Caudal peduncle slender. Intestine short, with a single S-shaped loop; peritoneum silvery with dark speckles. Breast and prepectoral area naked; belly scaled. Breeding males with unique nuptial tuberculation; tubercles present above upper lip and a cluster on lower jaw, with minute tubercles covering rest of head. An additional dense patch of tubercles occurs in the gular area, extending from the lower lip to slightly behind the jaw angle. Moderate-sized tubercles on dorsal surfaces of pectoral rays 2–6 or 7 that form a single row basad, 2 rows with about 3 tubercles per row per fin ray segment medially, and extending a short distance past major branching of rays as a single row on each branch (Etnier and Starnes 1993). The first pectoral fin ray has a poorly defined single row of smaller tubercles. Tubercles absent elsewhere. Maximum reported size around 2.6 inches (66 mm) TL, but largest Arkansas specimen collected by TMB was 2.76 inches (70 mm) TL from Lake Chicot.

LIFE COLORS   Body color silver to green dorsally and silver laterally and ventrally. Dorsal and lateral scales distinctly outlined with melanophores, producing a diamond-shaped or crosshatched appearance. A narrow, dark lateral band usually prominent, but its degree of development is variable and is most prominent in the young. The bases of 3 or 4 caudal rays at the end of the lateral band moderately to heavily pigmented. Breeding males with a large anterior and posterior black blotch in the dorsal fin with a clear "window" between them, giving the dorsal fin a strikingly bicolored appearance. Breeding males develop a pink or reddish coloration on the fins with tips of paired fins white, and the dorsal fin tends to be blackish.

SIMILAR SPECIES   No other Arkansas minnow has the combination of 9 dorsal rays, 0,5–5,0 pharyngeal teeth, and a tiny, upturned mouth. Although *Pteronotropis hubbsi* often has 9 dorsal rays and a small mouth, *O. emiliae* differs from it in having 0,5–5,0 pharyngeal teeth (vs. 0,4–4,0 pharyngeal teeth), and in having more than 9 pored lateral line scales (vs. 9 or fewer pored scales).

VARIATION AND TAXONOMY   *Opsopoeodus emiliae* was described by Hay (1881) from Horsehunter Creek, upper Tombigbee River system, Noxubee County, Mississippi. Gilbert and Bailey (1972) transferred it to genus *Notropis*. Campos and Hubbs (1973) argued in favor of retaining *Opsopoeodus* on the basis of chromosome differences, but Dimmick (1987) further supported placing this species in *Notropis* based on genetic evidence. Mayden et al. (2006a) supported its inclusion in genus *Pimephales*, but subsequent analysis of nuclear gene sequences was consistent with previously published morphological, chromosomal, and behavioral data that supported the recognition of *Opsopoeodus* as a genus (Blanton et al. 2011). Based on its reproductive behavior, its closest affinities probably are with *Pimephales* (Page and Johnston 1990a). Gilbert and Bailey (1972) recognized two subspecies, *O. e. peninsularis* in peninsular Florida, and the nominate form *O. e. emiliae* elsewhere. Intergrades were reported in southern Georgia and northern Florida.

Degree of lateral line development is variable in Arkansas populations. In most specimens, the lateral line ends on the caudal peduncle, or sometimes more anteriorly. In many specimens, the lateral line is interrupted in one or more places, and some individuals have both an incomplete and interrupted lateral line. Up to one-third of individuals in some Arkansas populations have a complete lateral line.

DISTRIBUTION   The Pugnose Minnow occurs in the Great Lakes and Mississippi River basins from Ontario, Canada, and southern Minnesota south to the Gulf of Mexico; in Atlantic Slope drainages from South Carolina to southern Florida; and in Gulf Slope drainages from Florida to the Nueces River, Texas. In Arkansas, it is widespread throughout all rivers of the Coastal Plain and the Arkansas River Valley (Map 84). It is found throughout the Red River in Arkansas, where it is uncommon in the main stem (Buchanan et al. 2003) but common in Red River oxbows (TMB, unpublished data). It penetrates some streams along the southern edge of the Ouachita Mountains ecoregion. A large population exists in Lake Nimrod (Fourche LaFave drainage). Recently, the first documented specimen from the Current River was taken in Randolph County at the edge of the Ozark Mountains.

HABITAT AND BIOLOGY   The Pugnose Minnow is primarily a lowland species and typically inhabits tannin-stained, vegetated, quiet regions of sluggish streams and borrow pits, sloughs, or oxbow lakes over mud and sand or debris substrates in or near vegetation. It is found in streams of different sizes, ranging from creeks to the largest rivers in the state. Gilbert and Bailey (1972) and Trautman (1981) believed that the presence of both emergent and submerged aquatic vegetation was strongly associated with stable populations. It is also found in clearwater habitats but is tolerant of some turbidity. It apparently adapts well to some reservoirs, and Buchanan (2005) reported it from 25 Arkansas impoundments.

Mouth morphology indicates a midwater or surface-feeding species (Gilbert and Bailey 1972), but Boschung and Mayden (2004) noted that the most commonly reported food items are benthic organisms. Our data from Arkansas populations support the conclusion of Boschung and Mayden (2004) that most food is ingested from the substrate or lower part of the water column. McLane (1955) reported that it is an omnivore, with the diet in Florida consisting of chironomid larvae, filamentous algae, fish eggs and larvae, and microcrustaceans (copepods and cladocerans). Becker (1983) reported a similar diet in Wisconsin. Pflieger (1997) believed that its structural adaptations suggest it is carnivorous. TMB examined gut contents of 77 individuals collected from Chicot, Clark, Garland, Faulkner, Mississippi, Monroe, and Yell counties in Arkansas from March into November. In Arkansas, it was less of an omnivore than reported for Florida and Wisconsin populations. The diet in all months was dominated by immature aquatic insects, primarily various types of midge larvae. Other insect food items consisted of coleopteran larvae, other dipteran larvae and pupae, and trichopteran and ephemeropteran nymphs. Eleven percent of the guts sampled in spring and summer contained winged adult insects taken near the water surface. Microcrustaceans (cladocerans, copepods, isopods, and ostracods) were an important dietary component, occurring in 27% of the guts, but were most important in individuals smaller than 60 mm TL. Food items other than insects and microcrustaceans made up only a small proportion of the diet, occurring in small amounts in 19% of the guts sampled, and consisted of detritus, filamentous algae, sand grains, and seeds. Guts sampled in October and November contained less food than those in spring and summer months.

A long spawning season was reported in Florida, extending from March into late summer (McLane 1955). Spawning in Alabama was believed to occur from mid-April to September (Boschung and Mayden 2004), and that agrees with our observations for Arkansas populations. Gilbert and Bailey (1972) reported finding spawning individuals in Arkansas in May and June. Based on examination of gonads by TMB, the Pugnose Minnow spawns in Arkansas from mid-April to at least mid-August. Females collected on 26 March from Indian Bayou in Monroe County contained enlarging ova that were still a few weeks away from reaching maturity. Adults in spawning condition were taken from Lake Hamilton in Garland County on 20 April. Ten ripe females collected on 29 July from Lake Chicot ranged from 64 to 70 mm TL and contained from 180 to 372 ($\bar{x} = 243$) mature ova. The presence of smaller developing egg sizes indicated multiple spawning. Ripe females and tuberculate males with enlarged testes were found on 6 August in Lake Nimrod in Yell County. The spawning mode of this minnow is egg clustering (Page and Johnston

1990a). During courtship, the territorial male tends to rapidly raise and lower the dorsal fin when excited, giving the white or clear area of the otherwise dark fin the appearance of a "flickering light" (Page and Johnston 1990a). Ross (2001) reported that the tips of dorsal rays 1–3 of the male may develop small, whitish-yellow knobs which function as egg mimics to induce the female to spawn. In aquaria, the partly inverted females deposited eggs on the underside of a flat object. The spawning act lasts about 1 second but is repeated numerous times. The nest site and eggs are defended by the male, who keeps the eggs aerated and free of fungus (Page and Johnston 1990a). Eggs hatched in 90 hours at 27°C and in 142 hours at 21°C, but temperatures above 27°C and below 21°C were lethal to the embryos (Ross 2001). Sexual maturity is likely reached in 1 year, and life span is not known to exceed 2 years (Boschung and Mayden 2004).

CONSERVATION STATUS   Although declines have been reported in the northern part of its range, the Pugnose Minnow is a common lowland species in Arkansas and its populatons are secure.

Genus *Phenacobius* Cope, 1867b— Suckermouth Minnows

The distinctive genus *Phenacobius* is composed of five species, only one of which (*P. mirabilis*) occurs in Arkansas. All members of this genus are slender fishes with eyes directed upward; mouth inferior and suckerlike; no barbels; 0,4–4,0 pharyngeal teeth that are hooked and without grinding surfaces; small scales, 42–69 in lateral series; lateral line complete; breast naked; deeply embedded scales on anterior belly; 7 anal fin rays; 8 pelvic fin rays; dorsal fin origin anterior to vertical from pelvic fin origin; intestine short; gill rakers about as long as their basal width; and gill membranes broadly joined to the isthmus.

Phenacobius mirabilis (Girard)
Suckermouth Minnow

Figure 15.103. *Phenacobius mirabilis*, Suckermouth Minnow. David A. Neely.

Map 85. Distribution of *Phenacobius mirabilis*, Suckermouth Minnow, 1988–2019.

CHARACTERS   A fairly large, terete minnow with a pronounced, blunt snout (snout length about twice the eye diameter) projecting beyond upper lip, an inferior suckerlike mouth, and a dusky lateral band extending from the head to the end of the caudal peduncle and terminating in a conspicuous horizontally elongate black blotch (Fig. 15.103). Eyes small (eye diameter going into head length 4.2–5 times). Compared with the other cyprinid species examined by Mayden (1989) the dentary, maxillary, and premaxillary are larger and heavier. Upper lip thick; lateral portions of the lower lip are fleshy and expanded, forming lobes; there is no cartilaginous ridge inside the lower lip. Barbels absent. Gill membranes broadly joined across isthmus. Gill rakers short and knoblike, usually 5–9 total. Breast naked; belly with embedded scales on anterior portion. Lateral line complete and nearly straight, with 42–50 scales. Scales around caudal peduncle 15–17. Anal fin rays 7. Pharyngeal teeth 0,4–4,0. Median and paired fins rounded. Dorsal fin rays 8; dorsal fin origin anterior to pelvic fin origin (posterior to it in *Campostoma* species). Intestine short, with S-shaped loop; peritoneum silvery with scattered speckles. Swim bladder reduced. Breeding males with dorsal and lateral body scales with marginal row of 3–5 tubercles; large and pointed tubercles on side of head and top of head extending onto snout. Males with nuptial pad on lower cheek and tubercles on dorsal surface of pectoral rays 2–8. Breeding female occasionally lightly tuberculate on top of head. Pectoral fins of male are large, broad, and fan shaped; pectoral fins of female are small, narrow, and elongate (Becker 1983). Maximum size about 4.8 inches (122 mm) (Trautman 1981).

LIFE COLORS   Body usually distinctly bicolored, olive above and silvery below with creamy white undersides. Basicaudal spot present. Thin dark stripe along back. Dorsolateral scales outlined by melanophores. A dusky lateral band is overlaid with silver and may not be prominent in live specimens. In preserved specimens, the dark lateral band is more distinct. It is about 2 scales wide, encircles the snout, extends across the opercles, and is most prominent on the caudal peduncle. No chromatic breeding colors.

SIMILAR SPECIES   The Suckermouth Minnow is similar to stonerollers, of the genus *Campostoma*, but can be distinguished from those species by lacking a cartilaginous ridge inside lower lip (vs. lower lip with cartilaginous ridge); dorsal fin origin anterior to pelvic fin origin (vs. posterior to it in *Campostoma*); intestine short, with S-shaped loop (vs. intestine long, with many coils); peritoneum silvery (vs. peritoneum black); and breast naked (vs. breast fully scaled). It differs from *Hybopsis*, *Erimystax*, *Platygobio*, *Macrhybopsis*, and *Nocomis* species in lacking barbels.

VARIATION AND TAXONOMY   *Phenacobius mirabilis* was described by Girard (1856) as *Exoglossum mirabile*. Surprisingly, for such an extremely rare fish in Arkansas, the type locality is the Arkansas River at Fort Smith. Mayden (1989) discussed several characters suggesting the monophyly of and species relationships within *Phenacobius*, and regarded *Erimystax* as its sister group. Dimmick and Burr (1999) corroborated this relationship in a detailed allozyme and DNA study. Coburn and Cavender (1992) suggested a close relationship with the Cutlips Minnow of the genus *Exoglossum*. In a phylogenetic analysis of 12S and 16S ribosomal RNA genes, Simons and Mayden (1999) supported the sister group relationship between *Phenacobius* and *Erimystax*. In a subsequent analysis of the mitochondrial cytochrome *b* gene, Mayden et al. (2006a) resolved *P. mirabilis* as sister to *P. catostomus*, with both species sister to a clade containing nine species of *Erimystax*.

DISTRIBUTION   *Phenacobius mirabilis* occurs in the Lake Erie drainage of Ohio and Michigan, and in the Mississippi River basin from Ohio and West Virginia to Wyoming and Colorado south to New Mexico, Oklahoma, and Alabama. It is also found in isolated populations of Gulf Slope drainages in the Sabine River, Louisiana and Texas,

the Trinity and Colorado rivers of Texas, and the Pecos River, New Mexico. This species is very rare in Arkansas. Until the 1980s, only 5 pre-1940 records were known from the state (Map A4.84), all from western Arkansas (Black 1940; Robison and Buchanan 1988). Those older records are: (1) Poteau River, one-half mile north of Waldron, Scott County, 1939 (Black 1940), (2) Horsehead Creek, 8 miles west of Clarksville, Johnson County, 1939 (Black 1940), (3) Arkansas River at Fort Smith, Sebastian County (Girard 1856), (4) Lee Creek near its mouth, Crawford County (Jordan and Gilbert 1886), and (5) Poteau River at Fort Smith, Sebastian County (Meek 1896).

Collections were made from the historical sites on several occasions from 1972 to 2010, but no specimens of *P. mirabilis* were found. However, on 16 July 1986, a single specimen of *P. mirabilis* was collected from Little Bay Ditch (St. Francis River drainage), 3 miles southeast of Jonesboro, Craighead County, Arkansas (UAFS 0431) (Keith et al. 1987). Buchanan et al. (2003) later reported the first record of this species from the Red River in Arkansas (UAFS 1558), a single specimen (59 mm TL) taken 6 miles upstream from the I-30 bridge on 18 August 1997 (Map 85). Although there are occasionally erroneous reports of this species from Arkansas, the seven collections listed above remain the only documented records of *Phenacobius mirabilis* from the state.

HABITAT AND BIOLOGY   The Suckermouth Minnow is primarily a northern and western Great Plains species and is a riffle inhabitant of sand and gravel-bottomed, permanent streams of moderate gradient (Rohde 1980a; Miller and Robison 2004). It is intolerant of reservoir conditions. In Kansas, it moved into shallower gravel riffles at night (Deacon 1961), and juveniles primarily occupied backwater habitats (Minckley 1959). Its distribution suggests that this species originated in the Ancestral Plains system (Metcalf 1966). *Phenacobius mirabilis* possesses several morphological features suited for benthic habitats in strong current, such as a reduced swim bladder, small anal fin, and supralateral eyes (Jenkins and McInich 1994). Because of its large, fleshy lips that surround an inferior mouth, the Suckermouth Minnow is morphologically and ecologically convergent with species of the family Catostomidae (suckers). Over most of its range it occurs in streams of all sizes and is tolerant of high turbidity, but it is most abundant in clearer streams. Because it is much less sensitive to fluctuating water levels and turbidity than other riffle fishes (based on observations in Oklahoma), its scarcity in western Arkansas streams is puzzling. In Illinois, Smith (1979) noted that it was probably most abundant in flowing streams of the natural prairie but absent or scarce in heavily forested areas where the tree canopy shades the streams. Most Arkansas streams are of the latter type, with only a few scattered remnants of natural prairie remaining in the state. Becker (1983) noted that in Wisconsin, populations are low or absent in the clearest streams, in streams of high gradients, and in normally turbid streams whose gradients are too low to prevent silt accumulations from covering the stones on the riffles. In that state, it reached its largest population densities in large streams and rivers, especially in habitats where competitive pressure from other riffle species was low.

Phenacobius mirabilis is morphologically specialized for benthic feeding, and it roots in the substrate with its snout (Pflieger 1997). Smith (1979) noted that in seeking food, the Suckermouth Minnow behaves much like a darter except that the food is sucked up rather than seized. It is predominantly an insectivore, and its foods consist mostly of aquatic dipteran and caddisfly larvae, mayfly nymphs, detritus, and plant materials (Haas 1977). It has also been reported to eat aquatic oligochaetes and snails (Starrett 1950a).

Cross (1967) reported a protracted breeding season from April through August in Kansas. In Wisconsin, tuberculate males have been taken from June to mid-September, and spawning occurs from at least early July to the end of August (Becker 1983). In Alabama, at latitudes similar to those of Arkansas, the reproductive season extends at least from May through June (Boschung and Mayden 2004). The breeding habitat is believed to be the same gravel riffles occupied by the species throughout the year. Haas (1977) and Pflieger (1997) also believed that spawning occurs in gravel riffles. Suckermouth Minnows prefer flowing waters and require high velocity areas during spring for spawning (Brewer et al. 2006). In Oklahoma, Cross (1950) collected *P. mirabilis* in breeding condition when water temperatures reached 57.2–77°F (22.4–25°C). Bestgen and Compton (2007) successfully spawned Suckermouth Minnows in the laboratory by using hormone injections. They spawned over interstitial spaces in gravel and cobble substrate at water temperatures of 17, 19, and 23°C. Only 1 or a few eggs were released in any single spawning bout. Viable, adhesive eggs were found on gravel and cobble substrate but not on sand. Suckermouth Minnows in Colorado streams first spawned in May when temperatures were 15–17°C. It has been found to spawn two or more times from April through August at water temperatures of 14–25°C (Cross 1950, 1967; Haas 1977; Pflieger 1997; Becker 1983). A tuberculate male was taken by TMB from the Red River in Miller County, Arkansas, on 18 August. Becker (1983) reported a 90 mm TL female with approximately 1,640 eggs, and Haas (1977)

reported an 88 mm female with 1,080 eggs. Bestgen and Compton (2007) suggested that Suckermouth Minnows probably ripen only a portion of their eggs at a time and spawn them in small batches. Sexual maturity is reached by age 2 when the fish exceeds about 60 mm SL (Boschung and Mayden 2004). Life expectancy is 4–5 years.

CONSERVATION STATUS   Approximately 80 years ago, Black (1940) stated, "In spite of the fact that sandy, swift waters abound in Arkansas this is a very rare species here." The discovery of only two records of Suckermouth Minnow in the past 30+ years in Arkansas is still puzzling today because there is presumably suitable habitat available for it, and because it is currently found in all six states bordering Arkansas. It has been found in the past 30 years in the Sulphur River of northeastern Texas (Carroll et al. 1977) (a stream that flows into Arkansas), in the mainstem Red River in Oklahoma 18 km upstream from the Arkansas state line (Oklahoma Department of Environmental Quality data provided by Jimmie Pigg); and in the lower Forked Deer River of western Tennessee, near its confluence with the Mississippi River across from Mississippi County, Arkansas (Boronow 1975). Waifs from any of these sites could possibly occur today in Arkansas. There are recent reports of specimens from Arkansas that have proven erroneous upon examination of the specimens. We consider the Suckermouth Minnow an endangered species in Arkansas.

Genus *Pimephales* Rafinesque, 1820a— Bluntnose Minnows

The North American cyprinid genus *Pimephales* contains four species, *P. notatus*, *P. promelas*, *P. tenellus*, and *P. vigilax*, all of which occur in Arkansas. This distinctive genus is characterized by a robust, cylindrical body, blunt snout, naked breast, crowded predorsal scales, 7 anal fin rays, 8 pelvic fin rays, 14–17 pectoral rays, 0,4–4,0 pharyngeal teeth, a detached (not fused), short splintlike anterior dorsal fin ray rudiment separated from the first principal fin ray by a membrane, and breeding males with fleshy nape pads, extremely large snout tubercles, and thickened pectoral and dorsal fin membranes. Bright colors are absent, but males darken during the breeding season. All *Pimephales* species have an extended spawning season from April to September. They are egg clusterers and have a complex reproductive behavior called cavity nesting which includes territorial defense of a nest cavity and parental care (McMillan and Smith 1974).

Mayden et al. (2006a) supported monophyly of the genus *Pimephales* only if *Opsopoeodus emiliae* is included.

Monophyly of the four species of *Pimephales* is supported by morphological (Mayden 1987b; Coburn and Cavender 1992) and chromosomal (Li and Gold 1991) characters. In a combined analysis of the cyt *b* and S7 genes, monophyly of *Pimephales* was strongly supported and *Notropis maculatus* was recovered as sister to *Pimephales* (Blanton et al. 2011). There was only marginal support for a sister relationship between *Opsopoeodus* and *Pimephales*. Relationships among the four species of *Pimephales* are not well resolved (Schmidt et al. 1994). Bielawski et al. (2002) in an analysis of the ND4L gene and part of the ND4 gene from the four species of *Pimephales*, recovered four different trees with the strict consensus providing no resolution of relationships among species.

Pimephales notatus (Rafinesque)
Bluntnose Minnow

Figure 15.104. *Pimephales notatus*, Bluntnose Minnow. *Uland Thomas.*

CHARACTERS   A slim, elongate, terete minnow with short, rounded fins, a somewhat subterminal, nearly horizontal mouth slightly overhung by a bluntly rounded snout, and a distinct black lateral band that extends around the snout (Fig. 15.104). Eyes round and more laterally directed and lower on the head than in *P. tenellus* and *P. vigilax*. Eye diameter going into head length 3.3–4.5 times. Upper lip approximately the same width at middle as on either side (but tuberculate males often have a medially swollen upper lip). Intestine long with several loops; peritoneum black. Lateral line complete, with 39–44 scales. Anal rays 7. Dorsal rays 8. Short ray at front of dorsal fin thickened, splintlike, and separated from first principal ray by a membrane. Predorsal area flattened, with scales conspicuously smaller and more crowded than adjacent scales on sides; predorsal scale rows 18–22. Breast naked and belly scaled. Body squarish in cross-section. Body depth going into SL 3.9–5.8 times. Dorsal fin origin slightly behind pelvic fin origin. Fewer than 32 scale rows around body. Pharyngeal teeth 0,4–4,0. Gill rakers 7–10. Nuptial males develop a distinct barbel at the top angle of the jaws

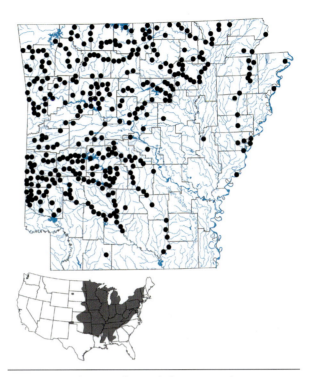

Map 86. Distribution of *Pimephales notatus*, Bluntnose Minnow, 1988–2019.

(unique among *Pimephales*) and 3 rows of large tubercles on snout; there are no chin tubercles. Maximum size 4.3 inches (108 mm) TL.

LIFE COLORS   Body color olivaceous above, white below. Dorsolateral scales outlined with fine, black lines, providing a diamond-shaped appearance. A narrow, well-developed dusky lateral band extends from snout to end of caudal peduncle, terminating in a distinct black spot at the caudal base. Development of a small black blotch in anterior rays of dorsal fin is highly variable in Arkansas populations. Most individuals, particularly the smaller specimens, lack the blotch which is present and most distinct during the breeding season in large individuals, especially the tuberculate males. It is present in some large females but absent in others in the same population. Breeding males become almost black, obscuring the lateral band, and develop a fleshy light-colored pad on front of back.

SIMILAR SPECIES   The Bluntnose Minnow differs from *P. promelas* in having a complete lateral line (vs. incomplete), a well-developed caudal spot (vs. caudal spot weak or absent), and a subterminal horizontal mouth (vs. mouth terminal and oblique). It differs from *P. tenellus* and *P. vigilax* in having a long, coiled intestine (vs. a short intestine with an S-shaped loop), a black peritoneum (vs. peritoneum silvery with scattered dark speckles), and round eyes more on the side of the head (vs. eyes somewhat elliptical

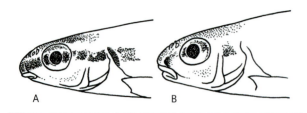

Figure 15.105. *Pimephales* heads:

  A—*P. notatus*
  B—*P. vigilax*

and placed higher on the head and directed upward) (Fig. 15.105A). It is distinguished from *Opsopoeodus emiliae* in having 0,4–4,0 pharyngeal teeth (vs. 0,5–5,0), a subterminal, horizontal mouth (vs. mouth extremely oblique, nearly vertical), and 8 dorsal rays (vs. 9).

VARIATION AND TAXONOMY *Pimephales notatus* was described by Rafinesque (1820a) as *Minnilus notatus* from the Ohio River (no specific locality given). It was subsequently placed in the monotypic genus *Hyborhynchus*, where it remained for more than seven decades until Bailey (1951) placed it in *Pimephales*. In the past, at least six nominal species have been synonymized with *P. notatus* (Schönhuth et al. 2016). Mayden (1987b) conducted a small-scale cladistic analysis of morphological characters and suggested that *P. notatus* was sister to *P. promelas*. Li and Gold (1991) hypothesized that *P. notatus* was sister to *P. vigilax*. Jacquemin et al. (2012) studied the morphology of *Pimephales notatus* and identified covariation of morphology with local environmental variation. Those authors also found sexually dimorphic shape variation in this species. Females had a distended abdomen, reduced dorsal and caudal fin base length, and the head shape of females differed from males. Jacquemin et al. (2012) interpreted the ability of the Bluntnose Minnow to respond morphologically to environmental variation as a character that enhances its ability to occupy multiple habitats. Beachum et al. (2016) found that *P. notatus* exhibits intraspecific morphological variation across its large geographic range. Schönhuth et al. (2016) studied the phylogeny and genetic variation of *Pimephales notatus* using one mitochondrial (*cytb*) and two nuclear (*S7* and *Rag1*) genes. Results suggested a largely drainage-independent genetic structure. Schönhuth et al. (2016) found (1) a well-supported lineage in the Ozark Highlands and adjacent areas in the Missouri River (Clade A), which is highly divergent from the remaining populations and may represent a long-term isolated lineage that should be considered a separate conservation and management unit, and (2) all remaining populations made up a widely distributed lineage,

covering most of the distribution for the species from the Red River and tributaries of the Great lakes in the north to the lower Mississippi and Mobile basin in the south, and from the Missouri River in the west to the Chesapeake and Albemarle Sound basins in the east (Clade B). Because their results from molecular analyses revealed a distinctive lineage in the Ozarks (Clade A), Schönhuth et al. (2016) suggested a need for additional morphological analysis within *P. notatus* to determine its taxonomic status. The high differentiation observed for Clade A supports the genetic isolation of this lineage in the Ozark Highlands.

DISTRIBUTION *Pimephales notatus* is widely distributed in the Great Lakes, Hudson Bay, and Mississippi River basins from Ontario to southern Manitoba, Canada, south to Louisiana; Atlantic Slope drainages from the St. Lawrence River, Canada, to the Roanoke River, Virginia; and Gulf Slope drainages from Mobile Bay, Alabama, to the Mississippi River. Page and Burr (2011) considered it the most common fish in eastern North America. In Arkansas, the Bluntnose Minnow is most abundant in the Interior Highlands, but it also occurs sporadically in appropriate clearwater habitats in the Coastal Plain (Map 86). There are several records from the Mississippi River.

HABITAT AND BIOLOGY The Bluntnose Minnow generally inhabits medium-sized to large, clear, permanent streams with sand or gravel bottoms and aquatic vegetation, but it can be found in a wide variety of stream and lake habitats throughout the state. Burr and Warren (1986b) noted that it is an ecologically labile species that inhabits a wide variety of riverine habitats in Kentucky. It is also sometimes found in slightly turbid to turbid environments, at varying depths in the water column, and over a variety of substrates. Although there are a few records from the Mississippi River in Arkansas, it typically avoids the largest rivers, where it is replaced by *P. vigilax*. It was one of the most widely distributed small species in Arkansas reservoirs, where it was found in large numbers in 29 impoundments (Buchanan 2005). Quiet pools and backwater areas are favored habitats, and it can often be seen schooling in midwater or benthic areas. Becker (1983) noted that in northern parts of its range, the Bluntnose Minnow lends itself well to pond cultivation and is used extensively as a bait minnow.

Pimephales notatus is a generalized omnivore, and foods consist primarily of plankton, algae, insect larvae, and small aquatic invertebrates. In the Illinois River, Oklahoma, it was primarily herbivorous, but an occasional individual had eaten benthic invertebrates (Felley and Hill 1983). In addition to algae and detritus, it is reported to feed on entomostracans, and midge larvae and pupae (Keast and Webb

1966; Moyle 1973). Larval and early juvenile Bluntnose Minnows feed mainly on microcrustaceans such as copepods and cladocerans (Keast 1985a), while larger individuals feed more in benthic areas and among aquatic plants (Moyle 1973). Moyle also noted seasonal changes in diet in a Minnesota lake, with diatoms and filamentous green algae most commonly eaten in winter and insects becoming more important in the diet in spring and summer. Walker and Hasler (1949) determined that the Bluntnose Minnow has a keen sense of smell and can discriminate between the odors of aquatic plants. Odor discrimination may play an important role in its feeding biology.

A rather long breeding season occurs in Arkansas from at least early May through August. Tuberculate males have been taken as early as 10 May in Columbia County and as late as August in the White River system. Ripe females were collected in the White River near Batesville on 15 May 2013 (data from Jeff Quinn). We have observed intense spawning activity in Lake Hamilton in early July. Breeding habits are similar to those of the Bullhead Minnow (Hubbs and Cooper 1936; Westman 1938). Spawning occurs in gravelly or sandy shoal regions (Van Cleave and Markus 1929) when water temperatures exceed 21°C (Hubbs and Cooper 1936). Like other *Pimephales*, the highly territorial male Bluntnose Minnow excavates a nest under stones or other objects on the stream or lake bottom. The swollen dorsal pad and the breeding tubercles are used to clean the roof of the nest after it has been excavated by sweeping movements of the body and caudal fin. Mating behavior was described by Page and Ceas (1989). Several females may spawn in turn in the nest, which may contain more than 5,000 eggs in several stages of development. Over a single spawning season, a female may produce 7–19 clutches of eggs, with a total annual production of 1,112–4,195 eggs (Gale 1983). The male guards the nest, fanning the eggs until hatching and for a short time thereafter (Figs. 15.106, 15.107). Eggs hatch in 6–10 days at water temperatures of 19–25°C (Westman 1938). Johnston and Johnson (2000) reported that the male produces sound during aggression associated with the breeding season. These sounds are pulsed and complex, usually consisting of several components, and are of low frequency. Because *Pimephales* males have black breeding coloration, defend cavity nests, and may need to attract mates from greater distances, acoustic signals that travel greater distances (i.e. lower frequency) may be important in cavity nesting species (Johnston and Johnson 2000). Maximum life span is around 3–4 years (Becker 1983).

CONSERVATION STATUS The Bluntnose Minnow is widespread and abundant in Arkansas, and its populations are secure.

Figure 15.106. *Pimephales notatus* male tending eggs attached to the underside of a large rock in a Flint Hills stream, Kansas. *Garold W. Sneegas.*

Figure 15.107. *Pimephales notatus* male rubbing eggs attached to the underside of a large rock in a Flint Hills stream, Kansas. *Garold W. Sneegas.*

Pimephales promelas Rafinesque
Fathead Minnow

Figure 15.108. *Pimephales promelas*, Fathead Minnow. *David A. Neely.*

CHARACTERS A robust, cylindrical-bodied minnow with an incomplete lateral line (42–48 scales in lateral series, with 7–40 pored scales), a small, oblique, terminal mouth, and a small rounded head (head length going 3.4–4 times into SL) (Fig. 15.108). Rudimentary anterior dorsal fin ray is short and thickened, splintlike, and separated from the first principal (unbranched) ray by a membrane. Body

Map 87. Distribution of *Pimephales promelas*, Fathead Minnow, 1988–2019. Inset shows native and introduced ranges in the contiguous United States.

depth going into SL 3.6–5.1 times. Eye small, round, and positioned laterally on head (eye diameter going into head length 4.3–5 times). Snout blunt, protruding only slightly, if at all, beyond mouth. Scales small and crowded in predorsal region, with 22–26 predorsal scale rows. Usually more than 39 scale rows around the body. Breast naked; belly scaled. Dorsal fin rounded, its origin over pelvic fin origin. Dorsal and pelvic fin rays 8. Anal fin rays 7. Pectoral rays 14–18. Pharyngeal teeth 0,4–4,0. Intestine long with several loops; peritoneum black. Barbels absent. Nuptial males develop a fleshy, fatty pad along the forward part of the back. Breeding males with large tubercles developed in 3 rows on snout and 4–6 tubercles on chin. Maximum length about 4 inches (102 mm).

LIFE COLORS Body color olivaceous above, grading to tan or silvery below. Dorsolateral scales not outlined by melanophores. A dusky, narrow, predorsal stripe is present. Dark spots at anterior dorsal and caudal fin bases diffuse or absent. Upper sides often having diagonal, chevron-like or "herring bone" markings. Lateral band may be present, but usually inconspicuous. Breeding males become dark orange brown or with 2 conspicuous broad, whitish transverse bands, one encircling the body behind head and a similar light band beneath the dorsal fin. Head noticeably darker than rest of body. Fins black.

SIMILAR SPECIES It differs from all other *Pimephales* species in having an incomplete lateral line (vs. complete). It is further distinguished from *P. notatus* in having a terminal oblique mouth (vs. mouth subterminal and horizontal), and in usually having more than 39 body circumferential scales (vs. 32 or fewer circumferential scales). The Fathead Minnow further differs from *P. tenellus* and *P. vigilax* in having a black peritoneum (vs. a silvery peritoneum often with dark specks), a long coiled intestine (vs. intestine short with a single S-shaped loop), and eyes round and more laterally placed (vs. eyes elliptical, higher on head, and more upwardly directed).

VARIATION AND TAXONOMY *Pimephales promelas* was described by Rafinesque (1820a) from a pond near Lexington, Kentucky. Vandermeer (1966) studied geographic variation and concluded that recognition of subspecies was unwarranted. Mayden (1987b) suggested that *P. notatus* was sister to *P. promelas* based on morphological characters. Li and Gold (1991) could not infer relationships of *P. promelas* and *P. tenellus*, as both species possess the presumably ancestral NOR character state. A mitochondrial DNA analysis by Bielawski et al. (2002) also provided no resolution of relationships.

Hatchery-raised Fathead Minnows occasionally exhibit polymorphism in coloration. A minute fraction of those produced at a hatchery may be red, yellow, or white. Presumably these deviants from the normal color pattern would not long survive predators in nature. In the early 1980s, Bobby Bland, an ornamental fish breeder from Taylor, Arkansas, developed a pure-breeding strain of red Fathead Minnow known as a rosy red (Fig. 15.109). One commercial breeder in Arkansas tested its potential as a bait minnow in the state, and it has been widely used for that purpose. It survives well in minnow buckets, ships well, and is more easily visible to gamefish in the water. The Arkansas breeders distribute rosy reds to several other states including Wisconsin and Texas. This species does well in an aquarium, and has become moderately popular among aquarium hobbyists. It is sometimes sold in pet stores as feeder fish for carnivorous fish kept in aquaria.

DISTRIBUTION The native range of *Pimephales promelas* is almost impossible to accurately determine because of its widespread introduction as a bait minnow. It is widely distributed over much of North America in the Great Lakes, Arctic, and Mississippi River basins from Quebec and Northwest Territories, Canada, south to Alabama, Louisiana, and Texas. It is absent from Atlantic Slope drainages south of the Potomac River and from Gulf Slope drainages to the Trinity River, Texas, except where introduced, but it is probably native throughout the Rio

Figure 15.109a. Rosy Reds (*Pimephales promelas*) baitfish school. *Keith Sutton.*

Figure 15.109b. Rosy Red bait minnow (*Pimephales promelas*). *Uland Thomas.*

Grande drainage, south into Mexico. It has also been introduced into the Colorado River drainage of New Mexico and Arizona, and into other western states. Etnier and Starnes (1993) believed that the native range excluded most of the southeastern states. In Arkansas, Meek (1894a, b) reported the Fathead Minnow from throughout the northern half of the state, but Lee and Shute (1980) considered it native to the state only in the headwaters of the Illinois and White rivers in northwestern Arkansas. Etnier and Starnes (1993) and Page and Burr (2011), however, suggested that all of Arkansas might be within the native range of this species. Etnier and Starnes (1993) speculated that because substantial reproducing populations are not known from natural habitats in Tennessee, the Fathead Minnow may not be indigenous to that state. The same may be true in Arkansas because natural, reproducing populations are also undocumented in this state. Black (1940) collected no *P. promelas* in Arkansas and concluded that many or all of Meek's reports of that species from northern Arkansas may have been misidentifications. It is currently widely introduced and sporadically collected throughout Arkansas (Map 87). The relatively few records from the natural waters of Arkansas are surprising considering its use in AGFC nursery ponds as a forage species and its widespread use as a bait minnow.

HABITAT AND BIOLOGY The Fathead Minnow is found in a variety of habitats in Arkansas, from clear to turbid streams, but it seems to favor sluggish streams, ditches, and ponds over a soft mud substrate. Becker (1983) believed this species probably exhibited the greatest ecological diversity of all cyprinids in North America, and Cross (1967) considered it a pioneer species. It is tolerant of high temperature, high turbidity, and low oxygen. It is not a commonly encountered species in Arkansas, and Smith (1979) noted that in Illinois it is most abundant in streams where *P. notatus* is absent, suggesting it is unable to compete successfully with that species. Starrett (1950b) considered it "intolerant to other species" in Iowa. Becker (1983) stated that it is unable to maintain itself in streams or lakes which have a heavy population of predaceous fish. Chiu and Abrahams (2010) found a strong preference for turbid habitats in the presence of predators, suggesting turbidity confers a benefit to feeding Fathead Minnows that more than compensates for the cost of predation risk. Despite its widespread fisheries and bait minnow use, it was found in only 1 out of 66 reservoirs sampled, including reservoirs with associated nursery ponds in which it is commonly raised (Buchanan 2005). This is probably indicative of its rapid consumption by predatory fishes when released into reservoirs.

Pimephales promelas is an omnivorous, opportunistic species over its range, but individual populations may feed predominantly on a single food category such as insects, algae, or detritus (Jenkins and Burkhead 1994). Food habits differ greatly from one habitat to another and in different months in the same habitat. It is reported to feed on detritus, plankton, insect larvae, and some plant material, but Boschung and Mayden (2004) considered its principal food to be algae, with mud and detritus being ingested incidentally during benthic feeding. In some areas, its diet consisted of several algal species in addition to detritus, with the algal species presumably selected for by the pharyngeal teeth (Coyle 1930). In North Dakota lakes, it fed primarily on small aquatic insects and zooplankton, especially cladocerans (Held and Peterka 1974). It is predominantly a benthic feeder.

Observation on reproduction in ponds at the Joe Hogan State Fish Hatchery in Lonoke, Arkansas, revealed that spawning usually begins in early April, reaches a peak in late May, and continues through June, ending in July (Berry Beavers, pers. comm.). Intense spawning activity was observed by TMB at the Lonoke hatchery on 23 April at a water temperature of 65°F (18.3°C). Breeding generally occurs in natural waters in the spring from May until summer. In Nebraska, Fathead Minnows begin to spawn when surface water temperatures reach 58–65°F (14.4–18.3°C) (McCarraher and Thomas 1968). Males become darkened, developing tubercles and a thick, fatty dorsal pad in front of the dorsal fin. Smith and Murphy (1974) investigated the functional morphology of this pad and determined that it secreted mucus that was later deposited on the spawning surface during contact movements performed by breeding males. They suggested that contact movements by breeding males may facilitate chemosensory sampling of the spawning site and eggs. Cole and Smith (1987, 1992) found that males produce waterborne chemical stimuli that attract females. Markus (1934) studied the life history of this species and found that it is quite prolific. Eggs are deposited on the underside of hard objects that are either submerged or floating in the stream. Males fertilize eggs and defend the nest site until hatching. The male is more defensive of its nest and more attentive to the eggs than are male Bluntnose Minnows (Smith 1979). The constant back and forth swimming of a nest-guarding male serves to drive away predators, aerate the eggs, and sweep the eggs free of sediment (McMillan and Smith 1974). Allopaternal care is known (Unger and Sargent 1988). In this behavior pattern, newly reproductive males adopt and care for the eggs of other males, presumably to increase their own chances of successful reproduction (females prefer males with eggs). Sargent (1989) found that adoptive eggs received less care and suffered heavier mortality than sired eggs. In experimental ponds, nest predation by conspecifics accounted for most of the egg mortality, and 25% of all nests failed to produce hatches (Divino and Tonn 2008). In that study, nests where parental care was observed were larger, lasted longer, and were more likely to produce hatchlings than nests where caregiving was not observed. The Fathead Minnow is a fractional spawner (Gale and Buynak 1982). Fractional spawning not only allows the female to increase fecundity by producing more eggs in a season than could be held at one time, but also decreases the chance that an entire generation will be eliminated by short-term environmental events (Gale and Buynak 1982). The total number of eggs spawned per female during the breeding season ranged from 6,803 to 10,164, but from 9 to 1,136 eggs were spawned per spawning session. Intervals between spawning sessions ranged from 2 to 16 days. Spawning often began before dawn and was usually finished before 10 a.m. Growth of young is rapid and maturity is quickly reached, so some individuals can reproduce later during the summer in which they are hatched. The AGFC often stocks Fathead

Minnows in its nursery ponds to serve as food for young bass and other fishes. Sometimes *P. promelas* is stocked in small impoundments to allow reproduction prior to bass stocking (Noble 1981). Bass typically eat them to extinction in small lakes.

The Fathead Minnow is raised commercially in Arkansas as a bait species, ranking only behind the Golden Shiner in its importance to the aquaculture industry. In 1982, Fathead Minnows valued at $1.83 million were produced in the state (891,600 pounds [404,430 kg]). By 2011, estimates for Fathead Minnow production were 1.4 million pounds valued at $2.9 million (Nathan Stone, pers. comm.).

CONSERVATION STATUS   The Fathead Minnow is a widely introduced baitfish across Arkansas and is found statewide. It is not known whether native populations currently exist.

Pimephales tenellus (Girard)
Slim Minnow

Figure 15.110. *Pimephales tenellus*, Slim Minnow. *David A. Neely.*

CHARACTERS   A small, terete minnow with a slender body, blunt snout, slightly oblique mouth with the upper lip considerably thickened at its middle, and a dark lateral band. Relationship of snout to upper lip variable depending on drainage, but mouth usually not noticeably overhung by snout. Eye moderately large (eye diameter going 3.5–4.5 times into head length) (Fig. 15.110). Eye somewhat elliptical, situated higher on head, and directed more upwardly than in *P. notatus* and *P. promelas*. Dorsal fin rounded, with 8 rays. Rudimentary anterior dorsal fin ray is short and thickened, splintlike, and separated from the first principal (unbranched) ray by a membrane. Anal fin rays 7. Predorsal scales small and crowded with fewer than 23 predorsal scale rows. Scale rows around the body 28–30. Pharyngeal teeth 0,4–4,0. Scales above lateral line 6 (rarely 5 or 7). Barbels absent. Lateral line complete with 37–41 scales. Intestine short with a single S-shaped loop; peritoneum silvery. Breeding males with 11–13 tubercles arranged in 3 rows on snout. It is the smallest *Pimephales* species with adults usually around 2 inches (51 mm). Maximum size 2.75 inches (70 mm).

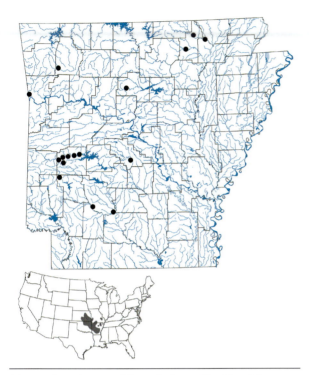

Map 88. Distribution of *Pimephales tenellus*, Slim Minnow, 1988–2019.

LIFE COLORS   Body color dark olivaceous above and creamy white below. Well-defined black lateral band terminating in a small distinctive basicaudal spot, a narrow, vertical black line extending above and below the spot. A distinct, narrow dark stripe present along underside of caudal peduncle. Dorsolateral scales outlined by melanophores. Middorsal stripe indistinct. A small dark spot often present on anterior membranes of the dorsal fin. Fin rays, especially in the caudal fin, often tinged with orange. Breeding males develop darker colors with black head and lateral band, and white on the leading edges of pectoral fins.

SIMILAR SPECIES   The Slim Minnow is similar to the Bullhead Minnow, but differs in having 6 scales (rarely 5 or 7) above the lateral line (vs. usually 7 or 8, rarely 6); dusky stripe on the underside of the caudal peduncle more distinct and narrower (vs. stripe weaker, broader, and more diffuse); upper lip considerably thickened at the middle (vs. upper lip little expanded at midline); anal region conspicuously marked with black pigment (vs. anal region almost devoid of pigment); spot at caudal base more or less vertically elongated (vs. caudal spot nearly round); and in lacking a dark crescent-shaped mark on snout between nostril and upper lip (vs. crescent-shaped mark usually present). It differs from *P. notatus* and *P. promelas* in having a silvery peritoneum (vs. black), a short intestine with a single S-shaped loop (vs. a long, coiled intestine), and eye

somewhat elliptical and directed upward (vs. eye round and directed laterally).

VARIATION AND TAXONOMY   *Pimephales tenellus* was originally described as *Hyborhynchus tenellus* by Girard (1856) from specimens "collected 20 miles west of Choctaw Agency," branch of San Bois Creek, Le Flore County, Oklahoma. Hubbs and Black (1947) placed the Slim Minnow in the genus *Ceratichthys* and recognized two subspecies, both of which occur in Arkansas: the Western Slim Minnow, *C. t. tenellus* (Girard) and the Eastern Slim Minnow, *C. t. parviceps* (Hubbs and Black). The eastern subspecies is more slender, with body depth going into SL more than 5 times, and the snout projects slightly beyond the upper lip, while the upper lip is terminal in the western subspecies. In the early 1950s, the Slim Minnow was placed in the genus *Pimephales* by Bailey (1951), where it has remained. Hubbs and Black (1947) considered *P. tenellus* to be most closely related to *P. vigilax* (which they also placed in genus *Ceratichthys*). This sister relationship was supported by Mayden (1989), but relationships of *P. tenellus* could not be inferred by Li and Gold (1991) in a study of chromosomal NORs. Bielawski et al. (2002) and Schönhuth et al. (2016) also concluded that additional collection and analysis of data are needed to resolve relationships among *Pimephales* species.

DISTRIBUTION   The Slim Minnow is found in only four states: Arkansas, Missouri, Kansas, and Oklahoma, in the St. Francis, White, Arkansas, Ouachita, and Little river systems, and is most common in the northwestern part of its range (Page and Burr 2011). In Arkansas, it is an uncommon species that occupies upland streams of the Interior Highlands (Map 88). The Western Slim Minnow, *P. t. tenellus*, is confined to extreme western Arkansas in lower Lee Creek in Crawford County, while the Eastern Slim Minnow, *P. t. parviceps*, occupies the White, Black, St. Francis, Ouachita, and Little river systems. Populations assumed to be intergrades were reported from the upper Poteau River and Horsehead Creek, Johnson County (Black 1940). The first records of *P. tenellus* from the Spring River were collected by HWR at Imboden, Lawrence County, on 16 June 1999 (Tumlison et al. 2015), and additional specimens were collected by TMB, Sam Henry, and Brett Timmons on 18 June 2014 (UAFS 2084).

HABITAT AND BIOLOGY   This schooling minnow frequents midwater to bottom areas of small to medium, clear, gravel and cobble-bottomed streams in quiet water and often in moderate current. It apparently does not adapt to reservoir conditions. In the upper Ouachita River, HWR collected this species only over fine gravel substrate in about 3–4 feet (0.9–1.2 m) of water in pool regions with slight to moderate current. Miller and Robison (2004) reported that unlike other species in the genus, it can often be found actively swimming in the current. In Missouri, the western subspecies of the Slim Minnow occurs in streams that are transitional between those of the Ozarks and the Prairies, and it avoids clear, spring-fed streams (Pflieger 1997). The eastern subspecies in Missouri inhabits clearer, swifter streams than the western subspecies (Pflieger 1997).

There is no information on feeding biology of *P. tenellus*, but mouth and gut morphologies indicate that it probably feeds largely on insects and other small invertebrates.

An extended breeding season occurs in Kansas from the end of April to the beginning of September (Tiemann 2007b), with spawning presumably occurring under rocks in swift riffles when water temperatures range from 24°C to 29°C (Cross 1967). *Pimephales* species have evolved reproductively associated morphological features and complex breeding behaviors (egg-clustering breeding strategy) (Page and Ceas 1989). Tiemann (2007a) observed aquarium spawning of *P. tenellus* and documented the egg-clustering strategy similar to the other species of *Pimephales*. Males vigorously defended the nest against intruders; however, males did not assume a banding pattern on the body as in other *Pimephales* species (Tiemann 2007b). Age 1 and older *P. tenellus* began showing signs of sexual maturity around the middle of April as photoperiod neared 15L:9D and water temperature approached 16°C. At that time, males shifted from schooling behavior to territorial behavior. After developing the secondary sexual characteristics (swollen, black head; breeding tubercles on snout; a thickened first dorsal ray; and a dorsal pad on the nape), a male established a nest under a cobble. Males courted females by using various visual and contact displays, such as approaching, lateral displaying, tapping, and leading. A receptive female entered the nest, and spawning ensued. Spawning behavior was described by Tiemann (2007a) as follows:

> When ready to spawn, a *P. tenellus* female entered and examined the nest of a male. She then turned on her side and allowed the male to juxtapose himself below her. The pair remained pressed together on their sides and simultaneously swam in circles while undulating. It was during this process that the female pressed adhesive eggs, which simultaneously were fertilized by the male, one-by-one in a single-layer cluster on the underside of cobble. To do so, the female released an egg from her genital papilla, rolled it along the upper side of her body, and pressed it to the underside of the submerged object while she was on her side and the male was pushing against her.

Females laid 2–3 clutches per female, and mean clutch sizes were 189 eggs for the first clutch, 151 for the second, and 89 for the third. After spawning, the female left the area and the male guarded the nest. Parental care by the male included rubbing his dorsal pad and thickened first dorsal ray over the eggs about once every 30 minutes. Although the number of eggs laid in a season is unknown, Tiemann (2007a) showed that three *P. tenellus* females laid 344, 375, and 390 eggs, respectively, in a spawning season. Eggs were spherical, transparent, and about 1.2 mm in diameter (Tiemann 2007b). Hatching occurred in about 6 days at 24°C, and fry began schooling 7 days after hatching. Apparently, there is no parental care of free-swimming fry, and fry are often eaten by their parents (McMillian and Smith 1974).

HWR and Chris McAllister collected an adult specimen from the Ouachita River at Rocky Shoals, Montgomery County on 13 October 2015 that possessed the parasitic copepod *Lernaea cyprinacea*. This is the first report of this parasite from *P. tenellus* (Tumlison et al. 2016).

CONSERVATION STATUS   Pflieger (1997) considered reservoir construction the likely cause of the apparent elimination of *P. tenellus* from the White and St. Francis rivers in Missouri. The Slim Minnow is a rarely encountered species in Arkansas. We consider the western subspecies, *Pimephales tenellus tenellus*, endangered and the eastern subspecies, *P. tenellus parviceps*, of special concern. Future monitoring of the populations of these forms is needed.

Pimephales vigilax (Baird and Girard)
Bullhead Minnow

Figure 15.111. *Pimephales vigilax*, Bullhead Minnow. *Uland Thomas.*

CHARACTERS   A stout minnow with a large head (head length going into SL 3.6–4.3 times), a blunt, rounded snout, a large conspicuous black spot on the anterior membranes of the dorsal fin, and short, rounded fins (Fig. 15.111). Mouth terminal, slightly oblique; upper lip only slightly thickened at the middle. Barbels absent. Eye moderately large, elliptical, situated higher on head, and directed more

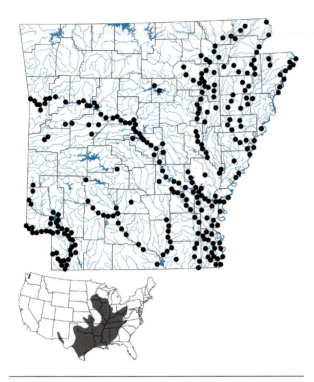

Map 89. Distribution of *Pimephales vigilax*, Bullhead Minnow, 1988–2019.

upwardly than in *P. notatus* and *P. promelas* (eye diameter of *P. vigilax* going into head length 3.2–4.0 times). Dorsal fin rounded with 8 rays, its origin approximately over pelvic fin origin. Anal fin with 7 rays. Pectoral rays 14–16. Lateral line complete, slightly decurved anteriorly, with 37–42 scales. Predorsal area flattened and covered by small crowded scales; predorsal scale rows 18–22. Breast naked; belly scaled. Intestine short with an S-shaped loop; peritoneum silvery with scattered dark speckles. Short ray at front of dorsal fin thickened and separated from first principal (unbranched) ray by a membrane. Pharyngeal teeth 0,4–4,0. Scales above lateral line 7 or 8. Breeding males lack a maxillary barbel, have 2 transverse rows of large tubercles on snout (5 in lower row, 4 in upper row), and have a fleshy white pad on top of the head and nape. Maximum size 3.8 inches (98 mm).

LIFE COLORS   Body color tan to light olive above with silvery sides. Lateral band narrow, dusky to black, and most prominent posteriorly when present; lateral band often absent in many individuals and is especially inconspicuous in live fish. Lateral line often outlined by a row of paired dark dots anteriorly. Dorsal fin with a black spot at its anterior base (absent only in small young). Dusky stripe along underside of caudal peduncle broad and indistinct. Caudal peduncle with a large, distinct wedge-shaped

caudal spot. Snout with a distinctive crescent-shaped dark mark on each side above the upper lip, but without a dark band around snout. Dorsolateral scales weakly outlined. Middorsal stripe absent or obscure. Breeding males are dark, and the head is mainly black. Crotches of dorsal and caudal rays are blackened, and the anterior dorsal fin spot intensifies. Pectoral fins pale, but with the leading edges dark.

SIMILAR SPECIES   *Pimephales vigilax* differs from *P. notatus* and *P. promelas* in having a silvery peritoneum with dark speckles (vs. black peritoneum), and a short intestine with a single S-shaped loop (vs. intestine long with several loops). It is similar to *P. tenellus*, but differs from that species in usually having 7 or 8 scales above the lateral line (vs. 6 scales above lateral line), upper lip only slightly thickened at middle (vs. upper lip decidedly thickened at middle), a dark crescent-shaped mark on snout between nostril and upper lip (vs. crescent-shaped mark absent), dusky stripe along underside of caudal peduncle broad and less distinct (vs. stripe on underside of caudal peduncle narrow and dark), and breeding males with 2 rows of tubercles on snout (vs. 3 rows of tubercles on snout).

VARIATION AND TAXONOMY   *Pimephales vigilax* was described by Baird and Girard (1853b) as *Ceraticthys* (subsequently changed to *Pimephales*) *vigilax* from Otter Creek, a tributary of the North Fork of the Red River (either Kiowa or Tillman County), southwestern Oklahoma. Variation in this species was studied by Hubbs and Black (1947). Cross (1953) supported the recognition of two allopatric subspecies. *Pimephales v. vigilax* (with a single row of 5 snout tubercles) is found in the headwaters of the Red River of Texas and Oklahoma, southward to the Brazos and Rio Grande. The remainder of the range (including Arkansas) is inhabited by *P. v. perspicuus*, which has 9 tubercles on the snout in 2 rows. Bielawski et al. (2002) found that ND4-ND4L gene sequence divergence between these subspecies was similar to or greater than that for several other pairs of cyprinids recognized as full species. Those authors suggested that morphological and ecological variation within *P. vigilax* warrants additional examination to determine whether separate species might exist. Mayden (1987b) considered *P. vigilax* most closely related to *P. tenellus*.

DISTRIBUTION   *Pimephales vigilax* occurs in the Mississippi River basin from Ohio to Minnesota, and south to the Gulf of Mexico, and in Gulf Slope drainages from Mobile Bay drainages of Georgia and Alabama, to the Rio Grande of Texas and Mexico. It is introduced upstream in the Rio Grande to New Mexico. In Arkansas, the Bullhead Minnow ranges throughout the Coastal

Plain and Arkansas River Valley, where it is common to abundant in the largest rivers (Map 89). It generally avoids upland streams, but penetrates farther upstream in the Ouachitas than in the Ozarks.

HABITAT AND BIOLOGY   Because of its tolerance of turbidity, siltation, high temperature, and low oxygen levels, the Bullhead Minnow is widespread and abundant in large sluggish streams and rivers with sand and mud bottoms throughout the Coastal Plain. It is the most abundant *Pimephales* species in the largest rivers of Arkansas, and it also occurs in medium-sized rivers. It is sometimes found in moderate to strong current but is most abundant in the large rivers in quiet backwater areas where it can be observed in schools. It is also abundant in some oxbow lakes and impoundments in Arkansas, and Buchanan (2005) reported it from 16 Arkansas reservoirs. It was the fourth-most abundant species in the Red River of Arkansas (Buchanan et al. 2003), and it is common to abundant throughout the Arkansas River navigation pools and the Mississippi River.

This schooling species feeds on or near the bottom of streams on a diet mostly of insects, with detritus and seeds seasonally abundant in the diet during winter and spring (Starrett 1950a). Although considered omnivorous by some authors, most available information on feeding indicates that it is largely carnivorous, with plant material being of minor importance in the diet. Eddy and Underhill (1974) reported that in Minnesota it feeds on algae, other plant material, snails, and other small bottom-dwelling animals. It apparently feeds to a much greater extent on aquatic insect larvae (mayflies, caddisflies, and midges) than *P. notatus* and *P. promelas*, and microcrustaceans (such as cladocerans) are a major food item in ponds (Ross 2001). In Oklahoma, the diet consisted mainly of immature and adult aquatic insects (primarily chironomids and psychodids) and small crustaceans (mainly cladocerans), and the volume of plant material was small (Parker 1964). A few individuals had consumed filamentous green and blue-green algae and diatoms, but their total volume was 5% or less. The young Bullhead Minnows feed primarily on bottom-ooze diatoms. Most feeding in adults occurs in schools during the day near the bottom, and schools disperse at night.

Pimephales vigilax breeds in late spring and summer (May into July) in Oklahoma in cavities excavated by males beneath stones, tree limbs, or other solid objects on the bottom (Parker 1964). Tiemann (2007b) reported an extended spawning season in Kansas from late April to September. A similar spawning season probably occurs in Arkansas. We have observed spawning in the Arkansas River at Fort Smith in late June and in Lake Hamilton in early July. *Pimephales* species are fractional spawners (Wynne-Edwards 1932; Gale 1983; Tiemann 2007a). Spawning occurs at water temperatures of 78°F (25.6°C) or higher (Parker 1964). Page and Ceas (1989) classified this species as an egg clusterer. Territorial males guard clutches of eggs attached to the upper surface of the nest cavity. The male keeps the eggs free of sediment by brushing them with the fleshy pad on his back; his constant swimming beneath the eggs also serves to aerate them. The males are very aggressive and active in guarding nests. Fecundity varies from 33 to 332 eggs ($\bar{x}$ = 124) (Tiemann 2007b). Egg size averaged 1.21 mm (Tiemann 2007b). The eggs hatch in 4–6 days at water temperatures of 79–83°F (26.1–28.3°C) (Parker 1964). Few individuals live more than 3 years, but maximum life span is around 5 years (Starrett 1951).

CONSERVATION STATUS   The Bullhead Minnow is widely distributed across Arkansas and is abundant in larger streams and rivers. Its populations are secure.

Genus *Platygobio* Gill, 1863—Flathead Chubs

Throughout most of the 20th century, *Platygobio* was considered a subgenus of the genus *Hybopsis*. Its elevation to full generic status (Coburn and Cavender 1992) is now widely accepted. *Platygobio* is a monotypic genus, including only *P. gracilis*. Genus characters are the same as those in the following species account.

Platygobio gracilis (Richardson)
Flathead Chub

Figure 15.112. *Platygobio gracilis*, Flathead Chub. *David A. Neely.*

CHARACTERS   A large, barbeled, silvery chub with a small eye, a pointed snout that scarcely protrudes beyond the mouth, and a broad, flattened, wedge-shaped head (Fig. 15.112). Eye diameter much less than snout length, and going into head length 4.4–6.2 times. Body depth going into SL 4.2–5.2 times. Mouth large, subterminal, upper jaw extending posteriorly to or past the anterior margin of the eye. Pharyngeal teeth 2,4–4,2. Barbels of moderate size. Lateral

Map 90. Distribution of *Platygobio gracilis*, Flathead Chub, 1988–2019.

line complete, with 42–56 scales. Dorsal rays 8; anal rays 8. Fins high and falcate. Anterior rays of dorsal fin extending beyond rear rays when fin is depressed. Pectoral fin long, pointed, and falcate, its tip when depressed nearly reaching pelvic fin base, but not extending behind it. Dorsal fin origin over or in front of pelvic fin origin. Anal and pelvic fins have compound taste buds in the interradial membranes. Body scales smooth, not keeled. Breast and belly well scaled. Intestine short, with a single S-shaped loop; peritoneum silvery, with scattered dark speckles. Breeding males with minute tubercles on top of head and snout, rows of tubercles on scales of nape and on pectoral fin rays 2–8, and fine tubercles on dorsal, pelvic, and anal fins. Females with tubercles on head and sometimes weakly developed elsewhere. Adults usually 3.7–7.5 inches (95–190 mm) in TL. Maximum size 12.5 inches (317 mm) TL (Scott and Crossman 1973).

LIFE COLORS   Body color of adults light brown, dusky olive, or black above with silvery sides and venter. All fins are clear, but the lower lobe of the caudal fin is darker than the upper lobe. Breeding males exhibit no bright colors.

SIMILAR SPECIES   *Platygobio gracilis* is similar to *Macrhybopsis meeki*, but it differs in having a well-scaled breast (vs. naked), in pectoral fins not extending behind

pelvic fin origins (vs. extending beyond pelvic fin origins), and in having 2,4–4,2 pharyngeal teeth (vs. 1,4–4,1). It differs from *M. gelida* in lacking keels on body scales (vs. keels present on scales), in having anterior dorsal rays extending past posterior rays when fin is depressed, and in having 2,4–4,2 pharyngeal teeth (vs. 1,4–4,1). The Flathead Chub can be distinguished from *Hybopsis amblops* and *M. storeriana* by its small eye and in having more than 40 lateral line scales (vs. eye large and fewer than 40 lateral line scales). It differs from *M. hyostoma*, *Erimystax harryi*, and *E. x-punctatus* in lacking spots or dark markings on body (vs. spots or dark markings present).

VARIATION AND TAXONOMY   *Platygobio gracilis* was described by Richardson (1836) as *Cyprinus gracilis* from the Saskatchewan River at Carleton House, Canada. It was subsequently assigned to genus *Platygobio* and later placed in genus *Hybopsis* by Bailey (1951). Mayden (1989) returned it to *Platygobio*. Olund and Cross (1961) studied geographic variation in *Platygobio* and recognized two subspecies. The nominate subspecies, *Platygobio gracilis gracilis*, is the form which occurs in Arkansas and most of the species range. A southwestern subspecies, *P. g. gulonella*, was also recognized. Analyses of mitochondrial and nuclear genes supported a sister relationship between *P. gracilis* and genus *Macrhybopsis* (Echelle et al. 2018).

DISTRIBUTION   *Platygobio gracilis* occurs in Arctic and Mississippi River drainages from northwestern Canada south through the Missouri River basin and lower Mississippi River through Louisiana, in the Arkansas River drainage from Colorado and New Mexico to central Oklahoma, and in the upper Rio Grande drainage (including the Pecos River) in New Mexico (Map 90). In Arkansas, this species is confined to the Mississippi River, and it has been collected on four occasions from three localities (Map A4.89). (1) Black (1940) collected five "young to half-grown" specimens from northern Mississippi County between Barfield and Hickman on 8 August 1939 (UMMZ 128573); (2) Buchanan (1976) collected one small specimen near the confluence of the White and Mississippi rivers in Desha County on 8 July 1974 (UAFS 0460); (3) five specimens were taken by T. M. Buchanan, Ken Shirley, and Drew Wilson at West Memphis in Crittenden County on 31 July 1980 (UAFS 2015); (4) and one specimen was collected by Buchanan and Shirley at West Memphis on 22 December 1980 (UAFS 2016). Etnier and Starnes (1993) reported two additional records from the Mississippi River from the Tennessee side. There have been no additional records for this species in Arkansas in almost 40 years.

HABITAT AND BIOLOGY   The Flathead Chub is a rare member of the Arkansas fish fauna, having been taken only in the Mississippi River from turbid, flowing waters over a firm sand or fine gravel substrate in the main channel. It is primarily a midwestern and Great Plains species, but sporadic records have been reported from the lower Mississippi River in Arkansas and Louisiana. This chub inhabits both clear and turbid streams over shifting sand bottoms or in murky pools with gravel or bedrock substrates (Olund and Cross 1961). Davis and Miller (1967) attributed the success of *P. gracilis* to its ability to capitalize on changing conditions, since no single sensory system is hyperdeveloped nor degenerated to an extent that utility is seriously impaired (Reno 1969). The ability to use either or both taste and sight in any environment gives *P. gracilis* a decided advantage over other fishes that must rely on one system regardless of environmental conditions (Davis and Miller 1967). Pflieger (1997) reported that it was one of the most abundant minnows in the Missouri River and the middle Mississippi River of Missouri in the 1940s, but by 1997 it was considered on the verge of extirpation in Missouri. Its decline in the Missouri River was attributed to the construction of six large reservoirs on that river upstream from Missouri that changed the natural flow regimen and reduced the bedload and turbidity all the way to its mouth (Pflieger 1997). Carter (1984) failed to find any specimens in his survey of the Mississippi River in Mississippi County, Arkansas, and Pennington et al. (1980, 1983) reported no specimens from the lower Mississippi River of Arkansas. No specimens were taken during extensive sampling of the Mississippi River by biologists from the Vicksburg Waterways Experiment Station in the 1990s and 2000s (data provided by K. J. Killgore). Extensive sampling of the middle and lower Mississippi River of Missouri using a Missouri trawl produced no records for *P. gracilis* (Herzog 2004). It is known from two localities in the Mississippi River of Louisiana (Douglas 1974). It remains relatively abundant only in northern and western parts of its range, particularly in South Dakota (Hayer et al. 2008).

The chemosensory barbels and buds of Flathead Chubs are considered adaptations to the characteristically turbid prairie rivers (Rahel and Thel 2004a), but Davis and Miller (1967) reported that vision is also an important sensory modality allowing this species to feed successfully in varying turbidity environments. They further suggested that it is an opportunistic feeder with a diet that includes mostly terrestrial insects supplemented by small invertebrates and plant material (Olund and Cross 1961). Terrestrial insects ingested include beetles, flies, grasshoppers, and ants; aquatic insects made up a much smaller proportion of the diet (Olund and Cross 1961). Fisher et al. (2002) found that Flathead Chubs smaller than 60 mm TL fed on hemipterans and copepods in high flow years and predominantly on coleopterans in average flow years. Pflieger (1997) speculated that in the highly turbid presettlement Missouri River it had little competition for food, but in the much clearer Missouri River of today, it is outcompeted by other sight-feeding specialists such as the Emerald Shiner.

In Iowa, *P. gracilis* spawned from mid-July to mid-August (Martyn and Schmulbach 1978), with mature females averaging 4,974 eggs per ovary. In South Dakota, spawning occurred in June and July (Hayer et al. 2008). Sexual maturity was reached in 2 years and a standard length of greater than 4 inches (105 mm). Although the precise spawning habitat of Flathead Chubs remains unknown, the timing of the spawning season appears to coincide with lower flows, reduced turbidity levels, and warmer water temperatures in summer (Olund and Cross 1961). Bonner and Wilde (2000) reported that the Flathead Chub belongs to a guild of prairie stream fishes that produce nonadhesive, semibuoyant eggs and spawn in response to floods that increase stream flows required to keep the semibuoyant eggs afloat until hatching occurs. Walters et al. (2014) showed that it migrates upstream to spawn during summer in Colorado. Those authors noted that the Flathead Chub requires long reaches of contiguous, flowing riverine habitat for drifting eggs or larvae to develop, and their declining populations have been attributed to habitat fragmentation or barriers to upstream movement such as dams. Newly hatched fry are weak swimmers, and strong currents are required to keep fry suspended so they do not settle to the bottom and become buried. Maximum reported ages were 6 years from the Belle Fourche River, South Dakota (Hayer et al. 2008), and 10 years from the Peace River, Canada (Bishop 1975).

CONSERVATION STATUS Pflieger (1997) documented the decline of this species in Missouri and considered it possibly on the verge of extirpation from that state. It is currently considered endangered in Missouri. This big-river chub is rarely encountered in Arkansas and should be considered endangered in our state.

Genus *Pteronotropis* Fowler, 1935— Flagfin Shiners

There are 9 species of *Pteronotropis*, one of which (*P. hubbsi*) occurs in Arkansas. Genus characters include body fairly deep to deep and compressed, the deepest part under or near the dorsal fin origin. No barbels present. Dorsal fin large; its origin posterior to pelvic fin origin; dorsal and anal fins of males relatively larger than those of other cyprinids. Pharyngeal teeth usually 2,4–4,2, but 1,4–4,1 in

P. welaka and 0,4–4,0 in *P. hubbsi*. Lateral line usually complete (except in *P. hubbsi* and *P. welaka*). A broad steel-blue to blackish lateral stripe present. Breeding males develop bright colors on head and body. Mayden et al. (2006a) did not resolve the genus *Pteronotropis* as monophyletic.

Pteronotropis hubbsi (Bailey and Robison)
Bluehead Shiner

Figure 15.113. *Pteronotropis hubbsi*, Bluehead Shiner. *Uland Thomas.*

Map 91. Distribution of *Pteronotropis hubbsi*, Bluehead Shiner, 1988–2019.

CHARACTERS   A distinctive, slab-sided minnow with 9 or 10 dorsal fin rays, and a broad, black lateral stripe extending from the tip of the snout to the caudal base, terminating in a deep caudal spot that extends onto the caudal fin (Fig. 15.113). Dorsal fin origin behind pelvic fin origin;

dorsal origin closer to caudal base than to tip of snout. Anal rays 9–11 (modally 10). Pectoral rays 12–14. Pharyngeal teeth 0,4–4,0. Mouth terminal, sharply upturned. Snout projecting slightly beyond upper lip. Eyes moderate in size, eye diameter about equal to length of lower jaw. Lateral line incomplete, with only 2–9 pored scales; 34–36 scales in the lateral series. Predorsal scales crossing the middorsal line 20–24. Body circumferential scales 26–29. Infraorbital canal interrupted, occurring in short sections. Intestine short, with a single lengthwise loop; peritoneum dusky. Nuptial tubercles well developed in males, but small, occurring over the head, and concentrated along the mandible, anterodorsal rim of orbit, lower edge of lachrymal, and with a rather uniform distribution over the dorsal surface. Tubercles also on the opercle and body. All fins with tubercles. Dorsal fin of males rounded and elongated, with middle rays longest. Anal fin also large and gently rounded. Maximum size 2.5 inches (64 mm).

LIFE COLORS   One of our most colorful fishes. Adult males have a dusky body above with a distinctive, broad black lateral stripe extending from the chin through the eye across the opercle backward to the caudal base, where it terminates in a deep, darker basicaudal spot that extends a short distance onto the caudal rays. Lower surface of head and belly white. Chin distinctly black. Dark pigment well developed on floor of mouth. Breeding males are dimorphic in coloration, with the larger flag males (Fig. 15.114) exhibiting different coloration from the smaller nonflag males. Fletcher and Burr (1992) provided the following information on breeding colors. The blue on top of the head of smaller flag males (42–45 mm SL) was faded or absent on the largest (>45 mm SL) individuals. The fins were olive yellow to cream gold. The dorsal fin was dark bluish black with 2 rows of cream-gold blotches at the bifurcations of the fin rays and with a metallic blue fringe on the distal border of the fin. The caudal fin was cream gold proximally and dark bluish black distally. Ephemeral dark vertical bars develop on the sides of flag males during aggressive and probably spawning encounters. The nonflag or secondary males had a patch of powder blue on top of the head, extending from the occiput to the tip of the snout. Dorsal, anal, pectoral, and pelvic fins were iridescent blue. The caudal fin was russet red proximally, black at the base of the fork, and blue to bluish black on the lobes. The russet coloration was observed only in nonflag males, and was most intense in the smallest (30–34 mm SL) individuals. The red russet formed a stripe above the black lateral body stripe, at the base of the caudal fin, and, occasionally, on the cheek and opercle. As the nonflag males grew, they gradually took on the coloration of a flag male. Breeding

Figure 15.114. *Pteronotropis hubbsi* "flag male" in spawning coloration. *Richard T. Bryant and Wayne C. Starnes.*

females had iridescent powder-blue heads as in the nonflag males, but no iridescent blue on any fins. Some females had a red-russet stripe above the black lateral stripe, around the caudal spot, and at the base of the caudal fin, but this coloration was not as intense as in the nonflag males. The name Bluehead Shiner is derived from the most commonly observed color features of the species; that is, the top of the head from occiput to between the anterior edges of nostrils is a deep azure blue with green iridescence in many individuals during the breeding season (Bailey and Robison 1978).

SIMILAR SPECIES   The combination of 9 or 10 dorsal fin rays and a short lateral line with only 2–9 pored scales distinguishes the Bluehead Shiner from all other Arkansas cyprinids.

VARIATION AND TAXONOMY   The type locality for *Pteronotropis hubbsi* is Locust Bayou at State Highway 4 bridge, 0.6 mile (1 km) west of Locust Bayou, Calhoun County, Arkansas. Bailey and Robison (1978) described it as *Notropis hubbsi* (subgenus *Pteronotropis*) and hypothesized a close relationship with *P. welaka* based on four morphological traits: reduced number of pored scales in lateral series; rounded and enlarged dorsal and anal fins; continuous, broad lateral stripe; and the coloration of nuptial males (intense blue color on the head of *P. hubbsi* and on the snout in *P. welaka*). They suggested that these taxa might be closely related to *Cyprinella*. Interestingly, some phylogenetic studies have found little evidence of relationship between *P. hubbsi* and *P. welaka* (Dimmick 1987; Mayden 1989). Mayden (1989) elevated *Pteronotropis* to genus level (leaving *P. hubbsi* in *Notropis*). Dimmick (1987) and Amemiya and Gold (1990) placed *N. hubbsi* within *Pteronotropis*, which had previously been regarded as a subgenus. Coburn and Cavender (1992) retained *Pteronotropis* as a subgenus, placing *N. hubbsi* within it. Simons et al. (2000) found no support for *Pteronotropis* as a natural group. Phylogenetic relationships among the taxa within

Pteronotropis remain unclear, and Mayden et al. (2006a) found no evidence for the monophyly of *Pteronotropis*. Suttkus and Mettee (2001) analyzed the subgenus *Pteronotropis* and considered *N. hubbsi* and *N. welaka* sufficiently different to warrant subgeneric placement. Simons et al. (2000) found strong support for the monophyly of the *P. hubbsi* plus *P. welaka* clade. *Pteronotropis welaka* and *P. hubbsi* are also both nest associates with centrarchids (Fletcher and Burr 1992; Johnston and Knight 1999) and spawn over the clean gravel substrate maintained by the male sunfish. Controversy over relationships among the taxa remains. We tentatively follow Page et al. (2013) in placing the Bluehead Shiner in the genus *Pteronotropis*.

DISTRIBUTION   *Pteronotropis hubbsi* has a disjunct distribution with one group of isolated populations occurring in the Red River drainage (the Red River and four of its major tributary rivers) west of the Mississippi River in Arkansas, Oklahoma, Louisiana, and Texas (Hargrave and Gary 2016), and a widely separated population (possibly introduced and now presumed extirpated) in Wolf Lake in southwestern Illinois. In Arkansas, the Bluehead Shiner is an inhabitant of tributaries of the Ouachita River system and occurs disjunctly in the Little River system of southwestern Arkansas (Map 91). Tumlison et al. (2015) provided additional locality records of *P. hubbsi* from the Saline and Ouachita rivers in Ashley County, Arkansas.

HABITAT AND BIOLOGY   The Bluehead Shiner is an inhabitant of quiet, backwater areas of small to medium-sized sluggish streams and oxbow lakes (Bailey and Robison 1978). Typically, the tannin-stained water where *P. hubbsi* occurs is heavily vegetated with such plants as *Proserpinaca palustris*, *Polygonum hydropiperoides*, and *Nelumbo pentapetala*. In a tributary of Bayou Bartholomew in Louisiana, *P. hubbsi* was often found around dense growths of hornwort, *Ceratophyllum demersum*, in water 0.5–1.2 m deep (Fletcher and Burr 1992). In Illinois, this species was almost always found in association with lotus beds (Smith 1979). The substrate is generally mud or a mixture of sand and mud. Buchanan (2005) reported the Bluehead Shiner from two Arkansas impoundments, Lake Millwood and an area of Champagnolle Creek impounded by a low-water dam. *Pteronotropis hubbsi* schools in backwater or side areas away from substantial current and seems to remain poised in midwater, just outside vegetation, into which it darts for protection if disturbed (Bailey and Robison 1978). Observations by HWR indicate that this species is migratory in Ouachita River drainage streams. At the type locality, adults are found only during the late spring (May) when they migrate upstream to spawn, but specimens have been taken in the main Ouachita River at various times of the year.

Most of the following information on the feeding biology of *P. hubbsi* is from the Louisiana study by Fletcher and Burr (1992). Most feeding occurs in the water column, but some individuals picked food items off vegetation and occasionally off the surface. The diet was diverse, and animal material, particularly microcrustaceans, dominated the diet of juveniles (<29 mm SL) and adults (>29 mm SL). The dominant food item of adults in April and May was Cladocera, followed by chironomid larvae and copepods. Juveniles also fed predominantly on cladocerans, copepods, and chironomids. Other foods consumed included ostracods, amphipods, ephemeropterans, odonates, hemipterans, other midges, oligochaetes, rotifers, and bryozoans. Some individuals had consumed seeds and unicellular and filamentous algae. TMB examined gut contents of 8 specimens collected in May from Tupelo Creek in Clark County, and 6 specimens collected in August from Champagnolle Creek in Ouachita County. In those months, it fed almost entirely at or near the water surface on terrestrial and aquatic insects and on microcrustaceans (mainly cladocerans).

Spawning occurs in Arkansas from the middle of May to July (tuberculate males have been taken from early to mid-May to June, and young-of-the-year have been collected in early June). We found ripe females and tuberculate males with enlarged testes on 23 May in Tupelo Creek, Arkansas. In southeastern Oklahoma, spawning occurred from May into July (Taylor and Norris 1992). From observations at the type locality, HWR discovered that two classes of males are present from May to July. The flag males are larger, have large dorsal fins, are socially dominant, and carry out most of the spawning. Younger, smaller nonflag males are not directly engaged in spawning activities (Robison and Buchanan 1988). Fletcher and Burr (1992) confirmed that *P. hubbsi* is unusual among cyprinids in having two phases of postjuvenile development in males and referred to those phases of postjuvenile development as terminal males (= flag males) and secondary males (= nonflag males). The two male phases exhibited different coloration (see Life Colors section). The transition from nonflag to flag males is apparently length-related and possibly age-related. The Bluehead Shiner spawns in Warmouth (*Lepomis gulosus*) nests in cavities formed by diverging roots of the trunks of bald cypress trees (*Taxodium distichum*) (Fletcher and Burr 1992). One nest was in water 15 cm deep, and water temperature was 24°C. Very little vegetation was present in the nest area. The floors of the nesting cavities differed

from the surrounding substrate. The floor of a nest cavity was covered with large deciduous leaves, cypress fronds, pine needles, and moss, and it was free of silt, even though the surrounding area was mixed silt and detritus. Fletcher and Burr (1992) found evidence that other habitats were used as oviposition sites in ponds in Louisiana where cavities of cypress trees were not available. Flag males establish and defend territories around the nest cavities using displays of aggression and dominance. Nonflag males are not territorial and may sneak into the territories of flag males and spawn. Spawning behavior of the Bluehead Shiner has not been described. Fletcher and Burr (1992) found the testes of flag males were only slightly heavier than those of nonflag males, but their somatic weights were dramatically higher. They hypothesized that upon transition into a flag male, more energy is allotted to increased somatic growth, fin development, and agonistic behavior than to gonad development. Another most unusual behavior of this cyprinid among nest-associating cyprinids is that only a single flag male defended the host nest. Females reach maturity between 36 and 40 mm SL and 1 year of age. In southeastern Oklahoma, females contained multiple size-classes of ova in May, suggesting that multiple clutches are spawned during the reproductive season (Taylor and Norris 1992). After hatching, larvae remain on the substrate for about 5 days, until inflation of the swim bladder. Larvae can be distinguished from other syntopic larval cyprinids by a distinctive caudal spot (Fletcher and Burr 1992). In Illinois, Burr and Warren (1986a) reported that females are sexually mature at 1 year of age and 1.9 inches (47 mm) SL. One female contained 781 mature ova averaging 0.8 mm (0.03 inch) in diameter. In Oklahoma, clutch size ranged from 172 to 1,129 mature ova in females ranging from 34.5 to 49.5 mm SL (Taylor and Norris 1992). Growth information from Illinois indicated that juveniles were 14–20 mm TL on 21 June, 18–26 mm on 25 July, and 22–38 mm on 30 August; one juvenile collected on 29 November was 44 mm TL. Two-year males probably die after spawning, and maximum life span appears to be 2 years. Cloutman (2011) described a new species of monogenoid parasite from *P. hubbsi*.

CONSERVATION STATUS   In Illinois, the Bluehead Shiner is presumed extinct (Ranvestel and Burr 2004), and Arkansas currently has more of the known populations of *P. hubbsi* than any other state. There has been little historical interest in protecting the lowland swamp habitat required by *P. hubbsi*. We consider the Bluehead Shiner a species of special concern in Arkansas due to a rather restricted distribution within the state and continued environmental degradation of lowland streams of the Ouachita River system, including gravel removal at the type locality.

Genus *Scardinius* Bonaparte, 1837—Rudds

This Eurasian genus, described by Bonaparte in 1837, contains about 10 species. Genus characters of *Scardinius* are included in the species account below.

Scardinius erythrophthalmus (Linnaeus)
Rudd

Figure 15.115. *Scardinius erythrophthalmus*, Rudd. *David A. Neely.*

Map 92. Distribution of *Scardinius erythrophthalmus*, Rudd, 1988–2019. Inset shows introduced range (outlined) in the contiguous United States.

CHARACTERS A stocky, deep, compressed-bodied cyprinid with a small head and terminal, oblique mouth, a protruding lower lip, and a forked tail (Fig. 15.115). Body depth going 2.7–3.2 times into SL. Eye moderate in size, eye diameter going into head length 3.6–4.7 times. A scaled bony keel runs along the belly from pelvic fins to anal fin. Lateral line complete and greatly decurved, with 37–45 scales. Usually 7 or 8 scale rows above the lateral line. Dorsal fin rays 9–11. Anal rays 10–14. Pectoral rays 15 or 16; pelvic rays 8 or 9. Gill rakers short and stout, 10–13. Intestine short, with a single S-shaped loop. Peritoneum silvery and thickly sprinkled with dark specks. Dorsal fin origin behind pelvic fin origin. Margins of dorsal and anal fins concave. Pharyngeal teeth are 3,5–5,3 (sometimes 2,5–5,2). Scales are conspicuously marked. Breeding males develop fine tubercles on head and body. Maximum size 19 inches (48 cm). Largest specimen from Nebraska weighed 3 pounds, 7 ounces (1.6 kg) (Hrabik et al. 2015).

LIFE COLORS Body brownish green above, brassy yellow laterally, and grading to a silvery-white belly. Anal, pelvic, and pectoral fins bright red in life; dorsal and caudal fins reddish brown. Eye color is gold with red spot (or fleck) at top.

SIMILAR SPECIES The Rudd is similar to the Golden Shiner, *Notemigonus crysoleucas*, but differs from it in having a scaled ventral keel (vs. a scaleless keel), 9–11 dorsal rays (vs. 7–9), and pharyngeal teeth 3,5–5,3 (vs. 0,5–5,0). In addition, the Rudd has blood-red to reddish-brown fins while *Notemigonus crysoleucas* has yellow fins.

VARIATION AND TAXONOMY *Scardinius erythrophthalmus* was originally described by Linnaeus (1758) as *Rutilus rutilus* from northern Europe. The Rudd is known to hybridize with the Golden Shiner (Burkhead and Williams 1991). The hybrid is apparently the first known nonsalmonid intergeneric cross of a North American native and an exotic (J. N. Taylor et al. 1984). The scaled ventral keel distinguishes the Rudd from native Arkansas cyprinids.

DISTRIBUTION The Rudd is native to Eurasia (western Europe to the Caspian and Aral sea basins), and was introduced into the United States in the early 1900s as a baitfish. It has been introduced into 20 states, and reproducing populations are known from Massachusetts, Nebraska, and South Dakota, with possibly other reproducing populations in other states not yet confirmed (Nico et al. 2018b). It was first found in Oklahoma in 1987–1988 (Pigg and Pham 1990). In Arkansas, records are known from Horseshoe Lake in Crittenden County (Robison and Buchanan 1993), Spring River in Lawrence County

(Rasmussen 1998), White River drainage, and Lonoke County (Map 92). The first two records are probably bait releases, and the Lonoke County record was an escapee from an aquaculture facility. It has also been reported from the lower Sulphur River in Red River County, Texas, adjacent to Arkansas. Other records are from southeastern Kansas and the Arkansas River drainage of eastern Oklahoma.

HABITAT AND BIOLOGY The Rudd has broad habitat tolerance and is most commonly found in weedy shoreline areas of lakes and sluggish pools of medium to large-sized rivers (Page and Burr 2011). It also occurs in pools and backwaters of streams and shallow margins of lakes and ponds (Pflieger 1997). Crossman et al. (1992) suggested that the Rudd should not be intentionally or accidentally transferred to other waters, in order to avoid damage to native cyprinid populations through competition and hybridization.

It is a benthic omnivore that can live up to 15 years (Nico et al. 2018b). Its diet consists of various macrophytes, bryophytes, filamentous algae, some animal material, and detritus (Nurminen et al. 2003). It consumes vegetation, which can potentially impact spawning areas of other species. In its native range, adults feed mainly on surface or aerial insects; young feed mainly on diatoms, algae, and copepods (Hensley and Courtenay 1980).

Reproduction occurs from April to August. Adhesive eggs are laid among vegetation. Males defend territories during the breeding period, including the plants on which the adhesive eggs are deposited. Males continue to discharge sperm for some time after the eggs have been deposited (Breder and Rosen 1966). Fecundity ranges from 3,500 to 232,000 eggs. The Rudd is known to hybridize under artificial conditions with the Golden Shiner, which it may encounter in nature (Burkhead and Williams 1991). Self-sustaining populations exist in three South Dakota lakes, where populations were dominated by individuals >200 mm TL (Blackwell et al. 2009). In those lakes, several year classes were present, natural mortality was low, and growth rates were fast. There is no evidence of self-sustaining populations in Arkansas.

CONSERVATION STATUS This introduced species does not warrant a conservation status in Arkansas.

Genus *Semotilus* Rafinesque, 1820a— Creek Chubs

The North American genus *Semotilus* consists of four described species. One species, *S. atromaculatus*, occurs in Arkansas. Characters of the genus include 2,5–4,2 pharyngeal teeth, and a triangular, flaplike barbel in the groove between the maxilla and snout (Fig. 15.5). Location of the barbel is different from most barbeled North American cyprinids in that it is preterminal, rather than terminal, with respect to the posterior end of the maxilla (Simons and Mayden 1997). Evans and Deubler (1955) documented pharyngeal tooth replacement in *Semotilus*. Affinities of *Semotilus* are with *Couesius plumbeus*, according to Lachner and Jenkins (1971). Cavender and Coburn (1987) treated *Margariscus* as a separate genus and sister group to *Phoxinus* (= *Chrosomus*), with *Semotilus* as closest relative to those two genera. *Semotilus* and *Hemitremia* were recovered consistently as sister-taxa in analysis of morphological (Coburn and Cavender 1992) and molecular (Simons and Mayden 1997; Simons et al. 2003; Bufalino and Mayden 2010a, b) data.

Semotilus atromaculatus (Mitchill)
Creek Chub

Figure 15.116. *Semotilus atromaculatus*, Creek Chub. *Uland Thomas.*

CHARACTERS A large, terete minnow with a large head, very large, oblique mouth (upper jaw extending behind front of eye), a conspicuous black spot present at the anterior base of the dorsal fin, and a small dark spot at the caudal base (Fig. 15.116). Head length going into TL 4.0–4.7 times. A small flaplike barbel present (sometimes absent) in the groove above the upper lip just ahead of the corner of the mouth (preterminal). Upper lip wider in the middle than on either side. Lateral line usually complete, with 52–63 scales. Anterior body scales smaller than posterior scales; anterodorsal scales are crowded. Predorsal scale rows 27–33. Body circumferential scales 43–46. Breast and belly fully scaled. Pharyngeal teeth usually 2,5–4,2 (Fig. A3.7). Gill rakers short and conical, 8–11 total. Dorsal rays 8. Dorsal fin origin situated behind pelvic fin origin. Anal rays 8. Pelvic rays 8. Pectoral fin rays 15–18. Intestine short and S-shaped; peritoneum silvery with a few dark

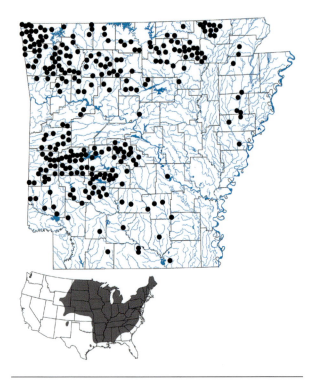

Map 93. Distribution of *Semotilus atromaculatus*, Creek Chub, 1988–2019.

speckles. Breeding males with 6–12 well-developed tubercles on snout and top of head and with smaller tubercles on the pectoral rays. The largest Arkansas specimen from Spring Creek in Benton County was 11.5 inches (295 mm) (Tumlison et al. 2017). Maximum size 12 inches (305 mm).

LIFE COLORS   Body color dark olive with a broad dusky lateral stripe well developed in young, but fading in large adults. Sides silver with green cast. Belly silvery white. A distinct middorsal stripe encircles the dorsal fin base. Fins often yellow. Breeding male develops an overall bronze-orange coloration with orange lower fins, orange at dorsal fin base, and often with a pink band along the side of the body.

SIMILAR SPECIES   It can be distinguished from minnows in the genera *Hybopsis, Macrhybopsis, Platygobio, Erimystax,* and *Nocomis* by its flaplike nonterminal barbel (vs. barbel at posterior end of upper jaw). All other native cyprinids lack barbels.

VARIATION AND TAXONOMY   *Semotilus atromaculatus* was described by Mitchill (1818) as *Cyprinus atromaculatus* from the Wallkill River, New York. Subsequent synonyms were *Leuciscus diplema* and *Semotilus corporalis* (Smith 1979). *Semotilus* was included in a group of 16 genera considered to form a chub clade (Coburn and Cavender 1992). Skalski et al. (2008) compared genetic variation in

small tributaries connected by a river with that of populations in small tributaries connected by a reservoir. Genetic diversity was higher in the river tributaries than in reservoir tributaries. Additional analysis of geographic variation over its large range is needed.

DISTRIBUTION   *Semotilus atromaculatus* is widely distributed in Atlantic Slope, Great Lakes, Hudson Bay, Mississippi River, and Gulf Slope drainages from Saskatchewan, Canada, and Wyoming to southeastern Canada, south to Georgia and Texas, with isolated populations in upper Pecos and Canadian river systems, New Mexico, and in northeastern Canada. In Arkansas, the Creek Chub occurs primarily in upland streams of the Interior Highlands, but it can be found wherever there is suitable habitat on the Coastal Plain and is common in streams of Crowley's Ridge (Map 93).

HABITAT AND BIOLOGY   The Creek Chub frequents the smallest clear, gravel-bottomed headwater streams throughout the Ozarks and Coastal Plain. The creeks where it is found usually have few other fishes, and in some small streams, it may be the only fish species present. Bart (1989) found that in Ozark streams it occurs primarily in pools and slow-moving side channels. It is reported to be somewhat tolerant of pollution and turbidity (Becker 1983; Pflieger 1997). It is apparently intolerant of reservoirs. Smith (1979) reported that it had increased in abundance in all parts of Illinois and attributed that increase to extensive habitat alteration, especially increased siltation and the elimination of competing species by human modification of streams. The Creek Chub usually avoids larger streams having a continuous strong flow and a variety of competing fishes, but we have sometimes found it in larger creeks, such as Lee Creek in Crawford County. Its distribution may also be limited in the lowlands by the availability of gravel bottoms required for spawning. Davey et al. (2011) evaluated the diel activity of juvenile Creek Chubs as a response to predator presence and habitat structural complexity. They found that Creek Chubs preferred structurally complex habitats during the day and moved to less complex habitat at night. In addition, Creek Chubs used diel shifts to avoid a predator. Etchison and Pyron (2014) found patch use by *Semotilus atromaculatus* changes from day to night. This species used coarse substrates during the day but not always at night, perhaps because of inability to distinguish sediment size at night. It showed directional movement into pools during periods of stream drying (Hodges and Magoulick 2011). Creek Chub movement in intermittent reaches of North Sylamore Creek, Arkansas, was greater than reported in previous studies (Walker and Adams 2016). Maximum distances moved were between 1,228 and

4,678 meters. The probability of Creek Chub movement decreased as habitat complexity and pool area increased.

It is a generalized carnivore that forages at all levels in the water column feeding on a variety of invertebrates, including crayfish, mollusks, diatoms, and insects, and also on small fishes. In the smallest headwaters, the Creek Chub diet consisted more of terrestrial insects than aquatic insects (Lotrich 1973), and it fed almost exclusively on flying insects in summer and autumn in the Illinois River, Oklahoma (Felley and Hill 1983). Tumlison et al. (2016) reported finding a cicada in the gut of a 130 mm TL male specimen from the Mulberry River in Johnson County. Small individuals (20–40 mm SL) feed more on larval and adult dipterans, mayflies, hymenopterans, spiders, and bugs; larger chubs (41–60 mm) add odonates, trichopterans, and mollusks to the diet; and large adults consume larger insect species, with fishes becoming increasingly important in the diet (Barber and Minckley 1971; Ross 2001). The largest adults have been known to consume salamanders and frogs (Keast 1966; Moshenko and Gee 1973; Newsome and Gee 1978). Boschung and Mayden (2004) noted that adults prefer to feed on drifting food. Barber and Minckley (1971) found that feeding activity was greatest in the early evening and least in the morning.

Breeding in Arkansas occurs in early spring (usually April–May). The male builds a pit-ridge nest unique to members of the genus *Semotilus* (Maurakis and Kahnke 1987). The nest consists of a small pit with a pile of pebbles (6.0–23.0 mm) at the upstream margin. Spawning occurs when a female enters the nest and approaches the male, who clasps her between his pectoral fins and body. A female may produce more than 7,500 eggs per year. After females spawn in the nest, the male continues digging in the downstream portion of the nest (pit) and depositing pebbles in the upstream portion to produce a long gravel ridge upstream of the stream pit, and the pit is displaced downstream (Woolcott and Maurakis 1988; Maurakis et al. 1990; Miller and Robison 2004). The male covers each spawning of eggs with substrate, thereby extending the pit downstream and eventually forming a long ridge of gravel. The developing eggs are buried within the ridge, having been deposited in successive spawnings. After all spawning activity is completed, the males abandon the nests. R. J. Miller (1967) observed spawning in Spavinaw Creek, Oklahoma, and noted that several males may work simultaneously in constructing a single nest. Up to 6 males at a time were observed picking up stones in their mouths and depositing them mainly upstream from the 18-inch (457 mm) diameter nest. HWR (pers. obs.) made similar observations involving 4 or 5 males working on a single nest in

Spavinaw Creek, Oklahoma, on 19 April 1972. Maurakis et al. (1995) observed and recorded nocturnal spawning behavior and suggested sensory systems other than vision are involved in communication among males and females. Male Creek Chubs are often aggressive toward other conspecific males during the reproductive season (Ross 1977) possibly due to a shortage of suitable nest sites. Challenges by similar-sized males often result in complex displays called parallel swims, first described in Creek Chubs by Reighard (1910). Many males do not build their own nests, but instead act as satellites, waiting to temporarily occupy and spawn in the nests of other territorial males (Johnston 1989b). Little aggression is shown toward other fish species, including potential egg predators. After the eggs hatch, the young Creek Chubs work their way up through the spaces between the stones. Maturity is reached in the second year. Few Creek Chubs live beyond 3 or 4 years (Rohde et al. 2009), and maximum life span is around 8 years. McAllister et al. (2013) reported blackspot disease in Arkansas Creek Chubs, and Richardson et al. (2013) documented the piscicolid fish leech, *Cystobranchus klemmi*.

CONSERVATION STATUS   The Creek Chub is widespread and abundant in Arkansas, and its populations are secure.

Family Catostomidae *Suckers*

The sucker family, Catostomidae, is the only family in suborder Catostomoidei of order Cypriniformes. It consists of 13 genera with about 78 extant species (Harris et al. 2014; Nelson et al. 2016). Suckers are distributed in freshwaters throughout North America, from northern Alaska to Guatemala, and a few species are found in the Arctic Circle, China, and eastern Siberia (Berra 2007). All sucker species are tetraploid, and the family possibly originated

Figure 15.117. *Moxostoma carinatum,* River Redhorse, in White River near Cotter, Arkansas. *Isaac Szabo.*

from genome duplication through hybridization from a cyprinid-like ancestor (Uyeno and Smith 1972). Many of the duplicate genes have been silenced, and suckers are functionally diploid. This family dates from the Eocene (55–35 mya) of North America and Asia (Smith 1981; Cavender 1986; Chang et al. 2001; Liu and Chang 2009). Catostomidae is divided into four subfamilies (Harris and Mayden 2001; Harris et al. 2002, 2014; Nelson et al. 2016): (1) Myxocyprininae, having a single species, *Myxocyprinus asiaticus*, in eastern China; (2) Ictiobinae, composed of two genera and eight species, *Carpiodes* (carpsuckers) with three species, and *Ictiobus* (buffalofishes) with five species; (3) Cycleptinae, containing a single genus, *Cycleptus*, the blue suckers with two species; and (4) Catostominae, with nine genera and 67 species (Nelson et al. 2016). Boschung and Mayden (2004), recognized two tribes within Catostominae, and Doosey et al. (2010) recognized four tribes: tribe Catostomini, with four genera and 34 species, including *Catostomus commersonii* in Arkansas; tribe Moxostomatini, with 22 species, including the redhorse suckers and jumprock suckers (*Moxostoma*); tribe Erimyzonini which includes five species, the Spotted Sucker (*Minytrema melanops*) and four species of chubsuckers (*Erimyzon*); and tribe Thoburnini, containing six species, three species of torrent suckers (*Thoburnia*) and three species of hog suckers (*Hypentelium*). Tan and Armbruster (2018) supported the above-listed recognition of subfamilies within Catostomidae and tribes within Catostominae.

Doosey et al. (2010) found the tribe Erimyzonini to be the earliest diverging lineage within the subfamily Catostominae and sister to remaining members of the tribe Moxostomatini + Catostomini. Miller (1959) had previously proposed a basal position for the Erimyzonini based on osteological evidence. This was later supported by Ferris and Whitt (1978) based on loss of duplicate gene expression patterns. Clements et al. (2012) examined phylogenetic relationships of suckers of the tribe Moxostomatini using mitochondrial cytochrome *b* gene sequences and nuclear intron sequences of the growth hormone gene. Based on their analysis, the genus *Scartomyzon* is not monophyletic. They also supported continued recognition of the tribe Erimynonini and recognition of the Thoburnini (*Hypentelium, Thoburnia*) as the sister clade to all other *Moxostoma*. There was also strong support for recognizing a V-lip species group (*M. anisurum, M. pappillosum, M. collapsum*); an *M. carinatum* species group (*M. carinatum, M. erythrurum, M.* sp. Sicklefin Redhorse); and a Shorthead Redhorse group (*M. macrolepidotum, M. pisolabrum, M. breviceps*).

Catostomids inhabit a wide range of lentic and lotic freshwater environments, with the greatest morphological and ecological diversity found in the Mississippi River basin and southeastern United States (Berra 2007). Eighteen native sucker species contained in six genera inhabit Arkansas, plus one presumed extinct catostomid, the Harelip Sucker (*Moxostoma lacerum* Jordan and Brayton), last collected in Arkansas in 1884 (Jordan and Gilbert 1886). Suckers are also possible indicator species for watershed change (Doherty et al. 2010). They often make considerable migrations to suitable spawning habitats; therefore, habitat connectivity is a key factor in their conservation (Cooke et al. 2005).

Catostomids are basically bottom-dwelling fishes with the mouth usually situated on the underside of the head with fleshy premaxillae and protrusible lips for sucking foods from the bottom. Most species have a long, coiled intestine. Dorsal fin ray counts are 9–17 in subfamily Catostominae and 23–35 in subfamilies Ictiobinae and Cycleptinae. Additional family characters include 20–300 teeth on a distinctive dental-pharyngeal arch with all teeth in 1 row, anal fin with fewer than 10 rays and positioned well posteriorly, cycloid scales, a scaleless head, a Weberian apparatus, paired fins attached low on the body, and fins lacking spines. The mouth is an extremely important character for identifying catostomids. Jenkins and Burkhead (1994) emphasized the importance of the shape and surface texture of the lips, and stated that with experience, it is possible to identify preserved specimens at least as small as 40 mm TL. The young and small juveniles have slightly smoother, less dissected lip surfaces than adults; large adults have more posteriorly developed, fuller lips than smaller fish (Jenkins and Burkhead 1994). Because genera and various species may be distinguished from one another by anatomical differences in the lips, it is imperative that the mouth of the specimen is closed when examining those features. This caveat of examining the lip features with the mouth closed applies to all catostomid genera.

Suckers are readily distinguished from the related minnows (Cyprinidae) by the position of the anal fin. In catostomids, the anal fin is positioned farther back on the body. The distance from the front of the anal fin base to the base of the caudal fin base goes into the distance from the anal fin base forward to the tip of the snout more than 2.5 times. In native minnows, it goes into that distance fewer than 2.5 times. In addition, there are 19 principal caudal rays in North American cyprinids, whereas catostomids have 18 principal caudal rays.

Arkansas suckers range in size from the Western Creek Chubsucker, which reaches a length of about 9 inches (229 mm) and a weight of 0.5 pound (0.2 kg), to the Smallmouth Buffalo, which set a state record at a length of 3 feet (914 mm) and a weight of 74 pounds (33.6 kg). Small suckers are valuable forage food for large predator species. Larger catostomids, such as buffalofishes, are important commercial species, as are carpsuckers and Blue Suckers. While the flesh of suckers is generally said to be quite tasty, it is very bony. Buffalofishes composed the most significant single segment of the annual commercial catch in Arkansas. From 1975 to 1985, an average of 6,543,52 pounds (2,968,144 kg) of buffalofishes per year were taken commercially. More recent commercial catch data for suckers in Arkansas are not available.

Key to the Suckers

1A Dorsal fin long-based, with more than 20 rays.   2

1B Dorsal fin shorter-based, with fewer than 20 rays.   8

2A Head small, its length going 5 or more times into standard length; lateral line scales small, 50 or more; body long and slender; lips papillose.
Cycleptus elongatus Page 391

2B Head larger, its length going into standard length fewer than 5 times; lateral line scales fewer than 50; body deep and compressed; lips weakly plicate or almost smooth.   3

3A Subopercle semicircular, broadest at middle (Fig. 15.118B); body coloration gray or blackish; pelvic fins densely speckled with black; peritoneum black; arrangement of intestine linear or S-shaped; anal rays usually 8–11.
Ictiobus 4

3B Subopercle subtriangular, broadest below middle (Fig. 15.118A); body gold to bronze-colored dorsally, grading to silver or white ventrally; pelvic fins scarcely or not at all speckled with black; peritoneum silvery; arrangement of intestines circular or coiled; anal rays usually 7, occasionally 8.   *Carpiodes* 6

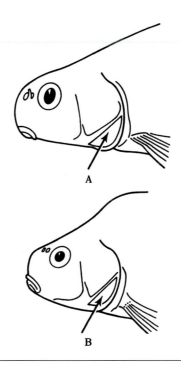

Figure 15.118. Diagrammatic views of the heads of *Carpiodes* and *Ictiobus*. *Miller and Robison (2004).*

A—*Carpiodes carpio*, showing the subtriangular shape of the subopercle bone.
B—*Ictiobus cyprinellus*, showing the semicircular shape of the subopercle bone.

4A Mouth large and oblique (Fig. 15.119A); front of upper lip on horizontal with lower margin of eye; upper jaw length about as long as snout length; lower dentitional pharyngeal arch thin with teeth weak.
Ictiobus cyprinellus Page 403

4B Mouth small, ventral or horizontal; front of upper lip far below margin of eye; upper jaw length shorter than snout length; lower dentitional pharyngeal arch somewhat thick with teeth strong.   5

5A Mouth small, almost horizontal, and inferior (Fig. 15.119C); body deep, body depth going into standard length 2.2–2.8 times; back in front of dorsal fin highly arched, thin and keeled; eye large, its width about equal to snout length.   *Ictiobus bubalus* Page 401

5B Mouth large, slightly oblique, almost terminal (Fig. 15.119B); body thick, body depth going 2.6–3.2 times into standard length; back in front of dorsal fin not highly arched, with a limited keel; eye smaller, its width less than snout length.   *Ictiobus niger* Page 404

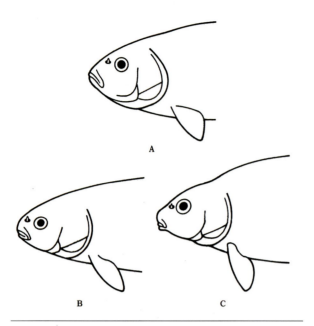

Figure 15.119. Lateral views of the heads of the *Ictiobus* species. *Smith (1979).*

A—*Ictiobus cyprinellus*
B—*Ictiobus niger*
C—*Ictiobus bubalus*

6A Middle of lower jaw without small, nipplelike knob at tip (Fig. 15.120A); lateral line scales 37–41; dorsal fin rays usually 28 or more; upper jaw not extending backward beyond front of eye. *Carpiodes cyprinus* Page 384

6B Middle of lower jaw with a small, nipplelike knob at tip (not well developed in young) (Fig. 15.120B); lateral line scales 33–37; dorsal fin rays usually 27 or fewer; upper jaw extending backward beyond front of eye.　7

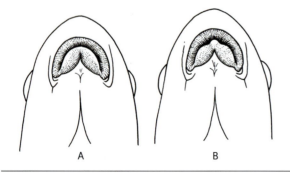

Figure 15.120. *Carpiodes* lips:

A—*C. cyprinus*
B—*C. carpio/velifer*

7A Anterior rays of dorsal fin longer than dorsal fin base; body quite deep (its depth going into standard length 2.6 times or less) and compressed in adults.
　　　　Carpiodes velifer Page 386

7B Anterior rays of dorsal fin not longer than half the length of dorsal fin base; body less deep (its depth usually going into standard length more than 2.6 times) and less compressed. *Carpiodes carpio* Page 382

8A Lateral line absent or poorly developed in adults; swim bladder with 2 chambers.　9

8B Lateral line present, well developed in adults; swim bladder with 3 chambers (2 chambers in *Catostomus* and *Hypentelium*).　11

9A Lateral line poorly developed; lateral scale rows usually 42 or more; mouth inferior, horizontal; each scale usually with an anterior black spot, forming parallel lines on sides. *Minytrema melanops* Page 407

9B Lateral line absent; lateral scale rows usually 41 or fewer; mouth slightly oblique; scales without dark spots, but sides with vertical bars, blotches, or plain in adults (in young, body with dark lateral stripe).　10

10A Lateral scale rows 34–37; dorsal fin rays 11 or 12; eye diameter going into snout length 2 times or less; head pointed. *Erimyzon sucetta* Page 396

10B Lateral scale rows 39–43; dorsal fin rays 9 or 10 (occasionally 11); eye diameter going into snout length more than 2 times; head bluntly rounded.
　　　　Erimyzon claviformis Page 394

11A Lips papillose (Fig. 15.121B); swim bladder with 2 chambers.　12

11B Lips plicate or somewhat papillose (Fig. 15.121C); swim bladder with 3 chambers. (*Moxostoma*)　13

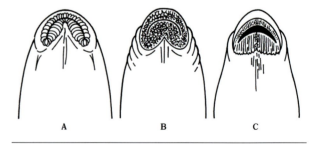

Figure 15.121. Characteristics of the lips of some suckers. *Miller and Robison (2004).*

A—Fleshy, plicate lips of *Erimyzon claviformis*.
B—Fleshy, papillose lips of *Hypentelium nigricans*.
C—Plicate lips of *Moxostoma* species.

12A Lateral line scales more than 50; head between eyes convex (Fig. 15.122B); eye near middle of head in adults; no conspicuous markings (except often 3 large spots or blotches laterally).
                                    Catostomus commersonii Page 388

12B Lateral line scales fewer than 50; head between eyes strongly concave (Fig. 15.122A); eye far behind middle of head in adults; body with conspicuous oblique bars.   *Hypentelium nigricans* Page 398

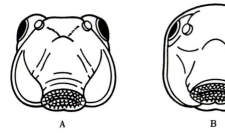

Figure 15.122. Frontal views of heads of two suckers. *Smith-Vaniz (1968)*.

A—*Hypentelium nigricans* (showing concave head).

B—*Catostomus commersonii* (showing rounded head).

13A Upper lip protractile (i.e., separated from snout by a groove); lower lip not divided into separate lobes; widespread.   14

13B Upper lip nonprotractile; lower lip completely divided into 2 lobes; probably extinct.
                                    Moxostoma lacerum Page 418

14A Pharyngeal-teeth arch very heavy, subtriangular in cross-section; teeth on lower half of arch greatly enlarged and molarlike (Fig. 15.123).
                                    Moxostoma carinatum Page 411

14B Pharyngeal-teeth arch moderately heavy, distinctly narrower than high in cross-section; teeth on lower half of arch not enlarged but compressed to form a comblike series.   15

Figure 15.123. Molarlike pharyngeal teeth of *Moxostoma carinatum. Robert E. Jenkins.*

15A Posterior margin of lower lip forming an acute angle (Fig. 15.124A); dorsal fin rays usually 14 or 15, occasionally 16; upper lip surface weakly papillose, but front of upper lip lacking papillae; lower lip semipapillose; posterior part of lower lip (from lip angle to just slightly anterior) abruptly narrower than rest of lip.
                                    Moxostoma anisurum Page 409

15B Posterior margin of lower lip nearly straight or forming a slight to moderate angle; dorsal fin rays usually 12 or 13, rarely 14 or 15; upper and lower lip surfaces basically plicate or lower lip plicae deeply dissected into moderate or large, somewhat ovoid papillae; posterior part of lower lip not abruptly narrower than rest of lip.   16

16A Caudal fin with a distinct black stripe on lower lobe, with the lowermost two rays of the lower lobe white; fins red in life.   *Moxostoma poecilurum* Page 423

16B Caudal fin without black stripe on lower lobe and without white border; fins red or not in life.   17

17A Posterior margin of lower lip forming almost a straight line; lower lip broad with plicae divided by grooves forming papillae; center of upper lip thickened with plicae obliterated at least in half-grown individuals and adults (Fig. 15.124C); head short; caudal fin bright red in life.   *Moxostoma pisolabrum* Page 420

17B Posterior margin of lower lip not forming a straight line, instead the halves forming a moderate to wide angle (Figs. 15.124D, 15.124E); lower lip relatively narrow and plicate, plicae not divided by transverse grooves forming papillae; center of upper lip not thickened; head longer; caudal fin not red in life.   18

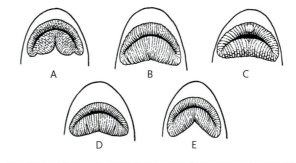

Figure 15.124. Ventral views of lips of *Moxostoma* species.

A—*Moxostoma anisurum*

B—*Moxostoma carinatum*

C—*Moxostoma pisolabrum*

D—*Moxostoma duquesnei*

E—*Moxostoma erythrurum*

18A Lateral line scales 44–48; pelvic rays usually 10 (often 9, rarely 8 or 11); body more slender; caudal peduncle long and slender, its least depth going 2 times or more into its length; eye smaller; breeding males without medium to large tubercles on head.

Moxostoma duquesnei Page 413

18B Lateral line scales 39–45; pelvic rays usually 9 (often 8, rarely 7 or 10); body stouter; caudal peduncle shorter, its least depth going fewer than 2 times into its length; eye larger; breeding males with medium to large tubercles on snout and cheek.

Moxostoma erythrurum Page 416

Genus *Carpiodes* Rafinesque, 1820a— Carpsuckers

Carpsuckers are golden to copper-colored dorsally, silvery to white laterally and ventrally, and have 2 fontanelles (openings) on the dorsal midline of the skull. The subopercle is subtriangular (Fig. 15.118A) and the swim bladder is divided into 2 chambers. The pharyngeal arch is paper thin and almost linear in cross section (Etnier and Starnes 1993). The intestine is coiled into whorls that form a helical pattern in ventral view (Berner 1948) (Fig. 15.125B). The peritoneum is silvery. Dorsal fin rays number 23 or more. Lateral line scales are 43 or fewer. Anal fin rays 8–11 and pelvic fins with few or no melanophores. All three species of genus *Carpiodes* inhabit Arkansas: *C. carpio, C. cyprinus,* and *C. velifer.*

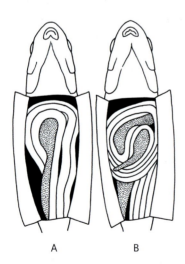

Figure 15.125. Intestinal coiling patterns of ictiobine suckers:

A—*Ictiobus*
B—*Carpiodes*

River Carpsucker

Figure 15.126. *Carpiodes carpio,* River Carpsucker. *Uland Thomas.*

Map 94. Distribution of *Carpiodes carpio,* River Carpsucker, 1988–2019.

CHARACTERS A large, deep-bodied, and compressed silvery fish with a slightly to moderately arched back and an almost flat ventral outline (Fig. 15.126). Subopercle bone subtriangular in shape (broadest anterior to middle). Body depth in adults usually going into SL 2.7 times or more (2.3–3.1). Body width going into body depth fewer than 2 times. Dorsal fin long, usually with 23–26 rays (22–29), and falcate, but with anterior dorsal rays distinctly shorter than the dorsal fin base length; the depressed first dorsal ray usually not extending posteriorly much beyond middle of fin in adults. Mouth small and subterminal. Upper jaw extending past front of eye. Lower lip thin with a small knob (nipple) present at its middle. Eyes large and closer to snout tip than to rear margin of the opercle. Snout almost square. Distance from snout tip to anterior nostril going into postorbital head length fewer than 3 times; total snout length less than distance from back of eye to upper end of gill opening. Pelvic and anal fins without black pigment (present in *Ictiobus* species). Anal fin rays usually 7 (6–8); pectoral fin rays usually 15–17; pelvic rays usually 9 (8–10). Lateral line complete, with 34–36 (modally 35) scales. Intestine long and coiled into circular loops; peritoneum silvery. As with all species of *Carpiodes, C. carpio* has two fontanelles (unossified openings) in the roof of the skull as opposed to a single fontanelle in *Ictiobus.* Breeding males develop numerous small tubercles on top and sides of head, snout, cheek, nape, and upper surfaces of paired fins. Maximum length about 25 inches (635 mm) (Carlander 1969). The largest specimen reported from Nebraska weighed 15 pounds, 12 ounces (7.1 kg) (Hrabik et al. 2015). Individuals weighing 1–4 pounds (0.5–1.8 kg) are common in the catches of commercial fishermen in Arkansas. The unrestricted-tackle Arkansas state record from the Arkansas River in 2017 was 3 pounds, 2 ounces (1.4 kg).

LIFE COLORS   Body color varies from faintly bronze to distinctly gold above, with silvery sides. Belly white. Medial fins dusky, paired fins colorless or tinged with pink or orange. Breeding colors lacking.

SIMILAR SPECIES   The River Carpsucker is most similar to the Highfin Carpsucker but differs (in adults) in lacking the long anterior rays of the dorsal fin, the first principal ray of the dorsal fin not reaching much beyond the middle of the fin when depressed (vs. first principal ray of dorsal fin very long, reaching to or beyond back of fin), and in being less deep-bodied, its greatest body depth usually going into SL 2.7 times or more (vs. body depth going into SL 2.6 times or less). From the Quillback, the River Carpsucker differs in having a nipple at the middle of the lower lip (vs. nipple absent), usually 23–26 dorsal rays (vs. usually 27 or 28), a rounded snout (vs. squarish), and a more elongate body. *Carpiodes carpio* differs from *Ictiobus* species in having a subtriangular subopercle (vs. a semicircular subopercle), pelvic fins scarcely or not at all speckled with black (vs. pelvic fins densely speckled with black), peritoneum silvery (vs. peritoneum black), and 2 fontanelles on the dorsal midline of the skull (vs. 1 fontanelle).

VARIATION AND TAXONOMY   *Carpiodes carpio* was described by Rafinesque (1820a) as *Catostomus carpio* from the falls of the Ohio River below Louisville, Kentucky. It was known by at least four other synonyms until Jordan (1878a) recognized it as *Carpiodes carpio.* The River

Carpsucker is generally divided into two subspecies: the widespread northern nominate form, *C. c. carpio,* which inhabits Arkansas, and a distinct southwestern subspecies, *C. c. elongatus,* in Gulf Slope drainages from eastern Texas to northwestern Mexico. Hubbs and Black (1940b) studied the systematics of *Carpiodes carpio* and differentiated *C. c. elongatus* on the basis of the semitubercular process at the top of the lower jaw (moderately diagnostic of *C. c. carpio*) being absent or weakly developed; the gape tending to be more semicircular and less four-sided; the anterior part of the back more arched; the crosshatching that outlines the scale pockets usually more conspicuous in adults; and finally the best differences involved the body depth, the length of the caudal peduncle, and length of the dorsal fin base. *Carpiodes c. elongatus* is more attenuate, especially toward the caudal fin, and the dorsal fin is generally shorter. The length of the dorsal base when projected forward usually falls far back of the eye in *elongatus* but reaches almost to or beyond the back edge of the eye in *C. c. carpio.*

Hubbs and Black (1940b) reported that while specimens from the Arkansas and Red rivers are referable to *C. c. carpio,* their features seem to approach those of *C. c. elongatus.* Two young specimens (UMMZ 109496) collected in the Mississippi River near the Louisiana-Arkansas border were assigned to *C. c. elongatus,* but there is some doubt about the identifications because the specimens were small. Northern Arkansas specimens seem to be easily recognizable as *C. c. carpio.* Suttkus and Bart (2002) reviewed the status of *Carpiodes carpio elongatus* Meek and concluded that specimens from the Red and lower Mississippi rivers (including those near the Louisiana-Arkansas border) were referable to *C. c. carpio.* A mitochondrial DNA analysis by Doosey et al. (2010) resolved *C. carpio* as sister to *C. velifer.*

DISTRIBUTION   *Carpiodes carpio* occurs in the Mississippi River basin from Pennsylvania to Montana, south to Louisiana, and in Gulf Slope drainages from the Mississippi River west to the Rio Grande drainage of Texas, New Mexico, and Mexico. The River Carpsucker is the most abundant carpsucker in Arkansas, occurring in all major river drainages. It generally avoids the mountainous regions of the state but is found in the northern Ozarks (Map 94). TMB and Jeff Quinn collected the first specimens verified from the Illinois River in Benton County on 2 August 2016.

HABITAT AND BIOLOGY   This schooling sucker occurs throughout the state in moderate to large streams, rivers, and reservoirs. Although it occurs in some Arkansas reservoirs, it is not an abundant species in that habitat. Its preferred habitat is the quiet, sand- or silt-bottomed pools, backwaters, and oxbows of large streams having moderate or low gradients (Pflieger 1997). Baker et al. (1991) found this species abundant along the backwaters and pools, sloughs, and oxbow lakes in the lower Mississippi River and common in the main river channel along revetted banks. Large individuals are often associated with *Cyprinus carpio* around brush piles and log jams (Smith 1979). Young *Carpiodes carpio* were commonly found in tributaries and shallow bays of Lake Texoma, Oklahoma (Riggs and Bonn 1959). It is abundant in the Arkansas River over both sand and silt substrates, and seems more tolerant of turbid waters than other carpsuckers.

Large schools of River Carpsuckers browse extensively over the substrate on attached filamentous algae, consuming large amounts of single-celled algae, protozoans, and small crustaceans with the attached algae (Brezner 1958), as well as aquatic oligochaetes, mollusks, and immature aquatic insects (Walberg and Nelson 1966). Buchholz (1957) described the food in the Des Moines River, Iowa, as bottom ooze consisting primarily of diatoms, green algae, bluegreen algae, desmids, immature dipterans, and other small invertebrates. Volumetric composition of foods in four Oklahoma reservoirs comsisted of 68% organic detritus, 16.2% plant and animal detritus, and 14.2% entomostracans, mainly ostracods and copepods (Summerfelt et al. 1972). Young-of-the-year feed mainly on the same materials as older fish, and little seasonal change in food habits has been reported. Several authors have reported sand in the gut contents (Brezner 1958; Walberg and Nelson 1966; Jester 1972). Aquarium observations by TMB of River Carpsuckers feeding over a sand substrate revealed that much sand was ingested along with the food, but most of the sand was apparently separated from the food items and ejected through the opercular openings as the fish vacuumed the substrate.

Spawning is from May to late July or August at water temperatures of 21–24°C (69.8–75.2°F) and may occur more than once per season (Behmer 1965). Young-of-the-year River Carpsuckers were present in the Arkansas River near Fort Smith on 4 June and were abundant there on 20 June (TMB, unpublished data). This carpsucker migrates upstream in May as water temperatures increase and moves downstream after spawning (Trautman 1981; Curry and Spacie 1984). Spawning occurs in the water column or on the bottom of rivers and some tributaries, over silt or sand substrate (Jester 1972; Fuiman 1982), and the spawning act is accompanied by much splashing. Cross (1967) observed River Carpsuckers in Kansas spawning over roots and fallen stems of rushes in current along an undercut bank. In Oklahoma, spawning occurred after dark over a firm

sand bottom in water 1 to 3 feet (0.3–0.9 m) deep; adhesive eggs were apparently broadcast loose in the water over the substrate (Walberg and Nelson 1966). Fuiman (1982) described eggs as adhesive and demersal, with a diameter of 1.7–2.1 mm. Fecundity ranges from about 4,000 mature ova in small females to 196,000 ova in large individuals (Behmer 1969). Maximum life span is 10–11 years.

CONSERVATION STATUS   The River Carpsucker is a common and often abundant sucker in Arkansas waters, and we consider its populations secure.

Carpiodes cyprinus (Lesueur)
Quillback

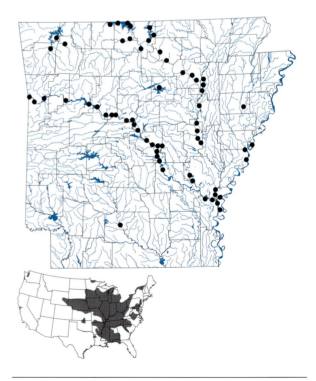

Figure 15.127. *Carpiodes cyprinus*, Quillback. *Uland Thomas.*

Map 95. Distribution of *Carpiodes cyprinus*, Quillback, 1988–2019.

CHARACTERS   A silvery, deep-bodied, laterally compressed carpsucker with a small head (head length going into SL 3.2–3.8 times) (Fig. 15.127). Subopercle asymmetrical and distinctly subtriangular (greatest width anterior to middle). Body depth going into SL 2.6–3.4 times. Body width usually going into body depth 2 times or less. Mouth positioned far forward on head, the upper jaw not extending backward past front of eye. Dorsal fin long and falcate, but as noted by Pflieger (1997) for Missouri specimens, the first principal dorsal ray of Arkansas specimens is variable in length. In some adult specimens, the first dorsal ray when fin is depressed is as long as or slightly longer than the dorsal fin base, while in other adult specimens, the depressed first dorsal ray reaches only to about the middle of the fin base. Snout longer than in other carpsuckers, its length about equal to distance from back of eye to upper end of gill opening. No median nipplelike knob on lower lip; lips semipapillose. Pharyngeal teeth comblike with about 150–180 teeth per arch. Dorsal fin rays 26–32. First dorsal ray and at least the first 8 or 9 pectoral rays are larger than in the River Carpsucker. Anal rays 7 or 8 (13 specimens from Arkansas River below Dam 2, 31 July 2001); pectoral rays 15 or 16. Lateral line complete, with 37–41 scales. Body circumferential scales 31–38; circumpeduncle scales 17–19 (Jenkins and Burkhead 1994). Intestine long and greatly coiled; peritoneum silvery. Breeding tubercles in males are widely scattered on the sides and ventral surface of the head, but mostly absent from the snout, top of head, and most of the body. Females occasionally with tubercles on side of head (Madsen 1971). Maximum length about 26 inches (660 mm). Largest specimen reported from Nebraska weighed 13 pounds, 10 ounces (6.2 kg) (Hrabik et al. 2015). The Arkansas unrestricted tackle record is 3 pounds, 3 ounces (1.4 kg) from the Arkansas River in 2016.

LIFE COLORS   Dorsal coloration of the body is gold to bronze or copper-colored and similar to the River Carpsucker, but the sides often have a tinge of golden yellow. Lower sides and belly silvery. Median fins are dusky, and the paired fins are whitish or orange.

SIMILAR SPECIES   The Quillback is distinguished from the River and Highfin carpsuckers by a lack of the nipplelike extension on the middle of lower lip (vs. nipplelike projection present at middle of lower lip), in usually having 37–41 lateral line scales (vs. 36 or fewer lateral line scales), and a rather long, bluntly rounded snout, its length about equal to distance from rear margin of eye to upper end of gill opening (vs. snout length less than distance from back of eye to upper end of gill opening). From the

buffalofishes (*Ictiobus* species) the Quillback differs by its silvery coloration (vs. largely gray or dark olive in buffalofishes), silvery peritoneum (vs. black), and by its subtriangular subopercle (vs. semicircular).

VARIATION AND TAXONOMY   *Carpiodes cyprinus* was described by Lesueur (1817b) as *Catostomus cyprinus* from the Elk River, tributary of Chesapeake Bay, Cecil County, Maryland. It was subsequently referred to by at least nine synonyms (Smith 1979). Trautman (1957) noted that few genera of North American freshwater fishes show such diverse morphological types as the carpsuckers and suggested that those types are presumably the result of environmental conditions. Because of these differences, subspecific and less often specific identification of some specimens is difficult or impossible morphologically. The great range of ontogenetic and individual variation is reflected by the complex synonymy of the Quillback (Smith 1979). Several subspecies have previously been recognized including *Carpiodes* c. *cyprinus*, *C. c. forbesi*, and *C. c. hinei*, but Bailey and Allum (1962) indicated subspecific recognition was unwarranted.

Carpiodes forbesi was described originally by Hubbs (1930) as a full species distinguished from *C. cyprinus* by its more slender, terete form, large head and mouth, and the lower and more posterior dorsal fin with the first depressed ray not extending to end of fin base. Bailey and Allum (1962) considered *C. forbesi* an environmental variant and placed *C. forbesi* in the synonymy of *C. cyprinus* based on the assumption that the *C. forbesi* phenotype was "found chiefly in prairie and plains areas where high turbidity and scanty food supplies in the rivers are characteristic." Pflieger (1971) later concurred with this placement, but noted that the *C. forbesi* phenotype was found in the Ozark Uplands of Missouri as well as in prairie regions. Pflieger's study revealed great variation in morphological and meristic characters among Missouri specimens of this phenotype. Similar variation in Quillback specimens in Arkansas has been noted, and we agree with Pflieger's conclusion that while it is possible that *C. forbesi* and *C. cyprinus* are sibling species, a combined morphological and genetic analysis of the Quillback throughout its range is needed to resolve this question. Analysis of mitochondrial DNA by Doosey et al. (2010) resolved *C. cyprinus* as sister to a clade composed of *C. carpio* and *C. velifer*.

DISTRIBUTION   *Carpiodes cyprinus* occurs in the St. Lawrence–Great Lakes, Hudson Bay, and Mississippi River basins from Quebec to Alberta, Canada, south to Louisiana; Atlantic Slope drainages from the Delaware River, New York, to the Savannah River, Georgia; and Gulf

Slope drainages from the Apalachicola River, Florida, west to the Pearl River, Louisiana. In Arkansas, the Quillback is found throughout the Arkansas River and sporadically in the White, Little Red, St. Francis, and Mississippi rivers, with records from the Ouachita River system (Map 95).

HABITAT AND BIOLOGY Pflieger (1997) reported that in Missouri the Quillback is characteristic of moderately clear, highly productive streams having large, permanent pools and stable, gravel bottoms. In Ohio, it mainly inhabited the lower gradient portions of large and medium-sized streams having bottoms of sandy silt or sandy muck and where the amount of clayey-silt was small (Trautman 1957). Our observations indicate that in Arkansas it occurs in both types of habitats reported by Pflieger (1997) and Trautman (1957), but it is more common in moderately turbid sandy-bottomed habitats of large rivers (although it is rare in the Mississippi River). In Indiana it was found in quiet water except when spawning, and it preferred water temperatures of 29–31°C (84.2–87.8°F) (Becker 1983). It was seldom found in vegetated areas. Beecher (1980) found that preferred habitat in northwestern Florida varied seasonally in response to river flow. During high water periods from winter to summer, Quillbacks occupied areas of reduced current downstream of sandbars, and during low flows (July through November) they preferred swifter water in midchannel areas or along outer edges of sandbars. It is known from Beaver (Rainwater and Houser 1982), Bull Shoals, and Norfork reservoirs in Arkansas, and from other reservoirs elsewhere (Rohde et al. 2009).

Quillbacks are food generalists, and in Wisconsin, they fed on debris in the bottom ooze, plant materials, and insect larvae (Becker 1983). Very often, gut contents contained little identifiable material. Insect remains were often represented only by larval cases, detached legs, and wings. Cahn (1927) reported that foods consisted of fragments of aquatic vegetation and algae, occasional larval chironomids, and a variety of snails and small clams. Smith (1979) noted that in Illinois streams it feeds on bottom ooze in pools and eddies of backwaters. Beecher (1979) and Ross (2001) reported that most feeding occurs during the day and listed major food items, including organic detritus, bivalve mollusks, larval insects, small crustaceans (copepods and ostracods), and algae (desmids and diatoms). Much sand was also ingested.

Little is known about the reproductive biology of this species in Arkansas. Adult Quillbacks in breeding condition were captured in Missouri from early April to late May (Pflieger 1997). William Keith (pers. comm.) reported that on 11 May 1986, seven *C. cyprinus* from War Eagle Creek were netted, with one male running milt and the

females all heavy with ripe eggs in water temperatures of 68–73.4°F (20–23°C). Spawning occurs from March to September in different parts of its range, and it is reported to spawn in different habitats in different areas. Coughlan et al. (2007) noted a spawning season from 14 April to 24 May for Quillback in South Carolina. Adults moved from a drainage main stem into tributaries, where they spawned over a sand bottom in calm water. In Canada, this schooling species also exhibited spawning migrations in a small prairie river (Parker and Franzin 1991). Harlan and Speaker (1956) reported that spawning occurred in quiet waters of streams over sand and mud bottoms in Iowa, and in a Canadian river, groups of 2–5 individuals spawned in a riffle over coarse to fine gravel (Parker and Franzin 1991). Pflieger (1997) also reported that spawning occurred in Missouri near the lower ends of deep, gravelly riffles. Spawning occurred in Ohio from late June through September (Woodward and Wissing 1976). Fecundity ranged from 15,235 eggs for age 6 fish to 63,779 for age 10 fish. Sexual maturity is reached by age 4 and maximum life span is about 11 years.

CONSERVATION STATUS The Quillback is a rather uncommon spcies in Arkansas, but we consider its populations secure. We believe that it merits future monitoring to detect any change in status.

Carpiodes velifer (Rafinesque)
Highfin Carpsucker

Figure 15.128. *Carpiodes velifer*, Highfin Carpsucker. *David A. Neely.*

CHARACTERS A small, silver carpsucker with a deep, highly compressed body (body depth going into SL fewer than 2.5 times) and a distinctive, long and falcate dorsal fin (22–27 rays). The first principal dorsal fin ray extremely long, when depressed, reaching beyond the back of the fin base in adults (Fig. 15.128). Subopercle subtriangular (widest anterior to middle). Body width going into body depth 2 or more times. Head small and conical, head length going into body depth 1.5 times or more. Snout short (shortest of all carpsuckers), blunt and rounded. Distance from snout tip

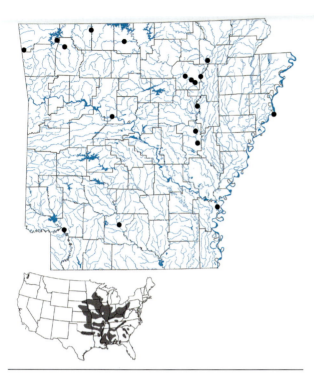

Map 96. Distribution of *Carpiodes velifer*, Highfin Carpsucker, 1988–2019.

to anterior nostril going more than 3 times into postorbital head length; total snout length less than distance from back of eye to upper end of gill opening. Eyes large, eye diameter going into head length 3.1–4.7 times. Mouth ventral and lower lip with a median nipple as in *C. carpio* (except in small young). Upper jaw extends back to behind the anterior edge of the eye. Anal rays 7 (sometimes 8); pectoral rays 15–17; pelvic rays 9 or 10. Lateral line complete, usually with 33–35 scales. Gill rakers 40–70. Intestine long and coiled; peritoneum silvery. Breeding males with large tubercles on snout, top and ventral regions of the head, and most of the body scales; tubercles present on rays and interradial membranes of all fins. Females lack tubercles. The smallest of the carpsucker species, with maximum size probably about 19.7 inches (500 mm) TL, and a weight of 3.3 pounds (1.5 kg). The Arkansas unrestricted tackle record is 1 pound, 12 ounces (0.79 kg) from the Arkansas River in 2017.

LIFE COLORS   Body with gold or bronze dorsally, grading to silver on the sides and white on the belly. Median fins dusky; paired fins are plain but may match the dorsal color.

SIMILAR SPECIES   *Carpiodes velifer* is quite similar to the Quillback; however, the Quillback lacks a tiny, nipplelike projection on the lower jaw. In addition, the Highfin Carpsucker has the deepest body of any of the carpsuckers and the anterior rays of the dorsal fin are more

produced than in the Quillback. The combination of the elongate anterior dorsal fin rays that reach to or past the posterior base of the fin and a nipplelike structure on the tip is diagnostic for adults (Etnier and Starnes 1993). Young carpsuckers are very difficult to identify with certainty. *Carpiodes velifer* and *C. carpio* smaller than 75–100 mm TL are especially difficult to separate since distinctive characters do not appear until later. Etnier and Starnes (1993) provided additional characters that may be useful in identifying subadults and juveniles: body width going 2 or more times into body depth in *C. velifer* (vs. fewer than 2 times in *C. carpio*), and body depth going 2.9 (young) to 2.4 (adults) times into SL in *C. velifer* (vs. 3.3 and 2.7 for young and adults, respectively, of *C. carpio*). It further differs (juveniles and adults) from *C. cyprinus* in scale and dorsal fin ray counts, having 33–35 lateral line scales (vs. 37–41) and 22–27 dorsal rays (vs. 26–32). *Carpiodes velifer* is similar to *I. bubalus*, but differs from that and other *Ictiobus* species in having a subtriangular subopercle (vs. semicircular), pelvic and anal fins unpigmented to lightly pigmented (vs. pelvic and anal fins well pigmented), a silvery peritoneum (vs. black), and life colors more silvery (vs. life colors gray or black).

VARIATION AND TAXONOMY   *Carpiodes velifer* was described by Rafinesque (1820a) as *Catostomus velifer* from the Ohio River between Louisville and Pittsburgh. It was recognized in genus *Carpiodes* by Jordan (1878a). G. R. Smith (1992) identified *C. velifer* as sister to *C. carpio*, and those two species formed a clade that was sister to *C. cyprinus*. A molecular analysis by Doosey et al. (2010) supported that relationship. A rangewide systematic study of the Highfin Carpsucker is needed.

DISTRIBUTION   *Carpiodes velifer* occurs in the Great Lakes (Lake Michigan) and Mississippi River basins from Pennsylvania to Minnesota, south to Louisiana; Atlantic Slope drainages from the Cape Fear River, North Carolina, to the Altamaha River, Georgia; and Gulf Slope drainages from the Apalachicola River, Georgia and Florida, to the Pearl River, Mississippi and Louisiana. This carpsucker is uncommon in Arkansas. It is found sporadically in the upper White and Black river systems, the lower White River (Buchanan 1976), and the Mississippi River (USACE biologists, pers. comm.) (Map 96). Recent isolated records are known from the Arkansas, Ouachita, and Red rivers (McAllister et al. 2010b). It has apparently declined in abundance in the upper White River in the Beaver Lake area (compare Maps 96 and A4.94).

HABITAT AND BIOLOGY   The Highfin Carpsucker inhabits relatively clear streams and rivers with firm substrates. It is more sensitive to high turbidity than other carpsuckers, possibly explaining its rarity in large Arkansas rivers. In Arkansas, it is most common in medium-sized rivers but is occasionally found in the state's largest rivers. It also occurs in small to modest numbers in the three large White River impoundments, Beaver, Bull Shoals, and Norfork reservoirs (Rainwater and Houser 1982; AGFC rotenone sample data). Although it is a riverine-adapted species, adults prefer quiet water areas of rivers. It occurs over substrates of sand and gravel and apparently avoids vegetation. Trautman (1957) reported that it has a curious habit of "skimming along near the water's surface with its dorsal fin and part of its back exposed, and it frequently jumps clear of the water." Becker (1983) noted that the Highfin Carpsucker has a tendency to inhabit "big water" (large lakes and rivers) in the northern portions of its range and "little water" (small rivers) in the southern portions.

It is a diurnal benthic feeder. It moves slowly upstream ingesting bottom materials. After moving 1–3 meters upstream, it drifts 1–2 meters downstream, then repeats the upstream feeding movements (Ross 2001). Foods of the Highfin Carpsucker consist of bottom ooze, algae, insect larvae, and bivalve mollusks (Beecher 1979; Boschung and Mayden 2004). In the Des Moines River, Iowa, gut contents of individuals 150–250 mm (6–10 inches) TL consisted of 93% bottom ooze and algae, and 7% insects (mainly bloodworms); in fish greater than 250 mm (10 inches) the percentages were 78% and 22%, respectively (Harrison 1950).

Pflieger (1997) collected adults in breeding condition in Missouri in late July over deep, gravelly riffles. Woodward and Wissing (1976) reported that spawning in Ohio occurred from late June through September, and individuals in breeding condition have been taken in Wisconsin from mid-May through July (Becker 1983). Breeding fish move to shallow areas and overflow ponds of streams, presumably to spawn (Harlan and Speaker 1956). The only information on spawning in Arkansas comes from William Keith (pers. comm.), who collected 2 males running milt on 11 May 1986 from War Eagle Creek in water temperatures of 68–73.4°F (20–23°C). Another ripe specimen was taken from the Kings River during the same sampling period at the same water temperature. Fecundity in Ohio specimens ranged from 41,644 eggs in an age 5 female to 62,355 in an age 7 female. Fertilized eggs are demersal and adhesive (Yeager 1980). The range in total body length was 2.6–12.4 inches (65–315 mm) at ages 1 through 8. In Tenkiller Reservoir, Oklahoma, *C. velifer* ranged from 4 to 13.5 inches (102–343 mm) at ages 1 through 5 (R. M. Jenkins et al. 1952). Maximum life span is about 8+ years.

CONSERVATION STATUS   The Highfin Carpsucker is uncommon in Arkansas, and its populations should be monitored to detect changes. We do not assign it a status of special concern at this time.

Genus *Catostomus* Lesueur, 1817b— Finescale Suckers

Catostomus contains 28 species. Genus characters include dorsal fin rays 16 or fewer, scales near head much smaller than those on the posterior part of the body, lateral line scales 60 or more, and the swim bladder with 2 chambers. There is a single species in Arkansas, *Catostomus commersonii*.

Catostomus commersonii (Lacépède)
White Sucker

Figure 15.129. *Catostomus commersonii*, White Sucker. *David A. Neely*.

Map 97. Distribution of *Catostomus commersonii*, White Sucker, 1988–2019.

CHARACTERS   An elongate and terete fish with a short dorsal fin base and large, fleshy, papillose lips (Fig. 15.129). Body depth going into SL 4.2–5.0 times. Head moderate in size. Mouth small, ventral, and extending posteriorly only as far as nostrils. Lips heavily papillose, with lower lip having a deep medial notch. Snout rounded. Eye situated about halfway between snout and rear margin of operculum. Area between eyes flat or convex, not concave. Pharyngeal teeth comblike. Intestine long, with 4 or 5 coils. Lateral line complete, with 55–76 scales. Scales on the body near the head are much smaller than scales on the posterior portion of the body. Scales around caudal peduncle 17–24. The edge of the dorsal fin is straight to slightly convex, and its height is approximately equal to the length of its base. Dorsal rays 11–13. Anal rays 7 or 8; pelvic rays 10 or 11. Swim bladder with 2 chambers. In breeding males, small tubercles develop on head, dorsum, and body, and medium to large tubercles are present on caudal and anal fins. Breeding females occasionally with weakly developed tubercles on anal and caudal fins. Maximum size 25 inches (635 mm) and 6 or 7 pounds (2.7–3.2 kg) (McPhail and Lindsey 1970; Lee and Kucas 1980), but Arkansas specimens rarely exceed 2 pounds (0.9 kg) and 12 inches (305 mm) in length.

LIFE COLORS   Body color olive or gray above with lighter sides and creamy white belly. Dorsal and caudal fins dusky, lower fins often yellow. Young fish with 3 dark blotches on each side; blotches disappear in older fish. Breeding males distinctly bicolored (black above, white below), and with a pink or red lateral band.

SIMILAR SPECIES   The White Sucker differs from all other suckers in Arkansas in having very small scales, with 55–76 in the lateral line (vs. fewer than 50 lateral line scales in all other suckers, except *C. elongatus*). It differs from *C. elongatus* in having a short dorsal fin, with 11–13 rays (vs. dorsal fin long, with 28–35 rays).

VARIATION AND TAXONOMY   *Catostomus commersonii* was described by Lacépède (1803) as *Cyprinus commersonnii* (type locality not given and holotype whereabouts unknown). The International Commission on Zoological Nomenclature clarified the disputed spelling of patronyms for species-group names ending in –*i* or -*ii* by ruling that original spelling shall prevail (Bailey et al. 2004). Therefore, the correct scientific name for the White Sucker was changed to *Catostomus commersonii* to reflect the original spelling of the specific epithet suffix (Eschmeyer et al. 2018). G. R. Smith (1992) identified *C. commersonii* as the sister species to all other species in the genus. This relationship was not supported in an analysis of mitochondrial DNA by Doosey et al. (2010).

DISTRIBUTION   *Catostomus commersonii* is widely distributed in Atlantic Slope, Great Lakes, Mississippi River and Arctic basins from Newfoundland to the Northwest Territories, Canada, south to the Savannah River drainage of Georgia, the Tennessee River drainage of northern Alabama, and the Arkansas River drainage of New Mexico. It also occurs in the upper Rio Grande drainage of New Mexico, in Pacific Slope drainages of British Columbia, Canada, and it has been introduced into the Colorado River drainage of Wyoming, Colorado, New Mexico, and Utah. The White Sucker scarcely penetrates into Arkansas along the northern boundary in the Arkansas (Illinois River drainage) and upper White river systems (Map 97). It has been documented from streams in seven counties in northern Arkansas, and McAllister et al. (2009b) reported that it occurs farther downstream in the White River than previously known.

HABITAT AND BIOLOGY   In Arkansas, the White Sucker is usually a schooling inhabitant of small streams, preferring spring pools, spring runs, and spring-fed feeder creeks with large amounts of submerged vegetation and gravel bottoms. It is also very common in the tailwater of Norfork Dam in Baxter County (Ken Shirley, pers. comm.). In more northern parts of its range (Lee and Kucas 1980; Becker 1983), where it is abundant, it occurs in a wide variety of habitats. Becker (1983) reported that the White Sucker is a common inhabitant of the most highly polluted and turbid waters in Wisconsin, and he considered it more tolerant of a wide range of environmental conditions than any other species of fish in that state. In streams, the largest White Suckers occurred in deep holes. McAllister et al. (2010b) reported a specimen (HSU 3331) from atypical habitat in Arkansas in Independence County, collected from Simpson Slough, an old White River channel and muddy lowland slough with high turbidity that drains cleared low, agricultural bottomlands. In adjacent Oklahoma, White Suckers are found only in clear Ozark streams (Miller and Robison 2004). Pflieger (1997) noted that in Missouri, the habitats where White Suckers are abundant are largely devoid of other suckers, possibly indicating that competition is a factor limiting its distribution. *Semotilus atromaculatus* is a common associated species in Missouri.

Larval White Suckers have a terminal mouth and initially feed near the surface for the first 10 or 11 days on true midges (Tendipedidae), microcrustaceans, protozoans, and other plankton (Stewart 1927). Rotifers contributed an important portion of the early diet, and copepods provided most of the remainder of the diet (Siefert 1972). Cladocerans increased in importance after the fish reached 14 mm. After the subterminal, horizontal mouth develops, food is then taken from the substrate. Juveniles feed mainly on bottom ooze and detritus, while adult White Suckers consume a variety of terrestrial and immature aquatic insects, particularly chironomid larvae and pupae, small crustaceans, mollusks, bottom ooze, and detritus (Stewart 1927; Eder and Carlson 1977; Rohde et al. 2009). The diet of adults can vary greatly from one locality to another. Stomach analyses have shown 100% insects in some collections, 100% higher plants in others, 95% mollusks in one collection, and 50% drift in other stomachs (Becker 1983). In Tennessee, gut contents included 50% benthic invertebrates and 50% unidentified matter (Etnier and Starnes 1993).

The White Sucker is potamodromous, has a high fecundity, is an iteroparous early spring spawner, and does not exhibit parental care of offspring (Winemiller and Rose 1992). In lakes that lack tributaries, White Suckers spawn in the lake littoral zone (Raney and Webster 1942; Geen et al. 1966; Page and Johnston 1990b). In tributaries of Lake Taneycomo, a cold-water reservoir of the White River in southwestern Missouri, White Suckers spawned from March through May, and larvae were collected from mid to late April in all tributaries of the lake (Wakefield and Beckman 2005). Apparently, no spawning occurred in the main channel of Lake Taneycomo. Doherty et al. (2010) reported White Suckers migrated on average 9.2 km and up to 40 km to reach spawning habitats, suggesting that White Suckers require distinct, specialized habitats to reproduce. Breeding adults congregate over clean, gravelly shoals near the lower ends of pools where the current begins to quicken (Pflieger 1997). Spawning temperatures range from 7.0°C to 19.0°C (Corbett and Powles 1983; Curry and Spacie 1984; Jenkins and Burkhead 1994; Boschung and Mayden 2004; Wakefield and Beckman 2005). McManamay et al. (2012) reported groups of 15–20 individuals engaged in spawning. Females move toward a riffle and are approached and sandwiched between two or more males. Females are typically upright with the pelvic area oriented toward the substrate, and males congregated on either side and on top of a female that was slightly vibrating. All fish had heads oriented in the upstream direction, with spawning following (McManamay et al. 2012). Fecundity is variable, ranging from 1,000 to 140,000 eggs and generally increasing with fish length (Raney and Webster 1942; Carlander 1969). Wakefield and Beckham (2005) found fecundity in females ranged from 5,000 to 59,000 eggs in Missouri. Eggs hatch in 5 days at 16°C and in 12–15 days at 10°C. (Rohde et al. 1994; Corbett and Powles 1983, 1986). Horak and Tanner (1964) reported that optimal temperature for growth and survival of White

Suckers was 19–24°C. In Lake Taneycomo tributaries, a small percentage of White Suckers attained maturity at age 3 and 260 mm TL (Wakefield and Beckham 2005). By age 4 and 300 mm TL, 90% were mature, and by age 6 all were mature. Females live longer and reach larger sizes than males. Trippel and Harvey (1991) reported that age and length at maturity were variable among northern populations, ranging 2–8 years and 180–390 mm fork length. Variation in age and length at maturity can be affected by several factors, including food availability, population density, water temperature, and water quality (Trippel and Harvey 1987a, b, 1989). While maximum age is 17 years, Wakefield and Beckham (2005) in their Missouri study found the most abundant age class for males was age 7 and for females age 8.

CONSERVATION STATUS   Petersen et al. (1996) collected a single specimen from the upper White River, and 1995 summer collections by ADPCE yielded 25 individuals from Osage and Spring creeks in the headwaters of the Illinois River drainage. These small, scattered populations in the Illinois River drainage are threatened by the progressive deterioration of that watershed's aquatic environment. Slight changes in one or more critical factors may have substantial effects upon a species near the edge of its distributional range (Petersen et al. 1996). The White Sucker is commonly collected in southern Missouri (Pflieger 1997), and Lake Taneycomo (White River) supports a viable population (Wakefield and Beckman 2005). Although one of the most abundant and widespread catostomids in North America, the White Sucker is uncommon in Arkansas. We consider it a species of special concern in this state.

Genus *Cycleptus* Rafinesque, 1819— Blue Suckers

Cycleptus is endemic to North America. It occupies a special place in the Catostomidae, having been regarded as both a basal taxon and as the sister genus to *Myxocyprinus*, the sucker complex indigenous to China (G. R. Smith 1992; Burr and Mayden 1999). The genus currently contains two species, but the population in the Rio Grande possibly represents an undescribed third species (Buth and Mayden 2001; Bessert 2006). *Cycleptus* is represented in Arkansas by the Blue Sucker, *C. elongatus*. Burr and Mayden (1999) described a second species of *Cycleptus*, *C. meridionalis*, occurring in Gulf Slope drainages in Louisiana, Mississippi, and Alabama. Genus characters include lips papillose, dorsal fin rays 23 or more, lateral line scales 50 or more, and swim bladder with 2 chambers.

Cycleptus elongatus (Lesueur)

Blue Sucker

Figure 15.130. *Cycleptus elongatus*, Blue Sucker. *David A. Neely*.

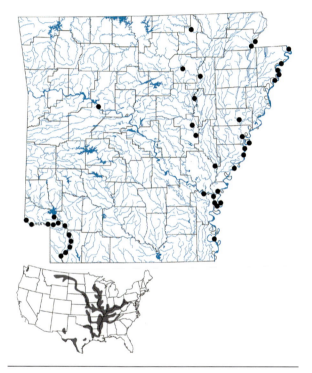

Map 98. Distribution of *Cycleptus elongatus*, Blue Sucker, 1988–2019.

Figure 15.131. Head of nuptial male Blue Sucker. *Photo courtesy of Paul Schumann.*

CHARACTERS A large, streamlined sucker having a long-based falcate dorsal fin with 25–34 rays (Fig. 15.130). The first and second principal dorsal rays are long, but rays quickly shorten at about the 7th or 8th ray, and remaining rays are all short. Dorsal fin base more than one-third of SL. Body oval in cross section; body depth going into SL 4–5 times. Head small and slender, head length going into SL 5–5.4 times. Eyes small, in posterior half of head. Snout rounded. Mouth small, inferior, and distinctly overhung by snout. Lips large and papillose; papillae short; papillae on upper lip generally longest. Pharyngeal arch moderately stout, symphysis short, with 32–45 teeth per arch (Eastman 1977); teeth relatively large, arranged in comblike fashion (Eastman 1977; Burr and Mayden 1999). Swim bladder with 2 chambers. Anal fin small with 7 or 8 rays. Pectoral fins large, falcate, with 16 or 17 rays. Pelvic fin large, with 10 or 11 rays. Caudal fin large, widely forked. Lateral line complete, nearly straight, usually with 49–58 scales. Peritoneum silvery. Breeding males and females covered with thousands of epidermal tubercles over most of the body and fins (Branson 1961). These tubercles range in size from 0.2 to 1.0 mm. In nuptial males, tubercles are prominent on the head (Fig. 15.131), snout, nearly all body scales, around the eyes, on the opercle, and on all rays of all fins. In breeding females the tubercles occur principally around the eyes, but are smaller and less numerous than those on males. Chromosomes 2n = 96–100 (Uyeno and Smith 1972). Maximum TL 39 inches (990 mm) (Page and Burr 2011). Trautman (1981) gave a maximum size of 40 inches TL (102 cm) and 12–15 pounds (5.4–6.8 kg). The largest specimen from Nebraska weighed 18 pounds, 14 ounces (8.6 kg) (Hrabik et al. 2015). Arkansas specimens are seldom over 25 inches (635 mm) and 6 pounds (10 kg).

LIFE COLORS Body with dorsum and sides blue black to dark gray with brassy reflections; belly bluish white. Lips white. All fins dark blue gray, dusky, or black. Breeding males becoming a much darker blue black. Forbes and Richardson (1920) noted that "spring males almost black." Burr and Mayden (1999) reported fall males are also blue black. They reported that adult females, exclusive of the spring spawning season, were indistinguishable from males in overall color patterns. Breeding females may be tan to light blue color (Burr and Mayden 1999). Lower lobe of caudal fin in juveniles and subadults black (Fig. 15.132). Coloration of young 62–80 mm SL described in detail by Cross (1967).

Figure 15.132. *Cycleptus elongatus* small juvenile, 32 mm SL, Mississippi River, Tennessee. *Richard T. Bryant and Wayne C. Starnes.*

SIMILAR SPECIES   The Blue Sucker does not resemble any other sucker species inhabiting Arkansas. Some redhorses (*Moxostoma*) and perhaps some minnows (chubs of genus *Macrhybopsis*) have similar body shapes and caudal fin pigmentation; however, all those species have short-based dorsal fins with 14 or fewer dorsal fin rays, while *C. elongatus* has 25–34 dorsal rays.

VARIATION AND TAXONOMY   *Cycleptus elongatus* was originally described by Lesueur (1817b) as *Catostomus elongatus* from a 2-foot-long dried specimen obtained from the Ohio River, presumably in western Pennsylvania (Burr and Mayden 1999). It was later placed in a different genus by Agassiz (1854, 1855). *Cycleptus elongatus* has been the subject of an in-depth osteological study (Branson 1962), morphological and molecular analyses (Smith 1992; Bessert 2006), and a few studies on aspects of its life history (Hogue et al. 1981; Rupprecht and Jahn 1980; Moss et al. 1983; Yeager and Semmens 1987). Analysis of mitochondrial DNA by Doosey et al. (2010) recovered *C. elongatus* as basal to a clade consisting of *Carpiodes* species, *Ictiobus* species, and *Myxocyprinus asiaticus*. Analysis of nuclear microsatellite genotypic data indicated distinct subpopulations of *C. elongatus* within the Mississippi River basin, while mitochondrial markers revealed a pattern of intermediate polyphyly with no gene flow between the two described *Cycleptus* species (Bessert 2006).

DISTRIBUTION   *Cycleptus elongatus* occurs in the Mississippi River basin from Pennsylvania to Montana, south to Louisiana, and in Gulf Slope drainages from the Sabine River, Louisiana, to the Rio Grande of Texas, New Mexico, and Mexico. In Arkansas, the Blue Sucker is found principally in the Mississippi and Red rivers (Map 98). A few to several records are known from the White River (Shirley et al. 2013), Arkansas River, and St. Francis River. Older records are known from the Little Red, Spring, and Black rivers (Map A4.96), with one recent record from the Spring River.

HABITAT AND BIOLOGY   The Blue Sucker is restricted to large riverine environments in Arkansas. It is morphologically well adapted for life on the bottom in deep, fast-moving rivers and channels of deep lakes (Miller and Robison 2004). Because of this habitat preference, it is difficult to capture, contributing to its rarity in collections. In the Red River of Arkansas, adult Blue Suckers inhabited deep, swift water in the main channel the entire year except during the breeding season (Layher 2007). Ross (2001) noted that adults prefer current speeds of at least 100 cm/s. Additional quantitative habitat data in the Neosho River were provided by Moss et al. (1983), where young Blue Suckers were captured over coarse substrates in shallow, swift currents. Juveniles were captured in broad riffle areas where channel constriction increased water velocity. Adults in the Neosho River were captured during the nonbreeding season over exposed limestone bedrock in deep riffles typically 1–2 m deep. The fastest current velocity measured was 2.6 m/s (at 0.6 m depth), with most adults captured in currents greater than 1 m/s. Moss et al. (1983) further concluded that the presence of adequate areas of swift, deep water over firm substrates, with sufficient flows over spawning riffles in the spring, were the most important factors in maintaining populations of Blue Suckers. Other substrates occupied are sand, cobble, gravel, and bedrock bottoms. Cross and Collins (1995) found young Blue Suckers in areas along gravel bars in shallower water and slower currents than adults. The larval stage requires nonflowing habitat areas, but after developing into juveniles, the young increase their use of flowing-water habitats and become more benthically oriented. Larval Blue Suckers have been captured in relatively low numbers in main channels, channel borders, tributary mouths, adjacent backwaters, and from intake screens in the Mississippi River basin (Adams et al. 2006). In the Mississippi River, Adams et al. (2006) found that slackwater areas associated with islands were the most important nursery areas for larval and young individuals. In that study, all sites where larvae were found were characterized by lack of flow, soft substrates (muds/silts), and low to moderate amounts of cover in the form of woody debris and flooded terrestrial vegetation. Neely et al. (2010) noted that habitat diversity was required by *C. elongatus* to complete its annual life history cycle in the middle Missouri River, as evidenced by interseasonal variation in habitat selection. Burr and Mayden (1999) noted that *C. elongatus* is often now associated with human-made structures that help constrict the channel, including the ends of wing dikes, the bases of dams, deep zones of reservoirs, and bridge abutments. Pennington et al. (1983) found this species associated with revetted banks in the lower Mississippi River, and Baker et al. (1991) reported that Blue Suckers were abundant to common in the main channel and along revetted and naturally steep banks of the lower Mississippi River. Riggs and Bonn (1959) recorded adults as deep as 15 feet (4.57 m) in Lake Texoma, Oklahoma.

Most feeding occurs over a firm substrate (Cowley and Sublette 1987). Foods of adults consist of aquatic insects, principally caddisfly larvae and pupae, a variety of other small invertebrates (hellgrammites and fingernail clams) from the bottom, and some filamentous algae (Rupprecht and Jahn 1980; Moss et al. 1983). In some environments, the diet was reportedly dominated by attached algae and zooplankton (Walburg et al. 1971). Eastman (1977) concluded that the pharyngeal apparatus of *C. elongatus* was adapted to permit moderate mastication and noted that the diet consisted principally of aquatic insect larvae, small, thin-shelled mollusks, and microcrustaceans. Miller and Evans (1965) reported that anatomical aspects of the brain and lip indicated a highly developed sense of taste. In the Mississippi River, larval and early juvenile fish 16–39 mm TL used diverse feeding modes in the shoreline areas, consuming benthic, nektonic, and neustonic forms (Adams et al. 2006). In that study, 18 invertebrate taxa, representing 13 major groups, were identified in larval Blue Sucker guts. Chironomid larvae, pupae, and adults (83.8% of larval guts), copepods (71.8%), and cladocerans (65.3%) occurred in the majority of guts examined. Other foods consumed by larvae were oligochaetes (16.1%), ostracods (11.3%), and bryozoan statoblasts (7.3%). In the Neosho River, YOY Blue Suckers fed on small insect larvae, primarily dipterans and caddisflies (Moss et al. 1983).

In Kansas, adult Blue Suckers winter in deep pools and move upriver in spring to spawn in riffles (Cross 1967). There is a massive upstream migration of adults to large riffles with strong current. Males migrate to spawning areas before females (Moss et al. 1983). Females are heavier than males for their length and can be sexed externally using size and shape characters (Beal 1967). Spawning females are heavy with eggs and have deeper, wider bodies. Pflieger (1975) reported adults from the Current River in Missouri in breeding condition as early as February and March. Hrabik et al. (2015) reported the collection of larval Blue Suckers from the Platte River, Nebraska, from early May to early June at water temperatures of 59–75°F (15–23.9°C). Cross (1967) found most spawning in Kansas occurred from late April through May at water temperatures ranging from 10°C to 20°C. Spawning occurred in the Red River of Arkansas in March and April (Layher 2007), but spawning possibly extends into May in northern parts of Arkansas. Spawning occurs in May and June in Arkansas, with tuberculation reaching its peak in April and May. Sex ratio of spawning adults is about 1:1 (Moss et al. 1983).

Spawning in the Mississippi River occurred over a 10–28 day period during spring and corresponded with rising water temperatures of 14–18°C from the middle of April to mid-May (Adams et al. 2006). In Indiana, Daugherty et al. (2008) captured ripe specimens in shallow water (0.3–3 m depth) over sand, gravel, and cobble substrates. Cross et al. (1983) reported spawning in Kansas in deep riffles (3–6 feet [1–2 m]) with cobble and bedrock substrates in May at water temperatures of 68–73.4°F (20–23°C). Moss et al. (1983) captured adults readily extruding eggs and milt during handling from a deep (1.4 m), swift (1.8 m/s) cobble riffle in the Neosho River, Kansas. Territoriality has not been observed in this species and apparently does not occur (Page and Johnston 1990b). Tuberculate males may be present from April until June. In the Grand River in Missouri, spawning aggregations were found in the fastest water available that was breaking over large cobbles and boulders and was between 0.5 and 1.0 m deep (Vokoun et al. 2003). Burr and Mayden (1999) suggested that in large rivers such as the Mississippi River, this species probably participates in the "flood-pulse" strategy of reproduction. Burr et al. (1996) observed young-of-the-year on both sides of the river in Illinois and Missouri in June after the major floods of 1993 and 1994. In the Red River of Arkansas, spawning occurred over coarse sand on the outside of inside bends in the river adjacent to strong currents but in slower water toward shore (Layher 2007). Rising water levels and temperatures of 66–68°F were required to trigger spawning. Two size classes of eggs have been reported in gravid females: opaque white eggs averaging 1.0 mm in diameter, and transparent eggs, averaging 0.4 mm in diameter (Rupprecht and Jahn 1980). Beal (1967) reported eggs of mature South Dakota females had a mean diameter of 1.7 mm. Eggs released from a ripe female in Kansas were opaque, slightly yellow, adhesive, and averaged 2.2 mm in diameter (Moss et al. 1983). In Indiana, mean fecundity of females was 150,704 and ranged from 26,829 to 267,471 eggs (Daugherty et al. 2008).

Adams et al. (2006) documented the importance of shorelines near the main channel as nursery habitat for the Blue Sucker. They also demonstrated that seining of shallow, nonflowing, shorelines of off-channel habitats in the Mississippi River was effective for sampling larval and early juvenile stages. Juveniles may reach 4.9 inches (125 mm) by the end of the first summer. Mature females reach greater maximum ages, weights, and lengths (9 years, 9 pounds [4.1 kg], and 30 inches [763 mm] TL) than males (7 years, 8.2 pounds [3.6 kg], and 29.5 inches [749 mm] TL) according to Moss et al. (1983). Males reach sexual maturity at ages 3 or 4, females at age 6. Sexual maturity was reached in the Red River at 4+ years (Layher 2007). *Cycleptus elongatus* lives at least 12 years (Elrod and Hassler 1971; Walburg et al. 1971). In the Mississippi (Rupprecht and Jahn 1980), Missouri (Beal 1967), and Neosho rivers (Moss et al. 1983), this species lived about 9 or 10 years. Maximum life span was reported to be 22 years (Vokoun et al. 2003).

CONSERVATION STATUS   Although the Blue Sucker is tolerant of high turbidity, its abundance has declined across the country since 1900. The decline is possibly due to impoundment, which has decreased or generally nullified current velocity and permitted siltation. The Arkansas River population studied by Buchanan (1976) was relatively stable and large, but a more recent assessment of its status is needed. Commercial fishermen in the 1980s reported that *C. elongatus* is frequently caught in the spring, sometimes as many as 100 fish a day, during presumed spawning migrations. Fishermen report that the Blue Sucker is highly valued as a food fish and preferred over other suckers. The Blue Sucker is considered jeopardized throughout much or all of its range (Boschung and Mayden 2004). It is uncommon in Arkansas and should be considered a species of special concern.

Genus *Erimyzon* Jordan, 1876b— Chubsuckers

The genus *Erimyzon* contains four species, two of which occur in Arkansas. *Erimyzon* is most closely related to *Minytrema* (Smith 1992), a conclusion supported by Tan and Armbruster (2018). Typically, these small suckers lack a lateral line, have 9–12 dorsal fin rays, 7 anal rays, and 18 principal caudal fin rays. The mouth is terminal to subterminal with fleshy, plicate lips, scales in lateral series 41 or fewer, and side with a single lateral stripe, blotched, or uniformly colored. Dorsal fin with a convex margin and never with a black blotch. Breast scales are large and slightly embedded, the swim bladder has 2 chambers, and there is only 1 ovary present (Etnier and Starnes 1993). Breeding males with large tubercles on head (usually 3 or 4 on each side of head between eye and upper lip).

Erimyzon claviformis (Girard)
Western Creek Chubsucker

CHARACTERS   A small, robust sucker with large fins and no lateral line (39–43 lateral scale rows) (Fig. 15.133). Body elongate, slightly compressed; body depth going into SL 3.2–4.2 times. Dorsal fin rounded with 10 (occasionally 11) rays. Predorsal scale rows 14–16. Distal edge of dorsal

Figure 15.133. *Erimyzon claviformis*, Western Creek Chubsucker. *Uland Thomas.*

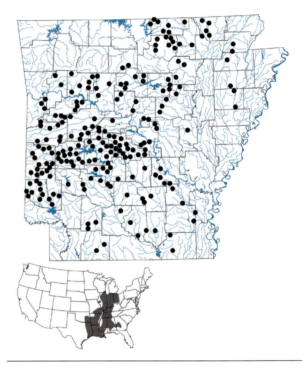

Map 99. Distribution of *Erimyzon claviformis*, Western Creek Chubsucker, 1988–2019.

fin convex rather than concave. Mouth slightly subterminal, oblique. Rear margin of lower lips forming an acute V-shaped angle. Pharyngeal teeth 56–76 (Eastman 1977). Eyes small, eye diameter going into head length 5.5–6.2 times. Eye diameter going 2.1–2.6 times into snout length in specimens greater than 70 mm (Wayne Starnes, pers. comm.). Anal rays 7; pelvic rays 8 or 9; pectoral rays 13–16. Swim bladder with 2 chambers. Intestine long with several coils; peritoneum silvery but may be darkened anteriorly in some individuals. Breeding males with 3 large tubercles on each side of snout (Fig. 15.134), and with a sickle-shaped, bilobed anal fin. Smallest Arkansas sucker with a maximum size in this state of 9 inches (229 mm) TL. Most adults in Arkansas are 4–7 inches (102–178 mm) TL. Maximum reported size is 14.8 inches (419 mm) TL (Ross 2001).

Figure 15.134. Head of *Erimyzon claviformis* male with 3 breeding tubercles on each side. *Renn Tumlison.*

LIFE COLORS   Color dark bronze above, grading to golden yellow on the sides. Scales of dorsum and upper sides prominently dark-edged. Belly and lower sides white or yellow. Dark stripe along midside continuous only in juveniles, broken into a series of somewhat connected blotches in adults. These blotches fade out in older adults. No parallel rows of dark spots along the sides. Paired and median fins are yellow orange to grayish; median fins tend to be darker than paired fins. Breeding males are dark brown above, pink yellow ventrally with orange paired fins and yellow median fins (Ross 2001). Juveniles with a whitish body and a dark continuous stripe along midside with two thin black stripes on the dorsum.

SIMILAR SPECIES   The Western Creek Chubsucker is similar to the Lake Chubsucker but usually has 10, occasionally 11 dorsal fin rays (vs. 11 or 12), 14–16 predorsal scale rows (vs. 12 or 13), and usually 39–43 lateral scale rows (vs. 34–37). It also has a less emarginate caudal fin than *E. sucetta.* Young Western Creek Chubsuckers have a black continuous stripe running along the midside and look very similar to minnow species having a black lateral stripe. The young chubsucker can be distinguished from those native cyprinids by lack of a lateral line (vs. lateral line at least partially developed in native minnows having a black lateral band), and by the distance from front of anal fin base to base of caudal fin going more than 2.5 times into distance from front of anal fin base to tip of snout (vs. going 2.5 times or less). It differs from *Minytrema melanops* in lacking rows of spots on lower sides (vs. rows of small spots present). It can be distinguished from all *Moxostoma* species in having the lateral line incomplete or absent (vs. complete), and air bladder with 2 chambers (vs. 3).

VARIATION AND TAXONOMY   *Erimyzon claviformis* was originally described by Girard (1856) as *Moxostoma claviformis* from Coal Creek, a tributary of the South Fork

of the Canadian River, Oklahoma. It was subsequently considered a junior synonym of *Erimyzon oblongus.* Hubbs (1930) studied geographic variation in *E. oblongus* and recognized three subspecies (including *E. oblongus claviformis*) and six total forms. For more than 50 years, two of those forms—*E. o. oblongus* of Atlantic Slope drainages from southern Maine to the Altamaha River, Georgia, and Lake Ontario drainage of New York, and *E. o. claviformis,* found in Arkansas and elsewhere in the Mississippi Valley—were considered valid subspecies. Bailey et al. (2004) noted a substantial geographical separation between populations of the subspecies; in addition, the subspecies differed notably in size (to 375 mm TL in *oblongus*; to about 178 mm TL, rarely 229 mm, in *claviformis*) and in dorsal fin ray counts (11–14, usually 12 in *oblongus*; 9–11, usually 10 in *claviformis*). Bailey et al. (2004) removed *E. o. claviformis* from the synonymy of *E. oblongus* and elevated it to full species level (*E. claviformis*) and provided a common name, Western Creek Chubsucker. We follow Bailey et al. (2004), Page and Burr (2011), and Page et al. (2013) in recognizing *E. claviformis* as a full species.

DISTRIBUTION   *Erimyzon claviformis* occurs in the lower Great Lakes and Mississippi River basins from southern Michigan and southeastern Wisconsin, south to the Gulf of Mexico; and in Gulf Slope drainages from the Apalachicola River drainage of Georgia west to the San Jacinto River, Texas. The Western Creek Chubsucker is widespread in Arkansas. It occurs in all major drainages but is absent from streams in the highest elevations of the Ozarks (Map 99).

HABITAT AND BIOLOGY   The Western Creek Chubsucker prefers small, clear creeks and streams of moderate gradient. It avoids medium to large rivers. In creeks, it is found in quiet waters in vegetation, over sand, gravel, or debris-laden substrates. It is often found in streams or ditches having submergent vegetation, and it has been reported from ponds (Ross 2001). It is sometimes found in habitats lacking vegetation. Buchanan (2005) found *E. claviformis* in two Arkansas impoundments, Blue Mountain Lake and the impounded area of Champagnolle Creek. Juveniles are typically found in small headwater creeks, and adults occupy pool areas when not breeding. During the breeding season, adults move onto riffles. Although usually associated with quiet water habitats, it is sometimes found in current. It is found in greatest abundance in Arkansas in small creeks of the Ouachitas, and it is widely distributed below the Fall Line in Coastal Plain streams having suitable habitat. Smith (1979) reported that the young are pioneering fish and among the first to ascend headwaters and previously dry stream courses.

Some authors have speculated that the food habits of *E. claviformis* are probably similar to those reported by Flemer and Woolcott (1966) and Gatz (1979) for *E. oblongus*: entomostracans, rotifers, small insects, small clams, algae, and plant fragments. Pflieger (1997) suggested that the nearly terminal mouth of *E. claviformis* may indicate that it feeds less on the bottom than many other suckers. That does not agree with our observations on the feeding biology of *E. claviformis* in Arkansas. Gut contents of 71 Arkansas specimens from Cross, Greene, and Logan counties were examined by TMB, who found that it is a benthic omnivore as its long, coiled intestine suggests. Gut contents were dominated by percent frequency of occurrence and volume at all sites in spring, summer, and fall in descending order by detritus, sand, algae (filamentous algae, diatoms), and benthic animal remains. Animal remains made up a slightly larger percentage of the diet at the Crowley's Ridge localities (Cross and Greene counties) than in an Arkansas River Valley stream (Big Shoal Creek in Logan County). Logan County specimens consumed proportionally more microcrustaceans than the Crowley's Ridge specimens. Most of the animals in the diet were probably ingested while sucking in detritus, algae, and other benthic materials rather than by individual selection. The most important animal components of the diet were chironomids, other midges, other dipterans, and larval coleopterans. Animals ingested in smaller amounts included microcrustaceans (copepods, ostracods, cladocerans), trichopterans, ephemeropterans, hemipterans, and rotifers. Feeding begins 7.4 days after hatching at a length of 7.7 mm (Kay et al. 1994). The diet of small Western Creek Chubsuckers consists of copepods, cladocerans, and small aquatic insect larvae. William J. Matthews (pers. comm.) observed young-of-the-year feeding on algae.

Erimyzon claviformis spawns in the spring in small creeks at water temperatures between 12°C and 24°C (Curry and Spacie 1984; Page and Johnston 1990b). In Arkansas, spawning occurs from March through May over a bed of fine gravel or sand in slow to moderately swift water. Adults in spawning condition were collected on 21 April and 17, 18 May from streams on Crowley's Ridge at a water temperature of approximately 20°C. Spawning streams in Illinois averaged 1 m wide and 20 cm deep, with a maximum depth of 50 cm (Page and Johnston 1990b). Males defend a territory, but no nest is constructed. Sometimes the territory is over a nest of *Campostoma* species or *Semotilus atromaculatus*. The male defends his territory against intraspecific rivals by pushing against other males and occasionally by butting another male or interlocking the tubercles. Johnston et al. (1996) documented interspecific parallel swims between *E. claviformis* and two cyprinids, *Semotilus atromaculatus* and *Luxilus chrysocephalus*. The female drifts downstream through the territory of the male when she is ready to spawn. The male presses her side as the eggs and milt are released (Page and Johnston 1990b). In the closely related *E. oblongus*, eggs are scattered at random (Whitworth et at. 1968). After spawning, adults migrate downstream into the larger creeks, where they remain through fall and winter (Becker 1983). Maturity is reached in 2 years, and maximum age is about 7 years (Rohde et al. 2009). Females live 6–7 years, while males live only 5 years (Carnes 1958; Wagner and Cooper 1963; Carlander 1969). Populations of this species in Arkansas tend to be small.

CONSERVATION STATUS   The Western Creek Chubsucker is a common and widespread species in Arkansas, and its populations are currently secure.

Erimyzon sucetta (Lacépède)
Lake Chubsucker

Figure 15.135. *Erimyzon sucetta*, Lake Chubsucker. *Uland Thomas*.

CHARACTERS   A small, moderately deep-bodied (in adults), slightly laterally compressed, olivaceous sucker lacking a lateral line and possessing a nearly terminal mouth (Fig. 15.135). Body depth in adults going into SL usually 2.9–3.1 times. Dorsal fin short and rounded, usually with 11 or 12 (10–13) rays. Head length going into SL 3.5–4.1 times. Eyes relatively small, eye diameter going into head length 5.5–6.6 times in adults and approximately 3.9 times in juveniles smaller than 3 inches (75 mm) TL. Lower lips plicate, forming an acute V-shaped angle along rear margins. Pharyngeal teeth short and fragile, with 80 or more per arch (Becker 1983). Intestine long with several coils (Goldstein and Simon 1999b). Swim bladder with 2 chambers. Anal rays 7; pelvic rays 9; pectoral rays 14 or 15. Predorsal scales usually 12 or 13. Lateral line absent; scales in lateral series usually 34–37. Scales around caudal

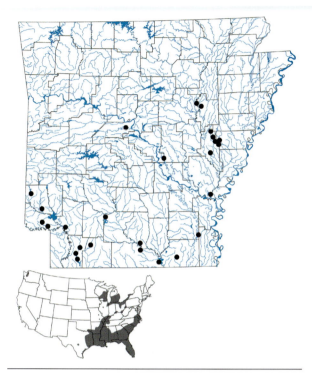

Map 100. Distribution of *Erimyzon sucetta*, Lake Chubsucker, 1988–2019.

peduncle usually 17 or 18. Breeding males with 3 or 4 large tubercles on each side of snout; other tubercles are along base of anal fin and on scales near urogenital opening. Anal fin of male distinctly bilobed. Adults usually 5–10 inches (127–254 mm) long; maximum size 16 inches (410 mm) and nearly 2.2 pounds (1.04 kg).

LIFE COLORS   Body color olive to bronze dorsally, grading to silver on sides and yellow ventrally. Fins dusky. Scale margins on back and upper sides are outlined in black. Caudal fin may have a reddish tinge. Dark stripe along midside continuous in juveniles and most adults; stripe often becomes indistinct in large adults.

SIMILAR SPECIES   The Lake Chubsucker is similar to the Western Creek Chubsucker (see account for *E. claviformis* for differences between the two species). It differs from *Minytrema melanops* in having fewer than 40 lateral scale rows (vs. more than 40), and lateral scales without dark spots (vs. each lateral scale with a dark spot in adults). It can be distinguished from all remaining catostomids in lacking a lateral line (vs. lateral line present).

VARIATION AND TAXONOMY   *Erimyzon sucetta* was described by Lacépède (1803) as *Cyprinus sucetta* from South Carolina (presumably in the vicinity of Charleston). Hubbs (1930) studied variation in *E. sucetta* and recognized two weakly defined subspecies. Bailey et al. (1954) considered the subspecies invalid, and subspecies are

not recognized by current workers. A molecular analysis by Harris et al. (2002) identified *F. sucetta* as sister to *E. tenuis*, with both of those species forming a clade sister to *E. oblongus* (no data for *E. claviformis*). A molecular analysis by Doosey et al. (2010) resolved *E. sucetta* as sister to *E. oblongus* (no data for *E. claviformis*). Larvae of *E. sucetta* were described by Fuiman (1979). Hybridization between *E. sucetta* and *E. oblongus* has been reported in North Carolina (Hanley 1977).

DISTRIBUTION   *Erimyzon sucetta* occurs in the Great Lakes and Mississippi River basin lowlands from southern Ontario, Canada, to Wisconsin, south to the Gulf of Mexico, and in Atlantic and Gulf slope drainages from southern Virginia to the Brazos River, Texas, with a disjunct population in the Guadalupe River, Texas. In Arkansas, this sucker is confined to the Coastal Plain, where most recent records are from the West Gulf Coastal Plain ecoregion (Connior et al. 2012) and the lower White River drainage (Map 100).

HABITAT AND BIOLOGY   In Arkansas, the Lake Chubsucker is strictly a lowland species occurring in heavily vegetated areas of oxbow lakes, sloughs of large rivers, and sluggish stream backwaters of varying clarity over soft bottoms of organic detritus, mud, and sand. It is frequently associated with dense aquatic vegetation and is often common in lakes having thick growths of cypress trees in their backwaters. In Arkansas, it is more common in oxbow lakes than in any other natural habitat, but AGFC rotenone population samples in lowland oxbows from 2000 to 2010 indicated that it may have declined in abundance in that habitat. It adapts well to some lowland reservoirs in southern Arkansas, such as Bois D'Arc, Erling, Felsenthal, Columbia, and Upper and Lower White Oak lakes. Lake Columbia probably has the largest reservoir population. An AGFC rotenone population sample in 2000 in that lake yielded a density of 458 individuals per hectare. In some states, it has been transplanted to reservoirs as a forage fish for Largemouth Bass and does well in those environments if the water is clear and vegetation is abundant (Smith 1979). Large adults are often found in deeper, more open waters than juveniles and small adults (Hill and Cichra 2005).

Foods consist primarily of detritus, algae, small clams, water mites, *Daphnia*, and other small crustaceans in addition to aquatic insects such as midge larvae (Ewers and Boesel 1935; Sheldon and Meffe 1993). Filamentous algae and other plant matter may sometimes compose more than 70% of the diet (Becker 1983; Shireman et al. 1978). In an Illinois lake, *E. sucetta* fed mainly on cladocerans (36.8%), chironomids (26.7%), and ostracods (22.5%) by volume;

other foods making up more than 1% of diet biomass were odonates (5.9%), gastropods (2.9%), and copepods (1.4%) (Eberts et al. 1998). Other reported foods include diatoms and adult insects. McLane (1955) observed predation on sunfish eggs in unguarded nests. Although it is a bottom feeder, its feeding does not roil the bottom nor increase the turbidity of the water.

Pflieger (1997) reported a May spawning in Missouri in ponds over submergent vegetation, and spawning occurred in Alabama from May to July (Mettee et al. 1996). Becker (1983) noted that a long spawning season begins in late March and probably extends into late July in Wisconsin. The demersal, adhesive eggs averaging about 2 mm in diameter are scattered over vegetation at random during spawning with no apparent prior preparation of the spawning site. In South Carolina, this species spawns primarily in the spring over a gravel substrate, sometimes in the nests of larger species in the family Centrarchidae (Marcy et al. 2005). Spawning behavior was described in a small Florida stream by McLane (1955). During the spawning act, the male draped his dorsal and caudal fins over the female and both fish vibrated intensely. The eggs were deposited over aquatic vegetation and filamentous algae. Up to 20,000 eggs may be produced per female. Mean fecundity for 14 Florida females (249–347 mm TL) was 18,478 (Shireman et al. 1978). Eggs hatch in 4 or 5 days at temperatures between 72°F and 85°F (22.2–29.4°C), and in 6 or 7 days at 68–72°F (20–22°C). Maximum age was reported as 6 years in South Carolina (Rohde et al. 2009), but Carlander (1969) stated that Lake Chubsuckers live about 8 years.

CONSERVATION STATUS Becker (1983) regarded the Lake Chubsucker as being tolerant of environmental stresses, but that does not agree with our observations in Arkansas. In Missouri, the abundance of *E. sucetta* has declined in the last 35 years. The same pattern is true in Arkansas, probably due to large-scale clearing of land throughout the Delta region. The Lake Chubsucker is uncommon, occurs sporadically in Arkansas, and appears to be declining in this state. We are not aware of any recent records from the St. Francis River drainage in Arkansas, although there are older records from that drainage. It also appears to be less common in some oxbow lakes than older data indicate, but stable populations are found in a few reservoirs. We consider it a species of special concern.

Genus *Hypentelium* Rafinesque, 1818c— Hog Suckers

There are three species in this genus, and one species, *H. nigricans*, inhabits Arkansas. These mottled suckers typically occur in clear, spring-fed streams with gravel substrates. Four to six conspicuous dorsal saddles are present. The head is large, flat, and concave between the eyes. Eyes are placed posteriorly behind the midpoint of the head length. Lips are fleshy and papillose. The lateral line is complete and well developed. Scales in lateral series fewer than 50. Body scales of uniform size. Dorsal fin rays 11 or fewer. The cylindrical body tapers to a narrow caudal peduncle. Swim bladder with 2 chambers. Anal rays 7; pelvic fin rays 9.

All molecular analyses to date support previous morphological evidence that the genus *Hypentelium* is closely related to genus *Thoburnia* (Clements et al. 2012). E. M. Nelson (1948; 1949) was the first to recognize the close affinity of *Hypentelium* with *Thoburnia*. Clements et al. (2012) always recovered *Thoburnia atripinnis* as the sister group to *Hypentelium* rather than to the other two species of *Thoburnia*. Although some recent authors assigned the genus *Hypentelium* to the tribe Moxostomatini (G. R. Smith 1992; Boschung and Mayden 2004), Harris et al. (2002), Doosey et al. (2010), and Tan and Armbruster (2018) grouped the genera *Thoburnia* and *Hypentelium* into the tribe Thoburnini.

Hypentelium nigricans (Lesueur)
Northern Hog Sucker

Figure 15.136. *Hypentelium nigricans*, Northern Hog Sucker. *David A. Neely.*

CHARACTERS A distinctly mottled and blotched, slender, terete sucker with a large head and large, fleshy, strongly papillose lips (Fig 15.136). Eyes small, located far back on the head and directed slightly upward. Snout long and strongly decurved. Head rather squarish in cross section; space between eyes broad and flat to distinctly concave (Fig. 15.122A). Mouth ventral, horizontal or oblique. Dorsal fin short-based, distal margin concave, and with 10–12 rays. Pectoral fins large and expansive; pectoral rays 15–18. Lateral line complete with 46–50 scales. Scales around caudal peduncle 14–17. Swim bladder with 2 reduced chambers. Breeding males with large tubercles on anal fin, pelvic fins, lower caudal lobe, and caudal peduncle

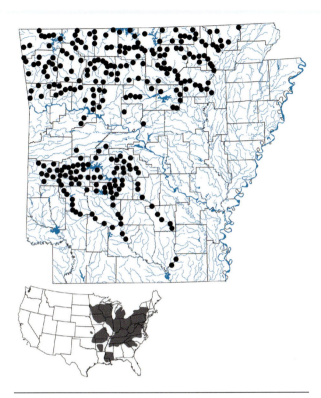

Map 101. Distribution of *Hypentelium nigricans*, Northern Hog Sucker, 1988–2019.

ventrally; minute tubercles occur on head and branchiostegal area. Breeding females often have small tubercles on the fins (Trautman 1957). Maximum size 24 inches (610 mm) TL and about 4.5 pounds (2 kg), but usually smaller than 18 inches (457 mm) TL. The Arkansas state angling record from the Spring River in 2011 is 1 pound, 2 ounces (0.5 kg).

LIFE COLORS   Body color reddish brown to olive above with yellow to white lower sides and a white belly. Body with 4–6 dark brown to black saddles extending obliquely forward onto the sides well below the lateral line. Dorsal fin plain or variably marked with dark pigment. Caudal fin with a dark basal bar and a thinner, lighter submarginal bar. Pectoral, pelvic, and anal fins olivaceous to dull orange.

SIMILAR SPECIES   The Northern Hog Sucker differs from *Moxostoma* species in having prominent dorsal saddles (vs. saddles absent), a 2-chambered swim bladder (vs. 3 chambers), and papillose lips (vs. plicate lips). Juveniles of the Northern Hog Sucker are more easily confused with redhorse suckers (*Moxostoma* species), and in addition to the above-listed differences, *H. nigricans* has a flat to concave head between the eyes, while the redhorses of Arkansas have a convex head. It differs from the Spotted Sucker in having a well-developed, complete lateral line

(vs. lateral line incomplete or absent, large adults having only about 4 unpored lateral scales).*

VARIATION AND TAXONOMY   *Hypentelium nigricans* was described by Lesueur (1817b) as *Catostomus nigricans* from Lake Erie. Buth's (1980) genetic study of systematic relationships in *Hypentelium* did not identify subspecies within *H. nigricans*. However, Berendzen et al. (2003) found significant intraspecific variation and hypothesized that the phylogenetic history of *Hypentelium* was shaped by old vicariant events associated with erosion of the Blue Ridge and separation of the Mobile and Mississippi river basins. Within *H. nigricans*, two clades existed prior to the Pleistocene: a widespread clade in the preglacial Teays-Mississippi River system, and a clade in the Cumberland and Tennessee rivers. Pleistocene events fragmented the Teays-Mississippi fauna. Following retreat of the glaciers, *H. nigricans* dispersed northward into previously glaciated regions. These patterns are replicated in other clades of fishes and are consistent with the predictions of Mayden's pre-Pleistocene vicariance hypothesis (Mayden 1988b). Berendzen et al. (2003) found two monophyletic groups within *H. nigricans*: clade A, containing haplotypes from the Eastern Highlands, and clade B, containing haplotypes from the Roanoke (Atlantic Slope) and New rivers, the Interior Highlands, and previously glaciated regions of the upper Mississippi River. Within clade B, two western haplotypes in the Interior Highlands and upper Mississippi River drainage were recovered in two monophyletic groups: a clade containing the northern Ozark Highlands and previously glaciated regions of the upper Mississippi River drainage, and a clade containing the southern Ozark and Ouachita highlands. A genetic analysis by Doosey et al. (2010) resolved *H. nigricans* as sister to *H. etowanum*, with the two forming a clade sister to *H. roanokense*.

DISTRIBUTION   *Hypentelium nigricans* occurs in the Great Lakes, Hudson Bay, and Mississippi River basins from New York and southern Ontario, Canada, west to Minnesota, and south to northern Alabama, southern Arkansas, and eastern Louisiana; Atlantic Slope drainages from the Hudson River drainage, New York, to the Oconee River in northern Georgia; and in Gulf Slope drainages from the Apalachicola River, Georgia (Chattahoochee River), Mobile Bay drainage of Tennessee, and Pascagoula River, Mississippi, to the Mississippi River, Louisiana. In Arkansas, the Northern Hog Sucker is widely distributed in the upland streams of the Ozark and Ouachita

---

* Lower lip of large juvenile to adult *Moxostoma pisolabrum* can be quite "pebbly" and that of *M. anisurum* not distinctly plicate.

mountains and sporadically distributed in streams of the Arkansas River Valley. Its range extends onto the Coastal Plain in the Saline and Ouachita rivers (Map 101).

HABITAT AND BIOLOGY The Northern Hog Sucker lives in clear, permanent streams with gravel or rocky bottoms. It is commonly found in medium to large creeks and small to medium-sized rivers in Arkansas. Adults generally prefer deep riffles, runs, raceways, or pools with a distinct current, whereas small juveniles are most often found in shallow pool areas. The broad bulky head, long slender body, reduced swim bladder, and orientation of paired fins are adaptations for maintaining position in swift water (Buth 1980; Lundberg and Marsh 1976). In the Current River, Missouri, habitat use changed seasonally from slower, deeper water and smaller substrates during winter to increasing use of faster, shallower water and larger substrates through warmer-water periods (Matheney and Rabeni 1995). Diel patterns of habitat use were also observed. In winter, fish used pool habitat with moderate flow during the day and riffle or edge habitat at night. In summer, fish used run habitat during the day and riffle or edge habitat at night. Northern Hog Suckers responded to flooding by moving onto adjacent floodplain areas until the high water receded (Matheney and Rabeni 1995). In Illinois, it overwinters in quiet water of large streams (Smith 1979). Although adapted primarily to habitats with considerable current, it is occasionally found in upland reservoirs of the Ozarks and Ouachitas near creek mouths. This sucker is intolerant of pollution, silt, and physical modification of stream channels (Smith 1979; Trautman 1981) and is often used as an indicator species of clean-water environments. The home ranges of *H. nigricans* in the Current River (276–812 m) were considerably larger than those reported for this species in smaller Indiana streams (91–122 m) by Gerking (1953) (Matheney and Rabeni 1995).

Northern Hog Suckers are often seen foraging alone or in small groups, but not in large schools. During summer in the Current River, Missouri, most feeding activity took place during the day (Matheney and Rabeni 1995). They feed on aquatic insects, plant material, small clams, and other invertebrates, often overturning rocks to secure food items. Feeding *H. nigricans* are often followed by schools of minnows that eagerly feed on foods dislodged when the hog suckers overturn stones. In Mississippi, dominant food items included diatoms, organic detritus, midge larvae, and small crustaceans (Greenwich 1979). In Wisconsin, *H. nigricans* was almost entirely carnivorous, feeding mainly on insects (coleopterans, and odonates), small clams, and shrimp; only a small proportion (12%) of the diet consisted of vegetable matter (Becker 1983). When

feeding, the mouth protrudes anteriorly and the oral and pharyngeal cavities expand, effecting a vacuum for sucking up food items. When not feeding, hog suckers rest on the bottom, where they are almost imperceptible because of their multitone cryptic coloration.

In Arkansas, spawning occurs in April and May in the lower ends of pools near gravel riffles. Spawning takes place at water temperatures of 15–19°C (59–66°F) at depths of 0.35–0.45 m (14–18 inches) in current speeds of 0.4–0.56 m/s (1.3–1.8 feet/s) (Greenwich 1979; Curry and Spacie 1984). In the Current River, Missouri, Northern Hog Suckers moved in late February and early March into pool-head or pool-tail habitats characterized by moderate current velocities and depths, with pebble and gravel as the dominant substrates (Matheney and Rabeni 1995). After water temperatures reached and remained above 10°C (March 21), circular areas of substrate about 0.3 m in diameter cleared of overlying silt were present, presumably in preparation for spawning. Raney and Lachner (1946) described the communal spawning behavior. The rapid body quivering of the spawning act cleans the substrate, allowing fine particulates to be current-carried downstream. Because successive/repeated spawning acts by the single female with at least two males occur at a particular spot, one could call this a nest or redd. When a ripe female moves into a spawning area, she attracts several males. When she stops, the males press their tails and caudal peduncles against her and quiver, resulting in a considerable disturbance of the substrate (Rohde et al. 2009). Eggs are broadcast and abandoned, often to be fed upon by minnows. The eggs hatch in 10 days at a mean temperature of 63.3°F (17.4°C) (Buynak and Mohr 1978a). Maturity is reached in about 3 years, and the maximum recorded age is 11 years. In Missouri, the Northern Hog Sucker is considered a popular game species usually taken by spearing (Matheney and Rabeni 1995).

CONSERVATION STATUS *Hypentelium nigricans* is widespread and common in Arkansas in upland streams, and its populations are secure.

Genus *Ictiobus* Rafinesque, 1820a— Buffalofishes

Buffalofishes are the largest and commercially most valuable of the suckers. These gray to black, robust-bodied fishes have only 1 fontanelle (opening) on the dorsal midline of the skull. The subopercle is semicircular, with its greatest depth in the middle (Fig. 15.118B). The pharyngeal arch is robust, with the edge opposite the teeth reinforced by thickening of the bone (Etnier and Starnes 1993). Swim bladder with 2 chambers. The intestine runs longitudinally and parallel to the sides (Fig. 15.125A). The peritoneum is black. Dorsal fin rays 23–32. Anal fin rays 7–11. Pelvic fins densely pigmented with melanophores. The genus *Ictiobus* contains five species, three of which inhabit Arkansas: *I. bubalus*, *I. cyprinellus*, and *I. niger*.

Ictiobus bubalus (Rafinesque)
Smallmouth Buffalo

Figure 15.137. *Ictiobus bubalus*, Smallmouth Buffalo. *David A. Neely*.

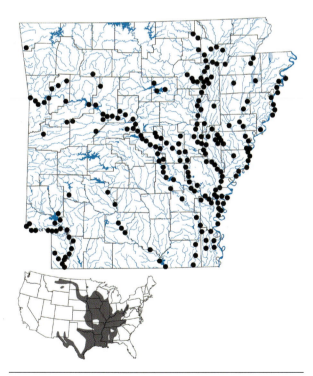

Map 102. Distribution of *Ictiobus bubalus*, Smallmouth Buffalo, 1988–2019.

CHARACTERS A deep-bodied and compressed fish compared with other species of *Ictiobus*, with a straight ventral contour and highly arched back (back strongly keeled) (Fig. 15.137). Body depth going into SL about 2.2–2.8 times; body width is less than the distance from the tip of the snout to the upper end of the operculum. Head small and conical, head length going into SL 3.4–4.1 times. Snout blunt. Mouth small, nearly horizontal and subterminal. Subopercle semicircular, broadest at middle, its outer margin gently rounded. Length of upper jaw much less than snout length. Lips thick and plicate. Front of upper lip well below level of lower margin of eye. Gill rakers fewer than 60; pharyngeal teeth 115–167 (Eastman 1977). Dorsal fin long-based and falcate, with 24–31 rays. Lateral line complete, nearly straight, and usually with 36–38 scales. Anal fin rays 9 or 10; pectoral rays 15–18; pelvic rays 9–11. Intestine long with many loops running parallel to body axis; peritoneum black. Small breeding tubercles covering the head and body of males and weakly developed on females (Sublette et al. 1990). Arkansas specimens usually 2–15 pounds (0.9–6.8 kg), and 15–30 inches (381–762 mm) TL. The Arkansas (and previous world) rod and reel angling record is 68 pounds, 8 ounces (31.1 kg) from Lake Hamilton in 1984. The Arkansas unrestricted tackle record from Millwood Lake in 2007 is 74 pounds (33.6 kg). Maximum size 35.8 inches (909 mm) TL and about 88 pounds (39.9 kg).

LIFE COLORS Body coloration dark; uniformly olive gray with strong coppery, lavender, or greenish reflections on back and sides. In breeding males, the upper sides and dorsum are metallic bronze. All fins are black. Belly white.

SIMILAR SPECIES The Smallmouth Buffalo most closely resembles the Black Buffalo (*I. niger*), differing from it in having a smaller head, a smaller and more horizontal mouth and, in the adult, a much deeper and more slab-sided body (young specimens are very difficult to separate). Body depth in *I. bubalus* goes into SL fewer than 2.8 times (vs. more than 2.9 times in *I. niger*). *I. bubalus* differs from *I. cyprinellus* in having a subterminal mouth (vs. a terminal mouth). Taxonomic work is needed on all *Ictiobus* species throughout their ranges. Although the characters given will differentiate most specimens, some individuals (especially the juveniles) are quite difficult to identify. *Ictiobus bubalus* has a single fontanelle in the roof of the skull (as do all buffalofishes), whereas *Carpiodes* has 2 fontanelles. It further differs from all *Carpiodes* species in having a medium to dark-colored body (vs. gold or coppery dorsally grading to silver or white ventrally in *Carpiodes*), pelvic fins densely speckled with black (vs. pelvic fins scarcely or not at all speckled with black), intestine with longitudinal folds (vs. circular folds in carpsuckers), black peritoneum (vs. silvery peritoneum), and in having a semicircular subopercle (vs. subtriangular). From the Common Carp, the Smallmouth Buffalo differs in lacking barbels and lacking a serrate spinelike ray at the dorsal and anal fin origins (vs. maxillary barbels present and front of dorsal and anal fins with a serrate spinelike ray).

VARIATION AND TAXONOMY *Ictiobus bubalus* was described by Rafinesque (1818b) as *Catostomus bubalus* from the Ohio River. Bart et al. (2010) described patterns of morphometric and DNA sequence variation across the geographic range of the species of *Ictiobus*. They showed that *Ictiobus* species form more or less discrete entities based on body morphometry consistent with current taxonomy; however, there is extensive sharing of alleles of nuclear and mitochondrial genes among species, and the species do not form reciprocally monophyletic groups in nuclear or mitochondrial gene trees. There has evidently been a long history of introgressive hybridization and gene flow among *Ictiobus* species inhabiting the Mississippi River basin. Smith (1992) found *I. bubalus* was sister to *I. cyprinellus* plus *I. labiosus*, suggesting an unusual pattern of vicariance involving a Mexican Plateau endemic and a species largely confined to the Mississippi basin. Analysis of mitochondrial DNA by Doosey et al. (2010) recovered *I. bubalus* as sister to *I. meridionalis*, with the two forming a clade sister to *I. niger*. Fossils identified as *I. cf. bubalus* are known from Lower Pliocene deposits of Oklahoma (C. L. Smith 1962).

DISTRIBUTION *Ictiobus bubalus* occurs in the lower Great Lakes, Hudson Bay, and Mississippi River basins from Pennsylvania to Montana, and south to the Gulf of Mexico, and in Gulf Slope drainages from Mobile Bay, Alabama, to the Rio Grande, Texas, New Mexico, and Mexico. It has been introduced into North Carolina, Arizona, and possibly elsewhere (Page and Burr 2011). The Smallmouth Buffalo is widely distributed in Arkansas, but most records come from eastern Arkansas (Map 102). It is abundant throughout the Arkansas, Red, lower White, St. Francis, and Mississippi rivers and also occurs in the Ouachita River.

HABITAT AND BIOLOGY The Smallmouth Buffalo is widespread, adaptable, and abundant in Arkansas and is found in streams, rivers, and reservoirs, where it has been collected over a variety of substrates. It prefers firm-bottomed channels but is sometimes taken in backwaters and in mouths of tributaries (Smith 1979). In the lower Mississippi River, it is abundant along revetted banks and other main channel habitats, and in oxbow lakes and borrow pits (Baker et al. 1991). It prefers slightly clearer water than the Bigmouth Buffalo and is found less often in strong current than the Black Buffalo (Pflieger 1997). It also occurs in smaller streams in Arkansas than the other two *Ictiobus* species, where it inhabits the deepest pools

available. Hrabik et al. (2015) noted that the Smallmouth Buffalo uses a wide variety of habitats, with the fry typically found in quiet backwaters, while the adults prefer areas in rivers with moderate current and low turbidity over firm substrates.

The adult Smallmouth Buffalo is an opportunistic bottom feeder ingesting whatever small organisms are abundant (McComish 1967) including zooplankton, attached algae, aquatic insects, crustaceans, bryozoans, snails, and detritus (Wrenn 1968; Tafanelli et al. 1971; Goldstein and Simon 1999; Walberg and Nelson 1999; Ross 2001). The diet can vary drastically from one season to another at the same locality (Minckley et al. 1970). In Arizona in January and February, the diet consisted mainly of diatoms; in March and April, cladocerans, copepods, and green algae were the dominant foods; in May and June, clams were the largest component of the diet; from July through October, blue-green algae dominated; and in November and December, diatoms and blue-green algae made up most of the diet. The larvae have a terminal mouth and feed near the surface (Wrenn 1968); young-of-the-year 35–64 mm TL continue feeding in the water column mainly on microcrustaceans (copepods and cladocerans made up 99% of the food volume). Feeding becomes more benthic-oriented at about 250 mm TL, and adults ingest considerable amounts of detritus and sand (Ross 2001). Individuals in the Tennessee River drainage fed mainly on bivalve mollusks (primarily *Corbicula*), copepods, and aquatic dipterans (Mettee et al. 1996).

Becker (1983) summarized information on the reproductive biology of *I. bubalus*. The length of the spawning season is variable over its range and often associated with periods of rising water. Spawning occurs in Wisconsin from April to early June at water temperatures of 60–65°F (15.6–18.3°C), and in New Mexico from early May to September at water temperatures between 71.6°F and 79.7°F (22–26.6°C) (Jester 1973). In Wheeler Reservoir (Tennessee River), Alabama, spawning occurs mainly in April and May when water temperatures reach 59–62°F (15–16°C) (Wrenn 1968). Spawning in Arkansas also occurs over a fairly long season beginning in early April and continuing well into June in quiet, shallow backwaters or on flooded lands during high-water periods. Spawning has also been reported in swift water below Lake Tuscaloosa Dam, Alabama (Mettee et al. 1996). This species is a non-guarder and an open substratum spawner (Simon 1999b). One female spawns with several males. The demersal, adhesive eggs are broadcast over almost all substrate types or on vegetation and are unattended by adults (Jester 1971, 1973; Padilla 1972). Spawning in a New Mexico reservoir occurred at depths of 4–8 feet (1.2–2.4 m) over recently inundated terrestrial vegetation (Jester 1973). Fecundity was estimated by Wrenn (1968) to be 290,000 eggs for fish 546–556 mm TL, whereas MacDonald (1978) estimated 98,630 and 501,360 eggs for two females 450 and 838 mm TL. Fertilized eggs average 1.6–2.1 mm in diameter (Yeager 1980) and 2.3–2.4 mm in diameter (Yeager and Baker 1982). Hatching occurs in 96–100 hours at 21.1°C (Wrenn 1968). Males mature in 4–5 years (minimum 411 mm TL), and females mature after 6 years (minimum size 444 mm TL) (Wrenn 1968). Maximum life span is reported to be 9–18 years (Carlander 1969; Jester 1973; MacDonald 1978).

Black (1940) noted that *I. bubalus* was probably the most common of the Arkansas buffalofishes. It is currently a valuable commercial species occurring in considerable numbers in the Arkansas River Navigation System (Buchanan 1976). Often called "razorback buffalo," it makes up the largest percentage by weight of all buffalofishes taken by commercial fishermen in Arkansas.

CONSERVATION STATUS   In Arkansas the Smallmouth Buffalo is the most common catostomid in the state. Its populations are secure.

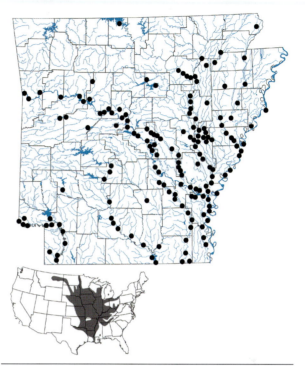

Map 103. Distribution of *Ictiobus cyprinellus*, Bigmouth Buffalo, 1988–2019.

Ictiobus cyprinellus (Valenciennes)
Bigmouth Buffalo

Figure 15.138. *Ictiobus cyprinellus*, Bigmouth Buffalo. David A. Neely.

CHARACTERS   A large, deep-bodied sucker with a large oblique, terminal mouth and a long-based, falcate dorsal fin with 23–32 rays (Fig. 15.138). Head large, ovoid. Lips thin, relatively smooth. Front of upper lip on level with lower margin of eye. Length of upper jaw about equal to snout length and much greater than diameter of eye. Eye small and closer to tip of snout than to rear of the operculum. Subopercle semicircular, widest in middle. Lateral line complete, nearly straight, and with 32–42 scales. Anal rays 7–10; pectoral rays 16–19; pelvic rays 9 or 10. Gill rakers 40–60. Pharyngeal teeth short, with 129–173 on each arch. Body depth going into SL about 2.5–3.3 times. Predorsal body surface rounded or weakly keeled. Intestine

very long with at least 4 longitudinal loops (Goldstein and Simon 1999). Peritoneum black. Breeding males with small tubercles on head, predorsal area, side of body, caudal peduncle, and fin rays. Twenty-pound (9 kg) specimens are not uncommon in Arkansas. The Arkansas rod and reel angling record from Lake Conway in 2007 is 50 pounds (22.7 kg), and the Arkansas unrestricted tackle record from Calion Lake in 2000 is 56 pounds, 4 ounces (25.5 kg). Maximum weight 80 pounds (36.3 kg) (Harlan and Speaker 1956). Maximum length around 40 inches (100 cm) (Page and Burr 2011).

LIFE COLORS   Body gray to olive brown above, becoming copper and green on sides, and white to pale yellow below. All fins usually dusky to black; lower fins often white-edged along distal margins. Specimens from turbid waters are often very pale and yellowish (Trautman 1957).

SIMILAR SPECIES   See account of the Smallmouth Buffalo, *I. bubalus*, for differences among the buffalofishes, carpsuckers (*Carpiodes* species), and the Common Carp, *Cyprinus carpio*.

VARIATION AND TAXONOMY   *Ictiobus cyprinellus* was originally described by Valenciennes *in* Cuvier and Valenciennes (1844) as *Sclerognathus cyprinella* from Lake Pontchartrain, Louisiana. It is known to hybridize with *I. bubalus* in nature. Fossil remains agreeing with

I. cyprinellus were identified from preglacial Early Pleistocene deposits of Nebraska (1.5–1 mya) (Smith and Lundberg 1972). Bart et al. (2010) described patterns of morphometric and DNA sequence variation across the geographic range of the species of *Ictiobus* (see *I. bubalus* account). Analysis of mitochondrial DNA by Doosey et al. (2010) resolved *I. cyprinellus* as sister to a clade containing all other *Ictiobus* species.

DISTRIBUTION *Ictiobus cyprinellus* occurs in the Great Lakes, Hudson Bay, and Mississippi River basins from Ontario to Saskatchewan, Canada, south to the Gulf of Mexico. Introduced populations are known in Mississippi, North Carolina, Arizona, and California. The Bigmouth Buffalo is widely distributed in Arkansas, occurring in all rivers of the Coastal Plain and throughout the Arkansas River (Map 103). It is rarely found in mountainous areas of the state.

HABITAT AND BIOLOGY The Bigmouth Buffalo, also called gourdhead by commercial fishermen, prefers quieter waters than the other two species of buffalofishes, contributing to its wide success in impoundments. It lives in deep pool sections of large streams and rivers, oxbow lakes, and reservoirs, and is raised commercially in ponds. Occasionally it enters small streams to spawn. Smith (1979) considered the Bigmouth Buffalo less of a bottom-dwelling fish than most other suckers. Pflieger (1997) reported that it is more tolerant of higher turbidity than other *Ictiobus* species. Trautman (1981) also considered it more tolerant of turbid conditions and rapid siltation of stream bottoms than *I. bubalus*, and concluded that greater turbidity tolerance was related to different feeding habits. It is also able to tolerate low oxygen tensions (0.9 ppm) and high water temperatures (Becker 1983).

Basically a midwater to bottom-dwelling, schooling species, the juveniles and adults feed mainly on zooplankton (usually more than 90% microcrustaceans), but sometimes ingest a small amount of plant material. The large terminal mouth and very fine gill rakers enable *I. cyprinellus* to operate as an efficient filter feeder, utilizing swarms of zooplankton found in quiet areas of lakes and rivers (Miller and Robison 2004). Larval fish (about 7 mm TL) first feed on algae and rotifers but soon add small cladocerans to the diet (Makeyeva 1980). Young-of-the-year fish in Lewis and Clark Lake fed almost entirely on zooplankton (cladocerans, copepods, and rotifers) (McComish 1967), but in another study in South Dakota, YOY fed primarily on benthic organisms (midge pupae and larvae and benthic cladocerans), and planktonic prey made up only 25% of the diet (Starostka and Applegate 1970). Older individuals feed mainly on small crustaceans and other organisms found near the bottom or that rest lightly upon the substrate. Swimming movements of feeding schools of adults create water swirls that bring benthic food items up into the water column where they can be strained by the filtering apparatus (Minckley et al. 1970; Pflieger 1997).

Spawning occurs in Arkansas in early to mid-spring (April–May) after spring floods raise water levels. Spawning is reported to occur at water temperatures of 15–27°C (59–80.6°F) (Johnson 1963). Burr and Heidinger (1983) observed no nest site preparation by spawning *I. cyprinellus* in Illinois. In that study, the adhesive eggs were deposited over vegetation and other objects on the substrate. Pflieger (1997) also observed this species spawning in late May over rock riprap in a quiet backwater of the Missouri River in water so shallow that the backs of the adults were exposed. Spawning is accomplished in groups of 3 or more individuals (usually 2 tuberculate males aligned alongside 1 female) and involves a "rush" along the water surface. Spawners then sink to the bottom, occasionally assuming a vertical posture with their tails exposed. Large females may produce up to 750,000 eggs. Adhesive eggs are scattered over shallow, vegetated areas and left unattended. Maturity is usually reached in 3 years, but in Arkansas, males were usually mature after 1 year and females after 2 years (Brady and Hulsey 1959). In most areas, the maximum reported life span is less than 10 years. Johnson (1963) reported a maximum life span in Canada of 19 or 20 years, and Paukert and Long (1999) reported a 26-year-old specimen from Keystone Reservoir (Arkansas River drainage), Oklahoma. Black (1940) considered it an important commercial species in the lower White, Arkansas, and Mississippi rivers and noted that it was marketed in "great quantities" at DeValls Bluff, St. Charles, and Lake Village. It is of less commercial importance than the Smallmouth Buffalo because it is not as desirable as a food fish in Arkansas (Buchanan 1976). Ross (2001) reported that in Mississippi it is considered an excellent food fish.

CONSERVATION STATUS Hrabik et al. (2015) speculated that Silver Carp and Bighead Carp may negatively affect *I. cyprinellus* populations. The Bigmouth Buffalo is common in its preferred habitat in Arkansas, and its populations are secure.

Ictiobus niger (Rafinesque)
Black Buffalo

CHARACTERS A large, heavy-bodied buffalo similar to the other buffalofishes but not as deep-bodied and with a less arched dorsal contour (Fig. 15.139). Predorsal region

Figure 15.139. *Ictiobus niger*, Black Buffalo. *David A. Neely.*

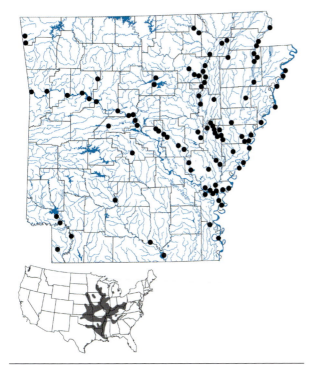

Map 104. Distribution of *Ictiobus niger*, Black Buffalo, 1988–2019.

of body rounded or only weakly keeled. Body depth going into SL about 2.6–3.2 times. Body width greater than distance from tip of snout to upper end of operculum. Mouth small (but larger than that of *I. bubalus*), subterminal, and slightly oblique. Length of upper jaw much less than snout length, equal to or slightly greater than eye diameter. Front edge of mouth much below lower edge of eye. Subopercle semicircular, widest in middle. Eyes small, eye diameter going into head length approximately 7.7 times. Upper lip thick and deeply grooved. Fewer than 60 gill rakers on first gill arch. Pharyngeal teeth about 195 per side; teeth short, narrow, and fragile. Intestine long with many longitudinal loops. Peritoneum black. Lateral line complete, nearly straight, and with 36–39 scales. Dorsal fin long-based and with 27–31 rays. Anal rays 8–10. Pectoral rays 15 or 16. Pelvic rays 9–11. Body of nuptial males covered by tiny to small whitish tubercles, most pronounced on cheek and opercle; tubercles smaller and less numerous on posterior half of body than on anterior half of body; body tuberculation most pronounced below lateral line, with individual scales containing 10–16 tubercles; above lateral line, tubercles are smaller and less numerous. Tubercles present on all fins, most pronounced on anal and caudal fins, with tubercles running along entire length of anterior rays (rays 1–6) but confined to distal portions of more posterior rays (Piller et al. 2003). The Arkansas rod and reel angling record from Lake Maumelle in 2001 is 92 pounds, 8 ounces (42 kg). The Arkansas unrestricted tackle record from Harris Brake Lake in 1994 is 105 pounds (47.6 kg). Maximum size more than 100 pounds (45.4 kg) and 37 inches (940 mm) TL.

LIFE COLORS   Body blue gray to dark brown above, light brown laterally, and yellowish white below (Miller and Robison 2004). Coloration similar to the other two *Ictiobus* species, but noticeably darker. All fins dark. In the spring, males are often blackish (Trautman 1981).

SIMILAR SPECIES   *Ictiobus niger* is similar to the carpsuckers (*Carpiodes*) in general appearance; however, in *Carpiodes* the body is more compressed and silvery, and the subopercle bone is asymmetrical, with its greatest depth anterior to the middle of the bone (vs. the subopercle in *Ictiobus* symmetrical and semicircular with its greatest depth at the middle of the bone) (Fig. 15.118B). Also, in *Ictiobus* the pelvic fins are densely pigmented with melanophores (vs. pelvic fins in *Carpiodes* with few or no melanophores), and the peritoneum is black (vs. peritoneum silvery). *Ictiobus niger* differs from *I. cyprinellus* in mouth being inferior and horizontal (vs. mouth terminal and oblique in *cyprinellus*). From *I. bubalus*, *I. niger* differs in having the eye diameter equal to or less than distance from fleshy posterior tip of upper lip to fleshy anterior tip of lower jaw (vs. eye diameter greater than distance from fleshy posterior tip of maxilla to fleshy anterior tip of lower jaw in *I. bubalus*). In addition, body depth in *I. niger* usually goes more than 2.9 times into SL and the back is rounded anterior to dorsal fin (vs. body deep and compressed, body depth usually going fewer than 2.7 times into SL, and back sharply ridged anterior to dorsal fin in *I. bubalus*). Specimens smaller than 12 inches (305 mm) TL are often very difficult to separate from *I. bubalus*.

VARIATION AND TAXONOMY   *Ictiobus niger* was described by Rafinesque (1819) as *Amblodon niger* from the Mississippi, Ohio, and Missouri rivers (no specific type locality given). A comprehensive phylogenetic analysis of the family Catostomidae involving osteological, allozymic, developmental, and external morphological characters found *I. niger* to be the most basal of extant *Ictiobus* species

(Smith 1992). A long history of introgressive hybridization and gene flow among *Ictiobus* species inhabiting the present-day Mississippi River basin is apparent (Bart et al. 2010). A molecular analysis by Doosey et al. (2010) recovered *I. niger* as sister to a clade composed of *I. bubalus* and *I. meridionalis*. The earliest fossil evidence of *I. niger* is from Late Pleistocene deposits (<0.2 mya) (Smith 1981).

DISTRIBUTION   *Ictiobus niger* occurs in the lower Great Lakes and Mississippi River basins from Michigan and Ohio to South Dakota, and south through Louisiana to the Gulf of Mexico. In Arkansas, the Black Buffalo is typically found in the larger rivers, primarily in the Coastal Plain and throughout the Arkansas River (Map 104). It is more common in eastern Arkansas than southern Arkansas.

HABITAT AND BIOLOGY   The Black Buffalo is rare to uncommon in Arkansas and is often described in those terms elsewhere within its range. Part of the explanation for fewer records of this species from Arkansas than for other buffalo species may be the difficulty in distinguishing it from *I. bubalus*, but there is also no doubt that it is more uncommon in this state than the other buffalofishes. Trautman (1981) noted that its habitat is often intermediate between that of *I. bubalus* and *I. cyprinellus*. It prefers flowing waters and is usually found in the stronger currents of large rivers, and in impoundments. It is also found in small numbers in backwater areas of the Arkansas River over sand and silt bottoms. In Oklahoma, Pigg and Gibbs (1995) noted that *I. niger* preferred a stronger current than other buffalofishes. In Kansas, Cross (1967) collected *Ictiobus niger* in deep, fast riffles where the channel narrowed. The Black Buffalo was the most numerous *Ictiobus* species taken in annual AGFC rotenone population samples in lentic areas of Greers Ferry Lake. About 20 individuals were taken each year from three sample areas totaling approximately 8 acres (3.2 hectares) in surface area (Carl Perrin, pers. comm.). This is the only locality in Arkansas where the Black Buffalo is known to outnumber the other *Ictiobus* species in collections.

The food of the Black Buffalo is similar to that of the Bigmouth Buffalo, according to Forbes and Richardson (1920). *Ictiobus niger* consumes vegetation, plankton, insect larvae, snails, and other small mollusks, often in large quantities (Shute 1980; Miller and Robison 2004). However, Minckley et al. (1970) found that *I. niger* and *I. bubalus* in Arizona reservoirs fed mainly on benthic macroinvertebrates, with mollusks being the most important group used. Diatoms, blue-green algae, and crustaceans were high in frequency of occurrence but made up only a minor part of diet volume. Black Buffaloes fed by searching the bottom, stirring the soft sediments, and probing beneath twigs, stones, and other debris (Minckley et al. 1970). Becker (1983) reported that 67% of its diet is animal matter, largely mollusks and insects. Vegetable foods eaten included duckweed and algae.

Spawning in Tennessee occurs in April and is often 4–5 days in duration (Piller et al. 2003). Becker (1983) suggested an April and May spawning period in Wisconsin, and captured a tuberculate male in mid-June. Yeager (1936) recorded an instance of spawning in Mississippi during April along the margin of a flooded swamp; several hundred adults participated, and the eggs were deposited over submerged trees and logs. Simon (1999b) classified *I. niger* as a lithopelagophil (rock and gravel spawner with pelagic larvae). They are nonguarders and open substrate spawners. In Citico Creek, Tennessee, Piller et al. (2003) described spawning habitat as mainly runs and pools (75% of observations), but sometimes riffles (25% of observations). Spawning occurred in the upper water column and adhesive eggs were broadcast over a variety of substrates, from gravel to bedrock. Eggs were found above the stream bottom, suggesting they were carried some distance downstream. Fecundity is more than 9,000 (Piller et al. 2003). Fertilized eggs are demersal, adhesive, and average 1.8–2.4 mm in diameter, with hatching occurring within 24–36 hours at 19–24°C (Ross 2001). Moore (1968) noted that *I. bubalus* and *I. niger* hybridize in some impoundments in Oklahoma, producing offspring impossible to identify morphologically. Black (1940) considered *I. niger* commercially important, noting that it was regularly marketed at DeValls Bluff and St. Charles and was a staple market fish throughout the lower White, Arkansas, and Mississippi rivers. It is currently the least captured of the buffalofishes in Arkansas and has little commercial significance. Commercial fishermen can readily identify this species they call the "blue rooter" or "rooter."

CONSERVATION STATUS   The Black Buffalo has declined in some areas such as the Ohio River (Trautman 1981), and it is a species of special concern in Mississippi (Ross 2001). Hrabik et al. (2015) speculated that diet overlap in feeding on mollusks between *I. niger* and the introduced Black Carp, *Mylopharyngodon piceus*, may adversely affect Black Buffalo populations. It is uncommon in Arkansas, and its populations may have declined somewhat since the 1940s. We do not believe it warrants a status of special concern.

Genus *Minytrema* Jordan, 1878b—
Spotted Suckers

Jordan described this monotypic genus in 1878. Genus characters are the same as those in the following species account.

Minytrema melanops (Rafinesque)
Spotted Sucker

Figure 15.140. *Minytrema melanops*, Spotted Sucker. *Uland Thomas.*

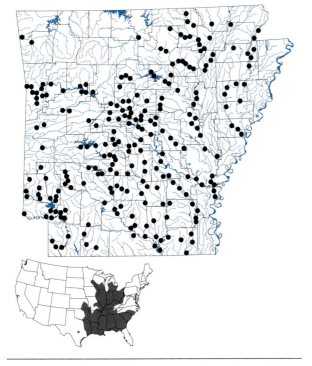

Map 105. Distribution of *Minytrema melanops*, Spotted Sucker, 1988–2019.

CHARACTERS   A moderately slender, terete sucker with a short-based dorsal fin; similar in body shape to the redhorses (*Moxostoma*), but differs in having the lateral line incomplete (adults) or absent (young), with 42–47 scales in lateral series; large adults may have a nearly complete lateral line with only a few unpored scales. Caudal peduncle scales 15 or 16. Swim bladder with two chambers. A conspicuous black, squarish spot or dash on the exposed anterior area of each scale on the side of the body, forming parallel rows. Dorsal fin with a straight to slightly concave margin (straight in young, slightly falcate in adults), with 11–13 rays. Anal rays 7; pectoral rays 16–18. Mouth inferior and horizontal. Lips fairly thin and plicate. Rear margin of lower lip forming an acute, V-shaped angle. Pharyngeal teeth 75–91 (Eastman 1977); R. E. Jenkins reported a tooth count of 101–104 (Ross 2001). Body depth going into SL 4–6 times. Intestine long. Medium and large breeding tubercles develop in males on head and on caudal and anal fins. Small tubercles present on anal fin of breeding females (McSwain and Gennings 1972). Maximum size 19.5 inches (495 mm) TL. Trautman (1957) listed maximum weight as about 3 pounds (1.4 kg). Largest Arkansas specimens we examined were a breeding male 17 inches (438.6 mm) TL taken from the lower Cossatot River by HWR, and a female 17.8 inches (453 mm) TL taken from Lake Fort Smith by Tim Smith in 2017 (examined by TMB). The Arkansas angling record from Caney Creek in 2015 is 1 pound, 15 ounces (0.88 kg) and 15.5 inches (394 mm) TL. The Arkansas unrestricted tackle record is 4 pounds, 6 ounces (2 kg) from Big Hill Creek in 2018 (the largest reported record for the species).

LIFE COLORS   Body olive to brown, turning to bronze or silver on sides. Scale bases along sides with a dark square or slightly rectangular spot forming longitudinal dashed stripes on sides except in very young specimens. The dark scale spot is never round, and it is often dashlike in juveniles (R. E. Jenkins, pers. comm.). Lateral spotting pattern is most evident on lower sides. Belly silver to white. Dorsal and caudal fins slate-colored, pectoral fins tinged with orange. Distal part of dorsal rays of juveniles darkly pigmented, often forming a black blotch. Lower fins typically white. Lower lobe of caudal fin has black edge. Breeding males with two dark lateral stripes separated by a pink band along midside.

SIMILAR SPECIES   *Minytrema melanops* is superficially similar to the redhorses (*Moxostoma*) in general shape; however, its sides are marked with a series of dark horizontal lines formed by dark square or rectangular spots at scale bases. Arkansas *Moxostoma* species lack distinct rows of dark spots or dashes on sides, but have dusky or dark crescents in the exposed, outer, half-moonlike portions (lunula) of scale pockets (R. E. Jenkins, pers. comm.). *Minytrema melanops* usually has about 16 scales around the caudal peduncle (vs. 12 in *Moxostoma* species), thicker lips (vs. thin lips), and lateral line incomplete or absent (vs. a complete lateral line). Juvenile Spotted Suckers (as well as adults) can be distinguished from juvenile *Moxostoma* by the two-chambered swim bladder (vs. three chambers in *Moxostoma*) and modally 16 caudal peduncle scales (vs. 12).

VARIATION AND TAXONOMY   *Minytrema melanops* was described by Rafinesque (1820a) as *Catostomus melanops* from the falls of the Ohio River at Louisville, Kentucky. It was placed in genus *Minytrema* by Jordan (1878b). The Spotted Sucker is the only species in the genus *Minytrema*. Morphological analyses (Nelson 1948; Miller 1959) and molecular analyses (Harris et al. 2002; Doosey et al. 2010) resolved the chubsuckers (*Erimyzon*) as its closest relatives. Tribal placement within subfamily Catostominae has been controversial. Smith (1992) placed *Minytrema* in the tribe Moxostomatini, but Harris et al. (2002) placed *Minytrema* and *Erimyzon* in the tribe Erimyzontini, a designation supported by Tan and Armbruster (2018).

DISTRIBUTION   *Minytrema melanops* occurs in the lower Great Lakes and Mississippi River basins from Pennsylvania to Minnesota, and south through Louisiana to the Gulf of Mexico; in Atlantic Slope drainages from the Cape Fear River, North Carolina through Georgia; and in Gulf Slope drainages from the Suwannee River, Florida, to the Brazos River, Texas, with an isolated population in the Colorado River drainage (Llano River), Texas. In Arkansas, the Spotted Sucker is widespread and common over the state, but very rare in the upper White River drainage. There is only one recent record from the upper White River drainage (Map 105), but a few older records exist (Map A4.103).

HABITAT AND BIOLOGY   The Spotted Sucker is primarily a lowland species preferring clear to slightly turbid, warm waters with submerged aquatic vegetation, lots of detritus, and soft substrates. It is found in small to moderate streams and in the backwaters of the large rivers in Arkansas, except for the Mississippi River, where there are surprisingly almost no records. It also occurs sporadically in streams in the Ouachitas and penetrates into the Ozarks, except for streams on the highest Ozark plateaus. It is also found in most Arkansas reservoirs, where it is the most widely distributed and often the most abundant catostomid. This species occurs typically in sluggish, somewhat turbid lowland ditches and quiet backwater areas away from notable current. In Kentucky, White and Haag (1977) observed adults in deep pools during the day, and over sandbars in early morning hours and at dusk; larvae and juveniles were found in shallow pools. Trautman (1957) considered it intolerant of high turbidity and industrial pollutants, but it is often found in moderately turbid waters in Arkansas. In northern areas of its range, it appears to be declining somewhat in numbers (P. W. Smith 1979; Gilbert and Burgess 1980b; Becker 1983).

Goldstein and Simon (1999) classified the Spotted Sucker as an invertivore. *Minytrema melanops* larvae (12–15 mm TL) first feed in midwater on small crustaceans and other zooplankton, but by 50 mm (2 inches) TL, the young shift more to feeding on the bottom. Adults are bottom foragers, predominantly consuming detritus, sand, zooplankton (cladocerans and copepods), diatoms, and benthic invertebrates such as midge larvae (White and Haag 1977). Other reported food items are ostracods, tendipedid larvae, ephemeropteran nymphs, amphipods, and filamentous algae. Its diet in the upper Mississippi River by percent volume consisted of copepods and ostracods (19.4%) and cladocerans (77.2%) (Bur 1976). Becker (1983) noted that most food organisms were smaller than 3 mm. Food items are obtained by sorting them from the bottom material sucked into the mouth. The finely spaced gill rakers are used to separate the food from the debris, which is then expelled from the mouth. White and Haag (1977) also observed seasonal changes in diet, with zooplankton consumed more in spring, summer, and fall, and organic detritus consumed more in winter.

In Oklahoma, the Spotted Sucker spawns in late April through May at 15–18°C (Jackson 1958). The spawning season in Alabama is from late February into April (Mettee et al. 1996). In Arkansas, *Minytrema melanops* spawns from early March to May. Three highly tuberculate males were taken on 9 March in Arkansas in Valley Creek, Hot Spring County (HSU 2545). An adult highly tuberculate male that was dying was collected in the Cossatot River on 25 March 1972 by HWR; the male was battered, bruised, and bleeding from fighting other males during a recent spawning event. Simon (1999b) classified this sucker as a nonguarder and open substrate spawner. Spawning behavior has been detailed by McSwain and Gennings (1972) and is similar to that of other suckers. In their Georgia study, spawning occurred in a riffle area 12–20 inches (0.3–0.5 m) deep over limestone rubble substrate at midday, with a flow rate of 1.4 m³/s and a surface velocity of 0.24 m/s. Spawning in Oklahoma also took place on shallow riffles having coarse substrates above large pools at water temperatures of 12–19°C (Jackson 1958). In Alabama, adults entered small stream tributaries of rivers and reservoirs where spawning occurred in long riffles having gravel and cobble substrates and moderate to swift current (Mettee et al. 1996). Males established loosely defined territories, which they defended against intruding males. When a female entered the riffle, two or more males swam back and forth over her, often prodding her with their snouts. The spawning act usually involved two males and one female. During spawning, the female settled to the bottom with a male on each side, all facing upstream. The males clasped the posterior half of the female's body between their posterior halves and vibrated vigorously. As they vibrated, the two rose to the surface and eggs and milt were shed. Most eggs drifted downstream below the spawning site. Fertilized eggs are demersal and adhesive, and hatching occurs in 108–156 hours at 61–68°F (16.1–20.0°C). Fertilized eggs prior to hatching measure 3.1 mm (White 1977). Larvae 7–8 days old formed a compact, integrated school when exposed to direct light but dispersed in a dark environment. Most juveniles move downstream when they reach 4–6 inches TL (Mettee et al. 1996). Individuals mature by the third year of life (age group 2) (Jackson 1958). Maximum life span is 6–8 years in Minnesota (Gilbert and Burgess 1980b), but southeastern fish may live only 5 years (Carlander 1969).

CONSERVATION STATUS   The Spotted Sucker is common and sometimes abundant in Arkansas waters, and its populations are secure.

Genus *Moxostoma* Rafinesque, 1820a— Redhorse Suckers

The redhorses and jumprock suckers are the most speciose of the catostomids with 22 described species. In Arkansas, there are 6 extant redhorse suckers: *M. anisurum*, *M. carinatum*, *M. duquesnei*, *M. erythrurum*, *M. pisolabrum*, and *M. poecilurum*. The head of a redhorse is convexly rounded between the eyes, with the eyes positioned near the midpoint of the head length, dorsal saddles and lateral blotches absent in adults, and a complete and well-developed lateral line. Scales in lateral line 37–51. Swim bladder with 3 chambers. Dorsal fin rays 12–16; anal fin rays 7; pectoral fin rays 15–19. It should be noted that species in the genus *Moxostoma* are often distinguished from one another by anatomical differences in the lips, and it is imperative that the mouth is closed when examining those features (but a strongly appressed or tight degree of mouth closure should be avoided, R. E. Jenkins, pers. comm.). Habitats of redhorse suckers in Arkansas range from big rivers to the swift-water areas of small streams. They are most abundant in small to medium-sized rivers. The adults migrate upstream into shallow gravel riffles and shoals for the spring spawning season. Robert E. Jenkins (pers. comm.) provided information on the sequence of initiation of spawning in

North Carolina and Virginia for four syntopic *Moxostoma* species that also occur in Arkansas. *Moxostoma anisurum* is the first to begin spawning, followed by *M. duquesnei*, then *M. erythrurum*, and finally *M. carinatum*.

According to Etnier and Starnes (1993), redhorse suckers are excellent food fish. Their firm, white, delicately flavored flesh compares favorably with the best of our gamefishes. Redhorse suckers are best when knife-scored and deep-fried in fat so the multitude of small bones soften, leaving a sweet-tasting morsel. Snagging of suckers in Ozark streams is a very popular spring sport, and Arkansas fishing regulations now include a daily limit of 20 suckers.

Moxostoma anisurum (Rafinesque)
Silver Redhorse

Figure 15.141. *Moxostoma anisurum*, Silver Redhorse. *Uland Thomas.*

Map 106. Distribution of *Moxostoma anisurum*, Silver Redhorse, 1988–2019.

CHARACTERS A deep-bodied redhorse with a short-based, convex to straight-edged dorsal fin, usually with 14–16 rays. Body depth going into SL 2.8 (adults) to 3.9 (young) times. Length of dorsal fin base about equal to the distance from the dorsal fin origin to the occiput (dorsal fin base usually shorter than in other sympatric redhorses). Lateral line complete, usually with 40–42 (38–46) scales. Scales around the caudal peduncle modally 12. Lower lip distinctly bilobed; posterior margin of lower lip forming an acute V-shaped angle, and lower lip usually distinctly narrowed just anterior to its junction with the upper lip (Fig. 15.124A) (the diagnostic thinning of the lower lip may not be evident if the mouth is appressed too tightly, R. E. Jenkins, pers. comm.). Lower lip surface semipapillose (any transections of ridges on the lower lip are shallow). Head large, its length going into SL 3.5–4.3 times. Swim bladder with 3 chambers. Anal rays 7. Caudal peduncle deep. Pharyngeal teeth numerous, bladelike, and lacking flattened grinding surfaces. Breeding males strongly tuberculate on caudal and anal fins with weaker tubercles scattered on pelvic fins. Females have weak tubercles on anal fin. Maximum size 28 inches TL (71 cm) and 8.25 pounds (3.75 kg) (Trautman 1957). The Arkansas angling record from the Spring River in 2016 weighed 3 pounds, 12 ounces (1.7 kg) and was 22 inches (558 mm) TL.

LIFE COLORS Body color usually silvery to grayish, but some specimens may be moderately gold, copper, or brassy overall (R. E. Jenkins, pers. comm.). Caudal and dorsal fins lack reddish color; caudal fin slate-colored. Pectoral, pelvic, and anal fins pale orange to orange red. Scales of back and upper sides without dark spots at bases, but usually with dusky to somewhat dark crescents (scale pockets). Young and small juveniles with a blotch-saddle pattern.

SIMILAR SPECIES All *Moxostoma* are similar, and *M. anisurum* is best distinguished from other species of Arkansas *Moxostoma* by the posterior margin of its lower lip forming an acute V-shaped angle (vs. rear margin of lower lip nearly straight or forming a slight to moderate angle), and by the subdivision of the lips into round or oblong papilla-like elements (vs. lower lips plicate). It further differs from other *Moxostoma* species in adults having the lower lip distinctly constricted just anterior to its junction with the upper lip (vs. lower lip not distinctly constricted). It is the only Arkansas *Moxostoma* with a convex dorsal fin margin, but adults often have a straight-margined dorsal fin. In addition, it usually has 14 or 15, occasionally 16 dorsal fin rays (vs. usually 12 or 13, rarely 14 dorsal rays in other *Moxostoma*).

VARIATION AND TAXONOMY *Moxostoma anisurum* was described by Rafinesque (1820a) as *Catostomus anisurus* from the Ohio River and its large tributaries. Jenkins (1970) studied variation and did not recognize subspecies, but considered the central and south Atlantic Slope form to be distinctive. The Atlantic Slope form is now recognized as a distinct species, *Moxostoma collapsum*, the Notchlip Redhorse (Page et al. 2013). Jenkins (1970) and Clements et al. (2012) presented evidence for a V-lip redhorse species group containing *M. anisurum*, *M. collapsum*, and *M. pappillosum*. Analyses of mitochondrial DNA recovered *M. anisurum* as sister to a *Moxostoma collapsum* clade (Harris et al. 2002; Doosey et al. 2010).

DISTRIBUTION *Moxostoma anisurum* occurs in the St. Lawrence–Great Lakes, Hudson Bay, and Mississippi River basins from Quebec to Alberta, Canada, and south to Georgia, Alabama, and Arkansas. The Silver Redhorse is rare in Arkansas. Only 12 records comprising 23 specimens from 5 rivers in the White River system (Buffalo, Current, Eleven Point, Strawberry, Black, and mainstem White rivers) were confirmed from the state (McAllister et al. 2009e) until 2 additional specimens were collected by HWR and Ken Shirley from the mainstem White River at Batesville, Independence County, on 12 March 2012. Another specimen was caught by a fisherman from the Spring River near Hardy in December 2016. It is currently documented from 18 localities in Arkansas (Map 106).

HABITAT AND BIOLOGY This medium-sized redhorse inhabits the larger, deeper, sluggish pools of moderate-sized streams to large rivers having relatively clear water over rocky or gravelly bottoms. In Illinois, it was found most commonly in deep, rather firm-bottomed pools that had undercut banks and tree roots protruding into the water (Smith 1979). In Alabama, it was most commonly found in the Tennessee River and in medium and large-sized streams having moderate to fast current and silt, sand, gravel, and bedrock substrates (Mettee et al. 1996). It avoids spring-fed streams with high gradients and excessively turbid waters, as well as the largest rivers. In the Des Moines River, Iowa, the adults were more common in slow water of deep pools (Meyer 1962). In Alabama, juveniles inhabited a reservoir until reaching sexual maturity, at which time they moved upstream into the Flint River (Hackney et al. 1971).

Foods in Iowa for age classes 1 and above consisted mainly of immature aquatic insects (primarily chironomids, ephemeropterans, and trichopterans) (Meyer 1962). In North Carolina, insects, small mollusks, and small crustaceans made up most of the diet (Gatz 1979). Jenkins and Burkhead (1994) reported that it also ingests micro-crustaceans, crayfish, algae, detritus, and even the fry of minnows. Other food items reported for this species are megalopterans, coleopterans, hemipterans, and a small amount of terrestrial insects. In Virginia, individuals were present in pools in the upper Roanoke River drainage but were not observed feeding in runs like other redhorse species (Jenkins and Burkhead 1994).

Moxostoma anisurum is one of the earliest-spawning species of *Moxostoma* (Jenkins 1970). Spawning in central Missouri occurs in early April (Pflieger 1997), in Alabama in late March and early April at water temperatures of 14–15°C (56–58°F) (Hackney et al. 1971), and in April and May in Wisconsin (Becker 1983). Spawning sites were characterized as shallow riffle areas with rocky and gravelly bottoms and moderate to swift currents (Hackney et al. 1971). In Iowa, spawning occurred in rather deep, clear riffles in the main channels of the Des Moines River (Meyer 1962). Males precede females to the spawning grounds. There is no overt defense of territory, but the males often nudge each other (R. E. Jenkins, pers. comm.). Tuberculate males and ripe females have been taken from the White River at Batesville, Arkansas, as early as 12 March (Robison et al. 2013). Fecundity of females (338–490 mm TL) in an Iowa population was about 15,000–36,000 eggs (Meyer 1962). Eggs hatch in 5–6 days at 17.8–21.0°C (64.0–69.8°F) (Kay et al. 1994). Fish (1932) described larvae. Sexual maturity is reached in the fifth or sixth year at total lengths of 507 mm (males) and 548 mm (females) in Alabama (Hackney et al. 1971). One age 0 *M. anisurum* was collected from the Strawberry River on 20 July 2007 (McAllister et al. 2009e). Maximum life span is 10–13 years, but R. E. Jenkins reported ages up to 26 years (Ross 2001). It is reportedly an excellent food fish (Hackney et al. 1971), but it must be thoroughly cooked because of numerous small intermuscular bones.

CONSERVATION STATUS   The White River drainage population of the Silver Redhorse in Arkansas is listed as Threatened and considered critically imperiled (S1) by the Nature Conservancy. The AGFC considers it a species of special concern, but we believe that it is threatened in Arkansas.

Moxostoma carinatum (Cope)
River Redhorse

CHARACTERS   A moderately heavy-bodied, terete redhorse possessing a short-based dorsal fin usually with 13 or 14 (12–15) rays and a complete lateral line usually with

Figure 15.142. *Moxostoma carinatum*, River Redhorse. *David A. Neely.*

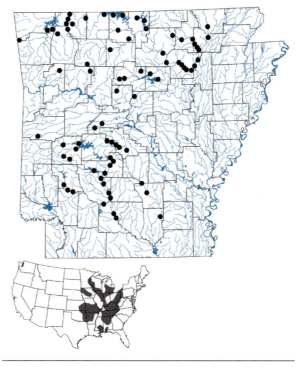

Map 107. Distribution of *Moxostoma carinatum*, River Redhorse, 1988–2019.

42–44 (41–46) scales. It is distinctive in having a very stout pharyngeal arch with large, molariform teeth, the crown of the lower teeth being rounded when first erupted but becoming beveled flat with wear (Fig. 15.123). Body depth going 3.5–4.5 times into SL. Mouth medium to large with thick plicate lips; upper lip without a pea-shaped swelling at middle; rear margin of lower lip forming an acute angle in juveniles that becomes a moderate to wide angle in adults (Fig. 15.124B). Head large, its length going into SL fewer than 4.5 times. Dorsal fin margin usually moderately to slightly concave, occasionally straight. Anal rays 7; pectoral rays usually 16–19; pelvic rays usually 9. Scales around caudal peduncle usually 12 or 13. Swim bladder with 3 chambers. Breeding male with medium to large tubercles on snout, cheek, and elsewhere on head (Jenkins and Burkhead 1994). Body tubercles, if present, minute and usually only on upper half of body. Minute tubercles may be present on unpaired fins. Females lack breeding tubercles. One of the larger redhorses, with maximum size about 30 inches (77 cm) (Becker 1983; Page and Burr 2011). The Arkansas angling record from Lake Ouachita in 2013 is 9 pounds (4.1 kg). A specimen weighing 17.1 pounds (7.8 kg) was caught by snagging in Missouri in 1966 (Ross 2001).

LIFE COLORS Body color silver to brassy to bronze on dorsum and upper sides. Scales of dorsum and upper sides with a crescent-shaped dark mark at base. Venter white. Caudal fin bright red distally; dorsal and anal fins a paler red distally; paired fins orange to orange red. Red coloration often obvious in older juveniles and females as well as males. Coloration becomes more intense in breeding males, and a dark stripe develops along the side.

SIMILAR SPECIES The River Redhorse is distinctive among *Moxostoma* species in having a very thick pharyngeal arch with large molarlike lower teeth. According to Jenkins and Burkhead (1994), the stoutness of the arch can be determined without dissection by probing with a fine-pointed tool into the gill chamber anteromediad from the inner edge of the cleithrum. The lateral edge of the pharyngeal arch in *M. carinatum* is very close to the cleithrum. In other species of *Moxostoma*, the lateral edge of the arch is well inward from the anterior edge of the cleithrum. Small *M. carinatum* can be distinguished from small *M. erythrurum* and *M. duquesnei* in having a tiny black spot at the base of each gill raker on the middle third of the first gill arch (Jenkins and Burkhead 1994). Juveniles smaller than 40 mm TL and adult *M. carinatum* lack this spot which *M. erythrurum* and *M. duquesnei* also usually lack. The combination of a moderate-sized head, entirely plicate lips, and dark scale bases further distinguishes *M. carinatum* from other *Moxostoma* species. It can be distinguished from *Catostomus commersonii* by its plicate lips (vs. papillose lips) and fewer than 50 lateral line scales (vs. more than 55). It differs from *Minytrema melanops* and *Erimyzon* species in having a complete lateral line (vs. incomplete or absent) and swim bladder with 3 chambers (vs. 2 chambers).

VARIATION AND TAXONOMY *Moxostoma carinatum* was described by Cope (1870) as *Placopharynx carinatus* from the Wabash River at Lafayette, Tippecanoe County, Indiana. Bailey (1951) and subsequent authors placed the River Redhorse in the subgenus *Moxostoma*. Jenkins (1970) studied variation and did not recognize subspecies. *Moxostoma carinatum* was considered most closely related to *M. erythrurum* (G. R. Smith 1992) or to the rediscovered "true" Robust Redhorse, *M. robustum* (Jenkins and Burkhead 1994). Harris et al. (2002) resolved this species as polyphyletic based on analysis of the mitochondrial

gene, cytochrome *b*. Doosey et al. (2010) considered *M. carinatum* sister to the undescribed *M.* sp. *cf. macrolepidotum* (Sicklefin Redhorse).

DISTRIBUTION *Moxostoma carinatum* occurs in the St. Lawrence–Great Lakes and Mississippi River basins from Quebec, Canada, to central Minnesota and western Iowa, south through eastern Oklahoma and Alabama, and in Gulf Slope drainages from the Escambia River, Florida, to the Pearl River, Mississippi and Louisiana. In Arkansas, it is widely distributed in the Ozark region (e.g., Spring River; McAllister et al. 2010b), at scattered localities in the Ouachita Mountains, and in some Arkansas River tributaries in the western half of the state (Map 107). A 1975 record of *M. carinatum* from the Petit Jean River at Danville (which had previously been misidentified as *M. erythrurum*) represents the first confirmed record from that Arkansas River tributary (R. E. Jenkins, pers. comm.). It is probably most abundant in the middle White River.

HABITAT AND BIOLOGY The River Redhorse generally inhabits pool and medium to swift water habitats of typically clear, medium to large upland streams and rivers with some gravel or rock bottoms. It is absent from the largest rivers and smallest creeks in Arkansas. It does best in streams with permanent flow and clean substrates and is seldom encountered in deeper waters with silt or sand bottoms. Specimens also have been collected from Beaver, Bull Shoals, and Norfork reservoirs (White River drainage) near stream mouths over rocky substrates and in Lake Ouachita. Adults avoid shallow portions of pools during daylight hours, but the young and small juveniles are often found in such habitats (Jenkins and Burkhead 1994). In the Kankakee River, Illinois, River Redhorse predominantly occupied deep runs (>1.5 m) with moderate to swift currents (>0.4 m/s) over gravel, cobble, or boulder substrates (Butler and Wahl 2017). Habitat use differed among seasons, with River Redhorse occupying faster current velocities during winter and spring than during summer and fall, and using deeper water over smaller substrates in winter than during summer. Butler and Wahl (2017) also found that movement patterns of River Redhorse varied across seasons, with the largest ranges and highest displacements occurring during spring, and the smallest ranges and lowest movement rates occurring during summer. It is more sensitive to pollution, turbidity, and intermittent flow than most other Arkansas redhorses.

Jenkins and Burkhead (1994) considered *M. carinatum* one of the most trophically divergent species of *Moxostoma*. The large molariform pharyngeal teeth adapt the River Redhorse to feed on snails and clams by crushing the shells, which pass through the digestive system and are ejected along with feces. Insects are also often taken in at least moderate numbers along with crayfish (Jenkins and Burkhead 1994). Ross (2001) reported *M. carinatum* has shifted to eating the widespread, nonnative Asian clam *Corbicula* where the native mussel fauna has declined. In the Cahaba River, Alabama, it fed largely on *Corbicula*, and other foods consumed in lesser amounts included ephemeropteran nymphs and larval chironomids and trichopterans (Hackney et al. 1968). Jenkins and Burkhead (1994) suggested that declines in River Redhorse populations in some parts of its range may, in part, be associated with documented declines in the molluscan fauna in many North American river basins.

Moxostoma carinatum assembles in spring spawning aggregations in swiftly flowing gravel riffles (Hackney et al. 1968). Spawning typically occurs in Alabama in mid-April in riffles and runs over a gravel to small cobble substrate when water temperatures are 22.2–24.4°C (Boschung and Mayden 2004). In Tennessee, spawning occurred from mid-April to early May, and this is presumably the peak spawning period in Arkansas. In the Ozarks, spawning was reported to be slightly earlier (beginning in early April) than that of the Black Redhorse, with spawning activities preceded by upstream migration (Pflieger 1997); however, in the Ohio River basin, *Moxostoma carinatum* is the last of the redhorses to begin spawning (R. E. Jenkins, pers. comm.). In the Kankakee River, Illinois, substantial upstream movements occurred in March, and River Redhorse were often found in areas of high current velocity ($\bar{x}$ = 0.75 m/s) during the spring spawning season (Butler and Wahl 2017). In that Illinois study, spawning aggregations were observed in a shallow (46–48 cm deep) gravel/cobble riffle with water velocities of 0.48–0.72 cm/s. After spawning, River Redhorse returned to a boulder-filled run and exhibited very little subsequent movement. Adult tuberculate males of *M. carinatum* were taken by HWR on 12 March 2012 in the White River at Batesville, Arkansas. Certain unique aspects of the River Redhorse's nesting and spawning behavior were given by R. E. Jenkins (*in* Jenkins and Burkhead 1994) from a study by P. A. Hackney (Hackney et al. 1968), and Hackney's unpublished manuscript of the Cahaba River, Alabama population; however, based on Jenkins's subsequent behavioral study of *M. carinatum* in North Carolina and Virginia, and videos from Ohio and the Cahaba River, *M. carinatum* does not purposely excavate nests nor exhibit overt courtship. Still, the two-male, one-female spawning trios of *M. carinatum* are distinctive in the family because of the relatively long duration of bodily alignment leading to gamete release (R. E. Jenkins, pers. comm.). The spawning act is similar to that

of the Golden Redhorse. One female joins two strongly territorial males, the three fish settle to or very near the substrate with the female pressed between the males, and the trio undulates, then quivers, and eggs are released, fertilized, and mostly buried in the gravel (Ross 2001; R. E. Jenkins, pers. comm.). The nest depression in the gravel substrate is created only by the forceful quivering of the repeated spawning acts by the trio. The demersal, nonadhesive fertilized eggs are large, ranging 3–4 mm in diameter, and hatching occurs in 3–4 days at 22°C. Fecundity ranges from 6,078 to 20,085 eggs in females 455–561 mm TL (Hackney et al. 1968). Based on age estimates from Wisconsin, Becker (1983) reported the smallest specimens of River Redhorse measuring 71–115 mm TL were about 1 year old. Retzer and Kowalik (2002) reported ages of 11–12 years in Illinois. In the James River in Missouri, Beckman and Hutson (2012) found the maximum age of 15 years, with ages 3 and 4 individuals composing 47% of the population. They validated opercles and otoliths as reliable estimators of age in the River Redhorse. This largest of our redhorses has a maximum reported life span in Ontario of 26 years (R. E. Jenkins, pers. comm.).

CONSERVATION STATUS   Although originally having a rather large range in upland areas of the Mississippi Basin, *Moxostoma carinatum* has declined substantially in range and abundance during the last century (Jenkins 1980c). It is recognized as imperiled in 12 U.S. states and 2 Canadian provinces (Butler and Wahl 2017). R. E. Jenkins (pers. comm.) considers siltation of spawning areas a likely major contributor to population declines. Boschung and Mayden (2004) attributed much of its decline to its dependence on mollusks for food, and in turn to the dependence of mollusks on a silt-free habitat. Additionally, where populations within a river system have been fragmented by impoundments, the isolated populations are especially vulnerable to catastrophic events. Its numbers in Arkansas have also been reduced by dams and reservoirs, but additional distributional records for this species were reported by McAllister et al. (2010a) and Robison et al. (2011a). In Arkansas, the River Redhorse is considered an uncommon species but does not appear to currently require a status of special concern.

Moxostoma duquesnei (Lesueur)
Black Redhorse

CHARACTERS   An elongate, terete sucker with a short-based, slightly concave dorsal fin usually with 12–14 rays (rarely 11 or 15), a complete lateral line usually with 44–47

Figure 15.143. *Moxostoma duquesnei*, Black Redhorse. *David A. Neely.*

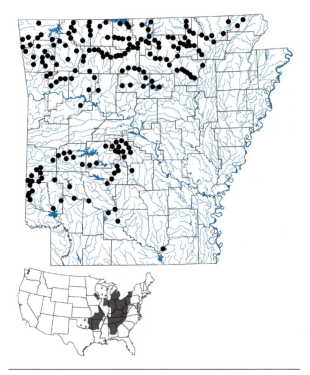

Map 108. Distribution of *Moxostoma duquesnei*, Black Redhorse, 1988–2019.

(43–51) scales, and a moderate-sized, horizontal mouth with moderately thick, deeply plicate lips (Fig. 15.124D). Rear margin of lower lip forming a moderate to very obtuse angle of usually 120–160 degrees. Snout truncate, pointed, or rounded. Caudal peduncle moderate in juveniles but noticeably slender in adults, its depth going into distance from caudal base to anal origin 2 or more times. Scales around caudal peduncle 12–14. Breast scales of adults are partly embedded and are noticeably smaller than the adjacent belly scales. Head length going into SL 3.5–4.6 times. Body depth going into SL 3.9–4.8 times. Anal rays 7; pectoral rays 16–18; pelvic fin rays usually 10, often 9 or 11. Swim bladder with 3 chambers. Pharyngeal teeth numerous (about 65 per arch), bladelike and fragile, without flattened grinding surfaces. Breeding males with tiny tubercles on head, and with tubercles on all fins except the

dorsal fin, the largest tubercles (medium-sized) occurring on anal fin rays and on rays of the lower lobe of the caudal fin. Females may have minute tubercles on head and some body scales. Maximum length usually 20 inches (510 mm), but a Missouri specimen was 25.9 inches (658 mm) TL and weighed 7 pounds (3.2 kg) (Pflieger 1975). Arkansas specimens are typically smaller than 15 inches (381 mm) TL and 1 pound (0.5 kg) in weight.

LIFE COLORS   Upper body dusky; body color gray, olive, or brown above, grading to mostly brassy or silvery on sides, and white below. Scales on back and upper sides without dark spots at base, but uniformly dusky. Dorsal and caudal fins slate-colored. Lower fins may have an orange tint. Young are olivaceous dorsally, grading to silver laterally.

SIMILAR SPECIES   The Black Redhorse can be distinguished as adults from other members of this family by differences in tail color, lip morphology, dorsal fin shape, and lateral line scale count (Bunt et al. 2013). It is most similar to, and often found with, *Moxostoma erythrurum* but differs from that species in having a more slender body, more lateral line scales, usually 44–48 (vs. 39–45), a shallower caudal peduncle, its least depth going into its length 2 or more times (vs. fewer than 2 times), and usually 10 pelvic rays (vs. usually 9). Breast scales of *M. duquesnei* are much smaller than the adjacent belly scales and are partially embedded, while those in *M. erythrurum* are conspicuous, exposed, and only slightly smaller than the adjacent belly scales. In smaller specimens of the Black Redhorse, the minimum caudal peduncle depth is less than two-thirds of peduncle length, whereas in smaller specimens of the Golden Redhorse, the minimum peduncle depth is more than two-thirds its length (Etnier and Starnes 1993). For most specimens greater than 50 mm SL, the anterior dorsolateral scales are fairly uniformly dusky or dark in *M. duquesnei*, but those scales in *M. erythrurum* have the posterior portion obviously darker than the median portion, the two pigment shades somewhat demarcated (Jenkins and Burkhead 1994). Nuptial males of these two species are easily differentiated by tubercle pattern differences. Black Redhorse males lack the large, conspicuous head tubercles, whereas in nuptial Golden Redhorse males, the large and prominent tubercles on the anterior portion of the head are quite conspicuous.

VARIATION AND TAXONOMY   *Moxostoma duquesnei* was described by Lesueur (1817b) as *Catostomus duquesnei* from the Ohio River at Fort Duquesne (Pittsburgh), Pennsylvania. Geographic variation occurs, and Jenkins (1970) divided the species into two races rather than subspecies: one in the Mobile Bay drainage, and the other race occurring elsewhere throughout its range. Smith (1992) considered *M. duquesnei* sister to the *M. macrolepidotum* clade with *M. poecilurum* and *M. lacerum*. Harris et al. (2002) identified two clades of *M. duquesnei* that formed a group sister to *M. valenciennesi*. Doosey et al. (2010) identified *M. duquesnei* as sister to a clade composed of *M. congestum* and *M. poecilurum*.

DISTRIBUTION *Moxostoma duquesnei* occurs in the lower Great Lakes and Mississippi River basins from southern Ontario, Canada, and New York to southeastern Minnesota, and south through eastern Oklahoma and Alabama, and in Gulf Slope drainages of upland areas of the Mobile Bay basin. In Arkansas, the Black Redhorse is one of the most common suckers in Ozark streams, and it also occurs in streams and rivers of the Ouachita Uplands (Map 108). It is found in some tributaries of the Arkansas River in the western half of the state, and it is occasionally found in reservoirs.

HABITAT AND BIOLOGY The Black Redhorse inhabits clear creeks to medium-sized rivers with gravelly or rocky bottoms and permanent flow and is generally absent from headwater streams. It is fairly sensitive to siltation and other forms of environmental disturbance (Jenkins and Burkhead 1994). Its abundance in Missouri streams was correlated with volume of flow (Bowman 1970). Streams with average annual discharges of 500–600 cfs (14–17 m³/s) had the largest populations of Black Redhorse. In summer, it was found mainly in pools, but by October and November it had moved into deeper wintering areas. When found syntopically with the Golden Redhorse (as it often is in Arkansas), the Black Redhorse tends to predominate in short, rocky pools with good current, whereas the Golden Redhorse prefers sluggish large pools and backwaters without noticeable current (Pflieger 1997). Bunt et al. (2013) reported larval and juvenile Black Redhorse occupied riffles, runs, pools, and backwater areas but showed a strong preference for runs. Although Bowman (1970) believed that it does not adapt well to reservoirs because of its low tolerance of siltation, Rainwater and Houser (1982) reported that it occurs in modest numbers in Beaver Reservoir. Based on AGFC rotenone population samples, it is commonly found in other Arkansas reservoirs in the Ozarks (Bull Shoals, Norfork, and Greers Ferry) in areas having rocky substrates.

Adults forage in the early evening in schools of 15–20 fish in pools or riffles ingesting primarily crustaceans and aquatic insect larvae (mainly Diptera) as well as small mollusks, aquatic oligochaetes, algae, and detritus (Bowman 1970). They suck in bottom material and are able to separate desirable food items from silt and other unsuitable

foods that are ejected. Bowman (1970) provided evidence that *M. duquesnei* does not compete directly with juvenile Smallmouth Bass for food. Feeding often occurs in pools under low light conditions, but Jenkins and Burkhead (1994) also observed adults feeding in a run. In Missouri, juveniles smaller than 65 mm SL have a more terminal mouth and fed mainly on phytoplankton, but also ingested cladocerans, copepods, and rotifers (Bowman 1970). As growth occurs, feeding activities shift to riffles where they mainly become benthic invertivores.

Spawning in Missouri takes place in late April or early May when breeding adults congregate in pools adjacent to the spawning riffles with water temperatures ranging from 56°F to 72°F (13.3–22.2°C). The start of the spawning season in Missouri varies as much as 2 weeks between years, but actual spawning usually lasts only about 4 days at a given locality (Bowman 1970). In Arkansas, we have taken adult tuberculate males as early as 12 March in the White River at Batesville, and a ripe female 382 mm TL was found at that locality on 5 April (Tumlison et al. 2017). Becker (1983) noted the narrow habitat requirements for spawning, which include proper substrate, water flow, and stream size. When ready to spawn, males establish and defend small territories in riffles. Black Redhorses selected spawning habitats with precise concentrations of fine and coarse gravel and sand at depths of 6–24 inches (152–6210 mm) (Bowman 1970). Spawning substrate was composed of approximately 70% fine rubble, 10% coarse rubble, and 20% sand and gravel. In Illinois, males exhibited no agonistic behavior during spawning, and spawning occurred in water that was slightly deeper, much swifter, and over coarser substrate than that used by the Golden Redhorse (Kwak and Skelly 1992). Spawning occurs both day and night after females drift into the spawning area (already occupied by males) from the pool above. Generally, two territorial males position themselves alongside a female. The males clasp the posterior half of the female's body between theirs and vibrate rapidly. At this time, the eggs are released and fertilized. The female returns to the upstream pool after spawning, and the males return to their territories to spawn again with different females. The fertilized eggs (demersal and nonadhesive) settle in the gravel of the riffle and are abandoned by the parents. The eggs hatch in 4 or 5 days at 22°C. Fecundity ranged from 6,078 to 23,085 eggs in females 455–561 mm TL (Hackney et al. 1968). Sexual maturity is reached in 2–5 years, and maximum life span is about 10 years.

CONSERVATION STATUS The Black Redhorse is common to abundant in Arkansas, and its populations are secure.

Moxostoma erythrurum (Rafinesque)
Golden Redhorse

Figure 15.144. *Moxostoma erythrurum*, Golden Redhorse. *David A. Neely.*

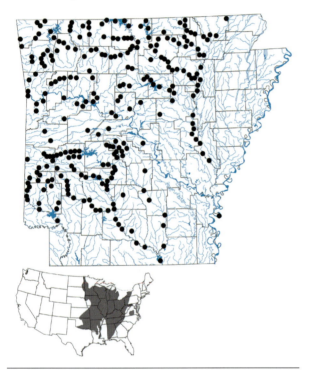

Map 109. Distribution of *Moxostoma erythrurum*, Golden Redhorse, 1988–2019.

CHARACTERS   A stout, terete sucker with a short-based, concave or straight-edged dorsal fin usually with 12–14 (10–15) rays, a complete lateral line usually with 39–42 scales, and usually 9 pelvic rays (Fig. 15.144). Snout blunt or slightly rounded. Mouth large, ventral, and horizontal. Upper and lower lips plicate. Rear margin of lower lip forming a moderately obtuse or slightly V-shaped angle (about 80–130°) (Fig. 15.124E). Scales on anterior upper body of preserved specimens greater than 50 mm SL are often bicolored, with the scale median distinctly paler than the dusky margin. Breast and belly scales are of approximately equal size, or breast scales only slightly smaller than adjacent belly scales. Anal rays 7; pectoral rays 16–18. Swim bladder with

3 chambers. Pharyngeal teeth numerous (53–79), bladelike, and lacking flattened grinding surfaces (Fig. A3.7). Caudal peduncle moderate in depth, its least depth going into its length fewer than 2 times. Scales around caudal peduncle usually 12 (11–14). Head moderately large; head length going into SL 3.6–4.1 times. Body depth going into SL 3.4–4.4 times. Breeding males with large tubercles on snout to operculum, sometimes also on top and bottom of head, and on all fins except dorsal fin. Small to tiny tubercles present elsewhere on the body. Females usually lack tubercles. The Arkansas angling record from the Spring River in 2011 is 1 pound, 2 ounces (0.5 kg). Maximum length about 30.5 inches (780 mm) (Page and Burr 2011), and weight about 11 pounds (5 kg).

LIFE COLORS   Body color olivaceous to coppery above, silvery brown to golden or brassy on sides and white below. Scales of sides and back usually have pigment concentrated on scale margins and without dark spots at scale bases. Caudal fin olive to slate-colored. Paired fins usually orangish. Dorsal fin blackish. Young-of-the-year with 3 diffuse, dusky saddle bands. Dorsal and caudal fins of young often tinted red.

SIMILAR SPECIES   The Golden Redhorse is most similar to the Black Redhorse and often occurs with that species in Arkansas. It differs from *M. duquesnei* in having 9 pelvic rays (vs. usually 10), and 39–42 lateral line scales (vs. 44–47). It is usually slightly more robust and has a shorter and deeper caudal peduncle, with depth:length ratio of the caudal peduncle typically greater than 2:3 (vs. depth:length ratio usually less than 2:3). In addition, breast scales of *M. erythrurum* are conspicuous, exposed, and only slightly smaller than the adjacent belly scales (vs. breast scales are noticeably smaller than the adjacent belly scales and are partially embedded).

VARIATION AND TAXONOMY   *Moxostoma erythrurum* was described by Rafinesque (1818b) as *Catostomus erythrurus* from the Ohio River. Jenkins (1970) studied variation and divided the species into two races rather than subspecies. One race occurs in the Talapoosa River system of the Mobile Bay drainage, and the other race occurs elsewhere. *Moxostoma erythrurum* is allied with several species of redhorses, including *M. poecilurum* (Buth 1978). Mitochondrial DNA analyses by Harris et al. (2002) and Doosey et al. (2010) did not agree on relationships of *M. erythrurum* with other *Moxostoma* species. Based on anatomy, coloration, and reproductive behavior, Clements et al. (2012) considered *M. erythrurum* most closely related to the undescribed *Moxostoma* sp., Carolina Redhorse.

DISTRIBUTION   *Moxostoma erythrurum* occurs in the Great Lakes, Hudson Bay, and Mississippi River basins from southern Ontario, Canada, and New York to Manitoba, Canada, and south through Oklahoma and Alabama, with an isolated population in southwestern Mississippi; Atlantic Slope drainages from the Potomac River, Pennsylvania, to the Roanoke River, North Carolina; and in the Gulf Slope drainage of Mobile Bay in Georgia, Alabama, and southeastern Tennessee. In Arkansas, the Golden Redhorse is one of the most commonly encountered redhorse suckers. It is common in upland streams in the Ozark and Ouachita mountains and is frequently found in upland reservoirs (Map 109). It is also common in Arkansas River tributaries in the western half of the state and occurs in the Ouachita and Saline rivers through the Coastal Plain. There is one reported post-1988 record from the Mississippi River (Lake Whittington, 1992) in Desha County, and a 1982 record from that river in Chicot County (R. E. Jenkins, pers. comm.), but Jenkins has not verified the identities of those specimens.

HABITAT AND BIOLOGY   The Golden Redhorse is found in a wider variety of habitats in Arkansas than any other *Moxostoma* species and is the most common and abundant redhorse in the upland regions of the state. It typically lives in clear, gravel- and cobble-bottomed streams like the Black Redhorse, but it seems to prefer slightly warmer water with less current. Specifically, it inhabits pools and slow raceways. It is also found in smaller creeks and is more tolerant of turbidity than any other Arkansas *Moxostoma*. Juveniles are often found in shallow backwater areas and are usually the only suckers found in intermittent streams in late summer. In Wisconsin, it occurred in medium to large rivers where it preferred pools in river bends with undercut banks and fallen trees where aquatic vegetation was almost nonexistent (Becker 1983). In Ohio, the Golden Redhorse ascends smaller streams in the spring, moves downstream during the summer and fall, and winters in the larger streams (Trautman 1957). The Golden Redhorse is commonly found in upland reservoirs in Arkansas, and it became the dominant redhorse in Beaver Reservoir shortly after impoundment (Rainwater and Houser 1982).

In Arkansas it feeds on bottom detritus, algae, insect larvae, and small mollusks (HWR, pers. obs.). Tumlison et al. (2016) reported aquatic mites (*Lebertia* sp.) in the gut of a Golden Redhorse specimen from Crooked Creek, Marion County. Jenkins (1980d) listed a similar diet composition with main food items including small mollusks, microcrustaceans, insects, detritus, and algae. Feeding apparently occurs during the day and at night, but the most intense feeding activity occurred at dawn and dusk in Ohio (Smith 1977). In the early life history stages, feeding

occurs more in the water column. Larvae first feed mainly on small crustaceans, but at between 2 and 4 inches (50–102 mm) TL, a shift to a more benthic pattern occurs and dominant foods in an Iowa study were immature midges, mayflies, and caddisflies (Meyer 1962). Foods in Wisconsin included trichopterans, Tendipedidae, ephemeropterans, Sphaeridae, copepods, coleopterans, hemipterans, and algae (Becker 1983).

Meyer (1962) reported that the Golden Redhorse spawns in Iowa at water temperatures between 15°C and 21°C (59–70°F). In April and May, adults congregate to spawn in smaller and higher-gradient streams over gravel shoals and shallow riffles in groups of from 3 to 5 up to 50 (R. E. Jenkins, pers. comm.). In Missouri, spawning usually begins in mid-April and sometimes continues into early June (Pflieger 1997). HWR collected adult tuberculate males of *M. erythrurum* as early as 12 March 2012 in the White River at Batesville, Arkansas, and observed spawning of *M. erythrurum* in a chute over a gravel substrate in the West Fork of the White River on 16 April 1992. Repeated localized spawning produces clear oviposition sites similar to nests (R. E. Jenkins, pers. comm.). Simon (1999b) classified this redhorse as a nonguarder, open substratum spawner and lithophil—rock and gravel spawner with benthic larvae that hide beneath stones. Kwak and Skelly (1992) found that the Golden Redhorse and Black Redhorse spawned simultaneously in Illinois, and they provided evidence that differences in spawning habitat and behavior served as reproductive isolating mechanisms between the two species. In that study, *M. erythrurum* spawned over finer substrates (sand and gravel) and in slower currents than *M. duquesnei*. Males of Golden Redhorse were also more territorial than those of Black Redhorse. Males defend territories on the spawning grounds, and females move into the spawning sites only briefly to spawn (Fig. 15.145). Most

Figure 15.145. Nuptial male *Moxostoma erythrurum* displaying over a spawning site in the Verdigris River, Chase County, Kansas. *Garold W. Sneegas.*

matings involve two males, one on either side of a female. The spawning trio vibrates, releasing eggs and sperm. Some of the fertilized eggs (which are demersal and nonadhesive) may be buried in the substrate by this activity, but others are swept downstream by the current (Kwak and Skelly 1992). Fecundity in Iowa was reported as 6,100–23,350 eggs in females ranging in TL from 292 to 399 mm (Meyer 1962). McAllister et al. (2009e) reported young-of-the-year Golden Redhorse (38.1–61.5 mm TL) from the Strawberry River, Arkansas, collected on 20 July 2007. Maximum life span is about 11 years (Jenkins and Burkhead 1994).

CONSERVATION STATUS   The Golden Redhorse is widespread and abundant in Arkansas, and its populations are secure.

Moxostoma lacerum (Jordan and Brayton)
Harelip Sucker

Figure 15.146. *Moxostoma lacerum*, Harelip Sucker. © *Joseph R. Tomelleri.*

CHARACTERS   Because this catostomid is extinct, the information for this species account is derived mainly from the detailed study of all 33 extant museum specimens by the catostomid authority, Robert E. Jenkins (1970, 1980b, 1994). A small sucker possessing a short head (head length going 4.1–5.1 times into SL), a short-based, slightly concave dorsal fin with 11 or 12 rays, and a complete lateral line with 42–46 scales. Mouth unique, with upper lip nonprotractile (frenum absent) and the lower lip fully divided into two separate lobes. Inner surfaces of lips finely plicate, finely papillate, or having both plicae and papillae. Snout well rounded in juveniles, proclivous to tip in adult. Eyes lateral and moderate to large in size. Pharyngeal arch very small and frail, the teeth comblike and very delicate. Swim bladder 4-chambered (fourth chamber very small). Predorsal scale rows 15–18. Body circumferential scales 31–34. Caudal peduncle scale rows 12. Anal rays 7. Pectoral fin rays 15–17.

Map 110. Distribution of *Moxostoma lacerum*, Harelip Sucker. The single 1884 locality record for *M. lacerum* in Arkansas is shown. Inset shows former range of this extinct species.

Pelvic fin rays 9. Caudal fin rays 18. Vertebrae 41–45 (Jenkins 1970). Maximum size of existing specimens is a gravid female 12.3 inches (313 mm) SL, or about 15.4 inches (390 mm) TL (Jenkins 1970). The Harelip Sucker was reported to reach 18 inches (400 mm) TL.

LIFE COLORS   Body with olivaceous to dark bluish-brown dorsum, grading to silver or white on sides and belly. Dorsal fin dusky and distally edged in black. Lower fins slightly orange and remaining fins are creamy to dusky (Jordan and Brayton 1877). Juveniles had 4 lateral blotches and 3 saddles (Jenkins and Burkhead 1994).

SIMILAR SPECIES   The Harelip Sucker is similar to other *Moxostoma* in body form and probably colored like the Black Redhorse and the Golden Redhorse; however, the lower lip structure of the Harelip Sucker is unique among all North American fishes (Etnier and Starnes 1993). The lower lip of *M. lacerum* is completely divided into two separate lobes (vs. lower lip not divided into two separate lobes in all other catostomids).

VARIATION AND TAXONOMY   *Moxostoma lacerum* was first collected in 1859 but was not formally described until 1877 when Jordan and Brayton (1877) described it as *Lagochila lacera* from the Chickamauga River, Georgia. Originally, *L. lacera* was compared with and considered related to *Moxostoma* by Jordan and Brayton (1877). Jenkins (1970) continued to recognize the genus *Lagochila* and hypothesized that the closest relatives of *Lagochila* were members of the subgenus *Moxostoma* of genus *Moxostoma*. G. R. Smith (1992) lumped the Harelip Sucker into *Moxostoma*, identifying *M. lacerum* as the sister species of *M. poecilurum,* in a monophyletic shortheaded redhorse group with *M. duquesnei* and *M. macrolepidotum*. R. E. Jenkins (pers. comm.) considers the sister species of *M. poecilurum* to be the Apalachicola Redhorse (*Moxostoma* sp. *cf. poecilurum*), and the phylogenetic position of the Harelip Sucker remains to be determined. However, Jenkins and Burkhead (1994) pointed out that if one accepts Smith's hypotheses that *Lagochila* belongs in the short-headed group and that the group stemmed transitionally within the genealogy of the subgenus *Moxostoma*, then retaining *Lagochila* as a monotypic genus renders *Moxostoma* an unnatural (paraphyletic) group. Harris et al. (2002) could not use molecular traits to examine the phylogenetic position of *M. lacerum* but found no evidence for the *M. poecilurum* plus Apalachicola Redhorse relationship. Fink and Humphries (2010) provided a morphological description of the Harelip Sucker based on high resolution x-ray computed tomography. We herein follow Page et al. (2013) in placing the Harelip Sucker in genus *Moxostoma*, but R. E. Jenkins (pers. comm.) continues to support its placement in *Lagochila*.

DISTRIBUTION   *Moxostoma lacerum* was last collected in 1893 in the Maumee River drainage of Lake Erie and is unanimously regarded as extinct. Etnier and Starnes (1993) reported that *M. lacerum* was collected at about 20 localities between 1859 and 1893. Records are widely scattered from the lower and middle Ohio River basin (including the Tennessee River) and the Maumee River system of western Lake Erie drainage east of the Mississippi River, and the upper White River drainage of Arkansas, west of the Mississippi River (Jenkins 1980b). In Arkansas, the Harelip Sucker was collected on a single date, 31 August 1884, by Jordan and Gilbert in collections from the White River drainage near Eureka Springs, Arkansas, in Carroll County (Map 110). One voucher specimen from those collections from the White River near Eureka Springs was deposited in the National Museum of Natural History of the Smithsonian Institution (USNM 36331.5006709). The only remaining trace of that USNM specimen is a radiograph. Another voucher specimen, 128 mm SL, from the 31 August 1884 collections was deposited in the University of Michigan Museum of Zoology (UMMZ 177430). The locality information for that UMMZ specimen was ambiguously given as "White River above the

narrows; Kings River, east of Eureka Springs; and brook trib. to White River." Therefore, its precise locality of collection is not known, and it is possible that *M. lacerum* was collected from all three listed areas near Eureka Springs, since Jordan and Gilbert (1886) considered it "not rare" on the date of collection. Black (1940) listed the 1884 records as from the White River. Seth Meek did not collect this species during his 1889–1894 survey of the White River, even though he seined in the same areas where Jordan and Gilbert had found it to be "not rare" in 1884. According to Black (1940), Meek apparently did not realize that by the time of his collections the Harelip Sucker was probably already rare or extinct in Arkansas. Meek (1894a) listed the Jordan and Gilbert record without comment. R. E. Jenkins (pers. comm.) regards the 1940 statement by Black (and also Black 1949) as possibly the first statement of its extinction. Black (1949) mentioned that he had made considerable effort to find the Harelip Sucker in the Arkansas Ozarks without success.

HABITAT AND BIOLOGY   As might be imagined, there is little information on the biology of the Harelip Sucker as only about 33 specimens, all juveniles except for one adult, are known. Jenkins (1970, 1980b, 1994) examined all extant specimens and presented pertinent information from which we are able to provide the following natural history information. It seems the Harelip Sucker normally occurred in clear creeks to medium rivers 10–70 m wide having moderate gradient and gravel or rock bottoms (Jenkins 1994). Jenkins (1994) reported a weak indication that small Harelip Suckers were taken most frequently from the smaller tributaries, while adults were from larger streams. It may have usually occupied areas of gentle current, as indicated by the small size of the cerebellum, a part of the brain associated with locomotion and balance (Jenkins 1994). Suckers with a large cerebellum tend to live in swift-water habitats (Miller and Evans 1965). Jordan and Gilbert (1886) did not give the specific habitat in which *M. lacerum* was found during their 31 August 1884 Arkansas collections, but R. E. Jenkins (pers. comm.) believes that it was found in the rivers and not in the brooks or springs. Other *Moxostoma* species collected with *M. lacerum* in 1884 were *M. carinatum, M. duquesnei,* and *M. pisolabrum.*

The unique modification of the lips and trophic specializations throughout the alimentary tract (e.g., short intestine) of the Harelip Sucker suggest that it had a very specialized diet. Examination of stomach contents revealed a diet of snails, limpets, fingernail clams, and microcrustaceans (Jenkins 1970). The Harelip Sucker probably ingested whole snails, then digested soft parts from the shell. Jenkins (1994) deemed the pharyngeal teeth of the

Harelip Sucker too delicate to crush snail shells. Mollusks numerically composed about 90% of their diet (Jenkins 1994).

Moxostoma lacerum was a spring spawner and migrated as do most other suckers (Jenkins 1994). Local fishermen often referred to *M. lacerum* as the May sucker (Etnier and Starnes 1993) due to its spawning migration at that time. Jordan and Brayton (1877) reported it was a valued food fish in the lower Tennessee River drainage.

How abundant the Harelip Sucker was in Arkansas is only speculation. The last time it was collected in Arkansas was in 1884 at one or two upper White River drainage sites in the Ozarks and it was noted as "not rare" (Jordan and Gilbert 1886). This remains the only capture west of the Mississippi River. Jenkins (1994) believed that Jordan and Evermann (1896) erred in claiming this species was "abundant only in the Ozark Mountains." It is unclear whether the Harelip Sucker was ever abundant. Eastward, in the southern bend of the Tennessee River region, Jordan and Brayton (1877) stated that fishermen of that region indicated that it was the most common of suckers. The last report of the Harelip Sucker was made in 1893 in northwest Ohio. Black (1940) listed it as probably extinct, and in the mid-1950s Lachner (1956) also pronounced the probable extinction of the Harelip Sucker.

The reason or reasons for its extinction remain a mystery. Miller and Evans (1965) studied the brain anatomy of *M. lacerum*. They found the relative size of the brain lobes, fairly large eyes, and a few labial taste buds indicated that it was a sight feeder, a major contrast with other suckers (Miller and Evans 1965). Jenkins (1970) cited this anatomical evidence suggesting it was a sight feeder and surmised that increasing turbidity may have inhibited its ability to find snails. Earlier, Black (1949) had also suggested that increased siltation was a major cause of its extinction. Increases in turbidity and sedimentation associated with the widespread deforestation and land cultivation that occurred through the 1800s probably contributed substantially to the demise of the Harelip Sucker (Jenkins 1994).

CONSERVATION STATUS   The Harelip Sucker was the first North American fish species documented as extinct (Williams et al. 1989; Fink and Humphries 2010). It was first and last collected in Arkansas in 1884 (Jordan and Gilbert 1896). Robert E. Jenkins (1994) pointed out that *Lagochila* (= *Moxostoma lacerum*) was not collected as frequently as the small-stream redhorses, *Moxostoma erythrurum* and *M. duquesnei*, but it was taken slightly more often than *Moxostoma carinatum* and *M. macrolepidotum*, both of which occur more often in big-river habitats. No one is sure whether *M. lacerum* persisted into the

20th century. Surveys of the White River by Keith (1964) and Cashner (1967), in addition to Black's earlier (1940) statewide survey, did not find this species. Dam-building in this country began in earnest in the mid-1930s; however, Etnier and Starnes (1993) and Jenkins (1994) suggested that the Harelip Sucker may have already disappeared or nearly disappeared by that time. It is extinct in Arkansas and elsewhere.

Moxostoma pisolabrum Trautman and Martin
Pealip Redhorse

Figure 15.147. *Moxostoma pisolabrum*, Pealip Redhorse. *David A. Neely.*

Map 111. Distribution of *Moxostoma pisolabrum*, Pealip Redhorse, 1988–2019.

CHARACTERS   A slender redhorse sucker with a short-based, slightly falcate dorsal fin usually with 12 or 13 (11–14) rays, a complete lateral line usually with 42–44 (40–46) scales, deeply plicate lips, and a short head, head length going into SL 4.3–5.4 times (Fig. 15.147). Mouth small; posterior margin of lower lip nearly straight (Fig. 15.124C). A distinctive pea-shaped thickening at middle of the upper lip of large juveniles and adults is unique among suckers. Often in large adults, the pea-shaped knob becomes cornified, often forcing the ventral surface of the lower lip to tilt so that its tip slips under the knob of the upper lip, giving the mouth a "sucked-in" appearance. Body depth going into SL 3.5–4.3 times. Pharyngeal arches moderately heavy; pharyngeal teeth numerous and bladelike. Intestine long with several loops. Swim bladder with 3 chambers. Anal rays 7; pectoral rays 15–18; pelvic rays 9. Breeding males with large tubercles on anal and caudal fins, and with small tubercles on pectoral and pelvic fins. Females occasionally with small tubercles on anal and caudal fins. Maximum size about 19 inches (483 mm) (Page and Burr 2011).

LIFE COLORS   Body color olivaceous above to golden yellow on sides. Scales of back and upper sides with a crescent-shaped dark mark at base. Belly white. Dorsal fin slate-colored; caudal fin bright red in life. Lower fins plain or with orange tinge.

SIMILAR SPECIES   It differs from other *Moxostoma* species and all other catostomids in having the upper lip with a distinctive pea-shaped swelling at middle that is best developed in adults and often not obvious in juveniles smaller than 37 mm SL (vs. upper lip not medially swollen). It can be further distinguished from Arkansas's only other red-tailed *Moxostoma* species, *M. carinatum*, in having a thin pharyngeal arch with slender, bladelike teeth (vs. pharyngeal arch thick with molarlike teeth), and in having the rear margin of the lower lip nearly straight (vs. rear margin of lower lip forming a definite V-shaped angle). It frequently occurs syntopically with *M. duquesnei* and *M. erythrurum* and can be further distinguished from those species in having a red caudal fin in life (vs. caudal fin not red), scales of back and upper sides each with a crescent-shaped dark spot at base (vs. scales without a dark spot at base), and in having a straight lower lip margin (vs. rear margin of lower lip forming a moderate to broadly acute V-shaped angle). It differs additionally from *M. anisurum* in usually having 12 or 13 dorsal rays (vs. 14 or 15), rear margin of lower lip straight (vs. V-shaped), and lower lip not constricted just before its junction with upper lip (vs. lower lip distinctly narrowed just anterior to its junction with upper lip).

VARIATION AND TAXONOMY   *Moxostoma pisolabrum* was described by Trautman and Martin (1951) as a subspecies of the Shorthead Redhorse, *Moxostoma aureolum pisolabrum*, from Coon Creek, tributary to the North Fork of the Spring River, 1 mile north of Jasper,

Jasper County, Missouri. *Moxostoma aureolum* was subsequently placed under synonymy of *Moxostoma macrolepidotum* (Hubbs and Lagler 1958), and the form in Arkansas was referred to as *M. macrolepidotum* in the first edition of *Fishes of Arkansas* (Robison and Buchanan 1988). Morphological analyses produced different interpretations of the status of the Ozarkian populations. Trautman and Martin (1951) treated populations from the Missouri River drainage and drainages to the south as typical *M. aureolum pisolabrum* and recognized a small zone of intergradation between *M. a. pisolabrum* and *M. a. aureolum* (= *M. macrolepidotum*) in Missouri in drainages north of the Missouri River. Cross (1967) included all of the Missouri River drainage in Kansas (except for the Osage drainage) in the zone of intergradation, and Pflieger (1971) supported a much larger zone of intergradation that included all of the Missouri River drainage in Missouri. All Arkansas specimens examined by us show the pealip trait, and we have not identified any specimens as intergrades. Robert E. Jenkins (pers. comm., 2019) informed us that he has not identified typical *M. macrolepidotum* from the White River or lower Mississippi River. Jenkins (1970) noted the high degree of separation between *M. pisolabrum* and *M. macrolepidotum* using indices based on the pealip character. Harris et al. (2002) sequenced the complete cytochrome *b* gene but removed *M. pisolabrum* from the synonymy of *M. macrolepidotum* and recognized it as a full species based largely on morphology. Clements et al. (2012) presented molecular evidence for a Shorthead Redhorse group comprising *M. macrolepidotum*, *M. pisolabrum*, and *M. breviceps*, with *M. hubbsi* as the sister species to the clade. Clements et al. (2012) did not include tissue samples of *M. pisolabrum* in their analysis and referred to it as "this questionable species." Harris et al. (2002) noted that resolution of its taxonomic status requires a more in-depth, population-level examination of genetic variation. Robert E. Jenkins (pers. comm., 2019) also believed that combined detailed morphological and molecular analyses are needed to clarify its status. We provisionally follow Harris et al. (2002) and Page et al. (2013) in considering *M. pisolabrum* a distinct species.

DISTRIBUTION   *Moxostoma pisolabrum* occurs in the Arkansas, White, and St. Francis river drainages of the Ozarks in Arkansas, Missouri, Kansas, and Oklahoma, and disjunctly in a short section of the Red River in Oklahoma. It occasionally occurs in the Mississippi River (Etnier and Starnes 1993). The known distribution of the Pealip Redhorse in Arkansas prior to 1988 was sporadic, probably because of inadequate big-river sampling rather than actual rarity (Map A4.108). The range of *Moxostoma aureolum* (prior to recognition of *M. pisolabrum*) was listed as "south to Arkansas" by Jordan and Evermann (1896) and Forbes and Richardson (1920). Black (1940) documented the first specimens of *M. aureolum* (almost certainly *M. pisolabrum*) from Arkansas, collected on 3 August 1939 from the White River at DeValls Bluff. About 18 scattered records were reported from the state during the next 50 years, but increased sampling efforts in large rivers after 1988 revealed that it is widely distributed throughout the Arkansas River and throughout the middle and lower White River (Map 111). There are also recent records from the St. Francis River in Greene County and the Illinois River in Benton County. There are no records from the Red River system in Arkansas, even though a single record from that drainage in Oklahoma was reported by Riggs and Moore (1963). McAllister et al. (2009e) reported the collection of two specimens (59.9 and 66.4 mm TL) from the Strawberry River in 2007. It was common in collections from the Spring River by TMB, Sam Henry, and Brett Timmons in 2014. McAllister et al. (2010b) reported one specimen from the Black River in Independence County, and there is a confirmed record from the Mississippi River of Arkansas in the early 1980s (R. E. Jenkins, pers. comm.).

HABITAT AND BIOLOGY The Pealip Redhorse inhabits a diversity of stream types over its range, but it is most abundant in moderately large rivers having a predominance of gravelly or rocky bottoms and permanent strong flow (Pflieger 1997). In Arkansas, it is not known from small creeks but is most common in the Arkansas and lower White rivers, where it usually occurs in moderate current over a sandy bottom. In the Spring River and middle portion of the White River, adults are most commonly found in deep runs with swift current over a gravel and cobble substrate. Juveniles are sometimes collected in backwater areas near creek mouths. It appears to be more tolerant of turbidity than any other *Moxostoma* in Arkansas. It prefers swifter stream sections near riffles in large streams, but it has also been taken from pools in small streams. Black (1949) indicated its abundance in the Arkansas River in the 1890s as follows:

> Fifty years ago *Placopharynx* [almost certainly *M. pisolabrum*] was so common in the Arkansas River near Fort Smith that farm boys frequently would fill a wagon bed with the fish in a short time of spearing and netting and haul them into the city to peddle out on the streets.

Moxostoma pisolabrum is mainly an aquatic insectivore, feeding predominantly on benthic larvae, pupae, and nymphs. Gut contents of 32 specimens collected in August and September by electrofishing and gill netting from the Arkansas River (Crawford, Franklin, Logan, Pope, Jefferson, and Lincoln counties), the lower White River (Independence and Monroe counties), and the Illinois River (Benton County) were examined by TMB. Summer diets were similar in the large-river environments of the Arkansas and White rivers, with chironomid and other midge larvae ingested most frequently and in greatest amounts. Also important in the diet in decreasing order of percent frequency of occurrence were coleopteran larvae, trichopterans, odonates, and annelids. The intestine in adults is long with several loops, and large adult gut contents included some organic debris, plant pieces, and filamentous algae. Juveniles (65–100 mm TL) have relatively shorter intestines than adults, and juvenile gut contents rarely included items other than benthic insects. Individuals collected from the Illinois River, an Ozark stream, in early August fed most heavily on a variety of coleopteran larvae (dyticids, psephenids, and others), megalopterans, and trichopterans. Also ingested were midges and other dipteran larvae and pupae, and plecopterans. Jenkins (1970) suggested that the enlarged, callous-like, pea-shaped thickening on the upper lip of *M. pisolabrum* is probably an adaptation for feeding among stones, which are generally more frequent in riffles than pools.

Its reproductive biology is presumably similar to that of the closely related *M. macrolepidotum*, and most available information comes from studies of that species. Spawning occurs during late April and May on gravelly riffles and raceways (Cross 1967; Sule and Skelly 1985) when a female moves into a riffle where males have congregated. HWR found adult tuberculate males of *M. pisolabrum* in Crooked Creek, Marion County, Arkansas, as early as 13 March. Burr and Morris (1977) observed spawning of more than 100 *M. macrolepidotum* near the edge of a sandbar in a high-gradient Illinois stream in mid-May at 60.8°F (16°C). No territoriality or aggressive displays were observed. Groups of 3–7 individuals, with the female in the middle or below the males, violently rolled and undulated until troughs and circular depressions were formed in the sand and gravel. The fertilized eggs are demersal and nonadhesive and hatch in 8 days at a temperature of 15.6°C (60°F); average fecundity is 18,000 eggs per female (Sule and Skelly 1985). Adults live 6–7 years, with females attaining slightly larger sizes than males after age 2. Miller and Robison (2004) found the flesh of this fish quite tasty, especially in the spring.

CONSERVATION STATUS The Pealip Redhorse is commonly encountered (and often misidentified) in Arkansas, and its populations are secure.

Moxostoma poecilurum Jordan
Blacktail Redhorse

Figure 15.148. *Moxostoma poecilurum*, Blacktail Redhorse. *Uland Thomas.*

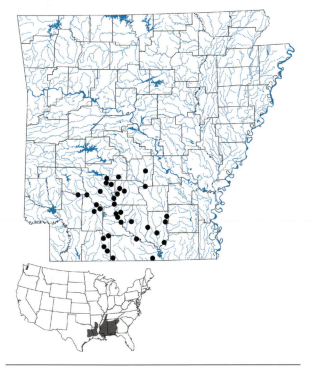

Map 112. Distribution of *Moxostoma poecilurum*, Blacktail Redhorse, 1988–2019.

CHARACTERS   A slender or moderate-bodied redhorse possessing a short-based, slightly concave or straight-margined dorsal fin usually with 11–13 (10–14) rays and a distinctive caudal fin with a black stripe on the lower caudal lobe. The dark caudal stripe is bordered ventrally by a sharply contrasting white marginal stripe (Fig. 15.148). Snout moderately short and rounded, its length going into head length 2–3 times. Head moderately short, its length going into SL 4.0–4.6 times. Eye moderate in size. Mouth subterminal; lips plicate, rear margin of lower lip forming a broad angle (slightly U-shaped). Pharyngeal teeth numerous and bladelike, without flattened grinding surfaces. Gill rakers in adults usually 23–26. Caudal peduncle long and shallow, its depth going into SL 9.3–10.7 times. Intestine

long, with several loops. Lateral line complete, usually with 41–44 (40–45) scales. Anal rays 7. Pectoral fin rays 15 or 16; pelvic rays 9. Breast usually fully scaled. Swim bladder with 3 chambers. Breeding males with well-developed tubercles on rays of caudal and anal fins. Dorsal fin and paired fins with minute tubercles. Head and body with scattered small tubercles. Females with minute tubercles on head and nape and a cornified layer on anal fin rays. Paired and dorsal fins of females with minute tubercles (Boschung and Mayden 2004). Maximum size is 20 inches (508 mm) TL (Carlander 1969). In Arkansas maximum length about 15 inches (381 mm).

LIFE COLORS   Body color olivaceous to greenish bronze dorsally grading to white ventrally. Scales have crescentic dusky to blackish marks. Body with faint patterns of dusky, longitudinal stripes along sides. The dusky stripes pass through the upper and lower edges of the scales but not through the midheight of the scales where the crescents are widest. Dorsal fin with a black submarginal band and occasionally a black marginal band, at least in the leading rays. Caudal and lower fins bright rust-colored, orange, or even red; lower lobe of caudal fin with a blackened, submarginal, horizontal stripe bordered by a white marginal stripe. The blackened area can be vague in some juveniles (Fig. 15.149), and the white stripe is often masked by red pigment in spawning adults.

SIMILAR SPECIES   The Blacktail Redhorse differs from all other redhorse (*Moxostoma*) suckers in Arkansas by the diagnostic blackened midlower caudal fin lobe, bordered ventrally by a white stripe (vs. lower caudal fin lobe without a black and white marginal pattern). It differs from *Minytrema melanops* in having a complete lateral line (vs. lateral line absent or incomplete), and in swim bladder with three chambers (vs. two chambers).

VARIATION AND TAXONOMY   *Moxostoma poecilurum* was described by Jordan (1877b) as *Myxostoma poecilura* from the Tangipahoa River, Mississippi. Jenkins (1970) presented strong morphological evidence that *M. poecilurum* is most closely related to *M.* sp. *cf. poecilurum*, the Apalachicola Redhorse, and speculated that they could constitute a species group. He cited 24 morphological,

Figure 15.149. *Moxostoma poecilurum* juvenile. *Renn Tumlison.*

osteological, coloration, and meristic characters that link these two species and distinguish them from other *Moxostoma* species. Jenkins (1970) did not recognize subspecies within *M. poecilurum*. Smith (1992) considered *M. poecilurum* sister to *M. macrolepidotum*, with both species forming a clade sister to *M. lacerum*. Based on analysis of the cytochrome *b* gene, Harris et al. (2002) considered *M. poecilurum* sister to a *Moxostoma congestum* clade. Doosey et al. (2010) also supported a sister relationship between *M. poecilurum* and *M. congestum*.

DISTRIBUTION   *Moxostoma poecilurum* occurs east of the Mississippi River in the lower Mississippi River basin from Kentucky and northern Georgia south to southeastern Louisiana; west of the Mississippi River from southern Arkansas south to central Louisiana and eastern Texas; and in Gulf Slope drainages from the Choctawhatchee River, Florida and Alabama, west to Galveston Bay, Texas. In Arkansas, the Blacktail Redhorse is limited to the lower Ouachita River system of the southcentral part of the state (Map 112).

HABITAT AND BIOLOGY   The Blacktail Redhorse occurs in small to large, tannin-stained, lowland streams with sand and gravel substrates. It seems to prefer sluggish pool regions in streams. During frequent but brief periods of flooding, Blacktail Redhorses moved onto the inundated floodplain (Slack 1996; Ross 2001). Seasonal movements have been reported in Louisiana, with adults inhabiting small tributaries from March to November and moving downstream into deeper water during winter (Gunning and Shoop 1964). In Village Creek, Texas, it was found in channel pools in summer and fall, and juveniles were found in backwater and sandbar habitats (Moriarty and Winemiller 1997). Jenkins (1980d) and Etnier and Starnes (1993) reported that this species does well in reservoirs, but we have records from only one reservoir in Arkansas (Calion Lake in Union County).

Foods of pond-raised Blacktail Redhorse in Alabama included caddisfly larvae, ostracods, midge larvae, cladocerans, rotifers, copepods, nematodes, protozoans, and detritus

(Kilgen 1972). Gut contents of 14 small to medium-sized (48–300 mm TL) specimens collected in March, July, August, and September from Clark, Drew, Grant, and Union counties in Arkansas were examined by TMB. It is predominantly a benthic invertivore, feeding mainly on immature aquatic insects. The dominant foods by volume and numbers in all guts examined were chironomids, other midge larvae, and other dipteran larvae and pupae. Ephemeropteran nymphs, coleopteran larvae, odonate nymphs, and small crustaceans (ostracods, isopods, copepods, and amphipods) were also important in the diet. Items ingested in small amounts were trichopterans, plecopterans, annelids, small clams, and limpets. All guts contained detritus and sand, indicating that invertebrate food items were not separated from the ingested benthic material. The intestine is long with 6–8 loops, suggesting that some nutrient value is obtained from the detritus.

Spawning occurs in Arkansas from late April through May. Spawning habitat is in the shoal areas of small streams (Gunning and Shoop 1964; Kilgen 1974). In Alabama, spawning commonly occurred in tailwaters below locks and dams (Mettee et al. 1996). Tuberculate males and ripe females have been taken in Arkansas as late as 21 May in Columbia County. Two or three males swim around a female, spawning intermittently in the manner described for several other species of redhorses (Kilgen 1974; Boschung and Mayden 2004). Fertilized eggs are demersal and nonadhesive with hatching occurring in 6–8 days at 20°C. Larvae move off the bottom into the water column about 6 days after hatching (Kilgen 1974). Two young-of-the-year specimens, 1.4 inches (34.7 mm) and 1.3 inches (33.1 mm) SL, were collected on 24 October 1976. Another YOY was collected in Union County from La Pere Creek on 17 October 1992 (Tumlison and Robison 2010).

CONSERVATION STATUS   The Blacktail Redhorse is an uncommon species confined in Arkansas to the southern portion of the state. We consider it a species of special concern, and its populations should be monitored to detect any future changes.

CHAPTER 16

Order Siluriformes
Catfishes

Siluriformes is a large, diverse clade of 40 extant families with about 490 genera and approximately 3,730 species, 2,053 of which live in the New World (Nelson et al. 2016). Sabaj et al. (2004) estimated 1,750 undescribed species. This order is characterized by a Weberian apparatus, in which the first five vertebrae are fused (vs. four fused vertebrae in the Weberian apparatus of the related order Cypriniformes), often with a single spinelike ray at the beginning of the dorsal and pectoral fins, usually with an adipose fin, a vomerine bone usually toothed (but mouth without large teeth), body naked or covered with bony plates, and 1–4 pairs of barbels on the snout. The dorsal and pectoral spines of catfishes are not homologous to the spines of perciform fishes; instead, they are fused and hardened rays. Principal caudal rays number 18 or fewer (most with 17), and eyes are usually small. Catfishes lack intermuscular bones. The cranium is a solid box, usually somewhat flattened, and the number of skull bones has been reduced (the symplectic, basihyal, parietal, and subopercle bones are missing). The pectoral girdle is modified to form a locking mechanism for the pectoral fin spine. Wiley and Johnson (2010) listed additional morphological synapomorphies for this order. Some species are venomous and can inflict painful wounds with their spines (primarily those of the pectoral fin) by injecting a poison produced by glandular cells or venom glands in the epidermal covering of the spines. Extreme variation occurs in this order as species vary in size from weighing only a few grams (e.g., species in the genus *Noturus* among the North American catfishes) to the giant European catfish (*Silurus glanis*), which can reach 16 feet (5 m) in length (Bruton 1996; Nelson 2006).

Catfishes are known from all continents, including Eocene and Oligocene fossils from Antarctica (Grande and Eastman 1986; Ferraris 2007). Thirteen families are endemic to South America (Roberts 1972) and only one family (Ictaluridae) normally occurs in the freshwaters of North America, including Arkansas. The trophic ecology of this diverse order ranges from pure herbivory to carnivory and even quasi-parasitism. Many questions remain about the classification of siluriforms, and disagreements exist on the interrelationships of some families (Nelson et al. 2016).

Family Ictaluridae *North American Catfishes*

The North American catfish family, Ictaluridae, is the largest freshwater fish family endemic to North America; it comprises roughly 50 extant and at least 14 fossil species distributed from Central America to Canada (Lundberg 1992). The oldest ictalurid fossils are Late Eocene members of *Ameiurus* and *Ictalurus* (Arce-H. et al. 2017). There is no consensus on the relationship of Ictaluridae to other cat-fish families. Arce-H. et al. (2017) found that Ictaluridae is sister to a clade formed by extant Schilbeidae, Bagridae, Siluridae, and Polyoxidae based on morphological and genetic evidence.

Ictaluridae has undergone major systematic studies by Taylor (1969), Lundberg (1982, 1992), LeGrande (1981), Grady (1987), Grady and LeGrande (1992), and Arce-H. et al. (2017). Sullivan et al. (2006) studied phylogenetic relationships of the major groups of catfishes using nuclear gene sequences. Monophyly of Ictaluridae is well estab-lished by morphological and molecular data (Lundberg 1970, 1975, 1982, 1992; Hardman and Page 2003; Hardman 2005; Egge 2010). Currently, there are seven recognized extant genera: *Ictalurus, Ameiurus, Trogloglanis, Noturus, Prietella, Satan,* and *Pylodictis* (Arce-H. et al. 2017; Nelson et al. 2016), four of which (*Ameiurus, Ictalurus, Pylodictis,* and *Noturus*) occur in Arkansas. *Noturus* (the madtoms) is the most speciose genus, with 29 described species and several species awaiting description (Burr and Stoeckel 1999). Nelson et al. (2004) changed the common name of the family Ictaluridae from Bullhead Catfishes to North American Catfishes because the latter name was more descriptive and similar to names used by other authors. Lundberg (1982) presented evidence that the bullheads (White Catfish, plus six other species formerly in the genus *Ictalurus*) are more closely related to several other currently recognized ictalurid genera than to the fork-tailed catfishes in the genus *Ictalurus*. We follow Lundberg (1982), Page et al. (2013), and Arce-H. et al. (2017) in using *Ameiurus* as the genus name for bullheads.

Catfishes are some of the most easily recognizable Arkansas fishes, characterized by 8 conspicuous barbels (2 on the snout, 2 on the end of the maxillae, and 4 on the chin) that are sensitive to touch and to chemical stimuli. Additional characters include a wide mouth, a scaleless body, hundreds of minute teeth arranged in bands on the roof of the mouth, a well-developed Weberian apparatus, and an adipose fin. In Arkansas, there are 19 species of icta-lurids in 4 distinct groups: the bullheads *(Ameiurus),* the madtoms *(Noturus),* the fork-tailed catfishes *(Ictalurus),* and the Flathead Catfish *(Pylodictis olivaris).* The bull-heads, fork-tailed catfishes, and Flathead Catfish have an adipose fin forming a free, flaplike lobe, while in the mad-toms, the adipose fin forms a low, keel-like ridge without the free posterior flaplike tip. The three native bullheads can be separated from the larger fork-tailed catfishes and Flathead Catfish by the square or rounded caudal fin and smaller size, rarely exceeding 3 pounds (1.4 kg). The

armature of the pectoral spine and the anal and caudal fin ray counts are especially useful characters in distinguishing species (a dissecting needle is useful to separate the soft tis-sue from the pectoral spine). In counts of anal and caudal rays, all rudimentary rays are counted.

Twenty of the 35 catfish families possess venom glands (Wright 2009), and all ictalurid species possess a deliv-ery system for the venom consisting of venomous spines at the origins of the dorsal and pectoral fins. The spines can inflict a painful wound. The stings of the madtoms (*Noturus*) seem to inflict the most pain on humans, but Birkhead (1967; 1972) found that the toxicity of the venom of 13 species of catfishes varied greatly (using *Gambusia* as a test subject). Catfish stings are almost certainly a defense mechanism to reduce predation and are not used to capture prey. In addition to toxins produced by venom glands, cri-notoxins are produced by specialized epidermal cells that are spread over the body of the fish (Egge and Simons 2011). There is no delivery apparatus associated with crinotoxins.

Taste buds are distributed over the body, allowing cat-fishes to forage successfully at night for food. Most spe-cies spend much of the day hiding in crevices, under sub-merged logs, in gravel shoals, or in deep pool regions, but Blue Catfish very often school in pelagic areas during the day. All ictalurids are summer breeders and construct nests under rocks or use natural cavities. Madtoms even spawn in discarded beer cans and similar containers (Taylor 1969). Although both sexes may contribute to nest site selection, construction, and guarding of eggs and young, parental care is often strictly a male function (Blumer 1985a, b). Following spawning, males typically guard and care for the eggs and young. In a few species, for example, *A. melas,* the female assists the male in guarding the eggs and young. Tiemann et al. (2011) summarized the ecological interac-tions between catfishes and freshwater mollusks in North America. Ictalurids are hosts for larvae of several native freshwater mussel species. The mussel larvae attach to the gills or fins to complete their development to the free-living juvenile stage. In turn, mussels serve as a food source for some catfish species, while other catfish species may use spent mussel shells for habitat and spawning sites.

In Arkansas, catfishes range from the diminutive Tadpole Madtom, 2 or 3 inches (51–76 mm) TL and weigh-ing only a few ounces, to the ever-popular food and sport fishing species, the Blue Catfish, and Flathead Catfish. The Blue Catfish can reach weights of approximately 132 pounds (60 kg) and lengths of more than 5 feet (1.5 m). The state angling record for Blue Catfish is 116 pounds, 12 ounces, and the state record for the Flathead Catfish is 80 pounds.

Unofficial reports by AGFC biologists are 139 pounds for the Blue Catfish and 132 pounds for the Flathead Catfish. As a group, the catfishes are among the most important fishes taken by commercial fishermen in Arkansas, ranking only behind the buffalofishes in number of pounds taken annually. From 1975 to 1985, an average of 3,804,846 pounds (1,725,878 kg) of wild catfish per year were taken commercially in Arkansas. More recent commercial catch data are unavailable. Catfishes are also important in the aquaculture industry in Arkansas. The Channel Catfish is the most commercially important species in Arkansas. Recently, a Channel Catfish × Blue Catfish hybrid has also become quite popular. Arkansas catfish farmers produced almost $33.42 million worth of catfish in 2011 and $40.11 million in 2012. Sales nationwide in 2012 were up 5% from the previous year at $423 million. As of 1 January 2011, Arkansas catfish producers were utilizing 13,200 water surface acres, 31% lower than in 2010 (Sider 2011). Arkansas is the third-largest catfish-producing state behind Mississippi and Alabama.

Channel Catfish and Blue Catfish stockings maintain popular fisheries for these species (Broach 1967). They are stocked primarily in lacustrine habitats where reproduction and recruitment are inadequate due to unfavorable habitat, predation, and heavy fishing pressure (Keith 1971a). Catchable-sized catfish (0.5–1.0 pound [0.2–0.5 kg] per fish) are produced by four state fish hatcheries and distributed according to management recommendations (Oliver and Rider 1986).

Key to the North American Catfishes

1A Adipose fin free posteriorly, separate from caudal fin (Fig. 16.1A).   *Ictalurus, Ameiurus,* and *Pylodictis* 2
1B Adipose fin adnate, continuous with the caudal fin (Fig. 16.1B). *Noturus*   7
2A Caudal fin deeply forked (Fig. 16.1D).   3
2B Caudal fin not forked, but emarginate or rounded (Fig. 16.1C).   4

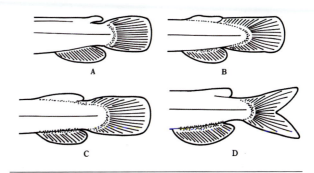

Figure 16.1. Characteristic caudal and adipose fins in catfishes.

A—Adipose fin with free lobe, caudal fin rounded.
B—Adipose fin adnate, a low keel-like ridge, caudal fin rounded.
C—Adipose fin with only a slight notch between caudal fin, caudal fin rounded.
D—Adipose fin tip entirely free, caudal fin deeply forked.

3A Anal fin with curved outer margin; anal rays 24–30; back not conspicuously humped at dorsal origin; body typically with dark spots. *Ictalurus punctatus* Page 438
3B Anal fin with straight outer margin; anal rays 30–35; back conspicuously humped at dorsal origin; body without dark spots.   *Ictalurus furcatus* Page 436
4A Premaxillary band of teeth with backward processes; head conspicuously flattened between eyes; caudal fin with upper tip light-colored; lower jaw projecting beyond upper jaw except in smallest young.
        Pylodictis olivaris Page 465
4B Premaxillary band of teeth without backward processes; head not conspicuously flattened between eyes; caudal fin without upper tip light-colored; lower jaw not projecting beyond upper jaw.   5
5A Anal fin long with 24–27 rays; anal base length going into standard length fewer than 3.8 times; chin barbels white or light-colored.   *Ameiurus natalis* Page 432
5B Anal fin shorter with 17–24 rays; anal base length going into standard length more than 3.8 times; chin barbels gray to black or dusky.   6
6A Anal rays 17–21; body not heavily mottled; teeth on rear margin of pectoral fin spine weakly developed or absent (Fig. 16.2A).   *Ameiurus melas* Page 430
6B Anal rays usually 21–24; body usually heavily mottled; teeth on rear margin of pectoral fin spine well developed (Fig. 16.2B).   *Ameiurus nebulosus* Page 434

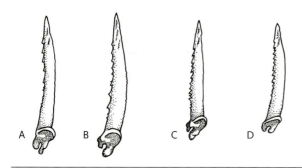

Figure 16.2. Pectoral fin spines of four *Ameiurus* species showing degree of serration.

A—*Ameiurus melas*

B—*Ameiurus nebulosus*

C—*Ameiurus natalis*

D—*Ameiurus catus*

7A Upper jaw not projecting beyond lower jaw, the two jaws about equal length (Fig. 16.3A).   8

7B Upper jaw projecting well beyond lower jaw (Fig. 16.3B).   10

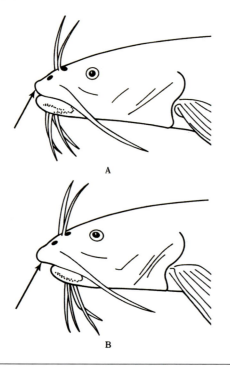

Figure 16.3. Lateral view of heads of catfishes. *Pflieger (1975)*.

A—Jaws equal

B—Upper jaw projecting beyond lower jaws

8A Posterior edge of pectoral spine with distinct serrae (Fig. 16.4A); anal, caudal, and dorsal fins dark-edged; typically 9 pectoral rays.   *Noturus exilis* Page 445

8B Posterior edge of pectoral spine without serrae (Fig. 16.4B, C, D); anal, caudal, and dorsal fins not dark-edged; typically 7 or 8 pectoral rays.   9

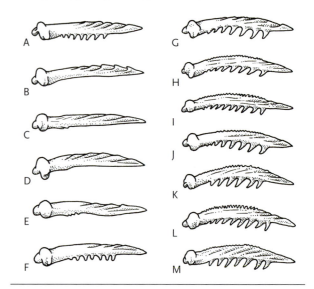

Figure 16.4. Pectoral fin spines showing degree of posterior serration in 13 *Noturus* species.

A—*Noturus exilis*

B—*Noturus flavus*

C—*Noturus gyrinus*

D—*Noturus lachneri*

E—*Noturus nocturnus*

F—*Noturus phaeus*

G—*Noturus albater*

H—*Noturus eleutherus*

I—*Noturus flavater*

J—*Noturus maydeni*

K—*Noturus miurus*

L—*Noturus gladiator*

M—*Noturus taylori*

9A Two pores present between the anterior and posterior nasal openings on each side of the head (Fig. 16.5B); anal rays 12–17 (modally 15); head length going into standard length 3–3.8 times.

Noturus gyrinus Page 453

9B One pore present between the anterior and posterior nasal openings on each side of the head (Fig. 16.5A); anal rays 16–19 (modally 17); head length going into standard length 3.7–4.2 times.

Noturus lachneri Page 455

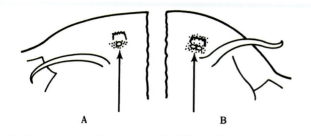

Figure 16.5. Dorsal view of one side of head showing internasal pores. *Smith-Vaniz (1968).*

A—*Noturus exilis*
B—*Noturus nocturnus*

10A Premaxillary tooth patch in upper jaw with a backward extension on each side (Fig. 16.6A).

Noturus flavus Page 449

10B Premaxillary tooth patch in upper jaw without a conspicuous backward extension on each side (Fig. 16.6B). 11

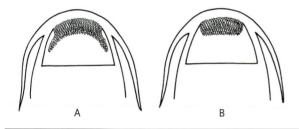

Figure 16.6. Premaxillary tooth patch in upper jaw of catfishes.

A—Backward extension on each side in *Noturus flavus*.
B—Without backward extension of tooth patch on each side.

11A Body unicolored, olive yellow or brown; pectoral spines with or without posterior serrae (Fig. 16.4A; 16.4F). 12

11B Body mottled or punctulate with black or brown; pectoral spines very strong with posterior serrae recurved (Fig. 16.4G, 16.4L). 13

12A Anal rays usually 18 or fewer; pectoral spine without distinct, well-developed posterior serrae (Fig. 16.4E); abdomen and lower surface of head mostly unpigmented, but narrow bands of pigment often present on chin and anterior to pelvic fins.

Noturus nocturnus Page 460

12B Anal rays usually 20–22; posterior serrae on pectoral spine distinct and well developed (Fig. 16.4F); abdomen and lower surface of head usually well pigmented with large chromatophores. *Noturus phaeus* Page 462

13A Typically 9 pectoral rays; head length going into standard length 3.7 times or more; outer third of dorsal fin without a large black blotch; posterior process of cleithrum shorter than the diameter of the pectoral spine shaft. 14

13B Typically 8 pectoral rays; head length going into standard length fewer than 3.7 times; outer third of dorsal fin with or without a large black blotch; posterior process of cleithrum longer than the diameter of the pectoral spine. 15

14A Confined to the Black River system (Strawberry, Spring, Eleven Point, Current, Black and tributaries) (cryptic species); 2N chromosomal count of 56–58; tends to be chunkier in body size; more mottled, with diffuse dorsal saddles; less slender.

Noturus maydeni Page 456

14B Confined to the White River and Little Red River systems (excluding the Black River system) (cryptic species); 2N chromosomal count of 65–72; tends to be larger in body size, has bolder and more distinct saddles and is more slender.

Noturus albater Page 441

15A Outer third of dorsal fin without a large black blotch; caudal fin rays generally 44–53. 16

15B Outer third of dorsal fin with a distinct, large black blotch; caudal fin rays generally more than 53. 17

16A Caudal fin with 3 vertical dark bands; pectoral fin sharply pointed; caudal fin rays usually 49–53; sides and belly prominently mottled with dark pigment; preoperculomandibular pores 11.

Noturus gladiator Page 451

16B Caudal fin with 1 or 2 dark vertical bands; pectoral fin rounded; caudal fin rays usually 43–49; sides and belly dimly mottled; preoperculomandibular pores 10.

Noturus eleutherus Page 443

17A Dark saddle or bar on adipose fin submarginal, extending dorsally only onto the basal half of the fin.

Noturus taylori Page 464

17B Broad dark saddle on adipose fin extending to the outer margin of the fin. 18

18A Basicaudal bar narrow, confined to posterior edge of caudal peduncle and base of caudal fin rays, not crossing caudal peduncle; 1 internasal pore; body with poorly defined dark bars and heavy mottling.

Noturus miurus Page 458

18B Basicaudal bar jet black, broad, extending from base of caudal fin rays across the caudal peduncle; 2 internasal pores; body with sharply defined dark bars and slight mottling. *Noturus flavater* Page 447

Genus *Ameiurus* Rafinesque, 1820a— Bullhead Catfishes

Ameiurus has at times been recognized as a genus, at other times as a subgenus of *Ictalurus*, and even as a synonym of *Ictalurus,* but for at least the past two decades, it has been universally accepted as a genus. Lundberg (1982, 1992) considered *Ameiurus* sister to a clade composed of the genera *Noturus*, *Prietella*, *Satan*, and *Pylodictis*. *Ameiurus* was recovered as monophyletic and supported by five morphological synapomorphies by Arce-H. et al. (2017). The genus was divided into two species groups by Lundberg (1975): the *natalis* group containing three species (all native to Arkansas), and the *catus* group with four species (one of which, *A. catus*, formerly occurred in Arkansas as a transplanted species). However, Arce-H. et al. (2017) divided extant species of the genus into a clade containing *A. natalis* and a second clade with all other species. Bullheads are medium-sized (rarely exceeding 1.4 kg [3 pounds]), heavy-bodied catfishes with dark coloration and a moderately forked, emarginate, or rounded caudal fin. All native Arkansas species have small eyes. Other genus characters include dorsal fin rays usually 6, pelvic fin rays 8, posterior tip of adipose fin free, and no posterior projections of the premaxillary tooth patch. There are three *Ameiurus* species in Arkansas: *A. melas*, *A. natalis*, and *A. nebulosus*.

Ameiurus melas (Rafinesque)
Black Bullhead

Figure 16.7. *Ameiurus melas*, Black Bullhead. *Uland Thomas.*

CHARACTERS   A dark, robust, small-eyed bullhead catfish with a free adipose fin and a truncate to slightly emarginate caudal fin (Fig. 16.7). Scaleless. Body moderately elongate. Upper jaw projects slightly beyond lower jaw; mouth slightly subterminal. Premaxillary tooth patch on upper jaw without backward extensions. Gray or black chin barbels present (barbel coloration similar to that of Brown Bullhead). Barbel at corner of mouth not reaching farther

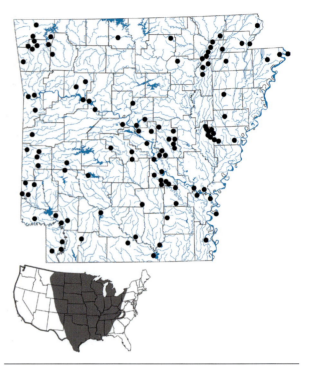

Map 113. Distribution of *Ameiurus melas*, Black Bullhead, 1988–2019. Inset shows native range (solid) and introduced range (outlined) in the contiguous United States.

posteriorly than base of pectoral fin. Gill rakers modally 18 or 19 (15–21). Branchiostegal rays 8 or 9. Dorsal fin with 1 spine and 6 rays. Pectoral fin with 1 spine and 7 or 8 rays; pectoral spine straight or slightly curved, with posterior serrae weakly developed or absent (Fig. 16.2A). Pelvic rays 8; Anal fin margin rounded; anal fin without spines and with 17–24 rays. Length of anal fin base less than head length. Usually less than 1 pound (0.5 kg) and 12–16 inches (305–406 mm) TL. The Arkansas state record from Point Remove Creek in 1986 weighed 4 pounds, 12 ounces (2.2 kg). Maximum length 24.5 inches (620 mm) TL and maximum weight 8 pounds (3.6 kg).

LIFE COLORS   Dorsal region of head, dorsum, and sides of adults solid brownish yellow to black, not mottled. Chin barbels pigmented, gray or black, but never uniformly white. Belly yellow to white. All fins dusky with dark edges, membranes darker than fin rays. A rectangular depigmented area often present at base of caudal fin in adults (Burkhead et al. 1980). Upper lobe of caudal fin never partly depigmented. Breeding male often jet black with bright yellow or white belly. Young individuals smaller than 2 inches (50 mm) TL are solid black; a whitish belly develops at about 50 mm.

SIMILAR SPECIES   The Black Bullhead is similar to *Ameiurus natalis* and *A. nebulosus*. It differs from the

Yellow Bullhead in having black chin barbels (vs. white) and usually 17–21 anal rays (vs. 24–27). It differs from the Brown Bullhead in having 17–20 gill rakers (vs. 13–15), a rectangular depigmented area often present at base of caudal fin in adults (vs. a uniformly pigmented caudal base), usually 17–21 anal rays (vs. 21–24), and in having no serrae on pectoral spine, or a roughened rear margin (vs. large serrae on posterior margin of pectoral spine). Small individuals differ from *Noturus* species in having a free adipose fin (vs. adipose fin adnate).

VARIATION AND TAXONOMY *Ameiurus melas* was described by Rafinesque (1820b) as *Silurus melas* from the Ohio River. It was subsequently placed in genus *Ameiurus*, then in *Ictalurus*, and was restored to *Ameiurus* by Lundberg (1982). Two subspecies are sometimes recognized: *A. m. catulus* along the Gulf Coast and in northern Mexico, and *A. m. melas* farther north. Some authors (Burr and Warren 1986b) felt that the recognition of subspecies was unwarranted, and most recent authors do not recognize subspecies. Hybridization between *A. melas* and *A. nebulosus* occurs in Ohio and Kentucky. Based on the molecular analysis of Hardman and Hardman (2008), *A. melas* was sister to *A. nebulosus*. A combined molecular and morphological analysis by Arce-H. et al. (2017) also supported *A. melas* as sister to *A. nebulosus*.

DISTRIBUTION *Ameiurus melas* is native to the Great Lakes, Hudson Bay, and Mississippi River basins from New York to Saskatchewan, Canada, and Montana, south to the Gulf of Mexico, and Gulf Slope drainages from Mobile Bay, Georgia and Alabama, to northern Mexico. It has been widely introduced elsewhere in the United States and southern Canada. In Arkansas, this species is found statewide in suitable habitats (Map 113).

HABITAT AND BIOLOGY In Arkansas, the Black Bullhead is found in a variety of habitats, from lowland streams to upland rivers of the Ozarks and Ouachitas, but it is more common below the Fall Line than above it. Rarely found in Arkansas's largest rivers, it is a common inhabitant of oxbow lakes, quiet mud-bottomed backwater areas, and pools of small streams away from strong currents. Snow et al. (2017) noted that *A. melas* is the most common *Ameiurus* species in Oklahoma, ranging across the state and inhabiting any aquatic ecosystem. Cross (1967) explained the presence of this species in many farm ponds in Kansas by its high vagility and predilection for headwater pools. After rains, bullheads rapidly invade formerly dry drainageways and gently sloping spillways when they overflow, allowing access of bullheads into many ponds. Pflieger (1997) reported the occurrence of this species in a variety of habitats with a preference for those having quiet, turbid water and a silt bottom. He also noted that in Missouri it is a typical inhabitant of waters having low diversity in the fish fauna. Trautman (1957) also reported its occurrence in a variety of habitats: base and low-gradient portions of small and moderate-sized streams, impoundments, farm ponds, backwaters, oxbows, and overflow ponds, particularly along the large rivers. Trautman noted its preference for silty water and soft mud bottoms, and considered it highly tolerant of many types of pollutants, as did Boschung and Mayden (2004). Jenkins and Burkhead (1994) also noted its tolerance for extreme environmental conditions, including high temperature, low dissolved oxygen, and considerable turbidity. Campbell and Branson (1978) studied the physiological ecology and population dynamics of the Black Bullhead in Kentucky.

Black Bullheads feed on a variety of animal and plant materials. Goldstein and Simon (1999) considered it an invertivore/carnivore, and Carlander (1969) noted that a variety of insects and fish made up the diet. Forney (1955) found that adult Black Bullheads principally ate immature aquatic insects, and young up to 1 inch (25 mm) TL fed almost exclusively on small crustaceans. The adults become active at night and search extensively for food. Darnell and Meierotto (1965) found that the young in a Wisconsin Stream had two primary feeding periods, one around dawn and the other around dusk. In Beaver Reservoir, Black Bullheads exhibited different food habitats during periods of relatively stable water levels than in periods of rapidly rising water levels (Applegate and Mullan 1967a). During stable water-level periods, entomostracans formed 72% of the volume of food of the young-of-the-year bullheads; filamentous algae, organic detritus, and crayfish made up 94% of the food volume found in 4.0–11.3-inch (102–287 mm) bullheads. During winter and spring months when water levels rose in the new reservoir, Black Bullheads longer than 4 inches (102 mm) TL predominantly ate terrestrial earthworms, insects, slugs, spiders, centipedes, millipedes, pill bugs, and a worm snake. Filamentous algae remained an important food item (26%). In Lake Carl Etling, Oklahoma, *A. melas* consumed a broad range of prey items, including 5 fish species, 5 crustacean species, 3 species of insects, and 3 plant species (Snow et al. 2017). Fish had the highest index of relative importance, with crustaceans having the lowest IRI. The most frequently consumed diet item was Gizzard Shad, and no significant differences in diet were found between seasons. Black Bullheads have the greatest digestive capacity during the summer months; individuals most likely exist at reduced activity levels on stored fat and glycogen during the winter months (Nordlie 1966).

In Arkansas, spawning begins in May and continues well into August. Black Bullheads nest in cleared holes or saucer-shaped depressions fanned out usually by the female under dense vegetation or logs in May or June in Oklahoma (Miller and Robison 2004). Boschung and Mayden (2004) suggested that both males and females usually participate in nest construction. Multiple spawnings by a pair occur over a single nest. Wallace (1967) described reproductive behavior of this species, noting that the spawning act resembled in some aspects that of Channel Catfish. Nests are guarded by both parents after eggs are deposited. Fecundity estimates range from 2,047 to 5,495 eggs. Hatching occurs in 5–10 days depending on water temperature. After hatching, the dark black young move about in dense schools for 2 weeks or longer accompanied by one or both adults. The fry swim actively in schools during most of the daylight hours, while the adults remain inactive and hidden in vegetation or other cover. When the young reach about 1 inch (25 mm) TL, the parents abandon them, but the young usually continue to school throughout the first summer of life. In Oklahoma, this species averaged 94 mm TL after 1 year, and 170, 229, 274, 312, and 351 mm TL after years 2–6, respectively (Houser and Collins 1962). Growth was faster in new populations, especially those in reservoirs, than for those in streams and overly populated ponds. Maturity is reached in 2 years. Individuals generally live less than 5 years, but some live 10 years or more.

CONSERVATION STATUS   The Black Bullhead is widespread and abundant throughout Arkansas. Its populations are secure.

Ameiurus natalis (Lesueur)
Yellow Bullhead

Figure 16.8. *Ameiurus natalis*, Yellow Bullhead. *Uland Thomas.*

CHARACTERS   A heavy-bodied, small-eyed catfish with a free adipose fin, deep caudal peduncle, and white to yellowish chin barbels (Fig. 16.8). Head large, rounded above, its length going about 3.1–3.8 times into SL. Mouth terminal. Barbel at corner of mouth not reaching much farther

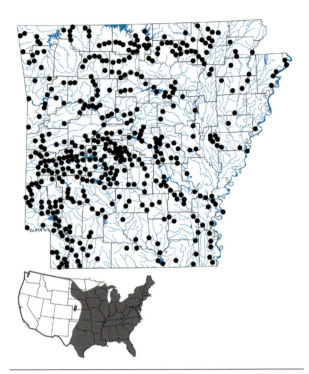

Map 114. Distribution of *Ameiurus natalis*, Yellow Bullhead, 1988–2019. Inset shows native range (solid) and introduced range (outlined) in the contiguous United States.

back than base of pectoral fin. Premaxillary band of teeth on upper jaw without lateral backward extensions. Gill rakers usually 12–16. Body scaleless. Dorsal fin with 1 spine and 6 rays. Pectoral fin with 1 spine; soft pectoral rays modally 9. Pectoral spine nearly straight with 5–8 large serrae on posterior edge (a few weakly developed serrae may be present on the anterior side) (Fig, 16.2C). Anal fin without spines and with 24–27 rays. Anal fin base length about equal to head length. Rear of caudal fin slightly emarginate, with midsection nearly straight. Caudal fin branched rays usually 15 or 16. Maximum size about 18.5 inches (470 mm) TL and about 5 pounds (2.3 kg). The Arkansas rod and reel angling record from a private lake in Texarkana in 2014 is 2 pounds, 7 ounces (1.1 kg). The Arkansas unrestricted tackle record from Blue Bayou in 2005 is 4 pounds, 9 ounces (2.1 kg).

LIFE COLORS   Adults have a solid yellowish to brownish or black body with yellowish or white belly. No mottling. Chin barbels unpigmented white or yellow, matching color of underside of head. All fins are dusky and usually with black margins. Young individuals smaller than 2 inches (50 mm) are black, except for whitish belly and underside of head.

SIMILAR SPECIES   *Ameiurus natalis* differs from the Black Bullhead and Brown Bullhead in having white chin barbels in individuals of all sizes (vs. chin barbels pigmented with gray or black) and in having 24–27 anal fin rays (vs. fewer than 24 anal rays). It differs from the Blue and Channel catfishes in having a straight or rounded caudal fin margin (vs. a forked caudal fin). It can be distinguished from *Pylodictis olivaris* in having upper and lower jaws almost equal (vs. lower jaw projecting beyond upper jaw). Small individuals can be distinguished from *Noturus* species by the free adipose fin (vs. adipose fin adnate to its tip).

VARIATION AND TAXONOMY *Ameiurus natalis* was described by Lesueur (1819) as *Pimelodus natalis*. The type locality was given by Jordan and Evermann (1896) as "North America." It was subsequently placed in genus *Ameiurus*, then in *Ictalurus*, and was restored to *Ameiurus* by Lundberg (1982). While some authors formerly recognized two subspecies, Bailey et al. (1954) concluded that subspecific recognition was not justified by the available evidence. A rangewide systematic study of *A. natalis* is still needed. Lundberg (1992) considered *A. natalis* sister to a clade containing *A. melas* and *A. nebulosus*. A molecular analysis by Hardman and Hardman (2008) identified *A. natalis* as basal to a clade containing all other *Ameiurus* species. Arce-H. et al. (2017) considered *A. catus* the most closely related living species to *A. natalis*.

DISTRIBUTION *Ameiurus natalis* is native to the St. Lawrence–Great Lakes and Mississippi River basins from southern Quebec, Canada, to central North Dakota, and south to the Gulf of Mexico; Atlantic Slope drainages from New York to southern Florida; and Gulf Slope drainages from Florida to Mexico. After the Channel Catfish, the Yellow Bullhead is the most widely distributed catfish in Arkansas and is found statewide (Map 114).

HABITAT AND BIOLOGY The Yellow Bullhead, although found in a variety of habitats in Arkansas, seems to prefer clear, gravel- and rocky-bottomed, permanent streams more than the Black Bullhead; however, it avoids strong current, is common in most Arkansas reservoirs, and is occasionally found in oxbow lakes. Rainwater and Houser (1982) considered the Yellow Bullhead uncommon in Beaver Reservoir, and Ross (2001) considered it uncommon or absent from reservoirs in general. An examination of AGFC rotenone population sample data from 2000 to 2010 by TMB revealed that *A. natalis* is indeed absent or uncommon in the large White River reservoirs, but it is very common in reservoirs elsewhere in Arkansas, sometimes exceeding the Channel Catfish in numbers and biomass per hectare at a given sampling locality. It is more common in Arkansas reservoirs, by far, than the other two *Ameiurus* species. This bullhead is sometimes associated with heavy

vegetation in shallow water (Glodek 1980b; Simon and Wallus 2004). Although not as tolerant of pollution as the Black Bullhead and Brown Bullhead, it is more tolerant of acidity than those species (Jenkins and Burkhead 1994). In a study of diel activity, Yellow Bullheads exhibited a nocturnal activity pattern (Reynolds and Casterlin 1977). In Missouri, Funk (1957) found that *A. natalis* was generally sedentary, with a tendency to travel greater distances upstream than downstream.

Yellow Bullheads are mainly carnivorous, feeding on insect larvae, mollusks (snails), crustaceans (amphipods, cladocerans, copepods, grass shrimp, and crayfish), and small fishes, which they locate by acute senses of smell and taste. Stomach contents sometimes include aquatic vegetation and organic detritus. The sense of smell in the Yellow Bullhead is important in social behavior as it enables an individual to recognize other individuals as well as their social status (Todd et al. 1967). Juveniles smaller than 100 mm SL feed mainly on microcrustaceans and immature insects (mainly midges and ephemeropterans), but adults are more omnivorous. Bullheads larger than 100 mm SL fed on crayfish, terrestrial arthropods, large immature aquatic insects (such as odonates), and fish (Sheldon and Meffe 1993). It apparently consumes more vegetation (Boschung and Mayden 2004) and is more selective and more nocturnal in its feeding habits (Harlan and Speaker 1956) than the other two Arkansas species of *Ameiurus*.

Spawning occurs in April to early June in Alabama (Boschung and Mayden 2004), and a similar spawning season is likely in Arkansas. Simon (1999b) classified this bullhead as a guarder and nest spawner. It is also classified as a speleophil, a hole nester (Wallace 1972; Simon 1999b). A saucer-shaped nest is excavated by both sexes at water depths of 1.5–4.0 feet (0.5–1.2 m) beside a bank, log, or tree root (Becker 1983). During spawning, male and female lie parallel, facing opposite directions, and each places the caudal fin over the head of the other (Wallace 1972). The male arches his ventral area toward the urogenital area of the female. This act is repeated several times and usually 300–700 yellowish-white eggs are deposited in a nest (Fowler 1917). Reported fecundities are between 1,650 and 7,000 eggs per female. A 282 mm TL female collected in the Ouachita River drainage in Louisiana contained 5,387 ovarian eggs; 243 of which were ripe (Simon and Wallus 2004). Fertilized eggs are 2.5–3.0 mm in diameter (Wallace 1972). Following spawning, the female is chased from the nest and the male returns to fan and guard the eggs (Wallace 1972). The incubation period is 5–10 days (Simon and Wallus 2004). After hatching, schools of several hundred black young feed in quiet waters, moving about in compact groups near the surface. The eggs, larvae, and juveniles are guarded by the male until the young reach about 1 or 2 inches (25–50 mm) in length (Mansueti and Hardy 1967). Sexual maturity is reached at age 2 or 3, and some individuals have attained an age of at least 7 years (Etnier and Starnes 1993). Maximum age was reported as 10 years (Ross 2001).

CONSERVATION STATUS   The Yellow Bullhead is widespread and abundant throughout Arkansas, and its populations are secure.

Ameiurus nebulosus (Lesueur)
Brown Bullhead

Figure 16.9. *Ameiurus nebulosus*, Brown Bullhead. *Uland Thomas.*

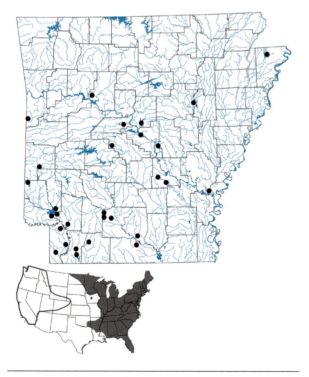

Map 115. Distribution of *Ameiurus nebulosus*, Brown Bullhead, 1988–2019. Inset shows native range (solid) and introduced range (outlined) in the contiguous United States.

CHARACTERS   A stout-bodied, small-eyed bullhead catfish with a mottled body, a free adipose fin, and a caudal fin margin that is slightly emarginate to rounded (Fig. 16.9). Head broad and long, its length going about 3.2–3.4 times into SL. Mouth terminal, jaws nearly equal in length, or upper jaw slightly longer than lower jaw. Premaxillary tooth patch without posterolateral extensions. Chin barbels darkly pigmented, at least at their bases. Barbel at corner of mouth reaching well behind base of pectoral fin in adults. Scaleless. Gill rakers usually 13–15 (11–15). Pectoral spine moderately to strongly serrate on posterior edge with 5–8 large serrae (Fig. 16.2B); pectoral rays 7–9. Dorsal fin with 1 spine and 6 or 7 rays. Anal fin rounded, without spines and with 20–24 rays. Maximum size 20.9 inches (532 mm) and about 7 pounds (3.2 kg), but seldom exceeding 1 pound (0.5 kg). Arkansas state angling record from Upper White Oak Lake in 2013 is 3 pounds, 3 ounces (1.45 kg).

LIFE COLORS   Adults with olive to gray body color above with strong black or brown mottling; however, degree of mottling is highly variable in Arkansas specimens. Sides mottled with yellow to brown blotches. Chin barbels darkly pigmented gray or black, but never uniformly white. Belly white to yellow. Anal fin slightly mottled, all fins dark-colored. Caudal base with uniform pigmentation, not depigmented. Young smaller than 2 inches (50 mm) TL are often black, except for a white belly and whitish underside of head.

SIMILAR SPECIES   Very similar to the Black Bullhead, but differs from it in having well-developed serrae on posterior edges of pectoral spines (vs. pectoral spine serrae absent or only weakly developed), in having 21–24 anal rays (vs. 17–21), and in having 11–15 gill rakers (vs. 15–21, but usually 18 or 19). It differs from the Yellow Bullhead in having chin barbels pigmented with gray or black (vs. chin barbels unpigmented, white). It further differs from other Arkansas *Ameiurus* and from *Ictalurus* in having a slightly emarginate caudal fin (vs. truncate in other *Ameiurus* and forked in *Ictalurus*). Small individuals differ from *Noturus* species in having a free adipose fin (vs. adipose fin adnate).

VARIATION AND TAXONOMY   *Ameiurus nebulosus* was described by Lesueur (1819) as *Pimelodus nebulosus* from the Delaware River near Philadelphia, Pennsylvania. It was subsequently placed in genus *Ameiurus*, then in *Ictalurus*, and was restored to *Ameiurus* by Lundberg (1982). The taxonomy of this species is confused and in need of revision. Most recent authors do not recognize subspecies, but some recognize a northern subspecies (*A. n. nebulosus*) and a southern subspecies (*A. n. marmoratus*), differing greatly in color pattern (Smith 1979). Hybridization between *A. nebulosus* and *A. melas* is known in Ohio and Kentucky (Burr and Warren 1986b). Molecular and morphological analyses identified *A. nebulosus* as sister species to *A. melas* (Hardman and Hardman 2008; Arce-H. et al. 2017).

DISTRIBUTION   *Ameiurus nebulosus* is native to the St. Lawrence–Great Lakes, Hudson Bay, and Mississippi River basins from Quebec west to southeastern Saskatchewan, Canada, south to Louisiana; Atlantic Slope drainages from Nova Scotia and New Brunswick, Canada, south to Florida; and Gulf Slope drainages from Florida to Mobile Bay, Alabama. The distribution of the Brown Bullhead in Arkansas is sporadic and confined mainly to southern and eastern parts of the state (Map 115). There are several valid records from the natural waters of Arkansas (some Red River oxbows, Bayou Bartholomew, the Black River, and a few lower Arkansas River tributaries), but the Brown Bullhead is uncommon in Arkansas. Although considered native to Arkansas (Black 1940; Glodek 1980c; Page and Burr 2011), *A. nebulosus* has also been stocked in several impoundments by the AGFC. A population of Brown Bullheads exists in Sugarloaf Lake in Sebastian County, and we have collected individuals as large as 4 pounds (1.8 kg) there. Other Arkansas reservoirs with small populations are Bois D'Arc (Hempstead County), and Upper and Lower White Oak lakes (Ouachita County).

HABITAT AND BIOLOGY   The Brown Bullhead seems to prefer quiet, clear waters with moderate to large amounts of aquatic vegetation (Pflieger 1997). In Alabama, it has been found in rivers, streams, oxbow lakes, swamps, and reservoirs in clear to turbid water, low to moderate flow, and over sand and mud substrates (Mettee et al. 1996). It was usually found near undercut banks and along the edges of aquatic vegetation. In Illinois, it occurs in clear, well-vegetated lakes (Smith 1979), and one of the largest Arkansas populations occurs in Sugarloaf Lake in Sebastian County, an impoundment that has a history of heavy growths of aquatic vegetation. Life history of this species is similar to the Black Bullhead and has been studied by Raney and Webster (1940). Trautman (1957) reported that the Brown Bullhead tended to remain in deeper water than other bullhead species, and it appeared to be far less tolerant of turbidity than the Black Bullhead; however, Becker (1983) considered it tolerant of high turbidity and of waters modified by domestic and industrial effluents. Mettee et al. (1996) noted that most Alabama records were from the eastern part of the state where rivers have not been dredged to maintain navigable channels.

Fry and fingerling Brown Bullheads feed on zooplankton and chironomids; adults feed on insects, fish, fish eggs, mollusks, and plants (Carlander 1969). Flemer and Woolcott (1966) reported that the diet included freshwater shrimp, crayfish, coleopterans, ephemeropterans, and fishes. Young bullheads smaller than 40–60 mm TL feed primarily on chironomid larvae, cladocerans, ostracods, and copepods (Keast 1985a, b). Larger fish are more benthic oriented in their feeding, ingesting amphipods, odonate larvae, chironomids, fish, and crayfish (Turner 1966). It is an opportunistic feeder, and availability usually determines what kinds of food items are eaten. Brown Bullheads primarily use their elaborate senses of smell and taste to find food, and they have been reported to feed at all times of the day and night but are most active at night (Harlan and Speaker 1956). Loeb (1964) reported a curious habit in *A. nebulosus* of burying the body in soft substrates, possibly as a feeding adaptation. The fish buried itself by tilting its head downward and swimming vigorously until only its mouth was exposed at the substrate surface and a funnel was formed leading into the mouth.

In a Michigan study, Blumer (1985a) reported spawning occurs in late May and early June. In the southern United States, spawning season is longer, beginning in April and continuing into late summer (Jenkins and Burkhead 1994) at water temperatures between 70 and 77°F (21–25°C). Nests (constructed usually by the female but occasionally by both parents) were generally in shallow water less than 3 feet (1 m) deep. Nest sites have been reported as holes excavated in banks, burrows under roots of aquatic plants, cavities in old stumps, and in a variety of discarded manmade objects. Breder and Rosen (1966) described spawning behavior that was similar to that of other bullhead species. With the male and female facing in opposite directions and their bodies in close contact, vibration occurs and eggs and milt are released. A female may deposit from 50 to 10,000 adhesive eggs in a nest. The eggs hatch in 6–9 days at water temperatures of 20.6–23.3°C (69–74°F). Brown Bullheads seem to be monogamous. The adults, principally males, care for the eggs (embryos), larvae, and juveniles as long as 20 days after oviposition. Females are more frequently involved in chasing away would-be predators, but they may also care for the eggs (Blumer 1985a). After leaving the nest, about 7–10 days after hatching, the young are herded and guarded by the male and sometimes the female. Parental care ceases soon thereafter, but the young bullheads often continue to travel in schools throughout their first summer. Maximum life span is 11+ years.

CONSERVATION STATUS   The Brown Bullhead is uncommon in Arkansas, but its populations are probably stable. Although we do not consider it in need of an official conservation status at this time, it is a species that deserves future monitoring to detect possible population changes.

Genus *Ictalurus* Rafinesque, 1820a— Forktail Catfishes

The monophyly of *Ictalurus* was supported by 10 morphological and four nucleotide synapomorphies (Arce-H. et al. 2017). The genus is morphologically characterized by a forked caudal fin, dorsal fin with 6 rays, anal rays 24–35, a supraoccipital process forming a bony ridge extending back to the dorsal fin origin, and no posterior projection of the premaxillary tooth patch. The posterior margin of the adipose fin is adnexed, that is, not attached to caudal peduncle. Lundberg (1992) investigated the phylogenetic systematics of *Ictalurus* and recognized two clades, an *I. furcatus* group and an *I. punctatus* group. However, Arce-H. et al. (2017) found that the most basal split was between *I. balsanus* and its sister clade formed by all other living and fossil species of *Ictalurus*. The genus contains 10 species (Page et al. 2013), two of which inhabit Arkansas, the Blue Catfish (*I. furcatus*) and Channel Catfish (*I. punctatus*).

Ictalurus furcatus (Lesueur)
Blue Catfish

Figure 16.10. *Ictalurus furcatus*, Blue Catfish. *Patrick O'Neil, Geological Survey of Alabama.*

CHARACTERS   A large, heavy-bodied catfish with a sloping, wedge-shaped head in profile, a prominent hump in front of the dorsal fin in adults, a deeply forked caudal fin, and a free adipose fin (Fig. 16.10). Anal fin straightedged, without spines and with 30–35 rays; length of anal fin base going into SL 2.9–3.3 times. The swim bladder is constricted in the middle, forming 2 approximately equal chambers. Scales absent. Dorsal fin with 1 spine and 6 soft rays. Pectoral fin with 1 sharp, robust spine and 8–10 soft rays; anterior margin of pectoral spine roughened; posterior

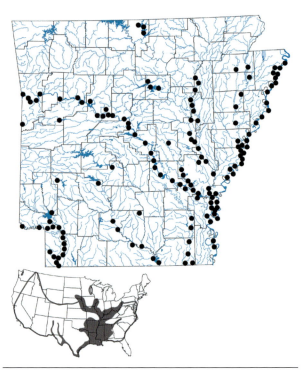

Map 116. Distribution of *Ictalurus furcatus*, Blue Catfish, 1988–2019. Inset shows native range (solid) and introduced range (outlined) in the contiguous United States.

margin with large, curved serrae (serrae are reduced in large fish). Pelvic fins without spines; pelvic rays 8. Gill rakers 14–21. Eye small, with large fish having proportionately smaller eyes. Mouth subterminal, upper jaw projecting well beyond lower jaw. Premaxillary tooth pad without backward extensions. Genital orifices of male and female are distinct. In the male, the papilla is more prominent and has a circular opening; in the female, it is more recessed and the opening is slitlike (Moyle 1976). Breeding male with a very wide head and bulging cheeks and opercles (Trautman 1957). Maximum length of 47 inches (1,194 mm) (Glodek 1980a); maximum weight well over 100 pounds (45.4 kg), but there are older reports of individuals greater than 200 pounds (91 kg). The Arkansas state angling record is 116 pounds, 12 ounces (53 kg) from the Mississippi River in 2001. In January 2005, a 139-pound (63.1 kg) Blue Catfish was caught by a commercial fisherman in the Mississippi River (Ken Shirley, pers. comm.).

LIFE COLORS   Body color bluish to gray above grading to grayish white on sides and belly, but never with prominent black spots. Fins light-colored, often with dusky outer margins. Chin barbels are usually unpigmented, nasal barbels unpigmented to dusky, and the long maxillary barbels are black. Juveniles have a yellowish-brown back and silvery sides. Breeding males dark blue (Moyle 1976).

SIMILAR SPECIES   The Blue Catfish is similar to the Channel Catfish, but the anal fin possesses a straight outer margin (vs. a curved outer margin in *I. punctatus*) and 30–35 anal rays (vs. 24–30). In addition, *I. furcatus* lacks black spots on the body and the back is conspicuously humped at the dorsal origin (vs. black spots usually present on body, especially in juveniles and young adults, and slope of head profile from dorsal fin to snout is less steep). Pflieger (1997) noted that the swim bladder of *I. furcatus* is constricted in the middle, dividing it into two parts, while the swim bladder of *I. punctatus* is not constricted in the middle. It can be distinguished from *Ameiurus* species by its deeply forked caudal fin (vs. caudal fin emarginate or rounded), and 30–35 anal rays (vs. anal rays fewer than 30). It differs from *Pylodictis olivaris* in having the upper jaw projecting beyond the lower jaw (vs. lower jaw projecting).

VARIATION AND TAXONOMY   *Ictalurus furcatus* was described by Lesueur *in* Cuvier and Valenciennes (1840) as *Pimelodus furcatus* from the Mississippi River, New Orleans, Louisiana. It was placed in genus *Ictalurus* by the 1880s, and several junior synonyms are known. The Blue Catfish is the type species of the *I. furcatus* species group, which is the sister group to the *I. punctatus* clade (Lundberg 1992). A molecular analysis by Hardman and Hardman (2008) resolved *I. furcatus* as sister to a clade formed by *I. lupus* and *I. punctatus*. A combined morphological and molecular analysis identified *I. furcatus* as sister to *I. meridionalis* (Arce-H. et al. 2017).

DISTRIBUTION   *Ictalurus furcatus* is considered native to the Mississippi River basin from western Pennsylvania to southern South Dakota and the Platte River of southwestern Nebraska, south to the Gulf of Mexico, and in Gulf Slope drainages from the Escambia River drainage of Florida and Alabama (where it is likely introduced) to the Rio Grande drainage of Texas and Mexico. It is introduced in Atlantic Slope drainages, western states, and Minnesota (Page and Burr 2011). In Arkansas, this species originally inhabited the larger rivers and their associated oxbow lakes, but it has been stocked by the AGFC into reservoirs throughout the state (Map 116).

HABITAT AND BIOLOGY   The Blue Catfish is mainly an inhabitant of the big rivers, where it is most abundant in swift, deep channels. It is usually found in moderate current over a sandy bottom. Small reservoir populations also exist in the state. A popular snagging fishery exists for this species in the swift tailwaters below locks and dams and power plants along the Arkansas River where fish are concentrated. Snagging rigs use deep-sea rods and reels that cast weights ranging from 6 ounces to 1 pound as far as 70 yards or more into the river. Above the weights are large

treble hooks that hook or snag fish in various parts of the fish's body. Snagging begins in the spring during heavy flows below the dams and continues for four or five months until the river attains its normal low summer flows. May is usually the peak month for snagging. Graham (1999) reviewed the biology and management of *I. furcatus*.

In the Ohio River prior to impoundment, most feeding by Blue Catfish occurred in the swiftly flowing chutes or rapids, and over bars or elsewhere in pools wherever there was a good current and where the bottom consisted of bedrock, boulders, gravel, or sand (Trautman 1957). This species feeds mainly near the bottom on fish, crayfish, aquatic insects, fingernail clams, and freshwater mussels (Brown and Dendy 1961). Foods vary widely depending on prey availability and age of the fish. Small individuals feed mainly on small invertebrates, and fish greater than 8–13 inches (203–330 mm) TL feed mainly on fishes and larger invertebrates (Brown and Dendy 1961). Eggleton and Schramm (2004) compared feeding habits of Blue Catfish in floodplain and main river channel habitats in the LMR. Differences in diets among habitats were strong. Caloric densities of consumed foods were generally greatest in floodplain lakes, least in the main river channel, and intermediate in the secondary river channels. There was also strong between-year variation in diet, with Blue Catfish feeding disproportionately on low-energy Zebra Mussels in the main river channel during 1997, and higher-energy chironomids and oligochaetes in floodplain lakes during 1998. William Keith (pers. comm.) reports it feeds heavily on the introduced clam *Corbicula* in the Arkansas and White rivers. Lewis (1999) found 52.9% of Blue Catfish sampled in Lake Dardanelle, Arkansas, from June 1996 to June 1998 consumed Zebra Mussels. Quinn and Limbird (2008) reported that biomass of *I. furcatus* had increased in the Arkansas River in association with the invasion of the river by Zebra Mussels. As with other catfishes, the senses of taste and smell are more important than sight in locating food.

Spawning occurs in late spring and early summer at water temperatures of 21–25°C (Sublette et al. 1990). Spawning in Louisiana is in April and May (Pflieger 1997) and in June in Illinois (Smith 1979). Pflieger (1997) reports nesting habits are similar to those of the Channel Catfish. Simon (1999b) classified the Blue Catfish as a guarder, nest spawner, and speleophil (hole nester). A nest is constructed by both parents in the late spring or early summer and guarded by the male until the young hatch. Spawning behavior is similar to that of the Channel Catfish. In Louisiana, males mature by the fourth year at lengths of 490 mm TL, and females mature during the fifth year at lengths of 590 mm

TL (Perry and Carver 1972). In Lake Texoma, Oklahoma, Blue Catfish reached 5.7 inches at the end of the first year, and averaged 10.0, 13.8, 17.4, 21, 25.8, 30.3, 34.3, 40.4, 42.1, and 44.0 inches at the end of the succeeding years (R. M. Jenkins 1956). In the Arkansas River prior to impoundment of the navigation system, average TL of Blue Catfish in years 0 through 4 were 7.2, 11.5, 16.7, 19.0, and 20.0 inches, respectively; the age 5 cohort was absent in the study area, and average TLs in age groups 6–8 were 24.4, 27.9, and 29.1 inches (Gray and Collins 1970). Growth of Blue Catfish was more rapid than that of Channel Catfish in Lake Texoma. This species migrates, and in the lower Mississippi River it moves farther down the river where water is warmest in winter, running upstream in summer (Jordan and Evermann 1916; Pflieger 1997). Ross (2001) and Smith (1979) noted that life expectancy is likely more than 20 years due to large sizes found. McAllister and McAllister (1988) reported an anomalous specimen of *I. furcatus* from Lake Overcup in Conway County, Arkansas, that lacked a caudal peduncle and fin.

CONSERVATION STATUS   Pflieger (1997) noted a general decline in abundance of this species in Missouri streams; however, good populations are present in the Mississippi and Arkansas rivers in Arkansas. The Blue Catfish is widespread and abundant throughout Arkansas, especially in the big rivers. Its populations are secure.

Ictalurus punctatus (Rafinesque)
Channel Catfish

Figure 16.11. *Ictalurus punctatus*, Channel Catfish. *Uland Thomas.*

CHARACTERS   A slender, elongate catfish with a deeply forked caudal fin and a free adipose fin (Fig. 16.11). Anal fin rounded, without spines and with 24–29 rays, and a short base (length of anal fin base usually going 3.4–3.7 times into SL). Swim bladder without elongate posterior extension and not constricted in the middle. Adults more robust than juveniles. Head broad but small, dorsal profile gently sloping from dorsal fin origin to snout. Head length going into SL 3.6–4.3 times. Mouth subterminal; upper jaw

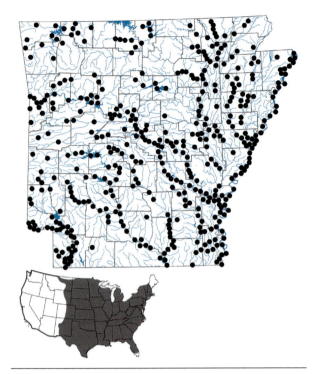

Map 117. Distribution of *Ictalurus punctatus*, Channel Catfish, 1988–2019. Inset shows native range (solid) and introduced range (outlined) in the contiguous United States.

protruding beyond the lower jaw. Scales absent. Dorsal fin with 1 spine and 6 rays. Pectoral fin with 1 spine and 7–9 rays; pectoral spine slightly dentate on anterior margin; posterior margin with well-developed, curved serrae that often become reduced in large fish. Pelvic rays 8 or 9. Gill rakers 13–15. Premaxillary tooth pad on upper jaw without backward extensions. Breeding males undergo drastic morphological change, and on several occasions, TMB has encountered fishermen who insisted that a breeding male was not the same species as the female. Nuptial males develop a blue-black enlarged head due to swelling of cranial musculature, thickened lips, and apparent increased width of mouth (Crawford 1957). Fin membranes are also thickened in breeding males (Sublette et al. 1990). In females the head is scarcely as wide as the body (Davis 1959). Males have a distinctive urogenital papilla extending posteriorly. In males there is one opening behind the vent, whereas females have two openings behind the vent (Moen 1959). The Arkansas state angling record from Lake Ouachita in 1989 is 38 pounds, 0 ounces (17.2 kg). Maximum size 50 inches (1,270 mm) TL (Glodek 1980d). Record weight was 58 pounds (26.3 kg) from Santee-Cooper Reservoir, South Carolina in 1964.

LIFE COLORS   Body color blue gray over the back and sides, grading to white on the belly. Sides with small dark, randomly scattered spots on all but the largest adults and small young. Adult Channel Catfish tend to lose their dark spots as they grow larger. Fingerlings develop black spots by about 60–70 mm SL; spots disappear gradually in fish older than 3–5 years (Canfield 1947) but sometimes persisting in adults. Breeding males with blue-black heads. Females lighter color than males. Barbels on lower jaw normally unpigmented, but dusky in large individuals. Large maxillary barbels usually dusky (Ross 2001).

SIMILAR SPECIES The Channel Catfish is similar to the Blue Catfish but differs in usually having scattered black spots on sides (vs. no spots), 24–29 anal rays (vs. 30–35), the outer margin of the anal fin is convex (vs. straight), and swim bladder without a constriction in the middle (vs. swim bladder constricted in middle forming 2 approximately equal chambers). It differs from *Ameiurus* species in having a deeply forked caudal fin (vs. emarginate or rounded caudal fin), and supraoccipital process forming a bony ridge extending back to the dorsal fin origin (vs. supraoccipital process forming a slightly projecting bony ridge that does not extend back to the dorsal fin origin). It differs from *Pylodictis olivaris* in having a sloping top of head (vs. top of head very flat), and in upper jaw projecting beyond lower jaw (vs. lower jaw projecting).

VARIATION AND TAXONOMY *Ictalurus punctatus* was described by Rafinesque (1818b) as *Silurus punctatus* from the Ohio River. The Channel Catfish was described as a new species several times by early workers. Bailey et al. (1954) presented data on geographic variation and did not recognize subspecies. It is the type species of the *I. punctatus* species group (Lundberg 1992) and was considered sister to the remaining species of that group by Arce-H. et al. (2017). A molecular analysis by Hardman and Hardman (2008) resolved *I. punctatus* sister to *I. lupus*. Extensive stockings of *I. punctatus* within and outside its natural range have probably obscured any noteworthy variation.

DISTRIBUTION The native range of *Ictalurus punctatus* is not precisely known because of widespread introductions throughout the United States, but it is probably native to the St. Lawrence–Great Lakes, Hudson Bay, and Mississippi River basins from southern Quebec to southern Manitoba, Canada, and Montana, south to the Gulf of Mexico; possibly native to some drainages along the Atlantic Slope from Canada to southern Florida; and Gulf Slope drainages from Florida to northern Mexico and eastern New Mexico. In Arkansas, the Channel Catfish occurs statewide (Map 117).

HABITAT AND BIOLOGY This extremely adaptable catfish seems to do equally well in farm ponds, reservoirs, streams, and rivers, but it is basically a stream fish. Adults seek out deep pools, submerged logs, and overhanging banks in the major rivers by day and at night feed in riffles and shallow pools. It reaches its greatest abundance in clear, fast-flowing, sand- or gravel-bottomed rivers of medium to large size (Smith 1979). Edds et al. (2002) found that when Channel and Blue catfish occur together, Channel Catfish tend to favor shallower portions of a lake, while blues occur in deeper sections; however, Channel Catfish inhabiting lakes without Blue Catfish are found in all habitats. The Channel Catfish is adversely affected by the hypolimnetic release of cold water below dams. Brown (1967) and Hoffman and Kilambi (1970) documented its elimination from the cold-water reaches of the White River below Bull Shoals Dam. Siegwarth (1992) provided evidence that the sparse Channel Catfish population in the Buffalo River may be partially attributed to reduced inputs from historic migratory stocks due to cold White River tailwaters. Several studies (summarized by Ross 2001) documented that Channel Catfish move extensively in rivers and reservoirs. Direction and degree of movement vary seasonally. In Beaver Reservoir, there was a trend for upstream migration in the spring, followed by downstream movement in the fall (Duncan and Myers 1978a). Siegwarth and Johnson (1994) cited several studies from a wide range of geographical locations showing that Channel Catfish exhibit extensive spring migrations from larger rivers and lakes into tributary streams. Those annual migrations appear to be in response to either a lack of overwintering habitat in the tributary or a lack of suitable spawning habitat in the confluence stream. Therefore, the Channel Catfish appears to require substantially different habitat areas for overwintering and spawning.

The Channel Catfish is an omnivore and feeds on a variety of foods, ranging from fish, insects, mollusks, and crayfish to occasional plant material and detritus, depending on availability. One measure of diet variability was provided by Bailey and Harrison (1948), who found 50 families of insects present in stomach contents of specimens from the Des Moines River, Iowa. It is also a scavenger, and Pflieger (1997) reported that it formerly consumed large quantities of slaughterhouse and rendering plant wastes along rivers when it was common practice to dispose of wastes in that manner. Dead fish and other animals are sometimes eaten. Numerous taste buds are located on the barbels and other areas of the fish and (along with the sense of smell) aid in the detection of prey items (Joyce and Chapman 1978). The larvae (called alevins) first feed on plankton but soon add small aquatic insect larvae to the diet (Jenkins and Burkhead 1994). In the Illinois River, Arkansas, alevins fed mainly on chironomid larvae and pupae (Armstrong and Brown 1983). Bailey and Harrison (1948) concluded that practically all the diet of individuals smaller than 102 mm TL consisted of small aquatic insects; fish larger than 102 mm switched mainly to larger aquatic insects, and with continued growth, additional food items were added to the diet, including fishes. Most studies indicate that there is much feeding activity at night, but several studies have also documented daytime feeding. Alevins in the Illinois River fed most actively in the hours just after dusk and just before dawn; during the day, they were inactive, remaining on the stream bottom or buried in it (Armstrong and Brown 1983). Those authors attributed the presence of alevins in nocturnal drift samples to displacement by river currents while feeding along the substrate.

Spawning begins in Arkansas and Oklahoma when the water warms to approximately 16–24°C (61–75°F) and continues at water temperatures up to 27°C (80°F) (Crawford 1957; Jearld and Brown 1971). Spawning occurs from May to August in Arkansas in a dark, natural cavity or a hole cleaned by the male in an undercut bank, or underneath a submerged log or pile of debris. Unlike other members of the genus *Ictalurus,* the female plays no part in nest site selection, nest construction, or subsequent guarding of the young (Clemens and Sneed 1957). Behavioral observations suggested that *I. punctatus* spawns in monogamous pairs, and genetic monogamy was confirmed by Tatarenkov et al. (2006). In that study, each male had mated with only one female in his nest. During spawning, the male and female were usually facing in opposite directions. Clemens and Sneed (1957) described spawning behavior as follows:

> Then he wrapped his tail around her head so that his caudal fin covered her eyes.... Then the male's body quivered, during which time his pectoral fins beat, but his pelvics remained rather motionless and pointed backwards, and in some cases slightly to the side. If the female responded, she usually did so within 5 seconds. When she participated she wrapped her tail around the head of the male and quivered in unison with him. With each reflex, a contraction of the abdominal muscles of the female moved the eggs posteriorly.... She then lunged forward about 3 to 5 inches as the eggs spurted out.... It was believed that the male released milt ... at the same moment that the female released eggs.

Following egg deposition and hatching, the male guards the fry until they leave the nest. After leaving the nest site, the male abandons his parental duties, but the young school for several weeks. After the young reach fingerling size, they tend to hide under shelters by day and become active at night. If adequate shelters are not available, the fingerlings form aggregations near the bottom and, when disturbed, flee together in a tight school (B. E. Brown et al. 1970). Fecundity of females weighing 1–4 pounds (0.45–1.81 kg) averaged about 8,800 eggs per kilogram of body weight (Clemens and Sneed 1957). Eggs hatch in 6 days at 25 °C and in 10 days at 15.6 °C. This species matures at an age of 4–5 years and a length of 12–15 inches (305–381 mm). The Channel Catfish may occasionally live more than 10 years, but the life span is usually 6–7 years or less. Maximum age in northern areas is 24 years, and in southern areas it is about 15 years (Mettee et al. 1996). Carlander (1969) reported a 40-year-old fish in Canada.

In an Arkansas study of heat tolerance of Channel Catfish, Allen and Strawn (1968) found that catfish 6 days old and 11 months old that were acclimated to temperatures of 79, 86, and 93°F (26, 30, and 34°C) had upper lethal temperatures of 97.9, 99.1, and 100.0°F (36.6, 37.3, and 37.8°C), respectively. Kilambi et al. (1970) found that the optimum conditions for raising Channel Catfish included a water temperature of 89.6°F (32°C) and a 14-hour photoperiod. The findings of Kilambi et al. (1970) were based on an evaluation of growth, food consumption, food conversion efficiency, and mortality. Stephens and Beadles (1979) studied the effects of cropping (periodic removal) on Channel Catfish growth in northeastern Arkansas.

The Channel Catfish, one of the foremost game and food fishes in Arkansas, is widely stocked in farm ponds and reservoirs throughout the state, and it is very important in the commercial fishing industry. Only Mississippi and Alabama produce more Channel Catfish than Arkansas.

CONSERVATION STATUS The Channel Catfish is widespread and abundant throughout Arkansas, and its status is secure.

Genus *Noturus* Rafinesque, 1818d—Madtoms

The monophyly of *Noturus* was supported by 20 morphological and 76 nucleotide synapomorphies, and based on 8 morphological and 23 nucleotide synapomorphies, *Noturus* formed a clade with the monotypic genus *Pylodictis* (Arce-H. et al. 2017). Lundberg (1970) provided morphological synapomorphies of the genus. Grady (1987) and Grady and LeGrande (1992) constructed a cladogram for 26 species using phylogenetic methods that incorporated allozymic, chromosomal, and morphological traits. Hardman (2004) used nucleotide sequence data to estimate genetic variation and infer phylogenetic relationships among species of *Noturus* and other ictalurids. Arce-H. et al. (2017) combined morphology, genes, and fossils to study phylogenetic relationships.

The genus *Noturus* contains 29 described species that range across biogeographic provinces in formerly glaciated and unglaciated areas of North America, with as many as 10 additional species awaiting description (Grady and LeGrande 1992; Burr and Stoeckel 1999). Some forms are undescribed because species of *Noturus* are cryptic, owing to limited variation of standard morphometric and meristic traits (Taylor 1969). As a result, subtle phenotypic variation often masks significant genetic divergence (Thomas and Burr 2004; Burr et al. 2005). *Noturus* is the most species-rich but fossil-poor genus of the Ictaluridae (Arce-H. et al. 2017). Within the genus, three subgenera have traditionally been recognized: *Rabida*, *Schilbeodes*, and *Noturus*. The subgenus *Rabida* contains species that are typically boldly patterned in black, white, and yellow, with curved pectoral fin spines (Fig. 16.4) heavily armed with recurved teeth or serrae. They generally inhabit swift, gravel-bottomed stream sections. Species in *Rabida* inhabiting Arkansas include *N. albater*, *N. eleutherus*, *N. flavater*, *N. gladiator*, *N. maydeni*, *N. miurus*, and *N. taylori*. Members of the subgenus *Schilbeodes* include *N. exilis*, *N. gyrinus*, *N. lachneri*, *N. nocturnus*, and *N. phaeus*. The monotypic subgenus *Noturus* contains only *N. flavus*. Arce-H. et al. (2017) concluded that the subgenera *Noturus* and *Schilbeodes* are not valid because their analysis recovered members of *Schilbeodes* nested within the *Noturus* group (paraphyletic). Those authors supported recognition of seven clades (species groups) within *Noturus*. In the following species accounts, we designate traditional subgenus assignment for each species.

A monograph of the genus *Noturus* by Taylor (1969) remains the best single source of pre-cladistic information on madtom catfishes. *Noturus* is characterized by a rounded or truncate caudal fin, short anal fin (12–27 rays), 8 pectoral fin rays, and usually 9 pelvic fin rays. The adipose fin is generally long and low and connected to the caudal peduncle, with the posterior margin adnate. Only *Noturus flavus* has posterior projections of the premaxillary tooth patch. Mayden (1983) reviewed the biology of the genus *Noturus*, and Burr and Stoeckel (1999) documented aspects of natural history. Boschung and Mayden (2004) commented on the ecological sensitivity of this group, calling them "ecological canaries" among groups of fishes because they are often the first to disappear as a result of environmental disturbance. Madtoms are small catfishes, with two of the largest species (*N. flavater* and *N. phaeus*) reaching lengths of 175–200 mm (7–8 inches). Trautman (1981) reported a specimen of *N. flavus* more than 1 foot in length.

Madtom catfishes are known for their painful stings.

Birkhead (1967, 1972) found that the toxicity is caused by poisonous secretions of cells in the integumentary sheath covering the dorsal and pectoral fin spines, rather than by secretions from axillary glands at the bases of the spines as some other ichthyologists had speculated. Egge and Simons (2011) studied venom gland and spine morphology of 29 species of *Noturus* and identified four basic sting morphologies: (1) smooth spine, no venom gland; (2) smooth spine with venom gland associated with shaft of spine; (3) serrated spine with venom gland associated with shaft of spine; and (4) serrated spine with venom gland associated with shaft of spine and posterior serrations. All species with serrated pectoral spines possessed venom glands and all but two of the species without serrae were venomous. The venom cells occurred along the shaft of the spine and frequently on the posterior serrae (when present). Venom cells were never associated with the anterior serrations present in some species. All 13 *Noturus* species found in Arkansas were included in the study, and all possessed venom glands (Egge and Simons 2011). *Noturus gyrinus* and *N. lachneri* had larger venom cells than other ictalurid species, with the venom gland extending nearly the entire length of the spine. In most species, the venom cells were most numerous on the distal third of the spine, and the venom gland did not extend more than half the length of the spine. The majority of *Noturus* species have serrated spines, and the serrations are thought to increase the mechanical damage caused by the sting and to facilitate the entry of venom into the wound (Egge and Simons 2011).

Noturus albater Taylor
Ozark Madtom

Figure 16.12. *Noturus albater*, Ozark Madtom. *David A. Neely.*

CHARACTERS A small, slender, elongate madtom of the subgenus *Rabida* with a proportionately small, rounded, or somewhat flattened head (head length going into SL 3.5 times or more), and the adipose fin forming a low, keel-like ridge and possessing a large dark bar (Fig. 16.12). Scaleless. Pectoral soft rays typically 9. *Noturus albater* and *N. maydeni* are the only members of the

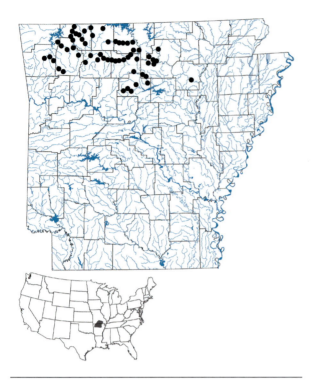

Map 118. Distribution of *Noturus albater*, Ozark Madtom, 1988–2019.

subgenus with 9 soft pectoral rays. Pectoral spine relatively short, with well-developed numerous posterior serrae with tips curved toward spine base (Fig. 16.4G), anterior serrae nearly obscure. Eye diameter going into snout length 1.5–2.4 times. Dorsal fin with 1 spine and 6 soft rays. Anal rays 13–16; pelvic rays typically 9; caudal rays 45–54. Normally 2 internasal and 11 preoperculomandibular pores. Jaws unequal, upper jaw extending beyond lower jaw. Premaxillary tooth patch with posterior corners rounded and without posterior extensions. Distance from tip of caudal fin to adipose fin notch going more than 2 times into distance from adipose fin notch to front of dorsal fin base. Maximum size about 5 inches (127 mm).

LIFE COLORS   Body color yellow with black or brown blotches. Body and fins mottled, and there are 4 indistinct dark bars crossing the dorsum. Caudal fin base without broad, dark bar. Dorsal fin not tipped with a dark blotch, but the spine and proximal portion are brown. Caudal fin with yellowish areas on the upper and lower procurrent rays.

SIMILAR SPECIES   The Ozark Madtom is endemic to the White River and is most similar to its sister species, the Black River Madtom, *N. maydeni*, a cryptic species of the Black River in Arkansas. See Variation and Taxonomy section below for more information on differences. Workers in Arkansas should remember that these species

are essentially morphologically indistinguishable and most reliably separated on the basis of their geographic distributions. *Noturus albater* is also superficially similar to the Mountain Madtom (*N. eleutherus*), Checkered Madtom (*N. flavater*), and Brindled Madtom (*N. miurus*) but differs from all three in having 9 pectoral rays (vs. 8), head length going into SL 3.5 times or more (vs. fewer than 3.5 times), and distance from tip of caudal fin to adipose fin notch going into distance from adipose fin notch to front of dorsal fin base more than 2 times (vs. fewer than 2 times).

VARIATION AND TAXONOMY   *Noturus albater* was described by Taylor (1969) from the White River at Forsyth, Missouri, in Taney County. It differs from *N. maydeni* in several very cryptic features. In the first edition of this book, we reported that William LeGrande (pers. comm.) found that *N. albater* specimens from the White River tributaries had relatively high chromosome numbers (65–72), and those from the Black River tributaries (Strawberry River) had distinctly lower numbers (56–58) (Robison and Buchanan 1988). Morphologically, LeGrande also concluded that White River specimens of *N. albater* reach larger body size, have bolder and more distinct dorsal saddles, and are more slender, while Black River counterparts have smaller body size, are usually more mottled with diffuse dorsal saddles, and appear somewhat more plump. However, Egge and Simons (2006) disagreed and could find no consistent morphological differentiation between the two species. Those authors described the Black and St. Francis form of *N. albater* as a distinct cryptic species, *N. maydeni*, based on fixed allele differences at five allozyme loci (Grady 1987), a lower chromosome count (Robison and Buchanan 1988), and differing base pairs in the mtDNA cytochrome *b* gene and the nuclear RAG 2 gene. Therefore, only karyotypes, allozyme variation, and DNA sequences provide reliable characters (along with geographic distribution) for distinguishing *N. albater* from *N. maydeni* (Egge and Simons 2006). Molecular analyses by Hardman and Hardman (2008) and Arce-H. et al. (2017) supported the sister species relationship between *N. albater* and *N. maydeni*.

DISTRIBUTION   *Noturus albater* is endemic to the upper White River drainage of Arkansas and Missouri (Map 118). In Arkansas, it is known only from the White River system upstream from its confluence with the Black River, and the Little Red River drainage above Greers Ferry Lake. It is locally abundant in the White River drainage and is replaced in the Black River drainage by *Noturus maydeni*.

HABITAT AND BIOLOGY   The Ozark Madtom occurs in clear, swift, moderate to large-sized streams with sand, gravel, rubble, or rock bottoms in riffles and rocky pools. *Noturus albater* prefers the swifter areas of shallow riffles. It is usually not found in headwater streams. Pflieger (1997) noted that it does not thrive in streams kept continuously cool by inflow from large springs. It occurs syntopically with *N. exilis*, *N. flavater*, and *N. miurus*.

Like all madtoms, *N. albater* hides by day under rocks and comes out to feed at night in riffles and shallow pools. Foods consist primarily of aquatic insects. Stomach contents of 38 specimens from six localities in Marion, Newton, Van Buren, and Washington counties were examined by TMB. Data from those samples (collected in April, May, July, and October) were pooled because of little dietary differences across sample sites and dates sampled. Ephemeropteran nymphs were, by far, the dominant food consumed. Other food items in decreasing order of percent frequency of occurrence were coleopteran larvae (especially riffle beetles), trichopterans, megalopterans, and ostracods. The presence of sand in most stomachs underscored the benthic nature of its feeding habits.

Based on examination of gonads by TMB, spawning occurs in Arkansas at least from mid-May through July. Ripe males and females were found in Kings River (Madison County) on 17 May at a water temperature of 19°C. The number of mature ova per female was rather small, with females 57–80 mm TL containing 5–59 mature ova. The greatest number of mature ova occurred in females taken in early July from Crooked Creek in Madison County. Size of the mature egg complement in females was similar to clutch sizes found in nests of *N. maydeni* by Mayden et al. (1980), but the mature egg complement per female in *N. maydeni* averaged more than twice the number found in *N. albater* by TMB. Other aspects of reproductive biology are also probably similar to *N. maydeni*.

CONSERVATION STATUS   The Ozark Madtom is widespread and common in the White River system of northern Arkansas. Its range does not appear to have changed perceptibly in this state in the past 30 years, and we consider its populations secure.

Noturus eleutherus Jordan
Mountain Madtom

CHARACTERS   A small, moderately elongate madtom with a very flattened head and a high adipose fin that is nearly free posteriorly from the caudal fin (Fig. 16.13). Head length going into SL fewer than 3.5 times. Eye diameter going into snout length 1.8–2.7 times. Scaleless. Mouth subterminal, lower jaw included. Premaxillary tooth patch

Figure 16.13. *Noturus eleutherus*, Mountain Madtom. *David A. Neely.*

Map 119. Distribution of *Noturus eleutherus*, Mountain Madtom, 1988–2019.

with obtusely angulate or rounded posterior corners, without posterior extensions. Adipose notch closer to distal margin of caudal fin than to front of dorsal fin base. Dorsal fin with 1 stout spine and 6 rays. Anal rays 12–16; pelvic rays 9; pectoral rays usually 8. Caudal fin short, usually with 43–49 (39–52) rays. Pectoral spine moderately long and slightly to moderately curved with 5–8 posterior serrae recurved toward base and anterior serrae also well developed (Fig. 16.4H). Two internasal pores and 10 or 11 preoperculomandibular pores. Maximum size about 5 inches (102 mm).

LIFE COLORS   Body color dull brown to yellowish brown with sides conspicuously mottled with dark brown. The venter is white to pale yellow. Dark blotch along the base of the adipose fin not extending to the outer fin margin, creating a wide, pale margin. Dorsal fin flecked with dark pigment but without a discrete, distal black blotch. A brown band crosses the lower part of the dorsal fin about three-fourths of the distance from the fin base but does not extend to fin margin. Dark brown bar crossing base of caudal peduncle (obscure in pale specimens). Development of 4 brown saddles on the back variable; saddles are absent in some individuals, indistinct in others, and well developed in a few. Front edge of first dorsal saddle at dorsal spine. Midcaudal crescent absent.

SIMILAR SPECIES   The Mountain Madtom is similar to *N. albater*, *N. maydeni*, and *N. miurus*. It differs from *N. albater* and *N. maydeni* in usually having 8 pectoral rays (vs. 9), a larger head with head length going into SL fewer than 3.5 times (vs. head length going into SL 3.5 times or more), and a high adipose fin, with a free posterior tip (vs. a more adnate adipose fin). It differs from *N. miurus* in having fewer caudal rays, usually 43–49 rays (vs. usually 57 or more), the dark bar at the base of the adipose fin extending only onto the basal half of the fin (vs. the dark bar at the base of the adipose fin extends to the margin of the fin); 2 internasal pores (vs. 1); 10 preoperculomandibular pores (vs. 11), and the dorsal fin not tipped with a black blotch (vs. dorsal fin tipped with a black blotch).

VARIATION AND TAXONOMY   *Noturus eleutherus* was described by Jordan (1877d) from Big Pigeon River, tributary to the French Broad River at Clifton (probably an error for near Newport, Cocke County), Tennessee. The type specimen was "taken alive from the mouth of a watersnake." It is a member of subgenus *Rabida* and has two population centers in Arkansas: one in southern Arkansas in the Little, Ouachita, and Saline rivers, and the other occurs disjunctly in northern Arkansas in the Strawberry, White, and Current rivers. According to W. R. Taylor (pers. comm. to HWR), specimens in the Strawberry River differ from typical *N. eleutherus* by having 36 vertebrae vs. usually 35. Also, there are 11 preoperculomandibular pores on both sides of the head, a condition Taylor considered rare in typical *N. eleutherus,* which usually has 10, frequently 11, on one side only. However, other characters, such as caudal ray counts, color pattern, and body structure, are typical of *N. eleutherus* elsewhere.

Various species relationships have been proposed for *N. eleutherus*, depending on the analysis used (LeGrande 1981; Grady 1987; Grady and LeGrande 1992). Analysis of mitochondrial and nuclear genes by Hardman (2004) and Hardman and Hardman (2008) placed *N. eleutherus* sister to a clade containing *N. placidus*, *N. stigmosus*, and *N. munitus*. This relationship was largely supported in a combined morphological and molecular analysis by Arce-H. et al. (2017).

DISTRIBUTION   *Noturus eleutherus* occurs in disjunct populations in the Mississippi River basin in Ohio River drainages from northwestern Pennsylvania to eastern Illinois, and south to northern Georgia and Alabama; the White and St. Francis river systems of Missouri and Arkansas; and the Ouachita and Red river systems of Arkansas and Oklahoma. David A. Etnier informed us of a single record from the main channel Mississippi River, Tennessee adjoining Arkansas. The Mountain Madtom has a curious disjunct distribution in Arkansas occurring in the Little, Ouachita (including the Caddo and Little Missouri rivers), and Saline rivers in southern Arkansas, and in the Strawberry, White, and Current river systems in northeastern Arkansas (Robison 1969) (Map 119). There are no records of this species from the St. Francis River drainage in Arkansas, but an older record exists from that river in Missouri (Pflieger 1997).

HABITAT AND BIOLOGY   The Mountain Madtom chiefly inhabits large, clear, swift to moderately flowing rivers and streams in Arkansas and avoids small creeks. It is most commonly found in river segments near the Fall Line. Specimens are frequently taken in association with dense vegetation and large stone, rock, and gravel bottoms. Smith (1979) reported that it is often found within the valves of long-dead mussels. In a study of the ecology and life history of *N. eleutherus* in Tennessee, Starnes and Starnes (1985) found this species associated with clean-swept, gravel-rubble riffles. Current velocities were estimated at 1.6–2.3 feet/s (0.5–0.7 m/s). Young-of-the-year were most frequently located in riffles less than 1.6 feet (0.5 m) deep. *Noturus eleutherus* was active at night, feeding under the edges of rocks or among *Podostemum* stems. During the daylight, it was found under large rocks or deep in substrate interstices (Fig. 16.14).

Starnes and Starnes (1985) found aquatic macroinvertebrates, consisting of immature ephemeropterans, dipterans, trichopterans, and plecopterans, were the dominant food items. Mountain Madtoms forage at night, seeking food under rocks and in vegetation in riffles.

The Mountain Madtom spawns during the first half of the summer (June and July) in Tennessee (Starnes and

Figure 16.14. *Noturus eleutherus* in gravel. Renn Tumlison.

Starnes 1985) and probably has a similar spawning season in Arkansas. Starnes and Starnes (1985) found a single nest in a shaded pool 2.3 feet (0.7 m) deep and about 13 feet (4 m) from shore on 2 July at a water temperature of 75.2°F (24°C). Substrate was primarily clean-swept, fine gravel and the nest site was under an elliptical rock 7.9 inches (20 cm) in diameter. Seventy eggs were guarded by a 2.6-inch (66 mm) SL male. After hatching, the larvae were guarded by the male until they dispersed. Maximum life span is 4 or 5 years.

CONSERVATION STATUS   The Mountain Madtom is uncommonly encountered in Arkansas and is usually not abundant at any locality; however, there are reports of hundreds of specimens collected in single samples from the Ouachita River (D. A. Neely, pers. comm.). Based on the relatively few locations in Arkansas from which this species is known, we consider it a species of special concern, and its status in Arkansas should be monitored periodically.

Noturus exilis Nelson
Slender Madtom

Figure 16.15a. *Noturus exilis*, Slender Madtom. *David A. Neely.*

Figure 16.15b. White River form of *Noturus exilis* with blackened margins of medial fins. *Uland Thomas.*

CHARACTERS   A slender, elongated, brownish madtom (body depth going into SL more than 4.5 times) of the subgenus *Schilbeodes* (Fig. 16.15) with a terminal mouth, jaws nearly equal, a single internasal pore, 10 preoperculomandibular pores, and 5–8 long, straight, conspicuous serrae along the rear margin of the pectoral spine; no anterior serrae are present (Fig. 16.4A). Scaleless. Head decidedly depressed and rather elongate. Eye small (eye diameter going into snout length 2.5–3.5 times). Premaxillary tooth patch without backward extensions. Adipose fin long, low, keel-like, closely united to the anterior caudal rays, and very

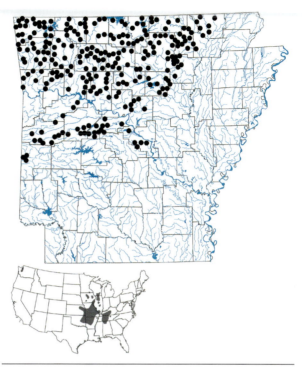

Map 120. Distribution of *Noturus exilis*, Slender Madtom, 1988–2019.

shallowly notched at the origin of the caudal fin. Notch in adipose fin closer to distal margin of caudal fin than to front of dorsal fin base. Dorsal fin with 1 spine and 6 rays. Caudal fin short (44–57 rays) and rounded or square. Anal rays 17–22; pelvic rays 8–10, usually 9. Generally 9 soft pectoral rays. Breeding males with enlarged head muscles, swollen lips, and enlarged genital papilla. Most adults are smaller than 4 inches (100 mm) TL. Largest Arkansas specimen 5.4 inches (137 mm) TL (Etnier and Starnes 1993). Maximum reported size from Missouri was 6 inches (152 mm) TL (Pflieger 1975).

LIFE COLORS   Body color gray to yellowish brown above grading to a white belly. Underside of head white or pale yellow. Body without saddles, dark blotches, or speckles. Dorsal, caudal, and anal fins often with blackened or darkened edges in some Arkansas populations.

SIMILAR SPECIES   The Slender Madtom is most similar to other brown madtoms—the Tadpole Madtom (*Noturus gyrinus*), Stonecat (*N. flavus*), Freckled Madtom (*N. nocturnus*), Brown Madtom (*N. phaeus*), and Ouachita Madtom (*N. lachneri*). It differs from the Stonecat, Brown Madtom, and Freckled Madtom in having a terminal mouth, with both jaws approximately equal, the upper jaw not projecting beyond the lower (vs. mouth inferior, the upper jaw projecting well beyond lower jaw). *Noturus exilis* differs from both the Tadpole and Ouachita madtoms in having the pectoral fin spine with well-developed posterior serrae (vs. pectoral fin spine without posterior serrae).

VARIATION AND TAXONOMY *Noturus exilis* was described by Nelson (1876) from Mackinaw Creek, McClean County, Illinois. Considerable genetic (Hardy et al. 2002; Blanton et al. 2013) and morphological (Hubbs and Raney 1944; Taylor 1969) variation exists among populations of *N. exilis* across its range. Differences in pigmentation are often striking. In Arkansas, there appear to be three forms of the Slender Madtom: a White River form, an Arkansas River form, and a Little Red River form. The White River form tends to be more elongate and the fins have very dark margins, whereas the Arkansas River form has more uniformly pigmented fins. The Little Red River form has 11 preoperculomandibular pores versus 10 in the other forms (Blanton et al. 2013). In addition, it is karyotypically distinct (William LeGrande, pers. comm.). Taylor (1969) documented meristic, pigmentation, and body shape differences among several Slender Madtom populations from the Interior Highlands but concluded that the observed variation did not warrant taxonomic recognition. Taylor also noted that, in general, the morphology of madtom catfishes (*Noturus*) was conserved, especially among closely related species, a result supported by several recent studies (Egge and Simons 2006, 2009). Hardy et al. (2002) found high levels of genetic structure and reduced levels of gene flow among populations and hypothesized that species divergence occurred prior to the Pleistocene based on a molecular clock rate of evolution applied to mtDNA sequences. Blanton et al. (2013) examined the genetic and morphological variation in *N. exilis* to explore the possibility of currently unrecognized species diversity. They recovered three well-supported and deeply divergent clades from analyses of genetic data: Little Red River (White River drainage) clade; Arkansas + Red River (Mississippi River) clade; and a large clade of populations from the rest of the range of *N. exilis*. The recovered clades showed few to no diagnostic morphological differences, which supported previous hypotheses of morphological conservatism in madtom catfishes. The Little Red River clade is the most morphologically distinct lineage of *N. exilis* with 11 POM pores (vs. 10 in other populations) and is also genetically distinctive. Blanton et al. (2013) did not treat it as a full species separate from *N. exilis* because they felt doing so would imply that the other major clades of *N. exilis* are more closely related to each other than they are to the Little Red River clade, which was not supported by their data. It is likely that new species will eventually be carved from the currently recognized *N. exilis*. The age of *N. exilis* was estimated as Late Miocene, 9.7 mya (Blanton et al. 2013).

DISTRIBUTION *Noturus exilis* has a disjunct range occurring in the Mississippi River basin in the Green, Cumberland, and Tennessee river drainages from central Kentucky to northern Alabama, and in the Mississippi River drainage of Wisconsin and Minnesota, south through Missouri, eastern Kansas, eastern Oklahoma, and Arkansas to the Ouachita Highlands. It had long been known that in Arkansas, the Slender Madtom occurred widely throughout the White River and Arkansas River systems (Map 120). In 1979, Robison and Winters reported the first specimens of *N. exilis* from the Mountain Fork River, a Red River drainage tributary. Five additional specimens were subsequently collected from Mill Creek (tributary of Mountain Fork River) (UF22742) by G. Burgess (pers. comm.). Robison et al. (2013) collected 53 individuals of *N. exilis* from the Mountain Fork River at State Highway 246, indicating that this species has established populations there.

HABITAT AND BIOLOGY The Slender Madtom inhabits small to moderate-sized, permanent, clear springfed creeks and rivers with cobble and gravel bottoms. It is absent from large-river habitats in Arkansas and apparently has little tolerance of turbidity. Adults generally prefer moderately flowing to swift riffles with some debris and large rocks for cover. Specimens are occasionally taken in gravel and rubble bottomed pools having current. In an Oklahoma study, 90% of the Slender Madtoms were found at depths less than 12 inches (300 mm), and small and large size classes did not differ in habitat use (Vives 1987). In Jacks Fork and Big Piney rivers in Missouri, it was the most common madtom catfish, occurring in all available macrohabitats and occupying a wide range of depths and current velocities; during summer, it exhibited increased use of backwater pools and emergent vegetation patches (Banks and DiStefano 2002). In winter, Slender Madtoms sometimes concealed themselves in leaf litter in deep, clear pools (Cross and Collins 1995). Buchanan (2005) reported this species from 7 Arkansas reservoirs, where 1,965 specimens were taken. Like most catfishes, it hides by day and is active at night. In Tyner Creek, Oklahoma, *N. exilis* was found under and among rocks in water 1–3 feet (0.3–0.9 m) deep during the day in May (Curd 1960). At the same locality after dark, Slender Madtoms were actively searching for food in shallow water less than 6 inches (0.15 m) deep, and 71 specimens were captured during 15 minutes of seining. Mayden and Burr (1981) reported densities of

Slender Madtoms in riffles in Southern Illinois streams of 549 individuals per 100 m², and in Arkansas, Brown and Lyttle (1992) reported densities of 78 per 100 m² in Clear Creek, 3 per 100 m² in Crooked Creek, and 2 per 100 m² in Kings River.

Foods include aquatic insect larvae (mayflies, caddisflies, stoneflies, and chironomids), crustaceans (especially ostracods and copepods), nematodes, and gastropods (Curd 1960; Mayden and Burr 1981; Vives 1987). Young fish feed more heavily on microcrustaceans (mainly copepods) and small dipteran larvae and pupae, while older fish feed more on larger crustaceans (isopods) and larger aquatic insects. Sullivan (1994) found a seasonal shift in diet, with ephemeropterans more important in August and dipterans dominating the diet in March. No crayfish were in the diet in March, but a small number occurred in August. Other food items were Zygoptera, Orthoptera, Psephenidae, and salamanders. Slender Madtoms feed nocturnally, sometimes in surprisingly shallow water, with a major feeding peak just before sunrise and a smaller peak just after sunset (Curd 1960; Mayden and Burr 1981).

The reproductive biology of the Slender Madtom has been studied in Illinois by Mayden and Burr (1981), and in northeastern Oklahoma by Vives (1987). Spawning occurred in Illinois from May through July and in Flint Creek, Oklahoma, from late April to early June. Vives (1987) found that oocyte sizes of *N. exilis* were larger, and clutch sizes smaller, in two northeastern Oklahoma populations than in the Illinois population studied by Mayden and Burr (1981). In Oklahoma, Vives (1987) reported a single egg cluster of *N. exilis* from Flint Creek that contained 44 embryos. Burr and Mayden (1984) reported an average clutch size of 72 embryos with a maximum clutch size of 124 in five clusters from the North Fork of the White River in Missouri, and 51 embryos with a maximum clutch size of 74 from southern Illinois (Mayden and Burr 1981). In Arkansas, we have collected ripe females during late April and May, but the spawning season extends into July. Connior et al. (2014) reported two embryo clusters of *N. exilis* from Flint Creek, Benton County (Arkansas River drainage) that had 86 embryos and 39 embryos with wet weights of 1.78 g and 0.84 g, respectively. Connior also found an embryo cluster with 154 embryos and a wet weight of 3.12 g from Water Creek (White River drainage) near Mull, Searcy County, Arkansas on 14 June. Ripe females were found in White Oak Creek and Fane Creek (both Arkansas River drainage) in Franklin County on 5 July 2015 (Tumlison et al. 2017). Nests were always found in cavities constructed by adult males in either pools or

riffles beneath large rocks and were guarded by a single male or by both parents. Males in the North Fork and Meramec rivers, Missouri, were also found guarding newly hatched fry during the first two weeks of July (Burr and Mayden 1984). Clutch size averaged 51 eggs. Eggs hatched in 187–210 hours in the laboratory at 77°F (25°C). Both sexes generally mature in 2 years although a small percentage matures at 1 year of age. Individual males live more than 4 years. McAllister et al. (2015c) reported helminth parasites of *Noturus exilis* in Arkansas.

CONSERVATION STATUS The Slender Madtom is widespread and common in northern Arkansas. Its populations are secure.

Noturus flavater Taylor
Checkered Madtom

Figure 16.16. *Noturus flavater*, Checkered Madtom. *Uland Thomas.*

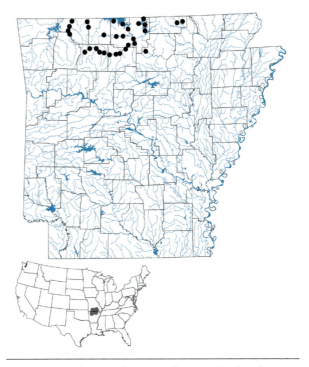

Map 121. Distribution of *Noturus flavater*, Checkered Madtom, 1988–2019.

CHARACTERS   A moderately large, prominently barred madtom with the upper jaw projecting beyond the lower jaw (Fig. 16.16). Scaleless. Head length going into SL fewer than 3.5 times. Caudal fin squarish or slightly rounded; adipose fin low and keel-like with a relatively free posterior tip. Premaxillary tooth patch without backward extensions. Pectoral spine with the posterior margin possessing large recurved serrae; the anterior margin with prominent long serrae that turn outward distally and inward proximal to the base (Fig. 16.41). Dorsal fin with 1 spine and 6 rays. Pectoral soft rays 8. Anal rays 14–17; pelvic rays 9. Caudal fin with 51–60 rays. Typically with 2 internasal pores, and 11 preoperculomandibular pores. Second-largest species of *Noturus*. with adults commonly 4–7 inches (102–180 mm) TL, but maximum reported size is 8.1 inches (206 mm). The largest Arkansas specimen we examined was 7 inches (178 mm) TL from the Buffalo River.

LIFE COLORS   The only Arkansas species of *Noturus* to have both a broad, black vertical bar at the base of the caudal fin and a large black blotch on the outer one-third of the dorsal fin; the blotch extends across all fin rays except the last. A broad, black bar extends from the base to the distal margin of the adipose fin. A narrow vertical bar in the middle of the caudal fin does not continue to margins. Caudal fin with a broad, basicaudal black bar and a prominent black border. Four prominent black saddles cross the yellow back, which has dusky mottling evident. Sides lightly mottled. Belly white.

SIMILAR SPECIES   The Checkered Madtom is somewhat similar to the Ozark Madtom, Black River Madtom, and Brindled Madtom. *Noturus flavater* differs from the Ozark and Black River madtoms in having 8 pectoral rays (vs. 9), a relatively free posterior margin of the adipose fin (vs. a moderately to broadly united posterior margin of the adipose fin), a broad, black, vertical basicaudal bar, and a large black blotch on the outer one-third of the dorsal fin (vs. no prominent basicaudal bar and no dark blotch on outer third of dorsal fin). *Noturus flavater* differs from *N. miurus* in having 2 internasal pores (vs. a single internasal pore) and a prominent basicaudal bar extending rather broadly across the caudal peduncle (vs. basicaudal bar seldom prominent on caudal peduncle in *N. miurus*).

VARIATION AND TAXONOMY   This White River endemic was described by Taylor (1969), who designated the type locality as Flat Creek, 12 miles northeast of Cassville, Barry County, Missouri. Taylor reported that populations of *N. flavater* from the Current and White rivers appeared to be identical morphologically, and no significant variational trends were found. The general similarity of color pattern and number of caudal rays appeared to align *N. flavater* with the *N. miurus* group (Taylor 1969). Based on a combined morphological, chromosomal, and allozymic analysis, Grady and LeGrande (1992) found that *N. flavater* was most closely related to *N. furiosus*, a relationship supported by an analysis of the cytochrome *b* and RAG2 genes by Hardman (2004). A combined morphological and molecular analysis by Arce-H. et al. (2017) found that *N. flavater* was most closely related to a clade consisting of *N. eleutherus*, *N. placidus*, *N. miurus*, *N. gladiator*, and *N. stigmosus*.

DISTRIBUTION   The Checkered Madtom is endemic to the White River system of Arkansas and Missouri. In Arkansas, it historically occurred throughout the White River system upstream of its confluence with the Black River, with a disjunct population in the Spring River (Map A4.119). Its distribution apparently skips the Strawberry River. It has likely been extirpated from the upper White River above Beaver Reservoir, and there are also no recent records from the White River main stem downstream of Baxter County (Map 121). There are no records from the Current and Eleven Point rivers in Arkansas, but it is known from those rivers in Missouri.

HABITAT AND BIOLOGY   The Checkered Madtom is restricted to cool, clear, high-gradient rivers. When not spawning, it inhabits deep, quiet backwater pool areas over bottoms of gravel, boulders, and flat rocks. Pflieger (1997) reported that it is often found in association with thick deposits of leaves, sticks, and other organic debris. In Jacks Fork River, Missouri, *N. flavater* was found at an average water depth of 0.67 m and was more often associated with boulders than two other madtom species in that river (Banks and DiStefano 2002). Fairly large populations are found in two Arkansas reservoirs, Lake Norfork and Bull Shoals Lake (Buchanan 2005).

Feeding habits are little known. Pflieger (1997) speculated that because of its large size, it probably consumes a greater proportion of crayfish and even small fishes than other Missouri madtoms. Stomach contents of 20 specimens collected from the White River in August were examined by TMB. The adult specimens, ranging from 78 to 116 mm TL, fed most heavily on ephemeropteran nymphs. Other food items consumed in small amounts were coleopterans, odonates, and megalopterans.

Burr and Mayden (1984) investigated the reproductive biology of *N. flavater*. Males are generally larger than females. No discernable sexual dimorphism in body color pattern or fin lengths was apparent. Males select a nest site and guard it. Nest-guarding males have enlarged cephalic epaxial muscles and swollen lips, giving their bodies a flattened appearance. The enlarged head of male madtoms may function in nest guarding as males typically face the opening of the nest site, blocking potential intruders (Burr and Mayden 1984). Females are ready to spawn (based on ovary size) by mid-April. They have two distinct classes of ova: mature (1.3–3.2 mm diameter, with a mean of 2.0 mm), and immature. Baker and Heins (1994) attributed this extreme range to the inclusion of females in widely varying stages within the clutch production cycle, with mean egg size therefore being biased downward. The number of mature ova in *N. flavater* ranged from 93 to 340 (Burr and Mayden 1984). Three nests were found in pools and one in a raceway about 3 feet deep (0.9 m) in early July under thick, flat rocks with siltfree gravel and some sand in the nests. Nest-guarding males were 3–5 years old. Males guarded nests 10–12 days post-hatching. The nests contained 250–350 larvae.

CONSERVATION STATUS   The Checkered Madtom is uncommon in Arkansas. Even though it has adapted somewhat to Norfork and Bull Shoals reservoirs, it apparently did not become established in Beaver Reservoir further upstream. Its populations have declined in Arkansas, and we consider it a species of special concern.

Noturus flavus Rafinesque
Stonecat

Figure 16.17. *Noturus flavus*, Stonecat. *Uland Thomas.*

CHARACTERS   A slender madtom with a unicolored dorsum representing the only member of the subgenus *Noturus* (Fig. 16.17). Distinguished from all other Arkansas madtoms by the posterior crescent-shaped extensions on each side of the premaxillary tooth patch on the upper jaw, and by the greater number of paired fin rays (usually 9 or 10 pelvic rays and 9–11, more often 10, soft pectoral rays). Scaleless. Pectoral spines reduced with rear margin nearly smooth (Fig. 16.4B). One dorsal spine and usually 6, sometimes 7 dorsal fin rays. Anal rays 15–18. Gill rakers 6 or 7. Lower jaw included; mouth subterminal. Eye small (eye diameter going into snout length 2.5–4.7 times). Adipose fin with notch, but not forming a free flaplike lobe. Caudal fin truncate to slightly rounded. Largest species of *Noturus*;

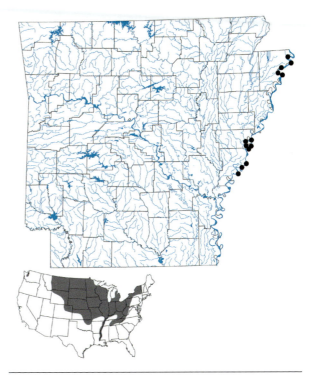

Map 122. Distribution of *Noturus flavus*, Stonecat, 1988–2019.

adult size 3–7.4 inches (76–188 mm) TL. Maximum reported size 12.3 inches (313 mm), and a weight of 1 pound, 1 ounce (0.48 kg) (Trautman 1957).

LIFE COLORS   Body brown, tan, or yellow, but without dark blotches, bars, or speckles. A slightly darker band of pigment runs through the caudal fin, with upper and lower rays lighter. Lower surface of head and abdomen immaculate. Mississippi River specimens in Arkansas have a distinct yellow coloration.

SIMILAR SPECIES   It differs from catfishes in the genera *Ictalurus, Ameiurus*, and *Pylodictis* in having an adnate adipose fin without a free posterior tip. It is distinguished from all other *Noturus* by the posterior extensions of the premaxillary tooth patch (vs. no posterior extensions).

VARIATION AND TAXONOMY   *Noturus flavus* was described by Rafinesque (1818c) from the falls of the Ohio River (no locality given). Taylor (1969) designated a neotype from Eagle Creek (tributary of lower Kentucky River), 3.5 miles east of Jonesville, Grant County, Kentucky. Taylor considered *N. flavus* unique among madtom catfishes based on a combination of morphological characters and placed it in the monotypic subgenus *Noturus*. LeGrande (1981) reported no convincing karyological differences between the subgenera *Noturus* and *Schilbeodes* and questioned the validity of their continued separate recognition. LeGrande and Cavender (1980) discovered evidence of a "chromosomal race" of *N. flavus* in the upper Tennessee River system. Burr and Warren (1986b) considered the Cumberland River drainage populations a separate species. Hardman (2004) identified divergence of mitochondrial DNA cytochrome *b* sequences within *N. flavus* similar to levels identified for some distinct species of *Noturus*. Faber et al. (2009) investigated intraspecific phylogeography of *N. flavus* and found two morphologically distinct forms, one in lowland drainages and another in the Eastern Highland drainages of North America. Etnier and Starnes (1993) noted that main channel Mississippi River populations and those found in the lower Missouri River differ noticeably from other populations in having tiny, nearly vestigial eyes. We have also noted that Mississippi River specimens from Arkansas have tiny eyes and a yellowish body coloration. Earlier, Taylor (1969) had also noted considerable variation in eye size, but found no consistent differences between populations. To further complicate matters, Faber et al. (2009) studied genetics of the two forms and found the small-eyed forms had the same mtDNA haplotypes as the adjacent large-eyed forms. Egge (2007) provided a complete osteological description of *N. flavus*. He confirmed *N. flavus* has several unique skeletal features compared with other madtom catfishes, including large posterior extensions of the premaxillary tooth patch, generally unfused pectoral radials, and a higher number of paired-fin rays. In addition, Egge (2007) discovered a coronomecklian bone associated with the dentary of *N. flavus* that had not been described previously. Egge (2007) concluded that the subgenus *Noturus* is monotypic and can be considered monophyletic as long as *N. flavus* is recognized as monophyletic.

Lundberg (1970, 1982, 1992) proposed *N. flavus* as a sister species in a clade consisting of *N. insignis* and *N. stigmosus*. More detailed relationships within the genus *Noturus* have been studied by LeGrande (1981), Grady (1987), Grady and LeGrande (1992), Bennetts et al. (1999), Hardman (2004), and Arce-H. et al. (2017). The relationship of *N. flavus* to other species is unresolved, and it has been placed by various authors as a sister species to all madtoms (Lundberg 1970; Grady and LeGrande 1992), sister to *Schilbeodes* (Taylor 1969), nested within *Schilbeodes* (LeGrande 1981; Hardmann 2004) and sister to *Rabida* (Grady and Le Grande 1992). A combined morphological and molecular analysis by Arce-H. et al. (2017) found that *N. flavus* was most closely related to a clade containing *N. nocturnus*, *N. gyrinus*, and *N. lachneri*.

DISTRIBUTION As currently recognized, *Noturus flavus* occurs in the St. Lawrence–Great Lakes, Hudson

Bay, and Mississippi River basins from Quebec to Alberta, Canada, south to northern Alabama, Mississippi, northeastern Oklahoma, and the Mississippi River main stem to southern Louisiana. It is also known from one Atlantic Slope drainage: the Hudson River, New York. Black (1940) did not collect *N. flavus* in Arkansas, and Taylor (1969) considered pre-1900 reports and other records from the Illinois River drainage in Benton and Washington counties erroneous (probably misidentifications of *N. exilis*). The first vouchered specimen reported from Arkansas was a female (TNHC 8347) collected from the mainstem Mississippi River in Phillips County (Buchanan 1973c). Later, Carter (1984) added a single specimen from Mississippi County. McAllister et al. (2012b) added five localities for the Stonecat in the Mississippi River from Mississippi County, and data from the Vicksburg Waterways Experiment Station (provided by K. J. Killgore, USACE) added several new records from more southern locations in Arkansas. The Stonecat has now been reported from 13 sites in the northern two-thirds of the Mississippi River in Arkansas, and it undoubtedly occurs throughout that river in this state, based on records of its occurrence in Louisiana (Map 122).

HABITAT AND BIOLOGY Although *N. flavus* has been found in rather clear, gravel-bottomed streams in other areas, the Arkansas specimens were collected in swift current in the turbid Mississippi River primarily from rock jetties and dikes and only rarely over a sand or gravel substrate. Given the scarcity of naturally occurring gravel and cobble substrates in the lower Mississippi River, stone dikes provide especially important habitat for filter-feeding insects (Payne and Miller 1996), thereby providing the most suitable habitat in the LMR for *N. flavus*. However, north of Arkansas in the main channels of the Mississippi, Missouri, and Ohio rivers, Stonecats are reportedly found mainly over a sand substrate in swift currents (Taylor 1969; Walsh and Burr 1985). LeGrande and Cavender (1980) reported that *N. flavus* inhabits moderate- to high-gradient streams, preferring riffles or rapids of moderate to large streams, often with a substrate of large rocks or cobbles under which it hides during the day. In the Gasconade River drainage, Missouri, it was distinctly rheophilic but associated with fairly shallow depths (0.03–0.46 m) and high current velocities (Banks and DiStefano 2002). Cross (1967) regarded it as the most adaptable madtom in terms of habitats occupied. Becker (1983) noted that the Stonecat is a solitary catfish compared with most other members of the family Ictaluridae.

Primarily nocturnal, *N. flavus* feeds on a wide variety of benthic organisms, principally aquatic larvae, insects, decapod crustaceans, and mollusks. Its diet is supplemented with an occasional darter or other small fish (Walsh and Burr 1985; Pflieger 1997). Tzilkowski and Stauffer (2004) found that it was a generalist feeder. The diet varies with size of the individual, with specimens smaller than 60 mm TL consuming mainly trichopteran and ephemeropteran nymphs; at between 60 and 120 mm TL, the diet included plecopterans and chironomids in addition to trichopterans and ephemeropterans; individuals greater than 120 mm TL fed mainly on ephemeropteran nymphs and crayfish. Specimens ranging between 50 and 107 mm TL collected from the Mississippi River in Mississippi County, Arkansas, on 26 July by TMB and Sam Barkley fed almost exclusively on ephemeropterans.

In Illinois and Missouri, *N. flavus* spawns in June and July when water temperatures exceed 77°F (25°C) (Walsh and Burr 1985). Etnier and Starnes (1993) reported a long spawning season in Tennessee, from April to July. Nuptial males have enlarged cephalic epaxial muscles and swollen genital papillae. Nests are found beneath very large rocks in pools and riffles of moderate current and slightly turbid water. Each nest is guarded by a male 3.4–4.1 inches (87–105 mm) TL. Clutch sizes range from 100 to 500 eggs, and total fecundity varies from 200 to 1,200 eggs (Walsh and Burr 1985; Boschung and Mayden 2004). Average clutch size in Pennsylvania was 261 eggs (Tzilkowski and Stauffer 2004). Males also protect the young for a short period after hatching. Individuals live 5–6 years (Walsh and Burr 1985).

CONSERVATION STATUS   The Stonecat is uncommon in the Mississippi River of Arkansas and is known from the northern two-thirds of that river in the state. Pflieger (1997) considered *N. flavus* rare in the lower Mississippi River in Missouri. Etnier and Starnes (1993) listed only three records from the Tennessee side of the mainstem Mississippi River. The Nature Conservancy lists *N. flavus* as critically imperiled (S1) in Arkansas. It should be considered a species of special concern, but it may be more common than presently thought in this state because of the difficulty in extracting specimens from its preferred Mississippi River habitat.

Noturus gladiator Thomas and Burr
Piebald Madtom

CHARACTERS   A medium-sized, robust, mottled madtom of the subgenus *Rabida* with an arched, slightly flattened head, included lower jaw, short body, and conspicuous pattern of black and pale yellow body coloration (Fig.

Figure 16.18. *Noturus gladiator*, Piebald Madtom. *David A. Neely.*

Map 123. Distribution of *Noturus gladiator*, Piebald Madtom, 1988–2019.

16.18). Scaleless. Eyes large, eye diameter going into snout length 1.7–2.4 times. Premaxillary tooth patch rounded and without posterior extensions. Anal rays 12–15; typically 9 pelvic fin rays. Caudal rays usually 48–55. Pectoral fin long, sickle-shaped, and when depressed, reaching beyond pelvic fin origin (except in very large specimens). Pectoral rays 7–9. Pectoral spine long and curved, with 5–10 well-developed posterior serrae and prominent anterior serrae (Fig. 16.4L). Internasal pores typically 2, preoperculomandibular pores typically 11. Dorsal fin with 1 spine and usually 6 (5–7) rays. Vertebrae 33 or 34. Adipose fin short and high, nearly free at posterior tip. Caudal fin truncate with slightly rounded edges. Slight sexual dimorphism between females and non-nuptial males. Nuptial condition in males includes pronounced development of head musculature and swollen lips. Sexes generally difficult to distinguish without internal examination of gonads. Adult size generally 1.5–3.9 inches (37–100 mm) SL. Maximum size 4.2 inches (106.5 mm) SL.

LIFE COLORS   Body color on dorsum and sides pale yellow or sandy to tan with a golden sheen. Four contrasting dark brown to black dorsal saddles, blotches, and bars on body. Ventral areas white to cream-colored. Abdomen and pectoral girdle speckled with discrete dark, round melanophores (often forming irregular clusters in larger specimens). Sides variously mottled with patches of darker melanophores. Two pale yellow spots occur posterior to the eyes, and a yellow saddle is located behind the occiput. Adipose blotch extends one-half to four-fifths the distance from the base toward the margin of the fin, but does not quite reach the fin margin. The adipose blotch may reach the fin margin in juveniles and subadults. Midcaudal crescent typically extends across the upper and lower procurrent caudal rays to the caudal peduncle.

SIMILAR SPECIES   Noturus gladiator is superficially similar to N. miurus and N. eleutherus. It differs from N. miurus in having 55 or fewer caudal rays (vs. 57 or more), lacking a black terminal blotch in the dorsal fin (vs. terminal blotch present), having a crescent-shaped midcaudal band (vs. midcaudal band absent), and the black bar in the adipose fin not reaching the fin margin (vs. dark adipose fin bar extends to fin margin). Noturus gladiator differs from N. eleutherus in having 11 preoperculomandibular pores (vs. 10), 48–55 caudal rays (vs. 43–49), and a midcaudal crescent (vs. no crescent).

VARIATION AND TAXONOMY   Noturus gladiator was described by Thomas and Burr (2004) from Pleasant Run (Hatchie River drainage), 2.5 km west of Bolivar, Hardeman County, Tennessee. Formerly part of the N. stigmosus complex, it is distinguished by its bold and contrasting pigment pattern, large body size, specific aspects of body shape, and more extensive armature on the anterior and posterior portions of the pectoral spine. Noturus gladiator shows no trenchant geographic variation in pigmentation, morphometric, or meristic characters among known populations (Thomas and Burr 2004). Noturus stigmosus is the sister taxon of N. gladiator, and both of those species are in the N. furiosus group (Thomas and Burr 2004; Arce-H. et al. 2017).

DISTRIBUTION   Noturus gladiator occurs only in three states of the lower Mississippi River basin from the Obion River system of western Tennessee, south to the Big Black River system of central Mississippi. Arkansas is included in the distribution of N. gladiator based on a single specimen (UT 48.243) taken from the main channel of the Mississippi River at a gravel island, 100 yards (91 m) from

a boat landing at Randolph, Tipton County, Tennessee, adjacent to Arkansas (Map A4.121) on 10 October 1976 by D. A. Etnier and W. C. Starnes (David A. Etnier, pers. comm.). There are no recent records from Arkansas (Map 123).

HABITAT AND BIOLOGY   Noturus gladiator typically occupies mainstem rivers of small to medium size and the lower reaches of their major tributaries. It is seldom found in the largest rivers, but there is one record from the Mississippi River. It is usually found over shifting sand and clay substrates and in water varying from clear to turbid with moderate current (Rohde 1980c; Thomas and Burr 2004). It occurs in sand-swept streams under debris piles (leaf packs, brush, and log jams) on the Coastal Plain. In Mississippi, Ross (2001) often found this species over sand and mud substrates in pools below riffles in streams 4–15 m wide at depths of 15–30 cm. It tends to avoid extreme silty situations, but it has often been found in turbid waters with poor visibility (Thomas and Burr 2004). There is no information on the diet of N. gladiator.

Spawning probably occurs from May through June. Conservation Fisheries Inc. (2007) spawned N. gladiator in 100-gallon aquaria in Knoxville, Tennessee. The following anecdotal observations on spawning are taken from Thomas and Burr (2004). Spawning adults used clay flowerpot saucers as nests. A 130 mm TL male was discovered guarding one nest on 21 May with about 100 metalarvae produced from a 100 mm TL female. A second nest with about 40 eggs was guarded on 19 June by a different male, and another male was guarding larvae close to the size (about 15 mm TL) at which they might disperse and leave the nest. Young, once becoming free-swimming, spent considerable time in the water column around floating objects. These observations are similar to what is known for N. stigmosus (Thomas and Burr 2004).

Thomas and Burr (2004) predicted that many of the life history features reported for N. stigmosus would be found similar to those of N. gladiator. Noturus gladiator has an equilibrium life history strategy because of its relatively small clutch size, large eggs, extreme parental care, and small body size (Winemiller and Rose 1992). Species of Noturus and ictalurid catfishes in general maximize juvenile survivorship by investing resources in small propagule size and parental care (Pianka 2000).

CONSERVATION STATUS   The Piebald Madtom is one of the rarest fish species in Arkansas, having been found only in the mainstem Mississippi River. Thomas and Burr (2004) considered N. gladiator vulnerable to local extinctions because of habitat deterioration over a significant portion of its small range, and because of its equilibrium life history strategy. It is considered vulnerable in Tennessee and a species of special concern in Mississippi (Ross 2001). We consider *N. gladiator* a species of special concern in Arkansas.

Noturus gyrinus (Mitchill)
Tadpole Madtom

Figure 16.19a. *Noturus gyrinus*, Tadpole Madtom, slender-bodied form. *David A. Neely.*

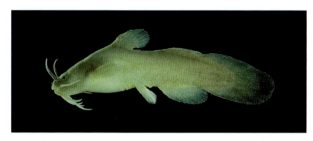

Figure 16.19b. *Noturus gyrinus*, Tadpole Madtom, hump-backed form. *Patrick O'Neil, Geological Survey of Alabama.*

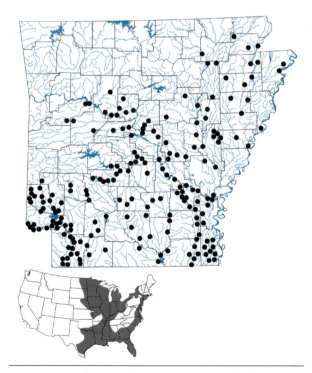

Map 124. Distribution of *Noturus gyrinus*, Tadpole Madtom, 1988–2019.

CHARACTERS A small, stout, uniformly brown-colored madtom with smooth pectoral spines and a terminal mouth (upper and lower jaws are nearly equal in length). Two internasal pores and 10 preoperculomandibular pores. Scaleless. Body shape variable in Arkansas populations (Fig. 16.19), usually short and chubby, deepest at or in front of dorsal fin, but body sometimes slender, elongated, and flattened anteriorly like that of *N. exilis*; therefore, the body depth into SL morphometric is not a reliable character for separating *N. gyrinus* and *N. exilis* in Arkansas. Head length going into SL 3.0–3.8 times. Eyes very small, eye diameter going into snout length 3.5–5.0 times. Pectoral spine small, nearly straight and deeply grooved, but without serrae (Fig. 16.4C). Rear margin of the caudal fin sometimes broadly rounded or truncate, but usually ending in an obtuse point formed by the extreme elongation of the middle caudal rays. Adipose fin high, widely united to the caudal rays, forming a low keel-like ridge; no free, flaplike lobe. Notch between adipose fin and caudal fin closer to front of dorsal fin base than to posterior tip of caudal fin. Dorsal fin with 1 spine and 6 or 7 rays. Caudal fin rays usually 53–63. Anal fin with 14–16 rays; 1 pectoral spine and 7 pectoral rays; pelvic rays typically 8. Premaxillary tooth patch with lateral edges and posterior corners rounded, without posterior extensions. Breeding males with enlarged muscles on top of head, swollen lips, and enlarged genital papillae. Maximum size 4 inches (102 mm).

LIFE COLORS Body uniformly tan to brown with a white belly. Taylor (1969) noted that coloration may vary from dull golden yellow to olive gray. Fins, barbels, and upper body surfaces are nearly uniformly pigmented. Underside of body variously unpigmented in young, lightly or irregularly pigmented in older specimens, but without dark blotches, bars, and speckles. A prominent, veinlike dark line usually present along midside.

SIMILAR SPECIES The Tadpole Madtom is similar to *N. exilis* and *N. lachneri* in having a uniform body coloration, but it differs from those species in having 2 internasal pores (vs. 1). It further differs from *N. exilis* and *N. phaeus* (which also have similar body coloration) in lacking posterior serrae on pectoral spine (vs. serrae present). It differs from another brown madtom, *N. nocturnus*, in having jaws equal (vs. lower jaw included).

VARIATION AND TAXONOMY *Noturus gyrinus* was described by Mitchill (1817) as *Silurus gyrinus* from the Wallkill River, New York. It is a member of subgenus *Schilbeodes*. Taylor (1969) described clinal and individual morphological variation in *N. gyrinus* throughout its range and recognized no subspecies. Smith (1979) noted that individual variation in color and body shape was extreme in Illinois, and we have noted that Arkansas populations of the Tadpole Madtom show local differentiation in certain characters, particularly body shape. Taylor (1969) believed that variation in body shape was related to nutrition during growth, but this has not been verified. Smith (1979) believed that the slender-bodied form was more strongly associated with high-gradient streams than the humpbacked form. Unusually large specimens have been collected from the Ouachita River near Malvern. Taylor (1969) considered *N. gyrinus* most closely related to *N. lachneri*. This sister species relationship was subsequently consistently supported by morphological and genetic analyses (Grady and LeGrande 1992; Hardman 2004; Hardman and Hardman 2008; Arce-H. et al. 2017).

DISTRIBUTION *Noturus gyrinus* is the widest-ranging madtom, occurring in the St. Lawrence–Great Lakes, Hudson Bay, and Mississippi River basins from southern Quebec to southern Saskatchewan, Canada, south to the Gulf of Mexico; in Atlantic Slope drainages from Massachusetts to southern Florida; and in Gulf Slope drainages from Florida west to the Nueces River, Texas. Introduced populations are found in the Snake River, Idaho and Oregon, and in the Merrimack River, New Hampshire (Page and Burr 2011). One of the most widely distributed and common madtoms in Arkansas, the Tadpole Madtom generally occurs below the Fall Line, ranging throughout the Coastal Plain province, with populations extending up the Arkansas Valley to Poteau, Oklahoma, in the Arkansas River drainage and into the Ouachita Mountains ecoregion in appropriate habitats. There are no recent records from the extreme western Arkansas River Valley in Arkansas (Map 124), even though a few older records are known (Map A4.122). It is nearly absent from the Ozarks, barely penetrating streams along the eastern edge.

HABITAT AND BIOLOGY The Tadpole Madtom is primarily a lowland species that inhabits clear to moderately turbid quiet water areas of small streams, ditches, bayous, oxbow lakes, and sluggish rivers where organic debris (dense branches, leaves) has accumulated, or where there are thick growths of aquatic vegetation over mud and sand bottoms. Abundant cover is an important requirement (Pflieger 1997). It has been found in accumulations of trash on the stream bottom as well as in empty mussel shells. Although found in backwaters of Arkansas' largest rivers (Mississippi, lower Arkansas, and Red rivers), it is not abundant in those environments. It is most abundant in habitats with little or no current, and it is commonly found in reservoirs within its area of occurrence in Arkansas. Buchanan (2005) reported 5,770 specimens from 36 of 66 reservoirs sampled.

Noturus gyrinus hides in leaf litter and decaying vegetation during the day and forages at night on the bottom and among aquatic plants feeding on small crustaceans (amphipods, ostracods, copepods), aquatic insects, oligochaetes, snails, algae, small fishes, and miscellaneous plant material and organic debris (Becker 1983). Smaller individuals fed more on crustaceans and oligochaetes, while larger ones fed more on immature insects. Whiteside and Burr (1986) reported dipteran larvae (chironomids) and small crustaceans (mainly isopods) formed the major portion of the diet of all size classes in all seasons in Illinois. In that study, individuals smaller than 35 mm SL fed mainly on copepods, cladocerans, and ostracods, while madtoms larger than 45 mm SL also consumed worms and grass shrimp. Stomach contents of 8 specimens, 51–72 mm TL, collected from Kiethley Lake in St. Francis County, Arkansas, by TMB on 28 May contained mainly ephemeropteran nymphs, amphipods, and isopods.

The Tadpole Madtom spawns in Missouri in June or July (Pflieger 1997) and from May to September in Mississippi (Ross 2001) and Alabama (Boschung and Mayden 2004). Based on examination of gonads by TMB, the reproductive season in Arkansas extends from May (possibly late April) at least through July. Ripe females between 56 and 68 mm TL contained from 38 to 47 mature ova in May and July. Scott and Crossman (1973) reported that average number of eggs per nest was 50 but numbering up to 117. Eggs are deposited in cavities or under flat rocks, if available, but it is also reported that hollow logs and crayfish holes are used as well. This madtom occasionally spawns in beer and soda cans and other discarded man-made objects. Egg masses are guarded by a male and a female (Adams and Hankinson 1928; Ross 2001), or by the male alone (Whiteside and Burr 1986) until they hatch. Post-hatching parental care has not been documented. Young-of-the-year have been collected in September in Illinois (Whiteside and Burr 1986). Small young between 12 and 15 mm TL were found by TMB in Kiethley Lake (St. Francis County) on 28 May; YOY between 17 and 29 mm TL were found in Lake Hamilton (Garland County) in early July. Maximum life span is about 4 years. This species has been collected syntopically with *N. nocturnus* at a few locations in Arkansas.

CONSERVATION STATUS The Tadpole Madtom is widespread and common in Arkansas. Its populations are secure.

Noturus lachneri Taylor
Ouachita Madtom

Figure 16.20. *Noturus lachneri*, Ouachita Madtom. *David A. Neely.*

Map 125. Distribution of *Noturus lachneri*, Ouachita Madtom, 1988–2019.

CHARACTERS   A slender, moderately elongate, uniformly colored brown madtom with an adnate adipose fin, terminal mouth, jaws about equal, and 10 preoperculomandibular pores (Fig. 16.20). A single internasal pore is present. Scaleless. Head short and flattened, head length going into SL 3.7–4.2 times. Eyes small, eye diameter going into snout length 2.3–2.9 times. Dorsal fin with 1 spine and 6 rays. Anal rays usually 16–18; pelvic rays 8 or 9; soft pectoral rays are typically 8. Pectoral spine nearly straight to slightly curved without serrations anteriorly or posteriorly (Fig. 16.4D). Caudal rays 56–61. Premaxillary tooth patch without backward extensions. Adipose fin relatively short and of moderate height, without a free posterior flap, broadly

connected to caudal fin. Adults range from 0.9 to 3.7 inches (23–80.3 mm) SL, but are usually 0.9–2.6 inches (23–66 mm) SL (Robison 1980b). Maximum size estimated at 3.2 inches (80.3 mm) SL by Tumlison and Hardage (2014), and reported as 3.7 inches (94 mm) TL by Gagen et al. (1998).

LIFE COLORS   Body color brown to gray (almost pink at night), grading to white on the belly. Median fins often with dark borders.

SIMILAR SPECIES   *Noturus lachneri* differs from *N. gyrinus* in having 1 internasal pore (vs. 2), 16–18 anal rays (vs. 14–16), and a shorter head, head length going into SL 3.7–4.2 times (vs. head length usually going into SL 3.7 times or less). It differs from *N. exilis* by lack of serrae on pectoral spine (vs. posterior serrae present), usually 56 or more caudal rays (vs. usually fewer than 56), and typically 8 pectoral rays (vs. 9). *Noturus lachneri* is sympatric with *N. nocturnus* but can be distinguished from that species by 1 internasal pore (vs. 2), and upper and lower jaws about equal (vs. lower jaw distinctly included).

VARIATION AND TAXONOMY   *Noturus lachneri* was described by Taylor (1969) from the Middle Fork of the Saline River at State Highway 7, 11.2 miles (18.1 km) north of Mountain Valley, Garland County, Arkansas. More recent measurements indicate that the type locality is actually about 6 miles (9.7 km) north of Mountain Valley. It is a member of subgenus *Schilbeodes* and has been resolved as sister species to *N. gyrinus* by morphological and molecular evidence (Hardman 2004; Hardman and Hardman 2008; Arce-H. et al. 2017).

DISTRIBUTION   The Ouachita Madtom is endemic to the Ouachita River drainage of Arkansas (Map 125). Originally thought to be restricted to the upper Saline River system (Ouachita River drainage) (Robison 1980b), *Noturus lachneri* was subsequently discovered by Larry Raymond in a small tributary of the main Ouachita River below Remmel Dam (Raymond 1975). It is now known to be fairly common in tributaries of the Ouachita River near Hot Springs.

HABITAT AND BIOLOGY   The Ouachita Madtom generally inhabits clear, high-gradient, small to moderate-sized gravel-bottomed streams, where it frequents rather quiet, backwater areas with substrates ranging from cobble to small gravel and sometimes softer substrates (Robison and Harp 1985). A few specimens have also been collected during the day in very shallow riffles under large rocks. Buchanan (2005) found *N. lachneri* in six small impoundments in the Saline River headwaters. Densities of *N. lachneri* were reported to be low by Robison and Harp (1985), who used seines and dip nets to collect specimens. Gagen et al. (1998) found higher densities with elctrofishing, averaging 95/100 m² (range 17.2–204/100 m²). Tumlison and Hardage (2014) usually found densities of 0–2/m², but up to 8/m² at four plots over 15 dates (60 samples—mean density 80/100 m²). They reported that densities differed at the two most productive sites from May to July, averaging 28.1/100 m² and 84.8/100 m². Much of the earlier perceived low densities of *N. lachneri* were probably due to less efficient sampling techniques.

Noturus lachneri began foraging 20–90 minutes after sunset. It consumed high numbers of a wide variety of organisms and was classified as a feeding generalist (Stoeckel et al. 2011). Foods consumed include primarily ephemeropterans, dipterans (Chironomidae), coleopterans, trichopterans, megalopterans, odonates, decapods, isopods, cladocerans, copepods, and gastropods (Robison and Harp 1985; Patton and Zornes 1991). Patton and Zornes (1991) found that *N. lachneri* was not highly selective in food preference in the pool and riffle habitats but was largely an opportunistic feeder. Stoeckel et al. (2011) studied food availability and diet of *N. lachneri* in pool and riffle habitats in three streams. In that two-year study, *N. lachneri* strongly selected for chironomids in all seasons in all habitats. Coleopterans, ephemeropterans, and odonates were avoided or unavailable compared with their abundance in the environment but did compose part of the diet along with isopods. Seasonally, ephemeropterans, zooplankton, and megalopterans made up a large proportion of the summer diet compared with winter and spring, with fall diets exhibiting intermediate values for those taxa.

Stoeckel et al. (2011) studied the reproductive biology of the Ouachita Madtom, and Tumlison and Hardage (2014) documented growth and reproduction at the periphery of its small range. The spawning season began in early June and terminated by mid-August. The nests were constructed at water depths of 1–42 cm in gravel or gravel/sand substrates located in glides above riffles underneath slab rocks that ranged in size from 100 to 1,200 cm² in area. The males guarded the bowl-shaped nests containing an average of 33 developing embryos. Tumlison and Hardage (2014) found breeding males with enlarged cephalic expaxial muscles from 6 June through 27 June at a water temperature of 17°C. Stoeckel et al. (2011) reported nests of *N. lachneri* from mid-June through July at water temperatures of 19–27°C. Fecundities of females ranged from 6 to 69 oocytes (Stoeckel et al. 2011). Females mature between 58 and 66 mm TL in the Saline River drainage populations. Most reproductive females are estimated to be 2 years old (Tumlison and Hardage 2014). Stoeckel et al. (2011) also reported that nesting could occur successfully in areas of low current velocity, but siltation apparently inhibits spawning activity. Six young-of-the-year specimens, 0.6–1.0 inch (16–25 mm) SL, were taken by HWR on 1 August 1980 from a small stream 2.0–3.9 feet (0.6–1.2 m) wide. Tumlison and Tumlison (1996) found hatchlings (20 mm SL) as early as 15 July but more commonly on 29 July (15 mm SL). Tumlison and Hardage (2014) first found YOY (17 mm TL) on 27 June and postulated that the young move to smaller and shallower reaches of the stream soon after hatching. They estimated growth of YOY during the warmer months at 0.20 mm/day. Smallest young measured 15 mm TL, juveniles averaged 61.4 mm TL (estimated 52.4 mm SL), and average size of adults was 73.4 mm TL (estimated 62.7 mm SL). Fiorillo et al. (1999) earlier reported 3 size classes in a February sample from the Saline River, based on a plot of SL versus body mass. Tumlison and Hardage (2014) reported longevity to be just over 2 years in the Cooper Creek population. They concluded that longevity and adult sizes for *N. lachneri* may differ among localities with maximum longevity of about 2+ years.

Fiorillo et al. (1999) studied the helminth parasite assemblage structure, prevalence, mean abundance, and mean intensity as well as size-dependent patterns of parasite load and helminth species richness in *N. lachneri*. McAllister et al. (2014b) also reported helminth parasites, and Fayton et al. (2018) described a new species of helminth from *N. lachneri*.

CONSERVATION STATUS   Robison and Harp (1985) considered *N. lachneri* a threatened species because of the multiple and complex environmental threats to its continued existence, coupled with a combination of small population size, and sporadic and restricted distribution. In the upper Saline River, habitat degradation from possible impoundment for a water supply, continued human population growth, commercial gravel operations, and channelization threaten this species. We consider it a threatened species.

Noturus maydeni Egge
Black River Madtom

CHARACTERS   A small, slender madtom of the subgenus *Rabida* with a small, rounded head (head length going into SL 3.5 times or more) with subcutaneous eyes directed dorsally (Fig. 16.21). Scaleless. Adipose fin adnate with a distinct notch at its junction with the caudal fin, with a low keel-like ridge, and possessing a large dark bar. Pectoral soft rays typically 9 instead of 8 as in all other *Rabida* except *N. albater*. Pectoral spine relatively short, with many small

Figure 16.21. *Noturus maydeni*, Black River Madtom. *Uland Thomas.*

Map 126. Distribution of *Noturus maydeni*, Black River Madtom, 1988–2019.

serrations on the posterior edge with tips curved toward spine base (Fig. 16.4J), Anterior serrae nearly obscure. Eye diameter going into snout length 1.5–2.4 times. Dorsal fin with 1 spine and 6 rays. Anal rays 15–17; pelvic rays typically 9; caudal rays 44–55. Normally 2 internasal and 11 preoperculomandibular pores. Upper jaw extending beyond lower jaw. Premaxillary tooth patch with posterior corners rounded and without posterior extensions. Distance from tip of caudal fin to adipose fin notch going more than two times into distance from adipose fin notch to front of dorsal fin base. Maximum size about 4 inches (102 mm).

LIFE COLORS   Body color gray to gray brown with pale yellow background and ventral surface. Body and fins mottled. Three distinct dorsal saddles cross the back. Caudal fin base with an area of dense pigment. Dorsal fin without a dark blotch on outer third of fin. Dorsal fin

spine well pigmented with some pigment extending from base onto rays and membranes. Adipose fin variously pigmented, but typically middle third of fin covered in dense pigment; pigment does not extend upward to the dorsal margin of the fin. Caudal fin with a dark margin posteriorly and with an immaculate region on the posterodorsal corner.

SIMILAR SPECIES   The Black River Madtom is most similar to its sister species, the Ozark Madtom, *N. albater*. It is also superficially similar to the Mountain Madtom (*N. eleutherus*), Checkered Madtom (*N. flavater*), and Brindled Madtom (*N. miurus*) but differs from the latter three madtoms in having 9 pectoral rays (vs. 8), head length going into SL 3.5 times or more (vs. fewer than 3.5 times), and distance from tip of caudal fin to adipose fin notch going into distance from adipose fin notch to front of dorsal fin base more than 2 times (vs. fewer than 2 times). From *N. albater*, *N. maydeni* differs in having a 2n chromosomal count of 56–58 (vs. 65–72). All other *Rabida* have lower chromosome counts of 2n = 40–52. It further differs from *N. albater* in its allopatric distribution, occurring only in the Black River drainage in Arkansas (vs. occurring only in the White and Little Red river drainages in Arkansas). Although Egge and Simons (2006) found no consistent morphological differences between *N. maydeni* and *N. albater*, W. H. LeGrande (pers. comm.) believed that there were subtle differences. LeGrande concluded that specimens of *N. albater* reach larger body size, have bolder and more distinct dorsal saddles, and are more slender, while *N. maydeni* specimens are usually more mottled with diffuse dorsal saddles and are somewhat more chunky. Egge and Simons (2006) found that *N. maydeni* exhibits variation in the adipose fin bar both in size and shape and that it also displays high variation in the relative intensity of the dorsal saddles. Some individuals have intense saddles, but others have less intense saddle coloration. They also concluded that mottling is variable.

VARIATION AND TAXONOMY   *Noturus maydeni* was described by Egge (*in* Egge and Simons 2006) from the Strawberry River at Simstown Public Access in Sharp County, Arkansas. Previously included under *N. albater*, fixed chromosomal and biochemical differences between populations of *N. albater* in the upper White River and those in the Black and St. Francis rivers led earlier workers to suggest that the Ozark Madtom represented two distinct species. The Black-St. Francis form was subsequently described as *N. maydeni*, based on substantial genetic evidence. Karyotypes, allozyme variation, and DNA sequences provided diagnostic characters for elevating the Black River populations to full specific status (Egge and Simons 2006). *Noturus maydeni* is the sister species to *N. albater* (Egge and Simons 2006; Arce-H. et al. 2017).

DISTRIBUTION The Black River Madtom, *N. maydeni*, is known only from the Black River drainage and St. Francis River in southeastern Missouri and northeastern Arkansas. In Arkansas, this species has been collected from the Strawberry, Spring, Eleven Point, Current, and Black rivers and their tributaries (Map 126). It is absent from the Arkansas portion of the St. Francis River.

HABITAT AND BIOLOGY The Black River Madtom is typically found in clear, swift, moderate to large-sized streams over gravel, rubble, or rock substrates in riffles and rocky pools. *Noturus maydeni* prefers the swifter areas of shallow riffles and shallow pool regions. It seems less common in headwater regions of creeks and streams. In Jacks Fork River, Missouri, a sixth-order Current River tributary, *N. maydeni* was largely rheophilic, occurring predominantly in riffles in depths of 0.06–0.57 m and current velocities ranging from 0 to 0.60 m/sec, but it was more commonly found at the higher velocities (Banks and DiStefano 2002).

Foods consist primarily of aquatic insects. In Big Creek, Missouri (St. Francis River drainage), 95.7% of the diet consisted of aquatic insects, with dipterans dominating the diet (70% of items ingested and found in 95% of the specimens) (Mayden et al. 1980). Ephemeropterans formed the second-largest portion of the diet, followed by megalopterans, trichopterans, coleopterans, and odonates.

Spawning occurs from June through July. On 8 July 2015, a 94 mm TL female full of ripe eggs was collected by HWR and Chris McAllister from Town Creek at Salem in Fulton County (Tumlison et al. 2016). Mayden et al. (1980) found nests located at the heads of riffles or in shallow pools 5.9–26.4 inches (15–67 cm) deep with current. The nests were under large, flat rocks varying in size from 7.1 to 10.2 inches (18–26 cm) wide and 11.8 to 17.7 inches (30–45 cm) long. Depth of the cavities ranged from 3.9 to 7.9 inches (10–20 cm). Each nest was guarded by a male, and clutch size averaged about 40 eggs per mass. Nest-guarding males probably do not feed while performing their parental duties. Mayden et al. (1980) found the number of mature ova averaged 111.6 per female. Individuals of both sexes live at least 2 years, probably becoming sexually mature by their second or third summer. McAllister et al. (2015a) reported helminth parasites of *Noturus maydeni* in Arkansas.

CONSERVATION STATUS Warren et al. (2000) listed populations of this species as currently stable, but because of its limited range, it is especially vulnerable to environmental disturbance. We consider it a species of special concern and recommend future monitoring of its populations.

Noturus miurus Jordan
Brindled Madtom

Figure 16.22. *Noturus miurus*, Brindled Madtom. *Uland Thomas.*

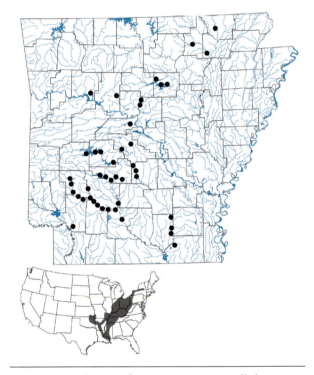

Map 127. Distribution of *Noturus miurus*, Brindled Madtom, 1988–2019.

CHARACTERS A small, moderately stout, weakly to conspicuously banded and conspicuously mottled member of the madtom subgenus *Rabida* (Fig. 16.22). Scaleless. Pectoral spine long, curved, and with 5–7 large, recurved posterior serrae and long, distinct anterior serrae (Fig. 16.4K). Adipose fin relatively high, forming a keel-like ridge that is only moderately notched at the caudal origin. Head length going into SL fewer than 3.5 times. Eyes moderately large, eye diameter usually going 2 times or less into snout length. Upper jaw projecting beyond the lower jaw. Premaxillary tooth patch with sharp or slightly rounded corners and lacking posterior extensions. Caudal fin rounded. Caudal rays 54–65, usually 57 or more. Dorsal fin with 1 spine and 6 or 7 rays. Anal rays 13–17, pelvic rays typically 9, soft pectoral rays 8. A single internasal pore, and 11 preoperculomandibular pores. A black adipose blotch extends to the margin of the fin. Maximum size usually around 3.5 inches (89 mm) SL. Largest Arkansas specimen from Lake Hamilton, Garland County, was 3.7 inches (93 mm) TL.

LIFE COLORS   Body color yellowish brown with heavy dusky mottling and 4 dark saddles crossing the dorsum but not extending downward onto the upper side. Under surfaces of head and abdomen with a few scattered chromatophores, but belly otherwise white. Black blotch in the distal third of the dorsal fin usually extending only across first 4 fin rays. Black blotch on adipose fin extends from base to margin of fin. Anal fin dusky near base, usually with a submarginal dark band. No black bar across the caudal base. Caudal fin with a black submarginal band and a white tip.

SIMILAR SPECIES   The Brindled Madtom is most similar to the Caddo Madtom from which it differs in having longer pectoral spines with larger serrae, a longer posterior process of the cleithrum, and a black blotch that extends to the margin of the adipose fin (vs. a submarginal adipose fin blotch). It differs from *N. eleutherus* by possessing a more adnate adipose fin, more prominent dorsal saddles, and 10 preoperculomandibular pores (vs. 11). It differs from *N. albater* and *N. maydeni* by having 8 pectoral rays (vs. 9), and a black-tipped dorsal fin (vs. outer third of dorsal fin without a black blotch). It is also superficially like *N. flavater* but differs in having a single internasal pore (vs. 2 pores), the absence of a black bar crossing the caudal peduncle (vs. a wide black bar present), and dorsal saddles poorly defined and not extending downward onto upper sides (vs. dorsal saddles prominent and extending downward onto upper sides). *Noturus miurus* differs from *N. gladiator* in usually having 57 or more caudal rays (vs. 49–53 caudal rays) and a dark bar on the adipose fin that extends to the margin (vs. dark bar does not extend to margin).

VARIATION AND TAXONOMY   *Noturus miurus* was described by Jordan (1877d) from the White River at Indianapolis, Marion County, Indiana. Taylor (1969) studied geographic variation in morphology of *N. miurus* and found remarkably little variation throughout its range. Grady and LeGrande (1992) considered *N. miurus* sister to

N. taylori, but some subsequent genetic analyses did not support this relationship. Arce-H. et al. (2017) resolved *N. miurus* as most closely related to *N. flavipinnis*.

In Arkansas, the ranges of *N. miurus* and *N. taylori* barely overlap in the Ouachita River in Montgomery County, and in the Caddo River in Clark County, and the potential for hybridization between the two species exists. Specimens collected from Lake Ouachita by TMB seemed to exhibit hybrid characteristics, but Neil Douglas examined the specimens and tentatively assigned them to *N. miurus*. A genetic analysis is needed to clarify the status of the Lake Ouachita population.

DISTRIBUTION   *Noturus miurus* occurs in lower Great Lakes drainages and the Mississippi River basin from New York and Ontario, Canada, southwest through most of the Ohio River drainage, west to southeastern Kansas and eastern Oklahoma in the Arkansas and Red river drainages and south through the lower Mississippi River basin to the Gulf of Mexico. It is known from one Atlantic Slope drainage, the Mohawk River (Hudson River system), New York, and from Lake Pontchartrain drainages on the Gulf Slope in Mississippi and Louisiana. In Arkansas, this species occurs in the Coastal Plain region below the Fall Line and penetrates a short distance above the Fall Line into the Ouachita and Ozark mountains (Map 127). It historically occurred sporadically throughout the upper Arkansas River Valley (Map A4.125), but there are no recent records from the extreme western part the ARV. There is one record from the Red River system (collected from Bois D'Arc Lake by TMB, Drew Wilson, and Les Claybrook). It is absent from most of the eastern Arkansas Delta and is most common in stream segments near the Fall Line boundary.

HABITAT AND BIOLOGY   The Brindled Madtom inhabits pools of low- to moderate-gradient streams with mud, sand, or fine gravel bottoms littered with twigs, leaves, and other cover such as large rocks. Although it has sometimes been found in riffles, it generally avoids strong current. Boschung and Mayden (2004) also noted its preference for still waters or slow-flowing pools in streams having ample shelter. Pflieger (1997) noted that it is almost entirely a lowland species in Missouri and is often found in habitats having considerable quantities of detritus, such as sticks and leaves. Although it is typically found in streams, it occasionally occurs in lakes, sometimes in large numbers. Buchanan (2005) collected 1,977 specimens from 7 Arkansas reservoirs. This species has been collected in moderately clear to turbid waters over a variety of substrates. In Arkansas, it is uncommon in the cool, clear upland streams, but is found mainly in that type of habitat in Oklahoma (Miller and Robison 2004). It is most active at night and seeks cover during the day.

Foods consist of aquatic dipteran larvae and pupae (predominantly chironomids), ephemeropteran nymphs, trichopteran larvae, and adult isopods (Burr and Mayden 1982c). Small individuals also fed heavily on microcrustaceans (mainly copepods) (Madding 1971). In an Indiana stream, individuals smaller than 40 mm SL fed predominantly on chironomid and trichopteran larvae and pupae, while larger individuals fed on a wider variety of foods (Whitaker 1981). Stomach contents of 50 adult specimens collected from five Arkansas counties (Cleburne, Garland, Lawrence, Pike, and Pope) from April through August were examined by TMB. The diet was dominated by ephemeropteran nymphs at all localities and on all dates of collection. Other food items ingested in decreasing order of percent frequency of occurrence were chironomid larvae, trichopterans, amphipods, odonates, and coleopterans. One individual had ingested fish eggs. Brindled Madtoms in lakes consumed more chironomids than those in streams.

In Illinois, spawning occurred from June through July at water temperatures of 24–27°C (Burr and Mayden 1982c). Males matured at 2 years, and during the breeding season their heads, lips, and genital papillae were swollen. The number of mature eggs produced by a female ranged from 42 to 143 (Taylor 1969). Based on examination of gonads by TMB, the breeding season in Arkansas extends at least from mid-May through early August. Gonads of females from Pope County contained enlarging ova on 21 April but no ripe eggs. Females with ripe ova were found in the Spring River in Lawrence County on 18 May. Peak spawning in Arkansas occurs in June and July, and ripe females have been found as late as 1 August in Lake Hamilton and 6 August in Greers Ferry Lake. Fecundity was lower than that reported for Illinois populations by Burr and Mayden (1982c). Females from 65 to 76 mm TL contained from 36 to 54 ripe ova in June and early to mid-July. Fecundity decreased later in the breeding season, with females 64–76 mm TL containing 25–38 ripe ova in late July and early August. Clutch size ranged from 56 to 81 yellow spherical eggs 3.4–3.8 mm in diameter (larger than those of most other Arkansas madtoms). The nest is usually a pitlike depression under a flat rock in a quiet pool and is guarded by the male. Man-made objects, such as discarded cans, bottles, and other materials, are also readily used as nesting sites. Small young have been found in early July in Arkansas. Maximum life span is about 3 years.

CONSERVATION STATUS　The Brindled Madtom

is sporadically distributed, and it is rarely abundant at any locality in Arkansas. Surprisingly, some of the largest Arkansas populations are found in reservoirs such as Lake Greeson, Lake Hamilton, Lake Ouachita, and DeGray Lake (Buchanan 2005). Its populations are currently secure, but future monitoring of its status is warranted.

Noturus nocturnus Jordan and Gilbert
Freckled Madtom

Figure 16.23. *Noturus nocturnus*, Freckled Madtom. *David A. Neely.*

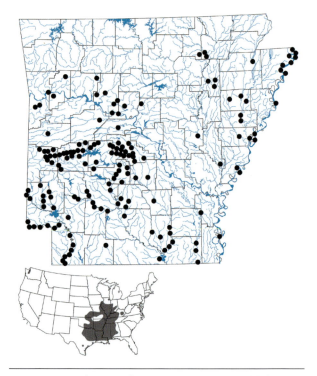

Map 128. Distribution of *Noturus nocturnus*, Freckled Madtom, 1988–2019.

CHARACTERS　A slender madtom of subgenus *Schilbeodes* with a unicolored dorsum, the upper jaw projecting distinctly beyond the lower jaw, and the underside of the head and body finely speckled (Fig. 16.23). Two internasal pores, and 10 or 11 preoperculomandibular pores. Scaleless. Adipose fin low, without a free, flaplike posterior lobe. Head rounded, slightly depressed. Eyes small, eye diameter going into snout length 2.0–3.5 times. Premaxillary tooth patch without posterior extensions. Pectoral fin spine relatively straight without anterior serrations; its posterior edge roughened, but usually without distinct serrae (sometimes with as many as 4 irregularly developed serrae) (Fig. 16.4E). Caudal fin rounded, with 55–64 rays. Dorsal fin with 1 spine and usually 6 rays. Anal rays 16–18. Pelvic rays typically 9. Pectoral soft rays 7–11, modally 9 or 10. Maximum size about 6.3 inches (161 mm) TL. Arkansas specimens generally smaller than 4 inches (101 mm); the largest specimens we have examined (from Vache Grasse Creek in Sebastian County) are about 5 inches (127 mm) TL.

LIFE COLORS Body color yellow to dark brown grading into a yellowish-white belly but without a prominent veinlike dark line along midside. Belly immaculate or sometimes with a few scattered chromatophores (dark pigmentation increases with age). Anterior part of chin heavily pigmented. Side and upper surface of body and head uniformly pigmented, without blotches or light markings. A moderate band of pigment across abdomen in front of pelvic fins. Median fins frequently with narrow, dark submarginal bands edged with narrow, clear to white margins. Large specimens, above 90 mm TL, almost uniformly and heavily pigmented over upper part of body, fins, and abdomen (Taylor 1969).

SIMILAR SPECIES The Freckled Madtom is most similar to Noturus exilis, N. gyrinus, N. phaeus, N. flavus, and N. lachneri in Arkansas. It differs from N. gyrinus, N. exilis, and N. lachneri in having the upper jaw projecting well beyond the lower jaw (vs. jaws about equal). It differs from N. flavus in having the pad of teeth in the upper jaw without backward extensions (vs. with backward extensions). It can be distinguished from N. phaeus in having a shorter anal fin, usually with fewer than 18 anal rays (vs. a longer anal fin with 18–25 rays), and lower surface of head and abdomen poorly pigmented (vs. profusely sprinkled with large chromatophores).

VARIATION AND TAXONOMY Noturus nocturnus was described by Jordan and Gilbert (1886), based on specimens from Arkansas and Oklahoma. Subsequent workers variously listed two of the mentioned localities as the type locality (Taylor 1969). The lectotype was designated in Jordan and Evermann (1900) as USNM 36461, with the type locality given as the Saline River at Benton, Saline County, Arkansas. Taylor (1969) noted that N. nocturnus is a variable species and appears to form distinctive localized populations. He considered the Red River populations noteworthy because of the shift in pore count inside the range of the species and also noted that specimens are often light gray to yellow and sometimes have extremely

short spines. In addition, preoperculomandibular pores are often 10 instead of 11. This may be indicative of two sibling species. Interestingly, speckled populations of N. nocturnus occur sporadically in the upper Saline and Little Missouri rivers. Various relationships of N. nocturnus to other Noturus species have been proposed (Taylor 1969; Grady and LeGrande 1992; Hardman 2004). Hardman (2004) resolved N. nocturnus as a complex of five clades, and Arce-H. et al. (2017) recovered N. nocturnus as sister to a clade containing N. gyrinus and N. lachneri.

DISTRIBUTION Noturus nocturnus occurs in the Mississippi River basin from eastern Kentucky, northern Illinois, and eastern Kansas south through Louisiana, and in Gulf Slope drainages from Mobile Bay, Alabama, to the Brazos River, Texas, with a disjunct population in the upper Guadalupe River, Texas. In Arkansas, it is found throughout the state in appropriate habitats, but is noticeably absent from the upper White River system in the Ozark Uplands (Map 128). It is common throughout the Red River in Arkansas (Buchanan et al. 2003) and the Ouachita River above Lake Ouachita.

HABITAT AND BIOLOGY The Freckled Madtom is most commonly an inhabitant of low- to moderate-gradient, medium-sized streams with slight to moderate current over gravel or sand bottoms but is frequently found in large rivers (we have numerous records from the Mississippi and Red rivers). In Arkansas, it is only rarely found in small streams, and it avoids headwater streams. It occurs in clear to moderately turbid waters, and adults are found in riffles and pools where leaves, twigs, and associated debris accumulate. The Freckled Madtom is sometimes found on silt and mud substrates. It is usually found in current, and unlike some other madtom species, does not adapt well to reservoirs. Buchanan (2005) found only 112 specimens in two impoundments (Felsenthal of the Ouachita River and Pool 9 of the Arkansas River Navigation System) out of 66 reservoirs sampled. Cross (1967) hypothesized that N. nocturnus may be a competitor of N. gyrinus and N. exilis as it is rarely found with either of those species; however, both Arkansas reservoirs in which N. nocturnus occurred also had N. gyrinus (Buchanan 2005).

Aquatic insect larvae are the predominant food items, with ephemeropterans, trichopterans, and chironomid larvae making up the bulk of the diet, and crustaceans, nematodes, oligochaetes, and arachnids making up smaller portions (Burr and Mayden 1982b). In that study, simuliids were important in the diet in autumn samples. Stomach contents of adult specimens (58–120 mm TL) collected in Pope, Johnson, Sebastian, Howard, and Little River counties, Arkansas, from April through August were examined by TMB. The diet at all localities was dominated by ephemeropteran nymphs. All other food items comprised other aquatic insect nymphs and larvae and included chironomids, odonates, simuliids, plecopterans, trichopterans, and coleopterans. Large individuals in southern Mississippi occasionally fed on other madtom species (Clark 1978). Feeding activity is greatest at night.

In Arkansas, spawning occurs from at least early May into mid-August, with peak spawning occurring in May and June (based on examination of gonads by TMB). In Vache Grasse Creek in Sebastian County, ripe females were found from 2 May to 10 August. Ripe females collected from four Arkansas counties (Pope, Johnson, Sebastian, and Howard) ranged from 69 to 133 mm TL and contained from 52 to 245 mature ova. Spawning occurs from late May through July in Illinois (Burr and Mayden 1982b) and Missouri (Pflieger 1997), and in the summer in southern Mississippi (Ross 2001). Nests are usually cavities under flat rocks, but beer cans and other man-made objects are also used as nesting sites. Nests are usually in shallow riffles (10–15 cm deep) with reduced flow at a water temperature of 77°F (25°C) (Burr and Mayden 1982b). Egg diameters ranged from 1.8 to 2.3 mm, and clutch size ranged from 47 to 154. The nest is guarded by the male alone. Eggs hatch within 139–161 hours at 77°F (25°C). Some females reach sexual maturity by one year of age, and males require two years. Individuals can live more than four years.

CONSERVATION STATUS   The Freckled Madtom is a commonly encountered species across Arkansas except in the Ozarks. Its populations are secure.

Noturus phaeus Taylor
Brown Madtom

Figure 16.24. *Noturus phaeus*, Brown Madtom. *David A. Neely.*

CHARACTERS   A heavy-bodied madtom of the subgenus *Schilbeodes* with a unicolored dorsum, with 18–25 (usually 20–22) anal rays, 9 pelvic rays, 11 preoperculomandibular pores, 2 internasal pores, and the lower jaw included (Fig. 16.24). Scaleless. Premaxillary tooth patch without

Map 129. Distribution of *Noturus phaeus*, Brown Madtom, 1988–2019.

backward extensions. Eyes small, eye diameter going into snout length 2.1–2.2 times. Pectoral spine short, stout, nearly straight, and with 3–8 well-developed serrae on posterior edge (Fig. 16.4F). Usually 8 or 9 soft pectoral rays. Dorsal fin pointed, with 1 spine and 6 rays; dorsal spine slender and usally flexible, but stiff in large specimens. Anal and caudal fins only slightly separated. Rear margin of caudal fin rounded or truncate. Adipose fin relatively low, connected to caudal fin. Average adult size 1.7–3.7 inches (43–95 mm) SL; however, Arkansas specimens are only 1.6 inches (41 mm). Maximum reported size from Mississippi was 6 inches (153 mm TL) (Chan 1995).

LIFE COLORS   Body color uniformly light or dark brown, the lower surfaces of the head and belly profusely sprinkled with large, distinct, brown chromatophores. All fins heavily pigmented, sometimes light-edged. Dorsal and anal fins occasionally with a distal dark band of pigment.

SIMILAR SPECIES   The Brown Madtom most closely resembles the Freckled Madtom; however, it differs from that species in having well-developed serrae on the posterior edge of the pectoral fin spine (vs. pectoral spine usually without posterior serrae). *Noturus phaeus* differs from *N. gyrinus* by the presence of 3–8 well-developed serrae on the posterior edge of the pectoral spine and in having an included lower jaw (vs. serrae absent on pectoral spine and lower jaw not included, the upper and lower jaws about equal). It can be distinguished from *N. exilis* by its included lower jaw (vs. jaws equal), and in having 2 internasal pores (vs. 1).

VARIATION AND TAXONOMY   *Noturus phaeus* was described by Taylor (1969) from the North Fork of the Obion River, Henry County, Tennessee. He noted little geographic variation in most morphological features, but pectoral ray counts differed between northern and southern populations. Taylor (1969) considered *N. phaeus* most closely related to *N. funebris*, with possible relationships to *N. insignis* and *N. nocturnus*. Subsequent genetic analyses supported various relationships, with some supporting a relationship of *N. phaeus* to a clade containing *N. funebris* and *N. leptacanthus* (Arce-H. et al. 2017).

DISTRIBUTION   *Noturus phaeus* is known from only five states, occurring in lower Mississippi River tributaries from the Obion River, Kentucky, to southwestern Mississippi; from southern Arkansas to central Louisiana; and in Tennessee River tributaries in southwestern Tennessee and northeastern Mississippi. It also occurs in two Gulf Slope drainages (Sabine River and Bayou Teche) in Louisiana, and one (Pearl River drainage) in Mississippi. Extremely rare in Arkansas, the Brown Madtom has been collected only twice from the state from Horsehead Creek in Columbia County (Ouachita River drainage) (Map 129). Three specimens were originally collected from Horsehead Creek in Columbia County in 1972, representing the first report from Arkansas (Robison 1974d), and one specimen was later taken from the same stream in 2001 (McAllister et al. 2009c).

HABITAT AND BIOLOGY   Strictly a lowland species, the Brown Madtom inhabits permanent springs and small streams (Rohde 1980b; Taylor 1969). In Kentucky, it is found in riffles and raceways of Coastal Plain creeks over a mixed sand and gravel bottom and around accumulations of sticks, logs, leaves, rocks, and other debris or in tree roots along undercut banks (Burr and Warren 1986b). Monzyk et al. (1997) reported that Brown Madtoms were always associated with woody debris (including root masses) during the day in Coastal Plain streams in Louisiana. Woody debris used by Brown Madtoms tended to have suspended leaves and higher flow immediately outside the structure than did unoccupied woody debris; therefore, structural complexity played a part in the presence of this madtom. In Mississippi, Brown Madtoms used woody debris and undercut banks as their primary daytime microhabitat, and stream flow was the best predictor of presence/absence, with madtoms being found most often in areas of complex (varying) flow (Chan and Parsons 2000). In that Mississippi study, *N. phaeus* was never collected in

areas lacking either debris, aquatic vegetation, or undercut banks. The four known Arkansas specimens were found associated with vegetation in moderate to fast, clear water over small gravel or coarse sand.

Most of the known life history information for *N. phaeus* is from a study by Chan (1995) of the Tallahatchie River, Mississippi population. Like other madtoms, most feeding occurred at night, with peaks at dawn and dusk, but feeding occurred continuously during the night. Major food items in fall, winter, and spring were chironomid and trichopteran larvae and crayfish. Trichopterans were not consumed in summer. Coleopteran larvae were eaten in the spring, and biting midges were consumed in spring and summer. Other dipterans consumed were ceratopogonids and simuliids. Additional identifiable foods included ephemeropterans, odonates, plecopterans, isopods, and arachnids (Chan and Parsons 2000).

Brown (1994) studied larval fishes of Coastal Plain streams in southcentral Louisiana and found that Brown Madtoms spawn during late summer. In Mississippi, spawning occurs from late May through August (Chan 1995), and a similar spawning season probably occurs in Arkansas. Ross (2001) described aquarium spawning and concluded that the female may initiate or at least share in nest construction. Clutch size is 41–130 eggs, depending on body size (Mayden and Walsh 1984), and hatching occurs in 8 or 9 days at a water temperature of 25 or 26°C (Ross 2001). The male cleans the nest site after spawning and guards the eggs. In Mississippi, both sexes reached sexual maturity in their second year, and minimum reproductive size was 40–50 mm TL (Ross 2001). It was common in Mississippi for individuals of both sexes to reach age 3, and some males were age 4 or older (Chan and Parsons 2000). Maximum life span may be up to 6 years (Ross 2001).

CONSERVATION STATUS   The greatest threat to *Noturus phaeus* in Arkansas and elsewhere is habitat degradation and destruction. Chan and Parsons (2000) concluded that natural channel shape (i.e., meandering channels with undercut banks) and woody debris are important components of its microhabitat. Monzyk et al. (1997) also found that Brown Madtoms selected for specific refuge qualities (greater cavity space, structural complexity, and suspended leaves) compared with unused debris sites. Chan and Parsons (2000) further suggested that managers should not assume that all instream debris is appropriate refuge for *N. phaeus* but should emphasize maintaining natural stream structure with a diversity of microhabitats. The Brown Madtom is one of the rarest fishes in Arkansas, where it is at the northern periphery of its range west of the Mississippi River. We consider it endangered in Arkansas.

Noturus taylori Douglas
Caddo Madtom

5 mm

Figure 16.25. *Noturus taylori*, Caddo Madtom.
David A. Neely.

Map 130. Distribution of *Noturus taylori*, Caddo Madtom, 1988–2019.

CHARACTERS   A moderately elongate, slender madtom of the subgenus *Rabida,* easily mistaken for *Noturus miurus* (Fig. 16.25). Scaleless. Dorsal fin with 1 spine and 6 rays. Anal rays 13–16. Pelvic rays 9, sometimes 10. Pectoral fin with 1 spine and 8 rays. Caudal rays 54–61. Adipose fin fully adnate with a well-defined notch. Internasal pores 1, infrequently 2. Preoperculomandibular pores 11. Pectoral fin spine is relatively short and slightly curved, with 5–9 recurved well-developed posterior serrae and numerous small but distinct anterior serrae (Fig. 16.4M). Premaxillary tooth patch without backward extensions. Maximum size 2.5 inches (64.7 mm) SL.

LIFE COLORS   Back light yellow or cream-colored, with prominent dark brown dorsal saddles. The anterior saddle originates anterior to the dorsal fin base and extends to near the middle of the fin and ventrally to midside. Another dorsal saddle lies between the dorsal and adipose fins, and a third saddle is at the base of the adipose fin (Douglas 1972). The adipose fin saddle is submarginal. Sides of body lightly pigmented gray or light brown. Sides and dorsum with uniformly scattered minute chromatophores. Ventral surface of the head, pelvic fins, and abdomen immaculate except for scattered chromatophores on the chin. Dorsal fin with black blotch on distal part of fin.

SIMILAR SPECIES   The Caddo Madtom is most similar to the Brindled Madtom but differs by possessing shorter spines with smaller serrae, a shorter posterior process of the cleithrum, and the adipose fin having dark pigment in the basal half, only rarely extending to the margin (vs. a black blotch extending to the margin of the adipose fin in *N. miurus*). It differs from *N. eleutherus* by possessing a more adnate adipose fin and more prominent saddles, and it differs from the allopatric *N. albater* and *N. maydeni* by having 8 soft pectoral rays (vs. 9), a submarginal adipose bar, and no prominent basicaudal bar (Douglas 1972).

VARIATION AND TAXONOMY   *Noturus taylori* was described by Douglas (1974) from the South Fork of the Caddo River, one mile (1.6 km) southeast of Hopper, Montgomery County, Arkansas. It is especially noteworthy in being the only madtom with heteromorphic sex chromosomes (LeGrande 1981). Turner and Robison (2006) assessed population structure and gene flow in *N. taylori* based on allozyme diversity. They found marked genetic divergence among rivers and low levels of genetic variation within sampling localities in *N. taylori* populations. Grady and LeGrande (1992) resolved *N. taylori* as sister to *N. miurus*. This relationship was not supported by analysis of the RAG2 gene by Hardman (2004) or in the analyses of Arce-H. et al. (2017).

DISTRIBUTION   The Caddo Madtom is endemic to the Caddo, upper Ouachita, and Little Missouri rivers of the Ouachita River drainage in Arkansas (Robison 1980c; Robison and Smith 1982) (Map 130 and Map A4.128). There are no recent records from the Little Missouri River drainage. Robison and Harris (1978) proposed two alternative hypotheses to account for the present distribution of the Caddo Madtom and, by extension, the dynamics of the individual movement among localities. They suggested that recent dispersal could account for the distribution because no morphological divergence was observed even among the most geographically distant localities. If this hypothesis is correct, then recent habitat fragmentation by dams might account for the high levels of genetic isolation observed in the Caddo Madtom in different rivers (Turner and Robison 2006). Alternatively, localities among rivers might have been isolated for a long time. In this case, higher order rivers, such as the reach formed by the confluence of the Caddo and Ouachita rivers, might have posed a barrier to gene flow for Caddo Madtoms even before the construction of dams on those rivers.

HABITAT AND BIOLOGY   The Caddo Madtom inhabits shallow, gravel-bottomed pools of clear upland streams. It prefers well-compacted gravel areas below gravel riffles where it lives under rocks, beneath large gravel, and in the interstices of rubble (Robison and Harris 1978). It is most frequently found in clear, shallow riffles near the shoreline and is not found in faster, deeper riffles (Douglas 1972). It has been taken with *N. eleutherus*, *N. miurus*, and *N. nocturnus*.

From unpublished studies (HWR), the Caddo Madtom feeds on snails, isopods, mayflies, dragonflies, caddisflies, stoneflies, aquatic lepidopterans, aquatic beetles, and dipterans. Ephemeropterans and dipterans are the dominant food items. Stomachs of seven specimens from the Ouachita River, collected in late May by TMB contained only ephemeropteran remains.

Our examination of the seasonal changes in ova size indicated that spawning likely occurs from late April through May, ending in early June. Nests have not been found. Ova counts from ripe females were as high as 48 mature eggs, but averaged 16 eggs per female in 18 specimens. Four ripe females, 50–58 mm TL, from the Ouachita River contained from 29–41 ripe ova on 31 May; the lack of smaller developing egg sizes in the ovaries indicated that the breeding season was almost over. McAllister et al. (2014b) reviewed helminth parasites of this madtom.

CONSERVATION STATUS   Turner and Robison (2006) suggested that species such as *N. taylori* that are adapted primarily to headwater habitats might be particularly vulnerable to local extirpation because natural recolonization from adjacent rivers is unlikely. The genetic results of Turner and Robison (2006) have implications for conservation and management of the Caddo Madtom. At the within-drainage scale, fragmentation by natural or manmade barriers (i.e., reservoirs) possibly affects individual movement and recolonization probabilities, resulting in genetic divergence (e. g., South Fork vs. Ouachita main stem). Fluker and McCall (2018) found *N. taylori* at nine localities in the Ouachita and Caddo river drainages and concluded that it was relatively abundant throughout its small, historic range. We consider this rare madtom threatened because of habitat vulnerability, restricted range, and possible extirpation from the Little Missouri River drainage.

Genus *Pylodictis* Rafinesque, 1819— Flathead Catfishes

Described by Rafinesque from a drawing by J. J. Audubon (Boschung and Mayden 2004). This monotypic genus is represented in Arkansas by *Pylodictis olivaris*, the Flathead Catfish, one of the largest catfishes in the state. Fossils of *Pylodictis* indistinguishable morphologically from *P. olivaris* are known from the Middle Miocene (Lundberg 1992). Genus characters are the same as those presented in the species account. Morphological analyses have generally placed *Pylodictis* as most closely related to the genera *Trogloglanis*, *Satan*, and *Prietella*, but combined datasets usually supported a close relationship with *Noturus* (Arce-H. et al. 2017).

Pylodictis olivaris (Rafinesque)
Flathead Catfish

Figure 16.26. *Pylodictis olivaris*, Flathead Catfish. *David A. Neely.*

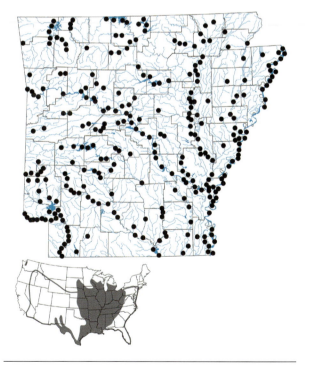

Map 131. Distribution of *Pylodictis olivaris*, Flathead Catfish, 1988–2019. Inset shows native range (solid) and introduced range (outlined) in the contiguous United States.

CHARACTERS A large, slender catfish with a large, broadly flattened head and a projecting lower jaw (Fig. 16.26). Eyes small. Body scaleless. Caudal fin slightly notched, but not forked, with about 17 rays. Adipose fin large and free posteriorly. Premaxillary tooth pad on upper jaw with a crescent-shaped backward extension on each side. Dorsal fin with 1 spine and 6 or 7 rays. Anal fin short, rounded, without spines and with 14–17 rays; anal fin base going into SL 5.3–6.3 times. Pectoral fin spine with well-developed serrae on anterior and posterior edges; pectoral fin rays 8–11. Pelvic rays 9. Internasal pores 2; preoperculomandibular pores modally 12. Gill rakers 8–10. The Arkansas state angling record from the Arkansas River in 1989 is 80 pounds (36.3 kg), but a 132-pound (59.9 kg) specimen was taken from the Mississippi River by a commercial fisherman. Among Arkansas catfishes, only the Blue Catfish attains a larger size than the Flathead Catfish.

LIFE COLORS Coloration varies with size and habitat. Body usually mottled with yellow to dark brown or black on back and sides, grading to white or yellow on the belly. Very large specimens becoming lighter yellow brown with mottling less noticeable. Chin barbels are white to yellow in most, but may be slightly dusky (never black) in large individuals. Nasal and maxillary barbels are dark. Dorsal lobe of caudal fin with white triangle in young individuals. Young individuals are darker and more boldly colored than adults (Fig. 16.27).

SIMILAR SPECIES A mottled body, head notably flattened between the eyes, somewhat square caudal fin, premaxillary band of teeth extending backward, large and free adipose fin, short anal fin (14–17 rays), and white dorsal and ventral edges of the caudal fin of small and medium-sized specimens combine to distinguish the Flathead Catfish from all other Arkansas catfishes. Specifically, the Flathead Catfish differs from *Ictalurus* species in having a rounded or slightly notched caudal fin (vs. forked), and from *Ameiurus* species in having a greatly flattened head (vs. head sloping), posterior extensions of the premaxillary tooth patch (vs. no extensions), and a projecting lower jaw (vs. lower jaw not projecting beyond upper jaw). Juvenile *P. olivaris* are long and slender and can be confused with madtoms. It differs from madtoms (*Noturus*) in having an adipose fin with a free, flaplike tip (vs. adipose fin adnate) and a projecting lower jaw (vs. jaws equal or upper jaw projecting).

VARIATION AND TAXONOMY *Pylodictis olivaris* was described by Rafinesque (1818b) as *Silurus olivaris* from the Ohio River. It was considered by Taylor (1969) and later by Lundberg (1982) to be most closely related to the Widemouth Blindcat, *Satan eurystomus*, a blind,

Figure 16.27. Juvenile Flathead Catfish. *Uland Thomas.*

hypogean species found only near San Antonio, Texas. However, Arce-H. et al. (2017) placed *Pylodictis* in a clade with *Noturus*, based on 8 morphological and 23 nucleotide synapomorphies. Morphological variation of the Flathead Catfish throughout its range has not been adequately studied, but a recent analysis of a mitochondrial DNA tandemly repeated sequence found that 95% of the samples from the Mississippi River and its tributaries as well as samples from the southwestern Gulf Coast drainages lacked the tandem repeat (Padhi 2014). The tandem repeat was present in 70% of individuals sampled from southeastern Gulf Coast drainages.

DISTRIBUTION *Pylodictis olivaris* is considered native to the lower Great Lakes and Mississippi River basins from western Pennsylvania to North Dakota, and south to Louisiana, and Gulf Slope drainages from the Mobile Bay drainage of Georgia and Alabama to Mexico. It is widely introduced elsewhere in the United States. In Arkansas, the Flathead Catfish is found statewide in appropriate habitats (Map 131).

HABITAT AND BIOLOGY The Flathead Catfish is a common ictalurid found in a variety of habitats, ranging from quiet to flowing waters. It is most abundant in the large sandy-bottomed, turbid rivers of the state, where it inhabits deep pools adjacent to strong current, but it is also found in the moderate to swift currents of those rivers. Commercial fishermen capture large numbers of Flathead Catfish (and other catfish) in hoop nets in strong current in deep water, primarily in spring during high flows. Buchanan (1976) reported *P. olivaris* to be common in all parts of the Arkansas River and found that it was second in abundance among the catfishes only to the Channel Catfish, with which it was frequently collected. Duncan and Myers (1978a) studied movements of tagged Flathead Catfish in Beaver Reservoir and found that large fish occasionally moved as much as 40–44 km from the point of tagging. Movement increased during times of higher river inflow. Vokoun and Rabeni (2006) used radiotracking to monitor round-the-clock movements in two Missouri streams. That study revealed an early afternoon activity period in addition to the nocturnal period of greatest movement. More large specimens of Flathead Catfish are caught than of any other catfish species. This species also occurs in reservoirs and in the clearer, moderate-sized rivers of Arkansas. In Oklahoma, it adapted well to newly impounded reservoirs (R. M. Jenkins 1952). Adults seem to prefer submerged logs, debris piles, rock riprap, and other cover, but the young live in much the same habitats as riffle-dwelling madtoms (Pflieger 1997).

The Flathead Catfish is a solitary carnivorous species feeding at night primarily on fish and crayfish (Minckley and Deacon 1959). Collections of younger individuals by HWR in riffles in the Mulberry River probably reflect their feeding tendencies. Trautman (1957) also noted that in streams it fed at night in riffles so shallow that its dorsal fin stuck out of the water. He also observed them at night lying on the bottom beside a log or other object in water less than 5 feet deep with the mouth wide open, and reported that fishermen have observed frightened fish dart into the open mouth to be swallowed immediately. The Flathead Catfish prefers live food and is not enticed by baits normally used for Channel Catfish. It is not a scavenger and rarely takes dead food. Ross (2001) summarized the changes in diet with increasing size and age: Flathead Catfish smaller than 102 mm TL feed mainly on aquatic insects (especially mayflies, caddisflies, and dipterans). Fish 102–254 mm TL continue to eat aquatic insects but add larger prey such as fish and crayfish to the diet. Fish 254–305 mm TL feed mainly on fish and crayfish, depending on availability, and fish larger than 480 mm TL are almost exclusively piscivorous. Fishes consumed by adult Flathead Catfish in streams include minnows, various sunfishes, darters, shad, Freshwater Drum, and other catfishes (Minckley and Deacon 1959). Turner and Summerfelt (1971) found that in Oklahoma reservoirs, fish made up more than 95% of the food volume, with Gizzard Shad making up 49.5–91.7% of the food volume, followed by Freshwater Drum and Common Carp. Crayfish populations are typically much smaller in reservoirs than in rivers, and crayfish made up less than 1% of the Flathead Catfish diet by both volume and numbers in the reservoirs. Feeding activity was greatest in reservoirs in September through October and April through May. Young *P. olivaris* smaller than 110 mm TL fed mainly on chironomids, but decapod crustaceans were the most important foods of fish 170–400 mm TL.

Spawning occurs from May through July at water temperatures of 22.2–23.9°C (Minckley and Deacon 1959), and spawning habits are similar to those of other large catfishes (Fontaine 1944). The male and female excavate a saucer-shaped depression in a natural cavity or in the stream bank. Up to 100,000 eggs are laid in a compact yellow mass, depending on the size of the female, but the usual clutch size of females 533–610 mm TL is 6,900–11,300 eggs (Minckley and Deacon 1959). Spawning may continue for 4 hours (Fontaine 1944). After spawning, the male guards the nest alone, agitating the egg mass by fin movements to keep silt off the eggs and provide oxygen. Several authors have reported that the males become very aggressive toward the females after spawning, viciously driving them away from the nest. Hatching reportedly occurs in 6–9 days at water temperatures between 23.9°C and 27.8°C (75–82°F). The young, still guarded by the male, form a compact school for a few days after hatching until the large yolk sac is absorbed and then disperse to eventually assume a solitary life. Sexual maturity is reached in 3 or 4 years and individuals may live up to 20 years (Smith 1979).

CONSERVATION STATUS The Flathead Catfish is widespread and abundant in Arkansas, and its populations are secure.

CHAPTER 17

Order Osmeriformes
Smelts

This order contains two suborders with 5 families, 20 genera, and about 47 species (Nelson et al. 2016). Smelts have sometimes been placed in the order Salmoniformes with the salmons, trouts, and their relatives (Wiley and Johnson 2010), but Begle (1991) suggested that they should be elevated to ordinal status. This elevation was subsequently followed by Bond (1996), Nelson (2006), Nelson et al. (2016), and Betancur-R. et al. (2017), who also recognized the smelts as a separate order. Smelts differ from salmoniforms in having larger scales and in lacking a pelvic axillary process. Other characters that distinguish Osmeriformes as a monophyletic group are loss of the orbitosphenoid and basisphenoid bones, and the presence of a cartilaginous vane ventrally on the first basibranchial (Begle 1991). The oldest fossil osmeriform is the Paleocene freshwater *Speirsaenigma lindoei* from Alberta, Canada (Wilson and Williams 1991). We follow Page et al. (2013) and Betancur-R. et al. (2017) in placing smelts in the order Osmeriformes, and we include them in superorder Protacanthopterygii, along with salmonids and esocids (Appendix 5).

Family Osmeridae *Smelts*

The osmerids are north circumpolar in distribution, with 6 genera and 15 species occurring in marine and freshwater habitats in North America, Europe, and Asia (Nelson et al. 2016). They are abundant in shallow seas and coastal areas. Some species are anadromous, and some of the anadromous species have landlocked populations. Six genera with 11 species inhabit North America (Page et al. 2013), with 1 species occurring in Arkansas. In areas where they are abundant, smelts are important commercial and sport fishes, but most species are rather small (usually smaller than 8 inches [20 cm] TL). They are pelagic carnivores, feeding on both zooplankton and small fish. Smelts are very fine food fishes, but the flesh is somewhat oily. These slender, silvery fishes are characterized by a terminal mouth with small teeth, a forked tail, an adipose fin, body with small easily shed cycloid scales, and gill membranes not joined across throat. Lopez et al. (2004) provided molecular evidence supporting the monophyly of Osmeridae.

Genus *Osmerus* Linnaeus, 1758— Rainbow Smelts

The taxonomy of this genus has been controversial. Page et al. (2013) recognized two North American species of *Osmerus*, *O. dentex* of Arctic and Pacific drainages of Asia and North America, and *O. mordax* of North Atlantic drainages of North America extending westward through the Great Lakes. There is also a European species, *O. eperlanus*. Genus characters include a small, slender body, silvery coloration, large mouth with well-developed teeth, and a small adipose fin. One species, *O. mordax*, occurs in Arkansas.

Osmerus mordax (Mitchill)
Rainbow Smelt

Figure 17.1. *Osmerus mordax*, Rainbow Smelt.
David A. Neely.

Map 132. Distribution of *Osmerus mordax*, Rainbow Smelt, 1988–2019. Inset shows native and introduced ranges in the contiguous United States.

CHARACTERS   A slender, elongate, silvery fish with an adipose fin, a pointed snout, and a large mouth, the upper jaw reaching middle of eye or beyond (Fig. 17.1). Large teeth on tongue and jaws; roof of mouth with 1–3 large canine teeth on each side of the tip of the vomer. Lower jaw projecting beyond upper jaw. Base of pelvic fin without axillary process found in some families having an adipose fin. Intestine short; peritoneum silvery and sprinkled with melanophores. Gill rakers long and slender, 26–35. Branchiostegal rays 6–10. Dorsal fin origin above or in front of pelvic fin origin and far in front of anal fin. Dorsal rays 9–11; anal rays 12–16; pectoral rays 11–14; pelvic rays 6–9. Caudal fin forked, with 17–20 rays. Scales large and cycloid. Lateral line incomplete, with 14–19 pored scales; scales in lateral series 56–72. Breeding males develop very small tubercles on the head, body, and fins. Individuals of landlocked populations reach 6–10 inches (152–254 mm) in length, while those of anadromous populations reach 12–14 inches (305–356 mm) and a weight of 5–8 ounces (142–227 g). All specimens collected in Arkansas have been juveniles, 1.3–3.5 inches (34–88 mm) TL.

LIFE COLORS   Predominantly silvery, but with dorsal half of head and body a steel blue or sometimes greenish gray. Undersides silvery white. Fins are usually clear. The young are translucent.

SIMILAR SPECIES   Juveniles closely resemble silversides (*Labidesthes* and *Menidia*), but differ from those species in having an adipose fin (vs. adipose fin absent), and a large mouth (with large teeth) extending to the back of the eye (vs. mouth not extending backward past front of eye, and without large teeth). The presence of an adipose fin also separates this species from the Mooneye, Goldeye, and the four Arkansas clupeid species (which lack an adipose fin). It differs from the trouts in lacking a pelvic axillary process (vs. pelvic axillary process present), and in having fewer than 72 scales in a lateral series (vs. more than 100 scales in a lateral series).

VARIATION AND TAXONOMY   *Osmerus mordax* was described by Mitchill (1814) as *Atherina mordax* from New York. Two subspecies are sometimes recognized, *O. m. mordax* of eastern North America (including Arkansas), and *O. m. dentex* of North Pacific–Arctic drainages, but Page et al. (2013) considered the latter form a distinct species. Baby et al. (1991) found significant geographical heterogeneity in the distribution of mtDNA genotypes among landlocked populations of smelt. A phylogenetic distinction was also revealed between landlocked and anadromous smelt in the St. Lawrence Estuary.

DISTRIBUTION   In North America, *Osmerus mordax* occurs in Atlantic, Great Lakes, Mississippi River, Arctic, and Pacific drainages from New Labrador, Canada, to Alaska, and is seasonally present in the Missouri, Mississippi, Illinois, and Ohio rivers from Kentucky and Montana, south to Louisiana. It also occurs in some Pacific drainages of Asia. Part of its current North American range is the result of introduction. In Arkansas, the Rainbow Smelt is found only in the Mississippi River, where it was first collected in December 1978 and January 1979 near Eudora (Pennington et al. 1982). It was subsequently collected in 1980 and 1981 near West Memphis and Blytheville (Buchanan et al. 1982; Carter 1984) (Map A4.130), and in Lake Neark off the Mississippi River near Osceola in March 1995 by TMB and Sam Henry (Map 132).

HABITAT AND BIOLOGY   Most populations over its range are anadromous, living in the cooler oceans but ascending freshwater streams to spawn. Some populations are landlocked and are strictly freshwater inhabitants. Landlocked populations spread throughout the Great Lakes during the first half of the twentieth century as a result of introductions (Bigelow and Schroeder 1963). It was introduced into Crystal Lake, Michigan, in 1912 (Van Oosten 1937) and from there colonized Lake Michigan (Foltz and Norden 1977). It entered southern Lake Michigan in the 1940s and 1950s (Smith 1979). From there this species probably spread into the upper Mississippi River via the Chicago sanitary canal and the Illinois River; it was first collected from the upper Mississippi River of Illinois in 1978 (Burr and Mayden 1980) and was found as far south as Louisiana by 1979 (Suttkus and Conner 1980). Buchanan et al. (1982) and Carter (1984) found juveniles abundant in the Mississippi River of northeastern Arkansas in the winter of 1980–81.

Previous reports speculated that the Rainbow Smelt used the Chicago Canal–Illinois River connector to disperse into the Mississippi River. Although this is the most likely source of the smelt found in Arkansas, Pennington et al. (1982) suggested another possible route via the Missouri River. Rainbow Smelt became established through stocking in Lake Sakakawea and Lake Oahe in the early 1970s, two mainstem reservoirs on the Missouri River in North Dakota. Smelt were captured in the Missouri River downstream from Lake Sakakawea and in the Missouri River in Kansas in 1978 (Cross and Collins 1995), and it is possible that this species gained access to the lower Mississippi River via the Missouri River. This view was supported by Mayden et al. (1987) and Pflieger (1997).

All *O. mordax* taken in Arkansas were juveniles of age group 0, ranging in size from 1.3 to 3.5 inches (34–88 mm). All collections of Rainbow Smelt in Arkansas occurred in winter or early spring, and its absence from summer collections from the Mississippi River probably reflects its preference for water temperatures below 60°F (15.6°C). In areas where it is abundant, such as Lake Michigan, it is an important forage species for larger predators. The adults, although not attaining a large size, are important sport and commercial fish. In inland lakes where landlocked populations have become established, the Rainbow Smelt has had negative impacts on native fish populations. In Wisconsin, invaded lakes showed significantly lower YOY Walleye densities than uninvaded lakes (Mercado-Silva et al. 2007).

Newly hatched fry absorbed the yolk sac by 7 mm TL and first fed on copepods, diatoms, and nonmotile green algae (Becker 1983). The food of juvenile Rainbow Smelt in marine and freshwater environments consisted primarily of small crustaceans, mainly copepods and cladocerans (Burbidge 1969). Burr and Mayden (1980) found juvenile diets in the Mississippi River were dominated by copepods and mayflies. Various feeding habits have been reported for the adults. Adult marine smelt feed on fish, shrimp, and other large crustaceans (Kendall 1927). In the Great Lakes, adults eat shrimp, insects, some fish, and small crustaceans (Schneberger 1937b). In Gull Lake, Michigan, smelt fed primarily on small crustaceans and dipteran larvae and did not vary their diet with age (Burbidge 1969). Out of four main zooplankton prey types present in Lake Winnipeg, Canada, Rainbow Smelt 70–130 mm TL preferred larger individuals of *Daphnia* spp. (Sheppard et al. 2012). In Lake Superior, Rainbow Smelt also selected larger taxa for prey, with *Mysis* being the most important food item (T. B. Johnson et al. 2004). In Lake Michigan, large smelt fed primarily on young-of-the-year Alewives (O'Gorman 1974). Greatest feeding activity occurred during the late spring, summer, and fall (Foltz and Norden 1977). Smelt fed at low levels in that lake during the winter and used gut and carcass energy reserves to meet their metabolic requirements.

In its native range, this anadromous species migrates from the ocean into streams to spawn in early spring. Landlocked populations move from lakes and large rivers into tributary streams. In the northern United States, spawning occurs in the spring (March–May) when temperatures exceed 40°F (4.4°C) and lasts about 2 weeks (Becker 1983). Spawning occurs at night in shallow water over a gravel bottom. The adults, which average about 7 or 8 inches TL (200 mm), gather in large schools over the spawning area, and each female accompanied by several males scatters thousands of eggs that sink to the bottom and adhere to the gravel (Rupp 1965). The eggs hatch in 19–29 days at water temperatures between 6°C and 10°C. Sexual maturity is reached in 2 or 3 years and maximum life span is about 5 or 6 years. Smelt generally spawn at the beginning of the third growing season in Michigan, and a pronounced mortality often occurs during or shortly after spawning runs (Foltz and Norden 1977). Maximum life span is about 8 years.

CONSERVATION STATUS   This species is very uncommon in Arkansas, occurring only in winter months in the Mississippi River.

CHAPTER 18

Order Salmoniformes
Trout, Salmon, and Whitefish

The order Salmoniformes is characterized by an adipose fin, the mesocoracoid bone is present in the pectoral girdle, the gill membranes extend far forward and are free from the isthmus, a pelvic axillary process is present, the last three vertebrae are turned up, there are 11–210 pyloric caeca, and parr marks are present on the sides of young of most species. The oldest fossil salmoniforms are from the Cretaceous (Bond 1996). Because no members of this order are native to Arkansas in historical times, the five species of trouts and chars found in this state have been introduced. This order comprises a single family, Salmonidae, that includes the whitefishes, graylings, salmons, chars, and trouts. Salmoniformes formerly included the smelts, family Osmeridae, but the osmerids are now generally recognized as a separate order, Osmeriformes (Page et al. 2013; Nelson et al. 2016).

Family Salmonidae *Trouts and Salmons*

This family is divided into 3 subfamilies—Coregoninae, Thymallinae, and Salmoninae (although some authors recognize 2 of those subfamilies as families)—with 10 genera and up to 223 species (Nelson et al. 2016). There is much controversy over the number of species, and Nelson (2006) reported that the biological diversity in the family is greater than recognized in our current taxonomy with its nomenclatural limits. There may be many species within the

family yet to be named (Nelson et al. 2016). Salmonidae, is a widely distributed family, with species occurring in North America, Europe, and Asia. These cold-water fishes are found in both freshwater and marine environments and are of great economic significance. Most species are anadromous or derived from anadromous forms; nonanadromous forms have repeatedly evolved when populations of anadromous fish became trapped above barriers (Moyle and Cech 2004). They are important gamefishes, and some salmonids, such as salmon, whitefish, and graylings, are of commercial importance in northern North America. The six species of Pacific salmon are especially important. Salmonids have been widely and successfully introduced into the cold freshwaters of the Southern Hemisphere (Moyle and Cech 2004).

No members of this family are native to Arkansas, but several species have been introduced into the state to take advantage of stream environments altered by the impoundment of large reservoirs in the mountainous regions. Prior to large-scale reservoir construction that began in the 1940s, there were sporadic attempts to introduce trout into some of the colder mountain streams of the Ozarks. The earliest known stockings in Arkansas were reported by Meek (1891, 1894b), who documented successful stocking of Rainbow Trout near Rogers in the early 1890s from the Neosho, Missouri Federal Hatchery. Meek also noted that a hatchery at Mammoth Spring was originally supplied with Brook and "English" (Brown) trout in addition to Rainbow Trout. Black (1940) listed several stockings of Rainbow Trout in Illinois and White river drainages by private landowners between 1917 and 1940 and considered it established in the Spring River near Mammoth Spring and in Marble Falls Creek in Boone County. The first small-scale experimental stocking of trout in cold tailwaters in Arkansas was made in 1948 when Rainbow Trout from the federal hatchery at Neosho, Missouri, were introduced into the North Fork of the White River below Norfork Dam (Baker 1970). In 1949 and 1950, larger numbers of Rainbow Trout and some Brown Trout were stocked there. The first trouts were stocked in the White River below Bull Shoals Dam in 1952, and by 1954 flourishing put-grow-and-take and put-and-take trout fisheries had become established. Rainbow Trout were introduced into reservoirs in the early 1960s. An example of the success of the Arkansas trout fishery can be found in the Beaver Lake tailwaters of the White River, where between 2008 and 2010 anglers caught an average of 71,939 Rainbow Trout each year (Kitterman et al. 2011). There is one record of Sockeye Salmon (*Oncorhynchus nerka*) from Bull Shoals Lake in 1969 (Buchanan 1973; Buchanan et al. 2000). That specimen probably came from Lake Taneycomo just upstream from Bull Shoals Lake as a result of stocking by the Missouri Department of Conservation in the mid-1960s. Dunham et al. (2002, 2004) described possible adverse impacts from the stocking of nonnative trouts, including the decline of native fish populations, amphibians, and invertebrates, alteration of ecosystem productivity and nutrient cycling, dispersal of pathogens and diseases, and additional indirect effects on the aquatic ecosystem. Few of those impacts apply to the Arkansas trout fisheries, which were created in response to already drastically altered aquatic environments caused by cold tailwaters below large dams.

Four species of salmonids currently form the basis of the fishery in Arkansas (although a fifth species, the Lake Trout, was stocked in some reservoirs in the mid-1980s). Natural reproduction of trout is limited by fluctuating tailwater conditions (R. F. Baker 1959; Aggus et al. 1977) and adverse environmental conditions, that is, warm water temperatures and unsuitable habitat in reservoirs (Stevenson and Hulsey 1961). While some natural reproduction occurs, particularly with Brown Trout, annual stockings are needed to maintain fishable populations (Oliver 1984). Most of the trout stocked are produced at the two federal hatcheries in Arkansas, but some are obtained from private sources. In 1985 a large private trout hatchery at Mammoth Spring was given to the state by the Kroger Company. This hatchery now enables the state to raise its own trout year-round. The Arkansas Game and Fish Commission developed a system to ensure a higher survival rate than observed in trout stocked directly from the hatchery. Net pens are used primarily to increase the size of trout stocked in reservoirs by protecting them from warmwater predators. Large floating net pens are used in some of the reservoirs such as Bull Shoals (Oliver and Rider 1986), Greers Ferry (small net pens, 20 × 20 × 12 feet, suspended from stalls in boat docks), and Ouachita (permanent net pen facility) to receive the small trout from the hatcheries. The fish are fed and retained in the pens until they reach a weight of 0.5–1 pound (0.2–0.5 kg); then they are dipped out and transported in tank trucks for stocking streams, or the net pens are towed to other parts of the lake where the fish are released. The Pot Shoals net pen in Bull Shoals Lake in 2006 produced 156,100 Rainbow Trout and 75,000 Cutthroat Trout. Unfortunately, the Bull Shoals Lake net pen facility was discontinued due to the presence of Zebra Mussels in that reservoir and the concern that trout from that net pen could introduce Zebra Mussels into other waters (Jeff Williams, pers. comm. 2014). More than 2 million trout were stocked annually in Arkansas in approximately 160 river miles (257 km) and 5 reservoirs (Hudy 1986) through the 1980s. Currently, Bull Shoals is the only Arkansas reservoir regularly stocked with trout (about 30,000 Rainbow Trout annually). Oliver (1984) reported that tourist surveys estimated $42 million was spent on fishing trips to the White River and North Fork of the White River in fiscal year 1982. A 1994 AGFC survey of trout anglers estimated the total expenditure of trout anglers in the state at $133.7 million, and by 2003, the estimate had increased to more than $175 million (Jeff Williams, pers. comm.).

There is an invasive species that may potentially adversely impact the Arkansas trout fishery. *Didymosphenia geminata* (usually referred to as didymo) is a cold-adapted, unicellular, circumboreal freshwater alga native to northern North America, Europe, and Asia. It is a diatom that thrives in cool, clear, nutrient-poor waters, where it forms thick brown mats that can cover large areas of the streambed. In some parts of the world, it covers 100% of the stream substrate with thicknesses of 20 cm over stream reaches greater than 1 km, where it greatly alters physical and biological conditions in the stream (Spaulding and Elwell 2007). Didymo attaches itself by stalks (extracellular mucopolysaccharides) to rocks, plants, and other materials in the water, where it develops flowing streamers that turn white at the ends and resemble tissue paper. These population explosions may last for several months or more during the year. This diatom has been expanding its geographical and ecological range and was first reported as a problem in Arkansas in 2005 from the Bull Shoals Dam tailwaters (Shelby 2006). It is now also found in Arkansas in the Beaver, Norfork, and Greers Ferry dam tailwaters. Didymo has the potential to alter food web structure by reducing or changing the available habitat for aquatic insects essential in trout diets. It may also reduce the habitat for baitfish or other trout forage.

Members of the salmonid family have a fleshy adipose fin and tiny cycloid scales on the body, with more than 70 scales in the lateral line (all species stocked in Arkansas have more than 110 scales in the lateral line). The fins are soft rayed, and an axillary process, a thin membranous flap, is present at the pelvic fin origin. The mouth is moderately large, extending to or behind the back of the eye. Many biologists refer to members of the genus *Salvelinus* as chars, applying the name trout only to members of the genera *Oncorhynchus* and *Salmo* (Stearley and Smith 1993). Esteve (2005) compared spawning behaviors of *Oncorhynchus*, *Salmo*, and *Salvelins*.

Key to the Trouts

1A Lateral line scales 185 or more; mouth large, upper jaw extending far behind eye; vomer with a troughlike, toothless shaft. *(Salvelinus)*   2

1B Lateral line scales usually fewer than 180; mouth moderate in size, upper jaw extending just past posterior edge of eye; vomer with a plane shaft bearing teeth.   3

2A Caudal fin slightly forked; lateral line scales usually more than 210; gill rakers 9–12; mandibular pores usually 7 or 8 on each side; pectoral, pelvic, and anal fins with a black stripe near anterior edge.
                               Salvelinus fontinalis Page 482

2B Caudal fin strongly forked; lateral line scales usually fewer than 210; gill rakers 12–14; mandibular pores usually 9 or 10 on each side; lower fins without black stripe.   *Salvelinus namaycush* Page 484

3A Caudal fin heavily spotted; dorsal fin rays 10–12; sides without red or orange spots in life.   4

3B Caudal fin without spots or with a few small spots along dorsal margin; dorsal fin rays 12–14; sides with or without red or orange spots present in life, if present, usually surrounded by blue halos.
                               Salmo trutta Page 480

4A Red dash (cutthroat mark) present on each side of throat in life; side without a pink or red midlateral stripe; small teeth present on midline of tongue; lateral line scales usually more than 150.
                               Oncorhynchus clarkii Page 475

4B Red cutthroat mark absent on each side of throat; side with broad pink to red lateral stripe in life; teeth absent on midline of tongue; lateral line scales usually fewer than 150.   *Oncorhynchus mykiss* Page 477

Genus *Oncorhynchus* Suckley, 1861— Pacific Trouts and Salmons

This genus, first described by Suckley in 1861 as a subgenus of *Salmo*, contains 12–14 species (depending on authority) and numerous subspecies of Pacific salmons and trouts of western North America and northeastern Asia. The western trouts were formerly included in the genus *Salmo* (along with species from the North Atlantic basin) but were transferred to *Oncorhynchus* by Smith and Stearley (1989) based on morphological and genetic evidence from several studies that indicated a more recent divergence from the anadromous Pacific salmons than from the *Salmo* of the North Atlantic basin (Vladykov 1963; Behnke 1979; Berg and Ferris 1984; and Kendall and Behnke 1984). The older of the two genus names (*Salmo*) was retained for the Atlantic group. Genus characters include dorsal fin rays 8–16, teeth on front and shaft of vomer, caudal fin with numerous black spots, and tiny (100–200) lateral scales. Two species of *Oncorhynchus* are found in Arkansas.

Oncorhynchus clarkii (Richardson)
Cutthroat Trout

Figure 18.1. *Oncorhynchus clarkii*, Cutthroat Trout. *Uland Thomas.*

CHARACTERS   A streamlined fish with a moderately large mouth, the upper jaw barely extending behind eye (Fig. 18.1). Teeth present on jaws, tongue, and palatine bones in roof of mouth; tongue with large teeth on tip and with minute teeth on midline. Lateral line complete; lateral line scales usually more than 150 (130–215). Dorsal rays 9–11 (usually 10); anal rays 8–10. Axillary process present at base of pelvic fin. Caudal fin slightly forked. Average TL is 6–16 inches (152–406 mm), but maximum reported size for the species is 39 inches (990 mm) (Page and Burr 2011). Maximum weight of the subspecies stocked in Arkansas is

Map 133. Distribution of *Oncorhynchus clarkii*, Cutthroat Trout, 1988–2019. Inset shows native range (solid) and introduced range (outlined) in the contiguous United States.

around 10 pounds (4.5 kg), and maximum TL is around 28 inches (710 mm). The Arkansas state record Cutthroat Trout weighed 10 pounds, 2 ounces (4.6 kg) and was taken from the White River in September 2018.

LIFE COLORS   The subspecies stocked in Arkansas has a profusion of small spots and specks over its body and on the dorsal, adipose, and caudal fins. The spotting is heaviest on the posterior half of the body. The cutthroat mark is red orange, and the pectoral, pelvic, and anal fins are bright red or orange. The yellow-brown body often has a silver-green or bronze sheen. Undersides white. Side of young with about 10 oval parr marks and small black spots. The characteristic cutthroat marks begin to develop when juveniles reach approximately 3–5 inches (75–127 mm) TL.

SIMILAR SPECIES   It can be distinguished from the Rainbow Trout and the Brown Trout by the red-orange cutthroat marks in life (vs. cutthroat mark absent), by the small teeth on the midline of the tongue (vs. teeth absent on midline of tongue), and by usually having more than 150 lateral line scales (vs. fewer than 150). It differs from the Brook Trout and Lake Trout in having dark spots on body and fins (vs. no dark spots on body and fins).

VARIATION AND TAXONOMY   *Oncorhynchus clarkii* was described by Richardson (1836) as *Salmo clarkii* from the North Fork of the Lewis River, Washington. It was transferred to *Oncorhynchus* by Smith and Stearley (1989). This highly variable species has 14 currently recognized subspecies (Page and Burr 2011). The distinctive Snake River Cutthroat (Finespotted Cutthroat Trout, *O. c. behnkei*), found in the Snake River drainage between Jackson Lake and the Palisades Reservoir in Wyoming, has been stocked in Arkansas. It is the most widely stocked Cutthroat subspecies outside its native range because it is extremely adaptable to different environments with different forage (Behnke 2002). This subspecies is unusual in that it has not hybridized in its native habitat with Rainbow Trout. Hybridization has often occurred where Rainbow Trout have been introduced into waters formerly inhabited only by Cutthroat Trout (Willers 1991). Phylogenetic analysis of 39 behavioral and life history traits failed to support a sister-group relationship between *O. clarkii* and *O. mykiss*, previously supported by nuclear and mitochondrial DNA evidence (Esteve and McLennan 2007).

DISTRIBUTION   *Oncorhynchus clarkii* occurs in Pacific drainages from the Kenai Peninsula, Alaska, to the Eel River, California, and eastward through the Rocky Mountains in Hudson Bay, Mississippi River, and Great Basin drainages from Alberta, Canada, to the Rio Grande in New Mexico. It has been widely introduced elsewhere in Canada and the United States. Cutthroat Trout eggs were first stocked in hatching boxes in the White River in Arkansas by private interests in February 1982. A stocking of 90,000 finespotted Snake River Cutthroat fingerlings from Wyoming occurred in September 1983, and 50,000 Cutthroat Trout (7–9 inches [178–229 mm]) were released in the spring of 1984 in the cold tailwaters of the White River below Bull Shoals Dam and in the North Fork of the White River below Norfork Dam. Stocking occurred in the Beaver Lake tailwaters in 1989. Subsequent stocking also occurred in the Little Red River. Cutthroat Trout are currently stocked in Arkansas trout waters in the Spring and White rivers and in the Norfork Dam tailwaters (Map 133).

HABITAT AND BIOLOGY   Anadromous forms are native to Pacific coastal streams from northern California to Alaska. Allopatric inland forms occur in a wide band in the Rocky Mountains from Alberta, Canada, south to Arizona and New Mexico. They inhabit small, gravel-bottomed rivers, mountain streams, and lakes. It is more of a headwater species than the Rainbow Trout and is less tolerant of higher water temperatures than that species. It requires well-oxygenated waters and can tolerate water temperatures up to 70–72°F (21–22°C). The Finespotted Cutthroat Trout subspecies stocked in Arkansas prefers large rivers more than the other subspecies, but it is also widely adaptable to streams and lakes. Cutthroat Trouts thrive in streams with complex habitat cover such as large wood, overhanging vegetation, undercut banks, boulders, and pools. In natural streams, tree canopy cover is important for maintaining shade to moderate stream temperatures. A well-vegetated riparian area contributes to channel stability and is an important source of insect prey. In the fall and winter, Cutthroat Trout seek habitats with abundant wood and/or rock cover, usually deep, slow-water pools. Juveniles overwinter in the interstitial spaces between rocks and boulders.

Its biology has been described by Carlander (1969), Scott and Crossman (1973), Moyle (1976), and Willers (1991). The diet of small Cutthroat Trout consists mostly of insects. Larger individuals also eat aquatic insects but often add minnows and other items to their diet. Finespotted Cutthroat Trout over 12 inches TL often feed on small fish and crayfish. Cutthroat Trout feed within or just downstream of riffles where aquatic invertebrates are abundant. The most common items in the diet are larval and pupal stages of mayflies, stoneflies, and caddisflies.

In its native range, the Cutthroat Trout is usually an early spring spawner. Spawning is similar to other trout species. It prefers to spawn in cold, clear, well-oxygenated headwater streams with loose, clean gravel bottoms. Preferred spawning sites have gravel smaller than 3.5 inches (89 mm), a water depth of 3.54–11.8 inches (90–299.7 mm), and water velocity of 6.2–23.6 in./s (15.8–59.9 cm/s). The eggs are deposited in depressions (redds) excavated by the female, fertilized by the male, and then covered with gravel. No natural reproduction of this introduced species has been documented in Arkansas. The Finespotted Cutthroat Trout is reportedly one of the easiest trouts to catch on hook and line, and Cutthroat Trout weighing 2–4 pounds (0.9–1.8 kg) are commonly taken in Arkansas; a few weighing almost 10 pounds (4.5 kg) have been caught (Hudy 1986). Its susceptibility to the whirling disease organism (*Myxobolus cerebralis*) was studied by Wagner et al. (2002).

CONSERVATION STATUS   The modest sport fishery for this species in Arkansas is maintained by stocking hatchery raised trout.

Oncorhynchus mykiss (Walbaum)
Rainbow Trout

CHARACTERS   A streamlined trout with a moderately large, terminal mouth, the upper jaw barely extending behind the eye (Fig. 18.2). Teeth are present on jaws, palatine bones in roof of mouth, and tip of tongue, but there

Figure 18.2. *Oncorhynchus mykiss*, Rainbow Trout. *Patrick O'Neil, Geological Survey of Alabama.*

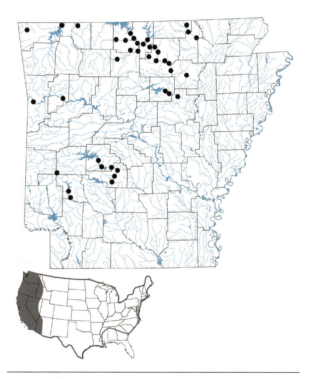

Map 134. Distribution of *Oncorhynchus mykiss*, Rainbow Trout, 1988–2019. Inset shows native range (solid) and introduced range (outlined) in the contiguous United States.

are no teeth present on midline of tongue. Lateral line complete, with 100–150 scales. Dorsal fin rays 10–13 (usually 11 or 12); anal fin rays 8–12 (usually 10 or 11); pectoral fin rays 11–17. Branchiostegal rays 9–13. Gill rakers 16–22. Axillary process present at base of pelvic fin. Caudal fin slightly forked. Breeding males develop a hook on lower jaw. Most Rainbow Trout caught in Arkansas are 10–16 inches (254–406 mm) long and weigh about 1 pound (0.4 kg), but individuals weighing 3–5 pounds (1.4–2.3 kg) are commonly caught. The state record is 19 pounds, 1 ounce (8.6 kg), taken from the White River in 1981. Maximum reported TL around 45 inches (1,140 mm) and weight of more than 35 pounds (15.9 kg).

LIFE COLORS Back and upper sides olive green, thickly speckled with black spots. Side with a broad pink or reddish longitudinal stripe, but without orange or red spots. Undersides silvery white. Dorsal, adipose, and caudal fins profusely spotted. Juveniles are less spotted and have 5–10 widely spaced oval dark parr marks. Anglers have occasionally caught solid-colored yellow-gold to orange Rainbow Trout in Arkansas.

SIMILAR SPECIES The Rainbow Trout differs from the Brown Trout in having dark spots on the caudal fin (vs. caudal fin without spots) and sides without orange or red spots in life (vs. sides with red or orange spots having blue halos). It is very similar in overall coloration to the Cutthroat Trout, but differs from that species in having fewer than 150 lateral line scales (vs. usually more than 150 lateral line scales), no teeth on midline of tongue (vs. teeth present on middle of tongue), and no red cutthroat mark (vs. red-orange cutthroat mark present on each side of throat). It can be distinguished from the Brook Trout and Lake Trout in having fewer than 150 lateral line scales (vs. more than 190), and in having distinct black spots on caudal fin (vs. caudal fin without distinct spots).

VARIATION AND TAXONOMY *Oncorhynchus mykiss* was described by Walbaum (1792) as *Salmo mykiss* from Kamchatka in eastern Siberia. For many decades, the Rainbow Trout was placed in the genus *Salmo*, and in the first edition of this book it was known as *Salmo gairdneri*. Since 1989, it has been assigned to the genus *Oncorhynchus* along with several other Pacific trouts (such as the Cutthroat Trout) and Pacific salmons (Smith and Stearley 1989). Because the Rainbow Trout was considered conspecific with the Kamchatkan Trout (then known as *Salmo mykiss*), the oldest available specific epithet (*mykiss*) was applied to the species when it was transferred to *Oncorhynchus*. This designation has subsequently been widely accepted (Nelson et al. 2004; Page et al. 2013). This is a highly variable species over its native range. Several races, varieties, and subspecies have been recognized, but much confusion still exists over the designation of these categories. Page and Burr (2011) listed six subspecies. Pacific coastal *O. mykiss,* called Steelhead, are very distinctive in color pattern and are anadromous.

DISTRIBUTION The Rainbow Trout is native to the Kamchatka Peninsula of Russia and the Pacific slope of North America from Alaska to northern Mexico and has been widely stocked throughout Canada, the United States, and elsewhere in the world (Page and Burr 2011). It

is stocked annually in Arkansas in the White River below Beaver Dam and Bull Shoals Dam, Bull Shoals Reservoir, North Fork of the White River below Norfork Dam, Blanchard Spring, Little Red River below Greers Ferry Lake, Spring River downstream from Mammoth Spring, Lake Hamilton, Lake Catherine, Ouachita River below Remmel Dam, and the Little Missouri River above and below Lake Greeson (Map 134). Strains of Rainbow Trout commonly stocked in Arkansas are the McConaughy, Arlee, Erwin, Arlee/Erwin, Fish Lake, and Shasta strains (Port et al. 2010). A Hot Creek strain of Rainbow Trout from California was stocked in the headwaters of the Illinois River (Spavinaw Creek) in the mid-1990s with some subsequent natural reproduction occurring. This California strain was selected for introduction because of its tolerance of higher water temperatures in the summer and its winter spawning characteristics. A golden-colored Rainbow Trout was stocked in White River drainage tailwaters in 2008 and 2019, with some subsequent reports of angler catches from the 2019 stocking. A put-and-take trout fishery also exists in the Illinois River below Tenkiller Dam in Oklahoma. This fishery is the likely source of occasional records of Rainbow Trout from the Arkansas River in western Arkansas. There are two records of *O. mykiss* from the Arkansas River at Fort Smith (the most recent one taken in 2007), and a record of a 4.2-pound specimen taken from the Arkansas River below the dam at Ozark in 1997. The Ozark specimen was wearing a tag from the Oklahoma Department of Wildlife Conservation. Rainbow Trout are commonly stocked by the Arkansas Game and Fish Commission in small lakes and ponds in various parts of Arkansas during cold months. These waters are not capable of supporting introduced trout for very long, but when stocked in winter or early spring, trout can provide a unique short-term fishing opportunity at nature centers, in urban areas, and for other youth-oriented events.

HABITAT AND BIOLOGY The Rainbow Trout has been widely introduced in suitable habitat worldwide. Anadromous populations occupy coastal streams, but resident nonmigratory populations are found farther inland. The Rainbow Trout occurs in streams, lakes and reservoirs in its native range. It always prefers well-oxygenated waters that remain below 70°F (21.3°C), but it can tolerate somewhat higher temperatures than most other trouts. Preferred water temperature was reported to be around 11.3°C (McCauley et al. 1977). It does well in Arkansas below the large dams in the cold tailwaters and occurs in the lower parts of small creeks (such as Piney Creek in Izard County) as strays from the White River (Matthews and Matthews

1978). In the Beaver Lake tailwaters, Rainbow Trout occupied the deepest habitats available and were randomly associated with cover at low flow (about 1 m³/s), but were strongly associated with velocity refugia near the river margins at high flow (about 215 m³/s) (Quinn and Kwak 2000). Quinn and Kwak also reported that instream and riparian habitat rehabilitation structures commonly used in small streams were a valid management technique for large tailwater rivers. The total detected linear range of movement of trout in a catch-and-release area of the Beaver Lake tailwater varied from 205 to 3,023 m (Quinn and Kwak 2011). Although it prefers swift, cool, gravel-bottomed streams where it is commonly found in sheltered water near structure (rocks, drop-offs, and logs) adjacent to currents carrying food, the Rainbow Trout also adapts to cool, deep Arkansas reservoirs. Successful experimental stocking in 1961 led to the development of a "two-story" fishery in Bull Shoals Reservoir (Baker and Mathis 1967). After the Brown Trout, the Rainbow Trout is the next most tolerant of high temperatures, with a tolerance range from about 32°F to 83°F (0–28.3°C) (Becker 1983).

Juveniles feed primarily on larval and emergent aquatic insects, while the adult Rainbow Trout feeds on aquatic and terrestrial insects as well as small fishes and snails. In Lake Taneycomo, Missouri, amphipod crustaceans made up 90% of its diet in the early 1970s (Pflieger 1977). Larger trout often feed at night, whereas juveniles smaller than 90 mm TL are usually inactive at night (Hill and Grossman 1993). In the Little Missouri River drainage, Arkansas, Rainbow Trout fed on a variety of invertebrates, with stonefly larvae dominating the diet in Long Creek and mayflies and dipteran pupae dominating trout diets in Blaylock Creek (Metcalf et al. 1997). Other food items making up trout diets in the Little Missouri River drainage were midge larvae, hymenopterans, caddisfly larvae, isopods, crayfish, coleopterans, and corn used as bait by fishermen. This study demonstrated the potential for introduced Rainbow Trout to compete with native juvenile Smallmouth Bass and other native insectivores for food resources in the oligotrophic Little Missouri River. Ebert and Filipek (1991) had previously suggested that Rainbow Trout had minimal impact on native fishes in the Little Missouri River because hatchery raised trout had low survival rates and were thought to exhibit little or no feeding after introduction. Flinders and Magoulick (2017) concluded that restrictive angling regulations in catch-and-release areas in Arkansas tailwaters may be unsuccessful if food availability limits energy for fish to grow. Based on bioenergetic simulations, Rainbow Trout

fed at submaintenance levels in all size classes (≤400 mm TL, >400 mm TL) throughout most seasons. In winter, the daily ration was substantially below the minimum required for maintenance, despite reduced metabolic costs associated with lower water temperatures.

Spawning success is limited in Arkansas but occurs in swift gravelly riffles from early winter to late spring. A wild population of Rainbow Trout was established in Collins Creek in Cleburne County after a continuous supply of cold water was diverted to that creek from Greers Ferry Dam in 2000 (Jeff Williams, pers. comm.). Spawning occurred in streams in the Missouri Ozarks from late December through early February (Turner 1979), and Pflieger (1997) reported that hatchery brood stock was selected to spawn in October and November. While the female is preparing the nest, the male drives away intruders. The female digs a shallow pit in which she lays the eggs, which are usually fertilized by one dominant male. The female covers the eggs with gravel and leaves them unguarded. Although the eggs and young receive no parental care, both male and female defend the nest against other fish while spawning. Average fecundity was reported by Carlander (1969) as 2,500–4,500 eggs. The eggs hatch in about 23–26 days at 12°C, and the newly hatched fry remain in the gravel until the yolk sac is absorbed. Reproducing populations are known from 14 spring branches and streams in the Missouri Ozarks (Turner 1979).

CONSERVATION STATUS   The only self-sustaining population in Arkansas occurs in Collins Creek in Cleburne County. The economically important sport fishery for this species is maintained by stocking hatchery-raised fish.

Genus *Salmo* Linnaeus, 1758— Atlantic Trouts and Salmons

This genus, described by Linnaeus in 1757 (but type species designated in 1758), is distributed mainly in Europe and western Asia, with one native anadromous species (Atlantic Salmon, *S. salar*) in northeastern North America. There is controversy over how many species and subspecies the genus contains, with some authors recognizing eight or nine species (Behnke 1990a, b) and other authors as many as 46 species (Turan et al. 2014). Most recognized species are endemic to single watersheds, but the Brown Trout (*Salmo trutta*) is broadly distributed and widely introduced throughout the world. Genus characters include dorsal rays 12–14, caudal fin lacking black spots, and males usually with red or orange spots when spawning.

Salmo trutta Linnaeus
Brown Trout

Figure 18.3. *Salmo trutta*, Brown Trout. *David A. Neely.*

Map 135. Distribution of *Salmo trutta*, Brown Trout, 1988–2019. Inset shows introduced range (outlined) in the contiguous United States.

CHARACTERS   A trout with a moderately large, terminal mouth, the upper jaw extending posteriorly to back of eye or beyond (Fig. 18.3). Teeth present on jaws, palatine bones in roof of mouth, and on tip of tongue, but no teeth are present on midline of tongue. Lateral line complete, with 120–140 scales. Caudal fin not deeply forked and usually without dark spots. Dorsal fin rays 10–14 (usually 11 or 12); anal fin rays 10–12; pectoral fin rays 13 or 14. Gill rakers 14–17. Axillary process present at base of pelvic fin. Breeding males develop a hook on lower jaw. Brown Trout reach a larger size in Arkansas than the other trouts except the Lake Trout. Most Brown Trout caught by anglers weigh 2–4 pounds, but individuals weighing 5–10 pounds (2.27–4.54 kg) are common. The Arkansas state angling record

is 40 pounds, 4 ounces (18.3 kg) from the Little Red River in 1992. It is the current 4-pound line-class world record.

LIFE COLORS   Back and upper sides a dark olive brown, with scattered dark spots on body. Large black spots present on operculum. Sides with red or orange spots usually having blue or gray halos; no lateral orange or red band present. Undersides yellowish white. Dorsal and adipose fins with black spots; caudal fin without spots or with only a few spots on dorsal portion. Adipose fin orange to orange red. Coloration and spotting patterns throughout North America can be highly diverse, with those inhabiting large lakes having a silvery coloration often with x-shaped markings dorsally (Bachman 1991). Juveniles have 9–14 short, narrow parr marks on side and a few red spots along the lateral line.

SIMILAR SPECIES   The Brown Trout differs from the Rainbow Trout in lacking dark spots on the caudal fin (vs. caudal fin with distinct dark spots) and in usually having red or orange spots on the side in life (vs. side without red or orange spots). It differs from the Cutthroat Trout in having fewer than 150 lateral line scales (vs. usually more than 150 lateral line scales), no teeth on midline of tongue (vs. small teeth present on midline of tongue), and no red cutthroat marks in life (vs. red cutthroat marks present). It differs from the Brook Trout and Lake Trout in having fewer than 150 lateral line scales (vs. more than 190).

VARIATION AND TAXONOMY   *Salmo trutta* was described by Linnaeus (1758) from "Europe." Throughout its native range, *Salmo trutta* occurs in a variety of habitats and is highly variable in its color pattern and appearance. Because of this great variation in appearance of populations, more than 50 species of trout were described that were eventually consolidated into *S. trutta* (Behnke 2002). Analysis of geographic variation in mitochondrial DNA identified five major evolutionary lineages that evolved in geographic isolation during the Pleistocene and have remained largely allopatric since then (Bernatchez 2001). Behnke (2002) reported that the Brown Trout has 80 chromosomes. Some sexual dimorphism occurs, with adult males having a rounded anal fin and females having the falcate anal fin characteristic of juveniles (Bachman 1991). Older breeding males may develop a hooked lower jaw.

DISTRIBUTION   The Brown Trout is not native to North America but occurs naturally in Europe, western Asia, and northern Africa. It was first stocked in North America in 1883. In Arkansas, it has been stocked in recent years primarily in the upper White River below Beaver Dam, Bull Shoals Dam, and Norfork Dam (Map 135). It is also stocked in the Spring River and in the Little Missouri River below Narrows Dam. Previous stockings have also occurred in Spavinaw Creek of the Illinois River drainage (Map A4.133). The most successful naturalized population is in the Little Red River.

HABITAT AND BIOLOGY   Carlander (1969), Belica (2007), and Lobon-Cervia and Sanz (2017) summarized biological information and the extensive literature on this species. It has been introduced around the world into at least 24 countries and has often established self-sustaining populations where introduced. In the United States, there are naturalized populations in 40 states (Behnke 2002). The Brown Trout inhabits mainly cold-water streams and lakes and is sometimes successfully stocked in reservoirs. In the Jocassee Reservoir, South Carolina, Brown Trout preferred water temperatures of 7.5–21.9°C during July and August, and temperatures of 8.8–12.0°C in September (Barwick et al. 2004). In that study, trout were found at oxygen concentrations of 4.2–10 mg/L and water depths of 14–54 m. Griffith and Smith (1993) studied habitat use by age 0 fish in the Snake River, Idaho, and found that young fish remained concealed during the day in substrates of river margins at depths less than 0.5 m. They estimated that between 61% and 66% of the fish emerged at night to swim in the water column near where they had been concealed during the day. In Ozark tailwaters, age 0 Brown Trout of the same size class were associated with different types of cover at different sites (Pender and Kwak 2002). Trout smaller than 65 mm were most frequently associated with woody debris at one site, cobble at another, and submerged vegetation at a third site. Cover preference changed with increasing trout size. The Brown Trout is most abundant in the White River below Bull Shoals Dam and the Little Red River, where it is often found around dense cover such as logs or undercut banks, or in deep water below riffles. Sigler and Miller (1963) reported that in Utah it is more tolerant of turbid water and pollution than other trouts. It prefers cool water, but it is more tolerant of higher temperatures than other trouts found in Arkansas. The critical maximum temperature is about 25°C (77°F) (Carlander 1969). Bachman (1984) studied social behavior in a Pennsylvania stream and found that Brown Trout did not have discrete, exclusive territories that were defended. Instead, they had overlapping home ranges with shared multiple feeding sites. The trout exhibited a linear dominance hierarchy in the wild, with older and larger fish having the highest ranks, and when an older fish disappeared, its range was taken over by YOY or yearlings (Bachman 1984). When hatchery-reared Brown Trout were introduced into the naturalized population, agonistic encounters between the two groups were observed within 20 minutes, but prior residence had no effect on the outcome of the encounters.

Brown Trout feed primarily on aquatic invertebrates, terrestrial insects, and small fish. The diet is diversified and varies broadly with size and age, spatial and temporal variability in food availability, behavior, and habitat characteristics (Belica 2007). Several studies have documented ontogenetic changes in feeding patterns. Recently emerged fry feed on zooplankton and other small invertebrates (Bachman 1991). As age 0 trout grow they begin to feed on terrestrial and aquatic insects, and the adults add fish and crayfish to the diet. Adults are largely opportunistic drift feeders in streams. Food habits of age 0 Brown Trout in White River tailwaters were similar at all sites sampled, with positive selection for dipterans of various life stages, particularly chironomid larvae (Pender and Kwak 2002). Large age 0 fish also positively selected isopods, amphipods, and gastropods at several localities and exhibited strong negative selection for rotifers, oligochaetes, and hydroids. Dunn (1999) also reported that age 0 Brown Trout larger than 75 mm predominantly fed on isopods in the Bull Shoals tailwaters. Diet overlap between the Knobfin Sculpin and age 0 Brown Trout differed significantly among four White River sites studied, with the greatest dietary overlap occurring in the Beaver Lake tailwaters (Pender and Kwak 2002). This suggests possible resource competition between the two species under conditions of low invertebrate food availability and low water fertility at that site. In the Little Red River, Arkansas, diet diversity was low in all size classes, and 70–80% of all prey consumed were isopods; other foods eaten were chironomid larvae, amphipods, gastropods, trout eggs, and age 0 trout (Johnson et al. 2006a, 2007). In those studies, piscivory (<0.5% of individuals sampled) and consumption of terrestrial invertebrates were rare. In areas where small fish are readily available as prey, smaller adult Brown Trout feed largely on aquatic invertebrates, but adults around 160 mm TL begin to incorporate fish into the diet; adults become predominantly piscivorous at around 350 mm TL (Mittelbach and Persson 1998). Behnke (2002) reported that few stream-dwelling Brown Trout transitioned to a piscivorous diet. Water fertility can influence food availability and fish diets (Pender and Kwak 2002). In the Little Red River, temperatures and food availability were less than required for maximal trout growth (Johnson and Harp 2005; Johnson et al. 2006a). In that river, Brown Trout growth was 54.8–57.0% of the maximum predicted by a bioenergetics model. The displacement of the native fish fauna and subsequent lack of establishment of cold-tolerant forage fish species due to the hypolimnetic water release from Greers Ferry Reservoir was considered the major barrier to Brown Trout growth in the Little Red

River (Johnson et al. 2006a). Most Brown Trout caught in Arkansas are taken by fishermen using live crayfish for bait, but this species is occasionally caught on artificial lures. Many fishermen have had excellent results using sculpins (*Uranidea*) for bait. Of the trout species found in Arkansas, the Brown Trout is reportedly the most difficult species for anglers to catch. Pflieger (1997) noted that it is more secretive than the Rainbow Trout, remaining under cover or in deep water during the day and feeding most actively at dusk and dawn. In a Pennsylvania stream, Brown Trout spent 86% of daylight time in a sit-and-wait state, searching the passing water column for food (Bachman 1984). In an Idaho stream, the proportion of active Brown Trout increased during crepuscular periods and at night (Young et al. 1997). Clapp et al. (1990) found that foraging of large Brown Trout varied seasonally and appeared to be related to light level, food availability, and water temperature.

The Brown Trout spawns in fall and early winter at water temperatures ranging from 35°F to 57°F (1.7–13.9°C) and is more successful in reproducing in Arkansas waters than the other four species of trout. The reproductive success of Brown Trout in White River tailwaters is highly variable, and the fecundity and condition factor of prespawning trout were significantly lower in Beaver Tailwater (a reach known for reproductive failure) than at other White River sites (Pender 1998). Significant natural reproduction occurs in the White River below Bull Shoals Dam and below Greers Ferry Dam in the Little Red River. The Brown Trout population in the Little Red River is sustained completely by natural reproduction. Seasonal fishing closures on the White River during the spawning season help maintain natural spawning by protecting the most actively used spawning areas from depletion of the spawning population; however, regular stocking is still sometimes required to maintain a fishery for Brown Trout in the state. Spawning occurs in White River tailwaters in October and November, with fry emergence beginning in late February (Pender and Kwak 2002). Spawning habitat in Beaver and Bull Shoals tailwaters was characterized by small cobble and coarse gravel substrates, water depths of 0.5–0.65 m, mean water velocities of 0.36–0.67 m/s, dissolved oxygen levels of 9.5–10.1 mg/L, and water temperatures of 9.5–13.4°C (Pender and Kwak 2002). Its spawning habitat is similar to that of other trouts: areas of swift current and gravel substrates. The female digs a shallow pit in the gravel substrate of a riffle. The eggs are deposited in it, are fertilized by the male, and are then buried in the gravel by the female. Both parents desert the eggs. Embryonic development takes place at temperatures between 1.4°C and 15.0°C but may occur at temperatures as low as 0°C (Elliot 1994).

Developmental time is rather lengthy, ranging from 34 to 151 days at temperatures from 11.2°C down to 2.0°C (Crisp 1981). Mean fecundities of 3 and 4-year-old Brown Trout in White River tailwaters ranged from 1,330 to 2,278 eggs (Pender and Kwak 2002). Carlander (1969) reported fecundity of larger females to be between 4,000 and 12,000 eggs. Predation by the Knobfin Sculpin, *Uranidea immaculata*, on trout eggs or on age 0 trout of any species was not found based on 418 sculpin stomachs examined (Pender and Kwak 2002). Dunn (1999) also reported that no trout eggs or fry were found in 131 *U. immaculata* stomachs examined from the Bull Shoals tailwater.

CONSERVATION STATUS   Although some natural reproduction occurs in Arkansas, the Brown Trout sport fishery is maintained by stocking hatchery-raised fish.

Genus *Salvelinus* Richardson, 1836—Chars

This genus, often referred to as chars, is distributed in the Northern Hemisphere from North America across Eurasia, occupying both marine and freshwater habitats. As with other salmonid genera, there is no consensus on the number of species it contains, but estimates range from 10 to around 51 species. Behnke (2002) provided biological information on the genus *Salvelinus* and discussed problems of char taxonomy. Three subgenera were recognized by Vladykov (1963) and Nelson (1984). There are five North American species of *Salvelinus* (Page et al. 2013), two of which are introduced into Arkansas. Genus characteristics include body with pale spots (lacking black spots), the presence of basibranchial teeth, tiny scales (100–200 in lateral series), and lower fins with a white anterior margin.

Salvelinus fontinalis (Mitchill)
Brook Trout

Figure 18.4. *Salvelinus fontinalis*, Brook Trout. *Patrick O'Neil, Geological Survey of Alabama.*

CHARACTERS   The Brook Trout has a large terminal mouth with the upper jaw extending well behind the eye (Fig. 18.4). Teeth are present on jaws, tongue, and palatine

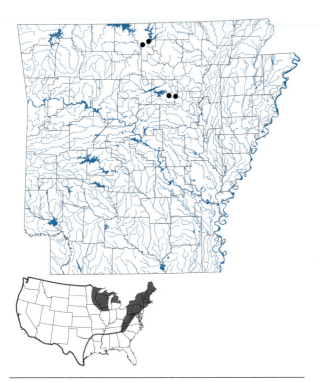

Map 136. Distribution of *Salvelinus fontinalis*, Brook Trout, 1988–2019. Inset shows native range (solid) and introduced range (outlined) in the contiguous United States.

bones in roof of mouth. Lateral line complete, with 210–244 scales. Dorsal fin rays 10–14; anal rays 9–13; pectoral rays 11–14. Gill rakers usually 16–21 (14–22). Pyloric caeca 23–55. Axillary process present at base of pelvic fin. Caudal fin very slightly forked to nearly straight-edged. Mandibular pores usually 7 or 8. Breeding male develops a hook on lower jaw. On average, a smaller trout than the other 4 introduced trout species. Adults are usually 10–16 inches (254–406 mm) long and weigh 1–2 pounds (0.5–0.9 kg). The Arkansas state record Brook Trout caught in the North Fork River in 2002 weighed 5 pounds (2.3 kg). Maximum size (from Canada) was 14 pounds, 8 ounces (6.6 kg).

LIFE COLORS   Back olive green with lighter green to cream-colored vermiculations; top of head also vermiculated. Dorsal and adipose fins with vermiculations or distinct dark spots. Caudal fin plain or with small spots along upper and sometimes lower margins. The dorsal vermiculate pattern fades down the upper body, and the sides are lighter with many bright red, yellow, or pink spots; the red spots often have blue halos. Lower sides with an orange or reddish band in breeding males. Eight to 10 parr marks are present on sides of young and are often still faintly evident in adults. Undersides usually white, but becoming brilliant orange or red in breeding males. Anterior rays of pectoral,

pelvic, and anal fins are a milky white, followed by a black stripe; posterior parts of those fins orange to red orange.

SIMILAR SPECIES   It differs from all trouts of the genera *Oncorhynchus* and *Salmo* in Arkansas in having more than 210 lateral line scales (vs. usually fewer than 210 lateral line scales), a larger mouth with the upper jaw extending well behind the eye (vs. mouth smaller, upper jaw barely extending past eye), and in having wormlike vermiculations on back (vs. back with spots but without vermiculations). It differs from the Lake Trout in having more than 210 lateral line scales (vs. usually fewer than 210 lateral line scales), 7 or 8 mandibular pores (vs. 9 or 10), and tail not deeply forked (vs. caudal fin forked).

VARIATION AND TAXONOMY   *Salvelinus fontinalis* was described by Mitchill (1814) as *Salmo fontinalis* from Long Island, New York. A subspecies, the Aurora Trout, *S. f. timagamiensis*, is a silvery form maintained only as hatchery stocks in some lakes in Ontario, Canada (Page and Burr 2011). Another subspecies, the Silver Trout, *S. f. agassizii*, formerly found in New Hampshire, is now extinct.

DISTRIBUTION   The Brook Trout is native to eastern Canada and eastern United States in the Great Lakes region and south through the Appalachian Mountains to Georgia. It has been widely introduced in North America and around the world. In the United States, it is now found in 41 states. Brook Trout have been stocked in the North Fork of the White River below Norfork Dam, the White River, and in the Little Red River (Map 136). It is much less common in Arkansas than the Rainbow and Brown trouts and is a very rare catch.

HABITAT AND BIOLOGY   Carlander (1969), Estes (1987), Willers (1991), and Karas (2015) summarized much of the literature on its biology. The Brook Trout inhabits cold, clear waters and in its native range is more numerous than other trouts in small creeks and ponds. It also inhabits medium rivers and some lakes, where it does best at water temperatures below 16°C (61°F). Its preferred temperature range is 11–13°C (52–55°F) (Mettee et al. 1996). Mortality usually occurs when temperatures exceed 24°C (Raleigh 1982). In Rocky Mountain streams, the Brook Trout and the Cutthroat Trout predominate in headwater areas, while the Brown and Rainbow trouts tend to inhabit mid- and lower-elevation stream segments or larger river habitats (Rahel and Nibbelink 1999). It is less tolerant of environmental fluctuations and warm water temperatures than other trouts, and in Arkansas, it is often found in the cold tailwaters just below dams. Like the Brown Trout, the Brook Trout is often found near cover such as rocks, logs, and undercut banks, but it feeds more in the daylight than that species.

The young feed on plankton and insects, the adults on insects and small fishes. Brook Trout fry in a Wisconsin creek fed mainly on chironomids, simuliids, and ephemeropterans, and diets of fry from different sections of the stream were similar (J. M. Miller 1974). Adults often feed on whatever is readily available, and reported food items include aquatic and terrestrial insects, crustaceans, worms, and fish (Meehan and Bjornn 1991; Behnke 2002). When Brook Trout occur together with Brown and/or Rainbow trout, Brook Trout tend to feed more on bottom-dwelling organisms, while Brown and Rainbow trout feed primarily on organisms in the water column and at the surface (Behnke 2002). During the winter months, Brook Trout reduce their feeding to sustain the minimal level of metabolic activity needed to survive (Raleigh 1982). It is reportedly the easiest trout for anglers to catch and is readily taken on a variety of artificial lures and natural baits (Etnier and Starnes 1993).

McPhail and Lindsey (1970) presented information on spawning. Although there is little successful reproduction in Arkansas waters, in its native range it spawns in late fall and winter. Brook Trout usually reach sexual maturity at age 2, but some males reach maturity at age 1 (Behnke 2002). Spawning is reportedly initiated by decreasing photoperiod, increased late fall flows, and drops in water temperature (Raleigh 1982; Meehan and Bjornn 1991). Spawning occurs at a water temperature of 55°F (12.8°C) or lower in gravelly riffles, where the female excavates a pit for depositing the eggs. Optimal spawning-substrate particle size was reported to be between 3 and 50 mm (Reiser and Wesche 1977). The male quickly releases milt over the eggs and then the female buries them. The eggs are deserted by the parents. The fertilized eggs hatch in 30 to 165 days at water temperatures of 11.2°C and 1.9°C, respectively (Raleigh 1982). The larvae remain in the redds for several weeks after hatching. Natural reproduction occurs successfully in Collins Creek, a tributary of the Little Red River below Greers Ferry Dam. Life expectancy of Brook Trout in most North American streams is about 3–5 years, but in large rivers and lakes, life expectancy may be up to 10 years (Meehan and Bjornn 1991; Behnke 2002).

The Brook Trout has been artificially crossed with the Brown Trout. The infertile hybrid is called a tiger trout because of its distinctive tiger-like markings. There have been some minor releases of tiger trout into Arkansas waters.

CONSERVATION STATUS   This species has been stocked in Arkansas in the White River drainage, and a modest sport fishery for Brook Trout is maintained in the state by stocking.

Salvelinus namaycush (Walbaum)
Lake Trout

Figure 18.5. *Salvelinus namaycush*, Lake Trout. *Rob Criswell.*

Map 137. Distribution of *Salvelinus namaycush*, Lake Trout, 1988–2019. Inset shows native range (solid) and introduced range (outlined) in the contiguous United States.

CHARACTERS   A large, elongate trout with a large, terminal mouth, the upper jaw extending far behind back of eye (Fig. 18.5). Teeth present on upper and lower jaws. Caudal fin deeply forked. Lateral line scales 185–210. Dorsal fin rays 8–10; anal fin rays 8–10. Axillary process present at base of pelvic fin. Gill rakers 12–24. Mandibular pores usually 9 or 10 on each side of jaw. Pyloric caeca 90–210. The Lake Trout is the largest of all North American trouts, averaging 15–20 inches (381–508 mm) in length, but attaining a length of more than 3 feet (914 mm) in Canada. The

Arkansas state angling record from Greers Ferry Lake in 1997 is 11 pounds, 5 ounces (5.1 kg).

LIFE COLORS One of the least colorful trouts, its back, sides, and head are silvery gray to greenish brown with many small, often bean-shaped cream or yellow spots or vermiculations; lacking dark spots on back; belly white. No orange or red on body. Dorsal and caudal fins with cream or yellow spots. Lower fins often orange with a white border, but lacking a black stripe. Breeding males usually develop a prominent black lateral band. Young with 7–12 dark, vertical parr marks.

SIMILAR SPECIES It differs from the Cutthroat, Rainbow, and Brown trouts in having a deeply forked tail (vs. tail slightly emarginate), and in lacking dark brown or black spots on back (vs. back with dark spots). It can be distinguished from Brook Trout in having fewer than 210 lateral scales (vs. more than 210 lateral line scales), no orange or red on body (vs. orange or red colors on body), and 9 or 10 mandibular pores (vs. 7 or 8).

VARIATION AND TAXONOMY *Salvelinus namaycush* was described by Walbaum (1792) as *Salmo namaycush* from "lakes far inland from Hudson Bay, Canada." A large-bodied form of Lake Trout from Lake Superior is called a Siscowet. The hybrid between the Lake Trout and the Brook Trout is called a splake. Splakes have been introduced into many parts of North America (Page and Burr 2011).

DISTRIBUTION *Salvelinus namaycush* is widely distributed in the Atlantic, Arctic, and Pacific basins from northern Canada and Alaska south to New England, the Great Lakes, and northern Montana. The Lake Trout has been widely stocked in deep reservoirs well outside its native range, where populations are managed for trophy fish because of its potential to reach large size and to live up to 37 years or more (B. M. Johnson and Martinez 2000). Ongoing stocking programs exist today in the adjoining states of Missouri and Tennessee. This native of northern North America was first stocked in Arkansas in 1986 in Bull Shoals and Greers Ferry lakes. Eggs from the "Jenny Lake strain" of Lake Trout were obtained from the federal fish hatchery at Jackson, Wyoming. The young trout were raised at the Norfork and Greers Ferry National Fish Hatcheries to a length of 6 or 7 inches (152–178 mm) and were moved to Bull Shoals Lake, where 100,000 young trout were eventually released into that lake from net pens and other stocking points. Fifty-five thousand young Lake Trout were released into Greers Ferry Lake (Map 137). Some Lake Trout were subsequently found in the Little Red River below Greers Ferry Lake (Carl Perrin, pers. comm.). It was hoped that the Lake Trout would thrive

in the cool 100–200-foot (30–60 m) depths of these lakes where little other fish life exists.

HABITAT AND BIOLOGY In its native range, the Lake Trout inhabits the deeper waters of cold lakes. Earlier studies suggested a temperature preference between 8°C and 12°C (McCauley and Tait 1970; Olson et al. 1988), but subsequent studies indicated a much broader temperature tolerance (Sellers et al. 1998) and a much cooler temperature preference of 6–8°C (Bergstedt et al. 2003; Mackenzie-Grieve and Post 2006). In a small Ontario lake during thermal stratification, trout were distributed over a broad range of temperatures from 2°C to 18°C (Plumb 2006). They occupied the highest temperatures and widest temperature range in the spring and fall. Plumb also found that seasonal and annual variations in fish depth were associated primarily with temperature-mediated changes in the timing and depth of thermal stratification. A combination of temperature (<12–15°C) and dissolved oxygen concentration (>4–6 mg/L) produced the most accurate estimates of habitat occupied by Lake Trout, with dissolved oxygen levels being the single most important factor determining the lower habitat boundary (trout were absent below 4–6 mg/L dissolved oxygen regardless of temperature) (Plumb 2006). Sellers et al. (1998) reported an absence of Lake Trout at water depths having <5.0 mg/L dissolved oxygen, supporting the avoidance thresholds of 4–6 mg/L summarized by Evans et al. (1991). It is an active fish that moves about extensively (Becker 1983). In the Great Lakes, it is most abundant at depths greater than 100 feet (30 m). It spends most of its time at or near the bottom but is occasionally found higher in the water column depending on water temperatures. In both small and large lakes, it moves into cooler and deeper water with the onset of thermal stratification (Plumb 2006).

Lake Trout fry feed initially on zooplankton but switch to larger crustaceans, insects, amphipods, and small fish as they grow (Elrod and O'Gorman 1991). Large Lake Trout feed on pelagic prey. Most of the time the adults feed almost exclusively on whatever fishes happen to be available in the immediate environment, but at certain times Lake Trout are opportunistic and take advantage of other abundant foods such as large hatches of insects (Becker 1983). They periodically consume fish such as Yellow Perch, suckers, and various minnows in the warm upper layer of a lake above the thermocline and in near-shore environments (Youngs and Oglesby 1972; Bronte 1993). MacLean et al. (1981) also reported brief feeding excursions into shallow (~3 m), warmer (>15°C) water during the stratified period, but Sellers et al. (1998) concluded that Lake Trout sometimes spend extended periods in shallow water at 19–21°C.

In the absence of pelagic prey fishes, Lake Trout typically feed on littoral prey fishes, and at the onset of thermal stratification, the diet shifts to primarily zooplankton and benthic invertebrates because of thermal restraints (Martin 1966; Konkle and Sprules 1986). In two Tennessee reservoirs, Lake Trout appeared to be spatially segregated from pelagic prey fishes during summer stratification (Russell and Bettoli 2013).

In its native range, spawning occurs in the fall from October to mid-December (Becker 1983; Smith 1979). It spawned in Otsego Lake, New York, from 20 October to 1 November when temperatures reached 10–12°C (Tibbits 2008). Spawning habitat in Lake Ontario and Lake Michigan consisted of large cobble substrate with no organic matter at depths of 4–6 m (13–19.7 feet) (Jude et al. 1981; Marsden and Krueger 1991; Marsden et al. 1995a). Spawning in Otsego Lake occurred over angular gravel and smooth pebble substrates at a depth of less than 1 m (3 feet) (Tibbits 2008). Spawning Lake Trout in New York were subject to predation by Bald Eagles and Great Blue Herons, and their eggs were observed being consumed by Common Carp (Tibbits 2008). Males are the first to arrive on the spawning grounds, where they can be seen swimming at any hour of the day or night, waiting for the females to appear (Esteve et al. 2008). Once the females arrive, the fish begin to congregate in groups, swimming slowly over the coarse gravel, rubble, or rock substrates. Lake Trout spawning differs from that of all other trouts, salmon, and chars (but is remarkably similar to that of the whitefish genus *Coregonus*) in that they (1) spawn without building a nest, (2) lack overt male-male competition, and (3) lack distinct sexual dimorphism. Although most studies have reported that Lake Trout limit their breeding activities almost entirely to periods of darkness (Gunn 1995; Watson 1999), Esteve et al. (2008), using a video camera, observed spawning during the day. Instead of establishing and defending a territory, females wander across large areas intermittently laying batches of eggs over selected substrate crevices. One or more females may spawn with several males in a compact group. The released eggs sink to the gravel substrate, where they fall into the interstitial spaces between rocks, but no attempt is made to bury them and, unlike some other species of *Salvelinus*, no female parental care of the eggs occurs. A detailed description of spawning behavior was provided by Esteve et al. (2008). The last stocking of Lake Trout in Greers Ferry Lake occurred in 1988, and there is evidence that successful spawning occurred in that lake in the late 1990s (Carl Perrin, pers. comm.). The last known angling records for Lake Trout in Arkansas occurred in the early 2000s. In Watauga Lake, Tennessee, 158 Lake Trout ranging in age from age-1 to age-20 were collected in 2009 and 2010 from stockings that began in the mid-1980s (Russell and Bettoli 2013). In South Holston Lake, Tennessee, which has been stocked with Lake Trout since 2006, the oldest trout taken were age-4 fish. The mean maximum life span in northern Canadian lakes was 34.3 years (McDermid et al. 2010).

Lake Trout supported an important commercial fishery in the Great Lakes during the first half of the 20th century. This commercial fishery declined steeply during the 1950s and early 1960s primarily because of predation by the Sea Lamprey and overfishing. Sharp restrictions on the commercial fishery of the Great Lakes have resulted in increased yields of Lake Trout each year since 1962 (Becker 1983).

CONSERVATION STATUS No recent attempts have been made to establish a put-and-take sport fishery in Arkansas for Lake Trout. Specimens weighing 8–11 pounds from the original stockings in the 1980s were occasionally taken by fishermen in Greers Ferry Lake into the late 1990s and early 2000s.

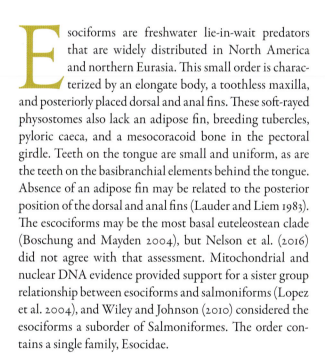

CHAPTER 19

Order Esociformes
Pikes and Mudminnows

Esociforms are freshwater lie-in-wait predators that are widely distributed in North America and northern Eurasia. This small order is characterized by an elongate body, a toothless maxilla, and posteriorly placed dorsal and anal fins. These soft-rayed physostomes also lack an adipose fin, breeding tubercles, pyloric caeca, and a mesocoracoid bone in the pectoral girdle. Teeth on the tongue are small and uniform, as are the teeth on the basibranchial elements behind the tongue. Absence of an adipose fin may be related to the posterior position of the dorsal and anal fins (Lauder and Liem 1983). The escociforms may be the most basal euteleostean clade (Boschung and Mayden 2004), but Nelson et al. (2016) did not agree with that assessment. Mitochondrial and nuclear DNA evidence provided support for a sister group relationship between esociforms and salmoniforms (Lopez et al. 2004), and Wiley and Johnson (2010) considered the esociforms a suborder of Salmoniformes. The order contains a single family, Esocidae.

Family Esocidae *Pikes and Mudminnows*

Controversy exists over whether the esociform fishes should be classified into one or two families. Until recently, the pikes and pickerels were grouped into the family Esocidae, and the mudminnows were considered to form a separate family, Umbridae. Several authors, using genetic evidence, argued that Umbridae constituted a paraphyletic grouping because *Esox* is nested within a clade with umbrid genera (Lopez et al. 2000, 2004); however, Betancur-R. et al. (2017) used molecular data to support recognition of Umbridae. Despite their obvious morphological differences, we follow Page and Burr (2011) and Page et al. (2013) in placing the species formerly in Umbridae in the family Esocidae. Nelson et al. (2016), however, recognized the two groups as separate families but moved two genera (*Novumbra* and *Dallia*) formerly placed in Umbridae into Esocidae.

Esocidae is a Holarctic family found in North America, Europe, and Asia and includes four genera with 11 species, 8 of which occur in North America. Following the classification of Page et al. (2013), the genus *Esox* contains five species, four of which are found in North America. There is one North American species of *Dallia* and another *Dallia* species in Russia. There is one North American species of the genus *Novumbra*, and two species of *Umbra*. A third *Umbra* species occurs in Europe. A combination of morphological and molecular data supported the monophyly of the genus *Esox*, and the monophyly of the subgenera *Esox* (pikes) and *Kenoza* (pickerels) (Grande et al. 2004).

The oldest fossils of this family (genera *Estesox*≠ and *Oldmanesox*≠) are from the Late Cretaceous (Nelson 2006). Three of the 8 extant North American species, the Redfin Pickerel, *E. americanus,* Chain Pickerel, *E. niger,* and Central Mudminnow, *Umbra limi,* are native to Arkansas. Two additional species, the Northern Pike, *E. lucius,* and the Muskellunge, *E. masquinongy,* have been introduced into the state but have not established reproducing populations. No records of the introduced *Esox* species in Arkansas have appeared in more than 30 years, but we have retained them in the keys and species accounts because they may be stocked in Arkansas waters again in the future. It is also possible that stragglers from northern populations might occasionally be found in the Mississippi River of Arkansas. Hybrids between *E. lucius* and *E. masquinongy,* called tiger muskies, have also been introduced into several reservoirs in Arkansas.

The three mudminnow genera include rather small species, usually 2.8–7.1 inches (70–180 mm) TL, with disjunct distributions in North America and Eurasia. They lack the ducklike beak of pikes and pickerels and have a blunt snout. Their fins are soft rayed, the pelvic fins are small and abdominal, the caudal fin is rounded, scales are cycloid, and the swim bladder is connected by a duct to the alimentary canal, allowing use of atmospheric oxygen. Mudminnows are lie-in-wait predators of invertebrates. Where abundant, mudminnows are often sold for bait and, in some waters, they are important forage species.

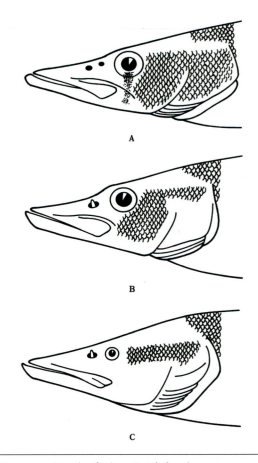

Figure 19.1. Heads of pikes. *Smith (1979).*

A—*Esox americanus*
B—*Esox lucius*
C—*Esox masquinongy*

Key to the Pikes and Mudminnows

1A Snout normal in appearance, not shaped like a duck's beak; 34–37 scales in a lateral series. *Umbra limi*
Page 496

1B Snout shaped like a duck's beak; more than 90 scales in a lateral series. 2

2A Opercle fully scaled (Fig. 19.1A); 4 pores on the lower surface of each lower jaw. 3

2B Opercle scaled only on upper half (Figs. 19.1B, 19.1C); 5 or more pores on the lower surface of each lower jaw. 4

3A Distance from tip of snout to center of eye distinctly greater than distance from center of eye to rear margin of operculum; branchiostegal rays usually 13 or 14.
Esox niger Page 494

3B Distance from tip of snout to center of eye equal to or less than distance from center of eye to rear margin of operculum; branchiostegal rays usually 11 or 12.
Esox americanus Page 489

4A Cheek fully scaled (Fig. 19.1B); branchiostegal rays 14–16; usually 5 pores on lower jaw; scales in lateral line usually fewer than 140. *Esox lucius* Page 491

4B Cheek scaled only on upper half (Fig. 19.1C); branchiostegal rays 17–19; 6 or more pores on lower jaw; scales in lateral line usually more than 140.
Esox masquinongy Page 492

Genus *Esox* Linnaeus, 1758—Pikes

This genus, described by Linnaeus in 1758, contains five species, three of which occur only in North America; one occurs across North America and northern Eurasia, and the fifth species occurs only in northeastern Asia. A distinctive snout shaped like a duck's beak is the most obvious feature of the genus. *Esox* is divided into two subgenera: subgenus *Esox*, containing *E. lucius*, *E. masquinongy*, and the east Asian *E. reicherti*, and subgenus *Kenoza*, which includes *E. americanus* and *E. niger*. All four North American species of *Esox* have been reported from Arkansas, but only *E. americanus* and *E. niger* are native to this state. Morphological and genetic data strongly support the monophyly of *Esox* (Grande et al. 2004). Additional genus characters include a long, cylindrical, green body, a large mouth with sharp teeth, a forked caudal fin, body covered with cycloid scales that are deeply scalloped on the front margins, and all fins with soft rays.

The three largest species of the genus, *E. lucius, E. masquinongy,* and *E. niger,* are popular gamefishes because of their fighting ability and willingness to take artificial lures. Even though the flavor of their flesh is considered good, they are not regarded as good food fishes because of their numerous intermuscular bones. They are also valuable as predators that help maintain balanced fish populations in reservoirs by preventing overcrowding and stunting of sunfishes and shad. Prey organisms are captured when the pike lunges from cover and engulfs the prey via ram-feeding, impaling a fish on its sharp teeth before usually swallowing it headfirst. All *Esox* in Arkansas are late winter and early spring spawners with similar spawning habits. No nest is prepared; the adhesive eggs are scattered over vegetation or debris on the substrate, and no parental care is given the eggs or young. Pikes and pickerels are sometimes referred to as "jackfish" by local fishermen.

Esox americanus Gmelin
Redfin Pickerel

Figure 19.2. *Esox americanus,* Redfin Pickerel. *Uland Thomas.*

CHARACTERS   An elongate fish with the snout formed into a ducklike beak and a large, terminal mouth with sharp canine teeth (Fig. 19.2). Snout short, snout length going into head length 2.4–2.7 times. Head length going into TL 3.2–3.8 times. Dorsal and anal fins far back on body. Undersurface of lower jaw with a row of 4 (3–5) sensory pores on each side. Branchiostegal rays usually 11 or 12. Gill rakers forming sharp, toothlike structures. Scales cycloid. Cheek and opercle fully scaled (Fig. 19.1A); top of head without scales. Scales in a lateral series 92–117 (usually fewer than 110). Although pored scales may be found along the lateral series, it is very rare that all scales in that series are pored. More commonly, there are numerous interruptions all along the lateral series where unpored scales are interspersed among the pored scales. Dorsal rays 17–20; anal rays 16–18; pectoral rays usually 14 or 15; pelvic rays 8–10. Arkansas's smallest member of the genus *Esox,* seldom exceeding 12–14 inches (305–356 mm) in length. Maximum length (*E. a. vermiculatus*) around 15 inches (380 mm) and maximum weight about 0.75 pound (0.34 kg).

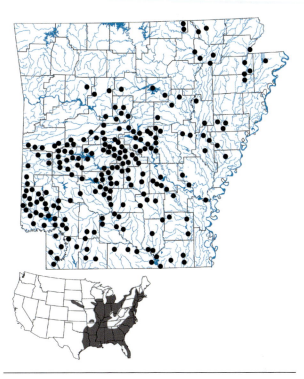

Map 138. Distribution of *Esox americanus,* Redfin Pickerel, 1988–2019.

LIFE COLORS   Back dark olive. Sides lighter olive with dark brown bars and vermiculations. Undersides yellow or white. Fins yellow green to dusky. A dark vertical bar extends downward and sometimes slightly backward from the eye. Juveniles with a light midlateral stripe.

SIMILAR SPECIES   Most often confused with the Chain Pickerel in Arkansas, the Redfin Pickerel differs from that species in having a shorter snout, the distance from tip of snout to center of eye equal to or less than the distance from center of eye to rear margin of operculum (vs. distance from tip of snout to center of eye greater than distance from center of eye to rear margin of operculum); in having 11 or 12 branchiostegal rays (vs. 13 or 14); and in usually having fewer than 110 scales in a lateral series (vs. more than 110). It differs from *E. lucius* and *E. masquinongy* in having a completely scaled operculum (vs. operculum with scales only on upper half), and in having 11–14 branchiostegal rays (vs. 14–19).

VARIATION AND TAXONOMY   *Esox americanus* was described by Gmelin (1789) from the vicinity of New York City. Variation was studied by Crossman (1966, 1978), who recognized two subspecies. The Redfin Pickerel, *E. a. americanus,* occupies the Atlantic Slope drainages to southern Georgia. The subspecies in Arkansas and throughout the Mississippi Valley is the Grass Pickerel, *Esox americanus vermiculatus* Lesueur 1846. A broad zone of intergradation between the two subspecies extends from Florida west through the Pascagoula River, Mississippi. Nelson et al. (2004) and Page et al. (2013) did not recognize subspecies and recommended the common name Redfin Pickerel for the species. We follow those authors in using Redfin Pickerel as the common name for the species, even though the subspecies found in Arkansas has dusky fins rather than red fins. The common name, Grass Pickerel, still applies to the western subspecies, *E. a. vermiculatus*. *Esox americanus* is the sister species of *E. niger* based on DNA sequence analysis (Lopez et al. 2000). Mitochondrial and nuclear DNA analysis indicated that hybridization and introgression may have occurred between pickerel species (Grande et al. 2004). Becker (1983) reported natural hybrids between *E. americanus* and *E. lucius*.

DISTRIBUTION   *Esox americanus* occurs in the Great Lakes and Mississippi River basins from Ontario, Canada, to Wisconsin and Nebraska, south to the Gulf of Mexico; in Atlantic Slope drainages from the St. Lawrence River, Quebec, to Florida; and in Gulf Slope drainages from Florida to the Brazos River, Texas. Introduced populations are known from Colorado, Washington, and California. This is Arkansas's most widely distributed *Esox* species. It is found statewide except in the extreme Ozark Uplands (Map 138). One valid upland record exists, from the headwaters of the White River in Madison County (Keith 1964) (Map A4.136).

HABITAT AND BIOLOGY   The Redfin Pickerel is most often found in quiet waters of small streams having heavy aquatic vegetation. In lowlands it also occurs in sloughs, bayous, swamps, and ditches having at least moderately clear water and little current and always with dense submerged vegetation. In the Ouachita Mountains it occurs in small to medium-sized creeks having permanent pools and is often found over gravel substrates. Ross et al. (1987) listed microhabitat characteristics in Mississippi as high amounts of vegetation and cover, fine substrata such as mud or silt, high litter amounts, and water depths greater than 0.5 meter. A stream habitat analysis in Indiana found that the Redfin Pickerel preferred habitats with aquatic macrophytes, logs and woody debris, and slow-moving water; it avoided riffle habitats (Cain et al. 2008). It is occasionally found in lakes in Arkansas, and Buchanan (2005) reported it from nine Arkansas reservoirs. Kwak (1988) found that *E. americanus* used floodplain habitat on a seasonal basis in Illinois, and that floodplain was an important habitat for juveniles. Trautman (1957) believed that there was competition between *E. americanus* and other species

of *Esox* in Ohio, noting that the Redfin Pickerel was rare or absent at any locality where another species of *Esox* was abundant. Becker (1983) also referred to possible competitive interactions between Redfin Pickerels and Northern Pike in Wisconsin. Competitive interactions of the two native *Esox* species in Arkansas have not been studied.

The Redfin Pickerel feeds mainly during the late afternoon or early evening on small fishes and occasionally on aquatic insects and crayfish (Crossman 1962). Young pickerels smaller than 50 mm (2 inches) feed on small crustaceans (cladocerans, amphipods, and isopods) and immature aquatic insects (odonates, mayflies, and caddisflies) and gradually become piscivorous as they grow. Between 50 and 100 mm (2–4 inches), the diet consists of invertebrates and small fishes, and pickerels larger than 100 mm feed almost entirely on fishes and crayfish (Ming 1968; Yeager 1990). In Oklahoma, sunfishes were the most frequently consumed food by pickerels greater than 7 inches (178 mm) TL, while smaller pickerels fed mainly on darters (Ming 1968).

Individuals are sexually mature at 1–2 years of age. Spawning is in February and March in Oklahoma (Ming 1968), but there is also evidence of additional spawning in late fall or early winter at water temperatures of about 50°F (10°C) (Crossman 1980a). Inundated floodplain is the preferred spawning habitat in lowland areas (Ross and Baker 1983; Kwak 1988). When this habitat is not available, spawning occurs in vegetated areas along stream or pond margins (Crossman 1962). Spawning occurs at water temperatures of 40–53°F (4.4–11.7°C). There is no territoriality, nest construction, or parental care associated with spawning. During spawning, one female is usually accompanied by several males. Ripe females in Wisconsin contained 843–4,584 mature eggs (Kleinert and Mraz 1966). Several hundred adhesive eggs are broadcast over submerged vegetation or debris on the bottom. Eggs hatch in 11–15 days at water temperatures of 46–48°F (7.8–8.9°C). Young-of-the-year *E. americanus* approximately 25 mm TL were found in Fourche Creek in Pulaski County, Arkansas, on 5 May 2015 by TMB and Jeff Quinn. In Oklahoma, few individuals live more than 3 years (Ming 1968), but Crossman (1980a) reported a maximum life span of 7–8 years in Canada.

The Redfin Pickerel is too small to be a gamefish. Although it is widespread and common in Arkansas, its populations are generally small. It is an excellent aquarium fish but must be provided live fish for food.

CONSERVATION STATUS   This is the most widespread and common esocid species in Arkansas, and its populations appear to be stable.

Esox lucius Linnaeus
Northern Pike

Figure 19.3. *Esox lucius*, Northern Pike. *Uland Thomas.*

Map 139. Distribution of *Esox lucius*, Northern Pike, 1988–2019. Inset shows native range (solid) and introduced range (outlined) in the contiguous United States.

CHARACTERS   A large, elongate fish with a snout shaped like a duck's bill (Fig. 19.3). Mouth large with sharp canine teeth. Dorsal and anal fins far back on the body. Snout long, snout length going into head length 2.1–2.4 times. Undersurface of lower jaw with a row of 4 or 5 sensory pores on each side. Branchiostegal rays 14–16. Cheek fully scaled; operculum scaled only on upper half (Fig. 19.1B). Top of head naked. Lateral line with 119–148 scales; the lateral line may be interrupted by unpored scales. Dorsal rays 15–19; anal fin rays 12–15. In its native range it can reach a length greater than 4 feet (1,219 mm) and a weight of more than 40 pounds (18.1 kg). The largest documented specimen was 68 pounds (30.8 kg) from Germany (Hrabik et al.

2015). The Arkansas state record of 16 pounds, 1 ounce (7.3 kg) is from DeGray Lake in 1973.

LIFE COLORS   Back olive or brownish green. Sides lighter olive with numerous small, oval white or yellow spots. Undersides yellowish white. All fins except pectorals orange yellow with black streaks or spots. No dark vertical bar beneath eye.

SIMILAR SPECIES   It differs from the Chain and Redfin pickerels in having the opercle scaled only on the upper half (vs. opercle fully scaled), and in lacking a dark suborbital bar (vs. dark vertical suborbital bar present). It can be distinguished from the Muskellunge in having a fully scaled cheek (vs. cheek scaled only on upper half), 14–16 branchiostegal rays (vs. 17–19), usually fewer than 130 lateral scales (vs. usually more than 140 scales in lateral series), and usually 5 sensory pores along each side of lower jaw (vs. 6–9).

VARIATION AND TAXONOMY   *Esox lucius* was described by Linnaeus (1758) from "Europe." Crossman (1978) summarized the systematics of this highly variable, widely distributed species and concluded that recognition of subspecies was unwarranted. Genetic analysis supported a sister group relationship between *E. lucius* and *E. reicherti* (Grande et al. 2004).

DISTRIBUTION   A Holarctic species, the Northern Pike occurs from northwestern Europe across northern Asia to northern North America. Although not native to Arkansas, it was stocked in some of the state's large reservoirs (Beaver, Norfork, DeGray, and Millwood) three decades ago (Map A4.137). There are no records of *E. lucius* from Arkansas since the 1980s, but in recent decades there has been an increase in the number of reports of Northern Pike from the Mississippi River in Missouri, leading Pflieger (1997) to speculate that limited reproduction might be occurring there (Map 139). There is the possibility that a straggler from that population might be found in the Mississippi River of Arkansas.

HABITAT AND BIOLOGY   In its native range it inhabits mainly lakes and marshes and, to a lesser extent, rivers. It prefers quiet waters with dense aquatic vegetation and is more tolerant of cold temperatures than any other esocid. It has been stocked in the large impoundments in Arkansas to provide an additional large sport fish and a predator of large prey species such as Gizzard Shad. The stockings were initially successful and fishermen regularly caught Northern Pike. Although some reproduction occurred in Norfork Lake tributaries after the initial stocking (Mark Oliver, pers. comm.), this species was apparently not able to establish reproducing populations in Arkansas.

The lack of successful reproduction in Arkansas is puzzling because of presumed similarities in spawning habitat requirements with the native esocids. Pflieger (1997) also noted a similar lack of spawning success in Missouri reservoirs where Northern Pike had been stocked. It is probably fortunate that no populations of Northern Pike became established in Arkansas, because in the Niobrara River, Nebraska, above Box Butte Reservoir, the introduced *E. lucius* became established and nearly eliminated the native fish community (Hrabik et al. 2015).

Young pike feed on plankton after hatching, and then on invertebrates. Fish (usually the fry of suckers or other species) are added to the diet when the young are about 50 mm TL. Adults are almost entirely piscivorous but occasionally eat other vertebrates. They are sight feeders, with feeding occurring almost entirely during the daytime.

Spawning is in early spring when water temperatures are around 39–51°F (4–11°C) (Crossman 1980b). In Minnesota, spawning runs into a slough occurred at water temperatures between 34°F and 40°F (1.1–4.4°C), but 36–37°F (2.2–2.8°C) is the preferred temperature range (Franklin and Smith 1963). Northern Pike congregate in the spawning areas a few days before spawning occurs. The spawning habitat is basically a flooded area with emergent vegetation (Becker 1983). Adults move into shallow water, where the adhesive eggs are scattered over submerged vegetation or debris on the bottom. No nest is constructed and the eggs are not guarded. A large female may have as many as 150,000 eggs (Threinen et al. 1966). Sexual maturity is reached in about 3–5 years. In nature, individuals are reported to live from 13 to 24 years, with specimens surviving in zoos for up to 75 years (Crossman 1980b; Pflieger 1997).

CONSERVATION STATUS   This introduced species did not adapt to the Arkansas reservoir environment as hoped. No recent attempts have been made to stock it in the state, and it warrants no conservation status.

Esox masquinongy Mitchill
Muskellunge

Figure 19.4. *Esox masquinongy*, Muskellunge. *David A. Neely.*

Map 140. Distribution of *Esox masquinongy*, Muskellunge, 1988–2019. Inset shows native range (solid) and introduced range (outlined) in the contiguous United States.

CHARACTERS   Like other pikes, it has an elongate body, dorsal and anal fins placed far back on body, and a snout shaped like a duck's bill, with many sharp teeth in the large mouth (Fig. 19.4). Snout long, snout length going into head length 2.1–2.3 times. Underside of lower jaw with a row of 6–9 sensory pores on each side. Branchiostegal rays 17–19. Cheek and operculum scaled only on upper half (Fig. 19.1C). Top of head unscaled. Lateral line with 130–167 (usually more than 140) scales; the lateral line may be interrupted by unpored scales. Dorsal rays 15–19; anal rays 14–16; pectoral rays 14–19. Largest esocid, reaching a maximum length and weight greater than 5 feet (1,524 mm) and 70 pounds (31.8 kg). No Arkansas state angling record exists for Muskellunge.

LIFE COLORS   Coloration can be variable but usually consists of dark markings on a light background. Back greenish yellow or brownish olive. Sides lighter yellow green with dark spots, blotches, or oblique dusky bars; sides sometimes without markings. Undersides white or yellow. No dark, vertical suborbital bar. Fins olive to reddish with dusky spots.

SIMILAR SPECIES   The Muskellunge differs from *E. lucius* in having the cheek scaled on the upper half only (vs. cheek fully scaled), 17–19 branchiostegal rays (vs.

14–16), and 6–9 sensory pores on each half of lower jaw (vs. 4 or 5). It differs from *E. americanus* and *E. niger* in having scales only on upper half of cheek and operculum (vs. cheek and operculum fully scaled) and in lacking a dark suborbital bar (vs. dark, vertical suborbital bar present). *Esox lucius × E. masquinongy* hybrids (tiger muskies) are intermediate in their features between the two parental species. Tiger muskies differ from the Northern Pike in having a color pattern of alternating light and dark vertical stripes (vs. side with light, horizontally elongated spots on a dark background), upper half to two-thirds of cheek with scales (vs. cheek fully scaled), and with 5–7 pores on each side of lower jaw (vs. lower jaw with 5 or fewer pores). Tiger muskies differ from the Muskellunge in having a color pattern of alternating light and dark vertical stripes (vs. sides with various dark markings on a light background), and in having lobes of caudal fin and tips of pectoral and pelvic fins rounded (vs. tips of those fins pointed).

VARIATION AND TAXONOMY   *Esox masquinongy* was described by Mitchill (1824). There is confusion about the type locality, and the traditional citation of the original description is not available. For almost 100 years, authors have noted the lack of access to Mitchill's original publication, and the oldest available firsthand account of that description is in DeKay (1842) (Crossman 1980c). Lake Erie is considered the possible type locality. Variation in this species was studied by Crossman (1978, 1980c), who considered the recognition of subspecies unwarranted. A silver color variant formerly thought to be a Muskellunge is now known to actually be a form of *E. lucius*. The Muskellunge hybridizes with *Esox lucius*. Some of the early work with *Esox* hybrids was done by John D. Black, a noted contributor to ichthyology in Arkansas (Black and Williamson 1946). The hybrid (tiger muskie) has been reared in hatcheries in Pennsylvania and stocked into several Arkansas reservoirs, including Lake Ashbaugh, Lake Frierson, and Spring River Lake at Mammoth Spring in 1990 and 2001. Stocking of the hybrids was suspended by 2005 to prevent the spread of viral hemorrhagic septicemia (VHS), a fish disease prevalent in areas east of the Mississippi River from which brood stock was obtained. Stocking of tiger muskies will likely be resumed in Arkansas when VHS-free fish can be obtained (Mark Oliver, pers. comm.). The hybrid is very hardy, grows rapidly, and has greater thermal resistance than either parental species (Becker 1983). Hybrid males are always sterile, but females are often fertile (Black and Williamson 1946). The state record for tiger muskie weighed 23 pounds, 12 ounces (10.8 kg) and was caught in the Spring River in 1995.

DISTRIBUTION   *Esox masquinongy* occurs in the St. Lawrence–Great Lakes, Hudson Bay, and Mississippi River basins from Quebec to southeastern Manitoba, Canada, and south in the Appalachian Mountains to Georgia and west to Iowa. It has been introduced into Atlantic Slope drainages as far south as southern Virginia. In Arkansas, the Muskellunge was stocked only in Norfork and DeGray reservoirs in the early 1970s (Map A4.138). There were a few subsequent angling records, and it did not become successfully established in those reservoirs. There are no recent records of *E. masquinongy* from Arkansas (Map 140). It has been stocked in several lakes in Missouri with good success, beginning in the 1960s and continuing into the 2000s (Boone 2007). The Missouri stocking program was expanded in 1995 into the southeastern part of the state to include Lake Girardeau in Cape Girardeau County, but there is no evidence of successful spawning in Missouri. Based on the successful stocking and management program, the long-range plan is to continue stocking Muskellunge in Missouri reservoirs (Boone 2007). Missouri currently stocks Muskellunge in five lakes and has a new multi-year management plan extending through 2027 (R. Hrabik, pers. comm. 2019). The possibility exists that occasional stragglers from populations north of Arkansas might be found in the Mississippi River of Arkansas.

HABITAT AND BIOLOGY   In its native range, the Muskellunge is most abundant in lakes and the backwaters of rivers. It prefers cool water temperatures from 33°F to 78°F (0.6–25.6°C) but can withstand temperatures up to 90°F (32.2°C) for short periods (Becker 1983). It is found more often in slow streams and larger rivers than the Northern Pike (Crossman 1980c). In Kentucky streams, optimum habitat was reported to be pools about 1 m deep with gradients of 0.6–1.2 m/km and with fallen trees (Axon and Kornman 1986). Adults are solitary and prefer heavily vegetated waters and are rarely found far from cover. Smaller individuals are often found in shallow, warmer waters; the adults are also shallow-water fish usually found in less than 15 feet (4.6 m) of water, but they occasionally inhabit rocky shorelines at depths of 20–30 feet (6.1–9.1 m). In Thornapple Lake, Michigan, Muskellunge preferentially selected vegetated habitats in shallow water, but moved into deeper offshore habitats when water temperatures exceeded 25°C (Eilers 2008). It reportedly thrives in reservoirs but often fails to reproduce in reservoir environments. Stocking of large esocids in Arkansas has now shifted to tiger muskies (*Esox masquinongy × E. lucius* hybrids).

Newly hatched young first feed on zooplankton but begin to eat small fish within 5 days of hatching. Adults are almost entirely piscivorous but will eat other vertebrates such as ducks and muskrats when the opportunity arises. In Wisconsin, fish made up 98% of the diet, and 31 fish species (mainly Yellow Perch and White Suckers) were consumed (Bozek et al. 1999). In that study the overall proportion of Muskellunge with food in the stomach differed among seasons, with the greatest proportion occurring in fall (69.0%), followed by summer (53.5%) and spring (25.4%).

Spawning is not known to occur in Arkansas. In northern areas, spawning occurs in the spring, slightly later than the Northern Pike when water temperatures are 48–59°F (9–15°C). The adults move to shallow spawning grounds, but no nest is constructed. Large numbers of small adhesive eggs are scattered over vegetation and debris and then abandoned. Maximum life span is about 25 years. Approximately 73% of North America's Muskellunge waters are sustained by natural reproduction, but the majority of areas rely on stocking to supplement natural reproduction or to maintain artificial fisheries (Kerr 2011).

CONSERVATION STATUS   Like the Northern Pike, the Muskellunge did not successfully adapt to reservoirs in Arkansas. No recent stocking attempts have been made.

Esox niger Lesueur
Chain Pickerel

Figure 19.5. *Esox niger*, Chain Pickerel. *Uland Thomas*.

CHARACTERS   Similar to other pikes in having an elongate body, a snout shaped like a duck's bill, a large mouth with many sharp teeth, and dorsal and anal fins placed far back on the body (Fig. 19.5). Adults easily recognized by dark, chainlike markings on the sides. Snout long, snout length going into head length fewer than 2.4 times. Underside of lower jaw with a row of 4 sensory pores on each side. Branchiostegal rays 13 or 14. Cheek and operculum fully scaled; top of head naked. Scales in a lateral series 110–138. Although pored scales may be found along the lateral series, it is very rare that all scales in that series are pored. More commonly, there are numerous interruptions all along the lateral series where unpored scales are interspersed among the pored scales. Dorsal rays 14 or 15; anal rays 11–13; pectoral rays 12–15. The Chain Pickerel is

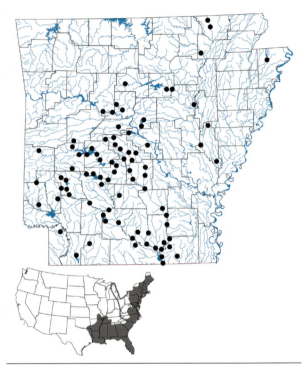

Map 141. Distribution of *Esox niger*, Chain Pickerel, 1988–2019. Inset shows native range (solid) and introduced range (outlined) in the contiguous United States.

Arkansas's largest native esocid, and 18–24-inch (457–610 mm) adults weighing 3–4 pounds (1.4–1.8 kg) are sometimes sought by anglers because of their fighting ability. The Arkansas state angling record is 7 pounds, 10.5 ounces (3.5 kg) from the Little Red River in 1979. Maximum reported size is 9.4 pounds (4.3 kg) and 39 inches (990 mm) TL.

LIFE COLORS   Back a dark olive or yellow green. Sides a lighter greenish yellow with dark chainlike reticulations in individuals greater than 10 inches (254 mm). A dark vertical bar present beneath the eye. Undersides white, sometimes tinged with yellow. Fins not pigmented. Young smaller than 10 inches (254 mm) lack the chainlike markings on the side, are barred and mottled, and are similar in color pattern to the Redfin Pickerel.

SIMILAR SPECIES   Similar to the smaller Redfin Pickerel, but differs from it in having a longer snout, the distance from the tip of the snout to the center of the eye distinctly greater than the distance from center of eye to the rear margin of the operculum (vs. distance from tip of snout to the center of the eye less than distance from center of eye to the rear margin of the operculum); 13 or 14 branchiostegal rays (vs. 11 or 12); and in usually having more than 110 scales in a lateral series (vs. fewer than 110 lateral scales). It differs from *E. lucius* and *E. masquinongy* in having a completely scaled operculum (vs. operculum with scales only on upper half), and in having 11–14 branchiostegal rays (vs. 14–19).

VARIATION AND TAXONOMY   *Esox niger* was described by Lesueur (1818b). Confusion exists over the correct type locality, with Jordan et al. (1930) listing it as South Carolina, and Crossman (1980d) giving it as Philadelphia. Eschmeyer et al. (2018) listed the type locality as Saratoga Lake, New York. Surprisingly, there has been no study of geographic variation in the Chain Pickerel. Genetic evidence indicated that *E. niger* is sister to *E. americanus* (Lopez et al. 2000), and mtDNA analysis indicated possible hybridization and introgression between the two pickerel species (Grande et al. 2004).

DISTRIBUTION   *Esox niger* occurs in Atlantic Slope drainages from Nova Scotia, Canada (introduced) to southern Florida; Gulf Slope drainages from Florida west to the Sabine River, Texas; and in the Mississippi River basin from the Gulf of Mexico north to Kentucky and Missouri. It has been introduced into Great Lakes drainages (Lake Ontario and Lake Erie) and some other areas of the United States. In Arkansas, the Chain Pickerel is most abundant in streams and lakes of the Coastal Plain lowlands in southern and eastern Arkansas (Map 141). It penetrates somewhat above the Fall Line in streams of the Ouachita Upland, and to a lesser extent in streams along the Ozark Upland boundary. It is absent from most of the Ozarks and the upper Arkansas River drainage, but Tumlison et al. (2015) reported a specimen from northern Arkansas in the lower Strawberry River in Lawrence County.

HABITAT AND BIOLOGY   Like other members of the genus *Esox*, the Chain Pickerel prefers clear, quiet waters with abundant aquatic vegetation but is found in lakes and larger streams more frequently than the Redfin Pickerel. It rarely occurs with the Redfin Pickerel, but when the two species are syntopic, one species usually replaces the other (Hrabik et al. 2015). It does not normally swim actively in search of prey, but lies in wait in shallow water near aquatic vegetation. In Missouri, it is found exclusively in mountain streams in the southeastern part of the state and avoids the lowlands (Pflieger 1997). In Arkansas, it occurs in both upland and lowland streams, but is more common below the Fall Line. The largest populations in the state are found in lakes in the southern part of the Coastal Plain. It is common in slightly alkaline waters in Arkansas, but it is also reported to be successful in waters with a pH as low as 3.8 (Crossman 1980d). It can also withstand salinities up to 22 ppt. It grows to a large size, but its popularity as a game species in Arkansas is limited by its rather small populations. It attains a larger size in the more northern areas of its range.

The young first feed on plankton, then on larger invertebrates (mainly immature aquatic insects), and finally on fish. Juveniles become primarily piscivorous at lengths greater than 100 mm (Meyers and Muncy 1962). Various studies have found that 38–50% of the adult diet consists of gamefishes, primarily centrarchids (McIlwain 1970). Crayfish are also sometimes an important component of the juvenile and small adult diets (Crossman 1980d). Other vertebrates are infrequently eaten by large adults (mice, salamanders, frogs).

Sexual maturity is usually reached after 2 or 3 years, but in southern populations maturity is achieved by some individuals at 1 year. Over its range, spawning occurs in late winter to spring, and some populations may spawn in the fall. The Chain Pickerel spawns in Arkansas in late February to early March when the water temperature reaches 45–50°F (7.2–10°C) (Keith 1972). Spawning in Alabama has been reported at water temperatures as high as 16°C (Boschung and Mayden 2004). Adults do not defend territories or build nests in the shallow spawning areas. Spawning occurs in inundated floodplain habitats or in the backwaters of streams. Large numbers of small adhesive eggs are scattered over aquatic vegetation or detritus on the bottom. No parental care is provided the eggs or young. Hatching occurs in 11 days at 10°C, and in 2–3 days at 17.2°C (Yeager 1990). Maximum life span in northern areas is 8–9 years, but 3–4 years are more common maximum ages in the southeastern U.S. (Ross 2001).

Keith (1972) reported that in the early 1970s the Arkansas Game and Fish Commission began a program of capturing and relocating adult *E. niger* to lakes that had an overabundance of small sunfish. Brood stock were often collected during the breeding season in February, especially from the large Lake Ouachita population, and transferred to other lakes. One of the largest successful populations established by stocking was in Lake Wilhelmina in Polk County.

CONSERVATION STATUS   This native gamefish species has been intentionally stocked in some Arkansas reservoirs with some success. It is widespread in natural waters across southern Arkansas, but its populations are small. Its populations in Arkansas, though small, are probably secure.

Genus *Umbra* Kramer, 1777—Mudminnows

The genus *Umbra* was described by Kramer *in* Scopoli (1777). This is a small genus consisting of two species in North America (*U. limi* and *U. pygmaea*) and one species in Europe (*U. krameri*). One species (*U. limi*) is native to

Arkansas. Genus characters are presented in the species account below.

Umbra limi (Kirtland)
Central Mudminnow

Figure 19.6. *Umbra limi*, Central Mudminnow. *Uland Thomas*.

Map 142. Distribution of *Umbra limi*, Central Mudminnow, 1988–2019.

CHARACTERS   A small fish with an oblong body and a terminal mouth (Fig. 19.6). Small villiform teeth present on premaxillary, lower jaws, and roof of mouth. All fins rounded. Dorsal fin short and situated posteriorly; dorsal fin origin distinctly in front of anal fin origin and just behind pelvic fin origin. Anal fin smaller than dorsal fin. Dorsal rays 13–15; anal rays 7–9; pectoral rays 14–16; pelvic rays 6 or 7. Gill rakers 13–15, short and stout. Frenum present, no groove separating tip of upper lip from snout. Scales rather large and cycloid. Top of head, cheeks, and opercles fully scaled. Lateral line absent; scales in lateral series 34–37. Adult male with long anal fin, its free tip almost reaching base of caudal fin; female with short anal fin, its free tip not reaching base of caudal fin. Adults usually 2–4 inches (51–102 mm) long, with a maximum reported size around 6 inches (150 mm).

LIFE COLORS   A dull-colored fish. Top of head, back, and upper sides a dark olive green or brown, with dark brown mottling. Mid and lower sides lighter with more distinct dark mottling; sides occasionally have 10–14 indistinct bars. Caudal fin base with a prominent dusky vertical bar, its width more than half the diameter of the eye. Undersides a yellowish white. All fins brownish. Breeding male with iridescent blue-green anal and pelvic fins.

SIMILAR SPECIES   Sometimes mistaken for members of the topminnow family (genus *Fundulus*), and the Western Mosquitofish, *Gambusia affinis*, but differs from those species in having no deep groove separating the upper lip and the tip of the snout (vs. deep groove separating upper lip and snout) and in having a terminal mouth (vs. superior mouth). It superficially resembles the young of Bowfins, but differs in having a short dorsal fin (vs. long dorsal fin) and in lacking a gular plate (vs. gular plate present). It can be distinguished from minnows and suckers by its rounded caudal fin and the large scales on top of the head (vs. forked tail and naked head). It differs from Pirate Perch in having the anus located immediately in front of the anal fin (vs. far in front), in lacking a dark suborbital bar (vs. suborbital bar present), and in having cycloid scales (vs. ctenoid).

VARIATION AND TAXONOMY   *Umbra limi* was described by Kirtland (1840a) as *Hydrargira limi* (genus misspelled as *Hydargira*) from Bull Creek, an upland tributary of Yellow Creek, upstream of Pine Lake, south of Poland, Mahoning County, Ohio. There have been no studies of geographic variation in the Central Mudminnow.

DISTRIBUTION   *Umbra limi* occurs in the St. Lawrence–Great Lakes, Hudson Bay, and Mississippi River basins from Quebec to Manitoba, Canada, south to central Ohio, western Tennessee, and northeastern Arkansas, and on the Atlantic Slope in the Hudson River drainage, New York (Map 142). An isolated population formerly occurred in the Missouri River drainage, South Dakota, and an introduced population is known from the Connecticut River, Massachusetts. The only Arkansas record for this species for more than 100 years was a specimen collected by Seth Meek in 1894 (Meek 1896) from a bayou near Greenway

(St. Francis River drainage) in Clay County (Map A 4.140). More recently, in either April or May of 1998 or 1999, a single specimen was collected in Village Creek upstream of State Highway 69 in Greene County by Hillary Hicks and Denver Dunn while seining fish for a biology class (George L. Harp, pers. comm.). The specimen was examined and identified by HWR, who noted that a baitfish farm was located immediately north of the collection locality. Unfortunately, the specimen was not cataloged into the ASU Fish Collection and was apparently discarded (G. L. Harp, pers. comm.). Whether this most recent record represents a bait introduction, an escapee from the nearby fish farm, or a natural record is not known, but its collection site is not far (approximately 25 miles SW) from the 1894 collection site of Meek. Northeastern Arkansas represents the southern limit of the range of this species, and extensive channelization and agricultural activities in Clay County and surrounding areas have greatly decreased the chances for its survival in the state; however, Becker (1983) reported it to be common in old and new channelization ditches in Wisconsin. Stragglers might occasionally be found today in the Mississippi River of northeastern Arkansas, as two individuals were found in that river near New Madrid, Missouri, in 1978, approximately 50 miles (80.5 km) upstream from the Arkansas border (Kofron and Schreiber 1983). Also, populations have been found in the southwestern corner of Kentucky, in Fulton County (Burr and Warren 1986b). Other populations are known to exist in northwestern Tennessee, especially around Reelfoot Lake (Etnier and Starnes 1993). There is a record of one specimen from Keystone Reservoir (Arkansas River) in Oklahoma collected in 1971 (Pigg et al. 1996). A survey of the fishes of the Mississippi River in Mississippi County, Arkansas, failed to produce any specimens of *Umbra limi* (Carter 1984). Our fish sampling with Sam Barkley and Sam Henry at more than 50 localities in the Mississippi River in Mississippi County, Arkansas, between 2000 and 2005 produced no *U. limi*.

HABITAT AND BIOLOGY   The Central Mudminnow is found in association with aquatic vegetation in quiet, clearer waters of sloughs, ditches, and ponds over bottoms of mud or organic debris. In Goose Pond, Missouri, the aquatic vegetation associated with mudminnow habitat included cattails, American lotus, sedges, pickerel weed, and watercress (Pflieger 1997). It avoids swift current and is most commonly found in water 0.5 m deep or less; it is less common in deeper water (Becker 1983). The soft bottom is necessary because of its alleged habit of burrowing tail first into the substrate to hide, rest, or estivate during droughts (Trautman 1957). Peckham and Dineen (1957) found no evidence of winter hibernation in Indiana despite previous speculation by others. Becker (1983) discounted both hibernation and reported substrate burrowing by this species. Eddy and Underhill (1974) found that it was a very difficult species to collect by seine and observed that *U. limi* dived into mud and detritus near the margins of ditches to avoid capture with seines and dip nets. It could be captured by seine more easily after the water had been stirred up to create intense turbidity. It is very tolerant of acid waters and low dissolved oxygen, and it can use its swim bladder as a lung. It is more common in the northern parts of its range because of its preference for cooler waters, but it is also able to survive high water temperatures (84°F [28.9°C]) in isolated pools (Scott and Crossman 1973). It is rather intolerant of high turbidity. Wisenden et al. (2008), based on field and laboratory studies, reported that Central Mudminnows exhibited antipredator behavioral responses to chemical alarm cues released by damaged conspecific tissue. J. R. Jenkins and Miller (2007) reported a tendency for shoaling behavior. Its ability to survive some environmental extremes may explain why it is often the only fish species inhabiting some swamps, shallow ponds, and bog ponds (Becker 1983).

Forbes and Richardson (1920) reported that mudminnows eat diverse foods, including aquatic and terrestrial insects, snails, and even algae and duckweed. Gunning and Lewis (1955) also found aquatic beetles and young Spring Cavefish in specimens from a spring-fed swamp in the Mississippi bottoms of southern Illinois. In Indiana, the Central Mudminnow was a carnivorous bottom feeder, with young-of-the-year feeding primarily on chironomids and small crustaceans and adults feeding on chironomids, mayflies, caddisflies, and mollusks (Peckham and Dineen 1957). Kofron and Schreiber (1983) found that in Missouri, *U. limi* fed mainly on small crustaceans and insects and to a lesser extent on arachnids and small clams. In Canada, mudminnows fed throughout the year, even in the winter under ice cover (Chilton et al. 1984). Large females often fed on small fishes during winter.

Spawning occurs at water temperatures between 55°F and 60°F (12.8–15.6°C), and the breeding season is correlated with the flooding of adjacent areas (Becker 1983). Spawning was reported in April in Illinois (Smith 1979) and Indiana (Peckham and Dineen 1957), March through May in Ohio (Trautman 1957), and in mid-March in Tennessee (Etnier and Starnes 1993). The adhesive eggs are deposited on aquatic vegetation in shallow water, and there is no parental care of eggs or young. Females may lay

220–1,500 eggs. Eggs hatch in about a week, depending on water temperature. Individuals usually live up to 4 years, but maximum life span was estimated to be 7–9 years (Scott and Crossman 1973).

CONSERVATION STATUS   We previously believed that the Central Mudminnow was possibly extirpated from Arkansas (Robison and Buchanan 1988). The discovery of a more recent record from the state (late 1990s) demonstrates that there is still a possibility of new records from bait releases and from stragglers from more northern populations. We consider it endangered in Arkansas.

CHAPTER 20

Order Percopsiformes
Trout-Perches

This small order of distinctive fishes contains 3 families, 7 genera, and at least 11 species (Nelson et al. 2016). Percopsiforms have reduced oral dentition and a nonprotractile premaxilla. There are two large bony plates (fused hypurals) in the caudal skeleton borne on two separate centra that support the caudal fin. Ctenoid scales are present in many species. There are 6 branchiostegal rays and 16 branched caudal rays. Pelvic fins, if present, are behind the pectorals (subthoracic) and have 3–8 soft rays. Spines of the dorsal fin are weakly developed. Percopsiform monophyly was previously questioned by Murray and Wilson (1999) but has since been supported by both morphological (Springer et al. 2004; Wiley and Johnson 2010) and molecular data (Dillman et al. 2011; Betancur-R. et al. 2017). The percopsiforms date back to the Cretaceous of North America.

The three families of Percopsiformes occur entirely in North America: Percopsidae (trout-perches, which do not occur in Arkansas); Aphredoderidae (pirate perches, a monotypic family of the eastern United States, including Arkansas); and Amblyopsidae (cavefishes, with three Arkansas species).

Family Aphredoderidae *Pirate Perches*

The Pirate Perch, confined to the freshwaters of the eastern United States, is the last surviving species of the monotypic North American family Aphredoderidae (Becker 1983). Several Oligocene and Miocene fossils of this species and related taxa have been found in North America and Europe (Uyeno and Miller 1963; Patterson 1981). Cavender (1986) reported fossil species from the Oligocene in Colorado.

The relationship of this family to other families has been controversial. Aphredoderidae was considered the sister group to the Amblyopsidae, the cavefish family, by Rosen (1962), Patterson (1981), and Patterson and Rosen (1989). Rosen (1962) considered both families related to the trout-perches, Percopsidae. In contrast, Patterson and Rosen (1989) noted that there is some evidence that Pirate Perches are not closely related to the trout-perch and cavefish clade. Wiley et al. (2000) provided morphological and molecular evidence for a sister group relationship between *Percopsis* and *Aphredoderus*.

Genus *Aphredoderus* Lesueur, 1833— Pirate Perches

This monotypic genus contains only the unique Pirate Perch, *Aphredoderus sayanus* Lesueur. Characters of the genus are presented below in the species account.

Aphredoderus sayanus (Gilliams)
Pirate Perch

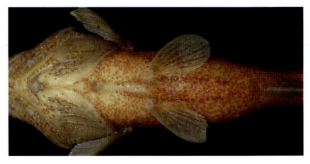

Figure 20.2. Jugular position of anus and urogenital opening in *Aphredoderus sayanus. Renn Tumlison.*

Figure 20.1. *Aphredoderus sayanus,* Pirate Perch. *Uland Thomas.*

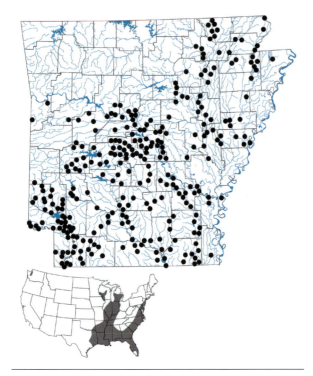

Map 143. Distribution of *Aphredoderus sayanus,* Pirate Perch, 1988–2019.

CHARACTERS   A small, dark, robust, large-mouthed fish with a single dorsal fin with 3 or 4 spines and 9–12 rays (Fig. 20.1). Lower jaw projecting beyond upper jaw. Back of upper jaw in adults reaching to anterior edge of eye. Adipose fin absent. The anus and urogenital opening are located between the pelvic fins in juveniles but shift forward to the throat (jugular) region just behind the isthmus in adults (Fig. 20.2). Body scales small and strongly ctenoid. Lateral line incomplete or sometimes absent, usually developed only anteriorly. Scales in lateral series 39–56 (usually

more than 45). Scale rows above lateral line 8–13; scale rows below lateral line 9–15. Cheeks and opercles fully scaled. Cheek with 4–8 scale rows. Scales around caudal peduncle 14–18. Anal fin with 2 or 3 spines and 6 or 7 rays. In young Pirate Perch, the third anal fin ray becomes transformed from a soft ray to a spine during growth (Nelson 2006). Pectoral fin rays 12 or 13. Pelvic fins subthoracic with 1 spine and 6 or 7 rays. Branchiostegal rays usually 6. Gill rakers usually 10–12 and blunt. Caudal fin margin slightly notched, not deeply forked. Twelve pyloric caeca. Swim bladder is simple, and a pneumatic duct to the pharynx is absent (physoclistous). Operculum with a sharp spine; rear margin of preopercle strongly serrate. Sensory canal system well developed on head, consisting of large canals; numerous cutaneous sensory organs on head and nape. Maximum size about 5.7 inches (144 mm).

LIFE COLORS   Body color varies from slate gray to black with a pink to iridescent purple cast on sides. Opercle usually with some pink to purple coloration. A dark bar beneath eye. Underside of head and body peppered with small melanophores. Caudal peduncle often with a dark vertical black bar. Fins dusky. Breeding colors are violet or purple. Young are dark to almost black, becoming lighter and often speckled ventrally.

SIMILAR SPECIES   Although the Pirate Perch superficially resembles a small bass or other sunfish, the jugular position of the anus, single dorsal fin with three spines, anal fin with 2 spines, and 1 pelvic fin spine render it so distinctive that it cannot be confused with any other species.

VARIATION AND TAXONOMY   *Aphredoderus sayanus* was described by Gilliams (1824) as *Scolopsis sayanus* from Harrogate near Philadelphia, Pennsylvania. Two subspecies are currently recognized: *A. s. sayanus* on the Atlantic Slope south to the Saltilla River, and *A. s. gibbosus* on the Gulf Slope west of the Mississippi River and in the Mississippi River Valley (Bailey et al. 1954; Boltz and

Stauffer 1993). Populations on the Gulf Slope are considered intergrades. *Aphredoderus s. gibbosus* is the form occupying Arkansas.

DISTRIBUTION   *Aphredoderus sayanus* occurs in the Great Lakes and Mississippi River basins from Michigan, Wisconsin, and southern Minnesota, south to the Gulf of Mexico, with an isolated population in the Lake Ontario drainage of New York; in Atlantic Slope drainages from southern New York to southern Florida; and in Gulf Slope drainages from Florida to the Brazos River, Texas. In Arkansas, the Pirate Perch occurs primarily throughout the Coastal Plain, extending into other regions above the Fall Line where local conditions are favorable (Map 143). It is also found throughout the Arkansas River Valley in Arkansas, and it is particularly abundant in Red River floodplain habitats of southwestern Arkansas. A small population exists in Spring Creek in Independence County in the Ozark Mountains (McAllister et al. 2010b), and Connior et al. (2014) reported the first record of this species from the Rolling Fork River. The Pirate Perch was found in 30 of 66 Arkansas impoundments surveyed (Buchanan 2005).

HABITAT AND BIOLOGY   The Pirate Perch is a solitary species typically inhabiting quiet ponds, oxbow lakes, swamps, ditches, and sluggish mud and sand-bottomed small rivers and streams. It also adapts well to reservoirs. and more than 2,300 were taken from 30 reservoirs in four ecoregions of Arkansas (Buchanan 2005). It is locally abundant over soft mud and silt bottoms with thick vegetation or organic debris and is found in both clear and turbid waters. It avoids current. Monzyk et al. (1997) found that woody debris (including root masses) was a requisite feature of diurnal cover for Pirate Perch during the summer in Louisiana. However, this is not always the case in Arkansas. A small population of *A. sayanus* inhabits Spring Creek in Independence County, a cold, spring-fed stream in the Ozark Mountains (McAllister et al. 2010b).

The Pirate Perch is a carnivore. In northcentral Arkansas, foods consisted mainly of grass shrimp (*Palaemonetes*), trichopterans, and chironomids; other items eaten included zygopterans, amphipods, and odonates (McCallum 2012). In Illinois (Forbes and Richardson 1920) and Virginia (Flemer and Woolcott 1966), it fed primarily on insects. In the Virginia study, foods in decreasing frequency of occurrence were dipteran larvae, ephemeropteran nymphs, hemipterans, amphipods, megalopterans, copepods, trichopterans, decapods, plant matter, plecopterans, dipteran pupae and adults, odonates, fishes, ostracods, and unidentifiable matter. In North Carolina, Shepherd and Huish

(1978) reported the diet consisted of cladocerans (particularly for young-of-the-year), dipteran larvae, isopods, and amphipods. Occasionally small fishes make up a small part of the diet. Resetarits and Binckley (2013) demonstrated the ability of *A. sayanus* to use chemical camouflage to resemble and more readily approach prey. Aquarium observations indicate that the Pirate Perch is a sedentary, nocturnal species remaining hidden during daylight hours and exhibiting peak activity at dusk and dawn (Abbott 1861; Parker and Simco 1975). Becker (1983) noted that in an aquarium, it will accept only live foods.

Spawning occurs primarily in May and early June in Arkansas, but it may begin as early as late April based on males with milt in Independence County (Tumlison et al. 2017). Speculation about the reproductive biology produced several hypotheses regarding egg deposition and fertilization. The function of the jugular anus and urogenital tract was unknown for many years. Tiemann (2004) proposed that one of the functions of the jugular anus is to allow the adult to leave its refuge partially and headfirst to defecate, avoiding pollution of the water around itself and avoiding undue exposure to predators. Originally, because of the jugular position of the genital opening in adults, it was presumed that *A. sayanus* was an oral incubator or branchial brooder similar to some species of cavefish (Amblyopsidae) (Eigenmann 1900; Smith 1979; Pflieger 1997). Boltz and Stauffer (1986) seemed to confirm this by providing the first report of eggs found in the branchial cavity (gill chamber) of a preserved adult female Pirate Perch. The source of those eggs was not confirmed, and it was not determined if the eggs had been fertilized. Brill (1977) and Fontenot and Rutherford (1999) reported that *A. sayanus* broadcast apparently aborted eggs on tank bottoms in aquaria. Katula (1987, 1992) described the deposition of viable eggs in or near a shallow depression swept by a female in the streambed, but Poly and Wetzel (2003) observed female Pirate Perch vigorously pushing their snouts into the gravel substrate to create furrows or pits. Tiemann (2004) reported that the eggs are broadcast or deposited in nests, based on finding a nest in an aquarium that also contained two sunfish species. Poly and Wetzel (2003) observed spawning behavior in aquaria and were the first to describe transbranchioral spawning. The male and female assumed parallel positions. Eggs were ejected from the female's urogenital pore directly into her buccal cavity and then through the mouth into the spawning substrate. The male then dispersed a cloud of sperm over the eggs through the same process. Fletcher et al. (2004) noted in field observations that the Pirate Perch hid the eggs in a cavity within dense submerged root masses of primarily woody riparian plants. The female thrust her head into, and released her eggs in, a small canal 1 cm in diameter in a root mass. Other females released their eggs into burrows made by insects and salamanders. Males congregated near these sites and subsequently entered the canal or burrow headfirst and then released their sperm. Although no parental care was observed, the male was observed plugging the entrance to the canal with his body. Fletcher et al. (2004) surmised that the forward-shifted urogenital pore may facilitate spawning under this special nesting circumstance. Such behaviors probably relate to selection pressures imposed by intense competition for fertilization success under these group-spawning conditions.

Discovery of spawning in the canals of root masses represents the first documentation of this behavior in any species of North American fish. During sequential spawning in the headfirst position, an extensive array of sensory pores in the head of the Pirate Perch (Moore and Burris 1956) may help adults locate eggs previously deposited in a dark canal. Mature eggs are yellow, adhesive, and measure 1.0–1.3 mm (Katula 1992). In the laboratory, eggs hatched in 5 or 6 days at water temperatures of 19–20°C (66–68°F) (Martin and Hubbs 1973). Hatchlings appeared 2 weeks after spawning. In Oklahoma, Hall and Jenkins (1954) studied age and growth and found that fish at the end of the first, second, third, and fourth years of life averaged 56, 84, 102, and 116 mm TL, respectively. The anus and urogenital pore are both located just anterior to the anal fin in specimens 9.5 mm TL but shift forward to the jugular position by the time the fish are 55 mm TL. Tiemann (2004) reported that ripe ovaries contained two sizes of eggs, mature and immature, with a mean number of 78 and 124 eggs, respectively. Most individuals live 2 or 3 years with a maximum up to 4 years (Shepherd and Huish 1978).

McAllister and Amin (2008) reported acanthocephalan parasites from Pirate Perch in the Caddo River, and McAllister et al. (2012a) reported a tapeworm, *Proteocephalus pearsei*, from the intestine of *A. sayanus* in northern Arkansas. Helminth parasites of *A. sayanus* were reviewed by McAllister et al. (2014b).

CONSERVATION STATUS   The Pirate Perch is common in Arkansas, and its populations are secure.

Family Amblyopsidae *Cavefishes*

Amblyopsid cavefishes form an unusual family occurring only in the eastern United States in regions south of former glacial advances. Phylogenetic relationships within the family have been the subject of considerable debate, and

Armbruster et al. (2016) reviewed the history and details of the various proposed amblyopsid phylogenies. The family, Amblyopsidae, contains six genera with nine described species. All species except the Swampfish, *Chologaster cornuta,* which lives in surface waters of the Atlantic Coastal Plain, are associated with limestone formations (Berra 1981). All species are small with the maximum length around 3.5 inches (90 mm). Cavefishes in the genera *Amblyopsis, Speoplatyrhinus, Troglichthys,* and *Typhlichthys* are blind, obligate cave-dwelling species living in complete darkness in environments having limited food resources (Chakrabarty et al. 2014). Three species, the Atlantic Coastal Plain *Chologaster cornuta,* and *Forbesichthys agassizii* and *Forbesichthys papilliferus* of the central Mississippi River drainage, unlike other cavefishes, have eyes and are usually found in springs and other surface waters, but *F. papilliferus* is also commonly found in caves of central Kentucky and southern Illinois.

In Arkansas, there are three species of cavefishes (two described and one undescribed species) representing two genera. All Arkansas species possess degenerate eyes and are depigmented, spending their entire lives in the pitch-black cave environment. Soares and Niemiller (2013) and Niemiller and Soares (2015) summarized our current understanding of nonvisual sensory modalities in cavefishes. Amblyopsid traits interpreted as adaptations to the subterranean environment are small size, long life span, infrequent reproduction, low fecundity, low metabolic rate, small population size, and high levels of population endemism (Poulson 1958, 1960, 1963, 1964; Noltie and Wicks 2001). Other general amblyopsid features are pelvic fins either absent (as in Arkansas species) or very reduced; the anus and urogenital opening are jugular; a strongly protruding lower jaw; a flattened head; scales cycloid and minute, producing a naked appearance; and the lateralis system is well developed, with prominent lines of superficial neuromasts on the head, body, and caudal fin. Armbruster et al. (2016) examined the skeletal morphology of all recognized amblyopsid species and found that monophyly of Amblyopsidae was supported by 34 synapomorphies. Those authors also found that the sister group to Amblyopsidae is Aphredoderidae based mainly on the presence of a unique set of upper jaw bones (which they termed lateromaxillae) in both families.

These small fishes are specialized for life in underground waters and are seldom seen. All Arkansas forms are rare with small populations, and the Ozark Cavefish, *Troglichthys rosae,* has been designated a Threatened species by the U.S. Fish and Wildlife Service. The fragile, energy-poor cave environments are extremely susceptible to human disturbances, and protective measures will be required to insure the future survival of the Arkansas cavefish species.

Key to the Cavefishes

1A   Sensory papillae on caudal fin in 0–2 rows (Fig. 20.3B).   *Typhlichthys eigenmanni* Page 507
1B   Sensory papillae on caudal fin in 4–6 rows (Fig. 20.3A, C).   2

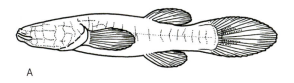

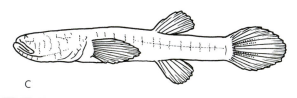

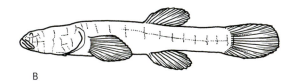

Figure 20.3. Sensory papillae on cavefish caudal fins:

A—Sensory papillae in 6 rows (*Troglichthys rosae*).
B—Sensory papillae in 2 rows (*Typhlichthys eigenmanni*).
C—Sensory papillae in 4 rows (*Typhlichthys* sp. *cf. eigenmanni*).

2A   Caudal fin rays 9–11; snout lacking in lateral outline; sensory papillae on caudal fin usually in 5 or 6 (sometimes 4) rows (Fig. 20.3A).
                                    Troglichthys rosae Page 503
2B   Caudal fin rays 14–20; head weakly depressed anteriorly into a snout; sensory papillae on caudal fin usually in 4 rows (Fig. 20.3C).
                              Typhlichthys sp. *cf. eigenmanni* Page 508

Genus *Troglichthys* Eigenmann, 1899a— Ozark Cavefishes

The Ozark Cavefish was originally described in the genus *Typhlichthys* by Eigenmann (1898a), who subsequently described it in more detail in several papers in 1898 and 1899. Eigenmann first referred to the Ozark Cavefish as *Troglichthys rosae* in 1898 and again referred to it by that name in papers in 1899 (Eigenmann 1898b, 1899a, b). The

1899a paper is usually considered the formal redescription (the same paper also appeared in *Proceedings of the Indiana Academy of Science* in 1899). The Ozark Cavefish remained in *Troglichthys* until Woods and Inger (1957) placed it in the genus *Amblyopsis,* where it has remained until recent times. Page and Burr (2011) placed the Ozark Cavefish back into *Troglichthys,* but Page et al. (2013) continued the long-standing placement of the Ozark Cavefish in *Amblyopsis.* Niemiller et al. (2013b) provided genetic evidence for placing the Ozark Cavefish in *Troglichthys.* Genetic evidence did not support the monophyly of the Ozark Cavefish with the two northern species of *Amblyopsis* (based on at least 10 different genes). We agree with Niemiller et al. (2013b) in returning the Ozark Cavefish to *Troglichthys.* Based on analysis of skeletal morphology, *Troglichthys* was supported as the sister to *Typhlichthys + Speoplatyrhinus* (Armbruster et al. 2016). This monotypic genus is characterized by a combination of morphological and genetic features (see species account for *T. rosae*).

Troglichthys rosae (Eigenmann)
Ozark Cavefish

Figure 20.4. *Troglichthys rosae*, Ozark Cavefish. *Dante Fenolio.*

CHARACTERS   A small, eyeless (vestigial eye tissue is present under the skin), depigmented fish with an elongate body, an elongate flattened head, and a rounded caudal fin (Fig. 20.4). Pelvic fins are absent. Dorsal fin single, with 7 or 8 (sometimes 9) rays; anal rays 8; branched caudal rays 9–11. Sensory papillae on caudal fin usually in 5 or 6 (sometimes 4) rows (Fig. 20.3A). Anus far in front of anal fin, jugular in adults. Newly hatched young have the anus just anterior to the anal fin, but as growth occurs, the anus gradually shifts forward until it occupies the jugular position of the adult (Woods and Inger 1957). Scales very small and imbedded, causing skin to appear naked. Mouth superior and snout lacking in lateral profile. Postcleithrum bone absent. Adults commonly around 2 inches (51 mm)

Map 144. Distribution of *Troglichthys rosae*, Ozark Cavefish, 1988–2019.

long. Maximum size 3 inches (76 mm) TL, but maximum reported size in Logan Cave, Arkansas, was 2.6 inches (65 mm) (Brown and Johnson 2001).

LIFE COLORS   Body depigmented and translucent with a pink coloration due to the blood flowing through superficial vessels. Preserved specimens are white.

SIMILAR SPECIES   Very similar to the Salem Plateau Cavefish, *Typhlichthys eigenmanni,* but differs in having 4–6 rows of sensory papillae on caudal fin (vs. 2 rows). It differs from the undescribed *Typhlichthys* sp. *cf. eigenmanni* in Stone County in having 9–11 caudal fin rays (vs. 14–23 caudal rays). The Ozark Cavefish further differs from both Arkansas species of *Typhlichthys* in lacking a postcleithrum bone (vs. postcleithrum bone present).

VARIATION AND TAXONOMY   *Troglichthys rosae* was originally described by Eigenmann (1898a) as *Typhlichthys rosae* "from the caves of Missouri" (see genus *Troglichthys* account). Cooper (1980) provided information indicating that the type locality was Day's Cave in Sarcoxie, Missouri. See genus account for additional details of its taxonomic history. Bergstrom et al. (1997) and Bergstrom (1997) identified at least four genetically distinct lineages of *T. rosae* based on mitochondrial DNA nucleotide sequence data and phylogenetic analyses. Those authors considered the differences sufficient to support recognition of the lineages as separate subspecies.

DISTRIBUTION   The Ozark Cavefish is limited to caves of the Springfield Plateau physiographic region of northwestern Arkansas, northeastern Oklahoma, and southwestern Missouri. Several surveys have been performed during the past three decades to monitor the status of this species (Brown 1985, 1991; Brown and Willis 1984; Willis and Brown 1985; Brown and Todd 1987; Brown et al. 1998a; Graening and Brown 1999, 2000a, b). This Ozark endemic species has been historically confirmed by observation or collection of specimens in Arkansas from 11 sites (caves, sinkholes, springs, and a nursery pond) in Benton County and from 4 sites (1 cave, 2 springs, and 1 creek) in Washington County (Map 144 and Map A4.142). There are other unconfirmed reports of cavefish in Boone, Carroll, Madison, and Newton counties, but those reports, if accurate, may not represent *T. rosae.*

HABITAT AND BIOLOGY   This obligatory cave species is only rarely found in surface waters when washed out of its cave habitat by a flood (Graening and Brown 2000b). It inhabits the permanently dark, constant temperature (55–60°F [12.8–15.6°C]) of underground passageways created by subsurface drainage, that has dissolved the highly soluble limestone bedrock. The rock strata of the Springfield Plateau in which Ozark Cavefish are found are riddled with completely or partially water-filled conduits, most (but not all) too small for human access (Noltie and Wicks 2001). Many of the passageways too small for humans may be occupied by cavefish, as evidenced by instances where cavefish have been found by landowners in hand-dug wells. Noltie and Wicks (2001) identified two types of subterranean sites in which Ozark Cavefish occur: (1) "Lateral sites" are cave streams that occupy relatively large passageways (penetrable by humans) that intersect the land surface near the base of either the Springfield Plateau bluff line or the river valley walls of the plateau escarpment. (2) "Vertical sites" (accessed through a sinkhole or well) where the downward opening penetrates the land surface and underlying rock units to intersect a conduit usually too small for human penetration. It is typically found in clear water in pools or very slow-moving water of small cave streams over a rubble bottom. Noltie and Wicks (2001) found that the stratigraphy of the Springfield Plateau causes the karst layer in which Ozark Cavefish are found to lie at or immediately adjacent to the land surface. The consequences of this near-surface occurrence are: (1) infiltration by surface waters into cavefish habitat tends to be rapid, (2) influx of allochthonous material is facilitated, and (3) greater risks to cavefish habitat from physical impacts such as well drilling and road construction (Noltie and Wicks 2001). The energy and materials that support the simple food web of a cave system are in low supply and were thought to be provided largely by the droppings of cave-dwelling bats and/or leaf litter (which enter via sinkholes). Aley and Aley (1979) estimated that the typical cave in northwestern Arkansas receives less than 1/1,000 of the energy input received by a surface ecosystem; however, Graening (2000) found that the oligotrophic Cave Springs Cave ecosystem received about 2% of the energy input received by the surface ecosystem above. Graening also provided the following information about the energy dynamics of Cave Springs Cave. Ninety-five percent of the annual organic matter input was in dissolved form, and Gray Bat guano represented only 1% of that amount. Particulate organic matter was scarce, emphasizing the importance of allochthonous inputs. Cave Springs Cave had an ecosystem efficiency of only 4%, a benthic organic matter turnover of 6 years, and a community respiration rate one-half that of surface streams, indicating that much organic matter input is exported and underused. The Ozark Cavefish is beautifully adapted to this energy-poor, fragile cave ecosystem (Poulson 1960, 1963). Its metabolism is very low, requiring less frequent ingestion of food than most fishes; it also spends most of its time motionless in the water (Poulson 1963, 1964; Boyd 1997). The neuromasts of the lateral line system project from the skin's surface, and their structure and arrangement are thought to enhance the detection of prey, conspecific, or predator movements, as well as currents and proximity to surroundings (Baker 1972; Noltie and Wicks 2001; Niemiller and Poulson 2010).

The Ozark Cavefish is a rare species not only in Arkansas, but also throughout its limited range in the Ozarks, and it was added to the Federal Threatened Species List in December 1984. All known populations are small; the species also has a very low reproductive rate (Means 1993; Means and Johnson 1995). The largest known populations of Ozark Cavefish occur in Cave Springs Cave and Logan Cave in Benton County, Arkansas (Graening et al. 2010). Those two caves are estimated to contain 61% of the known Ozark Cavefish numbers. Population counts conducted in 1983 and 1984 in this cave yielded sightings of 97 and 102 fish, respectively, but a more recent study indicated that cavefish populations are increasing in that cave and in Logan Cave (Graening et al. 2010). Cave Springs Cave was purchased by the Nature Conservancy and the Arkansas Natural Heritage Commission as a sanctuary for the Ozark Cavefish and the federally endangered Gray Bat (*Myotis grisescens*) in the 1980s, and Logan Cave has been designated a National Wildlife Refuge. Graening and Brown (2000b) found that Cave Springs Cave was

experiencing significant habitat degradation from surface pollution sources. Excessive bacterial and nutrient concentrations and the presence of toxic metals were found in the cave sediments and food web (Brown et al. 1998a; Graening and Brown 1999, 2000a).

One of the threats to the Ozark Cavefish is removal by collectors for sale to the aquarium trade or for museum specimens. It is easily collected if someone gains access to its subterranean waters because of its habit of remaining motionless in the water. The threatened status given by the U.S. Fish and Wildlife Service affords protection by the potential of a $20,000 fine. Although there is no evidence that the removal of small numbers of these fish from a cave can have long-lasting adverse effects on the cavefish population, the removal of large numbers can likely seriously affect the population for years and may even destroy it. In comparison, collecting fishes from a surface stream is usually of little or no consequence to the fish community. For example, 144 cavefish were removed from Cave Springs Cave in the 1950s and placed in the Tulane Museum of Natural History. Poulson (1960) counted 72 individuals in that cave in 1960, and it has taken another 40 years for this population to recover to its previous abundance (Graening and Brown 2000a). In Civil War Cave in Benton County, 30 cavefish were removed in one collection event in the 1930s and deposited at Tulane University (Aley and Aley 1979). Subsequent surveys have never reported more than 2 individuals (Graening et al. 2010). The primary threat to this species has now shifted from over-collection to habitat degradation (Graening et al. 2010). Groundwater pollution around Springfield, Missouri, may have contributed to the loss of six of seven formerly known cavefish populations near that city. Water in caves in northwestern Arkansas may experience low dissolved oxygen concentrations caused by surface land use (Aley and Aley 1979). The introduction of sewage, organic material, gasoline, and other substances into groundwater systems can seriously deplete dissolved oxygen concentrations. Reoxygenation of subsurface water is usually slower than reoxygenation of surface water. Toxic compounds and sediment can also adversely affect cavefish, as can certain agricultural activities. Aley and Aley (1979) described the damage to a cave ecosystem near Gravette in Benton County that had received a discharge of concentrated hog-farm wastes in early 1979. A study of six cavefish caves by Graening (2005) found that these sites had detectable contamination of water, sediments, and animal tissue by nutrients, toxic metals, and coliform bacteria, probably originating from septic systems and land application of concentrated animal wastes. Graening et al. (2010) reported that while the number of active cavefish sites throughout its range has decreased by more than 50% since 1990, the number of surveyed individuals has not.

The Ozark Cavefish feeds primarily on small crustaceans such as copepods, isopods, and amphipods but also consumes small crayfish, salamanders, and its own young (Poulson 1963). In Cave Springs Cave, *T. rosae* was the top predator, feeding mainly on isopods that ate imported organics including sewage (Graening 2000). Niemiller and Poulson (2010) reported that the gut contents of four individuals from Cave Springs Cave contained 65 copepods (100% frequency), 8 cladocerans (25%), 4 isopods (50%), and one 10 mm crayfish (25%). Based on stomach contents, extent of fat deposits, and annulus formation, *T. rosae* fed less in the fall than other times of the year (Poulson 1960). The Ozark Cavefish is more adept at sensing and capturing midwater prey, whereas the genus *Typhlichthys* is more adapted for capturing benthic prey (Noltie and Wicks 2001).

Breeding is thought to occur during high water from February through April (Poulson 1963); however, four females with ova visible in the body cavity were observed in Logan Cave in Benton County in late August, and five small cavefish (approximately 10 mm TL) likely less than a month old were observed in that cave on 20 June 1996 (Adams and Johnson 2001). Fewer than two-thirds of a population are sexually mature at any one time, and only about one-fifth spawn in each year (Poulson 1963). Each female lays about 25 eggs, and some authors have speculated that the eggs are possibly carried in the female's gill chamber until they hatch (although this has not been conclusively demonstrated for *T. rosae*). Branchial brooding was not observed by Adams and Johnson (2001), and Armbruster et al. (2016) considered it more likely that *T. rosae* exhibits transbranchioral spawning and is not a branchial brooder. Originally, the Ozark Cavefish was reported to reach sexual maturity in its fifth year and to have a maximum life span of 15 years (very long for such a small fish) (Poulson 1963); however, based on a discussion with A. V. Brown, Poulson revealed that he now believes that the life span of this cavefish is much longer than first reported by a factor of 2–4 times (Niemiller and Poulson 2010). Poulson believes the age of first reproduction could be as much as 16 years, but Niemiller and Poulson (2010) considered 6 years to be a more reasonable estimate of the age at first reproduction for *T. rosae*. A drastically different interpretation by J. Z. Brown and Johnson (2001) suggested that *T. rosae* has a maximum life span of only 4–5 years. Those authors found that growth averaged 0.6 mm per month, with maximum recorded growth of 6.0 mm per month.

CONSERVATION STATUS   The Ozark Cavefish has been listed as a federally Threatened species by the United States Fish and Wildlife Service and as an Endangered species by the Arkansas Game and Fish Commission. The rapid human population growth in the karst region of northwestern Arkansas since the first edition of this book has further degraded groundwater quality in that area and increased the threat to survival of *T. rosae*. We consider it endangered in Arkansas.

Genus *Typhlichthys* Girard, 1859— Southern Cavefishes

Analysis of molecular data has revealed considerable cryptic diversity in many taxonomic groups in recent decades, and this has been especially true in subterranean organisms (Niemiller et al. 2012, 2013b). There are numerous examples of multiple morphologically similar species being very distinct genetically. In the subterranean environment, similar selective pressures have operated to produce similar morphologies of the currently recognized hypogean cavefish species in the southeastern United States (Niemiller 2011). The Southern Cavefish, *Typhlichthys subterraneus*, as recognized by Page et al. (2013), has the largest range of any subterranean fish in the world (Niemiller and Poulson 2010). For at least three decades, several authors have suggested that *T. subterraneus* is a cryptic species complex (Swofford 1982; Barr and Holsinger 1985; Holsinger 2000; Niemiller and Fitzpatrick 2008; Niemiller and Poulson 2010). Niemiller et al. (2012), using multilocus genetic data, identified 10 cryptic lineages within that complex. Some of those lineages are believed to represent distinct species, and we are aware of confirmed populations in two caves in Stone County, Arkansas (and one unconfirmed report from Stone County by Graening and Brown 2000b) that we believe warrant recognition as a new species of *Typhlichthys*. Pending further study, Niemiller et al. (2012) identified at least 6 and possibly as many as 15 putative cryptic species within nominal *T. subterraneus* but did not provide formal descriptions for any new species. Niemiller et al. (2012) recognized all populations of *Typhlichthys* west of the Mississippi River in the Ozark Highlands of Arkansas and Missouri as specifically distinct from all other populations of *T. subterraneus*, assigning the name *Typhlichthys eigenmanni* to all Ozark populations. We herein treat the two confirmed Stone County populations as a new, undescribed species, *Typhlichthys* sp. *cf. eigenmanni*, and refer other known populations of *Typhlichthys* in Arkansas to *Typhlichthys eigenmanni* pending further analysis. Genus characters are given in the species account below.

Typhlichthys eigenmanni Charlton
Salem Plateau Cavefish

Figure 20.5. *Typhlichthys eigenmanni*, Salem Plateau Cavefish. *Dante Fenolio.*

Map 145. Distribution of *Typhlichthys eigenmanni*, Salem Plateau Cavefish, 1988–2019.

CHARACTERS   A small, eyeless (vestigial eye tissue present under skin), depigmented fish with an elongate body and an elongate, flattened head. Mouth superior. Caudal fin rounded (Fig. 20.5). Postcleithrum bone present. Pelvic fins absent. Dorsal fin single, usually with 8 or 9 rays (7–10). Sensory papillae on caudal fin usually in 2 rows (1 row on upper half, 1 row on lower half), but sometimes with 0, 1, or 2 rows (Fig. 20.3B). Anus far in front of anal fin in adults; anal rays 7–10. Caudal rays branched, 10–15. Scales very small and embedded, causing skin to appear naked. Maximum size just over 3 inches (76 mm).

LIFE COLORS   Body depigmented and translucent, but with a pinkish coloration created by the blood flowing through superficial vessels. Exposure to light causes the development of epidermal melanophores in some individuals (Woods and Inger 1957).

SIMILAR SPECIES   Closely resembles the Ozark Cavefish, *Troglichthys rosae,* and an undescribed Arkansas *Typhlichthys* species, but differs from both in having 2 rows of sensory papillae on caudal fin (vs. 4–6 rows).

VARIATION AND TAXONOMY   *Typhlichthys eigenmanni* was described by Charlton (1933) from Camden County, Missoui. It was synonymized, along with all other species recognized in *Typhlichthys*, under *T. subterraneus* by Woods and Inger (1957). Niemiller et al. (2012) resurrected the name *Typhlichthys eigenmanni* (Parenti 2006) for populations of this species in Arkansas and Missouri and assigned the common name Salem Plateau Cavefish. The stable cavefish populations in two caves in Stone County, formerly identified as *T. subterraneus,* are now considered an undescribed species (based on morphological and molecular evidence). Armbruster et al. (2016) reviewed the proposed phylogenetic relationships of *T. eigenmanni.*

DISTRIBUTION   *Typhlichthys eigenmanni,* as designated herein, inhabits the Salem Plateau of the Ozark Highlands in Missouri and Arkansas, and is much more widely distributed outside Arkansas (in Missouri) than the Ozark Cavefish. The recorded sites for *T. eigenmanni* in Arkansas are as follows: (1) A single specimen of this cavefish first collected in 1940 from a well in Randolph County (Woods and Inger 1957), (2) 20 individuals observed by Paige et al. (1981) in Richardson Cave in Fulton County, (3) Romero et al. (2010) reported an unusual collection of a specimen found outside the subterranean environment at Lake Norfork, Baxter County, Arkansas, on 20 February 2009 (ASU 13067), and (4) an identification by G. Graening of specimens from Riley's Springbox in Baxter County in 1996, based on a description by the landowner C. Riley (Map 145). About 20 cavefish were observed in 1975 in Alexander Cave in Stone County (Tim Ernst, pers. comm.), and Graening and Brown (2000b) listed it from three caves in Stone County: Alexander, Ennis, and Cave River caves. The taxonomic status of cavefish at all reported sites for *Typhlichthys* is undergoing morphological and molecular analysis.

HABITAT AND BIOLOGY   The habitat of the Salem Plateau Cavefish superficially appears similar to that of the Ozark Cavefish, but their ranges are not similar (Pflieger 1997; Brown and Willis 1984); however, Noltie and Wicks (2001) provided evidence for substantial differences in the subterranean habitats of the two species. It is found in limestone caves of the Salem Plateau, where it inhabits pools and the quieter water of small streams over a bedrock, rubble, or clay bottom. Coarser substrates are generally found where spring upwellings occur. Noltie and Wicks (2001) compared the geological differences of the Salem and Springfield plateaus and concluded that *T. eigenmanni* typically resides at appreciable depths, on the order of 100–200 meters below the land surface. They further speculated that most of our human perception of their habitat has been influenced (and misconstrued) by the nature of the sites in which we encounter the species near the surface. Instead, the near-surface sites (sinks) represent atypical habitats into which fish have been flushed, washed, carried, or transported. Most known sites for this species share a characteristic feature, the presence of a permanently effluent spring that either discharges at the surface into a spring pool, or contributes to a cave stream or the filling of a sinkhole (Noltie and Wicks 2001). Schubert and Noltie (1995) experimentally demonstrated that individuals are attracted to point sources of water efflux (small spring seeps), but the species exhibits less pronounced rheotaxis than the Ozark Cavefish. In a Missouri cave, all observed individuals occupied positions close to the bottom (within 0.5 cm) at depths of 1.0–2.0 m below the water surface; when disturbed, individuals moved to substrates of clean, loosely aggregated chert gravel (3–5 cm diameter), hiding in large interstitial spaces (Schubert et al. 1993). Like other Arkansas cavefishes, it is long lived and slow growing, and Noltie and Wicks (2001) suggested that its life span might be counted in decades rather than years. Populations are small and the reproductive capacity is low (Poulson 1963).

The diet of *T. eigenmanni* is dominated by midwater and substrate invertebrates (Poulson 1963). It feeds primarily on small crustaceans, especially copepods (Cooper and Beiter 1972), and Pflieger (1997) reported that its diet was more diverse than that of the Ozark Cavefish. Most amblyopsid cavefishes show little reaction to chemical stimuli of living or dead prey they find in their environments in a laboratory setting but will regularly react to mechanosensory stimuli (Niemiller and Poulson 2010; Niemiller and Soares 2015); however, *T. eigenmanni* is attracted to exudates of both conspecifics and prey (Aumiller and Noltie 2003).

Breeding probably occurs in late spring in association with rising water levels. Poulson (1985) reported that its eggs and clutches are larger than those of the Ozark Cavefish, and the females invest more energy in egg production and parental care. Only about 50% or fewer of the adult females breed each year, and fecundity is reportedly low with fewer than 50 ova per female (Poulson 1963). Niemiller and Poulson (2010) and Armbruster et al. (2016) believed that it is unlikely that *T. eigenmanni* broods its eggs in the female's gill chamber as documented for *Amblyopsis hoosieri* and *A. spelaea*. It is probably a transbranchial spawner.

CONSERVATION STATUS This species is very rare in Arkansas. Worldwide, *T. eigenmanni* was classified as Vulnerable (VU D2) by the World Conservation Monitoring Centre (WCMC) (Romero 1998). Its Global rank was listed as G4, indicating that this species is apparently secure. Populations of *T. eigenmanni* in Arkansas face several threats such as habitat degradation, groundwater pollution, and hydrological manipulations (Proudlove 2006; Niemiller and Poulson 2010). Romero et al. (2010) studied the population status of this species in Arkansas and found no indication that the populations of *Typhlichthys* in Arkansas needed urgent actions by either federal or state agencies (but that study included populations in caves in Stone County considered to be *T. subterraneus*). Niemiller et al. (2013a) considered *T. eigenmanni* the least threatened of 10 lineages, but most of the known secure populations are in Missouri. We consider *T. eigenmanni* endangered in Arkansas, and frequent monitoring of the populations should be continued.

Typhlichthys sp. *cf. eigenmanni*
Ghost Cavefish (or Ghost Fish)

Figure 20.6. *Typhlichthys* sp. *cf. eigenmanni*, Ghost Cavefish from Alexander Cave, Stone County. *Dante Fenolio.*

CHARACTERS A small, eyeless (vestigial eye tissue present under skin), depigmented fish with an elongate body and an elongate flattened head; head weakly depressed anteriorly into a snout. Mouth more terminal and less superior than that of *T. eigenmanni*. Caudal fin rounded (Fig. 20.6). Postcleithrum bone present. Pelvic fins absent. Pectoral fins long and pointed, with 9 or 10 rays. Dorsal fin single, usually with 6–8 rays. Sensory papillae on caudal

Map 146. Distribution of *Typhlichthys* sp. *cf. eigenmanni*, Ghost Cavefish, 1988–2019.

fin usually in 4 rows, 2 rows on upper half, and 2 rows on lower half, with 7–15 papillae per row (Fig. 20.3C). Anus far in front of anal fin in adults; anal rays 5–8. Caudal rays branched, 14–20. Scales very small and embedded, causing skin to appear naked. Largest specimen examined was 37 mm TL, but maximum size probably around 3 inches (76 mm).

LIFE COLORS   Body depigmented and translucent, but with a pinkish coloration caused by the blood flowing through superficial vessels.

SIMILAR SPECIES   This undescribed species superficially resembles the Ozark Cavefish, *Troglichthys rosae*, but differs from it in having 14–20 caudal fin rays (vs. 9–11 rays), and 4 rows of sensory papillae on the caudal fin (vs. usually 5 or 6 rows). It differs from *T. eigenmanni* in having 4 rows of sensory papillae on the caudal fin (vs. 0–2 rows).

VARIATION AND TAXONOMY   The cavefish populations in two (and possibly a third) Stone County caves, formerly identified as *T. subterraneus* or *T. eigenmanni*,

are now considered an undescribed species of *Typhlichthys* based on morphological and molecular evidence. Evidence from studies of mitochondrial DNA and nuclear DNA supports the specific distinctness and monophyly of *T.* sp. *cf. eigenmanni* (G. O. Graening, pers. comm.).

DISTRIBUTION   *Typhlichthys* sp. *cf. eigenmanni* has been confirmed from only two caves in Stone County, Arkansas: Alexander Cave and Ennis Cave (Romero et al. 2010) (Map 146). Graening and Brown (2000b) reported unconfirmed rumors of cavefish in Cave River Cave in Stone County, but Romero and Conner (2007) could not confirm those reported sightings.

HABITAT AND BIOLOGY   The habitat of *T.* sp. *cf. eigenmanni* is similar to that of *T. eigenmanni*, but their ranges do not overlap. It is found in two limestone caves of the Salem Plateau, where it inhabits pools and the quieter water of small subterranean streams of the middle White River drainage over a bedrock, rubble, or clay bottom. In 1975, T. Ernst (pers. comm.) reported seeing 23 individuals (mostly in one large pool) in one of the caves. In subsequent surveys, no more than 4 individuals were observed at one time (Romero and Conner 2007). Noltie and Wicks (2001) concluded that *T. eigenmanni* typically resides at appreciable depths, on the order of 100–200 meters below the land surface, and the same may be true of the Ghost Cavefish because of the similar geology of the two areas of the Salem Plateau occupied by those species. Because of the potential for cavefish to travel considerable distances in karst areas of the Salem Plateau, it is likely that the known cave occurrences of the Ghost Cavefish represent a single metapopulation rather than isolated groups. Like other Arkansas cavefishes, it is probably long lived and slow growing. Access to the caves where this species has been sighted is controlled by the cave owners, and the Alexander Cave entrance is gated (Romero et al. 2010).

Life history aspects for *T.* sp. *cf. eigenmanni* have not been studied. Food habits and reproductive biology are probably very similar to those of *T. eigenmanni*.

CONSERVATION STATUS   Because of its small range and population size, and because of the vulnerability of fragile subterranean environments to disturbance, this undescribed Arkansas endemic species should be considered endangered.

CHAPTER 21

Order Mugiliformes
Mullets

The order Mugiliformes contains only one family, the largely marine mullets; therefore, the characters for both order and family are identical. In mugiliforms, the two dorsal fins are widely separated, the first consists of only 4 spines and the second of 1 spine followed by soft rays. Pelvic fins have 1 spine and 5 rays and are subabdominal in position. The pelvic girdle does not articulate with the cleithra. The anal fin has 2 or 3 spines; the scales are large and cycloid (sometimes weakly ctenoid); the lateral line is absent or poorly developed; the gill rakers are long; the stomach is muscular, and the intestine is exceedingly long, indicating herbivory. The pseudobranchiae are large, the branchiostegal rays number 5 or 6, and the branchiostegal membranes are free from the isthmus. Some authors included Mugiliformes in series Mugilomorpha (Boschung and Mayden 2004) as we did in Appendix 5, but Nelson et al. (2016) placed Mugiliformes in series Percomorpha.

Family Mugilidae *Mullets*

The mullet family, Mugilidae, is a cosmopolitan marine family of about 20 genera and 75 or more species (Nelson et al. 2016). Mullets are mostly schooling fishes occurring in all tropical and temperate oceans, but most species are found in the northern Indian and western Pacific oceans (Eschmeyer et al. 1983). Adults range in size from 11.8 to 35.8 inches (300–910 mm), but rarely reach more than 20 inches (500 mm) in freshwater (Page and Burr 2011). Some are important food fishes in Africa and Asia, where they are often raised in ponds. Mugilid fossils from France date back to the Lower Oligocene (Berra 1981). Although it is a family of primarily coastal marine fishes, there are some species in Australia and New Guinea that spend their entire lives in freshwater (Berra 2001), and one species, the Striped Mullet (*Mugil cephalus*), ascends freshwaters far inland in North America.

Genus *Mugil* Linnaeus, 1758—Mullets

The genus *Mugil*, with 17 species, is characterized by prominent adipose eyelids and cycloid or very weakly ctenoid scales. Although five species occur in North American waters, only *Mugil cephalus* enters Arkansas waters occasionally from the Gulf of Mexico. Striped Mullets are found in many rivers in Texas and Louisiana, from as far inland as the Red River at Lake Texoma and the Colorado River near Austin, Texas (Hoese and Moore 1998). In Arkansas, the Striped Mullet has been taken at several sites and is generally regarded as a curiosity due to its uncommon occurrence. It is regarded as a food fish in Florida, but when taken from the muddier waters of the western Gulf of Mexico, this fish is not generally eaten because of an oily flavor.

Mugil cephalus Linnaeus
Striped Mullet

Figure 21.1. *Mugil cephalus*, Striped Mullet. *David A. Neely.*

Map 147. Distribution of *Mugil cephalus*, Striped Mullet, 1988–2019.

CHARACTERS The Striped Mullet, a stout, fusiform, somewhat elongate and rather distinctive member of the Arkansas fish fauna, has a blunt snout, a broad, flat, scaled head, two well-separated dorsal fins, highly placed pectoral fins, and adipose eyelids (Fig. 21.1). Upper end of pectoral fin base located near upper end of gill opening. Body depth going fewer than 5 times into SL. No teeth on vomer and palatine bones. Body covered with large scales; scales are cycloid in young, but become weakly ctenoid in adults. Lateral line absent; lateral scales 37–42. Mouth small and terminal with thin lips and shaped like an inverted V. Gill membranes separate; gill rakers are numerous and form a fine sievelike structure. There is a distinct gizzardlike stomach, and the intestine is extremely long (5–8 times the body length). Axillary process present at base of pectoral fin and

pelvic fin. First dorsal fin with 4 spines; second dorsal fin with 1 spine and 7–9 soft rays. Anal fin with 2 or 3 spines and 8 soft rays. Pectoral fin rays usually 16 or 17; pelvic fins with 1 spine and 5 soft rays. Usual size about 9–14 inches TL (230–360 mm). Maximum size 3 feet (910 mm) in India. In Arkansas, specimens are usually 12–24 inches (305–610 mm).

LIFE COLORS Body color generally olive green to bluish gray on back grading to silver on sides. Belly whitish to silvery. Large blue spot at base of pectoral fin. No lateral band present, but each scale on the lower-middle to upper side has a dark spot, creating the appearance of horizontal lines running the length of the body. Fish smaller than 6 inches (152 mm) lack lateral stripes and are a bright silvery color. Fins plain or slightly dusky. Bright golden spot on operculum in life.

SIMILAR SPECIES The Striped Mullet resembles silversides (*Labidesthes* and *Menidia*), but differs from them in having 7–9 soft anal rays (vs. 15 or more anal rays). It differs from Clupeidae, Hiodontidae, Cyprinidae, and Catostomidae in having 2 widely separated dorsal fins (vs. a single dorsal fin).

VARIATION AND TAXONOMY *Mugil cephalus* was described by Linnaeus in 1758 from "Europe." Campton and Mahmoudi (1991) studied allozyme variation in Florida and concluded that gene flow within and between the Gulf and Atlantic coasts was sufficient to maintain a genetically homogeneous population. A global study found high estimated gene diversities among widely separated populations, with an allele frequency variation among populations of 68% (Rossi et al. 1998).

DISTRIBUTION The Striped Mullet occurs in coastal waters in tropical and subtropical areas of all oceans. In North America, it is found along the Atlantic and Gulf coasts from Nova Scotia, Canada, to southern Mexico, and along the Pacific Coast from southern California south. It ascends the Mississippi River drainage as far inland as the Red River in central Oklahoma, the Mississippi River to the Missouri River, Missouri, and the lower Ohio River in Kentucky. Records of Striped Mullet from Arkansas are mainly from the main channels and backwater areas of large rivers (Map 147). Known localities of collection include the Sulphur/Red, upper Ouachita, lower Arkansas, and Mississippi rivers (Buchanan et al. 2003), and a 480 mm TL individual was collected from the middle White River. There is even a record of one large specimen from a Mississippi River oxbow lake, Lake Chicot, in 2000 (Jerry Smith, pers. comm.). Only pre-1988 records exist from the Red and Ouachita river drainages (Map A4.145). There is some evidence that individuals ascending rivers are mostly relatively large females (Rohde et al. 2009; D. Coughlan, pers. comm.). Reid and Ginny Adams (pers. comm.) reported observing a large school of Striped Mullet jumping in a backwater area off the old Arkansas River channel in Desha County in early October 2003. They collected two individuals as voucher specimens. In addition, Jeff Quinn (pers. comm.) observed numerous mullet from 2009 through 2011 in the lower Arkansas River below Dam 2 during electrofishing surveys.

HABITAT AND BIOLOGY   The Striped Mullet, widely distributed throughout the oceans of the world, is primarily a spring and summer visitor to Arkansas waters traveling up the Mississippi River and its tributaries from the Gulf of Mexico. Mullets are strong swimmers and can be seen jumping in open water. It has been reported as far inland in the Mississippi River as southern Illinois (Burr et al. 1990, 1996). It normally inhabits nearshore marine environments, where it is common in estuaries and salt marshes. In Arkansas, it has been found in large rivers, but there is one record from Lake Chicot (undoubtedly entering this oxbow lake from the nearby Mississippi River during high river flow) in the early 2000s.

This schooling species feeds on plant material, detritus, plankton, and algae, often taking mouthfuls from the bottom, and is therefore not generally caught by anglers. Fine gill rakers retain the larger food particles, which are then ground up in the gizzardlike stomach. The young feed on microcrustaceans and small invertebrates before switching to a more detritivorous diet as adults. Juveniles grow quickly in food-rich estuaries and begin their schooling behavior at a small size (Mettee et al. 1996). Adults also feed on plant material that they process in the gizzard.

Spawning occurs in the fall in the Gulf of Mexico near the edge of the continental shelf, along the Atlantic shore, and in winter along the Pacific Coast of California. Ripe females were found in South Carolina estuaries from October through February (McDonough et al. 2003). Spawning usually occurs well offshore (Eschmeyer et al. 1983) and fecundities are high, with females producing from 250,000 to 2.2 million eggs (Greeley et al. 1987). Eggs are small (about 0.72 mm), nonadhesive, and pelagic, and hatching occurs in about 48 hours. After hatching, the larvae and later the young drift or swim toward the coast, reaching estuarine areas when they are about 16–28 mm TL (Anderson 1958). One study suggested that Striped Mullets spawned in freshwater in the Colorado River of California (Johnson and McClendon 1970). There is no evidence that mullets spawn in Arkansas waters, but Ross (2001) reported that this species does attain sexual maturity in freshwater. Eggs and larvae apparently require nearly full-strength sea water to survive, and the juveniles are not fully tolerant of freshwater until they reach about 40–69 mm SL (Nordlie et al. 1982). Juveniles are collected for bait, and adults are important food fishes, exploited commercially for their flesh and roe (Etnier and Starnes 1993). Mullets can be seen exhibiting periodic jumping behavior. The causes of such behavior have not been determined but may include reaction to fright, ectoparasites, low dissolved-oxygen concentrations at certain positions within a school, or unknown factors (Rohde et al. 2009). Maximum life span is about 5 or 6 years.

CONSERVATION STATUS   This uncommon species is sporadically collected in Arkansas. The Nature Conservancy rank of S1 (critically imperiled in the state) is probably unwarranted, and we assign it a status of special concern in Arkansas.

CHAPTER 22

Order Atheriniformes
Silversides

This order consists of 8 families with 52 genera and about 351 species (Bloom et al. 2012; Werneke and Armbruster 2015; Nelson et al. 2016). Atheriniforms make up a clade of predominantly surface-dwelling fishes that occur throughout tropical and temperate regions (Bloom et al. 2012). They are widespread, abundant, and ecologically important forage fishes. Members of this order are primarily silvery fishes that usually have 2 separated dorsal fins, the first with weak, flexible spines, and the second with a single flexible spine followed by rays in most species. Pectoral fins are usually placed high on the body, and the pelvic fins are abdominal, subabdominal, or thoracic. The anal fin is much longer than the second dorsal fin and is generally preceded by a single weak spine. The lateral line is absent or very weak. Branchiostegal rays number 4–7.

Relationships of atheriniform fishes were reconsidered using molecular data (Setiamarga et al. 2008), morphological data (Saeed et al. 1994; Dyer and Chernoff 1996), and combined molecular and morphological approaches (Sparks and Smith 2004a). Above the order level, atheriniforms were grouped into the series Atherinomorpha with the Beloniformes and Cyprinodontiformes (Boschung and Mayden 2004), but Nelson et al. (2016) placed the atherinomorphs in the large series Percomorpha. We support retaining Atheriniformes in series Atherinomorpha (Appendix 5). White et al. (1984), Dyer (1998), and Betancur-R. et al. (2017) provided evidence for the monophyly of

Atheriniformes, but the monophyly and relationships among families within that order remain uncertain (Bloom et al. 2012). All silversides were formerly placed in the family Atherinidae, long known as the silverside family. That family was divided into several families by Dyer and Chernoff (1996), with Atherinidae as now constituted containing an assemblage of Atlantic and Indo-West Pacific species. The North American species are now placed (with very few exceptions) in the family Atherinopsidae, the New World Silversides (Dyer 1998).

Family Atherinopsidae
New World Silversides

Atherinopsidae is divided into two subfamilies consisting of about 13 genera and 110 species, of which about 50 are confined to freshwater (Nelson et al. 2016). Twelve species in six genera of this primarily marine family occur in coastal and freshwaters of the United States, and four of the species occur in freshwaters of North America (Page and Burr 2011; Werneke and Armbruster 2015). One of the best-known atherinopsids, the California Grunion, *Leuresthes tenuis,* has the peculiar adaptation of spawning on sandy beaches at night during high tides (Nelson 2006). There are no fossil records of atherinopsids in North America prior to the Late Tertiary (Cavender 1986). The genus *Menidia* is known from the Pliocene in Oklahoma

(Hubbs 1942). Three species of these small, slender fishes occur in Arkansas, the Brook Silverside, *Labidesthes sicculus*, the Hardy Silverside, *Labidesthes vanhyningi*, and the Mississippi Silverside, *Menidia audens*. All are placed in the silverside subfamily Atherinopsinae, tribe Menidiini, which was considered monophyletic by Bloom et al. (2012). Nelson et al. (2016) recognized Menidiini as a subfamily rather than a tribe.

New World silversides are important forage fishes characterized by a scaled head, two widely separated dorsal fins (the first with spines, the second with 1 thin spine and several soft rays), a broad silvery lateral band, pectoral fins placed high on the body, pelvic fins usually abdominal, and both pelvic and anal fins each with a small spine. Although the swim bladder is present, the pneumatic duct connection with the pharynx is absent (physoclistous). Silversides are common surface and midwater inhabitants of streams and lakes, generally preferring calm pools and backwater areas. Schools of silversides can be seen leaping from the water at times, especially when chased by a predator (or a net) (Miller and Robison 2004).

Key to the New World Silversides

1A Premaxilla, viewed from above, triangular and produced as a beak (Fig. 22.1A); predorsal scales small, more than 23; anal soft rays 20–26; scales in lateral series 74–87; lateral line not divided into two parts.
Labidesthes 2

1B Premaxilla, viewed from above, neither triangular nor produced as a beak (Fig. 22.1B); predorsal scales large, fewer than 23, usually 19; anal soft rays 15–20; scales in lateral series 38–46; lateral line divided into two parts, the anterior part higher on side than posterior part.
Menidia audens Page 522

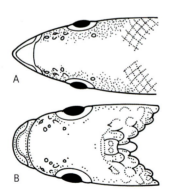

Figure 22.1. Dorsal view of silverside heads. *Moore (1968).*

A—*Labidesthes sicculus*
B—*Menidia audens*

2A Midlateral stripe narrows in front of first dorsal fin (Fig. 22.2A); ratio of thoracic length to abdominal length greater than 2; anterolateral process of the posttemporal is longer than its width at its base (Fig. 22.3B).
Labidesthes sicculus Page 517

2B Midlateral stripe expands in front of first dorsal fin (Fig. 22.2B); ratio of thoracic length to abdominal length less than 2; anterolateral process of the posttemporal is shorter than its width at its base (Fig. 22.3A).
Labidesthes vanhyningi Page 520

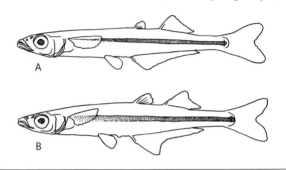

Figure 22.2. Midlateral stripe of *Labidesthes* species.

A—*Labidesthes sicculus*
B—*Labidesthes vanhyningi*

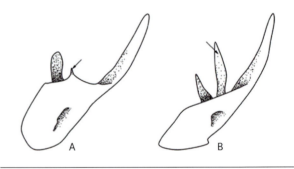

Figure 22.3. Morphological differences in posttemporal process of *Labidesthes* species. *Courtesy of David C. Werneke and Jonathan W. Armbruster.*

A—*Labidesthes vanhyningi*
B—*Labidesthes sicculus*

Genus *Labidesthes* Cope, 1870— Brook Silversides

The genus *Labidesthes* was first described by Cope (1870) to remove the Brook Silverside from the genus *Chirostoma* (where that species had originally been placed by Cope in 1865). *Labidesthes* is a member of the silverside tribe Menidiini within the subfamily Menidiinae of the Atherinopsidae (Chernoff 1986; Dyer and Chernoff 1996;

Bloom et al. 2012). Previous studies have also recovered it as a basal member of the tribe Menidiini (Echelle and Echelle 1984; White 1985; Chernoff 1986, Dyer 1997, 2006). Characters separating *Labidesthes* from *Menidia* in Arkansas are smaller scales, with more than 70 in a lateral series (vs. scales large, with fewer than 50 in a lateral series); a longer anal fin, with more than 20 rays (vs. anal fin shorter, with fewer than 20 rays), and a longer snout. There are two species of *Labidesthes*, both of which occur in Arkansas.

NOTE—Almost all points on our distribution maps for the two species of *Labidesthes* were originally plotted on our distribution map for *L. sicculus* prior to the description of *L. vanhyningi*. With a few exceptions, the specimens on which distribution points were based were not available for examination, and our distribution maps for *L. sicculus* and *L. vanhyningi* represent our best judgment of where those species occur in Arkansas. *Labidesthes sicculus* is distributed mainly above the Fall Line, with a few records from below it, while *L. vanhyningi* is known almost entirely below the Fall Line in Arkansas. Distribution points immediately above and below the Fall Line were especially difficult to plot, and we used the distributional information in Werneke and Armbruster (2015), along with an examination of some specimens, to make our decisions.

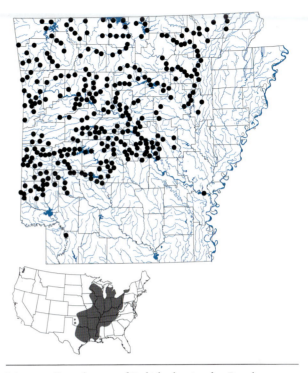

Map 148. Distribution of *Labidesthes sicculus*, Brook Silverside, 1988–2019.

Labidesthes sicculus (Cope)
Brook Silverside

Figure 22.4. *Labidesthes sicculus*, Brook Silverside. *David A. Neely.*

CHARACTERS   An extremely elongate, slender, translucent species with a long, beaklike, pointed snout with a weakly protrusible premaxillary, a small, flattened head, high pectoral fins, and 2 widely separated dorsal fins (Fig. 22.4). Midlateral stripe narrows distinctly in front of first dorsal fin. Mouth terminal and oblique, the elongate jaws protractile and with minute, posteriorly curved, pointed teeth; premaxillae appearing triangular when viewed from above. Eyes large, eye diameter going into snout length

about 1.5 times. Ratio of thoracic length to abdominal length in specimens over 55 mm SL greater than 2. The anterolateral process of the posttemporal bone is longer than its width at its base. First dorsal fin small and inconspicuous, with 3–5 (modally 4) thin, flexible spines; second dorsal fin with 1 thin, flexible spine and 8–12 soft rays. Anal fin long and falcate with 1 spine and 20–27 (modally 23) rays; anal fin origin approximately below the origin of the first dorsal fin. Pectoral fins set high on body, with 11–13 rays. Pelvic fin usually with 6 rays. Caudal fin forked with 16 or 17 rays. Scales small and cycloid. Predorsal scales 36–48. Lateral line absent or incomplete (sometimes with a few pores on caudal peduncle), with 72–91 scales in a lateral series. Opercle scaled; top of head with or without scales. Intestine short and S-shaped. Genital papilla short in males; genital papilla in females broad and round, with a large opening. Maximum size smaller than 4.4 inches (112 mm).

LIFE COLORS   Body color green to straw-colored above, white to silvery white below. Body translucent in life. Silvery midlateral stripe becoming narrower anteriorly in front of the first dorsal fin (Fig. 22.2A). Dorsal scales outlined with pigment. Head covered in melanophores, chin pigmented, dark spot on lower jaw at rear margin of mouth. Lower jaw and opercle a metallic blue. Dorsum with black stripe along middle; venter with pairs of black chromatophores from anus to origin of caudal fin, not forming a distinct stripe in Arkansas populations. Fins colorless, except for first dorsal fin of breeding male, which develops a black tip. Silver peritoneum is visible through body wall. Large males and females from Lake Fort Smith possessed distinct round or irregular dark spots on the silvery peritoneum (Fogle 1959).

SIMILAR SPECIES Labidesthes sicculus is most similar to L. vanhyningi (juvenile specimens of both species are more difficult to separate than adults) but differs in having the anterolateral process of the posttemporal longer than the width of its base (vs. shorter than width of base) (Fig. 22.3), a midlateral stripe that tapers anterior of the first dorsal fin (vs. lateral stripe maintaining a width or expanding in width anterior to first dorsal fin) (Fig. 22.2), and a ratio of thoracic length to abdominal length greater than 2 (vs. less than 2). Also similar to Menidia audens, but differs from that species in having smaller scales, with 72–87 scales in a lateral series (vs. 38–46 scales in a lateral series), anal fin longer with 20–27 rays (vs. 15–20 anal rays), and with no pored lateral line scales immediately behind the head above the lateral stripe (vs. up to 8 pored lateral line scales behind the head above the lateral stripe).

VARIATION AND TAXONOMY Labidesthes sicculus was described by Cope (1865b) as Chirostoma sicculum from Grosse Isle, Detroit River, Michigan, but a subsequent revision placed it in Labidesthes (Cope 1870). Previously, L. sicculus was believed to comprise two subspecies: the nominate form, Labidesthes s. sicculus, and L. s. vanhyningi; however, Werneke and Armbruster (2015) elevated L. s. vanhyningi to species level.

DISTRIBUTION The Brook Silverside is widely distributed throughout much of the St. Lawrence–Great Lakes and Mississippi River basins from Quebec, Canada, to eastern Minnesota, and south to Louisiana, and in Gulf Slope drainages from the Pearl River in Mississippi west to the Brazos River in Texas (Werneke and Armbruster 2015). In Arkansas, it is found largely along and above the Fall Line but also occurs sporadically below it (Map 148). It is likely that new, scattered distribution records will be reported for L. sicculus from the Coastal Plain of Arkansas. It occurs in appropriate habitats in all river drainages (with the possible exception of the St. Francis River) and is one of the most widely distributed nongame fish species in the state.

HABITAT AND BIOLOGY The Brook Silverside is a surface-dwelling fish, occurring in a variety of aquatic environments, including streams, small rivers, and lakes.

Occasionally, specimens are taken in the big rivers of the state. In an Ozark stream, populations of L. sicculus remained fairly isolated, stable, and showed little mobility, with marked individuals moving a distance of less than 0.5 km during a year (Dewey 1981). In new impoundments, it quickly becomes abundant, particularly in those with clear water and sandy-gravel bottoms (Smith 1979). Buchanan (2005) reported that L. sicculus was the most widely distributed small nongame species in Arkansas impoundments, and more than 20,000 specimens were taken from 57 of 66 reservoirs sampled (although some of the specimens were undoubtedly Labidesthes vanhyningi). Several authors have reported drastic declines in reservoir populations of L. sicculus after the introduction of Menidia audens (Dowell and Riggs 1958; Gomez and Lindsay 1972; Taylor et al. 2008; Simmons 2013). Within streams, this schooling species seems to prefer calm backwater regions, where it is often observed making short jumps out of the water; however, it is reported to prefer smaller, more lotic environments than Menidia audens (Strongin et al. 2011). Werneke and Armbruster (2015) considered L. sicculus to have broader habitat preferences than L. vanhyningi.

The Brook Silverside feeds mainly during daylight hours, but it is also known to feed on moonlit nights (Pflieger 1997). Foods consumed include chironomids, small insects, and zooplankton, depending on stage of life cycle (Zimmerman 1970; Keast 1985a). Mullan et al. (1968) provided the following information on the food habits of L. sicculus in two Arkansas reservoirs. This mobile surface feeder in Beaver Lake typically ingested microcrustaceans and aquatic insects throughout the year and terrestrial spiders in the spring. Brook Silversides in Bull Shoals Lake fed almost entirely on aquatic insects during the summer but added microcrustaceans and terrestrial insects to their diet at other times of the year. Young silversides 0.8–3.9 inches TL (20–100 mm) fed almost exclusively on microcrustaceans in both reservoirs. Late in their first summer, young silversides move closer to shore, and the diet shifts mainly to terrestrial and aquatic insects. Declines in populations of L. sicculus have been noted in several North American reservoirs after the introduction of M. audens (McComas and Drenner 1982; Pratt et al. 2002), and food competition has often been considered a major cause of the declines. Ramirez et al. (2006) found that populations of Labidesthes persist in several coves of Lake Texoma that have marinas with artificial lighting. They concluded that artificial lighting benefited Labidesthes by attracting terrestrial insects to these habitats. Menidia is considered a more efficient planktivore than Labidesthes, and increased abundance of dipterans near lighted marinas likely promotes the coexistence of these two species (Ramirez et al. 2006). In the Tennessee-Tombigbee Waterway (TTW), Mississippi, *Labidesthes sicculus* and *M. audens* had a high degree of dietary overlap, and neither species exhibited dietary shifts that would facilitate coexistence (Strongin et al. 2011). Because *Menidia* used a wider variety of prey items in that modified waterway environment and because of their previous interactions in other systems, Strongin et al. (2011) concluded that the native silverside, *L. sicculus*, is likely to be replaced by *M. audens* in lentic TTW habitats.

In Arkansas, the Brook Silverside spawns in late spring and summer from at least May into August when water temperatures reach 68–73°F (20–22.8°C). Spawning occurs in pools over aquatic vegetation or gravel beds. Spawning has been described by Cahn (1927), Hubbs (1921), J. S. Nelson (1968), and Pflieger (1997). Fogle (1959) found that spawning peaked in late May to early June in Lake Fort Smith, Arkansas, and spawning was essentially finished by late June. Ten ripe females collected on 13 August from Lake Nimrod in Yell County, Arkansas, by TMB ranged from 60 to 69 mm TL and contained from 100 to 130 mature ova. The scales of those females had no annuli, indicating that sexual maturity and spawning can occur in age 0 fish within 3–4 months after hatching. Spawning accounts of northern populations in Michigan and Wisconsin indicated that *L. sicculus* females were releasing eggs upon spawning (Hubbs 1921; Cahn 1927). Powles and Sandeman (2008) also concluded that eggs were released during mating and found no evidence of sperm or embryos in female *Labidesthes* from Ontario, Canada. Piteo et al. (2017) considered the presence of a genital papilla in males of *L. sicculus* (reported by Werneke and Armbruster 2015) to be evidence of internal fertilization. Those authors also cited the description of the reproductive behavior of *L. sicculus* by Hubbs (1921) as additional evidence for direct sperm transfer from males to females. Ripe females from several localities in the Ozarks of Arkansas examined by TMB contained mature ova but no obvious embryos. Pflieger (1997) noted that additional work is needed to determine whether *L. sicculus* has populations with internal fertilization (as reported for *L. vanhyningi*). Eggs (0.95–1.4 mm in diameter) with 1 or 2 long adhesive filaments attach to the vegetation or other objects as they sink to the pool bottom. Fogle (1959) reported that eggs from Lake Fort Smith specimens had 2 filaments. We do not believe that Fogle studied *L. vanhyningi* instead of *L. sicculus* in Lake Fort Smith as suggested by Piteo et al.

(2017) based on the presence of 2 egg filaments. Lake Fort Smith is an upland reservoir in the Boston Mountains Plateau, and TMB has found only *L. sicculus* in the Lake Fort Smith drainage. Becker (1983) noted that in a current, the filaments may act as flotation bodies transporting the egg for some distance before the sticky filaments attach the egg to the first thing with which it comes into contact. Cahn (1927) described spawning behavior and noted that hatching occurs in about a week at a temperature of 77°F (25°C). Becker (1983) reported that eggs hatch in 8 or 9 days at 22.8–24°C (73–75°F). During courtship, a male and female swim in alignment with the male above and slightly behind the female. As courtship proceeds, there is much darting and leaping by the female at the water surface until the female comes to rest, whereupon the male comes to lie beside her. Both fish then glide slowly down toward the substrate while making repeated abdominal contacts. It is at this time that eggs are emitted and presumably fertilized, and Cahn (1927) noted:

> During the descent, the eggs are extruded from the body of the female and may be seen slowly settling toward the bottom in the wake of the descending pair. Fertilization takes place in the water immediately after the eggs have left the female, the spermatozoa being extruded by the male coincident with the momentary contact with the abdomen of the female.

Brook Silversides have an extremely rapid growth rate, with individuals reaching adult size within 3 months after hatching (J. S. Nelson 1968). The young move to deeper water, where they form compact schools for the remainder of their first summer (Boschung and Mayden 2004). Young reach 2.4 inches (60 mm) TL before the end of the year and attain sexual maturity at that time or in the following spring and summer, when spawning occurs. Boschung and Mayden (2004) reported that few, if any, survive a severe winter; therefore, in Alabama, the Brook Silverside appears to be mostly an annual species leaving only one set of progeny. This view is consistent with previous perceptions that in most parts of its range, *L. sicculus* functions largely as an annual species, completing its entire life cycle within 1 year. Hubbs (1921) and Cahn (1936) reported that maximum life span is about 17 months; however, Fogle (1959) found a few males in Lake Fort Smith with 2 annuli. McAllister and Cloutman (2016) reported parasites of *L. sicculus* in Arkansas and Oklahoma.

CONSERVATION STATUS   The Brook Silverside is abundant in Arkansas, and its populations are secure.

Labidesthes vanhyningi Bean and Reid
Hardy Silverside

Figure 22.5. *Labidesthes vanhyningi*, Hardy Silverside. *Uland Thomas.*

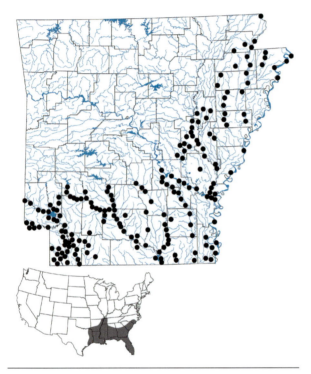

Map 149. Distribution of *Labidesthes vanhyningi*, Hardy Silverside, 1988–2019.

CHARACTERS A slender, elongate, translucent fish with elongate beaklike jaws. Mouth with moderately long, posteriorly curved, pointed teeth. Head short and flattened above, mouth terminal and oblique (Fig. 22.5). Eyes large, eye diameter going into snout length about 1.5 times. Midlateral silver band (dark in preserved specimens), narrowest on caudal peduncle, but broadening and fading anteriorly, especially in front of first dorsal fin. Ratio of thoracic length to abdominal length in specimens >55 mm SL less than 2. Length of the anterolateral process of the posttemporal bone shorter than the width of its base (Fig. 22.3A). Two widely separated dorsal fins; the first dorsal fin small and inconspicuous, with 4–6 (modally 5) thin, flexible spines; second dorsal fin with 1 thin, flexible spine

and 8–11 soft rays. Anal fin long and falcate with 1 spine and 19–25 (modally 22) rays; anal fin origin approximately below the origin of the first dorsal fin. Pectoral fins situated high on body, with 11–14 rays; pelvic fin with 6 rays. Caudal fin forked. Scales small and cycloid. Predorsal scales 31–44. Lateral line nearly absent; sometimes with a few pores on caudal peduncle, and with 68–88 scales in a lateral series. Opercle scaled; top of head with or without scales. Genital papilla short in males, absent in females (Grier et al. 1990). Maximum size just under 4 inches (102 mm) TL.

LIFE COLORS Body translucent, with green to yellow tint, white to silvery white below. Midlateral band silvery. Scales above midline and on dorsum outlined with melanophores, forming a weak black stripe along center of dorsum. Venter often with a black stripe from anus to origin of caudal fin in Arkansas populations. Chin pigmented, dark spot on lower jaw at rear margin of mouth. Iris silver. Inside of mouth with scattered melanophores along margin and on outside of cheeks. Peritoneum silvery. Breeding males in Florida usually have a reddish snout, a bright yellow-green body, and yellowish to orange second dorsal, anal, and caudal fins (Werneke and Armbruster 2015; Robins et al. 2018).

SIMILAR SPECIES Most similar to *Labidesthes sicculus* but differs in having the length of the anterolateral process of the posttemporal bone shorter than the width of its base (vs. longer than width of base) (this character is most reliable for separating specimens greater than 60 mm SL) (Fig. 22.3), a midlateral stripe that maintains its width or expands in width anterior to the first dorsal fin (vs. midlateral stripe tapering in advance of the first dorsal fin to the insertion of the pectoral fin) (Fig. 22.2), and a ratio of thoracic length to abdominal length less than 2 (vs. greater than 2), indicating a more anterior pelvic fin placement in *L. vanhyningi*. Also similar to *Menidia audens*, but differs from that species in having smaller scales, with 68–88 scales in a lateral series (vs. 38–46 scales in a lateral series), anal fin longer with 19–25 rays (vs. 15–20 anal rays), and with no pored lateral line scales immediately behind the head above the lateral stripe (vs. up to 8 pored lateral line scales behind the head above the lateral stripe).

VARIATION AND TAXONOMY Bean and Reid (1930) originally described *Labidesthes vanhyningi* from near Gainesville, Florida, calling it the Fiery-Finned Silverside. They distinguished *L. vanhyningi* from *L. sicculus* by its more slender, less compressed body and shorter snout, among other characters, but failed to define a range for the new species. Hubbs and Allen (1943) and Hubbs and Lagler (1947) considered *L. vanhyningi* a subspecies of *L. sicculus*, noting that it was a rather poorly defined geographical variant. Bailey et al. (1954) also placed *L. vanhyningi* in the synonymy of *L. sicculus* when they found putatively diagnostic characters *sensu* Bean and Reid (1930) of *L. vanhyningi* in populations of *Labidesthes* from Arkansas. Those authors recommended further investigation before recognizing *L. vanhyningi* as a full species. Later, Rasmussen (1980) summarized developmental differences between his findings and those of other workers, and noted that the differences in myomere counts and attachment filaments were clinal. He also cited the need for additional work on different populations to resolve the taxonomy of *Labidesthes*. Bloom et al. (2009) provided genetic evidence for the validity of *vanhyningi* as a species by finding a 14.7% sequence divergence between specimens from Florida and Minnesota in the mitochondrial NADH dehydrogenase subunit 2 (ND2) gene. Werneke and Armbruster (2015) examined morphometric, meristic, and osteological data among populations of *Labidesthes* rangewide to determine the taxonomic status of *L. vanhyningi*. Those authors redescribed both species of *Labidesthes*, defined their ranges, and discussed aspects of their distribution. Werneke and Armbruster (2015) also provided a new common name for the species, Golden Silverside. Piteo et al. (2017) noted that the Golden Silverside name was already assigned to *Menidia colei* and proposed a new common name of Stout Silverside, suggesting that live specimens of this species are "heartier, heftier, and tougher" than those of *L. sicculus*. We follow the suggestion of J. Armbruster and D. Werneke (pers. comm.) that a new common name of Hardy Silverside is more appropriate for *L. vanhyningi*. The name "hardy" refers to its greater ability to survive for a while after being out of water and to its ability to withstand handling more than other silverside species.

DISTRIBUTION *Labidesthes vanhyningi* occurs in Gulf of Mexico drainages from eastern Texas through Florida, and in Atlantic Slope drainages from Florida to North Carolina. It extends northward in the Mississippi Embayment to Reelfoot Lake in Tennessee. In Arkansas, it is known almost entirely from below the Fall Line, where its range overlaps broadly with that of *L. sicculus* (Map 149). It is largely absent from the Ouachita and Ozark mountains. Eleven lots of *Labidesthes* examined by Werneke and Armbruster (2015) contained both *L. sicculus* and *L. vanhyningi*, indicating that the two species were not only sympatric, but also occasionally syntopic.

HABITAT AND BIOLOGY *Labidesthes vanhyningi* usually occurs in streams, small rivers, and lakes of the lowlands of Arkansas having silt, sand, and detritus substrates, but it is also found in lower Mississippi River mainstem and floodplain habitats. It is primarily a Coastal Plain species

throughout its range. Substrate type is probably of little consequence, as long as water quality is suitable, because it spends almost its entire life near the surface, even remaining motionless near the surface on dark nights. It is common in the quiet waters of Red River oxbow lakes (TMB, unpublished data), and it adapts well to lowland reservoirs in Arkansas. Werneke and Armbruster (2015) considered it to have a narrower range of habitat preference than *L. sicculus*, even though *L. vanhyningi* is hardier than *L. sicculus*.

The following information on feeding biology is from a study by Piteo et al. (2017) in the Black Warrior River drainage near Tuscaloosa, Alabama. *Labidesthes vanhyningi* is a schooling species that feeds at or near the surface on a variety of prey organisms. In all months, cladocerans dominated the diet, making up 56–98% of the total diet. Aquatic insects made up 13% of the winter diet (collembolans 12%, chironomids 1%), but those two items constituted only 3% of the summer diet. Terrestrial insect stages (winged) accounted for 10% of the winter diet and 21% of the spring diet. Eggs of unknown origin represented 21% of the food items consumed in the fall. Less common food items included arachnids, copepods, larval fish, plant materials, and nematodes.

In the Black Warrior River drainage, Alabama, a prolonged breeding season extends from at least March through December (Piteo et al. 2017). In Arkansas, there is little information on spawning of *L. vanhyningi*, but its breeding season in this state extends at least from May into August. The latitude of extreme southern Arkansas is similar to that of the Alabama study site; therefore, a longer breeding season may occur along the southern tier of counties in Arkansas. It probably has internal fertilization with subsequent release of embryos in different stages of development. Mature ovarian eggs and embryos usually possess two long filaments that may serve as attachment structures when the eggs are released into the environment. Piteo et al. (2017) believed that egg filament number distinguished the two *Labidesthes* species, with *L. sicculus* eggs having one filament and *L. vanhyningi* having two filaments. Females in reproductive condition were collected on 23 May from Tupelo Creek in Clark County, Arkansas, by TMB. Ripe females from Tupelo Creek ranged from 50 to 65 mm TL and contained between 27 and 102 mature ova, but no obvious embryos were identified. Grier et al. (1990) found eyed embryos inside ovaries of *Labidesthes* females collected from southern Florida, demonstrating that fertilization was internal in *L. vanhyningi*. Those authors, using histological techniques and electron microscopy, also found sperm cells within the ovarian cavity and concluded that internal fertilization occurs shortly after ovulation. Because Rasmussen (1980) collected developing eggs with a plankton net in the Peace River, Florida, Grier et al. (1990) speculated that female *L. vanhyningi* may void embryos in different developmental stages in open water or, more likely, deposit them on aquatic vegetation and other substrates. Internal fertilization was also confirmed in *L. vanhyningi* in the Black Warrior River drainage of Alabama (Piteo et al. 2017). Fertilized ova were found in ovaries and were distinguishable by having orange-yellow yolk surrounded by clear albumin (unfertilized mature ova were orange or amber). Piteo et al. (2017) observed embryos in the ovaries in what appeared to be the beginning stages of blastulation, with clear signs of developed blastoderm, but found no eyed embryos as reported in Florida by Grier et al. (1990). Internal fertilization coupled with laying fertilized eggs is known in several atherinimorph fishes, but *L. vanhyningi* is the only atherinopsid species in which it has been documented. Grier et al. (1990) found that male *L. vanhyningi* possess a short genital palp immediately posterior to the anus that is lacking in females (an observation that we have also made in Arkansas specimens). Those authors concluded that during courtship, when the ventral surfaces of a breeding pair come into contact, the genital palp must serve as an intromittent organ for transfer of sperm to the female reproductive tract (since there is no modification of the anal fin for that purpose). If this reproductive pattern occurs in Arkansas populations (as seems likely), *L. vanhyningi* is the second Arkansas species (along with *Gambusia affinis*) known to exhibit internal fertilization. A collection of 224 juveniles from Beard's Lake in Hempstead County on 30 June 2017 ranged from 24 to 47 mm TL ($\bar{x}$ = 33.6 mm). McAllister and Cloutman (2016) provided information on the parasites of *L. vanhyningi* from Arkansas and Oklahoma.

CONSERVATION STATUS *Labidesthes vanhyningi* is widespread through the Coastal Plain of Arkansas (Werneke and Armbruster 2015), and its populations are currently secure.

Genus *Menidia* Bonaparte, 1836— Atlantic Silversides

The genus *Menidia* is a member of the silverside tribe Menidiini (Bloom et al. 2009). *Menidia* is a rather small genus of eight species that occur in fresh, salt, or brackish waters of the Atlantic and Gulf coasts of the United States, with one species, *M. audens*, occurring inland in the lower Mississippi River basin. Characters of the genus include a moderately compressed, elongate body, with a relatively short snout, and 2 dorsal fins, the first smaller

than the second. The origin of the first dorsal fin ranges from anterior to vertical from the anal fin origin. Pectoral fins are high on the side of the body, and the second dorsal fin and anal fin are both subfalcate. Side with a bright silvery lateral stripe. The lateral line is divided into two parts. The anterior one-quarter to one-third of the lateral line lies above the lateral midline with a vertical gap of 2–3 unpored scale rows between it and the posterior portion of the lateral line. The posterior two-thirds or so of the lateral line lies below the lateral midline. Hubbs and Lagler (2004) defined the lateral scale count in *Menidia* as from the base of the caudal fin to the first scale touching the shoulder girdle, and if some scales are not pored, the count includes the number of scales along the line in the position that would normally be occupied by a typical lateral line scale. Swift et al. (2014) noted that a count (from caudal base to shoulder girdle) along the lower pored lateral line segment is 2–4 scales fewer than a lateral count along the upper pored segment.

Menidia audens Hay
Mississippi Silverside

Figure 22.6. *Menidia audens*, Mississippi Silverside. *David A. Neely.*

CHARACTERS A slender, translucent species with a flattened head, long anal fin (15–20 rays), and 2 dorsal fins (Fig. 22.6). First dorsal fin small and inconspicuous, with 4 or 5 thin spines; first dorsal fin origin in front of anal fin origin. Second dorsal fin with 1 short, splintlike, flexible spine and 8 or 9 (rarely 10) soft rays, the last ray branched to the base, giving the false impression of 2 separate rays. Pectoral fin situated high on body, its origin about even with top of eye; pectoral rays 13–15. Pelvic rays 6. Predorsal scales usually fewer than 23 (Arkansas River specimens 18–25, $\bar{x}$ = 21.3). Scales between the two dorsal fins 4–6. Total gill rakers on first arch usually 22 or 23. Eyes large, eye diameter about equal to snout length. Snout rounded, not pointed, its length less than distance from back of eye to rear of operculum. Scales large, usually 40–43 (38–45) in lateral series. Lateral line interrupted and in two segments, the upper segment with about 8 pored scales is above the lateral stripe behind the head, and the lower segment continues below the lateral stripe from near the tip of the pectoral

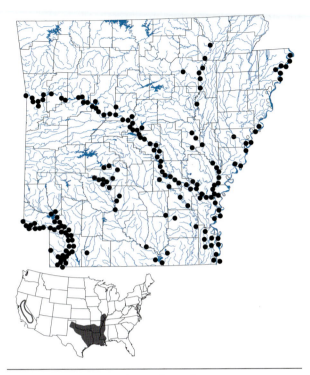

Map 150. Distribution of *Menidia audens*, Mississippi Silverside, 1988–2019. Inset shows native range (solid) and introduced range (outlined) in the contiguous United States.

fin to the caudal base; both upper and lower segments may show varying degrees of interruption. Size usually smaller than 4 inches (102 mm) SL, but maximum size around 6 inches (150 mm) TL. Females are usually larger than males.

LIFE COLORS   Background color is pale greenish yellow to tan on dorsum and generally straw-colored on sides. There is a broad silvery lateral band from the caudal base to above the pectoral fin. The venter is white. Scale margins outlined with melanophores, especially on the upper body. Fins are generally clear, with rows of small melanophores outlining some fin rays.

SIMILAR SPECIES   *Labidesthes sicculus* and *L. vanhyningi* are similar, but *M. audens* has a short, rounded snout (vs. a long, pointed snout) (Fig. 22.1). Scales are larger in *M. audens* with usually fewer than 23 predorsal scales (vs. more than 23), 38–45 scales in lateral series (vs. 74–87 scales in lateral series in *Labidesthes*), and there are usually 15–20 anal rays (vs. 20–26 anal rays in *Labidesthes*).

VARIATION AND TAXONOMY   Hay (1882) described *Menidia audens* and listed the type specimens as coming from the Mississippi River at Memphis (a few) and Vicksburg (many), one specimen from the Big Black River at Edwards, Mississippi, and a few from the Pearl River at Jackson, Mississippi. Suttkus and Thompson (2002)

provided justification for designating the Mississippi River at Vicksburg as the type locality. The taxonomic status of *Menidia audens* has been controversial for more than 40 years. *Menidia audens* was synonymized with *M. beryllina* (Cope 1869) by Chernoff et al. (1981), who considered them conspecific. However, Suttkus and Thompson (2002) and Suttkus et al. (2005) presented an alternative view of the systematic status of the two species and provided convincing morphological evidence that *M. audens* is a valid, unique taxon separate from *M. beryllina*. Ecologically, *M. audens* is basically a freshwater inhabitant, while *M. beryllina* is a brackish or tidewater inhabitant. Both species occur syntopically in the lower Mississippi River near its mouth and in the lower Sabine River, Texas-Louisiana, and there is no evidence of hybridization between them (Suttkus et al. 2005). In a subsequent phylogenetic analysis, Bloom et al. (2009) did not consider them separate species. We follow Suttkus and Thompson (2002), Suttkus et al. (2005), and Page et al. (2013) in recognizing *M. audens* as a separate species from *M. beryllina*. Suttkus et al. (2005) provided morphological data for Arkansas specimens of *M. audens* from the Arkansas River at Fort Smith.

DISTRIBUTION   *Menidia audens* is native to nine states in the lower Mississippi River basin, including the Red, Arkansas, and mainstem Mississippi rivers from southern Illinois, southeastern Missouri, and western Kentucky, south through western Tennessee and western Mississippi to Louisiana, and Gulf Slope drainages from the Pearl River, Mississippi, west to the Brazos River, Texas. It is considered native throughout southern Oklahoma and northern and eastern Texas. It likely spread into extreme westcentral Alabama after construction of the Tennessee-Tombigbee Waterway (although it is possibly native to the Tennessee River in extreme northwestern Alabama). Introduced populations are widely established in California (Swift et al. 2014). In Arkansas, the Mississippi Silverside is native to the large rivers of the state, including the Arkansas, lower White, Mississippi, Red, and Ouachita rivers (Map 150). Black (1940) first reported this species from Arkansas from Chicot and Miller counties, and speculated that it was probably more widely distributed in Coastal Plain lakes associated with large rivers in southern and eastern parts of the state. It has also been stocked in some large reservoirs. Originally present in Lake Chicot, it was, according to W. Keith (pers. comm.), probably accidentally introduced into several reservoirs in the early 1960s when Threadfin Shad were moved from Lake Chicot to other parts of the state.

HABITAT AND BIOLOGY   The Mississippi Silverside abounds in large rivers of the state, but it is also common in reservoirs and oxbow lakes. It is generally absent from small streams. In some states, it is abundant in reservoirs (e.g., Lake Texoma, Oklahoma), and Buchanan (2005) reported taking more than 21,000 individuals from 17 Arkansas impoundments, including the navigation pools of the Arkansas River. Some published descriptions of habitat use indicate that *M. audens* is more strongly associated with lentic habitats than the other two Arkansas species of atherinopsids, *L. sicculus* and *L. vanhyningi*, but we have frequently collected it in lotic habitats in Arkansas. Large numbers of *M. audens* were taken by TMB in current in the Arkansas, Mississippi, and Red rivers. Allopatric populations of *M. audens* and *L. sicculus* occurred at mean current velocities of 0.11 m/s and 0.62 m/s, respectively, in the Tennessee-Tombigbee Waterway (TTW) in Mississippi (Strongin et al. 2011). Although considered a freshwater species, *M. audens* can tolerate marine or near-marine salinities for at least 2 days (Hubbs et al. 1971). Swift et al. (2014) provided evidence that it is able to disperse from one freshwater environment to another through marine environments along the California coast, especially during the wet season when runoff makes coastal waters fresher for short periods.

Black (1940) found *M. audens* only in the Mississippi and Red rivers in Arkansas, but speculated that it probably also occurred in the lower Arkansas and White rivers. The first Oklahoma records for this species were reported from the Red River by Moore and Cross (1950), and it was first introduced into Lake Texoma in 1953 (Dowell and Riggs 1958). *Menidia audens* has apparently extended its range throughout the Arkansas River in Arkansas only in recent decades. Black (1940) and all previous collectors failed to report it from that river. Suttkus et al. (2005) reported that on 2 February 1967, R. D. Suttkus (RDS 4081) collected only a single specimen of *M. audens* from the impounded Arkansas River at Dardanelle, Yell County, Arkansas. Fourteen years later (31 October 1981), Suttkus made another collection (RDS 7719) from the Arkansas River just below Lock and Dam 13 at Fort Smith, Arkansas, that contained 2,152 specimens of *M. audens*. Sisk and Stephens (1964) reported the first records of *M. audens* from the Arkansas River in Oklahoma (from Boomer Lake in Payne County, collected in 1964) and speculated that it had been stocked there from Lake Texoma. Gomez and Lindsay (1972) reported that the first record of *M. audens* from the main stream of the Arkansas River below Keystone Reservoir was found in 1970, and an established population was found in that reservoir in 1971. Robison and Buchanan (1988) showed no pre-1960 records of *M. beryllina* (= *M. audens*) from the Arkansas River. Most collections from

the Arkansas River before 1970 were made in the upper portion, especially near Fort Smith; however, since 1971 the Mississippi Silverside has been the most abundant species in collections from the Arkansas River at Fort Smith, with some individual 10 × 30 m seine-hauls containing more than 1,000 individuals. If *M. audens* continues to spread throughout regions of Arkansas, particularly through intentional or unintentional introductions into impoundments, the abundance of *Labidesthes sicculus* would likely decrease. Strongin et al. (2011) noted the decline of *L. sicculus* in the TTW after the introduction of *M. audens*, and predicted that *L. sicculus* would likely be replaced by *M. audens* in this anthropogenic lentic habitat. Following invasion of the Tennessee River reservoirs by *M. audens*, mean abundance of *L. sicculus* linearly decreased with time, and 14 years after sympatry with *M. audens*, *L. sicculus* were undetectable or occurred at extremely low densities in TTW reservoirs (Simmons 2013). Previous studies have shown that once *M. audens* enters such an altered system, replacement of *L. sicculus* commonly follows (Riggs and Dowell 1956; Riggs and Bonn 1959; McComas and Drenner 1982; Bettoli et al. 1991). In a study of predation on silversides by Largemouth Bass in a pond, the Mississippi Silverside was more susceptible to capture by Largemouth Bass than the Brook Silverside, and smaller individuals of both species were more susceptible than larger individuals (Stoeckel and Heidinger 1992).

The Mississippi Silverside is currently one of the most abundant fishes in the Arkansas River, where it is most commonly found in moderate current along sandbars adjacent to the main channel. It is readily collected in this habitat at all times of the year and is an important forage species for predatory fishes. Echelle and Mense (1967) found that *M. audens* was the most important food for fishes that fed in surface waters or in littoral areas of Lake Texoma. It was a particularly important food for White Bass, gar, and young Largemouth Bass but was of little significance for the White Crappie, a species of deeper waters.

In a lower Mississippi River oxbow lake, *M. audens* fed mainly on chironomids, amphipods and zooplankton (Driscoll and Miranda 1999). In Lake Texoma, it was most abundant in the littoral zone, where it fed primarily on zooplankton and dipterans throughout the day and was less abundant in that habitat at night (Saunders 1959). Elston and Bachen (1976) investigated the effects of light on feeding intensity in Clear Lake, California, and found that *M. audens* fed most frequently in the pelagic zone at dawn and dusk. McComas and Drenner (1982), Pratt et al. (2002), and Strongin et al. (2011) considered *M. audens* a competitively superior zooplanktivore to *L. sicculus* in lentic habitats owing partly to its ability to capture a larger variety of prey items because of its more protrusible, tubular mouth. Ramirez et al. (2006) considered *M. audens* to forage less efficiently than *L. sicculus* on both terrestrial and aquatic insects because of differences in mouth morphology. The most frequently ingested food items of *M. audens* in the TTW were microcrustaceans (*Daphnia* sp., *Daphnia lumholtzi*, and *Cyclopoida*) and larval insects (Ceratopogonidae), and there was high dietary overlap with sympatric *L. sicculus* (Strongin et al. 2011).

Spawning is protracted and occurs from late March or April through July at water temperatures of 23.9–32.7°C (Mense 1967; Hubbs 1976). Length of the breeding season at a given locality varies with water temperature and population density (Stoeckel and Heidinger 1989). Mense (1967) reported that the average number of mature eggs per female in Lake Texoma was 984 (range 384–1,699). Subsequently, Hubbs (1982) found that female *M. audens* in Lake Texoma produced eggs daily, with large females producing up to 2,000 eggs per day during a three-month breeding period. Hubbs estimated that during the entire breeding season in Lake Texoma, an average-sized female can produce from 75,985 to 101,870 eggs, an egg complement equivalent to 6–8 times female body weight. Spawning in Lake Texoma occurred almost entirely during morning hours, with the greatest incidence of ripe females found at 9 a.m. CST in April and at 10 a.m. in June and July (Hubbs 1976). Other workers have reported that spawning occurs both day and night in shallow water over vegetation (Ross 2001). The females typically have ripe ova only during morning hours (Hubbs and Bryan 1974). During the breeding season, all adults collected from brushy areas of Lake Texoma were ripe, indicating a preferred spawning site (Hubbs et al. 1971). In those areas, *M. audens* eggs were found in algal growth on brush stems. Spawning was observed during midmorning on 11 May 1973 in Lexington Reservoir, California, in water 1–24 inches (25–610 mm) deep at a water temperature of 68°F (20°C) (Fisher 1973). The sloping spawning area was covered with rooted aquatic plants and some inundated terrestrial plants. The vegetation formed a mat 1–2 inches thick. Fisher (1973) described spawning behavior as follows:

A school of 25 to 150 individuals would approach parallel to the shoreline led by one or more females with visibly swollen abdomens. When sampled, one school contained 2 large females and 39 smaller males. When a school turned onto the slope, males began to swim vigorously around the female, nipping and prodding at her abdomen. Occasionally during this frenzied activity, a female would suddenly break free, closely accompanied by 3 to 5 males. She would dive, along with the males, into the rooted vegetation. There, both sexes began trembling violently. While lying on her side in close contact with the males, the female laid her eggs. Upon completion, she would rapidly swim away, still closely pursued by several males. The spawning group rejoined the larger school and left the shallow area. Examination of the vegetation showed each female deposited from 10 to 20 eggs. As each school passed, the females made a single spawning pass and were not observed to repeatedly broadcast eggs.

Fertilized eggs have a tuft of about 8 short gelatinous threads, in contrast to the eggs of *Labidesthes* species, which usually have 1 or 2 long filaments per egg. Eggs hatched in 6 days at a water temperature of 27.8°C and in 10 days at 25°C (Hildebrand 1922). Termination of the breeding season appears to be triggered when water temperatures exceed 86°F (30°C) (C. Hubbs and Bailey 1977). Mortality is temperature dependent, with minimal winter losses in normal years but high mortality rate in the summer. Large age 0 individuals are capable of reproducing by the end of their first summer. Reproduction occurs during the summer at the cost of substantially increased mortality (Hubbs 1982). Because of this extensive mortality, few adults survive to spawn a second year. Maximum life span is about 2 years.

CONSERVATION STATUS   The Mississippi Silverside is widespread and abundant throughout Arkansas in appropriate habitats and its populations are secure.

CHAPTER 23

Order Cyprinodontiformes
Killifishes, Topminnows, Livebearers, and Toothcarps

embers of this order inhabit fresh, marine, and brackish waters from the tropics through temperate North and Middle America, Eurasia, and Indo-Malaysia (Martin 1968 1972; Boschung and Mayden 2004). This order includes 10 families with 131 genera and about 1,257 species (Nelson et al. 2016). Parenti (1981), Wiley and Johnson (2010), Nelson et al. (2016), and Betancur-R. et al. (2017) summarized evidence for the monophyly of the order. Most recent classifications have usually placed Cyprinodontiformes in the series Atherinomorpha, but Nelson et al. (2016) placed it in the series Percomorpha.

Many species are among the world's most popular aquarium fishes and are often seen in pet stores. The order is also noted for the ability of many species to inhabit harsh environments, such as those with high salinity, high temperature, or extreme fluctuations in various attributes. Several morphological features unify the cyprinodontiforms (Parenti 1981): a symmetrical caudal skeleton; an unlobed caudal fin; low-set pectoral fins, each with a large, scalelike postcleithrum; sensory pores chiefly on the head; and the upper jaw bordered by a protractile premaxilla. Many of the species possess marked sexual differences, with males often more brightly colored than females and with enlarged pelvic, dorsal, and caudal fins. The pelvic fins are reduced or absent in many species, and fin spines are absent. In some groups, males possess a gonopodium used

in internal fertilization, whereas in others, fertilization is external. Scales are cycloid, but males of a few species have ctenoid scales (Moyle and Cech 2004). Cyprinodontiforms have a pattern of early sexual maturation and prolonged embryonic development. Among the killifishes are annual fishes whose eggs can withstand desiccation and develop at different rates. This is apparently an evolutionary adaptation that allows killifishes to survive unfavorable environments. The life cycle of cyprinodontiforms involves an annual, semiannual, or nonannual reproductive mode. Able (1984) provided a bibliography and summery of egg and larval development patterns of cyprinodontiforms.

Family Fundulidae *Topminnows*

This family occurs in the Western Hemisphere in North and Middle America. In North America, there are 45 species included in Fundulidae, with about 40 species in the genus *Fundulus*. Members of Fundulidae were previously included in the family Cyprinodontidae (killifishes), but Parenti (1981), in an extensive analysis of the order Cyprinodontiformes, reclassified the various groups of killifishes based primarily on osteological characters. Parenti's comprehensive analysis revealed that members of the genus *Fundulus* plus four other genera formed a distinct family, the Fundulidae. Able (1984), in a major work on the development of the Cyprinodontiformes, followed Parenti's conclusions, as have Wiley (1986), Burr and Warren (1986b), Cavender (1986), Etnier and Starnes (1993), Jenkins and Burkhead (1994), Miller and Robison (2004), Page et al. (2013), and Nelson et al. (2016). Based on Parenti's comprehensive work and general acceptance by the ichthyological community for almost four decades, we continue to recognize family Fundulidae for the six species of *Fundulus* inhabiting Arkansas.

All members of this family are small, surface-dwelling fishes, and the Arkansas species are referred to as topminnows because of their habit of feeding on insects and other small invertebrates at the water's surface. The flattened head and superior mouth, opening at the upper surface of the head, facilitate this feeding habit. Fine conical teeth are present on the edges of the jaws. Lewis (1970) presented evidence that the small, dorsally oriented mouth and dorsoventrally flattened head of fishes in this family and in Poeciliidae are also adaptations for utilizing a thin layer of oxygen-rich water at the atmosphere-water interface, and that these fishes are suited for occupying habitats characterized by periodic or continuous oxygen depletion. Other characteristics useful to distinguish the Arkansas members of this family include a deep groove separating the upper jaw from the snout, the origin of the pelvic fins closer to the tip of the snout than to the base of the caudal fin, the third ray of the anal fin branched, the caudal fin rounded, scales present on the top of the head, the fins soft-rayed, the body covered with cycloid scales, the dorsal fin single and placed far back over the anal fin, and lateral line lacking. All species have external fertilization.

In some natural environments, fundulids play a role in controlling mosquitoes and other aquatic insects, and some are important forage species. Some tropical forms of fundulid fishes are popular aquarium fishes in the United States, and native Arkansas *Fundulus* species also adapt well to aquaria. Members of this family have additional scientific value as the subjects of behavioral studies and genetics research.

Genus *Fundulus* Lacépède, 1803— Topminnows

This genus, described by Lacépède in 1803, contains about 40 species found in fresh and brackish waters of North America through Mexico. Phylogenetic relationships were discussed by Wiley (1986) and Cashner et al. (1992). *Fundulus* has been divided into four or five subgenera (depending on authority), two of which contain Arkansas species: subgenus *Xenisma* includes *Fundulus catenatus*,

and subgenus *Zygonectes* includes the remaining five species of *Fundulus* in Arkansas. All Arkansas species of *Fundulus* exhibit at least some sexual dimorphism. The dorsal and anal fins of the male are usually longer and more pointed than those of the female, and females have a scaly sheath over the anterior base of the anal fin. Some species exhibit sexual dimorphism in coloration. Genus characters include an elongate body with an upturned mouth and flattened dorsum, jaw teeth conical and in more than one series, and males usually smaller and more colorful than females.

Key to the Topminnows

1A Body with a broad dark lateral band extending from snout to base of caudal fin.   2
1B Body without a dark lateral band, sides uniformly colored or having streaks, spots, or vertical bars.   3
2A Spots above lateral band round, distinct, and same color as lateral band.   *Fundulus olivaceus* Page 539
2B Spots above lateral band absent or indistinct and diffuse, their color not as dark as lateral band.
Fundulus notatus Page 537
3A Dorsal fin origin directly over or in front of anal fin origin; scales in lateral series 40 or more; dorsal fin rays 13–16; anal rays 15–18.   *Fundulus catenatus* Page 530
3B Dorsal fin origin distinctly behind anal fin origin; scales in lateral series 38 or fewer; dorsal fin rays 6–9; anal rays 8–11.   4
4A No dark vertical bar under eye; sides without horizontal streaks or black or gray dots, but males often with vertical bars.   *Fundulus chrysotus* Page 533
4B Dark vertical bar under eye; sides with prominent horizontal streaks or black or gray dots, males with or without vertical bars.   5
5A Males with 3–12 dark vertical bars on sides, females without vertical bars and with little or no melanophore development between lateral stripes; young of both sexes usually with vertical bars on sides.
Fundulus dispar Page 535
5B Males without dark vertical bars on sides, females without vertical bars but with numerous dashes, dots, and less discrete melanophore development between lateral stripes; young of both sexes without vertical bars on sides.   *Fundulus blairae* Page 529

Fundulus blairae Wiley and Hall
Western Starhead Topminnow

Figure 23.1. *Fundulus blairae*, Western Starhead Topminnow. *David A. Neely.*

Map 151. Distribution of *Fundulus blairae*, Western Starhead Topminnow, 1988–2019.

CHARACTERS   A small, deep-bodied topminnow with a well-developed subocular teardrop (a prominent, dark wedge-shaped bar extending downward below the eye) (Fig. 23.1). Both sexes lack vertical bars on sides. Side of female with 7–9 horizontal stripes; side of male with numerous scattered irregular dots in Arkansas populations, but dots forming regular horizontal rows in more southern populations. Front of dorsal fin base behind front of anal fin base. Lateral line absent; scales in a lateral series usually 33 (30–36). Scale pattern on top of head having the large anterior scale overlapping the two scales posterior to it. Scales around body usually 23 or 24; scales around caudal peduncle usually 17 or 18 (16–20). Dorsal rays usually 8 (7–9);

pectoral rays 13 (11–14); pelvic rays 6; anal rays 10 or 11 (9–12). Breeding males of the *F. notti* species group develop nuptial tubercles on dorsal and anal fin rays (Etnier and Starnes 1993), but we have found no tubercles on breeding males of *F. blairae*. Maximum size usually around 2 inches (51 mm), but largest Arkansas specimen from Beard's Lake in Hempstead County was 2.8 inches (72 mm) TL.

LIFE COLORS   Back greenish yellow with a large golden spot on top of head. Sides yellowish tan with 7–9 horizontal brown stripes in the female and with scattered red dots in the male. Sides of female with numerous small dashes or spots between the horizontal stripes. A black wedge-shaped bar present beneath eye. Dorsal, caudal, and anal fins of male have black pigment along edges of fin rays and red spots on membranes of those fins. Dorsal and caudal fins of female with dark pigment bordering rays.

SIMILAR SPECIES   Very similar to its sibling species, the allopatric *Fundulus dispar*, but differs in that males lack vertical bars on sides and the females have numerous dashes and some melanophore development between the stripes on sides (vs. males with vertical bars on sides and females with little or no pigmentation between the lateral stripes in *F. dispar*); males have dots on sides in an irregular pattern (vs. males with dots in regular horizontal rows); usually 24 or fewer scales around the body (vs. 25 or more); and young of both sexes lack vertical bars on sides (vs. young of both sexes with vertical bars). It differs from *F. chrysotus* in having the dark vertical bar beneath eye (vs. dark teardrop absent). It can be distinguished from *F. catenatus* in having the dorsal fin origin behind the anal fin origin (vs. distinctly over or in front of anal fin origin); 38 or fewer lateral scales (vs. 40 or more) and in having 6–9 dorsal rays (vs. 13–16). It differs from *F. olivaceus* and *F. notatus* in lacking a wide dark lateral band (vs. dark lateral band present).

VARIATION AND TAXONOMY   *Fundulus blairae* was described by Wiley and Hall (1975) from Neville Bayou, Liberty County, Texas from populations formerly included in *Fundulus notti*. Arkansas populations of *F. blairae* in the Red River drainage differ somewhat in color pattern from more southern populations in Louisiana and Texas. The Red River population has a more solid subocular teardrop (vs. diffuse), and the dots on the side of the male are in irregular rows (vs. regular horizontal rows). Although some authors have considered *F. blairae* a subspecies of *F. dispar* (Ross 2001), Ghedotti and Grose (1997) recognized *F. blairae* as a sister species of *F. dispar* based on genetic evidence. We follow Boschung and Mayden (2004), Nelson et al. (2004), Miller and Robison (2004), and Page et al. (2013) in continuing to recognize *F. blairae* as a distinct species. The common name of this species was formerly Blair's Starhead Topminnow, but Nelson et al. (2004) designated it as the Western Starhead Topminnow.

DISTRIBUTION　*Fundulus blairae* occurs in the Red River drainage of the lower Mississippi River basin from southeastern Oklahoma, southwestern Arkansas, and northeastern Texas to eastern Louisiana, with a few records from the lower Mississippi River drainage of southwestern Mississippi. It is also sporadically distributed in Gulf Slope drainages from the Escambia River, Florida and Alabama, west to the Brazos River, Texas (Warren and Denette 1989; Gilbert 1992; Ross 2001; Boschung and Mayden 2004). It is more common west of the Mississippi River and is rare cast of that river (Page and Burr 2011). Within its range, populations are sporadically distributed and generally small. The Western Starhead Topminnow was first reported from Arkansas by Robison (1977b). It occurs in Arkansas in only six counties in the southwestern part of the state in the lower sections of the western Saline, Cossatot, and Rolling Fork rivers (Connior et al. 2014), the Little River in Sevier, Howard, Little River, and Hempstead counties, and oxbow lakes, small tributaries, and backwaters of the Red River in Miller and Lafayette counties (Map 151).

HABITAT AND BIOLOGY　The Western Starhead Topminnow is found in quiet, heavily vegetated waters of small creeks and swampy backwater overflows of rivers. It also occurs in a few Red River oxbow lakes and is especially common in one Arkansas reservoir, Millwood Lake (Buchanan 2005). It prefers clear water and a soft mud and detritus bottom and is often found near concentrations of *Myriophyllum*, *Potamogeton*, and duckweed. We have observed that *F. blairae* is most commonly found in association with patches of water primrose (*Ludwigia* sp.) in Arkansas. Although it may be locally common in this habitat, it should be considered rare in Arkansas because of its very restricted distribution. It is replaced eastward (Ouachita River drainage through eastern Arkansas) in the lowlands of Arkansas by its close relative, *F. dispar*.

Miller and Robison (2004) suggested that *F. blairae* was undoubtedly a surface feeding insectivore. Gut contents of 30 specimens collected from Hempstead, Little River, and Miller counties in southwestern Arkansas in July and August were examined by TMB. The diet was dominated almost equally by terrestrial dipterans (44% frequency) and aquatic larval and pupal dipterans (41%) that were taken near the surface. Other food items in decreasing order of importance were ants, hemipterans, arachnids, coleopteran adults, coleopteran larvae, and other terrestrial insects. One individual had eaten insect eggs, and another

individual had ingested filamentous green algae. Gut contents from six specimens taken in April in Sevier County contained mostly terrestrial and aquatic dipterans, with hemipterans, arachnids, coleopterans, and odonates also ingested in small amounts.

There is pronounced sexual dimorphism in color pattern (see Life Colors section of this account) and body size. The largest individuals examined were all males (60–72 mm TL). Based on examination of gonads by TMB, *F. blairae* has a long breeding season extending from April through August. Multiple spawning is indicated by the presence of several size classes of developing ova early in the breeding season. Ripe ova possess filaments for attaching eggs to vegetation. Ripe females were found in Sevier County as early as April 9, and in Miller County as late as August 12. Eight females containing mature ova revealed that clutch size is small, usually with 10–16 ripe ova per female. One large female (51 mm TL) contained 40 mature ova. The smallest female containing mature ova was 37 mm TL (12 ova). Courtship and spawning have not been observed, but are probably similar to those observed by Pflieger (1997) for *F. dispar*.

CONSERVATION STATUS　This species has a small range in Arkansas. Populations are generally small, but most of them are stable. Because of its very small range in Arkansas, we consider it a species of special concern, and its status should be monitored periodically.

Fundulus catenatus (Storer)
Northern Studfish

Figure 23.2a. *Fundulus catenatus*, Northern Studfish male. *Uland Thomas.*

Figure 23.2b. *Fundulus catenatus*, Northern Studfish female. *Uland Thomas.*

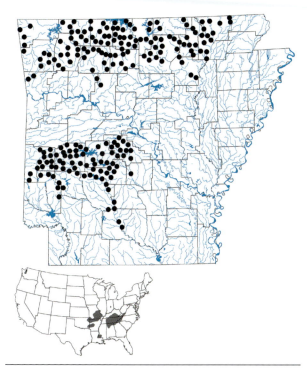

Map 152. Distribution of *Fundulus catenatus*, Northern Studfish, 1988–2019.

CHARACTERS   A small fish (but large *Fundulus*) with a slender, moderately compressed body, and a horizontal, terminal mouth (Fig. 23.2). Head large, broad, and flat; head length going into SL 3.1–3.4 times. Sides with horizontal streaks and without vertical bars. Front of dorsal fin situated directly over or slightly in front of anal fin origin. Lateral line absent; scales in lateral series usually 41–49. Scales around caudal peduncle 17–21. Dorsal rays 13–16; anal rays 15–18; pectoral rays 15 or 16; pelvic rays 5 or 6. Adult male with dorsal and anal fins elongate; mature female with an elongate anal fin. Breeding males with tubercles on side of head, body, and caudal peduncle and on dorsal, anal, and paired fins. Arkansas' largest topminnow, sometimes reaching 6 inches (152 mm) in length.

LIFE COLORS   Back yellowish brown. Sides bluish silver with 8–10 thin, brown horizontal lines that are often interrupted. A short gold stripe is present in front of dorsal fin. Dorsal and caudal fins dusky with small brown specks near bases; other fins usually unpigmented. Females and young are silvery with rows of small brown spots or dashes on sides, and with unmarked fins. Breeding males are brightly colored with bright blue sides with reddish-brown streaks; red spots also scattered on sides, fins, and especially on the operculum. Caudal fin with an orange margin and a black submarginal band in fewer than 50%

of males (Thomerson 1969). Paired fins of breeding males a lemon yellow.

SIMILAR SPECIES   It differs from all other Arkansas topminnows in having more than 40 lateral scales (vs. fewer than 40), 13 or more dorsal rays (vs. 10 or fewer), 15 or more anal rays (vs. 12 or fewer), and front of dorsal fin base situated over or in front of anal fin base (vs. front of dorsal fin base situated behind front of anal fin base).

VARIATION AND TAXONOMY   *Fundulus catenatus* was described by Storer (1846) as *Poecilia catenata* from Florence, Lauderdale County, Alabama. Variation was studied by Thomerson (1969), but no subspecies were recognized. Williams and Etnier (1982) diagnosed the subgenus *Xenisma* Jordan and placed *F. catenatus* in it. Cashner et al. (1992), based on an allozyme analysis, considered *F. catenatus* to be the sister species to *F. bifax*. An analysis of osteology and external morphology by Ghedotti et al. (2004) supported the monophyly of *Xenisma*, but concluded that *F. rathbuni* is sister to all other *Xenisma*. Hundt et al. (2017) used analyses of nuclear and mitochondrial genes to identify a cryptic, undescribed species within *F. catenatus*. According to that analysis, only populations in the Tennessee River drainage represent true *F. catenatus*. All other populations in the Ozark Highlands, Ouachita Highlands, White River of Indiana, Cumberland River, and southwestern Mississippi (Homochitto and Amite rivers) are an undescribed species, *Fundulus sp. cf. catenatus*. Within the undescribed species, Ouachita Highland populations represent a distinct clade, differing from remaining populations of *F. sp. cf. catenatus* by at least six nucleotide substitutions of the *cytb* gene. Lacking a formal description of that undescribed species, we herein continue to refer to Arkansas populations as *F. catenatus*.

DISTRIBUTION   *Fundulus catenatus*, as currently recognized, has a disjunct distribution, occurring in the Mississippi River basin east of the Mississippi River in the White River system of Indiana, upper Salt and Kentucky river drainages, Kentucky, upper Green, middle and lower Cumberland and Tennessee river drainages of Virginia, Kentucky, Tennessee, and northern Alabama, and in Mississippi River and Gulf Slope (Pearl River and Lake Pontchartrain) drainages in a small area of southwestern Mississippi. West of the Mississippi River, it is found primarily in the Ozark and Ouachita uplands of Arkansas, Missouri, and Oklahoma. In Arkansas, *F. catenatus* occurs in the Ouachita Mountains in the Red and Ouachita river drainages, in the Ozark Mountains in the White and Illinois river drainages, and in a few Arkansas River tributary creeks in western Arkansas (Map 152).

HABITAT AND BIOLOGY   The Northern Studfish is found in small sand- and gravel-bottomed clear mountain streams of moderate to high gradient and permanent flow as well as in medium-sized upland rivers, such as the Ouachita and White rivers. It is usually found in quiet, shallow waters along the margins of pools having rock and gravel bottoms, but it is sometimes found in slow to moderate current along stream margins and occasionally in shallow riffles. It is noted for its leaping ability to avoid predators, and Etnier and Starnes (1993) reported its aggressive nature toward other fishes when kept in aquaria. Despite its occurrence in a variety of stream sizes in Arkansas, and its reported tolerance of disturbance resulting from channelization, destabilization of the riparian corridor, and overgrazing (Pflieger 1997), it is apparently intolerant of reservoir conditions. Buchanan (2005) found the Northern Studfish in small numbers in only 2 (Bull Shoals and DeGray reservoirs) of 66 Arkansas reservoirs sampled.

Pflieger (1997) reported that the Northern Studfish feeds more from the stream bottom than other topminnows, a conclusion supported by our observations of Arkansas populations. McCaskill et al. (1972) also supported that observation and found that in Missouri its diet consisted of larval caddisflies and mayflies, adult beetles, and some mollusks. In Kentucky, Fisher (1981) found that most food items were taken from the substrate and water column of the stream, with peak feeding occurring in the morning and late afternoon. Immature and adult dipterans were the most important foods eaten by all age groups during autumn, spring, and summer months. Age 0 fish consumed crustaceans, *Hydracarina*, rotifers, gastropods, nematodes, and Hymenoptera. Major foods eaten by older fish were Ephemeroptera nymphs, Trichoptera, Coleoptera adults, nematodes, and Hymenoptera (Fisher 1981). Stomach contents of specimens from Ouachita Mountain streams (Saline and Little Missouri river drainages) in Arkansas did not support benthic feeding as an important contributor to the diet (Matthews et al. 2004). In that study, *F. catenatus* fed mainly at the water surface and in the water column, with terrestrial invertebrates making up an important percentage of the diet. Stomach contents of specimens from the Ouachitas were examined by TMB. During late fall and winter, the diet in the Saline River was dominated by benthic foods, with chironomids and isopods making up most of the diet; other items ingested were copepods, amphipods, ephemeropteran nymphs, dipteran pupae, and adult hemipterans and coleopterans. The summer diet in the Cossatot River was heavily dominated by trichopteran

larvae. The larval cases consisting of cemented sand grains (*Helicopsyche* and other genera) were ingested intact. Other food items in the Cossatot River specimens in decreasing order of percent frequency of occurrence were hemipteran adults, coleopteran larvae and adults, ephemeropteran nymphs, adult dipterans, and isopods. Specimens from the Ozarks examined by TMB from late spring and summer samples fed slightly more at the surface than those from the Ouachitas, they but were still largely benthic-oriented in their feeding habits. More than one-third of the stomachs examined contained sand grains, which indicated benthic feeding. The diets of Ozark specimens were dominated by chironomids and other dipteran larvae, ephemeropteran nymphs, adult coleopterans, and other terrestrial insects. Also present in small amounts were coleopteran larvae, odonates, trichopterans, other aquatic insects, amphipods, isopods, and fish remains (unidentified cyprinids). Tumlison et al. (2015) reported a small *Luxilus pilsbryi* from the gut of a *F. catenatus* from Crooked Creek in Marion County; other food items included crayfish (*Orconectes* sp.) and a terrestrial hemipteran (*Melanolestes picipes*).

This topminnow has a protracted spawning period in Arkansas, breeding from May to August. HWR collected nuptial males in Arkansas as early as 2 May, and females with ripening yellow eggs as late as 2 August. Specimens from the Spring River examined by TMB contained ripe ova on 8 May, and the presence of smaller maturing eggs indicated multiple spawning during the breeding season. Breeding also occurred from May to August in the Missouri Ozarks (Pflieger 1997), and from April into July in Kentucky (Fisher 1981). The males establish small territories in shallow, quiet waters. No nest is constructed, and the eggs are deposited on the gravel bottom. Pflieger (1997) observed a pair of studfish spawning in a Longear Sunfish nest. Fisher (1981) reported that the Northern Studfish probably spawns over extended periods of time and eggs reach 3+ mm in diameter when mature. Twenty-nine females sampled by Fisher had 28–245 ripening eggs 0.5–3.5 mm in diameter. *Fundulus catenatus* is the largest topminnow in Arkansas and the longest lived. Fisher (1981) found that it lives longer than other topminnows of its genus in Kentucky, with a maximum life span of at least 5 years. In Kentucky, this topminnow attained lengths of 85–115 mm. Helminth parasites of *F. catenatus* in Arkansas were studied by McAllister et al. (2015d).

CONSERVATION STATUS   Populations of this species are clustered in the Ozark and Ouachita mountains. It is widespread and common in those regions, and most populations are secure.

Fundulus chrysotus (Günther)
Golden Topminnow

Figure 23.3a. *Fundulus chrysotus*, Golden Topminnow male. *David A. Neely.*

Figure 23.3b. *Fundulus chrysotus*, Golden Topminnow. female. *Uland Thomas.*

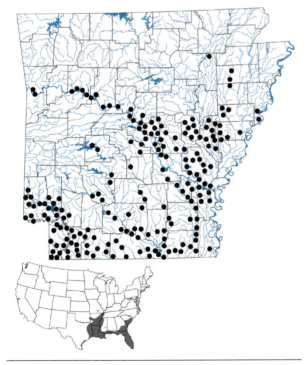

Map 153. Distribution of *Fundulus chrysotus*, Golden Topminnow, 1988–2019.

CHARACTERS A small, compressed topminnow with a deep caudal peduncle, a slightly arched back, and no lateral band or suborbital bar (Fig. 23.3). Mouth small and superior. Head length going into SL 3.4–3.6 times. Sides plain in females and with 6–10 narrow vertical bars in males. Front of dorsal fin base situated slightly behind front of anal fin base. Lateral line absent; scales in lateral series 31–35. Scales around caudal peduncle usually 18–20. Dorsal rays 7–10; anal rays 9–11; pectoral rays 14–16; pelvic rays 6. Typically 7 or 8 long gill rakers on first gill arch. During the breeding season, males develop tubercles on posterior dorsal fin rays, anal rays, outermost pelvic rays, and on many lateral scales. Maximum size around 2.8 inches (71 mm) (Foster 1967).

LIFE COLORS Males yellowish green with reddish-brown spots, smaller golden flecks, and large males typically with 8–11 green or dusky vertical bars on side. Fins of males white to yellow with reddish-brown spots. Underside of head and body light-colored. Breeding males becoming more intensely colored, with red spots on body and on unpaired fins, and with a red or red-orange caudal fin. Both sexes with a gold predorsal stripe and gold nasal spots. Nonbreeding females and young are more subdued in coloration, with plain, uniform olive-green sides; sides without the flecks, bars, and red spots of the male. Breeding females are yellowish with scattered small gold spots.

SIMILAR SPECIES It differs from *F. blairae* and *F. dispar* in lacking a dark suborbital bar (vs. dark teardrop present), and from *F. notatus* and *F. olivaceus* in lacking a wide, dark lateral band (vs. prominent, dark lateral stripe present). It differs from *F. catenatus* in having fewer than 40 lateral scales (vs. more than 40 lateral scales), usually fewer than 10 dorsal rays (vs. more than 10 dorsal rays), fewer than 12 anal rays (vs. 12 or more anal rays), and in lacking horizontal streaks on sides (vs. horizontal streaks present).

VARIATION AND TAXONOMY *Fundulus chrysotus* was originally described by Günther (1866) as *Haplochilus chrysotus* from Charleston, South Carolina. Because of strong sexual dimorphism, David Starr Jordan in 1880 erroneously described the female of *F. chrysotus* as a separate species, *Zygonectes henshalli*. The Golden Topminnow is placed in subgenus *Zygonectes*, and Wiley (1986) recognized *F. sciadicus* as the sister to the *F. notti* group, with *F. chrysotus*, *F. cingulatus,* and *F. luciae* forming an unresolved basal polyotomy in *Zygonectes*. Allozyme data support a reversal of the phylogenetic positions proposed by Wiley for these two clades, with the *F. chrysotus* clade sister to the *F. notti* group and *F. sciadicus* as basal (Cashner et al. 1992). A combined morphological and genetic analysis by

Ghedotti and Grose (1997) recognized *F. chrysotus* as the basal member of subgenus *Zygonectes*.

DISTRIBUTION *Fundulus chrysotus* occurs in the lower Mississippi River basin from southeastern Kentucky and southwestern Missouri, south through Louisiana to the Gulf of Mexico; Atlantic Slope drainages from the Waccamaw River, South Carolina to southern Florida; and Gulf Slope drainages from Florida to the Trinity River drainage, Texas. Although widely scattered in all major drainages of the Coastal Plain lowlands in southern and eastern Arkansas and extending through the Arkansas River Valley into Oklahoma, the Golden Topminnow is most common in the southern part of the state (Map 153). It is far more widespread and common in Arkansas than previously recognized, and McAllister et al. (2006) reported 98 new locality records for *F. chrysotus* from 27 counties. Additional records were reported from Grant County (McAllister et al. 2010b), Saline and Little River counties (Connior et al. 2014), and Lincoln and Desha counties (Tumlison et al. 2015).

HABITAT AND BIOLOGY The Golden Topminnow is strictly a lowland species inhabiting oxbow lakes, sluggish areas of creeks, and swampy backwater overflows of rivers. It is usually found in quiet, shallow water in or near aquatic vegetation over a soft mud and detritus bottom and often occurs with *F. dispar* or *F. blairae*. It is abundant in some Red River oxbow lakes, and Buchanan (2005) reported it from 11 Arkansas impoundments. *Fundulus chrysotus* has an upper salinity tolerance of 20–24 ppt (Griffith 1974), but there are only a few records from brackish waters along the Mississippi coast (Ross 2001).

Hunt (1953) reported that in Florida it fed mainly on insects and other aquatic invertebrates near the surface, but the diet also included crawling water beetles (Haliplidae) and midge larvae (Chironomidae). Goldstein and Simon (1999) classified it as an invertivore feeding mainly on drifting insects at the surface. In Reelfoot Lake, Tennessee, foods were reported as plants, small crustaceans (amphipods and isopods), mites, and gastropods (Sisk 1973). Gut contents of 48 specimens collected in July and August from four Arkansas counties were examined by TMB. Approximately 77% of the diet consisted of aquatic invertebrates (mostly immature insects) and 33% consisted of terrestrial insects (mostly dipterans). By percent frequency of occurrence, the diet was dominated by aquatic dipterans (31%), coleopteran larvae (25%), terrestrial dipterans (19%), and coleopteran adults (16%). Other food items ingested in smaller amounts were hemipterans, odonates, ephemeropterans, ostracods, cladocerans, copepods, and annelids. One specimen had ingested a small *Gambusia affinis*.

Like other species of *Fundulus*, *F. chrysotus* apparently has a long breeding season. Breeding has been reported from April into September throughout its range (Foster 1967; Hellier 1967; De Vlaming et al. 1978; Boschung and Mayden 2004). The breeding season in Arkansas occurs from April to September. Ripe females were found on 22 April in Union County (Tumlison et al. 2017). HWR observed spawning on 11 July in Cane Creek State Park, Lincoln County, Arkansas. Gonads of specimens collected between 25 July and 17 August from Arkansas, Hempstead, and Miller counties were examined by TMB. Eighteen females with ripe ova ranged in TL from 38 to 70 mm ($\bar{x}$ = 55.4). Clutch size ranged from 13 to 134 ripe ova, with a mean clutch size of 54.6 ripe ova per female. Smaller egg sizes were also present in the ovaries. Leitholf (1917) observed aquarium spawning and reported that eggs were released and fertilized one at a time and were deposited on submerged plants, stones, and the sides of the aquarium; spawning pairs seemed to prefer the roots of floating plants. The eggs have a yellow tinge and adhesive threads for attachment and hatched in 10–15 days at a water temperature of about 24°C (75°F) (Leitholf 1917). Fecundity is 10–20 fertilized eggs per day (Foster 1967). Boschung and Mayden (2004) provided the following description of spawning behavior in an aquarium as reported by Foster (1967):

> During courtship, the male displays to the female by looping and circling while he remains above her, making no contact at this point. When the female is ready to spawn, she nips the substrate (often *Ceratophyllum*) and immediately postures against it while pressing her anal fin against the same spot she nipped. The male, who has remained close above the female, maneuvers into a position beside her. Each spawning episode lasts about a second. Eggs are deposited a few at a time over a period of a week or more.

Newly hatched larvae are at first nonpelagic, resting either on vegetation or on the substrate. Growth of larvae is rapid, with individuals reaching 30 mm SL within three months, and the young soon shift to a pelagic existence and are sexually mature by 10 months after hatching. Maximum life span is about 2 years (Foster 1967).

CONSERVATION STATUS New records reported for the Golden Topminnow by McAllister et al. (2006) showed that it is known from 27 Arkansas counties. Populations appear to fluctuate from year to year in some Red River oxbows, but they can sometimes be quite large. Even though it has declined in some parts of its range (Boschung and Mayden 2004; Etnier and Starnes 1993;

Pflieger 1997), Arkansas currently has some of the largest, most stable populations. We consider this species secure in Arkansas.

Fundulus dispar (Agassiz)
Starhead Topminnow

Figure 23.4. *Fundulus dispar*, Starhead Topminnow. *Uland Thomas.*

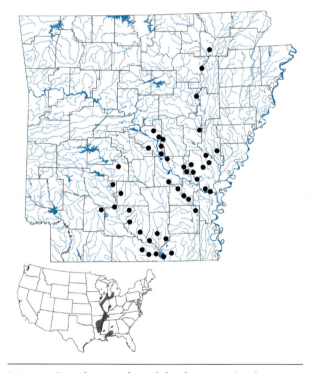

Map 154. Distribution of *Fundulus dispar*, Starhead Topminnow, 1988–2019.

CHARACTERS   A small, deep-bodied topminnow with a prominent, dark suborbital "tear-drop" (a wedge-shaped bar extending downward below each eye) (Fig. 23.4). Males with 3–13 dark vertical bars and 3–6 horizontal rows of dots on each side; females without vertical bars, but with 7–9 horizontal stripes on side. Front of dorsal fin base situated slightly behind front of anal fin origin. Mouth small, oblique, and opening dorsally. Head length going into SL 3.5–4.0 times. Top of head with scale pattern in which the

large anterior scale overlaps the two scales just posterior to it. Lateral line absent; scales in a lateral series usually 32 (31–34). Scales around body usually 25 or 26; scales around the caudal peduncle 18–20. Dorsal rays 7–9; pectoral rays 13 (12–14); pelvic rays 6; anal rays 10 (9–11). Intestine short. Breeding males of the F. *notti* species group develop tubercles on some dorsal and anal fin rays (Etnier and Starnes 1993), but we have found minute tubercle development only on the first anal ray of a few breeding males of *F. dispar* in Arkansas. Maximum size reportedly around 3 inches (78 mm) TL, but Arkansas specimens are much smaller, rarely reaching 2.5 inches (64 mm) TL.

LIFE COLORS   Back greenish yellow. A large golden spot on top of head. Sides yellow with 7–9 thin, brown horizontal stripes in the female; stripes usually continuous dorsally and broken ventrally; space between stripes with little or no pigment. Sides of males yellow with brown horizontal stripes on ventral half, the stripes often broken; upper half of sides with rows of red dots; sides with 3–12 (usually 6–9) narrow dark green vertical bars. Sides of females with vertical bars faint to well developed, but often absent in large females. Young of both sexes with dark vertical bars on sides. A black, wedge-shaped suborbital bar present in both sexes and well developed in adults and young. There is a narrow, dark, predorsal stripe. All fin rays are bordered with dark pigment. Pectoral and pelvic fins of both sexes without blotches. Dorsal, anal, and caudal fins of males with rows of dark spots.

SIMILAR SPECIES   The Starhead Topminnow, though closely related to *F. blairae*, differs from that species in the males having vertical bars on sides (vs. males of *F. blairae* without vertical bars on sides) and in the females having little or no pigmentation between the horizontal lateral stripes (vs. females having numerous dashes and dark pigment between lateral stripes); dots on upper sides only of males, those dots in regular horizontal rows (vs. dots on upper and lower sides in irregular rows); usually 25 or more scales around the body (vs. 24 or fewer); and young of both sexes with vertical bars on sides (vs. young of both sexes lacking vertical bars). It differs from all remaining topminnows in Arkansas in having the dark wedge-shaped vertical bar under the eye (vs. dark teardrop absent).

VARIATION AND TAXONOMY   *Fundulus dispar* was described by Agassiz (1854) as *Zygonectes dispar* from creeks near East St. Louis, Illinois and the Illinois River near Beardstown, Illinois. It was subsequently considered a subspecies of *Fundulus notti* until Wiley and Hall (1975) and Wiley (1977) recognized *F. dispar* as a distinct species. Wiley (1977) studied geographic variation in *F. dispar* and reported that Ouachita River drainage populations have lower numbers of scales around the body (mode of 25) and around the caudal peduncle (mode of 18) than more northern populations (26 and 20 scales, respectively). Also, males from the Ouachita drainage are more variable in number of vertical bars, having as few as 3 compared with 7 or more in northern populations. Wiley recognized no subspecies. Some recent authors do not recognize *F. dispar* and *F. blairae* as separate species, but treat both as subspecies of *F. dispar* (Ross 2001). We continue to recognize the specific distinctness of *F. dispar* and *F. blairae*, as do most other recent authors (Boschung and Mayden 2004; Nelson et al. 2004; Page and Burr 2011; Page et al. 2013).

DISTRIBUTION   *Fundulus dispar* occurs in the Great Lakes (Lake Michigan) and Mississippi River basins from Michigan and Wisconsin south to the Ouachita River drainage in Louisiana and the Big Black River, Mississippi, and in Gulf Slope drainages from Mobile Bay, Alabama, to the Pearl River, Mississippi. In Arkansas, this topminnow is primarily a Coastal Plain species found in the Ouachita, eastern Saline, lower Arkansas, lower White, Black (McAllister et al. 2010b), and St. Francis river drainages in southern and eastern Arkansas (Map 154). There are no recent records from the St. Francis River drainage and only one recent record from the lower Saline River. It is replaced in the Red River drainage of southwestern Arkansas by *F. blairae*.

HABITAT AND BIOLOGY   The Starhead Topminnow is a lowland species found in the quiet, sluggish waters of creeks, oxbow lakes, and swampy backwater overflows of rivers. Buchanan (2005) found it in small numbers in three Arkansas impoundments. It prefers clear water with a soft mud and detritus bottom and is almost always found near heavy growths of submerged aquatic vegetation such as *Myriophyllum* and *Potamogeton* and floating vegetation such as duckweed. In Arkansas, it is often found in association with *F. chrysotus*. Goodyear (1970) reported that when pursued by a bass or other predator, it sometimes jumps onto the bank, returning to the water when the predator leaves the area. Pflieger (1997) noted that it normally occurs singly or in pairs swimming just beneath the water surface. He also noted that it will not dive even when pursued. Its nongregarious tendencies may at least partly explain why it is not represented in large numbers in collections at an individual locality in Arkansas. Rotenone population sampling produced 111 specimens from three impoundments, Cane Creek, Champagnolle Creek, and Felsenthal (Buchanan 2005).

Forbes and Richardson (1920) found that in Illinois it fed mainly on insects near the water's surface, about half of which were terrestrial forms. Also in Illinois, Gunning and Lewis (1955) reported that in addition to insects it also feeds on snails, crustaceans, and algae. Etnier and Starnes (1993) found that stomachs of Tennessee specimens contained mainly terrestrial insects. In Reelfoot Lake, Tennessee, 31% of *F. dispar* stomachs contained mosquito larvae (*Anopheles*) that formed 10% by volume of the total food (Barnickol 1941). Twenty specimens collected from five Arkansas counties in July and August were examined by TMB. The diet was dominated by terrestrial dipterans and other insects (62% frequency of occurrence). Forty-six percent of the Arkansas specimens had ingested insect eggs, and the remainder of the diet comprised largely aquatic larvae of dipterans.

Spawning occurs in late spring to early summer among dense aquatic vegetation. Pflieger (1997) observed courtship behavior in early May in southeastern Missouri, where females were closely pursued by males for distances of 20–30 feet. Pflieger further noted that gravid females were observed being pursued by males in a pond-reared population near Columbia, Missouri, as late as 23 August, indicating an extended spawning period and probable maturation and spawning during the first summer of life. In Arkansas, spawning extends from April through August. Based on examination of gonads of Arkansas specimens by TMB, clutch size is small, ranging from 4 to 43 ripe ova per female, and multiple spawning occurs. Females with ripe eggs have been found as early as 22 April in Union County (Tumlison et al. 2017), and as late as mid-August in Lincoln County. Ripe eggs possess a filament for attaching them to vegetation. The smallest female (from Calhoun County) to contain ripe ova was 37 mm TL (4 ova), and the largest (from St. Francis County) was 56 mm TL (43 ova). In southern Illinois, peak reproductive readiness (based on gonadosomatic index values) occurred at water temperatures ranging from 17°C to 30°C (Taylor and Burr 1997); fertilized eggs hatched in 9–11 days at a water temperature of approximately 25°C. That study further revealed a multiple spawning strategy and other aspects of reproductive biology similar to other members of the genus *Fundulus*.

CONSERVATION STATUS   *Fundulus dispar* is sporadically distributed in the lowlands of Arkansas and is most common in the southern part of the state. Nowhere in Arkansas is it known to be abundant, but we consider its populations currently secure. Although there is no current evidence of a shrinking range in Arkansas, the Starhead Topminnow should be monitored to detect any future loss of populations.

Fundulus notatus (Rafinesque)
Blackstripe Topminnow

Figure 23.5. *Fundulus notatus*, Blackstripe Topminnow. *Uland Thomas.*

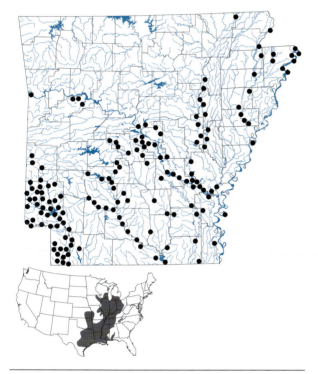

Map 155. Distribution of *Fundulus notatus*, Blackstripe Topminnow, 1988–2019.

CHARACTERS A slender, terete topminnow with a slender caudal peduncle and a wide, dark, horizontal lateral band extending the length of the body from the snout to the caudal fin base (Fig. 23.5). Front of dorsal fin base situated behind front of anal fin base. Upper sides without black spots or with spots that are irregular in outline and lighter in color than the black lateral stripe. Mouth small, oblique, and opening dorsally. Head length going 3.5–4.0 times into SL in adults. Lateral line absent; scales in lateral series usually 31–35 ($\bar{x}$ = 33.9) in Arkansas populations. Dorsal rays usually 9 (8–10); anal rays usually 12 or 13; pectoral rays 11–13; pelvic rays 6 or 7. Intestine short and

straight, or with a single loop anteriorly. Males are deeper-bodied than females, have larger dorsal and anal fins (the posterior rays of those fins become elongated), and have a serrate lateral band. Breeding males develop tubercles on anal fin rays and on some lateral body scales. Maximum size around 3 inches (76 mm).

LIFE COLORS Back and upper sides olive brown, sometimes with a few small, scattered, diffuse, and indistinct spots which, if present, are lighter in color than the blue-black lateral band. The lateral band is usually straight-edged in females but has vertical streaks extending from it in males. The width of the lateral band varies from 23% to 30% of the body depth, and it covers the upper 25–33% of the pectoral fin base. Undersides white but may be tinged with yellow in breeding males. Dorsal and caudal fins with dark dots; pectoral and pelvic fins plain or dusky, without dots.

SIMILAR SPECIES Often confused with the morphologically similar *F. olivaceus*, but differs from it in lacking prominent black spots on upper sides, or if present, spots are diffuse, irregular in shape, and lighter than the lateral band (vs. upper side with discrete, scattered small black spots). The wide, dark, lateral band distinguishes this species from all remaining Arkansas topminnows.

VARIATION AND TAXONOMY *Fundulus notatus* was described by Rafinesque (1820a) as *Semotilus notatus* from the Cumberland and Little rivers, Kentucky. It was later transferred to genus *Zygonectes* and then to genus *Fundulus*. This species is usually placed in the subgenus *Zygonectes* (Suttkus and Cashner 1981). Thomerson (1966) studied morphological variation throughout the ranges of *Fundulus notatus* and *F. olivaceus* and compared their biosystematics. He regarded the two as sibling species and found that the only morphological feature that will serve to separate the two species everywhere is the degree of development of dorsolateral spots, although other characters may be of use in separating these species at a given locality. Degree of development of dorsolateral spots is the only reliable morphological character for separating those two species in Arkansas. Hubbs and Burnside (1972), based on developmental and karyological data, concluded that *F. notatus* and *F. olivaceus* were sufficiently distinct from other *Fundulus* to warrant generic separation; however, this conclusion has not been followed by recent workers. Pflieger (1997) used lateral scale counts, in addition to dorsal spotting patterns, for separating *F. notatus* (usually 31–34 lateral scales) and *F. olivaceus* (usually 34–36 lateral scales) in Missouri. Arkansas *F. olivaceus* tend to have higher lateral scale counts ($\bar{x}$ = 34.7) than Arkansas *F. notatus*

($\bar{x}$ = 33.9), but there is much overlap in this character between the species; therefore, it is an unreliable tool for separating those species in this state. Welsh and Fuller (2015) found sexual dimorphism in the size and shape of anal and dorsal fins. Males have longer and more-pointed anal fins and longer, larger, and more-pointed dorsal fins than females (most noticeable in older age classes). Thomerson (1966) recognized no subspecies of *F. notatus*. Populations of *F. notatus* have a diploid chromosome number of 40, except for those in the Tombigbee River system, which have 44 chromosomes (Setzer 1970; Black and Howell 1978). The morphologically similar *F. olivaceus* has 48 chromosomes.

Boschung and Mayden (2004) considered the occurrence of natural hybrids between *F. notatus* and *F. olivaceus* to be rare, but they pointed out that the two species are interfertile in laboratory crosses. Despite the differences in chromosome numbers, hybridization between the two species has been widely reported in syntopic natural populations (Thomerson 1967; Setzer 1970; Gutierrez 2010; Schaefer et al. 2011b; Duvernell et al. 2013). Vigueira et al. (2008) assessed prezygotic (probability of spawning) and postzygotic (hatching success) reproductive isolation between the two species and found that prezygotic barriers were strong and postzygotic barriers were weak in crosses of nonhybrid heterospecifics. They also found that prezygotic barriers were weaker and postzygotic barriers stronger in crosses involving hybrid individuals. Duvernell et al. (2007) used nuclear and mitochondrial DNA to assess levels of hybridization and test for introgression in syntopic populations of these two species in four drainages in southern Illinois. They found that although hybridization was detected in all syntopic populations, an assessment of the proportion of hybrid individuals indicated a deficiency of hybrids relative to expectations under random mating. They further determined that although mtDNA introgression was prevalent and extended beyond the zones of contact, evidence of nuclear introgression was limited to the zone of sympatry. Schaefer et al. (2011b), Duvernell and Schaefer (2014), and Duvernell et al. (2013) found that local ecology clearly influenced contact zone structure and how the two species interacted. The system with the strongest ecological gradient (Pearl River) had no hybridization, while the system with the weakest ecological gradient (Sabine River) had the most individuals of hybrid origin.

DISTRIBUTION *Fundulus notatus* occurs in the Great Lakes (Erie, Huron, and Michigan) and Mississippi River basins from Ontario, Canada, to Michigan, Wisconsin, and Iowa, south through Kansas, Oklahoma, eastern Texas, and Louisiana, and in Gulf Slope drainages

from Mobile Bay, Alabama, west to San Antonio Bay, Texas. In Arkansas, it is most common below the Fall Line in Coastal Plain streams of southern and eastern parts of the state. It extends throughout the Arkansas River Valley and is also found to some extent in the Ouachita Mountains above the Fall Line (Map 155). It is absent from most of the Ozarks of northern Arkansas, but an older record exists from the Illinois River drainage in the northwestern corner of the state (Map A4.153). There are no recent verified records from northwestern Arkansas, but there are numerous records from the neighboring Neosho-Illinois drainages of Missouri, Kansas, and Oklahoma.

HABITAT AND BIOLOGY The Blackstripe Topminnow is found primarily in small to large low-gradient streams of moderate to high turbidity, in contrast to *F. olivaceus* which prefers clear waters. In the largest Coastal Plain rivers of Arkansas, it is more abundant than *F. olivaceus*. Even though the two species have similar ecological niches, they are seldom found in the same habitat even when both inhabit a single small stream (Thomerson and Wooldridge 1970). Pflieger (1997) reported that in areas where both species occur, one or the other usually predominates at any given locality. Pflieger believed that this apparent competitive exclusion was influenced by local habitat characteristics, and turbidity was considered very important in influencing the outcome of competitive interactions between the two species. However, Thomerson and Wooldridge (1970) believed that *F. notatus* may have a competitive advantage in feeding over *F. olivaceus* because of its more aggressive feeding behavior. Schaefer et al. (2011b) and Duvernell and Schaefer (2014) also believed that competition influenced observed distributional patterns of the two species. Those authors concluded that where the two species are sympatric, *F. olivaceus* may be excluding *F. notatus* from tributary streams, but competitive forces are not strong enough for *F. olivaceus* exclusion from rivers. They noted that river sites (in four Gulf Coast drainages) featuring larger and more complex habitats that included *F. olivaceus* always contained *F. notatus* as well. Furthermore, in allopatry, both species readily used the other's preferred habitat (Schaefer et al. 2011b). In headwater streams of the Saline River in southern Illinois, differences in habitat preferences between the two species were translated into measurable but relatively subtle differences in population genetic structure (Earnest et al. 2014). *Fundulus notatus* is found in small numbers in some clearer upland streams. It prefers the quiet backwaters and pool margins and can be found over a variety of bottom types. Becker (1983) noted that in Wisconsin it is rarely found in lakes, but Buchanan (2005) reported this species in small numbers from nine Arkansas impoundments. In

Cahokia Creek, Illinois, individuals moved an average of nearly 23 meters per day (Alldredge et al. 2011). Welsh et al. (2013) found that stream populations of age 1 *F. notatus* had larger body size than lake populations in Illinois. It is reportedly tolerant of high temperatures and low dissolved oxygen, and has a high critical thermal maximum of 41.6°C (Rutledge and Beitinger 1989). As with many other cyprinodontiform fishes, its ability to use oxygen in the thin surface layer of water along with its high temperature tolerance permits it to occupy isolated summer pools where few other fishes can survive (Ross 2001).

Fundulus notatus is a surface feeder (Thomerson and Wooldridge 1970), with terrestrial insects composing half its diet; the rest of its diet comprises aquatic insects, crustaceans, snails, and algae (Atmar and Stewart 1972; Shute 1980). Algal strands apparently pass through the intestine undigested, and filamentous algae is ingested presumably to obtain the invertebrates hidden in it (Atmar and Stewart 1972). In addition to terrestrial arthropods, young topminnows also feed on cladocerans and copepods. The species feeds most actively in early morning and late afternoon and functions as a third- or fourth-level predator. Spring and summer gut contents of 64 adult *F. notatus* from 7 Arkansas counties were examined by TMB, and the data were pooled because of diet similarities across localities and collection dates. In that study, the Blackstripe Topminnow fed almost exclusively at or near the surface, with only a very small percentage of food items taken near or from the bottom. The diet was dominated by percent frequency of occurrence and volume by terrestrial insects, especially mosquitoes, midges, flies, and other dipterans. Other important terrestrial insects consumed were ants, coleopterans, and orthopterans. Aquatic organisms consumed included hemipterans, cladocerans, copepods, dipteran larvae and pupae, late-instar odonate nymphs, and coleopteran larvae.

Breeding occurs from March to July in Mississippi (Cook 1959). Other authors have also referred to a late spring to summer breeding season (Etnier and Starnes 1993; Pflieger 1997). Based on examination of gonads by TMB, a long spawning season occurs in Arkansas at least from late April through August. Clutch size is typically small, ranging from 9 (in a 49 mm TL female) to 56 ripe eggs (in a 76 mm TL female). The presence of multiple egg sizes in ovaries from April through July indicated multiple spawning by females. A few ripe females in August samples also had smaller developing ova. Females with ripe eggs possessing filaments were found in Frazier Lake (Arkansas County) on 21 August and in the Sulphur River (Miller County) on 22 August, and some spawning activity may extend into September in some years. Spawning behavior was described by Carranza and Winn (1954) in Michigan. Both sexes defend a weakly defined territory near the shore. The male follows slightly below and behind the female as she swims along the shoreline. When ready to spawn, the female goes to the vegetation with the male following. The two spawn side by side, and the eggs are deposited singly on algae, other aquatic vegetation, or leaf litter and detritus when vegetation is absent. The eggs are attached by adhesive filaments to the vegetation. The female usually deposits 20–30 eggs at each spawning, and the eggs are not guarded by the parents. The life span is short, with few individuals surviving beyond their second summer (Nieman and Wallace 1974). Helminth and crustacean parasites of *Fundulus notatus* in Arkansas were reported by McAllister et al. (2016).

CONSERVATION STATUS   Accurate identification of this species is difficult because of its great morphological similarity to the much more common and abundant *F. olivaceus*. Misidentifications have frequently occurred. *Fundulus notatus* is locally abundant in the lowlands of southern Arkansas, and its populations are secure.

Fundulus olivaceus (Storer)
Blackspotted Topminnow

Figure 23.6. *Fundulus olivaceus*, Blackspotted Topminnow. *Uland Thomas.*

CHARACTERS   A slender, terete topminnow with a slender caudal peduncle and a wide, dark, horizontal, lateral band extending the length of the body from the snout to the base of the caudal fin (Fig. 23.6). Upper sides with a few to many small discrete black spots that are regular in outline and about as dark as the lateral stripe. Width of lateral stripe varies from 22% to 33% of body depth Front of dorsal fin base situated behind front of anal fin base. Head length going into SL 3.5–4.0 times. Lateral line absent; scales in lateral series 33–37 ($\bar{x}$ = 34.7) in Arkansas populations. Dorsal fin rays 9 or 10; anal rays usually 11–13; pectoral rays 13–15; pelvic rays 6. Dorsal and anal fins of males are more

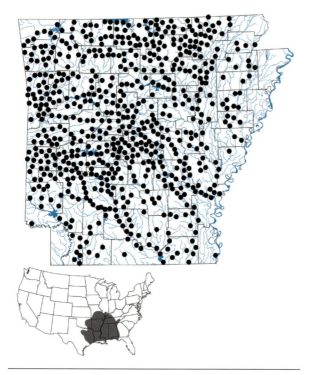

Map 156. Distribution of *Fundulus olivaceus*, Blackspotted Topminnow, 1988–2019.

elongate than those of females. Breeding males develop small tubercles on anal and dorsal fin rays and on some body scales. Adults are usually 2–3 inches (51–76 mm) long. Maximum size almost 4 inches (102 mm).

LIFE COLORS   Back and upper sides olive brown; upper sides with few to many small, discrete black spots. Spots generally more numerous in males than females; large males tend to have more spots than small males. Margins of the black lateral band are nearly straight in females and young, but margins are often uneven and with vertical streaks in males. Undersides white. Young with a faint predorsal stripe. Dorsal, anal, and caudal fins with small black spots; other fins plain or dusky. Breeding males have a bluish-black lateral stripe and yellowish dorsal, caudal, and anal fins; females lack coloration on fins.

SIMILAR SPECIES   Most like *F. notatus* but differs from it in having prominent small, black spots that are regular in outline on upper sides (vs. spots absent, or diffuse, irregular in shape, and lighter in color than the lateral band). It is distinguished from all remaining Arkansas topminnows by its wide, dark lateral band (vs. dark lateral band absent).

VARIATION AND TAXONOMY   *Fundulus olivaceus* was described by Storer (1845) as *Poecilia olivacea* (genus originally misspelled as *Paecilia*) from Florence,

Lauderdale County, Alabama. It is currently placed in the subgenus *Zygonectes* (Suttkus and Cashner 1981). Based on developmental and karyological data, Hubbs and Burnside (1972) concluded that *F. notatus* and *F. olivaceus* were sufficiently distinct from other *Fundulus* to warrant generic separation; however, this conclusion has not been followed by recent workers. Morphological variation in the Blackspotted Topminnow was studied by Thomerson (1966), who recognized no subspecies. *Fundulus olivaceus* has a diploid chromosome number of 48 while the similar *F. notatus* has 40 chromosomes; both species possess the same number of chromosome arms (Setzer 1970). Despite these genetic differences, Thomerson (1966) produced fertile hybrids between *Fundulus olivaceus* and *F. notatus* in the laboratory, and hybrids have been reported from natural populations in several states (Burr and Warren 1986b; Page and Burr 2011). In a study of syntopic populations of the two species in southern Illinois, Duvernell et al. (2007) found that introgression, based on nuclear DNA analysis, was limited to the zone of sympatry, while mtDNA introgression was prevalent and extended beyond the zones of contact. Specimens intermediate in appearance between these two species are often found in Arkansas.

DISTRIBUTION   *Fundulus olivaceus* occurs in the lower Mississippi River basin from Tennessee, western Kentucky, and southern Illinois, across the southern half of Missouri, south to the Gulf of Mexico, and in Gulf Slope drainages from the Apalachicola River drainage of Georgia and Florida, west to Galveston Bay, Texas. One of the most widespread and common fishes in Arkansas, the Blackspotted Topminnow is distributed statewide and occurs in all river drainages (Map 156). There are records of specimens from every Arkansas county except Miller County, but new records from that county would not be surprising.

HABITAT AND BIOLOGY   The Blackspotted Topminnow is found in a variety of habitats, including small headwater creeks to large rivers, impoundments, and oxbow lakes in both mountainous and lowland regions. In a study of small fish species of Arkansas impoundments, *F. olivaceus* was the third-most widely distributed species, occurring in 47 of 66 reservoirs studied (Buchanan 2005); however, it is more abundant in clear upland streams than in turbid lowland waters, and although apparently having wide ecological tolerances, it avoids extremely turbid waters. It is found primarily in quiet backwaters and pools and is often observed in small groups along stream margins near emergent vegetation such as water willow. It avoids swift current, but Etnier and Starnes (1993) noted that it prefers slightly faster-flowing streams than *F. notatus*. In the upper portion of Black Creek, Mississippi, it was found in habitats with an average current speed of 7.3 cm/s (2.9 inches/s) and an average depth of 53.5 cm (21 inches) (Ross et al. 1987). In Big Creek, Mississippi, *F. olivaceus* moved less than 1 meter per day (Alldredge et al. 2011). During floods, it moved out of the channel and foraged on the inundated floodplain (Ross and Baker 1983). Competition with *F. notatus* may play a role in determining the distribution and abundance of both species (Schaefer et al. 2011a) (see *F. notatus* species account). Earnest et al. (2014) summarized habitat differences between *F. olivaceus* and *F. notatus* when examined on a whole-stream landscape level. *Fundulus notatus* is typically found in downstream habitats with relatively low stream flow, mud or clay substrate, and with aquatic vegetation. *Fundulus olivaceus* in comparison is generally more abundant in typical headwater habitats with shallower and narrower channels, lower bank slope, larger substrate size, less emergent vegetation, more woody debris, and increased canopy cover.

It is a surface feeder primarily consuming terrestrial and aquatic insects, small crustaceans, and diatoms (Rice 1942). In a Mississippi creek, adult and immature aquatic insects composed most of the diet, and other foods eaten included beetles, amphipods, and spiders (Ross 2001). In Caney Bayou, Arkansas, the most frequently eaten items in summer in descending order were cladocerans, terrestrial insects, chironomids, detritus, ostracods, coleopterans, arachnids, algae, and gastropods (Neely and Pert 2000). In Ouachita Mountain streams of Arkansas, feeding occurred mainly at the surface and in the water column, with terrestrial invertebrates forming an important part of the diet (Matthews et al. 2004). Gut contents of 63 adults (54–81 mm TL) collected from March through September from nine Arkansas counties were examined by TMB. The diet consisted mainly of terrestrial insects and was also supplemented by aquatic insects taken near the surface. Very few benthic organisms were ingested. Adult dipterans and coleopterans were the most frequently consumed foods; also ingested in smaller amounts in decreasing order of importance were ants, larval and pupal dipterans, ephemeropterans, odonate nymphs, and isopods. Thomerson and Wooldridge (1970) believed that feeding is more leisurely in this topminnow than in the more aggressive *F. notatus*.

Pflieger (1997) reported that adults in breeding condition have been taken in May in Missouri. It apparently has a prolonged breeding season extending from May through August in Tennessee (Etnier and Starnes 1993). In Louisiana through Alabama, spawning season extended from March to September and peaked in May (Blanchard 1996; Boschung and Mayden 2004). Based on

examination of gonads by TMB, the breeding season in Arkansas extends from at least mid-April to mid-August. Ripe females were found in Pope County on 21 April and in Sebastian County on 23 April. Females (65–81 mm TL) from April through June contained from 47 to 130 mature ova. The number of mature ova in females decreased between July and mid-August, and by August, females contained few smaller developing ova. Courtship and spawning behavior in an aquarium were described by Ross (2001), and other aspects of its reproductive biology in the southeastern United States were presented by Blanchard (1996). Its reproductive habits are similar to those of *F. notatus*. Spawning occurs in vegetation, and fertilized eggs are attached to the vegetation by adhesive filaments (Carranza and Winn 1954). In aquaria, males spawned over a variety of spawning media and did not guard a limited territory (in contrast to the weak territorial behavior reported for *F. notatus*). One dominant male in a group sired all offspring (Gutierrez 2010). There is apparently sexual selection for male spotting pattern. In studies of female mate choice and predation pressure on spot phenotypes, Gutierrez (2010) and Schaefer et al. (2012) found that females preferred large males with a high number of dorsolateral spots. Larger males with more spots also mated more frequently than smaller males with fewer spots, but those larger, more spotted males suffered greater mortality due to predation. They further found that water clarity (turbidity) was the best predictor of spot density on the drainage scale, indicating that selection for this trait may be mediated by local light environments. This topminnow adapts well to aquarium life and readily accepts a wide range of food items (Thomerson and Wooldridge 1970). Maximum life span is about 3 years. Helminth parasites of *F. olivaceus* in Arkansas were reported by McAllister et al. (2015b).

CONSERVATION STATUS   This is the most widely distributed and abundant topminnow in Arkansas. Its populations are secure.

Family Poeciliidae *Livebearers*

This New World family, contains 42 genera and about 353 species (Nelson et al. 2016), with 93 species occurring in North America (Page and Burr 2011). It is especially rich in species in tropical and subtropical areas from Mexico southward into Central and South America (Schreiner 1989). Poeciliidae is represented in the United States by 14 native species, of which only one is native to Arkansas, the Western Mosquitofish (*Gambusia affinis*). The subfamily classification of this family has been changed based on Ghedotti (2000), and the species that are actually livebearers technically now form the subfamily Poeciliinae within the family Poeciliidae. This subfamily is unique among Arkansas fishes because of internal fertilization, sperm storage, and bearing of live young. Poeciliinae are sexually dimorphic (Bisazza et al. 2001; Evans et al. 2011), with males much smaller than females and possessing a gonopodium, a long slender modification of the anal fin, to transfer sperm to the female gonopore. Females are larger, can store sperm for several months (Constantz 1989), and are ovoviviparous (Parenti and Rauchenberger 1989). Females provision the oocyte with all needed resources prior to fertilization (lecithotrophic) and then retain the fertilized zygote internally without nutrient exchange until it hatches.

As males mature, the third, fourth, and fifth anal fin rays elongate to form the gonopodium on which distal armaments such as hooks, elbows, and serrae develop (Rosen and Gordon 1953; Chambers 1987). Despite the many studies investigating various structures of the gonopodium, little is known regarding variation and function of the more intricate structures (Clark et al. 1954; Stoops et al. 2013). These armaments are thought to be involved in either sexual conflict or as a lock-and-key isolating mechanism to avoid hybridization (Peden 1972; Eberhard 1985; Langerhans 2011; Stoops et al. 2013). Sperm storage by the female enables fertilization of successively developing clutches of ova from a single copulation. A single gravid mosquitofish may populate an entire fish pond. Peden (1973) demonstrated that the dark mark (gravid spot) on the posterior abdomen of females served as a thrusting target for males, with the mark increasing in size as ova mature. Females may produce several broods a year and the number of young increases as the female matures and gets larger. Egg diameters in poeciliids range from 0.8 to 3.6 mm (Thibault and Schultz 1978; Turner and Snelson 1984).

The largest poeciliid is *Belonesox belizanus*, which may attain a size of 20 cm, while Louisiana and other southeastern states are inhabited by the Least Killifish, *Heterandria formosa*, which may be the smallest vertebrate to bear live young (Robison 1968; Berra 1981). The livebearer family includes many tropical species popular in the aquarium trade such as guppies, mollies, platies, and swordtails. Livebearers closely resemble and are related to the topminnows of the family Fundulidae but differ in the shape of the anal fin of males, in having fewer scales along the mid-side, in the relative positions of the dorsal and anal fins, and in the structure of the sensory canals of the head. Although widely stocked throughout the world for mosquito control and considered an important invasive species across much of its current distribution (Schreiner 1989), the Western

Mosquitofish (and the Eastern Mosquitofish, *Gambusia holbrooki*) is sometimes less effective in controlling mosquitoes than the native fishes it often replaces (Pflieger 1997). Even though in some environments, it has proven to be a successful alternative to the use of pesticides for controlling mosquitoes, it often competes with or is a predator on various native species. For example, in the desert southwest of the United States, the introduction of this species caused the extirpation of several populations of rare and highly localized fishes that were unable to compete.

Genus *Gambusia* Poey, 1854—Gambusias

The large genus *Gambusia* contains about 45 species distributed from North America through Central America into northern South America, including several Caribbean islands. About 27 North American species occur in the southern United States extending westward and southward through Texas and eastern Mexico to Central America (Rosen and Bailey 1963; Page et al. 2013), 15 of which occur in Mexico (Miller et al. 2005). The genus is characterized by several features uniquely possessed by males, including the presence of knobby structures on segments of pectoral rays 3–6 and structures of the gonopodium (Etnier and Starnes 1993). A single species, *Gambusia affinis*, is known from Arkansas.

Gambusia affinis (Baird and Girard)
Western Mosquitofish

Figure 23.7a. *Gambusia affinis*, Western Mosquitofish male. *Uland Thomas.*

Figure 23.7b. *Gambusia affinis*, Western Mosquitofish female. *Uland Thomas.*

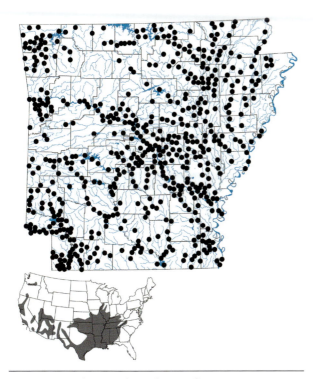

Figure 23.8. Gonopodium of male *Gambusia affinis*.
Uland Thomas.

Figure 23.9. Pregnant female *Gambusia affinis*.
Renn Tumlison.

Map 157. Distribution of *Gambusia affinis*, Western
Mosquitofish, 1988–2019. Inset shows native and introduced
populations.

CHARACTERS   The only livebearer native to Arkansas
(Fig. 23.7). It is a small, robust, surface-dwelling fish with
a terminal, upturned mouth resembling that of a topminnow, but easily distinguishable by its 6–8 dorsal rays (vs.
9 or more in topminnows), fewer than 10 anal rays (vs. 12
or more), and presence of a highly modified anal fin in the
male. Head flattened and wedge-shaped, with scales present
on top of head. Sensory canals on head are open. Eyes large,
eye diameter greater than length of snout. Body covered
with large cycloid scales. Belly of female often distended by
embryos. Pectoral fin rays 12–14; caudal fin rounded with
13–16 rays. Pelvic fin of male like that of female; usually
with 6 rays. Lateral line absent. Lateral scales large, 29–32
in lateral series. Third ray (counting rudimentary rays) of
anal fin unbranched in male and female. Males are small
and slender with the anal fin far forward and modified as
a gonopodium (Fig. 23.8), while females have much larger
and deeper body and normal anal fin in shape and position. Distal segments of dorsal edge of the first principal ray
of gonopodium of male without serrations. Reproductive
opening of female on end of large papilla within an open
urogenital sinus (Peden 1972, 1975). Females often have
a dark gravid spot on each side of belly above the vent.
Maximum total length of females about 2.5 inches (63
mm); males reach 1.2 inches (30 mm) TL.

LIFE COLORS   Color olive above, white below. A
blue-black suborbital bar present. Scales outlined with dark
pigment, giving a crosshatched appearance. Usually without discrete dark spots or stripes on sides. Dorsal and caudal fins usually with 1–3 rows of dark spots. Predorsal stripe
conspicuous, median fins with flecks of pigment forming
faint bars. Dark blotch (gravid spot) present near anus in
pregnant females (Fig. 23.9).

SIMILAR SPECIES   The Western Mosquitofish is
unique among Arkansas fishes in the males having an anal
fin modified into an elongate intromittent organ (gonopodium), and in pregnant females having a black gravid spot.
Fundulus dispar and *F. blairae* are sometimes confused
with *G. affinis* (both of which, like *G. affinis*, have a black
suborbital bar), but *G. affinis* can easily be separated from
those species by the lack of horizontal streaks along sides
(vs. horizontal streaks along the sides and with vertical
bars in males), a gonopodium in males (vs. males lacking
a gonopodium), and dorsal fin well behind anal fin origin
(vs. origin of the dorsal fin almost directly above rather
than well behind the anal fin origin). It is possible that the
Eastern Mosquitofish, *Gambusia holbrooki*, may occur in
Arkansas as a result of introduction. *Gambusia affinis* differs from *G. holbrooki* in having the distal segments of the
ventralmost major ray (anal ray 3) smooth dorsally (vs. distal segments of ventralmost major ray of the gonopodium
finely serrate dorsally) (Fig. 23.10), dorsal rays 6, sometimes

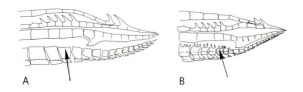

Figure 23.10. Gonopodia of:

A—*Gambusia affinis* (Rauchenberger 1989)
B—*G. holbrooki* (Etnier and Starnes 1993)

7 (vs. dorsal rays 7 or 8), and anal fin rays usually 9 (vs. anal fin rays usually10).*

VARIATION AND TAXONOMY *Gambusia affinis* was described by Baird and Girard (1853a) as *Heterandria affinis* from the Rio Medina and Rio Salado (San Antonio River drainage), Texas. It was subsequently placed in genus *Zygonectes* by Jordan (1878a) and in *Gambusia* by Hay (1882) and others. Rivas (1963) placed *Gambusia affinis* in the subgenus *Arthrophallus*. Two subspecies of *G. affinis* were previously recognized: *G. a. holbrooki,* which is found along the Atlantic Slope and peninsular Florida, and *G. a. affinis,* which is found in Arkansas and elsewhere through the Mississippi Valley to the Rio Grande. The Atlantic Slope subspecies was recognized as a full species (*G. holbrooki*) by Wooten et al. (1988), Rauchenberger (1989), M. H. Smith et al. (1989), and Lydeard et al. (1991) with a zone of hybridization occurring in Alabama. Recent authors have continued recognition of *G. affinis* and *G. holbrooki* as separate species (Page et al. 2013; Robins et al. 2018). *Gambusia affinis* does not share the same chromosomal sex determining mechanism with *G. holbrooki*. Females of *Gambusia affinis* have a large heteromorphic sex chromosome pair (WZ), but heteromorphic sex chromosomes are lacking in *G. holbrooki* (Black and Howell 1979). This does not prevent hybridization between the two species (Scribner and Avise 1993).

DISTRIBUTION The native range of *Gambusia affinis* is believed to be the Mississippi River basin from Kentucky, Illinois, Missouri, and southern Kansas, south to the Gulf of Mexico, and Gulf Slope drainages from Mobile Bay (with a population in the Chattahoochee River) to the Rio Grande of Texas, New Mexico, and Mexico. It is replaced in Atlantic Slope drainages and in Gulf Slope drainages east of Mobile Bay by *Gambusia holbrooki*, and the two mosquitofishes occur sympatrically in Mobile

Bay, Pascagoula, and coastal river drainages of southern Alabama and southeastern Mississippi. In Arkansas, the Western Mosquitofish occurs almost statewide in suitable habitats, but it is uncommon in the higher elevations of the Ozark Mountains (Map 157). *Gambusia affinis* and the Eastern Mosquitofish, *G. holbrooki,* have been introduced extensively throughout the United States and other parts of the world for mosquito control. In a great many instances, *Gambusia* introductions were made without any records being kept (Krumholz 1948). We have no information, records, or vouchered specimens indicating that *G. holbrooki* was ever stocked in Arkansas, but ichthyologists should be alert to the possibility that it might be found in this state. It would be easily overlooked, if present, because microscopic examination of the gonopodium of the male is necessary to accurately distinguish *G. affinis* from *G. holbrooki*.

HABITAT AND BIOLOGY The Western Mosquitofish occurs throughout the state in a wide variety of habitats, including swamps, ditches, farm ponds, streams, rivers, and lakes. It was the second-most widely distributed small fish species in Arkansas impoundments, occurring in 49 of 66 reservoirs sampled (Buchanan 2005). It prefers shallow, vegetated, backwater pool regions with little current, where it remains at the surface of the water, often moving about in groups. Mosquitofish exhibit an opportunistic life history strategy characterized by small body size, with early maturation, low fecundity per mating event, and low juvenile survivorship (Winemiller and Rose 1992; Hoeinghaus et al. 2007). Haynes and Cashner (1995) studied *G. affinis* in Louisiana and concluded that each population alters its life history (either genetically, via phenotypic plasticity, or by a combination of these mechanisms) to maximize fitness to a particular environment, and perhaps as a result, population dynamics are also altered. The ability of *G. affinis* to adapt to different, often harsh habitats by modifying its life history has allowed it to become one of the most successful vertebrate species on Earth (Haynes and Cashner 1995).

The Western Mosquitofish has broad ecological tolerances, surviving temperatures ranging from 42.8°F to 95°F (6–35°C) and up to 107.6°F (42°C) for brief periods, and a dissolved oxygen content as low as 0.18 ppm; given time to adapt, it may survive salinities more than twice that of seawater (Ahuja 1964; Pyke 2005). Temperature selection and heat resistance of Arkansas populations of Western Mosquitofish were studied by Bacon et al. (1968a). Female mosquitofish in an experimental temperature gradient spent most of their time between 77.9°F and 84.2°F (25.5–29°C), and the mean preferred temperature of their distribution was about 80.6°F (27°C).

---

* Anal fin rays of *Gambusia* species are more easily counted in females.

Introduced worldwide to control mosquitoes, this surface feeder eats a variety of terrestrial and aquatic insects, small crustaceans, and other invertebrates (snails, spiders), as well as plant material. Young *Gambusia* feed almost entirely on plankton (Barnickol 1941). In Caney Bayou, Arkansas, detritus (37%), terrestrial insects (34%) (mainly dipterans), and chironomids (16%) were the three major components of the summer diet; additional items consumed were algae, immature insects, and arachnids (Neely and Pert 2000). There was little dietary overlap with the syntopic *Fundulus olivaceus* in Caney Bayou. In Louisiana, surface prey and detritus were the dominant foods consumed (Daniels and Felley 1992). Feeding activity and growth are greatest at water temperatures between 25°C and 30°C, and mosquitofish can consume up to 75% of their body weight each day (Wurtsbaugh and Cech 1983; Haynes 1993). Chipps and Wahl (2004) reported that adults were capable of maximum food consumption rates of 42–167% of their body weight per day. Courtenay and Meffe (1989) argued that mosquitofish are ineffective mosquito control agents and that, generally, their introduction around the world has resulted in negative impacts on native fishes. Laboratory feeding experiments supported a high feeding capacity, and field-derived bioenergetics estimates of total food consumption found that a typical age 0 Western Mosquitofish consumed about 1,990 mosquito larvae from 22 August to 8 October 1998 in an Illinois population (Chipps and Wahl 2004). The almost worldwide introduction of *G. affinis* outside its native range has produced mixed results. In some environments, such as rice paddies and urban ponds, it has been a successful alternative to the use of pesticides for insect control (Moyle and Cech 2004). *Gambusia affinis* was first used in Arkansas rice fields in 1972 as a mosquito predator (Newton et al. 1977); however, mosquitofish may sometimes actually benefit mosquitoes by decreasing competitive and predation pressure from native zooplankton and predatory invertebrates (Blaustein and Karban 1990). It is an aggressive, competitive little fish that may be little or no better than native species as a mosquito destroyer. Myers (1965) noted that almost everywhere *G. affinis* has been introduced it has gradually eliminated most or all smaller native species of fishes. It has caused the elimination or decline of many populations of federally endangered and threatened fishes in the western United States and is responsible for the elimination of the Least Chub in several areas of Utah (Mills et al. 2004). In Nebraska, the disappearance of the Plains Topminnow (*Fundulus sciadicus*) from 11 localities on the Platte River was considered related to the establishment of *G. affinis* at those sites (Lynch 1991), and Haas

(2005) provided experimental evidence to support that conclusion. Matthews and Marsh-Matthews (2011) found that within its native range, *G. affinis* does not appear to negatively impact the assemblages in which it occurs, possibly because of fluctuations in its density. Those authors also cautioned that long-term studies of introduced mosquitofish populations are needed to adequately assess the threat to native fishes.

Gambusia affinis breeds throughout the summer in Arkansas. Hubbs (1971) hypothesized that photoperiod rather than temperature was the major factor structuring reproductive seasons of mosquitofish populations. Haynes and Cashner (1995) supported that hypothesis as they observed water temperatures were warm enough for reproduction to have continued throughout the winter of 1990–1991 in Louisiana, but instead, females ceased reproduction from October to April. Daniels and Felley (1992) found that females reproducing in late summer and fall in Louisiana had significantly smaller broods than females reproducing in spring and early summer. Internal fertilization is accomplished by the long intromittent organ (gonopodium) of modified anal rays in the male. Females store sperm in their reproductive tracts for up to several months and give birth to living young. Superfetation does not occur (broods do not overlap). One brood of young is born before any other ova have matured sufficiently to be fertilized (Krumholz 1948). Three or four broods are produced throughout the summer. Females mature at a length of about 1 inch (25 mm), often at 6 weeks of age, and reach a larger size than males. Because it is easily kept in aquaria, the interesting courtship and breeding habits of this species intrigue students of biology of all ages. The *Gambusia* spp. mating system is largely male driven, with males of all sizes forcing females to copulate (Pilastro et al. 1997). Hughes (1985) showed that small and large male Western Mosquitofish differ in mating behaviors, where small males coerced or force-copulated at higher rates than large males. Deaton (2008) studied the effects of male and female body size on mating behavior and found that while female body size was a significant predictor of male mating behavior, male size was not. Deaton also confirmed that factors other than female size are important predictors of male mating behavior in the Western Mosquitofish. Males constantly pursue and court females during the breeding season. The fertilized eggs hatch in 21–28 days and the young, numbering from a few to several hundred, are provided no parental protection. Average size of newborn young is 8 or 9 mm TL. Brood size increases with female body size, but also varies among water bodies and is affected by the age of the female and how many broods she has already produced

(Krumholz 1948). Females have fewer young per brood as the season progresses. Krumholz (1948) found 315 embryos in a 59 mm female that had overwintered, and 218 embryos in a 47 mm young-of-the-year female. Barney and Anson (1921) reported a maximum brood size of 226 young produced from a female of 43 mm SL (approximately 51 mm TL) in Louisiana. Bonham (1946) reported a single brood of 354 young produced by a female of unspecified size from a Texas pond. The life span is very short. Females that survive their first winter spawn the following spring and then die; male life spans are much shorter than those of females (Haynes and Cashner 1995).

CONSERVATION STATUS   The Western Mosquitofish is a very common inhabitant of Arkansas, occurring statewide, and is in no danger.

CHAPTER 24

Order Anabantiformes
Labyrinth Fishes

This order includes freshwater fishes native to Africa and southern and eastern Asia. Labyrinth fishes have traditionally been placed in the order Perciformes, but we follow Wiley and Johnson (2010), Nelson et al. (2016), and Betancur-R. et al. (2017) in placing them in the order Anabantiformes. There are several characters distinguishing this order, the most obvious of which is a suprabranchial chamber on each side of the head housing an air-breathing organ and the labyrinth apparatus (most species) (Tate et al. 2017), or lined with thin, highly vascularized respiratory epithelium (Channidae and some others). Other characters associated with Anabantiformes include distinctive circulation to and from the suprabranchial chamber, basioccipital with paired articular processes forming diarthrosis with the upper jaw, posterior extension of the swim bladder extending to the parhypural, and larva with a bilateral pair of oil vesicles for flotation (Nelson et al. 2016). Anabantiformes was divided into 2 suborders, 4 families, 21 genera, and about 207 species by Nelson et al. (2016). One family, Channidae, occurs in Arkansas.

Family Channidae *Snakeheads*

This is an air-breathing, as well as freshwater anabantiform family of the suborder Channoidei containing two genera: *Channa* (Scopoli 1777), with 34 species native to Asia, including Malaysia and Indonesia, and *Parachanna*

(Teugels and Daget 1984), with 3 species native to tropical Africa (Courtenay and Williams 2004; Nelson et al. 2016). The fossil record indicates the presence of channid fishes during the Upper Oligocene/Lower Miocene in western Switzerland and eastern France (Reichenbacher and Weidmann 1992). Size of channid species ranges from about 17 cm to 1.8 m in length. In the United States, members of this family are sold in the pet trade and in fish markets.

Members of this family are referred to as snakeheads because of the elongated, cylindrical body and the presence of large scales on the head of most species, reminiscent of the large epidermal scales (cephalic plates) on the head of many snakes (Courtenay and Williams 2004). In addition, the head is flattened, and the eyes are in a dorsolateral position on the anterior part of the head. Tubular nostrils are present, the dorsal and anal fins are elongated, and the caudal fin is rounded. All fins are supported only by soft rays. A few species lack pelvic fins (Nelson 2006; Berra 2001). Snakeheads have a large, terminal mouth with a protruding lower jaw that is toothed and often contains canine teeth. Scales are cycloid or ctenoid.

Snakeheads are non-ostariophysan primary freshwater fishes (Mirza 1975; 1995) and have little or no tolerance for saltwater. Most occur in streams and rivers, and all species can tolerate hypoxic conditions because they are air breathers after reaching late juvenile stages. All species are considered thrust predators, and most are piscivorous as adults

(Courtenay and Williams 2004). Snakeheads are highly valued as food fishes in their native ranges. The Northern Snakehead, *Channa argus*, is the market leader and was previously cultured in China and Korea. An abortive attempt was made to culture *C. argus* in Arkansas in 2002. Because of their air-breathing abilities, several species of snakeheads are capable of overland migration by wriggling motions (P. G. Lee and Ng 1991; Berra 2001), even though their pectoral fins lack spines. They possess a suprabranchial chamber located behind and above the gills for aerial respiration, and the ventral aorta is divided into two parts to permit bimodal (aquatic and aerial) respiration (Das and Saxena 1956; J. B. Graham 1997). Berg (1947) reported that the suprabranchial chambers are not labyrinthic but are lined with respiratory epithelium and first become functional during the juvenile stage (J. B. Graham 1997). Liem (1987) reported that *Channa* species burrow into mud to survive droughts.

Snakeheads build nests for reproduction. The nest is built by cleaning a generally circular area in aquatic vegetation, often weaving the removed vegetation around the centrally cleared area. This results in a vertical column of water surrounded by vegetation. Eggs are vigorously defended by one or both parents. There is one genus of Channidae in Arkansas.

Genus *Channa* Scopoli, 1877—Snakeheads

The genus *Channa* was described by Scopoli in 1777. This Asian genus contains approximately 34 species, and 1 species is established in Arkansas. *Channa argus* has so far been found only in eastern Arkansas. Characters of the genus are presented in the species account.

Channa argus (Cantor)
Northern Snakehead

Figure 24.1. *Channa argus*, Northern Snakehead. *Uland Thomas.*

CHARACTERS   A long, cylindrical body with a rounded caudal fin, and elongated dorsal (usually 49 or 50 rays) and anal (usually 31 or 32 rays) fins (Fig. 24.1). Head somewhat

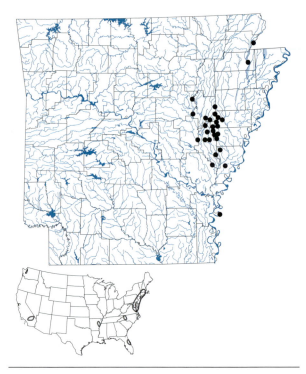

Map 158. Distribution of *Channa argus*, Northern Snakehead, 1988–2019. Inset shows introduced range (outlined) in the contiguous United States.

depressed anteriorly; interorbital area flat. Gular area without a patch of scales and without a gular plate (Fig. 24.2). Eye situated above middle of upper jaw. Scales present on top of head. Mouth large, terminal, and reaching behind eye. Villiform teeth present in bands, with some canine teeth on lower jaw and palatines (Fig. 24.3). Lateral line scales 60–67. Scale rows above lateral line to dorsal fin origin modally 8; 12–30 scale rows below lateral line to anal fin origin. Origin of pelvic fin beneath fourth dorsal fin ray. Tip of pectoral fin extending backward beyond base of pelvic fin. Suprabranchial chambers present for air breathing (but no labyrinth apparatus). Size up to 33 inches (85 cm) (Okada 1960); however, Courtenay and Williams (2004) reported a specimen from Russia approaching 59 inches (1.5 meters) TL.

LIFE COLORS   Body color is golden tan to pale brown with a series of dark blotches on the sides and saddlelike blotches across the back interrupted by the dorsal fin. Walter Courtenay (pers. comm.) observed that this species can darken its background coloration to the point of almost obscuring the dark blotches. Upper blotches on sides are typically separate anteriorly, but the more posterior blotches may coalesce with ventral blotches. Dark stripe present from just behind the eye to upper edge of operculum, and another dark stripe occurs below it from

Figure 24.2. Gular region of Northern Snakehead. *David A. Neely.*

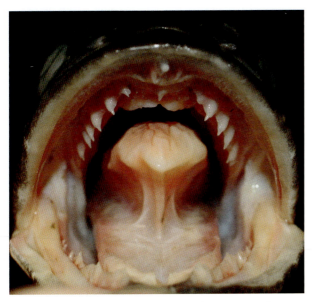

Figure 24.3. Northern Snakehead mouth. *Renn Tumlison.*

behind the orbit extending to the lower quadrant of the operculum. Fins are dark. Caudal fin without an ocellus. Juveniles are colored nearly the same as adults, a condition atypical for many snakehead species.

SIMILAR SPECIES The Bowfin is similar to the Northern Snakehead in having tubular nostrils, an elongated body, and a long dorsal fin; however, *C. argus* differs from the Bowfin in having thoracic or anterior-abdominal pelvic fins (vs. pelvic fins abdominal), an elongated anal fin usually with 31 or 32 rays (vs. a short anal fin with 9 or 10 rays), no gular plate (vs. a gular plate present), a rosette of enlarged scales on top of head (vs. no scales on top of head), and no ocellus on the caudal fin base (vs. ocellus usually present on upper base of caudal fin).

VARIATION AND TAXONOMY *Channa argus* was originally described by Cantor (1842) as *Ophicephalus argus*. The type locality is Chusan Island, China. The

diploid chromosome number is 48 (S. W. Lee and Lee 1986). Orrell and Weigt (2005) found seven unique mtDNA haplotypes in *C. argus* from the Potomac River system. Unique haplotypes found in Maryland, Massachusetts, and Pennsylvania support the conclusion that there were several independent introductions of the Northern Snakehead into these waters, with no two introductions coming from the same maternal source. That conclusion was also supported by Wegleitner et al. (2016). Zhu et al. (2014) studied genetic variation of microsatellite markers in four cultured populations in China and reported high genetic diversity.

DISTRIBUTION Native to Manchuria, Russia, China, and Korea. In the United States, introduced populations are established in the Potomac River drainage of Maryland and Virginia, and in Arkansas, with other nonindigenous records reported from California, Florida, North Carolina, Illinois, Delaware, Massachusetts, New York, and Pennsylvania (Fuller et al. 2019). In Arkansas, it is established in the lower White River system, probably through accidental introduction. A breeding population was confirmed in Big Piney Creek of the Big Creek drainage (a White River tributary) in Lee and Monroe counties in 2008. Despite a massive eradication attempt in 2008–2009, there were subsequent reports of fishermen catching snakeheads in the Big Creek drainage (Mark Oliver, pers. comm.). David A. Neely (pers. comm.) caught several snakeheads in a ditch tributary to Big Piney Creek at the northwest corner of Greenlee Reservoir in Monroe County on 16 March 2012 in just a few minutes of electrofishing. *Channa argus* has spread to drainages adjacent to the Big Creek drainage (the mainstem White River, Bayou DeView-Cache River, and the St. Francis River drainage), and records of specimens collected from 2008 to 2015 are now in the AFD from Cross, Lee, Monroe, Phillips, St. Francis, and Woodruff counties in Arkansas (Map 158). More than 1,060 individuals were collected in Arkansas between 2008 and 2015. On 5 June 2017, one *Channa argus* was taken by bow fishermen in Lake Whittington, a Mississippi River oxbow in Bolivar County, Mississippi (adjacent to Desha County, Arkansas), approximately 17 miles downstream from the mouth of the White River. A specimen taken in May 2018 from Ditch Number 8 on the St. Francis River near Lake City in Craighead County documented its continued upstream invasion in that system (Jimmy Barnett, pers. comm.). By April 2019, *C. argus* had spread northward throughout the St. Francis River into Dunklin County, Missouri (A. Asher and R. Hrabik, pers. comm.).

HABITAT AND BIOLOGY The Northern Snakehead prefers stagnant, shallow ponds or swamps with mud substrates and aquatic vegetation, and it is also found in muddy streams (Okada 1960). It avoids deep, flowing waters (Owens et al. 2008). This species also occurs in canals, lakes, reservoirs, and rivers (Dukravets and Machulin 1978; Dukravets 1992; Courtenay and Williams 2004). Odenkirk and Owens (2005) reported that Northern Snakeheads in the Potomac River occurred in water less than 2 m deep with submergent or emergent vegetation and near shorelines, channel edges, and boat slips. Lapointe et al. (2010) investigated seasonal habitat selection at meso- and microhabitat scales using radiotelemetry. In the pre-spawning season (April–June), fish moved to shallow waters of inner bays, creek channels, and embayment shore-lines, usually with macrophyte beds. In the spawning sea-son (June–September), snakeheads exhibited the strongest habitat selection and sought out macrophytes over coarse substrates along shorelines that may have been used for spawning (Lapointe et al. 2010). In the post-spawning fall period (September–November), snakeheads preferred off-shore vegetated beds with shallow water (about 115 cm) and soft substrates. In winter (November-April), they moved to deeper water (135 cm) having warmer temperatures. Because of its ability to breathe air, the Northern Snakehead may be able to move overland for short distances. However, Courtenay and Williams (2004) believed that because *C. argus* has a rounded body and lacks fin spines, it has very limited ability to move on land except as young, and only when some water is present during mild flooding. Lapointe et al. (2013) documented the great ability of this species to expand its distribution within and beyond the Potomac River. Since its discovery in the Potomac River in 2004, *Channa argus* has expanded its range at a rate of approxi-mately 2.7 subwatersheds per year in that drainage (Love and Newhard 2018). Those authors further noted that nat-ural expansion of its distribution was aided by high levels of precipitation. Odenkirk and Isel (2016) reported that population size in *Channa argus* appears to have declined in some streams of the Potomac River in Virginia, and Love and Newhard (2018) suggested that annually increas-ing levels of commercial harvest may be responsible for that decline. The latter authors reported that commercial har-vest of Northern Snakehead in Maryland increased from 5 kg in 2011 to 1,959 kg in 2016.

Foods of adult *Channa argus* consist of fishes, frogs, and aquatic insects, postlarvae feed on plankton, and juveniles eat small crustaceans and fish larvae (Courtenay and Williams 2004). Okada (1960) reported that this spe-cies is a voracious feeder. Dukravets and Machulin (1978) found that it fed on 17 different species of fishes, including young fish and fish up to 33% of the predator's body length.

Juveniles begin to feed on fishes at 4 cm TL, and fishes made up 64–70% of the diet of juveniles 13–15 cm TL. Juveniles up to 30 cm fed almost exclusively on fishes (90% of diet). Feeding activity reached its peak during June or July at temperatures of 20–27°C (Guseva 1990). Feeding declined by autumn as water temperatures declined to 12–18°C and ceased altogether when temperatures dropped below 10°C. Courtenay and Williams (2004) reported that this species feeds in schools with most activity at dusk or before dawn, typically in vegetation near shore. Adults actively feeding make grunting noises "like pigs" (Courtenay and Williams 2004). Soin (1960) also documented adults making click-ing sounds.

Northern Snakeheads construct and guard nests in patches of macrophytes (Courtney and Williams 2004). Nests must be constructed in areas that protect the eggs from current and wave action (Gascho Landis and Lapointe 2010). Soin (1960) provided information on spawning and development in its native range. Fertilized eggs about 1 day old were in an open nest (cleared of vegetation) about 1 m in diameter and 60–80 cm deep. This species builds a mostly circular nest next to pieces of aquatic plants in shallow aquatic vegetation. The male and female clean the water surface area above the nest, and spawning occurs at dawn or in early morning. The female rises near the surface and releases eggs which are then fertilized by the male. Eggs are pelagic, nonadhesive, spherical, yellow, and about 2 mm in diameter. After 24 hours, the embryo was 3.2 mm TL, and hatching occurred in 28 hours at 31°C, 45 hours at 25°C, and 120 hours at 18°C. Newly hatched larvae are black and about 4 mm TL (Courtenay and Williams 2004). Two days after hatching, each larva was about 4.5 mm long, and larvae approached 11 mm TL after about 2 weeks. The number of eggs released ranges from 1,300 to 15,000, with an average of 7,300. Frank (1970) reported total fecun-dity of about 50,000 oocytes, Nikol'skiy (1956) reported 22,000–51,000 eggs in the Amur basin, and Dukravets and Machulin (1978) gave a fecundity range from 28,600 to 115,000. Berg (1947) reported that the Northern Snakehead spawns five times per year in the Amur Basin, but Dukravets and Machulin (1978) noted that it spawns two to three times in the Syr Dar'ya basin from May to June at temperatures of 18–20°C. The larvae remain in the nest and are guarded by one or both parents until the yolk is absorbed and the body length is about 8 mm. Following yolk absorption, the larvae leave the nest as a group and start to feed on plankton. Parents guard the fry through the postlarval stage. Sexual maturity is reached in about 3 years at a length of 30–35 cm, but some individuals spawn during the second year (Dukravets and Machulin 1978).

In Japan, Okada (1960) reported spawning at 2 years at a length of 30 cm. Courtenay and Williams (2004) reported a specimen of *C. argus* that was almost 1.5 m in length, indicating a long-lived species.

CONSERVATION STATUS   The Northern Snakehead was imported for sale in live-food fish markets, and it was the most widely available snakehead species in the United States (Courtenay and Williams 2004). In 2001, an Arkansas fish farmer was discovered culturing Northern Snakeheads when possession of this species was legal. The original Arkansas aquaculturist had been approached by a live-food fish importer in New York and asked if he could culture snakeheads for sale in United States markets (Courtenay and Williams 2004). Later, two additional Arkansas aquaculturists decided that they too might profit from snakehead culture. By July 2002, three domestic sources of cultured Northern Snakehead were in operation, but by that time, the importation, culture, and possession of snakeheads in Arkansas were prohibited, and 14 states had banned possession of live snakeheads. The U.S. Fish and Wildlife Service published a proposal rule to list the family Channidae (snakehead fishes) as injurious wildlife in the Federal Register on 20 July 2002 (67 FR 48855) under the Lacey Act (18 U.S.C. 42). The final rule banning importation and interstate transport of live snakeheads was published in the Federal Register on 4 October 2002 (67 ER b62193).

The Northern Snakehead is a threat mainly because of its potential effects on native species. Because of its broad temperature tolerances (0–30°C) and varied diet, this species could potentially establish feral populations throughout most of the contiguous United States. Snakeheads may negatively impact native fish assemblages in Arkansas through direct predation, competition for food resources, and alteration of food webs (Courtenay and Williams 2004; Herborg et al. 2007). In addition, Northern Snakeheads may host diseases that could spread to native species, and they might even host some human parasites.

The economic cost of an eradication attempt was high. In 2008–2009, the Arkansas Game and Fish Commission spent more than $750,000 in an attempt to remove Northern Snakeheads from the Big Creek drainage of Lee and Monroe counties. Twenty-five miles of Big Piney Creek plus smaller tributaries and irrigation ditches were treated with rotenone to eradicate the snakehead population. In all, about 200 miles of waterway with a surface area of 4,000 acres were treated. The eradication attempt failed, and *Channa argus* has subsequently expanded its range in Arkansas. By 2015, it was documented from additional neighboring drainages, the mainstem White River, the Bayou DeView-Cache river drainage, and the St. Francis River drainage. Its occurrence in the St. Francis River drainage is especially troubling because that drainage is connected to the other known areas of occurrence in Arkansas only via the Mississippi River main stem. The St. Francis drainage specimens represent either a separate introduction from those in the White River drainage, or more likely, an invasion from the Mississippi River. Its documentation in a Mississippi River oxbow lake in 2017 indicates that it has spread to the Mississippi River. Because *Channa argus* has gained access to the Mississippi River, there is a strong probability that it will spread to other states in the Mississippi River basin.

CHAPTER 25

Order Scorpaeniformes
Mail-Cheeked Fishes

This large order of percomorph fishes consists of 6 suborders with 41 families, 398 genera, and approximately 2,092 species (Nelson et al. 2016), about 60 of which are confined to freshwater (Nelson 2006). Smith and Wheeler (2004), in an analysis of mitochondrial and nuclear DNA sequences, found the traditionally recognized scorpaeniform fishes to be polyphletic. Smith and Busby (2014) reviewed the classification history of the scorpaeniform suborder Cottoidei (sculpins, sandfishes, and snailfishes), one of the largest and most morphologically diverse teleostean suborders, using a combination of genetic and morphological characters to produce a revised cottoid taxonomy. Exactly where scorpaeniform fishes belong among the percomorphs has been disputed, with Gill and Mooi (2002) placing them in the order Perciformes as did Betancur-R. et al. (2017). The arrangement of families within Scorpaeniformes is also subject to much disagreement (see Nelson et al. 2016 for a summary). Most members of the order are marine, but a single family, Cottidae, is found in both marine and freshwater environments in Eurasia and North America (three families have freshwater species if Gasterosteidae and Indostomidae are also included in Scorpaeniformes, as in Nelson et al. 2016). The scorpaeniforms are referred to as the mail-cheeked fishes because of the suborbital stay, a posterior extension of the third infraorbital bone that extends across the cheek to the preopercle and is usually firmly attached to it (Nelson 2006). The head and body are usually spiny or have bony plates, or both (except in freshwater sculpins of North America). In the caudal skeleton, two platelike hypurals articulate with the terminal half of the centrum. Pectoral fins are usually large and rounded and sometimes have the lower interradial membranes incised. The caudal fin is usually rounded, occasionally truncate, and rarely forked.

Family Cottidae *Sculpins*

The sculpins are a distinctive, circumpolar, largely marine family of bottom-dwelling, cold-water fishes with 70 genera and about 282 species (Nelson et al. 2016). Cottidae as formerly constituted is not supported as monophyletic (Smith 2005), and a classification of Cottidae by Smith and Busby (2014) excluded most marine species from the family, recognizing 107 species of cottids in 18 genera. The family is distributed mainly in the Northern Hemisphere, with the greatest diversity occurring along the North Pacific coast. The only Southern Hemisphere cottids are four deep-water species found in eastern Australia, near New Guinea, and New Zealand (Nelson et al. 2016). In North America, four genera, *Cottus* (1 species), *Cottopsis* (10 species), *Uranidea* (23 species), and *Myoxocephalus* (2 species), have freshwater (as well as marine) species. In Arkansas, the family is represented by only two species of the genus *Uranidea* (*U. carolinae* and *U. immaculata*), both limited to cold, clear, often swift, highly oxygenated springs, spring-run

553

creeks, streams, and rivers of the northern mountainous part of the state.

Sculpins are easily recognized by the large, broad, flattened head; branchiostegal membranes fused to the isthmus; expanded pectoral fins; heavy body; large gape; and banded appearance. The body tapers abruptly from the large, broad head to a narrow caudal peduncle. The eyes are dorsal in position, the dorsal fin is divided into 2 distinct parts, teeth are well developed on the jaws, and the swim bladder is absent, as are scales; however, small prickles (modified ctenoid scales) are often found on the head and body. The preopercle is armored with bony spines. Care should be used when handling large individuals to avoid the sharp spines.

Sculpins feed primarily on microcrustaceans, crayfish, aquatic insects, and occasionally small fishes, including each other. They have been implicated in the consumption of trout eggs and young trout in streams, a charge that is likely untrue (Moyle 1977). The trout eggs eaten by sculpins were probably not buried in the nests. Becker (1983) noted that little evidence is available to show that sculpins limit trout numbers. In Arkansas, sculpins are sometimes used as bait for trout. Adams and Schmetterling (2007) reviewed the scientific literature on the biology of freshwater sculpins, including information on phylogenetics, behavior, and characteristics of sculpin movements, foraging, and reproduction.

Key to the Sculpins

1A Lateral line complete, extending to base of caudal fin; dorsal fins separate; dark bar beneath soft dorsal fin narrow; dark bar at caudal base angling anteroventrally to make a triangle. *Uranidea carolinae* Page 554
1B Lateral line incomplete, extending only to beneath soft dorsal fin; dorsal fins slightly connected; dark bar beneath soft dorsal fin wide; dark bar at caudal base narrow. *Uranidea immaculata* Page 556

Figure 25.1. Knobfin Sculpin on stream bottom in White River. *Isaac Szabo.*

Genus *Uranidea* DeKay, 1842— Uranidea Sculpins

The recognition of genera within Cottidae has dramatically changed in recent years, based largely on studies of mitochondrial and nuclear DNA. Kinziger (2003) and Kinziger et al. (2005) identified cottid clades that Smith and Busby (2014), Adams et al. (2015), and others recognized as genera. Smith and Busby (2014) provided an excellent taxonomic summary and justification for the generic change. All eastern North American species formerly in the genus *Cottus*, except for *C. ricei*, were placed in the genus *Uranidea*. Ten species from the western United States were placed in *Cottopsis* Girard (1850). *Cottus*, as now constituted, contains only European species plus *C. ricei*. Additional unpublished genetic evidence also provides strong support for recognition of *Uranidea* (D. Neely, pers. comm.). The genus *Uranidea* has a largely Nearctic distribution and is widely distributed across North America, where 23 species have evolved, including several species confined to one or two isolated springs. *Uranidea* species in Arkansas are primarily inhabitants of springs, spring-fed creeks, and cold rivers, but elsewhere they also occur in lakes and estuaries. The *Uranidea* fossil record dates to the Miocene in Nevada (G. R. Smith et al. 2002).

In Arkansas, the genus is characterized by relatively reduced armature of the head (usually 3 preopercular spines of which only the upper spine is prominent), villiform teeth present on the jaws, head of the vomer, and may be present or absent on the palatine bone (Etnier and Starnes 1993). This genus lacks a swim bladder, gill covers are separated by a wide isthmus, and there are 6 branchiostegal rays. Body shape is typically flattened with a large head and large, oblique pectoral fins.

Uranidea carolinae (Gill)
Banded Sculpin

CHARACTERS A distinctive fish with a large, flattened head, large eyes, large mouth, and scalelike prickles (modified scales) on head (Fig. 25.2). Lateral line usually complete, with 30 or more pores (usually 31–34) extending to the caudal base. Body scaleless. Body deepest at dorsal fin origin and tapering to a thin caudal peduncle; depth of caudal peduncle about 8% of SL. Head length going into SL 2.8–3.3 times. Preopercle with 3 sharp spines, the ventralmost 2 are covered by connective tissue. Palatine teeth present and usually contiguous with the vomerine tooth patch. Dorsal fins contiguous but separate. First dorsal fin

Figure 25.2. *Uranidea carolinae*, Banded Sculpin. *David A. Neely.*

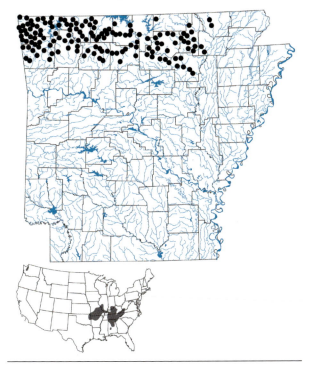

Map 159. Distribution of *Uranidea carolinae*, Banded Sculpin, 1988–2019.

small, with 7–9 weak spines; the second dorsal fin elongated, with 16–18 soft rays. Anal fin moderately long, with 12–14 rays. Pectoral fins large and greatly expanded; pectoral rays 15–18. Pelvic fin with 1 thin spine and 3 or 4 rays. Maximum length in Arkansas 8 inches (203 mm).

LIFE COLORS   Body color rusty brown or yellow, usually without strong mottling. Four or five dark brown or black dorsal saddles; the posterior three saddles extending down the sides as sharply defined bars and slanting obliquely forward; the first distinct saddle extending downward onto side is over the anal fin origin and the last saddle is at the base of the caudal fin, forming a well-defined, broad, dark vertical band crossing the body at the caudal base. Belly yellowish white and sprinkled with a faint to obvious dusting of dark specks. Chin strongly mottled. Fins yellow to

orange. Spots on median fins restricted to the surface of the rays. Pectoral fins faintly banded; pelvic fins white There is much color variation from very dark to quite pale even within a single population. This variation probably reflects substrate color occupied at time of capture. Breeding males develop very dark coloration, especially concentrated on the head.

SIMILAR SPECIES   The Banded Sculpin differs from *Uranidea immaculata* in having a complete lateral line (vs. lateral line incomplete, ending beneah posterior part of second dorsal fin). See the *U. immaculata* account for additional differences. The combination of a robust body, large flat head, large mouth, large expansive pectoral fins, and scales reduced to tiny prickles can be used to distinguish *U. carolinae* from all other Arkansas fishes.

VARIATION AND TAXONOMY   *Uranidea carolinae* was described by Gill (1861) as *Potamocottus carolinae* from near Maysville, Kentucky. It was long recognized in the genus *Cottus* (see genus *Uranidea* account for information on its placement in *Uranidea*), and is considered a species complex composed of up to 11 distinct species (Kinziger 2003; Kinziger et al. 2007). Five of those taxa occur in the Ozark Mountains, including two in Arkansas, the Midlands race (White and Strawberry rivers) and Black River race (Current and Eleven Point rivers). The two taxa are distinguished by pectoral fin ray number, modally 18 in the Black River race versus modally 16 in the Midlands race (Kinziger 2003). The two lineages can also be genetically distinguished (Kinziger et al. 2007). A distinct cave-dwelling species (*Uranidea specus*) sister to *U. carolinae* was identified in Missouri (G. L. Adams et al. 2013). *Uranidea carolinae* and *U. immaculata* are members of the Uranidea clade, but are distantly related to one another (Kinziger 2003).

DISTRIBUTION   *Uranidea carolinae* occurs in the Mississippi River basin from the upper Tennessee River drainage, West Virginia and Virginia, across Kentucky, Tennessee, southern Indiana and Illinois, south to Alabama, the Ozark Upland drainages of Arkansas, Missouri, Kansas, and Oklahoma, and in one Gulf Slope drainage, the eastern Mobile Bay basin in Georgia, Tennessee, and Alabama. In Arkansas, the Banded Sculpin inhabits the White and Black river systems and the Illinois River system of the Arkansas River drainage (Map 159). It is conspicuously absent from the Ouachita Mountains ecoregion.

HABITAT AND BIOLOGY   The Banded Sculpin is Arkansas's most widely distributed and abundant cottid. It occupies a variety of habitats but is found mainly in cold springs and spring-fed, clear, highly oxygenated gravel streams of the Ozark Uplands. Adults are most commonly found over gravel or rubble substrates in current; juveniles frequently occur in detritus in shallow, spring-fed pools. It has also been found in caves in Arkansas (Herbert 1994). Pflieger (1997) noted that it seems tolerant of higher temperatures than *U. immaculata* and is generally the most abundant sculpin in the larger and warmer Ozark streams. It is also found in the cold tailwaters of the White River immediately below Bull Shoals and Norfork dams. Cold water release from these dams has likely resulted in downstream range expansion. In Brawley's Fork, Tennessee, a large, spring-fed first order stream, adult *U. carolinae* were found almost exclusively in pools, YOY were found almost exclusively in riffles, and juveniles were found in both habitat types (Koczaja et al. 2005). None of the size classes exhibited a water velocity preference. The ontogenetic habitat shift from riffles to pools by juvenile sculpins may be explained by a change in predation risk as sculpins grow.

The Banded Sculpin is an ambush predator (Tumlison and Cline 2002). Greenberg and Holtzman (1987) reported that *U. carolinae* feeds primarily at night and seeks refuge under rocks during the day. It typically lives in current where it preys on immature insects (primarily stoneflies, mayflies, and caddisflies), small mollusks, small crustaceans, and small fish (Cross and Collins 1995). Larval feeding begins 5–7 days after hatching (Wallus and Grannemann 1979). Small sculpins feed mainly on immature aquatic insects, but adults switch to a diet dominated by crayfish, fish (including darters), large stoneflies, salamanders, and plecopterans (Etnier and Starnes 1993; Phillips and Kilambi 1996). Cooper (1975) reported foods consumed by 18 *Uranidea carolinae* from the North Fork River (Ozark County, Missouri). The most important food items were crayfish (*Orconectes*), followed by ephemeropterans (Baetidae), gastropods, fish (including other sculpins), trichopterans, plecopterans, elmid beetles, and a megalopteran and isopod. In Oklahoma, Tumlison and Cline (2002) found that Banded Sculpins fed primarily on ephemeropterans (Baetidae and Heptageniidae), dipterans (chironomids and black fly larvae), gastropods, amphipods, small fishes, and one salamander (*Eurycea tynerensis*). In the Illinois River, Arkansas, large sculpins (75–112 mm) fed mainly on isopods (17.4% by number), decapods (26.1%), and fish (43.5%), while small sculpins (37–59 mm) fed mainly on dipterans (61.8%) and ephemeropterans (18.1%) (Ebert et al. 1987).

Spawning occurs in late winter and early spring from January through March in Missouri and Arkansas. Spawning has been reported at water temperatures of 9–14°C (Craddock 1965; Wallus and Grannemann 1979).

Craddock (1965) found gravid females during the winter and early spring but did not observe spawning or fry. We have taken adults in breeding condition from the White River in mid-February. Wallus and Grannemann (1979) reported that *U. carolinae* uses the undersides of rocks and logs for spawning. About 100–300 adhesive eggs were found in nest cavities. The male guards the nest until the eggs hatch. Hatching occurs in 15–19 days at water temperatures of 16–19°C. Some authors suggested that *U. carolinae* may use underground spawning habitats to complete their life cycle (Craddock 1965; Herbert 1994; J. Z. Brown 1996). Herbert (1994) reported increased numbers of drifting larvae collected in effluent from Logan Cave, Arkansas during late winter compared to surface-stream drift, implying that spawning occurs primarily underground in that population. Although spawning was not observed in nearby surface populations, yolk-sac larvae of *U. specus* (formerly included in *U. carolinae*) were found drifting from a spring resurgence in a stream in southeastern Missouri in March (G. L. Adams, pers. comm.). Water temperature was 12°C in the main channel, but water temperatures on the edges of the stream ranged from 16°C to 18°C. Therefore, spawning appears to be strongly linked to water temperature, with cave temperatures reaching optimum level as early as three months before surface water temperatures. Most reproducing adults in Kentucky were in their second year (Craddock 1965). Maximum life span is around 6+ years (Etnier and Starnes 1993).

Uranidea carolinae is an important host fish for the larvae of numerous mussel species occupying the same riffle habitats (Yeager and Saylor 1995; Rogers et al. 2001). McAllister et al. (2014a) reviewed helminth parasites of *U. carolinae* in Arkansas.

CONSERVATION STATUS   The Banded Sculpin is common in appropriate upland habitats in Arkansas, and its populations are secure.

Uranidea immaculata (Kinziger and Wood)
Knobfin Sculpin

CHARACTERS   A moderate-sized, stout-bodied sculpin with a large head, eyes, and mouth, and naked body (Fig. 25.3). Body deep and rather strongly compressed posteriorly (body width under midpoint of second dorsal fin going into body depth at that point 2 or more times). Snout sharp, angle between gape and forehead smaller than 45°. Lateral line incomplete and rarely interrupted, usually with 21–25 (19–29) pores, ending beneath the posterior part of the soft dorsal fin. Dorsal fin union moderate to wide. First

Figure 25.3. *Uranidea immaculata*, Knobfin Sculpin. *David A. Neely.*

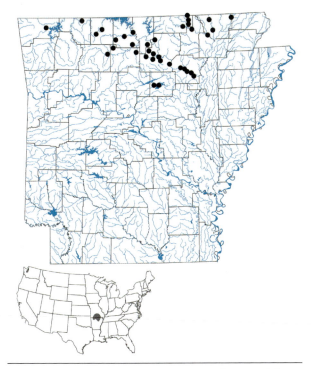

Map 160. Distribution of *Uranidea immaculata*, Knobfin Sculpin, 1988–2019.

dorsal fin usually with 8 or 9 spines (96% of specimens); second dorsal fin usually with 17 or 18 (15–19) soft rays; anal rays 13, rarely 12 or 14. Pectoral fins large, their tips usually extending back to second anal ray when pressed against body; pectoral rays usually 16 or 17 (84% of specimens). Pelvic fins well short of anus when pressed against body. Pelvic fin with 1 spine and modally 4 rays. Branchiostegal rays 6. Body prickling variable. Dorsal surface of head and rarely nape with papillae (Kinziger 2003). Palatine tooth patch moderately well developed. Premaxillary tooth patch short, not extending far laterally. Maximum size 3.5 inches (90 mm) SL.

LIFE COLORS   Coloration in living specimens is highly variable and depends to a large degree on substrate type and time of capture (Robins and Robison 1985). Generally, body is dusky overall and mottled with a background color of brown to olive. A horizontal row of 3–5 square blotches on light background or several pale dots (1–2 mm diameter) on dark background present postero-ventrad. Black spots present laterally and on head. Two faint suborbital bars usually present; bar 1 extending ventrally to maxilla; bar 2 extending posteriorly to uppermost preopercular spine. Chin dusky, or rarely, slightly mottled. Venter pallid. Three to five narrow to wide saddles crossing the dorsum. Caudal base band encircles fin base and extends anteriorly, forming a triangular blotch or sometimes connecting with saddle 5. Band at caudal base notched. Bar crossing caudal base narrow and indistinct. Basicaudal bar generally with straight posterior edge, usually forming a complete ring on caudal peduncle. Dorsal fin mottled and marbled. First dorsal fin membranes and spines black anteriorly between fin elements 1 and 3 and posteriorly between the last 3–4 spines, or uniformly dusky and pale margined. Males with tips of first dorsal spines forming thin knobs, 0.3–0.5 mm in diameter (Kinziger and Wood 2010). Second dorsal fin marbled or uniformly dusky. Anal fin rays unpigmented or tessellated, membranes uniformly dusky. Caudal fin banded or tessellated. Dark blotch usually present at pectoral fin base. Pectoral fin banded, tessellated, or uniformly dusky. Melanophores on prepectoral region evenly distributed and of nearly uniform size. Internally, pigmentation of the ventral surface of the peritoneum weak in 91% of specimens (Kinziger and Wood 2010). Juveniles (40 mm or less SL) sometimes with jet black bodies except for copper color at base of pectoral fin, and a U-shaped mark present on nape and on dorsal surface of caudal peduncle. Breeding males have a blue or black chin and belly. The body is dark grayish black, but not jet black. Blue-green or aqua coloration is present on the inside of the mouth and also marbles the chin, branchiostegal membranes, gular region, and belly to the anal fin insertion. Pelvic fin is blue or dusky black, first dorsal fin distal margin is orange to red, and all other fins are black.

SIMILAR SPECIES   *Uranidea carolinae* is the only *Uranidea* species known to occur sympatrically with *U. immaculata*. The Knobfin Sculpin differs from *U. carolinae* in having a blue or blue-green chin and belly in spawning males (vs. chin and belly not blue or blue green); an incomplete lateral line (vs. complete); 21–25 lateral line pores (vs. 31–34 pores); chin uniformly pigmented (vs. chin strongly mottled); and caudal base band notched posteriorly (vs. not notched) (Robins 1954; Jenkins and Burkhead 1994; Pflieger 1997; Kinziger and Wood 2010). If specimens suspected to be the allopatric *U. hypselura* are found in the Black River drainage of Arkansas, *U. immaculata* can be distinguished from that species in having 8 or 9 dorsal fin spines (vs. 6 or 7), 16 or 17 pectoral fin rays (vs. usually

13–15), pigmentation of ventral surface of peritoneum weak (vs. moderate to strong), and dorsal fins moderately to widely connected (vs. slightly to moderately connected).

VARIATION AND TAXONOMY *Uranidea immaculata* was originally included in *Cottus hypselurus*, described by Robins and Robison (1985) from the Ozarks of northern Arkansas and southern Missouri. Kinziger and Wood (2003) analyzed *hypselurus* populations in Arkansas and Missouri using cytochrome *b* sequences and found it to be a polytypic species composed of two monophyletic groups, one from the Osage, Gasconade, and Black river drainages (Osage-Black clade), and another from the Current, Eleven Point, and White river drainages (Current-White clade). Their data indicated that these clades represented two separate species. Because *C. hypselurus* was originally described by Robins and Robison (1985) from Bennett Springs, Missouri (Osage River drainage), the Osage-Black clade retained the name *C. hypselurus* (*Uranidea hypselura*), and the Current-White clade was elevated to new species status and named *C. immaculatus* (Kinziger and Wood 2010). The Osage-Black clade has not been found in Arkansas; therefore, *hypselurus* is no longer recognized in this state. The type locality for *C. immaculatus* is the Current River, 2.7 km east of Montauk State Park, Dent County, Missouri. These species formerly in genus *Cottus* are now in genus *Uranidea* (see genus *Uranidea* account). There is some variation within *U. immaculata*. Postpectoral prickling is weak (fewer than 10 prickles) in White River system individuals and moderate to strong (greater than 10 prickles) in the Current, Eleven Point, and Spring river systems specimens (Kinziger and Wood 2010).

DISTRIBUTION *Uranidea immaculata* is endemic to the White and Black river systems of Arkansas and Missouri. In Arkansas, it occurs in the White River and its tributaries upstream from its confluence with the Black River, including one recent record upstream from Beaver Dam. It is also found in the Black River system (Current, Eleven Point, and Spring rivers) of the Ozark Mountains ecoregion (Map 160) but is conspicuously absent from the Strawberry River of that drainage. It has been introduced into the Little Red River of the White River system where it is apparently established.

Uranidea immaculata was reported to be introduced in the Little Red River by Connior et al. (2013). However, Tom Bly (pers. comm. to TMB) had previously informed us that sculpins were introduced around 2008 by anglers in the Beech Island area, about 2 miles downstream of the Greers Ferry dam site. Large numbers of sculpins were found in 2013 trout samples from JFK Park downstream to Beech Island. Knobfin Sculpins have now extended their

Figure 25.4. Knobfin Sculpin in Cotter Spring, White River at Cotter, Arkansas. *Isaac Szabo.*

range downstream in the Little Red River as far as Cow Shoals and Jon's Pocket, but they are found there only in small numbers (Tom Bly, pers. comm.).

HABITAT AND BIOLOGY The Knobfin Sculpin inhabits cool, clear, highly oxygenated, swift streams and riffle portions of rivers over gravel or rock substrates. It is most abundant in cool to cold spring-fed creeks and rivers with cobble bottoms (Kinziger and Wood 2010). Those authors noted that it has not been found in the lower reaches of the Black River (between the confluences of the Current, Eleven Point, Spring, and White rivers) in Arkansas, presumably because of unsuitable habitat in this lowland faunal region. It is also abundant in the cold tailwaters of the White River immediately below Bull Shoals and Norfork lakes. Cold water release from those dams has likely resulted in downstream range expansion of this species. It is often collected in thick growths of watercress (*Nasturtium officinale*) (Pflieger 1997). At Mammoth Spring, this sculpin is abundant in a large spring run below the main spring area, where it lives in dense beds of *Potamogeton. Uranidea immaculata* seems to prefer cooler waters than *U. carolinae*, but the two are often found together.

Uranidea immaculata feeds on crayfish (*Orconectes*), Trichoptera larvae (Hydropsychidae), Plecoptera, Ephemeroptera, Diptera (*Chironomus* and *Simulium*), as well as gastropods, elmid beetles, and miscellaneous animal material (Cooper 1975), including other *U. immaculata* (Robins and Robison 1985).

Like other *Uranidea* species, *U. immaculata* is a territorial, cavity-nesting species that attaches eggs to the ceiling of its nesting cavity. Pflieger (1997) reported that egg masses of this species have been found attached to the undersides of rocks. Kinziger and Wood (2010) found that nuptial males possess enlarged knobs on the tips of the dorsal fin spines, a character previously unreported in genus *Uranidea*. They suggested that the knobs may serve as egg-mimics that prompt mating, or the knobs may protect eggs from puncture during nesting. Kinziger and Wood (2010) reported gravid females and nuptial males in collections from 4 March to 26 April in the Current, Eleven Point, Spring, and White river systems. We have collected nuptial males during the first week in February at Mammoth Spring. Robison et al. (2013) collected several hundred nuptial individuals on 11 March 2012 from the Spring River. Two different orange egg clusters of 50–80 eggs per mass were discovered in the swift, rocky riffle areas where these sculpins were taken. Breeding adults and an egg mass were found in the White River system on 29 March 1961 by Knapp and W. T. Richards (CU 37198) and on 6 March by Kinziger and Wood (2010). While these data suggest a February through April breeding season, Pflieger (1997) found an egg mass, presumably belonging to *U. immaculata*, on 2 January 1986 in the White River system in Missouri at a water temperature of 55.5°F (13.1°C). This suggests that spawning may begin as early as December. Spawning in *Uranidea* is triggered by rising water temperatures reaching a critical point. The time at which spawning temperature is reached varies from river to river and from one year to the next (Robins and Robison 1985).

CONSERVATION STATUS The Knobfin Sculpin is locally abundant and widespread in the White and Black river systems in Arkansas. We consider its populations secure.

CHAPTER 26

Order Perciformes
Perches and Relatives

Perciformes is the most diversified of all fish orders and is the largest order of vertebrate animals (Nelson 2006). Classification of the group was considered unsettled in 2006 (Nelson 2006), and controversy still envelops it today. This order, as constituted for the past several decades, was not monophyletic and could not be clearly defined cladistically (Boschung and Mayden 2004; Wiley and Johnson 2010). It has no unique specialized character or combination of specialized characters, and it has long been thought that several groups would eventually be split off as separate orders. Nelson et al. (2016) took that step by removing a number of taxa (mainly families) from Perciformes and either establishing them as new orders or placing them in other previously recognized orders as did Betancur-R. et al. (2017). To get an indication of the extent of those changes, Nelson (2006) recognized 20 suborders, 160 families, 1,539 genera, and 10,033 species in Perciformes, but Nelson et al. (2016) recognized 2 suborders, 62 families, about 365 genera, and about 2,248 species. Nelson et al. (2016) further stated that even after removal of many families to other groups of percomorphs, additional percoid families may be removed from the order in the future.

Several changes in Perciformes in Nelson et al. (2016) affect taxa found in Arkansas. The most significant change by those authors split off four perciform families found in Arkansas (Channidae, Moronidae, Sciaenidae, and Cichlidae) into new orders. We are unwilling at this time

to commit to those family changes except for removal of Channidae from Perciformes. We do not necessarily reject the other changes in Nelson et al. (2016), but we are not yet ready to incorporate them without evidence of widespread acceptance by the ichthyological community (which could be forthcoming). The systematic changes recognized by Nelson et al. (2016) and Betancur-R. et al. (2017) are noted at the appropriate places in our perciform family accounts. We recognize 82 perciform species in Arkansas (more than a third of the state's fish species), grouped into two suborders: (1) Percoidei (79 species), the principal suborder, containing families Moronidae, Centrarchidae, Elassomatidae, Percidae, and Sciaenidae; and (2) Labroidei (3 species) containing family Cichlidae.

Family Moronidae *Temperate Basses*

Temperate basses have had a controversial taxonomic history. They were originally placed in the family Serranidae, but Gosline (1966) separated these mainly freshwater percoids from the more specialized marine serranids, placing them in the family Percichthyidae. G. D. Johnson (1984) subsequently erected the family Moronidae based on lack of evidence of affinities with the rest of the Percichthyidae. Nelson et al. (2016), as noted in the Perciformes account above, removed Moronidae from Perciformes and placed it in the order Moroniformes, a change that we are hesitant to accept. We follow Johnson (1984) and other recent authors

(Page et al. 2013) in continuing to recognize the family Moronidae as a perciform group. Moronidae is represented by one genus, *Morone*, in North America containing four species, and by one genus, *Dicentrarchus*, in Europe and northern Africa with two species. Some authors place the Asian sea basses (*Lateolabrax* spp.) in this family, but E. P. Williams et al. (2012) found a unique genetic marker that grouped *Morone* and *Dicentrarchus* into a monophyletic family and excluded *Lateolabrax*, which is now in the family Polyprionidae. Additional genetic evidence reinforced the continued separation of *Morone* and *Dicentrarchus* as separate genera (Williams et al. 2012).

The temperate basses are popular gamefishes, especially the White Bass, which commonly exceeds 1 pound (0.5 kg) in weight, and the larger Striped Bass, commonly weighing 5–20 pounds (2.3–9 kg) in Arkansas. The family is characterized by 2 dorsal fins, the first with spines, the second with 1 spine and 11–14 soft rays; a large mouth with well-developed teeth; a body with ctenoid scales; the rear of the operculum with a sharp spine; and a forked caudal fin. Temperate basses superficially resemble sunfishes (family Centrarchidae) but are not closely related to them. Moronidae species, in contrast to the sunfishes, build no nests and provide no parental care for the eggs and young. Additionally, temperate basses possess well-developed pseudobranchiae (vs. pseudobranchiae small and concealed in sunfishes) and differ from the centrarchids in several other anatomical features.

Genus *Morone* Mitchill, 1814—Striped Basses

All four current species of *Morone* were formerly assigned to *Roccus* or *Lepibema* until the 1960s, when the priority of *Morone* was documented (Whitehead and Wheeler 1966). Three of the four species (all except *M. americana*) have prominent, horizontal dark stripes on the sides. Two species, the White Bass (*M. chrysops*) and the Yellow Bass (*M. mississippiensis*), are native to Arkansas, and two species are anadromous, euryhaline, nonindigenous forms. The Striped Bass (*M. saxatilis*) is an anadromous Atlantic Coast form that has been successfully introduced into the state. The White Perch (*M. americana*), another anadromous Atlantic Coast species, has also been documented from Arkansas (Buchanan et al. 2007). In addition, hybrids between Striped Bass and White Bass have been widely stocked in Arkansas. Natural hybridization among the four *Morone* species has been widely reported in North America, and the reproductive viability of F1 and later generation hybrids has been documented (Lueckenhoff 2011;

Harrell 2013). Harrell (2013) considered the frequency of occurrence of introgression in the wild to be low and discussed the implications of releasing hybrid *Morone* in natural waters with congeneric species. Characters of the genus *Morone* (in addition to family characters listed above) include body relatively deep and compressed, lower jaw projecting beyond upper jaw, preopercle serrate, a complete lateral line, 3 anal fin spines, pelvic fins with 1 spine and 5 rays, 7 branchiostegal rays, 17 principal caudal fin rays, and jaws with rows of villiform teeth. Sexual dimorphism in the male consists of a single urogenital opening behind the anus; in the female, there are two openings (genital and urinary) behind the anus (Moen 1959).

Key to the Temperate Basses

1A First and second dorsal fins slightly connected by a membrane; no tooth patches on back of tongue; anal rays 8 or 9; second anal spine thickened, as long as (or almost as long as) third spine (Fig. 26.1A).   2

1B First and second dorsal fins separate, not connected by a membrane; one or two tooth patches on back of tongue; anal rays 10–13; second anal spine not noticeably thickened and distinctly shorter than third spine (Fig. 26.1B).   3

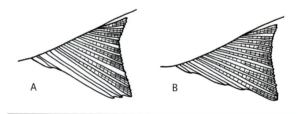

Figure 26.1. Anal fin spine lengths and size of second anal spine in *Morone* species.

A—*Morone americana* and *M. mississippiensis*
B—*Morone chrysops* and *M. saxatilis*

2A Lateral stripes absent, or if present, faint and not sharply offset on lower side in front of anal fin; greatest body depth occurring just in front of first dorsal fin; longest dorsal spine about half the head length; color in life silvery.   *Morone americana* Page 563

2B Lateral stripes dark, distinct and sharply offset on lower side in front of anal fin; greatest body depth occurring below first dorsal fin, the depth remaining fairly uniform below the entire fin; longest dorsal spine longer than half the head length; color in life usually brassy yellow.   *Morone mississippiensis* Page 568

3A One tooth patch on the tongue (near the midline toward the back of the tongue) (Fig. 26.2B).

Morone chrysops Page 565

3B Two distinct tooth patches toward the back of the tongue near the midline (Fig. 26.2A, C). 4

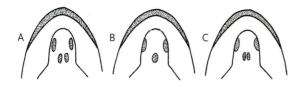

Figure 26.2. Tooth patches on tongue of *Morone* species:

A—*M. saxatilis*
B—*M. chrysops*
C—*M. chrysops* × *saxatilis* hybrid

4A Body depth going into standard length fewer than 3 times; most lateral stripes are broken.

Morone saxatilis × *Morone chrysops* hybrid.

4B Body depth going into standard length more than 3 times; lateral stripes distinct, usually only 1 or 2 are broken. *Morone saxatilis* Page 570

Morone americana (Gmelin)
White Perch

Figure 26.3. *Morone americana*, White Perch.
David A. Neely.

CHARACTERS A deep-bodied, laterally compressed fish with a distinctly arched back; greatest body depth occurs just anterior to or at the origin of the first dorsal fin, giving the body shape an appearance slightly reminiscent of *Aplodinotus grunniens* (Fig. 26.3). No distinct dark lines or stripes on back or sides of adults. Mouth rather small, length of upper jaw usually going 2.7–3.5 times into head length. A sharp spine present on posterior angle of opercle. First and second dorsal fins slightly connected. First dorsal fin usually with 9 spines; second dorsal fin with 1 spine and 11 or 12 rays. Anal fin with 3 spines and 8 or 9 rays. Anal

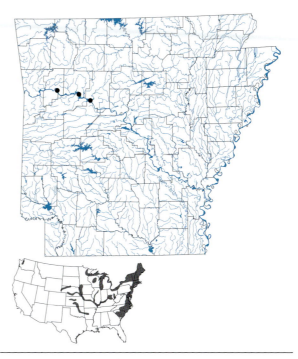

Map 161. Distribution of *Morone americana*, White Perch, 1988–2019. Inset shows native and introduced ranges in the contiguous United States.

spines not graduated, the first much shorter than the long, stout second anal spine, and the second spine just slightly shorter than the third anal spine (Fig. 26.1A). Pectoral fin rays 14–17; pelvic fin with 1 spine and 5 rays. Back of tongue without teeth. Lateral line complete, with 45–53 scales. Scales around caudal peduncle usually 20–22. A small species that rarely exceeds 9 or 10 inches (229–254 mm) and 0.5 pound (0.23 kg) outside its native range. Maximum TL 22.8 inches (580 mm) and maximum weight (from New Jersey) 3 pounds, 1 ounce (1.38 kg).

LIFE COLORS Back silvery gray to steel blue, sometimes with green reflections. Sides silvery. Belly white. No distinct dark stripes on back or sides of adults; if faint stripes or lines are present, they are not sharply offset on lower side anterior to anal fin. Juveniles have interrupted dark lines or bars on sides. Dorsal, caudal, and anal fins silvery near their bases, darker distally; pectoral and pelvic fins mostly silvery white or transparent (Trautman 1957).

SIMILAR SPECIES Despite its common name, the White Perch is not a member of the perch family. It can be distinguished from all true perches by its 3 anal spines (vs. 1 or 2 anal spines in Percidae). The White Perch differs from all other *Morone* species in adults usually lacking dark stripes on back and sides (vs. dark lateral stripes present). It further differs from the White Bass and Striped

Bass in having slightly connected first and second dorsal fins (vs. dorsal fins not connected by a membrane), back of tongue without teeth (vs. back of tongue with 1 or 2 patches of teeth), and anal rays usually 8 or 9 (vs. anal rays 10–13). The White Perch differs from the Yellow Bass in having its greatest body depth just in front of the first dorsal fin (vs. greatest body depth below first dorsal fin, and body depth remaining fairly uniform below the entire first dorsal fin), color in life silvery (vs. color in life yellow), and lateral stripes, if present, faint and not sharply offset in front of anal fin (vs. dark lateral stripes present and sharply offset on lower side just anterior to anal fin).

VARIATION AND TAXONOMY *Morone americana* was described by Gmelin *in* Linnaeus (1789) as *Perca americana* from Long Island, New York. There are few studies of morphological or genetic variation in White Perch. Mulligan and Chapman (1989) studied variation in mtDNA in tributaries of Chesapeake Bay and identified the existence of three White Perch stocks. Liu and Ely (2010) analyzed a nucleotide sequence of a fragment of the major histocompatibility complex (MHC) class Ia gene in all four *Morone* species and found high levels of allelic diversity and gene duplication in *M. americana*, *M. mississippiensis*, and *M. saxatilis*; extremely low levels of genetic diversity were found in *M. chrysops*. Natural hybridization between White Perch and White Bass was first noted in the Great Lakes region in the 1980s (Todd 1986), and hybrids between White Perch and Yellow Bass were found in 2000 in the middle Illinois River (Irons et al. 2002).

DISTRIBUTION The White Perch is native to freshwater coastal drainages and estuaries of the Atlantic Slope of North America from Nova Scotia, Canada, to South Carolina. It has been accidentally and intentionally introduced into the Mississippi River system for at least six decades (Fuller et al. 1999). It was found in Lake Ontario in the 1940s and subsequently spread throughout the Great Lakes. It entered the Mississippi River from Lake Michigan through the Chicago Sanitary Canal and has been reported from that river as far south as Cape Girardeau and Scott counties in Missouri in 1993 and 1994 (Pflieger 1997). Additional Missouri records are from the Missouri River resulting from introductions upstream in that system in Nebraska during the 1960s. In Oklahoma, it is established in Kaw, Sooner, and Keystone reservoirs. There are three known collection records for *M. americana* in Arkansas, all from Dardanelle Reservoir (Map 161): (1) The first records of *M. americana* from Arkansas were three adult specimens collected on 22 February 2006 from the Arkansas River (Dardanelle Reservoir) below Ozark Lock and Dam in Franklin County (Buchanan et al.

2007). (2) On 6 September 2006, one adult was taken in a rotenone population sample from the Cabin Creek arm of Dardanelle Reservoir in Johnson County (Buchanan et al. 2007). (3) One specimen was taken on 26 March 2018 from Dardanelle Reservoir in Yell County. The Arkansas specimens were probably the result of downstream movement from White Perch populations known to be established in the Arkansas River system of Oklahoma since 2000. It might also occasionally be found in the Mississippi River of Arkansas, based on its occurrence in that river in Cape Girardeau and Scott counties, Missouri, in 1993 and 1994 (Pflieger 1997).

HABITAT AND BIOLOGY The White Perch inhabits a variety of coastal environments in its native range but is most abundant in estuaries and other brackish-water habitats. It is more restricted in its seaward range than the Striped Bass, but it also thrives in many freshwater ponds, lakes, and rivers. In its native range, it is an important commercial and sport species. The White Perch is a gregarious, schooling species found over a variety of substrates, and it has little preference for vegetation or other cover. It preferred turbid water with high conductivity in Lake Champlain (Hawes and Parrish 2003). Large populations are established in reservoirs in Nebraska, and it also occurs in floodplain habitats along the Platte and Missouri rivers (Pflieger 1997). Introduced populations tend to become dominated by a high density of small-bodied, slow-growing individuals (Bethke et al. 2014). Five years after the White Perch was first recorded from Kaw Reservoir, Oklahoma, a small population had become established; however, there was no evidence that White Bass and White Crappie populations had been adversely affected based on catch rate data (Kuklinski 2007). Growth rates of White Perch were slow in Kaw Reservoir, with adults seldom exceeding 200 mm total length. White Perch abundance in Sooner Lake, Oklahoma, (a nonstunted population) rapidly increased after its establishment, and Largemouth Bass, White Bass, and Channel Catfish abundances decreased (Copeland 2016).

The White Perch occupies a wide trophic niche, which may aid in successfully establishing invasive populations (Feiner et al. 2013a). Its food in freshwater consists of microcrustaceans, immature aquatic insects, crayfish, and small fish (Jenkins and Burkhead 1994). In Lake Champlain, White Perch fed on benthic invertebrates, zooplankton assemblages dominated by *Daphnia*, native and invasive freshwater mussels, and small baitfish (Couture and Watzin 2008). The first foods of larvae from the Potomac River were mainly rotifers and copepod nauplii; late larval and early juvenile stages consumed mainly adult copepods and cladocerans (Stanley and Danie 1983). Subadult and adult White Perch prey mainly on benthic macroinvertebrates, but older adults become increasingly piscivorous. Food habit studies have shown that White Perch compete directly with White Bass throughout juvenile and adult stages (Parrish and Margraf 1991; Prout et al. 1990; Madenjian et al. 2000). There is also potential competition with White Crappie because the two species select similar prey items. White Perch and White Crappie both feed on *Daphnia* spp., *Cyclopoida* spp., and chironomids, especially during the first year of life (Kuklinski 2007). Schaeffer and Margraf (1987) and Roseman et al. (2006) found that White Perch also preyed upon the eggs of other fish species such as Walleye. In three North Carolina reservoirs, small and medium White Perch had the largest niche areas and high diet overlap with Bluegill, whereas large White Perch had moderate diet overlap with Striped Bass, Largemouth Bass, and Walleye (Feiner et al. 2013a). In Kaw Reservoir, Oklahoma, White Perch fed predominantly on insects, and no egg predation was found (Kuklinski 2007). In that study, there was moderate diet overlap between White Perch and White Bass smaller than 200 mm. In Sooner Lake, Oklahoma, a shift in diet occurred seasonally and by size from zooplankton to invertebrates and fish (Porta and Snow 2017). In that study, White Perch diets were dominated in spring by zooplankton, in summer by invertebrate prey, and in fall and winter by fish. Small fish ate mostly invertebrates and zooplankton, but as fish size increased, White Perch consumed larger amounts of fish, which became the predominant prey at quality and preferred size classes. Porta and Snow (2017) concluded that the shift in diet (seasonally and by size) from zooplankton to invertebrates and fish suggested that White Perch may compete with resident sport fishes for these resources in Sooner Lake. Feiner et al. (2013b) also found a high diet overlap between juvenile White Bass and all sizes of the recently introduced White Perch in Lake James, North Carolina. In Lake James, White Perch occupied lower trophic positions than White Bass, indicating that White Bass maintained a more piscivorous diet.

In its native range, the White Perch spawns in a variety of habitats, including rivers, lakes, and brackish tidal waters with up to 4.2 ppt salinity (Hardy 1978). It ascends coastal streams to spawn over fine gravel, rock, or sand substrates, but some landlocked populations exist. Movements, reproduction, and other ecological aspects were studied in Maryland by Mansueti (1961, 1964). The annual cycle of gametogenesis was studied by Jackson and Sullivan (1995). It ascends large rivers in spring to spawn when water temperatures are between 52°F and 61°F (11–16°C). Spawning

occurred in Nebraska in May and early June (Hrabik et al. 2015). The White Perch has adhesive eggs that sink and stick together in masses or attach to the substrate in quiet water, but eggs are pelagic in free-flowing streams and tidal waters (Lippson and Moran 1974). The eggs are not adhesive after water hardening, which occurs within the first half-hour after ovulation (Mansueti 1964). Hatching occurs in 6 days at 52°F (11°C). Fecundity ranges from 5,000 to 320,000 eggs per female, with an average fecundity of about 40,000 eggs (Hardy 1978; Carlander 1997). Males and females may reach sexual maturity by age 1, and maximum life span is 5+ years. In Arkansas, it is not known if White Perch use tributary streams for spawning at the same time as White Bass and Yellow Bass.

CONSERVATION STATUS   This nonindigenous species spread into Arkansas down the Arkansas River from Oklahoma in the early 2000s. Surprisingly, extensive electrofishing sampling in the upper three navigation pools of the Arkansas River in Arkansas between 2006 and 2016 produced no new records of *M. americana* (Frank Leone, pers. comm.). The 2018 record from Dardanelle Reservoir may represent a straggler from upstream populations or a record from a small established population. Lag times between introduction and range expansion are common with many invasive species. Based on its successful establishment in other areas of the United States, including Oklahoma, it is likely that the White Perch will become more widespread and established in Arkansas in the future.

Morone chrysops (Rafinesque)

White Bass

Figure 26.4. *Morone chrysops*, White Bass. *Uland Thomas.*

CHARACTERS   A moderately deep bodied, laterally compressed, silvery fish, usually with 6–12 dark longitudinal stripes on each side; the lower stripes are often discontinuous but not sharply offset above the front of the anal fin (Fig. 26.4). Back is arched in front of the first dorsal fin. Body depth going into SL fewer than 3 times (about

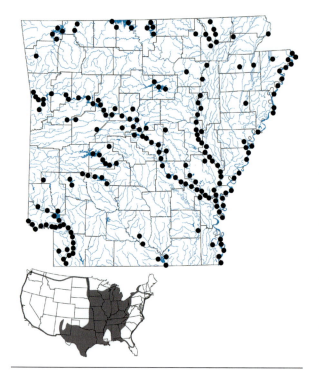

Map 162. Distribution of *Morone chrysops*, White Bass, 1988–2019. Inset shows native range (solid) and introduced range (outlined) in the contiguous United States.

2.7 times) in adults. Ratio of body depth to head length is greater than 1. Head small with an acute snout and a large mouth; lower jaw projects beyond upper jaw. Head length going into SL about 3 times. Jaws with minute needlelike teeth. Tongue with teeth near tip and along sides; teeth on back of tongue are usually in a single patch (sometimes in two patches). Gill rakers 20–25. Margin of preopercle serrate; a sharp spine present on posterior angle of opercle. First and second dorsal fins are entirely separate. Caudal fin is forked. Lateral line complete, nearly straight, with 50–58 scales. First dorsal fin with 9 spines; second dorsal fin with 1 spine and 13–15 soft rays. Anal fin with 3 graduated spines, the first shorter than the second, and the second shorter than the third (Fig. 26.1B); anal rays 11–13. Pectoral fin rays 15–17; pelvic fin with 1 spine and 5 rays. Adults are commonly 10–15 inches (254–381 mm) TL and weigh around 1 pound (0.5 kg). Maximum TL around 17.75 inches (450 mm), and weight of 6 pounds, 13 ounces (3.1 kg). The Arkansas state record, from the Mississippi River in 2005, is 5 pounds, 6 ounces (2.4 kg).

LIFE COLORS   Back blue gray with greenish, golden, and steel-blue reflections. Sides silvery white with 6–12 brown longitudinal stripes. In some individuals, the lateral stripes are faint. Undersides white. Dorsal, caudal, and anal fins dusky; pectoral and pelvic fins white to clear. Breeding adults brilliant blue white when taken from clear water and more olivaceous when taken from turbid waters (Trautman 1957).

SIMILAR SPECIES   It differs from the Yellow Bass and White Perch in having the first and second dorsal fins entirely separate (vs. slightly connected by a membrane), 11–13 soft rays in anal fin (vs. 10 or fewer), and second anal spine distinctly shorter than the third anal spine (vs. second and third anal spines about equal in length). It differs from the introduced Striped Bass in having a deeper body, its depth in adults going into SL fewer than 3 times (vs. body depth going into SL more than 3 times); in usually having a single patch of teeth on back of tongue (vs. teeth in two parallel patches); and in having the top of the head and forward part of the back concave in profile (vs. convex profile).

VARIATION AND TAXONOMY   *Morone chrysops* was described by Rafinesque (1820a) as *Perca chrysops* from the falls of the Ohio River at Louisville, Kentucky. It has, in the past, been included in the genera *Labrax*, *Roccus*, and *Lepibema* by various workers. It was considered the sister species of *M. saxatilis* (Boschung and Mayden 2004). A protein electrophoretic survey of White Bass found significant interpopulation genetic heterogeneity; however, the overall level of differentiation among populations was low (White 2000). Introduced Striped Bass can hybridize and produce fertile offspring with White Bass in the wild (Forshage et al. 1986; Avise and Van Den Avyle 1984), and natural hybridization between the two species has been reported from Lake Maumelle and Beaver Lake in Arkansas (Crawford et al. 1984). Artificial hybrids are readily produced in hatcheries and frequently stocked in Arkansas reservoirs. Natural hybrids have been reported between *M. chrysops* and *M. americana* (Boileau 1985; Todd 1986) and between *M. chrysops* and *M. mississippiensis* (Fries and Harvey 1989). In Toledo Bend Reservoir, Texas, three White Bass × Striped Bass hybrids were confirmed, but there was no evidence for either a hybrid swarm or substantial introgression among White, Yellow, and Striped Bass (based on analysis of microsatellite DNA genotypes) despite the introduction of more than 9 million Striped Bass over four decades (S. S. Taylor et al. 2013). The hybrids are reported to have a shorter life span and less reproductive success than White Bass (Bartley et al. 2000). Volitional tank spawning of female Striped Bass with male White Bass was reported by Woods et al. (1995), but those authors suggested that the Striped Bass females had a behavioral aversion to spawning with White Bass males. In other volitional hybridization trials, White Bass and White Perch hybridized with one another, but Striped Bass and White Bass did not (Salek et al. 2001).

Crawford et al. (1984) and Kerby (1979) provided morphological features for distinguishing F₁ White Bass × Striped Bass hybrids from both parental species, and Harrell and Dean (1988) provided features for distinguishing juvenile *Morone* hybrids, although accuracy in identification is often low (Muoneke et al. 1991). Back-crossed individuals can probably be correctly identified only by genetic analysis. In a Texas study, *Morone* spp. and hybrids >254 mm TL were identified with a 95% accuracy (Storey et al. 2000). Fulford and Rutherford (2000) investigated the use of geometric morphological shape differences to differentiate laboratory-reared larval (4–22 days post-hatching, <10 mm SL) *M. chrysops*, *M. saxatilis*, and their hybrids. Lueckenhoff (2011) reported that age 0 White Bass and age 0 hybrid Striped Bass (*M. saxatilis* × *M. chrysops*) could be distinguished from one another with only 71% or less accuracy using traditional morphological features (such as number of tooth patches on the base of the tongue, and broken lateral stripes in the hybrids). In that study, a 98.9% accuracy in identification of juveniles was achieved by using the ratio of caudal peduncle depth to standard length: if the standard length divided by the caudal peduncle depth is 7.30 or less, the fish should be classified as a White Bass; if the ratio is greater than 7.30, the fish should be classified as a hybrid Striped Bass.

DISTRIBUTION *Morone chrysops* occurs in the St. Lawrence–Great Lakes, Hudson Bay, and Mississippi River basins from Quebec, to Manitoba, Canada, and south to Louisiana, and in Gulf Slope drainages from the Apalachicola River, Florida, to the Rio Grande of Texas and New Mexico. The native range is uncertain, and most, if not all, Gulf Slope populations may be introduced. It has been widely introduced elsewhere in the United States. Widely distributed in Arkansas, the White Bass occurs in all major drainages (Map 162). It is most abundant in the large rivers and large impoundments, and is mostly absent from the more mountainous regions of the Ozarks and Ouachitas.

HABITAT AND BIOLOGY The White Bass, commonly called sand bass and (mistakenly) striped bass by some fishermen, is primarily an inhabitant of moderate-sized to large rivers, where it is found in both current and backwater areas in clear water over a firm sand or rock bottom. In impounded Alabama rivers, it was commonly collected along riprap-lined banks, downed trees, and other structures below dams (Mettee et al. 1996). In Nebraska reservoirs, the White Bass was most commonly found in open midwater areas (Hrabik et al. 2015). It is intolerant

of continuous, high turbidity and has become abundant through stocking in most large reservoirs of Arkansas. It thrives in clear, open waters of reservoirs and can usually be found in the deeper pools of rivers.

Adult White Bass are mobile, active, open-water sight feeders that travel in large schools in search of food. Olmsted and Kilambi (1969, 1971) and Pflieger (1997) described the food and feeding activities of this species in Beaver Reservoir. It fed most actively in early morning and late afternoon. The young ate primarily small crustaceans and insects, and adults fed mainly on small fishes. The dominant food was shad, followed by centrarchids and cyprinids. Aquatic insects and crustaceans were also important. Larval White Bass first begin feeding when the yolk sac is absorbed (at about 7 mm SL), and fish between 7 and 12 mm SL consume larval fishes as well as copepods, cladocerans, and midge larvae (Clark and Pearson 1979). Olmsted and Kilambi (1969) found that the introduction of Threadfin Shad into Beaver Lake did not appreciably alter overall food habits. In some studies, piscivory declined in individuals between 48 and 120 mm TL, and the bulk of the food consisted of zooplankton (primarily microcrustaceans) and insect larvae (Matthews et al. 1992). As adult size is reached (usually >260 mm TL), fishes increasingly dominate the diet, but in some northern areas the adults fed mainly on invertebrates (Priegel 1970). The success of White Bass in Arkansas reservoirs is dependent on an abundance of shad. Significant populations of White Bass exist throughout the Arkansas River largely because of the abundance of shad and Mississippi Silversides.

Morone chrysops exhibits impressive spawning migrations in late winter and early spring, moving into smaller tributary streams when water temperatures rise above 7–13°C (45–55°F). Anglers in Arkansas can often be observed crowding together in early March along surprisingly small streams to take advantage of the spawning runs. In reservoirs that lack suitable tributaries for spawning, White Bass spawn over windswept shores in water 0.6–2.1 m deep. In Arkansas, males usually begin the spawning run in late February, well ahead of the females. In the tributary streams, males form schools over the spawning areas, and the females arrive a few weeks later and form schools in deeper water not far away. When ready to spawn, the females enter the spawning areas occupied by the males. Just before and during spawning, males displayed a stereotypical circling behavior (Salek et al. 2001). Spawning takes place near the surface, usually in some current at water temperatures between 12°C and 20°C (54–68°F). The adhesive fertilized eggs settle to the bottom and become attached to the gravel substrate. Fertilized eggs hatch in about 2 days at 19–21°C (66–70°F) (Dorsa and Fritzsche 1979). Based on stages of oocyte development in mature females, Berlinsky et al. (1995) concluded that the White Bass is a multiple-clutch, group-synchronous spawner. The peak of spawning activity in Beaver Lake occurred in mid- to late March (Newton and Kilambi 1969). In that reservoir, White Bass apparently had a short spawning season and females reached sexual maturity at age 2 (most of the spawning females were in that age group). Fecundity ranged from 61,700 to 994,000 eggs. In DeGray Reservoir, Arkansas, spawning usually occurs during the last two weeks of March at water temperatures of 53.6–62.6°F (12–17°C) (Moen and Dewey 1980). Yellayi and Kilambi (1970) investigated early development, and Houser and Bryant (1970) studied age, growth, and maturity of White Bass in Bull Shoals Reservoir. White Bass YOY averaged 24 mm SL on 4 June 1997, and 40 mm SL on 20 June 1997 in the Arkansas River near Fort Smith (TMB, unpublished data). Growth of the young is rapid and the life span is short, with individuals rarely living more than 5 years (Yellayi and Kilambi 1976); however, Duncan and Myers (1978b) reported that three individuals tagged in Beaver Reservoir and later recaptured by anglers were 8 years old (1) and 9 years old (2), establishing a longevity record for the White Bass in the southern United States. During the spawning runs, White Bass are easily caught by fishermen using artificial lures or live bait.

CONSERVATION STATUS   This is a widely distributed gamefish. Its populations are currently secure, but it may be decreasing in abundance in some Arkansas River navigation pools such as Dardanelle Reservoir, where Yellow Bass populations have been increasing during the past two decades. It is also possible that its numbers could decrease in the Arkansas River as populations of the non-native White Perch increase. Hybridization of White Bass with White Perch has occurred in some areas (Boileau 1985; Todd 1986) and could adversely impact this species in Arkansas in the future.

Morone mississippiensis Jordan and Eigenmann
Yellow Bass

CHARACTERS   A moderately deep-bodied, laterally compressed, silvery-yellow fish with 5–8 dark longitudinal stripes on the side; the lower stripes are sharply broken and offset above the front of the anal fin (Fig. 26.5). Back somewhat arched in front of first dorsal fin; body depth going into SL fewer than 3 times (about 2.7 times). Head small, head length going into SL about 2.8 times. Snout acute

Figure 26.5. *Morone mississippiensis*, Yellow Bass. *David A. Neely.*

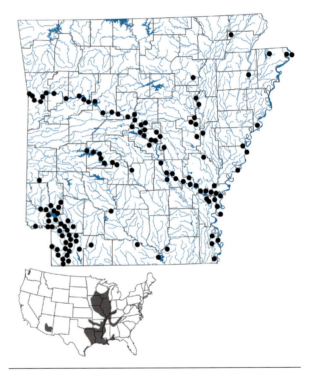

Map 163. Distribution of *Morone mississippiensis*, Yellow Bass, 1988–2019. Inset shows native and introduced ranges in the contiguous United States.

and mouth large. Lower jaw about equal in length to upper jaw. Margin of preopercle serrate; a sharp spine is present on rear margin of opercle. First and second dorsal fins are slightly connected by a membrane. Caudal fin forked. Back of tongue without tooth patches, but teeth are well-developed on anterior margin of tongue. Gill rakers 18–22. Lateral line complete, usually with 51–55 scales. First dorsal fin with 9 spines; second dorsal fin with 1 spine and 11 or 12 rays. Anal fin with 3 spines not graduated in length, the second and third spines nearly equal and much longer than the first spine (Fig. 26.1A); anal rays 9 or 10. Pectoral fin rays 15–17; pelvic fin with 1 spine and 5 rays. Arkansas's smallest native *Morone* species, reaching a maximum length of 18 inches (457 mm) and a maximum weight of 3 pounds, 2 ounces (1.42 kg) (Becker 1983). The Arkansas state record from Gillham Lake in 2009 is 2 pounds, 2 ounces (1.0 kg).

LIFE COLORS   Back olive green. Sides brassy yellow with 6–8 prominent brown or black longitudinal stripes. In some individuals, the overall body coloration is more silvery than yellow. Undersides white to yellow. The eye is often yellow. Dorsal, caudal, and anal fins blue or dusky; pectoral and pelvic fins are white to clear.

SIMILAR SPECIES   The Yellow Bass differs from the White Bass in having the first and second dorsal fins slightly connected by a membrane (vs. separate), 9 or 10 soft rays in anal fin (vs. 11–13), second and third anal spines about equal in length (vs. second anal spine distinctly shorter than the third), and in having the lateral stripes sharply offset above front of anal fin (vs. stripes not offset). It differs from the Striped Bass in having a deeper body, its depth going into SL fewer than 3 times (vs. body depth going into SL more than 3 times), in lacking patches of teeth on back of tongue (vs. 2 parallel patches), and in having the lateral stripes sharply offset above front of anal fin (vs. stripes not offset). This species differs from *M. americana* in having dark lateral stripes tht are sharply offset on the lower side in front of anal fin (vs. lateral stripes absent, or if present, faint and not offset in front of anal fin).

VARIATION AND TAXONOMY   *Morone mississippiensis* was described by Jordan and Eigenmann *in* Eigenmann (1887) from "St. Louis, Missouri, and New Orleans, Louisiana." Until the 1950s, it was often incorrectly referred to in the literature as *Morone interrupta* (or *Roccus interruptus*), a junior synonym of *Morone saxatilis* (Bailey 1956). Bailey's work resulted in the designation of *mississippiensis* as the specific epithet. Little information on geographic variation in morphological or genetic features is available. Natural hybridization between Yellow Bass and White Bass has been reported, and in hatchery crosses, the resulting offspring were 100% females (Wolters and DeMay 1996). A genetic analysis of *Morone* specimens from Toledo Bend Reservoir, Texas, identified only one Yellow Bass × Striped Bass hybrid (Taylor et al. 2013), leading those authors to conclude that introduced Striped Bass were not causing introgression in Yellow Bass.

DISTRIBUTION   *Morone mississippiensis* occurs in the Great Lakes drainage of Lake Michigan and the Mississippi River basin from Wisconsin and Minnesota south to the Gulf of Mexico, and in Gulf Slope drainages from Mobile Bay, Alabama, west to the Galveston Bay drainage, Texas. In Arkansas, the Yellow Bass is found primarily in the Coastal Plain lowlands in the Mississippi, St. Francis, White, Arkansas, Ouachita, and Red rivers, and it also extends upstream in the Arkansas River into Oklahoma

(Map 163). It has been accidentally or intentionally introduced into reservoirs upstream in the Ouachita River drainage, where it is well-established in Lake Catherine and Lake Hamilton and occurs in smaller numbers in Lake Ouachita (Brett Hobbs, pers. comm.).

HABITAT AND BIOLOGY   A lowland species, the Yellow Bass is found primarily in clear to turbid water in the backwaters of large rivers and especially in their associated oxbow lakes. It seems to be more of a lake-adapted species than the White Bass. Some authors have noted a preference for clear water and vegetation (Etnier and Starnes 1993; Ross 2001); those authors further suggest that the Yellow Bass is less tolerant of turbidity than the White Bass. It is an open-water, schooling fish, but generally not an important game species in Arkansas because of its small size and rather small, scattered populations. There is a small winter fishery for this species in Lake Hamilton, where anglers catch them in deep water, usually while fishing for Walleye (Brett Hobbs, pers. comm.). Some anglers report that it is a good food fish, and Becker (1983) noted that its flesh is white, firm, flaky, and delicious. Three of the largest Arkansas populations are found in Lake Chicot in Chicot County, Horseshoe Lake in Crittenden County, and Mallard Lake in Mississippi County. More recently, its numbers have drastically increased throughout the western part of the Arkansas River Navigation System in Arkansas, and over the past two decades (based on AGFC sampling data) it has become the numerically dominant *Morone* species in Dardanelle Reservoir. Etnier and Starnes (1993) noted a similar increase in abundance of Yellow Bass in main channel reservoirs formed by impoundment of the Tennessee River in eastern Tennessee. They attributed this increase to the creation of more lentic habitat favored by this species coupled with an increase in water clarity.

Adult Yellow Bass live in schools and feed from midwater to the surface. In Iowa lakes, the small young fed mostly on small crustaceans, midge larvae, and mayfly nymphs (Kutkuhn 1955; Collier 1959; Welker 1962). As the juveniles grow, aquatic insects become more important in the diet, but zooplankton may also continue to be important (Darnell 1961; Van Den Avyle et al. 1983). The large adults consumed mainly fish in Iowa (Kutkuhn 1955), but in coastal areas of Louisiana, large invertebrates (such as shrimp and crabs) were important in the adult diet (Darnell 1961; Lambou 1961). In Barren River Lake, Kentucky, the introduced Yellow Bass adults and juveniles fed heavily on chironomid larvae and pupae throughout the year, and YOY *Dorosoma cepedianum* dominated the diet in spring and summer (Zervas 2010). In a lower Mississippi River oxbow lake, age 1 Yellow Bass fed on chironomids, amphipods, and fish eggs; age 2+ fish consumed fish, fish eggs, amphipods, and benthic insects, and there was very little dietary overlap between age 1 and age 2 fish (Driscoll and Miranda 1999). Peak feeding activity occurred in Iowa at night (Kraus 1963), but we have observed considerable feeding activity during the day in Arkansas.

Yellow Bass spawn in April and May in Missouri (Pflieger 1997). In Mississippi, spawning occurs in April and May at water temperatures between 61°F and 72°F (16–22°C) (Ross 2001). Spawning occurred in the upper Barataria Estuary, Louisiana, from mid-February through April (Fox 2010). In that study, GSI values decreased as temperatures reached 64–72°F (18–22°C), indicating that spawning had been completed. Although movement into tributary streams occurs in some areas, the spectacular spawning runs of the White Bass are not as pronounced in this species in Arkansas. Brett Hobbs (pers. comm.) informed us that large numbers of Yellow Bass gather in the Little Mazarn Creek backwaters of Lake Hamilton during April when White Bass are also congregating in the shoals. As in all *Morone* species, no nest is constructed. The eggs of the Yellow Bass are smaller than those of the White Bass and are deposited over the substrate in water 2–3 feet (0.6–0.9 m) deep. Apparently, males and females segregate as the spawning season approaches, but they often form pairs at the time of spawning. Frequently, one female is accompanied by several males. The spawning fish swim slowly near the surface releasing eggs and sperm. The fertilized, demersal eggs sink to the substrate where they become attached to silt-free objects (Boschung and Mayden 2004). The Yellow Bass is reportedly an asynchronous batch spawner because ova in different stages of development have been found in reproducing females (Bulkley 1970; Fox 2010). Females produce 276–560 mature ova/mm of total body length. In Wisconsin, females 148–166 mm TL contained 14,300–39,400 ripe ova (Becker 1983). Bulkley (1970) noted that there was wide variation not only in fecundity among females of similar lengths, but also from year to year. Eggs hatch in 4–6 days at a water temperature of 21.1°C (70°F). The yolk sac is absorbed in about 4 days, and the young then swim toward the surface, where they school in shallow water. Males and females mature by age 2 (Fox 2010) or age 3 (Schoffman 1958), and maximum life span is about 8 years (Zervas 2010).

CONSERVATION STATUS The Yellow Bass is secure in its preferred Arkansas habitats, and it has distinctly increased in abundance westward through the Arkansas River during at least the past two decades based on AGFC fish population sampling data.

Morone saxatilis (Walbaum)
Striped Bass

Figure 26.6. *Morone saxatilis*, Striped Bass. *David A. Neely.*

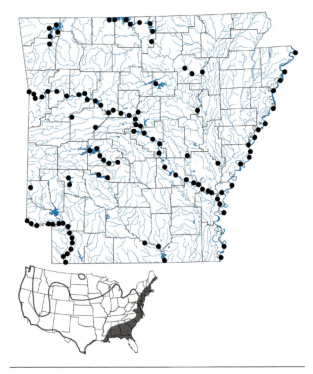

Map 164. Distribution of *Morone saxatilis*, Striped Bass, 1988–2019. Inset shows native range (solid) and introduced range (outlined) in the contiguous United States.

CHARACTERS A streamlined, moderately compressed, silvery fish, usually with 6–9 dark longitudinal stripes on sides; lower stripes not sharply broken or offset above front of anal fin (Fig. 26.6). Back not noticeably arched in front of first dorsal fin; body depth going into SL 3.5–4 times. Head small, with an acute snout and a large mouth; the lower jaw projects beyond the upper jaw. Head length going into SL 3.2–3.5 times. Ratio of body depth to head length less than 1. Margin of preopercle serrate; a sharp spine present on rear margin of opercle. First and second dorsal fins are entirely separate. Caudal fin forked. Teeth on back of tongue in two elongate parallel patches, each patch more than half as long as the patch of lateral tongue teeth; teeth at the tip of the tongue are weak or absent. Gill rakers 20–29. Lateral line complete, nearly straight, with 57–67 scales. Scales around caudal peduncle 25–30. First dorsal fin with 9 spines; second dorsal fin with 1 spine and usually 12 (8–14) soft rays. Anal fin with 3 graduated spines, the first shorter than the second, and the second shorter than the third (Fig. 26.1B); anal rays usually 11 (10–14). Pectoral fin rays 14–16; pelvic fin with 1 spine and 5 rays. A large species, with adults commonly weighing 5–20 pounds (2.3–9 kg) in freshwater. Maximum size in saltwater more than 100 pounds (45 kg) and a TL of 78 inches (1,981 mm). The Arkansas state record, from the Beaver Lake tailwater in 2000, is 64 pounds, 8 ounces (29.3 kg); the world record (from Alabama in 2013) for Striped Bass in freshwater was 69 pounds, 9 ounces (31.6 kg).

LIFE COLORS   Back olive green. Sides silver with 7 or 8 distinct black horizontal stripes. Undersides white. Dorsal, caudal, and anal fins dusky; pectoral and pelvic fins white to clear. Young less than 5 inches (127 mm) with several narrow, faint dusky vertical bars on sides.

SIMILAR SPECIES   The Striped Bass differs from White Bass, Yellow Bass, and White Perch in having a more slender body, its depth going into SL more than 3 times (vs. body depth going into SL fewer than 3 times); in having 2 parallel patches of teeth on back of tongue (vs. 1 patch or no patches of teeth on back of tongue); and in having a dorsal profile that is little arched (vs. well arched).

VARIATION AND TAXONOMY   *Morone saxatilis* was described by Walbaum (1792) as *Perca saxatilis* from "New York." Morphological and genetic studies have resulted in the recognition of distinct differences between Atlantic and Gulf Coast Striped Bass populations (Boschung and Mayden 2004). Couch et al. (2006) identified 149 novel microsatellite DNA markers for Striped Bass. A study of 12 microsatellites of Striped Bass across watersheds in the southeastern U.S. revealed that most populations were genetically diverse (Anderson et al. 2014). Hybrid Striped Bass have been produced in hatcheries since the early 1960s. One common cross, *M. saxatilis* × *chrysops* (often called the sunshine bass), is between a male Striped Bass and a female White Bass. The reciprocal cross (between a male White Bass and a female Striped Bass) is often called a palmetto bass, a wiper, or a whiterock bass. In some areas where the Striped Bass has been introduced (including Arkansas), it has hybridized with native White Bass and Yellow Bass, often producing fertile hybrid offspring (Forshage et al. 1986; Avise and Van Den Avyle 1984; Crawford et al. 1984; Taylor et al. 2013).

DISTRIBUTION   The Striped Bass is a marine species native to the Atlantic and Gulf coasts of North America, ascending freshwater rivers from the St. Lawrence River of Quebec, Canada, south to the St. Johns River, Florida, on the Atlantic Slope, and on the Gulf Slope from the Suwannee River, Florida, to Lake Pontchartrain, Louisiana. It has been widely stocked in coastal areas outside its native range. Stocking occurred along the Pacific Coast of California in 1879 and 1881, and within 20 years it had become an important commercial sport fish there (Raney 1952). It was first stocked in Arkansas in Lake Ouachita in 1956, when 27 adults were released. In 1957, more adults were stocked in Lake Greeson and again in Lake Ouachita. By the mid-1960s, the AGFC developed the ability to rear Striped Bass fry to the fingerling stage in nursery ponds. The fingerlings were eventually stocked in 12 large reservoirs in Arkansas, and a successful fishery has now become established in most of them. The largest populations are found today in Beaver Reservoir and Lake Ouachita. Significant fisheries also exist in the Arkansas and Red rivers (Map 164).

HABITAT AND BIOLOGY   This important commercial and sport species has been the subject of numerous studies. Much of the important biological literature was summarized by Scott and Crossman (1973), Burgess (1980c), and Bulak et al. (2013). The Striped Bass is an anadromous marine coastal and estuarine fish that moves inland up rivers to spawn. A landlocked population developed in the Santee-Cooper River system of South Carolina, and descendants of this stock have been widely introduced into reservoirs throughout the United States, including Arkansas.

In Arkansas reservoirs, it lives in moving schools similar to those of the White Bass. It is apparently less tolerant of warm water temperatures than our native *Morone* species, and large individuals avoid water temperatures above 72°F (22°C). This strongly influences summer distribution of Striped Bass in reservoirs, depending on reservoir temperature characteristics (Matthews 1985b; Matthews et al. 1985, 1988).

The Striped Bass is a piscivore, with diet varying depending on prey availability. The fry feed mainly on microcrustaceans, gradually adding small invertebrates and fish larvae as growth occurs. By the time juveniles reach 2–3 inches TL (50–75 mm), small fish become an important part of the diet, and individuals greater than 5–6 inches TL (152 mm) feed almost entirely on fish (Matthews et al. 1988, 1992). In Arkansas reservoirs, it feeds primarily on the abundant Threadfin Shad and Gizzard Shad. In a study of Striped Bass food habits in Lake Hamilton, Filipek and Claybrook (1984) found that 99% of the diet consisted of fish, with Threadfin Shad and Gizzard Shad composing 92.8% of the total food intake. Less than 1% of the diet consisted of crayfish and mayflies. Striped Bass in Beaver Reservoir fed primarily on Gizzard Shad (96.7% of the diet), and growth was similar to that of anadromous and freshwater populations (Kilambi and Zdinak 1981b).

Morone saxatilis is annually stocked in the major reservoirs of the state and survives well in them; however, it usually does not establish reproducing populations in reservoirs (unless those reservoirs have sizeable tributaries flowing into them) because of its special spawning requirements. Natural reproduction and recruitment have been documented in 10 of the more than 450 North American reservoirs stocked with Striped Bass, and Shelton and Snow (2017) documented reproduction and recruitment in Lake Texoma, Oklahoma. Natural spawning is successful throughout the Arkansas River in Arkansas. Striped Bass YOY ranged from 37 to 50 mm SL on 20 June 1997, 54 to 60 mm SL on 7 July 1994, and 55 to 75 mm SL on 14 August 1997 in the Arkansas River near Fort Smith (TMB, unpublished data). There is evidence that spawning also occurs in the Mississippi and Red rivers in Arkansas (juveniles averaging 73 mm TL were found in the Red River in Little River County on 18 August 1997 by TMB, Drew Wilson, and Les Claybrook). Spawning occurs in strong currents of large rivers in the spring when water temperatures exceed 58°F (14.4°C) (Pflieger 1997). In Arkansas, peak spawning activity occurs in April. In the Red and Washita rivers, Oklahoma, Striped Bass eggs were collected from early April to mid-May, with peak spawning apparently occurring in mid-April (Baker et al. 2009). The spawning act includes much splashing near the surface, with one female often accompanied by several males. In aquarium observations, the best predictor of imminent spawning was a significant increase in male attending behavior characterized by close and continuous following of the female, with males sometimes contacting the female's abdomen with the snout (Salek et al. 2001). The eggs are semibuoyant (specific gravity of approximately 1.0005) and are carried along by the current in freshwater until they hatch, which usually takes 33 hours at 70°F (21.1°C), 44 hours at 65°F (18.3°C), and 70 hours at 60°F (15.6°C) (Barkuloo 1970). Drift distance of the developing embryos until hatching has been reported to be from 30 to 160 km, depending on temperature and current velocity (Shelton and Snow 2017). Eggs have little chance for survival if they are not suspended (May and Fuller 1962; Albrecht 1964), and a minimum current velocity of 60 cm/s is required to keep water-hardened

Figure 26.7. *Morone chrysops* × *M. saxatilis* hybrid. *David A. Neely.*

eggs suspended. When there is no current, such as in reservoirs, the eggs sink to the bottom, become silted over, and die, but some hatching success has been reported in reservoirs when eggs settle onto coarse substrates with no subsequent siltation (Bayless 1968; Gustaveson et al. 1984). Sexual maturity is usually reached at 4 or 5 years of age, and fecundity is positively correlated with the size and age of females. Fecundity is reported to range from 14,000 to more than 40 million eggs (Lewis and Bonner 1966). Estimated fecundity in Beaver Reservoir ranged from 84,2320 to 1,086,540 ova in females 404–870 mm TL (Kilambi and Zdinak 1981b). Maximum life span is around 12–14 years.

Striped Bass females have been artificially crossed with White Bass males in AGFC hatcheries to produce hybrid offspring that are intermediate in features between the two parental species (some hybrids are the result of crossing White Bass females with Striped Bass males) (Fig. 26.7). These hybrids were first introduced into Lake DeGray in 1975. Twelve other Arkansas reservoirs now also have a hybrid fishery. The hybrids usually do best in lakes with hard water and have limited survival in environments having soft water (S. Lochmann, pers. comm.). Hybrids grow much larger than the White Bass, reaching a weight of around 20 pounds (9.1 kg), but they are easier to catch and grow faster than the Striped Bass. The hybrids also have the excellent fighting qualities of the White Bass. In a DeGray Lake study (Filipek and Claybrook 1984), 95.6% of the food eaten by hybrids consisted of fish, with shad composing 81.8% of the total. Crayfish made up 4.4% of the diet. The Arkansas record for a Striped Bass hybrid came from Greers Ferry Lake in 1997 and weighed 27 pounds, 5 ounces (12.4 kg).

CONSERVATION STATUS This introduced game species is maintained in Arkansas largely by stocking hatchery-raised fish. Natural reproduction occurs in the Arkansas River and probably in the Mississippi and Red rivers.

Family Centrarchidae *Sunfishes*

The sunfish family is native to North America and includes 8 genera with 33 species (Page et al. 2013). Nelson et al. (2016) placed the pygmy sunfishes in Centrarchidae (a view we do not support), based on molecular evidence, and recognized 9 genera of centrarchids with about 45 species. Betancur-R. et al. (2017) removed Centrarchidae from the order Perciformes and recognized the order Centrarchiformes. Morphologically, the sunfishes are characterized by a deep, compressed body covered with ctenoid scales (although scales lack ctenii in one species, *Acantharchus pomotis*) and by the presence of spines in the front part of the dorsal and anal fins. The spinous dorsal and soft dorsal fins are well connected with, at most, a deep notch between them. The pelvic fins are thoracic and have 1 spine and 5 soft rays. There are 5–7 branchiostegal rays, branchiostegal membranes are separate, not connected to the isthmus, pseudobranch small and concealed, and there are 17 principal caudal fin rays. Infraorbital bones are present in addition to the lachrymal (Nelson et al. 2016). Molecular divergence time estimates indicate that Centrarchidae originated in the Cenozoic and that the Oligocene and Miocene were times of peak diversification (Near et al. 2005, 2011). Fossils are known only from North America, with the oldest fossils dating to the Late Eocene (Near and Koppelman 2009).

Many of the species are important gamefishes, and some (such as the Largemouth Bass) are almost worldwide in distribution today because of intentional introductions. In Arkansas, there are five genera of sunfishes containing 19 species. Seventeen of those species are native to the state and two are introduced. The species account for a third introduced sunfish included in the first edition of this book has been removed from this edition. The Redeye Bass, *Micropterus coosae*, was stocked only in the Spring River in Arkansas in the 1970s, but no populations were successfully established and there have been no records for this species from Arkansas for at least 40 years.

Some of Arkansas's most popular gamefishes are in the sunfish family, including the black basses (*Micropterus*), crappies (*Pomoxis*), rock basses (*Ambloplites*), and bream (*Lepomis*). Members of the genus *Lepomis* are also incorrectly referred to as perch throughout Arkansas, but that common name applies exclusively to members of family Percidae. Most of the game species in the family are found statewide in lakes, streams, and ponds, but the rock basses and Smallmouth Bass are found mainly in clear, free-flowing mountain streams, a type of habitat that has declined in Arkansas during the past few decades.

Smallmouth Bass, and to a lesser extent the rock basses, have adapted to a few of the large reservoirs. Two small species of sunfishes, the Dollar Sunfish (*Lepomis marginatus*) and the Bantam Sunfish (*L. symmetricus*), are unfamiliar to most fishermen and are not as common in Arkansas as other members of the genus *Lepomis*.

One of the most distinctive characteristics of centrarchids is their reproductive behavior (Warren 2009). Breeding males of many sunfish species become quite colorful, and spawning habits are similar for most of the species. Nests consist of circular depressions in the substrate and are constructed by the males who, after spawning, guard the eggs and later the larvae. Only one centrarchid species, the Sacramento Perch (*Archoplites interruptus*) of western North America, does not build a nest. In a few species (*Micropterus*), extended parental care occurs, with the male herding the fry for up to a few weeks. Natural hybridization between various species of sunfishes is common. The hybrids, which are often misidentified, show some characteristics of both parental species. Although there is much literature reporting a high rate of natural hybridization, Bolnick (2009) emphasized that a relatively small number of species pairs of centrarchid hybrids have been documented. The number of centrarchid species whose ranges overlap could hypothetically produce 250 species pairs capable of natural hybridization, but only 31 of those 250 species pairs are known to hybridize in nature (Bolnick 2009). When hybridization occurs, the hybrids usually make up a small percentage of the local centrarchid population (1–5%), but there are localities where *Lepomis* hybrids make up 75% of the centrarchid population (Trautman 1957). Bolnick (2009) considered hybridization as posing significant risks to natural populations of centrarchids. Sunfishes become stunted (especially apparent in the genus *Lepomis*) when their populations are too large for the food supply.

Key to the Sunfishes

1A Anal fin spines 5–8.   2
1B Anal fin spines 3 (rarely 2 or 4).   7
2A Dorsal fin spines 10–13.   3
2B Dorsal fin spines 5–8. (*Pomoxis*).   6
3A Length of anal fin base almost equal to length of dorsal fin base; anal fin with 7 or 8 spines and 13–15 rays; gill rakers on first gill arch 25 or more.
                                    Centrarchus macropterus Page 582
3B Length of anal fin base distinctly less than length of dorsal fin base; anal fin with 6 spines and 10 or 11 rays; gill rakers on first arch fewer than 20. (*Ambloplites*) 4

4A   Anal fin of male with a black band in margin; anal fin elements (spines and rays) usually 16.

Ambloplites rupestris Page 580

4B   Anal fin of male without a black marginal band, but entire fin may be dusky or darkened; anal fin elements usually 17.   5

5A   Side with 4 or 5 large dark blotches that extend well down onto lower side; dark spots on side in regular rows, producing a pattern of parallel horizontal lines, typically every scale in the row having a spot; breast scale rows usually 18 or fewer.

Ambloplites ariommus Page 576

5B   Color pattern of side not dominated by 4 or 5 large dark blotches; dark spots on side in an irregular pattern, in any scale row usually only 1–3 scales (but never more than 6) in succession having spots; breast scale rows usually 20 or more.

Ambloplites constellatus Page 578

6A   Dorsal fin spines usually 6; length of dorsal fin base less than distance from dorsal fin origin to posterior margin of eye; markings on side arranged in definite vertical bars.   *Pomoxis annularis* Page 619

6B   Dorsal fin spines usually 7 or 8; length of dorsal fin base equal to or greater than distance from dorsal fin origin to posterior margin of eye; dark markings on side irregularly scattered, not arranged in definite vertical bars.   *Pomoxis nigromaculatus* Page 622

7A   Lateral line scales more than 55; body depth going into standard length 3 times or more (except in largest adults).   *(Micropterus)* 8

7B   Lateral line scales fewer than 55; body depth going into standard length fewer than 3 times.

(Lepomis) 10

8A   Deep notch present between spinous and soft dorsal fins, the 2 fins only slightly connected by a membrane; length of shortest posterior dorsal spine less than half the length of longest spine; scales on cheeks large, in 9–12 rows; dark lateral stripe usually present; caudal fin of young two-colored, its posterior part darker than its base; upper jaw extending far behind back of eye in adults.   *Micropterus salmoides* Page 615

8B   Shallow notch present between spinous and soft dorsal fins, the 2 fins well connected; length of shortest posterior dorsal spine more than half the length of longest spine; scales on cheeks small, in 13 or more rows; dark lateral stripe present or absent, if present, usually broken into blotches; caudal fin of young three-colored with a dark vertical bar separating yellow or orange fin base from white-fringed posterior margin; upper jaw not extending much behind back of eye in adults.   9

9A   Side plain or with a series of indistinct vertical bars; lower side without horizontal rows of small spots; young without prominent black spot at base of caudal fin; scales around narrowest part of caudal peduncle 28–32; scale rows above lateral line usually 12 or 13; scale rows below lateral line usually 20–23.

Micropterus dolomieu Page 609

9B   Side with a dark horizontal stripe usually broken into blotches; lower side with horizontal rows of small spots; young with a prominent black spot at base of caudal fin; scales around narrowest part of caudal peduncle usually 24 or 25; scale rows above lateral line usually 7–10; scale rows below lateral line usually 14–19.   *Micropterus punctulatus* Page 613

10A   Tongue with a patch of teeth; several distinct dark lines radiating backward from the eye; mouth large, upper jaw usually extending behind middle of eye.

Lepomis gulosus Page 589

10B   Tongue without teeth; no distinct dark lines radiating backward from eye (there may be light-colored lines behind the eye in some species); mouth small, upper jaw usually not extending behind middle of eye (except in adult Green Sunfish).   11

11A   Pectoral fin of adult long and pointed, extending to or beyond front of eye when bent forward toward eye; pectoral fin length going into standard length 3.5 times or less.   12

11B   Pectoral fin of adult short and rounded, not extending to front of eye when bent forward toward eye; pectoral fin length going into standard length more than 3.5 times.   14

12A   Soft dorsal fin with a distinct dark blotch near base of posterior rays; opercular flap dark to its margin, without a light-colored border.

Lepomis macrochirus Page 593

12B   Soft dorsal fin without a dark blotch near base of posterior rays; opercular flap with a light-colored border.   13

13A   Gill rakers on first arch long and slender, the length of each raker more than twice its width (Fig. 26.8); opercular flap without a red or orange spot in life; anal rays usually 7–9; male with scattered orange or red spots on side, female with scattered brown spots.

Lepomis humilis Page 591

13B   Gill rakers on first arch short, the length of each raker less than twice its width (Fig. 26.8); opercular flap of adult with a prominent red or orange spot in life; anal rays usually 10 or 11; male and female without orange, red, or brown spots on sides.

Lepomis microlophus Page 603

Figure 26.8. Gill rakers in *Lepomis* species.

Left—*L. macrochirus, L. humilis, L. miniatus*
Middle—*L. cyanellus, L. symmetricus*
Right—*L. microlophus, L. auritus, L. megalotis, L. marginatus*

14A Gill rakers on front of first arch long and slender, the length of each raker more than twice its width; rear margin of gill cover (not including membranous opercular flap) stiff (Fig. 26.9A), membranous opercular flap not greatly elongated. 15

14B Gill rakers on front of first arch short and stubby, the length of each raker less than twice its width; rear margin of gill cover thin and flexible (Fig. 26.9B), membranous opercular flap elongated in adults. 17

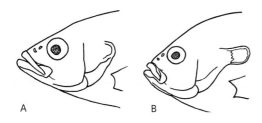

Figure 26.9. Differences in flexibility of opercles in sunfishes. *Cross (1967).*

A—Opercle inflexible posteriorly, its bony edge sharply defined where joined by the marginal gill membrane (as in *Lepomis gulosus, L. cyanellus,* and *L. microlophus*).
B—Opercle flexible posteriorly, attenuated as a thin, fimbriate, cartilaginous extension into the gill membrane (as in *L. macrochirus, L. megalotis,* and *L. humilis*).

15A Lateral line scales usually more than 40; body slender, its depth less than distance from tip of snout to front of dorsal fin base; snout long, its length going into distance from back of eye to rear margin of opercular flap fewer than 2 times. *Lepomis cyanellus* Page 587

15B Lateral line scales usually 40 or fewer; body deep, its depth greater than distance from tip of snout to front of dorsal fin base; snout short, its length going into distance from back of eye to rear margin of opercular flap about 2 times. 16

16A Lateral line incomplete or interrupted; soft dorsal fin with a prominent black blotch near base of posterior rays except in largest adults; lateral line scales usually fewer than 35. *Lepomis symmetricus* Page 607

16B Lateral line complete; soft dorsal fin without a black blotch near base of posterior rays; lateral line scales usually 35–40. *Lepomis miniatus* Page 605

17A Teeth present in roof of mouth on palatine bones (Fig. 26.10); cheek scale rows usually 6–9; width of opercular flap in adults less than eye diameter. *Lepomis auritus* Page 585

17B Palatine teeth absent; cheek scale rows 3–6; width of opercular flap in adults greater than eye diameter. 18

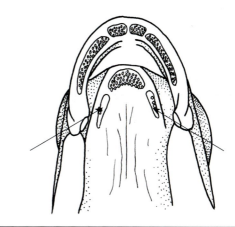

Figure 26.10. Roof of mouth showing palatine teeth (arrows) of *Lepomis auritus.*

18A Cheek scales usually in 5 or 6 rows (occasionally 7); pectoral fin rays 13–15 (usually 14); anal fin base nearly straight; membranous opercular flap of adults usually with a white (sometimes red) border. *Lepomis megalotis* Page 599

18B Cheek scales in 3 or 4 rows (occasionally 5); pectoral fin rays usually 12 (occasionally 13); anal fin base usually convex; membranous opercular flap of adults usually with a conspicuous green border. *Lepomis marginatus* Page 597

Genus *Ambloplites* Rafinesque, 1820a— Rock Basses

This genus, described by Rafinesque in 1820, is endemic to eastern North America and contains four species. Two of the species are native to Arkansas, and one species is introduced in this state. The four species consist of two sister group pairs: *A. ariommus* and *A. rupestris* form one sister pair, and *A. constellatus* and *A. cavifrons* the other (Warren

2009). Monophyly of the genus *Ambloplites* was supported by genetic evidence (Roe et al. 2002, 2008; Near et al. 2004). *Ambloplites* species apparently differ from other centrarchids (except for genus *Pomoxis*) in their nest construction methods and reproductive behaviors (summarized by Warren 2009). Males do not use caudal sweeping to construct a nest. Nest construction primarily involves undulations of the anal fin and sweeping and pushing of material with the pectoral fins. Males show no overt courtship of females and do not actively lead or guide females to the nest (as do species of *Lepomis* and *Micropterus*). Instead, males remain with the nest and eventually allow a persistent, submissive female to enter and spawn.

Genus characters include a large oblique mouth with large supramaxillae, and with the lower jaw projecting slightly beyond the upper jaw. The opercle has two flat projections, there are usually 5 or 6 anal fin spines, a short, rounded pectoral fin, an emarginate caudal fin, teeth present on tongue, vomer, and palatines, and a black or dusky teardrop (suborbital bar). The young are usually more highly patterned than the adults, with large, irregularly shaped dark blotches alternating with lighter areas. Young and adults are capable of rapidly changing the pigmentation intensity to blend in with their background.

Ambloplites ariommus Viosca
Shadow Bass

Figure 26.11a. *Ambloplites ariommus*, Shadow Bass. *Renn Tumlison.*

CHARACTERS A robust, deep-bodied rock bass, its body depth going into SL usually 2.1–2.4 times (Fig. 26.11). Eye large, eye diameter going into head length more than 3.3 times. Mouth large, upper jaw extending past middle of eye, lower jaw protruding beyond upper jaw. Side with 4 or

Figure 26.11b. *Ambloplites ariommus*, Shadow Bass juvenile. *Uland Thomas.*

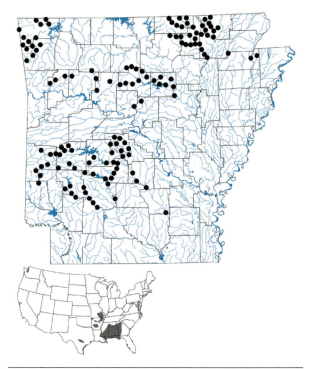

Map 165. Distribution of *Ambloplites ariommus*, Shadow Bass, 1988–2019.

5 wide, dark blotches extending the full width of the body and often obscuring horizontal lines on lower side formed by dusky spots on lateral scales. Breast scale rows (starting with scale at lower base of left pectoral fin insertion, counting downward and forward to ventral midline and then upward and backward to lower insertion of right pectoral fin) usually 18 or fewer (14–20). Cheek fully scaled; cheek scale rows usually 7, often 8 (5–9). Lateral line complete, usually with fewer than 41 scales (35–43). Scale rows above lateral line usually 6 or 7 (5–8). Diagonal scale rows usually 22 or 23. Dorsal spines 11; dorsal rays 11 or 12 (total dorsal

fin elements usually 22 or 23); anal spines 5 or 6; anal rays 9–11. Arkansas's smallest species of rock bass, the largest adults rarely exceeding 8 inches (203 mm) in length and 1 pound (0.5 kg) in weight. Arkansas state angling record from the Spring River in 1999 was 1 pound, 13 ounces (0.82 kg).

LIFE COLORS   Young with large, irregularly shaped dark brown blotches alternating with lighter areas on body; this juvenile pattern usually persists in the adult. Like all rock basses, both young and adults are capable of rapid pigmentation changes to provide camouflage (Warren 2009). Back and sides olive, with prominent dark brown lateral blotches. Lower sides with dusky spots forming horizontal lines, spots usually much lighter than lateral blotches and often obscured by the blotches. Undersides dusky white. Iris of eye red. Dorsal, caudal, and anal fins dusky or lightly mottled; anal fin lacks a black marginal band.

SIMILAR SPECIES   The Shadow Bass differs from the Rock Bass, *A. rupestris*, in color pattern dominated by 4 or 5 wide, dark lateral blotches extending full depth of body (vs. color pattern dominated by dark spots on lateral scales forming prominent horizontal lines; lateral blotches absent or weakly developed above midside); usually 17 anal fin elements (vs. 16); usually 18 or fewer breast scale rows (vs. more than 18); and no black marginal band in anal fin of male (vs. black margin present). It differs from the Ozark Bass, *A. constellatus*, in having a deeper body, its body depth usually going fewer than 2.38 times into standard length (vs. body depth usually going more than 2.38 times into SL); in having color pattern dominated by 4 or 5 wide, dark blotches with dark spots on scales of lower sides in regular rows (vs. color pattern dominated by irregular spotting, usually only 1–3 but never more than 6 lateral scales in succession having a dark spot); usually fewer than 41 lateral line scales (vs. 41 or more); and breast scale rows usually 18 or fewer (vs. 20 or more). It differs from the crappies (*Pomoxis*) in having more than 10 dorsal spines (vs. 5–8), from the Flier (*Centrarchus*) in having 5 or 6 anal spines (vs. 7 or 8), and from other sunfishes in having 5 or 6 anal spines (vs. 3).

VARIATION AND TAXONOMY   *Ambloplites ariommus* was described by Viosca (1936) from Little Bogue Falia Creek, Tchefuncte River drainage, 4.8 km north of Covington, St. Tammany Parish, Louisiana. This polytypic species was for many years regarded as a subspecies of *A. rupestris*, based largely on reported intergrades between *A. ariommus* and *A. rupestris* in northeastern Arkansas and southeastern Missouri (Cashner 1980a); however, Cashner and Suttkus (1977) referred the questionable specimens to *A. ariommus*, and elevated this form to full

specific rank. Later studies of allozyme-derived genetic data from Missouri River drainages in the Missouri Ozarks found that populations were intermediate between *A. ariommus* and *A. rupestris* (Koppelman et al. 2000). Despite the difficulty of field identification of *A. ariommus* from the Missouri Ozarks, Warren (2009) concluded that morphological differences could be used to accurately separate adults of *A. ariommus* and *A. rupestris*. Based largely on analysis of the mitochondrial gene, cytochrome *b*, Roe et al. (2008) concluded that Arkansas populations (Ozarks and Ouachitas) of Shadow Bass should be assigned to *A. rupestris*, a conclusion that we are not ready to accept. It is also possible that the lack of *cytb* distinctiveness between Arkansas *A. ariommus* and *A. rupestris* is due to introgressive hybridization between those species. *Ambloplites rupestris* has been widely introduced throughout North America, and intentional and suspected introductions into the range of *A. ariommus* have possibly affected the genetic integrity of the Shadow Bass (Warren 2009). Because a thorough rangewide morphological and genetic analysis (using nuclear DNA) is needed to clarify these taxonomic questions, we continue to recognize *A. ariommus* as a valid species in Arkansas as did Warren (2009), Page and Burr (2011), and Robins et al. (2018). Johnson and Cavenaugh (2003) studied mitochondrial DNA fragments of Shadow Bass in Arkansas using 14 restriction endonucleases. The dominant haplotype for the Shadow Bass was found in all three populations sampled. Unique restriction profiles between *A. ariommus* and *A. constellatus* were identified for six of the restriction endonucleases.

DISTRIBUTION   *Ambloplites ariommus* occurs in the lower Mississippi River basin from southeastern Missouri south through the uplands of Arkansas and in southwestern Louisiana and southeastern Mississippi, and in Gulf Slope drainages from the Apalachicola River, Georgia and Florida to Louisiana, and disjunctly in the upper Guadalupe River, Texas (where it is probably introduced). Arkansas's most widely distributed rock bass, the Shadow Bass occurs in the Red, Ouachita, Arkansas, Illinois, Little Red, Strawberry, Spring, Black, and St. Francis river drainages (Map 165). Some authors consider the Arkansas River drainage populations to be introduced. Individuals of the Illinois River population in that drainage are especially difficult to distinguish from the introduced *A. rupestris* (Koppelman et al. 2000).

HABITAT AND BIOLOGY   In Arkansas, the Shadow Bass is an inhabitant of clear mountain streams of moderate to high gradient above the Fall Line, where it is found in rocky-bottomed pools often near large boulders. It is intolerant of high turbidity and siltation and often occurs in association with Smallmouth Bass. Juveniles are often found in shallow water near water willows or other aquatic vegetation (Probst et al. 1984; McClendon and Rabeni 1987; Schieble 1998). In the Spring River and lower Eleven Point River, Arkansas, the Shadow Bass was the second-most abundant gamefish species (after Smallmouth Bass), and abundances were greatest upstream and declined significantly downstream in both rivers (Johnson et al. 2010). Downstream declines in stream slope and frequency of riffles were associated with this distribution trend. Outside Arkansas, it is usually a lowland form found below the Fall Line (Douglas 1974; Smith-Vaniz 1968). Pflieger (1997) reported that in the lowlands of southeastern Missouri, it inhabits the clearer ditches in strong current and dense beds of aquatic vegetation. It is only rarely taken from large reservoirs in Arkansas.

Various studies indicate that *A. ariommus* is primarily a benthic feeder. Viosca (1936) studied Shadow Bass diet in a Gulf Coastal Plain stream in Louisiana, a habitat quite different from that occupied by this species in Arkansas. He reported that foods consisted mainly of small fishes (darters, madtom catfish, minnows) and insects (dragonflies, stoneflies, caddisflies), with little predation on crayfish. In Missouri, crayfish were the most important prey of Shadow Bass larger than 100 mm TL, followed by fish and benthic invertebrates (Probst et al. 1984; Rabeni 1992). Midges and mayflies were the most important food items for juveniles smaller than 50 mm, and for individuals between 50 and 100 mm, crayfish became increasingly important food items. The diet of Shadow Bass in the Spring River, Arkansas, consisted mostly of crayfish (Johnson et al. 2010).

Viosca (1936) reported that spawning lasted from April until mid-August in Louisiana and Mississippi. More recent data indicate that the spawning period of southern populations in Louisiana can extend from January to June depending on water temperatures ranging from 15°C to 26°C (Schieble 1998). There is no information on nest construction, courtship, or other aspects of reproductive behavior. Spawning habits are probably similar to those described for *A. rupestris*. Pflieger (1997) reported a maximum life span in Missouri of between 6 and 9 years. It is a popular gamefish in Arkansas in the scattered mountain streams where it is common, and it will readily take artificial lures and natural baits.

CONSERVATION STATUS   This is the most widely distributed rock bass in Arkansas. Its populations tend to be small but stable.

Ambloplites constellatus Cashner and Suttkus
Ozark Bass

Figure 26.12. *Ambloplites constellatus*, Ozark Bass. *David A. Neely.*

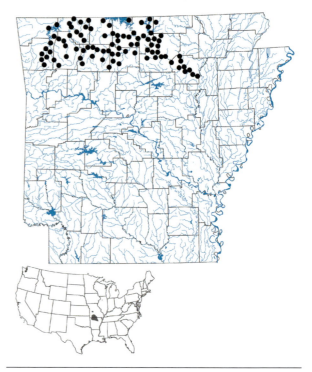

Map 166. Distribution of *Ambloplites constellatus*, Ozark Bass, 1988–2019. Inset shows native range and probable introduced population in the Osage River basin of Missouri.

CHARACTERS   A large, relatively slender-bodied rock bass, its body depth going into SL usually more than 2.38 times (Fig. 26.12). Eye large, eye diameter going into head length 3.7 times or less. Adults with an irregular, freckled, lateral color pattern, no unbroken horizontal rows of spots on side; in any scale row, usually 3, but no more than 6 scales in succession have dark spots; the spotting is more irregular above the lateral line and more continuous below it but does not extend downward to extreme lower side or belly. Some individuals have 4 or 5 wide blotches visible above lateral line, but blotches do not extend full depth of body and are not dark enough to obscure the freckled pattern. Mouth large, upper jaw extending past middle of eye, lower jaw protruding beyond upper jaw. Breast scale rows 20 or more (20–26). Cheek fully scaled; cheek scale rows usually 9, often 8 (6–11). Lateral line complete, usually with more than 41 scales (40–48). Scale rows above lateral line usually 8 or 9. Diagonal scale rows 22–24. Dorsal spines 11; dorsal rays 11–13; anal spines 5 or 6; anal rays 9–12. A large rock bass, often attaining a length of 10 inches (254 mm) and a weight of over 1 pound (0.5 kg). The Arkansas state record for Ozark Bass from Lake Norfork in 1982 is 1 pound, 8 ounces (0.68 kg).

LIFE COLORS   Young with 4 or 5 wide, dark blotches above lateral line, but if present, blotches never dark enough to obscure irregularly spotted pattern; this juvenile pattern does not persist in the adult. Adults with back and sides dark olive to tan and sides with irregularly scattered small, dark brown spots. If large blotches are visible in the adult, they occur above the lateral line, are saddlelike, and are never dark enough to obscure the spotted pattern on lateral scales. Dark spots also scattered on cheek, opercle, and preopercle. Fins usually olive green. Dorsal and caudal fins with dark spots. Anal fin of male without black marginal band. Iris of eye red-orange in adults.

SIMILAR SPECIES   Other rock basses lack the freckled color pattern of the Ozark Bass. It differs from both *A. ariommus* and *A. rupestris* in having color pattern dominated by irregular rows of spots on sides, with usually only 1–3 but never more than 6 scales in succession having a dark spot (vs. dark spots on sides in regular rows, with every scale in row typically having a spot); in having a more slender body, body depth usually going into SL more than 2.38 times (vs. usually fewer than 2.38 times); and in usually having 41 or more lateral line scales (vs. usually fewer than 41). It differs from the crappies (*Pomoxis*) in having more than 10 dorsal spines (vs. 5–8), from the Flier (*Centrarchus*) in having 5 or 6 anal spines (vs. 7 or 8), and from all remaining sunfishes in having 5 or 6 anal spines (vs. 3).

VARIATION AND TAXONOMY   *Ambloplites constellatus* was originally described from the Buffalo River at the mouth of Rush Creek in Marion County, Arkansas, by Cashner and Suttkus (1977). It was previously included in *A. rupestris*. Morphological and genetic evidence support its distinctiveness from congeners (Koppelman et al. 2000; Near et al. 2004; Bolnick and Near 2005). R. L. Johnson and Cavenaugh (2003) studied mtDNA from a Buffalo River population. A single haplotype was found,

and unique restriction profiles were identified between the Ozark Bass and Shadow Bass.

DISTRIBUTION   *Ambloplites constellatus* is endemic to the White River drainage in northern Arkansas and southern Missouri. Populations in the Sac and Pomme de Terre stream drainages of the Osage River basin (Missouri River drainage) in Missouri may have been introduced (Pflieger 1997). In Arkansas, the Ozark Bass is widely distributed in the White River system above its confluence with the Black River, and it is especially common in the Buffalo River (Map 166). It is replaced in the Black River drainage, and surprisingly in the Little Red River drainage, by *A. ariommus*. Before its specific distinctness was diagnosed, the state fish hatchery at Centerton took brood stock of this species from the upper White River and introduced the resulting fingerlings into the Illinois River drainage in the 1960s (Cashner and Suttkus 1977) (Map A4.162). Similar introductions of *A. constellatus* were made in the upper and middle Osage River (Missouri River drainage) in Missouri, where populations became established (Pflieger 1997).

HABITAT AND BIOLOGY   The Ozark Bass inhabits small to moderate-sized, clear Ozark streams with permanent flow, high dissolved oxygen, abundant aquatic vegetation, low turbidity, and silt-free substrates. It is usually found in pools with slow to moderate current, where it is often taken near boulders, logs, and beds of water willow. Juveniles are often found in shallow water in backwater areas along the shore, usually near aquatic vegetation or other cover. It is definitely a stream fish that does not adapt well to reservoirs. Only an occasional specimen was taken in AGFC rotenone population samples in Beaver and Bull Shoals reservoirs, but a small population of Ozark Bass exists in Norfork Reservoir.

Stomach contents of 47 Ozark Bass collected in January, May, June, August, and September from Baxter, Madison, Marion, and Newton counties were examined by TMB. The diet in all months was dominated by crayfish in both percent frequency of occurrence and percent volume. Other items eaten in decreasing order of importance were immature aquatic insects (including plecopteran nymphs, coleopteran pupae, ephemeropterans, and odonates), fishes (cyprinids), terrestrial insects, and adult hemipterans. Ozark Bass smaller than 75 mm TL fed almost exclusively on immature aquatic insects.

Pflieger (1997) observed males on nests in late May in Missouri. Ripe females were found as late as 30 June in the White River in Madison County, Arkansas, by TMB. The following information on reproduction was provided by Walters et al. (2000), who studied spawning behavior and other aspects of reproductive biology in the Buffalo River, Arkansas. Nest guarding and egg deposition occurred from mid-May to mid-June at water temperatures of 17–23.5°C. Nest guarding males ranged from 150 to 230 mm TL. New nests with embryos were found throughout this period. Nests were constructed in gravel and cobble substrates at depths of 0.5–2.9 m. Seventy-four percent of the nests were less than 1.0 m from a boulder or logs and were usually downstream from that cover. Noltie and Keenleyside (1987a) also noted that nests in streams were usually constructed downstream from obstructions to current. Courtship behavior of a spawning pair in the Buffalo River was described by Walters et al. (2000). At the time of spawning, males become black; after spawning, the typical coloration is restored. The male chased intruders away when they approached within a meter of the nest. During courtship, both sexes constantly waved their dorsal, caudal, and pectoral fins. The pair circled the nest several times prior to egg deposition, with the male occasionally nipping the caudal peduncle of the female. Walters et al. (2000) described the spawning act as follows:

> The Ozark Bass pair then made their way down to the nest, with the male usually arriving first. The male rarely directed or pushed the female into the nest. While broadside and facing the same direction, the pair pivoted in circles around the nest, usually with the female outside of, and pressing her body against the male. The male was angled headfirst toward the substrate and remained slightly posterior to the female while she remained perpendicular to the bottom. The female then began slow undulations of her body against the male, which increased in amplitude. She rolled to her side, with her vent towards the male, until nearly parallel with the substrate, and appeared to release eggs at this time. The male leaned towards the female slightly, but did not conspicuously undulate. He then began to drift above the nest and often darted away to chase nearby fish, while the female continued to vibrate for another second while moving slightly forward or in a quarter circle around the nest.

Eighty-eight spawning events occurred within 2 hours, with the pair drifting up from the nest between spawning bouts. Eggs hatched in 5 days or less at a mean temperature of 21°C, and larvae remained in the nest for 5–7 days. Dispersing young were gray, in contrast to the black young reported for *A. rupestris* by Noltie and Keenleyside (1987a). Young-of-the-year were caught in larger numbers during nighttime seining than during daytime seining. Young

were not observed during daytime snorkeling. This suggests that young-of-the-year are predominantly nocturnal. Predation and high water were major factors preventing successful Ozark Bass broods in the Buffalo River. Walters et al. (2000) observed Longear Sunfish preying upon Ozark Bass nests. No nests in the study area produced dispersing young in 1991, and 5 of 52 nests monitored in 1992 were successful. Asynchronous nesting and the ability to spawn over a wide range of water temperatures allow Ozark Bass to counter the effects of flooding and nest predation. Pflieger (1997) reported that Ozark Bass reached a length of about 1.4 inches in the first year of life and had a maximum life span of at least 11 years.

CONSERVATION STATUS   Most populations of this White River endemic are secure. In some areas, it is locally common. Because of its restricted range in Arkansas and its importance as a game species, periodic monitoring of its status in this state is warranted.

Ambloplites rupestris (Rafinesque)
Rock Bass

Figure 26.13. *Ambloplites rupestris*, Rock Bass. *Patrick O'Neil, Geological Survey of Alabama.*

CHARACTERS   A robust, deep-bodied rock bass, body depth going into SL approximately 2.2–2.4 times (Fig. 26.13). Eye large, eye diameter going into head length 3.3 times or fewer. Color pattern of side dominated by horizontal rows of dark spots; each scale in a row has a dark spot, producing a pattern of parallel lines. Horizontal rows most obvious below lateral line. Obscure dark blotches present on back in predorsal region. Mouth large, upper jaw extending backward past middle of eye; lower jaw protruding beyond upper jaw. Breast scale rows usually 20 or more (18–24). Cheek fully scaled; cheek scale rows usually 8, often 9 (6–10). Lateral line complete, usually with fewer than 41 scales (36–44). Scale rows above lateral line usually 7 or 8; diagonal scale rows 20–24. Dorsal spines 11–13;

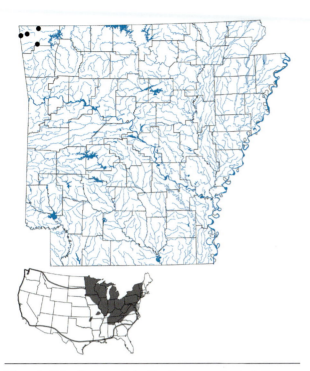

Map 167. Distribution of *Ambloplites rupestris*, Rock Bass, 1988–2019. Inset shows native range (solid) and introduced range (outlined, with some populations shown in solid) in the contiguous United States.

dorsal rays 11–13; anal spines 5 or 6; anal rays 9–11. Pectoral fin with 12–15 rays; pelvic fin with 1 spine and 5 rays. A large *Ambloplites* reaching a maximum size of more than 13.4 inches (340 mm) and a maximum weight of about 3 pounds, 12 ounces (1.7 kg) (Becker 1983). It is unusual for Rock Bass to exceed 0.5 pound (0.23 kg) in weight. No Arkansas state record has been established. (The previously published Arkansas record for "Rock Bass" was for *A. constellatus*).

LIFE COLORS   Young with well-developed dark brown blotches on sides; this juvenile pattern does not persist in the adult. Adults with back and upper sides dark green or brown, grading to a lighter shade below lateral line. Side with longitudinal rows of brown spots. Back of adult sometimes with 4 or 5 obscure brown blotches extending down to or just below the lateral line. If large blotches are visible in the adult, they are never dark enough to obscure the spotted pattern on lateral scales. Undersides dusky white. Iris of eye red. Dorsal, caudal, and anal fins mottled with brown; anal fin (and sometimes pelvic fins) of male with a distinct black marginal band. Paired fins olive to transparent and unspotted. Males become much darker during the breeding season.

SIMILAR SPECIES   The Rock Bass differs from *A. ariommus* in having a color pattern dominated by prominent horizontal rows of dark spots and absence of dark, broad lateral blotches (vs. color pattern dominated by wide, dark lateral blotches extending the full depth of the body); usually more than 18 breast scale rows (vs. 18 or fewer); and in having a black marginal band in anal fin of male (vs. anal fin without black marginal band). It differs from *A. constellatus* in having regular horizontal rows of dark spots on sides (vs. spots in irregular rows); a deeper body, body depth usually going into standard length 2.38 times or less (vs. usually more than 2.38 times); and in usually having fewer than 41 lateral line scales (vs. 41 or more). It differs from the crappies (*Pomoxis*) in having more than 10 dorsal spines (vs. 5–8 dorsal spines), from the Flier (*Centrarchus*) in having 5 or 6 anal spines (vs. 7 or 8), and from other sunfishes in having 5 or 6 anal spines (vs. 3).

VARIATION AND TAXONOMY   *Ambloplites rupestris* was originally described by Rafinesque (1817b) as *Bodianus rupestris* from "lakes of New York, Vermont, and Canada." Formerly all Rock Bass in Arkansas were included in *A. rupestris*, but now two other species of *Ambloplites* native to Arkansas are recognized (Cashner and Suttkus 1974). Little variation occurs in *A. rupestris* and recognition of subspecies is unwarranted. Analysis of allozyme and morphological variation of rock basses in the Missouri Ozarks did not permit reliable separation of the Rock Bass and Shadow Bass (Koppelman et al. 2000), but Warren (2009) considered morphological differences sufficient for separating adult specimens in that region.

DISTRIBUTION   *Ambloplites rupestris* is native to the St. Lawrence–Great Lakes, Hudson Bay, and Mississippi River basins from Quebec to Saskatchewan, Canada, south to Georgia, northern Alabama, and the Meramec River, Missouri. It is introduced into Atlantic Slope drainages from Quebec to Virginia and elsewhere in the United States. In Arkansas, the Rock Bass is an introduced species known only from the Illinois and Neosho river drainages in Benton County in the northwestern part of the state (Map 167). Cashner and Suttkus (1978) reported some of the early stocking attempts for Rock Bass west of its native range. It is known that *A. rupestris* was present in the Neosho drainage of southwestern Missouri as early as 1888, but the source of this stock is uncertain.

HABITAT AND BIOLOGY   In Arkansas, the Rock Bass is found in small to moderate-sized clear streams of the western Ozarks. It typically occurs in slight to moderate current in pools having rocky bottoms and is usually found near boulders or other cover such as submerged logs. Young Rock Bass are often found in backwaters and protected areas along the shore, often in beds of water willow or other aquatic vegetation (Pflieger 1997). It is intolerant of high turbidity and siltation and requires streams with permanent flow. In northern areas of its range, it is also found in small, cool, weedy lakes or littoral regions of larger lakes (Cashner 1980b). Some populations seek deeper water for wintering areas (Hatzenbeler et al. 2000). In small Virginia streams (similar to habitats occupied in Arkansas by this species), Rock Bass moved from headwaters to the deepest downstream pools available in the winter (Pajak and Neves 1987). It is a sedentary, solitary, secretive species.

Because of its large range in eastern North America, the biology of the Rock Bass has been extensively studied. Pflieger (1997) reported that in Missouri, smaller Rock Bass feed mainly on adult and immature aquatic insects; in addition to those items, larger Rock Bass also consume small minnows and crayfish. Studies in other areas show that the Rock Bass is predominantly a benthic feeder, with crayfish often constituting a major part of the diet (Johnson and Dropkin 1993; Roell and Orth 1993). Small fish are added to the diet in the second year of life, and fish become even more important in the diets of large adults (Keast 1985a, b). Young-of-the-year were reported to feed predominantly on cladocerans, isopods, amphipods, and chironomids (Keast 1977; George and Hadley 1979).

Pflieger (1997) found that breeding occurs in Missouri from April to early June when stream temperatures reach 55–60°F (12.8–15.6°C). The male fans out a depression in the gravel bottom of the stream in water 1–5 feet (0.3–1.5 m) deep, near a boulder or other cover in slight current. It is a solitary nester in streams, but synchronous spawning is known from lakes (Gross and Nowell 1980). Mating success was positively correlated with male size (Sabat 1994). Nolte and Keenleyside (1987a) compared the reproductive ecology of lake and stream-dwelling Rock Bass and gave the following information about spawning (Nolte and Keenleyside 1987b). There was usually much circling of the nest by the male and female before spawning. A single spawning bout lasted an average of 2 hours and involved an average of 120 separate egg releases. Females deposited between 400 and 500 eggs in a 2-hour spawning session and may produce 3,000–11,000 eggs per year (Scott and Crossman 1973). In streams, males and females appeared to be nearly monogamous. The eggs hatched in 5 days at a mean temperature of 22°C (72.5°F), and larvae dispersed after 9 days in the nest. After spawning, only the male remains to guard the eggs and larvae (for an average of 14 days). Most Rock Bass live only 5 or 6 years, but a few 10-year-old individuals have been reported.

CONSERVATION STATUS This introduced species is very uncommon in Arkansas, but small reproducing populations are still present in the northwest corner of the state.

Genus *Centrarchus* Cuvier, 1829—Fliers

This monotypic genus was described by Cuvier in 1829. Relationships to other centrarchid genera have not been universally agreed upon. *Centrarchus* has been considered most closely related to *Ambloplites*, *Archoplites*, *Enneacanthus*, or *Pomoxis* by various authors (Eaton 1956; Branson and Moore 1962; Mok 1981; Chang 1988; Wainwright and Lauder 1992; Roe et al. 2002). Near et al. (2004, 2005) considered *Centrarchus* sister to a clade comprising *Ambloplites*, *Archoplites*, *Enneacanthus*, and *Pomoxis*. Genus characters are the same as those given in the species account below.

Centrarchus macropterus (Lacépède)
Flier

Figure 26.14. *Centrarchus macropterus*, Flier. *Uland Thomas.*

CHARACTERS A deep-bodied, compressed sunfish, with body depth about half of SL (Fig. 26.14). Anal fin nearly as long and large as the dorsal fin. Dorsal fin origin distinctly in front of anal fin origin. Anterior profile of head concave over eye. Head small, snout pointed and short, mouth of moderate size, upper jaw reaching almost to (but not past) middle of eye; lower jaw projecting. Supramaxilla moderate in size. Eye large, its diameter equal to or greater than snout length. Rear margin of preopercle finely serrate. Two patches of teeth present on the tongue; teeth present on vomer and palatine bones. Spinous dorsal and soft dorsal fins broadly connected, without a notch between them. Pectoral fin long and pointed. Gill rakers long and slender, usually 30–36. Lateral line complete and highly arched anteriorly; lateral line scales 36–44. Cheek scale rows 4–7. Dorsal spines 11–13; dorsal rays 12–15; anal spines 7 or 8. Pectoral rays 12–14. Rarely exceeding 5 or 6 inches (127–152 mm) in length. Maximum length of 8 inches (203 mm) and weight of 1 pound (0.5 kg) in northern Louisiana (Douglas

Figure 26.15. Juvenile *Centrarchus macropterus*. *Uland Thomas.*

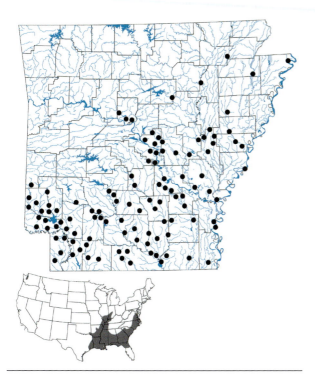

Map 168. Distribution of *Centrarchus macropterus*, Flier, 1988–2019.

1974). The Arkansas state angling record from the Saline River in 1985 is 14 ounces (0.4 kg).

LIFE COLORS   Back and upper sides olive green, becoming more yellow on lower sides and venter. Sides with longitudinal rows of brown spots. Dark wedge-shaped bar beneath eye continues vertically through eye. Dorsal, caudal, and anal fins with small light spots and reticulations. Small Fliers (to about 65 mm TL) with prominent black spot surrounded by red-orange coloration on posterior rays of dorsal fin; spot disappears with age. Small young with 4 or 5 broad dark bars on side (Fig. 26.15).

SIMILAR SPECIES   Most closely resembles the crappies, but differs from them in having 10–13 dorsal spines (vs. 6–8). It can be distinguished from other sunfishes by its 7 or 8 anal spines (vs. other sunfishes having fewer than 7).

VARIATION AND TAXONOMY   *Centrarchus macropterus* was described as *Labrus macropterus* from Charleston, South Carolina, by Lacépède in 1801. No extensive study of geographic variation is available. Natural hybridization between *C. macropterus* and *Pomoxis annularis* was reported by Burr (1974).

DISTRIBUTION   *Centrarchus macropterus* occurs in the Mississippi River basin from southern Indiana and southern Illinois, south to the Gulf of Mexico; Atlantic Slope drainages from the Potomac River, Maryland, to central Florida; and Gulf Slope drainages from central Florida west to the Trinity River, Texas. In Arkansas, the Flier is found in all major drainages of the Coastal Plain lowlands of the state. It is most common in southern Arkansas and sporadically distributed in eastern Arkansas (Map 168). One pre-1988 record exists from the upper Arkansas River Valley near Russellville. McAllister et al. (2004) provided several new locality records for the Little River drainage in southwestern Arkansas.

HABITAT AND BIOLOGY   The Flier is a lowland species inhabiting oxbow lakes, bayous, creeks, and swampy backwaters of large rivers. It prefers quiet, clear, heavily vegetated waters having mud and detritus bottoms. Fliers frequently seek cover in aquatic vegetation, around submerged tree roots, and in concentrations of rotting leaves and other plant matter (Mettee et al. 1996). The Flier appears to be more tolerant of acidic waters and hypoxic conditions than other sunfishes (Laerm and Freeman 1986; Rutherford et al. 2001). Fliers do poorly in neutral to alkaline waters (Ross 2001). Study of a North Carolina population indicated fidelity to small home ranges over extended periods (Whitehurst 1981). It is not widely used as a gamefish in Arkansas because of its small size, sporadic distribution, and generally small populations.

Gunning and Lewis (1955) found its diet in Illinois consisted mostly of insects, crustaceans, and filamentous algae (not digested). Conley (1966) found that food habits varied considerably in relation to size in southeastern Missouri. Individuals smaller than 1 inch (25 mm) TL fed entirely on copepod crustaceans. Small crustaceans made up the bulk of the diet of Fliers smaller than 7 inches (178 mm) TL, and larger individuals fed mainly on insects, with 7% of the diet consisting of fish and crustaceans. Long, slender gill rakers (more than any other centrarchid) make it possible for the Flier to feed on copepods throughout life. In western Tennessee, adult Fliers ate mainly terrestrial insects, indicating surface-oriented feeding (Etnier and

Starnes 1993). Conley (1966) reported that it is primarily a nocturnal feeder.

It is one of the earliest-spawning centrarchids, with spawning occurring in southeastern Missouri in April over a 2-week period (Conley 1966). Spawning has been reported as early as February and as late as May in other areas (Lee and Gilbert 1980). The few details of its reproductive biology that have been reported are similar to those of other sunfishes, and the following information on reproduction comes from Dickson (1949) and Conley (1966). The male establishes and defends a territory and prepares a saucer-shaped depression nest. Nest-building activity begins in early spring when water temperature reaches 14–17°C. Nests are constructed at depths of 0.3–1.2 m, often in colonies. The male leads a female to the nest and circles around her several times before spawning ensues. There is mutual biting during spawning. Mature ovarian eggs are the smallest of any centrarchid (Conley 1966), and the eggs hatch in 7–8 days at 19°C. Fecundity is high, with females producing from 4,412 eggs at 70 mm TL to 48,254 eggs at 205 mm TL (Dickson 1949; Conley 1966). Sexual maturity is reached at 1 year of age and life expectancy is about 5–7 years.

CONSERVATION STATUS   Most populations of this widely distributed lowland species are apparently stable or secure in Arkansas.

Genus *Lepomis* Rafinesque, 1819—Panfishes

This genus, described in 1819 by Rafinesque, is native to North America east of the Continental Divide. It contains 13 species, 10 of which are found in Arkansas (one species is introduced into Arkansas). For several decades, there was disagreement on the gender of the genus name (which would determine the correct latinized endings of the names of the species in the genus), but in 1992, the ICZN fixed the name *Lepomis* as masculine. During the past five decades, various authors have questioned the monophyly of *Lepomis*, but genetic analyses by Near et al. (2004, 2005) and Harris et al. (2005) supported its recognition as a monophyletic clade. Near et al. also considered *Lepomis* as sister to the genus *Micropterus*. *Lepomis* species frequently hybridize in nature because of similarities in habitats and reproductive behaviors (Boschung and Mayden 2004). Bolnick (2009) listed 22 known intrageneric hybrid combinations.

In the genus *Lepomis*, a combination of coloration, morphology, and orientation of the opercular flap is distinctive for each species and can play an important role in species recognition, mate choice, and aggressive encounters

between rival males. It should be noted that juvenile *Lepomis* smaller than about 40 mm TL are extremely difficult to correctly identify to species based solely on morphological features (Fig. 26.16). The small juveniles exhibit a similar barred color pattern on the sides (Etnier and Starnes 1993), and the species-specific shape and color pattern of the opercular flap have not yet developed. Genus characters include 3 anal fin spines, emarginate caudal fin, lateral line complete in all except *L. symmetricus*, fewer than 55 lateral scale rows, vomerine teeth present, edge of opercle smooth, not serrate, and opercle without a flat spinous projection.

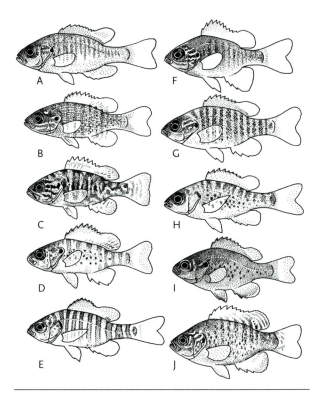

Figure 26.16. Small juveniles of *Lepomis* species:

A—*Lepomis auritus*
B—*Lepomis cyanellus*
C—*Lepomis gulosus*
D—*Lepomis humilis*
E—*Lepomis macrochirus*
F—*Lepomis marginatus*
G—*Lepomis megalotis*
H—*Lepomis microlophus*
I—*Lepomis miniatus*
J—*Lepomis symmetricus*

Lepomis auritus (Linnaeus)

Redbreast Sunfish

Figure 26.17. *Lepomis auritus*, Redbreast Sunfish. *Uland Thomas.*

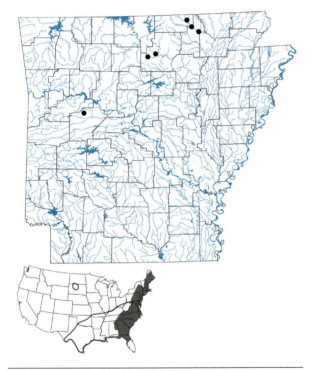

Map 169. Distribution of *Lepomis auritus*, Redbreast Sunfish, 1988–2019. Inset shows native range (solid) and introduced range (outlined) in the contiguous United States.

CHARACTERS A moderate-sized *Lepomis* with a long, narrow opercular flap (best developed in adults, but short or only slightly elongated in juveniles) that is horizontally oriented or pointing upward, and uniformly dark to its margin (Fig. 26.17). Opercular flap width usually equal to or narrower than eye diameter. Mouth small, upper jaw extending to or just past front of eye. Pectoral fin short and rounded, its tip usually not reaching past front of eye

when bent forward. Pectoral rays usually 14 (13–16) Gill rakers short but not reduced to blunt knobs, usually 9–12. Palatine teeth present in roof of mouth; villiform teeth also present on premaxillary, dentary, and vomer (Fig. 26.10). No teeth on tongue. Lateral line complete, arched anteriorly, usually with 42–51 (40–54) scales. Cheek scale rows usually 6–9. Scale rows above lateral line 7–9; scale rows below lateral line 14–16. Dorsal fin with 10 or 11 spines and 10–12 rays; soft dorsal fin with an acute posterior tip. Anal fin spines 3; anal rays 9 or 10. Most adults are around 6 inches (152 mm). Maximum size is about 10 inches (254 mm) and 2+ pounds (1 kg). No Arkansas state record has been established.

LIFE COLORS Back and upper sides olive or olive brown; lower sides grading into reddish orange or yellow of undersides. Dorsal, anal, and caudal fins are dusky. Dorsal fin without a posterior black blotch. The long opercular flap is black to its margin, and bordered above and below by blue lines. Wavy blue lines radiate across snout onto cheek and opercle (similar to *L. megalotis* and *L. marginatus*). Breeding male with bright orange or red-orange breast and belly, orange fins, and with much blue and orange on the sides.

SIMILAR SPECIES *Lepomis auritus* most closely resembles the Longear Sunfish and Dollar Sunfish, but differs from those species in having teeth on palatine bones in roof of mouth (vs. no palatine teeth), and adults with a long, narrow opercular flap not wider than the eye diameter (vs. opercular flap wider than eye diameter in adults). Another feature used by most authors to separate those species is the opercular flap dark to its margin in *L. auritus* (vs. opercular flap with a light-colored border in *L. megalotis* and *L. marginatus*). We have found this feature to be unreliable for separating *L. auritus* from some Arkansas populations of *L. megalotis*. It is common for some specimens of *L. megalotis* from the White River (and other areas of the state) to have the opercular flap dark to its margin and narrower than the eye diameter. Although there is overlap between the two species in scale counts, *L. auritus* usually has 43–50 lateral line scales and 6–9 cheek scale rows, while *L. megalotis* usually has fewer than 43 lateral line scales and 5 or 6 cheek scale rows. *Lepomis auritus* can be distinguished from all other Arkansas *Lepomis* by the long, thin opercular flap.

VARIATION AND TAXONOMY *Lepomis auritus* was described by Linnaeus (1758) as *Labrus auritus* from Philadelphia, Pennsylvania. A rangewide study of variation has not been done. Based on morphological characters, Branson and Moore (1962) hypothesized a close relationship with *L. megalotis* and *L. marginatus*. This was supported by Mabee (1993) who added *L. gibbosus*, *L. microlophus*, and *L. punctatus* as additional relatives. Early genetic analyses indicated a close relationship to *L. punctatus* (Avise and Smith 1974b), but subsequent analyses of DNA identified *L. auritus* as sister to a clade that included *L. marginatus*, *L. megalotis*, and *L. peltastes* (Near et al. 2004, 2005). Natural hybrids have been reported between *L. auritus* and four other *Lepomis* species (*L. cyanellus*, *L. gibbosus*, *L. gulosus*, and *L. macrochirus*) (Bolnick 2009), and we believe that hybridization with *L. megalotis* may have occurred in Arkansas (based on morphological evidence).

DISTRIBUTION   The Redbreast Sunfish is native to Atlantic and Gulf Slope drainages from New Brunswick, Canada, to central Florida, and west to the Apalachicola and Choctawhatchee river drainages of Alabama, Georgia, and Florida. It has been widely introduced elsewhere (including several drainages in Texas) but has become successfully established in only a few areas (Page and Burr 2011). The only Arkansas Game and Fish Commission records for stocking this species in Arkansas are from the 1960s, but museum records for *L. auritus* from Arkansas (unverified by us) exist prior to that time (Map 169). There is a 1925 record of 1 specimen from Phillips Bay in Lawrence County (United States National Museum, USNM 91788.5), and 7 specimens from Long Creek in Boone County (University of Arkansas Fayetteville, UAFC 2848) from 1958.

In late 1963, adult *L. auritus* were obtained from the state of Virginia in exchange for yearling *Pylodictis olivaris* and placed in a pond on the Joe Hogan State Fish Hatchery in Lonoke. In the spring of 1964, 87 adults weighing 0.33–0.5 pound (0.15–0.2 kg) were stocked in a small lake at Camp Hedges (Sylamore Creek, White River drainage) in Stone County (Keith 1971b). Keith (1971b) found that the Redbreast Sunfish stocked in the lake at Camp Hedges successfully reproduced within 7 months of their introduction. In early 1964, the Game and Fish Commission obtained 7,000 *L. auritus* from North Carolina and placed them in a hatchery pond at what is now the Andrew H. Hulsey State Fish Hatchery in Hot Springs. In December 1964, 7,300 mixed adult and juvenile *L. auritus* were harvested from the hatchery pond and stocked in the Spring River near Hardy in Sharp County. Brood stock was retained at the hatchery in Hot Springs, and on 13 January 1966, an additional 40,000 *L. auritus* juveniles were stocked in the Spring River. On 4 March 1966, 15,000 juveniles were stocked in Spring Lake, a small National Forest Service lake in Yell County. It is possible that undocumented stockings

of Redbreast Sunfish occurred elsewhere in Arkansas, particularly during the years when they were being reared at two state fish hatcheries.

Until recently, reported records for *L. auritus* from Arkansas (subsequent to the 1960s stockings) consisted of 7 specimens from the White River at St. Paul in Madison County in 1965 (UAFC 2040), 2 collections from the White River in Independence County (Academy of Natural Sciences at Philadelphia, 1 specimen ANSP 146277 in 1980; 1 specimen ANSP 168983 in 1991), 1 specimen from War Eagle Creek in Madison County in 1992 (Illinois Natural History Survey, INHS 28597), and 1 specimen from South Sylamore Creek in Stone County in 2002 (Tulane University, TU 194218 [specimen rechecked and identification confirmed by H. L. Bart, Jr. in 2016]). Sampling of fishes from the three Arkansas sites historically stocked with *L. auritus* was conducted by TMB in May and June 2014. That sampling indicated the possibility of introgressive hybridization (based on morphological features) between *L. auritus* and *L. megalotis* in all three drainages, and specimens that were closest to *L. auritus* were found in Spring Lake, Sylamore Creek, and the Spring River. A combined analysis of morphology and DNA is needed to clarify the extent of hybridization, if any, and the status of these populations. The Redbreast Sunfish may be more common in Arkansas than our current information indicates, because it is easily confused with the Longear Sunfish, which occurs statewide; however, in our attempt to track down records of *L. auritus* from Arkansas, we found 13 lots of *L. megalotis* in 6 curated fish collections that had been incorrectly identified as *L. auritus*.

HABITAT AND BIOLOGY   The Redbreast Sunfish is mainly a stream-adapted species in its native range, where it inhabits rocky, sandy, or mud-bottomed pools of small creeks to medium rivers (Warren 2009). Even though it is reportedly more of a stream species than most other *Lepomis* (Boschung and Mayden 2004), it also occurs in some lakes, ponds, and reservoirs. This relatively sedentary sunfish is usually associated with cover, such as instream wood, stumps, or undercut banks, and is often found near vegetated lake margins (Warren 2009; Page and Burr 2011). It prefers areas of quiet water, and its abundance is positively associated with increasing depth and cover (Meffe and Sheldon 1988). Because the Redbreast Sunfish was believed to be better adapted to cooler streams (Carlander 1977) than Arkansas's native *Lepomis* species, it was stocked most heavily in the Spring River in northern Arkansas, and in smaller numbers in two cool, spring-fed lakes in Stone and Yell counties (Keith 1971b); however, Saecker and Woolcott (1988) reported that *L. auritus* was one of the few fishes found in elevated water temperatures (to 39°C) downstream from a power plant in the James River, Virginia. Almost all post-stocking reports of *L. auritus* from Arkansas have come from the White River drainage, and no assessment of possible ecological effects or introgression with *L. megalotis* has been made. There is evidence that introduced *L. auritus* in the upper Tennessee River drainage may be replacing the native *L. megalotis* (Etnier and Starnes 1993).

The Redbreast Sunfish has been described as an opportunistic invertivore that feeds by day or night (Warren 2009). Fish are occasionally consumed but are not an important component of the diet. Young up to 100 mm TL feed benthically, mostly on microcrustaceans (ostracods) and small chironomids. Larger juveniles and adults feed predominantly on aquatic insects, with mayfly nymphs, dragonfly nymphs, caddisfly larvae, and dipteran larvae making up the bulk of the diet. Other invertebrates commonly consumed are beetles, terrestrial and emerging aquatic insects, and crayfish (Cooner and Bayne 1982; Murphy et al. 2005).

Sexual maturity is reached at ages of 1+ to 2+ years or a minimum size of about 90–114 mm TL (Davis 1972; Levine et al. 1986; Lukas and Orth 1993). Spawning habits of *L. auritus* are similar to those of other *Lepomis*. It is a colonial nester in quiet water, but in flowing water it usually constructs solitary nests. In Alabama, spawning occurred from April to August, and adults moved into shallow water (0.2–1.2 m deep) and began constructing nests when water temperatures reached 16–21°C and continued to 31°C; nests were about 50 mm (2 inches) deep and varied in diameter from 0.25 to 1.0 m (9.8–39.4 inches) (Boschung and Mayden 2004). Those authors also reported that *L. auritus* nests in flowing water more readily than other *Lepomis* species. Nests were generally constructed in coarse sand and gravel substrates and were typically located among plants near submerged logs, stumps, or on the sheltered side of protective rocks. Boschung and Mayden (2004) described reproductive behavior and noted that stone carrying by the males during nest construction was unusual among *Lepomis*. Depending on size, females can produce from 963 to 8,250 eggs in a single batch (Davis 1972). Fertilized eggs hatch in 3 days at 20–24°C, and the male defends the nest, eggs, and larvae from predators until the larvae become free-swimming. Young grow to about 60 mm (2.4 inches) TL during their first year and grow an additional 30 mm per year until age 6 (Buynak and Mohr 1978b). In early 1964, 7,000 juvenile *L. auritus* between 0.75 and 1.0 inch (19–25 mm) TL were placed in a hatchery pond in Hot Springs. When harvested several months later (December 1964), the fish from the original stocking were

between 5 and 6 inches (127–152 mm) TL, and a second size group of offspring was present averaging 2–3 inches (51–76 mm) TL.

CONSERVATION STATUS This is an introduced species that has no conservation status in Arkansas.

Lepomis cyanellus Rafinesque
Green Sunfish

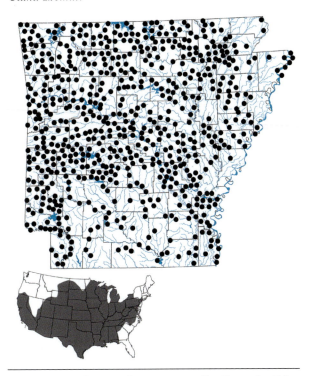

Figure 26.18. *Lepomis cyanellus*, Green Sunfish.
Uland Thomas.

Map 170. Distribution of *Lepomis cyanellus*, Green Sunfish, 1988–2019. Inset shows native and introduced ranges in the contiguous United States.

CHARACTERS A robust, elongate *Lepomis* with a large mouth, the upper jaw usually extending to about middle of eye or past middle of eye in large individuals (Fig. 26.18). Supramaxilla small. Body depth going into SL 2.2–2.7

times. Soft dorsal fin of large juveniles and adults with a prominent dark blotch at posterior base. Pharyngeal arches narrow and strong with conical, blunt teeth. Palatine teeth present. Gill rakers moderately long and slender, 11–14. Pectoral fin short and rounded, its tip usually not reaching eye when bent forward across cheek. Opercle stiff to its margin (rear margin lies within the base of the membranous ear flap); ear flap not elongated. Lateral line complete, arched anteriorly, with 41–52 scales. Scale rows above lateral line 8–10. Cheek scale rows 6–9. Dorsal spines 9–11; dorsal rays 10 or 11; anal spines 3; anal rays 13 or 14; pectoral rays 13–15; pelvic fin with 1 spine and 5 rays. Adults generally reaching 8–10 inches (203–254 mm) in length and 1 pound (0.5 kg) in weight. Maximum weight about 2 pounds, 7 ounces (1.1 kg). The Arkansas state angling record from a pond in Dierks in 1976 is 1 pound, 11 ounces (0.8 kg).

LIFE COLORS   Back and sides olive to bluish green. Sides sometimes with 7–12 faint to dark vertical bars. Undersides yellow orange. Side of head with iridescent blue or green mottling; ear flap black with white or yellow-orange margin. Black blotch usually present near posterior base of soft dorsal fin and often in the anal fin. Fins of breeding males become dark; margin of caudal, anal, and pelvic fins whitish orange. During actual spawning, females develop prominent dark vertical bars on sides. Young-of-the-year mostly dark gray and without vertical bands (unusual among YOY *Lepomis*); older juveniles often with closely spaced bars on sides. The black blotch in the posterior base of the soft dorsal fin begins to develop in individuals greater than 1 inch (26 mm) SL (Ross 2001).

SIMILAR SPECIES   The Green Sunfish differs from the Bantam Sunfish in having more than 40 lateral line scales (vs. fewer than 35); from the Redspotted, Redbreast, Longear, and Dollar sunfishes in having a black blotch in the soft dorsal fin (vs. black blotch absent); and from all remaining *Lepomis* in having a short, rounded pectoral fin (vs. long and pointed). We have also noted that live Green Sunfish are more slippery than other *Lepomis* species.

VARIATION AND TAXONOMY   *Lepomis cyanellus* was described by Rafinesque (1819) from the Ohio River. It was considered the sister species of *L. symmetricus* by Near et al. (2004, 2005). Little information exists on geographic variation. Natural hybridization has been reported between *L. cyanellus* and seven other *Lepomis* species: *L. auritus*, *L. gibbosus*, *L. gulosus*, *L. humilis*, *L. macrochirus*, *L. megalotis*, and *L. microlophus* (Bolnick 2009).

DISTRIBUTION   *Lepomis cyanellus* is native to the Great Lakes, Hudson Bay, and Mississippi River basins from New York and Ontario, Canada, to Minnesota, and south to the Gulf of Mexico, and Gulf Slope drainages from the Escambia River, Florida, and Mobile Bay drainages of Georgia and Alabama west to the Rio Grande drainage of Texas, New Mexico, and Mexico. It has been introduced into almost every other state. In Arkansas, the Green Sunfish occurs statewide in all drainages (Map 170).

HABITAT AND BIOLOGY   The Green Sunfish is a highly adaptable species and can be found in almost every type of aquatic habitat in Arkansas. Pflieger (1997) and Smale and Rabeni (1995a, b) noted its tolerance of a wide range of ecological conditions, particularly extremes of turbidity, dissolved oxygen, temperature, high alkalinity, and flow. It is aggressive and competitive, and it is also a pioneering species, readily populating new bodies of water. It is among the first fishes to repopulate streams after droughts (Pflieger 1997; Smith 1979). Although widely distributed in a variety of habitats in Arkansas, it is most abundant in small creeks and ponds that will not support most other sunfishes. Adults are very territorial, with a well-defined, small home range.

Carlander (1977) and Warren (2009) provided a summary of the extensive literature on the biology and life history of the Green Sunfish. Its large mouth allows predation on larger food items than by most other *Lepomis*. Even though its diet is similar to that of Bluegills, where the two species are found together, resource partitioning apparently allows the two to coexist. The Green Sunfish has a larger mouth than the Bluegill, and it selects food items of a larger average size than those in the Bluegill's diet. Its food consists mainly of insects (both aquatic and terrestrial), small fish, and crayfish. In Bull Shoals Reservoir, Green Sunfish smaller than 48 mm (1.9 inches) TL fed mainly on aquatic insects and small crustaceans (copepods and cladocerans), whereas those from 51 to 76 mm (2–3 inches) ate primarily large aquatic insects such as mayfly nymphs, and fish larger than 102 mm (4 inches) fed most heavily on crayfish (Applegate et al. 1967). In Beaver Lake, Green Sunfish ate more Largemouth Bass eggs and fry than other sunfish, but the amount was not large after mid-May (Mullan and Applegate 1968). By late June, stomachs of Green Sunfish 51–99 mm (2–3.9 inches) TL contained 33% fish larvae by volume, and they had turned primarily to Gizzard Shad and young sunfish for food. Stream populations of Green Sunfish usually consume a wider variety of foods at a given locality than reservoir populations, including crayfish, terrestrial insects, immature and adult aquatic insects, and a variety of fishes.

In Arkansas, an extended spawning period occurs from

April through August with the most intense activity occurring in June. Spawning activity begins as water temperatures approach 20°C, and peak spawning occurs between 20°C and 28°C (Hunter 1963). Colonial nesting and synchronous spawning sometimes occur. Sound production is apparently an important part of courtship (Gerald 1971). The males construct nests in shallow backwater areas. Gravel substrates are preferred, but clay and detritus bottoms can be used. The nests are circular depressions formed mainly by vigorous fanning with the tail. Hunter (1963) described spawning behavior as follows:

> The male and female circled in the nest side by side, paused momentarily and released sperm and eggs. The consummatory act took place when the female reclined on her side and vibrated while the male remained in an upright position. An isolated pair might circle and spawn in a nest for considerable periods of time but in crowded colonies the male frequently interrupted spawning to chase intruding fish. After spawning the male expelled the female from the nest with a nip. Both sexes usually spawned with more than one individual.

The eggs are attached to the substrate and are guarded by the male for about a week, until the fry become free-swimming. Fertilized eggs hatch in 2.1 days at 23.8°C (Childers 1967). Fecundity has been reported as 2,000–10,000 eggs depending on size of female (Moyle 2002). Sexual maturity is attained in 1 or 2 years, and maximum life span is about 7 years. Growth rates in Bull Shoals Reservoir averaged 47.7 mm TL after 1 year, and 71.1 mm, 91.4 mm, 109.2 mm, and 121.9 mm TL in years 2 through 5, respectively (Applegate et al. 1967). *Cyprinella lutrensis* and *Lythrurus umbratilis* sometimes spawn in Green Sunfish nests (Warren 2009).

The Green Sunfish is not a highly prized gamefish in Arkansas, and some anglers derisively refer to them as ricefield slicks. Stunting is often severe in farm ponds because of overpopulation. Its greatest importance to anglers is in small creeks that will not support other gamefish. The Arkansas Game and Fish Commission in 1986 began producing "hybrid bream" by crossing male Green Sunfish with female Bluegills. Fingerling hybrids have been stocked in some Arkansas reservoirs, but their greatest use is in ponds to create put-and-take fisheries. Ross (2001) noted that the hybrids are fast growing, aggressive, and easy to catch.

CONSERVATION STATUS The Green Sunfish is widespread and secure in Arkansas.

Lepomis gulosus (Cuvier)
Warmouth

Figure 26.19. *Lepomis gulosus*, Warmouth. *David A. Neely.*

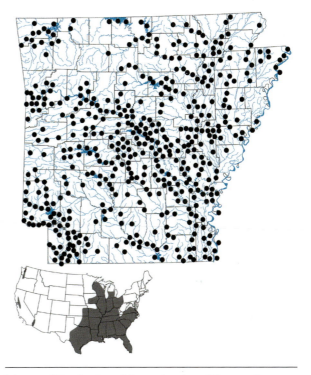

Map 171. Distribution of *Lepomis gulosus*, Warmouth, 1988–2019. Inset shows native and introduced ranges in the contiguous United States.

CHARACTERS A medium-sized, robust sunfish with a large head and the largest mouth of any *Lepomis* species, the upper jaw in adults reaching to or beyond middle of eye; lower jaw projecting slightly (Fig. 26.19). Body depth going into SL 2.0–2.5 times. Prominent dark lines radiating backward from eye to margin of gill cover. Tongue with a small patch of teeth. Villiform palatine teeth present. Supramaxilla large at all ages. Gill rakers long and slender, 9–13. Pectoral fin short and rounded, its tip usually not reaching past eye when bent forward across cheek. Opercular flap short and stiff with a black spot and light border; opercle stiff to rear margin. Lateral line complete, arched anteriorly, with 35–44 scales. Scale rows above lateral line 6–9. Cheek scale rows 5–7. Dorsal spines 10 or 11; dorsal rays 10, rarely 11; anal spines 3; anal rays 9 or 10; pectoral rays 12–14. Adults rarely exceed 8 inches (203 mm) in length or 0.5 pound (0.2 kg) in weight. The world angling record from Florida is 2 pounds, 6 ounces (1.1 kg). The Arkansas state record, from Black Dog Bayou in 1998 is 1 pound, 8 ounces (0.68 kg).

LIFE COLORS   Back and sides olive brown and mottled or barred with dark purplish brown. Undersides a light yellow. Four or five dark reddish-brown streaks radiating backward from eye; iris of eye red. Breeding male usually with red border on opercular flap. Fins dark and strongly mottled or banded with brown, especially the soft dorsal and anal fins; pelvic fins white-edged. Young-of-the-year and small juveniles are purple, darker than most other small Lepomis, and with dark lines radiating backward from eye.

SIMILAR SPECIES   Sometimes mistaken for Rock Bass, but differs from the 3 Arkansas species of Ambloplites in having 3 anal spines (vs. 5 or 6). Distinguished from other species of Lepomis by presence of small patch of teeth on tongue, 4 or 5 dark lines radiating backward from eye, and a very large supramaxilla at all sizes.

VARIATION AND TAXONOMY   Lepomis gulosus was described by Cuvier in Cuvier and Valenciennes (1829) as Pomotis gulosus from Lake Pontchartrain, New Orleans, Louisiana. Disagreement over whether to place the Warmouth in genus Chaenobryttus or genus Lepomis persisted through most of the 20th century, but recent authors have supported placement in Lepomis. Smith (1979) noted little morphological variation geographically in the Warmouth. Bermingham and Avise (1986) recognized distinct eastern and western clades based on mitochondrial DNA analysis. Lepomis gulosus populations from the Mobile Basin westward constituted a separate lineage from those to the east. Near et al. (2004, 2005) considered the Warmouth sister to the species pair L. cyanellus and L. symmetricus. Natural hybridization has been reported between L. gulosus and five other Lepomis species (L. auritus, L. cyanellus, L. gibbosus, L. macrochirus, and L. microlophus) (Bolnick 2009). A rangewide study of variation is lacking.

DISTRIBUTION   Lepomis gulosus is native to the Great Lakes (Lake Erie and southern Lake Michigan) and Mississippi River basins from Ontario, Canada, and Pennsylvania to Minnesota and south to the Gulf of Mexico; the Atlantic Slope from the James River, Virginia,

to southern Florida; and the Gulf Slope from Florida to the Rio Grande of Texas and Mexico. It has been introduced into Pacific Coast drainages from Oregon to Mexico. In Arkansas, the Warmouth is widely distributed over the state in all major drainages but is most abundant in the Coastal Plain lowlands (Map 171). In the more mountainous regions, it is most common in reservoirs and ponds.

HABITAT AND BIOLOGY   Primarily a lowland species, the Warmouth inhabits oxbow lakes, quiet waters of bayous and rivers, and swamps having mud and detritus bottoms. It is most abundant in habitats having little current. It prefers clear water and thick growths of aquatic vegetation and is often found near submerged stumps, logs, and cypress knees. It can tolerate moderate levels of turbidity, low oxygen levels, and moderately brackish water (Loftus and Kushlan 1987; Killgore and Hoover 2001). Becker (1983) considered it to have greater tolerance of these unfavorable conditions than most other sunfishes. In upland areas, it is found in sluggish streams and quiet pools and does well in impoundments. Juveniles are more closely associated with vegetation in shallow water throughout the year, while adults usually occupy deeper water. The Warmouth is a sedentary, secretive fish usually found in cover, and it avoids strong light. It does not form schools, even in winter. Pflieger (1997) noted that it generally occurs in low population densities, seldom achieving the abundance of other Lepomis species; therefore, it has no tendency to become dominant at the expense of other species.

Larimore (1957) studied its life history in Illinois, and Warren (2009) summarized published information. Its food habits are similar to those of the Green Sunfish, another Lepomis with a large mouth. Adults feed more on fishes than do most other Lepomis, but its diet also consists of crayfish, freshwater shrimp, larval aquatic insects, and isopods. In a Florida study, the largest adults fed almost exclusively on crayfish (Guillory 1978). The young feed initially on microcrustaceans and transition to midge and caddisfly larvae and other small invertebrates at about 75 mm TL (Larimore 1957; Guillory 1978). Most studies have shown that fish and crayfish are the most important foods in the adult diet. Individuals greater than 127 mm (5 inches) TL in Illinois consumed crayfish (29.7% by volume), fish (20.5%), damselflies (15.7%), dragonflies (5.2%), dipteran larvae (4.6%), mayflies (2.9%), entomostracans (0.7%), hemipterans (0.3%), miscellaneous insects (12.2%), and unidentifiable items (8.2%) (Larimore 1957). It is apparent that the Warmouth is somewhat of an opportunistic feeder, often consuming any items that are readily available.

The spawning period in Arkansas extends from April

into August, but peak activity occurs in June. Spawning activity begins as water temperatures approach 21°C, with peak spawning occurring at 27–29°C (Larimore 1957; Guillory 1978). Circular depression nests are constructed in the substrate by males near a stump, log, rock, vegetation, or other object at depths between 0.15 and 1.5 m (6–59 inches). Availability of cover is more important than substrate type as long as the chosen substrate can be swept free of silt. Nests are not found in areas that are completely exposed. Nests are usually solitary, but colonial nesting also occurs if spawning habitat is limited. Courtship behavior, as described by Larimore (1957), is typical of the genus, and nests are guarded by males until the fry are free-swimming, usually 5 or 6 days after spawning. After leaving the nest, the young do not school but hide in dense vegetation or other cover (Larimore 1957). Mature females contained two or more egg class sizes throughout the breeding season, and fecundity increased with female size. Batch fecundity estimates in Louisiana ranged from 6,825 to 20,238 eggs (Guillory 1978), but in Illinois, females 94–137 mm (3.7–5.4 inches) TL contained 17,200–63,200 eggs (Becker 1983). Eggs hatch in 35.5 hours at 25–26.4°C (77–79.5°F). Maximum life span is around 7+ years (Smith 1979). Cloutman (1974) reported parasites of Warmouth in Lake Fort Smith.

In Arkansas, the Warmouth has its greatest angling importance in lowland waters, where it is commonly known as a goggle-eye. It will readily strike natural or artificial baits but tires easily after an initially strong strike. It is considered an excellent food fish (Larimore 1957).

CONSERVATION STATUS The Warmouth is widespread and common in all areas of Arkansas except the Ozarks. Its populations in some environments are large and the species is secure in the state.

Lepomis humilis (Girard)
Orangespotted Sunfish

CHARACTERS A small sunfish with a compressed, moderately deep body, body depth going into SL 2.2–2.6 times (Fig. 26.20). Mouth moderately large, the upper jaw extending past front of eye. Supramaxilla absent. Pectoral fin moderately long and pointed, its length going into SL fewer than 3.5 times; when bent forward across cheek, pectoral fin reaches to about anterior edge of eye. Gill rakers moderately long and slender, 10–15. Two large sensory pits on top of head in skull between eyes, the width of each pit approximately equal to distance between pits. Sensory pores along preopercle elongated and very large, larger

Figure 26.20. *Lepomis humilis*, Orangespotted Sunfish. *Uland Thomas.*

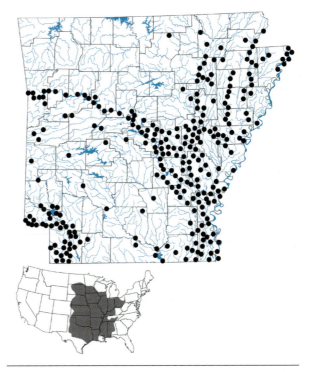

Map 172. Distribution of *Lepomis humilis*, Orangespotted Sunfish, 1988–2019.

than those of any other *Lepomis* species (Moore 1956). Opercular flap elongated, flexible, and angled upward in adults; rear margin of opercle (which lies within the base of the membranous ear flap) thin and flexible. Lateral line complete or incomplete, arched anteriorly, with 32–40 scales. Cheek scale rows usually 5. Palatine teeth present. No teeth on tongue. Lower pharyngeal arches narrow, with short, pointed teeth. Dorsal spines 10 or 11; dorsal rays 9 or 10; anal spines 3; anal rays 8 or 9; pectoral rays 13–15; pelvic fin with 1 spine and 5 rays. It is the second-smallest species of *Lepomis* (after *L. symmetricus*). Never large enough to have significance as gamefish, there is no Arkansas state angling record. Maximum reported length approximately 7 inches (177 mm) TL (Carlander 1977), but most adults range from 1.6 to 4 inches (40–102 mm) TL. Maximum weight about 5 ounces (0.142 kg).

LIFE COLORS   A brightly colored sunfish. Adults with back and sides greenish silver; sides with scattered reddish-brown to orange spots. Cheek with faint orange or brown wavy lines. Undersides white or yellow. Elongated opercular flap black with broad light-colored (usually white) margin. Fins plain. Breeding males with bright orange lateral spots; breeding females with reddish-brown spots. Males with belly and median fins bright reddish orange; sides of head iridescent blue with orange streaks; iris of eye red or orange. Nonbreeding adults with pastel versions of the breeding colors. Young with widely spaced dusky vertical bars on sides and no brown spots.

SIMILAR SPECIES   The Orangespotted Sunfish differs from other sunfishes in having very large, elongated sensory pores along the preopercle (vs. sensory pores small and circular) and in having a pair of large sensory pits on the top of the head (vs. pits on head small). It differs from the Warmouth in lacking teeth on tongue (vs. teeth present), and in lacking prominent dark lines radiating backward from eye (vs. prominent dark lines present behind eye). It differs from the Bluegill in lacking a black blotch in the soft dorsal fin (vs. black blotch present), and from the Redear Sunfish in having long, slender gill rakers (vs. short and stubby). It differs from remaining *Lepomis* in having a moderately long pectoral fin, its length going into SL fewer than 3.5 times (vs. more than 3.5 times). Other salient features include a crappie-like profile and a short, upturned opercular flap.

VARIATION AND TAXONOMY   *Lepomis humilis* was originally described as *Bryttus humilis* by Girard (1858b) from "the freshwaters of Arkansas" with no specific type locality given. Only one of the specimens listed by Girard as *Bryttus humilis* (ANSP 13154) is referable to *Lepomis humilis* (the other specimens are *L. cyanellus*), and that specimen was small and imperfectly preserved. Based on information from R. M. Bailey, Böhlke (1984) selected as the lectotype the specimen illustrated as figure 13, plate 7 in Girard (1858b) from Sugar Loaf Creek in Sebastian County, Arkansas. Smith (1979) noted considerable sexual and ontogenetic variation. A rangewide study of variation is lacking. Near et al. (2004, 2005) considered *L. humilis* and *L. macrochirus* as forming a sister pair. Natural hybrids have been reported between *L. humilis* and four other *Lepomis* species (*L. cyanellus*, *L. gibbosus*, *L. macrochirus*, and *L. megalotis*) (Bolnick 2009).

DISTRIBUTION   *Lepomis humilis* occurs in the lower Great Lakes (Erie and Michigan), Hudson Bay, and Mississippi River basins from Ohio to North Dakota, and south to Louisiana; Gulf Slope drainages from the Pearl River system of Alabama and Mississippi, to the Colorado River, Texas; and isolated populations in southern Texas. It has been introduced into the Mobile Bay drainage, and elsewhere in Canada and the United States. In Arkansas, the Orangespotted Sunfish is widely scattered over the state but is most common in the Coastal Plain lowlands, the Arkansas River Valley, and the Red River (Map 172). It was one of 10 species represented in the greatest number of collections in a Red River study (Buchanan et al. 2003). This species is nearly absent from the more mountainous regions of the Ozarks and Ouachitas.

HABITAT AND BIOLOGY   Several authors have noted the wide ecological tolerances of the Orangespotted Sunfish, and Boschung and Mayden (2004) considered its greater tolerance of pollution than other sunfishes a key factor in its expanding range and increasing numbers in Alabama. Cross (1967) reported that in Kansas it is indifferent to bottom type, occurring over rocky, sandy, or muddy substrates and that it tolerates high turbidity and extensive fluctuation in water level. Smith (1979) observed that in Illinois it occurs in small ponds, large lakes, creeks, and large rivers and prefers areas with silt or detritus on the bottom. Pflieger (1997) noted that it avoids streams with high gradient, clear or cool water, and continuous strong flow. It is tolerant of low dissolved oxygen (<1.0 ppm) and high water temperatures (Smale and Rabeni 1995a; Beitinger et al. 2000). In the Sulphur River, Texas, it was found in pool-rootwad, backwater, and backwater–bank snag habitats (Morgan 2002), and Burgess (2003) reported that it was the only invertivore associated with the channelized reach of the South Sulphur River. In Arkansas, it is primarily a stream fish most commonly found in turbid or clear lowland waters over a mud and detritus bottom in medium to large rivers. It is also common in some sand, mud, and gravel-bottomed small creeks that drain patches of tallgrass prairie in westcentral Arkansas, and in several small creeks of Crowley's Ridge. It is occasionally found in lowland oxbow lakes, several impoundments, and is abundant in backwater areas all along the Arkansas and Red rivers. It occurred in 16 of 66 Arkansas impoundments studied by Buchanan (2005). H. C. Miller (1964) described a distinctive sleeping posture, with the body from the chin to the pelvic fins touching the substrate and the back part of the body elevated.

Warren (2009) summarized life history information and described the Orangespotted Sunfish as an opportunistic insectivore feeding on both benthic and terrestrial insects.

Barney and Anson (1922) found that its diet consisted primarily of small crustaceans, larval aquatic insects (chironomids, caddisflies, mosquito larvae), hemipterans, and occasionally small fish. Foods reported from Wisconsin specimens included immature trichopterans, ephemeropterans, plecopterans, odonates, and dipterans, in addition to copepods and other small crustaceans (Becker 1983). Two Iowa studies also confirmed that insects composed 87% (Kutkuhn 1955) and 100% (Harrison 1950) of the Orangespotted Sunfish diet. In addition to food items reported from other studies, the Iowa specimens had also eaten homopterans, terrestrial coleopterans, lepidopterans, and traces of algae and debris. The range of foods ingested, including terrestrial insects, indicated both benthic and surface feeding (Etnier and Starnes 1993).

In Arkansas, spawning occurs from April through August. We observed aquarium spawning in May 1995. Spawning occurred all afternoon in typical sunfish fashion. Spawning begins when water temperatures reach 18°C, but peak spawning has been reported to occur between 24°C and 32°C (Cross 1967; Noltie 1990). The males construct small circular depression nests in the substrate, often in colonies at water depths of 0.3–0.61 m. Nests are somewhat smaller than those of most other *Lepomis* species (150–189 mm diameter), and 30–40 mm deep (Becker 1983). The male entices females into the nest by rushing out to them and then rapidly returning to the nest. Gerald (1971) provided evidence that sound production by the males is also a part of courtship behavior. Peak spawning activity occurs in June in this gregarious species. The eggs are very small and colorless and adhere to objects on the nest floor. After spawning, the female leaves the nest, and the male guards the embryos until hatching, usually about 5 days at 80°F (26.7°C). Fecundity ranges from 138 eggs at 30 mm TL to 5,776 eggs at 105 mm TL (Becker 1983). Maximum life span is about 4+ years.

Although not large enough to be important as a game species, *L. humilis* is occasionally an important forage fish for larger sunfishes. It might compete with young of game species for food. It is an attractive species that is sometimes kept in the home aquarium but, like all sunfishes, would not be a good community aquarium fish. Some consider *L. humilis* more difficult to maintain in an aquarium than most sunfishes, and H. C. Miller (1964) found that in captivity it would accept only live food. We do not consider it any more difficult to maintain in an aquarium than other *Lepomis* species.

CONSERVATION STATUS This small sunfish is widely distributed and sometimes locally abundant in Arkansas. Its populations are secure.

Lepomis macrochirus Rafinesque
Bluegill

Figure 26.21a. *Lepomis macrochirus*, Bluegill male. *Uland Thomas.*

Figure 26.21b. *Lepomis macrochirus*, Bluegill female. *David A. Neely.*

CHARACTERS A moderately large, laterally compressed, deep-bodied *Lepomis*; body depth going into SL 1.8–2.3 times in adults (Fig. 26.21). Mouth small and oblique, the upper jaw rarely reaching and not extending past front of eye. Supramaxilla absent. Soft dorsal fin with a black blotch near posterior base in adults and sometimes in juveniles; spot is least conspicuous or absent in specimens from turbid waters and in small young. Pectoral fin long and pointed, its tip usually reaching past front of eye when bent forward across cheek. Gill rakers long and slender, 13–16. The opercular flap is short in juveniles, becoming longer in adults (especially males), flexible and dark to its margin; rear margin of opercle (which lies within the base of the opercular flap) thin and flexible. Lateral line complete, arched anteriorly, with 39–46 scales. Scale rows above lateral line 7–9. Cheek scale rows 4–7. Palatine teeth present or absent. No teeth on tongue. Pharyngeal arches moderately wide, with thin, pointed teeth. Dorsal spines 10 (9–11); dorsal rays 11 or 12; anal spines 3; anal rays 10 or 11; pectoral rays 12–15; pelvic fin with 1 spine and 5 rays.

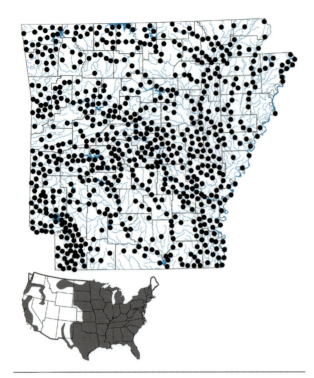

Map 173. Distribution of *Lepomis macrochirus*, Bluegill, 1988–2019. Inset shows native range (solid) and introduced range (outlined with some introduced populations shown in solid) in the contiguous United States.

Adults commonly reach a length of 6–10 inches (152–254 mm) and a weight of 0.5–1 pound (0.2–0.5 kg). Maximum weight about 4 pounds, 12 ounces (2.16 kg), and maximum TL around 16.25 inches (410 mm). The Arkansas state record from Fulton County in 1998 is 3 pounds, 4 ounces (1.47 kg).

LIFE COLORS   Back and upper sides dark olive green, with bluish iridescence; lower sides silver. Sides often with 6–8 dark olive bars, but large adults and fish from turbid waters generally lack bars; juveniles with 9–12 evenly spaced dark vertical bars on sides. Posterior rays of soft dorsal fin usually with a black blotch. Undersides yellow to reddish orange. Chin and lower part of operculum blue; opercular flap entirely black to its margin. In breeding males, all colors more intense, fins becoming densely pigmented, pelvic and anal fins becoming almost black.

SIMILAR SPECIES   The Bluegill differs from the Redear Sunfish and the Orangespotted Sunfish in having a black blotch in the soft dorsal fin and an opercular flap that is dark to its margin (vs. no black blotch in soft dorsal fin and a light-colored margin of opercular flap). It differs from other *Lepomis* in having a very long, pointed pectoral fin (vs. short and rounded).

VARIATION AND TAXONOMY   *Lepomis macrochirus* was described by Rafinesque (1819) from the "Ohio River." This polytypic species may eventually have some of its variant populations elevated to full species status (Near and Koppelman 2009). Three subspecies of the Bluegill are recognized, but their ranges are not clearly defined (Page and Burr 2011). *Lepomis m. macrochirus*, the form native to Arkansas, is found throughout the Mississippi River drainage and the Gulf Slope drainages east to Alabama. A second subspecies, often known as Coppernose Bluegill, was reported in coastal drainages from North Carolina to Florida (Hubbs and Allen 1943). The taxonomic status of the Coppernose Bluegill has not been determined, and there is considerable disagreement about its distribution and recognition. Avise and Smith (1974a) reported that it occurred only in the Florida peninsula and identified a broad intergrade zone to the north between the two subspecies. Felley (1980) also found the two subspecies genetically as well as morphologically distinguishable and recognized an intergrade zone between them. The name traditionally applied to this eastern subspecies was *Lepomis macrochirus purpurescens*, based on the assumption that this subspecies' range extended from Florida along the Atlantic Coast to North Carolina (where the type locality for the subspecies is located); however, some subsequent studies indicated that this subspecies is restricted to peninsular Florida and was originally described as a separate species, *Lepomis mystacalis* by Cope in 1877 (Near and Koppelman 2009). Therefore, according to those authors, the correct designation for the Florida subspecies (or Coppernose Bluegill) is *Lepomis macrochirus mystacalis*. *Lepomis macrochirus purpurescens* was considered a synonym of *L. m. macrochirus* by Gilbert (1998), but some recent authors (e.g., Page and Burr 2011; Robins et al. 2018) still use the subspecific name, *L. m. purpurescens*. Robins et al. (2018) noted that populations of Bluegills in Florida are highly variable and in need of taxonomic study. The Coppernose Bluegill has been widely introduced into Arkansas by the AGFC and by private landowners who obtain them from private aquaculture operations throughout the state. The Coppernose Bluegill can be distinguished from the native Arkansas Bluegill by its red or red-orange fins with white margins in breeding males (vs. clear or dusky to dark-colored fins), copper-colored patch on head in adults, usually 12 anal rays (vs. usually 10 or 11 anal rays), and broader lateral bars in young. Most specimens of Coppernose Bluegills cultured, introduced, or caught in Arkansas have a dark-colored body with a distinct purplish sheen (as indicated by the scientific name *L. m. purpurescens*). The Florida Bluegill shown in Fig. 26.22 may be taxonomically distinct from the true

Figure 26.22. Coppernose Bluegill, Fanning Springs, Suwannee River drainage, Florida. *Isaac Szabo.* Arkansas specimens of Coppernose Bluegill are much darker in overall body coloration than shown in the photograph.

Coppernose Blugill, or its overall lighter body coloration may be environmentally influenced; however, it does show the characteristic fin and snout coloration of a Coppernose.

The Coppernose Bluegill reportedly reaches a larger size than other Bluegill subspecies, a feature largely responsible for its widespread introduction outside its native range; however, Ross (2001) reported that stockings of Coppernose Bluegill have shown little advantage in fish weight and growth over native Bluegill. The third subspecies, *L. m. speciosus*, occurs in west Texas and Mexico and intergrades with *L. m. macrochirus* in Arkansas and Red river drainages of Arkansas, Oklahoma, and Texas. Differences in modal number of anal rays are often used to distinguish the three subspecies of Bluegill, with *L. m. speciosus* typically having 10 anal rays, *L. m. macrochirus* having 11, and *L. m. mystacalis/purpurescens* having 12.

Near et al. (2004, 2005) considered the Bluegill to form a sister pair with *L. humilis*. Artificial hybrids between *L. cyanellus* and *L. macrochirus*, often referred to as hybrid bream, have been stocked in a number of Arkansas reservoirs. Natural hybrids have been reported between *L. macrochirus* and eight other *Lepomis* species (*L. auritus*, *L. cyanellus*, *L. gibbosus*, *L. gulosus*, *L. humilis*, *L. megalotis*, *L. microlophus*, and *L. punctatus*) (Bolnick 2009).

DISTRIBUTION *Lepomis macrochirus* is considered native to the St. Lawrence–Great Lakes and Mississippi River basins from Quebec, Canada, and New York, west to Minnesota, and south to the Gulf of Mexico; Atlantic Slope drainages from the Cape Fear River, North Carolina, to southern Florida; and Gulf Slope drainages from Florida to the Rio Grande of Texas, New Mexico, and Mexico. It has been introduced throughout the United States, and also in Europe, Asia, and Africa. In Arkansas, the native

Bluegill, *L. m. macrochirus*, is found statewide in all drainages, and the Coppernose Bluegill, *L. m. mystacalis*, has been widely stocked (Map 173). Between 1991 and 2000, the AGFC stocked approximately 1 million Coppernose Bluegills in public waters in northeastern Arkansas (Greene and Mississippi counties), central Arkansas (Faulkner and Lonoke counties), western Arkansas (Sebastian County), and southern and southwestern Arkansas (Garland, Clark, Hempstead, Miller, and Lafayette counties). Based on information provided by Tommy Laird (AGFC), Coppernose Bluegills continue to be routinely stocked statewide in private waters with fish obtained from private hatcheries.

HABITAT AND BIOLOGY One of the most popular gamefishes in the state, the Bluegill has been widely stocked in farm ponds and reservoirs, and a tremendous volume of literature is available on its biology. In Arkansas, it is one of two species commonly referred to as bream. According to Carlander's 1977 summary of the literature on the biology and management of this species, its largest populations are found in clear, quiet, warm waters having at least some aquatic vegetation and other cover, but it is also found in streams and rivers. In an Illinois lake, juveniles moved away from vegetated habitat near the shore at night and toward vegetation during the day, apparently in response to reduced predator abundance and/or activity, combined with increases in abundance of zooplankton in open-water habitat at night (Shoup et al. 2014). Gaston and Lauer (2015) found that environment influenced morphology of Bluegills in Indiana. Male and female *L. macrochirus* (and female *L. cyanellus*) from lentic systems had a deeper body than those from lotic systems. They concluded that a deeper body promotes greater maneuverability desirable in lentic systems. In contrast, the more streamlined body of the Bluegills found in lotic systems reduces drag in flowing water, ultimately maximizing energy efficiency. It is sometimes found in turbid water, but it is intolerant of continuous high turbidity and siltation. It is among the most tolerant *Lepomis* species of low dissolved oxygen and high water temperatures (Smale and Rabeni 1995a, b). Pflieger (1997) noted that its ecological requirements are very similar to those of the Largemouth Bass, and it is often found with that species.

The Bluegill is a generalized sight feeder that selects food items individually. It is a gregarious species, often moving in loose schools, and it feeds at various depths depending on food availability. It is able to switch foraging habitats and quickly learn new foraging behaviors to exploit new prey in response to changes in prey abundance, competition, or predation risk (Warren 2009). Werner and Hall

(1976) found that Bluegills adjusted their feeding strategies depending on which competing species of *Lepomis* (Green Sunfish and Pumpkinseed) co-occurred with them. The young feed mainly on small crustaceans and the adults feed primarily on insects, but they also consume crayfish, snails, and small fish. Its small mouth limits the size of the food items taken. Occasionally, vegetation is ingested. In Bull Shoals and DeGray reservoirs, larger Bluegills fed mainly on terrestrial insects, planktonic crustaceans, filamentous algae, fish eggs, and bryozoans (Applegate et al. 1967; Bryant and Moen 1980a). Bluegill larvae begin feeding at about 5.0–5.9 mm TL on small rotifers and copepod nauplii; between 7 and 10 mm TL, larger zooplankton dominate the diet; between 14–17 mm TL, they seek more protected areas near shore and feed mainly on aquatic insects (especially midge larvae) and small crustaceans; and large fish consume mainly aquatic and terrestrial insects and other available prey small enough to ingest (Flemer and Woolcott 1966; Siefert 1972; Werner and Hall 1988). The Bluegill is one of the few freshwater fishes known to exhibit cleaning symbiosis (Sulak 1975; Powell 1984). Large Bluegills and Largemouth Bass have been observed to allow small Bluegills to approach and pick parasites off their bodies. Powell (1984) described cleaning of Striped Mullet (and human divers) by Bluegills in Crystal River, Florida, and TMB has observed cleaning behavior by Bluegills on humans in a clear Arkansas lake. Spall (1970) reported cleaning of Flathead Catfish by Bluegills in captivity.

Spawning in Arkansas takes place from April through August when water temperatures exceed 68°F (20°C), and spawning continues up to about 88°F (31°C). Peak spawning activity occurs in June and early July. The males fan out nests that are circular depressions usually 8–12 inches (203–305 mm) in diameter, in a variety of substrates. The nests are often closely spaced in water 1–4 feet (0.3–1.2 m) deep and are usually away from cover. Colonies sometimes contain hundreds of adjacent nests, and spawning events are synchronous, occurring at intervals of 10–14 days (Warren 2009). Several females may spawn in the same nest. Sound production by the male appears to play a role in courtship (Gerald 1971). Fecundity increases with female size, ranging from 17,990 mature eggs in females 165 mm TL to 45,575 eggs in females 216 mm TL (Warren 2009). Eggs hatch in 2.1 days at 23.8°C and 1.3 days at 27.1°C, and the larvae become free-swimming about 3–4 days after hatching (Childers 1967). The male guards the nest and larvae for about 7 days, but does not guard the free-swimming fry. The free-swimming young usually move from the littoral areas to the limnetic zone to feed on zooplankton and

subsequently return to the littoral zone prior to their first winter; in some populations, the free-swimming larvae remain in the littoral zone (DeVries et al. 2009).

Alternate mating strategies (or female mimicry) referred to as cuckoldry are well documented in male Bluegills (Gross 1979, 1982). Sneaker and satellite male behaviors have been documented in a number of studies (Dominey 1980; Ehlinger 1997; Drake et al. 1997). In such instances, reproductive males that have a color pattern resembling the female do not construct nests. These nonaggressive, sneaker males attempt to fertilize the eggs being deposited in the nest of a territorial male while that male is engaged in spawning activity with females. Satellite males also attempt to enter the nest of a territorial male to fertilize eggs, but in contrast to sneaker males, they are willing to engage in aggressive encounters, if necessary. In one study, these parasitic males were estimated to have fertilized 80% of the eggs when directly competing with territorial males (Fu et al. 2001); however, most studies have found that parasitic males fertilized 1.3–22.6% of the eggs (Neff and Knapp 2009). Neff and Clare (2008) found that cuckoldry rates peaked in the middle third of the breeding season in Bluegills.

Growth rates of Bluegills in Arkansas (as elsewhere) can be highly variable, depending on a number of environmental factors. Hogue and Kilambi (1975) studied age and growth of Bluegills in Lake Fort Smith, Sewell (1985) reported growth rates in the Big Creek watershed of Craighead and Greene counties, and Hulsey and Stevenson (1958) compared growth rates of Bluegills and other gamefishes in Lake Catherine, Lake Hamilton, and Lake Ouachita. Pflieger (1997) provided data showing that growth is most rapid in ponds and new reservoirs and slowest in old reservoirs and streams. Cross (1967) reported that stunting is most common in muddy ponds, but it can also occur in clear ponds that have extensive areas of shallow water overgrown by aquatic vegetation. Bluegills grow best and produce more fish of catchable size in ponds having clear water, steep shorelines, and at least one predaceous species such as Largemouth Bass to prevent overpopulation. It is an eagerly sought species and will readily take both natural and artificial bait. When hooked it is an excellent fighter, and its flesh is very flavorful. Maximum life span in northern areas of its range is about 11 years.

CONSERVATION STATUS   This is the most widely stocked gamefish in Arkansas. It is widespread and locally abundant statewide, and its populations are secure; however, it is likely that many, if not most, Arkansas populations contain genetic material from *L. m. mystacalis*.

Lepomis marginatus (Holbrook)

Dollar Sunfish

Figure 26.23. *Lepomis marginatus*, Dollar Sunfish. *David A. Neely.*

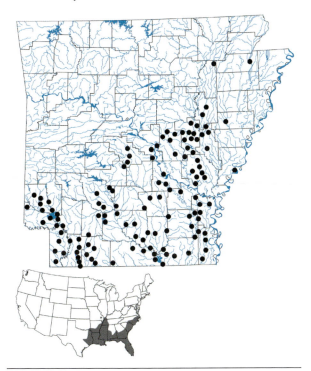

Map 174. Distribution of *Lepomis marginatus*, Dollar Sunfish, 1988–2019.

CHARACTERS A small sunfish with a short, compressed, deep body; body depth going into SL approximately 2.0 times (1.93–2.29 for Arkansas populations) (Fig. 26.23). Mouth moderately small and oblique, the upper jaw not extending much past anterior margin of eye. A small supramaxilla is present. Opercular flap usually considerably elongated in adults, angled upward, and with a light border; rear margin of opercle (which lies within the base of the opercular flap) thin and flexible. Pectoral fin short and rounded, its tip usually not reaching the eye when bent

forward across the cheek. Palatine teeth absent; no teeth on tongue. Gill rakers short and stubby, 11–14 (8–10 on lower limb of arch and 2–5 on upper limb). Cheek scales usually in 3 or 4 rows, with approximately 8% of Arkansas specimens examined having 5 cheek scale rows (B. H. Bauer, pers. comm.). Pectoral fin rays usually 12 (11–14). Anal fin base convex. Lateral line complete, arched anteriorly, usually with fewer than 40 scales (35–42, $\bar{x}$ = 38.4). Scale rows above lateral line 5 or 6. Dorsal spines usually 10 (9–11); dorsal rays usually 10 (9–12); anal spines 3; anal rays usually 9 (8–11). It is a smaller species than the Longear Sunfish, and most adults are around 3–4 inches (76–102 mm) TL. It reportedly reaches a maximum size of 5 inches (127 mm) TL in Louisiana (Douglas 1974).

LIFE COLORS Coloration variable but resembles that of Longear Sunfish. Back and sides blue green to olive, occasionally flecked with yellow or orange. Each lateral scale usually with a dark spot, causing a dusky general appearance (spots in preserved specimens are fairly discrete); lateral spotting pattern extends downward to about level of pectoral fins but not extending onto belly. Some have noted that scales along the lateral line of adults are often red and lighter in color than adjacent lateral scales, causing the appearance of a red line along the side of the body (Pflieger 1997). Undersides yellow or orange. Cheeks often with wavy blue-green lines (that are light-colored in preserved specimens). Eye is often red. Opercular flap black with a faint green or white margin; adults usually have silvery blotches on basal part of opercular flap. Fins without prominent pigmentation, but may be dusky. Young are very similar in coloration to the young of Longear Sunfish. Breeding males develop a yellowish body and a bright golden-orange venter; sides speckled with iridescent blue.

SIMILAR SPECIES Often confused with the Longear Sunfish, the Dollar Sunfish differs from it (and from *L. auritus*) in typically having cheek scales in 3 or 4 rows (vs. usually 5 or more rows), pectoral fin usually with 12 rays (vs. 13–15), opercular flap usually distinctly upturned (vs. usually not upturned), and adults with silvery blotches on basal part of dark ear flap (vs. silvery blotches absent). It differs from *L. cyanellus*, *L. symmetricus*, and *L. miniatus* in having short, stubby gill rakers (vs. long and slender); from *L. gulosus* in lacking teeth on tongue; and from all other *Lepomis* in having a short, rounded pectoral fin (vs. pectoral fin long and pointed).

VARIATION AND TAXONOMY *Lepomis marginatus* was originally described from the St. Johns River, Florida, by Holbrook in 1855 as *Pomotis marginatus*. Fowler (1945) considered *L. marginatus* a subspecies of *L. megalotis*, but the Dollar Sunfish has been universally recognized as a valid species for several decades. Jennings and Philipp (1992a) provided genetic evidence that it is specifically distinct from *L. megalotis*, as did Near et al. (2004). Based on a morphological analysis, Barlow (1980) considered *L. marginatus* most similar to the allopatric *L. peltastes*. Near et al. (2004, 2005) included the Dollar Sunfish in a clade with *L. peltastes* and *L. megalotis*. A rangewide study of variation is lacking, but limited data indicate that this species is polytypic (Warren 2009). Based on our data, specimens from the Red River drainage in Arkansas are slightly higher in some meristic counts than specimens from other drainages in the state. Forty percent of Red River drainage specimens have 40–42 lateral line scales (vs. 11% of specimens elsewhere), modally 6 scales above lateral line (vs. modally 5), and 6% have 3 cheek scale rows (vs. 42% elsewhere).

DISTRIBUTION *Lepomis marginatus* is native to the lower Mississippi River basin from western Kentucky and southeastern Missouri, south to the Gulf of Mexico; Atlantic Slope drainages from the Tar River, North Carolina, south to southern Florida; and Gulf Slope drainages from Florida to the Brazos River, Texas. The Dollar Sunfish is one of the least common sunfishes in Arkansas, restricted to the Coastal Plain lowlands at scattered localities. It is less widely distributed in eastern Arkansas than in the southern part of the state. Where it occurs, it is often abundant. It is known today primarily from the Red, Ouachita, lower Arkansas, and lower White river drainages (Map 174). Buchanan (2005) reported this species from nine reservoirs in southern Arkansas. It was formerly reported from the St. Francis River drainage by Black (1940), and there are only two known records (both recent) from that drainage in the past 75 years.

HABITAT AND BIOLOGY This lowland sunfish is found in swamps, oxbow lakes, and small sluggish creeks and bayous having clear water, usually moderate to heavy vegetation, and mud, sand, and detritus bottoms. When present in a natural environment, the Dollar Sunfish is considered an ecological indicator of relatively undisturbed lowland and wetland ecosystems (Warren 2009). It is almost always found in Arkansas in or near vegetation, and the removal of vegetation by Grass Carp in a Texas reservoir caused the elimination of the Dollar Sunfish (Bettoli et al. 1993). Paller et al. (1992) believed that the Dollar Sunfish is unable to maintain permanent populations in reservoirs having large populations of other centrarchids such as Largemouth Bass, Bluegill, and Redear Sunfish. In Arkansas, it occurs in at least nine reservoirs (Buchanan 2005), and AGFC rotenone population sampling data

indicate that small populations are maintained in those reservoirs despite sizeable populations of other centrarchids. Some of the most stable populations occur in Upper and Lower White Oak lakes (Ouachita County), and in Lake Columbia (Columbia County). Other Arkansas reservoirs where it is commonly found include Calion, Erling, Felsenthal, and Millwood. In a North Carolina impoundment, adults moved to deep water during winter and returned to shallow water the following spring (Lee and Burr 1985).

It is an opportunistic feeder, preying on available small invertebrates (Warren 2009). McLane (1955) reported that it fed mainly on insects in Florida. Lee and Burr (1985) studied life history aspects in a man-made North Carolina pond and found that *L. marginatus* fed on immature aquatic insects (mostly midge larvae) and small crustaceans. Other reported foods are terrestrial insects, snails, and aquatic oligochaetes (Sheldon and Meffe 1993). Ross and Baker (1983) reported that during floods, it leaves stream channels to forage on the floodplain. Etnier and Starnes (1993) believed the occurrence of large amounts of detritus, filamentous algae, and terrestrial insects in stomachs indicated that it feeds at all levels in the water column.

Spawning in Arkansas occurs from mid-May to early August. McLane (1955) reported a slightly longer spawning period in Florida from April to September. Nests are constructed by caudal sweeping by males on various substrates usually with little aquatic vegetation. Both solitary and colonial nesting are known. Lee and Burr (1985) reported nest densities of 3–5 nests/m². Reproductive behavior appears typical of other *Lepomis* (Winkelman 1996), and MacKiewicz et al. (2002) used genetic analysis to determine that males spawn on average with 2.5 females in a given spawning event; about 95% of offspring in a nest were sired by the guardian male. Males guard eggs and larvae for about 6 days until the larvae disperse from the nest. Nests produce around 150–200 larvae (Lee and Burr 1985). Minimum breeding size (2.4 inches [60 mm] SL) is reached in the second year of life, and maximum life span is about 6 years.

CONSERVATION STATUS This lowland species is probably more common than verified records indicate because of the difficulty most persons have in distinguishing it from the Longear Sunfish. Its populations in Arkansas appear to be secure.

Lepomis megalotis (Rafinesque)
Longear Sunfish

Figure 26.24. *Lepomis megalotis*, Longear Sunfish. *Uland Thomas.*

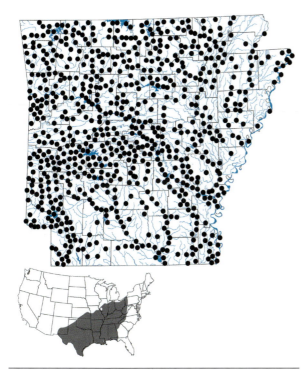

Map 175. Distribution of *Lepomis megalotis*, Longear Sunfish, 1988–2019.

CHARACTERS   A moderate-sized *Lepomis* with a deep, laterally compressed body; body depth going into SL approximately 2.3 times (1.62–2.89) (Fig. 26.24). It is polytypic in Arkansas, and some morphological characters vary by drainage (Table 14). Mouth moderately small, the upper jaw sometimes reaching to or past front of eye, but not

extending past middle of eye. Small supramaxilla present. Opercular flap considerably elongated in adults and usually wider than eye diameter, especially in large males, but it is short or only slightly elongated in juveniles. Opercular flap usually with a distinct light-colored border, but this feature is highly variable in Arkansas populations. Rear margin of opercle (which lies within the base of the opercular flap) is thin and flexible. Pectoral fin short and rounded, its tip usually not reaching past eye when bent forward across cheek. Soft dorsal fin usually rounded. Palatine teeth absent; no teeth on tongue. Lower pharyngeal arches narrow, with pointed teeth. Gill rakers short and stubby, usually 13 (11–15) total rakers on first arch (8–10 on lower part and 3–5 on upper part). Cheek scale rows usually 5 or 6 (rarely 4 or 7). Scales around caudal peduncle usually 19 or 20 (17–24). Pectoral fin rays usually 14 (12–15). Anal fin base nearly straight. Lateral line complete, arched anteriorly, usually with 38–42 (34–49 depending on drainage) scales. Scale rows above lateral line usually 6 or 7 (5–9); scale rows below lateral line usually 15 (12–18). Dorsal spines usually 10 (9–11); dorsal rays usually 11 (9–13); anal spines 3; anal rays 9 or 10 (8–12). Adults rarely exceed 5 inches (127 mm) in length. Maximum reported length around 7 inches (178 mm); maximum weight is about 1 pound, 12 ounces (0.8 kg). The Arkansas state record from Table Rock Lake in 1991 is 1 pound, 2 ounces (0.51 kg), but there is some question about the validity of that specimen's identification (Ken Shirley, pers. comm.). On 16 June 2018, a Longear Sunfish from Bull Shoals Lake measured 8.75 inches (222 mm) TL and weighed approximately 9 ounces (0.26 kg) (Paul Port, pers. comm.).

LIFE COLORS   Coloration in Arkansas specimens can be highly variable depending on habitat, stage of maturity, and drainage (Table 14). Back and sides olive, sometimes flecked with yellow or green. Undersides yellow to orange. Opercular flap black, usually with a distinct white margin; adults without silvery blotches on basal part of blackish opercular flap. The white margin varies in thickness among and within populations. Breeding males are among the most brilliantly colored fishes in Arkansas, particularly in the upper and middle White River drainage (Fig. 26.25), with iridescent green or turquoise above and bright reddish orange below; unpaired fins become reddish brown, pelvic fins becoming dark blue (bright orange in upper White River), eyes red, and sides of head have alternating orange and blue or blue-green wavy lines (which are light-colored in preserved specimens). Young uniformly olive; side speckled with yellow flecks, sometimes with chainlike bars; underside white or light yellow.

Table 14. Comparison of Some Characters of Five Forms of *Lepomis megalotis* in Arkansas. Data are mainly from Arkansas populations, but Ouachita River information includes 16 specimens from Louisiana.

Character	Form 1—Illinois River dr. N = 43	Form 2—White River dr.* N = 53	Form 3—Arkansas River dr.** N = 40	Form 4—Ouachita River dr. N = 57	Form 5—Little River dr. N = 56
Dorsal fin spines	$\bar{x}$ = 10 (9–11)	$\bar{x}$ = 10 (9–11)	$\bar{x}$ = 10 (9–11)	Usually 10 (10 or 11)	$\bar{x}$ = 10 (9–11)
Dorsal fin rays	Usually 11 (10–12)	Usually 11 (9–13)	Usually 11 (10–12)	Usually 11 (10–12)	Usually 11 (10–12)
Pectoral fin rays	Usually 13 (12–14)	Usually 14 (13–15)	13 or 14 (13–15)	Usually 14 (12–15)	Usually 14 (13–15)
Anal fin rays	Usually 9 (8–10)	Usually 9 (8–11)	Usually 9 (9–12)	Usually 9 (8–10)	Usually 10 (9 or 10)
Lateral line scales	34–41 ($\bar{x}$ = 37.7)	35–47 ($\bar{x}$ = 40.3)	35–43 ($\bar{x}$ = 38.3)	36–44 ($\bar{x}$ = 39.8)	40–49 ($\bar{x}$ = 45.4)
Scale rows above LL	Usually 6 (5–8)	Usually 7 (6–9)	Usually 6 (5–8)	Usually 6 (5–8)	Usually 7 (6–9)
Cheek scale rows	Usually 5 (5–7)	Usually 5 (4–6)	5 or 6, occasionally 7	Usually 5 (4–6)	Usually 5 (4–6)
Body depth into SL	$\bar{x}$ = 2.21 (1.97–2.56)	$\bar{x}$ = 2.18 (1.62–2.75)	$\bar{x}$ = 2.2 (1.74–2.65)	$\bar{x}$ = 2.22 (1.72–2.52)	$\bar{x}$ = 2.39 (1.81–2.89)
Brick-red nuchal stripe	Present	Absent	Absent	Absent	Absent
Opercular flap decurved	No	Yes, in upper White River drainage	No	No	No
Dark blotch in posterior base of soft dorsal fin	Absent	Absent	Present in some individuals	Usually Present	Absent
Round dark spots on lateral scales scattered or in broken rows	No	No	No	No	Yes

* There may be more than one form of *L. megalotis* in the White River drainage of Arkansas. Form 2 occurs in the upper White River drainage, specimens from the Spring and Strawberry rivers lack the decurved opercular flap, and Form 3 dominates the lower White River drainage.

** More than one form may occur in the Arkansas River drainage from Fort Smith to its mouth.

SIMILAR SPECIES Closely resembles the less common Dollar Sunfish but differs in having cheek scales usually in 5 or 6 rows (vs. 3 or 4 rows) and pectoral fin with 13–15 rays (vs. 12). The Longear Sunfish is often confused with the introduced Redbreast Sunfish, and frequent misidentifications have occurred in Arkansas. It differs from *L. auritus* in lacking palatine teeth (vs. palatine teeth present), in generally having fewer than 43 lateral line scales, except in Little River drainage and about one-third of Buffalo River specimens examined (vs. usually 43 or more lateral line scales), and in usually having 5 or 6 cheek scale rows (vs. 6–9 cheek scale rows). We have found that pigmentation of the opercular flap margin is not a completely reliable character for separating *L. megalotis* from *L. auritus* in Arkansas. *Lepomis megalotis* differs from *L. cyanellus*, *L. symmetricus*, and *L. miniatus* in having short stubby gill rakers (vs. long and slender), from *L. gulosus* in lacking teeth on tongue, and from all other *Lepomis* in having a short, rounded pectoral fin.

VARIATION AND TAXONOMY *Lepomis megalotis* was described by Rafinesque (1820a) as *Icthelis megalotis* from the Licking and Sandy rivers, Kentucky. The Longear Sunfish is the most polytypic member of the family Centrarchidae (Bauer 1980). Boschung and Mayden

Figure 26.25. Male *Lepomis megalotis* from the White River in breeding coloration. *Isaac Szabo.*

(2004) summarized previous designations of subspecies and races. Despite several analyses of morphological and genetic variation, there is still no consensus on recognition of subspecies, races, or other evolutionarily significant units, and Near and Koppelman (2009) speculated that new species (now included in *L. megalotis*) will eventually be recognized. Near et al. (2004, 2005) included *L. megalotis* in a clade with *L. marginatus* and *L. peltastes*. Within Arkansas, *L. megalotis* is polytypic and shows considerable geographic variation in body color pattern and opercular flap length, width, orientation, and coloration of margin. Within populations, body coloration can be variable in different size groups. Populations in the St. Francis River, Cache River, lower White River, and possibly the Lower Arkansas River drainage are the closest to the typical *L. megalotis* phenotypes in scale counts, body shape, and coloration, especially in having a distinct light-colored border on the wide opercular flap. Some White and Ouachita river individuals have very long, narrow opercular flaps that are sometimes dark to or almost to the margin (often causing them to be mistaken for *L. auritus*). Much of the White River system is populated by Longear Sunfish with consistently decurved opercular flaps, and this character was historically considered sufficient to recognize these populations as a new species (B. H. Bauer, pers. comm.). Specimens from the Ouachita River system usually have a posterior basidorsal spot in the soft dorsal fin, similar to Bluegills, and TMB has observed a similar soft dorsal spot in many specimens from the Arkansas River main stem near Fort Smith. Spring and Strawberry river populations are characterized by opercular flaps with a very thin light border or with no light border. Trautman (1957) indicated that Arkansas was within the range of the Northern Longear Sunfish, *L. m. peltastes* (now considered a full species, *L. peltastes*, the Northern Sunfish), but more recent workers determined that the range of *L. peltastes* is far to the north of Arkansas (Page and Burr 2011).

An ongoing rangewide morphological study of variation in *L. megalotis* by Bruce H. Bauer (pers. comm.) identified five distinct, largely allopatric forms of Longear Sunfish in Arkansas. Additional sampling is needed to more precisely define the areas of occurrence of those forms in Arkansas, particularly throughout the White and Arkansas river drainages. The White River drainage is especially problematic, and there may be three forms of *L. megalotis* in that drainage. Arkansas has more of the *L. megalotis* forms identified by Bauer than any other state. Taxonomic assignments for those forms are awaiting further morphological and molecular analyses. The following information on the approximate distribution of the five *L. megalotis* forms in Arkansas is used with permission of B. H. Bauer. Form 1—Found in the Illinois River drainage (Arkansas River drainage) in northwestern Arkansas in Benton and Washington counties, and possibly in western Arkansas. It may also occur down the Arkansas River main stem in Arkansas (along with Form 3), but there are currently no documented records. It is primarily a Great Plains form and is characterized by a brick-red nuchal stripe. It is also found in Missouri, Kansas, and Oklahoma. Form 2—Occurs primarily in the upper and middle White River drainage in Arkansas and Missouri, but it may also occur in the Black River drainage and the lower White River drainage in Arkansas. It may occur syntopically in the lower White River with Form 3. The White River form is characterized by a decurved opercular flap in many individuals and distinctive male breeding coloration. Form 3—This form is closest to the typical *L. megalotis*. In Arkansas, it is found primarily below the Fall Line in eastern Arkansas north of the Arkansas River, but it also extends down the Mississippi River main stem. It may occur throughout the Arkansas River drainage into Oklahoma. It also occurs in lower White River drainages. Elsewhere, it is the most widely distributed form of *L. megalotis*, occurring mostly east of the Mississippi River. Form 4—Found mainly in the Ouachita River drainage and the mainstem Red River in Arkansas. Outside Arkansas, it occurs throughout Louisiana, eastern Texas, and southern Oklahoma. It is usually characterized by a posterior basidorsal spot. Specimens resembling this form have also been found in the Arkansas River by TMB. Form 5—Little River Form. Considered by Bauer (pers. comm.) to be an undescribed species (*L.* sp. *cf. megalotis*) restricted to the Little River system of Oklahoma and Arkansas. It is distinguished from all other *L. megalotis* in Arkansas (based on data from B. H. Bauer and TMB) primarily by a more elongate body, higher lateral line scale counts, and lateral pigmentation pattern (Table 14). A formal description is planned by B. H. Bauer. It is possible that combined morphological and genetic analyses would support the elevation of some or all of the above-mentioned forms to full species status. Because their precise distributions in Arkansas are undetermined, we do not present a map showing the distributions of the five above-listed forms. Additional sampling is needed to determine the extent of sympatry and allopatry among the five forms in Arkansas and to identify any other possible forms of *L. megalotis* in this state.

Schanke et al. (2017) found that genetic connectivity of Longear Sunfish populations in streams in the Ouachita Mountains was reduced by stream drying and instream barriers. Natural hybridization has been reported between *L. megalotis* and four other species of *Lepomis*: *L. cyanellus*, *L. gibbosus*, *L. humilis*, and *L. macrochirus*) (Bolnick 2009).

DISTRIBUTION   As currently recognized, *Lepomis megalotis* is native to the Mississippi River basin from Pennsylvania to Illinois, Missouri, and Kansas, and south to the Gulf of Mexico, and in Gulf Slope drainages from the Apalachicola River, Georgia, to the Rio Grande of Texas, New Mexico, and Mexico. It has been introduced elsewhere in the United States. In Arkansas, the Longear Sunfish is found statewide, occurring in all drainages (Map 175).

HABITAT AND BIOLOGY   Although occurring in a variety of habitats statewide, the Longear Sunfish is most abundant in small, clear, upland streams with rocky bottoms and permanent or semipermanent flow. It is typically found in pools and does well in reservoirs and farm ponds. It is tolerant of low dissolved oxygen and high water temperatures (Smale and Rabeni 1995a, b). Individuals often have well-defined home ranges (<100 m) and good homing abilities, using olfactory cues over short distances (Gunning 1959, 1965). Home ranges are abandoned in winter by the larger individuals as they move to deeper water. In Bull Shoals Reservoir, Longear Sunfish were observed hiding singly or in small groups under rocks or stumps on the bottom during March, but as the water warmed, they became more active; aggregations of fish dispersed when the water temperature reached about 17.8°C (64°F) (Boyer and Vogele 1971). The Longear Sunfish has good ability to repopulate stream reaches affected by drought (Lonzarich et al. (1998, 2004). Neill (1967) and Neill et al. (1966) studied heat resistance in Longear Sunfish from the White River. The incipient upper lethal temperatures of young Longear Sunfish were estimated to be 95.9, 97.9, and 100.8°F (35.5, 36.6, and 38.2°C) for fish acclimated to 77, 86, and 95°F (25, 30, and 35°C), respectively. In general,

larger young sunfish were more resistant to upper lethal temperatures than their smaller siblings.

Lepomis megalotis feeds mainly on aquatic and terrestrial insects as well as other invertebrates and an occasional small fish. In Bull Shoals Reservoir, aquatic insects and Entomostraca made up 86% of the diet of Longear Sunfish smaller than 1.9 inches (48 mm). Aquatic insects (48%), fish eggs (23%), terrestrial insects (9%), and bryozoans (9%) were the main foods eaten by 1.9–3.9-inch (48–99 mm) sunfish; and individuals longer than 4 inches (102 mm) ate terrestrial insects (37%), fish (29%), and aquatic insects (15%) (Applegate et al. 1967). Large Longear Sunfish in Bull Shoals Lake ingested moderate to heavy quantities of young bass in May, and other newly hatched sunfish in June (Mullan and Applegate 1968). In DeGray Reservoir, Longear Sunfish consumed insects (52.6%), crayfish (12.5%), fish (7.4%), and plant material (6.7%) (Bryant and Moen 1980a). Pflieger (1997) reported that it is known to follow turtles and large suckers as they forage on the bottom, feeding on small invertebrates that are exposed. They are common nest predators of other sunfishes, feeding on eggs and larvae when the guardian male is distracted. Kwak et al. (1992) found that feeding occurred throughout a 24-hour period, with peak feeding at dawn and dusk. Cleaning behavior by Longear Sunfish on Flathead Catfish was observed in an aquarium (Spall 1970).

Sexual maturity is reached at 1 year of age and a minimum size of 60 mm TL in females and 100 mm TL for males (Jennings and Philipp 1992b). In Arkansas, spawning occurs from May to August. It is a colonial nester and begins to spawn when water temperatures exceed 75°F (23.9°C) (Cross 1967). Boyer and Vogele (1971) reported that spawning began in Beaver and Bull Shoals reservoirs near the end of May when water temperatures reached 71–73°F (21–23°C); spawning can continue up to temperatures of 29–31°C. In those reservoirs, spawning colonies were generally found in brush-free areas having a gradually sloping gravel substrate. Nests were constructed at depths ranging from 0.7 to 11.2 feet (0.2–3.4 m) with an average depth of 4.9 feet (1.5 m). The average nest diameter was approximately 1.6 feet (0.5 m). The males construct circular depression nests close together in the substrate, and spawning is asynchronous. The male courts an approaching female by swimming around and above her and displaying his bright red-orange belly. The coloration of the opercular flap may play a role in species recognition during courtship, and females prefer males with longer flaps (Goddard and Mathis 1997, 2000). Gerald (1971) provided evidence that sound production is a part of male courtship. Spawning occurs as the two fish circle above the nest. The male may later spawn again with the same female or with different ones. Females frequently spawn in more than one nest. Fecundity increases with female size, with estimates of numbers of spawned ova ranging from 1,417 to 4,213 eggs in females between 100 and 130 mm TL (Boyer 1969). Hatching occurs in 3–5 days at a water temperature of about 25°C (77°F). The male guards the nest (for 9–14 days) until the fry become free-swimming (Robison 1977d). Upon becoming free-swimming, fry from several nests initially form large schools in dense vegetation before separating into smaller groups or dispersing individually (Boyer 1969). *Lepomis megalotis* is one of four centrarchid species (all in the genus *Lepomis*) for which alternative reproductive tactics have been documented (Neff and Knapp 2009). Female mimicry has been reported for male Longear Sunfish, and sneaker males have been observed in Arkansas reservoirs (Boyer 1969; Boyer and Vogele 1971). Sneaker males, which resemble females in coloration, dart into a nest and release milt while the resident male is spawning with a female.

Bacon (1968) and Bacon and Kilambi (1968) studied age and growth of Longear Sunfish in the Kings River and Beaver Reservoir. The largest specimen collected was 6.1 inches (155 mm) long and 6 years of age. They reported that most Longear Sunfish reach a catchable size of 5 inches (127 mm) at the end of their fifth growing season, but some individuals attain this length at the end of the fourth year. Males grew faster than females after the second year of life. Maximum life span reported from northern areas is 9 years (Carlander 1977).

CONSERVATION STATUS   This is one of the most ubiquitous sunfishes in Arkansas, and its populations are secure. The form believed to be an undescribed species, *L.* sp. *cf. megalotis*, in the Little River drainage is also secure but may warrant future monitoring because of its restricted distribution.

Lepomis microlophus (Günther)
Redear Sunfish

CHARACTERS   A moderately large sunfish with a laterally compressed, rounded body; body depth going into SL 2.0–2.4 times (Fig. 26.26). Mouth relatively small, the upper jaw not extending posteriorly to pupil of eye. Small supramaxilla present. Rear margin of opercular flap with red or orange crescent. Pectoral fin very long and pointed, its tip extending past front of eye when bent forward across cheek. Pharyngeal arches very broad and heavy, with large, rounded, molariform teeth. Gill rakers short and stout,

Figure 26.26. *Lepomis microlophus*, Redear Sunfish. *Uland Thomas*.

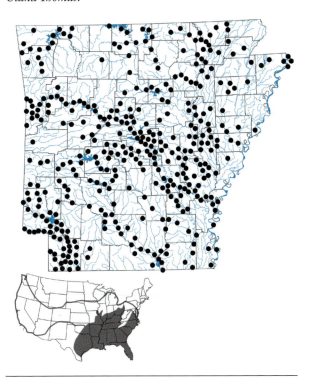

Map 176. Distribution of *Lepomis microlophus*, Redear Sunfish, 1988–2019. Inset shows native range (solid) and introduced range (outlined) in the contiguous United States.

9–11. Opercular flap short, moderately flexible; rear margin of opercle (which lies within the base of the opercular flap) thin and flexible. Palatine teeth present or absent; no teeth on tongue. Sensory pits in head small and poorly developed, pit diameter much less than distance between pits. Cheek usually with 3 or 4 (3–6) scale rows. Lateral line complete with 34–40 scales. Scale rows above lateral line 6–8. Dorsal spines usually 10 (9–11); dorsal rays 11 or 12; anal spines 3; anal rays 10 or 11; pectoral rays 13–16; pelvic fin with 1 spine and 5 rays. Arkansas's largest species of *Lepomis*, commonly reaching 8–10 inches (203–254 mm) in length and around 1 pound (0.5 kg) in weight in farm ponds and reservoirs; usually much smaller in streams and rivers. World angling record: 5 pounds, 7.5 ounces (2.48 kg) is from South Carolina. The Arkansas state record from Bois D'Arc Lake near Hope in 1985 is 2 pounds, 14 ounces (1.3 kg) and 15 inches (381 mm) TL.

LIFE COLORS   Back olive green. Sides silver, often with brown mottling that produces an almost checkered pattern; sides sometimes with a series of black vertical bars, especially prominent in smaller individuals. Short opercular flap black with prominent, crescent-shaped red or orange spot on posterior border in adults and larger young; red spot developed or not in small individuals. Dorsal fin dusky, without a posterior dark blotch. Undersides yellow. Breeding male a brassy gold with dusky pelvic fins. Breeding female less brilliantly colored; opercular border orange red or orange. Young with silvery body and about 8 (5–10) dark vertical bars on side.

SIMILAR SPECIES   It differs from the Bluegill and the Orangespotted Sunfish (both of which have long pectoral fins) in having a red spot on the opercular flap in life (vs. red spot absent), short gill rakers (vs. long), and molariform teeth on a broad lower pharyngeal arch (vs. lower pharyngeal arch thin and without molariform teeth). *Lepomis gulosus* sometimes has red pigment on the opercular margin, but it has short pectoral fins and lacks molariform pharyngeal teeth. The Redear Sunfish can be distinguished from other *Lepomis* by its red opercular spot and long, pointed pectoral fins.

VARIATION AND TAXONOMY   *Lepomis microlophus* was described by Günther (1859) as *Pomotis microlophus* from the St. Johns River, Florida. Two subspecies are recognized: *L. m. microlophus* in Georgia, Florida, and south Alabama, and an unnamed subspecies found throughout the remainder of the species range including Arkansas (Page and Burr 2011). Based on a genetic analysis, Near et al. (2004) considered *L. microlophus* sister to the species pair, *L. punctatus* and *L. miniatus*. Natural hybridization has been reported between *L. microlophus* and three other *Lepomis* species, *L. cyanellus*, *L. gulosus*, and *L. macrochirus* (Bolnick 2009).

DISTRIBUTION   *Lepomis microlophus* occurs in the Mississippi River basin from southwestern Pennsylvania, to Iowa and eastern Oklahoma, south to the Gulf of Mexico; Atlantic Slope drainages from northern Virginia to southern Florida; and Gulf Slope drainages from Florida to the Nueces River, Texas. It is widely introduced elsewhere in the United States. In Arkansas, the Redear Sunfish is native to all major drainages. It is widely stocked in ponds and reservoirs throughout the state (Map 176).

HABITAT AND BIOLOGY   The Redear Sunfish was probably much more common originally in the lowlands than in upland areas of Arkansas. It is still abundant today in the lowlands in oxbow lakes and the sluggish backwaters of rivers and streams. It prefers warm, clear waters without current and with an abundance of stumps, logs, brush, and other aquatic vegetation. It is usually found over a mud and detritus bottom and is one of two species commonly referred to in Arkansas as bream. Its affinity for logs, stumps, and other standing timber has led to another common name, stumpknocker (a local common name shared with *L. miniatus*). It is also called a shellcracker in reference to its food habits. The Redear Sunfish is today widely distributed in upland areas of the state, especially in ponds and reservoirs, but it is sometimes found in quiet pools of upland streams. Smith (1979) reported it to be more tolerant of silt than most other sunfishes. It is one of the most euryhaline centrarchids and is known from salinities up to 20 ppt (Rutherford et al. 2001). It can also acclimate quickly to salinity changes (Peterson 1988).

The Redear Sunfish is a specialized benthic feeder. Huish (1957), in a Florida study, found the food of the Redear consisted mainly of snails and other small mollusks. It is able to consume a wide variety of small, hard-bodied prey (Huckins 1997). Its broad, flat pharyngeal teeth are ideally suited for crushing the shells of those animals. Moyle (2002) reported that it fed on nonnative snails and clams in California, and it may have potential to control such nonindigenous populations elsewhere, including populations of snails that serve as intermediate hosts of parasites (Ledford and Kelly 2006). It also feeds on insect larvae and cladocerans taken near the bottom. We have observed that a small percentage (10–15%) of Redear stomachs contain vegetation or even detritus possibly incidentally obtained because of its bottom feeding behavior. The flesh of those individuals is not as tasty as the flesh of those that have not ingested such incidental materials. Young Redear Sunfish smaller than 25 mm feed mainly on microcrustaceans and other small, soft-bodied invertebrates, but they begin to shift dramatically from soft-bodied food to snails at a size between 25 and 75 mm TL (VanderKooy et al. 2000). Adults have also been reported to feed on larval dipterans, burrowing mayflies, amphipods, and larval odonates, particularly in environments where snails are scarce (Wilbur 1969; Desselle et al. 1978; VanderKooy et al. 2000).

In Arkansas, spawning occurs from April into August. In Crystal Lake in Benton County, the spawning season lasts from May through July, with females exhibiting multiple spawning (Adams and Kilambi 1979). Fecundity of 9–10-inch (230–260 mm) females ranged from 35,500 to 64,000 mature ova. An even longer spawning period occurred in Florida (late February to September), beginning when water temperature reached 21°C (Wilbur 1969). Peak spawning appears to occur between 23.8°C and 26°C. The nests are circular depressions fanned out in the substrate by the males. Spawning substrates can be variable, ranging from sand to mud, and nests are often located in submerged vegetation (Boschung and Mayden 2004). A colonial nester with synchronous spawning, the male Redear Sunfish guards the nest after spawning until the fry become free-swimming. Eggs hatch in 50.3 hours at 23.8°C, and 26.6–28.1 hours at 28.5°C, and the larvae become free-swimming about 3 days after hatching (Childers 1967; Yeager 1981). Reproductive behavior is presumably similar to that of other *Lepomis* species, and there is evidence that males produce distinctive popping sounds while courting females (Gerald 1971). Nests have been reported from water depths of less than 0.5 m to slightly less than 2 m (1.6–6.6 feet). Sexual maturity was attained by 50% of the females (151–213 mm TL) in the second summer in Crystal Lake, Arkansas, and all age 3 and older females were mature; maximum life span was around 6 or 7 years (Adams and Kilambi 1979). Roberg et al. (1986) reported 8 age groups in Bob Kidd Lake in northwestern Arkansas. Other studies have reported a much earlier age for attaining sexual maturity (Warren 2009).

The Redear Sunfish is a popular gamefish in Arkansas but a little more difficult to catch than most sunfishes because of its bottom-feeding habits. It is frequently stocked in combination with Largemouth Bass and Bluegills in ponds and reservoirs in Arkansas. The greatest angling success is with natural baits fished near the bottom. Under ideal conditions, the Redear attains a large size, and its flesh is of excellent flavor and quality.

CONSERVATION STATUS This is a widely distributed and locally abundant gamefish. Its populations may have increased in some parts of the Arkansas River as a result of the spread of Zebra Mussels in that system (Quinn and Limbird 2008). We consider its populations secure.

Lepomis miniatus Jordan
Redspotted Sunfish

CHARACTERS A moderately small sunfish with a compressed, short, deep body with small squarish spots forming longitudinal rows on sides (spots especially pronounced below the lateral line). Body depth going into SL 2.0–2.2

Figure 26.27a. *Lepomis miniatus*, Redspotted Sunfish male. *David A. Neely.*

Figure 26.27b. *Lepomis miniatus*, Redspotted Sunfish female. *David A. Neely.*

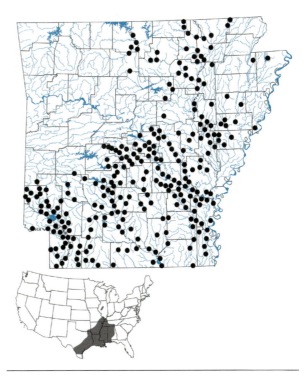

Map 177. Distribution of *Lepomis miniatus*, Redspotted Sunfish, 1988–2019.

times (Fig. 26.27). Mouth moderate-sized, the upper jaw extending to or just past front of eye. Small supramaxilla present. Pectoral fin short and rounded, its tip usually not reaching past front of eye when bent forward across cheek. Gill rakers on middle of first arch moderately long and slender (distinctly shorter than those of Bluegill but distinctly longer than those of Longear Sunfish), usually 8–11. Opercular flap short, stiff, distinctly upturned, and dark with a narrow light margin, widest below rear margin of operculum; rear margin of opercle (which lies within the base of the opercular flap) stiff, not flexible. Snout short, its length going into distance from back of eye to rear margin of opercular flap fewer than 2 times. Lateral line complete, arched anteriorly, usually with 35–39 scales. Scale rows above lateral line usually 6 or 7. Cheek scale rows 4–6. Palatine teeth present or absent; no teeth on tongue. Dorsal spines 10; dorsal rays usually 11; anal spines 3; anal rays usually 10; pectoral rays 13 or 14. Adults rarely exceed 5 or 6 inches (127–152 mm); maximum size around 8 inches (203 mm) in length and 7 ounces (0.2 kg) in weight. No Arkansas state record exists, even though this species is often caught by anglers (and is often referred to by them as stumpknocker) wherever it occurs.

LIFE COLORS   Back and sides dark blue or olive, often with a brassy sheen. Scales of sides with reddish-orange (males) or yellow (females) spots forming longitudinal rows; young without spots but with dark olive sides that are often strongly mottled. Face and head black with a pale red area just above the opercular flap. Ventral curvature of dark eye outlined with an iridescent turquoise crescent. Undersides yellow to dusky white. Short ear flap black, narrowly edged with white or pale yellow. Fins dusky or plain. Breeding males more intensely colored with much red orange on breast and belly; pelvic fins dark; and distal parts of soft dorsal, caudal, and anal fins with red-orange coloration.

SIMILAR SPECIES   Similar to the Bantam Sunfish, but differs in having a complete lateral line (vs. lateral line incomplete or interrupted) and in lacking a black blotch near the base of the soft dorsal fin (vs. black blotch present). It differs from the Green Sunfish in usually having 40 or fewer lateral line scales (vs. more than 40); from *L. auritus*, *L. megalotis*, and *L. marginatus* in having long, slender gill rakers (vs. short and stubby) and a short opercular flap (vs. opercular flap long in adults). It differs from *L. macrochirus*, *L. humilis*, and *L. microlophus* in having short, rounded pectoral fins (vs. pectoral fins long and pointed); and from *L. gulosus* in lacking teeth on tongue (vs. tooth patch present on tongue), and in lacking dark lines radiating backward from eye across cheek (vs. dark lines present).

VARIATION AND TAXONOMY   *Lepomis miniatus* was described by Jordan (1877b) as *Lepiopomis miniatus* from the Tangipahoa River, Mississippi. It was long recognized as one of two subspecies of *Lepomis punctatus*: *L. p. punctatus*, ranging from North Carolina to Florida, and *L. p. miniatus*, occurring in Arkansas and elsewhere in the lower Mississippi River Valley and Gulf Slope drainages. The subspecies occurring in Arkansas was elevated to full species status (*L. miniatus*) by Warren (1992), and that elevation was subsequently recognized by most authors. A broad zone of presumed hybridization occurs along the Gulf Coast from Alabama to the Florida panhandle (Boschung and Mayden 2004), and some consider this evidence for continued recognition of subspecies (Robins et al. 2018). Near and Koppelman (2009) pointed out that morphological differentiation between the sister species *L. miniatus* and *L. punctatus* is concordant with the mtDNA inferred phylogeographic breaks. Although the status of this taxon is not universally agreed upon, we follow Page et al. (2013) in recognizing specific status for *L. miniatus*. Natural hybridization has been reported between *L. miniatus* and *L. punctatus* (Bolnick 2009).

DISTRIBUTION   *Lepomis miniatus* occurs in the Mississippi River basin from eastern Illinois, south to the Gulf of Mexico, including the lowermost Ohio, Cumberland, and Tennessee river drainages; and in Gulf Slope drainages from the Mobile Bay basin, Alabama, west to the Rio Grande, Texas. In Arkansas, it occurs in all major drainages of the Coastal Plain lowlands, and it is occasionally found above the Fall Line in a few Ozark and Ouachita mountain streams (Map 177). It especially penetrates far above the Fall Line in the White River drainage.

HABITAT AND BIOLOGY   The Redspotted Sunfish is found in streams, rivers, and oxbow lakes in the lowlands, where it inhabits quiet or sluggish, clear waters usually with much aquatic vegetation and other cover over mud and detritus bottoms. In the few upland streams it penetrates, it occupies quiet pools near boulders or logs; the pools usually have heavy growths of water willow along the margins. It is a common species in the heavily vegetated upper Spring River in Fulton County. It also occurs in a spring off the White River near Cotter in Baxter County. It is sometimes found in moderately turbid waters and can tolerate periodic low dissolved oxygen concentrations (Killgore and Hoover 2001). The Redspotted Sunfish is able to adapt somewhat to reservoirs within its range if suitable habitat is present. Small populations persist in reservoirs in southern Arkansas, including Calion, Columbia, Erling, Felsenthal, and Upper and Lower White Oak lakes. Tolerance of small fluctuation in salinity in coastal habitats has been reported

(Desselle et al. 1978). Removal of aquatic vegetation by Grass Carp in a Texas reservoir resulted in almost complete elimination of *L. miniatus* (Bettoli et al. 1993).

Lepomis miniatus is primarily an invertivore that feeds throughout the water column in shallow water environments, and it forages most frequently in or near aquatic vegetation. In the San Marcos River, Texas, Redspotted Sunfish diets consisted (by volume) of 60% insect adults and larvae, 27% aquatic vegetation, 6% organic detritus, and the remaining 7% of crustaceans, clams, snails, arachnids, and fish remains (Buchanan 1971). In coastal habitats of varying salinity, adults fed on mud crabs, isopods, amphipods, dipteran larvae, caddisfly larvae, and terrestrial insects (Desselle et al. 1978). Juveniles smaller than 60 mm fed on copepods, midges, cladocerans, mysid shrimp, and mayfly nymphs.

Spawning has been reported in May in Illinois (Forbes and Richardson 1920), in July in Missouri (Pflieger 1997), and from early spring to November in Florida (Carr 1946). TMB noted spawning activity from early April to August in the San Marcos River, Texas, and HWR observed spawning of *L. miniatus* on 7 June 1994 over solitary nests in a clear backwater area of Dorcheat Bayou at U.S. Route 82 in Columbia County, Arkansas, at a water temperature of 24.4°C (76°F) and depth of 1.2 m (3.8 ft.) (Connior et al. 2011). Pflieger (1997) found that nests in stream environments were located in shallow water (a few centimeters deep) among stems of water willow over a sand and gravel bottom. The Redspotted Sunfish is mainly a solitary nester in the San Marcos River, Texas (Buchanan 1971), but nesting in small groups of two or three is also known (Pflieger 1997). Its courtship, spawning, and nest defense are typical of other sunfishes (Carr 1946). Gerald (1971) reported sound production by courting males. Eggs hatch in 36 hours at 26°C, and larvae become free-swimming about 4 or 5 days after hatching (Roberts et al. 2004). Maximum life span is about 4+ years.

CONSERVATION STATUS This species is widely distributed in eastern and southern Arkansas. No large populations are known, but most populations are probably secure.

Lepomis symmetricus Forbes
Bantam Sunfish

CHARACTERS A very small, robust, deep-bodied and less compressed sunfish; body depth going into SL approximately 1.9–2.1 times (Fig. 26.28). Mouth moderately large, the upper jaw extending well past front of eye.

Figure 26.28. *Lepomis symmetricus*, Bantam Sunfish. *David A. Neely.*

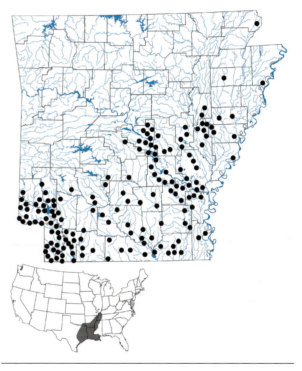

Map 178. Distribution of *Lepomis symmetricus*, Bantam Sunfish, 1988–2019.

Small supramaxilla present. Lateral line incomplete or interrupted as many as 6 times, with 1–18 unpored scales; lateral scale rows usually 32–36 (30–40). Scale rows above lateral line 5–7. Cheek scale rows 4–6. Pectoral fin short and rounded, its tip usually not reaching eye when bent forward across cheek. Palatine teeth present; no teeth on tongue. Gill rakers moderately long and slender, usually 12–15 on first arch. Opercular flap short and dark with a light margin; rear margin of opercle (which lies within the opercular flap) stiff, not flexible. Caudal fin slightly emarginate. Dorsal spines usually 10 (9–11); dorsal rays usually

10 (9–12); anal spines 3; anal rays usually 10 (9–12); pectoral rays 11–13. This is the smallest species of *Lepomis*, reaching a maximum size of 3 inches (76 mm). Largest Arkansas specimen examined by TMB was 2.8 inches (70 mm) and weighed about 0.31 ounce (8.8 g).

LIFE COLORS   Our least colorful *Lepomis* species. Back and sides dark green. Many dark brown spots on body, often one per scale, creating longitudinal rows or sometimes irregular vertical bands. Undersides yellowish brown with many tiny, dark speckles. All fins, except pectorals, dusky; soft dorsal and anal fins usually have several light spots. Cheeks and head dark. Juveniles and subadults with more distinct vertical bands and a prominent black spot with an orange halo near the posterior base of the soft dorsal fin; the spot becomes less intense with age in males and disappears altogether in large males. Adult females retain the posterior dark spot in the soft dorsal fin, but the orange halo disappears with increasing size (Wetzel 2007). Young with some red-orange pigmentation in soft dorsal and soft anal fins. Breeding males becoming very dark overall and iris of eye bright red.

SIMILAR SPECIES   Because of its small size, *L. symmetricus* is often mistaken for the young of other centrarchids. The Bantam Sunfish can be distinguished from all other Arkansas sunfishes by its incomplete or interrupted lateral line (vs. lateral line complete). It further differs from *L. macrochirus*, *L. humilis*, and *L. microlophus* in having short, rounded pectoral fins (vs. long and pointed); from *L. cyanellus* in having fewer than 40 lateral line scales (vs. more than 40); from *L. miniatus* in usually having a dark spot in the soft dorsal fin (vs. dark spot absent); from *L. auritus*, *L. megalotis*, and *L. marginatus* in having a short opercular flap (vs. adults with a long opercular flap) and long, slender gill rakers (vs. gill rakers short and stubby); and from *L. gulosus* in lacking teeth on tongue (vs. tooth patch present on tongue).

VARIATION AND TAXONOMY   *Lepomis symmetricus* was described by Forbes *in* Jordan and Gilbert (1883) from the Illinois River at Pekin, Tazewell County, Illinois. Burr (1977) studied morphological variation of *L. symmetricus* from throughout its range and recognized no subspecies. From the 1930s through the 1990s, it was generally considered most closely related to *Lepomis cyanellus* (Bailey 1938; Branson and Moore 1962; Wainwright and Lauder 1992), and Near et al. (2004, 2005) provided genetic evidence that it forms a sister pair with *L. cyanellus*.

DISTRIBUTION   *Lepomis symmetricus* occurs in the lower Mississippi River basin from southern Illinois to the Gulf of Mexico, and in Gulf Slope drainages from the Biloxi River, Mississippi, to the Colorado River, Texas. In Arkansas, this small sunfish is found in the Coastal Plain lowlands in all major drainages but is much more common in southern Arkansas than in the eastern part of the state (Map 178). It is especially abundant in the Red River drainage in southwestern Arkansas. Buchanan (2005) reported it from 6 Arkansas impoundments, and McAllister et al. (2008) reported 77 new locality records and 15 new county records for this sunfish in Arkansas. Connior et al. (2011) provided additional records from several localities in the Ouachita and Little river drainages.

HABITAT AND BIOLOGY   The Bantam Sunfish is a lowland species inhabiting swamps, oxbow lakes, and sluggish backwaters of creeks, sloughs, and bayous. It is always found in association with aquatic vegetation and usually in environments with submerged logs, standing timber, and stumps over a detritus and mud bottom in shallow water. It rarely ventures into deeper water away from shore. Vegetation commonly found in its habitat includes submerged plants (coontail, *Ceratophyllum demersum*, elodea, *Elodea canadensis*, giant elodea, *E. densa*, water milfoil, *Myriophyllum heterophyllum*) and floating plants (duckweeds of the family Lemnaceae) (Wetzel 2007). This species is also known from reservoirs (Bettoli et al. 1993; Buchanan 2005) and freshwater coastal marshes (Gelwick et al. 2001). Arkansas impoundments with known populations are Champagnolle Creek, Dierks, Erling, Merrisach, Millwood, and Pool 2 of the Arkansas River Navigation System. It is apparently very tolerant of high turbidity and siltation, but it is often found in clear waters. Killgore and Hoover (2001) reported that it is tolerant of low dissolved oxygen concentrations. Removal of aquatic vegetation from a Texas reservoir by Grass Carp resulted in the disappearance of the Bantam Sunfish (Bettoli et al. 1993). During major flood events, it apparently has good ability to move across flooded lowlands to establish founder populations in previously unoccupied areas (Burr et al. 1996). Warren (2009) considered its presence and abundance in natural habitats an indicator of functioning, relatively intact wetland ecosystems.

Burr (1977) studied the life history of the Bantam Sunfish in Illinois and also summarized what little previously published information was available for that species. It is an opportunistic insectivore feeding predominantly on odonate larvae, amphipods, hemipterans, dipteran larvae, mayflies, and gastropods. Juveniles smaller than 30 mm TL ate a higher percentage of microcrustaceans and midge larvae than adults, and did not feed on snails. Some surface feeding occurs in this species, as indicated by the presence of some terrestrial insects in the adult diet. Burr (1977) also noted some seasonal variation in diet, with heaviest consumption of gastropods occurring in winter and spring and hemipterans in summer.

Spawning occurred in southern Illinois from late April to late May, presumably taking place in shallow water over a soft mud bottom near vegetation. The typical spawning period in more southern areas of its range is from mid-April to early June. In Arkansas, spawning begins by late April and continues at least through May. Peak spawning activity in Illinois occurred in May when water temperatures ranged from 64.4°F to 71.6°F (18–22°C). Features of its reproductive biology differ from other *Lepomis* in that it consistently matures at a very young age and smaller size, in its earlier and shorter spawning period, in having relatively small mature ova, and in its low batch fecundity (Warren 2009). Nests are closely spaced and are excavated in substrates of fibrous root material in dense aquatic vegetation or over mud and leaf litter (Robison 1975; Zeman and Burr 2004). Although actual field observations are lacking, aquarium studies indicate that nest construction, courtship, and spawning behavior are similar to those of other sunfishes (Wetzel 2007). Males attempt to lead females to the nest. While a pair circles over the nest, courtship behavior by the male includes nudging the female with his snout and nipping at her caudal fin. Batch fecundity in Illinois ranged from 248 to 1,544 eggs. Eggs hatch in 26–36 hours at 22–26°C, and the larvae become free-swimming about 5 days after hatching (Wetzel 2007). Males defend the nest until the larvae disperse from it (Zeman and Burr 2004). Sexual maturity is reached at 1 year of age or less, and the maximum life span is about 3+ years.

CONSERVATION STATUS   Loss of wetlands has led to extirpation of some northern populations in Illinois and Indiana. It is considered critically imperiled in those states and in Missouri and Oklahoma (Warren 2009). This lowland species is more widely distributed in Arkansas than previously thought (McAllister et al. 2008). It is abundant in a few areas of suitable habitat in southern Arkansas. We currently consider the Bantam Sunfish secure in Arkansas, but periodic monitoring of its status is warranted.

Genus *Micropterus* Lacépède, 1802— Black Basses

This genus, described by Lacépède in 1802, is native to North America east of the Continental Divide and contains eight species, three of which are native to Arkansas. *Micropterus* species, the largest centrarchids, are referred to as black basses because of their black larvae (Boschung and Mayden 2004). Taxonomy of the genus was clarified largely through the work of Hubbs and Bailey (1940) and

Bailey and Hubbs (1949). Philipp and Ridgway (2002) presented information on ecology, conservation, and management of the black basses. The eight species form a monophyletic clade (Near et al. 2004, 2005), and hypotheses of intrageneric relationships were summarized by Boschung and Mayden (2004). Although hybridization between *Micropterus* species is reportedly not as common as in *Lepomis* species, recent genetic evidence suggests that natural hybridization is more common than previously believed. Bolnick (2009) reported five known hybrid combinations between *Micropterus* species. Genus characters include an elongate, slightly compressed body; mouth large, extending to at least below the center of the eye in adults; lower jaw projecting beyond upper jaw; opercle with 2 flat projections, the lower projection longer than the upper; an emarginate caudal fin; preopercle margin smooth, not serrate; lateral line complete with 55 or more lateral line scales; 3 anal fin spines; usually 6 or 7 branchiostegal rays; and dark brown lines radiating backward from snout and back of eye to edge of opercle. In addition, there is little sexual dimorphism and no bright breeding colors. There is extended parental care (male care from spawning through fry dispersal to the juvenile stage can last from 2 to 7 weeks or more).

Micropterus dolomieu Lacépède
Smallmouth Bass

Figure 26.29. *Micropterus dolomieu*, Smallmouth Bass. *Uland Thomas.*

CHARACTERS   A large, slender, elongate bass with body depth going into SL 3.6–5.6 times (Fig. 26.29). Mouth moderately large, the upper jaw usually extends to beneath center of eye in juveniles smaller than 6 inches (152 mm) TL, to beneath the posterior half of the eye in the smaller adults, and to slightly beyond the rear margin of the eye in some large adults. Supramaxilla large. Tongue usually without a patch of teeth. No lateral band present, sides plain or with several separate vertical bars. Spinous dorsal fin low and broadly joined to soft dorsal fin, with a shallow notch between them; the shortest dorsal spine typically

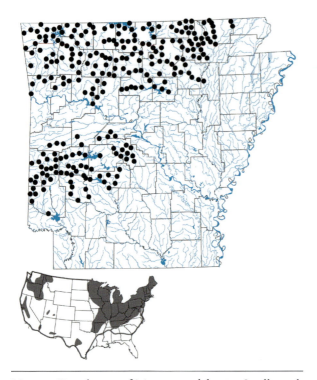

Map 179. Distribution of *Micropterus dolomieu*, Smallmouth Bass, 1988–2019. Inset shows native range (solid) and introduced range (outlined, with some introduced populations shown in solid) in the contiguous United States.

Figure 26.30. *Micropterus dolomieu* juvenile with tricolored caudal fin. *David A. Neely.*

more than half the length of the longest spine. Gill rakers on first gill arch 6–8. Pyloric caeca not forked, about 10–15. Lateral line complete, usually with 68–76 scales. Scale rows above lateral line usually 12 or 13. Cheek scale rows usually 15–18. Scales around caudal peduncle 29–32. Dorsal spines 9 or 10; dorsal rays 12–15; anal spines 3; anal rays usually 11; pectoral rays usually 16 or 17; pelvic fin with 1 spine and 5 rays. Most adults are 10–20 inches (254–508 mm) long and weigh 1–4 pounds (0.5–1.8 kg). Maximum weight around 12 pounds (5.4 kg), and maximum TL more than 27 inches (686 mm). The Arkansas state record, from Bull Shoals Lake in 1969, is 7 pounds, 5 ounces (3.3 kg).

LIFE COLORS   Back and sides uniform olive brown to bronze, giving rise to the local name of brown bass. Arkansas specimens from the Ozarks tend to be more yellowish than Smallmouth Bass from east of the Mississippi River and the upper Midwest (D. Neely, pers. comm.). Sides plain or with faint dark mottling and sometimes with 8–16 dark brown vertical bars; lower sides sometimes with dark spots irregularly arranged but not forming horizontal rows. Three conspicuous dark horizontal bars radiate across operculum (usually present in juveniles as well as adults). Eye becomes red in breeding males. Undersides white, often

with dusky pigment. Fins dusky. Young with various lateral patterns of vertical bars and blotches, or plain (Fig. 26.30). Tail of young tricolored, with yellow base, black middle, and white edge.

SIMILAR SPECIES   The Smallmouth Bass differs from *Micropterus salmoides* in having a smaller mouth, the upper jaw usually not reaching beyond back of eye except in large adults (vs. upper jaw usually extending distinctly behind eye) and an overall olive-brown coloration in life (vs. body more greenish in coloration, usually with a dark midlateral stripe). It further differs from Spotted Bass in having 12 or 13 scale rows above lateral line (vs. 7–10) and in lacking regular horizontal rows of spots on lower sides (vs. lower side with rows of small spots).

VARIATION AND TAXONOMY   *Micropterus dolomieu* was described by Lacépède (1802), with no type locality given. The spelling of the specific epithet has changed from *dolomieui* to *dolomieu* since the previous edition of this book because of changes in the rules of zoological nomenclature (Nelson et al. 2004). Hubbs and Bailey (1940) recognized two subspecies: *M. d. dolomieui*, occurring over most of the species range, and a southern form, *M. d. velox*, occurring in middle Arkansas River drainages. Intergrades between these two subspecies have been found in the White and Black river drainages in Arkansas and Missouri, and in the Ouachita River drainage in Arkansas. Widespread introductions have resulted in much mixing of the gene pools of these two forms, and Bailey (1956) ultimately recommended that *M. d. velox* no longer be recognized as a valid subspecies. This view was further supported by Hoyt (1965) who found that even though Smallmouth Bass in Arkansas represented two statistically distinct populations (one inhabiting the Arkansas River drainage in northwestern Arkansas, and the other in the White, Ouachita, and Red river systems), the two populations did not merit subspecific status. Castro (1963) compared meristic variations of wild and laboratory-raised Arkansas Smallmouth Bass. Stark and Echelle (1998) conducted an allozyme survey of genetic variation at 33 gene loci of Smallmouth Bass from 51 localities in the Ozark and Ouachita uplands. Their data supported the recognition of three clades from the Interior Highlands: (1) the previously recognized Neosho Smallmouth Bass in Ozark tributaries of the middle Arkansas River, (2) the Ouachita Smallmouth Bass in the Little and Ouachita river drainages, and (3) a clade that included populations from the White, Black, St. Francis, Meramec, and Missouri rivers, and other streams in the northern and eastern Ozarks. Near et al. (2004, 2005) considered *M. dolomieu* to form a sister pair with *M. punctulatus*.

Natural hybridization has been reported between *M. dolomieu* and four other species of *Micropterus*: *M. coosae*, *M. punctulatus*, *M. salmoides*, and *M. treculi* (Koppelman 1994; Bolnick 2009).

DISTRIBUTION *Micropterus dolomieu* is considered native to the St. Lawrence–Great Lakes, Hudson Bay, and Mississippi River basins from Quebec, Canada, to North Dakota, and south to northern Alabama, eastern Oklahoma, and southern Arkansas. It has been introduced into northern Atlantic Slope drainages, western states, and elsewhere. In Arkansas, the Smallmouth Bass is found in all major drainages of the Ozark and Ouachita uplands, including the White, Arkansas, Ouachita, and Red river drainages, but it is most widespread in the clear, upland streams of the Ozarks (Map 179).

HABITAT AND BIOLOGY In Arkansas, the Smallmouth Bass is mainly an inhabitant of cool, clear mountain streams with permanent flow and rocky bottoms. In the Illinois Bayou, Arkansas, Smallmouth Bass were consistently found at a median depth of 0.8 m, and boulder habitat was preferred when it was available (Hafs et al. 2010). Cobble, gravel, and bedrock substrates were also used. Walters and Wilson (1996) described habitat segregation between age 0 and older Smallmouth Bass in the Buffalo River, Arkansas, one of the best remaining riverine habitats for this species in the state. Runs and pools were occupied by both groups of *M. dolomieu*, but a higher proportion of age 0 fish were observed in pools. Significant-run microhabitat predictors for the presence of age 0 fish were bedrock, silt, sand, gravel, aquatic macrophytes, boulders, and light period. Significant predictors in pool microhabitats were cobble, undercut banks, depth, and light. In Ozark border streams in Missouri, abundance of Smallmouth Bass was inversely related to percent pool area and maximum summer water temperature (Sowa and Rabeni 1995). In Oklahoma, densities of age 0, age 1, and older Smallmouth Bass were approximately an order of magnitude greater in the Boston Mountains and Ozark Highlands streams than in Ouachita Mountains streams (Dauwalter et al. 2007). In a study of the effect of watershed land-use patterns on stream fish assemblages in

Arkansas River tributaries draining the Boston Mountains Plateau of the Arkansas Ozarks, Rambo (1998) found that Smallmouth Bass production was highest in a watershed with an intermediate proportion of agriculture and lower in areas with very high and very low agricultural proportions. Although annual production rates of Smallmouth Bass were low, four of the streams studied met the criteria for a blue ribbon or quality Smallmouth Bass stream (Rambo 1998). Lonzarich et al. (1998) found that recolonization rates were high in pools of Ouachita Mountain streams from which Smallmouth Bass had been experimentally removed. Interpool movement in Ouachita streams was high in summer, with 35% of marked individuals moving among adjacent pools during a 3-day observation period. It is more intolerant of habitat alteration than the other black basses, and especially intolerant of high turbidity and siltation. It also has lower tolerance than other centrarchids to low dissolved oxygen concentrations and low pH, with reproductive impairment occurring at pH below 6.0 (Warren 2009). Successful Smallmouth fisheries exist in a few Arkansas reservoirs such as Bull Shoals, Beaver, and Greers Ferry. These large Ozark reservoirs have clear, deep waters and a well-developed, cool, oxygenated hypolimnion. Bowman (1993) found that Smallmouth Bass habitat varied longitudinally in Beaver Reservoir. The eutrophic up-lake area was not good habitat, the oligotrophic down-lake area contained excellent habitat, and the mesotrophic mid-lake area was a transition zone that provided adequate habitat. Smallmouth Bass made up 5–10% of the total black bass catch by electrofishing in 1991 and 1992 (Bowman 1993). It has also been stocked in Lake Ouachita. Stocking has not been successful in Lake Greeson. Even though it has sufficient depth, Greeson has low dissolved oxygen and is probably not quite large enough to support a successful Smallmouth Bass fishery (Brett Hobbs, pers. comm.). Hubert and Lackey (1980) found that the major habitat variable determining Smallmouth distribution in a Tennessee reservoir was bottom relief. Drop-offs of 30–43° from the overbank into the original river channel or inundated creek channels were preferred. Smallmouth Bass also used all forms of submerged cover in that reservoir: rocks, stumps, sunken trees, and crevices in hard clay banks, without apparent preference for one type. The damming of optimal bass streams in Arkansas has drastically reduced the suitable habitat for this species. It is primarily a stream species in Arkansas that either does not survive in impoundments or survives only in small numbers. Bowman (1993) noted that Smallmouth Bass populations in upper reaches of the four main tributaries to Beaver Reservoir have become isolated from one another by the area of unsuitable, eutrophic habitat in the lower reaches of those tributaries where they enter the eutrophic upper reservoir. Stream habitats for Smallmouth Bass have also been eliminated for many miles below large dams by the cold tailwaters of those dams. Smallmouth populations have also declined in some Arkansas streams because of detrimental land-use practices, such as gravel mining and clearing of riparian vegetation. These practices have resulted in increased sedimentation and chemical inflow in numerous streams.

Warren (2009) referred to the Smallmouth Bass as an opportunistic, top carnivore that feeds from the surface to the bottom. Pflieger (1997) reported that the free-swimming fry of the Smallmouth Bass in Ozark streams of Missouri feed on midge larvae and microcrustaceans. Small fish are added to the diet when the young bass are smaller than 1 inch (25 mm) long, and fish remain important in the diet throughout life. Crayfish (if available) are of about equal importance with fish in the adults' diet, but insects are also frequently eaten. Others have also emphasized the great importance of crayfish in the diets of stream populations of Smallmouth Bass (Rabeni 1992; Roell and Orth 1993). In second-, third-, and fourth-order segments of the Kings River, Arkansas, the most important items in Smallmouth Bass diets in descending order were crayfish, forage fish (mainly *Campostoma* spp.), and aquatic insects (predominantly Ephemeroptera) (Roberg 1999). In the Buffalo River, Arkansas, fish made up the major food by weight for immature Smallmouth Bass (45–170 mm TL), and crayfish were the next most abundant food items except during summer, when insects became the second-most abundant food item (Kilambi et al. 1977). Adults (176–382 mm TL) fed most heavily on fishes and crayfish during all seasons, but crayfish were the dominant food in winter. Whisenant (1984) also found that adult Smallmouth Bass in the Buffalo River fed primarily on fish, followed by crayfish. Crayfish are often absent or present in small numbers in reservoirs, and in those environments the adult diet is made up almost entirely of fish. Aggus (1972) found that Smallmouth Bass in Bull Shoals Reservoir fed mainly on young-of-the-year shad from July to October, with centrarchids composing a significant portion of the diet during that time. Fish made up 64% of the weight of the annual diet. Crayfish were found in Smallmouth stomachs throughout the year but were dominant in the diet only during midwinter. Mayfly nymphs and terrestrial insects made up a small percentage of the diet in spring and summer.

Minimum age for attaining sexual maturity has been reported as 3–7 years for females and 2–5 years for males

(Vogele 1981; Ridgway et al. 1989; Dunlop et al. 2005). Many populations in streams, rivers, lakes, and reservoirs make pronounced spring migrations to spawning areas, with individuals often nesting in the same area over multiple years (Todd and Rabeni 1989; Kraai et al. 1991). Spawning reportedly occurs over a wide range of temperatures (12.0–26.7°C), but peak spawning apparently occurs between 15°C and 22°C (Warren 2009). Pflieger (1966a) described reproduction in Ozark streams. Spawning in Missouri and northern Arkansas begins in April and continues through most of June. Vogele and Rainwater (1975) found that peak Smallmouth spawning activity occurred in Bull Shoals Reservoir in northwestern Arkansas from late April through early May. In that reservoir, Smallmouth showed no preference for man-made brush shelters as nesting sites (unlike the other two black bass species) but instead appeared selective for a specific substrate of rocks and gravel. Nests were constructed in the reservoir at water depths ranging from 3.9 to 15.1 feet (1.2–4.6 m) and averaging 8.9 feet (2.7 m). In streams, males excavate circular nests in the gravel substrate, usually in some current just downstream from a boulder or other object. Males in Baron Fork Creek, Oklahoma, selected nest sites of intermediate water depths (0.24–1.8 m), low water velocity (<0.26 m/s mean velocity), and near cover (boulders and woody structure) (Dauwalter and Fisher 2007). Usually only one female spawns in the nest with the male, and from 2,000 to 10,000 eggs are deposited. Fecundity estimates in the Buffalo River ranged from 1,884 to 23,366 in females 229–408 mm TL (Kilambi et al. 1977). The spawning pair circles the nest rim with the male repeatedly nipping the female. The pair then settles to the nest bottom, where the female releases eggs while vibrating her body. The male releases milt while in an upright position. More than 100 subsequent spawning bouts by the pair may occur over a more than 2-hour period (Wiegmann et al. 1992, 1997). During the spawning act, the male's iris becomes bright red, and the female develops a series of dark vertical bars or mottlings that are lacking in the male (Warren 2009). The male guards the nest, fanning the eggs, until the eggs have hatched and even after the black, free-swimming fry have dispersed from it. Like all *Micropterus*, the male Smallmouth Bass vigorously defends the embryos and fry for 2–7 weeks, much longer than centrarchids in other genera (Cooke et al. 2006). Peek (1965a, b) studied age and growth of wild and laboratory-raised Arkansas Smallmouth Bass. Maximum growth rate occurred at 82.4°F and 84.2°F (28°C, 29°C). Buchanan (1967) and Buchanan and Strawn (1969) demonstrated that laboratory and hatchery-raised fish could readily be distinguished from stream-raised bass on the basis of first-year scale growth when stocked in streams where Smallmouth Bass were known to have slow rates of growth. Bare (2005) estimated the survival of adult Smallmouth Bass in the Buffalo River in a September–March period to be approximately 84%, and Hafs et al. (2010) estimated annual survival rates in the Illinois Bayou as 66–71%. Maximum life span is probably around 15 years.

CONSERVATION STATUS This gamefish is widespread and locally common in streams of the Ozark and Ouachita mountains. Gravel mining and clear-cutting have adversely impacted some populations in recent years; however, few new dams that affect Smallmouth streams have been constructed since the first edition of this book. Rambo (1998) found that upper stream reaches in Arkansas River tributaries of the Boston Mountains Plateau are important spawning and rearing areas for Smallmouth Bass. Most of its populations in Arkansas are stable if not secure. Consideration should be given to proper management practices for the three genetically distinct clades of Smallmouth Bass in Arkansas (Stark and Echelle 1998). Specifically, stocking individuals of one clade into habitats occupied by the other clades should be avoided.

Micropterus punctulatus (Rafinesque)
Spotted Bass

Figure 26.31. *Micropterus punctulatus*, Spotted Bass. *Uland Thomas.*

CHARACTERS A large, slender, elongate bass, with body depth going into SL 3.7–5.9 times (Fig. 26.31). Mouth large, the posterior end of the upper jaw usually extends to beneath posterior half of eye in young and many adults; in large adults, upper jaw usually extends to or slightly behind the rear margin of the eye. Supramaxilla large. Tongue with a patch of teeth. Midside with a broad, longitudinal dark band of more or less confluent blotches. Lower sides with horizontal rows of dark spots. Spinous dorsal fin broadly joined to soft dorsal fin, with a shallow notch between them, the shortest spine in dorsal fin typically more than half the length of the longest spine. Gill rakers on first gill

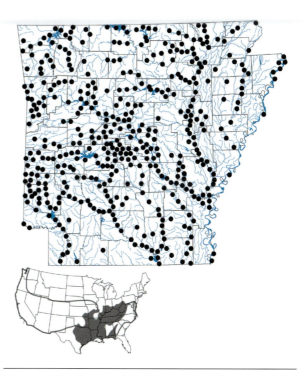

Map 180. Distribution of *Micropterus punctulatus*, Spotted Bass, 1988–2019. Inset shows native range (solid) and introduced range (outlined) in the contiguous United States.

arch 5–7. Pyloric caeca not forked, 10–13. Lateral line complete, usually with 59–68 scales. Scale rows above lateral line usually 7–9. Scale rows around caudal peduncle usually 24 or 25. Cheek scale rows usually 13–18. Dorsal spines usually 10 (9–11); dorsal rays usually 12 or 13; anal spines 3; anal rays usually 10 (9–11); pectoral rays 15–17; pelvic fin with 1 spine and 5 rays. Most adults 12–15 inches (305–381 mm) long and weighing 1–3 pounds (0.5–1.4 kg). Maximum size about 25 inches (640 mm) TL; maximum weight around 10 pounds, 4 ounces (4.65 kg). The Arkansas state record, from Bull Shoals Lake in 1983, is 7 pounds, 15 ounces (3.6 kg).

LIFE COLORS Back and upper sides olive green with darker mottlings and often with golden reflections; midside with a black band or a series of partly joined blotches; lower sides white with regular rows of dark brown or black spots. Cheek with 3 or 4 faint to dark horizontal bars. Undersides white. Base of caudal fin usually with a prominent black spot, especially in smaller individuals. Iris of eye red in breeding males. Fins dusky. Young smaller than 6 inches (152 mm) TL with a tricolored tail; a yellowish-orange base is separated from a white fringe by a black vertical bar. Sides of young with a series of vertically elongated dark blotches, but sometimes with a solid band. Rows of spots on lower sides less well developed or absent in young (Fig. 26.32).

Figure 26.32. *Micropterus punctulatus* juvenile with tricolored caudal fin. *Renn Tumlison.*

SIMILAR SPECIES The Spotted Bass differs from the Largemouth Bass in having a shallow notch between spinous and soft dorsal fins (vs. deep notch) and in having regular horizontal rows of spots on lower sides (vs. rows of spots absent, but lower sides irregularly spotted). Juvenile Spotted Bass have a tricolored tail, with yellow, black, and white coloration (vs. juvenile Largemouth Bass with a bicolored tail with a white base and wide black margin and lacking yellow-orange pigment). It differs from the Smallmouth Bass in having rows of spots on lower sides (vs. rows of spots absent) and in having 7–10 scale rows above the lateral line (vs. 12 or 13).

VARIATION AND TAXONOMY *Micropterus punctulatus* was described by Rafinesque (1819) as *Calliurus punctulatus* from the falls of the Ohio River. Three subspecies were formerly recognized, two of which had very restricted distributions (Hubbs and Bailey 1940). (1) *Micropterus p. henshalli* occurs in the Mobile Bay drainage and has been elevated to full species status (Page et al. 2013). (2) *M. p. wichitae*, restricted to West Cache Creek, Oklahoma, was based on hybrids between *M. punctulatus* and *M. dolomieu* and is not a valid subspecies (Cofer 1995). (3) All Spotted Bass currently belong to the former subspecies, *M. p. punctulatus* (subspecific recognition no longer valid), the widest-ranging form found throughout the central and lower Mississippi basin and along most of the Gulf Coast. Near et al. (2003, 2004, 2005) considered the Spotted Bass to form a sister pair with *M. dolomieu*. Natural hybridization is common among *Micropterus* species (especially in reservoirs). Lake Norman, North Carolina, a system with a historically strong Largemouth Bass fishery, experienced considerable hybridization between *M. salmoides* and *M. punctulatus* after the introduction of *M. punctulatus* (Godbout et al. 2009). Natural hybridization has also been reported between *M. punctulatus* and *M. dolomieu* (Bolnick 2009).

DISTRIBUTION *Micropterus punctulatus* occurs in the Mississippi River basin from southern Pennsylvania and West Virginia, west to southeastern Kansas, and south to the Gulf of Mexico; and in Gulf Slope drainages from the Apalachicola River drainage, Georgia, Alabama, and Florida, west to the Guadalupe River, Texas (but absent from the Mobile Bay basin). It has been widely introduced elsewhere in the United States. In Arkansas, the Spotted Bass is found statewide, occurring in all major drainages, but is most abundant in the Ozark and Ouachita uplands (Map 180).

HABITAT AND BIOLOGY Primarily an inhabitant of larger streams and rivers, *M. punctulatus* is especially abundant in the mountainous regions of the state. It has also adapted well to man-made impoundments of various sizes and occurs in the larger lowland streams. It is most abundant in streams having clear water, permanent flow, and gravel bottoms, where it is usually found in deep pools with cover. It is more tolerant of turbidity and warm water than the Smallmouth Bass (Layher et al. 1987), and it is also more common in reservoirs than that species. The Largemouth Bass was the numerically dominant *Micropterus* species in Beaver Reservoir from its filling in 1964 through 1974, but by 1980, the Spotted Bass had become the dominant black bass species in that reservoir (Rainwater and Houser 1982). Bowman (1993) also found that *M. punctulatus* was the numerically dominant *Micropterus* species in Beaver Reservoir in 1991 and 1992, making up 57% and 60% of the total black bass catch in those years, respectively. Olmsted (1974) found that *M. punctulatus* and *M. salmoides* in Lake Fort Smith demonstrated a high degree of ecological segregation as a result of different habitat preferences. Largemouth Bass were more abundant in areas having a mud substrate and abundant emergent aquatic vegetation, while Spotted Bass were more abundant in areas having a rocky substrate, steeply sloping shores, and little emergent vegetation. This same habitat segregation also occurs in Bull Shoals, Beaver, Greers Ferry, and Ouachita reservoirs (William Keith, pers. comm.). Other authors have also noted that in southern reservoirs the Spotted Bass is most abundant in oligo-mesotrophic reaches of reservoirs, with abundance decreasing as eutrophication increases (Buynak et al. 1989; Greene and Maceina 2000; Maceina and Bayne 2001).

The Spotted Bass is an opportunistic carnivore that feeds from the water's surface to the bottom (Warren 2009). Cross (1967) reported that crayfish and insects were the principal food items of Spotted Bass in Kansas. Smith and Page (1969) found that in Illinois, all sizes fed mainly on the immature stages of aquatic insects; young bass also ate small crustaceans, and larger bass fed on crayfish and fish. Crayfish were the dominant food item in Lake Fort Smith, and the Spotted Bass was more insectivorous than the Largemouth Bass (Olmsted 1974). The fry initially feed largely on cladocerans and copepods before transitioning to immature aquatic insects, and then to fish and crayfish. Mullan and Applegate (1968, 1970) reported adult Spotted Bass in Beaver Lake were largely piscivorous. In Bull Shoals Reservoir, Spotted Bass smaller than 2 inches (51 mm) ate aquatic insects (79%) and Entomostraca (21%), bass between 2.0 and 3.9 inches (51–99 mm) fed on fish (56%), insects (30%), and Entomostraca (14%), and bass over 4.0 inches (102 mm) were almost entirely piscivorous (85%) (Applegate et al. 1967). In a subsequent study, Aggus (1972) found that crayfish made up the bulk of the food of Spotted Bass in Bull Shoals Reservoir on an annual basis both by weight (79%) and by total number of food items ingested (49%). Young-of-the-year shad were the dominant food items only during summer and early autumn. Aquatic insects were eaten during the spring but contributed little to the total biomass ingested.

Spotted Bass reproductive habits are similar to those of the Smallmouth Bass, but spawning activity usually begins a week or more later. In Arkansas, spawning occurs from early April to mid-June, with peak activity occurring in Bull Shoals Reservoir from late April through May at water temperatures between 17.2°C and 25.6°C (Vogele 1975; Vogele and Rainwater 1975). In that reservoir, nests were most frequently found along steep bluffs on talus accumulations where the broken rock substrate had been swept free of silt and periphyton (Vogele 1975). Solid rock ledges, large flat rocks, and patches of rubble and gravel were also commonly used. In coves, nesting substrate was variable and appeared to be less important than cover for determining specific locations and depths of nests. Spawning began in Lake Fort Smith when water temperatures reached 15°C and occurred from late April through May (Olmsted 1974).

Spotted Bass exhibited the greatest preference of any of the black basses in Bull Shoals Lake for man-made brush shelters as nesting sites. Nests were found at water depths of 0.9–5.5 m (3–18 feet). In streams, the male fans out a circular depression nest in the substrate, usually in a pool with some current. The nest is usually located just downstream from a boulder, log, or other cover. In reservoirs, nests are located at depths of 2.3–3.7 m (7.6–12.1 feet). It is a solitary nester. Vogele (1975) described spawning behavior in Bull Shoals Reservoir, where spawning occurred between 11 a.m. and 7 p.m., with peak activity occurring in late afternoon. Courtship and spawning sequences were lengthy, requiring up to 3.5 hours for completion. Although spawning usually involved one pair of Spotted Bass in one nest (average nest diameter was 23.6 inches [60 cm]), other combinations were observed. In one instance, two females were actively involved in courtship in a nest with one male; in another, four females were seen circling a nest with one male. After spawning, the male guards the nest until the translucent fry become free-swimming, at which time the fry form compact schools that the male continues to guard for up to 4 weeks. Eggs hatched in 2 days at a water temperature of 21.1°C and in 5 days at 14.5°C. Larvae become free-swimming at 6.0–7.5 mm TL in 4 days after hatching at 25°C and 8 days after hatching at 15–18°C (Vogele 1975). The schools break up when the fry reach about 30 mm TL (Vogele 1975). In hatchery ponds, males exhibited a much shorter parental care period of about 8–10 days. Fecundity of Spotted Bass in Lake Fort Smith ranged from 1,727 to 9,552 eggs, depending on size of female (Olmsted 1974). In Bull Shoals Lake, the number of eggs produced varied with size and age of females, ranging from 3,249 eggs in females 311 mm TL (age 3 fish) to 30,586 eggs in 444 mm TL fish (age 6) (Vogele 1975). The average number of eggs in nests was 5,016, and the average number of larvae was 1,476. Sexual maturity is attained in 1–3 years.

Olmsted and Kilambi (1978) provided information on age and growth of Spotted Bass in Lake Fort Smith, Arkansas, a small oligotrophic impoundment. They found that the Spotted Bass had a relatively slow rate of growth, especially in the older age groups. This slow growth was attributed to the small size of the reservoir (438 acres [177 ha]), frequent periods of high turbidity, and the oligotrophy of the reservoir. The maximum life span in Lake Fort Smith appeared to be around 7 years. Maximum known life span is about 11+ years.

The Spotted Bass is a significant game species in streams and reservoirs of the uplands. Although it has not been as important as the Largemouth Bass to the state's anglers, it is probably now of greater importance than the Smallmouth Bass because of the latter species marked decline caused by damming of streams and other habitat alterations.

CONSERVATION STATUS   This is a widespread and important gamefish. Its populations are apparently secure in Arkansas.

Micropterus salmoides (Lacépède)
Largemouth Bass

CHARACTERS   A large, slender, elongate bass, with body depth going into SL 3.4–4.2 times (Fig. 26.33). Mouth very large, in adults the upper jaw extends well behind rear margin of eye; in specimens smaller than 5 inches (127 mm) TL, mouth extends to the posterior margin of eye. Supramaxilla large. Tongue usually without a patch of teeth. Midside

Figure 26.33. *Micropterus salmoides*, Largemouth Bass. *Uland Thomas.*

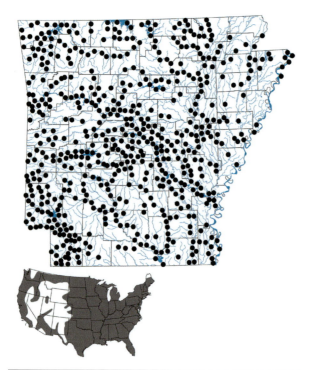

Map 181. Distribution of *Micropterus salmoides*, Largemouth Bass, 1988–2019. Inset shows native and introduced ranges in the contiguous United States.

with a broad, dark longitudinal band (often indistinct in large adults); lower sides with irregularly arranged spots that do not form continuous horizontal rows. Spinous and soft dorsal fins narrowly joined, with a deep notch almost completely separating them, the shortest dorsal spine less than half the length of the longest spine. Spinous dorsal fin highest at middle, low at rear. Gill rakers on first gill arch 7–9. Pyloric caeca branched at base, 12–45 (counting all branches). Lateral line complete, usually with 59–73 scales (when counts for the introduced subspecies *M. s. floridanus* are included). Scale rows above lateral line 7–9. Cheek scale rows 9–13. Dorsal spines usually 9 (9–11); dorsal rays usually 12–14; anal spines 3; anal rays usually 11 or 12; pectoral rays usually 14 or 15; pelvic fin with 1 spine and 5 rays.

Figure 26.34. *Micropterus salmoides* juvenile. *David A. Neely.*

Most adults are between 12 and 20 inches (305–508 mm) long and weigh from 1 to 5 pounds (0.5–2.3 kg). Maximum size about 40 inches (787 mm) TL. Maximum weight 22 pounds, 4 ounces (10.1 kg). The Arkansas state record from Mallard Lake in 1976 is 16 pounds, 4 ounces (7.4 kg).

LIFE COLORS   Back and upper sides olive to brassy green with faint to dark olive mottling. Midside with black lateral band formed from confluent blotches and often interrupted anteriorly, but almost straight-edged on caudal peduncle. Lower sides and undersides white, often with scattered dark brown spots. Caudal fin of young bicolored, with light-colored base and wide black margin and lacking yellow or orange pigment (Fig. 26.34).

SIMILAR SPECIES   The Largemouth Bass differs from other black basses in having a larger mouth in adults, the upper jaw extending well past rear margin of eye (vs. upper jaw extending to or slightly past rear margin of eye), in having a deep notch between spinous and soft dorsal fins (vs. the 2 dorsal fins more broadly joined and only slightly emarginate), and the shortest dorsal spine less than half the length of the longest spine (vs. shortest dorsal spine more than half the length of longest spine). It further differs from the Smallmouth Bass in having a broad lateral stripe (vs. side plain or with a series of vertically elongated blotches), and from the Spotted Bass in having scattered dark spots on lower sides (vs. lower side with dark spots in distinct rows).

VARIATION AND TAXONOMY   *Micropterus salmoides* was described by Lacépède (1802) as *Labrus salmoides* from Charleston, South Carolina. Two subspecies of Largemouth Bass were recognized by Bailey and Hubbs (1949), *M. s. salmoides*, occurring over most of the species range, including Arkansas, and *M. s. floridanus*, which is native to peninsular Florida and differs morphologically from the former subspecies in coloration, in having smaller and more numerous scales, and in attaining a larger maximum size. It is the last attribute that has caused the Florida subspecies to be introduced into several Arkansas reservoirs since the late 1970s, notably Mallard Lake, from which the state record Largemouth Bass was taken. Between 2003 and 2013, approximately 1 million Florida

Largemouth Bass fingerlings were stocked annually in Arkansas reservoirs, mainly in the southern half of the state (Lamothe and Johnson 2013). Morphological features for separating the two subspecies include *M. s. salmoides* having 14–35 (x̄ = 23) pyloric caeca (vs. 26–53, x̄ = 37 in *M. s. floridanus*), and 59–67 lateral line scales (vs. 69–73) (Buchanan 1973). Widespread introgression between the Florida Largemouth Bass and *M. s. salmoides* has been reported across the southern United States, including Arkansas (Johnson and Fulton 1999; Johnson and Staley 2001; Allen and Johnson 2009). Largemouth Bass stocks at three Arkansas state fish hatcheries had contamination from Florida Largemouth Bass alleles, as did stocks at seven private fish farms (Johnson and Staley 2001). Bass from those sources are routinely stocked throughout Arkansas in public and private waters, so there are probably few if any bodies of water in the state with pure strains of *M. s. salmoides*. Lamothe and Johnson (2013) used microsatellite genetic markers to analyze 148 trophy Largemouth Bass heavier than 5 pounds (2.3 kg) from 14 Arkansas reservoirs between 2006 and 2011 and found that 83.1% possessed Florida Largemouth Bass alleles. Fifty-six (37.8%) of the 148 bass sampled were identified as Florida Largemouth Bass, and 32 (64%) of the heaviest 50 fish sampled were genetically confirmed to be Florida Largemouth Bass. The dispersal of Florida Largemouth Bass alleles from reservoirs to rivers has been documented in Oklahoma (Gelwick et al. 1995), Texas (Ray et al. 2012), and Arkansas (Brockway and Johnson 2015). Presumed intergrades between the two subspecies (which occur naturally in northern Florida and southern Georgia) are intermediate in all features, including number of pyloric caeca. In recent years there has been considerable disagreement about the taxonomic status of *M. s. floridanus*. It has been considered a distinct species (*M. floridanus*) by Nedbal and Philipp (1994), Kassler et al. (2002), Near et al. (2003, 2004), and Warren (2009) based on genetic analysis of allozyme loci, mitochondrial DNA, and nuclear DNA; however, because no study has yet genetically examined populations within the zone of intergradation, we tentatively follow Page and Burr (2011), Page et al. (2013), and Robins et al. (2018) in continued subspecific recognition for this taxon. Natural hybridization has been reported between *M. salmoides* and two other *Micropterus* species (*M. dolomieu* and *M. treculi*) (Bolnick 2009).

DISTRIBUTION The native range of *Micropterus salmoides* is not precisely known but is thought to include the St. Lawrence–Great Lakes, Hudson Bay, and Mississippi River basins from southern Quebec, Canada, to Minnesota, and south to the Gulf of Mexico; Atlantic

Slope drainages from North Carolina to southern Florida; and Gulf Slope drainages from Florida to Texas and northern Mexico. It has been introduced throughout the United States and southern Canada in North America, and it has also been introduced into Europe, Asia, and Africa. In Arkansas, the Largemouth Bass is found statewide in all drainages and is commonly stocked throughout the state (Map 181).

HABITAT AND BIOLOGY Because it is probably the most important freshwater gamefish in the United States, the biology of the Largemouth Bass has been extensively studied throughout its range including Arkansas, and most of the following biological information is from Arkansas populations. Summaries of available information on *M. salmoides* can be found in Carlander (1977), Heidinger (1976), Philipp and Ridgway (2002), and Warren (2009). Cage culture in Arkansas was studied by Kilambi (1981) and Kilambi et al. (1978).

The Largemouth Bass occupies a variety of habitats, but it is adapted to warmer, more eutrophic waters than other *Micropterus* species. It is most commonly found in clear, quiet waters in natural and man-made lakes and ponds, and in the backwaters and pools of streams and rivers. In streams, both young and adults are most common in deep pools or low-velocity habitats near undercut banks, instream wood, overhanging and aquatic vegetation, or other cover (Sowa and Rabeni 1995). In reservoirs, young and adults are associated with shallow water habitats (<3 m deep) near aquatic vegetation, logs, or other cover, but the young also use gravel substrates and steep shoreline slopes if vegetation or other cover are not available (Annett et al. 1996). Stahr and Shoup (2015) found that American water willow (*Justicia americana*), an emergent macrophyte, resulted in significantly higher juvenile survival from predator attack in laboratory trials and provided a similar increase in survival compared with previous studies using submersed macrophytes. Those authors further concluded that water willow is an excellent candidate for establishment in reservoirs because it is easier to establish than many other macrophyte species yet still reduces predation risk on juvenile Largemouth Bass. It is somewhat tolerant of low dissolved oxygen levels but appears to avoid hypoxic conditions in the summer by seeking out densely vegetated areas in reservoirs and wetlands (Killgore and Hoover 2001). Smaller individuals are more tolerant of low dissolved oxygen than larger individuals (Burleson et al. 2001). It is also more tolerant of turbidity and salinity than other *Micropterus* species (Carlander 1977).

The Largemouth Bass exploits a variety of prey, feeding from the bottom to the water surface. The first food of the young consists of microcrustaceans and other zoo-plankton, and insects are later added to the diet. Adult bass feed primarily on fish, crayfish, and insects. In Lake Fort Smith, immature and adult bass were largely piscivorous (Olmsted 1974). Applegate and Mullan (1967b) compared the food habits of young Largemouth Bass in new (Beaver) and old (Bull Shoals) Arkansas reservoirs and found striking differences in stomach contents. Bull Shoals bass first fed on small microcrustaceans and later switched to larval shad and other fish fry. In Bull Shoals Lake, Largemouth Bass 0.8–1.9 inches (20–48 mm) TL consumed mainly Entomostraca (99% by volume), 2–3.9-inch (51–99 mm) fish fed on Entomostraca (50%) and aquatic insects (50%), and 4–23.5-inch (102–597 mm) Largemouth fed on fish (88–99%). Beaver Lake bass consumed larger volumes of food of a larger size, first feeding on large microcrustaceans and midge larvae and then switching to Gizzard Shad when the young bass reached 1.6 inches (40 mm). Bass growth was substantially faster in the new Beaver Lake than in Bull Shoals. Hodson (1966) found that the greatest use of entomostracans by Largemouth and Spotted bass fingerlings in Beaver Reservoir occurred in late summer and fall when the entomostracans were most abundant. Bryant and Moen (1980b) examined 748 Largemouth Bass from DeGray Reservoir and found that fish constituted 59% of the weight of the total diet and occurred in 81% of the stomachs, and crayfish made up 38% of the weight and occurred in 24% of the stomachs. Bettoli et al. (1992) found that sudden changes in aquatic plant density can lead to large dietary changes. The Largemouth Bass was a keystone species in Oklahoma streams (an Ozark stream and a prairie stream) profoundly affecting prey habitat use, community structure, and trophic level biomasses (Power and Matthews 1983; Matthews et al. 1994; Power et al. 1996).

Sexual maturity can be attained at 1 year of age but is attained most commonly at 2–4 years (Bryant and Houser 1971). Peak spawning occurs at water temperatures between 15°C and 21°C, and spawning occurs in Arkansas from April through June when water temperatures rise above 65°F (18.3°C). Spawning occurred in Lake Fort Smith when water temperatures reached 15°C (Olmsted 1974). Peak spawning activity occurred in Bull Shoals Reservoir from late April through early May (Vogele and Rainwater 1975). The male sweeps out a circular depression nest in whatever substrate is available (silt to boulders, stump tops, logs, clay), but coarse gravel bottoms are preferred (Annett et al. 1996; Hunt and Annett 2002). Males tend to select nest sites near simple cover such as logs and tree trunks (Annett et al. 1996). In Bull Shoals Reservoir, Largemouth Bass showed the greatest preference for man-made brush shelters

as nesting sites early in the spawning season (Vogele and Rainwater 1975). Nests were constructed at water depths ranging from 3 to 13 feet (0.9–4 m). It is a solitary nester, with average nest spacing in an Arkansas reservoir ranging from 6.2 to 9.4 m (15 nests/100 m transect) (Hunt and Annett 2002). Vogele and Rainwater (1975) reported much lower densities of nests in Bull Shoals Reservoir (fewer than 1–3 nests/100 m of shoreline). Nests are usually constructed at depths of 1–15 feet (0.3–4.6 m) where there is no current (Pflieger 1997). Carlander (1977) reported that fecundity ranges from 2,000 to 145,000 eggs depending on size of female. In Lake Fort Smith, fecundity ranged from 2,942 to 30,709 (Olmsted 1974). The male guards the nest until the fry become free-swimming and then remains with the schooling young until the school disperses. Total time of parental care by the male from fertilization to dispersal of the fry is usually between 3 and 5 weeks (Cooke et al. 2006). An unusual case of biparental care was reported from a population in a North Carolina stream, where more than half the nests examined were guarded by a male and a female (DeWoody et al. 2000).

Largemouth Bass stocked into the Arkansas River (an open system) contributed to year-class strength to a similar degree as fish stocked into closed reservoirs and lakes (Heitman et al. 2006). Colvin et al. (2008) studied the contribution of stocked fish 50 and 100 mm TL to the Largemouth Bass population in Arkansas River backwaters. Fingerlings 50 mm TL stocked at a density of 309 fish/ha in June 2003 contributed 13.2% to the year class in November 2003, and 17.6% in May 2004. Fingerlings 100 mm TL stocked at 62 fish/ha in August 2003 contributed 13.8% to the year class in November 2003, and 17.2% in May 2004. Growth rates of Largemouth Bass are extremely variable, depending on a number of environmental conditions. Zdinak et al. (1980) found good growth and standing crop in Lake Elmdale in Washington County. In contrast, bass in Lake Ashbaugh in northeastern Arkansas grew slowly and were in poor condition (Johnson and Davis 1997). A study of 12 southern Arkansas lakes to assess the effects of stocking *M. s. floridanus* on the condition of the bass populations was inconclusive (Allen et al. 2009). Bass in streams grow more slowly than bass in large lakes. Cross (1967) also noted that weather cycles and turbidity can affect abundance, reproduction, and growth. The greatest numbers of young were usually produced when adult bass were scarce, and weak year classes were more likely when large bass were abundant. Cross also noted that few lakes and ponds can support more than 100 pounds of bass per acre (112.1 kg / ha) and that the average standing crop (including adults and young fish) is considerably less than 100 pounds per acre even in good bass lakes. Eggleton et al. (2010) studied population characteristics of Largemouth Bass in the 11 navigation pools of the Arkansas River in Arkansas. Populations were young, with 94% of the individuals between 1 and 4 years of age, and the populations exhibited above-average condition and growth rates. The Largemouth Bass has a maximum life span in southern parts of its range of about 10–12 years.

A potential threat to Largemouth Bass populations throughout the range of this species emerged in the mid-1990s in the form of a virus first discovered to affect bass in Santee-Cooper Reservoir in South Carolina. Since that discovery in 1995, the Largemouth Bass virus (LMBV) has been found in a number of states including Arkansas. The LMBV has been documented in more than 10 reservoirs throughout Arkansas and was blamed for the die-off of more than 1,000 adult bass in Lake Monticello in 2000. Quinn and Limbird (2008) concluded that the decline in Largemouth Bass greater than 2.27 kg (5 pounds) or 533 mm (about 21 inches) TL in Lake Dardanelle between 1996 and 2000 was related to LMBV, based on data from AGFC rotenone and electrofishing samples, Arkansas Tournament Information Program data, and the date of confirmation of LMBV in that lake. Little is known about LMBV. It may be present in bass in some lakes without apparent negative effects. In lakes where the virus has caused die-offs, the bass populations seemingly recovered.

CONSERVATION STATUS   This is the most widely distributed and abundant black bass in Arkansas. Its populations are secure.

Genus *Pomoxis* Rafinesque, 1818c—Crappies

This genus, native to eastern North America, was described by Rafinesque in 1818 and has experienced little or no taxonomic turmoil since its inception. A fossil species of *Pomoxis* (*P. lanei*) is known from the Miocene of Kansas and Nebraska, dated at 6.6 mya, and an undescribed fossil *Pomoxis* was reported from Kansas dated at 12 mya (Warren 2009). *Pomoxis* contains two extant sister species, both of which are native to Arkansas. Phylogenetic analyses support the monophyly of *Pomoxis* and the sister group relationship between this genus and a clade containing *Archoplites* and *Centrarchus* (Wainwright and Lauder 1992; Near et al. 2004, 2005).

Genetic evidence indicates that hybridization and introgression between the two *Pomoxis* species are common, especially in reservoirs. Hybrids were found in four consecutive year classes in Weiss Lake, Alabama, and made up 22% of the crappie population (Smith et al. 1994). In a study of 22 southern Minnesota lakes, all but two lakes contained crappie hybrids (L. M. Miller et al. 2008). A study of both species in reservoirs in Alabama, Arkansas, and Tennessee, found that natural hybridization was common and the major source of genetic variation (Dunham et al. 1994). Hooe and Buck (1991) reported that F₁ hybrids between *P. annularis* and *P. nigromaculatus* and the reciprocal cross were fully fertile and capable of backcrossing with both parental species. In some environments, hybrids occur in frequencies as high as 54.5% of the crappie population, but Bolnick (2009) concluded that most sites with sympatric *Pomoxis* species have few or no hybrids.

Ecologically, the crappies are intermediate between *Lepomis* and *Micropterus* species (Aday et al. 2009). Size at age 1, later onset of piscivory, and early dates of spawning are similar to *Micropterus*, while duration of parental care, age at maturity, and planktivory as YOY are similar to *Lepomis*.

Genus characters include an oblong, strongly compressed body with a long and sloping anterior dorsal profile with a sharp dip in profile over the eye. The head is small and the eye is large, teeth are present on tongue, vomer, and palatine bones, the preopercle margin is serrate, there are 5–8 anal fin spines, the lateral line is complete, and there are 34–50 lateral line scales. The mouth is large and oblique, and the lower jaw projects beyond the upper jaw.

Pomoxis annularis Rafinesque
White Crappie

Figure 26.35. *Pomoxis annularis*, White Crappie. *Uland Thomas.*

CHARACTERS   A silver sunfish with a deep, laterally compressed body, body depth going into SL 2.1–3.0 times (sometimes less in unusually deep-bodied, large adults) (Fig. 26.35). Head small, mouth large and oblique, upper jaw extending well past middle of eye. Supramaxilla of moderate size. Preopercle finely serrated. Sides with faint vertical bars. Spinous and soft dorsal fins broadly joined, with no notch between them. Length of dorsal fin base

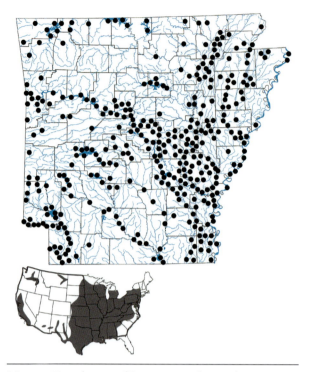

Map 182. Distribution of *Pomoxis annularis*, White Crappie, 1988–2019. Inset shows native range (solid) and introduced range (outlined, with some introduced populations shown in solid) in the contiguous United States.

much less than distance from front of dorsal fin base to back of eye. Lower jaw protruding beyond upper jaw. Gill rakers long and slender, usually more than 28 on first arch. Branchiostegal rays 7. Teeth on tongue in two patches. Lower pharyngeal arches long and narrow, with numerous fine teeth. Lateral line complete, arched anteriorly, usually with 38–45 (34–50) scales. Cheek scale rows 5 or 6. Dorsal spines usually 6 (sometimes 5); dorsal rays 13–15; anal spines 6 (sometimes 5) or 7; anal rays 16–19; pectoral rays 14–16; pelvic fin with 1 spine and 5 rays. Adults are usually 8–15 inches (203–381 mm) long and weigh 1–2 pounds (0.5–0.9 kg). Maximum length around 21 inches (530 mm) TL; maximum weight about 5 pounds, 6 ounces (2.35 kg). The Arkansas state angling record, from Mingo Creek in 1993 is 4 pounds, 7 ounces (2.01 kg).

LIFE COLORS   Back dark olive or brownish green. Sides silver, usually with 10 or fewer vague, dark vertical bars; bars most distinct on upper sides and absent on ventral surface of body. Undersides white. Dorsal, caudal, and anal fins banded and mottled with black; pectoral and pelvic fins usually colorless. Breeding males become much darker and more boldly marked, with sides of head and breast becoming nearly black. Specimens from turbid waters are often very pallid.

SIMILAR SPECIES   The White Crappie differs from the Black Crappie in having 6 dorsal spines (vs. 7 or 8) a shorter dorsal fin base, its length much less than the distance from the front of the dorsal fin base to back of eye (vs. length of dorsal fin base greater than the distance from the front of the dorsal fin base to back of eye), and in usually having definite vertical bars on sides (vs. side with irregularly arranged black speckles and blotches). Siefert (1969) provided characters for separating the larvae of the crappie species. It differs from other sunfishes in usually having 6 dorsal spines (all other sunfishes have 10 or more). *Elassoma zonatum* (Elassomatidae) has 4 or 5 dorsal spines.

VARIATION AND TAXONOMY   *Pomoxis annularis* was described by Rafinesque (1818c) from "falls of the Ohio River." Near et al. (2004, 2005) assigned *P. annularis* to a sister pair with *P. nigromaculatus*. Surprisingly for such an important game species, comparative studies of variation across its range are lacking.

DISTRIBUTION   *Pomoxis annularis* is native to the Great Lakes, Hudson Bay, and Mississippi River basins from New York and southern Ontario, Canada, west to Minnesota and North Dakota, and south to the Gulf of Mexico; and Gulf Slope drainages from the Mobile Bay basin of Georgia and Florida, west to the Nueces River, Texas. It has been widely introduced throughout the United States. In Arkansas, the White Crappie occurs almost statewide in all major drainages but is uncommon in streams of the more mountainous regions of the Ozarks and Ouachitas (Map 182).

HABITAT AND BIOLOGY   The White Crappie is found in natural and man-made lakes and ponds, and in streams and rivers. It has wide ecological tolerances but prefers quiet waters and is almost always found near brush piles, the tops of fallen trees, or standing timber. It is often found at water depths greater than 13 feet (4 m) and is reportedly more tolerant of turbidity than the Black Crappie (Miranda and Lucas 2004). In Ohio, the White Crappie was outnumbered by the Black Crappie in cool, clear waters having an abundance of submerged aquatic vegetation and a clean, hard substrate (Trautman 1957). In turbid systems containing both species, White Crappie often have higher relative abundance than Black Crappie (Andree and Wahl 2019). An explanation of the reported differences in turbidity tolerance is lacking, but some studies of growth and survival suggest that White Crappies can feed more efficiently in turbid waters than Black Crappies, or that White Crappies compete poorly in clear waters with other centrarchids (Ellison 1984; Barefield and Ziebell 1986). Growth in small ponds was similar in both crappie species across a range of turbidities (Spier and Heidinger

2002). Andree and Wahl (2019) found that turbidities of 0, 25, and 50 NTU did not impair foraging of juvenile White and Black crappies. White Crappie populations tend to become stunted in farm ponds and other small bodies of water, but many larger reservoirs and natural lakes produce a steady supply of large individuals for anglers. It is not a strongly schooling species but tends to congregate in loose aggregations near suitable cover.

Young White Crappies feed mainly during the day on small crustacean zooplankton (copepods and cladocerans) and small dipteran larvae and pupae, but adults feed primarily on fish, with insect larvae also making up a small percentage of the diet. The shift to a largely piscivorous diet occurred at a relatively large size (about 160 mm TL) in a Kansas reservoir (O'Brien et al. 1984). In turbid Lake Carl Blackwell, Oklahoma, ephemeropteran nymphs were the most important dietary component for White Crappie 150 mm TL or smaller, while crappies greater than 150 mm TL mainly consumed Gizzard Shad (Muoneke et al. 1992). Large White Crappies are known for their nocturnal feeding and high nocturnal activity (Warren 2009). White Crappies in Beaver Reservoir fed on fish all year-round, and Ball and Kilambi (1972) believed that this food habit caused the White Crappie to become more dominant than the Black Crappie. As that new reservoir aged, there was an apparent reduction in the number of insects and earthworms important in the Black Crappie's diet. As its long, slender gill rakers suggest, it is capable of feeding on zooplankton at all life stages. The retina of the White Crappie has a high density of cones along the eye's horizontal meridian, and this is interpreted as an adaptation for detecting open-water zooplankton (Browman et al. 1990; Browman and O'Brien 1992).

The White Crappie spawns in Arkansas from early April to about the middle of June, but the spawning season for both *Pomoxis* species is shorter than in other Arkansas centrarchids. At a given locality, the crappie population usually completes all spawning activities within a 3-week period. It is one of the earliest spawners among the centrarchids, and the testes and ovaries begin enlarging in the fall. In Beaver Reservoir, Thomas (1981) and Thomas and Kilambi (1981) reported White Crappie spawning from late April through May. Nesting activity begins when water temperatures rise to about 56°F (13.3°C). The males fan out nests (but apparently do not use caudal sweeping) that are often not as well developed as those of other sunfishes. It is a colonial nester, and sometimes as many as 35 nests are constructed in the same area. The nests may be located from less than 1 foot down to 20 feet (0.3–6.1 m) deep in silt-free substrates near a log, stump, or aquatic vegetation. Siefert (1968) found

that White Crappie in hatchery ponds in South Dakota showed no definite substrate preference for nest location but selected areas with some protective object or bottom vegetation. Siefert also found that nest size was variable (average diameter was approximately 11.8 inches [30 cm]) and that a well-defined nest depression was not constructed. Each nest simply consisted of an area swept clean of loose silt. In Table Rock Lake, Missouri, nesting invariably occurred in coves protected from wave action (Pflieger 1997). A large female may deposit from 10,000 to 160,000 adhesive eggs in a nest (Siefert 1968; Smith 1979). Siefert (1968) gave the following account of spawning behavior:

> The female, after being repulsed from the territory a number of times, would finally stop retreating from the territory when chased, and would be accepted by the male. After circling the nest several times, the female would position herself beside the male and face the same direction as the male. They would remain motionless for a few seconds, and then the sides of their bodies would touch. Both fish then would slowly move forward and upward with their bodies quivering. The female would slide under the male in the process, pushing him up and to the side, causing the pair to move in a curve as the sex products presumably were emitted. The male exerted a steady pressure on the female's abdomen. Each spawning act lasted from 2 to 5 seconds with most lasting 4.

In Beaver Reservoir, estimated fecundity ranged from 48,058 to 232,026 ripe ova in females ranging from 217 to 335 mm TL (Thomas and Kilambi 1981). During embryonic development, the male fans the eggs with the pectoral fins. The eggs hatch in about 4 days at 57.2°F (14°C) and in about 2 days at 66.2°F (19°C), and the fry remain in the nest for several days while being guarded by the male. Length of parental care by the male extends from egg deposition until the fry leave the nest, a period of 4–11 days depending on water temperature (Siefert 1968). After leaving the nest, the young move to the limnetic zone. Crappie fry are apparently more sensitive to unfavorable environmental conditions than the fry of most other centrarchids.

Stocking of White Crappies in Arkansas lakes has generally not produced an increase in the number of harvestable crappies available to anglers. Wright (2012), in a study of White Crappie recruitment and supplemental stocking in six Arkansas reservoirs, found that stocking made little contribution to crappie populations. The stocking contribution during the four months following stocking ranged from 0% to 2.1%, and was 0% for all six reservoirs 1 year after stocking. Stocking of fingerling White Crappie in

Lake Chicot, a Mississippi River oxbow, resulted in year-class contributions ranging from 0% to 3.1% (Racey and Lochmann 2002). Sewell (1979b) studied age and growth of the White Crappie in a Mississippi County flood-created pond. Sexual maturity was attained during the second or third summer of life (Thomas and Kilambi 1981) and maximum life span was around 8–10 years.

The White Crappie is one of the most important gamefishes in oxbow lakes of the lowlands and in large reservoirs statewide. It will readily take both natural and artificial baits. Most Arkansas fishermen rank the flavor of its flesh (and that of the Black Crappie) above that of any other species of fish.

CONSERVATION STATUS This popular gamefish is widespread and locally abundant in suitable habitat in Arkansas. Its populations are secure.

Pomoxis nigromaculatus (Lesueur)
Black Crappie

Figure 26.36. *Pomoxis nigromaculatus*, Black Crappie. *Uland Thomas.*

CHARACTERS A silvery sunfish (but darker overall than the White Crappie) with irregularly scattered black speckles and blotches on sides (Fig. 26.36). Body deep and laterally compressed, body depth going into SL 2.2–2.7 times. Head small, and mouth large, upper jaw extending well past middle of eye. Supramaxilla of moderate size. Preopercle finely serrated. Spinous and soft dorsal fins broadly joined, without a notch between them. Length of dorsal fin base equal to or greater than distance from front of dorsal fin base to back of eye. Lower jaw protruding beyond upper jaw. Branchiostegal rays 7. Gill rakers long and slender, usually more than 28 on first arch. Teeth on tongue in two patches. Pharyngeal arches long and narrow, with numerous fine teeth. Lateral line complete, arched anteriorly with 35–44 scales. Cheek scale rows 5 or 6. Dorsal spines 7 or 8; dorsal rays 14–16; anal spines 6; anal

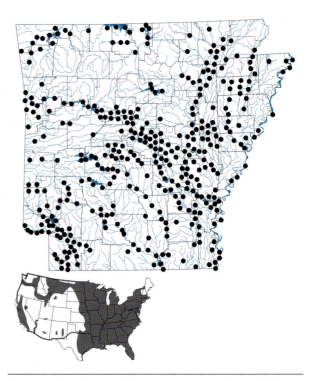

Map 183. Distribution of *Pomoxis nigromaculatus*, Black Crappie, 1988–2019. Inset shows native range (solid) and introduced range (outlined, with some introduced populations shown in solid) in the contiguous United States.

rays usually 18 (16–19); pectoral rays 13–15; pelvic fin with 1 spine and 5 rays. Adults usually 8–15 inches (203–381 mm) long and weigh 1–2 pounds (0.5–0.9 kg). Maximum length about 22 inches (560 mm) TL; maximum weight about 6 pounds (2.73 kg). The Arkansas state angling record, from Lake Wilhelmina in 2011, is 5 pounds, 0 ounces (2.1 kg).

LIFE COLORS Back dark olive with green reflections. Sides silvery with scattered black vermiform blotches that do not form vertical bars. Intensity of mottling varies greatly. Specimens from turbid waters are usually faintly marked, while those from clear, vegetated waters are brilliantly and strikingly pigmented (Trautman 1957). Undersides silvery white. Dorsal, caudal, and anal fins strongly mottled with black pigment. Breeding males become much darker, particularly on head and breast.

SIMILAR SPECIES The Black Crappie is similar to the White Crappie, but differs in having 7 or 8 dorsal spines (vs. 6); a longer dorsal fin base, its length equal to or greater than distance from front of dorsal fin to back of eye (vs. length of dorsal fin base distinctly less than distance from front of dorsal fin to back of eye); and a lateral color pattern of irregularly scattered black speckles and blotches (vs. sides with dark pigment arranged in vertical bars). It

Figure 26.37. Blacknose Crappie, *Pomoxis nigromaculatus*. M. E. Harness.

differs from other sunfishes in having 7 or 8 dorsal spines (all other sunfishes have 10 or more). *Elassoma zonatum* (Elassomatidae) has 4 or 5 dorsal spines.

VARIATION AND TAXONOMY   *Pomoxis nigromaculatus* was described by Lesueur *in* Cuvier and Valenciennes (1829) as *Cantharus nigromaculatus* from the Wabash River, Indiana. Some Black Crappie in Arkansas have a broad dark brown or black stripe along the midline of the back extending from just in front of the dorsal fin to the tip of the lower jaw and ventrally onto the throat and are commonly referred to as Blacknose Crappie by fishermen (Fig. 26.37).

A predorsal stripe was found on 28.1% of 2,192 Black Crappie from Beaver Reservoir (Buchanan and Bryant 1973). This trait has been observed in Black Crappie from Bull Shoals Reservoir and in crappie propagated at the state fish hatchery in Lonoke. We have noted its occurrence in populations in several eastern Arkansas lakes in St. Francis, Crittenden, Lee, and Phillips counties. It has also been reported from at least 13 other states. Buchanan and Bryant (1973) hypothesized that this polymorphic feature is inherited as a simple Mendelian recessive character and is expressed only by homozygous individuals. It occurs with equal frequency in males and females. Gomelsky et al. (2005) crossed crappies with and without the predorsal stripe in various combinations and concluded that the trait was controlled by a single dominant gene; therefore, the ratios found by Buchanan and Bryant (1973) in Beaver Reservoir probably reflected differential survival of the two forms rather than a Mendelian genetic ratio. In a study of genetic variation in crappies, Dunham et al. (1994) found that Black Crappies possessing the blacknose trait had allele frequencies similar to those of Black Crappies with normal coloration. Blacknose Crappie have occasionally

been stocked in reservoirs by the AGFC. *Pomoxis nigromaculatus* forms a sister pair with *P. annularis* (Near et al. 2004, 2005). Studies of variation across the range of this species are lacking.

DISTRIBUTION   The native range of *Pomoxis nigromaculatus* is not precisely known, but is thought to include the St. Lawrence–Great Lakes, Hudson Bay, and Mississippi River basins from Quebec to Manitoba, Canada, south to the Gulf of Mexico; Atlantic Slope drainages from Virginia to Florida; and Gulf Slope drainages from Florida through Texas. It has been widely introduced throughout the United States. In Arkansas the Black Crappie occurs in all major drainages, but it is uncommon in streams of the more mountainous regions of the Ozarks and Ouachitas (Map 183).

HABITAT AND BIOLOGY   The Black Crappie occupies a habitat similar to that of the White Crappie. It seems to be the dominant crappie in clear, vegetated, more acid waters (oligotrophic), while the White Crappie dominates in richer (more eutrophic), more turbid, alkaline waters. It is most abundant in large reservoirs, natural lowland lakes, and in the quiet backwaters of large rivers, where it is almost always found near fallen treetops, standing timber, logs, or other cover. At most localities it is less abundant than the White Crappie because it is less tolerant of turbidity and siltation than that species. It prefers cooler, deeper waters and does best in clear lakes with abundant aquatic vegetation. Pflieger (1997) reported that the Black Crappie is noticeably more abundant in the clear stream-fed arms of reservoirs. For example, in the Little North Fork arm of Bull Shoals Reservoir in Missouri, Black Crappie occurred almost to the exclusion of the White Crappie (Burress 1965). By day, except for the breeding season, Black Crappie form loose schools near submerged treetops or other cover 3–10 feet (0.9–3 m) deep. At night the schools often move out over deep water (Moyle 1969).

Most aspects of the feeding biology of the Black Crappie are similar to those of the White Crappie, but adult foraging patterns of the two species often diverge. White Crappies become increasingly piscivorous as they age. Adult Black Crappies also increase consumption of fishes but often utilize more macroinvertebrate prey than adult White Crappies (Ellison 1984; Andree and Wahl 2019). Young crappies first feed on zooplankton, gradually adding aquatic insects to their diet as they grow. The shift to piscivory occurs at a later age and larger size than for most centrarchids (Keast 1985a, b). Zooplankton often continues to form a part of the adult diet until a relatively large size is reached (160–200 mm TL) because of the numerous, long gill rakers (Warren 2009). Ellison (1984) found that energetics is largely behind the increasingly piscivorous diet, and as adult size increases, zooplankton alone cannot supply enough energy to sustain growth. Large Black Crappie are some of the most active nocturnal feeders among the centrarchids (Shoup et al. 2004). The adults feed mainly on fish, but the diet is also supplemented by aquatic insects, crustaceans, and occasionally terrestrial insects and zooplankton. The Black Crappie was initially more abundant than the White Crappie in the newly constructed Beaver Reservoir, and Ball and Kilambi (1972) attributed this to differences in their food habits. Black Crappie adults ate benthic insects in the spring and fish in other seasons, while White Crappies ate fish the year-round. Black Crappies in Beaver Reservoir occasionally ate terrestrial insects during the summer months. Ball and Kilambi (1972) believed that the apparent reduction in the number of insects and earthworms as the reservoir aged gave an advantage to the White Crappie. In contrast to some previous studies showing decreased growth of Black Crappies at high turbidities, Andree and Wahl (2019) found that juvenile crappies did not decrease prey consumption in turbid water and showed little change in prey type selection.

Sexual maturity is reached at 2–4 years of age and a minimum size of about 178 mm TL (Cooke et al. 2006). It is an early-spawning centrarchid, and ovaries begin developing in the fall. Spawning occurs in colonies in quiet water near cover, where the males defend territories and construct nests. The nests of Black Crappies often have more of a definite depression than those of the White Crappie (Hansen 1965). In a small urban impoundment in Illinois, Black Crappies nested in colonies, usually with more than 10 nests in close proximity (Phelps et al. 2009). In that Illinois study, Black Crappies selected spawning areas close to deep water with firm substrates and low vegetation height and density. Black Crappies in three Minnesota lakes were more likely to nest adjacent to undeveloped shoreline areas than along developed lakeshores (Reed and Pereira 2009). The optimum spawning temperature is between 64°F and 68°F (17.8–20.0°C). The eggs hatch in 2.3 days at a temperature of 17.4°C. Spawning behavior is presumably similar to that of the White Crappie. Females may spawn with several males and may produce eggs several times during the spawning period. A 246 mm TL female can potentially produce 54,225–82,751 mature eggs in a single batch (Baker and Heidinger 1994). The male guards the nest until the young become free-swimming; total length of male parental care is about 7–11 days (Cooke et al. 2006). Natural F₁ hybrids between *P. nigromaculatus* and *P. annularis* in Weiss Lake, Alabama, had superior growth and survival

and recruited into the fishery earlier than the parental species (Smith et al. 1994). The F₁ hybrids were 7 times more abundant than second generation and higher hybrids and were more likely to successfully mate and produce offspring with Black Crappie than with White Crappie. Life span in most areas is 6–8 years, but maximum life span is 13+ years. It is an important gamefish and a highly regarded food fish.

CONSERVATION STATUS This species is widely distributed and locally abundant in Arkansas. Its populations in the state are secure.

Family Elassomatidae *Pygmy Sunfishes*

Family Elassomatidae contains a single genus, *Elassoma*. The members of this North American family are restricted to the southern and southeastern United States, mainly in the Coastal Plain (one Alabama species occurs entirely above the Fall Line). There is no fossil record for the family, and phylogenetic relationships of this group above the level of genus have been controversial for more than 100 years. Some authors have placed the pygmy sunfishes in the sunfish family, Centrarchidae, and others have placed them in a separate family, Elassomatidae. Wiley and Johnson (2010) even considered them sufficiently different to form their own order, Elassomatiformes. Originally, Jordan (1877b) grouped the pygmy sunfishes into subfamily Elassominae without inclusion in any family group (Gilbert 2004). They were placed in Elassomidae (= Elassomatidae) by Hay (1881), and in 1896, Jordan and Evermann recognized Elassomatidae as a distinct family. For most of the 20th century, various authors treated pygmy sunfishes as either centrarchids (e.g., Berg 1947; Greenwood et al. 1966) or in a separate family that was not believed to be closely related to centrarchids (Branson and Moore 1962). Jones and Quattro (1999) supported recognition of Elassomatidae as a monophyletic sister group to the Centrarchidae based on mitochondrial DNA and ribosomal RNA evidence. A distinctive pattern of breeding (no nest is constructed) along with many morphological differences also support the placement of the pygmy sunfishes in a separate family; however, Near et al. (2012), using an analysis of nuclear DNA, concluded that *Elassoma* should be placed in Centrarchidae. Nelson et al. (2016) returned the pygmy sunfishes to Centrarchidae as subfamily Elassomatinae, a move that we do not support.

We believe that morphological and ecological differences provide sufficient justification for separating the sunfishes and pygmy sunfishes at the family level. Pygmy sunfishes can be distinguished from the centrarchids in having only 4 or 5 dorsal spines (vs. more than 5 dorsal spines), cycloid scales on the body (vs. ctenoid), in lacking a lateral line (vs. lateral line present), lacking a lateralis canal on the mandible (vs. mandibular lateralis canal present), lacking infraorbital bones except for the lachrymal (vs. infraorbital bones present including the lachrymal), and in having the branchiostegal membranes joined across the isthmus (vs. branchiostegal membranes separate). *Elassoma zonatum* possesses a spectacle covering the eye, providing a further difference from Centrarchidae (Moore and Sisk 1963; Taber 1965). Epidermal and dermal layers of the head skin pass, without folding, across the eyeball, with the eyeball able to move freely beneath the immovable spectacle. As the common name implies, these tiny dark fishes reach a maximum size of about 2 inches (50 mm). Most recent authors support recognition of the family Elassomatidae (Page and Burr 2011; Page et al. 2013; Betancur-R. et al. 2017; Robins et al. 2018), and some believe that it is not very closely related to the Centrarchidae.

Its phylogenetic relationships above the family level have also recently been questioned. Some authors suggested that it is not even a percomorph (series Percomorpha) group, but should be placed in a new series, Smegmamorpha, along with Synbranchiformes, Mugiliformes, Gasterosteiformes, and Atherinomorpha (Johnson and Patterson 1993; Johnson and Springer 1997). Near et al. (2012) strongly disagreed with this view and provided DNA evidence that Smegmamorpha is not a monophyletic group and opposed its recognition. There is also no published phylogenetic analysis of morphological data that supports monophyly of Smegmamorpha (Near et al. 2012). We follow Page et al. (2013) in supporting continued placement of the pygmy sunfishes in order Perciformes, but we support a different phylogenetic placement of family Elassomatidae. To reflect the recent genetic evidence supporting a close relationship with Centrarchidae (Near et al. 2012), we have abandoned recognition of suborder Elassomatoidei and placed Elassomatidae immediately after Centrarchidae in suborder Percoidei (Appendix 5).

Genus *Elassoma* Jordan, 1877b— Pygmy Sunfishes

The genus was established by Jordan in 1877. *Elassoma* contains seven species, one of which, the Banded Pygmy Sunfish (*E. zonatum*), is common in the lowlands of Arkansas, where it inhabits well-vegetated, sluggish swamps and backwaters of oxbow lakes. Genus characters are the same as those presented in the family account.

Elassoma zonatum Jordan
Banded Pygmy Sunfish

Figure 26.38. *Elassoma zonatum*, Banded Pygmy Sunfish. *Uland Thomas.*

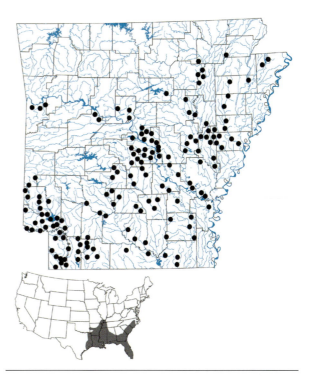

Map 184. Distribution of *Elassoma zonatum*, Banded Pygmy Sunfish, 1988–2019.

CHARACTERS   A tiny fish with a moderately elongate, laterally compressed body, a small terminal mouth, a slightly projecting lower jaw, and caudal fin rounded, not forked (Fig. 26.38). Body depth going into SL 3.7–4.4 times. Jaws with small, conical teeth. Head and body with cycloid scales; scales on cheeks and opercles mostly embedded, and no scales on top of head. Gill membranes broadly joined across throat. Gill rakers 6–8 on lower limb of first gill arch. Lateral line absent; scales in lateral series usually 31–36 (28–45). Diagonal scale rows 13 or 14; scale rows around caudal peduncle 21–24. Eye large, its diameter going into head length 3.0–4.5 times. Spinous and soft dorsal fins broadly joined, without a notch between them. Dorsal spines 4 or 5; dorsal rays 9 or 10 (8–11); anal spines 3; anal rays 5 or 6; pectoral rays usually 15 or 16; pelvic fin with 1 spine and usually 5 rays; principal caudal fin rays 13–16. Adults very small, about 1–1.5 inches (25–40 mm) in length. Maximum reported size 1.8 inches (44.6 mm) TL.

LIFE COLORS   Entire body a dark olive green to brown and stippled with many small dark specks. Side with 7–12 brown or black vertical bands and usually with 1 or 2 conspicuous, large black spots on side below front of dorsal fin. Soft dorsal, caudal, and anal fins banded with black. Breeding males develop dark jet-black vertical barring, or become almost solid black, with iridescent blue patches below the eye, on the opercle, and on base of pectoral fin; body sometimes with scattered gold flecks. Breeding females do not develop dark barring but become rather evenly dark purple brown in color (Taber 1965).

SIMILAR SPECIES   *Elassoma zonatum* superficially resembles very small sunfishes but differs from all members of the sunfish family in lacking a lateral line (vs. lateral line present), having 4 or 5 dorsal spines (vs. more than 5), and in having the body covered with cycloid scales (vs. ctenoid scales). It superficially resembles small Pirate Perch, but differs in having the anus situated immediately in front of anal fin (vs. anus jugular in position), 4 or 5 stiff dorsal spines (vs. 2 or 3 weak spines), and cycloid scales on head and body (vs. ctenoid scales). Conner (1979) provided characters for identifying larval specimens.

VARIATION AND TAXONOMY   *Elassoma zonatum* was described by Jordan (1877b) as *Elassoma zonata*. Jordan listed the type localities as Little Red River, White County, Arkansas, and Rio Brazos, Texas. Gilbert (2004) designated an Arkansas specimen (USNM 20496) as the lectotype and the Texas specimens as paralectotypes. No comprehensive study of geographic variation has been conducted. See Family Elassomatidae account for information on the controversy surrounding its taxonomic placement above the genus level.

DISTRIBUTION   *Elassoma zonatum* occurs in the Mississippi River basin from southern Illinois and the Wabash River floodplain of southern Indiana, south through the Mississippi Embayment to the Gulf of Mexico; in Atlantic Slope drainages from the Roanoke River, North Carolina, to the St. Johns River, Florida; and in Gulf Slope drainages from approximately the Suwannee

River drainage of Florida, west to the Brazos River, Texas. In Arkansas, the Banded Pygmy Sunfish is widely distributed in all major drainages of the Coastal Plain lowlands in eastern and southern Arkansas and is rarely found above the Fall Line (Map 184). It also occurs in floodplain habitats throughout the Arkansas River Valley in Arkansas; the first records from Sebastian County were found by Brian Wagner from three localities a few miles east of Fort Smith in 2007. Buchanan (2005) reported this species from nine Arkansas impoundments.

HABITAT AND BIOLOGY   This tiny fish is a lowland species inhabiting swamps, bayous, sluggish creeks, and oxbow lakes, It is found in quiet, clear, heavily vegetated waters over a mud and detritus bottom in shallow water with protective cover (logs, stumps, cypress knees, and overhanging banks). In Black Creek, Mississippi, it occurred in habitats with an average current speed of 2.3 cm/s (0.9 inches/s), an average depth of 49 cm (19.3 inches), moderate amounts of vegetation or other cover, and fine substrates (Ross 2001). It is a secretive, solitary species with no schooling tendencies. In Kentucky, it was usually taken alone or with species frequenting shoreline areas, such as sunfishes, topminnows, and mosquitofish (Walsh and Burr 1984). In that study of Mayfield Creek Swamp, *E. zonatum* was taken in all areas except open, deep-water habitats lacking vegetation or other cover. Estimates of densities of *E. zonatum* in Mayfield Creek Swamp ranged from a low of 3.8 fish/m^3 in October to a high of 30.4 fish/m^3 in June, and an intermediate density of 11.5 fish/m^3 in February.

Barney and Anson (1920) first studied the life history of *Elassoma zonatum* in Louisiana. Walsh and Burr (1984) investigated its life history in western Kentucky and summarized the pertinent literature. It is an opportunistic invertivore, feeding at all levels of the water column. It is a diurnal sight feeder, ingesting only living, moving prey. Food consists largely of microcrustaceans (mainly copepods and cladocerans), supplemented by midge larvae and pupae, other insects, and small snails and clams. Adults and juveniles have similar diets, but juveniles use smaller items such as copepod nauplii, rotifers, and small ostracods (Ross 2001). Other small prey are consumed when locally abundant. The presence of adult mosquitoes and springtails in the diet indicated that it occasionally feeds at the water's surface (Walsh and Burr 1984). Seasonal variation in diet probably reflects seasonal fluctuations in the invertebrate fauna.

Spawning occurs from mid-March to early May in Louisiana and Kentucky. Taber (1965) found spawning occurring on 2 March in southeastern Oklahoma, and spawning occurs in March and April in Alabama (Mettee et al. 1996). In Arkansas, spawning likely occurs from March through May. Taber (1965) and Walsh and Burr (1984) gave detailed descriptions of courtship and spawning behavior in aquaria. Aquarium spawning occurred at water temperatures between 21°C and 24°C (Taber 1965; Walsh and Burr 1984), but Walsh and Burr (1984) cited evidence that spawning occurs in natural environments at lower temperatures. Mettee (1974) also described aquarium spawning and provided information on embryology and larval development. Unlike true sunfishes (Centrarchidae), the Banded Pygmy Sunfish prepares no nest. The males defend territories and mate several times with the same or different females. Courtship behavior by the male involves erecting the fins, alternately erecting and lowering the pelvic fins, bobbing, and lateral fin undulations (Walsh and Burr 1984). A receptive female is then nudged near her vent area, and the expelled eggs are fertilized by the male. Spawning occurs in or above submerged vegetation, with the demersal, adhesive eggs becoming attached to the vegetation. Females spawn repeatedly with the same or different males at intervals of several hours to several days (Mettee 1974). Echelle and Echelle (2005) observed previously unreported spawning behaviors, including a striking change in male coloration during late courtship and spawning, stereotyped "pointing" behavior by the male that resembled nest showing in gasterosteid fishes, and a jerking behavior that occurs with sperm release. After spawning, the female swims away or is driven from the territory by the male, who guards the developing embryos for up to 48 hours. Eggs hatched in 7 days at a water temperature of 18.5°C (65°F) (Barney and Anson 1920), and in 4–4.8 days at 21°C (70°F) (Walsh and Burr 1984). Individual females may contain 96–970 eggs, and may spawn 6–76 eggs per spawning event (Barney and Anson 1920). Sexual maturity is attained at 1 year of age, and the maximum life span is 3 years in Louisiana, and just over 2 years in Kentucky. Both sexes in Kentucky reached sexual maturity at 10 to 12 months of age, and the vast majority of the population did not live to spawn a second year; just over 5% of the individuals in Kentucky were in their second year (Walsh and Burr 1984).

CONSERVATION STATUS   The populations of this widely distributed species are secure in Arkansas. Its preferred habitat is common in the lowlands, but large populations of Banded Pygmy Sunfishes are rare. Because of extensive draining of wetlands in the Coastal Plain, preservation of its wetland habitat is important to maintaining its widespread distributional pattern in the future.

Family Percidae *Perches and Darters*

The members of this family are distinguished by overlapping ctenoid scales on the body; two distinct dorsal fins that are separate or narrowly joined together, the first with 6 or more spines, the second with fewer than 23 soft rays; the pelvic fins thoracic in position; and 1 or 2 anal spines (the second anal spine is usually weak). The perches and darters are a large, relatively recent family, with most diversification occurring during the Late Oligocene and Pliocene (Near et al. 2011). Percids are widely distributed throughout the North Temperate zone in North America (more than 250 species) and Eurasia (16 species) in Atlantic and Arctic drainages. The family contains 11 genera, 7 found in Arkansas. Fifty-two percid species occur in Arkansas, and there are also at least three (possibly six or more) undescribed species in the state. Only two species native to Arkansas, the Walleye and Sauger, are large enough to be sought by anglers. Those species have counterparts in Europe. The Yellow Perch, a native of northern North America, but recently introduced and established in Arkansas, is also large enough to be considered a gamefish. The remaining 49 species (in five genera) inhabiting Arkansas are known as darters. Darters (Percidae: subfamily Etheostomatinae) represent a diverse clade of freshwater percomorph fishes endemic to eastern North America and estimated by Near et al. (2011) to contain about 250 species (including known undescribed forms).

The taxonomy of percids below the family level has been controversial, and various authors have supported different subfamilial groupings. Collette and Bănărescu (1977) presented proposed relationships of the subfamilies, tribes, and genera, and divided the family into two subfamilies: Percinae (with genera *Perca, Gymnocephalus, Percarina, Percina, Ammocrypta,* and *Etheostoma*) and Luciopercinae (with genera *Stizostedion, Zingel,* and *Romanichthys*). Page (1985) constructed a percid phylogeny based on reproductive behavior and also divided the Percidae into two subfamilies: Percinae and Etheostomatinae. Page placed *Perca, Percarina,* and *Gymnocephalus* in Percinae and divided Etheostomatinae into three tribes: Luciopercini (*Stizostedion*), Etheostomatini (*Percina, Ammocrypta,* and *Etheostoma*) and Romanichthyini (*Romanichthys* and *Zingel*). Wiley (1992) also divided the Percidae into two subfamilies, Percinae and Etheostomatinae, but placed *Perca* in the subfamily Percinae and all other percids into the subfamily Etheostomatinae. Wiley considered *Stizostedion* the sister group to *Zingel, Romanichthys, Crystallaria, Percina,* and *Etheostoma,* elevated the subgenus *Crystallaria* to a monotypic genus, and placed the remaining sand darters (*Ammocrypta*) in the genus *Etheostoma*. Near and Keck (2005) elevated the subgenus *Nothonotus* (of genus *Etheostoma*) to generic status, a designation that we support. Song et al. (1998) revised Percidae into three subfamilies on the basis of mitochondrial cytochrome *b* DNA sequence data: Percinae (*Perca, Percarina, Gymnocephalus*); Luciopercinae (*Stizostedion, Zingel, Romanichthys*); and Etheostomainae (*Ammocrypta, Crystallaria, Etheostoma* and *Percina*). Song et al. (1998) also removed *Ammocrypta* species from *Etheostoma* and placed them in their own genus separate from *Crystallaria*. Buth (2010) pointed out that mitochondrial DNA as a symbiont is extrinsic to its host nuclear DNA and is inappropriately used as a proxy for the latter (Bruner 2011). Buth also stated that mtDNA should be rechecked using intrinsic characters. Bruner (2011) divided the family Percidae into five subfamilies, with the focus on the Luciopercinae (Walleye, Sauger, Zander, and related species) mainly using osteological (intrinsic) characters, and Nelson et al. (2016) followed Bruner's classification. Bruner (2011) also provided a cogent explanation of why the use of Sander for *Stizostedion* by Kottelat (1997), Nelson et al. (2003, 2004), Page et al. (2013), and others is not correct (a view that we share).

We recognize the following five genera of darters (with approximate number of described species in parenthesis): *Ammocrypta* (6), *Crystallaria* (2), *Etheostoma* (145–150, depending on source), *Nothonotus* (20), and *Percina* (46–49). Nelson et al. (2016) recognized only three darter genera, *Ammocrypta* (8), *Etheostoma* (156), and *Percina* (46). The genus *Etheostoma* is the most diverse genus of freshwater fishes in North America (Page and Burr 2011). Species diversity of darters is concentrated in the Interior Highlands and Eastern Highlands, and darter communities often contain multiple species that occur in sympatry (Page 1983a; Mayden 1988b). Only the minnows (Cyprinidae), with about 320 species, are more diverse in North America. Darters partition habitat mostly by water column position, substrate type, and flow regime; however, those associations can change during breeding seasons (Winn 1958a, b; Greenberg 1991; Welsh and Perry 1998). Most species of darters are largely benthic-adapted and have secondarily lost the swim bladder. Although not all species of darters have been examined for the presence of a swim bladder, Evans and Page (2003) found that a swim bladder occurred in 15 of 17 species of *Percina* species, but was absent in the species of *Ammocrypta, Crystallaria*, and *Etheostoma* examined.

Darters exhibit considerable variation in reproductive strategies (Kelly et al. 2012; Harrington et al. 2012). Darter mating systems include simple behaviors such as egg scattering and clumping and more complex patterns of parental nest and egg guarding (Page 1985; Kelly et al. 2012). Three general egg deposition behaviors are known in darters: (1) egg buriers bury clumps of eggs in loose substrate with no subsequent parental care; (2) egg attachers attach 1–3 eggs at a time to various substrates with no subsequent parental care; and (3) the most complex behavior involves males defending a crevice under a suitable structure where females enter and deposit eggs and the male tends the fertilized eggs until hatching (Page 1985, 2000). Hybridization is common in darters, with more than 25% of species of darters hybridizing with at least one other species, compared with approximately 10% in other animal groups (Keck and Near 2009; Mallet 2005). Additionally, 12.5% of darter species have been identified as having introgressed mitochondrial genes (Near et al. 2011). Keck and Near (2009) noted that most hybrid crosses involved egg buriers. The mtDNA introgression is most often asymmetric, and there is typically little or no associated introgression of nuclear-encoded alleles (Keck and Near 2010). Arkansas darter species identified as containing genomes of heterospecific origin are *Etheostoma artesiae, E. asprigene, E. autumnale, E. caeruleum, E. collettei, E. fragi, E. mihileze, E. radiosum, E. uniporum*, and *E. whipplei* (Near et al. 2011).

Hubbs (1985) studied darter reproductive seasons and found that populations of a given species have longer breeding seasons in the south than in the north. Texas species often spawn for most of the year, whereas most Arkansas darters spawn for two to three months. Populations north of Arkansas rarely spawn more than two months. Within most species there is little difference in the date of reproductive termination in different geographic areas, but differences in date of initiation of reproduction are largely responsible for the different breeding season lengths at different latitudes. Darters living in stenothermal environments tend to have longer breeding seasons than those in eurythermal ones.

The males of many darter species develop intense breeding coloration, presumably as a result of sexual selection, but it is also generally assumed that there are antagonistic forces of natural selection favoring cryptic coloration. Bossu and Near (2015) concluded that the evolution of sexual dichromatism is coupled to habitat use in darter species. They found that midwater darter lineages exhibit a narrow distribution of dichromatism, possibly resulting from predator-mediated selection, while the transition to benthic habitats coincides with greater variability in the levels of dichromatism. Bossu and Near (2015) suggested a complex interaction of sexual selection with potentially two mechanisms of natural selection (predation and sensory drive) influencing the evolution of diverse male nuptial coloration in darters.

Most darters are benthic invertivores, with food habits within a species determined largely by availability of prey and size of the darter. Resource partitioning among species involving differences in preferred foraging location, jaw morphology, and microhabitat preferences often permit high levels of syntopy of darter species (Carlson et al. 2009; Ciccotto and Mendelson 2015).

Key to the Perches and Darters

1A Rear margin of preopercle strongly serrate; mouth large, upper jaw extending beyond middle of eye; branchiostegal rays usually 7 (sometimes 8).   2

1B Rear margin of preopercle smooth or very weakly serrate; mouth smaller, upper jaw usually not extending past middle of eye; branchiostegal rays usually 6.   4

2A Jaws without canine teeth; anal rays 6–8.
Perca flavescens Page 718

2B Jaws with prominent canine teeth; anal rays 10–14.
(*Stizostedion*) 3

3A Spinous dorsal fin with large black blotch on membranes at posterior base; membranes of spinous dorsal fin with dark streaks or mottlings; soft dorsal fin with 19–22 rays; cheeks with few or no scales.
Stizostedion vitreum Page 755

3B Spinous dorsal fin without large black blotch at posterior base; membranes of spinous dorsal fin with horizontal rows of distinct dark spots; soft dorsal fin with 17–20 rays; cheeks usually partially to well scaled.
Stizostedion canadense Page 753

4A Body very long and slender, its depth going into standard length 7 times or more; belly without scales; anal fin with 1 spine.   5

4B Body deeper, its depth going into standard length fewer than 7 times; belly partially or completely scaled; anal fin usually with 2 spines (2 species of *Etheostoma*, *E. chlorosoma* and *E. nigrum*, have only 1 anal spine). 7

5A Lateral line scales more than 80; premaxillary frenum present; dorsal spines usually 13–15; dorsal rays about 12–16; anal fin with 12–14 soft rays; back usually with 4 dark saddles that extend forward and downward on sides to the lateral line. *Crystallaria asprella* Page 638

5B Lateral line scales fewer than 80; premaxillary frenum absent; dorsal spines usually 7–11; dorsal rays 9–11; anal fin with 8–10 soft rays; back with a series of dark spots or narrow dusky blotches that do not extend onto sides.   6

6A Opercle with a long, prominent, pinlike spine near rear margin, length of spine much greater than its width at base; body with scales only on caudal peduncle and forward along midside; side without dark blotches, or with a row of indistinct, horizontally elongated blotches.
Ammocrypta clara Page 634

6B Opercle with a short, less prominent, triangular spine near its rear margin, length of spine about equal to its width at base; body well scaled, except for belly and breast; side with a row of well-developed vertically elongated dark blotches. *Ammocrypta vivax* Page 636

7A One or more enlarged and strongly toothed scales between pelvic fin bases (Fig. 26.39A); midline of belly usually without scales or with row of enlarged, toothed scales (females of some species have scales of normal size and shape); lateral line always complete to base of caudal fin; anal fin usually almost as large as soft dorsal fin.
(*Percina*) 8

7B No enlarged and modified scales between pelvic fin bases (Fig. 26.39B); midline of belly with normal scales which are not enlarged (belly sometimes partially naked); lateral line complete or incomplete; anal fin usually much smaller than soft dorsal fin.
(*Etheostoma* and *Nothonotus*) 21

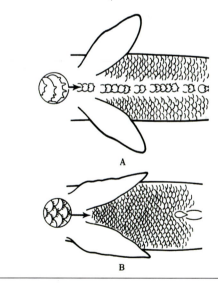

Figure 26.39. Differences in scales on the belly and width between bases of the pelvic fins in darters. *Cross (1967)*.

A—Belly with a medial row of modified scales that are sometimes lost, leaving a naked strip; pelvic fins separated by a space about as wide as the basal length of each pelvic fin (genus *Percina*).

B—Belly with scales like those on sides (sometimes partly naked anteriorly, but never with a medial scaleless strip); pelvic fins separated by a space less than the basal length of each fin (genera *Etheostoma* and *Nothonotus*).

8A Side with at least 15 dark, narrow, vertical bars alternating in length and producing a striped lateral pattern; total spinous and soft dorsal fin elements 30 or more. Logperches. 9

8B Side with various patterns but never with 15 vertical stripes; total spinous and soft dorsal fin elements 29 or fewer. 11

9A Snout only slightly fleshy, not protruding prominently beyond upper lip in juveniles or adults (Fig. 26.40A); usually 1 or more scales on breast; dark vertical suborbital bar usually absent.

Percina macrolepida Page 732

9B Snout conical and fleshy, protruding prominently beyond upper lip in individuals larger than 65 mm (Fig. 26.40B); scales absent on breast; dark vertical suborbital bar present. 10

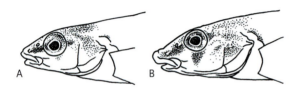

Figure 26.40. Snout differences between adult *Percina macrolepida* and *P. caprodes/P. fulvitaenia*. David A. Neely.

A—*Percina macrolepida*
B—*Percina caprodes/fulvitaenia*

10A Broad orange or yellow submarginal band present in first dorsal fin of adult males and often females.

Percina fulvitaenia Page 730

10B No orange or yellow submarginal band in first dorsal fin. *Percina caprodes* Page 723

11A Gill membranes moderately to broadly connected across throat (Fig. 26.41A); distance from front of upper lip to junction of gill covers equal to or greater than distance from junction of gill covers to back of pelvic fin base. 12

11B Gill membranes not connected or only slightly joined across throat (Fig. 26.41B); distance from front of upper lip to junction of gill covers much less than distance from junction of gill covers to back of pelvic fin base. 15

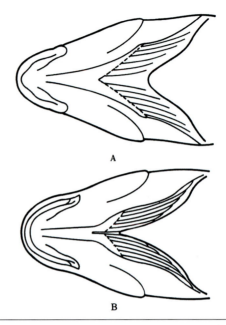

Figure 26.41. Gill membrane connections of darters.

A—Underside of head, showing gill membranes broadly connected.
B—Underside of head, showing gill membranes not broadly connected.

12A Base of caudal fin with a vertical row of 3 dark spots (lower 2 usually confluent); rear margin of preopercle finely serrate; first dorsal fin without orange submarginal band in life. *Percina sciera* Page 743

12B Base of caudal fin with 1 distinct medial black spot; rear margin of preopercle smooth, not serrate; first dorsal fin with orange submarginal band in life. 13

13A Snout depth greater than snout length; snout length less than 7.5% of standard length and less than 26% of head length; eye width more than 70% of snout length; lateral line scales usually fewer than 72.

Percina phoxocephala Page 741

13B Snout depth less than snout length; snout length more than 7.5% of standard length and more than 26% of head length; eye width less than 70% of snout length; lateral line scales usually more than 72. 14

14A Snout very long, its length 9% or more of standard length; snout length more than 5.7% of postorbital fish length; usually fewer than 7 lateral blotches between the insertion of the second dorsal fin and the hypural plate. *Percina nasuta* Page 737

14B Snout moderately long, its length less than 9% of standard length (7.5–9%); snout length less than 5.7% of postorbital fish length; usually 7 or more lateral blotches between the insertion of the second dorsal fin and the hypural plate.

Percina brucethompsoni Page 720

15A Upper lip bound to snout by a well-developed, wide frenum. 16

15B Upper lip separated from snout by a groove, or if present, frenum narrow and weakly developed. 18

16A Lateral line scales 81 or more; midside with a series of 10–14 round black spots, with smaller spots above; scales around caudal peduncle 28 or more.

Percina pantherina Page 739

16B Lateral line scales fewer than 81; midside with a series of 6–9 oval or rectangular blotches which are never round and without small spots above them, but with variously shaped markings; scales around caudal peduncle fewer than 28. 17

17A Dark blotches on side horizontally elongated and confined to a medial row; cheek partly or fully scaled; dorsal spines usually 13–15; base of caudal fin with small distinct black spot; underside of head not red orange in life. *Percina maculata* Page 735

17B Dark blotches on side vertically elongated and continuous with saddles on back; cheek naked; dorsal spines usually 12; base of caudal fin with dark spot indistinct or absent; underside of head red orange in life.

Percina evides Page 728

18A Anal fin rays usually 8 or 9 (rarely 7 or 10); scales around caudal peduncle 18 or fewer; anal fin of adult male not greatly elongated, similar in size to that of female. *Percina copelandi* Page 725

18B Anal fin rays 10–13; scales around caudal peduncle 19 or more; anal fin of adult male greatly elongated, extending nearly to base of caudal fin. 19

19A First dorsal fin with a small black blotch at anterior base and a large black blotch at posterior base; back without prominent dark saddles, but 5–9 obscure blotches are sometimes evident.

Percina shumardi Page 746

19B First dorsal fin with or without a small black blotch at anterior base but always without a black blotch at posterior base; back with 4 or 5 faint to dark saddles. 20

20A Back yellow to olive with 5 dark saddles that do not extend downward to connect with the midlateral row of dark blotches; 4th dorsal saddle separated by 1 or more scales from the keel-like upper margin of caudal fin; a small distinct dark spot usually present at base of caudal fin; first dorsal fin without a small black blotch at anterior base. *Percina vigil* Page 750

20B Back brown to olive with 4 dark saddles that extend downward and connect with dark midlateral blotches; 4th dorsal saddle touching keel-like upper margin of caudal fin; small dark spot at base of caudal fin indistinct or absent; first dorsal fin usually with a small, black blotch at anterior base.

Percina uranidea Page 748

21A Lateral line absent or with 8 or fewer pored scales anteriorly. 22

21B Lateral line complete or incomplete, but with more than 8 pored scales anteriorly. 23

22A Lateral line entirely absent or with 1 or 2 pored scales anteriorly; cheeks naked; opercles naked or with a few scales; upper jaw extending to or slightly behind front of eye; branchiostegal rays 5.

Etheostoma microperca Page 674

22B Lateral line very short, usually having only 2–4 pored scales anteriorly; cheeks fully scaled; opercles scaled; upper jaw extending distinctly behind front of eye; branchiostegal rays 6.

Etheostoma proeliare Page 685

23A Anal fin with 1 thin spine (first anal ray often unbranched and spinelike, but jointed). 24

23B Anal fin with 2 thick spines. 25

24A Lateral line incomplete, usually extending to below soft dorsal fin; dark stripe extending forward from each eye continuous around snout above upper lip (Fig. 26.42A); dorsal fins widely separated; cheeks scaled; soft dorsal fin usually with 10 or 11 rays.

Etheostoma chlorosoma Page 653

24B Lateral line complete or nearly so, extending to near base of caudal fin; dark stripe extending forward from each eye not continuous around snout but interrupted at midline (Fig. 26.42B); dorsal fins only slightly separated; cheeks usually naked; soft dorsal fin usually with 12 or 13 rays. *Etheostoma nigrum* Page 678

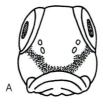 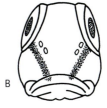

Figure 26.42. Differences in snout pigmentation between *Etheostoma chlorosoma* and *E. nigrum*.

A—*E. chlorosoma*

B—*E. nigrum*

25A Lateral line incomplete and distinctly arched upward anteriorly, usually with only 3 or 4 scale rows between lateral line and front of base of first dorsal fin.   26

25B Lateral line complete or incomplete but always straight and not arched upward anteriorly, usually with more than 4 scale rows between lateral line and front of base of first dorsal fin.   27

26A Breast and top of head naked or with a few scattered scales; infraorbital canal uninterrupted, usually with 8 pores; sides with vertical green bars or rectangles in life; first dorsal fin with a red medial band and a blue marginal band.   *Etheostoma gracile* Page 670

26B Breast and usually top of head fully scaled; infraorbital canal interrupted, usually with 4 or 5 pores; sides without vertical green bars or rectangles in life; first dorsal fin without red or blue bands.
    Etheostoma fusiforme Page 668

27A Lateral line complete, extending to caudal fin base or missing only 1 or 2 scales in front of the caudal fin base.   28

27B Lateral line incomplete, never extending to within 1 or 2 scales in front of the caudal fin base.   33

28A Gill membranes separate or only narrowly joined across throat, distance from membrane notch to tip of lower lip less than distance from notch to front of pelvic fin base; nape naked.
    Nothonotus moorei Page 714

28B Gill membranes broadly joined across throat, distance from membrane notch to tip of lower lip greater than distance from notch to front of pelvic fin base; nape scaled.   29

29A Midside with a row of 5–8 double bars forming H-, W-, or U-shaped markings; no groove separating skin of snout from back end of upper jaw.
    Etheostoma blennioides Page 648

29B Midside plain or with bars but never with H-, W-, or U-shaped markings; a long, horizontal groove separating skin of snout from back end of upper jaw.   30

30A Back with 1–4 distinct, dark brown saddles.   31

30B Back with 6 or 7 dark brown or green saddles, often indistinct.   32

31A Large dark yoke or saddle on back in front of spinous dorsal fin extending downward to pectoral fin base, followed on back by 3 less-prominent saddles; breast naked; anal fin rays 7 or 8.
    Nothonotus juliae Page 712

31B Four large, equally prominent dark saddles on back extending obliquely forward down side of body; breast at least partly scaled; anal fin rays 9–11.
    Etheostoma euzonum Page 662

32A Two large, dark brown blotches at base of caudal fin; breast and underside of head with large dark spots; cheeks, opercles, and front half of belly naked; spinous dorsal fin usually with 9 or 10 spines; pectoral fins long and expansive, extending backward far beyond tips of pelvic fins.
    Etheostoma histrio Page 672

32B Caudal fin base without 2 large dark blotches; breast and underside of head without large dark spots; cheeks, opercles, and belly well scaled; spinous dorsal fin usually with 11 or 12 spines; pectoral fins of moderate length, extending backward to or only slightly behind tips of pelvic fins.
    Etheostoma zonale Page 709

33A Premaxillary frenum absent, upper lip completely separated from snout by a continuous deep groove (Fig. 26.43A); upper side of body with X- or W-shaped markings.   *Etheostoma stigmaeum* complex 34

33B Premaxillary frenum present, upper lip joined to snout at midline by a bridge of skin (Fig. 26.43B); upper side of body without X- or W-shaped markings.   36

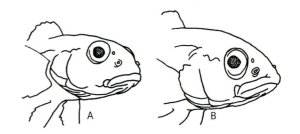

Figure 26.43. Darter frena.

A—Premaxillary frenum absent, e.g., *Percina shumardi*.
B—Premaxillary frenum present, e.g., *Etheostoma autumnale*.

34A Palatine teeth absent (>77% of specimens); nape fully scaled or nearly so; breeding male with a wide black submedial band in first dorsal fin; soft dorsal and caudal fins of breeding male with discrete orange spots.
    Etheostoma teddyroosevelt Page 703

34B Palatine teeth present (>85% of specimens) (Fig. 26.44); nape naked to fully scaled; breeding male with a wide blue or blue-green submedial band in first dorsal fin; soft dorsal and caudal fins of breeding male without orange spots.   35

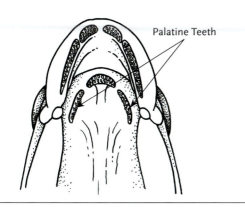

Figure 26.44. Palatine teeth in roof of mouth (arrows) of *Etheostoma stigmaeum*. David A. Neely.

35A   Preoperculomandibular pores 10 (>75% of specimens); anal rays 8; breeding males without a thin, dusky midlateral stripe running through lateral blue-green bars.   *Etheostoma stigmaeum* Page 701

35B   Preoperculomandibular pores 9 (>90% of specimens); anal rays 9; breeding males with a thin, dusky midlateral stripe running through lateral blue-green blotches.   *Etheostoma clinton* Page 655

36A   Gill membranes moderately to broadly joined across throat, distance from membrane notch to tip of lower lip greater than distance from notch to front of pelvic fin base.   37

36B   Gill membranes separate or narrowly joined across throat, distance from membrane notch to tip of lower lip less than distance from notch to front of pelvic fin base.   41

37A   Lateral line complete or nearly so, extending well behind second dorsal fin to near base of caudal fin; lateral line scales often without dark pigment near pores, producing a thin light-colored stripe along midside; no prominent black humeral spot.
   Etheostoma parvipinne Page 683

37B   Lateral line incomplete, usually ending beneath second dorsal fin, but never extending to near base of caudal fin; lateral line scales uniformly pigmented, not producing a pale stripe along midside; prominent black humeral spot present.   38

38A   First dorsal fin usually with fewer than 10 spines and less than half the height of second dorsal fin; side usually with dark horizontal streaks.
   Etheostoma flabellare Page 664

38B   First dorsal fin usually with 10 or more spines and nearly equal in height to second dorsal fin; side without dark horizontal streaks.   39

39A   Sides with red or yellow spots on a mottled or

reticulated background, sometimes with a few vertical bars posteriorly; opercle partly to fully scaled; pored lateral line scales usually 43 or more.   40

39B   Sides without red or yellow spots, but with a midlateral row of 8–12 dark blotches that usually form vertical bars posteriorly; opercle naked or with a few scales; pored lateral line scales usually fewer than 43.
   Etheostoma radiosum Page 694

40A   Usually 61 or fewer scales in lateral series; 26 or fewer caudal peduncle scale rows.
   Etheostoma artesiae Page 642

40B   More than 61 scales in lateral series; more than 26 caudal peduncle scale rows.
   Etheostoma whipplei Page 707

41A   Scales in lateral series 58 or more; second dorsal fin with 14–16 soft rays.   42

41B   Scales in lateral series fewer than 58; second dorsal fin with 10–13 (rarely 14) soft rays.   43

42A   Lateral scale rows usually 70–74; dorsal saddles 5 or 6; anal rays usually 9; infraorbital pores 8.
   Etheostoma autumnale Page 646

42B   Lateral scale rows usually 59–67; dorsal saddles 4; anal rays usually 8; infraorbital pores 9 or 10.
   Etheostoma mihileze Page 676

43A   Head and body usually sprinkled with small black flecks; lateral line usually ending beneath first dorsal fin and having fewer than 20 pored scales.   44

43B   Head and body not sprinkled with small black flecks; lateral line usually ending beneath second dorsal fin and having 20 or more pored scales.   45

44A   Back with a wide, medial, pale stripe; confined to Ouachita River system.
   Etheostoma pallididorsum Page 680

44B   Back dark brown, without a pale medial stripe; confined to Arkansas River system.
   Etheostoma cragini Page 659

45A   Infraorbital canal complete, uninterrupted; pectoral fin rays usually 13–15.   46

45B   Infraorbital canal interrupted; pectoral fin rays usually 11 or 12. (Orangethroat Darter Clade, *Ceasia* [Near et al. 2011])   48

46A   Anal fin usually with 6 or 7 soft rays; cheek usually naked, sometimes with a few small scales near eye; side usually with more than 8 vertical bars.
   Etheostoma caeruleum Page 650

46B   Anal fin usually with 8 or more soft rays; cheek fully scaled, scaled only on lower half, or naked in a few populations; side usually with 5–8 vertical bars.   47

47A   Back with 8 or 9 saddles, 3 or 4 saddles are much darker than the others; humeral spot small and dark;

prepectoral area naked; spinous dorsal fin of breeding male with a wide reddish-orange band; posterior side with 5–7 bars indistinct above lateral line.

Etheostoma collettei Page 657

47B Back with 8 or 9 dark saddles of about equal intensity; humeral spot faint or absent; prepectoral area scaled; spinous dorsal fin of breeding male with a narrow reddish-orange band; posterior side with 6–8 bars distinct above lateral line.

Etheostoma asprigene Page 644

48A 1 pore in the posterior segment of the infraorbital canal; supratemporal canal usually interrupted.

Etheostoma uniporum Page 705

48B Usually 3 pores in the posterior segment of the infra-orbital canal; supratemporal canal usually uninter-rupted.   49

49A Cheek with 5 or more rows of scales; breeding males with orange chevrons crossing the belly and without breeding tubercles.   *Etheostoma fragi* Page 667

49B Cheek with 1 or 2 rows of scales below or behind eye; breeding males without orange chevrons crossing the belly and with breeding tubercles on pectoral, pelvic, anal, and caudal fins and on ventral scales.   50

50A Blue is the prevalent color of dorsal fins, or blue and orange equal in area of development; restricted to upper White River drainage upstream from conflu-ence of White and Black rivers.

Etheostoma sp. *cf. spectabile* Page 696

50B Orange is the prevalent color of dorsal fins, suprabasal blue band of spinous dorsal thin and incomplete or absent; not found in upper White River drainage. 51

51A Top of head usually scaled; 3 dorsal saddles darker than others; vertical bars on sides usually less dense or lacking on upper side; distinct development of hori-zontal lines on upper and lower sides.

Etheostoma squamosum Page 698

51B Top of head usually naked; all dorsal saddles usually of equal intensity; vertical bars on sides usually of equal intensity on upper and lower sides; horizontal lines indistinct on sides, or if present, developed only near the lateral line (both above and below lateral line). 52

52A From Red River drainage of southwestern Arkansas.

Etheostoma pulchellum Page 687

52B Not from Red River drainage.   53

53A Breeding male with red or red-orange belly; from Arkansas River tributaries from Illinois Bayou west-ward into Oklahoma.

Etheostoma sp. *cf. pulchellum* 1 Page 689

53B Breeding male with blue belly; from Arkansas River tributaries from Illinois Bayou eastward and Des Arc Bayou of White River drainage.

Etheostoma sp. *cf. pulchellum* 2 Page 691

Genus *Ammocrypta* Jordan, 1877b— Sand Darters

Ammocrypta was first used as a genus name by Jordan in 1877. This genus contains six described species of small (less than 70 mm SL), slender, pallid or clear-bodied fishes usually associated with sandy substrates, giving them the common name of sand darters. They typically inhabit medium to large streams and are noted for burying them-selves in the sandy substrate with only their eyes and snout exposed. Controversy has existed about the relationships of sand darters to other darter genera, and various species rela-tionships within *Ammocrypta* have been proposed. Wiley and Hagen (1997) considered *Ammocrypta* a subgenus of *Etheostoma*. The subgeneric placement of *Ammocrypta* within *Etheostoma* was not widely accepted (Page et al. 2013), and a genetic analysis by Near et al. (2011) supported the monophyly of *Ammocrypta* as a taxon equivalent to other darter genera. We agree with the continued recogni-tion of *Ammocrypta* as a genus. There are two species of *Ammocrypta* in Arkansas.

Genus characters include a very slender, terete body, fre-num absent, caudal fin truncate or emarginate, and usually 6 branchiostegal rays. The sensory canals of the head are uninterrupted, the lateral line is complete, there is 1 anal fin spine, there are no modified scales on breast or belly, and the swim bladder is absent. Adults are smaller than 3 inches (76 mm) TL.

Ammocrypta clara Jordan and Meek
Western Sand Darter

Figure 26.45. *Ammocrypta clara*, Western Sand Darter. *David A. Neely.*

CHARACTERS   A pale, very slender darter with a large opercular spine and with lateral and middorsal blotches weakly developed or absent (Fig. 26.45). Opercular spine

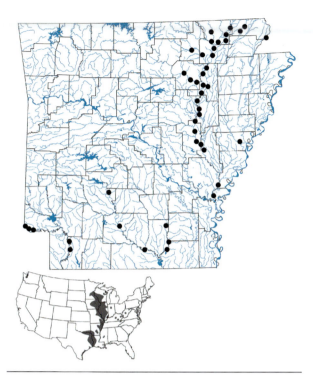

Map 185. Distribution of *Ammocrypta clara*, Western Sand Darter, 1988–2019.

long and prominent, its length much greater than the width of its base. Body almost cylindrical, body depth going into SL more than 7 times. Premaxillary frenum absent. Posterior margin of preopercle smooth. Body largely unscaled except for scales on the caudal peduncle and a narrow band of scales bordering the lateral line along the midside. Caudal fin only slightly forked. Cheek and opercle partly scaled; nape naked or nearly so; breast and belly naked. Gill membranes separate or narrowly joined across throat. Infraorbital canal complete. There is little sexual dimorphism, but males have longer dorsal and anal fins than females (Driver and Adams 2013). Lateral line complete, not arched anteriorly, and strongly deflected downward posteriorly (usually the last 3–5 scales) with 66–78 scales. Scale rows above lateral line 0–2; scale rows below lateral line 1–5. Diagonal scale rows usually 5 or 6 (3–12); scale rows around caudal peduncle (when completely scaled) 21–26. Dorsal spines 9–13; dorsal rays 9–13; anal spine 1; anal rays 8–11. Breeding males with tubercles on anal, caudal, and pelvic fins. Maximum length about 2.8 inches (70 mm), but most Arkansas specimens are smaller than 2 inches (50 mm) SL.

LIFE COLORS   A translucent fish with a faint yellow back and silvery white belly. Cheek and opercle may be

iridescent green. Twelve to 16 small dusky spots on mid-dorsum; sometimes with a row of small, dark rectangular blotches along the side, but midlateral coloration often lacking. Lateral scale margins above lateral line often outlined in black. Median fins sometimes with dark pigment along the fin elements. Pectoral and pelvic fins generally without markings.

SIMILAR SPECIES   The Western Sand Darter is most like the Scaly Sand Darter from which it is separated by the general lack of body scales, except for the caudal peduncle and along midside (vs. completely scaled sides), and by a long, prominent spine near rear margin of operculum, the length of the spine much greater than its width (vs. length of spine about equal to its width at base). It is distinguished from the Crystal Darter by the lack of a premaxillary frenum and a caudal fin that is only slightly forked (vs. frenum present and caudal fin forked), and from most remaining darters by a slender body, body depth going into SL 7 times or more (vs. fewer than 7 times).

VARIATION AND TAXONOMY   *Ammocrypta clara* was described by Jordan and Meek in 1885 with collection locality data provided as follows: "A few specimens taken in a sandy part of the river opposite Ottumwa. Specimens were also obtained in Red River at Fulton, Arkansas, and in the Sabine River at Longview, Texas." Because the Ottumwa, Iowa, locality in the Des Moines River was listed first, it is accepted as the type locality for the species. Variation was studied by J. D. Williams (1975) who recognized no subspecies and placed *A. clara* in the *A. beani* species group (*beani*, *bifascia*, and *clara*). Wiley and Hagen (1997) studied variation in mitochondrial DNA and favored a basal placement for *A. clara* (sister to other *Ammocrypta* species). A mitochondrial and nuclear DNA analysis by Near et al. (2011) basically supported that conclusion.

DISTRIBUTION   *Ammocrypta clara* occurs in the Mississippi River basin from Wisconsin and Minnesota south to Virginia, Mississippi, and Texas, including small disjunct populations in the Ohio River drainage. It is also found in Lake Michigan drainages and on the Gulf Slope in the Sabine and Neches river drainages of Texas. In Arkansas it occurs in southern and eastern parts of the state in the Red, Ouachita, eastern Saline, White, Black, and St. Francis rivers, where it has a sporadic distribution with small populations (Map 185). It is most common from the mouth of the White River upstream to Batesville and in the Black River system, including the Current River (Tumlison et al. 2015). Buchanan et al. (2003) reported it as uncommon in the Red River of Arkansas. Although several records are known from the Mississippi River in

Missouri, there are no records of *A. clara* from that river in Arkansas or farther south.

HABITAT AND BIOLOGY   The Western Sand Darter inhabits moderate-sized rivers, where it is usually found in slight to moderate current over a sandy bottom. There are no records from the two largest rivers of Arkansas, the Mississippi and Arkansas rivers, despite an abundance of sandy substrate. Page (1983a) attributed its spotty distribution and small population sizes rangewide to increased siltation and related stream degradation. Pflieger (1997) reported that it is intolerant of excessive siltation and turbidity and that it has steadily declined in abundance in Missouri during the last few decades. Moriarty and Winemiller (1997) also considered it to be intolerant of environmental alterations. Kuehne and Barbour (1983) and Ross (2001) reported an apparent trend toward decreasing abundance over much of its range. It is not known to what extent, if any, its range in Arkansas has decreased, but it is not a commonly collected species. Although it is occasionally found with *A. vivax* (such as in the Saline River), Black (1940) noted that it seems to prefer larger rivers than that species. Like all members of the genus *Ammocrypta*, it spends much of its time buried in the sand. In a study of *A. pellucida* to test three hypotheses on the function of burying in *Ammocrypta*, Daniels (1989) concluded that this behavior was used primarily to maintain its position on the relatively homogeneous sand substrate.

It feeds mainly on aquatic insects (Forbes and Richardson 1920; Miller and Robison 2004). Eddy and Underhill (1974) reported that it feeds on small or immature aquatic insects, such as mayflies and midge larvae, and on amphipods. In Powell River, Tennessee, it fed exclusively on midge larvae (Etnier and Starnes 1993). Page (1983a) reported that it feeds on a variety of immature aquatic insects.

Southern populations were reported to breed from June to August (Williams 1975; Kuehne and Barbour 1983). Driver (2009) and Driver and Adams (2013) reported its reproductive biology in the Black River system of northeastern Arkansas and found gravid females and mature males in the Current, Strawberry, and Black rivers from June to mid-September, making this Arkansas's latest-breeding darter. Males and females reached maturity at 35–40 mm SL, gravid females averaged 57 eggs per clutch, and individual clutches varied from 15 to 141 eggs per female. Driver and Adams (2013) also described aquarium spawning behavior as follows:

Males and females were observed undulating and vibrating vigorously in corners and along the side of the tank, creating depressions in the sand substrate. Spawning events varied in the number of individuals participating, from one male and one female, up to 8 individuals of unknown sex ratios.

Most eggs were buried singly in the sand, but some eggs were also found at the sand surface. Sneaker males were also observed to participate in aquarium spawning bouts. Driver and Adams (2013) found that *A. clara* specimens from Arkansas were smaller in size, had lower GSI values, smaller clutch sizes, and smaller egg sizes compared with populations in the northern part of the species range (Wisconsin). Maximum life span in Wisconsin was 3+ years (Lutterbie 1979), but it is probably less than that in Arkansas populations.

CONSERVATION STATUS   *Ammocrypta clara* is considered vulnerable throughout its range (Warren et al. 2000; Jelks et al. 2008). It is widely distributed but nowhere abundant in Arkansas. Its distribution in Arkansas does not appear to have changed during the past 30 years, and its populations are probably stable. Even though we do not assign it a conservation status of concern, its populations should be periodically monitored in this state because of population declines elsewhere in its range.

Ammocrypta vivax Hay
Scaly Sand Darter

Figure 26.46. *Ammocrypta vivax*, Scaly Sand Darter. *David A. Neely.*

CHARACTERS   A slender darter with opercular spine small or absent; margin of preopercle weakly serrated (Fig. 26.46). Body depth going more than 7 times into SL. A series of 9–16 distinct, dusky (usually vertically oriented) lateral blotches, and 10–15 irregularly shaped dark spots on the back. Body mostly scaled (the most completely scaled *Ammocrypta*), except for the breast and belly. There is no premaxillary frenum, and the caudal fin is only slightly forked. Gill membranes narrowly joined across throat. Infraorbital canal complete. Cheek and opercle scaled; nape partially scaled; breast and midline of belly naked. Lateral line complete, not arched anteriorly, and only slightly or not at all turned downward posteriorly, with

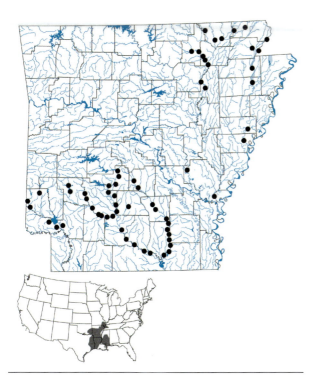

Map 186. Distribution of *Ammocrypta vivax*, Scaly Sand Darter, 1988–2019.

58–79 (usually 64–78) scales. Scale rows above lateral line usually 4–6 (1–7); scale rows below lateral line usually 8–10 (6–12). Diagonal scale rows usually 13–16; scale rows around caudal peduncle (when completely scaled) 21–25. Dorsal spines 8–14; dorsal rays 9–12; anal spine 1; anal rays 7–10 (usually 8 or 9). Pectoral rays usually 14 or 15. Breeding males with tubercles on spines and rays of the anal and pelvic fins. Largest Arkansas specimen examined was 2.76 inches (70 mm) TL from Lake Greeson in Pike County. Maximum reported length about 2.8 inches (72 mm).

LIFE COLORS   Similar to the Western Sand Darter, but the body is less translucent and more pigmented. Body light yellowish dorsally. Cheek and opercle often an iridescent green. Back with 10–15 irregular dark blotches. Lateral blotches usually darker, more distinct, and vertically elongated. Usually with thin dark bands in dorsal and caudal fins. Lateral scales often outlined in black. Males with black marginal and submarginal bands in spinous and soft dorsal fins and with black pigmentation in the pelvic fins. The anal, pectoral, and pelvic fins of males have melanophores along the rays.

SIMILAR SPECIES   It can be distinguished from *Ammocrypta clara* by fully scaled sides (vs. sides naked except for about 3 rows of scales along midside), a short

spine near rear margin of operculum, its length about equal to its base width (vs. opercular spine large), and by bolder blotches and pigmentation on side of body, with a midlateral row of round spots as large as or larger than pupil of eye (vs. side of body unpigmented or with midlateral spots much smaller than pupil of eye). It differs from *Crystallaria asprella* by the lack of a premaxillary frenum (vs. frenum present), only slightly forked tail (vs. tail noticeably forked), and lack of dorsal saddles (vs. 4 broad, dark dorsal saddles present). It differs from most other darters by its slender body, body depth going into standard length 7 times or more (vs. fewer than 7 times).

VARIATION AND TAXONOMY   *Ammocrypta vivax* was described by Hay (1882) from the Pearl River at Jackson, Mississippi. Variation was studied by Williams (1975), who recognized no subspecies and placed *A. vivax* in the *A. pellucida* species group along with *A. meridiana*. Wiley and Hagen (1997) studied variation in mitochondrial DNA and found that estimated nucleotide diversity among *A. vivax* was 4 times higher than for other *Ammocrypta* species studied. Their data further supported the hypothesis that *A. vivax* and *A. meridiana* form a monophyletic species pair, a conclusion also supported by the genetic analyses of Near et al. (2011).

DISTRIBUTION   The Scaly Sand Darter is found in only 9 states of the lower Mississippi River basin, from western Kentucky and southeastern Missouri south to southern Mississippi, and west to eastern Oklahoma; and in Gulf Slope drainages from the Pascagoula River drainage of southwestern Alabama, west to the San Jacinto River, Texas. In Arkansas, it is known from the Red, Ouachita, Arkansas, White, and St. Francis river drainages (Map 186). There are two pre-1960 records from Arkansas River tributaries in central Arkansas (Map A4.183), but it has been found in the Arkansas River drainage (Bayou Meto) of Arkansas only once during the past 60 years (Heckathorn 1993). Recent records are known from the St. Francis River, the lower White River (McAllister et al. 2012b), and the Cossatot River (Brook L. Fluker, pers. comm.).

HABITAT AND BIOLOGY   The Scaly Sand Darter is found in moderate-sized rivers but also occurs in the larger tributary streams of those rivers much more commonly than *A. clara* or *Crystallaria asprella*. Over most of its range, it is found only in Coastal Plain lowland streams. In Arkansas, the southern populations (Red and Ouachita drainages) are mostly confined to the Coastal Plain, but the northeastern populations (White River drainage) are concentrated just above the boundary between Coastal Plain and uplands. It is found most often in moderate current over a sandy bottom where, like all *Ammocrypta*, it buries itself in the substrate. It is sometimes taken over a mixed sand/fine gravel substrate. Surprisingly, it is commonly found in channelized ditches in southeastern Missouri (Kaszubski 1990), but it is generally considered intolerant of environmental alterations (Moriarty and Winemiller 1997). This species was routinely found in summer fish population samples taken by the AGFC from Lake Greeson (Buchanan 2005). It has also been collected from Millwood Lake.

Most of the following life history information is from a study in southeastern Missouri by Kaszubski (1990). The diet consisted of chironomid larvae (84.84%), oligochaetes (5.70%), zooplankton (3.25%), and ephemeropteran larvae (1.06%). Diet composition fluctuated throughout the year, and chironomid larvae made up the majority of the diet for all size classes. Stomachs from one Arkansas collection contained only midge larvae (Etnier and Starnes 1993).

Spawning has been reported from April through August (based mainly on the presence of ripe females in collections) in different parts of the species range (Kuehne and Barbour 1983; Miller and Robison 2004). Etnier and Starnes (1993) reported ripe females from Arkansas in late May with a fecundity of 60–70 ova per female. Hubbs (1985) found ripe individuals in east Texas in early April. Spawning was reported in Louisiana from April to August at water temperatures from 16.9°C to 26.0°C (Mitchell 1987). Kaszubski (1990) reported that spawning occurred in southeastern Missouri from early April to late June or early July; peak spawning occurred in April based on gonadosomatic indices and ovum diameters. The number of mature ova ranged from 87 to 325, and the smallest reproductive female was about 49 mm SL. In Louisiana, individuals attained sexual maturity by 39 mm SL, and the number of mature ova ranged from 43 to 108 (Mitchell 1987). Spawning behavior has not been reported but is probably similar to that reported for *Ammocrypta pellucida* by Johnston (1989a). Agonistic behaviors which may be involved in courtship were observed in aquaria. Those behaviors consisted of frontal and lateral displays, tail beating, head butting, and biting among males. A reproductive male appeared to guard the female by placing himself between the female and rival males, or by positioning himself on top of the female. Maximum life span in southeastern Missouri was 3 years (Kaszubski 1990).

CONSERVATION STATUS *Ammocrypta vivax* is occasionally taken in large numbers at some localities in Arkansas. We do not consider it currently in need of a conservation status, but its populations should be periodically monitored to detect future changes.

Genus *Crystallaria* Jordan and Gilbert, 1885—Crystal Darters

Jordan and Gilbert *in* Jordan (1885b) recognized *Crystallaria* as a monotypic genus containing only *C. asprella*. In 2008 a second species of *Crystallaria* (*C. cincotta* of the Ohio River drainage) was described (Welsh and Wood 2008). Bailey and Gosline (1955) included the Crystal Darter in the genus *Ammocrypta*, and it was usually assigned to that genus for more than three decades (as in the first edition of this book). Bruner (2004) reported a unique character found only in *Ammocrypta* and *Crystallaria* called the Spreitzer vertebrae, which he concluded is a generic character defining the genus *Ammocrypta* and supporting the inclusion of *Crystallaria* as a subgenus of *Ammocrypta*. Returning *Crystallaria* to *Ammocrypta* has not been widely accepted by most recent authors (Near et al. 2011; Page et al. 2013) except for Nelson et al. (2016). Simons (1991) hypothesized that *Crystallaria* and *Ammocrypta* are phylogenetically remote based on morphological comparisons and should be considered distinct genera, with *Crystallaria* sister to a lineage containing all other darters. Simon et al. (1992), examining characters of protolarvae of sand darters, confirmed Simons's earlier elevation of *Crystallaria* to genus status but found a sister relationship between the genera *Ammocrypta* and *Etheostoma*. Genetic analysis by Near et al. (2011) supported a sister relationship between *Ammocrypta* and *Crystallaria*. We follow Moore (1968), Miller and Robison (1973, 2004), Simons (1991), Wiley (1992), Nelson et al. (2004), Near et al. (2011), and Page et al. (2013) in recognizing the genus *Crystallaria*. There is one described species (and possibly one undescribed species) of *Crystallaria* in Arkansas. Morphologically, *Crystallaria* differs from *Ammocrypta* in the presence of palatine and prevomerine teeth, a narrow premaxillary frenum, deeply forked caudal fin, maximum SL of 130 mm, 12–15 anal rays, 4 large, dark dorsal saddles, up to 4 pored scales on the caudal fin, and other features (Page 1983a).

Crystallaria asprella (Jordan)
Crystal Darter

CHARACTERS   A large, very slender darter usually with 4 dark, wide dorsal saddles extending obliquely forward down upper side, and with a midlateral series of 8–10 dark oblong blotches (Fig. 26.47). Body depth going into SL 7.8–10.0 times. Caudal peduncle long and narrow. Head conical, flat and wide. Eyes closely set. Opercular

Figure 26.47. *Crystallaria asprella*, Crystal Darter.
David A. Neely.

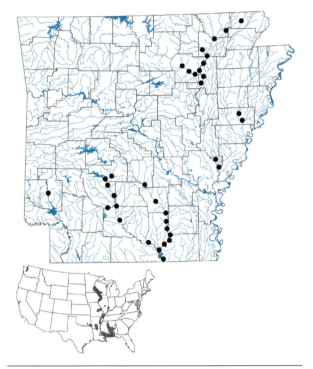

Map 187. Distribution of *Crystallaria asprella*, Crystal
Darter, 1988–2019.

spine present. Preopercle smooth on posterior edge. Gill
membranes separate or slightly connected across throat.
Infraorbital canal uninterrupted. Narrow premaxillary
frenum present. Mouth horizontal, lower jaw included.
Palatine teeth present. Body fully scaled except for breast
and belly. Caudal fin noticeably forked. Lateral line complete, not arched anteriorly, with 81–98 scales. Scale rows
above lateral line 6–8; scale rows below lateral line 9–15.
Diagonal scale rows 19–26; scales around caudal peduncle
25–30. Dorsal spines 12–15; dorsal rays 12–15; anal spine 1;
anal rays 12–15; pectoral rays 15–17. Swim bladder absent.
Breeding males with tubercles on anal and pelvic fin rays.
George et al. (1996) reported that females from the Saline
River, Arkansas, occasionally possess minute tubercles.
Males have longer soft dorsal and anal fins than females.
Maximum length in Arkansas around 5 inches (127 mm).

Pflieger (1997) reported a maximum length in Missouri of
6 inches (152 mm).

LIFE COLORS   Translucent, with yellow back and
upper sides and silver belly. Dorsal saddles dark brown and
connected to midlateral brown blotches. Dark brown mottling sometimes found on dorsum between saddles. Wide
dark brown stripe extending around snout from eye to eye.
Fins usually clear or with a few scattered melanophores. No
bright breeding colors.

SIMILAR SPECIES   The Crystal Darter can be distinguished from the two Arkansas species of *Ammocrypta*
by the well-developed premaxillary frenum (vs. frenum
absent), noticeably forked caudal fin (vs. slightly forked),
usually 4 dark dorsal saddles (vs. dorsal saddles absent),
usually more than 82 lateral line scales (vs. fewer than 82),
and dorsal rays modally 13 or 14 (vs. 10 or 11). It differs from
all remaining darters by its extremely slender body, body
depth usually going into SL 7 times or more (vs. fewer than
7 times).

VARIATION AND TAXONOMY   *Crystallaria asprella*
was described by Jordan (1878a) as *Pleurolepis asprellus* from
a small tributary of the Mississippi River, Hancock County,
Illinois. In Arkansas, there is some variation in color pattern, especially in Ouachita River drainage populations.
Genetic variation of the Crystal Darter was studied by
Wood and Raley (2000), who recognized the Saline River,
Arkansas, population as a highly differentiated, distinct
lineage, although they did not assign a formal name to this
population. This distinct lineage may warrant taxonomic
recognition in the future, and it certainly merits protection.

DISTRIBUTION   *Crystallaria asprella* occurs in
disjunct populations in the Mississippi River basin from
Minnesota to Louisiana, and in Gulf Slope drainages from
the Escambia River, Florida, to the Pearl River, Mississippi
and Louisiana. In Arkansas it is found in southern and
eastern parts of the state in the western Saline (Little River
drainage) (Robison et al. 2013), Little Missouri, Ouachita,
eastern Saline, St. Francis, Black (McAllister et al. 2010b),
and lower White rivers (Map 187). Although not previously
known from the St. Francis River in Arkansas, it was found
in that river in modest numbers in St. Francis County in
2013 by Jeff Quinn (AGFC).

HABITAT AND BIOLOGY   The Crystal Darter
inhabits the lower reaches of moderate-sized rivers, mainly
along and below the Fall Line, where it is usually found in
strong current over a sand or fine gravel substrate. It was
typically collected in the Saline River, Arkansas, at depths
of 114–148 cm and water velocities of 46–90 cm/s over substrates of gravel, small cobble, and patches of sand; it was

not collected from habitats containing mud, clay, or submerged aquatic vegetation (George et al. 1996). During that 1992–1993 study, dissolved oxygen levels ranged from 6.81 to 11.0 ppm, pH varied from 5.73 to 6.6, and conductivity was between 175 and 250 S. It occupies deeper water during the day and moves into shallower water at night. The most common species associates in the Saline River were *Percina caprodes* and *Cyprinella whipplei*. It buries itself in the substrate with only its eyes protruding and darts out after passing small prey (Miller and Robison 2004). It is intolerant of reservoir habitats.

In Wisconsin, the Crystal Darter fed mainly on insect larvae, including mayflies, craneflies, black flies, caddisflies, and midges (Becker 1983). In Minnesota, small juveniles fed mainly on microcrustaceans; larger juveniles and adults fed primarily on chironomid and hydropsychid larvae, but 11 taxa of insects were consumed (Hatch 1997). A comparison of the diets of *C. asprella* populations in the Tennessee-Tombigbee Waterway (TTW), in a remaining free-flowing tributary of the TTW, and museum specimens collected before construction of the TTW found that diets differed before and after construction of the TTW (Roberts et al. 2007). Simuliids were most common in historical diets, chironomids were most common in waterway diets, and tributary diets were intermediate between the two. Indicator species analysis isolated invertebrate taxa indicative of waterway specimens, as well as historical/tributary specimens. Waterway indicators were mayflies (Caenidae), cladocerans (Daphnidae), and ostracods. Shared indicators for the historical and tributary groups were mayflies (Baetidae and Heptageniidae), caddisflies (Hydropsychidae), and water mites (Hydracarina). The tributary and waterway groups shared the Chironomidae as indicators, and Tricorythidae were indicative of tributary diets (Roberts et al. 2007).

Little was known about the reproductive biology of this species until recently. It spawns in the Tallapoosa River of Alabama in late February when water temperatures reach approximately 12°C (Simon et al. 1992). Boschung and Mayden (2004) described its spawning behavior as follows:

> Crystal Darters leave the deeper riffles of the main stream and presumably spawn in meandering side-channel riffles where the water depth is 60 to 90 cm, the current is moderate to swift, and the substrate is composed of gravel. Females partially bury themselves in the substrate, where they remain stationary while being mounted by one to several males. The fertilized eggs are slightly straw-colored, about 1.8 mm in diameter, and are strongly adhesive. Left unguarded,

they become attached to coarse sand and small gravel. After an apparently short spawning period of about a week, Tallapoosa River adults return to the deeper and swifter runs of the main channel.

In the Saline River, Arkansas, Crystal Darters spawned when 1 year old from late January through mid-April over gravel without constructing a nest, and with clutch sizes ranging from 106 to 576 mature ova (George et al. 1996). Eggs are dispersed by the current shifting the fine gravel and sand substrates. A ripe female 91 mm TL collected from the White River in Independence County on 7 March examined by TMB contained 754 mature ova. Life expectancy in the Saline River is at least 2 years (George et al. 1996).

It was unintentionally, but successfully, cultured in an AGFC nursery pond just off the White River near Augusta in the spring of 1981. The pond had been filled with water from the White River in late winter 1981 and stocked with Walleye fry. When the pond was subsequently drained on 5 May 1981, no Walleye were present but hundreds of YOY Crystal Darters (1.2–1.5 inches [30–38 mm] SL) were found. Ruble et al. (2014) successfully propagated the closely related *Crystallaria cincotta* by developing specific methods for breeding, care of eggs, and rearing of larval darters in laboratory aquaria. McAllister et al. (2013) documented black-spot disease in *C. asprella* in Arkansas.

CONSERVATION STATUS   *Crystallaria asprella* has apparently declined in numbers over much of its range (Kuehne and Barbour 1983). It requires clean water and a silt-free substrate and is particularly vulnerable to habitat-altering activities such as channelization, dredging, and impoundment. Deacon et al. (1979) placed this species in a category of special concern. It is considered extirpated in Illinois, Indiana, Kentucky, and Tennessee. Arkansas may currently possess more stable populations than any other state, but *C. asprella* is rarely abundant at any locality. The current status of this species in Arkansas is unknown. It appears to be very vulnerable to environmental disturbance, and we consider it a species of special concern in the state. The genetically distinct population in the Saline River warrants protection and should be closely monitored.

Genus *Etheostoma* Rafinesque, 1819— Etheostoma Darters

The North American genus *Etheostoma* contains approximately 145 described species and a number of undescribed forms. Twenty-nine described species and at least three undescribed species are found in Arkansas, and three of the described species, *E. clinton*, *E. fragi*, and *E. pallidi-dorsum*, are endemic to the state. There are more species of *Etheostoma* than any other North American freshwater fish genus. *Etheostoma* species are almost entirely confined to Atlantic drainages of North America, with a single species (*Etheostoma pottsi*) found in a Pacific Slope stream system in Mexico (Miller et al. 2005).

Rafinesque (1819) established *Etheostoma* as a genus to include *E. blennioides*. For the next 130 years, various species were placed in the genus, but most darter species were distributed among several other genera. In the early 1950s, Bailey (1951) and Bailey et al. (1954) envisioned *Etheostoma* as an assemblage of all darters not assignable to *Percina* or *Ammocrypta* (Page 1983a). Bailey and Gosline (1955) also recognized this large, inclusive version of *Etheostoma*, and it has largely remained intact to the present time. There has been little stability or consensus in the recognition of subgenera of *Etheostoma*. Bailey and Gosline (1955) established 12 subgenera, and Page (1981, 2000) and Bailey and Etnier (1988) increased the number to 17. Near et al. (2011) summarized the traditional classification of darter subgenera. Molecular phylogenetic analyses using mitochondrial and nuclear DNA did not support 7 of the subgenera recognized by Page (2000) as monophyletic (Near 2002; Mayden et al. 2006b; Lang 2007; Lang and Mayden 2007). More recent genetic evidence has not resolved the problem of subgeneric recognition, with analyses of different genes often producing conflicting conclusions. Four of Page's subgenera were not monophyletic in phylogenies based on discretely coded morphological characters (Ayache and Near 2009). Although genetic analyses by Lang and Mayden (2007) did not support monophyly of the large *Etheostoma* subgenus, *Oligocephalus*, subsequent analyses of nuclear genes by Bossu and Near (2009) and Near et al. (2011) supported monophyly of *Oligocephalus* with some minor changes in its species composition. This new composition of *Oligocephalus* included several distinct clades, three of which are represented by Arkansas species: (1) the Rainbow Darter clade (*E. caeruleum*, *E. collettei*, and *E. asprigene*), (2) the Redfin Darter clade (*E. whipplei*, *E. artesiae*, and *E. radiosum*), and (3) the Orangethroat Darter clade. Near et al. (2011) named the Orangethroat Darter clade *Ceasia*, and recognized it as a rank-free clade consisting of 15 described and undescribed species nested within the subgenus *Oligocephalus*. *Ceasia* has also been commonly referred to as the *Etheostoma spectabile* complex and is currently considered to consist of 19 described and undescribed species (P. A. Ceas, pers. comm.). There are 7 species of *Ceasia* in Arkansas: (1) *E. fragi* (endemic to the Strawberry River), (2) *E. uniporum* (Spring, Current, and Black rivers), (3)

E. pulchellum (found in Arkansas only in the Red River drainage), (4) *E. squamosum* (found in Benton and Washington counties in the Neosho and Illinois river drainages), (5) *E.* sp. cf. *spectabile* (the undescribed Ozark Darter found in the upper White River drainage), (6) *E.* sp. cf. *pulchellum* 1, red belly form (an undescribed species found in Arkansas River tributaries west of Dover in Pope County into eastern Oklahoma), and (7) *E.* sp. cf. *pulchellum* 2, blue belly form (an undescribed species found in lower White and Arkansas river tributaries east of Dover and disjunctly in one stream in eastern Oklahoma). Near (2002) suggested abandonment of darter subgenera in favor of the recognition of monophyletic species clades, but Near et al. (2011) developed a phylogeny-based classification of darters that followed the principles of phylogenetic nomenclature. The clade names provided by Near et al. (2011) were unranked, but most of the clades were nested within traditionally recognized genera and would therefore be available as subgenus names under the ICZN rules. Despite controversial interpretations of the monophyly of darter subgenera, we consider subgenera useful for indicating past interpretations of phylogenetic relationships and list 11 traditionally recognized subgenera of *Etheostoma* with Arkansas species: *Boleosoma*, *Catonotus*, *Doration*, *Etheostoma*, *Fuscatelum*, *Hololepis*, *Microperca*, *Oligocephalus*, *Ozarka*, *Poecilichthys*, and *Vaillantia*. Subgeneric designations are also presented because of their extensive use in past taxonomic literature. The subgenus of each Arkansas species of *Etheostoma* is given in its species account. Boschung and Mayden (2004) provided a more thorough treatment of the characters and systematics of 15 traditionally recognized subgenera of *Etheostoma* in Alabama.

The males of most *Etheostoma* species become brightly colored during the breeding season. Some species may breed as early as February, others as late as June. Hubbs (1985) presented evidence that many *Etheostoma* have long reproductive seasons with multiple spawnings. Paine (1990) studied life history tactics of darters in the genus *Etheostoma* and found that larger species generally grow faster, mature at a larger size, produce bigger clutches of eggs, and have longer reproductive and life spans, and shorter spawning seasons than small darter species. Paine further found that rare species may not match the reproductive performance of more common and widely distributed species. There is considerable diversity in habitat preference in this group, but nearly all of them feed on insect larvae and small crustaceans. There is also a wide range in size, from *E. microperca*, which reaches a maximum length of around 1.5 inches (38 mm), to *E. blennioides*, which can exceed 5 inches (127 mm), but most species of *Etheostoma* are smaller than most species of *Percina*. The greatest number of *Etheostoma* species in Arkansas inhabit upland streams of the Ozarks and Ouachitas, but a few are strictly lowland forms, and some species occur in both upland and lowland waters. Most are rather sensitive to environmental disturbance. Morphological features of *Etheostoma* species are variable, but the genus differs from *Percina* in lacking enlarged and modified scales between the pelvic fin bases and in lacking modified scales along the midline of the belly. It differs from *Ammocrypta* and *Crystallaria* in having a deeper and well-scaled body. *Etheostoma* is distinguished from *Nothonotus* primarily by genetic differences, but a combination of morphological characters will distinguish *Nothonotus* from most *Etheostoma* species (see genus *Nothonotus* account).

In *Etheostoma,* genital papillae of females range from simple tubes to complex pleated and multilobed structures. Martin and Page (2015), in a study of 128 *Etheostoma* species, found correlations between papilla morphology and spawning behaviors and oviposition. A simple tube papilla was considered the ancestral condition and is characteristic of species that bury their eggs. Egg-clumping species have mound papillae, most species that attach eggs to objects above the substrate have tube papillae, and all egg-clustering species have complex, wide, pleated rosette papillae for attaching eggs to the undersides of rocks or logs.

Etheostoma artesiae (Hay)
Redspot Darter

Figure 26.48a. *Etheostoma artesiae*, Redspot Darter male. *Uland Thomas.*

Figure 26.48b. *Etheostoma artesiae*, Redspot Darter female. *Uland Thomas.*

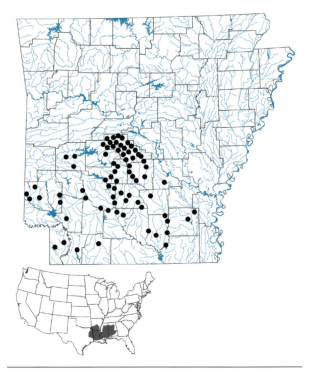

Map 188. Distribution of *Etheostoma artesiae*, Redspot Darter, 1988–2019.

CHARACTERS   A moderate-sized, fairly robust darter with a moderately pointed snout and a deep caudal peduncle (Fig. 26.48). Back with 8–12 small saddles that are sometimes faint or absent. Pigmentation pattern of side and back variable but usually dominated by dark mottlings and reticulations, with small scattered light spots. Dark vertical bars sometimes present on sides of preserved specimens and, if present, best developed posteriorly, especially on the caudal peduncle and on smaller individuals. Lateral bars and blotches usually not as prominent as in *E. radiosum*. Gill membranes usually moderately to broadly joined across throat. Premaxillary frenum present. Preopercle smooth or weakly serrate. Eye diameter less than snout length. Infraorbital canal uninterrupted. Belly fully scaled; nape and opercle partly to fully scaled; cheek varies from naked to fully scaled; breast naked to partly scaled. Lateral line incomplete, ending beneath second dorsal fin. Scales in lateral series 50–63 ($\bar{x}$ = 55.8); pored lateral line scales 35–53 ($\bar{x}$ = 42); scale rows around caudal peduncle 20–26 ($\bar{x}$ = 22.8); diagonal scale rows from origin of anal fin to first dorsal fin base 14–22 ($\bar{x}$ = 17). Dorsal spines 9–12 (usually 10 or 11); dorsal rays 11–16 (usually 12–14); anal spines 2; anal rays 6–9 (usually 7 or 8). Male with breeding tubercles on belly, along anal fin base, and on ventral caudal peduncle scales. Maximum length just over 3 inches (80 mm).

LIFE COLORS   Back and sides are a mottled olive brown; venter and underside of head are white to light brown. Back usually with 8–12 dark saddles; dark lateral bands sometimes present posteriorly. A small black humeral spot and a dark vertical suborbital bar are present. Adult males with bright red scattered spots on sides; females with scattered yellow spots. Coloration of breeding males very intense, with red spots often coalescing. First and second dorsal, caudal, and anal fins similarly colored with a bright blue marginal band, a narrow white submarginal band, and a red medial to basal band; pelvic fins dusky blue; pectoral fins clear. The venter is red orange. Breeding females less colorful but with scattered yellow spots on sides.

SIMILAR SPECIES   *Etheostoma artesiae* is most similar to *E. whipplei* and *E. radiosum*. It differs from *E. whipplei* in usually having 61 or fewer lateral scales (separates 97.9% of individuals) and 26 or fewer caudal peduncle scale rows (separates 95.4% of individuals), (vs. more than 61 lateral scales and more than 26 caudal peduncle scale rows in *E. whipplei*). No specimens of *E. artesiae* have the combination of more than 61 lateral scales *and* more than 26 caudal peduncle scales, and no specimens of *E. whipplei* have the combination of fewer than 61 lateral scales *and* fewer than 26 caudal peduncle scales (Piller et al. 2001). *Etheostoma artesiae* differs from *E. radiosum* in the presence of red or yellow spots on sides (vs. spots absent), exposed or embedded scales covering most of the opercle (vs. opercle unscaled or with scales only on the upper half), usually 43 or more pored lateral line scales (vs. fewer than 43 pored scales), and in having thin dark lines with many interruptions along the side of the body from head to caudal fin on the upper half of the body and from the middle of the side to the caudal fin on the lower half (vs. rarely with dark lines on sides, or if present, lines are wide and restricted to the area above the lateral line in *E. radiosum*). *Etheostoma artesiae* is sometimes syntopic with *E. asprigene* but differs from that species in having red or yellow spots on sides (vs. no red or yellow spots on sides), gill membranes moderately to broadly joined across throat (vs. gill membranes separate or only slightly joined across throat), and in usually having more than 51 lateral scales (vs. 51 or fewer lateral scales). Juveniles of *E. artesiae* and *E. parvipinne* are sometimes confused, but *E. artesiae* differs in having uninterrupted infraorbital and supratemporal canals (vs. interrupted in *E. parvipinne*).

VARIATION AND TAXONOMY   *Etheostoma artesiae* is a member of the subgenus *Oligocephalus*. Originally described as *Poecilichthys artesiae* by Hay in 1881 from Lowndes County, Mississippi, the Redspot Darter has for most of the past century been considered a subspecies of the Redfin Darter, *E. whipplei* (Hubbs and Black 1941; Moore and Rigney 1952; Retzer et al. 1986). It was elevated to full species status by Piller et al. (2001) based primarily on meristic differences with no intergrade populations known. Subsequent studies of mitochondrial and nuclear DNA supported the specific status of *E. artesiae* (Lang and Mayden 2007; Near et al. 2011). We examined a few specimens of presumptive *E. whipplei* from Rock Creek in Little Rock that appear to show possible intergradation of characters. This population needs further study, but we nevertheless support full species recognition for *E. artesiae*.

DISTRIBUTION   *Etheostoma artesiae* occurs in the lower Mississippi River basin from northern Alabama and Mississippi to southern Arkansas, south to eastern Texas and southern Mississippi, and in Gulf Slope drainages from the Chattahoochee River drainage, Alabama, west to the Neches River, Texas. In Arkansas, the Redspot Darter occurs in the Ouachita and Red river drainages (Tumlison et al. 2015), where it is found primarily below the Fall Line (Map 188). The similar *E. radiosum* is usually found in smaller creeks above the Fall Line; however, we have collected the two species syntopically in the Saline River drainage in habitats distinctly above the Fall Line.

HABITAT AND BIOLOGY   *Etheostoma artesiae* is primarily a lowland form widely distributed in small to moderate-sized streams of the Gulf Coastal Plain across southern Arkansas. Those streams typically have moderate flow and mixed sand and gravel substrates. It is especially common in the upper Saline River. It occurs syntopically with *E. radiosum* immediately above the Fall Line in the Red and Ouachita river drainages, but is replaced by *E. radiosum* in the smaller streams of the Ouachita Highlands having higher gradients, often with intermittent flow, and gravel-cobble substrates. *Etheostoma artesiae* was found in six impoundments in the upper Saline River above the Fall Line and in Lake Erling, a Gulf Coastal Plain reservoir below the Fall Line (Buchanan 2005). Tyrone (2007) found that timber harvest reduced population densities in the Kisatchie Bayou drainage of Louisiana (one of only three of the 26 species in the study area to exhibit a population decline).

In the Saline River, Arkansas, *E. artesiae* was a benthic invertivore, feeding primarily on dipterans, odonates, plecopterans, and coleopterans (Matthews et al. 2004). Ephemeropterans and trichopterans were also frequently consumed. A similar diet composition was found by TMB in stomach contents of 15 specimens (43–71 mm TL) from the upper Saline River in Saline County. Ephemeropterans, trichopterans, plecopterans, and chironomids dominated the diet, but small crustaceans (amphipods, isopods, and copepods) were also important.

Spawning occurs from late February to mid-May in Alabama and Mississippi (Heins and Machado 1993; Mettee et al. 1996; Boschung and Mayden 2004), where clutch sizes ranged from 31 to 207 eggs. Males apparently attained reproductive condition somewhat earlier than females, and mean male size was significantly larger than mean female size. A similar spawning season likely occurs in Arkansas. Individuals from the Saline River, Arkansas, nearing reproductive condition on 26 January were examined by TMB. Gonads of males and females were enlarged, but no ova nearing ripeness were present. Spawning in Arkansas probably begins in mid- to late February. Reproductive behavior has not been reported but is probably similar to that of *E. whipplei*.

CONSERVATION STATUS This lowland darter is widely distributed and common in its preferred habitat in southern Arkansas. Its populations are secure.

Etheostoma asprigene (Forbes)
Mud Darter

Figure 26.49a. *Etheostoma asprigene*, Mud Darter male. *Uland Thomas.*

Figure 26.49b. *Etheostoma asprigene*, Mud Darter female. *Uland Thomas.*

CHARACTERS A moderately slender darter usually with 8–10 subdued dorsal saddles of equal intensity, sides with 6–9 vertical bars, the posterior 5 or 6 bars distinct, the anterior ones faint or absent; anterior side often mottled brown (Fig. 26.49). Gill membranes separate or only

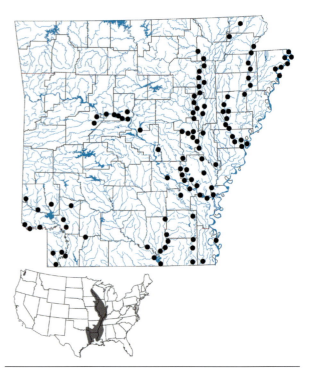

Map 189. Distribution of *Etheostoma asprigene*, Mud Darter, 1988–2019.

slightly connected across throat. Premaxillary frenum present. Greatest body depth usually occurring under middle or anterior half of first dorsal fin. Cheek, opercle, and nape fully scaled; belly fully scaled or naked anteriorly; breast naked or with a few scales. Infraorbital canal uninterrupted, usually with 8 pores. Lateral line incomplete, usually ending beneath second dorsal fin. Scales in lateral series 40–53; pored scales 33–44. Scale rows above lateral line 4–6; scale rows below lateral line 6–10. Diagonal scale rows 12–19; scale rows around caudal peduncle 17–23. Dorsal spines 10 or 11; dorsal rays 11–14; anal spines 2; anal rays 7–9; pectoral rays usually 14 (13–15). There is no development of breeding tubercles. Maximum length around 2.8 inches (71 mm).

LIFE COLORS Less colorful than other species of subgenus *Oligocephalus*. Back and anterior sides a mottled olive brown; 8–10 brown dorsal saddles of equal intensity (saddles are often indistinct or sometimes absent); lateral bars a dark greenish brown, spaces between bars become orange in breeding males. Usually 3 distinct spots vertically aligned at base of caudal fin, the medial spot more posteriorly located and often more distinct. Pre- and suborbital bars dark and prominent. Breast and lower surface of head pale and often evenly stippled with scattered, small melanophores. Small, dark humeral spot usually evident; humeral spot becomes very prominent in breeding males. Pelvic fin membranes usually with at least a few small melanophores. Lower sides of breeding males orange. First dorsal fin with a broad, black basal band, a narrow red-orange submarginal band, and a blue margin; the black basal band is usually expanded posteriorly, forming a distinct blotch at the rear of the first dorsal fin (usually faint in females). Second dorsal fin with a wide orange band or rows of brown spots, and caudal fin with rows of black or brown spots and often with some orange coloration. Remaining fins usually clear. In breeding males, the anal fin is blue green and the pelvic fins are blue black. Coloration of females remains the same throughout the year, without bright body colors, but with a single submarginal red band in the first dorsal fin.

SIMILAR SPECIES   Most like *E. collettei,* with which it is found syntopically in the lower eastern Saline River, the Mud Darter differs in having 8–10 dorsal saddles or blotches of about equal intensity (vs. 3 or 4 of its 8–10 saddles much darker than the others); usually a dark blotch in the posterior part of the first dorsal fin (vs. dark blotch absent); posterior lateral bars distinct above midside (vs. indistinct above midside); a deeper caudal peduncle (vs. narrower caudal peduncle); and no breeding tubercles in males (vs. breeding tubercles present). It differs from *E. caeruleum* in having fully scaled cheeks (vs. cheeks naked or with a few embedded scales behind eye), and in breeding males lacking tubercles and red pigmentation in anal fin (vs. tubercles present and red pigment present in anal fin). The Mud Darter is distinguished from species of the *E. spectabile* complex in having an uninterrupted infraorbital canal (vs. interrupted), and usually 13 or 14 pectoral fin rays (vs. usually 11 or 12). It differs from *E. artesiae* and *E. whipplei* in having gill membranes separate (vs. moderately to broadly joined across throat); usually 51 or fewer lateral scales (vs. more than 51); and in breeding males lacking red coloration in anal fin (vs. bright red coloration present in anal fin).

VARIATION AND TAXONOMY   *Etheostoma asprigene* was described by Forbes *in* Jordan (1878a) as *Poecilichthys asprigenis* from a small creek near Pekin, Illinois. Lang and Mayden (2007) studied mitochondrial and nuclear DNA and identified an *E. asprigene* species group (referred to as the Rainbow Darter species group by Bossu and Near 2009 and by Near et al. 2011) within the subgenus *Oligocephalus.* The only other Arkansas species in the *E. asprigene* group are *E. collettei* and *E. caeruleum.*

DISTRIBUTION   This lowland darter is found in the Mississippi River basin from Minnesota south to Louisiana and eastern Texas. In Arkansas, it is most abundant in the eastern part of the state in the St. Francis (Connior et al.

2012), lower White, lower Arkansas, and lower eastern Saline river drainages, and it is less abundant in the Gulf Coastal Plain. It is commonly found in the Mississippi River and is rare in the Red River (Buchanan et al. 2003; McAllister et al. 2010a). It extends upstream in the Arkansas River Valley to the Petit Jean River and Cove Lake near Paris in Logan County (Map 189).

HABITAT AND BIOLOGY   The Mud Darter inhabits a variety of lowland habitats. It is found primarily in low-gradient streams, rivers, sloughs, and bayous of small to moderate size, where it is usually found in current over a bottom of mud and detritus but sometimes over a mixed sand, gravel, and silt substrate. In Lake Creek, Illinois, adults typically occurred in shallow clay, gravel, detritus, and cypress-stump riffles and were occasionally found in shallow to deep pools directly upstream or downstream from the riffles (Cummings et al. 1984). Adults moved from riffles in the daytime to pools at night. Densities in riffle habitats were 4.5–8.3 fish/m^2 from August to November; density in pools was highest in February (1.5 fish/m^2). Juveniles occurred more frequently in pools or quiet water areas than in riffles. It appears to be moderately tolerant of siltation and turbidity but is most abundant in clearer lowland waters. It is occasionally taken in lowland oxbow lakes, reservoirs, and large rivers such as the Mississippi. Locally large populations exist in the lower eastern Saline River and in the relatively undisturbed lowland streams of the White River National Wildlife Refuge in Arkansas, Monroe, and Phillips counties. Buchanan (2005) reported this species from five Arkansas impoundments.

In Lake Creek, Illinois, the predominant food items were dipteran larvae (chironomids and simuliids), isopods, mayfly nymphs, and caddisfly larvae. Less important food items included amphipods and other small crustaceans, water mites, stoneflies, beetle larvae, aquatic hemipterans, and fish eggs (Cummings et al. 1984). Dipteran larvae made up the bulk of the diet in all size classes and in all seasons. Stomach contents of 54 specimens from Arkansas were examined by TMB. The diets of specimens from Monroe, Yell, Jackson, and Chicot counties in spring and early summer were about equally dominated by ephemeropterans and chironomids, with plecopterans and coleopterans also important. Trichopterans, megalopterans, and copepods were ingested in small amounts. The August diet of Mud Darters from the Mississippi River in Mississippi County was dominated by chironomids (83% frequency of occurrence), with small numbers of ephemeropterans, coleopterans, and copepods also consumed. Ten specimens collected on 26 September from the Saline River in Bradley County had empty stomachs despite an apparent availability of chironomid larvae and other foods (12 *Percina vigil* collected at the same locality on the same date had full stomachs).

Spawning occurred in Texas from mid-February to late March (Hubbs 1985) and in southern Illinois from early March to early May (Cummings et al. 1984). Examination of ovaries and testes by TMB indicated that peak spawning occurs in Arkansas in March and April, and the spawning season probably extends from late February to early May. Females with ripe ova were collected on 26 March in Big Cypress Bayou in Monroe County, and on 21 April in Smiley Bayou in Yell County. Females from 48 to 55 mm TL contained 97–304 ripe ova. Cummings et al. (1984) captured adults in breeding condition in shallow (7.9 inches [20 cm]) riffles that contained an abundance of leaves, sticks, and cypress tree stumps at water temperatures from 11°C to 15°C. Turbid water prevented field observations of spawning, but in aquaria, males did not establish territories. Males actively pursued and courted females by repeatedly swimming around them and displaying erect dorsal fins. The female selected the egg deposition site while the male followed close behind. Spawning occurred over sticks, leaves, or vegetation, with the male mounted vertically on the back of the female. The adhesive eggs attach to objects as they sink. No parental care has been observed. Eggs hatch in 5 or 6 days at temperatures of 20–24°C (Cummings et al. 1984). Individuals reach sexual maturity at 1 year of age, and maximum life span is around 3 years (Lutterbie 1979).

CONSERVATION STATUS   This is a widespread lowland species, but large populations are uncommon. Its populations in Arkansas are currently stable.

Etheostoma autumnale Mayden
Autumn Darter

CHARACTERS   A moderately stout darter with a large head, conical snout, and terminal mouth (Fig. 26.50). Dark suborbital bar present. Back usually with 5 or 6 dark saddles extending somewhat downward onto upper side. Lateral caudal peduncle with 3 large dark blotches arranged from beneath middle of second dorsal fin to caudal fin base. Gill membranes separate or slightly joined across throat. Premaxillary frenum present. Infraorbital canal uninterrupted, usually with 8 pores. Nape and belly fully scaled; cheek naked or with embedded scales; opercle naked; and breast usually with embedded scales but sometimes naked. Lateral line incomplete, extending to beneath (and sometimes past) second dorsal fin. Scales in lateral series

Figure 26.50a. *Etheostoma autumnale*, Autumn Darter male. *Uland Thomas.*

Figure 26.50b. *Etheostoma autumnale*, Autumn Darter female. *Uland Thomas.*

Map 190. Distribution of *Etheostoma autumnale*, Autumn Darter, 1988–2019.

65–76 (but usually 70–75); pored lateral line scales highly variable (51–75). Scale rows above lateral line 10–13; scale rows below lateral line usually 10–12. Diagonal scale rows usually 21–26; scale rows around caudal peduncle usually 30–34. Dorsal spines 10–13 (modally 11); dorsal rays 13–15 (usually 14 or 15); anal spines 2; anal rays 7–10 (usually 8 or 9); pectoral rays 12–14. Breeding males with tubercles on pelvic and anal fin spines and rays, on scales of the belly, and along the anal fin base. Maximum length around 3.5 inches (89 mm).

LIFE COLORS Mayden (2010) provided detailed descriptions of breeding and nonbreeding coloration. First dorsal fin of breeding male with a broad distal orange band, a broad bright blue medial band, and a broad dusky basal band without orange coloration. Second dorsal fin usually with dusky ovals or scattered melanophores. Pelvic and anal fins darkly pigmented. Bright orange coloration on venter extending entire length of body. Nonbreeding males, females, and juveniles with head, back, and sides light brown or tan; dorsal saddles dark brown; underside white to yellow orange. Body lightly stippled with dark melanophores; stippling most obvious in nonbreeding and freshly preserved individuals; venter and head not mottled. Dark vertical suborbital bar distinct. Humeral spot black and conspicuous.

SIMILAR SPECIES *Etheostoma autumnale* is most like the allopatric *E. mihileze*, but differs from it in usually having 70–75 lateral scales (vs. usually 59–66), a lightly stippled body (vs. body heavily stippled and mottled), usually 5 or 6 dorsal saddles (vs. usually 4), and in lacking vermiculated melanin coloration pattern on top and sides of head (vs. vermiculations present). For other differences in breeding male coloration, see Mayden (2010). It differs from other Arkansas members of the subgenus *Ozarka* (the allopatric *E. cragini* and *E. pallididorsum*) in having a longer lateral line extending well past the base of the first dorsal fin (vs. lateral line short, not extending past base of first dorsal fin), more than 65 lateral scales (vs. fewer than 57), and 13 or more soft dorsal rays (vs. 12 or fewer). It also differs from *E. pallididorsum* in lacking a wide, pale stripe down middle of back (vs. wide, pale middorsal stripe present).

VARIATION AND TAXONOMY Formerly included in *E. punctulatum* in the first edition of this book, but Mayden (2010) described White River populations as a new species, *E. autumnale*, based on morphological and allozyme characters. The type locality is Sawyer Creek at the junction of Missouri highways B and D just west of Henderson, Webster County, Missouri. *Etheostoma autumnale* is a member of the *Etheostoma punctulatum* species group within the subgenus *Ozarka*. Genetic evidence did not support monophyly of *Ozarka* (Near et al. 2011). The Autumn Darter forms a sister pair with *E. mihileze*, and that pair is sister to *E. punctulatum* (Mayden 2010).

DISTRIBUTION The Autumn Darter is endemic to the White River drainage of the Ozark Plateau in Missouri and Arkansas, where it occurs in Arkansas in the White, Kings, Buffalo, Little Red, and Strawberry rivers (Map 190). It has not been found in the Current and Eleven Point rivers in Arkansas, although it is known from a few localities in the headwaters of those rivers in Missouri.

HABITAT AND BIOLOGY The Autumn Darter is most common in Arkansas in the headwaters of the White River in the western Ozarks and in streams of the north-central Ozarks; it becomes less common eastward. Pflieger (1997) also reported it as rare in the southeastern Ozarks of Missouri. It is primarily an inhabitant of small headwater creeks and large springs and spring-runs but is also found in smaller numbers in large creeks and small rivers. Surprisingly, 716 specimens were taken from Beaver, Bull Shoals, Norfork, and Greers Ferry reservoirs (Buchanan 2005). It requires clear water, some permanent flow, and is intolerant of silt. During most of the year it is found in gravel- or rubble-bottomed pools in quieter waters in vegetation or detritus. While electrofishing, we have often taken this species next to large rocks scattered over a pool bottom. It is often associated with undercut banks with overhanging rooted vegetation. In March and April, it is usually found in shallow gravelly riffles with moderate to swift current; these riffles often have aquatic vegetation.

Stomach contents of 28 specimens, 39–80 mm TL, collected from White River headwaters and tributaries in Madison, Searcy, Fulton, and Newton counties from late May to mid-September, and 30 specimens, 42–92 mm TL, collected from middle White River tributaries in Baxter, Stone, Fulton, and Independence counties from mid-October to late November were examined by TMB. Full stomachs were not common at any season. Stomachs in May and June contained little food, mostly ephemeropteran remains. In August and September, the stomachs had more food, primarily ephemeropterans and chironomids, and in October through November, the diet was dominated by plecopterans, followed in importance by ephemeropterans and trichopterans.

It is believed that spawning occurs in riffles (Pflieger 1997). According to Hubbs (1985), spawning occurs in Arkansas from mid-February through mid-May in stenothermal spring environments, but examination of ovaries and testes of specimens taken in November by TMB indicated that the onset of spawning activity may occur earlier. Specimens from Brushy Creek in Fulton County had large testes, and ovaries had begun to enlarge. By 28 November, females from Spring Creek in Independence County had considerably enlarged ovaries containing developing eggs of various sizes. The largest developing cohort consisted of about 300 eggs. The males from Spring Creek also had greatly enlarged testes. Aquarium observations of spawning in other members of the *E. punctulatum* species group indicate that the eggs are buried in gravel (Page and Simon 1988; Simon and Garcia 1990). Males reach sexual maturity at 1 year of age, but only those yearling females at least 1.9 inches (49 mm) TL produced mature ova. Maximum life span is slightly more than 4 years (Hotalling and Taber 1987).

CONSERVATION STATUS   *Etheostoma autumnale* is common in favorable habitats and most of its populations are currently stable, although large numbers are rarely found at a single locality.

Etheostoma blennioides Rafinesque
Greenside Darter

Figure 26.51a. *Etheostoma blennioides*, Greenside Darter male in breeding coloration. *Uland Thomas.*

Figure 26.51b. *Etheostoma blennioides*, Greenside Darter, nonbreeding pigmentation. *Uland Thomas.*

CHARACTERS   Arkansas's largest species of *Etheostoma;* adults commonly 3–4 inches (76–102 mm) long, but reaching 5.5 inches (140 mm) TL in the Buffalo River (Fig. 26.51). A moderately slender darter with a blunt snout that overhangs the inferior mouth. Adults with a nipplelike extension or fleshy tab at middle of upper lip (nipple poorly developed in small individuals). Skin over maxilla fused with skin of snout, leaving a groove only above posterior end of maxilla; premaxillary frenum not evident, upper lip separated from snout at midline by a continuous groove. Lower side with a series of 5–8 dark V-, H-, or W-shaped blotches which during breeding season are often obscured by dark lateral bands extending from dorsum to venter. Back with 5–8 saddles of varying intensity. Upper side flecked with small dark blotches and irregular marks. Gill membranes broadly connected. Infraorbital canal uninterrupted,

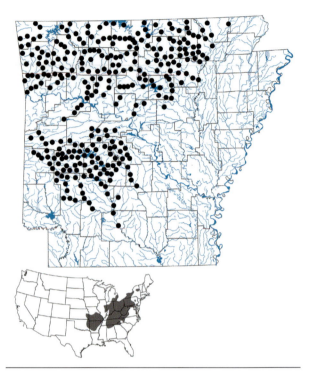

Map 191. Distribution of *Etheostoma blennioides*, Greenside Darter, 1988–2019.

usually with 8 pores. Opercle, cheek, nape, and belly scaled; breast naked to partly scaled. Lateral line complete; lateral line scale count highly variable in Arkansas populations with 61–86 scales (see Variation and Taxonomy section). Diagonal scale rows 17–24; scale rows around caudal peduncle 22–30. Dorsal spines 12–14; dorsal rays 12–15; anal spines 2; anal rays 7–9. Breeding males usually with weakly developed tubercles on ventral scales from midbelly area to caudal fin base, but not on the fins (males are often without tubercles). Breeding males and females with tips of pelvic fin spines and rays of pelvic, pectoral, and anal fins with thickened epithelium. Maximum reported size is 6.75 inches (170 mm) TL (Page and Burr 2011).

LIFE COLORS   Body overall yellow green with dark green to brown saddles. Prominent pre- and suborbital bars are usually present. Lateral V- or W-shaped blotches are dark green. Upper sides with brownish-red spots and small blotches. Both dorsal fins green with basal red bands. Other fins green. Underparts yellowish white to dusky. Breeding males bright green, developing intensely dark green lateral bands; sometimes the intense green color obscures other body markings. The first dorsal fin has a brick-red base and is otherwise green; other fins are green. Breeding females not brightly colored, having an overall yellow-green appearance.

SIMILAR SPECIES   The Greenside Darter can be distinguished from all other Arkansas darters by the fusion of skin over the maxilla with skin of the snout, and in having a nipplelike extension at middle of upper lip in all but smaller individuals. It differs from *E. zonale* (with which it is almost always found in Arkansas) and from *E. histrio* in its lack of an evident premaxillary frenum (vs. frenum present), lower sides with V- or W-shaped blotches (vs. sides usually with vertical bars), and usually more than 63 lateral line scales (vs. fewer than 63).

VARIATION AND TAXONOMY   *Etheostoma blennioides* was described by Rafinesque (1819) from the Ohio River (possibly near Louisville, Kentucky). The Greenside Darter has traditionally been placed in the subgenus *Etheostoma*. Near et al. (2011) supported recognition of a new clade (subgenus) name, *Neoetheostoma*, to avoid redundancy of the genus name *Etheostoma*. Controversy in interpreting the morphological variation within this species over its large range has existed for decades, and more recent genetic analyses have also resulted in conflicting interpretations (e.g., see Piller and Bart 2009, and a different interpretation by Stepien and Haponski 2010). In a rangewide study, R. V. Miller (1968) recognized four subspecies and several races based on morphological characters. The four subspecies of Miller were (1) *E. b. blennioides* in the upper Potomac, Genesee, and Ohio rivers, (2) *E. b. gutselli* in the Little Tennessee and Pigeon rivers, (3) *E. b. pholidotum* in the Great Lakes drainage and in the Wabash, Maumee, Mohawk, and Missouri systems, and (4) *E. b. newmanii* in most of the Tennesee and Cumberland river drainages and west of the Mississippi River in the Ozark and Ouachita uplands. Miller (1968) recognized two races of *E. b. newmanii* in Arkansas. Populations in the St. Francis, Current, Black, and upper White rivers were assigned to the Cumberland race of *newmanii*. Those in the Little Red, Arkansas, and Ouachita drainages were assigned to an Arkansas race of that subspecies. Based on morphological differences, Etnier and Starnes (2001) elevated *E. b. gutselli* to full species status (*E. gutselli*, the Tuckasegee Darter), a designation that was accepted by Nelson et al. (2004) and Page et al. (2013). Piller and Bart (2017) found additional morphological evidence for species status of *E. gutselli* and provided a rediagnosis of the species. Haponski and Stepien (2008) and Stepien and Haponski (2010), using analyses of one mitochondrial and one nuclear gene, supported the specific recognition of *E. gutselli*, and the Tennessee/Hiwassee River clade of *E. b. newmanii* was also supported as a full species (*E. newmanii*). This left a more restricted Greenside Darter clade consisting of the remaining subspecies of Miller (*blennioides*, *newmanii* west

of the Mississippi River, and *pholidotum*). However, none of the three subspecies were monophyletic in the nuclear DNA trees. Stepien and Haponski (2010) concluded that the putative subspecies *newmanii* and *pholidotum sensu* Miller (1968) are polyphyletic and do not merit continued taxonomic recognition; *E. b. blennioides* was paraphyletic in the nuclear DNA analyses. Near et al. (2011) supported the elevation of all four subspecies of Miller (1968) to full species, with *E. b. newmanii* divided into two new species, *Etheostoma newmanii* (Highlands Greenside Darter) in the Tennessee and Hiwassee rivers, and an undescribed species, *Etheostoma* sp. cf. *newmanii* (Arkansas Greenside Darter) in the Ozarks and Ouachitas. Page et al. (2013) recognized only the elevation of *Etheostoma gutselli*. Because additional study is needed to clarify the status of Arkansas populations of *E. blennioides*, we refrain from referring to those populations as the undescribed species, *E.* sp. cf. *newmanii sensu* Near et al. (2011) and continue to list all Arkansas populations of Greenside Darters under *E. blennioides*. Other genetic studies identified multiple unique lineages of *Etheostoma blennioides* (Haponski and Stepien 2008; Piller et al. 2008; Piller and Bart 2009); however, the lineages did not correspond to the subspecies identified by Miller (1968). Page and Burr (2011) recognized three subspecies of *E. blennioides*, and continued to assign all Arkansas populations to *E. b. newmanii*.

Meristic features of Arkansas populations exhibit considerable variation. Lateral line scale counts provided by Miller (1968) were Current-Black rivers 61–78 ($\bar{x}$ = 70.29), upper White River 68–86 ($\bar{x}$ = 77.05), Little Red River 67–86 ($\bar{x}$ = 76.08), Arkansas River 65–82 ($\bar{x}$ = 73.53), Ouachita River 63–79 ($\bar{x}$ = 69.86), and Saline River 61–72 ($\bar{x}$ = 66.48). The Little Red River population differs considerably in meristic characters from populations of the upper White, Current-Black, and St. Francis rivers. It also differs from those populations in the shape of the head and snout (lacking the bulging cheeks, the long, swollen snout, and flattened head). Miller (1968) noted that the Little Red River population most closely resembles Arkansas River populations and should be recognized as part of the Arkansas race.

DISTRIBUTION   A widely distributed upland darter east of the Mississippi River, found in the Great Lakes and Mississippi River basins from southern Canada to northern Alabama, and in some Atlantic Slope drainages from New York to Virginia. West of the Mississippi River, it is restricted primarily to the Ozark and Ouachita uplands. In Arkansas, the Greenside Darter is abundant throughout Ozark Upland drainages of the state, common in the Ouachita River system of the central Ouachitas, and sporadic in Arkansas River tributaries in the western half of the state. It is apparently more common in Arkansas River tributaries north of the river than in southern tributaries and is conspicuously absent from Red River tributaries of the Little River system in the Ouachita Upland (Map 191).

HABITAT AND BIOLOGY   One of Arkansas's most common darters, the Greenside Darter is found from small creeks (but usually not in headwater streams) to moderate-sized rivers with moderate to swift current and low turbidity. The largest adults usually occupy the swiftest and deepest areas of riffles (Fahy 1954; Winn 1958a). In Kansas, it requires permanent flow and is most abundant in streams having a large, stable volume of flow, allowing the growth of filamentous algae on riffles (Cross and Collins 1995). It is most commonly found in riffles over a gravel and rubble bottom, where it is sometimes collected in large numbers. In Missouri, the average current speed in Greenside Darter habitats was 64 cm/s, and water depth averaged 28.8 cm (White and Aspinwall 1984). Fullenkamp (2010) noted a preference for cobble and boulder substrate in Indiana streams. In another study of microhabitat selection in Indiana streams, Greenside Darters were most commonly found over substrates of intermediate particle size (cobble and boulder) and in deeper parts of riffles with high water velocities (Pratt and Lauer 2013). It is often found in riffles containing the alga *Cladophora* (Miller and Robison 1973) and the aquatic moss *Fontinalis*, and it is sometimes found in silt-free bedrock pools having a steady current (Kuehne and Barbour 1983). The young are commonly found in quiet water among growths of aquatic vegetation or in accumulations of leaves, sticks, and other organic debris (Pflieger 1997). Its numbers were greatly reduced in the Beaver Lake tailwaters 30 years after impoundment (Quinn and Kwak 2003), and Buchanan (2005) found Greenside Darters in small numbers in five Arkansas impoundments in areas of clear water and rocky substrates. Miller (1968) also noted that it occasionally occurs in a range of habitats, including relatively quiet lake shores.

Several early ecological studies of the Greenside Darter were summarized by Miller (1968). Pflieger (1997) reported that it often forages over boulders and logs in pools with little current, and Fahy (1954) found that food was picked from algal-covered rocks in riffles. Its food consists largely of riffle-dwelling insect larvae and nymphs, mainly midges, caddisflies, and mayflies (Turner 1921; Pflieger 1997). Fahy (1954) also listed immature coleopterans, megalopterans, and odonates as other frequently consumed items. In Ohio, adults preferred midge and simuliid larvae, with chironomids making up 62% of the diet by weight (Wynes and Wissing 1982; Hlohowskyj and White 1983). Juveniles ate primarily midge larvae and microcrustaceans, while adults ate midge larvae, black fly larvae, and immature mayflies and caddisflies. Etnier and Starnes (1993) reported that adults consumed large numbers of snails in the Little River, Tennessee.

Several investigators reported Greenside Darter spawning habits (Fahy 1954; Pflieger 1997; Winn 1958a, b). Its reproductive biology was studied in Ohio by Barron and Albin (2006) who reported a spawning season extending from mid-March to early June. Breeding began in Ohio when water temperatures reached 8–10°C, and breeding ceased at water temperatures of 18–20°C. Spawning occurred in Missouri in March and April (Pflieger 1997), in Alabama in April and May (Boschung and Mayden 2004), and in Arkansas from late February to late March (Hubbs 1985) in riffles over filamentous algae or water moss. McCormick and Aspinwall (1983) found that Greenside Darters used olfactory cues to locate aquatic vegetation and algae. Males actively defend territories. During spawning a female moves into algae and is followed by a male who mounts her. The two vibrate and 3–15 eggs are released and fertilized. A pair may spawn more than once at short time intervals. The eggs are attached to the vegetation, but aquarium observations indicated that algae was not a requisite for successful spawning (Schwartz 1965). Several females spawn with one male and a single female may lay 400–1,800 eggs in a season (Fahy 1954). Average clutch size in Ohio was 107.1 eggs, and clutch size was approximately 20–30% greater in the middle of the breeding season than at either end (Barron and Albin 2006). Females can potentially recruit, mature, and ripen a single clutch of eggs in as little as 2 or 3 days. Hatching occurs in 18 days at 55–58°F (12.8–14.4°C). No parental care is given the eggs. Age and growth were studied in Kentucky by Wolfe et al. (1978). Maximum life span is around 5 years. Fayton et al. (2017) described a new species of helminth parasite (*Plagioporus*) from *E. b. newmanii* in Arkansas.

CONSERVATION STATUS   This species is restricted almost entirely to the Ozark and Ouachita uplands in Arkansas. Its populations are widespread and currently secure.

Etheostoma caeruleum Storer
Rainbow Darter

CHARACTERS   A colorful, moderately deep-bodied darter with 6–12 dark dorsal saddles, 2 or 3 of which are darker than the rest (Fig. 26.52). Side with 6–12 dark

Figure 26.52a. *Etheostoma caeruleum*, Rainbow Darter male in breeding coloration. *Uland Thomas*.

Figure 26.52b. *Etheostoma caeruleum*, Rainbow Darter female. *Uland Thomas*.

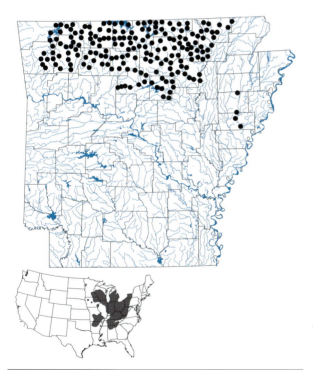

Map 192. Distribution of *Etheostoma caeruleum*, Rainbow Darter, 1988–2019.

narrow vertical bars best developed posteriorly and often meeting counterparts on belly and across back, especially in males; when anterior bars are indistinct, that area is darkly mottled. Lateral bars noticeably slanting backward. Gill membranes narrowly joined across throat. Premaxillary

frenum present. Infraorbital canal uninterrupted. Opercle and belly scaled; cheek and breast naked. Lateral line straight and incomplete, usually ending beneath second dorsal fin. Scales in lateral series 38–54; pored scales 16–39. Scale rows above lateral line 4–7; scale rows below lateral line 6–10. Diagonal scale rows 11–16; scale rows around caudal peduncle 16–23. Pectoral rays usually 13 or more; dorsal spines 9–11; dorsal rays 12–14; anal spines 2; anal rays 6–8; principal caudal rays usually 17. Breeding males with tubercles on ventral scales (on belly, along the base of the anal fin, and on caudal peduncle). Maximum length about 3 inches (76 mm).

LIFE COLORS   Back mottled brown or tan with dark brown saddles. Sides tan with blue-green vertical bars alternating with wider orange spaces. Adult males are colorful throughout the year, but colors become especially intense in breeding males. Small orange spot on each scale on side, forming horizontal rows. Cheeks blue, lower surface of head and throat bright orange; breast blue black. First dorsal fin orange red basally and blue distally. Second dorsal fin blue at margin and base, but broadly suffused with red. Pelvic fins blue; anal fin with mixture of blue and red. Females generally drab olive brown with thin red and blue bands in first dorsal fin, wavy black lines in second dorsal fin, and sometimes traces of red and blue in the anal fin. Females are also more mottled with less-distinct lateral bars.

SIMILAR SPECIES   *Etheostoma caeruleum* is often confused with species of the *E. spectabile* complex, and it is sympatric with three of those species (*E. fragi*, *E.* sp. *cf. spectabile*, and *E. uniporum*) in northern Arkansas. It is distinguished from those species by the presence of 13 or more pectoral fin rays (vs. fewer than 13), an uninterrupted infraorbital canal (vs. interrupted infraorbital canal), greatest body depth occurring beneath first dorsal fin (vs. just anterior to first dorsal fin), and males with red in anal fin in life (vs. males without red in anal fin). It differs from *E. asprigene* in having naked cheeks (vs. fully scaled), 2 or 3 dorsal saddles darker than others (vs. dorsal saddles of about equal intensity), and males with red in anal fin in life (vs. usually no red pigment in anal fin).

VARIATION AND TAXONOMY   *Etheostoma caeruleum* was described by Storer (1845) as *Etheostoma caerulea* from the Fox River, Illinois. It is a member of subgenus *Oligocephalus*. Its specific distinctness from *E. spectabile* was confirmed by Trautman (1930). Based on genetic evidence, Near et al. (2011) recognized a Rainbow Darter species group (referred to by some authors as the *E. asprigene* group) consisting of *E. caeruleum*, *E. asprigene*, and *E. collettei*. Morphological variation in this wide-ranging species was studied by Knapp (1964), who recognized three subspecies and several races. McCormick (1990) reported that there are at least two distinct forms of *E. caeruleum* in Arkansas: one undescribed form occurs in most of the White River drainage, and a second unnamed form, which may be specifically distinct, occurs in the Little Red River drainage and on Crowley's Ridge. Ray et al. (2006) studied mitochondrial DNA cytochrome *b* genetic variation and recognized four major clades of *E. caeruleum* with many unique haplotypes, few of which are shared across drainages. Two of the clades are represented in Arkansas: (1) the White-Little Red River clade, and (2) the Mississippi River corridor clade, which includes populations in the Spring and Strawberry rivers, and presumably the small populations on Crowley's Ridge. Based largely on analysis of nuclear DNA, Near et al. (2011) recognized populations from the White and Little Red river drainages as an undescribed species, *E.* sp. *cf. caeruleum*, the Ozark Rainbow Darter. Because additional genetic analysis of Arkansas populations of *E. caeruleum* is needed, particularly populations on Crowley's Ridge, we refrain from recognizing *E.* sp. *cf. caeruleum sensu* Near et al. (2011) as a separate species and provisionally continue to assign all Arkansas populations of Rainbow Darters to *E. caeruleum*.

Ray et al. (2008) found evidence for mitochondrial introgression between *E. caeruleum* and *E. uniporum* based on analysis of the cytochrome *b* gene. They speculated that the mechanism of introgression was likely asymmetric sneaking behavior by male *E. uniporum*, a mating tactic observed in related species. Bossu and Near (2009) provided additional evidence for mitochondrial introgression between *E. caeruleum* and other members of the *E. spectabile* species complex. Zhou and Fuller (2014) found evidence for reinforcement of prezygotic reproductive barriers between sympatric *E. caeruleum* and *E. spectabile* populations in Illinois and Michigan, and Zhou et al. (2014) showed that male color traits differed across species, populations, and body sizes, with size differences contributing the most to individual color variation.

DISTRIBUTION   The Rainbow Darter is found in the Great Lakes and Mississippi River drainages from southern Canada west to Minnesota and south to Alabama and Arkansas, with disjunct populations in the lower Mississippi River basin (Mississippi and Louisiana), Hudson Bay basin, and upper Potomac River drainage of the Atlantic Slope. In Arkansas, it occurs primarily across the northern part of the state in the White River and its tributary streams, including the Current, Spring, Strawberry, and Little Red rivers. A disjunct population occurs in eastern Arkansas in four small creeks of the

southern part of Crowley's Ridge in the St. Francis River drainage in Cross and St. Francis counties (Map 192).

HABITAT AND BIOLOGY   Typically, the Rainbow Darter is found in creeks and small rivers of high to moderate gradient where it often occurs in large numbers in clear, swift riffles having coarse gravel and rubble bottoms. It was most common in Indiana streams having cobble and boulder substrates, where it occurred in the deeper riffles (Pratt and Lauer 2013). It is reportedly intolerant of turbidity and intermittent flow. Weston et al. (2010) studied habitat partitioning between the Rainbow and Yellowcheek darters in the Middle Fork of the Little Red River, Arkansas. In riffles where Yellowcheek Darters were present, Rainbow Darters preferred the downstream areas of the riffles. Rainbow Darters also retreated to pools when riffles dried and were able to recolonize previously dry areas, whereas Yellowcheek Darters did not show recolonization ability. Rainbow Darters were at their greatest densities in upstream riffles where Yellowcheek Darters had not recolonized, suggesting niche partitioning (Weston et al. 2010). Similarly, Geihsler (1975) found that *E. caeruleum* was most abundant in the lower portions of riffles in the Buffalo River, Arkansas, while *Nothonotus juliae* occupied the upper sections of the same riffles. Commens and Matthis (1999) provided evidence for recognition of alarm pheromones by *E. caeruleum*. Rainbow Darters exposed to extracts from macerated skin of conspecifics and *N. juliae* responded by decreasing their activity. Rainbow Darter populations in the Beaver Lake tailwaters were greatly reduced 30 years after impoundment (Quinn and Kwak 2003), but Buchanan (2005) collected more than 1,500 specimens from four White River drainage reservoirs in northern Arkansas. The Rainbow Darter is not usually found in extreme headwaters, a habitat preferred by members of the Orangethroat Darter clade (*Ceasia*). In the scattered localities in Arkansas where the Rainbow Darter and members of the Orangethroat Darter clade are syntopic, the former is found in the swifter, deeper riffles, the latter in shallow riffles or near the margins of pools in slow current. Surprisingly, the disjunct populations of *E. caeruleum* of southern Crowley's Ridge (St. Francis River drainage) are found in habitat normally more suited for species of the Orangethroat Darter clade. The creeks there are small and have gravel and sandy bottoms and are subject to periodic siltation and turbidity. The largest of the Crowley's Ridge populations occurs in Village Creek of Village Creek State Park in Cross and St. Francis counties.

Larval Rainbow Darters up to 15 mm TL fed on zooplankton, primarily copepods, but soon added immature insects to their diet (Turner 1921). Above 15 mm TL, copepods decreased in importance. The stomachs of fish 15–20 mm TL contained 40% copepods, and stomachs of fish 35–40 mm long contained only 0.25% copepods. Foraging behavior of Rainbow Darters in an Ozark stream consisted of a roving search pattern characterized by frequent short moves and turns across the substrate while using head and eye movements to locate prey (Vogt and Coon 1990). Adult Rainbow Darters feed on a variety of aquatic insect larvae but have also been reported to eat snails, crayfish, and even fish eggs (Kuehne and Barbour 1983). In Ohio, its diet in fall and winter was dominated by chironomids; other important components were simuliid larvae, trichopteran larvae (*Cheumatopsyche* and *Hydropsyche*), and coleopteran larvae (*Stenelmis*) (Stewart 1988). In an Ozark stream in Missouri, adults fed primarily on chironomid larvae obtained by foraging in the open along the stream bottom (Vogt and Coon 1990). Stomach contents of 34 specimens collected in March and April from Village Creek in Cross and St. Francis counties were examined by TMB. The diet was dominated by chironomids (69% frequency of occurrence), amphipods (56%), and ephemeropteran nymphs (31%). Other items ingested in small amounts were coleopterans, plecopteran nymphs, isopods, and insect eggs, Twenty-seven specimens collected from White River tributaries in Madison County in May by TMB fed primarily on plecopterans, ephemeropterans, and fish eggs, with chironomids, simuliids, trichopterans, and arachnids composing a small percentage of the diet.

Hubbs (1985) reported that spawning occurs in Arkansas from late February into late May in swift gravel riffles when water temperatures rise above 59°F (15°C). Examination of gonads by TMB supported Hubbs's observations. Ripe males and females were found in the West Fork of the White River on 27 March, in Village Creek on Crowley's Ridge on 25 and 31 March at a water temperature of 15°C, and in Kings River on 17 May at a water temperature of 19°C. Ripe females of the Crowley's Ridge population (Village Creek) ranged from 42 to 63 mm TL and contained from 53 to 242 mature ova ($\bar{x}$ = 96.2) in March and April; the presence of smaller egg sizes in ovaries indicated multiple spawning. By mid to late May, ripe females in White River tributaries in Sharp and Madison counties were nearing the end of the spawning season. Those females (14 individuals from 48 to 55 mm TL) contained from 11 to 166 mature ova, and there were no apparent smaller developing ova present. Females normally avoid the riffles during the spawning season, except when actually spawning (Winn 1958a). Males have shifting, ill-defined territories on the riffles with fine to moderate-sized gravel. Females and younger males congregate immediately below the riffles.

When ready to spawn, a female enters the riffle and signals her readiness to spawn by performing nose digs (pushing her head into the gravel). She then buries the lower half of her body in the gravel and is mounted by the male. The eggs are fertilized and buried in the gravel as both vibrate their bodies rapidly. This act may be repeated several times as male and female move upstream. Occasionally, sneaker males release sperm next to a spawning pair and successfully fertilize some of the female's eggs (Fuller 1999). Group spawning was occasionally observed in Michigan, where two males mated simultaneously with a female, and male interactions, in the form of guarding, played a larger role in determining male mating success than did female choice (Fuller 2003). After spawning, there is no further parental care. Clutch size in Mississippi ranged from 16 to 60, and hatching occurred in 10 or 11 days at temperatures of 17–18.5°C (Heins et al. 1996). Fuller (1998) estimated average annual fecundity from aquarium spawning as 309 eggs. Winn (1958a) counted 508–1,462 eggs in females 34–50 mm TL, but it is not certain that all eggs were laid during the spawning season. Sexual maturity is attained at 1 year of age. Maximum life span is about 4 years, but few individuals survive that long.

CONSERVATION STATUS *Etheostoma caeruleum* is widespread and often abundant in its preferred habitat in the Ozarks, and those populations are currently secure. If the Little Red River and Crowley's Ridge populations are ultimately described as taxonomically distinct, those populations would warrant future monitoring. The Crowley's Ridge populations are especially vulnerable because of gravel mining and other land use practices. Rainbow Darters were first collected from Crowley's Ridge in Village Creek (St. Francis County) in 1937 by E. Moore (reported by Black 1940). Additional Ridge populations were found in Sugar Creek (Cross County), Harefarm Creek (Cross County, Fulmer and Harp 1977), Village Creek (Cross County), Willow Creek (a Village Creek tributary in St. Francis County), and Crow Creek (St. Francis County) in the 1960s and 1970s. In 2017, populations persisted in three of those creeks (Sugar, Village, and Crow creeks). We consider the Crowley's Ridge *E. caeruleum* populations to be of special concern if not threatened.

Etheostoma chlorosoma (Hay)
Bluntnose Darter

CHARACTERS A small, slender darter with a small head and blunt snout (Fig. 26.53). One of the least colorful *Etheostoma* species, but pigmentation patterns are

Figure 26.53. *Etheostoma chlorosoma*, Bluntnose Darter.
Uland Thomas.

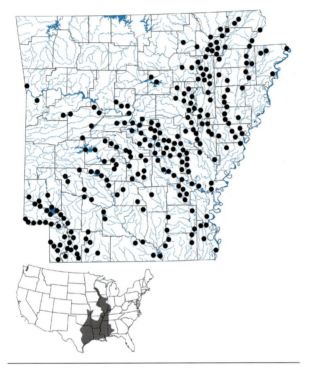

Map 193. Distribution of *Etheostoma chlorosoma*, Bluntnose
Darter, 1988–2019.

very useful for identification. Back with about 6 faint
dusky saddles. Midside with small dark blotches, often
in X- or M-shaped marks; upper sides with smaller X- or
M-shaped markings. Dark preorbital bar extending onto
snout and fusing with counterpart from other side, forming a continuous dark band around snout. Premaxillary
frenum absent, upper lip and snout separated by a continuous groove. Gill membranes separate or only slightly
connected. Infraorbital canal interrupted, with 2–4 anterior pores and 0 or 1 posterior pore. Cheek, opercle, and
prepectoral area fully scaled; breast partly to fully scaled;
belly fully scaled or scaled only posteriorly. Lateral line not
arched anteriorly, incomplete, and usually ending beneath
second dorsal fin. Scales in midlateral series 50–60; pored
scales 4–36. Scale rows above lateral line 3–7; scale rows
below lateral line 4–9. Diagonal scale rows 10–17; scale
rows around caudal peduncle 15–23. First and second dorsal

fins moderately to widely separated. Dorsal spines 9 or 10;
dorsal rays 10 or 11; anal spine 1; anal rays 7–9; pectoral rays
12–14. Breeding males with tubercles on pelvic and anal
fin rays. Breeding males and females with 2 rows of fleshy,
tubercle-like structures on interradial membranes of caudal
fin that may function as contact organs during spawning
(Bart and Cashner 1986). Maximum size around 2.4 inches
(61 mm).

LIFE COLORS   Basically a straw-colored darter with a
greenish-yellow back and sides having brown or black spots
and vermiculations. Back with about 6 dusky saddles that
may be faint or distinct. Confluent preorbital bars black,
forming a bridle around front of snout. Suborbital bar thin
and black. Dorsal and caudal fins with light brown bands;
other fins clear. Breeding males darker overall, with a dusky
edge on first dorsal fin but lacking bright colors; operculum
with a metallic green tint.

SIMILAR SPECIES   Sometimes confused with the
Johnny Darter, *E. nigrum,* the Bluntnose Darter differs
in having a dark stripe extending forward from each eye
continuous around snout above upper lip (vs. dark stripe
distinctly interrupted on midline of snout); lateral line
incomplete, usually extending only to below soft dorsal fin
(vs. lateral line complete or nearly so); and scaled cheeks
(vs. naked cheeks). It differs from *E. clinton, E. stigmaeum,*
and *E. teddyroosevelt* in having only 1 anal spine (vs. 2 anal
spines), breast partly to fully scaled (vs. naked), usually 9
or 10 dorsal spines (vs. usually 12 or 13), and lack of bright
green, red, and blue coloration in breeding males (vs. breeding males brightly colored).

VARIATION AND TAXONOMY   *Etheostoma chlorosoma* was described by Hay (1881) as *Vaillantia chlorosoma* from a tributary of the Tuscumbia River at Corinth,
Alcorn County, Mississippi. It is the only Arkansas member of subgenus *Vaillantia.* The long-time spelling of the
specific epithet (*chlorosomum*) was replaced with the
gender-correct *chlorosoma* (Nelson et al. 2004). The monophyly of *Vaillantia* was supported by genetic analysis (Near
et al. 2011). Bart and Cashner (1986) studied geographic
variation in Gulf Slope populations of *E. chlorosoma* and
found variation in meristic characters east and west of the
Mississippi River. They did not support the recognition of
subspecies.

DISTRIBUTION   The Bluntnose Darter occurs in
the Mississippi River basin from southern Minnesota
and Wisconsin, south through Louisiana, and in Gulf
Slope drainages from Alabama west through the San
Antonio River drainage of Texas. In Arkansas, it is found
in streams of the Coastal Plain lowlands in eastern and
southern Arkansas and throughout the Arkansas River

Valley. It is absent from the most mountainous regions of the Ozarks in northcentral and northwestern Arkansas, and mostly absent from the most mountainous regions of the Ouachitas; however, the Bluntnose Darter penetrates farther into the Ouachitas than into the Ozarks (Map 193).

HABITAT AND BIOLOGY   Primarily a lowland species, the Bluntnose Darter occurs in quiet water in sluggish streams, bayous, swamps, oxbow lakes, and sometimes moderate to large rivers. It is usually found over a bottom of mud and detritus or sand. Aquatic vegetation is apparently not a habitat requirement, and the removal of submerged aquatic vegetation from a Texas reservoir did not affect the abundance of *E. chlorosoma* (Bettoli et al. 1993). It is most abundant in Arkansas near the Fall Line and occasionally found in streams above the Fall Line, particularly in the Ouachitas, where it inhabits the quieter waters of pools with mainly sand and detritus bottoms. In Kansas, it is found in streams that become intermittent and turbid during severe droughts (Cross 1967). Because of its habitat preferences, it is sometimes the only darter found at a collection site. The Bluntnose Darter readily adapts to some reservoirs. Buchanan (2005) found it in 30 of 66 Arkansas impoundments sampled, and more than 2,400 specimens were collected in that study. *Etheostoma chlorosoma* is widely distributed and common in most of the lowlands of Arkansas, but there are reports of declines in northern areas of its range (Kuehne and Barbour 1983; Smith 1979). Pflieger (1997) reported its decline in the prairie streams of northeastern Missouri and attributed it to increased siltation.

In Tennessee, Bluntnose Darters fed on larval insects such as caddisflies, dytiscid beetles, and midge larvae (Etnier and Starnes 1993). Stomach contents of 51 *E. chlorosoma* collected in March, July, and October from six localities in Arkansas were examined by TMB. As expected in a small darter with a small mouth, the diet was dominated at all times of year by small insect larvae and nymphs, and small crustaceans. Specimens from Mississippi County in March fed mainly on small crustaceans (cladocerans and copepods), insect eggs, trichopterans, and chironomids. Specimens collected in July from Hempstead, Lafayette, and Garland counties ate primarily chironomids and other dipteran larvae and also consumed small crustaceans (primarily copepods). The diet in October in Lake Conway (Faulkner County) was not dominated by any one food item, but consisted mostly of small crustaceans (copepods, cladocerans, and isopods) along with ephemeropterans, chironomids, and other dipteran larvae.

Spawning occurred from early January to late March in Texas (Hubbs 1985), and Cross and Collins (1995) believed that spawning occurred in April or May in Kansas. The spawning period in Arkansas is not precisely known but probably occurs from March to May based on examination of ovaries and testes by TMB. Females collected from Mississippi County, Arkansas, on 14 March contained ripe ova, and males had greatly enlarged testes. The ripe females contained from 84 to 105 mature ova, and the presence of smaller egg sizes indicates multiple spawning. Ripe females were found on 11 April in Clark County (Tumlison et al. 2017). In Tennessee, tuberculate specimens were found in March and April (Etnier and Starnes 1993). Spawning temperatures range from 22°C to 26°C (Simon 1994). It is an egg attacher, and Winn (1958a) reported that its eggs are laid on plants or plant debris. Page et al. (1982) described aquarium spawning 1 day after specimens were collected on 30 April from an Illinois stream with a water temperature of 22°C. Males were aggressive toward one another and courted the females. Lateral displays by the males were the most important features of aggressive and courtship behaviors. After a brief courtship, the female selected spawning substrates such as filamentous algae, twigs, or other materials. The male mounted the female and both vibrated with the female releasing 1–3 adhesive eggs at a time. Maximum life span is about 3 years.

CONSERVATION STATUS   This widespread lowland species is common and sometimes abundant in southern states but is in danger of extirpation from some northern areas of its range. Its populations in Arkansas are currently secure.

Etheostoma clinton Mayden and Layman
Beaded Darter

Figure 26.54a. *Etheostoma clinton*, Beaded Darter male in breeding coloration. *David . Neely.*

Figure 26.54b. *Etheostoma clinton*, Beaded Darter female. *Uland Thomas.*

Map 194. Distribution of *Etheostoma clinton*, Beaded Darter, 1988–2019.

CHARACTERS   The smallest species of the subgenus *Doration*, *E. clinton* is a slender darter with a moderately blunt snout and a horizontal mouth (Fig. 26.54). Back with 6 dark saddles. Midside with a series of 9–11 dark quadrate blotches. Upper side and back with small scattered X-shaped markings and/or zigzag lines. Premaxillary frenum absent, upper lip and snout separated by a continuous deep groove. Gill membranes narrowly joined across throat. Vomerine teeth present; palatine teeth present. Infraorbital canal uninterrupted, with 7–9 pores. Opercle nearly fully scaled; nape squamation highly variable; cheek naked or with a few scales near eye; breast naked; belly mostly scaled. Lateral line incomplete, usually ending beneath second dorsal fin. Scales in lateral series usually 46–53; unpored lateral scales usually 11–21. Scale rows above lateral line usually 4 or 5; scale rows below lateral line usually 6–8. Diagonal scale rows usually 12–15; scale rows around caudal peduncle usually 15–18. Dorsal spines 10–13 (modally 11); dorsal rays 10–13 (modally 11); anal spines 2; anal rays 7–10 (modally 9); pectoral rays modally 13 or 14. Breeding male with tubercles on pelvic and anal fin rays and occasionally on posterior two-thirds of belly. Maximum length around 1.5 inches (35 mm).

LIFE COLORS   Layman and Mayden (2012) provided detailed descriptions of breeding and nonbreeding coloration. Breeding male: spinous dorsal fin with blue-green marginal and submarginal bands, a red-orange medial band, with basal band lacking bright orange pigment. Soft dorsal and anal fins with blue green in base of fin. Face and lower head gray with blue or blue green on operculum. Midside with iridescent turquoise-blue or blue-green quadrate blotches with a narrow continuous midlateral band of melanophores running through blotches. Upper side with small red-orange X-markings and spots. Nonbreeding and freshly preserved specimens with straw-colored back and upper sides; dorsal saddles and lateral blotches brown; underside light-colored. Preorbital dark bar extending onto snout, but not continuous around snout.

SIMILAR SPECIES   It is sometimes confused with *E. chlorosoma* and *E. nigrum*, with which it is occasionally syntopic, but *E. clinton* differs in having 2 anal spines (vs. 1), usually 11 or more dorsal spines (vs. fewer than 11), and breeding males with bright colors (vs. breeding males lacking bright colors). It is most similar to the closely related allopatric species *E. stigmaeum* and *E. teddyroosevelt*. It differs from *E. stigmaeum* in having 9 preoperculomandibular pores (vs. 10), modally 9 anal rays (vs. 8), and in breeding male with a thin, dusky midlateral stripe running through lateral blue-green blotches (vs. breeding male lacking a thin, dusky midlateral stripe). Both *E. clinton* and *E. stigmaeum* differ from *E. teddyroosevelt* in having palatine teeth present (vs. palatine teeth absent) and in breeding male lacking bright orange spots in soft dorsal and caudal fins (vs. breeding male usually with bright orange spots in soft dorsal and caudal fins). *Etheostoma clinton* has the shortest snout, narrowest body depth at the spinous and soft dorsal fin origins, shortest distance between the spinous and soft dorsal fin origins, narrowest trans-pelvic width, and narrowest body of all *Doration* species.

VARIATION AND TAXONOMY   Based on analysis of 34 discrete morphological and behavioral characters, Layman and Mayden (2012) revised the subgenus *Doration* and supported its monophyly. Formerly included in *Etheostoma stigmaeum*, *E. clinton* was elevated to full species status by Mayden and Layman (*in* Layman and Mayden 2012) based largely on variation in morphology and male breeding colors. The type locality is the Caddo River at Arkansas State Highway 182, 3.2 km north of Amity, Clark County, Arkansas. The species epithet was selected to honor William J. Clinton, the 42nd President of the United States of America and a native Arkansan.

Despite its small range, the Beaded Darter populations of the upper Ouachita and upper Caddo rivers vary in several meristic features such as pectoral rays and several scale counts (see Layman and Mayden 2012 for details). The sexes exhibit dimorphism in seven body proportions, with males having a longer upper jaw, taller spinous dorsal fin, and larger soft dorsal, anal, and pelvic fins. Layman and Mayden (2012) also found a significant difference ($P < 0.05$) in mean pored lateral line scales (males 34.1, females 32.7).

DISTRIBUTION   The Beaded Darter is endemic to the upper Ouachita River system of Arkansas, where it is known only from the upper Caddo and upper Ouachita rivers above the Fall Line in Clark, Montgomery, Pike, and Polk counties (Map 194). In the Caddo River, it is found upstream of DeGray Reservoir, and in the Ouachita River, it occurs upstream of Lake Ouachita in the upper Ouachita and South Fork Ouachita rivers.

HABITAT AND BIOLOGY   Descriptions of habitat of species of the subgenus *Doration* provided by Layman and Mayden (2012) included observations of *E. clinton*. *Doration* species are found in clear medium to large-sized creeks and small rivers of moderate gradient. Adults and juveniles occur in slow to moderate current over mixtures of sand, gravel, and occasionally silt substrates. They are typically found just downstream of riffles, in moderate runs, gentle riffles, or along margins of pools. Caldwell (2011) reported population densities of 2.42 fish/100 m² for *E. clinton* from the upper Ouachita River, with densities decreasing from upstream sampling sites downstream toward Lake Ouachita. *Etheostoma clinton* is replaced in the Coastal Plain portion of the Ouachita River system downstream from Remmel and DeGray dams by *E. stigmaeum*.

The life history features of the Beaded Darter are poorly known. No studies of this recently described species have yet been undertaken. Layman and Mayden (2012) reported that all *Doration* species move to gravelly runs and shallow riffles in the spring, presumably for spawning. Spawning probably occurs from March into May, and tuberculate males have been taken from the Caddo River between 25 March and 4 April.

CONSERVATION STATUS   Populations of this darter appear to be small. In 19 field samples in which this species occurred, the modal number of specimens in a sample was 1 (7 samples), and only 2 of the 19 samples had more than 4 specimens (based on collection data reported in Layman and Mayden 2012). The largest number of *E. clinton* in a sample was 10. A survey of the current status of this Arkansas endemic is needed, and its populations should be monitored in the future. Because of its small range and population size, we tentatively consider the Beaded Darter threatened.

Etheostoma collettei Birdsong and Knapp
Creole Darter

Figure 26.55a. *Etheostoma collettei*, Creole Darter male. *Uland Thomas.*

Figure 26.55b. *Etheostoma collettei*, Creole Darter male in breeding coloration. *Renn Tumlison.*

Figure 26.55c. *Etheostoma collettei*, Creole Darter female. *Renn Tumlison.*

CHARACTERS   A small, colorful darter with body deepest at origin of spinous dorsal fin (Fig. 26.55). Back with 8 or 9 brown saddles, 3 or 4 saddles usually much darker than the others. Side with 5–7 vertical bars developed posteriorly, bars indistinct above midside. Upper side with distinct longitudinal rows of small dark spots similar to those of some members of the Orangethroat Darter clade. Dark humeral spot present. Snout slightly decurved, blunt. Gill membranes separate or only slightly joined across throat. Premaxillary frenum present. Eye ovoid. Opercle scaled; upper half of cheek usually scaled, but cheek naked or with only a few scales under eye in eastern Saline River populations; breast and prepectoral area naked (the latter sometimes with a few embedded scales). Nape partly to fully scaled. Infraorbital canal uninterrupted, with 8 pores. Lateral line nearly straight and incomplete, usually ending beneath second dorsal fin. Scales in midlateral

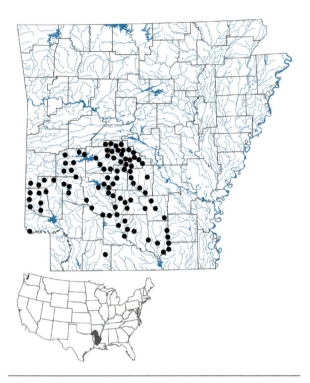

Map 195. Distribution of *Etheostoma collettei*, Creole Darter, 1988–2019.

series usually 47–52 (44–60); pored scales usually 32–42; unpored scales usually 8–18. Scale rows above lateral line modally 5 (4–6); scale rows below lateral line usually 8–10 (7–12). Diagonal scale rows usually 13–15 (11–16); scale rows around caudal peduncle usually 19–22 (18–24). Dorsal spines 8–12; dorsal rays 11–14; pectoral rays usually 12–14; anal spines 2; anal rays 6–8. Breeding males with tubercles on lower body scales, along midventral line of belly, and sometimes along the base of the anal fin. Maximum length is around 2.5 inches (64 mm).

LIFE COLORS   Usually 8 or 9 olive-brown dorsal saddles, with 3 or 4 of the saddles distinctly more prominent than the others. Small, horizontally aligned darks spots often found on scales of upper two-thirds of body. Dark vertical suborbital bar present. Distinct, dark humeral spot present. Most specimens with 3 faint or indistinct vertically aligned spots at the caudal fin base (similar to *E. asprigene*). Breast white or yellow; breast pigmentation variable, ranging from no melanophores to numerous small, scattered melanophores. Pigmentation of underside of head ranges from scattered melanophores to, more commonly, heavily stippled. Pelvic fins clear or with a few scattered melanophores in membranes. Breeding males brightly colored with dark blue or turquoise bars posteriorly on side; area between bars bright orange. Belly orange. Pelvic and anal

fins and head dark blue green. First dorsal fin with a narrow blue marginal band, a wide red submarginal band, and a blue basal band. Second dorsal fin red and blue. Female color more subdued with olive-brown back and sides; dorsal saddles dark brown, posterior lateral bands green and indistinct. First dorsal fin of female with a narrow red-orange submarginal band.

SIMILAR SPECIES   It is often confused with members of the Orangethroat Darter clade, which it resembles in overall coloration and body shape, but *E. collettei* occurs syntopically with only one member of that clade, *E. pulchellum*, in Little River tributaries of southwestern Arkansas. It differs from *E. pulchellum* and other species of *Ceasia* in having an uninterrupted infraorbital canal (vs. interrupted). The Creole Darter differs from *E. asprigene* in having 3 or 4 dorsal saddles much darker than the others (vs. dorsal saddles, if present, are all of equal intensity), no black blotch at the posterior base of the first dorsal fin (vs. black basal band of first dorsal fin usually expanded posteriorly to form a distinct blotch), posterior lateral bars indistinct above midside (vs. usually 1–3 posterior lateral bars distinct above midside), eye ovoid (vs. round), and breeding males with tubercles on lower body scales (vs. no breeding tubercles).

VARIATION AND TAXONOMY   *Etheostoma collettei* was described by Birdsong and Knapp (1969) from the Dugdemona River, Jackson Parish, Louisiana. It had previously been confused with *E. asprigene*, and Black (1940) referred to it as a subspecies of that species, *Poecilichthys jessiae striatus* Hubbs and Black, the Striped Mud Darter. Arkansas populations of the Creole Darter are variable, especially in cheek scalation. In the Saline River population, the cheek is generally naked, but it is clearly scaled in other Ouachita River populations. The status of the small Little River population in southwestern Arkansas and southeastern Oklahoma is undetermined and requires further study. Robison (1978b) studied variation in the caudal skeleton of *E. collettei* and found 69.9% had no hypural fusion; 22.8% had fusion between hypurals 2 and 3; 1.9% had fusion between hypurals 4 and 5; and 5.7% exhibited fusion between hypurals 2 and 3 and 4 and 5. Robison also found 55.7% of the 159 specimens of *E. collettei* examined had 15 branched caudal rays, 32.3% had 14, 10.1% had 13, 1.3% had 16, and 0.6% had 12. Karlin and Rickett (1990) studied microgeographic genetic variation of populations in the headwaters of the Saline River in Saline County and found little variation among populations at that small scale. An analysis of mitochondrial and nuclear DNA identified the Creole Darter as a member of the *E. asprigene* species group (sometimes referred to as the *E. caeruleum* group) of the subgenus *Oligocephalus* (Lang and Mayden 2007).

DISTRIBUTION   Found only in Arkansas, Oklahoma, and Louisiana, in the lower Mississippi River basin and a few Gulf Slope drainages, the Creole Darter is most abundant in southern Arkansas in the eastern Saline, Ouachita, Caddo, and Little Missouri rivers and is less abundant in the western Saline, Cossatot, and Rolling Fork rivers and some of their tributary creeks. Matthews and Robison (1982) considered it rare in the Little River system, but Brook L. Fluker (pers. comm.) provided several new records from that system in 2016 (Map 195). It is not usually taken in large numbers at any locality.

HABITAT AND BIOLOGY   Platania and Robison (1980) noted the occurrence of this species in several types of habitat. In the southern Ouachita Uplands it occurs in headwater creeks and small rivers in swift current over a gravel bottom or in rocky chutes with heavy submergent vegetation. In Coastal Plain streams it is often found over a hard clay or mud substrate in moderate current in vegetation and debris (Douglas 1974), and it has been reported from turbid waters above and below dam sites (Fruge 1971). Fernandez (1982) found adults occurring primarily along the shoreline in pools in January and February and mainly in riffles in March and April. It has been reported from impoundments (Wood 1981), and Buchanan (2005) found the Creole Darter in nine Arkansas reservoirs. It was reported as rare in the Red River main stem in Arkansas (Buchanan et al. 2003).

Unpublished data from HWR revealed that the Creole Darter feeds mainly on aquatic insects. Specimens collected in April and May by Fernandez (1982) contained mayflies as the most common food item, with small numbers of mosquito pupae and amphipods also ingested. Matthews et al. (2004) also reported that *E. collettei* consumed primarily immature dipterans, odonates, plecopterans, and coleopterans in the Saline and Little Missouri river drainages. Stomach contents of 20 specimens collected in late May in the tailwaters below Narrows Dam on the Little Missouri River in Pike County were examined by TMB. The diet was dominated by small crustaceans (amphipods 62% by frequency of occurrence, and isopods 23%). Other food items in decreasing order of importance were plecopterans, coleopterans, chironomids, ephemeropterans, and trichopterans. Specimens examined by TMB from the upper Saline River in Garland and Saline counties in July and August fed predominantly on plecopterans and trichopterans, with chironomids, ephemeropterans, isopods, and copepods ingested in small amounts.

Based on data from different workers, spawning of

E. collettei in Arkansas occurs from February through April. Fernandez (1982) reported that peak spawning activity occurred in February in the Saline River, Arkansas, based on peak gonadosomatic indices and genital papilla development in that month. Ripe individuals were found by TMB in Ten-Mile Creek (Saline County) on 22 February. Miller and Robison (2004) reported a spawning season of March and April in Arkansas. A ripe female collected from the Ouachita River on 6 March 2012 had 187 mature ova and almost no immature ova (TMB data). Fernandez (1982) reported rather low fecundities ranging from 146 to 998 total ova. The following details on reproductive biology, including spawning behavior, were taken from Fernandez (1982) and Robison and Voiers (1996). Riffle areas of low-gradient second- and third-order streams are preferred spawning sites. The males defend areas around prospective mates but do not defend permanent territories. Male courtship behavior involves following, nuzzling, and rapidly repeated head twitches. In an aquarium at a water temperature of 21°C, the two largest males defended a spawning site by driving the smaller males away. As a ripe female buried her ventral side in the sand substrate with a rapid lateral vibrating motion, a male positioned himself along her left side and then quickly placed himself above her so that both fish were facing the same direction. Once the male was situated on top of the female, both fish vibrated simultaneously to release eggs and milt. The female pushed the male aside and repeated her burying behavior about 3 inches in front of the first spawning site. The second dominant male then spawned with her in the same manner, and the female spawned two additional times, once with the first male and then with a new male. One to several eggs were deposited during each spawning bout. While courting one female, a male may leave to court a second female, returning to the first one if the second female is unreceptive to his overtures. Occasionally a second male joins a spawning pair and vibrates with them. No parental care of eggs is provided by either sex. Maximum life span is about 3 years.

CONSERVATION STATUS   This species is widespread, common, but not known to be locally abundant in Arkansas. Its populations in the state currently appear secure.

Etheostoma cragini Gilbert
Arkansas Darter

CHARACTERS   A small, moderately stout darter with a large head, head length going into SL about 3.2 times (Fig. 26.56). Snout moderately blunt. Body distinctly bicolored,

Figure 26.56a. *Etheostoma cragini*, Arkansas Darter male. *Uland Thomas.*

Figure 26.56b. *Etheostoma cragini*, Arkansas Darter female. *Uland Thomas.*

Map 196. Distribution of *Etheostoma cragini*, Arkansas Darter, 1988–2019.

dark above midside, light below. Back with 6–9 small dusky saddles, sometimes indistinct. Side with 9–12 dusky blotches that may be distinct or mostly obscured by mottling. Body and head thickly stippled with small black spots. Orbital bars, especially the suborbital, prominent. Gill membranes separate or only slightly joined across throat. Premaxillary frenum present. Breast scaled; cheek and opercle naked or with a few embedded scales. Lateral line incomplete and short, usually not extending behind base of first dorsal fin. Scales in midlateral series 44–57; usually 12–20 pored scales. Scale rows above lateral line modally 7 (6–8); scale rows below lateral line modally 7 (6–8). Diagonal scale rows 13–16; scale rows around caudal peduncle 22–24. Dorsal spines 8–10; dorsal rays 11–13; anal spines 2; anal rays 5–8. Breeding males with tubercles on pelvic and anal fins and sometimes on ventral body scales. Maximum size around 2.5 inches (64 mm). Taber et al. (1986) found that females achieved a larger maximum size than males.

LIFE COLORS   Body dark brown above to white below, and stippled with black specks. Cheek beneath eye with a dusky narrow, vertical bar. First dorsal fin with a thin dark margin and often a black basal band; male with a yellow-orange medial band. Other fins weakly spotted or lightly banded with brown. Breeding males develop bright orange on gill membranes and entire underside of body; median fins tinged with orange.

SIMILAR SPECIES   Closely related to the allopatric Paleback Darter of the Ouachita River drainage, but differs in having a dark brown dorsum (vs. a wide, pale stripe down middle of back) and a more robust body. It is sympatric in Arkansas with *E. mihileze,* and differs from that species in having a shorter lateral line which does not extend past base of first dorsal fin, with 57 or fewer lateral scales (vs. lateral line extending to beneath and sometimes past second dorsal fin and usually with more than 57 scales), and usually 11 or 12 soft dorsal rays (vs. usually 14 or 15). It occurs in habitat similar to that of *E. microperca* and can easily be distinguished from that species by the lateral line extending at least to beneath the first dorsal fin, with 12–20 pored scales (vs. lateral line usually absent or with at most 1 or 2 pored scales).

VARIATION AND TAXONOMY   *Etheostoma cragini* was described by Gilbert (1885) from a small brook leading from the "lake" to the Arkansas River at Garden City, Finney County, Kansas. It is a member of subgenus *Ozarka.* Genetic evidence did not support the monophyly of *Ozarka* (Near et al. 2011). Meek (1894c) described *Etheostoma pagei* as a new species based on two specimens from Neosho, Missouri, but the publication was erroneously titled "A new *Etheostoma* from Arkansas." Black (1940) reported that Meek's two specimens were actually *Poecilichthys* (= *Etheostoma*) *cragini,* and Collette and Knapp (1966) considered *E. pagei* a junior synonym of *E. cragini.* A genetic analysis resolved *E. cragini* as sister to the *E. punctulatum* complex (Near et al. 2011).

DISTRIBUTION *Etheostoma cragini* is endemic to Arkansas River drainages of five states, from eastern Colorado through southern Kansas and northwestern Oklahoma, with disjunct populations in the Ozarks of northeastern Oklahoma, northwestern Arkansas, southwestern Missouri, and southeastern Kansas. Despite its common name, the Arkansas Darter has been known from Arkansas only since 1979, when it was first found within the city limits of Fayetteville during construction of I-49 through that city. Four additional localities were reported by Harris and Smith (1985), but Hargrave and Johnson (2003) found it at only three of the historically known sites. Wagner and Kottmyer (2006) found *E. cragini* at three of the five historic locations but also reported it from five additional disjunct stream reaches within a 2 km radius of historic sites. A more concentrated sampling effort in 2009–2011 documented populations at 13 additional sites (Wagner et al. 2011), and Brian Wagner (pers. comm.) found it at 16 additional localities in 2016. All known sites in Arkansas for *E. cragini* are in the headwater drainages of the Illinois River in Benton and Washington counties (Map 196).

HABITAT AND BIOLOGY In Arkansas, this species inhabits small permanent-flow springs and spring-run creeks often less than 3 feet (0.9 m) wide and 1 foot (0.3 m) deep. It is usually found in association with multiple types of rooted aquatic vegetation over a substrate of silt, mud, gravel, and some sand (Hargrave and Johnson 2003; Wagner and Kottmyer 2006). It is commonly observed in accumulations of sticks and other plant debris. When frightened, it buries itself in the substrate. It is found in essentially the same habitat in Missouri (Pflieger 1997), Oklahoma (Miller and Robison 2004) and Colorado (D. L. Miller 1984). In Colorado populations, the strongest predictors of site occupancy by age 0 and age 1 Arkansas Darters were cool stream temperature and sufficient length of available habitat (Groce et al. 2012). Blair and Windle (1961) stated that in northeastern Oklahoma, *E. cragini* was never found in the same collection with *E. microperca* despite the apparent similarity of habitats. Those authors concluded that differences in microhabitat were responsible, with optimal microhabitat for *E. cragini* consisting of a spring or spring-run filled with watercress, while optimal microhabitat for *E. microperca* was a *Myriophyllum*-filled cutoff or backwater pool of a creek. In Kansas, *E. cragini* has been found in similar habitat (Cross 1967) but also in several turbid streams with sand and mud bottoms away from any nearby springs (F. B. Cross, pers. comm.; Matthews and McDaniel 1981). Wagner and Kottmyer (2006) found that 78% of the Arkansas sites with *E. cragini* had a

predominantly mud/silt substrate. Cross also believed that it is a very resilient species, and its ability to adapt to turbid environments could allow survival during droughts when springs become desiccated. Wagner and Kottmyer (2006), however, found that 100% of Arkansas sites with *E. cragini* had very clear water. Additional habitat attributes found by Wagner and Kottmyer at 100% of the *E. cragini* sites were (1) less than 25% of shoreline wooded and (2) stream width less than 10 meters. The Arkansas Darter is reported to be very tolerant of high temperatures and low dissolved oxygen (Labbe and Fausch 2000; Smith and Fausch 1997).

In Arkansas, *E. cragini* is replaced in larger springs and second- to third-order streams by *E. mihileze*, although *E. mihileze* sometimes occurs in springs and spring-fed creeks with *E. cragini*. Population size of *E. cragini* shows considerable variation. Harris and Smith (1985) reported that two out of the five known populations in the 1980s probably contained 500–1,000 individuals; the remaining three populations were much smaller (around 100 each). Mark-and-recapture population estimates for two populations by Hargrave and Johnson (2003) were 546 and 79 individuals. Wagner and Kottmyer (2006) provided the following population numbers (length of stream segment in parentheses): 19 (45 m), 43 (657 m), 28 (886 m), and 60 (2,467 m).

Distler (1972) noted that snails made up a large part of the diet of this darter in Kansas. In Missouri, *E. cragini* fed primarily on isopods (58% by volume), ephemeropterans (12%), and chironomids (8%) (Taber et al. 1986). Stomach contents of 20 specimens from northwestern Arkansas collected in March through June were examined by TMB. The diet during all four months was dominated by small crustaceans, with isopods occurring in 60% of stomachs. Other crustaceans consumed in decreasing order of importance were copepods, cladocerans, ostracods, and one small crayfish. The most important noncrustacean food items were chironomids (20% frequency of occurrence) and ephemeropterans (20%). Other items consumed in small amounts were trichopterans, dipteran larvae other than chironomids, and hemipterans.

A long spawning season occurs in Arkansas in the relatively stenothermal environments inhabited by this species. In southwestern Missouri, a long spawning period extended from mid-February to mid-July at water temperatures from 48.2°F to 62.6°F (9–17°C) (Taber et al. 1986), and a similar breeding season probably occurs in Arkansas. Based on examination of gonads of Arkansas specimens by TMB, ripe females and males were found as early as 19 March and as late as 7 June. It is a fractional spawner, as indicated by the presence of multiple egg sizes in females early in the breeding season, and Taber et al. (1986) also concluded that females spawned more than once in Missouri. In Arkansas, the smallest female with ripe ova was 34 mm TL, and the largest was 51 mm TL (x̄ = 41.2 mm). Clutch size ranged from 53 to 278 ripe ova, with an average clutch size of 127. This is less than reported in Missouri, where average-sized females contained 294–472 eggs (Taber et al. 1986). Distler (1972) observed spawning habits in aquaria and found that territories are not established; several males spawned simultaneously with one female. The female buries her body in the gravel substrate accompanied by one or more males. Males and females vibrate and the fertilized eggs are buried about 0.4 inch (1 cm) deep in the substrate. This spawning act was repeated at about 30 second intervals until the female expended her ripe eggs. Little or no aggression was observed among the males attempting to spawn with the same female, but the largest males were most successful in acquiring positions immediately next to the female. Both sexes became sexually mature at 1 year or less, and maximum life span was around 3 years.

CONSERVATION STATUS The Arkansas Darter has declined throughout its range in response to changes in flow regimes and habitat fragmentation (Groce et al. 2012). In Arkansas, this species is restricted to small spring habitats in Benton and Washington counties, and because of the rapid human population growth in that part of the state during the past two decades, we consider existing populations as endangered. Wagner et al. (2011) noted that the fragmented nature of existing populations provides little opportunity for movement between populations; therefore, local populations are highly susceptible to extirpation (as has occurred with one Arkansas population discovered in 2005).

Etheostoma euzonum (Hubbs and Black)
Arkansas Saddled Darter

CHARACTERS A large, stout *Etheostoma* with a large, gradually sloping head and a moderately blunt snout (Fig. 26.57). When viewed from the front, the head appears almost triangular. The only *Etheostoma* in Arkansas that exceeds it in length is *E. blennioides*. Back with 4 wide, conspicuous dark saddles, which extend obliquely forward down side of body. Side often mottled with small orange and green spots and blotches. Gill membranes broadly connected across throat. Broad premaxillary frenum present. Infraorbital canal uninterrupted, usually with 8 pores. Opercle, breast, and belly partly to fully scaled; cheek varies from naked to fully scaled; nape fully scaled. Lateral

Figure 26.57a. *Etheostoma euzonum*, Arkansas Saddled Darter male. *Uland Thomas.*

Figure 26.57b. *Etheostoma euzonum*, Arkansas Saddled Darter female. *Uland Thomas.*

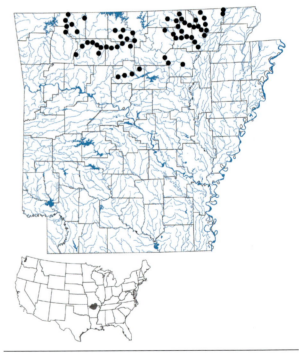

Map 197. Distribution of *Etheostoma euzonum*, Arkansas Saddled Darter, 1988–2019.

line straight and complete, with 54–71 scales (usually 59 or more). Scale rows above lateral line usually 7–9; scale rows below lateral line usually 9–11. Diagonal scale rows usually 17–20 (15–23); scale rows around caudal peduncle usually 23–27. Dorsal spines 12–15; dorsal rays 12–15; anal spines 2; anal rays 9–11. Breeding tubercles develop on both males and females on ventral scales from breast to base of anal fin and on lower half of caudal peduncle. Maximum length around 4.5 inches (114 mm).

LIFE COLORS Back and upper sides olive tan with 4 large dark brown saddles. Side mottled with red-orange spots and indistinct green blotches. Underside whitish yellow. First dorsal fin with a red-orange margin, a dark blue medial band, and dusky or dark green basally. Second dorsal, caudal, and pectoral fins with light orange spots and faint brown bands. Breeding males brightly colored, developing reddish-orange lower sides and dark blue breast; sides often develop a series of alternating blue-green and orange bars.

SIMILAR SPECIES No other *Etheostoma* species in Arkansas has color pattern of 4 wide, dark dorsal saddles extending obliquely forward down side of body. Smaller specimens are sometimes mistaken for *Percina uranidea* or *P. vigil*. It differs from those species in lacking a row of enlarged and modified scales between pelvic fin bases and on midline of belly (vs. enlarged, strongly toothed scales present between pelvic fin bases and along midline of belly), in usually having 59 or more lateral line scales (vs. fewer than 59), dorsal spines usually 13 or more (vs. fewer than 13), and gill membranes broadly joined across throat (vs. gill membranes separate or only narrowly joined across throat).

VARIATION AND TAXONOMY *Etheostoma euzonum* was originally described from the Buffalo River near St. Joe, Searcy County, Arkansas (Hubbs and Black 1940c) as *Poecilichthys euzonus*. It is the only Arkansas species in subgenus *Poecilichthys*. Hubbs and Black and later Page and Cordes (1983) recognized two subspecies: *Etheostoma e. euzonum,* which has no scales on the cheek, a rounded snout, and a large eye (eye length almost equal to snout length) and occurs in the White River system above Batesville (including the Little Red River), and *E. e. erizonum,* which has one or more scales on the cheek, a much more produced snout, and a small eye (eye length much shorter than snout) and occurs only in the Current River. Supposed intergrades between these two subspecies were reported from the Spring, Strawberry, and Black rivers. Near et al. (2011) found genetic evidence for recognizing the Current River form as a separate species, *Etheostoma erizonum*, the Current Saddled Darter. We tentatively continue to refer to all Arkansas populations as *E. euzonum*, but we believe that recognition of *E. erizonum* will be supported by additional analysis of morphological and molecular evidence.

DISTRIBUTION The Arkansas Saddled Darter is endemic to the White River drainage of Arkansas and Missouri. It occurs across northern Arkansas in the Current, Black, Spring, Strawberry, White, Buffalo, and Little Red rivers (Map 197). There is at least one record in the 1990s from the Eleven Point River in Missouri (R. Hrabik, pers. comm.), but there are only pre-1988 records from that river in Arkansas (Map A4.194). It has not been found in the upper White River above Beaver Reservoir in at least 30 years, but it still occurs in the King's River drainage of Carroll County.

HABITAT AND BIOLOGY *Etheostoma euzonum* inhabits moderate to large streams of the Ozark Uplands in deep, swift riffles having gravel and rubble bottoms. It is sometimes found in creeks having strong permanent flow, but it avoids extreme headwaters or other intermittent creeks. It is intolerant of reservoir conditions and appears to be sensitive to other types of environmental disturbance as well. Most upper White River drainage populations in Arkansas have been extirpated, and it has not been reported from the White River drainage of Missouri since 1968 (Pflieger 1997). It is still common at some localities, particularly in the Current, Spring, and Strawberry rivers. It is also common in the Buffalo River and occurs sporadically in the Kings River in the upper White River drainage. In the mainstem White River, it is found only in a short segment between Batesville and the mouth of the Black River. It is rare in the Little Red River. Most of the more upstream populations in the White River in northwestern Arkansas were extirpated by the impoundment of Beaver Lake. A preimpoundment survey of the Beaver Lake watershed found small numbers of the Arkansas Saddled Darter (Keith 1964). Two short-term postimpoundment surveys of the Beaver Lake tailwaters and a survey 30 years after impoundment produced no specimens of *E. euzonum* (Quinn and Kwak 2003). Greers Ferry Lake has undoubtedly had a similar effect on Little Red River drainage populations, although a recent survey found small numbers in the South Fork of the Little Red River (Quinn et al. 2012). Gravel mining is another threat. Brown et al. (1998b) documented the reduction in *E. euzonum* population density downstream from gravel mining sites in Crooked Creek and Kings River.

Stomach contents of 70 specimens, from 44 to 100 mm TL, collected between 17 and 20 May from the Spring, Strawberry, and White rivers were examined by TMB. The diet from all three drainages was dominated by ephemeropterans by percent volume and percent frequency, with plecopterans, chironomids, and trichopterans also eaten in large amounts. Other items ingested by decreasing order of importance were simuliids, coleopterans, megalopterans, and tipulids. Stomachs from Little Red River specimens in June were largely empty, and White and Buffalo river specimens collected in August and October, respectively, had eaten chironomids and plecopterans.

Little is known of the reproductive biology of this species, but the spawning season in Arkansas extends at least from late March through mid-May (Hubbs 1985). A ripe male was found in the Middle Fork of the Little Red River in Van Buren County on 24 April (Tumlison et al. 2017). Examination of ovaries and testes of specimens collected from the Spring River on 17 and 18 May by TMB supports a mid- to late May termination of breeding. Most females on those dates still contained mature ova, some had apparently just spawned, and a few possessed some atretic eggs. Mature ovum counts from 11 ripe females, 55–80 mm TL, ranged from 41 to 108 ($\bar{x}$ = 72), and there were no obvious smaller developing eggs.

CONSERVATION STATUS   Additional extirpation of populations by reservoir construction has not been a major threat since the first edition of this book; however, existing populations in Arkansas appear to be small and threatened by gravel mining and other environmental alterations such as nutrient enrichment. Because of its intolerance of habitat alteration, we consider it a species of special concern if not threatened. Future monitoring of this species should be a priority.

Etheostoma flabellare Rafinesque
Fantail Darter

Figure 26.58. *Etheostoma flabellare*, Fantail Darter. *David A. Neely.*

CHARACTERS   A moderately slender darter, with an unusually short and deep caudal peduncle. First dorsal fin very low, its height less than half the height of the second dorsal fin. Each lateral scale having a small black spot, causing side to have horizontal streaks; side otherwise plain or with 10–15 vertical bars (Fig. 26.58). Back sometimes with about 8 dark saddles that are often continuous with lateral bars. Gill membranes broadly joined across throat. Premaxillary frenum present. Lower jaw protruding beyond upper jaw. Infraorbital canal interrupted, usually with 4 anterior and 2 posterior pores. Cheek, opercle, breast, nape, and prepectoral area naked; belly fully

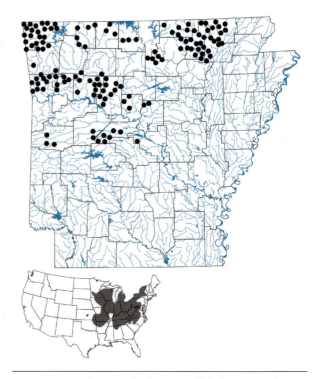

Map 198. Distribution of *Etheostoma flabellare*, Fantail Darter, 1988–2019.

scaled or naked anteriorly. Lateral line incomplete, ending beneath second dorsal fin. Scales in midlateral series usually 45–59 (41–62); pored lateral line scales usually 28–38 (24–40). Scale rows above lateral line 7–9; scale rows below lateral line 8–10. Diagonal scale rows 15–19; scale rows around caudal peduncle 24–30. Dorsal spines 6–10; dorsal rays 12–14; anal spines 2; anal rays 7–9; pectoral rays 12–14. Breeding tubercles are absent, but breeding males develop swollen, fleshy, nonkeratinized ridges on scales below the lateral line from 6 or 7 scale rows posterior to the pectoral fin insertion to the caudal peduncle (Mayden 1985b). Breeding males also develop fleshy knobs on tips of dorsal fin spines. Adult males lack scale ridges and fin knobs outside the reproductive season, and females show no development of those structures at any time of the year. Maximum length is around 3 inches (76 mm).

LIFE COLORS   Color patterns quite variable in Arkansas populations. Back olive brown, sides yellowish brown with dark horizontal lines in most Arkansas populations. Dorsal saddles and lateral markings, if present, black or dark brown. Black humeral spot present. Caudal and second dorsal fins with black bands or spots. First dorsal fin with a black base, clear medially and dusky marginally, with white or orange knobs on tips of spines especially prominent in breeding males. Other fins clear or dusky.

Breeding males become much darker overall and often have a yellowish cast to the body.

SIMILAR SPECIES   Rarely mistaken for any other Arkansas darter. It differs from *Nothonotus moorei* in having broadly connected gill membranes (vs. gill membranes separate or narrowly joined across throat), incomplete lateral line (vs. complete), and usually fewer than 10 dorsal spines (vs. more than 10). It differs from *N. juliae* in having a naked opercle (vs. partly scaled), incomplete lateral line (vs. complete), and usually fewer than 10 dorsal spines (vs. more than 10).

VARIATION AND TAXONOMY   *Etheostoma flabellare* was described by Rafinesque (1819) as *Etheostoma flabellaris* from tributaries of the Ohio River. Currently the only Arkansas species in subgenus *Catonotus*, *Etheostoma flabellare* was reported by Kuehne and Barbour (1983) to be the most variable of all darters, both within individual populations and geographically over its large range. McGeehan (1985), using a series of multivariate analyses, found that *Etheostoma flabellare* is a widespread polytypic species and recognized five allopatric forms. The long-recognized subspecies, *Etheostoma flabellare lineolatum* (Agassiz), which included all Arkansas populations except for a form discovered in the White River, was reduced to the synonymy of the nominate form *E. f. flabellare* due to clinal variation between the two forms. Page and Burr (2011) recognized three subspecies, including the most widespread form, *E. f. flabellare*. Pflieger (1971) first mentioned the morphologically distinct form of *E. flabellare* found in the White River in Missouri, and it was later found by HWR in the Arkansas portion of the White River system. The White River form has a more yellow body coloration than other populations and does not have bars or streaks well developed on the sides (Pflieger 1997). McGeehan (1985) recognized this questionable White River form but linked it to a population in the lower Tennessee River in the southern Appalachians, an interpretation we are hesitant to accept. Additional analysis is needed to clearly define the geographic distribution and relationships of the White River drainage populations. Recent mitochondrial and nuclear DNA evidence supports the designation of two forms of Fantail Darter in Arkansas, but it does not support allying the White River form with lower Tennessee River *E. flabellare* (Blanton 2007; Rebecca Blanton Johansen, pers. comm. 2014).

DISTRIBUTION   A widely distributed darter, *E. flabellare* occurs in Atlantic, Great Lakes, and Mississippi River basins from southern Canada to Minnesota, and south to South Carolina and westward into Oklahoma. The Fantail Darter has an unusual distributional pattern in Arkansas because of its absence from the White River headwaters in the western Ozarks of Washington and Madison counties upstream of Beaver Reservoir. Downstream of Beaver Reservoir, it is known from small White River tributaries in Arkansas in Madison, Carroll, Boone, Marion, Fulton, and Stone counties, but it is surprisingly absent from the Buffalo River drainage. There is one older record from the Little Red River drainage (Map A4.195), but there are no recent records from that drainage. It is most abundant in the Black River drainage (White River system) of northeastern Arkansas in the Current, Eleven Point, Black, Spring, and Strawberry rivers; the Illinois River system of extreme northwestern Arkansas in Benton and Washington counties; and in other Arkansas River tributaries draining the Boston Mountains Plateau north of the river. South of the Arkansas River, it is found in only a few scattered Arkansas River tributaries (Poteau, Fourche LaFave, and Little Maumelle river drainages) (Map 198).

HABITAT AND BIOLOGY   The Fantail Darter is found in clear, small creeks to small rivers where it is often abundant in swift, shallow riffles over a gravel and rubble bottom at depths of less than 1 foot (0.3 m). In Benton and Washington counties of northwestern Arkansas, it commonly occurs in springs and spring-run creeks. In Kansas, it preferred riffles less than 8 inches (40 cm) deep having slow current (Cross and Collins 1995). The young are sometimes found in pools and slackwater areas downstream from riffles. In the Frog Bayou (Clear Creek) drainage of Crawford County, Arkansas, population density estimates ranged from 0.14 to 0.48 fish/m^2, and in the Mulberry River drainage in Madison and Johnson counties, density estimates ranged from 0.0022 to 0.2 fish/m^2 (Rambo 1998). In Indiana streams, Fantail Darters were most strongly associated with cobble and bedrock substrates and occurred in shallow riffles of low velocity (Pratt and Lauer 2013). Moore and Paden (1950) attributed the absence of Fantail Darters in portions of the Illinois River, Oklahoma and Arkansas, to changes in the substrate.

Floyd (1948) studied life history of *E. flabellare* in Oklahoma and Arkansas, and Page (1983a) summarized ecological investigations. In Indiana, juveniles and adults fed mainly on ephemeropteran nymphs and chironomid larvae; plecopteran nymphs and small crustaceans were also seasonally important (Strange 1993). Additional food items reported from other studies include trichopterans, simuliid larvae, and isopods (Fuller and Hynes 1987; Matthews et al. 1982). Stomach contents of 75 specimens from 7 Arkansas counties were examined by TMB. From April through June, the diet of specimens from Arkansas River drainages was dominated by ephemeropteran nymphs (66%

frequency of occurrence) and plecopterans (24%). Other items consumed included simuliids, chironomids, odonates, amphipods, isopods, and hemipterans. Individuals from the Strawberry River drainage in May fed almost entirely on ephemeropterans. Specimens collected from spring and spring-run creeks of Benton and Washington counties from October to early December fed most heavily on ephemeropterans and plecopterans and also consumed small crustaceans (amphipods and isopods), chironomids, and trichopterans. Feeding occurs during the day, with peak activity occurring near dawn and dusk (Adamson and Wissing 1977), but some feeding may occur at night (Matthews et al. 1982).

Hubbs (1985) reported that Fantail Darter spawning occurs mainly in April and May in Arkansas, and examination of gonads by TMB indicated that spawning probably begins in mid-March and continues through May. However, Floyd (1948) reported that *E. flabellare* nests were first found in early April in Tahlequah Creek, Oklahoma (Illinois River drainage) but were most numerous in early June, and nests were found as late as 19 July. Females collected by TMB from Clear Creek in Crawford County on 18 February contained developing ovaries, but eggs appeared to be a few weeks away from maturity. Twenty-seven females collected in April and May from Crawford, Scott, Conway, Pulaski, Sharp, Benton, and Washington counties, Arkansas, contained ripe ova. *Etheostoma flabellare* has the largest mature ova of all darter species in Arkansas, with egg diameters averaging 2.5–2.7 mm. Large egg size is characteristic of the subgenus *Catonotus* (Page 1983a). Large egg size in the Fantail Darter is accompanied by small clutch size, and both features are associated with greater parental care than found in most darters. Clutch size in small females (33–42 mm TL) from Arkansas ranged from 5 to 23 ova ($\bar{x} = 15$), and large reproductive females (43–64 mm TL) contained between 23 and 78 ripe ova ($\bar{x} = 50$). Spawnng behavior was summarized by Page (1975, 1983a). The males establish territories centered around cavities beneath rocks in the riffles. When ready to spawn, a female enters a cavity guarded by a male, turns upside down, and begins laying eggs in a concentrated patch on the underside of the stone (Fig. 26.59). This may take up to 2 hours. During that time, the male briefly inverts and positions himself beside the female in a head-to-tail posture. Sperm are then released to fertilize the eggs. During spawning, both the males and females exhibited color changes, with the head, dorsal saddles, pelvic and pectoral fins becoming deep black (Moretz and Rogers 2004). The female leaves the nest after spawning,

Figure 26.59. *Etheostoma flabellare* eggs on underside of rock. *Renn Tumlison.*

and other females may enter and spawn in turn with the same male. The number of eggs in 27 nests in Tahlequah Creek, Oklahoma, ranged from 6 to 239, with a mean of 83.5 eggs (Floyd 1948). The male guards the eggs until they hatch. Some authors have suggested that the male uses the knobs of his dorsal fin to keep the eggs free of silt. Knouft et al. (2003) confirmed that *E. crossopterum* males, and presumably other species of the subgenus *Catonotus*, provide parental care by reducing the amount of microbial infestation of eggs, resulting in increased egg viability. Knapp and Sargent (1989), Page and Bart (1989), and Bart and Page (1991) suggested that the dorsal fin knobs function as egg mimics to entice females to deposit their eggs in the male's nest. A female may lay 200–400 eggs (about 45 per spawning) during the breeding season. The eggs hatch in 21 days at 70°F (21°C) (Lake 1936). In a South Carolina study, only the largest males were successful in mating with females (Moretz and Rogers 2004). Males are larger than females, and maximum life span is around 4 years (Karr 1964). Speares et al. (2011) reported sound production in *E. flabellare*, describing vocalizations during aggressive encounters.

CONSERVATION STATUS   Widely distributed and locally common in its preferred habitat, this largely upland species is currently secure in Arkansas.

Etheostoma fragi Distler
Strawberry Darter

Figure 26.60. *Etheostoma fragi*, Strawberry Darter. *Uland Thomas.*

Map 199. Distribution of *Etheostoma fragi*, Strawberry Darter, 1988–2019.

CHARACTERS A small to medium-sized member of the *Etheostoma spectabile* complex. It has a somewhat arched predorsal region with the greatest body depth occurring just in front of the first dorsal fin (Fig. 26.60). Back with 6–10 saddles that are often indistinct. Side usually with 10–12 narrow vertical bars that are sharply defined along the entire length of the body above and below the lateral line and continuous across venter; anterior bars are oblique with a slight forward slant along plane of diagonal scale rows. Horizontal lines on sides faint, present only below lateral line. Gill membranes narrowly joined across throat. Premaxillary frenum present. Supratemporal canal usually

uninterrupted. Infraorbital canal interrupted, with modally 3 posterior pores. Nape usually well scaled; breast varies from naked to completely scaled; belly fully scaled; cheek usually with 5 or more rows of scales; opercle usually more than 90% scaled. Lateral line incomplete, usually ending beneath second dorsal fin. Scales in lateral series usually 46–54 ($\bar{x}$ = 49.8); pored lateral line scales usually 28–37 (usually 29 or more). Pectoral fin rays usually 11 or 12; dorsal spines 10 or 11; dorsal rays usually 12 or 13; anal spines 2; anal rays 6 or 7. Scales around caudal peduncle usually 20–25. Diagonal scale rows from anal fin origin to first dorsal fin base usually 14–16. Breeding tubercles absent. Maximum length around 2 inches (51 mm).

LIFE COLORS Back and upper side olivaceous to rusty brown; dorsal saddles narrow, light brown and often indistinct. Venter usually light-colored and unmarked except for meeting lateral bars. Breeding males with turquoise and orange as dominant coloration. Throat red; lateral bars iridescent turquoise with the anterior interspaces reddish gray to reddish brown; the posterior interspaces orange, gradually increasing in intensity toward the caudal peduncle and forming orange chevrons across belly. First dorsal fin with 5 bands, with the medial orange band the widest. Anal fin turquoise. Breeding females with little coloration but with a thin, medial orange band in the first dorsal fin and prominent narrow lateral bars.

SIMILAR SPECIES The Strawberry Darter is similar to *E. uniporum* and other described and undescribed species of the *E. spectabile* complex in Arkansas. It is allopatric to all other members of the *E. spectabile* complex and can be distinguished from all of them by breeding males with orange chevrons crossing the belly and no breeding tubercles (vs. breeding males without orange chevrons crossing belly and breeding tubercles present on pectoral, pelvic, anal, and caudal fins and on ventral scales), usually 29 or more pored lateral line scales (vs. usually fewer than 29 pored lateral lines scales), and cheek usually with 5 or more rows of scales (vs. cheek with 1 or 2 rows of scales below or behind the eye). It further differs from *E. uniporum* in having an uninterrupted supratemporal canal and usually having 3 pores in the posterior segment of the infraorbital canal (vs. supratemporal canal interrupted and usually 1 posterior infraorbital pore). Like all members of the *E. spectabile* complex, *E. fragi* resembles *E. caeruleum*, with which it is sympatric. It can be distinguished from that species in having fewer than 13 pectoral rays, an interrupted infraorbital canal, greatest body depth occurring just anterior to first dorsal fin, and breeding males without red in anal fin (vs. 13 pectoral rays, infraorbital canal uninterrupted, greatest body depth occurring beneath first dorsal fin, and breeding males with red in anal fin in *E. caeruleum*).

VARIATION AND TAXONOMY   *Etheostoma fragi* is a member of subgenus *Oligocephalus*. The Strawberry Darter was originally described by Distler (1968) as a subspecies of Orangethroat Darter, *E. spectabile fragi*, even though he did not document hybrids or areas of intergradation. Wiseman et al. (1978) studied variation at the Ldh-1 gene locus in various *E. spectabile* populations and reported that *E. s. fragi* was fixed for an allele with a unique staining intensity. The Strawberry Darter was elevated to full specific status by Ceas and Page (1997) based on unique color patterns of breeding males, diagnostic modes for various meristic counts, and the lack of intergrade populations. The type locality is Spring Creek, a tributary of Big Creek (Strawberry River drainage), Sharp County, Arkansas. Analyses of six nuclear genes also supported *E. fragi* as nested in the Orangethroat Darter clade, and the mitochondrial gene *cytb* indicated that indeterminate mtDNA introgression with an unknown species had occurred (Bossu and Near 2009; Near et al. 2011). Those authors speculated that the mtDNA donor lineage may have become extinct after transfer of the mtDNA into *E. fragi*. There was no evidence of nuclear gene introgression. Bossu and Near (2009) found that *E. fragi* and *E. uniporum* are sister species. As expected, a species with a small range exhibits little geographic variation. Ceas and Page (1997) reported that specimens from Piney Fork modally have 6 anal rays, while other populations have 7. Big Creek and Bray Branch specimens tend to have more breast scales than other populations.

DISTRIBUTION   *Etheostoma fragi* is endemic to the Strawberry River (the source of the common name Strawberry Darter) and its upland tributaries in Fulton, Izard, and Sharp counties in northeastern Arkansas (Map 199). It is allopatric to all other species of the *E. spectabile* complex.

HABITAT AND BIOLOGY   The Strawberry Darter occurs in headwaters, tributary creeks, and the upper three-quarters of the main Strawberry River, where it is found in shallow riffles in slow to moderate current over gravel and rubble substrates. It is sometimes found in raceways and can survive in isolated pools when streams become intermittent. Additional habitat data provided by Brian Wagner showed that 81% of specimens collected in 2006 were taken over gravel/rubble substrate, 15% over boulder/bedrock substrate, and 4% over sand/gravel substrate. Most specimens taken in that study were found in spring-fed, clear tributary creeks less than 10 m wide. Aquatic vegetation was usually absent at collecting sites, but four

sites had watercress or other rooted aquatic vegetation. *Etheostoma fragi* can apparently occupy different habitats within its small range, and it was found syntopically with *E. caeruleum* at 5 of 12 sites where it occurred. Hecke (2017) found that it was most abundant in the upper part of the Strawberry River drainage in tributary and mainstem habitats, and both abundance and occurrence decreased drastically downstream. No Strawberry Darters were found at mainstem sites in the middle and lower reaches of the drainage. Hecke (2017) reported the following environmental measurements at sites where *E. fragi* was collected: mean dissolved oxygen 9.02 mg/L, mean current speed 0.25 m/s, mean pH 8.2, mean summer water temperature 21.6°C, and mean water depth 0.15 m.

Stomach contents of 48 specimens collected between 18 September and 20 October 2006 from 12 localities in the Strawberry River drainage in Fulton, Lawrence, and Sharp counties were examined by TMB. Seventy-nine percent of the stomachs contained identifiable food items. By percent frequency of occurrence, chironomids, ephemeropterans, and trichopterans dominated the diet, followed by ostracods, annelids, isopods, plecopterans, and cladocerans. Odonates and amphipods were also infrequently ingested.

Strawberry Darters in spawning condition have been collected from late February through April. Ripe males were found in the Strawberry River in Fulton County on 24 April (Tumlison et al. 2017). Breeding adults apparently congregate in shallow riffles to spawn. Its reproductive biology is probably similar to that of other well-studied species of the *E. spectabile* complex.

CONSERVATION STATUS   Hecke (2017) concluded that populations of *E. fragi* appear to be in decline. We consider the Strawberry Darter threatened because of its restricted range and habitat alterations, such as nutrient enrichment in the Strawberry River. Future monitoring of this species is needed, and some type of habitat protection will probably be required to ensure its long-term survival.

Etheostoma fusiforme (Girard)
Swamp Darter

CHARACTERS   A small darter with a small head and mouth and a slender body (Fig. 26.61). Back with indistinct markings or with about 9 small square saddles. Midside with 9–12 small dark rectangular blotches. Cheek, opercle, and lower sides often with small scattered black spots. Gill membranes slightly to moderately joined across throat. Premaxillary frenum present. Infraorbital canal interrupted, with 3 or 4 anterior pores and 1 posterior pore.

Figure 26.61. *Etheostoma fusiforme*, Swamp Darter. *Uland Thomas.*

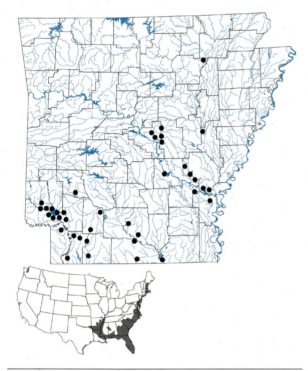

Map 200. Distribution of *Etheostoma fusiforme*, Swamp Darter, 1988–2019.

Cheek, opercle, breast, belly, nape, and top of head fully scaled. Margin of preopercle smooth or moderately serrate. Lateral line highly arched and incomplete, usually not extending past middle of first dorsal fin. Usually only 3 or 4 (2–5) scale rows between lateral line and front base of first dorsal fin; scale rows below lateral line 6–12. Scales in lateral series usually 46–56 (40–63), with 11–27 pored scales. Diagonal scale rows 12–15; scale rows around caudal peduncle 17–22. Dorsal spines 8–13; dorsal rays 8–13; anal spines 2; anal rays usually 7–9 (5–10); pectoral rays 12–15. Breeding males with tubercles on pelvic and anal fin rays. Maximum length is about 2.3 inches (58 mm).

LIFE COLORS   Coloration variable. Upper body tan; back with or without dark saddles; side with small dark rectangular blotches; lower side often spotted or stippled. Underside white with or without small scattered black spots. The short lateral line appears as a light stripe. Brown or black suborbital bar faint to distinct. There is a prominent black submedial basicaudal spot and a smaller supramedial spot. Spines of first dorsal fin outlined with black pigment; other fins weakly to strongly banded with dark brown. Breeding males become darker, but no bright colors develop.

SIMILAR SPECIES   The Swamp Darter differs from *E. gracile* in having the breast and top of head scaled (vs. naked), infraorbital canal interrupted, with 4 or 5 pores (vs. uninterrupted, with 8 pores), and no green bars on side in life (vs. green bars present on sides). It can be distinguished from *E. microperca* and *E. proeliare* in having more than 8 pored lateral line scales (vs. fewer than 8), more than 40 scales in a lateral series (vs. fewer than 40), and usually more than 12 pectoral rays (vs. fewer than 12).

VARIATION AND TAXONOMY   *Etheostoma fusiforme* was described by Girard (1854) as *Boleosoma fusiforme* from a tributary of the Charles River at Framingham, Massachusetts. It was placed in subgenus *Hololepis* by Collette (1962). That placement was supported by Bailey and Etnier (1988), even though Page (1981) had synonymized *Hololepis* and *Microperca* in subgenus *Boleichthys*. Collette (1962) recognized two subspecies: *E. f. fusiforme*, occurring in Atlantic Slope lowland drainages from Maine to Waccamaw River, North Carolina, and *E. f. barratti*, occurring in the rest of the range (including Arkansas) from the Pee Dee River, North Carolina, south and west to southeastern Oklahoma and eastern Texas, and extending northward through the Mississippi Embayment to southern Illinois. Near et al. (2011) supported elevation of the latter subspecies to full specific status as *E. barratti* (Holbrook), the Scalyhead Darter. Because additional study of morphological and molecular variation in *E. f. barratti* is needed (there may be more than one species hiding under that name), we herein refrain from recognizing *E. barratti* as a full species and continue to recognize all Arkansas populations as *E. fusiforme*.

DISTRIBUTION   The Swamp Darter is a Coastal Plain species found from Maine to Louisiana, and extending up the Mississippi Embayment to Kentucky, southern Illinois, and southeastern Missouri. A disjunct population occurs in the San Jacinto River drainage, Texas. It is found in Coastal Plain streams in southern and eastern Arkansas in the Red, Ouachita, Arkansas, White, St. Francis, and Mississippi river drainages. It is most common in oxbow lakes of the Red River and to a lesser extent in the lower Arkansas and Ouachita river drainages. There are few recent records from the Mississippi Alluvial Plain of eastern Arkansas (Map 200).

HABITAT AND BIOLOGY   Although widely scattered in the lowlands, *E. f. barratti* is not a common species in Arkansas and is never abundant at any locality; however, areas along the Atlantic Coast are often inhabited by large populations of *E. f. fusiforme* (Page 1983a). It is usually found in swamps, bayous, and oxbow lakes in areas of little or no current over a bottom of mud and detritus. It is almost always associated with dense aquatic vegetation. In 1974, it was found in Lake Merrisach, a heavily vegetated lake just off the Arkansas River in Arkansas County. Grass Carp were introduced into that lake in 1974 and within one year had virtually eliminated the aquatic vegetation. Efforts to collect *E. fusiforme* from the denuded lake in 1975 and 1976 were unsuccessful. It subsequently repopulated Lake Merrisach when the vegetation returned. It can adapt to lowland reservoirs if there is abundant aquatic vegetation and habitat similar to an oxbow lake, and Buchanan (2005) found the Swamp Darter in three Coastal Plain impoundments in southern Arkansas (Champagnolle Creek, Lake Merrisach, and Lake Millwood). Kuehne and Barbour (1983) reported that this species can tolerate a wide range of pH values, but it is commonly found in acid waters with lower pH values than those tolerated by most fishes. Hill and Cichra (2005) found that the Swamp Darter was tolerant of a variety of water chemistry features in Florida and occurred in several different habitat types in that state. Other lowland species often found with *E. fusiforme* in Arkansas are *Fundulus blairae, F. dispar, Centrarchus macropterus, Lepomis marginatus, L. symmetricus,* and *Elassoma zonatum.*

The Swamp Darter fed on small crustaceans (cladocerans, copepods, amphipods, and isopods), mosquito larvae, and midge larvae in Florida (McLane 1955). A similar diet was found for *E. f. fusiforme* in southern New England, with cladocerans, copepods, amphipods, ostracods, and midge larvae making up the bulk of foods eaten (Schmidt and Whitworth 1979). Stomach contents of 37 Swamp Darters from seven Arkansas counties were examined by TMB. The diet was dominated by small crustaceans in all months, at almost all localities, and in individuals of all sizes sampled (30–54 mm TL). Specimens collected June through August from Red River drainages in southwestern Arkansas fed mainly on copepods and cladocerans, supplemented by dipteran larvae, while specimens from Merrisach Lake in Arkansas County fed mainly on chironomids, ephemeropteran nymphs, and copepods in August. Thirteen specimens collected from Hickerson Lake in Monroe County on 10 November fed heavily on small crustaceans (copepods, cladocerans, and ostracods) with chironomids, other dipteran larvae, and coleopteran larvae also contributing to the diet.

Limited data from North Carolina, New Jersey, and Florida indicate that spawning may occur as early as December and as late as May (Collette 1962), and McLane (1955) concluded that spawning occurred year-round in the St. Johns River, Florida, based on collection of individuals in breeding condition in seven months including January, March, April, May, June, September, and December. Exact time of spawning in Arkansas is not known but probably occurs from February to April. Females collected from Crittenden County, Arkansas on 14 April by TMB contained ova that were nearly mature. Its spawning behavior, as described by Collette (1962), indicates no territoriality. It does not prepare a spawning site or defend the eggs. The male mounts the ripe female and beats her with his pelvic fins. The eggs are laid singly on plant leaves and fertilized. Eggs hatch in 6 days at temperatures of 21–25°C. The Swamp Darter is a short-lived species with most individuals not surviving beyond 1+ years and a maximum life span of about 2 years (Schmidt and Whitworth 1979). Sexual maturity is reached during the first year at 22–27 mm SL (Ross 2001), and a 34 mm TL female in Florida had 94 mature ova (McLane 1955).

CONSERVATION STATUS   This is an uncommon fish with small populations. We consider it a species of special concern, and its populations should be monitored to detect future changes.

Etheostoma gracile (Girard)
Slough Darter

Figure 26.62. *Etheostoma gracile*, Slough Darter. *Uland Thomas.*

CHARACTERS   A small, slender darter with a small head and mouth; head length going into SL 3.5–3.9 times (Fig. 26.62). Body markings highly variable. Back without saddles or with 9 or 10 small saddles of varying intensity. Side with dark reticulations and sometimes with 8–10 indistinct to distinct blotches or bars, best developed posteriorly. Gill membranes usually only slightly joined across

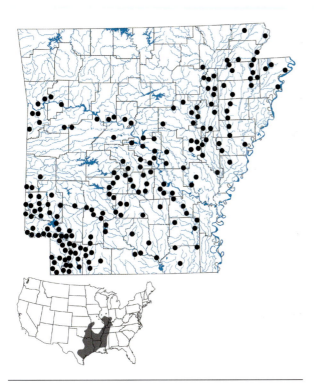

Map 201. Distribution of *Etheostoma gracile*, Slough Darter, 1988–2019.

throat but sometimes moderately joined. Premaxillary frenum present. Infraorbital canal uninterrupted, usually with 8 (7–9) pores. Usually with a faint suborbital vertical bar. Cheek and opercle fully scaled; belly fully scaled or naked anteriorly; breast and top of head naked. Lateral line highly arched anteriorly and incomplete, ending beneath first dorsal fin. Usually only 2–4 (2–6) scale rows between lateral line and front base of first dorsal fin; scale rows below lateral line 7–12. Scales in lateral series 40–55 (usually 46–51); pored scales usually 16–23. Diagonal scale rows 11–14; scale rows around caudal peduncle 16–23. Dorsal spines usually 8–10; dorsal rays 9–13 (usually 10–12); anal spines 2; anal rays 5–8 (usually 6 or 7); pectoral rays usually 13 (12–14). Breeding males with tubercles on lower jaw (2 rows on each side) and on pelvic and anal fins. Maximum length about 2 inches (52 mm).

LIFE COLORS   Back a mottled olive brown to yellowish, sometimes with dark greenish-brown saddles. Sides lighter brown with dark brown reticulations; female with a midlateral series of dark green rectangular blotches, male with bright green to turquoise vertical bars. A thin dark suborbital bar present in both sexes. The short lateral line appears as a light stripe. Breast and belly of male with scattered spots; those of female unspotted. First dorsal fin of breeding male with black or blue-black marginal and basal

bands and a broad, bright red submarginal to medial band; first dorsal fin of female with a narrow, light red-orange submarginal band. Second dorsal fin and caudal fin of male with light brown bands. Remaining fins usually clear.

SIMILAR SPECIES   Differs from *E. fusiforme* in having breast and top of head naked (vs. scaled), infraorbital canal uninterrupted (vs. interrupted), and green bars on side in life (vs. bright colors absent). It differs from *E. microperca* and *E. proeliare* in having more than 8 pored lateral line scales (vs. fewer than 8), more than 40 scales in lateral series (vs. fewer than 40), usually more than 12 pectoral rays (vs. fewer than 12), and green bars on side in life (vs. no green bars). It sometimes occurs syntopically with members of the *E. stigmaeum* complex (usually *E. stigmaeum* or *E. teddyroosevelt*), but it can be distinguished from those species in having a premaxillary frenum (vs. frenum absent), 16–23 pored lateral line scales (vs. 26 or more pored scales), and 8–10 spines in the first dorsal fin (vs. usually 11 or 12).

VARIATION AND TAXONOMY   *Etheostoma gracile* was described by Girard (1859) as *Boleosoma gracile* from the Rio Seco near Fort Inge, Texas. Variation in *Etheostoma gracile* was studied by Collette (1962), who placed it in subgenus *Hololepis* but recognized no subspecies. Subgeneric assignment has been controversial, with Page (1981, 1983a) merging the subgenera *Hololepis* and *Microperca* into subgenus *Boleichthys*. Most subsequent classifications retained the subgenus *Hololepis* (Etnier and Starnes 1993; Boschung and Mayden 2004).

DISTRIBUTION   The Slough Darter occurs in the Mississippi River basin from central Illinois and northeastern Missouri, south to Louisiana, and in Gulf Slope drainages from Mississippi to the Nueces River of southern Texas. It is widespread throughout the Coastal Plain lowlands of eastern and southern Arkansas in the St. Francis, Black, lower White, lower Arkansas, Ouachita, and Red river drainages. It is also found throughout the Arkansas River Valley in Arkansas but avoids the Ozark and Ouachita uplands (Map 201).

HABITAT AND BIOLOGY   The Slough Darter is a lowland species found in bayous, sloughs, oxbow lakes, and creeks in areas with sluggish or no current. It is occasionally found in moderate current, particularly in creeks of the lower Arkansas River Valley and in Red River floodplain habitats. It usually prefers a bottom of mud, covered with plant debris and some aquatic vegetation, but is sometimes found over a sand and fine gravel bottom. The Slough Darter is often found in habitats lacking vegetation (Collette 1962), and it tolerates occasional muddy water (Cross and Collins 1995). In Dismal Creek, Illinois, specimens were collected almost exclusively from mud-bottomed pools, the few exceptions being occasional strays or individuals overwintering in deep, sand-bottomed pools when shallower pools were frozen (Braasch and Smith 1967). Several authors noted that this species often occurs with the Bluntnose Darter (Kuehne and Barbour 1983; Page 1983a; Pflieger 1997). Buchanan (2005) found the Slough Darter in low numbers in 17 Arkansas impoundments.

In Illinois, Braasch and Smith (1967) found that Slough Darters fed mainly on midge and mayfly larvae and small crustaceans (copepods, cladocerans, ostracods, and amphipods). Within 2 weeks of hatching, young were able to feed on those items. The greatest volume of food of adults was consumed during spring months, with decreasing amounts consumed during winter, summer, and fall. Braasch and Smith (1967) also noted seasonal changes in types of food eaten and differences in food preference between adults and young. Midge larvae made up 64% by volume of the food eaten in February, with copepods composing 21%. By May, mayfly nymphs were the predominant food items (56%), and midge larvae made up 41%. Stomach contents of 62 specimens from seven Arkansas counties were examined by TMB. The diet from March through May was equally dominated by small crustaceans (copepods, amphipods, and isopods) and chironomids, with ephemeropterans also ingested in small amounts. In July, a similar diet composition was found in specimens from Lake Atkins in Pope County; specimens taken in July from the Red River (Little River County) fed mainly on chironomids and other larval dipterans, but cladocerans, copepods, and amphipods were also ingested.

Spawning occurs mainly in May and June in Illinois, April or May in Kansas (Cross and Collins 1995), but as early as March in Texas (Collette 1962). We have collected ripe individuals from early March through April in Arkansas, and spawning probably continues through May. A reproductive female 46 mm TL collected on 13 March in Mississippi County by TMB contained 177 ripe ova as well as smaller egg sizes; a 52 mm TL female from Franklin County contained 175 ripe ova on 17 April. Slough Darters collected from Crawford County, Arkansas, on 15 March, spawned the next day in an aquarium. The eggs were attached to aquatic vegetation and the aquarium glass. Braasch and Smith (1967) described reproductive behavior and egg deposition. Males are apparently territorial. When a female is ready to spawn, she enters a male's territory but is rather passive to the male's courtship attempts. Eventually the persistent male places himself on her back or alongside her with his head above hers. All stimulation by the male appeared to be tactile rather than visual, and no color displays were noted. The male rubs the female with his pectoral fins, breast, and chin tubercles. The female then moves to an object, usually a leaf or twig, and deposits several eggs in a line, which are fertilized by the male. Each female lays 30–50 eggs, but a single female can produce more than 2,500 eggs per ovary in one season (only about 20% ever reach maturity and only a few of those are actually laid). Hatching occurred in 5 days at a water temperature of 22.8°C. Sexual maturity occurs within 1 year of hatching (Collette 1962), and maximum reported life span is 4 years (Braasch and Smith 1967).

CONSERVATION STATUS   This is one of the most widespread and common largely lowland darters. Its apparent tolerance of turbidity (Etnier and Starnes 1993) probably contributes to its continued commonness. Populations tend to be small but are apparently secure in Arkansas.

Etheostoma histrio Jordan and Gilbert
Harlequin Darter

Figure 26.63. *Etheostoma histrio*, Harlequin Darter. *David A. Neely.*

CHARACTERS   A moderately robust *Etheostoma* with a short, rounded head, head length going 4.2 times or more into SL (Fig. 26.63). Snout blunt and mouth ventral. Pectoral fins expansive. Back with 6 or 7 dark saddles, the second and fourth saddles narrower than adjacent ones. Side with 7–11 dark vertical bars, some of which often fuse with dorsal saddles. Body at base of caudal fin with 2 large dark blotches. Underside of head and breast with dark spots or markings. Dark suborbital bar usually present. All fins conspicuously spotted. Gill membranes broadly joined across throat. Premaxillary frenum absent or very narrow. Infraorbital canal uninterrupted, with 7 pores. Nape fully scaled; cheek, opercle, and belly naked to partly scaled; breast naked. Lateral line complete with 45–58 scales. Scale rows above lateral line 4–8; scale rows below lateral line 6–9. Diagonal scale rows 12–18; scale rows around caudal peduncle 15–21. Dorsal spines 9–11; dorsal rays 11–14; anal spines 2; anal rays 6–8; pectoral rays usually 14 or 15. Males and females without breeding tubercles. Maximum length about 3 inches (76 mm).

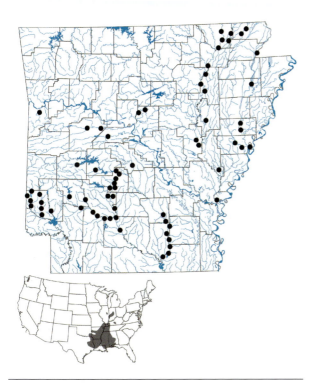

Map 202. Distribution of *Etheostoma histrio*, Harlequin Darter, 1988–2019.

LIFE COLORS   Back and upper sides a mottled green with dark brown or green saddles and dark brown or green lateral bars. Underparts yellowish white with scattered dark spots or blotches. All fins marked with conspicuous dark brown or black spots. Base of caudal fin with 2 large dark spots. First dorsal fin of male with a dark red submarginal band. Suborbital bar dark and prominent. Breeding males more boldly marked with contrasting bright green and brown colors, a wide red band in the spinous dorsal fin, dusky pelvic fins, and turquoise gill membranes.

SIMILAR SPECIES   Differs from all other Arkansas *Etheostoma* species in having 2 large, dark basicaudal blotches that extend onto caudal fin base. It is often confused with *E. zonale* and is syntopic with it in a few areas of the state, such as the Ouachita River. It differs from that species in having very long pectoral fins that extend backward past tips of pelvic fins (vs. tips of pectoral fins not extending backward past tips of pelvic fins), breast and underside of head with large, dark spots (vs. unspotted), 9 or 10 dorsal spines (vs. 11 or 12), and cheek, opercle, and front half of belly usually naked (vs. fully scaled).

VARIATION AND TAXONOMY   *Etheostoma histrio* was described by Jordan and Gilbert (*in* Gilbert 1887), giving the following type locality information: "Abundant in the Poteau River, near Hackett City, Ark. Found also in the Saline River at Benton, Ark., and in the Washita [*sic*] River at Arkadelphia." Collette and Knapp (1966) redescribed the species and designated the Poteau River, Oklahoma, as the locality for the lectotype. Tsai (1968) studied geographic variation and recognized no subspecies. *Etheostoma histrio* is traditionally placed in subgenus *Etheostoma*. Some molecular studies resulted in conflicting conclusions about its subgeneric relationships (Porterfield 1998; Porter 1999; Porter et al. 2002), but a combined mtDNA and nuclear gene phylogeny found that *Etheostoma histrio* and *E. baileyi* formed a significantly supported clade that was sister to a clade containing *E. cinereum*, all other greenside darters, and all other snubnose darter species (subgenera *Ulocentra* or *Nanostoma*) (Near et al. 2011).

DISTRIBUTION   The Harlequin Darter is found mainly in the Mississippi Embayment of the lower Mississippi River basin from Kentucky southward to Louisiana, and Gulf Slope drainages from the Escambia River, Florida and Alabama, west to the Neches River, Texas. Northern disjunct populations occur in Mississippi River drainages of Kentucky, Indiana, and Illinois. In Arkansas, *Etheostoma histrio* is found in Coastal Plain lowland streams and lowland-upland boundary streams above and below the Fall Line in eastern and southern parts of the state and in a few Arkansas River Valley streams such as the Vache Grasse Creek drainage in Sebastian County. It also occurs in the St. Francis, Black, lower White, Arkansas, Ouachita, and Red river drainages (Map 202).

HABITAT AND BIOLOGY   *Etheostoma histrio* occurs in Arkansas primarily along and below the Fall Line in large creeks and small to moderate-sized rivers, usually in moderate to swift current in riffles over fine to coarse gravel or broken shale. It is intolerant of reservoir conditions. Most populations throughout its range occupy streams on the Coastal Plain, but small populations occur well above the Fall Line in Illinois, Indiana, and Kentucky (Tsai 1968; Page 1983a). Pflieger (1997) noted that in Missouri it had a definite preference for sandy substrates where logs, sticks, and other organic debris were present. Etnier and Starnes (1993) reported that Harlequin Darters were regularly collected in the main channel of the Mississippi River in Tennessee, but we have only one record from the Mississippi main stem in Arkansas. In lowland streams of the White River National Wildlife Refuge it is found over a sand, mud, and detritus bottom in moderate current in habitat more like that described for this species in Missouri by Pflieger (1997), east Texas by Hubbs and Pigg (1972), and western Kentucky by Sisk and Webb (1976). Boschung and Mayden (2004) noted the variety of habitat features associated with this species and found that the unifying characteristic of its habitat was detritus in the form of dead leaves, brush, sticks and twigs, logs, and/or mats of rootlets. Etnier and Starnes (1993) noted that apparently considerable seasonal movement occurs in this species.

In Kentucky, Harlequin Darters fed mainly on midge, black fly, and caddisfly larvae; mayfly nymphs made up a small portion of the diet (Kuhajda and Warren 1989). Stomach contents of 53 specimens collected from Sebastian, Yell, St. Francis, and Monroe counties, Arkansas, from August through November were examined by TMB. Data were pooled because there were no obvious differences in diet among months or among localities. Ninety-eight percent of the stomachs contained food. Chironomids were the most important food by percent frequency of occurrence at all localities in all months, and dominated the diet (by percent frequency and percent volume) of individuals smaller than 40 mm TL. Chironomids and other dipteran larvae (mostly simuliids) dominated the diet of individuals of all sizes (76% frequency of occurrence), followed by trichopterans (55%), and ephemeropterans (38%). Also contributing to the diet in decreasing order of importance were coleopteran larvae, plecopteran nymphs, isopods, odonates, and megalopterans.

Hubbs and Pigg (1972) reported that spawning occurred in Texas in February, and Kuehne and Barbour (1983) collected individuals in spawning condition in mid-March in Mississippi. In Kentucky, spawning occurred in February and March (Kuhajda and Warren 1989). Females from the Ouachita River in Hot Spring County examined by TMB contained ripe ova as early as 22 February; one mature female 51 mm TL contained 70 ripe ova. On 17 March, TMB found ripe males and females spawning in deep riffles (46 cm deep) in swift current over a cobble and gravel substrate at a water temperature of 15°C in Vache Grasse Creek in Sebastian County, Arkansas. The eggs were attached to moss covering the cobbles. The breeding males were a brilliant green. Spawning behavior has not been observed in nature, but aquarium spawning was described by Steinberg et al. (2000). Agonistic displays involved lateral displays and sparring (males) and lateral displays and nipping (females). Egg deposition sites were chosen by females, and the eggs were attached to rocks covered by moss or algae (similar to the egg deposition sites in Vache Grasse Creek, Arkansas) in the aquarium. Based on the absence of adults in colder months from streams where they had been abundant in May through October, Etnier and Starnes (1993) believed that considerable movement occurred in Tennessee populations. They speculated that adults moved downstream into larger rivers and perhaps reservoirs during winter months. Kuhajda and Warren (1989) found that sexual maturity

occurred at age 1 and fecundity of females ranged from 90 to 450 eggs. Maximum life span was 4+ years.

CONSERVATION STATUS   This lowland darter is more widespread in Arkansas than previously thought. Its populations are generally small but secure.

Etheostoma microperca Jordan and Gilbert
Least Darter

Figure 26.64. *Etheostoma microperca*, Least Darter. *David A. Neely.*

Map 203. Distribution of *Etheostoma microperca*, Least Darter, 1988–2019.

CHARACTERS   Arkansas's smallest darter, reaching a maximum length of around 1.5 inches (38 mm) (Fig. 26.64). Lateral line usually entirely absent, but sometimes very short with only 1 or 2 pored scales. Back usually without saddles or blotches, but with small scattered dark spots. Side with a midlateral row of 7–10 small dark blotches. Gill membranes broadly joined across throat. Head and mouth small. Narrow premaxillary frenum present. Infraorbital canal interrupted, lacking posterior segment, with 2 or 3 pores in anterior segment. Nape, cheek, and breast naked; opercle with a few scales or naked; belly partly to fully scaled. Scales in a midlateral series 32–36. Scale rows above lateral line 3 or 4; scale rows below lateral line 4 or 5. Diagonal scale rows usually 9–11 (8–12); scale rows around caudal peduncle modally 14 (12–16). Dorsal spines 6–8; dorsal rays 8–11 (usually 9 or 10); anal spines 2 (occasionally 1); anal rays usually 5 or 6; pectoral rays 9–12. Breeding males with tubercles on pelvic and anal fins and unique among darters in having them on dorsal fins (in Ozark populations); the pelvic fin is long and has an expansive flap of skin.

LIFE COLORS   Back and sides olive green with scattered dark brown specks and zigzag markings. Lateral blotches dark green black. Undersides white or yellow with irregular spots or stippling. Prominent dark pre- and suborbital bars present. Dorsal and caudal fins splotched or banded with dark brown; other fins mostly clear. Breeding males much darker. First dorsal fin with a red-orange submarginal band; second dorsal and caudal fins milky white; pelvic and anal fins become dark red black. The pelvic fin is long and has an expansive flap of skin. Breeding colors poorly developed in females.

SIMILAR SPECIES   The Least Darter differs from all other Arkansas darters in lacking a lateral line (sometimes having 1 or rarely 2 pored scales). It is similar to *E. proeliare,* but differs in having naked cheeks (vs. fully scaled), and 5 branchiostegal rays (vs. 6).

VARIATION AND TAXONOMY   *Etheostoma microperca* was described by Jordan and Gilbert *in* Jordan (1888) from Michigan, Wisconsin, Illinois, and Alabama. It is placed in subgenus *Microperca* by most current workers. Genetic analyses by Near et al. (2011) supported a *Microperca* clade consisting of *E. fonticola, E. microperca,* and *E. proeliare.* Morphological variation in the Least Darter was studied by Burr (1978), who found that the most divergent populations occurred in the Ozarks. Those populations were different in several meristic characters, pigmentation features, tubercle patterns, and maximum size. The variation was sharply defined in several features, but Burr did not designate a distinct taxonomic status for the Ozark populations. Echelle et al. (2015), in a combined analysis of mitochondrial and nuclear DNA, found that *E. microperca* populations in the western Ozarks supported most of the genetic diversity in the species and could be divided into three markedly divergent clades based on the cyt*b* gene. Populations in the Illinois River drainage of Arkansas (which includes all known Arkansas populations of this species) were the most genetically divergent of all *E. microperca* populations studied, and probably represent an undescribed cryptic species. The Illinois River populations appeared to have been isolated for about 5 million years. Echelle et al. (2015) further rejected the long-held hypothesis that *E. microperca* did not occupy the Ozarks until glacially driven range displacement occurred in the Pleistocene. Instead, they concluded from their phylogeographic analysis that the Ozark region is the center of diversity for the species, and that the northern populations were derived from Ozark ancestors.

DISTRIBUTION   As currently recognized, most of the range of *E. microperca* is in the Great Lakes, Hudson Bay, and upper Mississippi River basins from southern Ontario, Canada, west to Minnesota, and south to Ohio, Indiana, and Illinois. Disjunct populations occur in the Ozarks of Missouri, Kansas, Arkansas, and Oklahoma, and in the Red River drainage (Blue River) of Oklahoma. The Least Darter is very rare in Arkansas. It was historically known from five springs in the headwaters of the Illinois River in Benton and Washington counties. Harris and Smith found it in only three of those springs in 1985, and Hargrave and Johnson (2003) found it at only two of the historic sites. In an extensive survey of 75 spring and spring-run habitats in the Illinois River drainage headwaters, Wagner and Kottmyer (2006) found *E. microperca* at eight sites. A subsequent survey determined that the Least Darter had been extirpated from Wildcat Creek, Clear Creek, and Elkhorn Springs. Previously undocumented populations were found in the Clear Creek and Flint Creek watersheds (Wagner et al. 2012). Wagner et al. (2012) also reported that populations persist in a spring-run tributary to Osage Creek and in Healing Spring Run, and additional populations were found in other nearby tributaries to Little Osage Creek and in vegetated backwaters along the creek itself (Map 203).

HABITAT AND BIOLOGY   In Arkansas, this tiny darter inhabits small clear springs and quiet pools of spring creeks having permanent flow and gravel bottoms, often with accumulations of silt and detritus and thick growths of rooted aquatic vegetation and filamentous algae. Johnson and Hatch (1991) noted that it appears to associate with dense aquatic vegetation, soft bottoms of sand and/or organic sediment, and quiet water regardless of geographic region. In northeastern Oklahoma, optimal microhabitat was described as a *Myriophyllum*-filled cutoff or backwater pool of a creek (Blair and Windle 1961). Hargrave and Johnson (2003) found that *E. microperca* in Arkansas always occurred in close association with coontail, *Ceratophyllum demersum.* Wagner et al. (2012) reported that *E. microperca* occurred in the same habitat type as *E. cragini*, with main habitat features of clear water, stream width less than 10 m, less than 25% of shoreline wooded, multiple types of rooted aquatic vegetation, mud/silt as dominant substrate, and slow current. It requires cool water temperatures in the summer (Burr 1978), thereby restricting its habitat in Arkansas to springs and spring-influenced streams (Echelle et al. 2015). Pflieger (1997) reported that it spends much of its time off the stream bottom moving over the leaves and stems of aquatic plants.

It feeds mainly on small crustaceans and midge larvae, with peak feeding occurring at midday (Page 1983a). In the Iroquois River, Illinois, small individuals fed mainly on small crustaceans (cladocerans, copepods, ostracods, and isopods) and chironomids, but gastropods, plecopterans, and several kinds of dipterans (mainly Tabanidae, Ephydridae, and Simuliidae) became increasingly important in the diets of large (22–30 mm TL) individuals (Burr and Page 1979). In that study, microcrustaceans and chironomids were important in the diet of individuals of all sizes. In Minnesota, the diet comprised chiefly cladocerans and copepods in spring and fall, and chironomids during the summer (Johnson and Hatch 1991).

Spawning occurs from April through June in Missouri (Pflieger 1997) and from February through April or May in Oklahoma (Etnier and Starnes 1993; Miller and Robison 2004). Its reproductive biology in Michigan was reported by Petravicz (1936) and Winn (1958a), and Burr and Page (1979) studied reproduction in Illinois. In Minnesota, *E. microperca* spent the winter in deep water (1.0–1.5 m) of pools, but as the spawning season approached, they moved to the margins of the pools and then to shallow, weedy margins of runs immediately downstream where spawning occurred (Johnson and Hatch 1991). The males are weakly territorial during the spawning season, defending small territories of about 30 cm diameter (Petravicz 1936). Spawning often occurs in a vertical position on the stems and leaves of aquatic plants. The male mounts the female and clasps her with his enlarged pelvic fins. The breeding tubercles aid the male in maintaining this position. The two fish vibrate and adhesive eggs are deposited one or a few at a time on the vegetation. More eggs are then laid at additional spots. Johnson and Hatch (1991) provided evidence for multiple clutches and estimated a mean of 88.6 ova for the first clutch, 88.3 ova for the second clutch, and 49.2 ova for the third clutch. This species was found to have a very short life span in Illinois, living only 18–20 months (Burr and Page 1979), but a few individuals in a Minnesota population had completed 3 growing seasons (Johnson and Hatch 1991). Both sexes mature and spawn at 1 year of age.

Population sizes in Arkansas were estimated at 500 individuals at one locality and 1,000 individuals at three other localities in 1985 (Harris and Smith 1985). Hargrave and Johnson (2003) estimated population size at two localities as 1,000 and 129.

CONSERVATION STATUS The Least Darter is a very rare species in Arkansas. The populations are mostly isolated, leaving them highly susceptible to local extirpation with little opportunity for natural recolonization. Least Darters are most common in the Osage Creek watershed, and those populations have a greater potential to move between spring runs than do the other more isolated Arkansas populations (Wagner et al. 2012). The Arkansas populations, which possibly represent an undescribed species, are at the very least an ancient lineage deserving especially high conservation priority (Echelle et al. 2015). Its continued existence in the state probably depends on protecting its critical habitat in northwestern Arkansas. As with the other darter species (*E. cragini*) restricted to a few springs and spring-runs in that part of the state, we consider it endangered in Arkansas.

Etheostoma mihileze Mayden
Sunburst Darter

Figure 26.65. *Etheostoma mihileze*, Sunburst Darter. *Uland Thomas*.

CHARACTERS A moderately stout darter with a large head, conical snout, and terminal mouth (Fig. 26.65). Body relatively deep and robust. Dark suborbital bar present. Back usually with 4 dark saddles extending somewhat downward onto upper side. Lateral caudal peduncle occasionally with 3 or 4 dark blotches arranged from beneath middle of second dorsal fin to caudal fin base. Head and body heavily stippled and mottled. Gill membranes separate or slightly joined across throat. Premaxillary frenum present. Infraorbital canal uninterrupted, usually with 9 or 10 pores. Nape and belly fully scaled; cheek, opercle, and breast scalation variable, ranging from scaled to naked or with embedded scales. Lateral line incomplete, extending to beneath (and sometimes past) second dorsal fin. Scales

Map 204. Distribution of *Etheostoma mihileze*, Sunburst Darter, 1988–2019.

in lateral series 55–76 (but usually 59–67); pored lateral line scales highly variable (34–57). Scale rows above lateral line usually 9–11; scale rows below lateral line usually 9 or 10 (8–11). Diagonal scale rows usually 19–23 (17–24); scale rows around caudal peduncle 25–32. Dorsal spines 10–12 (usually 10 or 11); dorsal rays 13–16 (usually 14 or 15); anal spines 2; anal rays usually 7–9; pectoral rays 12–14. Breeding males with tubercles on anal and pelvic spines and rays, on scales of the belly, and along the anal fin base. Maximum length around 3.5 inches (89 mm).

LIFE COLORS Mayden (2010) provided detailed descriptions of breeding and nonbreeding coloration. Breeding male: First dorsal fin with a broad distal orange band and a single basal dark band that extends to median of fin; first dorsal fin without a basal orange band. Second dorsal fin heavily mottled with up to 5 bars. All other fins with some type of pigmentation. Dorsum and upper side darkly pigmented. Bright orange coloration on venter restricted to abdominal region from pelvic fins to origin of anal fin and dorsally along flanks to below lateral line. Nonbreeding males, females, and juveniles with head, back, and sides brown to olive tan; dorsal saddles dark brown; underside white to cream-colored and with scattered melanophores. Body heavily stippled and mottled, especially around head; stippling most obvious in nonbreeding and

freshly preserved individuals. Dorsum and lateral areas of head with dark vermiculations. Dark vertical suborbital bar distinct. Humeral spot black and conspicuous.

SIMILAR SPECIES The Sunburst Darter is most like the allopatric Autumn Darter, but it differs from it in usually having 59–66 lateral scales (vs. usually 70–74), a heavily stippled and mottled body (vs. body lightly stippled and with no mottling), usually 4 dorsal saddles (vs. usually 5 or 6), and in having a vermiculated melanin coloration pattern on top and sides of head (vs. no vermiculations present). For other differences in breeding male coloration, see Mayden (2010). It differs from other Arkansas members of the subgenus *Ozarka* (the sympatric *E. cragini* and allopatric *E. pallididorsum*) in having a longer lateral line extending well past the base of the first dorsal fin (vs. lateral line short, not extending past base of first dorsal fin), more than 57 lateral scales (vs. fewer than 57), and 13 or more soft dorsal rays (vs. 12 or fewer). It also differs from *E. pallididorsum* in lacking a wide, pale stripe down middle of back (vs. pale middorsal stripe present).

VARIATION AND TAXONOMY *Etheostoma mihileze* is a member of the *Etheostoma punctulatum* species group within the subgenus *Ozarka*. Mayden (2010) described Arkansas River drainage populations formerly included under *E. punctulatum* as a new species, *E. mihileze*, based on morphological and allozyme characters. The type locality is the Spring River at U.S. Route 60, just south of Verona, Lawrence County, Missouri. The *Etheostoma punctulatum* species group was diagnosed as monophyletic, with *E. punctulatum* as the sister group to the *E. autumnale* and *E. mihileze* pair.

DISTRIBUTION *Etheostoma mihileze* is endemic to the Arkansas River drainage of the Ozark Plateau in Kansas, Missouri, Oklahoma, and Arkansas. In Arkansas, it is known from eight counties: Benton, Washington, Crawford, Johnson, Madison, Newton, Pope, and Franklin. It is most common in the Illinois-Neosho river drainages of Benton and Washington counties. It is also found in Arkansas River tributaries north of the river from Lee Creek in Crawford County eastward to Illinois Bayou in Pope County, but it is absent from Arkansas River tributaries south of the river (Map 204). According to Mayden (2010), *E. mihileze* is most abundant outside Arkansas.

HABITAT AND BIOLOGY The Sunburst Darter is most abundant in Arkansas in springs and spring-fed creeks in the karst regions of Benton County. It is found over gravel and cobble substrates and often occurs in association with watercress (*Nasturtium* spp.) in the spring runs. This darter becomes much less common in nonkarst areas to the south, where it is primarily an inhabitant of medium-sized to large creeks (e.g., Lee Creek) and small rivers (e.g., the Mulberry River) draining the southern slopes of the Ozarks. It requires clear water and some permanent flow and is intolerant of silt. During most of the year it is found in gravel- or rubble-bottomed pools in quieter waters in vegetation or detritus. Mayden (2010) noted that it is often associated with undercut banks with vegetation or overhanging trees and roots. Its population densities in Arkansas are usually lower than those of other syntopic *Etheostoma* species. In the Frog Bayou drainage (Clear Creek) of Crawford County, population density estimates ranged from 0.0025 to 0.03 fish/m^2, and in the Mulberry River drainage in Madison and Johnson counties, density estimates ranged from 0 to 0.05 fish/m^2 (Rambo 1998).

The major food items in Spring River, southwestern Missouri, were isopods (66.6% by volume), mayfly nymphs (12.4%), and caddisfly larvae (8%); other important food items were amphipods, crayfish, and earthworms. In Flint Creek, Oklahoma, stomachs of 205 *E. mihileze* contained 30 different prey taxa. Both juveniles and adults fed on chironomids in the coldest months (January and February) but relied much less on those food items in those months (and generally throughout the year) than did the syntopic *E. squamosum*; feeding by *E. mihileze* during those months was more focused on mayflies, isopods (*Asellus*), and amphipods (Todd and Stewart 1985). Feeding in spring was almost exclusively on the mayfly *Leptophlebia*, and during summer months (June–September) and into November, juvenile *E. mihileze* fed mainly on amphipods and the mayflies *Stenonema*, *Baetis*, and *Leptophlebia*. Mature *E. mihileze* showed little pattern in feeding, except in January to March, when the chironomids *Leptophlebia* and *Ephemerella* were ingested in high volumes.

The life history of *E. mihileze* was investigated in Missouri (Hotalling and Taber 1987) and Oklahoma (Vives 1987). During the breeding season it is usually found in shallow gravelly riffles with moderate to swift current where spawning may occur; these riffles often have aquatic vegetation. Spawning occurs in Arkansas from mid-February through mid-May (Hubbs 1985) in stenothermal spring environments, and Hotalling and Taber (1987) reported ripe females from Spring River, Missouri, from February through early May. Aquarium observations of spawning in other members of the *E. punctulatum* species group indicate that the eggs are buried in gravel (Page and Simon 1988; Simon and Garcia 1990). Males reach sexual maturity at 1 year of age, but only those yearling females at least 1.9 inches (49 mm) TL produced mature ova. Maximum life span is slightly more than 4 years (Hotalling and Taber 1987).

CONSERVATION STATUS The habitats favored by this darter are rapidly disappearing in Benton and Washington counties. It is still found in favorable habitats but may eventually warrant some sort of protection to keep it from becoming threatened in Arkansas. We consider it a species of special concern.

Etheostoma nigrum Rafinesque
Johnny Darter

Figure 26.66. *Etheostoma nigrum*, Johnny Darter. *Uland Thomas.*

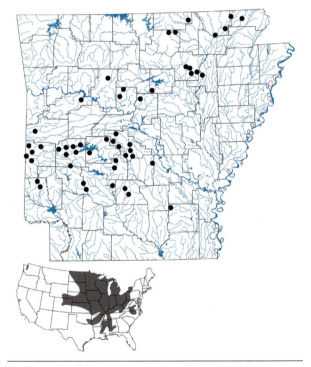

Map 205. Distribution of *Etheostoma nigrum*, Johnny Darter, 1988–2019.

CHARACTERS A small, slender, straw-colored darter with a small head, a short, moderately blunt snout, and a horizontal mouth (Fig. 26.66). Premaxillary frenum absent, upper lip and snout separated by a continuous groove. Head length going into SL 3.7 times or more. Back with about 6 saddles which may be faint or dark. Midside with a series of dark W- and X-shaped marks; upper side and back with scattered smaller W- and X-shaped markings and irregular vermiculations. Dark preorbital bars directed downward, not continuous around snout, but with a small, clear gap present on tip of snout (preorbital bars in some specimens may extend onto the upper lip but do not join on the midline). Gill membranes usually only slightly joined across throat. Squamation can be variable. Breast and nape naked; cheek naked or with a few scales behind eye; opercle scaled (but sometimes naked in Red River drainage). Lateral line complete or nearly so, extending well behind second dorsal fin, usually with fewer than 6 unpored scales. About 52% of Arkansas specimens examined by TMB had a complete lateral line, but the percentage varied among drainages. Scales in lateral series in Arkansas populations 49–59 ($\bar{x}$ = 53.7). Scale rows above lateral line usually 4 or 5 (2–7); scale rows below lateral line 4–9. Diagonal scale rows 11–16; scale rows around caudal peduncle 13–21. Infraorbital canal interrupted, usually with 4 or 5 anterior pores and 2 posterior pores. First and second dorsal fins scarcely to slightly separated. Dorsal spines 8–10; dorsal rays 11–13; anal spine 1; anal rays 7–9; pectoral rays usually 11 or 12 (10–14). Breeding males without tubercles. Maximum length is about 3 inches (76 mm).

LIFE COLORS   Back and upper sides yellow to light green, with dark brown saddles and markings. Underside yellowish white. Dorsal and caudal fins with faint brown bands; remaining fins usually clear. Breeding males much darker, especially head and fins, often with dark vertical bars on sides. A large anterior black blotch on first dorsal fin. Bright colors absent, except for metallic green on operculum. Breeding males develop white knobs on tips of dorsal spines and pelvic rays.

SIMILAR SPECIES   The Johnny Darter is often confused with *E. chlorosoma*, *E. clinton*, *E. stigmaeum*, and *E. teddyroosevelt*. It differs from the first species in not having dark preorbital bars continuous around the snout (vs. continuous), in usually having naked cheeks (vs. cheeks scaled), and in usually having a complete lateral line (vs. incomplete, usually with 23–45 unpored scales in lateral series). It differs from the latter three species in having only 1 anal spine (vs. 2), 10 or fewer dorsal spines (vs. usually 12 or 13), usually a complete lateral line (vs. incomplete), and lack of bright colors in breeding males (vs. breeding males with bright colors). It sometimes occurs with *E. proeliare* but can be distinguished from that species in having a complete or nearly complete lateral line (vs. lateral line very short, with only 2–7 pored scales).

VARIATION AND TAXONOMY   *Etheostoma nigrum* was described by Rafinesque (1820a) as *Etheostoma nigra*

from the Green River, Kentucky. It is the only Arkansas member of subgenus *Boleosoma*. Monophyly of *Boleosoma* was supported by genetic evidence (Near et al. 2011). Krotzer (1990) found morphological differences in populations of *E. nigrum* from direct tributaries of the Mississippi River in western Kentucky, Tennessee, and Mississippi compared with populations from the Ozarks, lower Tennessee River, and Mobile Basin. Two subspecies of *E. nigrum* were recognized by Page and Burr (2011). One subspecies, *E. n. susanae*, is endemic to the upper Cumberland River drainage of Kentucky and Tennessee, and the other subspecies, *E. n. nigrum*, is found throughout the remainder of its range, including Arkansas. The Cumberland River form was recognized as a full species, *E. susanae*, by Near et al. (2011) and Page et al. (2013).

DISTRIBUTION   *Etheostoma nigrum* is found in the St. Lawrence–Great Lakes, Hudson Bay, and Mississippi River basins from Canada south to Mississippi; in Atlantic Slope drainages of Virginia and North Carolina; and in Gulf Slope (Mobile Bay) drainages of Alabama and Mississippi. It is introduced into the Colorado River drainage of Colorado (Page and Burr 2011). The Johnny Darter has a sporadic distribution in Arkansas. It is most common in the Ouachita Mountains in the upper Ouachita and Saline rivers. It also occurs in upland tributaries of the Red River and in the Poteau River in westcentral Arkansas. Other populations are found in a few Arkansas River tributaries and in the White, Spring, Strawberry, Black, and Current rivers in northeastern Arkansas (Map 205).

HABITAT AND BIOLOGY   Although one of the most wide-ranging and abundant of all darters over much of its range north and east of Arkansas, the Johnny Darter is uncommon in Arkansas. It is found mainly in creeks and small rivers in the Ouachita Mountains, but the known localities for it there are widely scattered and the populations are small. Reynolds (1971), in a survey of the eastern Saline River, collected only 10 specimens from a single locality. A survey of the Caddo River by Fruge (1971) yielded only 20 specimens from four localities, and Dewey and Moen (1978) found no specimens in a subsequent study of that river. Etnier and Starnes (1993) reported that northern populations occur in shoreline areas of lakes as well as in streams. Buchanan (2005) found Johnny Darters in five Arkansas reservoirs, and the Lake Hinkle (Scott County) records were the first reported from the Poteau River drainage of Arkansas in more than 45 years. *Etheostoma nigrum* occurs in moderate to high gradient streams, where it is usually found in slow current near the edges of pools having a sand or mixed sand and gravel bottom. Fullenkamp (2010) and Pratt and Lauer (2013) also reported a preference for small substrate size (silt-sand), moderate water velocities, and deeper water than most darters its size. Pflieger (1997) reported that it is more tolerant of turbidity than most darters but avoids streams that are excessively turbid and silty. Becker (1983) considered it a pioneer species that quickly becomes established in disturbed areas. In Arkansas, it is found mainly in clear water. The transition zone where the Ouachita River transitions to Lake Ouachita provides a unique habitat for Johnny Darters, which are very rare in the riverine section of the upper Ouachita River (Stoeckel and Caldwell 2014). Pflieger speculated that the Johnny Darter may be excluded from the lowlands of Missouri by competition with the ecologically similar Bluntnose and Speckled darters. One or both of those species are almost always found syntopically with the Johnny Darter in Arkansas, perhaps contributing to its scarcity in this state.

Etheostoma nigrum has been the subject of a number of studies (summarized by Page 1983a). It feeds mainly on midge larvae, mayfly nymphs, and caddisfly larvae, as well as small crustaceans, especially ostracods and copepods (Roberts and Winn 1962; Karr 1963). Smart and Gee (1979) found it to be a daytime benthic feeder that locates food mainly by sight and, to a lesser extent, olfaction. The young feed mainly on entomostracans and small midge larvae, and these items are often important in the diet throughout life (Karr 1963). In Indiana, midges were the dominant prey, and there was little change in diet among seasons (Strange 1991). Becker (1983) provided a list of known foods of the Johnny Darter in Wisconsin, including chironomid, black fly, and other midge larvae, copepods, cladocerans, amphipods, and bottom ooze.

Winn (1958a) reported spawning habits of Johnny Darters in southern Michigan in detail, as did Pflieger (1997) for Missouri, who found spawning to occur in April and May. Spawning also occurred in April and May in Kansas (Cross and Collins 1995). At the southern end of its range in Mississippi, spawning occurred from mid-March to mid-May (Parrish et al. 1991). Spawning likely occurs in Arkansas from late March into May. HWR collected two highly tuberculate males from the Cossatot River on 25 March 1972. The males migrate to the spawning grounds slightly before the females. Spawning usually takes place at water temperatures between 11.7°C and 21.1°C (53–70°F). Breeding usually occurs in pools and slow raceways of streams and protected shallow waters of lakes which contain large rocks, logs, mussel shells, or other objects (Becker 1983). A male establishes a territory under a rock and cleans the underside of the rock with his fins by moving over it in an upside-down position. When a female enters

the territory, both sexes invert and spawning occurs in this upside-down position. Grant and Colgan (1983) reported that females preferred males that moved farther out from their nests to aggressively respond to intruders. Bart and Page (1991) suggested that the white knobs of the paired fins of males may be egg mimics used to induce females to spawn. The adhesive eggs are deposited on the underside of the rock. The female leaves the nest after spawning and the male guards the eggs. Success in defending the eggs against intruding crayfish was inversely related to crayfish size (Rahel 1989). Crayfish smaller than 15 mm carapace length were routinely evicted from the nest, but only 33% of large crayfish (16–32 mm) were evicted, resulting in some egg predation. Small size of nest entrances prevented larger crayfish from entering the nests. The male moves over the eggs in an inverted position, pulling them with his pelvic, anal, and caudal fins from 13 to 16 times per half hour. The male also fans them with his pectoral fins. The patterns of territorial defense by nest-guarding males were similar to those of other darter species exhibiting parental care (Grant and Colgan 1984). In Mississippi, clutch size varied from 54 to 192 eggs (Parrish et al. 1991). Both males and females are polygamous. One male spawns with more than one female, and a female may spawn with 4–6 different males and deposit between 30 and 200 eggs at each spawning (Winn 1958a). Winn counted 30–1,150 eggs in seven nests. The male becomes more aggressive immediately after spawning (Grant and Colgan 1983) and vigorously defends the nest, even against much larger intruders (Rahel 1989). Eggs hatch in 16 days at 12.8°C (55°F), in 10 days at 20°C (68°F), and in 6 days at 22.8°C (73°F). Maximum life span for southern populations is probably not more than 3 years, but reported life spans of 4 and 5 years are known for northern populations (Ross 2001).

CONSERVATION STATUS   This species is of sporadic occurrence in Arkansas and is nowhere abundant in the state. We list its populations as currently stable but not secure. It is a species of special concern.

Etheostoma pallididorsum Distler and Metcalf
Paleback Darter

CHARACTERS   A small, slender darter with a large head and a moderately rounded snout (Fig. 26.67). A wide, pale stripe extends along the middle of the back from the head to the base of the caudal fin; the pale dorsal stripe is crossed in many specimens by 1–6 poorly developed saddles. Side below dorsal stripe dark brown with 4 or 5 indistinct vertical bars. Lower side and venter light-colored with dark

Figure 26.67. *Etheostoma pallididorsum*, Paleback Darter. *David A. Neely.*

Map 206. Distribution of *Etheostoma pallididorsum*, Paleback Darter, 1988–2019.

spots or stippling. Gill membranes slightly joined across throat. Premaxillary frenum present. Infraorbital canal uninterrupted, with 6–9 pores. Cheek, opercle, breast, and anterior portion of belly naked. Lateral line straight, incomplete, and short, usually not extending behind base of first dorsal fin. Scales in lateral series 43–55; usually 10–16 pored scales. Scale rows above lateral line usually 7 or 8 (7–10); scale rows below lateral line modally 7 (6–8). Scales around caudal peduncle 20–25; diagonal scale rows 13–16 (counted from origin of second dorsal fin to anal fin base). Dorsal spines 8–11; dorsal rays 10–14; anal spines 2; anal rays 7–9; pectoral rays 10–12. Breeding males with tubercles on anal and pelvic fin spines and rays and on the ventral scales of the belly and caudal peduncle. Maximum length around 2 inches (51 mm).

LIFE COLORS   Middorsal stripe olive; side distinctly bicolored with upper portion dark brown, lower side and

venter orange (male) or white with black stippling (female). Head dark brown dorsally and orange or yellowish white ventrally; male with many dark spots on lower half. Black suborbital bar prominent. Dark humeral spot present. First dorsal fin black marginally and basally, and (in male) orange medially. Remaining fins with narrow dark bands. Breeding male with bright orange venter and a bright orange medial band in first dorsal fin. Males retain some breeding coloration through August, well after the spring breeding season (Hambrick and Robison 1979).

SIMILAR SPECIES   Similar to the closely related but allopatric Arkansas Darter. It differs in having a wide pale middorsal stripe (vs. dorsum brown, without pale middorsal stripe), a more slender body, and a naked anterior belly (vs. belly fully scaled).

VARIATION AND TAXONOMY   The type locality for this Arkansas endemic species is the Caddo River, 13.7 km west of Black Springs, Montgomery County, Arkansas (Distler and Metcalf 1962). Williams and Robison (1980) placed the Paleback Darter in the subgenus *Ozarka*, but some recent authors (Near et al. 2011) did not support monophyly of that subgenus. Near et al. (2011) resolved *E. pallididorsum* as sister to a clade containing *E. boschungi* and *E. tuscumbia*.

DISTRIBUTION   *Etheostoma pallididorsum* is endemic to Arkansas, where it occurs only in the upper Caddo River and in some small tributaries of the upper Ouachita River in the Ouachita Mountains of Polk, Montgomery, Pike, and Garland counties (Robison 1974e, 1980e) (Map 206). Prior to 1990, it had only been collected from tributaries of the upper Caddo River and one tributary of the Ouachita River below Lake Ouachita. A survey of 91 sites from 1990 to 1992 produced records of *E. pallididorsum* from 15 new sites in the Caddo River drainage and 11 new sites in the Ouachita River drainage of Polk County (Pardew et al. 1993). Fluker and McCall (2018) found it at 16 sites, two of which were previously undocumented localities.

HABITAT AND BIOLOGY   The preferred habitat of the Paleback Darter was described by Hambrick and Robison (1979) and Robison (1980e) as quiet shallow pools (15–30 cm deep) at the margins of small gravel-bottomed, spring-fed streams and rivulets. The backwater pools were clear with much subsurface percolation through small gravel-rubble substrate. There was usually abundant leaf litter covering the bottom, but mud substrate was occasionally noted. It usually avoids swift current and is occasionally found associated with vegetation. Distler and Metcalf (1962) also noted that considerable organic matter was present at localities where *E. pallididorsum* occurred. Hambrick and Robison (1979) found that this darter hides in rubble and gravel interstices in pool areas. The pale mid-dorsal stripe of this species disrupts the fish's outline and makes individuals difficult to detect among multicolored gravel, rubble, and leaf litter. Pardew et al. (1993) reported that during the spawning season, *E. pallididorsum* was occasionally found in fast-moving water 1 foot (0.3 m) in depth or was strongly associated with aquatic vegetation. Hambrick and Robison (1979) provided additional physicochemical data for Paleback Darter habitat as follows:

> The upper Caddo River where *E. pallididorsum* occurs is relatively clear and unpolluted. The following physicochemical data, while not intended to be indicative of parameter limits for *E. pallididorsum*, are suggestive of the general type of waters frequented. Stream temperatures range from 3°C during the winter to 41°C during the summer. Chemical characteristics at normal flow vary as follows: dissolved oxygen, 7.8–13.2 mg/l; conductivity, 51–120 umhos; pH, 7.01–7.90; total phosphorous, 0.01–0.09 mg/l; total hardness (CaCO3), 28–80 mg/l; and iron 17–671 ug/l.

The most common species associates of the Paleback Darter are *Etheostoma radiosum*, *Campostoma spadiceum*, *Semotilus atromaculatus*, *Notropis boops*, *Fundulus catenatus*, and *Lepomis megalotis*. Less frequent associates are *E. blennioides*, *Ameiurus natalis*, *Erimyzon claviformis*, *Pimephales notatus*, and *Lythrurus umbratilis*. The relatively low number of species associates reflects the headwater, small stream habitat of *E. pallididorsum*.

Hambrick and Robison (1979) reported food habits. The contents of alimentary tracts from 97 specimens collected during seven months consisted mostly of immature insects and small crustaceans, with the dominant items being cladocerans, ephemeropterans, and dipteran larvae. Cladocerans and dipteran larvae were the most abundant items in the diet during most months, whereas, by volume, ephemeropterans represented the major items. The high occurrence of cladocerans in the diet relates to the backwater habitat preference of *E. pallididorsum*. Other food items ingested in smaller amounts included trichopterans, coleopterans, megalopterans, gastropods, and isopods. Detritus and extraneous material (e.g., sand, twigs) were rarely found. Sand grains were probably inadvertently ingested during the consumption of dipteran larvae, as grains were most abundant in specimens that contained large numbers of immature benthic insects.

Most of the following information on reproductive biology is from Hambrick and Robison (1979) and Johnston (1995). Eggs begin to mature in late fall to early winter. The Paleback Darter reaches sexual maturity at 1 year of age and spawns mainly in February and March. Reproductive condition of females decreased during April and was minimal by early summer. Females outnumbered males by a 2:1 ratio during March, when peak spawning occurred (Hambrick and Robison 1979). Two ripe females on 20 March were 30 and 50 mm TL and contained 207 and 697 total ova, respectively. HWR found that spawning habitat was similar to that of the Slackwater Darter, *Etheostoma boschungi* (Boschung and Nieland 1986), in that breeding occurs in small seepage water in open pastures or wooded areas and differs from the nonbreeding habitat, which usually includes small creeks where current is slow. The following account of aquarium spawning behavior was provided by Johnston (1995). Males were aggressive but nonterritorial, and no elaborate courtship displays were observed. The male pursued the female until she chose a spawning substrate. After the male mounted the female, the pair vibrated and 1–3 eggs were attached to the spawning substrate. Although a variety of substrates were provided, most eggs were attached to an artificial turf. A few eggs were attached to the sides of the tank. Females probably spawn more than one clutch per breeding season. *Etheostoma pallididorsum* is the only Arkansas species of the subgenus *Ozarka* that is an egg attacher. The other three Arkansas species of *Ozarka* (*E. cragini* and *E. autumnale*, and *E. mihileze*) are egg buriers.

Hambrick and Robison (1979) found three age groups: age 0, age 1, and age 2. *Etheostoma pallididorsum* attained a standard length of less than 30 mm at the end of its first winter of life, and age 1 fish attained a length of about 40 mm at the end of their second winter. All fish longer than 40 mm were age 2. Females attain a larger size than males, and all specimens longer than 42 mm SL were females. Paleback Darters have a short life span, living a maximum of 2 years. Hambrick and Robison (1978) reported 2 hermaphroditic specimens (KU6 158) from Montgomery County.

CONSERVATION STATUS   Pardew et al. (1993) found that populations of Paleback Darters were small, with an average of 4 individuals taken from sites studied; the most found at one site was 12. Fluker and McCall (2018), however, found that it was relatively abundant throughout its historic range and concluded that populations were stable despite small population sizes. Kuehne and Barbour (1983) cited the small range of this species broken by two reservoirs, the small populations, and the potential for channel modification as serious threats to its continued survival. It is a rare species that will probably require future habitat protection for long-term survival. Because of its restricted range, small population size, and specialized habitat requirements, we consider this Arkansas endemic species threatened.

Etheostoma parvipinne Gilbert and Swain
Goldstripe Darter

Figure 26.68. *Etheostoma parvipinne*, Goldstripe Darter. *David A. Neely.*

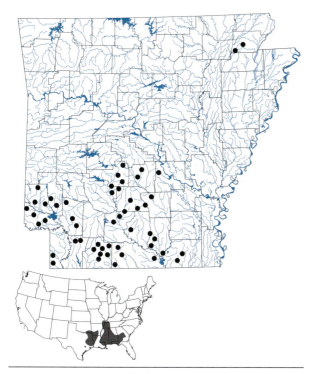

Map 207. Distribution of *Etheostoma parvipinne*, Goldstripe Darter, 1988–2019.

CHARACTERS   A small, moderately slender darter with a short, rounded snout (Fig. 26.68). Common name derived from pale stripe along lateral line, which contrasts with the otherwise darker side. Back plain, without definite saddles. Side often with 10–12 faint bars, which are bisected by the light stripe. Gill membranes moderately to broadly joined across throat. Premaxillary frenum present. Caudal peduncle relatively deep, its depth going into SL about 8.4

times. Infraorbital canal uninterrupted, usually with 7 or 8 pores. Opercle, nape, and belly fully scaled; breast and cheek partly to fully scaled. Lateral line nearly straight and almost complete, extending well behind second dorsal fin. Scales in lateral series usually 48–58; fewer than 10 unpored scales. Scale rows above lateral line 5–7; scale rows below lateral line 7–10. Diagonal scale rows 14–17; scale rows around caudal peduncle 20–27. Dorsal spines 8–10; dorsal rays 10–12; anal spines 2; anal rays 7–9; pectoral rays 14–16. Breeding tubercles present on anterior anal fin rays of male. Most individuals are usually 2 inches (51 mm) or less in length but reported to attain a maximum length of 3 inches (76 mm) (Pflieger 1997).

LIFE COLORS   One of the least colorful *Etheostoma* species. Back light brown to brownish gray; sides mottled darker brown, often with a midlateral series of brown bars. Anterior two-thirds of lateral line yellow or golden, especially in females and nonbreeding males. Underside white to yellow, sometimes stippled with melanophores. Suborbital bar dark. Dorsal and caudal fins with dark brown spots often forming rows. Anal fin dusky, paired fins are clear. Breeding males dusky with no bright colors.

SIMILAR SPECIES   Somewhat similar in body shape, coloration, and size to the allopatric species *E. cragini* and *E. pallididorsum*. It differs from those species in having gill membranes broadly joined across throat (vs. gill membranes separate), a fully scaled opercle (vs. naked), and a long lateral line extending past end of second dorsal fin (vs. short lateral line not extending past end of first dorsal fin). It differs from the sympatric *E. fusiforme* and *E. proeliare* in having an almost complete lateral line (vs. lateral line incomplete, not extending past second dorsal fin base) and in having a straight lateral line (vs. lateral line, if present, highly arched anteriorly).

VARIATION AND TAXONOMY   *Etheostoma parvipinne* was described by Gilbert and Swain *in* Gilbert (1887) from a small spring branch tributary of the Black Warrior River at Tuscaloosa, Alabama. It was in the synonymy of *E. squamiceps* (Jordan and Evermann 1896) until it was elevated to specific status by Hubbs and Black (1941). Originally placed in subgenus *Oligocephalus* by Bailey and Gosline (1955), it was later placed in the monotypic subgenus *Fuscatelum* by Page (1981). The subgenus was later expanded to include *E. phytophilum* (Bart and Taylor 1999). Monophyly of *Fuscatelum* was supported by genetic evidence (Near et al. 2011). Variation in Arkansas was studied by Robison (1977c), who found it was variable in several meristic characters, especially the number of lateral line scales. The species was redescribed by Bart and Taylor

(1999), who found high morphological variation throughout its range.

DISTRIBUTION *Etheostoma parvipinne* is practically confined to the Gulf Coastal Plain from the Florida panhandle to the Colorado River drainage, Texas, and extending up the Mississippi Embayment to southwestern Kentucky and southeastern Missouri. It also occurs in one Atlantic Slope drainage, the Ocmulgee River system of Georgia above the Fall Line. Almost all recent Arkansas records of the Goldstripe Darter are from the Gulf Coastal Plain in the southern part of the state in the Red and Ouachita river drainages (McAllister et al. 2007). Historically, there are few records from eastern Arkansas, and post-1988 records of this species from eastern Arkansas north of the Arkansas River are from Greene County from Betty's Spring (McAllister et al. 2007) and Poplar Creek (Map 207). Winston (2002) found 106 *E. parvipinne* at 15 sites on or near Crowley's Ridge in southeastern Missouri, but it is not as widely distributed in streams of the Arkansas portion of the Ridge.

HABITAT AND BIOLOGY Robison (1977c) described the habitat of the Goldstripe Darter in Arkansas as small, spring-fed, shallow (0.5–2 feet [0.2–0.6 m]) feeder streams or spring branches of low to moderate gradient. Such streams are typically about 2–8 feet (0.6–2.4 m) wide with a sand bottom and generally lack rooted aquatic vegetation because of heavy tree canopies overhead. It is usually found in sandy areas where fallen twigs, decaying leaves, and other detritus form protected areas in long shallow pools having slight to moderate current. Goldstripe Darter abundance in first-order Mississippi streams was positively correlated with canopy cover, water temperature, and sand substrate (Smiley et al. 2006). In that study, Goldstripe Darters exhibited the greatest associations with Creek Chub (*Semotilus atromaculatus*), Brown Madtom (*Noturus phaeus*), and Least Brook Lamprey (*Lampetra aepyptera*). We have only rarely found the Goldstripe Darter over a mud or gravel substrate or associated with aquatic vegetation in Arkansas, but this species has been reported to inhabit heavily vegetated springs and spring creeks in Alabama (Smith-Vaniz 1968) and Oklahoma (Moore and Cross 1950). Lowland species are often found in acidic waters, but there is little information on pH tolerance of *E. parvipinne*. On 6 July 2000, 405 Goldstripe Darters were taken from a 1,000 ft² (93.4 m²) sample area of Clearwater Lake (Saline River drainage) in Hot Spring County, Arkansas, at a pH of approximately 4.8 (TMB, unpublished data). It was the only fish species present. Clearwater Lake is an abandoned barite strip-mining pit (contributing to the low pH) with an outflow to

Rayburn Creek. A population was documented at an even lower pH ($\bar{x}$ = 3.7, range 2.9–4.0) in a small cattle pond in eastern Texas (Abdul 1987). That Texas pond still contained a reproducing population of *E. parvipinne* in 2000 (Robbins et al. 2003). Jeffers and Bacon (1979) reported a stable population in a shallow farm pond on the campus of the University of Arkansas at Monticello, Drew County.

Tennessee populations fed on a variety of small arthropods, including midge larvae, dipteran pupae, caddisfly larvae, dytiscid larvae, and small mayflies (Etnier and Starnes 1993). A similar diet composition was found in specimens from Calhoun and Union counties in southern Arkansas. Stomachs of 18 specimens (32–40 mm TL) collected on 10 and 11 July from those counties were examined by TMB. The diet was dominated by larval and pupal dipterans, primarily chironomids. Trichopteran larvae and ephemeropteran nymphs were also important food items, and larval coleopterans, plecopteran nymphs, and small crustaceans (cladocerans and copepods) were ingested in small amounts.

Peak breeding season occurs in March and April, possibly extending into May in Mississippi and Tennessee (Ross 2001). A similar March through May breeding season probably occurs in Arkansas. Individuals in spawning condition were found in Alabama near root masses, aquatic vegetation, and snags in swift chutes of Coastal Plain streams (Mettee et al. 1996). Aquarium spawning was observed by Johnston (1994b) in specimens taken in early to mid-April from Mississippi. Territoriality was not observed and no elaborate courtship displays occurred between males and females, but males were aggressive toward other males. Previously thought to be an egg burier, this species appeared to Johnston (1994b) to be an egg attacher. Males mounted the females in typical darter fashion, both vibrated, and eggs were attached singly to plant stems, leaves, and roots. Both males and females spawned with more than one partner, and no parental care was observed. Fecundity appears to be low, with females averaging about 66 eggs per year. Eggs hatched in 8 days at 20°C. Maximum life span is unknown, but Etnier and Starnes (1993) reported three age groups (0, 1, and 2) present in early spring collections in Tennessee.

CONSERVATION STATUS Although *E. parvipinne* is fairly widespread in the lowlands of southern Arkansas, it is usually not abundant. Ninety percent of all known specimens from the state have come from tributaries of the Ouachita River (Robison 1977c). The populations are very small. We currently consider it a species of special concern, and future monitoring of its status is recommended.

Etheostoma proeliare (Hay)
Cypress Darter

Figure 26.69. *Etheostoma proeliare*, Cypress Darter. *Uland Thomas.*

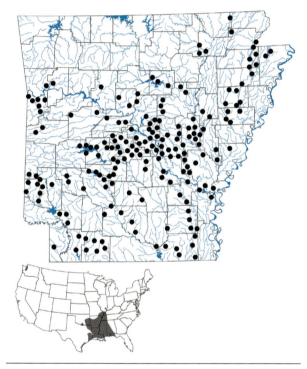

Map 208. Distribution of *Etheostoma proeliare*, Cypress Darter, 1988–2019.

CHARACTERS Arkansas's second-smallest darter, reaching a maximum length of about 1.6 inches (40 mm) (Fig. 26.69). Head and mouth small. Back without saddles or with 6–9 faint saddles. Midside with 7–12 dark horizontally elongated blotches. Gill membranes moderately joined across throat. A narrow premaxillary frenum present. Infraorbital canal interrupted, usually with 3 anterior pores and 1 posterior pore. Cheek and opercle scaled; breast and nape naked; belly usually scaled posteriorly. Lateral line very short, usually having only 2–7 pored scales; scales in lateral series 34–38. Scale rows above lateral line 2 or 3; scale rows below lateral line 5–8. Diagonal scale rows 8–12;

scale rows around caudal peduncle 13–18. Dorsal spines 7–9; dorsal rays 10–13; anal spines usually 2, but up to 43% of specimens of some Arkansas populations have only 1 anal spine (Burr 1978); anal rays 4–7; pectoral rays 9–11. Breeding males with tubercles on anal and pelvic fins. The pelvic fins of the nuptial male have expansive lateral flaps (forming cuplike appendages).

LIFE COLORS Back and upper sides olive brown, with dark brown lateral blotches. Numerous small brown spots, often distributed 1 per scale on sides, forming irregular longitudinal rows. Pre- and suborbital bars dark and prominent; there is usually a dark postorbital spot. Underside white with scattered large melanophores. Dusky mottlings on dorsal and caudal fins tend to form bars; other fins finely speckled to clear. Breeding males with some intensification of coloration. The body becomes dark brown and the lateral blotches darken even more. Suborbital and postorbital bars become an intense black. The lower half of the opercle becomes iridescent yellow. The eye becomes red orange. The first dorsal fin is black basally and red orange medially; second dorsal and caudal fins with orange bands. Anal and pelvic fins becoming dusky. Breeding females show little change in coloration, but they may develop some orange coloration in the iris of the eye.

SIMILAR SPECIES *Etheostoma proeliare* is very similar to the allopatric *E. microperca* but differs from it in usually having 2–7 pored scales in lateral line (vs. usually 0 or 1 pored scales), cheeks scaled (vs. naked); and 6 branchiostegal rays (vs. 5). It differs from *E. gracile* (with which it often occurs) in having an interrupted infraorbital canal (vs. uninterrupted), and in having 2–7 pored lateral line scales (vs. 16–23 pored lateral line scales). The Cypress Darter differs from *E. fusiforme* in having fewer than 40 lateral scales (vs. more than 40), and 2–7 pored lateral line scales (vs. more than 11 pored scales).

VARIATION AND TAXONOMY *Etheostoma proeliare* was described by Hay (1881) as *Microperca proeliaris* from a tributary of the Tuscumbia River at Corinth, Alcorn County, Mississippi. Subgeneric placement of *E. proeliare* differs among taxonomists, but it is usually recognized in the subgenus *Microperca*. The monophyly of *Microperca* was supported by genetic evidence (Near et al. 2011). Burr (1978) concluded that the recognition of subspecies was unwarranted based on an analysis of geographic variation throughout its range. Robison (1978b) studied variation in the caudal skeleton of *E. proeliare* and found that 58.3% of the 259 specimens examined had 11 branched caudal rays, 23.4% had 10, 6.2% had 12, 9.9% had 9, and one fish each had 13 and 8. None of the specimens had 15 branched caudal rays, which is the number occurring in most living perciform species (Greenwood et al. 1966). Lang and Echelle (2011) studied geographic variation in the ND2 gene and assigned populations from the Black and St. Francis rivers to a northern clade; the remaining Arkansas populations were assigned to one of three southern clades. Those authors also reported that finer-scale sampling combining genetic, morphological, and ecological data is needed to resolve phylogenetic structure and detect possible cryptic species within or among the Red and Arkansas river drainages. The mitochondrial gene studied by Lang and Echelle (2011) showed no evidence for hybridization in *Etheostoma proeliare*.

DISTRIBUTION The Cypress Darter occurs in the lower Mississippi River basin from southern Illinois through Louisiana, and in Gulf Slope drainages from the Florida panhandle to the San Jacinto River, Texas. In Arkansas, it is widely distributed throughout the Coastal Plain lowlands, where it occurs in all major river drainages. This species is mostly absent from the Ozark Uplands in northcentral and northwestern Arkansas except for records from Greers Ferry Lake in the Little Red River drainage and Norfork Lake in the White River drainage (Buchanan 2005). It is found in streams throughout the Arkansas River Valley and in several streams above the Fall Line in the Ouachitas (Map 208).

HABITAT AND BIOLOGY Found in several different habitats in Arkansas, the Cypress Darter is most abundant in the lowlands, where it occurs in bayous, oxbow lakes, swamps, and streams in little or no current over a soft mud and detritus bottom. It is frequently, but not always, found near aquatic vegetation. In Max Creek, Illinois, it occurred almost exclusively in habitats having leaves and/or vegetation at densities up to 5.5 darters/m² (Burr and Page 1978). Pools and riffles without leaves or vegetation yielded few or no *E. proeliare*, while those with leaves or vegetation typically yielded large numbers of *E. proeliare*. Collections by Burr and Page (1978) in five other states supported their observation of its affinity for leaves or vegetation in small to large streams and along the margins of lakes. In more upland areas it is found in moderate to low gradient creeks with gravel and sand bottoms, where it is usually taken in pools having silty bottoms with some vegetation and organic debris. It has sometimes been found over a gravel substrate. It was the second-most widely distributed darter species in Arkansas impoundments (Buchanan 2005). Even though it was found in all six ecoregions of Arkansas in 22 of 66 impoundments studied, its populations were small.

Burr and Page (1978) studied the life history of the Cypress Darter in Illinois and summarized the previous literature on this species. Its diet is similar to that of *E. microperca*, probably due to its small body size and small mouth. *Etheostoma proeliare* of all sizes fed mainly on small crustaceans (cladocerans, copepods, ostracods, isopods, and amphipods) and midge larvae. Isopods were consumed in significant amounts only by *E. proeliare* over 30 mm SL (Burr and Page 1978). In that study, chironomids and copepods were important diet items every month of the year. Cladocerans were most important in the fall, isopods from March to September, and amphipods from May to December. In Reelfoot Lake, Tennessee, stomach contents contained ephemeropterans (50%), copepods (35%), ostracods (10%), mosquito larvae (3%), and leafhoppers (2%) (Rice 1942). It is reportedly preyed upon by Largemouth Bass and White Crappie.

Spawning occurs in relatively shallow, sluggish water with accumulations of dead leaves, filamentous algae, or other aquatic plants (Burr and Page 1978). Breeding occurs from mid-March to early June in Illinois at water temperatures of 10–16°C and as early as January in Louisiana (Kuehne and Barbour 1983). Burr and Page (1978) reported the collection of individuals in breeding condition in Arkansas from the Red, Ouachita, Arkansas, and White river drainages between early February and early June. Jordan-Mathis (1994) studied reproductive biology in Bayou Bartholomew, Arkansas and Louisiana, and reported that spawning occurs from January through May. Reproductive females ranged in size from 21.1 to 36.4 mm SL and produced an average of 48 ova per clutch (range 9–138). Eggs hatched in 12.5 days at 15°C, 8–10 days at 20°C, and 5.5 days at 23°C. Burr and Ellinger (1980) reported that unfertilized eggs are nonspherical and indented on one side (unlike the eggs of most other North American fishes). Both sexes matured and spawned at 1 year of age; maximum life span in Arkansas was about 18 months. Aquarium spawning was described by Burr and Page (1978). Males exhibited no territoriality. The male closely follows a darting female and eventually mounts her, clasping her upper sides with his large pelvic fins and occasionally rubbing his chin on her cheeks and the top of her head. Apparently, the expanded cuplike appendages on the pelvic fins of the nuptial male aid in maintaining contact with the female during spawning. A spawning pair usually lays eggs at several sites, attaching the eggs to dead leaves, twigs, aquatic plants, or other objects. The spawning position is often vertical, but sometimes occurs upside down under leaves. A female may contain 26–110 mature ova. The eggs are not guarded. The Cypress Darter is a short-lived species with maximum reported life span between 15 and 18 months in different areas of its range.

CONSERVATION STATUS This is one of the most widely distributed and common darters in Arkansas. Its populations are usually not large, but we consider it a secure species.

Etheostoma pulchellum (Girard)
Plains Darter

Figure 26.70a. *Etheostoma pulchellum*, Plains Darter male. *David A. Neely.*

Figure 26.70b. *Etheostoma pulchellum*, Plains Darter female. *Brook L. Fluker.*

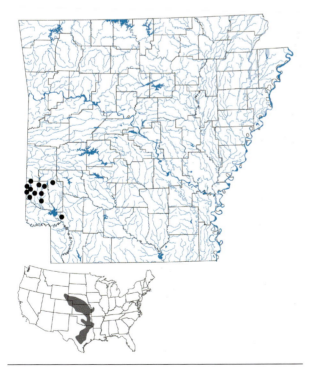

Map 209. Distribution of *Etheostoma pulchellum*, Plains Darter, 1988–2019.

CHARACTERS A small to medium-sized species of the Orangethroat Darter complex with a somewhat arched predorsal region; greatest body depth occurring just in front of the first dorsal fin (Fig. 26.70). Back usually with 3–9 dark saddles (which are sometimes indistinct) from occiput to base of caudal fin. Sides usually with 9 (6–11) dark vertical bars, usually more distinct posteriorly but well developed on both lower and upper sides. Horizontal lines on sides usually poorly developed or absent, but if present, are most evident just above and below the lateral line. Gill membranes narrowly joined across throat. Premaxillary frenum present. Infraorbital canal interrupted, usually with 4 pores in the anterior segment and 3 pores (2–4) in the posterior segment. Supratemporal canal uninterrupted. Cheek and nape naked to partly scaled; opercle with 1 to several scales; breast naked but sometimes with 1 to several scales; belly fully scaled. Lateral line incomplete, usually ending beneath second dorsal fin. Lateral scales 42–51 (usually around 48); pored lateral line scales 21–35 (usually around 28). Diagonal scale rows usually 13 (11–16); scales above lateral line 5–7. Pectoral fin rays usually 11 or 12; dorsal spines 9–11; dorsal rays 12 or 13; anal spines 2; anal rays 5–7; usually 15 or 16 principal caudal rays. Breeding males with tubercles on pectoral, pelvic, and anal fins and on some ventral scales. Maximum length around 2+ inches (51 mm).

LIFE COLORS Back and upper sides light tan to olivaceous with darker dorsal saddles. Venter usually unmarked except for meeting lateral bars. Nonbreeding females lack all chromatic colors. Breeding males are brightly colored. Throat orange. Sides with 9 (6–11) blue or blue-green vertical bars. Interspaces on upper sides reddish orange, or brownish orange; interspaces of posterior lower sides mostly orange. Belly light blue to blue gray. Anal fin blue to blue green, without red pigment. Spinous dorsal fin usually with a blue or greenish-black margin, a wide median bright orange to brick-red band, and a light blue basal band with orange to red spots. Soft dorsal fin mostly orange except for a narrow dusky-blue margin and a better-developed, but often incomplete, basal or suprabasal blue or blue-green band. Pelvic fins black suffused with blue or blue green. Membranes of caudal and pectoral fins clear to yellow, sometimes with dark pigment along rays. Breeding females with light coloration, the body sometimes having dull orange bands. Throat of female light orange; spinous and soft dorsal fins with a faint marginal light blue band and a supramedial light orange band. Anal fin and paired fins of female colorless, and caudal fin rays outlined with melanophores.

SIMILAR SPECIES *Etheostoma pulchellum* is most similar to two undescribed species, *E.* sp. *cf. pulchellum*

1 and *E.* sp. *cf. pulchellum* 2. The best nonmolecular characters for distinguishing these species are geographic distribution and coloration of breeding males. In Arkansas, *E. pulchellum* is found only in the Little River drainage, while the two undescribed species are found in western Arkansas River tributaries (*E.* sp. *cf. pulchellum* 1) and in central Arkansas and lower White river drainages (*E.* sp. *cf. pulchellum* 2). Breeding males exhibit several coloration differences. Most importantly, *E. pulchellum* has a silvery-gray to light blue belly, while *E.* sp. *cf. pulchellum* 1 has an orange to reddish-orange belly. It differs from *E.* sp. *cf. pulchellum* 2 (which has a blue belly) in having blue bands with orange interspaces well developed on anterior and posterior sides (vs. lateral bands poorly developed, usually confined to posterior part of body). It differs from *E. fragi* in having cheek with 1 or 2 rows of scales below or behind eye (vs. cheek with 5 or more rows of scales), breeding males with tubercles present on pectoral, pelvic, anal, and caudal fins and on ventral scales, and without orange chevrons crossing the belly (vs. breeding males without tubercles and with orange chevrons crossing the belly). It differs from *E. uniporum* in usually having 2 or 3 pores in the posterior segment of the infraorbital canal, supratemporal canal uninterrupted, and breeding males with the suprabasal band in the first dorsal fin not the widest band (vs. modally 1 posterior infraorbital pore, supratemporal canal interrupted, and breeding males with the suprabasal blue band in the first dorsal fin being the widest band). It differs from *E.* sp. *cf. spectabile* of the upper White River drainage in having predominantly orange dorsal fins (vs. dorsal fins dominated by blue coloration), and from *E. squamosum* in having cheek and top of head usually naked (vs. usually moderately to fully scaled) and in having dorsal saddles all of equal intensity (vs. 3 dorsal saddles darker than others). *Etheostoma pulchellum* does not occur sympatrically with *E. caeruleum* in Arkansas, but it differs from that species in having 11 or 12 pectoral rays, an interrupted infraorbital canal, greatest body depth occurring just anterior to first dorsal fin, and breeding males without prominent red pigment in anal fin (vs. 13 or more pectoral rays, infraorbital canal uninterrupted, greatest body depth occurring beneath first dorsal fin, and breeding males with prominent red pigment in anal fin). It also resembles *E. collettei*, with which it occurs syntopically in the Little River drainage of Arkansas, but differs in having an interrupted infraorbital canal and dorsal saddles of about equal intensity (vs. infraorbital canal uninterrupted and 4 dorsal saddles usually much darker than the others). It differs from *E. asprigene* in usually having fewer than 8 anal rays, usually fewer than 35 pored lateral line scales, interrupted infraorbital canal, fewer than 13 pectoral rays, and breeding males

with tubercles (vs. 8 anal rays, more than 35 pored lateral line scales, uninterrupted infraorbital canal, 13 or more pectoral rays, and breeding males without tubercles).

VARIATION AND TAXONOMY   One of seven species of the Orangethroat Darter complex in Arkansas, *Etheostoma pulchellum* was first described as *Oligocephalus pulchellus* by Girard (1859). Black (1940) regarded it as a subspecies of the Orangethroat Darter, *Poecilichthys spectabilis pulchellus*. *Etheostoma pulchellum* is in the subgenus *Oligocephalus*. Distler (1968) studied morphological variation and also regarded it as a subspecies of *Etheostoma spectabile*, *E. s. pulchellum*. Distler identified the correct type locality of Girard (1859) as Gypsum Creek, a tributary of the Washita River, Indian Territory, now Custer County, Oklahoma. Several recent authors, based on a combination of morphological and DNA evidence, elevated *E. s. pulchellum* to full species status (Bossu and Near 2009; Near et al. 2011; Bossu et al. 2013; Matthews and Turner 2019), a designation we also support. The elevation of *E. pulchellum* to specific status was not recognized by Page et al. (2013).

Arkansas populations in Arkansas and lower White river tributaries formerly assigned to *E. pulchellum* are now considered to represent two undescribed species (P. A. Ceas, pers. comm.).

DISTRIBUTION   *Etheostoma pulchellum*, as currently recognized, has one of the largest ranges with the greatest north-south distribution of any species in the Orangethroat Darter clade (Bossu et al. 2013). Its common name of Plains Darter is derived from its distribution primarily in the Great Plains in the Mississippi River basin from Nebraska to the Red River drainage of southern Oklahoma, and a few Gulf Slope drainages in central Texas. The population in the Blue River, Oklahoma, probably represents an undescribed species endemic to that river (Schwemm 2013). In Arkansas, it has a very restricted distribution, occurring only in Little River tributaries (Red River drainage) in southwestern Arkansas in Polk, Sevier, Howard, Little River, and Hempstead counties (Map 209). It is absent from the remainder of the Ouachita Mountains in Arkansas.

HABITAT AND BIOLOGY   The Plains Darter prefers small headwater creeks, where it is found in shallow riffles of slow to moderate current over a gravel to rubble substrate throughout the year. It is often found near the margins of rocky-bottomed pools having some current. It is not very silt-tolerant, and it is not known from reservoirs in Arkansas (Buchanan 2005). In Nebraska, it was considered tolerant of moderate turbidity but was most abundant in clear streams (Hrabik et al. 2015). Various authors have reported its avoidance of larger, deeper streams having swift, deep riffles, but in Arkansas, it is sometimes found in larger creeks and small rivers where it seeks out microhabitats similar to those typically preferred by most members of the Orangethroat Darter complex.

It is a benthic invertivore. The larvae feed on microcrustaceans and small insects; adults feed on immature dipterans, caddisflies, other insect larvae, and often on fish eggs (Cross 1967). Gillette (2012) studied the effects of variation among riffles on prey use and feeding selectivity of *E. pulchellum* in Brier Creek, Oklahoma. The following information on feeding is from that study. Ten riffles were sampled over a 3-day period during midsummer. The majority of prey items consisted of mayfly, stonefly, and dipteran nymphs and larvae, but a few other taxa (primarily ostracods) were also eaten. Size of prey consumed did not vary among riffles, but total number of prey consumed varied significantly. Prey selection varied greatly among riffles and appeared to be a function of habitat differences. Mean water depth, flow velocity, and substrate composition predicted selection for multiple prey items.

Sexual maturity is reached at 1 year of age, but most breeding adults are in their second or third summer. Breeding males averaged 40 mm SL in most areas of its range (Distler 1968). The reproductive season in Texas occurs from November through July (Hubbs 1961a) and in Kansas from March through May at 15–21°C (Cross 1967). Ripe adults were found in the Cossatot River, Arkansas, as early as 24 February. Ripe females were found in the Rolling Fork River in Sevier County on 4 March (Tumlison et al. 2016). During the spawning season, the males congregate on shallow riffles but are not territorial. Neither size nor breeding color correlated with spawning success under natural field conditions (Pyron 1995). Males and females do not differ significantly in size, and males showed no preference for larger females in Buckhorn Creek, Oklahoma (Pyron 1996a). Pyron speculated that *E. pulchellum* may lack sexual size dimorphism as a result of the lack of female choice for size and the ineffectiveness of male attempts to monopolize females. Spawning behavior was described in Pennington Creek, Oklahoma (Mendelson 2003). When ready to spawn, a female enters the riffle, burrows into the gravel substrate, and is mounted by a male. The male grasps the female's nape with his pelvic fins and both fish vibrate for about 8 seconds while eggs and sperm are released into the gravel. The spawning act may be repeated several times, and each male may spawn with several females. The buried eggs are left unguarded. After hatching, the larvae drift downstream into pools (Cross 1967). The young later leave the pools and move into shallow riffles along stream bank margins. Maximum life span is probably about 3+ years.

CONSERVATION STATUS   This is an uncommon species in Arkansas with a very restricted distribution. Populations appear to be small, and we consider it a threatened species in our state.

Etheostoma sp. *cf. pulchellum 1*
(Red Belly Form)

Figure 26.71. *Etheostoma* sp. *cf. pulchellum* 1 (Red Belly Form). *Uland Thomas.*

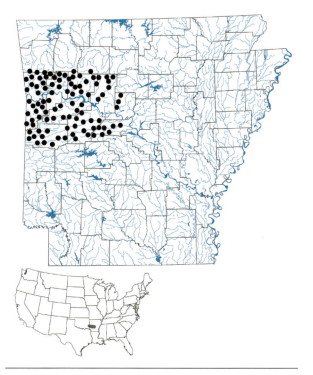

Map 210. Distribution of *Etheostoma* sp. *cf. pulchellum* 1 (Red Belly Form), 1988–2019.

CHARACTERS   A small, undescribed species of the Orangethroat Darter clade with a somewhat arched predorsal region; the greatest body depth occurring just in front of the first dorsal fin (Fig. 26.71). Back usually with 3–9 dark saddles (which are sometimes indistinct) from occiput to base of caudal fin. Sides usually with 9 (6–11) dark vertical bars, usually more distinct posteriorly but well developed on both lower and upper sides. Horizontal lines on sides usually poorly developed or absent, but if present, are most evident just above and below the lateral line. Gill membranes narrowly joined across throat. Premaxillary frenum present. Infraorbital canal interrupted, usually with 4 pores in the anterior segment and 2 or 3 pores in the posterior segment. Supratemporal canal uninterrupted. Cheek and nape naked to partly scaled; opercle partly to fully scaled; breast naked but sometimes with 1 to several scales; belly fully scaled. Lateral line incomplete, usually ending beneath second dorsal fin. Lateral scales 45–56 (usually around 50); pored lateral line scales 23–38 (usually around 31 or 32). Diagonal scale rows usually 13 (11–16); scales above lateral line 5–7. Pectoral fin rays usually 11 or 12; dorsal spines 9–11; dorsal rays 12 or 13; anal spines 2; anal rays usually 6 (5–7); usually 15 or 16 principal caudal rays. Breeding males with tubercles on pectoral, pelvic, and anal fins and on some ventral scales. Maximum length around 2+ inches (51 mm).

LIFE COLORS   Information on coloration was taken from Distler (1968), P. A. Ceas (pers. comm.), and TMB (unpublished data). Back and upper sides light tan to olivaceous with darker dorsal saddles. Venter usually unmarked except for meeting lateral bars. Nonbreeding females lack all chromatic colors. Breeding males brightly colored. Color pattern of body dominated by orange or red-orange colors. Throat orange to brick red. Belly and lower anterior sides orange to brick red. Sides with 9 (6–11) narrow blue or blue-green vertical bars separated by wider orange or reddish-orange bars; blue bars best developed posteriorly. Pelvic and anal fins blue to turquoise, usually without red pigment; however, some breeding males from Scott County examined by TMB had reddish-orange pigment invading the posterior one-third of the anal fin proximally. Spinous dorsal fin usually with a blue margin; lower two-thirds of fin bright orange to brick red. Specimens from Scott County exhibited five distinct pigment bands in the spinous dorsal fin from distal to proximal: margin of fin dark blue, followed by a thin white band, a wide medial orange or red-orange band (the dominant band), a narrow light blue to turquoise band, and a basal band consisting of round orange spots between the bases of the spines. Soft dorsal fin mostly orange except for a narrow dusky-blue margin and a better developed, but often incomplete, basal or suprabasal blue or blue-green band. Membranes of caudal and pectoral fins clear to yellow, sometimes with dark pigment along rays. Breeding females with little coloration, the body sometimes having dull orange bands and a thin yellow-orange band in the first dorsal fin. In both

males and females, preorbital and suborbital dark bars are prominent; the postorbital dark bar is short and spotlike; dark humeral spot absent or only faintly developed (TMB observations).

SIMILAR SPECIES   Etheostoma sp. cf. pulchellum 1 is most like E. pulchellum and E. sp. cf. pulchellum 2, but differs mainly in breeding males with reddish-orange or orange belly (vs. males with blue or blue-gray belly). It differs from E. fragi in having cheek with 1 or 2 rows of scales below or behind eye, breeding males with tubercles present on pectoral, pelvic, anal, and caudal fins and on ventral scales, and without orange chevrons crossing the belly (vs. cheek with 5 or more rows of scales, and breeding males with orange chevrons crossing the belly and without breeding tubercles). It differs from E. uniporum in usually having 2 or 3 pores in the posterior segment of the infraorbital canal, supratemporal canal uninterrupted, and breeding males with the suprabasal band in the first dorsal fin not the widest band (vs. modally 1 posterior infraorbital pore, supratemporal canal interrupted, and breeding males with the suprabasal blue band in the first dorsal fin being the widest band). It differs from E. sp. cf. spectabile (Ozark Darter) of the White River drainage in having predominantly orange dorsal fins (vs. dorsal fins dominated by blue coloration), and from E. squamosum in having cheek and top of head usually naked (vs. usually moderately to fully scaled) and in having dorsal saddles all of equal intensity (vs. 3 dorsal saddles darker than others). Etheostoma sp. cf. pulchellum 1 does not occur sympatrically with E. caeruleum, but it differs from that species in having 11 or 12 pectoral rays, an interrupted infraorbital canal, greatest body depth occurring just anterior to first dorsal fin, and breeding males usually without prominent red pigment in anal fin (vs. 13 or more pectoral rays, infraorbital canal uninterrupted, greatest body depth occurring beneath first dorsal fin, and breeding males with prominent red pigment in anal fin). It also resembles E. collettei, but differs in having an interrupted infraorbital canal and dorsal saddles of about equal intensity (vs. infraorbital canal uninterrupted and 4 dorsal saddles usually much darker than the others). It differs from E. asprigene in usually having fewer than 8 anal rays, usually fewer than 35 pored lateral line scales, interrupted infraorbital canal, fewer than 13 pectoral rays, and breeding males with tubercles (vs. 8 anal rays, more than 35 pored lateral line scales, uninterrupted infraorbital canal, 13 or more pectoral rays, and breeding males without tubercles).

VARIATION AND TAXONOMY   One of seven species of the Orangethroat Darter complex in Arkansas, Etheostoma sp. cf. pulchellum 1 is an undescribed species of the subgenus *Oligocephalus* (P. A. Ceas, pers. comm.). Distler (1968) noted that Orangethroat Darters occurring in the Poteau, Fourche LaFave, and Petit Jean rivers were smaller than specimens from other parts of the range of *Etheostoma spectabile pulchellum*, as it was then known. It is the smallest species of the Orangethroat Darter complex in Arkansas.

DISTRIBUTION   *Etheostoma* sp. *cf. pulchellum* 1 (red belly form) is found in Arkansas River tributaries from the Point Remove Creek drainage in Conway County and the Petit Jean River and Fourche LaFave River drainages in Yell, Perry, and Scott counties westward into extreme eastern Oklahoma (Poteau River, Lee Creek, and possibly other Arkansas River tributaries in Sequoyah and Le Flore counties) (Map 210). It occurs in streams draining the southern Ozark Plateau, the Arkansas River Valley, and the northern slopes of the Fourche Mountains of the Ouachitas. It is syntopic with another undescribed species of the Orangethroat Darter clade (*E.* sp. *cf. pulchellum* 2, blue belly form) in Illinois Bayou. These are the only two of the 19 species of the Orangethroat Darter clade in the United States known to occur syntopically.

HABITAT AND BIOLOGY   *Etheostoma* sp. *cf. pulchellum* 1 prefers small headwater creeks, where it is found in shallow riffles of slow to moderate current over a variety of substrates throughout the year. Habitat data, compiled from Brian Wagner's field notes on 13 western Arkansas streams in fall 2007, indicate that 77% of *E.* sp. *cf. pulchellum* 1 specimens were collected from streams less than 10 m wide and in slow current. Most specimens were found over a boulder-bedrock substrate (38%), with remaining specimens taken over gravel-rubble (24%), cobble (23%), and silt-clay (15%) substrates with no aquatic vegetation. In the Frog Bayou drainage (Clear Creek) of Crawford County, Arkansas, population density estimates ranged from 0.21 to 0.9 fish/m², and in the Mulberry River drainage in Madison and Johnson counties, density estimates ranged from 0.06 to 0.46 fish/m² (Rambo 1998). It does not appear to be very silt-tolerant, and *E.* sp. *cf. pulchellum* 1 was found in small numbers in one Arkansas reservoir, Lake Hinkle, Poteau River drainage (Buchanan 2005).

There is limited information about life history aspects of this undescribed species. It is a benthic invertivore like other species of *Ceasia*, feeding opportunistically on available benthic organisms. Stomach contents of 57 specimens collected from five western Arkansas counties between 13 November and 4 December 2007 were examined by TMB. The dominant food items by percent frequency of occurrence were chironomids (32%), copepods (32%), trichopterans (27%), ephemeropterans (25%), cladocerans (20%), plecopterans (18%), and isopods (16%). Other food items ingested less frequently were megalopterans (9%), arachnids (5%), coleopterans (5%), ostracods (2%), and odonates (2%). Stomach contents of 28 specimens collected on 10 April 2015 were dominated by trichopterans (58%) and chironomids (26%), with plecopterans, arachnids, and tardigrades making up small percentages.

Sexual maturity is reached in this small member of the Orangethroat Darter clade at 1 year of age, but most breeding adults are in their second or third summer. Breeding males averaged 30 mm SL in Arkansas populations, compared with 40 mm SL in *E. pulchellum* (Distler 1968). The smallest female containing ripe ova was 32 mm TL from Scott County, Arkansas (TMB collection). We observed spawning of *E.* sp. *cf. pulchellum* 1 in Vache Grasse Creek (Sebastian County) from mid-March to mid-April. Spawning was observed by TMB in Lee Creek in Crawford County on 26 April, and on 10 April; ripe individuals were found in Dirty Creek (Johnson County) on 28 April. Six females collected on 4 April in Sebastian County by TMB ranged in size from 38 to 45 mm TL and contained from 56 to 94 ripe ova; six females ranging 32 to 44 mm TL collected on 10 April from Mill Creek in Scott County had 15–40 ripe ova. Some spawning activity probably extends into early May, but spawning had apparently ceased by 5 May 2015 in West Fork Point Remove Creek in Conway County. During the spawning season, the males congregate on shallow riffles but are not territorial. Male breeding colors fade rapidly after the spawning season ends, making species identification difficult.

CONSERVATION STATUS   *Etheostoma* sp. *cf. pulchellum* 1 has a small range, most of which is in western Arkansas. It is common in some streams and its populations are secure, but because of its restricted range, its populations should periodically be monitored.

Etheostoma sp. *cf. pulchellum* 2 (Blue Belly Form)

CHARACTERS   A small undescribed species of the Orangethroat Darter complex with a somewhat arched predorsal region; greatest body depth occurring just in front of the first dorsal fin (Fig. 26.72). Back usually with 3–9 dark saddles (which are sometimes indistinct) from occiput to base of caudal fin. Sides usually with 9 (6–11) dark vertical bars, usually more distinct posteriorly but usually well developed on both lower and upper sides. Horizontal lines on sides usually poorly developed or absent, but if present, are most evident just above and below the lateral line. Gill

Figure 26.72. *Etheostoma* sp. *cf. pulchellum* 2 (Blue Belly Form). *Uland Thomas.*

Map 211. Distribution of *Etheostoma* sp. *cf. pulchellum* 2 (Blue Belly Form), 1988–2019.

membranes narrowly joined across throat. Premaxillary frenum present. Infraorbital canal interrupted, usually with 4 pores in the anterior segment and 2 or 3 pores in the posterior segment. Supratemporal canal uninterrupted. Cheek and nape naked to partly scaled; opercle partly to fully scaled; breast naked but sometimes with 1 to several scales; belly fully scaled. Lateral line incomplete, usually ending beneath second dorsal fin. Lateral scales 45–56 (usually around 50); pored lateral line scales 23–38 (usually around 31 or 32). Diagonal scale rows usually 13 (11–16); scales above lateral line 5–7. Pectoral fin rays usually 11 or 12; dorsal spines 9–11; dorsal rays 12 or 13; anal spines 2; anal rays usually 6 (5–7); usually 15 or 16 principal caudal rays. Breeding males with tubercles on pectoral, pelvic, and anal

fins and on some ventral scales. Maximum length around 2+ inches (51 mm).

LIFE COLORS Back and upper sides light tan to olivaceous with darker dorsal saddles. Venter usually unmarked except for meeting lateral bars. Dark humeral spot usually well developed in both sexes. Nonbreeding females lack all chromatic colors. Breeding males brightly colored. Color pattern of body dominated by dark blue colors. Throat orange to reddish orange. Cheek blue. Pre- and suborbital bars well developed; postorbital bar usually poorly developed, often just a spot or short horizontal streak. Belly and lower anterior sides with alternating dark and light blue coloration. Sides with 9 (6–11) wide blue vertical bars (lateral bars are darkest blue along midside); blue bars on posterior sides separated by narrow orange bars, the last orange bar divided into 2 round spots on base of caudal peduncle (TMB, pers. obs.). Pelvic and anal fins dark blue, without red pigment. Spinous dorsal fin usually with a blue margin, a thin white band, a bright orange to brick-red medial band, a narrow blue band, and a basal band consisting of round orange spots often suffused with black blotches. Soft dorsal fin mostly orange (orange band much wider than orange band in spinous dorsal fin) except for a narrow dusky-blue (or black) margin and a better-developed, but often incomplete, basal or suprabasal blue or blue-green and orange band. Membranes of caudal and pectoral fins clear to yellow, but caudal fin sometimes with dark blue or black pigment along rays and sometimes with splotches of orange. Breeding females with little coloration, the body sometimes having dull orange bands and a thin yellow-orange band in the first dorsal fin.

SIMILAR SPECIES *Etheostoma* sp. *cf. pulchellum* 2 differs from *E. fragi* in having cheek with 1 or 2 rows of scales below or behind eye, breeding males with tubercles present on pectoral, pelvic, anal, and caudal fins and on ventral scales, and without orange chevrons crossing the belly (vs. cheek with 5 or more rows of scales, and breeding males with orange chevrons crossing the belly and without breeding tubercles). It differs from *E. uniporum* in usually having 2 or 3 pores in the posterior segment of the infraorbital canal, supratemporal canal uninterrupted, and breeding males with the suprabasal band in the first dorsal fin not the widest band (vs. modally 1 posterior infraorbital pore, supratemporal canal interrupted, and breeding males with the suprabasal blue band in the first dorsal fin being the widest band). It differs from *E.* sp. *cf. spectabile* of the upper White River drainage in having predominantly orange dorsal fins (vs. dorsal fins dominated by blue coloration), and from *E. squamosum* in having cheek and top of head usually naked (vs. usually moderately to fully scaled) and in having dorsal saddles all of equal intensity (vs. 3 dorsal saddles darker than others). *Etheostoma* sp. *cf. pulchellum* 2 does not occur sympatrically with *E. caeruleum*, but it differs from that species in having 11 or 12 pectoral rays, an interrupted infraorbital canal, greatest body depth occurring just anterior to first dorsal fin, and breeding males without prominent red pigment in anal fin (vs. 13 or more pectoral rays, infraorbital canal uninterrupted, greatest body depth occurring beneath first dorsal fin, and breeding males with prominent red pigment in anal fin). It also resembles *E. collettei* but differs in having an interrupted infraorbital canal and dorsal saddles of about equal intensity (vs. infraorbital canal uninterrupted and 4 dorsal saddles usually much darker than the others). It differs from *E. asprigene* in usually having fewer than 8 anal rays, usually fewer than 35 pored lateral line scales, an interrupted infraorbital canal, fewer than 13 pectoral rays, and breeding males with tubercles (vs. 8 anal rays, more than 35 pored lateral line scales, uninterrupted infraorbital canal, 13 or more pectoral rays, and breeding males without tubercles). *Etheostoma* sp. *cf. pulchellum* 2 differs from *E.* sp. *cf. pulchellum* 1 in breeding males with color pattern of body dominated by blue coloration, especially the belly (vs. body coloration dominated by reddish-orange or orange), and in usually having a dark, well-developed humeral spot in both sexes (vs. humeral spot absent in females and usually absent or faintly developed in males). It differs from *E. pulchellum* in having blue lateral bands of breeding males poorly developed, except posteriorly (vs. blue lateral bands well developed along sides, even anteriorly).

VARIATION AND TAXONOMY One of seven species of the Orangethroat Darter clade in Arkansas, *Etheostoma* sp. *cf. pulchellum* 2 is an undescribed species of the subgenus *Oligocephalus* (P. A. Ceas, pers. comm.).

DISTRIBUTION *Etheostoma* sp. *cf. pulchellum* 2 has a smaller range than *E.* sp. *cf. pulchellum* 1 and is found in Arkansas River tributaries from Illinois Bayou in Pope County eastward in the Arkansas River Valley to the headwaters of Bayou Meto in Pulaski County, with a disjunct population in Des Arc Bayou (White River drainage) in White County (Map 211). Another disjunct population occurs westward in Sallisaw Creek, Sequoyah County, Oklahoma (Pat Ceas, pers. comm.). Most populations are known from tributaries north of the Arkansas River, and it is known only from the Maumelle River, Little Maumelle River, and Fourche Creek drainages south of the Arkansas River in Pulaski County (with possible occurrence in Perry County). It is syntopic with another undescribed species of

the Orangethroat Darter clade (*E.* sp. *cf. pulchellum* 1) in Illinois Bayou, the only known syntopic occurrence of two species of the Orangethroat Darter clade.

HABITAT AND BIOLOGY Habitat data for *Etheostoma* sp. *cf. pulchellum* 2 were taken from field notes of Brian Wagner (2001 and 2002) and TMB (2015) from localities in Pulaski County. It prefers small to medium-sized creeks, where it is found in shallow riffles in slow to swift current. Stream width at most localities (71%) was 10–25 m, and 29% of the localities were less than 10 m in width. It was found over a gravel and rubble substrate throughout the year at both low and high flows, and 100% of the collection sites had American water willow along the stream margins. This species prefers clear water but can apparently tolerate periods of varying turbidity. A good population exists in Fourche Creek in Little Rock, and it currently thrives in that stream in Hindman Park.

Like most darters, this undescribed species is a benthic invertivore. Stomach contents of 72 specimens collected on 5 May 2015 from three localities in Pulaski County were examined by TMB, Eighty-nine percent of the stomachs contained food, with trichopterans the dominant food item by percent frequency of occurrence (70.3%). The next most common food items were isopods (31.3%) and plecopterans (20.3%). Other food items ingested in smaller percentages were chironomids (9.4%), fish embryos (6.3%), ephemeropterans (4.7%), odonates (4.7%), megalopterans (3.1%), cladocerans (3.1%), amphipods (3.1%), ostracods (1.6%), arachnids (1.6%), and coleopterans (1.6%).

Aspects of its reproductive biology are probably similar to those of other species of the Orangethroat Darter clade. It apparently breeds from March into early May. Ripe adults were collected on 21 April in Pope County by TMB. Specimens collected in Pulaski County on 5 May 2015 by TMB and Jeff Quinn appeared to be nearing the end of the breeding season. Most males still possessed bright breeding colors and mature testes, but about half the females contained some atretic eggs. Ten females between 35 and 53 mm TL contained 5–92 mature ova. Distinctive breeding coloration of males fades rapidly when the breeding season ends.

CONSERVATION STATUS *Etheostoma* sp. *cf. pulchellum* 2 has a small range, almost all in central Arkansas. It is common in some streams, but because of its restricted range and the proximity of a large metropolitan area (Little Rock), we consider it a species of special concern. Its populations should periodically be monitored. The population in Fourche Creek in Little Rock is especially vulnerable to human impact, and its protection should be encouraged.

Etheostoma radiosum (Hubbs and Black)
Orangebelly Darter

Figure 26.73. *Etheostoma radiosum*, Orangebelly Darter. *David A. Neely.*

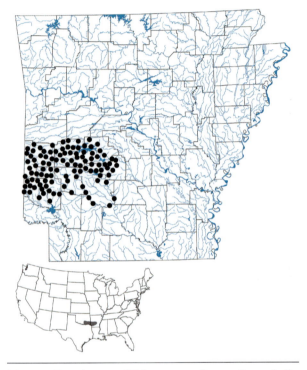

Map 212. Distribution of *Etheostoma radiosum*, Orangebelly Darter, 1988–2019.

CHARACTERS A colorful, moderately robust, medium-sized darter with much individual variation in body markings within populations (Fig. 26.73). Back usually with 8–10 dark saddles. Side with 8–12 lateral blotches darker below lateral line and often forming vertical bars on caudal peduncle. Dorsal saddles often connected to lateral blotches by dark vermiculations. In some individuals, the entire side of the body anterior to the caudal peduncle has a more vermiculated or mottled pattern. A narrow, dark marginal band on the caudal fin usually retained in preserved specimens. Gill membranes separate or slightly joined across throat. Premaxillary frenum present. Infraorbital canal uninterrupted. Breast naked; opercle naked to partly scaled; cheek partly to fully scaled; nape and belly fully scaled. Lateral line incomplete, usually ending beneath second dorsal fin. Scales in lateral series 47–66; Arkansas populations usually with 52–61 lateral scales (36–48 pored lateral line scales and 7–16 unpored scales). Scale rows above lateral line usually 7 or 8 (6–9); scale rows below lateral line 10–13. Diagonal scale rows 16–20; scale rows around caudal peduncle 23–29. Dorsal spines 9–12; dorsal rays usually 13–15; anal spines 2; anal rays 6–9; pectoral rays usually 12. Breeding males with tubercles on ventral belly scales, on several scale rows on either side of anal fin, and on several scale rows on ventral caudal peduncle. Maximum length around 3 inches (76 mm).

LIFE COLORS Back and upper sides olive with dark green dorsal saddles and dark greenish-black lateral blotches. Lower sides and venter yellowish orange from throat to caudal peduncle. Orbital bars dusky to black. Humeral spot large and black. Dorsal and caudal fins of male with blue margins, red-orange submarginal bands, and dusky brown basally; anal fin red with a blue margin; pelvic fins blue; pectoral fins yellow. Fins of female yellow to clear; all median fins have blue margins, first dorsal fin with orange submarginal band. Caudal fin with a narrow blue marginal band. Breeding males brightly colored, with previously described colors becoming brilliant.

SIMILAR SPECIES Similar to the Redfin Darter and the Redspot Darter, with which it is rarely syntopic in Arkansas. Live specimens are easily distinguished by absence of red or yellow spots on side of body in *E. radiosum* (vs. red spots present). It further differs from *E. artesiae* and *E. whipplei* in usually having fewer than 43 pored lateral line scales (vs. usually 43 or more). It also differs from *E. artesiae* in having the opercle unscaled or with scales only on upper half (vs. exposed or embedded scales over most of opercle). It differs from *E. asprigene* and *E. collettei* in having gill membranes broadly joined across throat (vs. separate or narrowly joined) and in breeding males with a broad red band in the anal fin and a blue marginal band in the caudal fin (vs. no red in anal fin or blue marginal band in caudal fin).

VARIATION AND TAXONOMY A member of subgenus *Oligocephalus*, *Etheostoma radiosum* was originally described as a subspecies of the Redfin Darter, *Poecilichthys whipplii radiosus*, from Sugar Loaf Creek, a tributary of the Caddo River (Ouachita River drainage) at U.S. Route 70, Hot Spring County, Arkansas (Hubbs and Black 1941). *Poecilichthys* became a subgenus of *Etheostoma* in the early 1950s, and Moore and Rigney (1952) elevated *P. w. radiosus* to full species status and recognized three subspecies, *E. r. cyanorum*, endemic to the Blue River, Oklahoma,

E. r. paludosum in the Kiamichi, Muddy Boggy, and Clear Boggy systems east of the Blue River drainage and in tributaries of the Red River west of the Blue River, Oklahoma, and *E. r. radiosum* in the Little River and Ouachita River drainages of Oklahoma and Arkansas east of the Kiamichi River drainage. Echelle et al. (1975) studied variation in two isozymes (lactate dehydrogenase LDH-1 and esterase ES-3) at multiple sites within each drainage in the range of *E. radiosum* and found a high degree of interdrainage differentiation in allelic frequencies. This suggested very little gene flow between populations in different drainages. Matthews and Gelwick (1988) studied variation throughout its range and recommended continued recognition of three subspecies pending further analysis of variation. Those authors found that *E. radiosum* exists as relatively isolated populations in at least 13 distinct upland streams in Arkansas and Oklahoma, with little extant gene flow among systems. The Blue River form differed substantially in morphology from all other populations of *E. radiosum*. Some authors subsequently supported recognition of the Blue River, Oklahoma, endemic subspecies as a full species (Near et al. 2011; April et al. 2011; Schwemm 2013). Matthews and Turner (2019) provided a formal description based on a combination of morphological and genetic evidence, recognizing full species status of *E. cyanorum*, the Blue River Orangebelly Darter. Natural hybridization between *E. cyanorum* and *E. pulchellum* (*E. spectabile* clade) was reported by Branson and Campbell (1969) and Echelle et al. (1974). Matthews et al. (2016) found that a local hybrid swarm between *E. radiosum paludosum* and *E. pulchellum* in Little Glasses Creek, Oklahoma, first discovered in 1985, had broken down by 2003, with *E. radiosum* completely replacing *E. pulchellum*. A genetic analysis by Lang and Mayden (2007) supported the inclusion of *E. radiosum* in an *E. whipplei* species group along with *E. artesiae*.

Despite previous studies indicating much temporal, spatial, and genetic isolation of populations, the mitochondrial gene trees of Matthews and Turner (2019) indicate recent, substantial gene flow across drainages between *E. r. radiosum* and *E. r. paludosum* (e.g., Ouachita and Kiamichi basins share identical haplotypes). Those authors noted that a broad geographic, molecular assessment is needed to determine the validity of the two remaining subspecies of *E. radiosum*.

DISTRIBUTION   The Orangebelly Darter occurs only in southwestern Arkansas and southeastern Oklahoma in Ouachita Mountain streams of the Red and Ouachita river drainages (Map 212). There is one record from the Red River drainage of Lamar County, Texas (Page and Burr

2011). It is found almost entirely above the Fall Line but is sometimes taken in waters immediately below it.

HABITAT AND BIOLOGY   The Orangebelly Darter is the most abundant darter in the southern Ouachita Mountains of Arkansas, often having large populations in optimal habitat. It is found in small high-gradient creeks to moderate-sized rivers with clear water, where it often occurs in large numbers in gravel- and rubble-bottomed riffles with moderate to swift current. It is sometimes found in smaller numbers in other habitats, such as sluggish streams and quiet backwaters, and it has been found in six Arkansas reservoirs (Buchanan 2005). It is able to adapt somewhat to habitat alteration, and it apparently has good ability to repopulate areas that have been environmentally disturbed after the disturbance has been removed. When exposed to laboratory thermal gradients, *E. radiosum* did not actively select a particular thermal regime but merely moved to avoid thermal extremes (Hill and Matthews 1980).

Jones and Maughan (1989) reported life history aspects of *E. radiosum* in Oklahoma, and Scalet (1972, 1973a, b) provided data for the Blue River form. The young feed primarily on crustaceans, the adults on mayfly nymphs and other insect larvae. Jones and Maughan (1989) studied the nominate subspecies, *E. r. radiosum*, which is the form that occurs in Arkansas. They found the diet consisted primarily of aquatic insect larvae, especially dipterans, but with a wide variety of ephemeropterans, plecopterans, and trichopterans occurring less frequently. Microcrustaceans (copepods and cladocerans) were also commonly consumed. Other taxa eaten included odonates, megalopterans, lepidopterans, annelids, ostracods, and gastropods. *Etheostoma r. radiosum* fed selectively and exhibited a distinct preference for particular food items such as dipterans. Juveniles generally ate the same foods as adults. Examination of stomach contents of 56 adult specimens from the Little Missouri River, Caddo River, and Hot Spring Creek by TMB supported the conclusions of previous investigators. The diet in spring was dominated by ephemeropteran nymphs and chironomids. Other food items ingested in decreasing order of percent frequency of occurrence were plecopterans, trichopterans, simuliids, coleopteran larvae, amphipods, snails, limpets, isopods, and megalopterans.

Most information on reproductive biology is based on studies of the Blue River, Oklahoma form, *E. cyanorum* (Scalet 1973a). In the Blue River, spawning occurs in gravel riffles from late February to May but peaks from mid-March to mid-April. Based on examination of gonads by TMB, *E. radiosum* in Arkansas spawns from at least late March through May. Ripe females collected from Hot Spring Creek on 31 March ranged from 48 to 64 mm TL and contained 59–85 mature ova. The presence of smaller maturing eggs in the ovaries in March indicated multiple spawning by females. Males and females in spawning condition were found in the Little Missouri River (Pike County) as late as 30 May; females 52–64 mm TL contained 47–59 mature ova, and the lack of smaller developing eggs indicated that the spawning season was almost over. Spawning behavior was described in the Blue River form by Scalet (1973a). Males defend territories in the riffles. When a female enters the territory, the male courts her by following and nipping her and by displaying his brightly colored fins. When ready to spawn, the female partly buries herself in the substrate. She is mounted by the male and both fish vibrate to release eggs and sperm. The spawning act is repeated with the same or with different males until as many as 270 eggs are laid. The eggs are not guarded. Young fish live in quieter portions of the stream but move into progressively faster water as they grow. Sexual maturity is reached at 1 year of age and individuals may live as long as 4 years.

CONSERVATION STATUS   Despite its rather small range, it is one of the most abundant darters in streams of the Ouachitas. We consider it secure in Arkansas.

Etheostoma sp. *cf. spectabile*
Ozark Darter

Figure 26.74a. *Etheostoma* sp. *cf. spectabile*, Ozark Darter. *Uland Thomas.*

Figure 26.74b. Ozark Darters in Cotter Spring, Arkansas. *Isaac Szabo.*

Map 213. Distribution of *Etheostoma* sp. *cf. spectabile*, Ozark Darter, 1988–2019.

CHARACTERS   A moderately robust small to medium-sized member of the Orangethroat Darter clade (*Ceasia*) with a somewhat arched predorsal region; greatest body depth occurring just in front of the first dorsal fin (Fig. 26.74). Back with 3–9 dark saddles which are sometimes indistinct. Pigmentation of sides highly variable with dark blotches or vertical bars which are usually more distinct posteriorly. Horizontal lines poorly developed on sides, usually present only on 1–4 scale rows below the lateral line and often reduced to dots, the intensity of the lines varying from faint to dark. Gill membranes narrowly joined across throat. Premaxillary frenum present. Infraorbital canal interrupted, usually with 3 pores (2–4) in the posterior segment. Supratemporal canal uninterrupted. Cheek naked to lightly scaled; opercle usually moderately scaled; nape and breast usually naked to partly scaled; belly fully scaled. Lateral line incomplete, usually ending beneath second dorsal fin. Lateral scales usually 38–56 ($\bar{x}$ = 52); pored lateral line scales usually 20–39. Diagonal scale rows modally 13 (10–16); scales above lateral line modally 6 (4–8). Dorsal spines 9–11; dorsal rays 12 or 13; anal spines 2; anal rays modally 7 (4–8); pectoral rays usually 11 or 12; usually 15 or 16 principal caudal rays. Breeding males with tubercles on pelvic, anal, and lower caudal fin rays and on ventral scales; sometimes with tubercles on lower pectoral rays. Maximum length around 2.5 inches (64 mm).

LIFE COLORS The following information on coloration is from Distler (1968). Color pattern variable in different parts of range, especially among breeding males. Back and upper side light tan to olivaceous; dorsal saddles brownish black. Vertical bars on sides indistinct or absent in nonbreeding individuals. Horizontal lines on sides not well developed, usually present on the 4 scale rows below the lateral line or reduced to rows of dots. Venter usually unmarked except for meeting lateral bars (if present). Breeding males brightly colored. Throat orange to reddish orange (TMB noted that males from Norfork Lake usually have a red throat). Lateral bars or blotches blue, well developed along entire side and interspersed with orange or deep red pigment. Belly usually with an extensive orange or reddish-orange patch. Predominant color of spinous dorsal fin blue. In White River headwaters, the spinous dorsal fin has marginal and suprabasal blue bands partially to completely fused and a narrow basal red band that is often incomplete. Downstream in the eastern part of the White River, the spinous dorsal fin has marginal and suprabasal blue or blue-black bands that are usually separated by a narrow red or red-and-white band; the blue-black marginal band is widest. In White River headwaters, the soft dorsal fin is predominantly blue (two-thirds or more), with red pigment limited to a narrow submarginal band. In the eastern part of the White River drainage, the soft dorsal fin has 4 bands, but either blue or orange may be the dominant color. Pelvic fins blue; caudal and pectoral fins clear to yellow, sometimes with dark pigment. Anal fin blue to blue green, without red pigment. Breeding females with little coloration, body having dull orange bands and a thin yellow-orange band in the first dorsal fin.

SIMILAR SPECIES *Etheostoma* sp. *cf. spectabile* differs from *E. fragi* in having cheek with 1 or 2 rows of scales below or behind eye, breeding males with tubercles present on pectoral, pelvic, anal, and caudal fins and on ventral scales, and without orange chevrons crossing the belly (vs. cheek with 5 or more rows of scales, breeding tubercles absent, and breeding males with orange chevrons crossing the belly). It differs from *E. uniporum* in usually having 3 pores in the posterior segment of the infraorbital canal, supratemporal canal uninterrupted, and breeding males with the suprabasal band in the first dorsal fin not the widest band (vs. modally 1 posterior infraorbital pore, supratemporal canal interrupted, and breeding males with the suprabasal blue band in the first dorsal fin being the widest band). The Ozark Darter differs from *E. pulchellum*, *E.* sp. *cf. pulchellum* 1 and 2, and *E. squamosum* in having predominantly blue coloration in dorsal fins (vs. predominantly orange coloration). It is sympatric with

E. caeruleum, but differs from that species in having 11 or 12 pectoral rays, an interrupted infraorbital canal, greatest body depth occurring just anterior to first dorsal fin, and breeding males without prominent red pigment in anal fin (vs. 13 or more pectoral rays, infraorbital canal uninterrupted, greatest body depth occurring beneath first dorsal fin, and breeding males with prominent red pigment in anal fin).

VARIATION AND TAXONOMY A member of subgenus *Oligocephalus*, populations formerly known as *Etheostoma spectabile spectabile* in the upper White River drainage were recognized by Distler (1968) as the White River race of that widespread subspecies. Ceas and Burr (2002) considered the upper White River populations to be an undescribed species, the Ozark Darter, *Etheostoma* sp. *cf. spectabile*. Bossu and Near (2009), Near et al. (2011), and Bossu et al. (2013), based largely on DNA analyses (including specimens from Thomas Creek in Newton County, Arkansas), supported its elevation to full species status. We agree that it is a distinct species of the Orangethroat Darter complex awaiting formal description. In a genetic analysis, Near et al. (2011) recovered the Ozark Darter as sister to a clade containing *Etheostoma burri* and *E. spectabile*.

Distler (1968) found that the Ozark Darter is not a homogeneous taxon. Based on coloration and squamation, populations in the headwaters of the White River above its confluence with Roaring River could be distinguished from populations in White River tributaries downstream from the mouth of Roaring River (see Life Colors section of this account for color differences of dorsal fins). The well-developed vertical bars on the sides in populations in the White River headwaters are usually reduced to dark blotches along the lateral line in eastern White River populations. Horizontal lines on sides near the lateral line are more evident in headwater populations and are reduced to rows of dots in eastern populations. Headwater populations also have more extensive tubercle development in breeding males, and a higher mean number of lateral scales (52 vs. 44) and pored lateral line scales (33 vs. 25) than eastern White River populations.

Coloration of breeding males and meristic features of populations in Millers Creek, a tributary of the White River near Batesville, Independence County, and in Curia and Data creeks, the only two tributary systems of the Black River between the Strawberry and White rivers, suggest that these populations warrant further study.

DISTRIBUTION This undescribed species is endemic to the upper and middle White River drainages of Arkansas and Missouri. In Arkansas, it is found from the White River headwaters downstream in tributaries of the

White River to its confluence with the Black River (Map 213). It is apparently rather evenly distributed over its area of occurrence in Arkansas.

HABITAT AND BIOLOGY   The Ozark Darter prefers small headwater creeks and spring runs, where it is found in shallow riffles of slow to moderate current over gravel to rubble substrate. It is often found near the margins of rocky-bottomed pools having some current. It is not very silt-tolerant. A survey of the fishes of the Beaver Lake tail-waters 30 years after impoundment produced no specimens of Ozark Darters (Quinn and Kwak 2003). *Etheostoma* sp. *cf. spectabile* was regularly found in Arkansas Game and Fish Commission population samples taken from Beaver, Bull Shoals, and Norfork reservoirs (Buchanan 2005). Kuehne and Barbour (1983) noted that its greatest abundance is in alkaline waters. Various authors have reported its avoidance of larger, deeper streams having swift, deep riffles. In those streams it tends to be replaced by the Rainbow Darter.

Few aspects of its life history have been studied. In first- through third-order streams of the upper White River, the Ozark Darter fed almost exclusively on chironomid larvae, with some predation on mayfly larvae (Phillips and Kilambi 1996). Diet breadth was narrow in terms of prey numbers and volume, and food habits were similar during all four seasons and among size classes. Stomach contents of 62 specimens collected by Brian Wagner between 31 October and 27 November 2006 from creeks in Baxter, Fulton, Izard, and Independence counties were examined by TMB. Eighty-five percent of the stomachs contained food, with chironomids dominating the diet (58% frequency of occurrence). Also important in the autumn diet were trichopterans, ephemeropterans, and small crustaceans (copepods, amphipods, cladocerans, isopods, and ostracods). Other food items consumed in smaller amounts were plecopterans, simuliids, coleopterans, arachnids, and annelids. One individual had eaten a small clam and another a small crayfish. Pflieger (1966b) reported that in Missouri streams the fry move into the nests of Smallmouth Bass in pools where they are protected by the male bass and where they feed on abundant microorganisms.

There is little information on its reproductive biology, but most aspects of reproduction are probably similar to those reported for other members of the Orangethroat Darter clade. Sexual maturity is reached at 1 year of age, but most breeding adults are in their second or third summer (Pflieger 1997). The reproductive season in northwestern Arkansas extends from March to May (Hubbs and Armstrong 1962). Members of the Orangethroat Darter clade are egg buriers, lacking parental care and territorial guarding. The species is sexually dimorphic for color but not for size. During the spawning season, the males congregate on shallow riffles. When ready to spawn, a female enters the riffle, burrows into the gravel substrate, and is mounted by a male. Both fish vibrate and eggs and sperm are released. The buried eggs are left unguarded. Embryos of Ozark Darters from Arkansas can survive developmental temperatures from 57.2°F to 82.4°F (14–28°C) and have higher developmental temperature maxima than populations of *E. pulchellum* in central Texas (Hubbs and Armstrong 1962).

CONSERVATION STATUS   Populations of this undescribed species are secure.

Etheostoma squamosum Distler
Plateau Darter

Figure 26.75a. *Etheostoma squamosum*, Plateau Darter. *Isaac Szabo.*

Figure 26.75b. *Etheostoma squamosum*, Plateau Darter. *David A. Neely.*

CHARACTERS   Arkansas's largest member of the Orangethroat Darter clade. A robust, medium-sized *Etheostoma* with a somewhat arched predorsal region; greatest body depth occurring just in front of the first dorsal fin (Fig. 26.75). Back with 6–10 saddles, 3 of which are darker than the rest. Sides with dark vertical bars which are more distinct posteriorly and less distinct on upper sides. Dark horizontal lines conspicuous on upper sides, more conspicuous in females. Gill membranes narrowly to moderately joined across throat. Premaxillary frenum present.

Map 214. Distribution of *Etheostoma squamosum*, Plateau Darter, 1988–2019.

Infraorbital canal interrupted, usually with 3 pores in the posterior segment; supratemporal canal uninterrupted. Anterior region of body moderately scaled. Cheek naked to well scaled, usually moderately scaled. Opercle naked to fully scaled, usually moderately scaled. Breast naked to well scaled, usually lightly scaled; belly fully scaled. Nape naked to well scaled, but usually well scaled. Lateral line incomplete, usually ending beneath second dorsal fin. Lateral scales 40–56 (usually fewer than 51); pored lateral line scales 20–41 ($\bar{x}$ = 30). Diagonal scale rows modally 13 (11–16); scales above the lateral line usually 6 (5–7). Pectoral fin rays usually 11 or 12; dorsal spines 9–11; dorsal rays 12 or 13; anal spines 2; anal rays usually 7 (6–8); usually 15 or 16 principal caudal rays. Breeding males with tubercles on pectoral, pelvic, anal, and caudal fins and on ventral scales. Maximum reported length around 2.5 inches (64 mm), but largest Arkansas specimen examined by TMB was 2.8 inches (72 mm) TL from Great House Spring in Washington County.

LIFE COLORS   Information on coloration is from Distler (1968). In nonbreeding males, the back and upper side are light tan to olivaceous, the dorsal saddles are dark brown, and there is faint orange coloration on sides, dorsal fins, and throat. The banded area of the caudal peduncle is a faint blue green. Venter usually unmarked except for

meeting lateral bars. Females devoid of chromatic colors even at peak of breeding season, occasionally with a pale submarginal orange band in dorsal fins. Females with more conspicuous broken horizontal lines on sides than males. Breeding males brightly colored. Throat orange. Anal fin dark blue to blue green, without red pigment. Lateral bars blue to blue green and best developed on lower sides and across venter on caudal peduncle, interspersed with orange pigment. Spinous and soft dorsal fins predominantly orange but with a narrow suprabasal, green band. Amorphic blotches of orange present on membranes of anal fin, principal rays of caudal fin, and distal margins of pectoral and pelvic fins. Pelvic fins dark blue; caudal and pectoral fins clear to yellow, sometimes with dark pigment.

SIMILAR SPECIES   *Etheostoma squamosum* differs from *E. fragi* in having cheek with 1 or 2 rows of scales below or behind eye (vs. 5 or more scale rows), breeding males with tubercles present on pectoral, pelvic, anal, and caudal fins and on ventral scales (vs. breeding tubercles absent), and without orange chevrons crossing the belly (vs. breeding males with orange chevrons crossing the belly). It differs from *E. uniporum* in usually having 3 pores in the posterior segment of the infraorbital canal, supratemporal canal uninterrupted, and breeding males with the suprabasal band in the first dorsal fin not the widest band (vs. modally 1 posterior infraorbital pore, supratemporal canal interrupted, and breeding males with the suprabasal blue band in the first dorsal fin being the widest band). It differs from *E. pulchellum*, *E*. sp. *cf. pulchellum* 1 and 2, and *E*. sp. *cf. spectabile* in having 3 dorsal saddles darker than the rest (vs. dorsal saddles all of equal intensity) and in usually having the cheek and breast lightly to moderately scaled (vs. cheek and breast usually naked). the Plateau Darter does not occur sympatrically with *E. caeruleum*, but it differs from that species in having 11 or 12 pectoral rays, an interrupted infraorbital canal, greatest body depth occurring just anterior to first dorsal fin, and breeding males without prominent red or orange pigment in anal fin (vs. 13 or more pectoral rays, infraorbital canal uninterrupted, greatest body depth occurring beneath first dorsal fin, and breeding males with prominent red pigment in anal fin).

VARIATION AND TAXONOMY   *Etheostoma squamosum* is a member of subgenus *Oligocephalus*. Black (1940) referred to this form as a subspecies of the Orangethroat Darter, *Poecilichthys spectabilis squamigenis* Hubbs, but that description was never published. Distler (1968) described *Etheostoma squamosum* as a subspecies of the Orangethroat Darter, *Etheostoma spectabile squamosum*, based on a morphological analysis. The type locality is Indian Creek, a tributary of Elk River, 25 miles southwest of Sweetwater,

Newton County, Missouri. Bossu and Near (2009), Near et al. (2011), and Bossu et al. (2013), based largely on DNA analysis, supported the elevation of *E. s. squamosum* to full species status. The common name given by Distler (1968), Arkansas River Scaly Orangethroat, was changed to Plateau Darter to more accurately reflect its geographic distribution. Although *E. squamosum* was not recognized by Page et al. (2013), we agree with those authors who recognize its elevation that there is substantial morphological and genetic evidence to support full species status for the Plateau Darter. Species tree analyses by Bossu and Near (2009), using mtDNA and nDNA, supported a sister relationship between *E. squamosum* and *E. pulchellum*.

DISTRIBUTION   *Etheostoma squamosum* is endemic to streams draining the Springfield Plateau of the Ozark Mountains west and southwest into the Arkansas River. It is found in southwestern Missouri, southeastern Kansas, northeastern Oklahoma, and northwestern Arkansas. In Arkansas, it occurs only in Benton and Washington counties in tributaries of the Neosho-Spring River system and in the Illinois River system (Map 214).

HABITAT AND BIOLOGY   *Etheostoma squamosum*, like other members of the Orangethroat Darter clade, occurs primarily in headwater creeks and spring runs, where it is usually found in shallow riffles of slow to moderate current over a gravel to rubble substrate; however, in Arkansas it is also often found in larger streams, such as the Illinois River, where it occupies stronger currents in deeper riffles, unlike other members of the Orangethroat Darter complex. We have also noted that it occupies swift riffles in Spavinaw Creek in Benton County. Our observations support Distler's (1968) statement that it often occupies the strong currents of habitats equivalent to those preferred by *E. caeruleum* (an allopatric species). Perhaps its larger body size (compared with other *Ceasia* species) permits the observed broader occupation of stream orders and habitats. It is not very silt-tolerant and apparently does not adapt well to reservoir conditions.

It is a benthic invertivore like other members of *Ceasia*. Phillips and Kilambi (1996) found that in Clear Creek, Arkansas (an Illinois River tributary), chironomids composed 99% of the diet throughout the year. The following information on feeding by *E. squamosum* is from a study in Flint Creek, Oklahoma, by Todd and Stewart (1985).

The stomachs of 244 *E. spectabile* [= *E. squamosum*] contained 35 different prey taxa. Both juveniles and adults fed heavily on chironomids during winter and early spring (January–March). Chironomids continued to constitute 16.1% to 59.5%, by dry weight, of juvenile fish diets in April to July, as they increased feeding on microcrustaceans and small mayflies. This shift to planktonic prey items coincided with recruitment of darter fry. In August, diets of smaller fish were exclusively composed of microcrustaceans, particularly Cladocera and Copepoda, and for the rest of the year microcrustaceans and amphipods continued to make up the major percentages by number and dry weight, into December. Mayfly nymphs (*Baetis*, *Stenonema*, *Ephemerella*, *Leptophlebia*, and *Caenis*) made up 49.1–77.3%, by dry weight, of mature *E. spectabile* [*squamosum*] diets in spring and early summer until June, corresponding with the emergence and recruitment of these insects. They were ingested at lower levels for the rest of the summer, and none were eaten by 12 fish in October. Unlike juveniles, mature fish fed on chironomids at levels less than 11% by dry weight from April to July, and microcrustaceans were not present in mature fish guts in late spring and summer (June, August) in amounts above 3.1% by dry weight. Amphipods appeared in diets of mature Orangethroat Darters [Plateau Darters] at levels of 42.5% and 91.0% dry weight in September and October, respectively, when ingestion of amphipods generally decreased. Chironomid feeding increased in November and December until winter, when chironomids again predominated in guts. Mature fish relied more heavily on larger items such as *Baetis*, *Ephemerella*, *Leptophlebia*, *Stenonema*, and amphipods as these items were available.

Differences in the diets of *E. squamosum*, *E. mihileze*, and *Uranidea carolinae*, the three most abundant small invertivores in Flint Creek, Oklahoma, were related to a complex of factors, including utilization of different habitats, prey size selection, and the selection of specific prey items (Todd and Stewart 1985). *Etheostoma squamosum* generally fed on smaller prey than *E. mihileze*, in part because of the larger gape size of the latter species.

The reproductive season in Arkansas extends from March to mid-May (Hubbs and Armstrong 1962). Ripe individuals were found in Great House Spring (Washington County) on 10 March by TMB. Females with mature eggs were collected on 15 May from Flint Creek in Benton County (Tumlison et al. 2017). Aspects of reproductive biology are probably similar to those of other *Ceasia* species, which are all egg buriers (Page 1985) lacking parental care and territorial guarding. The species is sexually dimorphic for color but not for size. During the spawning season, the males congregate on shallow riffles.

Males become increasingly aggressive as the spawning season approaches (Buchanan 1966). Plateau Darter embryos from Arkansas can survive developmental temperatures from 57.2°F to 82.4°F (14–28°C) and have higher developmental temperature maxima than do southern populations (of the *Ceasia* clade) in central Texas (Hubbs and Armstrong 1962). Fertilized eggs hatched in 4 days at 21°C, and the yolk sac was absorbed 3 days after hatching at that temperature (West 1966). West also found that maximum growth rate of Arkansas *E. squamosum* occurred at 26.0°C. Fayton et al. (2017) described a new species of helminth parasite (*Plagioporus*) from *E. squamosum* in Arkansas.

CONSERVATION STATUS   The Plateau Darter has a very restricted range in Arkansas. It is found only in Benton and Washington counties, where widespread loss of habitat has occurred due to rapid human population growth in recent decades. Its populations have undoubtedly decreased, and periodic monitoring should be done to insure its future survival in our state. We consider it a species of special concern.

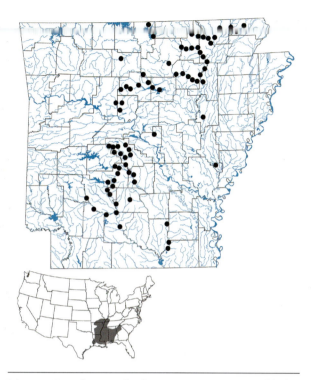

Map 215. Distribution of *Etheostoma stigmaeum*, Speckled Darter, 1988–2019.

Etheostoma stigmaeum (Jordan)
Speckled Darter

Figure 26.76. *Etheostoma stigmaeum*, Speckled Darter. *Patrick O'Neil, Geological Survey of Alabama.*

CHARACTERS   A small, slender darter with a moderately blunt snout and a horizontal mouth (Fig. 26.76). Back with 6 dark saddles. Midside with a series of 8–11 dark, almost square, blotches. Upper side and back with small scattered W- and X-shaped markings and/or zigzag lines. Premaxillary frenum absent, upper lip and snout separated by a continuous deep groove. Palatine teeth present. Gill membranes narrowly joined across throat. Infraorbital canal uninterrupted, usually with 8 pores. Opercle nearly fully scaled; nape partly to fully scaled; breast usually naked; cheek squamation variable, usually 30–70% scaled. Lateral line incomplete, usually ending beneath second dorsal fin. Scales in lateral series usually 44–51; usually 10–16 unpored scales. Scale rows above lateral line 3–6; scale rows below lateral line 5–9. Diagonal scale rows 10–15; scale rows around caudal peduncle usually 14–19. Dorsal spines 10–12 (modally 11); dorsal rays 10–12 (modally 11); anal spines 2; anal rays 7–9; pectoral rays 12–15 (modally 14). Breeding male with tubercles on pelvic and anal fin rays and occasionally on ventral body scales. Maximum length around 2 inches (51 mm).

LIFE COLORS   Layman and Mayden (2012) provided detailed descriptions of breeding and nonbreeding coloration. Breeding male: spinous dorsal fin with a thin gray to blue-green marginal band, a narrow white submarginal band, a wide red-orange medial band, a wide blue-green submedial band, and a narrow clear basal band with dark triangular areas in the posterior portion. Soft dorsal, pelvic, and anal fins with much black pigment and lacking distinct orange spots. Soft dorsal, anal, and caudal fins with blue green in base of fin. Lateral bars iridescent blue green. Sides with red-orange spots and X-markings between lateral bars extending to dorsum. Lateral belly yellow or orange yellow. Nonbreeding and freshly preserved specimens with straw-colored back and upper side, with dark brown saddles and lateral blotches, and with scattered smaller brown markings. Underside yellowish white. Preorbital dark bar extending onto snout, but not continuous around snout. Dorsal and caudal fins with faint brown bands; other fins clear or dusky.

SIMILAR SPECIES Sometimes confused with *E. chlorosoma* and *E. nigrum*, with which it is occasionally syntopic in Arkansas, *E. stigmaeum* differs from those species in having 2 anal spines (vs. 1), usually 11 or more dorsal spines (vs. fewer than 11), and breeding males with bright colors (vs. lacking bright colors). It is most similar to the allopatric *E. clinton* and *E. teddyroosevelt*, but differs from the former species in having modally 8 anal rays (vs. 9), usually naked or near-naked cheeks (vs. usually partly scaled), and in having modally 10 preoperculomandibular pores (vs. 9). *Etheostoma stigmaeum* differs from *E. teddyroosevelt* in having partly scaled cheeks (vs. usually naked cheeks), palatine teeth present (vs. usually absent) (Fig. 26.44), and breeding males having a blue-green submedial band in spinous dorsal fin (vs. a black submedial band) and no orange spots in soft dorsal, caudal, and anal fins (vs. orange spots present in those fins). *Etheostoma stigmaeum* differs from *E. gracile* in lacking a premaxillary frenum (vs. frenum present), in usually having more than 26 pored lateral line scales (vs. 16–23 pored scales), and in usually having 11 or 12 spines in the first dorsal fin (vs. usually 8–10 spines).

VARIATION AND TAXONOMY *Etheostoma stigmaeum* was originally described by Jordan (1877d) as *Boleosoma stigmaeum* from specimens from small tributaries of the Etowah and Oostanaula rivers near Rome, Floyd County, Georgia. A lectotype was subsequently designated from the Etowah River by Bailey et al. (1954). A revision of the subgenus *Doration* by Layman and Mayden (2012) resulted in a redescription of *E. stigmaeum*, redescriptions of three other nominal species, and the descriptions of five new species, two of which occur in Arkansas. The monophyly of *Doration* was supported by 10 derived character states, and the three *Doration* species in Arkansas (and *Doration* species elsewhere) are diagnosable on morphological (Layman and Mayden 2012) and genetic (Layman 1994) bases. Near et al. (2011) provided genetic evidence supporting the monophyly of *Doration*. Minor geographic variation in scale counts and degree of squamation occurs in Arkansas populations of the Speckled Darter. Howell (1967) provided meristic data by river system for Arkansas populations.

DISTRIBUTION The Speckled Darter is distributed in Gulf Coast drainages from Pensacola Bay in Florida and Alabama west to the Red-Atchafalaya and Sabine river systems in Louisiana and north up the Mississippi River Embayment to western Kentucky and southeastern Missouri (Layman and Mayden 2012). *Etheostoma stigmaeum* is the most widely distributed of three species of the subgenus *Doration* in Arkansas, even though some Arkansas populations were assigned to two new species

in 2012. It occurs in the White River drainage, where it is most common in the Ozark Uplands downstream of Bull Shoals Dam (it is replaced in the upper White River drainage by *E. teddyroosevelt*) and is rare in the Coastal Plain segment of that river system. In the Ouachita River drainage, it occurs most commonly in drainages downstream from Remmel Dam and is replaced in the upper Ouachita River drainage by *E. clinton*, with a few Coastal Plain records from that system (Buchanan 1997). The first reported record of *E. stigmaeum* from the Arkansas River drainage was from Bayou Meto (Heckathorn 1993), and the first records verified by us from that drainage are four specimens taken from West Fork Point Remove Creek in Conway County by TMB and Jeff Quinn on 5 May 2015 (Map 215). It is possible that additional records might be found in other lower Arkansas River tributaries. It is rare today in the Coastal Plain lowlands of eastern Arkansas but was probably more common there prior to extensive stream channelization and other habitat modifications now prevalent in that highly agricultural area. Older records exist from the St. Francis River drainage (Map A4.212), but there are no recent records from that drainage.

HABITAT AND BIOLOGY The Speckled Darter occurs in different habitats within the state, but it is most abundant in rivers and in clear, small to medium-sized streams of high to moderate gradient. It is found mainly in pools having some current but is sometimes taken in riffles, and it is almost invariably associated with a sandy substrate, either a pure sand bottom or a mixture of sand and gravel. The Speckled Darter is also found in moderate-sized rivers over a sandy substrate and has been collected in small numbers from Greers Ferry and Norfork reservoirs (Buchanan 2005). In southern Arkansas, it is found primarily in the eastern Saline, Ouachita, and Little Missouri rivers in slow to moderate current over a sand and gravel bottom. In the eastern lowlands it is known today from only a few moderately clear, sandy-bottomed streams of the White River National Wildlife Refuge and the Cache River National Wildlife Refuge. It has not been found in the St. Francis River drainage of Arkansas in the post-1988 period.

Food of Alabama specimens consisted mainly of midge larvae supplemented with microcrustaceans and mayfly nymphs (Boschung and Mayden 2004). Stomach contents of 48 specimens from 8 Arkansas counties were examined by TMB, and the results were pooled by season. Speckled Darters collected from midsummer to midfall (21 July to 23 October) had a high percentage of empty stomachs (44%), and food volume and diversity were low in stomachs having food. Summer and fall diets were dominated by chironomid larvae and ephemeropteran nymphs. Other dipteran larvae and small crustaceans (cladocerans) were present in small numbers. All Speckled Darter stomachs from late winter through spring collections (28 February through 9 June) contained food, usually in large amounts. The late winter and spring diets were dominated by chironomids (73% by frequency of occurrence), followed by small crustaceans (isopods and copepods), trichopteran nymphs, ephemeropterans, plecopterans, and annelids.

Etheostoma stigmaeum moves onto gravelly riffles to spawn in the spring. Hubbs (1985) reported that spawning in Arkansas occurs from mid-March through May at water temperatures of 14–17°C, and this largely agrees with our observations. Ripe individuals were found by TMB in Indian Bayou (Monroe County), Arkansas, on 26 March at a water temperature of 14°C. Based on examination of ovaries and testes of Speckled Darters from nine Arkansas counties by TMB, spawning begins by late February in southern Arkansas (Ouachita River) and continues to at least mid-May in northern Arkansas (Newton and Randolph counties). Females ranging from 36 to 45 mm TL collected from Conway County in April and May contained ripe ova, with clutch sizes ranging from 36 to 82 mature ova. Spawning behavior of *E. stigmaeum* has not been observed, but Winn (1958a, b) observed aquarium spawning of a closely related species and reported that the males are territorial. When a ripe female enters a male's territory, he pursues and mounts her. Vibration of the two fish causes the fertilized eggs to be buried in the substrate. Kuehne and Barbour (1983) reported that in nature the males have shifting territories over the breeding riffles, much like the males of *E. caeruleum*.

CONSERVATION STATUS   Populations of this darter are widely distributed and generally small. The largest Arkansas populations are currently found in the White and Ouachita river drainages. It was one of the most abundant darters in trawl samples from the Ouachita River in 2018 (Jeff Quinn, pers. comm.). We consider its populations secure.

Etheostoma teddyroosevelt Layman and Mayden
Highland Darter

CHARACTERS   A small, slender darter with a moderately blunt snout and a horizontal mouth (Fig. 26.77). Back with 6 dark saddles. Midside with a series of 8–10 dark, quadrate blotches. Upper side and back with small, scattered, W- and X-shaped markings and/or zigzag lines. Premaxillary frenum absent, upper lip and snout separated by a continuous, deep groove. Vomerine teeth present;

Figure 26.77. *Etheostoma teddyroosevelt*, Highland Darter. *Uland Thomas.*

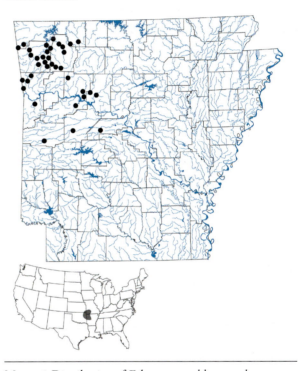

Map 216. Distribution of *Etheostoma teddyroosevelt*, Highland Darter, 1988–2019.

palatine teeth usually absent (present in 16% of specimens). Gill membranes narrowly to moderately joined across throat. Infraorbital canal uninterrupted, usually with 7–9 pores. Opercle squamation variable, but usually less than 60% scaled; nape usually 50–100% scaled; cheek naked or with a few scales near eye; breast usually naked; belly usually fully scaled. Lateral line incomplete, usually ending beneath second dorsal fin. Scales in lateral series usually 45–51; unpored lateral scales usually 12–19. Scale rows above lateral line 4–6; scale rows below lateral line usually 6–9. Diagonal scale rows usually 12–16; scale rows around caudal peduncle usually 15–19. Dorsal spines 10–13 (modally 11 or 12); dorsal rays 10–13 (modally 11 or 12); anal spines 2; anal rays 7–10 (modally 8); pectoral rays modally 14. Breeding male with tubercles on pelvic and anal fin rays and occasionally on posterior one-third to two-thirds of belly. Maximum length around 1.8 inches (45 mm).

LIFE COLORS Layman and Mayden (2012) provided detailed descriptions of breeding and nonbreeding coloration. Breeding male: spinous dorsal fin with a thin dusky to black marginal band, a narrow white to clear submarginal band, a wide bright orange medial band, a wide black submedial band, a clear basal band lacking bright orange pigment and with black pigment extending vertically through posterior portions of membranes. Soft dorsal, caudal, anal, and pectoral fins with subdued but discrete orange spots on rays. Side with 8–10 iridescent deep blue midlateral quadrate blotches; side with orange coloration on scales; dorsolateral area also with scattered dark black and blue markings; body often quite dark during peak breeding time. Nonbreeding and freshly preserved specimens with straw-colored back and upper sides; dorsal saddles and lateral blotches brown; underside light-colored. Preorbital dark bar on each side of head extending onto snout, but not continuous around snout.

SIMILAR SPECIES Sometimes confused with *E. chlorosoma*, with which it is occasionally syntopic in Arkansas, and *E. nigrum*, but *E. teddyroosevelt* differs from those species in having 2 anal spines (vs. 1), usually 11 or more dorsal spines (vs. fewer than 11), and breeding males with bright colors (vs. breeding males lacking bright colors). It is most similar to the closely related allopatric species *E. clinton* and *E. stigmaeum*, but differs from those species in breeding males having a black submedial band in the spinous dorsal fin (vs. a blue-green submedial band); orange spots on soft dorsal, caudal, and anal fins (vs. orange spots absent); deep blue lateral blotches (vs. blue-green lateral blotches); and in lacking blue pigment on the lips and mid-gular region (vs. blue pigment present on lips and mid-gular region). It further differs from those two species in having a more completely and consistently scaled nape. It also differs from *E. stigmaeum* in usually having naked cheeks (vs. partly scaled) and in usually lacking palatine teeth (vs. palatine teeth present); however, up to 29% of *E. teddyroosevelt* specimens from the upper White River drainage have palatine teeth (Layman and Mayden 2012). *Etheostoma teddyroosevelt* is relatively robust, with greater body width and body depth at the spinous dorsal fin origin than other Arkansas species of *Doration*. *Etheostoma teddyroosevelt* differs from *E. gracile* in lacking a premaxillary frenum (vs. frenum present), in usually having more than 26 pored lateral line scales (vs. 16–23 pored scales), and in usually having 11 or 12 spines in the first dorsal fin (vs. usually 8–10).

VARIATION AND TAXONOMY A member of the subgenus *Doration* and formerly included in *Etheostoma stigmaeum*, *E. teddyroosevelt* was elevated to full species

status by Layman and Mayden (2012) based on variation in morphology and male breeding colors. The type locality is the Spring River at Kansas State Highway 96, Cherokee County, Kansas. The species epithet was selected to honor Theodore Roosevelt, the 26th president of the United States of America and a noted conservationist. Some variation in nape squamation and the frequency of palatine teeth occurs (Layman and Mayden 2012). The nape is usually 90–100% scaled in Arkansas River specimens but varies widely, from 10% to 100%, in those from the upper White River. Palatine teeth are present in only 7% of Arkansas River specimens but occur in 29% of upper White River specimens.

DISTRIBUTION The Highland Darter is found only in the Arkansas and upper White river drainages of the Ozark Plateau in Arkansas, Missouri, Kansas, and Oklahoma. In Arkansas, it occurs in the upper White River drainage upstream of Bull Shoals Dam and is replaced by *E. stigmaeum* in the White River drainages downstream of Bull Shoals Dam (Layman and Mayden 2012). It also occurs in northern and southern Arkansas River tributaries in western Arkansas downstream to Illinois Bayou in Pope County, and the Fourche LaFave and Petit Jean river drainages in Perry and Yell counties (Map 216). There are no post-1988 records from the Petit Jean River drainage.

HABITAT AND BIOLOGY Pflieger (1997) reported that in the Ozarks of Missouri, it is largely restricted to rivers, occurring only rarely in small creeks and never in headwaters. In Kansas, the Highland Darter lives in rather large streams of moderate to steep gradient, where it prefers pools below riffles, often at depths of 2 feet (60 cm) or more, except when spawning (Cross and Collins 1995). That generally agrees with our observations of its habitat in Arkansas, but it is most commonly found in Arkansas in large creeks, such as Lee Creek and War Eagle Creek. It is typically found in pools with some current over sand and gravel substrates. Reservoir construction has apparently had a negative impact on its populations. Quinn and Kwak (2003) reported that specimens were not found in the Beaver Lake tailwaters shortly after impoundment or 30 years later. Buchanan (2005) collected no specimens in impoundments within its range, even though the closely related *E. stigmaeum* was found in small numbers in two White River drainage impoundments (Norfork and Greers Ferry reservoirs).

Stomach contents of 43 Highland Darters collected from Crawford, Johnson, Madison, and Washington counties, Arkansas, from March through July were examined by TMB. All stomachs examined contained food and the data were pooled for analysis. Chironomid larvae dominated the diet in all months in percent frequency and volume and occurred in 91% of the stomachs. Small crustaceans (copepods, isopods, and ostracods) made up the next most important part of the diet (occurring in 34% of stomachs), followed in importance by other dipteran larvae, ephemeropterans, and trichopterans. Coleopteran larvae, plecopterans, and megalopterans were also consumed in small amounts.

During the breeding season, males gather on broad, shallow riffles over clean, rounded gravel of uniformly small size, where the eggs are buried in the gravel (Cross and Collins 1995; Pflieger 1997). It spawns in April and May in Kansas and Missouri (Cross 1967; Pflieger 1997). Cross and Collins (1995) reported that the spawning activity of this species is similar to that of members of the Orangethroat Darter complex. Based on examination of ovaries and testes by TMB, spawning occurs in Arkansas from March to late May. Layman and Mayden (2012) reported finding tuberculate males in the Kings River in Madison County, Arkansas, as late as 22 May, but specimens collected from the White River in Madison County on 23 May by TMB contained only atretic eggs and regressing testes. We collected ripe adults from shallow gravel riffles in late April and early May in Johnson County, Arkansas, and spawning was observed by TMB in Lee Creek in Crawford County on 15 March and 10 April, and ripe individuals were taken from Vache Grasse Creek in Sebastian County on 11 April. Mature females from Crawford and Johnson counties ranging in TL from 34 to 40 mm contained clutch sizes of 48–62 ripe ova. The presence of different sizes of ova in the ovaries on 12 April in Dirty Creek in Johnson County indicates multiple spawning.

CONSERVATION STATUS   Populations of *E. teddyroosevelt* appear to be small based on collection data reported in Layman and Mayden (2012). The largest number of specimens reported from a single sample in Arkansas River drainages was 5, with most collections from that drainage containing only 1 specimen. The upper White River may have somewhat larger populations of the Highland Darter, with the largest number in a sample (12) taken from War Eagle Creek. The Kings River with 10 specimens in one sample and Richland Creek with two samples containing 8 and 9 individuals were the next-largest samples reported by Layman and Mayden (2012). The largest known series taken in Arkansas was from the White River in Madison County (17 specimens) on 23 May 1984 by TMB and Ralph Fourt (AGFC). A survey of the current status of this species is needed. We consider the Highland Darter a species of special concern in Arkansas.

Etheostoma uniporum Distler
Current Darter

Figure 26.78. *Etheostoma uniporum*, Current Darter. *Uland Thomas.*

Map 217. Distribution of *Etheostoma uniporum*, Current Darter, 1988–2019.

CHARACTERS   A small to medium-sized member of the *Etheostoma spectabile* complex, it has a somewhat arched predorsal region, with the greatest body depth occurring just in front of the first dorsal fin (Fig. 26.78). Back with 6–10 moderately pigmented saddles, but often with the saddles at the posterior insertion of the first dorsal fin and posterior insertion of the second dorsal fin darker than the others. Side usually with 8–10 narrow dark bars along the entire length of the body, but bars sharply defined only below lateral line; bars continuous across entire venter and anterior bars slanting forward. Horizontal lines on sides faint, absent below lateral line. Gill membranes narrowly joined across throat. Premaxillary frenum present.

Infraorbital canal interrupted, usually with 1 posterior pore (never with 3 pores). Supratemporal canal usually interrupted. Cheek with 1 or 2 rows of scales below or behind eye; opercle scalation variable, but usually less than 90% scaled; belly fully scaled. Lateral line incomplete, usually ending beneath second dorsal fin. Scales in lateral series usually 42–49 ($\bar{x}$ = 45.5); pored lateral line scales usually 16–28. Pectoral fin rays 11 or 12; dorsal spines 10 or 11; dorsal rays 11–13; anal spines 2; anal rays 6 or 7. Scale rows around caudal peduncle usually 16–20. Diagonal scale rows from anal fin origin to first dorsal fin base usually 12–15. Breeding males with tubercles on pectoral, pelvic, anal, and caudal fins and on ventral scales. Maximum length around 2 inches (50.8 mm).

LIFE COLORS   Back and upper side olivaceous to rusty brown. Dorsal saddles moderate to dark brown. Venter light-colored and unmarked except for lateral bars meeting ventrally. Breeding males with bright turquoise as the dominant coloration. Throat blue gray to orange. Transverse lateral bars iridescent turquoise; interspaces of bars blue gray to rusty gray, with red color increasing in intensity toward the caudal peduncle. Belly blue gray. Distinct blue-green humeral spot present. First dorsal fin with a five-band sequence, but the suprabasal blue band is the widest; a submarginal white band is prominent; medial orange pigment continuous, not reduced to a series of separated triangles. Second dorsal fin with typical five-banded pattern; medial orange band widest. Anal fin turquoise, without red or orange band but with occasional flecks of orange. Breeding females much more subdued in coloration, but with a thin medial orange band in the first dorsal fin.

SIMILAR SPECIES   The Current Darter is most similar to *E. fragi*, *E. pulchellum*, *E.* sp. *cf. pulchellum* 1, *E.* sp. *cf. pulchellum* 2, *E.* sp. *cf. spectabile*, and *E. squamosum* of the Orangethroat Darter clade, but is allopatric to those species. It differs from all other members of the *E. spectabile* complex in having only 1 pore in the posterior segment of the infraorbital canal (vs. usually 3 pores in the posterior segment of the infraorbital canal). *Etheostoma uniporum* also resembles *E. caeruleum* with which it is often sympatric. It can be distinguished from *E. caeruleum* in having fewer than 13 pectoral rays (vs. usually 13), an interrupted infraorbital canal (vs. uninterrupted), greatest body depth occurring just anterior to first dorsal fin (vs. beneath first dorsal fin), and breeding males without prominent red pigment in anal fin (vs. breeding males with red in anal fin).

VARIATION AND TAXONOMY   The Current Darter was originally recognized by Distler (1968) as a subspecies of *Etheostoma spectabile*, *E. s. uniporum*. Wiseman et al. (1978) studied variation at the Ldh-1 gene locus in various

E. spectabile populations and reported a unique fixed allele in the two *E. s. uniporum* populations examined. The Current Darter was elevated to full species status by Ceas and Page (1997) based on genetic evidence, unique color patterns of breeding males, diagnostic modes for various meristic counts, and the lack of intergrade populations. The type locality is Pigeon Creek, a headwater tributary of the Current River, Dent County, Missouri. The specific epithet, *uniporum*, refers to the single pore occurring in the posterior segment of the infraorbital canal. The common name, Current Darter, is a reference to the Current River, which runs through the range of this species. *Etheostoma uniporum* is a member of subgenus *Oligocephalus*. Although having a small range, the species has geographic variation in numerous characters, but little geographic concordance in character states (Ceas and Page 1997). Ray et al. (2008) concluded that introgression had occurred between *E. uniporum* and *E. caeruleum*, based on analysis of mitochondrial DNA. The mechanism of introgression was likely asymmetric sneaking behavior by male *E. uniporum*. The entire mitochondrial genome of *E. uniporum* has been replaced by that of *E. caeruleum*. MtDNA replacement may have occurred in *E. uniporum* due to drift fixation in a historically small female effective population.

DISTRIBUTION   *Etheostoma uniporum* is endemic to tributaries of the Black River system in Missouri and Arkansas. In Arkansas, it is most abundant in the Spring River in Fulton, Izard, and Sharp counties, but it is also found in the Eleven Point River and Fourche Creek in Randolph County. A small population occurs in Flint Creek in Lawrence County (Map 217).

HABITAT AND BIOLOGY   Like most members of the *E. spectabile* complex, the Current Darter is most abundant in headwater streams and tributary creeks in shallow riffles with slow to moderate current and gravel and cobble substrates. The following habitat data summary for 10 creeks in Randolph County having *E. uniporum* was obtained from field notes of Brian Wagner based on collections made on 18 and 19 September 2006. All 10 creeks were less than 10 m in width, and current ranged from none to moderate. Seven of the streams were spring-fed and clear, and three streams were moderately turbid. Most specimens (63%) were taken over a gravel/rubble substrate, 30% over a boulder/bedrock substrate, and 7% over a silt/mud substrate. There was no vegetation in 7 creeks, but 2 creeks had watercress and 1 had water willow along the margin. Like most members of the Orangethroat Darter clade, *E. uniporum* usually occupies a different habitat from that of *E. caeruleum*, and it occurred syntopically with that species in only 3 of the 10 Randolph County creeks.

Stomach contents of 43 individuals (ranging in TL from 27 to 55 mm) collected from 10 creeks in Randolph County on 18 and 19 September were examined by TMB. Seventy percent of the stomachs contained identifiable food items. Trichopterans were the main food consumed based on percent frequency of occurrence, followed by chironomids, ephemeropterans, copepods, amphipods, isopods, and ostracods. Other foods ingested in small amounts were cladocerans, odonates, and coleopterans.

There is no information on reproductive biology, but presumably it would be similar to other well-studied members of the *E. spectabile* complex.

CONSERVATION STATUS   Because of its restricted range and small populations, the Current Darter is a species of special concern. Its populations should be monitored for any future changes.

Map 218. Distribution of *Etheostoma whipplei*, Redfin Darter, 1988–2019.

Etheostoma whipplei (Girard)

Redfin Darter

Figure 26.79a. *Etheostoma whipplei*, Redfin Darter male. *Uland Thomas.*

Figure 26.79b. *Etheostoma whipplei*, Redfin Darter female. *David A. Neely*

CHARACTERS   A moderate-sized, fairly robust darter with a pointed snout and a deep caudal peduncle (Fig. 26.79). Back with 8–12 small saddles that are sometimes faint or absent. Color pattern of side and back variable but usually dominated by dark mottlings and reticulations, with small scattered light spots. Dark vertical bars sometimes present on sides, and if present, best developed posteriorly, especially on the caudal peduncle and on smaller individuals. Gill membranes usually moderately to broadly joined across throat. Premaxillary frenum present.

Infraorbital canal uninterrupted. Belly fully scaled but sometimes scales are embedded; nape and opercle partly to fully scaled; cheek varies from naked to fully scaled; breast naked to 75% scaled. Lateral line incomplete, usually ending beneath second dorsal fin but sometimes extending well past it. Scales in lateral series usually 61–71 ($\bar{x} = 65.6$); pored lateral line scales 45–58 ($\bar{x} = 51$); caudal peduncle scale rows 26–32 ($\bar{x} = 29.4$); diagonal scale rows from origin of anal fin to first dorsal fin base 17–25 ($\bar{x} = 20$). Dorsal spines 10–12; dorsal rays 12–15; anal spines 2; anal rays 7–9. Breeding males with tubercles on ventral scales of belly and caudal peduncle and on scales along the anal fin base. Maximum length around 3 inches (76 mm).

LIFE COLORS   Back and sides mottled olive brown; dorsal saddles, if present, dark brown. Venter and underside of head white to light brown. Dark vertical bands sometimes present posteriorly on sides. Small black humeral spot and dark vertical suborbital bar present. Adult males with dark red or orange spots on sides; females with yellow spots. Breeding males brightly colored with bright red spots on sides, blue breast, orange belly, and bright red and blue bands in dorsal, caudal, and anal fins; pectoral and pelvic fins blue. Breeding females with yellow spots on sides and a red-orange submarginal band in the first dorsal fin; second dorsal and caudal fins with brown bands; other fins clear.

SIMILAR SPECIES *Etheostoma whipplei* is very similar to *E. artesiae* and *E. radiosum* and is allopatric to both of those species. It can be separated with complete accuracy from *E. artesiae* by the combination of having more than 61 lateral scales *and* more than 26 caudal peduncle scale rows (no specimens of *E. artesiae* have the combination of more than 61 lateral scales *and* more than 26 caudal peduncle scale rows). It differs from *E. radiosum* in having red or yellow spots on sides (vs. no red or yellow spots on sides), usually 43 or more pored lateral line scales (vs. fewer than 43 pored lateral line scales), usually more than 61 lateral scales (vs. usually 61 or fewer lateral scales), and in having thin dark lines with many interruptions along the side of the body from head to caudal fin on the upper half of the body and from the middle of the side to the caudal fin on the lower half (vs. rarely with dark lines on sides). It can be distinguished from *E. asprigene* by having 61 or more lateral scales (vs. fewer than 55) and by gill membranes moderately to broadly joined across throat (vs. gill membranes separate or only slightly joined).

VARIATION AND TAXONOMY *Etheostoma whipplei* was originally described by Girard (1859) as *Boleichthys whipplii*. Girard listed the type locality as Coal Creek, Arkansas, but Hubbs and Black (1941) believed this locality was "Coal Creek, a southern tributary of the Arkansas River in eastern Oklahoma." This latter designation was accepted by Piller et al. (2001). It is a member of subgenus *Oligocephalus*. The Redfin Darter is variable over its entire range, and much variation in markings and color pattern is usually observed within a single population among individuals of different sizes, sexes, and reproductive states. Hubbs and Black (1941) originally separated *E. whipplei* into three subspecies, but much confusion resulted regarding the validity and distributions of those subspecies. One of those subspecies, *E. w. artesiae*, is found in Coastal Plain lowland streams of southern Arkansas and has now been elevated to full species status (Piller et al. 2001). The previously recognized subspecies, *E. w. montanus*, which Hubbs and Black (1941) described as endemic to the Clear Creek drainage of Crawford and Washington counties in Arkansas, is no longer considered a valid subspecies and does not have higher scale counts than specimens of *E. w. whipplei* (Retzer et al. 1986). Genetic and morphological analyses support the monophyly of an *E. whipplei* species group that also includes *E. artesiae* and *E. radiosum* (Lang and Mayden 2007).

DISTRIBUTION *Etheostoma whipplei* has a small range, occurring in only four states: Arkansas, Missouri, Kansas, and Oklahoma, in the White and Arkansas river systems of the Mississippi River basin. Most of the species'

range is in Arkansas and Oklahoma. In Arkansas, the Redfin Darter is primarily an upland species, occurring almost entirely above the Fall Line in the Arkansas River and its tributaries. It also occurs in the middle White River and historically in its major tributaries, the Black and Little Red rivers (Map 218). There are no recent records from the Black River drainage, and it is absent from the upper and lower White River. The only record from the lower Arkansas River downstream from Little Rock is from Jefferson County (Robison 2005b).

HABITAT AND BIOLOGY The Redfin Darter is apparently an adaptable species occurring in a variety of habitats in Arkansas. It is most abundant in small to medium-sized streams of high gradient, where it is found in shallow gravel-bottomed riffles. We have also found this species in small intermittent headwater creeks over bedrock and broken shale bottoms, in large rivers (including the Arkansas River) in quiet water over sand and silt substrates, and in small creeks of the Arkansas River floodplain over sand, clay, and detritus substrates. It was one of the most widely distributed darters in Arkansas impoundments, occurring in eight reservoirs in two ecoregions (Buchanan 2005). In Cypress Creek, Arkansas, *E. whipplei* was most abundant at sites characterized by gravel bottoms with pebble and cobble cover, small catchment sizes, and higher gradients (Stearman 2011). In first- to fourth-order streams, it appeared largely indifferent to riffle and pool habitat at most sites, occupying primarily riffles only in the lowest-gradient streams and pools only in the highest-gradient streams. In the Frog Bayou drainage (Clear Creek) of Crawford County, population density estimates ranged from 0.02 to 0.1 fish/m^2, and in the Mulberry River drainage in Madison and Johnson counties, density estimates ranged from 0.01 to 0.03 fish/m^2 (Rambo 1998). It seems somewhat tolerant of turbidity but not siltation. It is most abundant in clear waters and is absent from the extensively disturbed, highly agricultural lowlands of eastern Arkansas north of the Arkansas River; however, it does not occur in the relatively undisturbed lowland streams of the White River National Wildlife Refuge or other lowland refuges and wildlife management areas of the state.

Cross and Collins (1995) and Miller and Robison (2004) reported that it feeds on insects and other microinvertebrates. Food habits in Vache Grasse Creek in Sebastian County were studied by Brandy Ree and TMB (unpublished data). Stomachs removed from 64 specimens collected in October 2006 and January through April 2007 were examined. Chironomid larvae dominated the diet in both percent volume and percent frequency during every month except March, when isopods were the dominant food items. Other sporadically important foods included ephemeropterans, plecopterans, and amphipods. Rarely ingested foods included odonates, small snails, and coleopterans.

Spawning occurs in early spring in Oklahoma (Miller and Robison 2004) and in April in Kansas (Cross 1967). We have collected breeding adults from February through April over shallow gravel riffles in Vache Grasse Creek (Sebastian County) and in late April in Johnson County, Arkansas. In Vache Grasse Creek, mean GSI and mean number of mature eggs (68) per female peaked in February 2007 (Jade Ryles and TMB, unpublished data). Reproductive biology was studied in Cypress Creek, in Conway County, Arkansas (Stearman et al. 2015), and the following information is from that source. The potential reproductive season, based on presence of mature gonads, was from mid-January to mid-May, with ripe females occurring in March and April. Mean water temperature at time of peak spawning ranged from 12°C to 16°C. Spawning was observed only in runs from 15 to 30 cm deep over medium- to cobble-sized substrate, not in gravel riffles, as reported by some other workers. Males occupied territories averaging 0.5 m² in size with shifting boundaries. Stearman et al. (2015) provided the following description of spawning behavior in the stream:

Females entered runs from upstream and downstream riffle habitat. Upon entering a territory, the female immediately began exploring pebble and cobble-sized substrate. Males became immediately attentive and followed females closely, using their bodies, fin nips and fin displays to attempt to coerce the females against stones. Females selected suitable substrate by searching until finding a rock with a crevice narrower than their total height. Stones used for spawning ranged from 5 to 20 cm in diameter, though most were around 10 cm, and all provided protection from flow. Following substrate selection, the females turned sideways and edged into the crevice. Males followed, mounting the female dorsally while turned sideways. In 11 of the events, the female began vibrating and releasing eggs before the male began releasing sperm, while in the remaining nine events gamete release was synchronous. Upon completion of spawning, the female rapidly exited the crevice and began exploring for a new piece of substrate. Males exited the crevice slower than females. Pursuit followed, and we observed three repeat spawning events.

Redfin Darters were not observed spawning in gravel substrates by Stearman et al. (2015), and clutch size ranged

from 72 to 346 mature ova for females from 38.4 to 69.0 mm SL. It is one of the easiest darters to keep in an aquarium, where it thrives on live or frozen food; however, it will also accept commercially prepared dried foods.

CONSERVATION STATUS   Because the Redfin Darter is widespread, common, found in a variety of habitats, and somewhat adaptable to different environments, we consider it secure in Arkansas.

Etheostoma zonale (Cope)
Banded Darter

Figure 26.80a. *Etheostoma zonale*, Banded Darter breeding male. *Uland Thomas.*

Figure 26.80b. *Etheostoma zonale*, Banded Darter female. *Uland Thomas.*

CHARACTERS   A moderately slender darter with a small head and a decurved, moderately blunt snout (Fig. 26.80). Mouth small and subterminal. Premaxillary frenum present. Back with 6 or 7 dark saddles. Side with 9–13 dark blotches or vertical bars, which often extend the full depth of the body and sometimes encircle the venter on the caudal peduncle. Gill membranes broadly connected across throat. Supratemporal and infraorbital canals uninterrupted. Cheek, opercle, nape, and belly well scaled; breast naked to fully scaled. Lateral line complete, usually with 44–58 scales. Scale rows above lateral line usually 4–6 (3–8); scale rows below lateral line usually 7–9 (6–11). Diagonal scale rows 11–19; scale rows around caudal peduncle 15–23. Dorsal spines 10–12; dorsal rays 10–12; anal spines 2; anal rays 6–9; pectoral rays usually 13–15. Breeding tubercles absent in both sexes, but tips of pelvic,

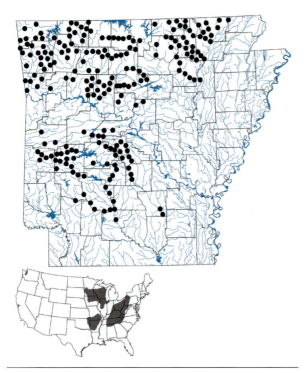

Map 219. Distribution of *Etheostoma zonale*, Banded Darter, 1988–2019.

lower pectoral, and anterior anal fin elements become swollen in breeding males (Etnier and Starnes 1993). Maximum length around 3 inches (78 mm).

LIFE COLORS   Back and upper sides a yellow green with dark green-brown dorsal saddles and dark green lateral blotches and vertical bars. Underside yellowish white. Dark pre- and suborbital bars present. Base of caudal fin with a vertical row of 3, sometimes 4, small, dark spots. First dorsal fin of male is dusky red medially and green marginally, that of the female usually with a thin red margin; other fins dusky or green in males and lightly spotted with brown in the females. Breeding males brilliantly colored with bright green vertical bars encircling the body and with dark green fins. Breeding females less colorful but with vertical bars on body which become a brighter green. Young individuals are poorly marked and easily confused in the field with a number of other species (Kuehne and Barbour 1983).

SIMILAR SPECIES   *Etheostoma zonale* closely resembles *E. histrio* but differs in lacking the 2 large dark basicaudal blotches and large dark spots on breast and underside of head, in having fully scaled cheeks, opercles, and belly (vs. naked cheeks, opercles, and front half of belly), usually 11 or 12 dorsal spines (vs. 9 or 10), and shorter pectoral fins, their tips extending backward only to about

tips of pelvic fins (vs. tips of pectoral fins extending backward past tips of pelvic fins). It differs from *E. blennioides* in lacking a nipplelike extension at middle of upper lip, lacking V- or W-shaped blotches on lower sides, and in having fewer than 63 lateral line scales (vs. more than 63).

VARIATION AND TAXONOMY   *Etheostoma zonale* was described by Cope (1868b) as *Poecilichthys zonalis* from the Holston River, Virginia. It is in the subgenus *Etheostoma*. Tsai and Raney (1974) examined geographic variation and recognized two subspecies (one of which is now recognized as a full species, *E. lynceum*). Banded Darters in Arkansas were recognized as *E. z. zonale*, and Tsai and Raney (1974) reported that two races occur in this state. A Black River race has a fully scaled breast and is found in the Black, Current, and Eleven Point rivers in northeastern Arkansas. Most of the rest of the range of *E. zonale* in Arkansas is occupied by the Arkansas River race, which has a naked or only partially scaled breast. Intergrades between these two races were found in the Strawberry and Spring rivers. More recently, in a rangewide analysis using mitochondrial and nuclear genes, Halas and Simons (2014) concluded that all *E. zonale* in Arkansas belong to a single clade that contains three genetically identifiable groups of populations or subclades: (1) Ouachita River and upper Arkansas River group, (2) White River group, and (3) Little Red River group. Molecular evidence supports a sister relationship between *E. zonale* and *E. lynceum* (Porterfield et al. 1999).

DISTRIBUTION   As currently recognized, *Etheostoma zonale* occurs in three disjunct areas: (1) Ohio River basin from southwestern New York to eastern Indiana and south to northern Georgia and Alabama; (2) Lake Michigan and upper Mississippi River basins from Michigan to Minnesota and south to northwestern Indiana and central Illinois; and (3) the Ozark-Ouachita highlands of southern Missouri, southeastern Kansas, Arkansas, and eastern Oklahoma. It has been introduced into some Atlantic Slope drainages (Page and Burr 2011). In Arkansas, it is an upland species occurring above the Fall Line in all major drainages of the Ozark Mountains (White and Arkansas river drainages). It also occurs in some Arkansas River tributaries draining the northern Ouachita Mountains (Poteau, Petit Jean, and Fourche LaFave rivers), and in the Saline and Ouachita river drainages in the Ouachita Mountains (Map 219). It is absent from the Little River drainage of the Ouachitas.

HABITAT AND BIOLOGY   The Banded Darter is found in upland streams of various sizes and in medium-sized rivers but is most abundant in moderate-sized streams of moderate to high gradient and clear water, where it lives in gravel and rubble riffles and in pools having some current. It often prefers the swifter, deeper riffles, and the largest populations are found where there is an abundance of filamentous algae (Lachner et al. 1950; Miller and Robison 2004). The young are often found in quiet water around aquatic plants or in accumulations of leaves, sticks, and other organic debris (Pflieger 1997). It is almost always found with *E. blennioides* in Arkansas. It is one of the most intolerant of reservoir conditions among the darters. In a survey of 66 Arkansas impoundments, Buchanan (2005) found no Banded Darters. It was found in moderate numbers in a preimpoundment fish survey of the Beaver Lake watershed (Keith 1964), but was not found in the Beaver Lake tailwaters shortly after impoundment or 30 years later (Quinn and Kwak 2003).

The diet of *E. zonale* in Ohio consisted mainly of immature aquatic insects (chironomids, black flies, and mayflies were the dominant food items) taken from the stream bottom (Adamson and Wissing 1977). In the Iroquois River, Illinois, feeding began after sunrise, peaked at midday, and decreased after sunset in this diurnal carnivore (Cordes and Page 1980). In that study, *E. zonale* fed most heavily on chironomids at all hours. Stomach contents of 84 specimens from three Arkansas localities were examined by TMB: 24 specimens, 37–58 mm TL, from the Spring River in Sharp County, collected on 17 May; 37 specimens, 34–60 mm TL, from the Caddo River in Clark County, collected on 31 May; and 23 specimens, 38–55 mm TL, from the Illinois River in Benton County, collected on 11 June. At all three sites, the diet was dominated by chironomids, trichopterans, ephemeropterans, and plecopterans by percent frequency and percent volume. Spring River specimens had the most diverse diet, with megalopterans, coleopterans, odonates, and simuliids also consumed in small amounts. Specimens from the Illinois River had the least diverse diets. We have noted that the Spring River is a particularly productive environment, with an abundance of benthic macroinvertebrate prey throughout the year. At all three sites there was a conspicuous lack of small crustacean and other noninsect prey in the diets.

The Banded Darter spawns in Missouri and Oklahoma from mid-April into May (Pflieger 1997; Miller and Robison 1973), but in Arkansas, spawning occurs from late March to mid-June (Hubbs 1985; Walters 1994). TMB observed spawning in Lee Creek in Crawford County on 10 April 1989, and ripe females containing 56–80 mature ova were collected from Point Remove Creek in Conway County on 5 May 2015. Spawning takes place in the riffles, and the eggs are attached to strands of filamentous algae and aquatic mosses. Pflieger noted that it spawns in slower current, over sparser vegetation, and slightly later than the Greenside Darter. In the Buffalo River, Arkansas, spawning occurred in runs approximately 0.6–1.6 m deep at water temperatures ranging from 11°C to 21°C. (Walters 1994). Males aggressively defended territories in the spawning area. Courtship and spawning behavior were described by Walters (1994) as follows:

> Most pairs courted and spawned in a 1 m² area. When a receptive female swam near a male, the male followed and repeatedly bobbed his head toward her, at times nipping or nudging her with his snout, but not always making contact. This motion was usually directed at the nape of the female. The number of head bobs/min ranged from 3–25. When the female moved about slowly or was stationary, the male generally had one pectoral fin across her back or in contact with her side. As female *E. zonale* moved about, they probed the substrate (patches of *Rhizoclonium* sp.) with their snout. Often only the caudal fin of the female remained visible, with the rest of her body buried in algae. Immediately before oviposition, probing intensified with more force behind each probe. At this time, the female began to undulate her caudal area from side to side, which may have been a cue to the male that she was ready to oviposit. Immediately after the female *E. zonale* began undulating, the male quickly mounted her with his head just posterior to hers; and they both quivered in unison while moving 2–3 cm forward, generally in an upstream direction. During this time, the female was pressed tightly against and burrowed into the algae. Time from the point at which the male mounted the female until they stopped quivering was no more than 1 second.

Walters (1994) observed up to 18 spawning bouts by a single pair, and females occasionally spawned with more than one male in succession; simultaneous spawning by two males with one female and sneaking behavior were not observed. Walters (1994) estimated that females deposited 1 or 2 eggs per spawning bout. Clutch and egg size variation were studied in three separated Arkansas populations (Guill and Heins 1996). Significant differences in standard length-adjusted mean clutch size and mean ovum mass were found. Clutch sizes ranged from 9 to 92 ova in Caddo River females, 4 to 75 ova in females from Martin Creek (Sharp County), and 11 to 63 ova in Strawberry River females. Guill and Heins (2000) also found significant variation among years in the reproductive life history traits of *E. zonale* from the Duck River drainage of Tennessee. Maximum life span is 3–4 years.

CONSERVATION STATUS This widespread and common upland darter does not usually occur in large numbers at a given locality. We consider its populations in Arkansas secure.

Genus *Nothonotus* Putnam, 1863— Lined Darters

Nothonotus was first used as a genus name by Putnam in 1863. Jordan (1877b) also used it as a genus, but it was subsequently treated as a subgenus of *Etheostoma* by Jordan and Evermann (1896), where it remained for more than 100 years (sometimes in genus *Poecilichthys*). Subgenus *Nothonotus* was morphologically diagnosed by Bailey (1959), Zorach (1972), and Page (1981). Monophyly of *Nothonotus* was supported in a combined allozymic, behavioral, and morphological analysis (Wood 1996). Some recent authors supported the elevation of *Nothonotus* to generic status (Near and Keck 2005; Bossu and Near 2009; Near et al. 2011; Keck and Near 2013), but others continued to treat *Nothonotus* as a subgenus of *Etheostoma* (Lang and Mayden 2007; Bruner 2011; T. A. Smith et al. 2011; Page et al. 2013). Near and Keck (2005) justified the elevation of *Nothonotus* to a genus on phylogenetic analyses of mtDNA sequences and lack of morphological synapomorphies to diagnose *Etheostoma*. Gene trees inferred from mtDNA sequences suggest that the genus *Etheostoma* (if *Nothonotus* is included) is not monophyletic (Bossu and Near 2009). In a combined mtDNA and nuclear gene analysis, *Nothonotus* was significantly supported as a clade equivalent in rank to *Ammocrypta*, *Crystallaria*, *Etheostoma*, and *Percina* (Near et al. 2011). All previous attempts to designate taxonomic ranks for darters (as a group) were based on morphological diagnoses that did not assess monophyly of the proposed taxa. The study by Near et al. (2011) was the first attempt to provide a comprehensive taxonomy for darters based on the results of explicit phylogenetic analysis of comparative data for a near-complete taxon sampling (245 of the known 248 darter species were sampled). Based on the above-cited genetic analyses, we support the elevation of *Nothonotus* to generic status.

No non-DNA synapomorphies have yet been discovered for *Nothonotus*, but the genus can be characterized by a suite of morphological features. Morphologically, the 20 described species of *Nothonotus* are characterized by a slab-sided body shape, deep caudal peduncle, the presence of thin alternating dark and light longitudinal lines on the side of the body in almost all species (therefore we use "lined darters" for the genus common name), lateral line complete or nearly so, no breeding tubercles, uninterrupted

head canals, frenum well developed, and darkened anterior interradial membranes of the first dorsal fin. Females of *Nothonotus* have short, mound, and mound-tube genital papillae and typically bury their entire body to deposit eggs (Martin and Page 2015).

Nothonotus juliae (Meek)
Yoke Darter

Figure 26.81. *Nothonotus juliae*, Yoke Darter. *Uland Thomas.*

Map 220. Distribution of *Nothonotus juliae*, Yoke Darter, 1988–2019.

CHARACTERS A stout darter with a wide, dark saddle (yoke) anterior to the first dorsal fin that extends down the side to the pectoral fin base; this yoke is followed by 3 less-distinct saddles (Fig. 26.81). Side with alternating light and dark horizontal lines and with 7–10 dark vertical bars (more developed in males). Gill membranes broadly connected across throat. Premaxillary frenum present. Caudal peduncle deep. Snout moderately pointed. Infraorbital canal uninterrupted, usually with 8 pores. Palatine teeth present. Nape and belly scaled; opercle partly scaled; breast and cheek naked. Lateral line complete or nearly so, extending behind second dorsal fin. Scales in lateral series 50–65; 0–2 unpored scales. Scale rows above lateral line 7–9; scale rows below lateral line 10 or 11. Diagonal scale rows 19–21; scale rows around caudal peduncle 26–30. Dorsal spines 11 or 12; dorsal rays 11 or 12; anal spines 2; anal rays 7 or 8. Breeding males and females without tubercles. Maximum total length around 3.3 inches (85 mm) (Geihsler 1975).

LIFE COLORS   Back and sides mottled orange brown with four dark brown saddles and 9 or 10 lateral bars or blotches; the first saddle is wide and dark, and the other three saddles are smaller and less prominent. Sides are thickly spotted with yellow; a large blue-green humeral spot is present. Undersides are yellowish white to light green. A suborbital bar is dark and distinct. The underside of the head is speckled with black. First dorsal fin is dusky on lower half with an orange marginal band; other fins are yellow or orange. No dark bands in caudal, soft dorsal, and anal fins. Breeding males develop black pigment on the head, throat, and breast (James and Taber 1986; HWR, pers. obs.); spinous dorsal fin becomes black basally and bright orange distally; other fins become bright orange. Females less colorful than males but more colorful than females of most darter species.

SIMILAR SPECIES   Most like the allopatric Yellowcheek Darter, Nothonotus moorei, the Yoke Darter differs in having broadly joined gill membranes (vs. separate or narrowly joined), fully scaled nape (vs. naked), and a dark, wide saddle in front of first dorsal fin (vs. dorsal saddles absent or if present, weakly developed).

VARIATION AND TAXONOMY   Meek (1891) described the Yoke Darter as Etheostoma juliae from the James River (White River drainage) near Springfield, Greene County, Missouri. It has sometimes been placed in the genus Poecilichthys (Black 1940) but was retained in genus Etheostoma, subgenus Nothonotus through most of the 20th century. We agree with several recent authors that Nothonotus should be recognized as a genus (see genus Nothonotus account). Etnier and Williams (1989) indicated that Nothonotus is monophyletic and that N. juliae is sister to all other Nothonotus. This relationship was supported by subsequent molecular studies (Near and Keck 2005; Near et al. 2011). Hill (1968) found significant differences in mean vertebral numbers among populations of Yoke Darters in northwestern Arkansas. Most downstream population counts were significantly lower than those of other populations, and differences were found within populations.

DISTRIBUTION   The Yoke Darter is endemic to the White River system of northern Arkansas and southern Missouri. In northwestern Arkansas it occurs in the upper White River and its major tributaries, War Eagle Creek and the Kings River. In northcentral Arkansas it is found throughout the Buffalo River and in the White River near the mouth of the Buffalo River (Map 220). It is not found in the White River drainage downstream of Stone County.

HABITAT AND BIOLOGY   The Yoke Darter inhabits only large, high-gradient creeks and small rivers of the Ozarks having clear water and permanent flow. It is most abundant in the swiftest, deepest riffles available, over a gravel and rubble bottom strewn with boulders. It is sometimes taken in shallow riffles with strong current, but it is usually found in the most turbulent water at depths of 1–20 inches (254–508 mm). It requires strong current, permanent flow, and silt-free substrates. It avoids pools. In the Buffalo River, it was found mainly at the heads of riffles, whereas E. caeruleum primarily occupied the riffle tails (Geihsler 1975). The Yoke Darter is most abundant in Arkansas in the Buffalo River, where it is still locally common, but much of its suitable habitat elsewhere was eliminated by the impoundment of Beaver, Bull Shoals, and Norfork reservoirs. It is intolerant of reservoir conditions. At least 15 upper White River localities reported for this species by Keith (1964) were inundated by Beaver Lake, and the cold tailwaters from large reservoirs have had a negative impact on this darter. Keith's preimpoundment survey found that N. juliae was the most abundant species (8,104 specimens composing 34% of the fish community) in the area of what is now the Beaver Lake tailwater. Postimpoundment surveys in 1965–1966 (Brown et al. 1967) and in 1968 (Bacon et al. 1968b) found only four Yoke Darters, and 30 years after impoundment, the tailwater fish assemblage was composed almost entirely of coldwater species with no Yoke Darters present (Quinn and Kwak 2003). Kuehne and Barbour (1983) reported large-scale pollution from animal wastes in the Kings River and in War Eagle Creek. Both areas were badly degraded by chicken wastes and possibly by chemicals used in the poultry industry. In 1984, we found moderate-sized populations of Yoke Darters in War Eagle Creek and small populations in the main White River above Beaver Reservoir. William Keith (pers. comm.) reported finding large populations in War Eagle and Kings River in the summer of 1985. Collections by TMB in 1992 and 2002 also produced records for N. juliae from the White River above Beaver Reservoir. Our fish sampling in War Eagle and Kings River between 2005 and 2010 revealed that small populations persisted in both systems. The most common species associates of the Yoke Darter in the Buffalo River riffles during all seasons were *Etheostoma caeruleum* and *Luxilus pilsbryi* (Gehsler 1975). Other species that were seasonally common in the same riffles were *Campostoma* spp., *Noturus albater*, *N. exilis*, *Uranidea carolinae*, and *Etheostoma zonale*.

Most information on feeding biology comes from Geihsler (1975). Yoke Darters fed almost exclusively on larvae, nymphs, and pupae of aquatic insects. Other food items ingested in very small amounts included water mites and fish eggs. Insect taxa represented in the diet were as follows: Plecoptera (Perlodidae, Nemouridae, and Perlidae), Ephemeroptera (Baetidae, Heptageniidae, and Siphlonuridae), Trichoptera (Hydropsychidae, Hydroptilidae, and Psychomyiidae), and Diptera (Chironomidae and Simuliidae). The most common and abundant immature insects in the diet were ephemeropteran and plecopteran nymphs and larvae of chironomids, simuliids, and trichopterans. There was some seasonal variation in the ingestion of insect taxa. Plecopteran nymphs were ingested mainly from October through May, trichopterans from May through November with a peak in July, and ephemeropteran nymphs from April into December. Male and female Yoke Darters basically ate the same foods, and the amounts eaten varied slightly. Feeding occurred during all daylight hours. James (1983) also found that Yoke Darters in the James River, Missouri, fed only on immature aquatic insects. In that study, chironomid larvae were the major food items ingested during all seasons, and trichopteran larvae were eaten throughout most of the year. Simuliids were consumed mainly in spring and summer, and plecopterans and ephemeropterans were most abundant in the diet in the fall.

Hill (1968) reported that breeding occurred in Arkansas between April and early July, and Hubbs (1985) provided data supporting this rather late breeding season. James and Taber (1986) reported the breeding season in Missouri extended from mid-May through mid-July at water temperatures between 18°C and 22°C. We have found this species breeding in the swift gravel riffles of the White River near Fayetteville in mid-May. Geihsler (1975) reported GSI values in the Buffalo River were lowest in August and gradually increased to a peak in April. During April and May, GSI values were highest followed by a sharp decrease in June, which Geihsler assumed to mark the end of the spawning period. Spawning occurred in swift, deep (11.8–23.6 inches [30–60 cm]) riffles. Five to ten males often follow a gravid female who seeks out small patches of fine gravel just downstream from large rocks. The female burrows headfirst into the gravel with violent

thrashing movements, leaving only her head and pectoral fins exposed. During this activity, the males make rapid darting movements around the female. One male, usually the largest, finally succeeds in positioning himself beside or on top of the female. This dominant male then drives the other males away and begins additional courtship of the female by darting around her, nudging her with his snout, and perching beside or on top of her. This courtship may last as long as thirty minutes until both fish vibrate to release eggs and milt. The adhesive eggs are buried about 0.8 inches (2 cm) in the substrate and left unguarded. Hill (1968) suggested *N. juliae* had a fall as well as a spring spawning period to explain intrayear class variation in vertebral numbers. Data from Geihsler (1975) and James and Taber (1986) do not support this, as those workers did not find a bimodality with peaks in spring and fall. James and Taber (1986) concluded that females produce multiple clutches with decreasing clutch size during the spawning season. Ova averaged 1.75 mm, were yellow to orange in color, and had a clearly defined oil droplet. A single clutch typically contained 70–80 eggs. In the Buffalo River, fecundity estimates ranged from 85 ova in a 48 mm female to 381 ova in a 78 mm female. Males and females spawn at 1 year of age provided they have reached lengths of 1.2 and 1.3 inches (30 mm and 32 mm), respectively. Some faster-growing females may mature and be able to spawn during their second year as several 2-year-old females showed class II or mature ova (Geihsler 1975). Maximum life span is approximately 39 months, but Geihsler (1975) reported slow growth rates and longer life spans in the Buffalo River population.

CONSERVATION STATUS   Even though some populations have been extirpated by reservoir construction, it is still widely found throughout most of its formerly known range in Arkansas; however, populations of this species appear to have declined in number and density (based on numbers reported from collection sites) during the past two decades. Its populations should be closely monitored, and we consider it a species of special concern.

Nothonotus moorei (Raney and Suttkus)
Yellowcheek Darter

CHARACTERS   A small darter with a moderately sharp snout, a compressed, deep body, and a deep caudal peduncle (Fig. 26.82). Back sometimes with 8 or 9 weakly developed saddles, but saddles are often absent. Side with about 13 dusky, often very faint, narrow vertical bars, last 1 or 2 bars often circling the caudal peduncle. Posterior half of

Figure 26.82a. *Nothonotus moorei*, Yellowcheek Darter male. *Uland Thomas.*

Figure 26.82b. *Nothonotus moorei*, Yellowcheek Darter female. *Uland Thomas.*

Map 221. Distribution of *Nothonotus moorei*, Yellowcheek Darter, 1988–2019.

body with thin, dark horizontal lines between scale rows. Gill membranes separate or only narrowly joined across throat. Premaxillary frenum present. Infraorbital canal uninterrupted, usually with 8 pores. Palatine teeth present. Opercle scaled; belly naked anteriorly, scaled posteriorly;

cheek with a few scales behind eye; nape and breast naked. Lateral line complete, with 51–60 scales. Scale rows above lateral line usually 6 (4–8); scale rows below lateral line usually 8 or 9 (7–11). Diagonal scale rows 15–19; scale rows around caudal peduncle 19–23. Dorsal spines 10–12; dorsal rays 9–12; anal spines 2; anal rays 6–8; pectoral rays usually 13 (12–14). Breeding males and females without tubercles. Maximum length approximately 2.5 inches (64 mm).

LIFE COLORS   Back and sides grayish brown, often with darker brown saddles and lateral bars. Dark brown or black horizontal stripes on posterior side. Body of female with reddish-orange to light yellow spots; males sometimes with a few light spots. Undersides lighter brown. Oblique dark suborbital bar present; preorbital bar dark. Median fins with black margins. Breeding males brightly colored with a brilliant turquoise-blue breast and throat; lower sides and belly green to blue green. Median fins of male with a thin, yellow submarginal band, a broad red-orange medial band, and green basally. Anal spines become an intense blue. Paired fins reddish orange. Breeding female not as brightly colored, but developing orange or red-orange spots on sides; second dorsal, caudal, and anal fins heavily spotted with black; paired fins tinged with yellowish orange.

SIMILAR SPECIES   Similar in appearance to *Nothonotus juliae* but differs in having separate to narrowly joined gill membranes (vs. gill membranes broadly joined), naked nape (vs. fully scaled), and lacking a wide, dark saddle anterior to the first dorsal fin that extends down each side to pectoral fin bases (vs. yokelike dark dorsal saddle present). It differs from *E. flabellare* in having separate or narrowly joined gill membranes (vs. broadly joined), a complete lateral line (vs. lateral line incomplete, ending beneath second dorsal fin), and usually more than 10 dorsal spines (vs. fewer than 10).

VARIATION AND TAXONOMY   *Nothonotus moorei* was described by Raney and Suttkus (1964) as *Etheostoma moorei* from the Devil's Fork of the Little Red River, 4 km SW of Woodrow and 9.7 km W of Drasco, Cleburne County, Arkansas. The type locality was subsequently inundated by impoundment of Greers Ferry Reservoir. One of only two species of the genus *Nothonotus* known to occur west of the Mississippi River (the other is *N. juliae*), *N. moorei* was considered a member of the *N. maculatus* species group (Williams and Etnier 1978; Etnier and Williams 1989). Etnier and Williams (1989) also reported that the *N. maculatus* species group is monophyletic as are its branches of *N. moorei* and *N. rubrus*; furthermore, *N. moorei* and *N. rubrus* are sister to the remaining members of the *maculatus* species group. More recent genetic analyses support *N. moorei* and *N. rubrus* as sister species (Keck and Near 2009; Near et al. 2011), but the relationship of the *N. moorei-N. rubrus* clade to other clades varies depending on whether the analyses are based on mitochondrial or nuclear genes (Keck and Near 2009).

Mitchell et al. (2002) demonstrated that Turkey Fork populations of *N. moorei* were markedly divergent genetically from both Middle Fork and South Fork populations. They found no statistically significant meristic differences among those populations. Genetic variation was also studied by Johnson et al. (2006b) and Johnson (2009), who found surprisingly high heterozygosity and polymorphism values, especially for a species with such restricted range and small population size. Of concern was the high degree of Hardy-Weinberg disequilibrium of polyallelic loci in some subpopulations. Those authors speculated that because this species commonly experiences extirpation and recolonization events in the headwater streams of the Little Red River (due to periodic droughts), the founder effect may have contributed to the genetic disequilibrium.

DISTRIBUTION   The Yellowcheek Darter is endemic to Arkansas and found only in the upper Little Red River drainage above Greers Ferry Lake in Cleburne, Searcy, Stone, and Van Buren counties (Map 221).

HABITAT AND BIOLOGY   *Nothonotus moorei* inhabits the high-gradient headwater forks of the Little Red River, having clear water, usually permanent flow, and moderate to strong current. It inhabits swift to moderate riffles over a gravel, rubble, and boulder substrate. Prime habitat is characterized by high levels of dissolved oxygen (7.2–13.0 ppm) and pH ranging from 6.8 to 7.8 (McDaniel 1984). Adults are commonly taken at depths of 10–20 inches (254–508 mm); small individuals are found in shallow riffles. Raney and Suttkus (1964) found the water plant *Podostemum* growing in the riffles at some localities. Weston et al. (2010) found evidence for niche partitioning between the Yellowcheek and Rainbow darters in the Middle Fork of the Little Red River. Yellowcheek Darters were more associated with cobble and gravel substrates than the Rainbow Darter. Where the two species occurred together, Yellowcheek Darters were found farther upstream in the riffles in or near crevices between gravel and cobble, whereas Rainbow Darters were found farther downstream in the riffles moving along the substrate surface. More importantly, Yellowcheek Darters did not retreat to pools during times of riffle drying like the Rainbow Darters and did not show the ability of the Rainbow Darter to recolonize previously dry stream segments. Yellowcheek Darters were not found in the hyporheic zone during riffle drying

(Weston et al. 2010). Wine et al. (2008) compared density dynamics of *N. moorei* in drought and nondrought periods. They found that upstream sites dried periodically during the drought of 1999–2001, and that Yellowcheek Darters occupying those sites were extirpated; at downstream sites, densities were significantly lower than historical levels. During normal precipitation in 2003–2004, densities increased significantly, yet several upstream sites and one complete stream remained extirpated. Weston (2006) provided density and population size estimates. A 2011 survey found that *N. moorei* was present at 7 of 12 sites sampled in the Middle and South forks of the Little Red River (Magoulick and Lynch 2015). Detection probability and population density were positively related to current velocity. Densities were highly variable within and among streams, and the highest single riffle density of 2.1 individuals/m^2 occurred at a site on the Middle Fork. The most common species associates of *N. moorei* were *Etheostoma blennioides*, *Noturus albater*, and *Etheostoma caeruleum* (McDaniel 1984).

Much of the habitat of *N. moorei*, including the type locality, was destroyed by the impoundment of Greers Ferry Lake in the early 1960s. Seasonal drought and stream drying are common features of streams of the Boston Mountains Plateau (Magoulick 2000), and have undoubtedly played a role in the decline of *N. moorei* populations. Populations still persist in tributaries above the lake, and it is common at a few scattered localities. It is intolerant of reservoir conditions, and its preferred habitat is now more fragmented, and suitable riffles are usually widely separated. A drastic decline in numbers has been documented over the past three decades. Robison and Harp (1981) provided a total estimate of 60,000 individuals for this species. Twenty years later (Wine et al. 2001), the total population was estimated at 10,300 individuals, an 83% decline. Protection of some of its remaining habitat will prove critical to its future survival.

All the following information on the feeding biology of *N. moorei* comes from McDaniel (1984). The Yellowcheek Darter was a very selective feeder, utilizing only a small portion of the benthic fauna available as a food source. Of 182 invertebrate taxa identified in the riffles from benthic samples, 37 taxa were found in *N. moorei* stomachs. Foods consisted almost entirely of immature aquatic insects. Dipteran larvae (mainly simuliids and chironomids) were the most important food source throughout the year. Trichopteran larvae were the next most commonly consumed foods, followed in decreasing order of importance by ephemeropterans, plecopterans, coleopterans, megalopterans, and arachnids. Although numerically low in the diet, stoneflies and mayflies by virtue of their large size were very important by volume in the diet. Seasonally, trichopterans made up the bulk of the diet in summer months and were low in importance in the winter. Although consumed year-round, chironomids were eaten less frequently in summer than in winter. Age 0 *N. moorei* fed almost exclusively on dipteran larvae, and the frequency at which larger benthic organisms appeared in the diet increased with age. No differences in feeding biology were detected between the sexes.

The Yellowcheek Darter may have a late, extended spawning period similar to that documented for *N. juliae*. McDaniel (1984) found that maximum GSI values for males and females occurred in May and June (but no gonadal information was obtained for March and April). He first found mature ova in females in May, and ovum diameters did not diminish through August. Peak spawning likely occurs from mid-May through June, and multiple spawnings may occur. McDaniel (1984) speculated that spawning occurred in the swifter portions of riffles apparently around or under the largest substrate particles available, but based on observations of aquarium spawning by W. D. Voiers, it is probable that *N. moorei* seeks out a fine gravel substrate just downstream of a boulder as a microhabitat for spawning similar to *N. juliae*. William D. Voiers (pers. comm.) confirmed that *N. moorei* is an egg burier and described aquarium spawning behavior, egg deposition, and agonistic behavior as follows:

> Both sexes have large display repertoires and are intensely agonistic. Following intense activity to clear a spawning area of conspecific competition, the female selects the spawning site (with minimal courtship by the male) and buries in the substrate in a near vertical to horizontal position. She is then joined by the male who vibrates simultaneously or successively with her while eggs are being deposited. The female may emerge and rebury several times, sometimes at more than one site. Several females may bury at more or less the same time and within the same general area. Sneak spawning by males is common but is deterred by dominant males who jerk the sneakers out of the substrate by their caudal fins. During courtship both sexes engage in lateral, frontal, tail-lashing, and appeasement displays.

McDaniel (1984) estimated that total fecundity over a spawning season was between 78 and 455 ova per female, depending on age. Simon et al. (1987) noted egg diameters ranged from 1.3 to 2.1 mm and hatchling length ranged from 3.8 to 7.2 mm in seven *Nothonotus* species studied. Individuals were removed from the Little Red River by Conservation Fisheries, Inc. (CFI) and transported to Knoxville, Tennessee, where an attempt was made to develop spawning and rearing techniques that would eventually allow reintroduction and restoration projects in Arkansas; however, Yellowcheek Darters proved to be one of the most difficult species to propagate of all the darters they have attempted to spawn. CFI (2007) reported eggs were about 1.8 mm in diameter and moderately adhesive, with small grains of sand and debris weakly attached. Egg clusters of 20–40 eggs were found. Older eggs were completely colorless and translucent; recently spawned eggs were pale amber with slight opacity (CFI 2007). Apparently, all members of *Nothonotus* have tiny, pelagic larvae, which present special culture difficulties (Rakes et al. 1999). CFI (2007) reported pelagic larvae tended to hold position in the current and feed on suspended fine live and prepared zooplankton-sized foods. Males were larger than females in all age groups (McDaniel 1984). Males and females reach sexual maturity at 1 year of age and maximum life span is around 4 years.

CONSERVATION STATUS   The already restricted range of this Arkansas endemic decreased even more during the past two decades. It is very vulnerable to disturbance (USFWS 2011, 2012). The Yellowcheek Darter was added to the Federal Endangered Species list on 8 September 2011. An attempt to develop a captive breeding program to potentially support reintroduction and restoration projects has produced mixed results (CFI 2007). Yellowcheek Darters were successfully raised to the juvenile stage at the University of Arkansas at Pine Bluff in quantities sufficient for restocking sites where that species had been extirpated (S. Lochmann, pers. comm.). Even if the Yellowcheek Darter can be propagated in large numbers, there will be little habitat left for reintroduction if present trends continue. The mixed success generally associated with reintroducing captive-bred animals is another concern. Resources should be directed toward protecting the remaining suitable habitat to allow natural reproduction to maintain populations in the future. Future monitoring of populations is imperative for continued survival of this endangered species.

Genus *Perca* Linnaeus, 1758—Perches

This genus was established by Linnaeus in 1758. There are three species, *P. fluviatilis* and *P. schrenki* of Eurasia, and *P. flavescens* of North America. Genus characters are given in the following species account.

Perca flavescens (Mitchill)
Yellow Perch

Figure 26.83. *Perca flavescens*, Yellow Perch. *Uland Thomas.*

Map 222. Distribution of *Perca flavescens*, Yellow Perch, 1988–2019. Inset shows native range (solid) and introduced range (outlined, with some introduced populations shown in solid) in the contiguous United States.

CHARACTERS Body fairly deep and compressed (Fig. 26.83). Body depth going into SL 3.5–4.2 times in small young, and 3.0–3.8 times in adults. First and second dorsal fins separated by a distinct gap. Caudal fin forked. Mouth large, upper jaw extending behind middle of eye. Jaws and roof of mouth with small, inconspicuous teeth and without prominent canine teeth. Rear margin of preopercle strongly serrate. Back with 6–9 dark saddles that extend downward onto sides as dark bars, bars most prominent in young and small adults. Gill rakers on first arch heavy, close-set, about 18–20. Gill membranes separate. Branchiostegal rays

usually 7 (rarely 8). Premaxillary frenum absent. Swim bladder well developed. Pyloric caeca 3, short and thick. Sensory canals of head uninterrupted. Opercular spine present. No modified scales on breast or along midline of belly. Nape, cheeks, breast, belly, and upper area of opercle scaled. Lateral line complete, with 51–65 scales. First dorsal fin with 13–15 spines; second dorsal fin with 1 or 2 spines and 12–15 rays; anal spines 2; anal rays 6–8. Breeding tubercles absent. Adult length usually around 10 inches; maximum length around 18 inches (400 mm). World record Yellow Perch is 18 inches (457 mm) and 4 pounds, 3 ounces (1.9 kg). Arkansas state angling record from Bull Shoals Lake in 2010 was 1 pound, 11 ounces (0.77 kg).

LIFE COLORS Dorsum brassy green, greenish olive, or golden yellow. Dorsal saddles and lateral bars dark green to brown. Sides a lighter yellow, venter white. Dorsal and caudal fins dusky; first dorsal fin with a posterior black blotch and sometimes with a smaller black anterior blotch. Pectoral, pelvic, and anal fins yellow orange to white. Breeding adults very brassy and golden; pelvic and anal fins a rich orange yellow; bands usually blackish (Trautman 1957). Very small young are more transparent or silvery.

SIMILAR SPECIES The Yellow Perch can be distinguished from the Walleye and Sauger by its lack of prominent canine teeth in upper and lower jaws (vs. prominent canine teeth present in upper and lower jaws) and by its 6–8 anal fin rays (vs. 11–14 anal rays). It differs from all members of the darter genus *Percina* in having normal scales along the midline of the belly and adults commonly reaching 10 inches or more in length (vs. midline of belly naked or with modified, strongly toothed scales, and usually smaller than 6 inches TL). It further differs from all other Arkansas darters in rear margin of preopercle strongly serrate and its large adult size (vs. rear margin of preopercle smooth or weakly serrate and usually smaller than 5 inches TL). The Yellow Perch superficially resembles *Morone* species, but differs from them in first dorsal fin with 13–15 spines (vs. 8–10 spines), and anal fin with 2 spines (vs. anal fin with 3 spines).

VARIATION AND TAXONOMY *Perca flavescens* was described by Mitchill (1814) as *Morone flavescens* from "New York." It is most closely related to *P. fluviatilis* of Europe and Asia, and some authors considered the two to be conspecific (McPhail and Lindsey 1970; Thorpe 1977). Most recent authors support the specific distinctness of the two species based on morphological (Collette and Bănărescu 1977) and genetic (Marsden et al. 1995b) evidence.

DISTRIBUTION Native to northern North America east of the Continental Divide, the Yellow Perch occurs in the Atlantic, Arctic, Great Lakes, and Mississippi River basins south to Nebraska, Illinois, Ohio, and the Savannah River of South Carolina and Georgia (Page and Burr 2011). Robins et al. (2018) considered the disjunct population in the Apalachicola basin of Florida to be native. It is not possible to precisely define its native range because of widespread introductions throughout the United States. Fuller et al. (1999) reported nonnative records of *P. flavescens* from every state outside its native range except Alaska, Hawaii, and Louisiana. It has become established in most states where it has been introduced, and it was intentionally stocked in Arkansas for food and sport fishing in 1918 (O'Malley 1919). No subsequent reports of its capture in Arkansas occurred until AGFC biologists collected one 72 mm specimen from the Trimble Creek arm of Bull Shoals Lake on 26 August 1999 (UAFS 1611, Buchanan et al. 2000). Previous reports of Yellow Perch from Bull Shoals Lake, Missouri, were confirmed in 1997, and the AGFC subsequently collected additional specimens from Bull Shoals Lake in Arkansas. The source of the Bull Shoals Lake specimens is unknown, but there have been no recent intentional introductions of this species in Arkansas by federal or state agencies. A record of *P. flavescens* was obtained from natural waters in Arkansas on 28 July 2003, when a single specimen (UAFS 1861) 70 mm TL was collected from the Mississippi River 2 miles above Barfield Landing in Mississippi County by Sam Barkley and TMB. At the time of collection, the Mississippi River was unusually high, and the specimen may represent a straggler from an established population farther north (Map 222).

HABITAT AND BIOLOGY   The Yellow Perch is primarily a lake-adapted species (Pflieger 1997), which probably accounts for its widespread introduction throughout the United States. Becker (1983) noted that it is adaptable to a wide variety of habitats. In its native range, it is commonly found in ponds, lakes, and quiet waters of creeks to large rivers, where it is often associated with aquatic vegetation. Trautman (1957) also noted a preference for low-gradient waters where there was an abundance of rooted aquatic vegetation and the substrate consisted of mud, organic debris, sand, or gravel. It is apparently intolerant of continuous turbidity and siltation, and Trautman (1957) reported that its numbers decreased drastically in environments where rooted aquatic vegetation disappeared. It is often found in schools or small groups near vegetation or other cover during the summer months, but in winter it retreats to deeper water (Jenkins and Burkhead 1994).

The Yellow Perch is mainly a diurnal sight feeder. It is adapted to a diet of small, live animals, and it will take whatever is available (Thorpe 1977). Cannibalism is common

and may begin at an early feeding stage. Free-swimming larvae are pelagic, feeding primarily on zooplankton (Etnier and Starnes 1993). First foods of Yellow Perch from a eutrophic lake were copepod nauplii and cyclopoid copepods; after fish reached 11 mm in length, *Bosmina coregoni* was the dominant food (Siefert 1972). Yellow Perch from an oligotrophic lake fed first on the rotifer *Polyarthra*, and to a lesser extent on copepod nauplii; after fish reached 11 mm, cyclopoid copepods became the most important food items. At about 25 mm TL, the juveniles become benthic oriented, feeding primarily on bottom-dwelling invertebrates. Turner (1920) found that juveniles approximately 25 mm TL fed mainly on copepods; between 30 and 40 mm TL, cladocerans became increasingly important and copepods dropped below 50% of the diet composition. Insect larvae and nymphs (chironomids, ephemeropterans, and others) constituted the main foods up to 120 mm TL. Large snails and crayfish entered the diet at about 100 mm TL, but some small snails were ingested by Yellow Perch smaller than 100 mm (Turner 1920). The adults are food generalists, feeding on small crustaceans, insects, and fish depending on food availability (Boschung and Mayden 2004). Some studies have shown considerable variety in the food habits of adults from year to year, even at the same locality. Most anglers in Kansas catch Yellow Perch on earthworms or minnows (Cross and Collins 1995).

Reproductive biology of *P. flavescens* has been well documented. Spawning occurs in winter and early spring from February to March in southern portions of its range in shallow water usually over aquatic vegetation or submerged brush at water temperatures of 6.7–19°C (44–66°F). No nest is constructed, and there is no guarding of eggs or young. Harrington (1947) described courtship and spawning behavior. Males precede females to the spawning areas, and a ripe female is pursued by 15–20 males when she enters the spawning area. The males prod the female with their snouts and the female lays eggs in gelatinous strands sometimes 2 m long. Two or more males simultaneously release sperm. Large females may contain more than 100,000 eggs. The fertilized ropes of eggs are draped over vegetation or debris but sometimes land on gravel or sand substrates, where their chances of successful development are greatly reduced. No additional parental care is provided. Maximum life span is around 8 years.

It is doubtful that the establishment of Yellow Perch populations in lakes and reservoirs in Arkansas is desirable. Although often an important game species in its native range, it is a rather small species, rarely exceeding 12 inches in length and 1 pound in weight. Northern lakes sometimes become overpopulated with stunted individuals.

Trautman (1957) reported that in Ohio lakes overpopulated by Yellow Perch, the species became dwarfed, with the adults breeding when smaller than 4 inches (102 mm) TL. A number of native Arkansas gamefish species are more desirable than Yellow Perch, and the establishment of *P. flavescens* in Arkansas lakes could negatively impact those native species populations. In addition, Yellow Perch were found to compete with trout for food, and they also preyed on young trout (Coots 1966). It is an important food and game fish in the Great Lakes Region and was once an important part of the commercial catch in that area (Hubbs and Lagler 2004). Although the flesh of Yellow Perch is reportedly very good, it does not provide much of a fight when hooked (Etnier and Starnes 1993). Jenkins and Burkhead (1994) also extolled the flavor of its flesh, remarking that "the fillet is as valued as that of its Walleye kin." Even though a population has become established in Bull Shoals Reservoir, it is unlikely that the Yellow Perch could establish large populations in Arkansas lakes because of its intolerance of warm water. In an Alabama reservoir, *P. flavescens* did not survive water temperatures above 32°C (Boschung and Mayden 2004).

CONSERVATION STATUS   The transfer of this introduced species to other Arkansas waters should be discouraged. The Bull Shoals Reservoir introduced population has apparently become established.

Genus *Percina* Haldeman, 1842— River Darters and Logperches

Haldeman originally described *Percina* as a subgenus of *Perca* in 1842. Many species currently in *Percina* were for many years distributed among several genera. Black (1940) recognized only one Arkansas species (*P. caprodes*) of *Percina* and placed other Arkansas species now in that genus in the genera *Cottogaster*, *Hadropterus*, or *Imostoma*. The genus *Percina* was enlarged by Bailey et al. (1954) to include species previously assigned to the nominal genera *Cottogaster*, *Hadropterus*, and others. The genus was subsequently divided into eight subgenera by Bailey and Gosline (1955), and Page (1974) raised the number to nine. Near (2002), using mtDNA sequencing, supported the monophyly of five of the subgenera but proposed that the use of subgenera in *Percina* taxonomy be abandoned in favor of the recognition of monophyletic species clades. Seven of the traditionally recognized subgenera are represented by Arkansas species: *Alvordius*, *Cottogaster*, *Ericosma*, *Imostoma*, *Hadropterus*, *Percina*, and *Swainia*. The subgenus of each Arkansas species of *Percina* is given in the species account. Near and Benard (2004), using mtDNA

analysis, concluded that speciation in the logperch clade (= subgenus *Percina*) was recent and rapid, with most speciation occurring allopatrically in the Pleistocene.

There are 49 described species in *Percina*, 14 of which are found in Arkansas. There is also at least one undescribed species in the state. One species, *Percina brucethompsoni* (the Ouachita Darter), is endemic to Arkansas. The genus *Percina* contains our largest darters. Several species reach 4 inches (102 mm) in length, and the logperches sometimes exceed 7 inches (178 mm); however, some species, such as the Channel Darter and Saddleback Darter, are small and do not exceed 2.5 inches (64 mm). *Percina* is considered the most generalized and least derived darter genus, both morphologically and behaviorally. All species for which data are available spawn on gravel riffles and bury their eggs in the substrate. There is no nest building or parental care of the eggs. It is possible that *Percina macrolepida* is an exception to this ancestral spawning pattern of egg burying (see account for that species). The best feature for distinguishing *Percina* species from other darters is the presence of 1 or more enlarged and modified (strongly toothed) scales between the pelvic fin bases of both sexes; the midline of the belly also has a row of enlarged and modified scales in the males, and the midline is usually naked in the females. The modified midventral scales of *Percina* males are believed to be used to induce tactile stimulation of the female during spawning (Wiley and Collette 1970; Page 1976a). The genus is also characterized by a general lack of bright breeding colors in Arkansas species (except for the colorful *Percina evides*), sensory canals of head uninterrupted, a complete lateral line, 2 anal spines, a large anal fin usually almost as large as the second dorsal fin, and breeding tubercles present in males of some species. Most species of *Percina* retain a small swim bladder and spend more time swimming off the bottom than other darters (Page and Burr 2011).

Percina brucethompsoni Robison, Cashner, and Near
Ouachita Darter

CHARACTERS   A medium-sized species of *Percina*, with a long head and a long snout, the snout length going into head length 3.8 times or fewer, and snout length less than 9% of SL (Fig. 26.84). Head length going into SL fewer than 3.7 times. Midside with a series of 12–15 dark blotches. No suborbital bar. Gill membranes moderately to broadly joined across throat. A wide premaxillary frenum present; branchiostegal rays typically 6. Infraorbital

Figure 26.84. *Percina brucethompsoni*, Ouachita Darter. *Uland Thomas.*

Map 223. Distribution of *Percina brucethompsoni*, Ouachita Darter, 1988–2019.

canal uninterrupted, usually having 8 pores. One or more enlarged and modified scales present between pelvic fin bases. There is a large modified scale at the center of the breast, often surrounded by 1–8 smaller scales, forming a cluster. The remainder of the breast is naked. Belly with a series of 15–30 enlarged and modified scales along midline (males) or with scales of normal size and shape (females). Nape, opercle, and cheek fully scaled. Lateral line complete, with 72–85 pored scales (modally 78). Diagonal scale rows 27–34 (modally 30); scales above the lateral line 9–13 (modally 12); caudal peduncle circumferential scale rows 27–32 (modally 30). Dorsal spines 11–14 (modally 12); dorsal rays 12–15 (modally 14); anal spines 2, the first thicker than the second; anal rays 8–10 (modally 9); pectoral rays 12–15 (modally 14). It is a more robust species than *P. nasuta*, and more similar in body shape to *P. phoxocephala*. Breeding

males and females without discrete breeding tubercles. Maximum length is around 3.6 inches (90 mm).

LIFE COLORS   Very similar to those of the Longnose Darter, *Percina nasuta*. The first dorsal fin has an orange submarginal band. Breeding males exhibit a blackening of the head and ventral fins. The basal half of the spinous dorsal fin is black, and the soft dorsal fin has the same coloration as in nonspawning individuals.

SIMILAR SPECIES   *Percina brucethompsoni* differs from the allopatric *P. nasuta* in having a shorter snout, snout length less than 9% of its SL (vs. 9% or more), and in having 7 or more lateral blotches between the origin of the second dorsal fin and the hypural plate (vs. fewer than 7). Its lateral blotches are not as vertically elongated, never forming stripes. It differs from *P. phoxocephala* in snout depth less than snout length (vs. snout depth greater than snout length), 72 or more pored lateral line scales (vs. usually fewer than 72 pored scales), and vertically elongated distinct lateral blotches (vs. round lateral blotches). It can be distinguished from other species of *Percina* by the same features used to separate *P. nasuta* from those species.

VARIATION AND TAXONOMY   Thompson (1977) first recognized the probable specific distinctness of the Ouachita Darter but did not provide a formal description. It was treated as an undescribed species by Robison and Buchanan (1988). Formerly included in *Percina nasuta*, *P. brucethompsoni* was elevated to species status in 2014, based on morphological features and phylogenetic analysis of mitochondrial DNA sequence data (Robison et al. 2014). The specific epithet was selected to honor the late Bruce A. Thompson. The type locality is the Ouachita River, Arkansas Highway 298, approximately 1.6 km south of Sims, Montgomery County, Arkansas. Variation in fin element and scale counts was presented in Robison et al. (2014). There is moderate sexual dimorphism, with males having a deeper caudal peduncle, longer pelvic fins, a deeper and wider head, and longer upper jaws than females; females have slightly larger eyes. Populations of uncertain taxonomic status in the lower White, Spring, and Strawberry rivers in northeast Arkansas, thought to be similar to *P. brucethompsoni*, were tentatively assigned to *P. phoxocephala* (B. A. Thompson, pers. comm.). For more than 60 years the subgenus *Swainia* included four species based mainly on morphological features: *P. nasuta*, *P. oxyrhynchus*, *P. phoxocephala*, and *P. squamata* (Bailey and Gosline 1955; Page 2000), and analyses of mitochondrial and nuclear DNA supported the monophyly of this clade (Near 2002; Near et al. 2011; Robison et al. 2014). *Percina brucethompsoni* represents the fifth species of *Swainia*.

DISTRIBUTION   The entire range of the Ouachita

Darter is within Arkansas in the Ouachita River system, where it is known from the upper Ouachita River mainly above Lake Ouachita, the Little Missouri River from below Lake Greeson to its confluence with the Ouachita River, and the lower Caddo River (Map 223). The lower Caddo River population was thought to have been extirpated by the tailwater effects of DeGray Dam (Buchanan 1984), but 3 specimens were collected in the tailwaters below DeGray Dam in 1994 (McAllister et al. 2009c), and Jeff Quinn (AGFC) collected 3 specimens from those tailwaters on 11 April 2012. Jeff Quinn also provided records of this species from three localities from the Ouachita River downstream from Remmel Dam in 2012 and 2013.

HABITAT AND BIOLOGY   The habitat and abundance of the Ouachita Darter were studied by Gagen et al. (2002), Caldwell (2011), and Stoeckel and Caldwell (2014), and almost all our information on those subjects comes from those studies. The Ouachita Darter occupies a habitat that is difficult to sample effectively by electrofishing or seining. Stoeckel and Caldwell (2014), however, found that juvenile and adult Ouachita Darters can be sampled effectively by snorkeling and trawling. Gagen et al. (2002) identified the preferred late-spring habitat as stream reaches with emergent macrophytes (primarily water willow) growing along the edges of runs. This differed somewhat from the preferred late-summer habitat, which was the upstream edges of runs having approximately 60% cobble and 40% gravel substrate. The microhabitats of this species also invariably included slight surface agitation, a mean depth of 22 cm, and mean water velocity of 0.15 m/s (during summer low-flow periods); however, Caldwell (2011) found Ouachita Darters at a mean depth of 0.58 m in run/glide and pool habitats and did not observe darters in less than 0.3 m of water depth; it preferred the upstream edges of runs (just below riffles) in late summer when the water level was low. The preferred cobble habitat of this species was always sediment-free, unlike some nearby habitats where no Ouachita Darters were found. Gagen et al. (2002) speculated that clean cobble substrate may constitute a critical summer habitat for this species. The number of Ouachita Darters per site where the darters occurred ranged from 2 to 18. A population estimate of 32 darters per 28 m^2 was obtained at sites of highest densities (low, but not extremely low for darters); however, only 13.3% of the river segment known to contain Ouachita Darters was riffle or run habitat. Caldwell (2011) found Ouachita Darters in much lower densities than previously reported. Ouachita Darter densities from 2010 visual surveys were much lower than those reported from visual surveys in 2001 (Caldwell 2011). In Caldwell's study, Ouachita Darters occupied moderately

deep pool and run/glide habitat with moderate to low average velocities, low bottom velocities, and gravel/cobble substrate. Darters selected for moderate current velocities, cobble substrates, and water depths greater than 0.61 m; habitat use did not differ throughout their range in the upper Ouachita River. Based on snorkeling surveys, adults occupy slightly shallower (mean of 0.37 vs. 0.56 m), faster (mean of 0.24 vs. 0.18 m/s) water than juveniles; trawl data indicated that Ouachita Darters were more abundant at depths less than 1.8 m and velocities of approximately 0.10 m/s (Stoeckel and Caldwell 2014). Highest densities (1.36 fish/100 m^2) were found in the 3 km transition reach where the Ouachita River transitions to Lake Ouachita (Caldwell 2011). In the riverine section, Ouachita Darter densities were somewhat sporadic and apparently decreased in an upstream direction (Stoeckel and Caldwell 2014). Caldwell (2011) also noted that the Ouachita Darter differed somewhat in habitat preference from that reported for the Longnose Darter, with the former species not occurring in deeper parts of pools, near vegetation, or over sandy substrates as much as the latter species. It is apparently intolerant of reservoir conditions, and the construction of large Ouachita River drainage impoundments has undoubtedly caused loss of populations. Stoeckel and Caldwell (2014) found that Ouachita Darters tolerate slight amounts of silt, but they are uncommon in moderately or heavily silted areas.

Stomach contents of 14 adults (75–95 mm TL) collected between March and July were examined by TMB. The diet in all months was dominated by large benthic insect nymphs, larvae, and pupae. Plecopterans and ephemeropterans occurred in all stomachs and composed more than 90% of foods eaten by volume. Chironomids and trichopterans were consumed in small amounts.

Spawning occurs from mid-March to late May, based on examination of ovaries and testes by TMB. Ovaries of two females collected on 6 March from the Ouachita River near the mouth of De Roche Creek in Hot Spring County contained eggs nearing maturity. One female contained 194 eggs nearing ripeness, and the eggs of the other female were less mature. Both females contained two distinct complements of eggs by size, indicating more than one spawning during the breeding season. Peak spawning apparently occurs in April. Two ripe females, 78 and 80 mm TL, collected on 11 April from the Caddo River below DeGray Dam in Clark County contained 141 and 184 mature ova, respectively. A third female with large ovaries from that collection contained no ripe ova and had apparently just spawned; that female contained a large complement of developing ova of a single size group. A female

95 mm TL collected on 13 May 2013 from the Ouachita River in Hot Spring County below Remmel Dam contained 199 ripe ova and appeared to be nearing the end of reproductive condition because the ovaries contained no smaller size-group of developing eggs. A female collected on 17 July from the Ouachita River had regressed ovaries. Caldwell (2011) found that YOY *P. brucethompsoni* collected from the Ouachita River from 4 June through 10 June 2010 ranged from 24 mm to 31 mm TL; generally, juveniles were observed on top of the substrate. Even though some apparently stable populations exist, the restricted range of this Arkansas endemic species should be cause for concern over its future status.

CONSERVATION STATUS   Most recent studies of the status of this species have either failed to find specimens or have found low population densities at known collection localities. Gagen et al. (2002) showed that the difficulty of collecting this species by traditional methods, such as seining or electrofishing, can be attributed to its preferred habitat of cobble-strewn runs as well as to its avoidance behavior of seeking shelter beneath cobbles when frightened. Caldwell (2011) found that visual surveys and trawling (with a mini Missouri trawl) were the most effective methods for detecting Ouachita Darters, and backpack electrofishing was ineffective (o captures at 16 sites). Because of small population sizes and low densities at individual sites, effective conservation of this species may require a protection status and management plan from appropriate state and federal governmental agencies (Robison et al. 2014). We provisionally consider it a species of special concern, but it may be threatened.

Percina caprodes (Rafinesque)
Logperch

Figure 26.85. *Percina caprodes*, Logperch. *Uland Thomas.*

CHARACTERS   A large darter, with adults commonly reaching 4–5 inches (100–125 mm) TL; larger specimens are occasionally found. A fleshy, conical snout projects well beyond the upper lip in adults (Fig. 26.85); specimens smaller than 70 mm may not have a projecting snout. Side with 12–19 narrow dark vertical stripes or bars; alternate

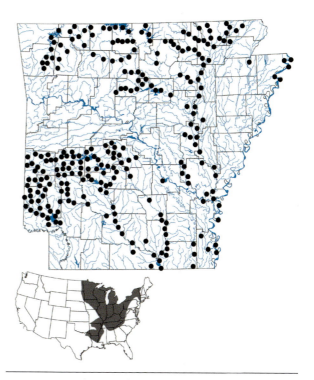

Map 224. Distribution of *Percina caprodes*, Logperch, 1988–2019.

bars (whole bars) are noticeably wider than bars on either side (half bars) in adults; usually 2–4 lateral bars from each side meeting ventrally. One or more enlarged and modified scales between pelvic fin bases; midline of belly naked (females) or with a row of 20–37 modified scales (males). Gill membranes usually separate. Infraorbital canal uninterrupted, usually with 7 or 8 pores. Lateral line complete, usually with 84–93 scales ($\bar{x}$ = 88.3). Diagonal scale rows from anal fin origin to first dorsal fin base usually 26–33 ($\bar{x}$ = 29.4). Dorsal spines 14–16 ($\bar{x}$ =15.2); dorsal rays 15–17 ($\bar{x}$ = 15.9); anal spines 2; anal rays 9–12 ($\bar{x}$ = 10.6). Males with breeding tubercles on ventral scales and caudal peduncle. Maximum size 7.25 inches (180 mm) TL.

LIFE COLORS   Back and sides yellow brown to yellow green with dark brown or black vertical stripes or bars. Venter white or cream-colored. There is a conspicuous black, round basicaudal spot. Usually a faint to distinct dark horizontal bar beneath orbit and a faint to distinct vertical suborbital bar (teardrop). First dorsal, second dorsal, and caudal fins often with light brown bands. First dorsal fin is black basally and usually distally, sometimes with scattered yellow chromatophores but without a distinct yellow or orange submarginal band (some specimens from the upper White River drainage have a yellow or orange submarginal band in the first dorsal fin and may not be

P. caprodes). Pectoral, pelvic, and anal fins usually clear. Breeding males becoming somewhat dusky.

SIMILAR SPECIES *Percina caprodes* differs from *P. fulvitaenia* in lacking an orange or yellow submarginal band in the first dorsal fin, and in having narrower lateral bars (vs. a distinct orange or yellow submarginal band in first dorsal fin, and wider lateral bars in *P. fulvitaenia*). *Percina caprodes* differs from *P. macrolepida* in having a fleshy, conical snout projecting well beyond upper lip in adults (vs. snout not projecting noticeably beyond upper lip), 26–33 ($\bar{x}$ = 29.4) diagonal scale rows (vs. 21–28 [$\bar{x}$ = 24.9] diagonal scale rows), and no scales on breast, prepectoral area, top of head, and along edge of preopercle (vs. usually with scales on breast, prepectoral area, edge of preopercle, and often on top of head).

VARIATION AND TAXONOMY *Percina caprodes* was described by Rafinesque (1818b) as *Sciaena caprodes* from the Ohio River (presumably in the vicinity of Louisville, Kentucky). It is in the subgenus *Percina*. Morris and Page (1981) recognized two subspecies of *P. caprodes* in Arkansas: (1) *P. c. fulvitaenia*, occurring in the Arkansas River and its tributaries, and (2) *P. c. caprodes*, occurring in the Ouachita River drainage and in Red River tributaries. Morris and Page (1981) considered logperch occurring in the White, Black, and St. Francis rivers intergrades between *P. c. fulvitaenia* and *P. c. caprodes* (based largely on variability in the development of an orange submarginal band in the first dorsal fin). Thompson (1997b) did not agree with this evaluation and recognized logperches from both the upper and lower White River as *P. c. caprodes*. Thompson also elevated *P. c. fulvitaenia* to full species status, a designation that we support (see species account for *P. fulvitaenia*). The presence or absence of an orange band in the first dorsal fin can be accurately judged only by examination of live or freshly preserved specimens. In Arkansas, Ouachita, and Red river drainages, *P. caprodes* does not have the orange submarginal band in the first dorsal fin. Specimens from the mainstem Mississippi River and the lower White River (downstream of Bull Shoals Dam) also lack the orange submarginal band. Specimens from several sites in the upper White River drainage examined by D. A. Neely (pers. comm.) lacked the orange band. TMB examined 138 live specimens taken from Greers Ferry Lake on 3 August 1987 and 56 specimens taken from Lake Norfork on 18 August 1987, and found that all lacked an orange band in the first dorsal fin. TMB also examined live specimens collected in mid-June 2014 from some of the other presumed intergrade populations of Morris and Page (1981) in Arkansas, and all specimens (5 specimens from Sylamore Creek in Stone County and 15 specimens from the Spring River in Sharp

County) showed no trace of orange band development in the first dorsal fin. The only documented occurrence of an orange band in the first dorsal fin in the upper White River was from Beaver Lake. TMB tabulated the occurrence of an orange band in the first dorsal fin of 68 freshly collected logperch specimens from an AGFC rotenone population sample on Beaver Lake. Eighty-seven percent of the males and 84% of the females possessed a submarginal band that was always orange. In males, 50% had a distinct orange band and 37% had a faint band; for females the same figures were 8% and 76%, respectively. A combined morphological and genetic analysis is needed to clarify the taxonomic status of the upper White River populations.

Near and Benard (2004) studied variation in mitochondrial gene sequences and reported high diversification rates in logperches. In another study, *P. caprodes* exhibited higher rates of gene flow than 14 other species of darters (Turner and Trexler 1998). This higher rate was attributed to life history features. Of the species examined, *P. caprodes* had the largest body size and clutch size and the smallest eggs. Larvae hatching from smaller eggs disperse over greater distances; as a result, populations of species producing smaller eggs are hypothesized to be less genetically isolated from one another.

DISTRIBUTION The most widely distributed logperch species, *Percina caprodes* occurs in the St. Lawrence–Great Lakes, Hudson Bay, and Mississippi River basins from Canada south to Louisiana. On the Atlantic Slope, it is found only in the Hudson River drainage of Vermont and New York. In Arkansas, *Percina caprodes* occurs throughout the White River drainage, including the Black and Little Red rivers. A few records are known from the St. Francis River. Populations in southern Arkansas are found in the Ouachita River drainage and upland tributaries of the Red River (Map 224). It is replaced in the Red River main stem by *P. macrolepida*. Although it is possible that specimens might be found in the Red River main stem, all logperch (more than 300 specimens) examined by TMB from throughout the Red River in Arkansas were *P. macrolepida*. In recent years we collected *P. caprodes* from the Mississippi River main stem at several localities in Mississippi and Chicot counties. It is absent from most of the Arkansas River main stem and its tributaries, where it is replaced by *P. fulvitaenia*; however, it has been found in the lower Arkansas River in the old river channel below Dam 2, Lake Merrisach off the Arkansas Post Canal, lower Bayou Meto, and in Jefferson County.

HABITAT AND BIOLOGY The Logperch is most abundant in streams of the Ozark and Ouachita uplands, but it also occurs at scattered localities throughout the

Coastal Plain. It is one of the most adaptable darters, occurring in a variety of habitats in Arkansas, including mountain streams, moderate and large rivers of medium to low gradient, numerous impoundments, and a few oxbow lakes; however, in those more unusual habitats, it seeks out areas of rocky substrate whenever possible. Excessive turbidity and siltation appear to be the main factors excluding this species from some lowland waters in eastern and southern Arkansas. In streams, it tends to be most abundant in deep, swift riffles over a gravel bottom, but it sometimes occurs in pools as well. It is rarely found over sand and mud substrates. The Logperch is common in some Arkansas reservoirs (Buchanan 2005), and annual rotenone population samples from a 4-acre (1.6 ha) area of Lake Hamilton between 1972 and 1981 yielded an average of 222 Logperch per sample. It is less tolerant of the conditions in cold tailwaters below dams. Its numbers were greatly reduced in fish samples from the Beaver Lake tailwaters shortly after impoundment as well as 30 years after impoundment (Quinn and Kwak 2003).

Percina caprodes feeds mainly on insect larvae and other small invertebrates. Pflieger (1997) reported that when searching for food it uses its conical snout to turn over rocks. Mullan et al. (1968) found the Logperch to be a sedentary, littoral bottom forager in two Arkansas reservoirs. In Bull Shoals Lake it had a mixed diet of benthic crustaceans and aquatic insects (primarily larvae), with 10% of its diet in the spring consisting of black bass eggs. In newly constructed Beaver Reservoir, Logperch fed almost exclusively on aquatic insects. In the Glover River, Oklahoma, Jones and Maughan (1987) found that dipterans (54%), principally chironomids (40%), were the most common items in the stomachs of Logperch, but ephemeropterans (21%) and trichopterans (10%) contributed to the diet as did gastropods (6%), cladocerans (3%), and plecopterans (2%). Stomach contents of 38 specimens collected from War Eagle Creek in Madison County and the White River in Washington County on 22 and 23 May were examined by TMB. The diet was dominated by plecopterans (89% frequency of occurrence), followed by ephemeropterans (50%), coleopterans (28%), and chironomids and other dipterans (28%). Odonates, trichopterans, hemipterans, and isopods were also ingested in small amounts. Thirty Logperch collected by TMB from the Caddo River in Clark County, Arkansas on 31 May fed predominantly on ephemeropterans (73%) and chironomids (67%), with plecopterans, coleopterans, trichopterans, and fish eggs ingested in small amounts.

Hubbs (1985) reported that Logperch spawning begins in Arkansas in mid-March and ends in May, and that finding agrees with our observations. Based on examination of ovaries and testes by TMB, most females had finished spawning by 31 May in the Caddo River, Arkansas. Only 2 out of 10 females collected on 31 May contained ripe ova (1 female 96 mm TL had 177 ripe ova, and 1 female 153 mm TL contained 285 ripe ova). One female collected by TMB in War Eagle Creek in Madison County on 22 May contained 764 ripe ova. Ripe females were found in the White River near Fayetteville at water temperatures from 57.2°F to 71.6°F (14–22°C) during the breeding season (Hubbs and Strawn 1963). Total fecundities are among the highest reported for any darter species, ranging from 1,000 to 3,000 eggs. Winn (1958a) provided information on spawning patterns in lakes and streams. In lakes, the males form schools above an area of sand or gravel parallel to the shoreline and are completely nonterritorial. The females feed in deeper water beyond the male school until ready to spawn. In streams, the males form schools over sand and gravel riffles and raceways. Each male protects a "moving territory" around a female which has come onto the breeding grounds. Spawning occurs when a ripe female settles to the substrate and a male assumes a position on top of her with his pelvic fins clasping her head and his tail at the side of hers. A rapid vibration of both fish excavates a small pit in the substrate. During this activity the eggs are emitted and fertilized. The female may spawn in many other pits. The eggs are buried in the substrate and left unguarded. Eggs hatch in about 8 days at 16.5°C and in 5 or 6 days at 22°C. Maximum life span is around 4 years.

CONSERVATION STATUS   The Logperch is widespread and common in Arkansas, and its populations are secure.

Percina copelandi (Jordan)
Channel Darter

Figure 26.86. *Percina copelandi*, Channel Darter. *Uland Thomas.*

CHARACTERS   One of the smaller members of the genus *Percina*, the Channel Darter has a blunt snout and a small horizontal mouth (Fig. 26.86). Premaxillary frenum usually absent, but 16% of Arkansas specimens have a frenum (Suttkus et al. 1994). A series of 9–12 dark

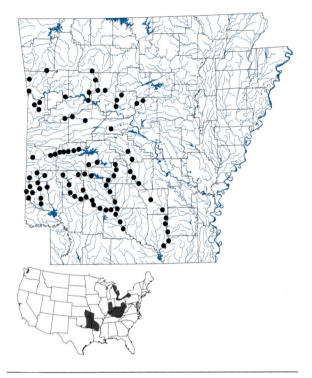

Map 225. Distribution of *Percina copelandi*, Channel Darter, 1988–2019.

oval, quadrate, or diamond-shaped, variously confluent blotches along midside, 5–8 faint middorsal blotches, and scattered X- or V-shaped markings on the back and upper side. Typically, a black basicaudal spot is present. Gill membranes separate, not broadly connected across throat. Infraorbital canal uninterrupted, usually with 8 or 9 pores. One or more enlarged and modified scales present between pelvic fin bases; midline of belly naked or scaled only posteriorly (females) or with a row of 3–16 enlarged and modified scales (males). Squamation variable in Arkansas populations. Breast and anterior nape usually naked; but nape varies from 25% to 100% scaled; opercle 66–100% scaled; cheek usually 25–100% scaled. Lateral line complete, with 50–66 scales (usually 51–61 in Arkansas River drainage, and 54–66 in Red-Ouachita river drainages). Diagonal scale rows 14–20 (usually 14–18 in Arkansas River drainage, and 15–19 in Red-Ouachita river drainages). Scales around caudal peduncle 16–22 (usually 16–21 in Arkansas River drainage, and 17–22 in Red-Ouachita river drainages). Dorsal spines usually 10–12; dorsal rays usually 11–13; anal spines 2; anal rays 7–9; pectoral rays modally 14. Breeding males with tubercles weakly developed on pelvic and anal fins. Maximum length about 2.5 inches (64 mm) TL, but Arkansas specimens rarely exceeding 2.2 inches (55 mm) TL.

LIFE COLORS   Back yellow to olive with brown specks or X-markings and sometimes about 8 faint brown blotches, but distinct saddles are absent. Sides yellow with 9–12 black blotches, which tend to be interconnected by a faint narrow band. Black basicaudal spot present. Belly whitish yellow. Black preorbital bar present; suborbital bar usually present. Breeding males heavily pigmented, with the throat, breast, and pelvic fins nearly black. First dorsal fin black basally and marginally. Other fins clear or with faint brown spots or bands.

SIMILAR SPECIES   The Channel Darter is sometimes mistaken for smaller individuals of other *Percina* species and even some *Etheostoma* species. It differs from *P. vigil*, *P. shumardi,* and *P. uranidea* in having 8 or 9 anal rays (vs. 10–13), 18 or fewer scales around caudal peduncle (vs. more than 18), and in having a more rounded snout (vs. snout more pointed). It can be distinguished from other *Percina* species by lack of a premaxillary frenum (vs. frenum present). It differs from darters of other genera in having 1 or more enlarged and modified scales present between pelvic fin bases, and midline of belly naked or with a row of modified scales.

VARIATION AND TAXONOMY   *Percina copelandi* was originally described by Jordan (1877b) as *Rheocrypta copelandi* from the White River, 8 km north of Indianapolis, Indiana. It is currently the only Arkansas species of subgenus *Cottogaster*. Suttkus et al. (1994) gave a detailed account of the tortuous taxonomic history of this species. A genetic analysis supported the monophyly of *Cottogaster* (Near et al. 2011). In the formal descriptions of *P. aurora* and *P. brevicauda*, morphological variability observed in *Percina copelandi* led Suttkus et al. (1994) to suggest that additional species may exist within subgenus *Cottogaster*. A recent mtDNA molecular phylogeny (Dugo et al. 2012) recovered two distinct clades of *P. copelandi*, including a nominal clade representing populations east of the Mississippi River inclusive of localities from the Arkansas River drainage west of the Mississippi River. The novel clade was recovered from localities of the Ouachita (Saline and Caddo rivers) and Red river drainages (Little/Glover River) (Todd Slack, pers. comm.). It is possible that the populations in the Ouachita and Red river drainages will eventually be elevated to full species status (also see Dugo et al. 2008, 2011).

DISTRIBUTION   As currently recognized, populations of *P. copelandi* east of the Mississippi River occur in the St. Lawrence–Great Lakes and Mississippi River basins from Canada south to Tennessee. West of the Mississippi River, it is found in Arkansas, Ouachita, and Red river drainages from Kansas, Missouri, and eastern Oklahoma south to northern Louisiana. In Arkansas, the Channel Darter is found mainly in the Red and Ouachita river drainages of the Ouachitas. It also occurs in tributaries of the Arkansas River in the western half of the state (both north and south of the Arkansas River). It is historically known from the Illinois River in northwestern Arkansas (Map A4.221), but there are no recent records from that drainage. It is absent from the White River drainage (Map 225).

HABITAT AND BIOLOGY   Various authors noted differences in preferred habitat in different parts of its range. In Arkansas, the Channel Darter is found primarily in small to moderate-sized rivers and large creeks, where it typically occurs in riffles of moderate to swift current over a gravel or rocky substrate. In Tennessee, it is often associated with sandy areas (Etnier and Starnes 1993), and Reid et al. (2005) found that riffles at most stream sites in the Lake Ontario basin having *P. copelandi* flowed into deep sand-bottomed run or pool habitats. In eastern Canada, summer habitat was characterized by a strong preference for environments with mean water velocities between 39 and 48 cm/s (Boucher et al. 2009). In that same study, the presence of periphyton, woody debris, and plant cover were linked to the occurrence of the Channel Darter. It prefers clear water and a silt-free bottom. It is most abundant in Arkansas in the larger streams of the Ouachita Uplands, and its greatest penetration into the Coastal Plain is in the eastern Saline River. In some areas of its range the Channel Darter prefers pools and quieter waters (Miller and Robison 1973; Cross and Collins 1995), and it is sometimes taken in those habitats in Arkansas. Branson (1967) described seasonal habitats for the Channel Darter in the Neosho River, where it was seldom found in swift riffles but was abundant over sand in slow current; late in the year it moved into quiet backwaters where it overwintered. Suttkus et al. (1994) summarized habitat data from numerous observations and collection records for populations in the southern U.S. and concluded that the Channel Darter prefers moderate to fast current over gravel, rubble, or rock-strewn bedrock (not sand) as spawning habitat. Post-spawning habitat may often be in slower current over sand substrate, as well as over gravel, rubble, and bedrock. It is sometimes found in lakes in northern parts of its range, and Buchanan (2005) found *P. copelandi* in five impoundments in the Arkansas River Valley and Ouachita Mountains ecoregions of Arkansas.

Percina copelandi fed mainly on midge larvae (chironomids) and other small insect larvae and nymphs (ephemeropterans), and microcrustaceans (copepods and cladocerans) in Illinois and Oklahoma (Thomas 1970; Jones and Maughan 1987). In Kansas, it fed on chironomids, trichopterans, and microcrustaceans (Cross and Collins 1995). Stomach contents of 41 specimens from Arkansas River tributaries in Arkansas were examined by TMB. From late February through May, ephemeropteran nymphs were the most frequently ingested food item, followed in descending frequency of occurrence by chironomids and other dipteran larvae, fish eggs, trichopterans, small crustaceans (copepods, and isopods), and plecopterans. From June through August, the diet in Arkansas River tributaries was dominated by chironomids, trichopterans, and ephemeropterans, with microcrustaceans ingested only infrequently. Stomach contents of 45 specimens collected in May from Red and Ouachita river tributaries by TMB contained predominantly ephemeropteran nymphs by percent frequency of occurrence (46%). Chironomids and other larval and pupal dipterans (Tipulidae) made up the next highest percentage of the diet (37%), followed in decreasing order of importance by trichopterans, small crustaceans (copepods, isopods, and cladocerans), coleopterans, plecopterans, fish eggs, odonates, and rotifers.

Individuals in spawning condition were found in late May in Tennessee (Etnier and Starnes 1993), and spawning was reported to occur in June in Oklahoma (Hubbs and Bryan 1975), April and May in Kansas (Cross and Collins 1995), and July in Michigan (Winn 1953). Suttkus et al. (1994) reported that spawning occurs from mid-March through mid-April in southwestern Arkansas. Based on examination of ovaries and testes of specimens from Red, Ouachita, and Arkansas river tributaries by TMB, the spawning season extends from mid-March through May in Arkansas. Females in spawning condition from Lee Creek in Crawford County in mid-April ranged from 39 to 50 mm TL and contained from 43 to 152 ripe ova. The presence of smaller egg sizes in the ovaries indicated multiple spawning. Females in spawning condition were collected from the Mountain Fork River in Polk County on 5 and 7 May, and the latest collection date in Arkansas for females containing ripe ova was 30 May from the cool tailwaters below Narrows Dam on the Little Missouri River in Pike County. Spawning occurs over a gravel substrate in moderate current. The males defend territories smaller than a meter in diameter in swift current over a sand and fine gravel substrate and spawn with females after burrowing into the substrate (Winn 1953). Suttkus et al. (1994) reported that southern populations spawn in moderate to fast current over gravel, rubble, or rock-strewn bedrock substrates, not sand, as was reported for some northern populations by Winn (1953). Each female may spawn with several males until around 400 eggs are laid. After spawning, the buried eggs receive no parental care. Fecundity ranges from 400 to 700 eggs, and hatching occurs in 6–10 days at 15–20°C. Maximum life span is believed to be about 3+ years.

CONSERVATION STATUS   This species is widespread and common in Arkansas but rarely abundant at any locality. It was, however, one of the most abundant darters in trawl samples from the Ouachita River in 2018 (Jeff Quinn, pers. comm.). We consider its populations secure.

Percina evides (Jordan and Copeland)
Gilt Darter

Figure 26.87. *Percina evides*, Gilt Darter. *David A. Neely.*

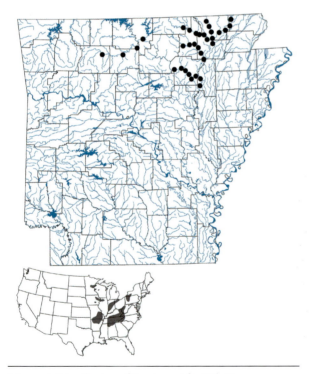

Map 226. Distribution of *Percina evides*, Gilt Darter, 1988–2019.

CHARACTERS   A stout-bodied *Percina* with a series of 7–9 dark blotches along midside, each blotch directly below a dorsal saddle and usually connected with the saddle to form a wide bar crossing the back (Fig. 26.87). A narrow dusky vertical bar beneath the eye. Preopercle smooth on

posterior edge. Gill membranes separate or barely connected across the throat; wide premaxillary frenum present. Infraorbital canal uninterrupted, usually with 8 pores. One or more enlarged and modified scales between pelvic fin bases; midline of belly naked (females) or with a row of 7–13 enlarged scales (males). Cheek and breast naked; nape and opercle scaled. Lateral line complete, with 64–76 scales. Dorsal spines 11–13; dorsal rays 12 or 13; anal spines 2; anal rays 7–10; pectoral rays 13–15. Breeding males exhibit the most extensive tubercle development of any species of darter. Tubercles occur on lower half of head, ventral scales, and pectoral, pelvic, and caudal fins. Tubercles are sometimes present on females. Maximum size is around 3.75 inches (96 mm) TL.

LIFE COLORS   Arkansas's most colorful *Percina* species. Breeding male brilliantly colored, developing fiery red orange on the cheeks, mouth, underside of head and throat; body and fins tinged with orange. Lateral blotches and dorsal saddles an intense dark green. Colors retained from early spring well into summer. Breeding females not developing bright colors, but having dull yellow lower body with dark green lateral blotches and dorsal saddles.

SIMILAR SPECIES   It can be distinguished from all other *Percina* in Arkansas by the broad bands that cross the back and join lateral blotches, and by its bright breeding colors. It further differs from the three logperch species (*P. caprodes*, *P. fulvitaenia*, and *P. maculata*) in lacking an elongate, terete body and narrow "tiger stripes" on the sides. It differs from *P. sciera* in having a single basicaudal spot (vs. 3, the lower 2 sometimes fused), and in having a preopercle with a smooth rear margin (vs. weakly serrate). It can be distinguished from *P. shumardi*, *P. uranidea*, and *P. vigil* in having a wide frenum (vs. narrow or no frenum), and an orange submarginal band in the first dorsal fin (vs. no orange band). It differs from darters in other genera in having 1 or more enlarged and modified scales between the pelvic fin bases and midline of belly naked or with a row of modified scales.

VARIATION AND TAXONOMY   *Percina evides* was described by Jordan and Copeland *in* Jordan (1877b) as *Alvordius evides* from the White River near Indianapolis, Indiana. It is the only Arkansas species of subgenus *Ericosma*. Genetic analysis by Near et al. (2011) did not support the monophyly of *Ericosma*. Subspecies are unnamed, but populations west of the Mississippi River differ morphologically (Denoncourt 1969; Page and Burr 2011) and genetically (Near et al. 2001) from populations elsewhere and may represent an undescribed species. The unnamed Ozark form occurs in the Ozark Uplands of Arkansas and Missouri and is characterized by a partly scaled cheek, an orange first dorsal fin, and a yellow belly in large males (Page and Burr 2011). Near et al. (2001) examined geographic variation in mitochondrial DNA and assigned Arkansas populations to a White River clade of a western clade (west of the Mississippi River). Near (2002) considered *P. evides* the sister species of *P. aurantiaca*. Near et al. (2011) considered populations in Minnesota, Wisconsin, Iowa (extirpated), Illinois (extirpated), Missouri, and Arkansas to be a separate, undescribed species, *Percina* sp. *cf. evides*, the Western Gilt Darter. Because additional study of Arkansas populations and other populations west of the Mississippi River is needed, we herein refrain from recognizing *P.* sp. *cf. evides sensu* Near et al. (2011) as a separate, undescribed species and provisionally continue to assign all Arkansas populations to *P. evides*.

DISTRIBUTION   As currently recognized, the Gilt Darter occurs in disjunct populations in the Mississippi River basin from New York to Minnesota, and south to northern Arkansas and Alabama. It is also known from one Great Lakes drainage (Maumee River system of Lake Erie) in Ohio and Indiana. In Arkansas, it is limited to the White, Buffalo, Strawberry, Spring, Current, and Black rivers in northern Arkansas (Map 226). Historically known populations in the upper White River of Benton, Carroll, and Washington counties (Map A4.222) have apparently been extirpated.

HABITAT AND BIOLOGY   An inhabitant of large creeks and small to medium-sized rivers in the Ozark Mountains, the Gilt Darter prefers clean, clear water with permanent flow and is found in moderate to swift current in deep riffles over a rubble and gravel bottom. Individuals are often taken in deep chutes and runs at the head of a riffle. It is intolerant of reservoir conditions. Skyfield (2006) and Skyfield and Grossman (2008) quantified aspects of microhabitat use by Gilt Darters in two North Carolina streams. Those authors found that *P. evides* preferred habitats with higher percent cobble and higher average velocities than randomly available in the environment. Larger darters tended to use microhabitats with more heterogeneous substrata and larger numbers of boulders than smaller individuals. In the same studies, 40% of population movements were within 5 meters of initial capture, and population density was estimated to be 0.31 darters/m².

Hickman and Fitz (1978) reported that Gilt Darters fed mainly on midge larvae and less often on other insects and snails in Tennessee. In Pennsylvania, Gray et al. (1997) reported that *P. evides* fed on larger prey (1–13 mm), fewer chironomids, and more fish eggs than syntopic *Etheostoma* species. Stomach contents of 36 specimens from three Arkansas rivers were examined by TMB. Thirteen

individuals, 41–76 mm TL, collected from the Strawberry River in Sharp County on 18 May, fed predominantly on ephemeropterans, with trichopterans, chironomids, and plecopterans consumed in small numbers. Fourteen specimens, 52–79 mm TL, collected on 18 May from the Spring River in Lawrence and Sharp counties, also fed most heavily on ephemeropterans. Other items ingested in decreasing percent frequency in the Spring River were chironomids, trichopterans, plecopterans, simuliids, and fish eggs. Nine individuals, 39–58 mm TL, collected from the White River in Independence County on 1 August fed almost exclusively on chironomids, with trichopterans and ephemeropterans eaten in small numbers. At all Arkansas sites, most individuals had ingested fine sand grains.

Based on information from Tennessee (Etnier and Starnes 1993) and Missouri (Pflieger 1997), the breeding season in Arkansas probably extends from April to July. Spawning was reported in Missouri from mid-April through June (Pflieger 1997). Hubbs (1985) reported the breeding season in Arkansas as April and May. We have found this species spawning in swift riffles in mid-May in the Spring and Strawberry rivers. Five ripe females, 61–70 mm TL, collected from those rivers on 17 and 18 May by TMB contained from 85 to 151 mature ova. Page et al. (1982) observed spawning in Little River, Tennessee, from June to early July at water temperatures of 17–20°C. Spawning occurs in the upper parts of riffles over a sand and gravel substrate having scattered cobbles and boulders; spawning occurred at depths of 30–60 cm. Fecundity ranged from 132 to 379 eggs. Hatch (1982, 1986) reported that males established specific territories (unusual among *Percina*), and spawning was similar to that of *P. caprodes*.

CONSERVATION STATUS   Stream degradation has severely reduced the range of this species outside of Arkansas, and it has been extirpated from all of Ohio, Illinois, and Iowa (Kuehne and Barbour 1983; Page 1983a). It is considered endangered in Wisconsin (Becker 1983). It has also suffered some range reduction in Arkansas, particularly in the western half of its range within the state (Buchanan 1974; Cloutman and Olmsted 1976). The construction of Beaver Reservoir has apparently eliminated most of the previously known populations from the upper White River in Benton, Carroll, Madison, and Washington counties, where it has not been reported since a preimpoundment study by Keith (1964). Postimpoundment fish surveys of the Beaver Lake tailwaters conducted a few years after impoundment and 30 years after impoundment yielded no Gilt Darters (Quinn and Kwak 2003). Stable populations still persist in the Buffalo, Strawberry, Black, and Spring rivers, and in a short section of the White

River in Independence County. It has apparently declined in numbers in Arkansas during the past two decades, and we consider it a species of special concern. Its populations should be periodically monitored.

Percina fulvitaenia Morris and Page
Ozark Logperch

Figure 26.88. *Percina fulvitaenia*, Ozark Logperch. *Uland Thomas.*

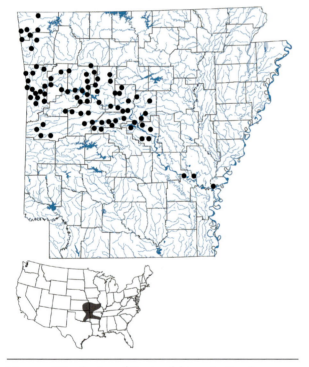

Map 227. Distribution of *Percina fulvitaenia*, Ozark Logperch, 1988–2019.

CHARACTERS   A large logperch with adults commonly reaching 5 inches (125 mm); occasionally even larger specimens are found. A rounded, fleshy, conical snout projects well beyond the small, inferior, horizontal mouth (Fig. 26.88); juvenile specimens smaller than 2–3 inches (70 mm) sometimes have a nonprotruding snout. First dorsal fin with a distinct orange or yellow-orange submarginal band. Usually 11–17 dark vertical bars on the side, usually with every other whole body bar somewhat expanded

(noticeably thicker), especially below the lateral line in adults. One or more pairs of vertical bars from each side meeting ventrally. Preopercle smooth on posterior edge. One or more enlarged and modified scales between pelvic fin bases; midline of belly naked (females) or with a row of 22–37 modified scales (males). Gill membranes usually separate. Infraorbital canal uninterrupted. Lateral line complete with 83–94 scales. Nape entirely scaled; breast, top of head, prepectoral area, and edge of preopercle without scales. Diagonal scale rows from anal fin origin to first dorsal fin base usually 25–33. Dorsal spines 14–16; dorsal rays 15–18; anal spines 2; anal rays usually 11 (9–12). Breeding males with tubercles on ventral scales. Maximum size around 7 inches (178 mm).

LIFE COLORS   First dorsal fin of adults with a narrow, dark marginal band, a broad orange or yellow-orange submarginal band (males) occupying at least 25% of the fin width and a dark basal band; females usually with a submarginal yellow to orange band. A dark horizontal bar present beneath eye and a dark vertical suborbital bar present. Back and upper sides light green to yellow with dark brown vertical bars. Venter cream-colored. Second dorsal and caudal fins clear or yellow with 4 or 5 bands of dark spots on rays. Distinct black spot at base of caudal fin. Pectoral, pelvic, and anal fins clear. Body and fins of breeding males becoming dusky.

SIMILAR SPECIES   Near (2008) noted broad overlap in meristic traits among many logperch species, but a combination of meristic characters, patterns of squamation, and pigmentation patterns will usually permit an accurate separation of logperch species in Arkansas. The Ozark Logperch differs from the other two Arkansas logperch species in having a broad orange or yellow submarginal band in the first dorsal fin of adult males and often in females (vs. no submarginal orange or yellow band in the first dorsal fin). It further differs from *P. caprodes* in having wider lateral bars and fewer lateral line scales. *Percina fulvitaenia* further differs from *P. macrolepida* in adults having a fleshy, conical snout protruding well beyond upper lip, and in lacking scales on the breast, prepectoral area, top of head, and along edge of preopercle (vs. pointed snout not usually protruding prominently beyond upper lip, and usually having scales on the breast, prepectoral area, top of head, and along edge of preopercle in *P. macrolepida*).

VARIATION AND TAXONOMY   *Percina fulvitaenia* was originally described by Morris and Page (1981) as a subspecies of *P. caprodes*, *P. c. fulvitaenia*, from Big Piney River (Gasconade River drainage), 5 km west of Houston, Texas County, Missouri. It is a member of subgenus *Percina*. There has been considerable disagreement about the status of this species. Morris and Page (1981), recognized two geographic races of Ozark Logperch. Arkansas River populations of *P. fulvitaenia* have a shorter snout and slightly wider, more even-edged vertical bars on the sides than Missouri River populations in the northern part of its range. Arkansas River drainage males also have a deep orange submarginal band in the first dorsal fin, whereas Missouri River drainage males have a yellow-orange band. Morris and Page (1981) regarded all populations of Ozark Logperch as subspecies (*P. c. fulvitaenia*) based on presumed intergrade populations. Thompson (1997b) disagreed with the interpretation of White River forms as intergrades and elevated the Ozark Logperch to full species status, *P. fulvitaenia*. Thompson's elevation of the Ozark Logperch to full species was recognized by Nelson et al. (2004) and several other authors. Studies of mitochondrial and nuclear DNA also supported its recognition as a species (Near 2008; Near et al. 2011, 2017). Near's 2008 genetic analysis included data from specimens from the Fourche LaFave River in Perry County, Arkansas. Despite the genetic evidence, Page et al. (2013) again assigned this taxon to *P. caprodes*. We examined more than 2,000 freshly collected logperch specimens from the Arkansas, Mississippi, and White rivers in Arkansas (either living or freshly preserved). There were no apparent intergrades in the Arkansas, Mississippi, or lower White rivers. Live Logperch examined from Sylamore Creek (White River drainage) and the Spring River on 17 and 18 June 2014 showed no trace of an orange band in the first dorsal fin. Except for specimens from Beaver Lake, logperch populations in the upper White River upstream from Bull Shoals Dam had no orange band in the first dorsal fin (see *P. caprodes* species account for more details on upper White River populations). All three logperch species in Arkansas (*P. caprodes*, *P. fulvitaenia*, and *P. macrolepida*) occur sympatrically without apparent morphological intergradation in the lower Arkansas River. We have not examined specimens from reported intergrade populations in the Missouri River drainage, but we interpret the Arkansas River race of Ozark Logperch to be a distinct evolutionary lineage worthy of species recognition pending further study. Near (2008) performed mitochondrial and nuclear DNA analysis on specimens from the Maries River, Missouri (Missouri River drainage) and assigned them to *P. fulvitaenia*. We follow Thompson (1997b), Scharpf (2008), Near (2008), and Near et al. (2011, 2017) in recognizing full species status for *Percina fulvitaenia*.

DISTRIBUTION   *Percina fulvitaenia* is found in the Missouri and Arkansas river drainages in Missouri, Arkansas, Kansas, and Oklahoma, and in the Red River drainage (Blue River) of Oklahoma. In Arkansas, the Ozark Logperch occurs throughout the Arkansas River and all its major tributaries (Map 227).

HABITAT AND BIOLOGY   The Ozark Logperch is most abundant in tributaries of the Arkansas River having permanent flow, where it prefers deep riffles, clear water, and gravel and rubble substrates. It is the only logperch found in Arkansas River tributaries. It occurs syntopically in the Arkansas River main stem with the presumably introduced *P. macrolepida* (Buchanan and Stevenson 2003; Buchanan 2005). The two species apparently maintain reproductive isolation in the Arkansas River, but there is some morphological evidence of hybridization between them in certain areas of Dardanelle Reservoir. In the Arkansas River, *P. fulvitaenia* is most abundant over rocky substrates, whereas *P. macrolepida* predominates over sand and silt substrates in quiet water. The Ozark Logperch is abundant in some impoundments on Arkansas River tributaries such as Blue Mountain, Hinkle, and Nimrod reservoirs (Buchanan 2005). It has better ability than most darters to repopulate an area affected by human activity. It demonstrated that ability by reestablishing a population in McKinney Creek in Johnson County in 1978 after a chemical spill killed all fishes along several miles of that stream in 1977.

In Kansas, Ozark Logperch fed on immature insects, crustaceans, and some algae (Cross and Collins 1995). Like *P. caprodes*, the Ozark Logperch often uses its conical snout to turn over small stones and other objects on the substrate to expose food items (TMB, pers. obs.). Stomach contents of 45 specimens, 70–115 mm TL, collected from Arkansas River tributaries (Fletcher Creek in Logan County, Lee Creek in Crawford County, and the Petit Jean River in Scott County) in late April to mid-May, and 30 specimens, 84–110 mm TL collected from the Arkansas River main stem near Fort Smith from mid-September to early October were examined by TMB. The diet in tributaries in spring was dominated by ephemeropterans, plecopterans, and chironomids by percent volume and percent frequency. Almost all specimens from Fletcher Creek had consumed fish embryos, with three specimens feeding on that food exclusively. Other items consumed in small amounts were trichopterans and isopods. The September-October diet in the Arkansas River was very different, with small crustaceans of equal importance with insect larvae in the stomach contents. Crustaceans consumed were mainly amphipods and isopods, along with a few small crayfish and shrimp. The dominant insect larvae were ephemeropterans and chironomids. Other items ingested in small amounts in the Arkansas River were limpets, coleopterans, and other dipterans.

Spawning begins in Arkansas in mid-March and ends in May (Hubbs 1985), and our observations agree with Hubbs's data. Ripe females were found in the Illinois River near Fayetteville at temperatures from 57.0°F to 71.6°F (14–22°C) during the breeding season (Hubbs and Strawn 1963). We have observed spawning in swift current over a gravel and cobble substrate in Vache Grasse Creek (Sebastian County) from as early as 10 March at a water temperature of 11°C to mid-April at a water temperature of 19°C. TMB found ripe individuals in Lee Creek in Crawford County on 10 April 1989, and in Clear Creek (Crawford County) on 30 April 1987. Three ripe females, 70, 95, and 107 mm TL, collected from Fletcher Creek in Logan County on 26 April by TMB contained 180, 373, and 633 mature ova, respectively. In the breeding season, males congregate on riffles, with the females gathering in nearby pools. Ripe females join the males on the riffles when ready to lay eggs. One male at a time spawns with a female, and fertilized eggs are buried in the gravel substrate by vigorous body vibrations of both parents. A female may spawn with more than one male. After burial of the eggs, no subsequent parental care is given. Maximum life span is about 3+ years.

CONSERVATION STATUS   *Percina fulvitaenia* has probably declined in the mainstem Arkansas River as a result of competition with *P. macrolepida*, which has spread through that navigation system in the past few decades (Buchanan and Stevenson 2003). *P. fulvitaenia* is still widespread and sometimes locally abundant in Arkansas River tributaries in the state, and we consider it currently secure.

Percina macrolepida Stevenson
Bigscale Logperch

Figure 26.89. *Percina macrolepida*, Bigscale Logperch. *David A. Neely.*

CHARACTERS   Snout pointed, not fleshy, and usually not protruding prominently beyond the front of the upper lip (Fig. 26.89). Twelve to 20 ($\bar{x}$ = 16) narrow vertical bars along the side, with the shorter bars nearly as long as the whole bars and with slight to moderate thickening of the whole body bars; usually 1 or 2 pairs of bars from each side

Map 228. Distribution of *Percina macrolepida*, Bigscale Logperch, 1988–2019. Inset shows native and introduced ranges.

meeting ventrally. Vertical suborbital bar usually absent (77%) or faint (15%); horizontal bar beneath orbit usually distinct (67%); preorbital bar usually distinct; postorbital bar diffuse. Preopercle smooth on posterior edge. One or more enlarged and modified scales between pelvic fin bases; midline of belly naked (females) or with a row of modified scales usually extending three-fourths the distance from the breast toward the vent (males). Gill membranes separate. Infraorbital canal uninterrupted. Lateral line complete, usually with 82–89 scales ($\bar{x}$ = 85.6). Scale rows above lateral line 7–10; scale rows below lateral line 10–14. One or more scales usually present on breast (83%); breast scales often embedded and difficult to see. One or more prepectoral scales usually present (79%); usually 1 or more scales present on edge of preopercle (74%); approximately 53% of specimens with 1 or more scales on top of head. Diagonal scale rows from origin of anal fin to first dorsal fin base usually 21–28 ($\bar{x}$ = 24.9); scale rows around caudal peduncle 27–32. Dorsal spines 13–15; dorsal rays 12–15; anal spines 2; anal rays 7–10; pectoral rays 12–14. Breeding males with tubercles on ventral scales. Arkansas's smallest logperch species, with adults commonly reaching 3–4 inches TL (75–100 mm). Largest Arkansas specimen (Arkansas River) 5 inches (125 mm).

LIFE COLORS   Ground color of head, back, and upper sides green to yellow green. Vertical lateral bars dark green to brown. Venter white to cream-colored. Small, black basicaudal spot present. First dorsal fin with melanophores in the membrane, usually more intense near the base. Yellow or orange chromatophores may also be scattered in the dorsal fin membrane but are never concentrated into a submarginal band. Second dorsal fin with dark pigment between the rays only on the posterior half of the membrane. Pectoral, pelvic, and anal fins dusky in breeding males and with very little pigment in females. Breeding male develops a dusky head and the body has a green or yellow tinge.

SIMILAR SPECIES   The Bigscale Logperch is morphologically very similar to the Logperch, *P. caprodes*, and the Ozark Logperch, *P. fulvitaenia*, but it can be distinguished from those species by its less fleshy snout which usually does not protrude prominently beyond its upper lip (vs. snout conical and fleshy, protruding prominently beyond upper lip in adult specimens) (Fig. 26.40). It also differs from those species in having 1 or more scales on the breast, prepectoral area, edge of preopercle, and usually on top of head (vs. scales absent in those areas), in usually having 24–26 diagonal scale rows (vs. usually 27–32), and by usually lacking a dark, vertical suborbital bar (vs. vertical suborbital bar present). *Percina macrolepida* has a smaller, more delicate head and dorsal fin elements than *P. caprodes* or *P. fulvitaenia* and generally reaches a smaller adult body size than those species. The Bigscale Logperch further differs from *P. caprodes* in having fewer lateral line scales ($\bar{x}$ = 85.6 vs. 89.1 in *P. caprodes*). It further differs from *P. fulvitaenia* in lacking an orange submarginal band in the first dorsal fin in life.

VARIATION AND TAXONOMY   *Percina macrolepida* was described by Stevenson (1971) from the Guadalupe River, below the dam at Kerrville State Park, Kerr County, Texas. It is a member of the subgenus *Percina*. Stevenson (1971), Sturgess (1976), and Buchanan et al. (1996) noted that *P. macrolepida* is a smaller species than *P. caprodes* and *P. fulvitaenia*. Red River populations in Arkansas reach a much smaller maximum size (approximately 4.0 inches TL) than Arkansas River populations (5.0 inches TL). Arkansas populations are also different in scale patterns and counts from populations in central Texas (Stevenson 1971; Thompson 1997a). Nearly 100% of Texas specimens have scales on the top of the head, versus 53% of Arkansas specimens. Stevenson and Thompson (1978) reported an eastward cline in that feature, with a reduction in number occurring from central Texas to the Sabine River. Arkansas specimens also have higher mean lateral line scale counts

(85.6) and a lower percentage of individuals (74%) with scales on the edge of the preopercle than Texas populations (82.8 and 100%, respectively). The presence of scales on the breast is unique among members of the subgenus *Percina*. Arkansas specimens rarely have a fully scaled breast; usually there are 1 or a few scattered, embedded or exposed scales with prominent ctenii (Buchanan et al. 1996). Thompson (1997a) placed *P. macrolepida* in a clade with *P. austroperca* and *P. suttkusi*. A genetic analysis by Near et al. (2011) supported the traditional morphology-based grouping into subgenus *Percina*, but those authors proposed a new subgenus name, *Pileoma*, in an attempt to avoid redundant group names.

DISTRIBUTION   The native range is considered to be the Red River drainage of Arkansas, Oklahoma, Texas, and Louisiana southwestward (including Gulf Slope drainages of Texas) to the Rio Grande drainage of Texas, New Mexico, and Mexico. Introduced populations have been reported from California (McKechnie 1966), Colorado (Platania 1990), New Mexico (Sublette et al. 1990), Oklahoma (Cashner and Matthews 1988), and Arkansas (Buchanan and Stevenson 2003). The Bigscale Logperch was first reported from Arkansas in Red River mainstem and floodplain habitats (Buchanan et al. 1996, 2003). It was subsequently reported from the Arkansas River main stem by Buchanan and Stevenson (2003), where it occurs throughout the Arkansas River Navigation System, including the lower White River portion of that system (Map 228). It occurs sympatrically with *P. caprodes* in the lower Little River downstream of Millwood Dam and in Millwood Lake. The Red River is considered within the native range of *P. macrolepida* (Fuller et al. 1999), but the Arkansas River populations are probably the result of introductions in western reaches of the drainage basin as early as the 1950s (Buchanan and Stevenson 2003).

HABITAT AND BIOLOGY   There is conflicting information from different parts of its range about the preferred habitat of *P. macrolepida*. Sublette et al. (1990) described its preferred habitat in New Mexico as deep rivers with a strong current and rubble and gravel substrates; however, they also noted that it was found in rivers with little flow and in impoundments. Davenport and Archdeacon (2008), in a survey of its status in New Mexico, collected the greatest number of specimens from reservoirs. Stevenson (1971) collected *P. macrolepida* from a variety of habitats in central Texas and noted that it avoided turbulent riffles in any stream in which it occurred. Jackson (1984) reported that juveniles in Lake Texoma, Oklahoma, had a strong midday affinity for littoral areas having relatively deep sand substrates, but night collections (between 8:00 p.m. and 6:00 a.m.) showed that juveniles occurred over a greater variety of substrates. Adults in Lake Texoma were most often collected from wave-swept locations having clay substrates with little sand (Jackson 1984). Buchanan et al. (1996) reported that Red River populations in Arkansas preferred quiet water of 1–2 m depth over a clay, silt, or sand substrate. In that study, 36.2% were captured over a sand substrate, 4.5% over a silt substrate, 58.9% over a mixed silt and sand substrate, and 0.4% over a gravel substrate. There was little or no aquatic vegetation or other cover at Red River sample sites, but it has been found in association with aquatic vegetation in Lake Millwood (Buchanan 2005). It is apparently the most turbidity-tolerant logperch species, with Secchi disk visibility ranging from 20 to 61 cm at Red River sampling sites (Buchanan et al. 1996). The largest Arkansas populations occurred in Red River oxbow lakes, particularly Fifty-one Cutoff Lake near Fulton. Our observations of habitat preference in Arkansas are similar to those reported by Moyle (1976) for Bigscale Logperch introduced into the Sacramento-San Joaquin system of California: lakes or slow-moving stretches of mud-bottomed, turbid sloughs of the Delta and lower Sacramento River.

The Bigscale Logperch occurs syntopically with Ozark Logperch throughout the Arkansas River main stem in Arkansas, but *P. fulvitaenia* is the only logperch found in Arkansas River tributaries. In the main Arkansas River, *P. macrolepida* prefers quiet water over sand or silt substrates, whereas *P. fulvitaenia* is more abundant over rocky substrates (Buchanan and Stevenson 2003). Environmental changes caused by the construction of the Arkansas River Navigation System probably increased the habitat favored by *P. macrolepida* at the expense of *P. fulvitaenia*.

California populations of *P. macrolepida* fed on a variety of benthic insect larvae and small crustaceans, occasionally rising from the bottom to snap up small, free-swimming organisms (Moyle 1976). Larval Bigscale Logperch become planktivorous after the mouth develops, and later switch to insects, crustaceans, and other small invertebrates. Stomach contents of 29 Arkansas specimens collected by seine and electrofishing were examined by TMB. Ten individuals, 50–88 mm TL, from the Arkansas River near Fort Smith contained little food in June (100% ephemeropterans), but specimens collected in September and October had full stomachs, with the diet dominated by chironomids and ephemeropterans. Other foods consumed in smaller amounts included isopods, recently hatched crayfish, and limpets. Ten specimens, 54–70 mm TL, collected on 25 July from the Red River in Little River County contained 90%

chironomids by volume, with some ephemeropterans also ingested. Nine individuals, 60–69 mm TL, collected on 20 August from a Red River oxbow lake in Lafayette County fed predominantly on chironomids, but odonates, ephemeropterans, coleopterans, and freshwater shrimp were also included in the diet.

There is little information on the reproductive biology of the Bigscale Logperch. It usually reaches sexual maturity in the second year of life in California, where it spawns from April to June (Moyle 1976). Simon and Kaskey (1992) described the eggs, larvae, and early juveniles from the Trinity River basin, Texas. Spawning is believed to occur from late February until mid-April in central Texas (Stevenson 1971) and between February and April in Oklahoma (Hubbs 1985). Larval Bigscale Logperch are pelagic and were found mainly at night in surface and mid-depth plankton tows in Eagle Mountain Lake, Texas, from March to May (Simon and Kaskey 1992). Bigscale Logperch juveniles taken from the Red River in Lafayette County by TMB, Drew Wilson, and Les Claybrook on 18 July 1997 ranged from 37 to 47 mm TL. The actual spawning of *P. macrolepida* has not been observed in nature, but all other members of the genus *Percina* bury their eggs in gravel or sand substrates in swift current. Our observations of populations in Red River oxbows lead us to speculate that spawning successfully occurs in habitats lacking the presumably suitable substrate and current conditions for egg burial. It is possible that *P. macrolepida* differs from all other members of its genus in that it attaches its eggs to vegetation or other objects above the substrate so that siltation does not kill the developing embryos. In support of this hypothesis, Moyle (1976) reported that Bigscale Logperch collected from California sloughs spawned in a vertical position in aquaria, attaching their eggs to the stems of aquatic plants. Ten to 20 eggs were laid at each spawning, and females spawned multiple times with different males over an extended period. Stevenson (1971) reported 186 mature eggs in a female 72 mm TL, and 365 eggs in a female 83 mm TL. The eggs averaged 1.4 mm in diameter. Based on length frequency distribution of specimens from the Arkansas River compiled by TMB, maximum life span is probably 3+ years.

CONSERVATION STATUS   The presumably native populations of Bigscale Logperch in the mainstem Red River and the large populations in some Red River oxbows are currently secure. It is now the numerically dominant logperch in most navigation pools throughout the Arkansas River main stem in the state. The introduced Arkansas River populations are secure.

Percina maculata (Girard)
Blackside Darter

Figure 26.90. *Percina maculata*, Blackside Darter. *Uland Thomas*.

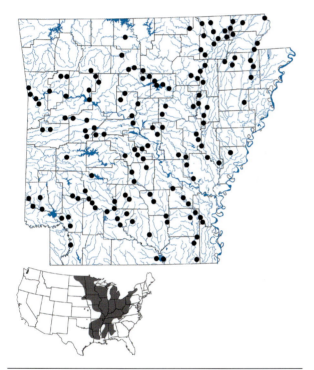

Map 229. Distribution of *Percina maculata*, Blackside Darter, 1988–2019.

CHARACTERS   A moderately slender, medium-sized darter with 6–12 (usually 6–9) dark brown oval to rectangular blotches along midside, the blotches somewhat connected by a narrow dark line along the lateral line (Fig. 26.90). A narrow dark vertical suborbital bar present beneath the eye, and a single small black basicaudal spot (darker than lateral blotches) present. Mouth terminal. Preopercle smooth on posterior edge. A well-developed premaxillary frenum present. Gill membranes separate. One or more enlarged and modified scales present between pelvic fin bases; anterior part of belly naked but the midline with 7–13 enlarged and modified scales (male), or belly scaled except along the midline (female). Breast naked;

opercle fully scaled; cheek scalation variable, naked to fully scaled; nape naked or with embedded scales. Lateral line complete, with 62–79 scales. Scale rows above lateral line 7–11; scale rows below lateral line 10–13. Diagonal scale rows 17–23; scale rows around caudal peduncle 20–27. Dorsal spines 13–15; dorsal rays 12–14; anal spines 2; anal rays usually 10 or 11; pectoral rays usually 13 or 14. Branchiostegal rays usually 6. Most authors report a lack of breeding tubercles (Page 1983a; Boschung and Mayden 2004), but Etnier and Starnes (1993) reported that breeding males have tuberculate ridges on rays and spines of anal and pelvic fins, and hardened swellings are found on the exposed portion of modified midventral scales. Maximum size around 4.5 inches (114 mm).

LIFE COLORS   Back and upper sides olive yellow to brown with black vermiculations and 8 or 9 irregular, inconspicuous dusky saddles on dorsum. Lateral blotches and median basicaudal spot black (basicaudal spot more distinct in smaller specimens). Basicaudal spot is usually darker than the lateral blotches, but this pigmentation will occasionally be sufficiently broad and diffuse to form a blotch. Underside of body a pale white or yellow, usually without melanophores. A dark subocular bar extends down and slightly back from the eye to the ventral margin of the preopercle. Dorsal and caudal fins dusky; lower fins clear or yellowish white. Pectoral fins with 4–6 pale bands formed by concentrations of melanophores in patches along the rays. Pelvic fins may be immaculate or dusky but do not have pigment arranged in bands. Spinous dorsal fin of males and females with a proximal concentration of melanophores on the first 2–4 interradial membranes (often obscured in breeding males by the overall darkening of the fin). Males (but not females) usually also have some concentration of pigment forming a proximal dark band on all interradial membranes of the spinous dorsal fin. Body and fins of breeding male becoming very dusky and head develops an iridescence. Colors of breeding female similar to those of male but less intense.

SIMILAR SPECIES   The Blackside Darter is most commonly confused in Arkansas with the Dusky Darter, but it differs from *P. sciera* in having separate gill membranes (vs. broadly joined), a single medial basicaudal spot (vs. vertical row of 3 basicaudal spots), and no serrae on preopercle (vs. rear margin of preopercle weakly serrate). It differs from *P. pantherina* in having fewer than 80 lateral line scales (vs. more than 80 lateral line scales) and 6–9 oval lateral blotches (vs. 10–15 round lateral blotches). It is distinguished from *P. nasuta, P. phoxocephala,* and *P. brucethompsoni* by its separate gill membranes (vs. moderately joined), shorter snout, and lack of an orange submarginal band in the first dorsal fin in life (vs. longer snout and an orange band present in first dorsal fin).

VARIATION AND TAXONOMY *Percina maculata* was described by Girard (1859) as *Alvordius maculatus* from Fort Gratiot, Lake Huron, Michigan. Bailey et al. (1954) assigned all *Hadropterus* species to genus *Percina*. Hubbs and Raney (1939) referred to *P. maculata* as "probably a complex of subspecies," and in subsequent decades, a few populations were recognized as new species. In 2002, Near concluded that genetic variation indicated that *P. maculata* was still a complex of undescribed species. Beckham (1980, 1986) studied morphological variation of the Blackside Darter and provided a redescription of the species.

During the past 100 years, various authors have placed *P. maculata* in the subgenus *Alvordius* with different combinations of other species. Based on genetic evidence, Near et al. (2011) found that the commonly recognized species composition of the subgenus was not monophyletic and recognized *Alvordius* as a clade containing *P. maculata*, *P. gymnocephala*, *P. notogramma*, and *P. pantherina*. Natural hybridization was reported between *Percina maculata* and *P. phoxocephala*, *P. caprodes*, and *Etheostoma gracile* (Page 1976b).

DISTRIBUTION *Percina maculata* is widely distributed in the Great Lakes, Hudson Bay, and Mississippi River basins from Canada south to Louisiana, and in Gulf Slope drainages from Alabama west to the Sabine River drainage, Texas. It is widely distributed in Arkansas and has been reported from all major drainages in the state (Map 229). Except for a single pre-1960 record (Map A4.225), it is absent from the upper White River. It is rarely found in the central Ouachitas and is commonly found in the Coastal Plain of eastern and southern Arkansas.

HABITAT AND BIOLOGY Although it occurs in widely scattered localities in Arkansas, the Blackside Darter is rarely abundant. It is not commonly found in the most mountainous regions of the Ozarks and Ouachitas, and it is practically absent from the largest rivers, with only a few records from the Arkansas River and the Red River (Buchanan et al. 2003). The Blackside Darter is most abundant in the state in medium-sized to large creeks and in small to medium-sized rivers near the Coastal Plain-upland boundary and in Arkansas River Valley boundary streams north of the Arkansas River and above the Fall Line. Thomas (1970) and Kuehne and Barbour (1983) reported that it occurs in a variety of habitats over its range, from shallow, swift water over a hard bottom, often of rubble or rock, to sluggish water over muddy bottoms. In Arkansas, it occurs in pools and in riffles of slow to moderate current having gravel or other rock substrates. Some authors

have noted that it often occurs in association with undercut banks, detritus, brush piles, and aquatic vegetation (Boschung and Mayden 2004). It has also been found in nine Arkansas impoundments (Buchanan 2005). A few observers (Kuehne and Barbour 1983; Larimore et al. 1959; Scott and Crossman 1973) reported a tolerance of turbidity and siltation by this species and a good ability to reinvade disturbed environments. This does not appear to be true of *P. maculata* in Arkansas.

Black (1940), Trautman (1957), and Pflieger (1997) noted that *P. maculata* is less of a bottom fish than most darters, often swimming in midwater and sometimes rising to the surface to feed (it is one of the few darters to have a small, well-developed swim bladder). Trautman (1957) observed it rising to the surface to feed on ovipositing dipterans, and even observed it jumping into the air to capture flying insects. Smart and Gee (1979) also noted its pelagic feeding habits. It also feeds on insect larvae and nymphs, such as chironomids, simuliids, and ephemeropterans, and on small crustaceans. The first foods of larvae are cladocerans (Labay et al. 2004); juveniles also feed mainly on microcrustaceans, and adults feed mainly on insects (Page 1983a). Fourteen specimens collected from the White River, Arkansas, on 7 March fed almost exclusively on benthic insect nymphs and larvae (stomachs examined by TMB). The diet was dominated by ephemeropterans, followed by chironomids, trichopterans, plecopterans, and simuliids. Three individuals had ingested fish eggs. Specimens from Arkansas River drainages in Perry, Scott, and Yell counties collected from June through August by TMB fed mainly on ephemeropterans, with odonates, copepods, chironomids, and coleopterans also ingested in small amounts. Seventeen specimens collected in September in Pope County had ingested comparatively small amounts of food, with ephemeropterans and cladocerans making up most of the diet; chironomids, simuliids, odonates, and coleopterans were also ingested in small amounts.

Spawning occurs in Illinois from April to June (Smith 1979) and in Kansas from April to early May at water temperatures of 15°C (60°F) or slightly warmer (Cross and Collins 1995). Spawning begins in Arkansas by early to mid-March and continues through May. Ripe females were found in Departe Creek in Independence County on 7 March by TMB. Various authors have reported different spawning habitats for this species. Winn (1958a, b) reported that *P. maculata* spawned in pools and raceways, where it laid its eggs in coarse sand or fine gravel in water depths of 305 mm or more. Petravicz (1938) found it spawning over sand and gravel in water 305 mm deep with moderate current. In the Kaskaskia River, Illinois, *P. maculata* was the first of four *Percina* species to enter the spawning riffles in large numbers in the spring (Thomas 1970). Winn (1958b) provided the following details of its spawning behavior. The males move first to the spawning riffles or raceways but do not establish territories. The females remain in deeper water until ready to spawn. When a female moves onto the spawning grounds, a male forms a moving territory around her. There is little courtship and no nest is constructed. The female comes to rest on the gravel substrate and the male mounts and clasps her with his pelvic fins. The two fish vibrate their bodies to release eggs and sperm. The fertilized eggs are buried in the substrate and are deserted by the parents. Year-old females produce 630–860 eggs; 2-year-old females produce 1,000–1,750 eggs. After hatching, the larvae are pelagic for about 3 weeks before assuming a more benthic life style. Maximum life span is around 4+ years.

CONSERVATION STATUS   The Blackside Darter is widely distributed throughout most of Arkansas and is common at a few localities. Its populations appear stable.

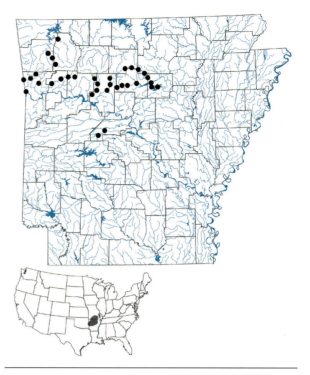

Map 230. Distribution of *Percina nasuta*, Longnose Darter, 1988–2019.

Percina nasuta (Bailey)

Longnose Darter

Figure 26.91. *Percina nasuta*, Longnose Darter.
Uland Thomas.

CHARACTERS   A slender darter with a long head and a long, pointed snout, with snout length going into head length 3.5 times or less (Fig. 26.91). Snout length greater than 9% of SL. Head length going into SL fewer than 3.5 times. Midside with a series of 12–15 vertically elongated dark blotches, usually forming bars; fewer than 7 lateral blotches between the origin of the second dorsal fin and the hypural plate. No suborbital bar. Gill membranes moderately to broadly connected across throat. A wide premaxillary frenum present. Usually 7 (occasionally 6) branchiostegal rays. Infraorbital canal uninterrupted. One or more enlarged and modified scales present between pelvic fin bases; breast naked except for 1 to a few enlarged and modified scales; belly with a series of 15–30 enlarged and modified scales along midline (males) or with scales of normal size and shape (females). Nape, opercle, and cheek fully scaled (lower part of cheek appears to be naked

because the few small cycloid scales are embedded). Lateral line complete, with 65–84 scales (usually 69–77). Diagonal scale rows usually 23–30. Scales above lateral line 9–13 (modally 11); scale rows below lateral line 12–16 (usually 15 or 16). Caudal peduncle circumferential scale rows 23–32 (modally 27). Dorsal spines 11–14 (modally 13); dorsal rays 12–14 (modally 13); anal spines 2; anal rays 8–10 (modally 9); pectoral rays 12–15 (modally 14). Breeding males and females without tubercles. Maximum size about 4.5 inches (110 mm).

LIFE COLORS   Back and upper sides yellowish brown, with dark midlateral blotches and a small black basicaudal spot. Upper sides and dorsum with irregular brown blotches; underside of body yellowish white. First dorsal fin dusky with bright orange or yellow submarginal band; second dorsal and caudal fins lightly banded with brown; other fins clear to dusky. Breeding males with dramatic secondary development of black pigment on fins and body (Thompson 1977).

SIMILAR SPECIES   The Longnose Darter differs from *P. brucethompsoni* and *P. phoxocephala* in having a longer snout, its length 9% or more of the SL (vs. snout length less than 9% of SL). It also differs from *P. phoxocephala* in that snout depth is less than snout length (vs. snout depth equal to or greater than snout length), and in usually having more than 72 lateral line scales (vs. usually fewer than 72 lateral line scales). It differs from *P. sciera* in having 1 basicaudal spot (vs. 3), lack of serrate preopercle (vs. rear margin of preopercle weakly serrate), and presence of orange submarginal band in first dorsal fin in life (vs. no orange submarginal band in first dorsal fin). It can be distinguished from remaining *Percina* species by moderately joined gill membranes (vs. separate).

VARIATION AND TAXONOMY *Percina nasuta* was originally described by Bailey (1941) as *Hadropterus nasutus* from the Middle Fork of the Little Red River near U.S. Route 64 bridge, 2.42 km southeast of Leslie, Searcy County, Arkansas. It is a member of subgenus *Swainia*. When *Hadropterus* was reduced to a subgenus of *Percina*, *P. nasuta* was moved to subgenus *Swainia* (Bailey and Gosline 1955). Variation was studied by Thompson (1977). Populations in the Ouachita River drainage formerly included under *P. nasuta* have now been described as a separate species, *Percina brucethompsoni* (Robison et al. 2014). Other populations in the lower White, Spring, and Strawberry rivers, thought possibly to be an undescribed species closely related to *P. nasuta*, have now been tentatively assigned to *P. phoxocephala*. A combined morphological and genetic analysis of those populations is needed to clarify their status.

There is considerable variation in *P. nasuta* in some characters, particularly in various scale counts. Arkansas River drainage populations have a wider range of lateral line scales with lower mean counts than those in the upper White and Little Red rivers. Longnose Darters from the Little Red River almost always have 7 branchiostegal rays, while populations from the upper White and Arkansas river drainages have 6 (55%) or 7 (45%) branchiostegal rays. The population in the Fourche LaFave River (Arkansas River drainage) is distinctive in color pattern, having more pronounced vertically elongated, narrow, stripelike bars along the side than other populations (striping is more pronounced in males than in females; B. A. Thompson, pers. comm.). Analysis of two mitochondrial genes revealed that *P. nasuta* was not monophyletic, and specimens were distributed among two clades: (1) a White River clade that was the sister lineage of *P. brucethompsoni*, and (2) specimens sampled from the Arkansas River drainage that were nested in the clade of Arkansas River *P. phoxocephala* (Robison et al. 2014). Those authors found that *P. nasuta* and *P. phoxocephala* sampled from the Arkansas River drainage form a clade that is the sister lineage of a clade containing *P. brucethompsoni* and *P. nasuta* from the White River drainage. Robison et al. (2014) also speculated that the similarity of the mtDNA haplotypes observed in the

Arkansas River *P. phoxocephala* and *P. nasuta* specimens indicated a possibility of mtDNA introgression from *P. phoxocephala* to *P. nasuta*.

DISTRIBUTION *Percina nasuta* occurs in only three states in the White River drainage of Missouri and Arkansas, the St. Francis River drainage of Missouri, and the Arkansas River drainage of Arkansas and Oklahoma. Most of its range is in Arkansas, where it inhabits the upper White River, the upper Little Red River, and a few tributary streams of the western half of the Arkansas River (Mulberry River, Big Piney Creek, Illinois Bayou, Point Remove Creek, and South Fork Fourche LaFave River) (Map 230). There has been little change in its distribution in Arkansas during the past 30+ years, but its abundance may have declined. A population persists in Lee Creek in Sequoyah County, Oklahoma, upstream from Lee Creek Reservoir (Gatlin and Long 2011). Holley and Long (2018) reported a specimen of *P. nasuta* from the Poteau River in Oklahoma near the Arkansas state line in 2015.

HABITAT AND BIOLOGY The Longnose Darter inhabits clear, silt-free upland streams. Large streams and small rivers with cobble and gravel bottoms are preferred. It is mainly a pool darter, and in the spring it inhabits raceway areas of pools having moderate to strong current at depths of 1–3 feet (0.3–0.9 m). During periods of low flow in late summer and early fall, it is found in the deeper parts of the pools in little or no current, often over a sandy bottom and frequently near aquatic vegetation. Pflieger (1997) reported it from a reservoir in Missouri, and in August 1984, one small specimen was taken from Lake Nimrod in Yell County, Arkansas. Buchanan (2005) reported seven specimens from Greers Ferry Lake. It has not been found in the Beaver Lake tailwaters since the impoundment of that reservoir (Quinn and Kwak 2003). Although known from several widely scattered localities in Arkansas, it is fairly rare in the state. Where it does occur, *P. nasuta* populations are very small and its range has been reduced by reservoir construction (Beaver Lake and Greers Ferry Lake). It appears to be very sensitive to environmental disturbance.

Little is known of the life history of this species. Stomach contents of eight specimens, 55–92 mm TL, collected in October from the South Fork of the Little Red River in Van Buren County were examined by TMB. Plecopterans occurred in all stomachs and dominated the diet by percent volume. Ephemeropterans and trichopterans were consumed in small amounts.

The breeding season probably extends from March into May, but there is little information about spawning habitat and behavior. Spawning was observed by TMB in a pool raceway in the upper White River in mid-May and in similar habitat in Lee Creek in Crawford County on 10 April. Thompson (1977) believed that peak spawning probably occurs in April.

CONSERVATION STATUS   Laboratory spawning of Longnose Darters was induced by manipulating temperature and photoperiod, but little success in raising the offspring beyond the larval stage was achieved (Labay 1992; Anderson et al. 1998). Populations of this uncommon darter have been periodically monitored during the past two decades, and the monitoring should continue in the future. This species appears to be especially vulnerable to environmental disturbance, and we consider it a species of special concern, if not threatened.

Percina pantherina (Moore and Reeves)
Leopard Darter

Figure 26.92. *Percina pantherina*, Leopard Darter. Uland Thomas.

Map 231. Distribution of *Percina pantherina*, Leopard Darter, 1988–2019.

CHARACTERS   A moderate-sized darter with a series of 10–14 round (sometimes square) black spots along midside connected by a narrow lateral band (Fig. 26.92). Gill membranes separate; premaxillary frenum present. Infraorbital canal uninterrupted. Cheek and opercle scaled (cheek scales embedded); nape naked or with a few embedded scales. One or more enlarged and modified scales between pelvic fin bases; midline of belly naked (females) or with a row of 11–15 modified scales (males). Lateral line complete, with 81–96 scales. Scale rows above lateral line usually 10–12 (9–13); scale rows below lateral line usually 16–18 (14–19). Diagonal scale rows 25–30; scale rows around caudal peduncle 28–33. Dorsal spines 12–16; dorsal rays 10–14; anal spines 2; anal rays 8–11; pectoral rays 13 or 14. Breeding males without tubercles. Maximum size about 3.5 inches (92 mm).

LIFE COLORS   A row of 10–14 round, black spots along midside. Midline of back with 11–13 irregular dark blotches; upper sides with many smaller scattered round or irregularly shaped dusky spots and blotches. Back and upper sides olive; lateral spots black; dorsal markings dusky. Underside of body white. Pre- and suborbital bars dark. Small black somewhat vertically elongated basicaudal spot present. Dorsal spines and soft rays often margined in black.

SIMILAR SPECIES   The Leopard Darter closely resembles the Blackside Darter. It differs in having 81 or more lateral line scales (vs. fewer than 80) and 10–14 round midlateral blotches (vs. 6–9 horizontal oblong blotches). It differs from *P. sciera* in having separate gill membranes (vs. broadly joined), a single basicaudal spot (vs. vertical row of 3 spots), and no serrae on preopercle (vs. posterior margin of preopercle finely serrate). It is distinguished from *P. brucethompsoni*, *P. nasuta*, and *P. phoxocephala* by its separate gill membranes, shorter snout, and lack of an orange band in the first dorsal fin in life (vs. gill membranes moderately joined, snout noticeably elongate, and an orange submarginal band present in first dorsal fin).

VARIATION AND TAXONOMY   *Percina pantherina* was described by Moore and Reeves (1955) as *Hadropterus pantherinus* from the Little River west of Pickens, Pushmataha County, Oklahoma. It was formerly included in *Hadropterus maculatus* (*Percina maculata*). *Hadropterus* was subsequently diagnosed as a subgenus of *Percina*, and *P. pantherina* was moved to subgenus *Alvordius* (Knapp 1964; Page 1974; Near et al. 2011). Echelle et al. (1999) studied allozyme variation in seven populations of *P. pantherina* and found fairly low average heterozygosity but a relatively high level of allele diversity, with two or more alleles detected at 21 of the 31 gene loci examined. An allele frequency analysis identified three primary clades: (1) populations in Little and Glover rivers, (2) populations from the Mountain Fork River drainage, and (3) populations in Robinson Fork and Cossatot rivers. Near (2002) and Near et al. (2011), based on phylogenetic analyses of the cytochrome *b* gene and nuclear DNA, recognized *P. maculata* as the sister taxon of *P. pantherina*. Schwemm and Echelle (2013) studied geographic variation of 8 microsatellite loci and mitochondrial DNA cytochrome *b*, and reported significant subdivision between tributaries in frequency of genetic markers. The Cossatot River population showed the lowest genetic diversity.

DISTRIBUTION   The Leopard Darter is endemic to the Little River system of Arkansas and Oklahoma (Robison 1980f). Most of the species' range is in Oklahoma, and it is a rare species in Arkansas, found only in the Mountain Fork, Rolling Fork, and Cossatot rivers in southwestern Arkansas (Map 231). It was federally listed as a threatened species in 1978. Schwemm (2013) interpreted a microsatellite genetic analysis of historical demography as being consistent with the hypothesis that effective population sizes in *P. pantherina* have declined in response to the construction of large-reservoir dams on the Little River system.

HABITAT AND BIOLOGY   Early reports of Leopard Darter habitat indicated that it preferred gravel- and cobble-bottomed riffles and raceways of large creeks and small rivers with high gradients (Moore and Reeves 1955; Eley et al. 1975; Page 1983a). Robison (1980f) reported its occurrence in swift riffles and pools in these streams. In a study of the Leopard Darter in Glover Creek, Oklahoma, over a three-year period, Jones et al. (1984) found that it is primarily a pool darter. It demonstrated a preference for pools during all seasons but was captured in riffle and run areas more frequently in winter and spring during the presumed breeding season. James and Maughan (1989) also provided evidence that it lives in pools during summer and fall and moves into riffles in late February when water temperature reaches about 10°C. Although it is adapted to stream pools, it is apparently intolerant of reservoir conditions. Swimming performance in relation to road culverts was studied by Toepfer et al. (1999). Its populations were characterized by small numbers and low densities. It was captured most frequently at depths of 7.9–31.1 inches (20–79 cm) in slow current usually over a rubble and boulder substrate.

Food habits have been reported as small insect larvae (Robison 1978c), mayfly nymphs, black fly larvae, midge larvae, coleopterans, chaoborids, and Chlorophyta (Miller and Robison 2004). Williams et al. (2006) found that, in general, Leopard Darters selected food items that were relatively common in their environment; therefore, food availability may not limit population density. In that study, nine families of macroinvertebrates representing six orders were found in the gut contents, and ephemeropterans (Baetidae and Heptageniidae) and chironomids were the most common food items eaten. Three of 18 Leopard Darter larvae collected in light traps from the Glover River, Oklahoma, contained cladoceran prey (Labay et al. 2004).

Leopard Darters begin moving into riffles in late February, and spawning occurs on the riffles from mid-March to mid-May at water temperatures of 12–20°C (James et al. 1991). Spawning behavior was described by James and Maughan (1989) as follows:

A gravid female, followed by one or more males, moved from the riffle tailwaters upstream into the riffle. The female moved slowly over the gravel and rubble and occasionally settled on the substrate. Males appeared to establish and defend moving territories around a gravid female and attempted to chase other males away from the female. One of the males, usually the largest, attempted to position himself directly on top of her. If a female was receptive, a male positioned himself with his pelvic fins on her spinous dorsal fin. With both fish oriented in the same direction, the male curved his body into an S-shape and the pair began to vibrate rapidly and presumably release gametes. The vibrating movements of the pair left the fertilized eggs buried in fine gravel, in the interstices of coarse gravel, and in rubble. The female and attendant males then selected another spawning site and repeated the spawning act.

James and Maughan (1989) found that spawning sites were at depths of 30–90 cm over gravel substrates at water velocities of 0–50 cm/s mainly in the riffle tailwaters. Egg predation by *Percina copelandi* was observed on two occasions. Berkhouse (1990) was able to maintain and spawn Leopard Darters in captivity. Life span is approximately 18 months, and most individuals spawn only once during their lifetime (James et al. 1991; Zale et al. 1994).

Schaefer et al. (2003) studied the effects of barriers and thermal refuges on local movement of Leopard Darters. At one study site where patches of preferred habitat were separated by a riffle, darters moved downstream across the riffle, but they also moved upstream into deeper water when water temperatures exceeded 29°C in the preferred habitat. At a second site where preferred habitat was separated by a road crossing with culverts, all documented movement was in a downstream direction.

CONSERVATION STATUS   In 1978, the U.S. Fish and Wildlife Service listed this species as Threatened. Its already small range in Arkansas has been further reduced and fragmented by the impoundment of reservoirs in Oklahoma and Gillham Lake on the upper Cossatot River in Arkansas, and Schwemm (2013) considered the Cossatot River population to be near extinction. Molecular estimates of population size along with severe barriers to gene flow caused Schwemm (2013) to conclude that, without human intervention, *P. pantherina* is a species among the "living dead" doomed to extinction. Measures that would reverse the processes of habitat loss and fragmentation, such as dam removal, are probably not feasible. Another threat to this species is silviculture, with the resultant road cutting, erosion, and increased siltation. Recent studies reveal that small populations still exist in Arkansas. A snorkeling survey in July 2017 documented the presence of Leopard Darters in two separate stream reaches of the Cossatot River (Jeff Quinn, pers. comm.). We consider this species endangered in Arkansas.

Percina phoxocephala (Nelson)
Slenderhead Darter

Figure 26.93. *Percina phoxocephala*, Slenderhead Darter. *Uland Thomas.*

CHARACTERS   A slender, terete darter with a slender head and a moderately long, pointed snout, the snout length going into head length usually more than 3.5 times; snout longer than eye diameter (Fig. 26.93). Snout length in adults less than 7.5% of SL. Head length going into SL 3.5 times or less. Preopercle smooth on posterior edge. A series of 10–15 often indistinct, dark, round to square blotches along midside, the anterior blotches typically longer and slimmer; back and upper sides with smaller dark blotches and vermiculations highly variable in size and intensity. Body markings varying with size, juveniles having numerous contrasting dorsal crossbars and lateral blotches, while adults are plainer and darker (Cross 1967). Gill membranes moderately to broadly joined across throat. A well-developed premaxillary frenum present. Branchiostegal rays 6. One or more enlarged and modified scales present between pelvic fin bases in both sexes; a few enlarged and modified scales on breast; posterior half of male's breast

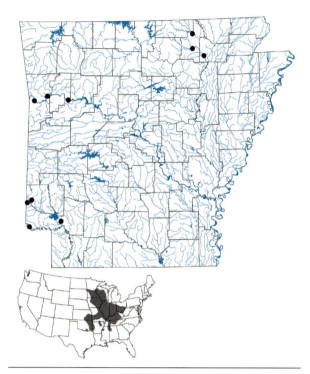

Figure 26.93. *Percina phoxocephala*, Slenderhead Darter. *Uland Thomas.*

is scaled and the anterior half is naked; breast of female is naked (except for 1 or 2 modified scales between pelvic fins). Belly with a series of 10–26 modified scales on midline (males) or with scales of normal size and shape or naked posteriorly (females). Nape and opercle fully scaled; cheek fully or partly scaled. Lateral line complete, with 61–73 (usually fewer than 72) scales. Scale rows above lateral line usually 8–10 (7–11); scale rows below lateral line 12–15. Diagonal scale rows 20–28; scale rows around caudal peduncle 23–28. Dorsal spines 11–14; dorsal rays 11–15; anal spines 2; anal rays 7–10 (usually 9); pectoral rays 13–15. Breeding males without discrete tubercles, but sometimes the pelvic and anal fins with weak tuberculate ridges. Maximum size about 3.75 inches (96 mm) TL, but adults in Little River noticeably smaller than elsewhere. Largest Arkansas specimen examined by TMB was 3.3 inches (85 mm) TL from the Arkansas River in Crawford County.

LIFE COLORS   Back and upper sides yellow brown with darker brown vermiculations and blotches; lower sides and venter white to yellow. Black preorbital bar prominent; a suborbital bar is sometimes present. A small black basicaudal spot present. First dorsal fin with a dusky basal band and an orange (males) or yellow (females) submarginal band; second dorsal and caudal fins with light brown bands. Breeding males becoming darker; breeding females show no color differences from nonbreeding females.

SIMILAR SPECIES The Slenderhead Darter differs from *P. nasuta* and *P. brucethompsoni* in having a shorter snout, its length less than 7.5% of standard length (vs. more than 7.5%); snout depth greater than snout length (vs. snout depth less than snout length); eye diameter more than 70% of snout length (vs. eye less than 70% of snout length); and lateral line scales usually fewer than 72 (vs. lateral line scales usually more than 72). It differs from *P. sciera* in having only 1 black basicaudal spot (vs. 3), an orange submarginal band in first dorsal fin in life (vs. no orange submarginal band in first dorsal fin), and lack of serrate preopercle (vs. rear margin of preopercle weakly serrate). Distinguished from remaining *Percina* species in Arkansas by moderately joined gill membranes (vs. gill membranes not joined or only slightly joined across throat).

VARIATION AND TAXONOMY *Percina phoxocephala* was described by Nelson (1876) as *Etheostoma phoxocephalum* from the Illinois River and its tributaries in Illinois. It is in the subgenus *Swainia*. Near (2002) in a genetic analysis identified *P. phoxocephala* as a sister species to *P. nasuta*. Based on analysis of two mitochondrial genes, *P. phoxocephala* was not monophyletic, and sampled specimens were distributed among three clades in the mtDNA gene tree: (1) an Arkansas River drainage clade, (2) a clade containing all specimens sampled from the Mississippi, Missouri, and Ohio river drainages, and (3) a clade of specimens sampled from the Red River drainage (Robison et al. 2014). Those authors also found that specimens of *P. phoxocephala* from the Red River drainage were resolved as the sister lineage of a clade containing all other sampled populations of *P. phoxocephala*, *P. brucethompsoni*, and *P. nasuta*. Therefore, Red River populations (and possibly other Arkansas populations) may represent an undescribed species. Robison et al. (2014) also provided evidence that indicated a possible mtDNA introgression from *P. phoxocephala* to *P. nasuta* in Arkansas River drainage populations. *Percina* specimens from the White River near Batesville (Buchanan 1973b), the Spring River, and the Strawberry River (reported by Meek 1894a) referred to in the first edition of this book as being of uncertain taxonomic status, were later tentatively assigned to *P. phoxocephala* (Bruce A. Thompson, pers. comm.); however, there has been no thorough genetic analysis of those populations. *Percina phoxocephala* is known to hybridize with *P. maculata* and *P. caprodes* (Etnier and Starnes 1993).

DISTRIBUTION *Percina phoxocephala* occurs in the upper portion of the Mississippi River basin from West Virginia to northeastern South Dakota, and south to northern Alabama and northeastern Texas. It is also found in the Lake Michigan drainage of Wisconsin.

There are reports of *P. phoxocephala* from Arkansas more than 120 years old. Prior to 1900, Jordan and Gilbert (1886) reported the first Arkansas records for this species from Lee Creek in Crawford County, remarking that it was "not very common." Jordan and Evermann (1896), however, stated that the Slenderhead Darter was "locally common, especially in Arkansas." Meek (1891, 1894a, b, 1896) listed it as "abundant" in the Ouachita River at Crystal Springs in Garland County and reported additional records from the Middle Fork of the Little Red River in Cleburne County, the Spring and Strawberry rivers in Lawrence County, the White River at Batesville in Independence County, and the Poteau River at Fort Smith in Sebastian County. Carl L. Hubbs examined most of Meek's specimens in The Field Museum of Natural History and confirmed their identification as *P. phoxocephala*. It is likely that the specimens from Crystal Springs and the Little Red River were *P. brucethompsoni* and *P. nasuta*, respectively, but the remaining pre-1900 records may well have been *P. phoxocephala*. The Slenderhead Darter was possibly less common in Arkansas by 1940, and Black (1940) stated, "We have never collected this species." The next verified collection of *P. phoxocephala* from Arkansas was by Larry Rider on 23 August 1973 from Blue Mountain Lake, a Petit Jean River impoundment in Logan County. The Slenderhead Darter was subsequently found in the Illinois, Arkansas, Little, and Red rivers in western Arkansas. It has also been reported from the White River near Batesville (Buchanan 1973b), and the Spring and Strawberry rivers. Buchanan (2005) reported this species from two navigation pools (three collection localities) of the Arkansas River. There are no post-1988 verified records from the Illinois and White rivers (Map 232).

HABITAT AND BIOLOGY Though sporadically distributed and uncommon in Arkansas, the Slenderhead Darter is often common in other parts of its range, where it most frequently inhabits gravel or rocky riffles with moderate to swift current and clear water in moderate-sized rivers and large creeks (Page 1983a). Population density estimates in Illinois were 1 darter/35.4 m² in the Embarras River (Page and Smith 1971), and 1 darter/24 m² in the Kaskaskia River (Thomas 1970). In Illinois, it demonstrated a preference for medium-sized rivers and was rare in downstream large-river habitats and in headwater habitats (Thomas 1970). In the Embarras River, Illinois, *P. phoxocephala* was generally found occupying the same habitat as *P. sciera*, the relatively shallow, gravel-bottomed portion of the river channel; however, it was somewhat less restricted to gravel than *P. sciera* and occasionally taken in the channel over sand. Although sporadically distributed in Arkansas, it has been found in a variety of habitats. It occurs in medium-sized rivers in the Illinois and Little rivers; however, it has also been found in impoundments (Blue Mountain Lake, Ozark Pool, and Dardanelle Reservoir), in the more turbid environment of the Arkansas River main stem near creek mouths over a sand and gravel bottom, and in the sandy Red River main stem. The largest collection known from Arkansas was 30 specimens taken from Dardanelle Reservoir on 6 September 2000 by TMB and Bob Limbird. Miller and Robison (2004) reported that in Oklahoma the Slenderhead Darter is somewhat more tolerant of turbidity than many darters, and it is often collected over a sandy substrate. Thompson (1980) noted that it is somewhat variable in its habitat preference but is moderately intolerant of siltation. Page and Smith (1971) also noted its plasticity in habitat choice in Illinois. Some authors reported that it moves downstream into deeper water in winter and returns to riffle areas in the spring (Etnier and Starnes 1993; Boschung and Mayden 2004).

In the Kaskaskia River, Illinois, the young fed mainly on chironomids, and larger individuals fed mainly on ephemeropterans, chironomids, and trichopterans (Thomas 1970). In the Embarras River, Illinois, 99% of the diet consisted of chironomids, simuliids, trichopterans, and ephemeropterans, with amphipods, fish eggs, and terrestrial insects consumed less frequently (Page and Smith 1971). Dipterans and ephemeropterans dominated the diet during all seasons, and trichopterans formed a large portion of the diet during the summer. Heaviest feeding occurred just prior to spawning, and the least feeding activity occurred in the months following spawning. As the darters grew, the diet composition changed from one dominated by dipteran larvae to a diet that included substantially greater proportions of larger immature mayflies and caddisflies. The diet in Iowa was similar to that reported from Illinois, with ephemeropterans the most common food item, and dipteran and caddisfly larvae making up most of the remainder of the food ingested (Karr 1963).

Spawning in Arkansas probably occurs from March to June. In Illinois and Missouri, it spawns in April and May in swift gravel riffles 6–24 inches (152–610 mm) deep (Page and Smith 1971; Page and Simon 1988; Pflieger 1997). Males move into the spawning habitat well before the females, possibly indicating territoriality during the breeding season. After spawning, the adults returned to deeper water (Page and Smith 1971). Juveniles remained in the spawning habitat (shallow riffles along gravel bars) for several weeks after hatching. At approximately 1 month of age, the young moved out into the deeper water of the channel (Page and Smith 1971). Spawning in Kansas occurred from late March

to early May in swift riffles over a substrate of rubble and gravel at depths of 45 cm (18 inches) or more and a water temperature of approximately 70°F (21.1°C) (Cross 1967; Cross and Collins 1995). Brewer et al. (2006) reported that spawning occurred between 12°C and 21°C, with most fish spawning between 19°C and 21°C. Aquarium spawning was described by Page and Simon (1988). Mating pairs selected a mixed gravel and sand substrate near rocks that sheltered them from the current. The male mounted the female in typical *Percina* fashion, and the vibrating pair produced transparent, nonadhesive fertilized eggs buried in the substrate. Eggs hatched in 120–124 hours at 22°C. Fecundity ranged from 100 to 750 mature ova per female in Iowa (Karr 1963), and Page and Smith (1971) estimated that each female laid from 50 to 1,000 eggs in one season in Illinois. Growth in Iowa populations averaged 1.4 inches (35.6 mm) during the first year of life, and 1.9, 2.0, and 2.3 inches (48.3, 50.8, and 58.4 mm, respectively) in subsequent years (Karr 1963). Sexual maturity was attained in Illinois at 1 year of age and minimum SLs of 40 mm (males) and 42 mm (females) (Page and Smith 1971). Maximum life span in different populations ranges from 2 to 4 years (Karr 1963, Thomas 1970, Page and Smith 1971).

CONSERVATION STATUS This is a rare species in Arkansas and is seldom found. We consider it threatened in this state.

Percina sciera (Swain)
Dusky Darter

Figure 26.94. *Percina sciera*, Dusky Darter. *David A. Neely.*

CHARACTERS A moderately large and deep-bodied *Percina* with a midlateral series of 8–10 dark blotches that tend to be connected by a lateral band (Fig. 26.94). Back with 7–10 small, indistinct dusky saddles; upper sides with dark vermiculations and zigzag lines. A vertical row of 3 round black spots, the lower 2 usually connected, on body at base of caudal fin. Gill membranes broadly joined across throat. Premaxillary frenum present. Rear margin of preopercle usually weakly serrate. Infraorbital canal uninterrupted, usually with 8 pores. Nape, cheek, opercle, and belly scaled. Breast usually naked in females and either

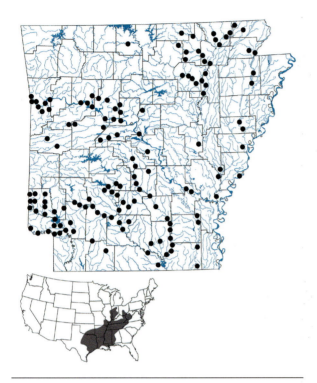

Map 233. Distribution of *Percina sciera*, Dusky Darter, 1988–2019.

partially or fully scaled in males. One or more (usually 2 or 3) enlarged and modified scales present between pelvic fin bases; midline of belly with a row of 10–20 enlarged and modified scales (males) or naked (females). Lateral line complete, with 59–72 scales. Scale rows above lateral line 7–10; scale rows below lateral line 12–16. Diagonal scale rows 20–26; scale rows around caudal peduncle 21–26. Dorsal spines 11–14; dorsal rays 11–13; anal spines 2; anal rays 7–10; pectoral rays 13–15. Breeding males without discrete breeding tubercles. Maximum size around 5.1 inches (130 mm).

LIFE COLORS   Bright colors are absent. Body olive green dorsally, with dark mottlings and dusky saddles; lateral blotches black. Side yellowish; undersides creamy white, sometimes with small, dark, irregular blotches in males. Dark preorbital bars meet on snout; postorbital bar dark and extends across the opercle; suborbital bar usually absent. Fins dusky and often barred with narrow brown lines. Breeding males become darkened and develop black vertical bars on sides, which extend over the dorsum. The top of the head, belly, breast, and fins also become darkened. Breeding females are mainly yellow and are heavily mottled with dark olive or black but do not become darkened like the male.

SIMILAR SPECIES   The Dusky Darter can be distinguished from most other Arkansas darters by its preopercle, which usually has a weakly serrate rear margin (vs. rear margin of preopercle not serrate). It differs from *P. maculata* in having broadly joined gill membranes (vs. separate) and a vertical row of 3 dark basicaudal spots (vs. a single median basicaudal spot). It differs from *P. brucethompsoni*, *P. nasuta*, and *P. phoxocephala* by its 3 basicaudal spots (vs. 1) and in lacking an orange submarginal band in first dorsal fin in life (vs. orange submarginal band present in first dorsal fin). It can be distinguished from *P. evides* by its broadly connected gill membranes (vs. gill membranes separate or slightly joined), scaled cheeks (vs. naked), and lateral blotches not connected to dorsal saddles (vs. dorsal saddles usually connected to lateral blotches). It resembles *P. vigil* but differs from that species in having 7–10 dorsal saddles (vs. 4 saddles) and a serrate preopercle (vs. rear margin of preopercle not serrate). The young of *P. sciera* are distinguished from the young of other *Percina* species by the fused blotches on the base of the caudal fin.

VARIATION AND TAXONOMY   *Percina sciera* was described by Swain (1883) as *Hadropterus scierus* from Bean Blossom Creek, tributary of the White River, Monroe County, Indiana. It is the only Arkansas species of subgenus *Hadropterus*. Hubbs and Black (1954) discussed its complex synonymy and reviewed Arkansas specimens in a study of the status of *P. sciera*. Two subspecies were recognized by C. Hubbs (1954): *P. s. apristis* of the Guadalupe River, Texas, and *P. s. sciera* in the remainder of the species range. The Guadalupe form has now been elevated to full species status (Robins and Page 2007). Based on analysis of the mitochondrial cytochrome *b* gene, Near (2002) found that subgenus *Hadropterus* was not monophyletic. Near proposed recognition of a *Percina sciera* clade comprising the sister taxa *P. sciera* and *P. aurolineata*. Natural hybridization has been reported between *P. sciera* and *P. caprodes*, *P. nigrofasciata*, and *Etheostoma spectabile* (Hubbs and Laritz 1961a, b; Suttkus and Ramsey 1967).

DISTRIBUTION   *Percina sciera* occurs in the Mississippi River basin from eastern Illinois to Ohio and West Virginia southwestward to western Alabama, most of Louisiana, and eastern Texas, and in Gulf Slope drainages from the Mobile Bay basin, Alabama, to the Colorado River, Texas. In Arkansas, the Dusky Darter is widely distributed in the White, St. Francis, Arkansas, Ouachita, and Red (Connior et al. 2011) river drainages but is almost absent from the Ozark Uplands in northcentral and northwestern Arkansas (Map 233). It is also rare in the more mountainous regions of the central Ouachitas.

HABITAT AND BIOLOGY The Dusky Darter is found in creeks and small to medium-sized rivers of low gradient with clear water and permanent current. It usually occurs in riffles in slow to moderate current over a gravel bottom and is often found in accumulations of sticks and leaves. Dusky Darters occurred in their optimal habitat in the Embarras River, Illinois, at an estimated density of 1 darter per 5.3 m² (Page and P. W. Smith 1970). This species is common in low-gradient Arkansas River Valley streams in western Arkansas, and it has been collected from the Arkansas River in sandy-bottomed backwater areas near creek mouths. The young are often found along the shallow, gravel edges of pools with moderate current and often enter smaller tributaries where adults are not found. Observations by Suttkus et al. (1994) found that the gravel raceway, midwater habitat described for this species by Page (1983a) is only seasonal; it overwinters in eddy pools or in slow current among organic debris lodged around snags or along the bank. In summer, Dusky Darters in Alabama moved to deep pools, where they hid in vegetation or debris (Boschung and Mayden 2004). It avoids high-gradient streams in the most mountainous regions of Arkansas and has probably been extirpated from many lowland streams in the Coastal Plain along the Mississippi River by channelization, agricultural practices, and other habitat altering activities. Buchanan (2005) reported that it occurred in small numbers in nine Arkansas impoundments, primarily Arkansas River navigation pools but also in Blue Mountain, Felsenthal, and Millwood reservoirs.

Dusky Darters in the Embarras River, Illinois, fed entirely on aquatic insects; juveniles fed largely on midge and black fly larvae, adults mainly on caddisfly and midge larvae (Page and P. W. Smith 1970). Other insects consumed were rhagionids (snipe flies), and nymphs of ephemeropterans and plecopterans. Trichopteran larvae dominated the diet during summer, and dipterans and ephemeropterans were the main components during other seasons, probably reflecting the relative availability of the different insect larvae throughout the year (Page and P. W. Smith 1970). Similar food habits were reported from Mississippi (Miller 1983), and Ross (2001) noted that small crustaceans, which are common in the diets of pool-inhabiting darters, were rare in the diet of the Dusky Darter. Stomach contents of specimens from eight Arkansas counties were examined by TMB, and it was also confirmed as a benthic insectivore in this state. Individuals from the White River collected in March fed primarily on chironomids, ephemeropterans, plecopterans, and adult dipterans. The diet in western Arkansas (Crawford, Sebastian, Scott, Pike, and Pope counties) in April, May, and June was dominated by chironomids and ephemeropterans, with plecopterans, simuliids, tipulids, coleopterans, and hemipterans also ingested in small amounts. Specimens collected in October from the St. Francis River in St. Francis County fed mainly on ephemeropterans and coleopterans, with dipterans (chironomids and simuliids) making up a small portion of the diet.

In Illinois, the dates of spawning and duration of the spawning period varied from year to year, primarily because of variation in climate and river conditions (Page and P. W. Smith 1970). Peak spawning in Illinois apparently occurred in June in most years. In central Texas, the Dusky Darter spawns from February through June (Hubbs 1961b). Based on examination of ovaries and testes by TMB, spawning occurs in Arkansas from March through May. Gonads of specimens collected in mid-June in Sebastian County had regressed and contained no ripe or maturing ova. We found Dusky Darters spawning in late March in Vache Grasse Creek (Sebastian County), late April in Horsehead Creek (Johnson County), and in mid-May in the Spring River, Arkansas. Mature females ranged from 63 to 100 mm TL and contained from 207 to 794 ripe ova ($\bar{x}$ = 342). In Illinois, females can produce 500–2,000 ova in a season, and females may contain 80–196 mature eggs at one time (Page and P. W. Smith 1970). Few details of its spawning habits are known, and spawning behavior has not actually been observed, but Pflieger (1997) suggested that it spawns over a gravelly substrate and then abandons the eggs. Boschung and Mayden (2004) believed that spawning occurs in the shallower portions (35–100 cm deep) of raceways in Alabama streams, where the eggs are scattered over a gravel substrate. The adhesive fertilized eggs probably settle into the gravel interstices, where development occurs without parental care. In the Pearl River, Louisiana, spawning occurred over sand and gravel bars (Suttkus and Ramsey 1967). Eggs hatch in 90–108 hours at temperatures between 23°C and 26°C (C. Hubbs 1961a). Larvae began to feed 4 days after hatching, and at 5 days the yolk sac was completely absorbed. Fry and small young were pelagic. Males are ready to breed at 1 year of age (55 mm SL), and females spawn when larger than 40 mm SL (Page and P. W. Smith 1970). The maximum life span is 4+ years.

CONSERVATION STATUS This widespread darter is common at a few localities. Its populations in Arkansas appear secure.

Percina shumardi (Girard)
River Darter

Figure 26.95. *Percina shumardi*, River Darter.
Uland Thomas.

Map 234. Distribution of *Percina shumardi*, River Darter,
1988–2019.

CHARACTERS A robust, medium-sized *Percina* with
8–13 dark, vertically elongated blotches along the side
of body, the anterior blotches more vertically elongated
than the posterior ones (Fig. 26.95). Two prominent black
blotches in spinous dorsal fin; blotch at anterior end of
fin small, posterior blotch large. Back usually without
saddles and having smaller obscure blotches; occasionally
some individuals have 7–12 indistinct blotches resembling
saddles. Black suborbital bar present. Small black basicaudal spot present, often partly joined with last lateral
blotch. Anal fin of adult male greatly elongated, extending
past end of soft dorsal fin. Eyes closely set high on head.
Mouth nearly terminal. Premaxillary frenum absent or

very narrowly developed. Preopercle smooth on posterior
edge. Gill membranes separate or slightly joined across
throat. Infraorbital canal uninterrupted, with 8 or 9 pores.
Cheek and opercle scaled; nape partly to fully scaled; breast
naked. One or more enlarged and modified scales present
between pelvic fin bases; midline of belly naked (females)
or with row of 1–18 enlarged and weakly modified scales
(males). Lateral line complete and straight, usually with
47–58 scales; pored scales on caudal fin 0–2. Scale rows
above lateral line 5–7; scale rows below lateral line 7–11.
Diagonal scale rows 14–17; scale rows around caudal
peduncle 20–25. Dorsal spines 9–12 (usually 9 or 10); dorsal rays usually 13–15 (11–16); anal spines 2; anal rays usually
11 or 12 (10–13); pectoral rays usually 13 or 14. Breeding
males highly tuberculate; tubercles occur on pelvic, anal,
and caudal fins, and sometimes on the underside of the
head and on other areas of the venter. Maximum length
about 3.7 inches (95 mm).

LIFE COLORS Back and upper sides brownish olive,
with brown blotches or saddles; sides light tan yellow and
lateral blotches are black. Undersides whitish yellow. Pre-
and suborbital bars black. First dorsal fin with 2 prominent
black blotches; second dorsal and caudal fins with faint
brown bands; other fins without dark pigment. Breeding
males darker overall and without bright colors.

SIMILAR SPECIES Most similar to other Arkansas
members of subgenus *Imostoma*, *P. vigil* and *P. uranidea*
(because all three species share the enlarged anal fin in
males), but it can be distinguished from those species and
other Arkansas *Percina* species by the large black blotch
at the posterior base of the first dorsal fin (vs. no black
blotch at posterior base of first dorsal fin). It further differs
from *P. uranidea* and *P. vigil* in lacking broad dark dorsal
saddles (vs. dorsal saddles well developed). It differs from
P. copelandi in having 10 or more anal rays (vs. fewer than
10) and 19 or more scales around the caudal peduncle (vs.
fewer than 19) and from remaining *Percina* species in usually having fewer than 59 lateral line scales (vs. usually more
than 60).

VARIATION AND TAXONOMY *Percina shumardi*
was described by Girard (1859) as *Hadropterus shumardi*
from the Arkansas River near Fort Smith, Sebastian
County, Arkansas. It was subsequently placed in the genus
Imostoma (Jordan 1877c) until Bailey and Gosline (1955)
recognized *Imostoma* as a subgenus of *Percina*. It currently remains in subgenus *Imostoma*. The monophyly of
Imostoma was supported by genetic evidence (Near et al.
2011). Near (2002) recognized *P. shumardi* as the sister
group to *P. vigil*. Boschung and Mayden (2004) noted that
the latitudinal range of nominal *P. shumardi* extends south

from 56°N to 28°N and includes several genetically distinct populatons. It is probable that future studies of this widely distributed species will result in the recognition of some populations as distinct species.

DISTRIBUTION Widely distributed from the Hudson Bay region of Ontario and Manitoba, Canada, southward in the Great Lakes and Mississippi River basins to Louisiana, and in a few Gulf Slope drainages from the Mobile Bay basin in Alabama to the Neches River, Texas. In Arkansas, the River Darter is found throughout the Arkansas River and in the Red, eastern Saline, Bayou Bartholomew, Mississippi, St. Francis, Black, and lower White (Robison et al. 2013) rivers (Map 234). It is the most commonly collected darter from the Mississippi River.

HABITAT AND BIOLOGY The River Darter is an inhabitant of the largest rivers and the lower reaches of medium-sized lowland rivers. In the lower eastern Saline River and middle section of the White River, it is found in the deeper, lower ends of riffles and raceways over a gravel and rubble substrate, often at depths of 2 feet (0.6 m) or more. This is typical of the habitat described by most observers (Miller and Robison 2004; Page 1983a; Pflieger 1997; Thomas 1970); however, in the lower White River, the Red River, and throughout the Arkansas River it is found in moderate to swift current over a sand bottom. In the Arkansas River it is most abundant during the day in 4 or more feet (1.2 m) of water, but at night it migrates shoreward and is commonly found at depths of 2–3 feet (0.6–0.9 m) along sandbars. Although it prefers strong current, it is occasionally taken in backwater areas adjacent to the main current at its type locality near Fort Smith. Buchanan et al. (2003) reported this species as uncommon in the Red River in Arkansas, but more than 1,700 P. shumardi were taken from seven Arkansas River navigation pools and Lake Millwood (Buchanan 2005). Several authors have noted its great tolerance of turbidity (Kuehne and Barbour 1983; Page 1983a; Pflieger 1997), which probably explains its success in the largest rivers. It is fairly common in the Mississippi River north of Arkansas (Gilbert 1980f; Pflieger 1997), and TMB frequently collected this species from the Mississippi River of Mississippi County, Arkansas, while sampling fishes with Sam Barkley and Sam Henry of the AGFC. Its populations do not appear to be large in most areas; however, Reynolds (1971) reported taking 491 specimens from one locality in the lower Saline River.

Thomas (1970) found that P. shumardi fed largely on chironomids and caddisflies in the Kaskaskia River, Illinois. Other food items consumed were other dipteran larvae, ephemeropteran nymphs, and small crustaceans (copepods and cladocerans). In the Ohio River, food also consisted mostly of chironomid larvae (Sanders and Yoder 1989). Some populations feed largely on snails (Thompson 1974). In Alabama, P. shumardi fed heavily on snails in two streams (Haag and Warren 2006). In that study, snail feeding was greatest in October (virtually 100% of darter food items), declined through the spring, nearly ceased in May, and increased again to high levels in July. Nonsnail food items were chironomid larvae, trichopteran larvae, and ephemeropteran nymphs. The Asian clam, Corbicula fluminea, was also rarely ingested. Haag and Warren (2006) suggested that snail feeding specialization may be characteristic for all species in the subgenus Imostoma. We have not documented snail feeding in Arkansas populations. Stomach contents of 63 specimens from three Arkansas rivers were examined by TMB. Fifteen individuals, 42–60 mm TL, from the Arkansas River near Fort Smith fed mainly on ephemeropterans and chironomids, with trichopterans and odonates also infrequently eaten. Specimens collected in June contained few food items, but those collected in October had full stomachs. Eight specimens collected in March from the White River in Independence County had consumed ephemeropterans, chironomids, trichopterans, and plecopterans. Stomachs of 40 specimens, 48–76 mm TL, collected from the St. Francis River in St. Francis County on 30 October were full and contained mainly larval coleopterans and chironomids. Other foods consumed in small amounts in the St. Francis River were trichopterans, ephemeropterans, simuliids, and plecopterans.

Little is known of its reproductive habits. Spawning in Illinois and Wisconsin occurs from April to June (Thomas 1970; Becker 1983). In Kansas, spawning occurred in April over a gravel substrate in current near shore at depths of 60 cm (2 feet) or less (Cross and Collins 1995). Hubbs (1985) reported the spawning season in Texas as extending from January to April. Spawning may occur as early as February in Tennessee, based on ripe specimens taken from the Little Tennessee River (Etnier and Starnes 1993). It probably spawns from March to May in Arkansas. We collected tuberculate males with well-developed testes from the Arkansas River near Fort Smith on 24 April at a water temperature of 17°C. Spawning behavior has not been observed in nature. Normal life span is 2–2.5 years; maximum life span 3–4 years (Thomas 1970).

CONSERVATION STATUS Although the River Darter has disappeared from some northern areas of its range where it was formerly common, it is widespread and sometimes common in its preferred large-river habitat in Arkansas. Despite extensive habitat modification, it is still common today at its type locality near Fort Smith and throughout the navigation pools of the Arkansas River. Its populations are currently secure.

Percina uranidea (Jordan and Gilbert)
Stargazing Darter

Figure 26.96. *Percina uranidea*, Stargazing Darter. *Uland Thomas.*

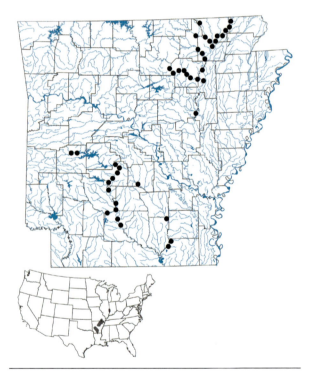

Map 235. Distribution of *Percina uranidea*, Stargazing Darter, 1988–2019.

CHARACTERS   A robust darter with 4 dark dorsal saddles, the posterior margin of the fourth saddle over the dorsal insertion of the caudal fin (Fig. 26.96). Dorsal saddles extending downward and connected to lateral row of 9–12 vertically oblong dark blotches (mostly below the lateral line). Eyes closely set high on head, snout moderately blunt, gill membranes separate to slightly joined, and premaxillary frenum absent or weakly developed. Anal fin of adult

male enlarged. One or more enlarged and modified scales between pelvic fin bases; midline of belly naked (females), but posteriorly in the male there is a row of 9–16 weakly modified scales. Opercle and nape fully scaled; breast usually naked. Lateral line complete, with 46–58 scales. Scale rows above lateral line 5 or 6; scale rows below lateral line 7–9. Diagonal scale rows usually 15–17 (14–19); scale rows around caudal peduncle usually 19 or 20 (18–22). Dorsal spines usually 11 or 12 (9–12); dorsal rays usually 14–16 (13–18); anal spines 2; anal rays usually 11 or 12 (10–13); pectoral rays 13–15. Breeding male with tubercles on the spine and at least the first 4 rays of pelvic fin, and often on the anal and caudal fins. Maximum size around 3.5 inches (89 mm).

LIFE COLORS   Back reddish brown with four distinct dark brown saddles; lateral blotches dark brown or black. Underside cream-colored. Head is heavily darkened between eyes, and pigment actually enters upper third of each eye (Kuehne and Barbour 1983). Chin, throat, and prepectoral area lightly stippled; breast lacks markings. Suborbital bar wide and black; a short preorbital bar extends somewhat downward to corner of mouth; no discrete basicaudal spot. All fins lightly banded. First dorsal fin usually with a small anterior black basal blotch. Breeding males become darker and develop gold cheeks.

SIMILAR SPECIES   Most similar to the Saddleback Darter, the Stargazing Darter is found in deeper riffles and has 4 dark dorsal saddles, with posterior margin of fourth saddle over dorsal insertion of caudal fin (vs. 5 faint to moderately dark saddles, with posterior margin of fourth saddle anterior to dorsal insertion of caudal fin in *P. vigil*). It is distinguished from *P. copelandi* by its 11 or 12 anal rays (vs. 8 or 9 anal rays), and by 4 prominent dorsal saddles (vs. saddles absent). It differs from *P. shumardi* in lacking a large posterior black blotch in first dorsal fin (vs. black blotch present at posterior base of first dorsal fin). It superficially resembles the Arkansas Saddled Darter, *Etheostoma euzonum,* but differs in having 1 or more enlarged and modified scales between pelvic fin bases (vs. no enlarged, modified scales between pelvic fin bases), gill membranes not connected or slightly connected across throat (vs. broadly connected), and fewer than 59 lateral line scales (vs. 59 or more).

VARIATION AND TAXONOMY   *Percina uranidea* was originally described by Jordan and Gilbert (*in* Gilbert 1887) as *Etheostoma (Cottogaster) uranidea* from the Ouachita River at Arkadelphia, Clark County, Arkansas. It is a member of subgenus *Imostoma*. Genetic evidence supported a monophyletic *Imostoma* (Near et al. 2011). Dorsal saddles of *P. uranidea* from the White River drainage are much narrower than those of specimens from the Ouachita River drainage (Etnier 1976). Phylogenetic analyses of the mitochondrial DNA cytochrome *b* and cytochrome *c* oxidase genes showed that the disjunct populations in Arkansas are genetically divergent, but the level of genetic diversity does not warrant classification of either group of populations as evolutionarily significant units (Rigsby 2009). Near et al. (2011) resolved *P. uranidea* as sister to *P. tanasi*, with both of those species most closely related to *P. vigil*.

DISTRIBUTION *Percina uranidea* was formerly found in the St. Francis, White, and Ouachita river drainages of Missouri, Arkansas, and Louisiana, and in the lower Wabash River (Ohio River drainage) of Indiana and Illinois. Most of the Stargazing Darter's range is in Arkansas. It is one of the rarest fishes in Missouri (Pflieger 1997) and is considered extirpated in Louisiana, Illinois, and Indiana. *Percina uranidea* has a disjunct distribution in Arkansas almost identical to that of *P. vigil* with which it often occurs (Map 235). It is found in the White (Robison et al. 2013), Strawberry, Spring, Current, and Black rivers in northeastern Arkansas. It was reported from the St. Francis River by Black (1940) but has not been taken from that river in many years. It historically occurred in the eastern Saline and Ouachita rivers in southern Arkansas, and Caldwell (2011) documented it from the upper Ouachita River above Lake Ouachita for the first time since 1978. There are recent records from the Ouachita River downstream from Remmel Dam (Robison et al. 2013), the lower Little Missouri River, and in 2013, it was found in small numbers in the eastern Saline River.

HABITAT AND BIOLOGY The Stargazing Darter occupies moderate-sized rivers in swift current of deep riffles having gravel bottoms. It prefers clear water and is intolerant of silt. Most of the range of this species now lies within Arkansas, unlike the closely related *P. vigil*, which is more widely distributed outside the state. The following information on habitat preferences was obtained from Rigsby (2009). Populations in the Black River drainage were found in slower bottom-water velocities and over a greater range of substrates than populations in the Ouachita River. In the Black River, Stargazing Darters occurred at depths between 1.0 and 2.0 meters (3.3–6.6 feet) and avoided depths outside that range. They selected mean water column velocities greater than 0.70 m/s (2.3 feet/s) and avoided velocities less than 0.5 m/s (1.64 feet/s); they selected bottom water velocities from 0.21 to 0.40 m/s (0.69–1.31 feet/s) and avoided bottom velocities less than 0.20 m/s (0.66 feet/s). They selected substrates in the

Black River ranging in composition from 99% sand and 1% gravel to 1% gravel and 99% cobble. Stargazing Darters in the Ouachita River preferred depths less than 1.5 meters (4.9 feet) and avoided depths greater than that. Also in the Ouachita River, they selected mean water column velocities greater than 0.70 m/s (2.3 feet/s) and bottom-water velocities greater than 0.30 m/s (0.98 feet/s). Substrates selected in the Ouachita River ranged from 100% gravel to 51% gravel and 49% cobble; they avoided coarser and finer substrates. No other benthic species in the study areas used the same habitat type as *P. uranidea*. The transition zone where the Ouachita River transitions to Lake Ouachita provided the most suitable habitat for Stargazing Darters, and 33 specimens were collected in that area (Stoeckel and Caldwell 2014).

Thompson (1974) reported that populations in the Current River fed almost exclusively on snails and limpets. Ten specimens from the White River in Independence County collected on 7 March (examined by TMB) had full stomachs, with the diet dominated by ephemeropterans. Other food items consumed in smaller amounts in decreasing order of importance were plecopterans, trichopterans, chironomids, coleopterans, and megalopterans. Eleven *P. uranidea* from the Saline River in Drew County (examined by TMB) fed predominantly on ephemeropterans, plecopterans, and trichopterans in July and November, and specimens larger than 50 mm TL fed almost exclusively on snails in December.

Little is known about reproductive biology. Thompson (1974) believed that *P. uranidea* spawned in late winter and early spring. Six ripe females, 53–57 mm TL (examined by TMB), collected from the White River near Jacksonport in Independence County on 7 March 2006 by Brian Wagner had abdominal cavities greatly distended with eggs. Those females contained from 149 to 404 mature ova (x̄ = 279). There were also two smaller size classes of developing eggs, indicating the probability of multiple spawning.

CONSERVATION STATUS Recent information on the status of the Stargazing Darter was provided by Rigsby (2009). Its distribution in the Black River was consistent with historical records, but it has apparently declined in the Ouachita River. Rigsby also found that populations in both the Black and Ouachita river drainages were larger than previously documented and judged them to be stable. Rigsby (2009) sampled seven segments of the Saline River and found no Stargazing Darters, but a 2013 survey found it in the Saline River in small numbers (data provided by Jeff Quinn). We tentatively list it as a species of special concern with some apparently stable populations.

Percina vigil (Hay)
Saddleback Darter

Figure 26.97. *Percina vigil*, Saddleback Darter. *Uland Thomas.*

Map 236. Distribution of *Percina vigil*, Saddleback Darter, 1988–2019.

CHARACTERS A slender delicate darter with 5 dorsal saddles of faint to moderate intensity. The first saddle is beneath first dorsal fin, the fourth saddle is separated by 1 or more scales from the keel-like upper margin of the caudal fin, and the fifth saddle is over the end of the caudal peduncle and usually indistinct (Fig. 26.97). Midside with 8–10 dark rectangular blotches usually not connected to the dorsal saddles but extending well downward. Crosshatching on upper side sometimes forming zigzag lines. Eyes closely set high on head. Snout moderately pointed. Gill membranes separate to slightly joined across throat. Premaxillary frenum absent or very narrowly developed. Infraorbital canal uninterrupted, usually with 8 pores. Anal fin of adult male greatly enlarged. One or

more enlarged and modified scales between pelvic fins bases; midline of belly naked (females) or with a row of 1–13 weakly modified, slightly enlarged scales posteriorly (males). Opercle and nape fully scaled; cheek scalation variable from a few partly embedded scales to fully scaled; breast usually with some scales, but not fully scaled. Lateral line straight and complete, usually with 47–56 (48–62) scales. Scale rows above lateral line 4–7; scale rows below lateral line 9–12. Diagonal scale rows 14–19; scale rows around caudal peduncle 19–22. Dorsal spines 9–11; dorsal rays 12–15; anal spines 2; anal rays 10–12; pectoral rays usually 14 or 15. Breeding males extremely tuberculate, with tubercles developing on the venter, all elements of the pelvic fins, anal and caudal fins, and sometimes on pectoral fins and head. One of the smaller members of its genus, rarely exceeding 2.5 inches (64 mm) in length. Maximum reported size is 3 inches (78 mm) TL (Page and Burr 2011).

LIFE COLORS Upper sides and back yellow to olive with brown or black lateral blotches, saddles, and crosshatching. Page (1983a) noted that individuals collected over gravel have distinct dorsal saddles and lateral blotches, and those collected over sand are more translucent and have much more diffuse saddles and more confluent lateral blotches. Underside white to cream-colored. Small black basicaudal spot usually present. First dorsal fin with dark basal and marginal bands; second dorsal, caudal, and usually pectoral fins faintly banded; pelvic and anal fins usually clear. The head is dark dorsally and light ventrally with prominent dark pre- and suborbital bars. There is a small dark patch on the chin and another at the junction of the gill membranes. Spinous dorsal fin of breeding males with a wide black basal band, a clear submarginal band, and a black marginal band.

SIMILAR SPECIES The Saddleback Darter has long been confused with *P. uranidea,* with which it is usually syntopic in Arkansas. They are segregated by habitat, with *P. vigil* found in shallow riffles and *P. uranidea* occurring in deeper, swifter riffles. *Percina vigil* also differs from *P. uranidea* in having 5 faint to moderately dark wide dorsal saddles, with the posterior margin of fourth saddle anterior to dorsal insertion of caudal fin (vs. 4 dark narrow dorsal saddles, with posterior margin of fourth saddle over dorsal insertion of caudal fin); a slender, more delicate body (vs. robust body), lighter in overall body coloration, moderately pointed snout, and presence of a dark basicaudal spot (vs. darker coloration, moderately blunt snout, and basicaudal spot faint or absent). It differs from *P. shumardi* in lacking a large posterior black spot in first dorsal fin (vs. black blotch present in posterior base of first dorsal fin). It occurs syntopically with *P. copelandi* in the Ouachita River drainage, but it can be distinguished from that species by its 10–12 anal rays and 5 dorsal saddles (vs. anal rays 8 or 9 and dorsal saddles absent or with 8 faint blotches).

VARIATION AND TAXONOMY *Percina vigil*, a member of subgenus *Imostoma*, was originally described by Hay (1882) as *Ioa vigil*. For many years the type locality for the Saddleback Darter (known as *Percina ouachitae*) was considered to be the Saline River at Benton, Saline County, Arkansas (Jordan and Gilbert *in* C. H. Gilbert 1887), but Hay (1882) had described this species earlier as *Ioa vigil* from the Pearl River at Jackson, Mississippi. Therefore, the type locality is not in Arkansas. Controversy has existed over the correct name of this fish, and in the first edition of this book, we continued to recognize it as *P. ouachitae* until a consensus was achieved regarding the correct scientific name. Hubbs and Black (1940c) placed *P. ouachitae* in the synonymy of *P. uranidea*, an action that was later supported by Collette and Knapp (1966); however, from their studies on the Current River in Arkansas, Thompson and Cashner (1975) recognized *Percina ouachitae* (Jordan and Gilbert) as being distinct from *P. uranidea*. This was corroborated by Robison in syntopic collections of the two species from the lower Saline River, Bradley County, in 1971. Page (1983b) examined the holotype of a form known as *Ioa vigil* Hay (USNM 32201, 23 mm SL) and reidentified it as *Percina shumardi*; however, Suttkus (1985) reexamined the single, small specimen and disagreed with Page's identification, declaring instead that it was actually *Percina ouachitae*. Suttkus recommended that *Ioa vigil* Hay be recognized as a valid species in the genus *Percina* (*Percina vigil*), replacing *P. ouachitae*. Burr and Warren (1986b) concurred, and we follow Page et al. (2013) and other recent authors in recognizing *P. vigil* as a senior synonym of *P. ouachitae*.

Rigsby (2009) studied mitochondrial DNA cytochrome *b* and cytochrome *c* oxidase genes and confirmed that *P. vigil* and *P. uranidea* are genetically distinct. Little genetic diversity was found in *P. vigil* within drainages or between the Black and Ouachita river drainages. Genetic variation was not sufficient to warrant recognition of different populations as evolutionarily significant units. Rigsby further noted that the number of Saddleback Darter haplotypes found in each drainage was relatively low compared with those found in the Stargazing Darter, and the *P. vigil* haplotypes were not unique to each drainage. A genetic analysis by Near et al. (2011) resolved *P. vigil* as sister to a clade containing *P. tanasi* and *P. uranidea*.

DISTRIBUTION The Saddleback Darter is more widely distributed than *P. uranidea* and is found in the lower Mississippi River basin from southwestern Kentucky and Indiana and southeastern Missouri south to Louisiana,

Mississippi, and Alabama. It also occurs in Gulf Slope drainages from the Escambia River of the Florida panhandle, west to Louisiana. *Percina vigil* has a disjunct distribution in Arkansas almost identical to that of *P. uranidea* (Map 236), but the two species are segregated by habitat preferences. In northeastern Arkansas it is found in the middle portion of the White River from near Sylamore to just downstream of its confluence with the Black River, and upstream in the Black River and its major tributaries, the Current, Spring, and Strawberry rivers. There are disjunct records from the lower White River. In southern Arkansas it is found in the eastern Saline and lower Little Missouri rivers, and in the Ouachita River upstream and downstream of Remmel Dam.

HABITAT AND BIOLOGY The Saddleback Darter is found in large creeks and small to moderate-sized rivers, where it occurs in shallow to medium-sized riffles over a fine gravel or sand bottom. It is also found where snags and log jams create enough current to sweep away silt and fine organic matter (Kuehne and Barbour 1983). Chutes with swift current bisecting extensive shoal areas were especially favored habitats in Alabama (Mettee et al. 1996). Saddleback Darters were usually found over a hard-packed clay substrate in Bayou Sara, Mississippi and Louisiana (Ross 2001). It is not very tolerant of silt (Thompson and Cashner 1980) and avoids pools and quiet waters, but Kuehne and Barbour (1983) reported that it is tolerant of turbidity. Although not widely distributed in Arkansas, it is sometimes abundant where it does occur.

Thompson (1974) found that Current River, Arkansas, populations fed mainly on insect larvae and microcrustaceans. G. L. Miller (1983) studied the trophic ecology of *P. vigil* in Mississippi and reported that it consumed primarily hydropsychids (Trichoptera), simuliids and chironomids (Diptera), and baetids (Ephemeroptera). Crustaceans were rarely eaten. In Little Escambia Creek, Alabama, it fed almost exclusively on trichopteran larvae in the fall, but the summer diet was more varied and included coleopterans, ephemeropterans, and plecopterans (Mettee et al. 1987). East of the Mississippi River, it feeds to some extent on snails (Etnier and Starnes 1993; Ross 2001; Boschung and Mayden 2004), but TMB found no snails in stomachs of 62 specimens collected between 6 July and 10 November from the Ouachita, Saline, and White rivers in Arkansas. Specimens from the White River drainage fed primarily on chironomids in summer and fall. The summer diet in the Ouachita and Saline rivers was dominated by ephemeropterans, followed by chironomids and other dipterans, coleopterans, plecopterans, trichopterans, and isopods. In the fall, ephemeropterans and chironomids dominated the diet in the Saline River, with hemipterans, odonates, cole-opterans, and isopods ingested in small amounts.

Reproduction of *Percina vigil* was studied by Heins and Baker (1989) in Mississippi, and the following information on its reproductive biology comes from that source. Spawning occurred from mid-February through April in Mississippi at water temperatures of 12–22°C, and it likely spawns during that time frame in Arkansas. Average clutch size in Mississippi was 142–296 eggs, with females apparently producing three or more clutches during a breeding season. Growth of YOY *P. vigil* was rapid, with individuals averaging 36.5 mm SL at 4 months of age, 41.2 mm SL at 6 months, and 46.2 mm SL at 1 year. Maximum life span was 32 months and substantial age structure differences occurred between years.

CONSERVATION STATUS   The current status of this uncommon species is uncertain, but populations in Arkansas appear to have declined in recent years. Rigsby (2009) judged populations to be stable. We consider the Saddleback Darter a species of special concern. It is hoped that future study will reveal that its populations are stable. Because of the potential for rapid change in its population status, it is a species that merits periodic monitoring.

Genus *Stizostedion* Rafinesque, 1820a— Pikeperches

Rafinesque described *Stizostedion* as a subgenus of *Perca* in 1820. There is disagreement about the correct generic name for Sauger, Walleye, and Eurasian pikeperches. In some early works, it is clearly indicated that Sander is the Latvian common name for *Stizostedion lucioperca* (*Perca lucioperca* of Bloch 1785, p. 66, and Fischer 1791, p. 246).

Vitins et al. (2001, p. 86) wrote of Fischer,

The book *Vesuch einer Naturgeschichte von Livland* [Livland = Latvia] by J. B. Fischer (1791) was the first scientific overview of local flora and fauna, including descriptions of 48 fish species *and presentations of their Latvian names* (italics ours). His work can be considered the beginning of ichthyology in Latvia.

Gill (1903, p. 967) wrote,

I [Gill 1894] was unable to find a Latinized generic name for the Pike-perches earlier than 1820, when Rafinesque published the name *Stizostedion*. The name Sander, published in the year 1817 [by Oken] as Cuvier's, must now be received and take its place.

The 1903 paper, in which Gill incorrectly believed Sander to be a Latin name, was mostly ignored. North American ichthyologists continued to use *Stizostedion* as the genus name for Walleye and Sauger (except for Jordan [1917, 1923, 1929], who recognized Sander as a valid genus but used genus *Cynoperca* for Sauger), and Europeans who continued to use *Lucioperca* for the European pikeperches until Collette (1963) and Collette and Bănărescu (1977) recombined them under the genus name *Stizostedion*. The long-used *Stizostedion* Rafinesque, 1820 was changed to Sander by Kottelat (1997), supposedly based on the Rule of Priority, because that name first appeared in Oken (1817). Nelson et al. (2003, 2004) made the same error as Gill (1903) when they assumed Sander was a Latin name and accepted Kottelat's (1997) conclusion. An 1816 encyclopedia by Oken was placed on the *Official Index of Rejected and Invalid Works in Zoological Nomenclature* in 1956 by the International Commission on Zoological Nomenclature, primarily because of Oken's arbitrary use of common names as scientific names (Hemming 1956). Unfortunately, the 1956 ICZN ruling did not include Oken's 1817 publication, which also was guilty of arbitrary and confusing use of common and scientific names and in which there was a single instance where the name Sander was mentioned. It is unclear whether Oken's 1817 publication intended Sander as a common or scientific name. Because of its origin and use as a common name, we reject the Latvian common name Sander as a generic name and support the use of *Stizostedion* for the genus name for Sauger, Walleye, and European pikeperches. In doing this, we follow Collette (1963), Collette and Bănărescu (1977), Miller and Robison (2004), Bruner (2011), and Nelson et al. (2016) in recognizing *Stizostedion*.

Stizostedion contains five species, two in North America and three in Eurasia. Genus characters include an elongate body that is moderately terete, caudal fin forked, mouth large, the upper jaw extending beyond middle of eye, and well-developed canine teeth present on jaws, vomer, and palatine bones. There is an opercular spine present and the preopercle is serrate. The lateral line is complete, with 75–91 scales, sensory canals on the head are uninterrupted, there are 2 anal fin spines, and a swim bladder is present. Mitochondrial DNA and allozyme analyses showed that the two North American species (*S. canadense* and *S. vitreum*) clustered in one group, while two European species (*S. lucioperca* and *S. volgense*) formed a second group (Billington et al. 1991). Haponski and Stepien (2013), using analyses of nuclear and mitochondrial DNA, concluded that North American *Stizostedion* species differentiated about 15.4 million years ago, while Eurasian species date to 13.8 mya. Ebbers et al. (1988) compiled a Walleye-Sauger bibliography, and B. A. Barton, ed. (2011) provided information on their biology, culture, and management.

Stizostedion canadense (Griffith and Smith)
Sauger

Figure 26.98. *Stizostedion canadense*, Sauger. *Uland Thomas.*

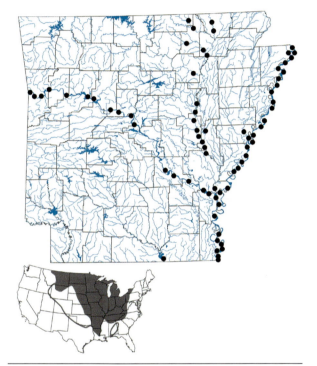

Map 237. Distribution of *Stizostedion canadense*, Sauger, 1988–2019. Inset shows native range (solid) and introduced range (outlined) in the contiguous United States.

CHARACTERS   One of two native Arkansas species in the perch family large enough to be a gamefish but smaller than the related Walleye (Fig. 26.98). Body slender and streamlined, usually with 3 or 4 dusky saddles that extend obliquely forward and downward on sides; saddles especially prominent in young. First dorsal fin with rows of distinct black spots and lacking a black blotch near bases of last few spines. Mouth large, upper jaw extending to rear margin of eye. Gill rakers on lower limb of first arch slender, about 8. Swim bladder well developed. Usually 4–8 pyloric caeca, each usually shorter than stomach. Strong canine teeth present on premaxillaries, lower jaw, and palatines.

Pharyngeal arches with short, recurved teeth. Opercle with a sharp spine. Rear margin of preopercle strongly serrate. Gill membranes separate or narrowly joined to one another, and narrowly joined to isthmus. Branchiostegal rays usually 7. Cheek partly to fully scaled; opercle, nape, and breast well scaled. Lateral line complete, with 78–91 scales. The 2 dorsal fins are well separated, the first with 12 or 13 spines, and the second with 1 or 2 short spines and 17–20 rays. Anal spines 2; anal rays usually 12 or 13. Pectoral rays 14–16. Breeding tubercles absent. Adults commonly around 18 inches (457 mm) long and weighing about 2.5 pounds (1 kg). Maximum length of 32.2 inches (818 mm) and weight of 8 pounds, 12 ounces (3.97 kg) (world angling record from North Dakota). Arkansas state angling record from the Arkansas River in 1976 is 6 pounds, 12 ounces (3 kg).

LIFE COLORS   Back and sides olive or brown with 3 or 4 dark brown saddles or blotches. Undersides creamy white. Spots in first dorsal fin black. Second dorsal and caudal fins dusky with narrow brown bands. Pectoral fin with a dark basal blotch; pelvic and anal fins mostly unpigmented.

SIMILAR SPECIES   The Sauger is most like the Walleye but differs in having rows of round black spots in first dorsal fin (vs. first dorsal fin without rows of dark spots), no large black blotch in posterior membranes of first dorsal fin (vs. black blotch present), cheeks partly to fully scaled (vs. cheeks usually naked), and 17–20 soft dorsal rays (vs. 19–22). Priegel (1969) separated very young Saugers from Walleyes by the absence of pigmentation on the dorsal surface of the midbrain area (visible through the top of the head) in the Sauger (vs. Walleye larvae have this area heavily pigmented, and it appears dark to black). The Sauger differs from the Yellow Perch in having well-developed canine teeth (vs. canine teeth absent), and in having 11 or more anal fin soft rays (vs. 8 or fewer). Young Saugers can be distinguished from all darters by the strongly serrate preopercle (*Percina sciera* preopercle has a weakly serrate rear margin) and a large mouth with upper jaw extending backward behind middle of eye (vs. upper jaw not extending backward past middle of eye).

VARIATION AND TAXONOMY   *Stizostedion canadense* was described by Griffith and Smith (1834) as *Lucioperca canadensis* from "Canada." We do not follow Nelson et al. (2004) and Page et al. (2013) in recognizing the genus name *Sander* for this species (see genus *Stizostedion* account). Two subspecies are sometimes recognized. *Stizostedion c. boreum* occurs in the upper Missouri River, and *S. c. canadense* occurs throughout the remainder of its range, including Arkansas. During the past two decades, a *Stizostedion canadense* × *S. vitreum* hybrid called a "saugeye" has been widely stocked in Arkansas impoundments. The hybrid grows larger than the Sauger and resembles the Sauger in appearance. It is often difficult to identify the saugeye, but Etnier and Starnes (1993) provided the following general features: saugeyes have the same pyloric caeca counts as Walleyes (3), longer gill rakers than Sauger, lack discrete black spots in the spinous dorsal fin, and have reduced cheek and opercular squamation compared to the Sauger. The hybrids and their backcrosses are fertile (Hearn 1986), and reproducing populations have become established in Ohio and West Virginia (Fuller et al. 1999). Mitochondrial genotypes of Walleye and Sauger are readily differentiated, and there is genetic evidence of introgression between those species in Lake Simcoe, Canada (Billington et al. 1988).

DISTRIBUTION   The Sauger is considered native to the St. Lawrence–Great Lakes, Hudson Bay, and Mississippi River basins of Canada (Quebec to Alberta) south to Alabama and Louisiana; it has been widely introduced into Atlantic Slope, Gulf Slope, and southern Mississippi river drainages with little success (Page and Burr 2011). The earliest report of this species from Arkansas was by Jordan and Gilbert (1886) from Lee Creek near Fort Smith. Meek (1894a, b, 1896) listed it from Illinois Bayou at Russellville and reported that it was common in the Poteau River at Fort Smith. It was considered by Barila (1980a) to be native only to extreme northern Arkansas in the upper White and Strawberry rivers and to be transplanted in all other areas of occurrence in the state. This interpretation of its range was not supported by Page and Burr (2011). It also occurs in the lower White River, the St. Francis River drainage, the lower Ouachita River drainage, and throughout the Mississippi and Arkansas rivers (Map 237).

HABITAT AND BIOLOGY   The Sauger is strictly a fish of moderate and large rivers in Arkansas, but it is also reported from reservoirs elsewhere (Etnier and Starnes 1993; Boschung and Mayden 2004). It is most widespread and abundant in the Mississippi and Arkansas rivers, where it is commonly taken by anglers in moderate to swift current near the ends of rock dikes and along riprap-lined banks. Its numbers increased substantially during the 1970s after the construction of the Arkansas River Navigation System. It is sometimes taken in dike-field backwaters over a sandy bottom, but it is usually found in strong current (in contrast to the Walleye, which prefers quieter waters). Becker (1983) noted that young Saugers sometimes frequent shallow mudflats. In early spring, more Saugers are caught in swift water close to the bases of the navigation dams on the Arkansas River. It is more tolerant of turbidity

than the Walleye. The tapetum lucidum (reflecting layer behind the retina of the eye) is more uniformly developed in the Sauger than in the Walleye (Collette et al. 1977) and appears to give the Sauger an advantage over the Walleye in turbid water. Saugers are also more adapted to darkness and feed in deeper water than Walleyes (Becker 1983).

The Sauger is a predator, with adults feeding on fish and the young on invertebrates and small fish. Its life history in the upper Missouri River drainage of North Dakota was studied by Carufel (1963). Kendall (1978) and Becker (1983) summarized other aspects of its biology, culture, and management. The larvae first feed on microcrustaceans, primarily cladocerans and copepods, but soon switch to a diet mainly of fish larvae and fry (W. R. Nelson 1968). In Wisconsin, young Saugers 12–50 mm TL fed most heavily on copepods, cladocerans, and chironomid larvae and pupae, but 47–65% of the stomachs also contained the fry of other fishes (Becker 1983). The YOY are pelagic during the early part of their first summer. Saugers 51–150 mm primarily consumed young fish, but when fish were not available in sufficient quantities, they turned to microcrustaceans and chironomids. In the Ohio River, fish were the primary food of adults both by weight (99.7%) and by frequency of occurrence (100%) (McBride and Tarter 1983). Emerald Shiners and Gizzard Shad composed 96% of all identifiable items, and invertebrates were of minor dietary importance. In reservoirs, the dominant fishes preyed on by adults include Emerald Shiners, Threadfin and Gizzard shad, and young Black Crappie (W. R. Nelson 1968; Fitz and Holbrook 1978). Stomach contents of specimens from the Arkansas River near Fort Smith (examined by TMB) contained mainly Gizzard Shad, Threadfin Shad, Mississippi Silversides, and Emerald Shiners.

In many parts of its range, it has been reported to spawn at night in early spring over gravel or rubble shoals (Scott and Crossman 1973; Smith 1979). Spawning began in the Missouri River, South Dakota, when water temperatures reached 42°F (5.6°C) in late April and was essentially completed within 2 weeks (W. R. Nelson 1968). Spawning occurs over rubble flats at depths of 2–12 feet (0.6–3.7 m). Males arrive at the spawning site first and are later joined by the females. No nest is built and spawning often occurs at night, the eggs being scattered over gravel or rocks. Fecundity ranges from 9,000 to 150,000 eggs, depending on fish size (Carufel 1963; Jenkins and Burkhead 1994). The incubation period was approximately 21 days at 47°F (8.3°C) and 9–14 days at 55°F (12.8°C) (W. R. Nelson et al. 1965). In the Arkansas River, Saugers congregate below dams, where they probably spawn over silt-free rock riprap in late February or early March. The scattered and abandoned eggs are not strongly adhesive and settle to the substrate until hatching occurs. In Wisconsin, male Saugers attained maturity at age 2 and females at age 4 (Becker 1983). Maximum life span is around 4–7 years in southern populations, but Scott and Crossman (1973) reported a maximum age of 13 years in Canada.

CONSERVATION STATUS Stable populations exist in the Arkansas and Mississippi rivers, and we list this species as currently secure.

Stizostedion vitreum (Mitchill)
Walleye

Figure 26.99. *Stizostedion vitreum*, Walleye. *David A. Neely.*

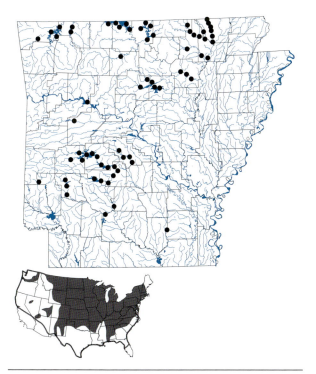

Map 238. Distribution of *Stizostedion vitreum*, Walleye, 1988–2019. Inset shows native range (solid) and introduced range (outlined) in the contiguous United States.

CHARACTERS One of two native gamefish species of the perch family in Arkansas. A slender, streamlined fish similar to the Sauger, but usually larger (Fig. 26.99). Back sometimes with 6–8 obscure saddles, but never with wide, oblique lateral blotches; young often with 5–12 vertical bars on side. First dorsal fin never with rows of round black spots (sometimes with dark streaks), but with a large posterior black blotch near bases of last few spines. Mouth large, upper jaw extending to about rear margin of eye. Gill rakers on lower limb of first arch long and thin, about 8. Pharyngeal arches with short, recurved teeth. Swim bladder well developed. Sharp canine teeth present on premaxillaries, lower jaw, and palatines. Pyloric caeca 3, each about as long as stomach. Opercle with a sharp spine. Rear margin of preopercle strongly serrate. Gill membranes narrowly joined to one another, and narrowly joined to isthmus. Branchiostegal rays usually 7. Cheeks usually naked, sometimes with a few scales. Opercle partially to fully scaled; breast and nape fully scaled. Lateral line complete, usually with more than 80 scales (77–90). The 2 dorsal fins are well separated, the first usually with 13 or 14 spines, the second with 1 or 2 short spines and 19–22 rays. Anal spines 2; anal rays 11–14. Pectoral rays 13–16. Breeding tubercles absent. Commonly reaching 15–30 inches (381–762 mm) in length and a weight of 4–10 pounds (1.8–4.5kg). Maximum size about 41 inches (1,041 mm) TL and 25 pounds (11.4 kg). The Arkansas state record is 22 pounds, 11 ounces (10.3 kg), from Greers Ferry Lake in 1982. It is the 12-pound line-class world record.

LIFE COLORS Body greenish yellow with darker mottlings and blotches. Underside white. First dorsal fin dusky with a large, black, basal posterior blotch; second dorsal and caudal fins with narrow brown bands. Pectoral fin with black spot at base. Pelvic and anal fin white to yellow; tip of anal fin and lower lobe of caudal fin white.

SIMILAR SPECIES The Walleye is similar to the Sauger but differs in lacking rows of round black spots in the first dorsal fin (vs. rows of round black spots present in Sauger); in having a large black blotch near the posterior base of first dorsal fin (vs. black blotch absent), cheeks naked or with a few scales (vs. cheeks fully scaled), and 19–22 soft dorsal rays (vs. 17–20). It differs from the Yellow Perch in having well-developed canine teeth (vs. canine teeth absent), and in having 11 or more anal fin soft rays (vs. 8 or fewer). Young Walleye can be distinguished from all darters by the strongly serrate preopercle and large mouth, with upper jaw extending well behind the middle of the eye.

VARIATION AND TAXONOMY *Stizostedion vitreum* was described by Mitchill (1818) as *Perca vitrea* from Cayuga Lake, New York. We do not follow Nelson et al. (2004) and Page et al. (2013) in recognizing the generic name *Sander* for this species (see genus *Stizostedion* account). Two subspecies have been recognized. *Stizostedion v. glaucum,* the

Blue Walleye, occurs in Lake Erie and Lake Ontario. It was on the Federal Endangered Species List and is now thought to be extinct (Nelson et al. 2004). The more wide-ranging subspecies, *S. v. vitreum,* occurs elsewhere over its range, including Arkansas. Hybrids between *S. canadense* and *S. vitreum* (called a saugeye) have been stocked in several Arkansas reservoirs (see *S. canadense* species account for characters to identify saugeye). The Arkansas state record for saugeye from Lake Frierson in 2012 is 9 pounds (4 kg).

DISTRIBUTION *Stizostedion vitreum* is native to the St. Lawrence–Great Lakes, Arctic, and Mississippi River basins from Canada (Quebec to Northwest Territories) south to Alabama and Arkansas; it is widely introduced into Atlantic and Pacific drainages of the United States (Page and Burr 2011). Black (1940) listed the Walleye from throughout the White, Arkansas, and Ouachita river systems, but Barila (1980b) considered it native to Arkansas only in the White River drainage of the northern part of the state (Map 238). The Walleye has been stocked in the larger reservoirs, primarily in the Ozark and Ouachita mountains. One of the largest stream populations in Arkansas is in the Eleven Point River (Johnson et al. 2009). Small, stable populations are also found in the Caddo, eastern Saline, and lower Ouachita rivers. Whether these represent original native or introduced populations is problematic, as reports of Walleye in these streams to the Arkansas Game and Fish Commission predate their stocking by that agency.

HABITAT AND BIOLOGY Much of the original habitat of this species in Arkansas was inundated by the impoundment of the upper White River lakes, but it has adapted to those reservoirs, and populations continue to survive in them. In northern parts of its range, the Walleye is most frequently found in the quiet water of lakes in clear water, over substrates of sand, gravel, mud, rubble, boulders, clay, silt, and detritus in decreasing order of occurrence (Becker 1983). The Walleye today is most abundant in Greers Ferry Lake and its headwater streams. When that lake was impounded in the early 1960s, enough suitable tributary streams remained to allow spawning success. The Walleye has been stocked in all large upland reservoirs of the state as a result of a Game and Fish Commission propagation program begun in the late 1960s. Bull Shoals and Norfork reservoirs have very good Walleye populations (Ken Shirley, pers. comm.). During the peak spawning months of March and April, ripe adults are collected as they begin their upstream spawning migration for use as brood stock for propagation of the species at state fish hatcheries. Game and Fish Commission biologists concentrate their efforts to collect mature fish in the South Fork,

Devils Fork, and Middle Fork of the Little Red River. The eggs and milt are stripped, and the fertilized eggs and later the young develop in nursery ponds until stocking size is reached. A moderately successful Walleye fishery has been established in Lake Ouachita through stocking. Stocking has been unsuccessful in some reservoirs, mainly in the Ouachitas, but there are Walleye fisheries in lakes Ouachita, Hamilton, Catherine, and Greeson. Walleye are occasionally taken in larger White River drainage streams in the Ozarks, such as the Kings River, but are usually not common there. In northeast Arkansas, good Walleye populations are found in the Spring, Current, and Eleven Point rivers. In the Eleven Point River, catch rates, growth rates, and size structure were high relative to other streams studied in North America (Johnson et al. 2009). In rivers, Walleyes tend to occur in deep pools but are sometimes found in deep runs above and below riffles. The Walleye is less tolerant of turbidity, silt, and high temperatures than the Sauger (Smith 1979).

Several authors have documented the nocturnal or crepuscular habits of this species (Cross 1967; Pflieger 1997; Smith 1979). It retreats to deeper waters by day, moving inshore to shoal areas to feed at dusk. Its eye has a reflective layer, the tapetum lucidum, which is an adaptation for vision in dim light, giving it the "walleyed" appearance. It is very sensitive to strong light and avoids it by seeking deeper water and sheltered areas.

There is much scientific literature on the feeding biology of this carnivore. In summary, it consumes food items that are most readily available, and it has been shown to feed on almost all species of fishes in its environment, as well as large invertebrates. The first foods of larval Walleyes (after absorption of the yolk sac) are small crustaceans, primarily cladocerans and copepods (Houde 1967), but other invertebrates are soon added to the diet, including larger zooplankton, insect larvae and pupae, crayfish, and mollusks (Mathias and Li 1982). Larval Walleyes are also known to feed on their siblings (Li and Mathias 1982). The diet switches from invertebrates to fish at about 34–80 mm TL, and adults feed almost exclusively on fish (Mathias and Li 1982). Piscivory of adults has been supported by various studies throughout its range, with the diet consisting of 85–100% fish. Walleyes in reservoirs feed on the most abundant forage species available, usually concentrating their efforts on only one or two dominant species. In rivers, where available forage species diversity is usually greater than in reservoirs, Walleyes feed on a greater diversity of fishes. In the Eleven Point River, Arkansas, a stream of high fish species diversity, *S. vitreum* fed on a variety of fishes (Johnson et al. 2009). The diet consisted predominantly of minnows or bottom dwelling species such as stonerollers, chubs, and sculpins, but also included Gizzard Shad. No centrarchids were found in Walleye stomachs despite their high abundance in the river.

In Arkansas, reservoir populations of Walleye ascend tributary streams to spawn when water temperatures approach 50°F (10°C). Spawning migrations occur in this state from early February through early April, depending on weather conditions. Peak spawning activity in the Greers Ferry Lake area usually occurs in mid-March (Carl Perrin, pers. comm.). Fish were tagged in Greers Ferry Lake in February 1996 to document spawning movements. Based on tag return data, Walleyes traveled an average of 12.2 miles (19.6 km) and a maximum of 27.0 miles (43.4 km) before recapture (Carl Perrin and Tom Bly, pers. comm.). The males arrive at the spawning sites first and are later joined by the females. Both sexes at this time occupy long, deep pools below swift riffles. Intense spawning activity is apparently triggered by a sudden rise in water level after a heavy rain (William E. Keith, pers. comm.). The ripe adults move into the shallow gravel-bottomed riffles to spawn at night. No nest is constructed, and the adhesive eggs are scattered over the gravel substrate and abandoned. Spawning behavior was described by Ellis and Giles (1965). Although Walleye are not territorial, courtship behavior occurs among groups of fish over the spawning grounds. There is much pushing and circling among the fish during the initial, promiscuous courtship. As courtship activity increased in intensity, one or more females and one or more males came closely together and the compact group rushed upward. At the surface, the group swam vigorously around releasing sperm and eggs. Fecundity ranges from 23,000 to 615,000 eggs (Hardy 1978). After spawning, the adults return to the reservoirs, where they remain in the deep, oxygen-rich waters. In some Arkansas reservoirs, such as Lake Norfork and Bull Shoals, a small amount of spawning occurs along rocky shorelines, but a stocking program is required to maintain a Walleye fishery in those reservoirs. The eggs hatch in 26 days at a water temperature of 4.4°C (40°F), in 21 days at 10–12.8°C (50–55°F), and in 7 days at around 13.9°C (57°F) (Becker 1983). The larvae quickly become pelagic and are dispersed by currents (W. R. Nelson 1968). Although few Walleye live more than 6–8 years, a maximum life span of around 10 years was reported by Hackney and Holbrook (1978). In the Eleven Point River, Arkansas, the oldest adults were 10+ years of age (Johnson et al. 2009).

CONSERVATION STATUS Native populations of this valued gamefish are supplemented by stocking hatchery raised fish. This species is secure in Arkansas.

Family Sciaenidae *Drums and Croakers*

Based largely on the molecular evidence presented in Betancur-R. et al. (2013), Nelson et al. (2016) removed Sciaenidae from Perciformes and placed it in the order Acanthuriformes (surgeonfishes and relatives) along with 17 other families, several of which were also removed from Perciformes. We are hesitant to support removal of Sciaenidae from Perciformes (see Appendix 5 for our classification), but it is possible that this change will be widely supported by most fish taxonomists in the future. The drum and croaker family, Sciaenidae, is a widely distributed, primarily marine or brackish water family found on the continental shelves of tropical and temperate seas worldwide. Of the approximately 283 species in about 67 genera recognized by Nelson et al. (2016) in Sciaenidae, only a few species enter freshwater in South America, and only one species does so in North America. That lone species, the Freshwater Drum, *Aplodinotus grunniens*, occurs in Arkansas and is the only strictly freshwater member of the family. Sciaenid fossils date back to the Upper Cretaceous, at least 70 mya (Berra 1981). This family is also known from marine deposits in North America from the Eocene (Romer 1966), and *Aplodinotus grunniens* has been reported from Pliocene to Recent sediments from the Mississippi basin and Gulf Coastal Plain (Cavender 1986).

The drum and croaker family got its common name because most species have a complicated swim bladder with special muscles and tendons capable of producing audible sounds when vibrated against the bladder (Becker 1983). The lower pharyngeal arches are large, sometimes fused, and have buttonlike teeth for crushing food. The spinous and soft dorsal fins are only slightly connected, and the anal fin has 1 or 2 spines. In addition, a distinctive, large, white otolith is located in the sacculus of the inner ear. The skull consists of heavy bones characterized by large cavities that are adaptations for the mucous glands of the lateral line system (Becker 1983). There are many small, sharp teeth in broad bands on the jaws. The lateral line, which extends past the hypural plate to the distal end of the caudal fin, is also distinctive.

The family contains several important marine sport fish species, such as the Red Drum, Spotted Seatrout, croakers, and kingfishes, and the Freshwater Drum is a favorite of many freshwater anglers because of its size; however, opinion is divided as to its quality as a food fish. In Arkansas, the Freshwater Drum is an important commercial species (ranking only behind buffalofishes, catfishes, and Common Carp).

Genus *Aplodinotus* Rafinesque, 1819— Freshwater Drums

The monotypic genus *Aplodinotus* occurs throughout central North America and south into Mexico. It is represented in Arkansas by the Freshwater Drum, *A. grunniens*, which is commonly found in large rivers, lakes, and reservoirs in the state. Characters of the genus are presented in the following species account.

Aplodinotus grunniens Rafinesque
Freshwater Drum

Figure 26.100. *Aplodinotus grunniens*, Freshwater Drum. *Uland Thomas.*

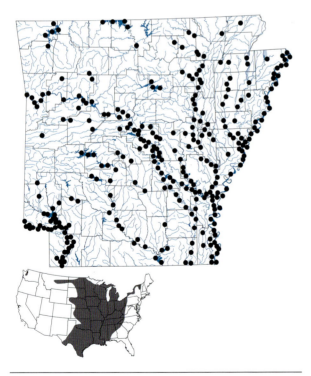

Map 239. Distribution of *Aplodinotus grunniens*, Freshwater Drum, 1988–2019.

CHARACTERS A distinctive gray or silvery fish with a blunt, rounded snout, subterminal mouth, a highly developed lateral line system extending far onto the caudal fin, a high arching back, and a deep, strongly compressed body (Fig. 26.100). The greatest body depth is near the dorsal fin origin, and the body tapers posteriorly to a somewhat slender caudal peduncle. Head with cavernous lateral line canals, particularly the preoperculomandibular and supraorbital regions. Jaws with small comblike teeth, but canine teeth are absent. Pharyngeal arches are heavy; pharyngeal teeth rounded, molariform, and closely set. Frenum absent. A deep notch occurs between the spinous and soft dorsal fins, but the two fins are conjoined. First part of dorsal fin usually with 9 or 10 stiff spines; the second part of the dorsal fin (posterior to the deep notch) with 1 spine and 26–32 soft rays. Base of spinous dorsal fin about half the length of soft dorsal fin base. Anal fin with 2 spines, the first small, the second greatly elongated and much larger than the first; anal rays 7. Pectoral fin with 15–17 rays. Pelvic fins thoracic in position with 1 spine and 5 rays, the outer ray elongated into a distinct slender filament. Lateral line arched anteriorly and complete, with 48–55 scales; the count extends posteriorly to the end of the hypural plate. Caudal fin is triangular. Body, top of head, cheeks, and opercles covered with ctenoid scales. Otolith of inner ear much enlarged (Priegel 1967a). Sexual dimorphism is apparent. In the male, there is a single urogenital opening behind the anus, but the female has two openings (genital and urinary) behind the anus (Moen 1959). Adults usually range from 12 to 20 inches (305–508 mm) and from 12 ounces to 5 pounds (0.3–2.3 kg). The Arkansas state angling record from Lake Wilson in 2004 weighed 45 pounds, 7 ounces (20.6 kg). Maximum weight near 60 pounds (27 kg) and maximum length about 35 inches (89 cm).

LIFE COLORS Body color basically silver with a gray dorsum, lighter-colored silvery sides, and a white belly. Lower region of head white. Pectoral fins lightly pigmented and other fins are dusky. Pectoral and pelvic fins often tinged with orange.

SIMILAR SPECIES The Freshwater Drum is similar to carpsuckers (*Carpiodes*) and buffaloes (*Ictiobus*) in having a deep body and silver coloration; however, carpsuckers and buffaloes differ in lacking sharp spines in the dorsal and anal fins and in having the pelvic fins in the abdominal position rather than a thoracic position. The Freshwater Drum also has the lateral line continuing to the end of the caudal fin, a feature that separates it from all other Arkansas fishes.

VARIATION AND TAXONOMY *Aplodinotus grunniens* was described by Rafinesque (1819) from the Ohio

River. The Freshwater Drum is in need of a thorough systematic study. Boschung and Mayden (2004) reported preliminary morphological data indicating that there may be more than one species. Jacquemin and Pyron (2013) concluded that morphological variation in *A. grunniens* in the Wabash River, Indiana, was the result of a combination of developmental, sexual, and environmental influences.

DISTRIBUTION *Aplodinotus grunniens* has the greatest latitudinal range of any North American freshwater fish, occurring from northern Canada south to the Rio Usumacinta, Guatemala (Page and Burr 2011). North of Mexico, it is found in the St. Lawrence–Great Lakes, Hudson Bay, and Mississippi River basins from Quebec to northern Manitoba and southern Saskatchewan, Canada, south to the Gulf of Mexico; and in Gulf Slope drainages from the Mobile Bay basin, Georgia and Alabama, west through Texas. The Freshwater Drum occurs throughout Arkansas primarily in the larger rivers, oxbow lakes, and large reservoirs (Map 239). It mostly avoids small tributary streams and the most mountainous regions.

HABITAT AND BIOLOGY This species, known to some commercial fishermen and recreational anglers as the gaspergou, is the only North American strictly freshwater representative of the family Sciaenidae. The drum derives its name from the peculiar ability of the members of the family Sciaenidae to make loud booming or drumming sounds by contracting special muscles along the walls of the swim bladder. This sound may signal other males and females to congregate in the spawning area. In Arkansas, drums inhabit deep pools of medium to large rivers, lakes, and large impoundments. Drums are benthic-oriented and prefer turbid waters with mud bottoms and are only occasionally found in clear water. An inflow of turbid water into an Oklahoma reservoir caused larval drums to become distributed throughout the water column, in contrast to their normal concentration near the bottom (Matthews 1984). Priegel (1967b) noted that individuals became distressed when water temperatures exceeded 25.6°C and when dissolved oxygen concentrations remained low for an extended period. This species also avoids heavily vegetated areas. Rypel et al. (2006) showed that hydraulic-based habitat was an important variable in determining Freshwater Drum condition and growth. In that Alabama study, growth rates of Freshwater Drum were significantly greater in rivers than in reservoirs, and the drums in lotic environments were more robust than those in lentic habitats. In Missouri streams, drums often moved distances greater than 161 km (100 miles) (Funk 1957). Freshwater Drums are schooling fish, and movement and activity are largely governed by water temperatures and water

movement (Nord 1967). Drums tend to move into shallow water in the spring and back to deeper water in late fall. Diel movements of drums during the summer in Alabama were described by Rypel and Mitchell (2007), who found that many drums shifted location to nearshore habitat after dark. Occasionally, drums are found in large concentrations in the tailwaters of dams, and in certain years, large die-offs occur due to gas bubble disease, as happened in the Arkansas River in Arkansas in December 2010 and January 2011.

This bottom-dwelling species feeds on a wide variety of prey, mostly benthic invertebrates, depending on availability. Feeding occurs at all hours of the day (Priegel 1967b). Rypel et al. (2006), based on examination of the literature on food habit studies, concluded that lotic habitats provided more of the macroinvertebrate foods Freshwater Drums are adapted to eat than lentic environments; therefore, studies reporting large amounts of fish in the diet are usually from reservoirs. For example, in the Ohio River, fish made up only 8.4% of the adult drum diet (Wahl et al. 1988), but in eutrophic Wheeler Lake, Alabama, fish made up >50% of the annual food consumption, and 97% of large adult drum stomachs contained fish (Wrenn 1968). In four Oklahoma reservoirs, fish made up 80% of the diet by volume (Summerfelt et al. 1972). Foods typically include bivalve mollusks (including introduced *Corbicula* and *Dreissena*), small fish, chironomids, small crustaceans, and other aquatic macroinvertebrates (Daiber 1952; Priegel 1967a; Summerfelt et al. 1972; Lewis 1999; Magoulick and Lewis 2002; Shields and Beckman 2015). Typically, larval drums feed on zooplankton, especially copepods and cladocerans (Ross 2001; Sullivan 2009), but in a Tennessee study, larval drums first fed on other larval fishes and switched to a diet of zooplankton at about 10–12 mm TL (Clark and Pearson 1979). Young drums feed on small crustaceans and aquatic insect larvae, and adults feed on snails, small clams, crayfish, small fishes, and insect larvae (Krumholz and Cavanah 1968; Swedberg 1968; Bur 1982). Age 0 drums averaging 4.6–8.4 mm TL in the Missouri River fed mainly on dipteran larvae and pupae, followed in decreasing order of importance by trichopteran larvae, copepods, ephemeropteran nymphs, cladocerans, amphipods, and odonate nymphs (Starks and Long 2017). Butler (1962) and Swedberg (1968) attributed an increase in growth and abundance of drums in the Mississippi and Missouri rivers to increases in abundance of mayflies (*Hexagenia*). Boschung and Mayden (2004) reported larger drums feed predominantly on bivalve mollusks (*Corbicula*) and fish, especially Gizzard Shad. Small Bluegills, Black Crappies, and Black Bullheads are also reportedly consumed by adult drums (Moen 1955). The Freshwater Drum has large pharyngeal teeth well suited for crushing the shells of mollusks (French 1997). Morrison et al. (1997) studied predation on nonnative Zebra Mussels (*Dreissena polymorpha*) by Freshwater Drums in western Lake Erie, and found that the mussels were fed upon when the drums reached about 150 mm TL. Zebra Mussel consumption increased with increasing drum size, composing up to 33% of estimated dry weight food volume in drums >350 mm TL in May, and mussels were absent from the diets of all drums in October; fish (mainly *Dorosoma cepedianum* and *Notropis* spp.) made up about 93% of drum diets in October. Based on Arkansas Game and Fish Commission population sampling data between 1970 and 2006, Freshwater Drum biomass increased in the Arkansas River after the invasion of the river by nonnative Zebra Mussels (Quinn and Limbird 2008). However, French and Love (1995) concluded that a size limitation on Zebra Mussels consumed by *A. grunniens* may preclude the effectiveness of drums as agents of biological control for that invasive species.

Spawning occurs in late spring in Arkansas (April or May) after adults move out of reservoirs and large rivers into tributary streams. It also spawns successfully in lakes (Cross and Collins 1995), and it spawns successfully in the large rivers of Arkansas. In the upper Mississippi River, spawning occurs from May to July at water temperatures of 66–72°F (18.9–22.2°C) (Butler 1965; Nord 1967). Sound production by males begins with the approach of the spring breeding season. The drumming sounds become longer and more frequent as breeding reaches its peak (Schneider and Hasler 1960). The drumming gradually diminishes through the summer and ceases in late August. The Freshwater Drum is a pelagic spawner, far away from the shoreline. In Wisconsin, schools of spawning fish were observed milling at the surface with their backs out of the water (Wirth 1958). Females produce an enormous number of 1 mm diameter pelagic eggs (43,000–508,000) that float at the surface until hatching (Daiber 1953). The eggs hatch in 24–48 hours (27 hours at 23°C) and the newly hatched 3.2 mm TL larvae remain attached to the surface film for 1 or 2 days (Swedberg and Walburg 1970). This characteristic of floating eggs and larvae is found in many saltwater fish and is rare among freshwater fishes in North America (Davis 1959; Becker 1983). After detaching from the surface film, the larvae drift near the surface, forming an important part of the larval drift in rivers (Muth and Schmulbach 1984), for about 2 weeks, until they are about 0.4 inch (10 mm) TL. Then they move to deeper water and finally assume their bottom-dwelling mode of life at a length of around 1 inch (25 mm) (Priegel 1967a). Freshwater Drum larvae were

abundant in the surface waters of the lower Mississippi River in June (Gallagher and Conner 1983). Scales first appear on Freshwater Drums at approximately 15 mm and the young are fully scaled by 22 mm (Priegel 1966). Life span may vary by region and habitat. Priegel (1967a) noted that few Freshwater Drums in Wisconsin live more than 10 years, but Becker (1983) reported records of individuals as old as 17 years. Rypel et al. (2006) found a 32-year-old drum (467 mm TL) in the Cahaba River, Alabama.

In the Mississippi River, the maturation status of young male Freshwater Drums increases with age. While a small percentage of the youngest mature males were age 2 (mean length of 310 mm), 63.4% of age 3 males were mature (mean length of 351 mm), and almost all the age 4 males (97%) had reached maturity (Butler and Smith 1950). Females first mature at age 4, at which time 46% were mature at an average length of 386 mm. All age 5 females were mature (Butler and Smith 1950).

CONSERVATION STATUS The Freshwater Drum is abundant and widespread in Arkansas and in no need of protection.

Family Cichlidae *Cichlids and Tilapias*

Wiley and Johnson (2010) supported the placement of Cichlidae in the order Labriformes. Nelson et al. (2016) also removed Cichlidae from the order Perciformes but placed it in the order Cichliformes along with the family Pholidichthyidae, the convict blennies. Although listed as a separate order by Nelson et al. (2016), no order account or detailed justification for recognition of Cichliformes was provided. Betancur-R. et al. (2017) also supported recognition of order Cichliformes based on molecular evidence. We feel that morphological (morphological synapomorphies are lacking) and molecular evidence are insufficient at this time to support removal of Cichlidae from the order Perciformes (see our classification in Appendix 5).

This large, widely distributed family includes about 202 genera and more than 1,762 described species (Nelson et al. 2016). There are believed to be many undescribed species, with estimates of total number of cichlid species ranging between 2,000 and 3,000 (Stiassny et al. 2007). The cichlid family has been divided into nine subfamilies; eight subfamilies were diagnosed on morphological features (Kullander 1998) and one subfamily was based on a genetic analysis (Sparks and Smith 2004b). Nelson et al. (2016) noted that there is disagreement about the recognition of cichlid genera, and additional work is needed to identify monophyletic groups at the genus level.

The cichlid family is found from southern Texas through

Mexico, Central America, and most of South America. It is also found throughout most of Africa, parts of the Middle East, southern India and some islands such as Cuba and Madagascar (Berra 1981, 2007). Only one cichlid species, the Rio Grande Cichlid, *Herichthys cyanoguttatus* Baird and Girard, is native to the United States, occurring in the Rio Grande drainage of southern Texas (but established in some other drainages in Texas, Florida, and Louisiana). The majority of cichlid species occur in the tropics and subtropics, and the East African Rift Lakes are noted for their cichlid species swarms. Many of those rift lake cichlids are endemic species. Some cichlid species resemble our North American centrarchids and are widely considered their ecological equivalents. Buchanan (1971) found evidence for resource partitioning between *H. cyanoguttatus* and centrarchid species in the San Marcos River, Texas. Montaña and Winemiller (2013) compared local assemblages of cichlids in Venezuela and Peru with an assemblage of centrarchids in Texas to reveal patterns of morphological and ecological convergence. Those authors found that with few exceptions, fishes with similar morphologies had similar trophic patterns. Cichlids and centrarchids had the same set of ecomorph types that corresponded to the same trophic niches, including substrate-sifting invertivores, epibenthic invertebrate gleaners, and piscivores (Montaña and Winemiller 2013).

Cichlids are distinguished from centrarchids by the presence of a single nostril on each side (vs. 2 nasal openings on each side of snout) and an interrupted lateral line consisting of two distinct parts, the front portion of the lateral line is higher on the body than the rear portion (vs. lateral line complete in all Arkansas centrarchids except *Lepomis symmetricus*, which has an interrupted lateral line). Cichlids are also noted for extended parental care of their eggs and young, with one or both parents often providing protection for their young far beyond the parental care usually provided by male centrarchids. Cichlids are very popular in the aquarium trade and a number of larger species are widely used in aquaculture. At least 44 species of cichlids have been found in lakes or streams of the United States because of intentional or unintentional introductions, and 20 species have become established in Florida (Robins et al. 2018). Those introductions were the result of aquarium releases, escapes from aquaculture operations, or from intentional stocking for vegetation control (Page and Burr 2011).

The common name "tilapia" refers to more than 30 cichlid species in three genera that are endemic to Africa and the Middle East. Various tilapia species are commercially farmed throughout the world, including Arkansas. As a

visit to almost any supermarket will attest, the market for tilapia has drastically expanded in recent years, and this expansion is attributed to tilapia's firm flesh and mild flavor (Popma and Masser 1999). Tilapia are particularly suited to warmwater aquaculture because they are easily spawned, eat a variety of natural and artificial foods, tolerate poor water quality (especially low dissolved oxygen), and most importantly for Arkansas's public waters, die at prolonged water temperatures below 48°F (9°C) (Popma and Masser 1999).

Three species in the genus *Oreochromis* (and their hybrids) are the only cichlid species approved by the Arkansas Game and Fish Commission for stocking in private ponds and lakes in Arkansas. Over a 22-year period beginning in 1987, the AGFC stocked tilapia in public reservoirs in Arkansas, Cleburne, Faulkner, Greene, Hempstead, Mississippi, Ouachita, Poinsett, and Randolph counties. Tilapia were also maintained at state fish hatcheries in Benton, Clay, Garland, and Lonoke counties. There is reason to believe that some attempts to stock tilapia in Arkansas occurred prior to the more recent documented efforts (Sam Henry, pers. comm.). In spring 2001, the AGFC stocked Nile Tilapia and their hybrids in Lake Hogue, a 250-acre impoundment in Poinsett County, in a pilot project to improve the forage fish base for predators such as Largemouth Bass (Henry 2005). By summer 2001, anglers were catching eating-size tilapia in large numbers, and in November when water temperatures dropped below 45°F (7°C) (a lethal temperature), fishermen could harvest the dying tilapia with dip nets. An AGFC creel survey in 2004 indicated that about 80% of the stocked fish were harvested by anglers, either by hook and line or by dipnets, during the die-off that almost always occurred during the last two weeks of November (Sam Henry, pers. comm.). The average net weight gain of tilapia during the 6–7-month growing season in Lake Hogue was 0.9 pounds (0.4 kg) per fish. The positive response from fishermen prompted the AGFC to stock Nile Tilapia (and their hybrids) in Lake Mallard. The put-and-take tilapia fishery was discontinued by 2010, primarily because of the cost of purchasing brood stock and the difficulty in overwintering tilapia at the state fish hatcheries (sufficient brood stock had to be maintained over winter in heated buildings). In addition, the expected increases in gamefish growth rates and condition (as a result of stocking tilapia) did not occur (Brett Timmons, pers. comm.). Tilapia species have been used as forage fish in bass nursery ponds at some state fish hatcheries; therefore, when bass from those hatcheries were stocked in lakes, it is possible that some tilapia individuals were inadvertently stocked as well. This could account for records of tilapia from public waters that were not intentionally stocked with tilapia (such as recent records from Pulaski and Columbia counties). Separate species angling records are not kept, and the state record for "tilapia" is 3 pounds, 8 ounces (1.6 kg) from a lake at Camp Robinson in Pulaski County on 26 October 2011. The world angling record caught in 2010 is 9.6 pounds (4.4 kg) (a Blue Tilapia) from Florida. Sam Henry (pers. comm.) noted that tilapia species are the fishes of choice for aquaponic systems. Their toughness, good growth rates, and ability to eat a wide variety of foods make them an excellent choice to grow in water recirculation systems. The AGFC discontinued propagation and use of tilapia at all state fish hatcheries by 2016.

Genus *Oreochromis* Günther, 1889—Tilapias

The genus *Oreochromis* was established by Günther in 1889, but it has sometimes been regarded as a subgenus, and placement of species in it has been problematic and controversial. All three cichlid species introduced into Arkansas were formerly placed in other genera, and Trewavas (1981, 1982) recognized *Oreochromis* as a genus and assigned several species formerly in *Tilapia* to *Oreochromis*. Miller et al. (2005) continued to recognize *Tilapia* as a genus for the species assigned to *Oreochromis* by Trewavas, but most recent authors (e.g., Page et al. 2013; Robins et al. 2018) support recognition of *Oreochromis*, with some species still assigned to *Tilapia*. It should be noted that although the three species introduced into Arkansas are no longer in genus *Tilapia*, they are all still referred to as tilapia and all have "Tilapia" in their common names. *Oreochromis* currently contains 32 species. Genus characters include first gill arch with 13–26 rakers on the lower limb, the outer jaw teeth are readily movable, scales on belly are usually much smaller than scales on sides, and the genital papilla of male and female is well developed.

All species of *Oreochromis* are maternal mouth brooders, and all are polygamous in their natural habitats. No lasting pair-bond is retained during the breeding season (Trewavas 1983). Males of *Oreochromis* have distinctive and conspicuous breeding colors and are generally larger than females. The male excavates a nest in the substrate, usually at depths of less than 1 m, and spawns with several females in succession; polyandry also occurs. After each mating, the female scoops up the fertilized eggs with her mouth and broods them in her buccal cavity for 7–17 days depending on water temperature. After hatching, the young remain in the mother's mouth through the yolk sac stage, and free-swimming fry are guarded for several more days, often retreating to the safety of the mother's mouth at night or when danger threatens. Total length of parental care may be up to 3 weeks. The entire breeding pattern that occurs in nature is often not completely observed in aquarium spawning (Trewavas 1983).

All *Oreochromis* species in Arkansas show strong sexual dichromatism in large, sexually mature adults. Determining the sex of smaller adults and juveniles is more difficult but can be done with a high degree of accuracy by examining the genital papilla located immediately behind the anal opening. In males, the genital papilla has a single opening through which both urine and milt pass. The genital papilla of the female has two openings, one for urine and one for eggs. In smaller individuals, it may be necessary to apply a drop of food coloring and use magnification to make the openings easier to see.

Tilapias are successful invasive species because of their wide salinity, oxygen, and temperature tolerances, ability to exist at high population densities, high reproductive potential, and ability to feed at multiple trophic levels (Canonico et al. 2005; Henson et al. 2018). The interactions of tilapias with native sport fishes are not well understood, and various studies have produced conflicting information. Hybrids between various *Oreochromis* species are frequently produced for aquaculture and fisheries purposes. The hybrids, which are mostly males, are fully fertile and capable of backcrossing with females of either parental species. The primary advantages of an all-male population in fish culture are the generally faster growth of males and the avoidance of dwarfing and overpopulation resulting from early mating (Trewavas 1983). Some of the tilapia stocked in Arkansas are hybrids.

Key to The Cichlids and Tilapias

1A Caudal fin without thin, black, vertical bands; lower limb of first gill arch usually with 20 or fewer (usually 17 or 18) gill rakers; length of lower jaw of mature male going into head length 2.0–2.63 times.

Oreochromis mossambicus Page 764

1B Caudal fin with thin, black, vertical bands developed to varying degrees; lower limb of first gill arch usually with more than 20 gill rakers; length of lower jaw of mature male going into head length 2.70–3.45 times. 2

2A Caudal fin with irregular, vague, narrow, vertical bands with bands usually most prominent in proximal part of fin; dorsal spines 14–17 (typically 15 or 16).

Oreochromis aureus Page 763

2B Caudal fin with distinct, narrow vertical black stripes or bars; dorsal spines usually 17 or 18 (modally 17).

Oreochromis niloticus Page 766

Oreochromis aureus (Steindachner)
Blue Tilapia

Figure 26.101. *Oreochromis aureus*, Blue Tilapia.
Patrick O'Neil, Geological Survey of Alabama.

Map 240. Distribution of *Oreochromis aureus*, Blue Tilapia, 1988–2019. Inset shows introduced range (outlined) in the contiguous United States.

CHARACTERS Body laterally compressed, deep and sunfish-like, with long dorsal fins (Fig. 26.101). Body depth going into SL 2.0–2.8 times. A single nasal opening on each side of snout. Mouth small. Length of lower jaw of mature male going into head length 2.70–3.45 times. Spinous and soft dorsal fins broadly connected. Caudal fin truncate. Lower limb of first gill arch with 18–26 gill rakers (usually more than 20). Teeth in 3–5 rows in jaws, bicuspid in outermost row, tricuspid in inner rows. Dorsal spines 14–17 (typically 15 or 16); dorsal rays 11–15; anal spines 3; anal rays 8–11. Lateral line interrupted, divided into 2 distinct parts. Scales in lateral series usually 30–33. Cheek scale rows usually 2, sometimes 3. Scale rows above lateral line usually 4 or 5. Preorbital bone with 5 sensory canal openings. Male genital papilla simple, or at most with a narrow flange. Maximum length about 17 inches (420 mm).

LIFE COLORS Juveniles usually with wide, dark, vertical bars on sides, the bars sometimes persisting in adults. Juveniles also have a black blotch anteriorly on the base of the second dorsal fin ("tilapia spot"); blotch fades with increasing size. Adults are olive to bluish gray dorsally, yellow olive to silver laterally, and white ventrally. Each lateral scale with a small dark spot, often causing sides to appear horizontally striped. Caudal fin with or without bars, but if present, bands are indistinct dusky reticulations or vague, interrupted bands. Breeding males an intense bluish gray, with intense metallic blue on head. Dorsal fin with a vermilion edge, and caudal fin margin with intense pink coloration; posterior half of dorsal fin with alternating dark and light spots. Breeding females lighter in coloration with a pale orange margin on dorsal and caudal fins.

SIMILAR SPECIES The Blue Tilapia can be distinguished from our native sunfishes by the presence of a single nostril on each side of head (vs. 2 nasal openings on each side) and by an interrupted lateral line consisting of two distinct parts (vs. lateral line complete in all native sunfishes except *Lepomis symmetricus*). It differs from the other two introduced *Oreochromis* species in usually having fewer dorsal spines (usually 15 or 16 vs. 17 or 18). It further differs from *O. mossambicus* in having a deeper body and in breeding males lacking the charcoal to black body and white to yellow-gold throat of *O. mossambicus*.

VARIATION AND TAXONOMY *Oreochromis aureus* was originally described by Steindachner (1864) as *Chromis aureus* from west Africa. The Blue Tilapia has also been assigned to the genera *Tilapia* and *Sarotherodon* by various workers, and it was eventually placed in *Oreochromis* (Trewavas 1983).

DISTRIBUTION *Oreochromis aureus* is native to tropical and subtropical Africa and the Jordan River system of the Middle East. It has been introduced into at least 15 states in the United States, including Arkansas. It is considered locally established or possibly established in 10 states, including Oklahoma and Texas (Nico et al. 2018a). The Blue Tilapia is commonly sold for bait and is raised in farm ponds for food in Arkansas (A. Carter, pers. comm.). It was used as a forage species in bass ponds at the Andrew H. Hulsey State Fish Hatchery in Hot Springs (B. Hobbs, pers. comm.). It has been stocked in Arkansas impoundments, and two specimens were collected from the Arkansas River below Dam 3 on 11 August 1998 (Buchanan et al. 2000) (Map 240).

HABITAT AND BIOLOGY   The Blue Tilapia is a warmwater species and cannot survive Arkansas winters outdoors. It generally stops feeding at water temperatures below 63°F (17°C) but tolerates the lowest environmental temperatures of any of the introduced *Oreochromis* species. It is a relatively cold-tolerant tilapia, but sustained temperatures below 48°F (9°C) are lethal (Popma and Masser 1999). It is also known to form schools. Optimal water temperatures for growth are 85–88°F (29–31°C). The Blue Tilapia grows well in brackish water up to 20 ppt salinity (Popma and Masser 1999). In Florida, where it has become the most widespread nonindigenous fish, it is considered a competitor with native species for spawning areas, food, and space (Courtenay and Robins 1975). Some Florida streams where this species is abundant have lost most vegetation and nearly all native fishes. The Blue Tilapia has also been associated with reduction of native fish populations in warm springs in Nevada (Scoppettone et al. 2005). There is evidence that Blue Tilapia may be associated with shad population declines in lakes and reservoirs in Florida and Texas through competition for food (Hendricks and Noble 1980; Kushlan 1986; Taylor et al. 1984; Zale and Gregory 1990). There is also evidence for competition with the native Rio Grande Cichlid, *Herichthys cyanoguttatus*, in Texas (Edwards and Contreras-Balderas 1991). Henson et al. (2018) found substantial diet overlap between Blue Tilapia and Bluegills in a North Carolina reservoir but found little evidence of tilapias influencing the relative abundance of native sport species. Schramm and Zale (1985) found that small Blue Tilapia can serve as the optimal and preferred prey of Largemouth Bass.

The adults are omnivorous and feed on algae, aquatic macrophytes, plankton, benthic invertebrates, and detritus. The adult diet in Alabama ponds consisted primarily of phytoplankton (McBay 1961). Other reported food items include protozoans, annelids, Formicidae, rotifers, nematodes, oligochaetes, trichopterans, cladocerans, copepods, ostracods, dipterans, fry of Blue Tilapia, and unidentified eggs (McBay 1961; Spataru and Zorn 1978). It is very efficient at harvesting phytoplankton using a secreted mucous to trap plankton rather than just physical filtration by the gill rakers. It is not considered piscivorous but larval fish are sometimes consumed. Young fish have a more varied diet than adults, consuming large quantities of copepods and cladocerans.

Sexual maturity is reached at lengths of 3–4 inches (75–100 mm), and maximum life span is 5+ years (Boschung and Mayden 2004). Spawning occurs from late March through May in Israel (Ben-Tuvia 1959) and begins in late April in ponds in Alabama (McBay 1961). Aureli and

Torrans (1988) studied spawning frequency and fecundity of Blue Tilapia in indoor facilities in Arkansas. Between 25°C and 28°C (77–82°F), females spawned in intervals as short as 2 weeks if they were not allowed to incubate the eggs, and fecundity ranged from 97 to 1,042 eggs. In Alabama ponds, females 128–168 mm TL produced 64–655 eggs per spawn (McBay 1961), and a single female can hold up to 2,000 eggs in her mouth (Ben-Tuvia 1959). Reproductive behavior of this mouth brooder is described in the genus *Oreochromis* account. Torrans and Lowell (1987) reported that a combination of Blue Tilapia and Channel Catfish grown together in ponds in Arkansas reduced off-flavor in the catfish by 54%. Where *O. aureus* has established breeding populations in the warm waters of Florida and Texas, native fish assemblages were unable to control tilapia population levels either by competition or by predation (Buntz and Manooch 1969). In those states, the Blue Tilapia has continued to expand its range (Courtenay and Robins 1973; Courtenay et al. 1991; Nico et al. 2018a). It has become so abundant in Florida that a substantial commercial fishery for Blue Tilapia exists in that state (Hale et al. 1995).

CONSERVATION STATUS   This cichlid has been introduced into some Arkansas lakes to provide a summer and fall sport fishery. Because of winter die-off, it is not likely to establish permanent populations in Arkansas. It is the only cichlid for which voucher specimens exist from the state's natural waters (Buchanan et al. 2000).

Oreochromis mossambicus (Peters)
Mozambique Tilapia

Figure 26.102. *Oreochromis mossambicus*, Mozambique Tilapia. *Robert H. Robins, Florida Museum.*

CHARACTERS   Body laterally compressed, moderately deep and sunfish-like, with long dorsal fins (Fig. 26.102). Body depth going into SL 2.02–2.78 times in adults. Spinous and soft dorsal fins broadly connected. Caudal fin

Map 241. Distribution of *Oreochromis mossambicus*, Mozambique Tilapia, 1988–2019. Inset shows introduced range (outlined) in the contiguous United States.

rounded with slight notch along edge. Lower limb of first gill arch with 14–20 gill rakers (usually 17 or 18). Dorsal spines 15–17 (modally 16); dorsal rays 10–13; anal spines 3 (sometimes 4); anal rays 9–12 (usually 10 or 11); pectoral rays 14 or 15. Lateral line interrupted, divided into two distinct parts. Scales in lateral series 29–33 (modally 31). Scale rows above lateral line 3–5. Cheek scale rows usually 3, occasionally 2. Preorbital bone with 5 sensory canal openings. A single nasal opening on each side of snout. Jaws of mature males enlarged, resulting in a concave upper profile. Upper jaw ending between nostril and eye in females and immature males, but ending below anterior edge of eye in breeding males. Length of lower jaw of mature male going into head length 2.0–2.63 times. Jaw teeth are typically bicuspid but gradually, by replacement, become unicuspid in large individuals; both upper and lower jaws have an inner row of smaller tricuspid teeth (Robins et al. 2018). Pharyngeal teeth are very fine. Genital papilla of male simple or with a shallow distal notch. Maximum length around 15 inches (390 mm).

LIFE COLORS   Sides of females and nonbreeding males silvery with 2–7 midlateral black blotches and sometimes with dorsal blotches that extend downward to the lateral blotches; sides of adults sometimes without distinct markings. Fins are clear to yellow with some dusky pigment. Caudal fin without narrow, dark bands or reticulations. Juveniles up to 60 mm SL usually with 6–8 wide, black vertical bars on silvery sides. Juveniles also have the "tilapia spot"—a black spot on the anterior base of the second dorsal fin. Breeding males becoming mostly charcoal gray to black with white to yellow-gold throat and cheeks; dorsal and caudal fins with reddish-pink to pink margins (pink margin widest on upper and lower caudal lobes); pectoral fins usually red. The striking contrast of the white lower parts of the head with the black general color in territorial males is characteristic of this species. Breeding females are much lighter in coloration.

SIMILAR SPECIES   The Mozambique Tilapia can be distinguished from our native sunfishes by the presence of a single nostril on each side of head (vs. 2 nasal openings on each side) and by an interrupted lateral line consisting of 2 distinct parts (vs. lateral line complete in all native sunfishes except *Lepomis symmetricus*). *Oreochromis mossambicus* differs from the other two introduced *Oreochromis* species in usually having 17 or 18 gill rakers on the lower limb of the first gill arch (vs. usually more than 20 gill rakers), in lacking dark vertical bands on the caudal fin (vs. dark bands present), and breeding males with a charcoal-gray to black body and a white to yellow-gold throat (vs. body of breeding male usually with a pink wash and without a yellow-gold throat). Additionally, *O. mossambicus* is not as deep bodied as the other two *Oreochromis* species.

VARIATION AND TAXONOMY   *Oreochromis mossambicus* was originally described by Peters (1852) as *Chromis mossambicus* from Tette, Sena, Quellimane, Lumbo, Inhambane, and Querimba in Mozambique. It was subsequently placed in *Tilapia* and *Sarotherodon* by later workers, and most recently in *Oreochromis* by Trewavas (1981).

DISTRIBUTION   The native range of *O. mossambicus* includes tropical and subtropical regions of Africa, where it occurs in both fresh and brackish water. It has been widely introduced for aquaculture and is also an important commercial, game, and aquarium species. Nonindigenous occurrences have been reported from 14 states in the United States (Nico and Neilson 2016) (Map 241). It is raised in pond culture in Arkansas and has been stocked in public reservoirs and private ponds in the state.

HABITAT AND BIOLOGY   Like our other introduced tilapias, it prefers quiet, warm waters and is tolerant of high salinity, low dissolved oxygen, and high ammonia concentrations (Popma and Masser 1999). The Mozambique Tilapia is the most salinity tolerant of the three *Oreochromis* species introduced into Arkansas, and it grows well at salinities near or at full-strength seawater. It is intolerant of cold water temperatures and cannot survive Arkansas winters outdoors. Prolonged water temperatures below 50°F (10°C) are lethal.

The consensus, based on studies of wild and stocked populations, is that juveniles are omnivorous and adults feed mainly on detritus (Trewavas 1983). Boschung and Mayden (2004) considered *O. mossambicus* an omnivore and noted that its long, coiled intestine indicated that it was at least a partial herbivore. The juveniles are at least partly carnivorous and the adults are omnivorous, feeding on algae, aquatic macrophytes, invertebrates, and detritus. This species is less efficient than the other two introduced *Oreochromis* species at harvesting phytoplankton. It opportunistically feeds on small fish.

Reproduction is most successful at temperatures above 80°F (27°C) and does not occur below 68°F (20°C). Sexual maturity is attained at lengths of 5–6 inches (120–140 mm) and at ages as young as 6 months. Reproductive behavior of this mouth brooder is described in the genus *Oreochromis* account. Buchanan (1971) studied reproduction in the San Marcos River, Texas, and found that males usually established spawning territories in colonies. Very little courtship preceded spawning, which occurred quickly. The sexes segregated immediately after spawning, with the brooding females seeking shelter in vegetation away from the spawning area. Amorim et al. (2003) documented sound production by males during all phases of courtship and spawning. Maximum age is around 11 years (Boschung and Mayden 2004).

CONSERVATION STATUS   This species is stocked in some reservoirs to provide a summer and fall sport fishery. It is not likely to establish populations in Arkansas because it does not survive the state's winter temperatures.

Oreochromis niloticus (Linnaeus)
Nile Tilapia

Figure 26.103. *Oreochromis niloticus*, Nile Tilapia. *Zachary Randall, Florida Museum.*

Map 242. Distribution of *Oreochromis niloticus*, Nile Tilapia, 1988–2019. Inset shows introduced range (outlined) in the contiguous United States.

CHARACTERS   Body laterally compressed, deep and sunfish-like, with long dorsal fins (Fig. 26.103). Body depth going into SL 1.79–2.94 times. Mouth small. Length of lower jaw of mature male going into head length 2.70–3.45 times. Spinous and soft dorsal fins broadly connected. Caudal fin truncate, but sometimes rounded, with many distinct, narrow, vertical stripes. Jaw teeth are bicuspid, with 2 or 3 smaller inner rows. Lower limb of first gill arch with 20–26 gill rakers. Dorsal spines 16–18 (modally 17); dorsal rays usually 12 or 13 (11–14); anal spines 3; anal rays 9–11. Lateral line interrupted, divided into 2 distinct parts. Scales in lateral series usually 31–33 (30–34). Scale rows above lateral line usually 4 or 5. Cheek scale rows usually 2 or 3. Preorbital bone with 5 sensory canal openings. A single nasal opening on each side of snout. Genital papilla of male short and conical or bluntly bifid at tip. Maximum length around 24 inches (620 mm) and weight of 8 pounds (3.65 kg) by 9 years of age.

LIFE COLORS   Sides of females and nonbreeding males silver gray, sometimes with a light violet wash; side usually with a series of dusky black bars. Throat region gray or pink. Distal margin of dorsal fin black or gray but never pink or orange, the melanin sometimes slightly mixed with red. Caudal fin with distinct, dark vertical stripes or bands. Other fins clear to yellowish and usually with dusky markings. Juveniles usually with wide dark vertical bars on sides, and opercle with a distinct black spot. Juveniles also have a black "tilapia spot" anteriorly on the base of the second dorsal fin. Breeding males become uniformly darker with a pink-red flush on the body, the lower part of the head, and sides. Dorsal fin with a black margin; caudal fin with a pinkish-red posterior margin. Breeding females are variable in coloration but are usually much lighter than males. Females develop a characteristic color pattern while brooding young: dark stripes across the forehead and a dark operculum, chin, and eye (Turner and Robinson 2000).

SIMILAR SPECIES The Nile Tilapia can be distinguished from our native sunfishes by the presence of a single nostril on each side of head (vs. 2 nasal openings on each side) and by an interrupted lateral line consisting of 2 distinct parts (vs. lateral line complete in all native sunfishes except *Lepomis symmetricus*). *Oreochromis niloticus* differs from *O. mossambicus* in having many distinct dark vertical bands on caudal fin (vs. caudal fin without dark bands), in usually having more than 20 gill rakers on the lower limb of the first gill arch (vs. fewer than 20 gill rakers), and distal margin of dorsal fin of breeding male black or gray (vs. distal margin of dorsal fin pink or red). It differs from *O. aureus* in having distinct black bands on caudal fin (vs. bands on caudal fin irregular and vague), in having modally 17 (16–18) dorsal fin spines (vs. typically 15 or 16 dorsal spines), and in dorsal fin of breeding male with a black margin (vs. a pink to pink-red margin).

VARIATION AND TAXONOMY *Oreochromis niloticus* was described by Linnaeus in 1758 as *Labrus niloticus* from Egypt. It has subsequently been assigned by various workers to the genera *Chromis*, *Tilapia*, and *Sarotherodon*, until its assignment to *Oreochromis* by Trewavas (1983). This is a variable species over its native range, and several populations originally described as separate species have been synonymized with *O. niloticus* and sometimes considered subspecies.

DISTRIBUTION The native range of the Nile Tilapia includes tropical and subtropical Africa and the Middle East. It has been widely introduced for aquaculture, and more than 90% of all commercially farmed tilapia species outside Africa are Nile Tilapia (Popma and Masser 1999). Nonindigenous occurrences have been reported from eight states in the United States (Nico et al. 2016a). It is raised in pond culture in Arkansas and has been stocked in at least two public reservoirs in the state. In 2013, the city of Blytheville, Arkansas, stocked Nile Tilapia in Walker Park City Lake. There is a record of a specimen of *O. niloticus* caught with a cast net in late September 2015 from Lake Columbia in Columbia County (Map 242).

HABITAT AND BIOLOGY The Nile Tilapia is similar in habitat requirements and environmental tolerances to the other two introduced *Oreochromis* species, but it is the least euryhaline of the three, tolerating salinities up to 15 ppt (Popma and Masser 1999). It thrives in warm water above 80°F (27°C), and prolonged temperatures below 50°F (10°C) are lethal. It cannot survive Arkansas winters outdoors and was thought to be established in the United States only in a single reservoir bordering Florida and Georgia; however, an established population was documented from southern Mississippi (Peterson et al. 2004; Grammer et al. 2012), and Godwin et al. (2016) confirmed that *O. niloticus* was reproducing in southern Alabama. Zambrano et al. (2006) studied the invasive potential of *O. niloticus* in American freshwater systems using an ecological niche modeling approach and concluded that its distributional potential is limited to tropical and coastal areas. It prefers shallow, quiet waters near the edges of lakes and wide rivers with sufficient vegetation.

Nile Tilapia exhibit trophic plasticity according to the environment and the other species with which they coexist (Bwanika et al. 2007). The fry are omnivorous, actively pursuing copepods, hydracarines, and various insects (both aquatic larvae and terrestrial insects that fall into the water); they also ingest aufwuchs and detritus (Trewavas 1983). Small juveniles are mainly carnivorous, feeding on microcrustaceans, chironomids, and other small invertebrates; adults are omnivorous but predominantly herbivorous, feeding mainly on filamentous algae (Boschung and Mayden 2004). Adults also feed on aquatic macrophytes, benthic invertebrates, larval fish, and detritus. They are also efficient planktivores, using a thick mucus secreted by the gills to trap plankton. Most nutrition is obtained by grazing on periphyton mats (Trewavas 1983). Nile Tilapia were stocked in a Blytheville, Arkansas, city lake (Walker Park Lake) in 2013 to improve water quality because of the propensity of that species to select blue-green algae as a food source (Sam Henry, pers. comm.). Prior to the introduction of *O. niloticus*, annual fish kills were common in Walker Park Lake due to overly fertile water and an abundance of blue-green algae caused by a "waterfowl overload." There was no fish die-off in that 5-acre lake in 2013, probably because of a combination of the introduction of Nile Tilapia and an attempt to scare away waterfowl. In Lake Hogue, Arkansas (a 101 ha closed-system reservoir), anglers harvested an estimated 83% of the stocked Nile Tilapia; 56% of stocked fish were captured by traditional pan fishing techniques and 27% were captured with dip nets during the winter die-off in late November (Henry et al. 2005).

Females can mature and produce eggs as early as 50 days

of age and length of 4 inches (100 mm). Large females can mouth-brood 2,000 eggs, about twice the number of most mouth-brooding tilapias (Boschung and Mayden 2004). Spawning begins when the water temperature reaches 24°C (75°F), but spawning has been observed at temperatures as low as 19°C (Trewavas 1983). Temperature-dependent sex determination is known to occur in *O. niloticus* (Baroiller et al. 1995). Reproductive behavior is described in the genus *Oreochromis* account. Maximum life span is 10+ years.

CONSERVATION STATUS This tilapia has been stocked in some Arkansas reservoirs to provide a summer and fall sport fishery. It is not likely to become established because it is killed by the winter temperatures in Arkansas.

APPENDIX 1

Arkansas Fish Collections

Collections of Arkansas fishes are available for study at several museums. One of the largest holdings of Arkansas fishes (30,000+ lots) was formerly housed at the University of Louisiana at Monroe Museum of Zoology, derived primarily from collections leading to masters' theses. Unfortunately, ULM decided to dispose of its entire fish collection, and in 2017, that collection was divided among Arkansas State University, the Mississippi Museum of Natural Science, Tulane University, Louisiana State University, Southeastern Louisiana, and the University of Texas Natural History Collection (with ASU receiving approximately 10,000 of the Arkansas lots). Tulane University also serves as a repository for many Arkansas specimens, including some lots obtained from the University of Arkansas. Significant older collections from Arkansas, examined and reported by Black (1940), are available at the University of Michigan Museum of Zoology, the United States National Museum, the Museum of Comparative Zoology at Harvard University, Academy of Natural Sciences at Philadelphia, and The Field Museum of Natural History in Chicago. The California Academy of Sciences houses some Arkansas collections made by Seth E. Meek and David Starr Jordan. More recently collected Arkansas material is housed at Arkansas State University Museum of Zoology (currently the largest collection of fishes located within Arkansas), University of Arkansas–Fort Smith, University of Arkansas Fayetteville, Texas Natural History Science Center (TNHC), University of Kansas Museum of Natural History, University of Florida (Florida State Museum), and Cornell University. In addition, small collections of Arkansas fishes are maintained by the University of Arkansas at Monticello, Henderson State University, Southern Arkansas University, University of Central Arkansas, Arkansas Tech University, and University of Arkansas at Pine Bluff. A search of online fish databases*, such as FishNet2, can be used to find cataloged lots of fishes from Arkansas in several other curated museum collections, including Sam Noble Oklahoma Museum of Natural History (University of Oklahoma), Illinois Natural History Survey, University of Tennessee–Etnier Ichthyological Research Collection, Mississippi Museum of Natural Science, University of Alabama Ichthyological Collection, Auburn University Museum of Natural History, Louisiana State University Museum of Zoology, The Ohio State University Museum, Los Angeles County Museum of Natural History, North Carolina State Museum of Natural Sciences, Texas A&M University Biodiversity Research and Teaching Collections, University of Minnesota Bell Museum of Natural History, and Yale University Peabody Museum. Information on the taxonomic status and history of all genera and species of Arkansas fishes can be found in Eschmeyer's Catalog of Fishes Online Database of the California Academy of Science (Eschmeyer et al. 2018).

---

* Even the most prestigious fish collections contain some cataloged specimens that have been misidentified. Caution should be used in accepting extralimital or other noteworthy or unusual records without examination of the specimens.

APPENDIX 2

Preservation of Specimens

In Arkansas, anyone collecting fishes other than by angling or seining minnows for bait should apply for a collection permit from the Arkansas Game and Fish Commission, No. 2 Natural Resources Drive, Little Rock, AR 72205. If the permit is granted, the collector may use a variety of methods to collect fishes, including seining, angling, gill netting, and traps; however, one may not use chemicals or electrofishing unless accompanied by an Arkansas Game and Fish Commission fisheries biologist. Sometimes the collecting permit allows the use of electrofishing with certain restrictions. It is also necessary to have the appropriate federal collecting permits if fish collections will be made in streams under federal jurisdiction, such as a national forest. All collections must be reported annually to the Arkansas Game and Fish Commission. Methods for collecting and preserving fishes have been described previously in several readily available works (e.g., Etnier and Starnes 1993; Pflieger 1997; Miller and Robison 2004); however, those methods are repeated here in an abbreviated form for the convenience of the reader. It is important to know something about the standard methods for preserving and maintaining specimens retained in local collections or sent to institutional museums.

The best containers for field collections are wide-mouthed plastic jars from one quart to one gallon capacity. Larger plastic containers are used for preserving large specimens. The use of glass containers in the field is discouraged because of the danger of breakage. Jars with some type of metal bail or clamp for securing the lid are ideal and save time; however, they can be difficult to obtain. While plastic jars have screw-on lids, which slow the opening and closing process, they have the advantage of being resistant to breakage. Specimens should be placed in a preservative while still alive. It is recommended that collectors anesthetize the specimens prior to preservation. The standard field preservative for fishes is a 10% solution of formalin (9 parts water, 1 part concentrated formalin). Full strength formalin (35–40%) can be purchased at chemical supply outlets. Formalin kills and fixes the tissues of the specimens, but it is a known carcinogen and caution should be exercised in its use and disposal. Specimens larger than 6 inches should be slit along the right side of the abdomen to ensure proper preservation of deeper tissues and internal organs. Extremely large specimens should also have formalin injected with a hypodermic needle into the larger muscle masses. Specimens and tissue samples to be used for DNA analysis are generally frozen or preserved in 95–100% nondenatured ethyl alcohol.

It is important that a label of good-quality, high rag-content paper be placed inside each container when the collection is made. Do not place labels only on the outside of jars, as these labels may be lost and the collection rendered worthless for distributional information. A technical pen (Micron Pigma Pen or Rapidograph) or a pencil may be used to make temporary tags in the field. Higgins Eternal Ink or Higgins Engrossing Ink is generally recommended for preparation of permanent labels. Many museums now use computer generated labels. The following information should be on the label: date; name of stream, lake, or pond where the collection was made; the precise locality (state, county, section, township, and range if known); the highway number of the access road; names of the collectors; and a collection number. It is now common practice to use a GPS instrument to get precise measurements of the latitude and longitude of a collecting locality. Most professionals keep notes in a field notebook for later examination. These contain more detailed information on the physical description of the collection site: temperature, depth, bottom type; weather conditions; collecting methods; time of collection; and notes on abundance, habitat, life colors, stage of maturity, etc.

Specimens should be left in the original formalin solution (10%) for at least a week to ensure proper preservation. After removal from the formalin preservative, they should be soaked in freshwater for two or three days, changing the water several times during this washing process. The specimens should then be permanently stored in 50% isopropyl alcohol or 70% ethyl alcohol. Most large museums currently use ethyl alcohol as the preferred medium for long-term storage, but isopropyl alcohol is much less expensive and does not evaporate as easily. Larochelle et al. (2016) found that long-term storage of specimens in isopropanol does not alter fish morphometrics. Any specimen shrinkage occurs shortly after preservation and does not exacerbate over decadal time spans. Clamp-top glass jars with rubber gaskets are the best containers for permanent storage; however, these are difficult to find and quite expensive; therefore, most collectors use screw-top lids. The collection should be checked periodically for evaporation and any fluid loss should be replaced.

APPENDIX 3

Aids to Identification of Fishes

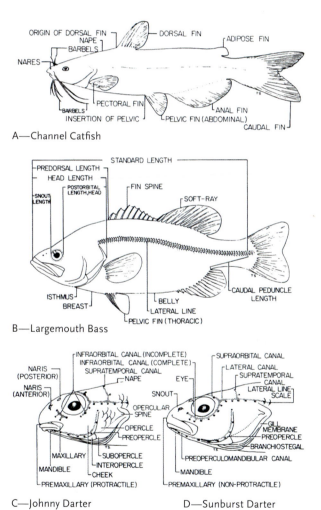

A—Channel Catfish

B—Largemouth Bass

C—Johnny Darter   D—Sunburst Darter

Figure A3.1. Structural features of fishes often used for identification *(Cross 1967)*.

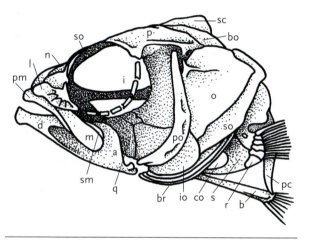

Figure A3.2. The major head bones of a generalized bony fish.

d—dentary	p—parietal
a—angular	sc—supraoccipital
co—coracoid	r—radials
pc—postcleithrum	s—scapula
pm—premaxilla	bo—basioccipital
m—maxilla	po—preopercle
sm—supramaxilla	o—opercle
l—lachrymal	io—interopercle
n—nasal	so—subopercle
br—branchiostegals	q—quadrate
f—frontal	so—supraorbital
b—basipterygium	

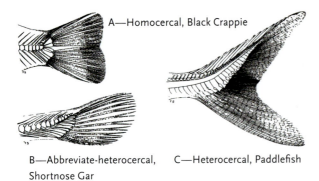

A—Homocercal, Black Crappie

B—Abbreviate-heterocercal,
Shortnose Gar

C—Heterocercal, Paddlefish

Figure A3.3. Kinds of caudal fins in fishes *(Cross 1967)*.

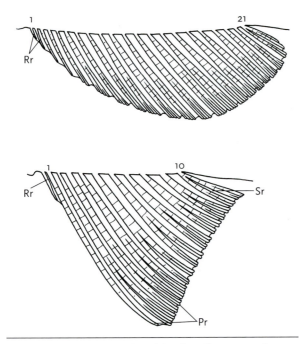

Figure A3.4. Hypothetical anal fins (in effect, inverted dorsal fins) showing how fin rays are counted *(Cross 1967)*.

Top: Total ray count, as taken in fins that slope gradually away from the body contour.

Bottom: Principal ray count, as taken in fins that have a straight leading edge (rudimentary [procurrent] rays contiguous anteriorly).

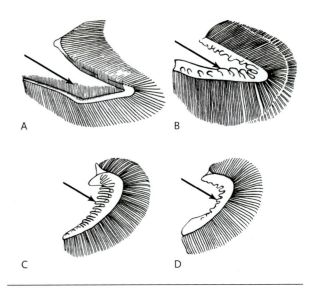

A   B

C   D

Figure A3.5. First gill arches, showing differences in number and shape of gill rakers (see arrows on concave side of each arch) *(Cross 1967)*.

A—Rakers numerous and slender, as in Clupeidae. (Gizzard Shad illustrated).

B—Rakers short and knoblike, as in Hiodontidae. (Goldeye illustrated).

C—Rakers slender, as in species of *Lepomis* other than the Longear and Redear sunfishes. (Green Sunfish illustrated).

D—Rakers short and knoblike, as in Longear and Redear sunfishes. (Longear Sunfish illustrated).

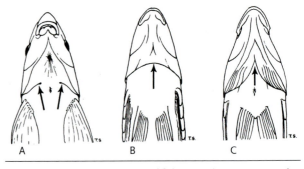

A   B   C

Figure A3.6. Gill membranes of fishes in relation to ventral body wall (note arrows) *(Cross 1967)*.

A—Right and left membranes bound down to isthmus, as in minnows and suckers (*Carpiodes* illustrated); a needle-tip slipped into the gill cleft on one side cannot be moved freely across to the opposite side.

B—Gill membranes broadly joined across (but free from) isthmus, as in Banded Darter.

C—Gill membranes separate (right and left sides not conjoined) and free from isthmus, as in Autumn Darter.

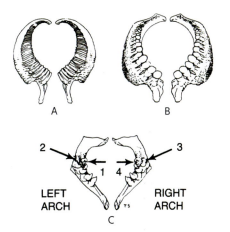

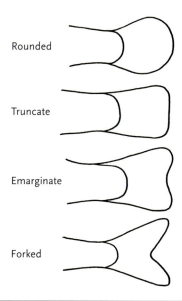

Figure A3.7. Pharyngeal arches of two suckers (A and B) and a minnow (C) on a horizontal surface with teeth projecting upward *(Cross 1967)*.

A—Golden Redhorse; teeth numerous and slender, in a single row (comblike).

B—River Redhorse; lower teeth stumplike or molariform, but numerous and uniserial.

C—Creek Chub; teeth few, confined to central part of arch, often hooked (2-rowed in this species). The sequence in which rows are counted is indicated by numerals 1, 2, 3, 4. (The count is 2,5–4,2).

Figure A3.9. Some caudal fin shapes in fishes. Top to bottom: rounded, truncate (straight margin), emarginate, and forked.

Figure A3.8. (*Right*) Areas and appendages of fishes. In the bottom fish, the spinous dorsal fin = first dorsal fin, and the rayed (soft) dorsal fin = second dorsal fin; these fins may be broadly joined or separate.

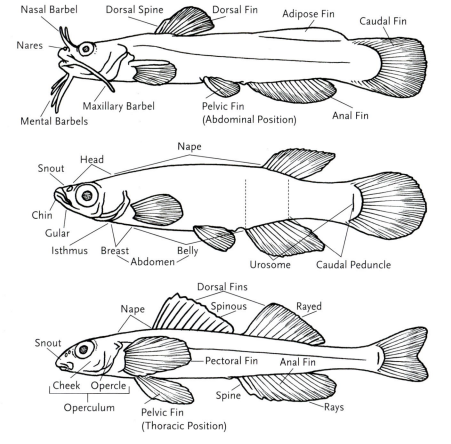

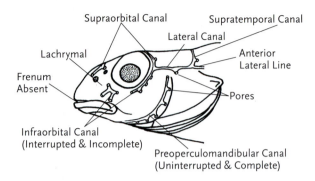

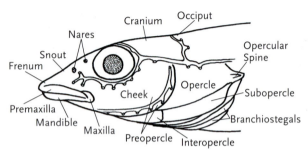

Figure A3.10. Areas and parts of the head, including the canals of the cephalic lateralis system.

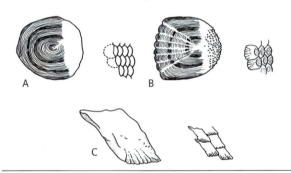

Figure A3.11. Types of scales:

    A—Cycloid
    B—Ctenoid
    C—Ganoid

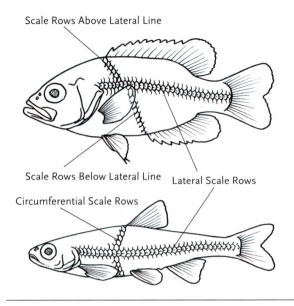

Figure A3.12. Counting lateral line scales and scale rows above and below the lateral line.

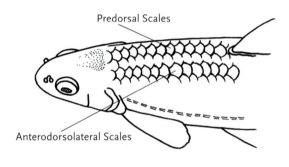

Figure A3.13. Predorsal and anterodorsal scales.

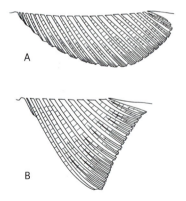

Figure A3.14. Counts of median fin rays:

A—Anal fin with all rays counted, as in catfishes; last 2 ray elements [Sr] counted as 1 ray. Types of rays: Rr—rudimentary, Pr—principal; Sr—split ray.

B—Anal fin with only principal rays counted; Sr counted as 1 ray.

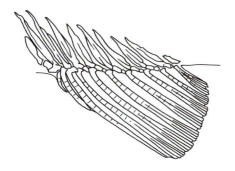

Figure A3.15. *Chrosomus* anal fin.

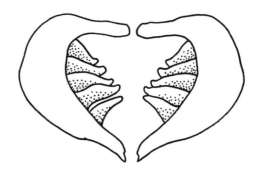

Figure A3.18. *Chrosomus erythrogaster* pharyngeal teeth.

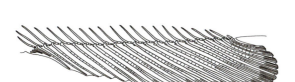

Figure A3.16. *Noturus* anal fin.

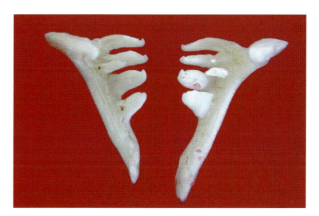

Figure A3.19. Dorsal view of pharyngeal teeth of minnow with arches rotated outward so teeth are visible. *Chad Thomas.*

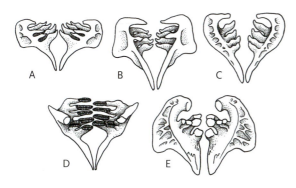

Figure A3.17. Anterior views of pharyngeal arches of cyprinids showing teeth configurations:

A—2,4–4,2 (*Notropis shumardi*)
B—1,4–4,1 (*Cyprinella*)
C—4–4 (*Nocomis*)
D—2,5–4,2 (*Ctenopharyngodon*)
E—Molariform teeth of *Cyprinus*

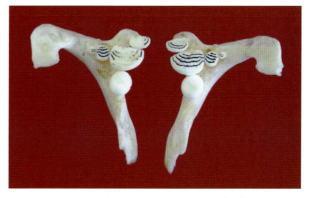

Figure A3.20. Dorsal view of pharyngeal teeth of Common Carp, *Cyprinus carpio*, with arches rotated outward so teeth are visible. *Chad Thomas.*

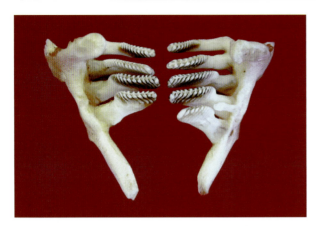

Figure A3.21. Dorsal view of pharyngeal teeth of Grass Carp, *Ctenopharyngodon idella*, with arches rotated outward so teeth are visible. *Chad Thomas.*

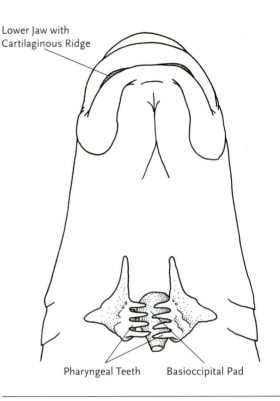

Figure A3.23. Stoneroller (*Campostoma*) pharyngeal arches, showing relative position of pharyngeal teeth and basioccipital pad.

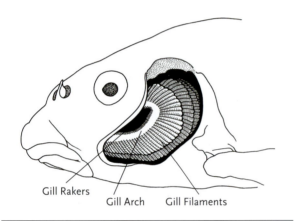

Figure A3.22. Stoneroller (*Campostoma*) with operculum removed to show gill arches, gill rakers, and gill filaments.

APPENDIX 4

Pre-1988 Distribution Maps for Arkansas Fishes

The pre-1988 distribution maps are primarily the maps from the first edition of this book. Additional maps have been added to show the pre-1988 distributions (when known) for species described since 1988 (such as *Notropis suttkusi* and *Etheostoma clinton*). These additions sometimes necessitated the revision of some of the other maps. For example, a new "old" map was required for *Etheostoma stigmaeum* because some populations in Arkansas formerly under that name have now been recognized as *E. clinton* or *E. teddyroosevelt*. Similarly, a new map was required for *Labidesthes sicculus* because most lowland records now fall under the recently described *L. vanhyningi*. Pre-1988 distribution maps are included for *Ameiurus catus* (Map A4.110) and *Micropterus coosae* (Map A4.175), two introduced species that

have been removed from the 2019 Arkansas checklist because they are believed to no longer occur in this state. Modifications were required only to the inset maps for some species. The old inset range map for *Menidia audens* (formerly *M. beryllina*) required an adjustment to reflect the removal of Atlantic Slope populations from that species' range. The Redspotted Sunfish and Western Mosquitofish distribution maps required similar adjustments. Maps are arranged by family according to our current concept of phylogenetic order (not the same order as in the first edition of this book). Species maps within a family are presented in alphabetical order by genus, and then by species. Species are listed by their 2019 scientific and common names.

Fish collections in Arkansas (Pre-1960)

Map A4.1. Fish collection localities in Arkansas (pre-1960).

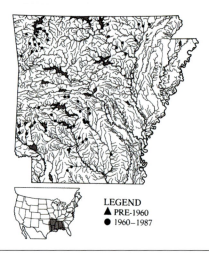

LEGEND
▲ PRE-1960
● 1960–1987

Map A4.4. Distribution of *Ichthyomyzon gagei*, Southern Brook Lamprey.

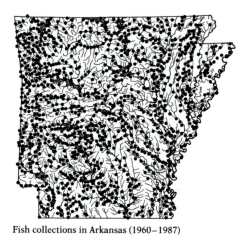

Fish collections in Arkansas (1960–1987)

Map A4.2. Fish collection localities in Arkansas (1960–1987)

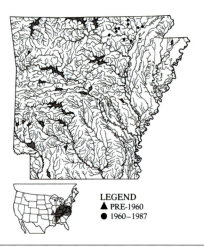

LEGEND
▲ PRE-1960
● 1960–1987

Map A4.5. Distribution of *Lampetra aepyptera*, Least Brook Lamprey.

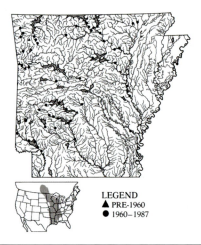

LEGEND
▲ PRE-1960
● 1960–1987

Map A4.3. Distribution of *Ichthyomyzon castaneus*, Chestnut Lamprey.

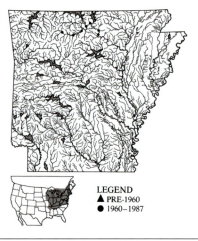

LEGEND
▲ PRE-1960
● 1960–1987

Map A4.6. Distribution of *Lethenteron appendix*, American Brook Lamprey.

Map A4.7. Distribution of *Acipenser fulvescens*, Lake Sturgeon. The single distribution point in the Little Missouri River is for *A. oxyrinchus*, Atlantic Sturgeon.

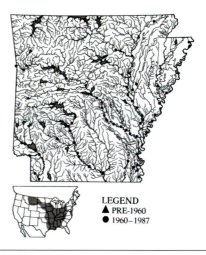

Map A4.10. Distribution of *Polyodon spathula*, Paddlefish.

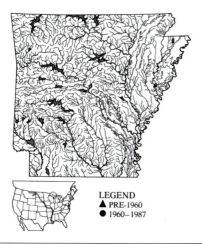

Map A4.8. Distribution of *Scaphirhynchus albus*, Pallid Sturgeon.

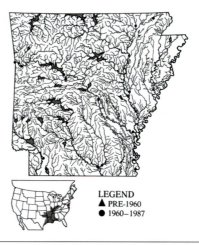

Map A4.11. Distribution of *Atractosteus spatula*, Alligator Gar.

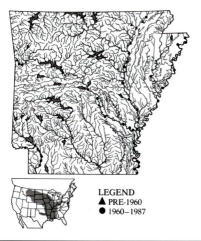

Map A4.9. Distribution of *Scaphirhynchus platorynchus*, Shovelnose Sturgeon.

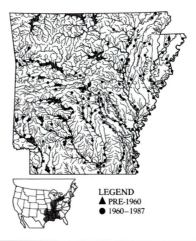

Map A4.12. Distribution of *Lepisosteus oculatus*, Spotted Gar.

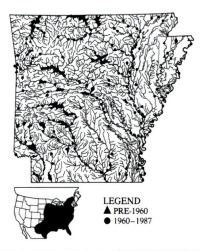

Map A4.13. Distribution of *Lepisosteus osseus*, Longnose Gar.

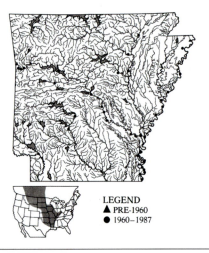

Map A4.16. Distribution of *Hiodon alosoides*, Goldeye.

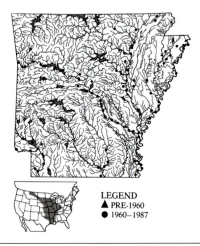

Map A4.14. Distribution of *Lepisosteus platostomus*, Shortnose Gar.

Map A4.17. Distribution of *Hiodon tergisus*, Mooneye.

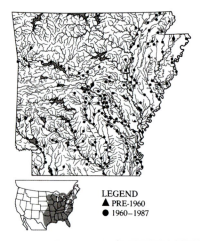

Map A4.15. Distribution of *Amia calva*, Bowfin.

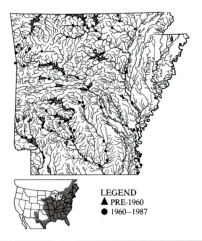

Map A4.18. Distribution of *Anguilla rostrata*, American Eel.

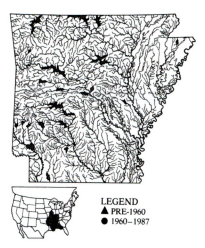

Map A4.19. Distribution of *Alosa alabamae*, Alabama Shad.

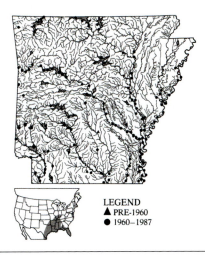

Map A4.22. Distribution of *Dorosoma petenense*, Threadfin Shad.

Map A4.20. Distribution of *Alosa chrysochloris*, Skipjack Herring.

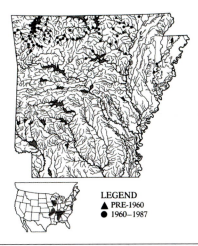

Map A4.23. Distribution of *Campostoma oligolepis*, Largescale Stoneroller.

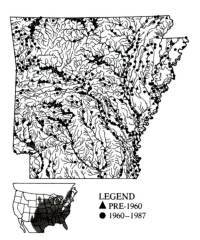

Map A4.21. Distribution of *Dorosoma cepedianum*, Gizzard Shad.

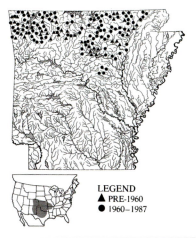

Map A4.24. Distribution of *Campostoma plumbeum*, Plains Stoneroller.

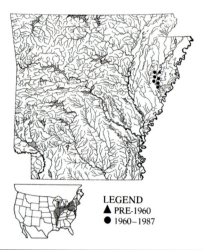

Map A4.25. Distribution of *Campostoma pullum*, Finescale Stoneroller.

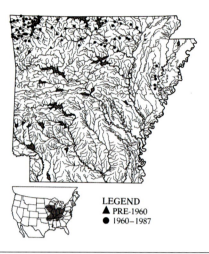

Map A4.28. Distribution of *Chrosomus erythrogaster*, Southern Redbelly Dace.

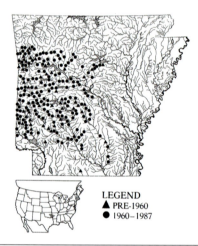

Map A4.26. Distribution of *Campostoma spadiceum*, Highland Stoneroller.

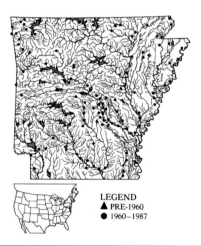

Map A4.29. Distribution of *Ctenopharyngodon idella*, Grass Carp.

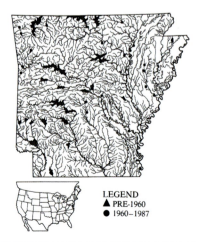

Map A4.27. Distribution of *Carassius auratus*, Goldfish.

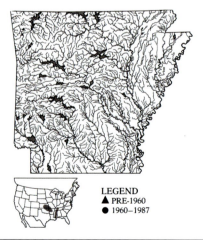

Map A4.30. Distribution of *Cyprinella camura*, Bluntface Shiner.

Map A4.31. Distribution of *Cyprinella galactura*, Whitetail Shiner.

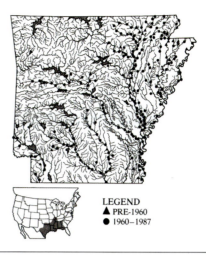

Map A4.34. Distribution of *Cyprinella venusta*, Blacktail Shiner.

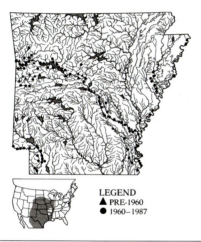

Map A4.32. Distribution of *Cyprinella lutrensis*, Red Shiner.

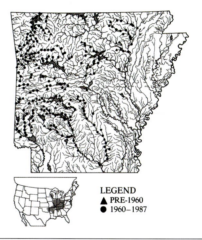

Map A4.35. Distribution of *Cyprinella whipplei*, Steelcolor Shiner.

Map A4.33. Distribution of *Cyprinella spiloptera*, Spotfin Shiner.

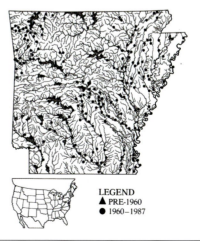

Map A4.36. Distribution of *Cyprinus carpio*, Common Carp.

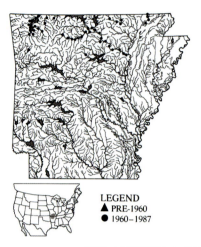

Map A4.37. Distribution of *Erimystax harryi*, Ozark Chub.

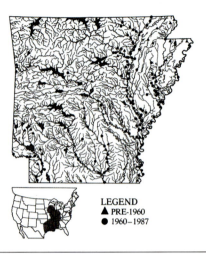

Map A4.40. Distribution of *Hybognathus nuchalis*, Mississippi Silvery Minnow.

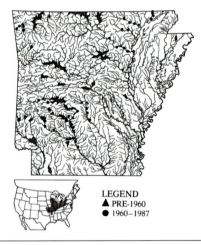

Map A4.38. Distribution of *Erimystax x-punctatus*, Gravel Chub.

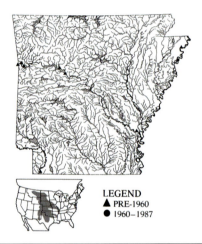

Map A4.41. Distribution of *Hybognathus placitus*, Plains Minnow.

Map A4.39. Distribution of *Hybognathus hayi*, Cypress Minnow.

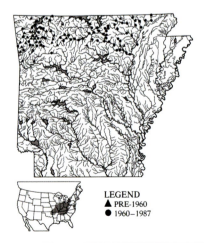

Map A4.42. Distribution of *Hybopsis amblops*, Bigeye Chub.

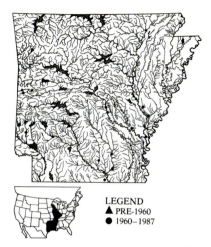

Map A4.43. Distribution of *Hybopsis amnis*, Pallid Shiner.

Map A4.46. Distribution of *Luxilus cardinalis*, Cardinal Shiner.

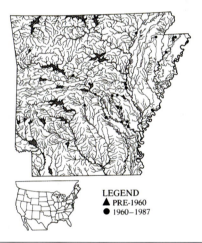

Map A4.44. Distribution of *Hypophthalmichthys molitrix*, Silver Carp.

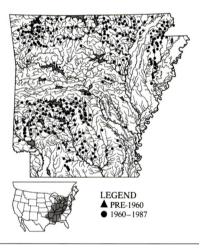

Map A4.47. Distribution of *Luxilus chrysocephalus*, Striped Shiner.

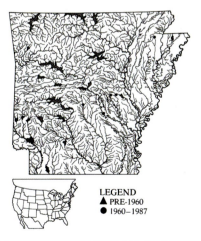

Map A4.45. Distribution of *Hypophthalmichthys nobilis*, Bighead Carp.

Map A4.48. Distribution of *Luxilus pilsbryi*, Duskystripe Shiner.

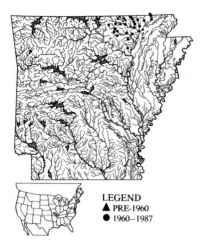

Map A4.49. Distribution of *Luxilus zonatus*, Bleeding Shiner.

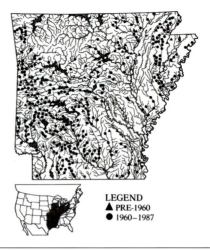

Map A4.52. Distribution of *Lythrurus umbratilis*, Redfin Shiner.

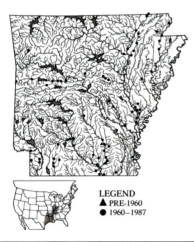

Map A4.50. Distribution of *Lythrurus fumeus*, Ribbon Shiner.

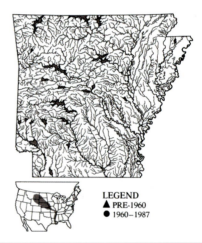

Map A4.53. Distribution of *Macrhybopsis gelida*, Sturgeon Chub.

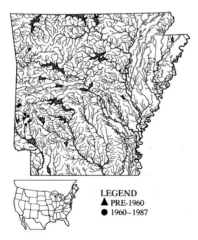

Map A4.51. Distribution of *Lythrurus snelsoni*, Ouachita Mountain Shiner.

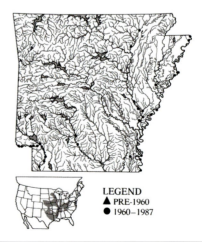

Map A4.54. Distribution of *Macrhybopsis hyostoma*, Shoal Chub.

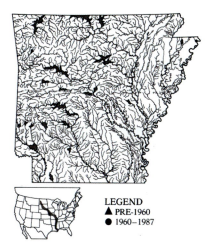

Map A4.55. Distribution of *Macrhybopsis meeki*, Sicklefin Chub.

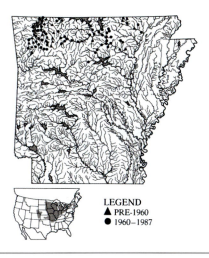

Map A4.58. Distribution of *Nocomis biguttatus*, Hornyhead Chub.

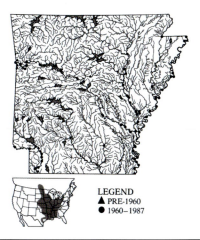

Map A4.56. Distribution of *Macrhybopsis storeriana*, Silver Chub.

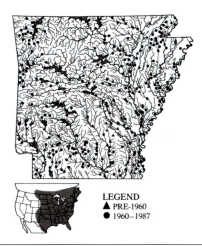

Map A4.59. Distribution of *Notemigonus crysoleucas*, Golden Shiner.

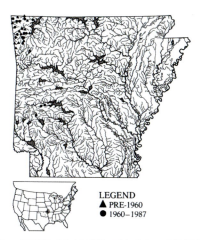

Map A4.57. Distribution of *Nocomis asper*, Redspot Chub.

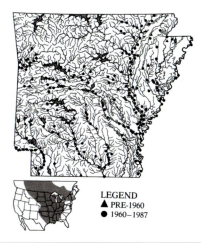

Map A4.60. Distribution of *Notropis atherinoides*, Emerald Shiner.

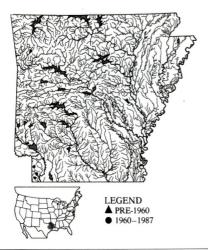

Map A4.61. Distribution of *Notropis atrocaudalis*, Blackspot Shiner.

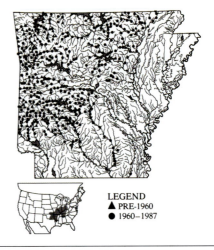

Map A4.64. Distribution of *Notropis boops*, Bigeye Shiner.

Map A4.62. Distribution of *Notropis bairdi*, Red River Shiner.

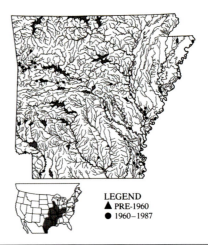

Map A4.65. Distribution of *Notropis buchanani*, Ghost Shiner.

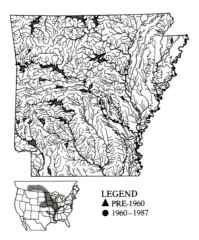

Map A4.63. Distribution of *Notropis blennius*, River Shiner.

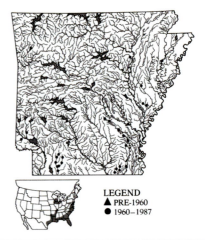

Map A4.66. Distribution of *Notropis chalybaeus*, Ironcolor Shiner.

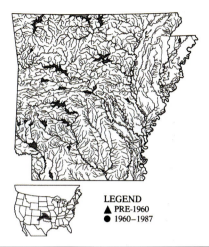

Map A4.67. Distribution of *Notropis girardi*, Arkansas River Shiner.

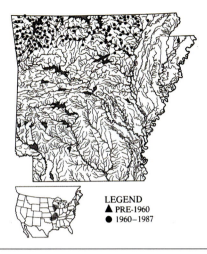

Map A4.70. Distribution of *Notropis nubilus*, Ozark Minnow.

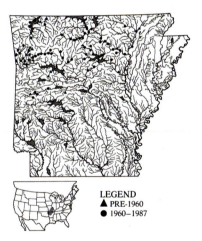

Map A4.68. Distribution of *Notropis greenei*, Wedgespot Shiner.

Map A4.71. Distribution of *Notropis ortenburgeri*, Kiamichi Shiner.

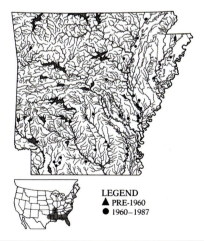

Map A4.69. Distribution of *Notropis maculatus*, Taillight Shiner.

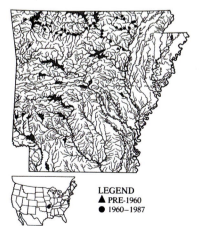

Map A4.72. Distribution of *Notropis ozarcanus*, Ozark Shiner.

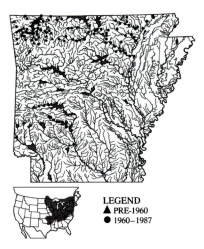

Map A4.73. Distribution of *Notropis percobromus*, Carmine Shiner.

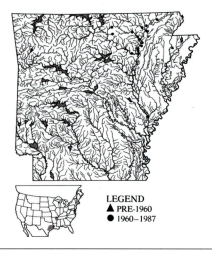

Map A4.76. Distribution of *Notropis sabinae*, Sabine Shiner.

Map A4.74. Distribution of *Notropis perpallidus*, Peppered Shiner.

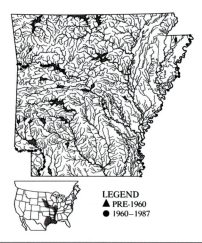

Map A4.77. Distribution of *Notropis shumardi*, Silverband Shiner.

Map A4.75. Distribution of *Notropis potteri*, Chub Shiner.

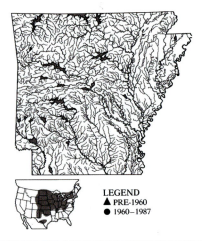

Map A4.78. Distribution of *Notropis stramineus*, Sand Shiner.

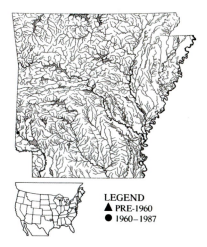

Map A4.79. Distribution of *Notropis suttkusi*, Rocky Shiner.

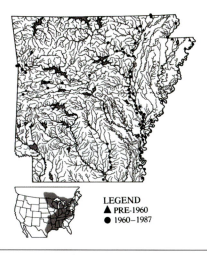

Map A4.82. Distribution of *Notropis volucellus*, Mimic Shiner. Some points represent the Channel Shiner.

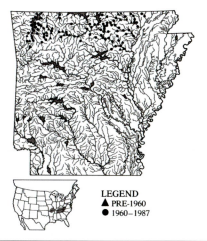

Map A4.80. Distribution of *Notropis telescopus*, Telescope Shiner.

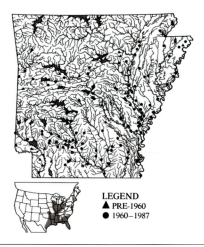

Map A4.83. Distribution of *Opsopoeodus emiliae*, Pugnose Minnow.

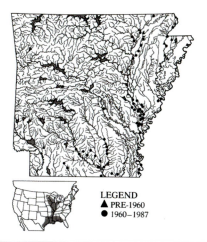

Map A4.81. Distribution of *Notropis texanus*, Weed Shiner.

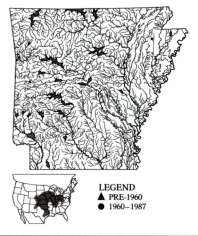

Map A4.84. Distribution of *Phenacobius mirabilis*, Suckermouth Minnow.

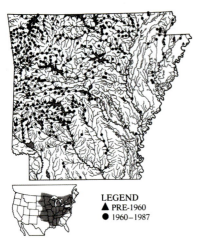

Map A4.85. Distribution of *Pimephales notatus*, Bluntnose Minnow.

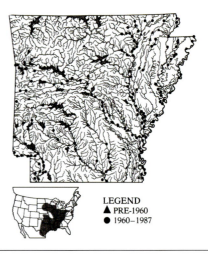

Map A4.88. Distribution of *Pimephales vigilax*, Bullhead Minnow.

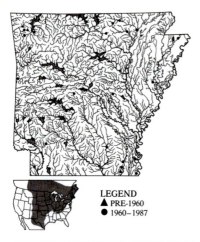

Map A4.86. Distribution of *Pimephales promelas*, Fathead Minnow.

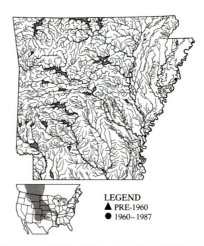

Map A4.89. Distribution of *Platygobio gracilis*, Flathead Chub.

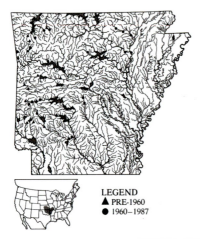

Map A4.87. Distribution of *Pimephales tenellus*, Slim Minnow.

Map A4.90. Distribution of *Pteronotropis hubbsi*, Bluehead Shiner.

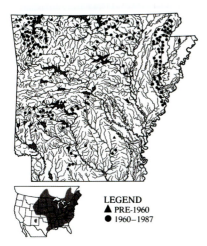

Map A4.91. Distribution of *Semotilus atromaculatus*, Creek Chub.

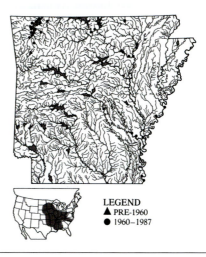

Map A4.94. Distribution of *Carpiodes velifer*, Highfin Carpsucker.

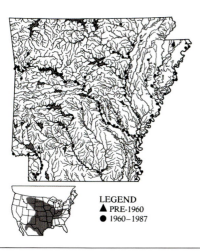

Map A4.92. Distribution of *Carpiodes carpio*, River Carpsucker.

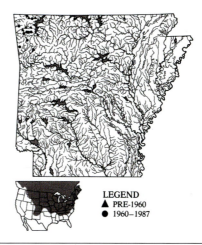

Map A4.95. Distribution of *Catostomus commersonii*, White Sucker.

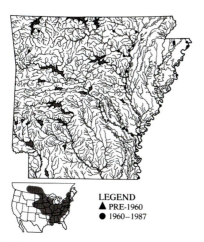

Map A4.93. Distribution of *Carpiodes cyprinus*, Quillback.

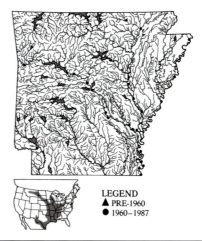

Map A4.96. Distribution of *Cycleptus elongatus*, Blue Sucker.

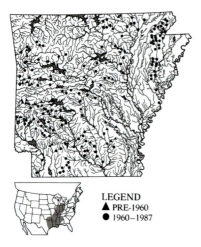

Map A4.97. Distribution of *Erimyzon claviformis*, Western Creek Chubsucker.

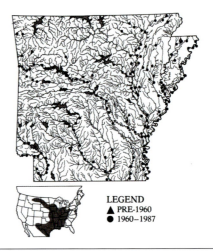

Map A4.100. Distribution of *Ictiobus bubalus*, Smallmouth Buffalo.

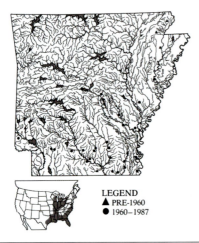

Map A4.98. Distribution of *Erimyzon sucetta*, Lake Chubsucker.

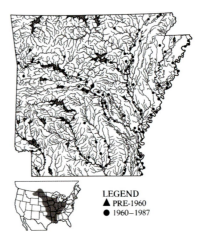

Map A4.101. Distribution of *Ictiobus cyprinellus*, Bigmouth Buffalo.

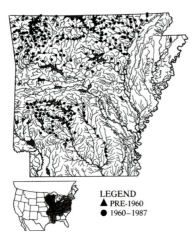

Map A4.99. Distribution of *Hypentelium nigricans*, Northern Hog Sucker.

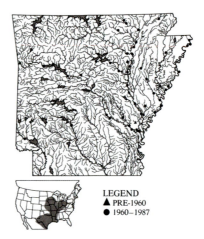

Map A4.102. Distribution of *Ictiobus niger*, Black Buffalo.

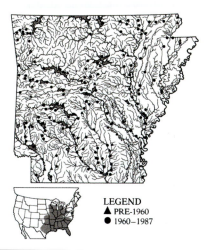

Map A4.103. Distribution of *Minytrema melanops*, Spotted Sucker.

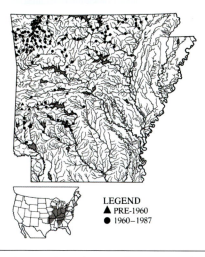

Map A4.106. Distribution of *Moxostoma duquesnei*, Black Redhorse.

Map A4.104. Distribution of *Moxostoma anisurum*, Silver Redhorse.

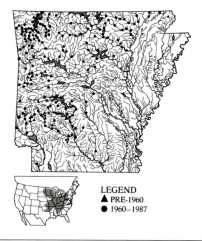

Map A4.107. Distribution of *Moxostoma erythrurum*, Golden Redhorse.

Map A4.105. Distribution of *Moxostoma carinatum*, River Redhorse.

Map A4.108. Distribution of *Moxostoma pisolabrum*, Pealip Redhorse.

Map A4.109. Distribution of *Moxostoma poecilurum*, Blacktail Redhorse.

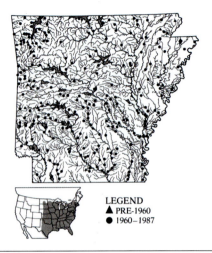

Map A4.112. Distribution of *Ameiurus natalis*, Yellow Bullhead.

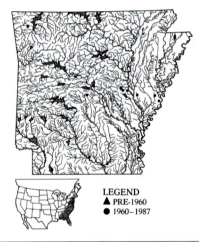

Map A4.110. Distribution of *Ameiurus catus*, White Catfish

Map A4.113. Distribution of *Ameiurus nebulosus*, Brown Bullhead.

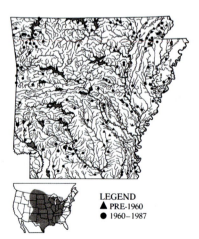

Map A4.111. Distribution of *Ameiurus melas*, Black Bullhead.

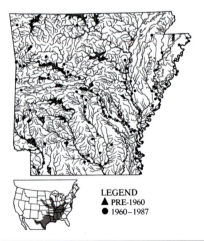

Map A4.114. Distribution of *Ictalurus furcatus*, Blue Catfish.

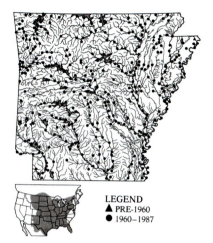

Map A4.115. Distribution of *Ictalurus punctatus*, Channel Catfish.

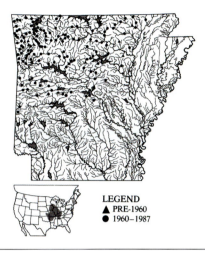

Map A4.118. Distribution of *Noturus exilis*, Slender Madtom.

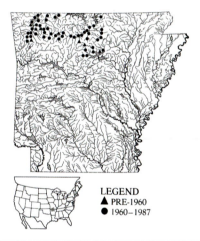

Map A4.116. Distribution of *Noturus albater*, Ozark Madtom.

Map A4.119. Distribution of *Noturus flavater*, Checkered Madtom.

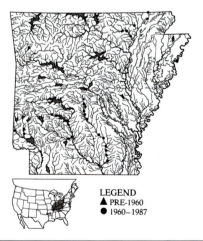

Map A4.117. Distribution of *Noturus eleutherus*, Mountain Madtom.

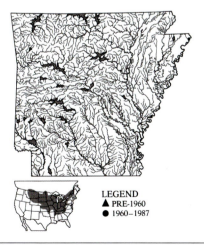

Map A4.120. Distribution of *Noturus flavus*, Stonecat.

Map A4.121. Distribution of *Noturus gladiator*, Piebald Madtom.

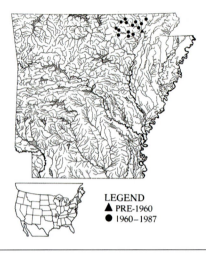

Map A4.124. Distribution of *Noturus maydeni*, Black River Madtom.

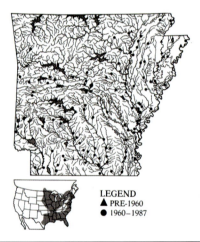

Map A4.122. Distribution of *Noturus gyrinus*, Tadpole Madtom.

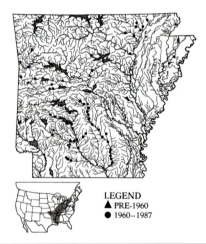

Map A4.125. Distribution of *Noturus miurus*, Brindled Madtom.

Map A4.123. Distribution of *Noturus lachneri*, Ouachita Madtom.

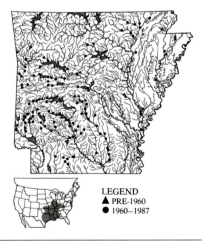

Map A4.126. Distribution of *Noturus nocturnus*, Freckled Madtom.

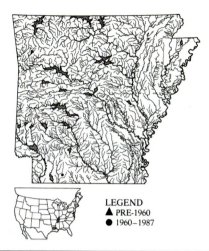

Map A4.127. Distribution of *Noturus phaeus*, Brown Madtom.

Map A4.130. Distribution of *Osmerus mordax*, Rainbow Smelt.

Map A4.128. Distribution of *Noturus taylori*, Caddo Madtom.

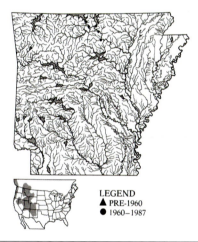

Map A4.131. Distribution of *Oncorhynchus clarkii*, Cutthroat Trout.

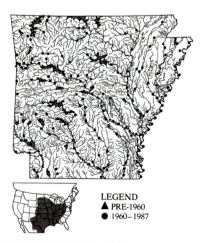

Map A4.129. Distribution of *Pylodictis olivaris*, Flathead Catfish.

Map A4.132. Distribution of *Oncorhynchus mykiss*, Rainbow Trout.

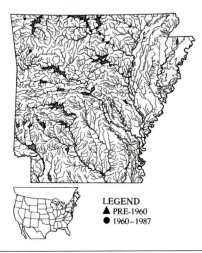

Map A4.133. Distribution of *Salmo trutta*, Brown Trout.

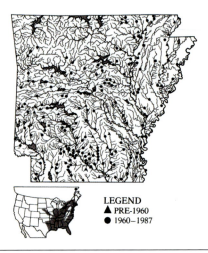

Map A4.136. Distribution of *Esox americanus*, Redfin Pickerel.

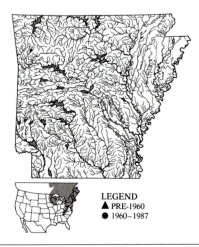

Map A4.134. Distribution of *Salvelinus fontinalis*, Brook Trout.

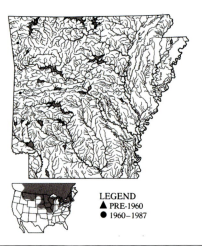

Map A4.137. Distribution of *Esox lucius*, Northern Pike.

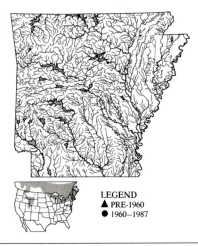

Map A4.135. Distribution of *Salvelinus namaycush*, Lake Trout.

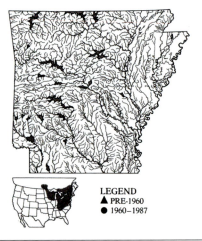

Map A4.138. Distribution of *Esox masquinongy*, Muskellunge

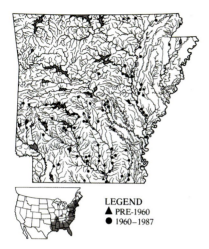

Map A4.139. Distribution of *Esox niger*, Chain Pickerel.

Map A4.142. Distribution of *Troglichthys rosae*, Ozark Cavefish.

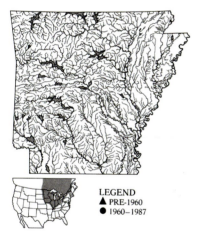

Map A4.140. Distribution of *Umbra limi*, Central Mudminnow.

Map A4.143. Distribution of *Typhlichthys eigenmanni*, Salem Plateau Cavefish.

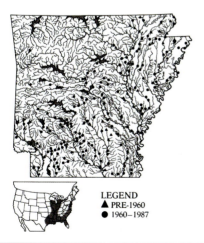

Map A4.141. Distribution of *Aphredoderus sayanus*, Pirate Perch.

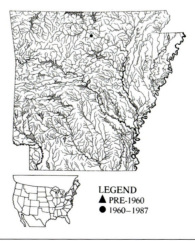

Map A4.144. Distribution of *Typhlichthys* sp. *cf. eigenmanni*, Ghost Cavefish.

Map A4.145. Distribution of *Mugil cephalus*, Striped Mullet.

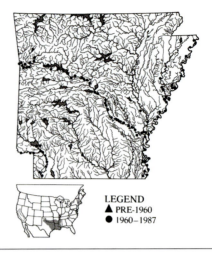

Map A4.148. Distribution of *Menidia audens*, Mississippi Silverside.

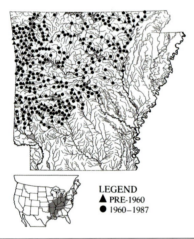

Map A4.146. Distribution of *Labidesthes sicculus*, Brook Silverside.

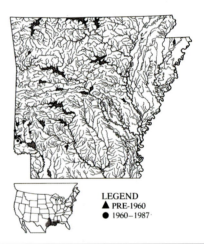

Map A4.149. Distribution of *Fundulus blairae*, Western Starhead Topminnow.

Map A4.147. Distribution of *Labidesthes vanhyningi*, Hardy Silverside.

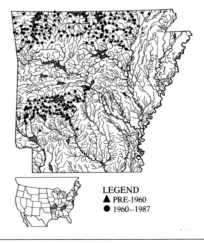

Map A4.150. Distribution of *Fundulus catenatus*, Northern Studfish.

Map A4.151. Distribution of *Fundulus chrysotus*, Golden Topminnow.

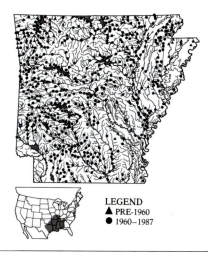

Map A4.154. Distribution of *Fundulus olivaceus*, Blackspotted Topminnow.

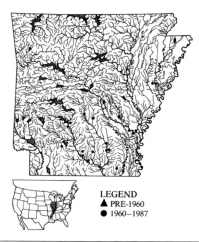

Map A4.152. Distribution of *Fundulus dispar*, Starhead Topminnow.

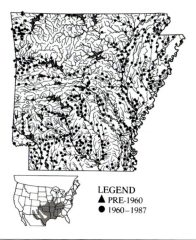

Map A4.155. Distribution of *Gambusia affinis*, Western Mosquitofish.

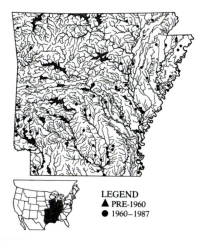

Map A4.153. Distribution of *Fundulus notatus*, Blackstripe Topminnow.

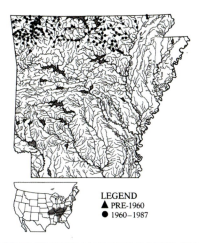

Map A4.156. Distribution of *Uranidea carolinae*, Banded Sculpin.

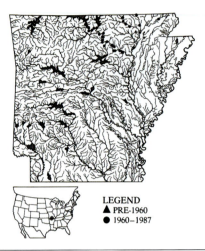

Map A4.157. Distribution of *Uranidea immaculata*, Knobfin Sculpin.

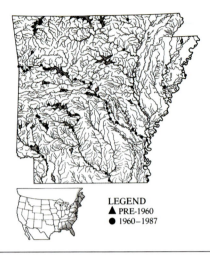

Map A4.160. Distribution of *Morone saxatilis*, Striped Bass.

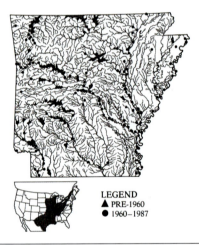

Map A4.158. Distribution of *Morone chrysops*, White Bass.

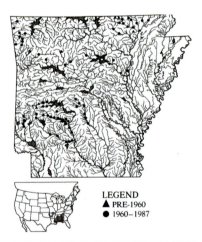

Map A4.161. Distribution of *Ambloplites ariommus*, Shadow Bass.

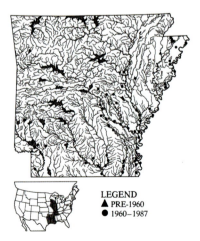

Map A4.159. Distribution of *Morone mississippiensis*, Yellow Bass.

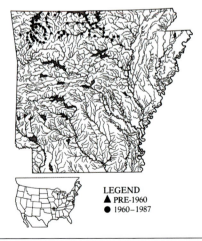

Map A4.162. Distribution of *Ambloplites constellatus*, Ozark Bass.

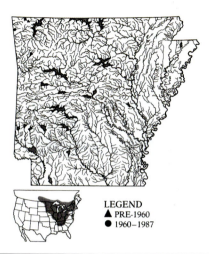

Map A4.163. Distribution of *Ambloplites rupestris*, Rock Bass.

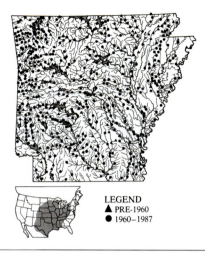

Map A4.166. Distribution of *Lepomis cyanellus*, Green Sunfish.

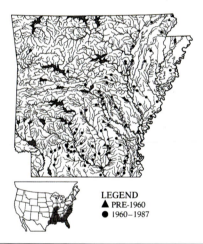

Map A4.164. Distribution of *Centrarchus macropterus*, Flier.

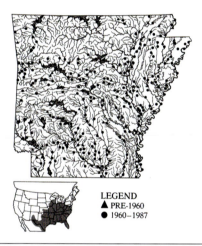

Map A4.167. Distribution of *Lepomis gulosus*, Warmouth.

Map A4.165. Distribution of *Lepomis auritus*, Redbreast Sunfish.

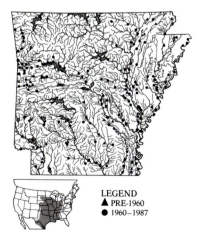

Map A4.168. Distribution of *Lepomis humilis*, Orangespotted Sunfish.

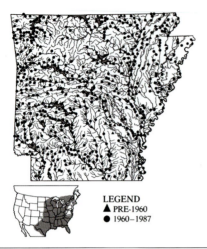

Map A4.169. Distribution of *Lepomis macrochirus*, Bluegill.

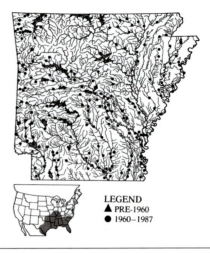

Map A4.172. Distribution of *Lepomis microlophus*, Redear Sunfish.

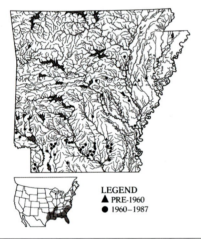

Map A4.170. Distribution of *Lepomis marginatus*, Dollar Sunfish.

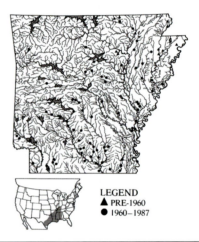

Map A4.173. Distribution of *Lepomis miniatus*, Redspotted Sunfish.

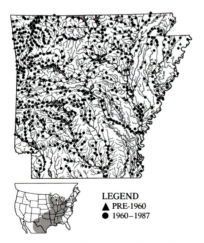

Map A4.171. Distribution of *Lepomis megalotis*, Longear Sunfish.

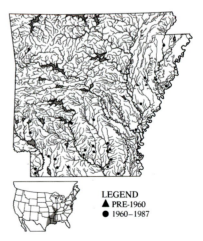

Map A4.174. Distribution of *Lepomis symmetricus*, Bantam Sunfish.

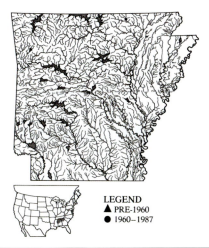

Map A4.175. Distribution of *Micropterus coosae*, Redeye Bass.

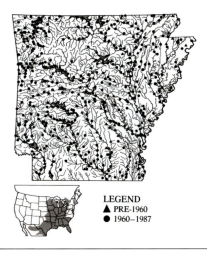

Map A4.178. Distribution of *Micropterus salmoides*, Largemouth Bass.

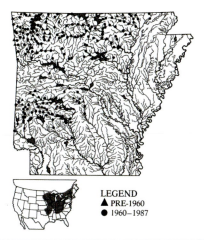

Map A4.176. Distribution of *Micropterus dolomieu*, Smallmouth Bass.

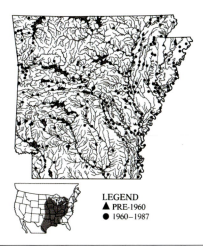

Map A4.179. Distribution of *Pomoxis annularis*, White Crappie.

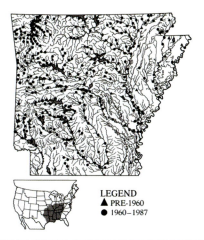

Map A4.177. Distribution of *Micropterus punctulatus*, Spotted Bass.

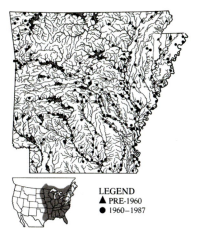

Map A4.180. Distribution of *Pomoxis nigromaculatus*, Black Crappie.

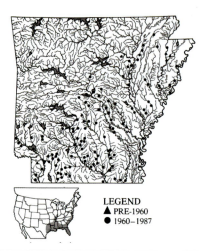

Map A4.181. Distribution of *Elassoma zonatum*, Banded Pygmy Sunfish.

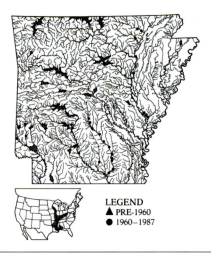

Map A4.184. Distribution of *Crystallaria asprella*, Crystal Darter.

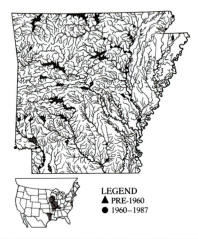

Map A4.182. Distribution of *Ammocrypta clara*, Western Sand Darter.

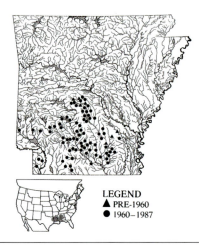

Map A4.185. Distribution of *Etheostoma artesiae*, Redspot Darter.

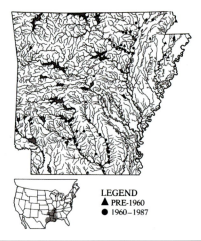

Map A4.183. Distribution of *Ammocrypta vivax*, Scaly Sand Darter.

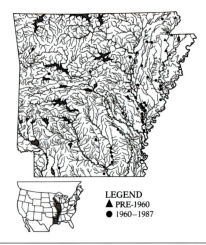

Map A4.186. Distribution of *Etheostoma asprigene*, Mud Darter.

Map A4.187. Distribution of *Etheostoma autumnale*, Autumn Darter.

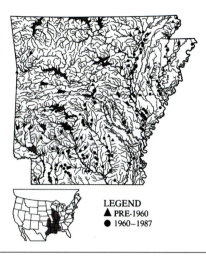

Map A4.190. Distribution of *Etheostoma chlorosoma*, Bluntnose Darter.

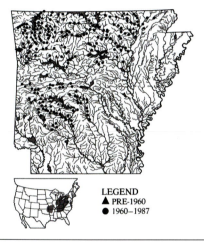

Map A4.188. Distribution of *Etheostoma blennioides*, Greenside Darter.

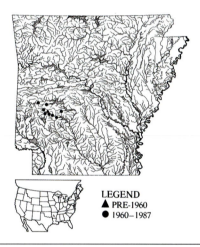

Map A4.191. Distribution of *Etheostoma clinton*, Beaded Darter.

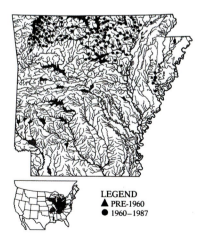

Map A4.189. Distribution of *Etheostoma caeruleum*, Rainbow Darter.

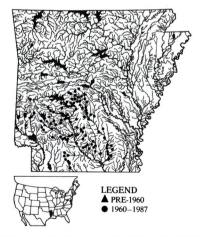

Map A4.192. Distribution of *Etheostoma collettei*, Creole Darter.

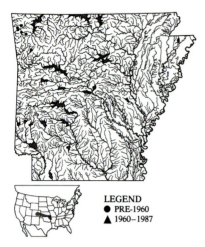

Map A4.193. Distribution of *Etheostoma cragini*, Arkansas Darter.

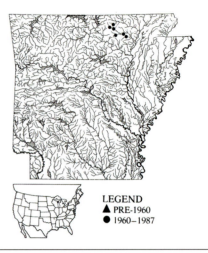

Map A4.196. Distribution of *Etheostoma fragi*, Strawberry Darter.

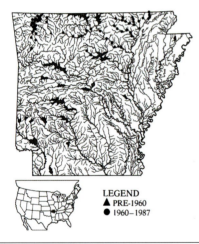

Map A4.194. Distribution of *Etheostoma euzonum*, Arkansas Saddled Darter.

Map A4.197. Distribution of *Etheostoma fusiforme*, Swamp Darter

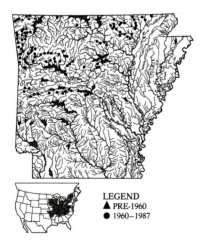

Map A4.195. Distribution of *Etheostoma flabellare*, Fantail Darter.

Map A4.198. Distribution of *Etheostoma gracile*, Slough Darter.

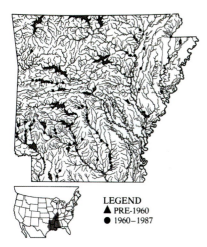

Map A4.199. Distribution of *Etheostoma histrio*, Harlequin Darter.

Map A4.202. Distribution of *Etheostoma nigrum*, Johnny Darter.

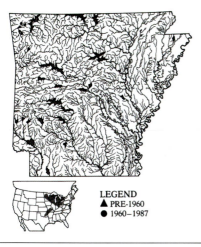

Map A4.200. Distribution of *Etheostoma microperca*, Least Darter.

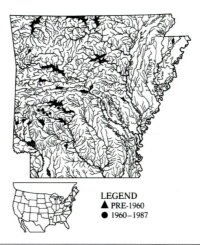

Map A4.203. Distribution of *Etheostoma pallididorsum*, Paleback Darter.

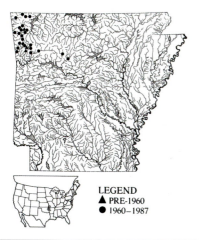

Map A4.201. Distribution of *Etheostoma mihileze*, Sunburst Darter.

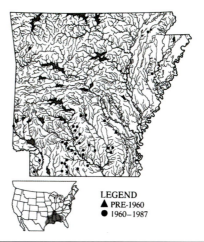

Map A4.204. Distribution of *Etheostoma parvipinne*, Goldstripe Darter.

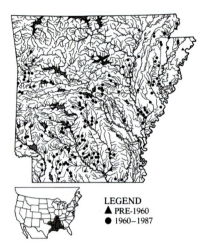

Map A4.205. Distribution of *Etheostoma proeliare*, Cypress Darter.

Map A4.208. Distribution of *Etheostoma* sp. *cf. pulchellum* 2, Blue Belly Form.

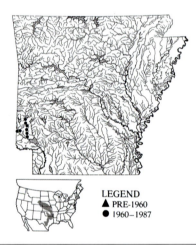

Map A4.206. Distribution of *Etheostoma pulchellum*, Plains Darter.

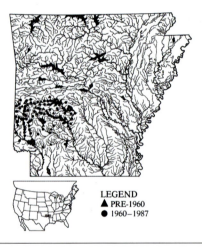

Map A4.209. Distribution of *Etheostoma radiosum*, Orangebelly Darter.

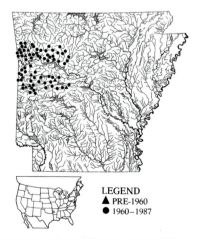

Map A4.207. Distribution of *Etheostoma* sp. *cf. pulchellum* 1, Red Belly Form

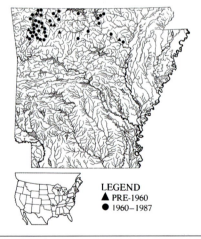

Map A4.210. Distribution of *Etheostoma* sp. *cf. spectabile*, Ozark Darter.

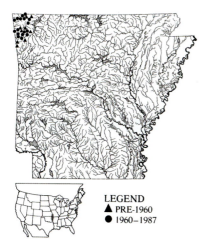

Map A4.211. Distribution of *Etheostoma squamosum*, Plateau Darter.

Map A4.214. Distribution of *Etheostoma uniporum*, Current Darter.

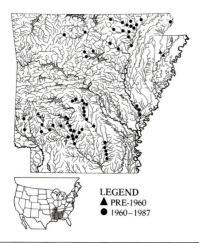

Map A4.212. Distribution of *Etheostoma stigmaeum*, Speckled Darter.

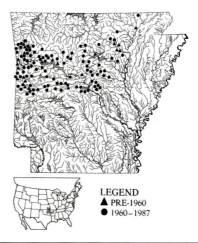

Map A4.215. Distribution of *Etheostoma whipplei*, Redfin Darter.

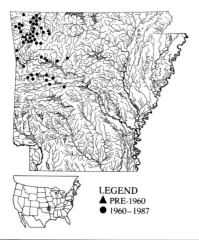

Map A4.213. Distribution of *Etheostoma teddyroosevelt*, Highland Darter.

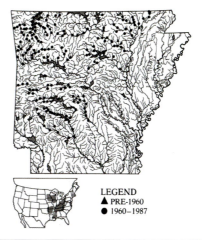

Map A4.216. Distribution of *Etheostoma zonale*, Banded Darter.

Map A4.217. Distribution of *Nothonotus juliae*, Yoke Darter.

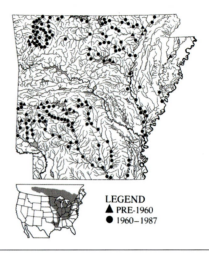

Map A4.220. Distribution of *Percina caprodes*, Logperch.

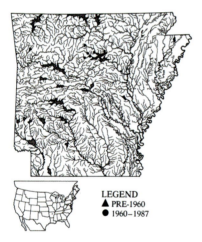

Map A4.218. Distribution of *Nothonotus moorei*, Yellowcheek Darter.

Map A4.221. Distribution of *Percina copelandi*, Channel Darter.

Map A4.219. Distribution of *Percina brucethompsoni*, Ouachita Darter.

Map A4.222. Distribution of *Percina evides*, Gilt Darter.

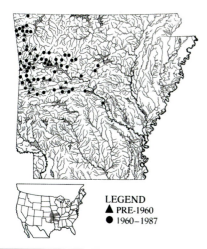

Map A4.223. Distribution of *Percina fulvitaenia*, Ozark Logperch.

Map A4.224. Distribution of *Percina macrolepida*, Bigscale Logperch.

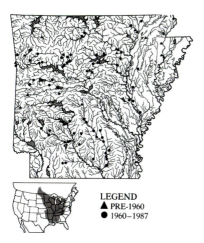

Map A4.225. Distribution of *Percina maculata*, Blackside Darter.

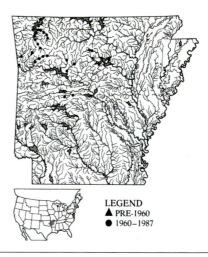

Map A4.226. Distribution of *Percina nasuta*, Longnose Darter.

Map A4.227. Distribution of *Percina pantherina*, Leopard Darter.

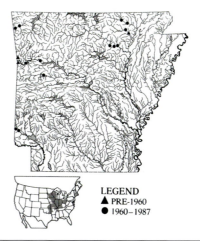

Map A4.228. Distribution of *Percina phoxocephala*, Slenderhead Darter.

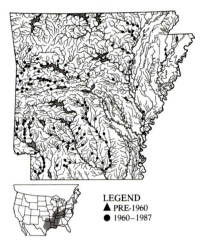

Map A4.229. Distribution of *Percina sciera*, Dusky Darter.

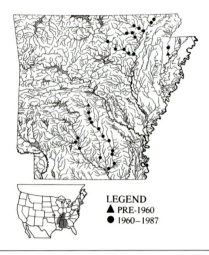

Map A4.232. Distribution of *Percina vigil*, Saddleback Darter.

Map A4.230. Distribution of *Percina shumardi*, River Darter.

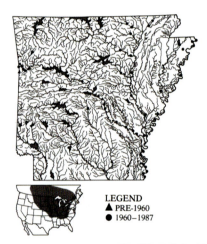

Map A4.233. Distribution of *Stizostedion canadense*, Sauger.

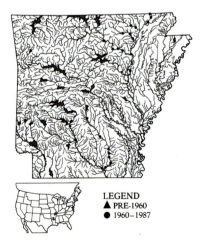

Map A4.231. Distribution of *Percina uranidea*, Stargazing Darter.

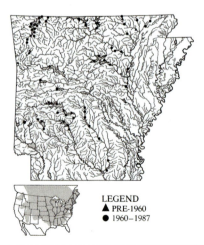

Map A4.234. Distribution of *Stizostedion vitreum*, Walleye.

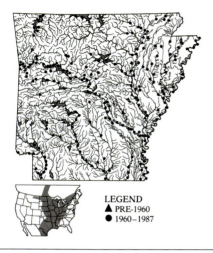

Map A4.235. Distribution of *Aplodinotus grunniens*, Freshwater Drum.

APPENDIX 5

Checklist of Species and List of Higher Taxonomic Categories of Arkansas Fishes

Checklist of the Fishes of Arkansas

This checklist of 243 species includes the 219 native fishes (one species is extinct) historically known from Arkansas, along with 24 nonnative species known to recently inhabit this state. Twenty-two species (indicated by asterisks) differ in scientific or common or both names from species listed in the most recent AFS list of common and scientific names (Page et al. 2013). The list is provided to enable the reader to quickly locate a particular species or group and to see what species are included in each of the 19 orders and 28 families that inhabit Arkansas waters.

Class Petromyzontida—Lampreys

Order Petromyzontiformes—Lampreys

Family Petromyzontidae—Lampreys
Ichthyomyzon castaneus Girard, 1858 | Chestnut Lamprey
Ichthyomyzon gagei Hubbs and Trautman, 1937 | Southern Brook Lamprey
Ichthyomyzon unicuspis Hubbs and Trautman, 1937 | Silver Lamprey
Lampetra aepyptera (Abbott, 1860) | Least Brook Lamprey
Lethenteron appendix (DeKay, 1842) | American Brook Lamprey

Class Actinopterygii—Ray-Finned Fishes

Order Acipenseriformes—Sturgeons and Paddlefishes

Family Acipenseridae—Sturgeons
Acipenser fulvescens Rafinesque, 1817 | Lake Sturgeon
Acipenser oxyrinchus Mitchill, 1815 | Atlantic Sturgeon
Scaphirhynchus albus (Forbes and Richardson, 1905) | Pallid Sturgeon

Scaphirhynchus platorynchus (Rafinesque, 1820) | Shovelnose Sturgeon

Family Polyodontidae—Paddlefishes
Polyodon spathula (Walbaum, 1792) | Paddlefish

Order Lepisosteiformes—Gars

Family Lepisosteidae—Gars
Atractosteus spatula (Lacépède, 1803) | Alligator Gar
Lepisosteus oculatus Winchell, 1864 | Spotted Gar
Lepisosteus osseus (Linnaeus, 1758) | Longnose Gar
Lepisosteus platostomus Rafinesque, 1820 | Shortnose Gar

Order Amiiformes—Bowfins

Family Amiidae—Bowfins
Amia calva Linnaeus, 1766 | Bowfin

Order Hiodontiformes—Mooneyes

Family Hiodontidae—Mooneyes
Hiodon alosoides (Rafinesque, 1819) | Goldeye
Hiodon tergisus Lesueur, 1818 | Mooneye

Order Anguilliformes—Eels

Family Anguillidae—Freshwater Eels
Anguilla rostrata (Lesueur, 1817) | American Eel

Order Clupeiformes—Herrings

Family Clupeidae—Herrings
Alosa alabamae Jordan and Evermann, 1896 | Alabama Shad
Alosa chrysochloris (Rafinesque, 1820) | Skipjack Herring
Dorosoma cepedianum (Lesueur, 1818) | Gizzard Shad
Dorosoma petenense (Günther, 1867) | Threadfin Shad

Order Cypriniformes—Minnows, Carps, Loaches, and Suckers

Family Cyprinidae—Carps and Minnows

Campostoma oligolepis Hubbs and Greene, 1935 | Largescale Stoneroller

Campostoma plumbeum (Girard, 1856) | Plains Stoneroller*

Campostoma pullum (Agassiz, 1854) | Finescale Stoneroller*

Campostoma spadiceum (Girard, 1856) | Highland Stoneroller

Carassius auratus (Linnaeus, 1758) | Goldfish

Chrosomus erythrogaster (Rafinesque, 1820) | Southern Redbelly Dace

Ctenopharyngodon idella (Valenciennes, 1844) | Grass Carp

Cyprinella camura (Jordan and Meek, 1884) | Bluntface Shiner

Cyprinella galactura (Cope, 1868) | Whitetail Shiner

Cyprinella lutrensis (Baird and Girard, 1853) | Red Shiner

Cyprinella spiloptera (Cope, 1867) | Spotfin Shiner

Cyprinella venusta Girard, 1856 | Blacktail Shiner

Cyprinella whipplei Girard, 1856 | Steelcolor Shiner

Cyprinus carpio Linnaeus, 1758 | Common Carp

Erimystax harryi (Hubbs and Crowe, 1956) | Ozark Chub

Erimystax x-punctatus (Hubbs and Crowe, 1956) | Gravel Chub

Hybognathus hayi Jordan, 1885 | Cypress Minnow

Hybognathus nuchalis Agassiz, 1855 | Mississippi Silvery Minnow

Hybognathus placitus Girard, 1856 | Plains Minnow

Hybopsis amblops (Rafinesque, 1820) | Bigeye Chub

Hybopsis amnis (Hubbs and Greene, 1951) | Pallid Shiner

Hypophthalmichthys molitrix (Valenciennes, 1844) | Silver Carp

Hypophthalmichthys nobilis (Richardson, 1845) | Bighead Carp

Luxilus cardinalis (Mayden, 1988) | Cardinal Shiner

Luxilus chrysocephalus Rafinesque, 1820 | Striped Shiner

Luxilus pilsbryi (Fowler, 1904) | Duskystripe Shiner

Luxilus zonatus (Agassiz, 1863) | Bleeding Shiner

Lythrurus fumeus (Evermann, 1892) | Ribbon Shiner

Lythrurus snelsoni (Robison, 1985) | Ouachita Mountain Shiner*

Lythrurus umbratilis (Girard, 1856) | Redfin Shiner

Macrhybopsis gelida (Girard, 1856) | Sturgeon Chub

Macrhybopsis hyostoma (Gilbert, 1884) | Shoal Chub

Macrhybopsis meeki (Jordan and Evermann, 1896) | Sicklefin Chub

Macrhybopsis storeriana (Kirtland, 1845) | Silver Chub

Mylopharyngodon piceus (Richardson, 1846) | Black Carp

Nocomis asper Lachner and Jenkins, 1971 | Redspot Chub

Nocomis biguttatus (Kirtland, 1840) | Hornyhead Chub

Notemigonus crysoleucas (Mitchill, 1814) | Golden Shiner

Notropis atherinoides Rafinesque, 1818 | Emerald Shiner

Notropis atrocaudalis Evermann, 1892 | Blackspot Shiner

Notropis bairdi Hubbs and Ortenburger, 1929 | Red River Shiner

Notropis blennius (Girard, 1856) | River Shiner

Notropis boops Gilbert, 1884 | Bigeye Shiner

Notropis buchanani Meek, 1896 | Ghost Shiner

Notropis chalybaeus (Cope, 1867) | Ironcolor Shiner

Notropis girardi Hubbs and Ortenburger, 1929 | Arkansas River Shiner

Notropis greenei Hubbs and Ortenburger, 1929 | Wedgespot Shiner

Notropis maculatus (Hay, 1881) | Taillight Shiner

Notropis nubilus (Forbes, 1878) | Ozark Minnow

Notropis ortenburgeri Hubbs, 1927 | Kiamichi Shiner

Notropis ozarcanus Meek, 1891 | Ozark Shiner

Notropis percobromus (Cope, 1871) | Carmine Shiner

Notropis perpallidus Hubbs and Black, 1940 | Peppered Shiner

Notropis potteri Hubbs and Bonham, 1951 | Chub Shiner

Notropis sabinae Jordan and Gilbert, 1886 | Sabine Shiner

Notropis shumardi (Girard, 1856) | Silverband Shiner

Notropis stramineus (Cope, 1865) | Sand Shiner

Notropis suttkusi Humphries and Cashner, 1994 | Rocky Shiner

Notropis telescopus (Cope, 1868) | Telescope Shiner

Notropis texanus (Girard, 1856) | Weed Shiner

Notropis volucellus (Cope, 1865) | Mimic Shiner

Notropis wickliffi Trautman, 1931 | Channel Shiner

Opsopoeodus emiliae Hay, 1881 | Pugnose Minnow

Phenacobius mirabilis (Girard, 1856) | Suckermouth Minnow

Pimephales notatus (Rafinesque, 1820) | Bluntnose Minnow

Pimephales promelas Rafinesque, 1820 | Fathead Minnow

Pimephales tenellus (Girard, 1856) | Slim Minnow

Pimephales vigilax (Baird and Girard, 1853) | Bullhead Minnow

Platygobio gracilis (Richardson, 1836) | Flathead Chub

Pteronotropis hubbsi (Bailey and Robison, 1978) | Bluehead Shiner

Scardinius erythrophthalmus (Linnaeus, 1758) | Rudd

Semotilus atromaculatus (Mitchill, 1818) | Creek Chub

Family Catostomidae—Suckers

Carpiodes carpio (Rafinesque, 1820) | River Carpsucker

Carpiodes cyprinus (Lesueur, 1817) | Quillback

Carpiodes velifer (Rafinesque, 1820) | Highfin Carpsucker

Catostomus commersonii (Lacépède, 1803) | White Sucker

Cycleptus elongatus (Lesueur, 1817) | Blue Sucker

Erimyzon claviformis (Girard, 1856) | Western Creek Chubsucker

Erimyzon sucetta (Lacépède, 1803) | Lake Chubsucker

---

*A species for which we use a different scientific or common name from the one listed in the 2013 AFS list.

Hypentelium nigricans (Lesueur, 1817) | Northern Hog Sucker
Ictiobus bubalus (Rafinesque, 1818) | Smallmouth Buffalo
Ictiobus cyprinellus (Valenciennes, 1844) | Bigmouth Buffalo
Ictiobus niger (Rafinesque, 1819) | Black Buffalo
Minytrema melanops (Rafinesque, 1820) | Spotted Sucker
Moxostoma anisurum (Rafinesque, 1820) | Silver Redhorse
Moxostoma carinatum (Cope, 1870) | River Redhorse
Moxostoma duquesnei (Lesueur, 1817) | Black Redhorse
Moxostoma erythrurum (Rafinesque, 1818) | Golden Redhorse
Moxostoma lacerum (Jordan and Brayton, 1877) | Harelip Sucker[†]
Moxostoma pisolabrum Trautman and Martin, 1951 | Pealip Redhorse
Moxostoma poecilurum Jordan, 1877 | Blacktail Redhorse

Order Siluriformes—Catfishes

Family Ictaluridae—North American Catfishes
Ameiurus melas (Rafinesque, 1820) | Black Bullhead
Ameiurus natalis (Lesueur, 1819) | Yellow Bullhead
Ameiurus nebulosus (Lesueur, 1819) | Brown Bullhead
Ictalurus furcatus (Lesueur, 1840) | Blue Catfish
Ictalurus punctatus (Rafinesque, 1818) | Channel Catfish
Noturus albater Taylor, 1969 | Ozark Madtom
Noturus eleutherus Jordan, 1877 | Mountain Madtom
Noturus exilis Nelson, 1876 | Slender Madtom
Noturus flavater Taylor, 1969 | Checkered Madtom
Noturus flavus Rafinesque, 1818 | Stonecat
Noturus gladiator Thomas and Burr, 2004 | Piebald Madtom
Noturus gyrinus (Mitchill, 1817) | Tadpole Madtom
Noturus lachneri Taylor, 1969 | Ouachita Madtom
Noturus maydeni Egge, 2006 | Black River Madtom
Noturus miurus Jordan, 1877 | Brindled Madtom
Noturus nocturnus Jordan and Gilbert, 1886 | Freckled Madtom
Noturus phaeus Taylor, 1969 | Brown Madtom
Noturus taylori Douglas, 1972 | Caddo Madtom
Pylodictis olivaris (Rafinesque, 1818) | Flathead Catfish

Order Osmeriformes—Smelts

Family Osmeridae—Smelts
Osmerus mordax (Mitchill, 1814) | Rainbow Smelt

Order Salmoniformes—Trout, Salmon, and Whitefish

Family Salmonidae—Trouts and Salmons
Oncorhynchus clarkii (Richardson, 1836) | Cutthroat Trout

Oncorhynchus mykiss (Walbaum, 1792) | Rainbow Trout
Salmo trutta Linnaeus, 1758 | Brown Trout
Salvelinus fontinalis (Mitchill, 1814) | Brook Trout
Salvelinus namaycush (Walbaum, 1792) | Lake Trout

Order Esociformes—Pikes and Mudminnows

Family Esocidae—Pikes and Mudminnows
Esox americanus Gmelin, 1789 | Redfin Pickerel
Esox lucius Linnaeus, 1758 | Northern Pike
Esox masquinongy Mitchill, 1824 | Muskellunge
Esox niger Lesueur, 1818 | Chain Pickerel
Umbra limi (Kirtland, 1840) | Central Mudminnow

Order Percopsiformes –Trout-Perches

Family Aphredoderidae—Pirate Perches
Aphredoderus sayanus (Gilliams, 1824) | Pirate Perch

Family Amblyopsidae—Cavefishes
Troglichthys rosae (Eigenmann, 1898) | Ozark Cavefish[*]
Typhlichthys eigenmanni Charlton, 1933 | Salem Plateau Cavefish[*]
Typhlichthys sp. *cf. eigenmanni* | Ghost Cavefish[**]

Order Mugiliformes—Mullets

Family Mugilidae—Mullets
Mugil cephalus Linnaeus, 1758 | Striped Mullet

Order Atheriniformes—Silversides

Family Atherinopsidae—New World Silversides
Labidesthes sicculus (Cope, 1865) | Brook Silverside
Labidesthes vanhyningi Bean and Reid, 1930 | Hardy Silverside[***]
Menidia audens Hay, 1882 | Mississippi Silverside

Order Cyprinodontiformes—Killifishes, Topminnows, Livebearers, and Toothcarps

Family Fundulidae—Topminnows
Fundulus blairae Wiley and Hall, 1975 | Western Starhead Topminnow
Fundulus catenatus (Storer, 1846) | Northern Studfish
Fundulus chrysotus (Günther, 1866) | Golden Topminnow
Fundulus dispar (Agassiz, 1854) | Starhead Topminnow

---

[*] A species for which we use a different scientific or common name from the one listed in the 2013 AFS list.

[**] An undescribed species that we recognize and for which we provide a species account (AFS list does not include undescribed species).

[***] A species that has been described since the publication of the most recent AFS list.

Fundulus notatus (Rafinesque, 1820) | Blackstripe
   Topminnow
Fundulus olivaceus (Storer, 1845) | Blackspotted Topminnow

Family Poeciliidae—Livebearers
Gambusia affinis (Baird and Girard, 1853) | Western
   Mosquitofish

Order Anabantiformes—Labyrinth Fishes

Family Channidae—Snakeheads
Channa argus (Cantor, 1842) | Northern Snakehead

Order Scorpaeniformes—Mail-Cheeked Fishes

Family Cottidae—Sculpins
Uranidea carolinae (Gill, 1861) | Banded Sculpin*
Uranidea immaculata (Kinziger and Wood, 2010) | Knobfin
   Sculpin*

Order Perciformes—Perches and Relatives

Family Moronidae—Temperate Basses
Morone americana (Gmelin, 1789) | White Perch
Morone chrysops (Rafinesque, 1820) | White Bass
Morone mississippiensis Jordan and Eigenmann, 1887 | Yellow
   Bass
Morone saxatilis (Walbaum, 1792) | Striped Bass

Family Centrarchidae—Sunfishes
Ambloplites ariommus Viosca, 1936 | Shadow Bass
Ambloplites constellatus Cashner and Suttkus, 1977 | Ozark
   Bass
Ambloplites rupestris (Rafinesque, 1817) | Rock Bass
Centrarchus macropterus (Lacépède, 1801) | Flier
Lepomis auritus (Linnaeus, 1758) | Redbreast Sunfish
Lepomis cyanellus Rafinesque, 1819 | Green Sunfish
Lepomis gulosus (Cuvier, 1829) | Warmouth
Lepomis humilis (Girard, 1858) | Orangespotted Sunfish
Lepomis macrochirus Rafinesque, 1819 | Bluegill
Lepomis marginatus (Holbrook, 1855) | Dollar Sunfish
Lepomis megalotis (Rafinesque, 1820) | Longear Sunfish
Lepomis microlophus (Günther, 1859) | Redear Sunfish
Lepomis miniatus Jordan, 1877 | Redspotted Sunfish
Lepomis symmetricus Forbes, 1883 | Bantam Sunfish
Micropterus dolomieu Lacépède, 1802 | Smallmouth Bass
Micropterus punctulatus (Rafinesque, 1819) | Spotted Bass
Micropterus salmoides (Lacépède, 1802) | Largemouth Bass

Pomoxis annularis Rafinesque, 1818 | White Crappie
Pomoxis nigromaculatus (Lesueur, 1829) | Black Crappie

Family Elassomatidae—Pygmy Sunfishes
Elassoma zonatum Jordan, 1877 | Banded Pygmy Sunfish

Family Percidae—Perches and Darters
Ammocrypta clara Jordan and Meek, 1885 | Western Sand
   Darter
Ammocrypta vivax Hay, 1882 | Scaly Sand Darter
Crystallaria asprella (Jordan, 1878) | Crystal Darter
Etheostoma artesiae (Hay, 1881) | Redspot Darter
Etheostoma asprigene (Forbes, 1878) | Mud Darter
Etheostoma autumnale Mayden, 2010 | Autumn Darter
Etheostoma blennioides Rafinesque, 1819 | Greenside Darter
Etheostoma caeruleum Storer, 1845 | Rainbow Darter
Etheostoma chlorosoma (Hay, 1881) | Bluntnose Darter
Etheostoma clinton Mayden and Layman, 2012 | Beaded
   Darter***
Etheostoma collettei Birdsong and Knapp, 1969 | Creole
   Darter
Etheostoma cragini Gilbert, 1885 | Arkansas Darter
Etheostoma euzonum (Hubbs and Black, 1940) | Arkansas
   Saddled Darter
Etheostoma flabellare Rafinesque, 1819 | Fantail Darter
Etheostoma fragi Distler, 1968 | Strawberry Darter
Etheostoma fusiforme (Girard, 1854) | Swamp Darter
Etheostoma gracile (Girard, 1859) | Slough Darter
Etheostoma histrio Jordan and Gilbert, 1887 | Harlequin
   Darter
Etheostoma microperca Jordan and Gilbert, 1888 | Least
   Darter
Etheostoma mihileze Mayden, 2010 | Sunburst Darter
Etheostoma nigrum Rafinesque, 1820 | Johnny Darter
Etheostoma pallididorsum Distler and Metcalf, 1962 | Paleback
   Darter
Etheostoma parvipinne Gilbert and Swain, 1887 | Goldstripe
   Darter
Etheostoma proeliare (Hay, 1881) | Cypress Darter
Etheostoma pulchellum (Girard, 1859) | Plains Darter*
Etheostoma sp. *cf. pulchellum* 1 | (Red Belly Form)**
Etheostoma sp. *cf. pulchellum* 2 | (Blue Belly Form)**
Etheostoma radiosum (Hubbs and Black, 1941) | Orangebelly
   Darter
Etheostoma sp. *cf. spectabile* | Ozark Darter**
Etheostoma squamosum Distler, 1968 | Plateau Darter*
Etheostoma stigmaeum (Jordan, 1877) | Speckled Darter
Etheostoma teddyroosevelt Layman and Mayden,
   2012 | Highland Darter***

---

* A species for which we use a different scientific or common name from the one listed in the 2013 AFS list.
** An undescribed species that we recognize and for which we provide a species account (AFS list does not include
   undescribed species).
*** A species that has been described since the publication of the most recent AFS list.
† Extinct throughout its range.

Etheostoma uniporum Distler, 1968 | Current Darter

Etheostoma whipplei (Girard, 1859) | Redfin Darter

Etheostoma zonale (Cope, 1868) | Banded Darter

Nothonotus juliae (Meek, 1891) | Yoke Darter*

Nothonotus moorei (Raney and Suttkus, 1964) | Yellowcheek Darter*

Perca flavescens (Mitchill, 1814) | Yellow Perch

Percina brucethompsoni Robison, Cashner, and Near, 2014 | Ouachita Darter***

Percina caprodes (Rafinesque, 1818) | Logperch

Percina copelandi (Jordan, 1877) | Channel Darter

Percina evides (Jordan and Copeland, 1877) | Gilt Darter

Percina fulvitaenia Morris and Page, 1981 | Ozark Logperch*

Percina macrolepida Stevenson, 1971 | Bigscale Logperch

Percina maculata (Girard, 1859) | Blackside Darter

Percina nasuta (Bailey, 1941) | Longnose Darter

Percina pantherina (Moore and Reeves, 1955) | Leopard Darter

Percina phoxocephala (Nelson, 1876) | Slenderhead Darter

Percina sciera (Swain, 1883) | Dusky Darter

Percina shumardi (Girard, 1859) | River Darter

Percina uranidea (Jordan and Gilbert, 1887) | Stargazing Darter

Percina vigil (Hay, 1882) | Saddleback Darter

Stizostedion canadense (Griffith and Smith, 1834) | Sauger*

Stizostedion vitreum (Mitchill, 1818) | Walleye*

Family Sciaenidae—Drums and Croakers

Aplodinotus grunniens Rafinesque, 1819 | Freshwater Drum

Family Cichlidae—Cichlids and Tilapias

Oreochromis aureus (Steindachner, 1864) | Blue Tilapia

Oreochromis mossambicus (Peters, 1852) | Mozambique Tilapia

Oreochromis niloticus (Linnaeus, 1758) | Nile Tilapia

---

* A species for which we use a different scientific or common name from the one listed in the 2013 AFS list.

*** A species that has been described since the publication of the most recent AFS list.

Classification of Arkansas Fishes at Higher Taxonomic Categories (from Domain to Family) in the Systematic Hierarchy

The major categories and some of the most commonly used intermediate categories in the systematic hierarchy in descending order from domain to family are:

Domain
  Kingdom
    Phylum
      Subphylum
        Infraphylum
          Superclass
            Class
              Subclass
                Division
                  Subdivision
                    Superorder
                      Series
                        Order
                          Suborder
                            Family

Fish classification is currently in a state of flux (even more than usual), and there is no consensus on assigning fishes to hierarchical categories. Drastic changes in how we view the Tree of Life of fishes are certain to develop in coming years. Our classification is based on morphological and molecular evidence from multiple sources. The list that follows shows how Arkansas fishes fit into the categories of the systematic hierarchy. The taxa above the class level include not only Arkansas fishes, but also all other living fishes. Three extant classes of fishes (Myxini, Chondrichthyes, and Sarcopterygii) are omitted from the list because they contain no species found in Arkansas. Below the class level, only the taxa found in Arkansas are listed. Recognition of classes, orders, and families, and their phylogenetic sequence follow Page et al. (2013), with two exceptions: (1) We follow Nelson et al. (2016) in removing Channidae from order Perciformes and placing it in order Anabantiformes. (2) We changed the phylogenetic placement of family Elassomatidae by placing it immediately after family Centrarchidae to reflect recent molecular evidence indicating a close relationship between those two families.

Domain Eukarya—Eukaryotes
  Kingdom Animalia—Animals
    Phylum Chordata—Chordates
      Subphylum Craniata—Craniates
        Infraphylum Myxinomorphi—Hagfishes
        Infraphylum Vertebrata—Vertebrates
          Superclass Petromyzontomorphi
            Class Petromyzontida

Order Petromyzontiformes—Lampreys
    Family Petromyzontidae—Lampreys
Superclass Gnathostomata—Jawed Fishes
(and other jawed vertebrates)
  Class Actinopterygii—Ray-Finned Fishes
    Subclass Chondrostei
      Order Acipenseriformes—Sturgeons and
      Paddlefishes
        Family Acipenseridae—Sturgeons
        Family Polyodontidae—Paddlefishes
    Subclass Neopterygii
     Division Ginglymodi
      Order Lepisosteiformes—Gars
        Family Lepisosteidae—Gars
     Division Halecomorphi
      Order Amiiformes—Bowfins
        Family Amiidae—Bowfins
     Division Teleostei—Teleosts
      Subdivision Osteoglossomorpha
        Order Hiodontiformes—Mooneyes
          Family Hiodontidae—Mooneyes
      Subdivision Elopomorpha
        Order Anguilliformes—Eels
          Family Anguillidae—Freshwater Eels
      Subdivision Clupeomorpha
        Order Clupeiformes—Herrings
          Family Clupeidae—Herrings
      Subdivision Euteleostei
       Superorder Ostariophysi
        Series Otophysi
        Order Cypriniformes—Minnows, Carps,
        Loaches and Suckers
          Family Cyprinidae—Carps and Minnows
          Family Catostomidae—Suckers
        Order Siluriformes—Catfishes
          Family Ictaluridae—North American
          Catfishes

Superorder Protacanthopterygii
  Order Osmeriformes—Smelts
    Family Osmeridae—Smelts
  Order Salmoniformes—Trout, Salmon, and
  Whitefish
    Family Salmonidae—Trouts and Salmons
  Order Esociformes—Pikes and Mudminnows
    Family Esocidae—Pikes and Mudminnows
Superorder Paracanthopterygii
  Order Percopsiformes—Trout-perches
    Family Aphredoderidae—Pirate Perches
    Family Amblyopsidae—Cavefishes
Superorder Acanthopterygii
  Series Mugilomorpha[†]
    Order Mugiliformes—Mullets
      Family Mugilidae—Mullets
  Series Atherinomorpha[†]
    Order Atheriniformes—Silversides
      Family Atherinopsidae—New World
      Silversides
    Order Cyprinodontiformes—Killifishes,
    Topminnows, Livebearers, and Toothcarps
      Family Fundulidae—Topminnows
      Family Poeciliidae—Livebearers
  Series Percomorpha
    Order Anabantiformes—Labyrinth Fishes
      Family Channidae—Snakeheads
    Order Scorpaeniformes—Mail-Cheeked Fishes
      Family Cottidae—Sculpins
    Order Perciformes—Perches and Relatives
     Suborder Percoidei
      Family Moronidae—Temperate Basses[‡]
      Family Centrarchidae—Sunfishes
      Family Elassomatidae—Pygmy Sunfishes
      Family Percidae—Perches and Darters
      Family Sciaenidae—Drums and Croakers[‡]
     Suborder Labroidei
      Family Cichlidae—Cichlids and Tilapias[†]

[†] A series that Nelson et al. (2016) placed in series Percomorphi.
[‡] A family that Nelson et al. (2016) removed from the order Perciformes and placed in a different order.

GLOSSARY

ABBREVIATE-HETEROCERCAL. Type of caudal fin found in gars in which the posterior end of the vertebral column is distinctly curved upward and partially enters the upper half of the fin.

ABDOMINAL. Pertaining to the belly or ventral surface of the body. Pelvic fins are abdominal when inserted far behind the base of the pectoral fins.

ADIPOSE FIN. A small, fleshy, rayless fin located on the midline of the back between the dorsal and caudal fins in catfishes, salmonids, and some other fishes.

ADNATE. Grown together (united); conjoined; keel-like.

ADNEXED. Flaglike. With a free edge, not united.

AGE. Age categories of fish are herein designated as follows: age 0 (young-of-the-year), age 1 (a fish past its first winter), age 2 (a fish past its second winter), etc.

ALLOMETRIC. Relative growth of a body part in relation to the entire organism. The proportion of that part, relative to body length, increases or decreases as the individual grows.

ALLOPATRIC. Occupying different geographical regions.

ALLOPATRIC SPECIATION. The hypothesis that new species are formed by dividing an ancestral species into geographically isolated subpopulations that evolve reproductive barriers between them through independent evolutionary divergence from their common ancestor.

ALLOZYMES. Variant forms of an enzyme coded by different alleles at the same gene locus.

ALLUVIUM. Sediments transported by streams and deposited on land.

ALPHA TAXONOMY. The description and naming of organisms

AMMOCOETE. Larval stage of lampreys.

AMPHIPODS. Crustaceans of the order Amphipoda (class Crustacea), characterized by a compressed body, first thoracic segment fused with the head, and no true carapace.

AMPLIFIED FRAGMENT LENGTH POLYMORPHISM. Developed in the early 1990s, AFLP uses restriction enzymes to digest genomic DNA, followed by ligation of adaptors to the sticky ends of the restriction fragments. A subset of the restriction fragments is then selected to be amplified. The amplified fragments are separated and visualized on denaturing polyacrylamide gels.

ANADROMOUS. A fish, such as the Alabama Shad, that spawns and spends its early life in freshwater but moves into the ocean, where it attains sexual maturity and spends most of its life span.

ANAL FIN. Median, ventral unpaired fin situated posterior to the anus.

ANCESTRAL. Denotes a character state inferred to have been present in the most recent common ancestral population of a group of organisms.

ANGULATE. Having definite angles or corners.

ANTERIOR (ANTERIAD). In the anterior direction; toward the front.

ANTRORSE. Directed forward or upward; opposite of retrorse.

ANTEROHYAL. Anterior bone to which branchiostegal rays attach; formerly ceratohyal.

ANUS. The exterior opening of the digestive tract; the vent.

AQUACULTURE. The science and practice of raising fish in artificially controlled ponds or pools.

AQUIFER. A subterranean layer of porous water-bearing rock, gravel, or sand.

ASYMMETRIC INTROGRESSIVE HYBRIDIZATION. Heterospecific spawning of the males of one species with the females of another species occurs more frequently than the reciprocal heterospecific spawning event.

ATLANTIC SLOPE. The land area of eastern North America where the rivers drain into the Atlantic Ocean.

ATRETIC EGGS. Eggs that are degenerating in an ovary and not capable of being fertilized.

AUFWUCHS. A film of organic detritus, algae, and small invertebrates attached to substrates in a body of water. Also called periphyton.

AXILLARY PROCESS. A long, thin, membranous flap or modified scale at the anterior base of the pelvic fins in the Hiodontidae and Salmonidae, and at the anterior bases of both the pectoral and pelvic fins in the Clupeidae and Mugilidae.

BAND. A broad, straight, vertical or diagonal color mark that crosses the body.

BAR. A short, straight, vertical color mark of varying width.

BARBEL. A slender, elongate, fleshy process located near the mouth, snout, and chin areas; tactile and gustatory in function.

BASE (OF FINS). That line along which a fin is attached to the body.

BASIBRANCHIALS. Three median bones on the floor of the gill chamber, joined to the ventral ends of the five gill arches.

BASICAUDAL. Referring to the area at the base of the caudal fin.

BASIN. A major group of drainages interconnected by a major river.

BASIOCCIPITAL. Posteriormost bone on the underside of the skull, articulating with centrum of first vertebra.

BAYESIAN ANALYSIS. A statistical procedure based on Bayes' theorem that attempts to estimate parameters of an underlying distribution based on the observed distribution. It is used to update the probability of a hypothesis as more evidence or information becomes available.

BELLY. Ventral surface posterior to the base of the pelvic fins and anterior to the anal fin.

BENTHIC. Pertaining to the bottom; bottom dwelling.

BENTHOS. Organisms living on the bottom of bodies of water.

BICOLORED. Of two colors.

BICUSPID. Having two points (referring to a two-pointed tooth on one base).

BIODIVERSITY. The number, distribution, and abundance of species within a given area.

BIOMASS. Living weight; the weight of a fish stock, or of some defined portion of it; sometimes expressed as weight per unit area.

BIOTA. The animal and plant life of a region.

BLACK-SPOT DISEASE. An encysted intermediate life stage of a flatworm parasite. The cyst is imbedded in the skin and covered by black pigment, causing it to appear as a black spot.

BOULDER. Used here to refer to rock substrate fragments greater than 10 inches (256 mm) in diameter.

BRAIDED. Division of a stream or river channel into multiple small channels that divide and recombine numerous times.

BRANCHED RAY. A soft ray which is forked or branched away from its base.

BRANCHIAL ARCHES. Bony or cartilaginous structures supporting the gills, filaments, and rakers; gill arches.

BRANCHIAL REGION. The pharyngeal region where branchial arches and gills develop.

BRANCHIOSTEGAL RAY. One of the elongate, flattened bones supporting the gill membranes, ventral to the operculum.

BREAST. Ventral surface in front of the pelvic fins; anterior to the belly.

BREEDING TUBERCLES. See nuptial tubercles.

CADUCOUS. Readily shed (as belly scales in genus *Percina*).

CAECUM (PLURAL, CAECA). Blind pouch or other saclike evagination of the digestive tract, especially at the pylorus (junction of the stomach and small intestine).

CANINE TEETH. Sharp, conical teeth in the front part of the jaws; conspicuously larger than the rest of the teeth.

CATADROMOUS. A life history in which a fish species hatches in the ocean, migrates to freshwater to grow and reach sexual maturity, then returns to the ocean to spawn. Example, the American Eel.

CAUDAL FIN. Tail fin.

CAUDAL PEDUNCLE. The narrow region of the body in front of the caudal fin from the posterior end of the base of the anal fin to the base of the caudal fin.

CEMENT GLANDS. Discrete or diffuse structures that allow a larva to adhere to a substrate.

CENTER OF ORIGIN. The geographic area occupied by a species or higher taxon during its initial evolution; contrasts to areas colonized by dispersal following a species evolutionary origin.

CENTRUM. Central body of a vertebra.

CEPHALIC. Pertaining to the head.

CF. Latin *confer*. Often used to mean "compare with."

CHALK. A loosely consolidated variety of limestone made up in part of biochemically derived calcite in the form of the skeletons or skeletal fragments of microscopic ocean plants and animals mixed with very fine-grained calcite deposits of either biochemical or inorganic chemical origin.

CHEEK. The area between the eye and the preopercle bone.

CHERT. A hard sedimentary rock composed of microcrystalline quartz (silica), usually light-colored (dark gray to black varieties are known as flint); not nearly as abundant as most other types of sedimentary rock, chert is common in the Springfield Plateau of the Ozarks in pebble beds and gravel deposits because its hardness and resistance to chemical decay enable it to survive rough treatment from streams and other erosional forces.

CHIN. The lower surface of the head between the mandibles.

CHIRONOMIDS. True flies of the midge family Chironomidae.

CHROMATOPHORES. Pigment cells capable of expansions and contractions that change their size, shape, and color.

CHUTE. A section of a river in which the water is rather deep and flows swiftly and in which its surface, although containing eddies or small whirlpools, is not conspicuously broken up by protruding rocks.

CIRCUMORAL TEETH. Innermost circle of horny teeth that surround the esophagus in lampreys.

CLADE. A monophyletic group of organisms; a taxon or other group consisting of an ancestral species and all its descendants, forming a distinct branch on a cladogram or phylogenetic tree.

CLADISTICS. A system of arranging taxa by analysis of evolutionarily derived characters so that the arrangement reflects phylogenetic relationships.

CLADOCERA. An order of small (0.2–3 mm) crustaceans, the water fleas, found in nearly all types of freshwater habitats.

CLADOGRAM. A branching diagram showing the pattern of sharing of evolutionarily derived characters among species or higher taxa. Cladograms focus on identifying the common ancestry of related groups.

CLASSIFICATION. The organization of living organisms.

CLAY. Finely crystalline particles that form as a result of the weathering of various silicate minerals; particle size is less than 0.00016 inch (0.004 mm) in diameter.

CLEITHRUM. Major bone of the pectoral girdle, extending upward from the fin base and forming the posterior margin of the gill chamber.

CLOACA. The common chamber into which the digestive tract and urogenital ducts discharge in some fishes and other vertebrates.

COASTAL PLAIN. A physiographic region extending from sea level to the Fall Line. Found in eastern and southern Arkansas.

COBBLE. See rubble.

COELOMIC. Pertaining to the body cavity.

COLEOPTERA. The largest order of insects with a small percentage wholly or partly aquatic in the adult or larval stages. Beetles.

COLIFORM BACTERIA. A type of bacterium occurring in the human gut; the degree of its presence is used as an index of stream or lake water contamination by human sewage.

COMPLEMENTARY DISTRIBUTION. A distribution pattern where two taxa occupy adjacent geographic areas, with their ranges having little or no overlap (e.g., *Fundulus blairae* and *F. dispar* in Arkansas).

COMPRESSED. Narrow from side to side (flattened laterally); deeper than broad.

CONCAVE. Curved inward.

CONGENER OR CONGENERIC. Species assigned to the same genus.

CONSPECIFIC. Of the same species.

CONVEX. Curved outward.

COPEPODS. Small crustaceans of the class Maxillopoda, subclass Copepoda, characterized by rigid, sclerotized, cylindrical segments, and a true head with five pairs of appendages.

CORACOID. One of a series of small bones in the pectoral girdle that support the pectoral fin. The coracoids from each pectoral girdle fuse anteriorly on the ventral surface of the fish in the gular region to form the coracoid symphysis (easily observed just under the skin of the isthmus posterior to the gill membranes in *Labidesthes* species).

COURTSHIP. The series of behaviors preliminary to mating.

CRANIATES (CRANIATA). A subphylum of the phylum Chordata. Often used interchangeably with "Vertebrata".

CREPUSCULAR. Active at dawn or dusk.

CRYPTIC SPECIES. Species that are morphologically indistinguishable yet represent independent evolutionary lineages, based on either molecular or phylogeographic evidence.

CRUSTACEANS. Members of the class Crustacea, phylum Arthropoda; all have a hard exoskeleton, usually with a carapace and telson; includes crayfish, crabs, barnacles, shrimp, water fleas, and others.

CTENOID SCALES. Thin scales that bear a patch of tiny spinelike prickles (ctenii) on the exposed (posterior) surface; found in most spiny-rayed fishes.

CULTERS. Ten genera of medium-sized Asian fishes in the subfamily Cultrinae that are sometimes considered a clade of the family Cyprinidae. Tan and Armbruster (2018) placed them in a new family, Xenocyprididae, along with Asian carps and their allies.

CYCLOID SCALES. More or less rounded scales that are flat and bear no ctenii; found in most soft-rayed fishes.

CYTOCHROME C OXIDASE SUBUNIT I (COI OR COX I). Is one of three mitochondrial DNA (mtDNA) encoded subunits of cytochrome *c* oxidase, a key enzyme in aerobic metabolism of prokaryotes and eukaryotes.

DDT. (dichloro-diphenyl-trichloroethane) a colorless, crystalline, tasteless and almost odorless human-made organochloride chemical widely used to control insects on agricultural crops and insects that carry diseases like malaria and typhus. Its use was banned in the United States in 1972.

DECIDUOUS SCALE. A loosely attached scale that is easily shed or knocked off.

DECURVED. Curving downward.

DEME. A local population of closely related animals.

DEMERSAL EGGS. Eggs that are heavy and sink.

DENTARY. Major bony element of the lower jaw, usually bearing teeth.

DENTATE. Having toothlike notches.

DEPRESSED. Flattened dorsoventrally (from top to bottom); wider than deep.

DERIVED CHARACTER STATE. Condition of a taxonomic character inferred by cladistic analysis to have arisen within a taxon being examined cladistically rather than having been inherited from the most recent common ancestor of all members of the taxon. A character not present in the ancestor of a taxon.

DETRITUS. Decomposing organic matter; organic debris such as accumulations of leaves on a lake bottom.

DIADROMOUS. Refers to a species that migrates between freshwater and marine environments as a regular part of its life cycle.

DIATOMS. Unicellular or colonial algae of the phylum Chrysophyta, having cell walls impregnated with silica.

DIMORPHISM. Two body forms in the same species, often referring to differences between male and female.

DIOXIN. An extremely toxic impurity occurring in the herbicide 2,4,5-T; it is suspected of causing birth defects and miscarriages in mammals.

DIPHYCERCAL. Having the caudal fin approximately symmetrical, with the vertebral column extending to the tip of the tail without any upturning.

DIPTERA. An order of insects including flies, mosquitoes, crane flies, midges, black flies, horse flies, and others; the adults are never aquatic, but many families of dipterans have members that have aquatic larvae and pupae.

DISC TEETH. Teeth on the roof of the mouth (buccal disc) of lampreys.

DISJUNCT. Separated or disjoined; populations of organisms geographically isolated from a species' main range.

DISPERSAL. Movement of organisms from their place of birth to a new geographic area for permanent residence.

DISTAL. Farthest from point of attachment (for example, the free edge of a fin, farthest from its base).

DIURNAL. Active by day.

DNA BARCODING. A technique for identifying organisms to species using sequence information of a standard gene present in all animals. The mitochondrial gene encoding cytochrome *c* oxidase I (*COI*) is often used.

DOLOMITE. A mineral composed of carbonate of calcium and magnesium, $CaMg(CO_3)_2$; also used as a rock name for formations composed largely of the mineral dolomite.

DORSAL. Referring to the back; used as an abbreviation for dorsal fin.

DORSAL FIN. Median, longitudinal, vertical fin located on the back.

DORSUM. The upper part of the body; the back.

DRAINAGE. Used here to refer to an interconnected group of streams and tributaries forming a major river basin.

EAR FLAP. A flattened, flexible extension of the posterior part of the operculum; well developed in some sunfishes.

ECOLOGY. The study of the interactions of organisms with their environment and other organisms.

ECOSYSTEM. An ecological unit consisting of both the biotic (living) communities and the abiotic (nonliving) environment, which interact to produce a stable system.

ECOTONE. A transitional area between two adjacent communities.

ECTOTHERMIC. Having a body temperature derived from heat acquired from the environment.

ELECTROFISHING. A fish collection method in which fish are temporarily stunned by a current of electricity that renders them immobile and easy to pick up.

ELECTROPHORESIS. A procedure for separating different proteins (and sometimes other molecules) based on their electric charge and size.

ELEVATED SCALES. Scales on sides that are much higher than they are long (distance from top to bottom of scales much greater than distance from front to rear).

ELONGATE. Long and narrow.

EMARGINATE. Having a distal margin notched, indented, or slightly forked.

EMBEDDED SCALES. Scales that are hidden in the skin and difficult to see.

ENDANGERED. In danger of extinction.

ENDEMIC. Found only in, or limited to, a specified geographic region or locality.

ENDOTHERMIC. Having a body temperature determined by heat derived from an animal's own oxidative metabolism.

ENTOMOSTRACA. A descriptive term used to designate collectively all small freshwater Crustacea.

EPAXIAL. Portion of the body dorsal to the horizontal or median myospetum.

EPHEMEROPTERA (MAYFLIES). An order of insects in which the delicate, transparent-winged, terrestrial adults are found only near bodies of freshwater; the immature stages (naiads) are aquatic.

EPILIMNION. The upper layer of water in a lake above the thermocline characterized by a temperature gradient of less than 1°C per meter of depth.

EPURALS. Modified vertebral elements that lie above the vertebrae and support part of the caudal fin.

ESTIVATION. Dormancy during summer.

EUPHOTIC ZONE. The layer of water receiving 1% or more of the incident radiation striking the water's surface.

EURYHALINE. Pertaining to organisms that can tolerate a wide range of salinities.

EURYTHERMAL. Pertaining to organisms that can tolerate a wide range of temperatures.

EURYTHERMAL ENVIRONMENT. An aquatic environment that experiences wide seasonal fluctuations in water temperature.

EUTROPHICATION. The enrichment of an aquatic ecosystem with nutrients (nitrates, phosphates) that promote biological productivity (growth of algae and other vegetation).

EUTROPHIC LAKE. A lake rich in dissolved nutrients but usually having a seasonal oxygen deficiency in the bottom layers of water.

EXOTIC. Foreign; used herein as not native to North America.

EXTANT. Opposite of extinct. Still existing in nature or maintained under artificial conditions.

EXTINCT. No longer represented by living individuals anywhere within its former range; refers to a species or other taxon.

EXTIRPATED. Exterminated on a local basis (political or geographic part of range); used here to refer to a species exterminated from Arkansas.

FALCATE. A fin is said to be falcate when its margin is deeply concave or sickle-shaped.

FALL LINE. A transition zone 8–16 km wide in which streams pass from the older and harder rocks of the Uplands to the loose unconsolidated young sediments of the Coastal Plain, generally considered the boundary between Uplands and Coastal Plain. The Fall Line is shown on Figure 3.2 as the boundary between the brown- and green-shaded areas.

FAMILY. A taxonomic group containing two or more closely related genera.

FECUNDITY. Used here to refer to the number of ripening eggs produced by a female fish prior to its next spawning period.

FILAMENT. A threadlike process usually associated with the fins.

FIMBRIATE. Fringed at the margin with slender, elongate processes.

FINGERLING. A young fish, usually late in the first year.

FIN INSERTION. Posterior point at which the fin attaches to the body.

FIN ORIGIN. Anteriormost point at which fin attaches to body.

FIN RAY. A bony or cartilaginous rod supporting the fin membrane. Soft rays are usually segmented (cross-striated), often branched and flexible near their tips, whereas spines are not segmented, never branched, and are usually stiff to their sharp distal tips.

FLUVIATILE. Of, found in, or produced by a river.

FONTANELLE. An aperture or opening in a bony surface, often covered by a membrane.

FRACTIONAL SPAWNING. Pertaining to eggs spawned in several clutches, with periods of rest between clutches.

FRENUM. A ridge or fold of tissue that connects the upper lip to the snout.

FRY. A young fish at the age when yolk has been consumed and the fish is actively feeding.

FULTON CONDITION FACTOR. An index intended to assess the condition, health, or plumpness of a fish based on a relationship between weight and length. The formula, $K = W/L^3 \times 100$, usually attributed to T. W. Fulton, is used where K = fish condition, W = fish weight, and L = fish length.

FUSIFORM. Tapering gradually at both ends; spindle shaped.

GANOID SCALES. Thick, strong rhombic scales with a covering of ganoin, as in gars.

GAPE. Refers to the mouth. In fish, width of gape is the transverse distance between the two ends of the mouth cleft when the mouth is closed; length of gape is the diagonal distance from the anterior (median) end of the lower lip to one end of the mouth cleft.

GAS BUBBLE DISEASE. Caused by the development of gases (often oxygen) in a fish's bloodstream. This usually occurs when a fish breathes water supersaturated with gases, such as in turbulent water immediately below a dam or waterfall. Either supersaturation of oxygen or nitrogen can cause this disease, but the total dissolved gas is more important than individual gases. Fish with gas bubble disease often exhibit loss of equilibrium, abnormal buoyancy, and may float at the water surface (in addition to other symptoms).

GASTROPODS. Mollusks of the class Gastropoda, characterized by usually having a spiral, one-piece shell (absent in some); snails, limpets, slugs, and relatives.

GENERALIST. A fish that does not specialize on any particular type of food item or prey or is not restricted to a particular habitat.

GENITAL PAPILLA. A fleshy protuberance surrounding an aperture between the anus and anal fin through which eggs and sperm are discharged and deposited (as in darters).

GENUS (PL. GENERA). A taxonomic category; subdivision of a family, including one or more species with certain characteristics in common.

GILL ARCHES. See branchial arches.

GILL COVER. Assemblage of various bones (opercle, preopercle, subopercle, interopercle, and branchiostegal rays) that cover the gills. It opens and closes at the rear during respiration. Also called the operculum.

GILL FILAMENTS. Respiratory structures projecting posteriorly from gill arches.

GILL MEMBRANES. Membranes that close the gill cavity ventrolaterally, supported by the branchiostegals.

GILL NET. A net with meshes that allow the head of the fish to pass and captures fish by entanglement; it is a single mesh wall with cork and lead lines keeping the net upright.

GILL RAKERS. Projections (knobby or comblike) from the concave anterior surface of the gill arches.

GLOBAL POSITIONING SYSTEM (GPS). A worldwide radio-navigation system formed by at least 24 satellites and their ground stations. This satellite-based navigational system provides latitude and longitude to a radio receiver.

GLOBOSE. Shaped like a globe or a ball.

GONADOSOMATIC INDEX (GSI). The calculation of the gonad mass as a proportion of the total body mass. It is usually represented by the formula: GSI = [gonad weight / total tissue weight] × 100.

GONOPODIUM. Modified, rodlike anal fin of male *Gambusia* used in the transfer of sperm to genital pore of female.

GRADIENT. As used in relation to streams, the rate of descent in feet per mile (or meters per kilometer) of streams.

GRAVEL. Rock substrate fragments 0.08–2.5 inches (2–64 mm) in diameter; subdivisions include: coarse gravel—1.3–2.5 inches (32–64mm), medium gravel—0.3–1.3 inches (8–32 mm), and fine gravel—0.08–0.3 inches (2–8 mm).

GRAVID. Used in this book to refer to a female fish containing ripe or nearly ripe eggs.

GROUNDWATER. Water that has accumulated in materials below the Earth's surface.

GULAR FOLD. Transverse fold of soft tissue across the throat.

GULAR PLATE. Large, median, dermal bone on the throat region of certain primitive fishes such as the bowfin.

GULAR REGION. Throat area.

GULF SLOPE. The land area where the rivers (usually excluding the Mississippi River) drain directly into the Gulf of Mexico.

HAPLOTYPE. A set of DNA variations, or polymorphisms, that tend to be inherited together. It usually refers to a set of single nucleotide polymorphisms (SNPs) found on the same chromosome.

HEAD LENGTH. Distance from the tip of the snout to the rear edge of the opercular membrane; in eels, measure to the nearest edge of the gill opening.

HEADWATER SITE. A site located in the upstream-most part of a basin.

HECTARE (HA). A metric unit of area equal to 2.471 acres or 10,000 square meters.

HEMIPTERA. A large order of insects containing the true bugs; most species are terrestrial, but a few species are adapted to aquatic environments.

HETEROCERCAL. A type of tail of fishes in which the vertebral column turns upward into the dorsal lobe; examples are sturgeons and sharks.

HOLOTYPE. A single specimen designated by the author or authors of a species to represent that species.

HOMOCERCAL. A type of tail in which the vertebrae terminate at the caudal fin base in a hypural plate; neither lobe of the caudal fin is invaded by the vertebral column; usually symmetrical.

HOOP NET. A more or less conical fish trap typically made of twine mesh hung on round frames (hoops) and having two funnel-shaped throats in series.

HORIZONTAL MYOSEPTUM (HORIZONTAL SEPTUM). A midlateral connective tissue separating the epaxial muscle mass from the hypaxial muscle mass.

HUMERAL "SPOT" OR "SCALE". Scalelike bone, often dark-colored, behind the gill opening and above the base of the pectoral fin (in darters).

HYBRID. Used here to refer to the offspring resulting from a cross between individuals of two different species.

HYOID TEETH. Teeth on the tongue.

HYPAXIAL. That portion of the body ventral to the horizontal myoseptum.

HYPOGEAN. Underground; subterranean.

HYPOLIMNION. The cool bottom layer of water in a stratified lake below the thermocline characterized by a temperature gradient of less than 1°C per meter of depth.

HYPURAL PLATE. Expanded terminal vertebral process at the caudal base in fishes having a homocercal tail; formed by fusion of the last few caudal vertebrae; end of hypural plate usually appearing as a crease or line across end of caudal peduncle.

ICHTHYOFAUNA. The assemblage of fishes inhabiting a specific body of water or geographic area.

ICHTHYOLOGY. The scientific study of fishes.

ICHTHYOLOGIST. A scientist who studies fishes.

IMBRICATE. Overlapping, as in fish scales.

IMMACULATE. Without spots or other pigmentation.

IMPOUNDMENT. A reservoir. A lentic body of water created by a dam.

INCLUDED. Used in reference to the upper or lower jaw when one is shorter than the other.

INDEX OF BIOTIC INTEGRITY (also called Index of Biological Integrity). An index used to assess the environmental health of a stream or fish community based on attributes of the fish assemblage. Various IBIs have been developed using a variety of metrics that are sensitive to changes in aquatic habitat for the specific watershed being evaluated.

INDIGENOUS. Pertaining to fishes native to a given area.

INFERIOR. Beneath, lower, or on the ventral side; for example, inferior mouth.

INFRAORBITAL CANAL. Segment of the lateral line canal in the suborbital bones that curves beneath the eye and extends forward onto the snout; canal may be complete or interrupted.

INSERTION. Anteriormost end of the bases of the paired fins.

INTERGRADE. Used here to refer to the offspring resulting from a cross between individuals of two different subspecies and found in a zone of intergradation.

INTERMITTENT. Pertaining to small streams whose flow is discontinuous in the dry season.

INTERMUSCULAR BONES. Fragile, branched bones isolated in the connective tissue between body muscles.

INTEROPERCLE. A small bone of the operculum situated between the preopercle and the subopercle.

INTERORBITAL. Region on top of the head between the eyes.

INTERORBITAL DISTANCE. The least bony distance (skull width) on the top of the head between the eye orbits.

INTERRADIAL MEMBRANES. Membranes between fin rays.

INTESTINE. The most posterior region of the digestive tube and site of most digestion and absorption of food.

INTRODUCED. Refers to nonnative species that have been transplanted in Arkansas waters.

INTROGRESSION. The spread of genes of one species into the gene pool of another species by hybridization and subsequent backcrossing of hybrid individuals with one or both parental forms.

INTROMITTENT ORGAN. Copulatory structure of the male used to inseminate females in species with internal fertilization.

INVERTIVORE. A fish that feeds primarily on invertebrate animals.

ISOPODS. Members of the crustacean order Isopoda, they are strongly flattened dorsoventrally. Most are terrestrial or marine, and only a small percentage of North American species live in freshwater.

ISOZYMES. Enzymes that catalyze the same chemical reaction but are coded by different genes that are not alleles.

ISTHMUS. The narrow part of the breast that projects forward between (and separates) the gill chambers.

ITEROPARITY. Repeated reproduction after reaching sexual maturity (characteristic of most fishes).

JUGULAR. Refers to the throat region; pelvic fins are jugular when they are inserted in front of the pectoral fin bases.

JUVENILE. Young fish after attainment of minimum adult fin-ray counts and complete absorption of the median fin-fold and before sexual maturation.

KARST TOPOGRAPHY. Irregular topography characterized by underground stream channels, eaves, and sinkholes developed by the action of surface and underground water in soluble rock such as limestone.

KARYOTYPE. A display of the chromosome pairs of an organism.

KEEL. Scales or tissue forming a sharp edge or ridge.

KERATIN. A protein that is the substance making up teeth in lampreys (class Petromyzontida). In mammals, this protein forms fingernails, claws, and hair.

LACHRYMAL. A dermal bone of the skull located in front of the orbit; the most anterior bone in the suborbital series and often the largest.

LACUSTRINE. Pertaining to lakes.

LARVA (PL. LARVAE). The early, immature form of an animal that changes structurally as it becomes an adult.

LATERAL LINE. System of sensory tubules communicating to the body surface by pores; refers most often to a longitudinal row of scales that bear tubules and visible pores; considered incomplete if only the anterior scales possess pores and complete if all scales in that row (to base of caudal fin) have pores.

LATERAL LINE SCALES. Pored or notched scales associated with the lateral line.

LATERAL SERIES. Referring to a longitudinal scale count on the side of the body when there is no lateral line.

LECITHOTROPHIC. During development, embryos are nourished by yolk deposited in the egg before fertilization.

LECTOTYPE. A single specimen designated from a series of syntypes by someone other than the original describer to function as the "type specimen" after publication of the original description.

LENTIC. Standing water situation such as a lake or pond.

LEPTOCEPHALUS. The transparent, ribbonlike stage of the larval eel.

LIMESTONE. A fine-grained sedimentary rock composed largely of the mineral calcite, $CaCO_3$, which has been formed either as a chemical precipitate or by the consolidation of shell fragments.

LIMNETIC ZONE. The region of open water in a lake away from the shore extending downward to the maximal depth at which sunlight is sufficient for photosynthesis.

LINGUAL LAMINA. Horny ridge on the "tongue" of a lamprey.

LITHOPHILOUS SPECIES. Species that spawn on a coarse substrate (gravel, rock, blocks) or in fast-flowing water.

LITTORAL. Refers to the area along the shore of bodies of water.

LOCALITY. A fish sampling site. Used herein to refer to a place in a stream or lake that would be included in a single typical collection or observational sample from a fish assemblage by an ichthyologist (Matthews 1998).

LOTIC. Running water situations such as a river or stream.

LMR (LOWER MISSISSIPPI RIVER). As used herein, the free-flowing portion of the Mississippi River from the mouth of the Ohio River to the Gulf of Mexico.

MACROPHYTES. Vascular plants and macroscopic algae that grow in or near water and can be either submergent, emergent, or floating. Macrophytes include flowering plants, ferns, and bryophytes.

MANDIBLE. The main bone of the lower jaw; comprises 3 bones: dentary, angular, and articular.

MANDIBULAR PORES. Small openings along a tube that traverses the underside of each lower jaw (part of the lateral line system).

MARGINAL. Refers to the portion along the edges of fins.

MARL. Porous sediments of marine origin that contain a mixture of clay and finely divided shell fragments.

MAXILLA (MAXILLARY). A bone of each upper jaw that lies immediately above (or behind) and parallel to the premaxilla.

MECKEL'S CARTILAGE. Embryonic cartilaginous axis of the lower jaw in bony fishes; forms the area of the jaw articulation in adults.

MEGALOPTERA. An order of insects that includes alderflies, dobsonflies, and fishflies. Megalopterans have an aquatic larval stage.

MELANOPHORE. Black pigment cell.

MENTAL. Pertaining to the chin.

MERISTIC. Pertaining to the number of serial parts; for example, fin rays or lateral line scales.

METAPOPULATIONS. Spatially isolated populations that function as independent populations but that can exchange occasional individuals.

MICROHABITAT. A small, defined area within the general habitat.

MICROSATELITE DNA. Repetitive segments of DNA usually 2–6 nucleotides in length, scattered throughout the genome in the noncoding regions between genes or within genes (introns), often used as markers for linkage analysis because of high variability in repeat number between individuals. These regions are inherently unstable and susceptible to mutations. Microsatellites are widely used in parentage analysis, gene mapping, and assessments of population structure.

MIDDORSAL. Pertaining to the midline of the back.

MILT. Sperm and associated liquids released by a spawning male fish.

MISSISSIPPI EMBAYMENT. An inland extension of the Coastal Plain up the Mississippi River Delta to southern Illinois. It represents an area covered by ocean during the Early Cenozoic.

MMR (MIDDLE MISSISSIPPI RIVER). The segment of the Mississippi River between the mouth of the Missouri River and the mouth of the Ohio River.

MOLAR TEETH. Teeth with broad, flat surfaces adapted for grinding.

MONOPHYLETIC. When a taxon consists of groups whose members are all derived from a common ancestor. A monophyletic grouping of organisms contains the most recent ancestor of the group and all its descendants. Such a grouping of organisms forms a clade (contrasts with paraphyletic and polyphyletic).

MONOPHYLY. The condition that a taxon or other group of organisms contains the most recent ancestor of the group and all its descendants; grouping together of an organism and all its descendants in a clade; contrasts with paraphyly and polyphyly.

MONOTYPIC. Having only one type; for example, a monotypic genus has only one species in the genus.

MORPHOLOGY. The form and structure of organisms or parts of organisms.

MORPHOMETRIC. Morphological characters that can be measured with a millimeter scale such as length of body parts; often expressed as proportions or as percent of a fish's standard length (SL).

MOTTLED. Marked with spots or blotches of different colors.

MULTIVARIATE ANALYSIS. A statistical technique for analyzing more than one dependent or independent variable simultaneously. It can include analysis of variance, multiple regression, principal components analysis, or discriminant function analysis.

MYOMERE. Serially arranged blocks of muscle; muscle segment.

MYOSEPTA. Connective tissue partitions separating myomeres.

NAKED. Without scales.

NAPE. Dorsal part of the body from the occiput to dorsal fin origin; area of back immediately behind the head.

NARES. Nostrils; in fish, each nostril usually has an anterior and a posterior narial opening, located above and in front of the eyes.

NATIVE. Naturally occurring, indigenous, or endemic to a specified geographic area.

NEOTYPE. A specimen selected as the type subsequent to the original description in cases in which the primary types are definitely known to have been destroyed.

NEST ASSOCIATE. A fish that uses the nests of other fish species for spawning.

NEURAL ARCH. Dorsal part of a vertebra that surrounds the spinal cord.

NOMENCLATURE. A standard, international system for naming organisms; it provides stability and the Rule of Priority.

NOMINATE SUBSPECIES. In a species, the first subspecies to be described (named) to science; nominate subspecies have the same scientific species and subspecies name (e.g. the nominate subspecies *Luxilus chrysocephalus chrysocephalus*).

NONPROTRACTILE. Not capable of being thrust out. Premaxillaries are nonprotractile if they are not fully separated from the snout by a continuous groove.

NOSTRIL. External opening of the olfactory organ.

NOTOCHORD. Longitudinal supporting axis of the body eventually replaced by the vertebral column in teleostome fishes.

NOVACULITE. A very hard, fine-grained rock of silica, used as an abrasive stone and as a silica source in manufacturing.

NUCHAL. Pertaining to the nape.

NUCLEOLUS ORGANIZER REGION (NOR). The part of a chromosome associated with the nucleolus after the nucleolus divides. The region contains several tandem copies of ribosomal DNA genes.

NUPTIAL TUBERCLES. Hardened protuberances seen in adult males of many fishes just before or during their breeding season; these structures help maintain contact between the spawning pair and may occur in males or in both sexes. Also called pearl organs or breeding tubercles.

OBTUSE. With a blunt or rounded end; an angle greater than 90 degrees.

OCCIPUT. The posterior dorsal extremity of the head, often marked by the line separating scaly and scaleless portions of the dorsum.

OCELLUS. An eyelike spot, usually round with a light or dark border.

ODONATES. Insects of the order Odonata, dragonflies and damselflies. Odonates have an aquatic larval stage.

OLD-GROWTH FOREST. Mature woodland ecosystem characterized by the presence of old trees, and the associated wildlife and smaller plants living with them.

OLIGOCHAETA. A class of segmented worms including terrestrial and aquatic earthworms and other freshwater forms such as tubifex worms.

OLIGOTROPHIC LAKE. A nutrient-poor lake characterized by low production of plankton and fish and having considerable dissolved oxygen in the bottom waters (due to a low organic content).

OPEN POSTERIOR MYODOME (OPM). A cavity in the posterior of the cranium that is the site of the origin of the lateral rectus muscles.

OPERCLE. The large posterior bone of the gill cover; may be spiny, serrate, or entire (smooth).

OPERCULAR FLAP. Posterior flaplike extension of the operculum, especially in sunfishes ("ear" flap in *Lepomis* species).

OPERCULUM. Bony flap covering the gills of fish, also called the gill cover.

ORAL DISC (HOOD). Circular, flattened area around the mouth of lampreys; equipped with horny epidermal teeth; used as a suction cup to attach to prey.

ORAL CAVITY. The space inside the mouth.

ORBIT. Eye socket.

ORIGIN (OF FINS). Anterior end of the base of a dorsal fin or anal fin; point at which fin begins, that is, the point at which the first ray is inserted.

OSTRACODS. Small crustaceans of the subclass Ostracoda, characterized by a bivalve shell enclosing an indistinctly segmented body.

OTOLITHS. Calcareous "earstones" found in the inner ear of bony fishes; the three otoliths in each inner ear respond to gravitational forces and changes in orientation of the fish as well as to sound-wave vibrations.

OVIPARITY. Reproduction in which eggs are released by the female; development of offspring occurs outside the maternal body; oviparous.

OVOVIVIPARITY. Reproduction in which eggs develop within the maternal body without additional nourishment from the parent and hatch within the parent, or immediately after laying; ovoviviparous.

OUTGROUP. In phylogenetic systematic studies, a species or group of species closely related to but not included within a taxon whose phylogeny is being studied and used to polarize variation of characters and to root the phylogenetic tree.

OUTGROUP COMPARISON. A method for determining the polarity of a character in cladistics analysis of a taxonomic group. Character states found within the group being studied are judged ancestral if they occur in related taxa outside the study group (= outgroups); character states that occur only within the taxon being studied but not in outgroups are judged to have been derived evolutionarily within the group being studied.

OXBOW LAKE. A natural lake formed from an abandoned former river meander which has become isolated or almost isolated from the main stream channel by erosion and deposition.

OXIDATIVE (AEROBIC) METABOLISM. Cellular respiration using molecular oxygen as the final electron acceptor.

PAIRED FINS. The pectoral and pelvic fins.

PALATINE TEETH. Teeth on the paired palatine bones that form the anterolateral roof of the mouth (Fig. 26.10).

PANMIXIA. Generally used to refer to fish species in which all individuals compose a single group of potentially mating individuals. In a panmictic population, all individuals are potential reproductive partners with those of the opposite sex.

PAPILLA (PL PAPILLAE). A small, soft, fleshy, rounded projection on the skin.

PAPILLOSE. Covered with papillae (as contrasted with plicate when applied to lips of suckers).

PARAPATRIC. Populations or species with contiguous, non-overlapping geographic ranges that meet along a common boundary.

PARAPHYLETIC. When a taxon or other group of organisms contains the most recent common ancestor of all members of a group but excludes some descendants of that ancestor; contrasts with monophyly and polyphyly.

PARATYPE(S). Name-bearing specimen(s), usually a series, designated by the original describer of a species to represent that species.

PARIETAL. A paired dermal bone of the roof of the skull, located between the frontal and occipital.

PARR MARKS. Nearly square or oblong blotches found along the sides of young trout before they have developed the adult color pattern.

PECTORAL FIN. The anteriormost paired fin on the side, or on the breast, behind the head.

PEDUNCLE. See caudal peduncle.

PELAGIC. Of open waters; in Arkansas, usually referring to lakes or ponds.

PELVIC FIN. The ventral-paired fin, lying below the pectoral fin or between it and the anal fin; these fins vary in position from on the belly just anterior to anal fin (abdominal) to under the pectoral fin (thoracic) to below a point in front of the base of the pectoral fins on the isthmus (jugular).

PEPPERED. Stippled with dark pigment.

PERIPHYTON. Refers to microfloral growth on an underwater substrate.

PERITONEUM. Membranous lining of the body cavity.

PERSONAL COMMUNICATION (PERS. COMM.). The acknowledgment of unpublished information received from a colleague.

PHARYNGEAL CAVITY. The space that contains the gills.

PHARYNGEAL TEETH. Bony projections from the fifth pharyngeal (gill) arch.

PHYLOGENY. The origin and diversification of any taxon, or the evolutionary history of its origin and diversification, usually presented in the form of a dendrogram.

PHYLOGENETIC TREE. A tree diagram whose branches represent current or past evolutionary lineages and which shows the hypothesized patterns of common descent among those lineages.

PHYLOGENETIC SYSTEMATICS. See CLADISTICS.

PHYLOGEOGRAPHY. The study of the historical processes that may be responsible for the contemporary geographic distributions of taxa using the genetic and geographic structure of populations and species.

PHYSIOGRAPHIC REGION. An area defined by the geological and general topography of the landscape.

PHYSOCLISTOUS. Having no connection between the esophagus and the pneumatic duct; typical of perciform fishes.

PHYSOSTOMOUS. Having the swim bladder connected to the esophagus by the pneumatic duct; typical of cypriniform fishes.

PHYTOLITHOPHILOUS SPECIES. Species that spawn on varied material beds, organic or not, with vegetation present.

PHYTOPHILIC SPAWNER (OR PHYTOPHILOUS). An obligatory plant spawner.

PHYTOPLANKTON. Microscopic algae that float or drift near the surface of a body of water; they are somewhat smaller than most zooplankton.

PISCIVOROUS. Fish eating.

PLANKTON. Very small or microscopic plants and animals that swim passively or float near the surface of a body of water.

PLECOPTERA. An order of insects known as stoneflies; the terrestrial adults are seldom found far from running water, but the nymphs are all strictly aquatic.

PLESIOMORPHY. In cladistics, refers to an ancestral trait. A symplesiomorphy or symplesiomorphic character is an ancestral trait shared by two or more taxa.

PLICATE. Having parallel folds of soft ridges; grooved lips, especially prominent on the lips of certain catostomids.

PNEUMATIC DUCT. The tube that connects the swim bladder to the digestive tube.

POLYANDRY. A reproductive strategy in which the female mates with more than one male.

POLYGYNY. Also known as polygamy; a reproductive strategy in which the male mates with more than one female.

POLYMERASE CHAIN REACTION (PCR). A biochemical technology in molecular biology used to amplify a single copy or a few copies of a piece of DNA across several orders of magnitude, generating thousands to millions of copies of a particular DNA sequence.

POLYMORPHISM. The presence in a population of two or more distinct forms.

POLYPHYLETIC. When members of a taxonomic group are derived from more than one ancestral group.

POLYTYPIC. A polytypic species is composed of two or more populations or groups of populations that are morphologically distinct in some feature or features. Variant populations were often formerly designated as subspecies, but that concept has fallen out of favor.

PORED SCALE. A scale with an opening into the acoustico-lateralis system (e.g., scales along the lateral line).

POSTERIAD. In a posterior direction.

POSTERO-HYAL. Posterior bone to which branchiostegal rays attach; formerly epihyal.

POTAMODROMOUS. Refers to a species that migrates entirely within freshwater environments.

PREDORSAL SCALES. Scales on the midline of the dorsum between the back edge of the head and the dorsal fin origin.

PREMAXILLA (PREMAXILLARY). A paired bone at the front of the upper jaw. The right and left premaxillae join anteriorly and form all or part of the border of the jaw.

PREOPERCLE. Sickle-shaped bone lying behind and below the eye; may be serrated or smooth.

PREOPERCULOMANDIBULAR CANAL. A cephalic portion of the lateral line system that extends along the preopercle and mandible.

PREORBITAL. Bone forming the anterior rim of the eye socket and extending forward on side of snout.

PRIMARY FEEDER (PRIMARY CONSUMER). An herbivorous fish occupying the second trophic level.

PRIMARY SEWAGE TREATMENT. A rudimentary sewage treatment that removes a substantial amount of the settleable solids and about 90% of the biological oxygen demand (BOD).

PRINCIPAL RAYS. Fin rays that extend to the distal margin of median fins; enumerated by counting only 1 unbranched ray anteriorly plus subsequent branched rays.

PRINCIPLE OF PARSIMONY. Choosing the simplest scientific explanation that fits the evidence. In terms of constructing evolutionary trees, this means that, all other things being equal, the best hypothesis is the one that requires the fewest evolutionary changes.

PROCURRENT (RUDIMENTARY) RAYS. Small, contiguous rays at the anterior bases of the dorsal, caudal, and anal fins of many fishes; not included in the count of principal fin rays.

PROTRACTILE. Capable of being thrust out. Used herein when the upper jaw is completely separated from the snout by a continuous groove (frenum absent).

PROXIMAL. Nearest the body; center or base of attachment.

PSEUDOBRANCH. Accessory gill-like structure on the inner surface of the operculum.

PTERYGIOPHORE. Bones of the internal skeleton supporting the dorsal and anal fins.

PUNCTULE. Fine dot.

PUNTULOSE. Having small blisterlike projections or elevations.

PYLORIC CAECA. Fingerlike blind tubes at the junction of the stomach and intestine in most fishes.

QUADRATE. Squarish or having four sides.

R-SELECTED (R-STRATEGISTS). Refers to fishes that are successful in a given environment largely because of their extremely high reproductive potential (r), as opposed to fishes that are successful because they are adapted for maximizing trophic level efficiency and competitive ability (k-strategists).

RADIALS. Part of the girdle supporting a paired fin to which the fin rays attach.

RADII (SING. RADIUS). Grooves on a fish scale that radiate outward from its central part.

RANGE. The geographical area throughout which an organism is distributed.

RAY. See fin ray; may be soft or spiny in a fin.

RAY-FINNED FISHES. Fishes of the class Actinopterygii, characterized by attachment of fin rays in paired fins directly to underlying girdles.

RECURVED. Curved upward and inward.

REDD. An excavated area or nest into which trout spawn.

REGENERATED SCALE. A scale that has replaced a lost scale; characterized by a large focus and irregular circuli. A regenerated scale cannot be used for age analysis.

RESERVOIR. Used here to refer to a man-made impoundment of approximately 500 surface acres (202 hectares) or more at normal pool level.

RESTRICTION ENDONUCLEASE. An enzyme having the property of cleaving DNA molecules at or near a specific sequence of bases (called a restriction site).

RESTRICTION FRAGMENT LENGTH POLYMORPHISM. RFLP is a technique that exploits variations in homologous DNA sequences. It refers to a difference between samples of homologous DNA molecules that come from differing locations of restriction enzyme sites, and to a related laboratory technique by which these segments can be illustrated. In RFLP analysis, the DNA sample is broken into pieces (i.e., digested) by restriction enzymes, and the resulting restriction fragments are separated according to their lengths by gel electrophoresis.

RETICULATE. Containing a network of lines in a chain-like pattern, such as the pattern on the sides of the Chain Pickerel.

RETRORSE. Pointing backward.

REVETMENTS. Structures installed along river banks to prevent bank caving and erosion. They generally consist of stone riprap.

RHAGIONIDS. Snipe flies of the dipteran family Rhagionidae.

RHEOTAXIS. Oriented movement of fish in response to a

water current. Many positively rheotactic fishes turn to face into an oncoming current.

RIFFLE. A section of stream in which the water is usually shallower than in the connecting pools and over which the water runs more swiftly than it does in the pools. A riffle is shallower than a chute and usually has at least some "white water" breaking over the substrate.

RIPARIAN. Living or located at the bank of a watercourse such as a river or stream. Also refers to a vegetated zone adjacent to a stream.

RIPE. Refers to a state of reproductive readiness in which light abdominal pressure causes female fish to exude mature eggs capable of being fertilized and males to exude sperm.

RIPRAP. Large rocks used for stabilizing the banks of rivers and reservoirs.

RIVERINE. Pertaining to a river. Generally used as a habitat reference for fishes adapted to rivers.

RIVER SYSTEM. A river and its included streams, rivers, and lakes.

ROBUST. Strongly or stoutly built; husky.

ROE. Fish eggs.

ROSTRUM. Snout.

ROTENONE. An organic compound ($C_{23}H_{22}O_6$) derived from certain tropical plants and used as a fish toxicant to achieve complete fish kills during fish population sampling, to remove invasive fish species, or to renovate ponds and lakes. It is only slightly toxic to humans when used according to EPA regulations and quickly breaks down into relatively harmless compounds in aquatic environments. The half-life of rotenone in California waters typically ranges from 0.5 day at 24°C to 3.5 days at 0°C (Finlayson et al. 2001), but in 2007, the EPA reported half-lives ranging from a few days to several weeks (EPA document 738-R-07–005). The AGFC discontinued routine rotenone population sampling after 2010.

ROTIFERS. Microscopic multicellular animals of the phylum Rotatoria, having an anterior corona of cilia and occurring in all aquatic habitats in Arkansas.

ROUGH FISH. Nongame fishes generally considered undesirable in certain aquatic habitats from a standpoint of angling or fisheries management.

RUBBLE (COBBLE). Used here to refer to rock substrate fragments 2.5–10 inches (64–256 mm) in diameter.

RUDIMENTARY RAYS. Short rays that occur at the leading edge of a median fin (dorsal or anal fin) and do not reach the distal fin margin. The rudimentary rays grade in length into the principal rays.

RUN. A stream habitat transitional between fast, shallow riffles and slow, deeper pools. It is characterized by a smooth or roiled surface, moderate or swift current, and shallow or moderate depth. Also termed a raceway or glide.

SACCULUS. A saclike chamber forming a small portion of the inner ear of a fish and containing an otolith.

SADDLE. Rectangular or linear bars or bands that cross the back and extend partially or entirely downward across the sides.

SAND. Fine clastic particles ranging in size from 0.062 to 2 mm in diameter and commonly but not always composed of the mineral quartz.

SANDBAR. A deposit of sand formed on the inside bend of a river.

SANDSTONE. A detrital sedimentary rock formed by the cementation of individual grains of sand and commonly composed of the mineral quartz.

SATELLITE MALE. An alternate reproductive strategy, known in some centrarchids, where a male with body coloration similar to a female slowly enters the nest of a territorial male spawning with a female in an attempt to fertilize the eggs. In contrast to "sneaker" males, satellites males are aggressive toward other fish of equal size (Neff and Knapp 2009).

SCUTE. A modified scale in the form of a horny or bony plate that is often spiny or keeled.

SECONDARY SEWAGE TREATMENT. An advanced type of sewage treatment that involves both mechanical and biological (bacterial action) phases; although superior to primary treatment, much of the phosphates and nitrates remain in the effluent.

SEMELPARITY. A reproductive pattern in which the adults spawn once in a lifetime and then die.

SERRAE. Teeth of a sawlike organ or structure.

SERRATE. Toothed or notched on the edge, like a saw.

SEXUAL DIMORPHISM. Males and females of the same species differ in shape, size, color, or other characteristics.

SHAGREEN. Condition of being covered with small round granulations providing a rough texture.

SHALE. A fine-grained, soft sedimentary rock made up of silt and clay-sized particles and usually occurring in thin layers.

SIBLING SPECIES. Two species that are the closest relative of each other and have not been distinguished from one another taxonomically.

SILT. Used here to refer to waterborne sediment with a particle size of 0.00016–0.00244 inch (0.004–0.062 mm) in diameter.

SIMULIIDS. Insects of the order Diptera, family Simuliidae (black flies). Black flies have an aquatic larval stage and often occur at great densities in streams where they are preyed upon by fish.

SINGLE NUCLEOTIDE POLYMORPHISM (SNP). A DNA sequence variation occurring commonly within a population or taxon in which a single nucleotide—A, T, C, or G—in the genome (or other shared sequence) differs between members of a biological species or paired chromosomes. For example, two sequenced DNA fragments from different individuals, AAGCCTA to AAGCTTA, contain a difference in a single nucleotide.

SISTER TAXA OR SPECIES. Two taxa derived from the same immediate common ancestor and therefore, each is the other's closest relative.

SNEAKER MALE. A reproductive male that has a color pattern resembling that of a female. The sneaker male attempts to fertilize eggs being deposited in the nest of a territorial male engaged in spawning activity with females. Sneakers use a rapid nest entry and exit during spawning and are nonaggressive (Neff and Knapp 2009). Among Arkansas fishes, sneaker males are known in some species of minnows, sunfishes, and darters.

SNOUT. Part of the head anterior to the eye but not including the lower jaw.

SOFT RAY. A flexible, generally segmented and branched fin ray.

SPATULATE. Paddle-shaped or spoon-shaped.

SPECIALIST. A fish restricted in its choice of diet or habitat.

SPECIES. The fundamental taxonomic category; subdivision of a genus. There are several definitions of a species. In the Biological Species Concept, a species is a group of organisms that naturally or potentially interbreed, are reproductively isolated from other such groups, and are usually morphologically separable from them. In the Evolutionary Species concept, a species is defined as a single lineage of ancestor-descendant populations that maintains its identity from other such lineages and that has its own evolutionary tendencies and historical fate. In the Phylogenetic Species Concept, a species is an irreducible grouping of organisms diagnosably distinct from other such groupings and within which there is a parental pattern of ancestry and descent.

SP. *cf.* Used herein to denote a distinct, undescribed species formerly included in another described species. In such cases, sp. *cf.* is placed between the genus and specific epithet names. For example, *Etheostoma* sp. *cf. spectabile* indicates an undescribed species of the Orangethroat Darter complex formerly recognized as *Etheostoma spectabile*.

SPECIES EPITHET. The second word in the Linnaean binomial system of nomenclature of species, never capitalized, and differentiates an individual species from other members of the same genus.

SPINE. A stiff, bony, supporting structure in the fin membrane. Unlike rays, spines are never branched or segmented.

SPIRAL VALVE. Spiral fold of mucous membrane projecting into the intestines.

SPIRACLE. Opening on the back part of the head (above and behind the eye) in paddlefish and some sturgeons, representing a primitive gill cleft.

SPRING RUN. A stream formed by an outflow of water from a spring.

SQUAMATION. The arrangement of scales.

STANDARD LENGTH (SL). Distance from the tip of the snout to the end of the hypural plate.

STELLATE. Star-shaped.

STENOTHERMAL ENVIRONMENT. An aquatic environment, such as a spring, that does not experience wide seasonal fluctuations in water temperature.

STIPPLED. Sparsely dotted or spotted with small melanophores.

STOMACH. Region of the digestive tube between the esophagus and the intestine that can expand to store food and the site where digestion of food begins. Not all fishes have a true stomach.

STREAM ORDER. A ranking of the relative positions of streams within a watershed. A first-order stream is found in the extreme headwaters and receives no tributaries. A second-order stream results from the confluence of two first-order streams, a third-order stream results from the confluence of two second-order streams, and so on.

STRIATE. Streaked or striped with narrow parallel lines or grooves.

SUBADULT. A juvenile about the size of an adult but not sexually mature.

SUBEQUAL. Almost the same length.

SUBGENUS. A named subdivision of a genus; groups of species within a genus sharing certain characteristics not shared by other groups within the genus.

SUBOPERCLE. Bone immediately below the opercle in the operculum.

SUBORBITAL. Thin bone forming the lower part of the orbital rim.

SUBSPECIES. A subdivision of a species; a group of local populations inhabiting a geographic subdivision of the species range, and differing morphologically from other populations of the species. The subspecies concept is infrequently used today because it has no meaningful definition as a taxon.

SUBSTRATE. Bottom materials in a body of water.

SUBTERMINAL MOUTH. A mouth positioned slightly ventrally, rather than straight forward and from the front of the head; upper jaw slightly exceeds the lower jaw.

SUBTRIANGULAR. Almost triangular in shape.

SUPERIOR MOUTH. Condition in which the lower jaw extends upward and the mouth opens dorsally.

SUPRAMAXILLA. Small wedge-shaped movable bone adherent to the upper edge of the maxilla near its posterior tip.

SUPRAORAL LAMINA. The tooth-bearing plate immediately above the mouth opening in a lamprey.

SUPRAORBITAL CANAL. A paired branch of the lateral line system that extends along the top of the head between the eyes and forward onto the snout.

SUPRATEMPORAL CANAL. A branch of the lateral line system that extends across the top of the head at the occiput, connecting the lateral canals.

SWIM BLADDER. Membranous, gas-filled sac in the upper part of the body cavity, beneath the vertebral column. Also called air bladder or gas bladder.

SYMPATRIC. Two or more populations occupying identical or broadly overlapping geographical areas.

SYMPHYSIS. Articulation of two bones in the median plane

of the body, especially that of the two halves of the lower jaw (mandibles) at the chin.

SYMPLESIOMORPHY. Sharing among species of ancestral characteristics, not necessarily indicative that the species form a monophyletic group.

SYNAPOMORPHY. Shared, evolutionarily derived character states that are used to recover patterns of common descent among two or more species.

SYNONYM. One of two or more names for the same taxon.

SYNONYMY. A collection or listing of all properly described names for the same species or other taxon.

SYNTYPE. Formerly, when a new species was described, multiple types or type series were sometimes designated called syntypes. This practice has largely been abandoned.

SYNTOPIC. Occurring at same locations.

SYSTEMATICS. The study of the evolutionary relationships of species or higher taxa such as families or orders; the study of biodiversity relevant to relationships among organisms.

TAILWATER. The part of a stream immediately downstream from a dam. Stream flow is regulated by water released from the dam.

TALUS. An accumulation of rock fragments at the foot of a cliff or slope.

TAXON. Used for any named group of two or more organisms; it can be at any level of the classification system.

TAXONOMY. That part of systematics dealing with the theory and practice of describing diversity and erecting classifications; the scientific field of classification and naming of organisms.

TEARDROP. A black, drop-shaped mark below the eye.

TELEOSTS. A subset (division or infraclass Teleostei) of the class Actinopterygii (the ray-finned fishes) that share a number of characters, including the support structures of the caudal fin, the lack of a spiral valve intestine, and scales (if present) composed of only bone. Teleosts comprise more than 12,000 living species of inland fishes and up to 28,000 fish species worldwide.

TERETE. Cylindrical and tapering with circular cross section; having a rounded body form, the width and depth about equal.

TERMINAL MOUTH. When the upper and the lower jaws form the extreme anterior tip of the head; condition when the lower and upper jaws are equal in length and the mouth opens terminally.

TESSELLATED. Checkered, with the markings or colors arranged in squares or forming a mosaic.

THERMOCLINE. The middle layer of water in a lake in summer characterized by a temperature gradient of more than 1°C per meter of depth.

THREATENED. A form or forms likely to become endangered within the foreseeable future if certain conditions continue to deteriorate.

THORACIC. Pertaining to the chest region in fishes; pelvic fins thoracic when inserted below the pectoral fins.

TIPULIDS. Insects of the order Diptera, family Tipulidae (crane flies). Many species have an aquatic larval stage and are preyed upon by fishes.

TRAMMEL NET. A net that captures fish by entanglement; it is a wall formed by three parallel pieces of mesh, the outer two of a wide mesh, the inner one finer.

TRANSBRANCHIORAL SPAWNING. A spawning pattern found in *Aphredoderus sayanus* and *Amblyopsis* species (and possibly in *Troglichthys rosae* and *Typhlichthys* species) in which ripe eggs are extruded from the jugularly situated urogenital opening of the female into the branchial cavity and through the mouth into the external spawning substrate to be fertilized by the male.

TRANSVERSE. To go from side to side across the long axis.

TRANSVERSE LINGUAL LAMINA. The anterior tongue plate of lampreys, which can be straight (linear) from side to side, or strongly bilobed, or divided at the middle.

TREE OF LIFE. A metaphor used to describe the phylogenetic relationships among organisms, both living and extinct.

TRICHOPTERA (CADDISFLIES). An order of insects in which the small to medium-sized mothlike adults are found near streams, ponds, and lakes; the larvae and pupae are aquatic and are found in all types of freshwater habitats.

TRICUSPID. Having three points (applicable especially to teeth).

TROPHIC LEVEL. A feeding level; for example, a herbivore or a first level carnivore.

TUBERCLE. A small projection or lump. Refers herein to keratinized or osseous structures developed during the breeding period (breeding tubercles).

TUBERCULATE. Having tubercles.

TURBIDITY. Refers to the decreased ability of water to transmit light caused by suspended particulate matter ranging in size from colloidal to coarse dispersions; turbidity may result from colloidal clay particles entering a body of water with runoff, colloidal organic matter originating from decay of vegetation, or from an abundance of plankton.

TYPE LOCALITY. The locality from which the specimens used in the original description of a species were collected.

UNICUSPID. A tooth with one projection.

UMR (UPPER MISSISSIPPI RIVER). The portion of the Mississippi River upstream from the mouth of the Missouri River.

VALID SPECIES. A species properly named and described and currently recognized as such by biologists.

VASCULARIZED. Supplied with blood vessels.

VENT. The anus; external opening of the alimentary canal.

VENTER. The belly or lower sides of a fish.

VENTRAL. Pertaining to the lower surface.

VERMICULATE. Wormlike; marked with irregular or wavy lines.

VERMIFORM. Wormlike in shape.

VICARIANCE. Existence of closely related taxa or biota in different geographical regions that have been separated by the formation of a natural barrier to dispersal and therefore to gene flow.

VILLIFORM TEETH. Teeth slender and crowded together, giving the appearance of a velvety band.

VIVIPAROUS. Live bearing; young receive nutrients from the mother during development.

VOMER. A median, unpaired bone, usually bearing teeth, at the anterior extremity of the roof of the mouth.

WAIF. A stray from the usual geographic area of occurrence.

WATER HARDENED. Pertaining to egg capsules that are expanded and toughened by the absorption of water into the perivitelline space.

WATER WILLOW. A shallow-water emergent plant (genus *Justicia*) with willowlike leaves often found along the margins of Arkansas creeks, rivers, and lakes.

WEBERIAN APPARATUS. A series of ossicles (small bones) that conduct vibrations or pressure changes from the swim bladder to the ear, involving the first four or five fused vertebrae behind the head, and restricted to species of the families Cyprinidae, Catostomidae, and Ictaluridae in Arkansas.

YOUNG-OF-THE-YEAR (YOY). A fish in its first growing season belonging to the age-group 0, which has usually reached the fingerling stage.

XANTHIC. Colored with some shade of yellow or gold.

ZOOPLANKTON. Minute animals such as protozoans, rotifers, and microcrustaceans that float or drift near the surface of a body of water; they are somewhat larger than most phytoplankton.

ZYGOTE. A cell formed by the union of sperm and egg; a fertilized egg.

LITERATURE CITED*

Abbott, C. C. 1860. Descriptions of new species of American fresh-water fishes. Proc. Acad. Nat. Sci. Phila. 12: 325–328.

Abbott, C. C. 1861. Notes on the habits of *Aphredoderus sayanus*. Proc. Acad. Nat. Sci. Phila. 13: 95–96.

Abdul, M. N. 1987. The Goldstripe Darter *Etheostoma parvipinne* Gilbert and Swain: its habitat, morphology, and feeding habits. M.S. thesis. Stephen F. Austin State Univ., Nacogdoches, TX. 102 pp.

Abdusamadov, A. S. 1987. Biology of White Amur, *Ctenopharyngodon idella*, Silver Carp, *Hypophthalmichthys molitrix*, and Bighead Carp, *Aristichthys nobilis*, acclimatized in the Terek region of the Caspian Basin. J. Ichthyol. 26(4): 41–49.

Abell, R., D. M. Olson, E. Dinerstein, P. Hurley, J. T. Diggs, W. Eichbaum, S. Walters, W. Wettengel, T. Allnutt, C. J. Loucks, P. Hedao, and C. Taylor. 2000. Freshwater ecoregions of North America: a conservation assessment. Island Press, Washington, DC. 368 pp.

Able, K. W. 1984. Cyprinodontiformes: development. Pages 362–368. *In*: H. G. Moser et al., eds. Ontogeny and systematics of fishes. American Society of Ichthyology and Herpetology, Special Publ. 1. 760 pp.

Adams, C. C. and T. L. Hankinson. 1928. The ecology and economics of Oneida Lake fish. Bull. N.Y. State Coll. Forest. Syracuse Univ., Roosevelt Wild Life Ann. 1(3–4): 235–548.

Adams, G. L. and J. E. Johnson. 2001. Metabolic rate and natural history of Ozark Cavefish, *Amblyopsis rosae*, in Logan Cave, Arkansas. Environ. Biol. Fishes 62(1–3): 97–105.

Adams, G. L., B. M. Burr, J. L. Day, and D. E. Starkey. 2013. *Cottus specus*, a new troglomorphic species of sculpin (Cottidae) from southeastern Missouri. Zootaxa 3609(5): 484–494.

Adams, J. C. and R. V. Kilambi. 1979. Maturation and fecundity of Redear Sunfish. Proc. Arkansas Acad. Sci. 33: 13–16.

Adams, L. A. 1942. Age determination and rate of growth in *Polyodon spathula*, by means of the growth rings of the otoliths and dentary bone. Am. Midl. Nat. 28(3): 617–630.

Adams, R., E. Kluender, and L. Lewis. 2013. Movements, habitat use, and reproduction of Alligator Gar in the Fourche LaFave River. Final Report, State Wildlife Grant Program, Arkansas Game and Fish Commission, Little Rock. 83 pp.

Adams, S. B. and D. A. Schmetterling. 2007. Freshwater sculpins: phylogenetics to ecology. Trans. Am. Fish. Soc. 136(6): 1736–1741.

Adams, S. B., D. A. Schmetterling, and D. A. Neely. 2015. Summer stream temperatures influence sculpin distributions and spatial partitioning in the upper Clark Fork River basin, Montana. Copeia 103(2): 416–428.

Adams, S. R., G. L. Adams, and J. J. Hoover. 2003. Oral grasping; a distinctive behavior of cyprinids for maintaining station in flowing water. Copeia 2003(4): 851–857.

Adams, S. R., J. J. Hoover, and K. J. Killgore. 1999. Swimming endurance of juvenile Pallid Sturgeon, *Scaphirhynchus albus*. Copeia 1999(3): 802–807.

Adams, S. R., M. B. Flinn, B. M. Burr, M. R. Whiles, and J. E. Garvey. 2006. Ecology of larval Blue Sucker (*Cycleptus elongatus*) in the Mississippi River. Ecol. Freshw. Fish 15(3): 291–300.

Adamson, S. W. and T. E. Wissing. 1977. Food habits and feeding periodicity of the Rainbow, Fantail, and Banded darters in Four Mile Creek. Ohio J. Sci. 77(4): 164–169.

Aday, D. D., J. J. Parkos III, and D. H. Wahl. 2009. Population and community ecology of Centrarchidae. Pages 39–69. *In*: S. J. Cooke and D. P. Philipp, eds. Centrarchid fishes: diversity, biology, and conservation. Wiley-Blackwell, Chichester.

ADEQ. 2014. Integrated water quality monitoring and assessment report. Arkansas Department of Environmental Quality, Little Rock. 408 pp.

ADEQ. 2016. Integrated water quality monitoring and assessment report. Arkansas Department of Environmental Quality, Office of Water Quality, North Little Rock. 524 pp.

Agassiz, L. 1854. Notice of a collection of fishes from the southern bend of the Tennessee River, in the state of Alabama. Am. J. Sci. Arts (Ser. 2) 17: 297–308, 353–365.

Agassiz, L. 1855. Article XII. Synopsis of the ichthyological fauna of the Pacific slope of North America, chiefly from the collections made by the U.S. Expl. Exped., under the command of Captain C. Wilkes, with recent additions and comparisons with eastern types. Am. J. Sci. Arts (Ser. 2) 19(55): 71–99; (56): 215–231.

---

* We follow Nelson et al. (2002) and Page et al. (2013) in capitalizing the common names of fishes in all chapters of this book; therefore, to maintain consistency, we also capitalized common names in the Literature Cited section even though the names may not have been capitalized in the original publications.

Agassiz, L. 1856. Young gar-pikes from Lake Ontario. Proc. Boston Soc. Nat. Hist. 6: 47–48.

Agassiz, L. 1870. Lake Superior: Its physical character, vegetation, and animals, compared with those of other and similar regions . . . with a narrative of the tour by J. Elliot Cabot and contributions by other scientific gentlemen. Gould, Kendall, and Lincoln, Boston. 488 pp.

Aggus, L. R. 1972. Food of angler harvested Largemouth, Spotted, and Smallmouth bass in Bull Shoals Reservoir. Proc. 26th Annu. Conf. Southeast. Assoc. Game Fish Comm.: 519–529.

Aggus, L. R., D. I. Morais, and R. F. Baker. 1977. Evaluation of the trout fishery in the tailwater of Bull Shoals Reservoir, Arkansas, 1971–73. Proc. 31st. Annu. Conf. Southeast. Assoc. Game Fish Comm.: 656–673.

Ahuja, S. K. 1964. Salinity tolerance of *Gambusia affinis*. Indian J. Exp. Biol. 2(1): 9–11.

Albers, J. L. 2014a. Sturgeon Chub: *Macrhybopsis gelida* (Jordan and Evermann 1896). Page 179. *In*: Kansas Fishes Committee. Kansas Fishes. Univ. Press of Kansas, Lawrence. 542 pp.

Albers, J. L. 2014b. Sicklefin Chub: *Macrhybopsis meeki* (Jordan and Evermann 1896). Pages 184–186. *In*: Kansas Fishes Committee. Kansas Fishes. Univ. Press of Kansas, Lawrence. 542 pp.

Albers, J. L. and M. L. Wildhaber. 2017. Reproductive strategy, spawning induction, spawning temperatures and early life history of captive Sicklefin Chub *Macrhybopsis meeki*. J. Fish Biol. 91(1): 58–79.

Albrecht, A. B. 1964. Some observations on factors associated with survival of Striped Bass eggs and larvae. California Fish Game 50(2): 100–113.

Alexander, A. M. and J. S. Perkin. 2013. Notes on the distribution and feeding ecology of a relict population of the Cardinal Shiner, *Luxilus cardinalis* (Teleostei: Cyprinidae), in Kansas. Trans. Kansas Acad. Sci. 116(1–2): 11–21.

Alexander, H. 1973. Outdoor recreation, fish and wildlife, and the White River plan. Arkansas Game and Fish 5(4): 20–23.

Alexander, M. L. 1914. The paddle-fish (*Polyodon spathula*). (Commonly called "spoonbill cat"). Trans. Am. Fish. Soc. 44(1): 73–78.

Aley, T. and C. Aley. 1979. Prevention of adverse impacts on endangered, threatened, and rare animal species in Benton and Washington counties, Arkansas. Ozark Underground Laboratory Report, 25 June 1979. 35 pp.

Alfaro, R. M., C. A. Gonzalez, and A. M. Ferrara. 2008. Gar biology and culture: status and prospects. Aquac. Res. 39(7): 748–763.

Alfermann, T. J. and L. E. Miranda. 2013. Centrarchid assemblages in floodplain lakes of the Mississippi Alluvial Valley. Trans. Am. Fish. Soc. 142(2): 323–332.

Allan, J. D., R. Abell, Z. Hogan, C. Revenga, B. W. Taylor, R. L. Welcomme, and K. Winemiller. 2005. Overfishing of inland waters. BioScience 55(12): 1041–1051.

Alldredge, P., M. Gutierrez, D. Duvernell, J. Schaefer, P. Brunkow, and W. Matamoros. 2011. Variability in movement dynamics of topminnow (*Fundulus notatus* and *F. olivaceus*) populations. Ecol. Freshw. Fish 20(4): 513–521.

Allen, K. O. and K. Strawn. 1968. Heat tolerance of Channel Catfish, *Ictalurus punctatus*. Proc. 21st Annu. Conf. Southeast. Assoc. Game Fish Comm. 21: 399–411.

Allen, M. S., M. V. Hoyer, and D. E. Canfield, Jr. 2000. Factors related to Gizzard Shad and the Threadfin Shad occurrence and abundance in Florida lakes. J. Fish Biol. 57(2): 291–302.

Allen, R., C. Cato, C. Denis, and R. L. Johnson. 2009. Condition relative to phenotype for bass populations in southern Arkansas lakes. J. Arkansas Acad. Sci. 63: 20–27.

Allen, R. M. and R. L. Johnson. 2009. Temporal changes of Largemouth Bass alleles in a northern Arkansas reservoir stocked with Florida Bass. J. Arkansas Acad. Sci. 63: 28–33.

Allen, T. C., Q. E. Phelps, R. D. Davinroy, and D. M. Lamm. 2007. A laboratory examination of substrate, water depth, and light use at two water velocity levels by individual juvenile Pallid (*Scaphirhynchus albus*) and Shovelnose (*Scaphirhynchus platorynchus*) sturgeon. J. Appl. Ichthyol. 23(4): 375–381.

Alpaugh, W. C. 1972. High lethal temperatures of Golden Shiners (*Notemigonus crysoleucas*). Copeia 1972(1): 185.

Al-Rawi, A. H. and F. B. Cross. 1964. Variation in the Plains Minnow, *Hybognathus placitus* Girard. Trans. Kansas Acad. Sci. 67(1): 154–168.

Amemiya, C. T. and J. R. Gold. 1990. Chromosomal NOR phenotypes of seven species of North American Cyprinidae, with comments on cytosystematic relationships of the *Notropis volucellus* species-group, *Opsopoeodus emiliae,* and the genus *Pteronotropis*. Copeia 1990(1): 68–78.

Amorim, M. C. P., P. J. Fonseca, and V. C. Almada. 2003. Sound production during courtship and spawning of *Oreochromis mossambicus*: male-female and male-male interactions. J. Fish Biol. 62(3): 658–672.

Anderson, A. P., M. R. Denson, and T. L. Darden. 2014. Genetic structure of Striped Bass in the southeastern United States and effects from stock enhancement. N. Am. J. Fish. Manag. 34(3): 653–667.

Anderson, K. A., P. M. Rosenblum, B. G. Whiteside, R. W. Standage, and T. M. Brandt. 1998. Controlled spawning of Longnose Darters. Prog. Fish-Cult. 60(2): 137–145.

Anderson, W. W. 1958. Larval development, growth, and spawning of Striped Mullet (*Mugil cephalus*) along the South Atlantic coast of the United States. USFWS Fish. Bull. 144: 501–519.

Andree, S. R. and D. H. Wahl. 2019. Effects of turbidity on foraging of juvenile Black and White crappies. Ecol. Freshw. Fish. 28: 123–131.

Angermeier, P. L. 1982. Resource seasonality and fish diets in an Illinois stream. Environ. Biol. Fishes 7(3): 251–264.

Ankley, G. T. and R. D. Johnson. 2004. Small fish models for identifying and assessing the effects of endocrine-disrupting chemicals. ILAR J. 45(4): 469–483.

Ankley, G. T., D. C. Bencic, M. S. Breen, T. W. Collette, R. B. Conolly, N. D. Denslow, S. W. Edwards, D. R. Ekman, N. Garcia-Reyero, K. M. Jensen, J. M. Lazorchak, D. Martinovic, D. H. Miller, E. J. Perkins, E. F. Orlando, D. L. Vileneuve, R. Wang, and K. H. Watanabe. 2009. Endocrine disrupting chemicals in fish: developing exposure indicators and predictive models of effects based on mechanism of action. Aquat. Toxicol. 92(3): 168–178.

Annett, C., J. Hunt, and E. D. Dibble. 1996. The compleat bass: habitat use patterns of all stages of the life cycle of Largemouth Bass. Pages 306–314. In: L. E. Miranda and D. R. DeVries, eds. Multidimensional approaches to reservoir fisheries management. Symp. 16. American Fisheries Society, Bethesda, MD.

Applegate, R. L. and J. W. Mullan. 1967a. Food of the Black Bullhead (Ictalurus melas) in a new reservoir. Proc. 20th Annu. Conf. Southeast. Assoc. Game Fish Comm. 20: 288–292.

Applegate, R. L. and J. W. Mullan. 1967b. Food of young Largemouth Bass, Micropterus salmoides in a new and old reservoir. Trans. Am. Fish. Soc. 96(1): 74–77.

Applegate, R. L., J. W. Mullan, and D. I. Morais. 1967. Food and growth of six centrarchids from shoreline areas of Bull Shoals Reservoir. Proc. 20th Annu. Conf. Southeast. Assoc. Game Fish Comm. 20: 469–482.

April, J., R. L. Mayden, R. H. Hanner, and L. Bernatchez. 2011. Genetic calibration of species diversity among North America's freshwater fishes. Proc. Natl. Acad. Sci. USA. 108(26): 10602–10607.

Arce-H., M., J. G. Lundberg, and M. A. O'Leary. 2017. Phylogeny of the North American catfish family Ictaluridae (Teleostei: Siluriformes) combining morphology, genes and fossils. Cladistics 33(4): 406–428.

Arkansas Department of Pollution Control and Ecology. 1976. Arkansas Water Quality Inventory Report. 1976. ADPCE, Little Rock.

Arkansas Department of Pollution Control and Ecology. 1984. Arkansas Water Quality Inventory Report. 1984. ADPCE, Little Rock. 495 pp.

Armbruster, J. W., M. L. Niemiller, and P. B. Hart. 2016. Morphological evolution of the cave-, spring-, and swamp-fishes of the Amblyopsidae (Percopsiformes). Copeia 104(3): 763–777.

Armstrong, D. L., Jr., D. R. Devries, C. Harman, and D. R. Bayne. 1998. Examining similarities and differences between congeners: do larval Gizzard Shad and Threadfin Shad act as ecologically equivalent units? Trans. Am. Fish. Soc. 127(6): 1006–1020.

Armstrong, M. L. and A. V. Brown. 1983. Diel drift and feeding of Channel Catfish alevins in the Illinois River, Arkansas. Trans. Am. Fish. Soc. 112(2b): 302–307.

Artyukhin, E. N. 1995. On biogeography and relationships within the genus Acipenser. Sturgeon Q. 3(2): 6–8.

Ashley, K. W. and R. T. Rachels. 1999. Food habits of Bowfin in the Black and Lumber rivers, North Carolina. Proc. Annu. Conf. Southeast. Assoc. Fish Wildl. Agencies 53: 50–60.

Atmar, G. L. and K. W. Stewart. 1972. Food, feeding selectivity and ecological efficiencies of Fundulus notatus. Am. Midl. Nat. 88(1): 76–89.

Auer, N. A. 1996. Importance of habitat and migration to sturgeons with emphasis on Lake Sturgeon. Can. J. Fish. Aquat. Sci. 53(S1): 152–160.

Auer, N. A. 2004. Conservation. Pages 252–276. In: G. T. O. LeBreton, F. W. H. Beamish, R. S. McKinley, eds. Sturgeons and Paddlefish of North America. Kluwer Academic. 323 pp.

Aumiller, S. R. and D. B. Noltie. 2003. Chemoreceptive responses of the Southern Cavefish Typhlichthys subterraneus Girard, 1860 (Pisces, Amblyopsidae) to conspecifics and prey. Subterr. Biol. 1: 79–92.

Aureli, T. J. and L. Torrans. 1988. Spawning frequency and fecundity of Blue Tilapia. Proc. Arkansas Acad. Sci. 42: 108.

Avise, J. C. 1974. Systematic value of electrophoretic data. Syst. Biol. 23(4): 465–481.

Avise, J. C. 2004. Molecular markers, natural history, and evolution. 2nd ed. Sinauer, Sunderland, MA. 684 pp.

Avise, J. C. and M. H. Smith. 1974a Biochemical genetics of sunfish. I. Geographic variation and subspecific intergradation in the Bluegill, Lepomis macrochirus. Evolution 28(1): 42–56.

Avise, J. C. and M. H. Smith. 1974b. Biochemical genetics of sunfish. II. Genic similarity between hybridizing species. Am. Nat. 108(962): 458–472.

Avise, J. C. and M. J. Van Den Avyle. 1984. Genetic analysis of reproduction of hybrid White Bass × Striped Bass in the Savannah River. Trans. Am. Fish. Soc. 113(5): 563–570.

Avise, J. C., G. S. Helfman, N. C. Saunders, and L. S. Hales. 1986. Mitochondrial DNA differentiation in North Atlantic eels: population genetic consequences of an unusual life history pattern. Proc. Natl. Acad. Sci. USA. 83(12): 4350–4354.

Avise, J. C., J. Arnold, R. M. Ball, E. Bermingham, T. Lamb, J. E. Neigel, C. A. Reeb, and N. C. Saunders. 1987. Intraspecific phylogeography: the mitochondrial DNA bridge between population genetics and systematics. Annu. Rev. Ecol. Syst. 18: 489–522.

Axon, J. R. and L. E. Kornman. 1986. Characteristics of native Muskellunge streams in eastern Kentucky. Pages 263–272. In: G. E. Hall, ed. Managing muskies: a treatise on the biology and propagation of Muskellunge in North America. American Fisheries Society, Special Publ. 15. 272 pp.

Ayache, N. C. and T. J. Near. 2009. The utility of morphological data in resolving phylogenetic relationships of darters as exemplified with Etheostoma (Teleostei: Percidae). Bull. Peabody Mus. Nat. Hist. 50(2): 327–346.

Baby, M.-C., L. Bernatchez, and J. J. Dodson. 1991. Genetic structure and relationships among anadromous and landlocked populations of Rainbow Smelt, *Osmerus mordax*, Mitchill, as revealed by mtDNA restriction analysis. J. Fish Biol. 39(SA): 61–68.

Bachman, R. A. 1984. Foraging behavior of free-ranging wild and hatchery Brown Trout in a stream. Trans. Am. Fish. Soc. 113(1): 1–32.

Bachman, R. A. 1991. Brown Trout (*Salmo trutta*). Pages 208–229. *In*: J. Stolz and J. Schnell, eds. Trout. Stackpole Books, Harrisburg, PA.

Bacon, E. J. 1968. Age and growth of the Longear Sunfish in Northwest Arkansas. M.S. thesis. Univ. Arkansas, Fayetteville.

Bacon, E. J. and R. V. Kilambi. 1968. Some aspects of the age and growth of the Longear Sunfish in Arkansas waters. Proc. Arkansas Acad. Sci. 22: 44–56.

Bacon, E. J., W. H. Neill, Jr., and R. V. Kilambi. 1968a. Temperature selection and heat resistance of the Mosquitofish, *Gambusia affinis*. Proc. 22nd Annu. Conf. Southeast. Assoc. Game Fish Comm.: 411–416.

Bacon, E. J., Jr., S. H. Newton, R. V. Kilambi, and C. E. Hoffman. 1968b. Changes in the ichthyofauna of the Beaver Reservoir tailwaters. Proc. 22nd Annu. Conf. Southeast. Assoc. Game Fish Comm.: 245–248.

Baglin, R. E., Jr. and R. V. Kilambi. 1968. Maturity and spawning periodicity of the Gizzard Shad, *Dorosoma cepedianum* (Lesueur), in Beaver Reservoir. Proc. Arkansas Acad. Sci. 22: 38–43.

Bailey, R. M. 1938. A systematic revision of the centrarchid fishes, with a discussion of their distribution, variations, and probable interrelationships. Ph.D. diss. Univ. Michigan, Ann Arbor. 512 pp.

Bailey, R. M. 1941. *Hadropterus nasutus*, a new darter from Arkansas. Occas. Pap. Mus. Zool. Univ. Michigan, 440: 1–8.

Bailey, R. M. 1951. A check-list of the fishes of Iowa, with keys for identification. Pages 185–238. *In*: J. R. Harlan and E. B. Speaker, eds. Iowa fish and fishing. 3rd ed. Iowa State Conserv. Comm., Des Moines.

Bailey, R. M. 1956. A revised list of the fishes of Iowa: with keys for identification. Pages 325–377. *In*: J. R. Harlan and E. B. Speaker, eds. Iowa fish and fishing. Iowa State Conserv. Comm., Des Moines.

Bailey, R. M. 1959. *Etheostoma acuticeps*, a new darter from the Tennessee River system, with remarks on the subgenus *Nothonotus*. Occas. Pap. Mus. Zool. Univ. Michigan 603. 10 pp.

Bailey, R. M. 1980. Comments on the classification and nomenclature of lampreys—an alternative view. Can. J. Fish. Aquat. Sci. 37(11): 1626–1629.

Bailey, R. M. and M. O. Allum. 1962. Fishes of South Dakota. Misc. Publ. Mus. Zool. Univ. Michigan 119: 1–131.

Bailey, R. M. and F. B. Cross. 1954. River sturgeons of the American genus *Scaphirhynchus*: characters, distribution, and synonymy. Papers Michigan Acad. Sci. Arts Let. 39: 169–208.

Bailey, R. M. and D. A. Etnier. 1988. Comments on the subgenera of darters (Percidae) with descriptions of two new species of *Etheostoma (Ulocentra)* from southeastern United States. Misc. Publ. Mus. Zool. Univ. Michigan 175. 48 pp.

Bailey, R. M. and W. A. Gosline. 1955. Variation and systematic significance of vertebral counts in the American fishes of the family Percidae. Misc. Publ. Mus. Zool. Univ. Michigan 93: 1–44.

Bailey, R. M. and H. M. Harrison, Jr. 1948. Food habits of the Southern Channel Catfish (*Ictalurus lacustris punctatus*) in the Des Moines River, Iowa. Trans. Am. Fish. Soc. 75(1): 110–138.

Bailey, R. M. and C. L. Hubbs. 1949. The black basses (*Micropterus*) of Florida, with description of a new species. Occas. Pap. Mus. Zool. Univ. Michigan 516. 40 pp.

Bailey, R. M. and H. W. Robison. 1978. *Notropis hubbsi*, a new cyprinid fish from the Mississippi River basin. Occas. Pap. Mus. Zool. Univ. Michigan 683. 21 pp.

Bailey, R. M., W. C. Latta, and G. R. Smith. 2004. An atlas of Michigan fishes with keys and illustrations for their identification. Misc. Publ. Mus. Zool. Univ. Michigan 192. 215 pp.

Bailey, R. M., H. E. Winn, and C. L. Smith. 1954. Fishes from the Escambia River, Alabama and Florida, with ecologic and taxonomic notes. Proc. Acad. Nat. Sci. Phila. 106: 109–164.

Bailey, W. M. and R. L. Boyd. 1970. A preliminary report on spawning and rearing of Grass Carp (*Ctenopharyngodon idella*) in Arkansas. Proc. 24th Annu. Conf. Southeast. Assoc. Game Fish Comm.: 560–569.

Bain, M. B., D. H. Webb, M. D. Tangedal, and L. N. Mangum. 1990. Movements and habitat use by Grass Carp in a large mainstream reservoir. Trans. Am. Fish. Soc. 119(3): 553–561.

Baird, S. F. and C. Girard. 1853a. Descriptions of new species of fishes collected by Mr. John H. Clark, on the U.S. and Mexican Boundary Survey, under Lt. Col. Jas. D. Graham. Proc. Acad. Nat. Sci. Phila. 6: 387–390.

Baird, S. F. and C. Girard. 1853b. Description of new species of fishes, collected by Captains R. B. Marcy, and Geo. B. McClellan, in Arkansas. Proc. Acad. Nat. Sci. Phila. 6: 390–392.

Baker, B. 1970. Profits from the river. Arkansas Game and Fish 3(2): 10–11.

Baker, C. D. 1972. The cephalic lateral line systems of amblyopsid blind cavefishes and other percopsiform fishes (Pisces: Percopsiformes). Ph.D. diss. Univ. Louisville, KY. 113 pp.

Baker, C. D. and E. H. Schmitz. 1971. Food habits of adult Gizzard Shad and Threadfin Shad in two Ozark reservoirs. Pages 3–11. *In*: G. E. Hall, ed. Reservoir fisheries and limnology. American Fisheries Society, Special Publ. 8.

Baker, C. D., D. W. Martin, and E. H. Schmitz. 1971. Separation of taxonomically identifiable organisms and detritus taken from shad foregut contents using density-gradient centrifugation. Trans. Am. Fish. Soc. 100(1): 138–139.

Baker, J. A. and S. A. Foster. 1994. Observations on a foraging association between two freshwater stream fishes. Ecol. Freshw. Fish 3(3): 137–139.

Baker, J. A. and D. C. Heins. 1994. Reproductive life history of the North American Madtom Catfish, *Noturus hildebrandi* (Bailey and Taylor 1950), with a review of data for the genus. Ecol. Freshw. Fish 3(4): 167–175.

Baker, J. A. and S. T. Ross. 1981. Spatial and temporal resource utilization by southeastern cyprinids. Copeia 1981(1): 178–189.

Baker, J. A., K. J. Killgore, and S. A. Foster. 1994. Population variation in spawning current speed selection in the Blacktail Shiner, *Cyprinella venusta* (Pisces: Cyprinidae). Environ. Biol. Fishes 39(4): 357–364.

Baker, J. A., K. J. Killgore, and R. L. Kasul. 1991. Aquatic habitats and fish communities in the lower Mississippi River. Rev. Aquat. Sci. 3(4): 313–356.

Baker, R. F. 1959. Historical review of the Bull Shoals Dam and Norfork Dam tailwater trout fishery. Proc. 13th Annu. Conf. Southeast. Assoc. Fish Wildl. Agencies: 229–236.

Baker, R. F. and W. P. Mathis. 1967. A survey of Bull Shoals Lake, Arkansas, for the possibility of an existing two-story lake situation. Proc. 21st Annu. Conf. Southeast. Assoc. Game Fish Comm.: 360–368.

Baker, S. C. and M. L. Armstrong. 1987. Recent collections of fishes from the Spring River drainage in northeast Arkansas. Proc. Arkansas Acad. Sci. 41: 96.

Baker, S. S. and R. C. Heidinger. 1994. Individual and relative fecundity of Black Crappie (*Pomoxis nigromaculatus*) in Baldwin Cooling Pond. Trans. Illinois State Acad. Sci. 87(3–4): 145–150.

Baker, W. P., J. Boxrucker, and K. E. Kuklinski. 2009. Determination of Striped Bass spawning locations in the two major tributaries of Lake Texoma. N. Am. J. Fish. Manag. 29(4): 1006–1014.

Baldwin, G. L. 1983. A taxonomic study of the fishes of the Little Red River, northcentral Arkansas. M.S. thesis, Northeast Louisiana Univ., Monroe. 118 pp.

Ball, R. L. and R. V. Kilambi. 1972. The feeding ecology of the Black and White crappies in Beaver Reservoir, Arkansas and its effect on the relative abundance of the crappie species. Proc. 26th Annu. Conf. Southeast. Assoc. Game Fish Comm.: 577–580.

Balon, E. K. 1975. Reproductive guilds of fishes: a proposal and definition. J. Fish. Res. Board Can. 32(6): 821–864.

Balon, E. K. 1981. Additions and amendments to the classification of reproductive styles in fishes. Environ. Biol. Fishes 6(3–4): 377–389.

Balon, E. K. 1995. Origin and domestication of the wild carp, *Cyprinus carpio*: from Roman gourmets to the swimming flowers. Aquaculture 129(1–4): 3–48.

Balon, E. K., M. N. Bruton, and D. L. G. Noakes, eds. 1994. Women in ichthyology: an anthology in honor of ET, Ro, and Genie. Kluwer Academic. Dordrecht. 456 pp.

Bănărescu, P. 1964. Fauna Republicii Populare Romine.

Pisces-Osteichthyes. Vol. 13. Editura Academiei Republicii Populare Romine, Bucharest, Romania. 963 pp.

Banks, S. M. and R. J. DiStefano. 2002. Diurnal habitat associations of the madtoms, *Noturus albater, N. exilis, N. flavater,* and *N. flavus* in Missouri Ozarks streams. Am. Midl. Nat. 148(1): 138–145.

Barber, W. E. and W. L. Minckley. 1971. Summer foods of the cyprinid fish *Semotilus atromaculatus*. Trans. Am. Fish. Soc. 100 (2): 283–289.

Barbin, G. P. and J. D. McCleave. 1997. Fecundity of the American Eel *Anguilla rostrata* at 45°N in Maine, U.S.A. J. Fish Biol. 51(4): 840–847.

Bardach, J. E., J. H. Ryther, and W. O. McLarney. 1972. Aquaculture: the farming and husbandry of freshwater and marine organisms. Wiley-Interscience, New York. 884 pp.

Bardack, D. and R. Zangerl. 1968. First fossil lamprey: a record from the Pennsylvanian of Illinois. Science 162(3859): 1265–1267.

Bardack, D. and R. Zangerl. 1971. Lampreys in the fossil record. Pages 67–84. *In:* M. W. Hardisty and I. C. Potter, eds. The biology of lampreys. Vol. I. Academic Press, New York. 305 pp.

Bare, C. M. 2005. Movement, habitat use, and survival of Smallmouth Bass in the Buffalo National River drainage of Arkansas. M.S. thesis. Univ. Arkansas, Fayetteville. 76 pp.

Barefield, R. L. and C. D. Ziebell. 1986. Comparative feeding ability of small White and Black crappie in turbid water. Iowa State J. Res. 61(1): 143–146.

Barila, T. Y. 1980a. *Stizostedion canadense*, Sauger. Page 745. *In:* D. S. Lee et al. Atlas of North American freshwater fishes. N.C. State Mus. Nat. Hist., Raleigh.

Barila, T. Y. 1980b. *Stizostedion vitreum*, Walleye. Pages 747–748. *In:* D. S. Lee et al. Atlas of North American freshwater fishes. N.C. State Mus. Nat. Hist., Raleigh.

Barkuloo, J. M. 1970. Taxonomic status and reproduction of Striped Bass (*Morone saxatilis*) in Florida. Tech. Paper 44, Bureau of Sport Fisheries and Wildlife, Washington, DC. 16 pp.

Barlow, J. A. 1980. Geographic variation in *Lepomis megalotis* (Rafinesque) (Osteichthyes: Centrarchidae). Ph.D. diss., Texas A&M Univ., College Station. 250 pp.

Barney, R. L. and B. J. Anson. 1920. Life history and ecology of the Pygmy Sunfish, *Elassoma zonatum*. Ecology 1(4): 241–256.

Barney, R. L. and B. J. Anson. 1921. Seasonal abundance of the mosquito destroying top-minnow, *Gambusia affinis*, especially in relation to male frequency. Ecology 2(1): 53–69.

Barney, R. L. and B. J. Anson. 1922. Life history and ecology of the Orange-spotted Sunfish, *Lepomis humilis*. Append. XV, Rep. U.S. Comm. Fish. 1922, Bur. Fish. Doc. 938. 16 pp.

Barnickol, P. G. 1941. Food habits of *Gambusia affinis* from Reelfoot Lake, Tennessee, with special reference to malaria control. Rep. Reelfoot Lake Biol. Sta. 5: 5–13.

Baroiller, J., D. Chourrout, A. Foster, and B. Jalabert. 1995.

Temperature and sex chromosomes govern sex ratios of the mouthbrooding cichlid fish *Oreochromis niloticus*. J. Exp. Zool. 273(3): 216–223.

Barr, T. C., Jr. and J. R. Holsinger, Jr. 1985. Speciation in cave faunas. Annu. Rev. Ecol. Syst. 16: 313–337.

Barron, J. N. and H. T. Albin. 2006. Multi-year reproduction in *Etheostoma blennioides* in south-central Ohio. Am. Midl. Nat. 155(1): 103–112.

Bart, H. L., Jr. 1989. Fish habitat association in an Ozark stream. Environ. Biol. Fishes 24(3): 173–186.

Bart, H. L., Jr. and R. C. Cashner. 1986. Geographic variation in Gulf Slope populations of the Bluntnose Darter, *Etheostoma chlorosomum* (Hay). Tulane Stud. Zool. Bot. 25(2): 151–170.

Bart, H. L., Jr. and L. M. Page. 1991. Morphology and adaptive significance of fin knobs in egg-clustering darters. Copeia 1991(1): 80–86.

Bart, H. L., Jr. and M. S. Taylor. 1999. Systematic review of subgenus *Fuscatelum* of *Etheostoma* with description of a new species from the upper Black Warrior River system, Alabama. Tulane Stud. Zool. Bot. 31: 23–50.

Bart, H. L., Jr., M. D. Clements, R. E. Blanton, K. R. Piller, and D. L. Hurley. 2010. Discordant molecular and morphological evolution in buffalofishes (Actinopterygii: Catostomidae). Mol. Phylogenet. Evol. 56(2): 808–820.

Bartley, D. M., K. Rana, and A. J. Immink. 2000. The use of inter-specific hybrids in aquaculture and fisheries. Rev. Fish Biol. Fish. 10(3): 325–337.

Barton, B. A., ed. 2011. Biology, management, and culture of Walleye and Sauger. American Fisheries Society. Bethesda, MD. 570 pp.

Barton, M. 2007. Bond's Biology of fishes. 3rd ed. Thomson Brooks/Cole. Belmont, CA. 891 pp.

Barwick, D. H., J. W. Foltz, and D. M. Rankin. 2004. Summer habitat use by Rainbow Trout and Brown Trout in Jocassee Reservoir. N. Am. J. Fish. Manag. 24(2): 735–740.

Battle, H. I. and W. M. Sprules. 1960. A description of the semi-buoyant eggs and early developmental stages of the Goldeye, *Hiodon alosoides* (Rafinesque). J. Fish. Res. Board Can. 17(2): 245–266.

Bauer, B. H. 1980. *Lepomis megalotis* (Rafinesque), Longear Sunfish. Page 600. *In*: D. S. Lee et al. Atlas of North American freshwater fishes. N.C. State Mus. Nat. Hist., Raleigh.

Bayless, J. D. 1968. Striped Bass hatching and hybridization experiments. Proc. Annu. Conf. Southeast. Assoc. Game Fish Comm. 21: 233–244.

Beach, M. L. 1974. Food habits and reproduction of the Taillight Shiner, *Notropis maculatus* (Hay), in central Florida. Fla. Sci. 37(1): 5–16.

Beachum, C. E., M. J. Michel, and J. H. Knouft. 2016. Differential responses of body shape to local and reach scale stream flow in two freshwater fish species. Ecol. Freshw. Fish 25(3): 446–454.

Beadles, J. K. 1974. The Spotfin Shiner in northeastern Arkansas. Southwest. Nat. 19(2): 219–220.

Beadles, J. K. 1979. The Goldeye in the Black River. Proc. Arkansas Acad. Sci. 33: 75.

Beal, C. D. 1967. Life history information on the Blue Sucker, *Cycleptus elongatus* (LeSueur) in the Missouri River. M.S. thesis. Univ. South Dakota, Brookings. 36 pp.

Beamish, F. W. H. and E. J. Thomas. 1983. Potential and actual fecundity of the "paired" lampreys, *Ichthyomyzon gagei* and *I. castaneus*. Copeia 1983(2): 367–374.

Beamish, F. W. H. and E. J. Thomas. 1984. Metamorphosis of the Southern Brook Lamprey, *Ichthyomyzon gagei*. Copeia 1984(2): 502–515.

Beamish, R. J. 1973. Determination of age and growth of populations of the White Sucker (*Catostomus commersoni*) exhibiting a wide range in size at maturity. J. Fish. Res. Board Can. 30(5): 607–616.

Bean, B. A. and E. D. Reid. 1930. On a new species of Brook Silverside, *Labidesthes vanhyningi*, from Florida. Proc. Biol. Soc. Washington 43: 193–194.

Bean, P. T., C. S. Williams, P. H. Diaz, and T. H. Bonner. 2010. Habitat associations, life history, and diet of the Blackspot Shiner, *Notropis atrocaudalis*. Southeast. Nat. 9(4): 673–686.

Becker, G. C. 1983. Fishes of Wisconsin. Univ. Wisconsin Press, Madison. 1052 pp.

Beckett, D. C. and C. H. Pennington. 1986. Water quality, macroinvertebrates, larval fishes, and fishes of the lower Mississippi River: a synthesis. Tech. Rep. E-86-12, U.S. Army Engineers, Waterways Experiment Station, Vicksburg, Miss. 139 pp.

Beckham, E. C. 1980. *Percina gymnocephala*, a new percid fish of the subgenus *Alvordius*, from the New River in North Carolina, Virginia, and West Virginia. Occas. Pap. Mus. Zool. Louisiana State Univ. 57: 1–11.

Beckham, E. C. 1986. Systematics and redescription of the Blackside Darter, *Percina maculata* (Girard), (Pisces: Percidae). Occas. Pap. Mus. Zool. Louisiana State Univ. 62: 1–11.

Beckman, D. W. and C. A. Hutson. 2012. Validation of aging techniques and growth of the River Redhorse, *Moxostoma carinatum*, in the James River, Missouri. Southwest. Nat. 57(3): 240–247.

Beecher, H. A. 1979. Comparative functional morphology and ecological isolating mechanisms in sympatric fishes of the genus *Carpiodes* in northwestern Florida. Ph.D. diss. Florida State Univ., Tallahassee. 416 pp.

Beecher, H. A. 1980. Habitat segregation of Florida carpsuckers (Osteichthyes: Catostomidae: *Carpiodes*). Fla. Sci. 43(2): 92–97.

Begle, D. P. 1991. Relationships of the osmeroid fishes and the use of reductive characters in phylogenetic analysis. Syst. Zool. 40(1): 33–53.

Behmer, D. J. 1965. Spawning periodicity of the River

Carpsucker, *Carpiodes carpio*. Proc. Iowa Acad. Sci. 72: 253–262.

Behmer, D. J. 1969. A method of estimating fecundity; with data on River Carpsuckers, *Carpiodes carpio*. Trans. Am. Fish. Soc. 98(3): 523–524.

Behnke, R. J. 1979. Monograph of the native trouts of the genus *Salmo* of western North America. U.S. Forest Service, Lakewood, CO. 163 pp.

Behnke, R. J. 1990a. Still a rainbow—by any other name. Trout (Winter 1990): 41–45.

Behnke, R. J. 1990b. About trout: how many species? Trout 31(3): 66–69.

Behnke, R. J. 2002. Trout and salmon of North America. The Free Press, Simon and Schuster, New York. 360 pp.

Beitinger, T. L., W. A. Bennett, and R.W. McCauley. 2000. Temperature tolerances of North American freshwater fishes exposed to dynamic changes in temperature. Environ. Biol. Fishes 58(3): 237–275.

Belica, L. 2007. Brown Trout (*Salmo trutta*): a technical conservation assessment. USDA Forest Service, Rocky Mountain Region. 118 pp.

Bemis, W. E. and L. Grande. 1992. Early development of the actinopterygian head. I. External development and staging of the Paddlefish *Polyodon spathula*. J. Morph. 213(1): 47–83.

Bemis, W. E. and B. Kynard. 1997. Sturgeon rivers: an introduction to acipenseriform biogeography and life history. Environ. Biol. Fishes 48(1–4): 167–183.

Bemis, W. E., E. K. Findeis, and L. Grande. 1997. An overview of Acipenseriformes. Environ. Biol. Fishes 48(1–4): 25–71.

Bennett, C., J. Giese, B. Keith, R. McDaniel, M. Maner, N. O'Shaughnessy, and B. Singleton. 1987. Physical, chemical and biological characteristics of least-disturbed reference streams in Arkansas' ecoregions. Arkansas Dept. Pollution and Control and Ecology, Little Rock.

Bennetts, R. Q., J. M. Grady, F. C. Rohde, and J. M. Quattro. 1999. Discordant patterns of morphological and molecular change in broadtail madtoms (genus *Noturus*). Mol. Ecol. 8(10): 1563–1569.

Ben-Tuvia, A. 1959. The biology of the cichlid fishes of Lake Tiberias and Huleh. Bull. Res. Counc. Israel. B. Zool. 8B(4): 153–188 (reprinted as Bull. Sea. Fish. Res. Sta. Israel No. 27, 1960).

Berendzen, P. B., J. F. Dugan, and T. Gamble. 2010. Postglacial expansion into the Paleozoic Plateau: evidence of an Ozarkian refugium for the Ozark Minnow *Notropis nubilus* (Teleostei: Cypriniformes). J. Fish Biol. 77(5): 1114–1136.

Berendzen, P. B., T. Gamble, and A. M. Simons. 2008a. Phylogeography of the Bigeye Chub *Hybopsis amblops* (Teleostei: Cypriniformes): Early Pleistocene diversification and post-glacial range expansion. J. Fish Biol. 73(8): 2021–2039.

Berendzen, P. B., A. M. Simons, R. M. Wood, T. E, Dowling, and C. L. Secor. 2008b. Recovering cryptic diversity and ancient drainage patterns in eastern North America:

historical biogeography of the *Notropis rubellus* species group (Teleostei: Cypriniformes). Mol. Phylogenet. Evol. 46(2): 721–737.

Berendzen, P. B., W. M. Olson, and S. M. Barron. 2009. The utility of molecular hypotheses for uncovering morphological diversity in the *Notropis rubellus* species complex (Cypriniformes: Cyprinidae). Copeia 2009(4): 661–673.

Berendzen, P. B., A. M. Simons, and R. M. Wood. 2003. Phylogeography of the Northern Hogsucker, *Hypentelium nigricans* (Teleostei: Cypriniformes): genetic evidence for the existence of the ancient Teays River. J. Biogeogr. 30(8): 1139–1152.

Berg, L. S. 1947. Classification of fishes both recent and fossil. Travaux Inst. Zool. de Acad. Sci. USSR. Engl. Transl., 1947, Edwards Press, Ann Arbor, MI. 517 pp.

Berg, L. S. 1964. Freshwater fishes of the USSR and adjacent countries. 4th ed. Vol. 2. Israel Program for Scientific Translations, Jerusalem.

Berg, W. J. and S. D. Ferris. 1984. Restriction endonuclease analysis of salmonid mitochondrial DNA. Can. J. Fish. Aquat. Sci. 41(7): 1041–1047.

Bergstedt, R. A., R. L. Argyle, J. G. Seelye, K. T. Scribner, and G. L. Curtis. 2003. In situ determination of the annual thermal habitat use by Lake Trout (*Salvelinus namaycush*) in Lake Huron. J. Great Lakes Res. 29(S1): 347–361.

Bergstrom, D. E. 1997. The phylogeny and historical biogeography of Missouri's *Amblyopsis rosae* (Ozark Cavefish) and *Typhlichthys subterraneus* (Southern Cavefish). M.S. thesis. Univ. Missouri, Columbia. 63 pp.

Bergstrom, D. E., D. B. Noltie, and T. P. Holtsford. 1997. Molecular phylogenetics and historical biogeography of the family Amblyopsidae. Pages 4–5. *In*: I. D. Sasowsky, D. W. Fong, and E. L. White, eds. Conservation and protection of the biota of karst. Karst Waters Institute Special Publ. 3. Charles Town, W.V. 118 pp.

Berkhouse, C. S. 1990. Laboratory spawning of the threatened Leopard Darter (*Percina pantherina*). M.S. thesis. Southwest Texas State Univ., San Marcos.

Berlinsky, D. L., L. F. Jackson, T. I. J. Smith, and C. V. Sullivan. 1995. The annual reproductive cycle of the White Bass, *Morone chrysops*. J. World Aquac. Soc. 26(3): 252–260.

Bermingham, E. and J. C. Avise. 1986. Molecular zoogeography of freshwater fishes in the southeastern United States. Genetics 113(4): 939–965.

Bernatchez, L. 2001. The evolutionary history of Brown Trout (*Salmo trutta* L.) inferred from phylogeographic, nested clade, and mismatch analyses of mitochondrial DNA variation. Evolution 55(2): 351–379.

Berner, L. M. 1948. The intestinal convolutions: new generic characters for the separation of *Carpiodes* and *Ictiobus*. Copeia 1948(2): 140–141.

Berra, T. M. 1981. An atlas of distribution of the freshwater fish families of the world. Univ. Nebraska Press, Lincoln. 197 pp.

Berra, T. M. 2001. Freshwater fish distribution. 2nd ed. Academic Press, San Diego, CA. 604 pp.

Berra, T.M. 2007. Freshwater fish distribution. Univ. Chicago Press, Chicago. 615 pp.

Berry, F. H. 1964. Review and emendation of: Family Clupeidae by Samuel F. Hildebrand. Pages 257–454. *In*: Fishes of the western North Atlantic. Copeia 1964(4): 720–730.

Berry, P. Y. and M. P. Low. 1970. Comparative studies on some aspects of the morphology and histology of *Ctenopharyngodon idellus*, *Aristichthys nobilis* and their hybrid (Cyprinidae). Copeia 1970(4): 708–726.

Bessert, M. L. 2006. Molecular systematics and population structure in the North American endemic fish genus *Cycleptus* (Teleostei: Catostomidae). Ph.D. diss. Univ. Nebraska, Lincoln. 219 pp.

Bestgen, K. R. and R. I. Compton. 2007. Reproduction and culture of Suckermouth Minnow. N. Am. J. Aquac. 69(4): 345–350.

Bestgen, K. R., S. P. Platania, J. E. Brooks, and D. L. Propst. 1989. Dispersal and life history traits of *Notropis girardi* (Cypriniformes: Cyprinidae), introduced into the Pecos River, New Mexico. Am. Midl. Nat. 122(2): 228–235.

Betancur-R., R., R. E. Broughton, E. O. Wiley, K. Carpenter, J. A. Lopez, C. Li, N. I. Holcroft, D. Arcila, M. Sanciangco, J. C. Cureton II, F. Zhang, T. Buser, M. A. Campbell, J. A. Ballesteros, A. Roa-Varon, S. Willis, W. C. Bordon, T. Rowley, P. C. Reneau, D. J. Hough, G. Lu, T. Grande, G. Arratia, and G. Orti. 2013. The tree of life and a new classification of bony fishes. PLOS Currents Tree of Life. 18 April 2013. Edition 1. Available:. https://doi.org/10.1371/currents.tol.53ba26640df0ccaee75bb165c8c26288.

Betancur-R., R., E. O. Wiley, G. Arratia, A. Acero, N. Bailly, M. Miya, G. Lecointre, and G. Orti. 2017. Phylogenetic classification of bony fishes. BMC Evol. Biol. 17(1): 162.

Bethke, B. J., J. A. Rice, and D. D. Aday. 2014. White Perch in small North Carolina reservoirs: what explains variation in population structure? Trans. Am. Fish. Soc. 143(1): 77–84.

Bettoli, P. W., J. E. Morris, and R. L. Noble. 1991. Changes in the abundance of two atherinid species after aquatic vegetation removal. Trans. Am. Fish. Soc. 120(1): 90–97.

Bettoli, P. W., J. A. Kerns, and G. D. Scholten. 2009. Status of Paddlefish in the United States. Pages 23–28. *In*: C. P. Paukert and G. D. Scholten, eds. Paddlefish management, propagation, and conservation in the 21st century: building from 20 years of research and management. Symp. 66. American Fisheries Society. Bethesda, MD. 443 pp.

Bettoli, P. W., M. J. Maceina, R. L. Noble, and R. K. Betsill. 1992. Piscivory in Largemouth Bass as a function of aquatic vegetation abundance. N. Am. J. Fish. Manag. 12(3): 509–516.

Bettoli, P. W., M. J. Maceina, R. L. Noble, and R. K. Betsill. 1993. Response of a reservoir fish community to aquatic vegetation removal. N. Am. J. Fish. Manag. 13(1): 110–124.

Beugly, J. and M. Pyron. 2010. Variation in fish and macroinvertebrate assemblages among seasonal and perennial headwater streams. Am. Midl. Nat. 163(1): 2–13.

Bevelander, G. 1934. The gills of *Amia calva* specialized for respiration in an oxygen deficient habitat. Copeia 1934(3): 123–127.

Bickford, D., D. J. Lohman, N. S. Sodhi, P. K. L. Ng, R. Meier, K. Winker, K. K. Ingram, and I. Das. 2007. Cryptic species as a window on diversity and conservation. Trends Ecol. Evol. 22(3): 148–155.

Bielawski, J. P. and J. R. Gold. 2001. Phylogenetic relationships of cyprinid fishes in subgenus *Notropis* inferred from nucleotide sequences of the mitochondrially encoded cytochrome *b* gene. Copeia 2001(3): 656–667.

Bielawski, J. P., A. Brault, and J. R. Gold. 2002. Phylogenetic relationships within the genus *Pimephales* as inferred from ND4 and ND4L nucleotide sequences. J. Fish Biol. 61(1): 293–297.

Bigelow, H. B. and W. C. Schroeder. 1963. Family Osmeridae. Pages 553–597. *In*: H. B. Bigelow, ed. Fishes of the Western North Atlantic. Mem. Sears Found. Mar. Res. No. 1, Part 3.

Billington, N., R. G. Danzmann, P. D. N. Hebert, and R. D. Ward. 1991. Phylogenetic relationships among four members of *Stizostedion* (Percidae) determined by mitochondrial DNA and allozyme analyses. J. Fish Biol. 39(sA): 251–258.

Billington, N., P. D. N. Hebert, and R. D. Ward. 1988. Evidence of introgressive hybridization in the genus *Stizostedion*: interspecific transfer of mitochondrial DNA between Sauger and Walleye. Can J. Fish. Aquat. Sci. 45(11): 2035–2041.

Birdsong, R. S. and L. W. Knapp. 1969. *Etheostoma collettei*, a new darter of the subgenus *Oligocephalus* from Louisiana and Arkansas. Tulane Stud. Zool. Bot. 15(3): 106–112.

Birkhead, W. S. 1967. The comparative toxicity of stings of the ictalurid catfish genera *Ictalurus* and *Schilbeodes*. Comp. Biochem. Physiol. 22(1): 101–111.

Birkhead, W. S. 1972. Toxicity of stings of ariid and ictalurid catfishes. Copeia 1972(4): 790–807.

Biro, P. 1999. *Mylopharyngodon piceus* (Richardson, 1846). Pages 347–365. *In*: P. M. Bănărescu, ed. The freshwater fishes of Europe, v. 5/I, Cyprinidae 2, Part I: *Rhodeus* to *Capoeta*. Wiebelshiem, Germany, AULU-Verlag. 426 pp.

Birstein, V. J. and W. E. Bemis. 1997. How many species are there within the genus *Acipenser*? Environ. Biol. Fishes 48(1–4): 157–163.

Birstein, V. J. and R. DeSalle. 1998. Molecular phylogeny of Acipenserinae. Mol. Phylogenet. Evol. 9(1): 141–155.

Birstein, V. J., P. Doukakis, and R. DeSalle. 2002. Molecular phylogeny of Acipenseridae: nonmonophyly of Scaphirhynchinae. Copeia 2002(2): 287–301.

Bisazza, A., G. Vaccari, and A. Pilastro. 2001. Female mate choice in a mating system dominated by male sexual coercion. Behav. Ecol. 12(1): 59–64.

Bishop, F. G. 1975. Observations on the fish fauna of the Peace River in Alberta. Can. Field Nat. 89(4): 423–430.

Black, A. and W. M. Howell. 1978. A distinctive chromosomal race of the cyprinodontid fish, *Fundulus notatus*, from the upper Tombigbee River system of Alabama and Mississippi. Copeia 1978(2): 280–288.

Black, D. A. and W. M. Howell. 1979. The North American Mosquitofish, *Gambusia affinis*: a unique case in sex chromosome evolution. Copeia 1979(3): 509–513.

Black, J. D. 1940. The distribution of the fishes of Arkansas. Ph.D. diss., Univ. Michigan, Ann Arbor. 243 pp.

Black, J. D. 1945. Natural history of the northern Mimic Shiner, *Notropis volucellus volucellus* Cope. Investig. Indiana Lakes Streams 2(18): 449–469.

Black, J. D. 1949. Changing fish populations as an index to pollution and soil erosion. Trans. Illinois State Acad. Sci. 42: 145–148.

Black, J. D. and L. O. Williamson. 1946. Artificial hybrids between Muskellunge and Northern Pike. Trans. Wisconsin Acad. Sci. Arts Lett. 38: 299–314.

Blackwell, B. G., T. M. Kaufman, and W. H. Miller. 2009. Occurrence of Rudd (*Scardinius erythrophthalmus*) and dynamics of three populations in South Dakota. J. Freshwater Ecol. 24(2): 285–291.

Blair, A. P. and J. Windle. 1961. Darter associates of *Etheostoma cragini* (Percidae). Southwest. Nat. 6(3–4): 201–202.

Blair, W. F., A. P. Blair, P. Brodkorb, F. R. Cagle, and G. A. Moore. 1968. Vertebrates of the United States. 2nd ed. McGraw-Hill, New York. 616 pp.

Blanchard, T. A. 1996. Ovarian cycles and microhabitat use in two species of topminnow, *Fundulus olivaceus* and *F. euryzonus*, from the southeastern United States. Environ. Biol. Fishes 47(2): 155–163.

Blanton, R. E. 2007. Evolution of the Fantail Darter, *Etheostoma flabellare* (Percidae: Catonotus): systematics, phylogeography, and population history. Ph.D. diss. Tulane Univ. New Orleans. 294 pp.

Blanton, R. E., L. M. Page, and B. M. Ennis. 2011. Phylogenetic relationships of *Opsopoeodus emiliae*, with comments on the taxonomic implications of discordance among datasets. Copeia 2011(1): 82–92.

Blanton, R. E., L. M. Page, and S. A. Hilber. 2013. Timing of clade divergence and discordant estimates of genetic and morphological diversity in the Slender Madtom, *Noturus exilis* (Ictaluridae). Mol. Phylogenet. Evol. 66(3): 679–693.

Blaustein, L. and R. Karban 1990. Indirect effects of the Mosquitofish *Gambusia affinis* on the mosquito *Culex tarsalis*. Limnol. Oceanogr. 35(3): 767–771.

Bleeker, P. 1860. Conspectus systematis Cyprinorum. Natuurkundig Tijdschrift voor Nederlandsch Indië v. 20 (no. 3): 421–441. [Date of 1860 given by Eschmeyer et al. 2018]

Bloch, M. E. 1785. Naturgeschichte der ausländischen Fische. Berlin Vol. 1: 1–136, Pls. 109–144.

Bloom, D. D., P. J. Unmack, A. E. Gosztonyi, K. R. Piller, and N. R. Lovejoy. 2012. It's a family matter: molecular phylogenetics of Atheriniformes and the polyphyly of the surf silversides (family Notocheiridae). Mol. Phylogenet. Evol. 62(3): 1025–1030.

Bloom, D. D., K. R. Piller, J. Lyons, N. Mercado-Silva, and M. Medina-Nava. 2009. Systematics and biogeography of the silverside tribe Menidiini (Teleostomi: Atherinopsidae) based on the mitochondrial ND2 gene. Copeia 2009(2): 408–417.

Blum, M. J., D. A. Neely, P. M. Harris, and R. L. Mayden. 2008. Molecular systematics of the cyprinid genus *Campostoma* (Actinopterygii: Cypriniformes): disassociation between morphological and mitochondrial differentiation. Copeia 2008(2): 360–369.

Blumer, L. S. 1985a. Reproductive natural history of the Brown Bullhead *Ictalurus nebulosus* in Michigan. Am. Midl. Nat. 114(2): 318–330.

Blumer, L. S. 1985b. The significance of biparental care in the Brown Bullhead, *Ictalurus nebulosus*. Environ. Biol. Fishes 12(3): 231–236.

Bodola, A. 1966. Life history of the Gizzard Shad, *Dorosoma cepedianum* (Lesueur), in western Lake Erie. USFWS, Fish. Bull. 65(2): 391–425.

Bodznick, D. and D. G. Preston. 1983. Physiological characterization of electroreceptors in the lampreys *Ichthyomyzon unicuspis* and *Petromyzon marinus*. J. Comp. Physiol. 152(2): 209–217.

Boesel, M. W. 1938. The food of nine species of fish from the western end of Lake Erie. Trans. Am. Fish. Soc. 67(1): 215–223.

Bogue, M. B. 2000. Fishing the Great Lakes: an environmental history, 1783–1933. Univ. Wisconsin Press, Madison. 456 pp.

Böhlke, E. B. 1984. Catalog of type specimens in the ichthyological collection of the Academy of Natural Sciences of Philadelphia. Acad. Nat. Sci. Phila. Special Publ. 14. 246 pp.

Boileau, M. G. 1985. The expansion of White Perch, *Morone americana*, in the lower Great Lakes. Fisheries 10(1): 6–10.

Bolnick, D. I. 2009. Hybridization and speciation in centrarchids. Pages 39–69. *In*: S. J. Cooke and D. P. Philipp, eds. Centrarchid fishes: diversity, biology, and conservation. Wiley-Blackwell, Chichester.

Bolnick, D. I. and T. J. Near. 2005. Tempo of hybrid inviability in centrarchid fishes (Teleostei: Centrarchidae). Evolution 59(8): 1754–1767.

Boltz, J. M. and J. R. Stauffer, Jr. 1986. Branchial brooding in the Pirate Perch, *Aphredoderus sayanus* (Gilliams). Copeia 1986(4): 1030–1031.

Boltz, J. M., and J. R. Stauffer, Jr. 1993. Systematics of *Aphredoderus sayanus* (Teleostei: Aphredoderidae). Copeia 1993(1): 81–98.

Bonaparte, C. L. 1836. Iconografia della fauna italica per le quattro classi degli animali vertebrati. Tomo III. Pesci. Roma. Fasc. 15–18, puntata 80–93, 10 pls.

Bonaparte, C. L. 1837. Iconografia della fauna italica per le quattro classi degli animali vertebrati. Tomo III. Pesci. Roma. Fasc. 19–21, puntata 94–103, 105–109, 5 pls.

Bond, C. E. 1996. Biology of fishes. 2nd ed. Brooks/Cole, Thompson Learning, Toronto. 750 pp.

Bonham, K. 1941. Food of gars in Texas. Trans. Am. Fish. Soc. 70(1): 356–362.

Bonham, K. 1946. Management of a small fish pond in Texas. J. Wildl. Manag. 10(1): 1–4.

Bonnaterre, J. P. 1788. Tableau encyclopédique et methodique des trois règnes de la nature. Ichthyologie, pp. 1–215. Panckoucke, Paris.

Bonner, T. H. 2000. Life history and reproductive ecology of the Arkansas River Shiner and Peppered Chub in the Canadian River, Texas and New Mexico. Ph.D. diss. Texas Tech Univ. Lubbock.

Bonner, T. H. and G. R. Wilde. 2000. Changes in the Canadian River fish assemblage associated with reservoir construction. J. Freshw. Ecol. 15(2): 189–198.

Bonner, T. H. and G. R. Wilde. 2002. Effects of turbidity on prey consumption by prairie stream fishes. Trans. Am. Fish. Soc. 131(6): 1203–1208.

Boone, M. F. 2007. Muskellunge in Missouri: a 10 year strategic plan. Environ. Biol. Fishes 79(1–2): 171–177.

Boronow, G. F. 1975. The fishes of the Forked Deer River system. M.S. thesis, Univ. Tennessee. Knoxville. 71 pp.

Bortone, S. A. 1989. *Notropis melanostomus*, a new species of cyprinid fish from the Blackwater-Yellow River drainage of northwest Florida. Copeia 1989(3): 737–741.

Boschung, H. T., Jr. and R. L. Mayden. 2004. Fishes of Alabama. Smithsonian Books. Washington, DC. 736 pp.

Boschung, H. T., Jr. and D. Nieland. 1986. Biology and conservation of the Slackwater Darter, *Etheostoma boschungi* (Pisces: Percidae). SFC Proc. 4(4): 1–4.

Boschung, H. and P. O'Neil. 1981. The effects of forest clear-cutting on fishes and macroinvertebrates in an Alabama stream. Pages 200–217. *In:* L. A. Krumholz, G. D. Pardue, and G. A. Hall, eds. Proc. Warmwater Streams Symposium. Southern Division American Fisheries Society. Bethesda, MD.

Bossu, C. M. and T. J. Near. 2009. Gene trees reveal repeated instances of mitochondrial DNA introgression in Orangethroat Darters (Percidae: *Etheostoma*). Syst. Biol. 58(1): 114–129.

Bossu, C. M. and T. J. Near. 2015. Ecological constraint and the evolution of sexual dichromatism in darters. Evolution 69(5): 1219–1231.

Bossu, C. M., J. M. Beaulieu, P. A. Ceas, and T. J. Near. 2013. Explicit tests of paleodrainage connections of southeastern North America and the historical biogeography of Orangethroat Darters (Percidae: *Etheostoma: Ceasia*). Mol. Ecol. 22(21): 5397–5417.

Bottrell, C. E., R. H. Ingersol, and R. W. Jones. 1964. Notes on the embryology, early development, and behavior of *Hybopsis aestivalis tetranemus* (Gilbert). Trans. Am. Microsc. Soc. 83(4): 391–399.

Boucher, J., P. Berube, and R. Cloutier. 2009. Comparison of the Channel Darter (*Percina copelandi*) summer habitat in two rivers from eastern Canada. J. Freshw. Ecol. 24(1): 19–28.

Boulenger, G. A. 1895. Catalogue of the perciform fishes in the British Museum. Taylor and Francis, London.

Bounds, S. M. 1977. Addendum to: "Fishes of the Fourche River in northcentral Arkansas." Proc. Arkansas Acad. Sci. 31: 112.

Bounds, S. M. and J. K. Beadles. 1976. Fishes of the Fourche River in northcentral Arkansas. Proc. Arkansas Acad. Sci. 30: 22–26.

Bounds, S. M., J. K. Beadles, and B. M. Johnson. 1977. Fishes of Randolph County, Arkansas. Proc. Arkansas Acad. Sci. 31: 21–25.

Bowen, B. R., B. R. Kreiser, P. F. Mickle, J. F. Schaefer, and S. B. Adams. 2008. Phylogenetic relationships among North American *Alosa* species (Clupeidae). J. Fish Biol. 72(5): 1188–1201.

Bowman, D. W. 1993. Black bass in Beaver Reservoir and its tributaries: distribution and abundance in relation to water quality. M.S. thesis, Univ. Arkansas. 113 pp.

Bowman, M. L. 1970. Life history of the Black Redhorse, *Moxostoma duquesnei* (Lesueur), in Missouri. Trans. Am. Fish. Soc. 99(3): 546–559.

Boyd, G. L. 1997. Metabolic rates and life history of aquatic organisms inhabiting Logan Cave stream in northwest Arkansas. M.S. thesis, Univ. Arkansas, Fayetteville. 107 pp.

Boyer, R. L. 1969. Aspects of the behavior and biology of the Longear Sunfish, *Lepomis megalotis* (Rafinesque) in two Arkansas reservoirs. M.S. thesis, Oklahoma State Univ., Stillwater.

Boyer, R. L. and L. E. Vogele. 1971. Longear Sunfish behavior in two Ozark reservoirs. Reservoir Fisheries and Limnology, Special Publ. 8, Am. Fish. Soc. pp. 13–25.

Bozek, M. A., T. M. Burri, and R. V. Frie. 1999. Diets of Muskellunge in northern Wisconsin lakes. N. Am. J. Fish. Manag. 19(1): 258–270.

Braasch, M. E. and P. W. Smith, 1967. The life history of the Slough Darter, *Etheostoma gracile* (Pisces: Percidae). Illinois Nat. Hist. Surv. Biol. Notes 58: 1–12.

Braaten, P. J. and D. B. Fuller. 2007. Growth rates of young-of-year Shovelnose Sturgeon in the upper Missouri River. J. Appl. Ichthyol. 23(4): 506–515.

Bradbury, J. W. and S. L. Vehrencamp. 2011. Principles of animal communication. 2nd ed. Sinauer, Sunderland, MA. 697 pp.

Brady, L. and A. Hulsey. 1959. Propagation of buffalofishes. Proc. Annu. Conf. Southeast. Assoc. Game Fish Comm. 13: 80–89.

Branner, J. C. 1891. Preface to the Annual Report of the Arkansas Geological Survey for 1889, pp. i–xix.

Branson, B. A. 1961. Observations on the distribution of nuptial tubercles in some catostomid fishes. Trans. Kansas Acad. Sci. 64(4): 360–372.

Branson, B. A. 1962. Comparative cephalic and appendicular

osteology of the fish family Catostomidae. Part 1, *Cycleptus elongatus* (Lesueur). Southwest. Nat. 7(2): 81–153.

Branson, B. A. 1963. The olfactory apparatus of *Hybopsis gelida* (Girard) and *Hybopsis aestivalis* (Girard) (Pisces: Cyprinidae). J. Morphol. 113(2): 215–229.

Branson, B. A. 1967. Fishes of the Neosho River system in Oklahoma. Am. Midl. Nat. 78(1): 126–154.

Branson, B. A. 1970. Measurements, counts, and observations on four lamprey species from Kentucky (*Ichthyomyzon*, *Lampetra*, and *Entosphenus*). Am. Midl. Nat. 84(1): 243–247.

Branson, B. A. and G. A. Moore. 1962. The lateralis components of the acoustico-lateralis system in the sunfish family Centrarchidae. Copeia 1962(1): 1–108.

Branson, B. A. and J. B. Campbell. 1969. Hybridization in the darters *Etheostoma spectabile* and *Etheostoma radiosum cyanorum*. Copeia 1969(1): 70–75.

Breder, C. M., Jr. and D. E. Rosen. 1966. Modes of reproduction in fishes. Natural History Press, New York. 941 pp.

Brenneman, W. M. 1992. Ontogenetic aspects of upper and lower stream reach cyprinid assemblages in a south Mississippi watershed. Ph.D. diss. Univ. of Southern Mississippi, Hattiesburg.

Bresnick, G. I. and D. C. Heins. 1977. The age and growth of the Weed Shiner, *Notropis texanus* (Girard). Am. Midl. Nat. 98(2): 491–495.

Brewer, S. K., D. M. Papoulias, and C. F. Rabeni. 2006. Spawning habitat associations and selection by fishes in a flow-regulated prairie river. Trans. Am. Fish. Soc. 135(3): 763–778.

Brezner, J. 1958. Food habits of the northern River Carpsucker in Missouri. Prog. Fish-Cult. 20(4): 170–174.

Briggs, J. C. 1974. Operation of zoogeographic barriers. Syst. Zool. 23(2): 248–256.

Briggs, J. C. 1986. Introduction to the zoogeography of North American fishes. Pages 1–16. *In*: C. H. Hocutt and E. O. Wiley, eds. The zoogeography of North American fishes. John Wiley and Sons, New York.

Briggs, J. C., ed. 1995. Global biogeography. Developments in Paleontology and Stratigraphy. Vol. 14. 451 pp.

Brigham, W. U. 1973. Nest construction of the lamprey, *Lampetra aepyptera*. Copeia 1973(1): 135–136.

Brill, J. S., Jr. 1977. Notes on abortive spawnings of the Pirateperch, *Aphredoderus sayanus*, with comments on sexual distinctions. Am. Currents 5(4): 10–16.

Brinkman, E. L. 2008. Contributions to the life history of Alligator Gar, *Atractosteus spatula*, (Lacépède), in Oklahoma. M.S. thesis, Oklahoma State Univ. Stillwater. 37 pp.

Britton, J. R., R. R. Boar, J. Grey, J. Foster, J. Lugonzo, and D. M. Harper. 2007. From introduction to fishery dominance: the initial impacts of the invasive Carp *Cyprinus carpio* in lake Naivasha, Kenya, 1999 to 2006. J. Fish Biol. 71(sd): 239–257.

Broach, R. W. 1967. Arkansas catchable Channel Catfish program. Proc. 21st Annu. Conf. Southeast. Assoc. Fish Wildl. Agencies: 445–452.

Brockway, W. J. and R. L. Johnson. 2015. Introduction of Florida bass alleles into Largemouth Bass inhabiting northeast Arkansas stream systems. J. Arkansas Acad. Sci. 69: 29–35.

Bronte, C. R. 1993. Evidence of spring spawning Lake Trout in Lake Superior. J. Great Lakes Res. 19(3): 625–629.

Broughton, R. E. and J. R. Gold. 2000. Phylogenetic relationships in the North American cyprinid genus *Cyprinella* (Actinopterygii: Cyprinidae) based on sequences of the mitochondrial ND2 and ND4L genes. Copeia 2000(1): 1–10.

Browman, H. I., W. C. Gordon, B. I. Evans, and W. J. O'Brien. 1990. Correlation between histological and behavioral measures of visual acuity in a zooplanktivorous fish, the White Crappie (*Pomoxis annularis*). Brain Behav. Evol. 35(2): 85–97.

Browman, H. I. and W. J. O'Brien. 1992. The ontogeny of search behavior in the White Crappie, *Pomoxis annularis*. Environ. Biol. Fishes 34(2): 181–195.

Brown, A. V. 1985. The Ozark Cavefish (*Amblyopsis rosae*). Arkansas Fish Game 16(2): 25–26.

Brown, A. V. 1991. Status survey of *Amblyopsis rosae* in Arkansas. A final report submitted to the Arkansas Game and Fish Commission, Little Rock.

Brown, A. V. and M. Lyttle. 1992. Impacts of gravel mining on Ozark stream ecosystems. Final Report to the Fisheries Division, Arkansas Game and Fish Commission. Little Rock. 116 pp.

Brown, A. V. and C. S. Todd. 1987. Status review of the threatened Ozark Cavefish (*Amblyopsis rosae*). Proc. Arkansas Acad. Sci. 41: 99–100.

Brown, A. V. and L. D. Willis. 1984. Cavefish (*Amblyopsis rosae*) in Arkansas: populations, incidence, habitat requirements and mortality factors. Final Report to Arkansas Game and Fish Comm., Little Rock. Fed. Aid Proj. E-1-6, P-L 93-205. 61 pp.

Brown, A. V., G. O. Graening, and P. Vendrell. 1998a. Monitoring cavefish populations and environmental quality in Cave Springs Cave, Arkansas. Misc. Publ. No. 214. Arkansas Water Resources Center, Univ. Arkansas, Fayetteville. 32 pp.

Brown, A. V., M. M. Lyttle, and K. B. Brown. 1998b. Impacts of gravel mining on gravel bed streams. Trans. Am. Fish. Soc. 127(6): 979–994.

Brown, B. E. and J. S. Dendy. 1961. Observations on the food habits of the Flathead and Blue catfish in Alabama. Proc. 15th Annu. Conf. Southeast. Assoc. Game Fish Comm.: 219–222.

Brown, B. E., I. Inman, and A. Jearold, Jr. 1970. Schooling and shelter seeking tendencies in fingerling Channel Catfish. Trans. Am. Fish. Soc. 99(3): 540–545.

Brown, J. A., H. W. Robison, and C. T. McAllister. 2016. Occurrence of Shoal Chub, *Macrhybopsis hyostoma* (Cypriniformes: Cyprinidae) in unusual habitat in the

Arkansas River system of Arkansas: could direct tributaries be refugia allowing persistence despite fragmentation of instream habitat? J. Arkansas Acad. Sci. 70: 260–262.

Brown, J. D. 1967. A study of the fishes of the tailwaters of three impoundments in northern Arkansas. M.S. thesis, Univ. of Arkansas. Fayetteville. 45 pp.

Brown, J. D., C. R. Liston, and R. W. Dennie. 1967. Some physico-chemical and biological aspects of three cold tailwaters in northern Arkansas. Proc. 21st Annu. Conf. Southeast. Assoc. Game Fish Comm.: 369–381.

Brown, J. R., K. Beckenbach, A. T. Beckenbach, and M. J. Smith. 1996. Length variation, heteroplasmy and sequence divergence in the mitochondrial DNA of four species of sturgeon (*Acipenser*). Genetics 142(2): 525–535.

Brown, J. Z. 1996. Population dynamics and growth of Ozark Cavefish in Logan Cave National Wildlife Refuge, Benton County, Arkansas. M.S. thesis. Univ. of Arkansas. Fayetteville. 105 pp.

Brown, J. Z. and J. E. Johnson. 2001. Population biology and growth of Ozark Cavefish in Logan Cave National Wildlife Refuge, Arkansas. Environ. Biol. Fishes 62(1–3): 161–169.

Bruch, R. M. and F. P. Binkowski. 2002. Spawning behavior of Lake Sturgeon (*Acipenser fulvescens*). J. Appl. Ichthyol. 18(4–6): 570–579.

Bruch, R. M., T. A. Dick, and A. Choudhury. 2001. A field guide for the identification of stages of gonad development in Lake Sturgeon, *Acipenser fulvescens* Rafinesque, with notes on Lake Sturgeon reproductive biology and management implications. Publ. Wisconsin Dept. Natural Resources. Oshkosh and Sturgeon for Tomorrow. Appleton, WI. 38 pp.

Brueggen-Bowman, T. R., S. Choi, and J. L. Bouldin. 2015. Response of water-quality indicators to the implementation of best-management practices in the upper Strawberry River watershed, Arkansas. Southeast. Nat. 14(4): 697–713.

Bruner, J. C. 2004. "Spreitzer" vertebrae, a unique character found only in *Ammocrypta* (crystal and sand darters). Pages 57–58. *In*: T. P. Barry and J. A. Malison, eds. Proceedings of PERCIS III: the Third International Percid Fish Symposium. Univ. Wisconsin Sea Grant Institute, Madison. 136 pp.

Bruner, J. C. 2011. A phylogenetic analysis of Percidae using osteology. Chapter 2. Pages 5–84. *In*: B. A. Barton, ed. Biology, management, and culture of Walleye and Sauger. American Fisheries Society, Bethesda, MD.

Brusca, R. C. and G. J. Brusca. 2003. Invertebrates. 2nd ed. Sinauer, Sunderland, MA. 936 pp.

Bruton, M. N. 1996. Alternative life-history strategies of catfishes. Aquat. Living Resour. 9(S1): 35–41.

Bryant, H. E. and A. Houser. 1968. Growth of Threadfin Shad in Bull Shoals Reservoir. Proc. 22nd Annu. Conf. Southeast. Assoc. Game Fish Comm.: 275–283.

Bryant, H. E. and A. Houser. 1971. Population estimates and growth of Largemouth Bass in Beaver and Bull Shoals reservoirs. Pages 349–357. *In:* G. E. Hall, ed. Reservoir fisheries

and limnology. Special Publ. 8, American Fisheries Society, Washington, DC. 511 pp.

Bryant, H. E. and T. E. Moen. 1980a. Food of Bluegill and Longear Sunfish in DeGray Reservoir, Arkansas, 1976. Proc. Arkansas Acad. Sci. 34: 31–33.

Bryant, H. E. and T. E. Moen. 1980b. Food of Largemouth Bass (*Micropterus salmoides*) in DeGray Reservoir, Arkansas, 1976. Proc. Arkansas Acad. Sci. 34: 34–37.

Buchanan, J. P. 1973. Separation of the subspecies of Largemouth Bass *Micropterus salmoides salmoides* and *M. s. floridanus* and intergrades by use of meristic characters. Proc. 27th Annu. Conf. Southeast. Assoc. Game Fish Comm.: 608–619.

Buchanan, J. P. and H. E. Bryant. 1973. The occurrence of a predorsal stripe in the Black Crappie, *Pomoxis nigromaculatus*. J. Alabama Acad. Sci. 44(4): 293–297.

Buchanan, T. M. 1966. A study of the "killing phenomenon" in isolated groups of *Etheostoma spectabile* (Agassiz). Proc. Arkansas Acad. Sci. 20: 54–58.

Buchanan, T. M. 1967. A field test of the use of scale size at the formation of the first annulus to permanently mass-mark Smallmouth Bass, *Micropterus dolomieui* Lacépède. M.S. thesis, Univ. Arkansas, Fayetteville. 31 pp.

Buchanan, T. M. 1971. The reproductive ecology of the Rio Grande Cichlid, *Cichlasoma cyanoguttatum* (Baird and Girard). Ph.D. diss., Univ. Texas, Austin. 226 pp.

Buchanan, T. M. 1973a. Checklist of Arkansas fishes. Proc. Arkansas Acad. Sci. 27: 27–29.

Buchanan, T. M. 1973b. Key to the fishes of Arkansas. Arkansas Game and Fish Comm., Little Rock. 68 pp. + 198 maps.

Buchanan, T. M. 1973c. First Arkansas record of *Noturus flavus* (Ictaluridae). Southwest. Nat. 18(1): 98–99.

Buchanan, T. M. 1974. Threatened native fishes of Arkansas. Pages 67–92. *In*: W. M. Shepherd, ed. Arkansas Natural Area Plan. Ark. Dept. Planning, Little Rock.

Buchanan, T. M. 1975. Fish. *In*: E. E. Dale, Jr. Environmental evaluation report on various completed channel improvement projects in eastern Arkansas. Arkansas Water Resources Research Center, Publ. 30. 42 pp.

Buchanan, T. M. 1976. An evaluation of the effects of dredging within the Arkansas River Navigation System. Vol. 5. The effects upon the fish fauna. Arkansas Water Resources Research Center, Publ. 47. 277 pp.

Buchanan, T. M, 1984. Status of the Longnose Darter, *Percina nasuta* (Bailey), in Arkansas. Arkansas Natural Heritage Comm. Report, June 1984. 29 pp.

Buchanan, T. M. 1997. The fish community of Indian Bayou, a coastal plain stream of remarkable species richness in the lower White River drainage of Arkansas. J. Arkansas Acad. Sci. 51: 55–65.

Buchanan, T. M. 2005. Small fish species of Arkansas reservoirs. J. Arkansas Acad. Sci. 59: 26–42.

Buchanan, T. M. and H. W. Robison. 1994. A recent record

of the Plains Minnow, *Hybognathus placitus* Girard, from Arkansas. Proc. Arkansas Acad. Sci. 48: 242.

Buchanan, T. M. and M. M. Stevenson. 2003. Distribution of Bigscale Logperch, *Percina macrolepida* (Percidae), in the Arkansas River basin. Southwest. Nat. 48(3): 454–460.

Buchanan, T. M. and K. Strawn. 1969. A field test of the use of scale size at the formation of the first annulus to permanently mass-mark Smallmouth Bass, *Micropterus dolomieui* Lacépède. Proc. 23rd Annu. Conf. Southeast. Assoc. Game Fish Comm.: 303–311.

Buchanan, T. M., R. L. Limbird, and F. J. Leone. 2007. First Arkansas records for the White Perch, *Morone americana* (Gmelin), (Teleostei: Moronidae). J. Arkansas Acad. Sci. 61: 123–124.

Buchanan, T. M., H. W. Robison, and K. Shirley. 1993. New distributional records for Arkansas sturgeons. Proc. Arkansas Acad. Sci. 47: 133.

Buchanan, T. M., A. Carter, S. Henry, and K. Shirley. 1982. Rainbow Smelt, *Osmerus mordax* (Pisces: Osmeridae), in the Mississippi River of Arkansas and Tennessee. Southwest. Nat. 27(2): 225–226.

Buchanan, T. M., W. G. Layher, C. T. McAllister, and H. W. Robison. 2012. The Alabama Shad, *Alosa alabamae* (Jordan and Evermann) (Clupeiformes: Clupeidae) in the White River, Arkansas. Southwest. Nat. 57(3): 352–354.

Buchanan, T. M., D. Wilson, L. G. Claybrook, and W. G. Layher. 2003. Fishes of the Red River in Arkansas. Proc. Arkansas Acad. Sci. 57: 18–26.

Buchanan, T. M., C. Hargrave, D. Wilson, L. G. Claybrook, and P. W. Penny, Jr. 1996. First Arkansas records for Bigscale Logperch, *Percina macrolepida* Stevenson (Pisces: Percidae), with comments on habitat preference and distinctive characters. Proc. Arkansas Acad. Sci. 50: 28–35.

Buchanan, T. M., J. D. Houston, J. F. Nix, R. L. Meyer, and E. H. Schmitz. 1978. A limnological study of Ricks Pond and the Gulpha Creek drainage in Garland County, Arkansas. Arkansas Water Resources Research Center, Publ. No. 62. 109 pp.

Buchanan, T. M., J. Nichols, D. Turman, C. Dennis, S. Wooldridge, and B. Hobbs. 1999. Occurrence and reproduction of the Alabama Shad, *Alosa alabamae* Jordan and Evermann, in the Ouachita River system of Arkansas. Proc. Arkansas Acad. Science 53: 21–26.

Buchanan, T. M., J. Smith, D. Saul, J. Farwick, T. Burnley, M. Oliver, and K. Shirley. 2000. New Arkansas records for two nonindigenous fish species, with a summary of previous introductions of nonnative fishes in Arkansas. Proc. Arkansas Acad. Science 54: 143–145.

Buchholz, M. 1957. Age and growth of River Carpsucker in Des Moines River, Iowa. Proc. Iowa Acad. Sci. 64(1): 589–600.

Buckmeier, D. L., N. G. Smith, and D. J. Daugherty. 2013. Alligator Gar movement and macrohabitat use in the lower Trinity River, Texas. Trans. Am. Fish. Soc. 142(4): 1025–1035.

Bufalino, A. P. and R. L. Mayden. 2010a. Molecular phylogenetics of North American phoxinins (Actinopterygii: Cypriniformes: Leuciscidae) based on RAG1 and S7 nuclear DNA sequence data. Mol. Phylogenet. Evol. 55(1): 274–283.

Bufalino, A. P. and R. L. Mayden. 2010b. Phylogenetic relationships of North American phoxinins (Actinopterygii: Cypriniformes: Leuciscidae) as inferred from S7 nuclear DNA sequences. Mol. Phylogenet. Evol. 55(1): 143–152.

Bufalino, A. P. and R. L. Mayden. 2010c. Phylogenetic evaluation of North American Leuciscidae (Actinopterygii: Cypriniformes: Cyprinoidea) as inferred from analyses of mitochondrial and nuclear DNA sequences. Syst. Biodivers. 8(4): 493–505.

Bulak, J. S. 1985. Distinction of larval Blueback Herring, Gizzard Shad, and Threadfin Shad from the Santee-Cooper drainage, South Carolina. J. Elisha Mitchell Sci. Soc. 101(3): 177–186.

Bulak, J. S., C. C. Coutant, and J. A. Rice, eds. 2013. Biology and management of inland Striped Bass and hybrid Striped Bass. Symp. 80. American Fisheries Society, Bethesda, MD. 588 pp.

Bulkley, R. V. 1970. Changes in Yellow Bass reproduction associated with environmental conditions. Iowa State J. Sci. 45(2): 137–180.

Bunt, C. M., N. E. Mandrak, D. C. Eddy, S. A. Choo-Wing, T. G. Heiman, and E. Taylor. 2013. Habitat utilization, movement and use of groundwater seepages by larval and juvenile Black Redhorse, *Moxostoma duquesnei*. Environ. Biol. Fishes 96(10–11): 1281–1287.

Buntz, J. and C. S. Manooch. 1969. *Tilapia aurea* (Steindachner), a rapidly spreading exotic in south central Florida. Proc. Annu. Conf. Southeast. Assoc. Game Fish Comm. 22: 495–501.

Bur, M. T. 1976. Age, growth and food habits of Catostomidae in Pool 8 of the upper Mississippi River. M.S. thesis. Univ. Wisconsin, LaCrosse. 107 pp.

Bur, M. T. 1982. Food of Freshwater Drum in western Lake Erie. J. Great Lakes Res. 8(4): 672–675.

Burbidge, R. G. 1969. Age, growth, length-weight relationship, sex ratio, and food habits of American Smelt, *Osmerus mordax* (Mitchill), from Gull Lake, Michigan. Trans. Am. Fish. Soc. 98(4): 631–640.

Burgess, C. C. 2003. Summer fish assemblages in channelized and unchannelized reaches of the South Sulphur River, Texas. M.S. thesis. Texas A&M Univ., College Station. 94 pp.

Burgess, G. H. 1980a. *Alosa alabamae* Jordan and Evermann, Alabama Shad. Page 62. *In*: D. S. Lee et al. Atlas of North American freshwater fishes. N.C. State Mus. Nat. Hist., Raleigh.

Burgess, G. H. 1980b. *Dorosoma petenense* (Günther), Threadfin Shad. Page 70. *In*: D. S. Lee et al. Atlas of North American freshwater fishes. N.C. State Mus. Nat. Hist., Raleigh.

Burgess, G. H. 1980c. *Morone saxatilis* (Walbaum), Striped

Bass. Page 576. *In:* D. S. Lee et al. Atlas of North American freshwater fishes. N.C. State Mus. Nat. Hist., Raleigh.

Burkhead, N. M. 2012. Extinction rates in North American freshwater fishes, 1900–2010. BioScience 62(9): 798–808.

Burkhead, N. M. and R. E. Jenkins. 1991. Fishes. Pages 321–409. *In:* K. A. Terwilliger, coordinator. Virginia's endangered species. McDonald and Woodward, Blacksburg, VA. 672 pp.

Burkhead, N. M., and J. D. Williams. 1991. An intergeneric hybrid of a native minnow, the Golden Shiner, and an exotic minnow, the Rudd. Trans. Am. Fish. Soc. 120(6): 781–795.

Burkhead, N. M., R. E. Jenkins, and E. G. Maurakis. 1980. New records, distribution and diagnostic characters of Virginia ictalurid catfishes with an adnexed adipose fin. Brimleyana 4: 75–93.

Burleson, M. L., D. R. Wilhelm, and N. J. Smatresk. 2001. The influence of fish size on the avoidance of hypoxia and oxygen selection by Largemouth Bass. J. Fish Biol. 59(5): 1336–1349.

Burnie, D. and D. E. Wilson, eds. 2011. Animal: the definitive visual guide to the world's wildlife. Smithsonian Collection, Dorling Kindersley, London. 624 pp.

Burns, J. W. 1966. Threadfin Shad. Pages 481–492. *In:* A. Calhoun, ed. Inland fisheries management. California Department of Fish and Game, Sacramento. 546 pp.

Burns, T. A., D. T. Stalling, and W. Goodger, 1981. Gar ichthyootoxin- its effect on crayfish, with notes on Bluegill sunfish. Southwest. Nat. 25(4): 513–515.

Burr, B. M. 1974. A new intergeneric hybrid combination in nature: *Pomoxis annularis × Centrarchus macropterus.* Copeia 1974(1): 269–271.

Burr, B. M. 1977. The Bantam Sunfish, *Lepomis symmetricus*: systematics and distribution, and life history in Wolf Lake, Illinois. Illinois Nat. Hist. Surv. Bull. 31(10): 437–465.

Burr, B. M. 1978. Systematics of the percid fishes of the subgenus *Microperca*, genus *Etheostoma*. Bull. Alabama Mus. Nat. Hist. 4: 1–53.

Burr, B. M. and R. C. Cashner. 1983. *Campostoma pauciradii*, a new cyprinid fish from southeastern United States, with a review of related forms. Copeia 1983(1): 101–116.

Burr, B. M. and W. W. Dimmick. 1983. Redescription of the Bigeye Shiner, *Notropis boops* (Pisces: Cyprinidae). Proc. Biol. Soc. Washington 96(1): 50–58.

Burr, B. M. and M. S. Ellinger. 1980. Distinctive egg morphology and its relationship to development in the percid fish *Etheostoma proeliare*. Copeia 1980(3): 556–559.

Burr, B. M. and R. C. Heidinger. 1983. Reproductive behavior of the Bigmouth Buffalo *Ictiobus cyprinellus* in Crab Orchard Lake, Illinois. Am. Midl. Nat. 110(1): 220–221.

Burr, B. M. and R. L. Mayden. 1980. Dispersal of Rainbow Smelt, *Osmerus mordax*, into the upper Mississippi River (Pisces: Osmeridae). Am. Midl. Nat. 104(1): 198–201.

Burr, B. M. and R. L. Mayden. 1982a. Status of the Cypress Minnow, *Hybognathus hayi* Jordan, in Illinois. Nat. Hist. Miscellanea, Chicago Acad. Sci. No. 215: 1–10.

Burr, B. M. and R. L. Mayden. 1982b. Life history of the Freckled Madtom, *Noturus nocturnus*, in Mill Creek, Illinois (Pisces: Ictaluridae). Occas. Pap. Mus. Nat. Hist. Univ. Kansas. 98: 1–15.

Burr, B. M. and R. L. Mayden. 1982c. Life history of the Brindled Madtom, *Noturus miurus* in Mill Creek, Illinois (Pisces: Ictaluridae). Am. Midl. Nat. 107(1): 25–41.

Burr, B. M. and R. L. Mayden. 1984. Reproductive biology of the Checkered Madtom (*Noturus flavater*) with observations on nesting in the Ozark (*N. albater*) and Slender (*N. exilis*) madtoms (Siluriformes: Ictaluridae). Am. Midl. Nat. 112(2): 408–414.

Burr, B. M. and R. L. Mayden. 1992. Phylogenetics and North American freshwater fishes. Pages 18–75. *In:* R. L. Mayden, ed. Systematics, historical ecology, and North American freshwater fishes. Stanford Univ. Press, Stanford, CA. 969 pp.

Burr, B. M. and R. L. Mayden. 1999. A new species of *Cycleptus* (Cypriniformes: Catostomidae) from Gulf slope drainages of Alabama, Mississippi, and Louisiana, with a review of the distribution, biology, and conservation status of the genus. Bull. Alabama Mus. Nat. Hist. 20: 19–57.

Burr, B. M. and M. A. Morris. 1977. Spawning behavior of the Shorthead Redhorse, *Moxostoma macrolepidotum*, in Big Rock Creek, Illinois. Trans. Am. Fish. Soc. 106(1): 80–82.

Burr, B. M. and L. M. Page. 1975. Distribution and life history notes on the Taillight Shiner, *Notropis maculatus* in Kentucky. Trans. Kentucky Acad. Sci. 36(3–4): 71–74.

Burr, B. M. and L. M. Page. 1978. The life history of the Cypress Darter, *Etheostoma proeliare*, in Max Creek, Illinois. Illinois Nat. Hist. Surv. Biol. Notes 106. 15 pp.

Burr, B. M. and L. M. Page. 1979. The life history of the Least Darter, *Etheostoma microperca*, in the Iroquois River, Illinois. Illinois Nat. Hist. Surv. Biol. Notes 112: 1–15.

Burr, B. M. and L. M. Page. 1986. Zoogeography of fishes of the lower Ohio–upper Mississippi Basin. Pages 363–412. *In:* C. H. Hocutt and E. O. Wiley, eds. The zoogeography of North American freshwater fishes. John Wiley and Sons, New York. 866 pp.

Burr, B. M. and P. W. Smith. 1976. Status of the Largescale Stoneroller, *Campostoma oligolepis*. Copeia 1976(3): 521–531.

Burr, B. M. and J. N. Stoeckel. 1999. The natural history of madtoms (Genus *Noturus*), North America's diminutive catfishes. Am. Fish. Soc. Symp. 24: 51–101.

Burr, B. M. and M. L. Warren, Jr. 1986a. Status of the Bluehead Shiner (*Notropis hubbsi*) in Illinois. Trans. Illinois State Acad. Sci. 79(1–2): 129–136.

Burr, B. M. and M. L. Warren, Jr. 1986b. A distributional atlas of Kentucky fishes. Kentucky Nature Preserves Comm., Sci. Tech. Ser. No. 4. 398 pp.

Burr, B. M., R. C. Cashner, and W. L. Pflieger. 1979. *Campostoma oligolepis* and *Notropis ozarcanus* (Pisces: Cyprinidae), two additions to the known fish fauna of the

Illinois River, Arkansas and Oklahoma. Southwest. Nat. 24(2): 381–403.

Burr, B. M., D. J. Eisenhour, and J. M. Grady. 2005. Two new species of *Noturus* (Siluriformes: Ictaluridae) from the Tennessee River drainage: description, distribution, and conservation status. Copeia 2005(4): 783–802.

Burr, B. M., M. L. Warren, G. K. Weddle, and R. R. Cicerello. 1990. Records of nine endangered, threatened, or rare Kentucky fishes. Trans. Kentucky Acad. Sci. 51(3–4): 188–190.

Burr, B. M., K. M. Cook, D. J. Eisenhour, K. R. Piller, W. J. Poly, R. W. Sauer, C. A. Taylor, E. R. Atwood, and G. L. Seegert. 1996. Selected Illinois fishes in jeopardy: new records and status evaluations. Trans. Illinois State Acad. Sci. 89(3–4): 169–186.

Burress, R. M. 1965. A quantitative creel census of two arms of Bull Shoals Reservoir, Missouri. Proc. 16th Annu. Conf. Southeast. Assoc. Game Fish Comm.: 387–398.

Burton, C. 1970. Alligator Gar-fish terror or fishery tool? Arkansas Game and Fish 3(1): 5–7.

Buth, D. G. 1978. Biochemical systematics of the Moxostomatini (Cypriniformes, Catostomidae). Ph.D. diss. Univ. Illinois, Urbana. 191 pp.

Buth, D. G. 1979. Biochemical systematics of the cyprinid genus *Notropis*. I. The subgenus *Luxilus*. Biochem. Syst. Ecol. 7(1): 69–79.

Buth, D. G. 1980. Evolutionary genetics and systematic relationships in the catostomid genus *Hypentelium*. Copeia 1980(2): 280–290.

Buth, D. G. 2010. Should mitochondrial DNA sequences be used in phylogenetic studies? Bull. South. California Acad. Sci. 109(2): 108–109.

Buth, D. G. and B. M. Burr. 1978. Isozyme variability in the cyprinid genus *Campostoma*. Copeia 1978(2): 298–311.

Buth, D. G. and R. L. Mayden. 1981. Taxonomic status and relationships among populations of *Notropis pilsbryi* and *N. zonatus* (Cypriniformes: Cyprinidae) as shown by the glucosephosphate isomerase, lactate dehydrogenase and phosphoglucomutase enzyme systems. Copeia 1981(3): 583–590.

Buth, D. G. and R. L. Mayden. 2001. Allozymic and isozymic evidence for polytypy in the North American catostomid genus *Cycleptus*. Copeia 2001(4): 899–906.

Butler, R. L. 1962. The life history and ecology of the Sheepshead, *Aplodinotus grunniens* Rafinesque, in the commercial fishery of the upper Mississippi River. Ph.D. diss. Univ. Minnesota, Minneapolis. 178 pp.

Butler, R. L. 1965. Freshwater Drum, *Aplodinotus grunniens*, in the navigational impoundments of the upper Mississippi River. Trans. Am. Fish. Soc. 94(4): 339–349.

Butler, R. L. and L. L. Smith, Jr. 1950. The age and rate of growth of the Sheepshead, *Aplodinotus grunniens* Rafinesque, in the upper Mississippi River navigation pools. Trans. Am. Fish. Soc. 79(1): 43–54.

Butler, R. S. and R. L. Mayden. 2003. Cryptic biodiversity. Endangered Species Bull. 28(2): 24–26.

Butler, S. E. and D. H. Wahl. 2017. Movements and habitat use of River Redhorse (*Moxostoma carinatum*) in the Kankakee River, Illinois. Copeia 2017(4): 734–742.

Buynak, G. L. and H. W. Mohr, Jr. 1978a. Larval development of the Northern Hog Sucker (*Hypentelium nigricans*) from the Susquehanna River. Trans. Am. Fish. Soc. 107(4): 595–599.

Buynak, G. L. and H. W. Mohr, Jr. 1978b. Larval development of the Redbreast Sunfish (*Lepomis auritus*) from the Susquehanna River. Trans. Am. Fish. Soc. 107(4): 600–604.

Buynak, G. L., L. E. Kornman, A. Surmont, and B. Mitchell. 1989. Longitudinal differences in electrofishing catch rates and angler catches of black bass in Cave Run Lake, Kentucky. N. Am. J. Fish. Manag. 9(2): 226–230.

Bwanika, G. N., D. J. Murie, and L. J. Chapman. 2007. Comparative age and growth of Nile Tilapia (*Oreochromis niloticus* L.) in lakes Nabugabo and Wamala, Uganda. Hydrobiologia 589(1): 287–301.

Bylinsky, G. 1971. The limited war on water pollution. *In:* T. R. Detwyler, ed. Man's impact on environment. McGraw-Hill, New York. 731 pp.

Byrkjedal, I., D. J. Rees, and E. Willassen. 2007. Lumping lumpsuckers: molecular and morphological insights into the taxonomic status of *Eumicrotremus spinosus* (Fabricius, 1776) and *Eumicrotremus eggvinii* Koefoed, 1956 (Teleostei: Cyclopteridae). J. Fish Biol. 71(SA): 111–131.

Cahn, A. R. 1927. An ecological study of southern Wisconsin fishes; the Brook Silversides (*Labidesthes sicculus*) and the Cisco (*Leucichthys artedi*) in their relations to the region. Illinois Biol. Monogr. 11(1): 1–151.

Cahn, A. R. 1936. Observations on the breeding of the Lawyer, *Lota maculosa*. Copeia 1936(3): 163–165.

Cain, M. L., T. E. Lauer, and J. K. Lau. 2008. Habitat use of Grass Pickerel *Esox americanus vermiculatus* in Indiana streams. Am. Midl. Nat. 160(1): 96–109.

Caldwell, J. M. 2011. Density, distribution, habitat use and the associated darter assemblage of the Ouachita Darter in the upper Ouachita River, Arkansas. M.S. thesis. Arkansas Tech Univ., Russellville. 112 pp.

Calkins, H. A., S. J. Tripp, and J. E. Garvey. 2012. Linking Silver Carp habitat selection to flow and phytoplankton in the Mississippi River. Biol. Invasions 14(5): 949–958.

Call, R. E. 1891. The geology of Crowley's Ridge. Annu. Rep. Arkansas Geol. Surv. 1889, 283 pp.

Campbell, J. S. and H. R. MacCrimmon. 1970. Biology of the Emerald Shiner, *Notropis atherinoides* Rafinesque in Lake Simcoe, Canada. J. Fish Biol. 2(3): 259–273.

Campbell, R. D. and B. A. Branson. 1978. Ecology and population dynamics of the Black Bullhead, *Ictalurus melas* (Rafinesque), in central Kentucky. Tulane Stud. Zool. Bot. 20(3–4): 99–136.

Campos, H. H. and C. Hubbs. 1973. Taxonomic implications

of the karyotype of *Opsopoeodus emiliae*. Copeia 1973(1): 161–163.

Campton, D. E. and B. Mahmoudi. 1991. Allozyme variation and population structure of Striped Mullet (*Mugil cephalus*) in Florida. Copeia 1991(2): 485–492.

Campton, D. E., A. L. Bass, F. A. Chapman, and B. W. Bowen. 2000. Genetic distinction of Pallid, Shovelnose, and Alabama sturgeon: emerging species and the US Endangered Species Act. Conserv. Genet. 1(1): 17–32.

Canfield, H. L. 1947. Artificial propagation of those Channel Cats. Prog. Fish-Cult. 9(1): 27–30.

Canonico, G. C., A. Arthington, J. K. McCrary, and M. L. Thieme. 2005. The effects of introduced tilapias on native biodiversity. Aquat. Conserv. Mar. Freshw. Ecosyst. 15(5): 463–483.

Cantor, T. E. 1842. General features of Chusan, with remarks on the flora and fauna of that island. Annu. Mag. Nat. Hist. (New Ser.) 9(58–60): 265–495.

Carlander, K. D. 1969. Handbook of freshwater fishery biology. Vol. 1. Iowa State Univ. Press, Ames. 752 pp.

Carlander, K. D. 1977. Handbook of freshwater fishery biology. Vol. 2. Life history data on the centrarchid fishes of the United States and Canada. Iowa State Univ. Press, Ames. 431 pp.

Carlander. K. D. 1997. Handbook of freshwater fishery biology. Vol. 3. Life history data on ichthyopercid and percid fishes of the United States and Canada. Blackwell Professional, Ames, IA. 397 pp.

Carlson, D. M., M. K. Kettler, S. E. Fisher, and G. S. Whitt. 1982. Low genetic variability in Paddlefish populations. Copeia 1982(3): 721–725.

Carlson, D. M., W. L. Pflieger, L. Trial, and P. S. Haverland. 1985. Distribution, biology and hybridization of *Scaphirhynchus albus* and *S. platorynchus* in the Missouri and Mississippi rivers. Environ. Biol. Fishes 14(1): 51–59.

Carlson, R. L., P. C. Wainwright, and T. J. Near. 2009. Relationship between species co-occurrence and rate of morphological change in *Percina* darters (Percidae: Etheostomatinae). Evolution 63(3): 767–778.

Carnes, W. C., Jr. 1958. Contributions to the biology of the Eastern Creek Chubsucker, *Erimyzon oblongus oblongus* (Mitchill). M.S. thesis, North Carolina State College, Raleigh. 69 pp.

Carpenter, C. C. 1975. Functional aspects of the notochordal appendage of young-of-the-year gar (*Lepisosteus*). Proc. Oklahoma Acad. Sci. 55: 57–64.

Carpenter, C. C. 1995. Notochordal filament in young gar: morphological characteristics. Southwest. Nat. 40(4): 427–428.

Carr, M. H. 1946. Notes on the breeding habits of the eastern stumpknocker, *Lepomis punctatus punctatus* (Cuvier). Q. J. Florida Acad. Sci. 9(2): 101–106.

Carranza, J. and H. E. Winn. 1954. Reproductive behavior of the Blackstripe Topminnow, *Fundulus notatus*. Copeia 1954(4): 273–278.

Carroll, J. H., Jr., D. Ingold, and M. Bradley. 1977. Distribution and species diversity of summer fish populations in two channelized rivers in northeast Texas. Southwest. Nat. 22(1): 128–134.

Carroll, R. L. 1988. Vertebrate paleontology and evolution. W. H. Freeman, New York. 698 pp.

Carter, F. A. 1984. Fishes collected from the Mississippi River and adjacent flood areas in Arkansas, river mile 770.0 to river mile 816.0. M.S. thesis. Arkansas State Univ., Jonesboro.

Carter, F. A. and J. K. Beadles. 1980. The fishes of Rock Creek, Sharp County, Arkansas. Proc. Arkansas Acad. Sci. 34: 38–40.

Carter, F. A. and J. K. Beadles. 1983a. Sicklefin Chub, *Hybopsis meeki,* in the Mississippi River bordering Arkansas. Proc. Arkansas Acad. Sci. 37: 80.

Carter, F. A. and J. K. Beadles. 1983b. Range extension of the Silver Carp, *Hypophthalmichthys molitrix.* Proc. Arkansas Acad. Sci. 37: 80.

Carufel, L. H. 1963. Life history of Saugers in Garrison Reservoir. J. Wildl. Manag. 27(3): 450–456.

Case, B. 1970. Spawning behaviour of the Chestnut Lamprey (*Ichthyomyzon castaneus*). J. Fish. Res. Board Can. 27(10): 1872–1874.

Cashner, M. F. and H. L. Bart, Jr. 2010. Reproductive ecology of nest associates: use of RFLPs to identify cyprinid eggs. Copeia 2010(4): 554–557.

Cashner, M. F., K. R. Piller, and H. L. Bart, Jr. 2011. Phylogenetic relationships of the North American cyprinid subgenus *Hydrophlox*. Mol. Phylogenet. Evol. 59(3): 725–735.

Cashner, R. C. 1967. A survey of the fishes of the cold tailwaters of the White River in northwestern Arkansas, and a comparison of the White River with selected warm-water streams. M.S. thesis, Univ. Arkansas, Fayetteville. 143 pp.

Cashner, R. C. 1980a. *Ambloplites ariommus* Viosca, Shadow Bass. Page 578. *In*: D. S. Lee et. al. Atlas of North American freshwater fishes. N.C. State Mus. Nat. Hist., Raleigh.

Cashner, R. C. 1980b. *Ambloplites rupestris*, Rock Bass. Page 581. *In*: D. S. Lee et al. Atlas of North American freshwater fishes. N.C. State Mus. Nat. Hist., Raleigh.

Cashner, R. C. and J. D. Brown. 1977. Longitudinal distribution of the fishes of the Buffalo River in northwestern Arkansas. Tulane Stud. Zool. Bot. 19(3–4): 37–46.

Cashner, R. C. and W. J. Matthews. 1988. Changes in the known Oklahoma fish fauna from 1973 to 1988. Proc. Oklahoma Acad. Sci. 68: 1–7.

Cashner, R. C. and R. D. Suttkus. 1974. A new rock bass (*Ambloplites*: Centrarchidae) from the Ozark Uplands of the White River in Missouri and Arkansas. ASB Bull. 21(2): 45.

Cashner, R. C. and R. D. Suttkus. 1977. *Ambloplites constellatus*, a new species of rock bass from the Ozark Uplands of Arkansas and Missouri with a review of western rock bass populations. Am. Midl. Nat. 98(1): 147–161.

Cashner, R. C. and R. D. Suttkus. 1978. The status of the

Rock Bass population in Blue Spring, New Mexico, with comments on the introduction of Rock Bass in the western United States. Southwest. Nat. 23(3): 463–472.

Cashner, R. C., J. S. Rogers, and J. M. Grady. 1992. Phylogenetic studies of the genus *Fundulus*. Pages 421–437. *In:* R. L. Mayden, ed. Systematics, historical ecology, and North American freshwater fishes. Stanford Univ. Press, Stanford, CA. 969 pp.

Cashner, R. C., W. J. Matthews, E. Marsh-Matthews, P. J. Unmack, and F. M. Cashner. 2010. Recognition and redescription of a distinctive stoneroller from the southern Interior Highlands. Copeia 2010(2): 300–311.

Casten, L. R. 2006. Life history plasticity of the Blacktail Shiner (*Cyprinella venusta*) across disturbance gradients in Alabama streams. M.S. thesis, Auburn University, Auburn, AL. 64 pp.

Castro, G. A. 1963. Meristic variations of wild and laboratory-raised Smallmouth Bass, *Micropterus dolomieui* Lacépède. M.S. thesis, Univ. Arkansas, Fayetteville. 69 pp.

Cavender, T. M. 1986. Review of the fossil history of North American freshwater fishes. Pages 699–724. *In:* C. H. Hocutt and E. O. Wiley, eds. The zoogeography of North American freshwater fishes. John Wiley and Sons, New York. 866 pp.

Cavender, T. M. 1991. The fossil record of the Cyprinidae. Pages 34–54. *In:* I. J. Winfield and J. S. Nelson, eds. Cyprinid fishes: systematics, biology, and exploitation. Chapman & Hall, London. 667 pp.

Cavender, T. M. and M. M. Coburn. 1985. Interrelationships of North American Cyprinidae, Part II [abstract]. 65th Annu. Meet. Am. Soc. Ichthyol. Herpetol., p 49.

Cavender, T. M. and M. M. Coburn. 1986. Cladistic analysis of eastern North American Cyprinidae. Ohio J. Sci. 86: 1.

Cavender, T. M. and M. M. Coburn. 1987. The *Phoxinus* group and its relationship with North American cyprinids [abstract]. 67th Annu. Meet. Am. Soc. Ichthyol. Herpetol., p. 38.

Cavender, T. M. and M. M. Coburn. 1989. Relationships of American Cyprinidae [abstract]. 69th Annu. Meet. Am. Soc. Ichthyol. Herpetol., p. 74.

Cavender, T. M. and M. M. Coburn. 1992. Phylogenetic relationships of North American cyprinids. Pages 293–327. *In:* R. L. Mayden, ed. Systematics, historical ecology, and North American freshwater fishes. Stanford Univ. Press, Stanford, CA. 969 pp.

Cavin, L. M. 1962. Natural history of the cyprinid fishes, *Notropis lutrensis* (Baird and Girard) and *Notropis camurus* (Jordan and Meek). M.S. thesis. Univ. Kansas, Lawrence.

Ceas, P. A. and B. M. Burr. 2002. *Etheostoma lawrencei*, a new species of darter in the *E. spectabile* species complex (Percidae: subgenus *Oligocephalus*), from Kentucky and Tennessee. Ichthyol. Explor. Freshw. 13(3): 203–216.

Ceas, P. A. and L. M. Page. 1997. Systematic studies of the *Etheostoma spectabile* complex (Percidae; subgenus *Oligocephalus*), with descriptions of four new species. Copeia 1997(3): 496–522.

Chakrabarty, P., J. A. Prejean, and M. L. Niemiller. 2014. The Hoosier Cavefish, a new and endangered species (Amblyopsidae, *Amblyopsis*) from the caves of southern Indiana. ZooKeys 412: 41–57.

Chambers, J. 1987. The cyprinodontiform gonopodium, with an atlas of the gonopodia of the fishes of the genus *Limia*. J. Fish Biol. 30(4): 389–418.

Chambers, S. R. 1971. Aspects of the life history of the Bleeding Shiner, *Notropis zonatus*, in Missouri. M.A. thesis. Univ. Missouri. Columbia.

Chan, M. D. 1995. Life history and bioenergetics of the Brown Madtom, *Noturus phaeus*. M.S thesis. Univ. Mississippi, Oxford.

Chan, M. D. and G. R. Parsons. 2000. Aspects of Brown Madtom, *Noturus phaeus*, life history in northern Mississippi. Copeia 2000(3): 757–762.

Chang, C-H. M. 1988. Systematics of the Centrarchidae (Perciformes: Percoidei) with notes on the haemal-anal-axial character complex. Ph.D. diss. City University of New York. 490 pp.

Chang, M., D. Miao, Y. Chen, J. Zhou, and P. Chen. 2001. Suckers (Fish, Catostomidae) from the Eocene of China account for the family's current disjunct distributions. Sci. China (Ser. D) 44(7): 577–586.

Chang, Y. F. 1966. Culture of freshwater fish in China. *In:* E. O. Gangstad, ed. 1980. Chinese fish culture, Report 1, Tech. Report A-79. Aquatic Plant Control Research Program. U.S. Army Corps of Engineers, Waterways Expt. Stn., Washington, DC. (Draft transl. by T. S. Y. Koo, 1980).

Chapman, D. C., D. Chen, J. J. Hoover, H. Du, Q. E. Phelps, L. Shen, C. Wang, Q. Wei, and H. Zhang. 2016. Bigheaded carps of the Yangtze and Mississippi rivers: biology, status, and management. Am. Fish. Soc. Symp. 84: 113–126.

Charlton, H. H. 1933. The optic tectum and its related fiber tracts in blind fishes. A. *Troglichthys rosae* and *Typhlichthys eigenmanni*. J. Comp. Neurol. 57(2): 285–325.

Chen, P., E. O. Wiley, and K. M. McNyset. 2007. Ecological niche modeling as a predictive tool: Silver and Bighead carps in North America. Biol. Invasions 9(1): 43–51.

Chen, W.-J. and R. L. Mayden. 2009. Molecular systematics of the Cyprinoidea (Teleostei: Cypriniformes), the world's largest clade of freshwater fishes: further evidence from six nuclear genes. Mol. Phylogenet. Evol. 52(2): 544–549.

Chen, W.-J. and R. L. Mayden. 2010. A phylogenomic perspective on the new era of ichthyology. BioScience 60(6): 421–432.

Chen, W.-J., G. Orti, and A. Meyer. 2004. Novel evolutionary relationship among four fish model systems. Trends Genet. 20(9): 424–431.

Chernoff, B. 1986. Phylogenetic relationships and reclassification of menidiine silverside fishes with emphasis on the tribe Membradini. Proc. Acad. Nat. Sci. Phila. 138(1): 189–249.

Chernoff, B., J. V. Conner, and C. F. Bryan. 1981. Systematics of the *Menidia beryllina* complex (Pisces: Atherinidae) from the Gulf of Mexico and its tributaries. Copeia 1981(2): 319–336.

Chew, R. L. 1974. Early life history of the Florida Largemouth Bass. Fish. Bull. Florida Game Freshw. Fish Comm. 7: 1–76.

Chiasson, W. B., D. L. G. Noakes, and F. W. H. Beamish. 1997. Habitat, benthic prey, and distribution of juvenile Lake Sturgeon (*Acipenser fulvescens*) in northern Ontario rivers. Can. J. Fish. Aquat. Sci. 54(12): 2866–2871.

Childers, W. F. 1967. Hybridization of four species of sunfishes (Centrarchidae). Bull. Illinois Nat. Hist. Surv. 29: 159–214.

Chilton, E. W. II and M. I. Muoneke. 1992. Biology and management of Grass Carp (*Ctenopharyngodon idella*, Cyprinidae) for vegetation control: a North American perspective. Rev. Fish Biol. Fish. 2(4): 283–320.

Chilton, G., K. A. Martin, and J. H. Gee. 1984. Winter feeding: an adaptive strategy broadening the niche of the Central Mudminnow, *Umbra limi*. Environ. Biol. Fishes 10(3): 215–219.

Chipps, S. R. and D. H. Wahl. 2004. Development and evaluation of a Western Mosquitofish bioenergetics model. Trans. Am. Fish. Soc. 133(5): 1150–1162.

Chiu, S. and M. V. Abrahams. 2010. Effects of turbidity and risk of predation on habitat selection decisions by Fathead Minnow (*Pimephales promelas*). Environ. Biol. Fishes 87(4): 309–316.

Choudhury, A. and T. A. Dick. 1998. The historical biogeography of sturgeons (Osteichthyes: Acipenseridae): a synthesis of phylogenetics, paleontology and paleogeography. J. Biogeogr. 25(4): 623–640.

Chu, K. H., M. Xu, and C. P. Li. 2009. Rapid DNA barcoding analysis of large datasets using the compositional vector method. BMC Bioinform. 10(S14): S8.

Ciccotto, P. J. and T. C. Mendelson. 2015. Evolution of the premaxillary frenum and substratum in snubnose darters and allies (Percidae: *Etheostoma*). J. Fish Biol. 87(4): 1090–1098.

Clapp, D. F., R. D. Clark, Jr., and J. S. Diana. 1990. Range, activity, and habitat of large, free-ranging Brown Trout in a Michigan stream. Trans. Am. Fish. Soc. 119(6): 1022–1034.

Clark, A. L. and W. D. Pearson. 1979. Early piscivory in larvae of the Freshwater Drum, *Aplodinotus grunniens*. Pages 31–59. *In*: R. Wallus and C. W. Voigtlander, eds. Proc. 2nd Annu. Larval Fish Workshop. TVA. Norris, TN. 241 pp.

Clark, E., L. R. Aronson, and M. Gordon. 1954. Mating behavior patterns in two sympatric species of xiphophorin fishes: their inheritance and significance in sexual isolation. Bull. Am. Mus. Nat. Hist. 103: 139–225.

Clark, K. E. 1978. Ecology and life history of the Speckled Madtom, *Noturus leptacanthus* (Ictaluridae). M.S. thesis. Univ. Southern Mississippi, Hattiesburg.

Clark-Kolaks, S. J., J. R. Jackson, and S. E. Lochmann. 2009. Adult and juvenile Paddlefish in floodplain lakes along the lower White River, Arkansas. Wetlands 29(2): 488–496.

Clay, T. A. 2009. Growth, survival, and cannibalism rates of Alligator Gar *Atractosteus spatula* in recirculating aquaculture systems. M.S. thesis. Nichols State Univ. Thibodaux, LA. 100 pp.

Clemens, H. P. and K. E. Sneed. 1957. The spawning behavior of the Channel Catfish *Ictalurus punctatus*. USFWS Special Sci. Rep. Fish. No. 219. 11 pp.

Clements, M. D., H. L. Bart, Jr., and D. L. Hurley. 2012. A different perspective on the phylogenetic relationships of the Moxostomatini (Cypriniformes: Catostomidae) based on cytochrome-b and growth hormone intron sequences. Mol. Phylogenet. Evol. 63(1): 159–167.

Clemmer, G. H. 1971. The systematics and biology of the *Hybopsis amblops* complex. Ph.D. diss. Tulane Univ., New Orleans. 155 pp.

Clemmer, G. H. 1980a. *Hybopsis amblops*, Bigeye Chub. Page 181. *In*: D. S. Lee et al. Atlas of North American freshwater fishes. N.C. State Mus. Nat. Hist., Raleigh.

Clemmer, G. H. 1980b. *Notropis amnis* (Hubbs and Greene), Pallid Shiner. Page 224. *In*: D. S. Lee et al. Atlas of North American freshwater fishes. N.C. State Mus. Nat. Hist., Raleigh.

Clemmer, G. H. and R. D. Suttkus. 1971. *Hybopsis lineapunctata*, a new cyprinid fish from the upper Alabama River system. Tulane Stud. Zool. Bot. 17: 21–30.

Cloutman D. G. 1974. Parasite community structure of Largemouth Bass, Warmouth, and Bluegill in Lake Fort Smith, Arkansas. Trans. Am. Fish. Soc. 104(2): 277–283.

Cloutman, D. G. 1976. Parasitism in relation to taxonomy of the sympatric sibling species of stonerollers, *Campostoma anomalum pullum* (Agassiz) and *C. oligolepis* Hubbs and Greene, in the White River, Arkansas. Southwest. Nat. 21(1): 67–70.

Cloutman, D. G. 2011. *Dactylogyrus robisoni* n. sp. (Monogenoidea: Dactylogyridae) from the Bluehead Shiner, *Pteronotropis hubbsi* (Bailey and Robison), 1978 (Pisces: Cyprinidae). Comp. Parasitol. 78(1): 1–3.

Cloutman, D. G. and L. L. Olmsted. 1974. A survey of the fishes of the Cossatot River in southwestern Arkansas. Southwest. Nat. 19(3): 257–266.

Cloutman, D. G. and L. L. Olmsted. 1976. The Fishes of Washington County, Arkansas. Arkansas Water Resources Research Center Publ. No. 39, 109 pp.

Coburn, M. M. 1982a. Anatomy and relationships of *Notropis atherinoides*. Ph.D. diss., Ohio State Univ., Columbus. 400 pp.

Coburn, M. M. 1982b. The systematic relationships of *Notropis oxyrhynchus* and *Notropis jemezanus* [abstract]. 62nd Annu. Meet. Am. Soc. Ichthyol. Herpetol.

Coburn, M. M. 1986. Egg diameter variation in eastern North American minnows (Pisces: Cyprinidae): correlation with vertebral number, habitat, and spawning behavior. Ohio J. Sci. 86(3): 110–120.

Coburn, M. M. and T. M. Cavender. 1992. Interrelationships of

North American cyprinid fishes. Pages 328–373. *In*: R. L. Mayden, ed. Systematics, historical ecology, and North American freshwater fishes. Stanford Univ. Press, Stanford, CA. 969 pp.

Cochran, P. A. 2014a. Observations on spawning by captive Sand Shiners (*Notropis stramineus*) from Minnesota. Am. Currents 39(2): 13–14.

Cochran, P. A. 2014b. Field and laboratory observations on the ecology and behavior of the Chestnut Lamprey, *Ichthyomyzon castaneus*. J. Freshw. Ecol. 29(4): 491–505.

Cochran, P. A. and R. E. Jenkins. 1994. Small fishes as hosts for parasitic lampreys. Copeia 1994(2): 499–504.

Cochran, P. A. and J. Lyons. 2004. Field and laboratory observations on the ecology and behavior of the Silver Lamprey (*Ichthyomyzon unicuspis*) in Wisconsin. J. Freshw. Ecol. 19(2): 245–253.

Cochran, P. A. and M. E. Sneen. 1995. The effect of preservation on urogenital papilla length in the Least Brook Lamprey, *Lampetra aepyptera*. SFC Proc. 31: 7–9.

Cochran, P. A., D. D. Bloom, and R. J. Wagner. 2008. Alternative reproductive behaviors in lampreys and their significance. J. Freshw. Ecol. 23(3): 437–444.

Cochran, P. A., J. Lyons, and M. R. Gehl. 2003. Parasitic attachments by overwintering Silver Lampreys, *Ichthyomyzon unicuspis*, and Chestnut Lampreys, *Ichthyomyzon castaneus*. Environ. Biol. Fishes 68(1): 65–71.

Cockerell, T. D. A. and E. M. Allison. 1909. The scales of some American Cyprinidae. Proc. Biol. Soc. Washington 22: 157–163.

Cofer, L. M. 1995. Invalidation of the Wichita Spotted Bass, *Micropterus punctulatus wichitae*, subspecies theory. Copeia 1995(2): 487–490.

Cohen, D. M. 1970. How many Recent fishes are there? Proc. California Acad. Sci. Ser. 4, 38(17): 341–346.

Coker, R. E. 1930. Studies of common fishes of the Mississippi River at Keokuk. Bull. U.S. Bur. Fish. 45: 141–225.

Cole, K. S. and R. J. F. Smith. 1987. Release of chemicals by prostaglandin-treated female Fathead Minnows, *Pimephales promelas,* that stimulate male courtship. Horm. Behav. 21(4): 440–456.

Cole, K. S. and R. J. F. Smith. 1992. Attraction of female Fathead Minnows, *Pimephales promelas*, to chemical stimuli from breeding males. J. Chem. Ecol. 18(7): 1269–1284.

Cole, L. C. 1971. Thermal pollution. *In*: Man's impact on environment. McGraw-Hill, New York. 731 pp.

Cole, R. A. 2006. Freshwater aquatic nuisance species impacts and management costs and benefits at federal water resources projects. Aquatic Nuisance Species Research Program. ANSRP Technical Notes Collection, ERDC/TN ANSRP-06-3. 14 pp.

Collette, B. B. 1962. The swamp darters of the subgenus *Hololepis* (Pisces: Percidae). Tulane Stud. Zool. 9(4): 115–211.

Collette B. B. 1963. The subfamilies, tribes, and genera of the Percidae (Teleostei). Copeia 1963(4): 615–623.

Collette, B. B. and P. Bănărescu. 1977. Systematics and zoogeography of the fishes of the family Percidae. J. Fish. Res. Board Can. 34(10): 1450–1463.

Collette, B. B. and L. W. Knapp. 1966. Catalog of type specimens of the darters (Pisces: Percidae, Etheostomatini). Proc. U.S. Natl. Mus. 119(3550): 1–88.

Collette, B. B., M. A. Ali, K. E. F. Hokanson, M. Nagiec, S. A. Smirnov, J. E. Thorpe, A. H. Weatherley, and J. Willemsen. 1977. Biology of the percids. J. Fish. Res. Board Can. 34(10): 1890–1899.

Collier, J. E. 1959. Changes in fish populations and food habits of Yellow Bass in North Twin Lake, 1956–1958. Proc. Iowa Acad. Sci. 66: 518–522.

Colvin, N. E., C. L. Racey, and S. E. Lochmann. 2008. Stocking contribution and growth of Largemouth Bass stocked at 50 and 100 mm into backwaters of the Arkansas River. N. Am. J. Fish. Manag. 28(2): 434–441.

Commens, A. M. and A. Mathis. 1999. Alarm pheromones of Rainbow Darters: responses to skin extracts of conspecifics and congeners. J. Fish Biol. 55(6): 1359–1362.

Conley, J. M. 1966. Ecology of the Flier, *Centrarchus macropterus* (Lacépède) in southeast Missouri. M.A. thesis. Univ. Missouri, Columbia. 119 pp.

Conner, J. V. 1977. Zoogeography of freshwater fishes in western Gulf Slope drainages between the Mississippi and the Rio Grande. Ph.D. diss. Tulane Univ., New Orleans. 280 pp.

Conner, J. V. 1979. Identification of larval sunfishes (Centrarchidae, Elassomatidae) from southern Louisiana. Pages 17–52. *In*: R. D. Hoyt, ed. Proceedings of the Third Symposium on Larval Fish. Western Kentucky Univ., Bowling Green. 236 pp.

Conner, J. V. and R. D. Suttkus. 1986. Zoogeography of freshwater fishes of the western Gulf Slope of North America. Pages 413–436. *In*: C. H. Hocutt and E. O. Wiley, eds. The zoogeography of North American freshwater fishes. John Wiley and Sons, New York.

Conner, J. V., R. P. Gallagher, and M. F. Chatry. 1980. Larval evidence for natural reproduction of the Grass Carp (*Ctenopharyngodon idella*) in the lower Mississippi River. Proc. 4th Annu. Larval Fish. Conf. Biol. Serv. Prog. National Power Plant Team. Ann Arbor, MI. FWS/OBS-80/43: 1–19.

Conner, J. V., C. H. Pennington, and T. R. Bosley. 1983. Larval fish of selected aquatic habitats on the lower Mississippi River. Tech. Report E-83-4. U.S. Army Corps of Engineers, Waterways Expt. Stn. Vicksburg, Miss. 48 pp.

Connior, M. B., R. Tumlison, and H. W. Robison. 2011. New records and notes on the natural history of vertebrates from Arkansas. J. Arkansas Acad. Sci. 65: 160–165.

Connior, M. B., R. Tumlison, and H. W. Robison. 2012. New vertebrate records and natural history notes from Arkansas. J. Arkansas Acad. Sci. 66: 180–184.

Connior, M. B., C. T. McAllister, and H. W. Robison. 2013. Status of an exotic salamander, *Desmognathus monticola*

(Caudata: Plethodontidae) and discovery of an introduced population of *Cottus immaculatus* (Perciformes: Cottidae), in Arkansas. J. Arkansas Acad. Sci. 67: 165–167.

Connior, M. B., R. Tumlison, H. W. Robison, C. T. McAllister, and D. A. Neely. 2014. Natural history notes and records of vertebrates from Arkansas. J. Arkansas Acad. Sci. 68: 140–145.

Conservation Fisheries, Inc. 2007. Development of propagation methods for the Yellowcheek Darter, *Etheostoma moorei*. Final report to U.S. Fish and Wildlife Service, Conway, AR Field Office. Summary of Services performed Agreement #1448–40181–02-G-070. August 1, 2002–December 31, 2006.

Constantz, G. D. 1989. Reproductive biology of poeciliid fishes. Pages 33–50. *In:* G. K. Meffe and F. F. Snelson, Jr., eds. Ecology and evolution of livebearing fishes (Poeciliidae). Prentice-Hall. Englewood Cliffs, NJ. 453 pp.

Cook, F. A. 1959. Freshwater fishes of Mississippi. Mississippi Game and Fish Commission, Jackson. 239 pp.

Cook, J. A., K. R. Bestgen, D. L. Propst, and T. L. Yates. 1992. Allozymic divergence and systematics of the Rio Grande Silvery Minnow, *Hybognathus amarus*, (Teleostei: Cyprinidae). Copeia 1992(1): 36–44.

Cooke, S. J. and D. P. Philipp. 2009. Centrarchid fishes: diversity, biology, and conservation. Blackwell Scientific Press, Cambridge. 560 pp.

Cooke, S. J., D. P. Philipp, D. H. Wahl, and P. J. Weatherhead. 2006. Energetics of parental care in six syntopic centrarchid fishes. Oecologia 148(2): 235–249.

Cooke, S. J., C. M. Bunt, S. J. Hamilton, C. A. Jennings, M. P. Pearson, M. S. Cooperman, and D. F. Markle. 2005. Threats, conservation strategies, and prognosis for suckers (Catostomidae) in North America: insights from regional case studies of a diverse family of nongame fishes. Biol. Conserv. 121(3): 317–331.

Cooke, S. L., W. R. Hill, and K. P. Meyer. 2009. Feeding at different plankton densities alters invasive Bighead Carp (*Hypophthalmichthys nobilis*) growth and zooplankton species composition. Hydrobiologia 625(1): 185–193.

Cooner, R. W. and D. R. Bayne. 1982. Diet overlap in Redbreast and Longear sunfishes from small streams in east central Alabama. Proc. Annu. Conf. Southeast. Assoc. Fish Wildl. Agencies. 36: 106–114.

Cooper, H. R. 1975. Food and feeding selectivity of two cottid species in an Ozark stream. M.S. thesis. Arkansas State Univ., Jonesboro. 45 pp.

Cooper, J. E. 1980. Ozark Cavefish, *Amblyopsis rosae*. Page 478. *In:* D. S. Lee et al. Atlas of North American freshwater fishes. N.C. State Mus. Nat. Hist., Raleigh.

Cooper, J. E. and D. P. Beiter. 1972. The Southern Cavefish, *Typhlichthys subterraneus* (Pisces: Amblyopsidae), in the eastern Mississippi Plateau of Kentucky. Copeia 1972(4): 879–881.

Coots, M. 1966. Yellow Perch. Pages 426–430. *In:* A. Calhoun, ed. Inland fisheries management. California Dept. Fish and Game.

Cope, E. D. 1865a. Partial catalogue of the cold-blooded vertebrata of Michigan. Part 1. Proc. Acad. Nat. Sci. Phila. 16: 276–285.

Cope, E. D. 1865b. Partial catalogue of the cold-blooded vertebrata of Michigan. Part 2. Proc. Acad. Nat. Sci. Phila. 17: 78–88.

Cope, E. D. 1867a. Synopsis of the Cyprinidae of Pennsylvania. Trans. Am. Philos. Soc. 13(13): 351–410.

Cope, E. D. 1867b. Description of a new genus of cyprinoid fishes from Virginia. Proc. Acad. Nat. Sci. Phila. 19: 95–97.

Cope, E. D. 1868a. On the genera of fresh-water fishes *Hypsilepis* Baird and *Photogenis* Cope, their species and distribution. Proc. Acad. Nat. Sci. Phila. (1867) 19: 156–166.

Cope, E. D. 1868b. On the distribution of fresh-water fishes in the Allegheny region of southwestern Virginia. J. Acad. Nat. Sci. Phila. (Ser. 2) 6: 207–247.

Cope, E. D. 1869. Synopsis of the Cyprinidae of Pennsylvania, with supplement on some new species of American and African fishes. Trans. Am. Philo. Soc. 13(3): 351–410.

Cope, E. D. 1870. A partial synopsis of the fishes of the fresh waters of North Carolina. Proc. Am. Philos. Soc. 11(84): 448–495.

Cope, E. D. 1871. Recent reptiles and fishes. Report on the reptiles and fishes obtained by the naturalists of the expedition. Pages 432–442. *In:* U.S. Geological Survey of Wyoming and contiguous territories (for 1870). Part IV. Special Reports.

Copeland, N. 2016. Predation of White Perch in Sooner Reservoir: is a biological control possible? Final Report: Project No. F15AF00550. Oklahoma Department of Wildlife Conservation, Norman.

Corbett, B. W. and P. M. Powles. 1983. Spawning and early-life ecological phases of the White Sucker in Jack Lake, Ontario. Trans. Am. Fish. Soc. 112(2B): 308–313.

Corbett, B. W. and P. M. Powles. 1986. Spawning and larva drift of sympatric Walleyes and White Suckers in an Ontario stream. Trans. Am. Fish. Soc. 115(1): 41–46.

Cordes, L. E. and L. M. Page. 1980. Feeding chronology and diet composition of two darters (Percidae) in the Iroquois River system, Illinois. Am. Midl. Nat. 104(1): 202–206.

Corkern, C. K. 1979. A comprehensive ichthyological survey of the Rolling Fork River, southwest, Arkansas. M.S. thesis. Northeast Louisiana Univ., Monroe. 27 pp.

Cote, C. L., P. A. Gagnaire, V. Bourret, G. Verreault, M. Castonquay, and L. Bernatchez. 2013. Population genetics of the American Eel (*Anguilla rostrata*): FST = 0 and North Atlantic oscillation effects on demographic fluctuations of a panmictic species. Mol. Ecol. 22(7): 1763–1776.

Couch, C. R., A. F. Garber, C. E. Rexroad, J. M. Abrams, J. A. Stannard, M. E. Westerman, and C. V. Sullivan. 2006. Isolation and characterization of 149 novel microsatellite DNA markers for Striped Bass, *Morone saxatilis*, and cross-species amplification in White Bass, *Morone chrysops*, and their hybrid. Mol. Ecol. Notes 6(3): 667–669.

Coughlan, D. J., B. K. Baker, D. H. Barwick, A. B. Garner, and W. R. Doby. 2007. Catostomid fishes of the Wateree River, South Carolina. Southeast. Nat. 6(2): 305–320.

Coulter, A. A., D. Keller, J. J. Amberg, E. J. Bailey, and R. R. Goforth. 2013. Phenotypic plasticity in the spawning traits of bigheaded carp (*Hypophthalmichthys* spp.) in novel ecosystems. Freshw. Biol. 58(5): 1029–1037.

Courtenay, W. R., Jr. and G. K. Meffe. 1989. Small fishes in strange places: a review of introduced poeciliids. Pages 319–331. *In*: G. K. Meffe and F. F. Snelson, Jr., eds. Ecology and evolution of livebearing fishes (Poeciliidae). Prentice Hall, Englewood Cliffs, NJ. 453 pp.

Courtenay, W. R., Jr. and C. R. Robins. 1973. Exotic aquatic organisms in Florida with emphasis on fishes: a review and recommendations. Trans. Am. Fish. Soc. 102(1): 1–12.

Courtenay, W. R., Jr. and C. R. Robins. 1975. Exotic organisms: an unsolved, complex problem. BioScience 25(5): 306–313.

Courtenay, W. R., Jr. and J. D. Williams. 2004. Snakeheads (Pisces, Channidae)—a biological synopsis and risk assessment. U. S. Geological Survey Circular 1251. Denver, CO. 143 pp.

Courtenay, W. R., Jr., D. A. Hensley, J. N. Taylor, and J. A. McCann. 1984. Distribution of exotic fishes in the continental United States. Pages 41–77. *In*: W. R. Courtenay, Jr. and J. R. Stauffer, Jr, eds. Distribution, biology, and management of exotic fishes. Johns Hopkins Univ. Press, Baltimore, MD. 430 pp.

Courtenay, W. R., Jr., D. P. Jennings, and J. D. Williams. 1991. Appendix 2: Exotic fishes. Pages 97–107. *In*: C. R. Robins, R. M. Bailey, C. E. Bond, J. R. Brooker, E. A. Lachner, R. N. Lea, and W. B. Scott. Common and scientific names of fishes from the United States and Canada, 5th edition. Special Publ. 20. American Fisheries Society, Bethesda, MD.

Couture, S. C. and M. C. Watzin. 2008. Diet of invasive adult White Perch (*Morone americana*) and their effects on the zooplankton community in Missisquoi Bay, Lake Champlain. J. Great Lakes Res. 34(3): 485–494.

Cowell, B. C. and B. S. Barnett. 1974. Life history of the Taillight Shiner, *Notropis maculatus*, in central Florida. Am. Midl. Nat. 91(2): 282–293.

Cowley, D. E. and J. E. Sublette. 1987. Food habits of *Moxostoma congestum* and *Cycleptus elongatus* (Catostomidae: Cypriniformes) in Black River, Eddy County, New Mexico. Southwest. Nat. 32(3): 411–413.

Cox, C. A. 2014. Population demographics and upstream migration of American Eels in the Ouachita, White, and Arkansas rivers. M.S. thesis. Univ. Central Arkansas, Conway. 123 pp.

Cox, C. A., J. W. Quinn, L. C. Lewis, S. R. Adams, and G. L. Adams. 2016. Population demographics of American Eels *Anguilla rostrata* in two Arkansas, U.S.A., catchments that drain into the Gulf of Mexico. J. Fish Biol. 88(3): 1088–1103.

Coyle, E. E. 1930. The algal food of *Pimephales promelas* (Fathead Minnow). Ohio J. Sci. 30(1): 23–35.

Craddock, J. R. 1965. Some aspects of the life history of the Banded Sculpin, *Cottus carolinae carolinae* in Doe Run, Meade County, Kentucky. Ph.D. diss. Univ. Louisville, Louisville, KY. 157 pp.

Crawford, B. 1957. Propagation of Channel Catfish (*Ictalurus punctatus*). Unpublished mimeo. 9 pp. Arkansas Game and Fish Commission, Centerton.

Crawford, T. 1982. The status of aquaculture in Arkansas, 1982. Annual Report for Federal Aid to Commercial Fisheries PL88-309 and Natl. Marine Fisheries Serv. Proj. 2-371-R, Job C. Arkansas Game and Fish Comm., Little Rock. 10 pp.

Crawford, T. and M. Freeze. 1982. Wild commercial fishery of Arkansas. Proc. Arkansas Acad. Sci. 36: 17–19.

Crawford, T., M. Freeze, R. Fourt, S. Henderson, G. O'Bryan, and D. Phillipp. 1984. Suspected natural hybridization of Striped Bass and White Bass in two Arkansas reservoirs. Proc. Annu. Conf. Southeast. Assoc. Fish Wildl. Agencies 38: 455–469.

Creaser, C. W. and C. L. Hubbs. 1922. A revision of the Holarctic lampreys. Occas. Pap. Mus. Zool. Univ. Michigan 120: 1–14.

Cremer, M. C. and R. O. Smitherman. 1980. Food habits and growth of Silver and Bighead carp in cages and ponds. Aquaculture 20(1): 57–64.

Crisp, D. T. 1981. A desk study of the relationship between temperature and hatching time for the eggs of five species of salmonid fishes. Freshw. Biol. 11(4): 361–368.

Croizat, L. 1958. Panbiogeography or an introductory synthesis of zoogeography, phytogeography, and geology, with notes on evolution, systematics, ecology, anthropology, etc. Published by the author. Caracas, Venezuela.

Croizat, L., G. Nelson, and D. E. Rosen. 1974. Centers of origin and related concepts. Syst. Zool. 23(2): 265–287.

Croneis, C. 1930. Geology of the Arkansas Paleozoic area. Arkansas Geological Survey, Bull. 3. 457 pp.

Cross, F. B. 1950. Effects of sewage and of a headwaters impoundment on the fishes of Stillwater Creek in Payne County, Oklahoma. Am. Midl. Nat. 43(1): 128–145.

Cross, F. B. 1953. Nomenclature in the Pimephalinae, with special reference to the Bullhead Minnow, *Pimephales vigilax perspicuus* (Girard). Trans. Kansas Acad. Sci. 56(1): 92–96.

Cross, F. B. 1967. Handbook of fishes of Kansas. Mus. Nat. Hist., Univ. of Kansas, Misc. Publ. 45: 1–357.

Cross, F. B. 1970. Occurrence of the Arkansas River Shiner, *Notropis girardi* Hubbs and Ortenburger, in the Red River system. Southwest. Nat. 14(3): 370.

Cross, F. B. and L. M. Cavin. 1971. Effects of pollution, especially from feedlots, on fishes in the upper Neosho River basin. Contribution 79. Kansas Water Resources Research Institute, Manhattan.

Cross, F. B. and J. T. Collins. 1995. Fishes in Kansas. 2nd ed. Kansas Natural History Museum, Lawrence. 315 pp.

Cross, F. B. and G. A. Moore 1952. The fishes of the Poteau River, Oklahoma and Arkansas. Am. Midl. Nat., 47(2): 396–412.

Cross, F. B., O. T. Gorman, and S. G. Haslouer. 1983. The Red River Shiner, *Notropis bairdi*, in Kansas with notes on depletion of its Arkansas River cognate, *Notropis girardi*. Trans. Kansas Acad. Sci. 86(2–3): 93–98.

Cross, F. B., R. L. Mayden, and J. D. Stewart. 1986. Fishes in the western Mississippi Basin (Missouri, Arkansas and Red rivers). Pages 362–412. *In*: C. H. Hocutt and E. O. Wiley, eds. The zoogeography of North American freshwater fishes. John Wiley and Sons, New York. 866 pp.

Cross, F. B., R. E. Moss, and J. T. Collins. 1985. Assessment of dewatering impacts on stream fisheries in the Arkansas and Cimarron rivers. Final Report to Kansas Fish and Game Comm. Contract No. 46 (1985), 161 pp.

Crossman, E. J. 1962. Predator-prey relationships in pikes (Esocidae). J. Fish. Res. Board Can. 19(5): 979–980.

Crossman, E. J. 1966. A taxonomic study of *Esox americanus* and its subspecies in eastern North America. Copeia 1966(1): 1–20.

Crossman, E. J. 1978. Taxonomy and distribution of North American esocids. Am. Fish. Soc. Special Publ. 11: 13–26.

Crossman, E. J. 1980a. *Esox americanus* Gmelin, Redfin Pickerel and Grass Pickerel. Page 131. *In*: D. S. Lee et al. Atlas of North American freshwater fishes. N.C. State Mus. Nat. Hist., Raleigh.

Crossman, E. J. 1980b. *Esox lucius* Linnaeus, Northern Pike. Pages 133–134. *In*: D. S. Lee et al. Atlas of North American freshwater fishes. N.C. State Mus. Nat. Hist., Raleigh.

Crossman, E. J. 1980c. *Esox masquinongy* Mitchill, Muskellunge. Pages 135–136. *In*: D. S. Lee et al. Atlas of North American freshwater fishes. N.C. State Mus. Nat. Hist., Raleigh.

Crossman, E. J. 1980d. *Esox niger* Lesueur, Chain Pickerel. Pages 137–138. *In*: D. S. Lee et al. Atlas of North American freshwater fishes. N.C. State Mus. Nat. Hist., Raleigh.

Crossman, E. J., E. Holm, R. Cholmondeley, and K. Tuininga. 1992. First record for Canada of the Rudd, *Scardinius erythrophthalmus*, and notes on the introduced Round Goby, *Neogobius melanostomus*. Can. Field Nat. 106(2): 206–209.

Crump. B. G. and H. W. Robison. 2000. A record of the Lake Sturgeon, *Acipenser fulvescens* Rafinesque, from the Caddo River (Ouachita River drainage), Arkansas. J. Arkansas. Acad. Sci. 54: 146.

Crumpton, J. 1971. Food habits of Longnose Gar (*Lepisosteus osseus*) and Florida Gar (*Lepisosteus platyrhincus*) collected from five central Florida lakes. Proc. Annu. Conf. Southeast. Assoc. Game Fish Comm. 24: 419–424.

Cucherousset, J. and J. D. Olden. 2011. Ecological impacts of nonnative freshwater fishes. Fisheries 36(5): 215–230.

Cudmore, B. and N. E. Mandrak. 2004. Biological synopsis of Grass Carp (*Ctenopharyngodon idella*). Canadian Manuscript Report of Fisheries and Aquatic Sciences 2705. v + 44 pp.

Cuhna, C., N. Mesquita, T. E. Dowling, A. Gilles, and M. M. Coelho. 2002. Phylogenetic relationships of Eurasian and American cyprinids using cytochrome *b* sequences. J. Fish Biol. 61(4): 929–944.

Cummings, K. S., J. M. Grady, and B. M. Burr. 1984. The life history of the Mud Darter, *Etheostoma asprigene*, in Lake Creek, Illinois. Illinois Nat. Hist. Surv. Biol. Notes 122, 16 pp.

Curd, M. R. 1960. On the food and feeding habits of the catfish *Schilbeodes exilis* (Nelson) in Oklahoma. Proc. Oklahoma Acad. Sci. 40: 26–29.

Curry, K. D. and A. Spacie. 1984. Differential use of stream habitat by spawning catostomids. Am. Midl. Nat. 111(2): 267–279.

Curtis, G. L., J. S. Ramsey, and D. L. Scarnecchia. 1997. Habitat use and movements of Shovelnose Sturgeon in Pool 13 of the upper Mississippi River during extreme low flow conditions. Environ. Biol. Fishes 50(2): 175–182.

Cuvier, G. A. 1829. Le Règne Animal, distribué d'après son organisation, pour servir de base à l'histoire naturelle des animaux et d'introduction à l'anatomie comparée. Edition 2. v. 2: i–xv + 1–406.

Cuvier, G. A. and M. A. Valenciennes. 1829. Histoire naturelle des poissons. Tome troisième. Suite du Livre troisième. Des percoïdes à dorsale unique à sept rayons branchiaux et à dents en velours ou en cardes. F. G. Levrault, Paris. v. 3: i–xxviii + 2 pp. + 1–500, Pls. 41–71.

Cuvier, G. A. and M. A. Valenciennes. 1840. Histoire naturelle des poissons. Tome quinzième. Suite du livre dix-septième. Siluroïdes. v. 15: i–xxxi + 1–540, Pls. 421–455.

Cuvier, G. A. and M. A. Valenciennes. 1844. Histoire naturelle des poissons. Tome dix-septième. Suite du livre dix-huitième. Cyprinoïdes. v. 17: i–xxiii + 1–497 + 2 pp., Pls. 487–519.

Daiber, F. C. 1952. The food and feeding relationships of the Freshwater Drum, *Aplodinotus grunniens* Rafinesque in western Lake Erie. Ohio J. Sci. 52(1): 35–46.

Daiber, F. C. 1953. Notes on the spawning population of the Freshwater Drum, *Aplodinotus grunniens* Rafinesque, in western Lake Erie. Am. Midl. Nat. 50(1): 159–171.

Dale, E. E., Jr. 1975. Environmental evaluation report on various completed channel improvement projects in eastern Arkansas. Arkansas Water Resources Research Center Publ. No. 30. 42 pp.

Daniels, G. L. and J. D. Felley. 1992. Life history and foods of *Gambusia affinis* in two waterways of southwestern Louisiana. Southwest. Nat. 37(2): 157–165.

Daniels, R. A. 1989. Significance of burying in *Ammocrypta pellucida*. Copeia 1989(1): 29–34.

Darlington, P. J., Jr. 1957. Zoogeography: the geographical distribution of animals. John Wiley & Sons, New York. 675 pp.

Darnell, R. M. 1961. Trophic spectrum of an estuarine

community, based on studies of Lake Pontchartrain, Louisiana. Ecology 42(3): 553–568.

Darnell, R. M. and R. R. Meierotto. 1965. Diurnal periodicity in the Black Bullhead, *Ictalurus melas* (Rafinesque). Trans. Am. Fish. Soc. 94(1): 1–8.

Das, S. M. and D. B. Saxena. 1956. Circulation of the blood in the respiratory region of the fishes *Labeo rohita* and *Ophicephalus striatus*. Copeia 1956(2): 100–109.

Daugherty, D. J., T. D. Bacula, and T. M. Sutton. 2008. Reproductive biology of Blue Sucker in a large midwestern river. J. Appl. Ichthyol. 24(3): 297–302.

Dauwalter, D. C. and W. L. Fisher. 2007. Spawning chronology, nest site selection and nest success of Smallmouth Bass during benign streamflow conditions. Am. Midl. Nat. 158(1): 60–78.

Dauwalter, D. C., D. K. Splinter, W. L. Fisher, and R. A. Marston. 2007. Geomorphology and stream habitat relationships with Smallmouth Bass (*Micropterus dolomieu*) abundance at multiple spatial scales in eastern Oklahoma. Can. J. Fish. Aquat. Sci. 64(8): 1116–1129.

Davenport, S. R. and T. Archdeacon. 2008. Status of Bigscale Logperch (*Percina macrolepida*) and Greenthroat Darter (*Etheostoma lepidum*) in New Mexico. Report to New Mexico Dept. of Game and Fish, Share With Wildlife Program, Albuquerque. 30 pp.

Davey, A. J. H., D. J. Booker, and D. J. Kelly. 2011. Diel variation in stream fish habitat suitability criteria: implications for instream flow assessment. Aquat. Conserv. Mar. Freshw. Ecosyst. 21(2): 132–145.

Davis, B. J. and R. J. Miller. 1967. Brain patterns in minnows of the genus *Hybopsis* in relation to feeding habits and habitat. Copeia 1967(1): 1–39.

Davis, C. C. 1959. A planktonic fish egg from fresh water. Limnol. Oceanog. 4(3): 352–355.

Davis, J. G. 2006. Reproductive biology, life history and population structure of a Bowfin, *Amia calva*, population in southeastern Louisiana. M.S. thesis. Nicholls State Univ., Thibodaux, LA. 83 pp.

Davis, J. R. 1972. The spawning behavior, fecundity rates, and food habits of the Redbreast Sunfish in southeastern North Carolina. Proc. Annu. Conf. Southeast. Assoc. Game Fish Comm. 25: 556–560.

Deacon, J. E. 1961. Fish populations, following a drought, in the Neosho and Marais des Cygnes rivers of Kansas. Mus. Nat. Hist. Univ. Kansas 13(9): 359–427.

Deacon, J. E., G. Kobetich, J. D. Williams, and S. Contreras. 1979. Fishes of North America: endangered, threatened, or of special concern, 1979. Fisheries 4(2): 29–44.

Deaton, R. 2008. Factors influencing male mating behaviour in *Gambusia affinis* (Baird & Girard) with a coercive mating system. J. Fish Biol. 72(7): 1607–1622.

DeGrandchamp, K. L., J. E. Garvey, and L. A. Csoboth. 2007. Linking adult reproduction and larval density of invasive carp in a large river. Trans. Am. Fish. Soc. 136(5): 1327–1334.

Dekar, M. P., C. McCauley, J. W. Ray, and R. S. King. 2014. Thermal tolerance, survival, and recruitment of cyprinids exposed to competition and chronic heat stress in experimental streams. Trans. Am. Fish. Soc. 143(4): 1028–1036.

DeKay, J. E. 1842. Zoology of New-York, or the New York fauna; comprising detailed descriptions of all the animals hitherto observed within the state of New-York, with brief notices of those occasionally found near its borders, and accompanied by appropriate illustrations. Part IV. Fishes. W. White, A. White, and J. Visscher, Albany, NY. 415 pp.

Delco, E. A., Jr. 1960. Sound discrimination by males of two cyprinid fishes. Texas J. Sci. 12(1–2): 48–54.

DeMont, D. J. 1982. Use of *Lepomis macrochirus* Rafinesque nests by spawning *Notemigonus crysoleucas* (Mitchill) (Pisces: Centrarchidae and Cyprinidae). Brimleyana 8:61–63.

Dence, W. A. 1933. Notes on a large Bowfin (*Amia calva*) living in a mud puddle. Copeia 1933(1): 35.

Dendy, J. S. and D. C. Scott. 1953. Distribution, life history, and morphological variations of the Southern Brook Lamprey, *Ichthyomyzon gagei*. Copeia 1953(3): 152–162.

Denoncourt, R. F. 1969. A systematic study of the Gilt Darter, *Percina evides* (Jordan and Copeland) (Pisces, Percidae). Ph.D. diss. Cornell Univ., Ithaca, NY. 209 pp.

Desbrow, C., E. J. Routledge, G. C. Brighty, J. P. Sumpter, and M. Waldock. 1998. Identification of estrogenic chemicals in STW effluent. I. Chemical fractionation and in vitro biological screening. Environ. Sci. Technol. 32(11): 1549–1558.

Desselle, W. J., M. A. Poirrier, J. S. Rogers, and R. C. Cashner. 1978. A discriminant functions analysis of sunfish (*Lepomis*) food habits and feeding niche segregation in the Lake Pontchartrain, Louisiana estuary. Trans. Am. Fish. Soc. 107(5): 713–719.

Deters, J. E., D. C. Chapman, and B. McElroy. 2013. Location and timing of Asian carp spawning in the lower Missouri River. Environ. Biol. Fishes 96(5): 617–629.

Dettmers, J. M., S. Gutreuter, D. H. Wahl, and D. A. Soluk. 2001a. Patterns in abundance of fishes in main channels of the upper Mississippi River system. Can. J. Fish. Aquat. Sci. 58(5): 933–942.

Dettmers, J. M., D. H. Wahl, D. A. Soluk, and S. Gutreuter. 2001b. Life in the fast lane: fish and food web structure in the main channel of large rivers. J. North Am. Benthol. Soc. 20(2): 255–265.

De Vlaming, V. L., A. Kuris, and F. R. Parker, Jr. 1978. Seasonal variation of reproduction and lipid reserves in some subtropical cyprinodontids. Trans. Am. Fish. Soc. 107(3): 464–472.

DeVries, D. R., J. E. Garvey, and R. A. Wright. 2009. Early life history and recruitment. Pages 105–133. *In*: S. J. Cooke and D. P. Philipp, eds. Centrarchid fishes: diversity, biology, and conservation. Wiley-Blackwell, Chichester.

DeWalt, R. E. 2011. DNA barcoding: a taxonomic point of view. J. North Am. Benthol. Soc. 30(1): 174–181.

Dewey, M. R. 1981. Seasonal abundance, movement and

diversity of fishes in an Ozark stream. Proc. Arkansas. Acad. Sci. 35: 33–39.

Dewey, M. R. and T. E. Moen. 1978. Fishes of the Caddo River, Arkansas after impoundment of DeGray Lake. Proc. Arkansas Acad. Sci. 32: 39–42.

DeWoody, J. A., D. E. Fletcher, S. D. Wilkins, W. S. Nelson, and J. C. Avise. 2000. Genetic monogamy and biparental care in an externally fertilizing fish, the Largemouth Bass (*Micropterus salmoides*). Proc. Royal Soc. London B 267(1460): 2431–2437.

Diana, M. 1966. The diet of Longnose Gar (*Lepisosteus osseus*) in Lake Griffin, before, during, and after a selective rotenone treatment with differences in male and female diets. Florida Game Freshwater Fish. Comm. 1–22 pp.

Diamond, J. 1999. Guns, germs, and steel: the fates of human societies. W. W. Norton, New York. 512 pp.

DiBenedetto, K. C. 2009. Life history characteristics of Alligator Gar *Atractosteus spatula* in the Bayou DuLarge area of southcentral Louisiana. M.S. thesis, Louisiana State Univ., Baton Rouge. 69 pp.

DiCenzo, V. J., M. J. Maceina, and M. R. Stimpert. 1996. Relations between reservoir trophic state and Gizzard Shad population characteristics in Alabama reservoirs. N. Am. J. Fish. Manag. 16(4): 888–895.

Dickson, F. J. 1949. The biology of the Round Flier, *Centrarchus macropterus* (Lacépède). M.S. thesis. Alabama Polytechnic Institute, Auburn. 56 pp.

Dieterman, D. J. 2000. Spatial patterns in phenotypes and habitat use of Sicklefin Chub, *Macrhybopsis meeki*, in the Missouri and lower Yellowstone rivers. Ph.D. diss. Univ. Missouri, Columbia.

Dieterman, D. J. and D. L. Galat. 2004. Large-scale factors associated with Sicklefin Chub distribution in the Missouri and lower Yellowstone rivers. Trans. Am. Fish. Soc. 133(3): 577–587.

Dieterman, D. J. and D. L. Galat. 2005. Variation in body form, taste buds, and brain patterns of the Sicklefin Chub, *Macrhybopsis meeki*, in the Missouri River and Lower Yellowstone River USA. J. Freshw. Ecol. 20(3): 561–573.

Dieterman, D. J., E. Roberts, P. J. Braaten, and D. L. Galat. 2006. Reproductive development in the Sicklefin Chub in the Missouri and lower Yellowstone rivers. Prairie Nat. 38(2): 113–130.

Dillman, C. B., D. E. Bergstrom, D. B. Noltie, T. P. Holtsford, and R. L. Mayden. 2011. Regressive progression, progressive regression or neither? Phylogeny and evolution of the Percopsiformes (Teleostei, Paracanthopterygii). Zool. Scripta 40(1): 45–60.

Dillman, C. B., R. M. Wood, B. R. Kuhajda, J. M. Ray, V. B. Salnikov, and R. L. Mayden. 2007. Molecular systematics of Scaphirhynchinae: an assessment of North American and central Asian freshwater sturgeon fishes. J. Appl. Ichthyol. 23(4): 290–296.

Dimmick, W. W. 1987. Phylogenetic relationships of *Notropis hubbsi, N. welaka* and *N. emiliae* (Cypriniformes: Cyprinidae). Copeia 1987(2): 316–325.

Dimmick, W. W. 1988. Ultrastructure of North American cyprinid maxillary barbels. Copeia 1988(1): 72–80.

Dimmick, W. W. 1993. A molecular perspective on the phylogenetic relationships of the barbeled minnows, historically assigned to the genus *Hybopsis* (Cyprinidae: Cypriniformes). Mol. Phylogenet. Evol. 2(3): 173–184.

Dimmick, W. W. and B. M. Burr. 1999. Phylogenetic relationships of the suckermouth minnows, genus *Phenacobius*, inferred from parsimony analyses of nucleotide sequence, allozymic and morphological data (Cyprinidae, Cypriniformes). Biochem. Syst. Ecol. 27(5): 469–485.

Dimmick, W. W. and A. Larson. 1996. A molecular and morphological perspective on the phylogenetic relationships of the otophysan fishes. Mol. Phylogenet. Evol. 6(1): 120–133.

Dingerkus, G. and W. M. Howell. 1976. Karyotypic analysis and evidence of tetraploidy in the North American Paddlefish, *Polyodon spathula*. Science 194(4267): 842–844.

Distler, D. A. 1968. Distribution and variation of *Etheostoma spectabile* (Agassiz) (Percidae. Teleostei). Univ. Kansas Sci. Bull. 48(5): 143–208.

Distler, D. A. 1972. Observations on the reproductive habits of captive *Etheostoma cragini* Gilbert. Southwest. Nat. 16(3–4): 439–441.

Distler, D. A. and A. L. Metcalf. 1962. *Etheostoma pallididorsum*, a new percid fish from the Caddo River system of Arkansas. Copeia 1962(3): 556–561.

Divers, S. J., S. S. Boone, J. J. Hoover, K. A. Boysen, K. J. Killgore, C. E. Murphy, S. G. George, and A. C. Camus. 2009. Field endoscopy for identifying gender, reproductive stage and gonadal anomalies in free-ranging sturgeon (*Scaphirhynchus*) from the lower Mississippi River. J. Appl. Ichthyol. 25(S2): 68–74.

Divers. S. J., S. S. Boone, A. Berliner, E. A. Kurimo, K. A. Boysen, D. R. Johnson, K. J. Killgore, S. G. George, and J. J. Hoover. 2013. Non-lethal acquisition of large liver samples from free-ranging river sturgeon (*Scaphirhynchus*) using single-entry endoscopic biopsy forceps. J. Wildl. Dis. 49(2): 321–331.

Divino, J. N. and W. M. Tonn. 2008. Importance of nest and paternal characteristics for hatching success in Fathead Minnow. Copeia 2008(4): 920–930.

Doan, K. H. 1938. Observations on Dogfish (*Amia calva*) and their young. Copeia 1938(4): 204.

Docker, M. F. and F. W. H. Beamish. 1991. Growth, fecundity, and egg size of Least Brook Lamprey, *Lampetra aepyptera*. Environ. Biol. Fishes 31(3): 219–227.

Docker, M. F., N. E. Mandrak, and D. D. Heath. 2012. Contemporary gene flow between "paired" Silver (*Ichthyomyzon unicuspis*) and Northern Brook (*I. fossor*) lampreys: implications for conservation. Conserv. Genet. 13(3): 823–835.

Docker, M. F., J. H. Youson, R. J. Beamish, and R. H. Devlin. 1999. Phylogeny of the lamprey genus *Lampetra* inferred from mitochondrial cytochrome *b* and ND3 gene sequences. Can. J. Fish. Aquat. Sci. 56(12): 2340–2349.

Doherty, C. A., R. A. Curry, and K. R. Munkittrick. 2010. Spatial and temporal movements of White Sucker: implications for use as a sentinel species. Trans. Am. Fish. Soc. 139(6): 1818–1827.

Dominey, W. J. 1980. Female mimicry in male Bluegill sunfish—a genetic polymorphism? Nature 284: 546–548.

Donabauer, S. B. 2007. Reproduction, habitat use, survival, and interpool movement of Paddlefish in the mid-reaches of the Arkansas River, Arkansas. M.S. thesis. Arkansas Tech Univ., Russellville. 88 pp.

Donabauer, S. B., J. N. Stoeckel, and J. W. Quinn. 2009. Exploitation, survival, reproduction, and habitat use of gravid female Paddlefish in Ozark Lake, Arkansas River, Arkansas. Pages 123–140. *In*: C. P. Paukert and G. D. Scholten, eds. Paddlefish management, propagation, and conservation in the 21st century: building from 20 years of research and management. Symp. 66. American Fisheries Society, Bethesda, MD. 443 pp.

Donald, D. B. and A. H. Kooyman. 1977. Food, feeding habits and growth of Goldeye, *Hiodon alosoides* (Rafinesque), in waters of the Peace-Athabasca Delta. Can. J. Zool. 55(6): 1038–1047.

Doosey, M. H., H. L. Bart, Jr., K. Saitoh, and M. Miya. 2010. Phylogenetic relationships of catostomid fishes (Actinopterygii: Cypriniformes) based on mitochondrial ND4/ND5 gene sequences. Mol. Phylogenet. Evol. 54(3): 1028–1034.

Doroshov, S. I. 1985. Biology and culture of sturgeon Acipenseriformes. Pages 251–274. *In:* J. F. Muir and R. J. Roberts, eds. Recent advances in aquaculture. Vol. 2. Westview Press, Boulder, CO.

Dorsa, W. J. and R. A. Fritzsche. 1979. Characters of newly hatched larvae of *Morone chrysops* (Pisces: Percichthyidae), from Yocona River, Mississippi. J. Mississippi Acad. Sci. 24: 37–41.

Douglas, N. H. 1972. *Noturus taylori*, a new species of madtom (Pisces: Ictaluridae) from the Caddo River, southwest Arkansas. Copeia 1972(4): 785–789.

Douglas, N. H. 1974. Freshwater fishes of Louisiana. Claitors Publ. Div., Baton Rouge. 443 pp.

Douglas, N. H. and J. L. Harris. 1977. Occurrence of the chub genus *Nocomis* (Cyprinidae) in the Ouachita River drainage, west-central Arkansas. ASB Bull. 24(2): 47.

Dowell, V. E. and C. D. Riggs. 1958. Further observations on *Astyanax fasciatus* and *Menidia audens* in Lake Texoma. Proc. Oklahoma Acad. Sci. 36: 52–53.

Dowling, T. E. and W. S. Moore. 1984. Level of reproductive isolation between two cyprinid fishes, *Notropis cornutus* and *N. chrysocephalus*. Copeia 1984(3): 617–628.

Dowling, T. E. and W. S. Moore. 1985a. Evidence for selection against hybrids in the family Cyprinidae (genus *Notropis*). Evolution 39(1): 152–158.

Dowling, T. E. and W. S. Moore. 1985b. Genetic variation and divergence of the sibling pair of cyprinid fishes, *Notropis cornutus* and *N. chrysocephalus*. Biochem. Syst. Ecol. 13(4): 471–476.

Dowling, T. E. and W. S. Moore. 1986. Absence of population subdivision in the Common Shiner, *Notropis cornutus* (family Cyprinidae). Environ. Biol. Fishes 15(2): 151–155.

Dowling, T. E. and G. J. P. Naylor. 1997. Evolutionary relationships of minnows in the genus *Luxilus* (Teleostei: Cyprinidae) as determined from cytochrome b sequences. Copeia 1997(4): 758–765.

Dowling, T. E., G. R. Smith, and W. M. Brown. 1989. Reproductive isolation and introgression between *Notropis cornutus* and *Notropis chrysocephalus* (family Cyprinidae): comparison of morphology, allozymes, and mitochondrial DNA. Evolution 43(3): 620–634.

Dowling, T. E., W. R. Hoeh, G. R. Smith, and W. M. Brown. 1992. Evolutionary relationships of shiners in the genus *Luxilus* (Cyprinidae) as determined by analysis of mitochondrial DNA. Copeia 1992(2): 306–322.

Drake, M. T., J. E. Claussen, D. P. Philipp, and D. L. Pereira. 1997. A comparison of Bluegill reproductive strategies and growth among lakes with different fishing intensities. N. Am. J. Fish. Manag. 17(2): 496–507.

Drenner, R. W., J. R. Mummert, and W. J. O'Brien. 1982. Filter-feeding rates of Gizzard Shad. Trans. Am. Fish. Soc. 111(2): 210–215.

Driscoll, M. P. and L. E. Miranda. 1999. Diet ecology of Yellow Bass, *Morone mississippiensis*, in an oxbow of the Mississippi River. J. Freshw. Ecol. 14(4): 477–486.

Driver, L. J. 2009. Reproductive life history of the Western Sand Darter, *Ammocrypta clara,* in Northeast Arkansas. M.S. thesis. Univ. Central Arkansas, Conway. 68 pp.

Driver, L. J. and G. L. Adams. 2013. Life history and spawning behavior of the Western Sand Darter (*Ammocrypta clara*) in northeast Arkansas. Am. Midl. Nat. 170(2): 199–212.

Dryer, M. P. and A. J. Sandvol. 1993. Recovery plan for the Pallid Sturgeon (*Scaphirhynchus albus*). USFWS. Denver, CO. 55 pp.

Dudely, R. K. and S. P. Platania. 2007. Flow regulation and fragmentation imperil pelagic-spawning riverine fishes. Ecol. Appl. 17(7): 2074–2086.

Dudgeon, D., A. H. Arthington, M. O. Gessner, Z.-I. Kawabata, D. J. Knowler, C. Leveque, R. J. Naiman, A. H. Prieur-Richard, D. Soto, M. L. J. Stiassny, and C. A. Sullivan. 2006. Freshwater biodiversity: importance, threats, status, and conservation challenges. Biol. Rev. Cambridge Philos. Soc. 81(2): 163–182.

Dugas, C. N., M. Konikoff, and M. F. Trahan. 1976. Stomach contents of Bowfin (*Amia calva*) and Spotted Gar (*Lepisosteus oculatus*) taken in Henderson Lake, Louisiana. Louisiana Acad. Sci. 39: 28–34.

Dugo, M. A., W. T. Slack, B. R. Kreiser. 2008. Conservation of the *Cottogaster* (*Percina copelandi* sp. clade): applying molecular phylogenetics and phylogeography towards the reconstruction of evolutionary history (Year II). Museum Technical Report, No. 137. Mississippi Museum of Natural Science, Jackson.

Dugo, M. A., W. T. Slack, and B. R. Kreiser. 2011. Phylogeography of the *Cottogaster* (*Percina copelandi* sp. clade). Abstract. 2011 Southeastern Fishes Council Annual Meeting, Chattanooga, TN.

Dugo, M. A., B. R. Kreiser, W. T. Slack, P. B. Tchounwou. 2012. Gulf coastal plain vicariance and speciation within the *Cottogaster* (*Percina copelandi* sp. clade). Abstract. 2012 Southeastern Fishes Council Annual Meeting, New Orleans, LA.

Dukravets, G. M. 1992. The Amur Snakehead, *Channa argus warpachowskii*, in the Talas and Chu River drainages. J. Ichthyol. 31(5): 147–151.

Dukravets, G. M. and A. L. Machulin. 1978. The morphology and ecology of the Amur Snakehead, *Ophiocephalus argus warpachowskii*, acclimatized in the Syr Dar'ya basin. J. Ichthyol. 18(2): 203–208.

Duncan, T. O. and M. R. Myers, Jr. 1978a. Movements of Channel Catfish and Flathead Catfish in Beaver Reservoir, northwest Arkansas. Proc. Arkansas Acad. Sci. 32: 43–45.

Duncan, T. O. and M. R. Myers, Jr. 1978b. Longevity of White Bass in Beaver Reservoir, Arkansas. Proc. Arkansas Acad. Sci. 32: 87–88.

Dunham, J. B., S. B. Adams, R. E. Schroeter, and D. C. Novinger. 2002. Alien invasions in aquatic ecosystems: toward an understanding of Brook Trout invasions and potential impacts on inland Cutthroat Trout in western North America. Rev. Fish Biol. Fish. 12(4): 373–391.

Dunham, J. B., D. S. Pilliod, and M. K. Young. 2004. Assessing the consequences of nonnative trout in headwater ecosystems in western North America. Fisheries 29(6): 18–26.

Dunham, R. A., K. G. Norgren, L. Robison, R. O. Smitherman, T. Steeger, D. C. Peterson, and M. Gibson. 1994. Hybridization and biochemical genetics of Black and White crappies in the southeastern USA. Trans. Am. Fish. Soc. 123(2): 141–149.

Dunlop, E. S., B. J. Shuter, and M. S. Ridgway. 2005. Isolating the influence of growth rate on maturation patterns in the Smallmouth Bass (*Micropterus dolomieu*). Can. J. Fish. Aquat. Sci. 62(4): 844–853.

Dunn, D. M. 1999. Determining the possible limitations by cottid species on Brown Trout populations in the Bull Shoals Lake cold tailwaters. M.S. thesis, Arkansas State Univ., Jonesboro. 44 pp.

Dunn, J. R. 1997. Charles Henry Gilbert (1859–1928): Pioneer ichthyologist of the American West. Pages 265–278. *In*: T. W. Pietsch and W. D. Anderson, Jr., eds. Collection building in ichthyology and herpetology. Am. Soc. Ichthyol. Herpetol Special Publ. No. 3. Lawrence, KS. 593 pp.

Durham, B. W. and G. R. Wilde. 2008a. Asynchronous and synchronous spawning by Smalleye Shiner *Notropis buccula* from the Brazos River, Texas. Ecol. Freshw. Fish 17(4): 528–541.

Durham, B. W. and G. R. Wilde. 2008b. Composition and abundance of drifting fish larvae in the Canadian River, Texas. J. Freshw. Ecol. 23(2): 273–280.

Duvernell, D. and J. F. Schaefer. 2014. Variation in contact zone dynamics between two species of topminnows, *Fundulus notatus* and *F. olivaceus*, across isolated drainage systems. Evol. Ecol. 28(1): 37–53.

Duvernell, D. D., J. F. Schaefer, D. C. Hancks, J. A. Fonoti, and A. M. Ravanelli. 2007. Hybridization and reproductive isolation among syntopic populations of the topminnows, *Fundulus notatus* and *F. olivaceus*. J. Evol. Biol. 20(1): 152–164.

Duvernell, D. D., S. L. Meier, J. F. Schaefer, and B. R. Kreiser. 2013. Contrasting phylogeographic histories between broadly sympatric topminnows in the *Fundulus notatus* species complex. Mol. Phylogenet. Evol. 69(3): 653–663.

Dyer, B. S. 1997. Phylogenetic revision of Atherinopsinae (Teleostei, Atherinopsidae), with comments on the systematics of the South American freshwater fish genus *Basilichthys* Girard. Misc. Publ. Mus. Zool. Univ. Michigan. 185: 1–64.

Dyer, B. S. 1998. Phylogenetic systematics and historical biogeography of the Neotropical silverside family Atherinopsidae (Teleostei: Atheriniformes). Pages 519–536. *In*: L. R. Malabarba, R. E. Reis, R. P. Vari, Z. M. S. Lucena, and C. A. S. Lucena, eds. Phylogeny and classification of Neotropical fishes. EDIPUCRS, Porto Alegre, Brazil. 603 pp.

Dyer, B. S. 2006. Systematic revision of the South American silversides (Teleostei, Atheriniformes). Biocell 30(1): 69–88.

Dyer, B. S. and B. Chernoff. 1996. Phylogenetic relationships among atheriniform fishes (Teleostei: Atherinomorpha). Zool. J. Linn. Soc. 117(1): 1–69.

Earnest, K., J. Scott, J. Schaefer, and D. Duvernell. 2014. The landscape genetics of syntopic topminnows (*Fundulus notatus* and *F. olivaceus*) in a riverine contact zone. Ecol. Freshw. Fish 23(4): 572–580.

Eastman, J. T. 1977. The pharyngeal bones and teeth of catostomid fishes. Am. Midl. Nat. 97(1): 68–88.

Eaton, T. H., Jr. 1956. Notes on the olfactory organs in Centrarchidae. Copeia 1956(3): 196–199.

Ebbers, M. A., P. J. Colby, and C. A. Lewis. 1988. Walleye-Sauger bibliography. Investigative Report No. 396. Minnesota Department of Natural Resources. 201 pp.

Eberhard, W. G. 1985. Sexual selection and animal genitalia. Harvard Univ. Press, Cambridge, MA. 244 pp.

Ebert, D. J. and S. P. Filipek. 1991. Evaluation of feeding and habitat competition between native Smallmouth Bass (*Micropterus dolomieui*) and Rainbow Trout (*Oncorhynchus mykiss*) in a coolwater stream. Pages 49–54. *In*: D. C. Jackson, ed. First International Smallmouth Bass Symposium. MAFES, Mississippi State Univ., Starkville.

Ebert, D. J., A. V. Brown, and C. B. Fielder. 1987. Distribution of fish within headwater riffles of the Illinois River system, Washington County, Arkansas. Proc. Arkansas Acad. Sci. 41: 38–42.

Eberts, R. C., Jr., V. J. Santucci, Jr., and D. H. Wahl. 1998. Suitability of the Lake Chubsucker as prey for Largemouth Bass in small impoundments. N. Am. J. Fish. Manag. 18(2): 295–307.

Echelle, A. A. 1968. Food habits of young-of-year Longnose Gar in Lake Texoma, Oklahoma. Southwest. Nat. 13(1): 45–50.

Echelle, A. A. and A. F. Echelle. 1984. Evolutionary genetics of a species flock: atherinid fishes on the Mesa Central of Mexico. Pages 93–110. *In*: A. A. Echelle and I. Kornfield, eds. Evolution of fish species flocks. Univ. Maine at Orono Press.

Echelle, A. A. and J. B. Mense. 1967. Forage value of Mississippi Silversides in Lake Texoma. Proc. Oklahoma Acad. Sci. 47: 394–396.

Echelle, A. A. and C. D. Riggs. 1972. Aspects of the early life history of gars (*Lepisosteus*) in Lake Texoma. Trans. Am. Fish. Soc. 101(1): 106–112.

Echelle, A. A., A. F. Echelle, and L. G. Hill. 1972. Interspecific interactions and limiting factors of abundance and distribution in the Red River Pupfish, *Cyprinodon rubrofluviatilis*. Am. Midl. Nat. 88(1): 109–130.

Echelle, A. A., J. R. Schenck, and L. G. Hill. 1974. *Etheostoma spectabile– E. radiosum* hybridization in Blue River, Oklahoma. Am. Midl. Nat. 91(1): 182–194.

Echelle, A. A., A. F. Echelle, M. H. Smith, and L. G. Hill. 1975. Analysis of genic continuity in a headwater fish, *Etheostoma radiosum* (Percidae). Copeia 1975(2): 197–204.

Echelle, A. A., A. F. Echelle, L. R. Williams, C. S. Toepfer, and W. L. Fisher. 1999. Allozyme perspective on genetic variation in a threatened percid fish, the Leopard Darter (*Percina pantherina*). Am. Midl. Nat. 142(2): 393–400.

Echelle, A. A., M. R. Schwemm, N. J. Lang, J. S. Baker, R. M. Wood, T. J. Near, and W. L. Fisher. 2015. Molecular systematics of the Least Darter (Percidae: *Etheostoma microperca*): historical biogeography and conservation implications. Copeia 2015(1): 87–98.

Echelle, A. A., M. R. Schwemm, N. J. Lang, B. C. Nagle, A. M. Simons, P. J. Unmack, W. L. Fisher, and C. W. Hoagstrom. 2014. Molecular systematics and historical biogeography of the *Nocomis biguttatus* species group (Teleostei: Cyprinidae): nuclear and mitochondrial introgression and a cryptic Ozark species. Mol. Phylogenet. Evol. 81: 109–119.

Echelle, A. A., N. J. Lang, W. C. Borden, M. R. Schwemm, C. W. Hoagstrom, D. J. Eisenhour, R. L. Mayden, and R. A. Van Den Bussche. 2018. Molecular systematics of the North American chub genus *Macrhybopsis* (Teleostei: Cyprinidae). Zootaxa 4375(4): 537–554.

Echelle, A. F. and A. A. Echelle. 2005. Reproductive behavior in Banded Pygmy Sunfish, *Elassoma zonatum* (Elassomatidae), with comments on the implications for relationships of the genus. Am. Currents 31(1): 1–6.

Edds, D. R., W. J. Matthews, and F. P. Gelwick. 2002. Resource use by large catfishes in a reservoir: is there evidence for interactive segregation and innate differences? J. Fish Biol. 60(3): 739–750.

Eddy, S. and J. C. Underhill. 1974. Northern fishes: with special reference to the upper Mississippi valley. Univ. Minnesota Press, Minneapolis. 436 pp.

Edeline, E. 2007. Adaptive phenotypic plasticity of eel diadromy. Mar. Ecol. Prog. Ser. 341: 229–232.

Eder, S. and C. A. Carlson. 1977. Food habits of Carp and White suckers in the South Platte and St. Vrain rivers and Goosequill Pond, Weld County, Colorado. Trans. Am. Fish. Soc. 106(4): 339–346.

Edwards, R. J. 1978. The effect of hypolimnion reservoir releases on fish distribution and species diversity. Trans. Am. Fish. Soc. 107(1): 71–77.

Edwards, R. J. and S. Contreras-Balderas. 1991. Historical changes in the ichthyofauna of the lower Rio Grande (Rio Bravo del Norte), Texas and Mexico. Southwest. Nat. 36(2): 201–212.

Egge, J. J. D. 2007. The osteology of the Stonecat, *Noturus flavus* (Siluriformes: Ictaluridae), with comparisons to other siluriforms. Bull. Alabama Mus. Nat. Hist. 25: 71–89.

Egge, J. J. D. 2010. Systematics of ictalurid catfishes: a review of the evidence. Pages 363–378. *In*: J. S. Nelson, H. P. Schultze, and M. V. H. Wilson, eds. Origin and phylogenetic interrelationships of teleosts. Verlag Dr. Friedrich Pfeil. München, Germany.

Egge, J. J. D. and A. M. Simons. 2006. The challenge of truly cryptic diversity: diagnosis and description of a new madtom catfish (Ictaluridae: *Noturus*). Zool. Scripta 35(6): 581–595.

Egge, J. J. D. and A. M. Simons. 2009. Molecules, morphology, missing data and the phylogenetic position of a recently extinct madtom catfish (Actinopterygii: Ictaluridae). Zool. J. Linn. Soc. 155(1): 60–75.

Egge, J. J. D. and A. M. Simons. 2011. Evolution of venom delivery structures in madtom catfishes (Siluriformes: Ictaluridae). Biol. J. Linn. Soc. 102(1): 115–129.

Eggleton, M. A. and H. L. Schramm, Jr. 2004. Feeding ecology and energetic relationships with habitat of Blue Catfish, *Ictalurus furcatus*, and Flathead Catfish, *Pylodictis olivaris*, in the lower Mississippi River, U.S.A. Environ. Biol. Fishes 70(2): 107–121.

Eggleton, M. A., B. G. Batten, and S. E. Lochmann. 2010. Largemouth Bass fishery characteristics in the Arkansas River, Arkansas. Proc. Annu. Conf. Southeast. Assoc. Fish Wildl. Agencies 64: 160–167.

Ehlinger, T. J. 1989. Foraging mode switches in the Golden Shiner (*Notemigonus crysoleucas*). Can. J. Fish. Aquat. Sci. 46(7): 1250–1254.

Ehlinger, T. J. 1997. Male reproductive competition and sex-specific growth patterns in Bluegill. N. Am. J. Fish. Manag. 17(2): 508–515.

Eichelberger, J. S., P. J. Braaten, D. B. Fuller, M. S. Krampe, and E. J. Heist. 2014. Novel single-nucleotide polymorphism markers confirm successful spawning of endangered Pallid Sturgeon in the upper Missouri River basin. Trans. Am. Fish. Soc. 143(6): 1373–1385.

Eigenmann, C. H. 1887. Notes on the specific names of certain North American fishes. Proc. Acad. Nat. Sci. Phila. 39: 295–296.

Eigenmann, C. H. 1897. The Amblyopsidae, the blind fish of America. Report of the British Association for the Advancement of Science 1897: 685–686.

Eigenmann, C. H. 1898a. On the Amblyopsidae. Science (n.s.) 7 (164): 227.

Eigenmann, C. H. 1898b. A new blind fish. Proc. Indiana Acad. Sci. 1897: 231.

Eigenmann, C. H. 1899a. A case of convergence. Science 9(217): 280–282.

Eigenmann, C. H. 1899b. The eyes of the blind vertebrates of North America. I. The eyes of the Amblyopsidae. Archiv. für Entwicklungsmechanik der Organismen 8(4): 545–617.

Eigenmann, C. H. 1900. The blind-fishes. Biological Lectures from the Marine Biological Laboratory of Woods Hole (for 1899) 8: 113–126.

Eilers, C. D. 2008. Movement, home-range, and habitat selection of Muskellunge (*Esox masquinongy* Mitchell) in Thornapple Lake, Michigan. M.S. thesis, Central Michigan Univ., Mount Pleasant. 109 pp.

Eisenhour, D. J. 1996. Distribution and systematics of *Notropis wickliffi* (Cypriniformes: Cyprinidae) in Illinois. Trans. Illinois State Acad. Sci. 90(1–2): 65–78.

Eisenhour, D. J. 1997. Systematics, variation and speciation of the *Macrhybopsis aestivalis* complex (Cypriniformes: Cyprinidae) west of the Mississippi River. Ph.D. diss., Southern Illinois Univ., Carbondale. 230 pp.

Eisenhour, D. J. 1999. Systematics of *Macrhybopsis tetranema* (Cypriniformes: Cyprinidae). Copeia 1999(4): 969–980.

Eisenhour, D. J. 2004. Systematics, variation and speciation of the *Macrhybopsis aestivalis* complex west of the Mississippi River. Bull. Alabama Mus. Nat. Hist. 23: 9–47.

Eldredge, N. and J. Cracraft. 1980. Phylogenetic patterns and the evolutionary process: method and theory in comparative biology. Columbia Univ. Press, New York. 349 pp.

Eley, R. L., J. C. Randolph, and R. J. Miller. 1975. Current status of the Leopard Darter, *Percina pantherina*. Southwest. Nat. 20(3): 343–354.

Elliot, J. M. 1994. Quantitative ecology and the Brown Trout. Oxford Univ. Press, New York. 298 pp.

Ellis, D. V. and M. A. Giles. 1965. The spawning behavior of the Walleye, *Stizostedion vitreum* (Mitchill). Trans. Am. Fish. Soc. 94(4): 358–362, and N. Am. J. Fish. Manag. 4(4A): 355–364.

Ellison, D. G. 1984. Trophic dynamics of a Nebraska Black Crappie and White Crappie population. N. Am. J. Fish. Manag. 4(4A): 355–364.

Elrod, J. H. and T. J. Hassler. 1971. Vital statistics of seven fish species in Lake Sharpe, South Dakota 1964–69. Pages 27–40. *In*: G. E. Hall, ed. Reservoir fisheries and limnology. American Fisheries Society, Special Publ. 8.

Elrod, J. H. and R. O'Gorman. 1991. Diet of juvenile Lake Trout in southern Lake Ontario in relation to abundance and size of prey fishes, 1979–1987. Trans. Am. Fish. Soc. 120(3): 290–302.

Elston, R. and B. Bachen. 1976. Diel feeding cycle and some effects of light on feeding intensity of the Mississippi Silverside, *Menidia audens*, in Clear Lake, California. Trans. Am. Fish. Soc. 105(1): 84–88.

Ely, P. C., S. P. Young, and J. J. Isely. 2008. Population size and relative abundance of adult Alabama Shad reaching Jim Woodruff Lock and Dam, Apalachicola River, Florida. N. Am. J. Fish. Manag. 28(3): 827–831.

Ensign, W. E., K. N. Leftwich, P. L. Angermeier, and C. A. Dolloff. 1997. Factors influencing stream fish recovery following a large-scale disturbance. Trans. Am. Fish. Soc. 126(6): 895–907.

Epifanio, J. M., J. B. Koppelman, M. A. Nedbal, and D. P. Philipp. 1996. Geographic variation of Paddlefish allozymes and mitochondrial DNA. Trans. Am. Fish. Soc. 125(4): 546–561.

Eschmeyer, W. N. and J. D. Fong. 2017. Species by family/subfamily in the catalog of fishes. Available: http://research archive.calacademy.org/research/ichthyology/catalog /SpeciesByFamily.asp [Accessed 11 February 2017]

Eschmeyer, W. N., E. S. Herald, and H. Hammann. 1983. A field guide to Pacific Coast fishes of North America. Houghton Mifflin, Boston. 336 pp.

Eschmeyer, W. N., R. Fricke, and R. van der Laan, eds. 2018. Catalog of fishes: genera, species, references. California Acad. Sci., San Francisco. [Subsequently renamed Eschmeyer's Catalog of Fishes]. Available: http://research archive.calacademy.org/research/ichthyology/catalog /fishcatmain.asp.

Estes, R. D. 1987. Selected bibliography, the eastern Brook Trout, *Salvelinus fontinalis* (Mitchill). Trout Commitee, Southern Division, American Fisheries Soc. 82 pp.

Esteve, M. 2005. Observations of spawning behaviour in Salmoninae: *Salmo*, *Oncorhynchus*, and *Salvelinus*. Rev. Fish Biol. Fish. 15(1–2): 1–21.

Esteve, M. and D. A. McLennan. 2007. The phylogeny of *Oncorhynchus* (Euteleostei: Salmonidae) based on behavioral and life history characters. Copeia 2007(3): 520–533.

Esteve, M., D. A. McLennan, and J. M. Gunn. 2008. Lake Trout (*Salvelinus namaycush*) spawning behaviour: the evolution of a new female strategy. Environ. Biol. Fishes 83(1): 69–76.

Etchison, L. J. and M. Pyron. 2014. Day and night substrate use in six minnow species. Am. Midl. Nat. 171(2): 321–327.

Etnier, D. A. 1976. *Percina* (*Imostoma*) *tanasi*, a new percid fish from the Little Tennessee River, Tennessee. Proc. Biol. Soc. Washington 88(44): 469–488.

Etnier, D. A. and H. W. Robison. 2004. An unusual *Hybognathus* (Osteichthyes, Cyprinidae) from lower White River, Arkansas. J. Arkansas Acad. Sci. 58: 109–110.

Etnier, D. A. and W. C. Starnes. 1993. The fishes of Tennessee. Univ. Tennessee Press, Knoxville. 681 pp.

Etnier, D. A. and W. C. Starnes. 2001. The fishes of Tennessee. Univ. Tennessee Press, Knoxville. 681 pp.

Etnier, D. A. and J. D. Williams. 1989. *Etheostoma* (*Nothonotus*) *wapiti* (Osteichthyes, Percidae), a new darter from the southern bend of the Tennessee River in Alabama and Tennessee. Proc. Biol. Soc. Washington 102(4): 987–1000.

Etnier, D. A., W. C. Starnes, and B. H. Bauer. 1979. Whatever happened to the Silvery Minnow (*Hybognathus nuchalis*) in the Tennessee River? SFC Proc. 2(3): 1–3.

Etheridge, G. G. 1974. A comprehensive ichthyological survey of the Cossatot River, southwest Arkansas, prior to the completion of Gillham Dam. M.S. thesis, Northeast Louisiana Univ., Monroe. 38 pp.

Evans, D. O., J. M. Casselman, and C. C. Wilcox. 1991. Effects of exploitation, loss of nursery habitat, and stocking on the dynamics and productivity of Lake Trout populations in Ontario lakes. Lake Trout Synthesis, Ontario Ministry of Natural Resources, Canada. 193 pp.

Evans, H. E. and E. E. Deubler, Jr. 1955. Pharyngeal tooth replacement in *Semotilus atromaculatus* and *Clinostomus elongatus*, two species of cyprinid fishes. Copeia 1955(1): 31–44.

Evans, J. D. and L. M. Page. 2003. Distribution and relative size of the swim bladder in *Percina*, with comparisons to *Etheostoma*, *Crystallaria*, and *Ammocrypta* (Teleostei: Percidae). Environ. Biol. Fishes 66(1): 61–65.

Evans, J. P., A. Pilastro, and I. Schlupp, eds. 2011. Ecology and evolution of poeciliid fishes. Univ. Chicago Press, Chicago. 424 pp.

Evans, J. W. and R. L. Noble. 1979. The longitudinal distribution of fishes in an east Texas stream. Am. Midl. Nat. 101(2): 333–343.

Evenhuis, B. L. 1970. Seasonal and daily food habits of Goldeye, *Hiodon alosoides* (Rafinesque), in the Little Missouri arm of Lake Sakakawea, North Dakota. M.S. thesis, Univ. North Dakota, Grand Forks. 41 pp.

Evermann, B. W. 1892. A report upon investigations made in Texas in 1891. Bull. U.S. Fish. Comm. 11 (for 1891): 61–90.

Evermann, B. W. 1896. Description of a new species of shad (*Alosa alabamae*) from Alabama. Rep. U.S. Comm. Fish and Fisheries 21 (for 1895): 203–205.

Ewers, L. A. and M. W. Boesel. 1935. The food of some Buckeye Lake fishes. Trans. Am. Fish. Soc. 65(1): 57–70.

Faber, J. E., J. Rybka, and M. M. White. 2009. Intraspecific phylogeography of the Stonecat madtom, *Noturus flavus*. Copeia 2009(3): 563–571.

Facey, D. E. and G. W. LaBar. 1981. Biology of American Eels in Lake Champlain, Vermont. Trans. Am. Fish. Soc. 110(3): 396–402.

Fahy, W. E. 1954. The life history of the Northern Greenside Darter, *Etheostoma blennioides blennioides* Rafinesque. J. Elisha Mitchell Sci. Soc. 70(2): 139–205.

Farr, M. D. 1996. Reproductive biology of *Cyprinella camura*, the Bluntface Shiner, in Morganfork Creek, southwestern Mississippi. Ecol. Freshw. Fish 5(3): 123–132.

Farringer, R. T., A. A. Echelle, and S. F. Lehtinen. 1979. Reproductive cycle of the Red Shiner, *Notropis lutrensis*, in central Texas and south-central Oklahoma. Trans. Am. Fish. Soc. 108(3): 271–276.

Fayton, T. J., A. Choudhury, C. T. McAllister, and H. W. Robison. 2017. Three new species of *Plagioporus* Stafford, 1904 from darters (Perciformes: Percidae), with a redescription of *Plagioporus boleosomi* (Pearse, 1924) Peters, 1957. Syst. Parasitol. 94(2): 159–182.

Fayton, T. J., C. T. McAllister, H. W. Robison, and M. B. Connior. 2018. Two new species of *Plagioporus* (Digenea: Opecoelidae) from the Ouachita Madtom, *Noturus lachneri*, and the Banded Sculpin, *Cottus carolinae*, from Arkansas. J. Parasitol. 104(2): 145–156.

Federal Register. 1990. Determination of endangered status for the Pallid Sturgeon; final rule. Sept. 6, 1990, 55: 36641–36647.

Feiner, Z. S., J. A. Rice, and D. D. Aday. 2013a. Trophic niche of invasive White Perch and potential interactions with representative reservoir species. Trans. Am. Fish. Soc. 142(3): 628–641.

Feiner, Z. S., J. A. Rice, A. J. Bunch, and D. D. Aday. 2013b. Trophic niche and diet overlap between invasive White Perch and resident White Bass in a southeastern reservoir. Trans. Am. Fish. Soc. 142(4): 912–919.

Felley, J. 1980. Analysis of morphology and asymmetry in Bluegill sunfish (*Lepomis macrochirus*) in the southeastern United States. Copeia 1980(1): 18–29.

Felley, J. D. 1984. Piscivorous habits of the Chub Shiner, *Notropis potteri* (Cyprinidae). Southwest. Nat. 29(4): 495–496.

Felley, J. D. and E. G. Cothran. 1981. *Notropis bairdi* (Cyprinidae) in the Cimarron River, Oklahoma. Southwest. Nat. 25(4): 564.

Felley, J. D. and S. M. Felley. 1987. Relationships between habitat selection by individuals of a species and patterns of habitat segregation among species: fishes of the Calcasieu drainage. Pages 61–68. *In:* W. J. Matthews and D. C. Heins, eds. Community and evolutionary ecology of North American stream fishes. Univ. Oklahoma Press, Norman.

Felley, J. D. and L. G. Hill. 1983. Multivariate assessment of environmental preferences of cyprinid fishes of the Illinois River, Oklahoma. Am. Midl. Nat. 109(2): 209–221.

Felsenstein, J. 2003. Inferring phylogenies. Sinauer, Sunderland, MA. 663 pp.

Fenneman, N. M. 1938. Physiography of the eastern United States. McGraw-Hill, New York. 714 pp.

Fernandez, A. G. 1982. A study comparing two populations

of the Creole Darter, *Etheostoma collettei*, in the Ouachita River drainage, with notes on distribution and life history aspects. M.S. thesis, Northeast Louisiana Univ., Monroe. 44 pp.

Ferrara, A. M. 2001. Life-history strategy of Lepisosteidae: implications for the conservation and management of Alligator Gar. Ph.D. diss., Auburn Univ., Auburn, AL. 252 pp.

Ferraris, C. J., Jr. 2007. Checklist of catfishes, recent and fossil (Osteichthyes: Siluriformes), and catalogue of siluriform primary types. Zootaxa 1418: 1–628.

Ferris, S. D. and G. S. Whitt. 1978. Phylogeny of tetraploid catostomid fishes based on the loss of duplicate gene expression. Syst. Biol. 27(2): 189–206.

Ficetola, G. F., C. Miaud, F. Pompanon, and P. Taberlet. 2008. Species detection using environmental DNA from water samples. Biol. Lett. 4(4): 423–425.

Fiegel, D. H. and M. Freeze. 1981. The aquaculture industry of Arkansas in 1979–1980. Proc. Arkansas. Acad. Sci. 35: 40–42.

Filipek, S. and L. Claybrook. 1984. Stripers and hybrids—what do they really eat? Arkansas Game and Fish 15(4): 8–9.

Findeis, E. K. 1997. Osteology and phylogenetic interrelationships of sturgeons (Acipenseridae). Environ. Biol. Fishes 48(1–4): 73–126.

Fingerman, S. W. and R. D. Suttkus. 1961. Comparison of *Hybognathus hayi* Jordan and *Hybognathus nuchalis* Agassiz. Copeia 1961(4): 462–467.

Fink, S. V. and W. L. Fink. 1981. Interrelationships of the ostariophysan fishes (Teleostei). Zool. J. Linn. Soc. 72(4): 297–353.

Fink, W. L. and J. H. Humphries. 2010. Morphological description of the extinct North American sucker *Moxostoma lacerum* (Ostariophysi: Catostomidae) based on high-resolution x-ray computed tomography. Copeia 2010(1): 5–13.

Finlayson, B. J., S. Siepmann, and J. Trumbo. 2001. Chemical residues in surface and ground waters following rotenone application to California lakes and streams. Pages 37–54. *In:* R. L. Cailteux, L. DeMong, B. J. Finlayson, W. Horton, W. McClay, R. A. Schnick, and C. Thompson, eds. Rotenone in fisheries: are the rewards worth the risks? Trends in Fisheries Science and Management 1. American Fisheries Society. Bethesda, MD. 124 pp.

Finnel, J. C., R. M. Jenkins, and G. E. Hall. 1956. The fishery resources of the Little River system, McCurtain County, Oklahoma. Oklahoma Fish. Res. Lab Rep. No. 55, Norman. 40 pp.

Fiorillo, R. A., R. B. Thomas, M. L. Warren, Jr., and C. M. Taylor. 1999. Structure of the helminth assemblage of an endemic madtom catfish (*Noturus lachneri*). Southwest. Nat. 44(4): 522–526.

Firehammer, J. A. and D. L. Scarnecchia. 2006. Spring migratory movements by Paddlefish in natural and regulated river segments of the Missouri and Yellowstone rivers, North Dakota and Montana. Trans. Am. Fish. Soc. 135(1): 200–217.

Fischer, J. B. 1791. Vesuch einer Naturgeschichte von Livland 2, Aufl. 24 Fridrich Nicolovins, Königsberg.

Fish, M. P. 1932. Contributions to the early life histories of sixty-two species of fishes from Lake Erie and its tributary waters. Bull. U.S. Bur. Fish. 47(10): 293–398.

Fisher, F. 1973. Observations on the spawning of the Mississippi Silversides, *Menidia audens* Hay. California Fish Game 59(4): 315–316.

Fisher, J. W. 1981. Ecology of *Fundulus catenatus* in three interconnected stream orders. Am. Midl. Nat. 106(2): 372–378.

Fisher, S. J., D. W. Willis, M. M. Olson, and S. C. Krentz. 2002. Flathead Chubs, *Platygobio gracilis,* in the upper Missouri River: the biology of a species at risk in an endangered habitat. Can. Field-Nat. 116(1): 26–41.

Fitz, R. B. 1968. Fish habitat and population changes resulting from impoundment of Clinch River by Melton Hill Dam. Tennessee Acad. Sci. 43(1): 7–15.

Fitz, R. B. and J. A. Holbrook II. 1978. Sauger and Walleye in Norris Reservoir, Tennessee. Pages 82–88. *In:* R. L. Kendall, ed. Selected coolwater fishes of North America. Special Publ. No. 11, American Fisheries Society, Washington, DC.

Flemer, D. A. and W. S. Woolcott. 1966. Food habits and distribution of the fishes of Tuckahoe Creek, Virginia, with special emphasis on the Bluegill, *Lepomis m. macrochirus* Rafinesque. Chesapeake Sci. 7(2): 75–89.

Fletcher, D. E. and B. M. Burr. 1992. Reproductive biology, larval description, and diet of the North American Bluehead Shiner, *Pteronotropis hubbsi* (Cypriniformes: Cyprinidae), with comments on conservation status. Ichthyol. Explor. Freshw. 3: 193–218.

Fletcher, D. E., E. E. Dakin, B. A. Porter, and J. C. Avise. 2004. Spawning behavior and genetic parentage in the Pirate Perch (*Aphredoderus sayanus*), a fish with an enigmatic reproductive morphology. Copeia 2004(1): 1–10.

Flinders, J. M. and D. D. Magoulick. 2017. Spatial and temporal consumption dynamics of trout in catch-and-release areas in Arkansas tailwaters. Trans. Am. Fish. Soc. 146(3): 432–449.

Flittner, G. A. 1964. Morphometry and life history of the Emerald Shiner *Notropis atherinoides* Rafinesque. Ph.D. diss., Univ. Michigan, Ann Arbor. 213 pp.

Floyd, E. P. 1948. Observations on the life history of the Fantailed Darter, *Catonotus flabellaris* Rafinesque. M.S. thesis, Univ. Arkansas, Fayetteville. 17 pp.

Fluker, B. L. and B. L. McCall. 2018. Status survey and population characteristics of the Paleback Darter (*Etheostoma pallididorsum*) and the Caddo Madtom (*Noturus taylori*). Final Report submitted to the Arkansas Game and Fish Commission, State Wildlife Grants Program, Mayflower. 56 pp.

Fogle, N. E. 1959. Some aspects of the life history of the Brook Silversides, *Labidesthes sicculus*, in Lake Fort Smith, Arkansas. M.S. thesis, Univ. Arkansas, Fayetteville. 25 pp.

Foltz, J. W. and C. R. Norden. 1977. Seasonal changes in food consumption and energy content of Smelt (*Osmerus mordax*) in Lake Michigan. Trans. Am. Fish. Soc. 106(3): 230–234.

Fontaine, P. A. 1944. Notes on the spawning of the Shovelhead Catfish, *Pylodictis olivaris* (Rafinesque). Copeia 1944(1): 50–51.

Fontenot, Q. C. and D. A. Rutherford. 1999. Observations on the reproductive ecology of Pirate Perch, *Aphredoderus sayanus*. J. Freshw. Ecol. 14(4): 545–549.

Forbes, S. A. 1884. Pages 60–89. *In:* A catalogue of the native fishes of Illinois. Report of the Illinois State Fish Commissioner, Springfield.

Forbes, S. A. and R. E. Richardson. 1905. On a new Shovelnose Sturgeon from the Mississippi River. Bull. Illinois State Lab. Nat. Hist. 7: 37–44.

Forbes, S. A. and R. E. Richardson. 1920. The fishes of Illinois. 2nd ed. Illinois Nat. Hist. Surv., Urbana. 357 pp.

Forey, P. L. and L. Grande. 1998. An African twin to the Brazilian *Calamopleurus* (Actinopterygii: Amiidae). Zool. J. Linn. Soc. 123(2): 179–195.

Forney, J. L. 1955. Life history of the Black Bullhead *Ameiurus melas* (Rafinesque), of Clear Lake, Iowa. Iowa State Coll. J. Sci. 30(1): 145–162.

Forshage, A. A., W. D. Harvey, K. E. Kulzer, and L. T. Fries. 1986. Natural reproduction of White Bass × Striped Bass hybrids in a Texas reservoir. Proc. Annu. Conf. Southeast. Assoc. Fish Wildl. Agencies. 40: 9–14.

Forrest, T. G., G. L. Miller, and J. R. Zagar. 1993. Sound propagation in shallow water: implications for acoustic communication by aquatic animals. Bioacoustics 4(4): 259–270.

Foster, N. R. 1967. Comparative studies on the biology of killifishes (Pisces, Cyprinodontidae). Ph.D. diss. Cornell Univ., Ithaca, NY. 369 pp.

Foti, T. L. 1974. Natural divisions of Arkansas. *In:* Arkansas natural areas plan. Arkansas Dept. of Planning, Little Rock. 247 pp.

Foti, T. L. 1976. Arkansas: Its land and people. Arkansas Dept. Education, Little Rock.

Fowler, C. L. and G. L. Harp. 1974. Ichthyofaunal diversification and distribution in Jane's Creek watershed, Randolph County, Arkansas. Proc. Arkansas Acad. Sci. 28: 13–18.

Fowler, H. W. 1904. Notes on fishes from Arkansas, Indian Territory, and Texas. Proc. Acad. Nat. Sci. Phila. 56(1): 242–249.

Fowler, H. W. 1906. Some new and little-known percoid fishes. Proc. Acad. Nat. Sci. Phila. 58(3): 510–528.

Fowler, H. W. 1917. Some notes on the breeding habits of local catfishes. Copeia 42: 32–36.

Fowler, H. W. 1935. Notes on South Carolina fresh-water fishes. Contributions from the Charleston Museum 7: 1–28.

Fowler, H. W. 1945. A study of the fishes of the southern Piedmont and Coastal Plain. Acad. Nat. Sci. Phila., Monogr. 7: 1–408.

Fowler, J. F. and C. A. Taber. 1985. Food habits and feeding periodicity in two sympatric stonerollers (Cyprinidae). Am. Midl. Nat. 113(2): 217–224.

Fowler, J. F., P. W. James, and C. A. Taber. 1984. Spawning activity and eggs of the Ozark Minnow, *Notropis nubilus*. Copeia 1984(4): 994–996.

Fox, C. N. 2010. Seasonal abundance, age structure, gonadosomatic index, and gonad histology of Yellow Bass *Morone mississippiensis* in the upper Barataria Estuary, Louisiana. M.S. thesis, Nicholls State Univ., Thibodaux, LA. 89 pp.

Frank, S. 1970. Acclimatization experiments with Amur Snakehead, *Ophiocephalus argus warpachowskii* (Berg, 1909) in Czechoslovakia. Vstnik Eskoslovenske Spolenosti Zoologicke 34: 277–283.

Franklin, D. R. and L. L. Smith, Jr. 1963. Early life history of the Northern Pike, *Esox lucius* L., with special reference to the factors influencing the numerical strength of year classes. Trans. Am. Fish. Soc. 92(2): 91–110.

Fraser, D. F. and R. D. Cerri. 1982. Experimental evaluation of predator-prey relationships in a patchy environment: consequences for habitat use patterns in minnows. Ecology 63(2): 307–313.

Frazer, K. S., H. T. Boschung, and R. L. Mayden. 1989. Diet of juvenile Bowfin *Amia calva* Linnaeus, in the Sipsey River, Alabama. SFC Proc. 20: 13–15.

Frazier, G. C. and J. K. Beadles. 1977. The fishes of Sylamore Creek, Stone County, Arkansas. Proc. Arkansas Acad. Sci. 31: 38–41.

Freeze, M. and S. Henderson. 1982. Distribution and status of the Bighead Carp and Silver Carp in Arkansas. N. Am. J. Fish. Manag. 2(2): 197–200.

Fremling, C. R., J. L. Rasmussen, R. E. Sparks, S. P. Cobb, C. F. Bryan, and T. O. Claflin. 1989. Mississippi River fisheries: a case history. Pages 309–351. *In:* D. P. Dodge, ed. Proceedings of the International Large River Symposium. Can. Special Publ. Fish. Aquat. Sci. 106.

French, J. R. P. 1997. Pharyngeal teeth of the Freshwater Drum (*Aplodinotus grunniens*) a predator of the Zebra Mussel (*Dreissena polymorpha*). J. Freshw. Ecol. 12(3): 495–498.

French, J. R. P. and J. G. Love. 1995. Size limitation on Zebra Mussels consumed by Freshwater Drum may preclude the effectiveness of drum as a biological controller. J. Freshw. Ecol. 10(4): 379–383.

Frenette, B. D. and R. Snow. 2016. Natural habitat conditions in a captive environment lead to spawning of Spotted Gar. Trans. Am. Fish. Soc. 145(4): 835–838.

Fries, L. T. and W. D. Harvey. 1989. Natural hybridization of White Bass with Yellow Bass in Texas. Trans. Am. Fish. Soc. 118(1): 87–89.

Frietsche, R. A. 1982. The warmwater fish community of a cool tailwater in Arkansas. Proc. Arkansas Acad. Sci. 36: 28–30.

Frothingham, K. M., B. L. Rhoads, and E. E. Herricks. 2001. Stream geomorphology and fish community structure in channelized and meandering reaches of an agricultural stream. Pages 105–118. *In:* J. M. Dorava, D. R. Montgomery,

B. B. Palcsak, and F. A. Fitzpatrick, eds. Geomorphic processes and riverine habitat. Water Science and Application. Volume 4. American Geophysical Union, Washington, DC.

Fruge, D. W. 1971. Fishes of the Caddo River, west-central Arkansas. M.S. thesis, Northeast Louisiana Univ., Monroe. 48 pp.

Fu, P., B. D. Neff, and M. R. Gross. 2001. Tactic-specific success in sperm competition. Proc. Royal Soc. London B Biol. Sci. 268(1472): 1105–1112.

Fuchs, E. H. 1967. Life history of the Emerald Shiner, *Notropis atherinoides* in Lewis and Clark Lake, South Dakota. Trans. Am. Fish. Soc. 96(3): 247–256.

Fuhrman, F. A., G. J. Fuhrman, D. L. Dull, and H. S. Mosher. 1969. Toxins from eggs of fishes and amphibia. J. Agric. Food Chem. 17(3): 417–424.

Fuiman, L. A. 1979. Descriptions and comparisons of catostomid fish larvae: northern Atlantic drainage species. Trans. Am. Fish. Soc. 108(6): 560–603.

Fuiman, L. A. 1982. Family Catostomidae, suckers. Pages 345–435. *In*: N. A. Auer, ed. Identification of larval fishes of the Great Lakes basin with emphasis on the Lake Michigan drainage. Great Lakes Fish. Comm., Ann Arbor, MI. Special Publ. 82-3: 744 pp.

Fulford, R. S. and D. A. Rutherford. 2000. Discrimination of larval *Morone* geometric shape differences with landmark-based morphometrics. Copeia 2000(4): 965–972.

Fullenkamp, A. E. 2010. Microhabitat selection among five congeneric darter species in two Indiana watersheds. M.S. thesis, Ball State Univ., Muncie, IN. 53 pp.

Fuller, D. B., M. E. Jaeger, M. P. Ruggles, P. J. Braaten, M. A. Webb, and K. M. Kappenman. 2007. Spawning and associated movement patterns of Pallid Sturgeon in the lower Yellowstone River. Report to Western Area Power Administration, Upper Basin Pallid Sturgeon Work Group, and U.S. Army Corps of Engineers. Montana Fish, Wildlife and Parks, Fort Peck. 22 pp.

Fuller, P. L., L. G. Nico, and J. D. Williams. 1999. Nonindigenous fishes introduced into inland waters of the United States. Special Publ. 27. American Fisheries Society, Bethesda, MD. 622 pp.

Fuller, P. L., A. J. Benson, G. Nunez, A. Fusaro, and M. Neilson. 2019. *Channa argus* (Cantor, 1842): U. S. Geological Survey, Nonindigenous Aquatic Species Database, Gainesville, FL. Available: https://nas.er.usgs.gov/queries/factsheet.aspx?SpeciesID=2265 [Revision Date: 12 Sept. 2018; Peer Review Date: 1 April 2016; Access Date: 6 April 2019].

Fuller, P. L., A. Benson, E. Maynard, M. E. Neilson, J. Larson, and A. Fusaro. 2018. *Neogobius melanostomus* (Pallas, 1814): U.S. Geological Survey, Nonindigenous Aquatic Species Database, Gainesville, FL. Available: https://nas.er.usgs.gov/queries/factsheet.aspx?SpeciesID=713 [Revision Date: 14 May 2018; Peer Review Date: 1 April 2016, Access Date: 24 August 2018].

Fuller, R. C. 1998. Fecundity estimates for Rainbow Darters, *Etheostoma caeruleum*, in southwestern Michigan. Ohio J. Sci. 98(2): 2–5.

Fuller, R. C. 1999. Costs of group spawning to guarding males in the Rainbow Darter, *Etheostoma caeruleum*. Copeia 1999(4): 1084–1088.

Fuller, R. C. 2003. Disentangling female mate choice and male competition in the Rainbow Darter, *Etheostoma caeruleum*. Copeia 2003(1): 138–148.

Fuller, R. L. and H. B. N. Hynes. 1987. Feeding ecology of three predacious aquatic insects and two fish in a riffle of the Speed River, Ontario. Hydrobiologia 150(3): 243–255.

Fulmer, R. F. and G. L. Harp. 1977. The fishes of Crowley's Ridge in Arkansas. Proc. Arkansas Acad. Sci. 31: 42–45.

Funk, J. L. 1957. Movement of stream fishes in Missouri. Trans. Am. Fish. Soc. 85(1): 39–57.

Fuselier, L. and D. Edds. 1996. Seasonal variation of riffle and pool fish assemblages in a short mitigated stream reach. Southwest. Nat. 41(3): 299–306.

Gagen, C. J., R. W. Standage, and J. N. Stoeckel. 1998. Ouachita Madtom (*Noturus lachneri*) metapopulation dynamics in intermittent Ouachita Mountain streams. Copeia 1998(4): 874–882.

Gagen, C. J., K. R. Moles, L. J. Hlass, and R. W. Standage. 2002. Habitat and abundance of the Ouachita Darter (*Percina* sp. nov.). J. Arkansas Acad. Sci. 56: 230–234.

Gale, W. F. 1983. Fecundity and spawning frequency of caged Bluntnose Minnows—fractional spawners. Trans. Am. Fish. Soc. 112(3): 398–402.

Gale, W. F. and G. L. Buynak. 1982. Fecundity and spawning frequency of the Fathead Minnow—a fractional spawner. Trans. Am. Fish. Soc. 111(1): 35–40.

Gale, W. F. and C. A. Gale. 1977. Spawning habits of Spotfin Shiner (*Notropis spilopterus*)—a fractional, crevice spawner. Trans. Am. Fish. Soc. 106(2): 170–177.

Gallagher, R. P. and J. V. Conner. 1983. Comparison of two ichthyoplankton sampling gears with notes on microdistribution of fish larvae in a large river. Trans. Am. Fish. Soc. 112(2B): 280–285.

Galloway, M. L. and R. V. Kilambi, 1984. Temperature preference and tolerance of Grass Carp (*Ctenopharyngodon idella*). Proc. Arkansas Acad. Sci. 38: 36–37.

Gannon, J. E. and R. P. Howmiller. 1973. Ecological notes on Paddlefish (*Polyodon spathula*) with short rostrums. Michigan Acad. Sci. 6: 217–222.

Garcia de Leon, F. J., L. Gonzalez-Garcia, J. M. Herrera-Castillo, K. O. Winemiller, and A. Banda-Valdes. 2001. Ecology of the Alligator Gar, *Atractosteus spathula*, in the Vicente Guerrero Reservoir, Tamaulipas, Mexico. Southwest. Nat. 46(2): 151–157.

Garman, H. 1898. Some notes on the brain and pineal structures of *Polyodon folium*. Bull. Illinois State Lab. Nat. Hist. 4: 298–309 + plates.

Garvey, J. E. 2007. Spatial assessment of Asian carp population dynamics: development of a spatial query tool for predicting relative success of life stages. USFWS Tech. Report. 45 pp. Available: http://fishdata.siu.edu/carptools/carprep.pdf [Accessed 1 May 2018].

Garvey, J. E. and R. A. Stein. 1998. Competition between larval fishes in reservoirs: the role of relative timing of appearance. Trans. Am. Fish. Soc. 127(6): 1021–1039.

Gascho Landis, A. M. and N. W. R. Lapointe. 2010. First record of a Northern Snakehead (*Channa argus* Cantor) nest in North America. Northeast. Nat. 17(2): 325–332.

Gaston, K. A. and T. E. Lauer. 2015. Morphometric variation in Bluegill *Lepomis macrochirus* and Green Sunfish *Lepomis cyanellus* in lentic and lotic systems. J. Fish Biol. 86(1): 317–332.

Gatz, A. J., Jr. 1979. Ecological morphology of freshwater stream fishes. Tulane Stud. Zool. Bot. 21: 91–124.

Gatlin, M. R and J. M. Long. 2011. Persistence of the Longnose Darter (*P. nasuta*) in Lee Creek, Oklahoma. Proc. Oklahoma Acad. Sci. 91: 11–14.

Geen, G. H., T. G. Northcote, G. F. Hartman, and C. C. Lindsey. 1966. Life histories of two species of catostomid fishes in Sixteenmile Lake, British Columbia, with particular reference to inlet stream spawning. J. Fish. Res. Board Can. 23(11): 1761–1788.

Geihsler, M. R. 1975. Life history of the Yoke Darter, *Etheostoma juliae* Meek, in the Buffalo River, Arkansas. M.S. thesis. Univ. Arkansas, Fayetteville. 65 pp.

Geihsler, M. R., E. D. Short, and P. D. Kittle. 1975. A preliminary checklist of the fishes of the Illinois River, Arkansas. Proc. Arkansas Acad. Sci. 29: 37–39.

Gelwick, F. P. and W. J. Matthews. 1992. Effects of an algivorous minnow on temperate stream ecosystem properties. Ecology 73(5): 1630–1645.

Gelwick, F. P. and W. J. Matthews. 1997. Effects of algivorous minnows (*Campostoma*) on spatial and temporal heterogeneity of stream periphyton. Oecologia 112(3): 386–392.

Gelwick, F. P., E. R. Gilliland, and W. J. Matthews. 1995. Introgression of the Florida Largemouth Bass genome into stream populations of northern Largemouth Bass in Oklahoma. Trans. Am. Fish. Soc. 124(4): 550–562.

Gelwick, F. P., S. Akin, D. A. Arrington, and K. O. Winemiller. 2001. Fish assemblage structure in relation to environmental variation in a Texas Gulf coastal wetland. Estuaries 24(2): 285–296.

Gentry, J. L., G. P. Johnson, B. T. Baker, C. T. Witsell, and J. D. Ogle, eds. 2013. Atlas of the vascular plants of Arkansas. Univ. Arkansas Herbarium. Fayetteville. 709 pp.

George, E. L. and W. F. Hadley. 1979. Food and habitat partitioning between Rock Bass (*Ambloplites rupestris*) and Smallmouth Bass (*Micropterus dolomieui*) young of the year. Trans. Am. Fish. Soc. 108(3): 253–261.

George, S. G., J. J. Hoover, and H. P. Brown. 1997. Paddlefish (*Polyodon spathula*) as samplers of riffle beetles (Coleoptera: Elmidae). Entomol. Sci. 108(3): 179–182.

George, S. G., W. T. Slack, and N. H. Douglas. 1996. Demography, habitat, reproduction, and sexual dimorphism of the Crystal Darter, *Crystallaria asprella*, from south-central Arkansas. Copeia 1996(1): 68–78.

George, S. G., W. T. Slack, and J. J. Hoover. 2012. A note on the fecundity of Pallid Sturgeon. J. Appl. Ichthyol. 28(4): 512–515.

Gerald, J. W. 1971. Sound production during courtship in six species of sunfish (Centrarchidae). Evolution 25(1): 75–87.

Gerken, J. E. and C. P. Paukert. 2009. Threats to Paddlefish habitat: implications for conservation. Pages 173–183. *In*: C. P. Paukert and G. D. Scholten, eds. Paddlefish management, propagation, and conservation in the 21st century: building from 20 years of research and management. Symp. 66. American Fisheries Society. Bethesda, MD. 443 pp.

Gerking, S. D. 1953. Evidence for the concepts of home range and territory in stream fishes. Ecology 34(2): 347–365.

Gerrity, P. C., C. S. Guy, and W. M. Gardner. 2006. Juvenile Pallid Sturgeon are piscivorous: a call for conserving native cyprinids. Trans. Am. Fish. Soc. 135(3): 604–609.

Ghedotti, M. J. 2000. Phylogenetic analysis and taxonomy of the poecilioid fishes (Teleostei, Cyprinodontiformes). Zool. J. Linn. Soc. 130(1): 1–53.

Ghedotti, M. J. and M. J. Grose. 1997. Phylogenetic relationships of the *Fundulus notti* species group (Fundulidae, Cyprinodontiformes) as inferred from the cytochrome *b* gene. Copeia 1997(4): 858–862.

Ghedotti, M. J., A. M. Simons, and M. P. Davis. 2004. Morphology and phylogeny of the studfish clade, subgenus *Xenisma* (Teleostei: Cyprinodontiformes). Copeia 2004(1): 53–61.

Gibbs, R. H., Jr. 1957a. Cyprinid fishes of the subgenus *Cyprinella* of *Notropis*. I. Systematic status of the subgenus *Cyprinella*, with a key to the species exclusive of the *lutrensis-ornatus* complex. Copeia 1957(3): 185–195.

Gibbs, R. H., Jr. 1957b. Cyprinid fishes of the subgenus *Cyprinella* of *Notropis*. II. Distribution and variation of *Notropis spilopterus*, with the description of a new subspecies. Lloydia 20(3): 186–211.

Gibbs, R. H., Jr. 1957c. Cyprinid fishes of the subgenus *Cyprinella* of *Notropis*. III. Variation and subspecies of *Notropis venustus* (Girard). Tulane Stud. Zool. 5(8): 175–203.

Gibbs, R. H., Jr. 1961. Cyprinid fishes of the subgenus *Cyprinella* of *Notropis*. IV. The *Notropis galacturus-camurus* complex. Am. Midl. Nat. 66(2): 337–354.

Gibbs, R. H., Jr. 1963. Cyprinid fishes of the subgenus *Cyprinella* of *Notropis*. The *Notropis whipplei-analostanus-chloristius* complex. Copeia 1963(3): 511–528.

Gido, K. B. 2001. Feeding ecology of three omnivorous fishes in Lake Texoma (Oklahoma–Texas). Southwest. Nat. 46(1): 23–33.

Gido, K. B. 2002. Interspecific comparisons and the potential importance of nutrient excretion by benthic fishes in a large reservoir. Trans. Am. Fish. Soc. 131(2): 260–270.

Gido, K. B. 2003. Effects of Gizzard Shad on benthic communities in reservoirs. J. Fish Biol. 62(6): 1392–1404.

Gilbert, C. H. 1884. A list of fishes collected in the east fork of White River, Indiana, with descriptions of two new species. Proc. U.S. Natl. Mus. 7 (423): 199–205.

Gilbert, C. H. 1885. Second series of notes on the fishes of Kansas. Bull. Washburn Coll. Lab. Nat. Hist. 1(3): 97–99.

Gilbert, C. H. 1887. Descriptions of new and little known etheostomoids. Proc. U. S. Natl. Mus. 10(607): 47–64.

Gilbert, C. H. 1889. A list of fishes from a small tributary of the Poteau River, Scott County, Arkansas. Proc. U. S. Natl. Mus. 11(1888): 609–610.

Gilbert, C. R. 1961. Hybridization versus intergradation: an inquiry into the relationship of two cyprinid fishes. Copeia 1961(2): 181–192.

Gilbert, C. R. 1964. The American cyprinid fishes of the subgenus *Luxilus* (genus *Notropis*). Bull. Florida State Mus. Biol. Sci. 8(2): 95–194.

Gilbert, C. R. 1969. Systematics and distribution of the American cyprinid fishes *Notropis ariommus* and *Notropis telescopus*. Copeia 1969(3): 474–492.

Gilbert, C. R. 1976. Composition and derivation of the North American freshwater fish fauna. Florida Sci. 39(2): 104–111.

Gilbert, C. R. 1978. Type catalogue of the North American cyprinid fish genus *Notropis*. Bull. Florida State Mus., Biol. Sci. 23(1): 1–104.

Gilbert, C. R. 1980a. *Hybopsis storeriana* (Kirtland), Silver Chub. Page 194. *In:* D. S. Lee et al. Atlas of North American freshwater fishes. N.C. State Mus. Nat. Hist., Raleigh. 854 pp.

Gilbert, C.R. 1980b. *Notropis atrocaudalis* (Evermann), Blackspot Shiner. Page 234. *In:* D. S. Lee et al. Atlas of North American freshwater fishes. N.C. State Mus. Nat. Hist., Raleigh. 854 pp.

Gilbert, C. R. 1980c. *Notropis chrysocephalus* (Rafinesque), Striped Shiner. Page 256. *In:* D. S. Lee et al. Atlas of North American freshwater fishes. N.C. State Mus. Nat. Hist., Raleigh. 854 pp.

Gilbert, C. R. 1980d. *Notropis cornutus* (Mitchill), Common Shiner. Page 258. *In:* D. S. Lee et al. Atlas of North American freshwater fishes. N.C. State Mus. Nat. Hist., Raleigh. 854 pp.

Gilbert, C. R. 1980e. *Notropis girardi* Hubbs and Ortenburger, Arkansas River Shiner. Page 268. *In:* D. S. Lee et al. Atlas of North American freshwater fishes. N.C. State Mus. Nat. Hist., Raleigh. 854 pp.

Gilbert, C. R. 1980f. *Percina shumardi* (Girard), River Darter. Page 741. *In:* D. S. Lee et al. Atlas of North American freshwater fishes. N. C. State Mus. Nat. Hist., Raleigh. 854 pp.

Gilbert, C. R. 1992. Southern Starhead Topminnow, *Fundulus dispar blairae*, Pages 63–67. *In:* C. R. Gilbert, ed. Rare and endangered biota of Florida. Vol. II. Fishes. Univ. Press of Florida, Gainesville. 336 pp.

Gilbert, C. R. 1998. Type catalogue of recent and fossil North American freshwater fishes: families Cyprinidae, Catostomidae, Ictaluridae, Centrarchidae, and Elassomatidae. Florida Museum of Natural History, Special Publ. 1: 1–284.

Gilbert, C. R. 2004. Family Elassomatidae Jordan 1877—pygmy sunfishes. Calif. Acad. Sci. Annotated Checklists of Fishes No. 33. 5 pp.

Gilbert, C. R. and R. M. Bailey. 1962. Synonymy, characters and distribution of the American cyprinid fish, *Notropis shumardi*. Copeia 1962(4): 807–819.

Gilbert, C. R. and R. M. Bailey. 1972. Systematics and zoogeography of the American cyprinid fish *Notropis (Opsopoeodus) emiliae*. Occas. Pap. Mus. Zool. Univ. Michigan. No. 664: 35 pp.

Gilbert, C. R. and G. H. Burgess. 1980a. *Notropis greenei* Hubbs and Ortenburger, Wedgespot Shiner. Page 269. *In:* D. S. Lee et al. Atlas of North American freshwater fishes. N.C. State Mus. Nat. Hist., Raleigh. 854 pp.

Gilbert, C. R. and G. H. Burgess. 1980b. *Minytrema melanops* (Rafinesque), Spotted Sucker. Page 408. *In:* D. S. Lee et al. Atlas of North American freshwater fishes. N.C. State Mus. Nat. Hist., Raleigh. 854 pp.

Gilbert, C. R. and G. H. Burgess. 1980c. *Notropis spilopterus* Cope, Spotfin Shiner. Page 312. *In:* D. S. Lee et al. Atlas of North American freshwater fishes. N.C. State Mus. Nat. Hist., Raleigh. 854 pp.

Gilbert, C. R. and J. D. Williams. 2002. National Audubon Society field guide to fishes: North America. Alfred A. Knopf, New York. 896 pp.

Gilbert, C. R., R. L. Mayden, and S. L. Powers. 2017. Morphological and genetic evolution in eastern populations of the *Macrhybopsis aestivalis* complex (Cypriniformes: Cyprinidae), with the descriptions of four new species. Zootaxa 4247(5): 501–555.

Gill, A. C. and R. D. Mooi. 2002. Phylogeny and systematics of fishes. Pages 15–42. *In:* P. J. B. Hart and J. D. Reynolds, eds. Handbook of fish biology and fisheries. Vol. 1. Fish Biology. Blackwell, London.

Gill, H. S., C. B. Renaud, F. Chapleau, R. L. Mayden, and I. C. Potter. 2003. Phylogeny of living parasitic lampreys (Petromyzontiformes) based on morphological data. Copeia 2003(4): 687–703.

Gill, T. N. 1861. Observations on the genus *Cottus*, and descriptions of two new species. Proc. Boston Soc. Nat. Hist. 8 (1861–1862): 40–42.

Gill, T. N. 1863. Fishes. Page 178. *In:* F. V. Hayden, ed. On the geology and natural history of the upper Missouri. [Chapter XVII. Reptiles, fishes, and recent shells (pp. 177–182)]. Trans. Am. Phil. Soc. (New Ser.) v. 12 (pt 1): 1–218. [Gill authorship assumed from literature; overall authorship is Hayden. (Eschmeyer et al. 2018)]

Gill, T. N. 1894. On the relations and nomenclature of *Stizostedion* or *Lucioperca*. Proc. U.S. Natl. Mus. 17(993): 123–128.

Gill, T. N. 1903. On some fish genera of the first edition of Cuvier's Regne Animal and Oken's names. Proc. U. S. Natl. Mus. 26(1346): 965–7.

Gill, T. N. 1907. Some noteworthy extra-European cyprinids. Smithson. Misc. Collect. 48(22): 297–340.

Gillen, A. L. and T. Hart. 1980. Feeding interrelationships between the Sand Shiner and the Striped Shiner. Ohio J. Sci. 80(2): 71–76.

Gillette, D. P. 2012. Effects of variation among riffles on prey use and feeding selectivity of the Orangethroat Darter *Etheostoma spectabile*. Am. Midl. Nat. 168(1): 184–201.

Gilliams, J. 1824. Description of a new species of fish of the Linnaean genus *Perca*. J. Acad. Sci. Phila. 4: 80–82.

Ginter, M. 2004. Devonian sharks and the origin of Xenacanthiformes. Pages 473–486. *In:* G. Arratia, M. V. H. Wilson, and R. Cloutier, eds. Recent advances in the origin and early radiation of vertebrates: honoring Hans-Peter Schultze. Verlag Dr. Friedrich Pfeil. Munchen.

Girard, C. F. 1850. A monograph of the fresh water *Cottus* of North America. Proc. Am. Assoc. Adv. Sci. 2: 409–411.

Girard, C. F. 1854. Description of some new species of fish from the state of Massachusetts. Proc. Boston Soc. Nat. Hist. 5: 40–43.

Girard, C. F. 1856. Researches upon the cyprinoid fishes inhabiting the fresh waters of the United States of America, west of the Mississippi Valley, from specimens in the museum of the Smithsonian Institution. Proc. Acad. Nat. Sci. Phila. 8: 165–213. [often cited as 1857].

Girard, C. F. 1858a. Part 4. General report upon the zoology of the several Pacific Railroad routes. Fishes. 10(4): 1–400. *In:* Reports of explorations and surveys, to ascertain the most practicable and economical route for a railroad from the Mississippi River to the Pacific Ocean, made under the direction of the Secretary of War, in 1853–4. Gov. Printing Office, 1855–61. Washington, DC.

Girard, C. F. 1858b. Notice upon new genera and new species of marine and fresh-water fishes from western North America. Proc. Acad. Nat. Sci. Phila. 9(15) (for 1857): 200–202.

Girard, C. F. 1859. Ichthyological notices. Proc. Acad. Nat. Sci. Phila. 11: 56–68; 100–104.

Glass, W. R., L. D. Corkum, and N. E. Mandrak. 2012. Spring and summer distribution and habitat use by adult threatened Spotted Gar in Rondeau Bay, Ontario, using radiotelemetry. Trans. Am. Fish. Soc. 141(4): 1026–1035.

Glazier, J. R. and C. A. Taber. 1980. Reproductive biology and age and growth of the Ozark Minnow, *Dionda nubila*. Copeia 1980(3): 547–550.

Gleason, C. A. and T. M. Berra. 1993. Demonstration of reproductive isolation and observation of mismatings in *Luxilus cornutus* and *L. chrysocephalus* in sympatry. Copeia 1993(3): 614–628.

Glenn, C. L. 1975a. Annual growth rates of Mooneye, *Hiodon tergisus*, in the Assiniboine River. J. Fish. Res. Board Can. 32(3): 407–410.

Glenn, C. L. 1975b. Seasonal diets of Mooneye, *Hiodon tergisus*, in the Assiniboine River. Can. J. Zool. 53(3): 232–237.

Glenn, C. L. 1978. Seasonal growth and diets of young-of-the-year Mooneye (*Hiodon tergisus*) from the Assiniboine River, Manitoba. Trans. Am. Fish. Soc. 107(4): 587–589.

Glenn, C. L. and R. R. G. Williams. 1976. Fecundity of Mooneye, *Hiodon tergisus*, in the Assiniboine River. Can. J. Zool. 54(2): 156–161.

Glodek, G. S. 1980a. *Ictalurus furcatus*, Blue Catfish. Page 439. *In:* D. S. Lee et al. Atlas of North American freshwater fishes. N.C. State Mus. Nat. Hist., Raleigh.

Glodek, G. S. 1980b. *Ictalurus natalis*, Yellow Bullhead. Page 442. *In:* D. S. Lee et al. Atlas of North American freshwater fishes. N.C. State Mus. Nat. Hist., Raleigh.

Glodek, G. S. 1980c. *Ictalurus nebulosus*, Brown Bullhead. Page 443. *In:* D. S. Lee et al. Atlas of North American freshwater fishes. N.C. State Mus. Nat. Hist., Raleigh.

Glodek, G. S. 1980d. *Ictalurus punctatus*, Channel Catfish. Page 446. *In:* D. S. Lee et al. Atlas of North American freshwater fishes. N.C. State Mus. Nat. Hist., Raleigh.

Gmelin, J. F., ed. 1789. Caroli a Linne Systema Naturae (13th ed.). Tomus I. Pars III: 1033–2224.

Godbout, J. D., D. D. Aday, J. A. Rice, M. R. Bangs, and J. M. Quattro. 2009. Morphological models for identifying Largemouth Bass, Spotted Bass, and Largemouth Bass × Spotted Bass hybrids. N. Am. J. Fish. Manag. 29(5): 1425–1437.

Goddard, K. and A. Mathis. 1997. Do opercular flaps of male Longear Sunfish (*Lepomis megalotis*) serve as sexual ornaments during female mate choice? Ethol. Ecol. Evol. 9(3): 223–231.

Goddard, K. and A. Mathis. 2000. Opercular flaps as sexual ornaments for male Longear Sunfish (*Lepomis megalotis*): male condition and male-male competition. Ethology 106(7): 631–643.

Godwin, J. C., D. A. Steen, D. Werneke, and J. W. Armbruster. 2016. Two significant records of exotic tropical freshwater fishes in southern Alabama. Southeast. Nat. 15(4): N57–N60.

Goff, G. P. 1984. Brood care of Longnose Gar (*Lepisosteus osseus*) by Smallmouth Bass (*Micropterus dolomieui*). Copeia 1984(1): 149–152.

Goldstein, R. M. and T. P. Simon. 1999. Toward a unified definition of guild structure for feeding ecology of North American freshwater fishes. Pages 123–202. *In:* T. P. Simon, ed. Assessing the sustainability and biological integrity of water resources using fish communities. CRC Press, Boca Raton, FL. 671 pp.

Gomelsky, B., S. D. Mims, R. J. Onders, and N. D. Novelo. 2005. Inheritance of predorsal black stripe in Black Crappie. N. Am. J. Aquac. 67(2): 167–170.

Gomez, R. and H. L. Lindsay, Jr. 1972. Occurrence of the Mississippi Silversides, *Menidia audens* Hay, in Keystone

Reservoir and the Arkansas River. Proc. Oklahoma Acad. Sci. 52: 16–18.

Gong, L. 1991. Systematics of two morphologically similar fish species, *Notropis volucellus* and *Notropis wickliffi* (Cyprinidae). M.S. thesis. Ohio State Univ. Columbus.

Gong, L. and T. M. Cavender. 1991. Systematics of *Notropis volucellus* and *Notropis wickliffi* (Cyprinidae: Pisces) from Ohio waters. Ohio J. Sci. 91(2): 23.

Goodman, M., J. Czelusniak, B. F. Koop, D. A. Tagle, and J. L. Slightom. 1987. Globins: a case study in molecular phylogeny. Cold Spring Harbor Symp. Quant. Biol. 52: 875–890.

Goodyear, C. P. 1967. Feeding habits of three species of gars, *Lepisosteus*, along the Mississippi Gulf Coast. Trans. Am. Fish. Soc. 96(3): 297–300.

Goodyear, C. P. 1970. Terrestrial and aquatic orientation in the Starhead Topminnow, *Fundulus notti*. Science 168(3931): 603–605.

Gorman, O. T. 1988. The dynamics of habitat use in a guild of Ozark minnows. Ecol. Monogr. 58(1): 1–18.

Gosline, W. A. 1960. Contributions toward a classification of modern isospondylous fishes. Bull. British Mus. (Nat. Hist.) Zool. 6(6): 325–365.

Gosline, W. A. 1966. The limits of the fish family Serranidae, with notes on other lower percoids. Proc. California Acad. Sci. 33(6): 91–112.

Gosline, W. A. 1973. Considerations regarding the phylogeny of cypriniform fishes, with special reference to structures associated with feeding. Copeia 1973(4): 761–776.

Gosline, W. A. 1978. Unbranched dorsal-fin rays and subfamily classification of the fish family Cyprinidae. Occas. Pap. Mus. Zool. Univ. Michigan. No. 684. 21 pp.

Gowanloch, J. N. 1933. Fishes and fishing in Louisiana. Bulletin 23, Department of Conservation, New Orleans. 638 pp.

Grady, J. M. 1987. Biochemical systematics and evolution of the ictalurid catfish genus *Noturus* (Pisces: Siluriformes). Ph.D. diss. Southern Illinois Univ. Carbondale. 184 pp.

Grady, J. M. and B. S. Elkington. 2009. Establishing and maintaining Paddlefish populations by stocking. Pages 385–396. *In*: C. P. Paukert and G. D. Scholten, eds. Paddlefish management, propagation, and conservation in the 21st century: building from 20 years of research and management. Symp. 66. American Fisheries Society, Bethesda, MD. 443 pp.

Grady. J. M. and W. H. LeGrande. 1992. Phylogenetic relationships, modes of speciation, and historical biogeography of the madtom catfishes, genus *Noturus* Rafinesque (Siluriformes: Ictaluridae). Pages 747–777. *In*: R. L. Mayden, ed. Systematics, historical ecology, and North American freshwater fishes. Stanford University Press, Stanford, CA. 969 pp.

Grady, J. M. and J. Milligan. 1998. Status of selected cyprinid species at historic Lower Missouri River sampling sites. U. S. Fish and Wildlife Service. Columbia, MO. 49 pp.

Graening, G. O. 2000. Ecosystem dynamics of an Ozark cave. Ph.D. diss. Univ. Arkansas, Fayetteville. 106 pp.

Graening, G. O. 2005. Trophic structure of Ozark cave streams containing endangered species. J. Oceanol. Hydrobiol. Stud. 34(3): 3–17.

Graening, G. O. and A. V. Brown. 1999. Cavefish population status and environmental quality in Cave Springs Cave, Arkansas. Report submitted to the Arkansas Natural Heritage Commission, Publication No. MSC-276, Arkansas Water Resources Center, Univ. Arkansas, Fayetteville.

Graening, G. O. and A. V. Brown. 2000a. Trophic dynamics and pollution effects in Cave Springs Cave, Arkansas. Report to the Arkansas Natural Heritage Commission. Publication No. MSC-285, Arkansas Water Resources Center, Univ. Arkansas, Fayetteville.

Graening, G. O. and A. V. Brown. 2000b. Status survey of aquatic cave fauna in Arkansas. Arkansas Water Resources Center Publication No. MSC-286, Univ. Arkansas, Fayetteville.

Graening, G. O., D. B. Fenolio, M. L. Niemiller, A. V. Brown, and J. B. Beard. 2010. The 30-year recovery effort for the Ozark Cavefish (*Amblyopsis rosae*): analysis of current distribution, population trends, and conservation status of this threatened species. Environ. Biol. Fishes 87(1): 55–88.

Graham, J. B. 1997. Air-breathing fishes: evolution, diversity, and adaptation. Academic Press, San Diego, CA. 299 pp.

Graham, K. 1997. Contemporary status of the North American Paddlefish, *Polyodon spathula*. Environ. Biol. Fishes 48(1–4): 279–289.

Graham, K. 1999. A review of the biology and management of Blue Catfish. Pages 37–49. *In*: E. R. Irwin, W. A. Hubert, C. F. Rabeni, H. L. Schramm, Jr., and T. Coon., eds. Catfish 2000: Proceedings of the International Ictalurid Symposium. Symp. 24. American Fisheries Society, Bethesda, MD. 532 pp.

Grammer, G. L., W. T. Slack, M. S. Peterson, and M. A. Dugo. 2012. Nile Tilapia *Oreochromis niloticus* (Linnaeus, 1758) establishment in temperate Mississippi, USA: multi-year survival confirmed by otolith ages. Aquat. Invasions 7(3): 367–376.

Grande, L. 1980. Paleontology of the Green River formation, with a review of the fish fauna. Geol. Survey Wyoming, Laramie. 333 pp.

Grande, L. 1982a. A revision of the fossil genus *Diplomystus*, with comments on the interrelationships of clupeomorph fishes. Am. Mus. Novit. 2728. 34 pp.

Grande, L. 1982b. A revision of the fossil genus *Knightia*, with a description of a new genus from the Green River formation (Teleostei, Clupeidae). Am. Mus. Novit. 2731. 22 pp.

Grande, L. 1984. Paleontology of the Green River Formation, with a review of the fish fauna. 2nd ed. Geol. Surv. Wyoming Bull. 63: 1–333.

Grande, L. 1985. Recent and fossil clupeomorph fishes with materials for revision of the subgroups of clupeoids. Bull. Am. Mus. Nat. Hist. 181(2): 231–372.

Grande, L. 1999. The first *Esox* (Esocidae: Teleostei) from the

Eocene Green River Formation, and a brief review of esocid fishes. J. Vert. Paleontol. 19(2): 271–292.

Grande, L. 2010. An empirical synthetic pattern study of gars (Lepisosteiformes) and closely related species, based mostly on skeletal anatomy. The resurrection of Holostei. Am. Soc. Ichthyol. Herpetol. Special Publ. 6: 1–871.

Grande, L. and W. E. Bemis. 1991. Osteology and phylogenetic relationships of fossil and recent Paddlefishes (Polyodontidae) with comments on the interrelationships of Acipenseriformes. J. Vert. Paleontol. 11 (Memoir 1). 212 pp.

Grande, L. and W. E. Bemis. 1996. Interrelationships of Acipenseriformes, with comments on "Chondrostei." Pages 85–115. *In*: M. L. J. Stiassny, L. R. Parenti, and G. D. Johnson, eds. Interrelationships of fishes. Academic Press, New York. 496 pp.

Grande, L. and W. E. Bemis. 1998. A comprehensive phylogenetic study of amiid fishes (Amiidae) based on comparative skeletal anatomy: an empirical search for interconnected patterns of natural history. Supplement to J. Vert. Paleontol. 18, Memoir 4. 696 pp.

Grande, L. and W. E. Bemis. 1999. Historical biogeography and historical paleoecology of Amiidae and other halecomorph fishes. Pages 413–424. *In*: G. Arratia and H-P. Schultze, eds. Mesozoic fishes 2—Systematics and fossil record. Pfeil. München. 604 pp.

Grande, L. and J. T. Eastman. 1986. A review of Antarctic ichthyofaunas in the light of new fossil discoveries. Paleontology 29: 113–137.

Grande, L., F. Jin, Y. Yabumoto, and W. E. Bemis. 2002. *Protopsephurus liui*, a well-preserved primitive paddlefish (Acipenseriformes: Polyodontidae) from the Lower Cretaceous of China. J. Vert. Paleontol. 22(2): 209–237.

Grande, T., H. Laten, and J. A. Lopez. 2004. Phylogenetic relationships of extant esocid species (Teleostei: Salmoniformes) based on morphological and molecular characters. Copeia 2004(4): 743–757.

Grant, J. W. A. and P. W. Colgan. 1983. Reproductive success and mate choice in the Johnny Darter, *Etheostoma nigrum* (Pisces: Percidae). Can. J. Zool. 61(2): 437–446.

Grant, J. W. A. and P. W. Colgan. 1984. Territorial behaviour of the male Johnny Darter, *Etheostoma nigrum*. Environ. Biol. Fishes 10(4): 261–269.

Gray, D. L. and R. A. Collins. 1970. Age and growth of the Blue Catfish, *Ictalurus furcatus*, in the Arkansas River. Proc. Arkansas Acad. Sci. 24: 62–65.

Gray, E. V. S., J. M. Boltz, K. A. Kellogg, and J. R. Stauffer, Jr. 1997. Food resource partitioning by nine sympatric darter species. Trans. Am. Fish. Soc. 126(5): 822–840.

Gray, S. M., L. J. Chapman, and N. E. Mandrak. 2012. Turbidity reduces hatching success in threatened Spotted Gar (*Lepisosteus oculatus*). Environ. Biol. Fishes 94(4): 689–694.

Greeley, M. S., Jr., D. R. Calder, and R. A. Wallace. 1987.

Oocyte growth and development in the Striped Mullet, *Mugil cephalus*, during seasonal ovarian recrudescence: relationship to fecundity and size at maturity. Fish. Bull. (U.S.) 85(2): 187–200.

Green, J. F. and J. K. Beadles. 1974. Ichthyofaunal survey of the Current River within Arkansas. Proc. Arkansas Acad. Sci. 28: 22–26.

Greene, J. C. and M. J. Maceina. 2000. Influence of trophic state on Spotted Bass and Largemouth Bass spawning time and age-0 population characteristics in Alabama reservoirs. N. Am. J. Fish. Manag. 20(1): 100–108.

Greenberg, L. A. 1991. Habitat use and feeding behavior of thirteen species of benthic stream fishes. Environ. Biol. Fishes 31(4): 389–401.

Greenberg, L. A. and D. A. Holtzman. 1987. Microhabitat utilization, feeding periodicity, home range and population size of the Banded Sculpin, *Cottus carolinae*. Copeia 1987(1): 19–25.

Greenfield, D. W. 1973. An evaluation of the advisability of the release of the Grass Carp, *Ctenopharyngodon idella*, into natural waters of the United States. Trans. Illinois State Acad. Sci. 66(1–2): 48–53.

Greenwich, W. W. 1979. Life style and habits of the Northern Hog Sucker, *Hypentelium nigricans*. M.S. thesis, Univ. Southern Mississippi, Hattiesburg.

Greenwood, P. H. 1973. Interrelationships of osteoglossomorphs. Pages 307–332. *In*: P. H. Greenwood, S. R. Miles, and C. Patterson, eds. Interrelationships of fishes. Linn. Soc. London Publ., Academic Press, London. 536 pp.

Greenwood, P. H., D. E. Rosen, S. H. Weitzman, and G. S. Myers. 1966. Phyletic studies of teleostean fishes, with a provisional classification of living forms. Bull. Am. Mus. Nat. Hist. 131(4): 339–456.

Grier, H. J., D. P. Moody, and B. C. Cowell. 1990. Internal fertilization and sperm morphology in the Brook Silverside, *Labidesthes sicculus* (Cope). Copeia 1990(1): 221–226.

Griffith, E. and C. H. Smith. 1834. The class Pisces, arranged by the Baron Cuvier, with supplementary additions, by Edward Griffith, F.R.S., &c. and Lieut.-Col. Charles Hamilton Smith, F.R., L.S.S., &c. &c. In: Cuvier, G: The animal kingdom, arranged in conformity with its organization, by the Baron Cuvier, member of the Institute of France, &c. &c. &c., with supplementary additions to each order, by Edward Griffith . . . and others. (2nd ed.) Whittaker & Co., London. 1–680, Pls. 1–62 + 3. [English treatment of Cuvier's "Régne Animal"; supplements by Griffith and Smith; authorship variously cited.]

Griffith, J. S. and R. W. Smith. 1993. Use of winter concealment cover by juvenile Cutthroat and Brown trout in the South Fork of the Snake River, Idaho. N. Am. J. Fish. Manag. 13(4): 823–830.

Griffith, R. W. 1974. Environment and salinity tolerance in the genus *Fundulus*. Copeia 1974(2): 319–331.

Grisak, G. G. 1996. The status and distribution of the Sicklefin

Chub in the middle Missouri River, Montana. M.S thesis. Montana State Univ., Bozeman. 77 pp.

Groce, M. C., L. L. Bailey, and K. D. Fausch. 2012. Evaluating the success of Arkansas Darter translocations in Colorado: an occupancy sampling approach. Trans. Am. Fish. Soc. 141(3): 825–840.

Groen, C. L. and J. C. Schmulbach. 1978. The sport fishery of the unchannelized and channelized middle Missouri River. Trans. Am. Fish. Soc. 107(3): 412–418.

Grohs, K. L., R. A. Klumb, S. R. Chipps, and G. A. Wanner. 2009. Ontogenetic patterns in prey use by Pallid Sturgeon in the Missouri River, South Dakota and Nebraska. J. Appl. Ichthyol. 25 (S2): 48–53.

Grose, M. J. and E. O. Wiley. 2002. Phylogenetic relationships of the *Hybopsis amblops* species group (Teleostei: Cyprinidae). Copeia 2002(4): 1092–1097.

Gross, M. R. 1979. Cuckoldry in sunfishes (*Lepomis*: Centrarchidae). Can. J. Zool. 57(7): 1507–1509.

Gross, M. R. 1982. Sneakers, satellites, and parentals: polymorphic mating strategies in North American sunfishes. Zeit. Tierpsy. 60(1): 1–26.

Gross, M. R. and W. A. Nowell. 1980. The reproductive biology of Rock Bass, *Ambloplites rupestris* (Centrarchidae), in Lake Opinicon, Ontario. Copeia 1980(3): 482–494.

Gruchy, C. G. and B. Parker. 1980. *Acipenser fulvescens* Rafinesque, Lake Sturgeon. Page 39. *In:* D. S. Lee et al. Atlas of North American freshwater fishes. N.C. State Mus. Nat. Hist., Raleigh. 854 pp.

Gudger, E. W. 1925. The fishes of Arkansas, a historical sketch. Arkansas Gazette (Little Rock), Jan. 11, 1925: part II, p. 4.

Gudger, E. W. 1932. The Shovel-Nosed Sturgeon in the Arkansas River. Science 76(1971): 323–324.

Guidroz, T. P. 1975. Fishes of the Buffalo River, White River system, Arkansas. M.S. thesis, Northeast Louisiana Univ., Monroe. 33 pp.

Guldin, J. M. 2004. Ouachita and Ozark mountains symposium: ecosystem management research. Gen. Tech. Rep. SRS-74. U.S. Dept. Agriculture, Forest Service, Southern Research Station, Asheville, NC. 321 pp.

Guill, J. M. and D. C. Heins. 1996. Clutch and egg size variation in the Banded Darter, *Etheostoma zonale*, from three sites in Arkansas. Environ. Biol. Fishes 46(4): 409–413.

Guill, J. M. and D. C. Heins. 2000. Interannual variation in clutch and egg size of the Banded Darter, *Etheostoma zonale*. Copeia 2000(1): 230–233.

Guillory, V. 1978. Life history of Warmouth in Lake Conway, Florida. Proc. Annu. Conf. Southeast. Assoc. Fish Wildl. Agencies 32: 490–501.

Guillory, V. 1980. *Ctenopharyngodon idella* (Valenciennes), Grass Carp. Page 151. *In:* D. S. Lee et al. Atlas of North American freshwater fishes. N.C. State Mus. Nat. Hist., Raleigh. 854 pp.

Guillory, V. and R. D. Gasaway. 1978. Zoogeography of the Grass Carp in the United States. Trans. Am. Fish. Soc. 107(1): 105–112.

Gunn, J. M. 1995. Spawning behavior of Lake Trout: effects on colonization ability. J. Great Lakes Res. 21(S1): 323–329.

Gunning, G. E. 1959. The sensory basis for homing in the Longear Sunfish, *Lepomis megalotis megalotis* (Rafinesque). Invest. Indiana Lakes Streams 5: 103–130.

Gunning, G. E. 1965. A behavioral analysis of the movement of tagged Longear Sunfish. Prog. Fish-Cult. 27: 211–215.

Gunning, G. E. and W. M. Lewis. 1955. The fish population of a spring-fed swamp in the Mississippi bottoms of southern Illinois. Ecology 36(4): 552–558.

Gunning, G. E. and C. R. Shoop. 1964. Stability in a headwater stream population of the Sharpfin Chubsucker. Prog. Fish-Cult. 26(2): 76–79.

Gunning, G. E. and R. D. Suttkus. 1990. Decline of the Alabama Shad, *Alosa alabamae*, in the Pearl River, Louisiana-Mississippi: 1963–1988. SFC Proc. 21: 3–4.

Gunter, G. 1938. Notes on invasion of fresh water by fishes of the Gulf of Mexico, with special reference to the Mississippi-Atchafalaya River system. Copeia 1938(2): 69–72.

Günther, A. 1859. Catalogue of the fishes in the British Museum. Catalogue of the acanthopterygian fishes in the collection of the British Museum. Gasterosteidae, Berycidae, Percidae, Aphredoderidae, Pristipomatidae, Mullidae, Sparidae. v. 1: i–xxxi + 1–524.

Günther, A. 1866. Catalogue of fishes in the British Museum. Catalogue of the Physotomi, containing the families Salmonidae, Percopsidae, Galaxidae, Mormyridae, Gymnarchidae, Esocidae, Umbridae, Scombresocidae, Cyprinodontidae, in the collection of the British Museum. v. 6: i–xv + 1–368.

Günther, A. 1867. On the fishes of the states of Central America, founded upon specimens collected in the fresh and marine waters of various parts of that country by Messrs. Salvin and Godman and Capt. J. M. Dow. Proc. Zool. Soc. London 1866 (part 3): 600–604.

Günther, A. 1889. On some fishes from Kilima-Njaro District. Proc. Zool. Soc. London 1889 (pt. 1): 70–72.

Gurgens, C., D. F. Russell, and L. A. Wilkens. 2000. Electrosensory avoidance of metal obstacles by the Paddlefish. J. Fish Biol. 57(2): 277–290.

Guseva, L. N. 1990. Food and feeding rations of the Amur Snakehead, *Channa argus warpachowskii*, in water bodies in the lower reaches of the Amu Darya. J. Ichthyol. 30(3): 439–446.

Gustaveson, A. W., T. D. Pettengill, J. E. Johnson, and J. R. Wahl. 1984. Evidence of in-reservoir spawning of Striped Bass in Lake Powell, Utah-Arizona. N. Am. J. Fish. Manag. 4(4B): 540–546.

Gutierrez, M. A. 2010. Reproductive behaviors of male and female Blackspotted Topminnows, *Fundulus olivaceus*. M.S. thesis, Univ. Southern Mississippi, Hattiesburg, 32 pp.

Gutreuter, S., J. M. Dettmers, and D. H. Wahl. 2003.

Estimating mortality rates of adult fish from entrainment through the propellers of river towboats. Trans. Am. Fish. Soc. 132(4): 646–661.

Gutreuter, S., J. M. Vallazza, and B. C. Knights. 2006. Persistent disturbance by commercial navigation alters the relative abundance of channel-dwelling fishes in a large river. Can. J. Fish. Aquat. Sci. 63(11): 2418–2433.

Gutreuter, S., J. M. Vallazza, and B. C. Knights. 2009. Lateral distribution of fishes in the main-channel trough of a large floodplain river: implications for restoration. River Res. Appl. 26(5): 619–635.

Haag, W. R. and M. L. Warren, Jr. 2006. Seasonal feeding specialization on snails by River Darters (*Percina shumardi*) with a review of snail feeding by other darter species. Copeia 2006(4): 604–612.

Haas, J. D. 2005. Evaluation of the impacts of the introduced Western Mosquitofish, *Gambusia affinis*, on native Plains Topminnow, *Fundulus sciadicus*, in Nebraska. M.S. thesis, Univ. Nebraska at Kearney. 105 pp.

Haas, M. A. 1977. Some aspects of the life history of the Suckermouth Minnow, *Phenacobius mirabilis* (Girard). M.S. thesis, Univ. Missouri, Columbia. 100 pp.

Haase, B. L. 1969. An ecological life history of the Longnose Gar, *Lepisosteus osseus* (Linnaeus), in Lake Mendota and in several other lakes of southern Wisconsin. Ph.D. diss. Univ. Wisconsin, Madison. 448 pp.

Hackney, P. A. and J. A. Holbrook. 1978. Sauger, Walleye, and Yellow Perch in the southeastern United States. Pages 74–81. *In*: R. L. Kendall, ed. A symposium on selected coolwater fishes of North America. Am. Fish. Soc. Special Publ. 11. 437 pp.

Hackney, P. A., G. R. Hooper, and J. F. Webb. 1971. Spawning behavior, age and growth, and sport fishery for the Silver Redhorse, *Moxostoma anisurum* (Rafinesque) in the Flint River, Alabama. Proc. 24th Annu. Conf. Southeast. Assoc. Game Fish Comm.: 569–576.

Hackney, P. A., W. M. Tatum, and S. L. Spencer. 1968. Life history study of the River Redhorse, *Moxostoma carinatum* (Cope) in the Cahaba River, Alabama, with notes on the management of the species as a sport fish. Proc. 21st Annu. Conf. Southeast. Assoc. Game Fish Comm.: 324–342.

Hafs, A. W., C. J. Gagen, and J. K. Whalen. 2010. Smallmouth Bass summer habitat use, movement, and survival in response to low flow in the Illinois Bayou, Arkansas. N. Am. J. Fish. Manag. 30(2): 604–612.

Hajibabaei, M., G. A. C. Singer, P. D. N. Hebert, and D. A. Hickey. 2007. DNA barcoding: how it complements taxonomy, molecular phylogenetics, and population genetics. Trends Genet. 23(4): 167–172.

Hajibabaei, M., M. A. Smith, D. H. Janzen, J. J. Rodriguez, J. B. Whitfield, and P. D. N. Hebert. 2006. A minimalist barcode can identify a specimen whose DNA is degraded. Mol. Ecol. Res. 6(4): 959–964.

Halas, D. and A. M. Simons. 2014. Cryptic speciation reversal in the *Etheostoma zonale* (Teleostei: Percidae) species group, with an examination of the effect of recombination and introgression on species tree inference. Mol. Phylogenet. Evol. 70: 13–28.

Haldeman, S. S. 1842. Description of two new species of the genus *Perca*, from the Susquehanna River. J. Acad. Nat. Sci. Phila. 8(pt. 2): 330.

Hale, M. C. 1963. A comparative study of the food of the shiners *Notropis lutrensis* and *Notropis venustus*. Proc. Oklahoma Acad. Sci. 43: 125–129.

Hale, M. M., J. E. Crumpton, and R. J. Schuler, Jr. 1995. From sportfishing bust to commercial fishing boon: a history of the Blue Tilapia in Florida. Am. Fish. Soc. Symp. 15: 425–430.

Hall, G. E. 1956. Additions to the fish fauna of Oklahoma with a summary of introduced species. Southwest. Nat. 1(1): 16–26.

Hall, G. E. and R. M. Jenkins. 1954. Notes on the age and growth of the Pirate Perch, *Aphredoderus sayanus*, in Oklahoma. Copeia 1954(1): 69.

Hall, G. E. and G. A. Moore. 1954. Oklahoma lampreys: their characterization and distribution. Copeia 1954(2): 127–135.

Hall, J. D. 1963. An ecological study of the Chestnut Lamprey, *Ichthyomyzon castaneus* (Girard), in the Manistee River, Michigan. Ph.D. diss. Univ. Michigan, Ann Arbor. 101 pp.

Hambrick, P. S. and R. G. Hibbs, Jr. 1976. Spring diel feeding activity of the Southern Striped Shiner, *Notropis chrysocephalus isolepis* Hubbs and Brown, in Bayou Sara, Louisiana. Proc. Louisiana Acad. Sci. 39: 16–18.

Hambrick, P. S. and R. G. Hibbs, Jr. 1977. Feeding chronology and food habits of the Blacktail Shiner, *Notropis venustus* (Cyprinidae), in Bayou Sara, Louisiana. Southwest. Nat. 22(4): 511–516.

Hambrick, P. S. and H. W. Robison. 1978. An hermaphroditic Paleback Darter, *Etheostoma pallididorsum*, with notes on other aberrant darters (Percidae). Southwest. Nat. 23(1): 170–171.

Hambrick, P. S. and H. W. Robison. 1979. Life history aspects of the Paleback Darter *Etheostoma pallididorsum* (Pisces: Percidae), in the Caddo River system, Arkansas. Southwest. Nat. 24(3): 475–484.

Hammond, S. D. 2003. Seasonal movements of yellow-phase American Eels (*Anguilla rostrata*) in the Shenandoah River, West Virginia. M.S. thesis. West Virginia Univ., Morgantown. 32 pp.

Hankinson, T. L. 1930. Breeding behavior of the Silverfin Minnow, *Notropis whipplii spilopterus* (Cope). Copeia 1930(3): 73–74.

Hanley, R. W. 1977. Hybridization between the chubsuckers *Erimyzon oblongus* and *E. sucetta* in North Carolina. Abstract, ASIH 57th Annu. Meeting. Gainesville, FL.

Hansen, D. F. 1965. Further observations on nesting of the White Crappie, *Pomoxis annularis*. Trans. Am. Fish. Soc. 94(2): 182–184.

Hansen, K. A. and C. P. Paukert. 2009. Current management of Paddlefish sport fisheries. Pages 277–290. *In*: C. P. Paukert and G. D. Scholten, eds. Paddlefish management, propagation, and conservation in the 21st century: building from 20 years of research and management. Symp. 66. American Fisheries Society, Bethesda, MD. 443 pp.

Haponski, A. E. and C. A. Stepien. 2008. Molecular, morphological, and biogeographic resolution of cryptic taxa in the Greenside Darter *Etheostoma blennioides* complex. Mol. Phylogenet. Evol. 49(1): 69–83.

Haponski, A. E. and C. A. Stepien. 2013. Phylogenetic and biogeographical relationships of the *Sander* pikeperches (Percidae: Perciformes): patterns across North America and Eurasia. Biol. J. Linn. Soc. 110(1): 156–179.

Hardisty, M. W. and I. C. Potter. 1971a. The behavior, ecology and growth of larval lampreys. Pages 85–125. *In:* M. W. Hardisty and I. C. Potter, eds. The biology of lampreys. Vol. 1. Academic Press, London. 422 pp.

Hardisty, M. W. and I. C. Potter. 1971b. The general biology of adult lampreys. Pages 127–206. *In*: M. W. Hardisty and I. C. Potter, eds. The biology of lampreys. Vol. 1. Academic Press, London. 422 pp.

Hardman, M. 2004. The phylogenetic relationships among *Noturus* catfishes (Siluriformes: Ictaluridae) as inferred from mitochondrial gene cytochrome *b* and nuclear recombination activating gene 2. Mol. Phylogenet. Evol. 30(2): 395–408.

Hardman, M. 2005. The phylogenetic relationships among non-diplomystid catfishes as inferred from mitochondrial cytochrome *b* sequences; the search for the ictalurid sister taxon (Otophysi: Siluriformes). Mol. Phylogenet. Evol. 37(3): 700–720.

Hardman, M. and L. M. Hardman. 2008. The relative importance of body size and paleoclimatic change as explanatory variables influencing lineage diversification rate: an evolutionary analysis of bullhead catfishes (Siluriformes: Ictaluridae). Syst. Biol. 57(1): 116–130.

Hardman, M. and L. M. Page. 2003. Phylogenetic relationships among bullhead catfishes of the genus *Ameiurus* (Siluriformes: Ictaluridae). Copeia 2003(1): 20–33.

Hardy, J. D., Jr. 1978. Development of fishes of the Mid-Atlantic Bight: an atlas of egg, larval, and juvenile stages. Vol. III. Aphredoderidae through Rachycentridae. U.S. Fish and Wildlife Service, Div. Biol. Surv. Prog., FWS/OBS-78/12. 394 pp.

Hardy, M. E., J. M. Grady, and E. J. Routman. 2002. Intraspecific phylogeography of the Slender Madtom: the complex evolutionary history of the Central Highlands of the United States. Mol. Ecol. 11(11): 2393–2403.

Hargrave, C. W. 2006. A test of three alternative pathways for consumer regulation of primary productivity. Oecologia 149(1): 123–132.

Hargrave, C. W. and K. P. Gary. 2016. Historical distribution of Bluehead Shiner (*Pteronotropis hubbsi*). Southeast. Nat. 15(SI 9): 110–116.

Hargrave, C. W. and K. B. Gido. 2004. Evidence of reproduction by exotic Grass Carp in the Red and Washita rivers, Oklahoma. Southwest. Nat. 49(1): 89–93.

Hargrave, C. W. and J. E. Johnson. 2003. Status of Arkansas Darter, *Etheostoma cragini*, and Least Darter, *E. microperca*, in Arkansas. Southwest. Nat. 48(1): 89–92.

Harkness, W. J. K. and J. R. Dymond. 1961. The Lake Sturgeon: the history of its fishery and problems of conservation. Fish and Wildlife Branch. Ontario Dept. Lands and Forests. Toronto, Ontario. 121 pp.

Harlan, J. R. and E. B. Speaker. 1956. Iowa fish and fishing. 3rd ed. Iowa Conserv. Comm, Des Moines. 377 pp.

Haro, A. J. and W. H. Krueger. 1991. Pigmentation, otolith rings, and upstream migration of juvenile American Eels (*Anguilla rostrata*) in a coastal Rhode Island stream. Can. J. Zool. 69(3): 812–814.

Haro, A., W. Richkus, K. Whalen, A. Hoar, W. Busch, S. Lary, T. Brush, and D. Dixon. 2000. Population decline of the American Eel: implications for research and management. Fisheries 25(9): 7–16.

Harp, G. L. and W. J. Matthews. 1975. First Arkansas records of *Lampetra* spp. (Petromyzontidae). Southwest. Nat. 20(3): 414–416.

Harrell, R. M. 2013. Releasing hybrid *Morone* in natural waters with congeneric species: implications and ethics. Pages 531–549. *In:* J. S. Bulak, C. C. Coutant, and J. A. Rice, eds. Biology and management of inland Striped Bass and hybrid Striped Bass. Symp. 80. American Fisheries Society, Bethesda, MD. 588 pp.

Harrell, R. M. and J. M. Dean. 1988. Identification of juvenile hybrids of *Morone* based on meristics and morphometrics. Trans. Am. Fish. Soc. 117(6): 529–535.

Harrington, R. C., E. Benavides, and T. J. Near. 2012. Phylogenetic inference of nuptial trait evolution in the context of asymmetrical introgression in North American darters (Teleostei). Evolution 67(2): 388–402.

Harrington, R. W., Jr. 1947. Observations on the breeding habits of the Yellow Perch, *Perca flavescens* (Mitchill). Copeia 1947(3): 199–200.

Harris, J. L. 1977. Fishes of the Mountain Province section of the Ouachita River from the headwaters to Remmel Dam. M.S. thesis, Northeast Louisiana Univ., Monroe. 101 pp.

Harris, J. L. 1986. Systematics, distribution, and biology of fishes currently allocated to *Erimystax* (Jordan), a subgenus of *Hybopsis* (Cyprinidae). Ph.D. diss., Univ. Tennessee, Knoxville. 335 pp.

Harris, J. L. and N. H. Douglas. 1978. Fishes of the mountain province section of the Ouachita River. Proc. Arkansas Acad. Sci. 32: 55–59.

Harris, J. L. and K. L. Smith. 1985. Distribution and status of *Etheostoma cragini* Gilbert and *E. microperca* Jordan and Gilbert in Arkansas. Proc. Arkansas Acad. Sci. 39: 135–156.

Harris, P. M. and R. L. Mayden. 2001. Phylogenetic relationships of major clades of Catostomidae (Teleostei:

Cypriniformes) as inferred from mitochondrial SSU and LSU rDNA sequences. Mol. Phylogenet. Evol. 20(2): 225–237.

Harris, P. M., G. Hubbard, and M. Sandel. 2014. Catostomidae: Suckers. Pages 451–501. *In*: M. L. Warren, Jr. and B. M. Burr, eds. Freshwater fishes of North America. Vol. 1: Petromyzontidae to Catostomidae. Johns Hopkins Univ. Press, Baltimore, MD. 644 pp.

Harris, P. M., K. J. Roe, and R. L. Mayden. 2005. A mitochondrial DNA perspective on the molecular systematics of the sunfish genus *Lepomis* (Actinopterygii: Centrarchidae). Copeia 2005(2): 340–346.

Harris, P. M., R. L. Mayden, H. S. Espinosa Perez, and F. Garcia de Leon. 2002. Phylogenetic relationships of *Moxostoma* and *Scartomyzon* (Catostomidae) based on mitochondrial cytochrome *b* sequence data. J. Fish Biol. 61(6): 1433–1452.

Harrison, A. B., W. T. Slack, and K. J. Killgore. 2014. Feeding habitats of young-of-year river sturgeon *Scaphirhynchus* spp. in the lower Mississippi River. Am. Midl. Nat. 171(1): 54–67.

Harrison, H. M. 1950. The foods used by some common fish of the Des Moines River drainage. Pages 31–44. *In*: Biology seminar held at Des Moines, Iowa, 11 July 1950. Iowa Conserv. Comm. Div. Fish Game.

Hartel, K. E., D. B. Halliwell, and A. E. Launer. 2002. Inland fishes of Massachusetts. Massachusetts Audubon Society, Lincoln, MA. 378 pp.

Hartfield, P. D., N. M. Kuntz, and H. L. Schramm, Jr. 2013. Observations on the identification of larval and juvenile *Scaphirhynchus* spp. in the Lower Mississippi River. Southeast. Nat. 12(2): 251–266.

Hartman, K. J., B. Vondracek, D. L. Parrish, and K. M. Muth. 1992. Diets of Emerald and Spottail shiners and potential interactions with other western Lake Erie planktivorous fishes. J. Great Lakes Res. 18(1): 43–50.

Harvill, M. L. 1989. The fishes of the St. Francis River in Arkansas. M.S. thesis. Arkansas State Univ., Jonesboro. 64 pp.

Hatch, J. T. 1982. Life history of the Gilt Darter, *Percina evides* (Jordan and Copeland), in the Sunrise River, Minnesota. Ph.D. diss., Univ. Minnesota, Minneapolis. 162 pp.

Hatch, J. T. 1986. Distribution, habitat, and status of the Gilt Darter (*Percina evides*) in Minnesota. J. Minnesota Acad. Sci. 51(2): 11–16.

Hatch, J. T. 1997. Resource utilization and life history of the Crystal Darter, *Crystallaria asprella* (Jordan), in the lower Mississippi River, Minnesota. Report to Minnesota Natural Heritage and Nongame Wildlife Research Program, Minnesota Department of Natural Resources, St. Paul. 27 pp.

Hatch, J. T. and E. E. Elias. 2002. Ovarian cycling, clutch characteristics, and oocyte size of the River Shiner *Notropis blennius* (Girard) in the upper Mississippi River. J. Freshw. Ecol. 17(1): 85–92.

Hatcher, H. R., M. J. Moore, and D. J. Orth. 2017. Spawning observations of Clinch Dace: comparison of *Chrosomus* spawning behavior. Am. Midl. Nat. 177(2): 318–326.

Hatfield, J. S., T. E. Wissing, S. I. Guttman, and M. P. Farrell. 1982. Electrophoretic analysis of Gizzard Shad from the lower Mississippi River and Ohio. Trans. Am. Fish. Soc. 111(6): 742–748.

Hatzenbeler, G. R., M. A. Bozek, M. J. Jennings, and E. E. Emmons. 2000. Seasonal variation in fish assemblage structure and habitat structure in the nearshore littoral zone of Wisconsin lakes. N. Am. J. Fish. Manag. 20(2): 360–368.

Hawes, E. J. and D. L. Parrish. 2003. Using abiotic and biotic factors to predict the range expansion of White Perch in Lake Champlain. J. Great Lakes Res. 29(2): 268–279.

Hay, O. P. 1881. On a collection of fishes from eastern Mississippi. Proc. U.S. Natl. Mus. 3(179): 488–515.

Hay, O. P. 1882. On a collection of fishes from the lower Mississippi Valley. Bull. U.S. Fish. Comm. 2: 57–75.

Hayer, C. A. and E. R. Irwin. 2008. Influence of gravel mining and other factors on detection probabilities of Coastal Plain fishes in the Mobile River basin, Alabama. Trans. Am. Fish Soc. 137(6): 1606–1620.

Hayer, C. A., N. L. Ahrens, and C. R. Berry, Jr. 2008. Biology of Flathead Chub, *Platygobio gracilis*, in three Great Plains rivers. Proc. South Dakota Acad. Sci. 87: 185–196.

Haynes, J. L. 1993. Annual reestablishment of Mosquitofish populations in Nebraska. Copeia 1993(1): 232–235.

Haynes, J. L. and R. C. Cashner. 1995. Life history and population dynamics of the Western Mosquitofish: a comparison of natural and introduced populations. J. Fish Biol. 46(6): 1026–1041.

He, C., R. L. Mayden, X. Wang, W. Wang, K. L. Tang, W. J. Chen, and Y. Chen. 2008. Molecular phylogenetics of the family Cyprinidae (Actinopterygii: Cypriniformes) as evidenced by sequence variation in the first intron of S7 ribosomal protein-coding gene: further evidence from a nuclear gene of the systematic chaos in the family. Mol. Phylogenet. Evol. 46(3): 818–829.

Hearn, M. C. 1986. Reproductive viability of Sauger-Walleye hybrids. Prog. Fish-Cult. 48(2): 149–150.

Hebert, P. D., A. Cywinska, S. L. Ball, and J. R. deWaard. 2003. Biological identifications through DNA barcodes. Proc. Royal Soc. London Series B: Biol. Sci. 270(1512): 313–321.

Hebert, P. D. N., E. H. Penton, J. M. Burns, D. H. Janzen, and W. Hallwachs. 2004. Ten species in one: DNA barcoding reveals cryptic species in the Neotropical skipper butterfly *Astraptes fulgerator*. Proc. Natl. Acad. Sci. USA 101(41): 14812–14817.

Heckathorn, W. D., Jr. 1993. Fishes of Bayou Meto and Wattensaw Bayou, two lowland streams in east central Arkansas. Proc. Arkansas Acad. Sci. 47: 44–53.

Hecke, K. 2017. Distribution and status of the Strawberry Darter *Etheostoma fragi* in the main stem and tributaries of the Strawberry River drainage. M.S. thesis, Univ. Arkansas at Pine Bluff. 44 pp.

Heckel, J. J. 1836. *Scaphirhynchus*, eine neue Fischgattung aus der Ordnung der Chondropterygier mit freien Kiemen. Annalen des Wiener Museums der Naturgeschichte v. 1: 69–78. [Possibly published in 1835.]

Hedrick, M. S., S. L. Katz, and D. R. Jones. 1994. Periodic air-breathing behaviour in a primitive fish revealed by spectral analysis. J. Exp. Biol. 197(1): 429–436.

Heidinger, R. C. 1976. Synopsis of biological data on the Largemouth Bass, *Micropterus salmoides* (Lacépède) 1802. FAO Fisheries Synopsis No. 115. 85 pp.

Heins, D. C. 1981. Life history pattern of *Notropis sabinae* (Pisces: Cyprinidae) in the lower Sabine River drainage of Louisiana and Texas. Tulane Stud. Zool. Bot. 22(2): 67–84.

Heins, D. C. 1990. Mating behaviors of the Blacktail Shiner, *Cyprinella venusta*, from southeastern Mississippi. SFC Proc. 21: 5–7.

Heins, D. C. and J. A. Baker. 1989. Growth, population structure, and reproduction of the percid fish *Percina vigil*. Copeia 1989(3): 727–736.

Heins, D. C. and J. A. Baker. 1992. Historical and recent influences on reproduction in North American stream fishes. Pages 573–599. *In:* R. L. Mayden, ed. Systematics, historical ecology, and North American freshwater fishes. Stanford Univ. Press, Stanford, CA.

Heins D. C. and D. Davis. 1984. The reproductive season of the Weed Shiner, *Notropis texanus* (Pisces: Cyprinidae), in southeastern Mississippi. Southwest. Nat. 29(1): 133–135.

Heins, D. C. and D. R. Dorsett. 1986. Reproductive traits of the Blacktail Shiner, *Notropis venustus* (Girard), in southeastern Mississippi. Southwest. Nat. 31(2): 185–189.

Heins, D. C. and M. D. Machado. 1993. Spawning season, clutch characteristics, sexual dimorphism and sex ratio in the Redfin Darter *Etheostoma whipplei*. Am. Midl. Nat. 129(1): 161–171.

Heins, D. C. and F. G. Rabito, Jr. 1986. Spawning performance in North American minnows: direct evidence of the occurrence of multiple clutches in the genus *Notropis*. J. Fish Biol. 28(3): 343–357.

Heins, D. C. and F. G. Rabito, Jr. 1988. Reproductive traits in populations of the Weed Shiner, *Notropis texanus*, from the Gulf Coastal Plain. Southwest. Nat. 33(2): 147–156.

Heins, D. C., J. A. Baker, and D. J. Tylicki. 1996. Reproductive season, clutch size, and egg size of the Rainbow Darter, *Etheostoma caeruleum*, from the Homochitto River, Mississippi, with an evaluation of data from the literature. Copeia 1996(4): 1005–1010.

Heiser, J. B. 2009. Fishes. Pages 581–584. *In:* M. Ruse and J. Travis, eds. Evolution: the first four billion years. Belknap Press of Harvard Univ. Press, Cambridge, MA. 979 pp.

Heist, E. J. and A. Mustapha. 2008. Rangewide genetic structure in Paddlefish inferred from DNA microsatellite loci. Trans. Am. Fish. Soc. 137(3): 909–915.

Heitman, N. E., C. L. Racey, and S. E. Lochmann. 2006. Stocking contribution and growth of Largemouth Bass in pools of the Arkansas River. N. Am. J. Fish. Manag. 26(1): 175–179.

Held, J. W. 1969. Some early summer foods of the Shovelnose Sturgeon in the Missouri River. Trans. Am. Fish. Soc. 98(3): 514–517.

Held, J. W. and J. J. Peterka. 1974. Age, growth, and food habits of the Fathead Minnow, *Pimephales promelas*, in North Dakota saline lakes. Trans. Am. Fish. Soc. 103(4): 743–756.

Helfman, G. S. 2007. Fish conservation: a guide to understanding and restoring global aquatic biodiversity and fishery resources. Island Press, Washington, DC. 584 pp.

Helfman, G. S., B. B. Collette, D. E. Facey, and B. W. Bowen. 2009. The diversity of fishes: biology, evolution, and ecology. 2nd ed. Wiley-Blackwell, Chichester. 736 pp.

Helfman, G. S., D. E. Facey, L. S. Hales, Jr., and E. L. Bozeman, Jr. 1987. Reproductive ecology of the American Eel. Pages 42–56. *In:* M. J. Dadswell, R. J. Klauda, C. M. Moffitt, R. L. Saunders, R. A. Rulifson, and J. E. Cooper, eds. Common strategies of anadromous and catadromous fishes. Am. Fish. Soc. Symp. 1.

Hellier, T. R., Jr. 1967. The fishes of the Santa Fe River system. Bull. Florida State Mus. Biol. 2(1): 1–46.

Hemming, F. 1956. Opinion 417. Rejection for nomenclatorial purposes of volume 3 (Zoologie) of the work by Lorenz Oken entitled Okens Lehrbuch der Naturgeschichte published in 1815–1816. Opinions and declarations rendered by the International Commission on Zoological Nomenclature Vol. 14, Part 1: 1–42.

Hempstead, F. 1890. A pictorial history of Arkansas from earliest times to the year 1890. N. D. Thompson Publ., St. Louis, MO. 1240 pp.

Henderson, S. 1976. Observations on the Bighead and Silver carp and their possible application in pond fish culture. Arkansas Game and Fish Comm., Little Rock. 18 pp.

Henderson, S. 1977. An evaluation of filter feeding fishes for water quality improvement. Arkansas Game and Fish Comm., Little Rock. 26 pp.

Henderson, S. 1979. Utilization of Silver and Bighead carp for water quality improvement. Arkansas Game and Fish Comm., Little Rock. 32 pp.

Henderson, S. 1983. An evaluation of filter feeding fishes for removing excessive nutrients and algae from wastewater, Proj. Summary EPA-600/S2-83-019. May 1983, U. S. Environmental Protection Agency. Robert S. Kerr Env. Res. Lab., Ada, OK. 5 pp.

Hendricks, M. K. and R. L. Noble. 1980. Feeding interactions of three planktivorous fishes in Trinidad Lake, Texas. Proc. Annu. Conf. Southeast. Assoc. Fish Wildl. Agencies 33(1979): 324–330.

Hennig, W. 1966. Phylogenetic systematics. Univ. Illinois Press, Urbana. 263 pp.

Henry, S. 2005. Time for *Tilapia*. Arkansas Wildlife 36(3): 6–9.

Henry, S. D., S. W. Barkley, and R. L. Johnson. 2005. Exploitation of Nile Tilapia in a closed-system public fishing reservoir in northern Arkansas. N. Am. J. Fish. Manag. 25(3): 853–860.

Hensley, D. A. and W. R. Courtenay, Jr. 1980. *Scardinius erythrophthalmus* (Linnaeus), Rudd. Page 360. *In:* D. S. Lee et al. Atlas of North American freshwater fishes. N.C. State Mus. Nat. Hist., Raleigh.

Henson, M. N., D. D. Aday, J. A. Rice, and C. A. Layman. 2018. Assessing the influence of tilapia on sport fish species in North Carolina reservoirs. Trans. Am. Fish. Soc. 147(2): 350–362.

Herald, E. S. 1961. Living fishes of the world. Doubleday, Garden City, NY. 304 pp.

Herbert, K. A. 1994. Drift of aquatic macrofauna in Logan Cave Stream, Benton County, Arkansas. M.S. thesis. Univ. Arkansas, Fayetteville. 175 pp.

Herborg, L. M., N. E. Mandrak, B. C. Cudmore, and H. J. MacIsaac. 2007. Comparative distribution and invasion risk of snakehead (Channidae) and Asian carp (Cyprinidae) species in North America. Can. J. Fish. Aquat. Sci. 64(12): 1723–1735.

Herman, P. A., A. Plauck, N. Utrup, and T. Hill. 2008. Three year summary age and growth report for Sicklefin Chub (*Macrhybopsis meeki*). Pallid Sturgeon Population Assessment Project and Associated Fish Community Monitoring for Missouri River. Columbia National Fish and Wildlife Conserv. Office. Columbia, MO. 64 pp.

Herrala, J. R., P. T. Kroboth, N. M. Kuntz, and H. L. Schramm, Jr. 2014. Habitat use and selection by adult Pallid Sturgeon in the lower Mississippi River. Trans. Am. Fish. Soc. 143(1): 153–163.

Herrington, S. J., K. N. Hettiger, E. J. Heist, and D. B. Keeney. 2008. Hybridization between Longnose and Alligator gars in captivity, with comments on possible gar hybridization in nature. Trans. Am. Fish. Soc. 137(1): 158–164.

Herting, G. E. and A. Witt, Jr. 1967. The role of physical fitness of forage fishes in relation to their vulnerability to predation by Bowfin (*Amia calva*). Trans. Am. Fish. Soc. 96(4): 427–430.

Herzog, D. P. 2004. Capture efficiency and habitat use of the Sturgeon Chub (*Macrhybopsis gelida*) and Sicklefin Chub (*Macrhybopsis meeki*) in the Mississippi River. M.S. thesis. Southeast Missouri State Univ., Cape Girardeau. 186 pp.

Herzog, D. P. and R. A. Hrabik. 2013. Missouri trawler. Missouri Conservationist 74(1): 16–21.

Herzog, D. P., D. E. Ostendorf, R. A. Hrabik, and V. A. Barko. 2009. The mini-Missouri trawl: a useful methodology for sampling small-bodied fishes in small and large river systems. J. Freshw. Ecol. 24(1): 103–108.

Herzog, D. P., V. A. Barko, J. S. Scheibe, R. A. Hrabik, and D. E. Ostendorf. 2005. Efficacy of a benthic trawl for sampling small-bodied fishes in large river systems. N. Am. J. Fish. Manag. 25(2): 594–603.

Hewitt, L. M., S. M. Smyth, M. G. Dube, C. I. Gilman, and D. L. MacLatchy. 2002. Isolation of compounds from bleached kraft mill recovery condensates associated with reduced levels of testosterone in Mummichog (*Fundulus heteroclitus*). Environ. Toxicol. Chem. 21(7): 1359–1367.

Hickman, C. P., Jr., L. S. Roberts, S. L. Keen, D. J. Eisenhour, A. Larson, and H. I'Anson. 2017. Integrated principles of zoology. 17th ed. McGraw-Hill, New York. 823 pp.

Hickman, G. D. and R. B. Fitz. 1978. A report on the ecology and conservation of the Snail Darter (*Percina tanasi* Etnier) 1975–1977. TVA Div. For. Fish. Wildl. Dev. Tech. Note B 28. Norris, TN. 212 pp.

Hilburn, D. C. 1987. A comparative study of the fishes of the Strawberry River. M.S. thesis. Arkansas State Univ. Jonesboro. 62 pp.

Hildebrand, H. 2005. Size, age composition, and upstream migration of American Eels at the Millville Dam eel ladder, Shenandoah River, West Virginia. M.S. thesis. West Virginia Univ. Morgantown. 54 pp.

Hildebrand, S. F. 1922. Notes on habits and development of eggs and larvae of the silversides, *Menidia menidia* and *Menidia beryllina*. Bull. U.S. Bur. Fish. 38: 113–120.

Hildebrand, S. F. and W. C. Schroeder. 1928. Fishes of Chesapeake Bay. Bull. U.S. Bur. Fish. 43(1): 1–366.

Hill, J. E. and C. E. Cichra. 2005. Biological synopsis of six selected Florida non-game, littoral fishes with an emphasis on the effects of water level fluctuations. Final Report to St. Johns River Water Management District, Palatka, FL. 108 pp.

Hill, J. and G. D. Grossman. 1993. An energetic model of microhabitat use for Rainbow Trout and Rosyside Dace. Ecology 74(3): 685–698.

Hill, L. G. 1968. Inter- and intrapopulational variation of vertebral numbers of the Yoke Darter, *Etheostoma juliae*. Southwest. Nat. 13(2): 175–191.

Hill, L. G. 1972. Social aspects of aerial respiration of young gars (*Lepisosteus*). Southwest. Nat. 16(3–4): 239–247.

Hill, L. G. and W. J. Matthews. 1980. Temperature selection by the darters *Etheostoma spectabile* and *Etheostoma radiosum* (Pisces: Percidae). Am. Midl. Nat. 104(2): 412–415.

Hill, L. G., J. L. Renfro, and R. Reynolds. 1972. Effects of dissolved oxygen tensions upon the rate of aerial respiration of young Spotted Gar, *Lepisosteus oculatus* (Lepisosteidae). Southwest. Nat. 17(3): 273–278.

Hill, L. G., G. D. Schnell, and A. A. Echelle. 1973. Effect of dissolved oxygen concentration on locomotory reactions of the Spotted Gar, *Lepisosteus oculatus* (Pisces: Lepisosteidae). Copeia 1973(1): 119–124.

Hillis, D. M., C. Moritz, and B. K. Mable, eds. 1996. Molecular systematics. 2nd ed. Sinauer, Sunderland, MA. 655 pp.

Hilton, E. J. 2011. Bony fish skeleton. Pages 434–448. *In:* A. P. Farrell, ed. Encyclopedia of fish physiology: from genome to environment. Vol. 1. Academic Press, San Diego, CA. 2,272 pp.

Hilton, E. J. and L. Grande. 2006. Review of the fossil record of sturgeons, family Acipenseridae (Actinopterygii: Acipenseriformes), from North America. J. Paleontol. 80(4): 672–683.

Hilton, E. J., W. E. Bemis, and L. Grande. 2014. Hiodontidae: Mooneyes. Pages 299–312. In: M. L. Warren, Jr. and B. M. Burr, eds. Freshwater fishes of North America. Vol. 1: Petromyzontidae to Catostomidae. Johns Hopkins Univ. Press, Baltimore, MD. 644 pp.

Hilton, W. A. 1900. On the intestine of Amia calva. Am. Nat. 34(405): 717–735.

Hintz, W. D., N. K. MacVey, A. M. Asher, A. P. Porreca, and J. E. Garvey. 2017. Variation in prey selection and foraging success associated with early-life ontogeny and habitat use of American Paddlefish (Polyodon spathula). Ecol. Freshw. Fish 26(2): 181–189.

Hitt, N. P., S. Eyler, and J. E. B. Wofford. 2012. Dam removal increases American Eel abundance in distant headwater streams. Trans. Am. Fish. Soc. 141(5): 1171–1179.

Hlass, L. J., W. L. Fisher, and D. J. Turton. 1998. Use of the index of biotic integrity to assess water quality in forested streams of the Ouachita Mountains ecoregion, Arkansas. J. Freshw. Ecol. 13(2): 181–192.

Hlohowskyj, C. P., M. M. Coburn, and T. M. Cavender. 1989. Comparison of a pharyngeal filtering apparatus in seven species of the herbivorous cyprinid genus, Hybognathus (Pisces: Cyprinidae). Copeia 1989(1): 172–183.

Hlohowskyj, I. and A. M. White. 1983. Food resource partitioning and selectivity by the Greenside, Rainbow, and Fantail darters (Pisces: Percidae). Ohio J. Sci. 83(4): 201–208.

Hoagstrom C. W. and J. E. Brooks. 2005. Distribution and status of Arkansas River Shiner Notropis girardi and Rio Grande Shiner Notropis jemezanus, Pecos River, New Mexico. Texas J. Sci. 57(1): 35–58.

Hoagstrom, C. W., N. D. Zymonas, S. R. Davenport, D. L. Propst, and J. E. Brooks. 2010. Rapid species replacements between fishes of the North American plains: a case history from the Pecos River. Aquatic Invasions 5(2): 141–153.

Hocutt, C. H., and E. O. Wiley, eds. 1986. The zoogeography of North American freshwater fishes. John Wiley and Sons, New York. 866 pp.

Hodges, S. W. and D. D. Magoulick. 2011. Refuge habitats for fishes during seasonal drying in an intermittent stream: movement, survival and abundance of three minnow species. Aquat. Sci. 73(4): 513–522.

Hodson, R. 1966. Growth and food of young-of-year Largemouth and Spotted Bass in Beaver Reservoir. M.S. thesis, Univ. Arkansas, Fayetteville. 55 pp.

Hoeinghaus, D. J., K. O. Winemiller, and J. S. Birnbaum. 2007. Local and regional determinants of stream fish assemblage structure: inferences based on taxonomic vs. functional groups. J. Biogeogr. 34(2): 324–338.

Hoese, H. D. and R. H. Moore. 1998. Fishes of the Gulf of Mexico: Texas, Louisiana, and adjacent waters. 2nd ed. Texas A&M Univ. Press, College Station. 422 pp.

Hoffman, C. E. and R. V. Kilambi. 1970. Environmental changes produced by cold-water outlets from three Arkansas reservoirs. Water Resour. Res. Center Publ. No. 005, Univ. of Arkansas, Fayetteville. 169 pp.

Hogue, J. J. and R. V. Kilambi. 1975. Age and growth of Bluegill, Lepomis macrochirus Rafinesque, from Lake Fort Smith, Arkansas. Proc. Arkansas Acad. Sci. 29: 43–46.

Hogue, J. J., J. V. Conner, and V. R. Kranz. 1981. Descriptions and methods for identifying larval Blue Sucker Cycleptus elongatus (LeSueur). J. Cons. Int. Explor. Mer 178: 585–587.

Hogue, J. J., R. Wallus, and L. K. Kay. 1976. Preliminary guide to the identification of larval fishes in the Tennessee River. TVA Technical Note B19. 66 pp.

Holbrook, J. E. 1855. An account of several species of fish observed in Florida, Georgia, etc. J. Acad. Nat. Sci. Phila. Second ser. 3: 47–58.

Holder, T. H. 1970. Disappearing wetlands in eastern Arkansas. Arkansas Planning Comm., Little Rock, 72 pp.

Holland, L. E. and J. R. Sylvester. 1983. Distribution of larval fishes related to potential navigation impacts on the upper Mississippi River, Pool 7. Trans. Am. Fish. Soc. 112(2b): 293–301.

Holland-Bartels, L. E., S. K. Littlejohn, and M. L. Huston. 1990. A guide to larval fishes of the upper Mississippi River. U.S. Fish and Wildlife Service, National Fisheries Research Center, LaCrosse, WI. 107 pp.

Holley, C. T. and J. M. Long. 2018. Potential Longnose Darter population in the Kiamichi River of Oklahoma. Proc. Oklahoma Acad. Sci. 98: 14–17.

Holmes, B., L. Whittington, L. Marino, A. Adrian, and B. Stallsmith. 2010. Reproductive timing of the Telescope Shiner, Notropis telescopus, in Alabama, USA. Am. Midl. Nat. 163(2): 326–334.

Holsinger, J. R. 2000. Ecological derivation, colonization, and speciation. Pages 399–415. In: H. Wilkins, D. C. Culver, and W. F. Humphreys, eds. Ecosystems of the World 30: Subterranean ecosystems. Elsevier, New York. 808 pp.

Holt, A. and G. L. Harp. 1993. Ichthyofauna of the Village Creek system. Proc. Arkansas Acad. Sci. 47: 54–60.

Holt, D. E. and C. E. Johnston. 2011. Can you hear the dinner bell? Response of cyprinid fishes to environmental acoustic cues. Animal Behav. 82(3): 529–534.

Hooe, M. L. and D. H. Buck. 1991. Evaluation of F₁ hybrid crappies as sport fish in small impoundments. N. Am. J. Fish. Manag. 11(4): 564–571.

Hoopes, D. T. 1960. Utilization of mayflies and caddisflies by some Mississippi River fishes. Trans. Am. Fish. Soc. 89(1): 32–34.

Hoover, J. J. and K. J. Killgore. 1998. Fish communities. Pages 237–260. In: M. G. Messina and W. H. Conner, eds. Southern forested wetlands ecology and management. CRC Press. 640 pp.

Hoover, J. J., S. G. George, and K. J. Killgore. 2007. Diet of Shovelnose Sturgeon and Pallid Sturgeon in the free-flowing Mississippi River. J. Appl. Ichthyol. 23(4): 494–499.

Hoover, J. J., A. Turnage, and K. J. Killgore. 2009a. Swimming performance of juvenile Paddlefish: quantifying risk of entrainment. Pages 141–155. *In*: C. P. Paukert and G. D. Scholten, eds. Paddlefish management, propagation, and conservation in the 21st century: building from 20 years of research and management. Symp. 66. American Fisheries Society, Bethesda, MD. 443 pp.

Hoover, J. J., K. A. Boysen, C. E. Murphy, and S. G. George. 2009b. Morphological variation in juvenile Paddlefish. Pages 157–171. *In*: C. P. Paukert and G. D. Scholten, eds. Paddlefish management, propagation, and conservation in the 21st century: building from 20 years of research and management. Symp. 66. American Fisheries Society, Bethesda, MD. 443 pp.

Hoover, J. J., A. W. Katzenmeyer, J. Collins, B. R. Lewis, W. T. Slack, and S. G. George. 2015. Age and reproductive condition of an unusually large Bighead Carp from the lower Mississippi River basin. Southeast. Nat. 14(4): N55–N60.

Hoover, J. J., K. J. Killgore, D. G. Clarke, H. Smith, A. Turnage, and J. Beard. 2005. Paddlefish and sturgeon entrainment by dredges: swimming performance as an indicator of risk. DOER Technical Notes Collection (ERDC TN-DOER-E22), U.S. Army Engineer Research and Development Center, Vicksburg, Miss. 13 pp.

Hopkins, G. S. 1895. On the enteron of American ganoids. J. Morphol. 11(2): 411–442.

Horak, D. L. and H. A. Tanner. 1964. The use of vertical gill nets in studying fish depth distribution, Horsetooth Reservoir, Colorado. Trans. Am. Fish. Soc. 93(2): 137–145.

Horn, M. H. and C. D. Riggs. 1973. Effects of temperature and light on the rate of air breathing of the Bowfin, *Amia calva*. Copeia 1973(4): 653–657.

Horwitz, R. J. 1982. The range and co-occurrence of the shiners *Notropis analostanus* and *N. spilopterus* in southeastern Pennsylvania. Proc. Acad. Nat. Sci. Phila. 134: 178–193.

Hotalling, D. R. and C. A. Taber. 1987. Aspects of the life history of the Stippled Darter *Etheostoma punctulatum*. Am. Midl. Nat. 117(2): 428–434.

Houde, E. D. 1967. Food of pelagic young of the Walleye, *Stizostedion vitreum vitreum*, in Oneida Lake, New York. Trans. Am. Fish. Soc. 96(1): 17–24.

Houser, A. and H. E. Bryant. 1970. Age, growth, sex composition, and maturity of White Bass in Bull Shoals Reservoir. Tech. Paper 49, Bureau of Sport Fisheries and Wildlife. 11 pp.

Houser, A. and C. Collins. 1962. Growth of Black Bullhead catfish in Oklahoma. Oklahoma Fisheries Res. Lab Publ. Rep. 79. 18 pp.

Houser, A. and J. E. Dunn. 1967. Estimating the size of the Threadfin Shad population in Bull Shoals Reservoir from midwater trawl catches. Trans. Am. Fish. Soc. 96(2): 176–184.

Howard, J. M. 1989. The geology of Arkansas: the natural state. Rocks and Minerals 64(4): 270–276.

Howell, J. M., M. J. Weber, and M. L. Brown. 2014. Evaluation of trophic niche overlap between native fishes and young-of-the-year Common Carp. Am. Midl. Nat. 172(1): 91–106.

Howell, W. M. 1967. Taxonomy and distribution of the percid fish, *Etheostoma stigmaeum* (Jordan), with the validation and redescription of *Etheostoma davisoni* Hay. Ph.D. diss., Univ. Alabama, Tuscaloosa. 77 pp.

Howells, R. G. 1992. Guide to identification of harmful and potentially harmful fishes, shellfishes and aquatic plants prohibited in Texas. Texas Parks and Wildlife Dept., Inland Fisheries Div. Special Publ., Austin. 182 pp.

Howes, G. J. 1981. Anatomy and phylogeny of the Chinese major carps *Ctenopharyngodon* Steind., 1866 and *Hypophthalmichthys* Blkr., 1860. Bull. British Mus. Nat. Hist. (Zool.) 41(1): 1–52.

Howes, G. J. 1985. A revised synonymy of the minnow genus *Phoxinus* Rafinesque, 1820 (Teleostei: Cyprinidae) with comments on its relationships and distribution. Bull. British Mus. Nat. Hist. (Zool.) 48(1): 57–74.

Howes, G. J. 1991. Systematics and biogeography: an overview. Pages 1–33. *In*: I. J. Winfield and J. S. Nelson, eds. Cyprinid fishes: systematics, biology, and exploitation. Chapman & Hall, London. 667 pp.

Hoyt, R. D. 1965. Meristic variation of Smallmouth Bass in Arkansas. M.S. thesis, Univ. Arkansas, Fayetteville. 62 pp.

Hrabik, R. A. 1996. Taxonomic and distributional status of *Notropis volucellus* and *Notropis wickliffi* in the Mississippi River drainage: a literature review. LTRMP 96-S001. National Biological Service, Environmental Management Technical Center, Onalaska, WI. 15 pp.

Hrabik, R. A. 1997. Meristic and morphometric variation of some populations of *Notropis volucellus* and *Notropis wickliffi* in the upper Mississippi River basin. M.S. thesis. Southeast Missouri State Univ. Cape Girardeau. 211 pp.

Hrabik, R. A., D. P. Herzog, D. E. Ostendorf, and M. D. Petersen. 2007. Larvae provide first evidence of successful reproduction by Pallid Sturgeon, *Scaphirhynchus albus*, in the Mississippi River. J. Appl. Ichthyol. 23(4): 436–443.

Hrabik, R. A., S. C. Schainost, R. H. Stasiak, and E. J. Peters. 2015. The fishes of Nebraska. Univ. Nebraska, Conservation and Survey Division. Lincoln. 542 pp.

Hrbek, T. and A. Larson. 1999. The evolution of diapause in the killifish family Rivulidae (Atherinomorpha, Cyprinodontiformes): a molecular phylogenetic and biogeographic perspective. Evolution 53(4): 1200–1216.

Hubbs, C. 1954. A new Texas subspecies, *apristis*, of the darter *Hadropterus scierus*, with a discussion of variation within the species. Am. Midl. Nat. 52(1): 211–220.

Hubbs, C. 1961a. Developmental temperature tolerances of four etheostomatine fishes occurring in Texas. Copeia 1961(2): 195–198.

Hubbs, C. 1961b. Differences in the egg complement of *Hadropterus scierus* from Austin and San Marcos. Southwest. Nat. 6(1): 9–12.

Hubbs, C. 1971. Competition and isolation mechanisms in the *Gambusia affinis* × *G. heterochir* hybrid swarm. Texas Meml. Mus. Bull. 19: 1–47.

Hubbs, C. 1976. The diel reproductive pattern and fecundity of *Menidia audens*. Copeia 1976(2): 386–388.

Hubbs, C. 1982. Life history dynamics of *Menidia beryllina* from Lake Texoma. Am. Midl. Nat. 107(1): 1–12.

Hubbs, C. 1985. Darter reproductive seasons. Copeia 1985(1): 56–68.

Hubbs, C. and N. E. Armstrong. 1962. Developmental temperature tolerance of Texas and Arkansas-Missouri *Etheostoma spectabile* (Percidae: Osteichthyes). Ecology 43(4): 742–744.

Hubbs, C. and H. H. Bailey. 1977. Effects of temperature on the termination of breeding season of *Menidia audens*. Southwest. Nat. 22(4): 544–547.

Hubbs, C. and C. Bryan. 1974. Effects of parental temperature experience on thermal tolerance of eggs of *Menidia audens*. Pages 431–435. *In*: J. H. S. Blaxter, ed. The early life history of fish. Springer-Verlag, New York. 765 pp.

Hubbs, C. and C. Bryan. 1975. Ontogenetic rates and tolerances of the Channel Darter, *Hadropterus copelandi*. Texas J. Sci. 26: 623–625.

Hubbs, C. and D. F. Burnside. 1972. Developmental sequences of *Zygonectes notatus* at several temperatures. Copeia 1972(4): 862–865.

Hubbs, C. and C. M. Laritz. 1961a. Natural hybridization between *Hadropterus scierus* and *Percina caprodes*. Southwest. Nat. 6(3–4): 188–192.

Hubbs, C. and C. M. Laritz. 1961b. Occurrence of a natural intergeneric etheostomatine fish hybrid. Copeia 1961(2): 231–232.

Hubbs, C. and J. Pigg. 1972. Habitat preferences of the Harlequin Darter, *Etheostoma histrio*, in Texas and Oklahoma. Copeia 1972(1): 193–194.

Hubbs, C. and J. Pigg. 1976. The effects of impoundments on threatened fishes of Oklahoma. *In*: L. G. Hill and R. C. Summerfelt, eds. Oklahoma reservoir resources. Ann. Oklahoma Acad. Sci. 5: 113–117.

Hubbs, C. and K. Strawn. 1956. Interfertility between two sympatric fishes, *Notropis lutrensis* and *Notropis venustus*. Evolution 10(4): 341–344.

Hubbs, C. and K. Strawn. 1963. Differences in the developmental temperature tolerance of central Texas and more northern stocks of *Percina caprodes* (Percidae: Osteichthyes). Southwest. Nat. 8(1): 43–45.

Hubbs, C., R. J. Edwards, and G. P. Garret. 1991. An annotated checklist of freshwater fishes of Texas, with keys to identification of species. Texas J. Sci., Suppl. 43(4): 1–56.

Hubbs, C., R. J. Edwards, and G. P. Garrett. 2008. An annotated checklist of the freshwater fishes of Texas, with keys to identification of species. 2nd ed. Texas Acad. Sci. 87 pp. Available: http://www.texasacademyofscience.org/

Hubbs, C., H. B. Sharp, and J. F. Schneider. 1971. Developmental rates of *Menidia audens* with notes on salt tolerance. Trans. Am. Fish. Soc. 100(4): 603–610.

Hubbs, C. L. 1921. An ecological study of the life-history of the fresh-water atherine fish *Labidesthes sicculus*. Ecology 2(4): 262–276.

Hubbs, C. L. 1926. A check-list of the fishes of the Great Lakes and tributary waters, with nomenclatorial notes and analytical keys. Univ. Michigan Mus. Zool. Misc. Publ. 15. 77 pp.

Hubbs, C. L. 1930. Materials for a revision of the catostomid fishes of eastern North America. Univ. Michigan Mus. Zool. Misc. Publ. 20. 47 pp.

Hubbs, C. L. 1940. Speciation of fishes. Am. Nat. 74(752): 198–211.

Hubbs, C. L. 1941. Relation of hydrological conditions to speciation in fishes. Pages 182–195. *In*: A symposium on hydrobiology. Univ. Wisconsin Press, Madison. 495 pp.

Hubbs, C. L. 1942. An atherinid fish from the Pliocene of Oklahoma. J. Paleontol. 16(3): 399–400.

Hubbs, C. L. 1951. *Notropis amnis*, a new cyprinid fish of the Mississippi fauna, with two subspecies. Occas. Pap. Mus. Zool., Univ. Michigan 530. 31 pp.

Hubbs, C. L. 1964. History of ichthyology in the United States after 1850. Copeia 1964(1): 42–60.

Hubbs, C. L. 1979. History of ichthyology in the United States after 1850. Pages 9–21. *In*: M. S. Love and G. M. Cailliet, eds. Readings in ichthyology. Goodyear Publ., Santa Monica, CA. 525 pp.

Hubbs, C. L. and E. R. Allen. 1943. Fishes of Silver Springs, Florida. Proc. Florida Acad. Sci. 6(3–4): 110–130.

Hubbs, C. L. and R. M. Bailey. 1940. A revision of the black basses (*Micropterus* and *Huro*) with descriptions of four new forms. Misc. Publ. Mus. Zool. Univ. Michigan 48: 51 pp.

Hubbs, C. L. and J. D. Black. 1940a. *Notropis perpallidus*, a new minnow from Arkansas. Copeia 1940(1): 46–49.

Hubbs, C. L. and J. D. Black. 1940b. Status of the catostomid fish, *Carpiodes carpio elongatus* Meek. Copeia 1940(4): 226–230.

Hubbs, C. L. and J. D. Black. 1940c. Percid fishes related to *Poecilichthys variatus*, with descriptions of three new forms. Occas. Pap. Mus. Zool. Univ. Michigan 416: 30 pp.

Hubbs, C. L. and J. D. Black. 1941. The subspecies of the American percid fish, *Poecilichthys whipplii*. Occas Pap. Mus. Zool. Univ. Michigan 429: 1–21.

Hubbs, C. L. and J. D. Black. 1947. Revision of *Ceratichthys*, a genus of American cyprinid fishes. Misc. Publ. Mus. Zool. Univ. Michigan 66. 56 pp.

Hubbs, C. L. and J. D. Black. 1954. Status and synonymy of the American percid fish *Hadropterus scierus*. Am. Midl. Nat. 52(1): 201–210.

Hubbs, C. L. and K. Bonham. 1951. New cyprinid fishes of the genus *Notropis* from Texas. Texas J. Sci. 3(1): 91–110.

Hubbs, C. L. and G. P. Cooper. 1936. Minnows of Michigan. Cranbrook Inst. Sci. Bull. 8: 1–84.

Hubbs, C. L. and W. R. Crowe. 1956. Preliminary analysis of the American cyprinid fishes, seven new, referred to the genus *Hybopsis*, subgenus *Erimystax*. Occas. Pap. Mus. Zool., Univ. Michigan 578. 8 pp.

Hubbs, C. L. and C. W. Greene. 1935. Two new subspecies of fishes from Wisconsin. Trans. Wisconsin Acad. Sci. Arts. Lett. 29: 89–101.

Hubbs, C. L. and K. F. Lagler. 1947. Fishes of the Great Lakes Region. Cranbrook Inst. Sci. Bull. 26. 186 pp.

Hubbs, C. L. and K. F. Lagler. 1958. Fishes of the Great Lakes Region. Cranbrook Inst. Sci. Bull. 26. 213 pp.

Hubbs, C. L. and K. F. Lagler. 2004. Fishes of the Great Lakes Region, Revised Edition. Revised and updated by G. R. Smith. Univ. Michigan Press, Ann Arbor. 276 pp.

Hubbs, C. L. and G. A. Moore. 1940. The subspecies of *Notropis zonatus*, a cyprinid fish of the Ozark upland. Copeia 1940(2): 91–99.

Hubbs, C. L. and A. I. Ortenburger. 1929a. Further notes on the fishes of Oklahoma with descriptions of new species of Cyprinidae. Publ. Univ. Oklahoma Biol. Surv. 1(2): 15–43.

Hubbs, C. L. and A. I. Ortenburger. 1929b. Fishes collected in Oklahoma and Arkansas in 1927. Univ. Oklahoma Bull. (N.S.) 434. Publ. Univ. Oklahoma Biol. Surv. 1(3): 47–112.

Hubbs, C. L. and I. C. Potter. 1971. Distribution, phylogeny, and taxonomy. Pages 1–65. *In*: M. W. Hardisty and I. C. Potter, eds. The biology of lampreys. Vol. I. Academic Press, New York.

Hubbs, C. L. and E. C. Raney. 1939. *Hadropterus oxyrhynchus*, a new percid fish from Virginia and West Virginia. Occas. Pap. Mus. Zool. Univ. Michigan 396: 1–9.

Hubbs, C. L. and E. C. Raney. 1944. Systematic notes on North American siluroid fishes of the genus *Schilbeodes*. Occas. Pap. Mus. Zool. Univ. Michigan 487: 1–36.

Hubbs, C. L. and M. B. Trautman. 1937. A revision of the lamprey genus *Ichthyomyzon*. Univ. Michigan Mus. Zool. Misc. Publ. 35: 1–109.

Hubert, W. A. and R. T. Lackey. 1980. Habitat of adult Smallmouth Bass in a Tennessee River reservoir. Trans. Am. Fish. Soc. 109(4): 364–370.

Huckins, C. J. F. 1997. Functional linkages among morphology, feeding performance, diet, and competitive ability in molluscivorous sunfish. Ecology 78(8): 2401–2414.

Hudson, L. and T. M. Buchanan. 2001. Life history of the River Shiner, *Notropis blennius* (Cyprinidae), in the Arkansas River of western Arkansas. J. Arkansas Acad. Sci. 55: 57–65.

Hudy, M. 1986. Natural state trout. Arkansas Game and Fish 17(2): 8–11.

Hughes, A. L. 1985. Male size, mating success, and mating strategy in the Mosquitofish, *Gambusia affinis* (Poeciliidae). Behav. Ecol. Sociobiol. 17(3): 271–278.

Huish, M. T. 1954. Life history of the Black Crappie of Lake George, Florida. Trans. Am. Fish. Soc. 83(1): 176–193.

Huish, M. T. 1957. Food habits of three Centrarchidae in Lake George, Florida. Proc. 11th Annu. Conf. Southeast. Assoc. Game Fish Comm.: 293–302.

Hulsey, A. H. and J. H. Stevenson. 1958. Comparison of growth rates of game fish in Lake Catherine, Lake Hamilton, and Lake Ouachita, Arkansas. Proc. Arkansas Acad. Sci. 12: 17–31.

Hulsey, C. D. and F. J. Garcia de Leon. 2005. Cichlid jaw mechanics: linking morphology to feeding specialization. Func. Ecol. 19(3): 487–494.

Humphries, J. M. and R. C. Cashner. 1994. *Notropis suttkusi*, a new cyprinid from the Ouachita Uplands of Oklahoma and Arkansas, with comments on the status of Ozarkian populations of *N. rubellus*. Copeia 1994(1): 82–90.

Humphries, P. and P. S. Lake. 2000. Fish larvae and the management of regulated rivers. Regulated Rivers: Res. Manag. 16(5): 421–432.

Hundt, P. J., P. B. Berendzen, and A. M. Simons. 2017. Species delimitation and phylogeography of the studfish *Fundulus catenatus* species group (Ovalentaria: Cyprinodontiformes). Zool. J. Linn. Soc. 180(2): 461–474.

Hunt, B. P. 1953. Food relationships between Florida Spotted Gar and other organisms in the Tamiami Canal, Dade County, Florida. Trans. Am. Fish. Soc. 82(1): 13–33.

Hunt, J. and C. A. Annett. 2002. Effects of habitat manipulation on reproductive success of individual Largemouth Bass in an Ozark reservoir. N. Am. J. Fish. Manag. 22(4): 1201–1208.

Hunter, J. R. 1963. The reproductive behavior of the Green Sunfish, *Lepomis cyanellus*. Zoologica 48(2): 13–24.

Hunter, J. R. and A. D. Hasler. 1965. Spawning association of the Redfin Shiner, *Notropis umbratilis* and the Green Sunfish, *Lepomis cyanellus*. Copeia 1965(3): 265–281.

Hunter, J. R. and W. J. Wisby. 1961. Utilization of the nests of Green Sunfish (*Lepomis cyanellus*) by the Redfin Shiner (*Notropis umbratilis cyanocephalus*). Copeia 1961(1): 113–115.

Hurley, S. T., W. A. Hubert, and J. G. Nickum. 1987. Habitats and movements of Shovelnose Sturgeons in the upper Mississippi River. Trans. Am. Fish. Soc. 116(4): 655–662.

Inebnit, T. E. III. 2009. Aspects of the reproductive and juvenile ecology of Alligator Gar in the Fourche la Fave River, Arkansas. M.S. thesis, Univ. Central Arkansas, Conway. 69 pp.

Ingram, T. R. 2007. Age, growth and fecundity of Alabama Shad (*Alosa alabamae*) in the Apalachicola River, Florida. M.S. thesis, Clemson Univ., Clemson, SC. 27 pp.

Inoue J. G., M. Miya, M. J. Miller, T. Sado, R. Hanel, K. Hatooka, J. Aoyama, Y. Minegishi, M. Nishida, and K. Tsukamoto. 2010. Deep-ocean origin of the freshwater eels. Biol. Lett. 6(3): 363–366.

Irons, K. S., T. M. O'Hara, M. A. McClelland, and M. A. Pegg. 2002. White Perch occurrence, spread, and hybridization in the middle Illinois River, upper Mississippi River system. Trans. Illinois State Acad. Sci. 95(3): 207–214.

Irwin, E. R., W. A. Hubert, C. F. Rabeni, H. L. Schramm, Jr., and T. Coon, eds. 1999. Catfish 2000: Proceedings of the International Ictalurid Symposium. Symp. 24. American Fisheries Society, Bethesda, MD. 532 pp.

Jackson, D. C. 1984. Substrate preference of the Bigscale Logperch, *Percina macrolepida* (Percidae) in Lake Texoma, Oklahoma. Southwest. Nat. 29(3): 351–353.

Jackson, L. F. and C.V. Sullivan. 1995. Reproduction of White Perch: the annual gametogenic cycle. Trans. Am. Fish. Soc. 124(4): 563–577.

Jackson, S. W., Jr. 1958. Comparison of the age and growth of four fishes from lower and upper Spavinaw lakes, Oklahoma. Proc. Annu. Conf. Southeast. Assoc. Game Fish Comm. 11(1957): 232–249.

Jackson, W. D. and G. L. Harp. 1973. Ichthyofaunal diversification and distribution in an Ozark stream in northcentral Arkansas. Proc. Arkansas Acad. Sci. 27: 42–46.

Jacoby, D. M. P., J. M. Casselman, V. Crook, M. DeLucia, H. Ahn, K. Kaifu, T. Kurwie, P. Sasal, A. M. C. Silfvergrip, K. G. Smith, K. Uchida, A. M. Walker, and M. J. Gollock. 2015. Synergistic patterns of threat and the challenges facing global anguillid eel conservation. Global Ecol. Conserv. 4: 321–333.

Jacquemin, S. J. and M. Pyron. 2013. Effects of allometry, sex, and river location on morphological variation of Freshwater Drum *Aplodinotus grunniens* in the Wabash River, USA. Copeia 2013(4): 740–749.

Jacquemin, S. J., E. Martin, and M. Pyron. 2012. Morphology of Bluntnose Minnow *Pimephales notatus* (Cyprinidae) covaries with habitat in a central Indiana watershed. Am. Midl. Nat. 169(1): 137–146.

Jagnandan, K. and C. P. J. Sanford. 2013. Kinematics of ribbon-fin locomotion in the Bowfin, *Amia calva*. J. Exp. Zool. 319A(10): 569–583.

James, D. A. and J. C. Neal. 1986. Arkansas birds: their distribution and abundance. Univ. Arkansas Press, Fayetteville. 402 pp.

James, P. W. 1983. The life history of the Yoke Darter, *Etheostoma juliae*. M.S. thesis. Southwest Missouri State Univ., Springfield. 37 pp.

James, P. W. and O. E. Maughan. 1989. Spawning behavior and habitat of the threatened Leopard Darter, *Percina pantherina*. Southwest. Nat. 34(2): 298–301.

James, P. W. and C. A. Taber. 1986. Reproductive biology and age and growth of the Yoke Darter, *Etheostoma juliae*. Copeia 1986(2): 536–540.

James, P. W., O. E. Maughan, and A. V. Zale. 1991. Life history of the Leopard Darter *Percina pantherina* in Glover River, Oklahoma. Am. Midl. Nat. 125(2): 173–179.

Jansen, C. R. 2012. Population characteristics and capture techniques for Shovelnose Sturgeon *Scaphirhynchus platorynchus* in the Arkansas River. M.S. thesis. Arkansas Tech Univ., Russellville. 65 pp.

Janvier, P. 1984. Cladistics: theory, purpose and evolutionary implications. Pages 39–75. *In:* J. W. Pollard, ed. Evolutionary theory: paths into the future. Wiley Interscience, New York. 294 pp.

Janvier, P. 1996. The dawn of the vertebrates: characters versus common ascent in the rise of current vertebrate phylogenies. Palaeontology 39(2): 259–287.

Janvier, P. 1999. Catching the first fish. Nature 402: 21–22.

Janvier, P. and R. Lund. 1983. *Hardistiella montanensis* n. gen. et sp. (Petromyzontida) from the Lower Carboniferous of Montana, with remarks on the affinities of lampreys. J. Vert. Paleontol. 2(4): 407–413.

Jarocki, F. P. 1822. Zoologiia czyli zwiérzetopismo ogólne podlug náynowszego systematu. Drukarni Lakiewicza, Warszawie (Warsaw). v. 4: 1–464.

Jearld, A., Jr. and B. E. Brown. 1971. Fecundity, age and growth, and condition of Channel Catfish in an Oklahoma reservoir. Proc. Oklahoma Acad. Sci. 51: 15–22.

Jeffers, C. D. and E. J. Bacon, Jr. 1979. A distributional survey of the fishes of Ten Mile Creek in southeastern Arkansas. Proc. Arkansas Acad. Sci. 33: 46–48.

Jelks, H. L., S. J. Walsh, N. M. Burkhead, S. Contreras-Balderas, E. Diaz-Pardo, D. A. Hendrickson, J. Lyons, N. E. Mandrak, F. McCormick, J. S. Nelson, S. P. Platania, B. A. Porter, C. B. Renaud, J. J. Schmitter-Soto, E. B. Taylor, and M. L. Warren, Jr. 2008. Conservation status of imperiled North American freshwater and diadromous fishes. Fisheries 33(8): 372–407.

Jenkins, J. R. and B. A. Miller. 2007. Shoaling behavior in the Central Mudminnow (*Umbra limi*). Am. Midl. Nat. 158(1): 226–232.

Jenkins, J. T. and G. L. Harp. 1971. Ichthyofaunal diversification and distribution in the Big Creek Watershed, Craighead and Green counties, Arkansas. Proc. Arkansas. Acad. Sci. 25: 80–87.

Jenkins, R. E. 1970. Systematic studies of the catostomid fish tribe Moxostomatini. Ph.D. diss. Cornell Univ., Ithaca, NY. 799 pp.

Jenkins, R. E. 1976. A list of undescribed freshwater fish species of continental United States and Canada, with additions to the 1970 checklist. Copeia 1976(3): 642–644.

Jenkins, R. E. 1980a. *Hybopsis gelida* (Girard), Sturgeon Chub. Page 185. *In:* D. S. Lee et al. Atlas of North American freshwater fishes. N.C. State Mus. Nat. Hist., Raleigh. 854 pp.

Jenkins, R. E. 1980b. *Lagochila lacera* Jordan and Brayton, Harelip Sucker. Page 407. *In:* D. S. Lee et al. Atlas of North American freshwater fishes. N.C. State Mus. Nat. Hist., Raleigh. 854 pp.

Jenkins, R. E. 1980c. *Moxostoma carinatum* (Cope), River Redhorse. Pages 415–416. *In:* D. S. Lee et al. Atlas of North American freshwater fishes. N.C. State Mus. Nat. Hist., Raleigh. 854 pp.

Jenkins, R. E. 1980d. *Moxostoma poecilurum* (Jordan), Blacktail Redhorse. Page 430. *In:* D. S. Lee et al. Atlas of North

American freshwater fishes. N.C. State Mus. Nat. Hist., Raleigh. 854 pp.

Jenkins, R. E. 1994. Harelip Sucker *Moxostoma lacerum* (Jordan and Brayton). Pages 519–523. *In*: R. E. Jenkins and N. M. Burkhead, Freshwater fishes of Virginia. American Fisheries Society, Bethesda, MD. 1079 pp.

Jenkins, R. E. and N. M. Burkhead. 1994. Freshwater fishes of Virginia. American Fisheries Society. Bethesda, MD. 1079 pp.

Jenkins, R. E. and E. A. Lachner. 1971. Criteria for analysis and interpretation of the American fish genera *Nocomis* Girard and *Hybopsis* Agassiz. Smithson. Contr. Zool. 90: 1–15.

Jenkins, R. E. and E. A. Lachner. 1980. *Nocomis biguttatus*, Hornyhead Chub. Page 211. *In*: D. S. Lee et al. Atlas of North American freshwater fishes. N.C. State Mus. Nat. Hist., Raleigh. 854 pp.

Jenkins, R. E. and S. P. McIninch. 1994. Suckermouth minnows, genus *Phenacobius* Cope. Pages 337–345. *In*: R. E. Jenkins and N. M. Burkhead, Freshwater fishes of Virginia. American Fisheries Society, Bethesda, MD. 1079 pp.

Jenkins, R. M. 1952. Growth of the Flathead Catfish, *Pylodictis olivaris* in Grand Lake (Lake O' the Cherokees), Oklahoma. Proc. Oklahoma Acad. Sci. 33: 11–20.

Jenkins, R. M. 1956. Growth of Blue Catfish *Ictalurus furcatus* in Lake Texoma. Southwest. Nat. 1(4): 166–173.

Jenkins, R. M., E. M. Leonard, and G. E. Hall. 1952. An investigation of the fisheries resources of the Illinois River and pre-impoundment study of Tenkiller Reservoir, Oklahoma. Oklahoma Fish. Res. Lab. Rep. 26. 136 pp.

Jennings, C. A. and S. J. Zigler. 2000. Ecology and biology of Paddlefish in North America: historical perspectives, management approaches, and research priorities. Rev. Fish Biol. Fish. 10(2): 167–181.

Jennings, C. A. and S. J. Zigler. 2009. Biology and life history of Paddlefish in North America: an update. Pages 1–22. *In*: C. P. Paukert and G. D. Scholten, eds. Paddlefish management, propagation, and conservation in the 21st century: building from 20 years of research and management. Symp. 66. American Fisheries Society, Bethesda, MD. 443 pp.

Jennings, M. J. and D. P. Philipp. 1992a. Genetic variation in the Longear Sunfish (*Lepomis megalotis*). Can. J. Zool. 70(9): 1673–1680.

Jennings, M. J. and D. P. Philipp. 1992b. Reproductive investment and somatic growth rates in Longear Sunfish. Environ. Biol. Fishes 35(3): 257–271.

Jessop, B. M. 1997. An overview of European and American eel stocks, fisheries, and management issues. Can. Tech. Rep. Fish. Aquat. Sci. 1296: 6–20.

Jessop, B. M. 2010. Geographic effects on American Eel (*Anguilla rostrata*) life history characteristics and strategies. Can. J. Fish. Aquat. Sci. 67(2): 326–346.

Jester, D. B. 1971. Effects of commercial fishing, species introductions, and drawdown control on fish populations in Elephant Butte Reservoir, New Mexico. Pages 265–286. *In*:

G. E. Hall, ed. Reservoir fisheries and limnology. Spec. Publ. No. 8, American Fisheries Society, Washington, DC.

Jester, D. B. 1972. Life history, ecology, and management of the River Carpsucker, *Carpiodes carpio* (Rafinesque), with reference to Elephant Butte Lake. Res. Rep. 243. New Mexico State Univ. Agric. Exp. Station, Las Cruces. 120 pp.

Jester, D. B. 1973. Life history, ecology, and management of the Smallmouth Buffalo, *Ictiobus bubalus* (Rafinesque), with reference to Elephant Butte Lake. Res. Rep. 261. New Mexico State Univ. Agric. Exp. Station. 11 pp.

Jobling, S. and C. R. Tyler. 2003. Endocrine disruption in wild freshwater fish. Pure Appl. Chem. 75(11–12): 2219–2234.

Johnsgard, P. A. 2005. Prairie dog empire: a saga of the shortgrass prairie. Univ. Nebraska Press, Lincoln. 243 pp.

Johnson, B. L. 1994. Migration and population demographics of stream-spawning Longnose Gar (*Lepisosteus osseus*) in Missouri. M.S. thesis. Univ. Missouri, Columbia. 280 pp.

Johnson, B. L. and D. B. Noltie. 1996. Migratory dynamics of stream-spawning Longnose Gar (*Lepisosteus osseus*). Ecol. Freshw. Fish 5(3): 97–107.

Johnson, B. L. and D. B. Noltie. 1997. Demography, growth, and reproductive allocation in stream-spawning Longnose Gar. Trans. Am. Fish. Soc. 126(3): 438–466.

Johnson, B. M. and J. K. Beadles. 1977. Fishes of the Eleven Point River within Arkansas. Proc. Arkansas Acad. Sci. 31: 58–61.

Johnson, B. M. and P. J. Martinez. 2000. Trophic economics of Lake Trout management in reservoirs of different productivity. N. Am. J. Fish. Manag. 20(1): 127–143.

Johnson, C. R. 2011. Recovery of a lowland fish assemblage following large-scale rotenone application in eastern Arkansas. M.S. thesis. Univ. Central Arkansas, Conway. 109 pp.

Johnson, D. H. 1963. The food habits of the Goldeye, *Hiodon alosoides*, of the Missouri River and Lewis and Clark Reservoir, South Dakota. M.S. thesis. Univ. South Dakota, Vermillion. 72 pp.

Johnson, D. W. and E. L. McClendon. 1970. Differential distribution of the Striped Mullet, *Mugil cephalus* Linnaeus. California Fish Game 56: 138–139.

Johnson, G. D. 1984. Percoidei: development and relationships. Pages 464–498. *In*: H. G. Moser et al., eds. Ontogeny and systematics of fishes. American Society of Ichthyology and Herpetology, Special Publ. 1. 760 pp.

Johnson, G. D. and C. Patterson. 1993. Percomorph phylogeny: a survey of acanthomorphs and a new proposal. Bull. Mar. Sci. 52(1): 554–626.

Johnson, G. D. and V. G. Springer. 1997. *Elassoma*: another look [abstract]. 77th Annu. Meet. Am. Soc. Ichthyol. Herpetol., 176.

Johnson, J. D. and J. T. Hatch. 1991. Life history of the Least Darter, *Etheostoma microperca*, at the northwestern limits of its range. Am. Midl. Nat. 125(1): 87–103.

Johnson, J. E., W. D. Heckathorn, Jr., and A. L. Thompson. 1996. Dispersal and persistence of 2,3,7,8-tetrachlorodibe nzo-*p*-dioxin (TCDD) in a contaminated aquatic ecosystem, Bayou Meto, Arkansas. Trans. Am. Fish. Soc. 125(3): 450–457.

Johnson, J. H. and D. S. Dropkin. 1991. Summer food habits of Spotfin Shiner, Mimic Shiner, and subyearling Fallfish in the Susquehanna River basin. J. Freshw. Ecol. 6(1): 35–42.

Johnson, J. H. and D. S. Dropkin. 1993. Diel variation in diet composition of a riverine fish community. Hydrobiologia 271(3): 149–158.

Johnson, P. D., A. E. Bogan, K. M. Brown, N. M. Burkhead, J. R. Cordeiro, J. T. Garner, P. D. Hartfield, D. A. Lepitzki, G. L. Mackie, E. Pip, T. A. Tarpley, J. S. Tiemann, N. V. Whelan, and E. E. Strong. 2013. Conservation status of freshwater gastropods of Canada and the United States. Fisheries 38(6): 247–282.

Johnson, R. L. 2009. A comparison of genetic structuring of Yellowcheek Darters (*Etheostoma moorei*) using AFLPs and allozymes. Biochem. Syst. Ecol. 37(4): 298–303.

Johnson, R. L. and C. F. Cavenaugh. 2003. Genetic similarity of Shadow and Ozark basses (*Ambloplites*) as determined by mitochondrial DNA analysis. J. Arkansas Acad. Sci. 57: 206–207.

Johnson, R. L. and R. M. Davis. 1997. Age, growth, and condition of Largemouth Bass, *Micropterus salmoides*, of Lake Ashbaugh, Arkansas. J. Arkansas Acad. Sci. 51: 95–102.

Johnson, R. L. and T. Fulton. 1999. Persistence of Florida Largemouth Bass alleles in a northern Arkansas population of Largemouth Bass, *Micropterus salmoides* Lacépède. Ecol. Freshw. Fish 8(1): 35–42.

Johnson, R. L. and G. L. Harp. 2005. Spatio-temporal changes of benthic macroinvertebrates in a cold Arkansas tailwater. Hydrobiologia 537(1–3): 15–24.

Johnson, R. L. and R. Staley. 2001. Identification of Florida Largemouth Bass alleles in Arkansas public and private aquaculture ponds. J. Arkansas Acad. Sci. 55: 82–85.

Johnson, R. L., S. C. Blumenshine, and S. M. Coghlan. 2006a. A bioenergetic analysis of factors limiting Brown Trout growth in an Ozark tailwater river. Environ. Biol. Fishes 77(2): 121–132.

Johnson, R. L., R. M. Mitchell, and G. L. Harp. 2006b. Genetic variation and genetic structuring of a numerically declining species of darter, *Etheostoma moorei* Raney and Suttkus, endemic to the upper Little Red River, Arkansas. Am. Midl. Nat. 156(1): 37–44.

Johnson, R. L., S. M. Coghlan, and T. Harmon. 2007. Spatial and temporal variation in prey selection of Brown Trout in a cold Arkansas tailwater. Ecol. Freshw. Fish 16(3): 373–384.

Johnson, R. L., S. D. Henry, and S. W. Barkley. 2009. Distribution and population characteristics of Walleye in the lower Eleven Point River, Arkansas. J. Arkansas Acad. Sci. 63: 99–105.

Johnson, R. L., S. D. Henry, and S. W. Barkley. 2010. Distribution and population characteristics of Shadow Bass in two Arkansas Ozark streams. N. Am. J. Fish. Manag. 30(6): 1522–1528.

Johnson, R. M. 1978. Fishes of the Saline River, southwest Arkansas. M.S. thesis. Northeast Louisiana Univ., Monroe. 96 pp.

Johnson, R. P. 1963. Studies on the life history and ecology of the Bigmouth Buffalo, *Ictiobus cyprinellus* (Valenciennes). J. Fish. Res. Board Can. 20(6): 1397–1429.

Johnson, T. B., W. P. Brown, T. D. Corry, M. H. Hoff, J. V. Scharold, and A. S. Trebitz. 2004. Lake Herring (*Coregonus artedi*) and Rainbow Smelt (*Osmerus mordax*) diets in western Lake Superior. J. Great Lakes Res. 30(S1): 407–413.

Johnston, C. E. 1989a. Spawning in the Eastern Sand Darter, *Ammocrypta pellucida* (Pisces: Percidae), with comments on the phylogeny of *Ammocrypta* and related taxa. Trans. Illinois State Acad. Sci. 82(3–4): 163–168.

Johnston, C. E. 1989b. Male minnows build spawning nests. Illinois Nat. Hist. Surv. Rep., Nov. 1989, No. 291.

Johnston, C. E. 1994a. The benefit to some minnows of spawning in the nests of other species. Environ. Biol. Fishes 40(2): 213–218.

Johnston, C. E. 1994b. Spawning behavior of the Goldstripe Darter (*Etheostoma parvipinne* Gilbert and Swain) (Percidae). Copeia 1994(3): 823–825.

Johnston, C. E. 1995. Spawning behavior of the Paleback Darter, *Etheostoma pallididorsum* (Percidae). Southwest. Nat. 40(4): 422–425.

Johnston, C. E. 1999. The relationship of spawning mode to conservation of North American minnows (Cyprinidae). Environ. Biol. Fishes 55(1–2): 21–30.

Johnston, C. E. and D. L. Johnson. 2000. Sound production in *Pimephales notatus* (Rafinesque) (Cyprinidae). Copeia 2000(2): 567–571.

Johnston, C. E. and C. L. Knight. 1999. Life-history traits of the Bluenose Shiner, *Pteronotropis welaka* (Cypriniformes: Cyprinidae). Copeia 1999(1): 200–205.

Johnston, C. E. and L. M. Page. 1992. The evolution of complex reproductive strategies in North American minnows (Cyprinidae). Pages 600–621. *In*: R. L. Mayden, ed. Systematics, historical ecology, and North American freshwater fishes. Stanford Univ. Press, Stanford, CA. 969 pp.

Johnston, C. E. and C. T. Phillips. 2003. Sound production in sturgeon *Scaphirhynchus albus* and *S. platorynchus* (Acipenseridae). Environ. Biol. Fishes 68(1): 59–64.

Johnston, C. E., J. W. Armbruster, and C. A. Laird. 1996. Parallel swims as a means of intra- and interspecific assessment in stream fishes. Environ. Biol. Fishes 46(4): 405–408.

Jones, E. B. D. III, G. S. Helfman, J. O. Harper, and P. V. Bolstad. 1999. Effects of riparian forest removal on fish assemblages in southern Appalachian streams. Conserv. Biol. 13(6): 1454–1465.

Jones, R. N. and O. E. Maughan. 1987. Food of two species of darters in Glover River, Oklahoma. Proc. Oklahoma Acad. Sci. 67: 73–74.

Jones, R. N. and O. E. Maughan. 1989. Food habits of the juvenile and adult Orangebelly Darter, *Etheostoma radiosum*, in Glover Creek, Oklahoma. Proc. Oklahoma Acad. Sci. 69: 39–43.

Jones, R. N., D. J. Orth, and O. E. Maughan. 1984. Abundance and preferred habitat of the Leopard Darter, *Percina pantherina*, in Glover Creek, Oklahoma. Copeia 1984(2): 378–384.

Jones, W. J. and J. M. Quattro. 1999. Phylogenetic affinities of pygmy sunfishes (*Elassoma*) inferred from mitochondrial DNA sequences. Copeia 1999(2): 470–474.

Jordan, D. S. 1876a. Class V.—Pisces. (The fishes.). Pages 199–362. *In*: Manual of the vertebrates of the northern United States, including the district east of the Mississippi River and north of North Carolina and Tennessee, exclusive of marine species. Jansen, McClurg & Co., Chicago.

Jordan, D. S. 1876b. Concerning the fishes of the Ichthyologia Ohiensis. Bull. Buffalo Soc. Nat. Sci. 3(3, art. 8): 91–97.

Jordan, D. S. 1877a. Contributions to North American ichthyology. Based primarily on the collections of the United States National Museum. I. Review of Rafinesque's memoirs on North American fishes. Bull. U.S. Natl. Mus. 9: 1–53.

Jordan, D. S. 1877b. Contributions to North American ichthyology based primarily on the collections of the United States National Museum. II. Part A. Notes on Cottidae, Etheostomatidae, Percidae, Centrarchidae, Aphredoderidae, Dorasomatidae, and Cyprinidae, with revisions of the genera and descriptions of new or little known species. Bull. U.S. Natl. Mus. 10: 1–68.

Jordan, D. S. 1877c. On the fishes of northern Indiana. Proc. Acad. Nat. Sci. Phila. 29: 42–82.

Jordan, D. S. 1877d. A partial synopsis of the fishes of upper Georgia; with supplementary papers on fishes of Tennessee, Kentucky, and Indiana. Ann. Lyceum Nat. Hist. N.Y. 11(11/12): 307–377.

Jordan, D. S. 1878a. A catalogue of the fishes of Illinois. Bull. Illinois State Lab. Nat. Hist. 1(2): 37–70.

Jordan, D. S. 1878b. Manual of the vertebrates of the northern United States, including the district east of the Mississippi River and north of North Carolina and Tennessee, exclusive of marine species. 2nd ed. Jansen, McClurg & Co., Chicago. 407 pp.

Jordan, D. S. 1880. Description of new species of North American fishes. Proc. U.S. Natl. Mus. 2(84): 235–241.

Jordan, D. S. 1882. Report on the fishes of Ohio. Report of the Geological Survey of Ohio 4 (pt 1, sect 4): 735–1002.

Jordan, D. S. 1885a. Description of a new species of *Hybognathus* (*Hybognathus hayi*) from Mississippi. Proc. U.S. Natl. Mus. 7(467): 548–550.

Jordan, D. S. 1885b. A catalogue of the fishes known to inhabit the waters of North America, north of the Tropic of Cancer, with notes on the species discovered in 1883 and 1884. United States Commission of Fish and Fisheries, Report of the Commissioner v. 13 (1885): 789–973.

Jordan, D. S. 1888. A manual of vertebrate animals of the northern United States, including the district north and east of the Ozark Mountains, south of the Laurentian hills, north of the southern boundary of Virginia, and east of the Missouri River; inclusive of marine species. 5th ed. McClurg & Co., Chicago. 375 pp.

Jordan, D. S. 1917. The genera of fishes. Part I. from Linnaeus to Cuvier, 1758–1833 seventy-five years with the accepted type of each. A contribution to the stability of scientific nomenclature. Leland Stanford Junior Univ. Publ., Stanford, CA. Univ. Ser. 27: 1–161 pp.

Jordan, D. S. 1923. A classification of fishes including families and genera as far as known. Stanford Univ. Publ., Univ. Ser., Biol. Sci. 3(2): 243 pp. [Reprinted in 1934 and 1963.]

Jordan, D. S. 1924. Concerning the American dace allied to the genus *Leuciscus*. Copeia 1924: 70–72.

Jordan, D. S. 1929. Manual of the vertebrate animals of the northeastern United States inclusive of marine species. 13th ed. World Book Company, Yonkers-on-Hudson, NY. 446 pp.

Jordan, D. S. and A. W. Brayton. 1877. On *Lagochila*, a new genus of catostomoid fishes. Proc. Acad. Nat. Sci. Phila. 29: 280–283.

Jordan, D. S. and A. W. Brayton. 1878. Contributions to North American Ichthyology. No. 3. A. On the distribution of the fishes of the Allegheny region of South Carolina, Georgia, and Tennessee, with descriptions of new or little known species. Bull. U.S. Natl. Mus. 12: 7–95.

Jordan, D. S. and B. W. Evermann. 1896. The fishes of North and Middle America: a descriptive catalogue of the species of fish-like vertebrates found in the waters of North America, north of the Isthmus of Panama. Part I. Bull. U.S. Natl. Mus. 47: 1–1240.

Jordan, D. S. and B. W. Evermann. 1900. The fishes of North and Middle America. Part 4. Bull. U.S. Natl. Mus. 47: 3137–3313.

Jordan, D. S. and B. W. Evermann. 1916. American food and game fishes: a popular account of all the species found in America north of the equator, with keys for ready identification, life histories and methods of capture. Doubleday, Page, and Co., Garden City, NY. 572 pp.

Jordan, D. S. and C. H. Gilbert. 1883. Synopsis of the fishes of North America. Bull. U.S. Natl. Mus. 16: 473–474. [Often cited as 1882]

Jordan, D. S. and C. H. Gilbert. 1886. List of fishes collected in Arkansas, Indian Territory, and Texas, in September, 1884, with notes and descriptions. Proc. U.S. Natl. Mus. 9(549): 1–25.

Jordan, D. S. and S. E. Meek. 1884. Description of four new species of Cyprinidae in the United States National Museum. Proc. U.S. Natl. Mus. 7(450): 474–477.

Jordan, D. S. and S. E. Meek. 1885. List of fishes collected in Iowa and Missouri in August, 1884, with descriptions of three new species. Proc. U.S. Natl. Mus. 8(470): 1–17.

Jordan, D. S., B. W. Evermann, and H. W. Clark. 1930.

Checklist of the fishes and fishlike vertebrates of North and Middle America north of the northern boundary of Venezuela and Columbia. Rep. U.S. Comm. Fish. Part 2: 1–670.

Jordan, G. R., E. J. Heist, P. J. Braaten, A. J. DeLonay, P. Hartfield, D. P. Herzog, K. M. Kappenman, and M. A. H. Webb. 2016. Status of knowledge of the Pallid Sturgeon (*Scaphirhynchus albus* Forbes and Richardson, 1905). J. Appl. Ichthyol. 32(S1): 191–207.

Jordan-Mathis, R. J. 1994. The reproductive biology of the Cypress Darter, *Etheostoma proeliare* (Hay), from the Bayou Bartholomew drainage. M.S. thesis, Northeast Louisiana Univ., Monroe.

Jorgensen, J. M., A. Flock, and J. Z. Wersäll. 1972. The Lorenzinian ampullae of *Polyodon spathula*. Z. Zellforsch. Mikrosk. Anat. 130(3): 362–377.

Joswiak, G. R. 1980. Genetic divergence within a genus of cyprinid fish (*Phoxinus*: Cyprinidae). Ph.D. dissertation. Wayne State Univ., Detroit, MI.

Joyce, E. C. and G. B. Chapman. 1978. Fine structure of the nasal barbel of the Channel Catfish, *Ictalurus punctatus*. J. Morphol. 158(2): 109–153.

Juanes, F., J. A. Buckel, and F. S. Scharf. 2002. Feeding ecology of piscivorous fishes. Pages 267–283. *In*: P. J. B. Hart and J. D. Reynolds, eds. Handbook of fish biology and fisheries. Blackwell, Malden, MA.

Jude, D. J., S. A. Klinger, and M. D. Enk. 1981. Evidence of natural reproduction by planted Lake Trout in Lake Michigan. J. Great Lakes Res. 7(1): 57–61.

Kaartinen, R., G. N. Stone, J. Hearn, K. Lohse, and T. Roslin. 2010. Revealing secret liaisons; DNA barcoding changes our understanding of food webs. Ecol. Entomol. 35(5): 623–638.

Kallemeyn, L. 1983. Status of the Pallid Sturgeon. Fisheries 8(1): 3–9.

Kammerer, C. F., L. Grande, and M. W. Westneat. 2006. Comparative and developmental functional morphology of the jaws of living and fossil gars (Actinopterygii: Lepisosteidae). J. Morphol. 267(9): 1017–1031.

Kappenman, K. M., M. A. H. Webb, and M. Greenwood. 2013. The effect of temperature on embryo survival and development in Pallid Sturgeon *Scaphirhynchus albus* (Forbes and Richardson 1905) and Shovelnose Sturgeon *S. platorynchus* (Rafinesque, 1820). J. Appl. Ichthyol. 29(6): 1193–1203.

Karas, N. 2015. Brook Trout: a thorough look at North America's great native trout—its history, biology, and angling possibilities. Skyhorse, New York. 492 pp.

Karlin, A. A. and J. D. Rickett. 1990. Microgeographic genetic variation in Creole (*Etheostoma collettei*) and Redfin (*Etheostoma whipplei*) darters (Percidae) in central Arkansas. Southwest. Nat. 35(2): 135–145.

Karr, J. R. 1963. Age, growth, and food habits of Johnny, Slenderhead and Blacksided darters of Boone County, Iowa. Proc. Iowa Acad. Sci. 70(1): 228–236.

Karr, J. R. 1964. Age, growth, fecundity and food habits of

Fantail Darters in Boone County, Iowa. Proc. Iowa Acad. Sci. 71(1): 274–280.

Kassler, T. W., J. B. Koppelman, T. J. Near, C. B. Dillman, J. M. Levengood, D. L. Swofford, J. L. VanOrman, J. E. Claussen, and D. P. Philipp. 2002. Molecular and morphological analyses of the black basses (*Micropterus*): implications for taxonomy and conservation. Am. Fish. Soc. Symp. 31: 291–322.

Kaszubski, J. L. 1990. The life history and ecology of the Scaly Sand Darter, *Ammocrypta vivax* (class: Osteichthyes, family: Percidae), from southeast Missouri. M.S. thesis, Southeast Missouri State Univ., Cape Girardeau. 180 pp.

Katano, O., T. Nakamura, and S. Yamamoto. 2006. Intraguild indirect effects through trophic cascades between stream-dwelling fishes. J. Animal Ecol. 75(1): 167–175.

Katechis, C. T., P. C. Sakaris, and E. R. Erwin. 2007. Population demographics of *Hiodon tergisus* (Mooneye) in the lower Tallapoosa River. Southeast. Nat. 6(3): 461–470.

Katula, R. 1987. Spawning of the Pirate Perch recollected. Am. Currents 20: 9.

Katula, R. 1992. The spawning mode of the Pirate Perch. Trop. Fish Hobbyist (August) 40: 156–159.

Katula, R. S. and L. M. Page. 1998. Nest association between a large predator, the Bowfin (*Amia calva*) and its prey, the Golden Shiner (*Notemigonus crysoleucas*). Copeia 1998(1): 220–221.

Kay, L. K., R. Wallus, and B. L. Yeager. 1994. Reproductive biology and early life history of fishes in the Ohio River drainage. Vol. 2. Catostomidae. Tennessee Valley Authority, Chattanooga, TN. 242 pp.

Ke, Z. X., P. Xie, and L. G. Guo. 2008. Controlling factors of spring–summer phytoplankton succession in Lake Taihu (Meiliang Bay, China). Hydrobiologia 607(1): 41–49.

Keast, A. 1966. Trophic interrelationships in the fish fauna of a small stream. Proc. 9th Conf. Great Lakes Res., Univ. Michigan: 51–79.

Keast, A. 1977. Mechanisms expanding niche width and minimizing intraspecific competition in two centrarchid fishes. Evol. Biol. 10: 333–395.

Keast, A. 1985a. Development of dietary specializations in a summer community of juvenile fishes. Environ. Biol. Fishes 13(3): 211–224.

Keast, A. 1985b. The piscivore feeding guild of fishes in small freshwater ecosystems. Environ. Biol. Fishes 12(2): 119–129.

Keast, A. and D. Webb. 1966. Mouth and body form relative to feeding ecology in the fish fauna of a small lake, Lake Opinicon, Ontario. J. Fish. Res. Board Can. 23(12): 1845–1874.

Keck, B. P. and T. J. Near. 2009. Patterns of natural hybridization in darters (Percidae: Etheostomatinae). Copeia 2009(4): 758–773.

Keck, B. P. and T. J. Near. 2010. Geographic and temporal aspects of mitochondrial replacement in *Nothonotus* darters (Teleostei: Percidae: Etheostomatinae). Evolution 64(5): 1410–1428.

Keck, B. P. and T. J. Near. 2013. A new species of *Nothonotus* darter (Teleostei: Percidae) from the Caney Fork in Tennessee, USA. Bull. Peabody Mus. Nat. Hist. 54(1): 3–21.

Keenlyne, K. D. 1997. Life history and status of the Shovelnose Sturgeon, *Scaphirhynchus platorynchus*. Environ. Biol. Fishes 48(1–4): 291–298.

Keenlyne, K. D. and L. G. Jenkins. 1993. Age at sexual maturity of the Pallid Sturgeon. Trans. Am. Fish. Soc. 122(3): 393–396.

Keenlyne, K. D., E. M. Grossman, and L. G. Jenkins. 1992. Fecundity of the Pallid Sturgeon. Trans. Am. Fish. Soc. 121(1):139–140.

Keenlyne, K. D., L. K. Graham, and B. C. Reed. 1994. Hybridization between the Pallid and Shovelnose sturgeon. Proc. South Dakota Acad. Sci. 73: 59–66.

Keith, W. E. 1964. A pre-impoundment study of the fishes, their distribution, and abundance in the Beaver Lake drainage of Arkansas. M.S. thesis. Univ. Arkansas, Fayetteville. 94 pp.

Keith, W. E. 1971a. Culture, stocking, and management of catfish in Arkansas. Proc. North Central Warmwater Fish Culture-Management Workshop: 164–170. Iowa Coop. Fish. Res. Unit.

Keith, W. E. 1971b. The Redbreast: cold-water bream. Arkansas Game and Fish 4(2): 10–11.

Keith, W. E. 1972. The biological aspect of the pickerel. Arkansas Game and Fish 5(1): 8–9.

Keith, W. E. 1987. Distribution of fishes in reference streams within Arkansas' ecoregions. Proc. Arkansas Acad. Sci. 41: 57–60.

Keith, W. E., T. M. Buchanan, and H. W. Robison. 1987. Rediscovery of the Suckermouth Minnow, *Phenacobius mirabilis* (Girard), in Arkansas. Proc. Arkansas Acad. Sci. 41: 110.

Kelly, H. A. 1924. *Amia calva* guarding its young. Copeia 1924(133): 73–74.

Kelly, N. B., T. J. Near, and S. H. Alonzo. 2012. Diversification of egg-deposition behaviours and the evolution of male parental care in darters (Teleostei: Percidae: Etheostomatinae). J. Evol. Biol. 25(5): 836–846.

Kendall, A. W. and R. J. Behnke. 1984. Salmonidae: development and relationships. Page 142–149. *In:* H. G. Moser, ed. Ontogeny and systematics of fishes. American Society of Ichthyologists and Herpetologists Special Publ. 1. Allen Press, Lawrence, KS.

Kendall, R. L., ed. 1978. Selected coolwater fishes of North America. Publ. 11, American Fisheries Society. 437 pp.

Kendall, W. C. 1927. The smelts. Bull. U.S. Bur. Fish. 42: 217–375.

Kennedy, W. A. and W. M. Sprules. 1967. Goldeye in Canada. Bulletin 161, Fisheries Research Board of Canada. 45 pp.

Kerby, J. H. 1979. Meristic characters of two *Morone* hybrids. Copeia 1979(3): 513–518.

Kerr, S. J. 2011. Distribution and management of Muskellunge in North America: an overview. Fisheries Policy Section, Biodiversity Branch, Ontario Ministry of Natural Resources, Peterborough. 22 pp. + appendices.

Kilambi, R. V. 1980. Food consumption, growth, and survival of Grass Carp *Ctenopharyngodon idella* Val at four salinities. J. Fish Biol. 17(6): 613–618.

Kilambi, R. V. 1981. Cage culture fish production and effects on resident Largemouth Bass. Proc. World Symp. on Aquaculture in Heated Effluents and Recirculation Systems, Stavanger, 28–30 May 1980. Vol. II, Berlin: 191–203.

Kilambi, R. V. and R. E. Baglin, Jr. 1969a. Fecundity of the Gizzard Shad, *Dorosoma cepedianum* (Lesueur), in Beaver and Bull Shoals reservoirs. Am. Midl. Nat. 82(2): 444–449.

Kilambi, R. V. and R. E. Baglin, Jr. 1969b. Fecundity of the Threadfin Shad, *Dorosoma petenense*, in Beaver and Bull Shoals reservoirs. Trans. Am. Fish. Soc. 98(2): 320–322.

Kilambi, R. V. and M. L. Galloway. 1985. Temperature preference and tolerance of hybrid carp (female Grass Carp, *Ctenopharyngodon idella* × male Bighead, *Aristichthys nobilis*) Environ. Biol. Fishes 12(4): 309–314.

Kilambi, R. V. and W. R. Robison. 1978. Age and growth of Carp from Beaver Reservoir, Arkansas. Proc. Arkansas Acad. Sci. 32: 91–92.

Kilambi, R. V. and W. R. Robison. 1979. Effects of temperature and stocking density on food consumption and growth of Grass Carp *Ctenopharyngodon idella* Val. J. Fish Biol. 15(3): 337–342.

Kilambi, R. V. and A. Zdinak. 1980. Food preference and growth of Grass Carp, *Ctenopharyngodon idella,* and hybrid carp, *C. idella* female × *Aristichthys nobilis* male. Proc. V. Int. Symp. Biol. Contr. Weeds, Brisbane, Australia, 1980: 281–286.

Kilambi, R. V. and A. Zdinak. 1981a. Comparison of early developmental stages and adults of Grass Carp, *Ctenopharyngodon idella,* and hybrid carp (female Grass Carp × male Bighead *Aristichthys nobilis*). J. Fish Biol. 19(4): 457–465.

Kilambi, R. V. and A. Zdinak. 1981b. The biology of Striped Bass, *Morone saxatilis,* in Beaver Reservoir, Arkansas. Proc. Arkansas Acad. Sci. 35: 43–45.

Kilambi, R. V., J. C. Adams, and W. A. Wickizer. 1978. Effects of cage culture on growth, abundance, and survival of resident Largemouth Bass (*Micropterus salmoides*). J. Fish. Res. Board Can. 35(1): 157–160.

Kilambi, R. V., J. Noble, and C. E. Hoffman. 1970. Influence of temperature and photoperiod on growth, food consumption and food conversion efficiency of Channel Catfish. Proc. 24th Annu. Conf. Southeast. Assoc. Game Fish Comm.: 519–531.

Kilambi, R. V., W. R. Robison, and J. C. Adams. 1977. Growth, mortality, food habits, and fecundity of the Buffalo River Smallmouth Bass. Proc. Arkansas Acad. Sci. 31: 62–63.

Kilgen, R. H. 1972. Food habits and growth of fingerling Blacktail Redhorse, *Moxostoma poecilurum* (Jordan) in ponds. Proc. Louisiana Acad. Sci. 35: 12–20.

Kilgen, R. H. 1974. Artificial spawning and hatching techniques for Blacktail Redhorse. Prog. Fish-Cult. 36(3): 174.

Killgore, K. J. and J. J. Hoover. 2001. Effects of hypoxia on fish assemblages in a vegetated waterbody. J. Aquat. Plant Manag. 39: 40–44.

Killgore, K. J., J. J. Hoover, J. P. Kirk, S. G. George, B. R. Lewis, and C. E. Murphy. 2007a. Age and growth of Pallid Sturgeon in the free-flowing Mississippi River. J. Appl. Ichthyol. 23(4): 452–456.

Killgore, K. J., J. J. Hoover, S. G. George, B. R. Lewis, C. E. Murphy, and W. E. Lancaster. 2007b. Distribution, relative abundance and movements of Pallid Sturgeon in the free-flowing Mississippi River. J. Appl. Ichthyol. 23(4): 476–483.

Killgore, K. J., L. E. Miranda, C. E. Murphy, D. M. Wolff, J. J. Hoover, T. M. Keevin, S. T. Maynord, and M. A. Cornish. 2011. Fish entrainment rates through towboat propellers in the Upper Mississippi and Illinois rivers. Trans. Am. Fish. Soc. 140(3): 570–581.

King-Heiden, T. C., V. Mehta, K. M. Xiong, K. A. Lanham, D. S. Antkiewicz, A. Ganser, W. Heideman, and R. E. Peterson. 2012. Reproductive and developmental toxicity of dioxin in fish. Mol. Cell. Endocrinol. 354(1–2): 121–138.

Kinney, E. C, Jr. 1954. A life history study of the Silver Chub, *Hybopsis storeriana* (Kirtland), in western Lake Erie, with notes on associated species. Ph.D. diss. Ohio State Univ., Columbus. 99 pp.

Kinziger, A. P. 2003. Evidence supporting two new forms and one previously described race within the *Cottus carolinae* species-complex from the Ozark Highlands. Am. Midl. Nat. 149(2): 418–424.

Kinziger, A. P. and R. M. Wood. 2003. Molecular systematics of the polytypic species *Cottus hypselurus* (Teleostei: Cottidae). Copeia 2003(3): 624–627.

Kinziger, A. P. and R. M. Wood. 2010. *Cottus immaculatus*, a new species of sculpin (Cottidae) from the Ozark Highlands of Arkansas and Missouri, USA. Zootaxa 2340: 50–64.

Kinziger, A. P., D. H. Goodman, and R. S. Studebaker. 2007. Mitochondrial DNA variation in the Ozark Highland members of the Banded Sculpin, *Cottus carolinae*, complex. Trans. Am. Fish. Soc. 136(6): 1742–1749.

Kinziger, A. P., R. M. Wood, and D. A. Neely. 2005. Molecular systematics of the genus *Cottus* (Scorpaeniformes: Cottidae). Copeia 2005(2): 303–311.

Kipp, R., B. Cudmore, and N. E. Mandrak. 2011. Updated (2006–early 2011) biological synopsis of Bighead Carp (*Hypophthalmichthys nobilis*) and Silver Carp (*H. molitrix*). Can. Manuscr. Rep. Fish. Aquat. Sci. 2962: v + 51 pp.

Kirtland, J. P. 1840a. Article III. Descriptions of four new species of fishes. Boston J. Nat. Hist. 3(1/2): 273–277.

Kirtland, J. P. 1840b. Article X. Descriptions of the fishes of the Ohio River and its tributaries. Boston J. Nat. Hist. 3: 338–352.

Kirtland, J. P. 1845. [Description of a new fish, from the continuation of Dr. Kirtland's paper on the fishes of Ohio]. Proc. Boston Soc. Nat. Hist. 1(21): 199–200.

Kitterman, C. L., J. S. Williams, P. Port, R. Moore, and J. Stein. 2011. Beaver tailwater creel survey 2008–2010 Final Report. AGFC Report TP-09-01. Arkansas Game and Fish Commission, Little Rock.

Klaassen, H. E. and K. L. Morgan. 1974. Age and growth of Longnose Gar in Tuttle Creek Reservoir, Kansas Trans. Am. Fish. Soc. 103(2): 402–405.

Kleckner, R. C. and J. D. McCleave. 1988. The northern limit of spawning by Atlantic eels (*Anguilla* spp.) in the Sargasso Sea in relation to thermal fronts and surface water masses. J. Mar. Res. 46(3): 647–667.

Kleinert, S. J. and D. Mraz. 1966. Life history of the Grass Pickerel (*Esox americanus vermiculatus*) in southeastern Wisconsin. Tech. Bull. 37, Wisconsin Conservation Dept., Madison. 41 pp.

Kluender, E. R. 2011. Seasonal habitat use of a leviathan, Alligator Gar, at multiple spatial scales in a river floodplain ecosystem. M.S. thesis. Univ. Central Arkansas, Conway. 76 pp.

Kluender, E. R., R. Adams, and L. Lewis. 2017. Seasonal habitat use of Alligator Gar in a river-floodplain ecosystem at multiple spatial scales. Ecol. Freshw. Fish 26(2): 233–246.

Knapp, F. T. 1951. Additional reports of lampreys from Texas. Copeia 1951(1): 87.

Knapp, L. W. 1964. Systematic studies of the Rainbow Darter, *Etheostoma caeruleum* (Storer) and the subgenus *Hadropterus* (Pisces, Percidae). Ph.D. diss. Cornell Univ., Ithaca, NY. 424 pp.

Knapp, R. A. and R. C. Sargent. 1989. Egg-mimicry as a mating strategy in the Fantail Darter, *Etheostoma flabellare*: females prefer males with eggs. Behav. Ecol. Sociobiol. 25(5): 321–326.

Knight, S. S., R. F. Cullum, F. D. Shields, Jr., and P. C. Smiley. 2012. Effects of channelization on fish biomass in river ecosystems. J. Environ. Sci. Eng. A1: 980–985.

Knouft, J. H., L. M. Page, and M. J. Plewa. 2003. Antimicrobial egg cleaning by the Fringed Darter (Perciformes: Percidae: *Etheostoma crossopterum*): implications of a novel component of parental care in fishes. Proc. Royal Soc. London Biol. Sci. 270(1531): 2405–2411.

Knowlton, N. 2010. Citizens of the sea: wondrous creatures from the census of marine life. National Geographic, Washington, DC. 216 pp.

Knowles, L. L. and L. S. Kubatko, eds. 2010. Estimating species trees: practical and theoretical aspects. Wiley-Blackwell, New Jersey. 232 pp.

Koch, B., R. C. Brooks, A. Oliver, D. Herzog, J. E. Garvey, R. Hrabik, R. Columbo, Q. Phelps, and T. Spier. 2012. Habitat selection and movement of naturally occurring Pallid Sturgeon in the Mississippi River. Trans Am. Fish. Soc. 141(1): 112–120.

Koch, J. D., M. C. Quist, K. A. Hansen, and G. A. Jones. 2009. Population dynamics and potential management of Bowfin (*Amia calva*) in the upper Mississippi River. J. Appl. Ichthyol. 25(5): 545–550.

Kočovský, P. M. 2019. Diets of endangered Silver Chub (*Macrhybopsis storeriana*, Kirtland, 1844) in Lake Erie and implications for recovery. Ecol. Freshw. Fish 28: 33–40.

Kočovský, P. M., D. C. Chapman, and J. E. McKenna. 2012. Thermal and hydrologic suitability of Lake Erie and its major tributaries for spawning of Asian carps. J. Great Lakes Res. 38(1): 159–166.

Koczaja, C., L. McCall, E. Fitch, B. Glorioso, C. Hanna, J. Kyzar, M. Niemiller, J. Spiess, A. Tolley, R. Wyckoff, and D. Mullen. 2005. Size-specific habitat segregation and intra-specific interactions in Banded Sculpin (*Cottus carolinae*). Southeast. Nat. 4(2): 207–218.

Koehn, R. K. 1965. Development and ecological significance of nuptial tubercles of the Red Shiner, *Notropis lutrensis*. Copeia 1965(4): 462–467.

Kofron, C. P. and A. A. Schreiber. 1983. The Central Mudminnow (*Umbra limi*) in Missouri, with an analysis of summer foods. Southwest. Nat. 28(3): 371–372.

Kohler, T. J., J. N. Murdock, K. B. Gido, and W. K. Dodds. 2011. Nutrient loading and grazing by the minnow *Phoxinus erythrogaster* shift periphyton abundance and stoichiometry in mesocosms. Freshw. Biol. 56(6): 1133–1146.

Kolar, C. S., D. C. Chapman, W. R. Courtenay, Jr., C. M. Housel, J. D. Williams, and D. P. Jennings. 2005. Asian carps of the genus *Hypophthalmichthys* (Pisces, Cyprinidae)—a biological synopsis and environmental risk assessment. U.S. Fish and Wildlife Service Report. 183 pp.

Kolar, C. S., D. C. Chapman, W. R. Courtenay, Jr., C. M. Housel, J. D. Williams, and D. P. Jennings. 2007. Bigheaded carps: a biological synopsis and environmental risk assessment. Special Publ. 33. American Fisheries Society, Bethesda, MD. 204 pp.

Konkle, B. R. and W. G. Sprules. 1986. Planktivory by stunted Lake Trout in an Ontario lake. Trans. Am. Fish. Soc. 115(4): 515–521.

Koppelman, J. B. 1994. Hybridization between Smallmouth Bass, *Micropterus dolomieu*, and Spotted Bass, *M. punctulatus*, in the Missouri River system, Missouri. Copeia 1994(1): 204–210.

Koppelman, J. B., C. M. Gale, and J. S. Stanovick. 2000. Allozyme and morphological variation among three nominal species of *Ambloplites* (Centrarchidae) inhabiting the Ozarks region. Trans. Am. Fish. Soc. 129(5): 1134–1149.

Kott, E., C. B. Renaud, and V. D. Vladykov. 1988. The urogenital papilla in the Holarctic lamprey (Petromyzontidae). Environ. Biol. Fishes 23(1–2): 37–44.

Kottelat, M. R. 1997. European freshwater fishes. An heuristic checklist of the freshwater fishes of Europe (exclusive of former USSR), with an introduction for non-systematists and comments on nomenclature and conservation. Biologia 52(S5); 1–271.

Kottelat, M. R., R. Britz, T. H. Hui, and K. E. Witte. 2006. *Paedocypris*, a new genus of southeast Asian cyprinid fish with a remarkable sexual dimorphism, comprises the world's smallest vertebrate. Proc. Royal Soc. B: Biol. Sci. 273(1589): 895–899.

Kozhova, O. M. and L. R. Izmest'eva. 1998. Lake Baikal: evolution and biodiversity. Backhuys Publ., Leiden. 447 pp.

Kraai, J. E., C. R. Munger, and W. E. Whitworth. 1991. Home range, movements, and habitat utilization of Smallmouth Bass in Meredith Reservoir, Texas. Pages 44–48. *In:* D. C. Jackson, ed. The First International Smallmouth Bass Symposium. Mississippi Agricultural and Forestry Experiment Station, Mississippi State Univ., Starkville. 177 pp.

Kramer, J. A. 1777. Page 450. *In:* J. A. Scopoli. Introductio ad historiam naturalem, sistens genera lapidum, plantarum et animalium hactenus detecta, caracteribus essentialibus donata, in tribus divisa, subinde ad leges naturae. Prague. i–x + 1–506.

Kramer, R. H. and L. L. Smith, Jr. 1960. Utilization of nests of Largemouth Bass, *Micropterus salmoides*, by Golden Shiners, *Notemigonus crysoleucas*. Copeia 1960(1): 73–74.

Kraus, R. 1963. Food habits of the Yellow Bass, *Roccus mississippiensis*, Clear Lake, Iowa, summer 1962. Proc. Iowa Acad. Sci. 70(1): 209–215.

Krieger, J., P. A. Fuerst, and T. M. Cavender. 2000. Phylogenetic relationships of the North American sturgeons (order Acipenseriformes) based on mitochondrial DNA sequences. Mol. Phylogenet. Evol. 16(1): 64–72.

Krieger, J., A. K. Hett, P. A. Fuerst, E. Artyukhin, and A. Ludwig. 2008. The molecular phylogeny of the order Acipenseriformes revisited. J. Appl. Ichthyol. 24(S1): 36–45.

Kristmundsottir, A. Y. and J. R. Gold. 1996. Systematics of the Blacktail Shiner (*Cyprinella venusta*) inferred from analysis of mitochondrial DNA. Copeia 1996(4): 773–783.

Krotzer, M. J. 1990. Variation and systematics of *Etheostoma nigrum* Rafinesque, the Johnny Darter (Pisces: Percidae). Ph.D. diss. Univ. Tennessee, Knoxville. 192 pp.

Krueger, W. H. and K. Oliveira. 1999. Evidence for environmental sex determination in the American Eel, *Anguilla rostrata*. Environ. Biol. Fishes 55(4): 381–89.

Krumholz, L. A. 1948. Reproduction in the Western Mosquitofish, *Gambusia affinis affinis* (Baird and Girard), and its use in mosquito control. Ecol. Monogr. 18(1): 1–43.

Krumholz, L. A. and H. S. Cavanah. 1968. Comparative morphometry of Freshwater Drum from two midwestern localities. Trans. Am. Fish. Soc. 97(4): 429–441.

Kuehne, R. A. and R. W. Barbour. 1983. The American darters. Univ. Press of Kentucky, Lexington. 177 pp.

Kuhajda, B. R. 2002. Systematics, taxonomy, and conservation status of sturgeon in the subfamily Scaphirhynchinae (Actinopterygii, Acipenseridae). Ph.D. diss. Univ. Alabama, Tuscaloosa. 291 pp.

Kuhajda, B. R. 2014. Acipenseridae: Sturgeons. Pages 160–206. *In*: M. L. Warren. Jr. and B. M. Burr, eds. Freshwater fishes of North America. Vol. 1: Petromyzontidae to Catostomidae. Johns Hopkins Univ. Press, Baltimore, MD. 644 pp.

Kuhajda, B. R. and M. L. Warren, Jr. 1989. Life history aspects of the Harlequin Darter, *Etheostoma histrio*, in western Kentucky. ASB Bull. 36(2): 66–67.

Kuhajda, B. R., R. L. Mayden, and R. M. Wood. 2007. Morphologic comparisons of hatchery-reared specimens of *Scaphirhynchus albus, Scaphirhynchus platorynchus,* and *S. albus × S. platorynchus* hybrids (Acipenseriformes: Acipenseridae). J. Appl. Ichthyol. 23(4): 324–347.

Kuklinski, K. E. 2007. Ecological investigation of the invasive White Perch in Kaw Lake, Oklahoma. Proc. Oklahoma Acad. Sci. 87: 77–84.

Kullander, S. O. 1998. A phylogeny and classification of the South American Cichlidae (Teleostei: Perciformes). Pages 461–498. *In:* L. R. Malabarba, R. E. Reis, R. P. Vari, Z. M. S. Lucena, and C. A. S. Lucena, eds. Phylogeny and classification of Neotropical fishes. EDIPUCRS, Porto Alegre, Brazil. 603 pp.

Kushlan, J. A. 1986. Exotic fishes of the Everglades: a reconsideration of proven impact. Environ. Conserv. 13(1): 67–69.

Kutkuhn, J. H. 1955. Food and feeding habits of some fishes in a dredged Iowa lake. Proc. Iowa Acad. Sci. 62: 576–588.

Kwak, T. J. 1988. Lateral movement and use of floodplain habitat by fishes of the Kankakee River, Illinois. Am. Midl. Nat. 120(2): 241–249.

Kwak, T. J. 1991. Ecological characteristics of a northern population of the Pallid Shiner. Trans. Am. Fish. Soc. 120(1): 106–115.

Kwak, T. J. and T. M. Skelly. 1992. Spawning habitat, behavior, and morphology as isolating mechanisms of the Golden Redhorse, *Moxostoma erythrurum,* and the Black Redhorse, *M. duquesnei,* two syntopic fishes. Environ. Biol. Fishes 34(2): 127–137.

Kwak, T. J., M. J. Wiley, L. L. Osbourne, and R. W. Larimore. 1992. Application of diel feeding chronology to habitat suitability analysis of warmwater fishes. Can. J. Fish. Aquat. Sci. 49(7): 1417–1430.

Kynard, B., E. Henyey, and M. Horgan. 2002. Ontogenetic behavior, migration, and social behavior of Pallid Sturgeon, *Scaphirhynchus albus,* and Shovelnose Sturgeon, *S. platorynchus,* with notes on the adaptive significance of body color. Environ. Biol. Fishes 63(4): 389–403.

Labay, A. A. 1992. Laboratory spawning and rearing of the Dusky Darter (*Percina sciera*) and the Longnose Darter (*P. nasuta*). M.S. thesis. Southwest Texas State Univ., San Marcos.

Labay, A. A., K. Collins, R. W. Standage, and T. M. Brandt. 2004. Gut content of first-feeding wild darters and captive-reared Dusky Darters. N. Am. J. Aquac. 66(2): 153–157.

Labbe, T. R. and K. D. Fausch. 2000. Dynamics of intermittent stream habitat regulate persistence of a threatened fish at multiple scales. Ecol. Appl. 10(6): 1774–1791.

Lacépède, B. G. E. 1797. Mémoire sur le polyodon feuille. Bulletin des Sciences, par la Société Philomathique de Paris. v. 1 (pt. 2, no. 7). 49.

Lacépède, B. G. E. 1798–1803. Historie Naturelle des Poissons. Vol. 3 (1801); Vol. 4 (1802); Vol. 5 (1803). 803 pp.

Lachner, E. A. 1950. The comparative food habits of the cyprinid fishes *Nocomis biguttatus* and *Nocomis micropogon* in western New York. J. Washington Acad. Sci. 40(7): 229–236.

Lachner, E. A. 1956. The changing fish fauna of the upper Ohio basin. Pages 64–78. *In*: Man and the waters of the upper Ohio basin. Special Publ. 1. Univ. Pittsburgh, Pymatuning Laboratory of Ecology. Pittsburgh, PA.

Lachner, E. A. and R. E. Jenkins. 1967. Systematics, distribution, and evolution of the chub genus *Nocomis* (Cyprinidae) in the southwestern Ohio River basin, with the description of a new species. Copeia 1967(3): 557–580.

Lachner, E. A. and R. E. Jenkins. 1971. Systematics, distribution and evolution of the *Nocomis biguttatus* species group (family Cyprinidae: Pisces) with a description of a new species from the Ozark Upland. Smithson. Contrib. Zool. 91: 1–28.

Lachner, E. A. and M. L. Wiley. 1971. Populations of the polytypic species *Nocomis leptocephalus* (Girard) with a description of a new subspecies. Smithson. Contrib. Zool. 92: 1–35.

Lachner, E. A., E. F. Westlake, and P. S. Handwerk. 1950. Studies on the biology of some percid fishes from western Pennsylvania. Am. Midl. Nat. 43(1): 92–111.

Laerm, J. and B. J. Freeman. 1986. Fishes of the Okefenokee Swamp. Univ. of Georgia Press, Athens. 128 pp.

Lagasse, P. F. 1986. River response to dredging. J. Waterway, Port, Coastal and Ocean Eng. Div., Am. Soc. Civil Eng. 112: 1–14.

Lagasse, P. F., B. R. Winkley, and D. B. Simmons. 1980. Impacts of gravel mining on river system stability. J. Waterway, Port, Coastal and Ocean Engineering Div., Am. Soc. Civil Eng. 106: 398–404.

Lagler, K. F. and F. V. Hubbs. 1940. Food of the Long-nosed Gar (*Lepisosteus osseus oxyurus*) and the Bowfin (*Amia calva*) in southern Michigan. Copeia 1940(4): 239–241.

Lagler, K. F., C. B. Obrecht, and G. V. Harry. 1942. The food habits of gars (*Lepisosteus* spp.) considered in relation to fish management. Invest. Indiana Lakes Streams 2: 117–135.

Lagler, K. F., J. E. Bardach, R. R. Miller, and D. R. M. Passino. 1977. Ichthyology. 2nd ed. John Wiley and Sons, New York. 506 pp.

LaHaye, M., A. Branchaud, M. Gendron, R. Verdon, and R. Fortin. 1992. Reproduction, early life history, and characteristics of the spawning grounds of the Lake Sturgeon (*Acipenser fulvescens*) in Des Prairies and L'Assomption rivers, near Montreal, Quebec. Can. J. Zool. 70(9): 1681–1689.

Laird, C. A. and L. M. Page. 1996. Non-native fishes inhabiting the streams and lakes of Illinois. Illinois Nat. Hist. Surv. Bull. 35: 1–51.

Lake, C. T. 1936. The life history of the Fan-tailed Darter, *Catonotus flabellaris flabellaris* (Rafinesque). Am. Midl. Nat. 17(5): 816–830.

Lambou, V. W. 1961. Utilization of macrocrustaceans for food by freshwater fishes in Louisiana and its effects on the determination of predator-prey relations. Prog. Fish-Cult. 23(1): 18–25.

Lambou, V. W. 1965. Observations on size distribution and spawning behavior of Threadfin Shad. Trans. Am. Fish. Soc. 94(4): 385–386.

Lamothe, K. A. and R. L. Johnson. 2013. Microsatellite analysis of trophy Largemouth Bass from Arkansas reservoirs. J. Arkansas Acad. Sci. 67: 71–80.

Lamson, H. M., J-C. Shiao, Y. Iizuka, W-N. Tzeng, and D. K. Cairns. 2006. Movement patterns of American Eels (*Anguilla rostrata*) between salt- and freshwater in a coastal watershed, based on otolith microchemistry. Mar. Biol. 149(6): 1567–1576.

Landolt, J. C. and L. G. Hill. 1975. Observations on the gross structure and dimensions of the gills of three species of gars (Lepisosteidae). Copeia 1975(3): 470–475.

Lang, N. J. 2007. Systematics of the subgenus *Oligocephalus* (Percidae: *Etheostoma*) with biogeographic investigations of constituent subspecies groups. Ph.D. diss. St. Louis Univ., St. Louis, MO.

Lang, N. J. and A. A. Echelle. 2011. Novel phylogeographic patterns in a lowland fish, *Etheostoma proeliare* (Percidae). Southeast. Nat. 10(1): 133–144.

Lang, N. J. and R. L. Mayden. 2007. Systematics of the subgenus *Oligocephalus* (Teleostei: Percidae: *Etheostoma*) with complete subgeneric sampling of the genus *Etheostoma*. Mol. Phylogenet. Evol. 43(2): 605–615.

Lang, N. J., K. J. Roe, C. B. Renaud, H. S. Gill, I. C. Potter, J. Freyhof, A. M. Naseka, P. Cochran, H. E. Perez, E. M. Habit, B. R. Kuhajda, D. A. Neely, Y. Reshetnikov, V. B. Salnikov, M. T. Stoumboudi, and R. L. Mayden. 2009. Novel relationships among lampreys (Petromyzontiformes) revealed by a taxonomically comprehensive molecular data set. Am. Fish. Soc. Symp. 72: 41–55.

Langerhans, R. B. 2011. Genital evolution. Pages 228–240. *In:* J. P. Evans, A. Pilastro, and I. Schlupp, eds. Ecology and evolution of poeciliid fishes. Univ. of Chicago Press, Chicago. 409 pp.

Lapointe, N. W. R., J. S. Odenkirk, and P. L. Angermeier. 2013. Seasonal movement, dispersal, and home range of Northern Snakehead *Channa argus* (Actinopterygii, Perciformes) in the Potomac River catchment. Hydrobiologia 709(1): 73–87.

Lapointe, N. W. R., J. T. Thorson, and P. L. Angermeier. 2010. Seasonal meso- and microhabitat selection by the Northern Snakehead (*Channa argus*) in the Potomac River system. Ecol. Freshw. Fish 19(10): 566–577.

Larimore, R. W. 1957. Ecological life history of the Warmouth (Centrarchidae). Bull. Illinois Nat. Hist. Surv. 27(1): 83 pp.

Larimore, R. W. and K. D. Carlander. 1971. Life history of Red Shiners, *Notropis lutrensis*, in the Skunk River, central Iowa. Iowa J. Sci. 45: 557–562.

Larimore, R. W., W. F. Childers, and C. Heckrotte. 1959. Destruction and re-establishment of stream fish and invertebrates affected by drought. Trans. Am. Fish. Soc. 88(4): 261–285.

Larochelle, C. R., F. A. T. Pickens, M. D. Burns, and B. L. Sidlauskas. 2016. Long-term isopropanol storage does not alter fish morphometrics. Copeia 2016(2): 411–420.

Larsson, D. G. J. and L. Forlin. 2002. Male-biased sex ratios of fish embryos near a pulp mill: temporary recovery after a short-term shutdown. Environ. Health Perspect. 110(8): 739–742.

Lau, J. K., T. E. Lauer, and M. L. Weinman. 2006. Impacts of channelization on stream habitats and associated fish assemblages in east central Indiana. Am. Midl. Nat. 156(2): 319–330.

Lauder, G. V. and K. F. Liem. 1983. The evolution and interrelationships of the actinopterygian fishes. Bull. Mus. Comp. Zool. 150(30): 95–197.

Laurence, G. C. and R. W. Yerger. 1966. Life history studies of the Alabama Shad, *Alosa alabamae*, in the Apalachicola River, Florida. Proc. 20th Annu. Conf. Southeast. Assoc. Game Fish. Comm.: 260–273.

Layher, W. G. 1998. Status and distribution of Pallid Sturgeon, Blue Sucker, and other large river fishes in the Red River. Report to Arkansas Game and Fish Comm., Little Rock. 31 pp.

Layher, W. G. 2007. Life history of the Blue Sucker in the Red River, Arkansas. Report to Arkansas Game and Fish Comm, Little Rock. 25 pp.

Layher, W. G. and J. W. Phillips. 2000. Status and distribution of Alligator Gar, *Lepisosteus spatula*, in the Red River. Report to Arkansas Game and Fish Comm., Little Rock. 32 pp.

Layher, W. G., O. E. Maughan, and W. D. Warde. 1987. Spotted Bass habitat suitability related to fish occurrence and biomass and measurements of physicochemical variables. N. Am. J. Fish. Manag. 7(2): 238–251.

Layman, S. R. 1994. Phylogenetic systematics and biogeography of darters of the subgenus *Doration* (Percidae: *Etheostoma*). Ph.D. diss., Univ. Alabama, Tuscaloosa. 299 pp.

Layman, S. R. and R. L. Mayden. 2012. Morphological diversity and phylogenetics of the darter subgenus *Doration* (Percidae: *Etheostoma*), with descriptions of five new species. Bull. Alabama Mus. Nat. Hist. 30: 1–83.

Leckvarcik, L. G. 2001. Life history of the Ironcolor Shiner *Notropis chalybaeus* (Cope) in Marshall's Creek, Monroe County, Pennsylvania. M.S. thesis, Pennsylvania State Univ. 75 pp.

Ledford, J. J. and A. M. Kelly. 2006. A comparison of Black Carp, Redear Sunfish, and Blue Catfish as biological controls of snail populations. N. Am. J. Aquac. 68(4): 339–347.

LeDuc, R. G. 1984. Morphological and biochemical variation in the Bluntface Shiner, *Notropis camurus* (Cyprinidae). M.S. thesis, Univ. Tennessee, Knoxville. 57 pp.

Lee, D. S. 1980a. *Scaphirhynchus albus* (Forbes and Richardson). Pallid Sturgeon. Page 43. *In:* D. S. Lee et al.

Atlas of North American freshwater fishes. N.C. State Mus. Nat. Hist., Raleigh. 854 pp.

Lee, D. S. 1980b. *Scaphirhynchus platorynchus*, Shovelnose Sturgeon. Page 44. *In*: D. S. Lee et al. Atlas of North American freshwater fishes. N.C. State Mus. Nat. Hist., Raleigh. 854 pp.

Lee, D. S. 1980c. *Anguilla rostrata*, American Eel. Page 59. *In*: D. S. Lee et al. Atlas of North American freshwater fishes. N.C. State Mus. Nat. Hist., Raleigh. 854 pp.

Lee, D. S. and B. M. Burr. 1985. Observations on life history of the Dollar Sunfish, *Lepomis marginatus* (Holbrook). ASB Bull. 32(2): 58.

Lee, D. S. and C. R. Gilbert. 1980. *Centrarchus macropterus* (Lacépède), Flier. Page 583. *In*: D. S. Lee et al. Atlas of North American freshwater fishes. N.C. State Mus. Nat. Hist., Raleigh. 854 pp.

Lee, D. S. and S. T. Kucas. 1980. *Catostomus commersoni* (Lacépède), White Sucker. Pages 375–376. *In*: D. S. Lee et al. Atlas of North American freshwater fishes. N.C. State Mus. Nat. Hist., Raleigh. 854 pp.

Lee, D. S. and J. R. Shute. 1980. *Pimephales promelas* Rafinesque, Fathead Minnow. Page 341. *In*: D. S. Lee et al. Atlas of North American freshwater fishes. N.C. State Mus. Nat. Hist., Raleigh. 854 pp.

Lee, D. S., S. P. Platania, and G. H. Burgess, eds. 1983. Atlas of North American freshwater fishes. 1983 Supplement. North Carolina State Mus. Nat. Hist., Raleigh. 854 pp.

Lee, D. S., C. R. Gilbert, C. H. Hocutt, R. E. Jenkins, D. E. McAllister, and J. R. Stauffer, Jr. 1980 et seq. Atlas of North American freshwater fishes. North Carolina State Mus. Nat. Hist., Raleigh. 854 pp.

Lee, P. G. and P. K. L. Ng. 1991. The snakehead fishes of the Indo-Malayan Region. Nature Malaysiana 16(4): 113–129.

Lee, S. S. 2004. The beauty of Bowfins. Wildlife in North Carolina 68(3): 22–27.

Lee, S. W. and Y. J. Lee. 1986. Karyotypes analysis of Korean Spotted Serpent Head (*Channa argus* (Cantor); Channiformes, Channidae). Korean J. Ichthyol. 29(2): 75–78.

LeGrande, W. H. 1981. Chromosomal evolution in North American catfishes (Siluriformes: Ictaluridae) with particular emphasis on the madtoms, *Noturus*. Copeia 1981(1): 33–52.

LeGrande, W. H. and T. M. Cavender. 1980. The chromosome complement of the Stonecat Madtom, *Noturus flavus* (Siluriformes: Ictaluridae), with evidence for the existence of a possible chromosomal race. Copeia 1980(2): 341–344.

Lehtinen, S. F. and A. A. Echelle. 1979. Reproductive cycle of *Notropis boops* (Pisces: Cyprinidae) in Brier Creek, Marshall County, Oklahoma. Am. Midl. Nat. 102(2): 237–243.

Lehtinen, S. F. and J. B. Layzer. 1988. Reproductive cycle of the Plains Minnow, *Hybognathus placitus* (Cyprinidae), in the Cimarron River, Oklahoma. Southwest. Nat. 33(1): 27–33.

Leitholf, E. 1917. *Fundulus chrysotus*. Aquat. Life 2(11): 141–142.

Lenhardt, M., G. Markovic, A. Hegedis, S. Maletin, M. Cirkovic, and Z. Markovic. 2011. Non-native and translocated fish species in Serbia and their impact on the native ichthyofauna. Rev. Fish Biol. Fish. 21(3): 407–421.

Leonard, N. J., W. W. Taylor, and C. Goddard. 2004. Multijurisdictional management of Lake Sturgeon in the Great Lakes and St. Lawrence River. Pages 231–252. *In*: G. T. O. LeBreton, F. W. H. Beamish, and R. S. McKinley, eds. Sturgeons and Paddlefishes of North America. Kluwer Academic Publ. 324 pp.

Leone, F. J. 2010. Population characteristics of Paddlefish in the Arkansas River, Arkansas: implications for the management of an exploited fishery. M.S. thesis, Arkansas Tech Univ., Russellville. 100 pp.

Leone, F. J., J. N. Stoeckel, and J. W. Quinn. 2012. Differences in Paddlefish populations among impoundments of the Arkansas River. N. Am. J. Fish. Manag. 32(4): 731–744.

Leslie, A. J., Jr., J. M. Van Dyke, L. E. Nall, and W. W. Miley II. 1982. Current velocity for transport of Grass Carp eggs. Trans. Am. Fish. Soc. 111(1): 99–101.

Lesueur, C. A. 1817a. A short description of five (supposed) new species of the genus *Muraena*, discovered by Mr. Le Sueur, in the year 1816. J. Acad. Nat. Sci. Phila. n. s. 1: 81–83.

Lesueur, C. A. 1817b. A new genus of fishes of the order Abdominales, proposed, under the name of *Catostomus*; and the characters of this genus, with those of its species, indicated, by the same. J. Acad. Nat. Sci. Phila. n. s. 1(5/6): 88–96; 102–111.

Lesueur, C. A. 1818a. Description of several new species of North American fishes. J. Acad. Nat. Sci. Phila. 1(2): 222–235; 359–368.

Lesueur, C. A. 1818b. Description of several new species of the genus *Esox*, of North America. J. Acad. Nat. Sci. Phila. 1(2): 413–417.

Lesueur, C. A. 1819. Notice de quelques poissons découverts dans les lacs du Haut-Canada, durant l'été de 1816. Mémoires du Muséum d'Histoire Naturelle, Paris v. 5: 148–161.

Lesueur, C. A. 1833. Page 445. *In*: G. Cuvier and A. Valenciennes. Histoire naturelle des poissons. Tome neuvième. Suite du livre neuvième. Des Scombéroïdes. v. 9: i–xxix + 3 pp. + 1–512, Pls. 246–279.

Lever, C. 1996. Naturalized fishes of the world. Academic Press, San Diego, CA. 408 pp.

Levine, D. S., A. G. Eversole, and H. A. Loyacano. 1986. Biology of Redbreast Sunfish in beaver ponds. Proc. Annu. Conf. Southeast. Assoc. Fish Wildl. Agencies 40: 216–226.

Lewis, L. C. 1999. Blue Catfish predation on zebra mussels in Lake Dardanelle: effects on predator and prey. M.S. thesis, Univ. Central Arkansas, Conway. 132 pp.

Lewis, R. M. and R. R. Bonner, Jr. 1966. Fecundity of the Striped Bass, *Roccus saxatilis* (Walbaum). Trans. Am. Fish. Soc. 95(3): 328–331.

Lewis, W. M., Jr. 1970. Morphological adaptations of

cyprinodontoids for inhabiting oxygen deficient waters. Copeia 1970(2): 319–326.

Li, G.-Q. 1987. A new genus of Hiodontidae from Luozigou Basin, east Jilin. Vert. PalAsiatica 25: 91–107.

Li, G.-Q. and M. V. H. Wilson. 1996. Phylogeny of Osteoglossomorpha. Pages 163–174. *In*: M. Stiassny, L. Parenti, and G. D. Johnson, eds. Interrelationships of fishes. Academic Press, San Diego, CA. 516 pp.

Li, G.-Q., M. V. H. Wilson., and L. Grande. 1997. Review of *Eohiodon* (Teleostei: Osteoglossomorpha) from western North America, with a phylogenetic reassessment of Hiodontidae. J. Paleontol. 71(6): 1109–1124.

Li., S. and J. A. Mathias. 1982. Causes of high mortality among cultured larval Walleyes. Trans. Am. Fish. Soc. 111(6): 710–721.

Li, S. F., J. W. Xu, Q. L. Yang, C. H. Wang, Q. Chen, D. C. Chapman, and G. Lu. 2009. A comparison of complete mitochondrial genomes of Silver Carp *Hypophthalmichthys molitrix* and Bighead Carp *Hypophthalmichthys nobilis*: implications for their taxonomic relationship and phylogeny. J. Fish Biol. 74(8): 1787–1803.

Li, Y. C. and J. R. Gold. 1991. Standard and NOR-stained karyotypes of three species of North American cyprinid fishes. Texas J. Sci. 43(2): 207–211.

Liem, K. F. 1987. Functional design of the air ventilation apparatus and overland excursions by teleosts. Fieldiana Zool. new ser. 37: 1–39.

Lienesch, P. W., W. I. Lutterschmidt, and J. F. Schaefer. 2000. Seasonal and long-term changes in the fish assemblage of a small stream isolated by a reservoir. Southwest. Nat. 45(3): 274–288.

Limp, W. F. and V. A. Reidhead. 1979. An economic evaluation of the potential of fish utilization in riverine environments. Am. Antiq. 44(1): 70–78.

Linck, H. F. 1790. Versuch einer Eintheilung der Fische nach den Zähnen. Magazin für das Neueste aus der Physik und Naturgeschichte, Gotha v. 6 (no. 3) (art. 3): 28–38. [Author also cited as H. F. Link. Date sometimes cited as 1789, but given as 1790 on journal cover page (Eschmeyer et al. 2018).]

Lindsey, H. L., J. C. Randolph, and J. Carroll. 1983. Updated survey of the fishes of the Poteau River, Oklahoma and Arkansas. Proc. Oklahoma Acad. Sci. 63: 42–48.

Linnaeus, C. 1758. Systema Naturae, 10th ed. (Systema naturae per regna tria naturae, secundum classes, ordines, genera, species, cum characteribus, differentiis, synonymis, locis. Tomus I. Editio decima, reformata.) Holmiae. 1: 1–824.

Linnaeus, C. 1766. Systema Naturae, 12th ed. (Systema naturae sive regna tria naturae, secundum classes, ordines, genera, species, cum characteribus, differentiis, synonymis, locis.) Laurentii Salvii, Holmiae. 1: 1–532.

Linnaeus, C. 1789. Systema Naturae, 13th ed. (Systema Naturae per regna tria naturae, secundum classes, ordines, genera, species; cum characteribus, differentiis, synonymis, locis.

Editio decimo tertia, aucta, reformata.) 3 vols. in 9 parts. Lipsiae, 1788–1793 v. 1 (pt. 3): 1126–1516.

Lippson, A. J. and R. L. Moran. 1974. Manual for identification of early developmental stages of fishes of the Potomac River Estuary. Power Plant Siting Program, Maryland Dept. Natural Resources, PPSP-MM-13. 282 pp.

Lipton, D. W. and I. E. Strand. 1997. Economic effects of pollution in fish habitats. Trans. Am. Fish. Soc. 126(3): 514–518.

Liu, J. and M.-M. Chang. 2009. A new Eocene catostomid (Teleostei: Cypriniformes) from northeastern China and early divergence of Catostomidae. Sci. China Ser. D, Earth Sci. 52: 189–202.

Liu, J.-X. and B. Ely. 2010. Evolution of an MHC class Ia gene fragment in four North American *Morone* species. J. Fish Biol. 76(8): 1984–1994.

Littrell, B. M. 2006. Can invasiveness of native cyprinids be predicted from life history traits? A comparison between a native invader and a regionally endemic cyprinid and status of an introgressed Guadalupe Bass population in a central Texas stream. M.S. thesis, Texas State Univ., San Marcos. 44 pp.

Lobon-Cervia, J. and N. Sanz, eds. 2017. Brown Trout: biology, ecology and management. John Wiley & Sons. 808 pp.

Loe, R. A. 1983. A subsequent study of the fishes of the Little Missouri River, west central Arkansas. M.S. thesis, Northeast Louisiana Univ., Monroe. 139 pp.

Loeb, H. A. 1964. Submergence of Brown Bullheads in bottom sediments. New York Fish and Game J. 11(2): 119–124.

Loftus, W. F. and J. A. Kushlan. 1987. Freshwater fishes of southern Florida. Bull. Florida State Mus. Biol. Sci. 31: 147–344.

Loh, W. K. W., P. Bond, K. J. Ashton, D. T. Roberts, and I. R. Tibbetts. 2014. DNA barcoding of freshwater fishes and the development of a quantitative qPCR assay for the species-specific detection and quantification of fish larvae from plankton samples. J. Fish Biol. 85(2): 307–328.

Lohmeyer, A. M. and J. E. Garvey. 2009. Placing the North American invasion of Asian carp in a spatially explicit context. Biol. Invasions 11(4): 905–916.

Long, J. M. and A. Nealis. 2011. Age estimation of a large Bighead Carp from Grand Lake, Oklahoma. Proc. Oklahoma Acad. Sci. 91: 15–18.

Long, J. M., J. D. Schooley, and C. P. Paukert. 2017. Long-term movement and estimated age of a Paddlefish (*Polyodon spathula*) in the Arkansas River basin of Oklahoma. Southwest. Nat. 62(3): 212–215.

Lonzarich, D. G., M. E. Lonzarich, and M. L. Warren, Jr. 2000. Effects of riffle length on the short-term movement of fishes among stream pools. Can. J. Fish. Aquat. Sci. 57(7): 1508–1514.

Lonzarich, D. G., M. L. Warren, Jr., and M. E. Lonzarich. 1998. Effects of habitat isolation on the recovery of fish assemblages in experimentally defaunated stream pools in Arkansas. Can. J. Fish. Aquat. Sci. 55(9): 2141–2149.

Lonzarich, D. G., M. L. Warren, Jr., and M. E. Lonzarich. 2004. Consequences of pool habitat isolation on fishes. Pages 246–252. *In:* J. M. Guldin Tech. Comp. SRS-74. U.S. Dept. Agriculture, Forest Service, Southern Research Station, Asheville, NC.

Lopez, J. A., P. Bentzen, and T. W. Pietsch. 2000. Phylogenetic relationships of esocoid fishes (Teleostei) based on partial cytochrome *b* and 16S mitochondrial DNA sequences. Copeia 2000(2): 420–431.

Lopez, J. A., W. Chen, and G. Orti. 2004. Esociform phylogeny. Copeia 2004(3): 449–464.

Lotrich, V. A. 1973. Growth, production, and community composition of fishes inhabiting a first-, second-, and third-order stream of eastern Kentucky. Ecol. Monogr. 43(3): 377–397.

Love, J. W. 2002. Sexual dimorphism in Spotted Gar *Lepisosteus oculatus* from southeastern Louisiana. Am. Midl. Nat. 147(2): 393–399.

Love, J. W. and J. J. Newhard. 2018. Expansion of Northern Snakehead in the Chesapeake Bay watershed. Trans. Am. Fish. Soc. 147(2): 342–349.

Ludwig, A., N. M. Belfiore, C. Pitra, V. Svirsky, and I. Jenneckens. 2001. Genome duplication events and functional reduction of ploidy levels in sturgeon (*Acipenser, Huso*, and *Scaphirhynchus*). Genetics 158(3): 1203–1215.

Lueckenhoff, R. W. 2011. Morphological variation between juvenile White Bass and juvenile hybrid Striped Bass. M.S. thesis, Univ. Nebraska, Lincoln. 149 pp.

Lukas, J. A. and D. J. Orth. 1993. Reproductive ecology of Redbreast Sunfish *Lepomis auritus* in a Virginia stream. J. Freshw. Ecol. 8(3): 235–244.

Lundberg, J. G. 1970. The evolutionary history of North American catfishes, family Ictaluridae. Ph.D. diss., Univ. Michigan, Ann Arbor. 524 pp.

Lundberg, J. G. 1975. The fossil catfishes of North America. Claude W. Hibbard Memorial Vol. II. Univ. Michigan Mus. Paleontol. Pap. 11: 1–51.

Lundberg, J. G. 1982. The comparative anatomy of the Toothless Blindcat, *Trogloglanis pattersoni* Eigenmann, with a phylogenetic analysis of the ictalurid catfishes. Misc. Publ. Mus. Zool., Univ. Michigan 163: 1–85.

Lundberg, J. G. 1992. The phylogeny of ictalurid catfishes: a synthesis of recent work. Pages 392–420. *In:* R. L. Mayden, ed. Systematics, historical ecology, and North American freshwater fishes. Stanford Univ. Press, Stanford, CA. 969 pp.

Lundberg, J. G. and E. Marsh. 1976. Evolution and functional anatomy of the pectoral fin rays in cyprinoid fishes, with emphasis on the suckers (family Catostomidae). Am. Midl. Nat. 96(2): 332–349.

Lundberg, J. G. and L. A. McDade. 1990. Systematics. Pages 65–108. *In:* C. B. Schreck and P. B. Moyle, eds. Methods for fish biology. American Fisheries Society, Bethesda, MD. 704 pp.

Lundberg, J. G., M. Kottelat, G. R. Smith, M. L. J. Stiassny, and A. C. Gill. 2000. So many fishes, so little time: an overview of recent ichthyological discovery in continental waters. Ann. Missouri Bot. Gard. 87(1): 26–62.

Lutterbie, G. W. 1979. Reproduction and age and growth in Wisconsin darters (Osteichthyes: Percidae). Univ. Wisconsin–Stevens Point Mus. Nat. Hist. Fauna Flora Rep. 10: 17–43.

Luttrell, G. R., A. A. Echelle, W. L. Fisher, and D. J. Eienhour. 1999. Declining status of two species of the *Macrhybopsis aestivalis* complex (Teleostei: Cyprinidae) in the Arkansas River basin and related effects of reservoirs as barriers to dispersal. Copeia 1999(4): 981–989.

Luttrell, G. R., D. M. Underwood, W. L. Fisher, and J. Pigg. 1995. Distribution of the Red River Shiner, *Notropis bairdi*, in the Arkansas River drainage. Proc. Oklahoma Acad. Sci. 75: 61–62.

Lydeard, C., M. C. Wooten, and M. H. Smith. 1991. Occurrence of *Gambusia affinis* in the Savannah and Chattahoochee drainages: previously undescribed geographic contacts between *G. affinis* and *G. holbrooki*. Copeia 1991(4): 1111–1116.

Lynch, J. D. 1991. A footnote to history: a tale of two fish species in Nebraska. Nebraskaland 60(6): 50–55.

Mabee, P. M. 1993. Phylogenetic interpretation of ontogenetic change: sorting out the actual and artefactual in an empirical case study of centrarchid fishes. Zool. J. Linn. Soc. 107(3): 175–291.

Mabee, P. M., G. Arratia, M. Coburn, M. Haendel, E. J. Hilton, J. G. Lundberg, R. L. Mayden, N. Rios, and M. Westerfield. 2007. Connecting evolutionary morphology to genomics using ontologies: a case study from Cypriniformes including Zebrafish. J. Exp. Zool. 308B(5): 655–668.

MacDonald, D. G. 1978. Life history studies of the Smallmouth Buffalo, *Ictiobus bubalus* Rafinesque, in Watts Bar and Chickamauga reservoirs, Tennessee. M.S. Thesis. Tennessee Technical Univ., Cookeville. 55 pp.

Maceina, M. J. and D. R. Bayne. 2001. Changes in the black bass community and fishery with oligotrophication in West Point Reservoir, Georgia. N. Am. J. Fish. Manag. 21(4): 745–755.

Machado, M. D., D. C. Heins, and H. L. Bart, Jr. 2002. Microgeographical variation in ovum size of the Blacktail Shiner, *Cyprinella venusta* Girard, in relation to streamflow. Ecol. Freshw. Fish 11(1): 11–19.

Mackenzie-Grieve, J. L. and J. R. Post. 2006. Thermal habitat use by Lake Trout in two contrasting Yukon Territory lakes. Trans. Am. Fish. Soc. 135(3): 727–738.

MacKiewicz, M., D. E. Fletcher, S. D. Wilkins, J. A. DeWoody, and J. C. Avise. 2002. A genetic assessment of parentage in a natural population of Dollar Sunfish (*Lepomis marginatus*) based on microsatellite markers. Mol. Ecol. 11(9): 1877–1883.

MacLean, J. A, D. O. Evans, N. V. Martin, and R. L. DesJardine. 1981. Survival, growth, spawning distribution, and movements of introduced and native Lake Trout

(*Salvelinus namaycush*) in two inland Ontario lakes. Can. J. Fish. Aquat. Sci. 38(12): 1685–1700.

Madding, R. S. 1971. Nutrition studies of the Brindled Madtom, *Noturus miurus* Jordan, based on stomach content analysis. M.S. thesis, Eastern Illinois Univ., Charleston.

Maddison, W. P. 1997. Gene trees in species trees. Syst. Biol. 46(3): 523–536.

Madenjian, C. P, R. L. Knight, M. T. Bur, and J. L. Forney. 2000. Reduction in recruitment of White Bass in Lake Erie after invasion of White Perch. Trans. Am. Fish. Soc. 129(6): 1340–1353.

Madsen, M. L. 1971. The presence of nuptial tubercles on female Quillback (*Carpiodes cyprinus*). Trans. Am. Fish. Soc. 100(1): 132–134.

Magoulick, D. D. 2000. Spatial and temporal variation in fish assemblages of drying stream pools: the role of abiotic and biotic factors. Aquat. Ecol. 34(1): 29–41.

Magoulick, D. D. and L. C. Lewis. 2002. Predation on exotic zebra mussels by native fishes: effects on predator and prey. Freshw. Biol. 47(10): 1908–1918.

Magoulick, D. D. and D. T. Lynch. 2015. Occupancy and abundance modeling of the endangered Yellowcheek Darter in Arkansas. Copeia 103(2): 433–439.

Maisey, J. G. 1996. Discovering fossil fishes. Henry Holt, New York. 233 pp.

Makeyeva, A. P. 1980. Early ontogenetic characteristics of the Bigmouth Buffalo, *Ictiobus cyprinellus* (Catostomidae). J. Ichthyol. 20: 73–89.

Mallet, J. 2005. Hybridization as an invasion of the genome. Trends Ecol. Evol. 20(5): 229–237.

Manley, T. A. 2012. Spotted Gar *Lepisosteus oculatus* diets in the upper Barataria Estuary. M.S. thesis, Nicholls State Univ., Thibodaux, LA. 93 pp.

Manning, T. 2005. Endocrine-disrupting chemicals: a review of the state of the science. Australian J. Ecotoxicol. 11: 1–52.

Mansueti, A. J. and J. D. Hardy, Jr. 1967. Development of fishes of the Chesapeake Bay region: an atlas of egg, larval, and juvenile stages—Part 1. Univ. Maryland, Baltimore Nat. Resour. Inst., Part I. 202 pp.

Mansueti, R. J. 1961. Movements, reproduction and mortality of the White Perch, *Roccus americanus*, in the Patuxent Estuary, Maryland. Chesapeake Sci. 2(3–4): 142–205.

Mansueti, R. J. 1964. Eggs, larvae, and young of the White Perch, *Roccus americanus*, with comments on its ecology in the estuary. Chesapeake Sci. 5(1): 3–45.

Marcy, B. C., Jr., D. E. Fletcher, F. D. Martin, M. H. Paller, and M. J. M. Reichert. 2005. Fishes of the middle Savannah River basin. Univ. of Georgia Press, Athens. 480 pp.

Markus, H. C. 1934. Life history of the Blackhead Minnow (*Pimephales promelas*). Copeia 1934(3): 116–122.

Marsden, J. E. and C. C. Krueger. 1991. Spawning by hatchery-origin Lake Trout (*Salvelinus namaycush*) in Lake Ontario: data from egg collections, substrate analysis, and diver observations. Can. J. Fish. Aquat. Sci. 48(12): 2377–2384.

Marsden, J. E., D. L. Perkins, and C. C. Krueger. 1995a. Recognition of spawning areas by Lake Trout: deposition and survival of eggs on small man-made rock piles. J. Great Lakes Res. 21(S1): 330–336.

Marsden, J. E., T. Kassler, and D. Philipp. 1995b. Allozyme confirmation that North American Yellow Perch (*Perca flavescens*) and Eurasian Yellow Perch (*Perca fluviatilis*) are separate species. Copeia 1995(4): 977–981.

Marshall, C. L. 1978. The distribution of *Notropis bairdi* along the Cimarron River in Logan County, Oklahoma. Proc. Oklahoma Acad. Sci. 58: 109.

Marshall. N. 1939. Annulus formation in scales of the Common Shiner, *Notropis cornutus chrysocephalus* (Rafinesque). Copeia 1939(3): 148–154.

Marshall, N. 1947. Studies on the life history and ecology of *Notropis chalybaeus* (Cope). Q. J. Florida Acad. Sci. 9(3–4): 163–188.

Marsh-Matthews, E., W. J. Matthews, K. B. Gido, and R. L. Marsh. 2002. Reproduction by young-of-year Red Shiner (*Cyprinella lutrensis*) and its implications for invasion success. Southwest. Nat. 47(4): 605–610.

Martin, F. D. 1968. Intraspecific variation in osmotic abilities of *Cyprinodon variegatus* Lacépède from the Texas coast. Ecology 49(6): 1186–1188.

Martin, F. D. 1972. Factors influencing local distribution of *Cyprinodon variegatus* (Pisces: Cyprinodontidae). Trans. Am. Fish. Soc. 101(1): 89–93.

Martin, F. D. and C. Hubbs. 1973. Observations on the development of Pirate Perch, *Aphredoderus sayanus* (Pisces: Aphredoderidae) with comments on yolk circulation patterns as a possible taxonomic tool. Copeia 1973(2): 377–379.

Martin, H. and M. M. White. 2008. Intraspecific phylogeography of the Least Brook Lamprey (*Lampetra aepyptera*). Copeia 2008(3): 579–585.

Martin, N. V. 1966. The significance of food habits in the biology, exploitation, and management of Algonquin Park, Ontario, Lake Trout. J. Fish. Res. Board Can. 95(4): 415–422.

Martin, R. 1970. Death of a watershed. Arkansas Game and Fish 3(2): 4–5.

Martin, Z. P. and L. M. Page. 2015. Comparative morphology and evolution of genital papillae in a genus of darters (Percidae: *Etheostoma*). Copeia 2015(1): 99–124.

Martyn, H. A. and J. C. Schmulbach. 1978. Bionomics of the Flathead Chub, *Hybopsis gracilis* (Richardson). Proc. Iowa. Acad. Sci. 85(2): 62–65.

Matamoros, W. A., C. W. Hoagstrom, J. F. Schaefer, and B. R. Kreiser. 2016. Fish faunal provinces of the conterminous United States of America reflect historical geography and familial composition. Biol. Rev. Cambridge Philos. Soc. 91(3): 813–832.

Mateus, C. S., R. Rodriguez-Munoz, B. R. Quintella, M. J. Alves, and P. R. Almeida. 2012. Lampreys of the Iberian

Peninsula: distribution, population status and conservation. Endangered Species Res. 16(2): 183–198.

Matheney, M. P., IV. and C. F. Rabeni. 1995. Patterns of movement and habitat use by Northern Hog Suckers in an Ozark stream. Trans. Am. Fish. Soc. 124(6): 886–897.

Mathias, J. A. and S. Li. 1982. Feeding habits of Walleye larvae and juveniles: comparative laboratory and field studies. Trans. Am. Fish. Soc. 111(6): 722–735.

Mathis, B. 1970. The rise and decline of commercial fishing on the White River. Arkansas Game and Fish 3(1): 2–4.

Matthews, M. M. and D. C. Heins. 1984. Life history of the Redfin Shiner, *Notropis umbratilis* (Pisces: Cyprinidae), in Mississippi. Copeia 1984(2): 385–390.

Matthews, W. J. 1977. Ingestion of sand by the Duskystripe Shiner, *Notropis pilsbryi* Fowler. Southwest. Nat. 22(4): 543–544.

Matthews, W. J. 1982. Small fish community structure in Ozark streams: structured assembly patterns or random abundance of species? Am. Midl. Nat. 107(1): 42–54.

Matthews, W. J. 1984. Influence of turbid inflows on vertical distribution of larval shad and Freshwater Drum. Trans. Am. Fish. Soc. 113(2): 192–198.

Matthews, W. J. 1985a. Distribution of midwestern fishes on multivariate environmental gradients, with emphasis on *Notropis lutrensis*. Am. Midl. Nat. 113(2): 225–237.

Matthews, W. J. 1985b. Summer mortality of Striped Bass in reservoirs of the United States. Trans. Am. Fish. Soc. 114(1): 62–66.

Matthews, W. J. 1986a. Fish faunal structure in an Ozark stream: stability, persistence and a catastrophic flood. Copeia 1986(2): 388–397.

Matthews, W. J. 1986b. Geographic variation in thermal tolerance of a widespread minnow *Notropis lutrensis* of the North American mid-west. J. Fish Biol. 28(4): 407–417.

Matthews, W. J. 1987a. Geographic variation in *Cyprinella lutrensis* (Pisces: Cyprinidae) in the United States, with notes on *Cyprinella lepida*. Copeia 1987(3): 616–637.

Matthews, W. J. 1987b. Physicochemical tolerance and selectivity of stream fishes as related to their geographic ranges and local distributions. Pages 111–120. *In*: W. J. Matthews and D. C. Heins, eds. Community and evolutionary ecology of North American stream fishes. Univ. of Oklahoma Press, Norman.

Matthews, W. J. 1995. Geographic variation in nuptial colors of Red Shiner (*Cyprinella lutrensis*; Cyprinidae) within the United States. Southwest. Nat. 40(1): 5–10.

Matthews, W. J. 1998. Patterns in freshwater fish ecology. Chapman & Hall, New York. 757 pp.

Matthews, W. J. 2015. Basic biology, good field notes, and synthesizing across your career. Copeia 2015(3): 495–501.

Matthews, W. J. and F. P. Gelwick. 1988. Variation and systematics of *Etheostoma radiosum*, the Orangebelly Darter (Pisces, Percidae). Copeia 1988(3): 543–554.

Matthews, W. J. and G. L. Harp. 1974. Preimpoundment ichthyofaunal survey of the Piney Creek watershed, Izard County, Arkansas. Proc. Arkansas Acad. Sci. 28: 39–43.

Matthews, W. J. and L. G. Hill. 1977. Tolerance of the Red Shiner, *Notropis lutrensis* (Cyprinidae) to environmental parameters. Southwest. Nat. 22(1): 89–98.

Matthews, W. J. and L. G. Hill. 1979a. Influence of physicochemical factors on habitat selection by Red Shiners, *Notropis lutrensis* (Pisces: Cyprinidae). Copeia 1979(1): 70–81.

Matthews, W. J. and L. G. Hill. 1979b. Age-specific differences in the distribution of Red Shiners, *Notropis lutrensis*, over physicochemical ranges. Am. Midl. Nat. 101(2): 366–372.

Matthews, W. J. and L. G. Hill. 1980. Habitat partitioning in the fish community of a southwestern river. Southwest. Nat. 25(1): 51–66.

Matthews, W. J. and J. D. Maness. 1979. Critical thermal maxima, oxygen tolerances and success of cyprinid fishes in a southwestern river. Am. Midl. Nat. 102(2): 374–377.

Matthews, W. J. and E. Marsh-Matthews. 2007. Extirpation of Red Shiner in direct tributaries of Lake Texoma (Oklahoma-Texas): a cautionary case history from a fragmented river-reservoir system. Trans. Am. Fish. Soc. 136(4): 1041–1062.

Matthews W. J. and E. Marsh-Matthews. 2011. An invasive fish species within its native range: community effects and population dynamics of *Gambusia affinis* in the central United States. Freshw. Biol. 56(12): 2609–2619.

Matthews, W. J. and E. Marsh-Matthews. 2015. Comparison of historical and recent fish distribution patterns in Oklahoma and western Arkansas. Copeia, 2015(1): 170–180.

Matthews, W. J. and E. Marsh-Matthews. 2016. Dynamics of an upland stream fish community over 40 years: trajectories and support for the loose equilibrium concept. Ecology 97(3): 706–719.

Matthews, W. J. and E. Marsh-Matthews. 2017. Stream fish community dynamics: a critical synthesis. Johns Hopkins Univ. Press, Baltimore, MD. 360 pp.

Matthews, W. J. and R. S. Matthews. 1978. Additions to the fish fauna of Piney Creek, Izard County, Arkansas. Proc. Arkansas Acad. Sci. 32: 92.

Matthews, W. J. and R. McDaniel. 1981. New locality records for some Kansas fishes, with notes on the habitat of the Arkansas Darter (*Etheostoma cragini*). Trans. Kansas Acad. Sci. 84(4): 219–222.

Matthews, W. J. and H. W. Robison. 1982. Addition of *Etheostoma collettei* (Percidae) to the fish fauna of Oklahoma and of the Red River drainage in Arkansas. Southwest. Nat. 27(2): 215–216.

Matthews, W. J. and H. W. Robison. 1988. The distribution of the fishes of Arkansas: a multivariate analysis. Copeia 1988(2): 358–374.

Matthews, W. J. and H. W. Robison. 1998. Influence of drainage connectivity, drainage area and regional species richness on fishes of the Interior Highlands in Arkansas. Am. Midl. Nat. 139(1): 1–19.

Matthews, W. J. and T. F. Turner. 2019. Redescription and recognition of *Etheostoma cyanorum* from Blue River, Oklahoma. Copeia 2019(2): 208–218.

Matthews, W. J., J. R. Bek, and E. Surat. 1982. Comparative ecology of the darters *Etheostoma podostemone*, *E. flabellare*, and *Percina roanoka* in the upper Roanoke River drainage, Virginia. Copeia 1982(4): 805–814.

Matthews, W. J., F. P. Gelwick, and J. J. Hoover. 1992. Food of and habitat use by juveniles of species of *Micropterus* and *Morone* in a southwestern reservoir. Trans. Am. Fish. Soc. 121(1): 54–66.

Matthews, W. J., B. C. Harvey, and M. E. Power. 1994. Spatial and temporal patterns in the fish assemblages of individual pools in a midwestern stream (U.S.A.). Environ. Biol. Fishes 39(4): 381–397.

Matthews, W. J., L. G. Hill, and S. M. Schellhaass. 1985. Depth distribution of Striped Bass and other fish in Lake Texoma (Oklahoma-Texas) during summer stratification. Trans. Am. Fish. Soc. 114(1): 84–91.

Matthews, W. J., M. E. Power, and A. J. Stewart. 1986. Depth distribution of *Campostoma* grazing scars in an Ozark stream. Environ. Biol. Fishes 17(4): 291–297.

Matthews, W. J., W. L. Shelton, and E. Marsh-Matthews. 2012. First-year growth of Longnose Gar (*Lepisosteus osseus*) from zygote to autumn juvenile. Southwest. Nat. 57(3): 335–337.

Matthews, W. J., W. D. Shepherd, and L. G. Hill. 1978. Aspects of the ecology of the Duskystripe Shiner, *Notropis pilsbryi* (Cypriniformes, Cyprinidae), in an Ozark stream. Am. Midl. Nat. 100(1): 247–252.

Matthews, W. J., T. F. Turner, and M. J. Osborne. 2016. Breakdown of a hybrid swarm between two darters (Percidae), *Etheostoma radiosum* and *Etheostoma spectabile*, with loss of one parental species. Copeia 2016(4): 873–878.

Matthews, W. J., E. Marsh-Matthews, G. L. Adams, and S. R. Adams. 2014. Two catastrophic floods: similarities and differences in effects on an Ozark stream fish community. Copeia 2014(4): 682–693.

Matthews, W. J., L. G. Hill, D. R. Edds, J. J. Hoover, and T. G. Heger. 1988. Trophic ecology of Striped Bass, *Morone saxatilis*, in a freshwater reservoir (Lake Texoma, U.S.A.). J. Fish. Biol. 33(2): 273–288.

Matthews, W. J., A. M. Miller-Lemke, M. L. Warren, Jr., D. Cobb, J. G. Stewart, B. Crump, and F. P. Gelwick. 2004. Context-specific trophic and functional ecology of fishes of small stream ecosystems in the Ouachita National Forest. Pages 221–230. *In*: J. M. Guilden, Ouachita and Ozark mountains symposium: ecosystem management research. Gen. Tech. Rep. SRS-74. U.S. Dept. Agriculture, Forest Service, Southern Research Station, Asheville, NC.

Mauney, M. and G. L. Harp. 1979. The effects of channelization on fish populations of the Cache River and Bayou DeView. Proc. Arkansas Acad. Sci. 33: 51–54.

Maurakis, E. G. and J. B. Kahnke. 1987. Construction of spawning nests by two recently described or redescribed

chub species from southeastern United States (Pisces: Cyprinidae). ASB Bull. 34(2): 86.

Maurakis, E. G. and W. Roston. 1998. Spawning behavior in *Nocomis asper* (Actinopterygii, Cyprinidae). Virginia J. Sci. 49(3): 199–202.

Maurakis, E. G. and W. S. Woolcott. 1989. Reproductive behavior of *Campostoma anomalum*. Video, RaukWool Productions. 14 min. Univ. of Richmond, VA.

Maurakis, E. G., W. S. Woolcott, and J. T. Magee. 1990. Pebble-nests of four *Semotilus* species. SFC Proc. 22: 7–13.

Maurakis, E. G., W. S. Woolcott, and W. R. McGuire. 1995. Nocturnal reproductive behaviors in *Semotilus atromaculatus* (Pisces: Cyprinidae). SFC Proc. 31: 1–3.

Maurakis, E. G., W. S. Woolcott, and M. H. Sabaj. 1991. Reproductive-behavioral phylogenetics of *Nocomis* species-groups. Am. Midl. Nat. 126(1): 103–110.

May, E. B. and A. A. Echelle. 1968. Young-of-year Alligator Gar in Lake Texoma, Oklahoma. Copeia 1968(3): 629–630.

May, O. D., Jr. and J. C. Fuller, Jr. 1962. A study on Striped Bass egg production in the Congaree and Wateree rivers. Proc. 16th Annu. Conf. Southeast. Assoc. Game Fish Comm.: 285–300.

Mayden, R. L. 1983. Madtoms, America's miniature catfishes. Trop. Fish Hobbyist 31: 66–73.

Mayden, R. L. 1985a. Biogeography of Ouachita Highland fishes. Southwest. Nat. 30(2): 195–211.

Mayden, R. L. 1985b. Nuptial structures in the subgenus *Catonotus*, genus *Etheostoma* (Percidae). Copeia 1985(3): 580–583.

Mayden, R. L. 1987a. Pleistocene glaciation and historical biogeography of North American central-highland fishes. Pages 141–151. *In*: W. C. Johnson, ed. Quaternary environments of Kansas. Kansas Geol. Surv. Guidebook Ser. 5.

Mayden, R. L. 1987b. Historical ecology and North American highland fishes: a research program in community ecology. Page 203–222. *In*: Community and evolutionary ecology of North American stream fishes. W. J. Matthews and D. C. Heins, eds. Univ. Oklahoma Press, Norman.

Mayden, R. L. 1988a. Systematics of the *Notropis zonatus* species group, with description of a new species from the Interior Highlands of North America. Copeia 1988(1): 153–173.

Mayden, R. L. 1988b. Vicariance biogeography, parsimony, and evolution in North American freshwater fishes. Syst. Zool. 37(4): 329–355.

Mayden, R. L. 1989. Phylogenetic studies of North American minnows: with emphasis on the genus *Cyprinella* (Teleostei, Cypriniformes). Misc. Publ. 80. Univ. of Kansas Museum of Natural History, Lawrence. 189 pp.

Mayden, R. L. 1991. Cyprinids of the New World. Pages 240–263. *In*: I. J. Winfield and J. S. Nelson, eds. Cyprinid fishes: systematics, biology, and exploitation. Fish and Fisheries Series 3, Chapman & Hall, London. 667 pp.

Mayden, R. L., ed. 1992. Systematics, historical ecology, and

North American freshwater fishes. Stanford Univ. Press, Stanford, CA. 969 pp.

Mayden, R. L. 1997. A hierarchy of species concepts: the denouement in the saga of the species problem. Pages 381–424. *In*: M. F. Claridge, H. A. Dawah, and M. R. Wilson, eds. Species: the units of biodiversity. Chapman & Hall, New York. 439 pp.

Mayden, R. L. 1999. Consilience and a hierarchy of species concepts: advances toward closure on the species puzzle. J. Nematol. 31(2): 95–116.

Mayden, R. L. 2010. Systematics of the *Etheostoma punctulatum* species group (Teleostei: Percidae), with descriptions of two new species. Copeia 2010(4): 716–734.

Mayden, R. L. and B. M. Burr. 1981. Life history of the Slender Madtom, *Noturus exilis*, in southern Illinois (Pisces: Ictaluridae). Occas. Pap. Mus. Nat. Hist., Univ. Kansas 93: 1–64.

Mayden, R. L. and W. J. Chen. 2010. The world's smallest vertebrate species of the genus *Paedocypris*: a new family of freshwater fishes and the sister group to the world's most diverse clade of freshwater fishes (Teleostei: Cypriniformes). Mol. Phylogenet. Evol. 57(1): 152–175.

Mayden, R. L. and C. R. Gilbert. 1989. *Notropis ludibundus* (Girard) and *Notropis tristis* (Girard), replacement names for *N. stramineus* (Cope) and *N. topeka* (Gilbert) (Teleostei: Cypriniformes). Copeia 1989(4): 1084–1089.

Mayden, R. L. and B. R. Kuhajda. 1989. Systematics of *Notropis cahabae*, a new cyprinid fish endemic to the Cahaba River of the Mobile Basin. Bull. Alabama Mus. Nat. Hist. 9: 1–16.

Mayden, R. L. and B. R. Kuhajda. 1996. Systematics, taxonomy and conservation status of the endangered Alabama Sturgeon, *Scaphirhynchus suttkusi* Williams and Clemmer (Actinopterygii, Acipenseridae). Copeia 1996(2): 241–273.

Mayden, R. L. and R. H. Matson. 1988. Evolutionary relationships of eastern North American cyprinids: an allozyme perspective [abstract]. 68th Annu. Meet. Am. Soc. Ichthyol. Herpetol., p. 138.

Mayden, R. L. and A. M. Simons. 2002. Crevice spawning behavior in *Dionda dichroma*, with comments on the evolution of spawning modes in North American shiners (Teleostei: Cyprinidae). Rev. Fish Biol. Fish. 12(2–3): 327–337.

Mayden, R. L. and S. J. Walsh. 1984. Life history of the Least Madtom *Noturus hildebrandi* (Siluriformes: Ictaluridae) with comparisons to related species. Am. Midl. Nat. 112(2): 349–368.

Mayden, R. L. and E. O. Wiley. 1992. The fundamentals of phylogenetic systematics. Pages 114–185. *In*: R. L. Mayden, ed. Systematics, historical ecology, and North American freshwater fishes. Stanford Univ. Press, Stanford, CA. 969 pp.

Mayden, R. L. and R. M. Wood. 1995. Systematics, species concepts, and the evolutionarily significant unit in biodiversity and conservation biology. Pages 58–113. *In*: J. L. Nielson, ed. Evolution and the aquatic ecosystem: defining unique units

in population conservation. Symp. 17, American Fisheries Society, Bethesda, MD. 435 pp.

Mayden, R. L., B. M. Burr, and S. L. Dewey. 1980. Aspects of the life history of the Ozark Madtom, *Noturus albater*, in southeastern Missouri (Pisces: Ictaluridae). Am. Mild. Nat. 104(2): 335–340.

Mayden, R. L., F. B. Cross, and O. T. Gorman. 1987. Distributional history of the Rainbow Smelt, *Osmerus mordax* (Salmoniformes: Osmeridae), in the Mississippi River basin. Copeia 1987(4): 1051–1055.

Mayden, R. L., R. H. Matson, and D. M. Hillis. 1992a. Speciation in the North American genus *Dionda* (Teleostei: Cypriniformes). Pages 710–746. *In*: R. L. Mayden, ed. Systematics, historical ecology, and North American freshwater fishes. Stanford Univ. Press, Stanford, CA. 996 pp.

Mayden, R. L., B. M. Burr, L. M. Page, and R. R. Miller. 1992b. The native freshwater fishes of North America. Pages 827–863. *In*: R. L. Mayden, ed. Systematics, historical ecology, and North American freshwater fishes. Stanford Univ. Press, Stanford, CA. 969 pp.

Mayden, R. L., A. M. Simons, R. M. Wood, P. M. Harris, and B. R. Kuhajda. 2006a. Molecular systematics and classification of North American notropin shiners and minnows (Cypriniformes: Cyprinidae). Pages 72–101. *In*: Ma. De Lourdes Lozano-Vilano and Armando J. Contreras-Balderas, eds. Studies of North American desert fishes in honor of E. P. (Phil) Pister, conservationist. Universidad Autonoma de Nuevo Leon, Mexico.

Mayden, R. L., R. M. Wood, N. J. Lang, C. B. Dillman, and J. F. Switzer. 2006b. Phylogenetic relationships of species of the darter genus *Etheostoma* (Perciformes: Percidae): evidence from parsimony and Bayesian analyses of mitochondrial cytochrome *b* sequences. Pages 20–39. *In*: Ma. De Lourdes Lozano-Vilano and Armando J. Contreras-Balderas, eds. Studies of North American desert fishes in honor of E. P. (Phil) Pister, conservationist. Universidad Autonoma de Nuevo Leon, Mexico.

Mayden, R. L., R. H. Matson, B. R. Kuhajda, J. M. Pierson, M. F. Mettee, and K. S. Frazer. 1989. The Chestnut Lamprey, *Ichthyomyzon castaneus* Girard, in the Mobile Basin. SFC Proc. 20: 10–13.

Mayden, R. L., K. L. Tang, K. W. Conway, J. Freyhof, S. Chamberlain, M. Haskins, L. Schneider, M. Sudkamp, R. M. Wood, M. Agnew, A. Bufalino, Z. Sulaiman, M. Miya, K. Saitoh, and S. He. 2007. Phylogenetic relationships of *Danio* within the order Cypriniformes: a framework for comparative and evolutionary studies of a model species. J. Exp. Zool. Mol. Dev. Evol. 308B(5): 642–654.

Mayden, R. L., K. L. Tang, R. M. Wood, W.-J Chen, M. K. Agnew, K. W. Conway, L. Yang, J. Li, X. Wang, K. Saitoh, M. Miya, S. He, H. Liu, Y. Chen, and M. Nishida. 2008. Inferring the tree of life of the order Cypriniformes, the earth's most diverse clade of freshwater fishes: implications of varied taxon and character sampling. J. Syst. Evol. 46(3): 424–438.

Mayden, R. L., W.-J. Chen, H. L. Bart, M. H. Doosey, A. M. Simons, K. L. Tang, R. M. Wood, M. K. Agnew, L. Yang, M. V. Hirt, M. D. Clements, K. Saitoh, T. Sado, M. Miya, and M. Nishida. 2009. Reconstructing the phylogenetic relationships of the earth's most diverse clade of freshwater fishes—order Cypriniformes (Actinopterygii: Ostariophysi): a case study using multiple nuclear loci and the mitochondrial genome. Mol. Phylogenet. Evol. 51(3): 500–514.

Mayr, E. 1942. Systematics and the origin of species. Harvard Univ. Press, Cambridge, MA. 334 pp.

Mayr, E. 1966. Animal species and evolution. Harvard Univ. Press, Cambridge, MA. 797 pp.

Mayr, E. and P. D. Ashlock. 1991. Principles of systematic zoology. 2nd ed. McGraw-Hill, New York. 475 pp.

McAllister, C. T. and O. Amin. 2008. Acanthocephalan parasites (Echinorhynchida: Heteracanthocephalidae: Pomphorhynchidae) from the Pirate Perch (Percopsiformes: Aphredoderidae), from the Caddo River, Arkansas. J. Arkansas Acad. Sci. 62: 151–152.

McAllister, C. T. and D. G. Cloutman. 2016. Parasites of Brook Silversides, *Labidesthes sicculus* and Golden Silversides, *L. vanhyningi* (Atheriniformes: Atherinopsidae), from Arkansas and Oklahoma, U.S.A. Comp. Parasitol. 83(2): 250–254.

McAllister, C. T. and J. T. McAllister., Jr. 1988. An unusual Blue Catfish, *Ictalurus furcatus* (Siluriformes: Ictaluridae) from Arkansas. Texas J. Sci. 40: 361–363.

McAllister, C. T., S. F. Barclay, and H. W. Robison. 2004. Geographic distribution records for the Flier, *Centrarchus macropterus* (Perciformes: Centrarchidae), from southwestern Arkansas. J. Arkansas Acad. Sci. 58: 131–132.

McAllister, C. T., C. R. Bursey, and H. W. Robison. 2012a. *Proteocephalus pearsei* (Cestoidea: Proteocephalidae) from the Pirate Perch, *Aphredoderus sayanus* (Percopsiformes: Aphredoderidae), in northern Arkansas, U.S.A. Comp. Parasitol. 79(2): 344–347.

McAllister, C. T., H. W. Robison, T. M. Buchanan, and D. A. Etnier. 2012b. Distributional records for four fishes from the Arkansas, Mississippi, and White rivers of Arkansas. Southwest. Nat. 57(2): 217–219.

McAllister, C. T., H. W. Robison, and T. M. Buchanan. 2006. Noteworthy geographic distribution records for the Golden Topminnow, *Fundulus chrysotus* (Cyprinodontiformes: Fundulidae), from Arkansas. J. Arkansas Acad. Sci. 60: 185–188.

McAllister, C. T., H. W. Robison, and T. M. Buchanan. 2009a. Distribution of the Pallid Shiner, *Hybopsis amnis* (Cypriniformes: Cyprinidae), in Arkansas. Texas J. Sci. 62(1): 15–24.

McAllister, C. T., H. W. Robison, and K. E. Shirley. 2009b. Two noteworthy geographic distribution records for the White Sucker, *Catostomus commersonii* (Cypriniformes:

Catostomidae), from northern Arkansas. Texas J. Sci. 62 (3): 232–240.

McAllister, C. T., R. Tumlison, and H. W. Robison. 2009c. Geographic distribution records for select fishes of central and southern Arkansas. Texas J. Sci. 61(1): 31–44.

McAllister, C. T., W. G. Layher, H. W. Robison, and T. M. Buchanan. 2009d. New geographic distribution records for three species of *Notropis* (Cypriniformes: Cyprinidae) from large rivers of Arkansas. J. Arkansas Acad. Sci. 63: 192–194.

McAllister, C. T., W. C. Starnes, H. W. Robison, R. E. Jenkins, and M. E. Raley. 2009e. Distribution of the Silver Redhorse, *Moxostoma anisurum* (Cypriniformes: Catostomidae), in Arkansas. Southwest. Nat. 54(4): 514–518.

McAllister, C. T., H. W. Robison, and R. Tumlison. 2007. Additional geographic records for the Goldstripe Darter, *Etheostoma parvipinne* (Perciformes: Percidae), from Arkansas. J. Arkansas Acad. Sci. 61: 125–127.

McAllister, C. T., H. W. Robison, and R. Tumlison. 2008. Distribution of the Bantam Sunfish, *Lepomis symmetricus* (Perciformes: Centrarchidae), in Arkansas. Texas J. Sci. 60(1): 23–32.

McAllister, C. T., M. C. Wooten, and T. L. King. 1981. Observations on size and fecundity of the Least Brook Lamprey, *Lampetra aepyptera* (Abbott), from northcentral Arkansas. Proc. Arkansas Acad. Sci. 35: 86–87.

McAllister, C. T., C. R. Bursey, H. W. Robison, and M. B. Connior. 2016. New records of helminth parasites (Trematoda, Cestoda, Nematoda) from fishes in the Arkansas and Red River drainages, Oklahoma 2016. Proc. Oklahoma Acad. Sci. 96: 83–92.

McAllister, C. T., M. B. Connior, W. F. Font, and H. W. Robison. 2014a. Helminth parasites of the Banded Sculpin, *Cottus carolinae* (Scorpaeniformes: Cottidae), from northern Arkansas. Comp. Parasitol. 81(2): 203–209.

McAllister, C. T., C. R. Bursey, H. W. Robison, D. A. Neely, M. B. Connior, and M. A. Barger. 2014b. Miscellaneous fish helminth parasite (Trematoda, Cestoidea, Nematoda, Acanthocephala) records from Arkansas. J. Arkansas Acad. Sci. 68: 78–86.

McAllister, C. T., W. Layher, H. W. Robison, and T. M. Buchanan. 2010a. Distributional records for fishes from five large rivers in Arkansas. Southwest. Nat. 55(4): 587–591.

McAllister, C. T., W. C. Starnes, M. E. Raney, and H. W. Robison. 2010b. Geographic distribution records for select fishes of central and northern Arkansas. Texas J. Sci. 62: 271–280.

McAllister, C. T., R. Tumlison, H. W. Robison, and S. E. Trauth. 2013. Black-spot disease (Digenea: Strigeoidea: Diplostomidae) in select Arkansas fishes. J. Arkansas Acad. Sci. 67: 200–203.

McAllister, C. T., M. A. Barger, T. J. Fayton, M. B. Connior, D. A. Neely, and H. W. Robison. 2015a. Acanthocephalan parasites of select fishes (Catostomidae, Centrarchidae,

Cyprinidae, Ictaluridae), from the White River drainage, Arkansas. J. Arkansas Acad. Sci. 69: 132–134.

McAllister, C. T., C. R. Bursey, T. J. Fayton, W. F. Font, H. W. Robison, M. B. Connior, and D. G. Cloutman. 2015b. Helminth parasites of the Blackspotted Topminnow, *Fundulus olivaceus* (Cyprinodontiformes: Fundulidae), from the Interior Highlands of Arkansas. J. Arkansas Acad. Sci. 69: 135–138.

McAllister, C. T., W. F. Font, M. B. Connior, C. R. Bursey, H. W. Robison, N. H. Stokes, and C. D. Criscione. 2015c. Trematode parasites (Digenea) of the Slender Madtom *Noturus exilis* and Black River Madtom *Noturus maydeni* (Siluriformes: Ictaluridae) from Arkansas, U.S.A. Comp. Parasitol. 82(1): 137–143.

McAllister, C. T., C. R. Bursey, W. F. Font, H. W. Robison, S. E. Trauth, D. G. Cloutman, and T. J. Fayton. 2015d. Helminth parasites of the Northern Studfish, *Fundulus catenatus* (Cyprinodontiformes: Fundulidae), from the Ouachitas and Ozarks of Arkansas. Comp. Parasitol. 83(1): 78–87.

McBay, L. G. 1961. The biology of *Tilapia nilotica* Linnaeus. Proc. Annu. Conf. Southeast. Assoc. Game Fish Comm. 15: 208–218.

McBride, S. I. and D. C. Tarter. 1983. Foods and feeding behavior of Sauger, *Stizostedion canadense* (Smith) (Pisces: Percidae) from Gallipolis Locks and Dam, Ohio River. Brimleyana 9: 123–134.

McCallum, M. L. 2012. Notes on the diet and egg clutches of the Pirate Perch (*Aphredoderus sayanus*) from central Arkansas. Southeast. Nat. 11(3): 543–545.

McCarraher, D. B. and R. Thomas. 1968. Some ecological observations on the Fathead Minnow, *Pimephales promelas*, in the alkaline waters of Nebraska. Trans. Am. Fish. Soc. 97(1): 52–55.

McCaskill, M. L., J. E. Thomerson, and P. R. Mills. 1972. Food of the Northern Studfish, *Fundulus catenatus*, in the Missouri Ozarks. Trans. Am. Fish. Soc. 101(2): 375–377.

McCauley, R. W. and J. S. Tait. 1970. Preferred temperature of yearling Lake Trout (*Salvelinus namaycush*). J. Fish. Res. Board Can. 27(10): 1729–1733.

McCauley, R. W., J. R. Elliott, and L. A. A. Read. 1977. Influence of acclimation temperature on preferred temperature in the Rainbow Trout *Salmo gairdneri*. Trans. Am. Fish. Soc. 106(4): 362–365.

McCaull, J. and J. Crossland. 1974. Water pollution. Harcourt Brace, New York. 206 pp.

McClendon, D. D. and C. F. Rabeni. 1987. Physical and biological variables useful for predicting population characteristics of Smallmouth Bass and Rock Bass in an Ozark stream. N. Am. J. Fish. Manag. 7(1): 46–56.

McComas, S. R. and R. W. Drenner. 1982. Species replacement in a reservoir fish community: silverside feeding mechanics and competition. Can. J. Fish. Aquat. Sci. 39(6): 815–821.

McComish, T. S. 1967. Food habits of Bigmouth and Smallmouth Buffalo in Lewis and Clark Lake and the Missouri River. Trans. Am. Fish. Soc. 96(1): 70–74.

McCormick, F. H. 1990. Systematics and relationships of the *Etheostoma caeruleum* species complex (Pisces: Percidae). Ph.D. diss. Univ. Oklahoma, Norman.

McCormick, F. H. and N. Aspinwall. 1983. Habitat selection in three species of darters. Environ. Biol. Fishes 8(3–4): 279–282.

McDaniel, R. E. 1984. Selected aspects of the life history of *Etheostoma moorei* Raney and Suttkus. M.S. thesis. Arkansas State Univ., Jonesboro. 124 pp.

McDermid, J. L., B. J. Shuter, and N. P. Lester. 2010. Life-history differences parallel environmental differences among North American Lake Trout (*Salvelinus namaycush*) populations. Can. J. Fish. Aquat. Sci. 67(2): 314–325.

McDonough, C. J., W. A. Roumillat, and C. A. Wenner. 2003. Fecundity and spawning season of Striped Mullet (*Mugil cephalus* L.) in South Carolina estuaries. Fish. Bull. 101(4): 822–834.

McGauhey, P. H. 1971. Manmade contamination hazards to groundwater. *In:* T. R. Detwyler, ed. Man's impact on environment. McGraw-Hill, New York. 731 pp.

McGeehan, L. T. 1985. Multivariate and univariate analyses of the geographic variation within *Etheostoma flabellare* (Pisces: Percidae) of eastern North America. Ph.D. diss. Ohio State Univ., Columbus. 156 pp.

McGrath, P. E. 2010. The life history of Longnose Gar, *Lepisosteus osseus*, an apex predator in the tidal waters of Virginia. Ph.D. diss., The College of William and Mary, Williamsburg, VA. 177 pp.

McGrath, P. E. and E. J. Hilton. 2011. Sexual dimorphism in Longnose Gar *Lepisosteus osseus*. J. Fish Biol. 80(2): 335–345.

McGrath, P. E., E. J. Hilton, and J. A. Musick. 2013. Temporal and spatial effects on the diet of an estuarine piscivore, Longnose Gar (*Lepisosteus osseus*). Estuaries and Coasts 36(6): 1292–1303.

McGrath, P. E., E. J. Hilton, and J. A. Musick. 2016. Population demographics of Longnose Gar, *Lepisosteus osseus*, from the tidal rivers of Virginia. Copeia 2016(3): 738–745.

McIlwain, T. D. 1970. Stomach contents and length-weight relationships of Chain Pickerel (*Esox niger*) in south Mississippi waters. Trans. Am. Fish. Soc. 99(2): 439–440.

McIlwain, T. D. and R. Waller. 1972. A xanthochroic gar, *Lepisosteus oculatus*, from Mississippi. Trans. Am. Fish. Soc. 101(2): 362.

McKechnie, R. J. 1966. Logperch. Pages 530–531. *In:* A. Calhoun, ed. Inland fisheries management. California Dept. Fish and Game, Sacramento.

McKee, J. E. 1956. Report on oily substances and their effects on the beneficial uses of water. California Water Polllution Control Board Publ. 16. 71 pp.

McKenna, J. E., Jr. and C. Castiglione. 2014. Model distribution of Silver Chub (*Macrhybopsis storeriana*) in western Lake Erie. Am. Midl. Nat. 171(2): 301–310.

McKenzie, D. J. and D. J. Randall. 1990. Does *Amia calva* aestivate? Fish Physiol. Biochem. 8(2): 147–158.

McKinley, S., G. Van Der Kraak, and G. Power. 1998. Seasonal migrations and reproductive patterns in the Lake Sturgeon, *Acipenser fulvescens*, in the vicinity of hydroelectric stations in northern Ontario. Environ. Biol. Fishes 51(3): 245–256.

McLane, W. M. 1955. The fishes of the St. John's River system. Ph.D. diss., Univ. Florida, Gainesville. 334 pp.

McMahon, T. E. and J. C. Tash. 1979. The use of chemosenses by Threadfin Shad, *Dorosoma petenense* to detect conspecifics, predators and food. J. Fish Biol. 14(3): 289–296.

McManamay, R. A., J. T. Young, and D. J. Orth. 2012. Spawning of White Sucker (*Catostomus commersoni*) in a stormwater pond inlet. Am. Midl. Nat. 168(2): 466–476.

McMillan, V. E. and R. J. F. Smith. 1974. Agonistic and reproductive behavior of the Fathead Minnow (*Pimephales promelas* Rafinesque). Z. Tierpsychol. 34(1): 25–58.

McNeely, D. L. 1987. Niche relations within an Ozark stream cyprinid assemblage. Environ. Biol. Fishes 18(3): 195–208.

McPhail, J. D. and C. C. Lindsay. 1970. Freshwater fishes of northwestern Canada and Alaska. Bulletin 173, Fisheries Research Board of Canada. 381 pp.

McQuire, W. R., W. S. Woolcott, and E. G.. Maurakis. 1996. Histomorphology of external and internal mandibular and cheek epidermis in four species of North American pebble nest-building minnows (Pisces: Cyprinidae). ASB Bull. 43(2): 37–43.

McSwain, L. E. and R. M. Gennings. 1972. Spawning behavior of the Spotted Sucker, *Minytrema melanops* (Rafinesque). Trans Am. Fish. Soc. 101(4): 738–740.

Meador, M. R. and A. O. Layher. 1998. Instream sand and gravel mining: environmental issues and regulatory process in the United States. Fisheries 23(11): 6–13.

Means, M. L. 1993. Population dynamics and movements of Ozark Cavefish in Logan Cave National Wildlife Refuge, Benton County, Arkansas with additional baseline water quality information. M.S. thesis. Univ. Arkansas, Fayetteville. 126 pp.

Means, M. L. and J. E. Johnson. 1995. Movement of threatened Ozark Cavefish in Logan Cave National Wildlife Refuge, Arkansas. Southwest. Nat. 40(3): 308–313.

Meehan, W. R. and T. C. Bjornn. 1991. Salmonid distributions and life histories: Brook Trout. Pages 78–79. *In:* W. R. Meehan, ed. Influences of forest and rangeland management on salmonid fishes and their habitats. American Fisheries Society Special Publ. 19. 622 pp.

Meek, S. E. 1891. Report of explorations made in Missouri and Arkansas during 1889, with an account of the fishes observed in each of the river basins examined. Bull. U.S. Fish. Comm. (for 1889) 9: 113–141.

Meek, S. E. 1894a. A catalog of the fishes of Arkansas. Annu. Rep. Arkansas Geol. Surv. for 1891, 2: 216–276.

Meek, S. E. 1894b. Report of investigations respecting the fishes of Arkansas, conducted during 1891, 1892, and 1893, with a synopsis of previous explorations in the same state. Bull. U.S. Fish. Comm. 14(1): 67–94.

Meek, S. E. 1894c. A new *Etheostoma* from Arkansas. Am. Nat. 28: 957.

Meek, S. E. 1896. A list of fishes and mollusks collected in Arkansas and Indian Territory in 1894. Bull. U.S. Fish. Comm. 15(for 1895): 341–349.

Meffe, G. K. and C. R. Carroll. 1997. Principles of conservation biology. 2nd ed. Sinauer, Sunderland, MA. 729 pp.

Meffe, G. K. and A. L. Sheldon. 1988. The influence of habitat structure on fish assemblage composition in southeastern blackwater streams. Am. Midl. Nat. 120(2): 225–240.

Mendelson, J. 1975. Feeding relationships among species of *Notropis* (Pisces: Cyprinidae) in a Wisconsin stream. Ecol. Monogr. 45(3): 199–230.

Mendelson, T. C. 2003. Evidence of intermediate and asymmetrical behavioral isolation between Orangethroat and Orangebelly darters (Teleostei: Percidae). Am. Midl. Nat. 150(2): 343–347.

Mendoza, R., C. Aguilera, G. Rodriguez, M. Gonzalez, and R. Castro. 2002. Morphophysiological studies on Alligator Gar (*Atractosteus spatula*) larval development as a basis for their culture and repopulation of their natural habitats. Rev. Fish Biol. Fish. 12(2–3): 133–142.

Mense, J. B. 1967. Ecology of the Mississippi Silversides, *Menidia audens* Hay, in Lake Texoma. Oklahoma Fish. Res. Lab. Bull. 6. 31 pp.

Menzel, B. W. and F. B. Cross. 1977. Systematics of the Bleeding Shiner species group (Cyprinidae: genus *Notropis* subgenus *Luxilus*). Abst. 1977 A.S.I.H. meeting. Gainesville, Florida.

Mercado-Silva, N., G. G. Sass, B. M. Roth, S. Gilbert, and M. J. Vander Zanden. 2007. Impact of Rainbow Smelt (*Osmerus mordax*) invasion on Walleye (*Sander vitreus*) recruitment in Wisconsin Lakes. Can. J. Fish. Aquat. Sci. 64(11): 1543–1550.

Metcalf, A. L. 1959. Fishes of Chautauqua, Cowley, and Elk counties, Kansas. Univ. Kansas Publ., Mus. Nat. Hist. 11(6): 345–400.

Metcalf, A. L. 1966. Fishes of the Kansas River system in relation to zoogeography of the Great Plains. Univ. Kansas Publ., Mus. Nat. Hist. 17(3): 25–189.

Metcalf, C., F. L. Pezold, and B. G. Crump. 1997. Food habits of introduced Rainbow Trout (*Oncorhynchus mykiss*) in the upper Little Missouri River drainage of Arkansas. Southwest. Nat. 42(2): 148–154.

Mettee, M. F. 1974. A study on the reproductive behavior, embryology, and larval development of the pygmy sunfishes of the genus *Elassoma*. Ph.D. diss., Univ. Alabama. Tuscaloosa. 130 pp.

Mettee, M. F. and P. E. O'Neil. 2003. Status of Alabama Shad and Skipjack Herring in Gulf of Mexico drainages. Pages 157–170. *In*: K. E. Limburg and J. R. Waldman, eds.

Biodiversity, status, and conservation of the world's shads. Symp. 35. American Fisheries Society, Bethesda, MD. 370 pp.

Mettee, M. F., P. E. O'Neil, and J. M. Pierson. 1996. Fishes of Alabama and the Mobile basin. Oxmoor House. Birmingham, AL. 820 pp.

Mettee, M. F., P. E. O'Neil, R. D. Suttkus, and J. M. Pierson. 1987. Fishes of the lower Tombigbee River system in Alabama and Mississippi. Alabama Geol. Surv. Bull. 107. 186 pp.

Meyer, W. H. 1962. Life history of three species of redhorse (*Moxostoma*) in the Des Moines River, Iowa. Trans. Am. Fish. Soc. 91(4): 412–419.

Meyers, C. D. and R. J. Muncy. 1962. Summer food and growth of Chain Pickerel, *Esox niger*, in brackish waters of the Severn River, Maryland. Chesapeake Sci. 3(2): 125–128.

Meyers, J. W. 1977. Fishes of the Little Missouri River, southwest Arkansas. M.S. thesis. Northeast Louisiana Univ., Monroe. 25 pp.

Michaletz, P. H., C. F. Rabeni, W. W. Taylor, and T. R. Russell. 1982. Feeding ecology and growth of young-of-the-year Paddlefish in hatchery ponds. Trans. Am. Fish. Soc. 111(6):700–709.

Mickle, P. F., J. F. Schaefer, S. B. Adams, and B. R. Kreiser. 2010. Habitat use of age 0 Alabama Shad in the Pascagoula River drainage, USA. Ecol. Freshw. Fish 19(1): 107–115.

Mickle, P. F., J. F. Schaefer, D. A. Yee, and S. B. Adams. 2013. Diet of juvenile Alabama Shad (*Alosa alabamae*) in two northern Gulf of Mexico drainages. Southeast. Nat. 12(1): 233–237.

Mickle, P. F., J. S. Franks, B. R. Kreiser, G. J. Gray, J. M. Higgs, and J. Havrylkoff. 2015. First molecular verification of a marine-collected specimen of *Alosa alabamae* (Teleostei: Clupeidae). Southeast. Nat. 14(3): 596–601.

Middleton, S., M. Perello, and T. P. Simon. 2013. Length-weight relationship of the Mimic Shiner *Notropis volucellus* (Cope 1865) in the western basin of Lake Erie. Ohio J. Sci. 112(2): 44–50.

Midwood, J. D., L. F. G. Gutowsky, B. Hlevca, R. Portiss, M. G. Wells, S. E. Doka, and S. J. Cooke. 2018. Tracking Bowfin with acoustic telemetry: insight into the ecology of a living fossil. Ecol. Freshw. Fish 27(1): 225–236.

Miles, G. W. 1913. A defense of the humble Dogfish. Trans. Am. Fish Soc. 42(1): 51–59.

Miller, D. L. 1984. Distribution, abundance, and habitat of the Arkansas Darter, *Etheostoma cragini* (Percidae) in Colorado. Southwest. Nat. 29(4): 496–499.

Miller, D. R. 1953. Two additions to Oklahoma's fish fauna from Red River in Bryan County. Proc. Oklahoma Acad. Sci. 34: 33–34.

Miller, G. L. 1983. Trophic resource allocation between *Percina sciera* and *P. ouachitae* in the Tombigbee River, Mississippi. Am. Midl. Nat. 110(2): 299–313

Miller, G. L. and W. R. Nelson. 1974. Goldeye, *Hiodon alosoides*, in Lake Oahe: abundance, age, growth, maturity, food, and the fishery, 1963–69. Tech. Papers U.S. Fish and Wildlife Service No. 79. 13 pp.

Miller, H. C. 1964. The behavior of the Pumpkinseed Sunfish, *Lepomis gibbosus* (Linnaeus), with notes on the behavior of other species of *Lepomis* and the Pygmy Sunfish, *Elassoma evergladei*. Behaviour 22: 88–151.

Miller, J. M. 1974. The food of Brook Trout *Salvelinus fontinalis* (Mitchill) fry from different subsections of Lawrence Creek, Wisconsin. Trans. Am. Fish. Soc. 103(1): 130–134.

Miller, L. M., M. C. McInerny, and J. Roloff. 2008. Crappie hybridization in southern Minnesota lakes and its effects on growth estimates. N. Am. J. Fish. Manag. 28(4): 1120–1131.

Miller, M. J. 2004. The ecology and functional morphology of feeding of North American sturgeon and Paddlefish. Pages 87–102. *In*: G. T. O. LeBreton, F. W. H. Beamish, and R. S. McKinley. 2004. Sturgeons and Paddlefish of North America. Kluwer Academic, Fish and Fisheries Series Vol. 27. Boston. 323 pp.

Miller, M. J. 2009. Ecology of anguilliform leptocephali: remarkable transparent fish larvae of the ocean surface layer. Aqua-BioSci. Monogr. 2(4): 1–94.

Miller, M. J., Y. Chikaraishi, N. O. Ogawa, Y. Yamada, K. Tsukamoto, and N. Ohkouchi. 2013. A low trophic position of Japanese eel larvae indicates feeding on marine snow. Biol. Lett. 9(1). Available: https://doi.org/10.1098/rsbl.2012.0826

Miller, M. J., T. Otake, J. Aoyama, S. Wouthuyzen, S. Suharti, H. Y. Sugeha, and K. Tsukamoto. 2011. Observations of gut contents of leptocephali in the North Equatorial Current and Tomini Bay, Indonesia. Coastal Mar. Sci. 35(1): 277–288.

Miller, R. F., R. Cloutier, and S. Turner. 2003. The oldest articulated chondrichthyian from the Early Devonian period. Nature 425(6957): 501–504.

Miller, R. J. 1962. Reproductive behavior of the Stoneroller Minnow, *Campostoma anomalum pullum*. Copeia 1962(2): 407–417.

Miller, R. J. 1964. Behavior and ecology of some North American cyprinid fishes. Am. Midl. Nat. 72(2): 313–357.

Miller, R. J. 1967. Nestbuilding and breeding activities of some Oklahoma fishes. Southwest. Nat. 12(4): 463–468.

Miller, R. J. 1968. Speciation in the Common Shiner: an alternate view. Copeia 1968(3): 640–647.

Miller, R. J. 1979. Relationship between habitat and feeding mechanisms in fishes. Pages 269–280. *In*: H. Clepper, ed. Predator-prey systems in fisheries management. Sport Fishing Institute, Washington, DC. 504 pp.

Miller, R. J. and H. E. Evans. 1965. External morphology of the brain and lips in catostomid fishes. Copeia 1965(4): 467–487.

Miller, R. J. and H. W. Robison. 1973. The fishes of Oklahoma. Oklahoma State Univ. Press, Stillwater. 246 pp.

Miller, R. J. and H. W. Robison. 2004. The fishes of Oklahoma. 2nd ed. Univ. of Oklahoma Press, Norman. 450 pp.

Miller, R. R. 1950. A review of the American clupeid fishes of the genus *Dorosoma*. Proc. U.S. Natl. Mus. 100(3267): 387–410.

Miller, R. R. 1959. Origin and affinities of the freshwater fish fauna of western North America. Pages 187–222. *In:* C. L. Hubbs, ed. Zoogeography, a symposium. Am. Assoc. Adv. Sci. Publ. 51. Washington, DC. 509 pp.

Miller, R. R. 1960. Systematics and biology of the Gizzard Shad (*Dorosoma cepedianum*) and related fishes. U.S. Fish Wildl. Serv. Fish. Bull. 60(173): 371–392.

Miller, R. R. 1965. Quaternary freshwater fishes of North America. Pages 569–581. *In:* H. E. Wright and D. J. Frey. eds. The Quaternary of the United States. Princeton Univ. Press, Princeton, NJ. 922 pp.

Miller, R. R. 1972. Threatened freshwater fishes of the United States. Trans. Am. Fish. Soc. 101(2): 239–252.

Miller, R. R., W. L. Minckley, and S. M. Norris. 2005. Freshwater fishes of Mexico. Univ. Chicago Press, Chicago. 490 pp.

Miller, R. R., J. D. Williams, and J. E. Williams. 1989. Extinctions of North American fishes during the past century. Fisheries 14(6): 22–38.

Miller, R. V. 1967. Food of the Threadfin Shad, *Dorosoma petenense*, in Lake Chicot, Arkansas. Trans. Am. Fish. Soc. 96(3): 243–246.

Miller, R. V. 1968. A systematic study of the Greenside Darter, *Etheostoma blennioides* Rafinesque (Pisces: Percidae). Copeia 1968(1): 1–40.

Mills, J. G. 1972. Biology of the Alabama Shad in northwest Florida. Tech Series No. 68. State of Florida Dept. of Natural Resources. 24 pp.

Mills, M. D., R. B. Rader, and M. C. Belk. 2004. Complex interactions between native and invasive fish: the simultaneous effects of multiple negative interactions. Oecologia 141(4): 713–721.

Mims, S. D., R. J. Onders, and W. L. Shelton. 2009. Propagation and culture of Paddlefish. Pages 357–383. *In:* C. P. Paukert and G. D. Scholten, eds. Paddlefish management, propagation, and conservation in the 21st century: building from 20 years of research and management. Symp. 66. American Fisheries Society, Bethesda, MD. 443 pp.

Minckley, W. L. 1959. Fishes of the Big Blue River basin, Kansas. Mus. Nat. Hist. Univ. Kansas 11(7): 401–442.

Minckley, W. L. 1972. Notes on the spawning behavior of Red Shiner, introduced into Burro Creek, Arizona. Southwest. Nat. 17(1): 101–103.

Minckley, W. L. and J. E. Deacon. 1959. Biology of the Flathead Catfish in Kansas. Trans. Am. Fish. Soc. 88(4): 344–355.

Minckley, W. L. and L. A. Krumholz. 1960. Natural hybridization between the clupeid genera *Dorosoma* and *Signalosa*, with a report on the distribution of *S. petenense*. Zoologica 45(4): 171–182.

Minckley, W. L., J. E. Johnson, J. N. Rinne, and S. E. Willoughby. 1970. Foods of buffalofishes, genus *Ictiobus*, in central Arizona reservoirs. Trans. Am. Fish. Soc. 99(2): 333–342.

Minder, M. and M. Pyron. 2018. Dietary overlap and selectivity among Silver Carp and two native filter feeders in the Wabash River. Ecol. Freshw. Fish 27(1): 506–512.

Ming, A. D. 1968. Life history of the Grass Pickerel, *Esox americanus vermiculatus*, in Oklahoma. Bull. Oklahoma Fish. Res. Lab. 8. 66 pp.

Miranda, L. E. and G. M. Lucas. 2004. Determinism in fish assemblages of floodplain lakes of the vastly disturbed Mississippi Alluvial Valley. Trans. Am. Fish. Soc. 133(2): 358–370.

Mirza, M. R. 1975. Freshwater fishes and zoogeography of Pakistan. Bijdragen tot de dierkunde 45(2): 143–180.

Mirza, M. R. 1995. Distribution of freshwater fishes in Pakistan and Kashmir. Proceedings of the Seminar on Aquatic Development of Pakistan, 1993: 1–15.

Mitchell, D. R. 1987. Reproductive cycle of the Scaly Sand Darter (*Ammocrypta vivax*) with notes on other life history characteristics. M.S. thesis. Northwestern State Univ. of Louisiana, Natchitoches. 40 pp.

Mitchell, R. M., R. L. Johnson, and G. L. Harp. 2002. Population structure of an endemic species of Yellowcheek Darter, *Etheostoma moorei* (Raney and Suttkus), of the upper Little Red River, Arkansas. Am. Midl. Nat. 148(1): 129–137.

Mitchill, S. L. 1814. Report, in part, of Samuel L. Mitchill, M.D., Professor of Natural History, etc., on the fishes of New York. D. Carlisle, New York. 28 pp.

Mitchill, S. L. 1815. The fishes of New York, described and arranged. Trans. Lit. Philos. Soc. N.Y. 1: 355–492.

Mitchill, S. L. 1817. Report on the ichthyology of the Wallkill, from the specimens of fishes presented to the society (Lyceum of Natural History) by Dr. B. Akerly. Am. Monthly Mag. Crit. Rev. 1(4): 289–290.

Mitchill, S. L. 1818. Memoir on ichthyology. The fishes of New York, described and arranged. Am. Monthly Mag. Crit. Rev. 2: 241–248; 321–328.

Mitchill, S. L. 1824. Masquinongy of the Great Lakes. Minerva, New York v. 1 (no. 16): 297. [Minerva is a supplement to the New York Mirror.]

Mittelbach, G. G. and L. Persson. 1998. The ontogeny of piscivory and its ecological consequences. Can. J. Fish. Aquat. Sci. 55(6): 1454–1465.

Miya, M., T. Sado, K. Saitoh, M. H. Doosey, H. L. Bart, Jr., I. Doadrio, Y. Keivany, J. Shrestha, V. Lheknim, R. Zardoya, M. Nishida, and R. L. Mayden. 2008. Higher-level relationships of the Cypriniformes (Actinopterygii: Ostariophysi) inferred from 238 whole mitochondrial genome sequences [abstract]. 88th Annu. Meet. Am. Soc. Ichthyol. Herpetol. 23–28 July 2008. Montreal, Quebec, Canada. pp. 272–273.

Modde, T. and J. C. Schmulbach. 1977. Food and feeding behavior of the Shovelnose Sturgeon, *Scaphirhynchus platorynchus*, in the unchannelized Missouri River, South Dakota. Trans. Am. Fish. Soc. 106(6): 602–608.

Moen, C. T., D. L. Scarnecchia, and J. S. Ramsey. 1992. Paddlefish movements and habitat use in Pool 13 of the upper

Mississippi River during abnormally low river stages and discharges. N. Am. J. Fish. Manag. 12(4): 744–751.

Moen, T. 1953. Food habits of the Carp in northwest Iowa lakes. Proc. Iowa Acad. Sci. 60(1): 665–686.

Moen, T. 1955. Food of the Freshwater Drum, *Aplodinotus grunniens* Rafinesque, in four Dickinson County, Iowa lakes. Proc. Iowa Acad. Sci. 62(1): 589–598.

Moen, T. E. 1959. Sexing of Channel Catfish. Trans. Am. Fish. Soc. 88(2): 149.

Moen, T. E. and M. R. Dewey. 1980. Growth and year class composition of White Bass (*Morone chrysops*) in DeGray Lake, Arkansas. Proc. Arkansas Acad. Sci. 34: 125–126.

Mok, H. K. 1981. The phylogenetic implications of centrarchid kidneys. Bull. Inst. Zool., Acad. Sinica 20(2): 59–67.

Monirian, J., Z. Sutphin, and C. Myrick. 2010. Effects of holding temperature and handling stress on the upper thermal tolerance of Threadfin Shad *Dorosoma petenense*. J. Fish Biol. 76(6): 1329–1342.

Montaña, C. G. and K. O. Winemiller. 2013. Evolutionary convergence in Neotropical cichlids and Nearctic centrarchids: evidence from morphology, diet, and stable isotope analysis. Biol. J. Linn. Soc. 109(1): 146–164.

Monzyk, F. R., W. E. Kelso, and D. A. Rutherford. 1997. Characteristics of woody cover used by Brown Madtoms and Pirate Perch in coastal plain streams. Trans. Am. Fish. Soc. 126(4): 665–675.

Mooi, R. D. and A. C. Gill. 2002. Historical biogeography of fishes. Pages 43–68. *In:* P. J. B. Hart and J. D. Reynolds, eds. Handbook of fish biology and fisheries. Vol. 1. Fish biology. Blackwell, London.

Moore, G. A. 1944. Notes on the early life history of *Notropis girardi*. Copeia 1944(4): 209–214.

Moore, G. A. 1950. The cutaneous sense organs of barbeled minnows adapted to life in the muddy waters of the Great Plains region. Trans. Am. Microsc. Soc. 69(1): 69–95.

Moore, G. A. 1956. The cephalic lateral-line system in some sunfishes (*Lepomis*). J. Comp. Neurol. 104(1): 49–55.

Moore, G. A. 1968. Fishes. Pages 22–165. *In:* W. F. Blair et al. Vertebrates of the United States, 2nd ed. McGraw Hill, New York.

Moore, G. A. 1973. Discovery of fishes in Oklahoma (1852–1972). Proc. Oklahoma Acad. Sci. 53: 1–26.

Moore, G. A. and W. E. Burris. 1956. Description of the lateral-line system of the Pirate Perch, *Aphredoderus sayanus*. Copeia 1956(1): 18–20.

Moore, G. A. and F. B. Cross. 1950. Additional Oklahoma fishes with validation of *Poecilichthys parvipinnis* (Gilbert and Swain). Copeia 1950(2): 139–148.

Moore, G. A. and R. C. McDougal. 1949. Similarity in the retinae of *Amphiodon alosoides* and *Hiodon tergisus*. Copeia 1949(4): 298.

Moore, G. A. and J. M. Paden. 1950. The fishes of the Illinois River in Oklahoma and Arkansas. Am. Midl. Nat. 44(1): 76–95.

Moore, G. A. and J. D. Reeves. 1955. *Hadropterus pantherinus*, a new percid fish from Oklahoma and Arkansas. Copeia 1955(2): 89–92.

Moore, G. A. and C. C. Rigney. 1952. Taxonomic status of the percid fish *Poecilichthys radiosus* in Oklahoma and Arkansas, with the descriptions of two new subspecies. Copeia 1952(1): 7–15.

Moore, G. A. and M. E. Sisk. 1963. The spectacle of *Elassoma zonatum* Jordan. Copeia 1963(2): 347–350.

Moore, G. A., M. B. Trautman, and M. R. Curd. 1973. A description of postlarval gar (*Lepisosteus spatula* Lacépède, Lepisosteidae), with a list of associated species from the Red River, Choctaw County, Oklahoma. Southwest. Nat. 18(3): 343–344.

Moore, J. W. and F. W. H. Beamish. 1973. Food of larval Sea Lamprey (*Petromyzon marinus*) and American Brook Lamprey (*Lampetra lamottei*). J. Fish. Res. Board Can. 30(1): 7–15.

Moretz, J. A. and W. Rogers. 2004. An ethological analysis of the breeding behavior of the Fantail Darter, *Etheostoma flabellare*. Am. Midl. Nat. 152(1): 140–144.

Morgan, M. N. 2002. Habitat associations of fish assemblages in the Sulphur River, Texas. M.S. thesis. Texas A&M Univ., College Station. 58 pp.

Moriarty, L. J. and K. O. Winemiller. 1997. Spatial and temporal variation in fish assemblage structure in Village Creek, Hardin County, Texas. Texas J. Sci. 49(3): 85–110.

Morris, M. A. and L. M. Page. 1981. Variation in western logperches (Pisces: Percidae), with description of a new subspecies from the Ozarks. Copeia 1981(1): 95–108.

Morrison, T. W., W. E. Lynch Jr., and K. Dabrowski. 1997. Predation on zebra mussels by Freshwater Drum and Yellow Perch in western Lake Erie. J. Great Lakes Res. 23(2): 177–189.

Morse, D. F. and P. A. Morse 1983. Archaeology of the Central Mississippi Valley. New World Archaeological Record Series, Academic Press, New York. 345 pp.

Moser, H. G., W. J. Richards, D. M. Cohen, M. P. Fahay, A. W. Kendall, Jr., and S. L. Richardson, eds. 1984. Ontogeny and systematics of fishes. Am. Soc. Ichthyol. Herpetol. Special Publ. 1:1–760.

Moshenko, R. W. and J. H. Gee. 1973. Diet, time and place of spawning, and environments occupied by Creek Chub (*Semotilus atromaculatus*) in the Mink River, Manitoba. J. Fish. Res. Board Can. 30(3): 357–362.

Moshin, A. K. M. and B. J. Gallaway. 1977. Seasonal abundance, distribution, food habits and condition of the Southern Brook Lamprey, *Ichthyomyzon gagei* Hubbs and Trautman, in an east Texas watershed. Southwest. Nat. 22(1): 107–114.

Moss, R. E., J. W. Scanlan, and C. S. Anderson. 1983. Observations on the natural history of the Blue Sucker (*Cycleptus elongatus* Lesueur) in the Neosho River. Am. Midl. Nat. 109(1): 15–22.

Moyer, G. R., B. L. Sloss, B. R. Kreiser, and K. A. Feldheim. 2009a. Isolation and characterization of microsatellite loci for Alligator Gar (*Atractosteus spatula*) and their variability in two other species (*Lepisosteus oculatus* and *L. osseus*) of Lepisosteidae. Mol. Ecol. Resour. 9(3): 963–966.

Moyer, G. R., R. K. Remington, and T. F. Turner. 2009b. Incongruent gene trees, complex evolutionary processes, and the phylogeny of a group of North American minnows (*Hybognathus* Agassiz 1855). Mol. Phylogenet. Evol. 50(3): 514–525.

Moyle, P. B. 1969. Ecology of the fishes of a Minnesota lake with special reference to the Cyprinidae. Ph.D. diss. Univ. Minnesota, Minneapolis. 169 pp.

Moyle, P. B. 1973. Ecological segregation among three species of minnows (Cyprinidae) in a Minnesota lake. Trans. Am. Fish. Soc. 102(4): 794–805.

Moyle, P. B. 1976. Inland fishes of California. Univ. California Press, Berkeley. 405 pp.

Moyle, P. B. 1977. In defense of sculpins. Fisheries 2(1): 20–23.

Moyle, P. B. 2002. Inland fishes of California. Revised and expanded. Univ. California Press, Berkeley. 517 pp.

Moyle, P. B. and J. J. Cech, Jr. 2004. Fishes, an introduction to ichthyology. 5th ed. Prentice-Hall, Englewood Cliffs, NJ. 726 pp.

Moyle, P. B. and R. A. Leidy. 1992. Loss of biodiversity in aquatic ecosystems: evidence from fish faunas. Pages 127–169. *In:* P. L. Fiedler, ed. Conservation biology: the theory and practice of nature conservation, preservation, and management. Chapman & Hall, New York. 507 pp.

Mueller, R. and M. Pyron. 2009. Substrate and current velocity preferences of Spotfin Shiner (*Cyprinella spiloptera*) and Sand Shiner (*Notropis stramineus*) in artificial streams. J. Freshw. Ecol. 24(2): 239–245.

Mullan, J. W. and R. L. Applegate. 1968. Centrarchid food habits in a new and old reservoir during and following bass spawning. Proc. 21st Annu. Conf. Southeast. Assoc. Game Fish Comm.: 332–342.

Mullan, J. W. and R. L. Applegate. 1970. Food habits of five centrarchids during filling of Beaver Reservoir 1965–66. Tech. Paper 50, Bureau of Sport Fisheries and Wildlife. 16 pp.

Mullan, J. W., R. L. Applegate, and W. C. Rainwater. 1968. Food of Logperch (*Percina caprodes*) and Brook Silverside (*Labidesthes sicculus*), in a new and old Ozark reservoir. Trans. Am. Fish. Soc. 97(3): 300–305.

Mulligan, T. J. and R. W. Chapman. 1989. Mitochondrial DNA analysis of Chesapeake Bay White Perch, *Morone americana*. Copeia 1989(3): 679–688.

Mundahl, N. D. and R. A. Sagan. 2005. Spawning ecology of the American Brook Lamprey, *Lampetra appendix*. Environ. Biol. Fishes 73(3): 283–292.

Mundahl, N. D., D. C. Melnytchuk, D. K. Spielman, J. P. Harkins, K. Funk, and A. M. Bilicki. 1998. Effectiveness of Bowfin as a predator on Bluegill in a vegetated lake. N. Am. J. Fish. Manag. 18(2): 286–294.

Mundahl, N. D., G. Sayeed, S. Taubel, C. Erickson, A. Zalatel, and J. Cousins. 2006. Densities and habitat of American Brook Lamprey (*Lampetra appendix*) larvae in Minnesota. Am. Midl. Nat. 156(1): 11–22.

Muoneke, M. I., O. E. Maughan, and M. E. Douglas. 1991. Multivariate morphometric analysis of Striped Bass, White Bass, and Striped Bass × White Bass hybrids. N. Am. J. Fish. Manag. 11(3): 330–338.

Muoneke, M. I., C. C. Henry, and O. E. Maughan. 1992. Population structure and food habits of White Crappie *Pomoxis annularis* Rafinesque in a turbid Oklahoma reservoir. J. Fish Biol. 41(4): 647–654.

Murphy, C. E., J. J. Hoover, S. G. George, B. R. Lewis, and K. J. Killgore. 2007a. Types and occurrence of morphological anomalies in *Scaphirhynchus* spp. of the middle and lower Mississippi River. J. Appl. Ichthyol. 23(4): 354–358.

Murphy, C. E., J. J. Hoover, S. G. George, and K. J. Killgore. 2007b. Morphometric variation among river sturgeons (*Scaphirhynchus* spp.) of the middle and lower Mississippi River. J. Appl. Ichthyol. 23(4): 313–323.

Murphy, G. W., T. J. Newcomb, D. J. Orth, and S. J. Reeser. 2005. Food habits of selected fish species in the Shenandoah River basin, Virginia. Proc. Annu. Conf. Southeast Assoc. Fish Wildl. Agencies 59: 325–335.

Murray, A. M. and M. V. H. Wilson. 1999. Contributions of fossils to the phylogenetic relationships of the percopsiform fishes (Teleostei: Paracanthopterygii): order restored. Pages 397–411. *In:* G. Arratia and H. P. Schultze, eds. Mesozoic fishes 2: systematics and the fossil record. Die Deutsche Bibliothek, Munich. 604 pp.

Muth, R. T. and J. C. Schmulbach. 1984. Downstream transport of fish larvae in a shallow prairie river. Trans. Am. Fish. Soc. 113(2): 224–230.

Myers, G. S. 1964. A brief sketch of the history of ichthyology in America to the year 1850. Copeia 1964(1): 33–41.

Myers, G. S. 1965. *Gambusia*, the fish destroyer. Trop. Fish Hobbyist 13(5): 31–32, 53–54.

Myers, G. S. 1979. A brief sketch of the history of ichthyology in America to the year 1850. Pages 1–8. *In:* M. S. Love and G. M. Cailliet eds. Readings in ichthyology. Goodyear, Santa Monica, CA. 525 pp.

Nachtrieb, H. F. 1910. The primitive pores of *Polyodon spathula* (Walbaum). J. Exp. Zool. 9(2): 455–468.

Nagle, B. C. and A. M. Simons. 2012. Rapid diversification in the North American minnow genus *Nocomis*. Mol. Phylogenet. Evol. 63(3): 639–649.

Narain, R. B., S. T. Kamble, T. O. Powers, and T. S. Harris. 2013. DNA barcode of thief ant complex (Hymenoptera: Formicidae). J. Entomol. Sci. 48(3): 234–242.

Near, T. J. 2002. Phylogenetic relationships of *Percina* (Percidae: Etheostomatinae). Copeia, 2002(1): 1–14.

Near, T. J. 2008. Rescued from synonymy: a redescription of *Percina bimaculata* Haldeman and a molecular phylogenetic analysis of logperch darters (Percidae: Etheostomatinae). Bull. Peabody Mus. Nat. Hist. 49(1): 3–18.

Near, T. J. and M. F. Benard. 2004. Rapid allopatric speciation in logperch darters (Percidae: *Percina*). Evolution 58(12): 2798–2808.

Near, T. J. and B. P. Keck. 2005. Dispersal, vicariance, and timing of diversification in *Nothonotus* darters. Mol. Ecol. 14(11): 3485–3496.

Near, T. J. and B. P. Keck. 2012. AFLPs do not support deep phylogenetic relationships among darters (Teleostei: Percidae: Etheostomatinae). Heredity 108(6): 647–648.

Near, T. J. and J. B. Koppelman. 2009. Species diversity, phylogeny, and phylogeography of Centrarchidae. Chapter 1. *In*: S. J. Cooke and D. P. Philipp, eds. Centrarchid fishes: diversity, biology, and conservation. Wiley-Blackwell. 560 pp.

Near, T. J., D. I. Bolnick, and P. C. Wainwright. 2004. Investigating phylogenetic relationships of sunfishes and black basses (Actinopterygii: Centrarchidae) using DNA sequences from mitochondrial and nuclear genes. Mol. Phylogenet. Evol. 32(1): 344–357.

Near, T. J., D. I. Bolnick, and P. C. Wainwright. 2005. Fossil calibrations and molecular divergence time estimates in centrarchid fishes (Teleostei: Centrarchidae). Evolution 59(8): 1768–1782.

Near, T. J., L. M. Page, and R. L. Mayden. 2001. Intraspecific phylogeography of *Percina evides* (Percidae: Etheostomatinae): an additional test of the Central Highlands pre-Pleistocene vicariance hypothesis. Mol. Ecol. 10(9): 2235–2240.

Near, T. J., J. C. Porterfield, and L. M. Page. 2000. Evolution of cytochrome *b* and the molecular systematics of *Ammocrypta* (Percidae: Etheostomatinae). Copeia 2000(3): 701–711.

Near, T. J., T. W. Kassler, J. B. Koppelman, C. B. Dillman, and D. P. Philipp. 2003. Speciation in North American black basses, *Micropterus* (Actinopterygii: Centrarchidae). Evolution. 57(7): 1610–1621.

Near, T. J., M. Sandel, K. L. Kuhn, P. J. Unmack, P. C. Wainwright, and W. L. Smith. 2012. Nuclear gene-inferred phylogenies resolve the relationships of the enigmatic pygmy sunfishes, *Elassoma* (Teleostei: Percomorpha). Mol. Phylogenet. Evol. 63(2): 388–395.

Near, T. J., C. M. Bossu, G. S. Bradburd, R. L. Carlson, R. C. Harrington, P. R. Hollingsworth, B. P. Keck, and D. A. Etnier. 2011. Phylogeny and temporal diversification of darters (Percidae: Etheostomatinae). Syst. Biol. 60(5): 565–595.

Near, T. J., J. W. Simmons, J. M. Mollish, M. A. Correa, E. Benavides, R. C. Harrington, and B. P. Keck. 2017. A new species of logperch endemic to Tennessee (Percidae: Etheostomatinae: *Percina*). Bull. Peabody Mus. Nat. Hist. 58(2): 287–309.

Neave, F. B., N. E. Mandrak, M. F. Docker, and D. L. Noakes. 2007. An attempt to differentiate sympatric *Ichthyomyzon* ammocoetes using meristic, morphological, pigmentation, and gonadal analyses. Can. J. Zool. 85(4): 549–560.

Nedbal, M. A. and D. P. Philipp. 1994. Differentiation of mitochondrial DNA in Largemouth Bass. Trans. Am. Fish. Soc. 123(4): 460–468.

Neely, A. R. and E. J. Pert. 2000. Feeding relationship between two syntopic, morphologically similar fishes, the Western Mosquitofish (*Gambusia affinis*) and the Blackspotted Topminnow (*Fundulus olivaceus*). J. Arkansas Acad. Sci. 54: 77–80.

Neely, B. C., M. A. Pegg, and G. E. Mestl. 2010. Seasonal resource selection by Blue Suckers *Cycleptus elongatus*. J. Fish Biol. 76(4): 836–851.

Neff, B. D. and E. L. Clare. 2008. Temporal variation in cuckoldry and paternity in two sunfish species (*Lepomis* spp.) with alternative reproductive tactics. Can. J. Zool. 86(2): 92–98.

Neff, B. D. and R. Knapp. 2009. Alternative reproductive tactics in the Centrarchidae. Pages 90–104. *In*: S. J. Cooke and D. P. Philipp, eds. Centrarchid fishes: diversity, biology, and conservation. Wiley-Blackwell, Chichester.

Neill, W. H., Jr. 1967. Factors affecting heat resistance and temperature selection in the Longear Sunfish, *Lepomis megalotis*. M.S. thesis, Univ. Arkansas, Fayetteville. 63 pp.

Neill, W. H., Jr., K. Strawn, and J. E. Dunn. 1966. Heat resistance experiments with the Longear Sunfish, *Lepomis megalotis* (Rafinesque). Proc. Arkansas Acad. Sci. 20: 39–49.

Neill, W. T. 1950. An estivating Bowfin. Copeia 1950(3): 240.

Nelson, E. M. 1948. The comparative morphology of the Weberian apparatus of the Catostomidae and its significance in systematics. J. Morphol. 83(2): 225–251.

Nelson, E. M. 1949. The opercular series of the Catostomidae. J. Morphol. 85(3): 559–567.

Nelson, E. W. 1876. A partial catalogue of the fishes of Illinois. Bull. Illinois Mus. Nat. Hist. 1: 33–52.

Nelson, G. J. 1969. Gill arches and the phylogeny of fishes, with notes on the classification of vertebrates. Bull. Am. Mus. Nat. Hist. 141: 475–552.

Nelson, G. J. and M. N. Rothman. 1973. The species of gizzard shads (Dorosomatinae) with particular reference to the Indo-Pacific region. Bull. Am. Mus. Nat. Hist. 150(2): 131–206.

Nelson, J. S. 1968. Life history of the Brook Silverside, *Labidesthes sicculus*, in Crooked Lake, Indiana. Trans. Am. Fish. Soc. 97(3): 293–296.

Nelson, J. S. 1994. Fishes of the World. 3rd ed. John Wiley and Sons, New York. 600 pp.

Nelson, J. S. 2006. Fishes of the World. 4th ed. John Wiley and Sons, New York. 601 pp.

Nelson, J. S., T. C. Grande, and M. V. H. Wilson. 2016. Fishes of the World. 5th ed. John Wiley and Sons, Hoboken, NJ. 707 pp.

Nelson, J. S., W. C. Starnes, and M. L. Warren, Jr. 2002. A capital case for common names of species of fishes—a white crappie or a White Crappie. Fisheries 27(7): 31–33.

Nelson, J. S., E. J. Crossman, H. Espinosa-Perez, L. T. Findley, C. R. Gilbert, R. N. Lea, and J. D. Williams. 2003. The "Names of Fishes" list, including recommended changes in fish names: Chinook salmon for chinook salmon, and *Sander* to replace *Stizostedion* for the Sauger and Walleye. Fisheries 28(7): 38–39.

Nelson, J. S., E. J. Crossman, H. Espinosa-Perez, L. T. Findley, C. R. Gilbert, R. N. Lea, and J. D. Williams. 2004. Common and scientific names of fishes from the United States, Canada, and Mexico. 6th ed. Special Publ. 29. American Fisheries Society, Bethesda, MD. 386 pp.

Nelson, W. R. 1968. Reproduction and early life history of Sauger, *Stizostedion canadense* in Lewis and Clark Lake. Trans. Am. Fish. Soc. 97(2): 159–166.

Nelson, W. R., N. R. Hines, and L. G. Beckman. 1965. Artificial propagation of Saugers and hybridization with Walleyes. Prog. Fish-Cult. 27(4): 216–218.

Netsch, N. F. 1964. Food and feeding habits of the Longnose Gar in central Missouri. Proc. 18th Annu. Conf. Southeast. Assoc. Game Fish. Comm.: 506–511.

Netsch, N. F. and A. Witt, Jr. 1962. Contributions to the life history of the Longnose Gar (*Lepisosteus osseus*) in Missouri. Trans. Am. Fish. Soc. 91(3): 251–262.

Netsch, N. F., G. M. Kersh, A. Houser, and R. V. Kilambi. 1971. Distribution of young Gizzard and Threadfin shad in Beaver Reservoir. Pages 95–105. *In:* G. E. Hall, ed. Reservoir fisheries and limnology. Special Publ. 8. American Fisheries Society.

Newsome, G. E. and J. H. Gee. 1978. Preference and selection of prey by Creek Chub (*Semotilus atromaculatus*) inhabiting the Mink River, Manitoba. Can. J. Zool. 56(12): 2486–2497.

Newton, S. H. and R. V. Kilambi. 1969. Determination of sexual maturity of White Bass from ovum diameters. Southwest. Nat. 14(2): 213–220.

Newton, S. H., A. J. Merkowsky, A. J. Handcock, and M. V. Meisch. 1977. Mosquitofish, *Gambusia affinis* (Baird and Girard) production in extensive polyculture systems. Proc. Arkansas Acad. Sci. 31: 77–78.

Niazi, A. D. and G. A. Moore. 1962. The Weberian apparatus of *Hybognathus placitus* and *H. nuchalis* (Cyprinidae). Southwest. Nat. 7(1): 41–50.

Nico, L. G. and P. L. Fuller. 1999. Spatial and temporal patterns of nonindigenous fish introductions in the United States. Fisheries 24(1): 16–27.

Nico, L. and M. Neilson. 2016. *Oreochromis mossambicus* (Peters, 1852). USGS Nonindigenous Aquatic Species Database, Gainesville, FL. Available: http://nas.er.usgs.gov/queries/factsheet.aspx?SpeciesID=466 [Revision Date: 28 July 2015].

Nico, L. G., and M. E. Neilson. 2018. *Mylopharyngodon piceus* (Richardson, 1846). USGS Nonindigenous Aquatic Species Database, Gainesville, FL. Available: https://nas.er.usgs.gov/queries/factsheet.aspx?SpeciesID=573 [Revision Date: 2 January 2018].

Nico, L., P. Fuller, and M. Neilson. 2018a. *Oreochromis aureus*. USGS Nonindigenous Aquatic Species Database, Gainesville, FL. Available: http://nas.er.usgs.gov/queries/factsheet.aspx?SpeciesID=463 [Revision Date: 19 June 2013].

Nico, L., P. Fuller, G. Jacobs, J. Larson, T. H. Makled, A. Fusaro, and M. Neilson. 2018b. *Scardinius erythrophthalmus*. USGS Nonindigenous Aquatic Species Database, Gainesville, FL. Available: http://nas.er.usgs.gov/queries/factsheet.aspx?SpeciesID=648 [Revision Date: 13 March 2015].

Nico, L., P. J. Schofield, and M. Neilson. 2016a. *Oreochromis niloticus*. USGS Nonindigenous Aquatic Species Database, Gainesville, FL. Available: http://nas.er.usgs.gov/queries/factsheet.aspx?SpeciesID=468 [Revision Date: 18 December 2013].

Nico, L. G., P. J. Schofield, J. Larson, T. H. Makled, and A. Fusaro. 2016b. *Carassius auratus*. USGS Nonindigenous Aquatic Species Database, Gainesville, FL. Available: http://nas.er.usgs.gov/queries/factsheet.aspx?SpeciesID=508 [Revision Date: 2 August 2013].

Nico, L. G., J. D. Williams, and H. L. Jelks. 2005. Black Carp: biological synopsis and risk assessment of an introduced fish. Special Publ. 32. American Fisheries Society. 337 pp.

Nieman, R. L. and D. C. Wallace. 1974. The age and growth of the Blackstripe Topminnow, *Fundulus notatus* Rafinesque. Am. Midl. Nat. 92(1): 203–205.

Niemiller, M. L. 2011. Evolution, speciation, and conservation of amblyopsid cavefishes. Ph.D. diss., Univ. Tennessee, Knoxville. 193 pp.

Niemiller, M. L. and B. M. Fitzpatrick. 2008. Phylogenetics of the Southern Cavefish (*Typhlichthys subterraneus*): implications for conservation and management. Pages 79–88. *In:* Proceedings of the 18th National Cave and Karst Management Symposium, St. Louis, MO.

Niemiller, M. L. and T. L. Poulson. 2010. Subterranean fishes of North America: Amblyopsidae. Pages 169–280. *In:* E. Trajano, M. E. Bichuette, and B. G. Kapoor, eds. Biology of subterranean fishes. CRC Press, Enfield, NH. 494 pp.

Niemiller, M. L. and D. Soares. 2015. Cave environments. Pages 161–191. *In:* R. Riesch, M. Tobler, and M. Plath, eds. Extremophile fishes: ecology, evolution, and physiology of teleosts in extreme environments. Springer International, Switzerland. 326 pp.

Niemiller, M. L., T. J. Near, and B. M. Fitzpatrick. 2012. Delimiting species using multilocus data: diagnosing cryptic diversity in the Southern Cavefish, *Typhlichthys subterraneus* (Teleostei: Amblyopsidae). Evolution 66(3): 846–866.

Niemiller, M. L., G. O. Graening, D. B. Fenolio, J. C. Godwin, J. R. Cooley, W. D. Pearson, B. M. Fitzpatrick, and T. J. Near. 2013a. Doomed before they are described? The need for conservation assessments of cryptic species complexes using an amblyopsid cavefish (Amblyopsidae: *Typhlichthys*) as a case study. Biodivers. Conserv. 22(8): 1799–1820.

Niemiller, M. L., B. M. Fitzpatrick, P. Shah, L. Schmitz, and T. J. Near. 2013b. Evidence for repeated loss of selective constraint in rhodopsin of amblyopsid cavefishes (Teleostei: Amblyopsidae). Evolution 67(3): 732–748.

Nikol'skiy, G. V. 1956. Ryby basseyna Amura [Fishes of the Amur basin]: Moscow. USSR Academy of Sciences.

Noble, R. L. 1981. Management of forage fishes in impoundments of the southern United States. Trans. Am. Fish. Soc. 110(6): 738–750.

Noltie, D. B. 1990. Status of the Orangespotted Sunfish, *Lepomis humilis*, in Canada. Can. Field Nat. 104(1): 69–86.

Noltie, D. B. and M. H. A. Keenleyside. 1987a. Breeding ecology, nest characteristics, and nest-site selection of stream- and lake-dwelling Rock Bass, *Ambloplites rupestris*. Can. J. Zool. 65(2): 379–390.

Noltie, D. B. and M. H. A. Keenleyside. 1987b. The breeding behaviour of stream-dwelling Rock Bass, *Ambloplites rupestris* (Centrarchidae). Biol. Behav. 12: 196–206.

Noltie, D. B. and C. M. Wicks. 2001. How hydrology has shaped the ecology of Missouri's Ozark Cavefish, *Amblyopsis rosae*, and Southern Cavefish, *Typhlichthys subterraneus*: insights on the sightless from understanding the underground. Environ. Biol. Fishes 62(1–3): 171–194.

Nord, R. C. 1967. A compendium of fishery information on the Upper Mississippi River. Upper Mississippi River Conservation Comm. 238 pp.

Nordlie, F. G. 1966. Thermal acclimation and peptic digestive capacity in the Black Bullhead, *Ictalurus melas* (Raf.). Am. Midl. Nat. 75(2): 416–424.

Nordlie, F. G., W. A. Szelistowski, and W. C. Nordlie. 1982. Ontogenesis of osmotic regulation in the Striped Mullet, *Mugil cephalus* L. J. Fish Biol. 20(1): 79–86.

Norris, H. W. 1923. On the function of the paddle of the Paddlefish. Proc. Iowa Acad. Sci. 30: 135–137.

Nuevo, M., R. J. Sheehan, and R. C. Heidinger. 2004a. Accuracy and precision of age determination techniques for Mississippi River Bighead Carp *Hypophthalmichthys nobilis* (Richardson 1845) using pectoral spines and scales. Arch. Hydrobiol. 160(1): 45–56.

Nuevo, M., R. J. Sheehan, and P. S. Willis. 2004b. Age and growth of the Bighead Carp *Hypophthalmichthys nobilis* (Richardson 1845) in the middle Mississippi River. Arch. Hydrobiol. 160(2): 215–230.

Nurminen, L., J. Horppila, J. Lappalainen, and T. Malinen. 2003. Implications of Rudd (*Scardinius erythrophthalmus*) herbivory on submerged macrophytes in a shallow eutrophic lake. Hydrobiologia 506(1–3): 511–518.

Nuttall, T. 1999. S. Lottinville, ed. A journal of travels into the Arkansas Territory during the year 1819. Univ. Arkansas Press, Fayetteville, 392 pp.

O'Brien, W. J., B. Loveless, and D. Wright. 1984. Feeding ecology of young White Crappie in a Kansas reservoir. N. Am. J. Fish. Manag. 4(4A): 341–349.

Odenkirk, J. S. and M. W. Isel. 2016. Trends in abundance of Northern Snakeheads in Virginia tributaries of the Potomac River. Trans. Am. Fish. Soc. 145(4): 687–692.

Odenkirk, J. S. and S. J. Owens. 2005. Northern Snakeheads in the tidal Potomac River system. Trans. Am. Fish. Soc. 134(6): 1605–1609.

Ogden, J. C. 1970. Relative abundance, food habits, and age of the American Eel, *Anguilla rostrata* (Lesueur), in certain New Jersey streams. Trans. Am. Fish. Soc. 99(1): 54–59.

O'Gorman, R. 1974. Predation by Rainbow Smelt (*Osmerus mordax*) on young-of-the-year Alewives (*Alosa pseudoharengus*) in the Great Lakes. Prog. Fish-Cult. 36(4): 223–224.

Okada, Y. 1960. Studies on the freshwater fishes of Japan, II. Special part: Prefectural Univ. of Mie J. Faculty of Fisheries 4(3): 1–860 pp.

Oken, L. 1816. Lehrbuch der Naturgeschichte: 3. Theil. Zool., Abth. 2, Fleischthiere. Jena. 1270 pp.

Oken, L. 1817. Cuvier's und Oken's zoologien naben einander gestellt. Isis, Encycl. Zeitung 8(148): 1179–1185.

O'Leary, M. A., R. Sarr, R. Malou, E. H. Sow, C. Lepre, and R. V. Hill. 2012. A new fossil amiid from the Eocene of Senegal and the persistence of extinct marine amiids after the Cretaceous-Paleogene boundary. Copeia 2012(4): 603–608.

Oliveira, K. 1999. Life history characteristics and strategies of the American Eel, *Anguilla rostrata*. Can. J. Fish. Aquat. Sci. 56(5): 795–802.

Oliver, J. R. 1986. Comparative reproductive biology of the Cahaba Shiner, *Notropis* sp., and the Mimic Shiner, *Notropis volucellus* (Cope), from the Cahaba River drainage, Alabama. M.S. thesis, Samford Univ., Birmingham, AL. 20 pp.

Oliver, M. L. 1984. The Rainbow Trout fishery in the Bull Shoals-Norfork tailwaters, Arkansas, 1971–81. Proc. 38th Annu. Conf. Southeast. Assoc. Fish Wildl. Agencies: 549–561.

Oliver, M. L. and L. L. Rider. 1986. Net-pen aquaculture in Bull Shoals Reservoir. *In:* R. H. Stroud, ed. Fish culture in fisheries management. Fish Culture Section and Fisheries Management Section, American Fisheries Society, Bethesda, MD. 481 pp.

Olmsted, L. L. 1974. The ecology of Largemouth Bass (*Micropterus salmoides*) and Spotted Bass (*Micropterus punctulatus*) in Lake Fort Smith, Arkansas. Ph.D. diss., Univ. Arkansas, Fayetteville. 134 pp.

Olmsted, L. L. and D. G. Cloutman. 1974. Repopulation after a fish kill in Mud Creek, Washington County, Arkansas following pesticide pollution. Trans. Am. Fish. Soc. 103(1): 79–87.

Olmsted, L. L. and R. V. Kilambi. 1969. Stomach contents analysis of White Bass (*Roccus chrysops*) in Beaver Reservoir, Arkansas. Proc. 23rd Annu. Conf. Southeast. Assoc. Game Fish Comm.: 244–250.

Olmsted, L. L. and R. V. Kilambi. 1971. Interrelationships

between environmental factors and feeding biology of White Bass of Beaver Reservoir, Arkansas. Pages 397–409. *In:* G. E. Hall, ed. Reservoir fisheries and limnology. Specieal Publ. 8. American Fisheries Society.

Olmsted, L. L. and R. V. Kilambi. 1978. Age and growth of Spotted Bass (*Micropterus punctulatus*) in Lake Fort Smith, Arkansas. Trans. Am. Fish. Soc. 107(1): 21–25.

Olmsted, L. L., G. D. Hickman, and D. G. Cloutman. 1972. A survey of the fishes of the Mulberry River, Arkansas. Arkansas Water Resources Research Center Publ. 10-B. 20 pp.

Olmsted, L. R., S. Krater, G. E. Williams, and R. G. Jaeger. 1979. Foraging tactics of the Mimic Shiner in a two-prey system. Copeia 1979(3): 437–441.

Olson, R. A., J. D. Winter, D. C. Nettles, and J. M. Haynes. 1988. Resource partitioning in summer by salmonids in south-central Lake Ontario. Trans. Am. Fish. Soc. 117(6): 552–559.

Olund, L. J. and F. B. Cross. 1961. Geographic variation in the North American cyprinid fish, *Hybopsis gracilis*. Mus. Nat. Hist. Univ. Kansas Publ. 13(7): 323–348.

O'Malley, H. 1919. The distribution of fish and fish eggs during the fiscal year 1918. Appendix I to the Report of the U.S. Commissioner of Fisheries for 1918, Bureau of Fisheries Document No. 863. Washington, DC. 82 pp.

Orrell, T. M. and L. Weigt. 2005. The Northern Snakehead *Channa argus* (Anabantomorpha: Channidae), a nonindigenous fish species in the Potomac River, U.S.A. Proc. Biol. Soc. Washington 118(2): 407–415.

Ortenburger, A. I. and C. L. Hubbs. 1927. A report on the fishes of Oklahoma, with descriptions of new genera and species. Proc. Oklahoma Acad. Sci. 6: 123–141.

Oshima, M. 1919. Contributions to the study of the fresh water fishes of the island of Formosa. Ann. Carnegie Mus. 12(2–4): 169–328.

Ostrand, K. G. and G. R. Wilde. 2001. Temperature, dissolved oxygen, and salinity tolerances of five prairie stream fishes and their role in explaining fish assemblage patterns. Trans. Am. Fish. Soc. 130(5): 742–749.

Ostrand, K. G., M. L. Thies, D. D. Hall, and M. Carpenter. 1996. Gar ichthyotoxin: Its effect on natural predators and the toxin's evolutionary function. Southwest. Nat. 41(4): 375–377.

Ostrand, K. G., G. R. Wilde, R. E. Strauss, and R. R. Young. 2001. Sexual dimorphism in Plains Minnow, *Hybognathus placitus*. Copeia 2001(2): 563–565.

Outten, L. M. 1958. Studies of the life history of the cyprinid fishes *Notropis galacturus* and *rubricroceus*. J. Elisha Mitchell Sci. Soc. 74(2): 122–134.

Outten, L. M. 1961. Observations on the spawning coloration and behavior of some cyprinid fishes. J. Elisha Mitchell Sci. Soc. 77: 18.

Owen, O. S. 1985. Natural resource conservation: an ecological approach. 4th ed. MacMillan, New York. 657 pp.

Owens, S. J., J. S. Odenkirk, and R. Greenlee. 2008. Northern Snakehead movement and distribution in the tidal Potomac River system. Proc. Annu. Conf. Southeast. Assoc. Fish Wildl. Agencies 62: 161–167.

Padhi, A. 2014. Geographic variation within a tandemly repeated mitochondrial DNA D-loop region of a North American freshwater fish, *Pylodictis olivaris*. Gene 538(1): 63–68.

Padilla, R. 1972. Reproduction of Carp, Smallmouth Buffalo, and River Carpsucker in Elephant Butte Lake. M.S. thesis. New Mexico State Univ., Las Cruces. 132 pp.

Page, L. M. 1974. The subgenera of *Percina* (Percidae: Etheostomatini). Copeia 1974(1): 66–86.

Page, L. M. 1975. Relations among the darters of the subgenus *Catonotus* of *Etheostoma*. Copeia 1975(4): 782–784.

Page, L. M. 1976a. The modified midventral scales of *Percina* (Osteichthyes; Percidae). J. Morphol. 148(2): 255–264.

Page, L. M. 1976b. Natural darter hybrids: *Etheostoma gracile* × *Percina maculata*, *Percina caprodes* × *Percina maculata*, and *Percina phoxocephala* × *Percina maculata*. Southwest. Nat. 21(2): 161–168.

Page, L. M. 1981. The genera and subgenera of darters (Percidae, Etheostomatini). Occas. Pap. Mus. Nat. Hist. Univ. Kansas. 90: 1–69.

Page, L. M. 1983a. Handbook of darters. TFH Publ., Neptune City, NJ. 271 pp.

Page, L. M. 1983b. Identification of the percids, *Boleosoma phlox* Cope and *Ioa vigil* Hay. Copeia 1983(4): 1082–1083.

Page, L. M. 1985. Evolution of reproductive behaviors in percid fishes. Illinois Nat. Hist. Surv. Bull. 33: 275–295.

Page, L. M. 2000. Etheostomatinae. Pages 225–253. *In:* J. F. Craig, ed. Percid fishes: systematics, ecology, and exploitation. Blackwell Scientific, Oxford. 368 pp.

Page, L. M. and H. L. Bart, Jr. 1989. Egg mimics in darters (Pisces: Percidae). Copeia 1989(2): 514–517.

Page, L. M. and B. M. Burr. 2011. Peterson field guide to freshwater fishes of North America north of Mexico. Second edition. Houghton Mifflin, Boston. 663 pp.

Page, L. M. and P. A. Ceas. 1989. Egg attachment in *Pimephales* (Pisces: Cyprinidae). Copeia 1989(4): 1074–1077.

Page, L. M. and L. E. Cordes. 1983. Variation and systematics of *Etheostoma euzonum*, the Arkansas Saddled Darter (Pisces: Percidae). Copeia 1983(4): 1042–1050.

Page, L. M. and C. E. Johnston. 1990a. The breeding behavior of *Opsopoeodus emiliae* (Cyprinidae) and its phylogenetic implications. Copeia 1990(4): 1176–1180.

Page, L. M. and C. E. Johnston. 1990b. Spawning in the Creek Chubsucker, *Erimyzon oblongus*, with a review of spawning behavior in suckers (Catostomidae). Environ. Biol. Fishes 27(4): 265–272.

Page, L. M. and T. P. Simon. 1988. Observations on the reproductive behavior and eggs of four species of darters, with comments on *Etheostoma tippecanoe* and *E. camurum*. Trans. Illinois State Acad. Sci. 81: 205–210.

Page, L. M. and P. W. Smith. 1970. The life history of the Dusky Darter, *Percina sciera*, in the Embarras River, Illinois. Illinois Nat. Hist. Surv. Biol. Notes 69. 15 pp.

Page, L. M. and P. W. Smith. 1971. The life history of the Slenderhead Darter, *Percina phoxocephala*, in the Embarras River, Illinois. Illinois Nat. Hist. Surv. Biol. Notes 74. 14 pp.

Page, L. M. and R. L. Smith. 1970. Recent range adjustments and hybridization of *Notropis lutrensis* and *Notropis spilopterus* in Illinois. Trans. Illinois State Acad. Sci. 63(3): 264–272.

Page, L. M., M. E. Retzer, and R. A. Stiles. 1982. Spawning behavior in seven species of darters (Pisces: Percidae). Brimleyana 8: 135–143.

Page, L. M., H. Espinosa-Perez, L. T. Findley, C. R. Gilbert, R. N. Lea, N. E. Mandrak, R. L. Mayden, and J. S. Nelson. 2013. Common and scientific names of fishes from the United States, Canada, and Mexico. 7th ed. Special Publ. 14. American Fisheries Society, Bethesda, MD. 254 pp.

Page, R. D. M. and E. C. Holmes. 1998. Molecular evolution: a phylogenetic approach. Blackwell Science, London. 346 pp.

Paige, K. N., C. R. Tumlison, and V. R. McDaniel. 1981. A second record of *Typhlichthys subterraneus* (Pisces: Amblyopsidae) from Arkansas. Southwest. Nat. 26(1): 67–92.

Paine, M. D. 1990. Life history tactics of darters (Percidae: Etheostomatiini) and their relationship with body size, reproductive behaviour, latitude and rarity. J. Fish Biol. 37(3): 473–488.

Pajak, P. and R. J. Neves. 1987. Habitat suitability and fish production: a model evaluation for Rock Bass in two Virginia streams. Trans. Am. Fish. Soc. 116(6): 839–850.

Paller, M. H., J. B. Gladden, and J. H. Heuer. 1992. Development of the fish community in a new South Carolina reservoir. Am. Midl. Nat. 128(1): 95–114.

Panek, F. M. 1987. Biology and ecology of Carp. Pages 1–5. *In*: E. L. Cooper, ed. Carp in North America. American Fisheries Society, Bethesda, MD.

Papoulias, D. M., D. C. Chapman, and D. E. Tillitt. 2006. Reproductive condition and occurrence of intersex in Bighead Carp and Silver Carp in the Missouri River. Hydrobiologia 571(1): 355–360.

Pardew, M. G., B. G. Cochran, and W. R. Posey II. 1993. Range extension of the Paleback Darter. Proc. Arkansas Acad. Sci. 47: 86–88.

Parenti, L. R. 1981. A phylogenetic and biogeographic analysis of cyprinodontiform fishes (Teleostei, Atherinomorpha). Bull. Am. Mus. Nat. Hist. 168(4): 335–557.

Parenti, L. R. 2006. *Typhlichthys eigenmanni* Charlton, 1933, an available name for a blind cavefish (Teleostei: Amblyopsidae), differentiated on the basis of characters of the central nervous system. Zootaxa 1374: 55–59.

Parenti, L. R. and M. Rauchenberger. 1989. Systematic overview of the poeciliines. Pages 3–12. *In:* G. K. Meffe and F. F. Snelson, Jr., eds. Ecology and evolution of livebearing fishes

(Poeciliidae). Prentice-Hall, Englewood Cliffs, New Jersey. 453 pp.

Parker, B. R. and W. G. Franzin. 1991. Reproductive biology of the Quillback, *Carpiodes cyprinus*, in a small prairie river. Can. J. Zool. 69(8): 2133–2139.

Parker, G. A. 1992. The evolution of sexual size dimorphism in fish. J. Fish Biol. 41(SB): 1–20.

Parker, H. L. 1964. Natural history of *Pimephales vigilax* (Cyprinidae). Southwest. Nat. 8(4): 228–235.

Parker, N. C. and B. A. Simco. 1975. Activity patterns, feeding and behavior of the Pirate Perch, *Aphredoderus sayanus*. Copeia 1975(3): 572–574.

Parks, T. P., M. C. Quist, and C. L. Pierce. 2014. Historical changes in fish assemblage structure in Midwestern non-wadeable rivers. Am. Midl. Nat. 171(1): 27–53.

Parrish, D. L. and F. J. Margraf. 1991. Prey selectivity by age-0 White Perch (*Morone americana*) and Yellow Perch (*Perca flavescens*) in laboratory experiments. Can. J. Fish. Aquat. Sci. 48(4): 607–610.

Parrish, J. D., D. C. Heins, and J. A. Baker. 1991. Reproductive season, clutch parameters and oocyte size of the Johnny Darter, *Etheostoma nigrum*, from southwestern Mississippi. Am. Midl. Nat. 125(2): 180–186.

Patterson, C. 1973. Interrelationships of holosteans. Pages 233–305. *In:* P. H. Greenwood, R. S. Miles, and C. Patterson, eds. Interrelationships of fishes. Academic Press, London. 536 pp.

Patterson, C. 1981. The development of the North American fish fauna—a problem of historical biogeography. Pages 265–681. *In:* P. L. Forey and P. H. Greenwood, eds. The evolving biosphere: chance, change, and challenge. Cambridge Univ. Press. 320 pp.

Patterson, C. 1982a. Morphology and interrelationships of primitive actinopterygian fishes. Am. Zool. 22(2): 241–259.

Patterson, C. 1982b. Cladistics and classification. New Sci. 94(1303): 303–306.

Patterson, C. and A. E. Longbottom. 1989. An Eocene amiid fish from Mali, West Africa. Copeia 1989(4): 827–836.

Patterson, C. and D. E. Rosen. 1977. Review of ichthyodectiform and other Mesozoic teleost fishes, and the theory and practice of classifying fossils. Bull. Am. Mus. Nat. Hist. 158: 81–172.

Patterson, C. and D. E. Rosen. 1989. The Paracanthopterygii revisited: order and disorder. Pages 5–36. *In:* D. M. Cohen, ed. Papers on the systematics of gadiform fishes. Nat. Hist. Mus. Los Angeles Cty. Sci. Ser. 32. 262 pp.

Patterson, G. 1985. Commercial fishermen: caught up in tradition. Arkansas Game and Fish 16(5): 10–12.

Patton, T. and C. Tackett. 2012. Status of Silver Carp (*Hypophthalmichthys molitrix*) and Bighead Carp (*Hypophthalmichthys nobilis*) in southern Oklahoma. Proc. Oklahoma Acad. Sci. 92: 53–58.

Patton, T. M. and M. L. Zornes. 1991. An analysis of stomach contents of the Ouachita Madtom (*Noturus lachneri*) in three streams of the upper Saline River drainage, Arkansas. Proc. Arkansas Acad. Sci. 45: 78–80.

Paukert, C. P. and J. M. Long. 1999. New maximum age of Bigmouth Buffalo, *Ictiobus cyprinellus*. Proc. Oklahoma Acad. Sci. 79: 85–86.

Payne, B. S. and A. C. Miller. 1996. Life history and production of filter-feeding insects on stone dikes in the lower Mississippi River. Hydrobiologia 319(2): 93–102.

Peckham, R. S. and C. F. Dineen. 1957. Ecology of the Central Mudminnow, *Umbra limi* (Kirtland). Am. Midl. Nat. 58(1): 222–231.

Peden, A. E. 1972. The function of gonopodial parts and behavioral pattern during copulation by *Gambusia* (Poeciliidae). Can. J. Zool. 50(7): 955–968.

Peden, A. E. 1973. Variation in anal spot expression of gambusiin females and its effect on male courtship. Copeia 1973(2): 250–263.

Peden, A. E. 1975. Differences in copulatory behavior as partial isolating mechanisms in the poeciliid fish *Gambusia*. Can. J. Zool. 53(9): 1290–1296.

Peek, F. W. 1965a. Growth studies of laboratory and wild population samples of Smallmouth Bass, *Micropterus dolomieui* Lacépède, with application to mass-marking of fishes. M.S. thesis, Univ. Arkansas, Fayetteville. 116 pp.

Peek, F. W. 1965b. Age and growth of the Smallmouth Bass, *Micropterus dolomieui* Lacépède in Arkansas. Proc. 19th Annu. Conf. Southeast. Assoc. Game Fish Comm.: 422–431.

Pegg, M. A., J. H. Chick, and B. M. Pracheil. 2009. Potential effects of invasive species on Paddlefish. Pages 185–201. *In*: C. P. Paukert and G. D. Scholten, eds. Paddlefish management, propagation, and conservation in the 21st century: building from 20 years of research and management. Symp. 66. American Fisheries Society, Bethesda, MD. 443 pp.

Pell, B. 1983. The natural divisions of Arkansas: a revised classification and description. Nat. Areas J. 3(2): 12–23.

Pender, D. R. 1998. Factors influencing Brown Trout reproductive success in Ozark tailwater rivers. M.S. thesis, Univ. Arkansas, Fayetteville. 146 pp.

Pender, D. R. and T. J. Kwak. 2002. Factors influencing Brown Trout reproductive success in Ozark tailwater rivers. Trans. Am. Fish. Soc. 131(4): 698–717.

Pennington, C. H., J. A. Baker, and M. E. Potter. 1983. Fish populations along natural and revetted banks on the lower Mississippi River. N. Am. J. Fish. Manag. 3(2): 204–211.

Pennington, C. H., G. J. Dahl, and H. L. Schramm, Jr. 1982. Occurrence of the Rainbow Smelt in the lower Mississippi River. J. Mississippi Acad. Sci. 27: 41–42.

Pennington, C. H., S. S. Knight, and M. P. Farrell. 1985. Response of fishes to revetment placement. Proc. Arkansas Acad. Sci. 39: 95–97.

Pennington, C. H., H. L. Schramm, Jr., M. E. Potter, and M. P. Farrell. 1980. Aquatic habitat studies on the lower Mississippi River, river mile 480 to 530. Rep. 5, Fish Studies—Pilot Report. Environmental and Water Quality Operational Studies. Misc. Paper E-80-1. U.S. Army Corps of Engineers, Vicksburg. 45 pp.

Pennock, C. A. and K. B. Gido. 2017. Density dependence of herbivorous Central Stoneroller *Campostoma anomalum* in stream mesocosms. Ecol. Freshw. Fish 26(2): 313–321.

Peoples, B. K. and E. A. Frimpong. 2013. Evidence of mutual benefits of nest association among freshwater cyprinids and implications for conservation. Aquat. Conserv. Mar. Freshw. Ecosyst. 23(6): 911–923.

Pendleton, R. M., J. J. Pritt, B. K. Peoples, and E. A. Frimpong. 2012. The strength of *Nocomis* nest association contributes to patterns of rarity and commonness among New River, Virginia cyprinids. Am. Midl. Nat. 168(1): 202–217.

Perkin, J. S. and K. B. Gido. 2011. Stream fragmentation thresholds for a reproductive guild of Great Plains fishes. Fisheries 36(8); 371–383.

Perkin, J. S., Z. R. Shattuck, and T. H. Bonner. 2012a. Life history aspects of a relict Ironcolor Shiner *Notropis chalybaeus* population in a novel spring environment. Am. Midl. Nat. 167(1): 111–126.

Perkin, J. S., Z. R. Shattuck, and T. H. Bonner. 2012b. Reproductive ecology of a relict population of Ironcolor Shiner (*Notropis chalybaeus*) in the headwaters of the San Marcos River, Texas. Am. Currents 37(2): 11–14, 18–21.

Perkin, J. S., C. S. Williams, and T. H. Bonner. 2009. Aspects of Chub Shiner *Notropis potteri* life history with comments on native distribution and conservation status. Am. Midl, Nat. 162(2): 276–288.

Perry, W. G. and D. C. Carver. 1972. Length at maturity and total length: collarbone length conversions for Channel Catfish, *Ictalurus punctatus*, and Blue Catfish, *Ictalurus furcatus*, collected from the marshes of southwest Louisiana. Proc. Annu. Conf. Southeast. Assoc. Game Fish Comm. 26: 541–553.

Peters, W. (C. H.) 1852. Diagnosen von neuen Flussfischen aus Mossambique. Monatsberichte der Königlichen Preussischen Akademie der Wissenschaften zu Berlin 1852: 275–276; 681–685.

Peters, W. (C. H.) 1881. Über die von der chinesischen Regierung zu der internationalen Fischerei-Austellung gesandte Fischsammlung aus Ningpo. Monatsberichte der Königlichen Preussischen Akademie der Wissenschaften zu Berlin 1880, v. 45: 921–927.

Petersen, J. C. 2004. Fish communities of the Buffalo River basin and nearby basins of Arkansas and their relation to selected environmental factors, 2001–2002. U.S. Dept. of the Interior, U.S. Geological Survey Sci. Invest. Rep. 2004-5119. 93 pp.

Petersen, J. C. and B. G. Justus. 2005. The fishes of Buffalo National River, Arkansas, 2001–2003. U.S. Dept. of the Interior, U. S. Geological Survey, Sci. Invest. Rep. 2005-5130. Little Rock, Arkansas. 41 pp.

Petersen, J. C., F. D. Usrey, W. E. Keith, and J. A. Wise. 1996. A recent record of the White Sucker, *Catostomus commersoni*, in the White River system, Arkansas. Proc. Arkansas Acad. Sci. 50: 141–142.

Petersen, J. C., B. G. Justus., and B. J. Meredith. 2014. Effects of land use, stream habitat, and water quality on biological communities of wadeable streams in the Illinois River basin of Arkansas, 2011 and 2012. U.S. Geological Survey Scientific Investigations Report 2014-5009, 89 pp.

Peterson, D. L., P. Vecsei, and C. A. Jennings. 2007. Ecology and biology of the Lake Sturgeon: a synthesis of current knowledge of a threatened North American Acipenseridae. Rev. Fish Biol. Fish. 17(1): 59–76.

Peterson, M. S. 1988. Comparative physiological ecology of centrarchids in hyposaline environments. Can. J. Fish. Aquat. Sci. 45(5): 827–833.

Peterson, M. S., W. T. Slack, N. J. Brown-Peterson, and J. L. McDonald. 2004. Reproduction in non-native environments: establishment of Nile Tilapia, *Oreochromis niloticus*, in coastal Mississippi watersheds. Copeia 2004(4): 842–849.

Petravicz, J. J. 1936. The breeding habits of the Least Darter, *Microperca punctulata* Putnam. Copeia 1936(2): 77–82.

Petravicz, W. P. 1938. The breeding habits of the Black-sided Darter, *Hadropterus maculatus* Girard. Copeia 1938(1): 40–44.

Pfeiffer, R. A. 1955. Studies on the life history of the Rosyface Shiner, *Notropis rubellus*. Copeia 1955(2): 95–104.

Pflieger, W. L. 1965. Reproductive behavior of the minnows, *Notropis spilopterus* and *Notropis whipplii*. Copeia 1965(1): 1–8.

Pflieger, W. L. 1966a. Reproduction of the Smallmouth Bass (*Micropterus dolomieui*) in a small Ozark stream. Am. Midl. Nat. 76(2): 410–418.

Pflieger, W. L. 1966b. Young of the Orangethroat Darter (*Etheostoma spectabile*) in nests of the Smallmouth Bass (*Micropterus dolomieui*). Copeia 1966(1): 139–140.

Pflieger, W. L. 1971. A distributional study of Missouri fishes. Univ. Kansas Mus. Nat. Hist., Publ. 20(3): 225–570.

Pflieger, W. L. 1975. The fishes of Missouri. Missouri Department of Conservation, Jefferson City. 343 pp.

Pflieger, W. L. 1977. Food habits of Rainbow Trout in Lake Taneycomo. Missouri Dept. of Conservation D-J Project F-1-R-25, Study I-12, Job Number 2. 21 pp.

Pflieger, W. L. 1978. Distribution and status of the Grass Carp (*Ctenopharyngodon idella*) in Missouri streams. Trans. Am. Fish. Soc. 107(1): 113–118.

Pflieger, W. L. 1997. The fishes of Missouri. 2nd ed. Missouri Dept. of Conservation, Jefferson City. 372 pp.

Pflieger, W. L. and T. M. Grace. 1987. Changes in the fish fauna of the lower Missouri River, 1940–1983. Pages 166–177. *In:* W. J. Matthews and D. C. Heins, eds. Community and evolutionary ecology of North American stream fishes. Univ. Oklahoma Press, Norman.

Phelps, Q. E., A. M. Lohmeyer, N. C. Wahl, J. M. Zeigler, and G. W. Whitledge. 2009. Habitat characteristics of Black Crappie nest sites in an Illinois impoundment. N. Am. J. Fish. Manag. 29(1): 189–195.

Phelps, Q. E., S. J. Tripp, W. D. Hintz, J. E. Garvey, D. P. Herzog, D. E. Ostendorf, J. W. Ridings, J. W. Crites, and R. A. Hrabik. 2010a. Water temperature and river stage influence mortality and abundance of naturally occurring Mississippi River *Scaphirhynchus* sturgeon. N. Am. J. Fish. Manag. 30(3): 767–775.

Phelps, Q. E., S. J. Tripp, J. E. Garvey, D. P. Herzog, D. E. Ostendorf, J. W. Ridings, J. W. Crites, and R. A. Hrabik. 2010b. Habitat use during early life history infers recovery needs for Shovelnose Sturgeon and Pallid Sturgeon in the middle Mississippi River. Trans. Am. Fish. Soc. 139(4): 1060–1068.

Phelps, Q. E., K. Baerwaldt, D. Chen, H. Du, L. Shen, C. Wang, Q. Wei, H. Zhang, and J. J. Hoover. 2016. Paddlefishes and sturgeons of the Yangtze and Mississippi rivers: status, biology, and management. Am. Fish. Soc. Symp. 84: 93–112.

Phelps, S. R. and F. W. Allendorf. 1983. Genetic identity of Pallid and Shovelnose sturgeon (*Scaphirhynchus albus* and *S. platorynchus*). Copeia 1983(3): 696–700.

Philipp, D. P. and M. S. Ridgway, eds. 2002. Black bass: ecology, conservation and management. Symp. 31. American Fisheries Society. 724 pp.

Phillips, C. T. and C. E. Johnston. 2008a. Sound production and associated behaviors in *Cyprinella galactura*. Environ. Biol. Fishes 82(3): 265–275.

Phillips, C. T. and C. E. Johnston. 2008b. Geographical divergence of acoustic signals in *Cyprinella galactura*, the Whitetail Shiner (Cyprinidae). Animal Behav. 75(2): 617–626.

Phillips, C. T. and C. E. Johnston. 2009. Evolution of acoustic signals in *Cyprinella*: degree of similarity in sister species. J. Fish Biol. 74(1): 120–132.

Phillips, E. C. and R. V. Kilambi. 1996. Food habits of four benthic fish species (*Etheostoma spectabile*, *Percina caprodes*, *Noturus exilis*, and *Cottus carolinae*) from northwest Arkansas streams. Southwest. Nat. 41(1): 69–73.

Phillips, G. L. 1969. Diet of minnow *Chrosomus erythrogaster* (Cyprinidae) in a Minnesota stream. Am. Midl. Nat. 82(1): 99–109.

Pianka, E. R. 2000. Evolutionary ecology. 6th ed. Benjamin Cummings, San Francisco. 512 pp.

Pietsch, T. W. and W. D. Anderson, Jr., eds. 1997. Collection building in ichthyology and herpetology. Am. Soc. Ichthyol. Herpetol. Special Publ. 3. 593 pp.

Pietsch, T. W. and D. B. Grobecker. 1987. Frogfishes of the world: systematics, zoogeography, and behavioral ecology. Stanford University Press, Stanford, CA. 420 pp.

Pigg, J. 1977. A survey of the fishes of the Muddy Boggy River in south central Oklahoma. Proc. Oklahoma Acad. Sci. 57: 68–87.

Pigg, J. 1991. Decreasing distribution and current status of the Arkansas River Shiner, *Notropis girardi*, in the rivers of Oklahoma and Kansas. Proc. Oklahoma Acad. Sci. 71: 5–15.

Pigg, J. 1998. Melanism in Longnose Gar, *Lepisosteus osseus* Linnaeus (Lepisosteidae). Proc. Oklahoma Acad. Sci. 78: 123.

Pigg, J. and R. Gibbs. 1995. Occurrence of catostomid fishes (suckers) in the North Canadian River and Lake Eufaula, Oklahoma. Proc. Oklahoma Acad. Sci. 75: 7–12.

Pigg, J. and T. Pham. 1990. The Rudd, *Scardinius erythrophthalamus*, a new fish in Oklahoma waters. Proc. Oklahoma Acad. Sci. 70: 37.

Pigg, J., R. Gibbs, and H. Weeks. 1991. Recent increases in number of Skipjack Herring, *Alosa chrysochloris* (Rafinesque), in the Arkansas River, Oklahoma. Proc. Oklahoma Acad. Sci. 71: 49–50.

Pigg, J., R. Gibbs, and T. Beard. 1996. Observations on two exotic fish species in Oklahoma. Proc. Oklahoma Acad. Sci. 76: 90.

Pigg, J., R. Gibbs, and K. K. Cunningham. 1999. Decreasing abundance of the Arkansas River Shiner in the South Canadian River, Oklahoma. Proc. Oklahoma Acad. Sci. 79: 7–12.

Pilastro, A., E. Giacomello, and A. Bisazza. 1997. Sexual selection for small size in male Mosquitofish (*Gambusia holbrooki*). Proc. Royal Soc. London B 264: 1125–1129.

Piller, K. R. and H. L. Bart, Jr. 2009. Incomplete sampling, outgroups, and phylogenetic inaccuracy: a case study of the Greenside Darter complex (Percidae: *Etheostoma blennioides*). Mol. Phylogenet. Evol. 53(1): 340–344.

Piller, K. R. and H. L. Bart, Jr. 2017. Rediagnosis of the Tuckasegee Darter, *Etheostoma gutselli* (Hildebrand), a Blue Ridge endemic. Copeia 2017(3): 569–574.

Piller, K. R., H. L. Bart, Jr., and D. L. Hurley. 2008. Phylogeography of the Greenside Darter complex, *Etheostoma blennioides* (Teleostomi: Percidae): a wide-ranging polytypic taxon. Mol. Phylogenet. Evol. 46(3): 974–985.

Piller, K. R., H. L. Bart, Jr., and J. A. Tipton. 2003. Spawning in the Black Buffalo, *Ictiobus niger* (Cypriniformes: Catostomidae). Ichthyol. Explor. Freshw. 14(2): 145–150.

Piller, K. R., H. L. Bart, Jr., and C. A. Walser. 2001. Morphological variation of the Redfin Darter, *Etheostoma whipplei*, with comments on the status of the subspecific populations. Copeia 2001(3): 802–807.

Pinion, A. K. and K. W. Conway. 2018. Tuberculation of *Macrhybopsis hyostoma* (Teleostei: Cyprinidae). Ichthyol. Explor. Freshw. 1095: 1–11.

Piteo, M. S., M. R. Kendrick, and P. M. Harris. 2017. Life history of *Labidesthes vanhyningi* (Atheriniformes: Atherinopsidae; Stout Silverside) in the Black Warrior River drainage, Alabama. Southeast. Nat. 16(3): 451–463.

Pittman, K. J. 2011. Population genetics, phylogeography, and morphology of *Notropis stramineus*. Ph.D. diss., Univ. Kansas, Lawrence. 143 pp.

Platania, S. P. 1990. Reports and verified occurrence of

logperches (*Percina caprodes* and *Percina macrolepida*) in Colorado. Southwest. Nat. 35(1): 87–88.

Platania, S. P. and C. S. Altenbach. 1998. Reproductive strategies and egg types of seven Rio Grande basin cyprinids. Copeia 1998(3): 559–569.

Platania, S. P. and H. W. Robison. 1980. *Etheostoma collettei* Birdsong and Knapp, Creole Darter. Page 636. *In*: D. S. Lee et al. Atlas of North American freshwater fishes. N.C. State Mus. Nat. Hist., Raleigh.

Plumb, J. M. 2006. Climate-mediated changes in habitat use by Lake Trout (*Salvelinus namaycush*). M.S. thesis, Univ. Manitoba, Winnipeg. 207 pp.

Poey, F. 1851–54. Memorias sobre la historia natural de la Isla de Cuba, acompañadas de sumarios Latinos y extractos en Francés. La Habana. v. 1: 1–463, Pls. 1–34.

Polivka, K. M. 1999. The microhabitat distribution of the Arkansas River Shiner, *Notropis girardi*: a habitat-mosaic approach. Environ. Biol. Fishes 55(3): 265–278.

Polivka, K. M. and W. J. Matthews. 1997. Habitat requirements of the Arkansas River Shiner, *Notropis girardi*: August 1, 1994–August 7, 1997. Final Report, Federal Aid Project No. E-33. Oklahoma Dept. Wildl. Conserv., Oklahoma City. 13 pp.

Poly, W. J. and J. E. Wetzel. 2003. Transbranchioral spawning: novel reproductive strategy observed for the Pirate Perch *Aphredoderus sayanus* (Aphredoderidae). Ichthyol. Explor. Freshw. 14(2): 151–158.

Ponder, M. D. 1983. A taxonomic survey of the fishes of Terre Noir Creek in southcentral Arkansas. M.S. thesis, Northeast Louisiana Univ., Monroe. 102 pp.

Popma, T. and M. Masser. 1999. Tilapia: life history and biology. Southern Regional Aquaculture Center Publ. SRAC-283. 4 pp.

Popper, A. N. and R. R. Fay. 2010. Rethinking sound detection by fishes. Hearing Res. 273(1–2): 25–36.

Port, P., J. Williams, T. Bly, K. Kitterman, and M Schroeder. 2010. Greers Ferry Tailwater Fish Population Sampling Report 2010. Trout Management Program Report TP-11-03, Arkansas Game and Fish Commission, Little Rock.

Porta, M. J. and R. A. Snow. 2017. Diet of invasive White Perch in Sooner Lake, Oklahoma. Proc. Oklahoma Acad. Sci. 97: 67–74.

Porter, B. A. 1999. Phylogeny, evolution, and biogeography of the darter subgenus *Ulocentra* (genus *Etheostoma*, family Percidae). Ph.D. diss., Ohio State Univ., Columbus. 176 pp.

Porter, B. A., T. M. Cavender, and P. A. Fuerst. 2002. Molecular phylogeny of the snubnose darters, subgenus *Ulocentra* (genus *Etheostoma*, family Percidae). Mol. Phylogenet. Evol. 22(3): 364–374.

Porter, H. T. and P. J. Motta. 2004. A comparison of strike and prey capture kinematics of three species of piscivorous fishes: Florida Gar (*Lepisosteus platyrhincus*), Redfin Needlefish (*Strongylura notata*), and Great Barracuda (*Sphyraena barracuda*). Mar. Biol. 145(5): 989–1000.

Porterfield, J. C. 1998. Phylogenetic systematics of snubnose daters (Percidae: *Etheostoma*), with discussion of reproductive behavior, sexual selection, and the evolution of male breeding color. Ph.D. diss., Univ. Illinois, Urbana. 154 pp.

Porterfield, J. C., L. M. Page, and T. J. Near. 1999. Phylogenetic relationships among Fantail Darters (Percidae: *Etheostoma*: *Catonotus*): total evidence analysis of morphological and molecular data. Copeia. 1999(3): 551–564.

Posey, B. 2006. A report of the commercial roe harvest in Arkansas, November 2005–May 2006. Arkansas Game and Fish Commission Report, Little Rock. 17 pp.

Potter, G. E. 1923. Food of the Short-nosed Gar-pike (*Lepidosteus platystomus*) in Lake Okoboji, Iowa. Proc. Iowa Acad. Sci. 30: 167–170.

Potter, G. E. 1926. Ecological studies of the Short-nosed Gar pike (*Lepidosteus platystomus*). Univ. Iowa Stud. Nat. Hist. 11(9): 17–27.

Potter, I. C. 1980a. Ecology of larval and metamorphosing lampreys. Can. J. Fish. Aquat. Sci. 37(11): 1641–1657.

Potter, I. C. 1980b. The Petromyzoniformes with particular reference to paired species. Can. J. Fish. Aquat. Sci. 37(11): 1595–1615.

Potter, I. C. and H. S. Gill. 2003. Adaptive radiation of lampreys. J. Great Lakes Res. 29(S1): 95–112.

Pough, F. H., C. M. Janis, and J. B. Heiser. 2013. Vertebrate life. 9th ed. Prentice Hall, Upper Saddle River, NJ. 634 pp.

Poulson, T. L. 1958. Cavefishes (Amblyopsidae). Bloomington Indiana Grotto Newsletter 1: 29–37.

Poulson, T. L. 1960. Cave adaptation in amblyopsid fishes. Ph.D. diss. Univ. Michigan, Ann Arbor. 185 pp.

Poulson, T. L. 1963. Cave adaptation in amblyopsid fishes. Am. Midl. Nat. 70(2): 257–290.

Poulson, T. L. 1964. Animals in aquatic environments: animals in caves. Pages 749–771. *In:* D. B. Dill, E. F. Adolph, and C. G. Wilbur, eds. Handbook of physiology: a critical, comprehensive presentation of physiological knowledge and concepts, Section 4: Adaptation to the environment. American Physiological Society, Washington, DC.

Poulson, T. L. 1985. Evolutionary reduction by neutral mutations: plausibility arguments and data from amblyopsid fishes and linyphiid spiders. Nat. Speleol. Soc. Bull. 47(2): 109–117.

Powell, J. A. 1984. Observations of cleaning behavior in the Bluegill (*Lepomis macrochirus*), a centrarchid. Copeia 1984(4): 996–998.

Powell, M. 1991. A rare golden gar. Alabama Conserv. 20: 7.

Power, M. E. and W. J. Matthews. 1983. Algae-grazing minnows (*Campostoma anomalum*), piscivorous bass (*Micropterus* spp.), and the distribution of attached algae in a small prarie-margin stream. Oecologia 60(3): 328–332.

Power, M. E., W. J. Matthews, and A. J. Stewart. 1985. Grazing minnows, piscivorous bass and stream algae: dynamics of a strong interaction. Ecology 66(5): 1448–1456.

Power, M. E., A. J. Stewart, and W. J. Matthews. 1988a. Grazer

control of algae in an Ozark mountain stream: effects of short-term exclusion. Ecology 69(6): 1894–1898.

Power, M. E., R. J. Stout, C. E. Cushing, P. P. Harper, F. R. Hauer, W. J. Matthews, P. B. Moyle, B. Statzner, and I. R. Wais De Badgen. 1988b. Biotic and abiotic controls in river and stream communities. J. N. Am. Benthol. Soc. 7(4): 456–479.

Power, M. E., D. Tilman, J. A. Estes, B. A. Menge, W. J. Bond, L. S. Mills, G. Daily, J. C. Castilla, J. Lubchenco, and R. T. Paine. 1996. Challenges in the quest for keystones. BioScience 46(8): 609–620.

Powers, P. K. and J. R. Gold. 1992. Cytogenetic studies in North American minnows (Cyprinidae). XX. Chromosomal NOR variation in the genus *Luxilus*. Copeia 1992(2): 332–342.

Powles, P. M. and I. M. Sandeman. 2008. Growth, summer cohort output, and observations on the reproduction of Brook Silverside, *Labidesthes sicculus* (Cope) in the Kawartha Lakes, Ontario. Environ. Biol. Fishes 82(4): 421–431.

Pramuk, J. B., M. J. Grose, A. L. Clarke, E. Greenbaum, E. Bonaccorso, J. M. Guayasamin, A. H. Smith-Pardo, B. W. Benz, B. R. Harris, E. Siegfried, Y. R. Reid, N. Holcroft-Benson, and E. O. Wiley. 2007. Phylogeny of finescale shiners of the genus *Lythrurus* (Cypriniformes: Cyprinidae) inferred from four mitochondrial genes. Mol. Phylogenet. Evol. 42(2): 287–297.

Pratt, A. E. and T. E. Lauer. 2013. Habitat use and separation among congeneric darter species. Trans. Am. Fish. Soc. 142(2): 568–577.

Pratt, A. E., C. W. Hargrave, and K. B. Gido. 2002. Rediscovery of *Labidesthes sicculus* (Atherinidae) in Lake Texoma (Oklahoma-Texas). Southwest. Nat. 47(1): 142–147.

Pratt, K. E. 2000. Life history traits of the Rocky Shiner, *Notropis suttkusi*. M.S. thesis, Univ. Oklahoma, Norman. 50 pp.

Priegel, G. R. 1966. Early scale development in the Freshwater Drum, *Aplodinotus grunniens* Rafinesque. Trans. Am. Fish. Soc. 95(4): 434–436.

Priegel, G. R. 1967a. Food of the Freshwater Drum, *Aplodinotus grunniens*, in Lake Winnebago, Wisconsin. Trans. Am. Fish. Soc. 96(2): 218–220.

Priegel, G. R. 1967b. The Freshwater Drum: its life history, ecology and management. Wisconsin Dept. Nat. Resour. Publ. 236. 15 pp.

Priegel, G. R. 1969. The Lake Winnebago Sauger: age, growth, reproduction, food habits, and early life history. Wisconsin Dept. Nat. Resour. Tech. Bull. 43. 63 pp.

Priegel, G. R. 1970. Food of the White Bass, *Roccus chrysops*, in Lake Winnebago, Wisconsin. Trans. Am. Fish. Soc. 99(2): 440–443.

Priegel, G. R. and T. L. Wirth. 1971. The Lake Sturgeon: its life history, ecology, and management. Publ. 240-7. Wisconsin Dept. Nat. Resour., Madison. 20 pp.

Probst, W. E., C. F. Rabeni, W. G. Covington, and R. E.

Marteney. 1984. Resource use by stream-dwelling Rock Bass and Smallmouth Bass. Trans. Am. Fish. Soc. 113(3): 283–294.

Prosek, J. 2010. Eels: an exploration from New Zealand to the Sargasso, of the world's most mysterious fish. Harper Collins, New York. 287 pp.

Proudlove, G. S. 2006. Subterranean fishes of the world: an account of the subterranean (hypogean) fishes described up to 2003 with a bibliography 1541–2004. International Society for Subterranean Biology, Moulis, France. 300 pp.

Prout, M. W., E. L. Mills, and J. L. Forney. 1990. Diet, growth, and potential competitive interactions between age-0 White Perch and Yellow Perch in Oneida Lake, New York. Trans. Am. Fish. Soc. 119(6): 966–975.

Purdom, C. E., P. A. Hardiman, V. J. Bye, N. C. Eno, C. R. Tyler, and J. P. Sumpter. 1994. Estrogenic effects of effluents from sewage treatment works. J. Chem. Ecol. 8(4): 275–285.

Purkett, C. A., Jr. 1961. Reproduction and early development of the Paddlefish. Trans. Am. Fish. Soc. 90(2): 125–129.

Putnam, F. W. 1863. List of the fishes sent by the museum to different institutions, in exchange for other specimens, with annotations. Bull. Mus. Comp. Zool. 1: 2–16.

Pyke, G. H. 2005. A review of the biology of Gambusia affinis and G. holbrooki. Rev. Fish Biol. Fish. 15(4): 339–365.

Pyron, M. 1995. Mating patterns and a test for female mate choice in Etheostoma spectabile (Pisces, Percidae). Behav. Ecol. Sociobiol. 36(6): 407–412.

Pyron, M. 1996a. Male Orangethroat Darters, Etheostoma spectabile, do not prefer larger females. Environ. Biol. Fishes 47(4): 407–410.

Pyron, M. 1996b. Sexual size dimorphism and phylogeny in North American minnows. Biol. J. Linn. Soc. 57(4): 327–341.

Quinn, J. W. 2009. Harvest of Paddlefish in North America. Pages 203–221. In: C. P. Paukert and G. D. Scholten, eds. Paddlefish management, propagation, and conservation in the 21st century: building from 20 years of research and management. Symp. 66. American Fisheries Society, Bethesda, MD. 443 pp.

Quinn, J. W. and T. J. Kwak. 2000. Use of rehabilitated habitat by Brown Trout and Rainbow Trout in an Ozark tailwater river. N. Am. J. Fish. Manag. 20(3): 737–751.

Quinn, J. W. and T. J. Kwak. 2003. Fish assemblage changes in an Ozark river after impoundment: a long-term perspective. Trans. Am. Fish. Soc. 132(1): 110–119.

Quinn, J. W. and T. J. Kwak. 2011. Movement and survival of Brown Trout and Rainbow Trout in an Ozark tailwater river. N. Am. J. Fish. Mngt. 31(2): 299–304.

Quinn, J. W. and R. L. Limbird. 2008. Trends over time for the Arkansas River fishery: a case history. Pages 169–191. In: M. S. Allen, S. Sammons, and M. J. Maceina, eds. Balancing fisheries management and water uses for impounded river systems. Symp. 62. American Fisheries Society. 697 pp.

Quinn, J. W., W. R. Posey II, F. J. Leone, and R. L. Limbird. 2009. Management of the Arkansas River commercial Paddlefish fishery with check stations and special seasons. Pages 261–275. In: C. P. Paukert and G. D. Scholten, eds. Paddlefish management, propagation, and conservation in the 21st century: building from 20 years of research and management. Symp. 66. American Fisheries Society, Bethesda, MD. 443 pp.

Quinn, J. W., S. Filipek, B. Wagner, S. O'Neal, C. Perrin, M. Schroeder, M. Horton, and T. Bly. 2012. South Fork of the Little Red River Gulf Mountain WMA monitoring sample report 2009–2011. Arkansas Game and Fish Commission Stream Sample Report STP2012-01. 9 pp.

Quist, M. C., J. S. Tillma, M. N. Burlingame, and C. S. Guy. 1999. Overwinter habitat use of Shovelnose Sturgeon in the Kansas River. Trans. Am. Fish. Soc. 128(3): 522–527.

Rabeni, C. F. 1992. Trophic linkage between stream centrarchids and their crayfish prey. Can. J. Fish. Aquat. Sci. 49(8): 1714–1721.

Rabito, F. G., Jr. and D. C. Heins. 1985. Spawning behaviour and sexual dimorphism in the North American cyprinid fish Notropis leedsi, the Bannerfin Shiner. J. Nat. Hist. 19(6): 1155–1163.

Racey, C. L. and S. E. Lochmann. 2002. Year-class contribution of White Crappies stocked into Lake Chicot, Arkansas. N. Am. J. Fish. Manag. 22(4): 1409–1415.

Rafinesque, C. S. 1817a. Additions to the observations on the sturgeons of North America. Am. Monthly Mag. Crit. Rev. 1: 288.

Rafinesque, C. S. 1817b. First decade of new North American fishes. Am. Monthly Mag. Crit. Rev. 2(2): 120–121.

Rafinesque, C. S. 1818a. Description of two new genera of North American fishes, Opsanus and Notropis. Am. Monthly Mag. Crit. Rev. 2(3): 203–204.

Rafinesque, C. S. 1818b. Discoveries in natural history, made during a journey through the western region of the United States. Am. Monthly Mag. Crit. Rev. 3(5): 354–356.

Rafinesque, C. S. 1818c. Description of three new genera of fluviatile fish, Pomoxis, Sarchirus and Exoglossum. J. Acad. Nat. Sci. Phila. 1(part 2): 417–422, Pl. 17. [possibly published in 1819].

Rafinesque, C. S. 1818d. Further account of discoveries in natural history, in the western states. Am. Monthly Mag. Crit. Rev. 4 (no. 1) (art. 5): 39–42.

Rafinesque, C. S. 1819. Prodrome de 70 nouveaux genres d'animaux découverts dans l'intérieur des États-Unis d'Amérique, durant l'année 1818. Journal de Physique, de Chimie et d'Histoire Naturelle 88: 417–429. [Fishes on pp. 418–422.]

Rafinesque, C. S. 1820a. Fishes of the Ohio River. [Ichthyologia Ohiensis, Part 8]. Western Revue and Miscellaneous Magazine: a monthly publ., devoted to literature and science, Lexington, Kentucky 3 (3): 165–173.

Rafinesque, C. S. 1820b. Description of the silures or catfishes of the River Ohio. Q. J. Sci. Lit. Arts 9: 48–52.

Rahel, F. J. 1989. Nest defense and aggressive interactions

between a small benthic fish (the Johnny Darter *Etheostoma nigrum*) and crayfish. Environ. Biol. Fishes 24(4): 301–306.

Rahel, F. J. 2000. Homogenization of fish faunas across the United States. Science 288(5467): 854–856.

Rahel, F. J. 2002. Homogenization of freshwater faunas. Annu. Rev. Ecol. Syst. 33: 291–315.

Rahel, F. J. and N. P. Nibbelink. 1999. Spatial patterns in relations among Brown Trout (*Salmo trutta*) distribution, summer air temperature, and stream size in Rocky Mountain streams. Can. J. Fish. Aquat. Sci. 56(S1): 43–51.

Rahel, F. J. and L. A. Thel. 2004a. Flathead Chub (*Platygobio gracilis*): a technical conservation assessment. USDA Forest Service, Rocky Mtn. Region. 53 pp. Available: http://www.fs.fed.us/r2/projects/scp/assessments/flatheadchub.pdf [Accessed 1 May 2018].

Rahel, F. J. and L. A. Thel. 2004b. Sturgeon Chub (*Macrhybopsis gelida*): a technical conservation assessment. USDA Forest Service, Rocky Mtn. Region. 48 pp. Available: http://www.fs.fed.us?r2/projects/scp/assessment/sturgeon chub.pdf. [Accessed 1 May 2018]

Rahman, M. M., M. Y. Hossain, Q. Jo, S. Kim, J. Ohtomi, and C. Meyer. 2009. Ontogenetic shift in dietary preference and low dietary overlap in Rohu (*Labeo rohita*) and Common Carp (*Cyprinus carpio*) in semi-intensive polyculture ponds. Ichthyol. Res. 56(1): 28–36.

Rainboth, W. J., Jr. and G. S. Whitt. 1974. Analysis of evolutionary relationships among shiners of the subgenus *Luxilus* (Teleostei, Cypriniformes, *Notropis*) with the lactate dehydrogenase and malate dehydrogenase isozyme systems. Comp. Biochem. Physiol. 49(2): 241–252.

Rainwater, W. C. and A. Houser. 1982. Species composition and biomass of fish in selected coves in Beaver Lake, Arkansas, during the first 18 years of impoundment (1963–1980). N. Am. J. Fish. Manag. 2(4): 316–325.

Rakes, P. L., J. R. Shute, and P. W. Shute. 1999. Reproductive behavior, captive breeding, and restoration ecology of endangered fishes. Environ. Biol. Fishes 55(1–2): 31–42.

Rakocinski, C. F. 1977. Evolutionary interactions of two sympatric related species of minnows in a creek in northwestern Illinois: hybridization and isolating mechanisms between *Campostoma oligolepis* and *Campostoma anomalum pullum*. M.S. thesis. Northern Illinois Univ., DeKalb. 148 pp.

Rakocinski, C. F. 1980. Hybridization and introgression between *Campostoma oligolepis* and *C. anomalum pullum* (Cypriniformes: Cyprinidae). Copeia 1980(4): 584–594.

Rakocinski, C. F. 1984. Aspects of reproductive isolation between *Campostoma oligolepis* and *Campostoma anomalum pullum* (Cypriniformes, Cyprinidae) in northern Illinois. Am. Midl. Nat. 112(1): 138–145.

Raleigh, R. F. 1982. Habitat suitability index models: Brook Trout. U.S. Fish and Wildlife Service Biological Report 82 (10.24), Fort Collins, CO. 42 pp.

Raley, M. E. and R. M. Wood. 2001. Molecular

systematics of members of the *Notropis dorsalis* species group (Actinopterygii: Cyprinidae). Copeia 2001(3): 638–645.

Raley, M. E., W. C. Starnes, and G. M. Hogue. 2004. Genetics of *Lythrurus umbratilis* in the upper Ouachita River system, Arkansas, with emphasis on a novel form [abstract]. 84th Annu. Meet. Am. Soc. Ichthyol. Herpetol., Norman, OK.

Raley, M. E., W. C. Starnes, G. M. Hogue, and H. W. Robison. 2005. Evidence of extensive mitochondrial introgression in the Redfin Shiner, *Lythrurus umbratilis* [abstract]. 85th Annu. Meet. Am. Soc. Ichthyol. Herpetol., Tampa, FL.

Rambo, R. D. 1998. Ozark stream fish assemblages and black bass population dynamics associated with watersheds of varying land use. M.S. thesis. Univ. Arkansas, Fayetteville. 98 pp.

Ramirez, R., E. R. Johnson, and K. B. Gido. 2006. Effects of artificial lighting and presence of *Menidia beryllina* on growth and diet of *Labidesthes sicculus*. Southwest. Nat. 51(4): 510–513.

Raney, E. C. 1939. The breeding habits of the Silvery Minnow, *Hybognathus regius* Girard. Am. Midl. Nat. 21(3): 674–680.

Raney, E. C. 1940. The breeding behavior of the Common Shiner, *Notropis cornutus* (Mitchill). Zoologica 25(1): 1–14.

Raney, E. C. 1942. Alligator Gar feeds upon birds in Texas. Copeia 1941(1): 50.

Raney, E. C. 1952. The life history of the Striped Bass, *Roccus saxatilis* (Walbaum). Bull. Bingham Oceanogr. Collect. 14: 5–97.

Raney, E. C. and E. A. Lachner. 1946. Age, growth, and habits of the Hog Sucker, *Hypentelium nigricans* (Lesueur) in New York. Am. Midl. Nat. 36(1): 76–86.

Raney, E. C. and R. D. Suttkus. 1964. *Etheostoma moorei*, a new darter of the subgenus *Nothonotus* from the White River system, Arkansas. Copeia 1964(1): 130–139.

Raney, E. C. and D. A. Webster. 1940. The food and growth of the young of the Common Bullhead, *Ameiurus nebulosus nebulosus* (Lesueur), in Cayuga Lake, New York. Trans. Am. Fish. Soc. 69(1): 205–209.

Raney, E. C. and D. A. Webster. 1942. The spring migration of the common White Sucker, *Catostomus c. commersonii* (Lacépède), in Skaneateles Lake Inlet, New York. Copeia 1942(3): 139–148.

Ranvestel, A. W. and B. M. Burr. 2004. Conservation assessment for Bluehead Shiner (*Pteronotropis hubbsi*). Am. Currents 30(1): 17–25.

Rasmussen, J. L. 1998. Aquatic nuisance species of the Mississippi River basin. 60th Midwest Fish and Wildlife Conference, Aquatic Nuisance Species Symposium, Dec. 7, 1998, Cincinnati, OH.

Rasmussen, R. P. 1980. Egg and larva development of Brook Silversides from the Peace River, Florida. Trans. Am. Fish. Soc. 109(4): 407–416.

Rauchenberger, M. 1989. Systematics and biogeography of the genus *Gambusia* (Cyprinodontiformes; Poeciliidae). Am. Mus. Novit. 2951: 1–74.

Ray, J. M., R. M. Wood, and A. M. Simons. 2006. Phylogeography and post-glacial colonization patterns of the Rainbow Darter, *Etheostoma caeruleum* (Teleostei: Percidae). J. Biogeogr. 33(9): 1550–1558.

Ray, J. M., N. J. Lang, R. M. Wood, and R. L. Mayden. 2008. History repeated: recent and historical mitochondrial introgression between the Current Darter, *Etheostoma uniporum* and Rainbow Darter, *Etheostoma caeruleum* (Teleostei: Percidae). J. Fish Biol. 72(2): 418–434.

Ray, J. M., C. B. Dillman, R. M. Wood, B. R. Kuhajda, and R. L. Mayden. 2007. Microsatellite variation among river sturgeons of the genus *Scaphirhynchus* (Actinopterygii: Acipenseridae): a preliminary assessment of hybridization. J. Appl. Ichthyol. 23(4): 304–312.

Ray, J. W., M. Husemann, R. S. King, and P. D. Danley. 2012. Genetic analysis reveals dispersal of Florida bass haplotypes from reservoirs to rivers in central Texas. Trans. Am. Fish. Soc. 141(5): 1269–1273.

Raymond, L. R. 1975. Fishes of the Hill Province section of the Ouachita River, from Remmel Dam to the Arkansas-Louisiana line. M.S. thesis. Northeast Louisiana Univ., Monroe. 38 pp.

Redmond, L. C. 1964. Ecology of the Spotted Gar (*Lepisosteus oculatus* Winchell) in southeastern Missouri. M.A. thesis, Univ. Missouri, Columbia. 144 pp.

Reece, J. B., M. R. Taylor, E. J. Simon, J. L. Dickey, and K. A. Hogan. 2015. Campbell biology: concepts and connections. 8th ed. Pearson Benjamin Cummings, London. 928 pp.

Reed, J. R. and D. L. Pereira. 2009. Relationships between shoreline development and nest site selection by Black Crappie and Largemouth Bass. N. Am. J. Fish. Manag. 29(4): 943–948.

Reed, R. J. 1957. Phases of the life history of the Rosyface Shiner, *Notropis rubellus*, in northwestern Pennsylvania. Copeia 1957(4): 286–290.

Regan, C. T. 1926. Organic evolution. Rep. Br. Assoc. Adv. Sci. 1925: 75–86.

Reichenbacher, V. B. and M. Weidmann. 1992. Fisch-otolithen aus der oligo-miozaenen molasses der west-Schweiz und der Haute-savoie (Frankreich). Stuttgarter Beitrage zur Naturkunde 184: 1–83.

Reid, S. M., L. M. Carl, and J. Lean. 2005. Influence of riffle characteristics, surficial geology, and natural barriers on the distribution of the Channel Darter, *Percina copelandi*, in the Lake Ontario basin. Environ. Biol. Fishes 72(3): 241–249.

Reigh, R. C. and D. S. Elsen. 1979. Status of the Sturgeon Chub (*Hybopsis gelida*) and Sicklefin Chub (*Hybopsis meeki*) in North Dakota. Prairie Nat. 11: 49–52.

Reighard, J. 1903. The natural history of *Amia calva* Linnaeus. Pages 57–109. *In:* G. H. Parker, ed. Mark anniversary volume. Henry Holt & Co., New York.

Reighard, J. 1910. Methods of studying the habits of fishes, with an account of the breeding habits of the Horned Dace. Bull. U.S. Bur. Fish. 28(2): 1111–1136.

Reighard, J. and J. Phelps. 1908. The development of the adhesive organ and head mesoblast of *Amia*. J. Morphol. 19(2): 469–496.

Reiser, D. W. and T. A. Wesche. 1977. Determination of physical and hydraulic preferences of Brown and Brook trout in the selection of spawning locations. Res. Ser. 64. Water Resource Research Institute, Univ. Wyoming. 112 pp.

Renaud, C. B. 1997. Conservation status of Northern Hemisphere lampreys (Petromyzontidae). J. Appl. Ichthyol. 13(3): 143–148.

Renaud, C. B. 2011. Lampreys of the world. An annotated and illustrated catalogue of lamprey species known to date. FAO Species Catalogue for Fishery Purposes No. 5. FAO, Rome. 109 pp.

Renfro, J. L. and L. G. Hill. 1970. Factors influencing the aerial breathing and metabolism of gars (*Lepisosteus*). Southwest. Nat. 15(1): 45–54.

Reno, H. W. 1966. The infraorbital canal, its lateral-line ossicles and neuromasts, in the minnows *Notropis volucellus* and *N. buchanani*. Copeia 1966(3): 403–413.

Reno, H. W. 1969. Cephalic lateral-line systems of the cyprinid genus *Hybopsis*. Copeia 1969(4): 736–773.

Resetarits, W. J., Jr. and C. A. Binckley. 2013. Is the Pirate really a ghost? Evidence for generalized chemical camouflage in an aquatic predator, Pirate Perch *Aphredoderus sayanus*. Am. Nat. 181(5): 690–699.

Resh, V. H., R. D. Hoyt, and S. E. Neff. 1971. The status of the Common Shiner, *Notropis cornutus chrysocephalus* (Rafinesque), in Kentucky. Proc. 25th Annu. Conf. Southeast. Assoc. Game Fish Comm.: 550–556.

Retzer, M. E. and C. R. Kowalik. 2002. Recent changes in the distribution of River Redhorse (*Moxostoma carinatum*) and Greater Redhorse (*Moxostoma valenciennesi*) (Cypriniformes: Catostomidae) in Illinois and comments on their natural history. Trans. Illinois State Acad. Sci. 95(4): 327–333.

Retzer, M. E., L. M. Page, and D. L. Swofford. 1986. Variation and systematics of *Etheostoma whipplei*, the Redfin Darter (Pisces: Percidae). Copeia 1986(3): 631–641.

Reynolds, J. T. 1971. The fishes of the Saline River, south central Arkansas. M.S. thesis, Northeast Louisiana Univ., Monroe. 36 pp.

Reynolds, W. W. and M. E. Casterlin. 1977. Diel activity in the Yellow Bullhead. Prog. Fish-Cult. 39(3): 132–133.

Reynolds, W. W., M. E. Casterlin, and S. T. Millington. 1978. Circadian rhythm of preferred temperature in the Bowfin, *Amia calva*, a primitive holostean fish. Comp. Biochem. Physiol. Part A 60(1): 107–109.

Rice, L. A. 1942. The food of seventeen Reelfoot Lake fishes in 1941. J. Tennessee Acad. Sci. 17(1): 4–13.

Richardson, D. J., R. E. Tumlison, W. E. Moser, C. T. McAllister, S. E. Trauth, and H. W. Robison. 2013. New host records for the fish leech *Cystobranchus klemmi* (Hirudinida: Piscicolidae) on cyprinid fishes from Arkansas and Oklahoma. J. Arkansas Acad. Sci. 67: 211–213.

Richardson, J. 1836. The Fish. *In:* Fauna Boreali-Americana; or the zoology of the northern parts of British America: containing descriptions of the objects of natural history collected on the late northern land expeditions, under the command of Sir John Franklin, R.N. J. Bentley, London. Part 3. 327 pp.

Richardson, J. 1845. Ichthyology. Pages 51–150. *In:* R. B. Hinds, ed. The zoology of the voyage of H.M.S. "Sulphur", under the command of Captain Sir Edward Belcher, during the years 1836–1842. Smith, Elder & Co., London.

Richardson, J. 1846. Report on the ichthyology of the seas of China and Japan. Rep. 15th Meeting [1845] of Brit. Assoc. Adv. Sci.: 187–320. Held at Cambridge in June 1845. John Murray, Albemarle Street, London.

Ridenour, C. J., W. J. Doyle, and T. D. Hill. 2011. Habitats of age-0 sturgeon in the lower Missouri River. Trans. Am. Fish. Soc. 140(5): 1351–1358.

Ridenour, C. J., A. B. Starostka, W. J. Doyle, and T. D. Hill. 2009. Habitat used by *Macrhybopsis* chubs associated with channel modifying structures in a large regulated river: implications for river modification. River Res. Appl. 25(4): 472–485.

Ridgway, M. S., G. P. Goff, and M. H. A. Keenleyside. 1989. Courtship and spawning behavior in Smallmouth Bass (*Micropterus dolomieui*). Am. Midl. Nat. 122(2): 209–213.

Riggs, C. D. and E. W. Bonn. 1959. An annotated list of the fishes of Lake Texoma, Oklahoma and Texas. Southwest. Nat. 4(4): 157–168.

Riggs, C. D. and V. E. Dowell. 1956. Some recent changes in the fish fauna of Lake Texoma. Proc. Oklahoma Acad. Sci. 35: 37–39.

Riggs, C. D. and G. A. Moore. 1963. A new record of *Moxostoma macrolepidotum pisolabrum*, and a range extension for *Percina shumardi*, in the Red River, Oklahoma and Texas. Copeia 1963(2): 451–452.

Rigsby, J. M. 2009. Status and genetics of the Stargazing Darter, *Percina uranidea*, in Arkansas. M.S. thesis, Arkansas Tech Univ., Russellville. 196 pp.

Rivas, L. R. 1963. Subgenera and species groups in the poeciliid fish genus, *Gambusia* Poey. Copeia 1963(2): 331–347.

Robbins, M., C. Rein, and M. Volkin. 2003. The Goldstripe Darter (*Etheostoma parvipinne*) and its tolerance to low pH in an east Texas pond. Texas J. Sci. 55(1): 87–89.

Roberg, R. R. 1999. Preference, importance, and relationships to stream order of food types available to Smallmouth Bass (*Micropterus dolomieui*) in the Kings River, Arkansas, USA. M.S. thesis, Univ. Arkansas, Fayetteville. 168 pp.

Roberg, R. R., T. T. Prabhakaran, and R. V. Kilambi. 1986. Age and growth of Redear Sunfish, *Lepomis microlophus* (Gunthur [sic]) from Bob Kidd Lake. Proc. Arkansas Acad. Sci. 40: 40–41.

Roberts, M. E., J. E. Wetzel II, R. C. Brooks, and J. E. Garvey. 2004. Daily increment formation in otoliths of the Redspotted Sunfish. N. Am. J. Fish. Manag. 24(1): 270–274.

Roberts, M. E., C. S. Schwedler, and C. M. Taylor. 2007. Dietary shifts in the Crystal Darter (*Crystallaria asprella*) after large-scale river fragmentation. Ecol. Freshw. Fish 16(2): 250–256.

Roberts, N. J. and H. E. Winn. 1962. Utilization of the senses in feeding behavior of the Johnny Darter, *Etheostoma nigrum*. Copeia 1962(3): 567–570.

Roberts, T. R. 1972. Ecology of fishes in the Amazon and Congo basins. Bull. Mus. Comp. Zool. 143(2): 117–147.

Roberts, T. R. 1973. Osteology and relationships of the Prochilodontidae, a South American family of characoid fishes. Bull. Mus. Comp. Zool. 145(4): 213–235.

Robertson, C. R., S. C. Zeug, and K. O. Winemiller. 2008. Associations between hydrological connectivity and resource partitioning among sympatric gar species (Lepisosteidae) in a Texas river and associated oxbows. Ecol. Freshw. Fish 17(1): 119–129.

Robins, C. R. 1954. A taxonomic revision of the *Cottus bairdi* and *Cottus carolinae* species groups in eastern North America. Ph.D. diss. Cornell Univ., Ithaca, NY. 248 pp.

Robins, C. R. and H. W. Robison. 1985. *Cottus hypselurus*, a new cottid fish from the Ozark Uplands, Arkansas and Missouri. Am. Midl. Nat. 114(2): 360–373.

Robins, R. H. and L. M. Page. 2007. Taxonomic status of the Guadalupe Darter, *Percina apristis* (Teleostei: Percidae). Zootaxa 1618(1): 51–60.

Robins, R. H., L. M. Page, J. D. Williams, Z. S. Randall, and G. E. Sheehy. 2018. Fishes in the fresh waters of Florida. Univ. Florida Press, Gainesville. 467 pp.

Robinson, M. R. and M. M. Ferguson. 2004. Genetics of North American Acipenseriformes. Pages 217–230. *In:* G. T. O. LeBreton, F. W. H. Beamish, and R. S. McKinley. Sturgeons and Paddlefish of North America. Kluwer Academic, Fish and Fisheries Series, Vol. 27. Boston. 323 pp.

Robison, H. W. 1968. The mini-livebearer. Aquarium 2(2): 8–9, 77–78.

Robison, H. W. 1969. An extension of the known range for *Noturus eleutherus* Jordan (Ictaluridae). Southwest. Nat. 14(2): 256–257.

Robison, H. W. 1974a. Environmental changes threaten some species of fish in Arkansas. Arkansas Gazette, June 9, 1974. p. 24.

Robison, H. W. 1974b. Threatened fishes of Arkansas. Proc. Arkansas Acad. Sci. 28: 59–64.

Robison, H. W. 1974c. New distributional records of some Arkansas fishes with addition of three species to the state ichthyofauna. Southwest. Nat. 19(2): 220–223.

Robison, H. W. 1974d. First record of the ictalurid catfish, *Noturus phaeus* from Arkansas. Southwest. Nat. 18(4): 475.

Robison, H. W. 1974e. An additional population of *Etheostoma pallididorsum* Distler and Metcalf in Arkansas. Am. Midl. Nat. 91(2): 478–479.

Robison, H. W. 1975a. A checklist of Arkansas fishes. *In:*

A partial inventory of the Arkansas biota. Special Publ., 9th Annu. ABCD Conf., Hendrix College, Conway, AR.

Robison, H. W. 1975b. New distributional records of fishes from the Ouachita River system of Arkansas. Proc. Arkansas Acad. Sci. 29: 54–56.

Robison, H. W. 1976. A preliminary survey of the fishes of the Petit Jean River system, Arkansas. ASB Bull. 23(2): 91.

Robison, H. W. 1977a. Distribution and habitat notes on the Ironcolor Shiner *Notropis chalybaeus* (Cope) in Arkansas. Proc. Arkansas Acad. Sci. 31: 92–94.

Robison, H. W. 1977b. *Fundulus blairae* Wiley and Hall (Cyprinodontidae) in Arkansas. Southwest. Nat. 22(4): 544.

Robison, H. W. 1977c. Distribution, habitat, variation and status of the Goldstripe Darter, *Etheostoma parvipinne* Gilbert and Swain, in Arkansas. Southwest. Nat. 22(4): 435–442.

Robison, H. W. 1977d. Nest building in the Ozarks. Ozark Soc. Bull. 11(1): 6.

Robison, H. W. 1978a. Distribution and habitat of the Taillight Shiner, *Notropis maculatus* (Hay), in Arkansas. Proc. Arkansas Acad. Sci. 32: 68–70.

Robison, H. W. 1978b. Variation in the caudal skeleton of two etheostomatine fishes. Southwest. Nat. 23(1): 165–167.

Robison, H. W. 1978c. The Leopard Darter (a status report). Endangered Species Report 3, U.S. Fish and Wildlife Service, Albuquerque, N.M. 28 pp.

Robison, H. W. 1979. Additions to the Strawberry River ichthyofauna. Proc. Arkansas Acad. Sci. 33: 89–90.

Robison, H. W. 1980a. *Notropis ortenburgeri* Hubbs, Kiamichi Shiner. Page 290. *In*: D. S. Lee et al. Atlas of North American freshwater fishes. N.C. State Mus. Nat. Hist., Raleigh.

Robison, H. W. 1980b. *Noturus lachneri* Taylor, Ouachita Madtom. Page 462. *In*: D. S. Lee et al. Atlas of North American freshwater fishes. N.C. State Mus. Nat. Hist., Raleigh.

Robison, H. W. 1980c. *Noturus taylori* Douglas, Caddo Madtom. Page 470. *In*: D. S. Lee et al. Atlas of North American freshwater fishes. N.C. State Mus. Nat. Hist., Raleigh.

Robison, H. W. 1980d. *Etheostoma moorei* Raney and Suttkus, Yellowcheek Darter. Page 669. *In*: D. S. Lee et al. Atlas of North American freshwater fishes. N.C. State Mus. Nat. Hist., Raleigh.

Robison, H. W. 1980e. *Etheostoma pallididorsum* Distler and Metcalf, Paleback Darter. Page 679. *In*: D. S. Lee et al. Atlas of North American freshwater fishes. N.C. State Mus. Nat. Hist., Raleigh.

Robison, H. W. 1980f. *Percina pantherina* (Moore and Reeves), Leopard Darter. Page 735. *In*: D. S. Lee et al. Atlas of North American freshwater fishes. N.C. State Mus. Nat. Hist., Raleigh.

Robison, H. W. 1985. *Notropis snelsoni*, a new cyprinid from the Ouachita Mountains of Arkansas and Oklahoma. Copeia 1985(1): 126–134.

Robison, H. W. 1986. Zoogeographic implications of the Mississippi River basin. Pages 267–283. *In*: C. H. Hocutt and E. O. Wiley, eds. The zoogeography of North American freshwater fishes. John Wiley and Sons, New York. 866 pp.

Robison, H. W. 1997. Distribution and status of the Ozark Shiner, *Notropis ozarcanus* Meek, in Arkansas. J. Arkansas Acad. Sci. 51: 150–158.

Robison, H. W. 1998. Status survey of the Strawberry River Darter, *Etheostoma fragi* (Distler), in Arkansas. Final Report to U.S. Fish and Wildlife Service.

Robison, H. W. 2000. An inventory of the fishes of the Pine Bluff Arsenal, Jefferson County, Arkansas. Final Report to The Nature Conservancy. 73 pp + 62 maps.

Robison, H. W. 2001. Distribution and status of the Kiamichi Shiner, *Notropis ortenburgeri* Hubbs (Cyprinidae). Final Report to USDA Forest Service, Ouachita National Forest. 50 pp.

Robison, H. W. 2005a. Distribution and status of the Kiamichi Shiner, *Notropis ortenburgeri* Hubbs (Cyprinidae). J. Arkansas Acad. Sci. 59: 137–147.

Robison, H. W. 2005b. Fishes of the Pine Bluff Arsenal, Jefferson County, Arkansas. J. Arkansas Acad. Sci. 59: 148–157.

Robison, H. W. 2006a. Status survey of the Peppered Shiner, *Notropis perpallidus* Hubbs and Black, in Arkansas and Oklahoma. J. Arkansas Acad. Sci. 60: 101–107.

Robison, H. W. 2006b. Fishes of the White River system of Arkansas and Missouri—Final Report. U.S. Army Corps of Engineers. Vicksburg, MS. 106 pp.

Robison, H. W. and R. T. Allen. 1995. Only in Arkansas: a study of the endemic plants and animals of the state. Univ. Arkansas Press, Fayetteville. 121 pp.

Robison, H. W. and J. K. Beadles. 1974. Fishes of the Strawberry River system in northcentral Arkansas. Proc. Arkansas Acad. Sci. 28: 65–70.

Robison, H. W. and T. M. Buchanan. 1988. Fishes of Arkansas. Univ. Arkansas Press, Fayetteville. 536 pp.

Robison, H. W. and T. M. Buchanan. 1993. Changes in the nomenclature and composition of the Arkansas fish fauna from 1988 to 1993. Proc. Arkansas Acad. Sci. 47: 145–148.

Robison, H. W. and T. M. Buchanan. 1994. First record of the Channel Shiner, *Notropis wickliffi* Trautman, in Arkansas and comments on the Current River population of *Notropis volucellus*. Proc. Arkansas Acad. Sci. 48: 264–265.

Robison, H. W. and G. L. Harp. 1971. A pre-impoundment limnological study of the Strawberry River in northeastern Arkansas. Proc. Arkansas Acad. Sci. 25: 70–79.

Robison, H. W. and G. L. Harp. 1981. A study of four endemic Arkansas threatened fishes. Project E-I-3. Office of Endangered Species. Washington, DC. Unpublished. Unnumbered.

Robison, H. W. and G. L. Harp. 1985. Distribution, habitat, and food of the Ouachita Madtom, *Noturus lachneri*, a Ouachita River drainage endemic. Copeia 1985(1): 216–220.

Robison, H. W. and J. L. Harris. 1978. Notes on the habitat and zoogeography of *Noturus taylori* (Pisces: Ictaluridae). Copeia 1978(3): 548–550.

Robison, H. W. and R. J. Miller. 1972. A new intergeneric cyprinid hybrid (*Notropis pilsbryi* × *Chrosomus erythrogaster*) from Oklahoma. Southwest. Nat. 16(3/4): 442–444.

Robison, H. W. and K. L. Smith. 1982. The endemic flora and fauna of Arkansas. Proc. Arkansas Acad. Sci. 36: 50–54.

Robison, H. W. and K. L. Smith. 1984. Aquatic ecosystems. Pages 97–113. *In*: W. Shepherd, ed. Arkansas's natural heritage. August House, Little Rock. 116 pp.

Robison, H. W. and B. Voiers. 1996. Reproductive behavior of the Creole Darter *Etheostoma collettei*. Pages 264–265. *In*: Program and Abstracts of the 76th Annual Meeting of the American Society of Ichthyologists and Herpetologists. New Orleans.

Robison, H. W. and S. W. Winters. 1978. The fishes of Moro Creek, a lower Ouachita tributary, in southern Arkansas. Proc. Arkansas Acad. Sci. 23: 71–75.

Robison, H. W. and S. W. Winters. 1979. Occurrence of the Slender Madtom, *Noturus exilis* Nelson (Osteichthyes: Ictaluridae), in the Little River system, Arkansas. Southwest. Nat. 23(4): 688–689.

Robison, H. W., E. Laird, and D. Koym. 1983. Fishes of the Antoine River, Ouachita River drainage, in southwestern Arkansas. Proc. Arkansas Acad. Sci. 37: 74–77.

Robison, H. W., C. T. McAllister, and K. E. Shirley. 2011a. The fishes of Crooked Creek (White River drainage) in north-central Arkansas, with new records and a list of species. J. Arkansas Acad. Sci. 65: 111–116.

Robison, H. W., S. G. George, W. T. Slack, and C. T. McAllister. 2011b. First record of the Silver Lamprey, *Ichthyomyzon unicuspis* (Petromyzontiformes: Petromyzontidae), from Arkansas. Am. Midl. Nat. 166(2): 458–461.

Robison, H. W., R. Tumlison, and J. C. Petersen. 2006. New distributional records of lampreys from Arkansas. J. Arkansas Acad. Sci. 60: 194–196.

Robison, H. W., R. C. Cashner, M. E. Raley, and T. J. Near. 2014. A new species of darter from the Ouachita Highlands in Arkansas related to *Percina nasuta* (Percidae: Etheostomatinae). Bull. Peabody Mus. Nat. Hist. 55(2): 237–252.

Robison, H. W., L. G. Henderson, M. L. Warren, Jr., and J. S. Rader. 2004. Computerization of the Arkansas fishes database. Pages 257–264. *In*: J. M. Guilden, Ouachita and Ozark mountains symposium: ecosystem management research. Gen. Tech. Rep. SRS-74. U.S. Dept. Agriculture, Forest Service, Southern Research Station, Asheville, NC.

Robison, H. W., D. A. Neely, U. Thomas, C. T. McAllister, K. E. Shirley, and J. K. Whalen. 2013. New distributional records and natural history notes on selected fishes from Arkansas. J. Arkansas Acad. Sci. 67: 115–120.

Robison, W. R. 1978. Food preference, food conversion and growth of Grass Carp *Ctenopharyngodon idella* at different temperatures. M.S. thesis, Univ. Arkansas, Fayetteville. 43 pp.

Roe, K. J., P. M. Harris, and R. L. Mayden. 2002. Phylogenetic relationships of the genera of North American sunfishes and basses (Percoidei: Centrarchidae) as evidenced by the mitochondrial cytochrome *b* gene. Copeia 2002(4): 897–905.

Roe, K. J., R. L. Mayden, and P. M. Harris. 2008. Systematics and zoogeography of the rock basses (Centrarchidae: *Ambloplites*). Copeia 2008(4): 858–867.

Roell, M. J. and D. J. Orth. 1993. Trophic basis of production of stream-dwelling Smallmouth Bass, Rock Bass, and Flathead Catfish in relation to invertebrate bait harvest. Trans. Am. Fish. Soc. 122(1): 46–62.

Rogers, S. O., B. T. Watson, and R. J. Neves. 2001. Life history and population biology of the endangered tan riffleshell (*Epioblasma florentina walkeri*) (Bivalvia: Unionidae). J. N. Am. Benthol. Soc. 20(4): 582–594.

Rohde, F. C. 1980a. *Phenacobius mirabilis* (Girard), Suckermouth Minnow. Page 332. *In*: D. S. Lee et al. Atlas of North American freshwater fishes. N.C. State Mus. Nat. Hist., Raleigh.

Rohde, F. C. 1980b. *Noturus phaeus* Taylor, Brown Madtom. Page 469. *In*: D. S. Lee et al. Atlas of North American freshwater fishes. N.C. State Mus. Nat. Hist., Raleigh.

Rohde, F. C. 1980c. *Noturus stigmosus* Taylor, Northern Madtom. Page 467. *In*: D. S. Lee et al. Atlas of North American freshwater fishes. N.C. State Mus. Nat. Hist., Raleigh.

Rohde, F. C., R. G. Arndt, and C. S. Wang. 1976. Life history of the freshwater lampreys, *Okkelbergia aepyptera* and *Lampetra lamottenii* (Pisces: Petromyzonidae), on the Delmarva Peninsula (East Coast, United States). Bull. South. Calif. Acad. Sci. 75(2): 99–111.

Rohde, F. C., R. G. Arndt, D. G. Lindquist, and J. F. Parnell. 1994. Freshwater fishes of the Carolinas, Virginia, Maryland, and Delaware. Univ. North Carolina Press, Chapel Hill. 228 pp.

Rohde, F. C., R. G. Arndt, J. W. Foltz, and J. M. Quatro. 2009. Freshwater fishes of South Carolina. Univ. South Carolina Press, Columbia. 430 pp.

Rohm, C. M., J. W. Giese, and C. C. Bennett. 1987. Evaluation of an aquatic ecoregion classification of streams in Arkansas. J. Freshw. Ecol. 4(1): 127–140.

Romer, A. S. 1966. Vertebrate paleontology. 3rd ed. Univ. Chicago Press. 468 pp.

Romero, A., Jr. 1998. Threatened fishes of the world: *Typhlichthys subterraneus* Girard, 1860 (Amblyopsidae). Environ. Biol. Fishes 53(1): 74.

Romero, A., Jr. and M. Conner 2007. Status report for the Southern Cavefish, *Typhlichthys subterraneus* in Arkansas. Final Report to Arkansas Game and Fish Commission, Little Rock. 37 pp.

Romero, A., Jr., M. S. Conner, and G. L. Vaughan. 2010.

Population status of the Southern Cavefish, *Typhlichthys subterraneus* in Arkansas. J. Arkansas Acad. Sci. 64: 106–110.

Roseman, E. F., W. W. Taylor, D. B. Hayes, A. L. Jones, and J. T. Francis. 2006. Predation on Walleye eggs by fish on reefs in western Lake Erie. J. Great Lakes Res. 32(3): 415–423.

Rosen, D. E. 1962. Comments on the relationships of the North American cave fishes of the family Amblyopsidae. Am. Mus. Novit. 2109. 35 pp.

Rosen, D. E. and R. M. Bailey. 1963. The poeciliid fishes (Cyprinodontiformes): their structure, zoogeography, and systematics. Bull. Am. Mus. Nat. Hist. 126: 1–176.

Rosen, D. E. and M. Gordon. 1953. Functional anatomy and evolution of male genitalia in poeciliid fishes. Zoologica: N. Y. Zool. Soc. 38(1): 1–47.

Rosen, R. A. and D. C. Hales. 1980. Occurrence of scarred Paddlefish in the Missouri River, South Dakota-Nebraska. Prog. Fish-Cult. 42(2): 82–85.

Rosen, R. A. and D. C. Hales. 1981. Feeding of Paddlefish, *Polyodon spathula*. Copeia 1981(2): 441–455.

Rosen, R. A., D. C. Hales, and D. G. Unkenholz. 1982. Biology and exploitation of Paddlefish in the Missouri River below Gavins Point Dam. Trans. Am. Fish. Soc. 111(2): 216–222.

Ross, M. R. 1977. Aggression as a social mechanism in the Creek Chub (*Semotilus atromaculatus*). Copeia 1977(2): 393–397.

Ross, S. T. 2001. The inland fishes of Mississippi. Univ. Press of Mississippi, Jackson. 624 pp.

Ross, S. T. 2013. The ecology of North American freshwater fishes. Univ. California Press, Berkeley. 480 pp.

Ross, S. T. and J. A. Baker. 1983. The response of fishes to periodic spring floods in a southeastern stream. Am. Midl. Nat. 109(1): 1–14.

Ross, S. T. and W. J. Matthews. 2014. Evolution and ecology of North American freshwater fish assemblages. Pages 1–49. *In:* M. L. Warren, Jr. and B. M. Burr, eds. Freshwater fishes of North America, Vol. 1. Johns Hopkins Univ. Press, Baltimore, MD. 644 pp.

Ross, S. T., J. A. Baker, and K. E. Clark. 1987. Microhabitat partitioning of southeastern stream fishes: temporal and spatial predictability. Pages 42–51. *In:* W. J. Matthews and D. C. Heins, eds. Community and evolutionary ecology of North American stream fishes. Univ. Oklahoma Press, Norman.

Ross, S. T., M. S. Peterson, and J. R. Brent. 1984. Fishery potential of American Eels in the northern Gulf of Mexico. Gulf and South Atlantic Fisheries Development Foundation, Tampa, FL. 68 pp.

Rossi, A. R., M. Capula, D. Crosetti, L. Sola, and D. E. Campton. 1998. Allozyme variation in global populations of Striped Mullet, *Mugil cephalus* (Pisces: Mugilidae). Mar. Biol. 131(2): 203–212.

Routledge, E. J., D. Sheahan, C. Desbrow, G. C. Brighty, M. Waldock, and J. P. Sumpter. 1998. Identification of estrogenic chemicals in STW effluent. 2. In vivo responses in trout and roach. Environ. Sci. Technol. 32(11): 1559–1565.

Ruble, C. L., P. L. Rakes, J. R. Shute, and S. A. Welsh. 2014. Captive propagation, reproductive biology, and early life history of the Diamond Darter (*Crystallaria cincotta*). Am. Midl. Nat. 172(1): 107–118.

Ruelle, R. and P. L. Hudson. 1977. Paddlefish (*Polyodon spathula*): growth and food of young of the year and a suggested technique for measuring length. Trans. Am. Fish. Soc. 106(6): 609–613.

Rupp, R. S. 1965. Shore-spawning and survival of eggs of the American Smelt. Trans. Am. Fish. Soc. 94(2): 160–168.

Rupprecht, R. J. and L. A. Jahn. 1980. Biological notes on Blue Suckers in the Mississippi River. Trans. Am. Fish. Soc. 109(3): 323–326.

Russell, D. and P. W. Bettoli. 2013. Population attributes of Lake Trout in Tennessee reservoirs. Southeast. Nat. 12(1): 217–232.

Russell, D. F., L. A. Wilkens, and F. Moss. 1999. Use of behavioural stochastic resonance by Paddlefish for feeding. Nature 402(6759): 291–294.

Rutherford, D. A., A. A. Echelle, and O. E. Maughan. 1987. Changes in the fauna of the Little River drainage, southeastern Oklahoma, 1948–1955 to 1981–1982: a test of the hypothesis of environmental degradation. Pages 178–183. *In:* W. J. Matthews and D. C. Heins, eds. Community and evolutionary ecology of North American stream fishes. Univ. Oklahoma Press, Norman.

Rutherford, D. A., A. A. Echelle, and O. E. Maughan. 1992. Drainage-wide effects of timber harvesting on the structure of stream fish assemblages in southeastern Oklahoma. Trans. Am. Fish. Soc. 121(6): 716–728.

Rutherford, D. A., K. R. Gelwicks, and W. E. Kelso. 2001. Physicochemical effects of the flood pulse on fishes in the Atchafalaya River basin, Louisiana. Trans. Am. Fish. Soc. 130(2): 276–288.

Rutledge, C. J. and T. L. Beitinger. 1989. The effects of dissolved oxygen and aquatic surface respiration on the critical thermal maxima of three intermittent-stream fishes. Environ. Biol. Fishes 24(2): 137–143.

Ryles, J. 2012. The effects of consecutive road crossings on fish movement and community structure in a Ouachita Mountain stream. M.S. thesis, Arkansas Tech Univ., Russellville. 53 pp.

Rypel, A. L. and J. B. Mitchell. 2007. Summer nocturnal patterns in Freshwater Drum (*Aplodinotus grunniens*). Am. Midl. Nat. 157(1): 230–234.

Rypel, A. L., D. R. Bayne, and J. B. Mitchell. 2006. Growth of Freshwater Drum from lotic and lentic habitats in Alabama. Trans. Am. Fish. Soc. 135(4): 987–997.

Sabaj, M. H. 1992. Spawning clasps and gamete deposition in pebble nest-building minnows (Pisces: Cyprinidae). M.S. thesis, Univ. of Richmond, VA. 85 pp.

Sabaj, M. H., L. M. Page, J. G. Lundberg, C. J. Ferraris, Jr.,

J. W. Armbruster, J. P. Friel, and P. J. Morris. 2004. All Catfish Species Inventory Website, http://clade.acnatsci.org/allcatfish.

Sabaj-Perez, M. H. 2009. Photographic atlas of fishes of the Guiana Shield. Bull. Biol. Soc. Washington 17(1): 52–59.

Sabat, A. M. 1994. Mating success in brood-guarding male Rock Bass, *Ambloplites rupestris*: the effect of body size. Environ. Biol. Fishes 39(4): 411–415.

Saecker, J. R. and W. S. Woolcott. 1988. The Redbreast Sunfish (*Lepomis auritus*) in a thermally influenced section of the James River, Virginia. Virginia J. Sci. 39: 1–17.

Saeed, B., W. Ivantsoff, and L. E. L. M. Crowley. 1994. Systematic relationships of atheriniform families within division 1 of the series Atherinomorpha (Actinopterygii) with relevant historical perspectives. J. Ichthyol. 34: 1–32.

Saitoh, K., T. Sado, R. L. Mayden, N. Hanzawa, K. Nakamura, M. Nishida, and M. Miya. 2006. Mitogenomic evolution and interrelationships of the Cypriniformes (Actinopterygii: Ostariophysi): the first evidence toward resolution of higher-level relationships of the world's largest freshwater fish clade based on 59 whole mitogenome sequences. J. Mol. Evol. 63(6): 826–841.

Salek, S. J., J. Godwin, C. V. Sullivan, and N. E. Stacey. 2001. Courtship and tank spawning behavior of temperate basses (genus *Morone*). Trans. Am. Fish. Soc. 130(5): 833–847.

Salinger, J. M. 2016. Aspects of host usage of the Chestnut Lamprey (*Ichthyomyzon castaneus*) in Arkansas. M.S. thesis, Arkansas State Univ., Jonesboro. 244 pp.

Saltzgiver, M. J., E. J. Heist, and P. W. Hedrick. 2012. Genetic evaluation of the initiation of a captive population: the general approach and a case study in the endangered Pallid Sturgeon (*Scaphirhynchus albus*). Conserv. Genet. 13(5): 1381–1391.

Sampson, S. J., J. H. Chick, and M. A. Pegg. 2009. Diet overlap among two Asian carp and three native fishes in backwater lakes on the Illinois and Mississippi rivers. Biol. Invasions 11(3): 483–496.

Sanders, L. G., J. A. Baker, C. L. Bond, and C. H. Pennington. 1985. Biota of selected aquatic habitats of the McClellan-Kerr Arkansas River Navigation System. Tech. Rep. E-85-6. U.S. Army Corps of Engineers Waterways Experiment Station, Vicksburg, MS. 89 pp.

Sanders, R. E. and C. O. Yoder. 1989. Recent collections and food items of River Darters, *Percina shumardi* (Percidae) in the Markland Dam pool of the Ohio River. Ohio J. Sci. 89(1): 33–35.

Sanderson, S. L., J. J. Cech, Jr., and A. Y. Cheer. 1994. Paddlefish buccal flow velocity during ram suspension feeding and ram ventilation. J. Exp. Biol. 186(1): 145–156.

Sargent, R. C. 1989. Allopaternal care in the Fathead Minnow, *Pimephales promelas*: stepfathers discriminate against their adopted eggs. Behav. Ecol. Sociobiol. 25(6): 379–385.

Saunders, R. P. 1959. A study of the food of the Mississippi Silversides, *Menidia audens* Hay, in Lake Texoma. M.S. thesis, Univ. Oklahoma, Norman. 42 pp.

Scalet, C. G. 1972. Food habits of the Orangebelly Darter, *Etheostoma radiosum cyanorum* (Osteichthyes: Percidae). Am. Midl. Nat. 87(2): 515–522.

Scalet, C. G. 1973a. Reproduction of the Orangebelly Darter, *Etheostoma radiosum cyanorum* (Osteichthyes: Percidae). Am. Midl. Nat. 89(1): 156–165.

Scalet, C. G. 1973b. Stream movements and population density of the Orangebelly Darter, *Etheostoma radiosum cyanorum* (Osteichthyes: Percidae). Southwest. Nat. 17(4): 381–387.

Scarnecchia, D. L. 1992. A reappraisal of gars and Bowfins in fishery management. Fisheries 17(5): 6–12.

Scarnecchia, D. L., L. F. Ryckman, Y. Lim, G. J. Power, B. J. Schmitz, and J. A. Firehammer. 2007. Life history and the costs of reproduction in Northern Great Plains Paddlefish (*Polyodon spathula*) as a potential framework for other acipenseriform fishes. Rev. Fish. Sci. 15(3): 211–263.

Scarnecchia, D. L., B. D. Gordon, J. D. Schooley, L. F. Ryckman, B. J. Schmitz, S. E. Miller, and Y. Lim. 2011. Southern and Northern Great Plains (United States) Paddlefish stocks within frameworks of acipenseriform life history and the metabolic theory of ecology. Rev. Fish. Sci. 19(3): 279–298.

Schaefer, J. F. 2001. Riffles as barriers to interpool movement by three cyprinids (*Notropis boops, Campostoma anomalum*, and *Cyprinella venusta*). Freshw. Biol. 46(3): 379–388.

Schaefer, J. F., D. Duvernell, and B. Kreiser. 2011a. Shape variability in topminnows (*Fundulus notatus* species complex) along the river continuum. Biol. J. Linn. Soc. 103(3): 612–621.

Schaefer, J. F., D. D. Duvernell, and B. R. Kreiser. 2011b. Ecological and genetic assessment of spatial structure among replicate contact zones between two topminnow species. Evol. Ecol. 25(5): 1145–1161.

Schaefer, J. F., E. Marsh-Matthews, D. E. Spooner, K. Gido, and W. J. Matthews. 2003. Effects of barriers and thermal refugia on local movement of the threatened Leopard Darter, *Percina pantherina*. Environ. Biol. Fishes 66(4): 391–400.

Schaefer, J. F., D. Duvernell, B. Kreiser, C. Champagne, S. R. Clark, M. Gutierrez, L. Stewart, and C. Coleman. 2012. Evolution of a sexually dimorphic trait in a broadly distributed topminnow (*Fundulus olivaceus*). Ecol. Evol. 2(7): 1371–1381.

Schaefer, S. A. and T. M. Cavender. 1986. Geographic variation and subspecific status of *Notropis spilopterus* (Pisces: Cyprinidae). Copeia 1986(1): 122–130.

Schaeffer, J. S. and F. J. Margraf. 1987. Predation on fish eggs by White Perch, *Morone americana*, in western Lake Erie. Environ. Biol. Fishes 18(1): 77–80.

Schaffler, J. J., S. P. Young, S. Herrington, T. Ingram, and J. Tannehill. 2015. Otolith chemistry to determine within-river origins of Alabama Shad in the Apalachicola-Chattahoochee-Flint river basin. Trans. Am. Fish. Soc. 144(1): 1–10.

Schanke, K. L. 2013. Effects of movement barriers on the gene flow of Longear Sunfish *Lepomis megalotis* and Highland Stoneroller *Campostoma spadiceum* within the Ouachita Mountains, Arkansas. M.S. thesis, Arkansas Tech Univ., Russellville. 81 pp.

Schanke, K. L, T. Yamashita, and C. J. Gagen. 2017. Reduced gene flow in two common headwater fishes in the Ouachita Mountains: a response to stream drying and in-stream barriers. Copeia 2017(1): 33–42.

Scharpf, C. 2008. Annotated checklist of North American freshwater fishes, including subspecies and undescribed forms. Part IV: Cottidae . . . [through] Percidae. Am. Currents 34(4): 1–43.

Schaus, M. H., M. J. Vanni, and T. E. Wissing. 2002. Biomass-dependent diet shifts in omnivorous Gizzard Shad: implications for growth, food web, and ecosystem effects. Trans. Am. Fish. Soc. 131(1): 40–54.

Scheidegger, K. J. and M. B. Bain. 1995. Larval fish distribution and microhabitat use in free-flowing and regulated rivers. Copeia 1995(1): 125–135.

Schieble, C. S. 1998. Life history of the Shadow Bass, *Ambloplites ariommus* Viosca, with emphasis on age and growth, reproduction, and habitats. M.S. thesis, Univ. New Orleans. 168 pp.

Schiemer, F., M. Zalewski, and J. E. Thorpe. 1995. Land/inland water ecotones: intermediate habitats critical for conservation and management. Hydrobiologia 303(1–3): 259–264.

Schmidt, J. 1923. The breeding places of the eel. Phil. Trans. Royal Soc. London, Ser. B. 211: 179–208.

Schmidt, R. E. and W. R. Whitworth. 1979. Distribution and habitat of the Swamp Darter (*Etheostoma fusiforme*) in southern New England. Am. Midl. Nat. 102(2): 408–413.

Schmidt, R. E., C. M. O'Reilly, and D. Miller. 2009. Observations of American Eels using an upland passage facility and effects of passage on the population structure. N. Am. J. Fish. Manag. 29(3): 715–720.

Schmidt, T. R. 1994. Phylogenetic relationships of the genus *Hybognathus* (Teleostei: Cyprinidae). Copeia 1994(3): 622–630.

Schmidt, T. R., J. P. Bielawski, and J. R. Gold. 1998. Molecular phylogenetics and evolution of the cytochrome *b* gene in the cyprinid genus *Lythrurus* (Actinopterygii: Cypriniformes). Copeia 1998(1): 14–22.

Schmidt, T. R., T. E. Dowling, and J. R. Gold. 1994. Molecular systematics of the genus *Pimephales* (Teleostei: Cyprinidae). Southwest. Nat. 39(3): 241–248.

Schneberger, E. 1937a. The food of small Dogfish, *Amia calva*. Copeia 1937(1): 61.

Schneberger, E. 1937b. The biological and economic importance of the smelt in Green Bay. Trans. Am. Fish. Soc. 66(1): 139–142.

Schneider, H. and A. D. Hasler. 1960. Laute und lauterzeugung beim Susswasser-Trommler *Aplodinotus grunniens* Rafinesque (Sciaenidae, Pisces). Zeitschr. Vergleich. Physiol., Berlin 43(5): 499–517.

Schneider, T. 2015. Shoal Chub: made for the Mississippi. Our Mississippi Magazine, Summer: p. 9.

Schoffman, R. J. 1955. Age and rate of growth of the Yellow Bullhead in Reelfoot Lake, Tennessee. J. Tennessee Acad. Sci. 30: 4–7.

Schoffman, R. J. 1958. Age and rate of growth of the Yellow Bass in Reelfoot Lake, Tennessee, for 1955 and 1957. J. Tennessee Acad. Sci. 33(1): 101–105.

Schönhuth, S. and I. Doadrio. 2003. Phylogenetic relationships of Mexican minnows of the genus *Notropis* (Actinopterygii, Cyprinidae). Biol. J. Linn. Soc. 80(2): 323–337.

Schönhuth, S. and R. L. Mayden. 2010. Phylogenetic relationships in the genus *Cyprinella* (Actinopterygii: Cyprinidae) based on mitochondrial and nuclear gene sequences. Mol. Phylogenet. Evol. 55(1): 77–98.

Schönhuth, S., C. E. Beachum, J. H. Knouft, and R. L. Mayden. 2016. Phylogeny and genetic variation within the widely distributed Bluntnose Minnow, *Pimephales notatus* (Cyprinidae), in North America. Zootaxa 4168(1): 38–60.

Schönhuth, S., M. J. Blum, L. Lozano-Vilano, D. A. Neely, A. Varela-Romero, H. Espinosa, A. Perdices, and R. Mayden. 2011. Inter-basin exchange and repeated headwater capture across the Sierra Madre Occidental inferred from the phylogeography of Mexican stonerollers. J. Biogeogr. 38(7): 1406–1421.

Schramm, H. L., Jr. 2017. The fishery resources of the Mississippi River: a model for conservation and management. Fisheries 42(11): 574–585.

Schramm, H. L., Jr. and A. V. Zale. 1985. Effects of cover and prey size on preferences of juvenile Largemouth Bass for Blue Tilapias and Bluegills in tanks. Trans. Am. Fish. Soc. 114(5): 725–731.

Schramm, H. L., Jr., R. B. Minnis, A. B. Spencer, and R. T. Theel. 2008. Aquatic habitat change in the Arkansas River after the development of a lock-and-dam commercial navigation system. River Res. Appl. 24(3): 237–248.

Schrank, F. von P. 1798. Fauna Boica. Durchgedachte Geschichte der in Baiern einheimischen und zahmen Thiere. Nürnberg. v. 1: i–xii + 1–720 pp.

Schrank, S. J. and C. S. Guy. 2002. Age, growth, and gonadal characteristics of adult Bighead Carp, *Hypophthalmichthys nobilis*, in the lower Missouri River. Environ. Biol. Fishes 64(4): 443–450.

Schrank, S. J., P. J. Braaten, and C. S. Guy. 2001. Spatiotemporal variation in density of larval Bighead Carp in the lower Missouri River. Trans. Am. Fish. Soc. 130(5): 809–814.

Schrank, S. J., C. S. Guy, and J. F. Fairchild. 2003. Competitive interactions between age-0 Bighead Carp and Paddlefish. Trans. Am. Fish. Soc. 132(6): 1222–1228.

Schreiner, I. 1989. Biological control introductions in the

Caroline and Marshall Islands. Proc. Hawaiian Entomol. Soc. 29: 57–69.

Schrey, A. W. and E. J. Heist. 2007. Stock structure of Pallid Sturgeon analyzed with microsatellite loci. J. Appl. Ichthyol. 23(4): 297–303.

Schrey, A. W., R. Boley, and E. J. Heist. 2011. Hybridization between Pallid Sturgeon *Scaphirhynchus albus* and Shovelnose Sturgeon, *Scaphirhynchus platorynchus*. J. Fish Biol. 79(7): 1828–1850.

Schrey, A. W., B. L. Sloss, R. J. Sheehan, R. C. Heidinger, and E. J. Heist. 2007. Genetic discrimination of middle Mississippi River *Scaphirhynchus* sturgeon into Pallid, Shovelnose, and putative hybrids with microsatellite loci. Conserv. Genet. 8(3): 683–693.

Schubert, A. L. S. and D. B. Noltie. 1995. Laboratory studies of substrate and microhabitat selection in the Southern Cavefish (*Typhlichthys subterraneus* Girard). Ecol. Freshw. Fish 4(4): 141–151.

Schubert, A. L. S., C. D. Nielsen, and D. B. Noltie. 1993. Habitat use and gas bubble disease in Southern Cavefish (*Typhlichthys subterraneus*). Int. J. Speleol. 22(1–4): 131–143.

Schwartz, F. J. 1965. The distribution and probable postglacial dispersal of the percid fish, *Etheostoma b. blennioides*, in the Potomac River. Copeia 1965(3): 285–290.

Schwemm, M. R. 2013. Zoogeography of Ouachita Highland fishes. Ph.D. diss. Oklahoma State Univ., Stillwater. 129 pp.

Schwemm, M. R. and A. A. Echelle. 2013. Development and characterization of eight polymorphic tetra-nucleotide microsatellite markers for the threatened Leopard Darter (*Percina pantherina*). Conserv. Genet. Res. 5(1): 73–75.

Schwemm, M. R., A. A. Echelle, and R. A. Van Den Bussche. 2014a. Isolation and characterization of 10 polymorphic microsatellite markers for the Ouachita Highlands endemic *Notropis suttkusi* (Teleostei: Cyprinidae). Conserv. Genet. Res. 6(1): 209–210.

Schwemm, M. R., A. A. Echelle, R. A. Van Den Bussche, and J. D. Schooley. 2014b. Development of diploid microsatellite markers for the North American Paddlefish (*Polyodon spathula*). Conserv. Genet. Res. 6(1): 217–218.

Scopoli, G. A. 1777. Introductio ad historiam naturalem, sistens genera lapidum, plantarum et animalium hactenus detecta, characteribus essentialibus donate, in tribus divisa, subinde ad leges naturae. Gerle, Prague. 506 pp.

Scoppettone, G. G., J. A. Salgado, and M. B. Nielsen. 2005. Blue Tilapia (*Oreochromis aureus*) predation on fishes in the Muddy River system, Clark County, Nevada. West. N. Am. Nat. 65(3): 410–414.

Scott, M. C. and G. S. Helfman. 2001. Native invasions, homogenization, and the mismeasure of the integrity of fish assemblages. Fisheries 26(11): 6–15.

Scott, W. 1938. Food of *Amia* and *Lepidosteus*. Investig. Indiana Lakes Streams 1: 111–115.

Scott, W. B. and E. J. Crossman. 1973. Freshwater fishes of Canada. Bulletin 184, Fisheries Research Board of Canada, Ottawa. 966 pp.

Scribner, K. T. and J. C. Avise. 1993. Cytonuclear genetic architecture in mosquitofish populations and the possible roles of introgressive hybridization. Mol. Ecol. 2(3): 139–149.

Seagle, H. H., Jr. and J. W. Nagel. 1982. Life cycle and fecundity of the American Brook Lamprey, *Lampetra appendix*, in Tennessee. Copeia 1982(2): 362–366.

Sechler, D. R., Q. E. Phelps, S. J. Tripp, J. E. Garvey, D. P. Herzog, D. E. Ostendorf, J. W. Ridings, J. W. Crites, and R. A. Hrabik. 2012a. Effects of river stage height and water temperature on diet composition of year-o sturgeon (*Scaphirhynchus* spp.): a multi-year study. J. Appl. Ichthyol. 29(1): 44–50.

Sechler, D. R., Q. E. Phelps, S. J. Tripp, J. E. Garvey, D. P. Herzog, D. E. Ostendorf, J. W. Ridings, J. W. Crites, and R. A. Hrabik. 2012b. Habitat for age-o Shovelnose Sturgeon and Pallid Sturgeon in a large river: interactions among abiotic factors, food, and energy intake. N. Am. J. Fish. Manag. 32(1): 24–31.

Secor, D. H., P. J. Anders, W. Van Winkel, and D. Dixon. 2002. Can we study sturgeons to extinction? What we do and don't know about the conservation of North American sturgeons. Am. Fish. Soc. Symp. 28: 3–10.

Seibert, J. R., Q. E. Phelps, S. J. Tripp, and J. E. Garvey. 2011. Seasonal diet composition of adult Shovelnose Sturgeon in the middle Mississippi River. Am. Midl. Nat. 165(2): 355–363.

Seidensticker, E. P. 1987. Food selection of Alligator Gar and Longnose Gar in a Texas reservoir. Proc. Annu. Conf. Southeast. Assoc. Fish Wildl. Agencies 41: 100–104.

Sellers, T. J., B. R. Parker, D. W. Schindler, and W. M. Tonn. 1998. Pelagic distribution of Lake Trout (*Salvelinus namaycush*) in small Canadian Shield lakes with respect to temperature, dissolved oxygen, and light. Can. J. Fish. Aquat. Sci. 55(1): 170–179.

Setiamarga, D. H. E., M. Miya, Y. Yamanoue, K. Mabuchi, T. P. Saitoh, J. G. Inoue, and M. Nishida. 2008. Interrelationships of Atherinomorpha (medakas, flyingfishes, killifishes, silversides, and their relatives): the first evidence based on whole mitogenome sequences. Mol. Phylogenet. Evol. 49(2): 598–605.

Settles, W. H. and R. D. Hoyt. 1976. Age structure, growth patterns, and food habits of the Southern Redbelly Dace *Chrosomus erythrogaster* in Kentucky. Trans. Kentucky Acad. Sci. 37(1–2): 1–10.

Settles, W. H. and R. D. Hoyt. 1978. The reproductive biology of the Southern Redbelly Dace, *Chrosomus erythrogaster* Rafinesque, in a spring-fed stream in Kentucky. Am. Midl. Nat. 99(2): 290–298.

Setzer, P. Y. 1970. An analysis of a natural hybrid swarm by means of chromosome morphology. Trans. Am. Fish. Soc. 99(1): 139–146.

Seversmith, H. F. 1953. Distribution, morphology and

life history of *Lampetra aepyptera*, a brook lamprey, in Maryland. Copeia 1953(4): 225–232.

Sewell, S. A. 1979a. The systematic status of the fishes of genus *Campostoma* (Cyprinidae) in the White, Black, and St. Francis rivers of northern Arkansas. M.S. thesis, Arkansas State Univ., Jonesboro. 63 pp.

Sewell, S. A. 1979b. Age and growth of White Crappie, *Pomoxis annularis* Rafinesque, from a flood-created pond in Mississippi County, Arkansas. Proc. Arkansas Acad. Sci. 33: 90–91.

Sewell, S. A. 1981. Preliminary report on the fishes of the upper Saline River, Polk and Howard counties, Arkansas, and observations on their relationships with land use and physicochemical conditions. Proc. Arkansas Acad. Sci. 35: 60–65.

Sewell, S. A. 1985. Age and growth of the Bluegill *Lepomis macrochirus* Rafinesque from an unmanaged watershed lake in northeast Arkansas with observations on lake ecology. Proc. Arkansas Acad. Sci. 39: 103–106.

Sewell, S. A., J. K. Beadles, and V. R. McDaniel. 1980a. The systematic status of the fishes of genus *Campostoma* (Cyprinidae) inhabiting the major drainages of northern Arkansas. Proc. Arkansas Acad. Sci. 34: 97–100.

Sewell, S. A., F. A. Carter, and C. T. McAllister. 1980b. Implications and considerations concerning the status, habitat, and distribution of the Least Brook Lamprey, *Lampetra aepyptera* (Abbott) (Pisces: Petromyzontidae) in Arkansas. Proc. Arkansas Acad. Sci. 34: 132–133.

Shane, B. S. 1994. Introduction to ecotoxicology. *In*: L. G. Cockerham and B. S. Shane, eds. Basic environmental toxicology. CRC Press, Boca Raton, FL. 640 pp.

Sharov, A., M. Wilberg, and J. Robinson. 2014. Developing biological reference points and identifying stock status for management of Paddlefish (*Polyodon spathula*) in the Mississippi River basin. Final Report to the Association of Fish and Wildlife Agencies. 210 pp.

Shaw, K., E. O. Wiley, and T. A. Titus. 1995. Phylogenetic relationships among members of the *Hybopsis amblops* species group (Teleostei: Cyprinidae). Occas. Pap. Nat. Hist. Mus. Univ. Kansas 172: 1–28.

Shelby, E. L. 2006. An assessment and analysis of benthic macroinvertebrate communities associated with the appearance of *Didymosphenia geminata* in the White River below Bull Shoals Dam. Arkansas Dept. Environmental Quality, Water Planning Division Report. 42 pp.

Sheldon, A. L. and G. K. Meffe. 1993. Multivariate analysis of feeding relationships of fishes in blackwater streams. Environ. Biol. Fishes 37(2): 161–171.

Shelton, W. L. 1972. Comparative reproductive biology of the Gizzard Shad, *Dorosoma cepedianum* (Lesueur) and the Threadfin Shad, *Dorosoma petenense* (Günther) in Lake Texoma, Oklahoma. Ph.D. diss., Univ. Oklahoma, Norman. 232 pp.

Shelton, W. L. and B. G. Grinstead. 1972. Hybridization between *Dorosoma cepedianum* and *D. petenense* in Lake Texoma, Oklahoma. Proc. 26th Annu. Conf. Southeast. Assoc. Game Fish Comm.: 506–510.

Shelton, W. L. and S. D. Mims. 2012. Evidence for female heterogametic sex determination in Paddlefish *Polyodon spathula* based on gynogenesis. Aquaculture 356–357: 116–118.

Shelton W. L. and R. A. Snow. 2017. Recruitment of two non-native river-spawning fishes in Lake Texoma, Oklahoma and Texas. Proc. Oklahoma Acad. Sci. 97: 61–66.

Shepherd, M. E. and M. T. Huish. 1978. Age, growth, and diet of the Pirate Perch in a coastal plain stream of North Carolina. Trans. Am. Fish. Soc. 107(3): 457–459.

Sheppard, K. T., A. J. Olynyk, G. K. Davoren, and B. J. Hann. 2012. Summer diet analysis of the invasive Rainbow Smelt (*Osmerus mordax*) in Lake Winnipeg, Manitoba. J. Great Lakes Res. 38(S3): 66–71.

Shields, F. D., Jr., S. S. Knight, and C. M. Cooper. 1995. Use of the index of biotic integrity to assess physical habitat degradation in warmwater streams. Hydrobiologia 312(3): 191–208.

Shields, R. C. and D. W. Beckman. 2015. Assessment of variation in age, growth, and prey of Freshwater Drum (*Aplodinotus grunniens*) in the lower Missouri River. Southwest. Nat. 60(4): 360–365.

Shireman, J. V. and C. R. Smith. 1983. Synopsis of biological data on the Grass Carp, *Ctenopharyngodon idella* (Cuvier and Valenciennes, 1844). FAO Fish. Synop. 135. 92 pp.

Shireman, J. V., R. L. Stetler, and D. E. Colle. 1978. Possible use of the Lake Chubsucker as a baitfish. Prog. Fish-Cult. 40(1): 33–34.

Shirley, K., C. T. McAllister, H. W. Robison, and T. M. Buchanan. 2013. The Blue Sucker, *Cycleptus elongatus* (Lesueur) (Cypriniformes: Catostomidae) from the transition zone between upper and lower White River, Arkansas. Texas J. Sci. 65(2): 41–47.

Shoup, D. E., R. E. Carlson, and R. T. Heath. 2004. Diel activity levels of centrarchid fishes in a small Ohio lake. Trans. Am. Fish. Soc. 133(5): 1264–1269.

Shoup, D. E., K. M. Boswell, and D. H. Wahl. 2014. Diel littoral-pelagic movements by juvenile Bluegills in a small lake. Trans. Am. Fish. Soc. 143(3): 796–801.

Shu, D.-G., H.-L. Luo, S. C. Morris, X.-L. Zhang, S.-X. Hu, L. Chen, J. Han, M. Zhu, Y. Li, and L.-Z. Chen. 1999. Lower Cambrian vertebrates from South China. Nature 402(6757): 42–46.

Shu, D.-G., H.-L. Luo, S. C. Morris, X.-L. Zhang, S.-X. Hu, L. Chen, J. Han, M. Zhu, Y. Li, and L.-Z. Chen. 2003. Head and backbone of the Early Cambrian vertebrate *Haikouichthys*. Nature 421(6922): 526–529.

Shute, J. R. 1980. *Fundulus notatus* (Rafinesque), Blackstripe Topminnow. Page 521. *In*: D. S. Lee et al. Atlas of North American freshwater fishes. N.C. State Mus. Nat. Hist., Raleigh.

Sider, A. 2011. Catfish acreage in state fell 31%. Arkansas Democrat-Gazette, Section D (3 February 2011). p. 1 D.

Siefert, R. E. 1968. Reproductive behavior, incubation and

mortality of eggs, and postlarval food selection in the White Crappie. Trans. Am. Fish. Soc. 97(3): 252–259.

Siefert, R. E. 1969. Characteristics for separation of White and Black crappie larvae. Trans. Am. Fish. Soc. 98(2): 326–328.

Siefert, R. E. 1972. First food of larval Yellow Perch, White Sucker, Bluegill, Emerald Shiner, and Rainbow Smelt. Trans. Am. Fish. Soc. 101(2): 219–225.

Siegwarth, G. L. 1992. Channel Catfish of the Buffalo National River, Arkansas: population abundance, reproductive output, and assessment of stocking catchable size fish. M.S. thesis, Univ. Arkansas, Fayetteville.

Siegwarth, G. L. and J. E. Johnson. 1994. Pre-spawning migration of Channel Catfish into three warmwater tributaries—effects of a cold tailwater. Proc. Arkansas Acad. Sci. 48: 168–173.

Sigler, W. F. and R. R. Miller. 1963. Fishes of Utah. Utah State Dept. Fish and Game, Salt Lake City. 203 pp.

Sillman, A. J. and D. A. Dahlin. 2004. Photoreceptor topography in the duplex retina of the Paddlefish (*Polyodon spathula*). J. Exp. Zool. 301(8): 674–681.

Simmons, B. R. 1999. Age, growth, reproduction, and population structure of the Striped Shiner (*Luxilus chrysocephalus*) and Duskystripe Shiner (*Luxilus pilsbryi*) in the James River, Missouri. M.S. thesis. Southwest Missouri State Univ., Springfield. 41 pp.

Simmons, B. R. and D. W. Beckman. 2012. Age determination, growth, and population structure of the Striped Shiner and Duskystripe Shiner. Trans. Am. Fish. Soc. 141(3): 846–854.

Simmons, J. W. 2013. Chronology of the invasion of the Tennessee and Cumberland river systems by the Mississippi Silverside, *Menidia audens*, with analysis of the subsequent decline of the Brook Silverside, *Labidesthes sicculus*. Copeia 2013(2): 292–302.

Simon, T. P. 1994. Ontogeny and systematics of darters (Percidae) with discussion of ecological effects of larval morphology. Ph.D. diss., Univ. Illinois, Chicago. 672 pp.

Simon, T. P., ed. 1999a. Assessing the sustainability and biological integrity of water resources using fish communities. CRC Press, Boca Raton, FL. 671 pp.

Simon, T. P. 1999b. Assessment of Balon's reproductive guilds with application to midwestern North American freshwater fishes. Pages 97–121. *In*: T. P. Simon, ed. Assessing the sustainability and biological integrity of water resources using fish communities. CRC Press, Boca Raton, FL. 671 pp.

Simon, T. P. and N. J. Garcia. 1990. Descriptions of eggs, larvae, and early juveniles of the Stippled Darter, *Etheostoma punctulatum* (Agassiz) from a tributary of the Osage River, Missouri. Southwest. Nat. 35(2): 123–129.

Simon, T. P. and J. B. Kaskey. 1992. Description of eggs, larvae, and early juveniles of the Bigscale Logperch, *Percina macrolepida* Stevenson, from the West Fork of the Trinity River basin, Texas. Southwest. Nat. 37(1): 28–34.

Simon, T. P. and R. Wallus. 1989. Contributions to the early life histories of gar (Actinopterygii: Lepisosteidae) in the Ohio

and Tennessee river basins with emphasis on larval development. Trans. Kentucky Acad. Sci. 50(1–2): 59–74.

Simon, T. P. and R. Wallus. 2004. Reproductive biology and early life history of fishes in the Ohio River drainage. Vol. 3: Ictaluridae—catfish and madtoms. CRC Press, New York. 204 pp.

Simon, T. P., R. D. Wallus, and K. B. Floyd. 1987. Descriptions of protolarvae of seven species of the subgenus *Nothonotus* (Percidae: Etheostomatini) with comments on intrasubgeneric characteristics. Pages 179–190. *In*: R. D. Hoyt, ed. 10th Annual Larval Fish Conference. Symp. 2. American Fisheries Society, Bethesda, MD.

Simon, T. P., E. J. Tyberghein, K. J. Scheiddeger, and C. E. Johnston. 1992. Descriptions of protolarvae of sand darters (Percidae: *Ammocrypta* and *Crystallaria*) with comments on systematic relationships. Ichthyol. Explor. Freshw. 3(4): 347–358.

Simons, A. M. 1991. Phylogenetic relationships of the Crystal Darter, *Crystallaria asprella* (Teleostei: Percidae). Copeia 1991(4): 927–936.

Simons, A. M. 2004. Phylogenetic relationships in the genus *Erimystax* (Actinopterygii: Cyprinidae) based on the cytochrome *b* gene. Copeia 2004(2): 351–356.

Simons, A. M. and R. L. Mayden. 1997. The phylogenetic relationships of the creek chubs and the spine-fins: an enigmatic group of North American cyprinid fishes (Actinopterygii: Cyprinidae). Cladistics 13(3): 187–205.

Simons, A. M. and R. L. Mayden. 1998. Phylogenetic relationships of the western North American phoxinins (Actinopterygii: Cyprinidae) as inferred from mitochondrial 12S and 16S ribosomal RNA sequences. Mol. Phylogenet. Evol. 9(2): 308–329.

Simons, A. M. and R. L. Mayden. 1999. Phylogenetic relationships of North American cyprinids and assessment of homology of the open posterior myodome. Copeia 1999(1): 13–21.

Simons, A. M., P. B. Berendzen, and R. L. Mayden. 2003. Molecular systematics of North American phoxinin genera (Actinopterygii: Cyprinidae) inferred from mitochondrial 12S and 16S ribosomal RNA sequences. Zool. J. Linn. Soc. 139(1): 63–80.

Simons, A. M., K. E. Knott, and R. L. Mayden. 2000. Assessment of monophyly of the minnow genus *Pteronotropis* (Teleostei: Cyprinidae). Copeia 2000(4): 1068–1075.

Sisk, M. E. 1966. Unusual spawning behavior of the Northern Creek Chub, *Semotilus atromaculatus* (Mitchill). Trans. Kentucky Acad. Sci. 27(1–2): 3–4.

Sisk, M. E. 1973. Six additions to the known piscine fauna of Kentucky. Trans. Kentucky Acad. Sci. 34: 49–50.

Sisk, M. E. and R. R. Stephens. 1964. *Menidia audens* (Pisces: Atherinidae) in Boomer Lake, Oklahoma, and its possible spread in the Arkansas River system. Proc. Oklahoma Acad. Sci. 44: 71–73.

Sisk, M. E. and D. H. Webb. 1976. Distribution and habitat

preference of *Etheostoma histrio* in Kentucky. Trans. Kentucky Acad. Sci. 37(1–2): 33–34.

Skalski, G. T., J. B. Landis, M. J. Grose, and S. P. Hudman. 2008. Genetic structure of Creek Chub, a headwater minnow, in an impounded river system. Trans. Am. Fish. Soc. 137(4): 962–975.

Skyfield, J. P. 2006. Microhabitat use and movements of Gilt Darters (*Percina evides*) in two southeastern streams. M.S. thesis, Univ. Georgia, Athens. 76 pp.

Skyfield, J. P. and G. D. Grossman. 2008. Microhabitat use, movements and abundance of Gilt Darters (*Percina evides*) in southern Appalachian (USA) streams. Ecol. Freshw. Fish 17(2): 219–230.

Slack, W. T. 1996. Fringing flood plains and assemblage structure of fishes in the DeSoto National Forest, Mississippi. Ph.D. diss., Univ. Southern Mississippi, Hattiesburg.

Slack, W. T., M. T. O'Connell, T. L. Peterson, J. A. Ewing III, and S. T. Ross. 1995. Status of the Southern Redbelly Dace, *Phoxinus erythrogaster*, in Hatcher Bayou and streams of the Yazoo drainage, Mississippi. SFC Proc. 32: 1–9.

Sliger, W. A. 1967. The embryology, egg structure, micropyle, and egg membranes of the Plains Minnow, *Hybognathus placitus* (Girard). M.S. thesis, Oklahoma State Univ., Stillwater. 55 pp.

Sloss, B. L., N. Billington, and B. M. Burr. 2004. A molecular phylogeny of the Percidae (Teleostei, Perciformes) based on mitochondrial DNA sequence. Mol. Phylogenet. Evol. 32(2): 545–562.

Smale, M. A. and C. F. Rabeni. 1995a. Hypoxia and hyperthermia tolerances of headwater stream fishes. Trans. Am. Fish. Soc. 124(5): 698–710.

Smale, M. A. and C. F. Rabeni. 1995b. Influences of hypoxia and hyperthermia on fish species composition in headwater streams. Trans. Am. Fish. Soc. 124(5): 711–725.

Smart, H. J. and J. H. Gee. 1979. Coexistence and resource partitioning in two species of darters (Percidae), *Etheostoma nigrum* and *Percina maculata*. Can. J. Zool. 57(10): 2061–2071.

Smiley, P. C., Jr., E. D. Dibble, and S. H. Schoenholtz. 2006. Spatial and temporal variation of Goldstripe Darter abundance in first-order streams in north-central Mississippi. Am. Midl. Nat. 156(1): 23–36.

Smith, A. J., J. H. Howell, and G. W. Piavis. 1968. Comparative embryology of five species of lampreys of the upper Great Lakes. Copeia 1968(3): 461–469.

Smith, B. D. 1978. Variation in Mississippian settlement patterns. Pages 479–503. *In*: B. D. Smith, ed. Mississippian settlement patterns. Academic Press, New York. 536 pp.

Smith, B. D. 1986. Archaeology of the southeastern United States: from Dalton to DeSoto 10,500–500 B.P. *In*: F. Wendorf and A. E. Close, eds. Advances in world archaeology, Vol. 5. Academic Press, New York. 374 pp.

Smith, B. G. 1908. The spawning habits of *Chrosomus erythrogaster* Rafinesque. Biol. Bull. 14(6): 9–18.

Smith, C. D., T. E. Neebling, and M. C. Quist. 2010. Population dynamics of the Sand Shiner (*Notropis stramineus*) in non-wadeable rivers of Iowa. J. Freshw. Ecol. 25(4): 617–626.

Smith, C. G. 1977. The biology of three species of *Moxostoma* (Pisces: Catostomidae) in Clear Creek, Hocking and Fairfield counties, Ohio, with emphasis on the Golden Redhorse, *M. erythrurum* (Rafinesque). Ph.D. diss. Ohio State Univ., Columbus. 158 pp.

Smith, C. L. 1962. Some Pliocene fishes from Kansas, Oklahoma, and Nebraska. Copeia 1962(3): 505–520.

Smith, C. L. 1985. The inland fishes of New York State. New York State Dept. Environmental Conservation, Albany. 522 pp.

Smith, C. L. 1988. Minnows first, then trout. Fisheries 13(4): 4–8.

Smith, D. M., S. A. Welsh, and P. J. Turk. 2011. Selection and preference of benthic habitat by small and large ammocoetes of the Least Brook Lamprey (*Lampetra aepyptera*). Environ. Biol. Fishes 91(4): 421–428.

Smith, G. R. 1981. Late Cenozoic freshwater fishes of North America. Annu. Rev. Ecol. Syst. 12: 163–193.

Smith, G. R. 1992. Phylogeny and biogeography of the Catostomidae, freshwater fishes of North America and Asia. Pages 778–826. *In*: R. L. Mayden, ed. Systematics, historical ecology, and North American freshwater fishes. Stanford University Press, Stanford, CA. 969 pp.

Smith, G. R. and J. G. Lundberg. 1972. The Sand Draw fish fauna. Pages 40–54. *In*: M. F. Skinner and C. W. Hubbard, eds. Early Pleistocene pre-glacial and glacial rocks and faunas of north-central Nebraska. Bull. Am. Mus. Nat. Hist. 148.

Smith, G. R. and R. F. Stearley. 1989. The classification and scientific names of Rainbow and Cutthroat trouts. Fisheries 14(1): 4–10.

Smith, G. R., C. Badgley, T. P. Eiting, and P. S. Larson. 2010. Species diversity gradients in relation to geological history in North American freshwater fishes. Evol. Ecol. Res. 12(6): 693–726.

Smith, G. R., T. E. Dowling, K. W. Gobalet, T. Lugaski, D. K. Shiozawa, and R. P. Evans. 2002. Biogeography and timing of evolutionary events among Great Basin fishes. Pages 175–234. *In*: R. Hershler, D. B. Madsen, and D. R. Currey, eds. Great Basin aquatic systems history. Smithsonian Inst. Press, Washington, DC.

Smith, J. J., S. Kuraku, C. Holt, T. Sauka-Spengler, N. Jiang, M. S. Campbell, M. D. Yandell, T. Manousaki, A. Meyer, O. E. Bloom, J. R. Morgan, J. D. Buxbaum, R. Sachidanandam, C. Sims, A. S. Garruss, M. Cook, R. Krumlauf, L. M. Wiedemann, S. A. Sower, W. A. Decatur, J. A. Hall, C. T. Amemiya, N. R. Saha, K. M. Buckley, J. P. Rast, S. Das, M. Hirano, N. McCurley, P. Guo, N. Rohner, C. J. Tabin, P. Piccinelli, G. Elgar, M. Ruffier, B. L. Aken, S. M. J. Searle, M. Muffato, M. Pignatelli, J. Herrero, M. Jones, C. T. Brown, Y.-W. Chung-Davidson, K. G. Nanlohy,

S. V. Libants, C.-Y. Yeh, D. W. McCauley, J. A. Langeland, Z. Pancer, B. Fritzsch, P. J. de Jong, B. Zhu, L. L. Fulton, B. Theising, P. Flicek, M. E. Bronner, W. C. Warren, S. W. Clifton, R. K. Wilson, and W. Li. 2013. Sequencing of the Sea Lamprey (*Petromyzon marinus*) genome provides insights into vertebrate evolution. Nat. Genet. 45(4): 415–421.

Smith, K. L., W. F. Pell, J. H. Rettig, R. H. Davis, and H. W. Robison. 1984. Arkansas's natural heritage. August House, Little Rock. 116 pp.

Smith, M. H., K. T. Scribner, J. D. Hernandez, and M. C. Wooten. 1989. Demographic, spatial, and temporal genetic variation in *Gambusia*. Pages 235–257. *In:* G. K. Meffe and F. F. Snelson, Jr., eds. Ecology and evolution of livebearing fishes (Poeciliidae). Prentice Hall, Englewood Cliffs, NJ.

Smith, N. A., R. E. Condrey, and B. C. Reed. 2009. The feeding ecology of Paddlefish in the Mermentau River, Louisiana. Pages 51–62. *In:* C. P. Paukert and G. D. Scholten, eds. Paddlefish management, propagation, and conservation in the 21st century: building from 20 years of research and management. Symp. 66, American Fisheries Society, Bethesda, MD. 443 pp.

Smith, O. A. 2008. Reproductive potential and life history of Spotted Gar *Lepisosteus oculatus* in the upper Barataria Estuary, Louisiana. M.S. thesis, Nicholls State Univ., Thibodaux, LA. 104 pp.

Smith, P. J., S. M. McVeagh, and D. Steinke. 2008. DNA barcoding for the identification of smoked fish products. J. Fish Biol. 72(2): 464–471.

Smith, P. W. 1979. The fishes of Illinois. Univ. Illinois Press, Urbana. 314 pp.

Smith, P. W. and L. M. Page. 1969. The food of Spotted Bass in streams of the Wabash River drainage. Trans. Am. Fish. Soc. 98(4): 647–651.

Smith, R. K. and K. D. Fausch. 1997. Thermal tolerance and vegetation preference of Arkansas Darter and Johnny Darter from Colorado plains streams. Trans. Am. Fish. Soc. 126(4): 676–686.

Smith, R. J. F. and B. D. Murphy. 1974. Functional morphology of the dorsal pad in Fathead Minnows (*Pimephales promelas* Rafinesque). Trans. Am. Fish. Soc. 103(1): 65–72.

Smith, S. M., M. J. Maceina, and R. A. Dunham. 1994. Natural hybridization between Black Crappie and White Crappie in Weiss Lake, Alabama. Trans. Am. Fish. Soc. 123(1): 71–79.

Smith, T. A., T. C. Mendelson, and L. M. Page. 2011. AFLPs support deep relationships among darters (Percidae: Etheostomatinae) consistent with morphological hypotheses. Heredity 107(6): 579–588.

Smith, W. L. 2005. The limits and relationships of mail-cheeked fishes (Teleostei: Percomorpha) and the evolution of venom in fishes. Ph.D. diss., Columbia Univ., New York. 249 pp.

Smith, W. L. and W. C. Wheeler. 2004. Polyphyly of the mail-cheeked fishes (Teleostei: Scorpaeniformes): evidence from mitochondrial and nuclear sequence data. Mol. Phylogenet. Evol. 32(2): 627–646.

Smith, W. L. and M. S. Busby. 2014. Phylogeny and taxonomy of sculpins, sandfishes, and snailfishes (Perciformes: Cottoidei) with comments on the phylogenetic significance of their early-life-history specializations. Mol. Phylogenet. Evol. 79(1): 332–352.

Smith-Vaniz, W. F. 1968. Freshwater fishes of Alabama. Auburn Univ. Agricultural Experiment Station. 211 pp.

Smith-Vaniz, W. F. and R. M. Peck. 1997. Contributions of Henry Weed Fowler (1878–1965), with a brief early history of ichthyology at the Academy of Natural Sciences of Philadelphia. Pages 377–389. *In:* T. W. Pietsch and W. D. Anderson, Jr., eds. Collection building in ichthyology and herpetology. Special Publ. 3. American Society of Ichthyologists and Herpetologists, Lawrence, KS. 593 pp.

Smylie, M., V. Shervette, and C. McDonough. 2015. Prey composition and ontogenetic shift in coastal populations of Longnose Gar *Lepisosteus osseus*. J. Fish Biol. 87(4): 895–911.

Smylie, M., V. Shervette, and C. McDonough. 2016. Age, growth, and reproduction in two coastal populations of Longnose Gars. Trans. Am. Fish. Soc. 145(1): 120–135.

Snedden, G. A., W. E. Kelso, and D. A. Rutherford. 1999. Diel and seasonal patterns of Spotted Gar movement and habitat use in the lower Atchafalaya River basin, Louisiana. Trans. Am. Fish. Soc. 128(1): 144–154.

Snelson, F. F., Jr. 1968. Systematics of the cyprinid fish *Notropis amoenus*, with comments on the subgenus *Notropis*. Copeia 1968(4): 776–802.

Snelson, F. F., Jr. 1972. Systematics of the subgenus *Lythrurus*, genus *Notropis* (Pisces: Cyprinidae). Bull. Florida State Mus. 17(1): 1–92.

Snelson, F. F., Jr. 1973. Systematics and distribution of the Ribbon Shiner, *Notropis fumeus* (Cyprinidae), from the central United States. Am. Midl. Nat. 89(1): 166–191.

Snelson, F. F., Jr. 1980. *Notropis fumeus* Evermann, Ribbon Shiner. Page 265. *In:* D. S. Lee et al. Atlas of North American freshwater fishes. N.C. State Mus. Nat. Hist., Raleigh.

Snelson, F. F., Jr. 1991. Phylogenetic studies of North American minnows, with emphasis on the genus *Cyprinella* (Teleostei Cypriniformes) by R. L. Mayden. Review. Copeia 1991(1): 258–260.

Snelson, F. F., Jr. and R. E. Jenkins. 1973. *Notropis perpallidus*, a cyprinid fish from south-central United States: description, distribution, and life history aspects. Southwest. Nat. 18(3): 291–304.

Snelson F. F., Jr. and W. L. Pflieger. 1975. Redescription of the Redfin Shiner, *Notropis umbratilis*, and its subspecies in the central Mississippi River basin. Copeia 1975(2): 231–249.

Snow, R. A. and J. M. Long. 2015. Estimating spawning times of Alligator Gar (*Atractosteus spatula*) in Lake Texoma, Oklahoma. Proc. Oklahoma Acad. Sci. 95: 46–53.

Snow, R. A., M. J. Porta, and A. L. Robison. 2017. Seasonal diet composition of Black Bullhead (*Ameiurus melas*) in Lake Carl Etling, Oklahoma. Proc. Oklahoma Acad. Sci. 97: 54–60.

Snow, R. A., M. J. Porta, R. W. Smmons, Jr., and J. B. Bartnicki. 2018. Early life history characteristics and contribution of stocked juvenile Alligator Gar in Lake Texoma, Oklahoma. Proc. Oklahoma Acad. Sci. 98: 46–54.

Snyder, D. E. 2002. Pallid and Shovelnose sturgeon larvae: morphological description and identification. J. Appl. Ichthyol. 18(4–6): 240–265.

Soares, D. and M. L. Niemiller. 2013. Sensory adaptations of fishes to subterranean environments. BioScience 63(4): 274–283.

Soin, S. G. 1960. Reproduction and development of the Snakehead *Ophiocephalus argus warpachowskii* (Berg). USSR Acad. Sci. Issues Ichthyol. 15: 127–137 [in Russian].

Solomon, L. E., Q. E. Phelps, D. P. Herzog, C. J. Kennedy, and M. S. Taylor. 2013. Juvenile Alligator Gar movement patterns in a disconnected floodplain habitat in southeast Missouri. Am. Midl. Nat. 169(2): 336–344.

Song, C. B., T. J. Near, and L. M. Page. 1998. Phylogenetic relations among percid fishes as inferred from mitochondrial cytochrome *b* DNA sequence data. Mol. Phylogenet. Evol 10(3): 343–353.

South, E. J. and W. E. Ensign. 2013. Life history of *Campostoma oligolepis* (Largescale Stoneroller) in urban and rural streams. Southeast. Nat. 12(4): 781–789.

Southall, P. D. and W. A. Hubert. 1984. Habitat use by adult Paddlefish in the upper Mississippi River. Trans. Am. Fish. Soc. 113(2): 125–131.

Sowa, S. P. and C. F. Rabeni. 1995. Regional evaluation of the relation of habitat to distribution and abundance of Smallmouth Bass and Largemouth Bass in Missouri streams. Trans. Am. Fish. Soc. 124(2): 240–251.

Spall, R. D. 1970. Possible cases of cleaning symbiosis among freshwater fishes. Trans. Am. Fish. Soc. 99(3): 599–600.

Sparks, J. S. and W. L. Smith. 2004a. Phylogeny and biogeography of the Malagasy and Australasian rainbowfishes (Teleostei: Melanotaenioidei): Gondwanan vicariance and evolution in freshwater. Mol. Phylogenet. Evol. 33(3): 719–734.

Sparks, J. S. and W. L. Smith. 2004b. Phylogeny and biogeography of cichlid fishes (Teleostei: Perciformes: Cichlidae). Cladistics 20(6): 501–517.

Spataru, P. and M. Zorn. 1978. Food and feeding habits of *Tilapia aurea* (Steindachner) (Cichlidae) in Lake Kinneret (Israel). Aquaculture 13(1): 67–79.

Spaulding, S. A. and L. Elwell. 2007. Increase in nuisance blooms and geographic expansion of the freshwater diatom *Didymosphenia geminata*. U.S. Geological Survey Open File Report 2007-1425. 38 pp.

Speares, P., D. Holt, and C. E. Johnston. 2011. The relationship between ambient noise and dominant frequency of vocalizations in two species of darters (Percidae: *Etheostoma*). Environ. Biol. Fishes 90(1): 103–110.

Spier, T. W. and R. C. Heidinger. 2002. Effect of turbidity on growth of Black Crappie and White Crappie. N. Am. J. Fish. Mngt. 22(4): 1438–1441.

Sprague, J. W. 1960. Report of fisheries investigations during the fifth year of impoundment of Gavin's Point Reservoir, 1959. Dingell-Johnson Project. F-1-R-9. 47 pp.

Springer, V. G., G. D. Johnson, T. M. Orrell, and K. Darrow. 2004. Study of the dorsal gill-arch musculature of teleostome fishes, with special reference to the Actinopterygii. Bull. Biol. Soc. Washington 11: 353–365.

Stackhouse, R. A. 1982. A subsequent study of the fishes of the Saline River, south central Arkansas. M.S. thesis, Northeast Louisiana Univ., Monroe. 142 pp.

Stahl, M. T. 2008. Reproductive physiology of Shovelnose Sturgeon from the middle Mississippi River in relation to seasonal variation in plasma sex steroids, vitellogenin, calcium, and oocyte diameters. M.S. thesis, Southern Illinois Univ. Carbondale. 71 pp.

Stahl, M. T., G. W. Whitledge, and A. M. Kelly. 2009. Reproductive biology of middle Mississippi River Shovelnose Sturgeon: insights from seasonal and age variation in plasma sex steroid and calcium concentrations. J. Appl. Ichthyol. 25(S2): 75–82.

Stahr, K. J. and D. E. Shoup. 2015. American water willow mediates survival and antipredator behavior of juvenile Largemouth Bass. Trans. Am. Fish. Soc. 144(5): 903–910.

Standage, R. W. 1999. Management indicator species. Pages 171–176. *In:* Chapter 2, USDA, Forest Service. Ozark-Ouachita Highlands assessment: aquatic conditions. Report 3 of 5. Gen. Tech. Rep. SRS-33. U.S. Dept. Agriculture, Forest Service, Southern Research Station, Asheville, NC. 317 pp.

Stanley, J. G. 1976. Reproduction of the Grass Carp (*Ctenopharyngodon idella*) outside its native range. Fisheries 1(3): 7–10.

Stanley, J. G. and D. S. Danie. 1983. Species profile: life histories and environmental requirements of coastal fishes and invertebrates (North Atlantic): White Perch. U.S. Fish and Wildlife Service, Div. Biol. Serv., FWS/OBS-82/11.7. 12 pp.

Stanley, J. G., W. W. Miley II, and D. L. Sutton. 1978. Reproductive requirements and likelihood for naturalization of escaped Grass Carp in the United States. Trans. Am. Fish. Soc. 107(1): 119–128.

Stark, W. J. and A. A. Echelle. 1998. Genetic structure and systematics of Smallmouth Bass, with emphasis on Interior Highlands populations. Trans. Am. Fish. Soc. 127(3): 393–416.

Starks, T. A. and J. M. Long. 2017. Diet composition of age-0 fishes in created habitats of the lower Missouri River. Am. Midl. Nat. 178(1): 112–122.

Starks, T. A., M. L. Miller, and J. M. Long. 2016. Early life history of three pelagic-spawning minnows *Macrhybopsis* spp. in the lower Missouri River. J. Fish Biol. 88(4): 1335–1349.

Starnes, L. B. and W. C. Starnes. 1985. Ecology and life history of the Mountain Madtom, *Noturus eleutherus* (Pisces: Ictaluridae). Am. Midl. Nat. 114(2): 331–341.

Starnes, W. C. and D. A. Etnier. 1980. Fishes. Pages B1–B134. *In:* D. C. Eager and R. M. Hatcher, eds. Tennessee's rare wildlife. Tennessee Dept. Conservation, Nashville.

Starostka, V. J. and R. L. Applegate. 1970. Food selectivity of Bigmouth Buffalo, *Ictiobus cyprinellus*, in Lake Poinsett, South Dakota. Trans. Am. Fish. Soc. 99(3): 571–576.

Starrett, W. C. 1950a. Food relationships of the minnows of the Des Moines River, Iowa. Ecology 31(2): 216–233.

Starrett, W. C. 1950b. Distribution of the fishes of Boone County, Iowa, with special reference to the minnows and darters. Am. Midl. Nat. 43(1): 112–127.

Starrett, W. C. 1951. Some factors affecting the abundance of minnows in the Des Moines River, Iowa. Ecology 32(1): 13–27.

Starrett, W. C., W. J. Harth, and P. W. Smith. 1960. Parasitic lampreys of the genus *Ichthyomyzon* in the rivers of Illinois. Copeia 1960(4): 337–346.

Stearley, R. F. and G. R. Smith. 1993. Phylogeny of the Pacific trouts and salmons (*Oncorhynchus*) and genera of the family Salmonidae. Trans. Am. Fish. Soc. 122(1): 1–33.

Stearman, L. W. 2011. Fish assemblage structure and the life history and ecology of *Etheostoma whipplei* in the Fayetteville shale natural gas basin. M.S. thesis, Univ. Central Arkansas, Conway. 125 pp.

Stearman, L. W., G. Adams, and R. Adams. 2015. Ecology of the Redfin Darter and a potential emerging threat to its habitat. Environ. Biol. Fish. 98(2): 623–635.

Steele, K. F. 1985. Groundwater in northwest Arkansas. Arkansas Nat. 3(7): 5–10.

Steffensen, K. D., D. A. Shuman, and S. Stukel. 2014. The status of fishes in the Missouri River, Nebraska: Shoal Chub (*Macrhybopsis hyostoma*), Sturgeon Chub (*M. gelida*), Sicklefin Chub (*M. meeki*), Silver Chub (*M. storeriana*), Flathead Chub (*Platygobio gracilis*), Plains Minnow (*Hybognathus placitus*), Western Silvery Minnow (*H. argyritis*), and Brassy Minnow (*H. hankinsoni*). Trans. Nebraska Acad. Sci. Affil. Soc. 34: 49–67.

Steffensen, K. D., L. A. Powell, and M. A. Pegg. 2017. Using the robust design framework and relative abundance to predict the population size of Pallid Sturgeon *Scaphirhynchus albus* in the lower Missouri River. J. Fish Biol. 91(5): 1378–1391.

Stein, D. W., J. S. Rogers, and R. C. Cashner. 1985. Biochemical systematics of the *Notropis roseipinnis* complex (Cyprinidae: subgenus *Lythrurus*). Copeia 1985(1): 154–163.

Steinberg, R., L. M. Page, and J. C. Porterfield. 2000. The spawning behavior of the Harlequin Darter, *Etheostoma histrio* (Osteichthyes: Percidae). Ich. Explor. Freshwaters 11(2): 141–148.

Steindachner, F. 1864. Ichthyologische Mittheilungen. VII. Verhandlungen der K.-K. Zoologisch-Botanischen Gesellschaft in Wien 14: 223–232.

Steindachner, F. 1866. Ichthyologische Mittheilungen. IX. Verhandlungen der K.-K. Zoologisch-Botanischen Gesellschaft in Wien 16: 761–796.

Stell, E. G., J. J. Hoover, B. A. Cage, D. Hardesty, and G. R. Parsons. 2018. Long-distance movements of four *Polyodon spathula* (Paddlefish) from a remote oxbow lake in the lower Mississippi River basin. Southeast. Nat. 17(2): 230–238.

Stephens, R. R. 1985. The lateral-line system of the Gizzard Shad, *Dorosoma cepedianum* Lesueur (Pisces: Clupeidae). Copeia 1985(3): 540–556.

Stephens, R. R. 2016. Description and comparison of the canals and canal branching pattern of the lateral line canal system of *Dorosoma petenense* (Günther) [Clupeomorpha: Clupeoidei]. Copeia 2016(2): 387–392.

Stephens, W. W. and J. K. Beadles. 1979. Effects of cropping on growth of Channel Catfish. Proc. 33rd Annu. Conf. Southeast. Assoc. Fish Wildl. Agencies: 572–583.

Stepien, C. A. and A. E. Haponski. 2010. Systematics of the Greenside Darter *Etheostoma blennioides* complex: consensus from nuclear and mitochondrial DNA sequences. Mol. Phylogenet. Evol. 57(1): 434–447.

Stevenson, J. H. 1965. Observations on Grass Carp in Arkansas. Prog. Fish-Cult. 27(4): 203–206.

Stevenson, J. H. and A. H. Hulsey. 1961. Vertical distribution of dissolved oxygen and water temperatures in Lake Hamilton with special reference to suitable Rainbow Trout habitat. Proc. 15th Annu. Conf. Southeast. Assoc. Fish Wildl. Agencies: 245–255.

Stevenson, M. M. 1971. *Percina macrolepida* (Pisces, Percidae, Etheostomatinea), a new percid fish of the subgenus *Percina* from Texas. Southwest. Nat. 16(1): 65–83.

Stevenson, M. M. and B. A. Thompson. 1978. Further distribution records for the Bigscale Logperch, *Percina macrolepida* (Osteichthyes: Percidae) from Oklahoma, Texas, and Louisiana with notes on its occurrence in California. Southwest. Nat. 23(2): 309–313.

Stewart, C. A. 1988. Brief note: Diet of the Rainbow Darter (*Etheostoma caeruleum*) in Rock Run, Clark County, Ohio. Ohio J. Sci. 88(5): 198–200.

Stewart, K. W. and D. A. Watkinson. 2004. The freshwater fishes of Manitoba. Univ. Manitoba Press, Winnipeg. 276 pp.

Stewart, N. H. 1927. Development, growth, and food habits of the White Sucker, *Catostomus commersonii* Lesueur. Bull. U.S. Bur. Fish. 42(1): 147–184.

Stiassny, M. L. J. 1996. An overview of freshwater biodiversity: with some lessons from African fishes. Fisheries 21(9): 7–13.

Stiassny, M. L. J., G. G. Teugels, and C. D. Hopkins. 2007. The fresh and brackish water fishes of lower Guinea, west-central Africa—Vol. 2. Musée Royal de l'Afrique Centrale. 269 pp.

Stockard, C. R. 1907. Observations on the natural history of *Polyodon spathula*. Am. Nat. 41(492): 753–766.

Stoeckel, J. N. and J. M. Caldwell. 2014. Ouachita Darter habitat use and distribution in the upper Ouachita River. Final Report to the U.S. Forest Service. 87 pp.

Stocckel, J. N. and R. C. Heidinger. 1989. Reproductive biology of the Inland Silverside, *Menidia beryllina* in southern Illinois. Trans. Illinois State Acad. Sci. 82: 59–70.

Stoeckel, J. N. and R. C. Heidinger. 1992. Relative susceptibility of Inland and Brook silversides to capture by Largemouth Bass. N. Am. J. Fish. Manag. 12(3): 499–503.

Stoeckel, J. N., C. J. Gagen, and R. W. Standage. 2011. Feeding and reproductive biology of Ouachita Madtom. Pages 267–279. *In:* P. H. Michaletz and V. H. Travnichek, eds. Conservation, ecology, and management of catfish: the second international symposium. Symp. 77. American Fisheries Society, Bethesda, MD. 800 pp.

Stone, N., E. Park, L. Dorman, and H. Thomforde. 1997. Baitfish culture in Arkansas: Golden Shiners, Goldfish, and Fathead Minnows. Publication MP 396, Arkansas Cooperative Extension Service, Univ. Arkansas at Pine Bluff. 68 pp.

Stoops, S. B., P. Fleming, G. P. Garrett, and R. Deaton. 2013. Gonopodial structures revisited: variation in genital morphology within and across four populations of the Western Mosquitofish (*Gambusia affinis*) in Texas. Southwest. Nat. 58(1): 97–101.

Storer, D. H. 1845. [No title. Original description of *Fundulus olivaceus* and *Etheostoma caeruleum*.] Proc. Boston Soc. Nat. Hist. 2 (1845–1848): 47–51.

Storer, D. H. 1846. A synopsis of the fishes of North America. Mem. Am. Acad. Arts Sci. 2(7): 253–550. [Reprinted by A. Asher and Co., Amsterdam, 1972.]

Storey, K. W., J. W. Schlechte, and L. T. Fries. 2000. Field identification accuracy for White Bass and hybrid Striped Bass. Proc. Annu. Conf. Southeast. Assoc. Fish Wildl. Agencies 54: 97–106.

Stout, J. F. 1975. Sound communication during the reproductive behavior of *Notropis analostanus* (Pisces: Cyprinidae). Am. Midl. Nat. 94(2): 296–325.

Strange, R. M. 1991. Diet selectivity in the Johnny Darter, *Etheostoma nigrum*, in Stinking Fork, Indiana. J. Freshw. Ecol. 6(4): 377–381.

Strange, R. M. 1993. Seasonal feeding ecology of the Fantail Darter, *Etheostoma flabellare*, from Stinking Fork, Indiana. J. Freshwater Ecol. 8(1): 13–18.

Strange, R. M. and B. M. Burr. 1997. Intraspecific phylogeography of North American highland fishes: a test of the Pleistocene vicariance hypothesis. Evolution 51(3): 885–897.

Strange, R. M. and R. L. Mayden. 2009. Phylogenetic relationships and a revised taxonomy for North American cyprinids currently assigned to *Phoxinus* (Actinopterygii: Cyprinidae). Copeia 2009(3): 494–501.

Strawn, K. 1965. Resistance of Threadfin Shad to low temperatures. Proc. 17th Annu. Conf. Southeast. Assoc. Game Fish Comm.: 290–293.

Strayer, D. L. 2006. Challenges for freshwater invertebrate conservation. J. N. Am. Benthol. Soc. 25(2): 271–287.

Strayer D. L. and D. Dudgeon. 2010. Freshwater biodiversity conservation: recent progress and future challenges. J. N. Am. Benthol. Soc. 29(1): 344–358.

Strong, D. R. and R. W. Pemberton. 2000. Biological control of invading species-risk and reform. Science 288(5473): 1969–1970.

Strongin, K., C. M. Taylor, M. E. Roberts, W. H. Neill, and F. Gelwick. 2011. Food habits and dietary overlap of two silversides in the Tennessee-Tombigbee waterway: the invasive *Menidia audens* versus the native *Labidesthes sicculus*. Am. Midl. Nat. 166(1): 224–233.

Stukel, E. D. 2001. Sturgeon and Sicklefin chubs. South Dakota Conserv. Digest (May/June): p. 25.

Sturgess, J. A. 1976. Taxonomic status of *Percina* in California. California Fish Game 62(1): 79–81.

Sublette, J. E., M. D. Hatch, and M. Sublette. 1990. The fishes of New Mexico. Univ. New Mexico Press, Albuquerque. 393 pp.

Suckley, G. 1861. Notices of certain new species of North American Salmonidae, chiefly in the collection of the N. W. Boundary Commission, in charge of Archibald Campbell, Esq., Commissioner of the United States, collected by Doctor C. B. R. Kennerly, naturalist to the … Ann. Lyceum Nat. Hist. N. Y. 7(art. 30): 306–313.

Sulak, K. J. 1975. Cleaning behaviour in the centrarchid fishes, *Lepomis macrochirus* and *Micropterus salmoides*. Animal Behav. 23, Part 2: 331–334.

Sule, M. J. and T. M. Skelly. 1985. The life history of the Shorthead Redhorse, *Moxostoma macrolepidotum*, in the Kankakee River drainage, Illinois. Illinois Nat. Hist. Surv. Biol. Notes 123: 1–15.

Sullivan, C. L. 2009. Zooplankton, Gizzard Shad, and Freshwater Drum: interactions in a Great Plains irrigation reservoir. M.S. thesis, Univ. Nebraska at Kearney. 139 pp.

Sullivan, J. J. 1994. Comparative food habits of *Noturus exilis* and *Cottus carolinae* in the Illinois River drainage. M.S. thesis, Univ. Arkansas, Fayetteville. 67 pp.

Sullivan, J. P., J. G. Lundberg, and M. Hardman. 2006. A phylogenetic analysis of the major groups of catfishes (Teleostei: Siluriformes) using *rag1* and *rag2* nuclear gene sequences. Mol. Phylogenet. Evol. 41(3): 636–662.

Summerfelt, R. C. and C. O. Minckley. 1969. Aspects of the life history of the Sand Shiner, *Notropis stramineus* (Cope), in the Smoky Hill River, Kansas. Trans. Am. Fish. Soc. 98(3): 444–453.

Summerfelt, R. C., P. E. Mauck, and G. Mensinger. 1972. Food habits of River Carpsucker and Freshwater Drum in four Oklahoma reservoirs. Proc. Oklahoma Acad. Sci. 52: 19–26.

Suttkus, R. D. 1958. Status of the nominal cyprinid species

Moniana deliciosa Girard and *Cyprinella texana* Girard. Copeia 1958(4): 307–318.

Suttkus, R. D. 1963. Order Lepisostei. Pages 61–88. *In:* H. B. Bigelow, C. M. Cohen, G. W. Mead, D. Merriam, Y. H. Olsen, W. C. Schroeder, L. P. Schultz, and J. Tee-Van, eds. Fishes of the Western North Atlantic Memoir 1 (part 3) of the Sears Foundation for Marine Research. Yale Univ., New Haven, CT.

Suttkus, R. D. 1980. *Notropis candidus*, a new cyprinid fish from the Mobile Bay basin, and a review of the nomenclatural history of *Notropis shumardi* (Girard). Bull. Alabama Mus. Nat. Hist. 5: 1–15.

Suttkus, R. D. 1985. Identification of the percid, *Ioa vigil* Hay. Copeia 1985(1): 225–227.

Suttkus, R. D. 1991. *Notropis rafinesquei*, a new cyprinid fish from the Yazoo River system in Mississippi. Bull. Alabama Mus. Nat. Hist. 10: 1–9.

Suttkus, R. D. and R. M. Bailey. 1990. Characters, relationships, distribution, and biology of *Notropis melanostomus*, a recently named cyprinid fish from southeastern United States. Occas. Pap. Mus. Zool. Univ. Michigan 722: 1–15.

Suttkus, R. D. and H. L. Bart, Jr. 2002. A preliminary analysis of the River Carpsucker, *Carpiodes carpio* (Rafinesque), in the southern portion of its range. Libro Jubilar en Honor al Dr. Salvador Contreras Balderas. Universidad Autonoma de Nuevo Leon, pp. 209–221.

Suttkus, R. D. and R. C. Cashner. 1981. A new species of cyprinodontid fish, genus *Fundulus* (*Zygonectes*), from Lake Pontchartrain tributaries in Louisiana and Mississippi. Bull. Alabama Mus. Nat. Hist. 6: 1–17.

Suttkus, R. D. and G. H. Clemmer. 1968. *Notropis edwardraneyi*, a new cyprinid fish from the Alabama and Tombigbee river systems and a discussion of related species. Tulane Stud. Zool. Bot. 15: 18–39.

Suttkus, R. D. and J. V. Conner. 1980. The Rainbow Smelt, *Osmerus mordax*, in the lower Mississippi River near St. Francisville, Louisiana. Am. Midl. Nat. 104(2): 394.

Suttkus, R. D. and M. F. Mettee. 2001. Analysis of four species of *Notropis* included in the subgenus *Pteronotropis* Fowler, with comments on relationships, origin, and dispersion. Geol. Surv. Alabama Bull. 170: 50 pp.

Suttkus, R. D. and J. S. Ramsey. 1967. *Percina aurolineata*, a new percid fish from the Alabama River system and a discussion of the ecology, distribution, and hybridization of darters of the subgenus *Hadropterus*. Tulane Stud. Zool. 13(4): 129–145.

Suttkus, R. D. and B. A. Thompson. 2002. The rediscovery of the Mississippi Silverside, *Menidia audens*, in the Pearl River drainage in Mississippi and Louisiana. SFC Proc. 44: 6–10.

Suttkus, R. D., B. A. Thompson, and H. L. Bart, Jr. 1994. Two new darters, *Percina* (*Cottogaster*), from the southeastern United States, with a review of the subgenus. Occas. Pap. Tulane Univ. Mus. Nat. Hist. 4: 1–46.

Suttkus, R. D., B. A. Thompson, and J. K. Blackburn. 2005. An analysis of the *Menidia* complex in the Mississippi River valley and in two nearby minor drainages. SFC Proc. 48: 1–9.

Sutton, K. 1986. The age of discovery and settlement. Arkansas Game and Fish 17(3): 2–12.

Swain, J. 1883. A description of a new species of *Hadropterus* (*Hadropterus scierus*) from southern Indiana. Proc. U.S. Natl. Mus. 6(379): 252.

Swedberg, D. V. 1968. Food and growth of the Freshwater Drum in Lewis and Clark Lake, South Dakota. Trans. Am. Fish. Soc. 97(4): 442–447.

Swedberg, D. V. and C. H. Walburg. 1970. Spawning and early life history of the Freshwater Drum in Lewis and Clark Lake, Missouri River. Trans. Am. Fish. Soc. 99(3): 560–570.

Swift, C. C. 1970. A review of the eastern North American cyprinid fishes of the *Notropis texanus* species group (subgenus *Alburnops*), with a definition of the subgenus *Hydrophlox*, and materials for a revision of the subgenus *Alburnops*. Ph.D. diss., Florida State Univ., Tallahassee. 515 pp.

Swift, C. C., R. W. Yerger, and P. R. Parrish. 1977. Distribution and natural history of the fresh and brackish water fishes of the Ochlockonee River, Florida and Georgia. Bull. Tall Timbers Res. Sta. 20: 1–111.

Swift, C. C., C. R. Gilbert, S. A. Bortone, G. H. Burgess, and R. W. Yerger. 1986. Zoogeography of the freshwater fishes of the southeastern United States: Savannah River to Lake Pontchartrain. Pages 213–265. *In:* C. H. Hocutt and E. O. Wiley, eds. The zoogeography of North American freshwater fishes. John Wiley and Sons, New York.

Swift, C. C., S. Howard, J. Mulder, D. J. Pondella II, and T. P. Keegan. 2014. Expansion of the non-native Mississippi Silverside, *Menidia audens* (Pisces, Atherinopsidae), into fresh and marine waters of coastal southern California. Bull. South. California Acad. Sci. 113(3): 153–164.

Swofford, D. L. 1982. Genetic variability, population differentiation, and biochemical relationships in the family Amblyopsidae. M.S. thesis, Eastern Kentucky Univ., Richmond.

Taber, C. 1965. Spectacle development in the Pygmy Sunfish, *Elassoma zonatum*, with observations on spawning habits. Proc. Oklahoma Acad. Sci. 45: 73–78.

Taber, C. A., B. A. Taber, and M. S. Topping. 1986. Population structure, growth and reproduction of the Arkansas Darter, *Etheostoma cragini* (Percidae). Southwest. Nat. 31(2): 207–214.

Tafanelli, R., P. E. Mauck, and G. Mensinger. 1971. Food habits of Bigmouth and Smallmouth Buffalo from four Oklahoma reservoirs. Proc. Annu. Conf. Southeast. Assoc. Game Fish Comm. 24: 649–658.

Tan, M. and J. W. Armbruster. 2018. Phylogenetic classification of extant genera of fishes of the order Cypriniformes (Teleostei: Ostariophysi). Zootaxa 4476(1): 006–039.

Tanyolac, J. 1973. Morphometric variation and life history of the cyprinid fish *Notropis stramineus* (Cope). Occas. Pap. Mus. Nat. Hist., Univ. Kansas 12: 1–28.

Tao W., M. Zou, X. Wang, X. Gan, R. L. Mayden, and H. Shunping. 2010. Phylogenomic analysis resolves the formerly intractable adaptive diversification of the endemic clade of East Asian Cyprinidae (Cypriniformes). PLOS ONE 5(10): e13508. Available: https://doi.org/10.1371/journal.pone.0013508.

Tarplee, W. H., Jr., D. E. Louder, and A. J. Weber. 1971. Evaluation of the effects of channelization on fish populations in North Carolina's Coastal Plain streams. North Carolina Wildlife Res. Comm., Raleigh. 26 pp.

Tatarenkov, A., F. Barreto, D. L. Winkelman, and J. C. Avise. 2006. Genetic monogamy in the Channel Catfish, *Ictalurus punctatus*, a species with uniparental nest guarding. Copeia 2006(4): 735–741.

Tate, M., R. E. McGoran, C. R. White, and S. J. Portugal. 2017. Life in a bubble: the role of the labyrinth organ in determining territory, mating and aggressive behaviours in anabantoids. J. Fish Biol. 91(3): 723–749.

Taylor, C. A. and B. M. Burr. 1997. Reproductive biology of the Northern Starhead Topminnow, *Fundulus dispar* (Osteichthyes: Fundulidae), with a review of data for freshwater members of the genus. Am. Midl. Nat. 137(1): 151–164.

Taylor, C. M. 1994. Distribution and abundance of *Lythrurus snelsoni* (Robison), an endemic species from the Ouachita Mountain uplift. USDA Forest Service Final Report. Hot Springs, AR. 29 pp.

Taylor, C. M. 2000. A large-scale comparative analysis of riffle and pool fish communities in an upland stream system. Environ. Biol. Fishes 58(1): 89–95.

Taylor, C. M. and P. W. Lienesch. 1995. Environmental correlates of distribution and abundance for *Lythrurus snelsoni*: a range-wide analysis of an endemic fish species. Southwest. Nat. 40(4): 373–378.

Taylor, C. M. and P. W. Lienesch. 1996. Regional parapatry of the congeneric cyprinids *Lythrurus snelsoni* and *L. umbratilis*: species replacement along a complex environmental gradient. Copeia 1996(2): 493–497.

Taylor, C. M. and R. J. Miller. 1990. Reproductive ecology and population structure of the Plains Minnow, *Hybognathus placitus* (Pisces: Cyprinidae), in central Oklahoma. Am. Midl. Nat. 123(1): 32–39.

Taylor, C. M. and S. M. Norris. 1992. Notes on the reproductive cycle of *Notropis hubbsi* (Bluehead Shiner) in southeastern Oklahoma. Southwest. Nat. 37(1): 89–92.

Taylor, C. M. and M. L. Warren, Jr. 2001. Dynamics in species composition of stream fish assemblages: environmental variability and nested subsets. Ecology 82(8): 2320–2330.

Taylor, C. M., D. S. Millican, M. E. Roberts, and W. T. Slack. 2008. Long-term change to fish assemblages and the flow regime in a southeastern U.S. river system after extensive aquatic ecosystem fragmentation. Ecography 31(6): 787–797.

Taylor, J. N., W. R. Courtenay, Jr., and J. A. McCann. 1984. Known impacts of exotic fishes in the continental United States. Pages 322–373. *In*: W. R. Courtenay, Jr. and J. R. Stauffer, Jr., eds. Distribution, biology, and management of exotic fishes. Johns Hopkins Univ. Press, Baltimore.

Taylor, S. S., S. Woltmann, A. Rodriguez, and W. E. Kelso. 2013. Hybridization of White, Yellow, and Striped bass in the Toledo Bend Reservoir. Southeast. Nat. 12(3): 514–522.

Taylor, W. R. 1969. Revision of the catfish genus *Noturus* Rafinesque, with an analysis of higher groups in the Ictaluridae. Bull. U.S. Natl. Mus. 282: 1–315.

Teletchea, F. 2009. Molecular identification methods of fish species: reassessment and possible applications. Rev. Fish Biol. Fish. 19(3): 265–293.

Teletchea, F. 2010. After 7 years and 1000 citations: comparative assessment of the DNA barcoding and the DNA taxonomy proposals for taxonomists and nontaxonomists. Mitochondrial DNA 21(6): 206–226.

Tesch, F. W. and J. E. Thorpe. 2003. The eel. 5th ed. Translated from the German by R. J. White and edited by J. E. Thorpe. Blackwell Science, Oxford. 408 pp.

Teugels, G. G. and J. Daget. 1984. *Parachanna* nom. nov. for the African snakeheads and rehabilitation of *Parachanna insignis* (Sauvage, 1884) (Pisces, Channidae). Cybium 8(4): 1–7.

Thibault, R. E. and R. J. Schultz. 1978. Reproductive adaptations among viviparous fishes (Cyprinodontiformes: Poeciliidae). Evolution 32(2): 320–333.

Thomas, C. E. 1976. Fishes of Bayou Bartholomew of southeast Arkansas and northeast Louisiana. M.S. thesis, Northeast Louisiana Univ., Monroe. 44 pp.

Thomas, D. L. 1970. An ecological study of four darters of the genus *Percina* (Percidae) in the Kaskaskia River, Illinois. Illinois Nat. Hist. Surv. Biol. Notes No. 70: 1–18.

Thomas, J. L. 1981. Maturation and fecundity of the White Crappie (*Pomoxis annularis*) in Beaver Reservoir, Arkansas. M.S. thesis, Univ. Arkansas, Fayetteville. 31 pp.

Thomas, J. L. and R. V. Kilambi. 1981. Maturation, spawning period, and fecundity of the White Crappie, *Pomoxis annularis* Rafinesque, in Beaver Reservoir, Arkansas. Proc. Arkansas Acad. Sci. 35: 70–73.

Thomas, M. R. and B. M. Burr. 2004. *Noturus gladiator*, a new species of madtom (Siluriformes: Ictaluridae) from Coastal Plain streams of Tennessee and Mississippi. Ichthyol. Explor. Freshw. 15(4): 351–368.

Thomerson, J. E. 1966. A comparative biosystematic study of *Fundulus notatus* and *Fundulus olivaceus* (Pisces: Cyprinodontidae). Tulane Stud. Zool. 13(1): 29–48.

Thomerson, J. E. 1967. Hybrids between the cyprinodontid fishes, *Fundulus notatus* and *Fundulus olivaceus* in southern Illinois. Trans. Illinois State Acad. Sci. 60(4): 375–379.

Thomerson, J. E. 1969. Variation and relationship of the studfishes, *Fundulus catenatus* and *Fundulus stellifer* (Cyprinodontidae, Pisces). Tulane Stud. Zool. Bot. 16(1): 1–21.

Thomerson, J. E. and D. P. Wooldridge. 1970. Food habits of allotopic and syntopic populations of the topminnows

Fundulus olivaceus and *Fundulus notatus*. Am. Midl. Nat. 84(2): 573–576.

Thomerson, J. E., T. B. Thorson, and R. L. Hempel. 1977. The Bull Shark, *Carcharhinus leucas*, from the upper Mississippi River, near Alton, Illinois. Copeia 1977(1): 166–168.

Thompson, B. A. 1974. An analysis of sympatric populations of two closely related species of *Percina*, with notes on food habits of the subgenus *Imostoma*. ASB Bull. 21(2): 87.

Thompson, B. A. 1977. An analysis of three subgenera (*Hypohomus*, *Odontopholis*, and *Swainia*) of the genus *Percina* (tribe Etheostomatini, family Percidae). Ph.D. diss., Tulane Univ., New Orleans. 409 pp.

Thompson, B. A. 1980. *Percina phoxocephala*, Slenderhead Darter. Page 737. *In*: D. S. Lee et al. Atlas of North American freshwater fishes. N.C. State Mus. Nat. Hist., Raleigh.

Thompson, B. A. 1997a. *Percina suttkusi* a new species of logperch (subgenus *Percina*) from Louisiana, Mississippi, and Alabama (Perciformes, Percidae, Etheostomatini). Occas. Pap. Mus. Nat. Sci. Louisiana State Univ. 72: 1–27.

Thompson, B. A. 1997b. *Percina kathae*, a new logperch endemic to the Mobile Basin in Mississippi, Alabama, Georgia, and Tennessee (Percidae, Etheostomatini). Occas. Pap. Mus. Nat. Sci. Louisiana State Univ. 73: 1–34.

Thompson, B. A. and R. C. Cashner. 1975. Systematics of the subgenus *Imostoma* (genus *Percina*) with comments on the status of *Percina ouachitae* and *P. uranidea* (family Percidae, tribe Etheostomatini) [abstract]. 55th Annu. Meet. Am. Soc. Ichthyol. Herpetol., p. 124.

Thompson, B. A. and R. C. Cashner. 1980. *Percina ouachitae* (Jordan and Gilbert), Yellow Darter. Page 732. *In*: D. S. Lee et al. Atlas of North American freshwater fishes. N.C. State Mus. Nat. Hist., Raleigh.

Thornbury, W. D. 1965. Regional geomorphology of the United States. John Wiley and Sons, New York. 609 pp.

Thorpe, J. E. 1977. Morphology, physiology, behavior, and ecology of *Perca fluviatilis* L. and *P. flavescens* Mitchill. J. Fish. Res. Board Can. 34(10): 1504–1514.

Threinen, C. W., C. Wistrom, B. Apelgren, and H. Snow. 1966. The Northern Pike: its life history, ecology, and management. Wisconsin Conservation Dept. Publ. 235. 16 pp.

Tibbits, W. T. 2008. The behavior of Lake Trout, *Salvelinus namaycush* (Walbaum, 1792) in Otsego Lake: a documentation of the strains, movements and the natural reproduction of Lake Trout under present conditions. Occasional Paper Number 42, Biological Field Station, Cooperstown, NY, State University College at Oneonta. 65 pp.

Tidwell, J. H. and S. D. Mims. 1990. Survival of Paddlefish fingerlings stocked with large Channel Catfish. Prog. Fish-Cult. 52(4): 273–274.

Tiemann, J. S. 2004. Observations of the Pirate Perch, *Aphredoderus sayanus* (Gilliams), with comments on sexual dimorphism, reproduction, and unique defecation behavior. J. Freshw. Ecol. 19(1): 115–121.

Tiemann, J. S. 2007a. Spawning behavior of the Slim Minnow, *Pimephales tenellus*. Southwest. Nat. 52(1): 137–141.

Tiemann, J. S. 2007b. Reproductive traits in the Slim Minnow (*Pimephales tenellus*) and the Bullhead Minnow (*Pimephales vigilax*). Trans. Kansas Acad. Sci. 110(3/4): 282–284.

Tiemann, J. S., S. E. McMurray, M. C. Barnhart, and G. T. Watters. 2011. A review of the interactions between catfishes and freshwater mollusks in North America. Am. Fish. Soc. Symp. 77: 733–743.

Timmons, T. J., W. L. Shelton, and W. S. Davies. 1977. Initial fish population changes following impoundment of West Point Reservoir, Alabama-Georgia. Proc. 31st Annu. Conf. Southeast. Assoc. Fish Wildl. Agencies: 312–317.

Todd, B. L. and C. F. Rabeni. 1989. Movement and habitat use by stream-dwelling Smallmouth Bass. Trans. Am. Fish. Soc. 118(3): 229–242.

Todd, C. S. and K. W. Stewart. 1985. Food habits and dietary overlap of nongame insectivorous fishes in Flint Creek, Oklahoma, a western Ozark foothills stream. Great Basin Nat. 45(4): 721–733.

Todd, J. H., J. Atema, and J. E. Bardach. 1967. Chemical communication in social behavior of a fish, the Yellow Bullhead (*Ictalurus natalis*). Science 158(3801): 672–673.

Todd, T. N. 1986. Occurrence of White Bass–White Perch hybrids in Lake Erie. Copeia 1986(1): 196–199.

Toepfer, C. S., W. L. Fisher, and J. A. Haubelt. 1999. Swimming performance of the threatened Leopard Darter in relation to road culverts. Trans. Am. Fish. Soc. 128(1): 155–161.

Toole, J. E. 1971. Food study of the Bowfin and gars in eastern Texas. Tech. Ser. 6, Texas Parks and Wildlife Dept., Austin. 35 pp.

Torrans, L. and F. Lowell. 1987. Effects of Blue Tilapia/ Channel Catfish polyculture on production, food conversion, water quality and Channel Catfish off-flavor. Proc. Arkansas Acad. Sci. 41: 82–86.

Tranah, G. J., H. L. Kincaid, C. C. Krueger, D. E. Campton, and B. May. 2001. Reproductive isolation in sympatric populations of Pallid and Shovelnose Sturgeon. N. Am. J. Fish. Manag. 21(2): 367–373.

Trautman, M. B. 1930. The specific distinctness of *Poecilichthys coeruleus* (Storer) and *Poecilichthys spectabilis* Agassiz. Copeia 1930(1): 12–13.

Trautman, M. B. 1931. *Notropis volucellus wickliffi*, a new subspecies of cyprinid fish from the Ohio and upper Mississippi rivers. Ohio J. Sci. 31(6): 468–474.

Trautman, M. B. 1957. The fishes of Ohio. Ohio State Univ. Press, Columbus. 683 pp.

Trautman, M. B. 1981. The fishes of Ohio, 2nd edition. Ohio State University Press, Columbus, 782 pp.

Trautman, M. B. and R. G. Martin. 1951. *Moxostoma aureolum pisolabrum*, a new subspecies of sucker from the Ozarkian streams of the Mississippi River system. Occas. Pap. Mus. Zool. Univ. Michigan 534: 1–10.

Tremblay, V. 2009. Reproductive strategy of female American Eels among five subpopulations in the St. Lawrence River watershed. Am. Fish. Soc. Symp. 58: 85–102.

Trewavas, E. 1981. Addendum to 'Tilapia and Sarotherodon?'. Buntbarsche Bull. (1065) 87: 12.

Trewavas, E. 1982. Generic groupings of Tilapiini used in aquaculture. Aquaculture 27(1): 79–81.

Trewavas, E. 1983. Tilapiine fishes of the genera Sarotherodon, Oreochromis and Danakilia. British Mus. Nat. Hist., Publ. 878. Comstock Publ., Ithaca, NY. 583 pp.

Tripp, S. J., Q. E. Phelps, R. E. Columbo, J. E. Garvey, B. M. Burr, D. P. Herzog, and R. A. Hrabik. 2009. Maturation and reproduction of Shovelnose Sturgeon in the middle Mississippi River. N. Am. J. Fish. Manag. 29(3): 730–738.

Trippel, E. A. and H. H. Harvey. 1987a. Abundance, growth, and food supply of White Suckers (Catostomus commersoni) in relation to lake morphometry and pH. Can. J. Zool. 65(3): 558–564.

Trippel, E. A. and H. H. Harvey. 1987b. Reproductive responses of five White Sucker (Catostomus commersoni) populations in relation to lake acidity. Can. J. Fish. Aquat. Sci. 44(5): 1018–1023.

Trippel, E. A. and H. H. Harvey. 1989. Missing opportunities to reproduce: an energy dependent or fecundity gaining strategy in White Sucker (Catostomus commersoni)? Can. J. Zool. 67(9): 2180–2188.

Trippel, E. A. and H. H. Harvey. 1991. Comparison of methods used to estimate age and length of fishes at sexual maturity using populations of White Sucker (Catostomus commersoni). Can. J. Fish. Aquat. Sci. 48(8): 1446–1459.

Trombulak, S. C. and C. A. Frissell. 2000. Review of ecological effects of roads on terrestrial and aquatic communities. Conserv. Biol. 14(1): 18–30.

Trubowitz, N. L. 1984. The aboriginal cemetery. Pages 97–108. In: N. L. Trubowitz, ed. Cedar Grove: an interdisciplinary investigation of a late Caddo farmstead in the Red River Valley. Arkansas Archeological Survey Research Series 23. 208 pp.

Tsai, C. 1968. Distribution of the Harlequin Darter, Etheostoma histrio. Copeia, 1968(1): 178–181.

Tsai, C. and E. C. Raney. 1974. Systematics of the Banded Darter, Etheostoma zonale (Pisces: Percidae). Copeia 1974(1): 1–24.

Tsehaye, I., M. Catalano, G. Sass, D. Glover, and B. Roth. 2013. Prospects for fishery-induced collapse of invasive Asian carp in the Illinois River. Fisheries 38(10): 445–454.

Tsukamoto, K. S., I. Nakai, and W. V. Tesch. 1998. Do all freshwater eels migrate? Nature 396(6712): 635–636.

Tsukamoto K., S. Chow, T. Otake, H. Kurogi, N. Mochioka, M. J. Miller, J. Aoyama, S. Kimura, S. Watanabe, T. Yoshinaga, A. Shinoda, M. Kuroki, M. Oya, T. Watanabe, K. Hata, S. Ijiri, Y. Kazeto, K. Nomura, and H. Tanaka. 2011. Oceanic spawning ecology of freshwater eels in the western North Pacific. Nat. Commun. 2(1): 179.

Tucker, J. K., F. A. Cronin, R. A. Hrabik, M. D. Petersen, and D. P. Herzog. 1996. The Bighead Carp (Hypophthalmichthys nobilis) in the Mississippi River. J. Freshw. Ecol. 11(2): 241–243.

Tumlison, R. and G. R. Cline. 2002. Food habits of the Banded Sculpin (Cottus carolinae) in Oklahoma with reference to predation on the Oklahoma Salamander (Eurycea tynerensis). Proc. Oklahoma Acad. Sci. 82: 111–113.

Tumlison, R. and J. O. Hardage. 2014. Growth and reproduction in the Ouachita Madtom (Noturus lachneri) at the periphery of its distribution. J. Arkansas Acad. Sci. 68: 110–116.

Tumlison, R. and H. W. Robison. 2010. New records and notes on the natural history of selected vertebrates from southern Arkansas. J. Arkansas Acad. Sci. 64: 145–150.

Tumlison, R. and C. Tumlison. 1996. A survey of the fishes in streams draining the Jack Mountain area, Hot Spring and Garland counties, Arkansas, with notes on the Ouachita Madtom (Noturus lachneri). Proc. Arkansas Acad. Sci. 50: 154–159.

Tumlison, R. and C. Tumlison. 1999. An extralimital population of Lampetra appendix (Petromyzontidae) in southwestern Arkansas. Southwest. Nat. 44(1): 106–108.

Tumlison, R., M. Connior, H. W. Robison, and C. T. McAllister. 2015. Notes on vertebrate natural history. J. Arkansas Acad. Sci. 69: 106–115.

Tumlison, R., C. T. McAllister, H. W. Robison, M. B. Connior, D. B. Sasse, D. G. Cloutman, L. A. Durden, C. R. Bursey, T. J. Fayton, S. Schratz, and M. Buckley. 2017. Vertebrate natural history notes from Arkansas, 2017. J. Arkansas Acad. Sci. 71: 7–16.

Tumlison, R., C. T. McAllister, H. W. Robison, M. B. Connior, D. B. Sasse, D. A. Saugey, and S. Chordas III. 2016. Vertebrate natural history notes from Arkansas, 2016. J. Arkansas Acad. Sci. 70: 248–254.

Turan, D., M. Kottelat, and S. Engin. 2014. Two new species of trouts from the Euphrates drainage, Turkey (Teleostei: Salmonidae). Ichthyol. Explor. Freshw. 24(3): 275–287.

Turner, C. L. 1920. Distribution, food and fish associates of young perch in the Bass Island region of Lake Erie. Ohio J. Sci. 20(5): 137–152.

Turner, C. L. 1921. Food of the common Ohio darters. Ohio J. Sci. 22(2): 41–62.

Turner, G. F. and R. L. Robinson. 2000. Reproductive biology, mating systems and parental care. Pages 33–58. In: M. C. M. Beveridge and B. J. McAndrew, eds. Tilapias: biology and exploitation. Kluwer Academic, Great Britain. 508 pp.

Turner, J. L. 1966. Distribution and food habits of ictalurid fishes in the Sacramento–San Joaquin Delta. Part II. Fish Bull. California Dept. Fish Game 136: 130–143.

Turner, J. S. and F. F. Snelson, Jr. 1984. Population structure, reproduction and laboratory behavior of the introduced Belonesox belizanus (Poeciliidae) in Florida. Environ. Biol. Fishes 10(1–2): 89–100.

Turner, P. R. and R. C. Summerfelt. 1971. Food habits of adult Flathead Catfish, *Pylodictis olivaris* (Rafinesque), in Oklahoma reservoirs. Proc. 24th Annu. Conf. Southeast. Assoc. Game Fish. Comm.: 387–401.

Turner, S. E. 1979. Life history of wild Rainbow Trout in Missouri. Missouri Dept. Conservation, D-J Project F-1-R-27, Study 22, Job 1. 14 pp.

Turner, T. F. and H. W. Robison. 2006. Genetic diversity of the Caddo Madtom, *Noturus taylori*, with comments on factors that promote genetic divergence in fishes endemic to the Ouachita Highlands. Southwest. Nat. 51(3): 338–345.

Turner, T. F. and J. C. Trexler. 1998. Ecological and historical associations of gene flow in darters (Teleostei: Percidae). Evolution 52(6): 1781–1801.

Turner, T. F., J. C. Trexler, D. N. Kuhn, and H. W. Robison. 1996. Life-history variation and comparative phylogeography of darters (Pisces: Percidae) from the North American central highlands. Evolution 50(5): 2023–2036.

Tyler, J. D. 1994. Albinistic Spotted Gar, *Lepisosteus oculatus*, in Oklahoma. Proc. Oklahoma Acad. Sci. 74: 39.

Tyler, J. D. and M. N. Granger. 1984. Notes on food habits, size, and spawning behavior of Spotted Gar in Lake Lawtonka, Oklahoma. Proc. Oklahoma Acad. Sci. 64: 8–10.

Tyler, J. D., J. R. Webb, T. R. Wright, J. D. Hargett, K. J. Mask, and D. R. Schucker. 1994. Food habits, sex ratios, and size of Longnose Gar in southwestern Oklahoma. Proc. Oklahoma Acad. Sci. 74: 41–42.

Tyrone, R. 2007. Effects of upland timber harvest and road construction on headwater stream fish assemblages in a southeastern forest. M.S. thesis, Texas State Univ., San Marcos. 41 pp.

Tzilkowski, C. J. and J. R. Stauffer, Jr. 2004. Biology and diet of the Northern Madtom (*Noturus stigmosus*) and Stonecat (*Noturus flavus*) in French Creek, Pennsylvania. J. Pennsylvania Acad. Sci. 78(1): 3–11.

Underwood, D. M., A. A. Echelle, D. J. Eisenhour, M. D. Jones, A. F. Echelle, and W. L. Fisher. 2003. Genetic variation in western members of the *Macrhybopsis aestivalis* complex (Teleostei: Cyprinidae), with emphasis on those of the Red and Arkansas river basins. Copeia 2003(3): 493–501.

Unger, L. M. and R. C. Sargent. 1988. Allopaternal care in the Fathead Minnow, *Pimephales promelas*: females prefer males with eggs. Behav. Ecol. Sociobiol. 23(1): 27–32.

USFWS. 1998: Final rule to list the Arkansas River basin population of the Arkansas River Shiner (*Notropis girardi*) as threatened. U.S. Federal Register 63(225): 64777–64799.

USFWS. 2007. Pallid Sturgeon (*Scaphirhynchus albus*) 5-year review summary and evaluation. USFWS Pallid Sturgeon Recovery Coordinator, Billings, MT. 120 pp.

USFWS. 2010. Endangered and threatened wildlife and plants: threatened status for Shovelnose Sturgeon under the Similarity of Appearance provisions of the Endangered Species Act. U.S. Federal Register 75(169): 53598–53606.

USFWS. 2011. Endangered status for the Cumberland Darter, Rush Darter, Yellowcheek Darter, Chucky Madtom, and Laurel Dace. U.S. Federal Register 76(153): 48722–48741.

USFWS. 2012. Proposed designation of critical habitat for the Cumberland Darter, Rush Darter, Yellowcheek Darter, Chucky Madtom, and Laurel Dace. Final Rule. U.S. Federal Register 77FR: 63603-63668.

USFWS. 2014. Revised recovery plan for the Pallid Sturgeon (*Scaphirhynchus albus*). U.S. Fish and Wildlife Service, Denver, CO. 115 pp.

Uyeno, T. and G. R. Smith. 1972. Tetraploid origin of the karyotype of catostomid fishes. Science 175(4022): 644–646.

Uyeno, T. and R. R. Miller. 1963. Summary of Late Cenozoic freshwater fish records for North America. Occas. Pap. Mus. Zool. Univ. Michigan 631: 1–34.

Van Cleave, H. J. and H. C. Markus. 1929. Studies on the life history of the Blunt-nosed Minnow. Am. Nat. 63(689): 530–539.

Van Den Avyle, M. J., B. J. Higginbotham, B. T. James, and F. J. Bulow. 1983. Habitat preferences and food habits of young-of-the-year Striped Bass, White Bass, and Yellow Bass in Watts Bar Reservoir, Tennessee. N. Am. J. Fish. Manag. 3(2): 163–170.

VanderKooy, K. E., C. F. Rakocinski, and R. W. Heard. 2000. Trophic relationships of three sunfishes (*Lepomis* spp.) in an estuarine bayou. Estuaries 23(5): 621–632.

Vandermeer, J. H. 1966. Statistical analysis of geographic variation of the Fathead Minnow *Pimephales promelas*. Copeia 1966(3): 457–466.

Van Oosten, J. 1937. The dispersal of Smelt, *Osmerus mordax* (Mitchill), in the Great Lakes region. Trans. Am. Fish. Soc. 66(1): 160–171.

Varble, K. A., J. J. Hoover, S. G. George, C. E. Murphy, and K. J. Killgore. 2007. Floodplain wetlands as nurseries for Silver Carp, *Hypophthalmichthys molitrix*: a conceptual model for use in managing local populations. ANSRP Technical Notes Collection (ERDC/TN ANSRP-07-4). U.S. Army Engineer Research and Development Center, Vicksburg, MS. 14 pp.

Verigin, B. V., A. P. Makeeva, and M. I. Zaki-Mohamed. 1978. Natural spawning of the Silver Carp *Hypophthalmichthys molitrix*, the Bighead Carp *Aristichthys nobilis*, and the Grass Carp *Ctenopharyngodon idella*, in the Syr-Dar'ya River. J. Ichthyol. 18(1): 143–146.

Victor, B. C., R. Hanner, M. Shivji, J. Hyde, and C. Caldow. 2009. Identification of the larval and juvenile stages of the Cubera Snapper, *Lutjanus cyanopterus*, using DNA barcoding. Zootaxa 2215: 24–36.

Vigueira, P. A., J. F. Schaefer, D. D. Duvernell, and B. R. Kreiser. 2008. Tests of reproductive isolation among species in the *Fundulus notatus* (Cyprinodontiformes: Fundulidae) species complex. Evol. Ecol. 22(1): 55–70.

Viosca, P., Jr. 1936. A new rock bass from Louisiana and Mississippi. Copeia 1936(1): 37–45.

Vitins, M., R. Gaumiga, and A. Mitāns. 2001. History of Latvian fisheries research. Proc. Estonian Acad. Sci. Biol. Ecol. 50(2): 85–109.

Vitousek, P. M., H. A. Mooney, J. Lubchenco, and J. M. Melillo. 1997. Human domination of Earth's ecosystems. Science 277(5325): 494–499.

Vitule, J. R. S., C. A. Freire, and D. Simberloff. 2009. Introduction of non-native freshwater fish can certainly be bad. Fish and Fisheries 10(1): 98–108.

Vives, S. P. 1987. Aspects of the life history of the Slender Madtom, *Noturus exilis*, in northeastern Oklahoma (Pisces: Ictaluridae). Am. Midl. Nat. 117(1): 167–176.

Vives, S. P. 1988. The reproductive behavior of minnows (Pisces: Cyprinidae) in two reproductive guilds. Ph.D. diss., Univ. Wisconsin, Madison. 174 pp.

Vives, S. P. 1990. Nesting ecology and behavior of Hornyhead Chub, *Nocomis biguttatus*, a keystone species in Allequash Creek, Wisconsin. Am. Midl. Nat. 124(1): 46–56.

Vives, S. P. 1993. Choice of spawning substrate in Red Shiner with comments on crevice spawning in *Cyprinella*. Copeia 1993(1): 229–232.

Vladykov, V. D. 1955. A comparison of Atlantic sea sturgeon with a new subspecies from the Gulf of Mexico (*Acipenser oxyrhynchus desotoi*) [oxyrinchus misspelled]. J. Fish. Res. Bd. Can. 12(5): 754–761.

Vladykov, V. D. 1963. A review of salmonid genera and their broad geographical distribution. Trans. Royal Soc. Can. (S4)1: 459–504.

Vladykov, V. D. 1985. Record of 61 parasitic lampreys (*Ichthyomyzon unicuspis*) on a single sturgeon (*Acipenser fulvescens*) netted in the St. Lawrence River (Quebec). Naturaliste Canadien 112: 435–436.

Vladykov, V. D. and E. Kott. 1976. Is *Okkelbergia* Creaser and Hubbs, 1922 (Petromyzonidae) a distinct taxon? Can. J. Zool. 54(3): 421–425.

Vladykov, V. D. and E. Kott. 1979. List of Northern Hemisphere lampreys (Petromyzonidae) and their distribution. Misc. Special Publ. 42. Dept. Fisheries and. Oceans, Ottawa, Ontario, Canada. 38 pp.

Vogele, L. E. 1975. Reproduction of Spotted Bass, *Micropterus punctulatus*, in Bull Shoals Reservoir, Arkansas. U.S. Fish and Wildlife Service Tech. Pap. 84. 21 pp.

Vogele, L. E. 1981. Reproduction of Smallmouth Bass, *Micropterus dolomieui*, in Bull Shoals Lake, Arkansas. U.S. Fish and Wildlife Service Tech. Pap. 106. 15 pp.

Vogele, L. E. and W. C. Rainwater. 1975. Use of brush shelters as cover by spawning black basses (*Micropterus*) in Bull Shoals Reservoir. Trans. Am. Fish. Soc. 104(2): 264–269.

Vogt, G. F., Jr. and T. G. Coon. 1990. A comparison of the foraging behavior of two darter (*Etheostoma*) species. Copeia 1990(1): 41–49.

Vokoun, J. C. and C. F. Rabeni. 2006. Summer diel activity and movement paths of Flathead Catfish (*Pylodictis olivaris*) in two Missouri streams. Am. Midl. Nat. 155(1): 113–122.

Vokoun, J. C., T. L. Guerrant, and C. F. Rabeni. 2003. Demographics and chronology of a spawning aggregation of Blue Sucker (*Cycleptus elongatus*) in the Grand River, Missouri, USA. J. Freshw. Ecol. 18(4): 567–575.

Volff, J. N. 2005. Genome evolution and biodiversity in teleost fish. Heredity 94(3): 280–294.

Wagner, B. A., A. A. Echelle, and O. E. Maughan. 1987. Abundance and habitat use of an uncommon fish, *Notropis perpallidus* (Cyprinidae): comparison with sympatric congeners. Southwest. Nat. 32(2): 251–260.

Wagner, B. K. and M. D. Kottmyer. 2006. Status and distribution of the Arkansas Darter (*Etheostoma cragini*) in Arkansas. J. Arkansas Acad. Sci. 60: 137–143.

Wagner, B. K., M. D. Kottmyer, and M. E. Slay. 2011. Summary of previous and new records of the Arkansas Darter (*Etheostoma cragini*) in Arkansas. J. Arkansas Acad. Sci. 65: 138–142.

Wagner, B. K., M. D. Kottmyer, and M. E. Slay. 2012. Summary of previous and new records of the Least Darter (*Etheostoma microperca*) in Arkansas. J. Arkansas Acad. Sci. 66: 173–179.

Wagner, C. C. and E. L. Cooper. 1963. Population density, growth, and fecundity of the Creek Chubsucker, *Erimyzon oblongus*. Copeia 1963(2): 350–357.

Wagner, E., R. Arndt, M. Brough, and D. W. Roberts. 2002. Comparison of susceptibility of five Cutthroat Trout strains to *Myxobolus cerebralis* infection. J. Aquat. Animal Health 14(1): 84–91.

Wagner, G. 1908. Notes on the fish fauna of Lake Pepin. Trans. Wisconsin Acad. Sci. Arts Lett. 16: 23–27.

Wahl, D. H., K. Bruner, and L. A. Nielson. 1988. Trophic ecology of Freshwater Drum in large rivers. J. Freshw. Ecol. 4(4): 483–491.

Wainwright, P. C. and G. V. Lauder. 1992. The evolution of feeding biology in sunfishes (Centrarchidae). Pages 472–491. *In*: R. L. Mayden, ed. Systematics, historical ecology and North American freshwater fishes. Stanford Univ. Press, Stanford, CA.

Wakefield, C. K. and D. W. Beckman. 2005. Life history attributes of White Sucker (*Catostomus commersoni*) in Lake Taneycomo and associated tributaries in southwestern Missouri. Southwest. Nat. 50(4): 423–434.

Walbaum, J. J. 1792. Petri Artedi sueci genera piscium. In quibus systema totum ichthyologiae proponitur cum classibus, ordinibus, generum characteribus, specierum differentiis, observationibus plurimis. Redactis speciebus 242 ad genera 52. Ichthyologiae pars III. Ant. Ferdin. Rose, Grypeswaldiae [Greifswald]. Part 3: [i–viii] + 1–723, Pls. 1–3. [Reprint 1966 by J. Cramer.]

Walburg, C. H. and W. R. Nelson. 1966. Carp, River Carpsucker, Smallmouth Buffalo, and Bigmouth Buffalo in Lewis and Clark Lake, Missouri River. U.S. Fish and Wildlife Service Research Rep. 69. U.S. Dept. of Interior, Washington, DC. 30 pp.

Walburg, C. H., G. L. Kaiser, and P. L. Hudson. 1971. Lewis

and Clark Lake tailwater biota and some relations of the tail-water and reservoir fish populations. Pages 449–467. *In:* G. E. Hall, ed. Reservoir fisheries and limnology. Am. Fish. Soc. Spec. Publ. No. 8. Washington, DC.

Walker, M. M. and L. H. Keith. 1992. EPA's pesticide fact sheet database. Lewis Publ., Chelsea, MI. 32 pp.

Walker, R. H. 2011. Movement patterns of Southern Redbelly Dace, *Chrosomus erythrogaster*, in a headwater reach of an Ozark stream. M.S. thesis, Univ. Central Arkansas, Conway. 84 pp.

Walker, R. H. and G. L. Adams. 2016. Ecological factors influencing movement of Creek Chub in an intermittent stream of the Ozark Mountains, Arkansas. Ecol. Freshw. Fish 25(2): 190–202.

Walker, R. H., G. L. Adams, and S. R. Adams. 2013a. Movement patterns of Southern Redbelly Dace, *Chrosomus erythrogaster*, in a headwater reach of an Ozark stream. Ecol. Freshw. Fish 22(2): 216–227.

Walker, R. H., E. R. Kluender, T. E. Inebnit, and S. R. Adams. 2013b. Differences in diet and feeding ecology of similar-sized Spotted (*Lepisosteus oculatus*) and Shortnose (*Lepisosteus platostomus*) gars during flooding of a southeastern US river. Ecol. Freshw. Fish 22(4): 617–625.

Walker, T. J. and A. D. Hasler. 1949. Detection and discrimination of odors of aquatic plants by the Bluntnose Minnow (*Hyborhynchus notatus*). Physiol. Zool. 22(1): 45–63.

Wallace, A. R. 1876. The geographical distribution of animals; with a study of the relations of living and extinct faunas as elucidating the past changes of the earth's surface. Vols. I and II. Harper & Brothers, New York.

Wallace, C. R. 1967. Observations on the reproductive behavior of the Black Bullhead (*Ictalurus melas*). Copeia 1967(4): 852–853.

Wallace, C. R. 1972. Spawning behavior of *Ictalurus natalis* (Lesueur). Texas J. Sci. 24(3): 307–310.

Wallus, R. 1986. Paddlefish reproduction in the Cumberland and Tennessee river systems. Trans. Am. Fish. Soc. 115(3): 424–428.

Wallus, R. and J. P. Buchanan. 1989. Contributions to the reproductive biology and early life ecology of Mooneye in the Tennessee and Cumberland rivers. Am. Midl. Nat. 122(1): 204–207.

Wallus, R. and K. L. Grannemann. 1979. Spawning behavior and early development of the Banded Sculpin, *Cottus carolinae* (Gill). Pages 200–235. *In*: R. Wallus and C. W. Voigtlander, eds. Proceedings of a workshop on freshwater larval fishes. Tennessee Valley Authority, Norris, TN.

Wallus, R. and L. K. Kay. 1990. Family Clupeidae. Pages 107–150. *In*: R. Wallus, T. P. Simon, and B. L. Yeager, eds. Reproductive biology and early life history of fishes in the Ohio River drainage. Vol. I. Acipenseridae through Esocidae. Tennessee Valley Authority, Chattanooga, TN.

Wallus, R., T. P. Simon, and B. L. Yeager. 1990. Reproductive biology and early life history of fishes in the Ohio River

drainage. Vol. I. Acipenseridae through Esocidae. Tennessee Valley Authority, Chattanooga, TN. 273 pp.

Walser, C. A. and H. L. Bart, Jr. 1999. Influence of agriculture on in-stream habitat and fish community structure in Piedmont watersheds of the Chattahoochee River system. Ecol. Freshw. Fish 8(4): 237–246.

Walsh, S. J. and B. M. Burr. 1981. Distribution, morphology and life history of the Least Brook Lamprey, *Lampetra aepyptera* (Pisces: Petromyzontidae), in Kentucky. Brimleyana 6: 83–100.

Walsh, S. J. and B. M. Burr. 1984. Life history of the Banded Pygmy Sunfish, *Elassoma zonatum* Jordan (Pisces: Centrarchidae), in western Kentucky. Bull. Alabama Mus. Nat. Hist. 8: 31–52.

Walsh, S. J. and B. M. Burr. 1985. Biology of the Stonecat, *Noturus flavus* (Siluriformes: Ictaluridae), in central Illinois and Missouri streams, and comparisons with Great Lakes populations and congeners. Ohio J. Sci. 85(3): 85–96.

Walsh, S. J., H. L. Jelks, and N. M. Burkhead. 2011. The decline of North American freshwater fishes. Am. Currents 36(4): 10–17.

Walters, D. M., R. E. Zuellig, H. J. Crockett, J. F. Bruce, P. M. Lukacs, and R. M. Fitzpatrick. 2014. Barriers impede upstream spawning migration of Flathead Chub. Trans. Am. Fish. Soc. 143(1): 17–25.

Walters, J., C. Annett, and G. Siegwarth. 2000. Breeding ecology and behavior of Ozark Bass, *Ambloplites constellatus*. Am. Midl. Nat. 144(2): 423–427.

Walters, J. P. 1994. Spawning behavior of *Etheostoma zonale* (Pisces: Percidae). Copeia 1994(3): 818–821.

Walters, J. P. and J. R. Wilson. 1996. Intraspecific habitat segregation by Smallmouth Bass in the Buffalo River, Arkansas. Trans. Am. Fish. Soc. 125(2): 284–290.

Wang, C. H. and W. N. Tzeng. 2000. The timing of metamorphosis and growth rates of American and European eel leptocephali: a mechanism of larval segregative migration. Fish. Res. 46(1–3): 191–205.

Wang, X., X. Gan, J. Li, R. L. Mayden, and S. He. 2012. Cyprinid phylogeny based on Bayesian and maximum likelihood analyses of partitioned data: implications for Cyprinidae systematics. Sci. China, Life Sci. 55(9): 761–773.

Ward, D. D., D. Dunn, H. Worley, M. Ruane, V. E. Hoffman, G. Troutman, and S. E. Trauth. 1999. The first report of *Lampetra appendix* (DeKay) from the Black River drainage of Arkansas. J. Arkansas Acad. Sci. 53: 156.

Ward, R. D., R. Hanner, and P. D. N. Hebert. 2009. The campaign to DNA barcode all fishes, FISH-BOL. J. Fish Biol. 74(2): 329–356.

Warren, M. A. and P. E. Denette. 1989. New records and comments on the distribution of Blair's Starhead Topminnow, *Fundulus blairae* (Fundulidae). SFC Proc. 20: 7–9.

Warren, M. L., Jr. 1992. Variation of the Spotted Sunfish, *Lepomis punctatus* complex (Centrarchidae): meristics, morphometrics, pigmentation and species limits. Bull. Alabama Mus. Nat. Hist. 12: 1–47.

Warren, M. L., Jr. 2009. Centrarchid identification and natural history. Pages 375–533. *In*: S. J. Cooke and D. P. Philipp, eds. Centrarchid fishes: diversity, biology, and conservation. Wiley-Blackwell, Chichester.

Warren, M. L., Jr. and B. M. Burr. 1989. Distribution, abundance, and status of the Cypress Minnow, *Hybognathus hayi*, an endangered Illinois species. Nat. Areas J. 9(3): 163–168.

Warren, M. L., Jr. and B. M. Burr. 1994. Status of freshwater fishes of the United States: overview of an imperiled fauna. Fisheries 19(1): 6–18.

Warren, M. L., Jr. and L. Hlass. 1999. Diversity of fishes. *In:* Chapter 2, USDA, Forest Service. Ozark-Ouachita Highlands assessment: aquatic conditions. Report 3 of 5. Gen. Tech. Rep. SRS-33. U.S. Dept. Agriculture, Forest Service, Southern Research Station, Asheville, NC. 317 pp.

Warren, M. L., Jr. and M. G. Pardew. 1998. Road crossings as barriers to small-stream fish movement. Trans. Am. Fish. Soc. 127(4): 637–644.

Warren, M. L., Jr. and K. Tinkle. 1999. Endangered, threatened, and other aquatic species of special concern. *In:* Chapter 2, USDA, Forest Service. Ozark-Ouachita Highlands assessment: aquatic conditions. Report 3 of 5. Gen. Tech. Rep. SRS-33. U.S. Dept. Agriculture, Forest Service, Southern Research Station, Asheville, NC. 317 pp.

Warren, M. L., Jr., B. M. Burr, and J. M. Grady. 1994. *Notropis albizonatus*, a new cyprinid fish endemic to the Tennessee and Cumberland river drainages, with a phylogeny of the *Notropis procne* species group. Copeia 1994(4): 868–886.

Warren, M. L., Jr., B. M. Burr, S. J. Walsh, H. L. Bart, Jr., R. C. Cashner, D. A. Etnier, B. J. Freeman, B. R. Kuhajda, R. L. Mayden, H. W. Robison, S. T. Ross, and W. C. Starnes. 2000. Diversity, distribution, and conservation status of the native freshwater fishes of the southern United States. Fisheries 25(10): 7–29.

Watson, R. 1999. Salmon, trout and char of the world: a fisherman's natural history. Swan Hill Press, England. 320 pp.

Webb, P. W. 1994. The biology of fish swimming. Pages 45–62. *In*: L. Maddock, Q. Bone, and J. M. V. Rayner, eds. The mechanics and physiology of animal swimming. Cambridge Univ. Press, Cambridge.

Weber, M. J. and M. L. Brown. 2009. Effects of Common Carp on aquatic ecosystems 80 years after "Carp as a dominant": ecological insights for fisheries management. Rev. Fish. Sci. 17(4): 524–537.

Weber, M. J. and M. L. Brown. 2013. Spatiotemporal variation of juvenile Common Carp foraging patterns as inferred from stable isotope analysis. Trans. Am. Fish. Soc. 142(5): 1179–1191.

Wegleitner, B. J., A. Tucker, W. L. Chadderton, and A. R. Mahon. 2016. Identifying the genetic structure of introduced populations of Northern Snakehead (*Channa argus*) in eastern USA. Aquat. Invasions 11(2): 199–208.

Weisel, G. F. 1975. The integument of the Paddlefish, *Polyodon spathula*. J. Morphol. 145(2): 143–150.

Weisel, G. F. 1978. The integument and caudal filament of the Shovelnose Sturgeon, *Scaphirhynchus platorynchus*. Am. Midl. Nat. 100(1): 179–189.

Welcomme, R. L. 1988. International introductions of inland aquatic species. FAO Fisheries Tech. Paper 294. FAO, Rome. 318 pp.

Welker, B. D. 1962. Summer food habits of Yellow Bass and Black Bullheads in Clear Lake. Proc. Iowa Acad. Sci. 69(1): 286–295.

Welker, T. L. and D. L. Scarnecchia. 2004. Habitat use and population structure of four native minnows (family Cyprinidae) in the upper Missouri and lower Yellowstone rivers, North Dakota (USA). Ecol. Freshw. Fish 13(1): 8–22.

Welker, T. L. and D. L. Scarnecchia. 2006. River alteration and niche overlap among three native minnows (Cyprinidae) in the Missouri River hydrosystem. J. Fish Biol. 68(5): 1530–1550.

Welsh, D. P. and R. C. Fuller. 2015. Influence of sex and habitat on the size and shape of anal and dorsal fins of the Blackstripe Topminnow *Fundulus notatus*. J. Fish Biol. 86(1): 217–227.

Welsh, D. P., M. Zhou, S. M. Mussmann, L. G. Fields, C. L. Thomas, S. P. Pearish, S. L. Kilburn, J. L. Parker, L. R. Stein, J. A. Bartlett, C. R. Bertram, T. J. Bland, K. L. Laskowski, B. C. Mommer, X. Zhuang, and R. C. Fuller. 2013. The effects of age, sex, and habitat on body size and shape of the Blackstripe Topminnow, *Fundulus notatus* (Cyprinodontiformes Fundulidae) (Rafinesque 1820). Biol. J. Linn. Soc. 108(4): 784–789.

Welsh, S. A. and H. L. Liller. 2013. Environmental correlates of upstream migration of yellow-phase American Eels in the Potomac River drainage. Trans. Am. Fish. Soc. 142(2): 483–491.

Welsh, S. A. and S. A. Perry. 1998. Habitat partitioning in a community of darters in the Elk River, West Virginia. Environ. Biol. Fishes 51(4): 411–419.

Welsh S. A. and R. M. Wood. 2008. *Crystallaria cincotta*, a new species of darter (Teleostei: Percidae) from the Elk River of the Ohio River drainage, West Virginia. Zootaxa 1680: 62–68.

Werneke, D. C. and J. W. Armbruster. 2015. Silversides of the genus *Labidesthes* (Atheriniformes: Atherinopsidae). Zootaxa 4032(5): 535–550.

Werner, E. E. and D. J. Hall 1976. Niche shifts in sunfishes: experimental evidence and significance. Science 191(4225): 404–406.

Werner, E. E. and D. J. Hall. 1988. Ontogenetic habitat shifts in Bluegill: the foraging rate–predation risk trade-off. Ecology 69(5): 1352–1366.

West, B. W. 1966. Growth rates at various temperatures of the Orange-throat Darter *Etheostoma spectabile* (Agassiz). Proc. Arkansas Acad. Sci. 20: 50–53.

Westman, J. R. 1938. Studies on the reproduction and growth of the Bluntnosed Minnow, *Hyborhychus notatus* (Rafinesque). Copeia 1938(2): 57–61.

Weston, M. R. 2006. Density, population size estimates, and movement patterns of *Etheostoma moorei* Raney and Suttkus. M.S. thesis, Arkansas State Univ., Jonesboro.

Weston, M. R., R. L. Johnson, and A. D. Christian. 2010. Niche partitioning of the sympatric Yellowcheek Darter *Etheostoma moorei* and Rainbow Darter *Etheostoma caeruleum* in the Little Red River, Arkansas. Am. Midl. Nat. 164(2): 187–200.

Wetzel, J. E. 2007. Spawning and raising the Bantam Sunfish, *Lepomis symmetricus*. Am. Currents 33(1): 11–15.

Wheeler, Q. and S. Pennak. 2013. What on Earth?: 100 of our planet's most amazing new species. Penguin Group, New York. 304 pp.

Whisenant, K. A. 1984. Characteristics of Smallmouth Bass and Ozark Bass populations in Buffalo National River, Arkansas. M.S. thesis, Oklahoma State Univ., Stillwater. 394 pp.

Whitaker, J. O., Jr. 1977. Seasonal changes in food habits of some cyprinid fishes from the White River at Petersburg, Indiana. Am. Midl. Nat. 97(2): 411–418.

Whitaker, J. O., Jr. 1981. Food habits of the Brindled Madtom, *Noturus miurus* Jordan, and the Redfin Shiner, *Notropis umbratilis* Girard, from west central Indiana. Proc. Indiana Acad. Sci. 91: 247–250.

White, B. N. 1985. Evolutionary relationships of the Atherinopsinae (Pisces: Atherinidae). Nat. Hist. Mus. Los Angeles Cty. Contr. Sci. 368: 1–20.

White, B. N., R. J. Lavenberg, and G. E. McGowen. 1984. Atheriniformes: development and relationships. Pages 355–362. *In:* H. G. Moser et al., eds. Ontogeny and systematics of fishes. Special Publ. 1, American Society of Ichthyology and Herpetology. 760 pp.

White, D. S. 1977. Early development and pattern of scale formation in the Spotted Sucker, *Minytrema melanops* (Catostomidae). Copeia 1977(2): 400–403.

White, D. S. and K. H. Haag. 1977. Foods and feeding habits of the Spotted Sucker, *Minytrema melanops* (Rafinesque). Am. Midl. Nat. 98(1): 137–146.

White, M. J. D. 1973. Animal cytology and evolution. Cambridge Univ. Press, London. 468 pp.

White, M. M. 2000. Genetic variation in White Bass. Trans. Am. Fish. Soc. 129(3): 879–885.

White, M. M. 2014. Intraspecific phylogeography of the American Brook Lamprey, *Lethenteron appendix* (DeKay, 1842). Copeia 2014(3): 513–518.

White, M. M. and N. Aspinwall. 1984. Habitat partitioning among five species of darters (Percidae: *Etheostoma*). Pages 55–56. *In:* D. G. Lindquist and L. M. Page, eds. Environmental biology of darters. Dr. D. W. Junk Publ., The Hague, Netherlands.

Whitehead, P. J. P. 1985. King herring: his place amongst the clupeoids. Can. J. Fish. Aquat. Sci. 42(S1): 3–20.

Whitehead, P. J. P. and A. C. Wheeler. 1966. The generic names used for the sea basses of Europe and N. America (Pisces: Serranidae). Ann. Mus. Civ. Storia Nat. Genova 76: 23–41.

Whitehurst, D. K. 1981. Seasonal movements of fishes in an eastern North Carolina swamp stream. Pages 182–190. *In:* L. A. Krumholz, ed. The Warmwater Streams Symposium, a National Symposium on Fisheries Aspects of Warmwater Streams. Southern Division, American Fisheries Society, Bethesda, MD. 90 pp.

Whiteside, L. A. and B. M. Burr. 1986. Aspects of the life history of the Tadpole Madtom, *Noturus gyrinus* (Siluriformes: Ictaluridae), in southern Illinois. Ohio J. Sci. 86(4): 153–160.

Whittier, T. R., D. B. Halliwell, and R. A. Daniels. 2000. Distributions of lake fishes in the northeast: II. The minnows (Cyprinidae). Northeast. Nat. 7(2): 131–156.

Whitworth, W. R., P. L. Berrien, and W. T. Keller. 1968. Freshwater fishes of Connecticut. State Geol. Nat. Hist. Surv. Conn. Bull. 101. 134 pp.

Wiegmann, D. D., J. R. Baylis, and M. H. Hoff. 1992. Sexual selection and fitness variation in a population of Smallmouth Bass, *Micropterus dolomieui* (Pisces: Centrarchidae). Evolution 46(6): 1740–1753.

Wiegmann, D. D., J. R. Baylis, and M. H. Hoff. 1997. Male fitness, body size, and timing of reproduction in Smallmouth Bass, *Micropterus dolomieui*. Ecology 78(1): 111–128.

Wilamovski, A. 1972. Structure of the gill apparatus and suprabranchial organ of *Hypophthalmichthys molitrix* Val. (Silver Carp). Bamidgeh 24: 87–98.

Wilbur, R. L. 1969. The Redear Sunfish in Florida. Fish. Bull. Florida Game Freshw. Fish Comm. 5: 1–64.

Wilde, G. R. 2002. Threatened fishes of the world: *Notropis girardi* Hubbs & Ortenburger, 1929 (Cyprinidae). Environ. Biol. Fishes 65(1): 98.

Wilde, G. R. 2005. Population dynamics and habitat use of the Arkansas River Shiner and Peppered Chub in the Canadian River, New Mexico and Texas. Final Report submitted to U.S. Bureau of Reclamation. 686 pp.

Wilde, G. R., T. H. Bonner, and P. J. Zwank. 2001. Diets of the Arkansas River Shiner and Peppered Chub in the Canadian River, New Mexico and Texas. J. Freshw. Ecol. 16(3): 403–410.

Wilder, B. G. 1876. Notes on the North American ganoids, *Amia*, *Lepidosteus*, *Acipenser*, and *Polyodon*. Proc. Am. Assoc. Adv. Sci. 24: 151–195.

Wiley, E. O. 1976. The phylogeny and biogeography of fossil and recent gars (Actinopterygii: Lepisosteidae). Univ. Kansas Mus. Nat. Hist. Misc. Publ. 64: 1–111.

Wiley, E. O. 1977. The phylogeny and systematics of the *Fundulus notti* species group (Teleostei: Cyprinodontidae). Occas. Pap. Mus. Nat. Hist. Univ. Kansas. 66: 1–31.

Wiley, E. O. 1978. The evolutionary species concept reconsidered. Syst. Zool. 27(1): 17–26.

Wiley, E. O. 1981. Phylogenetics: the theory and practice of phylogenetic systematics. John Wiley and Sons, New York. 439 pp.

Wiley, E. O. 1986. A study of the evolutionary relationships of *Fundulus* topminnows (Teleostei: Fundulidae). Am. Zool. 26(1): 121–130.

Wiley, E. O. 1992. Phylogenetic relationships of the Percidae (Teleostei: Perciformes): a preliminary hypothesis. Pages 247–267. *In*: R. L. Mayden, ed. Systematics, historical ecology, and North American freshwater fishes. Stanford Univ. Press, Stanford, CA. 969 pp.

Wiley, E. O. and R. H. Hagen. 1997. Mitochondrial DNA sequence variation among the sand darters (Percidae: Teleostei). Pages 75–96. *In*: T. D. Kocher and C. A. Stepien, eds. Molecular systematics of fishes. Academic Press, New York.

Wiley, E. O. and D. D. Hall. 1975. *Fundulus blairae*, a new species of the *Fundulus notti* complex (Teleostei, Cyprinodontidae). Am. Mus. Novit. No. 2577, pp. 1–13.

Wiley, E. O. and G. D. Johnson. 2010. A teleost classification based on monophyletic groups. Pages 123–182. *In*: J. S. Nelson, H. P. Schultze, and M. V. H. Wilson, eds. Origin and phylogenetic interrelationships of teleosts. Verlag Dr. Friedrich Pfeil, Munich, Germany.

Wiley, E. O. and B. S. Lieberman. 2011. Phylogenetics: theory and practice of phylogenetic systematics. 2nd ed. Wiley-Blackwell, New Jersey. 406 pp.

Wiley, E. O. and R. L. Mayden. 1985. Species and speciation in phylogenetic systematics, with examples from the North American fish fauna. Ann. Missouri Bot. Gar. 72(4): 596–635.

Wiley, E. O. and R. L. Mayden. 2000. The evolutionary species concept. Pages 70–89. *In*: Q. D. Wheeler and R. Meier, eds. Species concepts and phylogenetic theory: a debate. Columbia Univ. Press, New York. 256 pp.

Wiley, E. O. and H. P. Schultze. 1984. Family Lepisosteidae (gars) as living fossils. Pages 160–165. *In*: N. Eldredge and S. M. Stanley, eds. Living fossils. Springer-Verlag, New York. 291 pp.

Wiley, E. O. and D. Siegel-Causey. 1994. A phylogenetic analysis of the *Lythrurus roseipinnis* species complex (Teleostei: Cyprinidae) with comments on the relationships of other *Lythrurus*. Occas. Pap. Nat. Hist. Mus. Univ. Kansas 171: 1–20.

Wiley, E. O., G. D. Johnson, and W. W. Dimmick. 2000. The interrelationships of acanthomorph fishes: a total evidence approach using molecular and morphological data. Biochem. Syst. Ecol. 28(4): 319–350.

Wiley, M. L. and B. B. Collette. 1970. Breeding tubercles and contact organs in fishes: their occurrence, structure, and significance. Bull. Am. Mus. Nat. Hist. 143: 145–216.

Wilkens, L. A. and M. H. Hofmann. 2007. The Paddlefish rostrum as an electrosensory organ: a novel adaptation for plankton feeding. BioScience 57(5): 399–407.

Wilkens, L. A., M. H. Hofmann, and W. Wojtenek. 2002. The electric sense of the Paddlefish: a passive system for the detection and capture of zooplankton prey. J. Physiol. Paris 96(5–6): 363–377.

Wilkinson, C. D. and D. R. Edds. 2001. Spatial pattern and environmental correlates of a midwestern stream fish community: including spatial autocorrelation as a factor in community analyses. Am. Midl. Nat. 146(2): 271–289.

Willers, W. B. 1991. Trout biology: a natural history of trout and salmon. Univ. Wisconsin Press, Madison. 273 pp.

Williams, C. S. 2011. Life history characteristics of three obligate riverine species and drift patterns of lower Brazos River fishes. Ph.D. Dissertation. Texas State Univ. San Marcos. 166 pp.

Williams, C. S. and T. H. Bonner. 2006. Habitat associations, life history and diet of the Sabine Shiner, *Notropis sabinae* in an east Texas drainage. Am. Midl. Nat. 155(1): 84–102.

Williams, E. P., A. C. Peer, T. J. Miller, D. H. Secor, and A. R. Place. 2012. A phylogeny of the temperate seabasses (Moronidae) characterized by a translocation of the mt-nd6 gene. J. Fish. Biol. 80(1): 110–130.

Williams, J. C. 1963. The biology of the Silver Chub, *Hybopsis storeriana* (Kirkland), in the Ohio River basin. Ph.D. diss., Univ. Louisville, Louisville, KY. 130 pp.

Williams, J. D. 1975. Systematics of the percid fishes of the subgenus *Ammocrypta*, genus *Ammocrypta*, with descriptions of two new species. Bull. Alabama Mus. Nat. Hist. 1: 1–56.

Williams, J. D. and G. H. Clemmer. 1991. *Scaphirhynchus suttkusi*, a new sturgeon (Pisces: Acipenseridae) from the Mobile Basin of Alabama and Mississippi. Bull. Alabama Mus. Nat. Hist. 10: 17–31.

Williams, J. D. and D. A. Etnier. 1978. *Etheostoma aquali*, a new percid fish (subgenus *Nothonotus*) from the Duck and Buffalo rivers, Tennessee. Proc. Biol. Soc. Washington 91(2): 463–471.

Williams, J. D. and D. A. Etnier. 1982. Description of a new species, *Fundulus julisia*, with a redescription of *Fundulus albolineatus* and a diagnosis of the subgenus *Xenisma* (Teleostei: Cyprinodontidae). Occas. Pap. Mus. Nat. Hist. Univ. Kansas 102: 1–20.

Williams, J. D. and H. W. Robison. 1980. *Ozarka*, a new subgenus of *Etheostoma* (Pisces: Percidae). Brimleyana 4: 149–156.

Williams, J. D., M. L. Warren, Jr., K. S. Cummings, J. L. Harris, and R. J. Neves. 1993. Conservation status of freshwater mussels of the United States and Canada. Fisheries 18(9): 6–22.

Williams, J. E., J. E. Johnson, D. A. Hendrickson, S. Contreras-Balderas, J. D. Williams, M. Navarro-Mendoza, D. E. McAllister, and J. E. Deacon. 1989. Fishes of North America endangered, threatened, or of special concern: 1989. Fisheries 14(6): 2–20.

Williams, L. R., C. M. Taylor, M. L. Warren, Jr., and J. A.

Clingenpeel. 2002. Large-scale effects of timber harvesting on stream systems in the Ouachita Mountains, Arkansas, USA. Environ. Mngt. 29(1): 76–87.

Williams, L. R., M. G. Williams, A. R. Grubh, E. E. Swinehart, and R. W. Standage. 2006. Food habits of the federally threatened Leopard Darter (*Percina pantherina*). Am. Midl. Nat. 156(1): 208–211.

Williamson, C. J. and J. E. Garvey. 2005. Growth, fecundity, and diets of newly established Silver Carp in the middle Mississippi River. Trans. Am. Fish. Soc. 134(6): 1423–1430.

Willis, L. D. 1984. Distribution and habitat requirements of the Ozark Cavefish, *Amblyopsis rosae*. M.S. thesis. Univ. Arkansas, Fayetteville. 25 pp.

Willis, L. D. and A. V. Brown. 1985. Distribution and habitat requirements of the Ozark Cavefish, *Amblyopsis rosae*. Am. Midl. Nat. 114(2): 311–317.

Wills, P. S., R. J. Sheehan, R. Heidinger, B. L. Sloss, and R. Clevenstine. 2002. Differentiation of Pallid Sturgeon and Shovelnose Sturgeon using an index based on meristics and morphometrics. Pages 249–258. *In*: W. Van Winkle, P. J. Anders, D. H. Secor, and D. A. Dixon, eds. Biology, management, and protection of North American sturgeon. Symposium 28, American Fisheries Society, Bethesda, MD. 274 pp.

Wilson, J. A. and R. S. McKinley. 2004. Distribution, habitat, and movements. Pages 40–72. *In*: G. T. O. LeBreton, F. W. H. Beamish, and R. S. McKinley, eds. Sturgeons and Paddlefish of North America. Kluwer Academic Publ. Fish and Fisheries Series Vol. 27. Boston. 323 pp.

Wilson, J. L. and K. D. Cottrell. 1979. Catchability and organoleptic evaluation of Grass Carp in east Tennessee ponds. Trans. Am. Fish. Soc. 108(1): 97–99.

Wilson, M. V. H. 1982. A new species of the fish *Amia* from the Middle Eocene of British Columbia. Palaeo. 25(2): 413–424.

Wilson, M. V. H. and R. R. G. Williams. 1991. New Paleocene genus and species of smelt (Teleostei: Osmeridae) from freshwater deposits of the Paskapoo Formation, Alberta, Canada, and comments on osmerid phylogeny. J. Vert. Paleo. 11(4): 434–451.

Wilson, M. V. H. and R. R. G. Williams. 1992. Phylogenetic, biogeographic, and ecological significance of early fossil records of North American freshwater teleostean fishes. Pages 224–244. *In*: R. L. Mayden, ed. Systematics, historical ecology, and North American freshwater fishes. Stanford Univ. Press, Stanford, CA.

Winchell, A. 1864. Description of a gar-pike, supposed to be new: *Lepidosteus* (*Cylindrosteus*) *oculatus*. Proc. Acad. Nat. Sci. Philadelphia 16(4): 183–185.

Wine, M. S., M. R. Weston, and R. L. Johnson. 2008. Density dynamics of a threatened species of darter at spatial and temporal scales. Southeast. Nat. 7(4): 665–678.

Wine, M., S. Blumenshine, and G. L. Harp. 2001. Status survey of the Yellowcheek Darter (*Etheostoma moorei*), in the Little Red River basin. U. S. Fish and Wildlife Report. Conway, AR. 17 pp.

Winemiller, K. O. and K. A. Rose. 1992. Patterns of life-history diversification in North American fishes: implications for population regulation. Can. J. Fish. Aquatic Sci. 49(10): 2196–2218.

Winfield, I. J. and J. S. Nelson, eds. 1991. Cyprinid fishes: systematics, biology, and exploitation. Chapman & Hall, New York. 667 pp.

Winkelman, D. L. 1996. Reproduction under predatory threat: trade-offs between nest guarding and predator avoidance in male Dollar Sunfish (*Lepomis marginatus*). Copeia 1996(4): 845–851.

Winn, H. E. 1953. Breeding habits of the percid fish *Hadropterus copelandi* in Michigan. Copeia 1953(1): 26–30.

Winn, H. E. 1958a. Comparative reproductive behavior and ecology of fourteen species of darters (Pisces: Percidae). Ecol. Monogr. 28(2): 155–191.

Winn, H. E. 1958b. Observation on the reproductive habits of darters (Pisces: Percidae). Am. Midl. Nat. 59(1): 190–212.

Winn, H. E. and J. F. Stout. 1960. Sound production by the Satinfin Shiner, *Notropis analostanus*, and related fishes. Science 132(3421): 222–223.

Winston, M. R. 2002. Distribution and abundance of the Goldstripe Darter (*Etheostoma parvipinne*) in Missouri. Southwest. Nat. 47(2): 187–194.

Winston, M. R., C. M. Taylor, and J. Pigg. 1991. Upstream extirpation of four minnow species due to damming of a prairie stream. Trans. Am. Fish. Soc. 120(1): 98–105.

Winters, S. A. 1985. Taxa and occurrences of fishes within the Spring River subbasin (Black River drainage) of south central Missouri and northeast Arkansas. M.S. thesis, Northeast Louisiana Univ., Monroe. 143 pp.

Wirth, T. L. 1958. Lake Winnebago Freshwater Drum. Wisconsin Conserv. Bull. 23(5): 30–32.

Wirth, T. and L. Bernatchez. 2003. Decline of North Atlantic eels: a fatal synergy? Proc. Biol. Soc. Lond. 270(1516): 681–688.

Wiseman, E. D., A. A. Echelle, and A. F. Echelle. 1978. Electrophoretic evidence for subspecific differentiation and intergradation in *Etheostoma spectabile* (Teleostei: Percidae). Copeia 1978(2): 320–327.

Wisenden, B. D., J. Karst, J. Miller, S. Miller, and L. Fuselier. 2008. Anti-predator behaviour in response to conspecific chemical alarm cues in an esociform fish, *Umbra limi* (Kirtland 1840). Environ. Biol. Fishes 82(1): 85–92.

Wohlfarth, G. and M. Laharan. 1963. Genetic improvement of Carp VI. Leather and Line carp in fish ponds of Israel. Bamidgeh 15(1): 3–8.

Wolfe, G. W., B. H. Bauer, and B. A. Branson. 1978. Age and growth, length-weight relationships, and condition factors of the Greenside Darter from Silver Creek, Kentucky. Trans. Kentucky Acad. Sci. 39(3–4): 131–134.

Wolfe, J. C. 1969. Biological studies of the Skipjack Herring, *Alosa chrysochloris*, in the Apalachicola River, Florida. M.S. thesis, Florida State Univ., Tallahassee. 68 pp.

Wolters, W. R. and R. DeMay. 1996. Production characteristics of Striped Bass × White Bass and Striped Bass × Yellow Bass hybrids. J. World Aquacul. Soc. 27(2): 202–207.

Wong, E. H. K. and R. H. Hanner. 2008. DNA barcoding detects market substitution in North American seafood. Food Res. Int. 41(8): 828–837.

Wood, M. G. 1981. A taxonomic survey of the fishes of Bayou D'Arbonne after impoundment. M.S. thesis, Northeast Louisiana Univ., Monroe. 115 pp.

Wood, R. M. 1996. Phylogenetic systematics of the darter subgenus *Nothonotus* (Teleostei: Percidae). Copeia 1996(2): 300–318.

Wood, R. M. and R. L. Mayden. 1997. Phylogenetic relationships among selected darter subgenera (Teleostei: Percidae) as inferred from analysis of allozymes. Copeia 1997(2): 265–274.

Wood, R. M. and M. E. Raley. 2000. Cytochrome *b* sequence variation in the Crystal Darter *Crystallaria asprella* (Actinopterygii: Percidae). Copeia 2000(1): 20–26.

Wood, R. M., R. L. Mayden, R. H. Matson, B. R. Kuhajda, and S. R. Layman. 2002. Systematics and biogeography of the *Notropis rubellus* species group (Teleostei: Cyprinidae). Bull. Alabama Mus. Nat. Hist. 22: 37–80.

Woods, A. J., T. L. Foti, S. S. Chapman, J. M. Omernik, J. A. Wise, E. O. Murray, W. L. Prior, J. B. Pagan, Jr., J. A. Comstock, and M. Radford. 2004. Ecoregions of Arkansas. U.S. Geological Survey, Reston, VA. Scale 1:1,000,000.

Woods, L. C. III, C. C. Kohler, R. J. Sheehan, and C. V. Sullivan. 1995. Volitional tank spawning of female Striped Bass with male White Bass produces hybrid offspring. Trans. Am. Fish. Soc. 124(4): 628–632.

Woods, L. P. and R. F. Inger. 1957. The cave, spring, and swamp fishes of the family Amblyopsidae of central and eastern United States. Am. Midl. Nat. 58(1): 232–256.

Woodward, R. L. and T. E. Wissing. 1976. Age, growth, and fecundity of the Quillback (*Carpiodes cyprinus*) and Highfin (*C. velifer*) carpsuckers in an Ohio Stream. Trans. Am. Fish. Soc. 105(3): 411–415.

Woolcott, W. S. and W. L. Kirk. 1976. Melanism in *Lepisosteus osseus* from the James River, Virginia. Copeia 1976(4): 815–817.

Woolcott, W. S. and E. G. Maurakis. 1988. A need for clarification of the concept of nest building among cyprinid minnows. SFC Proc. 1(18): 4–6.

Wooley, C. M. and E. J. Crateau. 1985. Movement, microhabitat, exploitation and management of Gulf of Mexico sturgeon, Apalachicola River, Florida. N. Am. J. Fish. Manag. 5(4): 590–605.

Wooten, M. C., K. T. Scribner, and M. H. Smith. 1988. Genetic variability and systematics of *Gambusia* in the southeastern United States. Copeia 1988(2): 283–289.

Wrenn, W. B. 1968. Life history aspects of Smallmouth Buffalo and Freshwater Drum in Wheeler Reservoir, Alabama. Proc. Annu. Conf. Southeast. Assoc. Game Fish Comm. 22: 479–495.

Wright, J. J. 2009. Diversity, phylogenetic distribution, and origins of venomous catfishes. BMC Evol. Biol. 9(1): 282.

Wright, J. J., S. R. David, and T. J. Near. 2012. Gene trees, species trees, and morphology converge on a similar phylogeny of living gars (Actinopterygii: Holostei: Lepisosteidae), an ancient clade of ray-finned fishes. Mol. Phylogenet. Evol. 63(3): 848–856.

Wright, L. D. 2012. An evaluation of crappie recruitment and supplemental stocking in Arkansas reservoirs. M.S. thesis, Arkansas Tech Univ., Russellville. 162 pp.

Wurtsbaugh, W. A. and J. J. Cech, Jr. 1983. Growth and activity of juvenile Mosquitofish: temperature and ration effects. Trans. Am. Fish. Soc. 112(5): 653–660.

Wynne-Edwards, V. C. 1932. The breeding habits of the Black-headed Minnow (*Pimephales promelas* Raf.). Trans. Am. Fish. Soc. 62(1): 382–383.

Wynes, D. L. and T. E. Wissing. 1981. Effects of water quality on fish and macroinvertebrate communities of the Little Miami River. Ohio J. Sci. 81(6): 259–267.

Wynes, D. L. and T. E. Wissing. 1982. Resource sharing among darters in an Ohio stream. Am. Midl. Nat. 107(2): 294–304.

Xie, P. 1999. Gut contents of Silver Carp, *Hypophthalmichthys molitrix*, and the disruption of a centric diatom, *Cyclotella*, on passage through the esophagus and intestine. Aquaculture 180(3–4): 295–305.

Yang, L., R. L. Mayden, T. Sado, S. He, K. Saitoh, and M. Miya. 2010. Molecular phylogeny of the fishes traditionally referred to Cyprinini *sensu stricto* (Teleostei: Cypriniformes). Zool. Scripta 39(6): 527–550.

Yeager, B. E. and J. K. Beadles. 1976. Fishes of the Cane Creek watershed in southeast Missouri and northeast Arkansas. Proc. Arkansas Acad. Sci. 30: 100–104.

Yeager, B. L. 1979. Larval and early juvenile development of the Striped Shiner, *Notropis chrysocephalus* (Rafinesque). Pages 62–91. *In*: R. Wallus and C. W. Voigtlander, eds. Proceedings of a workshop on freshwater larval fishes. Tennessee Valley Authority, Norris, TN.

Yeager, B. L. 1980. Early development of the genus *Carpiodes* (Osteichthyes: Catostomidae). M.S. thesis, Univ. Tennessee, Knoxville. 80 pp.

Yeager, B. L. 1981. Early development of the Longear Sunfish, *Lepomis megalotis* (Rafinesque). J. Tennessee Acad. Sci. 56(3): 84–88.

Yeager, B. L. 1990. Family Esocidae. Pages 225–255. *In*: R. Wallus, T. P. Simon, and B. L. Yeager, eds. Reproductive biology and early life history of fishes in the Ohio River drainage. Vol. 1. Acipenseridae through Esocidae. Tennessee Valley Authority, Chattanooga, TN.

Yeager, B. L. and J. M. Baker. 1982. Early development of the genus *Ictiobus* (Catostomidae). Pages 63–69. *In*: C. F. Bryan, J. V. Conner, and F. M. Truesdale, eds. Proc. 5th Annual Larval Fish Conference. March 2–3, 1981. Louisiana State Univ. Press, Baton Rouge.

Yeager, B. L. and K. J. Semmens. 1987. Early development of the Blue Sucker, *Cycleptus elongatus*. Copeia 1987(2): 312–316.

Yeager, B. L. and C. F. Saylor. 1995. Fish hosts for four species of freshwater mussels (Pelecypoda: Unionidae) in the upper Tennessee River drainage. Am. Midl. Nat. 133(1): 1–6.

Yeager, L. E. 1936. An observation on spawning buffalofish in Mississippi. Copeia 1936(4): 238–239.

Yellayi, R. R. and R. V. Kilambi. 1970. Observations on early development of White Bass, *Roccus chrysops* (Rafinesque). Proc. 23rd Annu. Conf. Southeast. Assoc. Game Fish Comm.: 261–265.

Yellayi, R. R. and R. V. Kilambi. 1976. Population dynamics of White Bass in Beaver Reservoir, Arkansas. Proc. 29th Annu. Conf. Southeast. Assoc. Game Fish Comm.: 172–184.

Yi, B., Z. Liang, Z. Yu, R. Lin, and M. He. 2006. A study of the early development of Grass Carp, Black Carp, Silver Carp, and Bighead Carp in the Yangtze River, China. Pages 15–51. *In:* D. C. Chapman, ed. Early development of four cyprinids native to the Yangtze River, China. U.S. Geological Survey Data Ser. 239. 51 pp.

Young, M. D. 2011. Changes in discharge and morphology of a Boston Mountain stream following European settlement. M.S. thesis, Arkansas Tech Univ. Russellville. 168 pp.

Young, M. K., R. A. Wilkison, J. M. Phelps III, and J. S. Griffith. 1997. Contrasting movement and activity of large Brown Trout and Rainbow Trout in Silver Creek, Idaho. Great Basin Nat. 57(3): 238–244.

Young, S. P., T. R. Ingram, J. E. Tannehill, and J. J. Isely. 2012. Passage of spawning Alabama Shad at Jim Woodruff Lock and Dam, Apalachicola River, Florida. Trans. Am. Fish. Soc. 141(4): 881–889.

Youngs, W. D. and R. T. Oglesby. 1972. Cayuga Lake: effects of exploitation and introductions on the salmonid community. J. Fish. Res. Board Can. 29(6): 787–794.

Yu, S. L. and E. J. Peters. 2002. Diel and seasonal habitat use by Red Shiner (*Cyprinella lutrensis*). Zool. Stud. 41(3): 229–235.

Zachos, F. E. 2011. Linnean ranks, temporal banding, and time-clipping: why not slaughter the sacred cow? Biol. J. Linn. Soc. 103(3): 732–734.

Zale, A. V. and R. W. Gregory. 1990. Food selection by early life stages of Blue Tilapia, *Oreochromis aureus*, in Lake George, Florida: overlap with sympatric shad larvae. Florida Sci. 53(2): 123–129.

Zale, A. V., S. C. Leon, M. Lechner, O. E. Maughan, M. T. Ferguson, S. O'Donnell, B. James, and P. W. James. 1994. Distribution of the threatened Leopard Darter, *Percina pantherina* (Osteichthyes: Percidae). Southwest. Nat. 39(1): 11–20.

Zambrano, L., E. Martinez-Meyer, N. Menezes, and A. T. Peterson. 2006. Invasive potential of Common Carp (*Cyprinus carpio*) and Nile Tilapia (*Oreochromis niloticus*) in American freshwater systems. Can. J. Fish. Aquat. Sci. 63(9): 1903–1910.

Zdinak, A., Jr., R. V. Kilambi, M. Galloway, J. D. McClanahan, and C. Duffe. 1980. Estimated growth and standing crop of Largemouth Bass (*Micropterus salmoides*) from Lake Elmdale. Proc. Arkansas Acad. Sci. 34: 101–103.

Zeman, D. K and B. M. Burr. 2004. Conservation assessment for Bantam Sunfish (*Lepomis symmetricus*). Report of Dept. Zool., Southern Illinois Univ. Carbondale, to Shawnee National Forest, USDA Forest Service, Harrisburg, IL. 24 pp.

Zervas, P. G. 2010. Age, reproduction, growth, condition and diet of the introduced Yellow Bass, *Morone mississippiensis*, in Barren River Lake, Kentucky. M.S. thesis, Western Kentucky Univ., Bowling Green. 39 pp.

Zeug, S. C., D. Peretti, and K. O. Winemiller. 2009. Movement into floodplain habitats by Gizzard Shad (*Dorosoma cepedianum*) revealed by dietary and stable isotope analysis. Environ. Biol. Fishes 84(3): 307–314.

Zhou, M. and R. C. Fuller. 2014. Reproductive isolation between two darter species is enhanced and asymmetric in sympatry. J. Fish Biol. 84(5): 1389–1400.

Zhou, M., A. M. Johnson, and R. C. Fuller. 2014. Patterns of male breeding color variation differ across species, populations, and body size in Rainbow and Orangethroat darters. Copeia 2014(2): 297–308.

Zhu, S-R, J-L Lee, N. Xie, L-M Zhu, Q. Wang, and G-H Yue. 2014. Genetic diversity based on SSR analysis of the cultured snakehead fish, *Channa argus*, (Channidae) in China. Genet. Mol. Res. 13(3): 8046–8054.

Zigler, S. J., M. R. Dewey, and B. C. Knights. 1999. Diel movement and habitat use by Paddlefish in navigation pool 8 of the upper Mississippi River. N. Am. J. Fish. Manag. 19(1): 180–187.

Zigler, S. J., M. R. Dewey, B. C. Knights, A. L. Runstrom, and M. T. Steingraeber. 2003. Movement and habitat use by radio-tagged Paddlefish in the upper Mississippi River and tributaries. N. Am. J. Fish. Manag. 23(1): 189–205.

Zimmerman, C. J. 1970. Growth and food of the Brook Silverside, *Labidesthes sicculus*, in Indiana. Trans. Am. Fish. Soc. 99(2): 435–438.

Zorach, T. 1972. Systematics of the percid fishes, *Etheostoma camurum* and *E. chlorobranchium* new species, with a discussion of the subgenus *Nothonotus*. Copeia 1972(3): 427–447.

Zuckerkandl, E. and L. B. Pauling. 1962. Molecular disease, evolution and genetic heterogeneity. Pages 189–225. *In*: M. Kasha and B. Pullman, eds. Horizons in biochemistry. Academic Press, New York. 604 pp.

INDEX TO COMMON AND SCIENTIFIC NAMES

A

Acipenser, 129
   desotoi, 94, 132, 133
   fulvescens, 94, 129
   oxyrinchus, 129, 132
Acipenseridae, 106, 128
Acipenseriformes, 127
aepyptera, Lampetra, 115, 122
affinis, Gambusia, 542
Alabama Shad, 181
alabamae, Alosa, 180, 181
albater, Noturus, 429, 441
albus, Scaphirhynchus, 129, 133
Alligator Gar, 150
Alosa, 180
   alabamae, 180, 181
   chrysochloris, 180, 183
alosoides, Hiodon, 168
Ambloplites, 575
   ariommus, 574, 576
   constellatus, 574, 578
   rupestris, 574, 580
Ameiurus, 430
   catus, 95, 96
   melas, 427, 430
   natalis, 427, 432
   nebulosus, 427, 434
amblops, Hybopsis, 197, 257
Amblyopsidae, 106, 502
Amblyopsis rosae, xxvi
American Brook Lamprey, 124
American Eel, 174
americana, Morone, 562, 563
americanus, Esox, 488, 489
Amia calva, 162
Amiidae, 107, 161
Amiiformes, 161
Ammocrypta, 634
   clara, 629, 634
   vivax, 629, 636
amnis, Hybopsis, 203, 259
amnis, Notropis, 260

Anabantiformes, 547
Anguilla rostrata, 174
Anguillidae, 106, 173
Anguilliformes, 173
anisurum, Moxostoma, 381, 409
annularis, Pomoxis, 574, 619
anomalum, Campostoma, 204, 205, 209
Aphredoderidae, 107, 499
Aphredoderus sayanus, 500
Aplodinotus grunniens, 758
appendix, Lethenteron, 115, 124
argus, Channa, 548
argyritis, Hybognathus, 199, 250
ariommus, Ambloplites, 576
Arkansas Darter, 659
Arkansas River Shiner, 318
Arkansas Saddled Darter, 662
artesiae, Etheostoma, 633, 642
asper, Nocomis, 196, 297
asprella, Crystallaria, 629, 638
asprigene, Etheostoma, 634, 644
Atherinidae, 515
Atheriniformes, 515
atherinoides, Notropis, 202, 304
Atherinopsidae, 109, 515
Atlantic Sturgeon, 132
Atractosteus spatula, 149, 150
atrocaudalis, Notropis, 200, 306
atromaculatus, Semotilus, 196, 375
audens, Menidia, 516, 522
auratus, Carassius, 195, 221
aureus, Oreochromis, 762, 763
auritus, Lepomis, 575, 585
Autumn Darter, 646
autumnale, Etheostoma, 633, 646

B

bairdi, Notropis, 200, 308
Banded Darter, 709
Banded Pygmy Sunfish, 625
Banded Sculpin, 554
Bantam Sunfish, 607

Bass
   Florida Largemouth, 95, 617
   Largemouth, 615
   Ozark, 578
   Redeye, 95, 96, 573
   Rock, 580
   Shadow, 576
   Smallmouth, 609
   Spotted, 613
   Striped, 570
   White, 565
   Yellow, 568
Beaded Darter, 655
Bigeye Chub, 257
Bigeye Shiner, 312
Bighead Carp, 265
Bigheaded Carps, 261
Bigmouth Buffalo, 403
Bigscale Logperch, 732
biguttatus, Nocomis, 196, 299
Black Buffalo, 404
Black Bullhead, 430
Black Carp, 294
Black Crappie, 622
Blacknose Crappie, 623
Black Redhorse, 413
Black River Madtom, 456
Blackside Darter, 735
Blackspot Shiner, 306
Blackspotted Topminnow, 539
Blackstripe Topminnow, 537
Blacktail Redhorse, 423
Blacktail Shiner, 239
blairae, Fundulus, 528, 529
Blair's Starhead Topminnow, 530
Bleeding Shiner, 274
blennioides, Etheostoma, 632, 648
blennius, Notropis, 203, 310
Blue Belly Form, 691
Blue Catfish, 436
Bluegill, 593
Bluehead Shiner, 371

Blue Rooter, 406
Blue Sucker, 391
Blue Tilapia, 763
Bluntface Shiner, 229
Bluntnose Darter, 653
Bluntnose Minnow, 359
boops, Notropis, 203, 312
Bowfin, 162
Bream, 573, 595, 604
Brindled Madtom, 458
Brook Silverside, 517
Brook Trout, 482
Brown Bass, 610
Brown Bullhead, 434
Brown Madtom, 462
Brown Trout, 480
brucethompsoni, Percina, 631, 720
bubalus, Ictiobus, 379, 401
buchanani, Notropis, 201, 314
Buffalo
    Bigmouth, 403
    Black, 404
    Smallmouth, 401
Bullhead
    Black, 430
    Brown, 434
    Yellow, 432
Bullhead Catfishes, 430
Bullhead Minnow, 366

C
Caddo Madtom, 464
caeruleum, Etheostoma, 633, 650
calva, Amia, 162
Campostoma, 203
    anomalum, 204, 205, 209
    oligolepis, 197, 206
    plumbeum, 198, 211
    pullum, 198, 214
    spadiceum, 198, 217
camura, Cyprinella, 200, 229
canadense, Stizostedion, 629, 753
caprodes, Percina, 630, 723
Carassius auratus, 195, 221
Cardinal Shiner, 268
cardinalis, Luxilus, 203, 268
carinatum, Moxostoma, 381, 411
carolinae, Cottus, xxvi, 44
carolinae, Uranidea, 554
Carp
    Asian, 192
    Common, 243
    Bighead, 265
    Black, 294

Grass, 225
Silver, 262
carpio, Carpiodes, 380, 382
Carpiodes, 382
    carpio, 380, 382
    cyprinus, 380, 384
    velifer, 380, 386
castaneus, Ichthyomyzon, 115
catenatus, Fundulus, 528, 530
Catfish
    Blue, 436
    Channel, 438
    Flathead, 465
    White, 95, 96, 426
Catostomidae, 109, 377
Catostomus commersonii, 381, 388
Cavefishes, 502
    Ghost, 508
    Ozark, 503
    Salem Plateau, 507
    Southern, 506
Ceasia, 633, 641
Central Mudminnow, 496
Central Stoneroller, 204
Centrarchidae, 110, 573
Centrarchus macropterus, 573, 582
cepedianum, Dorosoma, 180, 185
cephalus, Mugil, 512
Chain Pickerel, 494
chalybaeus, Notropis, 203, 316
Channa argus, 548
Channel Catfish, 438
Channel Darter, 725
Channel Shiner, 351
Channidae, 108, 547
Checkered Madtom, 447
Chestnut Lamprey, 115
chlorosoma, Etheostoma, 631, 653
Chrosomus erythrogaster, 199, 223
chrysocephalus, Luxilus, 203, 270
chrysochloris, Alosa, 180, 183
chrysops, Morone, 563, 565
chrysotus, Fundulus, 528, 533
Chub
    Bigeye, 257
    Creek, 375
    Flathead, 368
    Gravel, 247
    Hornyhead, 299
    Redspot, 297
    Shoal, 286
    Sicklefin, 289
    Silver, 292

Speckled, 288
Sturgeon, 284
Chub Shiner, 334
Chubsucker
    Lake, 396
    Western Creek, 394
Cichlidae, 110, 760
Cichlids and Tilapias, 760, 762
clara, Ammocrypta, 629, 634
clarkii, Salmo, 476
clarkii, Oncorhynchus, 475
claviformis, Erimyzon, 380, 394
clinton, Etheostoma, 633, 655
Clupeidae, 108, 179
Clupeiformes, 179
collettei, Etheostoma, 634, 657
commersonii, Catostomus, 381, 388
Common Carp, 243
Common Shiner, 271
constellatus, Ambloplites, 574, 578
coosae, Micropterus, xxv, 95, 96, 573
copelandi, Percina, 631, 725
Coppernose Bluegill, 594, 595
Cottidae. 109, 553
Cottus, 553, 554
    carolinae, xxvi, 44
    hypselurus, xxvi, 44, 558
    immaculatus, xxvi, 44, 47
cragini, Etheostoma, 633, 659
Crappie
    Black, 622
    Blacknose, 623
    White, 619
Creek Chub, 37
Creek Chubsucker
    Western, 394
Creole Darter, 657
crysoleucas, Notemigonus, 198, 302
Crystal Darter, 638
Crystallaria asprella, 629, 638
Ctenopharyngodon idella, 195, 225
Current Darter, 705
Cutthroat Trout, 475
cyanellus, Lepomis, 575, 587
Cycleptus elongatus, 379, 391
Cypress Darter, 685
Cypress Minnow, 250
Cyprinella, 228
    camura, 200, 229
    galactura, 200, 231
    lutrensis, 200, 201, 233
    spiloptera, 200, 236
    venusta, 200, 239
    whipplei, 200, 241

cyprinellus, Ictiobus, 379, 403
Cyprinidae, 109, 192
Cypriniformes, 191
Cyprinodontidae, 527
Cyprinodontiformes, 527
Cyprinus carpio, 195, 243
cyprinus, Carpiodes, 380, 384

D
Dace
  Southern Redbelly, 223
Darter
  Arkansas, 659
  Arkansas Saddled, 662
  Autumn, 646
  Banded, 709
  Beaded, 655
  Blackside, 735
  Blue Belly Form, 691
  Bluntnose, 653
  Channel, 725
  Creole, 657
  Crystal, 638
  Current, 705
  Cypress, 685
  Dusky, 743
  Fantail, 664
  Gilt, 728
  Goldstripe, 683
  Greenside, 648
  Harlequin, 672
  Highland, 703
  Johnny, 678
  Least, 674
  Leopard, 739
  Longnose, 737
  Mud, 644
  Orangebelly, 694
  Ouachita, 720
  Ozark, 696
  Paleback, 680
  Plains, 687
  Plateau, 698
  Rainbow, 650
  Red Belly Form, 689
  Redfin, 707
  Redspot, 642
  River, 746
  Saddleback, 750
  Scaly Sand, 636
  Slenderhead, 741
  Slough, 670
  Speckled, 701
  Stargazing, 748

  Strawberry, 667
  Sunburst, 676
  Swamp, 668
  Western Sand, 634
  Yellowcheek, 714
  Yoke, 712
desotoi, Acipenser, 94, 132, 133
Dionda nubila, 325
dispar, Fundulus, 528, 535
dissimilis, Hybopsis, 247
Dollar Sunfish, 597
dolomieu, Micropterus, 574, 609
Dorosoma, 185
  cepedianum, 180, 185
  petenense, 180, 188
Drum
  Freshwater, 758
duquesnei, Moxostoma, 382, 413
Dusky Darter, 743
Duskystripe Shiner, 272

E
Eel
  American, 174
eigenmanni, Typhlichthys, 503, 507
eigenmanni, Typhlichthys sp. cf., 503, 508
Elassoma zonatum, 625
Elassomatidae, 110, 624
eleutherus, Noturus, 429, 443
elongatus, Cycleptus, 379, 391
Emerald Shiner, 304
emiliae, Opsopoeodus, 199, 354
Erimystax, 246
  harryi, 197, 246
  x-punctatus, 197, 247
Erimyzon, 394
  claviformis, 380, 394
  oblongus, 395
  sucetta, 380, 396
erythrogaster, Chrosomus, 199, 223
erythrogaster, Phoxinus, 194
erythrophthalmus, Scardinius, 198, 374
erythrurum, Moxostoma, 382, 416
Esocidae, 108, 487
Esociformes, 487
Esox, 488
  americanus, 488, 489
  lucius, 488, 491
  masquinongy, 488, 492
  niger, 488, 494
Etheostoma, 640
  artesiae, 633, 642
  asprigene, 634, 644
  autumnale, 633, 646

  blennioides, 632, 648
  caeruleum, 633, 650
  chlorosoma, 631, 653
  clinton, 633, 655
  collettei, 634, 657
  cragini, 633, 659
  euzonum, 632, 662
  flabellare, 633, 664
  fragi, 634, 667
  fusiforme, 632, 668
  gracile, 632, 670
  histrio, 632, 672
  juliae, 713
  microperca, 631, 674
  mihileze, 633, 676
  moorei, 715
  nigrum, 631, 678
  pallididorsum, 633, 680
  parvipinne, 633, 683
  proeliare, 631, 685
  pulchellum, 634, 687
  sp. cf. *pulchellum* 1, 634, 689
  sp. cf. *pulchellum* 2, 634, 691
  punctulatum, 647, 677
  radiosum, 633, 694
  spectabile, 688, 706, 744
  sp. cf. *spectabile,* 634, 696
  squamosum, 634, 698
  stigmaeum, 633, 701
  teddyroosevelt, 632, 703
  uniporum, 634, 705
  whipplei, 633, 707
  zonale, 632, 709
euzonum, Etheostoma, 632, 662
evides, Percina, 631, 728
exilis, Noturus, 428, 445

F
Fantail Darter, 664
Fathead Minnow, 361
Finescale Stoneroller, 214
flabellare, Etheostoma, 633, 664
Flathead Catfish, 465
Flathead Chub, 368
flavater, Noturus, 429, 447
flavus, Noturus, 429, 449
Flier, 582
fontinalis, Salvelinus, 475, 482
fragi, Etheostoma, 634, 667
Freckled Madtom, 460
Freshwater Drum, 758
Freshwater Eels, 174
fulvescens, Acipenser, 94, 129
fulvitaenia, Percina, 630, 730

fumeus, Notropis, 278
fumeus, Lythrurus, 202, 277
Fundulidae, 108, 527
Fundulus, 528
    blairae, 528, 529
    catenatus, 528, 530
    chrysotus, 528, 533
    dispar, 528, 535
    notatus, 528, 537
    olivaceus, 528, 539
furcatus, Ictalurus, 427, 436
fusiforme, Etheostoma, 632, 668

G

gagei, Ichthyomyzon, 115, 118
galactura, Cyprinella, 200, 231
Gambusia, 542
    affinis, 542
    holbrooki, 543, 544
Gar
    Alligator, 150
    Longnose, 156
    Shortnose, 159
    Spotted, 154
gelida, Macrhybopsis, 197, 284
Ghost Cavefish, 508
Ghost Shiner, 314
Gilt Darter, 728
girardi, Notropis, 200, 318
Gizzard Shad, 185
gladiator, Noturus, 429, 451
Gobiidae, 3, 9, 96, 192
Golden Redhorse, 416
Golden Shiner, 302
Golden Silverside, 521
Golden Topminnow, 533
Goldeye, 168
Goldfish, 221
Goldstripe Darter, 683
gracile, Etheostoma, 632, 670
gracilis, Platygobio, 197, 368
Grass Carp, 225
Grass Pickerel, 489, 490
Gravel Chub, 247
Green Sunfish, 587
greenei, Notropis, 199, 321
Greenside Darter. 648
Grinnel, 82, 162
grunniens, Aplodinotus, 758
Gulf Sturgeon, 94, 132, 133
gulosus, Lepomis, 574, 589
gyrinus, Noturus, 428, 453

H

Harelip Sucker, 418
Hardy Silverside, 520, 521
Harlequin Darter, 672
harryi, Erimystax, 197, 246
hayi, Hybognathus, 199, 250
Herring
    Skipjack, 183
Highfin Carpsucker, 386
Highland Darter, 703
Highland Stoneroller, 217
Hiodon, 168
    alosoides, 168
    tergisus, 168, 170
Hiodontidae, 109, 167
Hiodontiformes, 167
histrio, Etheostoma, 632, 672
Hornyhead Chub, 299
hubbsi, Notropis, 372
hubbsi, Pteronotropis, 199, 371
humilis, Lepomis, 574, 591
Hybognathus, 249
    argyritis, 199, 250
    hayi, 199, 250
    nuchalis, 199, 252
    placitus, 199, 254
Hybopsis, 257
    aestivalis, xxvi, 287
    amblops, 197, 257
    amnis, 203, 259
    dissimilis, 247
    meeki, 290
    storeriana, 284
hyostoma, Macrhybopsis, 197, 286
Hypentelium nigricans, 381, 398
Hypophthalmichthys, 261
    molitrix, 196, 262
    nobilis, 196, 265
hypselurus, Cottus, xxvi, 44, 558

I

Ichthyomyzon, 115
    castaneus, 115
    gagei, 115, 118
    unicuspis, 115, 120
Ictaluridae, 107, 425
Ictalurus, 436
    furcatus, 427, 436
    punctatus, 427, 438
Ictiobus, 400
    bubalus, 379, 401
    cyprinellus, 379, 403
    niger, 379, 404

idella, Ctenopharyngodon, 195, 225
immaculata, Uranidea, 554, 556
Ironcolor Shiner, 316

J

Johnny Darter, 678
juliae, Etheostoma, 713
juliae, Nothonotus, 632, 712

K

Kiamichi Shiner, 326
Killifishes, 527
Knobfin Sculpin, 556

L

Labidesthes, 516
    sicculus, 516, 517
    vanhyningi, 516, 520
lacerum, Moxostoma, 381, 418
lachneri, Noturus, 428, 455
Lake Chubsucker, 396
Lake Sturgeon, 129
Lake Trout, 484
Lampetra, 122
    aepyptera, 115, 122
    appendix, 125
Lamprey
    American Brook, 124
    Chestnut, 115
    Least Brook, 122
    Southern Brook, 118
    Silver, 120
Largemouth Bass, 615
Largescale Stoneroller, 206
Least Brook Lamprey, 122
Least Darter, 674
Leopard Darter, 739
Lepisosteidae, 106, 147
Lepisosteiformes, 147
Lepisosteus, 154
    oculatus, 149, 154
    osseus, 149, 156
    platostomus, 149, 159
Lepomis, 584
    auritus, 575, 585
    cyanellus, 575, 587
    gulosus, 574, 589
    humilis, 574, 591
    macrochirus, 574, 593
    macrochirus mystacalis, 95, 594, 595, 596
    macrochirus purpurescens, 594, 595
    marginatus, 575, 597

megalotis, 575, 599
microlophus, 574, 603
miniatus, 575, 605
punctatus, 606
symmetricus, 575, 607
Lethenteron appendix, 115, 124
Leuciscidae, 192
limi, Umbra, 488, 496
Livebearers, 527, 541
Logperch, 723
Longear Sunfish, 599
Longnose Darter, 737
Longnose Gar, 156
lucius, Esox, 488, 491
lutrensis, Cyprinella, 200, 201, 233
Luxilus, 268
cardinalis, 203, 268
chrysocephalus, 203, 270
pilsbryi, 203, 272
zonatus, 203, 274
Lythrurus, 276
fumeus, 202, 277
snelsoni, 202, 279
umbratilis, 202, 281
sp. cf. *umbratilis*, 17

M
macrochirus, Lepomis, 574, 593
Macrhybopsis, 284
gelida, 197, 284
hyostoma, 197, 286
meeki, 197, 289
storeriana, 197, 292
macrolepida, Percina, 630, 732
macrolepidotum, Moxostoma, 421
macropterus, Centrarchus, 573, 582
maculata, Percina, 631, 735
maculatus, Notropis, 200, 322
Madtom
Black River, 456
Brindled, 458
Brown, 462
Caddo, 464
Checkered, 447
Freckled, 460
Mountain, 443
Ouachita, 455
Ozark, 441
Piebald, 451
Slender, 445
Tadpole, 453
marginatus, Lepomis, 575, 597
masquinongy, Esox, 488, 492

maydeni, Noturus, 429, 456
meeki, Hybopsis, 290
meeki, Macrhybopsis, 197, 289
megalotis, Lepomis, 575, 599
melanops, Minytrema, 380, 407
melanostomus, Neogobius, 96
melas, Ameiurus, 427, 430
Menidia, 522
audens, 516, 522
beryllina, 523, 524
microlophus, Lepomis, 574, 603
microperca, Etheostoma, 631, 674
Micropterus, 609
coosae, xxv, 95, 96, 573
dolomieu, 574, 609
punctulatus, 574, 613
salmoides, 574, 615
salmoides floridanus, 95, 616, 617, 618
mihileze, Etheostoma, 633, 676
Mimic Shiner, 348
miniatus, Lepomis, 575, 605
Minnow
Bluntnose, 359
Bullhead, 366
Cypress, 250
Fathead, 361
Mississippi Silvery, 252
Ozark, 324
Plains, 254
Pugnose, 354
Slim, 364
Suckermouth, 356
Minytrema melanops, 380, 407
mirabilis, Phenacobius, 199, 356
Mississippi Silverside, 522
Mississippi Silvery Minnow, 252
mississippiensis, Morone, 562, 568
miurus, Noturus, 429, 458
molitrix, Hypophthalmichthys, 196, 262
Mooneye, 170
moorei, Etheostoma, 715
moorei, Nothonotus, 632, 714
mordax, Osmerus, 470
Morone, 562
americana, 562, 563
chrysops, 563, 565
mississippiensis, 562, 568
saxatilis, 563, 570
Moronidae, 110, 561
Mosquitofish
Eastern, 542, 543, 544
Western, 542
mossambicus, Oreochromis, 762, 764

Mountain Madtom, 443
Moxostoma, 409
anisurum, 381, 409
carinatum, 381, 411
duquesnei, 382, 413
erythrurum, 382, 416
lacerum, 381, 418
macrolepidotum, 421
pisolabrum, 381, 420
poecilurum, 381, 423
Mozambique Tilapia, 764
Mud Darter, 644
Mudminnow
Central, 496
Mugil cephalus, 512
Mugilidae, 109, 511
Mugiliformes, 511
Mullet
Striped, 512
Muskellunge, 492
mykiss, Oncorhynchus, 475, 477
Mylopharyngodon piceus, 195, 294

N
namaycush, Salvelinus, 475, 484
nasuta, Percina, 630, 737
natalis, Ameiurus, 427, 432
nebulosus, Ameiurus, 427, 434
Neogobius melanostomus, 96
niger, Esox, 488, 494
niger, Ictiobus, 379, 404
nigricans, Hypentelium, 381, 398
nigromaculatus, Pomoxis, 574, 622
nigrum, Etheostoma, 631, 678
Nile Tilapia, 766
niloticus, Oreochromis, 762, 766
nobilis, Hypophthalmichthys, 196, 265
Nocomis, 296
asper, 196, 297
biguttatus, 196, 299
nocturnus, Noturus, 429, 460
Northern Hog Sucker, 398

Northern Pike, 491
Northern Snakehead, 548

Northern Studfish, 530
notatus, Fundulus, 528, 537
notatus, Pimephales, 198, 359
Notemigonus crysoleucas, 198, 302
Nothonotus, 712
juliae, 632, 712
moorei, 632, 714

Notropis, 303
  amnis, 260
  atherinoides, 202, 304
  atrocaudalis, 200, 306
  bairdi, 200, 308
  blennius, 203, 310
  boops, 203, 312
  buchanani, 201, 314
  chalybaeus, 203, 316
  fumeus, 278
  girardi, 200, 318
  greenei, 199, 321
  hubbsi, 372
  maculatus, 200, 322
  nubilus, 199, 324
  ortenburgeri, 201, 326
  ozarcanus, 200, 328
  perpallidus, 203, 332
  pilsbryi, 269, 273, 274
  potteri, 200, 334
  rubellus, 330, 343
  sabinae, 200, 336
  shumardi, 203, 338
  snelsoni, 279
  stramineus, 200, 340
  telescopus, 203, 344
  texanus, 203, 346
  volucellus, 201, 348
  wickliffi, 201, 351
notti, *Fundulus*, 529, 535
Noturus, 440
  albater, 429, 441
  eleutherus, 429, 443
  exilis, 428, 445
  flavater, 429, 447
  flavus, 429, 449
  gladiator, 429, 451
  gyrinus, 428, 453
  lachneri, 428, 455
  maydeni, 429, 456
  miurus, 429, 458
  nocturnus, 429, 460
  phaeus, 429, 462
  stigmosus, 452
  taylori, 429, 464
nubila, *Dionda*, 325
nubilus, *Notropis*, 199, 324
nuchalis, *Hybognathus*, 199, 252

O

oblongus, *Erimyzon*, 395
oculatus, *Lepisosteus*, 149, 154

oligolepis, *Campostoma*, 197, 206
olivaceus, *Fundulus*, 528, 539
olivaris, *Pylodictis*, 427, 465
Oncorhynchus, 475
  clarkii, 475
  mykiss, 475, 477
Opsopoeodus emiliae, 199, 354
Orangebelly Darter, 694
Orangespotted Sunfish, 591
Orangethroat Darter Complex
  (or Clade), 633, 641
Oreochromis, 762
  aureus, 762, 763
  mossambicus, 762, 764
  niloticus, 762, 766
ortenburgeri, *Notropis*, 201, 326
Osmeridae, 107, 469
Osmeriformes, 469
Osmerus mordax, 470
osseus, *Lepisosteus*, 149, 156
Ouachita Darter, 720
Ouachita Madtom, 455
Ouachita Mountain Shiner, 279
ouachitae, *Percina*, 751
oxyrinchus, *Acipenser*, 129, 132
ozarcanus, *Notropis*, 200, 328
Ozark Bass, 578
Ozark Cavefish, 503
Ozark Chub, 246
Ozark Darter, 696
Ozark Logperch, 730
Ozark Madtom, 441
Ozark Minnow, 324
Ozark Sculpin, 44
Ozark Shiner, 328

P

Paddlefish, 141
Paleback Darter, 680
Pallid Shiner, 259
Pallid Sturgeon, 133
pallididorsum, *Etheostoma*, 633, 680
pantherina, *Percina*, 631, 739
parvipinne, *Etheostoma*, 633, 683
Pealip Redhorse, 420
Peppered Shiner, 332
Perca flavescens, 629, 718
Perch
  White, 563
  Yellow, 718
  Pirate, 500
Percichthyidae, 561

Percidae, 111, 627
Perciformes, 561
Percina, 720
  brucethompsoni, 631, 720
  caprodes, 630, 723
  copelandi, 631, 725
  evides, 631, 728
  fulvitaenia, 630, 730
  macrolepida, 630, 732
  maculata, 631, 735
  nasuta, 630, 737
  ouachitae, 751
  pantherina, 631, 739
  phoxocephala, 630, 741
  sciera, 630, 743
  shumardi, 631, 746
  uranidea, 631, 748
  vigil, 631, 750
Percopsiformes, 499
perpallidus, *Notropis*, 203, 332
petenense, *Dorosoma*, 180, 188
Petromyzontidae, 106, 114
Petromyzontiformes, 113
phaeus, *Noturus*, 429, 462
Phenacobius mirabilis, 199, 356
Phoxinus erythrogaster, 194
phoxocephala, *Percina*, 630, 741
piceus, *Mylopharyngodon*, 195, 294
Pickerel
  Chain, 494
  Grass, 489, 490
  Redfin, 489
Piebald Madtom, 451
Pike
  Northern, 491
pilsbryi, *Luxilus*, 203, 272
pilsbryi, *Notropis*, 269, 273, 274
Pimephales, 358
  notatus, 198, 359
  promelas, 198, 361
  tenellus, 198, 364
  vigilax, 198, 366
Pirate Perch, 500
pisolabrum, *Moxostoma*, 381, 420
placitus, *Hybognathus*, 199, 254
Plains Darter, 687
Plains Minnow, 254
Plains Stoneroller, 211
Plateau Darter, 698
platorynchus, *Scaphirhynchus*, 129,
  137
platostomus, *Lepisosteus*, 149, 159

Platygobio gracilis, 197, 368
plumbeum, Campostoma, 198, 211
Poeciliidae, 108, 541
poecilurum, Moxostoma, 381, 423
Polyodon spathula, 141
Polyodontidae, 106, 140
Pomoxis, 619
    annularis, 574, 619
    nigromaculatus, 574, 622
potteri, Notropis, 200, 334
proeliare, Etheostoma, 631, 685
promelas, Pimephales, 198, 361
Pteronotropis hubbsi, 199, 371
Pugnose Minnow, 354
pulchellum, Etheostoma, 634, 687
pullum, Campostoma, 198, 214
punctatus, Ictalurus, 427, 438
punctatus, Lepomis, 606
punctulatum, Etheostoma, 647, 677
punctulatus, Micropterus, 574, 613
Pygmy Sunfish
    Banded, 625
Pylodictis olivaris, 427, 465

Q
Quillback, 384

R
radiosum, Etheostoma, 633, 694
Rainbow Darter, 650
Rainbow Smelt, 470
Rainbow Trout, 477
Razorback Buffalo, 403
Redbreast Sunfish, 585
Red Belly Form, 689
Redear Sunfish, 603
Redeye Bass, 95, 96, 573
Redfin Darter, 707
Redfin Pickerel, 489
Redfin Shiner, 281
Redhorse
    Black, 413
    Blacktail, 423
    Golden, 416
    River, 411
    Pealip, 420
    Shorthead, 378, 421
    Silver, 409
Red River Shiner, 308
Red Shiner, 233
Redspot Chub, 297
Redspot Darter, 642

Redspotted Sunfish, 605
Ribbon Shiner, 277
River Carpsucker, 382
River Darter, 746
River Redhorse, 411
River Shiner, 310
Rock Bass, 580
Rooter, Blue, 406
rosae, Amblyopsis, xxvi
rosae, Troglichthys, 503
rostrata, Anguilla, 174
Round Goby, 96
rubellus, Notropis, 330, 343
Rudd, 374
rupestris, Ambloplites, 574, 580

S
sabinae, Notropis, 200, 336
Sabine Shiner, 336
Saddleback Darter, 750
Salem Plateau Cavefish, 507
Salmo, 479
    clarkii, 476
    gairdneri, 478
    trutta, 475, 480
salmoides, Micropterus, 574, 615
Salmonidae, 107, 473
Salmoniformes, 473
Salvelinus, 482
    fontinalis, 475, 482
    namaycush, 475, 484
Sand Shiner, 340
Sauger, 753
saxatilis, Morone, 563, 570
sayanus, Aphredoderus, 500
Scaly Sand Darter, 636
Scaphirhynchus, 129, 133
    albus, 129, 133
    platorynchus, 129, 137
Scardinius erythrophthalmus, 198, 374
Sciaenidae, 111, 757
sciera, Percina, 630, 743
Scorpaeniformes, 553
Sculpin
    Banded, 554
    Knobfin, 556
    Ozark, 44
Semotilus atromaculatus, 196, 375
Shad
    Alabama, 181
    Gizzard, 185
    Threadfin, 188

Shadow Bass, 576
Shiner
    Arkansas River, 318
    Bigeye, 312
    Blackspot, 306
    Blacktail, 239
    Bleeding, 274
    Bluehead, 371
    Bluntface, 229
    Cardinal, 268
    Channel, 351
    Common, 271
    Duskystripe, 272
    Emerald, 304
    Ghost, 314
    Ironcolor, 316
    Kiamichi, 326
    Mimic, 348
    Ouachita Mountain, 279
    Ozark, 328
    Pallid, 259
    Peppered, 332
    Red, 233
    Redfin, 281
    Red River, 308
    Ribbon, 277
    River, 310
    Sabine, 336
    Sand, 340
    Silverband, 338
    Spotfin, 236
    Steelcolor, 241
    Striped, 270
    Taillight, 322
    Telescope, 344
    Wedgespot, 321
    Weed, 346
    Whitetail, 231
Shoal Chub, 286
Shorthead Redhorse, 378, 421
Shortnose Gar, 159
Shovelnose Sturgeon, 137
shumardi, Notropis, 203, 338
shumardi, Percina, 631, 746
sicculus, Labidesthes, 516, 517
Sicklefin Chub, 289
Siluriformes, 425
Silverband Shiner, 338
Silver Carp, 262
Silver Chub, 292
Silver Lamprey, 120
Silver Redhorse, 409

Silverside
    Brook, 517
    Golden, 521
    Hardy, 520, 521
    Mississippi, 522
    Stout, 521
Skipjack Herring, 183
Slenderhead Darter, 741
Slender Madtom, 445
Slim Minnow, 364
Slough Darter, 670
Smallmouth Bass, 609
Smallmouth Buffalo, 401
Smelts, 469
    Rainbow, 470
Snakehead, Northern, 548
snelsoni, Notropis, 279
snelsoni, Lythrurus, 202, 279
Southern Brook Lamprey, 118
Southern Cavefish, 506
Southern Redbelly Dace, 223
spadiceum, Campostoma, 198, 217
spathula, Polyodon, 141
spatula, Atractosteus, 149, 150
Speckled Chub, 288
Speckled Darter, 701
spectabile, Etheostoma, 688, 706, 744
spiloptera, Cyprinella, 200, 236
Spoonbill Catfish, 140
Spotfin Shiner, 236
Spotted Bass, 613
Spotted Gar, 154
Spotted Sucker, 407
squamosum, Etheostoma, 634, 698
Stargazing Darter, 748
Starhead Topminnow, 535
Steelcolor Shiner, 241
stigmaeum, Etheostoma, 633, 701
stigmosus, Noturus, 452
Stizostedion, 752
    canadense, 629, 753
    vitreum, 629, 755
Stonecat, 449
Stoneroller
    Central, 204
    Largescale, 206
    Finescale, 214
    Highland, 217
    Plains, 211
storeriana, Hybopsis, 284
storeriana, Macrhybopsis, 197, 292
stramineus, Notropis, 200, 340

Strawberry Darter, 667
Striped Bass, 570
Striped Mullet, 512
Striped Shiner, 270
Studfish, Northern, 530
Sturgeon
    Atlantic, 132
    Gulf, 132, 133
    Lake, 129
    Pallid, 133
    Shovelnose, 137
Sturgeon Chub, 284
subterraneus, Typhlichthys, 506
sucetta, Erimyzon, 380, 396
Sucker
    Blue, 391
    Harelip, 418
    Northern Hog, 398
    Spotted, 407
    White, 388
Suckermouth Minnow, 356
Sunburst Darter, 676
Sunfish
    Banded Pygmy, 625
    Bantam, 607
    Dollar, 597
    Green, 587
    Longear, 599
    Orangespotted, 591
    Redbreast, 585
    Redear, 603
    Redspotted, 605
Swamp Darter, 668
symmetricus, Lepomis, 575, 607

T
Tadpole Madtom, 453
Taillight Shiner, 322
taylori, Noturus, 429, 464
teddyroosevelt, Etheostoma, 632, 703
Telescope Shiner, 344
telescopus, Notropis, 203, 344
Temperate Basses, 561
tenellus, Pimephales, 198, 364
tergisus, Hiodon, 168, 170
texanus, Notropis, 203, 346
Threadfin Shad, 188
Tilapia
    Blue, 763
    Mozambique, 764
    Nile, 766
Toothed Herrings, 167

Topminnow
    Blair's Starhead, 530
    Blackspotted, 539
    Blackstripe, 537
    Golden, 533
    Starhead, 535
    Western Starhead, 529
Troglichthys rosae, 503
Trout
    Brook, 482
    Brown, 480
    Cutthroat, 475
    Lake, 484
    Rainbow, 477
trutta, Salmo, 475, 480
Typhlichthys eigenmanni, 503, 507
Typhlichthys sp. cf. *eigenmanni*, 503, 508
Typhlichthys subterraneus, 506

U
Umbra limi, 488, 496
Umbridae, 487
umbratilis, Lythrurus, 202, 281
unicuspis, Ichthyomyzon, 115, 120
Uranidea, 554
    carolinae, 554
    immaculata, 554, 556
uniporum, Etheostoma, 634, 705
uranidea, Percina, 631, 748

V
vanhyningi, Labidesthes, 516, 520
velifer, Carpiodes, 380, 386
venusta, Cyprinella, 200, 239
vigil, Percina, 631, 750
vigilax, Pimephales, 198, 366
vitreum, Stizostedion, 629, 755
vivax, Ammocrypta, 629, 636
volucellus, Notropis, 201, 348

W
Walleye, 755
Warmouth, 589
Wedgespot Shiner, 321
Weed Shiner, 346
Western Creek Chubsucker, 394
Western Mosquitofish, 542
Western Sand Darter, 634
Western Starhead Topminnow, 529
whipplei, Cyprinella, 200, 241

whipplei, Etheostoma, 633, 707
White Bass, 565
White Catfish, 95, 96, 426
White Crappie, 619
White Perch, 563
White Sucker, 388
Whitetail Shiner, 231
wickliffi, Notropis, 201, 351

X
Xenocyprididae, 192
x-punctatus, Erimystax, 197, 247

Y
Yellow Bass, 568
Yellow Bullhead, 432
Yellow Perch, 718

Yellowcheek Darter, 714
Yoke Darter, 712

Z
zonale, Etheostoma, 632, 709
zonatum, Elassoma, 625
zonatus, Luxilus, 203, 274
zonatus, N., 275